KU-326-136

6167 4851

Handbook of Pharmaceutical Excipients

Handbook of Pharmaceutical Excipients

SEVENTH EDITION

Edited by

Raymond C Rowe BPharm, PhD, DSC, FRPharmS, FRSC, CPhys, MInstP

Chief Scientist

Intelligensys Ltd, Stokesley, North Yorkshire, UK

Paul J Sheskey BSc, RPh

Principal Research Scientist

The Dow Chemical Company, Midland, MI, USA

Walter G Cook BSc, PhD

Research Fellow

Materials Science group of Pharmaceutical R&D, Pfizer, Sandwich, Kent, UK

Marian E Fenton BSc, MSc

Development Editor

Royal Pharmaceutical Society of Great Britain, London, UK

London • Philadelphia

Published by Pharmaceutical Press
1 Lambeth High Street, London SE1 7JN, UK

and the American Pharmacists Association
2215 Constitution Avenue, NW, Washington, DC 20037-2985, USA

Pharmaceutical Press is the publishing division of the Royal Pharmaceutical Society of Great Britain

First published 1986
Second edition published 1994
Third edition published 2000
Fourth edition published 2003
Fifth edition published 2006
Sixth edition published 2009
Seventh edition published 2012
Reprinted 2013

Typeset by Data Standards Ltd, Frome, Somerset
Printed in Great Britain by CPI Group (UK) Ltd, Croydon, CR0 4YY

ISBN 978 0 85711 027 5 (UK)
ISBN 978 1 58212 169 7 (USA)

A catalogue record for this book is available from the British Library

Contents

Preface

Pharmaceutical dosage forms contain both pharmacologically active compounds and excipients added to aid the formulation and manufacture of the subsequent dosage form for administration to patients. Indeed, the properties of the final dosage form (i.e. its bioavailability and stability) are, for the most part, highly dependent on the excipients chosen, their concentration and interaction with both the active compound and each other. No longer can excipients be regarded simply as inert or inactive ingredients, and a detailed knowledge not only of the physical and chemical properties but also of the safety, handling and regulatory status of these materials is essential for formulators throughout the world. In addition, the growth of novel forms of delivery has resulted in an increase in the number of the excipients being used and suppliers of excipients have developed novel coprocessed excipient mixtures and new physical forms to improve their properties. Some excipient monographs in the *Handbook* describe materials no longer in common use and a comment is included where this is the case. These monographs are retained in the *Handbook* as a resource for users who may need to understand or reproduce the performance and properties of an old product. This database has been conceived as a systematic, comprehensive resource of information on all of these topics.

The first edition of the *Handbook* was published in 1986 and contained 145 monographs. Subsequent editions have contained more monographs, as well as revised existing content. The data is available in print and online. This new edition contains 380 excipient monographs with enhanced online features, compiled by over 140 experts in pharmaceutical formulation or excipient manufacture from Australia, Europe, India, Japan, and the US. All the monographs have been reviewed and revised in the light of current knowledge. There has been a greater emphasis on including published data from primary sources although some data from laboratory projects included in previous editions have been retained where relevant. Variations in test methodology can have significant effects on the data generated (especially in the case of the compactability of an excipient), and thus cause confusion. As a consequence, the editors have been more selective in including data relating to the physical properties of an excipient. However, comparative data that show differences between either source or batch of a specific excipient have been retained as this was considered relevant to the behavior of a material in practice. Over the past few years, there has been an increased emphasis on the harmonization of excipients. For information on the current status for each excipient selected for harmonization, the reader is directed to the General Information Chapter <1196> in the USP35–NF30, the General Chapter 5.8 in PhEur 7.0, along with the 'State of Work' document on the PhEur EDQM website (www.edqm.eu), and also the General Information Chapter 8 in the JP XV. The Suppliers Directory (Appendix I) has also been completely updated with many more international suppliers included.

In a systematic and uniform manner, the *Handbook of Pharmaceutical Excipients* collects essential data on the physical properties of excipients such as: boiling point, bulk and tap density, compression characteristics, hygroscopicity, flowability, melting point, moisture content, moisture-absorption isotherms, particle size distribution, rheology, specific surface area, and solubility. Scanning electron microphotographs (SEMs) are also included for many of the excipients along with over 130 near-infrared (NIR) spectra specifically generated for this publication. In addition, the current edition includes over 150 infrared (IR) spectra. The *Handbook* contains information from various international sources and personal observation and comments from monograph authors, steering committee members, and the editors.

All of the monographs in the *Handbook* are thoroughly cross-referenced and indexed so that excipients may be identified by either a chemical, a nonproprietary, or a trade name. Most monographs list related substances to help the formulator to develop a list of possible materials for use in a new dosage form or product. Related substances are not directly substitutable for each other but, in general, they are excipients that have been used for similar purposes in various dosage forms.

The *Handbook of Pharmaceutical Excipients* is a comprehensive, uniform guide to the uses, properties, and safety of pharmaceutical excipients, and is an essential reference source for those involved in the development, production, control, or regulation of pharmaceutical preparations. Since many pharmaceutical excipients are also used in other applications, the *Handbook of Pharmaceutical Excipients* will also be of value to persons with an interest in the formulation or production of confectionery, cosmetics, and food products.

Arrangement

The information consists of monographs that are divided into 22 sections to enable the reader to find the information of interest easily. Although it was originally intended that each monograph contain only information about a single excipient, it rapidly became clear that some substances or groups of substances should be discussed together. This gave rise to such monographs as 'Coloring Agents' and 'Hydrocarbons'. In addition, some materials have more than one monograph depending on the physical characteristics of the material, e.g. Starch versus Pregelatinized Starch. Regardless of the complexity of the monograph they are all divided into 22 sections as follows:

1 Nonproprietary Names
2 Synonyms
3 Chemical Name and CAS Registry Number
4 Empirical Formula and Molecular Weight
5 Structural Formula
6 Functional Category
7 Applications in Pharmaceutical Formulation or Technology
8 Description
9 Pharmacopeial Specifications
10 Typical Properties
11 Stability and Storage Conditions
12 Incompatibilities
13 Method of Manufacture
14 Safety
15 Handling Precautions
16 Regulatory Status
17 Related Substances
18 Comments
19 Specific References
20 General References
21 Authors
22 Date of Revision

Descriptions of the sections appear below with information from an example monograph if needed.

Section 1, Nonproprietary Names

Lists the excipient names used in the current British Pharmacopoeia, European Pharmacopeia, Japanese Pharmacopeia, and the United States Pharmacopeia/National Formulary.

Section 2, Synonyms

Lists other names for the excipient, including trade names used by suppliers (shown in italics). The inclusion of one supplier's trade name and the absence of others should in no way be interpreted as an endorsement of one supplier's product over the other. The large number of suppliers internationally makes it impossible to include all the trade names.

Section 3, Chemical Name and CAS Registry Number

Indicates the unique Chemical Abstract Services number for an excipient along with the chemical name, e.g., Acacia [9000-01-5].

Sections 4 and 5, Empirical Formula and Molecular Weight and Structural Formula

Are self-explanatory. Many excipients are not pure chemical substances, in which case their composition is described either here or in Section 8.

Section 6, Functional Category

Lists the function(s) that an excipient is generally thought to perform, e.g., diluent, emulsifying agent, etc. For the purpose of consistency, the functional categories have been thoroughly revised and updated for the current edition of the *Handbook*; *see* Table I.

Note that the use of the general term 'stabilizing agent' or 'stabilizer' has been replaced with terms specific to the type of stability issue addressed:

Physical stability
Antiadherent
Anticaking agent
Dispersing agent
Emulsion stabilizing agent
Foam stabilizing agent
Gelling agent
Humectant
Suspending agent
Viscosity-increasing agent

Microbiological stability
Antimicrobial preservative

Chemical stability
Acidulant
Air displacement
Alkalizing agent
Antioxidant
Buffering agent
Complexing agent
Opacifier

Section 7, Applications in Pharmaceutical Formulation or Technology

Describes the various applications of the excipient. Therapeutic applications and experimental studies are included in Section 18, Comments.

Section 8, Description

Includes details of the physical appearance of the excipient, e.g., white or yellow flakes, etc.

Section 9, Pharmacopeial Specifications

Briefly presents the compendial standards for the excipient. Information included is obtained from the British Pharmacopoeia (BP), European Pharmacopeia (PhEur), Japanese Pharmacopeia (JP), and the United States Pharmacopeia/National Formulary (USP–NF). Information from the JP, PhEur, and USP–NF are included if the substance is in those compendia. If the excipient is not in the PhEur but is included in the BP, information is included from the BP. Pharmacopeias are continually updated with most now being produced as annual editions. However, although efforts were made to include up-to-date information at the time of publication of this edition, the reader is advised to consult the most current pharmacopeias or supplements.

Section 10, Typical Properties

Describes the physical properties of the excipient which are not shown in Section 9. All data are for measurements made at 20°C unless otherwise indicated. Where the solubility of the excipient is

Table I: Functional categories used in the *Handbook*.

Functional category	Alternative term	Definition
Acidulant	Acidifying agent	Agent added to make a system more acid, decreasing the value of pH
Adsorbent	Adsorbing agent; sorbent; sorbing agent	Agent used to bind another component from within a formulation, acting as a carrier, reservoir or sequestrant
Aerosol propellant	Propellant	Agent used to provide an energy source within a formulation for generation of an aerosol on actuation of a valve
Air displacement	Gas flushing agent; sparging agent	Agent used to replace air in a product or pack with a gas phase of known composition
Alcohol denaturant	Bittering agent; denaturant	Agent added to make alcohol unfit to drink
Alkalizing agent		Agent added to make a system more alkaline, increasing the value of pH
Anionic surfactant	Detergent; surface active agent; wetting agent	Agent carrying an overall negative charge, added to reduce surface and interfacial tension
Antiadherent	Antiadhesive agent	Agent added to reduce the tendency of materials to remain attached to other surfaces
Anticaking agent	Flow aid; glidant	Agent added to reduce the tendency of materials to form non-redispersible masses
Antifoaming agent	Foam preventing agent	Agent added to reduce the stability of foams formed during processing or use of formulated products
Antimicrobial preservative	Disinfectant; preservative	Agent used to prevent spoilage due to microbial growth within a formulation
Antioxidant		Agent used to stabilize a system against oxidative degradation
Bioadhesive material	Mucoadhesive membrane	Agent used to promote adhesion to biomembranes
Biocompatible material		An agent that can be used in a parenteral implant product without producing an immune or imflammatory response
Biodegradable material		Used in products that can be degraded to nontoxic components while implanted over time
Buffering agent	Buffer	Agent used to stabilize pH within a defined range
Cationic surfactant	Detergent; surface active agent; wetting agent	Agent carrying an overall positive charge, added to reduce surface and interfacial tension
Coating agent	Enteric coating agent; film-coating agent; modified-release coating agent; sugar-coating agent	Agent used to produce a cosmetic or functional layer on the outer surface of a dosage form
Colorant	Color; colored lake; dye	An agent that imparts colour to a formulation
Complexing agent	Chelating agent; sequestering agent	Agent added to combine with another component, commonly to maintain or improve solubility or chemical stability
Cryoprotectant		Agent added to prevent cell damage during freeze-drying
Desiccant	Water-absorbing agent	Agent used to adsorb or absorb water
Direct compression excipient		Agent used to produce powder blends with flow and compaction properties suitable for tablet making without intermediate granulation steps
Dispersing agent		Agent added to prevent aggregation in liquid formulations
Dry powder inhaler carrier		Agent used in dry powder inhalation blends as a diluent providing suitable uniformity, flow and dispersion properties
Emollient		Agent added to topical formulations to promote softening of the skin
Emulsifying agent	Emulsifier	Agent added to promote mixing of immiscible phases
Emulsion stabilizing agent	Emulsion stabilizer	Agent added to improve stability against phase separation
Film-forming agent	Film-former; polymer film-former	A material which forms a thin film with some mechanical strength when applied to dosage form or other surfaces
Flavor enhancer	Flavor-enhancing agent	Agent added to enhance flavor
Flavoring agent	Flavor	Agent added to impart flavor to a product
Foam stabilizing agent	Foam stabilizer	Agent added to improve physical stability of foam
Gelling agent	Gel thickening agent	Agent added to produce a gel texture in a product
Glidant	Flow aid	Agent added to improve powder flow
Humectant	Moisture retention agent	Agent added to retain water within a product
Lubricant	Friction-reducing agent; lubricating agent	Agent added to reduce friction effects during processing or use
Lyophilization aid	Freeze-drying agent	Agent added to produce suitable physical properties in a freeze-dried product
Membrane-forming agent	Membrane former	A material which forms a thin film with defined permeability properties when applied to a surface or dosage form
Microencapsulating agent	Encapsulating agent	Agent used to form microcapsules with desirable physical properties

Functional category	Alternative term	Definition
Modified-release agent	Controlled-release agent; extended-release agent; release-modifying agent; sustained-release agent	Agent used to control the release rate of active ingredient from a dosage form
Nonionic surfactant	Detergent; surface active agent; wetting agent	Agent containing no ionisable functional groups added to reduce surface and interfacial tension
Ointment base		A nonaqueous vehicle for topical products
Oleaginous vehicle		An oil-based vehicle for topical products
Opacifier	Opacifying agent	Agent added to reduce light transmission in a product
Penetration enhancer	Penetration agent; penetration promoter; skin penetrant	Agent used to increase permeability of active ingredient through skin tissues
Pigment	Colored lake	An insoluble coloring agent
Plasticizing agent	Plasticizer	Agent added to promote flexibilty of films or coatings
Solubilizing agent	Solubilizer	Agent added to promote solubility of an active ingredient
Solvent	Cosolvent	Component used as the vehicle for dissolved ingredients
Stiffening agent		Agent added to increase stiffness of creams and ointments
Suppository base		Agent used as the carrier for other ingredients in suppository formulations
Suspending agent		Agent added to improve dispersion stability of solids in liquids
Sweetening agent	Sweetener	Agent used to produce a sweet taste
Tablet and capsule diluent	Tablet and capsule filler; tablet diluent; tablet filler	Material used to produce appropriate dosage form size, performance and processing properties for tablets and/or capsules
Tablet and capsule disintegrant	Tablet disintegrant	Agent used to promote break-up of tablet and/or capsule formulations after ingestion
Tablet and capsule lubricant	Tablet lubricant	Agent added to reduce friction effects during tablet and/or capsule processing
Tablet and capsule binder	Tablet binder	Agent used to promote granule formulation during tablet and/or capsule processing
Taste-masking agent		Agent added to improve and/or disguise taste
Tonicity agent		Agent added to alter osmotic potential of liquid formulations
Transdermal delivery component		Component used specifically in transdermal devices
Vaccine adjuvant		Agent added to activate antibody response in vaccines
Viscosity-increasing agent	Rheology modifier; thickening agent; viscosity-controlling agent	Agent added to increase the viscosity of liquid semi-solid products
Water-repelling agent	Water repellent	Agent used to make formulations hydrophobic products

described in words, the following terms describe the solubility ranges:

Very soluble	1 part in less than 1
Freely soluble	1 part in 1–10
Soluble	1 part in 10–30
Sparingly soluble	1 part in 30–100
Slightly soluble	1 part in 100–1000
Very slightly soluble	1 part in 1000–10 000
Practically insoluble or insoluble	1 part in more than 10 000

Near-infrared (NIR) reflectance spectra of samples as received (i.e. the samples were not dried or reduced in particle size) were measured using a FOSS NIRSystems 6500 spectrophotometer (FOSS NIRSystems Inc., Laurel, MD, USA) fitted with a Rapid Content Analyser against a ceramic standard supplied with the instrument. The instrument was controlled by Vision (version 2.22) software. Spectra were recorded over the wavelength range 1100–2498 nm (700 data points) and each saved spectrum was the average of 32 scans. Solid powdered samples were measured in glass vials of approximately 20 mm diameter. Each sample was measured in triplicate and the mean spectrum taken. When more than one batch of a material was available, the mean of all the batches is presented. Liquid samples were measured by transflectance using a gold reflector (2×0.5 mm optical path-length, FOSS) placed in a 45 mm silica reflectance cell against air as the reference. Spectra are presented as plots of (a) $\log(1/R)$ *vs* wavelength (dashed line, scale on right-hand side) and (b) second-derivative $\log(1/R)$ *vs* wavelength (solid line, scale on left-hand side). R is the reflectance and $\log(1/R)$ represents the apparent absorbance. Second-derivative spectra were calculated from the $\log(1/R)$ values using an 11 point Savitzky-Golay filter with second-order polynomial smoothing. Note, peak positions and amplitudes in the second-derivative spectrum are very sensitive to the method used to calculate the second-derivative.

For this edition, infrared (IR) spectra have been adapted with permission from Informa Healthcare — *Pharmaceutical Excipients: Characterisation by IR, Raman, and NMR Spectroscopy* by David E Bugay and W Paul Findlay eds, Marcel Dekker, Vol 94, 1999. All samples conformed to the USP–NF for identity and purity, and were used as received. The IR spectra were acquired on a Nicolet model 740 Fourier transform (FT) IR spectrophotometer equipped with a water-cooled globar source, Ge/KBr beamsplitter, and a deuterated triglycine sulfate (DTGS) detector. Interferograms of 16K data points were collected at a spectral resolution of 2 cm^{-1}. The number of scans acquired for each sample varied so that a minimum signal-to-noise ratio of 5000:1 was achieved. For each data set, a phase angle was calculated and then a Happ-Genzel apodization function applied. The Fourier transform was performed and the phase was corrected on the real portion of the data using the calculated phase angle to produce the single-beam spectrum. Subsequent ratioing of the single beam spectrum to the reference spectrum produced a frequency domain IR spectrum. The intensity of the IR spectra have been normalized such that the most intense absorption band equals 1 intensity unit (absorbance or Kubelka-Munk (K-M) units) and all other band intensities are relative to that band. The Spectra-Tech Collector sampling accessory was used to acquire the DR IR spectra of the solid powdered excipients. Samples were diluted in KCl (~1–5% w/w sample to KCl), ratioed against a KCl background, and the resultant spectra displayed in K-M units. IR spectra of liquid and waxy solid samples were acquired as neat smears (capillary film) between two 25– x 2–mm KBr transmission windows. A single beam data file was used as the background file for subsequent ratioing.

Where practical, data typical of the excipient or comparative data representative of different grades or sources of a material are included, the data being obtained from either the primary or the manufacturers' literature. In previous editions of the *Handbook* a laboratory project was undertaken to determine data for a variety of excipients and in some instances this data has been retained. For a description of the specific methods used to generate the data readers should consult the appropriate previous edition(s) of the *Handbook*.

Section 11, Stability and Storage Conditions

Describes the conditions under which the bulk material as received from the supplier should be stored. In addition some monographs report on storage and stability of the dosage forms that contain the excipient.

Section 12, Incompatibilities

Describes the reported incompatibilities for the excipient either with other excipients or with active ingredients. If an incompatibility is not listed it does not mean it does not occur but simply that it has not been reported or is not well known. Every formulation should be tested for incompatibilities prior to use in a commercial product.

Section 13, Method of Manufacture

Describes the common methods of manufacture and additional processes that are used to give the excipient its physical characteristics. In some cases the possibility of impurities will be indicated in the method of manufacture.

Section 14, Safety

Describes briefly the types of formulations in which the excipient has been used and presents relevant data concerning possible hazards and adverse reactions that have been reported. Relevant animal toxicity data are also shown.

Section 15, Handling Precautions

Indicates possible hazards associated with handling the excipient and makes recommendations for suitable containment and protection methods. A familiarity with current good laboratory practice (GLP) and current good manufacturing practice (GMP) and standard chemical handling procedures is assumed.

Section 16, Regulatory Status

Describes the accepted uses in foods and licensed pharmaceuticals where known. However, the status of excipients varies from one nation to another, and appropriate regulatory bodies should be consulted for guidance.

Section 17, Related Substances

Lists excipients similar to the excipient discussed in the monograph.

Section 18, Comments

Includes additional information and observations relevant to the excipient. Where appropriate, the different grades of the excipient available are discussed. Also includes therapeutic applications and experimental studies. Comments are the opinion of the listed author(s) unless referenced or indicated otherwise.

Section 19, Specific References

Is a list of references cited within the monograph.

Section 20, General References

Lists references which have general information about this type of excipient or the types of dosage forms made with these excipients.

Section 21, Authors

Lists the current authors of the monograph in alphabetical order. Authors of previous versions of the monograph are shown in previous printed editions of the text.

Section 22, Date of Revision

Indicates the date on which changes were last made to the text of the monograph.

Acknowledgments

A publication containing so much detail could not be produced without the help of a large number of pharmaceutical scientists based world-wide. The voluntary support of over 140 authors has been acknowledged as in previous editions, but the current editors would like to thank them all personally for their contribution. Grateful thanks also go to the members of the International Steering Committee who advised the editors and publishers on all aspects of the *Handbook* project. Many authors and Steering Committee members have been involved in previous editions of this text. For others, this was their first edition although not, we hope, their last. We extend our thanks to all for their support. Thanks are also extended to Roger Jee, Kelly Palmer, and Tony Moffat at The School of Pharmacy, University of London for supplying the NIR spectra, to Pfizer PGRD, Sandwich, UK for supplying SEMs, and to excipient manufacturers and suppliers who provided helpful information on their products. IR spectra have been adapted with permission from Informa Healthcare.

Thanks are also gratefully extended to the editorial and production staff of the Pharmaceutical Press and American Pharmacists Association, as well as to the team of dedicated copyeditors and proofreaders, who were involved in the production of this resource.

Two contributors, Stanley L Hem and Atul J Shukla, sadly passed away since the publication of the 6th print edition.

Raymond C Rowe, Paul J Sheskey, Walter Cook, Marian E Fenton
March 2012

Notice to Readers

The *Handbook of Pharmaceutical Excipients* is a reference work containing a compilation of information on the uses and properties of pharmaceutical excipients, and the reader is assumed to possess the necessary knowledge to interpret the information that this resource contains. The *Handbook of Pharmaceutical Excipients* has no official status and there is no intent, implied or otherwise, that any of the information presented should constitute standards for the substances. The inclusion of an excipient, or a description of its use in a particular application, is not intended as an endorsement of that excipient or application. Similarly, reports of incompatibilities or adverse reactions to an excipient, in a particular application, may not necessarily prevent its use in other applications. Formulators should perform suitable experimental studies to satisfy themselves and regulatory bodies that a formulation is efficacious and safe to use.

While considerable efforts were made to ensure the accuracy of the information presented in the *Handbook*, neither the publishers nor the compilers can accept liability for any errors or omissions. In particular, the inclusion of a supplier within the Suppliers Directory is not intended as an endorsement of that supplier or its products and, similarly, the unintentional omission of a supplier or product from the directory is not intended to reflect adversely on that supplier or its product.

Although diligent effort was made to use the most recent compendial information, compendia are frequently revised and the reader is urged to consult current compendia, or supplements, for up-to-date information, particularly as efforts are currently in progress to harmonize standards for excipients.

Data presented for a particular excipient may not be representative of other batches or samples.

Relevant data and constructive criticism are welcome and may be used to assist in the preparation of any future editions or digital versions of the *Handbook*. The reader is asked to send any comments to the Editor, Handbook of Pharmaceutical Excipients, Royal Pharmaceutical Society, 1 Lambeth High Street, London SE1 7JN, UK, or Editor, Handbook of Pharmaceutical Excipients, American Pharmacists Association, 2215 Constitution Avenue, NW, Washington, DC 20037-2985, USA.

Disclaimer and Liability

International Steering Committee

Editorial and Production Staff

Editorial Staff of the Pharmaceutical Press

Amy Cruse
Samuel Driver
Marian E Fenton
Elizabeth S Foan
Rebecca E Garner
Jennifer M Sharp
Nilufer Virani
Jo Watts

Production Staff of the Pharmaceutical Press

Tamsin Cousins
Simon Dunton
Karl Parsons
Linda Paulus
John Wilson

Editorial Staff of the American Pharmacists Association

Julian Graubart

Contributors

O AbuBaker
GlaxoSmithKline Inc
Collegeville, PA, USA

KS Alexander
University of Toledo
Toledo, OH, USA

LV Allen Jr
International Journal of Pharmaceutical Compounding
Edmond, OK, USA

FA Alvarez-Nunez
Amgen Inc
Thousand Oaks, CA, USA

GE Amidon
University of Michigan
Ann Arbor, MI, USA

GP Andrews
The Queen's University of Belfast
Belfast, UK

VD Antle
CyDex Pharmaceuticals Inc
Lenexa, KS, USA

NA Armstrong
Harpenden, Hertfordshire, UK

TI Armstrong
Pfizer Consumer Healthcare
Havant, Hampshire, UK

A Balasundaram
Pfizer Pharmatherapeutics
Sandwich, Kent, UK

BA Barner
The Dow Chemical Company
Midland, MI, USA

AC Bentham
Pfizer Global R&D
Sandwich, Kent, UK

C Bhugra
Boehringer-Ingelheim Pharmaceuticals Inc
Ridgefield, CT, USA

M Bond
Danisco (UK) Ltd
Redhill, Surrey, UK

K Boxell
Pfizer Global R&D
Sandwich, Kent, UK

S Brown
Pfizer Global R&D
Sandwich, Kent, UK

CG Cable
Royal Pharmaceutical Society
Edinburgh, UK

S Cantor
Pfizer Inc
Mount Airy, MD, USA

WG Chambliss
University of Mississippi
University, MS, USA

RK Chang
Supernus Pharmaceutical Inc
Rockville, MD, USA

R Chen
Pfizer Inc
Groton, CT, USA

D Chiappetta
Boehringer-Ingelheim Pharmaceuticals Inc
Ridgefield, CT, USA

WG Cook
Pfizer Global R&D
Sandwich, Kent, UK

JR Creekmore
AstraZeneca Pharmaceuticals LP
Wilmington, DE, USA

N Culver
Pfizer Inc
Groton, CT, USA

TC Dahl
Gilead Sciences
Foster City, CA, USA

PD Daugherity
Pfizer Inc
Groton, CT, USA

R Deanne
Boehringer-Ingelheim Pharmaceuticals Inc
Ridgefield, CT, USA

W Deng
University of Mississippi
University, MS, USA

E Draganoiu
Lubrizol Advanced Materials Inc
Cleveland, OH, USA

D Dubash
Pine Brook, NJ, USA

M Dudhedia
Boehringer-Ingelheim Pharmaceuticals Inc
Ridgefield, CT, USA

S Edge
Novartis Pharma AG
Basel, Switzerland

C Egger
Rockwood Pigments
Turin, Italy

H Ehlers
Åbo Akademi University
Turku, Finland

ME Fenton
Royal Pharmaceutical Society
London, UK

RT Forbes
University of Bradford
Bradford, UK

SO Freers
Grain Processing Corporation
Muscatine, IA, USA

B Fritzsching
BENEO-Palatinit GmbH
Mannheim, Germany

GP Frunzi
Time-Cap Labs Inc
Farmingdale, NY, USA

S Fulzele
Cephalon Inc
Brooklyn Park, MN, USA

LY Galichet
International Agency for Research on Cancer
Lyons, France

PL Goggin
Pfizer Global R&D
Sandwich, Kent, UK

S Gold
AstraZeneca
Macclesfield, Cheshire, UK

SR Goskonda
Sunnyvale, CA, USA

M Grachet
Pfizer Global R&D
Sandwich, Kent, UK

L Grove
General Chemical Corporation
Parsippany, NJ, USA

RT Guest
Lonza Biologics plc
Slough, Berkshire, UK

J Gunn
Boehringer-Ingelheim Pharmaceuticals Inc
Ridgefield, CT, USA

RT Gupta
SNDT Women's University
Mumbai, India

VK Gupta
CorePharma
Middlesex, NJ, USA

F Guth
BASF SE
Limburgerhof, Germany

E Hamed
Cephalon Inc
Brooklyn Park, MN, USA

BC Hancock
Pfizer Inc
Groton, CT, USA

BA Hanson
Schering-Plough Research Institute
Summit, NJ, USA

KP Hapgood
Monash University
Clayton, Victoria, Australia

O Häusler
Roquette Europe
Lestrem, France

X He
Boehringer-Ingelheim Pharmaceuticals Inc
Ridgefield, CT, USA

J Heinämäki
University of Tartu
Tartu, Estonia

P Heljo
University of Helsinki
Helsinki, Finland

SL Hem
Formerly, Purdue University
West Lafayette, IN, USA

L Hendricks
Innophos Inc
Cranbury, NJ, USA

S Hoag
University of Maryland at Baltimore
Baltimore, MD, USA

JC Hooton
AstraZeneca
Macclesfield, Cheshire, UK

WL Hulse
University of Bradford
Bradford, UK

M Isreb
University of Bradford
Bradford, UK

BR Jasti
University of the Pacific
Stockton, CA, USA

BA Johnson
Pfizer Inc
Groton, CT, USA

J Johnson
University of Tennessee
Memphis, TN, USA

DS Jones
The Queen's University Belfast
Belfast, UK

S Jones
Eastman Chemical Company
Kingsport, TN, USA

M Julien
Gattefossé SAS
Saint-Priest, France

B Kadri
PharmaForm, LLC
Austin, TX, USA

JS Kaerger
Aeropharm GmbH
Rudolstadt, Germany

AS Kearney
GlaxoSmithKline Inc
King of Prussia, PA, USA

VL Kett
The Queen's University of Belfast
Belfast, UK

AH Kibbe
Wilkes University School of Pharmacy
Wilkes-Barre, PA, USA

DD Ladipo
Pfizer Global R&D
Groton, CT, USA

N Ladyzhynsky
Boehringer-Ingelheim Pharmaceuticals Inc
Ridgefield, CT, USA

WJ Lambert
MedImmune
Gaithersburg, MD, USA

BA Langdon
Pfizer Global R&D
Groton, CT, USA

D Law
Abbott Laboratories
North Chicago, IL, USA

CM Lee
Boehringer-Ingelheim Pharmaceuticals Inc
Ridgefield, CT, USA

MG Lee
MHRA
London, UK

CS Leopold
University of Hamburg
Hamburg, Germany

J-X Li
Prinston Pharmaceutical Inc
Cranbury, NJ, USA

X Li
University of the Pacific
Stockton, CA, USA

H-P Lim
University of Maryland at Baltimore
Baltimore, MD, USA

EB Lindblad
Brenntag Biosector
Frederikssund, Denmark

O Luhn
BENEO-Palatinit GmbH
Mannheim, Germany

PE Luner
Boehringer-Ingelheim Pharmaceuticals Inc
Ridgefield, CT, USA

BR Matthews
Consultant
Surrey, UK

JS Maximilien
Pfizer Global R&D
Sandwich, Kent, UK

CP McCoy
The Queen's University of Belfast
Belfast, UK

C Medina
Amgen Inc
Thousand Oaks, CA, USA

RC Moreton
FinnBrit Consulting
Waltham, MA, USA

GL Mosher
Verrow Pharmaceuticals Inc
Lenexa, KS, USA

G Moss
Keele University
Keele, Staffordshire, UK

C Mroz
Colorcon Ltd
Dartford, Kent, UK

S Mujumdar
Boehringer-Ingelheim Pharmaceuticals Inc
Ridgefield, CT, USA

MP Mullarney
Pfizer Inc
Groton, CT, USA

S Murdande
Pfizer Inc
Groton, CT, USA

S Nema
Pfizer Global R&D
Chesterfield, MO, USA

S Obara
Shin-Etsu Chemical Co. Ltd
Niigata, Japan

A Palmieri
University of Florida
Gainesville, FL, USA

MA Pellett
PCS Ltd
Chichester, West Sussex, UK

L Peltonen
University of Helsinki
Helsinki, Finland

Y Peng
AstraZeneca Pharmaceuticals
Wilmington, DE, USA

M Penning
Mainz, Germany

JD Pipkin
CyDex Pharmaceuticals Inc
Lenexa, KS, USA

A Pirjanian
Amgen Inc
Thousand Oaks, CA, USA

F Podczeck
University College London
London, UK

P Pople
SNDT Women's University
Mumbai, India

W Qu
University of Tennessee
Memphis, TN, USA

A Rajabi-Siahboomi
Colorcon Inc
West Point, PA, USA

RD Reddy
Pfizer Inc
Groton, CT, USA

MA Repka
University of Mississippi
University, MS, USA

J Rexroad
Pacira Pharmaceuticals Inc
San Diego, CA, USA

B Richard
Pacira Pharmaceuticals Inc
San Diego, CA, USA

TL Rogers
The Dow Chemical Company
Bomlitz, Germany

RC Rowe
Intelligensys Ltd
Stokesley, UK

N Sandler
Åbo Akademi University
Turku, Finland

B Sarsfield
Bristol-Myers Squibb
New Brunswick, NJ, USA

P Satturwar
Apotex Inc
Toronto, ON, Canada

M Savolainen
University of Copenhagen
Copenhagen, Denmark

T Schmeller
BASF SE
Limburgerhof, Germany

A Schoch
BENEO-Palatinit GmbH
Mannheim, Germany

DR Schoneker
Colorcon Inc
West Point, PA, USA

J Schrier
Pacira Pharmaceuticals Inc
San Diego, CA, USA

CJ Sciarra
Sciarra Laboratories
Hickesville, NY, USA

JJ Sciarra
Sciarra Laboratories
Hickesville, NY, USA

HC Shah
Signet Chemical Corporation Pvt Ltd
Mumbai, India

SA Shah
ICON Development Solutions
Phoenix, AZ, USA

U Shah
Novartis Vaccines & Diagnostics
Holly Springs, NC, USA

RM Shanker
Pfizer Inc
Groton, CT, USA

PJ Sheskey
The Dow Chemical Company
Midland, MI, USA

J Shur
University of Bath
Bath, UK

S Sienkiewicz
Boehringer-Ingelheim Pharmaceuticals Inc
Ridgefield, CT, USA

D Simon
Roquette Frères
Lestrem, France

A Singh
University of Mississippi
University, MS, USA

KK Singh
SNDT Women's University
Mumbai, India

JLP Soh
Pfizer Global R&D
Sandwich, Kent, UK

RA Storey
AstraZeneca
Macclesfield, Cheshire, UK

C Subra
Gattefossé SAS
Saint-Priest, France

C Sun
University of Minnesota College of Pharmacy
Minneapolis, MN, USA

M Tanenbaum
Boehringer-Ingelheim Pharmaceuticals Inc
Ridgefield, CT, USA

M Tatipalli
University of Florida
Gainesville, FL, USA

AK Taylor
Baton Rouge
LA, USA

J Teckoe
Colorcon Ltd
Dartford, Kent, UK

C Telang
Boehringer-Ingelheim Pharmaceuticals Inc
Ridgefield, CT, USA

MS Tesconi
Pfizer Inc
Groton, CT, USA

S Tiwari
Colorcon Inc
West Point, PA, USA

D Traini
University of Sydney
Sydney, NSW, Australia

N Trivedi
Covidien
Webster Groves, MO, USA

BF Truitt
Xcelience, LLC
Tampa, FL, USA

CK Tye
Bristol-Myers Squibb
Princeton, NJ, USA

HM Unvala
Bayer Corp
Myerstown, PA, USA

S Wallace
Monash University
Hawthorn, VIC, Australia

D Wallick
The Dow Chemical Company
Midland, MI, USA

PJ Weller
BMJ Group
London, UK

W Yang
Astrazeneca
Waltham, MA, USA

P Ying
Pacira Pharmaceuticals Inc
San Diego, CA, USA

W Yu
Pfizer Inc
Groton, CT, USA

About the Editors

Raymond C Rowe

BPharm, PhD, DSc, FRPharmS, FRSC, MInstP

Raymond Rowe has been involved in the *Handbook of Pharmaceutical Excipients* since the first edition was published in 1986, initially as an author then as a Steering Committee member. He is currently Chief Scientist at Intelligensys, UK. He was formerly Senior Principal Scientist at AstraZeneca, UK, and Professor of Industrial Pharmaceutics at the School of Pharmacy, University of Bradford, UK. In 1998 he was awarded the Chiroscience Industrial Achievement Award, and in 1999 he was the British Pharmaceutical Conference Science Chairman. He has contributed to over 400 publications in the pharmaceutical sciences including a book and eight patents.

Paul J Sheskey

BSc, RPh

Paul Sheskey has been involved in the *Handbook of Pharmaceutical Excipients* as an author and member of the Steering Committee since the third edition. He is a Principal Research Scientist at The Dow Chemical Company in Midland, Michigan, USA. Paul received his BSc degree in pharmacy from Ferris State University. Previously, he has worked as a research pharmacist in the area of solid dosage form development at the Perrigo Company and the Upjohn (Pfizer) Company. Paul has authored numerous journal articles in the area of pharmaceutical technology. He is a member of the AAPS and the Controlled Release Society.

Walter G Cook

BSc, PhD

Walter Cook has been involved in the *Handbook of Pharmaceutical Excipients* as an author and member of the Steering Committee since the fourth edition. He is a Research Fellow in the Materials Science group of Pharmaceutical R&D at Pfizer, Sandwich, Kent, UK. Previously, Walter has held positions in formulation research and development at companies including AstraZeneca, UK and Abbott Laboratories, UK. He obtained a BSc in Pharmacy from the University of Strathclyde, UK and a PhD in Pharmaceutical Sciences from the University of Nottingham, UK.

Marian E Fenton

BSc, MSc

Marian Fenton (née Quinn) joined the publications department of the Royal Pharmaceutical Society of Great Britain in 2007 for the sixth edition of the *Handbook of Pharmaceutical Excipients*, having previously worked on the 34th and 35th editions of *Martindale: The Complete Drug Reference*. She has also previously worked at the National Institute for Medical Research, Blackwell Publishing, and Elsevier. Marian received her BSc (Hons) degree in microbiology from the University of Surrey, and her MSc in molecular genetics from the University of Leicester.

New Monographs

The following new monographs have been added to the *Handbook* since the publication of the 6th print edition.

Acetone Sodium Bisulfite
Butyl Stearate
Cellaburate
Cetyl Palmitate
Colophony
Corn Syrup Solids
Decyl Oleate
Dextran
Diethyl Sebacate
Diethylene Glycol Monoethyl Ether
Dipropylene Glycol
Dodecyl Gallate
Ethylene Glycol and Vinyl Alcohol Grafted Copolymer
Fructose and Pregelatinized Starch
Galactose
Hydroxyethylpiperazine Ethane Sulfonic Acid
Isopropyl Isostearate
Lysine Acetate
Lysine Hydrochloride
Mannitol and Sorbitol
D-Mannose
Octyl Gallate
Palm Kernel Oil
Palm Oil
Palm Oil, Hydrogenated
Polyvinyl Acetate Dispersion
Potassium Metaphosphate
Potassium Nitrate
Potassium Phosphate, Dibasic
Propylene Glycol Dilaurate
Propylene Glycol Monolaurate
Pullulan
Pyroxylin
Sodium Stearate
Squalane
Starch, Modified
Sucrose Palmitate
Sucrose Stearate
Tromethamine
Zinc Oxide

Bibliography

A selection of publications and websites which contain useful information on pharmaceutical excipients is listed below:

Ash M, Ash I. *Handbook of Pharmaceutical Additives*, 3rd edn. Endicott, NY: Synapse Information Resources, 2007.

Aulton ME, ed. *Aulton's Pharmaceutics: The Design and Manufacture of Medicines*, 3rd edn. Edinburgh: Churchill Livingstone, 2007.

Banker GS, Rhodes CT, eds. *Modern Pharmaceutics*, 4th edn. New York: Marcel Dekker, 2002.

British Pharmacopoeia 2012. London: The Stationery Office, 2012.

Bugay DE, Findlay WP. *Pharmaceutical Excipients Characterization by IR, Raman, and NMR Spectroscopy*. New York: Marcel Dekker, 1999.

European Pharmacopoeia, 7th edn. Strasbourg: Council of Europe, 2010.

Fiedler Encyclopedia of Excipients. Stuttgart, Germany: Wissenschaftliche Verlagsgesellschaft, 2010.

Florence AT, Salole EG, eds. *Formulation Factors in Adverse Reactions*. London: Butterworth, 1990.

Food and Drug Administration. *Inactive Ingredients Database*. www.accessdata.fda.gov/scripts/cder/iig/index.cfm (accessed 28 November 2011).

Food Chemicals Codex, 7th edn. Bethesda, MD: United States Pharmacopeia, 2010.

Health and Safety Executive. *EH40/2005: Workplace Exposure Limits*. Sudbury: HSE Books, 2011. www.hse.gov.uk/pubns/priced/eh40.pdf (accessed 12 April 2012).

Health Canada. *Canadian Natural Health Products Ingredients Database*. http://webprod.hc-sc.gc.ca/nhpid-bdipsn/search-rechercheReq.do (accessed 28 November 2011).

Japan Pharmaceutical Excipients Council. *Japanese Pharmaceutical Excipients 2004*. Tokyo: Yakuji Nippo, 2004.

Japanese Pharmacopeia, 15th edn. Tokyo: Yakuji Nippo, 2006.

Kemper FH *et al.* eds. *Blue List Cosmetic Ingredients*. Aulendorf, Germany: Editio Cantor, 2000.

Lewis RJ, ed. *Sax's Dangerous Properties of Industrial Materials*, 11th edn. New York: John Wiley, 2004.

Lund W, ed. *The Pharmaceutical Codex: Principles and Practice of Pharmaceutics*, 12th edn. London: Pharmaceutical Press, 1994.

Matthews BR, ed. *Pharmaceutical Excipients: A Manufacturer's Handbook*. Bethesda, MD: PDA Books, 2005.

National Library of Medicine. HSDB Database: ChemIDplus. Bethesda, MD, 2011. chem2.sis.nlm.nih.gov/chemidplus (accessed 28 November 2011).

National Library of Medicine. *TOXNET*. toxnet.nlm.nih.gov (accessed 28 November 2011).

O'Neil MJ *et al.* eds. *The Merck Index: an Encyclopedia of Chemicals, Drugs, and Biologicals*, 14th edn. Whitehouse Station, NJ: Merck, 2006.

Rowe RC, Sheskey PJ, Cook W, Fenton ME. Handbook of excipients: 25 years of marking developments in dosage forms. *Pharmaceutical Journal* 2011; 287: 31–32.

Smolinske SC. *Handbook of Food, Drug and Cosmetic Excipients*. Boca Raton, FL: CRC Press, 1992.

Swarbrick J, Boylan JC, eds. *Encyclopedia of Pharmaceutical Technology*, 2nd edn. New York: Marcel Dekker, 2002.

Sweetman SC, ed. *Martindale: the Complete Drug Reference*, 37th edn. London: Pharmaceutical Press, 2011.

United States Pharmacopeia 35 and National Formulary 30. Rockville, MD: United States Pharmacopeial Convention, 2012.

University of the Sciences in Philadelphia. *Remington: the Science and Practice of Pharmacy*, 21st edn. Philadelphia, PA: Lippincott Williams and Wilkins, 2006.

Weiner M, Bernstein IL. *Adverse Reactions to Drug Formulation Agents: A Handbook of Excipients*. New York: Marcel Dekker, 1989.

Weiner ML, Kotkoskie LA, eds. *Excipient Toxicity and Safety*. New York: Marcel Dekker, 2000.

Abbreviations

Some units, terms, and symbols are not included in this list as they are defined in the text. Common abbreviations have been omitted. The titles of journals are abbreviated according to the general style of the *Index Medicus*.

≈	approximately.
Ad	Addendum.
ADI	acceptable daily intake.
approx	approximately.
atm	atmosphere.
BAN	British Approved Name.
bp	boiling point.
BP	British Pharmacopoeia.
BS	British Standard (specification).
BSI	British Standards Institution.
cal	calorie(s).
CAS	Chemical Abstract Service.
CFC	chlorofluorocarbon.
cfu	colony-forming unit
cm	centimeter(s).
cm^2	square centimeter(s).
cm^3	cubic centimeter(s).
cmc	critical micelle concentration.
CNS	central nervous system.
cP	centipoise(s).
cSt	centistoke(s).
CTFA	Cosmetic, Toiletry, and Fragrance Association.
d	particle diameter (d_{10} at 10 percentile; d_{50} at 50 percentile; d_{90} at 90 percentile).
D&C	designation applied in USA to dyes permitted for use in drugs and cosmetics.
DoH	Department of Health (UK).
DSC	differential scanning calorimetry.
EC	European Community.
EC/SCF	European Commission Science Committee on Food.
e.g.	*exemplit gratia*, 'for example'.
EINECS	European Inventory of Existing Commercial Chemical Substances.
et al	*et alii*, 'and others'.
EU	European Union.
FAO	Food and Agriculture Organization of the United Nations.
FAO/WHO	Food and Agriculture Organization of the United Nations *and the* World Health Organization.
FCC	Food Chemicals Codex.
FDA	Food and Drug Administration of the USA.
FD&C	designation applied in USA to dyes permitted for use in foods, drugs, and cosmetics.
FFBE	Flat face beveled edge.
g	gram(s).
GMP	Good Manufacturing Practice.
GRAS	generally recognized as safe by the Food and Drug Administration of the USA.
HC	hydrocarbon.
HCFC	hydrochlorofluorocarbon.
HFC	hydrofluorocarbon.
HIV	human immunodeficiency virus.
HLB	hydrophilic–lipophilic balance.
HSE	Health and Safety Executive (UK).
i.e.	*id est*, 'that is'.
IM	intramuscular.
INN	International Nonproprietary Name.
IP	intraperitoneal.
IR	infrared.
ISO	International Organization for Standardization.
IU	International Units.
IV	intravenous.
J	joule(s).
JECFA	Joint FAO/WHO Expert Committee on Food Additives.
JP	Japanese Pharmacopeia.
JPE	Japanese Pharmaceutical Excipients
kcal	kilocalorie(s).
kg	kilogram(s).
kJ	kilojoule(s).
KM	Kubelka–Munk (K–M) units
kPa	kilopascal(s).
L	liter(s).
LAL	*Limulus* amoebocyte lysate.
LC_{50}	a concentration in air lethal to 50% of the specified animals on inhalation.
LD_{50}	a dose lethal to 50% of the specified animals or microorganisms.
LD_{Lo}	lowest lethal dose for the specified animals or microorganisms.
m	meter(s).
m^2	square meter(s).
m^3	cubic meter(s).
M	molar.
max	maximum.
MCA	Medicines Control Agency (UK).
mg	milligram(s).
MIC	minimum inhibitory concentration.
min	minute(s) *or* minimum.
mL	milliliter(s).
mm	millimeter(s).
mM	millimolar.
mm^2	square millimeter(s).
mm^3	cubic millimeter(s).
mmHg	millimeter(s) of mercury.
mmol	millimole(s).
mN	millinewton(s).
mol	mole(s).
mp	melting point.
mPa	millipascal(s).
MPa	megapascal(s).
μg	microgram(s).
μm	micrometer(s).
N	newton(s) *or* normal (concentration).
NIR	near-infrared.
NLC	nanostructured lipid carriers.
nm	nanometer(s).
o/w	oil-in-water.
o/w/o	oil-in-water-in-oil.
Pa	pascal(s).
pH	the negative logarithm of the hydrogen ion concentration.
PhEur	European Pharmacopeia.
pK_a	the negative logarithm of the dissociation constant.
PLGA	polylactide-*co*-glycolide.
pph	parts per hundred.

ppm	parts per million.
psia	pounds per square inch absolute.
R	reflectance.
RDA	recommended dietary allowance (USA).
rpm	revolutions per minute.
s	second(s).
SC	subcutaneous.
SEM	scanning electron microscopy *or* scanning electron microphotograph.
SI	Statutory Instrument *or* Système International d'Unites (International System of Units).
SLN	solid lipid nanoparticles.
TC_{Lo}	lowest toxic concentration for the specified animals or microorganisms.
TD_{Lo}	lowest toxic dose for the specified animals or microorganisms.
TPN	total parental nutrition.
TWA	time weighted average.
UK	United Kingdom.
US *or* USA	United States of America.
USAN	United States Adopted Name.
USP	The United States Pharmacopeia.
USP–NF	The United States Pharmacopeia National Formulary.
UV	ultraviolet.
v/v	volume in volume.
v/w	volume in weight.
WHO	World Health Organization.
w/o	water-in-oil.
w/o/w	water-in-oil-in-water.
w/v	weight in volume.
w/w	weight in weight.

Units of Measurement

The information below shows imperial to SI unit conversions for the units of measurement most commonly used in this reference work. SI units are used throughout with, where appropriate, imperial units reported in parentheses.

Area

1 square inch (in^2) = 6.4516×10^{-4} square meter (m^2)
1 square foot (ft^2) = 9.29030×10^{-2} square meter (m^2)
1 square yard (yd^2) = 8.36127×10^{-1} square meter (m^2)

Density

1 pound per cubic foot (lb/ft^3) = 16.0185 kilograms per cubic meter (kg/m^3)

Energy

1 kilocalorie (kcal) = 4.1840×10^3 joules (J)

Force

1 dyne (dynes) = 1×10^{-5} newton (N)

Length

1 angstrom (Å) = 10^{-10} meter (m)
1 inch (in) = 2.54×10^{-2} meter (m)
1 foot (ft) = 3.048×10^{-1} meter (m)
1 yard (yd) = 9.144×10^{-1} meter (m)

Pressure

1 atmosphere (atm) = 0.101325 megapascal (MPa)
1 millimeter of mercury (mmHg) = 133.322 pascals (Pa)
1 pound per square inch (psi) = 6894.76 pascals (Pa)

Surface tension

1 dyne per centimeter (dyne/cm) = 1 millinewton per meter (mN/m)

Temperature

Celsius (°C) = (1.8 × °C) + 32 Fahrenheit (°F)
Fahrenheit (°F) = (0.556 × °F) –17.8 Celsius (°C)

Viscosity (dynamic)

1 centipoise (cP) = 1 millipascal second (mPa s)
1 poise (P) = 0.1 pascal second (Pa s)

Viscosity (kinematic)

1 centistoke (cSt) = 1 square millimeter per second (mm^2/s)

Volume

1 cubic inch (in^3) = 1.63871×10^{-5} cubic meter (m^3)
1 cubic foot (ft^3) = 2.83168×10^{-2} cubic meter (m^3)
1 cubic yard (yd^3) = 7.64555×10^{-1} cubic meter (m^3)
1 pint (UK) = 5.68261×10^{-4} cubic meter (m^3)
1 pint (US) = 4.73176×10^{-4} cubic meter (m^3)
1 gallon (UK) = 4.54609×10^{-3} cubic meter (m^3)
1 gallon (US) = 3.78541×10^{-3} cubic meter (m^3)

Acacia

1 Nonproprietary Names

BP: Acacia
JP: Acacia
PhEur: Acacia
USP–NF: Acacia

2 Synonyms

Acaciae gummi; acacia gum; arabic gum; E414; gum acacia; gum arabic; gummi africanum; gummi arabicum; gummi mimosae; talha gum.

3 Chemical Name and CAS Registry Number

Acacia [9000-01-5]

4 Empirical Formula and Molecular Weight

Acacia is a complex, loose aggregate of sugars and hemicelluloses with a molecular weight of approximately 240 000–580 000. The aggregate consists essentially of an arabic acid nucleus to which are connected calcium, magnesium, and potassium along with the sugars arabinose, galactose, and rhamnose.

5 Structural Formula

See Section 4.

6 Functional Category

Emulsifying agent; modified-release agent; suspending agent; tablet and capsule binder; viscosity-increasing agent.

7 Applications in Pharmaceutical Formulation or Technology

Acacia is mainly used in oral and topical pharmaceutical formulations as a suspending and emulsifying agent, often in combination with tragacanth. It is also used in the preparation of pastilles and lozenges, and as a tablet binder, although if used incautiously it can produce tablets with a prolonged disintegration time. Acacia has also been evaluated as a bioadhesive;[1] and has been used in novel tablet formulations,[2] and modified release tablets.[3]

See Table I.

Table I: Uses of acacia.

Use	Concentration (%)
Emulsifying agent	10–20
Pastille base	10–30
Suspending agent	5–10
Tablet binder	1–5

Acacia is also used in cosmetics, confectionery, food products, and spray-dried flavors.[4]

See also Section 18.

8 Description

Acacia is available as white or yellowish-white thin flakes, spheroidal tears, granules, powder, or spray-dried powder. It is odorless and has a bland taste.

9 Pharmacopeial Specifications

The PhEur 7.4 provides monographs on acacia and spray-dried acacia, while the USP35–NF30 describes acacia in a single monograph that encompasses tears, flakes, granules, powder, and spray-dried powder. The USP35–NF30 also has a monograph on acacia syrup. The JP XV has monographs on acacia and powdered acacia; *see* Table II.

Table II: Pharmacopeial specifications for acacia.

Test	JP XV	PhEur 7.4	USP35–NF30
Identification	+	+	+
Characters	+	+	+
Microbial limit	–	≤10^4 cfu/g	+
Water	≤17.0%	≤15.0%	≤15.0%
	≤15.0%[a]	≤10.0%[b]	–
Total ash	≤4.0%	≤4.0%	≤4.0%
Acid-insoluble ash	≤0.5%	–	≤0.5%
Insoluble residue	≤0.2%	≤0.5%	≤50 mg/5 g
Arsenic	–	–	≤3 ppm
Lead	–	–	≤10 ppm
Heavy metals	–	–	≤40 ppm
Starch, dextrin, and agar	–	+	+
Tannin-bearing gums	+	+	+
Tragacanth	–	+	–
Sterculia gum	–	+	–
Glucose and fructose	+	+	–
Solubility and reaction	–	–	+

(a) Powdered acacia.
(b) Spray-dried acacia.

10 Typical Properties

Acidity/alkalinity pH = 4.5–5.0 (5% w/v aqueous solution)

Acid value 2.5

Hygroscopicity At relative humidities of 25–65%, the equilibrium moisture content of powdered acacia at 25°C is 8–13% w/w, but at relative humidities above about 70% it absorbs substantial amounts of water.

IR spectra This is a test graphic.

Solubility Soluble 1 in 20 of glycerin, 1 in 20 of propylene glycol, 1 in 2.7 of water; practically insoluble in ethanol (95%). In water, acacia dissolves very slowly, although almost completely

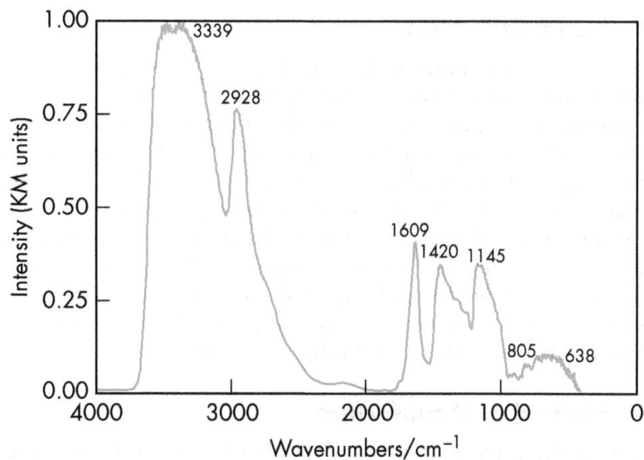

Figure 1: Infrared spectrum of acacia measured by diffuse reflectance. Adapted with permission of Informa Healthcare.

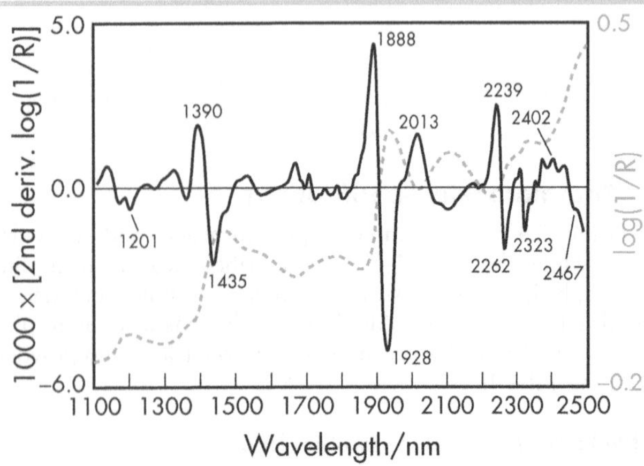

Figure 2: Near-infrared spectrum of acacia measured by reflectance.

after two hours, in twice the mass of water leaving only a very small residue of powder. The solution is colorless or yellowish, viscous, adhesive, and translucent. Spray-dried acacia dissolves more rapidly, in about 20 minutes.

Specific gravity 1.35–1.49

Spectroscopy

IR spectra *see* Figure 1.

NIR spectra *see* Figure 2.

Viscosity (dynamic) 100 mPa s for a 30% w/v aqueous solution at 20°C. The viscosity of aqueous acacia solutions varies depending upon the source of the material, processing, storage conditions, pH, and the presence of salts. Viscosity increases slowly up to about 25% w/v concentration and exhibits Newtonian behavior. Above this concentration, viscosity increases rapidly (non-Newtonian rheology). Increasing temperature or prolonged heating of solutions results in a decrease of viscosity owing to depolymerization or particle agglomeration.

See also Section 12.

11 Stability and Storage Conditions

Aqueous solutions are subject to bacterial or enzymatic degradation but may be preserved by initially boiling the solution for a short time to inactivate any enzymes present; microwave irradiation can also be used.(5) Aqueous solutions may also be preserved by the addition of an antimicrobial preservative such as 0.1% w/v benzoic acid, 0.1% w/v sodium benzoate, or a mixture of 0.17% w/v methylparaben and 0.03% propylparaben. Powdered acacia should be stored in an airtight container in a cool, dry place.

12 Incompatibilities

Acacia is incompatible with a number of substances including amidopyrine, apomorphine, cresol, ethanol (95%), ferric salts, morphine, phenol, physostigmine, tannins, thymol, and vanillin.

An oxidizing enzyme present in acacia may affect preparations containing easily oxidizable substances. However, the enzyme may be inactivated by heating at 100°C for a short time; *see* Section 11.

Many salts reduce the viscosity of aqueous acacia solutions, while trivalent salts may initiate coagulation. In aqueous solutions acacia carries a negative charge and will form coacervates with gelatin and other substances. In the preparation of emulsions, solutions of acacia are incompatible with soaps.

13 Method of Manufacture

Acacia is the dried gummy exudate obtained from the stems and branches of *Acacia senegal* (Linné) Willdenow or other related species of *Acacia* (Fam. Leguminosae) that grow mainly in the Sudan and Senegal regions of Africa.

The bark of the tree is incised and the exudate allowed to dry on the bark. The dried exudate is then collected, processed to remove bark, sand, and other particulate matter, and graded. Various acacia grades differing in particle size and other physical properties are thus obtained. A spray-dried powder is also commercially available.

14 Safety

Acacia is used in cosmetics, foods, and oral and topical pharmaceutical formulations. Although it is generally regarded as an essentially nontoxic material, there have been a limited number of reports of hypersensitivity to acacia after inhalation or ingestion.(6,7) Severe anaphylactic reactions have occurred following the parenteral administration of acacia and it is now no longer used for this purpose.(6)

The WHO has not set an acceptable daily intake for acacia as a food additive because the levels necessary to achieve a desired effect were not considered to represent a hazard to health.(8)

LD_{50} (hamster, oral): >18 g/kg(9)

LD_{50} (mouse, oral): >16 g/kg

LD_{50} (rabbit, oral): 8.0 g/kg

LD_{50} (rat, oral): >16 g/kg

15 Handling Precautions

Observe normal precautions appropriate to the circumstances and quantity of material handled. Acacia can be irritant to the eyes and skin and upon inhalation. Gloves, eye protection, and a dust respirator are recommended.

16 Regulatory Status

GRAS listed. Accepted for use as a food additive in Europe. Included in the FDA Inactive Ingredients Database (oral preparations and buccal or sublingual tablets). Included in the Canadian Natural Health Products Ingredients Database. Included in nonparenteral medicines licensed in the UK.

17 Related Substances

Ceratonia; guar gum; octenyl succinic anhydride modified gum acacia; tragacanth

Octenyl succinic anhydride modified gum acacia

CAS number [445885-22-0]

Synonyms talha gum; OSA modified gum acacia.

Comments Produced from gum acacia by the introduction of lipophilic groups to the polysaccharide in gum acacia by a controlled esterification process. It has been investigated for use as an emulsifier and has been found to be as safe as acacia.(13)

LD_{50} (oral, rat): >2g/kg

18 Comments

Concentrated aqueous solutions are used to prepare pastilles since on drying they form solid rubbery or glass-like masses depending upon the concentration used. Foreign policy changes and politically unstable conditions in Sudan, which is the principal supplier of acacia, has created a need to find a suitable replacement.(10) Poloxamer 188 (12–15% w/w) can be used to make an oil/water emulsion with similar rheological characteristics to a standard 4-2-1 acacia emulsion. Other natural by-products of foods can also be used.(11) Acacia is also used in the food industry as an emulsifier, stabilizer, and thickener. A specification for acacia is contained in the *Food Chemicals Codex* (FCC).(12)

The EINECS number for acacia is 232-519-5.

19 Specific References

1 Attama AA, *et al.* Studies on bioadhesive granules. *STP Pharma Sci* 2003; **13**(3): 177–181.
2 Streubel A, *et al.* Floating matrix tablets based on low density foam powder: effects of formulation and processing parameters on drug release. *Eur J Pharm Sci* 2003; **18**: 37–45.
3 Bahardwaj TR, *et al.* Natural gums and modified natural gums as sustained-release carriers. *Drug Dev Ind Pharm* 2000; **26**(10): 1025–1038.
4 Buffo R, Reineccius G. Optimization of gum acacia/modified starch/maltodextrin blends for spray drying of flavors. *Perfumer & Flavorist* 2000; **25**: 45–54.
5 Richards RME, Al Shawa R. Investigation of the effect of microwave irradiation on acacia powder. *J Pharm Pharmacol* 1980; **32**: 45P.
6 Maytum CK, Magath TB. Sensitivity to acacia. *J Am Med Assoc* 1932; **99**: 2251.
7 Smolinske SC. *Handbook of Food, Drug, and Cosmetic Excipients.* Boca Raton, FL: CRC Press, 1992: 7–11.
8 FAO/WHO. Evaluation of certain food additives and contaminants. Thirty-fifth report of the joint FAO/WHO expert committee on food additives. *World Health Organ Tech Rep Ser* 1990; No. 789.
9 Lewis RJ, ed. *Sax's Dangerous Properties of Industrial Materials*, 11th edn. New York: Wiley, 2004; 289.
10 Scheindlin S. Acacia – a remarkable excipient: the past, present, and future of gum arabic. *JAMA* 2001; **41**(5): 669–671.
11 I-Achi A, *et al.* Experimenting with a new emulsifying agent (tahini) in mineral oil. *Int J Pharm Compound* 2000; **4**(4): 315–317.
12 *Food Chemicals Codex*, 7th edn. Bethesda, MD: United States Pharmacopeia, 2010; 460.
13 European Food Safety Authority. Scientific opinion on the use of gum acacia modified with octenyl succinic anhydride (OSA) as a food additive. *The EFSA Journal* 2010; **8**(3): 1539.

20 General References

Anderson DMW, Dea ICM. Recent advances in the chemistry of acacia gums. *J Soc Cosmet Chem* 1971; **22**: 61–76.
Anderson DM, *et al.* Specifications for gum arabic (*Acacia Senegal*): analytical data for samples collected between 1904 and 1989. *Food Add Contam* 1990; **7**: 303–321.
Aspinal GO. Gums and mucilages. *Adv Carbohydr Chem Biochem* 1969; **24**: 333–379.
Whistler RL. *Industrial Gums.* New York: Academic Press, 1959.

21 Author

AH Kibbe.

22 Date of Revision

1 March 2012.

 Acesulfame Potassium

1 Nonproprietary Names

BP: Acesulfame Potassium
PhEur: Acesulfame Potassium
USP–NF: Acesulfame Potassium

2 Synonyms

Acesulfame K; ace K; acesulfamum kalicum; E950; 6-methyl-3,4-dihydro-1,2,3-oxathiazin-4(3*H*)-one-2,2-dioxide potassium salt; potassium 6-methyl-2,2-dioxo-oxathiazin-4-olate; *Sunett*; *Sweet One.*

3 Chemical Name and CAS Registry Number

6-Methyl-1,2,3-oxathiazin-4(3*H*)-one-2,2-dioxide potassium salt [55589-62-3]

4 Empirical Formula and Molecular Weight

$C_4H_4KNO_4S$ 201.24

5 Structural Formula

6 Functional Category

Sweetening agent.

7 Applications in Pharmaceutical Formulation or Technology

Acesulfame potassium is used as an intense sweetening agent in cosmetics, foods, beverage products, table-top sweeteners, vitamin and pharmaceutical preparations, including powder mixes, tablets, and liquid products. It is widely used as a sugar substitute in compounded formulations,[1] and as a toothpaste sweetener.[2]

The approximate sweetening power is 180–200 times that of sucrose, similar to aspartame, about one-third as sweet as sucralose, one-half as sweet as sodium saccharin, and about 4-5 times sweeter than sodium cyclamate.[3] It enhances flavor systems and can be used to mask some unpleasant taste characteristics.

8 Description

Acesulfame potassium occurs as a colorless to white-colored, odorless, crystalline powder with an intensely sweet taste.

9 Pharmacopeial Specifications

See Table I.

A

SEM 1: Excipient: acesulfame potassium; magnification: 150×; voltage: 5 kV.

Table I: Pharmacopeial specifications for acesulfame potassium.

Test	PhEur 7.4	USP35–NF30
Characters	+	—
Identification	+	+
Appearance of solution	+	—
Acidity or alkalinity	+	+
Acetylacetamide	0.125%	—
Impurity B[a]	≤20 ppm	—
Unspecified impurities	≤0.1%	≤0.002%
Total impurities	≤0.1%	—
Fluorides	≤3 ppm	≤3 ppm
Heavy metals	≤5 ppm	≤10 ppm
Loss on drying	≤1.0%	≤1.0%
Assay (dried basis)	99.0–101.0%	99.0–101.0%

(a) Impurity B is 5-chloro-6-methyl-1,2,3-oxathiazin-4(3*H*)-one 2,2-dioxide.

10 Typical Properties

Acidity/alkalinity pH = 5.5–7.5 (1% w/v aqueous solution)
Bonding index 0.007[4]
Brittle fracture index 0.08[4]
Density (bulk) 1.04 g/cm^3 [4]
Density (tapped) 1.28 g/cm^3 [4]
Elastic modulus 4000 MPa[4]
Flowability 19% (Carr compressibility index)[4]
Melting point 250°C
Solubility *see* Table II.

Table II: Solubility of acesulfame potassium.[3]

Solvent	Solubility at 20°C unless otherwise stated
Ethanol	1 in 1000
Ethanol (50%)	1 in 10
Ethanol (15%)	1 in 4.5
Water	1 in 6.7 at 0°C
	1 in 3.7 at 20°C
	1 in 0.77 at 100°C

Specific volume 0.538 cm^3/g[5]

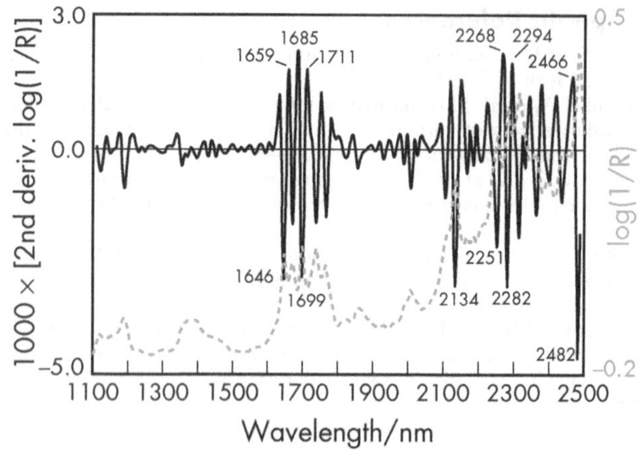

Figure 1: Near-infrared spectrum of acesulfame potassium measured by reflectance.

Spectroscopy
NIR spectra *see* Figure 1.
Surface tension 73.2 mN/m[6] (1% w/v aqueous solution at 20°C
Tensile strength 0.5 MPa[4]
Viscoelastic index 2.6[4]

11 Stability and Storage Conditions

Acesulfame potassium possesses good stability. In the bulk form it shows no sign of decomposition at ambient temperature over many years. In aqueous solutions (pH 3.0–3.5 at 20°C) no reduction in sweetness was observed over a period of approximately 2 years. Stability at elevated temperatures is good, although some decomposition was noted following storage at 40°C for several months. Sterilization and pasteurization do not affect the taste of acesulfame potassium.[7]

The bulk material should be stored in a well-closed container in a cool, dry place and protected from light.

12 Incompatibilities

—

13 Method of Manufacture

Acesulfame potassium is synthesized from acetoacetic acid *tert*-butyl ester and fluorosulfonyl isocyanate. The resulting compound is transformed to fluorosulfonyl acetoacetic acid amide, which is then cyclized in the presence of potassium hydroxide to form the oxathiazinone dioxide ring system. Because of the strong acidity of this compound, the potassium salt is produced directly.[8]

An alternative synthesis route for acesulfame potassium starts with the reaction between diketene and amidosulfonic acid. In the presence of dehydrating agents, and after neutralization with potassium hydroxide, acesulfame potassium is formed.

14 Safety

Acesulfame potassium is widely used in beverages, cosmetics, foods, and pharmaceutical formulations, and is generally regarded as a relatively nontoxic and nonirritant material. Pharmacokinetic studies have shown that acesulfame potassium is not metabolized and is rapidly excreted unchanged in the urine. Long-term feeding studies in rats and dogs showed no evidence to suggest acesulfame potassium is mutagenic or carcinogenic.[9]

The WHO has set an acceptable daily intake for acesulfame potassium of up to 15 mg/kg body-weight.[9] The Scientific Committee for Foods of the European Union has set a daily intake value of up to 9 mg/kg of body-weight.[3]

LD_{50} (rat, IP): 2.2 g/kg[7]
LD_{50} (rat, oral): 6.9–8.0 g/kg

15 Handling Precautions

Observe normal precautions appropriate to the circumstances and quantity of material handled. Eye protection, gloves, and a dust mask are recommended.

16 Regulatory Status

Included in the FDA Inactive Ingredients Database for oral and sublingual preparations. Included in the Canadian Natural Health Products Ingredients Database. Accepted for use as a food additive in Europe. It is also accepted for use in certain food products in the USA and several countries in Central and South America, the Middle East, Africa, Asia, and Australia.

17 Related Substances

Alitame.

18 Comments

The perceived intensity of sweeteners relative to sucrose depends upon their concentration, temperature of tasting, and pH, and on the flavor and texture of the product concerned.

Intense sweetening agents will not replace the bulk, textural, or preservative characteristics of sugar, if sugar is removed from a formulation.

Synergistic effects for combinations of sweeteners have been reported, e.g. acesulfame potassium with aspartame or sodium cyclamate; *see also* Aspartame. A ternary combination of sweeteners that includes acesulfame potassium and sodium saccharin has a greater decrease in sweetness upon repeated tasting than other combinations.[10]

Note that free acesulfame acid is not suitable for use as a sweetener.

A specification for acesulfame potassium is contained in the *Food Chemicals Codex* (FCC).[11]

19 Specific References

1 Kloesel L. Sugar substitutes. *Int J Pharm Compound* 2000; **4**(2): 86–87.
2 Schmidt R, *et al.* Evaluating toothpaste sweetening. *Cosmet Toilet* 2000; **115**: 49–53.
3 Wilson R, ed. *Sweeteners*, 3rd edn. Oxford, UK: Blackwell Publishing, 2007: 3–19.
4 Mullarney MP, *et al.* The powder flow and compact mechanical properties of sucrose and three high-intensity sweeteners used in chewable tablets. *Int J Pharm* 2003; **257**: 227–236.
5 Birch GG, *et al.* Apparent specific volumes and tastes of cyclamates, other sulfamates, saccharins and acesulfame sweeteners. *Food Chem* 2004; **84**: 429–435.
6 Hutteau F, *et al.* Physiochemical and psychophysical characteristics of binary mixtures of bulk and intense sweeteners. *Food Chem* 1998; **63**(1): 9–16.
7 Lipinski G-WvR, Huddart BE. Acesulfame K. *Chem Ind* 1983; **11**: 427–432.
8 Shetty K, ed. *Functional Foods and Biotechnology.* Boca Raton, FL: CRC Press, 2007: 327–344.
9 FAO/WHO. Evaluation of certain food additives and contaminants. Thirty-seventh report of the joint FAO/WHO expert committee on food additives. *World Health Organ Tech Rep Ser* 1991; No. 806.
10 Schiffman SS, *et al.* Effect of repeated presentation on sweetness intensity of binary and tertiary mixtures of sweetness. *Chem Senses* 2003; **28**: 219–229.
11 *Food Chemicals Codex*, 7th edn (Suppl. 1). Bethesda, MD: United States Pharmacopeia, 2010: 9.

20 General References

Anonymous. Artificial sweetners. *Can Pharm J* 1996; **129**: 22.
Lipinski G-WvR, Lück E. Acesulfame K: a new sweetener for oral cosmetics. *Manuf Chem* 1981; **52**(5): 37.
Marie S. Sweeteners. In: Smith J, ed. *Food Additives User's Handbook.* Glasgow: Blackie, 1991: 47–74.
Celanese Corp. Nutrinova technical literature: The *Sunett* guide to sweetness, 2008.

21 Author

BA Johnson.

22 Date of Revision

1 March 2012.

 Acetic Acid, Glacial

1 Nonproprietary Names

BP: Glacial Acetic Acid
JP: Glacial Acetic Acid
PhEur: Acetic Acid, Glacial
USP–NF: Glacial Acetic Acid

2 Synonyms

Acidum aceticum glaciale; E260; ethanoic acid; ethylic acid; methane carboxylic acid; vinegar acid.

See also Section 17 and Section 18.

3 Chemical Name and CAS Registry Number

Ethanolic acid [64-19-7]

4 Empirical Formula and Molecular Weight

$C_2H_4O_2$ 60.05

5 Structural Formula

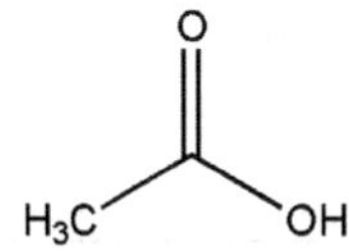

6 Functional Category

Acidulant; buffering agent.

A

7 Applications in Pharmaceutical Formulations or Technology

Glacial and diluted acetic acid solutions are widely used as acidulants in a variety of pharmaceutical formulations and food preparations. Acetic acid is used in pharmaceutical products as a buffer system when combined with an acetate salt such as sodium acetate.

8 Description

Glacial acetic acid occurs as a crystalline mass or a clear, colorless volatile solution with a pungent odor.

9 Pharmacopeial Specifications

See Table I.

Table I: Pharmacopeial specifications for glacial acetic acid.

Test	JP XV	PhEur 7.4	USP35–NF30
Identification	+	+	+
Characters	+	+	–
Freezing point	≥14.5°C	≥14.8°C	≥15.6°C
Nonvolatile matter	≤1.0 mg	≤0.01%	≤1.0 mg
Sulfate	+	+	+
Chloride	+	+	+
Heavy metals	≤10 ppm	≤5 ppm	≤5 ppm
Iron	–	≤5 ppm	–
Readily oxidizable impurities	+	+	+
Assay	≥99.0%	99.5–100.5%	99.5–100.5%

10 Typical Properties

Acidity/alkalinity

pH = 2.4 (1 M aqueous solution);

pH = 2.9 (0.1 M aqueous solution);

pH = 3.4 (0.01 M aqueous solution).

Boiling point 118°C

Dissociation constant $pK_a = 4.76$

Flash point 39°C (closed cup); 57°C (open cup).

Melting point 17°C

Refractive index $n_D^{20} = 1.3718$

Solubility Miscible with ethanol, ether, glycerin, water, and other fixed and volatile oils.

Specific gravity 1.045

Spectroscopy

IR spectra *see* Figure 1.

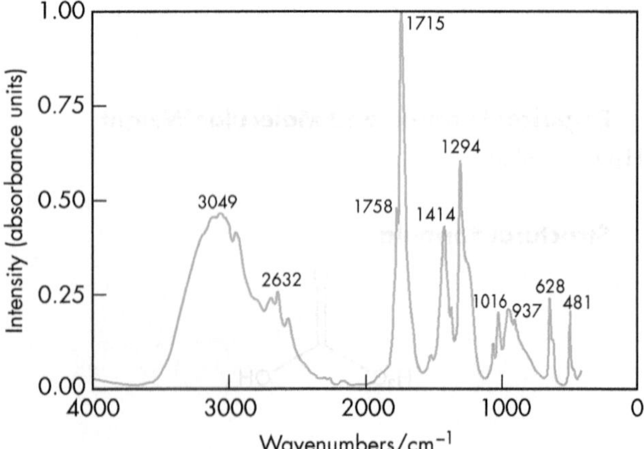

Figure 1: Infrared spectrum of acetic acid measured by transmission. Adapted with permission of Informa Healthcare.

11 Stability and Storage Conditions

Acetic acid should be stored in an airtight container in a cool, dry place.

12 Incompatibilities

Acetic acid reacts with alkaline substances.

13 Method of Manufacture

Acetic acid is usually made by one of three routes: acetaldehyde oxidation, involving direct air or oxygen oxidation of liquid acetaldehyde in the presence of manganese acetate, cobalt acetate, or copper acetate; liquid-phase oxidation of butane or naphtha; methanol carbonylation using a variety of techniques.

14 Safety

Acetic acid is widely used in pharmaceutical applications primarily to adjust the pH of formulations and is thus generally regarded as relatively nontoxic and nonirritant. However, glacial acetic acid or solutions containing over 50% w/w acetic acid in water or organic solvents are considered corrosive and can cause damage to skin, eyes, nose, and mouth. If swallowed, glacial acetic acid causes severe gastric irritation similar to that caused by hydrochloric acid.[1]

Dilute acetic acid solutions containing up to 10% w/w of acetic acid have been used topically following jellyfish stings.[2] Dilute acetic acid solutions containing up to 5% w/w of acetic acid have also been applied topically to treat wounds and burns infected with *Pseudomonas aeruginosa*.[3]

The lowest lethal oral dose of glacial acetic acid in humans is reported to be 1470 μg/kg.[4] The lowest lethal concentration on inhalation in humans is reported to be 816 ppm.[4] Humans, are, however, estimated to consume approximately 1 g/day of acetic acid from the diet.

LD_{50} (mouse, IV): 0.525 g/kg[4]

LD_{50} (rabbit, skin): 1.06 g/kg

LD_{50} (rat, oral): 3.31 g/kg

15 Handling Precautions

Observe normal precautions appropriate to the circumstances and quantity of material handled. Acetic acid, particularly glacial acetic acid, can cause burns on contact with the skin, eyes, and mucous membranes. Splashes should be washed with copious quantities of water. Protective clothing, gloves, and eye protection are recommended.

16 Regulatory Status

GRAS listed. Accepted as a food additive in Europe. Included in the FDA Inactive Ingredients Database (injections, nasal, ophthalmic, and oral preparations). Included in parenteral and nonparenteral preparations licensed in the UK.

17 Related Substances

Acetic acid; artificial vinegar; dilute acetic acid.

Acetic acid

Comments A diluted solution of glacial acetic acid containing 30–37% w/w of acetic acid. *See* Section 18.

Artificial vinegar

Comments A solution containing 4% w/w of acetic acid.

Dilute acetic acid

Comments A weak solution of acetic acid which may contain between 5.7–6.3% w/w of acetic acid. *See* Section 18.

18 Comments

In addition to glacial acetic acid, many pharmacopeias contain monographs for diluted acetic acid solutions of various strengths. For example, the USP33–NF28 S1 has monographs for acetic acid, which is defined as an acetic acid solution containing 36.0–37.0% w/w of acetic acid, and for diluted acetic acid, which is defined as an acetic acid solution containing 5.7–6.3% w/w of acetic acid. Similarly, the BP 2010 contains separate monographs for glacial acetic acid, acetic acid (33%), and acetic acid (6%). Acetic acid (33%) BP 2010 contains 32.5–33.5% w/w of acetic acid. Acetic acid (6%) BP 2010 contains 5.7–6.3% w/w of acetic acid. The JP XV also contains a monograph for acetic acid that specifies that it contains 30.0–32.0% w/w of acetic acid.

Acetic acid is also claimed to have some antibacterial and antifungal properties.

A specification for glacial acetic acid is contained in the *Food Chemicals Codex* (FCC).[5]

The EINECS number for acetic acid is 200-580-7. The PubChem Compound ID (CID) for glacial acetic acid is 176.

19 Specific References

1 Sweetman SC, ed. *Martindale: The Complete Drug Reference*, 36th edn. London: Pharmaceutical Press, 2009; 2244–2245.
2 Fenner PJ, Williamson JA. Worldwide deaths and severe envenomation from jellyfish stings. *Med J Aust* 1996; **165**: 658–661.
3 Milner SM. Acetic acid to treat *Pseudomonas aeruginosa* in superficial wounds and burns. *Lancet* 1992; **340**: 61.
4 Lewis RJ, ed. Sax's Dangerous Properties of Industrial Materials, 11th edn. New York: Wiley, 2004; 15–16.
5 *Food Chemicals Codex*, 7th edn. Bethesda, MD: United States Pharmacopeia, 2010; 12.

20 General References

—

21 Author

WG Chambliss.

22 Date of Revision

1 March 2012.

Acetone

1 Nonproprietary Names

BP: Acetone
PhEur: Acetone
USP–NF: Acetone

2 Synonyms

Acetonum; dimethylformaldehyde; dimethyl ketone; β-ketopropane; pyroacetic ether.

3 Chemical Name and CAS Registry Number

2-Propanone [67-64-1]

4 Empirical Formula and Molecular Weight

C_3H_6O 58.08

5 Structural Formula

H_3C–C(=O)–CH_3

6 Functional Category

Solvent.

7 Applications in Pharmaceutical Formulation or Technology

Acetone is used as a solvent or cosolvent in topical preparations up to a concentration of 13%, and as an aid in wet granulation.[1,2] It has also been used when formulating tablets with water-sensitive active ingredients, or to solvate poorly water-soluble binders in a wet granulation process.

Acetone has also been used in the formulation of microspheres to enhance drug release.[3]

8 Description

Acetone is a colorless volatile, flammable, transparent liquid, with a sweetish odor and pungent sweetish taste.

9 Pharmacopeial Specifications

See Table I. *See also* Section 18.

Table I: Pharmacopeial specifications for acetone.

Test	PhEur 7.4	USP35–NF30
Identification	+	+
Characters	+	–
Appearance of solution	+	–
Acidity or alkalinity	+	–
Relative density	0.790–0.793	≤0.789
Related substances	+	–
Matter insoluble in water	+	–
Reducing substances	+	+
Residue on evaporation	≤50 ppm	≤0.004%
Water	≤3 g/L	+
Assay	–	≥99.0%

10 Typical Properties

Boiling point 56.2°C
Flash point −20°C
Melting point 94.3°C
Refractive index $n_D^{20} = 1.359$
Solubility Soluble in water; freely soluble in ethanol (95%).

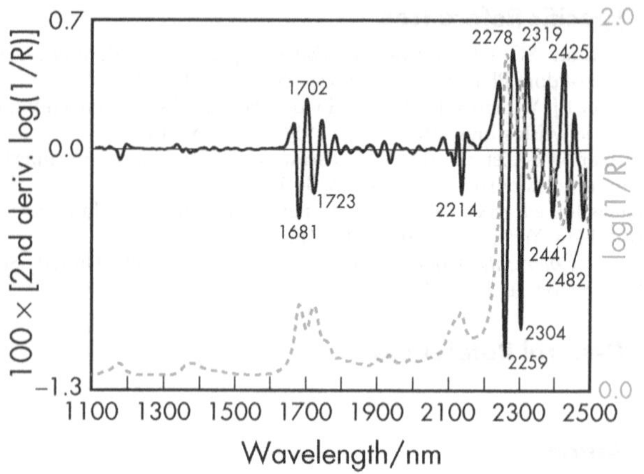

Figure 1: Near-infrared spectrum of acetone measured by transflectance (1 mm path-length).

Spectroscopy
NIR spectra *see* Figure 1.
Vapor pressure 185 mmHg at 20°C

11 Stability and Storage Conditions

Acetone should be stored in a cool, dry, well-ventilated place out of direct sunlight.

12 Incompatibilities

Acetone reacts violently with oxidizing agents, chlorinated solvents, and alkali mixtures. It reacts vigorously with sulfur dichloride, potassium *t*-butoxide, and hexachloromelamine. Acetone should not be used as a solvent for iodine, as it forms a volatile compound that is extremely irritating to the eyes.[4]

13 Method of Manufacture

Acetone is obtained by fermentation as a by-product of *n*-butyl alcohol manufacture, or by chemical synthesis from isopropyl alcohol; from cumene as a by-product in phenol manufacture; or from propane as a by-product of oxidation-cracking.

14 Safety

Acetone is considered moderately toxic, and is a skin irritant and severe eye irritant. It is important to remove it completely from the finished dosage if it was used as a solvent during manufacture. Skin irritation has been reported due to its defatting action, and prolonged inhalation may result in headaches. Inhalation of acetone can produce systemic effects such as conjunctival irritation, respiratory system effects, nausea, and vomiting.[5]

LD_{50} (mouse, oral): 3.0 g/kg[5]
LD_{50} (mouse, IP): 1.297 g/kg
LD_{50} (rabbit, oral): 5.340 g/kg
LD_{50} (rabbit, skin): 0.2 g/kg
LD_{50} (rat, IV): 5.5 g/kg
LD_{50} (rat, oral): 5.8 g/kg

15 Handling Precautions

Observe normal precautions appropriate to the circumstances and quantity of material handled. Acetone is a skin and eye irritant (*see* Section 14); therefore gloves, eye protection, and a respirator are recommended. In the UK, the long-term (8-hour TWA) workplace exposure limit for acetone is 1210 mg/m^3 (500 ppm). The short-term (15-minute) exposure limit is 3620 mg/m^3 (1500 ppm).[6]

16 Regulatory Status

Included in the FDA Inactive Ingredients Database (inhalation solution; oral tablets; topical preparations). Included in the Canadian Natural Health Products Ingredients Database. Included in nonparenteral medicines licensed in the UK.

17 Related Substances

—

18 Comments

The rate of removal of acetone from finished dosage forms may be a function of humidity.[7] Owing to its low boiling point, acetone has been used to extract thermolabile substances from crude drugs.[4]

A specification for acetone is included in the *Japanese Pharmaceutical Excipients* (JPE).[8]

The EINECS number for acetone is 200-662-2. The PubChem Compound ID (CID) for acetone is 180.

19 Specific References

1 Ash M, Ash I. *Handbook of Pharmaceutical Additives*, 3rd edn. Endicott, NY: Synapse Information Resources, 2007; 430.
2 Tang ZG, *et al.* Surface properties and biocompatibility of solvent-cast poly[ε-caprolactone] films. *Biomaterials* 2004; 25(19): 4741–4748.
3 Ruan G, Feng SS. Preparation and characterization of poly(lactic acid)–poly(ethylene glycol)–poly(lactic acid) (PLA-PEG-PLA) microspheres for controlled release of paclitaxel. *Biomaterials* 2003; 24(27): 5037–5044.
4 Todd RG, Wade A, eds. *The Pharmaceutical Codex*, 11th edn. London: Pharmaceutical Press, 1979; 6.
5 Lewis RJ, ed. *Sax's Dangerous Properties of Industrial Materials*, 11th edn. New York: Wiley, 2004; 22–23.
6 Health and Safety Executive. *EH40/2005: Workplace Exposure Limits*. Sudbury: HSE Books, 2011. http://www.hse.gov.uk/pubns/priced/eh40.pdf (accessed 28 February 2012).
7 Wang W, *et al.* A novel method for removing residual acetone from gelatin microspheres. *Pharm Dev Technol* 2002; 7(2): 169–180.
8 Japan Pharmaceutical Excipients Council. *Japanese Pharmaceutical Excipients 2004*. Tokyo: Yakuji Nippo, 2004; 35–36.

20 General References

—

21 Author

AH Kibbe.

22 Date of Revision

1 March 2012.

Acetone Sodium Bisulfite

1 Nonproprietary Names

None adopted.

2 Synonyms

Acetone sulfite; sodium acetone bisulfite.

3 Chemical Name and CAS Registry Number

2-Hydroxy-2-propanesulfonic acid, sodium salt [540-92-1]

4 Empirical Formula and Molecular Weight

$C_3H_7NaO_4S$ 162.14

5 Structural Formula

$$
\begin{array}{c}
CH_3 \\
| \\
H_3C - C - OH \\
| \\
O = S = O \\
| \\
ONa
\end{array}
$$

6 Functional Category

Antimicrobial preservative; antioxidant.

7 Applications in Pharmaceutical Formulation or Technology

Acetone sodium bisulfite is used in injections and solution inhalers. It has been used in nonaqueous solutions of doxorubicin as an antioxidant at concentrations of 0.1–20.0 mg/mL,[1] and may also be used in antifungal parenteral products.[2]

8 Description

Acetone sodium bisulfite occurs as a crystalline white powder with a slight odor of sulfur dioxide. It can feel waxy to the touch.

9 Pharmacopeial Specifications

—

10 Typical Properties

Solubility Freely soluble in water; sparingly soluble in ethanol (95%).

11 Stability and Storage Conditions

Acetone sodium bisulfite should be kept refrigerated at 2–8°C.

12 Incompatibilities

Acetone sodium bisulfite decomposes in the presence of acids.

13 Method of Manufacture

Acetone sodium bisulfite is prepared by the reaction of acetone with sodium bisulfite.

14 Safety

Acetone sodium bisulfite, being a sulfite, may cause allergic-type reactions in certain susceptible people.[3–5] Sulfite sensitivity is seen more frequently in asthmatics.[6]

15 Handling Precautions

Observe normal precautions appropriate to the circumstances and quantity of the material handled. Acetone sodium bisulfite is combustible.

16 Regulatory Status

Included in the FDA Inactive Ingredients Database (injections and inhalation solutions).

17 Related Substances

Sodium sulfite.

18 Comments

The EINECS number for acetone sodium bisulfite is 208-761-2. The PubChem Compound ID (CID) for acetone sodium bisulfite is 23673833.

19 Specific References

1 Kaplan MA, *et al.* Doxorubicin solutions. European Patent 372889; 1990.
2 Stogniew M. Antifungal parenteral products.United States Patent 6,991,800; 2006.
3 Settipane GA. Adverse reactions to sulfites in drugs and foods. *J Am Acad Dermatol* 1984; **10**: 1077–1080.
4 Dalton-Bunnow MF. Review of sulfite sensitivity. *Am J Hosp Pharm* 1985; **42**: 2220–2226.
5 Yang WH, Purchase EC. Adverse reactions to sulfites. *CMAJ* 1985; **133**: 880.
6 Gunnison AF, Jacobsen DW. Sulfite hypersensitivity. A critical review. *CRC Crit Rev Toxicol* 1987; **17**: 185–214.

20 General References

—

21 Authors

ME Fenton, RC Rowe.

22 Date of Revision

1 October 2010.

Acetyltributyl Citrate

1 Nonproprietary Names

BP: Tributyl Acetylcitrate
PhEur: Tributyl Acetylcitrate
USP–NF: Acetyltributyl Citrate

2 Synonyms

Acetylbutyl citrate; acetylcitric acid, tributyl ester; ATBC; *Citrofol AII*; *Citroflex A-4*; tributylis acetylcitras; tributyl *O*-acetylcitrate; tributyl citrate acetate.

3 Chemical Name and CAS Registry Number

1,2,3-Propanetricarboxylic acid, 2-acetyloxy, tributyl ester [77-90-7]

4 Empirical Formula and Molecular Weight

$C_{20}H_{34}O_8$ 402.5

5 Structural Formula

6 Functional Category

Plasticizing agent.

7 Applications in Pharmaceutical Formulation or Technology

Acetyltributyl citrate is used to plasticize polymers in formulated pharmaceutical coatings,[1–6] including capsules, tablets, beads, and granules for taste masking, immediate release, sustained-release and enteric formulations.

8 Description

Acetyltributyl citrate is a clear, odorless, practically colorless, oily liquid.

9 Pharmacopeial Specifications

See Table I.

Table I: Pharmacopeial specifications for acetyltributyl citrate.

Test	PhEur 7.4	USP35–NF30
Identification	+	+
Appearance	+	—
Characters	+	—
Specific gravity	—	1.045–1.055
Refractive index	1.442–1.445	1.4410–1.4425
Sulfated ash	≤0.1%	—
Acidity	+	+
Water	≤0.25%	≤0.25%
Heavy metals	≤10 ppm	≤10 ppm
Related substances	+	—
Assay (anhydrous basis)	99.0–101.0%	≥99.0%

10 Typical Properties

Acid value 0.02
Boiling point 326°C (decomposes)
Flash point 204°C
Pour point −59°C
Solubility Miscible with acetone, ethanol, and vegetable oil; practically insoluble in water.
Spectroscopy
IR spectra *see* Figure 1.
Vapor density 14.1 (air = 1) for *Citroflex A-4*
Viscosity (dynamic) 33 mPa s (33 cP) at 25°C

11 Stability and Storage Conditions

Acetyltributyl citrate should be stored in a well-closed container in a cool, dry location at temperatures not exceeding 38°C. When stored in accordance with these conditions, acetyltributyl citrate is a stable product.

12 Incompatibilities

Acetyltributyl citrate is incompatible with strong alkalis and oxidizing materials.

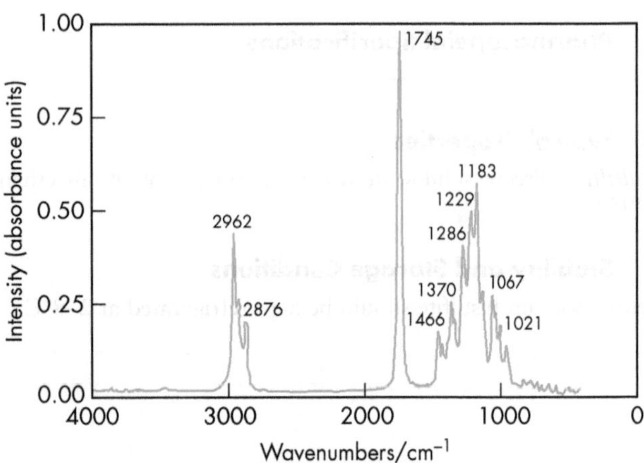

Figure 1: Infrared spectrum of acetyltributyl citrate measured by transmission. Adapted with permission of Informa Healthcare.

13 Method of Manufacture

Acetyltributyl citrate is prepared by the esterification of citric acid with butanol followed by acylation with acetic anhydride.

14 Safety

Acetyltributyl citrate is used in oral pharmaceutical formulations and films intended for direct food contact. It is also used in self-adhesive thin films used for topical delivery systems.(7) It is generally regarded as a relatively nontoxic and nonirritating material. However, ingestion of large quantities may be harmful.

LD_{50} (cat, oral): >50 mL/kg(8)

LD_{50} (mouse, IP): >4 g/kg

LD_{50} (rat, oral): >31.5 g/kg

15 Handling Precautions

Observe normal precautions appropriate to the circumstances and quantity of material handled. Acetyltributyl citrate is slightly irritating to the eyes and may be irritating to the respiratory system as a mist or at elevated temperatures. Gloves and eye protection are recommended for normal handling, and a respirator is recommended when using acetyltributyl citrate at elevated temperatures.

16 Regulatory Status

Included in FDA Inactive Ingredients Database (oral capsules and tablets). Included in the Canadian Natural Health Products Ingredients Database. Included in nonparenteral medicines licensed in the UK. Approved in the USA for direct food contact in food films.

17 Related Substances

Acetyltriethyl citrate; tributyl citrate; triethyl citrate.

18 Comments

Acetyltributyl citrate is used as a plasticizer in food contact films, although it has been known to migrate from food-grade PVC films into high-fat foods such as olive oil.(9)

Polylactide plasticized with acetyltributyl citrate has been investigated as a biodegradable barrier for use in guided-tissue regeneration therapy.(10)

The EINECS number for acetyltributyl citrate is 201-067-0. The PubChem Compound ID (CID) for acetyltributyl citrate is 6505.

19 Specific References

1 Gutierrez-Rocca JC, McGinity JW. Influence of water soluble and insoluble plasticizer on the physical and mechanical properties of acrylic resin copolymers. *Int J Pharm* 1994; **103**: 293–301.

2 Lehmann K. Chemistry and application properties of polymethacrylate coating systems. In: McGinity JW, ed. *Aqueous Polymeric Coatings for Pharmaceutical Dosage Forms*. New York: Marcel Dekker, 1989: 153–245.

3 Steurnagel CR. Latex emulsions for controlled drug delivery. In: McGinity JW, ed. *Aqueous Polymeric Coatings for Pharmaceutical Dosage Forms*. New York: Marcel Dekker, 1989: 1–61.

4 Gutierrez-Rocca JC, McGinity JW. Influence of aging on the physical-mechanical properties of acrylic resin films cast from aqueous dispersions and organic solutions. *Drug Dev Ind Pharm* 1993; **19**(3): 315–332.

5 Repka MA, *et al.* Influence of plasticisers and drugs on the physical-mechanical properties of hydroxypropylcellulose films prepared by hot melt extrusion. *Drug Dev Ind Pharm* 1999; **25**(5): 625–633.

6 Thumma S, *et al.* Influence of plasticizers on the stability and release of a prodrug of Delta(9)-tetrahydrocannabinol incorporated in poly(-ethylene oxide) matrices. *Eur J Pharm Biopharm* 2008; **70**(2): 605–614.

7 Lieb S, *et al.* Self-adhesive thin films for topical delivery of 5-aminolevulinic acid. *Eur J Pharm Biopharm* 2002; **53**(1): 99–106.

8 Lewis RJ, ed. *Sax's Dangerous Properties of Industrial Materials*, 11th edn. New York: Wiley, 2004: 3512.

9 Goulas AE, *et al.* Effect of high-dose electron beam irradiation on the migration of DOA and ATBC plasticizers from food-grade PVC and PVDC/PVC films, respectively, into olive oil. *J Food Prot* 1998; **61**(6): 720–724.

10 Dorfer CE, *et al.* Regenerative periodontal surgery in interproximal intrabony defects with biodegradable barriers. *J Clin Peridontol* 2000; 27(3): 162–168.

20 General References

Vertellus Specialties Inc. Material safety data sheet: *Citroflex A-4*, December 2009.

21 Authors

ME Fenton, PJ Sheskey.

22 Date of Revision

2 March 2012.

Acetyltriethyl Citrate

1 Nonproprietary Names

USP–NF: Acetyltriethyl Citrate

2 Synonyms

ATEC; *Citroflex A-2*; *Citrofol BII*; triethyl acetylcitrate; triethyl O-acetylcitrate; triethyl citrate acetate.

3 Chemical Name and CAS Registry Number

1,2,3-Propanetricarboxylic acid, 2-acetyloxy, triethyl ester [77-89-4]

4 Empirical Formula and Molecular Weight

$C_{14}H_{22}O_8$ 318.3

5 Structural Formula

6 Functional Category

Plasticizing agent.

7 Applications in Pharmaceutical Formulation or Technology

Acetyltriethyl citrate is used to plasticize polymers in formulated pharmaceutical coatings.[1] The coating applications include capsules, tablets, beads and granules for taste masking, immediate release, sustained-release and enteric formulations.[2–5] It is also used in diffusion-controlled release drug delivery systems.[6]

8 Description

Acetyltriethyl citrate occurs as a clear, odorless, practically colorless oily liquid.

9 Pharmacopeial Specifications

See Table I.

Table I: Pharmacopeial specifications for acetyltriethyl citrate.

Test	USP35–NF30
Identification	+
Specific gravity	1.135–1.139
Refractive index	1.432–1.441
Acidity	+
Water	≤0.3%
Heavy metals	≤0.001%
Assay (anhydrous basis)	≥99.0%

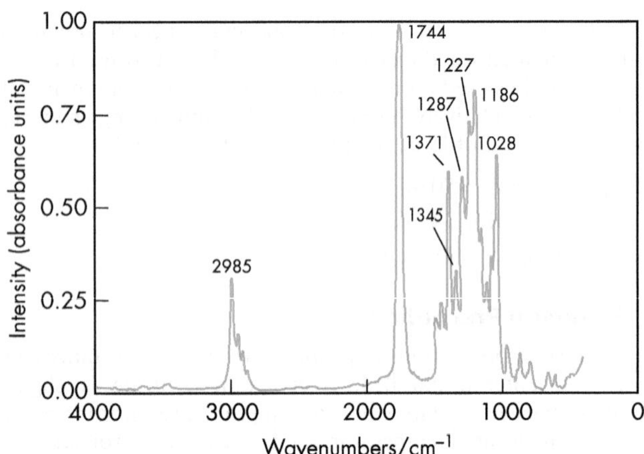

Figure 1: Infrared spectrum of acetyltriethyl citrate measured by transmission. Adapted with permission of Informa Healthcare.

10 Typical Properties

Acid value 0.02
Boiling point 294°C (decomposes)
Flash point 188°C
Pour point −43°C
Solubility Soluble 1 in 140 of water; miscible with acetone, ethanol, and propan-2-ol.
Spectroscopy
IR spectra *see* Figure 1.
Vapor density 14.1 (air = 1) for *Citroflex A-2*
Viscosity (dynamic) 54 mPa s (54 cP) at 25°C.

11 Stability and Storage Conditions

Acetyltriethyl citrate should be stored in dry, closed containers at temperatures not exceeding 38°C. When stored in accordance with these conditions, acetyltriethyl citrate is a stable product.

12 Incompatibilities

Acetyltriethyl citrate is incompatible with strong alkalis and oxidizing materials.

13 Method of Manufacture

Acetyltriethyl citrate is prepared by the esterification of citric acid with ethanol followed by acylation with acetic anhydride.

14 Safety

Acetyltriethyl citrate is used in oral pharmaceutical formulations and is generally regarded as a nontoxic and nonirritating material. However, ingestion of large quantities may be harmful.

LD_{50} (cat, oral): 8.5 g/kg[7]
LD_{50} (mouse, IP): 1.15 g/kg
LD_{50} (rat, oral): 7 g/kg

15 Handling Precautions

Observe normal precautions appropriate to the circumstances and quantity of material handled. Acetyltriethyl citrate may be irritating to the eyes or the respiratory system as a mist or at elevated temperatures. Gloves and eye protection are recommended for normal handling and a respirator is recommended if used at elevated temperatures.

16 Regulatory Status

Approved in the USA for direct food contact in food films.

17 Related Substances

Acetyltributyl citrate; tributyl citrate; triethyl citrate.

18 Comments

The EINECS number for acetyltriethyl citrate is 201-066-5. The PubChem Compound ID (CID) for acetyltriethyl citrate is 6504.

19 Specific References

1 Jensen JL, *et al.* Variables that affect the mechanism of drug release from osmotic pumps coated with acrylate/methacrylate copolymer latexes. *J Pharm Sci* 1995; **84**: 530–533.

2 Gutierrez-Rocca JC, McGinity JW. Influence of water soluble and insoluble plasticizer on the physical and mechanical properties of acrylic resin copolymers. *Int J Pharm* 1994; **103**: 293–301.

3 Lehmann K. Chemistry and application properties of polymethacrylate coating systems. In: McGinity JW, ed. *Aqueous Polymeric Coatings for Pharmaceutical Dosage Forms.* New York: Marcel Dekker, 1989: 153–245.

4 Steurnagel CR. Latex emulsions for controlled drug delivery. In: McGinity JW, ed. *Aqueous Polymeric Coatings for Pharmaceutical Dosage Forms.* New York: Marcel Dekker, 1–61.

5 Gutierrez-Rocca JC, McGinity JW. Influence of aging on the physical-mechanical properties of acrylic resin films cast from aqueous dispersions and organic solutions. *Drug Dev Ind Pharm* 1993; **19**(3): 315–332.

6 Siepmann J, *et al.* Diffusion-controlled drug delivery systems: calculation of the required composition to achieve desired release profiles. *J Control Release* 1999; **60**(2–3): 379–389.

7 Lewis RJ, ed. *Sax's Dangerous Properties of Industrial Materials*, 11th edn. New York: Wiley, 2004: 58–59.

20 General References

Vertellus Specialties Inc. Material safety data sheet: *Citroflex A-2*, July 2010.

21 Authors

ME Fenton, PJ Sheskey.

22 Date of Revision

10 March 2012.

Adipic Acid

1 Nonproprietary Names

PhEur: Adipic Acid

USP–NF: Adipic Acid

2 Synonyms

Acidum adipicum; acifloctin; acinetten; adilactetten; asapic; 1,4-butanedicarboxylic acid; E355; 1,6-hexanedioic acid; *Inipol DS.*

3 Chemical Name and CAS Registry Number

Hexanedioic acid [124-04-9]

4 Empirical Formula and Molecular Weight

$C_6H_{10}O_4$ 146.14

5 Structural Formula

HO, OH, O, O

6 Functional Category

Acidulant; buffering agent; flavoring agent; modified-release agent.

7 Applications in Pharmaceutical Formulation or Technology

Adipic acid is used as an acidifying and buffering agent in intramuscular, intravenous, and vaginal formulations. It is also used in food products as a leavening, pH-controlling, or flavoring agent.

Adipic acid has been incorporated into controlled-release formulation matrix tablets to obtain pH-independent release for both weakly basic[1,2] and weakly acidic drugs.[3,4] It has also been incorporated into the polymeric coating of hydrophilic monolithic systems to modulate the intragel pH, resulting in zero-order release of a hydrophilic drug.[5] The disintegration at intestinal pH of the enteric polymer shellac has been reported to improve when adipic acid was used as a pore-forming agent without affecting release in the acidic media.[6] Other controlled-release formulations have included adipic acid with the intention of obtaining a late-burst release profile.[7]

8 Description

Adipic acid occurs as a white or almost white, odorless nonhygroscopic crystalline powder. The crystal structure of adipic acid is monoclinic holohedral.

9 Pharmacopeial Specifications

See Table I.

Table I: Pharmacopeial specifications for adipic acid.

Test	PhEur 7.4	USP35–NF30
Identification	+	+
Characters	+	–
Melting range	151–154°C	151–154°C
Appearance of solution	+	–
Loss on drying	≤0.2%	≤0.2%
Residue on ignition	–	≤0.1%
Sulfated ash	≤0.1%	–
Chlorides	≤200 ppm	≤0.02%
Nitrates	≤30 ppm	≤30 ppm
Sulfates	≤500 ppm	≤0.05%
Iron	≤10 ppm	≤10 ppm
Heavy metals	≤10 ppm	≤10 ppm
Related substances	+	–
Assay (anhydrous)	99.0–101.0%	99.0–101.0%

10 Typical Properties

Acidity/alkalinity

pH = 2.7 (saturated solution at 25°C);

pH = 3.2 (0.1% w/v aqueous solution at 25°C)

Boiling point 337.5°C

Density 1.360 g/cm^3

Dissociation constant

pK_{a1}: 4.418 at 25°C;

pK_{a2}: 5.412 at 25°C.

Flash point 196°C (closed cup)

Heat of combustion 17 653.9 kJ/mol (4219.28 kcal/mol) at 25°C

Heat of solution 33.193 kJ/mol (7.9 kcal/mol) at 25°C

Melting point 152.1°C

Solubility *see* Table II.

Table II: Solubility of adipic acid.

Solvent	Solubility at 20°C unless otherwise stated
Acetone	Soluble
Benzene	Practically insoluble
Cyclohexane	Slightly soluble
Ethanol (95%)	Freely soluble
Ether	1 in 157.8 at 19°C
Ethyl acetate	Soluble
Methanol	Freely soluble
Petroleum ether	Practically insoluble
Water	1 in 71.4
	1 in 0.6 at 100°C

Spectroscopy

IR spectra *see* Figure 1.

Vapor pressure 133.3 Pa (1 mmHg) at 159.5°C

Viscosity (dynamic) 4.54 mPa s (4.54 cP) at 160°C for molten adipic acid.

11 Stability and Storage Conditions

Adipic acid is normally stable but decomposes above boiling point. It should be stored in a tightly closed container in a cool, dry place, and should be kept away from heat, sparks, and open flame.

12 Incompatibilities

Adipic acid is incompatible with strong oxidizing agents as well as strong bases and reducing agents. Contact with alcohols, glycols, aldehydes, epoxides, or other polymerizing compounds can result in violent reactions.

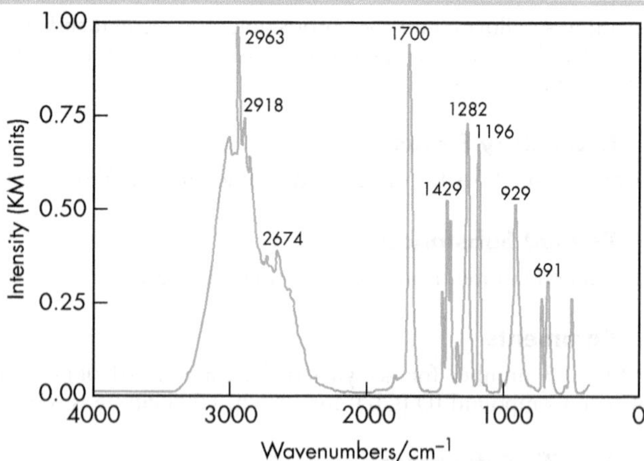

Figure 1: Infrared spectrum of adipic acid measured by diffuse reflectance. Adapted with permission of Informa Healthcare.

13 Method of Manufacture

Adipic acid is prepared by nitric acid oxidation of cyclohexanol or cyclohexanone or a mixture of the two compounds. Recently, oxidation of cyclohexene with 30% aqueous hydrogen peroxide under organic solvent- and halide-free conditions has been proposed as an environmentally friendly alternative for obtaining colorless crystalline adipic acid.(8)

14 Safety

Adipic acid is used in pharmaceutical formulations and food products. The pure form of adipic acid is toxic by the IP route, and moderately toxic by other routes. It is a severe eye irritant, and may cause occupational asthma.

LD_{50} (mouse, IP): 0.28 g/kg(9)

LD_{50} (mouse, IV): 0.68 g/kg

LD_{50} (mouse, oral): 1.9 g/kg

LD_{50} (rat, IP): 0.28 g/kg

LD_{50} (rat, oral): >11 g/kg

15 Handling Precautions

Observe normal precautions appropriate to the circumstances and quantity of the material handled.

Adipic acid is combustible and can react with oxidizing materials when exposed to heat and flame. It emits acrid smoke and fumes when heated to decomposition. Dust explosion is possible if in powder or granular form, mixed with air.

Adipic acid irritates the eyes and respiratory tract. Protective equipment such as respirators, safety goggles, and heavy rubber gloves should be worn when handling adipic acid.

16 Regulatory Status

GRAS listed. Included in the FDA Inactive Ingredients Database (IM, IV, and vaginal preparations). Accepted for use as a food additive in Europe. Included in an oral pastille formulation available in the UK. Included in the Canadian Natural Health Products Ingredients Database.

17 Related Substances

—

18 Comments

A specification for adipic acid is contained in the *Food Chemicals Codex* (FCC).(10)

The EINECS number for adipic acid is 204-673-3. The PubChem Compound ID (CID) for adipic acid is 196.

19 Specific References

1 Guthmann C, *et al.* Development of a multiple unit pellet formulation for a weakly basic drug. *Drug Dev Ind Pharm* 2007; **33**(3): 341–349.
2 Streubel A, *et al.* pH-independent release of a weakly basic drug from water-insoluble and -soluble matrix tablets. *J Control Release* 2000; **67**(1): 101–110.
3 Pillay V, Fassihi R. In situ electrolyte interactions in a disk-compressed configuration system for up-curving and constant drug delivery. *J Control Release* 2000; **67**(1): 55–65.
4 Merkli A, *et al.* The use of acidic and basic excipients in the release of 5-fluorouracil and mitomycin C from a semi-solid bioerodible poly(ortho ester). *J Control Release* 1995; **33**(3): 415–421.
5 Pillay V, Fassihi R. Electrolyte-induced compositional heterogeneity: a novel approach for rate-controlled oral drug delivery. *J Pharm Sci* 1999; **88**(11): 1140–1148.
6 Pearnchob N, *et al.* Improvement in the disintegration of shellac-coated soft gelatin capsules in simulated intestinal fluid. *J Control Release* 2004; **94**(2–3): 313–321.
7 Freichel OL, Lippold BC. A new oral erosion controlled drug delivery system with a late burst in the release profile. *Eur J Pharm Biopharm* 2000; **50**(3): 345–351.
8 Sato K, *et al.* A 'green' route to adipic acid: direct oxidation of cyclohexenes with 30 percent hydrogen peroxide. *Science* 1998; **281**: 1646–1647.
9 Lewis RJ, ed. *Sax's Dangerous Properties of Industrial Materials*, 11th edn. New York: Wiley, 2004; 83–84.
10 *Food Chemicals Codex*, 7th edn. Bethesda, MD: United States Pharmacopeia, 2010; 26.

20 General References

Grant DJW, York P. A disruption index for quantifying the solid state disorder induced by additives or impurities. II. Evaluation from heat of solution. *Int J Pharm* 1986; **28**(2–3): 103–112.

21 Author

D Law.

22 Date of Revision

1 March 2012.

Agar

1 Nonproprietary Names

JP: Agar
PhEur: Agar
USP–NF: Agar

2 Synonyms

Agar-agar; agar-agar flake; agar-agar gum; Bengal gelatin; Bengal gum; Bengal isinglass; Ceylon isinglass; Chinese isinglass; E406; gelosa; gelose; Japan agar; Japan isinglass; layor carang.

3 Chemical Name and CAS Registry Number

Agar [9002-18-0]

4 Empirical Formula and Molecular Weight

See Section 5.

5 Structural Formula

Agar is a dried, hydrophilic, colloidal polysaccharide complex extracted from the agarocytes of algae of the Rhodophyceae. The structure is believed to be a complex range of polysaccharide chains having alternating α-(1$\rightarrow$3) and β-(1$\rightarrow$4) linkages. There are three extremes of structure noted: namely neutral agarose; pyruvated agarose having little sulfation; and a sulfated galactan. Agar can be separated into a natural gelling fraction, agarose, and a sulfated nongelling fraction, agaropectin.

6 Functional Category

Emulsifying agent; gelling agent; modified-release agent; suppository base; suspending agent; tablet and capsule binder; viscosity-increasing agent.

7 Applications in Pharmaceutical Formulation or Technology

Agar is widely used in food applications as a stabilizing agent. In pharmaceutical applications, agar is used in a handful of oral tablet and topical formulations. It has also been investigated in a number of experimental pharmaceutical applications including as a sustained-release agent in gels, beads, microspheres, and tablets.(1–4) It has been reported to work as a disintegrant in tablets.(5) Agar has been used in a floating controlled-release tablet; the buoyancy in part being attributed to air entrapped in the agar gel network.(6) It can be used as a viscosity-increasing agent in aqueous systems. Agar can also be used as a base for nonmelting, and nondisintegrating suppositories.(7) Agar has an application as a suspending agent in pharmaceutical suspensions.(8)

8 Description

Agar occurs as transparent, odorless, tasteless strips or as a coarse or fine powder. It may be weak yellowish-orange, yellowish-gray to pale-yellow colored, or colorless. Agar is tough when damp, brittle when dry.

9 Pharmacopeial Specifications

See Table I.

A

Table I: Pharmacopeial specifications for agar.

Test	JP XV	PhEur 7.4	USP35–NF30
Identification	+	+	+
Characters	+	+	−
Swelling index	−	+	−
Arsenic	−	−	≤3 ppm
Lead	−	−	≤10 ppm
Sulfuric acid	+	−	−
Sulfurous acid and starch	+	−	−
Gelatin	−	+	+
Heavy metals	−	−	≤40 ppm
Insoluble matter	≤15.0 mg	≤1.0%	≤15.0 mg
Water absorption	≤75 mL	−	≤75 mL
Loss on drying	≤22.0%	≤20.0%	≤20.0%
Microbial contamination	−	$\leq 10^3$ cfu/g[a]	+
Total ash	≤4.5%	≤5.0%	≤6.5%
Acid-insoluble ash	≤0.5%	−	≤0.5%
Foreign organic matter	−	−	≤1.0%
Limit of foreign starch	−	−	+

(a) Total viable aerobic count, determined by plate-count.

10 Typical Properties

Solubility Soluble in boiling water to form a viscous solution; practically insoluble in ethanol (95%), and cold water. A 1% w/v aqueous solution forms a stiff jelly on cooling.

Spectroscopy

IR spectra *see* Figure 1.

NIR spectra *see* Figure 2.

11 Stability and Storage Conditions

Agar solutions are most stable at pH 4–10.

Agar should be stored in a cool, dry, place. Containers of this material may be hazardous when empty since they retain product residues (dust, solids).

12 Incompatibilities

Agar is incompatible with strong oxidizing agents. Agar is dehydrated and precipitated from solution by ethanol (95%). Tannic acid causes precipitation; electrolytes cause partial dehydration and decrease in viscosity of sols.[9]

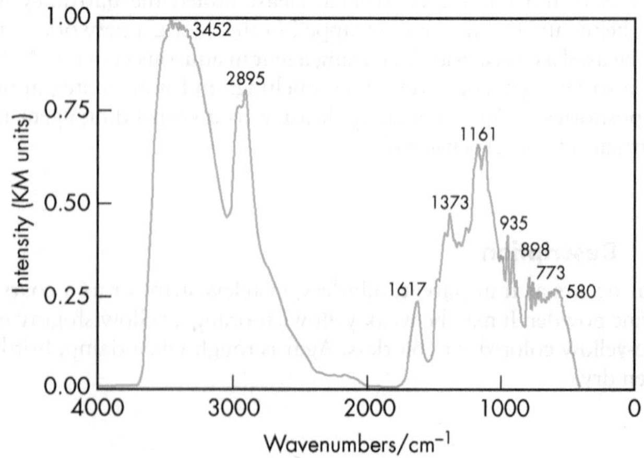

Figure 1: Infrared spectrum of agar measured by diffuse reflectance. Adapted with permission of Informa Healthcare.

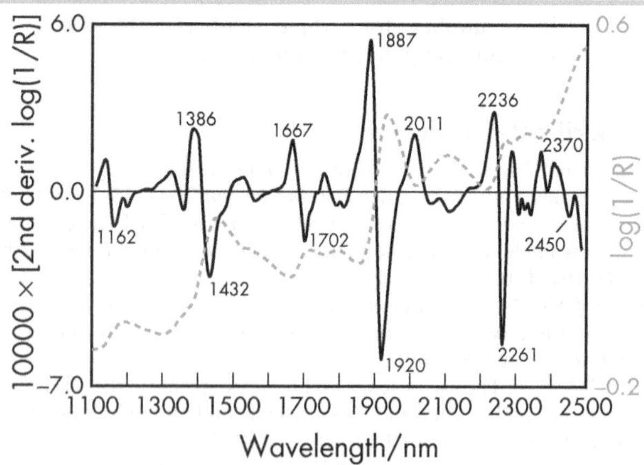

Figure 2: Near-infrared spectrum of agar measured by reflectance.

13 Method of Manufacture

Agar is obtained by freeze-drying a mucilage derived from *Gelidium amansii* Lamouroux, other species of the same family (Gelidiaceae), or other red algae (Rhodophyta).

14 Safety

Agar is widely used in food applications and has been used in oral and topical pharmaceutical applications. It is generally regarded as relatively nontoxic and nonirritant when used as an excipient.

LD_{50} (hamster, oral): 6.1 g/kg[10]

LD_{50} (mouse, oral): 16.0 g/kg

LD_{50} (rabbit, oral): 5.8 g/kg

LD_{50} (rat, oral): 11.0 g/kg

15 Handling Precautions

Observe normal precautions appropriate to the circumstances and quantity of the material handled. When heated to decomposition, agar emits acrid smoke and fumes.

16 Regulatory Status

GRAS listed. Accepted for use as a food additive in Europe. Included in the FDA Inactive Ingredients Database (oral tablets). Included in the Canadian Natural Health Products Ingredients Database. Included in nonparenteral medicines licensed in the UK.

17 Related Substances

—

18 Comments

The drug release mechanism of agar spherules of felodipine has been studied and found to follow Higuchi kinetics.[11] Agar has also been used to test the bioadhesion potential of various polymers.[12]

The EINECS number for agar is 232-658-1.

19 Specific References

1 Bhardwaj TJ, *et al.* Natural gums and modified natural gums as sustained release carriers. *Drug Dev Ind Pharm* 2000; **26**(10): 1025–1038.

2 Sakr FM, *et al.* Design and evaluation of a dry solidification technique for preparing pharmaceutical beads. *STP Pharma Sci* 1995; **5**(4): 291–295.

3 Boraie NA, Naggar VF. Sustained release of theophylline and aminophylline from agar tablets. *Acta Pharm Jugosl* 1984; **34**(Oct–Dec): 247–256.

4 Nakano M, *et al.* Sustained release of sulfamethizole from agar beads. *J Pharm Pharmacol* 1979; **31**: 869–872.

5 Fassihi AR. Characteristics of hydrogel as disintegrant in solid dose technology. *J Pharm Pharmacol* 1989; **41**(12): 853–855.
6 Desai S, Boston S. A floating controlled-release drug delivery system: *in vitro–in vivo* evaluation. *Pharm Res* 1993; **10**: 1321–1325.
7 Singh KK, *et al.* Studies on suppository bases: design and evaluation of sodium CMC and agar bases. *Indian Drugs* 1994; **31**(April): 149–154.
8 Kahela P, *et al.* Effect of suspending agents on the bioavailability of erythromycin ethylsuccinate mixtures. *Drug Dev Ind Pharm* 1978; 4(3): 261–274.
9 Gennaro AR, ed. *Remington: The Science and Practice of Pharmacy*, 20th edn. Baltimore: Lippincott Williams & Wilkins, 2000; 1030.
10 Lewis RJ, ed. *Sax's Dangerous Properties of Industrial Materials*, 11th edn. New York: Wiley, 2004; 90–91.
11 Patil AK, *et al.* Preparation and evaluation of agar spherules of felodipine. *J Pure Appl Microbiol* 2007; **1**(2): 317–322.
12 Bertram U, Bodmeier R. *In situ* gelling, bioadhesive nasal inserts for extended drug delivery: *in vitro* characterization of a new nasal dosage form. *Eur J Pharm* 2006; **27**(1): 62–71.

20 General References

—

21 Author

VK Gupta.

22 Date of Revision

1 March 2012.

Albumin

1 Nonproprietary Names

BP: Albumin Solution

PhEur: Human Albumin Solution

USP–NF: Albumin Human

2 Synonyms

Alba; *Albuconn*; *Albuminar*; albumin human solution; albumini humani solutio; *Albumisol*; *Albuspan*; *Albutein*; *Buminate*; human serum albumin; normal human serum albumin; *Octalbin*; *Plasbumin*; plasma albumin; *Pro-Bumin*; *Proserum*; *Zenalb*.

3 Chemical Name and CAS Registry Number

Serum albumin [9048-49-1]

4 Empirical Formula and Molecular Weight

Human serum albumin has a molecular weight of about 66 500 and is a single polypeptide chain consisting of 585 amino acids. Characteristic features are a single tryptophan residue, a relatively low content of methionine (6 residues), and a large number of cysteine (17) and of charged amino acid residues of aspartic acid (36), glutamic acid (61), lysine (59), and arginine (23).

5 Structural Formula

Primary structure Human albumin is a single polypeptide chain of 585 amino acids and contains seven disulfide bridges.

Secondary structure Human albumin is known to have a secondary structure that is about 55% α-helix. The remaining 45% is believed to be divided among turns, disordered, and β structures.[1]

Albumin is the only major plasma protein that does not contain carbohydrate constituents. Assays of crystalline albumin show less than one sugar residue per molecule.

6 Functional Category

Adsorbent; cryoprotectant; dispersing agent.

7 Applications in Pharmaceutical Formulation or Technology

Albumin is primarily used as an excipient in parenteral pharmaceutical formulations, where it is used as a stabilizing agent for formulations containing proteins and enzymes.[2,3] Albumin has also been used to prepare microspheres and microcapsules for experimental drug-delivery systems.[4–9]

As a stabilizing agent, albumin has been employed in protein formulations at concentrations as low as 0.003%, although concentrations of 1–5% are commonly used. Albumin has also been used as a cosolvent[10] for parenteral drugs, as a cryoprotectant during lyophilization,[11,12] and to prevent adsorption of other proteins to surfaces.

8 Description

The USP 33 S1 describes albumin human as a sterile nonpyrogenic preparation of serum albumin obtained from healthy human donors; *see* Section 13. It is available as a solution containing 4, 5, 20, or 25 g of serum albumin in 100 mL of solution, with not less than 96% of the total protein content as albumin. The solution contains no added antimicrobial preservative but may contain sodium acetyltryptophanate with or without sodium caprylate as a stablizing agent.

The PhEur 6.6 similarly describes albumin solution as an aqueous solution of protein obtained from human plasma; *see* Section 13. It is available as a concentrated solution containing 150–250 g/L of total protein or as an isotonic solution containing 35–50 g/L of total protein. Not less than 95% of the total protein content is albumin. A suitable stabilizer against the effects of heat, such as sodium caprylate (sodium octanoate) or *N*-acetyltryptophan or a combination of these two at a suitable concentration, may be added, but no antimicrobial preservative is added.

Aqueous albumin solutions are slightly viscous and range in color from almost colorless to amber depending upon the protein concentration. In the solid state, albumin appears as brownish amorphous lumps, scales, or powder.

9 Pharmacopeial Specifications

See Table I.

Table I: Pharmacopeial specifications for albumin.

Test	PhEur 7.4	USP35–NF30
Identification	+	–
Characters	+	–
Production	+	–
Protein composition	+	–
Molecular size distribution	+	–
Heat stability	–	+
pH (10 g/L solution)	6.7–7.3	+
Potassium	≤0.05 mmol/g	–
Sodium	≤160 mmol/L	130–160 mEq/L
Heme	+	+
Aluminum	≤200 μg/L	–
Sterility	+	+
Hepatitis B surface antigen	–	+
Pyrogens	+	+
Total protein	95–105%	≥96%
for 4 g in 100 mL	–	93.75–106.25%
for 5 to 25 g in 100 mL	–	94.0–106.0%
Prekallikrein activator	≤35 IU/mL	–

10 Typical Properties

Acidity/alkalinity pH = 6.7–7.3 for a 1% w/v solution, in 0.9% w/v sodium chloride solution, at 20°C.

Osmolarity A 4–5% w/v aqueous solution is isoosmotic with serum.

Solubility Freely soluble in dilute salt solutions and water. Aqueous solutions containing 40% w/v albumin can be readily prepared at pH 7.4. The high net charge of the peptide contributes to its solubility in aqueous media. The seven disulfide bridges contribute to its chemical and spatial conformation. At physiological pH, albumin has a net electrostatic charge of about −17. Aqueous albumin solutions are slightly viscous and range in color from almost colorless to amber depending on the protein concentration.

Spectroscopy

NIR spectra *see* Figure 1.

11 Stability and Storage Conditions

Albumin is a protein and is therefore susceptible to chemical degradation and denaturation by exposure to extremes of pH, high salt concentrations, heat, enzymes, organic solvents, and other chemical agents.

Albumin solutions should be protected from light and stored at a temperature of 2–25°C or as indicated on the label.

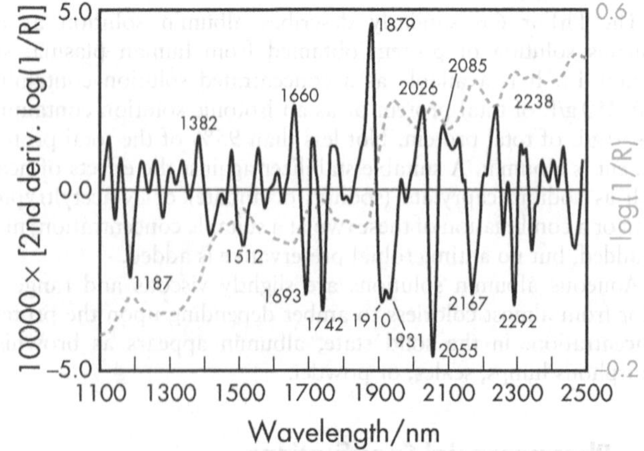

Figure 1: Near-infrared spectrum of albumin measured by reflectance.

12 Incompatibilities

See Section 11.

13 Method of Manufacture

Albumin human (USP35–NF30) Albumin human is a sterile nonpyrogenic preparation of serum albumin that is obtained by fractionating material (source blood, plasma, serum, or placentas) from healthy human donors. The source material is tested for the absence of hepatitis B surface antigen. It is made by a process that yields a product safe for intravenous use.

Human albumin solution (PhEur 7.4) Human albumin solution is an aqueous solution of protein obtained from plasma. Separation of the albumin is carried out under controlled conditions so that the final product contains not less than 95% albumin. Human albumin solution is prepared as a concentrated solution containing 150–250 g/L of total protein or as an isotonic solution containing 35–50 g/L of total protein. A suitable stabilizer against the effects of heat such as sodium caprylate (sodium octanoate) or *N*-acetyltryptophan or a combination of these two at a suitable concentration, may be added, but no antimicrobial preservative is added at any stage during preparation. The solution is passed through a bacteria-retentive filter and distributed aseptically into sterile containers, which are then closed so as to prevent contamination. The solution in its final container is heated to 60 ± 1.0°C and maintained at this temperature for not less than 10 hours. The containers are then incubated at 30–32°C for not less than 14 days or at 20–25°C for not less than 4 weeks and examined visually for evidence of microbial contamination.

14 Safety

Albumin occurs naturally in the body, comprising about 60% of all the plasma proteins. As an excipient, albumin is used primarily in parenteral formulations and is generally regarded as an essentially nontoxic and nonirritant material. Adverse reactions to albumin infusion rarely occur but include nausea, vomiting, increased salivation, chills, and febrile reactions. Urticaria and skin rash have been reported. Allergic reactions, including anaphylactic shock, can occur. Albumin infusions are contraindicated in patients with severe anemia or cardiac failure. Albumin solutions with aluminum content of less than 200 μg/L should be used in dialysis patients and premature infants.[13]

LD_{50} (monkey, IV): >12.5 g/kg[14]

LD_{50} (rat, IV): >12.5 g/kg

15 Handling Precautions

Observe handling precautions appropriate for a biologically derived blood product.

16 Regulatory Status

Included in the FDA Inactive Ingredients Database (IV injections, IV infusions and subcutaneous injectables). Included in parenteral products licensed in the UK. Included in the Canadian Natural Health Products Ingredients Database.

17 Related Substances

Albumins derived from animal sources are also commercially available, e.g. bovine serum albumin.

Recombinant albumin, human.

Recombinant albumin, human

Synonyms rAlbumin human; rHA

Description It is presented as a sterile and nonpyrogenic aqueous liquid consisting of a 10% or 20% solution in water for injection. It contains no added antimicrobial agents, but it may contain appropriate stabilizing agents. Clear, slightly viscous, and colorless to yellow-amber in color.

Stability and storage conditions Store in airtight glass containers at a temperature of 2–8°C. Do not allow to freeze.

Comments Listed in USP33–NF28 S1. Produced by recombinant DNA expression in *Saccharomyces cerevisiae*. Structural equivalence (primary, secondary and tertiary) between rHA and human serum albumin (HSA) has been demonstrated.

18 Comments

A 100 mL aqueous solution of albumin containing 25 g of serum albumin is osmotically equivalent to 500 mL of normal human plasma.

Therapeutically, albumin solutions have been used parenterally for plasma volume replacement and to treat severe acute albumin loss. However, the benefits of using albumin in such applications in critically ill patients has been questioned.[13]

The EINECS number for albumin is 310-127-6.

19 Specific References

1 Bramanti E, Benedetti E. Determination of the secondary structure of isomeric forms of human serum albumin by a particular frequency deconvolution procedure applied to Fourier transform IR analysis. *Biopolymers* 1996; **38**(5): 639–653.

2 Wang JUC, Hanson MA. Parenteral formulations of proteins and peptides: stability and stabilizers. *J Parenter Sci Technol* 1988; **42**(S): S1–S26.

3 Hawe A, Friess W. Stabilization of a hydrophobic recombinant cytokine by human serum albumin. *J Pharm Sci* 2007; **96**(11): 2987–2999.

4 Arshady R. Albumin microspheres and microcapsules: methodology of manufacturing techniques. *J Control Release* 1990; **14**: 111–131.

5 Callewaert M, *et al.* Albumin-alginate-coated microsopheres: resistance to steam sterilization and to lyophilization. *Int J Pharm* 2007; **344**(1–2): 161–164.

6 Dreis S, *et al.* Preparation, characterisation and maintenance of drug efficacy of doxorubicin-loaded human serum albumin (HSA) nanoparticles. *Int J Pharm* 2007; **341**(1–2): 207–214.

7 Mathew ST, *et al.* Formulation and evaluation of ketorolac tromethamine-loaded albumin microspheres for potential intramuscular administration. *AAPS Pharm Sci Tech* 2007; **8**(1): 14.

8 Zensi A, *et al.* Albumin nanoparticles targeted with Apo E enter the CNS by transcytosis and are delivered to neurones. *J Control Release* 2009; **137**: 78–86.

9 Hawkins MJ, *et al.* Protein nanoparticles as drug carriers in clinical medicine. *Adv Drug Deliv Rev* 2008; **60**: 876–885.

10 Olson WP, Faith MR. Human serum albumin as a cosolvent for parenteral drugs. *J Parenter Sci Technol* 1988; **42**: 82–85.

11 Hawe A, Friess W. Physicochemical characterization of the freezing behavior of mannitol-human serum albumin formulations. *AAPS Pharm Sci Tech* 2006; 7(4): 94.

12 Hawe A, Friess W. Physico-chemical lyophilization behaviour of mannitol, human serum albumin formulations. *Eur J Pharm Sci* 2006; **28**(3): 224–232.

13 Quagliaro DA, *et al.* Aluminum in albumin for injection. *J Parenter Sci Technol* 1988; **42**: 187–190.

14 Lewis RJ, ed. *Sax's Dangerous Properties of Industrial Materials*, 11th edn. New York: Wiley, 2004; 1970.

15 Cochrane Injuries Group Albumin Reviewers. Human albumin administration in critically ill patients: systematic review of randomised controlled trials. *Br Med J* 1998; **317**: 235–240.

20 General References

Chaubal MV. Human serum albumin as a pharmaceutical excipient. *Drug Deliv Technol* 2005; **5**: 22–23.

Kratz F. Albumin as a drug carrier: design of prodrugs, drug conjugates and nanoparticles. *J Control Release* 2008; **132**: 171–183.

Kragh-Hansen U. Structure and ligand properties of human serum albumin. *Danish Med Bull* 1990; **37**(1): 57–84.

Putnam FW, ed. *The Plasma Proteins, Structure, Function and Genetic Control.* London: Academic Press, 1975.

21 Author

RT Guest.

22 Date of Revision

1 March 2012.

Alcohol

1 Nonproprietary Names

BP: Ethanol (96 per cent)
JP: Ethanol
PhEur: Ethanol (96 per cent)
USP: Alcohol

2 Synonyms

Ethanolum (96 per centum); ethyl alcohol; ethyl hydroxide; grain alcohol; methyl carbinol.

3 Chemical Name and CAS Registry Number

Ethanol [64-17-5]

4 Empirical Formula and Molecular Weight

C_2H_6O 46.07

5 Structural Formula

6 Functional Category

Antimicrobial preservative; penetration enhancer; solvent.

7 Applications in Pharmaceutical Formulation or Technology

Ethanol and aqueous ethanol solutions of various concentrations (*see* Sections 8 and 17) are widely used in pharmaceutical formulations and cosmetics; *see* Table I. Although ethanol is primarily used as a solvent, it is also employed as a disinfectant, and in solutions as an antimicrobial preservative.[1,2] Topical ethanol solutions are used in the development of transdermal drug delivery systems as penetration enhancers.[3–11] Ethanol has also been used in the development of transdermal preparations as a co-surfactant.[12–14]

Table I: Uses of alcohol.

Use	Concentration (% v/v)
Antimicrobial preservative	≥10
Disinfectant	60–90
Extracting solvent in galenical manufacture	Up to 85
Solvent in film coating	Variable
Solvent in injectable solutions	Variable
Solvent in oral liquids	Variable
Solvent in topical products	60–90

8 Description

In the BP 2012, the term 'ethanol' used without other qualification refers to ethanol containing ≥99.5% v/v of C_2H_6O. The term 'alcohol', without other qualification, refers to ethanol 95.1–96.9% v/v. Where other strengths are intended, the term 'alcohol' or 'ethanol' is used, followed by the statement of the strength.

In the PhEur 7.4, anhydrous ethanol contains not less than 99.5% v/v of C_2H_6O at 20°C. The term ethanol (96%) is used to describe the material containing water and 95.1–96.9% v/v of C_2H_6O at 20°C.

In the USP35–NF30, the term 'dehydrated alcohol' refers to ethanol ≥99.5% v/v. The term 'alcohol' without other qualification refers to ethanol 94.9–96.0% v/v.

In the JP XV, ethanol (alcohol) contains 95.1–96.9% v/v (by specific gravity) of C_2H_6O at 15°C.

In the *Handbook of Pharmaceutical Excipients,* the term'alcohol' is used for either ethanol 95% v/v or ethanol 96% v/v.

Alcohol is a clear, colorless, mobile, and volatile liquid with a slight, characteristic odor and burning taste.

See also Section 17.

9 Pharmacopeial Specifications

The pharmacopeial specifications for alcohol have undergone harmonization of many attributes for JP, PhEur, and USP–NF.

See Table II. *See also* Sections 17 and 18.

Table II: Pharmacopeial specifications for alcohol.

Test	JP XV	PhEur 7.4	USP35–NF30
Identification	+	+	+
Characters	—	+	—
Specific gravity	0.809–0.816	0.805–0.812	0.812–0.816
Acidity or alkalinity	+	+	+
Clarity and color of solution[a]	+	+	+
Nonvolatile residue	≤2.5 mg	≤25 ppm	≤2.5 mg
Volatile impurities	+	+	+
Absorbance	+	+	+
at 240 nm	≤0.40	≤0.40	≤0.40
at 250–260 nm	≤0.30	≤0.30	≤0.30
at 270–340 nm	≤0.10	≤0.10	≤0.10
Assay	95.1–96.9%	95.1–96.9%	92.3–93.8% by weight 94.9–96.0% by volume

(a) These tests have not been fully harmonized at the time of publication.

10 Typical Properties

Antimicrobial activity Ethanol is bactericidal in aqueous mixtures at concentrations between 60% and 95% v/v; the optimum concentration is generally considered to be 70% v/v. Antimicrobial activity is enhanced in the presence of edetic acid or edetate salts.[1] Ethanol is inactivated in the presence of nonionic surfactants and is ineffective against bacterial spores.

Boiling point 78.15°C

Flammability Readily flammable, burning with a blue, smokeless flame.

Flash point 14°C (closed cup)

Solubility Miscible with chloroform, ether, glycerin, and water (with rise of temperature and contraction of volume).

Specific gravity 0.8119–0.8139 at 20°C

Spectroscopy

IR spectra *see* Figure 1.

NIR spectra *see* Figures 2 and 3.

Note The above typical properties are for alcohol (ethanol 95% or 96% v/v). *See* Section 17 for typical properties of dehydrated alcohol.

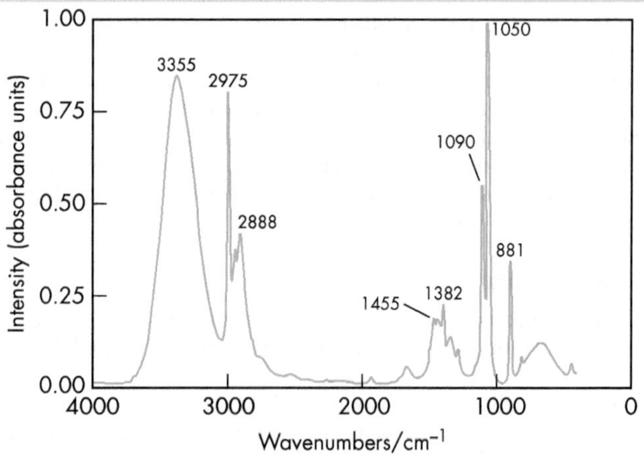

Figure 1: Infrared spectrum of alcohol measured by transmission. Adapted with permission of Informa Healthcare.

11 Stability and Storage Conditions

Aqueous ethanol solutions may be sterilized by autoclaving or by filtration and should be stored in airtight containers in a cool place.

12 Incompatibilities

In acidic conditions, ethanol solutions may react vigorously with oxidizing materials. Mixtures with alkali may darken in color owing to a reaction with residual amounts of aldehyde. Organic

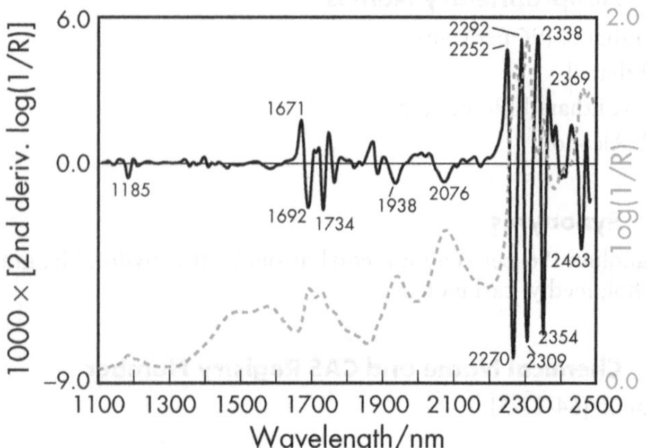

Figure 2: Near-infrared spectrum of alcohol (96%) measured by transflectance (1 mm path-length).

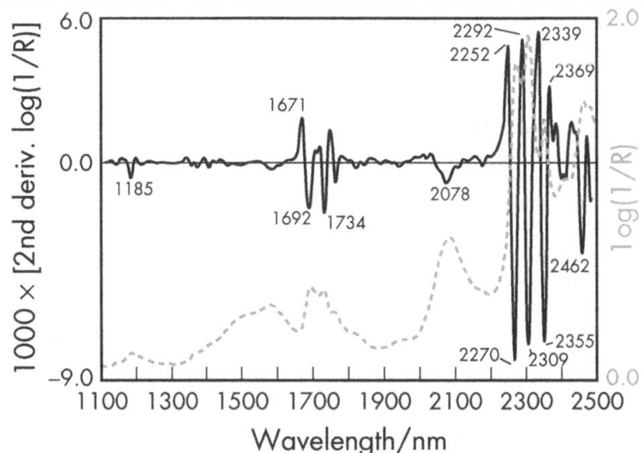

Figure 3: Near-infrared spectrum of alcohol (absolute) measured by transflectance (1 mm path-length).

salts or acacia may be precipitated from aqueous solutions or dispersions. Ethanol solutions are also incompatible with aluminum containers and may interact with some drugs.

13 Method of Manufacture

Ethanol is manufactured by the controlled enzymatic fermentation of starch, sugar, or other carbohydrates. A fermented liquid is produced containing about 15% ethanol; ethanol 95% v/v is then obtained by fractional distillation. Ethanol may also be prepared by a number of synthetic methods.

14 Safety

Ethanol and aqueous ethanol solutions are widely used in a variety of pharmaceutical formulations and cosmetics. It is also consumed in alcoholic beverages.

Ethanol is rapidly absorbed from the gastrointestinal tract and the vapor may be absorbed through the lungs; it is metabolized, mainly in the liver, to acetaldehyde, which is further oxidized to acetate.

Ethanol is a central nervous system depressant and ingestion of low to moderate quantities can lead to symptoms of intoxication including muscle incoordination, visual impairment, slurred speech, etc. Ingestion of higher concentrations may cause depression of medullary action, lethargy, amnesia, hypothermia, hypoglycemia, stupor, coma, respiratory depression, and cardiovascular collapse. The lethal human blood-alcohol concentration is generally estimated to be 400–500 mg/100 mL.

Although symptoms of ethanol intoxication are usually encountered following deliberate consumption of ethanol-containing beverages, many pharmaceutical products contain ethanol as a solvent, which, if ingested in sufficiently large quantities, may cause adverse symptoms of intoxication. In the USA, the maximum quantity of alcohol included in OTC medicines is 10% v/v for products labeled for use by people of 12 years of age and older, 5% v/v for products intended for use by children aged 6–12 years of age, and 0.5% v/v for products for use by children under 6 years of age.[15]

Parenteral products containing up to 50% of alcohol (ethanol 95 or 96% v/v) have been formulated. However, such concentrations can produce pain on intramuscular injection and lower concentrations such as 5–10% v/v are preferred. Subcutaneous injection of alcohol (ethanol 95% v/v) similarly causes considerable pain followed by anesthesia. If injections are made close to nerves, neuritis and nerve degeneration may occur. This effect is used therapeutically to cause anesthesia in cases of severe pain, although the practice of using alcohol in nerve blocks is controversial. Doses of 1 mL of absolute alcohol have been used for this purpose.[16]

Preparations containing more than 50% v/v alcohol may cause skin irritation when applied topically.

LD_{50} (mouse, IP): 0.93 g/kg[17]
LD_{50} (mouse, IV): 1.97 g/kg
LD_{50} (mouse, oral): 3.45 g/kg
LD_{50} (mouse, SC): 8.29 g/kg
LD_{50} (rat, IP): 3.75 g/kg
LD_{50} (rat, IV): 1.44 g/kg
LD_{50} (rat, oral): 7.06 g/kg

15 Handling Precautions

Observe normal precautions appropriate to the circumstances and quantity of material handled. Ethanol and aqueous ethanol solutions should be handled in a well-ventilated environment. In the UK, the long-term 8-hour TWA workplace exposure limit for ethanol is 1920 mg/m^3 (1000 ppm).[18] Ethanol may be irritant to the eyes and mucous membranes, and eye protection and gloves are recommended. Ethanol is flammable and should be heated with care. Fixed storage tanks should be electrically grounded to avoid ignition from electrostatic discharges when ethanol is transferred.

16 Regulatory Status

Included in the FDA Inactive Ingredients Database (dental preparations; inhalations; IM, IV, and SC injections; nasal and ophthalmic preparations; oral capsules, solutions, suspensions, syrups, and tablets; rectal, topical, and transdermal preparations). Included in the Canadian Natural Health Products Ingredients Database. Included in nonparenteral and parenteral medicines licensed in the UK.

17 Related Substances

Dehydrated alcohol; denatured alcohol; dilute alcohol; isopropyl alcohol.

Dehydrated alcohol

Synonyms Absolute alcohol; anhydrous ethanol; ethanol.
Autoignition temperature 365°C
Boiling point 78.5°C
Explosive limits 3.5–19.0% v/v in air
Flash point 12°C (closed cup)
Melting point −112°C
Moisture content Absorbs water rapidly from the air.
Refractive index n_D^{20} = 1.361
Specific gravity 0.7904–0.7935 at 20°C
Surface tension 22.75 mN/m at 20°C (ethanol/vapor)
Vapor density (relative) 1.59 (air = 1)
Vapor pressure 5.8 Pa at 20°C
Viscosity (dynamic) 1.22 mPa s (1.22 cP) at 20°C
Comments Dehydrated alcohol is ethanol ≥99.5% v/v. *See* Section 8. Dehydrated alcohol is one of the materials that have been selected for harmonization by the Pharmacopeial Discussion Group. For further information see the General Information Chapter <1196> in the USP35–NF30, the General Chapter 5.8 in PhEur 7.4, along with the 'State of Work' document on the PhEur EDQM website, and also the General Information Chapter 8 in the JP XV.

Denatured alcohol

Synonyms Industrial methylated spirit; surgical spirit.
Comments Denatured alcohol is alcohol intended for external use only. It has been rendered unfit for human consumption by the addition of a denaturing agent such as methanol or methyl isobutyl ketone.

Dilute alcohol

Synonyms Dilute ethanol.
Specific gravity *see* Table III.

A

Table III: Specific gravity of alcohol.

Strength of alcohol (% v/v)	Specific gravity at 20°C
90	0.8289–0.8319
80	0.8599–0.8621
70	0.8860–0.8883
60	0.9103–0.9114
50	0.9314–0.9326
45	0.9407–0.9417
25	0.9694–0.9703
20	0.9748–0.9759

Comments The term 'dilute alcohol' refers to a mixture of ethanol and water of stated concentration. The USP35–NF30 lists diluted alcohol. The BP 2012 lists eight strengths of dilute alcohol (dilute ethanol) containing 90%, 80%, 70%, 60%, 50%, 45%, 25%, and 20% v/v respectively of ethanol.

18 Comments

Alcohol is one of the materials that have been selected for harmonization by the Pharmacopeial Discussion Group. For further information see the General Information Chapter <1196> in the USP35–NF30, the General Chapter 5.8 in PhEur 7.4, along with the 'State of Work' document on the PhEur EDQM website, and also the General Information Chapter 8 in the JP XV.

Possession and use of nondenatured alcohols are usually subject to close control by excise authorities.

A specification for alcohol is contained in the *Food Chemicals Codex* (FCC).[19]

The EINECS number for alcohol is 200-578-6. The PubChem Compound ID (CID) for alcohol is 702.

19 Specific References

1 Chiori CO, Ghobashy AA. A potentiating effect of EDTA on the bactericidal activity of lower concentrations of ethanol. *Int J Pharm* 1983; **17**: 121–128.

2 Karabit MS, *et al.* Studies on the evaluation of preservative efficacy. IV. The determination of antimicrobial characteristics of some pharmaceutical compounds in aqueous solutions. *Int J Pharm* 1989; **54**: 51–56.

3 Liu P, *et al.* Quantitative evaluation of ethanol effects on diffusion and metabolism of β-estradiol in hairless mouse skin. *Pharm Res* 1991; **8**(7): 865–872.

4 Verma DD, Fahr A. Synergistic penetration enhancement of ethanol and phospholipids on the topical delivery of cyclosporin A. *J Control Release* 2004; **97**(1): 55–66.

5 Gwak SS, *et al.* Transdermal delivery of ondansetron hydrochloride: effects of vehicles and penetration enhancers. *Drug Dev Ind Pharm* 2004; **30**(2): 187–194.

6 Williams AC, Barry BW. Penetration enhancers. *Adv Drug Delivery Rev* 2004; **56**(5): 603–618.

7 Heard CA, *et al.* Skin penetration enhancement of mefenamic acid by ethanol and 1,8-cineole can be explained by the 'pull' effect. *Int J Pharm* 2006; **321**: 167–170.

8 Rhee YS, *et al.* Effects of vehicles and enhancers on transdermal delivery of clebopride. *Arch Pharm Res* 2007; **30**: 1155–1161.

9 Fang C, *et al.* Synergistically enhanced transdermal permeation and topical analgesia of tetracaine gel containing menthol and ethanol in experimental and clinical studies. *Eur J Pharm Biopharm* 2008; **68**: 735–740.

10 Krishnaiah YS, *et al.* Penetration-enhancing effect of ethanolic solution of menthol on transdermal permeation of ondansetron hydrochloride across rat epidermis. *Drug Deliv* 2008; **15**: 227–234.

11 Mutalik S, *et al.* A combined approach of chemical enhancers and sonophoresis for the transdermal delivery of tizanidine hydrochloride. *Drug Deliv* 2009; **16**(2): 82–91.

12 Kweon JH, *et al.* Transdermal delivery of diclofenac using microemulsions. *Arch Pharmacol Res* 2004; **27**(3): 351–356.

13 Huang YB, *et al.* Transdermal delivery of capsaicin derivative-sodium nonivamide acetate using microemulsions as vehicles. *Int J Pharm* 2008; **349**: 206–211.

14 El Maghraby GM. Transdermal delivery of hydrocortisone from eucalyptus oil microemulsion: effects of cosurfactants. *Int J Pharm* 2008; **355**: 285–292.

15 Jass HE. Regulatory review. *Cosmet Toilet* 1995; **110**(5): 21–22.

16 Lloyd JW. Use of anaesthesia: the anaesthetist and the pain clinic. *Br Med J* 1980; **281**: 432–434.

17 Lewis RJ, ed. *Sax's Dangerous Properties of Industrial Materials*, 11th edn. New York: Wiley, 2004: 1627–1628.

18 Health and Safety Executive. *EH40/2005: Workplace Exposure Limits.* Sudbury: HSE Books, 2011. http://www.hse.gov.uk/pubns/priced/eh40.pdf (accessed 29 February 2012).

19 *Food Chemicals Codex*, 7th edn. Bethesda, MD: United States Pharmacopeia, 2010: 346.

20 General References

European Directorate for the Quality of Medicines and Healthcare (EDQM). European Pharmacopoeia – State Of Work Of International Harmonisation. *Pharmeuropa* 2011; 22(4): 583–584. www.edqm.eu/en/International-Harmonisation-614.html (accessed 2 December 2011).

Lund W, ed. *The Pharmaceutical Codex: Principles and Practice of Pharmaceutics*, 12th edn. London: Pharmaceutical Press, 1994: 694–695.

Spiegel AJ, Noseworthy MN. Use of nonaqueous solvents in parenteral products. *J Pharm Sci* 1963; **52**: 917–927.

Wade A, ed. *Pharmaceutical Handbook*, 19th edn. London: Pharmaceutical Press, 1980: 227–230.

21 Author

ME Fenton.

22 Date of Revision

1 March 2012.

Alginic Acid

1 Nonproprietary Names

BP: Alginic Acid
PhEur: Alginic Acid
USP–NF: Alginic Acid

2 Synonyms

Acidum alginicum; E400; *Kelacid*; L-gulo-D-mannoglycuronan; polymannuronic acid; *Protacid*; *Satialgine H8*.

3 Chemical Name and CAS Registry Number

Alginic acid [9005-32-7]

4 Empirical Formula and Molecular Weight

Alginic acid is a linear glycuronan polymer consisting of a mixture of β-(1→4)-D-mannosyluronic acid and α-(1→4)-L-gulosyluronic acid residues, of general formula $(C_6H_8O)_n$. The molecular weight is typically 20 000–240 000.

5 Structural Formula

The PhEur 7.4 describes alginic acid as a mixture of polyuronic acids $[(C_6H_8O_6)_n]$ composed of residues of D-mannuronic and L-glucuronic acid, and obtained mainly from algae belonging to the Phaeophyceae. A small proportion of the carboxyl groups may be neutralized.

See also Section 4.

6 Functional Category

Modified-release agent; suspending agent; tablet and capsule binder; tablet and capsule disintegrant; taste-masking agent; viscosity-increasing agent.

7 Applications in Pharmaceutical Formulation or Technology

Alginic acid is used in a variety of oral and topical pharmaceutical formulations. In tablet and capsule formulations, alginic acid is used as both a binder and disintegrating agent at concentrations of 1–5% w/w.[1,2] Alginic acid is widely used as a thickening and suspending agent in a variety of pastes, creams, and gels, and as a stabilizing agent for oil-in-water emulsions.

Alginic acid has been used to improve the stability of levosimendan.[3]

See also Section 18.

8 Description

Alginic acid is a tasteless, practically odorless, white to yellowish-white, fibrous powder.

9 Pharmacopeial Specifications

See Table I.

10 Typical Properties

Acidity/alkalinity pH = 1.5–3.5 for a 3% w/v aqueous dispersion.

Crosslinking Addition of a calcium salt, such as calcium citrate or calcium chloride, causes crosslinking of the alginic acid polymer resulting in an apparent increase in molecular weight. Films crosslinked with triphosphate (tripolyphosphate) and calcium chloride were found to be insoluble but permeable to water vapor. Drug permeability varies with pH and the extent of crosslinking.[4]

SEM 1: Excipient: alginic acid; magnification: 100×; voltage: 25 kV.

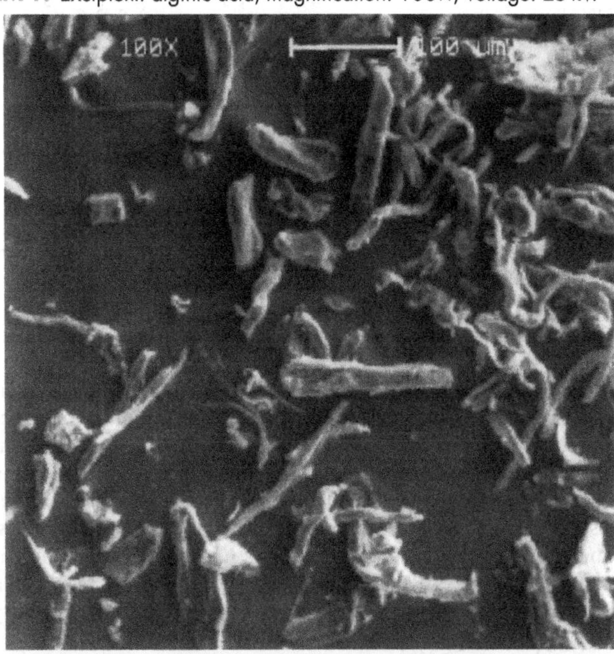

SEM 2: Excipient: alginic acid; magnification: 500×; voltage: 25 kV.

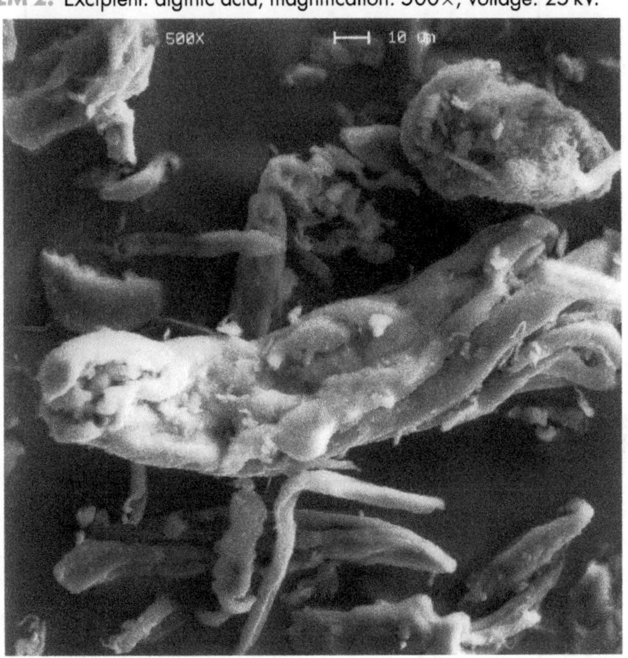

Density (true) 1.601 g/cm^3

Moisture content 7.01%

Solubility Soluble in alkali hydroxides, producing viscous solutions; very slightly soluble or practically insoluble in ethanol (95%) and other organic solvents. Alginic acid swells in water but does not dissolve; it is capable of absorbing 200–300 times its own weight of water.

Table I: Pharmacopeial specifications for alginic acid.

Test	PhEur 7.4	USP35–NF30
Identification	+	+
Characters	+	–
Microbial limits	≤10^2 cfu/g	≤200 cfu/g
pH (3% dispersion)	–	1.5–3.5
Loss on drying	≤15.0%	≤15.0%
Ash	–	≤4.0%
Sulfated ash	≤8.0%	–
Arsenic	–	≤3 ppm
Chloride	≤1.0%	–
Lead	–	≤10 ppm
Heavy metals	≤20 ppm	≤40 ppm
Acid value (dried basis)	–	≥230
Assay (of COOH groups)	19.0–25.0%	–

Spectroscopy

IR spectra *see* Figure 1.

NIR spectra *see* Figure 2.

Viscosity (dynamic) Various grades of alginic acid are commercially available that vary in their molecular weight and hence viscosity. Viscosity increases considerably with increasing concentration; typically a 0.5% w/w aqueous dispersion will have a viscosity of approximately 20 mPa s, while a 2.0% w/w aqueous dispersion will have a viscosity of approximately 2000 mPa s. The viscosity of dispersions decreases with increasing temperature. As a general rule, a 1°C increase in temperature results in a 2.5% reduction in viscosity. At low concentrations, the viscosity of an alginic acid dispersion may be increased by the addition of a calcium salt, such as calcium citrate. *See also* Section 11 and Section 18.

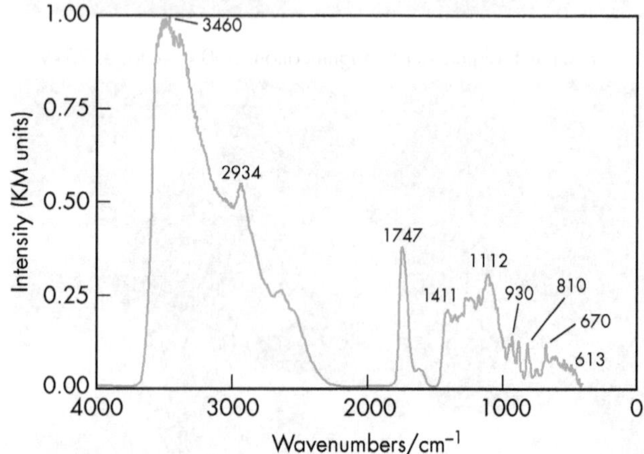

Figure 1: Infrared spectrum of alginic acid measured by diffuse reflectance. Adapted with permission of Informa Healthcare.

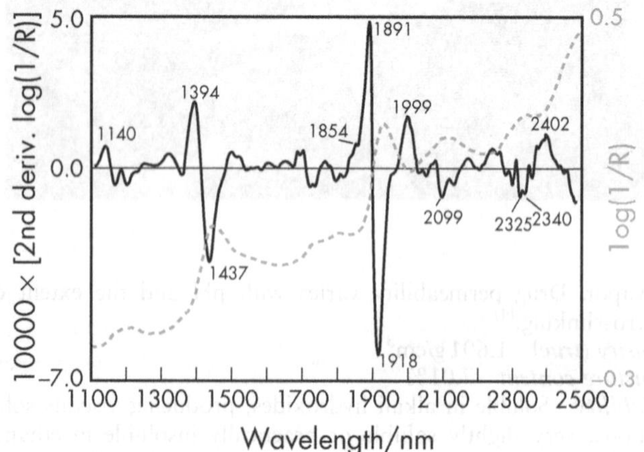

Figure 2: Near-infrared spectrum of alginic acid measured by reflectance.

11 Stability and Storage Conditions

Alginic acid hydrolyzes slowly at warm temperatures producing a material with a lower molecular weight and lower dispersion viscosity. When heated, alginic acid initially undergoes dehydration at approximately 80°C followed by decomposition at approximately 200°C.

Alginic acid dispersions are susceptible to microbial spoilage on storage, which may result in some depolymerization and hence a decrease in viscosity. Dispersions should therefore be preserved with an antimicrobial preservative such as benzoic acid; potassium sorbate; sodium benzoate; sorbic acid; or paraben. Concentrations of 0.1–0.2% are usually used.

Alginic acid dispersions may be sterilized by autoclaving or filtration through a 0.22 μm filter. Autoclaving may result in a decrease in viscosity which can vary depending upon the nature of any other substances present.[5]

Alginic acid should be stored in a well-closed container in a cool, dry place.

12 Incompatibilities

Incompatible with strong oxidizing agents; alginic acid forms insoluble salts in the presence of alkaline earth metals and group III metals with the exception of magnesium.

13 Method of Manufacture

Alginic acid is a hydrophilic colloid carbohydrate that occurs naturally in the cell walls and intercellular spaces of various species of brown seaweed (Phaeophyceae). The seaweed occurs widely throughout the world and is harvested, crushed, and treated with dilute alkali to extract the alginic acid.

14 Safety

Alginic acid is widely used in food products and topical and oral pharmaceutical formulations. It is generally regarded as a nontoxic and nonirritant material, although excessive oral consumption may be harmful. Inhalation of alginate dust may be an irritant and has been associated with industrially related asthma in workers involved in alginate production. However, it appears that the cases of asthma were linked to exposure to unprocessed seaweed dust rather than pure alginate dust.[6] An acceptable daily intake of alginic acid and its ammonium, calcium, potassium, and sodium salts was not set by the WHO because the quantities used, and the background levels in food, did not represent a hazard to health.[7]

LD_{50} (rat, IP): 1.6 g/kg[8]

15 Handling Precautions

Observe normal precautions appropriate to the circumstances and quantity of material handled. Alginic acid may be irritant to the eyes or respiratory system if inhaled as dust; *see* Section 14. Eye protection, gloves, and a dust respirator are recommended. Alginic acid should be handled in a well-ventilated environment.

16 Regulatory Status

GRAS listed. Accepted in Europe for use as a food additive. Included in the FDA Inactive Ingredients Database (ophthalmic preparations, oral capsules, and tablets). Included in the Canadian Natural Health Products Ingredients Database. Included in nonparenteral medicines licensed in the UK.

17 Related Substances

Ammonium alginate; calcium alginate; potassium alginate; propylene glycol alginate; sodium alginate.

18 Comments

Alginic acid has been investigated for use in an ocular formulation of carteolol.[9]

In the area of controlled release, the preparation of indomethacin sustained-release microparticles from alginic acid (alginate)–gelatin hydrocolloid coacervate systems has been investigated.[10] In addition, as controlled-release systems for liposome-associated macromolecules, microspheres have been produced encapsulating liposomes coated with alginic acid and poly-L-lysine membranes.[11] Alginate gel beads capable of floating in the gastric cavity have been prepared, the release properties of which were reported to be applicable for sustained release of drugs, and for targeting the gastric mucosa.[12]

Mechanical properties, water uptake, and permeability properties of a sodium salt of alginic acid have been characterized for controlled-release applications.[4] In addition, sodium alginate has been incorporated into an ophthalmic drug delivery system for pilocarpine nitrate.[13] Sodium alginate has been used to improve pelletization due to polyelectrolyte complex formation between cationic polymers such as chitosan.[14] Alginic acid has also been shown to be beneficial in the development of alginate gel-encapsulated, chitosan-coated nanocores, where the alginates act as a protective agent for sensitive macromolecules such as proteins and peptides for prolonged release.[15] In addition, the crosslinking of dehydrated paracetamol sodium alginate pellets has been shown to successfully mask the drug's unpleasant taste by an extrusion/spheronization technique.[16]

It has also been reported that associated chains of alginic acid complexed with cations can bind to cell surfaces and exert pharmacological effects, which depend on the cell type and the complexed cation. These complexes can be used to treat rheumatic disorders, diseases associated with atopic diathesis and liver diseases.[17]

An alginic oligosaccharide, obtained from a natural edible polysaccharide, has been shown to suppress Th2 responses and IgE production by inducing IL-12 production, and was found to be a useful approach for preventing allergic disorders.[18] Chemically modified alginic acid derivatives have also been researched for their anti-inflammatory, antiviral, and antitumor activities.[19] Alginate/antacid antireflux preparations have been reported to provide symptomatic relief by forming a physical barrier on top of the stomach contents in the form of a raft.[20]

Alginic acid gels for use in drug delivery systems may be prepared by adding D-glucono-D-lactone, which hydrolyzes in the presence of water to produce gluconic acid with a continuous lowering of pH.[21] Alginic acid forms a gel in the presence of di- or trivalent cations, and may be used as a carrier for treating joint disorders.[22,23]

Alginic acid dispersions are best prepared by pouring the alginic acid slowly and steadily into vigorously stirred water. Dispersions should be stirred for approximately 30 minutes. Premixing the alginic acid with another powder, such as sugar, or a water-miscible liquid such as ethanol (95%) or glycerin, aids dispersion.

Therapeutically, alginic acid has been used as an antacid.[24] In combination with an H_2-receptor antagonist, it has also been utilized for the management of gastroesophageal reflux.[25]

When using alginic acid in tablet formulations, the alginic acid is best incorporated or blended using a dry granulation process.

A specification for alginic acid is contained in the *Food Chemicals Codex* (FCC).[26]

The EINECS number for alginic acid is 232-680-1.

19 Specific References

1 Shotton E, Leonard GS. Effect of intragranular and extragranular disintegrating agents on particle size of disintegrated tablets. *J Pharm Sci* 1976; **65**: 1170–1174.

2 Esezobo S. Disintegrants: effects of interacting variables on the tensile strengths and disintegration times of sulfaguanidine tablets. *Int J Pharm* 1989; **56**: 207–211.

3 Larma I, Harjula M. Stable compositions comprising levosimendan and alginic acid. Patent No: WO9955337; 1999.

4 Remunan-Lopez C, Bodmeier R. Mechanical, water uptake and permeability properties of crosslinked chitosan glutamate and alginate films. *J Control Release* 1997; **44**: 215–225.

5 Vandenbossche GMR, Remon J-P. Influence of the sterilization process on alginate dispersions. *J Pharm Pharmacol* 1993; **45**: 484–486.

6 Henderson AK, *et al.* Pulmonary hypersensitivity in the alginate industry. *Scott Med J* 1984; **29**(2): 90–95.

7 FAO/WHO. Evaluation of certain food additives and naturally occurring toxicants. Thirty-ninth report of the joint FAO/WHO expert committee on food additives. *World Health Organ Tech Rep Ser* 1993; No. 837.

8 Lewis RJ, ed. *Sax's Dangerous Properties of Industrial Materials*, 11th edn. New York: Wiley, 2004; 101–102.

9 Tissie G, *et al.* Alginic acid effect on carteolol ocular pharmacokinetics in the pigmented rabbit. *J Ocul Pharmacol Ther* 2002; **18**(1): 65–73.

10 Joseph I, Venkataram S. Indomethacin sustained release from alginate-gelatin or pectin-gelatin coacervates. *Int J Pharm* 1995; **125**: 161–168.

11 Machluf M, *et al.* Characterization of microencapsulated liposome systems for the controlled delivery of liposome-associated macromolecules. *J Control Release* 1997; **43**: 35–45.

12 Murata Y, *et al.* Use of floating alginate gel beads for stomach-specific drug delivery. *Eur J Pharm Biopharm* 2000; **50**(2): 221–226.

13 Cohen S, *et al.* Novel *in situ*-forming opthalmic drug delivery system from alginates undergoing gelation in the eye. *J Control Release* 1997; **44**: 201–208.

14 Charoenthai N, *et al.* Use of chitosan-alginate as an alternative pelletization aid to microcrystalline cellulose in extrusion/spheronization. *J Pharm Sci* 2007; **96**: 2469–2484.

15 Rawat M, *et al.* Development and *in vitro* evaluation of alginate gel-encapsulated, chitosan-coated ceramic nanocores for oral delivery of enzyme. *Drug Dev Ind Pharm* 2008; **34**: 181–188.

16 Kulkami RB, Amin PD. Masking of unpleasant gustatory sensation by cross-linking of dehydrated paracetamol alginate pellets produced by extrusion-spheronization. *Drug Dev Ind Pharm* 2008; **34**: 199–205.

17 Gradl T. Use of alginic acid and/or its derivatives and salts for combating or preventing diseases. Patent No: DE19723155; 1998.

18 Tadashi Y, *et al.* Alginic acid oligosaccharide suppresses Th2 development and IgE production by inducing IL-12 production. *Int Arch Allergy Imm* 2004; **133**(3): 239–247.

19 Boisson-Vidal C, *et al.* Biological activities of polysaccharides from marine algae. *Drugs Future* 1995; **20**(Dec): 1247–1249.

20 Hampson FC, *et al.* Alginate rafts and their characterization. *Int J Pharm* 2005; **294**: 137–147.

21 Draget KI, *et al.* Similarities and differences between alginic acid gels and ionically crosslinked alginate gels. *Food Hydrocolloids* 2006; **20**: 170–175.

22 Fragonas E, *et al.* Articular cartilage repair in rabbits by using suspensions of allogenic chondrocytes in alginate. *Biomaterials* 2000; **21**: 795–801.

23 Iwasaki *et al.* Compositions for treating arthritic disorder. United States Patent 20100048506; 2010.

24 Vatier J, *et al.* Antacid drugs: multiple but too often unknown pharmacological properties. *J Pharm Clin* 1996; **15**(1): 41–51.

25 Stanciu C, Bennett JR. Alginate/antacid in the reduction of gastro-oesophageal reflux. *Lancet* 1974; i: 109–111.

26 *Food Chemicals Codex*, 7th edn. Bethesda, MD: United States Pharmacopeia, 2010; 29.

20 General References

Marshall PV, *et al.* Methods for the assessment of the stability of tablet disintegrants. *J Pharm Sci* 1991; **80**: 899–903.

Soares JP, *et al.* Thermal behavior of alginic acid and its sodium salt. *Ecletica Quimica* 2004; **29**(2): 57–63.

21 Authors

W Deng, MA Repka, A Singh.

22 Date of Revision

1 March 2012.

Aliphatic Polyesters

1 Nonproprietary Names

None adopted.

2 Synonyms

DL-PCL; DL-PLA; DL-PLGA; poly-ε-caprolactone; poly(lactic acid); poly(DL-lactic-*co*-glycolic acid); poly(L-lactide); poly(DL-lactide); poly(L-lactide-*co*-ε-caprolactone); poly(DL-lactide-*co*-ε-caprolactone); poly(DL-lactide-*co*-glycolide); *Resomer*.

See also Table I.

3 Chemical Name and CAS Registry Number

See Table I.

4 Empirical Formula and Molecular Weight

Aliphatic polyesters are synthetic homopolymers or copolymers of lactic acid, glycolic acid, lactide, glycolide and ε-hydroxycaproic acid. Typically, the molecular weights of homopolymers and copolymers range from 2 000 to >100 000 Da.

Co-monomer ratios of lactic acid and glycolic acid (or lactide and glycolide) for poly(DL-lactide-*co*-glycolide) range from 85 : 15 to 50 : 50.Table 1 shows the chemical and trade names of different commercially available aliphatic polyesters.

5 Structural Formula

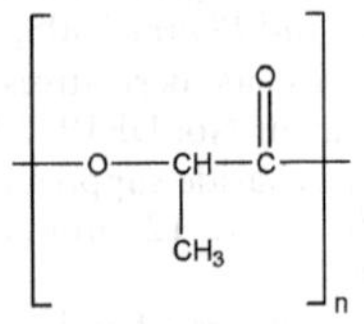

Poly(lactide)

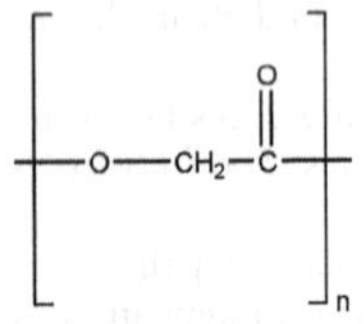

Poly(glycolide)

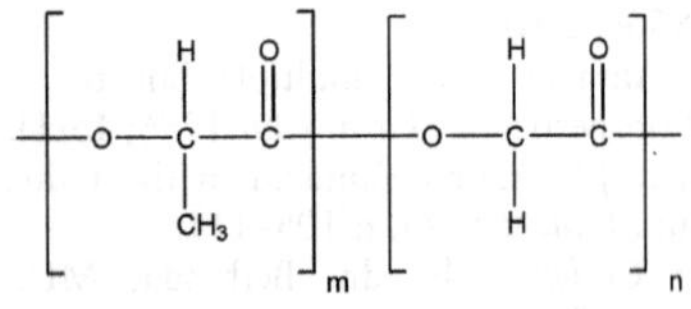

Poly(lactide-co-glycolide)

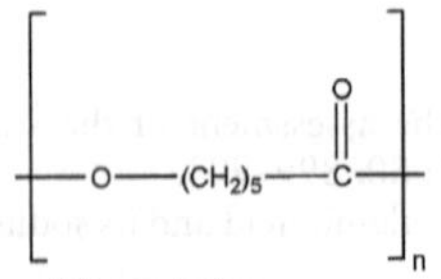

Polycaprolactone

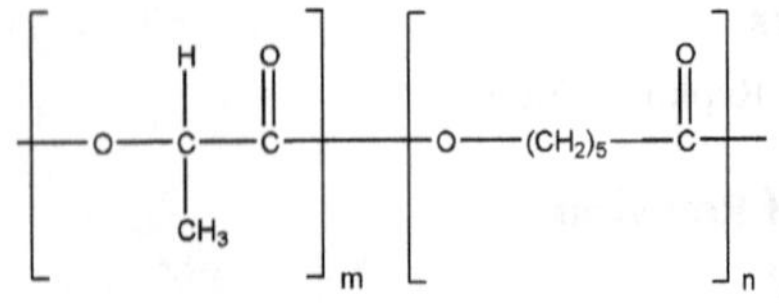

Poly(lactide-co-caprolactone)

6 Functional Category

Biocompatible material; biodegradable material; microencapsulating agent; modified-release agent.

7 Applications in Pharmaceutical Formulation or Technology

Owing to their reputation as safe materials and their biodegradability, aliphatic polyesters are primarily used as biocompatible and biodegradable carriers in many types of sustained-release implantable or injectable drug-delivery systems for both human and veterinary use.(1) Examples of implantable drug delivery systems include rods,(2) cylinders, tubing, films,(3) fibers, pellets, and beads.(4) Examples of injectable drug-delivery systems include microcapsules,(5) microspheres,(6) nanoparticles, and liquid injectable controlled-release systems such as gel formulations.(7)

See also Section 18.

8 Description

Aliphatic polyesters occur as white to off-white or light gold solid powders, pellets or granules, and are almost odorless.

9 Pharmacopeial Specifications

—

10 Typical Properties

For typical physical and mechanical properties of the aliphatic polyesters, *see* Tables II, III, IV, V, VI, and VII.

Polymer composition and crystallinity play important roles in the solubility of these aliphatic polyesters. The crystalline homopolymers of glycolide or glycolic acid are soluble only in strong solvents, such as hexafluoroisopropanol. The crystalline homopolymers of lactide or lactic acid also do not have good solubility in most organic solvents. However, amorphous polymers of DL-lactide or DL-lactic acid and copolymers of lactide or lactic acid with a low glycolide or glycolic acid content are soluble in many organic solvents (Table VII). Aliphatic polyesters are slightly soluble or insoluble in water, methanol, ethylene glycol, heptane, and hexane.

11 Stability and Storage Conditions

The aliphatic polyesters are easily susceptible to hydrolysis in the presence of moisture. Hence, they should be packaged under high-purity dry nitrogen and properly stored in airtight containers with desiccant, preferably refrigerated at –10°C or below. It is necessary to allow the polymers to reach room temperature in a dry environment before opening the container. After the original package has been opened, it is recommended to re-purge the package with high-purity dry nitrogen prior to resealing.

12 Incompatibilities

—

13 Method of Manufacture

Aliphatic polyesters are mainly synthesized via polycondensation of hydroxycarboxylic acids and catalytic ring-opening polymerization of lactones. Ring-opening polymerization is preferred because polyesters with high molecular weights and specific end groups can be produced.

Enzymatic polymerization is an alternative method carried out under mild conditions avoiding the use of toxic reagents, with the

Table I: Chemical names and CAS registry numbers of the aliphatic polyesters.

Generic name	Composition (%)			Synonyms	Trade name	Manufacturer	CAS name	CAS number
	Lactide	Glycolide	Caprolactone					
Poly(L-lactide)	100	0	0	L-PLA	*Lactel L-PLA* *100 L* *Resomer L 206 S, 207 S, 209 S, 210, 210 S*	Durect Lakeshore BI	Poly[oxy[(1*S*)-1-methyl-2-oxo-1,2-ethanediyl]]	[26161-42-2]
Poly(DL-lactide)	100	0	0	DL-PLA	*Lactel DL-PLA* *Purasorb PDL 02A, 02, 04, 05* *Resomer R 202 S, 202 H, 203 S, 203 H* *100 DL 7E*	Durect Purac BI Lakeshore	1,4-Dioxane-2,5-dione, 3,6-dimethyl-, homopolymer	[26680-10-4]
Poly(L-lactide-*co*-glycolide)	85	15	0		*Resomer LG 855 S, 857 S*	BI	1,4-Dioxane-2,5-dione, 3,6-dimethyl-, (3*S*,6*S*)-, polymer with 1,4-dioxane-2,5-dione	[30846-39-0]
Poly(L-lactide-*co*-glycolide)	82	18	0		*Resomer LG 824 S*	BI	1,4-Dioxane-2,5-dione, 3,6-dimethyl-, (3S,6S)-, polymer with 1,4-dioxane-2,5-dione	[30846-39-0]
Poly(L-lactide-*co*-glycolide)	10	90	0		*Resomer GL 903*	BI	1,4-Dioxane-2,5-dione, 3,6-dimethyl-, (3*S*,6*S*)-, polymer with 1,4-dioxane-2,5-dione	[30846-39-0]
Poly(DL-lactide-*co*-glycolide)	85	15	0	Polyglactin;DL-PLGA (85 : 15)	*Lactel 85 : 15 DL-PLG* *8515 DLG 7E* *Resomer RG 858 S*	Durect Lakeshore BI	1,4-Dioxane-2,5-dione, 3,6-dimethyl-, polymer with 1,4-dioxane-2,5-dione	[26780-50-7]
Poly(DL-lactide-*co*-glycolide)	75	25	0	Polyglactin;DL-PLGA (75 : 25)	*Lactel 75:25 DL-PLG* *Purasorb PDLG 7502A, 7502, 7507* *Resomer RG 752 H, 752 S, 753 S, 755 S, 756 S* *7525 DLG 7E*	Durect Purac BI Lakeshore	1,4-Dioxane-2,5-dione, 3,6-dimethyl-, polymer with 1,4-dioxane-2,5-dione	[26780-50-7]
Poly(DL-lactide-*co*-glycolide)	65	35	0	Polyglactin;DL-PLGA (65 : 35)	*Lactel 65:35 DL-PLG* *6535 DLG 7E* *Resomer RG 653 H*	Durect Lakeshore BI	1,4-Dioxane-2,5-dione, 3,6-dimethyl-, polymer with 1,4-dioxane-2,5-dione	[26780-50-7]
Poly(DL-lactide-*co*-glycolide)	50	50	0	Polyglactin;DL-PLGA (50 : 50)	*Lactel 50:50 DL-PLG* *5050 DLG 7E, 5E, 1A, 2A, 3A, 4A, 4.5A* *Purasorb PDLG 5002A, 5002, 5004A, 5004, 5010* *Resomer RG 502, 502H, 503, 503H, 504, 504H, 509S*	Durect Lakeshore Purac BI	1,4-Dioxane-2,5-dione, 3,6-dimethyl-, polymer with 1,4-dioxane-2,5-dione	[26780-50-7]
Poly-ε-caprolactone	0	0	100	PCL	*Lactel PCL* *100 PCL*	Durect Lakeshore	2-Oxepanone, homopolymer	[24980-41-4]
Poly(DL-lactide-*co*-ε-caprolactone)	85	0	15		*8515 DL/PCL*	Lakeshore	1,4-Dioxane-2,5-dione,3,6-dimethyl-, polymer with 2-oxepanone	[70524-20-8]
Poly(DL-lactide-*co*-ε-caprolactone)	80	0	20	DL-PLCL (80 : 20)	*Lactel 80 : 20 DL-PLCL*	Durect	1,4-Dioxane-2,5-dione, 3,6-dimethyl-, polymer with 2-oxepanone	[70524-20-8]
Poly(DL-lactide-*co*-ε-caprolactone)	25	0	75	DL-PLCL (25 : 75)	*Lactel 25 : 75 DL-PLCL*	Durect	1,4-Dioxane-2,5-dione, 3,6-dimethyl-, polymer with 2-oxepanone	[70524-20-8]
Poly(L-lactide-*co*-ε-caprolactone)	70	0	30		*Resomer LC 703 S*	BI	1,4-Dioxane-2,5-dione, 3,6-dimethyl-, (3*S*,6*S*)-, polymer with 2-oxepanone	[65408-67-5]
Poly(L-lactide-*co*-ε-caprolactone)	85	0	15		*8515 L/PCL*	Lakeshore	1,4-Dioxane-2,5-dione, 3,6-dimethyl-, (3*S*,6*S*)-, polymer with 2-oxepanone	[65408-67-5]
Poly(L-lactide-*co*-ε-caprolactone)	75	0	25		*7525 L/PCL*	Lakeshore	1,4-Dioxane-2,5-dione, 3,6-dimethyl-, (3*S*,6*S*)-, polymer with 2-oxepanone	[65408-67-5]

(a) BI, Boehringer Ingelheim; Durect, Durect Corporation; Lakeshore, Lakeshore Biomaterials; Purac, Purac America.

Table II: Typical resorption duration of certain aliphatic polyesters.[a]

Polymer	End group	Inherent viscosity (dL/g)[b]	Degradation time
Poly(DL-lactide-co-glycolide) (100 : 0)	Ester	0.60 – 0.80	12 – 16 months
Poly(DL-lactide-co-glycolide) (85 : 15)	Ester	0.60 – 0.80	5 – 6 months
Poly(DL-lactide-co-glycolide) (75 : 25)	Ester	0.60 – 0.80	4 – 5 months
Poly(DL-lactide-co-glycolide) (65 : 35)	Ester	0.60 – 0.80	3 – 4 months
Poly(DL-lactide-co-glycolide) (50 : 50)	Ester	0.60 – 0.80	1 – 2 months
Poly(DL-lactide-co-glycolide) (50 : 50)	Ester	0.45 – 0.55	1 – 2 months
Poly(DL-lactide-co-glycolide) (50 : 50)	Acid	0.05 – 0.15	1 – 2 weeks
Poly(DL-lactide-co-glycolide) (50 : 50)	Acid	0.15 – 0.25	2 – 4 weeks
Poly(DL-lactide-co-glycolide) (50 : 50)	Acid	0.25 – 0.35	3 – 4 weeks
Poly(DL-lactide-co-glycolide) (50 : 50)	Acid	0.35 – 0.45	3 – 4 weeks
Poly(DL-lactide-co-glycolide) (50 : 50)	Acid	0.40 – 0.50	3 – 4 weeks
Poly(L-lactide)	–	0.9 – 1.2	> 24 months
Poly(glycolide)	–	1.4 – 1.8[c]	6 – 12 months
Poly-ε-caprolactone	–	1.0 – 1.3	> 24 months

(a) Specifications obtained from DURECT Corp and SurModics Pharmaceuticals.
(b) 0.5% w/v in chloroform at 30°C with a size 25 Cannon-Fenske glass.
(c) Hexafluoroisopropanol.

Table III: General mechanical properties of selected aliphatic polyesters.[a]

Property	Polymers			
	Poly(DL-lactide-co-glycolide) (50:50)	Poly(DL-lactide)	Poly(L-lactide)	Poly(L-lactide-co-DL-lactide) (90:10)
Break stress (psi)	8296	6108	7323	7614
Strain at break (%)	5.2	5.0	5.5	5.2
Yield stress (psi)	8371	6666	7678	8414
Strain at yield (%)	5.1	3.7	4.9	4.5
Modulus (psi)	189 340	207 617	182 762	210 680

(a) Specifications obtained from SurModics Pharmaceuticals.

Table IV: Mechanical properties of Poly(L-lactide/caprolactone) [a]

Property	Poly(L-lactide/caprolactone) grade				
	50/50	75/25	85/15	90/10	95/05
Tensile Strength (psi)					
At max.	80	1488	3254	6232	6900
At 100%	79	400	1822	-	-
At 300%	44	950	2615	-	-
Elongation (%)					
To Yield	>1000	>400	>6.4	8.1	1.6
To Failure	>1000	>400	>500	8.1	1.6
Modulus (Kpsi)	0.1	5.3	84	167	185
Shore D-hardness	5	52	87	91	95
Specific gravity	1.2	1.2	1.23	1.25	1.26
Compression molding temperature (°C)	73 - 130	130 ± 15	140 ± 10	165 ± 5	165 ± 5

(a) Specifications obtained from SurModics Pharmaceuticals.

Table V: Mechanical properties of Poly(DL-lactide/caprolactone) [a]

Property	Poly(DL-lactide/caprolactone) grade				
	60/40	75/25	85/15	90/10	95/05
Tensile Strength (psi)					
At max.	65	1300	1555	4453	5493
At 100%	65	224	1555	-	-
At 300%	43	332	1041	-	-
Elongation (%)					
To Yield	-	-	-	5.6	-
To Failure	> 400	> 600	> 500	5.6	7.2
Modulus (Kpsi)	0.1	1.05	6.04	106	135
Shore D-hardness	0	42	79	88	95
Specific gravity	-	1.20	1.22	1.24	-
Compression molding temperature (°C)	-	82 - 140	82 - 140	82 - 140	120

(a) Specifications obtained from SurModics Pharmaceuticals.

Table VI: Glass transition temperature and melting point of selected aliphatic polyesters [(a)]

Polymer	Glass Transition Temperature (°C)	Melting point (°C)
Poly(glycolic acid)	35 – 40	225 – 230
Poly(L-lactide)	56 – 60	173 – 178
Poly(L-lactide-co-glycolide) (10:90)	35 – 45	180 – 200
Poly(DL-lactide)	50 – 55	Amorphous[(b)]
Poly(DL-lactide-co-glycolide) (85:15)	50 – 55	Amorphous[(b)]
Poly(DL-lactide-co-glycolide) (75:25)	48 – 53	Amorphous[(b)]
Poly(DL-lactide-co-glycolide) (65:35)	45 – 50	Amorphous[(b)]
Poly(DL-lactide-co-glycolide) (50:50)	43 – 48	Amorphous[(b)]
Poly(DL-lactide-co-caprolactone) (85:15)	20 – 25	Amorphous[(b)]
Poly(DL-lactide-co-caprolactone) (80:20)	17 – 23	Amorphous[(b)]
Poly(DL-lactide-co-caprolactone) (25:75)	(-50) – (-40)	Amorphous[(b)]
Poly(L-lactide-co-caprolactone) (85:15)	20 – 25	Amorphous[(b)]
Poly(L-lactide-co-caprolactone) (75:25)	13 – 20	Amorphous[(b)]
Poly(caprolactone)	(-60) – (-65)	60

(a) Specifications obtained from DURECT and SurModics Pharmaceuticals.
(b) Amorphous polymers process temperature range: 140 – 160°C

Table VII: Solubility of selected aliphatic polyesters [(a)(b)]

Polymer	Solvent						
	Ethyl acetate	Methylene chloride	Chloroform	Acetone	Dimethyl formamide (DMF)	Tetrahydrofuran (THF)	Hexafluoro-Isopropanol (HFIP)
Poly(L-lactide)	NS	S	S	NS	NS	NS	S
Poly(DL-lactide)	S	S	S	S	S	S	S
Poly(DL-lactide-co-glycolide) (85:15)	S	S	S	S	S	S	S
Poly(DL-lactide-co-glycolide) (75:25)	S	S	S	S	S	S	S
Poly(DL-lactide-co-glycolide) (65:35)	S	S	S	S	S	S	S
Poly(DL-lactide-co-glycolide) (50:50)	SS	S	S	SS	S	SS	S
Poly(caprolactone)	S	S	S	S	S	S	S
Poly(L-lactide-co-caprolactone) (75:25)	S	S	S	S	S	S	S
Poly(DL-lactide-co-caprolactone) (80:20)	S	S	S	S	S	S	S
Poly(glycolic acid)	NS	NS	NS	NS	NS	NS	S

(a) Specifications obtained from SurModics Pharmaceuticals.
(b) NS, not soluble; SS, slightly soluble (degree of solubility is dependent on molecular weight or inherent viscosity); S, soluble.

possibility to recycle the catalyst. Regional and stereo selectivity of enzymes provides attractive possibilities for the direct synthesis of functional polyesters, avoiding the use of protected monomers. Block copolymers may also be synthesized using enzymatic polymerization. However, the major drawback of this method is the relatively low molecular weight of the polymers obtained.(1)

14 Safety

Poly(lactic acid) or poly(lactide), poly(glycolic acid) or poly(glycolide), poly (lactic-*co*-glycolic acid) or poly(lactide-*co*-glycolide), and polycaprolactone are used in parenteral pharmaceutical formulations and are regarded as biodegradable, biocompatible, and bioabsorbable materials. Their biodegradation products are nontoxic, noncarcinogenic, and nonteratogenic. In general, these polyesters exhibit very little hazard.

15 Handling Precautions

Observe normal precautions appropriate to the circumstances and quantity of material handled. Contact with eyes, skin, and clothing, and breathing the dust of the polymers should be avoided. Aliphatic polyesters produce acid materials such as hydroxyacetic and/or lactic acid in the presence of moisture; thus, contact with materials that will react with acids, especially in moist conditions, should be avoided.

16 Regulatory Status

GRAS listed. Included in the FDA Inactive Ingredients Database (IM powder for injections, lyophilized suspensions; periodontal drug delivery system). Poly(lactide) and poly(lactide-*co*-glycolide) have been used in medical products and medical devices approved by the FDA. Included in the Canadian Natural Health Products Ingredients Database.

17 Related Substances

Glycolic acid; lactic acid.

Glycolic acid

Empirical formula $C_2H_4O_3$
Molecular weight 76.05
CAS number [79-14-1]
Synonyms 2-Hydroxyacetic acid; hydroxyethanoic acid.
Appearance Odorless, somewhat hygroscopic crystals.
Acidity/alkalinity pH of aqueous solution

2.5 (0.5%);
2.33 (1.0%);
2.16 (2.0%);
1.91 (5.0%);
1.73 (10.0%).

Dissociation constant 3.83 at 25°C
Melting point 80°C

Solubility Soluble in water, methanol, alcohol, acetone, acetic acid, and ether.
Safety LD_{50} (rat, oral): 1.95 g/kg
Comments A constituent of sugar cane juice. Manufactured by the action of sodium hydroxide on monochloroacetic acid; also by the electrolytic reduction of oxalic acid.

18 Comments

Aliphatic polyesters are a group of synthesized, nontoxic, biodegradable polymers. In an aqueous environment, the polymer backbone undergoes hydrolytic degradation, through cleavage of the ester linkages, into nontoxic hydroxycarboxylic acids. Aliphatic polyesters are eventually metabolized to carbon dioxide and water, via the citric acid cycle.

The rate of biodegradation and drug-release characteristics from injectable drug-delivery systems formulated with the aliphatic polyesters can be controlled by changing the physicochemical properties of the polymers, such as crystalline versus amorphous character, monomer ratio, end group, molecular weight, porosity and geometry of the formulation.

19 Specific References

1 Seyednejad H, *et al.* Functional aliphatic polyesters for biomedical and pharmaceutical applications. *J Control Release* 2011; **152**(1): 168–176.
2 Shim IK, *et al.* Healing of articular cartilage defects treated with a novel drug-releasing rod-type implant after microfracture surgery. *J Control Release* 2008; **129**: 187–191.
3 Aviv M, *et al.* Gentamicin-loaded bioresorbable films for prevention of bacterial infections associated with orthopedic implants. *J Biomed Mater Res A* 2007; **83**: 10–19.
4 Wang G, *et al.* The release of cefazolin and gentamicin from biodegradable PLA/PGA beads. *Int J Pharm* 2004; **273**: 203–212.
5 Snider C, *et al.* Microenvironment-controlled encapsulation (MiCE) process: effects of PLGA concentration, flow rate, and collection method on microcapsule size and morphology. *Pharm Res* 2008; **25**: 5–15.
6 Giovagnoli S, *et al.* Physicochemical characterization and release mechanism of a novel prednisone biodegradable microsphere formulation. *J Pharm Sci* 2008; **97**: 303–317.
7 Zare M, *et al.* Effect of additives on release profile of leuprolide acetate in an in situ forming controlled-release system: in vitro study. *J Appl Polym Sci* 2008; **107**: 3781–3787.

20 General References

Ahmed AR, *et al.* Reduction in burst release of PLGA microparticles by incorporation into cubic phase-forming systems. *Eur J Pharm Biopharm* 2008; **70**(3): 765–769.

Allison SD. Analysis of initial burst in PLGA microparticles. *Expert Opin Drug Delivery* 2008; **5**(6): 615–628.

Evonik Industries. *Resomer.* http://resomer.evonik.com (accessed 20 February 2012).

Determan AS, *et al.* Protein stability in the presence of polymer degradation products: Consequences for controlled release formulations. *Biomaterials* 2006; **27**(17): 3312–3320.

Durect Corporation. *Lactel* absorbable polymers. http://www.absorbables.com/ (accessed 20 February 2012).

Giteau A, *et al.* How to achieve sustained and complete protein release from PLGA-based microparticles? *Int J Pharm* 2008; **350**: 14–26.

Kim MS, *et al.* Preparation of methoxy poly(ethylene glycol)/polyester diblock copolymers and examination of the gel-to-sol transition. *J Polym Sci, Part A: Polym Chem* 2004; **42**(22): 5784–5793.

Kulkarni RK, *et al.* Biodegradable poly(lactic acid) polymers. *J Biomed Mater Res* 1971; **5**: 169–181.

Li M, *et al.* Microencapsulation by solvent evaporation: state of the art for process engineering approaches. *Int J Pharm* 2008; **363**(1-2): 26–39.

Patel RB, *et al.* Characterization of formulation parameters affecting low molecular weight drug release from *in situ* forming drug delivery systems. *J Biomed Mater Res, Part A* 2010; **94A**(2): 476–484.

Purac Biomaterials. Product information: Polymers for drug delivery. http://www.purac.com/Purac-Biomaterials/Products/Polymers-for-drug-delivery.aspx (accessed 21 December 2011).

SurModics Pharmaceuticals. Product information: Lakeshore Biomaterials. http://www.lakeshorebio.com/products.html (accessed 21 December 2011).

van der Walle CF, *et al.* Current approaches to stabilising and analysing proteins during microencapsulation in PLGA. *Expert Opin Drug Delivery* 2009; **6**(2): 177–186.

Wiggins JS, *et al.* Hydrolytic degradation of poly(D,L-lactide) as a function of end group: Carboxylic acid vs. hydroxyl. *Polymer* 2006; **47**(6): 1960–1969.

Zurita R, *et al.* Copolymerization of glycolide and trimethylene carbonate. *J Polym Sci, Part A: Polym Chem* 2005; **44**(2): 993–1013.

21 Authors

RK Chang, W Qu, N Trivedi, J Johnson.

22 Date of Revision

21 December 2011.

Alitame

1 Nonproprietary Names

None adopted.

2 Synonyms

Aclame; L-aspartyl-D-alanine-*N*-(2,2,4,4-tetramethylthietan-3-yl)amide; 3-(L-aspartyl-D-alaninamido)-2,2,4,4-tetramethylthietane.

3 Chemical Name and CAS Registry Number

Lα-Aspartyl-*N*-(2,2,4,4-tetramethyl-3-thietanyl)-D-alaninamide anhydrous [80863-62-3]
L-α-Aspartyl-*N*-(2,2,4,4-tetramethyl-3-thietanyl)-D-alaninamide hydrate [99016-42-9]

4 Empirical Formula and Molecular Weight

$C_{14}H_{25}N_3O_4S$ 331.44 (for anhydrous)
$C_{14}H_{25}N_3O_4S \cdot 2\frac{1}{2}H_2O$ 376.50 (for hydrate)

5 Structural Formula

6 Functional Category

Sweetening agent.

7 Applications in Pharmaceutical Formulation or Technology

Alitame is an intense sweetening agent developed in the early 1980s and is approximately 2000 times sweeter than sucrose. It has an insignificant energy contribution of 6 kJ (1.4 kcal) per gram of alitame.

Alitame is currently primarily used in a wide range of foods and beverages at a maximum level of 40–300 mg/kg.[1,2]

8 Description

Alitame is a white nonhygroscopic crystalline powder; odorless or having a slight characteristic odor.

9 Pharmacopeial Specifications

—

10 Typical Properties

Acidity/alkalinity pH = 5–6 (5% w/v aqueous solution)
Isoelectric point pH 5.6
Melting point 136–147°C
Solubility *see* Table I.

Table I: Solubility of alitame.

Solvent	Solubility at 20°C unless otherwise stated
Chloroform	1 in 5000 at 25°C
Ethanol	1 in 1.6 at 25°C
n-Heptane	Practically insoluble
Methanol	1 in 2.4 at 25°C
Propylene glycol	1 in 1.9 at 25°C
Water	1 in 8.3 at 5°C
	1 in 7.6 at 25°C
	1 in 3.3 at 40°C
	1 in 2.0 at 50°C

Specific rotation $[\alpha]_D^{25}$ = +40° to +50° (1% w/v aqueous solution)

11 Stability and Storage Conditions

Alitame is stable in dry, room temperature conditions but undergoes degradation at elevated temperatures or when in solution at low pH. Alitame can degrade in a one-stage process to aspartic acid and alanine amide (under harsh conditions) or in a slow two-stage process by first degrading to its β-aspartic isomer and then to aspartic acid and alanine amide. At pH 5–8, alitame solutions at 23°C have a half-life of approximately 4 years. At pH 2 and 23°C the half-life is 1 year.

Alitame should be stored in a well-closed container in a cool, dry place.

12 Incompatibilities

Alitame may be incompatible with oxidizing and reducing substances or strong acids and bases.

13 Method of Manufacture

Alitame may be synthesized by a number of routes.[3,4] For example, 3-(D-alaninamido)-2,2,4,4-tetramethylthietane is dissolved in water and L-aspartic acid *N*-thiocarboxyanhydride is then added in portions with vigorous stirring, maintaining the pH of 8.5–9.5. The pH is then adjusted to 5.5 and *p*-toluenesulfonic acid monohydrate is added over a period of one hour. The precipitated crystalline *p*-toluenesulfonate salt is collected by filtration. To obtain alitame from its salt, a mixture of *Amberlite LA-1* (liquid anion exchange resin), dichloromethane, deionized water, and the salt is stirred for one hour, resulting in two clear layers. The aqueous layer is treated with carbon, clarified by filtration, and cooled to crystallize alitame.

Alternatively, tetramethylthietane amine is condensed with an *N*-protected form of D-alanine to give alanyl amide. This is then coupled to a protected analogue of L-aspartic acid to give a crude form of alitame. The crude product is then purified.

The major impurities are L-β-aspartyl-*N*-(2,2,4,4-tetramethyl-3-thietanyl)-D-alaninamide hydrate (2 : 5) (referred to a β-isomer), formed after rearrangement of the aspartyl unit; and *N*-(2,2,4,4-tetramethyl-3-thietanyl)-D-alaninamide (referred to as alanine amide), formed after hydrolysis of the L-aspartic/D-alanine bond.

14 Safety

Alitame is a relatively new intense sweetening agent used primarily in foods and confectionary. It is generally regarded as a relatively nontoxic and nonirritant material.

A

Chronic animal studies in mice, rats, and dogs carried out for a minimum of 18 months at concentrations >100 mg/kg per day exhibited no toxic or carcinogenic effects. In people, no evidence of untoward effects were observed following ingestion of 15 mg/kg per day for two weeks.

Following oral administration, 7–22% of alitame is unabsorbed and excreted in the feces. The remaining amount is hydrolyzed to aspartic acid and alanine amide. The aspartic acid is metabolized normally and the alanine amide excreted in the urine as a sulfoxide isomer, as the sulfone, or conjugated with glucuronic acid.

The WHO has set an acceptable daily intake of alitame at up to 1 mg/kg body-weight.[1,2]

LD_{50} (mouse, oral): >5 g/kg
LD_{50} (rabbit, skin): >2 g/kg
LD_{50} (rat, oral): >5 g/kg

15 Handling Precautions

Observe normal precautions appropriate to the circumstances and quantity of material handled. Eye protection and gloves are recommended. Alitame should be stored in tightly closed containers, and protected from exposure to direct sunlight and higher than normal room temperatures.

16 Regulatory Status

Alitame is approved for use in food applications in a number of countries worldwide including Australia, Chile, China, Mexico, and New Zealand.

17 Related Substances

Acesulfame potassium; aspartame; saccharin; saccharin sodium; sodium cyclamate.

18 Comments

A specification for alitame is contained in the *Food Chemicals Codex* (FCC).[5]

The PubChem Compound ID (CID) for alitame is 64763.

19 Specific References

1 FAO/WHO. Evaluation of certain food additives and contaminants. Fifty-ninth report of the joint FAO/WHO expert committee on food additives. *World Health Organ Tech Rep Ser* 2002; No. 913.
2 FAO/WHO. Evaluation of certain food additives and contaminants. Forty-sixth report of the joint FAO/WHO expert committee on food additives. *World Health Organ Tech Rep Ser* 1997; No. 868.
3 Sklavounos C. Process for preparation, isolation and purification of dipeptide sweeteners. United States Patent No. 4,375,430; 1 Mar, 1983.
4 Brennan TM, Hendrick ME. Branched amides of L-aspartyl-D-amino acid dipeptides. United States Patent No. 4,411,925; 25 Oct, 1983.
5 *Food Chemicals Codex*, 7th edn. Suppl. 3. Bethesda, MD: United States Pharmacopeia, 2011; 1673.

20 General References

Anon. Use of nutritive and nonnutritive sweeteners—position of ADA. *J Am Diet Assoc* 1998; **98**: 580–587.
Hendrick ME. Alitame. In: Nabors L, Gelardi R, eds. *Alternative Sweeteners*. New York: Marcel Dekker, 1991; 29–38.
Hendrick ME. Grenby TH, ed. *Advances in Sweeteners*. Glasgow: Blackie, 1996; 226–239.
Renwick AG. The intake of intense sweeteners – an update review. *Food Addit Contam* 2006; **23**: 327–338.

21 Authors

LY Galichet.

22 Date of Revision

1 October 2010.

Almond Oil

1 Nonproprietary Names

BP: Virgin Almond Oil
PhEur: Almond Oil, Virgin
USP–NF: Almond Oil

2 Synonyms

Almond oil, bitter; amygdalae oleum virginale; artificial almond oil; bitter almond oil; expressed almond oil; huile d'amande; oleo de amêndoas; olio di mandorla; sweet almond oil; virgin almond oil.

3 Chemical Name and CAS Registry Number

Almond oil [8007-69-0]

4 Empirical Formula and Molecular Weight

Almond oil consists chiefly of glycerides of oleic acid, with smaller amounts of linoleic and palmitic acids.

5 Structural Formula

See Section 4.

6 Functional Category

Emollient; oleaginous vehicle; solvent.

7 Applications in Pharmaceutical Formulation or Technology

Almond oil is used as a vehicle in parenteral preparations,[1] such as oily phenol injection. It is also used in nasal spray,[2] and topical preparations.[3–6]

8 Description

Almond oil occurs as a clear, colorless, or pale-yellow colored oil with a bland, nutty taste.

The PhEur 7.4 describes almond oil as the fatty oil obtained by cold expression from the ripe seeds of *Prunus dulcis* (Miller) DA Webb var. *dulcis* or *Prunus dulcis* (Miller) DA Webb var. *amara* (DC) Buchheim or a mixture of both varieties. A suitable antioxidant may be added.

The USP35–NF30 describes almond oil as the refined fixed oil obtained by expression from the kernels of varieties of *Prunus dulcis* (Miller) DA Webb (formerly known as *Prunus amygdalus* Batsch) (Fam. Rosaceae) except for *Prunus dolcis* (Miller) DA Webb var. *amara* (De Candolle) Focke).

9 Pharmacopeial Specifications

See Table I.

Table I: Pharmacopeial specifications for almond oil.

Test	PhEur 7.4	USP35–NF30[a]
Identification	+	+
Absorbance	+	—
Acid value	≤2.0	≤0.5
Characters	+	—
Peroxide value	≤15.0	≤5.0
Saponification value	—	—
Specific gravity	—	0.910–0.915
Unsaponifiable matter	≤0.9%	≤0.9%
Water content	≤0.1%	—
Composition of fatty acids	+	+
Saturated fatty acids < C_{16}	≤0.1%	≤0.1%
Arachidic acid (20:0)	≤0.2%	≤0.2%
Behenic acid (22:0)	≤0.2%	≤0.2%
Eicosenoic acid (20:1)	≤0.3%	≤0.3%
Erucic acid (22:1)	≤0.1%	≤0.1%
Linoleic acid (18:2)	20.0–30.0%	20.0–30.0%
Linolenic acid (18:3)	≤0.4%	≤0.4%
Margaric acid (17:0)	≤0.2%	≤0.2%
Oleic acid (18:1)	62.0–86.0%	62.0–76.0%
Palmitic acid (16:0)	4.0–9.0%	4.0–9.0%
Palmitoleic acid (16:1)	≤0.8%	≤0.8%
Stearic acid (18:0)	≤3.0%	≤3.0%
Sterols	+	+
Δ^5-Avenasterol	≥10.0%	≥5.0%
Δ^7-Avenasterol	≤3.0%	≤3.0%
Brassicasterol	≤0.3%	≤0.3%
Cholesterol	≤0.7%	≤0.7%
Campesterol	≤4.0%	≤5.0%
Stigmasterol	≤3.0%	≤4.0%
β-Sitosterol	73.0–87.0%	73.0–87.0%
Δ^7-Stigmasterol	≤3.0%	≤3.0%

(a) The USP35–NF30 monograph refers to refined almond oil.

10 Typical Properties

Flash point 320°C
Melting point −18°C
Refractive index n_D^{40} = 1.4630–1.4650
Smoke point 220°C
Solubility Miscible with chloroform, and ether; slightly soluble in ethanol (95%).

11 Stability and Storage Conditions

Almond oil should be stored in a well-closed container in a cool, dry place away from direct sunlight and odors. It may be sterilized by heating at 150°C for 1 hour. Almond oil is relatively resistant to oxidation.

12 Incompatibilities

—

13 Method of Manufacture

Almond oil is expressed from the seeds of the bitter or sweet almond, *Prunus dulcis (Prunus amygdalus; Amygdalus communis)* var. *amara* or var. *dulcis* (Rosaceae).[7] *See also* Section 4.

14 Safety

Almond oil is widely consumed as a food and is used both therapeutically and as an excipient in topical and parenteral pharmaceutical formulations, where it is generally regarded as a nontoxic and nonirritant material. However, there has been a single case reported of a 5-month-old child developing allergic dermatitis attributed to the application of almond oil for 2 months to the cheeks and buttocks.[8]

15 Handling Precautions

Observe normal precautions appropriate to the circumstances and quantity of material handled.

16 Regulatory Status

Included in nonparenteral and parenteral medicines licensed in the UK. Included in the FDA Inactive Ingredient Database (topical, emulsion and cream). Widely used as an edible oil. Included in the Canadian Natural Health Products Ingredients Database.

17 Related Substances

Canola oil; corn oil; cottonseed oil; peanut oil; refined almond oil; sesame oil; soybean oil.

Refined almond oil

Synonyms Amygdalae oleum raffinatum.
Comments Refined almond oil is defined in some pharmacopeias such as the PhEur 7.4. Refined almond oil is a clear, pale yellow colored oil with virtually no taste or odor. It is obtained by expression of almond seeds followed by subsequent refining. It may contain a suitable antioxidant.

18 Comments

A 100 g quantity of almond oil has a nutritional energy value of 3700 kJ (900 kcal) and contains 100 g of fat of which 28% is polyunsaturated, 64% is monounsaturated and 8% is saturated fat.

Almond oil is used therapeutically as an emollient and to soften ear wax.[9,10] Studies have suggested that almond consumption is associated with health benefits, including a decreased risk of colon cancer.[11]

A specification for bitter almond oil is contained in the *Food Chemicals Codex* (FCC)[12] and in the *Japanese Pharmaceutical Excipients* (JPE).[13]

19 Specific References

1 Van Hoogmoed LM, *et al.* Ultrasonographic and histologic evaluation of medial and middle patellar ligaments in exercised horses following injection with ethanolamine oleate and 2% iodine in almond oil. *Am J Vet Res* 2002; **63**(5): 738–743.
2 Cicinelli E, *et al.* Progesterone administration by nasal spray in menopausal women: comparison between two different spray formulations. *Gynecol Endocrinol* 1992; **6**(4): 247–251.
3 Christen P, *et al.* Stability of prednisolone and prednisolone acetate in various vehicles used in semi-solid topical preparations. *J Clin Pharm Ther* 1990; **15**(5): 325–329.
4 Almeida IF, Bahia MF. Evaluation of the physical stability of two oleogels. *Int J Pharm* 2006; **327**(1-2): 73–77.

5 Almeida IF, *et al.* Moisturizing effect of oleogel/hydrogel mixtures. *Pharm Dev Technol* 2008; **13**(6): 487–494.
6 Ammar HO, *et al.* Design of a transdermal delivery system for aspirin as an antithrombotic drug. *Int J Pharm* 2006; **327**(1-2): 81–88.
7 Evans WC. *Trease and Evans' Pharmacognosy*, 14th edn. London: WB Saunders, 1996: 184.
8 Guillet G, Guillet M-H. Percutaneous sensitization to almond in infancy and study of ointments in 27 children with food allergy. *Allerg Immunol* 2000; **32**(8): 309–311.
9 Sweetman SC, ed. *Martindale: The Complete Drug Reference*, 37th edn. London: Pharmaceutical Press, 2011: 2463.
10 Ahmad Z. The uses and properties of almond oil. *Complement Ther Clin Pract* 2010; **16**(1): 10–12.
11 Davis PA, Iwahashi CK. Whole almonds and almond fractions reduce aberrant crypt foci in a rat model of colon carcinogenesis. *Cancer Lett* 2001; **165**(1): 27–33.
12 *Food Chemicals Codex*, 7th edn. Bethesda, MD: United States Pharmacopeia, 2010: 41.
13 Japan Pharmaceutical Excipients Council. *Japanese Pharmaceutical Excipients 2004*. Tokyo: Yakuji Nippo, 2004: 57.

20 General References

Allen LV. Oleaginous vehicles. *Int J Pharm Compound* 2000; 4(6): 470–472.
Anonymous. Iodine 2% in oil injection. *Int J Pharm Compound* 2001; **5**(2): 131.
Brown JH, *et al.* Oxidative stability of botanical emollients. *Cosmet Toilet* 1997; **112**(Jul): 87–9092, 94, 96–98.
Shaath NA, Benveniste B. Natural oil of bitter almond. *Perfum Flavor* 1991; **16**(Nov–Dec): 17, 19–24.

21 Authors

SL Cantor, SA Shah.

22 Date of Revision

1 March 2012.

Alpha Tocopherol

1 Nonproprietary Names

BP: *RRR*-Alpha-Tocopherol
JP: Tocopherol
PhEur: *RRR*-α-Tocopherol
USP–NF: Vitamin E
See also Sections 3, 9, and 17.

2 Synonyms

Copherol F1300; (±)-3,4-dihydro-2,5,7,8-tetramethyl-2-(4,8,12-trimethyltridecyl)-2*H*-1-benzopyran-6-ol; E307; *RRR*-α-tocopherolum; synthetic alpha tocopherol; all-*rac*-α-tocopherol; *dl*-α-tocopherol; 5,7,8-trimethyltocol.

3 Chemical Name and CAS Registry Number

(±)-(2*RS*,4'*RS*,8'*RS*)-2,5,7,8-Tetramethyl-2-(4',8',12'-trimethyltridecyl)-6-chromanol [10191-41-0]

Note that alpha tocopherol has three chiral centers, giving rise to eight isomeric forms. The naturally occurring form is known as *d*-alpha tocopherol or (2*R*,4'*R*,8'*R*)-alpha-tocopherol. The synthetic form, *dl*-alpha tocopherol or simply alpha tocopherol, occurs as a racemic mixture containing equimolar quantities of all the isomers.

Similar considerations apply to beta, delta, and gamma tocopherol and tocopherol esters.

See Section 17 for further information.

4 Empirical Formula and Molecular Weight

$C_{29}H_{50}O_2$ 430.72

5 Structural Formula

R1 R2 HO R3 O CH3 * * * CH3 CH3 CH3 CH3

Alpha tocopherol: $R^1 = R^2 = R^3 = CH_3$
Beta tocopherol: $R^1 = R^3 = CH_3$; $R^2 = H$
Delta tocopherol: $R^1 = CH_3$; $R^2 = R^3 = H$
Gamma tocopherol: $R^1 = R^2 = CH_3$; $R^3 = H$
* Indicates chiral centers.

6 Functional Category

Antioxidant; nonionic surfactant; plasticizer.

7 Applications in Pharmaceutical Formulation or Technology

Alpha tocopherol is primarily recognized as a source of vitamin E, and the commercially available materials and specifications reflect this purpose. While alpha tocopherol also exhibits antioxidant properties, the beta, delta, and gamma tocopherols are considered to be more effective as antioxidants.

Alpha tocopherol is a highly lipophilic compound, and is an excellent solvent for many poorly soluble drugs.[1–4] Of widespread regulatory acceptability, tocopherols are of value in oil- or fat-based pharmaceutical products and are normally used in the concentration range 0.001–0.05% v/v. There is frequently an optimum concentration; thus the autoxidation of linoleic acid and methyl linolenate is reduced at low concentrations of alpha tocopherol, and is accelerated by higher concentrations. Antioxidant effectiveness can be increased by the addition of oil-soluble synergists such as lecithin and ascorbyl palmitate.[4]

Alpha tocopherol may be used as an efficient plasticizer.[5] It has been used in the development of deformable liposomes as topical formulations.[6]

d-Alpha tocopherol has also been used as a nonionic surfactant in oral and injectable formulations.[3]

8 Description

Alpha tocopherol is a natural product. The PhEur 7.4 describes alpha tocopherol as a clear, colorless or yellowish-brown, viscous, oily liquid. *See also* Section 17.

9 Pharmacopeial Specifications

See Table I.

Table I: Pharmacopeial specifications for alpha tocopherol.

Test	JP XV	PhEur 7.4	USP35–NF30
Identification	+	+	+
Characters	–	+	–
Acidity	–	–	+
Optical rotation	–	+0.05° to +0.10°	+
Heavy metals	≤20 ppm	–	–
Related substances	–	+	–
Absorbance	+	–	–
at 292 nm	71.0–76.0	–	–
Refractive index	1.503–1.507	–	–
Specific gravity	0.947–0.955	–	–
Clarity and color of solution	+	–	–
Assay	96.0–102.0%	94.5–102.0%	96.0–102.0%

Note that the USP35–NF30 describes vitamin E as comprising *d*- or *dl*-alpha tocopherol, *d*- or *dl*-alpha tocopheryl acetate, or *d*- or *dl*-alpha tocopheryl acid succinate. However, the PhEur 7.4 describes alpha tocopherol and alpha tocopheryl acetate in separate monographs.
The diversity of the tocopherols described in the various pharmacopeial monographs makes the comparison of specifications more complicated; *see* Section 17.

10 Typical Properties

Boiling point 235°C
Density 0.947–0.951 g/cm^3
Flash point 240°C
Ignition point 340°C
Refractive index n_D^{20} = 1.503–1.507
Solubility Practically insoluble in water; freely soluble in acetone, ethanol, ether, and vegetable oils.
Spectroscopy

IR spectra *see* Figure 1.

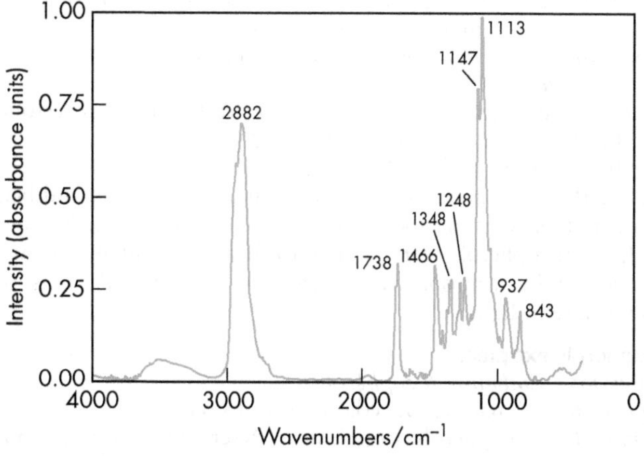

Figure 1: Infrared spectrum of alpha tocopherol measured by transmission. Adapted with permission of Informa Healthcare.

11 Stability and Storage Conditions

Tocopherols are oxidized slowly by atmospheric oxygen and rapidly by ferric and silver salts. Oxidation products include tocopheroxide, tocopherylquinone, and tocopherylhydroquinone, as well as dimers and trimers. Tocopherol esters are more stable to oxidation than the free tocopherols but are in consequence less effective antioxidants. *See also* Section 17.

Tocopherols should be stored under an inert gas, in an airtight container in a cool, dry place and protected from light.

12 Incompatibilities

Tocopherols are incompatible with peroxides and metal ions, especially iron, copper, and silver. Tocopherols may be absorbed into plastic.[7]

13 Method of Manufacture

Naturally occurring tocopherols are obtained by the extraction or molecular distillation of steam distillates of vegetable oils; for example, alpha tocopherol occurs in concentrations of 0.1–0.3% in corn, rapeseed, soybean, sunflower, and wheat germ oils.[8] Beta and gamma tocopherol are usually found in natural sources along with alpha tocopherol. Racemic synthetic tocopherols may be prepared by the condensation of the appropriate methylated hydroquinone with racemic isophytol.[9]

14 Safety

Tocopherols (vitamin E) occur in many food substances that are consumed as part of the normal diet. The daily nutritional requirement has not been clearly defined but is estimated to be 3.0–20.0 mg. Absorption from the gastrointestinal tract is dependent upon normal pancreatic function and the presence of bile. Tocopherols are widely distributed throughout the body, with some ingested tocopherol metabolized in the liver; excretion of metabolites is via the urine or bile. Individuals with vitamin E deficiency are usually treated by oral administration of tocopherols, although intramuscular and intravenous administration may sometimes be used.

Tocopherols are well tolerated, although excessive oral intake may cause headache, fatigue, weakness, digestive disturbance, and nausea. Prolonged and intensive skin contact may lead to erythema and contact dermatitis.

The use of tocopherols as antioxidants in pharmaceuticals and food products is unlikely to pose any hazard to human health since the daily intake from such uses is small compared with the intake of naturally occurring tocopherols in the diet.

The WHO has set an acceptable daily intake of tocopherol used as an antioxidant at 0.15–2.0 mg/kg body-weight.[10]

15 Handling Precautions

Observe normal precautions appropriate to the circumstances and quantity of material handled. Gloves and eye protection are recommended.

16 Regulatory Status

GRAS listed. Accepted for use as a food additive in Europe. Included in the FDA Inactive Ingredients Database (IV injections, powder, lyophilized powder for liposomal suspension; oral capsules, tablets, and topical preparations). Included in the Canadian Health Products Ingredients Database. Included in nonparenteral medicines licensed in the UK.

17 Related Substances

d-Alpha tocopherol; *d*-alpha tocopheryl acetate; *dl*-alpha tocopheryl acetate; *d*-alpha tocopheryl acid succinate; *dl*-alpha toco-

pheryl acid succinate; beta tocopherol; delta tocopherol; gamma tocopherol; tocopherols excipient.

d-Alpha tocopherol

Empirical formula $C_{29}H_{50}O_2$
Molecular weight 430.72
CAS number [59-02-9]
Synonyms Natural alpha tocopherol; (+)-(2*R*,4'*R*,8'*R*)-2,5,7,8-tetramethyl-2-(4',8',12'-trimethyltridecyl)-6-chromanol; *d*-α-tocopherol; vitamin E.
Appearance A practically odorless, clear, yellow, or greenish-yellow viscous oil.
Melting point 2.5–3.5°C
Solubility Practically insoluble in water; soluble in ethanol (95%). Miscible with acetone, chloroform, ether, and vegetable oils.
Specific gravity 0.95
Comments *d*-Alpha tocopherol is the naturally occurring form of alpha tocopherol.

d-Alpha tocopheryl acetate

Empirical formula $C_{31}H_{52}O_3$
Molecular weight 472.73
CAS number [58-95-7]
Synonyms (+)-(2*R*,4'R,8'*R*)-2,5,7,8-Tetramethyl-2-(4',8',12'-trimethyltridecyl)-6-chromanyl acetate; *d*-α-tocopheryl acetate; vitamin E.
Appearance A practically odorless, clear, yellow, or greenish-yellow colored viscous oil that may solidify in the cold.
Melting point 28°C
Solubility Practically insoluble in water; soluble in ethanol (95%). Miscible with acetone, chloroform, ether, and vegetable oils.
Specific rotation $[\alpha]_D^{25} = +0.25°$ (10% w/v solution in chloroform)
Comments Unstable to alkalis.

dl-Alpha tocopheryl acetate

Empirical formula $C_{31}H_{52}O_3$
Molecular weight 472.73
CAS number [7695-91-2]
Synonyms (±)-3,4-Dihydro-2,5,7,8-tetramethyl-2-(4,8,12-trimethyltridecyl)-2*H*-1-benzopyran-6-ol acetate; (±)-(2*RS*,4'*RS*,8'*RS*)-2,5,7,8-tetramethyl-2-(4',8',12'-trimethyltridecyl)-6-chromanyl acetate; (±)-α-tocopherol acetate; α-tocopheroli acetas; all-rac-α-tocopheryl acetate; *dl*-α-tocopheryl acetate; vitamin E.
Appearance A practically odorless, clear, yellow, or greenish-yellow viscous oil.
Density 0.953 g/cm^3
Melting point −27.5°C
Refractive index $n_D^{20} = 1.4950–1.4972$
Solubility Practically insoluble in water; freely soluble in acetone, chloroform, ethanol, ether, and vegetable oils; soluble in ethanol (95%).
Comments Unstable to alkali. However, unlike alpha tocopherol, the acetate is much less susceptible to the effects of air, light, or ultraviolet light. Alpha tocopherol acetate concentrate, a powdered form of alpha tocopherol acetate, is described in the PhEur 7.4. The concentrate may be prepared by either dispersing alpha tocopherol acetate in a suitable carrier such as acacia or gelatin, or by adsorbing alpha tocopherol acetate on silicic acid.

d-Alpha tocopheryl acid succinate

Empirical formula $C_{33}H_{54}O_5$
Molecular weight 530.8
CAS number [4345-03-3]
Synonyms (+)-α-Tocopherol hydrogen succinate; *d*-α-tocopheryl acid succinate; vitamin E.
Appearance A practically odorless white powder.
Melting point 76–77°C
Solubility Practically insoluble in water; slightly soluble in alkaline solutions; soluble in acetone, ethanol (95%), ether, and vegetable oils; very soluble in chloroform.
Comments Unstable to alkalis.

dl-Alpha tocopheryl acid succinate

Empirical formula $C_{33}H_{54}O_5$
Molecular weight 530.8
CAS number [17407-37-3]
Synonyms (±)-α-Tocopherol hydrogen succinate; *dl*-α-tocopheryl acid succinate; *dl*-α-tocopherol succinate; vitamin E.
Appearance A practically odorless, white crystalline powder.
Solubility Practically insoluble in water; slightly soluble in alkaline solutions; soluble in acetone, ethanol (95%), ether, and vegetable oils; very soluble in chloroform.
Comments Unstable to alkalis.

Beta tocopherol

Empirical formula $C_{28}H_{48}O_2$
Molecular weight 416.66
CAS number [148-03-8]
Synonyms Cumotocopherol; (±)-3,4-dihydro-2,5,8-trimethyl-2-(4,8,12-trimethyltridecyl)-2*H*-1-β-benzopyran-6-ol; 5,8-dimethyltocol; neotocopherol; *dl*-β-tocopherol; vitamin E; *p*-xylotocopherol.
Appearance A pale yellow-colored viscous oil.
Solubility Practically insoluble in water; freely soluble in acetone, chloroform, ethanol (95%), ether, and vegetable oils.
Specific rotation $[\alpha]_D^{20} = +6.37°$
Comments Less active biologically than alpha tocopherol. Obtained along with alpha tocopherol and gamma tocopherol from natural sources. Beta tocopherol is very stable to heat and alkalis, and is slowly oxidized by atmospheric oxygen.

Delta tocopherol

Empirical formula $C_{27}H_{46}O_2$
Molecular weight 402.64
CAS number [119-13-1]
Synonyms (±)-3,4-Dihydro-2,8-dimethyl-2-(4,8,12-trimethyltridecyl)-2*H*-1-benzopyran-6-ol; E309; 8-methyltocol; *dl*-δ-tocopherol; vitamin E.
Appearance A pale yellow-colored viscous oil.
Solubility Practically insoluble in water; freely soluble in acetone, chloroform, ethanol (95%), ether, and vegetable oils.
Comments Occurs naturally as 30% of the tocopherol content of soybean oil. Delta tocopherol is said to be the most potent antioxidant of the tocopherols.

Gamma tocopherol

Empirical formula $C_{28}H_{48}O_2$
Molecular weight 416.66
CAS number [7616-22-0]
Synonyms (±)-3,4-Dihydro-2,7,8-trimethyl-2-(4,8,12-trimethyltridecyl)-2*H*-1-benzopyran-6-ol; 7,8-dimethyltocol; E308; *dl*-γ-tocopherol; vitamin E; *o*-xylotocopherol.
Appearance A pale yellow-colored viscous oil.
Melting point −30°C
Solubility Practically insoluble in water; freely soluble in acetone, chloroform, ethanol (95%), ether, and vegetable oils.
Specific rotation $[\alpha]_D^{20} = -2.4°$ (in ethanol (95%))
Comments Occurs in natural sources along with alpha and beta tocopherol. Gamma tocopherol is biologically less active than alpha tocopherol. Very stable to heat and alkalis; slowly oxidized by atmospheric oxygen and gradually darkens on exposure to light.

Tocopherols excipient

Synonyms *Embanox tocopherol.*
Appearance A pale yellow-colored viscous oil.
Comments Tocopherols excipient is described in the USP35–NF30 as a vegetable oil solution containing not less than 50.0% of total tocopherols, of which not less than 80.0% consists of varying amounts of beta, delta, and gamma tocopherols.

18 Comments

Note that most commercially available tocopherols arc used as sources of vitamin E, rather than as antioxidants in pharmaceutical formulations.

Various mixtures of tocopherols, and mixtures of tocopherols with other excipients, are commercially available, and individual manufacturers should be consulted for specific information on their products.

Molecularly imprinted polymers for use in the controlled release of alpha tocopherol in gastrointestinal simulating fluids have been investigated.[11]

The EINECS numbers for alpha tocopherol, *d*-alpha tocopherol, and *dl*-alpha tocopherol are 215-798-8, 200-412-2, and 233-466-0, respectively. The PubChem Compound ID (CID) for alpha tocopherol includes 14 985 and 1 548 900.

19 Specific References

1 Nielsen PB, *et al.* The effect of α-tocopherol on the *in vitro* solubilisation of lipophilic drugs. *Int J Pharm* 2001; **222**: 217–224.

2 Constantinides PP, *et al.* Tocol emulsions for drug solubilization and parenteral delivery. *Adv Drug Delivery* 2004; **56**(9): 1243–1255.

3 Strickley RG. Solubilizing excipients in oral and injectable formulations. *Pharm Res* 2004; **21**(2): 201–230.

4 Johnson DM, Gu LC. Autoxidation and antioxidants. In: Swarbrick J, Boylan JC, eds. *Encyclopedia of Pharmaceutical Technology*. 1: New York: Marcel Dekker, 1988; 415–450.

5 Kangarlou S, *et al.* Physico-mechanical analysis of free ethyl cellulose films comprised with novel plasticizers of vitamin resources. *Int J Pharm* 2008; **356**: 153–166.

6 Gallarate M, *et al.* Deformable liposomes as topical formulations containing alpha-tocopherol. *J Dispers Sci Technol* 2006; **27**: 703–713.

7 Allwood MC. Compatibility and stability of TPN mixtures in big bags. *J Clin Hosp Pharm* 1984; **9**: 181–198.

8 Buck DF. Antioxidants. In: Smith J, ed. *Food Additive User's Handbook*. Glasgow: Blackie, 1991; 1–46.

9 Rudy BC, Senkowski BZ. *dl*-Alpha-tocopheryl acetate. In: Florey K, ed. *Analytical Profiles of Drug Substances*. 3: New York: Academic Press, 1974; 111–126.

10 FAO/WHO. Evaluation of certain food additives and contaminants. Thirtieth report of the joint FAO/WHO expert committee on food additives. *World Health Organ Tech Rep Ser* 1987; No. 751.

11 Puoci F, *et al.* Molecularly imprinted polymers for alpha-tocopherol delivery. *Drug Deliv* 2008; **15**: 253–258.

20 General References

US National Research Council Food and Nutrition Board. *Recommended Dietary Allowances*, 10th edn. Washington DC: National Academy Press, 1989; 99–105.

21 Author

ME Fenton.

22 Date of Revision

1 March 2012.

Aluminum Hydroxide Adjuvant

1 Nonproprietary Names

PhEur: Aluminium Hydroxide, Hydrated, for Adsorption

2 Synonyms

Alhydrogel; aluminii hydroxidum hydricum ad adsorptionem; aluminium hydroxide adjuvant; aluminium oxyhydroxide; poorly crystalline boehmite; pseudoboehmite; *Rehydragel*.

3 Chemical Name and CAS Registry Number

Aluminum oxyhydroxide [21645-51-2]

4 Empirical Formula and Molecular Weight

AlO(OH) 59.99

5 Structural Formula

Structural hydroxyl groups form hydrogen bonds between AlO(OH) octahedral sheets, where hydroxyl groups are exposed at the surface. The surface hydroxyl groups produce a pH-dependent surface charge by accepting a proton to produce a positive site, or donating a proton to produce a negative site. The pH-dependent surface charge is characterized by the point of zero charge, which is equivalent to the isoelectric point in protein chemistry. The surface hydroxyl groups may also undergo ligand exchange with fluoride, phosphate, carbonate, sulfate, or borate anions.

6 Functional Category

Adsorbent; vaccine adjuvant.

7 Applications in Pharmaceutical Formulation or Technology

Aluminum hydroxide adjuvant is used in parenteral human and veterinary vaccines.[1] It activates Th2 immune responses, including IgG and IgE antibody responses.

8 Description

Aluminum hydroxide adjuvant is a white hydrogel that sediments slowly and forms a clear supernatant.

9 Pharmacopeial Specifications

See Table I. Note that the USP35–NF30 includes a monograph for aluminum hydroxide gel, which is a form of aluminum hydroxide that is used as an antacid, in which there is a partial substitution of carbonate for hydroxide.

See Section 17.

10 Typical Properties

Acidity/alkalinity pH = 5.5–8.5

Particle size distribution Primary particles are fibrous with average dimensions of 4.5 × 2.2 × 10 nm. The primary particles form aggregates of 1–10 μm.

Point of zero charge pH = 11.4

Protein binding capacity >0.5 mg BSA/mg equivalent Al_2O_3

Table I: Pharmacopeial specifications for aluminum hydroxide adjuvant.

Test	PhEur 7.4
Identification	+
Characters	+
Solution	+
pH	5.5–8.5
Adsorption power	+
Sedimentation	+
Chlorides	≤0.33%
Nitrates	≤100 ppm
Sulfates	≤0.5%
Ammonium	≤50 ppm
Arsenic	≤1 ppm
Iron	≤15 ppm
Heavy metals	≤20 ppm
Bacterial endotoxins	+
Assay	90.0–110.0%

Solubility Soluble in alkali hydroxides and mineral acids. Heat may be required to dissolve the aluminum hydroxide adjuvant.

Specific surface area 500 m^2/g.[2]

X-ray diffractogram Exhibits characteristic x-ray diffraction pattern having diffraction bands at 6.46, 3.18, 2.35, 1.86, 1.44 and 1.31 Å.

11 Stability and Storage Conditions

Aluminum hydroxide adjuvant is stable for at least 2 years when stored at 4–30°C in well-sealed inert containers. It must not be allowed to freeze as the hydrated colloid structure will be irreversibly damaged.

12 Incompatibilities

When exposed to phosphate, carbonate, sulfate, or borate anions, the point of zero charge for aluminum hydroxide adjuvant decreases.

13 Method of Manufacture

Aluminum hydroxide adjuvant is prepared by the precipitation of a soluble aluminum salt by an alkali hydroxide, or the precipitation of an alkali aluminate by acid.

14 Safety

Aluminum hydroxide adjuvant is intended for use in parenteral vaccines and is generally regarded as nontoxic. It may cause mild irritation, dryness, and dermatitis on skin contact. On eye contact, aluminum hydroxide adjuvant may also cause redness, conjunctivitis, and short-term mild irritation. Ingestion of large amounts may cause gastrointestinal irritation with nausea, vomiting, and constipation. Inhalation of the dried product may cause respiratory irritation and cough. Type I hypersensitivity reactions following parenteral administration have been reported.[3]

15 Handling Precautions

Observe normal precautions appropriate to the circumstances and quantity of material handled. Eye protection and gloves are recommended.

16 Regulatory Status

GRAS listed. Accepted for use in human and veterinary parenteral vaccines in Europe and the USA. The limits for use in human vaccines are 0.85 mg aluminum/dose (FDA) and 1.25 mg aluminum/dose (WHO). There are no established limits for use in veterinary vaccines. Reported in the EPA TSCA Inventory.

17 Related Substances

Aluminum phosphate adjuvant.

18 Comments

Aluminum hydroxide adjuvant is used for the isolation of certain serum components such as blood clotting factors.[4]

Different grades of aluminum hydroxide adjuvant with various concentrations, protein binding capacities, and points of zero charge are available.

The impurity limits at 2% equivalent Al_2O_3 are Cl <0.5%; SO_4 <0.5%; PO_4 <0.1%; NO_3 <0.1%; NH_4 <0.1%; Fe <20 ppm; As <0.6 ppm; and heavy metals <20 ppm.

The aluminum hydroxide gel referred to in the USP 33 S1 is used in cosmetics as an emollient, filler, humectant, a mild astringent, and viscosity controlling agent. In pharmaceutical preparations it is used as an adsorbent, and as a protein binder.[5] It is also used therapeutically as an antacid, and as an abrasive in dentrifrices. It is not, however, used as a vaccine adjuvant.

19 Specific References

1 Shirodkar S, *et al.* Aluminum compounds used as adjuvants in vaccines. *Pharm Res* 1990; 7: 1282–1288.

2 Johnston CT, *et al.* Measuring the surface area of aluminum hydroxide adjuvant. *J Pharm Sci* 2002; **91**: 1702–1706.

3 Goldenthal KL, *et al.* Safety evaluation of vaccine adjuvants. *AIDS Res Hum Retroviruses* 1993; 9(Suppl. 1): S47–S51.

4 Prowse CV, *et al.* Changes in factor VIII complex activities during the production of a clinical intermediate purity factor VIII concentrate. *Thromb Haemost* 1981; **46**: 597–601.

5 Ash M, Ash I. *Handbook of Pharmaceutical Additives*, 3rd edn. Endicott, NY: Synapse Information Resources, 2007; 446.

20 General References

Gupta RK *et al.* Adjuvant properties of aluminum and calcium compounds. In: Powell MF, Newman MJ, eds. *Vaccine Design*. New York: Plenum, 1995; 229–248.

Hem SL, Hogenesch H. Aluminum-containing adjuvants: properties, formulation, and use. In: Singh M, ed. *Vaccine Adjuvants and Delivery Systems*. New York: Wiley, 2007; 81–114.

Lindblad EB. Aluminum adjuvants – in retrospect and prospect. *Vaccine* 2004; 22: 3658–3668.

Lindblad EB. Aluminum adjuvants. In: Stewart-Tull DES, ed. *The Theory and Practical Application of Adjuvants*. New York: Wiley, 1995; 21–35.

Vogel FR, Powell MF. A compendium of vaccine adjuvants and excipients. In: Powell MF, Newman MJ, eds. *Vaccine Design*. New York: Plenum, 1995; 229–248.

Vogel FR, Hem SL Immunogenic adjuvants. In: Plotkin SA *et al.* eds. *Vaccines*, 5th edn. New York: W.B. Saunders, 2008; 59–71.

White JL, Hem SL Characterization of aluminum-containing adjuvants. In: Brown F *et al.* eds. *Physico-Chemical Procedures for the Characterization of Vaccines*. IABS Symposia Series, Development in Biologicals, 103: New York: Karger, 2000; 217–228.

21 Authors

SL Hem, EB Lindblad, L Grove.

22 Date of Revision

1 March 2012.

Aluminum Monostearate

1 Nonproprietary Names

JP: Aluminum Monostearate
USP–NF: Aluminum Monostearate

2 Synonyms

Aluminum stearate; aluminum, dihydroxy (octadecanoate-O-); dihydroxyaluminum monostearate; octadecanoic acid aluminum salt; stearic acid aluminum salt; stearic acid aluminum dihydroxide salt; *Synpro*.

3 Chemical Name and CAS Registry Number

Aluminum monostearate [7047-84-9]

4 Empirical Formula and Molecular Weight

$C_{18}H_{37}AlO_4$ 344.50

5 Structural Formula

$[CH_3(CH_2)_{16}COO]Al(OH)_2$

6 Functional Category

Emollient; emulsion stabilizing agent; gelling agent; opacifier; viscosity-increasing agent.

7 Applications in Pharmaceutical Formulation or Technology

Aluminum monostearate is mainly used in microencapsulation[1-3] and in the manufacture of ointments. Aluminum monostearate produces a high gel strength and is used as a thixotropic agent and viscosity-increasing agent in nonaqueous cosmetic and pharmaceutical formulations.

8 Description

Aluminum monostearate is an aluminum compound of stearic acid and palmitic acid. The USP35–NF30 states that aluminum monostearate contains the equivalent of not less than 14.5% and not more than 16.5% of Al_2O_3, calculated on the dried basis. The JP XV states that it contains not less than 7.2% and not more than 8.9% of aluminum.

Aluminum monostearate occurs as a white, fine, bulky powder with a slight odor of fatty acid. It is a solid material.

9 Pharmacopeial Specifications

See Table I. *See also* Section 18.

Table I: Pharmacopeial specifications for aluminum monostearate.

Test	JP XV	USP35–NF30
Identification	+	+
Description	+	—
Loss on drying	≤3.0%	≤2.0%
Arsenic	≤2 ppm	≤4 ppm
Heavy metals	≤50 ppm	≤50 μg/g
Acid value for fatty acid	+	—
Free fatty acid	+	—
Water-soluble salts	≤10 mg	—
Assay of Al (dried basis)	7.2–8.9%	14.5–16.5%

10 Typical Properties

See Table II.

Melting point 220–225°C

Solubility Practically insoluble in water. Soluble in ethanol (95%) and benzene.

Specific gravity 1.14

Table II: Typical physical properties of selected commercially available aluminum monostearates.

Grade	Assay (as Al_2O_3) (%)	Loss on drying (%)	Median particle size (μm)
Synpro Aluminum Monostearate NF	15.5	0.8	7.0
Synpro Aluminum Monostearate NF Gellant	15.3	1.6	—

11 Stability and Storage Conditions

Aluminum monostearate should be stored in a well-closed container in a cool, dry, place. It is stable under ordinary conditions of use and storage.

12 Incompatibilities

—

13 Method of Manufacture

Aluminum monostearate is prepared by reacting aluminum with stearic acid.

14 Safety

Aluminum monostearate is generally regarded as relatively nontoxic and nonirritant when used as an excipient.

15 Handling Precautions

Observe normal precautions appropriate to the circumstances and quantity of the material handled. When heated to decomposition, aluminum monostearate emits acrid smoke and irritating vapors.

16 Regulatory Status

Aluminum monostearate and aluminum stearate are included in the FDA Inactive Ingredients Database (oral capsules and tablets, topical creams and ointments). Included in nonparenteral medicines licensed in the UK. Included in the Canadian Natural Health Products Ingredients Database.

17 Related Substances

Aluminum distearate; aluminum tristearate.

Aluminum distearate

Empirical formula $C_{36}H_{37}AlO_5$

Molecular weight 877.39

CAS number [300-92-5]

Synonyms Hydroxyaluminum distearate; aluminum stearate; aluminum monobasic stearate.

Description Aluminum distearate occurs as a fine white to off-white colored powder with a slight odor of fatty acid.

A

Melting point 150–165°C
Specific gravity 1.01
Solubility Soluble in benzene, and in ethanol (95%); practically insoluble in water.
Comments The EINECS number for aluminum distearate is 206-101-8.

Aluminum tristearate
Empirical formula $C_{54}H_{105}AlO_6$
Molecular weight 610.9
CAS number [637-12-7]
Synonyms Hydroxyaluminum tristearate; aluminum stearate.
Description Aluminum tristearate occurs as a fine white to off-white colored powder with a slight odor of fatty acid.
Melting point 117–120°C
Specific gravity 1.01
Solubility Practically insoluble in water. Soluble in ethanol (95%), benzene, turpentine oil, and mineral oils when freshly prepared.
Comments The EINECS number for aluminum tristearate is 211-279-5.

18 Comments

A specification for aluminum stearate, described as consisting mainly of the distearate, is included in the *Japanese Pharmaceutical Excipients* (JPE).[4]

It should be noted that unless otherwise specified, aluminum stearate could refer to the monostearate (CAS number 7047-84-9), distearate (CAS number 300-92-5) or tristearate (CAS number 637-12-7). The distearate exhibits the same excipient properties as the tristearate and is used in similar pharmaceutical applications. However, the monostearate is more widely used in both cosmetic and pharmaceutical preparations.

The EINECS number for aluminum monostearate is 230-325-5.

Aluminum monostearate can be used as an emulsion stabilizer in cosmetic emulsions and is used in cosmetics such as mascara, moisturizers, and sunscreens. It is also used to improve the flow properties and adherence of cosmetic powders.

19 Specific References

1 Horoz BB, *et al.* Effect of different dispersing agents on the characteristics of *Eudragit* microspheres prepared by a solvent evaporation method. *J Microencapsul* 2004; **21**: 191–202.
2 Wu PC, *et al.* Preparation and evaluation of sustained release microspheres of potassium chloride prepared with ethylcellulose. *Int J Pharm* 2003; **260**: 115–121.
3 Wu PC, *et al.* Design and evaluation of sustained release microspheres of potassium chloride prepared by *Eudragit*. *Eur J Pharm Sci* 2003; **19**: 115–122.
4 Japan Pharmaceutical Excipients Council. *Japanese Pharmaceutical Excipients 2004*. Tokyo: Yakuji Nippo, 2004: 74–75.

20 General References

Ferro Corporation. Material safety data sheet: *Synpro*, 2010.

21 Author

J Shur.

22 Date of Revision

1 March 2012.

Aluminum Oxide

1 Nonproprietary Names

USP–NF: Aluminum Oxide

2 Synonyms

Activated alumina; activated aluminum oxide; alpha aluminum oxide; alumina; alumina, calcined; alumina, tabular; aluminum oxide alumite; aluminum trioxide; gamma aluminum oxide.

3 Chemical Name and CAS Registry Number

Aluminum oxide [1344-28-1]

4 Empirical Formula and Molecular Weight

Al_2O_3 101.96

5 Structural Formula

Aluminum oxide occurs naturally as the minerals bauxite, bayerite, boehmite, corundum, diaspore, and gibbsite.

6 Functional Category

Adsorbent; dispersing agent.

7 Applications in Pharmaceutical Formulation or Technology

Aluminum oxide is used mainly in tablet formulations.[1] It is used for decoloring powders and is particularly widely used in antibiotic formulations. It is also used in suppositories, pessaries, and urethral inserts.

Aluminum oxide in the form of nanoparticles has been investigated for glidant properties in tablet formulation. It was shown to be effective but produced a large reduction in tablet tensile strength.[2]

Hydrated aluminum oxide (*see* Section 18) is used in cosmetic formulations.

8 Description

Aluminum oxide occurs as a white crystalline powder. Aluminum oxide occurs as two crystalline forms: α-aluminum oxide is composed of colorless hexagonal crystals, and γ-aluminum oxide is composed of minute colorless cubic crystals that are transformed to the α-form at high temperatures.

9 Pharmacopeial Specifications

See Table I. *See also* Section 18.

Table I: Pharmacopeial specifications for aluminum oxide.

Test	USP35–NF30
Identification	+
Arsenic	≤4 ppm
Heavy metals	≤60 ppm
Chloride	≤10 000 ppm
Sulfate	≤10 000 ppm
Microbial limits	
Bacteria	≤1000 cfu/g
Molds and yeasts	≤100 cfu/g
Clarity of solution	+
Alkaline impurities	+
Neutralizing capacity	+
Assay	47.0–60.0%

10 Typical Properties

Boiling point 2977°C
Density (bulk) 0.9–1.1 g/cm^3
Flammability Nonflammable.
Hardness (Mohs) 8.8
Hygroscopicity Very hygroscopic.
Melting point 2050°C
Solubility Slowly soluble in aqueous alkaline solutions with the formation of hydroxides; practically insoluble in nonpolar organic solvents, diethyl ether, ethanol (95%), and water.
Specific gravity 2.8 (becomes 4.0 at 800°C)
Vapor pressure 133.3 Pa at 2158°C

11 Stability and Storage Conditions

Aluminum oxide should be stored in a well-closed container in a cool, dry place. It is very hygroscopic.

12 Incompatibilities

Aluminum oxide should be kept well away from water. It is incompatible with strong oxidizers and chlorinated rubber. Aluminum oxide also reacts with chlorine trifluoride, ethylene oxide, sodium nitrate, and vinyl acetate. Exothermic reactions above 200°C with halocarbon vapors produce toxic hydrogen chloride and phosgene fumes.

13 Method of Manufacture

Most of the aluminum oxide produced commercially is obtained by the calcination of aluminum hydroxide.

14 Safety

Aluminum oxide is generally regarded as relatively nontoxic and nonirritant when used as an excipient. Inhalation of finely divided particles may cause lung damage (Shaver's disease).[3]

15 Handling Precautions

Observe normal precautions appropriate to the circumstances and quantity of the material handled.[4] In the UK, the workplace exposure limits for aluminum oxide are 10 mg/m^3 long-term (8-hour TWA) for total inhalable dust and 4 mg/m^3 for respirable dust.[5] In the USA, the OSHA limit is 15 mg/m^3 total dust, 5 mg/m^3 respirable fraction for aluminium oxide.[6]

16 Regulatory Status

Included in the FDA Inactive Ingredients Database (oral tablets and topical sponge). Included in nonparenteral medicines licensed in the UK. Included in the Canadian Natural Health Products Ingredients Database.

17 Related Substances

—

18 Comments

Aluminum oxide is used therapeutically as an antacid, and is reported to have therapeutic application as an abrasive in topical preparations.[7]

The PhEur 7.4 includes a specification for hydrated aluminum oxide that contains the equivalent of 47.0–60.0% of Al_2O_3. Hydrated aluminum oxide is used in mordant dyeing to make lake pigments. It has been shown that use of hydrated aluminum oxide as an adsorbent for dyes can improve the stability of tablet formulations, probably through separation of interacting ingredients.[8]

A specification for aluminum oxide is included in the *Japanese Pharmaceutical Excipients* (JPE);[9] *see* Table II. A specification for light aluminum oxide is also included.

The EINECS number for aluminum oxide is 215-691-6.

Table II: JPE specification for aluminum oxide.[7]

Test	JPE 2004
Identification	+
Water-soluble substances	+
Heavy metals	≤30 ppm
Lead	≤30 ppm
Arsenic	≤5 ppm
Loss on drying	≤1.5%
Loss on ignition	≤2.5%
Assay	≥96.0%

19 Specific References

1 Rupprecht H. Processing of potent substances with inorganic supports by imbedding and coating. *Acta Pharm Technol* 1980; **26**: 13–27.
2 Meyer K, Zimmerman I. Effect of glidants in binary powder mixtures. *Powder Technol* 2004; **139**: 40.
3 Lewis RJ, ed. *Sax's Dangerous Properties of Industrial Materials*, 11th edn. New York: Wiley, 2004; 136.
4 National Poisons Information Service 1997. Aluminium oxide. http://www.inchem.org/documents/ukpids/ukpids/ukpid33 (accessed 19 July 2011).
5 Health and Safety Executive. *EH40/2005: Workplace Exposure Limits*. Sudbury: HSE Books, 2011. http://www.hse.gov.uk/pubns/priced/eh40.pdf (accessed 27 February 2012).
6 JT Baker. Material safety data sheet: Aluminium oxide, 2009.
7 Sweetman SC, ed. *Martindale: The Complete Drug Reference*, 37th edn. London: Pharmaceutical Press, 2011: 1726.
8 Lobo M, *et al.* Interaction of omapatrilat with FD&C Blue No. 2 lake during dissolution of modified release tablets. *Int J Pharm* 2007; **339**: 168–174.
9 Japan Pharmaceutical Excipients Council. *Japanese Pharmaceutical Excipients 2004*. Tokyo: Yakuji Nippo, 2004: 67–68.

20 General References

—

21 Author

W Cook.

22 Date of Revision

1 March 2012.

A 

Aluminum Phosphate Adjuvant

1 Nonproprietary Names

None adopted.

2 Synonyms

Adju-Phos; aluminium hydroxyphosphate; aluminum hydroxyphosphate; *Rehydraphos*.

3 Chemical Name and CAS Registry Number

Aluminum phosphate [7784-30-7]

4 Empirical Formula and Molecular Weight

$Al(OH)_x (PO_4)_y$

The molecular weight is dependent on the degree of substitution of phosphate groups for hydroxyl groups.

5 Structural Formula

Aluminum phosphate adjuvant occurs as a precipitate of amorphous aluminum hydroxide in which some sites contain phosphate groups instead of hydroxyl groups. Both hydroxyl and phosphate groups are exposed at the surface. The hydroxyl groups produce a pH-dependent surface charge by accepting a proton to produce a positive site, or donating a proton to produce a negative site. The pH-dependent surface charge is characterized by the point of zero charge, which is equivalent to the isoelectric point in protein chemistry. The surface hydroxyl groups may also undergo ligand exchange with fluoride, phosphate, carbonate, sulfate, or borate anions.

Aluminum phosphate adjuvant is not a stoichiometric compound. Rather, the degree of phosphate group substitution for hydroxyl groups depends on the precipitation recipe and conditions.

6 Functional Category

Adsorbent; vaccine adjuvant.

7 Applications in Pharmaceutical Formulation or Technology

Aluminum phosphate adjuvant is used in parenteral human and veterinary vaccines.[1] It activates Th2 immune responses, including IgG and IgE antibody responses.

8 Description

Aluminum phosphate adjuvant is a white hydrogel that sediments slowly and forms a clear supernatant.

9 Pharmacopeial Specifications

—

10 Typical Properties

Acidity/alkalinity pH = 6.0–8.0
Al : P atomic ratio 1.0–1.4 : 1.0
Aluminum (%) 0.5–0.75
Particle size distribution Primary particles are platy with an average diameter of 50 nm. The primary particles form aggregates of 1–10 μm.
Point of zero charge pH = 4.6–5.6, depending on the Al : P atomic ratio.
Protein binding capacity >0.6 mg lysozyme/mg equivalent Al_2O_3
Solubility Soluble in mineral acids and alkali hydroxides.
X-ray diffractogram Amorphous to x-rays.

11 Stability and Storage Conditions

Aluminum phosphate adjuvant is stable for at least 2 years when stored at 4–30°C in well-sealed inert containers. It must not be allowed to freeze as the hydrated colloid structure will be irreversibly damaged.

12 Incompatibilities

The point of zero charge is related directly to the Al : P atomic ratio. Therefore, the substitution of additional phosphate groups for hydroxyl groups will lower the point of zero charge. Substitution of carbonate, sulfate, or borate ions for hydroxyl groups will also lower the point of zero charge.

13 Method of Manufacture

Aluminum phosphate adjuvant is formed by the reaction of a solution of aluminum chloride and phosphoric acid with alkali hydroxide.

14 Safety

Aluminum phosphate adjuvant is intended for use in parenteral vaccines and is generally regarded as safe. It may cause mild irritation, dryness, and dermatitis on skin contact. It may also cause redness, conjunctivitis, and short-term mild irritation on eye contact. Ingestion of large amounts of aluminum phosphate adjuvant may cause respiratory irritation with nausea, vomiting, and constipation. Inhalation is unlikely, although the dried product may cause respiratory irritation and cough. Type I hypersensitivity reactions following parenteral administration have also been reported.[2]

15 Handling Precautions

Observe normal precautions appropriate to the circumstances and quantity of material handled. Eye protection and gloves are recommended.

16 Regulatory Status

GRAS listed. Accepted for use in human and veterinary vaccines in Europe and the USA. The limits for use in human vaccines are 0.85 mg aluminum/dose (FDA) and 1.25 mg aluminum/dose (WHO). There are no established limits for use in veterinary vaccines. Reported in the EPA TSCA Inventory.

17 Related Substances

Aluminum hydroxide adjuvant.

18 Comments

The USP35–NF30 monograph for aluminum phosphate ($AlPO_4$) gel describes aluminum phosphate, which is used as an antacid, not as a vaccine adjuvant.

19 Specific References

1 Shirodkar S, *et al.* Aluminum compounds used as adjuvants in vaccines. *Pharm Res* 1990; 7: 1282–1288.
2 Goldenthal KL, *et al.* Safety evaluation of vaccine adjuvants. *AIDS Res Hum Retroviruses* 1993; 9: S47–S51.

20 General References

Hem SL, Hogenesch H. Aluminum-containing adjuvants: properties, formulation, and use. In: Singh M, ed. *Vaccine Adjuvants and Delivery Systems*. New York: Wiley, 2007; 81–114.

Gupta RK *et al.* Adjuvant properties of aluminum and calcium compounds. In: Powell MF, Newman MJ, eds. *Vaccine Design*. New York: Plenum, 1995; 229–248.

Lindblad EB. Aluminum adjuvants – in retrospect and prospect. *Vaccine* 2004; 22: 3658–3668.

Lindblad EB. Aluminum adjuvants. In: Stewart-Tull DES, ed. *The Theory and Practical Application of Adjuvants*. New York: Wiley, 1995; 21–35.

Vogel FR, Hem SL. Immunogenic adjuvants. In: Plotkin SA *et al.* eds. *Vaccines*, 5th edn. New York: W.B. Saunders, 2008; 59–71.

Vogel FR, Powell MF. A compendium of vaccine adjuvants and excipients. In: Powell MF, Newman MJ, eds. *Vaccine Design*. New York: Plenum, 1995; 142.

White JL, Hem SL. Characterization of aluminum-containing adjuvants. In: Brown F *et al.* eds. *Physico-Chemical Procedures for the Characterization of Vaccines*. IABS Symposia Series: Developments in Biologicals, 103: New York: Karger, 2000; 217–228.

21 Authors

SL Hem, EB Lindblad, L Grove.

22 Date of Revision

1 October 2010.

 Ammonia Solution

1 Nonproprietary Names

BP: Strong Ammonia Solution
PhEur: Ammonia Solution, Concentrated
USP–NF: Strong Ammonia Solution

2 Synonyms

Ammoniaca; ammoniacum; ammoniae solution concentrata; aqua ammonia; concentrated ammonia solution; spirit of hartshorn; stronger ammonia water.

3 Chemical Name and CAS Registry Number

Ammonia [7664-41-7]

4 Empirical Formula and Molecular Weight

NH_3 17.03

5 Structural Formula

See Section 4.

6 Functional Category

Alkalizing agent.

7 Applications in Pharmaceutical Formulation or Technology

Ammonia solution is typically not used undiluted in pharmaceutical applications. Generally, it is used as a buffering agent or to adjust the pH of solutions. Most commonly, ammonia solution (the concentrated form) is used to produce more dilute ammonia solutions.

8 Description

Strong ammonia solution occurs as a clear, colorless liquid having an exceedingly pungent, characteristic odor. The PhEur 7.4 states that concentrated ammonia solution contains not less than 25.0% and not more than 30.0% w/w of ammonia (NH_3). The USP35–NF30 states that strong ammonia solution contains not less than 27.0% and not more than 31.0% w/w of ammonia (NH_3).

See also Section 17.

9 Pharmacopeial Specifications

See Table I.

Table I: Pharmacopeial specifications for ammonia solution.

Test	PhEur 7.4	USP35–NF30
Identification	+	+
Characters	+	—
Appearance of solution	+	—
Oxidizable substances	+	+
Pyridine and related substances	≤2 ppm	—
Carbonates	≤60 ppm	—
Chlorides	≤1 ppm	—
Sulfates	≤5 ppm	—
Iron	≤0.25 ppm	—
Heavy metals	≤1 ppm	≤13 ppm
Residue on evaporation	≤20 mg/L	—
Limit of nonvolatile residue	—	≤0.05%
Assay (of NH_3)	25.0–30.0%	27.0–31.0%

10 Typical Properties

Solubility Miscible with ethanol (95%) and water.
Specific gravity 0.892–0.910

11 Stability and Storage Conditions

On exposure to the air, ammonia solution rapidly loses ammonia. Ammonia solution should be stored in a well-closed container, protected from the air, in a cool, dry place. The storage temperature should not exceed 20°C.

12 Incompatibilities

Ammonia solution reacts vigorously with sulfuric acid or other strong mineral acids and the reaction generates considerable heat; the mixture boils.

13 Method of Manufacture

Ammonia is obtained commercially chiefly by synthesis from its constituent elements, nitrogen and hydrogen, which are combined under high pressure and temperature in the presence of a catalyst. Ammonia solution is produced by dissolving ammonia gas in water.

A

14 Safety

Ingestion of strong solutions of ammonia is very harmful and causes severe pain in the mouth, throat, and gastrointestinal tract as well as severe local edema with cough, vomiting, and shock. Burns to the esophagus and stomach may result in perforation. Inhalation of the vapor causes sneezing, coughing, and, in high concentration, pulmonary edema. Asphyxia has been reported. The vapor is irritant to the eyes. Strong solutions are harmful when applied to the conjunctiva and mucous membranes. Topical application of even dilute ammonia solutions, used to treat insect bites, has caused burns, particularly when used with a subsequent dressing.[1–3]

When used as an excipient, ammonia solution is generally present in a formulation in a highly diluted form.

15 Handling Precautions

Observe normal precautions appropriate to the circumstances and quantity of material handled. Care should be used in handling strong or concentrated ammonia solutions because of the caustic nature of the solution and the irritating properties of its vapor. Before containers are opened, they should be well cooled. The closure should be covered with a cloth or similar material while opening. Ammonia solution should not be tasted and inhalation of the vapor should be avoided. Ammonia solution should be handled in a fume cupboard. Eye protection, gloves, and a respirator are recommended.

16 Regulatory Status

Included in the FDA Inactive Ingredients Database (oral suspensions, topical preparations). Included in nonparenteral medicines licensed in the UK. Included in the Canadian Natural Health Products Ingredients Database.

17 Related Substances

Dilute ammonia solution.

Dilute ammonia solution

Synonyms Ammonia water

Specific gravity 0.95–0.96

Comments Several pharmacopeias include monographs for dilute ammonia solution. The JP XV, for example, states that ammonia water contains not less than 9.5% and not more than 10.5% w/v of ammonia (NH_3).

18 Comments

Therapeutically, dilute ammonia solution is used as a reflex stimulant in 'smelling salts', as a rubefacient, and as a counter-irritant to neutralize insect bites or stings.[4]

Where 'ammonia solution' is prescribed therapeutically, dilute ammonia solution should be dispensed or supplied.

The EINECS number for ammonia solution is 231-635-3.

19 Specific References

1 Beare JD, *et al.* Ammonia burns of the eye: an old weapon in new hands. *Br Med J* 1988; **296**: 590.

2 Payne MP, Delic JI. Ammonia. In: *Toxicity Review 24*. London: HMSO, 1991; 1–12.

3 Leduc D, *et al.* Acute and long term respiratory damage following inhalation of ammonia. *Thorax* 1992; **47**: 755–757.

4 Frohman IG. Treatment of physalia stings. *J Am Med Assoc* 1996; **197**: 733.

20 General References

—

21 Author

PJ Sheskey.

22 Date of Revision

2 March 2012.

Ammonium Alginate

1 Nonproprietary Names

None adopted.

2 Synonyms

Alginic acid, ammonium salt; ammonium polymannuronate; E404; *Keltose*.

3 Chemical Name and CAS Registry Number

Ammonium alginate [9005-34-9]

4 Empirical Formula and Molecular Weight

$(C_6H_{11}NO_6)_n$ 193.16 (calculated)
217 (actual, average)

Ammonium alginate is the ammonium salt of alginic acid.

5 Structural Formula

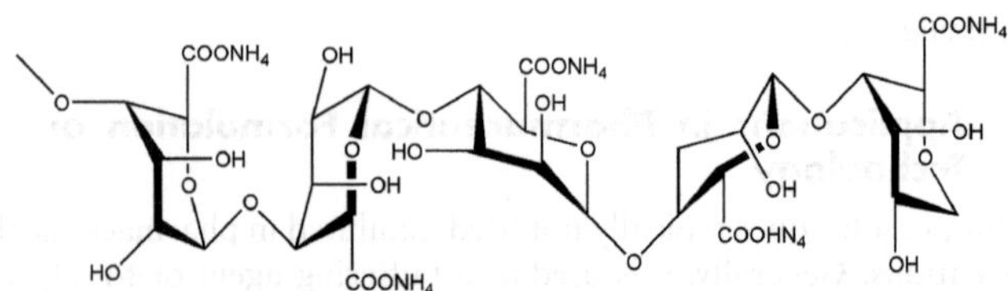

The number and sequence of the mannuronate and glucuronate residues shown above vary in the naturally occurring alginate. The associated water molecules are not shown.

6 Functional Category

Emulsifying agent; film-forming agent; humectant; viscosity-increasing agent.

7 Applications in Pharmaceutical Formulation or Technology

Ammonium alginate is used in pharmaceutical preparations as an emulsifier, film-former, and humectant.

A

8 Description

Ammonium alginate occurs as white to yellowish brown filamentous, grainy, granular, or powdered forms.

9 Pharmacopeial Specifications

See Section 18.

10 Typical Properties

Moisture content Not more than 15% at 105°C for 4 hours.

Solubility Dissolves slowly in water to form a viscous solution; insoluble in ethanol and in ether.

11 Stability and Storage Conditions

Ammonium alginate is a hygroscopic material, although it is stable if stored at low relative humidities and cool temperatures.

12 Incompatibilities

Incompatible with oxidizing agents and strong acids and alkalis.

13 Method of Manufacture

Ammonium alginate is prepared from alginic acid, which is a polysaccharide extracted from the giant brown seaweed species of Phaeophyceae (*see* Alginic Acid), by an ion-exchange reaction whereby the alginic acid is neutralized with ammonium hydroxide and sufficient water. Final preparation includes a drying stage and milling to the desired particle size.

14 Safety

Ammonium alginate is widely used in cosmetics and food products, and also in pharmaceutical formulations such as tablets. It is generally regarded as a nontoxic and nonirritant material, although excessive oral consumption may be harmful.

15 Handling Precautions

Observe normal precautions appropriate to the circumstances and quantity of the material handled. Eye protection, gloves, and a dust respirator are recommended.

16 Regulatory Status

GRAS listed. Accepted in Europe for use as a food additive. Included in the FDA Inactive Ingredients Database (oral, tablets). Included in the Canadian Natural Health Products Ingredients Database.

17 Related Substances

Alginic acid; calcium alginate; potassium alginate; propylene glycol alginate; sodium alginate.

18 Comments

Ammonium alginate and other alginate salts bind water very strongly due to the large number of carboxylate anions they contain.

Hydrophilic matrix tablets based on an alginate system have been investigated for modified drug release.[1] Alginates have also been investigated for use in microparticle and hydrogel systems.[2–8] Alginate microspheres have been produced by internal gelation using emulsification methods.[9,10]

Therapeutically, alginates provide an ideal moist healing environment, which makes them ideal for use in wound dressings.[11,12] Chitosan and alginates have been used together to produce sponges for use as wound dressings, or matrices for tissue engineering.[13]

Ammonium alginate is also widely used in foods as a thickener and emulsifier.

Although not included in any pharmacopeias, a specification for ammonium alginate is included in the *Food Chemicals Codex* (FCC),[14] *see* Table I.

Table I: FCC specification for ammonium alginate.[14]

Test	FCC 7[14]
Identification	+
Arsenic	≤3 mg/kg
Residue on ignition	≤7.0%
Lead	≤5 mg/kg
Loss on drying	≤15.0%
Assay	18.0–21.0% of CO_2, corresponding to 88.7–103.6% ammonium alginate

19 Specific References

1 Sriamornsak P, *et al.* Swelling, erosion and release behavior of alginate-based matrix tablets. *Eur J Pharm Biopharm* 2007; **66** (3): 435–450.

2 Matricardi P, *et al.* Recent advances and perspectives on coated alginate microspheres for modified drug delivery. *Expert Opin Drug Deliv* 2008; 5(4): 417–425.

3 Hori Y, *et al.* Modular injectable matrices based on alginate solution/microsphere mixtures that gel *in situ* and co-deliver immunomodulatory factors. *Acta Biomater* 2009; 5(4): 969–982.

4 Wittaya-areekul S, *et al.* Preparation and *in vitro* evaluation of mucoadhesive properties of alginate/chitosan microparticles containing prednisolone. *Int J Pharm* 2006; **312**(1–2): 113–118.

5 Xin J, *et al.* Study of branched cationic beta-cyclodextrin polymer/indomethacin complex and its release profile from alginate hydrogel. *Int J Pharm* 2010; **386**(1–2): 221–228.

6 George M, Abraham TE. pH sensitive alginate-guar gum hydrogel for the controlled delivery of protein drugs. *Int J Pharm* 2007; **335**(1–2): 123–129.

7 Notara M, *et al.* Cytocompatibility and hemocompatibility of a novel chitosan-alginate gel system. *J Biomed Mater Res A* 2009; **89**(4): 854–864.

8 George M, Abraham TE. Polyionic hydrocolloids for the intestinal delivery of protein drugs: alginate and chitosan – a review. *J Control Rel* 2006; **114**(1): 1–14.

9 Chan LW, *et al.* Production of alginate microspheres by internal gelation using an emulsification method. *Int J Pharm* 2002; **242**(1–2): 259–262.

10 Reis CP, *et al.* Review and current status of emulsion/dispersion technology using an internal gelation process for the design of alginate particles. *J Microencapsul* 2006; **23**(3): 245–257.

11 Morgan D. Wounds—what should a dressing formulary include? *Hosp Pharm* 2002; 9(9): 261–266.

12 Qin Y. The gel swelling properties of alginate fibers and their applications in wound management. *Polym Adv Technol* 2008; **19**: 6–14.

13 Lai HL, *et al.* The preparation and characterisation of drug-loaded alginate and chitosan sponges. *Int J Pharm* 2003; **251**(1–2): 175–181.

14 *Food Chemicals Codex*, 7th edn. Bethesda, MD: United States Pharmacopeia, 2010: 47.

20 General References

—

21 Authors

SL Cantor, SA Shah.

22 Date of Revision

1 November 2011.

A

Ammonium Chloride

1 Nonproprietary Names

BP: Ammonium Chloride
PhEur: Ammonium Chloride
USP–NF: Ammonium Chloride

2 Synonyms

Ammonii chloridum; ammonium muriate; E510; sal ammoniac; salmiac.

3 Chemical Name and CAS Registry Number

Ammonium chloride [12125-02-9]

4 Empirical Formula and Molecular Weight

NH_4Cl 53.49

5 Structural Formula

See Section 4.

6 Functional Category

Acidulant.

7 Applications in Pharmaceutical Formulation or Technology

Ammonium chloride is used as an acidifying agent in oral formulations and as a food additive.

8 Description

Ammonium chloride occurs as colorless, odorless crystals or crystal masses. It is a white, granular powder with a cooling, saline taste. It is hygroscopic and has a tendency to cake.

9 Pharmacopeial Specifications

See Table I.

Table I: Pharmacopeial specifications for ammonium chloride.

Test	PhEur 7.4	USP35–NF30
Identification	+	+
Characters	+	—
Appearance of solution	+	—
Acidity or alkalinity	+	+
Loss on drying	≤1.0%	≤0.5%
Residue on ignition	—	≤0.1%
Thiocyanate	—	+
Bromides and iodides	+	—
Sulfates	≤150 ppm	—
Sulfated ash	≤0.1%	—
Calcium	≤200 ppm	—
Iron	≤20 ppm	—
Heavy metals	≤10 ppm	≤0.001%
Assay (dried basis)	99.0–100.5%	99.5–100.5%

10 Typical Properties

Acidity/alkalinity pH = 4.5–5.5 (5.5% w/w aqueous solutions at 25°C)
Density (bulk) 0.6–0.9 g/cm^3
Hygroscopicity Hygroscopic with potential to cake.
Melting point Decomposes at 338°C; sublimes without melting.[1]
Solubility Soluble in water; hydrochloric acid and sodium chloride decrease its solubility in water. Also soluble in glycerin; sparingly soluble in methanol and ethanol. Almost insoluble in acetone, ether, and ethyl acetate.
Specific gravity 1.527 g/cm^3
Vapor pressure 133.3 Pa (1 mmHg) at 160°C

11 Stability and Storage Conditions

Ammonium chloride is chemically stable. It decomposes completely at 338°C to form ammonia and hydrochloric acid. Store in airtight containers in a cool, dry place.

12 Incompatibilities

Ammonium chloride is incompatible with strong acids and strong bases. It reacts violently with ammonium nitrate and potassium chlorate, causing fire and explosion hazards. It also attacks copper and its compounds.

13 Method of Manufacture

Ammonium chloride is prepared commercially by reacting ammonia with hydrochloric acid.

14 Safety

Ammonium chloride is used in oral pharmaceutical formulations. The pure form of ammonium chloride is toxic by SC, IV, and IM routes, and moderately toxic by other routes. Potential symptoms of overexposure to fumes are irritation of eyes, skin, respiratory system (cough, dyspnea, and pulmonary sensitization).[2] Ammonium salts are an irritant to the gastric mucosa and may induce nausea and vomiting.

LD_{50} (mouse, IP): 1.44 g/kg[3]
LD_{50} (mouse, oral): 1.3 g/kg
LD_{50} (rat, IM): 0.03 g/kg[4]
LD_{50} (rat, oral): 1.65 g/kg[5]

15 Handling Precautions

Observe normal precautions appropriate to the circumstances and quantity of the material handled.

All grades of ammonium chloride must be kept well away from nitrites and nitrates during transport and storage. They must be stored in a dry place, and effluent must not be discharged into the drains without prior treatment.

Ammonium chloride decomposes on heating, producing toxic and irritating fumes (nitrogen oxides, ammonia, and hydrogen chloride).

16 Regulatory Status

GRAS listed. Included in the FDA Inactive Ingredients Database (oral syrup and tablets). Accepted for use as a food additive in Europe. Included in medicines licensed in the UK (eye drops and oral syrup). Included in the Canadian Natural Health Products Ingredients Database.

17 Related Substances

Ammonia solution.

18 Comments

Ammonium chloride has the ability to cross the red blood cell membrane, and a solution that is isotonic to blood will still cause hemolytic rupture because it acts as a hypotonic solution.[6] Therapeutically, ammonium chloride is used in the treatment of severe metabolic alkalosis to maintain the urine at an acid pH in the treatment of some urinary tract disorders or in forced acid diuresis.[7–9] It is also used as an antiseptic agent[10] and as an expectorant in cough medicines.[11]

A specification for ammonium chloride is contained in the *Food Chemicals Codex* (FCC).[12]

The EINECS number for ammonium chloride is 235-186-4. The PubChem Compound ID (CID) for ammonium chloride is 25517.

19 Specific References

1 Zhu RS, *et al.* Sublimation of ammonium salts: a mechanism revealed by a first-principles study of the NH_4Cl system. *J Phys Chem* 2007; **111**: 13831–13838.
2 *NIOSH Pocket Guide to Chemical Hazards* (DHHS/NIOSH 97-140) 1997: 16.
3 Lewis RJ, ed. *Sax's Dangerous Properties of Industrial Materials*, 11th edn. New York: Wiley, 2004: 231.
4 Boyd EM, Seymour KGW. Ethylene diamine dihydrochloride. II. Untoward toxic reactions. *Exp Med Surg* 1946; **4**: 223–227.
5 Smeets P. Ammonium chloride [and water treatment]. *Tribune de l'Eau* 1994; 47(570): 26–29.
6 Jausel-Hüsken S, Deuticke B. General and transport properties of hypotonic and isotonic preparations of resealed erythrocyte ghosts. *J Membr Biol* 1981; **63**: 61–70.
7 Mainzer F. Acid therapy with neutral salts. *Klin Wochenschr* 1927; **6**: 1689–1691.
8 Portnoff JB, *et al.* Control of urine pH and its effect on drug excretion in humans. *J Pharm Sci* 1961; **50**: 890.
9 Davies HE. Rise in urine pH and in ammonium excretion during a water diuresis. *J Physiol* 1968; **194**: 79–80P.
10 Gottardi W, *et al.* *N*-Chlorotaurine and ammonium chloride: an antiseptic preparation with strong bactericidal activity. *Int J Pharm* 2007; **335**: 32–40.
11 Coleman W. Expectorant action of ammonium chloride. *Am J Med Sci* 1916; **152**: 569–574.
12 *Food Chemicals Codex*, 7th edn. Bethesda, MD: United States Pharmacopeia, 2010: 49.

20 General References

Ingham JW. The apparent hydration of ions. III. The densities and viscosities of saturated solutions of ammonium chloride in hydrochloric acid. *J Chem Soc* 1929; 2059–2067.

Kumaresan R, *et al.* Simultaneous heat and mass transfer studies in drying of ammonium chloride in fluidized bed dryer. *Process Plant Eng* 2007; 25(3): 60–66.

21 Author

W Cook.

22 Date of Revision

1 March 2012.

Ascorbic Acid

1 Nonproprietary Names

BP: Ascorbic Acid
JP: Ascorbic Acid
PhEur: Ascorbic Acid
USP–NF: Ascorbic Acid

2 Synonyms

Acidum ascorbicum; *C-97*; cevitamic acid; 2,3-didehydro-L-threo-hexono-1,4-lactone; E300; 3-oxo-L-gulofuranolactone, enol form; vitamin C.

3 Chemical Name and CAS Registry Number

L-(+)-Ascorbic acid [50-81-7]

4 Empirical Formula and Molecular Weight

$C_6H_8O_6$ 176.13

5 Structural Formula

6 Functional Category

Acidulant; antioxidant.

7 Applications in Pharmaceutical Formulation or Technology

Ascorbic acid is used as an antioxidant in aqueous pharmaceutical formulations at a concentration of 0.01–0.1% w/v. Ascorbic acid has been used to adjust the pH of solutions for injection, and as an adjunct for oral liquids. It is also widely used in foods as an antioxidant. Ascorbic acid has also proven useful as a stabilizing agent in mixed micelles containing tetrazepam.[1]

SEM 1: Excipient: ascorbic acid USP (fine powder); manufacturer: Pfizer Ltd; lot no.: 9A-3/G92040-CO 146; magnification: 120×; voltage: 20 kV.

SEM 2: Excipient: ascorbic acid USP (fine powder); manufacturer: Pfizer Ltd; lot no.: 9A-3/G92040-CO 146; magnification: 600×; voltage: 20 kV.

8 Description

Ascorbic acid occurs as a white to light-yellow-colored, nonhygroscopic, odorless, crystalline powder or colorless crystals with a sharp, acidic taste. It gradually darkens in color upon exposure to light.

9 Pharmacopeial Specifications

See Table I.

10 Typical Properties

Acidity/alkalinity pH = 2.1–2.6 (5% w/v aqueous solution)

Density (bulk)

0.7–0.9 g/cm^3 for crystalline material;

0.5–0.7 g/cm^3 for powder.

Density (particle) 1.65 g/cm^3

SEM 3: Excipient: ascorbic acid USP (fine granular); manufacturer: Pfizer Ltd; lot no.: 9A-2/G01280-CO 148; magnification: 120×; voltage: 20 kV.

Table I: Pharmacopeial specifications for ascorbic acid.

Test	JP XV	PhEur 7.4	USP35–NF30
Identification	+	+	+
Characters	–	+	–
Specific rotation (10% w/v solution)	+ 20.5° to + 21.5°	+ 20.5° to + 21.5°	+ 20.5° to + 21.5°
Residue on ignition	≤0.1%	–	≤0.1%
pH	2.2–2.5	2.1–2.6	–
Sulfated ash	–	≤0.1%	–
Copper	–	≤5 ppm	–
Heavy metals	≤20 ppm	≤10 ppm	≤20 ppm
Loss on drying	≤0.20%	–	–
Iron	–	≤2 ppm	–
Oxalic acid	–	+	–
Related substances	–	+	–
Appearance of solution	+	+	–
Assay	≥99.0%	99.0–100.5%	99.0–100.5%

Table II: Solubility of ascorbic acid.

Solvent	Solubility at 20°C
Chloroform	Practically insoluble
Ethanol	1 in 50
Ethanol (95%)	1 in 25
Ether	Practically insoluble
Fixed oils	Practically insoluble
Glycerin	1 in 1000
Propylene glycol	1 in 20
Water	1 in 3.5

Density (tapped)

1.0–1.2 g/cm^3 for crystalline material;

0.9–1.1 g/cm^3 for powder.

Density (true) 1.688 g/cm^3

Dissociation constant

$pK_{a1} = 4.17$;

$pK_{a2} = 11.57$.

Melting point 190°C (with decomposition)

Moisture content 0.1% w/w

Solubility see Table II.

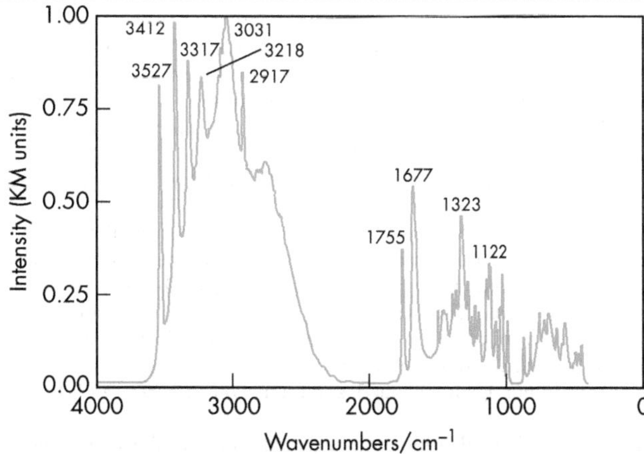

Figure 1: Infrared spectrum of ascorbic acid measured by diffuse reflectance. Adapted with permission of Informa Healthcare.

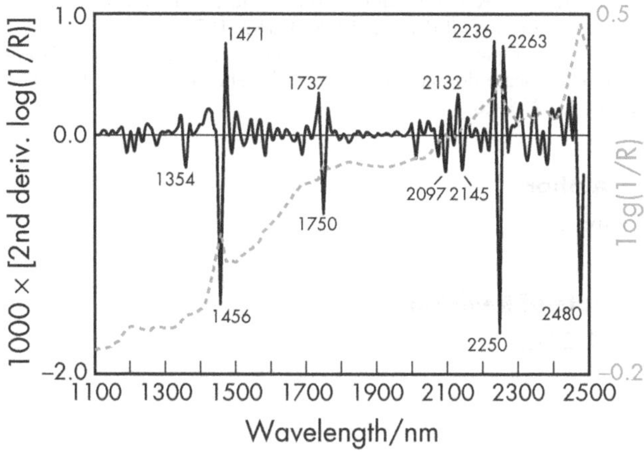

Figure 2: Near-infrared spectrum of ascorbic acid measured by reflectance.

Spectroscopy

IR spectra *see* Figure 1.

NIR spectra *see* Figure 2.

11 Stability and Storage Conditions

In powder form, ascorbic acid is relatively stable in air. In the absence of oxygen and other oxidizing agents it is also heat stable. Ascorbic acid is unstable in solution, especially alkaline solution, readily undergoing oxidation on exposure to the air.[2,3] The oxidation process is accelerated by light and heat and is catalyzed by traces of copper and iron. Ascorbic acid solutions exhibit maximum stability at about pH 5.4. Solutions may be sterilized by filtration.

The bulk material should be stored in a well-closed nonmetallic container, protected from light, in a cool, dry place.

12 Incompatibilities

Incompatible with alkalis, heavy metal ions, especially copper and iron, oxidizing materials, methenamine, phenylephrine hydrochloride, pyrilamine maleate, salicylamide, sodium nitrite, sodium salicylate, theobromine salicylate, and picotamide.[4,5] Additionally, ascorbic acid has been found to interfere with certain colorimetric assays by reducing the intensity of the color produced.[6]

13 Method of Manufacture

Ascorbic acid is prepared synthetically or extracted from various vegetable sources in which it occurs naturally, such as rose hips, blackcurrants, the juice of citrus fruits, and the ripe fruit of *Capsicum annuum* L. A common synthetic procedure involves the hydrogenation of D-glucose to D-sorbitol, followed by oxidation using *Acetobacter suboxydans* to form L-sorbose. A carboxyl group is then added at C1 by air oxidation of the diacetone derivative of L-sorbose and the resulting diacetone-2-keto-L-gulonic acid is converted to L-ascorbic acid by heating with hydrochloric acid.

14 Safety

Ascorbic acid is an essential part of the human diet, with 40 mg being the recommended daily dose in the UK[7] and 60 mg in the USA.[8] However, these figures are controversial, with some advocating doses of 150 or 250 mg daily. Megadoses of 10 g daily have also been suggested to prevent illness although such large doses are now generally considered to be potentially harmful.[9–11]

The body can absorb about 500 mg of ascorbic acid daily with any excess immediately excreted by the kidneys. Large doses may cause diarrhea or other gastrointestinal disturbances. Damage to the teeth has also been reported.[12] However, no adverse effects have been reported at the levels employed as an antioxidant in foods, beverages,[13] and pharmaceuticals. The WHO has set an acceptable daily intake of ascorbic acid, potassium ascorbate, and sodium ascorbate, as antioxidants in food, at up to 15 mg/kg body-weight in addition to that naturally present in food.[14]

LD_{50} (mouse, IV): 0.52 g/kg[15]

LD_{50} (mouse, oral): 3.37 g/kg

LD_{50} (rat, oral): 11.9 g/kg

15 Handling Precautions

Ascorbic acid may be harmful if ingested in large quantities and may be irritating to the eyes. Observe normal precautions appropriate to the circumstances and quantity of material handled. Eye protection and rubber or plastic gloves are recommended.

16 Regulatory Status

GRAS listed. Accepted for use as a food additive in Europe. Included in the FDA Inactive Ingredients Database (inhalations, injections, oral capsules, suspensions, tablets, topical preparations, and suppositories). Included in medicines licensed in the UK. Included in the Canadian Natural Health Products Ingredients Database.

17 Related Substances

Ascorbyl palmitate; erythorbic acid; sodium ascorbate.

18 Comments

Many dosage forms for ascorbic acid have been developed for its administration to patients, including microencapsulation.[16]

A specification for ascorbic acid is contained in the *Food Chemicals Codex* (FCC).[17]

The EINECS number for ascorbic acid is 200-066-2. The PubChem Compound ID (CID) for ascorbic acid is 5785.

19 Specific References

1 Hammad MA, Muller BW. Solubility and stability of tetrazepam in mixed micelles. *Eur J Pharm Sci* 1998; 7: 49–55.
2 Hajratwala BR. Stability of ascorbic acid. *STP Pharma* 1985; **1**: 281–286.
3 Touitou E, *et al.* Ascorbic acid in aqueous solution: bathochromic shift in dilution and degradation. *Int J Pharm* 1992; **78**: 85–87.
4 Botha SA, *et al.* DSC screening for drug–drug interactions in polypharmaceuticals intended for the alleviation of the symptoms of colds and flu. *Drug Dev Ind Pharm* 1987; **13**: 345–354.
5 Mura P, *et al.* Differential scanning calorimetry in compatibility testing of picotamide with pharmaceutical excipients. *Thermochim Acta* 1998; **321**: 59–65.

6 Krishnan G, *et al.* Estimation of phenylephrine hydrochloride in multicomponent pharmaceutical preparations. *Eastern Pharmacist* 1990; **33**: 143–145.
7 Department of Health. Dietary reference values for food energy and nutrients for the United Kingdom: report of the panel on dietary reference values of the committee on medical aspects of food policy. *Report on Health and Social Subjects 41*. London: HMSO, 1991.
8 Subcommittee on the tenth edition of the RDAs, Food and Nutrition Board, Commission on Life Sciences. National Research Council. *Recommended Dietary Allowances*, 10th edn. Washington, DC: National Academy Press, 1989.
9 Ovesen L. Vitamin therapy in the absence of obvious deficiency: what is the evidence? *Drugs* 1984; **27**: 148–170.
10 Bates CJ. Is there a maximum safe dose of vitamin C (ascorbic acid)? *Br Med J* 1992; **305**: 32.
11 Mason P. Vitamin C. In: *Dietary Supplements*, 2nd edn. London: Pharmaceutical Press, 2001; 227–233.
12 Giunta JL. Dental erosion resulting from chewable vitamin C tablets. *J Am Dent Assoc* 1983; **107**: 253–256.
13 Food and Drug Administration. CFSAN/Office of Food Additives Safety. Data on benzene in soft drinks and other beverages, May 2007.
14 FAO/WHO. Toxicological evaluation of certain food additives with a review of general principles and of specifications. Seventeenth report of the joint FAO/WHO expert committee on food additives. World Health Organ Tech Rep Ser 1974; No. 539.
15 Lewis RJ, ed. *Sax's Dangerous Properties of Industrial Materials*, 11th edn. New York: Wiley, 2004; 309–310.
16 Esposito E, *et al.* Spray-dried *Eudragit* microparticles as encapsulation devices for vitamin C. *Int J Pharm* 2002; **242**: 329–334.
17 *Food Chemicals Codex*, 7th edn. Bethesda, MD: United States Pharmacopeia, 2010; 71.

20 General References

Abramovici B, *et al.* [Comparative study of the tabletability of different grades of vitamin C]. *STP Pharma* 1987; **3**: 16–22 [in French].

Allwood MC. Factors influencing the stability of ascorbic acid in total parenteral nutrition infusions. *J Clin Hosp Pharm* 1984; **9**: 75–85.

Bhagavan HN, Wolkoff BI. Correlation between the disintegration time and the bioavailability of vitamin C tablets. *Pharm Res* 1993; **10**: 239–242.

Davies MB *et al. Vitamin C—Its Chemistry and Biochemistry*. London: Royal Society of Chemistry, 1991.

Hu F, *et al.* Effects of different adhesives on the stability of vitamin C buccal tablets. *Zhejiang Yike Daxue Xuebao* 1997; **26**: 108–110.

Krishna G, *et al.* Development of a parenteral formulation of an investigational anticancer drug, 3-aminopyridine-2-carboxaldehyde thiosemicarbazone. *Pharm Dev Technol* 1999; **4**: 71–80.

Nebuloni M, *et al.* Thermal analysis in preformulation studies of a lyophilized form of an antibiotic. *Boll Chim Farm* 1996; **135**: 94–100.

Pinsuwan S, *et al.* Degradation kinetics of 4-dedimethylamino sancycline, a new anti-tumor agent, in aqueous solutions. *Int J Pharm* 1999; **181**: 31–40.

Saleh SI, Stamm A. Evaluation of some directly compressible L-ascorbic acid forms. *STP Pharma* 1988; **4**: 10–14.

Saleh SI, Stamm A. Contribution to the preparation of a directly compressible L-ascorbic acid granular form: comparison of granules prepared by three granulation methods and evaluation of their corresponding tablets. *STP Pharma* 1988; **4**: 182–187.

Seta Y, *et al.* Preparation and pharmacological evaluation of Captopril sustained-release dosage forms using oily semisolid matrix. *Int J Pharm* 1988; **41**: 255–262.

21 Author

AH Kibbe.

22 Date of Revision

1 March 2012.

Ascorbyl Palmitate

1 Nonproprietary Names

BP: Ascorbyl Palmitate
PhEur: Ascorbyl Palmitate
USP–NF: Ascorbyl Palmitate

2 Synonyms

L-Ascorbic acid 6-palmitate; ascorbylis palmitas; E304; 3-oxo-L-gulofuranolactone 6-palmitate; Vitamin C ester; Vitamin C palmitate.

3 Chemical Name and CAS Registry Number

L-Ascorbic acid 6-hexadecanoate [137-66-6]

4 Empirical Formula and Molecular Weight

$C_{22}H_{38}O_7$ 414.54

5 Structural Formula

O H H_3C OH O 14 O O HO OH

6 Functional Category

Antioxidant.

7 Applications in Pharmaceutical Formulation or Technology

Ascorbyl palmitate is primarily used alone or in combination with alpha tocopherol as a stabilizer for oils in oral pharmaceutical formulations and food products; generally, a concentration of 0.05% w/v is used. It may also be used in oral and topical preparations as an antioxidant for drugs unstable to oxygen. The combination of ascorbyl palmitate with alpha tocopherol shows marked synergism, which increases the effect of the components and allows the amount used to be reduced. The solubility of ascorbyl palmitate in alcohol (freely soluble) permits it to be used in both nonaqueous and aqueous systems and emulsions.

8 Description

Ascorbyl palmitate is a practically odorless, white to yellowish powder.

9 Pharmacopeial Specifications

See Table I .

Table I: Pharmacopeial specifications for ascorbyl palmitate.

Test	PhEur 7.4	USP35–NF30
Identification	+	+
Appearance of solution	+	—
Melting range	—	107–117°C
Specific rotation (10% w/v in methanol)	+21° to +24°	+21° to +24°
Loss on drying	≤1.0%	≤2.0%
Residue on ignition	—	≤0.1%
Sulfated ash	≤0.1%	—
Heavy metals	≤10 ppm	≤10 ppm
Assay (dried basis)	98.0–100.5%	95.0–100.5%

10 Typical Properties

Solubility *see* Table II.

Table II: Solubility of ascorbyl palmitate.

Solvent	Solubility at 20°C unless otherwise stated[1]
Acetone	1 in 15
Chloroform	1 in 3300
	1 in 11 at 60°C
Cottonseed oil	1 in 1670
Ethanol	1 in 8
	1 in 1.7 at 70°C
Ethanol (95%)	1 in 9.3
Ethanol (50%)	1 in 2500
Ether	1 in 132
Methanol	1 in 5.5
	1 in 1.7 at 60°C
Olive oil	1 in 3300
Peanut oil	1 in 3300
Propan-2-ol	1 in 20
	1 in 5 at 70°C
Sunflower oil	1 in 3300
Water	Practically insoluble
	1 in 500 at 70°C
	1 in 100 at 100°C

Spectroscopy

IR spectra *see* Figure 1.

NIR spectra *see* Figure 2.

11 Stability and Storage Conditions

Ascorbyl palmitate is stable in the dry state, but is gradually oxidized and becomes discolored when exposed to light and high humidity. It has a shelf life of at least 12 months when stored in a sealed container in a cool place (8–15°C) and protected from light. During processing, temperatures greater than 65°C should be avoided.

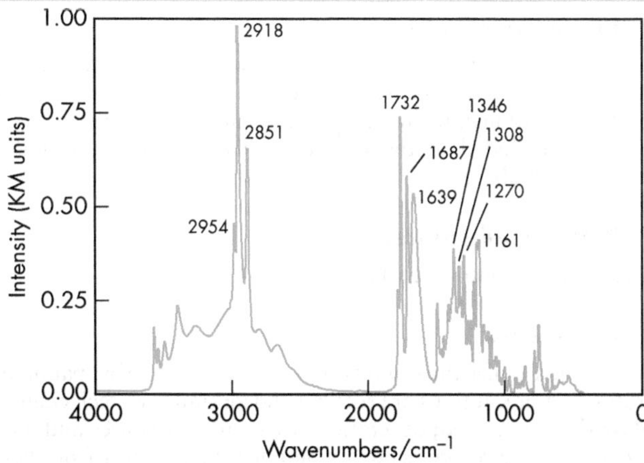

Figure 1: Infrared spectrum of ascorbyl palmitate measured by diffuse reflectance. Adapted with permission of Informa Healthcare.

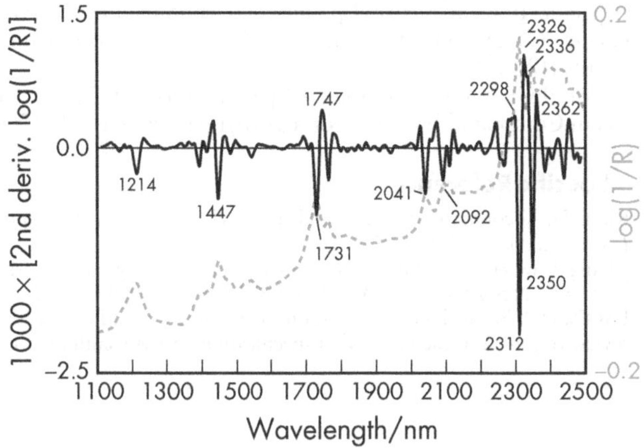

Figure 2: Near-infrared spectrum of ascorbyl palmitate measured by reflectance.

12 Incompatibilities

Incompatibilities are known with oxidizing agents; e.g. in solution oxidation is catalyzed by trace metal ions such as Cu^{2+} and Fe^{3+}.

13 Method of Manufacture

Ascorbyl palmitate is prepared synthetically by the reaction of ascorbic acid with sulfuric acid followed by reesterification with palmitic acid. Enzymatic catalysis is used to increase the production yield at lower cost.[2]

14 Safety

Ascorbyl palmitate is used in oral pharmaceutical formulations and food products, and is generally regarded as an essentially nontoxic and nonirritant material. The WHO has set an estimated acceptable daily intake for ascorbyl palmitate at up to 1.25 mg/kg body-weight.[3]

LD_{50} (mouse, oral): 25 g/kg[4]

LD_{50} (rat, oral): 10 g/kg

15 Handling Precautions

Observe normal precautions appropriate to the circumstances and quantity of material handled. Ascorbyl palmitate dust may cause irritation to the eyes and respiratory tract. Eye protection is recommended.

16 Regulatory Status

GRAS listed. Accepted for use as a food additive in Europe. Included in the FDA Inactive Ingredients Database (oral, rectal, topical preparations). Included in nonparenteral medicines licensed in the UK. Included in the Canadian Natural Health Products Ingredients Database.

17 Related Substances

Ascorbic acid; sodium ascorbate.

18 Comments

In order to maximize the stability and efficacy of ascorbyl palmitate the following precautions are recommended: stainless steel, enamel, or glass should be used; deaeration (vacuum) procedures and inert gas treatment are recommended where feasible; protect from light and radiant energy.

The potential of ascorbyl palmitate-cholesterol vesicles (Aspasomes) as a novel drug delivery platform has been investigated.[5,6]

A specification for ascorbyl palmitate is contained in the *Food Chemicals Codex* (FCC)[7] and the *Japanese Pharmaceutical Excipients* (JPE).[8]

The EINECS number for ascorbyl palmitate is 205-305-4. The PubChem Compound ID (CID) for ascorbyl palmitate is 5282566.

19 Specific References

1 Kläui H. Tocopherol, carotene and ascorbyl palmitate. *Int Flavours Food Addit* 1976; 7(4): 165–172.
2 Chang S-W, *et al.* Optimized synthesis of lipase-catalyzed l-ascorbyl laurate by *Novozym 435*. *J Mol Catal B Enzym* 2009; **56**: 7–12.
3 FAO/WHO. Toxicological evaluation of certain food additives with a review of general principles and of specifications. Seventeenth report of the joint FAO/WHO expert committee on food additives. *World Health Organ Tech Rep Ser* 1974; No. 539.
4 Sweet DV, ed. *Registry of Toxic Effects of Chemical Substances*. Cincinnati: US Department of Health, 1987.
5 Gopinath D, *et al.* Ascorbyl palmitate vesicles (Aspasomes): formation, characterization and applications. *Int J Pharm* 2004; **271**: 95–113.
6 Ojewole E, *et al.* Exploring the use of novel drug delivery systems for antiretroviral drugs. *Eur J Pharm Biopharm* 2008; **70**(3): 697–710.
7 *Food Chemicals Codex*, 7th edn. (Suppl. 1). Bethesda, MD: United States Pharmacopeia, 2010: 1434.
8 Japan Pharmaceutical Excipients Council. *Japanese Pharmaceutical Excipients 2004*. Tokyo: Yakuji Nippo, 2004: 85–86.

20 General References

Austria R, *et al.* Stability of vitamin C derivatives in solution and topical formulations. *J Pharm Biomed Anal* 1997; **15**: 795–801.
Daniel JW. Metabolic aspects of antioxidants and preservatives. *Xenobiotica* 1986; **16**(10–11): 1073–1078.
Pongracz G. Antioxidant mixtures for use in food. *Int J Vitam Nutr Res* 1973; **43**: 517–525.
Špiclin P, *et al.* Stability of ascorbyl palmitate in topical microemulsions. *Int J Pharm* 2001; **222**: 271–279.
Weller PJ, *et al.* Stability of a novel dithranol ointment formulation, containing ascorbyl palmitate as an anti-oxidant. *J Clin Pharm Ther* 1990; **15**: 419–423.

21 Author

JLP Soh.

22 Date of Revision

1 March 2012.

Aspartame

1 Nonproprietary Names

BP: Aspartame
PhEur: Aspartame
USP–NF: Aspartame

2 Synonyms

(3*S*)-3-Amino-4-[[(1*S*)-1-benzyl-2-methoxy-2-oxoethyl]amino]-4-oxobutanoic acid; 3-amino-*N*-(α-carboxyphenethyl)succinamic acid *N*-methyl ester; 3-amino-*N*-(α-methoxycarbonylphenethyl)-succinamic acid; APM; aspartamum; aspartyl phenylamine methyl ester; *Canderel*; E951; *Equal*; methyl *N*-L-α-aspartyl-L-phenylalaninate; *NatraTaste*; *NutraSweet*; *Pal Sweet*; *Pal Sweet Diet*; *Sanecta*; SC-18862; *Tri-Sweet*.

3 Chemical Name and CAS Registry Number

N-L-α-Aspartyl-L-phenylalanine 1-methyl ester [22839-47-0]

4 Empirical Formula and Molecular Weight

$C_{14}H_{18}N_2O_5$ 294.30

5 Structural Formula

6 Functional Category

Sweetening agent.

7 Applications in Pharmaceutical Formulation or Technology

Aspartame is used as an intense sweetening agent in beverage products, food products, and table-top sweeteners, and in pharma-

SEM 1: Excipient: aspartame; magnification: 70×; voltage: 3 kV.

ceutical preparations including tablets,[1,2] powder mixes, and vitamin preparations. It enhances flavor systems and can be used to mask some unpleasant taste characteristics; the approximate sweetening power is 180–200 times that of sucrose.

8 Description

Aspartame occurs as an off white, almost odorless crystalline powder with an intensely sweet taste.

9 Pharmacopeial Specifications

See Table I.

Table I: Pharmacopeial specifications for aspartame.

Test	PhEur 7.4	USP35–NF30
Identification	+	+
Characters	+	−
Appearance of solution	+	−
Conductivity	≤30 μS/cm	−
Specific optical rotation	+14.5° to +16.5°	+14.5° to +16.5°
Related substances	+	−
Heavy metals	≤10 ppm	≤10 ppm
Loss on drying	≤4.5%	≤4.5%
Residue on ignition	−	≤0.2%
Sulfated ash	≤0.2%	−
Impurities	+	−
Transmittance	−	+
Limit of 5-benzyl-3,6-dioxo-2-piperazineacetic acid	−	≤1.5%
Chromatographic purity	−	+
Assay	98.0–102.0%	98.0–102.0%

10 Typical Properties

Acidity/alkalinity pH = 4.5–6.0 (0.8% w/v aqueous solution)
Brittle fracture index 1.05[3]
Bonding index
0.8×10^2 (worst case)[3]
2.3×10^2 (best case)[3]
Flowability 44% (Carr compressibility index)[3]

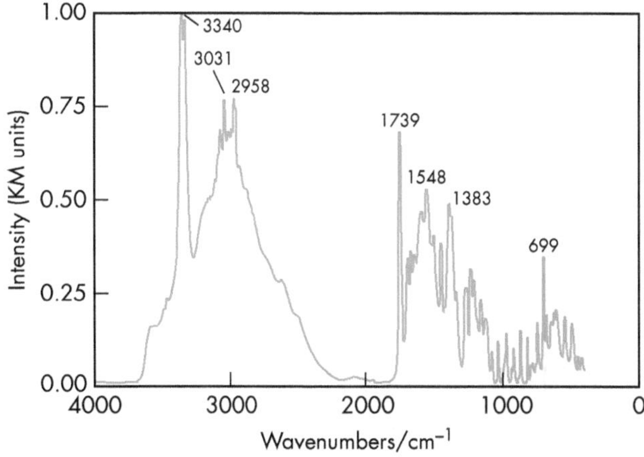

Figure 1: Infrared spectrum of aspartame measured by diffuse reflectance. Adapted with permission of Informa Healthcare.

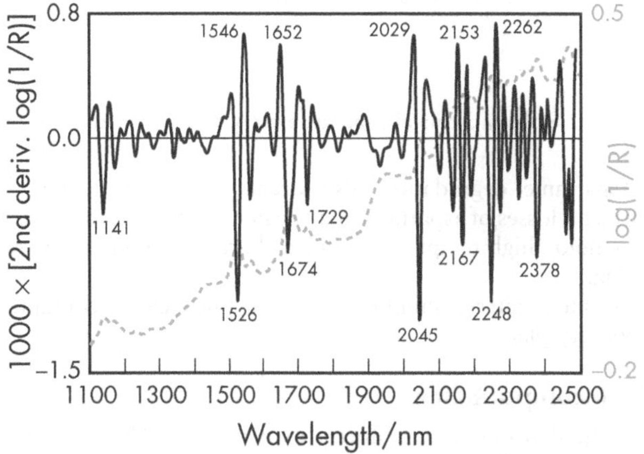

Figure 2: Near-infrared spectrum of aspartame measured by reflectance.

Density (bulk)
0.5–0.7 g/cm^3 for granular grade;
0.2–0.4 g/cm^3 for powder grade;
0.17 g/cm^3 (Spectrum Quality Products).[3]
Density (tapped) 0.29 g/cm^3 (Spectrum Quality Products)[3]
Density (true) 1.347 g/cm^3
Effective angle of internal friction 43.0°[3]
Melting point 246–247°C
Solubility Slightly soluble in ethanol (95%); sparingly soluble in water. At 20°C the solubility is 1% w/v at the isoelectric point (pH 5.2). Solubility increases at higher temperature and at more acidic pH, e.g., at pH 2 and 20°C solubility is 10% w/v.
Specific rotation $[\alpha]_D^{22} = -2.3°$ in 1 N HCl
Spectroscopy
IR spectra *see* Figure 1.
NIR spectra *see* Figure 2.

11 Stability and Storage Conditions

Aspartame is stable in dry conditions. In the presence of moisture, hydrolysis occurs to form the degradation products L -aspartyl-L-phenylalanine and 3-benzyl-6-carboxymethyl-2,5-diketopiperazine with a resulting loss of sweetness. A third-degradation product is also known, β-L-aspartyl-L-phenylalanine methyl ester. For the stability profile at 25°C in aqueous buffers, *see* Figure 3.

Stability in aqueous solutions has been enhanced by the addition of cyclodextrins,[4,5] and by the addition of polyethylene glycol 400 at pH 2.[6] However, at pH 3.5–4.5 stability is not enhanced by the replacement of water with organic solvents.[7]

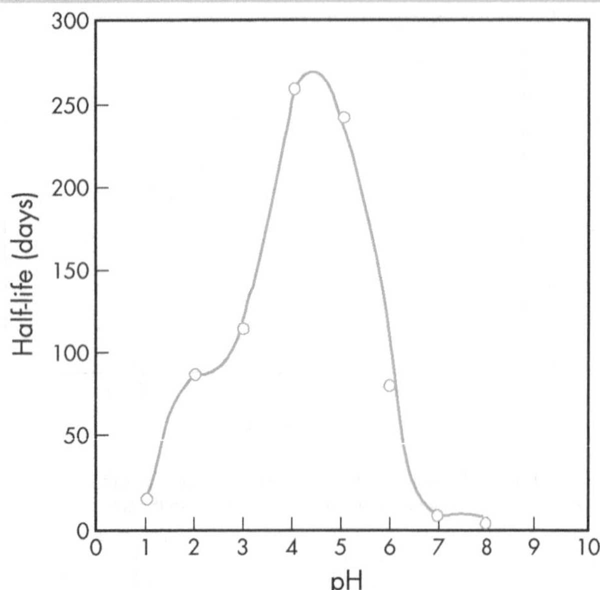

Figure 3: Stability profile of aspartame in aqueous buffers at 25°C.[8]

Aspartame degradation also occurs during prolonged heat treatment; losses of aspartame may be minimized by using processes that employ high temperatures for a short time followed by rapid cooling.

The bulk material should be stored in a well-closed container, in a cool, dry place.

12 Incompatibilities

Differential scanning calorimetry experiments with some directly compressible tablet excipients suggests that aspartame is incompatible with dibasic calcium phosphate and also with the lubricant magnesium stearate.[9] Reactions between aspartame and sugar alcohols are also known.

13 Method of Manufacture

Aspartame is produced by coupling together L-phenylalanine (or L-phenylalanine methyl ester) and L-aspartic acid, either chemically or enzymatically. The former procedure yields both the sweet α-aspartame and nonsweet β-aspartame from which the α-aspartame has to be separated and purified. The enzymatic process yields only α-aspartame.

14 Safety

Aspartame is widely used in oral pharmaceutical formulations, beverages, and food products as an intense sweetener, and is generally regarded as a nontoxic material. However, the use of aspartame has been of some concern owing to the formation of the potentially toxic metabolites methanol, aspartic acid, and phenylalanine. Of these materials, only phenylalanine is produced in sufficient quantities, at normal aspartame intake levels, to cause concern. In the normal healthy individual any phenylalanine produced is harmless; however, it is recommended that aspartame be avoided or its intake restricted by those persons with phenylketonuria.[10]

The WHO has set an acceptable daily intake for aspartame at up to 40 mg/kg body-weight.[11] Additionally, the acceptable daily intake of diketopiperazine (an impurity found in aspartame) has been set by the WHO at up to 7.5 mg/kg body-weight.[12]

A number of adverse effects have been reported following the consumption of aspartame,[10,12] particularly in individuals who drink large quantities (up to 8 liters per day in one case) of aspartame-sweetened beverages. Reported adverse effects include: headaches;[13] grand mal seizure;[14] memory loss;[15] gastrointestinal symptoms; and dermatological symptoms. However, scientifically controlled peer-reviewed studies have consistently failed to produce evidence of a causal effect between aspartame consumption and adverse health events.[16,17] Controlled and thorough studies have confirmed aspartame's safety and found no credible link between consumption of aspartame at levels found in the human diet and conditions related to the nervous system and behavior, nor any other symptom or illness. Aspartame is well documented to be nongenotoxic and there is no credible evidence that aspartame is carcinogenic.[18]

Although aspartame has been reported to cause hyperactivity and behavioral problems in children, a double-blind controlled trial of 48 preschool-age children fed diets containing a daily intake of 38 ± 13 mg/kg body-weight of aspartame for 3 weeks showed no adverse effects attributable to aspartame, or dietary sucrose, on children's behavior or cognitive function.[19]

15 Handling Precautions

Observe normal precautions appropriate to the circumstances and quantity of material handled. Measures should be taken to minimize the potential for dust explosion. Eye protection is recommended.

16 Regulatory Status

Accepted for use as a food additive in Europe and in the USA. Included in the FDA Inactive Ingredients Database (oral powder for reconstitution, buccal patch, granules, syrups, and tablets). Included in nonparenteral medicines licensed in the UK. Included in the Canadian Natural Health Products Ingredients Database.

17 Related Substances

Alitame; aspartame acesulfame; neotame.

Aspartame acesulfame
Empirical formula $C_{18}H_{23}O_9N_3S$
Molecular weight 457.46
CAS number [106372-55-8]
Comments A compound of aspartame and acesulfame approx. 350 times sweeter than sucrose. Aspartame acesulfame is listed in the USP35–NF30.

18 Comments

The intensity of sweeteners relative to sucrose depends upon their concentration, temperature of tasting, and pH, and on the flavor and texture of the product concerned. Unlike some other intense sweeteners, aspartame is metabolized in the body and consequently has some nutritive value: 1 g provides approximately 17 kJ (4 kcal). However, in practice, the small quantity of aspartame consumed provides a minimal nutritive effect.

Intense sweetening agents will not replace the bulk, textural, or preservative characteristics of sugar, if sugar is removed from a formulation.

Synergistic effects for combinations of sweeteners have been reported, e.g. aspartame with acesulfame potassium.

Aspartame can cause browning when used at high temperatures.

A specification for aspartame is contained in the *Food Chemicals Codex* (FCC).[20]

The PubChem Compound ID (CID) for aspartame includes 2242 and 21462246.

19 Specific References

1 Joachim J, *et al.* [The compression of effervescent aspartame tablets: the influence of particle size on the strain applied on the punches during compression.] *J Pharm Belg* 1987; 42: 17–28[in French].

2 Joachim J, *et al.* [The compression of effervescent aspartame tablets: the influence of particle size and temperature on the effervescence time and carbon dioxide liberation kinetics.] *J Pharm Belg* 1987; **42**: 303–314[in French].
3 Mullarney MP, *et al.* The powder flow and compact mechanical properties of sucrose and three high-intensity sweeteners used in chewable tablets. *Int J Pharm* 2003; **257**(1–2): 227–236.
4 Brewster ME, *et al.* Stabilization of aspartame by cyclodextrins. *Int J Pharm* 1991; **75**: R5–R8.
5 Prankerd RJ, *et al.* Degradation of aspartame in acidic aqueous media and its stabilization by complexation with cyclodextrins or modified cyclodextrins. *Int J Pharm* 1992; **88**: 189–199.
6 Yalkowsky SH, *et al.* Stabilization of aspartame by polyethylene glycol 400. *J Pharm Sci* 1993; **82**: 978.
7 Sanyude S, *et al.* Stability of aspartame in water: organic solvent mixtures with different dielectric constants. *J Pharm Sci* 1991; **80**: 674–676.
8 The NutraSweet Company. Technical literature: *NutraSweet* technical bulletin, 1991.
9 El-Shattawy HE, *et al.* Aspartame-direct compression excipients: preformulation stability screening using differential scanning calorimetry. *Drug Dev Ind Pharm* 1981; **7**: 605–619.
10 Golightly LK, *et al.* Pharmaceutical excipients: adverse effects associated with inactive ingredients in drug products (part II). *Med Toxicol* 1988; **3**: 209–240.
11 FAO/WHO. Evaluation of certain food additives and contaminants. Twenty-fifth report of the joint FAO/WHO expert committee on food additives. *World Health Organ Tech Rep Ser* 1981; No. 669.
12 Butchko HH, Kotsonis FN. Aspartame: review of recent research. *CommentsToxicol* 1989; **3**(4): 253–278.
13 Schiffman SS, *et al.* Aspartame and susceptibility to headache. *N Engl J Med* 1987; **317**: 1181–1185.
14 Wurtman RJ. Aspartame: possible effect on seizure susceptibility [letter]. *Lancet* 1985; **ii**: 1060.
15 Anonymous. Sweetener blamed for mental illnesses. *New Scientist* 1988; **February 18**: 33.
16 O'Donnell K. Aspartame and neotame. In: Mitchell H, ed. *Sweeteners and Sugar Alternatives in Food Technology.* Oxford, UK: Blackwell Publishing, 2006: 86–102.
17 European Commission. Opinion of the Scientific Committee on Food: update on the safety of aspartame, 2002. Available at: ec.europa.eu/food/fs/sc/scf/out155_en.pdf (accessed 2 December 2010).
18 Magnuson BA, *et al.* Aspartame: a safety evaluation based on current use levels, regulations, and toxicological and epidemiological studies. *Crit Rev Toxicol* 2007; **37**: 629–727.
19 Wolraich ML, *et al.* Effects of diets high in sucrose or aspartame on the behavior and cognitive performance of children. *N Engl J Med* 1994; **330**: 301–307.
20 *Food Chemicals Codex*, 7th edn. Bethesda, MD: United States Pharmacopeia, 2010: 73.

20 General References

Marie S. Sweeteners. In: Smith J, ed. *Food Additives User's Handbook.* Glasgow: Blackie, 1991: 47–74.

Roy GM. Taste masking in oral pharmaceuticals. *Pharm Technol Eur* 1994; 6(6): 24, 26–2830–3234, 35.

Stegink LD, Filer LJ, eds. *Aspartame, Physiology and Biochemistry.* New York: Marcel Dekker, 1984.

21 Author

S Brown.

22 Date of Revision

1 March 2012.

Attapulgite

1 Nonproprietary Names

BP: Attapulgite

2 Synonyms

Actapulgite; *Attaclay*; *Attacote*; *Attagel*; attapulgus; palygorscite; palygorskite; *Pharmasorb Colloidal*; *Pharmasorb Regular.*

3 Chemical Name and CAS Registry Number

Attapulgite [12174-11-7]

4 Empirical Formula and Molecular Weight

Attapulgite is a purified native hydrated magnesium aluminum silicate consisting of the clay mineral palygorskite, with the empirical formula $Mg(Al_{0.5-1}Fe_{0-0.5})Si_4O_{10}(OH)\cdot 4H_2O$.

5 Structural Formula

See Section 4.

6 Functional Category

Adsorbent; tablet and capsule binder; suspending agent; viscosity-increasing agent.

7 Applications in Pharmaceutical Formulation or Technology

Attapulgite is widely used as an adsorbent in solid dosage forms. Colloidal clays (such as attapulgite) absorb considerable amounts of water to form gels and in concentrations of 2–5% w/v usually form oil-in-water emulsions.

8 Description

Attapulgite occurs as a light cream colored, very fine powder. Particle size ranges depend on the grade and manufacturer.

9 Pharmacopeial Specifications

See Table I. *See also* Section 17.

10 Typical Properties

Acidity/alkalinity pH = 9.5 (5% w/v aqueous suspension)
Angle of repose 37.2–45.2°[1]
Density 2.2 g/cm^3
Density (tapped) 0.33 g/cm^3 [1]
Flowability 20.9–29.6% (Carr compressibility index)[1]
Particle size distribution

<2 μm in size for powder;

2–5 μm in size for aggregate.[1]

A

Table I: Pharmacopeial specifications for attapulgite.

Test	BP 2012
Identification	+
Characters	+
Acidity or alkalinity (5% w/v aqueous suspension)	7.0–9.5
Adsorptive capacity	5–14%
Arsenic	≤8 ppm
Heavy metals	≤20 ppm
Acid-soluble matter	+
Water-soluble matter	+
Loss on drying	≤17.0%
Loss on ignition	15.0–27.0%

11 Stability and Storage Conditions

Attapulgite can adsorb water. It should be stored in an airtight container in a cool, dry, location.

12 Incompatibilities

Attapulgite may decrease the bioavailability of some drugs such as loperamide[2] and riboflavin.[3] Oxidation of hydrocortisone is increased in the presence of attapulgite.[4]

13 Method of Manufacture

Attapulgite occurs naturally as the mineral palygorskite.

14 Safety

Attapulgite is widely used in pharmaceutical formulations and is generally regarded as an essentially nontoxic and nonirritant material. It is not absorbed following oral administration. In oral preparations, activated attapulgite up to 9 g is used in daily divided doses as an adjunct in the management of diarrhea.[5]

The Cosmetic Ingredient Review (CIR) Expert Panel have assessed attapulgite and concluded that the material is safe as used in cosmetics and personal care products.[6]

LD_{50} (rat, IP): 0.34 g/kg

15 Handling Precautions

Observe normal precautions appropriate to the circumstances and quantity of material handled. Eye protection, gloves, and a dust mask are recommended. Attapulgite should be handled in a well-ventilated environment and dust generation should be minimized. When heated to decomposition, attapulgite emits acrid smoke and irritating fumes.

16 Regulatory Status

Included in the FDA Inactive Ingredients Database (oral; powder). Included in nonparenteral medicines licensed in a number of countries worldwide including the UK and USA. Included in the Canadian Natural Health Products Ingredients Database.

17 Related Substances

Activated attapulgite; magnesium aluminum silicate.

Activated attapulgite

Comments Activated attapulgite is a processed native magnesium aluminum silicate that has been carefully heated to increase its adsorptive capacity. Monographs for activated attapulgite are included in the BP 2012, USP35–NF30, and other pharmacopeias. The USP35–NF30 also includes a monograph for colloidal activated attapulgite. Activated attapulgite is used therapeutically as an adjunct in the management of diarrhea.

18 Comments

The EINECS number for attapulgite is 302-243-0.

19 Specific References

1 Viseras C, López-Galindo A. Characteristics of pharmaceutical grade phyllosilicate powders. *Pharm Dev Technol* 2000; 5(1): 47–52.
2 Mboya SA, Bhargava HN. Adsorption and desorption of loperamide hydrochloride by activated attapulgites. *Am J Health Syst Pharm* 1995; 52: 2816–2818.
3 Khalil SAH, *et al*. Effect of attapulgite on the bioavailability of a model low dose drug (riboflavine) in humans. *Drug Dev Ind Pharm* 1987; **13**: 369–382.
4 Cornejo J, *et al.* Oxidative degradation of hydrocortisone in the presence of attapulgite. *J Pharm Sci* 1980; **69**: 945–948.
5 Sweetman SC, ed. *Martindale: The Complete Drug Reference*, 36th edn. London: Pharmaceutical Press, 2009: 1709.
6 Elmore AR. Final report on the safety assessment of aluminum silicate, calcium silicate, magnesium aluminum silicate, magnesium silicate, magnesium trisilicate, sodium magnesium silicate, zirconium silicate, attapulgite, bentonite, Fuller's earth, hectorite, kaolin, lithium magnesium silicate, lithium magnesium sodium silicate, montmorillonite, pyrophyllite, and zeolite. *Int J Toxicol* 2003; **22**(Suppl. 1): 37–102.

20 General References

Anonymous. The silicates: attapulgite, kaolin, kieselguhr, magnesium trisilicate, pumice, talc. *Int J Pharm Compound* 1998; 2(2): 162–163.
BASF. Product information sheet: Attapulgite, 2007.
Viseras C, *et al.* Characteristics of pharmaceutical grade phyllosilicate compacts. *Pharm Dev Technol* 2000; 5(1): 53–58.

21 Authors

A Palmieri, M Tatipalli.

22 Date of Revision

1 March 2012.

Bentonite

1 Nonproprietary Names

BP: Bentonite
JP: Bentonite
PhEur: Bentonite
USP–NF: Bentonite

2 Synonyms

Albagel; bentonitum; E558; mineral soap; *Polargel*; soap clay; taylorite; *Veegum HS*; wilkinite.

3 Chemical Name and CAS Registry Number

Bentonite [1302-78-9]

4 Empirical Formula and Molecular Weight

$Al_2O_3{\cdot}4SiO_2{\cdot}H_2O$ 359.16

Bentonite is a native colloidal hydrated aluminum silicate consisting mainly of montmorillonite, $Al_2O_3{\cdot}4SiO_2{\cdot}H_2O$; it may also contain calcium, magnesium, and iron. The average chemical analysis is expressed as oxides, *see* Table I, in comparison with magnesium aluminum silicate.

Table I: Average chemical analysis of bentonite expressed as oxides in comparison with magnesium aluminum silicate.

	Bentonite	Magnesium aluminum silicate
Silicon dioxide	59.92%	61.1%
Aluminum oxide	19.78%	9.3%
Magnesium oxide	1.53%	13.7%
Ferric oxide	2.96%	0.9%
Calcium oxide	0.64%	2.7%
Sodium oxide	2.06%	2.9%
Potassium oxide	0.57%	0.3%

5 Structural Formula

The PhEur 7.4 describes bentonite as a natural clay containing a high proportion of montmorillonite, a native hydrated aluminum silicate in which some aluminum and silicon atoms may be replaced by other atoms such as magnesium and iron.

The USP35–NF30 describes bentonite, purified benonite, and bentonite magma in three separate monographs. Bentonite is described as a native, colloidal, hydrated aluminum silicate; and purified bentonite is described as a colloidal montmorillonite that has been processed to remove grit and nonswellable ore compounds.

See also Section 4.

6 Functional Category

Adsorbent; suspending agent; viscosity-increasing agent

7 Applications in Pharmaceutical Formulation or Technology

Bentonite is a naturally occurring hydrated aluminum silicate used primarily in the formulation of suspensions, gels, and sols, for topical pharmaceutical applications. It is also used to suspend powders in aqueous preparations and to prepare cream bases containing oil-in-water emulsifying agents.

Bentonite may also be used in oral pharmaceutical preparations, cosmetics, and food products, *see* Section 18. In oral preparations, bentonite, and other similar silicate clays, can be used to adsorb cationic drugs and so retard their release.[1–3] Adsorbents are also used to mask the taste of certain drugs. *See* Table II.

Table II: Uses of bentonite.

Use	Concentration (%)
Adsorbent (clarifying agent)	1.0–2.0
Emulsion stabilizer	1.0
Suspending agent	0.5–5.0

8 Description

Bentonite is a crystalline, claylike mineral, and is available as an odorless, pale buff, or cream to grayish-colored fine powder, which is free from grit. It consists of particles about 50–150 μm in size along with numerous particles about 1–2 μm. Microscopic examination of samples stained with alcoholic methylene blue solution reveals strongly stained blue particles. Bentonite may have a slight earthy taste.

9 Pharmacopeial Specifications

See Table III.

Table III: Pharmacopeial specifications for bentonite.

Test	JP XV	PhEur 7.4	USP35–NF30
Identification	+	+	+
Characters	+	+	—
Alkalinity	—	+	—
Microbial limit	—	$\leq 10^3$ cfu/g	+
Coarse particles	—	≤0.5%	—
pH (2% w/v suspension)	9.0–10.5	—	9.5–10.5
Loss on drying	5.0–10.0%	≤15%	5.0–8.0%
Arsenic	≤2 ppm	—	≤5 ppm
Lead	—	—	≤40 ppm
Heavy metals	≤50 ppm	≤50 ppm	—
Gel formation	+	—	+
Sedimentation volume	—	≤2 mL	—
Swelling power	≥20 mL	≥22 mL	≥24 mL
Fineness of powder	+	—	+

The USP35–NF30 also contains specifications for bentonite magma and purified bentonite. *See* Section 17.

10 Typical Properties

Acidity/alkalinity pH = 9.5–10.5 for a 2% w/v aqueous suspension.

Flowability No flow.

Hygroscopicity Bentonite is hygroscopic.[4] *See also* Figure 1.

Moisture content 5–12%.

Solubility Practically insoluble in ethanol, fixed oils, glycerin, propan-2-ol, and water. Bentonite swells to about 12 times its original volume in water, to form viscous homogeneous suspensions, sols, or gels depending upon the concentration. Bentonite does not swell in organic solvents. Sols and gels may be conveniently prepared by sprinkling the bentonite on the surface of hot water and allowing to stand for 24 hours, stirring occasionally when the bentonite has become thoroughly wetted.

SEM 1: Excipient: bentonite; manufacturer: American Colloid Co.; lot no.: NMD 11780; magnification: 600×; voltage: 10 kV.

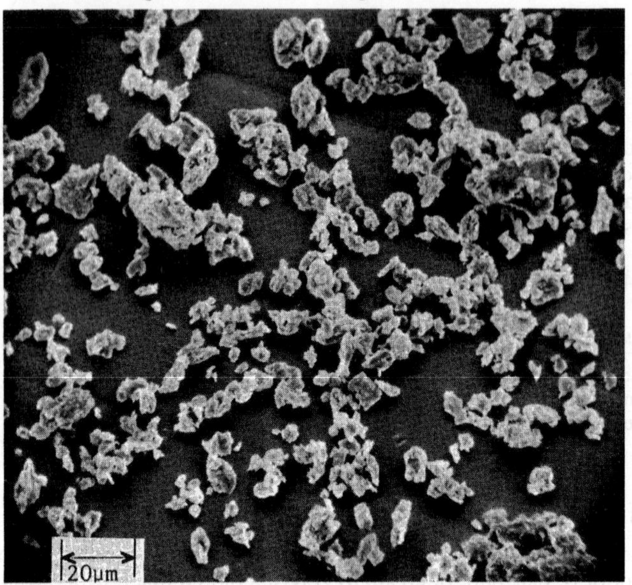

SEM 2: Excipient: bentonite; manufacturer: American Colloid Co.; lot no: NMD 11780; magnification: 2400×; voltage: 20 kV.

Water should not be added to bentonite alone, but bentonite may be satisfactorily dispersed in water if it is first triturated with glycerin or mixed with a powder such as zinc oxide. A 7% w/v aqueous suspension of bentonite is just pourable. *See also* Section 12.

Spectroscopy

NIR spectra *see* Figure 2.

Viscosity (dynamic) 75–225 mPa s (75–225 cP) for a 5.5% w/v aqueous suspension at 25°C. Viscosity increases with increasing concentration.

11 Stability and Storage Conditions

Bentonite is hygroscopic, and sorption of atmospheric water should be avoided.

Aqueous bentonite suspensions may be sterilized by autoclaving. The solid material may be sterilized by maintaining it at 170°C for 1 hour after drying at 100°C.

Bentonite should be stored in an airtight container in a cool, dry place.

12 Incompatibilities

Aqueous bentonite suspensions retain their viscosity above pH 6, but are precipitated by acids. Acid-washed bentonite does not have suspending properties. The addition of alkaline materials, such as magnesium oxide, increases gel formation.

Addition of significant amounts of alcohol to aqueous preparations will precipitate bentonite, primarily by dehydration of the lattice structure; *see also* Section 18.

Bentonite particles are negatively charged and flocculation occurs when electrolytes or positively charged suspensions are added. Bentonite is thus said to be incompatible with strong electrolytes, although this effect is sometimes used beneficially to clarify turbid liquids.

The antimicrobial efficacy of cationic preservatives may be reduced in aqueous bentonite suspensions, but nonionic and anionic preservatives are unaffected.[5]

Bentonite is incompatible with acriflavine hydrochloride.

13 Method of Manufacture

Bentonite is a native, colloidal, hydrated aluminum silicate, found in regions of Canada and the USA. The mined ore is processed to remove grit and nonswelling materials so that it is suitable for pharmaceutical applications.

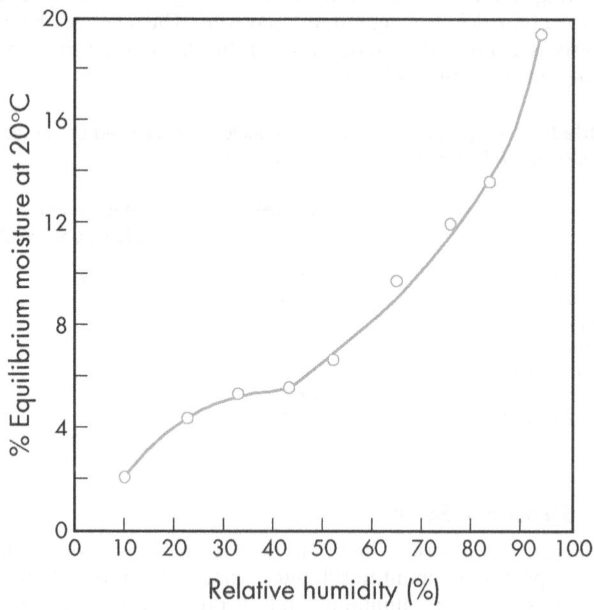

Figure 1: Equilibrium moisture content of bentonite USP–NF.

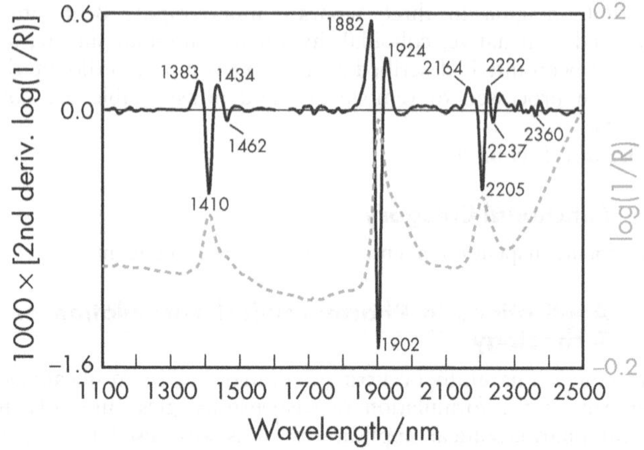

Figure 2: Near-infrared spectrum of bentonite measured by reflectance.

14 Safety

Bentonite is mainly used in topical pharmaceutical formulations but has also been used in oral pharmaceutical preparations, food products, and cosmetics.

Following oral administration, bentonite is not absorbed from the gastrointestinal tract. Bentonite is generally regarded as a nontoxic and nonirritant material.

LD_{50} (rat, IV): 0.035 g/kg[6]

15 Handling Precautions

Observe normal precautions appropriate to the circumstances and quantity of material handled. Eye protection, gloves, and a dust mask are recommended. Bentonite should be handled in a well-ventilated environment and dust generation minimized.

16 Regulatory Status

Accepted in Europe as a food additive in certain applications. Included in the FDA Inactive Ingredients Database (oral capsules, tablets and suspensions, topical suspensions, controlled release transdermal films and vaginal suppositories). Included in nonparenteral medicines licensed in the UK. Included in the Canadian Natural Health Products Ingredients Database.

17 Related Substances

Bentonite magma; kaolin; magnesium aluminum silicate; magnesium trisilicate; purified bentonite; talc.

Bentonite magma

Comments A 5% w/w suspension of bentonite in purified water appears in some pharmacopeias, such as the USP35–NF30.

Purified bentonite

Acidity/alkalinity pH = 9.0–10.0 for a 5% w/w aqueous suspension.

Viscosity (dynamic) 40–200 mPa s (40–200 cP) for a 5% w/w aqueous suspension.

Comments Specifications for purified bentonite occur in some pharmacopeias such as the USP35–NF30. Purified bentonite is bentonite that has been processed to remove grit and nonswellable ore components.

18 Comments

Bentonite may be used with concentrations of up to 30% ethanol or propan-2-ol; 50% glycerin; 30% propylene glycol; or high molecular weight polyethylene glycols. The EINECS number for bentonite is 215-108-5.

Bentonite is used in the food industry as a processing aid as a clarifying or filter agent. A specification for bentonite is contained in the *Food Chemicals Codex* (FCC).[7]

Bentonite has been investigated as a diagnostic agent for magnetic resonance imaging.[8]

Therapeutically, bentonite has been investigated as an adsorbent for lithium poisoning.[9]

19 Specific References

1 Stul MS, *et al. In vitro* adsorption-desorption of phenethylamines and phenylimidazoles by a bentonite and a resin. *J Pharm Sci* 1984; **73**: 1372–1375.

2 Shrivastava R, *et al.* Dissolution dialysis studies of metronidazole–montmorillonite adsorbates. *J Pharm Sci* 1985; **74**: 214–216.

3 Forni F, *et al.* Effect of montmorillonite on drug release from polymeric matrices. *Arch Pharm* 1989; **322**: 789–793.

4 Callahan JC, *et al.* Equilibrium moisture content of pharmaceutical excipients. *Drug Dev Ind Pharm* 1982; **8**: 355–369.

5 Harris WA. The inactivation of cationic antiseptics by bentonite suspensions. *Aust J Pharm* 1961; **42**: 583–588.

6 Lewis RJ, ed. *Sax's Dangerous Properties of Industrial Materials*, 11th edn. New York: Wiley, 2004: 351.

7 *Food Chemicals Codex*, 7th edn (Suppl. 1). Bethesda, MD: United States Pharmacopeia, 2010: 85.

8 Listinsky JJ, Bryant RG. Gastrointestinal contrast agents: a diamagnetic approach. *Magn Reson Med* 1988; **8**(3): 285–292.

9 Ponampalam R, Otten EJ. *In vitro* adsorption of lithium by bentonite. *Singapore Med J* 2002; **43**(2): 86–89.

20 General References

Altagracia M, *et al.* A comparative mineralogical and physico-chemical study of some crude Mexican and pharmaceutical grade montmorillonites. *Drug Dev Ind Pharm* 1987; **13**: 2249–2262.

Sadik F, *et al.* X-Ray diffraction analysis for identification of kaolin NF and bentonite USP. *J Pharm Sci* 1971; **60**: 916–918.

21 Authors

A Palmieri, M Tatipalli.

22 Date of Revision

1 March 2012.

Benzalkonium Chloride

1 Nonproprietary Names

BP: Benzalkonium Chloride
JP: Benzalkonium Chloride
PhEur: Benzalkonium Chloride
USP–NF: Benzalkonium Chloride

2 Synonyms

Alkylbenzyldimethylammonium chloride; alkyl dimethyl benzyl ammonium chloride; benzalkonii chloridum; BKC; *Hyamine 3500*; *Pentonium*; *Zephiran*.

3 Chemical Name and CAS Registry Number

Alkyldimethyl(phenylmethyl)ammonium chloride [8001-54-5]

4 Empirical Formula and Molecular Weight

The USP33–NF28 S1 describes benzalkonium chloride as a mixture of alkylbenzyldimethylammonium chlorides of the general formula $[C_6H_5CH_2N(CH_3)_2R]Cl$, where R represents a mixture of alkyls, including all or some of the group beginning with n-C_8H_{17} and extending through higher homologs, with n-$C_{12}H_{25}$, n-$C_{14}H_{29}$, and n-$C_{16}H_{33}$ comprising the major portion.

The average molecular weight of benzalkonium chloride is 360.

5 Structural Formula

$$\left[C_6H_5-\underset{H}{\overset{H}{C}}-\underset{CH_3}{\overset{CH_3}{N^+}}-R \right] Cl^-$$

R = mixture of alkyls: n-C_8H_{17} to n-$C_{18}H_{37}$; mainly n-$C_{12}H_{25}$ (dodecyl), n-$C_{14}H_{29}$ (tetradecyl), and n-$C_{16}H_{33}$ (hexadecyl).

6 Functional Category

Antimicrobial preservative; cationic surfactant; solubilizing agent.

7 Applications in Pharmaceutical Formulation or Technology

Benzalkonium chloride is a quaternary ammonium compound used in pharmaceutical formulations as an antimicrobial preservative in applications similar to other cationic surfactants, such as cetrimide.

In ophthalmic preparations, benzalkonium chloride is one of the most widely used preservatives,[1] at a concentration of 0.01–0.02% w/v. Often it is used in combination with other preservatives or excipients, particularly 0.1% w/v disodium edetate, to enhance its antimicrobial activity against strains of *Pseudomonas*.

In nasal[2] and otic formulations a concentration of 0.002–0.02% w/v is used, sometimes in combination with 0.002–0.005% w/v thimerosal. Benzalkonium chloride 0.01% w/v is also employed as a preservative in small-volume parenteral products. Benzalkonium chloride was also shown to enhance the topical penetration of lorazepam.[3]

Benzalkonium chloride is additionally used as a preservative in cosmetics.

8 Description

Benzalkonium chloride occurs as a white or yellowish-white amorphous powder, a thick gel, or gelatinous flakes. It is hygroscopic, soapy to the touch, and has a mild aromatic odor and very bitter taste.

9 Pharmacopeial Specifications

See Table I.

Table I: Pharmacopeial specifications for benzalkonium chloride.

Test	JP XV	PhEur 7.4	USP35–NF30
Identification	+	+	+
Characters	+	+	–
Acidity or alkalinity	–	+	+
Appearance of solution	+	+	–
Water	≤15.0%	≤10.0%	≤15.0%
Residue on ignition	≤0.2%	–	≤2.0%
Sulfated ash	–	≤0.1%	–
Water-insoluble matter	–	–	+
Foreign amines	–	+	–
Amines and amine salts	–	+	0.1 mmol/g
Ratio of alkyl components	–	+	+
Petroleum ether-soluble substances	≤1.0%	–	–
Benzyl alcohol	–	≤0.5%	–
Benzaldehyde	–	≤0.15%	–
Chloromethylbenzene	–	≤0.05%	–
Assay (dried basis)			
of n-$C_{12}H_{25}$	–	–	≥40.0%
of n-$C_{14}H_{29}$	–	–	≥20.0%
of n-$C_{12}H_{25}$ and n-$C_{14}H_{29}$	–	–	≥70.0%
for total alkyl content	95.0–105.0%	95.0–104.0%	97.0–103.0%

10 Typical Properties

Acidity/alkalinity pH = 5–8 for a 10% w/v aqueous solution.

Antimicrobial activity Benzalkonium chloride solutions are active against a wide range of bacteria, yeasts, and fungi. Activity is more marked against Gram-positive than Gram-negative bacteria and minimal against bacterial endospores and acid-fast bacteria, *see* Table II. The antimicrobial activity of benzalkonium chloride is significantly dependent upon the alkyl composition of the homolog mixture.[4] Benzalkonium chloride is ineffective against some *Pseudomonas aeruginosa* strains, *Mycobacterium tuberculosis*, *Trichophyton interdigitale*, and *T. rubrum*. However, combined with disodium edetate (0.01–0.1% w/v), benzyl alcohol, phenylethanol, or phenylpropanol, the activity against *Pseudomonas aeruginosa* is increased.[5] Antimicrobial activity may also be enhanced by the addition of phenylmercuric acetate, phenylmercuric borate, chlorhexidine, cetrimide, or *m*-cresol.[6,7] In the presence of citrate and phosphate buffers (but not borate), activity against *Pseudomonas* can be reduced. *See also* Sections 11 and 12. Benzalkonium chloride is relatively inactive against spores and molds, but is active against some viruses, including HIV.[8] Inhibitory activity increases with pH, although antimicrobial activity occurs at pH 4–10.

Table II: Minimum inhibitory concentrations (MICs) of benzalkonium chloride.

Microorganism	MIC (μg/mL)
Aerobacter aerogenes	64
Clostridium histolyticum	5
Clostridium oedematiens	5
Clostridium tetani	5
Clostridium welchii	5
Escherichia coli	16
Pneumococcus II	5
Proteus vulgaris	64
Pseudomonas aeruginosa	30
Salmonella enteritidis	30
Salmonella paratyphi	16
Salmonella typhosa	4
Shigella dysenteriae	2
Staphylococcus aureus	1.25
Streptococcus pyrogenes	1.25
Vibrio cholerae	2

Density ≈0.98 g/cm^3 at 20°C

Melting point ≈40°C

Partition coefficients The octanol : water partition coefficient varies with the alkyl chain length of the homolog: 9.98 for C_{12}, 32.9 for C_{14}, and 82.5 for C_{16}.

Solubility Practically insoluble in ether; very soluble in acetone, ethanol (95%), methanol, propanol, and water. Aqueous solutions of benzalkonium chloride foam when shaken, have a low surface tension and possess detergent and emulsifying properties.

Spectroscopy

IR spectra *see* Figure 1.

NIR spectra *see* Figure 2.

11 Stability and Storage Conditions

Benzalkonium chloride is hygroscopic and may be affected by light, air, and metals.

Solutions are stable over a wide pH and temperature range and may be sterilized by autoclaving without loss of effectiveness. Solutions may be stored for prolonged periods at room temperature. Dilute solutions stored in polyvinyl chloride or polyurethane foam containers may lose antimicrobial activity.

The bulk material should be stored in an airtight container, protected from light and contact with metals, in a cool, dry place.

12 Incompatibilities

Incompatible with aluminum, anionic surfactants, citrates, cotton, fluorescein, hydrogen peroxide, hypromellose,[9] iodides, kaolin, lanolin, nitrates, nonionic surfactants in high concentration, permanganates, protein, salicylates, silver salts, soaps, sulfonamides, tartrates, zinc oxide, zinc sulfate, some rubber mixes, and some plastic mixes.

Benzalkonium chloride has been shown to be adsorbed by various filtering membranes, especially those that are hydrophobic or anionic.[10]

13 Method of Manufacture

Benzalkonium chloride is formed by the reaction of a solution of *N*-alkyl-*N*-methylbenzamine with methyl chloride in an organic solvent suitable for precipitating the quaternary compound as it is formed.

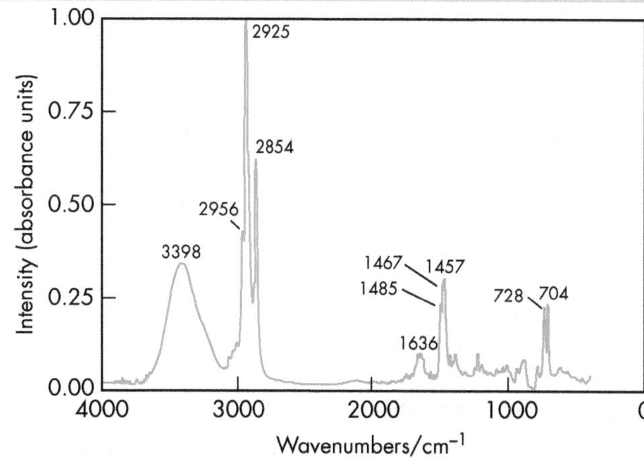

Figure 1: Infrared spectrum of benzalkonium chloride measured by transmission. Adapted with permission of Informa Healthcare.

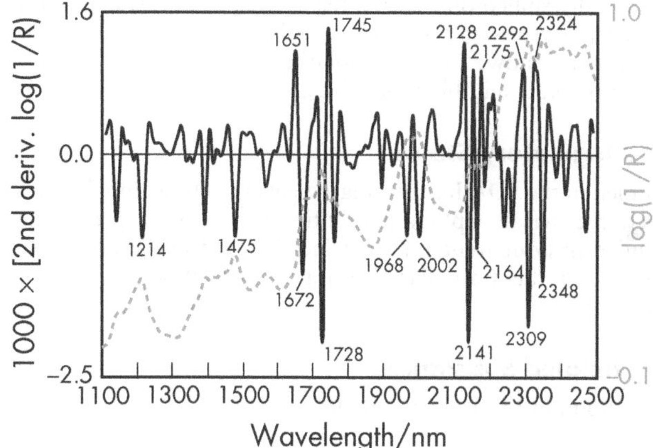

Figure 2: Near-infrared spectrum of benzalkonium chloride measured by reflectance.

14 Safety

Benzalkonium chloride is usually nonirritating, nonsensitizing, and is well tolerated in the dilutions normally employed on the skin and mucous membranes. However, benzalkonium chloride has been associated with adverse effects when used in some pharmaceutical formulations.[11]

Ototoxicity can occur when benzalkonium chloride is applied to the ear[12] and prolonged contact with the skin can occasionally cause irritation and hypersensitivity. Benzalkonium chloride is also known to cause bronchoconstriction in some asthmatics when used in nebulizer solutions.[13–17]

Toxicity experiments with rabbits have shown benzalkonium chloride to be harmful to the eye in concentrations higher than that normally used as a preservative. However, the human eye appears to be less affected than the rabbit eye and many ophthalmic products have been formulated with benzalkonium chloride 0.01% w/v as the preservative.

Benzalkonium chloride is not suitable for use as a preservative in solutions used for storing and washing hydrophilic soft contact lenses, as the benzalkonium chloride can bind to the lenses and may later produce ocular toxicity when the lenses are worn.[18] Solutions stronger than 0.03% w/v concentration entering the eye require prompt medical attention.

The use of benzalkonium chloride as a preservative in intranasal solutions at concentrations of between 0.00045% and 0.1% in treatments for up to one year has been found to demonstrate no toxic effects.[19]

Local irritation of the throat, esophagus, stomach, and intestine can occur following contact with strong solutions (>0.1% w/v). The fatal oral dose of benzalkonium chloride in humans is estimated to be 1–3 g. Adverse effects following oral ingestion include vomiting, collapse, and coma. Toxic doses lead to paralysis of the respiratory muscles, dyspnea, and cyanosis.

LD_{50} (mouse, oral): 150 mg/kg[20]
LD_{50} (rat, IP): 14.5 mg/kg
LD_{50} (rat, IV): 13.9 mg/kg
LD_{50} (rat, oral): 300 mg/kg
LD_{50} (rat, skin): 1.42 g/kg

15 Handling Precautions

Observe normal precautions appropriate to the circumstances and quantity of material handled. Benzalkonium chloride is irritant to the skin and eyes and repeated exposure to the skin may cause hypersensitivity. Concentrated benzalkonium chloride solutions accidentally spilled on the skin may produce corrosive skin lesions with deep necrosis and scarring, and should be washed immediately with water, followed by soap solutions applied freely. Gloves, eye protection, and suitable protective clothing should be worn.

16 Regulatory Status

Included in the FDA Inactive Ingredients Database (inhalations, IM injections, nasal, ophthalmic, otic, and topical preparations). Included in nonparenteral medicines licensed in the UK. It is also included in the Canadian Natural Health Products Ingredients Database.

17 Related Substances

Benzethonium chloride; cetrimide.

18 Comments

Benzalkonium chloride has been used in antiseptic wipes and has been shown to produce significantly less stinging or burning than isopropyl alcohol and hydrogen peroxide.[21]

The EINECS numbers for benzalkonium chloride are 264-151-6; 260-080-8; 269-919-4; 270-325-2; 287-089-1. The PubChem Compound ID (CID) for benzalkonium chloride is 3014024.

19 Specific References

1 Sklubalova Z. Antimicrobial substances in ophthalmic drops. *Ceska Slov Form* 2004; 53(3): 107–116.
2 Pisal SS, *et al.* Pluronic gels for nasal delivery of vitamin B. *Int J Pharm* 2004; **270**(1–2): 37–45.
3 Nokodchi A, *et al.* The enhancement effect of surfactants in the penetration of lorazepam through rat skin. *Int J Pharm* 2003; **250**(2): 359–369.
4 Euerby MR. High performance liquid chromatography of benzalkonium chlorides – variation in commercial preparations. *J Clin Hosp Pharm* 1985; **10**: 73–77.
5 Richards RME, McBride RJ. Enhancement of benzalkonium chloride and chlorhexidine acetate activity against *Pseudomonas aeruginosa* by aromatic alcohols. *J Pharm Sci* 1973; **62**: 2035–2037.
6 Hugbo PG. Additivity and synergism *in vitro* as displayed by mixtures of some commonly employed antibacterial preservatives. *Can J Pharm Sci* 1976; **11**: 17–20.
7 McCarthy TJ, *et al.* Further studies on the influence of formulation on preservative activity. *Cosmet Toilet* 1977; 92(3): 33–36.
8 Chermann JC, *et al.* HIV inactivation by a spermicide containing benzalkonium. *AIDS Forsch* 1987; **2**: 85–86.
9 Richards RME. Effect of hypromellose on the antibacterial activity of benzalkonium chloride. *J Pharm Pharmacol* 1976; **28**: 264.
10 Bin T, *et al.* Adsorption of benzalkonium chloride by filter membranes: mechanisms and effect of formulation and processing parameters. *Pharm Dev Technol* 1999; **4**(2): 151–165.
11 Smolinske SC. *Handbook of Food, Drug, and Cosmetic Excipients.* Boca Raton, FL: CRC Press, 1992; 31–39.
12 Honigman JL. Disinfectant ototoxicity [letter]. *Pharm J* 1975; **215**: 523.
13 Beasley CR, *et al.* Bronchoconstrictor properties of preservatives in ipratropium bromide (Atrovent) nebuliser solution. *Br Med J* 1987; **294**: 1197–1198.
14 Miszkiel KA, *et al.* The contribution of histamine release to bronchoconstriction provoked by inhaled benzalkonium chloride in asthma. *Br J Clin Pharmacol* 1988; **25**: 157–163.
15 Miszkiel KA, *et al.* The influence of ipratropium bromide and sodium cromoglycate on benzalkonium chloride-induced bronchoconstriction in asthma. *Br J Clin Pharmacol* 1988; **26**: 295–301.
16 Worthington I. Bronchoconstriction due to benzalkonium chloride in nebulizer solutions. *Can J Hosp Pharm* 1989; **42**: 165–166.
17 Boucher M, *et al.* Possible association of benzalkonium chloride in nebulizer solutions with respiratory arrest. *Ann Pharmacother* 1992; **26**: 772–774.
18 Gasset AR. Benzalkonium chloride toxicity to the human cornea. *Am J Ophthalmol* 1977; **84**: 169–171.
19 Marple B, *et al.* Safety review of benzalkonium chloride used as a preservative in intranasal solutions: an overview of conflicting data and opinions. *Otolaryngol Head Neck Surg* 2004; **130**(1): 131–141.
20 Lewis RJ, ed. *Sax's Dangerous Properties of Industrial Materials*, 11th edn. New York: Wiley, 2004; 104.
21 Pagnoni A, *et al.* Lack of burning and stinging from a novel first-aid formulation applied to experimental wounds. *J Cosmet Sci* 2004; **55**(2): 157–162.

20 General References

Cowen RA, Steiger B. Why a preservative system must be tailored to a specific product. *Cosmet Toilet* 1977; **92**(3): 15–20.
El-Falaha BM, *et al.* Surface changes in *Pseudomonas aeruginosa* exposed to chlorhexidine diacetate and benzalkonium chloride. *Int J Pharm* 1985; **23**: 239–243.
El-Falaha BM, *et al.* Activity of benzalkonium chloride and chlorhexidine diacetate against wild-type and envelope mutants of *Escherichia coli* and *Pseudomonas aeruginosa*. *Int J Pharm* 1985; **25**: 329–337.
Karabit MS, *et al.* Studies on the evaluation of preservative efficacy III: the determination of antimicrobial characteristics of benzalkonium chloride. *Int J Pharm* 1988; **46**: 141–147.
Lien EJ, Perrin JH. Effect of chain length on critical micelle formation and protein binding of quaternary ammonium compounds. *J Med Chem* 1976; **19**: 849–850.
Martin AR. Anti-infective agents. In: Doerge RF, ed. *Wilson and Gisvold's Textbook of Organic, Medicinal and Pharmaceutical Chemistry.* Philadelphia: JB Lippincott, 1982; 141–142.
Pensé AM, *et al.* Microencapsulation of benzalkonium chloride. *Int J Pharm* 1992; **81**: 111–117.
Prince HN, *et al.* Drug resistance studies with topical antiseptics. *J Pharm Sci* 1978; **67**: 1629–1631.
Wallhäusser KH. Benzalkonium chloride. In: Kabara JJ, ed. *Cosmetic and Drug Preservation Principles and Practice.* New York: Marcel Dekker, 1984; 731–734.

21 Author

AH Kibbe.

22 Date of Revision

1 March 2012.

Benzethonium Chloride

1 Nonproprietary Names

BP: Benzethonium Chloride
JP: Benzethonium Chloride
PhEur: Benzethonium Chloride
USP–NF: Benzethonium Chloride

2 Synonyms

Benzethonii chloridum; benzyldimethyl-[2-[2-(p-1,1,3,3-tetramethylbutylphenoxy) ethoxy]ethyl]ammonium chloride; BZT; diisobutylphenoxyethoxyethyl dimethyl benzyl ammonium chloride; *Hyamine 1622*.

3 Chemical Name and CAS Registry Number

N,N-Dimethyl-*N*-[2-[2-[4-(1,1,3,3-tetramethylbutyl)phenoxy]ethoxy]ethyl]benzene-methanaminium chloride [121-54-0]

4 Empirical Formula and Molecular Weight

$C_{27}H_{42}ClNO_2$ 448.10

5 Structural Formula

6 Functional Category

Antimicrobial preservative; stabilizing agent.

7 Applications in Pharmaceutical Formulation or Technology

Benzethonium chloride is a quaternary ammonium compound used in pharmaceutical formulations as an antimicrobial preservative. Typically, it is used for this purpose in injections, ophthalmic and otic preparations at concentrations 0.01–0.02% w/v. Benzethonium chloride may also be used as a wetting and solubilizing agent, and as a topical disinfectant.(1,2)

In cosmetics such as deodorants, benzethonium chloride may be used as an antimicrobial preservative in concentrations up to 0.5% w/v.

The physical properties and applications of benzethonium chloride are similar to those of other cationic surfactants such as cetrimide.

8 Description

Benzethonium chloride occurs as a white crystalline material with a mild odor and very bitter taste.

9 Pharmacopeial Specifications

See Table I.

Table I: Pharmacopeial specifications for benzethonium chloride.

Test	JP XV	PhEur 7.4	USP35–NF30
Identification	+	+	+
Characters	–	+	–
Appearance of solution	–	+	–
Acidity or alkalinity	–	+	–
Melting range	158–164°C	158–164°C	158–163°C
Loss on drying	≤5.0%	≤5.0%	≤5.0%
Residue on ignition	≤0.1%	–	≤0.1%
Sulfated ash	–	≤0.1%	–
Ammonium compounds	+	≤50 ppm	+
Assay (dried basis)	≥97.0%	97.0–103.0%	97.0–103.0%

10 Typical Properties

Acidity/alkalinity pH = 4.8–5.5 for a 1% w/v aqueous solution.

Antimicrobial activity Optimum antimicrobial activity occurs between pH 4–10. Preservative efficacy is enhanced by ethanol and reduced by soaps and other anionic surfactants. For typical minimum inhibitory concentrations (MICs) *see* Table II.(3)

Table II: Minimum inhibitory concentration (MIC) for benzethonium chloride.

Microorganism	MIC (μg/mL)
Aspergillus niger	128
Candida albicans	64
Escherichia coli	32
Penicillium notatum	64
Proteus vulgaris	64
Pseudomonas aeruginosa	250
Pseudomonas cepacia	250
Pseudomonas fluorescens	250
Staphylococcus aureus	0.5
Streptococcus pyogenes	0.5

Solubility Soluble 1 in less than 1 of acetone, chloroform, ethanol (95%), and water; soluble 1 in 6000 of ether. Dissolves in water to produce a foamy, soapy solution.

Spectroscopy

IR spectra *see* Figure 1.
NIR spectra *see* Figure 2.

11 Stability and Storage Conditions

Benzethonium chloride is stable. Aqueous solutions may be sterilized by autoclaving.

The bulk material should be stored in an airtight container protected from light, in a cool, dry place.

12 Incompatibilities

Benzethonium chloride is incompatible with soaps and other anionic surfactants and may be precipitated from solutions greater than 2% w/v concentration by the addition of mineral acids and some salt solutions.

13 Method of Manufacture

p-Diisobutylphenol is condensed in the presence of a basic catalyst with β, β'-dichlorodiethyl ether to yield 2-[2-[4-(1,1,3,3-tetramethylbutyl)phenoxy]ethoxy]ethyl chloride. Alkaline dimethylami-

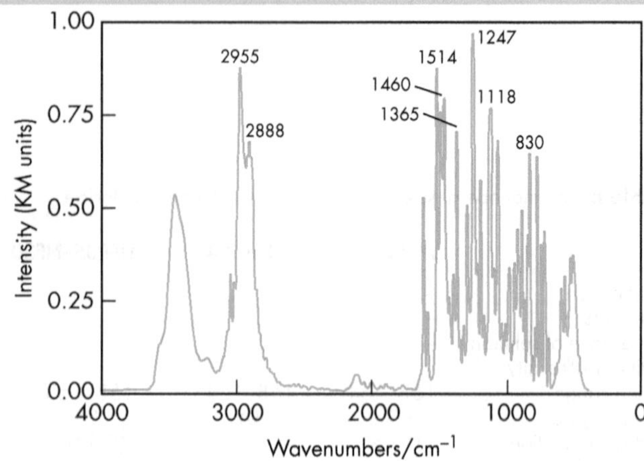

Figure 1: Infrared spectrum of benzethonium chloride measured by diffuse reflectance. Adapted with permission of Informa Healthcare.

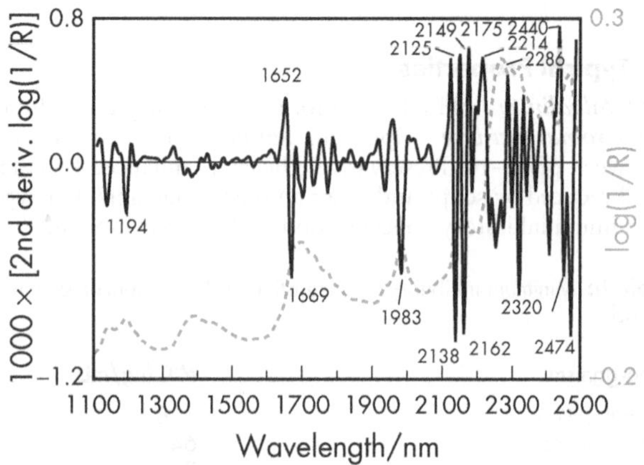

Figure 2: Near-infrared spectrum of benzethonium chloride measured by reflectance.

nation then produces the corresponding tertiary amine which, after purification by distillation, is dissolved in a suitable organic solvent and treated with benzyl chloride to precipitate benzethonium chloride.[4]

14 Safety

Benzethonium chloride is readily absorbed and is generally regarded as a toxic substance when administered orally. Ingestion may cause vomiting, collapse, convulsions, and coma. The probable lethal human oral dose is estimated to be 50–500 mg/kg bodyweight.

The topical use of solutions containing greater than 5% w/v benzethonium chloride can cause irritation although benzethonium chloride is not regarded as a sensitizer. The use of 0.5% w/v benzethonium chloride in cosmetics is associated with few adverse effects. A maximum concentration of 0.02% w/v benzethonium chloride is recommended for use in cosmetics used in the eye area and this is also the maximum concentration generally used in pharmaceutical formulations such as injections and ophthalmic preparations.[5]

See also Benzalkonium Chloride.

LD_{50} (mouse, IP): 15.5 mg/kg[6]
LD_{50} (mouse, IV): 30 mg/kg
LD_{50} (mouse, oral): 338 mg/kg
LD_{50} (rat, IP): 16.5 mg/kg
LD_{50} (rat, IV): 19 mg/kg
LD_{50} (rat, oral): 368 mg/kg
LD_{50} (rat, SC): 119 mg/kg

15 Handling Precautions

Observe normal precautions appropriate to the circumstances and quantity of material handled. Eye protection and gloves are recommended.

16 Regulatory Status

Included in the FDA Inactive Ingredients Database (IM and IV injections; nasal, ophthalmic, and otic preparations). Included in the Canadian Natural Health Products Ingredients Database.

17 Related Substances

Benzalkonium chloride; cetrimide.

18 Comments

Benzethonium chloride has been used therapeutically as a disinfectant and topical anti-infective agent. However, its use in these applications has largely been superseded by other more effective antimicrobials and it is now largely used solely as a preservative in a limited number of pharmaceutical and cosmetic formulations.

The EINECS number for benzethonium chloride is 204-479-9. The PubChem Compound ID (CID) for benethonium chloride includes 8478 and 547429.

19 Specific References

1 Shintre MS, *et al.* Efficacy of an alcohol-based healthcare hand rub containing synergistic combination of farnesol and benzethonium chloride. *Int J Hyg Environ Health* 2006; **209**: 477–487.
2 Bearden DT, *et al.* Comparative *in vitro* activities of topical wound care products against community-associated methicillin-resistant *Staphylococcus aureus*. *J Antimicrob Chemother* 2008; **62**: 769–772.
3 Wallhäusser KH. Benzethonium chloride. In: Kabara JJ, ed. *Cosmetic and Drug Preservation Principles and Practice.* New York: Marcel Dekker, 1984; 734–735.
4 Gennaro AR, ed. *Remington: The Science and Practice of Pharmacy,* 20th edn. Baltimore: Lippincott Williams and Wilkins, 2000; 1508.
5 The Expert Panel of the American College of Toxicology. Final report on the safety assessment of benzethonium chloride and methylbenzethonium chloride. *J Am Coll Toxicol* 1985; 4: 65–106.
6 Lewis RJ, ed. *Sax's Dangerous Properties of Industrial Materials,* 11th edn. New York: Wiley, 2004; 407.

20 General References

Lonza. Product data sheet: *Hyamine 1622* crystals, April 2005.

21 Author

ME Fenton.

22 Date of Revision

1 March 2012.

Benzoic Acid

1 Nonproprietary Names

BP: Benzoic Acid
JP: Benzoic Acid
PhEur: Benzoic Acid
USP–NF: Benzoic Acid

2 Synonyms

Acidum benzoicum; benzenecarboxylic acid; benzeneformic acid; carboxybenzene; dracylic acid; E210; phenylcarboxylic acid; phenylformic acid; *Purox B*.

3 Chemical Name and CAS Registry Number

Benzoic acid [65-85-0]

4 Empirical Formula and Molecular Weight

$C_7H_6O_2$ 122.12

5 Structural Formula

6 Functional Category

Antimicrobial preservative.

7 Applications in Pharmaceutical Formulation or Technology

Benzoic acid is widely used in cosmetics, foods, and pharmaceuticals (*see* Table I), as an antimicrobial preservative.[1–3] Greatest activity is seen at pH values between 2.5–4.5; *see* Section 10.

Table I: Uses of benzoic acid.

Use	Concentration (%)
IM and IV injections	0.17
Oral solutions	0.01–0.1
Oral suspensions	0.1
Oral syrups	0.15
Topical preparations	0.1–0.2
Vaginal preparations	0.1–0.2

8 Description

Benzoic acid occurs as feathery, light, white or colorless crystals or powder. It is essentially tasteless and odorless or with a slight characteristic odor suggestive of benzoin.

9 Pharmacopeial Specifications

See Table II.

Table II: Pharmacopeial specifications for benzoic acid.

Test	JP XV	PhEur 7.4	USP35–NF30
Identification	+	+	+
Characters	–	+	–
Melting point	121–124°C	121–124°C	121–123°C
Water	≤0.5%	–	≤0.7%
Residue on ignition	≤0.05%	≤0.1%	≤0.05%
Readily carbonizable substances	+	+	+
Readily oxidizable substances	+	+	+
Heavy metals	≤20 ppm	≤10 ppm	≤10 μg/g
Halogenated compounds and halides	+	≤300 ppm	–
Appearance of solution	–	+	–
Phthalic acid	+	–	–
Assay (anhydrous basis)	≥99.5%	99.0–100.5%	99.5–100.5%

10 Typical Properties

Acidity/alkalinity pH = 2.8 (saturated aqueous solution at 25°C)

Antimicrobial activity Only the undissociated acid shows antimicrobial properties; the activity therefore depends on the pH of the medium. Optimum activity occurs at pH values below 4.5; at values above pH 5, benzoic acid is almost inactive.[4] It has been reported that antimicrobial activity is enhanced by the addition of protamine, a basic protein.[5]

Bacteria Moderate bacteriostatic activity against most species of Gram-positive bacteria. Typical MIC is 100 μg/mL. Activity is less, in general, against Gram-negative bacteria. MIC for Gram-negative bacteria may be up to 1600 μg/mL.

Molds Moderate activity. Typical MICs are 400–1000 μg/mL at pH 3; 1000–2000 μg/mL at pH 5.

Spores Inactive against spores.

Yeasts Moderate activity. Typical MIC is 1200 μg/mL. The addition of propylene glycol may enhance the fungistatic activity of benzoic acid.

Autoignition temperature 570°C

Boiling point 249.2°C

Density

1.311 g/cm^3 for solid at 24°C;

1.075 g/cm^3 for liquid at 130°C.

Dissociation constant The dissociation of benzoic acid in mixed solvents is dictated by specific solute–solvent interactions as well as by relative solvent basicity. Increasing the organic solvent fraction favors the free acid form.[6]

$pK_a = 4.19$ at 25°C;

$pK_a = 5.54$ in methanol 60%.

Flash point 121–131°C

Melting point 122°C (begins to sublime at 100°C)

Moisture content 0.17–0.42% w/w

Partition coefficients

Benzene : water = 0.0044;[7]

Cyclohexane : water = 0.30;[8]

Octanol : water = 1.87.[9]

Refractive index
n_D^{15} = 1.5397 for solid;
n_D^{132} = 1.504 for liquid.

Solubility Apparent aqueous solubility of benzoic acid may be enhanced by the addition of citric acid or sodium acetate to the solution; *see* Table III.

Table III: Solubility of benzoic acid.

Solvent	Solubility at 25°C unless otherwise stated
Acetone	1 in 2.3
Benzene	1 in 9.4
Carbon disulfide	1 in 30
Carbon tetrachloride	1 in 15.2
Chloroform	1 in 4.5
Cyclohexane	1 in 14.6[8]
Ethanol	1 in 2.7 at 15°C
	1 in 2.2
Ethanol (76%)	1 in 3.72[10]
Ethanol (54%)	1 in 6.27[10]
Ethanol (25%)	1 in 68[10]
Ether	1 in 3
Fixed oils	Freely soluble
Methanol	1 in 1.8
Toluene	1 in 11
Water	1 in 300

Spectroscopy
IR spectra *see* Figure 1.
NIR spectra *see* Figure 2.

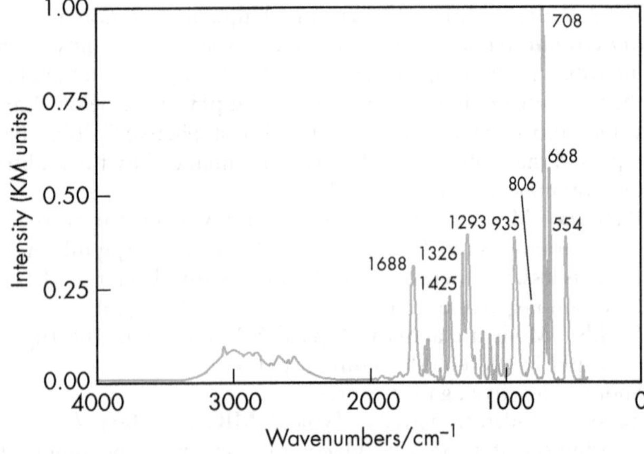

Figure 1: Infrared spectrum of benzoic acid measured by diffuse reflectance. Adapted with permission of Informa Healthcare.

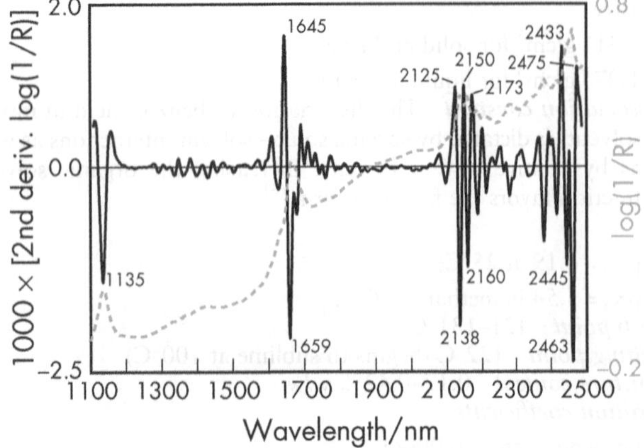

Figure 2: Near-infrared spectrum of benzoic acid measured by reflectance.

11 Stability and Storage Conditions

Aqueous solutions of benzoic acid may be sterilized by autoclaving or by filtration.

A 0.1% w/v aqueous solution of benzoic acid has been reported to be stable for at least 8 weeks when stored in polyvinyl chloride bottles, at room temperature.[11]

When added to a suspension, benzoic acid dissociates, with the benzoate anion adsorbing onto the suspended drug particles. This adsorption alters the charge at the surface of the particles, which may in turn affect the physical stability of the suspension.[12–14] The addition of sodium azide has been shown to increase the stability of benzoic acid in skin permeation experiments.[15]

The bulk material should be stored in a well-closed container in a cool, dry place.

12 Incompatibilities

Undergoes typical reactions of an organic acid, e.g. with alkalis or heavy metals. Preservative activity may be reduced by interaction with kaolin.[16]

13 Method of Manufacture

Although benzoic acid occurs naturally, it is produced commercially by several synthetic methods. One process involves the continuous liquid-phase oxidation of toluene in the presence of a cobalt catalyst at 150–200°C and 0.5–5.0 MPa (5.0–50.0 atm) pressure to give a yield of approximately 90% benzoic acid.

Benzoic acid can also be produced commercially from benzotrichloride or phthalic anhydride. Benzotrichloride, produced by chlorination of toluene, is reacted with 1 mole of benzoic acid to yield 2 moles of benzoyl chloride. The benzoyl chloride is then converted to 2 moles of benzoic acid by hydrolysis. Yield is 75–80%.

In another commercial process, phthalic anhydride is converted to benzoic acid, in about an 85% yield, by hydrolysis in the presence of heat and chromium and disodium phthalates.

Crude benzoic acid is purified by sublimation or recrystallization.

14 Safety

Ingested benzoic acid is conjugated with glycine in the liver to yield hippuric acid, which is then excreted in the urine;[17] care should be taken when administering benzoic acid to patients with chronic liver disease.[18] Benzoic acid is a gastric irritant, and a mild irritant to the skin.[19–22] It is also a mild irritant to the eyes and mucous membranes.[23] Allergic reactions to benzoic acid have been reported, although a controlled study indicated that the incidence of urticaria in patients given benzoic acid is no greater than in those given a lactose placebo.[24] It has been reported that asthmatics may become adversely affected by benzoic acid contained in some antiasthma drugs.[25]

The WHO acceptable daily intake of benzoic acid and other benzoates, calculated as benzoic acid, has been set at up to 5 mg/kg body-weight.[26,27] The minimum lethal human oral dose of benzoic acid is 500 mg/kg body-weight.[28,29]

LD_{50} (cat, oral): 2 g/kg[28]
LD_{50} (dog, oral): 2 g/kg
LD_{50} (mouse, IP): 1.46 g/kg
LD_{50} (mouse, oral): 1.94 g/kg
LD_{50} (rat, oral): 1.7 g/kg

See also Sodium benzoate.

15 Handling Precautions

Observe normal precautions appropriate to the circumstances and quantity of material handled. Benzoic acid may be harmful by inhalation, ingestion, or skin absorption and may be irritant to the eyes, skin, and mucous membranes. Benzoic acid should be handled in a well-ventilated environment; eye protection, gloves, and a dust mask or respirator are recommended. Benzoic acid is flammable.

16 Regulatory Status

GRAS listed. Accepted as a food additive in Europe. Included in the FDA Inactive Ingredients Database (IM and IV injections, irrigation solutions, oral solutions, suspensions, syrups and tablets, rectal, topical, and vaginal preparations). Included in nonparenteral medicines licensed in the UK. Included in the Canadian Health Products Ingredients Database.

17 Related Substances

Potassium benzoate; sodium benzoate.

18 Comments

Benzoic acid is known to dimerize in many nonpolar solvents. This property, coupled with pH-dependent dissociation in aqueous media, comprises a classic textbook example of the effects of dissociation and molecular association on apparent partitioning behavior. The principles involved may be practically applied in determination of the total concentration of benzoate necessary to provide a bacteriostatic level of benzoic acid in the aqueous phase of an oil-in-water emulsion.

Benzoic acid also has a long history of use as an antifungal agent[30] in topical therapeutic preparations such as Whitfield's ointment (benzoic acid 6% and salicylic acid 3%).

A specification for benzoic acid is contained in the *Food Chemicals Codex* (FCC).[31]

The EINECS number for benzoic acid is 200-618-2. The PubChem Compound ID (CID) for benzoic acid is 243.

19 Specific References

1 Buazzi MM, Marth EH. Characteristics of sodium benzoate injury of *Listeria monocytogenes*. *Microbios* 1992; **70**: 199–207.
2 Elder DJ, Kelly DJ. The bacterial degradation of benzoic acid and benzenoid compounds under anaerobic conditions: unifying trends and new perspectives. *FEMS Microbiol Rev* 1994; **13**(4): 441–468.
3 Hwang CA, Beuchat LR. Efficacy of a lactic acid/sodium benzoate wash solution in reducing bacterial contamination in raw chicken. *Int J Food Microbiol* 1995; **27**(1): 91–98.
4 Hurwitz SJ, McCarthy TJ. The effect of pH and concentration on the rates of kill of benzoic acid solutions against *E. coli*. *J Clin Pharm Ther* 1987; **12**: 107–115.
5 Boussard P, *et al*. *In vitro* modification of antimicrobial efficacy by protamine. *Int J Pharm* 1991; **72**: 51–55.
6 Ghosh SK, Hazra DK. Solvent effects on the dissociation of benzoic acid in aqueous mixtures of 2-methoxyethanol and 1,2-dimethoxyethane at 25°C. *J Chem Soc Perkin Trans* 1989; **2**: 1021–1024.
7 Pawlowski W, Wieckowska E. Hydration of benzoic acid in benzene solution II: calculation of hydration constant. *Z Phys Chem* 1990; **168**: 205–215.
8 Dearden JC, Roberts MJ. Cyclohexane–water partition coefficients of some pharmaceuticals. *J Pharm Pharmacol* 1989; **41**(Suppl.): 102P.
9 Yalkowsky SH, *et al*. Solubility and partitioning VI: octanol solubility and octanol–water partition coefficients. *J Pharm Sci* 1983; **72**: 866–870.
10 Pal A, Lahiri SC. Solubility and the thermodynamics of transfer of benzoic acid in mixed solvents. *Indian J Chem* 1989; **28A**: 276–279.
11 The Pharmaceutical Society of Great Britain, Department of Pharmaceutical Sciences. Plastic medicine bottles of rigid PVC. *Pharm J* 1973; **210**: 100.
12 Gallardo V, *et al*. Effect of the preservatives antipyrin, benzoic acid and sodium metabisulfite on properties of the nitrofurantoin/solution interface. *Int J Pharm* 1991; **71**: 223–227.
13 Wei L, *et al*. Preparation and evaluation of SEDDS and SMEDDS containing carvedilol. *Drug Dev Ind Pharm* 2005; **31**: 785–794.
14 Han J, Washington C. Partition of antimicrobial additives in an intravenous emulsion and their effect on emulsion physical stability. *Int J Pharm* 2005; **288**: 263–271.
15 Bode S, *et al*. Stability of the OECD model compound benzoic acid in receptor fluids of Franz diffusion cells. *Pharmazie* 2007; **62**: 470–471.
16 Clarke CD, Armstrong NA. Influence of pH on the adsorption of benzoic acid by kaolin. *Pharm J* 1972; **209**: 44–45.
17 Tremblay GC, Qureshi IA. The biochemistry and toxicology of benzoic acid metabolism and its relationship to the elimination of waste nitrogen. *Pharmacol Ther* 1993; **60**(1): 63–90.
18 Yamada S, *et al*. Clinical significance of benzoate-metabolizing capacity in patients with chronic liver disease: pharmacokinetic analysis. *Res Commun Chem Pathol Pharmacol* 1992; **76**(1): 53–62.
19 Downward CE, *et al*. Topical benzoic acid induces the increased biosynthesis of PGD_2 in human skin *in vivo*. *Clin Pharmacol Ther* 1995; **57**(4): 441–445.
20 Lahti A, *et al*. Immediate irritant reactions to benzoic acid are enhanced in washed skin areas. *Contact Dermatitis* 1995; **33**(3): 177–182.
21 Munoz FJ, *et al*. Perioral contact urticaria from sodium benzoate in a toothpaste. *Contact Dermatitis* 1996; **35**(1): 51.
22 Jacob SE, Stechschulte S. Eyelid dermatitis associated with balsam of Peru constituents: benzoic acid and benzyl alcohol. *Contact Dermatitis* 2008; **58**(2): 111–112.
23 Takeichi Y, Kimura T. Improvement of aqueous solubility and rectal absorption of 6-mercaptopurine by addition of sodium benzoate. *Biol Pharm Bull* 1994; **17**(10): 1391–1394.
24 Lahti A, Hannuksela M. Is benzoic acid really harmful in cases of atopy and urticaria? *Lancet* 1981; **ii**: 1055.
25 Balatsinou L, *et al*. Asthma worsened by benzoate contained in some antiasthmatic drugs. *Int J Immunopathol Pharmacol* 2004; **17**: 225–226.
26 FAO/WHO. Toxicological evaluation of certain food additives with a review of general principles and of specifications. Seventeenth report of the joint FAO/WHO expert committee on food additives. *World Health Organ Tech Rep Ser* 1974; No. 539.
27 FAO/WHO. Evaluation of certain food additives and contaminants. Twenty-seventh report of the joint FAO/WHO expert committee on food additives. *World Health Organ Tech Rep Ser* 1983; No. 696.
28 Lewis RJ, ed. *Sax's Dangerous Properties of Industrial Materials*, 11th edn. New York: Wiley, 2004; 379.
29 Nair B. Final report on the safety assessment of benzyl alcohol, benzoic acid and sodium benzoate. *Int J Toxicol* 2001; **20**(Suppl.3): 23–50.
30 Burlini N, *et al*. Metabolic effects of benzoate and sorbate in the yeast *Saccharomyes cerevisiae* at neutral pH. *Arch Microbiol* 1993; **159**(3): 220–224.
31 *Food Chemicals Codex*, 7th edn. Bethesda, MD: United States Pharmacopeia, 2010; 89.

20 General References

Garrett ER, Woods OR. The optimum use of acid preservatives in oil–water systems: benzoic acid in peanut oil–water. *J Am Pharm Assoc (Sci)* 1953; **42**: 736–739.

21 Author

ME Fenton.

22 Date of Revision

1 March 2012.

Benzyl Alcohol

1 Nonproprietary Names

BP: Benzyl Alcohol
JP: Benzyl Alcohol
PhEur: Benzyl Alcohol
USP–NF: Benzyl Alcohol

2 Synonyms

Alcohol benzylicus; benzenemethanol; α-hydroxytoluene; phenylcarbinol; phenylmethanol; α-toluenol.

3 Chemical Name and CAS Registry Number

Benzenemethanol [100-51-6]

4 Empirical Formula and Molecular Weight

C_7H_8O 108.14

5 Structural Formula

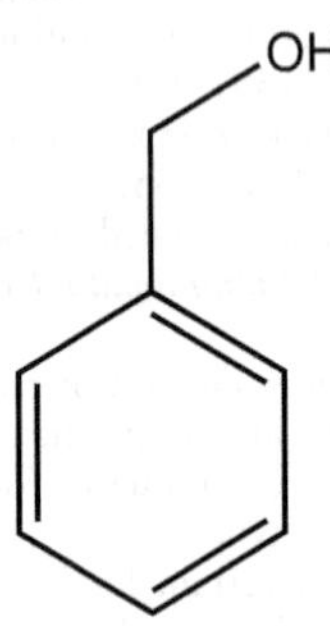

6 Functional Category

Antimicrobial preservative; solvent.

7 Applications in Pharmaceutical Formulation or Technology

Benzyl alcohol is an antimicrobial preservative used in cosmetics, foods, and a wide range of pharmaceutical formulations,[1–4] including oral and parenteral preparations, at concentrations up to 2.0% v/v. The typical concentration used is 1% v/v, and it has been reported to be used in protein, peptide and small molecule products, although its frequency of use has fallen from 48 products in 1996, 30 products in 2001, to 15 products in 2006.[5] In cosmetics, concentrations up to 3.0% v/v may be used as a preservative. Concentrations of 5% v/v or more are employed as a solubilizer, while a 10% v/v solution is used as a disinfectant.

Although widely used as an antimicrobial preservative, benzyl alcohol has been associated with some fatal adverse reactions when administered to neonates. It is now recommended that parenteral products preserved with benzyl alcohol, or other antimicrobial preservatives, should not be used in newborn infants if at all possible; *see* Section 14.

8 Description

A clear, colorless, oily liquid with a faint aromatic odor and a sharp, burning taste.

9 Pharmacopeial Specifications

The pharmacopeial specifications for benzyl alcohol have undergone harmonization of many attributes for JP, PhEur, and USP–NF.

See Table I. *See also* Section 18.

Table I: Pharmacopeial specifications for benzyl alcohol.

Test	JP XV	PhEur 7.4	USP35–NF30
Identification	+	+	+
Characters	+	+	−
Solubility	+	+	−
Acidity	+	+	+
Clarity and color of solution[a]	+	+	+
Specific gravity	1.043–1.049	1.043–1.049	−
Refractive index	1.538–1.541	1.538–1.541	1.538–1.541
Residue on evaporation	≤0.05%	≤0.05%	≤0.05%
Related substances	+	+	+
Benzaldehyde	+	+	0.05–0.15
Peroxide value	≤5	≤5	≤5
Assay	98.0–100.5%	98.0–100.5%	98.0–100.5%

(a) These tests have not been fully harmonized at the time of publication.

10 Typical Properties

Acidity/alkalinity Aqueous solutions are neutral to litmus.

Antimicrobial activity Benzyl alcohol is bacteriostatic and is used as an antimicrobial preservative against Gram-positive bacteria, molds, fungi, and yeasts, although it possesses only modest bactericidal properties. Optimum activity occurs at pH below 5; little activity is shown above pH 8. Antimicrobial activity is reduced in the presence of nonionic surfactants, such as polysorbate 80. However, the reduction in activity is less than is the case with either hydroxybenzoate esters or quaternary ammonium compounds. The activity of benzyl alcohol may also be reduced by incompatibilities with some packaging materials, particularly polyethylene; *see* Section 12.

See Table II for reported minimum inhibitory concentrations (MICs).

Table II: Minimum inhibitory concentrations (MICs) of benzyl alcohol.[4]

Microorganism	MIC (μg/mL)
Aspergillus niger	5000
Candida albicans	2500
Escherichia coli	2000
Pseudomonas aeruginosa	2000
Staphylococcus aureus	25

Bacteria Benzyl alcohol is moderately active against most Gram-positive organisms (typical MICs are 3–5 mg/mL), although some Gram-positive bacteria are very sensitive (MICs 0.025–0.05 mg/mL). In general, benzyl alcohol is less active against Gram-negative organisms.

Fungi Benzyl alcohol is effective against molds and yeasts; typical MICs are 3–5 mg/mL.

Spores Benzyl alcohol is inactive against spores, but activity may be enhanced by heating. Benzyl alcohol 1% v/v, at pH 5–6, has been claimed to be as effective as phenylmercuric nitrate 0.002% w/v against *Bacillus stearothermophilus* at 100°C for 30 min.

Autoignition temperature 436.5°C
Boiling point 204.7°C
Flammability Flammable. Limits in air 1.7–15.0% v/v.
Flash point
 100.6°C (closed cup);
 104.5°C (open cup).
Freezing point −15°C
Partition coefficients
 Liquid paraffin : water = 0.2;
 Octanol : water = 1.10;
 Peanut oil : water = 1.3.
Solubility *see* Table III.

Table III: Solubility of benzyl alcohol.

Solvent	Solubility at 20°C unless otherwise stated
Chloroform	Miscible in all proportions
Ethanol	Miscible in all proportions
Ethanol (50%)	1 in 1.5
Ether	Miscible in all proportions
Fixed and volatile oils	Miscible in all proportions
Water	1 in 25 at 25°C
	1 in 14 at 90°C

Spectroscopy
 IR spectra *see* Figure 1.
Surface tension 38.8 mN/m (38.8 dynes/cm)
Vapor density (relative) 3.72 (air = 1)
Vapor pressure
 13.3 Pa (0.1 mmHg) at 30°C;
 1.769 kPa (13.3 mmHg) at 100°C.
Viscosity (dynamic) 6 mPa s (6 cP) at 20°C

11 Stability and Storage Conditions

Benzyl alcohol oxidizes slowly in air to benzaldehyde and benzoic acid; it does not react with water. Aqueous solutions may be sterilized by filtration or autoclaving; some solutions may generate benzaldehyde during autoclaving.

Benzyl alcohol may be stored in metal or glass containers. Plastic containers should not be used; exceptions to this include polypropylene containers or vessels coated with inert fluorinated polymers such as Teflon; *see* Section 12.

Benzyl alcohol should be stored in an airtight container, protected from light, in a cool, dry place.

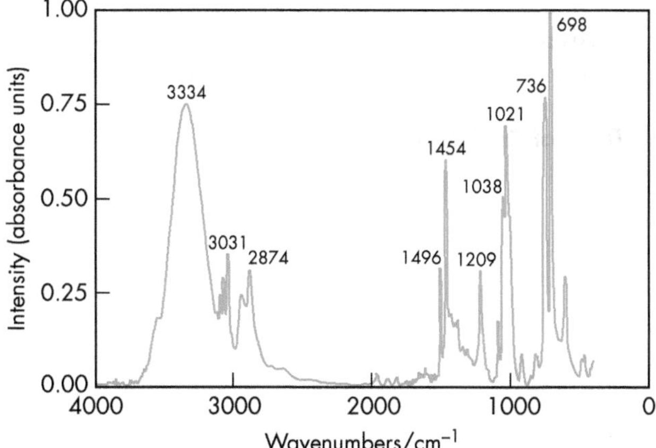

Figure 1: Infrared spectrum of benzyl alcohol measured by transmission. Adapted with permission of Informa Healthcare.

12 Incompatibilities

Benzyl alcohol is incompatible with oxidizing agents and strong acids. It can also accelerate the autoxidation of fats.

Although antimicrobial activity is reduced in the presence of nonionic surfactants, such as polysorbate 80, the reduction is less than is the case with hydroxybenzoate esters or quaternary ammonium compounds.

Benzyl alcohol is incompatible with methylcellulose and is only slowly sorbed by closures composed of natural rubber, neoprene, and butyl rubber closures, the resistance of which can be enhanced by coating with fluorinated polymers.(6) However, a 2% v/v aqueous solution in a polyethylene container, stored at 20°C, may lose up to 15% of its benzyl alcohol content in 13 weeks.(7) Losses to polyvinyl chloride and polypropylene containers under similar conditions are usually negligible. Benzyl alcohol can damage polystyrene syringes by extracting some soluble components.(8)

13 Method of Manufacture

Benzyl alcohol is prepared commercially by the distillation of benzyl chloride with potassium or sodium carbonate. It may also be prepared by the Cannizzaro reaction of benzaldehyde and potassium hydroxide.

14 Safety

Benzyl alcohol is used in a wide variety of pharmaceutical formulations. It is metabolized to benzoic acid, which is further metabolized in the liver by conjugation with glycine to form hippuric acid, which is excreted in the urine.

Ingestion or inhalation of benzyl alcohol may cause headache, vertigo, nausea, vomiting, and diarrhea. Overexposure may result in CNS depression and respiratory failure. However, the concentrations of benzyl alcohol normally employed as a preservative are not associated with such adverse effects.

Reports of adverse reactions to benzyl alcohol(9,10) used as an excipient include toxicity following intravenous administration;(11–13) neurotoxicity in patients administered benzyl alcohol in intrathecal preparations;(14,15) hypersensitivity,(16–18) although relatively rare; and a fatal toxic syndrome in premature infants.(19–22)

The fatal toxic syndrome in low-birth-weight neonates, which includes symptoms of metabolic acidosis and respiratory depression, was attributed to the use of benzyl alcohol as a preservative in solutions used to flush umbilical catheters. As a result of this, the FDA has recommended that benzyl alcohol should not be used in such flushing solutions and has advised against the use of medicines containing preservatives in the newborn.(23–25)

The WHO has set the estimated acceptable daily intake of the benzyl/benzoic moiety at up to 5 mg/kg body-weight daily.(26)

LD_{50} (mouse, IV): 0.32 g/kg(27)
LD_{50} (mouse, oral): 1.36 g/kg
LD_{50} (rat, IP): 0.4 g/kg
LD_{50} (rat, IV): 0.05 g/kg
LD_{50} (rat, oral): 1.23 g/kg

15 Handling Precautions

Observe normal precautions appropriate to the circumstances and quantity of material handled. Benzyl alcohol (liquid and vapor) is irritant to the skin, eyes, and mucous membranes. Eye protection, gloves, and protective clothing are recommended. Benzyl alcohol should be handled in a well-ventilated environment; a self-contained breathing apparatus is recommended in areas of poor ventilation. Benzyl alcohol is flammable.

16 Regulatory Status

Included in the FDA Inactive Ingredients Database (dental injections; oral capsules, solutions and tablets; topical and vaginal preparations). Included in parenteral and nonparenteral medicines licensed in the UK. Included in the Canadian Natural Health Products Ingredients Database.

17 Related Substances

—

18 Comments

Benzyl alcohol has undergone harmonization of many attributes for JP, PhEur, and USP–NF by the Pharmacopeial Discussion Group. For further information see the General Information Chapter <1196> in the USP35–NF30, the General Chapter 5.8 in PhEur 7.4, along with the 'State of Work' document on the PhEur EDQM website, and also the General Information Chapter 8 in the JP XV.

The EINECS number for benzyl alcohol is 202-859-9. The PubChem Compound ID (CID) for benzyl alcohol is 244.

Benzyl alcohol 10% v/v solutions also have some local anesthetic properties, which are exploited in some parenterals, cough products, ophthalmic solutions, ointments, and dermatological aerosol sprays.

19 Specific References

1 Croshaw B. Preservatives for cosmetics and toiletries. *J Soc Cosmet Chem* 1977; **28**: 3–16.

2 Karabit MS, *et al.* Studies on the evaluation of preservative efficacy II: the determination of antimicrobial characteristics of benzyl alcohol. *J Clin Hosp Pharm* 1986; **11**: 281–289.

3 Shah AK, *et al.* Physical, chemical, and bioavailability studies of parenteral diazepam formulations containing propylene glycol and polyethylene glycol 400. *Drug Dev Ind Pharm* 1991; **17**: 1635–1654.

4 Wallhäusser KH. Benzyl alcohol. In: Kabara JJ, ed. *Cosmetic and Drug Preservation Principles and Practice*. New York: Marcel Dekker, 1984; 627–628.

5 Meyer BK, *et al.* Antimicrobial preservative use in parenteral products: past and present. *J Pharm Sci* 2007; **96**(12): 3155–3167.

6 Royce A, Sykes G. Losses of bacteriostats from injections in rubber-closed containers. *J Pharm Pharmacol* 1957; **9**: 814–823.

7 Roberts MS, *et al.* The storage of selected substances in aqueous solution in polyethylene containers: the effect of some physicochemical factors on the disappearance kinetics of the substances. *Int J Pharm* 1979; **2**: 295–306.

8 Doull J *et al.* eds. *Casarett and Doull's Toxicology: The Basic Science of Poisons*. New York: Macmillan, 1980.

9 Reynolds RD. Nebulizer bronchitis induced by bacteriostatic saline [letter]. *J Am Med Assoc* 1990; **264**: 35.

10 Smolinske SC. *Handbook of Food, Drug, and Cosmetic Excipients*. Boca Raton, FL: CRC Press, 1992; 47–54.

11 Evens RP. Toxicity of intravenous benzyl alcohol [letter]. *Drug Intell Clin Pharm* 1975; **9**: 154–155.

12 López-Herce J, *et al.* Benzyl alcohol poisoning following diazepam intravenous infusion [letter]. *Ann Pharmacother* 1995; **29**: 632.

13 Chang YS, *et al. In vitro* benzyl alcohol cytotoxicity: implications for intravitreal use of triamcinolone acetonide. *Exp Eye Res* 2008; **86**(6): 942–950.

14 Hahn AF, *et al.* Paraparesis following intrathecal chemotherapy. *Neurology* 1983; **33**: 1032–1038.

15 Hetherington NJ, Dooley MJ. Potential for patient harm from intrathecal administration of preserved solution. *Med J Aust* 2000; **173**(3): 141–143.

16 Grant JA, *et al.* Unsuspected benzyl alcohol hypersensitivity [letter]. *N Engl J Med* 1982; **306**: 108.

17 Wilson JP, *et al.* Parenteral benzyl alcohol-induced hypersensitivity reaction. *Drug Intell Clin Pharm* 1986; **20**: 689–691.

18 Amado A, Jacob SE. Benzyl alcohol preserved saline used to dilute injectables poses a risk of contact dermatitis in fragrance-sensitive patients [letter]. *Dermatol Surg* 2007; 33(11): 1396–1397.

19 Brown WJ, *et al.* Fatal benzyl alcohol poisoning in a neonatal intensive care unit [letter]. *Lancet* 1982; **i**: 1250.

20 Gershanik J, *et al.* The gasping syndrome and benzyl alcohol poisoning. *N Engl J Med* 1982; **307**: 1384–1388.

21 McCloskey SE, *et al.* Toxicity of benzyl alcohol in adult and neonatal mice. *J Pharm Sci* 1986; **75**: 702–705.

22 Weissman DB, *et al.* Benzyl alcohol administration in neonates. *Anesth Analg* 1990; **70**(6): 673–674.

23 Anonymous. Benzyl alcohol may be toxic to newborns. *FDA Drug Bull* 1982; **12**: 10–11.

24 CDC. Neonatal deaths associated with the use of benzyl alcohol. *MMWR* 1982; 31(22): 290-291.

25 Belson JJ. Benzyl alcohol questionnaire. *Am J Hosp Pharm* 1982; **39**: 1850, 1852.

26 FAO/WHO. Evaluation of certain food additives. Twenty-third report of the joint FAO/WHO expert committee on food additives. *World Health Organ Tech Rep Ser* 1980; No. 648.

27 Lewis RJ, ed. *Sax's Dangerous Properties of Industrial Materials*, 11th edn. New York: Wiley, 2004; 398–399.

20 General References

Akers MJ. Considerations in selecting antimicrobial preservative agents for parenteral product development. *Pharm Technol* 1984; **8**(5): 36–4043, 44, 46.

Bloomfield SF. Control of microbial contamination part 2: current problems in preservation. *Br J Pharm Pract* 1986; **8**: 72, 74–7678, 80.

Carter DV, *et al.* The preparation and the antibacterial and antifungal properties of some substituted benzyl alcohols. *J Pharm Pharmacol* 1958; **10**(Suppl.): 149T–159T.

European Directorate for the Quality of Medicines and Healthcare (EDQM). European Pharmacopoeia – State Of Work Of International Harmonisation. *Pharmeuropa* 2011; 23(4): 713–714. www.edqm.eu/site/-614.html (accessed 1 December 2011).

Harrison SM, *et al.* Benzyl alcohol vapour diffusion through human skin: dependence on thermodynamic activity in the vehicle. *J Pharm Pharmacol* 1982; 34(Suppl.): 36P.

Russell AD, *et al.* The inclusion of antimicrobial agents in pharmaceutical products. *Adv Appl Microbiol* 1967; **9**: 1–38.

Schnuch A, *et al.* Contact allergy to preservatives. Analysis of IVDK data 1996-2009 . *Br J Dermatol* 2011; **164**: 1316–1325.

Sklubalova Z. Antimicrobial substances in ophthalmic drops. *Ceska Slov Form* 2004; 53(3): 107–116.

21 Author

RA Storey.

22 Date of Revision

1 March 2012.

Benzyl Benzoate

B

1 Nonproprietary Names

BP: Benzyl Benzoate
JP: Benzyl Benzoate
PhEur: Benzyl Benzoate
USP–NF: Benzyl Benzoate

2 Synonyms

Benzoic acid benzyl ester; benzylbenzenecarboxylate; benzylis benzoas; benzyl phenylformate; phenylmethyl benzoate.

3 Chemical Name and CAS Registry Number

Benzoic acid phenylmethyl ester [120-51-4]

4 Empirical Formula and Molecular Weight

$C_{14}H_{12}O_2$ 212.24

5 Structural Formula

6 Functional Category

Plasticizing agent; solubilizing agent; solvent.

7 Applications in Pharmaceutical Formulation or Technology

Benzyl benzoate is used as a solubilizing agent and nonaqueous solvent in intramuscular injections at concentrations of 0.01–46.0% v/v,[1] and as a solvent and plasticizer for cellulose and nitrocellulose. It is also used in the preparation of spray-dried powders using nanocapsules.[2]

Benzyl benzoate is also used as a solvent and fixative for flavors and perfumes in cosmetics and food products.

8 Description

Benzyl benzoate is a clear, colorless, oily liquid with a slightly aromatic odor. It produces a sharp, burning sensation on the tongue. At temperatures below 17°C it exists as clear, colorless crystals.

9 Pharmacopeial Specifications

See Table I.

Table I: Pharmacopeial specifications for benzyl benzoate.

Test	JP XV	PhEur 7.4	USP35–NF30
Identification	+	+	+
Characters	+	+	–
Specific gravity	≈ 1.123	1.118–1.122	1.116–1.120
Congealing temperature	≈ 17°C	≥17.0°C	≥18.0°C
Boiling point	≈ 323°C	≈320°C	–
Refractive index	1.568–1.570	1.568–1.570	1.568–1.570
Aldehyde	–	–	≤0.05%
Acidity	+	+	+
Sulfated ash	≤0.05%	≤0.1%	–
Assay	≥99.0%	99.0–100.5%	99.0–100.5%

10 Typical Properties

Autoignition temperature 481°C
Boiling point 323–324°C
Flash point 148°C
Freezing point 17°C
Partition coefficient Octanol : water $\log k_{ow} = 3.97$
Refractive index $n_D^{21} = 1.5681$
Solubility Soluble in acetone and benzene; practically insoluble in glycerin and water; miscible with chloroform, ethanol (95%), ether, and with fatty acids and essential oils.
Specific gravity 1.12
Spectroscopy IR spectra *see* Figure 1.
Vapor density (relative) 7.3 (air = 1)

11 Stability and Storage Conditions

Benzyl benzoate is stable when stored in tight, well-filled, light-resistant containers. Exposure to excessive heat (above 40°C) should be avoided.

12 Incompatibilities

Benzyl benzoate is incompatible with alkalis and oxidizing agents.

13 Method of Manufacture

Benzyl benzoate is a constituent of Peru balsam and occurs naturally in certain plant species. Commercially, benzyl benzoate is produced

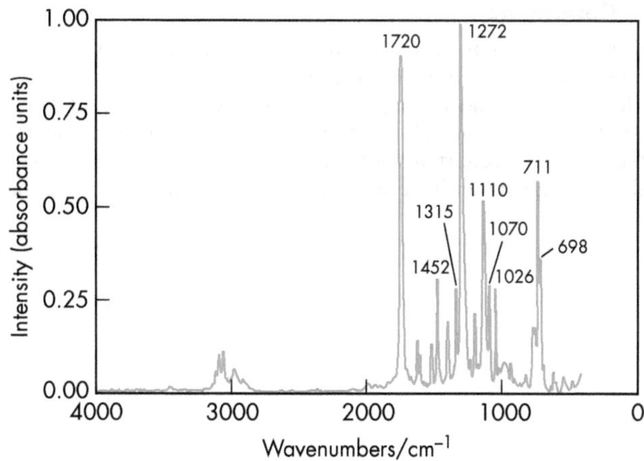

Figure 1: Infrared spectrum of benzyl benzoate measured by transmission. Adapted with permission of Informa Healthcare.

synthetically by the dry esterification of sodium benzoate and benzoyl chloride in the presence of triethylamine or by the reaction of sodium benzylate with benzaldehyde.

B

14 Safety

Benzyl benzoate is metabolized by rapid hydrolysis to benzoic acid and benzyl alcohol. Benzyl alcohol is then further metabolized to hippuric acid, which is excreted in the urine.

Benzyl benzoate is widely used as a 25% v/v topical application in the treatment of scabies and as an excipient in intramuscular injections and oral products. Adverse reactions to benzyl benzoate include skin and eye irritation, and hypersensitivity reactions. Oral ingestion may cause harmful stimulation of the CNS and convulsions. Benzyl benzoate should be avoided by people with perfume allergy.(3)

Benzyl benzoate has been reported as moderately toxic to humans with a probable lethal dose of 0.5–5.0 g/kg.(4) It has been reported that cats are particularly sensitive to benzyl benzoate if applied over too wide an area or too frequently.(5)

LD_{50} (cat, oral): 2.24 g/kg(6–9)
LD_{50} (dog, oral): 22.44 g/kg
LD_{50} (guinea pig, oral): 1.0 g/kg
LD_{50} (mouse, oral): 1.4 g/kg
LD_{50} (rabbit, oral): 1.68 g/kg
LD_{50} (rabbit, skin): 4.0 g/kg
LD_{50} (rat, oral): 0.5 g/kg
LD_{50} (rat, skin): 4.0 g/kg

15 Handling Precautions

Benzyl benzoate may be harmful if ingested, and is irritating to the skin, eyes, and mucous membranes. Observe normal precautions appropriate to the circumstances and quantity of material handled. Eye protection, gloves, and a respirator are recommended. It is recommended that benzyl benzoate is handled in a fume cupboard. Benzyl benzoate is flammable.

16 Regulatory Status

Included in the FDA Inactive Ingredients Database (IM injections and oral capsules). Included, as an active ingredient, in nonparenteral medicines licensed in the UK. Included in the Canadian Natural Health Products Ingredients Database.

17 Related Substances

—

18 Comments

Benzyl benzoate has been shown to have an inhibitory effect on angiotensin II-induced hypertension.(10)

However, the most widespread pharmaceutical use of benzyl benzoate is as a topical therapeutic agent in the treatment of scabies,(11) being reported as the most cost-effective treatment when used consecutively with ivermectin(12) and also more effective than ivermectin.(13) Additionally, benzyl benzoate is also used as a pediculicide.

Benzyl benzoate is also used therapeutically as a parasiticide in veterinary medicine.(14)

The EINECS number for benzyl benzoate is 204-402-9. The PubChem Compound ID (CID) for benzyl benzoate is 2345.

19 Specific References

1 Spiegel AJ, Noseworthy MM. Use of nonaqueous solvents in parenteral products. *J Pharm Sci* 1963; 52: 917–927.
2 Guterres SS, *et al.* Influence of benzyl benzoate as oil core on the physicochemical properties of spray-dried powders from polymeric nanocapsules containing indomethacin. *Drug Deliv* 2000; 7(4): 195–199.
3 Gilman AG *et al.* eds. *Goodman and Gilman's The Pharmacological Basis of Therapeutics*, 8th edn. New York: Pergamon Press, 1990; 1630.
4 Bachewar NP, *et al.* Comparison of safety, efficacy, and cost effectiveness of benzyl benzoate, permethrin, and ivermectin in patients of scabies. *Indian J Pharmacol* 2009; **41**(11): 914.
5 Ly F, *et al.* Ivermectin versus benzyl benzoate applied once or twice to treat human scabies in Dakar, Senegal: a randomized controlled trial. *Bull World Health Organ* 2009; **87**(6): 424–430.
6 Bishop Y, ed. *The Veterinary Formulary*, 6th edn. London: Pharmaceutical Press, 2005; 56.
7 Anderson DN *et al.* (eds). Danish Ministry of the Environment. Survey and health assessment of chemical substances in massage oils. Survey of Chemical Substances in Consumer Products, No. 78, 2006. http://www2.mst.dk/udgiv/publications/2006/87-7052-278-2/pdf/87-7052-279-0.pdf (accessed 27 February 2012).
8 Gosellin RE *et al. Clinical Toxicology of Commercial Products*, 4th edn. Baltimore: Williams and Wilkins, 1976; 137.
9 Humphreys DJ. *Veterinary Toxicology*, 3rd edn. London: Baltier Tindall, 1988; 131.
10 Graham BE, Kuizenga MH. Toxicity studies on benzyl benzoate and related benzyl compounds. *J Pharmacol Exp Ther* 1945; 84: 358–362.
11 Draize JH, *et al.* Toxicological investigations of compounds proposed for use as insect repellents. *J Pharmacol Exp Ther* 1948; 93: 26–39.
12 Sweet DV, ed. *Registry of Toxic Effects of Chemical Substances*. Cincinnati: US Department of Health, 1987; 965.
13 Hayes WJ, Jr, Laws ER, Jr, eds. *Handbook of Pesticide Toxicology*. 3. Classes of Pesticides: New York, NY: Academic Press Inc, 1991; 1505.
14 Ohno O, *et al.* Inhibitory effects of benzyl benzoate and its derivatives on angiotensin II-induced hypertension. *Bioorg Med Chem* 2008; **16**(16): 7843–7852.

20 General References

Gupta VD, Ho HW. Quantitative determination of benzyl benzoate in benzyl benzoate lotion NF. *Am J Hosp Pharm* 1976; 33: 665–666.
Hassan MMA, Mossa JS. Benzyl benzoate. In: Florey K, ed. *Analytical Profiles of Drug Substances*. 10: New York: Academic Press, 1981; 55–74.

21 Author

RA Storey.

22 Date of Revision

1 March 2012.

Boric Acid

1 Nonproprietary Names

BP: Boric Acid
JP: Boric Acid
PhEur: Boric Acid
USP–NF: Boric Acid

2 Synonyms

Acidum boricum; boracic acid; boraic acid; *Borofax*; boron trihydroxide; E284; orthoboric acid; trihydroxyborene.

3 Chemical Name and CAS Registry Number

Orthoboric acid [10043-35-3]
Metaboric acid [13460-50-9]

4 Empirical Formula and Molecular Weight

HBO_2 43.82 (for monohydrate)H_3BO_3 61.83 (for trihydrate)

5 Structural Formula

See Section 4.

6 Functional Category

Antimicrobial preservative; buffering agent.

7 Applications in Pharmaceutical Formulation or Technology

Boric acid is used as an antimicrobial preservative[1] in eye drops, cosmetic products, ointments, and topical creams. It is also used as an antimicrobial preservative in foods.

Boric acid and borate have good buffering capacity and are used to control pH; they have been used for this purpose in external preparations such as eye drops.[2]

In dilute concentrations it is used as a mild antiseptic, with weak bacteriostatic and fungistatic properties, although it has generally been superseded by more effective and less toxic disinfectants.[3] *See* Section 14.

8 Description

Boric acid occurs as a hygroscopic, white crystalline powder, colorless shiny plates, or white crystals.

9 Pharmacopeial Specifications

See Table I.

10 Typical Properties

Acidity/alkalinity pH = 3.5–4.1 (5% w/v aqueous solution)
Density 1.435 g/cm^3
Melting point 170.9°C. When heated slowly to 181.0°C, boric acid loses water to form metaboric acid (HBO_2); tetraboric acid ($H_2B_4O_7$) and boron trioxide (B_2O_3) are formed at higher temperatures.[4]

SEM 1: Excipient: boric acid; manufacturer: Alfa Aesar; lot no.: 23672; magnification: 100×; voltage: 5 kV.

SEM 2: Excipient: boric acid; manufacturer: Aldrich Chemical Company Inc.; lot no.: 01559BU; magnification: 100×; voltage: 5 kV.

Table I: Pharmacopeial specifications for boric acid.

Test	JP XV	PhEur 7.4	USP35–NF30
Identification	+	+	+
Characters	–	+	–
Appearance of solution	+	+	–
Loss on drying	≤0.50%	–	≤0.50%
Sulfate	–	≤450 ppm	–
Heavy metals	≤10 ppm	≤15 ppm	≤0.002%
Organic matter	–	+	–
Arsenic	≤5 ppm	–	–
pH	3.5–4.1	3.8–4.8	–
Solubility in ethanol (96%)	–	+	+
Completeness of solution	–	–	+
Assay	≥99.5%	99.0–100.5%	99.5–100.5%

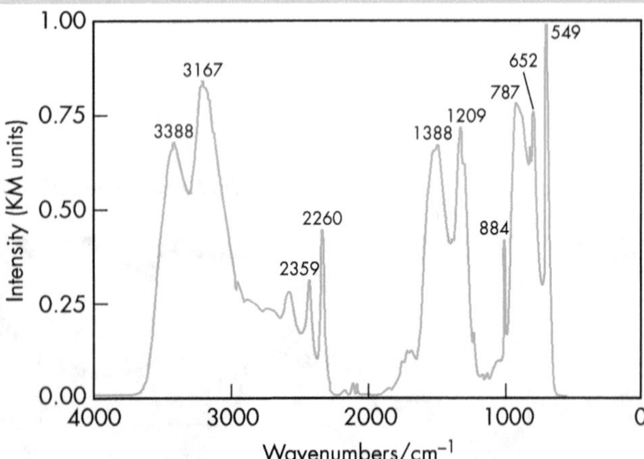

Figure 1: Infrared spectrum of boric acid measured by diffuse reflectance. Adapted with permission of Informa Healthcare.

Solubility Soluble in ethanol, ether, glycerin, water, and other fixed and volatile oils. Solubility in water is increased by addition of hydrochloric, citric, or tartaric acids.

Specific gravity 1.517

Spectroscopy

IR spectra *see* Figure 1.

11 Stability and Storage Conditions

Boric acid is hygroscopic and should therefore be stored in an airtight, sealed container. The container must be labeled 'Not for Internal Use'.

12 Incompatibilities

Boric acid is incompatible with water, strong bases, and alkali metals. It reacts violently with potassium and acid anhydrides. It also forms a complex with glycerin, which is a stronger acid than boric acid.

13 Method of Manufacture

Boric acid occurs naturally as the mineral sassolite. However, the majority of boric acid is produced by reacting inorganic borates with sulfuric acid in an aqueous medium. Sodium borate and partially refined calcium borate (colemanite) are the principal raw materials. When boric acid is made from colemanite, the fine-ground ore is vigorously stirred with mother liquor and sulfuric acid at about 90°C. The by-product calcium sulfate is removed by filtration, and the boric acid is crystallized by cooling the filtrate.

14 Safety

Boric acid is a weak bacteriostatic and antimicrobial agent, and has been used in topical preparations such as eye lotions, mouthwashes and gargles. It has also been used in US- and Japanese-approved intravenous products. Solutions of boric acid were formerly used to wash out body cavities, and as applications to wounds and ulcers, although the use of boric acid for these purposes is now regarded as inadvisable owing to the possibility of absorption.[3] Boric acid is not used internally owing to its toxicity. It is poisonous by ingestion and moderately toxic by skin contact. Experimentally it has proved to be toxic by inhalation and subcutaneous routes, and moderately toxic by intraperitoneal and intravenous routes.

Boric acid is absorbed from the gastrointestinal tract and from damaged skin, wounds, and mucous membranes, although it does not readily permeate intact skin. The main symptoms of boric acid poisoning are abdominal pain, diarrhea, erythematous rash involving both skin and mucous membrane, and vomiting. These symptoms may be followed by desquamation, and stimulation or depression of the central nervous system. Convulsions, hyperpyrexia, and renal tubular damage have been known to occur.[5]

Death has occurred from ingestion of less than 5 g in young children, and of 5–20 g in adults. Fatalities have occurred most frequently in young children after the accidental ingestion of solutions of boric acid, or after the application of boric acid powder to abraded skin.

The permissible exposure limit (PEL) of boric acid is 15 mg/m^3 total dust, and 5 mg/m^3 respirable fraction for nuisance dusts.[6]

LD_{Lo} (man, oral): 429 mg/kg[7]
LD_{Lo} (woman, oral): 200 mg/kg
LD_{Lo} (infant, oral): 934 mg/kg
LD_{Lo} (man, skin): 2.43 g/kg
LD_{Lo} (infant, skin): 1.20 g/kg
LD_{50} (mouse, oral): 3.45 g/kg
LD_{50} (mouse, IV): 1.24 g/kg
LD_{50} (mouse, SC): 1.74 g/kg
LD_{50} (rat, oral): 2.660 g/kg
LD_{50} (rat, IV): 1.33 g/kg
LD_{50} (rat, SC): 1.4 g/kg

15 Handling Precautions

Observe normal precautions appropriate to the circumstances and quantity of material handled. Boric acid is irritating to the skin and is potentially toxic by inhalation. Gloves, eye protection, protective clothing, and a respirator are recommended.

16 Regulatory Status

Accepted for use as a food additive in Europe. Included in the FDA Inactive Ingredients Database (IV injections; ophthalmic preparations; (auricular) otic solutions; topical preparations). Reported in the EPA TSCA Inventory. In the UK, the use of boric acid in cosmetics and toiletries is restricted. Included in the Canadian Natural Health Products Ingredients Database.

17 Related Substances

Sodium borate.

18 Comments

Boric acid has been used experimentally as a model oxo-acid to retard mannitol crystallization in the solid state.[8]

Boric acid has also been used therapeutically in the form of suppositories to treat yeast infections.[9,10]

The EINECS number for boric acid is 233-139-2. The PubChem Compound ID (CID) for boric acid includes 7628 and 24492.

19 Specific References

1 Borokhov O, Schubert D. Antimicrobial properties of boron derivatives. *ACS Symposium Series* 2007; **967**: 412–435.
2 Kodym A, *et al.* Technology of eye drops containing aloe (*Aloe arborescens* M–Liliaceae) and eye drops containing both aloe and neomycin sulphate. *Acta Pol Pharm* 2003; **60**(1): 31–39.
3 Prutting SM, Cerveny JD. Boric acid vaginal suppositories: a brief review. *Infect Dis Obstet Gynecol* 1998; **6**: 191–194.
4 Sobel JD. Current treatment options for vulvovaginal candidiasis. *Women's Health* 2005; **1**(2): 253–261.
5 Sweetman SC, ed. *Martindale: The Complete Drug Reference*, 36th edn. London: Pharmaceutical Press, 2009; 2268.

6 Lund W, ed. *The Pharmaceutical Codex: Principles and Practice of Pharmaceutics*, 12th edn. London: Pharmaceutical Press, 1994; 109.
7 Hubbard SA. Comparative toxicology of borates. *Biol Trace Elem Res* 1998; **66**: 343–357.
8 Dean JA, ed. *Lang's Handbook of Chemistry*, 13th edn. New York: McGraw-Hill, 1985; 4–57.
9 Lewis RJ, ed. *Sax's Dangerous Properties of Industrial Materials*, 11th edn. New York: Wiley, 2004; 536.
10 Yoshinari T, *et al*. Crystallisation of amorphous mannitol is retarded using boric acid. *Int J Pharm* 2003; **258**: 109–120.

20 General References

—

21 Authors

DD Ladipo, AC Bentham.

22 Date of Revision

1 March 2012.

Bronopol

1 Nonproprietary Names

BP: Bronopol

2 Synonyms

2-Bromo-2-nitro-1,3-propanediol; β-bromo-β-nitrotrimethyleneglycol; *Myacide.*

3 Chemical Name and CAS Registry Number

2-Bromo-2-nitropropane-1,3-diol [52-51-7]

4 Empirical Formula and Molecular Weight

$C_3H_6BrNO_4$ 200.00

5 Structural Formula

Br, NO_2, HO, OH

6 Functional Category

Antimicrobial preservative.

7 Applications in Pharmaceutical Formulation or Technology

Bronopol 0.01–0.1% w/v is used as an antimicrobial preservative either alone or in combination with other preservatives in topical pharmaceutical formulations, cosmetics, and toiletries; the usual concentration is 0.02% w/v.

8 Description

Bronopol is a white or almost white crystalline powder; odorless or with a faint characteristic odor.

9 Pharmacopeial Specifications

See Table I.

10 Typical Properties

Antimicrobial activity Bronopol is active against both Gram-positive and Gram-negative bacteria including *Pseudomonas aeruginosa*, with typical minimum inhibitory concentrations (MICs) between 10–50 μg/mL;[1–8] *see also* Table II. At room temperature, a 0.08% w/v aqueous solution may reduce the viability of culture collection strains of *Escherichia coli* and *Pseudomonas aeruginosa* by 100-fold or more in 15 minutes. Antimicrobial activity is not markedly influenced by pH in the range 5.0–8.0, nor by common anionic and nonionic surfactants, lecithin, or proteins.[2,5,6] Bronopol is less active against yeasts and molds, with typical MICs of 50–400 μg/mL or more, and has little or no useful activity against bacterial spores. *See also* Section 12.

Melting point 128–132°C

Partition coefficients

Mineral oil : water = 0.043 at 22–24°C;

Peanut oil : water = 0.11 at 22–24°C.

Solubility *see* Table III.

Table I: Pharmacopeial specifications for bronopol.

Test	BP 2012
Identification	+
Characters	+
Acidity or alkalinity (1% w/v solution)	5.0–7.0
Related substances	+
Sulfated ash	≤0.1%
Water	≤0.5%
Assay (anhydrous basis)	99.0–101.0%

Table II: Minimum inhibitory concentrations (MICs) of bronopol.[2,9]

Microorganism	MIC (μg/mL)
Aspergillus niger	3200
Bacillus subtilis	12.5
Burkholderia (Pseudomonas) cepacia	25
Candida albicans	1600
Escherichia coli	12.5–50
Klebsiella aerogenes	25
Legionella pneumophilia	50
Penicillium roqueforti	400
Penicillium funiculosum	1600
Pityrosporum ovale	125
Proteus mirabilis	25–50
Proteus vulgaris	12.5–50
Pseudomonas aeruginosa	12.5–50
Saccharomyces cerevisiae	3200
Salmonella gallinarum	25
Staphylococcus aureus	12.5–50
Staphylococcus epidermidis	50
Streptococcus faecalis	50
Trichophyton mentagrophytes	200
Trichoderma viride	6400

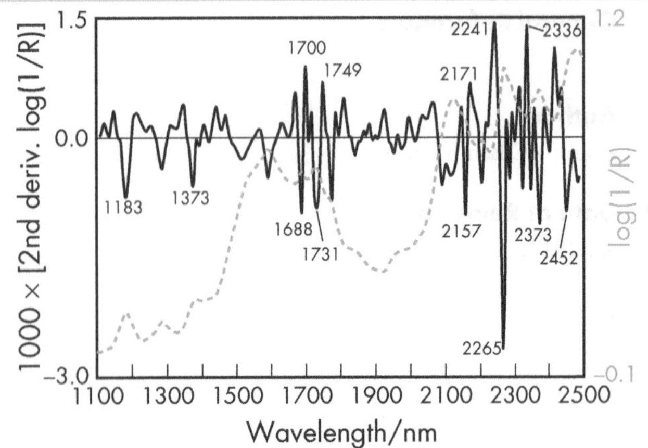

Figure 1: Near-infrared spectrum of bronopol measured by reflectance.

Table III: Solubility of bronopol.

Solvent	Solubility at 20°C
Cottonseed oil	Slightly soluble
Ethanol (95%)	1 in 2
Glycerol	1 in 100
Isopropyl myristate	1 in 200
Mineral oil	Slightly soluble
Propan-2-ol	1 in 4
Propylene glycol	1 in 2
Water	1 in 4

Spectroscopy

NIR spectra *see* Figure 1.

11 Stability and Storage Conditions

Bronopol is stable and its antimicrobial activity is practically unaffected when stored as a solid at room temperature and ambient relative humidity for up to 2 years.[3]

The pH of a 1.0% w/v aqueous solution is 5.0–6.0 and falls slowly during storage; solutions are more stable in acid conditions. Half-lives of bronopol in buffered aqueous solutions at 0.03% w/v are shown in Table IV.[9]

Microbiological assay results indicate longer half-lives than those obtained by HPLC and thus suggest that degradation products may contribute to antimicrobial activity. Formaldehyde and nitrites are among the decomposition products, but formaldehyde arises in such low concentrations that its antimicrobial effect is not likely to be significant. On exposure to light, especially under alkaline conditions, solutions become yellow or brown-colored but the degree of discoloration does not directly correlate with loss of antimicrobial activity.

The bulk material should be stored in a well-closed, non-aluminum container protected from light, in a cool, dry place.

Table IV: Half-lives of bronopol under different storage conditions.

Temperature (°C)	pH 4	pH 6	pH 8
5	>5 years	>5 years	6 months
25	>5 years	>5 years	4 months
40	2 years	4 months	8 days
60	2 weeks	<2 days	<1 day

12 Incompatibilities

Sulfhydryl compounds cause significant reductions in the activity of bronopol, and cysteine hydrochloride may be used as the deactivating agent in preservative efficacy tests; lecithin/polysorbate combinations are unsuitable for this purpose.[5] Bronopol is incompatible with sodium thiosulfate, with sodium metabisulfite, and with amine oxide or protein hydrolysate surfactants. Owing to an incompatibility with aluminum, the use of aluminum in the packaging of products that contain bronopol should be avoided.

13 Method of Manufacture

Bronopol is synthesized by the reaction of nitromethane with paraformaldehyde in an alkaline environment, followed by bromination. After crystallization, bronopol powder may be milled to produce a powder of the required fineness.

14 Safety

Bronopol is used widely in topical pharmaceutical formulations and cosmetics as an antimicrobial preservative.

Although bronopol has been reported to cause both irritant and hypersensitivity adverse reactions following topical use,[10–14] it is generally regarded as a nonirritant and nonsensitizing material at concentrations up to 0.1% w/v. At a concentration of 0.02% w/v, bronopol is frequently used as a preservative in 'hypoallergenic' formulations.

Animal toxicity studies have shown no evidence of phototoxicity or tumor occurrence when bronopol is applied to rodents topically or administered orally; and there is no *in vitro* or *in vivo* evidence of mutagenicity;[1] this is despite the demonstrated potential of bronopol to liberate nitrite on decomposition, which in the presence of certain amines may generate nitrosamines. Formation of nitrosamines in formulations containing amines may be reduced by limiting the concentration of bronopol to 0.01% w/v and including an antioxidant such as 0.2% w/v alpha tocopherol or 0.05% w/v butylated hydroxytoluene;[15] other inhibitor systems may also be appropriate.[16]

LD_{50} (dog, oral): 250 mg/kg [17]
LD_{50} (mouse, IP): 15.5 mg/kg
LD_{50} (mouse, IV): 48 mg/kg
LD_{50} (mouse, oral): 270 mg/kg
LD_{50} (mouse, SC): 116 mg/kg
LD_{50} (mouse, skin): 4.75 g/kg
LD_{50} (rat, IP): 26 mg/kg
LD_{50} (rat, IV): 37.4 mg/kg
LD_{50} (rat, oral): 180 mg/kg
LD_{50} (rat, SC): 170 mg/kg
LD_{50} (rat, skin): 1.6 g/kg

15 Handling Precautions

Observe normal precautions appropriate to the circumstances and quantity of material handled. Bronopol may be harmful upon inhalation and the solid or concentrated solutions can be irritant to the skin and eyes. Eye protection, gloves, and dust respirator are recommended. Bronopol burns to produce toxic fumes.

16 Regulatory Status

Included in topical pharmaceutical formulations licensed in Europe. Included in the Canadian Natural Health Products Ingredients Database.

17 Related Substances

—

18 Comments

Bronopol owes its usefulness as a preservative largely to its activity against *Pseudomonas aeruginosa*, and its affinity for polar solvents, which prevents the loss of preservative into the oil phase of emulsions that is seen with some other preservatives. Other advantages include a low incidence of microbial resistance; low concentration exponent;[18] and good compatibility with most surfactants, other excipients, and preservatives, with which it can therefore be used in combination. Bronopol is used therapeutically as an antiseptic.

The major disadvantages of bronopol are relatively poor activity against yeasts and molds, instability at alkaline pH, and the production of formaldehyde and nitrite on decomposition, although there is no evidence of serious toxicity problems associated with bronopol that are attributable to these compounds.

The EINECS number for bronopol is200-143-0. The PubChem Compound ID (CID) for bronopol is 2450.

19 Specific References

1 Croshaw B, *et al.* Some properties of bronopol, a new antimicrobial agent active against *Pseudomonas aeruginosa. J Pharm Pharmacol* 1964; **16**(Suppl.): 127T–130T.
2 Anonymous. Preservative properties of bronopol. *Cosmet Toilet* 1977; **92**(3): 87–88.
3 Bryce DM, *et al.* The activity and safety of the antimicrobial agent bronopol (2-bromo-2-nitropropane-1,3-diol). *J Soc Cosmet Chem* 1978; **29**: 3–24.
4 Moore KE, Stretton RJ. A method for studying the activity of preservatives and its application to bronopol. *J Appl Bacteriol* 1978; **45**: 137–141.
5 Myburgh JA, McCarthy TJ. Effect of certain formulation factors on the activity of bronopol. *Cosmet Toilet* 1978; **93**(2): 47–48.
6 Moore KE, Stretton RJ. The effect of pH, temperature and certain media constituents on the stability and activity of the preservative bronopol. *J Appl Bacteriol* 1981; **51**: 483–494.
7 Sondossi M. The effect of fifteen biocides on formaldehyde resistant strains of *Pseudomonas aeruginosa. J Ind Microbiol* 1986; **1**: 87–96.
8 Kumanova R, *et al.* Evaluating bronopol. *Manuf Chem* 1989; **60**(9): 36–38.
9 BASF Corp. Technical literature: *Bronopol products*, 2000.
10 Maibach HI. Dermal sensitization potential of 2-bromo-2-nitropropane-1,3-diol (bronopol). *Contact Dermatitis* 1977; **3**: 99.
11 Elder RL. Final report on the safety assessment for 2-bromo-2-nitropropane-1,3-diol. *J Environ Pathol Toxicol* 1980; **4**: 47–61.
12 Storrs FJ, Bell DE. Allergic contact dermatitis to 2-bromo-2-nitropropane-1,3-diol in a hydrophilic ointment. *J Am Acad Dermatol* 1983; **8**: 157–170.
13 Grattan CEH, Harman RRM. Bronopol contact dermatitis in a milk recorder. *Br J Dermatol* 1985; **113**(Suppl. 29): 43.
14 Choudry K, *et al.* Allergic contact dermatitis from 2-bromo-2-nitropropane-1,3-diol in Metrogel. *Contact Dermatitis* 2002; **46**: 60–61.
15 Dunnett PC, Telling GM. Study of the fate of bronopol and the effects of antioxidants on *N*-nitrosamine formation in shampoos and skin creams. *Int J Cosmet Sci* 1984; **6**: 241–247.
16 Challis BC, *et al.* Reduction of nitrosamines in cosmetic products. *Int J Cosmet Sci* 1995; **17**: 119–131.
17 Lewis RJ, ed. *Sax's Dangerous Properties of Industrial Materials*, 11th edn. New York: Wiley, 2004: 566.
18 Denyer SP, Wallhäusser KH. Antimicrobial preservatives and their properties. In: Denyer SP, Baird RM, eds. *Guide to Microbiological Control in Pharmaceuticals.* London: Ellis Horwood, 1990: 251–273.

20 General References

Croshaw B. Preservatives for cosmetics and toiletries. *J Soc Cosmet Chem* 1977; **28**: 3–16.
Rossmore HW, Sondossi M. Applications and mode of action of formaldehyde condensate biocides. *Adv Appl Microbiol* 1988; **33**: 223–273.
Schnuch A, *et al.* Contact allergy to preservatives. Analysis of IVDK data 1996-2009 . *Br J Dermatol* 2011; **164**: 1316–1325.
Shaw S. Patch testing bronopol. *Cosmet Toilet* 1997; **112**(4): 67, 68, 71–73.
Toler JC. Preservative stability and preservative systems. *Int J Cosmet Sci* 1985; **7**: 157–164.
Wallhäusser KH. Bronopol. In: Kabara JJ, ed. *Cosmetic and Drug Preservation Principles and Practice.* New York: Marcel Dekker, 1984: 635–638.

21 Authors

ME Fenton, PJ Sheskey.

22 Date of Revision

1 March 2012.

Butylated Hydroxyanisole

1 Nonproprietary Names

BP: Butylated Hydroxyanisole
PhEur: Butylhydroxyanisole
USP–NF: Butylated Hydroxyanisole

2 Synonyms

BHA; butylhydroxyanisolum; tert-butyl-4-methoxyphenol; 1,1-dimethylethyl-4-methoxyphenol; E320; *Nipanox BHA*; *Nipantiox 1-F*; *Tenox BHA.*

3 Chemical Name and CAS Registry Number

2-tert-Butyl-4-methoxyphenol [25013-16-5]

4 Empirical Formula and Molecular Weight

$C_{11}H_{16}O_2$ 180.25

The PhEur 7.4 describes butylated hydroxyanisole as 2-(1,1-dimethylethyl)-4-methoxyphenol containing not more than 10% of 3-(1,1-dimethylethyl)-4-methoxyphenol.

5 Structural Formula

OH
$C(CH_3)_3$
OCH_3

6 Functional Category

Antioxidant; antimicrobial preservative.

7 Applications in Pharmaceutical Formulation or Technology

Butylated hydroxyanisole is an antioxidant (*see* Table I) with some antimicrobial properties.[1–3] It is used in a wide range of cosmetics, foods, and pharmaceuticals. When used in foods, it is used to delay or prevent oxidative rancidity of fats and oils and to prevent loss of activity of oil-soluble vitamins.

Table I: Antioxidant uses of butylated hydroxyanisole.

Antioxidant use	Concentration (%)
β-Carotene	0.01
Essential oils and flavoring agents	0.02–0.5
IM injections	0.03
IV injections	0.0002–0.0005
Oils and fats	0.02
Topical formulations	0.005–0.02
Vitamin A	10 mg per million units

Butylated hydroxyanisole is frequently used in combination with other antioxidants, particularly butylated hydroxytoluene and alkyl gallates, and with sequestrants or synergists such as citric acid.

FDA regulations direct that the total content of antioxidant in vegetable oils and direct food additives shall not exceed 0.02% w/w (200 ppm) of fat or oil content or essential (volatile) oil content of food.

USDA regulations require that the total content of antioxidant shall not exceed 0.01% w/w (100 ppm) of any one antioxidant or 0.02% w/w combined total of any antioxidant combination in animal fats.

Japanese regulations allow up to 1 g/kg in animal fats.

8 Description

Butylated hydroxyanisole occurs as a white or almost white crystalline powder or a yellowish-white waxy solid with a faint, characteristic aromatic odor.

9 Pharmacopeial Specifications

See Table II.

Table II: Pharmacopeial specifications for butylated hydroxyanisole.

Test	PhEur 7.4	USP35–NF30
Identification	+	+
Characters	+	–
Appearance of solution	+	–
Residue on ignition	–	≤0.01%
Sulfated ash	≤0.1%	–
Related substances	+	–
Heavy metals	≤10 ppm	≤10 ppm
Assay	–	≥98.5%

10 Typical Properties

Antimicrobial activity Activity is similar to that of the *p*-hydroxybenzoate esters (parabens). The greatest activity is against molds and Gram-positive bacteria, with less activity against Gram-negative bacteria.

Boiling point 264°C at 745 mmHg

Density (true) 1.117 g/cm^3

Flash point 130°C

Melting point 47°C (for pure 2-*tert*-butyl-4-methoxyphenol); *see also* Section 18.

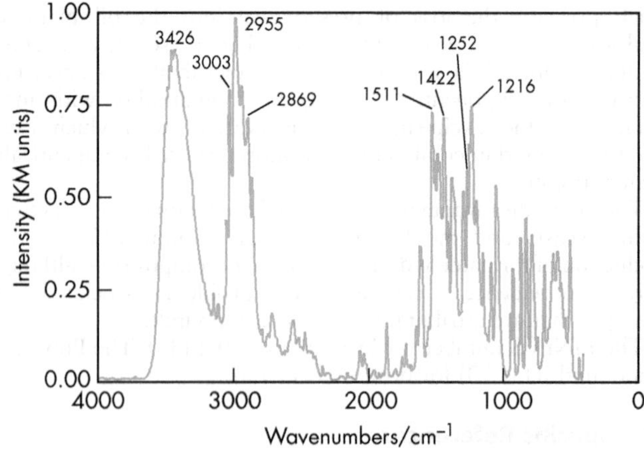

Figure 1: Infrared spectrum of butylated hydroxyanisole measured by diffuse reflectance. Adapted with permission of Informa Healthcare.

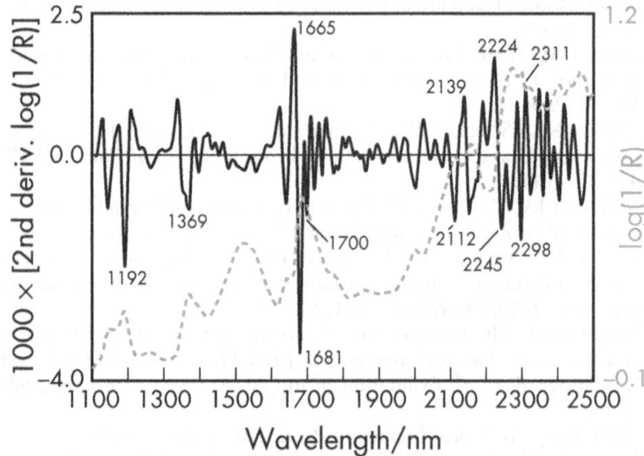

Figure 2: Near-infrared spectrum of butylated hydroxyanisole measured by reflectance.

Solubility Practically insoluble in water; soluble in methanol; freely soluble in ≥50% aqueous ethanol, propylene glycol, chloroform, ether, hexane, cottonseed oil, peanut oil, soybean oil, glyceryl monooleate, and lard, and in solutions of alkali hydroxides.

Spectroscopy

IR spectra *see* Figure 1.

NIR spectra *see* Figure 2.

Viscosity (kinematic) 3.3 mm^2/s (3.3 cSt) at 99°C.

11 Stability and Storage Conditions

Exposure to light causes discoloration and loss of activity. Butylated hydroxyanisole should be stored in a well-closed container, protected from light, in a cool, dry place.

12 Incompatibilities

Butylated hydroxyanisole is phenolic and undergoes reactions characteristic of phenols. It is incompatible with oxidizing agents and ferric salts. Trace quantities of metals and exposure to light cause discoloration and loss of activity.

13 Method of Manufacture

Prepared by the reaction of *p*-methoxyphenol with isobutene.

14 Safety

Butylated hydroxyanisole is absorbed from the gastrointestinal tract and is metabolized and excreted in the urine with less than 1% unchanged within 24 hours of ingestion.[4] Although there have been some isolated reports of adverse skin reactions to butylated hydroxyanisole,[5,6] it is generally regarded as nonirritant and nonsensitizing at the levels employed as an antioxidant.

Concern over the use of butylated hydroxyanisole has occurred following long-term animal feeding studies. Although previous studies in rats and mice fed butylated hydroxyanisole at several hundred times the US-permitted level in the human diet showed no adverse effects, a study in which rats, hamsters, and mice were fed butylated hydroxyanisole at 1–2% of the diet produced benign and malignant tumors of the forestomach, but in no other sites. However, humans do not have any region of the stomach comparable to the rodent forestomach and studies in animals that also do not have a comparable organ (dogs, monkeys, and guinea pigs) showed no adverse effects. Thus, the weight of evidence does not support any relevance to the human diet where butylated hydroxyanisole is ingested at much lower levels.[7] The WHO acceptable daily intake of butylated hydroxyanisole has been set at 500 μg/kg body-weight.[7]

LD_{50} (mouse, oral): 1.1–2.0 g/kg[8]
LD_{50} (rabbit, oral): 2.1 g/kg
LD_{50} (rat, IP): 0.88 g/kg
LD_{50} (rat, oral): 2.0 g/kg

15 Handling Precautions

Observe normal precautions appropriate to the circumstances and quantity of material handled. Butylated hydroxyanisole may be irritant to the eyes and skin, and on inhalation. It should be handled in a well-ventilated environment; gloves and eye protection are recommended. On combustion, toxic fumes may be given off.

16 Regulatory Status

GRAS listed. Accepted as a food additive in Europe. Included in the FDA Inactive Ingredients Database (IM and IV injections, nasal sprays, oral capsules and tablets, and sublingual, rectal, topical, and vaginal preparations). Included in nonparenteral medicines licensed in the UK. Included in the Canadian Natural Health Products Ingredients Database.

17 Related Substances

Butylated hydroxytoluene.

18 Comments

The commercially available material can have a wide melting point range (47–57°C) owing to the presence of varying amounts of 3-*tert*-butyl-4-methoxyphenol.

Tenox brands contain 0.1% w/w citric acid as a stabilizer.

A specification for butylated hydroxyanisole is contained in the *Food Chemicals Codex* (FCC).[9]

The EINECS number for butylated hydroxyanisole is 246-563-8. The PubChem Compound ID (CID) for butylated hydroxyanisole includes 8456 and 11954184.

19 Specific References

1 Lamikanra A, Ogunbayo TA. A study of the antibacterial activity of butyl hydroxy anisole (BHA). *Cosmet Toilet* 1985; **100**(10): 69–74.
2 Felton LA, *et al.* A rapid technique to evaluate the oxidative stability of a model drug. *Drug Dev Ind Pharm* 2007; **33**(6): 683–689.
3 Stein D, Bindra DS. Stabilization of hard gelatine capsule shells filled with polyethylene glycol matrices. *Pharm Dev Technol* 2007; **12**(1): 71–77.
4 El-Rashidy R, Niazi S. A new metabolite of butylated hydroxyanisole in man. *Biopharm Drug Dispos* 1983; **4**: 389–396.
5 Roed-Peterson J, Hjorth N. Contact dermatitis from antioxidants: hidden sensitizers in topical medications and foods. *Br J Dermatol* 1976; **94**: 233–241.
6 Juhlin L. Recurrent urticaria: clinical investigation of 330 patients. *Br J Dermatol* 1981; **104**: 369–381.
7 FAO/WHO. Evaluation of certain food additives and contaminants. Thirty-third report of the joint FAO/WHO expert committee on food additives. *World Health Organ Tech Rep Ser* 1989; No. 776.
8 Lewis RJ, ed. *Sax's Dangerous Properties of Industrial Materials*, 11th edn. New York: Wiley, 2004; 609.
9 *Food Chemicals Codex*, 7th edn. Bethesda, MD: United States Pharmacopeia, 2010; 101.

20 General References

Babich H, Borenfreund E. Cytotoxic effects of food additives and pharmaceuticals on cells in culture as determined with the neutral red assay. *J Pharm Sci* 1990; **79**: 592–594.

Verhagen H. Toxicology of the food additives BHA and BHT. *Pharm Weekbl Sci* 1990; **12**: 164–166.

21 Author

RT Guest.

22 Date of Revision

1 March 2012.

Butylated Hydroxytoluene

1 Nonproprietary Names

BP: Butylated Hydroxytoluene
PhEur: Butylhydroxytoluene
USP–NF: Butylated Hydroxytoluene

2 Synonyms

Agidol; BHT; 2,6-bis(1,1-dimethylethyl)-4-methylphenol; butylhydroxytoluene; butylhydroxytoluenum; *Dalpac*; dibutylated hydroxytoluene; 2,6-di-tert-butyl-p-cresol; 3,5-di-tert-butyl-4-hydroxytoluene; E321; *Embanox BHT*; *Impruvol*; *Ionol CP*; *Nipanox BHT*; *OHS28890*; *Sustane*; *Tenox BHT*; *Topanol*; *Vianol*.

3 Chemical Name and CAS Registry Number

2,6-Di-tert-butyl-4-methylphenol [128-37-0]

4 Empirical Formula and Molecular Weight

$C_{15}H_{24}O$ 220.35

5 Structural Formula

OH
$(H_3C)_3C$ $C(CH_3)_3$
CH_3

6 Functional Category

Antioxidant.

7 Applications in Pharmaceutical Formulation or Technology

Butylated hydroxytoluene is used as an antioxidant (*see* Table I) in cosmetics, foods, and pharmaceuticals.[1–4] It is mainly used to delay or prevent the oxidative rancidity of fats and oils and to prevent loss of activity of oil-soluble vitamins.

Table I: Antioxidant uses of butylated hydroxytoluene.

Antioxidant use	Concentration (%)
β-Carotene	0.01
Edible vegetable oils	0.01
Essential oils and flavoring agents	0.02–0.5
Fats and oils	0.02
Fish oils	0.01–0.1
Inhalations	0.01
IM injections	0.03
IV injections	0.0009–0.002
Topical formulations	0.0075–0.1
Vitamin A	10 mg per million units

8 Description

Butylated hydroxytoluene occurs as a white or pale yellow crystalline solid or powder with a faint characteristic phenolic odor.

9 Pharmacopeial Specifications

See Table II.

Table II: Pharmacopeial specifications for butylated hydroxytoluene.

Test	PhEur 7.4	USP35–NF30
Identification	+	+
Characters	+	–
Appearance of solution	+	–
Congealing temperature	–	≥69.2°C
Freezing point	69–70°C	–
Residue on ignition	–	≤0.002%
Sulfated ash	≤0.1%	–
Heavy metals	–	≤10 ppm
Related substances	+	+
Assay	–	≥99.0%

10 Typical Properties

Boiling point 265°C
Density (bulk) 0.48–0.60 g/cm^3
Density (true) 1.031 g/cm^3
Flash point 127°C (open cup)
Melting point 70°C
Moisture content ≤0.05%
Partition coefficient Octanol : water = 4.17–5.80
Refractive index n_D^{75} = 1.4859
Solubility Practically insoluble in water, glycerin, propylene glycol, solutions of alkali hydroxides, and dilute aqueous mineral acids. Freely soluble in acetone, benzene, ethanol (95%), ether, methanol, toluene, fixed oils, and mineral oil. More soluble than butylated hydroxyanisole in food oils and fats.
Specific gravity
1.006 at 20°C;
0.890 at 80°C;
0.883 at 90°C;
0.800 at 100°C.
Specific heat
1.63 J/g/°C (0.39 cal/g/°C) for solid;
2.05 J/g/°C (0.49 cal/g/°C) for liquid.
Spectroscopy
IR spectra *see* Figure 1.
NIR spectra *see* Figure 2.
Vapor density (relative) 7.6 (air = 1)
Vapor pressure
1.33 Pa (0.01 mmHg) at 20°C;
266.6 Pa (2 mmHg) at 100°C.
Viscosity (kinematic) 3.47 mm^2/s (3.47 cSt) at 80°C.

11 Stability and Storage Conditions

Exposure to light, moisture, and heat causes discoloration and a loss of activity. Butylated hydroxytoluene should be stored in a well-closed container, protected from light, in a cool, dry place.

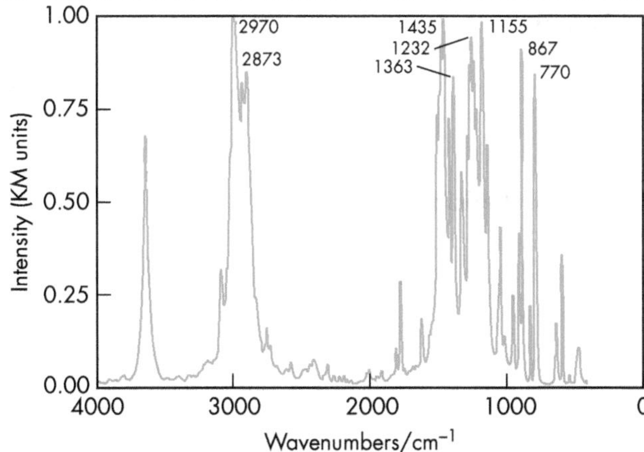

Figure 1: Infrared spectrum of butylated hydroxytoluene measured by diffuse reflectance. Adapted with permission of Informa Healthcare.

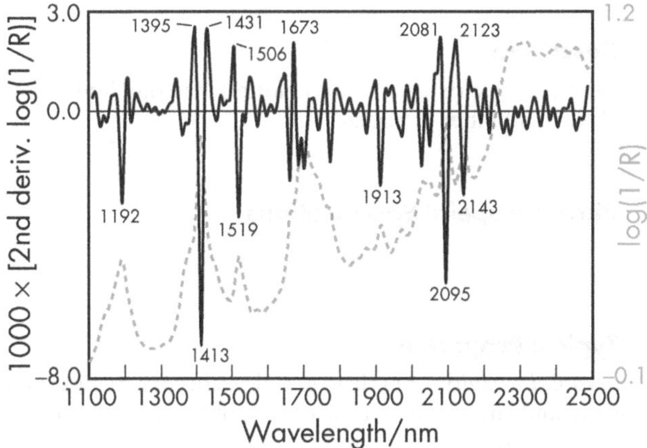

Figure 2: Near-infrared spectrum of butylated hydroxytoluene measured by reflectance.

12 Incompatibilities

Butylated hydroxytoluene is phenolic and undergoes reactions characteristic of phenols. It is incompatible with strong oxidizing agents such as peroxides and permanganates. Contact with oxidizing agents may cause spontaneous combustion. Iron salts cause discoloration with loss of activity. Heating with catalytic amounts of acids causes rapid decomposition with the release of the flammable gas isobutene.

13 Method of Manufacture

Prepared by the reaction of *p*-cresol with isobutene.

14 Safety

Butylated hydroxytoluene is readily absorbed from the gastro-intestinal tract and is metabolized and excreted in the urine mainly as glucuronide conjugates of oxidation products. Although there have been some isolated reports of adverse skin reactions, butylated hydroxytoluene is generally regarded as nonirritant and nonsensitizing at the levels employed as an antioxidant.[5,6]

The WHO has set a temporary estimated acceptable daily intake for butylated hydroxytoluene at up to 125 μg/kg body-weight.[7]

Ingestion of 4 g of butylated hydroxytoluene, although causing severe nausea and vomiting, has been reported to be nonfatal.[8]

LD_{50} (guinea pig, oral): 10.7 g/kg[9]
LD_{50} (mouse, IP): 0.14 g/kg
LD_{50} (mouse, IV): 0.18 g/kg
LD_{50} (mouse, oral): 0.65 g/kg
LD_{50} (rat, oral): 0.89 g/kg

15 Handling Precautions

Observe normal precautions appropriate to the circumstances and quantity of material handled. Butylated hydroxytoluene may be irritant to the eyes and skin, and on inhalation. It should be handled in a well-ventilated environment; gloves and eye protection are recommended. Closed containers may explode owing to pressure build-up when exposed to extreme heat.

16 Regulatory Status

GRAS listed. Accepted as a food additive in Europe. Included in the FDA Inactive Ingredients Database (IM and IV injections, nasal sprays, oral capsules and tablets, rectal, topical, and vaginal preparations). Included in nonparenteral medicines licensed in the UK. Included in the Canadian Natural Health Products Ingredients Database.

17 Related Substances

Butylated hydroxyanisole.

18 Comments

Butylated hydroxytoluene is also used at 0.5–1.0% w/w concentration in natural or synthetic rubber to provide enhanced color stability.

Butylated hydroxytoluene has some antiviral activity[10] and has been used therapeutically to treat herpes simplex labialis.[11]

A specification for butylated hydroxytoluene is contained in the *Food Chemicals Codex* (FCC).[12]

The EINECS number for butylated hydroxytoluene is 204-881-4. The PubChem Compound ID (CID) for butylated hydroxytoluene is 31404.

19 Specific References

1 Skiba M, *et al.* Stability assessment of ketoconazole in aqueous formulations. *Int J Pharm* 2000; **198**: 1–6.
2 Puz MJ, *et al.* Use of the antioxidant BHT in asymmetric membrane tablet coatings to stabilize the core to the acid catalysed peroxide oxidation of a thioether drug. *Pharm Dev Technol* 2005; **10**(1): 115–125.
3 Felton LA, *et al.* A rapid technique to evaluate the oxidative stability of a model drug. *Drug Dev Ind Pharm* 2007; **33**(6): 683–689.
4 Ramadan A, *et al.* Surface treatment: a potential approach for enhancement of solid-state photostability. *Int J Pharm* 2006; **307**(2): 141–149.
5 Roed-Peterson J, Hjorth N. Contact dermatitis from antioxidants: hidden sensitizers in topical medications and foods. *Br J Dermatol* 1976; **94**: 233–241.
6 Juhlin L. Recurrent urticaria: clinical investigation of 330 patients. *Br J Dermatol* 1981; **104**: 369–381.
7 FAO/WHO. Evaluation of certain food additives and contaminants. Thirty-seventh report of the joint FAO/WHO expert committee on food additives. *World Health Organ Tech Rep Ser* 1991; No. 806.
8 Shlian DM, Goldstone J. Toxicity of butylated hydroxytoluene. *N Engl J Med* 1986; **314**: 648–649.
9 Lewis RJ, ed. *Sax's Dangerous Properties of Industrial Materials*, 11th edn. New York: Wiley, 2004; 430.
10 Snipes W, *et al.* Butylated hydroxytoluene inactivates lipid-containing viruses. *Science* 1975; **188**: 64–66.

11 Freeman DJ, *et al*. Treatment of recurrent herpes simplex labialis with topical butylated hydroxytoluene. *Clin Pharmacol Ther* 1985; **38**: 56–59.
12 *Food Chemicals Codex*, 7th edn. Bethesda, MD: United States Pharmacopeia, 2010; 102.

20 General References

Verhagen H. Toxicology of the food additives BHA and BHT. *Pharm Weekbl (Sci)* 1990; **12**: 164–166.

21 Author

RT Guest.

22 Date of Revision

1 March 2012.

Butylene Glycol

1 Nonproprietary Names

None adopted.

2 Synonyms

Butane-1,3-diol; 1,3-butylene glycol; β-butylene glycol; 1,3-dihydroxybutane; methyltrimethylene glycol.

3 Chemical Name and CAS Registry Number

1,3-Butanediol [107-88-0]

4 Empirical Formula and Molecular Weight

$C_4H_{10}O_2$ 90.14

5 Structural Formula

HO OH

CH_3

6 Functional Category

Antimicrobial preservative; humectant; solvent; transdermal delivery component.

7 Applications in Pharmaceutical Formulation or Technology

Butylene glycol is used as a solvent and cosolvent for injectables.[1] It is used in topical ointments, creams, and lotions,[2–4] and it is also used as a vehicle in transdermal patches. Butylene glycol is a good solvent for many pharmaceuticals, especially estrogenic substances.[5]

In an oil-in-water emulsion, butylene glycol exerts its best antimicrobial effects at approximately 8% concentration.[6] Higher concentrations above 16.7% are required to inhibit fungal growth.[7]

8 Description

Butylene glycol occurs as a clear, colorless, viscous liquid with a sweet flavor and bitter aftertaste.

9 Pharmacopeial Specifications

—

10 Typical Properties

Antimicrobial activity Butylene glycol is effective against Gram-positive and Gram-negative bacteria, molds, and yeast, though it is not sporicidal.[6]
Density 1.004–1.006 (at 20°C)
Flash point 115–121°C (open cup)
Hygroscopicity Absorbs 38.5% w/w of water in 144 hours at 81% RH.
Melting point −77°C
Refractive index $n_D^{20} = 1.440$
Solubility Miscible with acetone, ethanol (95%), castor oil, dibutyl phthalate, ether, water; practically insoluble in mineral oil, linseed oil, ethanolamine, aliphatic hydrocarbons; dissolves most essential oils and synthetic flavoring substances.
Specific heat 2.34 J/g (0.56 cal/g) at 20°C
Surface tension 37.8 mN/m (37.8 dyne/cm) at 25°C
Vapor density (relative) 3.1 (air = 1)
Vapor pressure 8 Pa (0.06 mmHg) at 20°C
Viscosity (dynamic) 104 mPa s (104 cP) at 25°C

11 Stability and Storage Conditions

Butylene glycol is hygroscopic and should be stored in a well-closed container in a cool, dry, well-ventilated place. When heated to decomposition, butylene glycol emits acrid smoke and irritating fumes.

12 Incompatibilities

Butylene glycol is incompatible with oxidizing reagents.

13 Method of Manufacture

Butylene glycol is prepared by catalytic hydrogenation of aldol using Raney nickel.

14 Safety

Butylene glycol is used in a wide variety of cosmetic formulations and is generally regarded as a relatively nontoxic material. It is mildly toxic by oral and subcutaneous routes.

In topical preparations, butylene glycol is regarded as minimally irritant. Butylene glycol can cause allergic contact dermatitis, with local sensitivity reported in patch tests.[3,9–12] Some local irritation is produced on eye contact.

LD_{50} (guinea pig, oral): 11.0 g/kg[8]
LD_{50} (mouse, oral): 12.98 g/kg
LD_{50} (rat, oral): 18.61 g/kg
LD_{50} (rat, SC): 20.0 g/kg

15 Handling Precautions

Observe normal precautions appropriate to the circumstances and quantity of the material handled. Butylene glycol should be handled in a well-ventilated environment; eye protection is recommended. Butylene glycol is combustible when exposed to heat or flame.

16 Regulatory Status

GRAS listed. Included in the FDA Inactive Ingredients Database (transdermal patches). Included in licensed medicines in the UK (topical gel patches/medicated plasters).

17 Related Substances

Propylene glycol.

18 Comments

Butylene glycol is used in shaving lather preparations and cosmetics, where it can be used to replace glycerin.[2] Because of its high viscosity at low temperatures, heating may be required for pumping.

A specification for butylene glycol is included in the *Food Chemicals Codex* (FCC); *see* Table I.

Table I: FCC specification for butylene glycol.[13]

Test	FCC 7
Distillation range	200–215°C
Lead	≤2 mg/kg
Specific gravity	1.004–1.006 at 20°C
Assay	≥99.0%

The EINECS number for butylene glycol is 203-529-7. The PubChem Compound ID (CID) for butylene glycol is 7896.

19 Specific References

1 Anschel J. Solvents and solubilisers in injections. *Pharm Ind* 1965; **27**: 781–787.
2 Harb NA. 1:3 Butylene glycol as a substitute in shave lathers. *Drug Cosmet Ind* 1977; **121**: 38–40.
3 Shelanski MV. Evaluation of 1,3-butylene glycol as a safe and useful ingredient in cosmetics. *Cosmet Perfum* 1974; **89**: 96–98.
4 Comelles F, *et al.* Transparent formulations of a liposoluble sunscreen agent in an aqueous medium. *Int J Cosmet Sci* 1990; **12**: 185–196.
5 Hoeffner EM *et al.* eds. *Fiedler Encyclopedia of Excipients for Pharmaceuticals, Cosmetics and Related Areas*, 5th edn. Munich: Editio Cantor Verlag Aulendorf, 2002; 318.
6 Harb NA, Toama MA. Inhibitory effect of 1,3-butylene glycol on microorganisms. *Drug Cosmet Ind* 1976; **118**: 40–41136–137.
7 Osipow LI. 1,3-Butylene glycol in cosmetics. *Drug Cosmet Ind* 1968; **103**: 54–55167–168.
8 Lewis RJ, ed. *Sax's Dangerous Properties of Industrial Materials*, 11th edn. New York: Wiley, 2004; 585.
9 Sugiura M, Hayakawa R. Contact dermatitis due to 1,3-butylene glycol. *Contact Dermatitis* 1997; **37**: 90.
10 Diegenant C, *et al.* Allergic dermatitis due to 1,3-butylene glycol. *Contact Dermatitis* 2000; **43**: 234–235.
11 Oiso N, *et al.* Allergic contact dermatitis due to 1,3-butylene glycol in medicaments. *Contact Dermatitis* 2004; **51**: 40–41.
12 Tamagawa-Mineoka R, *et al.* Allergic contact dermatitis due to 1,3-butylene glycol and glycerol. *Contact Dermatitis* 2007; **56**: 297–298.
13 *Food Chemicals Codex*, 7th edn. Bethesda, MD: United States Pharmacopeia, 2010; 126.

20 General References

Dominguez-Gil A, Cadorniga R. [Stabilization of procaine hydrochloride with butanediols. Part II]. *Farmaco [Prat]* 1971; **26**: 405–420 [in Spanish].

Dominguez-Gil A, *et al.* [Solubilization of phenylethylbarbituric (phenobarbital) acid with polyols]. *Cienc Ind Farm* 1974; **6**: 53–57 [in Spanish].

21 Authors

ME Fenton, RC Rowe.

22 Date of Revision

1 October 2010.

Butylparaben

1 Nonproprietary Names

BP: Butyl Hydroxybenzoate
JP: Butyl Parahydroxybenzoate
PhEur: Butyl Parahydroxybenzoate
USP–NF: Butylparaben

2 Synonyms

Benzoic acid, 4-hydroxy-, butyl ester; *Butoben*; *Butyl Chemosept*; butyl *p*-hydroxybenzoate; butyl parahydroxybenzoate; *Butyl Parasept*; butylis parahydroxybenzoas; *CoSept B*; 4-hydroxybenzoic acid butyl ester; *Nipagin*; *Lexgard B*; *Nipabutyl*; *Tegosept B*; *Trisept B*; *Uniphen P-23*; *Unisept B*.

3 Chemical Name and CAS Registry Number

Butyl-4-hydroxybenzoate [94-26-8]

4 Empirical Formula and Molecular Weight

$C_{11}H_{14}O_3$ 194.23

5 Structural Formula

H_3C — O — O — OH

6 Functional Category

Antimicrobial preservative.

7 Applications in Pharmaceutical Formulation or Technology

Butylparaben is widely used as an antimicrobial preservative in cosmetics and pharmaceutical formulations; *see* Table I. It may be used either alone or in combination with other paraben esters or with other antimicrobial agents. In cosmetics, it is the fourth most frequently used preservative.[1]

As a group, the parabens are effective over a wide pH range and have a broad spectrum of antimicrobial activity, although they are most effective against yeasts and molds; *see* Section 10.

Owing to the poor solubility of the parabens, paraben salts, particularly the sodium salt, are frequently used in formulations. However, this may raise the pH of poorly buffered formulations.

See Methylparaben for further information.

Table I: Uses of butylparaben.

Use	Concentration (%)
Oral suspensions	0.006–0.05
Topical preparations	0.02–0.4

8 Description

Butylparaben occurs as colorless crystals or a white, crystalline, odorless or almost odorless, tasteless powder.

SEM 1: Excipient: butylparaben; magnification: 240×.

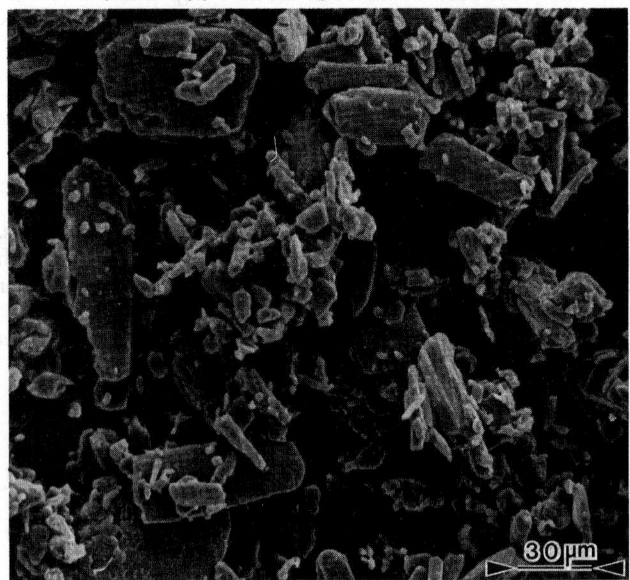

SEM 2: Excipient: butylparaben; magnification: 2400×.

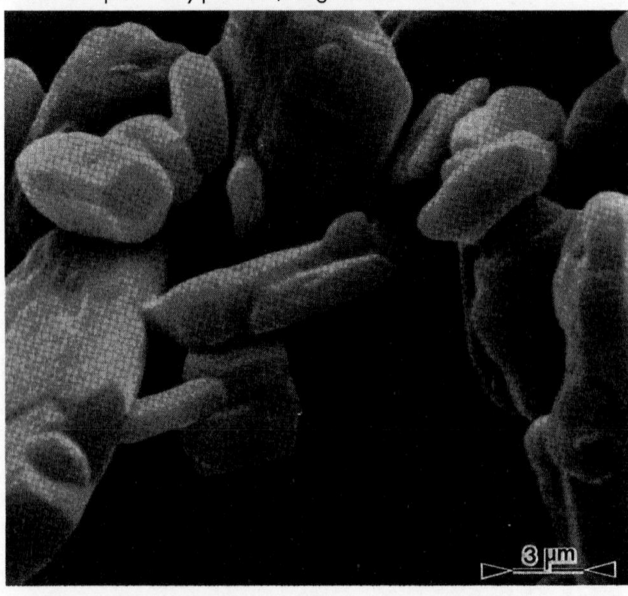

9 Pharmacopeial Specifications

The pharmacopeial specifications for butylparaben have undergone harmonization of many attributes for JP, PhEur, and USP–NF.

See Table II. *See also* Section 18.

Table II: Pharmacopeial specifications for butylparaben.

Test	JP XV	PhEur 7.4	USP35–NF30
Identification	+	+	+
Characters	+	+	—
Appearance of solution	+	+	+
Melting range	68–71°C	68–71°C	68–71°C
Acidity	+	+	+
Residue on ignition	≤0.1%	—	≤0.1%
Sulfated ash	—	≤0.1%	—
Related substances	+	+	+
Heavy metals[a]	≤20 ppm	—	—
Assay (dried basis)	98.0–102.0%	98.0–102.0%	98.0–102.0%

(a) These tests have not been fully harmonized at the time of publication.

10 Typical Properties

Antimicrobial activity Butylparaben exhibits antimicrobial activity between pH 4–8. Preservative efficacy decreases with increasing pH owing to the formation of the phenolate anion. Parabens are more active against yeasts and molds than against bacteria. They are also more active against Gram-positive than against Gram-negative bacteria; *see* Table III.[2]

The activity of the parabens increases with increasing chain length of the alkyl moiety, but solubility decreases. Butylparaben is thus more active than methylparaben. Activity may be improved by using combinations of parabens since synergistic effects occur. Activity has also been reported to be improved by the addition of other excipients; *see* Methylparaben for further information.

Table III: Minimum inhibitory concentrations (MICs) for butylparaben in aqueous solution.[2]

Microorganism	MIC (μg/mL)
Aerobacter aerogenes ATCC 8308	400
Aspergillus niger ATCC 9642	125
Aspergillus niger ATCC 10254	200
Bacillus cereus var. *mycoides* ATCC 6462	63
Bacillus subtilis ATCC 6633	250
Candida albicans ATCC 10231	125
Enterobacter cloacae ATCC 23355	250
Escherichia coli ATCC 8739	5000
Escherichia coli ATCC 9637	5000
Klebsiella pneumoniae ATCC 8308	250
Penicillium chrysogenum ATCC 9480	70
Penicillium digitatum ATCC 10030	32
Proteus vulgaris ATCC 13315	125
Pseudomonas aeruginosa ATCC 9027	>1000
Pseudomonas aeruginosa ATCC 15442	>1000
Pseudomonas stutzeri	500
Rhizopus nigricans ATCC 6227A	63
Saccharomyces cerevisiae ATCC 9763	35
Salmonella typhosa ATCC 6539	500
Serratia marcescens ATCC 8100	500
Staphylococcus aureus ATCC 6538P	125
Staphylococcus epidermidis ATCC 12228	250
Trichophyton mentagrophytes	35

Density (bulk) 0.731 g/cm^3

Density (tapped) 0.819 g/cm^3

Melting point 68–71°C

Partition coefficients Values for different vegetable oils vary considerably and are affected by the purity of the oil; *see* Table IV.[3]

Table IV: Partition coefficients for butylparaben between oils and water.[3]

Solvent	Partition coefficient oil : water
Mineral oil	3.0
Peanut oil	280
Soybean oil	280

Solubility *see* Table V.

Table V: Solubility of butylparaben.

Solvent	Solubility at 20°C unless otherwise stated
Acetone	Freely soluble
Ethanol	1 in 0.5
Ethanol (95%)	Freely soluble
Ether	Freely soluble
Glycerin	1 in 330
Methanol	1 in 0.5
Mineral oil	1 in 1000
Peanut oil	1 in 20
Propylene glycol	1 in 1
Water	1 in >5000
	1 in 670 at 80°C

Spectroscopy

NIR spectra *see* Figure 1.

11 Stability and Storage Conditions

Aqueous butylparaben solutions at pH 3–6 can be sterilized by autoclaving, without decomposition.[4] At pH 3–6, aqueous solutions are stable (less than 10% decomposition) for up to about 4 years at room temperature, while solutions at pH 8 or above are subject to rapid hydrolysis (10% or more after about 60 days at room temperature).[5]

Butylparaben should be stored in a well-closed container, in a cool, dry place.

12 Incompatibilities

The antimicrobial activity of butylparaben is considerably reduced in the presence of nonionic surfactants as a result of micellization.[6] Absorption of butylparaben by plastics has not been reported but appears probable given the behavior of other parabens. Some pigments, e.g. ultramarine blue and yellow iron oxide, absorb butylparaben and thus reduce its preservative properties.[7]

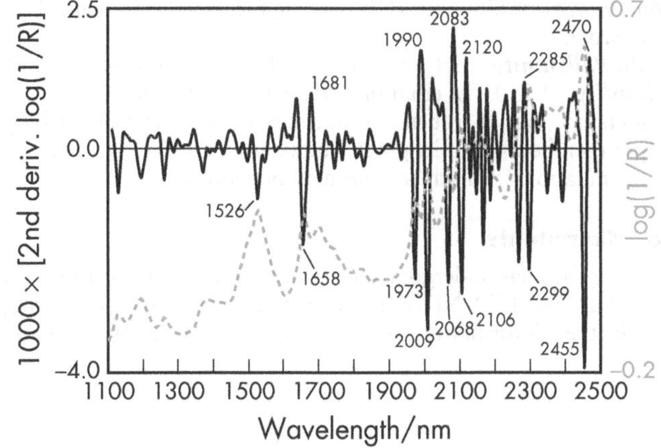

Figure 1: Near-infrared spectrum of butylparaben measured by reflectance.

Butylparaben is discolored in the presence of iron and is subject to hydrolysis by weak alkalis and strong acids.

See also Methylparaben.

B

13 Method of Manufacture

Butylparaben is prepared by esterification of *p*-hydroxybenzoic acid with *n*-butanol.

14 Safety

Butylparaben and other parabens are widely used as antimicrobial preservatives in cosmetics and oral and topical pharmaceutical formulations.

Systemically, no adverse reactions to parabens have been reported, although they have been associated with hypersensitivity reactions generally appearing as contact dematitis. Immediate reactions with urticaria and bronchospasm have occurred rarely.

Some possible estrogenic effects have been reported.[8–10]

A report has been published on the safety assessment of parabens including butylparaben in cosmetic products.[11]

See Methylparaben for further information.

LD_{50} (mouse, IP): 0.23 g/kg[12]

LD_{50} (mouse, oral): 13.2 g/kg

15 Handling Precautions

Observe normal precautions appropriate to the circumstances and quantity of material handled. Butylparaben may be irritant to the skin, eyes, and mucous membranes, and should be handled in a well-ventilated environment. Eye protection, gloves, and a dust mask or respirator are recommended.

16 Regulatory Status

Included in the FDA Inactive Ingredients Database (injections; oral capsules, solutions, suspensions, syrups and tablets; rectal, and topical preparations). Included in nonparenteral medicines licensed in the UK. Included in the Canadian Natural Health Products Ingredients Database.

17 Related Substances

Butylparaben sodium; ethylparaben; methylparaben; propylparaben.

Butylparaben sodium

Empirical formula $C_{11}H_{13}NaO_3$

Molecular weight 216.23

CAS number [36457-20-2]

Synonyms Butyl-4-hydroxybenzoate sodium salt; sodium butyl hydroxybenzoate.

Appearance White, odorless or almost odorless, hygroscopic powder.

Acidity/alkalinity pH = 9.5–10.5 (0.1% w/v aqueous solution)

Solubility 1 in 10 of ethanol (95%); 1 in 1 of water.

Comments Butylparaben sodium may be used instead of butylparaben because of its greater aqueous solubility. In unbuffered formulations, pH adjustment may be required.

18 Comments

Butylparaben has undergone harmonization for many attributes for JP, PhEur, and USP–NF by the Pharmacopeial Discussion Group. For further information see the General Information Chapter <1196> in the USP35–NF30, the General Chapter 5.8 in PhEur 7.4, along with the 'State of Work' document on the PhEur EDQM website, and also the General Information Chapter 8 in the JP XV.

See Methylparaben for further information and references.

The EINECS number for butylparaben is 202-318-7. The PubChem Compound ID (CID) for butylparaben is 7184.

19 Specific References

1 Decker RL, Wenninger JA. Frequency of preservative use in cosmetic formulas as disclosed to FDA—1987. *Cosmet Toilet* 1987; **102**(12): 21–24.

2 Haag TE, Loncrini DF. Esters of *para*-hydroxybenzoic acid. In: Kabara JJ, ed. *Cosmetic and Drug Preservation.* New York: Marcel Dekker, 1984; 63–77.

3 Wan LSC, *et al.* Partition of preservatives in oil/water systems. *Pharm Acta Helv* 1986; **61**: 308–313.

4 Aalto TR, *et al. p*-Hydroxybenzoic acid esters as preservatives I: uses, antibacterial and antifungal studies, properties and determination. *J Am Pharm Assoc (Sci)* 1953; **42**: 449–457.

5 Kamada A, *et al.* Stability of *p*-hydroxybenzoic acid esters in acidic medium. *Chem Pharm Bull* 1973; **21**: 2073–2076.

6 Aoki M, *et al.* [Application of surface active agents to pharmaceutical preparations I: effect of Tween 20 upon the antifungal activities of *p*-hydroxybenzoic acid esters in solubilized preparations.] *J Pharm Soc Jpn* 1956; **76**: 939–943[in Japanese].

7 Sakamoto T, *et al.* Effects of some cosmetic pigments on the bactericidal activities of preservatives. *J Soc Cosmet Chem* 1987; **38**: 83–98.

8 Vo TT, *et al.* Potential estrogenic effect(s) of parabens at the prepubertal stage of a postnatal female rat model. *Reprod Toxicol* 2010; **29**(3): 306–316.

9 Vo TT, Jeung EB. An evaluation of estrogenic activity of parabens using uterine calbindin-d9k gene in an immature rat model. *Toxicol Sci* 2009; **112**(1): 68–77.

10 Shaw J, deCatanzaro D. Estrogenicity of parabens revisited: impact of parabens on early pregnancy and an uterotrophic assay in mice. *Reprod Toxicol* 2009; **28**(1): 26–31.

11 Anonymous. Final amended report on the safety assessment of methylparaben, ethylparaben, propylparaben, isopropylparaben, butylparaben, isobutylparaben, and benzylparaben as used in cosmetic products. *Int J Toxicol* 2008; **27**(Suppl. 4): 1–82.

12 Lewis RJ, ed. *Sax's Dangerous Properties of Industrial Materials*, 11th edn. New York: Wiley, 2004: 637.

See also Methylparaben.

20 General References

Ballesta Claver J, *et al.* Analysis of parabens in cosmetics by low pressure liquid chromatography with monolithic column and chemiluminescent detection. *Talanta* 2009; **79**(2): 499–506.

European Directorate for the Quality of Medicines and Healthcare (EDQM). European Pharmacopoeia – State Of Work Of International Harmonisation. *Pharmeuropa* 2011; 23(4): 713–714. www.edqm.eu/site/-614.html (accessed 2 December 2011).

Golightly LK, *et al.* Pharmaceutical excipients associated with inactive ingredients in drug products (part I). *Med Toxicol* 1988; **3**: 128–165.

Schnuch A, *et al.* Contact allergy to preservatives. Analysis of IVDK data 1996-2009. *Br J Dermatol* 2011; **164**: 1316–1325.

See also Methylparaben.

21 Author

S Gold.

22 Date of Revision

2 December 2011.

Butyl Stearate

1 Nonproprietary Names

None adopted.

2 Synonyms

Butyl octadecanoate; *Crodamol BS*; *n*-butyl octadecanoate; *n*-butyl stearate; *Histar BS*; octadecanoic acid butyl ester.

3 Chemical Name and CAS Registry Number

Butyl octadecanoate [123-95-5]

4 Empirical Formula and Molecular Weight

$C_{22}H_{44}O_2$ 340.58

5 Structural Formula

See below.

6 Functional Category

Emollient; plasticizing agent.

7 Applications in Pharmaceutical Formulation or Technology

Butyl stearate is used as an emollient for topical emulsions, lotions and creams, normally up to a concentration of 3.7%.[1] It can be used as a plasticizer in film-coating formulations of ethylcellulose.[2]

Butyl stearate is also used as a flavoring agent in foods.[3]

8 Description

Butyl stearate at room temperature is a practically odorless, colorless or yellowish liquid, at lower temperatures it is an oily solid.

9 Pharmacopeial Specifications

—

10 Typical Properties

Acid value ≤0.5
Boiling point 343–360°C
Density 0.86–0.88 at 20°C
Flash point Approx 160°C (closed cup); 196°C (open cup)
Iodine value ≤1.0
Melting point 19–24°C
Refractive index $n_D^{20} = 1.442$
Saponification value 165–175
Solubility Soluble in acetone, ethanol, ether, mineral and vegetable oils; practically insoluble in water and propylene glycol.
Specific gravity 0.853–0.860 at 20°C
Surface tension 30 mN/m at 20°C
Vapor pressure 1.467 kPa at 150°C
Viscosity (dynamic) 8 cP at 20°C
Water content ≤0.5%

11 Stability and Storage Conditions

Butyl stearate solidifies at approx 19°C. Store in tightly closed containers. Keep away from heat and sources of ignition.

12 Incompatibilities

Butyl stearate is incompatible with oxidizing agents and alkalis.

13 Method of Manufacture

Butyl stearate is prepared by reacting silver stearate and *n*-butyl iodide or stearic acid and *n*-butanol.[4]

14 Safety

Butyl stearate is only mildly toxic by ingestion. May cause adverse reproductive effects based on animal test data, but no human data available. May cause irritation of the digestive tract if swallowed in large amounts. Prolonged or repeated skin contact may cause skin irritation.

LD_{50} (rat, oral): 32 g/kg[5]

15 Handling Precautions

Observe normal precautions appropriate to the circumstances and quantity of the material handled. When heated, butyl stearate emits acrid smoke and irritating fumes. Wear suitable protective clothing. Avoid contact with skin and eyes.

16 Regulatory Status

GRAS listed. Included in the FDA Inactive Ingredients Database (topical; emulsions, creams, lotions).

17 Related Substances

Ethyl oleate.

18 Comments

A specification for butyl stearate is included in the *Food Chemicals Codex* (FCC);[3] *see* Table I.

Table I: FCC specification for butyl stearate.

Test	FCC 7
Iodine value	≤1
Melting range	17–21°C
Saponification value	165–180
Solubility in 95% ethanol	1 in 6

The EINECS number for butyl stearate is 204-665-5. The PubChem Compound ID (CID) is 31278.

19 Specific References

1 Lower ES. Butyl stearate. *Manufacturing Chemist* 1982; 53: 57, 59–60.
2 Bodea A, Leucuta SE. Optimization of propranolol hydrochloride sustained release pellets using Box-Behnken design and desirability function. *Drug Dev Ind Pharm* 1998; 24: 145–155.

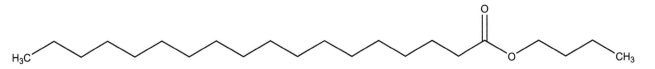

O
H_3C O CH_3

3 *Food Chemicals Codex*, 7th edn. Bethesda, MD: United States Pharmacopoeia, 2010; 125.
4 O'Neil MJ, ed. *The Merck Index: An Encyclopedia of Chemicals, Drugs, and Biologicals*, 14th edn. Whitehouse Station, NJ: Merck, 2006; 259.
5 Lewis RJ, ed. *Sax's Dangerous Properties of Industrial Materials*, 11th edn. New York: John Wiley, 2004: 643.

20 General References

Delta Education. Material safety data sheet: Butyl stearate, August 2005.
ScienceLab.com, Inc. Material safety data sheet: Butyl stearate, June 2008.
Oh Sung Chemical IND Co Ltd. Technical data sheet: *Histar BS*, 2006. http://www.oschem.co.kr/eng/product/index01.html (accessed 29 February 2012).

21 Authors

ME Fenton, RC Rowe.

22 Date of Revision

1 October 2010.

 Calcium Acetate

1 Nonproprietary Names

BP: Calcium Acetate
PhEur: Calcium Acetate, Anhydrous
USP–NF: Calcium Acetate

2 Synonyms

Acetate of lime; acetic acid, calcium salt; brown acetate; calcii acetas; calcium diacetate; E263; gray acetate; lime acetate; lime pyrolignite; vinegar salts.

3 Chemical Name and CAS Registry Number

Calcium acetate [62-54-4]
Calcium acetate monohydrate [5743-26-0]

4 Empirical Formula and Molecular Weight

$C_4H_6CaO_4$ 158.18
$C_4H_6CaO_4{\cdot}H_2O$ 176.17 (for monohydrate)

5 Structural Formula

6 Functional Category

Antimicrobial preservative; complexing agent.

7 Applications in Pharmaceutical Formulation or Technology

Calcium acetate is used as a preservative in oral and topical formulations.

Calcium acetate is also used in the food industry as a stabilizer, buffer, and sequestrant.

8 Description

Calcium acetate occurs as a white or almost white, odorless or almost odorless, hygroscopic powder. Calcium acetate monohydrate occurs as needles, granules, or powder.

9 Pharmacopeial Specifications

See Table I.

Table I: Pharmacopeial specifications for calcium acetate.

Test	PhEur 7.4	USP35–NF30
Identification	+	+
Readily oxidizable substances	+	+
pH	7.2–8.2	6.3–9.6
Nitrates	+	+
Chlorides	≤330 ppm	≤0.05%
Sulfates	≤600 ppm	≤0.06%
Heavy metals	≤10 ppm	≤25 ppm
Magnesium	≤500 ppm	≤0.05%
Arsenic	≤3 ppm	≤3 ppm
Aluminum	≤1 ppm	≤2 ppm
Barium	≤50 ppm	+
Potassium	≤500 ppm	≤0.05%
Sodium	≤500 ppm	≤0.5%
Strontium	≤500 ppm	≤0.05%
Water	≤7.0%	≤7.0%
Fluoride	≤50 ppm	≤50 ppm
Lead	–	≤10 ppm
Assay (anhydrous substance)	98.0–102.0%	99.0–100.5%

10 Typical Properties

Acidity/alkalinity pH = 6.3–9.6 (5% solution); pH = 7.6 (0.2 M aqueous solution) for monohydrate
Density: 1.50 g/cm^3
Solubility Soluble in water; slightly soluble in methanol; practically insoluble in acetone, ethanol (dehydrated alcohol), and benzene. The monohydrate is soluble in water, slightly soluble in alcohol.

11 Stability and Storage Conditions

Calcium acetate is stable although very hygroscopic, and so the monohydrate is the common form. It decomposes on heating (above 160°C) to form calcium carbonate and acetone.

Store in well-closed airtight containers.

12 Incompatibilities

Calcium acetate is incompatible with strong oxidizing agents and moisture.[1]

13 Method of Manufacture

Calcium acetate is manufactured by the reaction of calcium carbonate or calcium hydroxide with acetic acid or pyroligneous acid.[2]

14 Safety

Calcium acetate is used in oral and topical formulations. The pure form of calcium acetate is toxic by IP and IV routes.

LD_{50} (mouse, IP): 0.075 g/kg[3]
LD_{50} (mouse, IV): 0.052 g/kg[3]
LD_{50} (rat, oral): 4.28 g/kg[1]

15 Handling Precautions

Observe normal precautions appropriate to the circumstances and quantity of the material handled. Although regarded as safe during normal industrial handling, calcium acetate may cause eye and respiratory tract irritation.[1] It is combustible and when heated to decomposition, it emits acrid smoke and fumes. Avoid contact with eyes, skin, and clothing. Avoid breathing dust. Gloves, eye protection, respirator, and other protective clothing should be worn.

16 Regulatory Status

GRAS listed. Accepted for use as a food additive in Europe. Included in the FDA Inactive Ingredients Database (oral suspensions and tablets; topical emulsions, lotions, and creams). Included in nonparenteral medicines (oral tablets) licensed in the UK.

17 Related Substances

Sodium acetate.

18 Comments

Calcium acetate is used in the chemical industry for the manufacture of acetic acid, acetates, and acetone, and for the precipitation of oxalates.

Therapeutically, parenteral calcium acetate acts as a source of calcium ions for hypocalcemia or electrolyte balance.[4] Oral calcium acetate is used as a complexing agent for hyperphosphatemia in dialysis patients.[5,6]

A specification for calcium acetate is contained in the *Food Chemicals Codex* (FCC).[7]

The EINECS number for calcium acetate is 200-540-9. The PubChem Compound ID (CID) for calcium acetate is 6116.

19 Specific References

1 Mallinckrodt Baker Inc. *Material Safety Data Sheet C0264: Calcium Acetate*, 2007.
2 Speight JG. *Chemical and Process Design Handbook*. New York: McGraw-Hill, 2002; 121.
3 Lewis RJ, ed. *Sax's Dangerous Properties of Industrial Materials*, 11th edn. New York: Wiley, 2004; 667.
4 Todd RG, Wade A, eds. *The Pharmaceutical Codex*, 11th edn. London: Pharmaceutical Press, 1979; 124.
5 Almirall J, *et al*. Calcium acetate versus carbonate for the control of serum phosphorus in hemodialysis patients. *Am J Nephrol* 1994; **14**: 192–196.
6 Qunibi WY, *et al*. Treatment of hyperphosphatemia in hemodialysis patients: the Calcium Acetate Renagel Evaluation (CARE Study). *Kidney Int* 2004; **65**: 1914–1926.
7 *Food Chemicals Codex*, 7th edn. Bethesda, MD: United States Pharmacopeia, 2010: 130.

20 General References

—

21 Author

TI Armstrong.

22 Date of Revision

1 March 2012.

Calcium Alginate

1 Nonproprietary Names

None adopted.

2 Synonyms

Alginato calcico; algin; alginic acid, calcium salt; *CA33*; calc algin; calcium polymannuronate; *Calginate*; E404; *Kaltostat*.

3 Chemical Name and CAS Registry Number

Calcium alginate [9005-35-0]

4 Empirical Formula and Molecular Weight

$[(C_6H_7O_6)_2Ca]_n$ 195.16 (calculated)
219.00 (actual, average)

Each calcium ion binds with two alginate molecules. The molecular weight of 195.16 relates to one alginate molecule, and the equivalent of half a calcium ion, therefore $n = \frac{1}{2}$.

Calcium alginate is a polyuronide made up of a sequence of two hexuronic acid residues, namely D-mannuronic acid and L-guluronic acid. The two sugars form blocks of up to 20 units along the chain, with the proportion of the blocks dependent on the species of seaweed and also the part of the seaweed used. The number and length of the blocks are important in determining the physical properties of the alginate produced; the number and sequence of the mannuronate and guluronate residues varies in the naturally occurring alginate.

It has a typical macromolecular weight between 10 000 and 600 000.

5 Structural Formula

See Section 4.

6 Functional Category

Emulsifying agent; modified-release agent; tablet and capsule disintegrant; viscosity-increasing agent.

7 Applications in Pharmaceutical Formulation or Technology

In pharmaceutical formulations, calcium alginate and calcium-sodium alginate have been used as tablet disintegrants.[1] The use of a high concentration (10%) of calcium-sodium alginate has been reported to cause slight speckling of tablets.[1]

A range of different types of delivery systems intended for oral administration have been investigated. These exploit the gelling properties of calcium alginate.[2] Calcium alginate beads have been used to prepare floating dosage systems[3–7] containing amoxicillin,[8] furosemide,[9] meloxicam,[10] and barium sulfate,[11] and as a means of providing a sustained or controlled-release action for sulindac,[12] diclofenac,[13,14] tiaramide,[15] insulin,[16] and ampicillin.[17] The effect of citric acid in prolonging the gastric retention of calcium alginate floating dosage forms has been reported.[18,19] Impregnating meloxicam in calcium alginate beads may reduce the risk of ulceration and mucosal inflammation following oral administration.[20] The use of calcium alginate beads, reinforced with chitosan, has been shown to slow the release of verapamil,[21] and may be useful for the controlled release of protein drugs to the gastrointestinal tract.[22] The release rate of bovine serum albumin of two molecular weights loaded into calcium alginate beads, which were embedded in silk fibroin to form a three-dimensional dual release scaffold, was found to be dependent on the molecular weight of the protein.[23] The bioadhesive properties,[24] swelling and drug release[25] of calcium alginate beads have also been investigated. The mechanical properties of calcium alginate hydrogels have been investigated,[26] and a novel injectable and *in situ*-forming gel composite composed of calcium alginate gel and nanohydroxyapatite/collagen has been described.[27]

A series of studies investigating the production,[28] formulation,[29] and drug release[30] from calcium alginate matrices for oral administration have been published. The release of diltiazem hydrochloride from a polyvinyl alcohol matrix was shown to be controlled by coating with a calcium alginate membrane; the drug release profile could be modified by increasing the coating thickness of the calcium alginate layer.[31] The microencapsulation of live attenuated Bacillus Calmette–Guérin (BCG) cells within a calcium alginate matrix has also been reported.[32]

It has been shown that a modified drug release can be obtained from calcium alginate microcapsules,[33] pellets,[34,35] and microspheres.[36] When biodegradable bone implants composed of calcium alginate spheres and containing gentamicin were introduced into the femur of rats, effective drug levels in bone and soft tissue were obtained for 30 days and 7 days, respectively.[37] The incorporation of radioactive particles into calcium alginate gels may be useful for the localized delivery of radiation therapy to a wide range of organs and tissues.[38]

8 Description

Calcium alginate is an odorless or almost odorless, tasteless, white to pale yellowish-brown powder or fibers.

9 Pharmacopeial Specifications

See Section 18.

10 Typical Properties

Moisture content Loses not more than 22% of its weight on drying.

Solubility Practically insoluble in chloroform, ethanol, ether, water, and other organic solvents. Soluble in dilute solutions of sodium citrate and of sodium bicarbonate and in sodium chloride solution. Soluble in alkaline solutions or in solutions of substances that combine with calcium.

11 Stability and Storage Conditions

Calcium alginate can be sterilized by autoclaving at 115°C for 30 minutes or by dry heat at 150°C for 1 hour. Calcium alginate should be stored in airtight containers.

12 Incompatibilities

Calcium alginate is incompatible with alkalis and alkali salts. Propranolol hydrochloride has been shown to bind to alginate molecules, suggesting that propranolol and calcium ions share common binding sites in the alginate chains; the formation of the calcium alginate gel structure was impeded in the presence of propranolol molecules.[39] It has been reported that commonly used topical antimicrobials could potentially be incompatible with calcium alginate dressings.[40]

13 Method of Manufacture

Calcium alginate can be obtained from seaweed, mainly species of *Laminaria*.

Solutions of sodium alginate interact with an ionized calcium salt, resulting in the instantaneous precipitation of insoluble calcium alginate, which can then be further processed. Introducing varying proportions of sodium ions during manufacture can produce products having different absorption rates.

14 Safety

Calcium alginate is widely used in oral and topical formulations, and in foods.

In 1974, the WHO set an estimated acceptable daily intake of calcium alginate of up to 25 mg, as alginic acid, per kilogram bodyweight.[41]

When heated to decomposition, it emits acrid smoke and irritating fumes.

LD_{50} (rat, IP): 1.41 g/kg[42]
LD_{50} (rat, IV): 0.06 g/kg

15 Handling Precautions

Observe normal precautions appropriate to the circumstances and quantity of the material handled.

16 Regulatory Status

GRAS listed. Accepted for use as a food additive in Europe. Included in the FDA Inactive Ingredients Database (oral tablets). Included in nonparenteral medicines licensed in the UK.

17 Related Substances

Alginic acid; potassium alginate; propylene glycol alginate; sodium alginate.

18 Comments

Therapeutically, the gelling properties of calcium alginate are utilized in wound dressings in the treatment of leg ulcers, pressure sores, and other exuding wounds. These dressings are highly absorbent and are suitable for moderately or heavily exuding wounds. Calcium alginate dressings also have hemostatic properties, with calcium ions being exchanged for sodium ions in the blood; this stimulates both platelet activation and whole blood coagulation. A mixed calcium–sodium salt of alginic acid is used as fibers in dressings or wound packing material.

Sterile powder consisting of a mixture of calcium and sodium alginates has been used in place of talc in glove powders.

In foods, calcium alginate is used as an emulsifier, thickener, and stabilizer.

Although not included in any pharmacopeias, a specification for calcium alginate is contained in the *Food Chemicals Codex* (FCC),[43] and has been included in the *British Pharmaceutical Codex* (BPC);[44] *see* Table I.

Table I: FCC[43] and BPC[44] specifications for calcium alginate.

Test	FCC 7	BPC 1973
Arsenic	≤3 mg/kg	≤3 ppm
Iron	–	≤530 ppm
Lead	≤5 mg/kg	≤10 ppm
Loss on drying	≤15%	22.00%
Sulfated ash	–	31.0–34.0%
Assay	89.6–104.5%	–

19 Specific References

1 Khan KA, Rhodes CT. A comparative evaluation of some alginates as tablet disintegrants. *Pharm Acta Helv* 1972; **47**: 41–50.
2 Tonnesen HH, Karlsen J. Alginate in drug delivery systems. *Drug Dev Ind Pharm* 2002; **28**(6): 621–630.
3 Iannuccelli V, *et al.* Air compartment multiple-unit system for prolonged gastric residence. Part 1. Formulation study. *Int J Pharm* 1998; **174**: 47–54.

4 Whitehead L, *et al.* Floating dosage forms: *in vivo* study demonstrating prolonged gastric retention. *J Control Release* 1998; **55**: 3–12.
5 Iannuccelli V, *et al.* Oral absorption of riboflavin dosed by a floating multiple-unit system in different feeding conditions. *STP Pharma Sci* 2004; **14**(2): 127–133.
6 Tang YD. Sustained release of hydrophobic and hydrophilic drugs from a floating dosage form. *Int J Pharm* 2007; **336**(1): 159–165.
7 Stops F, *et al.* Floating dosage forms to prolong gastro-retention—the characterisation of calcium alginate beads. *Int J Pharm* 2008; **350**(1–2): 301–311.
8 Whitehead L, *et al.* Amoxicillin release from a floating dosage form based on alginates. *Int J Pharm* 2000; **210**: 45–49.
9 Iannuccelli V, *et al.* PVP solid dispersions for the controlled release of frusemide from a floating multiple-unit system. *Drug Dev Ind Pharm* 2000; **26**(6): 595–603.
10 Sharma S, Pawar A. Low density multiparticulate system for pulsatile release of meloxicam. *Int J Pharm* 2006; **313**: 150–158.
11 Iannuccelli V, *et al.* Air compartment multiple-unit system for prolonged gastric residence. Part 2. *In vivo* evaluation. *Int J Pharm* 1998; **174**: 55–62.
12 Abd-Elmageed A. Preparation and evaluation of sulindac alginate beads. *Bull Pharm Sci Assiut Univ* 1999; **22**(1): 73–80.
13 Mirghani A, *et al.* Formulation and release behavior of diclofenac sodium in Compritol 888 matrix beads encapsulated in alginate. *Drug Dev Ind Pharm* 2000; **26**(7): 791–795.
14 Turkoglu M, *et al.* Effect of aqueous polymer dispersions on properties of diclofenac/alginate beads and *in vivo* evaluation in rats. *STP Pharma Sci* 1997; 7(2): 135–140.
15 Fathy M, *et al.* Preparation and evaluation of beads made of different calcium alginate compositions for oral sustained release of tiaramide. *Pharm Dev Tech* 1998; 3(3): 355–364.
16 Rasmussen MR, *et al.* Numerical modelling of insulin and amyloglucosidase release from swelling Ca-alginate beads. *J Control Release* 2003; **91**(3): 395–405.
17 Torre ML, *et al.* Formulation and characterization of calcium alginate beads containing ampicillin. *Pharm Dev Tech* 1998; 3(2): 193–198.
18 Stops F, *et al.* The use of citric acid to prolong the *in vitro* gastro-retention of a floating dosage form in the fasted state. *Int J Pharm* 2006; **308**(1–2): 8–13.
19 Stops F, *et al.* Citric acid prolongs the gastro-retention of a floating dosage form and increases bioavailability of riboflavin in the fasted state. *Int J Pharm* 2006; **308**(1–2): 14–24.
20 Fathy M. Ca-alginate beads loaded with meloxicam: effect of alginate chemical composition on the properties of the beads and the ulcerogenicity of the drug. *J Drug Deliv Sci Technol* 2006; **16**(3): 183–189.
21 Pasparakis G, Bouropoulos N. Swelling studies and *in vitro* release of verapamil from calcium alginate and calcium alginate–chitosan beads. *Int J Pharm* 2006; **323**(1–2): 34–42.
22 Anal AK, *et al.* Chitosan-alginate multilayer beads for gastric passage and controlled intestinal release of protein. *Drug Dev Ind Pharm* 2003; **29**(6): 713–724.
23 Mandal BB, Kundu SC. Calcium alginate beads embedded in silk fibroin as 3D dual releasing scaffolds. *Biomaterials* 2009; **30**(28): 5170–5177.
24 Gaserod O, *et al.* Enhancement of the bioadhesive properties of calcium alginate gel beads by coating with chitosan. *Int J Pharm* 1998; **175**: 237–246.
25 Sriamornsak P, Kennedy RA. Development of polysaccharide gel-coated pellets for oral administration 2. Calcium alginate. *Eur J Pharm Sci* 2006; **29**(2): 139–147.
26 Grassi M, *et al.* Structural characterisation of calcium alginate matrices by means of mechanical and release tests. *Molecules* 2009; **14**(8): 3003–3017.
27 Tan R, *et al.* Preparation and characterisation of an injectable composite. *J Mater Sci Mater Med* 2009; **20**(6): 1245–1253.
28 Ostberg T, Graffner C. Calcium alginate matrices for oral multiple unit administration. Part 1. Pilot investigations of production method. *Acta Pharm Nord* 1992; **4**(4): 201–208.
29 Ostberg T, *et al.* Calcium alginate matrices for oral multiple unit administration. Part 2. Effect of process and formulation factors on matrix properties. *Int J Pharm* 1993; **97**: 183–193.
30 Ostberg T, *et al.* Calcium alginate matrices for oral multiple unit administration. Part 4. Release characteristics in different media. *Int J Pharm* 1994; **112**: 241–248.
31 Coppi G, *et al.* Polysaccharide film-coating for freely swellable hydrogels. *Pharm Dev Tech* 1998; 3(3): 347–353.
32 Esquisabel A, *et al.* Production of BCG alginate-PLL microcapsules by emulsification/internal gelation. *J Microencapsul* 1997; **14**(5): 627–638.
33 El-Gibaly I, Anwar MM. Development, characterization and *in vivo* evaluation of polyelectrolyte complex membrane gel microcapsules containing melatonin–resin complex for oral use. *Bull Pharm Sci Assiut Univ* 1998; **21**(2): 117–139.
34 Pillay V, Fassihi R. *In vitro* modulation from cross-linked pellets for site-specific drug delivery to the gastrointestinal tract. Part 1. Comparison of pH-responsive drug release and associated kinetics. *J Control Release* 1999; **59**: 229–242.
35 Pillay V, Fassihi R. *In vitro* release modulation from cross-linked pellets for site-specific drug delivery to the gastrointestinal tract. Part 2. Physicochemical characterization of calcium-alginate, calcium-pectinate and calcium-alginate-pectinate pellets. *J Control Release* 1999; **59**: 243–256.
36 Chickering DE, *et al.* Bioadhesive microspheres. Part 3. *In vivo* transit and bioavailability study of drug loaded alginate and poly (fumaric–co-sebacic anhydride) microspheres. *J Control Release* 1997; **48**: 35–46.
37 Iannuccelli V, *et al.* Biodegradable intraoperative system for bone infection treatment. Part 2. *In vivo* evaluation. *Int J Pharm* 1996; **143**: 187–194.
38 Holte O, *et al.* Preparation of a radionuclide/gel formulation for localised radiotherapy to a wide range of organs and tissues. *Pharmazie* 2006; **61**(5): 420–424.
39 Lim LY, Wan LSC. Propranolol hydrochloride binding in calcium alginate beads. *Drug Dev Ind Pharm* 1997; **23**(10): 973–980.
40 Goh CH, *et al.* Interactions of antimicrobial compounds with cross-linking agents of alginate dressings. *J Antimicrob Chemother* 2008; **62**(1): 105–108.
41 FAO/WHO. Toxicological evaluation of certain food additives with a review of general principles and of specifications. Seventeenth report of the joint FAO/WHO expert committee on food additives. *World Health Organ Tech Rep Ser* 1974; **539**: 1–40.
42 Lewis RJ, ed. *Sax's Dangerous Properties of Industrial Materials*, 11th edn. New York: Wiley, 2004; 668.
43 *Food Chemicals Codex*, 7th edn. Bethesda, MD: United States Pharmacopeia, 2010; 131.
44 *British Pharmaceutical Codex*. London: Pharmaceutical Press, 1973; 66.

20 General References

—

21 Author

CG Cable.

22 Date of Revision

1 October 2010.

Calcium Carbonate

1 Nonproprietary Names

BP: Calcium Carbonate

JP: Precipitated Calcium Carbonate

PhEur: Calcium Carbonate

USP: Calcium Carbonate

2 Synonyms

Balcarb; *Cal-Carb*; calcii carbonas; calcii carbonas praecipitatus; calcium carbonate (1 : 1); *Calcipress*; *Calopake*; carbonic acid calcium salt (1 : 1); creta preparada; *Destab*; E170; *MagGran CC*; *Pharma-Carb*; precipitated calcium carbonate; precipitated carbonate of lime; precipitated chalk; *Sturcal*; *Vicality*; *Vivapress*; *Witcarb*; *Vivapress Ca*.

3 Chemical Name and CAS Registry Number

Carbonic acid, calcium salt (1 : 1) [471-34-1]

4 Empirical Formula and Molecular Weight

$CaCO_3$ 100.09

5 Structural Formula

See Section 4.

6 Functional Category

Buffering agent; opacifier; tablet and capsule binder; tablet and capsule diluent.

7 Applications in Pharmaceutical Formulation or Technology

Calcium carbonate, employed as a pharmaceutical excipient, is mainly used in solid-dosage forms as a diluent.[1–5] It is also used as a base for medicated dental preparations,[6] as a buffering agent, and as a dissolution aid in dispersible tablets. Calcium carbonate is used as a bulking agent in tablet sugar-coating processes and as an opacifier in tablet film-coating.

Calcium carbonate is also used as a food additive.

8 Description

Calcium carbonate occurs as an odorless and tasteless white powder or crystals.

9 Pharmacopeial Specifications

See Table I. *See also* Section 18.

10 Typical Properties

Acidity/alkalinity pH = 9.0 (10% w/v aqueous dispersion)
Density (bulk) 0.8 g/cm^3
Density (tapped) 1.2 g/cm^3
Flowability Cohesive.
Melting point Decomposes at 825°C.
Moisture content *see* Figure 1.
Particle size *see* Figure 2.
Refractive index 1.59
Solubility Practically insoluble in ethanol (95%) and water. Solubility in water is increased by the presence of ammonium salts or carbon dioxide. The presence of alkali hydroxides reduces solubility.
Specific gravity 2.7
Specific surface area 6.21–6.47 m^2/g
Spectroscopy

IR spectra *see* Figure 3.

NIR spectra *see* Figure 4.

11 Stability and Storage Conditions

Calcium carbonate is stable and should be stored in a well-closed container in a cool, dry place.

SEM 1: Excipient: calcium carbonate; manufacturer: Whittaker, Clark & Daniels; lot no.: 15A-3; magnification: 600×; voltage: 20 kV.

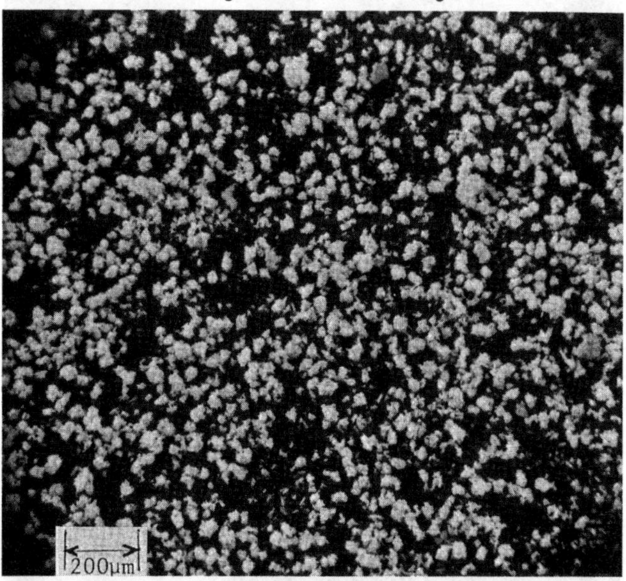

SEM 2: Excipient: calcium carbonate; manufacturer: Whittaker, Clark & Daniels; lot no.: 15A-3; magnification: 2400×; voltage: 20 kV.

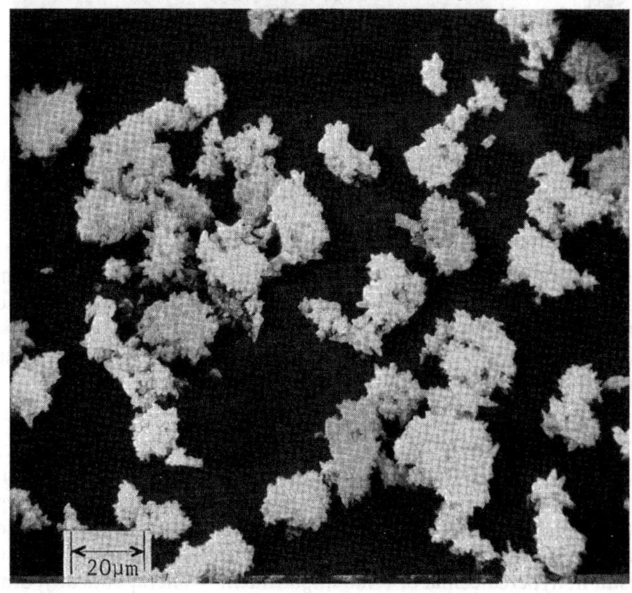

SEM 3: Excipient: calcium carbonate; manufacturer: Whittaker, Clark & Daniels; lot no.: 15A-4; magnification: 600×; voltage: 20 kV.

SEM 4: Excipient: calcium carbonate; manufacturer: Whittaker, Clark & Daniels; lot no.: 15A-4; magnification: 2400×; voltage: 20 kV.

SEM 5: Excipient: calcium carbonate; manufacturer: Whittaker, Clark & Daniels; lot no.: 15A-2; magnification: 600×; voltage: 20 kV.

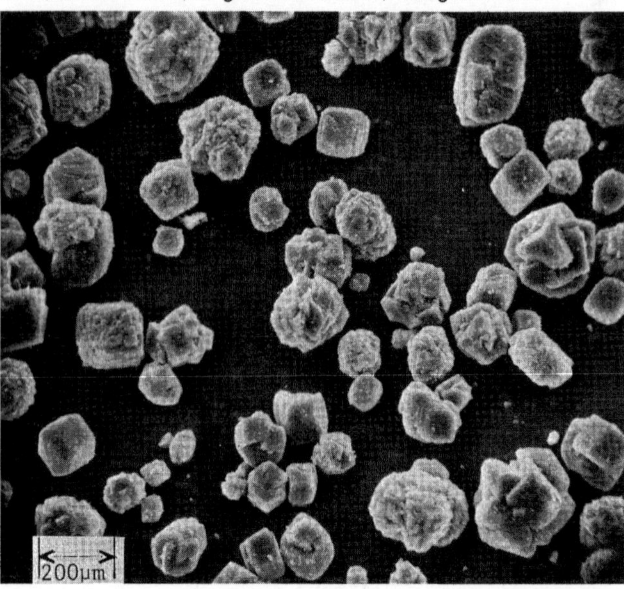

SEM 6: Excipient: calcium carbonate; manufacturer: Whittaker, Clark & Daniels; lot no.: 15A-2; magnification: 2400×; voltage: 20 kV.

12 Incompatibilities

Incompatible with acids and ammonium salts (*see also* Sections 10 and 18).

13 Method of Manufacture

Calcium carbonate is prepared by double decomposition of calcium chloride and sodium bicarbonate in aqueous solution. Density and fineness are governed by the concentrations of the solutions. Calcium carbonate is also obtained from the naturally occurring minerals aragonite, calcite, and vaterite.

14 Safety

Calcium carbonate is mainly used in oral pharmaceutical formulations and is generally regarded as a nontoxic material. However, calcium carbonate administered orally may cause constipation and flatulence. Consumption of large quantities (4–60 g daily) may also result in hypercalcemia or renal impairment.[7] Therapeutically, oral doses of up to about 1.5 g are employed as an antacid. In the treatment of hyperphosphatemia in patients with chronic renal failure, oral daily doses of 2.5–17 g have been used. Calcium carbonate may interfere with the absorption of other drugs from the gastrointestinal tract if administered concomitantly.

LD_{50} (rat, oral): 6.45 g/kg

15 Handling Precautions

Observe normal precautions appropriate to the circumstances and quantity of material handled. Calcium carbonate may be irritant to the eyes and on inhalation. Eye protection, gloves, and a dust mask are recommended. Calcium carbonate should be handled in a well-ventilated environment. In the UK, the long-term (8-hour TWA) workplace exposure limit for calcium carbonate is 10 mg/m^3 for total inhalable dust and 4 mg/m^3 for respirable dust.[8]

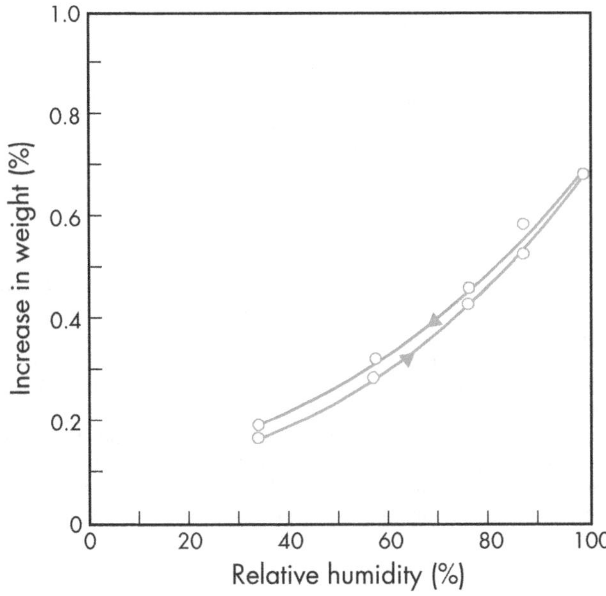

Figure 1: Moisture sorption–desorption isotherm of calcium carbonate.

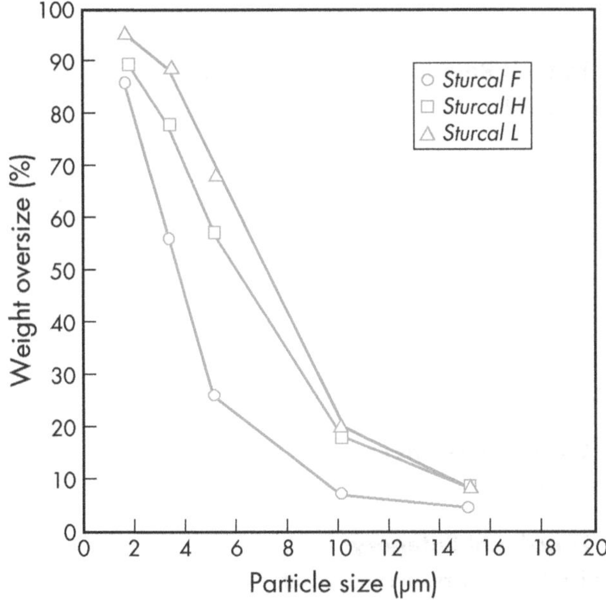

Figure 2: Particle-size distribution of calcium carbonate (*Sturcal*, Specialty Minerals Inc.).

16 Regulatory Status

GRAS listed. Accepted for use as a food additive in Europe. Included in the FDA Inactive Ingredients Database (buccal chewing gum, oral capsules and tablets; otic solutions; respiratory inhalation solutions). Included in nonparenteral medicines licensed in the UK. Included in the Canadian Natural Health Products Ingredients Database.

17 Related Substances

—

18 Comments

Calcium carbonate is one of the materials that have been selected for harmonization by the Pharmacopeial Discussion Group. For further information see the General Information Chapter <1196> in the USP35–NF30, the General Chapter 5.8 in PhEur 7.4, along with the 'State of Work' document on the PhEur EDQM website, and also the General Information Chapter 8 in the JP XV.

When calcium carbonate is used in tablets containing aspirin and related substances, traces of iron may cause discoloration. This may be overcome by inclusion of a suitable chelating agent. Grades with reduced lead levels are commercially available for use in antacids and calcium supplements.

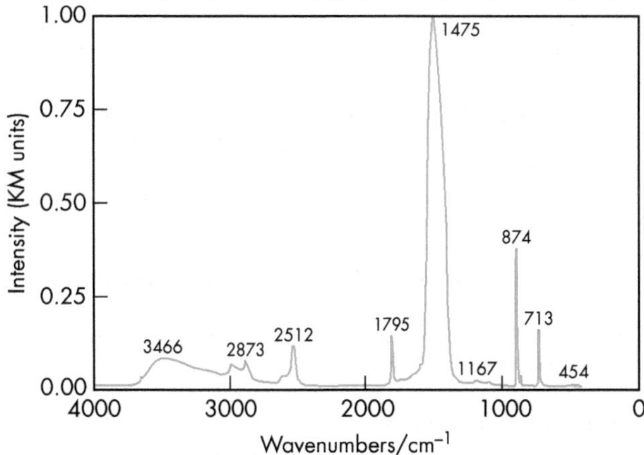

Figure 3: Infrared spectrum of calcium carbonate measured by diffuse reflectance. Adapted with permission of Informa Healthcare.

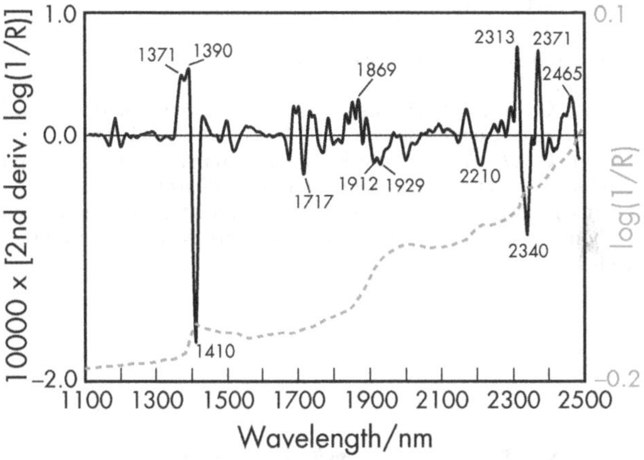

Figure 4: Near-infrared spectrum of calcium carbonate measured by reflectance.

Table I: Pharmacopeial specifications for calcium carbonate.

Test	JP XV	PhEur 7.4	USP35–NF30
Identification	—	+	+
Characters	+	+	—
Loss on drying	≤1.0%	≤2.0%	≤2.0%
Acid-insoluble substances	≤0.2%	≤0.2%	≤0.2%
Fluoride	—	—	≤50 ppm
Arsenic	≤5 ppm	≤4 ppm	≤3 ppm
Barium	+	+	+
Chlorides	—	≤330 ppm	—
Lead	—	—	≤3 ppm
Iron	—	≤200 ppm	≤0.1%
Heavy metals	≤20 ppm	≤20 ppm	≤20 ppm
Magnesium and alkali (metals) salts	≤0.5%	≤1.5%	≤1.0%
Sulfates	—	≤0.25%	—
Mercury	—	—	≤0.5 ppm
Assay (dried basis)	≥98.5%	98.5%–100.5%	98.0%–100.5%

Directly compressible grades containing only calcium carbonate are commercially available, such as *MagGran CC* (Magnesia GmbH). *Barcroft CS90* (SPI Pharma) is a directly compressible grade containing 10% starch and *Calci-Press MD* (Particle Dynamics Inc.) is a directly compressible blend of calcium carbonate and maltodextrin.

A specification for calcium carbonate is contained in the *Food Chemicals Codex* (FCC).[8]

The EINECS number for calcium carbonate is 207-439-9. The PubChem Compound ID (CID) for calcium carbonate includes 10112 and 516889.

Calcium carbonate is used therapeutically as an antacid and calcium supplement.

19 Specific References

1 Allen LV. Featured excipient: capsule and tablet diluents. *Int J Pharm Compound* 2000; 4(4): 306–310324–325.
2 Serra MD, Robles LV. Compaction of agglomerated mixtures of calcium carbonate and microcrystalline cellulose. *Int J Pharm* 2003; 258(1–2): 153–164.
3 Gorecki DKJ, *et al.* Dissolution rates in calcium carbonate tablets: a consideration in product selection. *Can J Pharm* 1989; **122**: 484–487508.
4 Bacher C, *et al.* Improving the compaction properties of roller compacted calcium carbonate. *Int J Pharm* 2007; **342**: 115–123.
5 Bacher C, *et al.* Compressibility and compactibility of granules produced by wet and dry granulation. *Int J Pharm* 2008; **358**: 69–74.
6 Carmargo IM, *et al.* Abrasiveness evaluation of silica and calcium carbonate used in the production of dentifrices. *J Cos Sci* 2001; 52: 163–167.
7 Orwoll ES. The milk-alkali syndrome: current concepts. *Ann Intern Med* 1982; 97: 242–248.
8 Health and Safety Executive. *EH40/2005: Workplace Exposure Limits.* Sudbury: HSE Books, 2011. http://www.hse.gov.uk/pubns/priced/eh40.pdf (accessed 29 February 2012).
9 *Food Chemicals Codex*, 7th edn. Bethesda, MD: United States Pharmacopeia, 2010: 133.

20 General References

Armstrong NA. Tablet manufacture. In: Swarbrick J, Boylan JC, eds. *Encyclopedia of Pharmaceutical Technology*, 2nd edn, 3. New York: Marcel Dekker, 2002: 2713–2732.
Ciancio SG. Dental products. In: Swarbrick J, Boylan JC, eds. *Encyclopedia of Pharmaceutical Technology*, 2nd edn, 3. New York: Marcel Dekker, 2002: 691–701.
European Directorate for the Quality of Medicines and Healthcare (EDQM). European Pharmacopoeia – State Of Work Of International Harmonisation. *Pharmeuropa* 2010; 22(4): 583–584. www.edqm.eu/en/International-Harmonisation-614.html (accessed 2 December 2010).
Specialty Minerals Inc. Technical data sheet: *Sturcal*, 1998.

21 Author

NA Armstrong.

22 Date of Revision

1 March 2012.

Calcium Chloride

1 Nonproprietary Names

BP: Calcium Chloride Dihydrate
Calcium Chloride Hexahydrate
JP: Calcium Chloride Hydrate
PhEur: Calcium Chloride Dihydrate
Calcium Chloride Hexahydrate
USP–NF: Calcium Chloride

Note that the JP XV and USP35–NF30 monographs list the dihydrate form.

2 Synonyms

Calcii chloridum dihydricum; calcii chloridum hexahydricum.

3 Chemical Name and CAS Registry Number

Calcium chloride anhydrous [10043-52-4]
Calcium chloride dihydrate [10035-04-8]
Calcium chloride hexahydrate [7774-34-7]

4 Empirical Formula and Molecular Weight

$CaCl_2$ 110.98 (for anhydrous)
$CaCl_2{\cdot}2H_2O$ 147.0 (for dihydrate)
$CaCl_2{\cdot}6H_2O$ 219.1 (for hexahydrate)

5 Structural Formula

See Section 4.

6 Functional Category

Antimicrobial preservative; desiccant.

7 Applications in Pharmaceutical Formulation or Technology

The main applications of calcium chloride as an excipient relate to its dehydrating properties and, therefore, it has been used as an antimicrobial preservative, and as a desiccant.

8 Description

Calcium chloride occurs as a white or colorless crystalline powder, granules, or crystalline mass, and is hygroscopic (deliquescent).

9 Pharmacopeial Specifications

See Table I.

10 Typical Properties

Acidity/alkalinity pH = 4.5–9.2 (5% w/v aqueous solution)
Boiling point >1600°C (anhydrous)
Density (bulk) 0.835 g/cm^3 (dihydrate)
Melting point
772°C (anhydrous);

Table I: Pharmacopeial specifications for calcium chloride.

Test	JP XV	PhEur 7.4	USP35–NF30
Identification	+	+	+
Characters	–	+	–
Appearance of solution	+	+	–
Acidity or alkalinity	4.5–9.2	+	4.5–9.2
Sulfates			
dihydrate	≤0.024%	≤300 ppm	–
hexahydrate	–	≤200 ppm	–
Aluminum	–	+	–
Aluminum (for hemodialysis only)			
dihydrate	–	≤1 ppm	≤1 ppm
hexahydrate	–	≤1 ppm	–
Iron, aluminum, and phosphate	+	–	+
Barium	+	+	–
Iron			
dihydrate	–	≤10 ppm	–
hexahydrate	–	≤7 ppm	–
Heavy metals			
dihydrate	≤10 ppm	≤20 ppm	≤10 ppm
hexahydrate	–	≤15 ppm	–
Magnesium and alkali salts			
dihydrate	–	≤0.5%	≤1.0% (residue)
hexahydrate	–	≤0.3%	–
Hypochlorite	+	–	–
Arsenic	≤2 ppm	–	–
Assay			
dihydrate	96.7–103.3%	97.0–103.0%	99.0–107.0%
hexahydrate	–	97.0–103.0%	–

176°C (dihydrate);
30°C (hexahydrate).

Solidification temperature 28.5–30°C (hexahydrate)

Solubility Freely soluble in water and ethanol (95%); insoluble in diethyl ether.

11 Stability and Storage Conditions

Calcium chloride is chemically stable; however, it should be protected from moisture. Store in airtight containers in a cool, dry place.

12 Incompatibilities

Calcium chloride is incompatible with soluble carbonates, phosphates, sulfates, and tartrates.[1] It reacts violently with bromine trifluoride, and a reaction with zinc releases explosive hydrogen gas. It has an exothermic reaction with water, and when heated to decomposition it emits toxic fumes of chlorine.

13 Method of Manufacture

Calcium chloride is a principal byproduct of the Solvay process.

14 Safety

Calcium chloride is used in topical, ophthalmic, and injection preparations. The pure form of calcium chloride is toxic by intravenous, intramuscular, intraperitoneal, and subcutaneous routes, and moderately toxic by ingestion, causing stomach and heart disturbances. It is a severe eye irritant and can cause dermatitis.

LD_{50} (mouse, IP): 0.21 g/kg[2]
LD_{50} (mouse, IV): 0.042 g/kg
LD_{50} (mouse, oral): 1.94 g/kg
LD_{50} (mouse, SC): 0.82 g/kg
LD_{50} (rat, IM): 0.025 g/kg
LD_{50} (rat, IP): 0.26 g/kg
LD_{50} (rat, oral): 1.0 g/kg
LD_{50} (rat, SC): 2.63 g/kg

15 Handling Precautions

Observe normal precautions appropriate to the circumstances and quantity of the material handled. Calcium chloride is irritating to eyes, the respiratory system, and skin. Gloves, eye protection, respirator, and other protective clothing should be worn.

16 Regulatory Status

GRAS listed. Included in the FDA Inactive Ingredients Database (injections, ophthalmic preparations, suspensions, creams). Included in medicines licensed in the UK (eye drops; intraocular irrigation; vaccines; injection powders for reconstitution; nebulizer solution; oral suspension).

17 Related Substances

—

18 Comments

The dissolution of calcium chloride in water is an exothermic reaction and, along with other excipients such as sodium sulfate, sodium acetate, and water, it has a potential application in hot packs.[3] Calcium chloride has been used to control the release of active ingredients from solid oral dosage forms by crosslinking pectin,[4] or by its interaction with chitosan.[5]

Therapeutically, calcium chloride injection 10% (as the dihydrate form) is used to treat hypocalcemia.[6] It has also been used as an astringent in eye lotions.

A specification for calcium chloride is contained in the *Food Chemicals Codex* (FCC).[7]

The EINECS number for calcium chloride is 233-140-8. The PubChem Compound ID (CID) for calcium chloride is 5284359.

19 Specific References

1 Todd RG, Wade A, eds. *The Pharmaceutical Codex*, 11th edn. London: Pharmaceutical Press, 1979; 125.
2 Lewis RJ, ed. *Sax's Dangerous Properties of Industrial Materials*, 11th edn. New York: Wiley, 2004; 670.
3 Donnelly WR. Exothermic composition and hot pack. United States Patent 4203418; 1980.
4 Wei X, *et al.* Sigmoidal release of indomethacin from pectin matrix tablets: effect of in situ crosslinking by calcium cations. *Int J Pharm* 2006; **318**: 132–138.
5 Rege PR, *et al.* Chitinosan-drug complexes: effect of electrolyte on naproxen release in vitro. *Int J Pharm* 2003; **250**: 259–272.
6 Joint Formulary Committee. *British National Formulary*, No. 55. London: British Medical Association and Royal Pharmaceutical Society of Great Britain, 2008.
7 *Food Chemicals Codex*, 7th edn. Bethesda, MD: United States Pharmacopeia, 2010; 134.

20 General References

Wenninger JA, McEwen JD Jr, eds. *CTFA Cosmetic Ingredient Handbook*. Washington DC: CTFA, 1992.

21 Author

MA Pellett.

22 Date of Revision

1 March 2012.

Calcium Hydroxide

1 Nonproprietary Names

BP: Calcium Hydroxide
JP: Calcium Hydroxide
PhEur: Calcium Hydroxide
USP–NF: Calcium Hydroxide

2 Synonyms

Calcii hydroxidum; calcium hydrate; E526; hydrated lime; slaked lime.

3 Chemical Name and CAS Registry Number

Calcium hydroxide [1305-62-0]

4 Empirical Formula and Molecular Weight

$Ca(OH)_2$ 74.1

5 Structural Formula

See Section 4.

6 Functional Category

Alkalizing agent.

7 Applications in Pharmaceutical Formulation or Technology

Calcium hydroxide is a strong alkali and is used as a pharmaceutical pH adjuster/buffer and antacid in topical medicinal ointments, creams, lotions, and suspensions, often as an aqueous solution (lime water).(1,2) It forms calcium soaps of fatty acids, which produce water-in-oil emulsions (calamine liniment).(3,4)

Calcium hydroxide is a common cosmetic ingredient in hair-straightening and hair-removal products, and in shaving preparations.(1)

8 Description

Calcium hydroxide occurs as a white or almost white, crystalline or granular powder. It has a bitter, alkaline taste. Calcium hydroxide readily absorbs carbon dioxide to form calcium carbonate.

9 Pharmacopeial Specifications

See Table I.

Table I: Pharmacopeial specifications for calcium hydroxide.

Test	JP XV	PhEur 7.4	USP35–NF30
Identification	+	+	+
Acid-insoluble substances	≤25 mg	≤0.5%	≤0.5%
Carbonates	–	≤5.0%	+
Chlorides	–	≤330 ppm	–
Sulfates	–	≤0.4%	–
Heavy metals	≤40 ppm	≤20 ppm	≤20 μg/g
Arsenic	≤4 ppm	≤4 ppm	–
Magnesium and alkali (metals) salts	≤24 mg	≤4.0%	≤4.8%
Assay	≥90.0%	95.0–100.5%	95.0–100.5%

10 Typical Properties

Acidity/alkalinity pH = 12.4 (saturated solution at 25°C)
Density 2.08–2.34 g/cm^3
Melting point When heated above 580°C, it dehydrates forming the oxide.
Solubility Soluble in glycerol and ammonium chloride solutions; dissolves in sucrose solutions to form calcium saccharosates;(2) soluble in acids with the evolution of heat; soluble 1 in 600 water (less soluble in hot water); insoluble in ethanol (95%).

11 Stability and Storage Conditions

Calcium hydroxide should be stored in an airtight container, in a cool, dry, well-ventilated place. Calcium hydroxide powder may be sterilized by heating for 1 hour at a temperature of at least 160°C.(2)

12 Incompatibilities

Incompatible with strong acids, maleic anhydride, phosphorus, nitroethane, nitromethane, nitroparaffins, and nitropropane. Calcium hydroxide can be corrosive to some metals.

13 Method of Manufacture

Calcium hydroxide is manufactured by adding water to calcium oxide, a process called slaking.

14 Safety

Calcium hydroxide is used in oral and topical pharmaceutical formulations. It is mildly toxic by ingestion. In the pure state, calcium hydroxide is a severe skin, eye, and respiratory irritant, and it is corrosive, causing burns. Typical exposure limits are TLV 5 mg/m^3 in air.(5)

LD_{50} (mouse, oral): 7.3 g/kg
LD_{50} (rat, oral): 7.34 g/kg(6)

15 Handling Precautions

Observe normal precautions appropriate to the circumstances and quantity of the material handled. Avoid contact with eyes, skin, and clothing. Avoid breathing the dust. Gloves, eye protection, respirator, and other protective clothing should be worn. In the USA, the OSHA permissible exposure limit is 15 mg/m^3 for total dust and 5 mg/m^3 respirable fraction for calcium hydroxide.(7)

16 Regulatory Status

GRAS listed. Accepted for use as a food additive in Europe. Included in the FDA Inactive Ingredients Database (intravenous and subcutaneous injections; oral suspensions and tablets; topical emulsions and creams). Included in parenteral preparations licensed in the UK.

17 Related Substances

Potassium hydroxide; sodium hydroxide.

18 Comments

Therapeutically, calcium hydroxide has been used as a topical astringent.[3,4]

In dentistry, it is used as a filling agent and in dental pastes to encourage deposition of secondary dentine.[8] Calcium hydroxide was traditionally used as an escharotic in Vienna Paste.[9]

A specification for calcium hydroxide is contained in the *Food Chemicals Codex* (FCC).[10]

The EINECS number for calcium hydroxide is 215-137-3. The PubChem Compound ID (CID) for calcium hydroxide is 14777.

19 Specific References

1 Wenninger JA, McEwen JD Jr, eds. *CTFA Cosmetic Ingredient Handbook*. Washington DC: CTFA, 1992; 53.
2 Todd RG, Wade A, eds. *The Pharmaceutical Codex*, 11th edn. London: Pharmaceutical Press, 1979; 127.
3 Allen L Jr *et al.* eds. *Ansel's Pharmaceutical Dosage Forms and Drug Delivery*. Philadelphia: Lippincott Williams and Wilkins, 2005; 411.
4 Allen L Jr *et al.* eds. *Ansel's Pharmaceutical Dosage Forms and Drug Delivery*. Philadelphia: Lippincott Williams and Wilkins, 2005; 368.
5 Ash M, Ash I, eds. *Handbook of Pharmaceutical Additives*, 3rd edn. Endicott, NY: Synapse Information Resources, 2007; 501.
6 Lewis RJ, ed. *Sax's Dangerous Properties of Industrial Materials*, 11th edn. New York: Wiley, 2004; 675.
7 Mallinckrodt Baker Inc. *Material safety data sheet: Calcium hydroxide*, 2007.
8 Foreman PC, Barnes IE. A review of calcium hydroxide. *Int Endod J* 1990; 23: 283–297.
9 Sweetman S, ed. *Martindale: The Complete Drug Reference*, 36th edn. London: Pharmaceutical Press, 2009; 2272.
10 *Food Chemicals Codex*, 7th edn. Bethesda, MD: United States Pharmacopeia, 2010; 139.

20 General References

—

21 Author

TI Armstrong.

22 Date of Revision

1 March 2012.

Calcium Lactate

1 Nonproprietary Names

BP: Calcium Lactate Pentahydrate

JP: Calcium Lactate Hydrate

PhEur: Calcium Lactate Pentahydrate

USP–NF: Calcium Lactate

2 Synonyms

Calcii lactas pentahydricus; calcium bis(2-hydroxypropanoate) pentahydrate; calcium dilactate; calcium lactate (1 : 2) hydrate; calcium lactate (1 : 2) pentahydrate; E327; 2-hydroxypropanoic acid, calcium salt; lactic acid, calcium salt; mixture of calcium (2*R*)-, (2*S*)- and (2*RS*)-2-hydroxypropanoates pentahydrates; propanoic acid, 2-hydroxy-, calcium salt (2 : 1), hydrate; *Puracal.*

3 Chemical Name and CAS Registry Number

Calcium lactate anhydrous [814-80-2]

Calcium lactate monohydrate and trihydrate [41372-22-9]

Calcium lactate pentahydrate [5743-47-5] and [63690-56-2]

4 Empirical Formula and Molecular Weight

$C_6H_{10}CaO_6$	218.2 (anhydrous)
$C_6H_{10}CaO_6{\cdot}H_2O$	236.0 (monohydrate)
$C_6H_{10}CaO_6{\cdot}3H_2O$	272.3 (trihydrate)
$C_6H_{10}CaO_6{\cdot}5H_2O$	308.3 (pentahydrate)

5 Structural Formula

$[CH_3CH(OH)COO^-]_2\ Ca^{2+}\ xH_2O$

6 Functional Category

Antimicrobial preservative; buffering agent; tablet and capsule binder; tablet and capsule diluent; viscosity-increasing agent.

7 Applications in Pharmaceutical Formulation or Technology

Calcium lactate is used as a bioavailability enhancer and nutrient supplement in pharmaceutical formulations.

A spray-dried grade of calcium lactate pentahydrate has been used as a tablet diluent in direct compression systems,[1] and has been shown to have good compactability. The properties of the pentahydrate form have been considered superior to those of calcium lactate trihydrate when used in direct compression tablet formulations.[2] Tablet properties may be affected by the hydration state of the calcium lactate and particle size of the material: reducing particle size increased crushing strength, whereas storage of tablets at elevated temperature resulted in dehydration accompanied by a reduction in crushing strength.[3]

Calcium lactate has also been used as the source of calcium ions in the preparation of calcium alginate microspheres for controlled-release delivery of active agents. It has been shown to result in lower calcium concentrations in the finished microspheres when compared with calcium acetate.[4]

C

8 Description

Calcium lactate occurs as white or almost white, crystalline or granular powder. It is slightly efflorescent.

9 Pharmacopeial Specifications

See Table I. *See also* Section 18.

Table I: Pharmacopeial specifications for calcium lactate.

Test	JP XV	PhEur 7.4	USP35–NF30
Identification	+	+	+
Characters	–	+	–
Appearance of solution	+	+	–
Acidity or alkalinity	+	+	≤0.45% lactic acid
Chlorides	–	≤200 ppm	–
Sulfates	–	≤400 ppm	–
Barium	–	+	–
Iron	–	≤50 ppm	–
Magnesium and alkali salts	≤5 mg	≤1.0%	≤1.0%
Heavy metals	≤20 ppm	≤10 ppm	≤20 ppm
Loss on drying			
Anhydrous	–	≤3.0%	≤3.0%
Monohydrate	–	5.0–8.0%	5.0–8.0%
Trihydrate	–	15.0–20.0%	15.0–20.0%
Pentahydrate	25–30%	22.0–27.0%	22.0–27.0%
Arsenic	≤4 ppm	–	–
Volatile fatty acid	+	–	+
Assay (dried basis)	≥97.0%	98–102%	98.0–101.0%

10 Typical Properties

Acidity/alkalinity pH = 6.0–8.5 for a 10% aqueous solution for *Puracal PP*[5]

Density (bulk) 0.56 g/cm^3;[2] 0.3–0.5 g/cm^3 for *Puracal PP*[5]

Density (tapped) 0.67 g/cm^3[2]

Density (true) 1.494 g/cm^3[2]

Hygroscopicity The pentahydrate form is nonhygroscopic (*see* Section 11).

Melting point >200°C for *Puracal PP*[5]

Solubility Soluble in water, freely soluble in boiling water; very slightly soluble in ethanol (95%).

11 Stability and Storage Conditions

Calcium lactate can exist in a number of hydration states, which are characterized as anhydrous, monohydrate, trihydrate, and pentahydrate. Dehydration of the pentahydrate form is rapid at temperatures of 55°C and above. Dehydration is reported to be accompanied by some loss of crystallinity.[6] Tablet crushing strength has been reported to be reduced following dehydration of calcium lactate pentahydrate.[3]

12 Incompatibilities

Calcium salts, including the lactate, can display physical incompatibility with phosphate in the diet or therapeutic preparations, for example in enteral feed mixtures.[7]

13 Method of Manufacture

Calcium lactate is prepared commercially by neutralization, with calcium carbonate or calcium hydroxide, of lactic acid obtained from fermentation of dextrose, molasses, starch, sugar, or whey.[8]

14 Safety

Calcium lactate was found to have no toxic or carcinogenic effects when dosed at levels of 0%, 2.5%, and 5% in drinking water to male and female rats for 2 years.[9]

15 Handling Precautions

Observe normal precautions appropriate to the circumstances and quantity of the material handled.

16 Regulatory Status

GRAS listed except for infant foods/formulas.[10] Accepted as a food additive in Europe. Calcium lactate (anhydrous) is included in the FDA Inactive Ingredients Database (vaginal, tablet). It is used in oral dosage forms. Included in vaginal pessary formulations licensed in the UK.

17 Related Substances

Lactic acid; sodium lactate.

18 Comments

Calcium lactate is available in a number of grades with respect to hydration state, purity, and particle size. Care should be taken to understand the hydration state of the material in use. The USP35–NF30 states that on product labeling, "calcium lactate" should be understood to be an amount of calcium equivalent to that contained in the stated amount of calcium lactate pentahydrate. Each 1.0 g of calcium lactate pentahydrate contains 130 mg (3.2 mmol) of calcium.

The use of calcium lactate in film coatings as an alternative white pigment to titanium dioxide has been reported.[11] The white coloration may be due to interactions between the hypromellose polymer and calcium ions in the film. The use of films containing calcium lactate as edible coatings for food products has also been reported. Milk proteins have been used as the film former, crosslinked by the calcium salt.[12]

Lactate salts, including calcium lactate, have been reported as having antimicrobial properties and have been applied as preservatives in foods.[13]

Therapeutically, calcium lactate has been used in preparations for the treatment of calcium deficiency.

The USP35–NF30 monograph for calcium lactate covers the anhydrous and hydrous forms. The PhEur 7.4 lists separate monographs for calcium lactate, anhydrous, calcium lactate monohydrate, calcium lactate pentahydrate, and calcium lactate trihydrate. The calcium in calcium lactate is bioavailable when administered orally; there are monographs for calcium lactate tablets in both BP 2012 and USP35–NF30. A specification for calcium lactate is contained in the *Food Chemicals Codex* (FCC).[14]

The optically active L-form of calcium lactate has a higher solubility in water than the racemic DL-form.[15]

The EINECS number for anhydrous calcium lactate is 212-406-7. The PubChem Compound ID (CID) for anhydrous calcium lactate is 521805 and for calcium lactate pentahydrate it is 165341.

19 Specific References

1 Bolhuis GK, Armstrong NA. Excipients for direct compaction – an update. *Pharm Dev Technol* 2006; **11**: 111–124.

2 Bolhuis GK, *et al.* DC calcium lactate, a new filler-binder for direct compaction of tablets. *Int J Pharm* 2001; **221**: 77–86.

3 Sakata Y, *et al.* Effect of pulverization and dehydration on the pharmaceutical properties of calcium lactate pentahydrate tablets. *Colloids Surf B Biointerfaces* 2006; **51**: 149–156.
4 Heng PWS, *et al.* Formation of alginate microspheres produced using emulsification technique. *J Microencapsul* 2003; **20**: 401–413.
5 Purac. Material safety data sheet. No. 2001/58/EC: *Puracal PP*, 26 October 2004.
6 Sakata Y, *et al.* Characterization of dehydration and hydration behavior of calcium lactate pentahydrate and its anhydrate. *Colloids Surf B Biointerfaces* 2005; **46**: 135–141.
7 Wong KK, Secker D. Calcium and phosphate precipitation – a medication and food interaction. *Can J Hosp Pharm* 1996; **49**: 119.
8 Inskeep GC, *et al.* Lactic acid from corn sugar. A staff-industry collaborative report. *Ind Eng Chem* 1952; **44**: 1955–1966.
9 Maekawa A, *et al.* Long-term toxicity/carcinogenicity study of calcium lactate in F344 rats. *Food Chem Toxicol* 1991; **29**: 589–594.
10 FDA Code of Federal Regulations. Direct food substances affirmed as generally recognised as safe – Calcium Lactate. Title 21, Section 184: 1207.
11 Sakate Y, *et al.* A novel white film for pharmaceutical coating formed by interaction of calcium lactate pentahydrate with hydroxypropyl methylcellulose. *Int J Pharm* 2006; **317**: 120–126.
12 Mei Y, Zhao Y. Barrier and mechanical properties of milk-protein based edible films containing neutraceuticals. *J Agric Food Chem* 2003; **51**: 1914–1918.
13 Shelef LA. Antimicrobial effects of lactates: a review. *J Food Prot* 1994; **57**: 444–450.
14 *Food Chemicals Codex*, 7th edn. Bethesda, MD: United States Pharmacopeia, 2010; 140.
15 Apelblat A, *et al.* Solubilities and vapour pressures of water over saturated solutions of magnesium-L-lactate, calcium-L-lactate, zinc-L-lactate, ferrous-L-lactate and aluminium-L-lactate. *Fluid Phase Equilib* 2005; **236**: 162–168.

20 General References

—

21 Author

W Cook.

22 Date of Revision

1 March 2012.

Calcium Phosphate, Dibasic Anhydrous

1 Nonproprietary Names

BP: Anhydrous Calcium Hydrogen Phosphate
JP: Anhydrous Dibasic Calcium Phosphate
PhEur: Calcium Hydrogen Phosphate, Anhydrous
USP–NF: Anhydrous Dibasic Calcium Phosphate

2 Synonyms

A-TAB; calcii hydrogenophosphas anhydricus; calcium monohydrogen phosphate; calcium orthophosphate; *Di-Cafos AN*; dicalcium orthophosphate; E341; *Emcompress Anhydrous*; *Fujicalin*; phosphoric acid calcium salt (1 : 1); secondary calcium phosphate.

3 Chemical Name and CAS Registry Number

Dibasic calcium phosphate [7757-93-9]

4 Empirical Formula and Molecular Weight

$CaHPO_4$ 136.06

5 Structural Formula

See Section 4.

6 Functional Category

Tablet and capsule diluent.

7 Applications in Pharmaceutical Formulation or Technology

Anhydrous dibasic calcium phosphate is used both as an excipient and as a source of calcium in nutritional supplements. It is used particularly in the nutritional/health food sectors. It is also used in pharmaceutical products because of its compaction properties, and the good flow properties of the coarse-grade material.[(1–5)] The predominant deformation mechanism of anhydrous dibasic calcium phosphate coarse-grade is brittle fracture and this reduces the strain-rate sensitivity of the material, thus allowing easier transition from the laboratory to production scale. However, unlike the dihydrate, anhydrous dibasic calcium phosphate when compacted at higher pressures can exhibit lamination and capping. This phenomenon can be observed when the material represents a substantial proportion of the formulation, and is exacerbated by the use of deep concave tooling. This phenomenon also appears to be independent of rate of compaction.

Anhydrous dibasic calcium phosphate is abrasive and a lubricant is required for tableting, for example 1% w/w magnesium stearate or 1% w/w sodium stearyl fumarate.

Two particle-size grades of anhydrous dibasic calcium phosphate are used in the pharmaceutical industry. Milled material is typically used in wet-granulated or roller-compacted formulations. The 'unmilled' or coarse-grade material is typically used in direct-compression formulations.

Anhydrous dibasic calcium phosphate is nonhygroscopic and stable at room temperature. It does not hydrate to form the dihydrate.

Anhydrous dibasic calcium phosphate is used in toothpaste and dentifrice formulations for its abrasive properties.

8 Description

Anhydrous dibasic calcium phosphate is a white, odorless, tasteless powder or crystalline solid. It occurs as triclinic crystals.

9 Pharmacopeial Specifications

The pharmacopeial specifications for anhydrous dibasic calcium phosphate have undergone harmonization of many attributes for JP, PhEur, and USP–NF.

See Table I. *See also* Section 18.

SEM 1: Excipient: *Emcompress Anhydrous*; manufacturer: JRS Pharma LP; magnification: 50×; voltage: 5 kV.

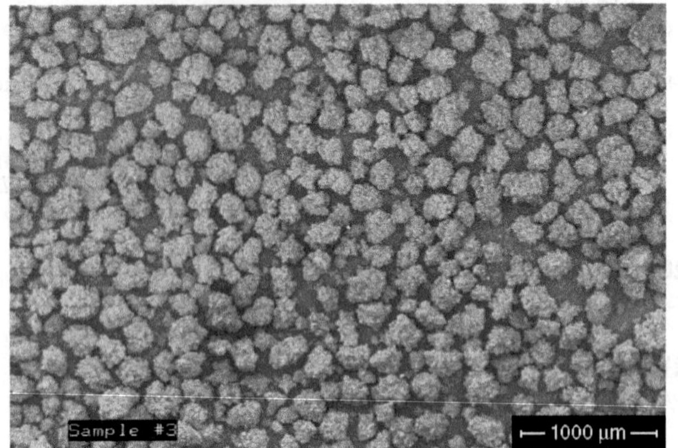

SEM 2: Excipient: *Emcompress Anhydrous*; manufacturer: JRS Pharma LP; magnification: 200×; voltage: 5 kV.

Table I: Pharmacopeial specifications for calcium phosphate, dibasic anhydrous.

Test	JP XV S1	PhEur 7.4	USP35–NF30
Identification	+	+	+
Characters	+	+	—
Loss on ignition	6.6–8.5%	6.6–8.5%	6.6–8.5%
Acid-insoluble substances	≤0.2%	≤0.2%	≤0.2%
Heavy metals[(a)]	≤31 ppm	≤40 ppm	≤30 ppm
Chloride	≤0.25%	≤0.25%	≤0.25%
Fluoride	—	≤100 ppm	≤50 ppm
Sulfate	≤0.5%	≤0.5%	≤0.5%
Carbonate	+	+	+
Barium	+	+	+
Arsenic	≤2 ppm	≤10 ppm	≤3 µg/g
Iron	—	≤400 ppm	—
Assay (dried basis)	98.0–103.0%	98.0–103.0%	98.0–103.0%

(a) These tests have not been fully harmonized at the time of publication.

10 Typical Properties

Acidity/alkalinity pH = 7.3 (20% slurry);
pH = 5.1 (20% slurry of *A-TAB*);
pH = 6.1–7.2 (5% slurry of *Fujicalin*).

Angle of repose 32° (for *Fujicalin*)

Density 2.89 g/cm^3

Density (bulk) 0.78 g/cm^3 for *A-TAB*;
0.45 g/cm^3 for *Fujicalin*.

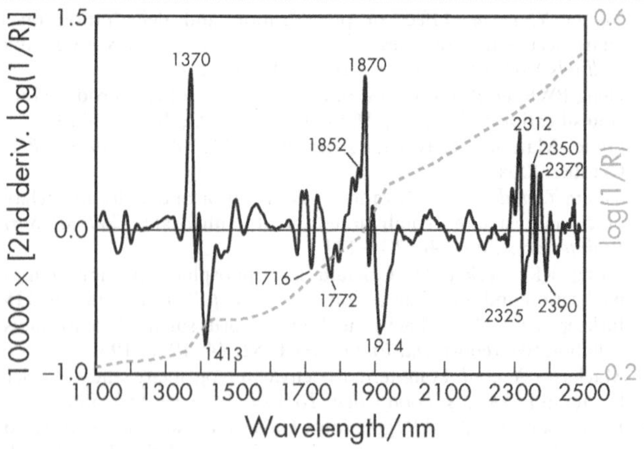

Figure 1: Near-infrared spectrum of anhydrous dibasic calcium phosphate measured by reflectance.

Density (tapped) 0.82 g/cm^3 for *A-TAB*;
0.46 g/cm^3 for *Fujicalin*.

Melting point Does not melt; decomposes at ≈425°C to form calcium pyrophosphate.

Moisture content Typically 0.1–0.2%. The anhydrous material contains only surface-adsorbed moisture and cannot be rehydrated to form the dihydrate.

Particle size distribution *A-TAB*: average particle diameter 180 µm;
Emcompress Anhydrous: average particle diameter 136 µm;
Fujicalin: average particle diameter 94 µm;
Powder: average particle diameter: 15 µm.

Solubility Practically insoluble in ether, ethanol, and water; soluble in dilute acids.

Specific surface area 20–30 m^2/g for *A-TAB*;
35 m^2/g for *Fujicalin*.

Spectroscopy NIR spectra *see* Figure 1.

11 Stability and Storage Conditions

Dibasic calcium phosphate anhydrous is a nonhygroscopic, relatively stable material. Under conditions of high humidity it does not hydrate to form the dihydrate.

The bulk material should be stored in a well-closed container in a dry place.

12 Incompatibilities

Dibasic calcium phosphate should not be used to formulate tetracyline antibiotics.[6]

The surface of milled anhydrous dibasic calcium phosphate is alkaline[2] and consequently it should not be used with drugs that are sensitive to alkaline pH. However, reports[7,8] suggest there are differences in the surface alkalinity/acidity between the milled and unmilled grades of anhydrous dibasic calcium phosphate; the unmilled form has an acidic surface environment. This difference has important implications for drug stability, particularly when reformulating from, e.g. roller compaction to direct compression, when the particle size of the anhydrous dibasic calcium phosphate might be expected to change.

Dibasic calcium phosphate dihydrate has been reported to be incompatible with a number of drugs and excipients, and many of these incompatibilities are expected to occur with dibasic calcium phosphate, anhydrous; *see* Calcium phosphate, dibasic dihydrate.

13 Method of Manufacture

Calcium phosphates are usually prepared by reacting very pure phosphoric acid with calcium hydroxide, $Ca(OH)_2$ obtained from limestone, in stoichiometric ratio in aqueous suspension[2] followed by drying at a temperature that will allow the correct hydration state to be achieved. After drying, the coarse-grade material is obtained by means of a classification unit; the fine particle-size material is obtained by milling. Dibasic calcium phosphate, anhydrous, may also be prepared by spray-drying.[5,9]

14 Safety

Dibasic calcium phosphate anhydrous is widely used in oral pharmaceutical products, food products, and toothpastes, and is generally regarded as a relatively nontoxic and nonirritant material.

15 Handling Precautions

Observe normal precautions appropriate to the circumstances and quantity of material handled. The fine-milled grades can generate nuisance dusts and the use of a respirator or dust mask may be necessary.

16 Regulatory Status

GRAS listed. Accepted as a food additive in Europe. Included in the FDA Inactive Ingredients Database (oral capsules and tablets). Included in nonparenteral medicines licensed in Europe. Included in the Canadian Natural Health Products Ingredients Database.

17 Related Substances

Calcium phosphate, dibasic dihydrate; calcium phosphate, tribasic; calcium sulfate.

18 Comments

Anhydrous dibasic calcium phosphate has undergone harmonization of many attributes for JP, PhEur, and USP–NF by the Pharmacopeial Discussion Group. For further information see the General Information Chapter <1196> in the USP35–NF30, the General Chapter 5.8 in PhEur 7.4, along with the 'State of Work' document on the PhEur EDQM website, and also the General Information Chapter 8 in the JP XV.

Grades of anhydrous dibasic calcium phosphate available for direct compression include *A-TAB* (Innophos), *Di-Cafos AN* (Chemische Fabrik Budenheim), *Emcompress Anhydrous* (JRS Pharma LP), and *Fujicalin* (Fuji Chemical Industry Co. Ltd.). A study has examined the use of calcium phosphate in reducing microbial contamination during direct compression in tableting.[10]

The EINECS number for calcium phosphate is 231-837-1. The PubChem Compound ID (CID) for anhydrous dibasic calcium phosphate is 24441.

19 Specific References

1 Fischer E. Calcium phosphate as a pharmaceutical excipient. *Manuf Chem* 1992; **64**(6): 25–27.
2 Schmidt PC, Herzog R. Calcium phosphates in pharmaceutical tableting 1: physico-pharmaceutical properties. *Pharm World Sci* 1993; **15**(3): 105–115.
3 Schmidt PC, Herzog R. Calcium phosphates in pharmaceutical tableting 2: comparison of tableting properties. *Pharm World Sci* 1993; **15**(3): 116–122.
4 Hwang R-C, Peck GR. A systematic evaluation of the compression and tablet characteristics of various types of lactose and dibasic calcium phosphate. *Pharm Technol* 2001; **25**(6): 54, 56, 58, 60, 62, 64, 66, 68.
5 Schlack H, *et al.* Properties of Fujicalin, a new modified anhydrous dibasic calcium phosphate for direct compression: comparison with dicalcium phosphate dihydrate. *Drug Dev Ind Pharm* 2001; **27**(8): 789–801.
6 Weiner M, Bernstein IL. *Adverse Reactions to Drug Formulation Agents: A Handbook of Excipients.* New York: Marcel Dekker, 1989; 93–94.
7 Dulin WA. Degradation of bisoprolol fumarate in tablets formulated with dicalcium phosphate. *Drug Dev Ind Pharm* 1995; **21**(4): 393–409.
8 Glombitza BW, *et al.* Surface acidity of solid pharmaceutical excipients I. Determination of the surface acidity. *Eur J Pharm Biopharm* 1994; 40(5): 289–293.
9 Kiyoshi T, *et al.* Novel preparation of free-flowing spherically granulated dibasic calcium phosphate anhydrous for direct tabletting. *Chem Pharm Bull* 1996; **44**(4): 868–870.
10 Ayorinde JO, *et al.* The survival of *B. subtilis* spores in dicalcium phosphate, lactose, and corn starch. *Pharmaceutical Technology* 2005; **29**(12): 56–67.

20 General References

Bryan JW, McCallister JD. Matrix forming capabilities of three calcium diluents. *Drug Dev Ind Pharm* 1992; **18**(19): 2029–2047.
Carstensen JT, Ertell C. Physical and chemical properties of calcium phosphates for solid state pharmaceutical formulations. *Drug Dev Ind Pharm* 1990; **16**(7): 1121–1133.
European Directorate for the Quality of Medicines and Healthcare (EDQM). European Pharmacopoeia – State Of Work Of International Harmonisation. *Pharmeuropa* 2011; 23(4): 713–714. www.edqm.eu/site/-614.html (accessed 5 December 2011).
Fuji Chemical Industry Co. Ltd. Technical literature: *Fujicalin*, 1998.
Innophos Inc. Product datasheet: Calcium Phosphates, 2008.

21 Author

RC Moreton.

22 Date of Revision

1 March 2012.

Calcium Phosphate, Dibasic Dihydrate

1 Nonproprietary Names

BP: Calcium Hydrogen Phosphate
JP: Dibasic Calcium Phosphate Hydrate
PhEur: Calcium Hydrogen Phosphate Dihydrate
USP–NF: Dibasic Calcium Phosphate Dihydrate

2 Synonyms

Calcii hydrogenophosphas dihydricus; calcium hydrogen orthophosphate dihydrate; calcium monohydrogen phosphate dihydrate; *Di-Cafos*; dicalcium orthophosphate; *DI-TAB*; E341; *Emcompress*; phosphoric acid calcium salt (1 : 1) dihydrate; secondary calcium phosphate.

3 Chemical Name and CAS Registry Number

Dibasic calcium phosphate dihydrate [7789-77-7]

4 Empirical Formula and Molecular Weight

$CaHPO_4{\cdot}2H_2O$ 172.09

5 Structural Formula

See Section 4.

6 Functional Category

Tablet and capsule diluent.

7 Applications in Pharmaceutical Formulation or Technology

Dibasic calcium phosphate dihydrate is widely used in tablet formulations both as an excipient and as a source of calcium and phosphorus in nutritional supplements.[1–8] It is one of the more widely used materials, particularly in the nutritional/health food sectors. It is also used in pharmaceutical products because of its compaction properties, and the good flow properties of the coarse-grade material. The predominant deformation mechanism of dibasic calcium phosphate coarse-grade is brittle fracture and this reduces the strain-rate sensitivity of the material, thus allowing easier transition from the laboratory to production scale. However, dibasic calcium phosphate dihydrate is abrasive and a lubricant is required for tableting, for example about 1% w/w of magnesium stearate or about 1% w/w of sodium stearyl fumarate is commonly used.

Two main particle-size grades of dibasic calcium phosphate dihydrate are used in the pharmaceutical industry. The milled material is typically used in wet-granulated, roller-compacted or slugged formulations. The 'unmilled' or coarse-grade material is typically used in direct-compression formulations.

Dibasic calcium phosphate dihydrate is nonhygroscopic and stable at room temperature. However, under certain conditions of temperature and humidity, it can lose water of crystallization below 100°C. This has implications for certain types of packaging and aqueous film coating since the loss of water of crystallization appears to be initiated by high humidity and by implication high moisture vapor concentrations in the vicinity of the dibasic calcium phosphate dihydrate particles.[8]

Dibasic calcium phosphate dihydrate is also used in toothpaste and dentifrice formulations for its abrasive properties.

8 Description

Dibasic calcium phosphate dihydrate is a white, odorless, tasteless powder or crystalline solid. It occurs as monoclinic crystals.

9 Pharmacopeial Specifications

The pharmacopeial specifications for dibasic calcium phosphate dihydrate have undergone harmonization of many attributes for JP, PhEur, and USP–NF.

See Table I. *See also* Section 18.

SEM 1: Excipient: dibasic calcium phosphate dihydrate, coarse grade; manufacturer: JRS Pharma LP; lot no.: W28C; magnification: 100×.

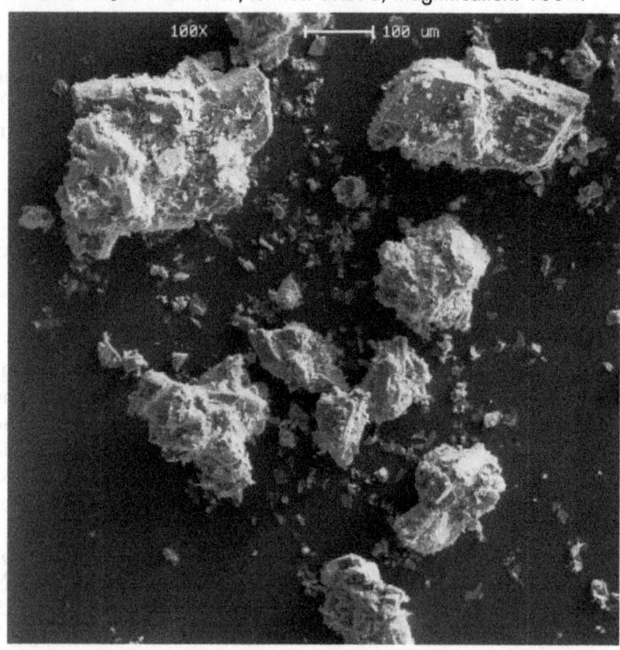

SEM 2: Excipient: dibasic calcium phosphate dihydrate, coarse grade; manufacturer: JRS Pharma LP; lot no.: W28C; magnification: 300×.

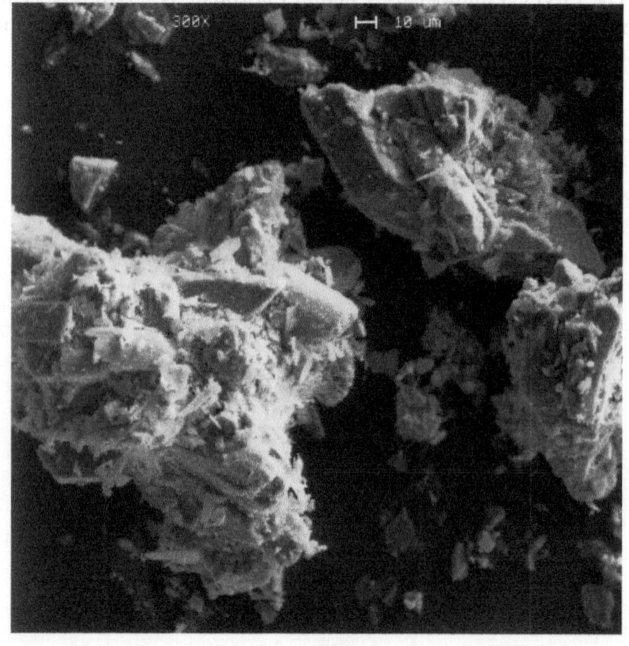

SEM 3: Excipient: dibasic calcium phosphate dihydrate; manufacturer: Innophos; lot no.: 16A-1 (89); magnification: 120×.

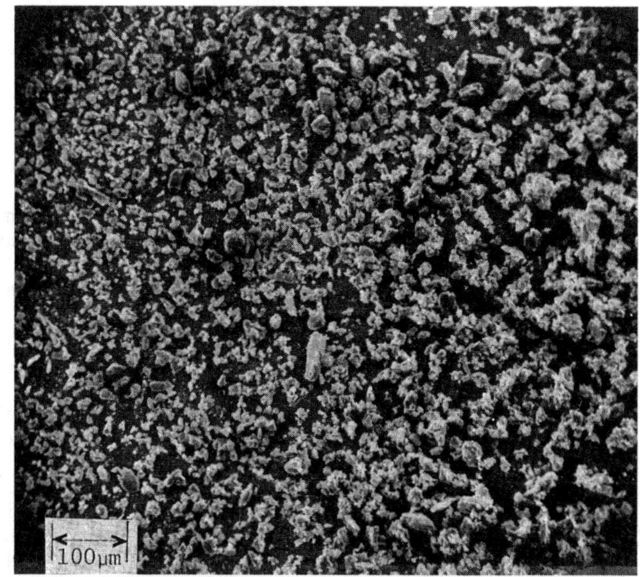

SEM 4: Excipient: dibasic calcium phosphate dihydrate, coarse grade; manufacturer: Innophos; lot no.: 16A-1 (89); magnification: 600×.

Table I: Pharmacopeial specifications for calcium phosphate, dibasic dihydrate.

Test	JP XV S1	PhEur 7.4	US35–NF30
Identification	+	+	+
Characters	+	+	—
Loss on ignition	24.5–26.5%	24.5–26.5%	24.5–26.5%
Acid-insoluble substances	≤0.2%	≤0.2%	≤0.2%
Heavy metals[a]	≤31 ppm	≤40 ppm	≤30 ppm
Chloride	≤0.25%	≤0.25%	≤0.25%
Fluoride	—	≤100 ppm	≤50 ppm
Sulfate	0.5%	≤0.5%	≤0.5%
Carbonate	+	+	+
Barium	+	+	+
Arsenic	≤2 ppm	≤10 ppm	≤3 μg/g
Iron	—	≤400 ppm	—
Assay	98.0–105.0%	98.0–105.0%	98.0–105.0%

(a) These tests have not been fully harmonized at the time of publication.

10 Typical Properties

Acidity/alkalinity pH = 7.4 (20% slurry of *DI-TAB*)
Angle of repose 28.3° for *Emcompress.*[9]
Density (bulk) 0.915 g/cm^3
Density (tapped) 1.17 g/cm^3
Density (true) 2.389 g/cm^3
Flowability 27.3 g/s for *DI-TAB*;11.4 g/s for *Emcompress.*[9]
Melting point Dehydrates below 100°C.
Moisture content Dibasic calcium phosphate dihydrate contains two molecules of water of crystallization, which can be lost at temperatures well below 100°C.
Particle size distribution *DI-TAB*: average particle diameter 180 μm; fine powder: average particle diameter 9 μm.
Solubility Practically insoluble in ethanol, ether, and water; soluble in dilute acids.
Specific surface area 0.44–0.46 m^2/g for *Emcompress.*
Spectroscopy

IR spectra *see* Figure 1.

NIR spectra *see* Figure 2.

11 Stability and Storage Conditions

Dibasic calcium phosphate dihydrate is a nonhygroscopic, relatively stable material. However, under certain conditions the dihydrate can lose water of crystallization. This has implications for both storage of the bulk material and coating and packaging of tablets containing dibasic calcium phosphate dihydrate.

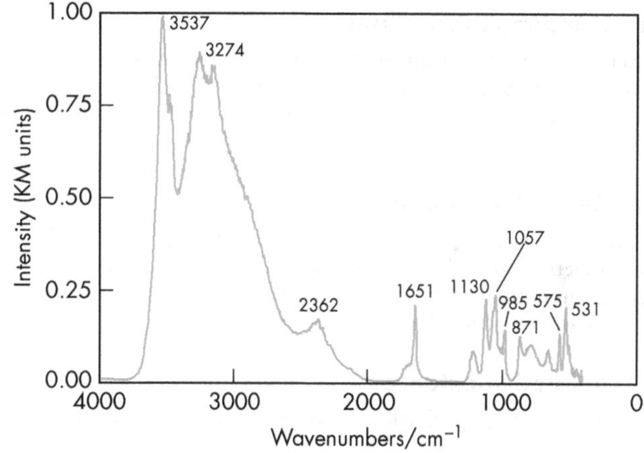

Figure 1: Infrared spectrum of dibasic calcium phosphate dihydrate measured by diffuse reflectance. Adapted with permission of Informa Healthcare.

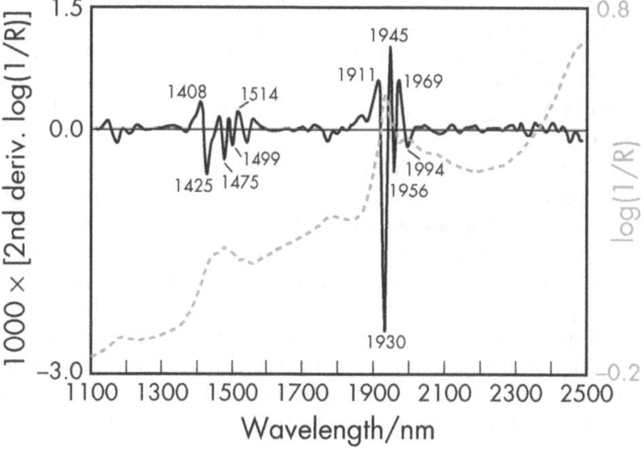

Figure 2: Near-infrared spectrum of dibasic calcium phosphate dihydrate measured by reflectance.

The bulk material should be stored in a well-closed container in a cool, dry place.

12 Incompatibilities

Dibasic calcium phosphate dihydrate should not be used to formulate tetracycline antibiotics.[10] Dibasic calcium phosphate dihydrate has been reported to be incompatible with indomethacin,[11] aspirin,[12] aspartame,[13] ampicillin,[14] cephalexin,[15] and erythromycin.[16] The surface of dibasic calcium phosphate dihydrate is alkaline[16] and consequently it should not be used with drugs that are sensitive to alkaline pH.

13 Method of Manufacture

Calcium phosphates are usually manufactured by reacting very pure phosphoric acid with calcium hydroxide, $Ca(OH)_2$ obtained from limestone, in stoichiometric ratio in aqueous suspension followed by drying at a temperature that will allow the correct hydration state to be achieved. After drying, the coarse-grade material is obtained by means of a classification unit; the fine particle-size material is obtained by milling.

14 Safety

Dibasic calcium phosphate dihydrate is widely used in oral pharmaceutical products, food products, and toothpastes, and is generally regarded as a nontoxic and nonirritant material. However, oral ingestion of large quantities may cause abdominal discomfort.

15 Handling Precautions

Observe normal precautions appropriate to the circumstances and quantity of material handled. The fine-milled grades can generate nuisance dusts and the use of a respirator or dust mask may be necessary.

16 Regulatory Status

GRAS listed. Accepted as a food additive in Europe. Included in the FDA Inactive Ingredients Database (oral capsules and tablets). Included in nonparenteral medicines licensed in Europe. Included in the Canadian Natural Health Products Ingredients Database.

17 Related Substances

Calcium phosphate, dibasic anhydrous; calcium phosphate, tribasic.

18 Comments

Dibasic calcium phosphate dihydrate has undergone harmonization of many attributes for JP, PhEur, and USP–NF by the Pharmacopeial Discussion Group. For further information see the General Information Chapter <1196> in the USP35–NF30, the General Chapter 5.8 in PhEur 7.4, along with the 'State of Work' document on the PhEur EDQM website, and also the General Information Chapter 8 in the JP XV.

Grades of dibasic calcium phosphate dihydrate available for direct compression include *Di-Cafos* (Chemische Fabrik Budenheim), *DI-TAB* (Innophos), and *Emcompress* (JRS Pharma LP).

Accelerated stability studies carried out at elevated temperatures on formulations containing significant proportions of dibasic calcium phosphate dihydrate can give erroneous results owing to irreversible dehydration of the dihydrate to the anhydrous form. Depending on the type of packaging and whether or not the tablet is coated, the phenomenon can be observed at temperatures as low as 40°C after 6 weeks of storage. As the amount of dibasic calcium phosphate dihydrate in the tablet is reduced, the effect is less easy to observe.

The EINECS number for calcium phosphate is 231-837-1. The PubChem Compound ID (CID) for dibasic calcium phosphate dihydrate is 104805.

19 Specific References

1 Lausier JM, *et al.* Aging of tablets made with dibasic calcium phosphate dihydrate as matrix. *J Pharm Sci* 1977; **66**(11): 1636–1637.
2 Carstensen JT, Ertell C. Physical and chemical properties of calcium phosphates for solid state pharmaceutical formulations. *Drug Dev Ind Pharm* 1990; **16**(7): 1121–1133.
3 Bryan JW, McCallister JD. Matrix forming capabilities of three calcium diluents. *Drug Dev Ind Pharm* 1992; **18**(19): 2029–2047.
4 Schmidt PC, Herzog R. Calcium phosphates in pharmaceutical tableting I: physico-pharmaceutical properties. *Pharm World Sci* 1993; **15**(3): 105–115.
5 Schmidt PC, Herzog R. Calcium phosphates in pharmaceutical tableting II: comparison of tableting properties. *Pharm World Sci* 1993; **15**(3): 116–122.
6 Landín M, *et al.* The effect of country of origin on the properties of dicalcium phosphate dihydrate powder. *Int J Pharm* 1994; **103**: 9–18.
7 Landín M, *et al.* Dicalcium phosphate dihydrate for direct compression: characterization and intermanufacturer variability. *Int J Pharm* 1994; **109**: 1–8.
8 Landín M, *et al.* Structural changes during the dehydration of dicalcium phosphate dihydrate. *Eur J Pharm Sci* 1994; **2**: 245–252.
9 Çelik M, Okutgen E. A feasibility study for the development of a prospective compaction functionality test and the establishment of a compaction data bank. *Drug Dev Ind Pharm* 1993; **19**(17–18): 2309–2334.
10 Weiner M, Bernstein IL. *Adverse Reactions to Drug Formulation Agents: A Handbook of Excipients.* New York: Marcel Dekker, 1989: 93–94.
11 Eerikäinen S, *et al.* The behaviour of the sodium salt of indomethacin in the cores of film-coated granules containing various fillers. *Int J Pharm* 1991; **71**: 201–211.
12 Landín M, *et al.* Chemical stability of acetyl salicylic acid in tablets prepared with different commercial brands of dicalcium phosphate dihydrate. *Int J Pharm* 1994; **107**: 247–249.
13 El-Shattawy HH, *et al.* Aspartame direct compression excipients: preformulation stability screening using differential scanning calorimetry. *Drug Dev Ind Pharm* 1981; **7**(5): 605–619.
14 El-Shattaway HH. Ampicillin direct compression excipients: preformulation stability screening using differential scanning calorimetry. *Drug Dev Ind Pharm* 1982; **8**(6): 819–831.
15 El-Shattaway HH, *et al.* Cephalexin I direct compression excipients: preformulation stability screening using differential scanning calorimetry. *Drug Dev Ind Pharm* 1982; **8**(6): 897–909.
16 El-Shattaway HH, *et al.* Erythromycin direct compression excipients: preformulation stability screening using differential scanning calorimetry. *Drug Dev Ind Pharm* 1982; **8**(6): 937–947.

20 General References

European Directorate for the Quality of Medicines and Healthcare (EDQM). European Pharmacopoeia – State Of Work Of International Harmonisation. *Pharmeuropa* 2011; 23(4): 713–714. www.edqm.eu/site/-614.html (accessed 5 December 2011).

Green CE, *et al.* R-P trials calcium excipient. *Manuf Chem* 1996; **67**(8): 55, 57.

Innophos Inc. Product data sheet: Calcium Phosphates, 2008.

21 Author

RC Moreton.

22 Date of Revision

1 March 2012.

Calcium Phosphate, Tribasic

1 Nonproprietary Names

BP: Calcium Phosphate
PhEur: Calcium Phosphate
USP–NF: Tribasic Calcium Phosphate

2 Synonyms

Calcium orthophosphate; E341(iii); hydroxylapatite; phosphoric acid calcium salt (2 : 3); precipitated calcium phosphate; tertiary calcium phosphate; *Tri-Cafos*; tricalcii phosphas; tricalcium diorthophosphate; tricalcium orthophosphate; tricalcium phosphate; *TRI-CAL WG*; *TRI-TAB*.

3 Chemical Name and CAS Registry Number

Tribasic calcium phosphate is not a clearly defined chemical entity but is a mixture of calcium phosphates. Several chemical names, CAS Registry Numbers, and molecular formulas have therefore been used to describe this material. Those most frequently cited are shown below.

Calcium hydroxide phosphate [12167-74-7]

Tricalcium orthophosphate [7758-87-4]

See also Sections 4 and 8.

4 Empirical Formula and Molecular Weight

$Ca_3(PO_4)_2$	310.20
$Ca_5(OH)(PO_4)_3$	502.32

5 Structural Formula

See Sections 3 and 4.

6 Functional Category

Anticaking agent; buffering agent; tablet and capsule diluent.

7 Applications in Pharmaceutical Formulation or Technology

Tribasic calcium phosphate is widely used as a capsule diluent and tablet filler/binder in either direct-compression or wet-granulation processes. The primary bonding mechanism in compaction is plastic deformation. As with dibasic calcium phosphate, a lubricant and a disintegrant should usually be incorporated in capsule or tablet formulations that include tribasic calcium phosphate. In some cases tribasic calcium phosphate has been used as a disintegrant.(1) It is most widely used in vitamin and mineral preparations(2) as a filler and as a binder.

In food applications, tribasic calcium phosphate powder is widely used as an anticaking agent. *See* Section 18.

See also Calcium phosphate, dibasic dihydrate.

8 Description

The PhEur 7.4 states that tribasic calcium phosphate consists of a mixture of calcium phosphates. It contains not less than 35.0% and not more than the equivalent of 40.0% of calcium. The USP35–NF30 specifies that tribasic calcium phosphate consists of variable mixtures of calcium phosphates having the approximate composition $10CaO{\cdot}3P_2O_5{\cdot}H_2O$. This corresponds to a molecular formula of $Ca_5(OH)(PO_4)_3$ or $Ca_{10}(OH)_2(PO_4)_6$.

Tribasic calcium phosphate is a white, odorless and tasteless powder.

9 Pharmacopeial Specifications

See Table I.

Table I: Pharmacopeial specifications for tribasic calcium phosphate.

Test	PhEur 7.4	USP35–NF30
Identification	+	+
Characters	+	—
Loss on ignition	≤8.0%	≤8.0%
Water-soluble substances	—	≤0.5%
Acid-insoluble substances	≤0.2%	≤0.2%
Carbonate	—	+
Chloride	≤0.15%	≤0.14%
Fluoride	≤75 ppm	≤75 ppm
Nitrate	—	+
Sulfate	≤0.5%	≤0.8%
Arsenic	≤4 ppm	≤3 ppm
Barium	—	+
Iron	≤400 ppm	—
Dibasic salt and calcium oxide	—	+
Heavy metals	≤30 ppm	≤30 ppm
Assay (as Ca)	35.0–40.0%	34.0–40.0%

10 Typical Properties

Acidity/alkalinity pH = 6.8 (20% slurry in water)

Density 3.14 g/cm^3

Density (bulk)

0.3–0.4 g/cm^3 for powder form;

0.80 g/cm^3 for granular *TRI-TAB*.(3)

Density (tapped) 0.95 g/cm^3 for granular *TRI-TAB*.(3)

Flowability 25.0 g/s for granular *TRI-TAB*.(3)

Melting point 1670°C

Moisture content Slightly hygroscopic. A well-defined crystalline hydrate is not formed, although surface moisture may be picked up or contained within small pores in the crystal structure. At relative humidities between about 15% and 65%, the equilibrium moisture content at 25°C is about 2.0%. At relative humidities above about 75%, tribasic calcium phosphate may absorb small amounts of moisture.

Particle size distribution

Tribasic calcium phosphate powder: typical particle diameter 5–10 μm; 98% of particles <44 μm.

TRI-CAL WG: average particle diameter 180 μm; 99% of particles <420 μm, 46% <149 μm, and 15% <44 μm.

TRI-TAB: average particle diameter 350 μm; 97% of particles <420 μm, and 2% <149 μm.

Solubility Soluble in dilute mineral acids; very slightly soluble in water; practically insoluble in acetic acid and alcohols.

Specific surface area 70–80 m^2/g(4)

Spectroscopy

IR spectra *see* Figure 1.

NIR spectra *see* Figure 2.

11 Stability and Storage Conditions

Tribasic calcium phosphate is a chemically stable material, and is also not liable to cake during storage.

The bulk material should be stored in a well-closed container in a cool, dry place.

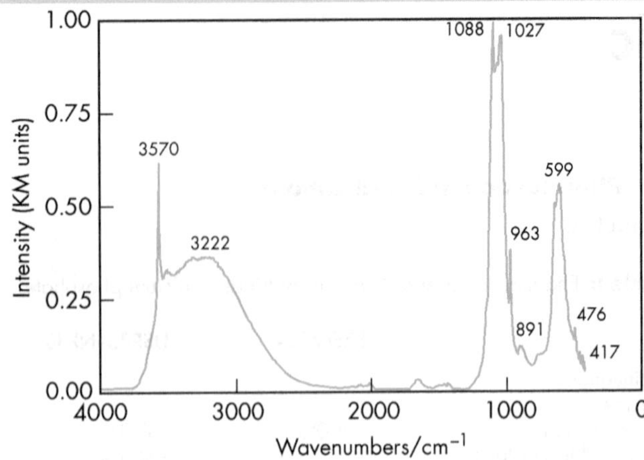

Figure 1: Infrared spectrum of tribasic calcium phosphate measured by diffuse reflectance. Adapted with permission of Informa Healthcare.

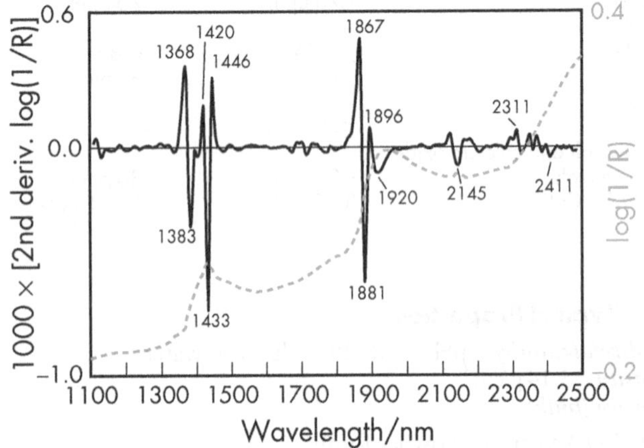

Figure 2: Near-infrared spectrum of tribasic calcium phosphate ($Ca_5(OH)(PO_4)_3$) measured by reflectance.

12 Incompatibilities

All calcium salts are incompatible with tetracycline antibiotics. Tribasic calcium phosphate is incompatible with tocopheryl acetate (but not tocopheryl succinate). Tribasic calcium phosphate may form sparingly soluble phosphates with hormones.

13 Method of Manufacture

Tribasic calcium phosphate occurs naturally as the minerals hydroxylapatite, voelicherite, and whitlockite. Commercially, it is prepared by treating phosphate-containing rock with sulfuric acid. Tribasic calcium phosphate powder is then precipitated by the addition of calcium hydroxide. Tribasic calcium phosphate is alternatively prepared by treating calcium hydroxide from limestone with purified phosphoric acid. It may also be obtained from calcined animal bones.[4] Some tribasic calcium phosphate products may be prepared in coarser, directly compressible forms by granulating the powder using roller compaction or spray drying.

14 Safety

Tribasic calcium phosphate is widely used in oral pharmaceutical formulations and food products, and is generally regarded as nontoxic and nonirritant at the levels employed as a pharmaceutical excipient.

Ingestion or inhalation of excessive quantities may result in the deposition of tribasic calcium phosphate crystals in tissues. These crystals may lead to inflammation and cause tissue lesions in the areas of deposition.

Oral ingestion of very large quantities of tribasic calcium phosphate may cause abdominal discomfort such as nausea and vomiting.

No teratogenic effects were found in chicken embryos exposed to a dose of 2.5 mg of tribasic calcium phosphate.[5]

LD_{50} (rat, oral): >1 g/kg[3]

15 Handling Precautions

Observe normal precautions appropriate to the circumstances and quantity of material handled. Eye protection and gloves are recommended. Handle in a well-ventilated environment since dust inhalation may be an irritant. For processes generating large amounts of dust, the use of a respirator is recommended.

16 Regulatory Acceptance

GRAS listed. Accepted for use as a food additive in Europe. Included in the FDA Inactive Ingredients Database (oral capsules and tablets). Included in nonparenteral medicines licensed in the UK. Included in the Canadian Natural Health Products Ingredients Database.

17 Related Substances

Calcium phosphate, dibasic anhydrous; calcium phosphate, dibasic dihydrate.

18 Comments

One gram of tribasic calcium phosphate represents approximately 10.9 mmol of calcium and 6.4 mmol of phosphate; 38% calcium and 17.3% phosphorus by weight.[3] Tribasic calcium phosphate provides a higher calcium load than dibasic calcium phosphate and a higher Ca/P ratio. Granular and fine powder forms of tribasic calcium phosphate are available from various manufacturers.

A specification for calcium phosphate tribasic is contained in the *Food Chemicals Codex* (FCC).[6]

The EINECS number for calcium phosphate is 231-837-1. The PubChem Compound ID (CID) for tribasic calcium phosphate includes 24456 and 516943.

Tribasic calcium phosphate is a source of both calcium and phosphorus, the two main osteogenic minerals for bone health. The bioavailability of the calcium is well known to be improved by the presence of cholecalciferol. Recent research reports that combinations of tribasic calcium phosphate and vitamin D3 are a cost-effective advance in bone fracture prevention.[7]

19 Specific References

1 Delonca H, *et al.* [Effect of excipients and storage conditions on drug stability I: acetylsalicylic acid-based tablets.] *J Pharm Belg* 1969; **24**: 243–252[in French].
2 Magid L. Stable multivitamin tablets containing tricalcium phosphate. United States Patent No. 3,564,097; 1971.
3 Rhodia. Technical literature: Calcium phosphate pharmaceutical ingredients, 1995.
4 Magami A. Basic pentacalcium triphosphate production. Japanese Patent 56 022 614; 1981.
5 Verrett MJ, *et al.* Toxicity and teratogenicity of food additive chemicals in the developing chicken embryo. *Toxicol Appl Pharmacol* 1980; **56**: 265–273.
6 *Food Chemicals Codex*, 7th edn (Suppl. 1). Bethesda, MD: United States Pharmacopeia, 2010: 150.
7 Lilliu H, *et al.* Calcium-vitamin D3 supplementation is cost-effective in hip fractures prevention. *Muturitas* 2003; **44**(4): 299–305.

20 General References

Bryan JW, McCallister JD. Matrix forming capabilities of three calcium diluents. *Drug Dev Ind Pharm* 1992; **18**: 2029–2047.

Chowhan ZT, Amaro AA. The effect of low- and high-humidity aging on the hardness, disintegration time and dissolution rate of tribasic calcium phosphate-based tablets. *Drug Dev Ind Pharm* 1979; **5**: 545–562.

Fischer E. Calcium phosphate as a pharmaceutical excipient. *Manuf Chem* 1992; **64**(6): 25–27.

Innophos Inc. Material safety data sheet: *TRI-TAB*, 2007.

Kutty TRN. Thermal decomposition of hydroxylapatite. *Indian J Chem* 1973; **11**: 695–697.

Molokhia AM, *et al*. Effect of storage conditions on the hardness, disintegration and drug release from some tablet bases. *Drug Dev Ind Pharm* 1982; **8**: 283–292.

Pontier C, Viana M. Energetic yields in apatitic calcium phosphate compression: influence of the Ca/P molar ratio. *Polymer International* 2003; **52**(4): 625–628.

Schmidt PC, Herzog R. Calcium phosphates in pharmaceutical tableting 1: physico-pharmaceutical properties. *Pharm World Sci* 1993; **15**(3): 105–115.

Schmidt PC, Herzog R. Calcium phosphates in pharmaceutical tableting 2: comparison of tableting properties. *Pharm World Sci* 1993; **15**(3): 116–122.

21 Author

L Hendricks.

22 Date of Revision

1 March 2012.

Calcium Silicate

1 Nonproprietary Names

USP–NF: Calcium Silicate

2 Synonyms

Calcium hydrosilicate; calcium metasilicate; calcium monosilicate; calcium polysilicate; *Micro-Cel*; okenite; silicic acid, calcium salt; tobermorite.

3 Chemical Name and CAS Registry Number

Calcium silicate [1344-95-2]

4 Empirical Formula and Molecular Weight

$CaSiO_3$ 116.2

5 Structural Formula

See Section 4.

6 Functional Category

Adsorbent; anticaking agent; opacifier; tablet and capsule diluent.

7 Applications in Pharmaceutical Formulation or Technology

Calcium silicate is used as a filler aid for oral pharmaceuticals. It has also been used in pharmaceutical preparations as an antacid.

The main applications of calcium silicate relate to its anticaking properties, and it has therefore been used in dusting powders and a range of different cosmetic products (e.g. face powders, eye shadow).[1]

8 Description

Calcium silicate occurs as a crystalline or amorphous white or off-white material, and often exists in different hydrate forms.

9 Pharmacopeial Specifications

The USP35–NF30 describes the material as containing not less than 4% of calcium oxide and not less than 35% of silicon dioxide.

See Table I.

Table I: Pharmacopeial specifications for calcium silicate.

Test	USP35–NF30
Identification	+
pH	8.4–11.2
Loss on ignition	≤20%
Heavy metals	≤20 ppm
Fluoride	≤50 ppm
Lead	≤10 ppm
Assay for silicon dioxide	90–110%
Assay for calcium oxide	90–110%
Ratio of silicon dioxide to calcium oxide	0.5 : 20
Sum of calcium oxide, silicon dioxide and loss on ignition	≥90%

10 Typical Properties

Acidity/alkalinity pH = 8.4–10.2 (5% w/v aqueous solution)

Density 2.10 g/cm^3

Melting point 1540°C

Solubility Practically insoluble in water; forms a gel with mineral acids. It can absorb up to 2.5 times its weight of liquids and still remain a free-flowing powder.

11 Stability and Storage Conditions

Calcium silicate is chemically stable and nonflammable, but it should be protected from moisture. Store in airtight containers in a cool, dry place.

12 Incompatibilities

—

13 Method of Manufacture

Calcium silicate is a naturally occurring mineral, but for commercial applications it is usually prepared from lime and diatomaceous earth under carefully controlled conditions.

14 Safety

When used in oral formulations, calcium silicate is practically nontoxic. Inhalation of the dust particles may cause respiratory tract irritation.

15 Handling Precautions

Observe normal precautions appropriate to the circumstances and quantity of the material handled. In large quantities, calcium silicate is irritating to eyes, the respiratory system, and skin. Gloves, eye protection, a respirator, and other protective clothing should be worn. In the UK, the long-term (8-hour TWA) workplace exposure standards for calcium silicate are 10 mg/m^3 for total inhalable dust and 4 mg/m^3 for respirable dust.[2]

C

16 Regulatory Status

GRAS listed. Included in the FDA Inactive Ingredients Database (oral dosage forms). Included in nonparenteral (oral, orodispersible, effervescent and enteric-coated tablets) formulations licensed in the UK.

17 Related Substances

Calcium diorthosilicate; calcium trisilicate.

Calcium diorthosilicate
Empirical formula Ca_2SiO_4
Molecular weight 172.2
CAS number [10034-77-2]
Synonyms Belite; dicalcium silicate
Comments The EINECS number for calcium diorthosilicate is 233-107-8.

Calcium trisilicate
Empirical formula Ca_3SiO_5
Molecular weight 228.3
CAS number [12168-85-3]
Synonyms Tricalcium silicon pentaoxide
Comments The EINECS number for calcium trisilicate is 235-336-9.

18 Comments

Studies utilizing the porous properties of calcium silicate granules have shown their ability to form floating structures, giving rise to potentially gastroretentive formulations.[3,4]

A specification for calcium silicate is contained in the *Food Chemicals Codex* (FCC).[5]

The EINECS number for calcium silicate is 215-710-8. The PubChem Compound ID (CID) for calcium silicate is 518851.

19 Specific References

1 Gottschalck TE, McEwen GN, eds. *International Cosmetic Ingredient Dictionary and Handbook*, 11th edn. Washington, DC: The Cosmetic, Toiletry, and Fragrance Association, 2006; 325.
2 Health and Safety Executive. *EH40/2005: Workplace Exposure Limits*. Sudbury: HSE Books, 2011. http://www.hse.gov.uk/pubns/priced/eh40.pdf (accessed 29 February 2012).
3 Jain AK, *et al*. Controlled release calcium silicate based floating granular delivery system of ranitidine hydrochloride. *Curr Drug Deliv* 2006; **3**(4): 367–372.
4 Jain SK, *et al*. Calcium silicate based microspheres of repaglinide for gastroretentive floating drug delivery: preparation and in vitro characterization. *J Control Release* 2005; **107**(2): 300–309.
5 *Food Chemicals Codex*, 7th edn. Bethesda, MD: United States Pharmacopeia, 2010; 154.

20 General References

Sigma-Aldrich. *Material safety data sheet, version 1.2: Calcium silicate*, 13 March 2004.

21 Author

MA Pellett.

22 Date of Revision

1 March 2012.

Calcium Stearate

1 Nonproprietary Names

BP: Calcium Stearate
JP: Calcium Stearate
PhEur: Calcium Stearate
USP–NF: Calcium Stearate

2 Synonyms

Calcii stearas; calcium distearate; calcium octadecanoate; *Ceasit PC*; *Kemistab EC-F*; octadecanoic acid, calcium salt; stearic acid, calcium salt; *Synpro*.

3 Chemical Name and CAS Registry Number

Octadecanoic acid calcium salt [1592-23-0]

4 Empirical Formula and Molecular Weight

$C_{36}H_{70}CaO_4$ 607.03 (for pure material)

The USP35–NF30 describes calcium stearate as a compound of calcium with a mixture of solid organic acids obtained from fats, and consists chiefly of variable proportions of calcium stearate and calcium palmitate. It contains the equivalent of 9.0–10.5% of calcium oxide.

The PhEur 7.4 describes calcium stearate as a mixture of calcium salts of different fatty acids consisting mainly of stearic acid [$(C_{17}H_{35}COO)_2Ca$] and palmitic acid [$(C_{15}H_{31}COO)_2Ca$] with minor proportions of other fatty acids. It contains a minimum of 40.0% stearic acid in the fatty acid fraction. The sum of stearic acid and palmitic acid in the fatty acid fraction is a minimum of 90.0% It contains the equivalent of 9.0–10.5% of calcium oxide.

5 Structural Formula

$$\left[H-CH_2-(CH_2)_{16}-C(=O)-O^- \right]_2 \; Ca^{2+}$$

6 Functional Category

Tablet and capsule lubricant.

7 Applications in Pharmaceutical Formulation or Technology

Calcium stearate is primarily used in pharmaceutical formulations as a lubricant in tablet and capsule manufacture at concentrations up to 1.0% w/w. Although it has good antiadherent and lubricant properties, calcium stearate has poor glidant properties.

Calcium stearate is employed as an emulsifier, stabilizing agent, and suspending agent, used in cosmetics and food products.

8 Description

Calcium stearate occurs as a fine, white to yellowish-white, bulky powder having a slight, characteristic odor. It is unctuous and free from grittiness.

SEM 1: Excipient: calcium stearate (standard); manufacturer: Durham Chemicals; lot no.: 0364; voltage: 20 kV.

SEM 2: Excipient: calcium stearate (precipitated); manufacturer: Witco Corporation; lot no.: 0438; voltage: 12 kV.

SEM 3: Excipient: calcium stearate (fused); manufacturer: Witco Corporation; voltage: 15 kV.

9 Pharmacopeial Specifications

See Table I.

Table I: Pharmacopeial specifications for calcium stearate.

Test	JP XV	PhEur 7.4	USP35–NF30
Identification	+	+	+
Characters	+	+	–
Microbial limit	–	10^3 cfu/g	–
Acidity or alkalinity	–	+	–
Loss on drying	≤4.0%	≤6.0%	≤4.0%
Arsenic	≤2 ppm	–	–
Heavy metals	≤20 ppm	–	≤10 ppm
Chlorides	–	≤0.1%	–
Sulfates	–	≤0.3%	–
Cadmium	–	≤3 ppm	–
Lead	–	≤10 ppm	–
Nickel	–	≤5 ppm	–
Assay (as CaO)	–	–	9.0–10.5%
Assay (as Ca)	6.4–7.1%	6.4–7.4%	–

10 Typical Properties

Acid value 191–203

Ash

9.9–10.3%;

9.2% for *Synpro*.

Chloride <200 ppm

Density (bulk and tapped) *see* Table II.

Density (true)

1.064–1.096 g/cm^3;

1.03 g/cm^3 for *Kemistab EC-F*.

Flash point 176°C

Flowability 21.2–22.6% (Carr compressibility index)

Free fatty acid

0.3–0.5%;

0.3% for *Synpro*.

Melting point

147–160°C;

130–156°C for *Kemistab EC-F*;

155°C for *Synpro*.

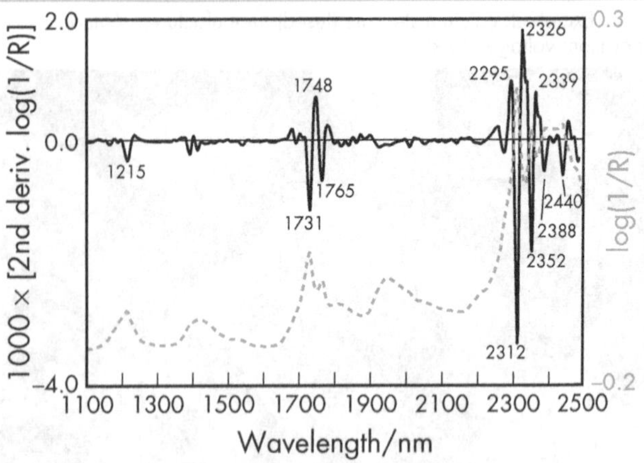

Figure 1: Near-infrared spectrum of calcium stearate measured by reflectance.

Table II: Density (bulk and tapped) of calcium stearate.

Manufacturer/Grade	Bulk density (g/cm^3)	Tapped density (g/cm^3)
Durham Chemicals		
Standard	–	0.26
A	–	0.45
AM	–	0.33
Ferro Corporation		
Synpro	0.2	–
Witco Corporation		
EA	0.21	0.27
Fused	0.38	0.48
Precipitated	0.16	0.20
Undesa		
Kemistab EC-F	0.3	–

Moisture content
<3.5%;
2.7% for *Synpro*.
Particle size distribution 1.7–60 μm; 100% through a 73.7 μm (#200 mesh); 99.5% through a 44.5 μm (#325 mesh).
Shear strength 14.71 MPa
Solubility Practically insoluble or insoluble in ethanol (95%), ether, chloroform, acetone, and water. Slightly soluble in hot alcohol, and hot vegetable and mineral oils. Soluble in hot pyridine.
Specific surface area 4.73–8.03 m^2/g
Spectroscopy
NIR spectra *see* Figure 1.
Sulfate <0.25%

11 Stability and Storage Conditions

Calcium stearate is stable and should be stored in a well-closed container in a cool, dry place.

12 Incompatibilities

Calcium stearate is incompatible with strong oxidizing agents, mineral acids, alkalis, organic acids, and amines of low molecular weight.

13 Method of Manufacture

Calcium stearate is prepared by the reaction of calcium chloride with a mixture of the sodium salts of stearic and palmitic acids. The calcium stearate formed is collected and washed with water to remove any sodium chloride.

14 Safety

Calcium stearate is used in oral pharmaceutical formulations and is generally regarded as a relatively nontoxic and nonirritant material.

15 Handling Precautions

Observe normal precautions appropriate to the circumstances and quantity of material handled. Calcium stearate should be used in a well-ventilated environment; eye protection, gloves, and a respirator are recommended.

16 Regulatory Status

GRAS listed. Included in the FDA Inactive Ingredients Database (oral capsules and tablets). Included in nonparenteral medicines licensed in the UK. Included in the Canadian Natural Health Products Ingredients Database.

17 Related Substances

Magnesium stearate; stearic acid; zinc stearate.

18 Comments

Calcium stearate exhibits interesting properties when heated: softening between 120–130°C, and exhibiting a viscous consistency at approximately 160°C. At approximately 100°C, it loses about 3% of its weight, corresponding to 1 mole of water of crystallization. The crystalline structure changes at this point, leading to the collapse of the crystal lattice at a temperature of about 125°C.[1]

Calcium stearate was studied as a component of a 'cushioning pellet' during compression of enteric-coated pellets to protect the enteric coating. The cushioning pellets were composed of stearic acid/microcrystalline cellulose (4 : 1 w/w) and were successful in avoiding rupture of the enteric coating during the compression process.[2]

Calcium stearate was hot-melt extruded with testosterone in a study to characterize testosterone solid lipid microparticles to be applied as a transdermal delivery system. The results showed good release of the drug from the matrix.[3]

Calcium stearate has been investigated in the preparation and evaluation of novel directly compressed fast-disintegrating furosemide tablets.[4] It has also been studied as a component of nanostructured lipid carriers containing minoxidil for topical formulations.[5]

See Magnesium Stearate for further information and references.

A specification for calcium stearate is contained in the *Food Chemicals Codex* (FCC).[6]

The EINECS number for calcium stearate is 216-472-8. The PubChem Compound ID (CID) for calcium stearate is 15324.

19 Specific References

1 Baerlocher (2010). *Calcium stearates*. http://www.baerlocher.com/en/products/product-groups/metallic-stearates/calcium-stearates/ (accessed 29 February 2012).
2 Qi XL, *et al*. Preparation of tablets containing enteric-coated diclofenac sodium pellets. *Yao Xue Xue Bao* 2008; **43**(1): 97–101.
3 El-Kamel AH. Testosterone solid lipid microparticles for transdermal drug delivery. Formulation and physicochemical characterization. *J Microencapsul* 2007; **24**(5): 457–475.
4 Koseki T, *et al*. Preparation and evaluation of novel directly-compressed fast-disintegrating furosemide tablets with sucrose stearic acid ester. *Biol Pharm Bull* 2009; **32**(6): 1126–1130.
5 Silva AC, *et al*. Minoxidil-loaded nanostructured lipid carriers (NLC): characterization and rheological behaviour of topical formulations. *Pharmazie* 2009; **64**(3): 177–182.
6 *Food Chemicals Codex*, 7th edn. Bethesda, MD: United States Pharmacopeia, 2010; 159.

20 General References

Büsch G, Neuwald F. [Metallic soaps as water-in-oil emulsifiers]. *J Soc Cosmet Chem* 1973; **24**: 763–769 [in German].

Phadke DS, Sack MJ. Evaluation of batch-to-batch and manufacturer-to-manufacturer variability in the physical and lubricant properties of calcium stearate. *Pharm Technol* 1996; 20(Mar): 126–140.

21 Authors

LV Allen, Jr.

22 Date of Revision

1 March 2012.

Calcium Sulfate

1 Nonproprietary Names

BP: Calcium Sulphate Dihydrate

PhEur: Calcium Sulfate Dihydrate

USP–NF: Calcium Sulfate

2 Synonyms

Calcium sulfate anhydrous anhydrite; anhydrous gypsum; anhydrous sulfate of lime; *Destab*; *Drierite*; E516; karstenite; muriacite; *Snow White*.

Calcium sulfate dihydrate alabaster; calcii sulfas dihydricus; *Cal-Tab*; *Compactrol*; *Destab*; E516; gypsum; light spar; mineral white; native calcium sulfate; precipitated calcium sulfate; satinite; satin spar; selenite; terra alba; *USG Terra Alba*.

3 Chemical Name and CAS Registry Number

Calcium sulfate [7778-18-9]

Calcium sulfate dihydrate [10101-41-4]

4 Empirical Formula and Molecular Weight

$CaSO_4$	136.14
$CaSO_4{\cdot}2H_2O$	172.17

5 Structural Formula

See Section 4.

6 Functional Category

Desiccant; tablet and capsule diluent.

7 Applications in Pharmaceutical Formulation or Technology

Calcium sulfate dihydrate is used in the formulation of tablets and capsules. In granular form it has good compaction properties and moderate disintegration properties.[1,2]

Anhydrous calcium sulfate is hygroscopic and is used as a desiccant. The uptake of water can cause the tablets to become very hard and to fail to disintegrate on storage. Therefore, anhydrous calcium sulfate is not recommended for the formulation of tablets, capsules, or powders for oral administration.

8 Description

Both calcium sulfate and calcium sulfate dihydrate are white or off-white, fine, odorless, and tasteless powder or granules.

9 Pharmacopeial Specifications

See Table I. *See also* Section 18.

Table I: Pharmacopeial specifications for calcium sulfate.

Test	PhEur 7.4	USP35–NF30
Identification	+	+
Characters	+	—
Acidity or alkalinity	+	—
Arsenic	≤10 ppm	—
Chlorides	≤300 ppm	—
Heavy metals	≤20 ppm	≤10 ppm
Iron	≤100 ppm	≤100 ppm
Loss on drying		
Anhydrous	—	≤1.5%
Dihydrate	—	19.0–23.0%
Loss on ignition	18.0–22.0%	—
Assay	98.0–102.0%	98.0–101.0%

10 Typical Properties

Acidity/alkalinity pH = 7.3 (10% slurry) for dihydrate; pH = 10.4 (10% slurry) for anhydrous material.

Angle of repose 37.6° for *Compactrol*.[2]

Compressibility *see* Figure 1.

Density (bulk) 0.94 g/cm³ for *Compactrol*;[2]
0.67 g/cm³ for dihydrate;
0.70 g/cm³ for anhydrous material.

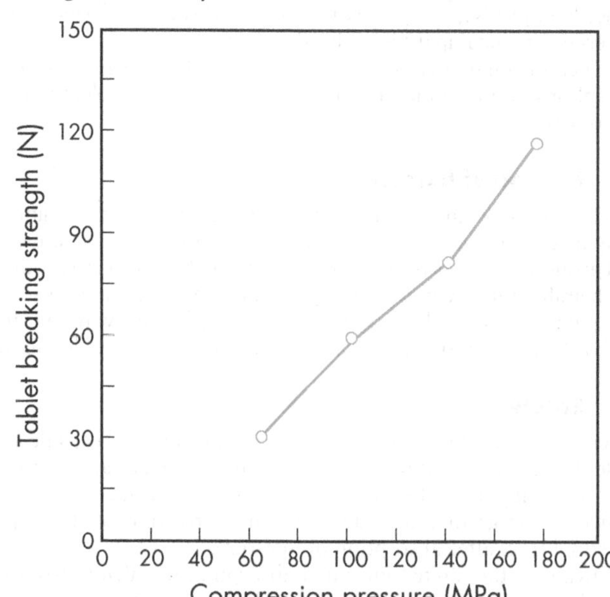

Figure 1: Compression characteristics of calcium sulfate dihydrate. Tablet weight: 700 mg.

Density (tapped) 1.10 g/cm³ for *Compactrol*;[2]
1.12 g/cm³ for dihydrate;
1.28 g/cm³ for anhydrous material.
Density (true) 2.308 g/cm³
Flowability 48.4% (Carr compressibility index);
5.2 g/s for *Compactrol*.[2]
Melting point 1450°C for anhydrous material.
Particle size distribution 93% less than 45 μm in size for the dihydrate (*USG Terra Alba*); 97% less than 45 μm in size for the anhydrous material (*Snow White*). Average particle size is 17 μm for the dihydrate and 8 μm for the anhydrous material. For *Compactrol*, not less than 98% passes through a #40 screen (425 μm), and not less than 85% is retained in a #140 screen (100 μm).
Solubility *see* Table II.

Table II: Solubility of calcium sulfate dihydrate.

Solvent	Solubility at 20°C unless otherwise stated
Ethanol (95%)	Practically insoluble
Water	1 in 375
	1 in 485 at 100°C

Specific gravity 2.32 for dihydrate;
2.96 for anhydrous material.
Specific surface area 3.15 m²/g (Strohlein apparatus)
Spectroscopy IR spectra *see* Figures 2 and 3.
NIR spectra *see* Figure 4.

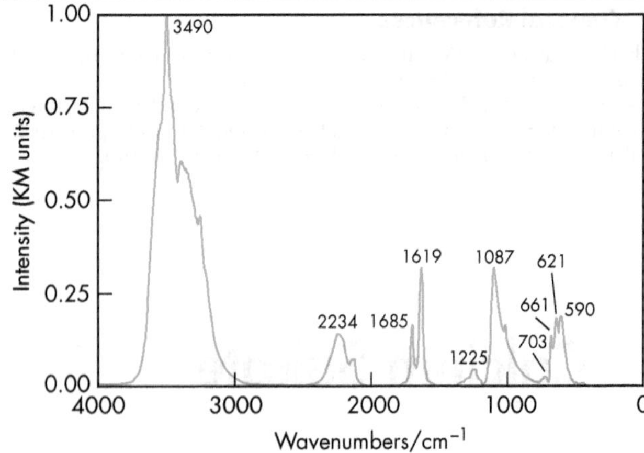

Figure 2: Infrared spectrum of anhydrous calcium sulfate measured by diffuse reflectance. Adapted with permission of Informa Healthcare.

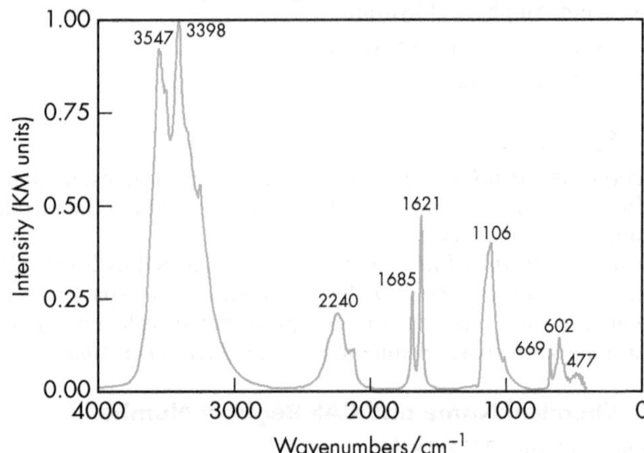

Figure 3: Infrared spectrum of calcium sulfate dihydrate measured by diffuse reflectance. Adapted with permission of Informa Healthcare.

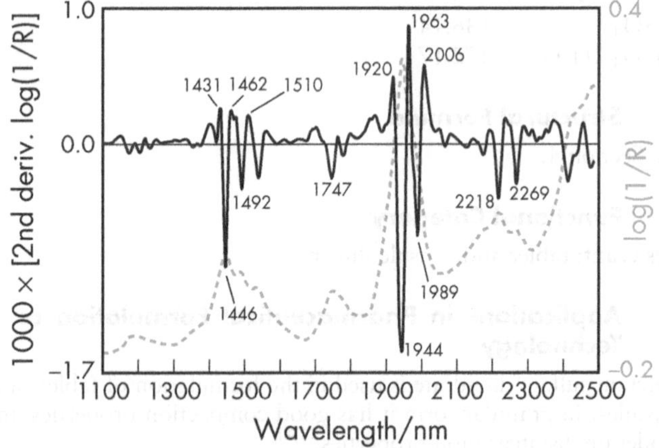

Figure 4: Near-infrared spectrum of calcium sulfate dihydrate measured by reflectance.

11 Stability and Storage Conditions

Calcium sulfate is chemically stable. Anhydrous calcium sulfate is hygroscopic and may cake on storage. Store in a well-closed container in a dry place, avoiding heat.

12 Incompatibilities

In the presence of moisture, calcium salts may be incompatible with amines, amino acids, peptides, and proteins, which may form complexes. Calcium salts will interfere with the bioavailability of tetracycline antibiotics.[3] It is also anticipated that calcium sulfate would be incompatible with indomethacin,[4] aspirin,[5] aspartame,[6] ampicillin,[7] cephalexin,[8] and erythromycin[9] since these materials are incompatible with other calcium salts.

Calcium sulfate may react violently, at high temperatures, with phosphorus and aluminum powder; it can react violently with diazomethane.

13 Method of Manufacture

Anhydrous calcium sulfate occurs naturally as the mineral anhydrite. The naturally occurring rock gypsum may be crushed and ground for use as the dihydrate or calcined at 150°C to produce the hemihydrate. A purer variety of calcium sulfate may also be obtained chemically by reacting calcium carbonate with sulfuric acid or by precipitation from calcium chloride and a soluble sulfate.

14 Safety

Calcium sulfate dihydrate is used as an excipient in oral capsule and tablet formulations. At the levels at which it is used as an excipient, it is generally regarded as nontoxic. However, ingestion of a sufficiently large quantity can result in obstruction of the upper intestinal tract after absorption of moisture.

Owing to the limited intestinal absorption of calcium from its salts, hypercalcemia cannot be induced even after the ingestion of massive oral doses.

Calcium salts are soluble in bronchial fluid. Pure salts do not induce pneumoconiosis.

15 Handling Precautions

Observe normal precautions appropriate to the circumstances and quantity of material handled. The fine-milled grades can generate nuisance dusts that may be irritant to the eyes or on inhalation. The use of a respirator or dust mask is recommended to prevent excessive powder inhalation since excessive inhalation may saturate the bronchial fluid, leading to precipitation and thus blockage of the air passages.

16 Regulatory Status

GRAS listed. Accepted for use as a food additive in Europe. Included in the FDA Inactive Ingredients Database (oral capsules, sustained release, tablets). Included in nonparenteral medicines licensed in the UK and Europe. Included in the Canadian Natural Health Products Ingredients Database.

17 Related Substances

Calcium phosphate, dibasic anhydrous; calcium phosphate, dibasic dihydrate; calcium phosphate, tribasic; calcium sulfate hemihydrate.

Calcium sulfate hemihydrate

Empirical formula $CaSO_4 \cdot \frac{1}{2}H_2O$
Molecular weight 145.14
CAS number [26499-65-0]
Synonyms annalin; calcii sulfas hemihydricus; calcined gypsum; dried calcium sulfate; dried gypsum; E516; exsiccated calcium sulfate; plaster of Paris; sulfate of lime; yeso blanco.
Appearance A white or almost white, odorless, crystalline, hygroscopic powder.
Solubility Practically insoluble in ethanol (95%); slightly soluble in water; more soluble in dilute mineral acids.
Comments The BP 2011 defines dried calcium sulfate as predominantly the hemihydrate, produced by drying powdered gypsum ($CaSO_4 \cdot 2H_2O$) at about 150°C, in a controlled manner, such that minimum quantities of the anhydrous material are produced. Dried calcium sulfate may also contain suitable setting accelerators or decelerators.

Calcium sulfate hemihydrate is used in the preparation of plaster of Paris bandage, which is used for the immobilization of limbs and fractures; it should not be used in the formulation of tablets or capsules.

18 Comments

Calcium sulfate will absorb moisture and therefore should be used with caution in the formulation of products containing drugs that easily decompose in the presence of moisture.

Therapeutically, calcium sulfate is used in dental and craniofacial surgical procedures.[10,11]

A specification for calcium sulfate is contained in the *Food Chemicals Codex* (FCC)[12] and the *Japanese Pharmaceutical Excipients*.[13]

The EINECS number for calcium sulfate is 231-900-3. The PubChem Compound ID (CID) for calcium sulfate is 24497.

19 Specific References

1 Bergman LA, Bandelin FJ. Effects of concentration, ageing and temperature on tablet disintegrants in a soluble direct compression system. *J Pharm Sci* 1965; **54**: 445–447.
2 Çelik M, Okutgen E. A feasibility study for the development of a prospective compaction functionality test and the establishment of a compaction data bank. *Drug Dev Ind Pharm* 1993; **19**: 2309–2334.
3 Weiner M, Bernstein IL. *Adverse Reactions to Drug Formulation Agents: A Handbook of Excipients.* New York: Marcel Dekker, 1989: 93–94.
4 Eerikäinen S, *et al.* The behaviour of the sodium salt of indomethacin in the cores of film-coated granules containing various fillers. *Int J Pharm* 1991; **71**: 201–211.
5 Landín M, *et al.* Chemical stability of acetylsalicylic acid in tablets prepared with different commercial brands of dicalcium phosphate dihydrate. *Int J Pharm* 1994; **107**: 247–249.
6 El-Shattawy HH, *et al.* Aspartame – direct compression excipients: preformulation stability screening using differential scanning calorimetry. *Drug Dev Ind Pharm* 1981; **7**(5): 605–619.
7 El-Shattawy HH. Ampicillin – direct compression excipients: preformulation stability screening using differential scanning calorimetry. *Drug Dev Ind Pharm* 1982; **8**(6): 819–831.
8 El-Shattawy HH, *et al.* Cephalexin 1 – direct compression excipients: preformulation stability screening using differential scanning calorimetry. *Drug Dev Ind Pharm* 1982; **8**(6): 897–909.
9 El-Shattawy HH, *et al.* Erythromycin – direct compression excipients: preformulation stability screening using differential scanning calorimetry. *Drug Dev Ind Pharm* 1982; **8**(6): 937–947.
10 Cho BC, *et al.* Clinical application of injectable calcium sulfate on early bone consolidation in distraction osteogenesis for the treatment of craniofacial microsomia. *J Craniofac Surg* 2002; **13**(3): 465–474.
11 Deporter DA, Todescan R. A possible 'rescue' procedure for dental implants with a textured surface geometry: a case report. *J Periodontol* 2001; **72**(10): 1420–1423.
12 *Food Chemicals Codex*, 7th edn. Bethesda, MD: United States Pharmacopeia, 2010: 159.
13 Japan Pharmaceutical Excipients Council. *Japanese Pharmaceutical Excipients 2004*. Tokyo: Yakuji Nippo, 2004; 124–125.

20 General References

Bryan JW, McCallister JD. Matrix forming capabilities of three calcium diluents. *Drug Dev Ind Pharm* 1992; **18**: 2029–2047.

21 Author

LV Allen Jr, RC Moreton.

22 Date of Revision

1 March 2012.

Canola Oil

1 Nonproprietary Names

BP: Refined Rapeseed Oil
PhEur: Rapeseed Oil, Refined
USP–NF: Canola Oil

2 Synonyms

Canbra oil; *Colzao CT*; huile de colza; *Lipex 108*; *Lipex 204*; *Lipovol CAN*; low erucic acid colza oil; low erucic acid rapeseed oil; rapae oleum raffinatum; tower rapeseed oil.

3 Chemical Name and CAS Registry Number

Canola oil [120962-03-0]

4 Empirical Formula and Molecular Weight

Canola oil contains approximately 6% saturated acids, 62% monounsaturated acids, and 32% polyunsaturated acids. Additionally, sulfur-containing fatty acids may also be present as minor constituents.

Unrefined canola oil is said to contain low levels of sulfur-containing fatty acids, resulting in the presence of sulfur in the oil in the stable form of triglycerides. These triglycerides resist refining procedures.[2] *See* Table I for the sulfur content of crude, refined, and deodorized canola oils.[3]

Table I: Total sulfur content in crude, refined, and bleached and deodorized canola oil.[a]

Oil sample	Range (mg/kg)	Mean	Standard deviation
Crude	23.6–24.1	23.8	1.0
Refined	19.1–20.2	19.7	2.85
Bleached and deodorized	15.6–16.5	16.2	2.7

(a) Determined using five replicates of each sample analyzed by ion chromatography.

5 Structural Formula

See Section 4.

6 Functional Category

Emollient; lubricant; oleaginous vehicle.

7 Applications in Pharmaceutical Formulation or Technology

Canola oil is a refined rapeseed oil obtained from particular species of rapeseed that have been genetically selected for their low erucic acid content.[4] In pharmaceutical formulations, canola oil is used mainly in topical preparations such as soft soaps and liniments. It is also used in cosmetics.

8 Description

A clear, light yellow-colored oily liquid with a bland taste.

9 Pharmacopeial Specifications

See Table II.

Table II: Pharmacopeial specifications for canola oil.

Test	PhEur 7.4	USP35–NF30
Identification	+	+
Characters	+	—
Specific gravity	—	0.906–0.920
Acid value	≤0.5	≤6.0
Alkaline impurities	+	—
Iodine value	—	110–126
Peroxide value	≤10.0	≤10.0
Saponification value	—	178–193
Unsaponifiable matter	≤1.5%	≤1.5%
Refractive index	—	1.465–1.467
Heavy metals	—	≤10 ppm
Fatty acid composition	+	+
Carbon chain length <14	—	0.1%
Eicosenoic acid	≤5.0%	<2.0%
Erucic acid	≤2.0%	≤2.0%
Linoleic acid	16.0–30.0%	<40%
Linolenic acid	6.0–14.0%	<14%
Oleic acid	50.0–67.0%	>50%
Palmitic acid	2.5–6.0%	<6.0%
Stearic acid	≤3.0%	<2.5%

10 Typical Properties

Boiling point 313°C
Density 0.913–0.917 g/cm^3
Flash point 290–330°C
Free fatty acid ≤0.05% as oleic acid
Freezing point −10 to −2°C
Solubility Soluble in chloroform and ether; practically insoluble in ethanol (95%); miscible with fixed oils.
Viscosity (dynamic) 77.3–78.3 mPa s (77.3–78.3 cP) at 20°C

11 Stability and Storage Conditions

Canola oil is stable and should be stored in an airtight, light-resistant container in a cool, dry place. The USP35–NF30 specifies that contact between canola oil and metals should be avoided. Containers should be filled to the top, while partially filled containers should be flushed with nitrogen. During storage, grassy, paintlike, or rancid off-flavors can develop.

Flavor deterioration has been attributed mainly to secondary oxidation products of linolenic acid, which normally makes up 6–14% of the fatty acids in canola oil. Storage tests of canola oil showed sensory changes after 2–4 days at 60–65°C in comparison to 16 weeks at room temperature. Canola oil seems to be more stable to storage in light than cottonseed oil and soybean oils, but is less stable than sunflower oil.[5] In addition, the effects of various factors on sediment formation in canola oil have been reported.[6]

It has been reported that oils stored at 2°C showed the highest rate of sediment formation, followed by those stored at 6°C.[5] All samples showed little sediment formation, as measured by turbidity, during storage at 12°C. Removal of sediment from canola oil prior to storage by cold precipitation and filtration did not eliminate this phenomenon, which still developed rapidly at 2°C.

A study on the effect of heating on the oxidation of low linolenic acid canola oil at frying temperatures under nitrogen and air clearly showed that a significantly lower development of oxidation was evident for the low linolenic acid canola oil. Reduction in the linolenic acid content of canola oil reduced the development of room odor at frying temperatures.

The thermal oxidation of canola oil studied during oven heating revealed an increase in peroxide values of pure and antioxidant-treated oils. Peroxide values were shown to differ between pure and antioxidant-treated canola oil during the initial stages of microwave heating (6 minutes). Formation of secondary products of oxidation, which contribute to off-flavors, were also observed.[7]

12 Incompatibilities

Canola oil is incompatible with strong oxidizing agents.

13 Method of Manufacture

Canola oil is obtained by mechanical expression or *n*-hexane extraction from the seeds of *Brassica napus (Brassica campestris)* var. *oleifera* and certain other species of *Brassica* (Cruciferae). The crude oil thus obtained is refined, bleached, and deodorized to substantially remove free fatty acids, phospholipids, color, odor and flavor components, and miscellaneous nonoil materials.

14 Safety

Canola oil is generally regarded as an essentially nontoxic and nonirritant material, and has been accepted by the FDA for use in cosmetics, foods, and pharmaceuticals.

Rapeseed oil has been used for a number of years in food applications as a cheap alternative to olive oil. However, there are large amounts of erucic acid and glucosinolates in conventional rapeseed oil, both substances being toxic to humans and animals.[7] Canola oil derived from genetically selected rapeseed plants that are low in erucic acid content has been developed to overcome this problem. The FDA specifies 165.55 mg as the maximum amount for each route or dosage form containing the ingredient.

Feeding studies in rats have suggested that canola oil is nontoxic to the heart, although it has also been suggested that the toxicological data may be unclear.[8,9]

15 Handling Precautions

Observe normal precautions appropriate to the circumstances and quantity of material handled. Spillages of this material are very slippery and should be covered with an inert absorbent material prior to disposal. Canola oil poses a slight fire hazard.

16 Regulatory Status

Accepted for use by the FDA in cosmetics and foods. Included in the FDA Inactive Ingredients Database (oral capsules). Included in the Canadian Natural Health Products Ingredients Database.

17 Related Substances

Almond oil; corn oil; cottonseed oil; peanut oil; rapeseed oil; sesame oil; soybean oil.

Rapeseed oil

CAS number [8002-13-9]

Synonyms *Calchem H-102*; colza oil; rape oil.

Appearance A clear, yellow to dark yellow-colored oily liquid.

Iodine number 94–120

Peroxide value <5

Saponification value 168–181

Comments Rapeseed oil contains 40–55% erucic acid. It is an edible oil and has been primarily used as an alternative, in foods and some pharmaceutical applications, to the more expensive olive oil. However, the safety of rapeseed oil as part of the diet has been questioned; *see* Section 14.

18 Comments

Canola oil has the lowest level of saturated fat compared to all other oils on the market at present and canola is now second only to soybean as the most important source of vegetable oil in the world. Canola has both a high protein (28%) and a high oil content (40%). When the oil is extracted, a high-quality and highly palatable feed concentrate of 37% protein remains. Canola oil is also high in the monounsaturated fatty acid oleic acid; *see* Table III.

The content of tocopherol, a natural antioxidant in canola oil, is comparable to those of peanut and palm oil. This is an important factor for oils with high linolenic acid content, which can reduce the shelf-life of the product, while the natural antioxidant, if present, can prevent oxidation during storage and processing.

The sulfur-containing compounds have been held responsible for the unpleasant odors from heated rapeseed oil. It has been suggested that the sulfur compounds in rapeseed oil are of three types: volatile, thermolabile, and nonvolatile.[1]

A specification for canola oil is contained in the *Food Chemicals Codex* (FCC).[10]

The EINECS number for canola oil is 232-313-5.

Table III: Comparison of the composition of crude soybean, canola, palm, and peanut oils.

Components	Canola	Palm	Peanut	Soybean
Fatty acid (%)	0.4–1.0	4.6	0.5–1.0	0.3–0.7
Phosphatides (gum) (%)	3.6	0.05–0.1	0.3–0.4	1.2–1.5
Sterols/triterpene alcohol (%)	0.53	0.1–0.5	0.2	0.33
Tocopherols (%)	0.06	0.003–0.1	0.02–0.06	0.15–0.21
Carotenoids (mg/kg)	25–50	500–1600	>1	40–50
Chlorophyll/ pheophytins (ppm)	5–25	—	—	1–2
Sulfur (ppm)	—	—	—	12–17
Iodine value	112–131	44–60	84–100	123–139

19 Specific References

1 Devinat G, *et al.* Sulfur-compounds in the rapeseed oils. *Rev Fr Corps Gras* 1980; **27**: 229–236.
2 Wijesundera RC, Ackman RG. Evidence for the probable presence of sulfur-containing fatty-acids as minor constituents in canola oil. *J Am Oil Chem Soc* 1988; **65**: 959–963.
3 Abraham V, de Man JM. Determination of total sulfur in canola oil. *J Am Oil Chem Soc* 1987; **64**: 384–387.
4 Hiltunen R, *et al.* Breeding of a zero erucic spring turnip-rape cultivar, *Brassica campestris* L. adapted to Finnish climatic conditions. *Acta Pharm Fenn* 1979; **88**: 31–34.
5 Przybylski R, *et al.* Formation and partial characterization of canola. *J Am Oil Chem Soc* 1993; **70**: 1009–1016.
6 Liu H, *et al.* Effects of crystalization conditions on sediment. *J Am Oil Chem Soc* 1994; **71**: 409–418.
7 Vieira T, *et al.* Canola oil thermal oxidation during oven test and microwave heating. *Lebensm-Wiss Technol* 2001; **34**: 215–221.
8 Anonymous. Rapeseed oil revisited. *Lancet* 1974; **ii**: 1359–1360.
9 Anonymous. Rapeseed oil and the heart. *Lancet* 1973; **ii**: 193.
10 *Food Chemicals Codex*, 7th edn. Bethesda, MD: United States Pharmacopeia, 2010: 162.

20 General References

Koseoglu SS, Iusas EW. Recent advances in canola oil hydrogenations. *J Am Oil Chem Soc* 1990; **67**: 3947.

Malcolmson LJ, *et al.* Sensory stability of canola oil: present status. *J Am Oil Chem Soc* 1994; **71**: 435–440.

Raymer PL. Canola: an emerging oilseed crop. In: Janick J, Whipkey A, eds. *Trends in New Crops and New Uses*. Alexandria, VA: ASHS Press, 2002: 122–126.

21 Author

KS Alexander.

22 Date of Revision

1 March 2012.

Carbomer

1 Nonproprietary Names

BP: Carbomers

PhEur: Carbomers

USP–NF: Carbomer

2 Synonyms

Acrypol; *Acritamer*; acrylic acid polymer; carbomera; *Carbopol*; carboxy polymethylene; carboxyvinyl polymer; *Pemulen*; polyacrylic acid; *Tego Carbomer*.

3 Chemical Name and CAS Registry Number

Carbomer [9003-01-4]

Note that alternative CAS registry numbers have been used: carbomer 934 ([9007-16-3]); carbomer 940 and carbomer homopolymer Type C ([9007-17-4]); and 941 ([9062-04-08]). The CAS registry number [9007-20-9] has also been used for carbomer (carboxypolymethylene).

4 Empirical Formula and Molecular Weight

Carbomers are synthetic high-molecular-weight polymers of acrylic acid that are crosslinked with either allyl sucrose or allyl ethers of pentaerythritol. They contain between 52% and 68% of carboxylic acid (COOH) groups calculated on the dry basis. The BP 2012 and PhEur 7.4 have a single monograph describing carbomer; the USP35–NF30 contains several monographs describing individual carbomer grades that vary in aqueous viscosity, polymer type, and polymerization solvent. The molecular weight of carbomer is theoretically estimated at 7×10^5 to 4×10^9.

5 Structural Formula

$$\left[-\overset{\displaystyle H}{\underset{\displaystyle H}{C}} - \overset{\displaystyle H}{\underset{\displaystyle \overset{\displaystyle |}{C}(=O)OH}{C}} - \right]_n$$

Acrylic acid monomer unit in carbomer polymers.

Carbomer polymers are formed from repeating units of acrylic acid. The monomer unit is shown above. The polymer chains are crosslinked with allyl sucrose or allyl pentaerythritol. *See also* Section 4.

6 Functional Category

Bioadhesive material; emulsifying agent; emulsion stabilizing agent; modified-release agent; suspending agent; viscosity-increasing agent.

7 Applications in Pharmaceutical Formulation or Technology

Carbomers are used in liquid or semisolid pharmaceutical and cosmetic formulations as rheology modifiers and emulsifying agents in the preparation of oil-in-water emulsions for external administration. Formulations include creams, gels, lotions and ointments for use in ophthalmic,[1–3] rectal,[4–6] topical[7–12] and vaginal[13–16] preparations; *see* Table I.

In tablet formulations, carbomers are used as controlled-release agents either alone[17–26] or in combination with other polymers such as hypromellose and polyvinyl acetate phthalate.[27–29] In contrast to linear polymers, higher viscosity does not result in slower drug release with carbomers. Lightly crosslinked carbomers (lower viscosity) are generally more efficient in controlling drug release than highly crosslinked carbomers (higher viscosity).

Carbomers are also used as binders in wet granulation using water, organic solvents, or their mixtures as the granulating fluid.

Carbomer polymers have also been studied in the preparation of multiparticulate systems for oral delivery[30–34] and in oral mucoadhesive controlled drug delivery systems.[35–39]

Table I: Uses of carbomers.

Use	Typical concentration (%)
Emulsifying agent	0.1–0.5
Gelling agent	0.5–2.0
Suspending agent	0.5–1.0
Tablet binder	0.75–3.0
Controlled-release agent	5.0–30.0

8 Description

Carbomers are white-colored, ‘fluffy’, acidic, hygroscopic powders with a characteristic slight odor. A granular carbomer is also available (*Carbopol 71G*).

9 Pharmacopeial Specifications

The USP354–NF30 has several monographs for different carbomer grades, while the BP 2012 and PhEur 7.4 have only a single monograph.

The USP–NF lists three umbrella monographs, carbomer copolymer, carbomer homopolymer and carbomer interpolymer, which separate carbomer products based on polymer structure and apply to products not polymerized in benzene. The differentiation within each umbrella monograph is based on viscosity characteristics (Type A, Type B and Type C).

The USP–NF also lists monographs for carbomer 934, 934P, 940 and 941, which are manufactured using benzene. Effective since January 1 2011, products manufactured without the use of benzene are officially titled Carbomer Homopolymer provided they comply with the carbomer homopolymer monograph. The USP–NF also includes carbomer 1342, which applies to carbomer copolymers manufactured using benzene.

Carbomer polymers are also covered either individually or together in other pharmacopeias.

See Table II. *See also* Section 18.

Note that unless otherwise indicated, the test limits shown above apply to all grades of carbomer.

SEM 1: Excipient: *Carbopol 971P*; manufacturer: Lubrizol Advanced Materials, Inc.; magnification: 2000×; voltage: 25 kV.

SEM 2: Excipient: *Carbopol 971P*; manufacturer: Lubrizol Advanced Materials, Inc.; magnification: 6000×; voltage: 25 kV.

Table II: Pharmacopeial specifications for carbomers.

Test	PhEur 7.4	USP35–NF30
Identification	+	+
Characters	+	−
Aqueous viscosity (mPa s)	300–115 000	+
Carbomer 934 (0.5% w/v)	−	30 500–39 400
Carbomer 934P (0.5% w/v)	−	29 400–39 400
Carbomer 940 (0.5% w/v)	−	40 000–60 000
Carbomer 941 (0.5% w/v)	−	4 000–11 000
Carbomer 1342 (1.0% w/v)	−	9 500–26 500
Carbomer copolymer (1% w/v)		
Type A	−	4 500–13 500
Type B	−	10 000–29 000
Type C	−	25 000–45 000
Carbomer homopolymer (0.5% w/v)		
Type A	−	4 000–11 000
Type B	−	25 000–45 000
Type C	−	40 000–60 000
Carbomer interpolymer		
Type A (0.5% w/v)	−	45 000–65 000
Type B (1% w/v)	−	47 000–77 000
Type C (0.5% w/v)	−	8 500–16 500
Loss on drying	≤3.0%	≤2.0%
Sulfated ash	≤4.0%	−
Residue on ignition	−	≤4.0%[(a)]
Heavy metals	≤20 ppm	≤0.002%
Benzene	≤2 ppm	+
Carbomer 934	−	≤0.5%
Carbomer 934P	−	≤0.01%
Carbomer 940	−	≤0.5%
Carbomer 941	−	≤0.5%
Carbomer 1342	−	≤0.2%
Carbomer copolymer	−	≤0.0002%
Carbomer homopolymer	−	≤0.0002%
Carbomer interpolymer	−	≤0.0002%
Free acrylic acid	≤0.25%	≤0.25%[(b)]
Ethylacetate	−	+
Carbomer copolymer	−	≤0.5%
Carbomer homopolymer	−	≤0.5%
Carbomer interpolymer	−	≤0.35%
Cyclohexane	−	+
Carbomer copolymer	−	≤0.3%
Carbomer homopolymer	−	≤0.3%
Carbomer interpolymer	−	≤0.15%
Assay (COOH content)	56.0–68.0%	56.0–68.0%[(c)]

(a) For carbomer homopolymer only.
(b) For carbomer copolymer, carbomer homopolymer and carbomer interpolymer only.
(c) Except for carbomer 1342, carbomer copolymer, and carbomer interpolymer, where the limits are 52.0–62.0%.

10 Typical Properties

Acidity/alkalinity

pH = 2.5–4.0 for a 0.2% w/v aqueous dispersion

Density (bulk)

0.2 g/cm^3 (powder);

0.4 g/cm^3 (granular).

Density (tapped)

0.3 g/cm^3 (powder);

0.4 g/cm^3 (granular).

Density (true) 1.41 g/cm^3

Dissociation constant pK_a = 6.0±0.5

Glass transition temperature 100–105°C

Melting point Decomposition occurs within 30 minutes at 260°C. *See* Section 11.

Moisture content Typical water content is up to 2% w/w. However, carbomers are hygroscopic and a typical equilibrium moisture content at 25°C and 50% relative humidity is 8–10% w/w. The moisture content of a carbomer does not affect its thickening efficiency, but an increase in the moisture content makes the carbomer more difficult to handle because it is less readily dispersed.

Particle size distribution Primary particles average about 0.2 μm in diameter. The flocculated powder particles average 2–7 μm in diameter and cannot be broken down into the primary particles. A granular carbomer has a particle size in the range 150–425 μm.

Solubility Swellable in water and glycerin and, after neutralization, in ethanol (95%). Carbomers do not dissolve but merely swell to a remarkable extent, since they are three-dimensionally crosslinked microgels.

Spectroscopy

IR spectra *see* Figure 1.

NIR spectra *see* Figure 2.

Viscosity (dynamic) Carbomers disperse in water to form acidic colloidal dispersions that, when neutralized, produce highly viscous gels. Neutralized aqueous gels are more viscous at pH 6–11. The viscosity is considerably reduced at pH values less than 3 or greater than 12, or in the presence of strong electrolytes.[(40,41)]

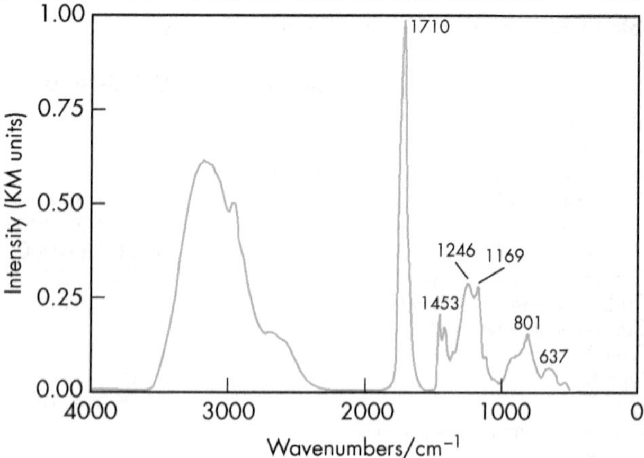

Figure 1: Infrared spectrum of carbomer measured by diffuse reflectance. Adapted with permission of Informa Healthcare.

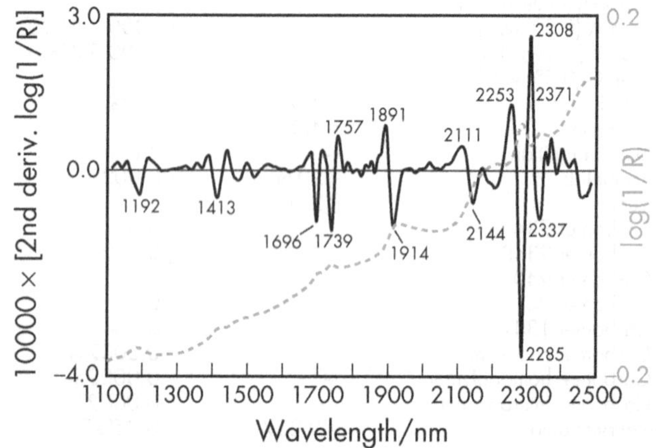

Figure 2: Near-infrared spectrum of carbomer measured by reflectance.

11 Stability and Storage Conditions

Carbomers are stable, hygroscopic materials that may be heated at temperatures below 104°C for up to 2 hours without affecting their thickening efficiency. However, exposure to excessive temperatures can result in discoloration and reduced stability. Complete decomposition occurs with heating for 30 minutes at 260°C. Dry powder forms of carbomer do not support the growth of molds and fungi. In contrast, microorganisms grow well in unpreserved aqueous dispersions, and therefore an antimicrobial preservative such as 0.1% w/v chlorocresol, 0.18% w/v methylparaben–0.02% w/v propylparaben, or 0.1% w/v thimerosal should be added. The addition of certain antimicrobials, such as benzalkonium chloride or sodium benzoate, in high concentrations (0.1% w/v) can cause cloudiness and a reduction in viscosity of carbomer dispersions. Aqueous gels may be sterilized by autoclaving[3] with minimal changes in viscosity or pH, provided care is taken to exclude oxygen from the system, or by gamma irradiation, although this technique may increase the viscosity of the formulation.[42,43] At room temperature, carbomer dispersions maintain their viscosity during storage for prolonged periods. Similarly, dispersion viscosity is maintained, or only slightly reduced, at elevated storage temperatures if an antioxidant is included in the formulation or if the dispersion is stored protected from light. Exposure to light causes oxidation that is reflected in a decrease in dispersion viscosity. Stability to light may be improved by the addition of 0.05–0.1% w/v of a water-soluble UV absorber such as benzophenone-2 or benzophenone-4 in combination with 0.05–0.1% w/v edetic acid.

Carbomer powder should be stored in an airtight, corrosion-resistant container and protected from moisture. The use of glass, plastic, or resin-lined containers is recommended for the storage of formulations containing carbomer.

12 Incompatibilities

Carbomers are discolored by resorcinol and are incompatible with phenol, cationic polymers, strong acids, and high levels of electrolytes. Certain antimicrobial adjuvants should also be avoided or used at low levels, *see* Section 11. Trace levels of iron and other transition metals can catalytically degrade carbomer dispersions.

Certain amino-functional actives form complexes with carbomer; often this can be prevented by adjusting the pH of the dispersion and/or the solubility parameter by using appropriate alcohols and polyols.

Carbomers also form pH-dependent complexes with certain polymeric excipients. Adjustment of pH and/or solubility parameter can also work in this situation.

13 Method of Manufacture

Carbomers are synthetic, high-molecular-weight, crosslinked polymers of acrylic acid. These acrylic acid polymers are crosslinked with allyl sucrose or allyl pentaerythritol. The polymerization solvent used previously was benzene; however, some of the newer commercially available grades of carbomer are manufactured using either ethyl acetate or a cyclohexane–ethyl acetate cosolvent mixture. The *Carbopol ETD* and *Carbopol Ultrez* polymers are produced in the cosolvent mixture with a proprietary polymerization aid.

14 Safety

Carbomers are used extensively in nonparenteral products, particularly topical liquid and semisolid preparations. Grades polymerized in ethyl acetate may also be used in oral formulations; *see* Section 18. There is no evidence of systemic absorption of carbomer polymers following oral administration.[44] Acute oral toxicity studies in animals indicate that carbomer 934P has a low oral toxicity, with doses up to 8 g/kg being administered to dogs without fatalities occurring. Carbomers are generally regarded as essentially nontoxic and nonirritant materials;[45] there is no evidence in humans of hypersensitivity reactions to carbomers used topically.

LD_{50} (guinea pig, oral): 2.5 g/kg for carbomer 934[46]
LD_{50} (guinea pig, oral): 2.5 g/kg for carbomer 934P
LD_{50} (guinea pig, oral): 2.5 g/kg for carbomer 940
LD_{50} (mouse, IP): 0.04 g/kg for carbomer 934P
LD_{50} (mouse, IP): 0.04 g/kg for carbomer 940
LD_{50} (mouse, IV): 0.07 g/kg for carbomer 934P
LD_{50} (mouse, IV): 0.07 g/kg for carbomer 940
LD_{50} (mouse, oral): 4.6 g/kg for carbomer 934P
LD_{50} (mouse, oral): 4.6 g/kg for carbomer 934
LD_{50} (mouse, oral): 4.6 g/kg for carbomer 940
LD_{50} (rat, oral): 10.25 g/kg for carbomer 910
LD_{50} (rat, oral): 2.5 g/kg for carbomer 934P
LD_{50} (rat, oral): 4.1 g/kg for carbomer 934
LD_{50} (rat, oral): 2.5 g/kg for carbomer 940
LD_{50} (rat, oral): > 1 g/kg for carbomer 941

No observed adverse effect level (NOAEL) (rat, dog, oral): 1.5 g/kg for carbomer homopolymer type B.[47]

15 Handling Precautions

Observe normal precautions appropriate to the circumstances and quantity of material handled. Excessive dust generation should be minimized to avoid the risk of explosion (lowest explosive concentration is 130 g/m^3). Carbomer dust is irritating to the eyes, mucous membranes, and respiratory tract. In the event of eye contact with carbomer dust, saline should be used for irrigation purposes. Gloves, eye protection, and a dust respirator are recommended during handling.

A solution of electrolytes (sodium chloride) or alkaline detergent is recommended for cleaning equipment after processing carbomers.

16 Regulatory Acceptance

Included in the FDA Inactive Ingredients Database (oral suspensions and tablets; ophthalmic, rectal, topical, and transdermal preparations; vaginal suppositories). Included in nonparenteral medicines licensed in Europe. Included in the Canadian Natural Health Products Ingredients Database.

17 Related Substances

Polycarbophil.

18 Comments

Carbomers have been investigated for formulations for nasal and pulmonary delivery routes.[48–51] The polymeric gels may be used as a vehicle for liposomes, niosomes, microemulsions, and nanoparticles.[52–57] Carbomers have also been investigated as inhibitors of intestinal proteases in peptide-containing dosage forms.[58–60] The polymers have been studied in dosage forms for colonic drug delivery,[61,62] in magnetic granules for site-specific drug delivery to the esophagus,[63] as a bioadhesive for a cervical patch[64] or in oral care formulations.[65,66]

A number of different carbomer grades are commercially available that vary in their chemical structure, degree of crosslinking, and residual components. These differences account for the specific rheological, handling, and use characteristics of each grade. Carbomer grades that have the polymer backbone modified with long-chain alkyl acrylates are used as polymeric emulsifiers or in formulations requiring increased resistance to ions.

Polycarbophils, poly(acrylic acid) polymers crosslinked with divinyl glycol, are available for bioadhesive or medicinal applications. In general, carbomers designated with the letter 'P', e.g. *Carbopol 971P*, are the pharmaceutical grade polymers for oral or mucosal contact products.

In an effort to measure the molecular weight between crosslinks, M_C, researchers have extended the network theory of elasticity to swollen gels and have utilized the inverse relationship between the elastic modulus and M_C.[67–69] Estimated M_C values of 237 600 g/mol for *Carbopol 941* and of 104 400 g/mol for *Carbopol 940* have been reported.[70] In general, carbomer polymers with lower viscosity and lower rigidity will have higher M_C values. Conversely, higher-viscosity, more rigid carbomer polymers will have lower M_C values.

Carbomer copolymer (or homopolymer or interpolymer) obtained from different manufacturers or produced in different solvents with different manufacturing processes may not have identical properties with respect to their use for specific pharmaceutical purposes, e.g. as tablet controlled-release agents, bioadhesives, topical gellants, etc. Therefore, types of carbomer copolymer (or homopolymer or interpolymer) should not be interchanged unless performance equivalency has been ascertained.

In the preparation of gels, carbomer powders should first be dispersed into vigorously stirred water, taking care to avoid the formation of indispersible agglomerates, then neutralized by the addition of a base. The *Carbopol ETD* and *Ultrez* series of carbomers have been introduced to overcome some of the problems of dispersing the powder into aqueous solvents. These carbomers wet quickly yet hydrate slowly, while possessing a lower unneutralized dispersion viscosity. Agents that may be used to neutralize carbomer polymers include amino acids, potassium hydroxide, sodium bicarbonate, sodium hydroxide, and organic amines such as triethanolamine. One gram of carbomer is neutralized by approximately 0.4 g of sodium hydroxide. During preparation of the gel, the solution should be agitated slowly with a broad, paddlelike stirrer to avoid introducing air bubbles.

Therapeutically, carbomer formulations have proved efficacious in improving symptoms of moderate-to-severe dry eye syndrome.[71,72]

A specification for carbomer (carboxyvinyl polymer) is contained in the *Japanese Pharmaceutical Excipients* (JPE).[73]

19 Specific References

1 Amin PD, *et al.* Studies on gel tears. *Drug Dev Ind Pharm* 1996; **22**(7): 735–739.

2 Ünlü N, *et al.* Formulation of carbopol 940 ophthalmic vehicles, and *in vitro* evaluation of the influence of simulated lacrimal fluid on their physico-chemical properties. *Pharmazie* 1991; **46**: 784–788.

3 Deshpande SG, Shirolkar S. Sustained release ophthalmic formulations of pilocarpine. *J Pharm Pharmacol* 1989; **41**: 197–200.

4 Dal Zotto M, *et al.* Effect of hydrophilic macromolecular substances on the drug release rate from suppositories with lipophilic excipient. Part 1: use of polyacrylic acids. *Farmaco* 1991; **46**: 1459–1474.

5 Morimoto K, Morisaka K. *In vitro* release and rectal absorption of barbital and aminopyrine from aqueous polyacrylic acid gel. *Drug Dev Ind Pharm* 1987; **13**(7): 1293–1305.

6 Green JT, *et al.* Nicotine carbomer enemas–pharmacokinetics of a revised formulation. *Ital J Gastroenterol Hepatol* 1998; **30**: 260–265.

7 Tamburic S, Craig DQM. Investigation into the rheological, dielectric and mucoadhesive properties of poly(acrylic acid) gel systems. *J Control Release* 1995; **37**: 59–68.

8 Amsellem E, *et al. In vitro* studies on the influence of carbomers on the availability and acceptability of estradiol gels. *Arzneimittelforschung* 1998; **48**: 492–496.

9 Jimenez-Kairuz A, *et al.* Mechanism of lidocaine release from carbomer-lidocaine hydrogels. *J Pharm Sci* 2002; **91**: 267–272.

10 Tanna S, *et al.* Covalent coupling of concanavalin to a carbopol 934P and 941P carrier in glucose-sensitive gels for delivery of insulin. *J Pharm Pharmacol* 2002; **54**: 1461–1469.

11 Tas C, *et al. In vitro* and *ex vivo* permeation studies of chlorpheniramine maleate gels prepared by carbomer derivatives. *Drug Dev Ind Pharm* 2004; **30**: 637–647.

12 A-Sasutjarit R, *et al.* Viscoelastic properties of *Carbopol 940* gels and their relationships to piroxicam diffusion coefficients in gel bases. *Pharm Res* 2005; **22**(12): 2134–2140.

13 Fiorilli A, *et al.* Successful treatment of bacterial vaginosis with a policarbophil-carbopol acidic vaginal gel: results from a randomised double-blind, placebo-controlled trial. *Eur J Obstet Gynecol Reprod Biol* 2005; **120**(2): 202–205.

14 Baloğlu E, *et al.* Bioadhesive controlled release systems of ornidazole for vaginal delivery. *Pharm Dev Technol* 2006; **11**(4): 477–484.

15 Cranage MP, *et al.* Repeated vaginal administration of trimeric HIV-1 clade C gp140 induces serum and mucosal antibody responses. *Mucosal Immunol* 2010; **3**(1): 57–68.

16 Bachhav YG, Patravale VB. Microemulsion based vaginal gel of fluconazole: formulation, *in vitro* and *in vivo* evaluation. *Int J Pharm* 2009; **365**(1–2): 175–179.

17 Wahlgren M, *et al.* Oral-based controlled release formulations using poly(acrylic acid) microgels. *Drug Dev Ind Pharm* 2009; **35**(8): 922–929.

18 Efentakis M, Peponaki C. Formulation study and evaluation of matrix and three-layer tablet sustained drug delivery systems based on Carbopols with isosorbite mononitrate. *AAPS PharmSciTech* 2008; **9**(3): 917–923.

19 Bermúdez JM, *et al.* A ciprofloxacin extended release tablet based on swellable drug polyelectrolyte matrices. *AAPS PharmSciTech* 2008; **9**(3): 924–930.

20 Huang LL, Schwartz JB. Studies on drug release from a carbomer tablet matrix. *Drug Dev Ind Pharm* 1995; **21**(13): 1487–1501.

21 Khan GM, Zhu JB. Studies on drug release kinetics from ibuprofen–carbomer hydrophilic matrix tablets: influence of co-excipients on release rate of the drug. *J Control Rel* 1999; **57**(2): 197–203.

22 Khan GM, Jiabi Z. Formulation and *in vitro* evaluation of ibuprofen-*Carbopol* 974P-NF controlled release matrix tablets III: influence of co-excipients on release rate of the drug. *J Control Rel* 1998; **54**(2): 185–190.

23 Parojcic J, *et al.* An investigation into the factors influencing drug release from hydrophilic matrix tablets based on novel carbomer polymers. *Drug Deliv* 2004; **11**(1): 59–65.

24 Samani SM, *et al.* The effect of polymer blends on release profiles of diclofenac sodium from matrices. *Eur J Phar Biopharm* 2003; **55**: 351–355.
25 Efentakis M, *et al.* Dimensional changes, gel layer evolution and drug release studies in hydrophilic matrices loaded with drugs of different solubility. *Int J Pharm* 2007; **339**(1–2): 66–75.
26 El-Malah Y, Nasal S. Hydrophilic matrices: application of Placket-Burman screening design to model the effect of *Polyox-Carbopol* blends on drug release. *Int J Pharm* 2006; **309**(1–2): 163–170.
27 Petrovic A, *et al.* Application of mixture experimental design in the formulation and optimization of matrix tablets containing carbomer and hydroxy-propylmethylcellulose. *Arch Pharm Res* 2009; **32**(12): 1767–1774.
28 Tiwari SB, Rajabi-Siahboomi AR. Applications of complementary polymers in HPMC hydrophilic extended release matrices. *Drug Deliv Technol* 2009; 9(7): 20–27.
29 Tiwari SB, Rajabi-Siahboomi AR. Modulation of drug release from hydrophilic matrices. *Pharm Technol Eur* 2008. http://pharmtech.findpharma.com/pharmtech/article/articleDetail.jsp?id=547891.
30 Ivic B, *et al.* Optimization of drug release from compressed multi unit particle system (MUPS) using generalized regression neural network (GRNN). *Arch Pharm Res* 2010; **33**: 103–113.
31 Paker-Leggs S, Neau SH. Propranolol forms affect properties of Carbopol-containing extruded-spheronized beads. *Int J Pharm* 2008; **361**(1–2): 169–176.
32 Bommareddy GS, *et al.* Extruded and spheronized beads containing *Carbopol* 974P to deliver nonelectrolytes and salts of weakly basic drugs. *Int J Pharm* 2006; **321**(1–2): 62–71.
33 Gomez-Carracedo A, *et al.* Extrusion-spheronization of blends of *Carbopol* 934 and microcrystalline cellulose. *Drug Dev Ind Pharm* 2001; **27**(5): 381–391.
34 Mezreb N, *et al.* Production of *Carbopol 974P* and *Carbopol 971P* pellets by extrusion-spheronization: optimization of the processing parameters and water content. *Drug Dev Ind Pharm* 2004; **30**(5): 481–490.
35 Singla AK, *et al.* Potential applications of carbomer in oral mucoadhesive controlled drug delivery system: a review. *Drug Dev Ind Pharm* 2000; **26**(9): 913–924.
36 Llabot JM, *et al.* Drug release from carbomer:carbomer sodium salt matrices with potential use as mucoadhesive drug delivery system. *Int J Pharm* 2004; **276**(1–2): 59–66.
37 Ikinci G, *et al.* Development and *in vitro/in vivo* evaluations of bioadhesive buccal tablets for nicotine replacement therapy. *Pharmazie* 2006; **61**(3): 203–207.
38 Surapaneni MS, *et al.* Effect of excipient and processing variables on adhesive properties and release profile of pentoxifylline from mucoadhesive tablets. *Drug Dev Ind Pharm* 2006; **32**(3): 377–387.
39 Jones DS, *et al.* Rheological, mechanical and mucoadhesive properties of thermoresponsive, bioadhesive binary mixtures composed of poloxamer 407 and carbopol 974P designed as platforms for implantable drug delivery systems for use in the oral cavity. *Int J Pharm* 2009; **372**(1–2): 49–58.
40 Neau SH, Chow MY. Fabrication and characterization of extruded and spheronized beads containing *Carbopol 974P-NF* resin. *Int J Pharm* 1996; **131**: 47–55.
41 Charman WN, *et al.* Interaction between calcium, a model divalent cation, and a range of poly(acrylic acid) resins as a function of solution pH. *Drug Dev Ind Pharm* 1991; **17**(2): 271–280.
42 Adams I, Davis SS. Formulation and sterilization of an original lubricant gel base in carboxypolymethylene. *J Pharm Pharmacol* 1973; **25**: 640–646.
43 Adams I, *et al.* Formulation of a sterile surgical lubricant. *J Pharm Pharmacol* 1972; **24**(Suppl.): 178P.
44 Riley RG, *et al.* The gastrointestinal transit profile of C14 labelled poly(acrylic acids): an *in vitro* study. *Biomaterials* 2001; **22**: 1861–1867.
45 Anonymous. Final report on the safety assessment of carbomers. *J Am Coll Toxicol* 1982; **1**(2): 109–141.
46 Lewis RJ, ed. *Sax's Dangerous Properties of Industrial Materials*, 11th edn. New York: Wiley, 2004: 71.
47 Lubrizol Advanced Materials, Inc. Technical data sheet TDS-328, *Carbopol* 971P NF polymer & *Carbopol* 71G NF polymer: October 2007.
48 Mahajan HS, Gattani S. *In situ* gels of metoclopramide hydrochloride for intranasal delivery: *in vitro* evaluation and *in vivo* pharmacokinetic study in rabbits. *Drug Deliv* 2010; **17**(1): 19–27.
49 Rathnam G, *et al.* Carbopol-based gels for nasal delivery of progesterone. *AAPS PharmSciTech* 2008; **9**(4): 1078–1082.
50 Vidgren P, *et al. In vitro* evaluation of spray-dried mucoadhesive microspheres for nasal administration. *Drug Dev Ind Pharm* 1992; **18**(5): 581–597.
51 Alhusban FA, Seville PC. Carbomer-modified spray-dried respirable powders for pulmonary delivery of salbutamol sulphate. *J Microencapsul* 2009; **26**(5): 444–455.
52 Hosny KM. Ciprofloxacin as ocular liposomal hydrogel. *AAPS PharmSciTech* 2010; **11**(1): 241–246.
53 Puglia C, *et al.* Evaluation of percutaneous absorption of naproxen from different liposomal formulations. *J Pharm Sci* 2010; **99**(6): 2819–2829.
54 Azeem A, *et al.* Nanocarrier for the transdermal delivery of an antiparkinsonian drug. *AAPS PharmSciTech* 2009; **10**(4): 1093–1103.
55 Dragicevic-Curic N, *et al.* Stability evaluation of temoporfin-loaded liposomal gels for topical application. *J Liposome Res* 2010; **20**(1): 38–48.
56 Dragicevic-Curic N, *et al.* Temoporfin-loaded liposomal gels: viscoelastic properties and *in vitro* skin penetration. *Int J Pharm* 2009; **373**(1–2): 77–84.
57 Liu W. Investigation of the carbopol gel of solid lipid nanoparticles for the transdermal iontophoretic delivery of triamcinolone acetonide acetate. *Int J Pharm* 2008; **364**(1): 135–141.
58 Luessen HL, *et al.* Mucoadhesive polymers in peroral peptide drug delivery. Part 1: influence of mucoadhesive excipients on the proteolytic activity of intestinal enzymes. *Eur J Pharm Sci* 1996; **4**: 117–128.
59 Luessen HL, *et al.* Mucoadhesive polymers in peroral peptide drug delivery. Part 2: carbomer and polycarbophil are potent inhibitors of the intestinal proteolytic enzyme trypsin. *Pharm Res* 1995; **12**: 1293–1298.
60 Vaidya AP, *et al.* Protective effect of *Carbopol* on enzymatic degradation of a peptide-like substrate I: effect of various concentrations and grades of *Carbopol* and other reaction variables on trypsin activity. *Pharm Dev Technol* 2007; **12**(1): 89–96.
61 McGirr ME, *et al.* The use of the InteliSite companion device to deliver mucoadhesive polymers to the dog colon. *Eur J Pharm Sci* 2009; **36**(4–5): 386–391.
62 Ali Asghar LF, *et al.* Design and *in vitro* evaluation of formulations with pH and transit time controlled sigmoidal release profile for colon-specific delivery. *Drug Deliv* 2009; **16**(6): 295–303.
63 Ito R, *et al.* Magnetic granules: a novel system for specific drug delivery to esophageal mucosa in oral administration. *Int J Pharm* 1990; **61**: 109–117.
64 Woolfson AD, *et al.* Bioadhesive patch cervical drug delivery system for the administration of 5-fluorouracil to cervical tissue. *J Control Release* 1995; **35**: 49–58.
65 Pader M. Product components: nontherapeutic ingredients. In *Oral Hygiene Products and Practice*. New York: Marcel Dekker, Inc. 1988: 299–302.
66 McConnell MD, *et al.* Bacterial plaque retention on oral hard materials: effect of surface roughness, surface composition, and physisorbed polycarboxylate. *J Biomed Mater Res A* 2010; **92**(4): 1518–1527.
67 Taylor A, Bagley A. Tailoring closely packed gel-particles systems for use as thickening agents. *J Appl Polym Sci* 1977; **21**: 113–122.
68 Taylor A, Bagley A. Rheology of dispersions of swollen gel particles. *J Polym Sci* 1975; **13**: 1133–1144.
69 Nae HN, Reichert WW. Rheological properties of lightly crosslinked carboxy copolymers in aqueous solutions. *Rheol Acta* 1992; **31**: 351–360.
70 Carnali JO, Naser MS. The use of dilute solution viscosity to characterize the network properties of carbopol microgels. *Colloid Polym Sci* 1992; **270**: 183–193.
71 Sullivan LJ, *et al.* Efficacy and safety of 0.3% carbomer gel compared to placebo in patients with moderate-to-severe dry eye syndrome. *Ophthalmology* 1997; **104**: 1402–1408.
72 Wang TJ, *et al.* Comparison of the clinical effects of carbomer-based lipid-containing gel and hydroxypropyl-guar gel artificial tear formulations in patients with dry eye syndrome: a 4-week, prospective, open-label, randomized, parallel-group, noninferiority study. *Clin Ther* 2010; **32**(1): 44–52.
73 Japanese Pharmaceutical Excipients Council. *Japanese Pharmaceutical Excipients 2004*. Tokyo: Yakuji Nippo, 2004: 138–141.

20 General References

Alexander P. Organic rheological additives. *Manuf Chem* 1986; **57**: 81, 83–84.
Corel Pharm Chem. Product data sheet: *Acrypol*. www.corelpharmachem.-com/acrypol.htm (accessed 9 June 2011).
Evonik Goldschmidt GmbH. Product data sheet: *TEGO Carbomer 134*, *TEGO Carbomer 140*, *TEGO Carbomer 141*, 2008.
Islam MT, *et al*. Rheological characterization of topical carbomer gels neutralized to different pH. *Pharm Res* 2004; **21**: 1192–1199.
Jimenez-Kairuz AF, *et al*. Swellable drug-polyelectrolyte matrices (SDPM). Characterization and delivery properties. *Int J Pharm* 2005; **288**: 87–99.
Lubrizol Advanced Materials, Inc. Technical literature: *Carbopol* and *Pemulen* polymers, 2008.
Pérez-Marcos B, *et al*. Interlot variability of carbomer 934. *Int J Pharm* 1993; **100**: 207–212.
Secard DL. Carbopol pharmaceuticals. *Drug Cosmet Ind* 1962; **90**: 28–30113, 115–116.

21 Authors

E Draganoiu, A Rajabi-Siahboomi, S Tiwari.

22 Date of Revision

1 March 2012.

Carbon Dioxide

1 Nonproprietary Names

BP: Carbon Dioxide
JP: Carbon Dioxide
PhEur: Carbon Dioxide
USP: Carbon Dioxide

2 Synonyms

Carbonei dioxidum; carbonic acid gas; carbonic anhydride; E290.

3 Chemical Name and CAS Registry Number

Carbon dioxide [124-38-9]

4 Empirical Formula and Molecular Weight

CO_2 44.01

5 Structural Formula

See Section 4.

6 Functional Category

Aerosol propellant; air displacement; solvent.

7 Applications in Pharmaceutical Formulation or Technology

Carbon dioxide and other compressed gases such as nitrogen and nitrous oxide are used as propellants for topical pharmaceutical aerosols. They are also used in other aerosol products that work satisfactorily with the coarse aerosol spray that is produced with compressed gases, e.g. cosmetics, furniture polish, and window cleaners.(1–3)

The advantages of compressed gases as aerosol propellants are that they are less expensive; are of low toxicity; and are practically odorless and tasteless. Also, in comparison to liquefied gases, their pressures change relatively little with temperature. However, the disadvantages of compressed gases are that there is no reservoir of propellant in the aerosol and pressure consequently decreases as the product is used. This results in a change in spray characteristics. Additionally, if a product that contains a compressed gas as a propellant is actuated in an inverted position, the vapor phase, rather than the liquid phase, is discharged. Most of the propellant is contained in the vapor phase and therefore some of the propellant will be lost and the spray characteristics will be altered. Also, sprays produced using compressed gases are very wet. Valves, such as the vapor tap or double dip tube, are currently available and will overcome these problems.

Carbon dioxide is also used to displace air from pharmaceutical products by sparging and hence to inhibit oxidation. As a food additive it is used to carbonate beverages and to preserve foods such as bread from spoilage by mold formation, the gas being injected into the space between the product and its packaging.(4,5)

Solid carbon dioxide is also widely used to refrigerate products temporarily, while liquid carbon dioxide, which can be handled at temperatures up to 31°C under high pressure, is used as a solvent for flavors and fragrances, primarily in the perfumery and food manufacturing industries.

8 Description

Carbon dioxide occurs naturally as approximately 0.03% v/v of the atmosphere. It is a colorless, odorless, noncombustible gas with a faint acid taste. Solid carbon dioxide, also known as dry ice, is usually encountered as white-colored pellets or blocks.

9 Pharmacopeial Specifications

See Table I.

10 Typical Properties

Density
0.714 g/cm^3 for liquid at 25°C;
0.742 g/cm^3 for vapor at 25°C.
Flammability Nonflammable.
Solubility 1 in about 1 of water by volume at normal temperature and pressure.
Vapor density (absolute) 1.964 g/m^3
Vapor density (relative) 1.53 (air = 1)
Viscosity (kinematic) 0.14 mm^2/s (0.14 cSt) at −17.8°C.

11 Stability and Storage Conditions

Extremely stable and chemically nonreactive. Store in a tightly sealed cylinder. Avoid exposure to excessive heat.

12 Incompatibilities

Carbon dioxide is generally compatible with most materials although it may react violently with various metal oxides or reducing metals such as aluminum, magnesium, titanium, and

Table I: Pharmacopeial specifications for carbon dioxide.

Test	JP XV	PhEur 7.4	USP35–NF30
Characters	+	+	–
Production	–	+	–
Total sulfur	–	≤1 ppm	–
Water	–	≤67 ppm	≤150 mg/m^3
Identification	+	+	+
Carbon monoxide	+	≤5 ppm	≤0.001%
Sulfur dioxide	–	≤2 ppm	≤5 ppm
Nitrogen monoxide and nitrogen dioxide	–	≤2 ppm	≤2.5 ppm
Impurities	–	+	–
Limit of ammonia	–	–	≤0.0025%
Limit of nitric oxide	–	–	≤2.5 ppm
Acid	+	–	–
Hydrogen phosphide, hydrogen sulfide or reducing organic substances	+	≤1 ppm	≤1 ppm
Oxygen and nitrogen	+	–	–
Assay	≥99.5%	≥99.5%	≥99.0%

zirconium. Mixtures with sodium and potassium will explode if shocked.

13 Method of Manufacture

Carbon dioxide is obtained industrially in large quantities as a by-product in the manufacture of lime; by the incineration of coke or other carbonaceous material; and by the fermentation of glucose by yeast. In the laboratory it may be prepared by dropping acid on a carbonate.

14 Safety

In formulations, carbon dioxide is generally regarded as an essentially nontoxic material.

15 Handling Precautions

Handle in accordance with standard procedures for handling metal cylinders containing liquefied or compressed gases. Carbon dioxide is an asphyxiant, and inhalation in large quantities is hazardous. It should therefore be handled in a well-ventilated environment equipped with suitable safety devices for monitoring vapor concentration.

It should be noted that carbon dioxide is classified as a greenhouse gas responsible for global warming. At the present time there are no restrictions on its use for aerosols and other pharmaceutical applications.

In the UK, the workplace exposure limits for carbon dioxide are 9150 mg/m^3 (5000 ppm) long-term (8-hour TWA) and 27 400 mg/m^3 (15 000 ppm) short-term (15-minute).[6] In the USA, the permissible exposure limits are 9000 mg/m^3 (5000 ppm) long-term and the recommended exposure limits are 18 000 mg/m^3 (10 000 ppm) short-term and 54 000 mg/m^3 (30 000 ppm) maximum, short-term.[7]

Solid carbon dioxide can produce severe burns in contact with the skin and appropriate precautions, depending on the circumstances and quantity of material handled, should be taken. A face shield and protective clothing, including thick gloves, are recommended.

16 Regulatory Status

GRAS listed. Accepted for use in Europe as a food additive. Included in the FDA Inactive Ingredients Database (aerosol formulation for nasal preparations; IM and IV injections). Included in the Canadian Natural Health Products Ingredients Database.

17 Related Substances

Nitrogen; nitrous oxide.

18 Comments

Supercritical carbon dioxide has been used in the formation of fine powders of stable protein formulations.[8,9]

Carbon dioxide has also been investigated for its suitability in Aerosol Solvent Extraction Systems (ASES), to generate microparticles of proteins suitable for aerosol delivery from aqueous based solutions.[10]

A specification for carbon dioxide is contained in the *Food Chemicals Codex* (FCC).[11]

The EINECS number for carbon dioxide is 204-696-9. The PubChem Compound ID (CID) for carbon dioxide is 280.

19 Specific References

1 Haase LW. Application of carbon dioxide in cosmetic aerosols. *Cosmet Perfum* 1975; 90(8): 31–32.

2 Sanders PA. Aerosol packaging of pharmaceuticals. In: Banker GS, Rhodes CT, eds. *Modern Pharmaceutics*. New York: Marcel Dekker, 1979: 591–626.

3 Anonymous. CO_2/acetone propellant kinder to ozone layer. *Manuf Chem* 1992; 63(1): 14.

4 King JS, Mabbitt LA. The use of carbon dioxide for the preservation of milk. In: Board RG *et al.* eds. *Preservatives in the Food, Pharmaceutical and Environmental Industries*. Oxford: Blackwell Scientific, 1987: 35–43.

5 Anonymous. Carbon dioxide breaks the mould. *Chem Br* 1992; 28: 506.

6 Health and Safety Executive. *EH40/2005: Workplace Exposure Limits*. Sudbury: HSE Books, 2011. http://www.hse.gov.uk/pubns/priced/eh40.pdf (accessed 29 February 2012).

7 National Institute for Occupational Safety and Health. Recommendations for occupational safety and health. *MMWR* 1988; 37(Suppl. S-7): 1–29.

8 Bettini R, *et al.* Solubility and conversion of carbamazepine polymorphs in supercritical carbon dioxide. *Eur J Pharm Sci* 2001; 13(3): 281–286.

9 Sellers SP, *et al.* Dry powders of stable protein formulations from aqueous solutions prepared using supercritical CO_2-assisted aerosolization. *J Pharm Sci* 2001; 90: 785–797.

10 Bustami RT, *et al.* Generation of microparticles of proteins for aerosol delivery using high pressure modified carbon dioxide. *Pharm Res* 2000; 17: 1360–1366.

11 *Food Chemicals Codex*, 7th edn. (Suppl. 1), Bethesda, MD: United States Pharmacopeia, 2010: 174.

20 General References

Johnson MA. *The Aerosol Handbook*, 2nd edn. Mendham, NJ: WE Dorland Co., 1982: 361–372.

Sanders PA. *Handbook of Aerosol Technology*, 2nd edn. New York: Van Nostrand Reinhold, 1979: 44–54.

Sciarra JJ, Sciarra CJ. Pharmaceutical and cosmetic aerosols. *J Pharm Sci* 1974; 63: 1815–1837.

Sciarra CJ, Sciarra JJ. Aerosols. In: *Remington: The Science and Practice of Pharmacy*, 21st edn. Philadelphia, PA: Lippincott, Williams and Wilkins, 2005: 1000–1017.

Sciarra CJ, Sciarra JJ. Pressurized dispensers. In: Schlossman ML. *The Chemistry and Manufacture of Cosmetics*, 4th edn, vol. 1. Carol Stream, IL: Allured Publishing Corporation, 2009: 451–478.

Sciarra JJ. Pharmaceutical aerosols. In: Banker GS, Rhoes CT, eds. *Modern Pharmaceutics*, 3rd edn. New York: Marcel Dekker, 1996: 547–574.

Sciarra JJ, Stoller L. *The Science and Technology of Aerosol Packaging*. New York: Wiley, 1974: 137–145.

21 Authors

CJ Sciarra, JJ Sciarra.

22 Date of Revision

1 March 2012.

Carboxymethylcellulose Calcium

1 Nonproprietary Names

BP: Carmellose Calcium
JP: Carmellose Calcium
PhEur: Carmellose Calcium
USP–NF: Carboxymethylcellulose Calcium

2 Synonyms

Calcium carboxymethylcellulose; calcium cellulose glycolate; carmellosum calcium; CMC calcium; *ECG 505*; Nymcel ZSC.

3 Chemical Name and CAS Registry Number

Cellulose, carboxymethyl ether, calcium salt [9050-04-8]

4 Empirical Formula and Molecular Weight

The USP35–NF30 describes carboxymethylcellulose calcium as the calcium salt of a polycarboxymethyl ether of cellulose.

5 Structural Formula

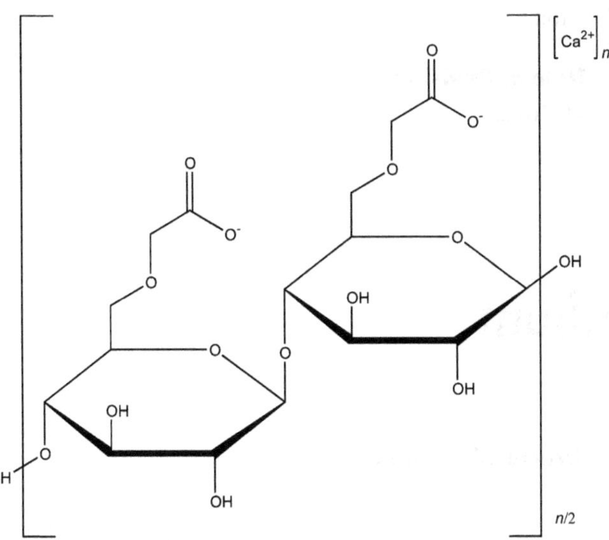

Structure shown with a degree of substitution (DS) of 1.0.

6 Functional Category

Adsorbent; coating agent; suspending agent; tablet and capsule binder; tablet and capsule disintegrant; viscosity-increasing agent.

7 Applications in Pharmaceutical Formulation or Technology

The main use of carboxymethylcellulose calcium is in tablet formulations (*see* Table I), where it is used as a binder, diluent, and disintegrant.[1–4] Although carboxymethylcellulose calcium is insoluble in water, it is an effective tablet disintegrant as it swells to several times its original bulk on contact with water. Concentrations up to 15% w/w may be used in tablet formulations; above this concentration, tablet hardness is reduced.

Carboxymethylcellulose calcium is also used in other applications similarly to carboxymethylcellulose sodium; for example, as a suspending or viscosity-increasing agent in oral and topical pharmaceutical formulations. Carboxymethylcellulose calcium is also used in modern wound dressings for its water absorption, retention and hemostatic properties.

Table I: Uses of carboxymethylcellulose calcium.

Use	Concentration (%)
Tablet binder	5–15
Tablet disintegrant	1–15

8 Description

Carboxymethylcellulose calcium occurs as a white to yellowish-white, hygroscopic powder.

9 Pharmacopeial Specifications

The pharmacopeial specifications for carboxymethylcellulose calcium have undergone harmonization of many attributes for JP, PhEur, and USP–NF.

See Table II. *See also* Section 18.

Table II: Pharmacopeial specifications for carboxymethylcellulose calcium.

Test	JP XV	PhEur 7.4	USP35–NF30
Identification	+	+	+
Characters	—	+	—
Alkalinity	+	+	+
pH	4.5–6.0	—	—
Loss on drying	≤10.0%	≤10.0%	≤10.0%
Residue on ignition	10.0–20.0%	—	10.0–20.0%
Chloride	≤0.36%	≤0.36%	≤0.36%
Sulfate	≤1.0%	≤1.0%	≤1.0%
Heavy metals[a]	≤20 ppm	≤20 ppm	≤20 ppm

(a) These tests have not been fully harmonized at the time of publication.

10 Typical Properties

Acidity/alkalinity pH = 4.5–6.0 for a 1% w/v aqueous dispersion.

Solubility Practically insoluble in acetone, chloroform, ethanol (95%), toluene, and ether. Insoluble in water, but swells to form a suspension.

11 Stability and Storage Conditions

Carboxymethylcellulose calcium is a stable, though hygroscopic material. It should be stored in a well-closed container in a cool, dry place.

See also Carboxymethylcellulose Sodium.

12 Incompatibilities

See Carboxymethylcellulose Sodium.

13 Method of Manufacture

Cellulose, obtained from wood pulp or cotton fibers, is carboxymethylated, followed by conversion to the calcium salt. It is then graded on the basis of its degree of carboxymethylation and pulverized.

14 Safety

Carboxymethylcellulose calcium is used in oral and topical pharmaceutical formulations, similarly to carboxymethylcellulose sodium, and is generally regarded as a nontoxic and nonirritant material. However, as with other cellulose derivatives, oral consumption of large amounts of carboxymethylcellulose calcium may have a laxative effect.

See also Carboxymethylcellulose Sodium.

15 Handling Precautions

Observe normal precautions appropriate to the circumstances and quantity of material handled. Carboxymethylcellulose calcium may be irritant to the eyes; eye protection is recommended.

16 Regulatory Status

Included in the FDA Inactive Ingredients Database (oral capsules and tablets). Included in nonparenteral medicines licensed in the UK. Included in the Canadian Natural Health Products Ingredients Database.

17 Related Substances

Carboxymethylcellulose sodium; croscarmellose sodium.

18 Comments

Carboxymethylcellulose calcium has undergone harmonization of many attributes for JP, PhEur, and USP–NF by the Pharmacopeial Discussion Group. For further information see the General Information Chapter <1196> in the USP35–NF30, the General Chapter 5.8 in PhEur 7.4, along with the 'State of Work' document on the PhEur EDQM website, and also the General Information Chapter 8 in the JP XV.

19 Specific References

1 Khan KA, Rooke DJ. Effect of disintegrant type upon the relationship between compressional pressure and dissolution efficiency. *J Pharm Pharmacol* 1976; **28**(8): 633–636.

2 Kitamori N, Makino T. Improvement in pressure-dependent dissolution of trepibutone tablets by using intragranular disintegrants. *Drug Dev Ind Pharm* 1982; **8**(1): 125–139.

3 Roe TS, Chang KY. The study of Key-Jo clay as a tablet disintegrator. *Drug Dev Ind Pharm* 1986; **12**(11–13): 1567–1585.

4 Ozeki T, *et al*. Design of rapidly disintegrating oral tablets using acid-treated yeast cell wall: a technical note. *AAPS Pharm Sci Tech* 2003; **4**(4): E70.

20 General References

Doelker E. Cellulose derivatives. *Adv Polym Sci* 1993; **107**: 199–265.

European Directorate for the Quality of Medicines and Healthcare (EDQM). European Pharmacopoeia – State Of Work Of International Harmonisation. *Pharmeuropa* 2011; 23(4): 713–714. www.edqm.eu/site/-614.html (accessed 5 December 2011).

21 Author

JC Hooton.

22 Date of Revision

1 March 2012.

Carboxymethylcellulose Sodium

1 Nonproprietary Names

BP: Carmellose Sodium
JP: Carmellose Sodium
PhEur: Carmellose Sodium
USP–NF: Carboxymethylcellulose Sodium

2 Synonyms

Akucell; *Aqualon CMC*; *Blanose*; *Carbose D*; carmellosum natricum; cellulose gum; CMC sodium; *Dynacel*; E466; *Finnfix*; *Nymcel ZSB*; SCMC; sodium carboxymethylcellulose; sodium cellulose glycolate; *Sunrose*; *Tylose CB*; *Tylose MGA*; *Walocel C*; *Walocel CRT*; *Xylo-Mucine*.

3 Chemical Name and CAS Registry Number

Cellulose, carboxymethyl ether, sodium salt [9004-32-4]

4 Empirical Formula and Molecular Weight

The USP35–NF30 describes carboxymethylcellulose sodium as the sodium salt of a polycarboxymethyl ether of cellulose.

5 Structural Formula

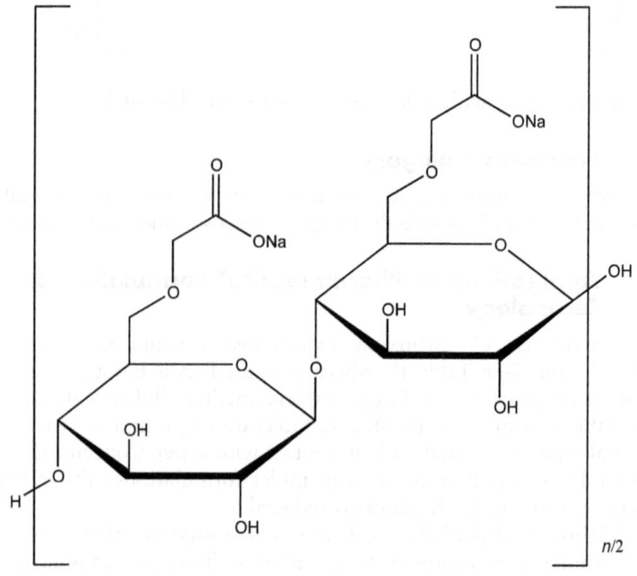

Structure shown with a degree of substitution (DS) of 1.0.

6 Functional Category

Adsorbent; emulsifying agent; suspending agent; tablet and capsule binder; tablet and capsule disintegrant; viscosity-increasing agent.

7 Applications in Pharmaceutical Formulation or Technology

Carboxymethylcellulose sodium is widely used in oral and topical pharmaceutical formulations, primarily for its viscosity-increasing properties. Viscous aqueous solutions are used to suspend powders intended for either topical application or oral and parenteral administration.[1,2] Carboxymethylcellulose sodium may also be used as a tablet binder and disintegrant,[3–6] and to stabilize emulsions.[7,8]

Encapsulation with carboxymethylcellulose sodium can affect drug protection and delivery.[6,9]

Higher concentrations, usually 3–6%, of the medium-viscosity grade are used to produce gels that can be used as the base for applications and pastes; glycols are often included in such gels to prevent them drying out.

Carboxymethylcellulose sodium is also used in cosmetics, toiletries,[10] personal hygiene, and food products.

See Table I.

Table I: Uses of carboxymethylcellulose sodium.

Use	Concentration (%)
Emulsifying agent	0.25–1.0
Gel-forming agent	3.0–6.0
Injections	0.05–0.75
Oral solutions	0.1–1.0
Tablet binder	1.0–6.0

8 Description

Carboxymethylcellulose sodium occurs as a white to almost white, odorless, tasteless, granular powder. It is hygroscopic after drying.

See also Section 18.

9 Pharmacopeial Specifications

See Table II. *See also* Section 18.

Table II: Pharmacopeial specifications for carboxymethylcellulose sodium.

Test	JP XV	PhEur 7.4	USP35–NF30
Identification	+	+	+
Characters	+	+	—
pH (1% w/v solution)	6.0–8.0	6.0–8.0	6.5–8.5
Appearance of solution	+	+	—
Viscosity	—	+	+
Loss on drying	≤10.0%	≤10.0%	≤10.0%
Heavy metals	≤20 ppm	≤20 ppm	≤20 ppm
Chloride	≤0.640%	≤0.25%	—
Arsenic	≤10 ppm	—	—
Sulfate	≤0.960%	—	—
Silicate	≤0.5%	—	—
Sodium glycolate	—	≤0.4%	—
Starch	+	—	—
Sulfated ash	—	20.0–33.3%	—
Assay (of sodium)	6.5–8.5%	6.5–10.8%	6.5–9.5%

10 Typical Properties

Density (bulk) 0.52 g/cm^3

Density (tapped) 0.78 g/cm^3

Dissociation constant pK_a = 4.30

SEM 1: Excipient: carboxymethylcellulose sodium; manufacturer: Buckeye Cellulose Corp.; lot no.: 9247 AP; magnification: 120×; voltage: 10 kV.

SEM 2: Excipient: carboxymethylcellulose sodium; manufacturer: Ashland Aqualon Functional Ingredients; lot no.: 21 A-1 (44390); magnification: 600×; voltage: 10 kV.

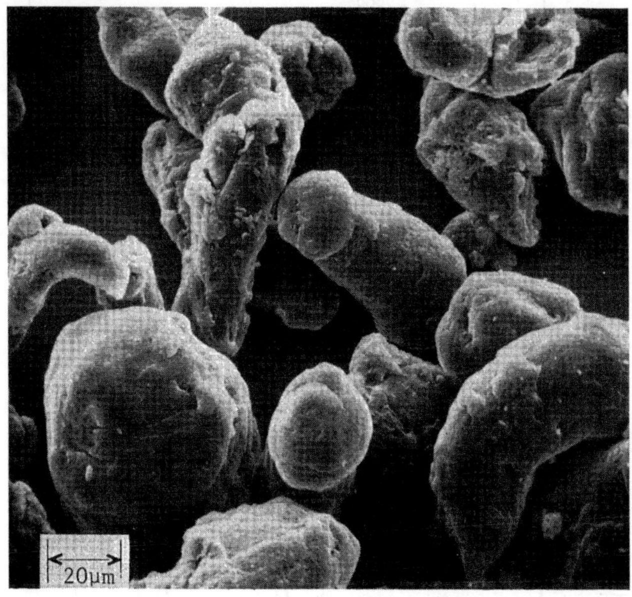

Melting point Browns at approximately 227°C, and chars at approximately 252°C.

Moisture content Typically contains less than 10% water. However, carboxymethylcellulose sodium is hygroscopic and absorbs significant amounts of water at temperatures up to 37°C at relative humidities of about 80%.

See Section 11. *See also* Figure 1.

Solubility Practically insoluble in acetone, ethanol (95%), ether, and toluene. Easily dispersed in water at all temperatures, forming clear, colloidal solutions. The aqueous solubility varies with the degree of substitution (DS). *See* Section 18.

Spectroscopy

IR spectra *see* Figure 2.

NIR spectra *see* Figure 3.

Viscosity Various grades of carboxymethylcellulose sodium are commercially available that have differing aqueous viscosities;

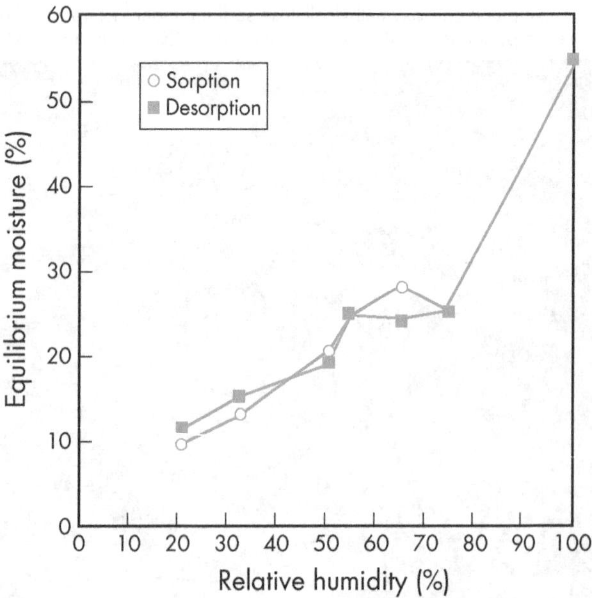

Figure 1: Sorption–desorption isotherm of carboxymethylcellulose sodium.

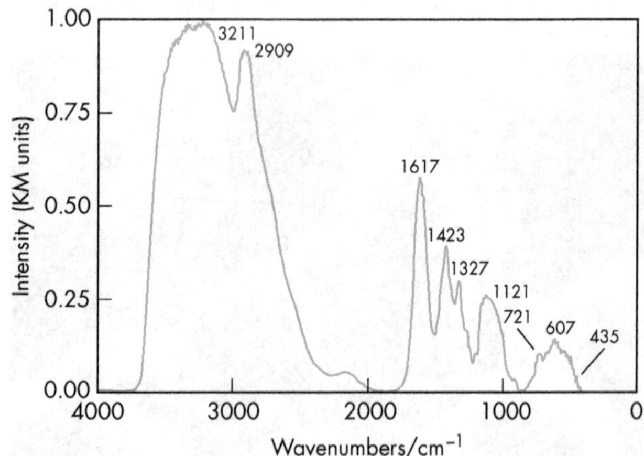

Figure 2: Infrared spectrum of carboxymethylcellulose sodium measured by diffuse reflectance. Adapted with permission of Informa Healthcare.

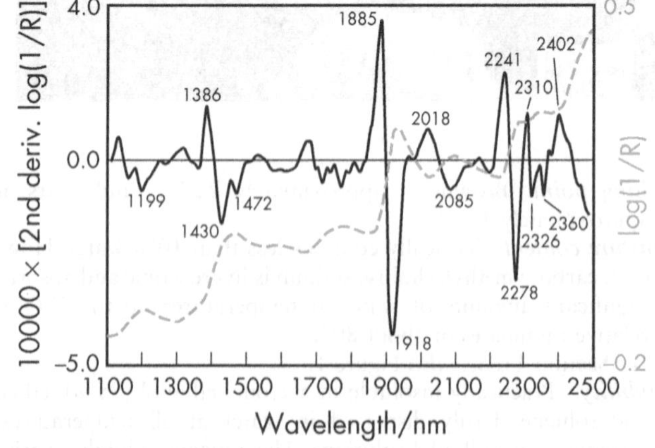

Figure 3: Near-infrared spectrum of carboxymethylcellulose sodium measured by reflectance.

see Table III. Aqueous 1% w/v solutions with viscosities of 5–2000 mPa s (5–2000 cP) may be obtained. An increase in concentration results in an increase in aqueous solution viscosity.[10] Prolonged heating at high temperatures will depolymerize the gum and permanently decrease the viscosity. The viscosity of sodium carboxymethylcellulose solutions is fairly stable over a pH range of 4–10. The optimum pH range is neutral.

See Section 11.

Table III: Viscosity of aqueous carboxymethylcellulose sodium 1% w/v solutions. (Measurements made with a Brookfield LVT viscometer at 25°C.)

	Grade	Viscosity (mPa s)	Spindle	Speed
Low viscosity	*Akucell AF 0305*	10–15	#1	60 rpm
Medium viscosity	*Akucell AF 2785*	1500–2500	#3	30 rpm
High viscosity	*Akucell AF 3085*	8000–12000	#4	30 rpm

11 Stability and Storage Conditions

Carboxymethylcellulose sodium is a stable, though hygroscopic material. Under high-humidity conditions, carboxymethylcellulose sodium can absorb a large quantity (>50%) of water. In tablets, this has been associated with a decrease in tablet hardness and an increase in disintegration time.[11]

Aqueous solutions are stable at pH 2–10; precipitation can occur below pH 2, and solution viscosity decreases rapidly above pH 10. Generally, solutions exhibit maximum viscosity and stability at pH 7–9.

Carboxymethylcellulose sodium may be sterilized in the dry state by maintaining it at a temperature of 160°C for 1 hour. However, this process results in a significant decrease in viscosity and some deterioration in the properties of solutions prepared from the sterilized material.

Aqueous solutions may similarly be sterilized by heating, although this also results in some reduction in viscosity. After autoclaving, viscosity is reduced by about 25%, but this reduction is less marked than for solutions prepared from material sterilized in the dry state. The extent of the reduction is dependent on the molecular weight and degree of substitution; higher molecular weight grades generally undergo a greater percentage reduction in viscosity.[12] Sterilization of solutions by gamma irradiation also results in a reduction in viscosity.

Aqueous solutions stored for prolonged periods should contain an antimicrobial preservative.[13]

The bulk material should be stored in a well-closed container in a cool, dry place.

12 Incompatibilities

Carboxymethylcellulose sodium is incompatible with strongly acidic solutions and with the soluble salts of iron and some other metals, such as aluminum, mercury, and zinc. It is also incompatible with xanthan gum. Precipitation may occur at pH <2, and also when it is mixed with ethanol (95%).

Carboxymethylcellulose sodium forms complex coacervates with gelatin and pectin. It also forms a complex with collagen and is capable of precipitating certain positively charged proteins.

13 Method of Manufacture

Alkali cellulose is prepared by steeping cellulose obtained from wood pulp or cotton fibers in sodium hydroxide solution. The alkaline cellulose is then reacted with sodium monochloroacetate to produce carboxymethylcellulose sodium. Sodium chloride and sodium glycolate are obtained as by-products of this etherification.

14 Safety

Carboxymethylcellulose sodium is used in oral, topical, and some parenteral formulations. It is also widely used in cosmetics, toiletries, and food products, and is generally regarded as a nontoxic and nonirritant material. However, oral consumption of large amounts of carboxymethylcellulose sodium can have a laxative effect; therapeutically, 4–10 g in daily divided doses of the medium- and high-viscosity grades of carboxymethylcellulose sodium have been used as bulk laxatives.[(14)]

The WHO has not specified an acceptable daily intake for carboxymethylcellulose sodium as a food additive since the levels necessary to achieve a desired effect were not considered to be a hazard to health.[(15–18)] However, in animal studies, subcutaneous administration of carboxymethylcellulose sodium has been found to cause inflammation, and in some cases of repeated injection fibrosarcomas have been found at the site of injection.[(19)]

Hypersensitivity and anaphylactic reactions have occurred in cattle and horses, which have been attributed to carboxymethylcellulose sodium in parenteral formulations such as vaccines and penicillins.[(20–23)]

LD_{50} (guinea pig, oral): 16 g/kg[(24)]

LD_{50} (rat, oral): 27 g/kg

15 Handling Precautions

Observe normal precautions appropriate to the circumstances and quantity of material handled. Carboxymethylcellulose sodium may be irritant to the eyes. Eye protection is recommended.

16 Regulatory Status

GRAS listed. Accepted as a food additive in Europe. Included in the FDA Inactive Ingredients Database (dental preparations; intra-articular, intrabursal, intradermal, intralesional, and intrasynovial injections; oral drops, solutions, suspensions, syrups and tablets; topical preparations). Included in nonparenteral medicines licensed in the UK. Included in the Canadian Natural Health Products Ingredients Database.

17 Related Substances

Carboxymethylcellulose calcium.

18 Comments

Carboxymethylcellulose sodium is one of the materials that have been selected for harmonization by the Pharmacopeial Discussion Group. For further information see the General Information Chapter <1196> in the USP35–NF30, the General Chapter 5.8 in PhEur 7.4, along with the 'State of Work' document on the PhEur EDQM website, and also the General Information Chapter 8 in the JP XV.

A number of grades of carboxymethylcellulose sodium are commercially available, such as *Accelerate*. These have a degree of substitution (DS) in the range 0.7–1.2. The DS is defined as the average number of hydroxyl groups substituted per anhydroglucose unit and it is this that determines the aqueous solubility of the polymer. Thermal crosslinking reduces solubility while retaining water absorption, therefore producing materials suitable for water absorption.

Grades are typically classified as being of low, medium, or high viscosity. The degree of substitution and the maximum viscosity of an aqueous solution of stated concentration should be indicated on any carboxymethylcellulose sodium labeling.

Carboxymethylcellulose sodium has been reported to give false positive results in the LAL test for endotoxins.[(25)]

Carboxymethylcellulose sodium is used in self-adhesive ostomy, wound care,[(26)] and dermatological patches as a mucoadhesive and to absorb wound exudate or transepidermal water and sweat. The mucoadhesive property is used in products designed to prevent post-surgical tissue adhesions;[(27–29)] to localize and modify the release kinetics of active ingredients applied to mucous membranes; and for bone repair. There have also been reports of its use as a cyto-protective agent.[(30,31)] Carboxymethylcellulose sodium is also used in concentrations of up to 1% as artificial saliva preparations for the management of dry mouth, and in eye drops for the management of dry eye. It has also been used in surgical prosthetics,[(32)] and in the treatment of constipation and incontinence.

The PubChem Compound ID (CID) for carboxymethylcellulose sodium is 23706213.

19 Specific References

1 Hussain MA, *et al.* Injectable suspensions for prolonged release nalbuphine. *Drug Dev Ind Pharm* 1991; **17**(1): 67–76.
2 Chang JH, *et al.* Radiographic contrast study of the upper gastro-intestinal tract of eight dogs using carboxymethylcellulose mixed with a low concentration of barium sulphate. *Vet Rec* 2004; **154**(7): 201–204.
3 Khan KA, Rhodes CT. Evaluation of different viscosity grades of sodium carboxymethylcellulose as tablet disintegrants. *Pharm Acta Helv* 1975; **50**: 99–102.
4 Shah NH, *et al.* Carboxymethylcellulose: effect of degree of polymerization and substitution on tablet disintegration and dissolution. *J Pharm Sci* 1981; **70**(6): 611–613.
5 Singh J. Effect of sodium carboxymethylcelluloses on the disintegration, dissolution and bioavailability of lorazepam from tablets. *Drug Dev Ind Pharm* 1992; **18**(3): 375–383.
6 Dabbagh MA, *et al.* Release of propanolol hydrochloride from matrix tablets containing sodium carboxymethylcellulose and hydroxypropylmethylcellulose. *Pharm Dev Technol* 1999; **4**(3): 313–324.
7 Oza KP, Frank SG. Microcrystalline cellulose stabilized emulsions. *J Disper Sci Technol* 1986; **7**(5): 543–561.
8 Adeyeye MC, *et al.* Viscoelastic evaluation of topical creams containing microcrystalline cellulose/sodium carboxymethylcellulose as stabilizer. *AAPS Pharm Sci Tech* 2002; **3**(2): E8.
9 Marschutz MK, *et al.* Design and *in vivo* evaluation of an oral delivery system for insulin. *Pharm Res* 2000; **17**(12): 1468–1474.
10 Mombellet H, Bale P. Sodium carboxymethylcellulose toothpaste. *Manuf Chem* 1988; **59**(11): 47, 49, 52.
11 Khan KA, Rhodes CT. Water-sorption properties of tablet disintegrants. *J Pharm Sci* 1975; **64**(6): 447–451.
12 Chu PI, Doyle D. Development and evaluation of a laboratory-scale apparatus to simulate the scale-up of a sterile semisolid and effects of manufacturing parameters on product viscosity. *Pharm Dev Technol* 1999; **4**(4): 553–559.
13 Banker G, *et al.* Microbiological considerations of polymer solutions used in aqueous film coating. *Drug Dev Ind Pharm* 1982; **8**(1): 41–51.
14 Wapnir RA, *et al.* Cellulose derivatives and intestinal absorption of water and electrolytes: potential role in oral rehydration solutions. *Proc Soc Exp Biol Med* 1997; **215**(3): 275–280.
15 FAO/WHO. Evaluation of certain food additives and contaminants. Thirty-fifth report of the joint FAO/WHO expert committee on food additives. *World Health Organ Tech Rep Ser* 1990; No. 789.
16 Til HP, Bar A. Subchronic (13-week) oral toxicity study of gamma-cyclodextrin in dogs. *Regul Toxicol Pharmacol* 1998; **27**(2): 159–165.
17 Diebold Y, *et al.* Carbomer- versus cellulose-based artificial-tear formulations: morphological and toxicologic effects on a corneal cell line. *Cornea* 1998; **17**(4): 433–440.
18 Ugwoke MI, *et al.* Toxicological investigations of the effects of carboxymethylcellulose on ciliary beat frequency of human nasal epithelial cells in primary suspension culture and *in vivo* on rabbit nasal mucosa. *Int J Pharm* 2000; **205**(1–2): 43–51.
19 Teller MN, Brown GB. Carcinogenicity of carboxymethylcellulose in rats. *Proc Am Assoc Cancer Res* 1977; **18**: 225.
20 Schneider CH, *et al.* Carboxymethylcellulose additives in penicillins and the elucidation of anaphylactic reactions. *Experientia* 1971; **27**: 167–168.
21 Aitken MM. Induction of hypersensitivity to carboxymethylcellulose in cattle. *Res Vet Sci* 1975; **19**: 110–113.

22 Bigliardi PL, *et al.* Anaphylaxis to the carbohydrate carboxymethylcellulose in parenteral corticosteroid preparations. *Dermatology* 2003; 207(1): 100–103.
23 Montoro J, *et al.* Anaphylactic shock after intra-articular injection of carboxymethylcellulose. *Allergol Immunopathol* 2000; **28**(6): 332–333.
24 Lewis RJ, ed. *Sax's Dangerous Properties of Industrial Materials*, 11th edn. New York: Wiley, 2004; 3236.
25 Tanaka S, *et al.* Activation of a limulus coagulation factor G by (1→3)-β-D-glucans. *Carbohydr Res* 1991; **218**: 167–174.
26 Fletcher J. The benefits if using hydrocolloids. *Nurs Times* 2003; 99(21): 57.
27 Yelimlies B, *et al.* Carboxymethylcellulose coated on visceral face of polypropylene mesh prevents adhesion without impairing wound healing in incisional hernia model in rats. *Hernia* 2003; 7(3): 130–133.
28 Hay WP, *et al.* One percent sodium carboxymethylcellulose prevents experimentally induced adhesions in horses. *Vet Surg* 2001; **30**(3): 223–227.
29 Liu LS, Berg RA. Adhesion barriers of carboxymethylcellulose and polyethylene oxide composite gels. *J Biomed Mater Res* 2002; **63**(3): 326–332.
30 Ahee JA, *et al.* Decreased incidence of epithelial defects during laser *in situ* keratomileusis using intraoperative nonpreserved carboxymethylcellulose sodium 0.5% solution. *J Cateract Refract Surg* 2002; **28**(9): 1651–1654.
31 Vehige JG, *et al.* Cytoprotective properties of carboxymethylcellulose (CMC) when used prior to wearing contact lenses treated with cationic disinfecting agents. *Eye Contact Lens* 2003; **29**(3): 177–180.
32 Valeriani M, *et al.* Carboxymethylcellulose hydrogel mammary implants: our experience. *Acta Chir Plast* 2002; **44**(3): 77–79.

20 General References

Doelker E. Cellulose derivatives. *Adv Polym Sci* 1993; **107**: 199–265.
European Directorate for the Quality of Medicines and Healthcare (EDQM). European Pharmacopoeia – State Of Work Of International Harmonisation. *Pharmeuropa* 2011; 23(4): 713–714. www.edqm.eu/site/-614.html (accessed 5 December 2011).

21 Author

JC Hooton.

22 Date of Revision

1 March 2012.

Carrageenan

1 Nonproprietary Names

BP: Carrageenan
PhEur: Carrageenan
USP–NF: Carrageenan

2 Synonyms

Carrageenanum; chondrus extract; E407; *Gelcarin*; *Genu*; *Grindsted*; *Hygum TP-1*; Irish moss extract; *Marine Colloids*; *SeaSpen PF*; *Viscarin*.

3 Chemical Name and CAS Registry Number

Carrageenan [9000-07-1]
ι-Carrageenan [9062-07-1]
κ-Carrageenan [11114-20-8]
λ-Carrageenan [9064-57-7]

4 Empirical Formula and Molecular Weight

The USP35–NF30 describes carrageenan as the hydrocolloid obtained by extraction with water or aqueous alkali from some members of the class Rhodophyceae (red seaweeds). It consists chiefly of potassium, sodium, calcium, magnesium, and ammonium sulfate esters of galactose and 3,6-anhydrogalactose copolymers. These hexoses are alternately linked at the α-1,3 and β-1,4 sites in the polymer.

5 Structural Formula

Kappa

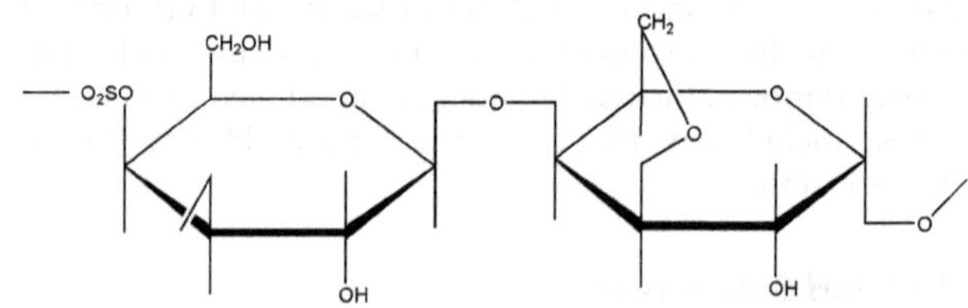

Iota

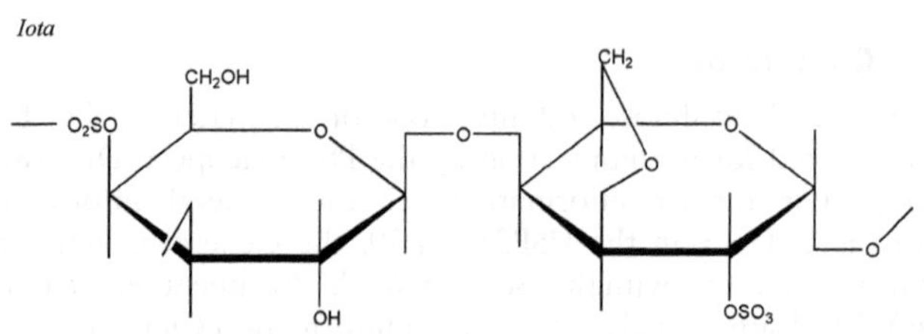

Lambda

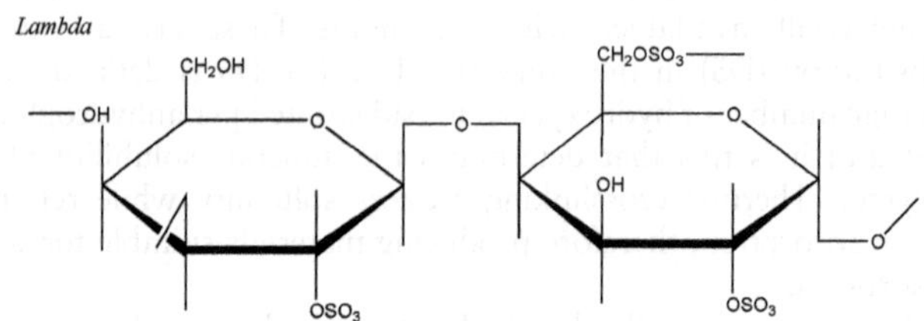

Three commercially important carrageenans exist currently and are divided according to the number and position of ester sulfate groups and the percentage of 3,6-anhydrogalactose.

κ-Carrageenan (kappa-carrageenan) is mainly the alternating polymer of D-galactose-4-sulfate and 3,6-anhydro-D-galactose. It contains approximately 25% ester sulfate and 34% 3,6-anhydrogalactose.

ι-Carrageenan (iota-carrageenan) is similar except that 3,6-anhydrogalactose is sulfated at carbon 2. It contains approximately 32% ester sulfate and 30% 3,6-anhydrogalactose.

λ-Carrageenan (lambda-carrageenan) has alternating monomeric units, which are mostly D-galactose-2-sulfate (1,3-linked) and D-galactose-2,6-disulfate (1,4-linked). It contains approximately 35% ester sulfate by weight and little or no 3,6-anhydrogalactose.

6 Functional Category

Emulsion stabilizing agent; gelling agent; microencapsulating agent; modified-release agent; suspending agent; viscosity-increasing agent.

7 Applications in Pharmaceutical Formulation or Technology

Carrageenan is used in a wide variety of nonparenteral dosage forms, including suspensions (wet and reconstitutable), emulsions, gels, creams, lotions, eye drops, suppositories, tablets, and capsules; *see* Table I.

In suspension formulations, usually only the ι-carrageenan and λ-carrageenan fractions are used. λ-Carrageenan is generally used at levels of 0.7% w/v or less, and provides viscosity to the liquid. With pure ι-carrageenan, about 0.4% w/v is required for most suspensions, plus the addition of calcium ions to establish a gel network. However, if *SeaSpen PF* is used, it should be used at about 0.75% w/v level, although no additional calcium is required as it is already present in the product to control the rate of gelation. Carrageenan has been shown to mask the chalkiness of antacid suspensions when used as a suspending agent in these preparations.[1] When used in concentrations of 0.1–0.5%, carrageenan gives stable emulsions.

Transdermal patches of cubic gels using carrageenan as the matrix have been formulated.[2,3] Carrageenan has also been used in the preparation of hard and soft capsule shells.[4]

ι-Carrageenan and κ-Carrageenan are claimed to provide a creamy mouthfeel in chewable tablet formulations. κ-Carrageenan, along with other excipients, is also used for oral instant-release formulations.[5] Carrageenan-based chewing gums are reported to be useful in the treatment of xerostomia.[6]

κ-Carrageenan is known as a novel pelletization aid in the manufacture of pellets by extrusion/spheronization and has been shown to exhibit better disintegration behavior and improved bioavailability for poorly water-soluble drugs at high doses.[7–9]

Due to its mucoadhesive properties, carrageenan has been evaluated for oral,[10] buccal,[11] nasal[12] and vaginal drug delivery.[13,14] It was found to prolong the local residence time of a poloxamer 407-based *in situ* vaginal gel, and also showed a synergistic bioadhesive effect with acrylic acid polymers.[15]

Different carrageenans along with various drugs and other excipients in tablet matrices have been shown to retard drug release in oral controlled-release formulations,[16–20] nasal inserts,[21] and occular films and microspheres.[22] Furthermore, the inclusion of calcium or potassium salts in the tablet creates a microenvironment for gelation to occur, which further controls drug release.

Carrageenan is used in toothpastes and cosmetic preparations such as hair conditioners, shampoos, moisturizers and other skin lotions. It is also widely used in food applications such as dairy products and dessert gels and glazes.

Table I: Typical uses of commercially available grades of carrageenan (FMC BioPolymer).

Carrageenan type	Trade name	Use concentration (%)	Use examples
Iota	*Gelcarin GP-379*	0.3–1.0	Creams, suspensions
Iota	*SeaSpen PF*	0.5–1.0	Suspensions, topical lotions/creams, reconstitutable suspensions
Kappa	*Gelcarin GP-812*	0.3–1.0	Gels
Kappa	*Gelcarin GP-911*	0.25–2.0	Encapsulation/delivery systems
Lambda	*Viscarin GP-109*	0.1–1.0	Creams, lotions
Lambda	*Viscarin GP-209*	0.1–1.0	Creams, lotions

8 Description

Carrageenan occurs as a yellowish or tan to white colored, coarse to fine powder that is odorless with a mucilaginous taste.

9 Pharmacopeial Specifications

See Table II.

Table II: Pharmacopeial specifications for carrageenan.

Test	PhEur 7.4	USP35–NF30
Identification	+	+
Characters	+	—
Acid insoluble matter	≤2.0%	≤2.0%
Arsenic	—	≤3 ppm
Heavy metals	≤20 ppm	≤40 ppm
Lead	—	≤10 ppm
Loss on drying	≤12%	≤12.5%
Total ash	≤40.0%	≤35.0%
Viscosity (at 75°C)	≥5 mPa s	≥5 mPa s
Microbial limits	—	≤200 cfu/g[a]

(a) Tests for *Salmonella* and *Escherichia coli* are negative.

10 Typical Properties

Because of the vast differences in the material that can be referred to as carrageenan, it is difficult to give descriptions of typical properties. *See* Table III.

Gelation ι-Carrageenan forms elastic gels and thixotropic fluids in the presence of calcium ions. κ-Carrageenan forms firm gels in the presence of potassium ions. λ-Carrageenan forms viscous, nongelling solutions. *See* Table IV

pH 7.0–10.5 (in solution)

Rheology Kappa gels cannot be sheared to form a flowable liquid; however, iota gels display thixotropic behavior when sheared. λ-Carrageenan displays pseudoplastic behavior under all conditions. *See* Table IV.

Solubility Soluble in water at 80°C. *See also* Tables III and IV.

Spectroscopy

IR spectra *see* Figure 1.

NIR spectra *see* Figures 2, 3, 4 and 5.

Viscosity (dynamic) 5 mPa s (5 cP) at 75°C. *See also* Table III.

11 Stability and Storage Conditions

Carrageenan is a stable, though hygroscopic, polysaccharide and should be stored in an airtight container, in a dry place at room temperature (approx. 25°C).

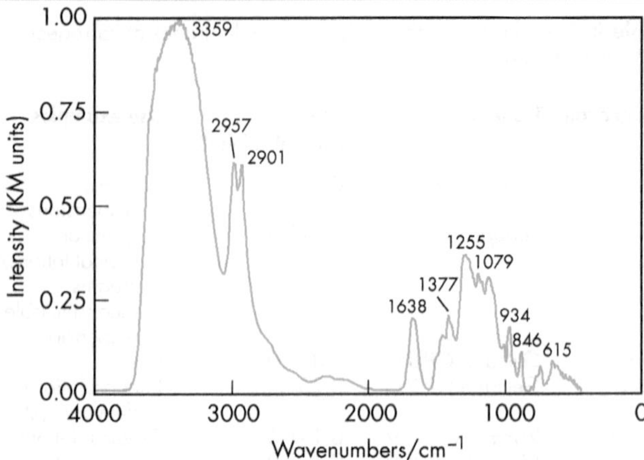

Figure 1: Infrared spectrum of carrageenan measured by diffuse reflectance. Adapted with permission of Informa Healthcare.

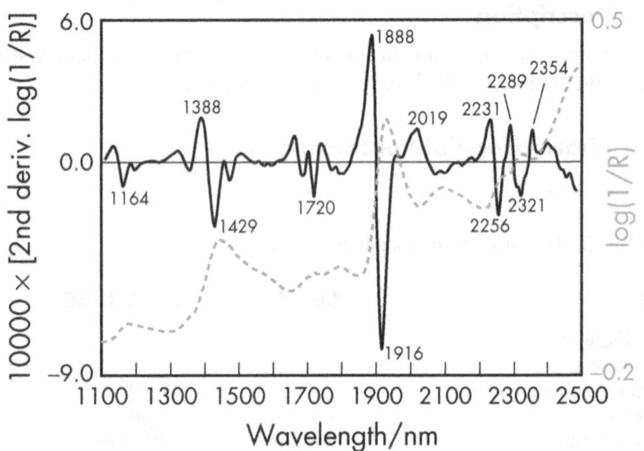

Figure 2: Near-infrared spectrum of carrageenan measured by reflectance.

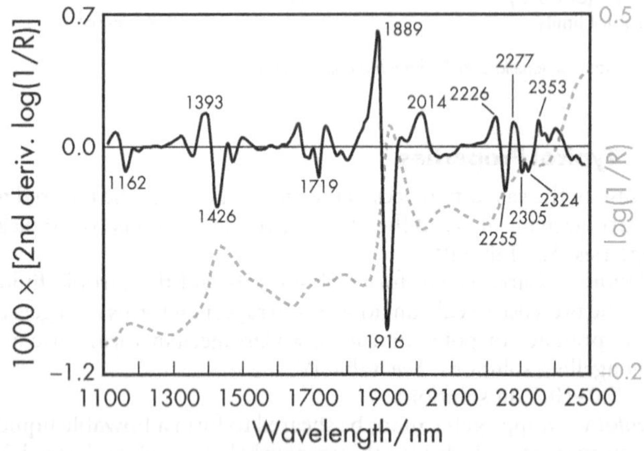

Figure 3: Near-infrared spectrum of carrageenan (iota) measured by reflectance.

All carrageenans are generally most stable at neutral and alkaline pH. Carrageenan solutions will lose viscosity and gel strength in systems below pH values of about 4.3 due to autohydrolysis. The rate of autohydrolysis increases at elevated temperatures and at low cation levels. Kappa and iota carrageenans in the gelled state are stable at low pH (pH $\geq$ 3). Lambda solutions are relatively more acid stable than kappa or iota solutions; *see* Table V.

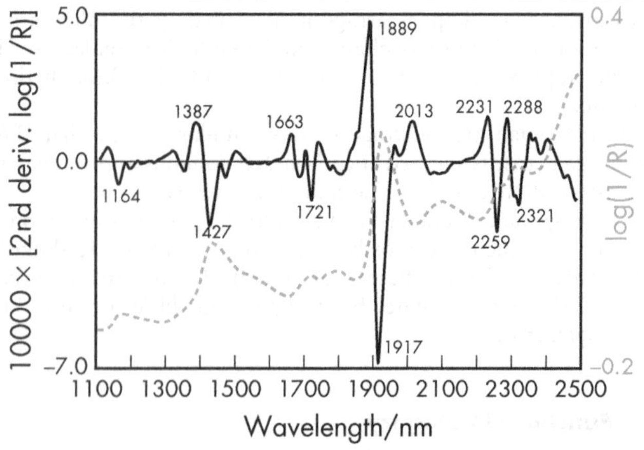

Figure 4: Near-infrared spectrum of carrageenan (kappa) measured by reflectance.

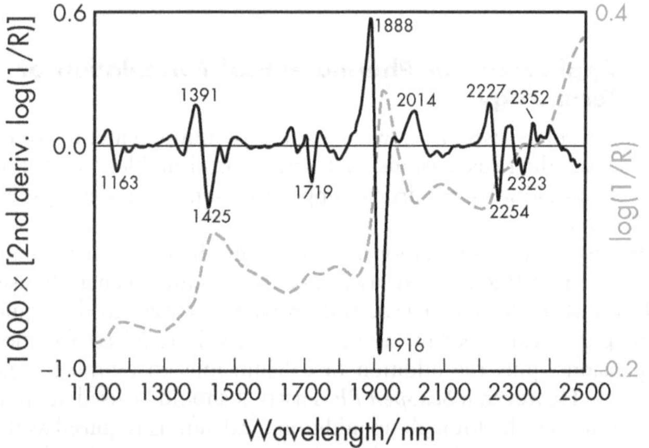

Figure 5: Near-infrared spectrum of carrageenan (lambda) measured by reflectance.

Table III: Typical properties of commercially available grades of carrageenan (FMC BioPolymer).

Carrageenan type	Trade name	Solubility in water	Viscosity
Iota	*Gelcarin GP-379*	Hot	High, thixotropic
Iota	*SeaSpen PF*	Cold, delayed gel formation	Medium, thixotropic
Kappa	*Gelcarin GP-812*	Hot	Low
Kappa	*Gelcarin GP-911*	Hot, partial in cold	Low
Lambda	*Viscarin GP-109*	Partial cold, full in hot	Medium
Lambda	*Viscarin GP-209*	Partial cold, full in hot	High

12 Incompatibilities

Carrageenan can form complexes with cationic materials.

Carrageenan may interact with other charged macromolecules, e.g. proteins, to give various effects such as viscosity increase, gel formation, stabilization, or precipitation.

13 Method of Manufacture

The main species of seaweed from which carrageenan is manufactured are *Eucheuma*, *Chondrus*, and *Gigartina*. The weed is dried quickly to prevent degradation. The seaweed is washed and then undergoes a hot alkali extraction process, releasing the carrageenan from the cell. It is then clarified and concentrated.

Table IV: Solubility and gelation properties of ι-, κ-, and λ-carrageenans.

	Iota	Kappa	Lambda
Solubility in water			
20°C	Na^+ salt only; Ca^{2+} salt swells to form thixotropic dispersion	Na^+ salt only; K^+, Ca^+, and NH_4^+ salts swell	Yes
80°C	Yes	Yes	Yes
Solubility in 5% salt solution			
Hot	Swell	Swell	Soluble
Cold	No	No	Soluble
Gelation			
Ions necessary	Ca^{2+}	K^+	No gel
Gel texture	Elastic	Brittle	No gel
Re-gelation after shear	Yes	No	No
Acid stability	>pH 3.8	>pH 3.8	—
Syneresis	No	Yes	No
Freeze/thaw stability	Yes	No	Yes
Synergism with other gums	No	Yes	No

Table V: Stability of different grades of carrageenan.

Grade	Stability at neutral and alkaline pH	Stability at acid pH
Iota	Stable	Hydrolyzed in solution. Stable in gelled form.
Kappa	Stable	Hydrolyzed in solution when heated. Stable in gelled form.
Lambda	Stable	Hydrolyzed

Three processes may be used to prepare dried carrageenan. The 'freeze–thaw' technique gels the solution with various salts before freezing. After thawing, the water is removed and the resultant mass of carrageenan and salt is ground to the desired particle size.

The alcohol precipitation method adds the concentrated solution of carrageenan to alcohol, causing precipitation. The precipitated carrageenan is recovered by drum drying.

The 'KCl precipitation' process uses evaporation to reduce the filtrate volume. The filtrate is then extruded through spinnerets into a cold 1.0–1.5% solution of potassium chloride. The resulting gel threads are washed with potassium chloride solution before drying and milling.

14 Safety

Carrageenan is widely used in numerous food applications and is increasingly being used in pharmaceutical formulations. Carrageenan is generally regarded as a relatively nontoxic and nonirritating material when used in nonparenteral pharmaceutical formulations.

However, carrageenan is known to induce inflammatory responses in laboratory animals, and for this reason it is frequently used in experiments for the investigation of anti-inflammatory drugs.[23–25] Animal studies suggest that degraded carrageenan (which is not approved for use in food products) may be associated with cancer in the intestinal tract, although comparable evidence does not exist in humans.[26,27]

The WHO has set an acceptable daily intake of carrageenan of 'not specified' as the total daily intake was not considered to represent a hazard to health.[28] In the UK, the Food Advisory Committee has recommended that carrageenan should not be used as an additive for infant formulas.[29]

LD_{50} (rat, oral): >5 g/kg

LD_{50} (rabbit, skin): >2 g/kg

LC_{50} (rat, inhalation): >0.93 mg/L (4 h)[30]

15 Handling Precautions

Observe normal precautions appropriate to the circumstances and quantity of material handled. On heating, carrageenan decomposes to produce oxides of sulfur.

16 Regulatory Status

GRAS listed. Accepted as a food additive in Europe, USA, and Canada. Included in the FDA Inactive Ingredients Database (dental pastes; oral capsules, granules, powders for suspension, and syrups; topical lotions; transdermal preparations; and controlled-release film preparations). Included in the Canadian Natural Health Products Ingredients Database. Included in nonparenteral medicines (oral granules for suspensions, capsules (shells), prolonged release and orodispersible tablets; powder for rectal suspension) licensed in the UK.

17 Related Substances

—

18 Comments

ι-Carrageenan and κ-carrageenan form thermoreversible gels. The gels become fluid when heated above their melting temperature and reset upon cooling with minimal to no loss of their original gel strength. κ-Carrageenan is synergistic with locust bean gum and konjac flour. The interaction significantly increases gel strength, improves moisture binding capabilities, and modifies gel texture to be more elastic. ι-Carrageenan increases the viscosity of starch systems as compared with the viscosity of starch alone. Carrageenans have been studied alone or in combination for use as gelling agents as alternative drug carrier systems for topical formulations.[31,32] κ-Carrageenan has been investigated as a component of complex gels for oral delivery of probiotic bacteria.[33]

Carrageenan has been shown to minimize polymorphic or pseudopolymorphic transitions from amorphous to crystalline forms in a tablet matrix[34] and in spray-dried co-precipitates.[35]

Carrageenan has been investigated in the preparation of nanospheres,[36,37] microcapsules,[38] encapsulating proteins,[39] and probiotic bacteria.[40] Enzyme loaded in κ-carrageenan beads was found to improve the stability of the formulation.[41] Hydrogel beads based on κ-carrageenan and sodium alginate/chitosan have been used as new carriers for floating and controlled delivery systems.[42–44]

Studies have shown that carrageenan compounds block infections by the herpes simplex[45] and human papilloma viruses.[46] Other investigations have revealed that carrageenan fails to prevent the sexual transmission of HIV.[47] However, the combination of carrageenan with a nonnucleoside reverse transcriptase inhibitor is more efficacious against HIV-1 and HIV-2.[48]

Poly(*N*-vinyl-2–pyrrolidone)-κ-carrageenan hydrogels (PVP-KC) have been evaluated for their usability in wound dressing applications.[49]

A specification for carrageenan is included in the *Food Chemicals Codex* (FCC)[50] and in the *Japanese Pharmaceutical Excipients* (JPE).[51]

The EINECS number for carrageenan is 232-524-2.

19 Specific References

1 FAO Corporate Document Repository. *Training manual on Gracilaria culture and seaweed processing in China*, 1990. http://www.fao.org/docrep/field/003/ab730e/AB730E03.htm (accessed 9 June 2011).

2 Valenta C, *et al.* Skin permeation and stability studies of 5-aminolevulinic acid in a new gel and patch preparation. *J Control Release* 2005; 107(3): 495–501.

3 Kählig H, *et al.* Rheology and NMR self-diffusion experiments as well as skin permeation of diclofenac-sodium and cyproterone acetate of new gel preparations. *J Pharm Sci* 2005; **94**(2): 288–296.

4 Tuleu C, *et al.* A scintigraphic investigation of the disintegration behaviour of capsules in fasting subjects: a comparison of hypromellose capsules containing carrageenan as a gelling agent and standard gelatin capsules. *Eur J Pharm Sci* 2007; **30**(3–4): 251–255.

5 Reher M *et al.* Oral rapid release pharmaceutical formulation for pyridylmethylsulfinyl-benzimidazoles. United States Patent 20090068261A1, 2009.

6 Alberto S. Carrageenan based chewing gum. European Patent 1946751A1, 2008.

7 Kranz H, *et al.* Drug release from MCC- and carrageenan-based pellets: experiment and theory. *Eur J Pharm Biopharm* 2009; **73**(2): 302–309.

8 Thommes M, *et al.* Improved bioavailability of darunavir by use of kappa-carrageenan versus microcrystalline cellulose as pelletisation aid. *Eur J Pharm Biopharm* 2009; **72**(3): 614–620.

9 Dukić-Ott A, *et al.* Production of pellets via extrusion-spheronisation without the incorporation of microcrystalline cellulose: a critical review. *Eur J Pharm Biopharm* 2009; **71**(1): 38–46.

10 Hanawa T, *et al.* Development of patient-friendly preparations: preparation of a new allopurinol mouthwash containing polyethylene(-oxide) and carrageenan. *Drug Dev Ind Pharm* 2004; **30**(2): 151–161.

11 Khairnar GA, Sayyad FJ. Development of buccal drug delivery system based on mucoadhesive polymers. *Int J PharmTech Res* 2010; **2**(1): 719–735.

12 Bertram U, Bodmeier R. *In situ* gelling, bioadhesive nasal inserts for extended drug delivery: *in vitro* characterization of a new nasal dosage form. *Eur J Pharm Sci* 2006; **27**(1): 62–71.

13 Valenta C. The use of mucoadhesive polymers in vaginal delivery. *Adv Drug Deliv Rev* 2005; **57**(11): 1692–1712.

14 Acartürk F. Mucoadhesive vaginal drug delivery systems. *Recent Pat Drug Deliv Formul* 2009; **3**(3): 193–205.

15 Liu Y, *et al.* Effect of carrageenan on poloxamer-based *in situ* gel for vaginal use: Improved *in vitro* and *in vivo* sustained-release properties. *Eur J Pharm Sci* 2009; **37**(3–4): 306–312.

16 Bonferoni MC, *et al.* Development of oral controlled-release tablet formulations based on diltiazem-carrageenan complex. *Pharm Dev Technol* 2004; **9**(2): 155–162.

17 Nerurkar J, *et al.* Controlled-release matrix tablets of ibuprofen using cellulose ethers and carrageenans: effect of formulation factors on dissolution rates. *Eur J Pharm Biopharm* 2005; **61**(1–2): 56–68.

18 Elmowafy EM, *et al.* Release mechanisms behind polysaccharides-based famotidine controlled release matrix tablets. *AAPS PharmSciTech* 2008; **9**(4): 1230–1239.

19 Prado HJ, *et al.* Preparation and characterization of a novel starch-based interpolyelectrolyte complex as matrix for controlled drug release. *Carbohydr Res* 2009; **344**(11): 1325–1331.

20 Nanaki S, *et al.* Miscibility study of carrageenan blends and evaluation of their effectiveness as sustained release carriers. *Carbohydr Res* 2010; **79**(4): 1157–1167.

21 Bertram U, Bodmeier R. Parameters affecting the drug release from *in situ* gelling nasal inserts. *Eur J Pharm Biopharm* 2006; **63**(3): 310–319.

22 Bonferoni MC, *et al.* Carrageenan-gelatin mucoadhesive systems for ion-exchange based ophthalmic delivery: *in vitro* and preliminary *in vivo* studies. *Eur J Pharm Biopharm* 2004; **57**(3): 465–472.

23 Vigil SV, *et al.* Efficacy of tacrolimus in inhibiting inflammation caused by carrageenan in a murine model of air pouch. *Transpl Immunol* 2008; **19**(1): 25–29.

24 Mazzon E, Cuzzocrea S. Role of TNF-α in lung tight junction alteration in mouse model of acute lung inflammation. *Respir Res* 2007; **8**: 75.

25 Igbe I, *et al.* Anti-inflammatory activity of aqueous fruit pulp extract of *Hunteria umbellata* K. Schum in acute and chronic inflammation. *Acta Pol Pharm* 2010; **67**(1): 81–85.

26 Tobacman JK. Review of harmful gastrointestinal effects of carrageenan in animal experiments. *Environ Health Perspect* 2001; **109**(10): 983–994.

27 Benard C, *et al.* Degraded carrageenan causing colitis in rats induces TNF secretion and ICAM-1 upregulation in monocytes through NF-kappaB activation. *PLoS One* 2010; **5**(1): e8666.

28 FAO/WHO. Safety evaluation of certain food additives (WHO Food Additives Series 42). Fifty-first report of the joint FAO/WHO expert committee on food additives. http://www.inchem.org/documents/jecfa/jecmono/v042je08.htm (accessed 10 June 2011).

29 MAFF. Food Advisory Committee: report on the review of the use of additives in foods specially prepared for infants and young children. FdAC/REP/12. London: HMSO, 1992.

30 Weiner ML. Toxicological properties of carrageenan. *Agents and Action* 1991; **32**(1-2): 46–51.

31 Valenta C, Schultz K. Influence of carrageenan on the rheology and skin permeation of microemulsion formulations. *J Control Release* 2004; **95**(2): 257–265.

32 Mangione MR, *et al.* Relation between structural and release properties in a polysaccharide gel system. *Biophys Chem* 2007; **129**(1): 18–22.

33 Jonganurakkun B, *et al.* DNA-based gels for oral delivery of probiotic bacteria. *Macromol Biosci* 2006; **6**(1): 99–103.

34 Schmidt AG, *et al.* Potential of carrageenans to protect drugs from polymorphic transformation. *Eur J Pharm Biopharm* 2003; **56**(1): 101–110.

35 Dhumal RS, *et al.* Development of spray-dried co-precipitate of amorphous celecoxib containing storage and compression stabilizers. *Acta Pharm* 2007; **57**(3): 287–300.

36 Daniel-da-Silva AL, *et al. In situ* synthesis of magnetite nanoparticles in carrageenan gels. *Biomacromolecules* 2007; **8**(8): 2350–2357.

37 Daniel-da-Silva AL, *et al.* Biofunctionalized magnetic hydrogel nanospheres of magnetite and kappa-carrageenan. *Nanotechnology* 2009; **20**(35): 355602.

38 Devi N, Maji TK. Effect of crosslinking agent on neem (*Azadirachta Indica A. Juss.*) seed oil (NSO) encapsulated microcapsules of κ-carrageenan and chitosan polyelectrolyte complex. *J Macromol Sci* 2009; **46**(11): 1114–1121.

39 Briones AV, Sato T. Encapsulation of glucose oxidase (GOD) in polyelectrolyte complexes of chitosan-carrageenan. *React Funct Polym* 2010; **70**(1): 19–27.

40 Ding WK, Shah NP. Effect of various encapsulating materials on the stability of probiotic bacteria. *J Food Sci* 2009; **74**(2): M100–107.

41 Sankalia MG, *et al.* Physicochemical characterization of papain entrapped in ionotropically cross-linked kappa-carrageenan gel beads for stability improvement using Doehlert shell design. *J Pharm Sci* 2006; **95**(9): 1994–2013.

42 Ishak RA, *et al.* Preparation, *in vitro* and *in vivo* evaluation of stomach-specific metronidazole-loaded alginate beads as local anti-*Helicobacter pylori* therapy. *J Control Release* 2007; **119**(2): 207–214.

43 Mohamadnia Z, *et al.* Ionically cross-linked carrageenan-alginate hydrogel beads. *J Biomater Sci Polym Ed* 2008; **19**(1): 47–59.

44 Piyakulawat P, *et al.* Preparation and evaluation of chitosan/carrageenan beads for controlled release of sodium diclofenac. *AAPS PharmSciTech* 2007; **8**(4): E97.

45 Carlucci MJ, *et al.* Protective effect of a natural carrageenan on genital herpes simplex virus infection in mice. *Antiviral Res* 2004; **64**(2): 137–141.

46 Buck CB, *et al.* Carrageenan is a potent inhibitor of papillomavirus infection. *PLoS Pathog* 2006; **2**(7): E69.

47 Skoler-Karpoff S, *et al.* Efficacy of *Carraguard* for prevention of HIV infection in women in South Africa: a randomised, double-blind, placebo-controlled trial. *Lancet* 2008; **372**(9654): 1977–1987.

48 Fernández-Romero JA, *et al.* Carrageenan/MIV-150 (PC-815), a combination microbicide. *Sex Transm Dis* 2007; **34**(1): 9–14.

49 Sen M, Avci EN. Radiation synthesis of poly(*N*-vinyl-2-pyrrolidone)-kappa-carrageenan hydrogels and their use in wound dressing applications. I. Preliminary laboratory tests. *J Biomed Mater Res A* 2005; **74**(2): 187–196.

50 *Food Chemicals Codex*, 7th edn. Bethesda, MD: United States Pharmacopeia; 2010: 181.

51 Japan Pharmaceutical Excipients Council. *Japanese Pharmaceutical Excipients 2004*. Tokyo: Yakuji Nippo, 2004: 149–150.

20 General References

CP Kelco. Genu Carrageenan. http://www.cpkelco.com (accessed 10 June 2011).

FMC BioPolymer. Technical literature: *Marine colloids, carrageenan application bulletins*, 2004.

21 Authors

HC Shah, KK Singh.

22 Date of Revision

1 March 2012.

Castor Oil

1 Nonproprietary Names

BP: Virgin Castor Oil
JP: Castor Oil
PhEur: Castor Oil, Virgin
USP–NF: Castor Oil

2 Synonyms

Crystal; *EmCon CO*; *Lipovol CO*; oleum ricini; ricini oleum virginale; ricinoleum; ricinus communis; ricinus oil; tangantangan.

3 Chemical Name and CAS Registry Number

Castor oil [8001-79-4]

4 Empirical Formula and Molecular Weight

Castor oil is a triglyceride of fatty acids. The fatty acid composition is approximately ricinoleic acid (87%); oleic acid (7%); linoleic acid (3%); palmitic acid (2%); stearic acid (1%) and trace amounts of dihydroxystearic acid.

5 Structural Formula

See Section 4.

6 Functional Category

Emollient; oleaginous vehicle; solvent.

7 Applications in Pharmaceutical Formulation or Technology

Castor oil is widely used in cosmetics, food products, and pharmaceutical formulations. In pharmaceutical formulations, castor oil is most commonly used in topical creams and ointments at concentrations of 5–12.5%. However, it is also used in oral tablet and capsule formulations, ophthalmic emulsions, and as a solvent in intramuscular injections.[1–3]

See also Section 18.

8 Description

Castor oil is a clear, almost colorless or pale yellow-colored viscous oil. It has a slight odor and a taste that is initially bland but afterwards slightly acrid.

9 Pharmacopeial Specifications

See Table I.

Table I: Pharmacopeial specifications for castor oil.

Test	JP XV	PhEur 7.4	USP35–NF30
Identification	+	+	—
Characters	+	+	—
Specific gravity	0.953–0.965	≈0.958	0.957–0.961
Heavy metals	—	—	≤10 ppm
Iodine value	80–90	—	83–88
Saponification value	176–187	—	176–182
Hydroxyl value	155–177	≥150	160–168
Acid value	≤1.5	≤2.0	—
Peroxide value	—	≤10.0	—
Refractive index	—	≈1.479	—
Optical rotation	—	+3.5° to +6.0°	—
Water	—	≤0.3%	—
Absorbance	+	≤1.0	—
Composition of fatty acids	—	+	—
Purity	+	—	—
Distinction from most other fixed oils	—	—	+
Free fatty acids	—	—	+
Unsaponifiable matter	—	≤0.8%	—

10 Typical Properties

Autoignition temperature 449°C
Boiling point 313°C
Density 0.955–0.968 g/cm^3 at 25°C
Flash point 229°C
Melting point −12°C
Moisture content ≤0.25%
Refractive index

n_D^{25} = 1.473–1.477;

n_D^{40} = 1.466–1.473.

Solubility Miscible with chloroform, diethyl ether, ethanol, glacial acetic acid, and methanol; freely soluble in ethanol (95%) and petroleum ether; practically insoluble in water; practically insoluble in mineral oil unless mixed with another vegetable oil. *See also* Section 11.

Spectroscopy

IR spectra *see* Figure 1.

Surface tension

39.0 mN/m at 20°C;

35.2 mN/m at 80°C.

Viscosity (dynamic)

1000 mPa s (1000 cP) at 20°C;

200 mPa s (200 cP) at 40°C.

11 Stability and Storage Conditions

Castor oil is stable and does not turn rancid unless subjected to excessive heat. On heating at 300°C for several hours, castor oil polymerizes and becomes soluble in mineral oil. When cooled to 0°C, it becomes more viscous.

Castor oil should be stored at a temperature not exceeding 25°C in well-filled airtight containers protected from light.

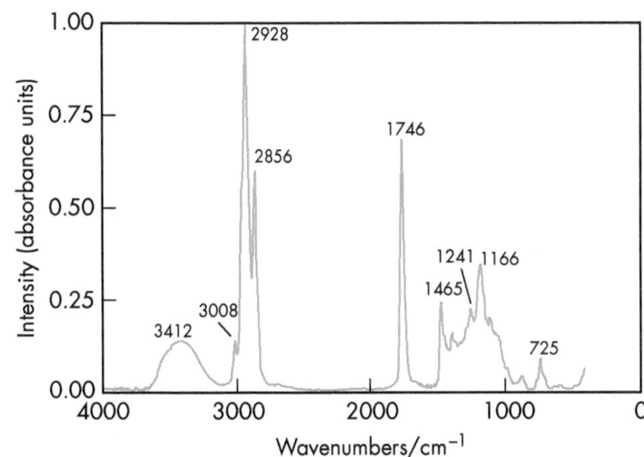

Figure 1: Infrared spectrum of castor oil measured by transmission. Adapted with permission of Informa Healthcare.

12 Incompatibilities

Castor oil is incompatible with strong oxidizing agents.

13 Method of Manufacture

Castor oil is the fixed oil obtained by cold-expression of the seeds of *Ricinus communis* Linné (Fam. Euphorbiaceae). No other substances are added to the oil.

14 Safety

Castor oil is used in cosmetics and foods and orally, parenterally, and topically in pharmaceutical formulations. It is generally regarded as a relatively nontoxic and nonirritant material when used as an excipient.[4]

Castor oil has been used therapeutically as a laxative and oral administration of large quantities may cause nausea, vomiting, colic, and severe purgation. It should not be given when intestinal obstruction is present.

Although widely used in topical preparations, including ophthalmic formulations, castor oil has been associated with some reports of allergic contact dermatitis, mainly to cosmetics such as lipsticks.[5–8]

15 Handling Precautions

Observe normal precautions appropriate to the circumstances and quantity of material handled. Castor oil may cause mild irritation to the skin and eyes. Castor oil is flammable when exposed to heat. Spillages are slippery and should be covered with an inert absorbant before collection and disposal.

16 Regulatory Status

GRAS listed. Included in the FDA Inactive Ingredients Database (IM injections; ophthalmic emulsions; oral capsules and tablets; topical creams, emulsions, ointments, and solutions). Included in nonparenteral medicines licensed in the UK. Included in the Canadian Natural Health Products Ingredients Database.

17 Related Substances

Castor oil, hydrogenated.

18 Comments

Castor oil has been described as a component in the formulation of microemulsions used for the solubilization of poorly soluble drug substances.[9] Studies into the development of nanolipidic formulations as drug delivery systems using castor oil have been reported.[10,11]

Therapeutically, castor oil has been administered orally for its laxative action, but such use is now obsolete.

A specification for castor oil is contained in the *Food Chemicals Codex* (FCC).[12]

The EINECS number for castor oil is 232-293-8. The PubChem Compound ID (CID) for castor oil is 6850719.

19 Specific References

1 Rifkin C, *et al.* Castor oil as a vehicle for parenteral administration of steroid hormones. *J Pharm Sci* 1964; **53**: 891–895.

2 Jumaa M, Müller BW. Development of a novel parenteral formulation for tetrazepam using a lipid emulsion. *Drug Dev Ind Pharm* 2001; 27(10): 1115–1121.

3 Yamaguchi M, *et al.* Formulation of an ophthalmic lipid emulsion containing an anti-inflammatory steroidal drug, difluprednate. *Int J Pharm* 2005; **301**(1–2): 121–128.

4 Irwin R. NTP technical report on the toxicity studies of castor oil (CAS no 8001-79-4) in F344/N rats and B6C3F1 mice (dosed feed studies). *Toxic Rep Ser* 1982; **12**: 1–B5.

5 Fisher LB, Berman B. Contact allergy to sulfonated castor oil. *Contact Dermatitis* 1981; **7**(6): 339–340.

6 Sai S. Lipstick dermatitis caused by castor oil. *Contact Dermatitis* 1983; **9**(1): 75.

7 Andersen KE, Nielsen R. Lipstick dermatitis related to castor oil. *Contact Dermatitis* 1984; **11**(4): 253–254.

8 Smolinske SC. *CRC Handbook of Food, Drug and Cosmetic Excipients*. Boca Raton, FL: CRC Press, 1992: 69–70.

9 Nazar MF, *et al.* Microemulsion system with improved loading of piroxicam: a study of microstructure. *AAPS PharmSciTech* 2009; **10**: 1286–1294.

10 Kelmann RG, *et al.* Carbamazepine parenteral nanoemulsions prepared by spontaneous emulsification process. *Int J Pharm* 2007; **342**(1–2): 231–239.

11 Lacoeuille F, *et al. In vivo* evaluation of lipid nanocapsules as a promising colloidal carrier for paclitaxel. *Int J Pharm* 2007; **344**(1–2): 143–149.

12 *Food Chemicals Codex*, 7th edn. Bethesda, MD: United States Pharmacopeia, 2010: 201.

20 General References

—

21 Author

W Cook.

22 Date of Revision

1 March 2012.

 Castor Oil, Hydrogenated

1 Nonproprietary Names

BP: Hydrogenated Castor Oil
PhEur: Castor Oil, Hydrogenated
USP–NF: Hydrogenated Castor Oil

2 Synonyms

Castorwax; *Castorwax MP 70*; *Castorwax MP 80*; *Cremophor*; *Croduret*; *Cutina HR*; *Fancol*; ricini oleum hydrogenatum.

3 Chemical Name and CAS Registry Number

Glyceryl-tri-(12-hydroxystearate) [8001-78-3]

4 Empirical Formula and Molecular Weight

$C_{57}O_9H_{110}$ 939.50

The USP35–NF30 describes hydrogenated castor oil as refined, bleached, hydrogenated, and deodorized castor oil, consisting mainly of the triglyceride of hydroxystearic acid.

5 Structural Formula

```
             O              HO
             ‖              |
CH2—O——C——(CH2)10——CH——(CH2)5——CH3
|            O              HO
|            ‖              |
CH2—O——C——(CH2)10——CH——(CH2)5——CH3
|            O              HO
|            ‖              |
CH2—O——C——(CH2)10——CH——(CH2)5——CH3
```

6 Functional Category

Coating agent; modified-release agent; stiffening agent; tablet and capsule lubricant.

7 Applications in Pharmaceutical Formulation or Technology

Hydrogenated castor oil is a hard wax with a high melting point used in oral and topical pharmaceutical formulations; *see* Table I.

Table I: Uses of hydrogenated castor oil.

Use	Concentration (%)
Coating agent (delayed release)	5.0–20.0
Delayed release drug matrix	5.0–10.0
Tablet die lubricant	0.1–2.0

In topical formulations, hydrogenated castor oil is used to provide stiffness to creams and emulsions.[1] In oral formulations, hydrogenated castor oil is used to prepare sustained-release tablet and capsule preparations;[2–4] hydrogenated castor oil may be used as a coat or to form a solid matrix. Studies have shown that hydrogenated castor oil may be used to enhance the stability of moisture-sensitive drug products.[5]

Hydrogenated castor oil is additionally used to lubricate the die walls of tablet presses,[6,7] and is similarly used as a lubricant in food processing.

Hydrogenated castor oil is also used in cosmetics.

8 Description

Hydrogenated castor oil occurs as a fine, almost white or pale yellow powder or flakes. The PhEur 7.4 describes hydrogenated castor oil as the oil obtained by hydrogenation of virgin castor oil. It consists mainly of the triglyceride of 12-hydroxystearic acid.

9 Pharmacopeial Specifications

See Table II.

Table II: Pharmacopeial specifications for hydrogenated castor oil.

Test	PhEur 7.4	USP35–NF30
Characters	+	–
Identification	+	–
Acid value	≤4.0	–
Hydroxyl value	145–165	154–162
Iodine value	≤5.0	≤5.0
Saponification value	–	176–182
Alkaline impurities	+	–
Composition of fatty acids	+	–
Palmitic acid	≤2.0%	–
Stearic acid	7.0–14.0%	–
Arachidic acid	≤1.0%	–
12-Oxostearic acid	≤5.0%	–
12-Hydroxystearic acid	78.0–91.0%	–
Any other fatty acid	≤3.0%	–
Free fatty acids	–	+
Nickel	≤1 ppm	–
Heavy metals	–	≤0.001%
Melting range	83–88°C	85–88°C

10 Typical Properties

Acid value ≤5
Density 0.98–1.10 g/cm^3
Flash point 316°C (open cup)
Moisture content ≤0.1%
Particle size distribution 97.7% ≥1000 μm in size for flakes.
Solubility Practically insoluble in water; soluble in acetone, chloroform, and methylene chloride.

11 Stability and Storage Conditions

Hydrogenated castor oil is stable at temperatures up to 150°C.

Clear, stable, chloroform solutions containing up to 15% w/v of hydrogenated castor oil may be produced. Hydrogenated castor oil may also be dissolved at temperatures greater than 90°C in polar solvents and mixtures of aromatic and polar solvents, although the hydrogenated castor oil precipitates out on cooling below 90°C.

Hydrogenated castor oil should be stored in a well-closed container in a cool, dry place.

12 Incompatibilities

Hydrogenated castor oil is compatible with most natural vegetable and animal waxes.

13 Method of Manufacture

Hydrogenated castor oil is prepared by the hydrogenation of castor oil using a catalyst.

14 Safety

Hydrogenated castor oil is used in oral and topical pharmaceutical formulations and is generally regarded as an essentially nontoxic and nonirritant material.

Acute oral toxicity studies in animals have shown that hydrogenated castor oil is a relatively nontoxic material. Irritation tests with rabbits show that hydrogenated castor oil causes mild, transient irritation to the eye.

LD_{50} (rat, oral): >10 g/kg

15 Handling Precautions

Observe normal precautions appropriate to the circumstances and quantity of material handled.

16 Regulatory Status

Accepted in the USA as an indirect food additive. Included in the FDA Inactive Ingredients Database (oral capsules, tablets, sublingual tablets, gels, and emulsion creams).

Included in nonparenteral medicines licensed in the UK. Included in the Canadian Natural Health Products Ingredients Database.

17 Related Substances

Castor oil; vegetable oil, hydrogenated.

18 Comments

Various different grades of hydrogenated castor oil are commercially available, the composition of which may vary considerably. *Sterotex K* (Karlshamns Lipid Specialities), for example, is a mixture of hydrogenated castor oil and hydrogenated cottonseed oil. *See* Vegetable Oil, Hydrogenated for further information.

The EINECS number for hydrogenated castor oil is 232-292-2.

19 Specific References

1 Kline CH. Thixcin R-thixotrope. *Drug Cosmet Ind* 1964; **95**(6): 895–897.
2 Yonezawa Y, *et al.* Release from or through a wax matrix system. III: Basic properties of release through the wax matrix layer. *Chem Pharm Bull (Tokyo)* 2002; **50**(6): 814–817.
3 Vergote GJ, *et al.* An oral controlled release matrix pellet formulation containing microcrystalline ketoprofen. *Int J Pharm* 2002; **219**: 81–87.
4 Nagaraju R, *et al.* Core-in-cup tablet design of metoprolol succinate and its evaluation for controlled release. *Curr Drug Discov Technol* 2009; **6**: 299–305.
5 Kowalski J, *et al.* Application of melt granulation technology to enhance stability of a moisture sensitive immediate-release drug product. *Int J Pharm* 2009; **381**: 56–61.
6 Danish FQ, Parrott EL. Effect of concentration and size of lubricant on flow rate of granules. *J Pharm Sci* 1971; **60**: 752–754.
7 Hölzer AW, Sjögren J. Evaluation of some lubricants by the comparison of friction coefficients and tablet properties. *Acta Pharm Suec* 1981; **18**: 139–148.

20 General References

Bose S, Bogner RH. Solventless pharmaceutical coating processes: a review. *Pharm Dev Technol* 2007; **12**: 115–131.
Jannin V, *et al.* Approaches for the development of solid and semi-solid lipid-based formulations. *Adv Drug Deliv Rev* 2008; **60**: 734–746.

21 Author

RT Guest.

22 Date of Revision

1 March 2012.

Cellaburate

1 Nonproprietary Names

BP: Cellulose Acetate Butyrate
PhEur: Cellulose Acetate Butyrate
USP–NF: Cellaburate

2 Synonyms

Acetylbutyrylcellulose; cellaburatum; cellulose acetate butanoate; cellulose butyrate acetate; cellulosi acetas butyras.

3 Chemical Name and CAS Registry Number

Cellulose acetate butanoate [9004-36-8]

4 Empirical Formula and Molecular Weight

Cellaburate is a cellulose derivative in which between 1% and 41% of the hydroxyl groups are acetylated, and between 5% and 56% of hydroxyl groups are butyrylated.

The PhEur 7.4 describes cellaburate as a partly or completely O-acetylated and O-butyrylated cellulose containing not less than 2.0% and not more than 30.0% acetyl groups, and not less than 16.0% and not more than 53.0% of butyryl groups, calculated with reference to the dried substance.

5 Structural Formula

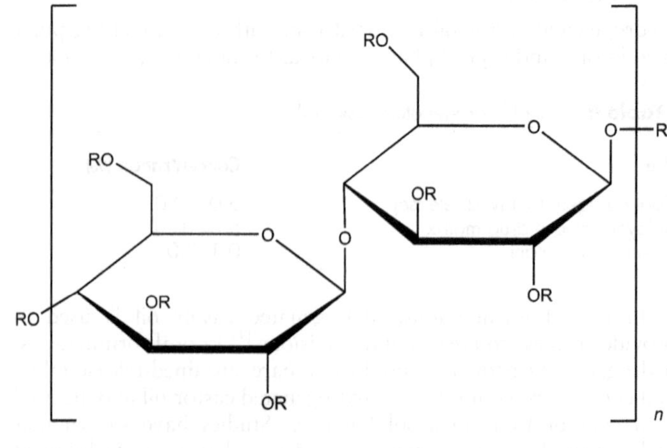

6 Functional Category

Coating agent; modified-release agent; tablet and capsule diluent.

7 Applications in Pharmaceutical Formulation or Technology

Cellaburate has been used in the formulation of oral modified-release drug delivery systems.[1] It is an alternative to cellulose acetate as the semipermeable barrier layer in osmotic pump devices.[2]

8 Description

Cellaburate occurs as a white to off-white, slightly hygroscopic powder or granules.

9 Pharmacopeial Specifications

See Table I.

Table I: Pharmacopeial specifications for cellaburate.

Test	PhEur 7.4	USP35–NF30
Identification	+	+
Residue on ignition	≤0.1%	≤0.1%
Heavy metals	≤20 ppm	≤0.002%
Free acid	–	≤0.1%
Acidity	+	–
Water determination	–	≤5.0%
Loss on drying	≤2.0%	–
Assay		
C_2H_3O groups	2.0–30.0%	1.0–41.0%
C_4H_7O groups	16.0–53.0%	5.0–56.0%

10 Typical Properties

Density 1.26 g/cm^3
Density (bulk) 0.224 g/cm^3
Density (tapped) 0.256 g/cm^3
Dielectric strength 784–984 kV/cm
Glass transition temperature (T_g) 151°C
Hardness 27 Knoops
Melting point 230–240°C
Refractive index 1.475
Solubility Practically insoluble in water and in ethanol (95%); soluble in acetone, formic acid, and in a mixture of equal volumes of methanol and dichloromethane.
Specific gravity 1.26

11 Stability and Storage Conditions

Slow hydrolysis of cellaburate will occur under prolonged adverse conditions such as high temperatures and high humidity, with a resultant increase in free acid content and water content. Store in airtight containers in a cool, dry place.

12 Incompatibilities

Cellaburate is incompatible with oxidizing materials. Mixing cellulose esters in a nonpolar hydrocarbon, such as toluene or xylene, may result in the buildup of static electricity, which can cause a flash fire or an explosion. When adding cellulose ester to any flammable liquid, an inert gas atmosphere should be maintained within the vessel.

13 Method of Manufacture

Cellaburate is produced by reacting cellulose with acetic anhydride and *n*-butyric anhydride in the presence of a tertiary organic base or a strong acid.

14 Safety

Cellaburate has been used in oral pharmaceutical formulations, and is generally regarded as a safe material. It may cause slight skin irritation but no skin sensitization has been observed in guinea pigs.

LD_{50} (guinea pig, skin): > 1.0 g/kg (highest dose tested)[3]
LD_{50} (rat, oral): > 6.4 g/kg (highest dose tested)[3]

15 Handling Precautions

Observe normal precautions appropriate to the circumstances and quantity of the material handled. Precautions to prevent dust explosions should be taken.

16 Regulatory Status

Included in licensed medicines marketed in Europe. Included in the Canadian Natural Health Products Ingredients Database.

17 Related Substances

Cellulose acetate; cellulose acetate phthalate.

18 Comments

Cellaburate has been used in the manufacture of hydrophobic contact lens materials.

19 Specific References

1 Yuan J *et al.* Tailoring permeability of coating films with cellulose esters polymers. IPCOM Patent 000181516D; 3 April 2009.
2 Shanbhag A, *et al.* Application of cellulose acetate butyrate-based membrane for osmotic drug delivery. *Cellulose* 2007; **14**(1): 65–71.
3 Eastman Chemical Company. Material safety data sheet: cellulose acetate butyrate, 2007. http://ws.eastman.com/ProductCatalogApps/PageControllers/MSDSShow_PC.aspx (accessed 6 July 2011).

20 General References

Eastman Chemical Company. Cellulose Esters for Pharmaceutical Drug Delivery. Kingsport, TN, USA, 2005. http://www.eastman.com/Literature_Center/P/PCI105.pdf (accessed 6 July 2011).

21 Authors

S Jones, RC Moreton.

22 Date of Revision

1 March 2012.

Cellulose, Microcrystalline

1 Nonproprietary Names

BP: Microcrystalline Cellulose
JP: Microcrystalline Cellulose
PhEur: Cellulose, Microcrystalline
USP–NF: Microcrystalline Cellulose

2 Synonyms

Avicel PH; *Cellets*; *Celex*; cellulose gel; hellulosum microcristallinum; *Celphere*; *Ceolus KG*; crystalline cellulose; *Cyclocel*; E460; *Emcocel*; *Ethispheres*; *Fibrocel*; *MCC Sanaq*; *Microcel*; *Pharmacel*; *Tabulose*; *Vivapur*.

3 Chemical Name and CAS Registry Number

Cellulose [9004-34-6]

4 Empirical Formula and Molecular Weight

$(C_6H_{10}O_5)_n$ ≈36 000 where $n \approx 220$.

5 Structural Formula

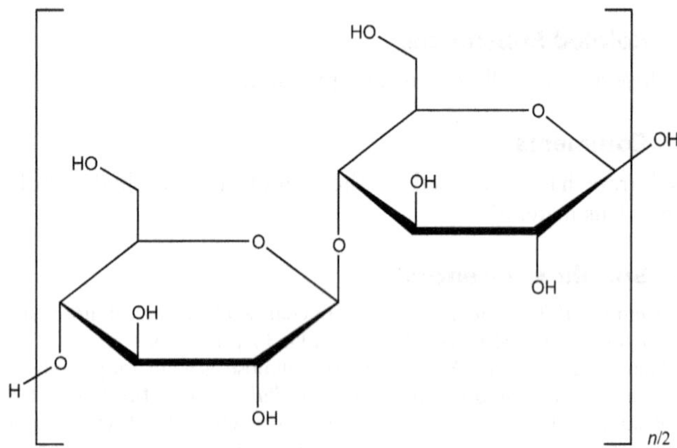

6 Functional Category

Adsorbent; direct compression excipient; suspending agent; tablet and capsule diluent.

7 Applications in Pharmaceutical Formulation or Technology

Microcrystalline cellulose is widely used in pharmaceuticals, primarily as a binder/diluent in oral tablet and capsule formulations where it is used in dry-granulation, wet-granulation, and direct-compression processes.[1–7] In addition to its use as a binder/diluent, microcrystalline cellulose may also reduce friction during tablet ejection[8] and facilitate tablet disintegration. Certain spherical-shaped grades may be used for drug layering.

Microcrystalline cellulose is also used in cosmetics and food products; *see* Table I.

Table I: Uses of microcrystalline cellulose.

Use	Concentration (%)
Adsorbent	20–90
Antiadherent	5–20
Capsule binder/diluent	20–90
Tablet disintegrant	5–15
Tablet binder/diluent	20–90

8 Description

Microcrystalline cellulose is a purified, partially depolymerized cellulose that occurs as a white, odorless, tasteless, powder composed of porous particles.

9 Pharmacopeial Specifications

The pharmacopeial specifications for microcrystalline cellulose have undergone harmonization of many attributes for JP, PhEur, and USP–NF.

See Table II. *See also* Section 18.

SEM 1: Excipient: microcrystalline cellulose; manufacturer: JRS Pharma LP; lot no.: 98662; magnification: 100×.

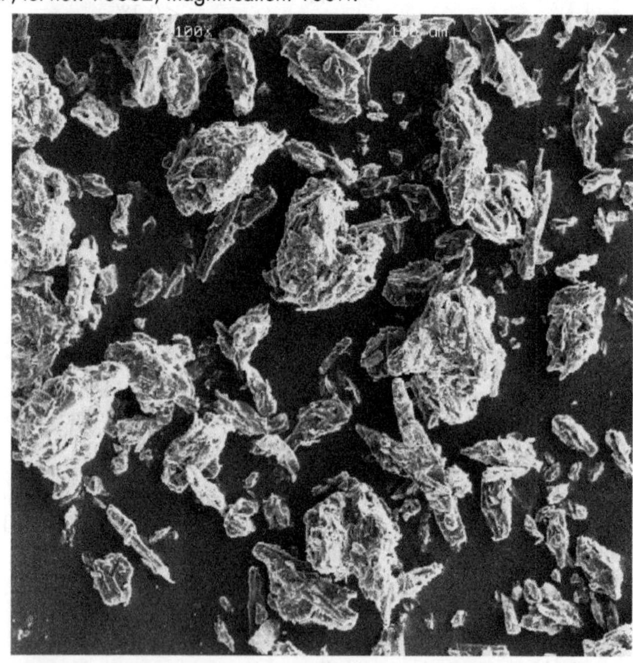

SEM 2: Excipient: microcrystalline cellulose (*Avicel PH-101*); manufacturer: FMC Biopolymer. magnification: 200×; voltage: 3 kV.

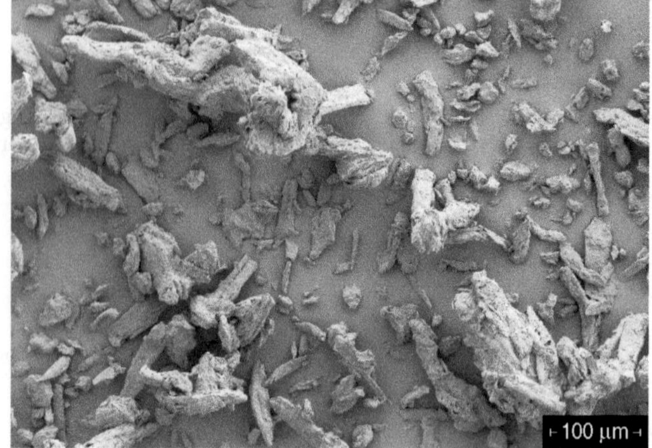

SEM 3: Excipient: microcrystalline cellulose (*Avicel PH-102*); manufacturer: FMC Biopolymer. magnification: 200×; voltage: 3 kV.

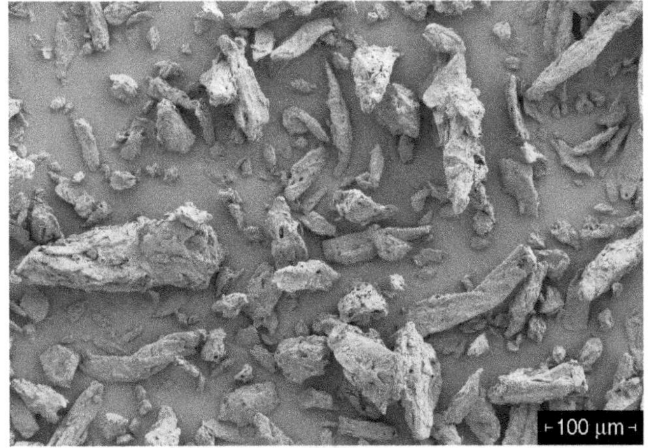

SEM 4: Excipient: microcrystalline cellulose (*Avicel PH-105*); manufacturer: FMC Biopolymer. magnification: 500×; voltage: 3 kV.

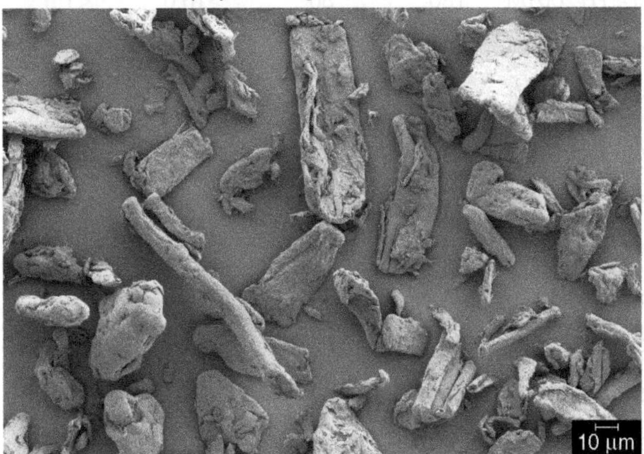

SEM 5: Excipient: microcrystalline cellulose (*Avicel PH-200*); manufacturer: FMC Biopolymer. magnification: 200×; voltage: 3 kV.

SEM 6: Excipient: microcrystalline cellulose (*Avicel PH-302*); manufacturer: FMC Biopolymer. magnification: 200×; voltage: 3 kV.

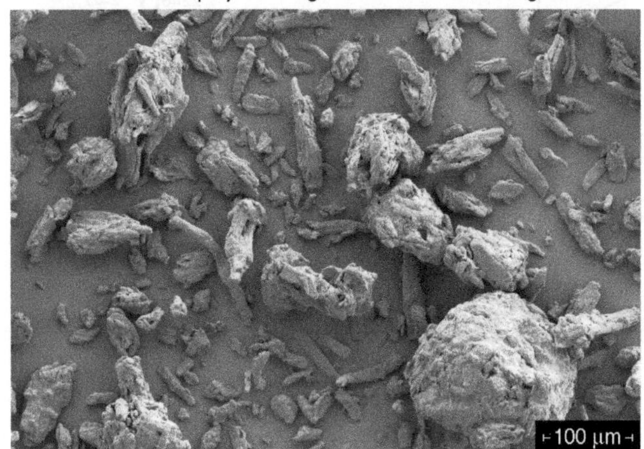

Table II: Pharmacopeial specifications for microcrystalline cellulose.

Test	JP XV	PhEur 7.4	USP35–NF30
Identification	+	+	+
Characters	+	+	—
pH	5.0–7.5	5.0–7.5	5.0–7.5
Bulk density	+	—	+
Loss on drying	≤7.0%	≤7.0%	≤7.0%
Residue on ignition	≤0.1%	—	≤0.1%
Conductivity	+	+	+
Sulfated ash	—	≤0.1%	—
Ether-soluble substances	≤0.05%	≤0.05%	≤0.05%
Water-soluble substances	+	≤0.25%	≤0.25%
Heavy metals[a]	≤10 ppm	≤10 ppm	≤10 ppm
Microbial limits	+	+	+
Aerobic	$\leq 10^3$ cfu/g	$\leq 10^3$ cfu/g	$\leq 10^3$ cfu/g
Molds and yeasts	$\leq 10^2$ cfu/g	$\leq 10^2$ cfu/g	$\leq 10^2$ cfu/g
Solubility	—	+	—
Particle size distribution	—	+	+

(a) These tests have not been fully harmonized at the time of publication.

10 Typical Properties

Angle of repose

49° for *Ceolus KG*;

34.4° for *Emcocel 90M*.[9]

Density (bulk)

0.337 g/cm^3;

0.32 g/cm^3 for *Avicel PH-101*;[10]

0.80 ± 0.05 g/cm^3 for *Cellets 100, 200, 350, 500, 700, 1000*;

0.29 g/cm^3 for *Emcocel 90M*;[9]

0.26–0.31 g/cm^3 for *MCC Sanaq 101*;

0.28–0.33 g/cm^3 for *MCC Sanaq 102*;

0.29–0.36 g/cm^3 for *MCC Sanaq 200*;

0.34–0.45 g/cm^3 for *MCC Sanaq 301*;

0.35–0.46 g/cm^3 for *MCC Sanaq 302*;

0.13–0.23 g/cm^3 for *MCC Sanaq UL-002*;

0.29 g/cm^3 for *Vivapur 101*.

Density (tapped)

0.478 g/cm^3;

0.45 g/cm^3 for *Avicel PH-101*;

0.35 g/cm^3 for *Emcocel 90M*.[9]

Density (true)

1.512–1.668 g/cm^3;

1.420–1.460 g/cm^3 for *Avicel PH-102*.[11]

Flowability 1.41 g/s for *Emcocel 90M*.[9] *Avicel PH102* exhibits flow properties adequate for high-speed tableting.[12]

Melting point Chars at 260–270°C.

Moisture content Typically less than 5% w/w. However, different grades may contain varying amounts of water. Microcrystalline cellulose is hygroscopic.[13]

See Table III.

Particle size distribution Different grades may have a different nominal mean particle size; *see* Table III.

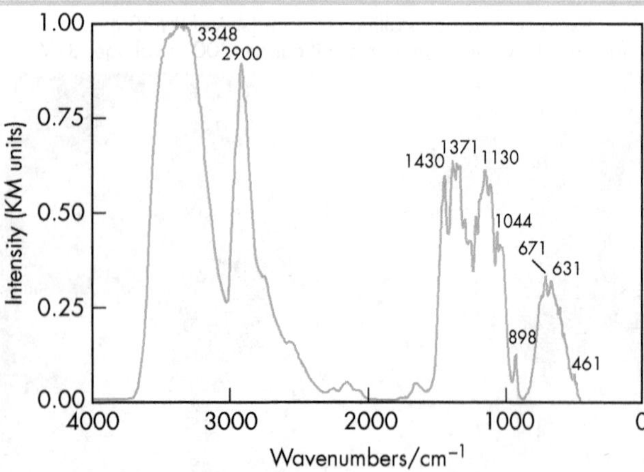

Figure 1: Infrared spectrum of cellulose, microcrystalline measured by diffuse reflectance. Adapted with permission of Informa Healthcare.

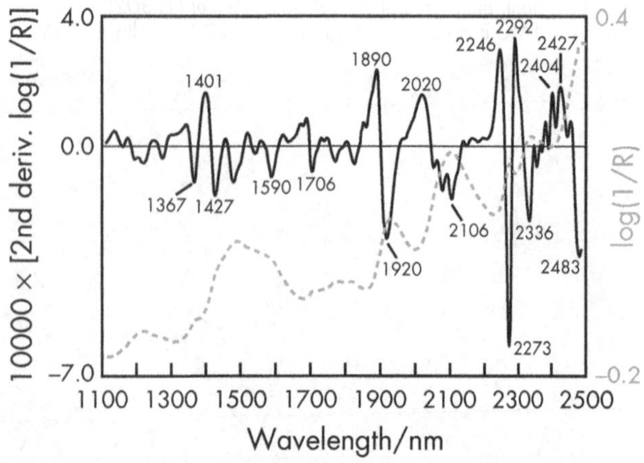

Figure 2: Near-infrared spectrum of cellulose, microcrystalline measured by reflectance.

Solubility Slightly soluble in 5% w/v sodium hydroxide solution; practically insoluble in water, dilute acids, and most organic solvents.

Specific surface area

1.06–1.12 m^2/g for *Avicel PH-101*;

1.21–1.30 m^2/g for *Avicel PH-102*;

0.78–1.18 m^2/g for *Avicel PH-200*.

Spectroscopy

IR spectra *see* Figure 1.

NIR spectra *see* Figure 2.

11 Stability and Storage Conditions

Microcrystalline cellulose is stable, physically and chemically, in ambient conditions. The bulk material is hygroscopic and should be stored in well-closed containers in a cool, dry place.

12 Incompatibilities

Microcrystalline cellulose is incompatible with strong oxidizing agents.

13 Method of Manufacture

Microcrystalline cellulose is typically manufactured by controlled hydrolysis with dilute mineral acid solutions of α-cellulose, obtained as a pulp from fibrous plant materials. Following hydrolysis, the hydrocellulose is washed with water and filtered. The aqueous slurry is then spray-dried to form a dry powder.

Table III: Properties of selected commercially available grades of microcrystalline cellulose.

Grade	Nominal mean particle size (μm)	Particle size analysis		Moisture content (%)
		Mesh size	Amount retained (%)	
Avicel PH-101 [a]	50	60	≤1.0	≤5.0
		200	≤30.0	
Avicel PH-102 [a]	100	60	≤8.0	≤5.0
		200	≥45.0	
Avicel PH-103 [a]	50	60	≤1.0	≤3.0
		200	≤30.0	
Avicel PH-105 [a]	20	400	≤1.0	≤5.0
Avicel PH-112 [a]	100	60	≤8.0	≤1.5
Avicel PH-113 [a]	50	60	≤1.0	≤1.5
		200	≤30.0	
Avicel PH-200 [a]	180	60	≥10.0	≤5.0
		100	≥50.0	
Avicel PH-200 LM[a]	180	60	≥10.0	≤1.5
		100	≥50.0	
Avicel PH-301 [a]	50	60	≤1.0	≤5.0
		200	≤30.0	
Avicel PH-302 [a]	100	60	≤8.0	≤5.0
		200	≥45.0	
Celex 101 [b]	75	60	≤1.0	≤5.0
		200	≥30.0	
Ceolus KG-802 [c]	50	60	≤0.5	≤6.0
		200	≤30.0	
Emcocel 50M [d]	50	60	≤0.25	≤5.0
		200	≤30.0	
Emcocel 90M [d]	91	60	≤8.0	≤5.0
		200	≥45.0	
MCC Sanaq 101[e]	50	60	≤1.0	≤6.0
		200	≤30.0	
MCC Sanaq 102[e]	100	60	≤8.0	≤6.0
		200	≥45.0	
MCC Sanaq 200[e]	180	60	≥10.0	≤6.0
		100	≥50.0	
MCC Sanaq 301[e]	50	60	≤1.0	≤6.0
		200	≥30.0	
MCC Sanaq 302[e]	100	60	≤8.0	≤6.0
		200	≥45.0	
MCC Sanaq UL-002[e]	50	60	<0.5	≤6.0
		100	<5.0	
		200	<5.0–30.0	
Vivapur 101 [d]	50	60	≤1.0	≤5.0
		200	≤30.0	
Vivapur 102 [d]	90	60	≤8.0	≤5.0
		200	≥45.0	
Vivapur 12 [d]	160	38	≤1.0	≤5.0
		94	≤50.0	

Suppliers:
(a) FMC Biopolymer
(b) International Specialty Products
(c) Asahi Kasei Corporation
(d) JRS Pharma
(e) Pharmatrans Sanaq AG

14 Safety

Microcrystalline cellulose is widely used in oral pharmaceutical formulations and food products and is generally regarded as a relatively nontoxic and nonirritant material.

Microcrystalline cellulose is not absorbed systemically following oral administration and thus has little toxic potential. Consumption of large quantities of cellulose may have a laxative effect, although this is unlikely to be a problem when cellulose is used as an excipient in pharmaceutical formulations.

Deliberate abuse of formulations containing cellulose, either by inhalation or by injection, has resulted in the formation of cellulose granulomas.[14]

15 Handling Precautions

Observe normal precautions appropriate to the circumstances and quantity of material handled. Microcrystalline cellulose may be irritating to the eyes. Gloves, eye protection, and a dust mask are recommended. In the UK, the workplace exposure limits for cellulose have been set at 10 mg/m^3 long-term (8-hour TWA) for total inhalable dust and 4 mg/m^3 for respirable dust; the short-term limit for total inhalable dust has been set at 20 mg/m^3.[15]

16 Regulatory Status

GRAS listed. Accepted for use as a food additive in Europe. Included in the FDA Inactive Ingredients Database (inhalations; oral capsules, powders, suspensions, syrups, and tablets; topical and vaginal preparations). Included in nonparenteral medicines licensed in the UK. Included in the Canadian Natural Health Products Ingredients Database.

17 Related Substances

Microcrystalline cellulose and carrageenan; microcrystalline cellulose and carboxymethylcellulose sodium; microcrystalline cellulose and dibasic calcium phosphate; microcrystalline cellulose and guar gum; microcrystalline cellulose and mannitol; powdered cellulose; silicified microcrystalline cellulose.

Microcrystalline cellulose and carrageenan

Synonyms *Lustre Clear.*

Comments *Lustre Clear* (FMC Biopolymer) is an aqueous film coating combining microcrystalline cellulose and carrageenan.

Microcrystalline cellulose and dibasic calcium phosphate

Synonyms *Avicel DG*

Comments *Avicel DG* (FMC Biopolymer) is a coprocessed mixture of microcrystalline cellulose and dibasic calcium phosphate used in roller compaction.

Microcrystalline cellulose and guar gum

Synonyms *Avicel CE-15.*

Comments *Avicel CE-15* (FMC Biopolymer) is a coprocessed mixture of microcrystalline cellulose and guar gum used in chewable tablet formulations.

Microcrystalline cellulose and mannitol

Synonyms *Avicel HFE-102*

Comments *Avicel HFE-102* (FMC Biopolymer) is a coprocessed mixture of microcrystalline cellulose and mannitol (anhydrous) used in direct compression and formation of orally disintegrating tablets.

18 Comments

Microcrystalline cellulose has undergone harmonization of many attributes for JP, PhEur, and USP–NF by the Pharmacopeial Discussion Group. For further information see the General Information Chapter <1196> in the USP35–NF30, the General Chapter 5.8 in PhEur 7.4, along with the 'State of Work' document on the PhEur EDQM website, and also the General Information Chapter 8 in the JP XV.

Several different grades of microcrystalline cellulose are commercially available that differ in their method of manufacture,[16,17] particle size, moisture, flow, and other physical properties.[18–30] The larger-particle-size grades generally provide better flow properties in pharmaceutical machinery. Low-moisture grades are used with moisture-sensitive materials. Higher-density grades have improved flowability.

Microcrystalline cellulose is generally regarded as a plastic material, with its plasticity affected by moisture content. The tabletability of raw powders and granulated powders depends on particle size, porosity, and moisture content.[31–33]

Microcrystalline cellulose exhibits birefringence when observed under polarized light microscopy but is X-ray amorphous probably due to the small size (nanometers) of crystalline domains.

Several coprocessed mixtures of microcrystalline cellulose with other excipients such as carrageenan, carboxymethylcellulose sodium, dibasic calcium phosphate, mannitol, and guar gum are commercially available; *see* Section 17.

Celphere (Asahi Kasei Corporation) is a pure spheronized microcrystalline cellulose available in several different particle size ranges. *Balocel Sanaq* (Pharmatrans Sanaq AG) is an excipient used mainly in the production of pellets and granulates in direct tableting, which contains lactose, microcrystalline cellulose, and sodium carboxymethylcellulose.

According to PhEur 7.4, microcrystalline cellulose has certain functionality related characteristics that are recognised as being relevant control parameters for one or more functions of the substance when used as an excipient. Non-mandatory testing procedures have been described for particle size distribution (2.9.31 or 2.9.38) and powder flow (2.9.36).

A specification for microcrystalline cellulose is contained in the *Food Chemicals Codex* (FCC).[34]

The PubChem Compound ID (CID) for microcrystalline cellulose is 14055602.

19 Specific References

1 Enézian GM. [Direct compression of tablets using microcrystalline cellulose.] *Pharm Acta Helv* 1972; **47**: 321–363[in French].

2 Lerk CF, Bolhuis GK. Comparative evaluation of excipients for direct compression I. *Pharm Weekbl* 1973; **108**: 469–481.

3 Lerk CF, *et al.* Comparative evaluation of excipients for direct compression II. *Pharm Weekbl* 1974; **109**: 945–955.

4 Lamberson RF, Raynor GE. Tableting properties of microcrystalline cellulose. *Manuf Chem Aerosol News* 1976; **47**(6): 55–61.

5 Lerk CF, *et al.* Effect of microcrystalline cellulose on liquid penetration in and disintegration of directly compressed tablets. *J Pharm Sci* 1979; **68**: 205–211.

6 Chilamkurti RN, *et al.* Some studies on compression properties of tablet matrices using a computerized instrumented press. *Drug Dev Ind Pharm* 1982; **8**: 63–86.

7 Wallace JW, *et al.* Performance of pharmaceutical filler/binders as related to methods of powder characterization. *Pharm Technol* 1983; 7(9): 94–104.

8 Omray A, Omray P. Evaluation of microcrystalline cellulose as a glidant. *Indian J Pharm Sci* 1986; **48**: 20–22.

9 Celik M, Okutgen E. A feasibility study for the development of a prospective compaction functionality test and the establishment of a compaction data bank. *Drug Dev Ind Pharm* 1993; **19**: 2309–2334.

10 Parker MD, *et al.* Binder–substrate interactions in wet granulation 3: the effect of excipient source variation. *Int J Pharm* 1992; **80**: 179–190.

11 Sun CC. True density of microcrystalline cellulose. *J Pharm Sci* 2005; 94(10): 2132–2134.

12 Sun CC. Setting the bar for powder flow properties in successful high speed tableting. *Powder Technol* 2010; **201**: 106–108.

13 Sun CC. Mechanism of moisture induced variations in true density and compaction properties of microcrystalline cellulose. *Int J Pharm* 2008; **346**: 93–101.

14 Cooper CB, *et al.* Cellulose granulomas in the lungs of a cocaine sniffer. *Br Med J* 1983; **286**: 2021–2022.

15 Health and Safety Executive. *EH40/2005: Workplace Exposure Limits.* Sudbury: HSE Books, 2011. www.hse.gov.uk/pubns/priced/eh40.pdf (accessed 12 April 2012).

16 Jain JK, *et al.* Preparation of microcrystalline cellulose from cereal straw and its evaluation as a tablet excipient. *Indian J Pharm Sci* 1983; **45**: 83–85.

17 Singla AK, *et al.* Evaluation of microcrystalline cellulose prepared from absorbent cotton as a direct compression carrier. *Drug Dev Ind Pharm* 1988; **14**: 1131–1136.

18 Doelker E, *et al.* Comparative tableting properties of sixteen microcrystalline celluloses. *Drug Dev Ind Pharm* 1987; **13**: 1847–1875.

19 Bassam F, *et al.* Effect of particle size and source on variability of Young's modulus of microcrystalline cellulose powders. *J Pharm Pharmacol* 1988; **40**: 68P.
20 Dittgen M, *et al.* Microcrystalline cellulose in direct tabletting. *Manuf Chem* 1993; **64**(7): 17, 19, 21.
21 Landin M, *et al.* Effect of country of origin on the properties of microcrystalline cellulose. *Int J Pharm* 1993; **91**: 123–131.
22 Landin M, *et al.* Effect of batch variation and source of pulp on the properties of microcrystalline cellulose. *Int J Pharm* 1993; **91**: 133–141.
23 Landin M, *et al.* Influence of microcrystalline cellulose source and batch variation on tabletting behavior and stability of prednisone formulations. *Int J Pharm* 1993; **91**: 143–149.
24 Podczeck F, Révész P. Evaluation of the properties of microcrystalline and microfine cellulose powders. *Int J Pharm* 1993; **91**: 183–193.
25 Rowe RC, *et al.* The effect of batch and source variation on the crystallinity of microcrystalline cellulose. *Int J Pharm* 1994; **101**: 169–172.
26 Hasegawa M. Direct compression: microcrystalline cellulose grade 12 versus classic grade 102. *Pharm Technol* 2002; **26**(5): 50, 52, 54, 56, 58, 60.
27 Kothari SH, *et al.* Comparative evaluations of powder and mechanical properties of low crystallinity celluloses, microcrystalline celluloses, and powdered celluloses. *Int J Pharm* 2002; **232**: 69–80.
28 Levis SR, Deasy PB. Production and evaluation of size-reduced grades of microcrystalline cellulose. *Int J Pharm* 2001; **213**: 13–24.
29 Wu JS, *et al.* A statistical design to evaluate the influence of manufacturing factors on the material properties and functionalities of microcrystalline cellulose. *Eur J Pharm Sci* 2001; **12**: 417–425.
30 Suzuki T, Nakagami H. Effect of crystallinity of microcrystalline cellulose on the compactability and dissolution of tablets. *Eur J Pharm Biopharm* 1999; **47**: 225–230.
31 Sun CC, Himmelspach MW. Reduced tabletability of roller compacted granules as a result of granule size enlargement. *J Pharm Sci* 2006; **95**: 200–206.
32 Patel S, *et al.* Understanding size enlargement and hardening of granules on tabletability of unlubricated granules prepared by dry granulation. *J Pharm Sci* 2011; **100**: 758–766.
33 Shi L, *et al.* Origin of profound changes in powder properties during wetting and nucleation stages of high shear wet granulation. *Powder Technol* 2011; **208**: 663–668.
34 *Food Chemicals Codex*, 7th edn. Bethesda, MD: United States Pharmacopeia, 2010.

20 General References

Asahi Kasei Corporation. *Ceolus KG, Celphere*. www.ceolus.com (accessed 6 December 2011).

Doelker E. Comparative compaction properties of various microcrystalline cellulose types and generic products. *Drug Dev Ind Pharm* 1993; **19**: 2399–2471.

European Directorate for the Quality of Medicines and Healthcare (EDQM). European Pharmacopoeia – State Of Work Of International Harmonisation. *Pharmeuropa* 2011; 23(4): 713–714. www.edqm.eu/site/-614.html (accessed 6 December 2011).

FMC Biopolymer. Problem Solver: *Avicel PH*, 2000.

FMC Biopolymer. Product literature: *Avicel DG*, *Avicel HFE-102*.

International Specialty Products. Material Safety data sheet: *Celex 101*, 2003.

JRS Pharma LP. Technical literature: *Emcocel*, 2003.

Pharmatrans Sanaq AG. Product literature: *Cellets*. www.cellets.com (accessed 6 December 2011).

Pharmatrans Sanaq AG. Product literature: *MCC Sanaq*. www.pharmatrans-sanaq.com/prod.html (accessed 6 December 2011).

Smolinske SC. *Handbook of Food, Drug, and Cosmetic Excipients*. Boca Raton, FL: CRC Press, 1992: 71–74.

Staniforth JN, *et al.* Effect of addition of water on the rheological and mechanical properties of microcrystalline celluloses. *Int J Pharm* 1988; **41**: 231–236.

21 Author

CC Sun.

22 Date of Revision

12 April 2012.

Cellulose, Microcrystalline and Carboxymethylcellulose Sodium

1 Nonproprietary Names

BP: Dispersible Cellulose

PhEur: Microcrystalline Cellulose and Carmellose Sodium

USP–NF: Microcrystalline Cellulose and Carboxymethylcellulose Sodium

2 Synonyms

Avicel CL-611; *Avicel RC-501*; *Avicel RC-581*; *Avicel RC-591*; *Avicel RC/CL*; cellulosum microcristallinum et carmellosum natricum; colloidal cellulose; *Vivapur MCG 591 PCG*; *Vivapur MCG 611 PCG*.

3 Chemical Name and CAS Registry Number

See Section 8.

4 Empirical Formula and Molecular Weight

See Section 8.

5 Structural Formula

See Section 8.

6 Functional Category

Dispersing agent; emulsion stabilizing agent; suspending agent; viscosity-increasing agent.

7 Applications in Pharmaceutical Formulation or Technology

Microcrystalline cellulose and carboxymethylcellulose sodium is used to produce thixotropic gels suitable as suspending vehicles in pharmaceutical and cosmetic formulations. The sodium carboxymethylcellulose aids dispersion and serves as a protective colloid.

Concentrations of less than 1% solids produce fluid dispersions, while concentrations of more than 1.2% solids produce thixotropic gels. When properly dispersed, it imparts emulsion stability, opacity, and suspension to a variety of products, and is used in nasal sprays, topical sprays and lotions, oral suspensions, emulsions, creams[1] and gels.

8 Description

Microcrystalline cellulose and carboxymethylcellulose sodium occurs as a white or off-white odorless and tasteless hygroscopic powder containing 5–22% sodium carboxymethylcellulose. It is a water-dispersible organic hydrocolloid.

9 Pharmacopeial Specifications

See Table I. *See also* Section 18.

Table I: Pharmacopeial specifications for microcrystalline cellulose and carboxymethylcellulose sodium.

Test	PhEur 7.4	USP35–NF30
Identification	+	+
pH	6.0–8.0	6.0–8.0
Solubility	+	–
Loss on drying	≤8.0%	≤8.0%
Sulfated ash	≤7.4%	–
Residue on ignition	–	≤5.0%
Apparent viscosity of nominal value	60–140%	60–140%
Heavy metals	–	≤0.001%
Assay (dried basis)	75–125%	75–125%

10 Typical Properties

Acidity/alkalinity pH 6–8 for a 1.2% w/v aqueous dispersion.
Density (bulk) 0.6 g/cm^3
Microbial content Total aerobic microbial count ≤100 cfu/g for *Avicel RC/CL* (*Escherichia coli*, *Staphylococcus aureus*, *Pseudomonas aeruginosa* and *Salmonella* species absent); total yeast and mold count ≤20 cfu/g for *Avicel RC/CL*.
Moisture content Not more than 6.0% w/w
Particle size distribution

Avicel CL-611: ≤0.1% retained on a #60 mesh and ≤50% retained on a #325 mesh;

Avicel RC-581: ≤0.1% retained on a #60 mesh and ≤35% retained on a #200 mesh;

Avicel RC-591: ≤0.1% retained on a #60 mesh and ≤45% retained on a #325 mesh;

Vivapur MCG 591 PCG: ≤5% retained on a #30 mesh and ≤50% retained on a #60 mesh;

Vivapur MCG 611 PCG: ≤5% retained on a #30 mesh and ≤50% retained on a #60 mesh.

Sodium content 0.8% for *Avicel RC-581* and *Avicel RC-591*; 1.2% for *Avicel CL-611*.
Solubility Practically insoluble in dilute acids and organic solvents. Partially soluble in dilute alkali and water (carboxymethylcellulose sodium fraction).
Viscosity (dynamic) (1.2% w/v aqueous dispersion)

50–118 mPa s (50–118 cP) for *Avicel CL-611*;

72–168 mPa s (72–168 cP) for *Avicel RC-581*;

39–91 mPa s (39–91 cP) for *Avicel RC-591*;

39–91 mPa s (39–91 cP) for *Vivapur MCG 591 PCG*;

50–118 mPa s (50–118 cP) for *Vivapur MCG 611 PCG*.

11 Stability and Storage Conditions

Microcrystalline cellulose and carboxymethylcellulose sodium is hygroscopic and should not be exposed to moisture. It is stable over a pH range of 3.5–11. Store in a cool, dry place. Avoid exposure to excessive heat.

12 Incompatibilities

Microcrystalline cellulose and carboxymethylcellulose sodium is incompatible with strong oxidizing agents. *See* Cellulose, Microcrystalline and Carboxymethylcellulose Sodium.

13 Method of Manufacture

Microcrystalline cellulose and carboxymethylcellulose sodium is a spray- or bulk-dried blend of microcrystalline cellulose and sodium carboxymethylcellulose. It is prepared by the chemical depolymerization of highly purified wood pulp. The original crystalline areas of the pulp fibers are combined with sodium carboxymethylcellulose, which serves as a protective colloid and also facilitates dispersion of the product; it is then either spray- or bulk-dried.

14 Safety

Microcrystalline cellulose and carboxymethylcellulose sodium is used in a wide range of pharmaceutical formulations and has low oral, dermal, and inhalation toxicity. It is nonirritating to the eyes and skin, and nonsensitizing to the skin. No significant acute toxicological effects are expected.

LD_{50} (skin, rabbit): >2.0 g/kg

LD_{50} (oral, rat): >5.0 g/kg

For further safety information, *see* Cellulose, Microcrystalline and Carboxymethylcellulose Sodium.

15 Handling Precautions

Observe normal precautions appropriate to the circumstances and quantity of material handled. Eye protection is recommended. *See also* Cellulose, Microcrystalline and Carboxymethylcellulose Sodium.

16 Regulatory Status

Included in the FDA Inactive Ingredients Database (inhalations; oral capsules, powders, suspensions, and tablets). Included in nonparenteral medicines licensed in the UK.

Microcrystalline cellulose and carboxymethylcellulose sodium is a mixture of two materials both of which are generally regarded as nontoxic:

Microcrystalline cellulose GRAS listed. Accepted for use as a food additive in Europe. Included in the FDA Inactive Ingredients Database (inhalations; oral capsules, powders, suspensions, syrups, and tablets; topical and vaginal preparations). Included in nonparenteral medicines licensed in the UK. Included in the Canadian Natural Health Products Ingredients Database.
Carboxymethylcellulose sodium GRAS listed. Accepted for use as a food additive in Europe. Included in the FDA Inactive Ingredients Database (dental preparations; intra-articular, intrabursal, intradermal, intralesional, and intrasynovial injections; oral drops, solutions, suspensions, syrups, and tablets; topical preparations). Included in nonparenteral medicines licensed in the UK. Included in the Canadian Natural Health Products Ingredients Database.

17 Related Substances

Carboxymethylcellulose sodium; cellulose, microcrystalline.

18 Comments

Microcrystalline cellulose is one of the materials that have been selected for harmonization by the Pharmacopeial Discussion Group. For further information see the General Information Chapter <1196> in the USP35–NF30, the General Chapter 5.8 in PhEur 7.4, along with the 'State of Work' document on the PhEur EDQM website, and also the General Information Chapter 8 in the JP XV.

The properties of preparations containing microcrystalline cellulose and carboxymethylcellulose sodium depend on the development of maximum colloidal dispersions in water.[2,3] *Avicel RC/CL* dispersions yield a highly thixotropic vehicle, which is primarily the result of the large number of colloidal microcrystal particles that result from full dispersion in aqueous media. The network establishes a weak gel structure with a measurable yield point that prevents drug particles from settling in a formulation. This gel structure is easily broken by mild shaking to yield a readily pourable liquid. Upon removal of shear, the gel structure re-establishes, providing a suspension medium with long-term stability against phase separation. *Avicel RC-591* has been found to have optimal formulation properties compared with other suspending agents in metronidazole benzoate suspensions.[4]

A specification for microcrystalline cellulose and carmellose sodium is contained in the *Japanese Pharmaceutical Excipients* (JPE).[5]

19 Specific References

1 Adeyeye MC, *et al.* Viscoelastic evaluation of topical creams containing microcrystalline cellulose/sodium carboxymethyl cellulose as stabilizer. *AAPS Pharm Sci Tech* 2002; 3(2): E8.

2 Rudraraju VS, Wyandt CM. Rheological characterization of microcrystalline cellulose/sodiumcarboxymethyl cellulose hydrogels using a controlled stress rheometer: part I. *Int J Pharm* 2005; **292**(1–2): 53–61.

3 Rudraraju VS, Wyandt CM. Rheological characterization of microcrystalline cellulose/sodiumcarboxymethyl cellulose hydrogels using a controlled stress rheometer: part II. *Int J Pharm* 2005; **292**(1–2): 63–73.

4 Zietsman S, *et al.* Formulation development and stability studies of aqueous metronidazole benzoate suspensions containing various suspending agents. *Drug Dev Ind Pharm* 2007; **33**(2): 191–197.

5 Japan Pharmaceutical Excipients Council. *Japanese Pharmaceutical Excipients 2004*. Tokyo: Yakuji Nippo, 2004; 551–554.

20 General References

Dell SM, Colliopoulos MS. FMC Biopolymer Problem Solver, Section 14: *Avicel RC/CL*, 2001.

Deorkar N. High-functionality excipients: a review. *Tablets and Capsules* 2008; 22–26. http://www.tabletscapsules.com.

European Directorate for the Quality of Medicines and Healthcare (EDQM). European Pharmacopoeia – State Of Work Of International Harmonisation. *Pharmeuropa* 2009; **21**(1): 142–143. http://www.edqm.eu/site/-614.html.

Farma International. Technical literature: *Avicel RC/CL*. http://www.farmainternational.com (accessed 29 February 2012).

FMC Biopolymer. Product literature: *Avicel RC-591*, 1994.

FMC Biopolymer. Specifications and analytical methods: *Avicel RC/CL*, 1995.

FMC Biopolymer. Material safety datasheet: *Avicel RC/CL*, 31 January 2008.

Gohel MC, Jogani PD. A review of co-processed directly compressible excipients. *J Pharm Pharm Sci* 2005; **8**(1): 76–93.

JRS Pharma. Technical literature: *Vivapur MCG*. http://jrspharma.de (accessed 29 February 2012).

Nachaegari SK, Bansal AK. Coprocessed excipients for solid dosage forms. *Pharm Tech* 2004; **28**: 52–64.

21 Authors

ME Fenton, RC Rowe.

22 Date of Revision

1 March 2012.

Cellulose, Powdered

1 Nonproprietary Names

BP: Powdered Cellulose
JP: Powdered Cellulose
PhEur: Cellulose, Powdered
USP–NF: Powdered Cellulose

2 Synonyms

Alpha-cellulose; *Arbocel*; cellulosi pulvis; E460; Elcema; *JustFiber*; *KC Flock*; *Microcel 3E-150*; *Sanacel*; *Sanacel Pharma*; *Sancel-W*; *Sanacel Wheat*; *Solka-Floc*.

3 Chemical Name and CAS Registry Number

Cellulose [9004-34-6]

4 Empirical Formula and Molecular Weight

$(C_6H_{10}O_5)_n$ ≈243 000 where $n \approx 500$.

Since cellulose is derived from a natural polymer, it has variable chain length and thus variable molecular weight.

See also Sections 8 and 13.

5 Structural Formula

HO HO OH OH O O OH OH H O OH n/2

6 Functional Category

Adsorbent; glidant; suspending agent; tablet and capsule diluent; tablet and capsule disintegrant; viscosity-increasing agent.

7 Applications in Pharmaceutical Formulation or Technology

Powdered cellulose is used as a tablet diluent and filler in two-piece hard capsules; *see* Table I. In both contexts it acts as a bulking agent to increase the physical size of the dosage form for formulations containing a small amount of active substance.

Powdered cellulose has acceptable compression properties, although the flow properties of most brands are poor. However, low-crystallinity powdered cellulose has exhibited properties that are different from standard powdered cellulose materials, and has shown potential as a direct-compression excipient.[1]

In soft gelatin capsules, powdered cellulose may be used to reduce the sedimentation rate of oily suspension fills. It is also used as the powder base material of powder dosage forms, and as a suspending agent in aqueous suspensions for peroral delivery. It may also be used to reduce sedimentation during the manufacture of suppositories.

Powdered cellulose has been investigated as an alternative to microcrystalline cellulose as an agent to assist the manufacture of pellets by extrusion/spheronization.[2,3] However, powdered cellulose alone requires too much water and due to water movement during extrusion cannot be used as an extrusion/spheronization aid on its own.[4]

Powdered cellulose is also used widely in cosmetics and food products as an adsorbent and thickening agent.

Table I: Uses of powdered cellulose.

Use	Concentration (%)
Capsule filler	5–30
Tablet and capsule binder	5–40 (wet granulation) 10–30 (dry granulation)
Tablet disintegrant	5–20
Tablet glidant	1–2

8 Description

Powdered cellulose occurs as a white or almost white, odorless and tasteless powder of various particle sizes, ranging from a free-flowing fine or granular dense powder, to a coarse, fluffy, nonflowing material.

9 Pharmacopeial Specifications

The pharmacopeial specifications for powdered cellulose have undergone harmonization of many attributes for JP, PhEur, and USP–NF. *See* Table II.

See also Section 18.

Table II: Pharmacopeial specifications for powdered cellulose.

Test	JP XV	PhEur 7.4[a]	USP35–NF30
Identification[b]	+	+	+
Characters	+	+	−
Microbial limits			
Aerobic	≤10^3 cfu/g	≤10^3 cfu/g	≤10^3 cfu/g
Fungi and yeast	≤10^2 cfu/g	≤10^2 cfu/g	≤10^2 cfu/g
pH (10% w/w suspension)	5.0–7.5	5.0–7.5	5.0–7.5
Loss on drying	≤6.5%	≤6.5%	≤6.5%
Residue on ignition	≤0.3%	≤0.3%	≤0.3%
Solubility	+	+	−
Ether-soluble substances	≤15.0 mg	≤0.15%	≤15.0 mg
Water-soluble substances	≤15.0 mg	≤1.5%	≤15.0 mg
Heavy metals[c]	≤10 ppm	≤10 ppm	≤10 ppm

(a) The PhEur 7.4 also includes crystallinity, particle size distribution and powder flow under functionality-related characteristics.
(b) Degree of polymerization is ≥440 for JP XV, PhEur 7.4 and USP35–NF30.
(c) These tests have not been fully harmonized at the time of publication.

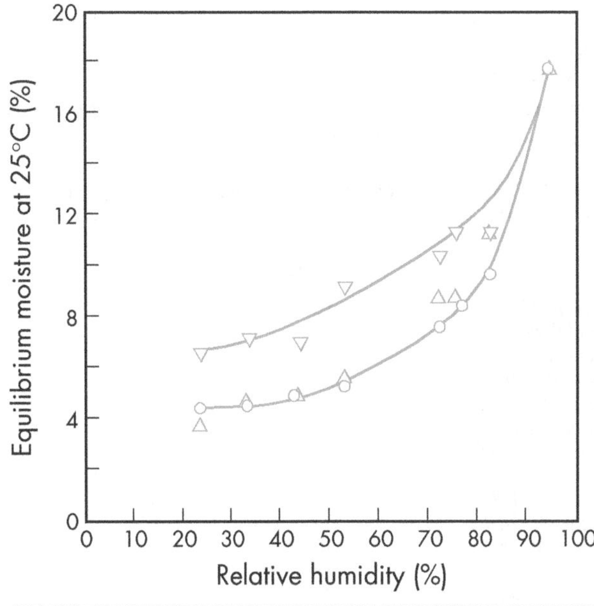

Figure 1: Equilibrium moisture content of powdered cellulose at 25°C.

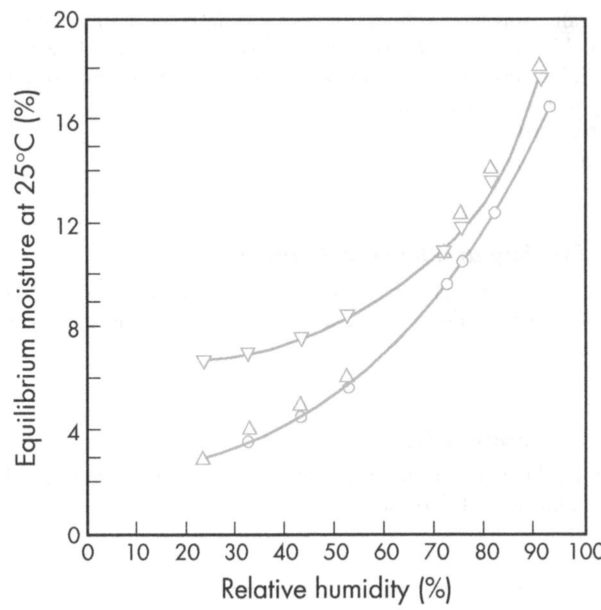

Figure 2: Equilibrium moisture content of powdered cellulose at 25°C.

10 Typical Properties

Density (bulk) 0.15–0.41 g/cm^3, depending on the source and grade.

Density (tapped) 0.21–0.48 g/cm^3, depending on the source and grade.

Density (true)

1.47–1.51 g/cm^3 [5] depending on source and grade;

1.27–1.61 g/cm^3 [6] depending on source and grade.

Moisture content Powdered cellulose is slightly hygroscopic;[7] *see* Figures 1 and 2.

Particle size distribution

Powdered cellulose is commercially available in several different particle sizes; *see* Table III.

Table III: Particle size distribution of selected commercially available powdered cellulose.

Grade	Average particle size (μm)
Arbocel M80	60
Arbocel P 290	80
Arbocel A 300	250
JustFiber BF 200	35
JustFiber WWF 40	60
JustFiber BVF 65	55
JustFiber WWF 200	35
KC Flock W–50	45
KC Flock W–100G	37
KC Flock W–200G	32
KC Flock W–250	30
KC Flock W–300G	28
KC Flock W–400G	24
KC Flock W-10MG2	10
Solka–Floc 900 NF	110
Solka–Floc 20 NF	100
Solka–Floc 40 NF	60

Solubility Practically insoluble in water, dilute acids, and most organic solvents, although it disperses in most liquids. Slightly soluble in 5% w/v sodium hydroxide solution. Powdered cellulose does not swell in water, but does so in dilute sodium hypochlorite (bleach).

Spectroscopy

NIR spectra *see* Figure 3.

11 Stability and Storage Conditions

Powdered cellulose is a stable, slightly hygroscopic material.[7] The bulk material should be stored in a well-closed container in a cool, dry place.

12 Incompatibilities

Incompatible with strong oxidizing agents, bromine pentafluoride, sodium nitrite and fluorine.[6]

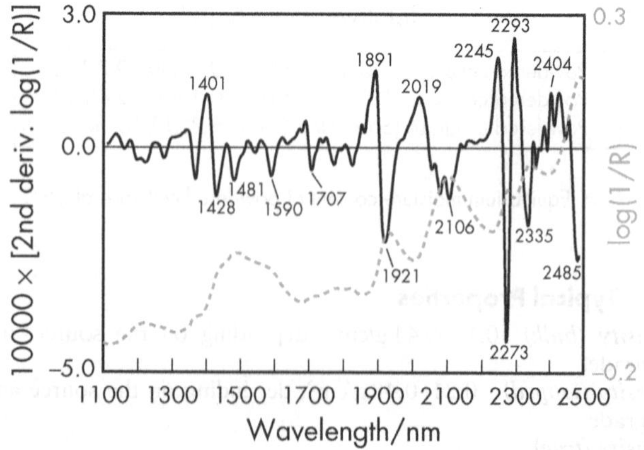

Figure 1: Near-infrared spectrum of powdered cellulose measured by reflectance.

13 Method of Manufacture

Powdered cellulose is manufactured by the purification and mechanical size reduction of α-cellulose obtained as a pulp from fibrous plant materials.

14 Safety

Powdered cellulose is widely used in oral pharmaceutical formulations and food products and is regarded as a nontoxic and nonirritant material. However, allergic reactions when inhaled, ingested or in contact with the skin are possible.[6]

Powdered cellulose is not absorbed systemically following peroral administration and thus has little toxic potential. Consumption of large quantities of cellulose may, however, have a laxative effect, although this is unlikely to be a problem when cellulose is used as an excipient in pharmaceutical formulations.

Deliberate abuse of formulations containing cellulose, either by inhalation or by injection, has resulted in the formation of cellulose granulomas.[8]

LD_{50} (rat, oral): >5 g/kg [6]

LD_{50} (rat, inhalation): 5.8 g/m^3/4 h [6]

LD_{50} (rabbit, skin): >2 g/kg [6]

15 Handling Precautions

Observe normal precautions appropriate to the circumstances and quantity of material handled. Powdered cellulose may be an irritant to the eyes. The material emits toxic fumes under fire conditions and when heated to decomposition.[6] Gloves, eye protection, and a dust mask (NIOSH approved), and engineering controls such as exhaust ventilation are recommended. In the UK, the workplace exposure limits for cellulose have been set at 10 mg/m^3 long-term (8-hour TWA) for total inhalable dust and 4 mg/m^3 for respirable dust; the short-term limit for total inhalable dust has been set at 20 mg/m^3.[9] In the US, the TWA exposure values are defined by NIOSH as 5 mg/m^3 (respirable fraction) and 10 mg/m^3 (total dust).[6]

16 Regulatory Status

GRAS listed. Accepted for use as a food additive in Europe (except for infant food in the UK). Included in nonparenteral medicines licensed in the UK. Included in the Canadian Natural Health Products Ingredients Database.

17 Related Substances

Cellulose, microcrystalline.

18 Comments

Powdered cellulose has undergone harmonization of many attributes for JP, PhEur, and USP–NF by the Pharmacopeial Discussion Group. For further information see the General Information Chapter <1196> in the USP35–NF30, the General Chapter 5.8 in PhEur 7.4, along with the 'State of Work' document on the PhEur EDQM website, and also the General Information Chapter 8 in the JP XV.

When powdered cellulose was used as a tablet disintegrant, tablets could be stored up to 78% RH without losing their rapid disintegration properties.[10] Tablets disintegrating rapidly in the oral cavity were also successfully prepared with powdered cellulose.[11] Coprocessing of powdered cellulose with magnesium carbonate by roller compaction has resulted in a promising excipient for direct tableting.[12] Highly porous matrices from powdered cellulose may be suitable for stabilization and handling of liquid drug substances.[13]

A specification for powdered cellulose is contained in the *Food Chemicals Codex* (FCC).[14]

The EINECS number for powdered cellulose is 232-674-9.

19 Specific References

1 Kothari SH, *et al.* Comparative evaluations of powder and mechanical properties of low crystallinity celluloses, microcrystalline celluloses, and powdered celluloses. *Int J Pharm* 2002; **232**: 69–80.
2 Alvarez L, *et al.* Powdered cellulose as excipient for extrusion-spheronization pellets of a cohesive hydrophobic drug. *Eur J Pharm Biopharm* 2003; **55**: 291–295.
3 Lindner H, Kleinebudde P. Use of powdered cellulose for the production of pellets by extrusion spheronization. *J Pharm Pharmacol* 1994; **46**: 2–7.
4 Fechner PM, *et al.* Properties of microcrystalline cellulose and powder cellulose after extrusion/spheronization as studied by Fourier transform Raman spectroscopy and environmental scanning electron microscopy. *AAPS PharmSci* 2003; **5**: Article 31.
5 Podczeck F, Révész P. Evaluation of the properties of microcrystalline and microfine cellulose powders. *Int J Pharm* 1993; **91**: 183–193.
6 United States Pharmacopeial Convention Inc. Material safety data sheet: Powdered cellulose, 2005.
7 Callahan JC, *et al.* Equilibrium moisture content of pharmaceutical excipients. *Drug Dev Ind Pharm* 1982; **8**: 355–369.
8 Cooper CB, *et al.* Cellulose granulomas in the lungs of a cocaine sniffer. *Br Med J* 1983; **286**: 2021–2022.
9 Health and Safety Executive. *EH40/2005: Workplace Exposure Limits.* Sudbury: HSE Books, 2011. www.hse.gov.uk//pubns/priced/eh40.pdf (accessed 12 April 2012).
10 Uhumwangho MU, Okor RS. Humidity effect on the disintegrant property of alpha-cellulose and the implication for dissolution rates in paracetamol tablets. *Pakistan J Sci Ind Res* 2005; **48**: 8–13.
11 Yamada Y, *et al.* Preparation and evaluation of rapidly disintegrating tablets in the oral cavity by the dry compression method – availability of powdered cellulose as an excipient. *J Pharm Sci Technol* 2006; **66**: 473–481.
12 Freitag F, *et al.* Coprocessing of powdered cellulose and magnesium carbonate: direct tabletting versus tabletting after roll compaction/dry granulation. *Pharm Dev Technol* 2005; **10**: 353–362.
13 Mihranyan A, *et al.* Sorption of nicotine to cellulose powders. *Eur J Pharm Sci* 2004; **22**: 279–286.
14 *Food Chemicals Codex*, 7th edn. Bethesda, MD: United States Pharmacopeia; 2010: 206.

See also Cellulose, microcrystalline.

20 General References

Allen LV. Featured excipient: capsule and tablet diluents. *Int J Pharm Compound* 2000; **4**: 306–310324–325.
Belda PM, Mielck JB. The tabletting behavior of cellactose compared with mixtures of celluloses with lactoses. *Eur J Pharm Biopharm* 1996; **42**: 325–330.
Blanver Europe. Product literature: *Microcel 3E-150*. www.blanver.com.br/microce2.htm (accessed 5 December 2011).
CFF GmbH. Product literature: *Sanacel*; *Sanacel Pharma*. www.cff.de/de/branch/pharmazie.htm (accessed 5 December 2011).
CFF GmbH. Product literature: *Sanacel Wheat*. www.prochem.ch/html/forum/thema_0407/BroschSANACELwheat.pdf (accessed 5 December 2011).
European Directorate for the Quality of Medicines and Healthcare (EDQM). European Pharmacopoeia – State Of Work Of International Harmonisation. *Pharmeuropa* 2011; 23(4): 713–714. www.edqm.eu/site/-614.html (accessed 5 December 2011).
International Fibre Corporation. Product literature: *Solka-Floc*; *JustFiber*. www.ifcfiber.com/applications/productData.php#pharm (accessed 5 December 2011).
JRS Pharma GmbH. Product literature: *Arbocel*. www.jrspharma.de/Pharma/wEnglisch/produktinfo/productinfo_arbocel.shtml (accessed 5 December 2011).
Kimura M, *et al.* The evaluation of powdered cellulose as a pharmaceutical excipient. *J Pharm Sci Tech Yakuzaigaku* 2002; **62**: 113–123.
NB Entrepreneurs. Product literature: Powdered cellulose - *Sancel-W*. www.nbent.com/SancelW.htm (accessed 5 December 2011).
Nippon Paper Chemicals. Product literature: *KC Flock*. www.npchem.co.jp/english/product/kcflock/index.html (accessed 5 December 2011).
Smolinske SC. *Handbook of Food, Drug, and Cosmetic Excipients*. Boca Raton, FL: CRC Press, 1992: 71–74.

21 Author

F Podczeck.

22 Date of Revision

5 December 2011.

Cellulose, Silicified Microcrystalline

1 Nonproprietary Names

USP–NF: Silicified Microcrystalline Cellulose

2 Synonyms

ProSolv.

3 Chemical Name and CAS Registry Number

See Section 8.

4 Empirical Formula and Molecular Weight

See Section 8.

5 Structural Formula

See Section 8.

6 Functional Category

Tablet and capsule diluent.

7 Applications in Pharmaceutical Formulation or Technology

Silicified microcrystalline cellulose is used as a diluent in the formulation of capsules and tablets. It has improved compaction properties in both wet granulation and direct compression compared to conventional microcrystalline cellulose.[1–5] Silicified microcrystalline cellulose was specifically developed to address the loss of compactability that occurs with microcrystalline cellulose after wet granulation. Silicified microcrystalline cellulose also appears to have beneficial properties for use in the formulation of powder-filled capsules.[6,7]

8 Description

Silicified microcrystalline cellulose is a synergistic, intimate physical mixture of two components: microcrystalline cellulose and colloidal silicon dioxide (for further information *see* Cellulose, Microcrystalline and Colloidal Silicon Dioxide). Silicified microcrystalline cellulose contains 2% w/w colloidal silicon dioxide.

SEM 1: Excipient: silicified microcrystalline cellulose; manufacturer: JRS Pharma LP; lot no.: CSD5866; magnification: 100×.

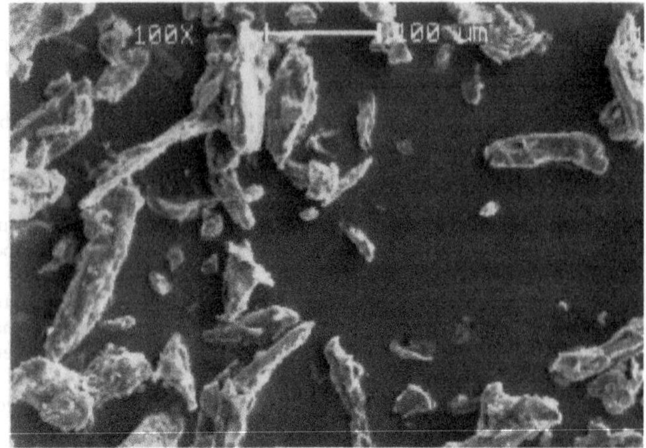

SEM 2: Excipient: silicified microcrystalline cellulose; manufacturer: JRS Pharma LP; lot no.: CSD5866; magnification: 300×.

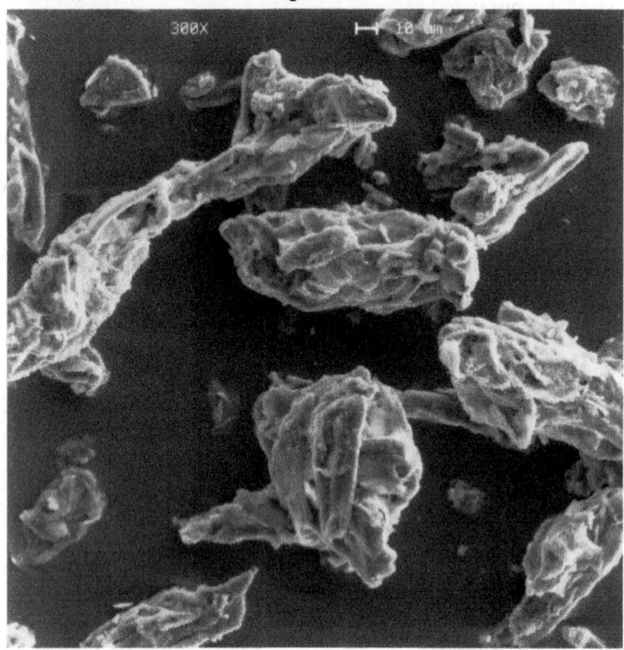

9 Pharmacopeial Specifications

See Table I. *See also* Section 18.

Table I: Pharmacopeial specifications for silicified microcrystalline cellulose.

Test	USP35–NF30
Identification	+
Microbial contamination	
Aerobic bacteria	≤1000 cfu per g
Fungi	≤100 cfu per g
Conductivity	+
pH	5.0–7.5
Loss on drying	7.0%
Residue on ignition	1.8–2.2%
Bulk density	+
Degree of polymerization	≤350
Particle size and distribution	+
Water-soluble substances	≤12.5 mg
Ether-soluble substances	≤5.0 mg
Heavy metals	≤0.001%

SEM 3: Excipient: silicified microcrystalline cellulose; manufacturer: JRS Pharma LP; lot no.: CSD5866; magnification: 500×.

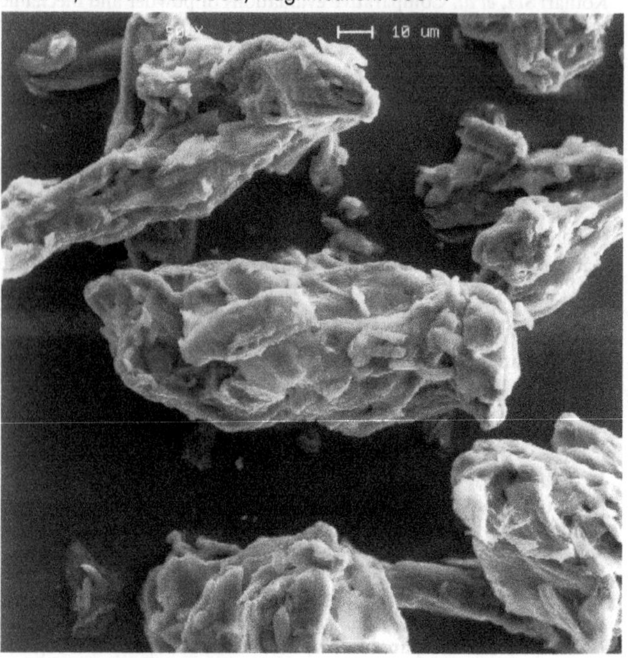

10 Typical Properties

Acidity/alkalinity pH = 5.0–7.5 (10% w/v suspension)

Density 1.58 g/cm[5]

Density (bulk) 0.31 g/cm^3

Density (tapped) 0.39 g/cm[5]

Melting point The microcrystalline cellulose component chars at 260–270°C.

Moisture content Typically less than 6% w/w.

Particle size distribution Typical particle size is 20–200 μm. Different grades may have a different normal mean particle size.

Solubility Practically insoluble in water, dilute acids, and most organic solvents. The microcrystalline cellulose component is slightly soluble in 5% w/w sodium hydroxide solution.

Spectroscopy

NIR spectra *see* Figure 1.

11 Stability and Storage Conditions

Silicified microcrystalline cellulose is stable when stored in a well-closed container in a cool, dry place.

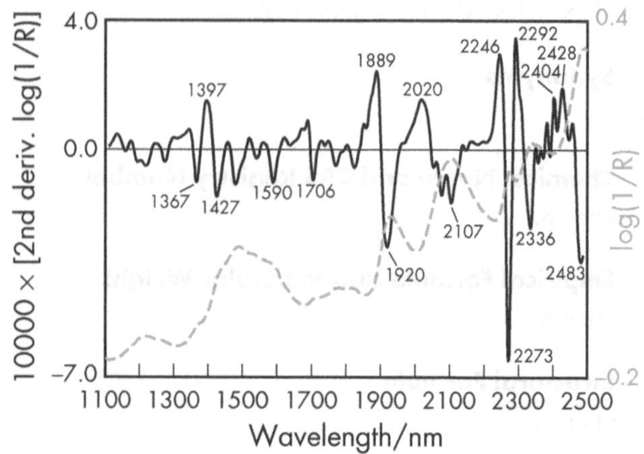

Figure 1: Near-infrared spectrum of silicified microcrystalline cellulose measured by reflectance.

12 Incompatibilities

See Cellulose, Microcrystalline and Colloidal Silicon Dioxide.

13 Method of Manufacture

Silicified microcrystalline cellulose is manufactured by co-drying a suspension of microcrystalline cellulose particles and colloidal silicon dioxide such that the dried finished product contains 2% w/w colloidal silicon dioxide.

The colloidal silicon dioxide appears physically bound onto the surface and inside the silicified microcrystalline cellulose particles. Extensive studies using different spectroscopic methods have failed to show any form of chemical interaction.[4,8,9]

14 Safety

See Cellulose, Microcrystalline and Colloidal Silicon Dioxide.

15 Handling Precautions

Observe normal precautions appropriate to the circumstances and quantity of material handled. Handling of silicified microcrystalline cellulose can generate nuisance dusts and the use of a respirator or dust mask may be necessary.

Microcrystalline cellulose may be irritant to the eyes. Gloves, eye protection, and a dust mask are recommended. In the UK the long-term workplace exposure limits (8-hour TWA) have been set at 10 mg/m^3 for total inhalable dust and 4 mg/m^3 for respirable dust; short-term limit for total inhalable dust has been set at 20 mg/m^3.[10]

Since the colloidal silicon dioxide is physically bound to the microcrystalline cellulose the general recommendations of gloves, eye protection, and a dust mask should be followed when handling silicified microcrystalline cellulose.

16 Regulatory Status

Silicified microcrystalline cellulose is a physical mixture of two materials both of which are generally regarded as nontoxic:

Microcrystalline cellulose GRAS listed. Included in the FDA Inactive Ingredients Database (inhalations, oral capsules, powders, suspensions, syrups, and tablets). Included in nonparenteral medicines licensed in Europe and the USA. Included in the Canadian Natural Health Products Ingredients Database.

Colloidal silicon dioxide GRAS listed. Included in the FDA Inactive Ingredients Database (oral capsules and tablets). Included in nonparenteral medicines licensed in Europe and the USA. Included in the Canadian Natural Health Products Ingredients Database.

17 Related Substances

Cellulose, microcrystalline; colloidal silicon dioxide.

18 Comments

Colloidal silicon dioxide and microcrystalline cellulose are two of the materials that have been selected for harmonization by the Pharmacopeial Discussion Group. For further information see the General Information Chapter <1196> in the USP35–NF30, the General Chapter 5.8 in PhEur 7.4, along with the 'State of Work' document on the PhEur EDQM website, and also the General Information Chapter 8 in the JP XV.

Silicified microcrystalline cellulose has greater tensile strength and requires lower compression pressures than regular grades of microcrystalline cellulose. Furthermore, silicified microcrystalline cellulose maintains its compactability after wet granulation; the compacts exhibit greater stiffness and they require considerably more energy for tensile failure to occur.[4,11,12]

19 Specific References

1 Sherwood BE, *et al.* Silicified microcrystalline cellulose (SMCC): a new class of high functionality binders for direct compression. *Pharm Res* 1996; **13**(9): S197.

2 Staniforth JN, *et al.* Towards a new class of high functionality tablet binders. II: silicified microcrystalline cellulose (SMCC). *Pharm Res* 1996; **13**(9): S197.

3 Tobyn MJ, *et al.* Compaction studies on a new class of high functionality binders: silicified microcrystalline cellulose (SMCC). *Pharm Res* 1996; **13**(9): S198.

4 Habib SY, *et al.* Is silicified wet-granulated microcrystalline cellulose better than original wet-granulated microcrystalline cellulose? *Pharm Dev Technol* 1999; 4(3): 431–437.

5 Tobyn MJ, *et al.* Physiochemical comparison between microcrystalline cellulose and silicified microcrystalline cellulose. *Int J Pharm* 1998; **169**: 183–194.

6 Guo M, *et al.* Evaluation of the plug information process of silicified microcrystalline cellulose. *Int J Pharm* 2002; **233**: 99–109.

7 Guo M, Augsburger LL. Potential application of silicified microcrystalline cellulose in direct-fill formulations for capsule-filling machines. *Pharm Dev Technol* 2003; **8**: 47–59.

8 Edge S, *et al.* The mechanical properties of compacts of microcrystalline cellulose and silicified microcrystalline cellulose. *Int J Pharm* 2000; **200**: 67–72.

9 Buckton G, Yonemochi E. Near IR spectroscopy to quantify the silica content and differences between silicified miicrocrystalline cellulose and physical mixtures of microcrystalline cellulose and silica. *Eur J Pharm Sci* 2000; **10**(1): 77–80.

10 Health and Safety Executive. *EH40/2005: Workplace Exposure Limits.* Sudbury: HSE Books, 2011. http://www.hse.gov.uk/pubns/priced/eh40.pdf (accessed 29 February 2012).

11 Buckton G, *et al.* Water sorption and near IR spectroscopy to study the differences between microcrystalline cellulose and silicified microcrystalline cellulose after wet granulation. *Int J Pharm* 1999; **181**: 41–47.

12 Mužíková J, Nováková P. A study of the properties of compacts from silicified microcrystalline cellulose. *Drug Dev Ind Pharm* 2007; **33**: 775–781.

20 General References

European Directorate for the Quality of Medicines and Healthcare (EDQM). European Pharmacopoeia – State Of Work Of International Harmonisation. *Pharmeuropa* 2010; 22(4): 583–584. www.edqm.eu/site/-614.html (accessed 13 October 2010).

Li JX, *et al.* Characterization of wet masses of pharmaceutical powders by triaxial compression test. *J Pharm Sci* 2000; **89**(2): 178–190.

Staniforth JN *et al.* Pharmaceutical excipient having improved compressibility. US Patent 5,585,115, 1996.

21 Author

RC Moreton.

22 Date of Revision

1 March 2012.

Cellulose Acetate

1 Nonproprietary Names

BP: Cellulose Acetate
PhEur: Cellulose Acetate
USP–NF: Cellulose Acetate

2 Synonyms

Acetic acid, cellulose ester; acetyl cellulose; cellulose diacetate; cellulose triacetate; cellulosi acetas.

3 Chemical Name and CAS Registry Number

Cellulose acetate [9004-35-7]
Cellulose diacetate [9035-69-2]
Cellulose triacetate [9012-09-3]

4 Empirical Formula and Molecular Weight

Cellulose acetate is cellulose in which a portion or all of the hydroxyl groups are acetylated. Cellulose acetate is available in a wide range of acetyl levels and chain lengths and thus molecular weights and viscosities. It contains not less than 29.0% and not more than 44.8% by weight of acetyl groups, calculated on the dried basis; *see* Table I.

5 Structural Formula

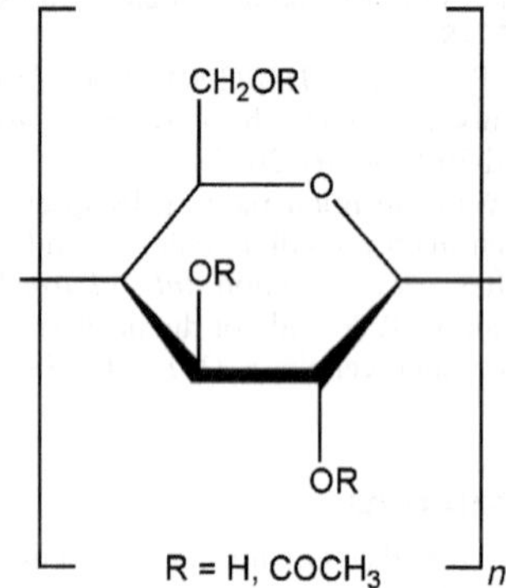

6 Functional Category

Coating agent; modified-release agent; tablet and capsule diluent; transdermal delivery component.

7 Applications in Pharmaceutical Formulation or Technology

Cellulose acetate is widely used in pharmaceutical formulations both in sustained-release applications and for taste masking.

Cellulose acetate is used as a semipermeable coating on tablets, especially on osmotic pump-type tablets and implants. This allows for controlled, extended release of actives.(1–5) Cellulose acetate films, in conjunction with other materials, also offer sustained release without the necessity of drilling a hole in the coating as is typical with osmotic pump systems. Cellulose acetate and other cellulose esters have also been used to form drug-loaded microparticles with controlled-release characteristics.(6–8)

Cellulose acetate films are used in transdermal drug delivery systems(9,10) and also as film coatings on tablets(11,12) or granules for taste masking. For example, acetaminophen granules have been coated with a cellulose acetate-based coating before being processed to provide chewable tablets. Extended-release tablets can also be formulated with cellulose acetate as a directly compressible matrix former.(2) The release profile can be modified by changing the ratio of active to cellulose acetate and by incorporation of plasticizer, but was shown to be insensitive to cellulose acetate molecular weight and particle size distribution.

8 Description

Cellulose acetate occurs as a hygroscopic white to off-white, free-flowing powder, pellet, or flake. It is tasteless and odorless, or may have a slight odor of acetic acid.

9 Pharmacopeial Specifications

The pharmacopeial specifications for cellulose acetate have undergone harmonization of many attributes for PhEur, and USP–NF.

See Table II. *See also* Section 18.

SEM 1: Excipient: cellulose acetate, CA-398-10; manufacturer: Eastman Chemical Co.; lot no.: AC65280; magnification: 60×; voltage: 3 kV.

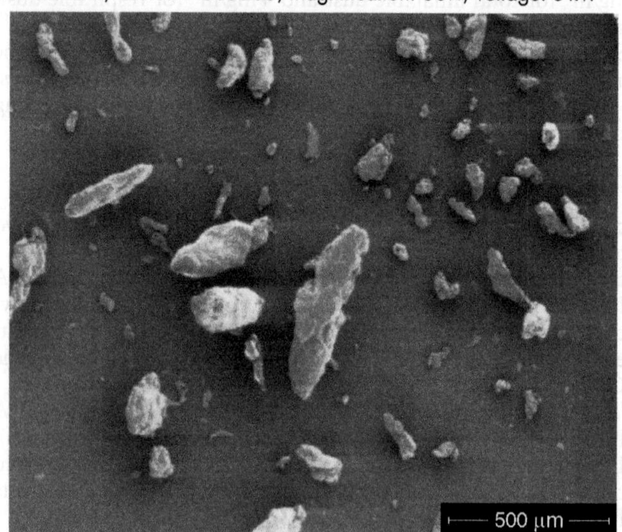

SEM 2: Excipient: cellulose acetate, CA-398-10; manufacturer: Eastman Chemical Co.; lot no.: AC65280NF; magnification: 600×; voltage: 2 kV.

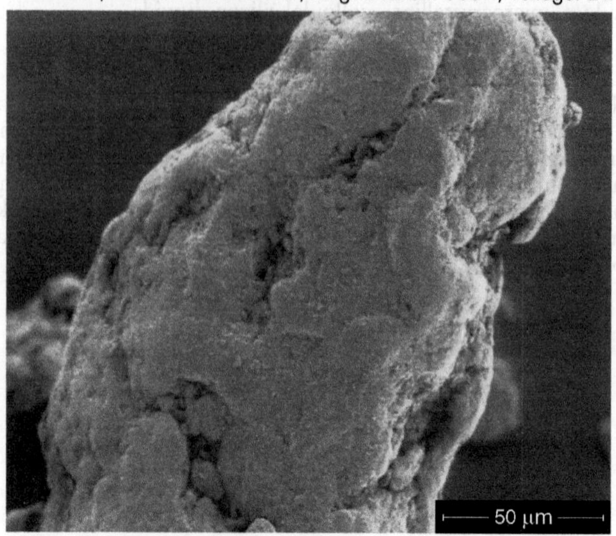

Table I: Comparison of different types of cellulose acetate.[2]

Type	Acetyl (%)	Viscosity (mPa s)[a]	Hydroxyl (%)	Melting range (°C)	T_g [b] (°C)	Density (g/cm^3)	MW_n [c]
CA-320S NF/EP	32.0	2.1	8.7	230–250	180	1.31	38 000
CA-398-3	39.8	11.4	3.5	230–250	180	1.31	30 000
CA-398-6	39.8	22.8	3.5	230–250	182	1.31	35 000
CA-398-10 NF/EP	39.8	38.0	3.5	230–250	185	1.31	40 000
CA-398-30	39.7	114.0	3.5	230–250	189	1.31	50 000
CA-394-60S	39.5	228.0	4.0	240–260	186	1.32	60 000

(a) ASTM D 817 (formula A) and D 1343.
(b) Glass transition temperature.
(c) Number average molecular weight in polystyrene equivalents determined using size-exclusion chromatography.

Table II: Pharmacopeial specifications for cellulose acetate.

Test	PhEur 7.4	USP35–NF30
Identification	+	+
Characters	+	—
Loss on drying	≤5.0%	≤5.0%
Residue on ignition	≤0.1%	≤0.1%
Free acid	≤0.1%	≤0.1%
Heavy metals[a]	≤10 ppm	≤10 ppm
Microbial contamination	+	—
Aerobic	≤10^3 cfu/g	—
Fungi and yeast	≤10^2 cfu/g	—
Assay (of acetyl groups)	29.0–44.8%	29.0–44.8%

(a) These tests have not been fully harmonized at the time of publication.

10 Typical Properties

Density (bulk) Typically 0.4 g/cm^3 for powders.
Glass transition temperature 170–190°C
Melting point Melting range 230–300°C
Solubility The solubility of cellulose acetate is greatly influenced by the level of acetyl groups present. The cellulose acetates of higher acetyl level are generally more limited in solvent choice than are the lower-acetyl materials. Practically insoluble in water and ethanol (96%); soluble in acetone, dioxane, dimethylformamide, formic acid, and in a mixture of equal volumes of methyl alcohol and dichloromethane.
Spectroscopy
NIR spectra *see* Figure 1.
Viscosity (dynamic) Various grades of cellulose acetate are commercially available that differ in their acetyl content and degree of polymerization. They can be used to produce 10% w/v solutions in organic solvents with viscosities of 10–230 mPa s (10–230 cP). Blends of cellulose acetates may also be prepared with intermediate viscosity values.
See also Table I.

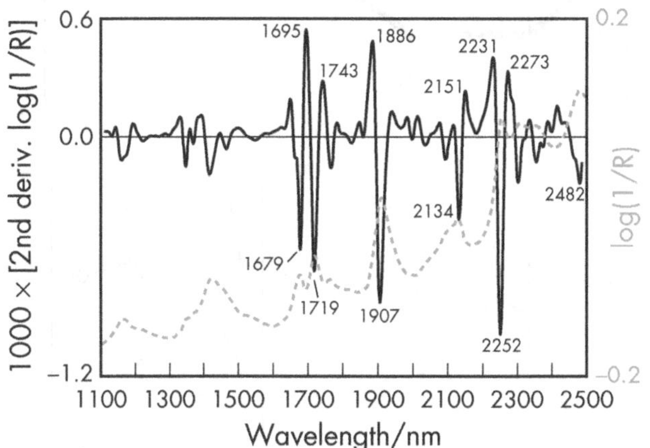

Figure 1: Near-infrared spectrum of cellulose acetate measured by reflectance.

11 Stability and Storage Conditions

Cellulose acetate is stable if stored in a well-closed container in a cool, dry place. Cellulose acetate hydrolyzes slowly under prolonged adverse conditions such as high temperature and humidity, with a resultant increase in free acid content and odor of acetic acid.

12 Incompatibilities

Cellulose acetate is incompatible with strongly acidic or alkaline substances. Cellulose acetate is compatible with the following plasticizers: diethyl phthalate, polyethylene glycol, triacetin, and triethyl citrate.

13 Method of Manufacture

Cellulose acetate is prepared from highly purified cellulose by treatment with acid catalysis and acetic anhydride.

14 Safety

Cellulose acetate is widely used in oral pharmaceutical products and is generally regarded as a nontoxic and nonirritant material.

15 Handling Precautions

Observe normal precautions appropriate to the circumstances and quantity of material handled. Dust may be irritant to the eyes and eye protection should be worn. Like most organic materials in powder form, these materials are capable of creating dust explosions. Cellulose acetate is combustible.

16 Regulatory Status

Included in the FDA Inactive Ingredients Database (oral tablets). Included in the Canadian Natural Health Products Ingredients Database.

17 Related Substances

Cellulose acetate phthalate.

18 Comments

Cellulose acetate has undergone harmonization of many attributes for PhEur, and USP–NF by the Pharmacopeial Discussion Group. For further information see the General Information Chapter <1196> in the USP35–NF30, the General Chapter 5.8 in PhEur 7.4, along with the 'State of Work' document on the PhEur EDQM website, and also the General Information Chapter 8 in the JP XV.

Therapeutically, cellulose acetate has been used to treat cerebral aneurysms, and also for spinal perimedullary arteriovenous fistulas.[13]

When solutions are being prepared, cellulose acetate should always be added to the solvent, not the reverse. Various grades of cellulose acetate are available with varying physical properties; *see* Table I.

19 Specific References

1 Meier MA, *et al.* Characterization and drug-permeation profiles of microporous and dense cellulose acetate membranes: influence of plasticizer and pre-forming agent. *Int J Pharm* 2004; **278**(1): 99–110.
2 Yum J, Hong-Wei Wu S. A feasibility study using cellulose acetate and cellulose acetate butyrate. *Pharm Technol* 2000; **24**(10): 92–106.
3 Theeuwes F. Elementary osmotic pump. *J Pharm Sci* 1975; **64**(12): 1987–1991.
4 Santus G, Baker RW. Osmotic drug delivery: review of the patent literature. *J Control Release* 1995; **35**: 1–21.
5 Van Savage G, Rhodes CT. The sustained release coating of solid dosage forms: a historical review. *Drug Dev Ind Pharm* 1995; **21**(1): 93–118.
6 Soppimath KS, *et al.* Development of hollow microspheres as floating controlled-release systems for cardiovascular drugs: preparation and release characteristics. *Drug Dev Ind Pharm* 2001; **27**(6): 507–515.
7 Soppimath KS, *et al.* Cellulose acetate microspheres prepared by o/w emulsification and solvent evaporation method. *J Microencapsul* 2001; **18**(6): 811–817.
8 Magdassi S *et al.* Organic nanoparticles obtained from microemulsions by solvent evaporation. International Patent WO 032327; 2008.
9 Rao PR, Diwan PV. Drug diffusion from cellulose acetate-polyvinyl pyrrolidine free films for transdermal administration. *Indian J Pharm Sci* 1996; 58(6): 246–250.
10 Rao PR, Diwan PV. Permeability studies of cellulose acetate free films for transdermal use: influences of plasticizers. *Pharm Acta Helv* 1997; 72: 47–51.
11 Wheatley TA. Water soluble cellulose acetate: a versatile polymer for film coating. *Drug Dev Ind Pharm* 2007; **33**: 281–290.
12 Yuan J, *et al.* Formulation effects on the thermomechanical properties and permeability of free films and coating films: characterization of cellulose acetate films. *Pharm Tech* 2009; **33**(3): 88–100.
13 Sugiu K, *et al.* Successful embolization of a spinal perimedullary arteriovenous fistula with cellulose acetate polymer solution: technical case report. *Neurosurgery* 2001; **49**(5): 1257–1260.

20 General References

Doelker E. Cellulose derivatives. *Adv Polym Sci* 1993; **107**: 199–265.

European Directorate for the Quality of Medicines and Healthcare (EDQM). European Pharmacopoeia – State Of Work Of International Harmonisation. *Pharmeuropa* 2011; 23(4): 713–714. www.edqm.eu/site/-614.html (accessed 2 December 2011).

21 Author

PD Daugherity.

22 Date of Revision

2 December 2011.

Cellulose Acetate Phthalate

1 Nonproprietary Names

BP: Cellacefate

JP: Cellacefate

PhEur: Cellulose Acetate Phthalate

USP–NF: Cellacefate

2 Synonyms

Acetyl phthalyl cellulose; *Aquacoat cPD*; CAP; cellacephate; cellulose acetate benzene-1,2-dicarboxylate; cellulose acetate hydrogen 1,2-benzenedicarboxylate; cellulose acetate hydrogen phthalate; cellulose acetate monophthalate; cellulose acetophthalate; cellulose acetylphthalate; cellulosi acetas phthalas.

3 Chemical Name and CAS Registry Number

Cellulose, acetate, 1,2-benzenedicarboxylate [9004-38-0]

4 Empirical Formula and Molecular Weight

Cellulose acetate phthalate is a cellulose in which some of the hydroxyl groups are acetylated and some are phthalylated.

5 Structural Formula

Cellulose acetate phthalate is a reaction product of phthalic anhydride and a partial acetate ester of cellulose. It contains 21.5–26.0 % of acetyl (C_2H_3O) groups and 30.0–36.0 % of phthalyl (*o*-carboxybenzoyl, $C_8H_5O_3$) groups, calculated on the anhydrous acid-free basis.

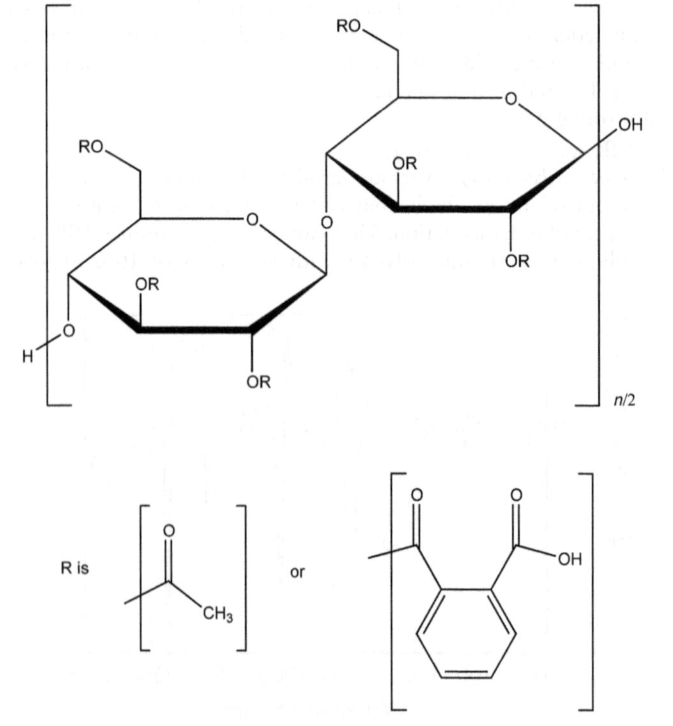

6 Functional Category

Coating agent.

Table I: Pharmacopeial specifications for cellulose acetate phthalate.

Test	JP XV	PhEur 7.4	USP35–NF30
Identification	+	+	+
Characters	+	+	−
Free acid	≤3.0%	≤3.0%	≤3.0%
Heavy metals[a]	≤10 ppm	≤10 ppm	≤0.001%
Phthaloyl groups	—	+	+
Residue on ignition	≤0.1%	≤0.1%	≤0.1%
Viscosity (15% w/v solution)	45–90 mPa s	45.0–90.0 mPa s	45.0–90.0 mPa s
Water	≤5.0%	≤5.0%	≤5.0%
Assay	+	+	+
Acetyl groups	21.5–26.0%	21.5–26.0%	21.5–26.0%
Carboxybenzoyl groups	30.0–36.0%	30.0–36.0%	30.0–36.0%

(a) These tests have not been fully harmonized at the time of publication.

7 Applications in Pharmaceutical Formulation or Technology

Cellulose acetate phthalate (CAP) is used as an enteric film coating material, or as a matrix binder for tablets and capsules.[1–8] Such coatings resist prolonged contact with the strongly acidic gastric fluid, but dissolve in the mildly acidic or neutral intestinal environment.

Cellulose acetate phthalate is commonly applied to solid-dosage forms either by coating from organic or aqueous solvent systems, or by direct compression. Concentrations generally used are 0.5–9.0% of the core weight. The addition of plasticizers improves the water resistance of this coating material, and formulations using such plasticizers are more effective than when cellulose acetate phthalate is used alone.

Cellulose acetate phthalate is compatible with many plasticizers, including acetylated monoglyceride; butyl phthalybutyl glycolate; dibutyl tartrate; diethyl phthalate; dimethyl phthalate; ethyl phthalylethyl glycolate; glycerin; propylene glycol; triacetin; triacetin citrate; and tripropionin. It is also used in combination with other coating agents such as ethyl cellulose in controlled-release preparations.

8 Description

Cellulose acetate phthalate is a hygroscopic, white to off-white, free-flowing powder, granule, or flake. It is tasteless and odorless, or might have a slight odor of acetic acid.

9 Pharmacopeial Specifications

The pharmacopeial specifications for cellulose acetate phthalate have undergone harmonization of many attributes for JP, PhEur, and USP–NF.

See Table I. *See also* Section 18.

10 Typical Properties

Density (bulk) 0.260 g/cm^3
Density (tapped) 0.266 g/cm^3
Melting point 192°C. Glass transition temperature is 160–170°C.[9]
Moisture content Cellulose acetate phthalate is hygroscopic and precautions are necessary to avoid excessive absorption of moisture. Equilibrium moisture content has been reported as 2.2%, but moisture content is a function of relative humidity.[10] *See also* Figure 1.
Solubility Practically insoluble in water, alcohols, and chlorinated and nonchlorinated hydrocarbons. Soluble in a number of ketones, esters, ether alcohols, cyclic ethers, and in certain solvent mixtures. It can be soluble in certain buffered aqueous solutions as low as pH 6.0. Cellulose acetate phthalate has a solubility of ≤10% w/w in a wide range of solvents and solvent mixtures; *see* Table II and Table III.

Table II: Examples of solvents with which cellulose acetate phthalate has ≤10% w/w solubility.

Acetone
Diacetone alcohol
Dioxane
Ethoxyethyl acetate
Ethyl glycol monoacetate
Ethyl lactate
Methoxyethyl acetate
β-Methoxyethylene alcohol
Methyl acetate
Methyl ethyl ketone

Table III: Examples of solvent mixtures with which cellulose acetate phthalate has ≤10% w/w solubility.

Acetone : ethanol (1 : 1)
Acetone : water (97 : 3)
Benzene : methanol (1 : 1)
Ethyl acetate : ethanol (1 : 1)
Methylene chloride : ethanol (3 : 1)

Spectroscopy
IR spectra *see* Figure 2.
NIR spectra *see* Figure 3.

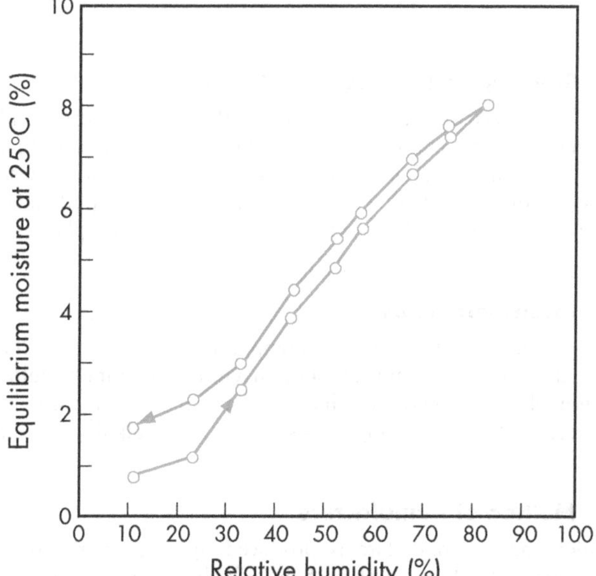

Figure 1: Sorption–desorption isotherm of cellulose acetate phthlate.

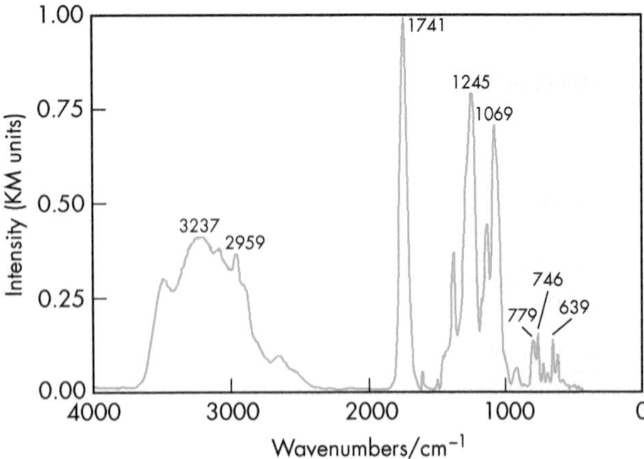

Figure 2: Infrared spectrum of cellulose acetate phthalate measured by diffuse reflectance. Adapted with permission of Informa Healthcare.

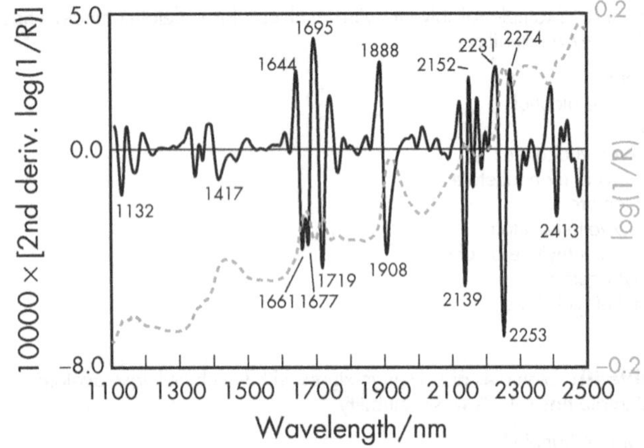

Figure 3: Near-infrared spectrum of cellulose acetate phthalate measured by reflectance.

Viscosity (dynamic) A 15% w/w solution in acetone with a moisture content of 0.4% has a viscosity of 50–90 mPa s (50–90 cP). This is a good coating solution with a honey-like consistency, but the viscosity is influenced by the purity of the solvent.

11 Stability and Storage Conditions

Slow hydrolysis of cellulose acetate phthalate will occur under prolonged adverse conditions such as high temperatures and high humidity, with a resultant increase in free acid content, viscosity, and odor of acetic acid. However, cellulose acetate phthalate is stable if stored in a well-closed container in a cool, dry place.

12 Incompatibilities

Cellulose acetate phthalate is incompatible with ferrous sulfate, ferric chloride, silver nitrate, sodium citrate, aluminum sulfate, calcium chloride, mercuric chloride, barium nitrate, basic lead acetate, and strong oxidizing agents such as strong alkalis and acids.

13 Method of Manufacture

Cellulose acetate phthalate is produced by reacting the partial acetate ester of cellulose with phthalic anhydride in the presence of a tertiary organic base such as pyridine, or a strong acid such as sulfuric acid.

14 Safety

Cellulose acetate phthalate is widely used in oral pharmaceutical products and is generally regarded as a nontoxic material, free of adverse effects.

Results of long-term feeding in rats and dogs have indicated a low oral toxicity. Rats survived daily feedings of up to 30% in the diet for up to 1 year without showing a depression in growth. Dogs fed 16 g daily in the diet for 1 year remained normal.

15 Handling Precautions

Observe normal precautions appropriate to the circumstances and quantity of material handled. Cellulose acetate phthalate may be irritant to the eyes, mucous membranes, and upper respiratory tract. Eye protection and gloves are recommended. Cellulose acetate phthalate should be handled in a well-ventilated environment; use of a respirator is recommended when handling large quantities.

16 Regulatory Status

Included in the FDA Inactive Ingredients Database (oral tablets). Included in nonparenteral medicines licensed in the UK. Included in the Canadian Natural Health Products Ingredients Database.

17 Related Substances

Cellulose acetate; hypromellose phthalate; polyvinyl acetate phthalate.

18 Comments

Cellulose acetate phthalate has undergone harmonization of many attributes for JP, PhEur, and USP–NF by the Pharmacopeial Discussion Group. For further information see the General Information Chapter <1196> in the USP35–NF30, the General Chapter 5.8 in PhEur 7.4, along with the 'State of Work' document on the PhEur EDQM website, and also the General Information Chapter 8 in the JP XV.

Any plasticizers that are used with cellulose acetate phthalate to improve performance should be chosen on the basis of experimental evidence. The same plasticizer used in a different tablet base coating may not yield a satisfactory product.

In using mixed solvents, it is important to dissolve the cellulose acetate phthalate in the solvent with the greater dissolving power, and then to add the second solvent. Cellulose acetate phthalate should always be added to the solvent, not the reverse.

Cellulose acetate phthalate films are permeable to certain ionic substances, such as potassium iodide and ammonium chloride. In such cases, an appropriate sealer subcoat should be used.

A reconstituted colloidal dispersion of latex particles rather than solvent solution coating material of cellulose acetate phthalate is also available. This white, water-insoluble powder is composed of solid or semisolid submicrometer-sized polymer spheres with an average particle size of 0.2 μm. A typical coating system made from this latex powder is a 10–30% solid-content aqueous dispersion with a viscosity in the 50–100 mPa s (50–100 cP) range.

Therapeutically, cellulose acetate phthalate has recently been reported to exhibit experimental microbicidal activity against sexually transmitted disease pathogens, such as the HIV-1 retrovirus.[11,12]

19 Specific References

1 Spitael J, *et al.* Dissolution rate of cellulose acetate phthalate and Brönsted catalysis law. *Pharm Ind* 1980; 42: 846–849.
2 Takenaka H, *et al.* Preparation of enteric-coated microcapsules for tableting by spray-drying technique and *in vitro* simulation of drug release from the tablet in GI tract. *J Pharm Sci* 1980; 69: 1388–1392.
3 Takenaka H, *et al.* Polymorphism of spray-dried microencapsulated sulfamethoxazole with cellulose acetate phthalate and colloidal silica, montmorillonite, or talc. *J Pharm Sci* 1981; 70: 1256–1260.

4 Stricker H, Kulke H. [Rate of disintegration and passage of enteric-coated tablets in gastrointestinal tract.] *Pharm Ind* 1981; **43**: 1018–1021[in German].
5 Maharaj I, *et al.* Simple rapid method for the preparation of enteric-coated microspheres. *J Pharm Sci* 1984; **73**: 39–42.
6 Beyger JW, Nairn JG. Some factors affecting the microencapsulation of pharmaceuticals with cellulose acetate phthalate. *J Pharm Sci* 1986; **75**: 573–578.
7 Lin SY, Kawashima Y. Drug release from tablets containing cellulose acetate phthalate as an additive or enteric-coating material. *Pharm Res* 1987; **4**: 70–74.
8 Thoma K, Heckenmüller H. [Effect of film formers and plasticizers on stability of resistance and disintegration behaviour. Part 4: pharmaceutical-technological and analytical studies of gastric juice resistant commercial preparations.] *Pharmazie* 1987; **42**: 837–841[in German].
9 Sakellariou P, *et al.* The thermomechanical properties and glass transition temperatures of some cellulose derivatives used in film coating. *Int J Pharm* 1985; **27**: 267–277.
10 Callahan JC, *et al.* Equilibrium moisture content of pharmaceutical excipients. *Drug Dev Ind Pharm* 1982; **8**: 355–369.
11 Neurath AR, *et al.* Cellulose acetate phthalate, a common pharmaceutical excipient, inactivates HIV-1 and blocks the coreceptor binding site on the virus envelope glycoprotein gp120. *BMC Infect Dis* 2001; **1**(1): 17.
12 Neurath AR, *et al.* Anti-HIV-1 activity of cellulose acetate phthalate: synergy with soluble CD4 and induction of 'dead-end' gp41 six-helix bundles. *BMC Infect Dis* 2002; **2**(1): 6.

20 General References

Doelker E. Cellulose derivatives. *Adv Polym Sci* 1993; **107**: 199–265.
European Directorate for the Quality of Medicines and Healthcare (EDQM). European Pharmacopoeia – State Of Work Of International Harmonisation. *Pharmeuropa* 2011; 23(4): 713–714. www.edqm.eu/site/-614.html (accessed 2 December 2011).
Obara S, Mc Ginity JW. Influence of processing variables on the properties of free films prepared from aqueous polymeric dispersions by a spray technique. *Int J Pharm* 1995; **126**: 1–10.
O'Connor RE, Berryman WH. Evaluation of enteric film permeability: tablet swelling method and capillary rise method. *Drug Dev Ind Pharm* 1992; **18**: 2123–2133.
Raffin F, *et al.* Physico-chemical characterization of the ionic permeability of an enteric coating polymer. *Int J Pharm* 1995; **120**(2): 205–214.
Wyatt DM. Cellulose esters as direct compression matrices. *Manuf Chem* 1991; **62**(12): 20, 21, 23.

21 Author

PD Daugherity.

22 Date of Revision

2 December 2011.

Ceratonia

1 Nonproprietary Names

None adopted.

2 Synonyms

Algaroba; carob bean gum; carob flour; ceratonia gum; ceratonia siliqua; ceratonia siliqua gum; Cheshire gum; E410; gomme de caroube; locust bean gum; *Meyprofleur*; St. John's bread.

3 Chemical Name and CAS Registry Number

Carob gum [9000-40-2]

4 Empirical Formula and Molecular Weight

Ceratonia is a naturally occurring plant material that consists chiefly of a high molecular weight hydrocolloidal polysaccharide, composed of D-galactose and D-mannose units combined through glycosidic linkages, which may be described chemically as galactomannan. The molecular weight is approximately 310 000.

5 Structural Formula

See Section 4.

6 Functional Category

Modified-release agent; suspending agent; tablet and capsule binder; viscosity-increasing agent.

7 Applications in Pharmaceutical Formulation or Technology

Ceratonia is a naturally occurring material generally used as a substitute for tragacanth or other similar gums. A ceratonia mucilage that is slightly more viscous than tragacanth mucilage may be prepared by boiling 1.0–1.5% of powdered ceratonia with water. As a viscosity-increasing agent, ceratonia is said to be five times as effective as starch and twice as effective as tragacanth. Ceratonia has also been used as a tablet binder[1] and is used in oral controlled-release drug delivery systems approved in Europe and the US.

8 Description

Ceratonia occurs as a yellow-green or white colored powder. Although odorless and tasteless in the dry powder form, ceratonia acquires a leguminous taste when boiled in water.

9 Pharmacopeial Specifications

See Section 18.

10 Typical Properties

Acidity/alkalinity pH = 5.3 (1% w/v aqueous solution)

Solubility Ceratonia is dispersible in hot water, forming a sol having a pH 5.4–7.0 that may be converted to a gel by the addition of small amounts of sodium borate. In cold water, ceratonia hydrates very slowly and incompletely. Ceratonia is practically insoluble in ethanol.

Spectroscopy

NIR spectra *see* Figure 1.

Viscosity (dynamic) 1200–2500 mPa s (1200–2500 cP) for a 1% w/v aqueous dispersion at 25°C. Viscosity is unaffected by pH within the range pH 3–11. Viscosity is increased by heating: if heated to 95°C then cooled, practically clear solutions may be obtained that are more viscous than prior to heating.

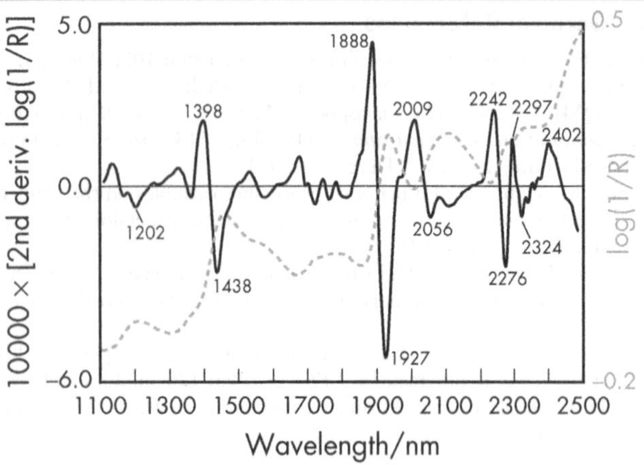

Figure 1: Near-infrared spectrum of ceratonia measured by reflectance.

11 Stability and Storage Conditions

The bulk material should be stored in a well-closed container in a cool, dry place. Ceratonia loses not more than 15% of its weight on drying.

12 Incompatibilities

The viscosity of xanthan gum solutions is increased in the presence of ceratonia.[2] This interaction is used synergistically in controlled-release drug delivery systems.

13 Method of Manufacture

Ceratonia is a naturally occurring material obtained from the ground endosperms separated from the seeds of the locust bean tree, *Ceratonia siliqua* (Leguminosae). The tree is indigenous to southern Europe and the Mediterranean region.

14 Safety

Ceratonia is generally regarded as an essentially noncarcinogenic,[3] nontoxic and nonirritant material. Therapeutically, it has been used in oral formulations for the control of vomiting and diarrhea in adults and children; 20–40 g daily in adults has been used dispersed in liquid.[4] As an excipient, ceratonia is used in oral controlled-release formulations approved in Europe and the US.

Ceratonia is also widely used in food products. The WHO has not specified an acceptable total daily intake for ceratonia as the total daily intake arising from its use at the levels necessary to achieve the desired effect, and from its acceptable background in food, was not considered to represent a hazard to health.[5] Ceratonia hypersensitivity has been reported, in a single case report, in an infant.[6] However, ceratonia is said to be nonallergenic in children with known allergy to peanuts.[7]

LD_{50} (hamster, oral): 10.0 g/kg[8]
LD_{50} (mouse, oral): 13.0 g/kg
LD_{50} (rabbit, oral): 9.1 g/kg
LD_{50} (rat, oral): 13.0 g/kg

15 Handling Precautions

Observe normal precautions appropriate to the circumstances and quantity of material handled. When heated to decomposition ceratonia emits acrid smoke and irritating fumes.

16 Regulatory Status

GRAS listed. Accepted for use in Europe as a food additive. In Europe and the US, ceratonia has been used in oral tablet formulations. Included in the Canadian Natural Health Products Ingredients Database.

17 Related Substances

Acacia; ceratonia extract; tragacanth; xanthan gum.

Ceratonia extract

Synonyms Ceratonia siliqua extract; extract of carob; locust tree extract.

CAS number [84961-45-5]

Comments Ceratonia extract is used as an emollient. The EINECS number for ceratonia extract is 284-634-5.

18 Comments

Therapeutically, ceratonia mucilage is used orally in adults and children to regulate intestinal function; *see* Section 14.

Ceratonia is widely used as a binder, thickening agent, and stabilizing agent in the cosmetics and food industry. In foods, 0.15–0.75% is used.

The EINECS number for ceratonia is 232-541-5.

Although not included in any pharmacopeias, a specification for ceratonia is contained in the *Food Chemicals Codex* (FCC); *see* Table I.[9]

Table I: Food Chemicals Codex specifications for ceratonia (locust bean gum).[9]

Test	FCC 7
Identification	+
Acid-insoluble matter	≤4.0%
Arsenic	≤3 mg/kg
Ash	≤1.2%
Galactomannans	≥75%
Lead	≤5 mg/kg
Loss on drying	≤14.0%
Protein	≤7.0%
Starch	+

19 Specific References

1 Georgakopoulos PP, Malamataris S. Locust bean gum as granulating and binding agent for tablets. *Pharm Ind* 1980; 42(6): 642–646.
2 Kovacs P. Useful incompatibility of xanthan gum with galactomannans. *Food Technol* 1973; 27(3): 26–30.
3 National Toxicology Program. Carcinogenesis bioassay of locust bean gum (CAS No. 9000-40-2) in F344 rats and B6C3F1 mice (feed study). *Natl Toxicol Program Tech Rep Ser* 1982; **221**(Feb): 1–99.
4 Wade A, ed. *Martindale: The Extra Pharmacopoeia*, 27th edn. London: Pharmaceutical Press, 1977: 921.
5 FAO/WHO. Evaluation of certain food additives. Twenty-fifth report of the FAO/WHO expert committee on food additives. *World Health Organ Tech Rep Ser* 1981; No. 669.
6 Savino F, *et al.* Allergy to carob gum in an infant. *J Pediatr Gastroenterol Nutr* 1999; **29**(4): 475–476.
7 Fiocchi A, *et al.* Carob is not allergenic in peanut-allergic subjects. *Clin Exp Allergy* 1999; **29**(3): 402–406.
8 Lewis RJ, ed. *Sax's Dangerous Properties of Industrial Materials*, 11th edn. New York: Wiley, 2004: 2247.
9 *Food Chemicals Codex*, 7th edn. Bethesda, MD: United States Pharmacopeia, 2010: 593.

20 General References

Bhardwaj TR, *et al.* Natural gums and modified gums as sustained-release carriers. *Drug Dev Ind Pharm* 2000; **26**(10): 1025–1038.

Griffiths C. Locust bean gum: a modern thickening agent from a biblical fruit. *Manuf Chem* 1949; **20**: 321–324.

Hoepfner E *et al.* eds. *Fiedler Encyclopedia of Excipients for Pharmaceuticals, Cosmetics and Related Areas*, 5th edn. Aulendorf: Editio Cantor Verlag, 2002: 358–359.

Knight WA, Dowsett MM. Ceratoniae gummi: carob gum. An inexpensive substitute for gum tragacanth. *Pharm J* 1936; **82**: 35–36.
Sujja-areevath J, *et al*. Release characteristics of diclofenac sodium from encapsulated natural gum mini-matrix formulations. *Int J Pharm* 1996; **139**: 53–62.
Vijayaraghavan C, *et al*. *In vitro* and *in vivo* evaluation of locust bean gum and chitosan combination as a carrier for buccal drug delivery. *Pharmazie* 2008; **63**: 342–347.
Woodruff J. Ingredients for success in thickening. *Manuf Chem* 1998; **69**(9): 49, 50, 52.
Yousif AK, Alghzawi HM. Processing and characterization of carob powder. *Food Chem* 2000; **69**: 283–287.

21 Author

PJ Weller.

22 Date of Revision

1 November 2011.

Ceresin

1 Nonproprietary Names

None adopted.

2 Synonyms

Cera microcrystallina; cera mineralis alba; ceresin refined; ceresine; ceresine wax; ceresine wax white; ceresin wax; cerin; cerosin; *Cirashine CS*; earth wax; *GS-Ceresin*; *Koster Keunen Ceresine*; microcrystalline wax; mineral wax; purified ozokerite; *Ross Ceresine Wax*; white ceresin wax; white ozokerite wax. *See also* Section 13.

3 Chemical Name and CAS Registry Number

Ceresin wax [8001-75-0]

4 Empirical Formula and Molecular Weight

Ceresin is a mineral wax composed of a wide and complex range of long-chain, high-molecular-weight, saturated and unsaturated hydrocarbons related to methane, ranging from C_{20} to C_{32}.

5 Structural Formula

See Section 4.

6 Functional Category

Coating agent; emulsion stabilizing agent; gelling agent; opacifier; stiffening agent; viscosity-increasing agent.

7 Applications in Pharmaceutical Formulation or Technology

Ceresin is used as a stiffening agent in creams[1,2] and ointments,[3,4] and as an emulsion stabilizer, opacifier, viscosity-increasing agent, gelling agent, and thickener in pharmaceutical protective, topical, and vaginal creams.[7] It is also used in cosmetics[5–7] and in a wide variety of personal care products[8,9] (*see* Section 18).

Ceresin is often used as a substitute for ozokerite wax, beeswax, and paraffin wax. It acts as a rheological modifier at low concentrations (2–3%) and has the ability to create very small crystallites, which crosslink and establish a network structure that does not allow flow in practical conditions.[10] Ceresin produces stable mixtures with oils and prevents bleeding or sweating of oil, and it produces a lighter cream that is less greasy.

Ceresin is also used for pharmaceutical coating applications of medicaments, for example, protective coatings,[11] enteric coatings,[12] and sustained-release coatings.[13] It has been used in the formulation of multivesicular emulsion topical delivery systems.[14]

8 Description

Ceresin is a natural, white-to-yellow waxy mixture of hydrocarbons of high molecular weight obtained by purification of ozokerite. It occurs as odorless, tasteless, solid, amorphous (noncrystalline) brittle, waxy cakes or pastilles.

Ceresin waxes are blends of complex hydrocarbons, sold in various grades, differing from one another chiefly in melting point and in the distribution of the hydrocarbons (*see* Table 1).

9 Pharmacopeial Specifications

See Section 18.

10 Typical Properties

Acid value *See* Table 1
Boiling point 343°C[15]
Congealing point 60–69°C; *see* Table 1
Density 0.91–0.92 g/cm^3
Flash point ≥204.4°C[15]
Iodine value 7–9[16]
Melting point 54–78°C; *see* Table 1
Saponification value *See* Table 1
Solubility Soluble in benzene, chloroform, naphtha, hot oils, petroleum ether, 30 parts absolute ethanol, turpentine, carbon disulfide, and most organic solvents. Insoluble in water.
Specific gravity ~0.880–0.935[17]
Viscosity (kinematic) 4.0 mm^2/s (4.0 cSt) at 100°C[18]

Table I: Typical properties of various commercial grades of ceresin (Koster Keunen).

Grade	Acid value	Congealing point °C	Melting point °C	Saponification value
Ceresin 130–135	<1	53.8–57.2	54.4–57.2	<1
Ceresin 140–145	<1	59.4–65.0	60.0–65.6	<1
Ceresin 155	<1	66.7–69.4	67.2–70.0	<1
Ceresin 1556	<1	72.8–78.3	74.0–79.4	<1
Ceresin 145	<1	59.4–62.2	60.0–62.8	<1
Ceresin 150–160	<1	65.0–70.6	65.6–71.1	<1

11 Stability and Storage Conditions

Ceresin is a stable material and should be stored in well-closed containers in a cool, dry, well-ventilated place, away from extreme heat, ignition sources and strong oxidizing agents.

12 Incompatibilities

Ceresin is incompatible with strong oxidizing agents. It is compatible with most animal, vegetable, and mineral waxes, as well as mineral oil and petrolatum.

13 Method of Manufacture

Ceresin is a complex combination of hydrocarbons prepared by extraction and purification of the native mineral fossil wax ozokerite, which is derived from coal and shale, with sulfuric acid and filtration through bone black to form waxy cakes. Ozokerite is mined from deposits in various localities around the world. It is found as irregular mineral veins or as a black mass in clay strata. Mined ozokerite is heated to melt it, and any earth or rock is removed. If necessary, it is heated to 115–120°C to remove any moisture and then treated with sulfuric acid or fuming sulfuric acid. After neutralization, it is decolorized using activated charcoal, bone black or silica gel, and filtered. If decolorizing is not sufficient, it is repeatedly treated with sulfuric acid and subjected to adsorption filtration to produce more refined ceresin.

Another method of producing ceresin involves dissolving ozokerite in ligroin, treating it with activated clay, and then removing the high-boiling-point fraction.

14 Safety

Ceresin is nontoxic, nonhazardous, and safe for use in personal care and cosmetic ingredients in the present practices of concentration and use. The Cosmetic Ingredient Review Expert Panel has concluded that ceresin does not result in dermal sensitization. When formulations containing these ingredients were tested, they produced no skin irritation and the formulations were not phototoxic.[19]

Ceresin may be slightly hazardous on ingestion and inhalation and may cause allergic reactions.[20] No definitive information is available on carcinogenicity, mutagenicity, target organs, or developmental toxicity. The FDA has established a cumulative estimated daily intake of ceresin of 0.00035 mg/kg body weight, and a cumulative dietary concentration in food of not more than 7 ppb.[21]

15 Handling Precautions

Observe normal precautions appropriate to the circumstances and quantity of the material handled. Ceresin should be handled in areas with adequate ventilation. Inhalation of vapors and contact with eyes, skin, and clothes should be avoided. Eye protection and gloves are recommended. Wash hands thoroughly after handling.

Ceresin should be kept away from heat and sources of ignition as it may be combustible at high temperature. It forms oxides of carbon as a hazardous decomposition product. The recommended operation temperatures are in the range of 75–95°C.

16 Regulatory Status

Included in the FDA Inactive Ingredients Database (topical ointments; vaginal emulsions and creams). Included in the Canadian Natural Health Products Ingredients Database. Included in nonparenteral medicines (topical creams) licensed in the UK. Accepted for use in cosmetics and personal care products marketed in Europe.

17 Related Substances

Wax, microcrystalline; wax, white; wax, yellow.

18 Comments

A specification for ceresin is included in *Japanese Pharmaceutical Excipients* (JPE); *see* Table II.[22]

Ceresin is used in many types of cosmetics and personal care products[23–25] including lipsticks,[26] lip salve, baby products, eye and facial makeup, as well as nail care, skin care, suntan, and sunscreen preparations,[27] deodorant sticks, fragrance, perfumes, pomades, and noncoloring hair preparations.[28] Ceresin is used in formulas that do not use animal products. It lessens the brittleness of cosmetic stick products, adding strength and stability, and it has been used in hair products as a waxy carrier.[29]

Ceresin is used in dentistry as one of the primary components of dental wax compounds[30] along with beeswax and microcrystalline and paraffin wax, and it is used in dental impressions.[31] It has also been used as a biodegradable wax in a sprayable, controlled-release insect control formulation.[32]

The EINECS number for ceresin is 232-290-1. The PubChem Substance ID (SID) for ceresin is 204276.

Table II: JPE specification for ceresin.[22]

Test	JPE 2004
Description	+
Melting point	61–95°C
pH	5.0–8.0
Sulfur compounds	+
Heavy metals	≤30 ppm
Residue on ignition	≤0.05%

19 Specific References

1 Morosawa K, *et al.* Study on the differential thresholds of sensory 'firmness" and 'viscousness" of cream base substances. *J Soc Cosmet Chem* 1974; 25(9): 481–494.
2 Bequette RJ *et al.* Stabilized estradiol cream composition. United States Patent No. 4,436,738; 1984.
3 Niazi SK. *Handbook of Pharmaceutical Manufacturing Formulations: Semisolid Products.* Vol. 4: New York, London: Informa Health Care, 2004; 156.
4 Strakosch EA. Studies on ointments I. Penetration of various ointment bases. *J Pharmacol Exp Ther* 1943; 78(1): 65–71.
5 Russell LW, Welch AE. Analysis of lipsticks. *Forensic Sci Int* 1984; 25(2): 105–116.
6 Engasser PG. Lip Cosmetics. *Dermatol Clin* 2000; 18(4): 641–649.
7 Takeo M. Raw materials of cosmetics. *New Cosmetic Science* 1997; 121–147.
8 Takeo M. Hair care cosmetics. *New Cosmetic Science* 1997; 406–438.
9 Yoshizumil H *et al.* Microbial hair tonic composition. United States Patent No. 4,565,698; 1986.
10 Marudova M, Jilov N. Creating a yield stress in liquid oils by the addition of crystallisable modifiers. *J Food Eng* 2002; 51(3): 235–237.
11 Coffey CR, Tesdahl TC. Protective coating compositions. United States Patent No. 3,700,013; 1972.
12 Bagaria SC, Lordi NG. Aqueous dispersions of waxes and lipids for pharmaceutical coating. United States Patent No. 5,023,108; 1991.
13 Takashima Y *et al.* Method of preparing sustained-release pharmaceutical/preparation. United States Patent No. 4,853,249; 1989.
14 Espinoza R. *Multivesicular emulsion drug delivery systems.* United States Patent No. 6,709,663; 2004.
15 Strahl & Pitsch Inc. Material safety data sheet: Ceresine wax, 2007.
16 Lambent Technologies – A Petroferm Company. Technical literature No. 1120398-2: *Personal care and Pharmaceuticals – Ceresin wax.*
17 Science stuff Inc. Material safety data sheet: Ceresin wax. http://www.sciencestuff.com/msds/C1470.html (accessed on 14 November 2011).
18 Akrochem Corporation. Technical Literature No. JH0102: *Ceresin wax.*
19 Cosmetic Ingredient Review Expert Panel. Final report on the safety assessment of fossil and synthetic waxes. *Int J Toxicol* 1984; 33: 43–99.
20 ScienceLab.com Inc. Material safety data sheet: Ceresin wax, 2005.
21 US FDA Center for Food Safety and Applied Nutrition: *CEDI/ADI Table. CFSAN/Office of Food Additive Safety,* 2007. http://www.accessdata.fda.gov/scripts/sda/sdNavigation.cfm?sd=edisrev (accessed 14 November 2011).
22 Japan Pharmaceutical Excipients Council. *Japanese Pharmaceutical Excipients 2004.* Tokyo: Yakuji Nippo, 2004: 155.

23 Hans M, Arnold G. *Cosmetics in Formulation Technology*. New York: Wiley-Interscience, 2001; 327–350.
24 Yoshida M *et al*. Oil-based composition for external use on skin for enhancing percutaneous absorption. United States Patent 20080124367; 2008.
25 Hasegawa Y *et al*. Powders coated with specific lipoamino acid composition and cosmetics containing the same. United States Patent No.7,374,783; 2008.
26 Szweda JA, Lotrario CA. A long-wearing lipstick. United States Patent No.5,667,770; 1996.
27 Luana P, *et al*. Sunscreen immobilization on ZnAl-hydrotalcite for new cosmetic formulations. *Microporous Mesoporous Mater* 2008; **107**: 180–189.
28 Walter A *et al*. Non-fluid hair treatment product comprising hair fixative absorbed on waxy carrier. United States Patent 20080089855; 2008.
29 Walter A, Birkel S. Use of di- or oligosaccharide polyester in hair styling products. United States Patent 20070184007; 2007.
30 Craig RG, *et al*. Properties of natural waxes used in dentistry. *J Dent Res* 1965; 44(6): 1308–1316.
31 Carl TO. Relining complete dentures. *J Prosthet Dent* 1961; **11**(2): 204–213.
32 Delwiche M *et al*. Aqueous emulsion comprising biodegradable carrier for insect pheromones and methods for controlled release thereof. United States Patent No.6,001,346; 1999.

20 General References

Ikuta K, *et al*. Scanning electron microscopic observation of oil/wax/water/ surfactant system. *Int J Cosmet Sci* 2005; **27**(2): 133.

Koster Keunen. Product information: Ceresin and ozokerite waxes. http://www.kosterkeunen.com (accessed 14 November 2011).

Lee RS, Blazynski TZ. Mechanical properties of a composite wax model material simulating plastic flow of metals. *J Mech Work Tech* 1984; **9**(3): 301–312.

Peris-Vicente J, *et al*. Characterization of waxes used in pictorial artworks according to their relative amount of fatty acids and hydrocarbons by gas chromatography. *J Chromatogr A* 2006; **1101**(1–2): 254–260.

21 Author

KK Singh.

22 Date of Revision

14 November 2011.

Cetostearyl Alcohol

1 Nonproprietary Names

BP: Cetostearyl Alcohol
PhEur: Cetostearyl Alcohol
USP–NF: Cetostearyl Alcohol

2 Synonyms

Alcohol cetylicus et stearylicus; cetearyl alcohol; cetyl stearyl alcohol; *Crodacol CS90*; *Lanette O*; *Speziol C16-18 Pharma*; *Tego Alkanol 1618*; *Tego Alkanol 6855*.

3 Chemical Name and CAS Registry Number

Cetostearyl alcohol [67762-27-0] and [8005-44-5]

4 Empirical Formula and Molecular Weight

Cetostearyl alcohol is a mixture of solid aliphatic alcohols consisting mainly of stearyl ($C_{18}H_{38}O$) and cetyl ($C_{16}H_{34}O$) alcohols. The proportion of stearyl to cetyl alcohol varies considerably, but the material usually consists of about 50–70% stearyl alcohol and 20–35% cetyl alcohol, with limits specified in pharmacopeias. The combined stearyl alcohol and cetyl alcohol comprise at least 90% of the material. Small quantities of other alcohols, chiefly myristyl alcohol, make up the remainder of the material.

5 Structural Formula

See Section 4.

6 Functional Category

Emollient; emulsifying agent; viscosity-increasing agent.

7 Applications in Pharmaceutical Formulation or Technology

Cetostearyl alcohol is used in cosmetics and topical pharmaceutical preparations. In topical pharmaceutical formulations, cetostearyl alcohol will increase the viscosity and act as an emulsifier in both water-in-oil and oil-in-water emulsions. Cetostearyl alcohol will stablize an emulsion and also act as a co-emulsifier, thus decreasing the total amount of surfactant required to form a stable emulsion. Cetostearyl alcohol is also used in the preparation of nonaqueous creams and sticks, and in nonlathering shaving creams.[1] *See also* Section 18.

8 Description

Cetostearyl alcohol occurs as white or cream-colored unctuous masses, flakes, pellets or granules. It has a faint, characteristic sweet odor. On heating, cetostearyl alcohol melts to a clear, colorless or pale yellow-colored liquid free of suspended matter.

9 Pharmacopeial Specifications

See Table I. *See also* Section 18.

10 Typical Properties

Boiling point $\approx$ 300-360°C (degradation temperature)
Density (bulk) $\approx 0.8\,g/cm^3$ at 20°C.
Solubility Soluble in ethanol (95%), ether, and oil; practically insoluble in water.
Spectroscopy
NIR spectra *see* Figure 1.

11 Stability and Storage Conditions

Cetostearyl alcohol is stable under normal storage conditions. Cetostearyl alcohol should be stored in a well-closed container in a cool, dry place.

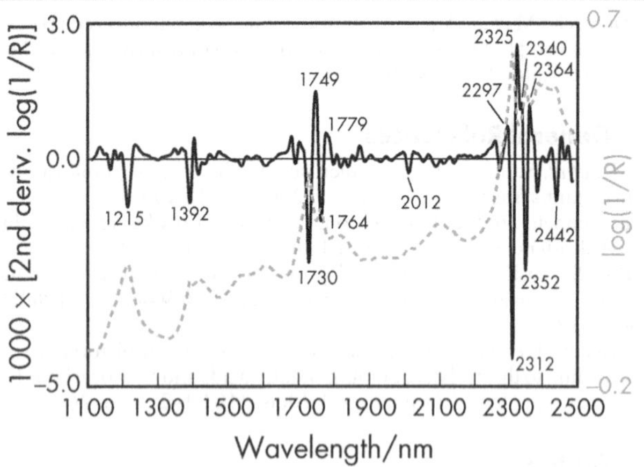

Figure 1: Near-infrared spectrum of cetostearyl alcohol measured by reflectance.

Table I: Pharmacopeial specifications for cetostearyl alcohol.

Test	PhEur 7.4	USP35–NF30
Identification	+	+
Characters	+	–
Appearance of solution	+	–
Melting range	49–56°C	48–55°C
Acid value	≤1.0	≤2.0
Iodine value	≤2.0	≤4
Hydroxyl value	208–228	208–228
Saponification value	≤2.0	–
Assay		
of $C_{18}H_{38}O$	≥40.0%	≥40.0%
of $C_{16}H_{34}O$ and $C_{18}H_{38}O$	≥90.0%	≥90.0%

12 Incompatibilities

Incompatible with strong oxidizing agents and metal salts.

13 Method of Manufacture

Cetostearyl alcohol is prepared by the reduction of the appropriate fatty acids from vegetable and animal sources. Cetostearyl alcohol can also be prepared directly from hydrocarbon sources.

14 Safety

Cetostearyl alcohol is mainly used in topical pharmaceutical formulations and topical cosmetic formulations.

Cetostearyl alcohol is generally regarded as a nontoxic material.[2] Although it is essentially nonirritating, sensitization reactions to cetostearyl, cetyl, and stearyl alcohols[3–8] have been reported.

Gamma radiation has been shown to be feasible for sterilization of petrolatum containing cetostearyl alcohol resulting in low levels of radiolysis products, which are of low toxicity.[9]

15 Handling Precautions

Observe normal precautions appropriate to the circumstances and quantity of material handled. Eye protection and gloves are recommended. Cetostearyl alcohol is flammable and on combustion may produce fumes containing carbon monoxide.

16 Regulatory Status

Accepted as an indirect food additive and as an adhesive and a component of packaging coatings in the US. Included in the FDA Inactive Ingredients Database (oral tablets; topical emulsions, lotions, ointments; vaginal suppositories). Included in nonparenteral medicines licensed in the UK. Included in the Canadian Natural Health Products Ingredients Database.

17 Related Substances

Anionic emulsifying wax; cetyl alcohol; sodium lauryl sulfate; stearyl alcohol.

18 Comments

The composition of cetostearyl alcohol from different sources may vary considerably. The composition of the minor components, typically straight-chain and branched-chain alcohols, varies greatly depending upon the source, which may be animal, vegetable, or synthetic. This has been reported in the literature to impart differences in emulsification behavior, particularly with respect to emulsion consistency or stability.[10–13]

Cetostearyl alcohol has also been reported to control or slow the dissolution rate of tablets or microspheres containing water-soluble drugs,[14–18] or poorly water-soluble drugs,[19–23] as well as to stabilize amorphous systems.[24] In combination with other surfactants, cetostearyl alcohol forms emulsions with very complex microstructures, which can include liquid crystals, lamellar structures, and gel phases.[10–13,25–33]

The PhEur 7.4 contains specifications for cetostearyl alcohol, emulsifying Type A, and Type B, respectively. Each contains at least 7% surfactant, with Type A containing sodium cetostearyl sulfate and Type B containing sodium lauryl sulfate. *See also* Wax, Anionic Emulsifying.

The EINECS number for cetostearyl alcohol is 267-008-6.

19 Specific References

1 Realdon N, *et al.* Influence of processing conditions in the manufacture of O/W creams II. Effect on drug availability. *Il Farmaco* 2002; **57**: 349–353.
2 Final report on the safety assessment of ceteryl alcohol, cetyl alcohol, isostearyl alcohol, myristyl alcohol, and behenyl alcohol. *J Am Coll Toxicol* 1988; 7(3): 359–413.
3 Tosti A, *et al.* Prevalence and sources of sensitization to emulsifiers: a clinical study. *Contact Dermatitis* 1990; **23**: 68–72.
4 Pasche-Koo F, *et al.* High sensitization rate to emulsifiers in patients with chronic leg ulcers. *Contact Dermatitis* 1994; **31**: 226–228.
5 Wilson CL, *et al.* High incidence of contact dermatitis in leg-ulcer patients – implications for management. *Clin Exp Dermatol* 1991; **16**: 250–253.
6 Pecegueiro M, *et al.* Contact dermatitis to hirudoid cream. *Contact Dermatitis* 1987; **17**: 290–293.
7 Hannuksela M. Skin contact allergy to emulsifiers. *Int J Cosmet Sci* 1988; **10**: 9–14.
8 Hannuksela M. Skin reactions to emulsifiers. *Cosmet Toilet* 1988; **10**: 81–86.
9 Hong L, Altorfer H. Radiolysis characterization of cetostearyl alcohol by gas chromatography-mass spectrometry. *J Pharm Biomed Anal* 2003; **31**: 753–766.
10 Rowe RC, McMahon J. The stability of oil-in-water emulsions containing cetrimide and cetostearyl alcohol. *Int J Pharm* 1986; **31**: 281–282.
11 Patel HK, *et al.* A comparison of the structure and properties of ternary gels containing cetrimide and cetostearyl alcohol obtained from both natural and synthethic sources. *Acta Pharm Technol* 1985; **31**(4): 243–247.
12 Fukushim S, Yamaguchi M. The effect of cetostearyl alcohol in cosmetic emulsions. *Cosmet Toilet* 1983; **98**: 89–102.
13 Ballmann C, Mueller BW. Stabilizing effect of cetostearyl alcohol and glyceryl monostearate as co-emulsifiers on hydrocarbon-free O/W glyceride creams. *Pharm Dev Technol* 2008; **13**(5): 433–445.
14 Al-Kassas RS, *et al.* Processing factors affecting particle size and *in vitro* drug release of sustained-release ibuprofen microspheres. *Int J Pharm* 1993; **94**: 59–67.
15 Lashmar UT, Beesley J. Correlation of rheological properties of an oil-in-water emulsion with manufacturing procedures and stability. *Int J Pharm* 1993; **91**: 59–67.
16 Wong LP, *et al.* Preparation and characterization of sustained-release ibuprofen-cetostearyl alcohol spheres. *Int J Pharm* 1992; **83**: 95–114.
17 Ahmed M, Enever RP. Formulation and evaluation of sustained release paracetamol tablets. *J Clin Hosp Pharm* 1981; **6**: 27–38.

18 Waters LJ, Pavlakis E. *In vitro* controlled drug release from loaded microspheres–dose regulation through formulation. *J Pharm Pharm Sci* 2007; **10**(4): 464–472.
19 Gowda DV, Shivakumar HG. Comparative bioavailability studies of indomethacin from two controlled release formulations in healthy albino sheep. *Indian J Pharm Sci* 2006; **68**(6): 760–763.
20 Gowda DV, Shivakumar HG. Preparation and evaluation of waxes/fat microspheres loaded with lithium carbonate for controlled release. *Indian J Pharm Sci* 2007; **69**(2): 251–256.
21 Karasulu E, *et al.* Extended release lipophilic indomethacin microspheres: formation factors and mathematical equations fitted drug release rates. *Eur J Pharm Sci* 2003; **19**: 99–104.
22 Pandit SS, Patil AT. Formulation and *in vitro* evaluation of buoyant controlled release lercanidipine lipospheres. *J Microencapsul* 2009; **26**(7): 635–641.
23 Swamy KM, *et al.* Matrix embedded microspherules containing indomethacin as controlled drug delivery systems. *Curr Drug Deliv* 2008; **5**(4): 248–255.
24 Urbanetz NA. Stabilization of solid dispersions of nimodipine and polyethylene glycol 2000. *Eur J Pharm Sci* 2006; **28**: 67–76.
25 Forster T, *et al.* Production of fine disperse and long-term stable oil-in-water emulsions by the phase inversion temperature method. *Disper Sci Technol* 1992; **13**(2): 183–193.
26 Niemi L, Laine E. Effect of water content on the microstructure of an O/W cream. *Int J Pharm* 1991; **68**: 205–214.
27 Eccleston GM, Beattie L. Microstructural changes during the storage of systems containing cetostearyl alcohol; polyoxyethylene alkyl ether surfactants. *Drug Dev Ind Pharm* 1988; **14**(15–17): 2499–2518.
28 Schambil F, *et al.* Interfacial and colloidal properties of cosmetic emulsions containing fatty alcohol and fatty alcohol polyglycol ethers. *Progr Colloid Polym Sci* 1987; **73**: 37–47.
29 Rowe RC, Bray D. Water distribution in creams prepared using cetostearyl alcohol and cetrimide. *J Pharm Pharmacol* 1987; **39**: 642–643.
30 Eros I, Kedvessy G. Applied rheological research on ointment bases. *Acta Chim Hung* 1984; **115**(4): 363–375.
31 Tsugita A, *et al.* Stable emulsion regions of surfactant-oil-water and surfactant-oil-water-long chain alcohol systems. *Yukagaku* 1980; **29**(4): 227–234.
32 Eccleston GM. Structure and rheology of cetomacrogol creams: the influence of alcohol chain length and homologue composition. *J Pharm Pharmacol* 1997; **29**: 157–162.
33 Fukushima S, *et al.* Effect of cetostearyl alcohol on stabilization of oil-in-water emulsion, II. Relation between crystal form of the alcohol and stability of the emulsion. *J Colloid Interface Sci* 1977; **59**(1): 159–165.

20 General References

Cognis. Product literature: *Lanette O*, 2009.
Cognis. Product literature: *Speziol C16-18 Pharma*, June 2006.
European Commission – European Chemicals Bureau. IUCLID Chemicals Datasheet: Substance ID 67762-27-0, Alcohols C16-18, February 2000.

21 Authors

S Shah, S Cantor.

22 Date of Revision

1 March 2012.

Cetrimide

1 Nonproprietary Names

BP: Cetrimide
PhEur: Cetrimide

2 Synonyms

Bromat; *Cetab*; *Cetavlon*; *Cetraol*; cetrimidum; *Lissolamine V*; *Micol*; *Morpan CHSA*; *Morphans*; *Quammonium*; *Sucticide*.

3 Chemical Name and CAS Registry Number

Cetrimide [8044-71-1]

Note that the above name, CAS Registry Number, and synonyms refer to the PhEur 7.4 material which, although it consists predominantly of trimethyltetradecylammonium bromide, may also contain smaller amounts of other bromides; *see* Section 4.

There is some confusion in the literature regarding the synonyms, CAS Registry Number, and molecular weight applied to cetrimide. Chemical Abstracts has assigned [8044-71-1] to cetrimide and describes that material as a mixture of alkyltrimethylammonium bromides of different alkyl chain lengths. Different CAS Registry Numbers have been assigned to the individual pure components. While these numbers should not be interchanged, it is common to find the molecular weight and CAS Registry Number of trimethyltetradecylammonium bromide [1119-97-7] used for cetrimide, as this is the principal component, defined in both the BP 2012 and PhEur 7.4. It should be noted however, that the original BP 1953 described the principal component of cetrimide as hexadecyltrimethylammonium bromide.

The CAS Registry Number for hexadecyltrimethylammonium hydroxide [505-86-2] has also been widely applied to cetrimide. Therefore, careful inspection of experimental details and suppliers' specifications in the literature is encouraged to determine the specific nature of the'cetrimide' material used in individual studies.

See Section 17 for further information.

4 Empirical Formula and Molecular Weight

Cetrimide consists mainly of trimethyltetradecylammonium bromide ($C_{17}H_{38}BrN$), and may contain smaller amounts of dodecyltrimethylammonium bromide ($C_{15}H_{34}BrN$) and hexadecyltrimethylammonium bromide ($C_{19}H_{42}BrN$).

$C_{17}H_{38}BrN$ 336.40

See also Section 17.

5 Structural Formula

$$H_3C-(CH_2)_n-\overset{CH_3}{\underset{CH_3}{N^+}}-CH_3\ Br^-$$

where
n = 11 for dodecyltrimethylammonium bromide
n = 13 for trimethyltetradecylammonium bromide
n = 15 for hexadecyltrimethylammonium bromide

6 Functional Category

Antimicrobial preservative; cationic surfactant.

7 Applications in Pharmaceutical Formulation or Technology

Cetrimide occurs as a quaternary ammonium compound that is used in cosmetics and pharmaceutical formulations as an antimicrobial preservative; *see* Section 10. It may also be used as a cationic surfactant. In eye-drops, it is used as a preservative at a concentration of 0.005% w/v.

Cetrimide is also used as a cleanser and disinfectant for hard contact lenses, although it should not be used on soft lenses. It is an ingredient of cetrimide emulsifying wax, and is used in o/w creams (e.g. cetrimide cream).

8 Description

Cetrimide is a white to creamy white, free-flowing powder, with a faint but characteristic odor and a bitter, soapy taste.

9 Pharmacopeial Specifications

See Table I.

Table I: Pharmacopeial specifications for cetrimide.

Test	PhEur 7.4
Identification	+
Characters	+
Acidity or alkalinity	+
Appearance of solution	+
Amines and amine salts	+
Loss on drying	≤2.0%
Sulfated ash	≤0.5%
Assay (as $C_{17}H_{38}BrN$, dried basis)	96.0–101.0%

10 Typical Properties

Acidity/alkalinity pH = 5.0–7.5 (1% w/v aqueous solution)

Antimicrobial activity Cetrimide has good bactericidal activity against Gram-positive species but is less active against Gram-negative species. *Pseudomonas* species, particularly *Pseudomonas aeruginosa*, may exhibit resistance. Cetrimide is most effective at neutral or slightly alkaline pH values, with activity appreciably reduced in acidic media and in the presence of organic matter. The activity of cetrimide is enhanced in the presence of alcohols. The activity of cetrimide against resistant strains of *Pseudomonas aeruginosa*, *Aspergillus niger*, and *Candida albicans* is significantly increased by the addition of edetic acid.[1] The combination of maleic acid with cetrimide has been shown to be significantly more effective in increasing activity against *Enterococcus faecalis* biofilm than combinations of cetrimide with edectic acid or citric acid.[2] Cetrimide has variable antifungal activity,[3,4] is effective against some viruses, and is inactive against bacterial spores. Typical minimum inhibitory concentrations (MICs) are shown in Table II.

Table II: Minimum inhibitory concentrations (MIC) of cetrimide.

Microorganism	MIC (μg/mL)
Escherichia coli	30
Pseudomonas aeruginosa	300
Staphylococcus aureus	10
Camphylobacter jejuni	8[5]
Staphylococcus aureus (NCTC-8325-4)	0.25[6]
Staphylococcus aureus (SH1000)	0.63[7]
Pseudomonas aeruginosa (PAO1 (ATCC 15692))	36[8]
Streptococcus pneumoniae R919	1.0[9]

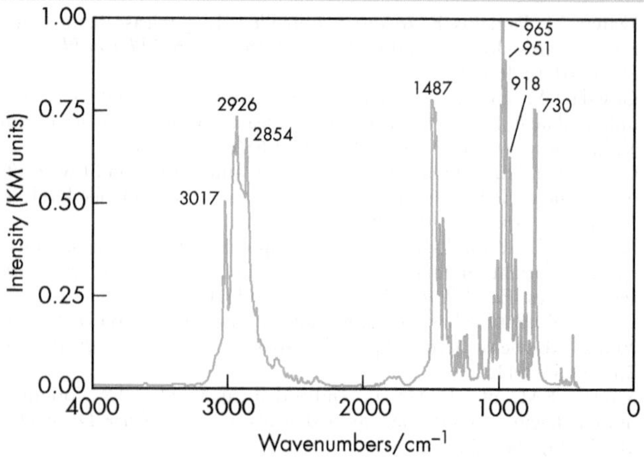

Figure 1: Infrared spectrum of cetrimide measured by diffuse reflectance. Adapted with permission of Informa Healthcare.

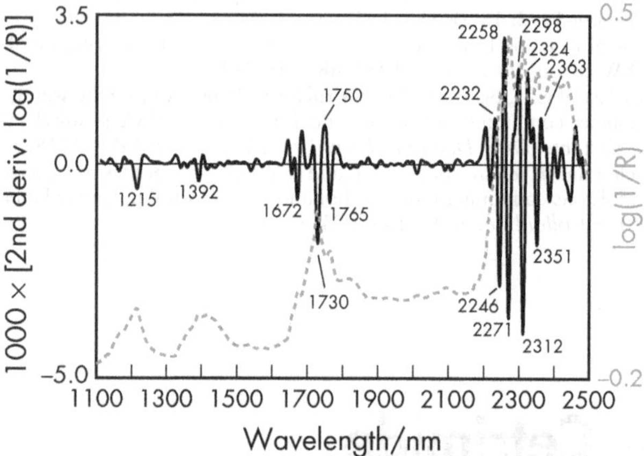

Figure 2: Near-infrared spectrum of cetrimide measured by reflectance.

Critical micelle concentration 3.08 mmol/kg[10] (in water)

Melting point 232–247°C

Moisture content At 40–50% relative humidity and 20°C, cetrimide absorbs sufficient moisture to cause caking and retard flow properties.

Partition coefficients

Liquid paraffin : water = <1;

Vegetable oil : water = <1.

Solubility Freely soluble in chloroform, ethanol (95%), and water; practically insoluble in ether. A 2% w/v aqueous solution foams strongly on shaking.

Spectroscopy

IR spectra *see* Figure 1.

NIR spectra *see* Figure 2.

11 Stability and Storage Conditions

Cetrimide is chemically stable in the dry state, and also in aqueous solution at ambient temperatures. Aqueous solutions may be sterilized by autoclaving. Water containing metal ions and organic matter may reduce the antimicrobial activity of cetrimide.

The bulk material should be stored in a well-closed container in a cool, dry place.

12 Incompatibilities

Incompatible with soaps, anionic surfactants, high concentrations of nonionic surfactants, bentonite, iodine, phenylmercuric nitrate, alkali hydroxides, and acid dyes. Aqueous solutions react with metals.

13 Method of Manufacture

Cetrimide is prepared by the condensation of suitable alkyl bromides and trimethylamine.

14 Safety

Most adverse effects reported relate to the therapeutic use of cetrimide. If ingested orally, cetrimide and other quaternary ammonium compounds can cause nausea, vomiting, muscle paralysis, CNS depression, and hypotension; concentrated solutions may cause esophageal damage and necrosis. The fatal oral human dose is estimated to be 1.0–3.0 g.[11]

At the concentrations used topically, solutions do not generally cause irritation, although concentrated solutions have occasionally been reported to cause burns. Cases of hypersensitivity have been reported following repeated application.[12,13]

Adverse effects that have been reported following irrigation of hydatid cysts with cetrimide solution include chemical peritonitis,[14] methemoglobinemia with cyanosis,[15] and metabolic disorders.[16]

15 Handling Precautions

Observe normal precautions appropriate to the circumstances and quantity of material handled. Cetrimide powder and concentrated cetrimide solutions are irritant; avoid inhalation, ingestion, and skin and eye contact. Eye protection, gloves, and a respirator are recommended.[17]

16 Regulatory Status

Included in nonparenteral medicines licensed in the UK. Included in the Canadian Natural Health Products Ingredients Database.

Cetrimide is on the list of 'Existing Active Substances' on the market in the Europe, and is registered according to REACH regulation. Cetrimide is not present in any approved product in the US.

17 Related Substances

Benzalkonium chloride; benzethonium chloride; dodecyltrimethylammonium bromide; hexadecyltrimethylammonium bromide; trimethyltetradecylammonium bromide.

Dodecyltrimethylammonium bromide
Empirical formula $C_{15}H_{34}BrN$
Molecular weight 308.35
CAS number [1119-94-4]
Synonyms DTAB; *N*-lauryl-*N,N,N*-trimethylammonium bromide; *N,N,N*-trimethyldodecylammonium bromide.
Safety
LD_{50} (mouse, IV): 5.2 mg/kg[18]
LD_{50} (rat, IV): 6.8 mg/kg

Hexadecyltrimethylammonium bromide
Empirical formula $C_{19}H_{42}BrN$
Molecular weight 364.48
CAS number [57-09-0]
Synonyms Cetrimide BP 1953; cetrimonium bromide; cetyltrimethylammonium bromide; CTAB; *N,N,N*-trimethylhexadecylammonium bromide.
Appearance A white to creamy-white, voluminous, free-flowing powder, with a characteristic faint odor and bitter, soapy taste.
Melting point 237–243°C
Safety
LD_{50} (guinea pig, SC): 100 mg/kg[19]
LD_{50} (mouse, IP): 106 mg/kg
LD_{50} (mouse, IV): 32 mg/kg
LD_{50} (rabbit, IP): 125 mg/kg
LD_{50} (rabbit, SC): 125 mg/kg
LD_{50} (rat, IV): 44 mg/kg
LD_{50} (rat, oral): 410 mg/kg
Solubility Freely soluble in ethanol (95%); soluble 1 in 10 parts of water.
Comments The original cetrimide BP 1953 consisted largely of hexadecyltrimethylammonium bromide, with smaller amounts of analogous alkyltrimethylammonium bromides. It contained a considerable proportion of inorganic salts, chiefly sodium bromide, and was less soluble than the present product.

Trimethyltetradecylammonium bromide
Empirical formula $C_{17}H_{38}BrN$
Molecular weight 336.40
CAS number [1119-97-7]
Synonyms Myristyltrimethylammonium bromide; tetradecyltrimethylammonium bromide; *N,N,N*-trimethyl-1-tetradecanaminium bromide.
Safety
LD_{50} (mouse, IV): 12 mg/kg[20]
LD_{50} (rat, IV): 15 mg/kg

18 Comments

Therapeutically, cetrimide is used in relatively high concentrations, generally as 0.1–1.0% w/v aqueous solutions, cream or spray as a topical antiseptic for skin, burns, and wounds.[21] Solutions containing up to 10% w/v cetrimide are used as shampoos to remove the scales in seborrheic dermatitis.

As a precaution against contamination with *Pseudomonas* species resistant to cetrimide, stock solutions may be further protected by adding at least 7% v/v ethanol or 4% v/v propan-2-ol.

The EINECS number for cetrimide is 214-291-9. The PubChem Compound ID (CID) for cetrimide includes 68166 (trimethylhexadecylammonium hydroxide) and 14250 (trimethyltetradecylammonium bromide).

19 Specific References

1 Esimone CO, *et al.* The effect of ethylenediamine tetraacetic acid on the antimicrobial properties of benzoic acid and cetrimide. *J Pharm Res Devel* 1999; **4**(1): 1–8.
2 Ferrer-Luque CM, *et al.* Antimicrobial activity of maleic acid and combinations of cetrimide with chelating agents against *Enterococcus faecalis* biofilm. *J Endodontics* 2010; **36**(10): 1673–1675.
3 Mahmoud YA-G. *In vitro* and *in vivo* antifungal activity of cetrimide (cetyltrimethyl ammonium bromide) against fungal keratitis caused by *Fusarium solani*. *Mycoses* 2007; **50**(1): 64–70.
4 Gupta AK, *et al.* Fungicidal activities of commonly used disinfectants and antifungal pharmaceutical spray preparations against clinical strains of *Aspergillus* and *Candida* species. *Med Mycol* 2002; **40**: 201–208.
5 Pumbwe L, *et al.* Evidence for multiple-antibiotic resistance in *Campylobacter jejuni* not mediated by CmeB or CmeF. *Antimicrob Agents Chemother* 2005; **49**(4): 1289–1293.
6 Kaatz GW, *et al.* Multidrug resistance in *Staphylococcus aureus* due to overexpression of a novel multidrug and toxin extrusion (MATE) transport protein. *Antimicrob Agents Chemomother* 2005; **49**(5): 1857–1864.
7 Kaatz GW, Seo SM. Effect of substrate exposure and other growth condition manipulations on norA expression. *J Antimicrob Chemother* 2004; **54**(2): 364–369.
8 Loughlin MF, *et al.* *Pseudomonas aeruginosa* cells adapted to benzalkonium chloride show resistance to other membrane-active agents but not to clinically relevant antibiotics. *J Antimicrob Chemother* 2002; **49**(4): 631–639.
9 Coyle EA, *et al.* Activities of newer fluoroquinolones against ciprofloxacin-resistant *Streptococcus pneumoniae*. *Antimicrob Agents Chemother* 2001; **45**(6): 1654–1659.
10 Attwood D, Patel HK. Composition of mixed micellar systems of cetrimide and chlorhexidine digluconate. *Int J Pharm* 1989; **49**(2): 129–134.
11 Arena JM. Poisonings and other health hazards associated with the use of detergents. *JAMA* 1964; **190**: 56–58.

12 Weiner M, Bernstein IL. *Adverse Reactions to Drug Formulation Agents: A Handbook of Excipients*. New York: Marcel Dekker, 1989.
13 Tomar J, *et al*. Contact allergies to cosmetics: testing with 52 cosmetic ingredients and personal products. *J Dermatol* 2005; **32**(12): 951–955.
14 Gilchrist DS. Chemical peritonitis after cetrimide washout in hydatid-cyst surgery [letter]. *Lancet* 1979; **2**: 1374.
15 Baraka A, *et al*. Cetrimide-induced methaemoglobinaemia after surgical excision of hydatid cyst [letter]. *Lancet* 1980; **2**: 88–89.
16 Momblano P, *et al*. Metabolic acidosis induced by cetrimonium bromide [letter]. *Lancet* 1984; **2**: 1045.
17 Jacobs JY. Work hazards from drug handling. *Pharm J* 1984; **233**: 195–196.
18 Lewis RJ, ed. *Sax's Dangerous Properties of Industrial Materials*, 11th edn. New York: Wiley, 2004; 1550.
19 Lewis RJ, ed. *Sax's Dangerous Properties of Industrial Materials*, 11th edn. New York: Wiley, 2004; 1925.
20 Lewis RJ, ed. *Sax's Dangerous Properties of Industrial Materials*, 11th edn. New York: Wiley, 2004; 3385–3386.
21 Langford JH, *et al*. Topical antimicrobial prophylaxis in minor wounds. *Ann Pharmacother* 1997; **31**(5): 559–563.

20 General References

August PJ. Cutaneous necrosis due to cetrimide application. *Br Med J* 1975; **1**: 70.

Eccleston GM. Phase transitions in ternary systems and oil-in-water emulsions containing cetrimide and fatty alcohols. *Int J Pharm* 1985; **27**: 311–323.

Eccleston GM, *et al*. Synchrotron X-ray investigations into the lamellar gel phase formed in pharmaceutical creams prepared with cetrimide and fatty alcohols. *Int J Pharm* 2000; **203**(1-2): 127–139.

European Commission, REACH Regulation (EC) 1907/2006, 2006. ec.europa.eu/environment/chemicals/reach/reach_intro.html (accessed 30 November 2011).

Evans BK, *et al*. The disinfection of silicone-foam dressings. *J Clin Hosp Pharm* 1985; **10**: 289–295.

Gilbert PM, Moore LE. Cationic antisepctics: diversity of action under a common epithet. *J Appl Microbial* 2005; **99**(4): 703–715.

Louden JD, Rowe RC. A quantitative examination of the structure of emulsions prepared using cetostearyl alcohol and cetrimide using Fourier transform infrared microscopy. *Int J Pharm* 1990; **63**: 219–225.

Rowe RC, Patel HK. The effect of temperature on the conductivity of gels and emulsions prepared from cetrimide and cetostearyl alcohol. *J Pharm Pharmacol* 1985; **37**: 564–567.

Rowe RC, *et al*. The stability of oil-in-water emulsions containing cetrimide and cetostearyl alcohol. *Int J Pharm* 1986; **31**: 281–282.

Smith ARW, *et al*. The differing effects of cetyltrimethylammonium bromide and cetrimide BP upon growing cultures of *Escherichia coli* NCIB 8277. *J Appl Bacteriol* 1975; **38**: 143–149.

21 Author

W Cook.

22 Date of Revision

1 March 2012.

Cetyl Alcohol

1 Nonproprietary Names

BP: Cetyl Alcohol
JP: Cetanol
PhEur: Cetyl Alcohol
USP–NF: Cetyl Alcohol

2 Synonyms

Alcohol cetylicus; *Avol*; *Cetanol*; *Cachalot*; *Crodacol C70*; *Crodacol C90*; *Crodacol C95*; ethal; ethol; *HallStar CO-1695*; 1-hexadecanol; *n*-hexadecyl alcohol; *Hyfatol 16-95*; *Hyfatol 16-98*; *Kessco CA*; *Lanette 16*; *Lipocol C*; *Nacol 16-95*; palmityl alcohol; *Rita CA*; *Speziol C16 Pharma*; *Tego Alkanol 16*; *Vegarol 1695*; *Vegarol 1698*.

3 Chemical Name and CAS Registry Number

Hexadecan-1-ol [36653-82-4]

4 Empirical Formula and Molecular Weight

$C_{16}H_{34}O$ 242.44 (for pure material)

Cetyl alcohol, used in pharmaceutical preparations, is a mixture of solid aliphatic alcohols comprising mainly 1-hexadecanol ($C_{16}H_{34}O$). The USP35–NF30 specifies not less than 90.0% of cetyl alcohol, the remainder consisting chiefly of related alcohols.

Commercially, many grades of cetyl alcohol are available as mixtures of cetyl alcohol (60–70%) and stearyl alcohol (20–30%), the remainder being related alcohols.

5 Structural Formula

$$H-\overset{\displaystyle H}{\underset{\displaystyle H}{\overset{|}{\underset{|}{C}}}}-(CH_2)_{14}-\overset{\displaystyle H}{\underset{\displaystyle H}{\overset{|}{\underset{|}{C}}}}-OH$$

6 Functional Category

Coating agent; emollient; emulsifying agent; stiffening agent.

7 Applications in Pharmaceutical Formulation or Technology

Cetyl alcohol is widely used in cosmetics and pharmaceutical formulations such as suppositories, modified-release solid dosage forms, emulsions, lotions, creams, and ointments.

In suppositories cetyl alcohol is used to raise the melting point of the base, and in modified-release dosage forms it may be used to form a permeable barrier coating. In lotions, creams, and ointments cetyl alcohol is used because of its emollient, water-absorptive, and emulsifying properties. It enhances stability, improves texture, and increases consistency. The emollient properties are due to absorption and retention of cetyl alcohol in the epidermis, where it lubricates and softens the skin while imparting a characteristic 'velvety' texture.

Cetyl alcohol is also used for its water absorption properties in water-in-oil emulsions. For example, a mixture of petrolatum and cetyl alcohol (19 : 1) will absorb 40–50% of its weight of water.

Cetyl alcohol acts as a weak emulsifier of the water-in-oil type, thus allowing a reduction of the quantity of other emulsifying agents used in a formulation. Cetyl alcohol has also been reported to increase the consistency of water-in-oil emulsions.

In oil-in-water emulsions, cetyl alcohol is reported to improve stability by combining with the water-soluble emulsifying agent. The combined mixed emulsifier produces a close packed, monomolecular barrier at the oil–water interface which forms a mechanical barrier against droplet coalescence.

In semisolid emulsions, excess cetyl alcohol combines with the aqueous emulsifier solution to form a viscoelastic continuous phase that imparts semisolid properties to the emulsion and also prevents droplet coalescence. Therefore, cetyl alcohol is sometimes referred to as a 'consistency improver' or a 'bodying agent', although it may be necessary to mix cetyl alcohol with a hydrophilic emulsifier to impart this property.

It should be noted that pure or pharmacopeial grades of cetyl alcohol may not form stable semisolid emulsions and may not show the same physical properties as grades of cetyl alcohol that contain significant amounts of other similar alcohols. *See* Section 4.

See Table I.

Table I: Uses of cetyl alcohol.

Use	Concentration (%)
Emollient	2–5
Emulsifying agent	2–5
Stiffening agent	2–10
Water absorption	5

8 Description

Cetyl alcohol occurs as waxy, white flakes, granules, cubes, or castings. It has a faint characteristic odor and bland taste.

9 Pharmacopeial Specifications

See Table II.

Table II: Pharmacopeial specifications for cetyl alcohol.

Test	JP XV	PhEur 7.4	USP35–NF30
Identification	—	+	+
Characters	—	+	—
Melting range	47–53°C	46–52°C	—
Residue on ignition	≤0.05%	—	—
Ester value	≤2.0	—	—
Alkali	+	—	—
Acid value	≤1.0	≤1.0	≤2
Iodine value	≤2.0	≤2.0	≤5
Hydroxyl value	210–232	218–238	218–238
Saponification value	—	≤2.0	—
Clarity and color of solution	+	+	—
Assay	—	≥95.0%	≥90.0%

10 Typical Properties

Boiling point

316–344°C;

300–320°C for *Nacol 16-95*;

310–360°C for *Speziol C16 Pharma*;

344°C for pure material.

Density

0.908 g/cm^3;

0.805-0.815 g/cm^3 for *Speziol C16 Pharma.*

Flash point 165°C

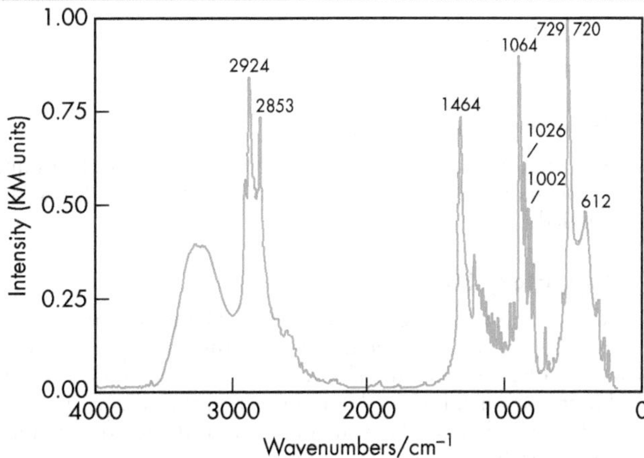

Figure 1: Infrared spectrum of cetyl alcohol measured by diffuse reflectance. Adapted with permission of Informa Healthcare.

Melting point

45–52°C;

49°C for pure material.

Refractive index n_D^{79} = 1.4283 for pure material.

Solubility Freely soluble in ethanol (95%) and ether, solubility increasing with increasing temperature; practically insoluble in water. Miscible when melted with fats, liquid and solid paraffins, and isopropyl myristate.

Specific gravity ≈ 0.81 at 50°C

Spectroscopy

IR spectra *see* Figure 1.

Viscosity (dynamic)

≈ 7 mPa s (7 cP) at 50°C;

8 mPa s (8 cP) at 60°C for *Nacol 16-95.*

11 Stability and Storage Conditions

Cetyl alcohol is stable in the presence of acids, alkalis, light, and air; it does not become rancid. It should be stored in a well-closed container in a cool, dry place.

12 Incompatibilities

Incompatible with strong oxidizing agents. Cetyl alcohol is responsible for lowering the melting point of ibuprofen, which results in sticking tendencies during the process of film coating ibuprofen crystals.[1]

13 Method of Manufacture

Cetyl alcohol may be manufactured by a number of methods such as esterification and hydrogenolysis of fatty acids or by catalytic hydrogenation of the triglycerides obtained from coconut oil or tallow. Cetyl alcohol may be purified by crystallization and distillation.

14 Safety

Cetyl alcohol is mainly used in topical formulations, although it has also been used in oral and rectal preparations.

Cetyl alcohol has been associated with allergic delayed-type hypersensitivity reactions in patients with stasis dermatitis.[2] Cross-sensitization with cetostearyl alcohol, lanolin, and stearyl alcohol has also been reported.[3,4] It has been suggested that hypersensitivity may be caused by impurities in commercial grades of cetyl alcohol since highly refined cetyl alcohol (99.5%) has not been associated with hypersensitivity reactions.[5]

LD_{50} (mouse, IP): 1.6 g/kg[6]

LD_{50} (mouse, oral): 3.2 g/kg

LD_{50} (rat, IP): 1.6 g/kg
LD_{50} (rat, oral): 5 g/kg

15 Handling Precautions

Observe normal precautions appropriate to the circumstances and quantity of material handled. Eye protection and gloves are recommended.

16 Regulatory Status

Included in the FDA Inactive Ingredients Database (ophthalmic preparations, oral capsules and tablets, otic and rectal preparations, topical aerosols, creams, emulsions, ointments and solutions, and vaginal preparations). Included in nonparenteral medicines licensed in the UK. Included in the Canadian Natural Health Products Ingredients Database.

17 Related Substances

Cetostearyl alcohol; stearyl alcohol.

18 Comments

The EINECS number for cetyl alcohol is 253-149-0. The PubChem Compound ID (CID) for cetyl alcohol is 2682.

19 Specific References

1 Schmid S, *et al.* Interactions during aqueous film coating of ibuprofen with Aquacoat ECD. *Int J Pharm* 2000; **197**: 35–39.
2 Smolinske SC. *Handbook of Food, Drug and Cosmetic Excipients*. Boca Raton, FL: CRC Press, 1992: 75–77.
3 van Ketel WG, Wemer J. Allergy to lanolin and 'lanolin-free' creams. *Contact Dermatitis* 1983; **9**(5): 420.
4 Degreef H, Dooms-Goossens A. Patch testing with silver sulfadiazine cream. *Contact Dermatitis* 1985; **12**: 33–37.
5 Hannuksela M, Salo H. The repeated open application test (ROAT). *Contact Dermatitis* 1986; **14**(4): 221–227.
6 Lewis RJ, ed. *Sax's Dangerous Properties of Industrial Materials*, 11th edn. New York: Wiley, 2004: 1923.

20 General References

Cognis. Material safety data sheet: *Speziol C16 Pharma*, 2007.
Eccleston GM. Properties of fatty alcohol mixed emulsifiers and emulsifying waxes. In: Florence AT, ed. *Materials Used in Pharmaceutical Formulation: Critical Reports on Applied Chemistry*. 6. Oxford: Blackwell Scientific, 1984: 124–156.
Mapstone GE. Crystallization of cetyl alcohol from cosmetic emulsions. *Cosmet Perfum* 1974; **89**(11): 31–33.
Ribeiro HM, *et al.* Structure and rheology of semisolid o/w creams containing cetyl alcohol/non-ionic surfactant mixed emulsifier and different polymers. *Int J Cosmet Sci* 2004; **26**(2): 47–59.
Sasol. Material safety data sheet: *Nacol 16-95*, 2005.

21 Author

HM Unvala.

22 Date of Revision

1 March 2012.

Cetyl Palmitate

1 Nonproprietary Names

BP: Cetyl Palmitate
PhEur: Cetyl Palmitate
USP–NF: Cetyl Palmitate

2 Synonyms

Cetaceum; cetyl esters wax; *Crodamol CP*; *Dynacerin CP*; *Estol 3694*; *Hallstar 653*; hexadecanoic acid hexadecyl ester; hexadecyl palmitate; N-hexadecyl palmitate; *Kessco CP*; *Palmil* C; palmitic acid N-hexadecyl ester; palmityl palmitate; *Pelemol CP*; *Sabowax CP*; spermaceti (synthetic); *Stepan 653*.

3 Chemical Name and CAS Registry Number

Hexadecyl hexadecanoate [540-10-3]

4 Empirical Formula and Molecular Weight

$C_{32}H_{64}O_2$ 480.87

The USP35–NF30 describes cetyl palmitate as consisting of esters of cetyl alcohol and saturated high-molecular-weight fatty acids, principally palmitic acid. The PhEur 7.4 describes cetyl palmitate as a mixture of C_{14}-C_{18} esters of lauric (dodecanoic), myristic (tetradecanoic), palmitic (hexadecanoic), and stearic (octadecanoic) acids ("cetyl esters wax").

5 Structural Formula

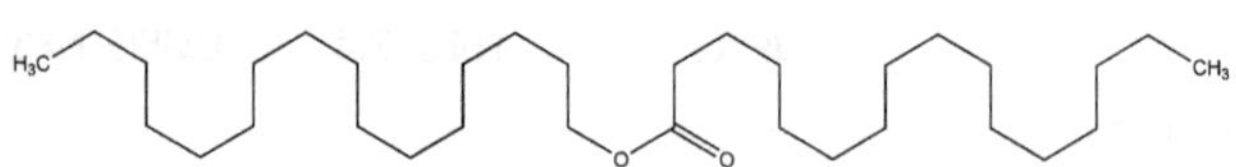

6 Functional Category

Emollient; emulsifying agent; emulsion stabilizing agent; stiffening agent; viscosity-increasing agent.

7 Applications in Pharmaceutical Formulation or Technology

Cetyl palmitate is used as an emollient for topical products. It is also used as an emulsion stabilizing agent in shampoos, and as an emulsifying agent and viscosity-increasing agent in various lotions and creams. It also helps add texture and body to various personal care and cosmetic products, including sunscreens and anti-aging treatments.

8 Description

Cetyl palmitate occurs as white or almost white waxy plates, flakes, or powder.

9 Pharmacopeial Specifications

See Table I.

Table I: Pharmacopeial specifications for cetyl palmitate.

Test	PhEur 7.4	USP35–NF30
Characters	+	–
Identification	+	+
Alkaline impurities	+	–
Melting point	45–52°C	46–53°C
Acid value	≤4	≤1
Appearance of solution	+	–
Hydroxyl value	≤20.0	≤6
Iodine value	≤2	≤1
Loss on drying	–	≤3.0%
Residue on ignition	–	≤0.05%
Saponification value	105–120	110–130
Heavy metals	–	≤0.002%
Nickel	≤1 ppm	–
Total ash	≤0.2%	–
Water	≤0.3%	–
Assay	+	+

10 Typical Properties

Boiling point 360–534°C[1–3]
Density 0.989 g/cm^3 [2,3]
Flash point >238°C for *Hallstar 653* (closed cup)[4]; 269.8°C[1]
Iodine value 0.5 for *Hallstar 653*[4]
Melting point 43–56°C;
approx. 50°C for *Pelemol CP*[5];
53°C for *Hallstar 653*[4].
Refractive index 1.4425 (at 49.8°C)[2]
Solubility Practically insoluble in water; soluble in boiling anhydrous ethanol and in methylene chloride; slightly soluble in light petroleum; practically insoluble in anhydrous ethanol.
Vapor pressure <0.1 at 20°C for *Pelemol C*[5]

11 Stability and Storage Conditions

Cetyl palmitate should be stored in tightly closed containers in a cool, dry place away from direct sunlight and odors. Avoid exposure to excessive heat.

12 Incompatibilities

Cetyl palmitate is incompatible with strong oxidizing agents.

13 Method of Manufacture

Cetyl palmitate is an ester produced from the reaction of cetyl alcohol and palmitic acid, a naturally occurring fatty acid found in both plants and animals. If manufactured from plants, cetyl esters may be used in cosmetics and personal care products marketed in Europe according to the general provisions of the Cosmetics Directive of the European Union.[6] If components of cetyl esters are manufactured from animal sources, they must comply with the EU's animal by-products regulations.

Cetyl palmitate also occurs naturally as a chief constituent of spermaceti (wax derived from sperm whale oil). In the past, spermaceti was extracted from sperm whale oil under pressure using a solution of caustic alkali, followed by crystallization at 6°C. However, the Endangered Species Act of 1973 in the US outlawed the killing of whales and the use of materials derived from those animals.

14 Safety

Cetyl palmitate is widely used in cosmetic and personal care products. The Cosmetic Ingredient Review (CIR) Expert Panel reviewed the safety of this ingredient and assessed it as nontoxic and nonsensitizing, deeming it safe for use in cosmetic formulations in the present practices of use and concentration.[7]

Slightly irritating to eyes and skin. While some palmitates have been reported to cause contact dermatitis, human skin tests of moisturizers containing between 2.5% and 2.7% of cetyl palmitate were found to be minimally irritating to the skin and showed no systemic toxicity or photosensitization. May cause irritation of the respiratory tract and digestive tract.LD_{50} (oral, rat): >14.4 g/kg[7]

15 Handling Precautions

Observe normal precautions appropriate to the circumstances and quantity of the material handled. Cetyl palmitate may be combustible at high temperatures. Protective clothing, gloves, and safety goggles are recommended. Wear suitable respiratory equipment.

16 Regulatory Status

Included in the FDA Inactive Ingredients Database (topical creams, ointments, emulsions, and solutions; vaginal emulsions and creams). Included in nonparenteral medicines licensed in the UK. Included in the Canadian Natural Health Products Ingredients Database.

17 Related Substances

Cetostearyl alcohol.

18 Comments

Cetyl palmitate has recently been used in the development of several new drug delivery systems such as nanostructured lipid carriers (NLC) and solid lipid nanoparticles (SLN). A NLC containing the endogenous cellular antioxidant coenzyme Q_{10} has been evaluated[8] and SLN formulations containing both nitrendipine[9] and insulin[10] have been shown to have superior properties in terms of stability[9] and bioavailability[10] compared with conventional formulations. An SLN dispersion prepared using 10% cetyl palmitate, 1.5% sucrose stearate, and 88.5% water incorporated into a cream has been shown to exhibit increased skin hydration.[11]

Cetyl palmitate is a cetylated fatty acid and it has been used in combination with other members of this class of chemical compounds (i.e. cetyl myristoleate) in the formulation of topical creams for the treatment of rheumatoid arthritis, osteoarthritis, fibromyalgia, and a variety of other syndromes.[12,13] Cetylated fatty acids may also be an alternative to the use of nonsteroidal anti-inflammatory drugs for the treatment of osteoarthritis.[14]

A botanical alternative to spermaceti wax is a derivative of jojoba oil, namely jojoba esters. Jojoba esters is a solid wax that has very similar physico-chemical properties to spermaceti and may be used in similar pharmaceutical applications.

A specification for cetyl palmitate is included in the *Japanese Pharmaceutical Excipients* (JPE).[15]

The EINECS number for cetyl palmitate is 208-736-6. The PubChem Compound ID (CID) for cetyl palmitate is 10889.

19 Specific References

1 LookChem. Product information: *Cetyl palmitate*. http://www.lookchem.com/(accessed 1 August 2011).
2 Chemexper.net. Product information: *Cetyl palmitate*, 2011. http://www.chemexper.net/specification_d/chemicals/supplier/cas/Cetyl%20-palmitate.asp (accessed 1 August 2011).
3 ScienceLab.com. Material safety data sheet: *Cetyl palmitate*, January 2010. http://www.sciencelab.com/msds.php?msdsId=9923365 (accessed 1 August 2011).
4 Hallstar Co. Product information sheet: *Hallstar 653*, 2010. http://www.hallstar.com/pis.php?product=1H031. (accessed 1 August 2011)
5 Phoenix Chemical Inc. Material safety data sheet: *Pelemol CP*, March 2012. http://www.cornelius.co.uk/Documents/MSDS/Pelemol_CP/$File/scan.pdf (accessed 1 August 2011).
6 European Commission. Cosmetics Directive of the European Union 76/768/EEC. http://ec.europa.eu/consumers/sectors/cosmetics/documents/directive/index_en.htm#h2-consolidated-version-of-cosmetics-directive-76/768/eec (accessed 1 August 2011).

7 Cosmetic Ingredient Review. Final Report on the Safety Assessment of Octyl Palmitate, Cetyl Palmitate and Isopropyl Palmitate. *Int J Toxicol* 1982; **1**(2): 13–35.
8 Teeranachaideekul V, *et al.* Cetyl palmitate-based NLC for topical delivery of Coenzyme Q_{10}-development, physicochemical characterization and *in vitro* release studies. *Eur J Pharm Biopharm* 2007; **67**(1): 141–148.
9 Kumar VV, *et al.* Development and evaluation of nitrendipine loaded solid lipid nanoparticles: influence of wax and glyceride lipids on plasma pharmacokinetics. *Int J Pharm* 2007; **335**(1–2): 167–175.
10 Sarmento B, *et al.* Oral insulin delivery by means of solid lipid nanoparticles. *Int J Nanomedicine* 2007; **2**(4): 743–749.
11 Wissing SA, Muller RH. The influence of solid lipid nanoparticles on skin hydration and viscoelasticity – *in vivo* study. *Eur J Pharm Biopharm* 2003; **56**: 67–72.
12 Kraemer WJ, *et al.* A cetylated fatty acid topical cream with menthol reduces pain and improves functional performance in individuals with arthritis. *J Strength Cond Res* 2005; **19**(2): 475–480.
13 Kraemer WJ, *et al.* Effect of a cetylated fatty acid topical cream on functional mobility and quality of life of patients with osteoarthritis. *J Rheumatol* 2004; **31**(4): 767–774.
14 Hesslink RJr, *et al.* Cetylated fatty acids improve knee function in patients with osteoarthritis. *J Rheumatol* 2002; **29**(8): 1708–1712.
15 Japan Pharmaceutical Excipients Council. *Japanese Pharmaceutical Excipients 2004*. Tokyo: Yakuji Nippo, 2004; 169.

20 General References

Croda International Plc. Product information: *Crodamol CP*, 2010.

Julius Hoesch Düren. Product information: *Kessco CP*. http://www.julius-hoesch.de/en/99.php (accessed 2 August 2011).

Phoenix Chemical Inc. 2002 Product information: *Cetyl palmitate*. http://www.cornelius.co.uk/Documents/MSDS/Pelemol_CP/$File/scan.pdf. (accessed 2 August 2011).

Polygon Chemical Ltd. Product information: *Sabowax CP*. http://www.polygon.ch/dateien/en/KosmGrd-en.pdf (accessed 2 August 2011).

Sasol. Product information: *Dynacerin CP*. http://www.sasoltechdata.com/MarketingBrochures/Excipients_Pharmaceuticals.pdf (accessed 2 August 2011).

Union Derivan SA. Product information: *Palmil C*. http://www.undesa.com/english/products/esters/monoalcohol_esters.htm (accessed 2 August 2011).

21 Authors

SL Cantor, SA Shah.

22 Date of Revision

1 March 2012.

Cetylpyridinium Chloride

1 Nonproprietary Names

BP: Cetylpyridinium Chloride
PhEur: Cetylpyridinium Chloride
USP-NF: Cetylpyridinium Chloride

2 Synonyms

C16-alkylpyridinium chloride; *Cepacol*; *Cepacol Chloride*; *Cetamiun*; cetylpridinii chloridum; cetyl pyridium chloride; *Dobendan*; hexadecylpyridinium chloride; 1-hexadecylpyridinium chloride; *Medilave*; *Pristacin*; *Pyrisept*.

3 Chemical Name and CAS Registry Number

1-Hexadecylpyridinium chloride [123-03-5]
1-Hexadecylpyridinium chloride monohydrate [6004-24-6]

4 Empirical Formula and Molecular Weight

$C_{21}H_{38}ClN$ 339.9 (for anhydrous)
$C_{21}H_{38}ClN \cdot H_2O$ 358.1 (for monohydrate)

5 Structural Formula

$N^+—(CH_2)_{15}CH_3 \quad Cl^- \cdot H_2O$

6 Functional Category

Antimicrobial preservative; cationic surfactant; solubilizing agent.

7 Applications in Pharmaceutical Formulation or Technology

Cetylpyridinium chloride is a quaternary ammonium cationic surfactant, used in pharmaceutical and cosmetic formulations as an antimicrobial preservative; *see* Section 10. Cetylpyridinium chloride is used in nonparenteral formulations licensed in the UK, and in oral and inhalation preparations at concentrations of 0.02–1.5 mg (*see* Section 16).

8 Description

Cetylypyridinium chloride is a white powder with a characteristic odor. It is slightly soapy to the touch.

9 Pharmacopeial Specifications

See Table I.

Table I: Pharmacopeial specifications for cetylpyridinium chloride.

Test	PhEur 7.4	USP 35–NF30
Absorbance	+	–
Acidity	+	+
Amines and amine salts	+	–
Appearance of solution	+	–
Characters	+	–
Heavy metals	–	≤0.002%
Identification	+	+
Melting range	–	80.0–84.0°C
Pyridine	–	+
Residue on ignition	–	≤0.2%
Sulfated ash	≤0.2%	–
Water	4.5–5.5%	4.5–5.5%
Assay	96.0–101.0%	99.0–102.0%

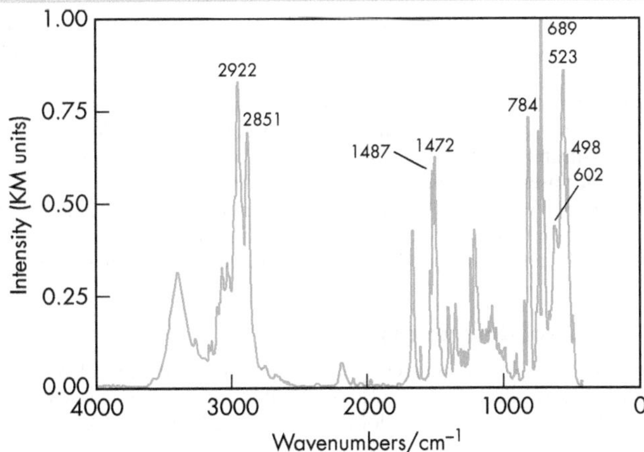

Figure 1: Infrared spectrum of cetylpyridinium chloride (anhydrous) measured by diffuse reflectance. Adapted with permission of Informa Healthcare.

Table II: Minimum inhibitory concentrations (MICs) for cetylpyridinium chloride.[3]

Microorganism	MIC (μg/mL)
Staphylococcus aureus	<2.0
Bacillus subtilis	<2.0
Salmonella typhimurium	8.0
Pseudomonas aeruginosa	16.0
Streptococcus pyogenes	<2.0

10 Typical Properties

Antibacterial activity Bactericidal to Gram-positive bacteria; relatively ineffective against some Gram-negative bacteria.[1] Cetylpyridinium chloride is also antibacterial against a number of oral bacteria;[2] *see* Table II.[3]

Critical micelle concentration 0.34 g/L (water, 25°C).[4,5]

Melting point 80–83°C

Solubility Freely soluble in water; very soluble in chloroform; very slightly soluble in ether; insoluble in acetone, acetic acid, and ethanol.

Spectroscopy

IR spectra *see* Figure 1.

11 Stability and Storage Conditions

Cetylpyridinium chloride is stable under normal conditions. It should be stored in well-closed containers.

12 Incompatibilities

Incompatible with strong oxidizing agents and bases. It is also incompatible with methylcellulose.

Magnesium stearate suspensions in cetylpyridinium chloride have been shown to significantly reduce its antimicrobial activity. This is due to the absorption of cetylpyridinium chloride on magnesium stearate.[6] The cetylpyridinium chloride ion also interacts with gelatin, resulting in reduced bioavailability.[7]

13 Method of Manufacture

Cetylpyridinium chloride is prepared from cetyl chloride by treatment with pyridine.

14 Safety

Cetylpyridinium chloride is used widely in mouthwashes as a bactericidal antiseptic. It is generally regarded as a relatively nontoxic material when used at a concentration of 0.05% w/v, although minor side effects such as mild burning sensations on the tongue have been reported.[8]

At higher concentrations, cetylpyridinium chloride may damage the mucous membranes in the mouth. It is harmful when ingested or inhaled. It can cause eye irritation, and is irritant to the respiratory system and the skin.

LD_{50} (mouse, IP): 0.01 g/kg[9]
LD_{50} (mouse, oral): 0.108 g/kg
LD_{50} (rabbit, IV): 0.036 g/kg
LD_{50} (rabbit, oral): 0.4 g/kg
LD_{50} (rat, IP): 0.006 g/kg
LD_{50} (rat, IV): 0.03 g/kg
LD_{50} (rat, oral): 0.2 g/kg
LD_{50} (rat, SC): 0.25 g/kg

15 Handling Precautions

Observe normal precautions appropriate to the circumstances and quantity of the material handled. When significant quantities are being handled, the use of a respirator with an appropriate gas filter is advised. When heated to decomposition, cetylpyridinium chloride emits very toxic fumes of NO_x and Cl^-. Eye protection, gloves and adequate ventilation are recommended.

16 Regulatory Status

Included in nonparenteral formulations licensed in the UK. Included in the FDA Inactive Ingredients Database, for use in inhalation and oral preparations. Reported in the EPA TSCA Inventory. It is not approved for use in Japan. Included in the Canadian Natural Health Products Ingredients Database.

17 Related Substances

Cetylpyridinium bromide.

Cetylpyridinium bromide

Empirical formula $C_{21}H_{38}BrN$

Molecular weight 384.45

CAS number [140-72-7]

Synonyms *Aceloquat CPB*; *Bromocet*; *Cetapharm*; *Cetasol*; N-cetylpyridinium bromide; hexadexylpyridinium bromide; *Nitrogenol*; *Seprison*; *Sterogenol.*

18 Comments

Cetylpyridinium chloride is used therapeutically as an antiseptic agent. It has been studied for use as an antimicrobial preservative for meat[10] and vegetables.[11] However, the residual levels after treatment are considered excessive for human consumption; *see* Section 14.

It is used alone or in combination with other drugs for oral and throat care. Mouthwashes containing cetylpyridinium chloride have been shown to inhibit plaque formation,[12–14] although efficacy is variable owing to limited published data.[15,16]

The EINECS number for cetylpyridinium chloride is 204-593-9. The PubChem Compound ID (CID) for cetylpyridinium chloride monohydrate is 22324.

19 Specific References

1 Baker Z, *et al.* The bactericidal action of synthetic detergents. *J Exp Med* 1941; 74: 611–620.
2 Smith RN, *et al.* Inhibition of intergeneric coaggregation among oral bacteria by cetylpyridinium chloride, chlorhexidine digluconate and octenidine dihydrochloride. *J Periodontal Res* 1991; 26(5): 422–428.
3 Bodor N, *et al.* Soft drugs. 1. Labile quaternary ammonium salts as soft antimicrobials. *J Med Chem* 1980; 23: 469–474.

4 Harada T, *et al.* Effect of surfactant micelles on the rate of reaction of tetranitromethane with hydroxide ion. *Bull Chem Soc Jpn* 1981; **54**: 2592–2597.
5 Wang K, *et al.* Aggregation behaviour of cationic fluorosurfactants in water and salt solutions. A cryoTEM survey. *J Phys Chem B* 1999; **103**(43): 9237–9246.
6 Richards RM, *et al.* Excipient interaction with cetylpyridinium chloride activity in tablet based lozenges. *Pharm Res* 2003; **13**(8): 1258–1264.
7 Ofner CM 3rd, Schott H. Swelling studies of gelatin. II: Effect of additives. *J Pharm Sci* 1987; **76**(9): 715–723.
8 Ciano SG, *et al.* Clinical evaluation of a quaternary ammonium containing mouthrinse. *J Periodontol* 1975; **46**: 397–401.
9 Lewis RJ, ed. *Sax's Dangerous Properties of Industrial Materials*, 11th edn. New York: Wiley, 2004; 737.
10 Cutter CN, *et al.* Antimicrobial activity of cetylpyridinium chloride washes against pathogenic bacteria on beef surfaces. *J Food Prot* 2000; **63**(5): 593–600.
11 Wang H, *et al.* Efficacy of cetylpyridinium chloride in immersion treatment reducing populations of pathogenic bacteria on fresh-cut vegetables. *J Food Prot* 2001; **64**(12): 2071–2074.
12 Volpe AR, *et al.* Antimicrobial control of bacterial plaque and calculus, and the effects of these agents on oral flora. *J Dent Res* 1969; **48**: 832–841.
13 Holbeche JD, *et al.* A clinical trial of the efficacy of a cetylpyridinium chloride-based mouth-wash. 1. Effect on plaque accumulation and gingival condition. *Aust Dent J* 1975; **20**: 397–404.
14 Ashley FP, *et al.* The effects of a 0.1% cetylpyridinium chloride mouthrinse on plaque and gingivitis in adult subjects. *Br Dent J* 1984; **157**: 191–196.
15 Bonosvoll P, Gjermo P. A comparison between chlorhexedine and some quaternary ammonium compounds with regard to retention, salivary concentration and plaque-inhibiting effect in the human mouth after mouth rinses. *Arch Oral Biol* 1978; **23**: 289–294.
16 Sheen S, Addy M. An *in vitro* evaluation of the availability of cetylpyridinium chloride and chlorhexidine in some commercially available mouthrinse products. *Br Dent J* 2003; **194**(4): 207–210.

20 General References

Huyck CL. Cetylpyridinium chloride. *Am J Pharm* 1944; **116**: 50–59.
Radford JR, *et al.* Effect of use of 0.005% cetylpyridinium chloride mouthwash on normal oral flora. *J Dent* 1997; **25**(1): 35–40.

21 Author

CP McCoy.

22 Date of Revision

1 March 2012.

Chitosan

1 Nonproprietary Names

BP: Chitosan Hydrochloride
PhEur: Chitosan Hydrochloride
USP–NF: Chitosan

2 Synonyms

2-Amino-2-deoxy-(1,4)-β-D-glucopyranan; *Chitopharm*; chitosani hydrochloridum; deacetylated chitin; deacetylchitin; β-1,4-poly-D-glucosamine; poly-D-glucosamine; poly-(1,4-β-D-glucopyranosamine); *Protasan*; *Protasan UP CL 113*.

3 Chemical Name and CAS Registry Number

Poly-β-(1,4)-2-Amino-2-deoxy-D-glucose [9012-76-4]

4 Empirical Formula and Molecular Weight

Partial deacetylation of chitin results in the production of chitosan, which is a polysaccharide comprising copolymers of glucosamine and *N*-acetylglucosamine. Chitosan is the term applied to deacetylated chitins in various stages of deacetylation and depolymerization and it is therefore not easily defined in terms of its exact chemical composition. A clear nomenclature with respect to the different degrees of *N*-deacetylation between chitin and chitosan has not been defined,[1–3] and as such chitosan is not one chemical entity but varies in composition depending on the manufacturer. In essence, chitosan is chitin sufficiently deacetylated to form soluble amine salts. The degree of deacetylation necessary to obtain a soluble product must be greater than 80–85%. Chitosan is commercially available in several types and grades that vary in molecular weight by 10 000–1 000 000, and vary in degree of deacetylation and viscosity.[4]

5 Structural Formula

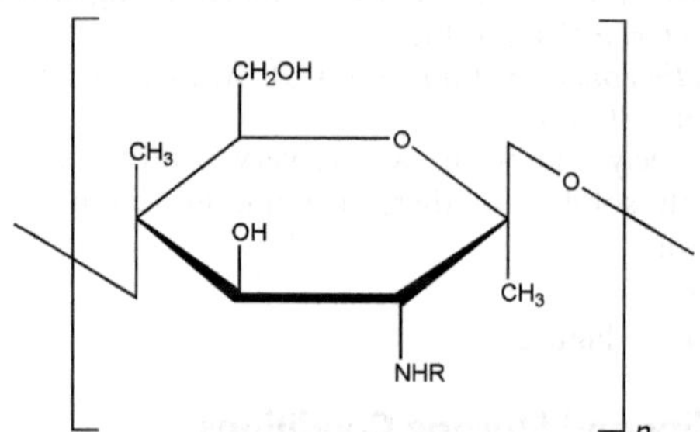

R = H or $COCH_3$

6 Functional Category

Coating agent; tablet and capsule disintegrant; bioadhesive material; film-forming agent; tablet and capsule binder; viscosity-increasing agent.

7 Applications in Pharmaceutical Formulation or Technology

Chitosan is used in cosmetics and is under investigation for use in a number of pharmaceutical formulations. The suitability and performance of chitosan as a component of pharmaceutical formulations for drug delivery applications has been investigated in numerous studies.[3,5–8] These include controlled drug delivery applications,[9–14] use as a component of mucoadhesive dosage forms,[15,16] rapid release dosage forms,[17,18] improved peptide delivery,[19,20] colonic drug delivery systems,[21,22] and use for gene delivery.[23] Chitosan has been processed into several pharmaceutical forms including gels,[24,25] films,[11,12,26,27] beads,[28,29] microspheres,[30,31] tablets,[32,33] and coatings for liposomes.[34] Furthermore, chitosan may be processed into drug delivery systems using several techniques including spray-drying,[15,16] coacervation,[35] direct compression,[32] and conventional granulation processes.[36]

8 Description

Chitosan occurs as odorless, white or creamy-white powder or flakes. Fiber formation is quite common during precipitation and the chitosan may look ‘cottonlike’.

9 Pharmacopeial Specifications

See Table I.

Table I: Pharmacopeial specifications for chitosan.

Test	PhEur 7.4
Identification	+
Characters	+
Appearance of solution	+
Matter insoluble in water	≤0.5%
pH (1% w/v solution)	4.0–6.0
Viscosity	+
Degree of deacetylation	+
Chlorides	10.0–20.0%
Heavy metals	≤40 ppm
Loss on drying	≤10%
Sulfated ash	≤1.0%

10 Typical Properties

Chitosan is a cationic polyamine with a high charge density at pH <6.5, and so adheres to negatively charged surfaces and chelates metal ions. It is a linear polyelectrolyte with reactive hydroxyl and amino groups (available for chemical reaction and salt formation).[7] The properties of chitosan relate to its polyelectrolyte and polymeric carbohydrate character. The presence of a number of amino groups allows chitosan to react chemically with anionic systems, which results in alteration of physicochemical characteristics of such combinations. The nitrogen in chitosan is mostly in the form of primary aliphatic amino groups. Chitosan therefore undergoes reactions typical of amines: for example, *N*-acylation and Schiff reactions.[3] Almost all functional properties of chitosan depend on the chain length, charge density, and charge distribution.[8] Numerous studies have demonstrated that the salt form, molecular weight, and degree of deacetylation as well as pH at which the chitosan is used all influence how this polymer is utilized in pharmaceutical applications.[7]

Acidity/alkalinity pH = 4.0–6.0 (1% w/v aqueous solution)

Density 1.35–1.40 g/cm^3

Glass transition temperature 203°C[37]

Moisture content Chitosan adsorbs moisture from the atmosphere, the amount of water adsorbed depending upon the initial moisture content and the temperature and relative humidity of the surrounding air.[38]

Particle size distribution <30 μm

Solubility Sparingly soluble in water; practically insoluble in ethanol (95%), other organic solvents, and neutral or alkali solutions at pH above approximately 6.5. Chitosan dissolves readily in dilute and concentrated solutions of most organic acids and to some extent in mineral inorganic acids (except phosphoric and sulfuric acids). Upon dissolution, amine groups of the polymer become protonated, resulting in a positively charged polysaccharide (RNH_3^+) and chitosan salts (chloride, glutamate, etc.) that are soluble in water; the solubility is affected by the degree of deacetylation.[7] Solubility is also greatly influenced by the addition of salt to the solution. The higher the ionic strength, the lower the solubility as a result of a salting-out effect, which leads to the precipitation of chitosan in solution.[39] When chitosan is in solution, the repulsions between the deacetylated units and their neighboring glucosamine units cause it to exist in an extended conformation. Addition of an electrolyte reduces this effect and the molecule possesses a more random, coil-like conformation.[40]

Viscosity (dynamic) A wide range of viscosity types is commercially available. Owing to its high molecular weight and linear, unbranched structure, chitosan is an excellent viscosity-enhancing agent in an acidic environment. It acts as a pseudo-plastic material, exhibiting a decrease in viscosity with increasing rates of shear.[7] The viscosity of chitosan solutions increases with increasing chitosan concentration, decreasing temperature, and increasing degree of deacetylation; *see* Table II.[40]

Table II: Typical viscosity (dynamic) values for chitosan 1% w/v solutions in different acids.[40]

Acid	1% acid concentration		5% acid concentration		10% acid concentration	
	Viscosity (mPa s)	pH	Viscosity (mPa s)	pH	Viscosity (mPa s)	pH
Acetic	260	4.1	260	3.3	260	2.9
Adipic	190	4.1	—	—	—	—
Citric	35	3.0	195	2.3	215	2.0
Formic	240	2.6	185	2.0	185	1.7
Lactic	235	3.3	235	2.7	270	2.1
Malic	180	3.3	205	2.3	220	2.1
Malonic	195	2.5	—	—	—	—
Oxalic	12	1.8	100	1.1	100	0.8
Tartaric	52	2.8	135	2.0	160	1.7

11 Stability and Storage Conditions

Chitosan powder is a stable material at room temperature, although it is hygroscopic after drying. Chitosan should be stored in a tightly closed container in a cool, dry place. The PhEur 7.4 specifies that chitosan should be stored at a temperature of 2–8°C.

12 Incompatibilities

Chitosan is incompatible with strong oxidizing agents.

13 Method of Manufacture

Chitosan is manufactured commercially by chemically treating the shells of crustaceans such as shrimps and crabs. The basic manufacturing process involves the removal of proteins by treatment with alkali and of minerals such as calcium carbonate and calcium phosphate by treatment with acid.[3,40] Before these treatments, the shells are ground to make them more accessible. The shells are initially deproteinized by treatment with an aqueous sodium hydroxide 3–5% solution. The resulting product is neutralized and calcium is removed by treatment with an aqueous hydrochloric acid 3–5% solution at room temperature to precipitate chitin. The chitin is dried so that it can be stored as a stable intermediate for deacetylation to chitosan at a later stage. *N*-Deacetylation of chitin is achieved by treatment with an aqueous sodium hydroxide 40–45% solution at elevated temperature (110°C), and the precipitate is washed with water. The crude sample is dissolved in acetic acid 2% and the insoluble material is removed. The resulting clear supernatant solution is neutralized with aqueous sodium hydroxide solution to give a purified white precipitate of chitosan. The product can then be further purified and ground to a fine uniform powder or granules.[1] The animals from which chitosan is derived must fulfil the requirements for the health of animals suitable for human consumption to the satisfaction of the competent authority. The method of production must consider inactivation or removal of any contamination by viruses or other infectious agents.

14 Safety

Chitosan is being investigated widely for use as an excipient in oral and other pharmaceutical formulations. It is also used in cosmetics. Chitosan is generally regarded as a nontoxic and nonirritant material. It is biocompatible[41] with both healthy and infected skin.[42] Chitosan has been shown to be biodegradable.[3,41]

LD_{50} (mouse, oral): >16 g/kg[43]

15 Handling Precautions

Observe normal precautions appropriate to the circumstances and quantity of material handled. Chitosan is combustible; open flames should be avoided. Chitosan is temperature-sensitive and should not be heated above 200°C. Airborne chitosan dust may explode in the presence of a source of ignition, depending on its moisture content and particle size. Water, dry chemicals, carbon dioxide, sand, or foam fire-fighting media should be used.

Chitosan may cause skin or eye irritation. It may be harmful if absorbed through the skin or if inhaled, and may be irritating to mucous membranes and the upper respiratory tract. Eye and skin protection and protective clothing are recommended; wash thoroughly after handling. Prolonged or repeated exposure (inhalation) should be avoided by handling in a well-ventilated area and wearing a respirator.

16 Regulatory Status

Chitosan is registered as a food supplement in some countries. Included in the Canadian Natural Health Products Ingredients Database.

17 Related Substances

See Section 18.

18 Comments

Chitosan derivatives are easily obtained under mild conditions and can be considered as substituted glucens.[3]

The PubChem Compound ID (CID) for chitosan includes 439300 and 3086191.

19 Specific References

1 Muzzarelli RAA, ed. *Natural Chelating Polymers*. New York: Pergamon Press, 1973: 83–227.
2 Zikakis JP, ed. *Chitin, Chitosan and Related Enzymes*. New York: Academic Press, 1974.
3 Kumar MNVR. A review of chitin and chitosan applications. *React Funct Polym* 2000; **46**: 1–27.
4 Genta I, *et al*. Different molecular weight chitosan microspheres: influence on drug loading and drug release. *Drug Dev Ind Pharm* 1998; 24: 779–784.
5 Illum L. Chitosan and its use as a pharmaceutical excipient. *Pharm Res* 1998; **15**: 1326–1331.
6 Paul W, Sharma CP. Chitosan, a drug carrier for the 21st century: a review. *STP Pharma Sci* 2000; **10**: 5–22.
7 Singla AK, Chawla M. Chitosan: some pharmaceutical and biological aspects – an update. *J Pharm Pharmacol* 2001; **53**: 1047–1067.
8 Dodane V, Vilivalam VD. Pharmaceutical applications of chitosan. *Pharm Sci Technol Today* 1998; **1**: 246–253.
9 Muzzarelli RAA, ed. *Chitin*. London: Pergamon Press, 1977: 69.
10 Nakatsuka S, Andrady LA. Permeability of vitamin-B-12 in chitosan membranes: effect of crosslinking and blending with poly(vinyl alcohol) on permeability. *J Appl Polym Sci* 1992; **44**: 7–28.
11 Kubota N, *et al*. Permeability properties of glycol chitosan membrane modified with thiol groups. *J Appl Polym Sci* 1991; **42**: 495–501.
12 Li Q, *et al*. Application and properties of chitosan. *J Bioact Compat Polym* 1992; **7**: 370–397.
13 Miyazaki S, *et al*. Sustained release and intragastric floating granules of indomethacin using chitosan in rabbits. *Chem Pharm Bull* 1988; **36**: 4033–4038.
14 Sawayangi Y, *et al*. Use of chitosan for sustained-release preparations of water soluble drugs. *Chem Pharm Bull* 1982; **30**: 4213–4215.
15 He P, *et al*. *In vitro* evaluation of the mucoadhesive properties of chitosan microspheres. *Int J Pharm* 1998; **166**: 75–88.
16 He P, *et al*. Sustained release chitosan microsphere produced by novel spray drying methods. *J Microencapsul* 1999; **16**: 343–355.
17 Sawayanagi Y, *et al*. Enhancement of dissolution properties of griseofulvin from ground mixtures with chitin or chitosan. *Chem Pharm Bull* 1982; **30**(12): 4464–4467.
18 Shirashi S, *et al*. Enhancement of dissolution rates of several drugs by low molecular weight chitosan and alginate. *Chem Pharm Bull* 1990; **38**: 185–187.
19 Luessen HL, *et al*. Bioadhesive polymers for the peroral delivery of drugs. *J Control Release* 1994; **29**: 329–338.
20 Luessen HL, *et al*. Mucoadhesive polymers in peroral peptide drug delivery, IV: polycarbophil and chitosan are potent enhancers of peptide transport across intestinal mucosae *in vitro*. *J Control Release* 1997; **45**: 15–23.
21 Tozaki H, *et al*. Validation of a pharmacokinetic model of colon-specific drug delivery and the therapeutic effects of chitosan capsules containing 5-aminosalicylic acid on 2,4,6-trinitrobenzene sulphonic acid-induced ulcerative colitis in rats. *J Pharm Pharmacol* 1999; **51**: 1107–1112.
22 Tozaki H, *et al*. Colon specific delivery of R 68070, a new thromboxane synthase inhibitor using chitosan capsules: therapeutic effects against 2,4,6-trinitrobenzene sulphonic acid-induced ulcerative colitis in rats. *Life Sci* 1999; **64**: 1155–1162.
23 Leong KW, *et al*. DNA-polycation nanospheres as non-viral gene delivery vehicles. *J Control Release* 1998; **53**: 183–193.
24 Kristl J, *et al*. Hydrocolloids and gels of chitosan as drug carriers. *Int J Pharm* 1993; **99**: 13–19.
25 Tasker RA, *et al*. Pharmacokinetics of an injectable sustained-release formulation of morphine for use in dogs. *J Vet Pharmacol Ther* 1997; **20**: 362–367.
26 Remunan-Lopez C, *et al*. Design and evaluation of chitosan/ethylcellulose mucoadhesive bilayered devices for buccal drug delivery. *J Control Release* 1998; **55**: 143–152.
27 Senel S, *et al*. Chitosan films and hydrogels of chlorhexidine gluconate for oral mucosal delivery. *Int J Pharm* 2000; **193**: 197–203.
28 Kofuji K, *et al*. Preparation and drug retention of biodegradable chitosan gel beads. *Chem Pharm Bull* 1999; **47**: 1494–1496.
29 Sezer AD, Akbuga J. Release characteristics of chitosan-treated alginate beads, II: sustained release of a macromolecular drug from chitosan treated alginate beads. *J Microencapsul* 1999; **193**: 197–203.
30 Ganza-Gonzalez A, *et al*. Chitosan and chondroitin microspheres for oral administration controlled release of metoclopromide. *Eur J Pharm Biopharm* 1999; **48**: 149–155.
31 Huang RG, *et al*. Microencapsulation of chlorpheniramine maleate-resin particles with crosslinked chitosan for sustained release. *Pharm Dev Technol* 1999; **4**: 107–115.
32 Yomota C, *et al*. Sustained-release effect of the direct compressed tablet based on chitosan and Na alginate. *Yakugaku Zasshi* 1994; **114**: 257–263.
33 Sabnis S, *et al*. Use of chitosan in compressed tablets of diclofenac sodium: inhibition of drug release in an acidic environment. *Pharm Dev Technol* 1997; **2**: 243–255.
34 Takeuchi H, *et al*. Enteral absorption of insulin in rats from mucoadhesive chitosan-coated liposomes. *Pharm Res* 1996; **13**: 896–901.
35 Bayomi MA, *et al*. Preparation of casein-chitosan microspheres containing diltiazem hydrochloride by an aqueous coacervation technique. *Pharma Acta Helv* 1998; **73**: 187–192.
36 Miyazaki S, *et al*. Drug release from oral mucosal adhesive tablets of chitosan and sodium alginate. *Int J Pharm* 1995; **118**: 257–263.
37 Sakurai K, *et al*. Glass transition temperature of chitosan and miscibility of chitosan/poly(*N*-vinyl pyrrolidone) blends. *Polymer* 2000; **41**: 7051–7056.
38 Gocho H, *et al*. Effect of polymer chain end on sorption isotherm of water by chitosan. *Carbohydr Polym* 2000; **41**: 87–90.
39 Errington N, *et al*. Hydrodynamic characterization of chitosans varying in degree of acetylation. *Int J Biol Macromol* 1993; **15**: 113–117.
40 Skaugrud O. Chitosan – new biopolymer for cosmetics and drugs. *Drug Cosmet Ind* 1991; **148**: 24–29.

41 Gebelein CG, Dunn RL, eds. *Progress in Biomedical Polymers*. New York: Plenum Press, 1990: 283.
42 Gooday GW *et al.* eds. *Chitin in Nature and Technology*. New York: Plenum Press, 1986: 435.
43 Arai K, *et al.* Toxicity of chitosan. *Bull Tokai Reg Fish Res Lab* 1968; 43: 89–94.

20 General References

Brine CJ *et al.* eds. *Advances in Chitin and Chitosan*. London: Elsevier Applied Science, 1992.
Skjak-Braek G *et al.* eds. *Chitin and Chitosan: Sources, Chemistry, Biochemistry, Physical Properties and Applications*. Amsterdam: Elsevier, 1992.

21 Author

DS Jones.

22 Date of Revision

1 March 2012.

Chlorhexidine

1 Nonproprietary Names

BP: Chlorhexidine Acetate
Chlorhexidine Gluconate Solution
Chlorhexidine Hydrochloride
JP: Chlorhexidine Gluconate Solution
Chlorhexidine Hydrochloride
PhEur: Chlorhexidine Diacetate
Chlorhexidine Digluconate Solution
Chlorhexidine Dihydrochloride
USP–NF: Chlorhexidine Acetate
Chlorhexidine Gluconate Solution
Chlorhexidine Hydrochloride

Chlorhexidine is usually encountered as the acetate, gluconate, or hydrochloride salt, and a number of pharmacopeias contain monographs for such materials. *See* Sections 9 and 17.

2 Synonyms

1,6-bis[*N*′-(*p*-Chlorophenyl)-N^5-biguanido]hexane; *N,N*″-bis(4-chlorophenyl)-3,12-diimino-2,4,11,13-tetraazatetradecanediimidamide; chlorhexidini diacetas; chlorhexidini digluconatis solutio; chlorhexidini dihydrochloridum; 1,6-di(4′-chlorophenyldiguanido)hexane; 1,1'-hexamethylene-bis[5-(*p*-chlorophenyl)biguanide].

3 Chemical Name and CAS Registry Number

1*E*-2-[6-[[Amino-[[amino-[(4-chlorophenyl)amino]methylidene]amino]methylidene]amino]hexyl]-1-[amino-[(4-chlorophenyl)amino]methylidene]guanidine [55-56-1]

4 Empirical Formula and Molecular Weight

$C_{22}H_{30}Cl_2N_{10}$ 505.48

5 Structural Formula

6 Functional Category

Antimicrobial preservative.

7 Applications in Pharmaceutical Formulation or Technology

Chlorhexidine salts are widely used in pharmaceutical formulations in Europe and Japan for their antimicrobial properties.[1,2] Although mainly used as disinfectants, chlorhexidine salts are also used as antimicrobial preservatives.

As excipients, chlorhexidine salts are mainly used for the preservation of eye-drops at a concentration of 0.01% w/v; generally the acetate or gluconate salt is used for this purpose. Solutions containing 0.002–0.006% w/v chlorhexidine gluconate have also been used for the disinfection of hydrophilic contact lenses.

8 Description

Chlorhexidine occurs as an odorless, bitter tasting, white crystalline powder. *See* Section 17 for information on chlorhexidine salts.

9 Pharmacopeial Specifications

See Table I.

SEM 1: Excipient: chlorhexidine; manufacturer: SST Corp.; magnification: 600×.

SEM 2: Excipient: chlorhexidine; manufacturer: SST Corp.; magnification: 2400×.

Table I: Pharmacopeial specifications for chlorhexidine.

Test	JP XV	PhEur 7.4	USP35–NF30
Identification	+	+	+
Characters	—	+	—
pH			
Chlorhexidine gluconate solution	5.5–7.0	5.5–7.0	5.5–7.0
Relative density			
Chlorhexidine gluconate solution	1.06–1.07	1.06–1.07	1.06–1.07
4-Chloroaniline			
Chlorhexidine acetate	—	500 ppm	≤500 ppm
Chlorhexidine gluconate solution	+	≤0.25%	≤500 μg/mL
Chlorhexidine hydrochloride	+	≤500 ppm	≤500 ppm
Related substances	—	+	+
Loss on drying			
Chlorhexidine acetate	—	≤3.5%	≤3.5%
Chlorhexidine hydrochloride	≤2.0%	≤1.0%	≤1.0%
Sulfated ash			
Chlorhexidine acetate	—	≤0.15%	≤0.15%
Chlorhexidine gluconate solution	≤0.1%	—	—
Chlorhexidine hydrochloride	≤0.1%	≤0.1%	≤0.1%
Heavy metals	≤10 ppm	—	—
Arsenic			
Chlorhexidine hydrochloride	≤2 ppm	—	—
Assay			
Chlorhexidine acetate	—	98.0–101.0%	98.0–101.0%
Chlorhexidine gluconate solution	19.0–21.0%	—	19.0–21.0%
Chlorhexidine hydrochloride	≥98.0%	98.0–101.0%	98.0–101.0%

See also Section 17.

10 Typical Properties

Antimicrobial activity Chlorhexidine and its salts exhibit antimicrobial activity against Gram-positive and Gram-negative microorganisms.[3] At the low concentrations normally used for preservation and antisepsis, chlorhexidine salts are rapidly bactericidal. However, species of *Proteus* and *Pseudomonas* are less susceptible to chlorhexidine, which is also inactive against acid-fast bacilli, bacterial spores, and some fungi. Chlorhexidine salts are effective against some lipophilic viruses such as adenovirus, herpes virus, and influenza virus. Optimum antimicrobial activity occurs at pH 5–7. Above pH 8, the chlorhexidine base may precipitate from aqueous solutions.

Bacteria (Gram-positive) Chlorhexidine salts are active against most species; the minimum inhibitory concentration (MIC) is normally in the range 1–10 μg/mL, although much higher concentrations are necessary for *Streptococcus faecalis*. Typical MIC values are shown in Table II.

Table II: Typical minimum inhibitory concentrations (MIC) of chlorhexidine against Gram-positive bacteria.

Microorganism	MIC (μg/mL)
Bacillus spp.	1.0–3.0
Clostridium spp.	1.8–70.0
Corynebacterium spp.	5.0–10.0
Staphylococcus spp.	0.5–6.0
Streptococcus faecalis	2000–5000
Streptococcus spp.	0.1–7.0

Bacteria (Gram-negative) Chlorhexidine salts are less active against Gram-negative species than against Gram-positive species. Typical MICs are 1–15 μg/mL, but pseudomonads, particularly *Pseudomonas aeruginosa*, may be more resistant. *Serratia marcescens* may also be resistant. Combinations of chlorhexidine acetate with the following substances have shown enhanced or more than additive activity towards *Pseudomonas aeruginosa*: benzalkonium chloride; benzyl alcohol; bronopol; edetic acid; phenylethanol, and phenylpropanol.[4,5] Typical MIC values are shown in Table III.

Table III: Typical MIC values of chlorhexidine against Gram-negative bacteria.

Microorganism	MIC (μg/mL)
Escherichia coli	2.5–7.5
Klebsiella spp.	1.5–12.5
Proteus spp.	3–100
Pseudomonas spp.	3–60
Serratia marcescens	3–75
Salmonella spp.	1.6–15

Fungi Chlorhexidine salts are slowly active against molds and yeasts, although they are generally less potent in their inhibitory activity against fungi than against bacteria. Typical MIC values are shown in Table IV.

Table IV: Typical MIC values of chlorhexidine against fungi.

Microorganism	MIC (μg/mL)
Aspergillus spp.	75.0–500.0
Candida albicans	7.0–15.0
Microsporum spp.	12.0–18.0
Penicillium spp.	150.0–200.0
Saccharomyces spp.	50.0–125.0
Trichophyton spp.	2.5–14.0

Spores Chlorhexidine salts are inactive against spores at normal room temperature.[6] At 98–100°C there is some activity against mesophilic spores.

Critical micelle concentration ≈0.6% w/v (depends on other ions in solution).[7]

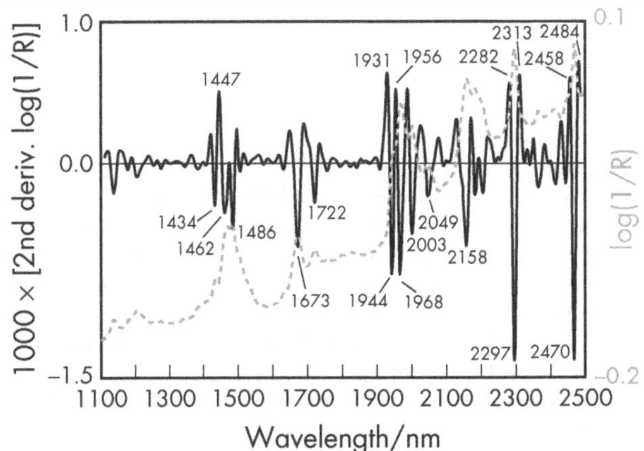

Figure 1: Near-infrared spectrum of chlorhexidine measured by reflectance.

Melting point

132–134°C

See also Section 17 for additional information.

Spectroscopy

NIR spectra *see* Figure 1.

11 Stability and Storage Conditions

Chlorhexidine and its salts are stable at normal storage temperatures when in the powdered form. However, chlorhexidine hydrochloride is hygroscopic, absorbing significant amounts of moisture at temperatures up to 37°C and relative humidities up to 80%.

Heating to 150°C causes decomposition of chlorhexidine and its salts, yielding trace amounts of 4-chloroaniline. However, chlorhexidine hydrochloride is more thermostable than the acetate and can be heated at 150°C for 1 hour without appreciable formation of 4-chloroaniline.

In aqueous solution, chlorhexidine salts may undergo hydrolysis to form 4-chloroaniline, catalyzed by heating and an alkaline pH. Following autoclaving of a 0.02% w/v chlorhexidine gluconate solution at pH 9 for 30 minutes at 120°C, it was found that 1.56% w/w of the original chlorhexidine content had been converted into 4-chloroaniline; for solutions at pH 6.3 and 4.7 the 4-chloroaniline content was 0.27% w/w and 0.13% w/w, respectively, of the original gluconate content.[8] In buffered 0.05% w/v chlorhexidine acetate solutions, maximum stability occurs at pH 5.6.

When chlorhexidine solutions were autoclaved at various time and temperature combinations, the rate of hydrolysis increased markedly above 100°C, and as pH increased or decreased from pH 5.6. At a given pH, chlorhexidine gluconate produced more 4-chloroaniline than the acetate.

It was predicted that in an autoclaved solution containing 0.01% w/v chlorhexidine, the amount of 4-chloroaniline formed would be about 0.00003%. At these low concentrations there would be little likelihood of any toxic hazard as a result of the increase in 4-chloroaniline content in the autoclaved solution.

Chlorhexidine solutions and aqueous-based products may be packaged in glass and high-density polyethylene or polypropylene bottles provided that they are protected from light. If not protected from light, chlorhexidine solutions containing 4-chloroaniline discolor owing to polymerization of the 4-chloroaniline.[9–11]

Cork-based closures or liners should not be used in packaging in contact with chlorhexidine solutions as chlorhexidine salts are inactivated by cork.

As a precaution against contamination with *Pseudomonas* species resistant to chlorhexidine, stock solutions may be protected by the inclusion of 7% w/v ethanol or 4% w/v propan-2-ol.

Chlorhexidine salts, and their solutions, should be stored in well-closed containers, protected from light, in a cool, dry place.

12 Incompatibilities

Chlorhexidine salts are cationic in solution and are therefore incompatible with soaps and other anionic materials. Chlorhexidine salts are compatible with most cationic and nonionic surfactants, but in high concentrations of surfactant chlorhexidine activity can be substantially reduced owing to micellar binding.

Chlorhexidine salts of low aqueous solubility are formed and may precipitate from chlorhexidine solutions of concentration greater than 0.05% w/v, when in the presence of inorganic acids, certain organic acids, and salts (e.g. benzoates, bicarbonates, borates, carbonates, chlorides, citrates, iodides, nitrates, phosphates, and sulfates).[12] At chlorhexidine concentrations below 0.01% w/v precipitation is less likely to occur.

In hard water, insoluble salts may form owing to interaction with calcium and magnesium cations. Solubility may be enhanced by the inclusion of surfactants such as cetrimide.

Other substances incompatible with chlorhexidine salts include viscous materials such as acacia, sodium alginate, sodium carboxymethylcellulose, starch, and tragacanth.[13,14] Also incompatible are brilliant green, chloramphenicol, copper sulfate, fluorescein sodium, formaldehyde, silver nitrate, and zinc sulfate.

Interaction has been reported between chlorhexidine gluconate and the hydrogel poly(2-hydroxyethyl methacrylate), which is a component of some hydrophilic contact lenses.[15,16]

13 Method of Manufacture

Chlorhexidine may be prepared either by condensation of polymethylene bisdicyandiamide with 4-chloroaniline hydrochloride or by condensation of 4-chlorophenyl dicyandiamine with hexamethylenediamine dihydrochloride. Chlorhexidine may also be synthesized from a series of biguanides.[17]

14 Safety

Chlorhexidine and its salts are widely used, primarily as topical disinfectants. As excipients, chlorhexidine salts are mainly used as antimicrobial preservatives in ophthalmic formulations.

Animal studies suggest that the acute oral toxicity of chlorhexidine is low, with little or no absorption from the gastrointestinal tract. However, although humans have consumed up to 2 g of chlorhexidine daily for 1 week, without untoward symptoms, chlorhexidine is not generally used as an excipient in orally ingested formulations.

Reports have suggested that there may be some systemic effects in humans following oral consumption of chlorhexidine.[18–20] Similarly, the topical application of chlorhexidine or its salts produced evidence of very slight percutaneous absorption of chlorhexidine, although the concentrations absorbed were insufficient to produce systemic adverse effects.[21]

Severe hypersensitivity reactions, including anaphylactic shock, have been reported following the topical administration of chlorhexidine,[22–26] although such instances are rare given the extensive use of chlorhexidine and it salts.

In ophthalmic preparations, irritation of the conjunctiva occurs with chlorhexidine solutions of concentration stronger than 0.1% w/v. Accidental eye contact with 4% w/v chlorhexidine gluconate solution may result in corneal damage.[27]

The aqueous concentration of chlorhexidine normally recommended for contact with mucous surfaces is 0.05% w/v. At this concentration, there is no irritant effect on soft tissues, nor is healing delayed. The gluconate salt (1% w/v) is frequently used in creams, lotions, and disinfectant solutions.

Direct instillation of chlorhexidine into the middle ear can result in ototoxicity;[28] when used in dental preparations, staining of teeth and oral lesions may occur.[29,30]

Use of chlorhexidine on the brain or meninges is extremely dangerous.

LD_{50} (mouse, IP): 0.04 g/kg[31]
LD_{50} (mouse, oral): 2.52 g/kg
LD_{50} (rat, IP): 0.06 g/kg
LD_{50} (rat, IV): 0.02 g/kg
LD_{50} (rat, oral): 9.2 g/kg

15 Handling Precautions

Observe normal precautions appropriate to the circumstances and quantity of material handled. The dust of chlorhexidine and its salts may be irritant to the skin, eyes, and respiratory tract. Gloves, eye protection, and a respirator are recommended.

16 Regulatory Status

Chlorhexidine salts are included in nonparenteral and parenteral medicines licensed in the UK. Included in the Canadian Natural Health Products Ingredients Database.

17 Related Substances

Chlorhexidine acetate; chlorhexidine gluconate; chlorhexidine hydrochloride.

Chlorhexidine acetate

Empirical formula $C_{22}H_{30}Cl_2N_{10}{\cdot}2C_2H_4O_2$
Molecular weight 625.64
CAS number [56-95-1]
Synonyms Chlorhexidini acetas; chlorhexidine diacetate; 1,1′-hexamethylenebis[5-(4-chlorophenyl)biguanide] diacetate; *Hibitane diacetate*; *Nolvasan*.
Appearance A white or almost white, microcrystalline powder.
Melting point 154°C
Moisture content Chlorhexidine acetate is hygroscopic, absorbing significant amounts of moisture at relative humidities up to about 80% and temperatures up to 37°C.
Partition coefficients

Mineral oil : water = 0.075;
Peanut oil : water = 0.04.

Solubility Soluble 1 in 15 of ethanol (95%), 1 in 55 of water; slightly soluble in glycerin, propylene glycol and polyethylene glycols.
Safety

LD_{50} (mouse, IP): 0.04 g/kg[31]
LD_{50} (mouse, IV): 0.03 g/kg
LD_{50} (mouse, oral): 2 g/kg
LD_{50} (mouse, SC): 0.33 g/kg

Comments Aqueous solutions may be sterilized by autoclaving; the solutions should not be alkaline or contain other ingredients that affect the stability of chlorhexidine. See Sections 11 and 12.
The EINECS number for chlorhexidine acetate is 200-302-4.

Chlorhexidine gluconate

Empirical formula $C_{22}H_{30}Cl_2N_{10}{\cdot}2C_6H_{12}O_7$
Molecular weight 897.88
CAS number [18472-51-0]
Synonyms Chlorhexidine digluconate; chlorhexidini digluconatis; *Corsodyl*; 1,1′-hexamethylenebis[5-(4-chlorophenyl)biguanide] digluconate; *Hibiclens*; *Hibiscrub*; *Hibitane*; *Unisept*.
Appearance Chlorhexidine gluconate is usually used as an almost colorless or pale yellow-colored aqueous solution.
Acidity/alkalinity pH = 5.5–7.0 for a 5% w/v aqueous dilution.
Solubility Miscible with water; soluble in acetone and ethanol (95%).
Safety

LD_{50} (mouse, IV): 0.02 g/kg[31]
LD_{50} (mouse, oral): 1.8 g/kg
LD_{50} (mouse, SC): 1.14 g/kg
LD_{50} (rat, IV): 0.02 g/kg
LD_{50} (rat, oral): 2 g/kg
LD_{50} (rat, SC): 3.32 g/kg

Comments The commercially available 5% w/v chlorhexidine gluconate solution contains a nonionic surfactant to prevent precipitation and is not suitable for use in body cavities or for the disinfection of surgical instruments containing cemented glass components. Aqueous dilutions of commercially available chlorhexidine gluconate solutions may be sterilized by autoclaving. *See* Sections 11 and 12.
The EINECS number for chlorhexidine gluconate is 242-354-0.

Chlorhexidine hydrochloride

Empirical formula $C_{22}H_{30}Cl_2N_{10}{\cdot}2HCl$
Molecular weight 578.44
CAS number [3697-42-5]
Synonyms Chlorhexidine dihydrochloride; chlorhexidini hydrochloridum; 1,1′-hexamethylenebis[5-(4-chlorophenyl)biguanide]dihydrochloride.
Appearance A white or almost white, crystalline powder.
Melting point 261°C, with decomposition.
Solubility Sparingly soluble in water; very slightly soluble in ethanol (95%); soluble 1 in 50 of propylene glycol.
Safety

LD_{50} (mouse, SC): >5 g/kg[31]

Comments Chlorhexidine hydrochloride may be sterilized by dry heat. *See* Sections 11 and 12.
The EINECS number for chlorhexidine hydrochloride is 223-026-6.

18 Comments

For skin disinfection, chlorhexidine has been formulated as a 0.5% w/v solution in 70% v/v ethanol and, in conjunction with detergents, as a 4% w/v surgical scrub. Chlorhexidine salts may also be used in topical antiseptic creams, mouthwashes, dental gels, and in urology for catheter sterilization and bladder irrigation.[1,2,32,33]

Chlorhexidine salts have additionally been used as constituents of medicated dressings, dusting powders, sprays, and creams.

The EINECS number for chlorhexidine is 200-238-7. The PubChem Compound ID (CID) for chlorhexidine is 9552079.

19 Specific References

1 Juliano C, *et al.* Mucoadhesive alginate matrices containing sodium carboxymethyl starch for buccal drug delivery in *in vitro* and *in vivo* studies. *STP Pharma Sci* 2004; **14**(2): 159–163.
2 Lupuleasa D, *et al.* Bucoadhesive dosage form with chlorhexidine dichlorhydrate. *Farmacia* 2003; **51**(5): 49–55.
3 Prince HN. Drug resistance studies with topical antiseptics. *J Pharm Sci* 1978; **67**: 1629–1631.
4 Richards RME, McBride RJ. Enhancement of benzalkonium chloride and chlorhexidine activity against *Pseudomonas aeruginosa* by aromatic alcohols. *J Pharm Sci* 1973; **62**: 2035–2037.
5 Russell AD, Furr JR. Comparitive sensitivity of smooth, rough and deep rough strains of *Escherichia coli* to chlorhexidine, quaternary ammonium compounds and dibromopropamidine isethionate. *Int J Pharm* 1987; **36**: 191–197.
6 Shaker LA, *et al.* Aspects of the action of chlorhexidine on bacterial spores. *Int J Pharm* 1986; **34**: 51–56.
7 Heard DD, Ashworth RW. The colloidal properties of chlorhexidine and its interaction with some macromolecules. *J Pharm Pharmacol* 1968; **20**: 505–512.
8 Jaminet F, *et al.* [Influence of sterilization temperature and pH on the stability of chlorhexidine solutions.] *Pharm Acta Helv* 1970; **45**: 60–63[in French].

9 Goodall RR, *et al.* Stability of chlorhexidine in solutions. *Pharm J* 1968; **200**: 33–34.
10 Dolby J, *et al.* Stability of chlorhexidine when autoclaving. *Pharm Acta Helv* 1972; **47**: 615–620.
11 Myers JA. Hospital infections caused by contaminated fluids [letter]. *Lancet* 1972; **ii**: 282.
12 Oelschläger H, Canenbley R. Clear indication of chlorhexidine dihydrochloride precipitate in isotonic eye-drops: report based on experience on the use of chlorhexidine as a preservative. *Pharm Ztg* 1983; **128**: 1166–1168.
13 Yousef RT, *et al.* Effect of some pharmaceutical materials on the bactericidal activities of preservatives. *Can J Pharm Sci* 1973; **8**: 54–56.
14 McCarthy TJ, Myburgh JA. The effect of tragacanth gel on preservative activity. *Pharm Weekbl* 1974; **109**: 265–268.
15 Plaut BS, *et al.* The mechanism of interaction between chlorhexidine digluconate and poly(2-hydroxyethyl methacrylate). *J Pharm Pharmacol* 1981; **33**: 82–88.
16 Stevens LE, *et al.* Analysis of chlorhexidine sorption in soft contact lenses by catalytic oxidation of [^{14}C]chlorhexidine and by liquid chromatography. *J Pharm Sci* 1986; **75**: 83–86.
17 Rose FL, Swain G. Bisguanides having antibacterial activity. *J Chem Soc* 1956; 4422–4425.
18 Massano G, *et al.* Striking aminotransferase rise after chlorhexidine self-poisoning [letter]. *Lancet* 1982; **i**: 289.
19 Emerson D, Pierce C. A case of a single ingestion of 4% *Hibiclens*. *Vet Hum Toxicol* 1988; **30**: 583.
20 Quinn MW, Bini RM. Bradycardia associated with chlorhexidine spray [letter]. *Arch Dis Child* 1989; **64**: 892–893.
21 Alder VG, *et al.* Comparison of hexachlorophane and chlorhexidine powders in prevention of neonatal infection. *Arch Dis Child* 1980; **55**: 277–280.
22 Lockhart AS, Harle CC. Anaphylactic reactions due to chlorhexidine allergy [letter]. *Br J Anaesth* 2001; **87**(6): 940–941.
23 Wahlberg JE, Wennersten G. Hypersensitivity and photosensitivity to chlorhexidine. *Dermatologica* 1971; **143**: 376–379.
24 Okano M, *et al.* Anaphylactic symptoms due to chlorhexidine gluconate. *Arch Dermatol* 1989; **125**: 50–52.
25 Evans RJ. Acute anaphylaxis due to topical chlorhexidine acetate. *Br Med J* 1992; **304**: 686.
26 Chisholm DG, *et al.* Intranasal chlorhexidine resulting in an anaphylactic circulatory arrest. *Br Med J* 1997; **315**: 785.
27 Tabor E, *et al.* Corneal damage due to eye contact with chlorhexidine gluconate [letter]. *J Am Med Assoc* 1989; **261**: 557–558.
28 Honigman JL. Disinfectant ototoxicity [letter]. *Pharm J* 1975; **215**: 523.
29 Addy M, *et al.* Extrinsic tooth discoloration by metals and chlorhexidine I: surface protein denaturation or dietary precipitation? *Br Dent J* 1985; **159**: 281–285.
30 Addy M, Moran J. Extrinsic tooth discoloration by metals and chlorhexidine II: clinical staining produced by chlorhexidine, iron and tea. *Br Dent J* 1985; **159**: 331–334.
31 Lewis RJ, ed. *Sax's Dangerous Properties of Industrial Materials*, 11th edn. New York: Wiley, 2004: 471.
32 Leyes Borrajo JL, *et al.* Efficacy of chlorhexidine mouthrinses with and without alcohol: a clinical study. *J Peridontol* 2002; **73**(3): 317–321.
33 Alaki SM, *et al.* Preventing the transfer of *Streptococcus mutans* from primary molars to permanent first molars using chlorhexidine. *Pediatr Dent* 2002; **24**(2): 103–108.

20 General References

Davies GE, *et al.* 1,6-Di-4′-chlorophenyldiguanidohexane (*Hibitane*): laboratory investigation of a new antibacterial agent of high potency. *Br J Pharmacol Chemother* 1954; **9**: 192–196.
McCarthy TJ. The influence of insoluble powders on preservatives in solution. *J Mond Pharm* 1969; **12**: 321–328.
Schnuch A, *et al.* Contact allergy to preservatives. Analysis of IVDK data 1996-2009. *Br J Dermatol* 2011; **164**: 1316–1325.
Senior N. Some observations on the formulation and properties of chlorhexidine. *J Soc Cosmet Chem* 1973; **24**: 259–278.
Sweetman SC, ed. *Martindale: the Complete Drug Reference*, 36th edn. London: Pharmaceutical Press, 2009: 1635.

21 Author

L Peltonen.

22 Date of Revision

1 March 2012.

Chlorobutanol

1 Nonproprietary Names

BP: Chlorobutanol
JP: Chlorobutanol
PhEur: Chlorobutanol, Anhydrous
USP–NF: Chlorobutanol

2 Synonyms

Acetone chloroform; anhydrous chlorbutol; chlorbutanol; chlorobutanolum anhydricum; chlorbutol; chloretone; *Coliquifilm*; methaform; sedaform; trichloro-*tert*-butanol; β,β,β-trichloro-*tert*-butyl alcohol; trichloro-*t*-butyl alcohol.

3 Chemical Name and CAS Registry Number

1,1,1-Trichloro-2-methyl-2-propanol [57-15-8]
1,1,1-Trichloro-2-methyl-2-propanol hemihydrate [6001-64-5]

4 Empirical Formula and Molecular Weight

$C_4H_7Cl_3O$	177.46 (for anhydrous)
$C_4H_7Cl_3O \cdot \frac{1}{2}H_2O$	186.46 (for hemihydrate)

5 Structural Formula

Cl Cl
H_3C
Cl
HO CH_3

6 Functional Category

Antimicrobial preservative; plasticizer.

7 Applications in Pharmaceutical Formulation or Technology

Chlorobutanol is primarily used in ophthalmic or parenteral dosage forms as an antimicrobial preservative at concentrations up to 0.5% w/v; *see* Section 10. It is commonly used as an antibacterial agent for epinephrine solutions, posterior pituitary extract solutions, and ophthalmic preparations intended for the treatment of miosis. It is especially useful as an antibacterial agent in nonaqueous formulations. Chlorobutanol is also used as a preservative in cosmetics (*see* Section 16); as a plasticizer for cellulose esters and ethers.

8 Description

Volatile, colorless or white crystals with a musty, camphoraceous odor.

9 Pharmacopeial Specifications

See Table I.

Table I: Pharmacopeial specifications for chlorobutanol.

Test	JP XV	PhEur 7.4	USP35–NF30
Identification	+	+	+
Characters	+	+	–
Appearance of solution	–	+	–
Melting point	≥76°C	+	–
Anhydrous	–	≈95°C	–
Hemihydrate	–	≈78°C	–
Acidity	+	+	+
Water (anhydrous form)	≤6.0%	≤1.0%	≤1.0%
Hemihydrate	–	4.5–5.5%	≤6.0%
Chloride	≤0.071%	+	≤0.07%
Anhydrous	–	≤300 ppm	–
Hemihydrate	–	≤100 ppm	–
Residue on ignition	≤0.10%	–	–
Sulfated ash	–	≤0.1%	–
Assay (anhydrous basis)	≥98.0%	98.0–101.0%	98.0–100.5%

Note: the JP XV and USP35–NF30 allow either the anhydrous form or the hemihydrate; the PhEur includes them as separate monographs.

10 Typical Properties

Antimicrobial activity Chlorobutanol has both antibacterial and antifungal properties. It is effective against Gram-positive and Gram-negative bacteria and some fungi, e.g. *Candida albicans*, *Pseudomonas aeruginosa*, and *Staphylococcus albus*. Antimicrobial activity is bacteriostatic, rather than bactericidal, and is considerably reduced above pH 5.5. In addition, activity may also be reduced by increasing heat and by incompatibilities between chlorobutanol and other excipients or packaging materials; *see* Sections 11 and 12. However, activity may be increased by combination with other antimicrobial preservatives; *see* Section 18. Typical minimum inhibitory concentrations (MICs) are: Gram-positive bacteria 650 μg/mL; Gram-negative bacteria 1000 μg/mL; yeasts 2500 μg/mL; fungi 5000 μg/mL.

Boiling point 167°C

Melting point

76–78°C for the hemihydrate;

95–97°C for the anhydrous form.

Partition coefficient Octanol : water log k_{ow} = 2.03

Refractive index n_D^{25} = 1.4339

Solubility *see* Table II.

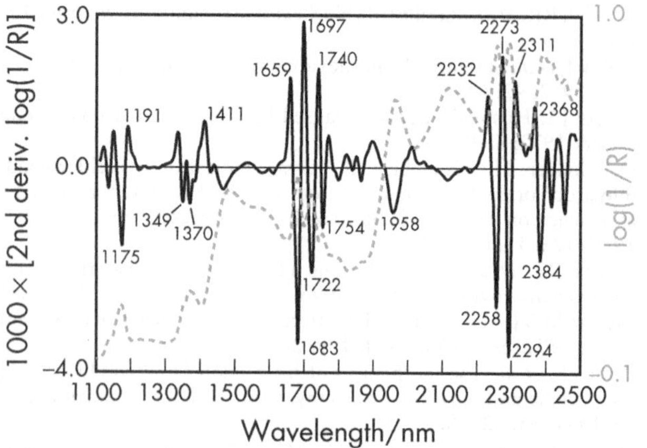

Figure 1: Near-infrared spectrum of chlorobutanol measured by reflectance.

Table II: Solubility of chlorobutanol.

Solvent	Solubility at 20°C unless otherwise stated
Acetic acid, glacial	Freely soluble
Acetone	Freely soluble
Chloroform	Freely soluble
Ethanol (95%)	Freely soluble (1 in 1)
Ether	Freely soluble
Glycerin	Freely soluble (1 in 10)
Methanol	Freely soluble
Volatile oils	Freely soluble
Water	Slightly soluble (1 in 125) Freely soluble in hot water

Spectroscopy

NIR spectra *see* Figure 1.

11 Stability and Storage Conditions

Chlorobutanol is volatile and readily sublimes. In aqueous solution, degradation to carbon monoxide, acetone and chloride ion is catalyzed by hydroxide ions. Stability is good at pH 3 but becomes progressively worse with increasing pH.[1] The half-life at pH 7.5 for a chlorobutanol solution stored at 25°C was determined to be approximately 3 months.[2] In a 0.5% w/v aqueous chlorobutanol solution at room temperature, chlorobutanol is almost saturated and may crystallize out of solution if the temperature is reduced.

Losses of chlorobutanol also occur owing to its volatility, with appreciable amounts being lost during autoclaving; at pH 5 about 30% of chlorobutanol is lost.[3] Porous containers result in losses from solutions, and polyethylene containers result in rapid loss. Losses of chlorobutanol during autoclaving in polyethylene containers may be reduced by pre-autoclaving the containers in a solution of chlorobutanol; the containers should then be used immediately.[4] There is also appreciable loss of chlorobutanol through stoppers in parenteral vials.

The bulk material should be stored in an airtight container at a temperature of 8–15°C.

12 Incompatibilities

Owing to problems associated with sorption, chlorobutanol is incompatible with plastic vials,[4–8] rubber stoppers, bentonite,[9] magnesium trisilicate,[9] polyethylene, and polyhydroxyethylmethacrylate, which has been used in soft contact lenses.[10] To a lesser extent, carboxymethylcellulose and polysorbate 80 reduce antimicrobial activity by sorption or complex formation.

13 Method of Manufacture

Chlorobutanol is prepared by condensing acetone and chloroform in the presence of solid potassium hydroxide.

14 Safety

Chlorobutanol is widely used as a preservative in a number of pharmaceutical formulations, particularly ophthalmic preparations. Although animal studies have suggested that chlorobutanol may be harmful to the eye, in practice the widespread use of chlorobutanol as a preservative in ophthalmic preparations has been associated with few reports of adverse reactions. A study of the irritation potential of a local anesthetic on the murine cornea indicated significant corneal surface damage in the presence of 0.5% w/v chlorobutanol, which may be related to the preservative's effective concentration.(11) Reported adverse reactions to chlorobutanol include: cardiovascular effects following intravenous administration of heparin sodium injection preserved with chlorobutanol;(12) neurological effects following administration of a large dose of morphine infusion preserved with chlorobutanol;(13) and hypersensitivity reactions, although these are regarded as rare.(14–16)

The lethal human dose of chlorobutanol is estimated to be 50–500 mg/kg.(17)

LD_{Lo} (dog, oral): 0.24 g/kg(18,19)
LD_{50} (mouse, oral): 0.15 g/kg(18)
LD_{Lo} (rabbit, oral): 0.21 g/kg(18,19)
LD_{50} (rat, oral): 0.51 g/kg(18)

15 Handling Precautions

Observe normal precautions appropriate to the circumstances and quantity of material handled. Chlorobutanol may be irritant to the skin, eyes, and mucous membranes. Eye protection and gloves are recommended along with a respirator in poorly ventilated environments. There is a slight fire hazard on exposure to heat or flame.

16 Regulatory Status

Included in the FDA Inactive Ingredients Database (IM, IV, and SC injection; inhalations; nasal, otic, ophthalmic, and topical preparations). Labeling must state 'contains chlorobutanol up to 0.5%.' Included in nonparenteral and parenteral medicines licensed in the UK. Included in the Canadian Natural Health Products Ingredients Database.

In the UK, the maximum concentration of chlorobutanol permitted for use in cosmetics, other than foams, is 0.5%. It is not suitable for use in aerosols.

17 Related Substances

Phenoxyethanol; phenylethyl alcohol.

18 Comments

It has been reported that a combination of chlorobutanol and phenylethanol, both at 0.5% w/v concentration, has shown greater antibacterial activity than either compound alone. An advantage of the use of this combination is that chlorobutanol dissolves in the alcohol; the resulting liquid can then be dissolved in an aqueous pharmaceutical preparation without the application of heat.

Chlorobutanol has been used therapeutically as a mild sedative and local analgesic in dentistry.

The EINECS number for chlorobutanol is 200-317-6. The PubChem ID Compound (CID) for chlorobutanol is 5977.

19 Specific References

1 Patwa NV, Huyck CL. In aqueous solutions—stability of chlorobutanol. *J Am Pharm Assoc* 1966; **NS6**: 372–373.
2 Nair AD, Lach JL. The kinetics of degradation of chlorobutanol. *J Am Pharm Assoc Am Pharm Ass (Baltim)* 1959; **48**: 390–395.
3 Lang JC *et al.* Design and evaluation of ophthalmic pharmaceutical products. In: Banker GS, Rhodes CT, eds. *Modern Pharmaceutics*, 4th edn. New York: Marcel Dekker, 2002: 415–478.
4 Blackburn HD, *et al.* The effect of container pre-treatment on the interaction between chlorbutol and polyethylene during autoclaving. *Aust J Hosp Pharm* 1983; **13**: 153–156.
5 Lachman L, *et al.* Stability of antibacterial preservatives in parenteral solutions I: factors influencing the loss of antimicrobial agents from solutions in rubber-stoppered containers. *J Pharm Sci* 1962; **51**: 224–232.
6 Friesen WT, Plein EM. The antibacterial stability of chlorobutanol stored in polyethylene bottles. *Am J Hosp Pharm* 1971; **28**: 507–512.
7 Blackburn HD, *et al.* Preservation of ophthalmic solutions: some observations on the use of chlorbutol in plastic containers [letter]. *J Pharm Pharmacol* 1978; **30**: 666.
8 Holdsworth DG, *et al.* Fate of chlorbutol during storage in polyethylene dropper containers and simulated patient use. *J Clin Hosp Pharm* 1984; **9**: 29–39.
9 Yousef RT, *et al.* Effect of some pharmaceutical materials on the bactericidal activities of preservatives. *Can J Pharm Sci* 1973; **8**: 54–56.
10 Richardson NE, *et al.* The interaction of preservatives with polyhydroxyethylmethacrylate (polyHEMA). *J Pharm Pharmacol* 1978; **30**: 469–475.
11 Kalin P, *et al.* Determination of the influence of preservatives on the irritation potential of a local anaesthetic on murine cornea. *Eur J Pharm Biopharm* 1996; **42**: 402–404.
12 Bowler GMR, *et al.* Sharp fall in blood pressure after injection of heparin containing chlorbutol [letter]. *Lancet* 1986; **i**: 848–849.
13 DeChristoforo R, *et al.* High-dose morphine infusion complicated by chlorobutanol-induced somnolence. *Ann Intern Med* 1983; **98**: 335–336.
14 Dux S, *et al.* Hypersensitivity reaction to chlorobutanol-preserved heparin [letter]. *Lancet* 1981; **i**: 149.
15 Itabashi A, *et al.* Hypersensitivity to chlorobutanol in DDAVP solution [letter]. *Lancet* 1982; **i**: 108.
16 Hofmann H, *et al.* Anaphylactic shock from chlorobutanol-preserved oxytocin. *Contact Dermatitis* 1986; **15**: 241.
17 Gosselin RE *et al. Clinical Toxicology of Commercial Products*, 4th edn. Baltimore: Williams & Wilkins, 1976: II-119.
18 Sweet DV, ed. *Registry of Toxic Effects of Chemical Substances*. Cincinnati: US Department of Health, 1987: 3838.
19 Lewis RJ, ed. *Sax's Dangerous Properties of Industrial Materials*, 11th edn. New York: Wiley, 2004: 784.

20 General References

Summers QA, *et al.* A non-bronchoconstrictor, bacteriostatic preservative for nebuliser solutions. *Br J Clin Pharmacol* 1991; **31**: 204–206.

21 Author

BA Hanson.

22 Date of Revision

1 March 2012.

Chlorocresol

1 Nonproprietary Names

BP: Chlorocresol

PhEur: Chlorocresol

USP–NF: Chlorocresol

2 Synonyms

Aptal; *Baktol*; chlorocresolum; 4-chloro-*m*-cresol; *p*-chloro-*m*-cresol; 1-chloro-4-hydroxy-2-methylbenzene; 2-chloro-5-hydroxytoluene; 6-chloro-3-hydroxytoluene; 4-chloro-3-methylphenol; 3-methyl-4-chlorophenol; *Nipacide PC*; parachlorometacresol; PCMC.

3 Chemical Name and CAS Registry Number

4-Chloro-3-methylphenol [59-50-7]

4 Empirical Formula and Molecular Weight

C_7H_7ClO 142.58

5 Structural Formula

OH

CH_3

Cl

6 Functional Category

Antimicrobial preservative.

7 Applications in Pharmaceutical Formulation or Technology

Chlorocresol is used as an antimicrobial preservative in cosmetics and pharmaceutical formulations. It is generally used in concentrations up to 0.2% in a variety of preparations except those intended for oral administration or that contact mucous membrane. Chlorocresol is effective against bacteria, spores, molds, and yeasts; it is most active in acidic media. Preservative efficacy may be reduced in the presence of some other excipients, particularly nonionic surfactants; *see* Sections 10 and 12.

In higher concentrations, chlorocresol is an effective disinfectant. *See* Table I.

Table I: Uses of chlorocresol.

Use	Concentration (%)
Eye drops	0.05
Injections	0.1
Shampoos and other cosmetics	0.1–0.2
Topical creams and emulsions	0.075–0.2

8 Description

Colorless or almost colorless, dimorphous crystals or crystalline powder with a characteristic phenolic odor.

9 Pharmacopeial Specifications

See Table II. *See also* Section 18.

Table II: Pharmacopeial specifications for chlorocresol.

Test	PhEur 7.4	USP35–NF30
Identification	+	+
Characters	+	−
Appearance of solution	+	−
Completeness of solution	−	+
Melting range	64–67°C	63–66°C
Nonvolatile residue	≤0.1%	≤0.1%
Acidity	+	−
Related substances	≤1.0%	−
Assay	98.0–101.0%	99.0–101.0%

10 Typical Properties

Acidity/alkalinity pH = 5.6 for a saturated aqueous solution

Antimicrobial activity Chlorocresol has bactericidal activity against both Gram-positive and Gram-negative organisms (including *Pseudomonas aeruginosa*), spores, molds, and yeasts. It is most active in acidic solutions, with antimicrobial effectiveness decreasing with increasing pH; it is inactive above pH 9. Antimicrobial activity may also be reduced by loss of chlorocresol from a formulation due to incompatibilities with packaging materials or other excipients, such as nonionic surfactants; *see* Section 12. Synergistic antimicrobial effects between chlorocresol and other antimicrobial preservatives, such as 2-phenylethanol, have been reported.[1,2] Reported minimum inhibitory concentrations (MICs) for chlorocresol are shown in Table III.[3] Like most antimicrobials, chlorocresol has a non-linear dose response.[4,5]

Table III: Minimum inhibitory concentrations (MICs) for chlorocresol.[3]

Microorganism	MIC (μg/mL)
Aspergillus niger	2500
Candida albicans	2500
Escherichia coli	1250
Klebsiella pneumoniae	625
Pseudomonas aeruginosa	1250
Pseudomonas fluorescens	1250
Staphylococcus aureus	625

Bacteria Concentrations of approximately 0.08%, with a contact time of 10 minutes, are bactericidal. A typical MIC is 0.02%.

Fungi Chlorocresol is active against molds and yeasts. Fungicidal concentrations (after 24 hours of contact) are in the range 0.01–0.04%.

Spores At temperatures of 80°C or above and in concentrations greater than 0.012%, chlorocresol is active against spores. It is much less active at room temperature. Heating at 98–100°C for 30 minutes in the presence of 0.2% chlorocresol has previously been used as a compendial method for the sterilization of solutions of substances that would not withstand autoclaving.

Boiling point 235°C

Dissociation constant $pK_a = 9.55$

Flash point 117.8°C

Melting point Dimorphous crystals with a melting point of 55.5°C and 65°C.
Partition coefficients

Cyclohexane/water = 0.15;

Hexane : water = 0.34;

Liquid paraffin : water = 1.53;

Octanol : water = 3.10;

Peanut oil : water = 117.

Refractive index 1.5403
Solubility *see* Table IV.

Table IV: Solubility of chlorocresol.

Solvent	Solubility at 20°C unless otherwise stated
Acetone	Soluble
Alkali hydroxide solutions	Soluble
Chloroform	Soluble
Ethanol	1 in 0.4
Ether	Soluble
Fixed oils	Soluble
Glycerin	Soluble
Terpenes	Soluble
Water	1 in 260[a]
	1 in 50 at 100°C[a]

(a) Aqueous solubility is decreased in the presence of electrolytes, particularly sodium chloride, potassium chloride, and potassium sulfonate.[6]

Specific gravity 1.37 at 20°C
Spectroscopy

NIR spectra *see* Figure 1.

Vapor pressure

0.008 kPa at 20°C;

0.67 kPa at 100°C.

11 Stability and Storage Conditions

Chlorocresol is stable at room temperature but is volatile in steam. Aqueous solutions may be sterilized by autoclaving. On exposure to air and light, aqueous solutions may become yellow colored. Solutions in oil or glycerin may be sterilized by heating at 160°C for 1 hour. The bulk material should be stored in a well-closed container, protected from light, in a cool, dry place.

12 Incompatibilities

Chlorocresol can decompose on contact with strong alkalis, evolving heat and fumes that ignite explosively. It is also incompatible with oxidizing agents, copper, and with solutions of calcium chloride, codeine phosphate, diamorphine hydrochloride, papaveretum, and quinine hydrochloride.[7] Chlorocresol is corrosive to metals and forms complex compounds with transition metal ions; discoloration occurs with iron salts. Chlorocresol also exhibits strong sorption or binding tendencies to organic materials such as rubber, certain plastics, and nonionic surfactants.[8–11]

Chlorocresol may be lost from solutions to rubber closures, and in contact with polyethylene may initially be rapidly removed by sorption and then by permeation, the uptake being temperature dependent. Presoaking of components may reduce losses due to sorption, but not those by permeation.[12,13] Chlorocresol may also be taken up by polymethylmethacrylate and by cellulose acetate. Losses to polypropylene or rigid polyvinyl chloride are usually small.[14]

At a concentration of 0.1%, chlorocresol may be completely inactivated in the presence of nonionic surfactants, such as polysorbate 80.[9] However, other studies have suggested an enhancement of antimicrobial properties in the presence of surfactants.[15,16] Bactericidal activity is also reduced, due to binding, by cetomacrogol, methylcellulose, pectin, or cellulose derivatives.[9,11] In emulsified or solubilized systems, chlorocresol readily partitions into the oil phase, particularly into vegetable oils, and higher concentrations will be required for efficient preservation.[10,17]

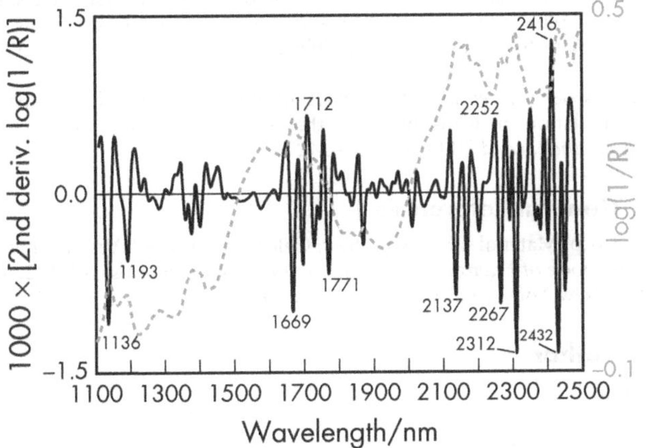

Figure 1: Near-infrared spectrum of chlorocresol measured by reflectance.

13 Method of Manufacture

Chlorocresol is prepared by the chlorination of *m*-cresol.

14 Safety

Chlorocresol is used primarily as a preservative in topical pharmaceutical formulations but has also been used in nebulized solutions[18] and ophthalmic and parenteral preparations. It should not, however, be used in formulations for intrathecal, intracisternal, or peridural injection.

Chlorocresol is metabolized by conjugation with glucuronic acid and sulfate and is excreted in the urine, mainly as the conjugate, with little chlorocresol being excreted unchanged.

Although less toxic than phenol, chlorocresol may be irritant to the skin, eyes, and mucous membranes, and has been reported to cause some adverse reactions when used as an excipient.[19,20]

Sensitization reactions may follow the prolonged application of strong solutions to the skin, although patch tests have shown that chlorocresol is not a primary irritant at concentrations up to 0.2%. Chlorocresol is recognized as a rare cause of allergic contact dermatitis.[21] Cross sensitization with the related preservative chloroxylenol has also been reported.[22,23] At concentrations of 0.005% w/v, chlorocresol has been shown to produce a reversible reduction in the ciliary movement of human nasal epithelial cells *in vitro*, and at concentrations of 0.1% chlorocresol produces irreversible ciliostasis; therefore it should be used with caution in nasal preparations.[24] However, a clinical study in asthma patients challenged with chlorocresol or saline concluded that preservative might be used safely in nebulizer solution.[18]

Chlorocresol is approved as safe for use in cosmetics in Europe at a maximum concentration of 0.2%, although not in products intended to come in contact with mucous membranes.[25]

Chlorocresol at a concentration as low as 0.05% produces ocular irritation in rabbits.[20] Despite such reports, chlorocresol has been tested in ophthalmic preparations.[26,27]

When used systemically, notably in a heparin injection preserved with chlorocresol 0.15%, delayed irritant and hypersensitivity reactions attributed to chlorocresol have been reported.[28,29] *See also* Section 18.

LD_{50} (mouse, IV): 0.07 g/kg[30]

LD_{50} (mouse, oral): 0.6 g/kg

LD_{50} (mouse, SC): 0.36 g/kg

LD_{50} (rabbit, dermal): >5 g/kg
LD_{50} (rat, dermal): >2 g/kg
LD_{50} (rat, oral): 1.83 g/kg
LD_{50} (rat, SC): 0.4 g/kg

C

15 Handling Precautions

Observe normal precautions appropriate to the circumstances and quantity of material handled. Chlorocresol can be irritant to the skin, eyes, and mucous membranes. Eye protection, gloves, and protective clothing are recommended. Chlorocresol presents a slight fire hazard when exposed to heat or flame. It burns to produce highly toxic fumes containing phosgene and hydrogen chloride.

16 Regulatory Status

Included in the FDA Inactive Ingredients Database (topical creams and emulsions). Included in nonparenteral and parenteral medicines licensed in the UK. Included in the Canadian Natural Health Products Ingredients Database.

In Europe, chlorocresol is approved for use in cosmetics at a maximum concentration of 0.2%; however, it is prohibited for use in products intended to come into contact with mucous membranes. In Japan, use of chlorocresol in cosmetics is restricted to a level of 0.5 g/100 g.

17 Related Substances

Cresol; chloroxylenol.

18 Comments

A specification for chlorocresol is contained in the *Japanese Pharmaceutical Excipients* (JPE).[31] The *Japanese Pharmaceutical Excipient Directory* (JPED) states a maximum concentration of 1 mg/g of chlorocresol in external pharmaceutical preparations.[32]

Chlorocresol has a characteristic odor which is difficult to mask in formulations, even at concentrations of 0.05–0.1%.

Although used in Europe, chlorocresol is not used in the US in parenteral formulations. Chlorocresol has also been used as an experimental *in vitro* diagnostic agent for the diagnosis of hyperthermia.[33,34]

The EINECS number for chlorocresol is 200-431-6.

19 Specific References

1 Denyer SP, *et al.* The biochemical basis of synergy between the antibacterial agents, chlorocresol and 2-phenylethanol. *Int J Pharm* 1986; **29**: 29–36.
2 Abdelaziz AA, El-Nakeeb MA. Sporicidal activity of local anaesthetics and their binary combinations with preservatives. *J Clin Pharm Ther* 1988; **13**: 249–256.
3 Wallhäusser KH. *p*-Chloro-*m*-cresol. In: Kabara JJ, ed. *Cosmetic and Drug Preservation Principles and Practice.* New York: Marcel Dekker, 1984: 683–684.
4 Lambert RJW, *et al.* Membrane damage to bacteria caused by single and combined biocides. *J Appl Microbiol* 2003; **94**: 1015–1023.
5 Lambert RJW, *et al.* Theory of antimicrobial combinations: biocide mixtures—synergy or additions? *J Appl Microbiol* 2003; **94**: 747–759.
6 Gadalla MAF, *et al.* Effect of electrolytes on the solubility and solubilization of chlorocresol. *Pharmazie* 1974; **29**: 105–107.
7 McEwan JS, Macmorran GH. The compatibility of some bactericides. *Pharm J* 1947; **158**: 260–262.
8 Yousef RT, *et al.* Effect of some pharmaceutical materials on the bactericidal activities of preservatives. *Can J Pharm Sci* 1973; **8**: 54–56.
9 McCarthy TJ. Dissolution of chlorocresol from various pharmaceutical formulations. *Pharm Weekbl* 1975; **110**: 101–106.
10 Kazmi SJA, Mitchell AG. Preservation of solubilized and emulsified systems I: correlation of mathematically predicted preservative availability with antimicrobial activity. *J Pharm Sci* 1978; **67**: 1260–1266.
11 Kazmi SJA, Mitchell AG. Preservation of solubilized and emulsified systems II: theoretical development of capacity and its role in antimicrobial activity of chlorocresol in cetomacrogol-stabilized systems. *J Pharm Sci* 1978; **67**: 1266–1271.
12 McCarthy TJ. Interaction between aqueous preservative solutions and their plastic containers III. *Pharm Weekbl* 1972; **107**: 1–7.
13 Roberts MS, *et al.* The storage of selected substances in aqueous solution in polyethylene containers: the effect of some physicochemical factors on the disappearance kinetics of the substances. *Int J Pharm* 1979; **2**: 295–306.
14 McCarthy TJ. Interaction between aqueous preservative solutions and their plastic containers. *Pharm Weekbl* 1970; **105**: 557–563.
15 Kurup TRR, *et al.* Preservative requirements in emulsions. *Pharma Acta Helv* 1992; **67**: 204–208.
16 Kurup TRR, *et al.* Effect of surfactants on the antimicrobial activity of preservatives. *Pharma Acta Helv* 1991; **66**: 274–280.
17 Sznitowska M, *et al.* Physicochemical screening of antimicrobial agents as potential preservatives for submicron emulsions. *Eur J Pharm Sci* 2002; **15**: 489–495.
18 Summers QA, *et al.* A non-bronchoconstrictor, bacteriostatic preservative for nebuliser solutions. *Br J Clin Pharmacol* 1991; **31**: 204–206.
19 Smolinske SC. *Handbook of Food, Drug and Cosmetic Excipients.* Boca Raton, FL: CRC Press, 1992: 87–90.
20 Zondlo M. Final report on the safety assessment of *p*-chloro-*m*-cresol. *J Toxicol* 1997; **16**: 235–268.
21 Jong CT, *et al.* Contact sensitivity to preservatives in the UK, 2004–2005: results of multicenter study. *Contact Dermatitis* 2007; **57**: 165–168.
22 Burry JN, *et al.* Chlorocresol sensitivity. *Contact Dermatitis* 1975; **1**: 41–42.
23 Andersen KE, Hamann K. How sensitizing is chlorocresol? Allergy tests in guinea pigs vs. the clinical experience. *Contact Dermatitis* 1984; **11**: 11–20.
24 Agu RU, *et al.* Effects of pharmaceutical compounds on ciliary beating in human nasal epithelial cells: a comparative study of cell culture models. *Pharm Res* 1999; **16**: 1380–1385.
25 Andersen A. Final report on the safety assessment of sodium *p*-chloro-*m*-cresol, *p*-chloro-*m*-cresol, chlorothymol, mixed cresols, *m*-cresol, *o*-cresol, *p*-cresol, isopropyl cresols, thymol, *o*-cymen-5-ol, and carvacrol. *Int J Toxicol* 2006; **25**(Suppl. 1): 29–127.
26 Palanichamy S, *et al.* Preservation of sodium chloride eye lotion BPC against contamination with Pseudomonas aeruginosa. *Indian Drugs* 1982; **19**: 153–155.
27 Palanichamy S, *et al.* Preservation of compound zinc sulfate eye lotion BPC 1963 against contamination with *Pseudomonas aeruginosa. Indian J Hosp Pharm* 1982; **19**: 64–65.
28 Hancock BW, Naysmith A. Hypersensitivity to chlorocresol-preserved heparin. *Br Med J* 1975; **3**: 746–747.
29 Ainley EJ, *et al.* Adverse reaction to chlorocresol-preserved heparin [letter]. *Lancet* 1977; **i**: 705.
30 Lewis RJ, ed. *Sax's Dangerous Properties of Industrial Materials*, 11th edn. New York: Wiley, 2004: 790.
31 Japanese Pharmaceutical Excipients Council. *Japanese Pharmaceutical Excipients 2004.* Tokyo: Yakuji Nippo, 2004: 171.
32 Japanese Pharmaceutical Excipients Council. *Japanese Pharmaceutical Excipient Directory.* Tokyo: Yakuji Nippo, 2005: 96.
33 Wappler F, *et al.* Multicenter evaluation of *in vitro* contracture testing with bolus administration of 4-chloro-*m*-cresol for diagnosis of malignant hyperthermia susceptibility. *Eur J Anaesth* 2003; **20**: 528–536.
34 Anetseder M, *et al.* The impact of 4-chloro-*m*-cresol in heparin formulation on malignant hyperthermia: *in vitro* and *in vivo*. *Acta Anaesthes Scand* 2000; **44**: 338–342.

20 General References

Aldon Corp. Material safety data sheet: Chlorocresol, 1999. Verschueren K. *Handbook of Environmental Data on Organic Chemicals*, vol. 1–2. 4th edn. New York: John Wiley & Sons, 2001: 491.

21 Author

S Nema.

22 Date of Revision

1 March 2012.

Chlorodifluoroethane (HCFC)

1 Nonproprietary Names

None adopted.

2 Synonyms

1,1-Difluoro-1-chloroethane; *Dymel 142b*; *Genetron 142b*; HCFC 142b; P-142b; propellant 142b; refrigerant 142b; *Solkane 142b*.

3 Chemical Name and CAS Registry Number

1-Chloro-1,1-difluoroethane [75-68-3]

4 Empirical Formula and Molecular Weight

$C_2H_3ClF_2$ 100.50

5 Structural Formula

```
      F     H
      |     |
Cl----C-----C----H
      |     |
      F     H
```

6 Functional Category

Aerosol propellant.

7 Applications in Pharmaceutical Formulation or Technology

Chlorodifluoroethane is a hydrochlorofluorocarbon (HCFC) aerosol propellant previously used in topical pharmaceutical formulations. However, it is no longer permitted for use in pharmaceutical formulations because of its harmful effects on the environment. It was also generally used in conjunction with difluoroethane to form a propellant blend with a specific gravity of 1. Chlorodifluoroethane was also used in combination with chlorodifluoromethane and hydrocarbon propellants.

8 Description

Chlorodifluoroethane is a liquefied gas and exists as a liquid at room temperature when contained under its own vapor pressure, or as a gas when exposed to room temperature and atmospheric pressure. The liquid is practically odorless and colorless. Chlorodifluoroethane is noncorrosive and nonirritating.

9 Pharmacopeial Specifications

—

10 Typical Properties

Autoignition temperature 632°C
Boiling point −9.8°C
Critical temperature 137.1°C
Density
1.11 g/cm^3 for liquid at 25°C;
1.03 g/cm^3 for liquid at 54.5°C.
Flammability Flammable. Limits of flammability 6.2–17.9% v/v in air.
Melting point −131°C
Solubility Soluble 1 in 715 parts of water at 20°C.
Vapor density (absolute) 4.487 g/m^3 at standard temperature and pressure.
Vapor density (relative) 3.48 (air = 1)
Vapor pressure
339 kPa (49.2 psia) at 25°C (29.1 psig at 21.1°C);
772 kPa (112.0 psia) at 54.5°C.
Viscosity (dynamic) 0.33 mPa s (0.33 cP) for liquid at 21°C.

11 Stability and Storage Conditions

Chlorodifluoroethane is a nonreactive and stable material. The liquefied gas is stable and should be stored in a metal cylinder in a cool, dry place.

12 Incompatibilities

Compatible with the usual ingredients used in the formulation of pharmaceutical aerosols. Chlorodifluoroethane can react vigorously with oxidizing materials.

13 Method of Manufacture

Chlorodifluoroethane is prepared by the chlorination of difluoroethane in the presence of a suitable catalyst; hydrochloric acid is also formed. The chlorodifluoroethane is purified to remove all traces of water and hydrochloric acid, as well as traces of the starting and intermediate materials.

14 Safety

Chlorodifluoroethane is no longer permitted for use as an aerosol propellant in topical pharmaceutical formulations. It is generally regarded as an essentially nontoxic and nonirritant material.

Deliberate inhalation of excessive quantities of chlorofluorocarbon propellant may result in death, and the following 'warning' statements must appear on the label of all aerosols:

> WARNING: Avoid inhalation. Keep away from eyes or other mucous membranes.

In the USA, the Environmental Protection Agency (EPA) additionally requires the following information on all aerosols containing chlorofluorocarbons as the propellant:

> WARNING: Contains a chlorofluorocarbon that may harm the public health and environment by reducing ozone in the upper atmosphere.

15 Handling Precautions

Chlorodifluoroethane is usually encountered as a liquefied gas and appropriate precautions for handling such materials should be taken. Eye protection, gloves, and protective clothing are recommended. Chlorodifluoroethane should be handled in a well-ventilated environment. Chlorofluorocarbon vapors are heavier than air and do not support life; therefore, when cleaning large tanks that have contained chlorofluorocarbons, adequate provisions for oxygen supply in the tanks must be made in order to protect workers cleaning the tanks.

Chlorodifluoroethane is flammable; *see* Section 10. When heated to decomposition, chlorodifluoroethane emits toxic fumes.

16 Regulatory Status

—

17 Related Substances

—

18 Comments

Although chlorodifluoroethane is no longer available for use in pharmaceutical formulation, the monograph remains in the *Handbook* to assist formulators in their understanding of old products, whose properties they may need to reproduce. It may also still be present in some commercial products.

For a discussion of the numerical nomenclature applied to this aerosol propellant, *see* Chlorofluorocarbons.

19 Specific References

1 Health and Safety Executive. *EH40/2005: Workplace Exposure Limits.* Sudbury: HSE Books, 2011. http://www.hse.gov.uk/pubns/priced/eh40.pdf (accessed 1 March 2012).

20 General References

Johnson MA. *The Aerosol Handbook*, 2nd edn. Caldwell: WE Dorland, 1982: 305–335.

Johnson MA. Flammability aspects of dimethy ether, p22, p-142b, p-1152a. *Aerosol Age* 1988; 33(8): 32, 34, 36, 38–39.

Sanders PA. *Handbook of Aerosol Technology*, 2nd edn. New York: Van Nostrand Reinhold, 1979: 19–35.

Sciarra JJ. Aerosols. In: *Remington: The Science and Practice of Pharmacy*, 21st edn. Philadelphia, PA: Lippincott Williams & Wilkins, 2006: 1000–1017.

Sciarra CJ, Sciarra JJ. Pressurized dispensers. In: Schlossman ML. *The Chemistry and Manufacture of Cosmetics*, vol. 1, 4th edn. Carol Stream, IL: Allured Publishing Corporation, 2009: 451–478.

Sciarra JJ Aerosol suspensions and emulsions. In: Lieberman H *et al.* eds. *Pharmaceutical Dosage Forms: Disperse Systems*, 2nd edn. New York: Marcel Dekker, 1996: 319–356.

Sciarra JJ. Pharmaceutical aerosols. In: Banker GS, Rhodes CT, eds. *Modern Pharmaceutics*, 3rd edn. New York: Marcel Dekker, 1996: 547–574.

Sciarra JJ, Stoller L. *The Science and Technology of Aerosol Packaging*. New York: Wiley, 1974: 137–145.

21 Authors

CJ Sciarra, JJ Sciarra.

22 Date of Revision

1 February 2011.

Chlorofluorocarbons (CFC)

(a) Dichlorodifluoromethane (Propellant 12)
(b) Dichlorotetrafluoroethane (Propellant 114)
(c) Trichloromonofluoromethane (Propellant 11)

1 Nonproprietary Names

(a) USP–NF: Dichlorodifluoromethane
(b) USP–NF: Dichlorotetrafluoroethane
(c) USP–NF: Trichloromonofluoromethane

2 Synonyms

Arcton; Dymel; Freon; Frigen; Genetron; Halon; Isceon; Isotron.

3 Chemical Name and CAS Registry Number

(a) Dichlorodifluoromethane [75-71-8]
(b) 1,2-Dichloro-1,1,2,2-tetrafluoroethane [76-14-2]
(c) Trichlorofluoromethane [75-69-4]

4 Empirical Formula and Molecular Weight

(a) CCl_2F_2	120.91
(b) $C_2Cl_2F_4$	170.92
(c) CCl_3F	137.37

5 Structural Formula

(a) CCl_2F_2: Cl–C(F)(F)–Cl

(a)

(b) F–C(Cl)(F)–C(Cl)(F)–F

(b)

(c) F–C(Cl)(Cl)–Cl

(c)

6 Functional Category

Aerosol propellant.

7 Applications in Pharmaceutical Formulation or Technology

Dichlorodifluoromethane, dichlorotetrafluoroethane, and trichloromonofluoromethane are chlorofluorocarbon (CFC) aerosol propellants used in pharmaceutical formulations. They are no longer used in metered-dose inhaler (MDI) formulations, with few exceptions for existing MDIs; *see also* Section 18.

Dichlorodifluoromethane is used as an aerosol propellant in MDIs, either as the sole propellant or in combination with dichlorotetrafluoroethane, trichloromonofluoromethane, or mixtures of these chlorofluorocarbons. Dichlorodifluoromethane may also be used as a propellant in an aerosolized sterile talc used for intrapleural administration and is also used alone in some MDIs containing a steroid.

Dichlorotetrafluoroethane is used in combination with dichlorodifluoromethane, and in several cases with dichlorodifluoromethane and trichloromonofluoromethane, as the propellant in metered-dose oral and nasal aerosols.

Trichloromonofluoromethane was used in combination with dichlorodifluoromethane as the propellant in metered-dose inhaler aerosols. It is also used in combination with dichlorotetrafluoroethane and dichlorodifluoromethane.

These three propellants have been blended to obtain suitable solubility characteristics for MDIs when formulated as solutions. They will produce suitable vapor pressures so that optimum particle-size distribution as well as suitable respiratory fractions may be achieved.

Blends of trichloromonofluoromethane and dichlorodifluoromethane (propellant 11/12) or propellant 11/114/12 produce vapor pressures of 103–484 kPa (15–70 psig) at 25°C, which adequately cover the range of pressures required to produce the proper particle-size distribution for satisfactory aerosol products. Trichloromonofluoromethane is unique among the chlorofluorocarbon propellants in that it is a liquid at room temperature and atmospheric pressure, and can be used to prepare a slurry with insoluble medicinal agents. These CFC propellants have been replaced by more environmentally accepted hydrofluoroalkane (HFA) propellants (HFA 134a and HFA 227) for use with the metered-dose inhalers.

8 Description

Dichlorodifluoromethane is a liquefied gas and exists as a liquid at room temperature when contained under its own vapor pressure, or as a gas when exposed to room temperature and atmospheric pressure. The liquid is practically odorless and colorless. The gas in high concentrations has a faint etherlike odor. Dichlorodifluoromethane is noncorrosive, nonirritating, and nonflammable.

Dichlorotetrafluoroethane is a colorless, nonflammable liquefied gas with a faint, ethereal odor.

Trichloromonofluoromethane is a clear, volatile liquid at room temperature and atmospheric pressure. It has a characteristic carbon tetrachloride-like odor and is nonirritating and nonflammable.

9 Pharmacopeial Specifications

See Table I.

Table I: Pharmacopeial specifications from USP35–NF30.

Test	Propellant 12	Propellant 114	Propellant 11
Identification	+	+	+
Boiling temperature	–30°C	4°C	24°C
Water	≤0.001%	≤0.001%	≤0.001%
High-boiling residues	≤0.01%	≤0.01%	≤0.01%
Inorganic chlorides	+	+	+
Chromatographic purity	+	+	+
Assay	99.6–100.0%	99.6–100.0%	99.6–100.0%

10 Typical Properties

See Table II for selected typical properties.

11 Stability and Storage Conditions

Chlorofluorocarbon propellants are nonreactive and stable at temperatures up to 550°C. The liquefied gas is stable when used as a propellant and should be stored in a metal cylinder in a cool, dry place.

12 Incompatibilities

The presence of greater than 5% water in solutions that contain trichloromonofluoromethane may lead to hydrolysis of the propellant and the formation of traces of hydrochloric acid, which may be irritant to the skin or cause corrosion of metallic canisters. Trichloromonofluoromethane may also react with aluminum, in the presence of ethanol, to cause corrosion within a cylinder with the formation of hydrogen gas. Similarly, alcohols in the presence of trace amounts of oxygen, peroxides, or other free-radical catalysts may react with trichloromonofluoromethane to form trace quantities of hydrochloric acid.

Both dichlorodifluoromethane and dichlorotetrafluoroethane are compatible with most ingredients used in pharmaceutical aerosols. Because of their poor miscibility with water, most MDIs are formulated as suspensions. However, solution MDIs can be prepared through the use of ethanol as a cosolvent for water and propellant, resulting in a clear solution (provided the water content is less than 5%).

13 Method of Manufacture

Dichlorodifluoromethane is prepared by the reaction of hydrogen fluoride with carbon tetrachloride in the presence of a suitable catalyst, such as polyvalent antimony. The dichlorodifluoromethane formed is further purified to remove all traces of water and hydrochloric acid as well as traces of the starting and intermediate materials.

Trichloromonofluoromethane is also obtained by this process.

Dichlorotetrafluoroethane is prepared by the reaction of hydrogen fluoride with chlorine and perchloroethylene in the presence of a suitable catalyst such as polyvalent antimony.

14 Safety

Dichlorodifluoromethane, dichlorotetrafluoroethane, and trichloromonofluoromethane have been used for over 50 years as propellants in topical, oral, and nasal aerosol formulations, and are generally regarded as nontoxic and nonirritant materials when used as directed.

The propellants used for metered-dose inhalant aerosol products generally vaporize quickly and most of the vapors escape and are not inhaled. However, a small amount of the propellant may be inhaled with the active ingredient and be carried to the respiratory system. These amounts of propellant do not present a toxicological problem and are quickly cleared from the lungs. Deliberate inhalation of excessive quantities of fluorocarbon propellant may result in death, and the following 'warning' statements must appear on the label of all aerosols:

> WARNING: Avoid inhalation. Keep away from eyes or other mucous membranes.

(Aerosols designed specifically for oral inhalation need not contain this statement).

> WARNING: Do not inhale directly; deliberate inhalation of contents can cause death.

or

> WARNING: Use only as directed; intentional misuse by deliberately concentrating and inhaling the contents can be harmful or fatal.

Additionally, the label should contain the following information:

C

Table II: Selected typical properties for chlorofluorocarbon propellants.

Test	Propellant 12	Propellant 114	Propellant 11
Boiling point	−29.8°C	4.1°C	23.7°C
Critical pressure	4.01 MPa (39.6 atm)	3268 kPa (474 psia)	4.38 MPa (43.2 atm)
Critical temperature	111.5°C	145.7°C	198°C
Density			
Liquid at 21°C	1.325 g/cm^3	1.468 g/cm^3	1.485 g/cm^3
Liquid at 54.5°C	1.191 g/cm^3	1.360 g/cm^3	1.403 g/cm^3
Flammability	Nonflammable	Nonflammable	Nonflammable
Freezing point	−158°C	−94°C	−111°C
Kauri-butanol value	18	12	60
Solubility at 20°C (unless otherwise stated)			
Ethanol (95%)	Soluble	Soluble	Soluble
Ether	Soluble	Soluble	Soluble
Water	1 in 3570 at 25°C	1 in 7690 at 25°C	1 in 909 at 25°C
Surface tension at 25°C	9 mN/m (9 dynes/cm)	13 mN/m (13 dynes/cm)	19 mN/m (19 dynes/cm)
Vapor density			
Absolute	5.398 g/m^3	7.63 g/m^3	6.133 g/m^3
Relative	4.19 (air = 1)	5.92 (air = 1)	5.04 (air = 1)
Vapor pressure			
At 21°C	585.4 kPa (84.9 psia)	190.3 kPa (27.6 psia)	92.4 kPa (13.4 psia)
At 54.5°C	1351.4 kPa (196.0 psia)	506.8 kPa (73.5 psia)	268.9 kPa (39.0 psia)
Viscosity (dynamic)			
Liquid at 21°C	0.262 mPa s (0.262 cP)	0.386 mPa s (0.386 cP)	0.439 mPa s (0.439 cP)
Liquid at 54.5°C	0.227 mPa s (0.227 cP)	0.296 mPa s (0.296 cP)	0.336 mPa s (0.336 cP)

WARNING: Contents under pressure. Do not puncture or incinerate container. Do not expose to heat or store at room temperature above 120°F (49°C). Keep out of the reach of children.

In the USA, the Environmental Protection Agency (EPA) additionally requires the following information on all aerosols containing chlorofluorocarbons as the propellant:

WARNING: Contains a chlorofluorocarbon that may harm the public health and environment by reducing ozone in the upper atmosphere.

(Metered-dose inhalers are exempt from this regulation.)

15 Handling Precautions

Dichlorodifluoromethane and dichlorotetrafluoroethane are usually encountered as a liquefied gas and appropriate precautions for handling such materials should be taken. Eye protection, gloves, and protective clothing are recommended. These propellants should be handled in a well-ventilated environment. Chlorofluorocarbon vapors are heavier than air and do not support life; therefore, when cleaning large tanks that have contained chlorofluorocarbons, adequate provisions for supply of oxygen in the tanks must be made in order to protect workers cleaning the tanks.

Although nonflammable, when heated to decomposition chlorofluorocarbons emit toxic fumes containing phosgene and fluorides. Although not as volatile as dichlorodifluoroethane or dichlorotetrafluoroethane, trichloromonofluoromethane should be handled as indicated above. Since it is a liquid at room temperature, caution should be exercised in handling this material to prevent spillage onto the skin. It is an irritant to the eyes.

The long-term workplace exposure limit (8-hour TWA) for dichlorodifluoromethane and dichlorotetrafluoroathane is 7110 mg/m^3 (1000 ppm). The short-term workplace exposure limit (15-minute) for both compounds is 8890 mg/m^3 (1250 ppm).[(1)]

16 Regulatory Status

Included in the FDA Inactive Ingredients Database (aerosol formulations for inhalation, nasal, oral, and topical applications). With few exceptions for existing MDIs, the FDA and EPA banned the use of CFCs in the US after 31st December 2008, with all CFCs to be phased out by 2010–2013.

The FDA has published its timeline for removing MDIs containing CFCs from the shelves of US pharmacies and hospitals. Most of the inhalers containing CFCs will no longer be sold after the year 2010, and any other CFC-containing MDIs will not be sold after mid-2011. The final two MDIs containing CFCs will no longer be available after 2013. Following this date, CFC-containing MDIs will no longer be available in the US and only those MDIs that contain an HFA as the propellant will be available.

17 Related Substances

—

18 Comments

Although chlorofluorocarbons are no longer available for use in pharmaceutical formulation, the monograph remains in the *Handbook* to assist formulators in their understanding of old products, whose properties they may need to reproduce. They may still be present in some commercial products.

Fluorocarbon (FC) aerosol propellants may be identified by a standardized numbering nomenclature; for example, dichlorodifluoromethane is known as propellant 12, while dichlorotetrafluoroethane is known as propellant 114.

Usually, three digits are used to describe the propellant, except when the first digit would be zero, in which case only two digits are used. The first digit is one less than the number of carbon atoms in the molecule. Thus, if the molecule is a methane derivative the first digit would be zero (1 − 1) and is ignored, so that only two digits are used in the propellant description; e.g. propellant 12. For an ethane derivative, the first digit would be a one (2 − 1); e.g. propellant 114.

The second digit is one more than the number of hydrogen atoms in the molecule, while the third digit represents the number of fluorine atoms in the molecule. The difference between the sum of the fluorine and hydrogen atoms and the number of atoms required to saturate the carbon chain is the number of chlorine atoms in the molecule. Isomers of a compound have the same identifying number and an additional letter; a, b, c, and so on. Cyclic derivatives are indicated by the letter C before the identifying number. With unsaturated propellants, the number 1 is used as the fourth digit from the right to indicate an unsaturated double bond.

Thus for dichlorodifluoromethane (propellant 12):

(a) First digit = 0 signifies number of C atoms = 1

(b) Second digit = 1 signifies number of H atoms = 0

(c) Third digit = 2 signifies number of F atoms = 2

(d) Number of Cl atoms = 4 − (2 − 0) = 2

Under the terms of the Montreal Protocol, aimed at reducing damage to the ozone layer, the use of chlorofluorocarbons, including dichlorodifluoromethane, dichlorotetrafluoroethane, and trichloromonofluoromethane, has been prohibited from January 1996.[2–6] However, this prohibition does not apply to essential uses such as existing pharmaceutical formulations for which no alternative chlorofluorocarbon-free product is available. The EPA and FDA approved essential-use status for dichlorodifluoromethane for a sterile aerosol talc used in the treatment of malignant pleural effusion in patients with lung cancer.

19 Specific References

1 Health and Safety Executive. *EH40/2005: Workplace Exposure Limits.* Sudbury: HSE Books, 2011. http://www.hse.gov.uk/pubns/priced/eh40.pdf (accessed 1 March 2012).

2 Fischer FX, *et al.* CFC propellant substitution: international perspectives. *Pharm Technol* 1989; **13**(9): 44, 48, 50, 52.

3 Kempner N. Metered dose inhaler CFC's under pressure. *Pharm J* 1990; 245: 428–429.

4 Dalby RN. Possible replacements for CFC-propelled metered-dose inhalers. *Med Device Technol* 1991; 2(4): 21–25.

5 CFC-free aerosols; the final hurdle. *Manuf Chem* 1992; **63**(7): 22–23.

6 Mackenzie D. Large hole in the ozone agreement. *New Scientist* 1992; Nov 28: 5.

20 General References

Amin YM, *et al.* Fluorocarbon aerosol propellants XII: correlation of blood levels of trichloromonofluormethane to cardiovascular and respiratory responses in anesthetized dogs. *J Pharm Sci* 1979; **68**: 160–163.

Byron PR, ed. *Respiratory Drug Delivery.* Boca Raton, FL: CRC Press, 1990.

Johnson MA. *The Aerosol Handbook*, 2nd edn. Caldwell: WE Dorland, 1982: 305–335.

Niazi S, Chiou WL. Fluorocarbon aerosol propellants XI: pharmacokinetics of dichlorodifluoromethane in dogs following single dose and multiple dosing. *J Pharm Sci* 1977; **66**: 49–53.

Sanders PA. *Handbook of Aerosol Technology*, 2nd edn. New York: Van Nostrand Reinhold, 1979: 19–35.

Sawyer E, *et al.* Microorganism survival in non-CFC propellant P134a and a combination of CFC propellants P11 and P12. *Pharm Technol* 2001; **25**(3): 90–96.

Sciarra JJ Pharmaceutical aerosols. In: Lachman L *et al.* eds. *The Theory and Practice of Industrial Pharmacy*, 3rd edn. Philadelphia: Lea and Febiger, 1986: 589–618.

Sciarra CJ, Sciarra JJ. Aerosols. In: *Remington: The Science and Practice of Pharmacy*, 21st edn. Philadelphia, PA: Lippincott, Williams and Wilkins, 2006: 1000–1017.

Sciarra CJ, Sciarra JJ. Pressurized dispensers. In: Schlossman ML. *The Chemistry and Manufacture of Cosmetics,* vol. 1, 4th edn. Carol Stream, IL: Allured Publishing Corporation, 2009: 451–478.

Sciarra JJ. Inc: Banker GS, Rhodes C, eds. *Modern Pharmaceutics*, 3rd edn. New York: Marcel Dekker, 1996: 547–574.

Sciarra JJ, Stoller L. *The Science and Technology of Aerosol Packaging.* New York: Wiley, 1974: 97–130.

Strobach DR. Alternatives to CFCs. *Aerosol Age* 1988; 32–3342–43.

21 Authors

CJ Sciarra, JJ Sciarra.

22 Date of Revision

1 February 2011.

Chloroxylenol

1 Nonproprietary Names

BP: Chloroxylenol

USP-NF: Chloroxylenol

2 Synonyms

4-Chloro-3,5-dimethylphenol; 2-chloro-5-hydroxy-1,3-dimethylbenzene; 4-chloro-1-hydroxy-3,5-dimethylbenzene; 2-chloro-5-hydroxy-*m*-xylene; 2-chloro-*m*-xylenol; 3,5-dimethyl-4-chlorophenol; *Nipacide PX*; parachlorometaxylenol; *p*-chloro-*m*-xylenol; PCMX.

3 Chemical Name and CAS Registry Number

4-Chloro-3,5-xylenol [88-04-0]

4 Empirical Formula and Molecular Weight

C_8H_9ClO 156.61

5 Structural Formula

OH

H_3C CH_3

Cl

6 Functional Category

Antimicrobial preservative.

7 Applications in Pharmaceutical Formulation or Technology

As a pharmaceutical excipient, chloroxylenol is commonly used in low concentrations, typically at concentrations of 0.1–0.8%, as an antimicrobial preservative in topical formulations such as creams and ointments. Chloroxylenol is also used in a number of cosmetic formulations.

C

8 Description

White or cream-colored crystals or crystalline powder with a characteristic phenolic odor. Volatile in steam.

9 Pharmacopeial Specifications

See Table I.

Table I: Pharmacopeial specifications for chloroxylenol.

Test	BP 2012	USP35-NF30
Identification	+	+
Characters	+	—
Residue on ignition	—	≤0.1%
Water	—	≤0.5%
Iron	—	≤0.01%
Melting range	114–116°C	114–116°C
Related substances	+	+
Assay	98.0–103.0%	≥98.5%

10 Typical Properties

Antimicrobial activity Chloroxylenol is effective against Gram-positive bacteria but less active against Gram-negative bacteria. The activity of chloroxylenol against Gram-negative bacilli can be increased by the addition of a chelating agent such as edetic acid.[1] Chloroxylenol is inactive against bacterial spores. Antimicrobial activity may be reduced by loss of chloroxylenol from a formulation due to incompatibilities with packaging materials or other excipients, such as nonionic surfactants.[2] Solution pH does not have a marked effect on the activity of chloroxylenol.[3] Minimum inhibitory concentration (MIC) determinations may be affected by growth conditions and the exact strain of organism tested. Reported ranges for chloroxylenol MIC against common microorganisms are: *Pseudomonas aeruginosa* 200-1000 ppm; *Escherichia coli* 125-500 ppm; *Staphylococcus aureus* 100-500 ppm; *Aspergillus niger* 50-100 ppm.[4]

Acidity $pK_a = 9.7$[5]

Boiling point 246°C

Density 0.89 g/cm^3 at 20°C[5]

Melting point 115.5°C

Partition coefficient Octanol : water $\log k_{ow} = 3.27$[5]

Solubility Freely soluble in ethanol (95%); soluble in ether, terpenes, and fixed oils; very slightly soluble in water. Dissolves in solutions of alkali hydroxides.

See also Table II.

Table II: Solubility of chloroxylenol.[6]

Solvent	Solubility at 15°C unless otherwise stated (g/100 mL)
Petroleum ether	0.5
Benzene	6.0
Acetone	58.0
Toluene	7.0
Chloroform	6.2
Isopropanol	38.0
Water (15°C)	0.03
Water (100°C)	0.5

11 Stability and Storage Conditions

Chloroxylenol is stable at normal room temperature, but is volatile in steam. Contact with natural rubber should be avoided. Aqueous solutions of chloroxylenol are susceptible to microbial contamination and appropriate measures should be taken to prevent contamination during storage or dilution. Chloroxylenol should be stored in polyethylene, mild steel or stainless steel containers, which should be well-closed and kept in a cool, dry place. Chloroxylenol does not absorb at wavelengths >290 nm and has been reported to be stable to sunlight for up to 24 hours.[5]

12 Incompatibilities

Chloroxylenol has been reported to be incompatible with nonionic surfactants and methylcellulose.

13 Method of Manufacture

Chloroxylenol is prepared by treating 3,5-dimethylphenol with chlorine or sulfuryl chloride (SO_2Cl_2).

14 Safety

Chloroxylenol is generally regarded as a relatively nontoxic and nonirritant material when used as an excipient in topical products. However, chloroxylenol has been placed in Toxicity Category I for eye irritation effects.[6] In addition, allergic skin reactions have been reported.[7–11] Taken orally, chloroxylenol is mildly toxic. However, ingestion of a chloroxylenol-containing disinfectant product has been associated with reports of fatal[12] or severe instances of self-poisoning.[13,14]

LD_{50} (mouse, IP): 0.115 g/kg[15]
LD_{50} (rat, dermal): >2.0 g/kg[5]
LD_{50} (rat, oral): 3.83 g/kg[15]

15 Handling Precautions

Observe normal precautions appropriate to the circumstances and quantity of material handled. Chloroxylenol is an eye irritant, and eye protection is recommended. When heated to decomposition, chloroxylenol emits toxic fumes.

16 Regulatory Status

Included in the FDA Inactive Ingredients Database (otic preparations; topical creams and emulsions). Included in nonparenteral medicines licensed in the UK. Approved in Europe as a preservative in cosmetics with a maximum authorized concentration of 0.5%.[16]

17 Related Substances

Chlorocresol.

18 Comments

Therapeutically, chloroxylenol has been investigated as a treatment for acne vulgaris,[17,18] for treating infected root canals,[19] and as an antifungal agent when impregnated into medical devices.[20] Chloroxylenol is included in drug products approved for topical acne, treatment of superficial fungal infections, and dandruff/seborrheic dermatitis/psoriasis. However the data supporting safety and effectiveness of chloroxylenol in the treatment of topical acne is stated to be inadequate.[21] Chloroxylenol is a common constituent of many proprietary disinfectants used for skin and wound disinfection, typically at concentrations of 2-5%.

The EINECS number for chloroxylenol is 201-793-8. The PubChem Compound ID (CID) for chloroxylenol described in this monograph is 2723.

19 Specific References

1 Ayliffe GAJ *et al.* eds. *Control of Hospital Infection*, 4th edn. London: Arnold, 2000; 78.
2 Kazmi SJA, Mitchell AG. Interaction of preservatives with cetomacrogol. *J Pharm Pharmacol* 1971; **23**: 482–489.
3 Judis J. Studies on the mechanism of action of phenolic disinfectants I. *J Pharm Sci* 1962; **51**: 261–265.
4 Anonymous. Final report on the safety of chloroxylenol. *Int J Toxicol* 1985; **4**(5): 147–169.
5 US National Library of Medicine (2008).*The Hazardous Substances Data Bank*. toxnet.nlm.nih.gov/cgi-bin/sis/htmlgen?HSDB (accessed 30 November 2011).
6 USEPA. *Reregistration Eligibility Decision (RED) Database for Chloroxylenol (88-04-0)*. EPA 738-R-94-032 (September 1994). www.epa.gov/pesticides/reregistration/status.htm (accessed 30 November 2011).
7 Mowad C. Chloroxylenol causing hand dermatitis in a plumber. *Am J Contact Dermatitis* 1998; **9**: 128–129.
8 Malakar S, Panda S. Post-inflammatory depigmentation following allergic contact dermatitis to chloroxylenol [letter]. *Br J Dermatol* 2001; **144**(6): 1275–1276.
9 Jong CT, *et al.* Contact sensitivity to preservatives in the UK, 2004-2005: Results of multicenter study. *Contact Dermatitis* 2007; **57**(3): 165–168.
10 Wilson M, Mowad C. Chloroxylenol. *Dermatitis* 2007; **18**(2): 120–121.
11 Berthelot C, Zirwas MJ. Allergic contact dermatitis to chloroxylenol. *Dermatitis* 2006; **17**(3): 156–159.
12 Meek D, *et al.* Fatal self-poisoning with Dettol. *Postgrad Med J* 1977; **53**: 229–231.
13 Joubert P, *et al.* Severe Dettol (chloroxylenol and terpineol) poisoning. *Br Med J* 1978; **1**: 890.
14 Chan TYK, *et al.* Chemical gastro-oesophagitis, upper gastrointestinal haemorrhage and gastroscopic findings following Dettol poisoning. *Hum Exp Toxicol* 1995; **14**: 18–19.
15 Lewis RJ, ed. *Sax's Dangerous Properties of Industrial Materials*, 11th edn. New York: Wiley, 2004; 906–907.
16 European Commission (2008). Consolidated version of Cosmetics Directive 76/768/EECEU: Annex VI. List of preservatives which cosmetic products may contain, 2008; 112. ec.europa.eu/enterprise/cosmetics/html/consolidated_dir.htm (accessed 30 November 2011).
17 Papageorgiou PP, Chu AC. Chloroxylenol and zinc oxide containing cream (Nels cream) vs. 5% benzoyl peroxide cream in the treatment of acne vulgaris. A double-blind, randomized, controlled trial. *Clin Exp Dermatol* 2000; **25**(1): 16–20.
18 Boutli F, *et al.* Comparison of chloroxylenol 0.5% plus salicylic acid 2% cream and benzoyl peroxide 5% gel in the treatment of acne vulgaris: a randomized double-blind study. *Drugs Exp Clin Res* 2003; **29**(3): 101–105.
19 Schafer E, Bossmann K. Antimicrobial efficacy of chloroxylenol and chlorhexidine in the treatment of infected root canals. *Am J Dent* 2001; **14**(4): 233–237.
20 Darouiche RO, *et al.* Antifungal activity of antimicrobial-impregnated devices. *Clin Microbiol Infect* 2006; **12**(4): 397–399.
21 US National Archives and Records Administration (2008). *Electronic Code of Federal Regulations*, 21 CFR 310.545(a)(1,7,22). www.gpoaccess.gov/ecfr (accessed 30 November 2011).

20 General References

Chapman DG. Preservatives available for use. In: Board RG *et al.* eds. *Preservatives in the Food, Pharmaceutical and Environmental Industries*. Oxford: Blackwell Scientific, 1987.

Gatti R, *et al.* HPLC-fluorescence determination of chlorocresol and chloroxylenol in pharmaceuticals. *J Pharm Biomed Anal* 1997; **16**: 405–412.

Paulson DS. Topical antimicrobials.*ACS Symposium Series, 967 (New Biocides Development)*, 2007: 124–150.

21 Author

W Cook.

22 Date of Revision

1 March 2012.

Cholesterol

1 Nonproprietary Names

BP: Cholesterol

JP: Cholesterol

PhEur: Cholesterol

USP–NF: Cholesterol

2 Synonyms

Cholesterin; cholesterolum.

3 Chemical Name and CAS Registry Number

Cholest-5-en-3β-ol [57-88-5]

4 Empirical Formula and Molecular Weight

$C_{27}H_{46}O$ 386.67

5 Structural Formula

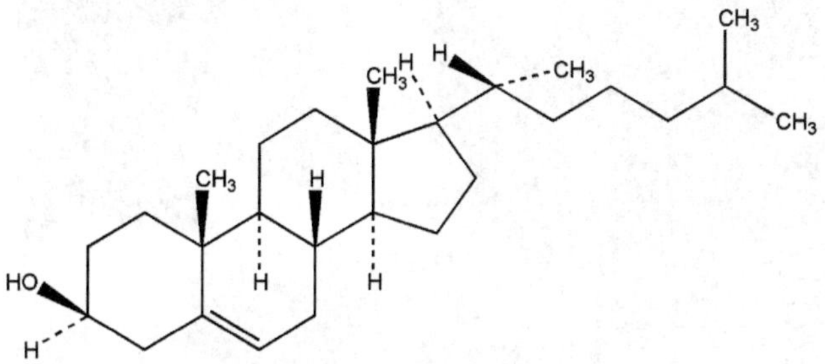

6 Functional Category

Emollient; emulsifying agent.

7 Applications in Pharmaceutical Formulation or Technology

Cholesterol is used in cosmetics and topical pharmaceutical formulations at concentrations of 0.3–5.0% w/w as an emulsifying agent. It imparts water-absorbing power to an ointment and has emollient activity.

8 Description

Cholesterol occurs as white or faintly yellow, almost odorless, pearly leaflets, needles, powder, or granules. On prolonged exposure to light and air, cholesterol acquires a yellow to tan color.

9 Pharmacopeial Specifications

See Table I.

10 Typical Properties

Boiling point 360°C (some decomposition)
Density 1.052 g/cm^3 for anhydrous form.
Dielectric constant $D^{20} = 5.41$
Melting point 147–150°C

SEM 1: Excipient: cholesterol; manufacturer: Pflatz & Bauer, Inc.; magnification: 240×.

SEM 2: Excipient: cholesterol; manufacturer: Pfaltz & Bauer, Inc.; magnification: 2400×.

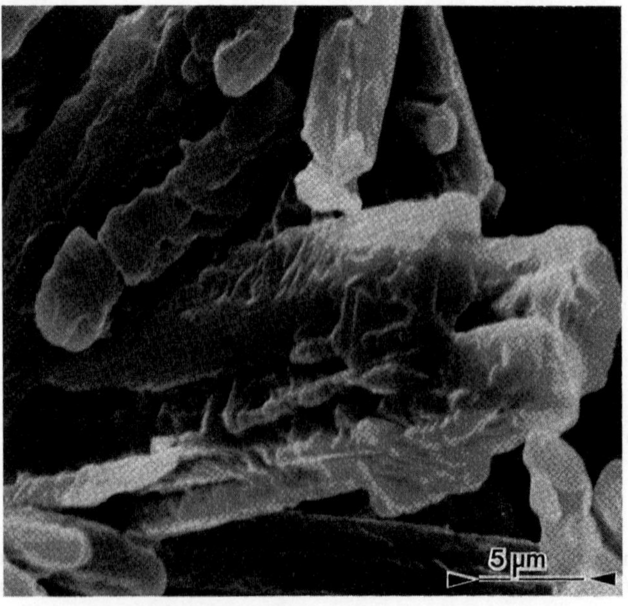

Table I: Pharmacopeial specifications for cholesterol.

Test	JP XV	PhEur 7.4	USP35–NF30
Identification	+	+	+
Acidity	+	+	+
Characters	—	+	—
Clarity of solution	+	—	—
Loss on drying	≤0.3%	≤0.3%	≤0.3%
Melting range	147–150°C	147–150°C	147–150°C
Residue on ignition	≤0.1%	—	≤0.1%
Solubility in alcohol	—	+	+
Specific rotation	–34° to –38°	—	–34° to –38°
Sulfated ash	—	≤0.1%	—
Assay (dried substance)			
Cholesterol	—	≥95.0%	—
Total sterols	—	97.0-103.0%	—

Solubility *see* Table II.[1–3]
Specific rotation
$[\alpha]_D^{20} = -39.5°$ (2% w/v solution in chloroform);
$[\alpha]_D^{20} = -31.5°$ (2% w/v solution in ether).

Table II: Solubility of cholesterol.

Solvent	Solubility at 20°C[1,2,3] unless otherwise stated
Acetone	Soluble
Benzene	1 in 7
Chloroform	1 in 4.5
Ethanol	1 in 147 at 0°C
	1 in 78 at 20°C
	1 in 29 at 40°C
	1 in 19 at 50°C
	1 in 13 at 60°C
Ethanol (95%)	1 in 78 (slowly)
	1 in 3.6 at 80°C
Ether	1 in 2.8
Hexane	1 in 52
Isopropyl myristate	1 in 19
Methanol	1 in 294 at 0°C
	1 in 153 at 20°C
	1 in 53 at 40°C
	1 in 34 at 50°C
	1 in 23 at 60°C
Vegetable oils	Soluble
Water	Practically insoluble

11 Stability and Storage Conditions

Cholesterol is stable and should be stored in a well-closed container, protected from light.

12 Incompatibilities

Cholesterol is precipitated by digitonin.

13 Method of Manufacture

The commercial material is normally obtained from the spinal cord of cattle by extraction with petroleum ethers, but it may also be obtained from wool fat. Purification is normally accomplished by repeated bromination. Cholesterol may also be produced by entirely synthetic means.[4]

Cholesterol produced from animal organs will always contain cholestanol and other saturated sterols.

See also Section 14.

14 Safety

Cholesterol is generally regarded as an essentially nontoxic and nonirritant material at the levels employed as an excipient.[3] It has, however, exhibited experimental teratogenic and reproductive effects, and mutation data have been reported.[5]

Cholesterol is often derived from animal sources and this must be done in accordance with the regulations for human consumption. The risk of bovine spongiform encephalopathy (BSE) contamination has caused some concern over the use of animal-derived cholesterol in pharmaceutical products.[6] However, synthetic methods of cholesterol manufacture have been developed.[4]

15 Handling Precautions

Observe normal precautions appropriate to the circumstances and quantity of material handled. Rubber or plastic gloves, eye protection, and a respirator are recommended.

May be harmful following inhalation or ingestion of large quantities, or over prolonged periods of time, owing to the possible involvement of cholesterol in atherosclerosis and gallstones. May be irritant to the eyes. When heated to decomposition, cholesterol emits acrid smoke and irritating fumes.

16 Regulatory Status

Included in the FDA Inactive Ingredients Database (injections; ophthalmic, topical, and vaginal preparations).

Included in nonparenteral medicines licensed in the UK. Included in the Canadian Natural Health Products Ingredients Database.

17 Related Substances

Lanolin; lanolin alcohols; lanolin hydrous.

18 Comments

A novel cholesterol-based cationic lipid has been developed that promotes DNA transfer in cells.[7,8] Cholesterol monohydrate becomes anhydrous at 70–80°C.

Cholesterol also has a physiological role. It is the major sterol of the higher animals, and it is found in all body tissues, especially in the brain and spinal cord. It is also the main constituent of gallstones.

The EINECS number for cholesterol is 200-353-2. The PubChem Compound ID (CID) for cholesterol includes 304 and 5997.

19 Specific References

1 Harwood RJ, Cohen EM. Solubility of cholesterol in isopropyl myristate. *J Soc Cosmet Chem* 1977; **28**: 79–82.
2 Flynn GL, *et al.* Cholesterol solubility in organic solvents. *J Pharm Sci* 1979; **68**: 1090–1097.
3 Cosmetic, Toiletry and Fragrance Association. Final report on the safety assessment of cholesterol. *J Am Coll Toxicol* 1986; **5**(5): 491–516.
4 Carmichael H. Safer by synthesis? *Chem Br* 2001; **37**(2): 40–42.
5 Lewis RJ, ed. *Sax's Dangerous Properties of Industrial Materials*, 11th edn. New York: Wiley, 2004: 912.
6 Anonymous. Beefing about contamination. *Pharm Dev Technol Eur* 1997; **9**(11): 12, 14.
7 Percot A, *et al.* Hydroxyethylated cholesterol-based cationic lipid for DNA delivery: effect of conditioning. *Int J Pharm* 2004; **278**(1): 143–163.
8 Reynier P, *et al.* Modifications in the head group and in the spacer of cholesterol-based cationic lipids promote transfection in melanoma B16-F10 cells and tumours. *J Drug Target* 2004; **12**(1): 25–38.

20 General References

Bogardus JB. Unusual cholesterol solubility in water/glyceryl-1-monooctanoate solutions. *J Pharm Sci* 1982; **71**: 370–372.

Cadwallader DE, Madan DK. Effect of macromolecules on aqueous solubility of cholesterol and hormone drugs. *J Pharm Sci* 1981; **70**: 442–446.

Feld KM, *et al.* Influence of benzalkonium chloride on the dissolution behavior of several solid-phase preparations of cholesterol in bile acid solutions. *J Pharm Sci* 1982; **71**: 182–188.

Singh VS, Gaur RC. Dispersion of cholesterol in aqueous surfactant solutions: interpretation of viscosity data. *J Disper Sci Technol* 1983; **4**: 347–359.

Udupa N, *et al.* Formulation and evaluation of methotrexate niosomes. *Drug Dev Ind Pharm* 1993; **19**: 1331–1342.

21 Author

L Peltonen.

22 Date of Revision

1 February 2011.

Citric Acid Monohydrate

1 Nonproprietary Names

BP: Citric Acid Monohydrate

JP: Citric Acid Hydrate

PhEur: Citric Acid Monohydrate

USP–NF: Citric Acid Monohydrate

2 Synonyms

Acidum citricum monohydricum; E330; 2-hydroxypropane-1,2,3-tricarboxylic acid monohydrate.

3 Chemical Name and CAS Registry Number

2-Hydroxy-1,2,3-propanetricarboxylic acid monohydrate [5949-29-1]

4 Empirical Formula and Molecular Weight

$C_6H_8O_7{\cdot}H_2O$ 210.14

5 Structural Formula

HO O HO OH OH O O $\cdot H_2O$

6 Functional Category

Acidulant; antioxidant; buffering agent; complexing agent; flavor enhancer.

7 Applications in Pharmaceutical Formulation or Technology

Citric acid (as either the monohydrate or anhydrous material) is widely used in pharmaceutical formulations and food products, primarily to adjust the pH of solutions. Citric acid monohydrate is used in the preparation of effervescent granules, while anhydrous citric acid is widely used in the preparation of effervescent tablets.[1–3] Citric acid has also been shown to improve the stability of spray-dried insulin powder in inhalation formulations.[4]

In food products, citric acid is used as a flavor enhancer for its tart, acidic taste. Citric acid monohydrate is used as a sequestering agent and antioxidant synergist; *see* Table I. It is also a component of anticoagulant citrate solutions.

Table I: Uses of citric acid monohydrate.

Use	Concentration (%)
Buffer solutions	0.1–2.0
Complexing agent	0.3–2.0
Flavor enhancer for liquid formulations	0.3–2.0

8 Description

Citric acid monohydrate occurs as colorless or translucent crystals, or as a white crystalline, efflorescent powder. It is odorless and has a strong acidic taste. The crystal structure is orthorhombic.

9 Pharmacopeial Specifications

The pharmacopeial specifications for citric acid monohydrate have undergone harmonization of many attributes for JP, PhEur, and USP–NF.

See Table II. *See also* Sections 17 and 18.

Table II: Pharmacopeial specifications for citric acid monohydrate (and anhydrous).

Test	JP XV	PhEur 7.4	USP35–NF30
Identification	+	+	+
Characters	—	+	—
Appearance of solution	+	+	+
Water			
(hydrous form)	7.5–9.0%	7.5–9.0%	7.5–9.0%
(anhydrous form)	≤1.0%	≤1.0%	≤1.0%
Sterility	—	—	+
Bacterial endotoxins[b]	—	+	+
Residue on ignition	≤0.1%	—	≤0.1%
Sulfated ash	—	≤0.1%	—
Calcium	+	—	—
Aluminum	—	≤0.2 ppm	≤0.2 ppm[a]
Oxalic acid	+	≤360 ppm	≤0.036%
Sulfate	+	≤150 ppm	≤0.015%
Heavy metals[b]	≤10 ppm	≤10 ppm	≤10 ppm
Readily carbonizable substances	+	+	+
Assay (anhydrous basis)	99.5–100.5%	99.5–100.5%	99.5–100.5%

(a) Where it is labeled as intended for use in dialysis.
(b) These tests have not been fully harmonized at the time of publication.

Note that the JP XV, PhEur 7.4 and USP35–NF30 have separate monographs for the monohydrate and anhydrous material.

SEM 1: Excipient: citric acid monohydrate; manufacturer: Pfizer Ltd; magnification: 60×.

SEM 2: Excipient: citric acid monohydrate; manufacturer: Pfizer Ltd; magnification: 600×.

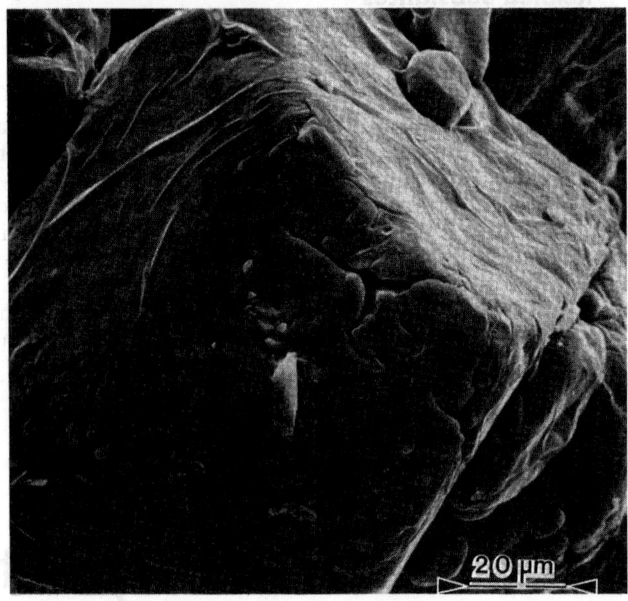

10 Typical Properties

Acidity/alkalinity pH = 2.2 (1% w/v aqueous solution)

Dissociation constant

pK_{a1}: 3.128 at 25°C;

pK_{a2}: 4.761 at 25°C;

pK_{a3}: 6.396 at 25°C.

Density 1.542 g/cm^3

Heat of combustion −1972 kJ/mol (−471.4 kcal/mol)

Heat of solution −16.3 kJ/mol (−3.9 kcal/mol) at 25°C

Hygroscopicity At relative humidities less than about 65%, citric acid monohydrate effloresces at 25°C, the anhydrous acid being formed at relative humidities less than about 40%. At relative humidities between about 65% and 75%, citric acid monohydrate absorbs insignificant amounts of moisture, but under more humid conditions substantial amounts of water are absorbed.

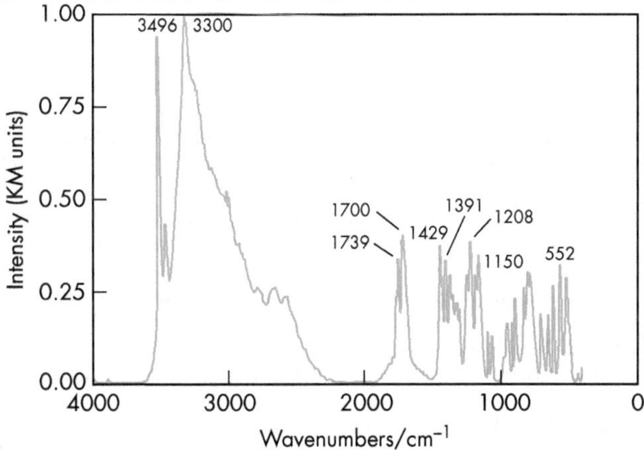

Figure 1: Infrared spectrum of citric acid monohydrate measured by diffuse reflectance. Adapted with permission of Informa Healthcare.

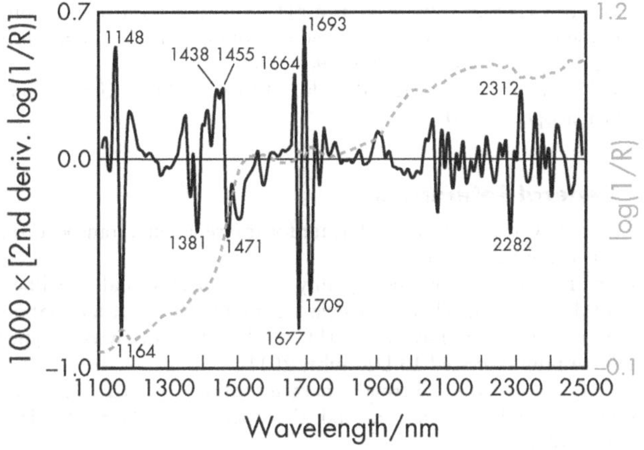

Figure 2: Near-infrared spectrum of citric acid monohydrate measured by reflectance.

Melting point ≈100°C (softens at 75°C)

Particle size distribution Various grades of citric acid monohydrate with different particle sizes are commercially available.

Solubility Soluble 1 in 1.5 parts of ethanol (95%) and 1 in less than 1 part of water; sparingly soluble in ether.

Spectroscopy

IR spectra *see* Figure 1.

NIR spectra *see* Figure 2.

Viscosity (dynamic)

6.5 mPa s (6.5 cP) for a 50% w/v aqueous solution at 25°C.

See also Section 17.

11 Stability and Storage Conditions

Citric acid monohydrate loses water of crystallization in dry air or when heated to about 40°C. It is slightly deliquescent in moist air. Dilute aqueous solutions of citric acid may ferment on standing.

The bulk monohydrate or anhydrous material should be stored in airtight containers in a cool, dry place.

12 Incompatibilities

Citric acid is incompatible with potassium tartrate, alkali and alkaline earth carbonates and bicarbonates, acetates, and sulfides. Incompatibilities also include oxidizing agents, bases, reducing agents, and nitrates. It is potentially explosive in combination with metal nitrates. On storage, sucrose may crystallize from syrups in the presence of citric acid.

13 Method of Manufacture

Citric acid occurs naturally in a number of plant species and may be extracted from lemon juice, which contains 5–8% citric acid, or pineapple waste. Anhydrous citric acid may also be produced industrially by mycological fermentation of crude sugar solutions such as molasses, using strains of *Aspergillus niger.* Citric acid is purified by recrystallization; the anhydrous form is obtained from a hot concentrated aqueous solution and the monohydrate from a cold concentrated aqueous solution.

14 Safety

Citric acid is found naturally in the body, mainly in the bones, and is commonly consumed as part of a normal diet. Orally ingested citric acid is absorbed and is generally regarded as a nontoxic material when used as an excipient. However, excessive or frequent consumption of citric acid has been associated with erosion of the teeth.[5]

Citric acid and citrates also enhance intestinal aluminum absorption in renal patients, which may lead to increased, harmful serum aluminum levels. It has therefore been suggested that patients with renal failure taking aluminum compounds to control phosphate absorption should not be prescribed citric acid or citrate-containing products.[6]

See Section 17 for anhydrous citric acid animal toxicity data.

15 Handling Precautions

Observe normal precautions appropriate to the circumstances and quantity of material handled. Eye protection and gloves are recommended. Direct contact with eyes can cause serious damage. Citric acid should be handled in a well-ventilated environment or a dust mask should be worn. It is combustible.

16 Regulatory Status

GRAS listed. The anhydrous form is accepted for use as a food additive in Europe. Included in the FDA Inactive Ingredients Database (inhalations; IM, IV, and other injections; ophthalmic preparations; oral capsules, solutions, suspensions and tablets; topical and vaginal preparations). Included in nonparenteral and parenteral medicines licensed in Japan and the UK. Included in the Canadian Natural Health Products Ingredients Database.

17 Related Substances

Anhydrous citric acid; fumaric acid; malic acid; sodium citrate dihydrate; tartaric acid.

Anhydrous citric acid

Empirical formula $C_6H_8O_7$

Molecular weight 192.12

CAS number [77-92-9]

Synonyms Acidum citricum anhydricum; citric acid; E330; 2-hydroxy-β-1,2,3-propanetricarboxylic acid; 2-hydroxypropane 1,2,3-tricarboxylic acid.

Appearance Odorless or almost odorless, colorless crystals or a white crystalline powder. Crystal structure is monoclinic holohedral.

Dissociation constants

pK_{a1}: 3.128 at 25°C;

pK_{a2}: 4.761 at 25°C;

pK_{a3}: 6.396 at 25°C.

Density 1.665 g/cm^3

Heat of combustion –1985 kJ/mol (–474.5 kcal/mol)

Hygroscopicity At relative humidities between about 25–50%, anhydrous citric acid absorbs insignificant amounts of water at 25°C. However, at relative humidities between 50% and 75%, it absorbs significant amounts, with the monohydrate being formed at relative humidities approaching 75%. At relative humidities greater than 75% substantial amounts of water are absorbed by the monohydrate.

Melting point 153°C

Solubility Soluble 1 in 1 part of ethanol (95%) and 1 in 1 of water; sparingly soluble in ether.

Safety

LD_{50} (mouse, IP): 0.9 g/kg[7]
LD_{50} (mouse, IV): 0.04 g/kg
LD_{50} (mouse, oral): 5.04 g/kg
LD_{50} (mouse, SC): 2.7 g/kg
LD_{50} (rabbit, IV): 0.33 g/kg
LD_{50} (rat, IP): 0.88 g/kg
LD_{50} (rat, oral): 3.0 g/kg
LD_{50} (rat, SC): 5.5 g/kg

Comments Anhydrous citric acid is listed in the PhEur 7.4 and USP35–NF30. Anhydrous citric acid is one of the materials that have been selected for harmonization by the Pharmacopeial Discussion Group. For further information see the General Information Chapter <1196> in the USP35–NF30, the General Chapter 5.8 in PhEur 7.4, along with the 'State of Work' document on the PhEur EDQM website, and also the General Information Chapter 8 in the JP XV.

The EINECS number for anhydrous citric acid is 201-069-1.

18 Comments

Citric acid monohydrate has undergone harmonization of many attributes for JP, PhEur, and USP–NF by the Pharmacopeial Discussion Group. For further information see the General Information Chapter <1196> in the USP35–NF30, the General Chapter 5.8 in PhEur 7.4, along with the 'State of Work' document on the PhEur EDQM website, and also the General Information Chapter 8 in the JP XV.

Citric acid monohydrate has been used experimentally to adjust the pH of tablet matrices in enteric-coated formulations for colon-specific drug delivery.[8]

Therapeutically, preparations containing citric acid have been used to dissolve renal calculi.

A specification for citric acid monohydrate is contained in the *Food Chemicals Codex* (FCC).[9]

The EINECS number for citric acid monohydrate is 201-069-1. The PubChem Compound ID (CID) for citric acid monohydrate is 22230.

19 Specific References

1 Anderson NR, *et al.* Quantitative evaluation of pharmaceutical effervescent systems II: stability monitoring by reactivity and porosity measurements. *J Pharm Sci* 1982; **71**(1): 7–13.
2 Yanze FM, *et al.* A process to produce effervescent tablets: fluidized bed dryer melt granulation. *Drug Dev Ind Pharm* 2000; **26**(11): 1167–1176.
3 Nykänen P, *et al.* Citric acid as excipient in multiple-unit enteric-coated tablets for targeting drugs on the colon. *Int J Pharm* 2001; **229**(1–2): 155–162.
4 Todo H, *et al.* Improvement of stability and absorbability of dry insulin powder for inhalation by powder-combination technique. *Int J Pharm* 2004; **271**(1–2): 41–52.
5 Anonymous. Citric acid: tooth enamel destruction. *Clin Alert* 1971; 151.
6 Main I, Ward MK. Potentiation of aluminium absorption by effervescent analgesic tablets in a haemodialysis patient. *Br Med J* 1992; **304**: 1686.
7 Lewis RJ, ed. *Sax's Dangerous Properties of Industrial Materials*, 11th edn. New York: Wiley, 2004: 955.
8 Nykaenen P, *et al.* Citric acid as a pH-regulating additive in granules and the tablet matrix in enteric-coated formulations for colon-specific drug delivery. *Pharmazie* 2004; **59**(4): 268–273.
9 *Food Chemicals Codex*, 7th edn. Bethesda, MD: United States Pharmacopeia, 2010.

20 General References

Cho MJ, *et al.* Citric acid as an adjuvant for transepithelial transport. *Int J Pharm* 1989; 52: 79–81.

European Directorate for the Quality of Medicines and Healthcare (EDQM). European Pharmacopoeia – State Of Work Of International Harmonisation. *Pharmeuropa* 2011; 23(4): 713–714. www.edqm.eu/site/-614.html (accessed 13 December 2011).

Timko RJ, Lordi NG. Thermal characterization of citric acid solid dispersions with benzoic acid and phenobarbital. *J Pharm Sci* 1979; 68: 601–605.

21 Author

GE Amidon.

22 Date of Revision

13 December 2011.

Coconut Oil

1 Nonproprietary Names

BP: Coconut Oil
JP: Coconut Oil
PhEur: Coconut Oil, Refined
USP–NF: Coconut Oil

2 Synonyms

Aceite de coco; cocois oleum raffinatum; coconut butter; copra oil; oleum cocois; *Pureco 76*; refined coconut oil.

3 Chemical Name and CAS Registry Number

Coconut oil [8001-31-8]

4 Empirical Formula and Molecular Weight

Coconut oil contains triglycerides, the fatty acid constituents of which are mainly lauric and myristic acids with smaller proportions of capric, caproic, caprylic, oleic, palmitic and stearic acids.

The PhEur 7.4 and USP33–NF28 S2 state that the fatty acid composition for coconut oil is caproic acid (≤1.5%), caprylic acid (5.0–11.0%), capric acid (4.0–9.0%), lauric acid (40.0–50.0%), myristic acid (15.0–20.0%), palmitic acid (7.0–12.0%), stearic acid (1.5–5.0%), arachidic acid (≤0.2%), palmitoleic acid (≤1.0%), oleic acid (4.0–10.0%), linoleic acid (1.0–3.0%), linolenic acid (≤0.2%), and eicosenoic acid (≤0.2%).

5 Structural Formula

See Section 4.

6 Functional Category

Emollient; ointment base.

7 Applications in Pharmaceutical Formulation or Technology

Coconut oil has traditionally been used in ointments where it forms a readily absorbable base. It has been used particularly in preparations intended for application to the scalp, where it could be applied as a solid but would liquefy when applied to the skin. Coconut oil is readily saponified by strong alkalis even in the cold and, as the soap produced is not readily precipitated by sodium chloride, it has been used in the making of 'marine' soap.

Coconut oil may be used in the formulation of a range of other preparations including emulsions[1,2] and nanoemulsions,[3,4] intranasal solutions,[5] and rectal capsules[6] and suppositories.[7]

See Table I.

Table I: Uses of coconut oil.

Use	Concentration (%)
Liquid soaps	4–20
Shampoos	1–20
Soaps	60–75
Topical ointments	50–70

8 Description

Coconut oil generally occurs as a white to light-yellow mass or colorless or light-yellow clear oil, with a slight odor characteristic of coconut and a mild taste. Refined coconut oil is a white or almost white unctuous mass.

The form that coconut oil takes depends on temperature; it occurs as a pale yellow to colorless liquid between 28°C and 30°C, as a semisolid at 20°C , and as a hard brittle crystalline solid below 15°C.

9 Pharmacopeial Specifications

See Table II.

Table II: Pharmacopeial specifications for coconut oil.

Test	JP XV	PhEur 7.4	USP35–NF30
Identification	—	+	+
Characters	+	+	—
Melting point	20–28°C	23–26°C	23–26°C
Acid value	≤0.2	≤0.5	≤0.5
Peroxide value	—	≤5.0	≤5.0
Saponification value	246–264	—	—
Unsaponifiable matter	≤1.0%	≤1.0%	≤1.0%
Iodine value	7–11	—	—
Alkaline impurities in fatty oils	—	+	+
Composition of fatty acids	—	+	+
Water	—	≤0.1%	≤0.1%
Arsenic	—	—	≤0.5 μg/g

Note: both the USP35–NF30 and PhEur 7.4 specify the fatty acid composition for coconut oil. In addition, the USP35–NF30 includes a specification for palmitoleic acid (≤0.1%).

10 Typical Properties

Boiling point >450°C
Flash point 216°C (closed cup)
Iodine number 8–9.5
Melting point 23–26°C
Refractive index n_D^{40} = 1.448–1.450[18]
Saponification number 255–258
Specific gravity 0.918–0.923
Solubility Practically insoluble in water; freely soluble in dichloromethane and in light petroleum (bp: 65–70°C); soluble in ether, carbon disulfide, and chloroform; soluble at 60°C in 2 parts of ethanol (95%) but less soluble at lower temperatures.
Surface tension
33.4 mN/m (33.4 dyne/cm) at 20°C;
28.4 mN/m (28.4 dyne/cm) at 80°C.

11 Stability and Storage Conditions

Coconut oil remains edible, and mild in taste and odor, for several years under ordinary storage conditions. However, on exposure to air, the oil readily oxidizes and becomes rancid, acquiring an unpleasant odor and strong acid taste.

Store in a tight, well-filled container, protected from light at a temperature not exceeding 25°C. Coconut oil may be combustible at high temperature, and may spontaneously heat and ignite if stored under hot and wet conditions.

12 Incompatibilities

Coconut oil reacts with oxidizing agents, acids and alkalis. Polyethylene is readily permeable to coconut oil.

It has been shown that the increased force required to expel coconut oil from plastic syringes was due to uptake of the oil into the rubber plunger; this resulted in swelling of the rubber plunger and an increased resistance to movement down the syringe barrel.[9]

13 Method of Manufacture

Coconut oil is the fixed oil obtained from the seeds of *Cocos nucifera* Linn. (Palmae). This oil is then refined to produce refined coconut oil, which is referred to in the coconut industry as RBD (refined, bleached, and deodorized) coconut oil.

14 Safety

When administered orally, coconut oil is essentially nontoxic, although ingestion of large amounts may cause digestive or gastrointestinal irritation or upset. Coconut oil can act as an irritant when applied to the skin and when in contact with the eyes; it may be absorbed through the skin. Inhalation of mist or vapor may cause respiratory tract irritation.

15 Handling Precautions

Observe normal precautions appropriate to the circumstances and quantity of the material handled. Coconut oil should be kept away from heat and sources of ignition, and contact with oxidizing agents, acids, and alkalis should be avoided.

If in the solid form, large spillages of coconut oil should be dealt with by shoveling the material into a waste disposal container. For liquid spillages, the oil should be absorbed with an inert material before removal for disposal.

16 Regulatory Status

Included in the FDA Inactive Ingredients Database (oral capsules and tablets; topical creams, solutions, and ointments). Included in scalp ointments and therapeutic shampoos licensed in the UK.

17 Related Substances

Almond oil; canola oil; castor oil; castor oil, hydrogenated; corn oil; cottonseed oil; medium-chain triglycerides; olive oil; peanut oil; sesame oil; soybean oil; sunflower oil.

18 Comments

In addition, coconut oil has been reported to have antifungal activity against a range of *Candida* species.[10]

Coconut oil has been used therapeutically in a lotion for the eradication of head lice,[11] and was included in a regime used to treat a patient who had ingested 16.8 g aluminum phosphide.[12] Virgin coconut oil may have anti-inflammatory, analgesic and antipyretic properties.[13]

Concern has been expressed at the potential use of coconut oil as a suntan lotion as it does not afford any protection against ultraviolet light.[14]

A specification for coconut oil (unhydrogenated) is contained in the *Food Chemicals Codex* (FCC).[15] The USP35–NF30 also includes a monograph for Hydrogenated Coconut Oil.

19 Specific References

1 Hung CF, *et al.* The effect of oil components on the physicochemical properties and drug delivery of emulsions: tocol emulsion versus lipid emulsion. *Int J Pharm* 2007; **335**(1–2): 193–202.
2 Garti N, Arkad O. Preparation of cloudy coconut oil emulsions containing dispersed TiO_2 using atomizer. *J Dispersion Sci Technol* 1986; **7**(5): 513–523.
3 Fang JY, *et al.* Lipid nano/submicron emulsions as delivery vehicles for topical flurbiprofen delivery. *Drug Deliv* 2004; **11**(2): 97–105.
4 Al-Edresi S, Baie S. Formulation and stability of whitening VCO-in-water nano-cream. *Int J Pharm* 2009; **373**: 174–178.
5 Nielsen HW, *et al.* Intranasal administration of different liquid formulations of bumetanide to rabbits. *Int J Pharm* 2000; **204**(1–2): 35–41.
6 Tanabe K, *et al.* Effect of different suppository bases on release of indomethacin. *J Pharm Sci Technol Jpn* 1984; **44**: 115–120.
7 Broda H, *et al.* Preparation and testing of suppositories with cephradine – evaluation of pharmaceutical availability and biological availability. *Farm Pol* 1993; **49**(11–12): 1–8.
8 Todd RG, Wade A, eds. *The Pharmaceutical Codex*, 11th edn. London: Pharmaceutical Press, 1979: 214.
9 Dexter MB, Shott MJ. The evaluation of the force to expel oily injection vehicles from syringes. *J Pharm Pharmacol* 1979; **31**: 497–500.
10 Ogbolu DO, *et al. In vitro* antimicrobial properties of of coconut oil on *Candida* species in Ibadan, Nigeria. *J Med Food* 2007; **10**(2): 384–387.
11 Anonymous. Nitlotion. *Pharm J* 2000; **265**: 345.
12 Pajoumand A, *et al.* Survival following severe aluminium phosphide poisoning. *J Pharm Pract Res* 2002; **32**(4): 297–299.
13 Intahphuak S, *et al.* Anti-inflammatory, analgesic, and antipyretic activities of virgin coconut oil. *Pharm Biol* 2010; **48**: 151–157.
14 Patel HR. Sales of coconut oil. *Pharm J* 1992; **249**: 252.
15 *Food Chemicals Codex*, 7th edn (Suppl. 1). Bethesda, MD: United States Pharmacopiea, 2010: 240.

20 General References

—

21 Author

CG Cable.

22 Date of Revision

1 March 2012.

Colloidal Silicon Dioxide

1 Nonproprietary Names

BP: Colloidal Anhydrous Silica
JP: Light Anhydrous Silicic Acid
PhEur: Silica, Colloidal Anhydrous
USP–NF: Colloidal Silicon Dioxide

2 Synonyms

Aerosil ; *Aeroperl*; *Cab-O-Sil*; colloidal silica; fumed silica; fumed silicon dioxide; hochdisperses silicum dioxid; SAS; silica colloidalis anhydrica; silica sol; silicic anhydride; silicon dioxide colloidal; silicon dioxide fumed; synthetic amorphous silica; *HDK*.

3 Chemical Name and CAS Registry Number

Colloidal silicon dioxide [112945-52-5]

4 Empirical Formula and Molecular Weight

SiO_2 60.08

5 Structural Formula

See Section 4.

6 Functional Category

Adsorbent; anticaking agent; emulsion stabilizing agent; glidant; suspending agent; tablet and capsule disintegrant; viscosity-increasing agent.

7 Applications in Pharmaceutical Formulation or Technology

Colloidal silicon dioxide is widely used in pharmaceuticals, cosmetics, and food products; *see* Table I. Its small particle size and large specific surface area give it desirable flow characteristics that are exploited to improve the flow properties of dry powders[1] in a number of processes such as tableting[2–4] and capsule filling.[5]

Colloidal silicon dioxide is also used to stabilize emulsions and as a thixotropic thickening and suspending agent in gels,[6] micoemulsions,[7] and semisolid preparations.[6] With other ingredients of similar refractive index, transparent gels may be formed. The degree of viscosity increase depends on the polarity of the liquid (polar liquids generally require a greater concentration of colloidal silicon dioxide than nonpolar liquids).[8] Viscosity is largely independent of temperature but may be affected by pH;[8] *see* Section 11.

In aerosols, other than those for inhalation, colloidal silicon dioxide is used to promote particulate suspension, eliminate hard settling, and minimize the clogging of spray nozzles. Colloidal silicon dioxide is also used as a tablet disintegrant and as an adsorbent dispersing agent for liquids in powders.[9] Colloidal silicon dioxide is frequently added to suppository formulations containing lipophilic excipients to increase viscosity, prevent sedimentation during molding, and decrease the release rate.[10,11]

Colloidal silicon dioxide is also used as an adsorbent during the preparation of wax microspheres;[12] as a thickening agent for topical preparations;[13] as a freeze-drying aid for nanocapsules and nanosphere suspensions.[14] as an inert core in solid dispersions;[15] and as a gel-former for transdermal systems.[8]

Table I: Uses of colloidal silicon dioxide.

Use	Concentration (wt %)
Aerosols	0.5–3.0
Emulsion stabilizing agent	1.0–5.0
Glidant	0.1–1.0
Suppositories	0.5–2.0
Suspending and thickening agent	2.0–10.0

8 Description

Colloidal silicon dioxide is a submicroscopic fumed silica with a particle size of about 15 nm. It is a light, loose, bluish-white-colored, odorless, tasteless, amorphous powder.

9 Pharmacopeial Specifications

See Table II. *See also* Section 18.

10 Typical Properties

Acidity/alkalinity pH = 3.8–4.2 (4% w/v aqueous dispersion) and 3.5–4.0 (10% w/v aqueous dispersion) for *Cab-O-Sil M-5P*

Density (bulk) 0.029–0.042 g/cm^3

Density (tapped) *see* Tables III, IV, and V.

Melting point 1600°C

Moisture content *see* Figure 1.[12,13]

Particle size distribution Primary particle size is 7–16 nm. *Aerosil* forms loose agglomerates of 10–200 μm. *See also* Figure 2.

Refractive index 1.48[8]

Solubility Practically insoluble in organic solvents, water, and acids, except hydrofluoric acid; soluble in hot solutions of alkali hydroxide. Forms a colloidal dispersion with water. For *Aerosil*, solubility in water is 150 mg/L at 25°C (pH 7).

Specific gravity 2.2

Specific surface area

100–400 m^2/g depending on grade. *See also* Tables III, IV,and V.

Several grades of colloidal silicon dioxide are commercially available, which are produced by modifying the manufacturing process. The modifications do not affect the silica content,

SEM 1: Excipient: colloidal silicon dioxide (*Aerosil A-200*); manufacturer: Evonik Degussa Corporation lot no.: 87A-1 (04169C); magnification: 600×; voltage: 20 kV.

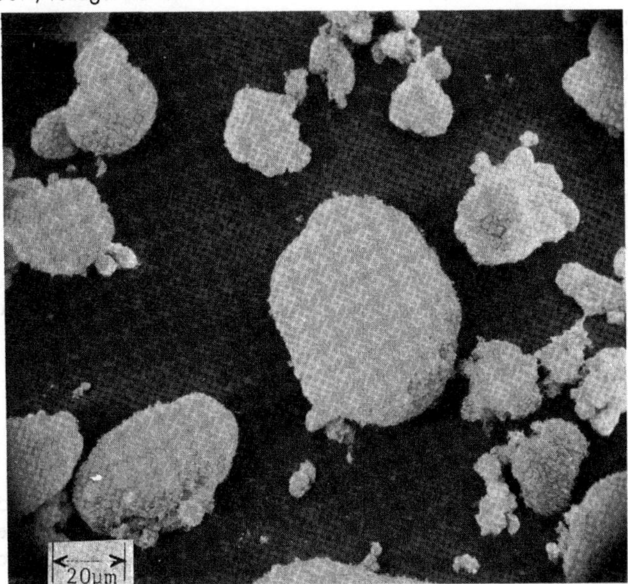

SEM 2: Excipient: colloidal silicon dioxide (*Aerosil A-200*); manufacturer: Evonik Degussa Corporation lot no.: 87A-1 (04169C); magnification: 2400×; voltage: 20 kV.

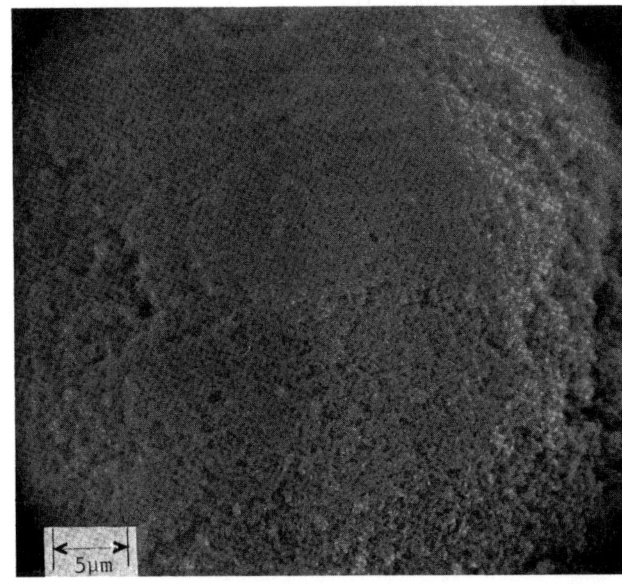

Table II: Pharmacopeial specifications for colloidal silicon dioxide.

Test	JP XV	PhEur 7.4	USP35–NF30
Identification	+	+	+
Characters	—	+	—
pH (4% w/v dispersion)	—	3.5–5.5	3.5–5.5
Arsenic	≤5 ppm	—	≤8 ppm
Chloride	≤0.011%	≤250 ppm	—
Heavy metals	≤40 ppm	≤25 ppm	—
Aluminum	+	—	—
Calcium	+	—	—
Iron	≤500 ppm	—	—
Loss on drying	≤7.0%	—	≤2.5%
Loss on ignition	≤12.0%	≤5.0%	≤2.0%
Volume test (5 g sample)	≥70 mL	—	—
Assay (on ignited sample)	≥98.0%	99.0–100.5%	99.0–100.5%

Table III: Physical properties of *Aerosil* and *Aeroperl*.

Grade	Specific surface area[(a)] (m^2/g)	Density (tapped) (g/cm^3)
Aerosil 130	130 ± 25	0.05
Aerosil 130v	130 ± 25	0.12
Aerosil 200	200 ± 25	0.05
Aerosil 200v	200 ± 25	0.20
Aerosil 300	300 ± 30	0.05
Aerosil 300v	300 ± 30	0.12
Aeroperl 300 Pharma	300 ± 30	0.28
Aerosil 380	380 ± 30	0.05
Aeorosil 380v	380 ± 30	0.12

(a) BET method.

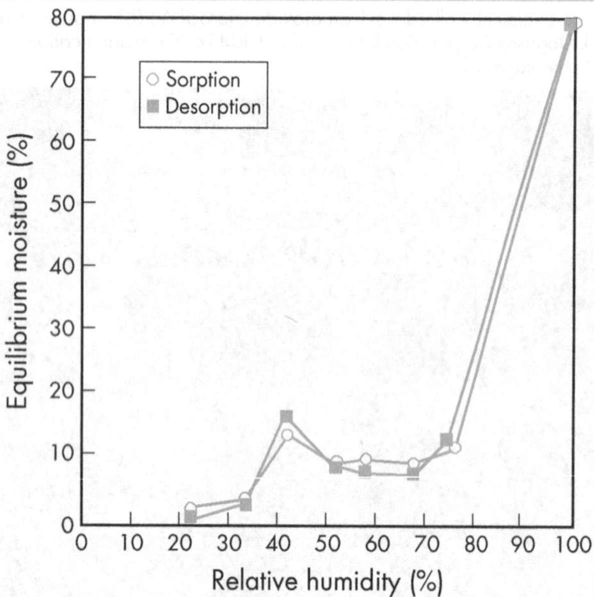

Figure 1: Sorption–desorption isotherm for colloidal silicon dioxide.

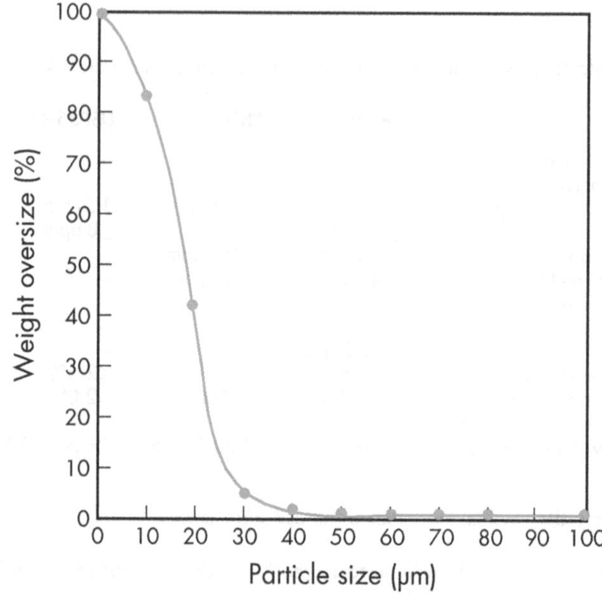

Figure 2: Particle size distribution of colloidal silicon dioxide (*Aerosil A-200*, Evonik Degussa Corp.).

Table IV: Physical properties of *Cab-O-Sil*.

Grade	Specific surface area[a] (m^2/g)	Density (tapped) (g/cm^3)
M-5P	200 ± 25	0.04
M-5DP	200 ± 25	0.04

(a) BET method.

specific gravity, refractive index, color, or amorphous form. However, particle size, surface areas, and densities are affected. The physical properties of three commercially available colloidal silicon dioxides, *Aerosil* (Evonik Degussa Corporation), *Cab-O-Sil* (Cabot Corporation), and *HDK* (Wacker-Chemie GmbH) are shown in Tables III, IV and V, respectively.

Spectroscopy

IR spectra *see* Figure 3.

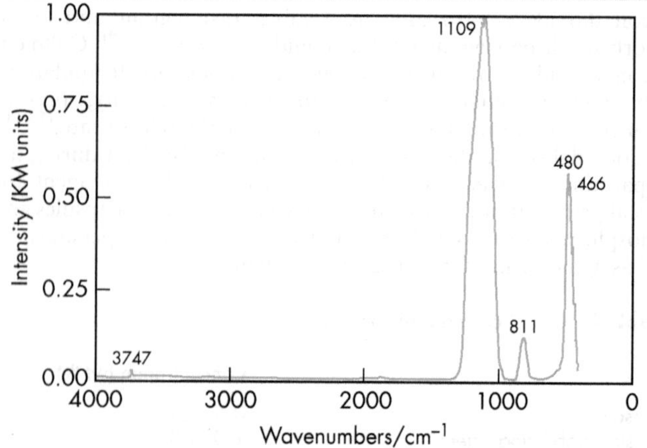

Figure 3: Infrared spectrum of colloidal silicon dioxide measured by diffuse reflectance. Adapted with permission of Informa Healthcare.

Table V: Physical properties of *HDK*.

Grade	Specific surface area[a] (m^2/g)	Density (tapped) (g/cm^3)
V15	150 ± 20	0.05
V15P	150 ± 20	0.05
N20	200 ± 30	0.04
N20P	200 ± 30	0.04

(a) BET method.

11 Stability and Storage Conditions

Colloidal silicon dioxide is hygroscopic but adsorbs large quantities of water without liquefying. When used in aqueous systems at a pH 0–7.5, colloidal silicon dioxide is effective in increasing the viscosity of a system. However, at a pH greater than 7.5 the viscosity-increasing properties of colloidal silicon dioxide are reduced; and at a pH greater than 10.7 this ability is lost entirely since the silicon dioxide dissolves to form silicates.[18] Colloidal silicon dioxide powder should be stored in a well-closed container.

12 Incompatibilities

Incompatible with diethylstilbestrol preparations.[19]

13 Method of Manufacture

Colloidal silicon dioxide is prepared by the flame hydrolysis of chlorosilanes, such as silicon tetrachloride, at 1800°C using a hydrogen–oxygen flame. Rapid cooling from the molten state during manufacture causes the product to remain amorphous.[20]

14 Safety

Colloidal silicon dioxide is widely used in oral and topical pharmaceutical products and is generally regarded as an essentially nontoxic and nonirritant excipient. However, intraperitoneal and subcutaneous injection may produce local tissue reactions and/or granulomas. Colloidal silicon dioxide should therefore not be administered parenterally.

LD_{50} (rat, IV): 0.015 g/kg[21]
LD_{50} (rat, oral): 3.16 g/kg

15 Handling Precautions

Observe normal precautions appropriate to the circumstances and quantity of material handled. Eye protection and gloves are recommended. Considered a nuisance dust, precautions should be taken to avoid inhalation of colloidal silicon dioxide. In the absence of suitable containment facilities, a dust mask should be worn when handling small quantities of material. For larger quantities, a dust respirator is recommended.[22] Colloidal silica may build up static charge due to friction during pneumatic conveying and some engineering risk controls may be required.[23]

Inhalation of amorphous colloidal silicon dioxide dusts may cause irritation to the respiratory tract but it is not associated with fibrosis of the lungs (silicosis), which can occur upon exposure to crystalline silica.[8,20,21]

16 Regulatory Acceptance

GRAS listed. Included in the FDA Inactive Ingredients Database (oral capsules, suspensions, and tablets; transdermal, rectal, and vaginal preparations). Also approved by the FDA as a food additive and for food contact. Included in nonparenteral medicines licensed in the UK. Included in the Canadian Natural Health Products Ingredients Database.

17 Related Substances

Hydrophobic colloidal silica.

18 Comments

Colloidal silicon dioxide is one of the materials that have been selected for harmonization by the Pharmacopeial Discussion Group. For further information see the General Information Chapter <1196> in the USP35–NF30, the General Chapter 5.8 in PhEur 7.4, along with the 'State of Work' document on the PhEur EDQM website, and also the General Information Chapter 8 in the JP XV.

The PhEur 7.4 also contains a specification for hydrated colloidal silicon dioxide (SiO_2,xH_2O). The incidence of microbial contamination of colloidal silicon dioxide is low[24] due to the high production temperatures and inorganic precursor materials.

Note that porous silica gel particles may also be used as a glidant, thickener, dispersant and to adsorb moisture, which may be an advantage for some formulations. *Syloid 244FP* meets the USP–NF requirements for silicon dioxide, and *Syloid 244 FP-BU* meets the PhEur and JP requirements for silicon dioxide.[25]

The former CAS number for colloidal silicon dioxide was 7631-86-9.[8]

Listed on the Japanese Ministry of International Trade and Industry (MITI) list of Existing and New Chemical Substances (ENCS) as 1-548.

The EINECS number for colloidal silicon dioxide is 231-545-4. The PubChem Compound ID (CID) for colloidal silicon dioxide is 24261.

19 Specific References

1 York P. Application of powder failure testing equipment in assessing effect of glidants on flowability of cohesive pharmaceutical powders. *J Pharm Sci* 1975; **64**: 1216–1221.
2 Lerk CF, *et al.* Interaction of lubricants and colloidal silica during mixing with excipients I: its effect on tabletting. *Pharm Acta Helv* 1977; **52**: 33–39.
3 Lerk CF, Bolhuis GK. Interaction of lubricants and colloidal silica during mixing with excipients II: its effect on wettability and dissolution velocity. *Pharm Acta Helv* 1977; **52**: 39–44.
4 Jonat S, *et al.* Influence of compacted hydrophobic and hydrophilic colloidal silicon dioxide on tabletting properties of pharmaceutical excipients. *Drug Dev Ind Pharm* 2005; **31**: 687–696.
5 Ito K, *et al.* Studies on hard gelatin capsules. II. The capsule filling on powders and effects of glidant by ring filling method-machine. *Chem Pharm Bull* 1969; **17**(6): 1138–1145.
6 Daniels R, *et al.* The stability of drug absorbates on silica. *Drug Dev Ind Pharm* 1986; **12**: 2127–2156.
7 Rozman B, *et al.* Dual influence of colloidal silica on skin deposition of vitamins C and E simultaneously incorporated in topical microemulsions. *Drug Dev Ind Pharm* 2010; **36**(7): 852–860.
8 Evonik Degussa Corporation. Technical Literature TI 1281: *Aerosil* colloidal silicon dioxide for pharmaceuticals. 2010.
9 Sherriff M, Enever RP. Rheological and drug release properties of oil gels containing colloidal silicon dioxide. *J Pharm Sci* 1979; **68**: 842–845.
10 Tukker JJ, De Blaey CJ. The addition of colloidal silicon dioxide to suspension suppositories II. The impact on *in vitro* release and bioavailability. *Acta Pharm Technol* 1984; **30**: 155–160.
11 Realdon N, *et al.* Effects of silicium dioxide on drug release from suppositories. *Drug Dev Ind Pharm* 1997; **23**: 1025–1041.
12 Mani N, *et al.* Microencapsulation of a hydrophilic drug into a hydrophobic matrix using a salting-out procedure. II. Effects of adsorbents on microsphere properties. *Drug Dev Ind Pharm* 2004; **30**(1): 83–93.
13 Gallagher SJ, *et al.* Effects of membrane type and liquid/liquid phase boundary on *in vitro* release of ketoprofen from gel formulations. *J Drug Target* 2003; **11**(6): 373–379.
14 Schaffazick SR, *et al.* Freeze-drying polymeric colloidal suspensions: nanocapsules, nanospheres and nanodispersion. A comparative study. *Eur J Pharm Biopharm* 2003; **56**(3): 501–505.
15 Tran HTT, *et al.* Preparation and characterization of pH-independent sustained release tablet containing solid dispersion granules of a poorly water-soluble drug. *Int J Pharm* 2011; **415**(1–2): 83–88.
16 Ettlinger M, *et al.* [Adsorption at the surface of fumed silica.] *Arch Pharm* 1987; **320**: 1–15[in German].
17 Callahan JC, *et al.* Equilibrium moisture content of pharmaceutical excipients. *Drug Dev Ind Pharm* 1982; **8**: 355–369.
18 Cabot Corporation. Technical literature: *Cab-O-Sil* fumed silicas, the performance additives, 1995.
19 Johansen H, M§ller N. Solvent deposition of drugs on excipients II: interpretation of dissolution, adsorption and absorption characteristics of drugs. *Arch Pharm Chem Sci Ed* 1977; **5**: 33–42.
20 Waddell WW. Silica amorphous. In: *Kirk-Othmer Encyclopedia of Chemical Technology*, 5th edn, 22. New York: John Wiley & Sons, 2001: 380–406.
21 Lewis RJ, ed. *Sax's Dangerous Properties of Industrial Materials*, 11th edn. New York: Wiley, 2004: 3205.
22 Evonik Industries. Technical bulletin: Fine Particles T28 – Handling of synthetic silica and silicate, May 2011.
23 Evonik Industries, Technical bulletin: Fine Particles T62 – Synthetic silica and electrostatic charges, June 2011.
24 Kienholz M. [The behaviour of bacteria in highly pure silicic acid.] *Pharm Ind* 1970; **32**: 677–679[in German].
25 WR Grace & Co. Technical literature: *Syloid 244FP*, 2002.

20 General References

Cabot Corporation. Technical literature PDS171: *Cab-O-Sil M-5P*, 2006.

Evonik Degussa Corporation. Technical bulletin–Fine Particles No. 11 TB0011-1: Basic characteristics of *Aerosil* fumed silica, 2006.

Evonik Degussa Corporation. Product information: Aeroperl, 2007.

European Directorate for the Quality of Medicines and Healthcare (EDQM). European Pharmacopoeia – State Of Work Of International Harmonisation. *Pharmeuropa* 2011; 23(4): 713–714. http://www.edqm.eu/site/-614.html (accessed 28 September 2011)

Meyer K, Zimmermann I. Effect of glidants in binary powder mixtures. *Powder Technol* 2004; **139**: 40–54.

Wacker-Chemie AG. Technical literature: *HDK*–Pyrogenic silica, May 2010.

Wagner E, Brunner H. *Aerosil*: its preparation, properties and behaviour in organic liquids. *Angew Chem* 1960; **72**: 744–750.

Yang KY, *et al.* Effects of amorphous silicon dioxides on drug dissolution. *J Pharm Sci* 1979; **68**: 560–565.

Zimmermann I, *et al.* Nanomaterials as flow regulators in dry powders. *Z Phys Chem* 2004; **218**: 51–102.

21 Author

KP Hapgood.

22 Date of Revision

1 March 2012.

Colophony

1 Nonproprietary Names

BP: Colophony
JP: Rosin
PhEur: Colophony

2 Synonyms

Abietic anhydride; colophane; colophonium; colophony; disproportionated rosin; gum rosin; pine rosin; rosin; wood rosin; yellow resin.

3 Chemical Name and CAS Registry Number

(2*R*,3*S*,4*S*,5*R*,6*R*)-2-(Hydroxymethyl)-6-[(*E*)-3-phenylprop-2-enoxy]oxane-3,4,5-triol [8050-09-7] and [8050-10-0]

4 Empirical Formula and Molecular Weight

Colophony is a mixture composed of approximately 90% acids. Rosin acids are monocarboxylic and have a typical molecular formula of $C_{20}H_{30}O_2$. The acids are diterpenes of two main types: abietic with conjugated double bonds and pimaric with nonconjugated double bonds. The molecular weight of abietic and primaric acids is 302.45. The 10% other components include a mixture of dihydroabietic acid ($C_{20}H_{32}O_2$) with a molecular weight of 302.45 and dehydroabietic acid ($C_{20}H_{28}O_2$) with a molecular weight of 300.44.

5 Structural Formula

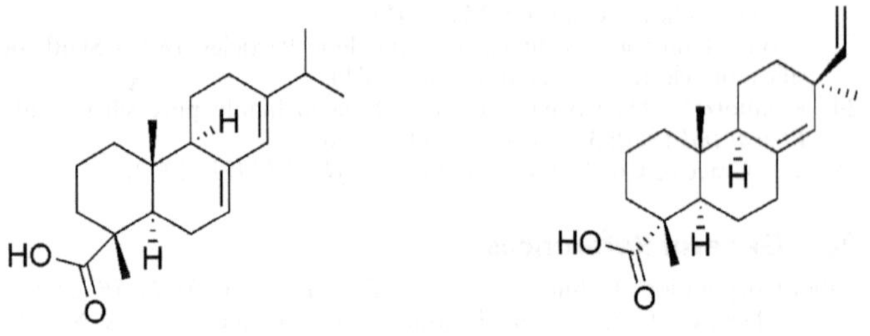

Abietic acid

Pimaric acid

6 Functional Category

Coating agent; film-forming agent; microencapsulating agent; modified-release agent; tablet and capsule binder.

7 Applications in Pharmaceutical Formulation or Technology

Colophony is used as a binder and coating agent in both capsules and tablets for oral delivery. It has also been used to prepare microcapsules for the controlled release of both aspirin[1] and diltiazem.[2].

In addition, colophony is widely used in cosmetics such as lipstick, mascara, depilatories, hair tonics, and shampoo formulations.[3–5] Food-related applications of colophony esters include use as coating agents on fruits and can linings, as polishing agents for roasted coffee beans, and for density adjustment of citrus oils in beverages.[5]

8 Description

Colophony occurs as pale yellow to dark amber chunks or fragments, and has a slight turpentine odor and taste.

Colophony has been classified into various color grades starting with D (very dark red) through E, F, G, H, I, K, M, N (golden amber), WG, WW, and finally X (pale yellow), according to the United States Department of Agriculture's standard method.

See also Sections 10 and 17.

9 Pharmacopeial Specifications

Table I: Pharmacopeial specifications for colophony.

Test	JP XV	PhEur 7.4
Characters	+	+
Identification	–	+
Acid value	150–177	145–180
Total ash	≤0.1%	≤0.2%

10 Typical Properties

Acid value
160 for grades N, WG, WW, X;
155 for grades D, H, K, M

Ash content
0.05% for grades N, WG, WW, X, D;
0.2% for grades H, K, M

Boiling point 318°C (100 mmHg)

Flash point 187°C

Insoluble matter
0.1% for grades N, WG, WW, X, D;
0.4% for grades H, K, M.

Melting point 100–150°C

Softening point 70–78°C

Solubility Insoluble in water; soluble in ethanol (96%), benzene, ether, methanol, glacial acetic acid, chloroform, oils, and carbon disulfide. Also soluble in dilute solutions and fixed alkali hydroxides.

Specific gravity 1.07–1.09

Unsaponifiable matter 6.0%

Volatile matter ≤2.0%

11 Stability and Storage Conditions

Stable under normal temperature and pressure. Colophony is combustible and is not stable in the presence of strong oxidizing agents and excess heat. It is biodegradable and biocompatible.[6]

Store in airtight containers in a dry, cool, well-ventilated area, away from heat or sources of ignition and oxidation. The PhEur 7.4 specifies that colophony should not be reduced to powder.

12 Incompatibilities

Colophony is incompatible with strong oxidizing agents.

13 Method of Manufacture

Colophony is the resinous constituent of the oleoresin exuded by various species of pine, known in commerce as crude turpentine. Natural sources include *Pinus soxburghi*, *Pinus lognifolium*, *Pinus palustris*, *Pinus pinaceae*, *Pinus caribaea*, and *Pinus toeda*.[7]

Colophony is obtained by tapping of the living pine tree to collect oleoresin, which is distilled to yield a vitreous resin. Wood rosin is derived by solvent extraction of aged and ground pine stump wood. The trunk of the pine tree is used as raw material in the production of tall oil rosin. Owing to their common source, gum, wood, and tall oil rosins are similar in composition. The crude colophony is treated by chemical and physical means to produce refined colophony with desired characteristics and applications.[8,9]

14 Safety

Colophony is generally considered to be a nontoxic material. However, it may be harmful if ingested in large quantities. Colophony may be irritating to the skin upon topical exposure and may cause allergic contact dermatitis.[10,11] It may also be an irritant to the respiratory system if inhaled as a dust.

LD_{50} (guinea pig, oral): 4.1 g/kg[12]
LD_{50} (mouse, oral): 0.002 g/kg[13]
LD_{50} (rat, oral): 0.003 g/kg[13]
LD_{50} (rat, inhalation): 0.11 g/kg[13]

15 Handling Precautions

Observe normal precautions appropriate to the circumstances and quantity of material handled.

Eye protection and gloves are recommended. Avoid the generation and accumulation of colophony dust. A respirator is not normally required in well-ventilated areas but is indicated for prolonged contact with heavy dust concentrations.

16 Regulatory Status

Included in the FDA Inactive Ingredients Database (oral capsules and oral coated, sustained- action, and repeat-action tablets). Included in the Canadian Natural Health Products Ingredients Database.

17 Related Substances

Glycerol ester of gum rosin; hydrogenated rosin; pentaerythritol ester of gum rosin; polymerized rosin.

Glycerol ester of gum rosin

Molecular weight ≈700–750 g/mol
CAS number [8050-31-5]
Synonyms *FloraRez G85, G85-L*; *Dertoline*; ester gum; glycerol-modified gum rosin; glycerol abietate; glycerol triabietate; rosin glycerol ester
Solubility Soluble in acetone and toluene; insoluble in water.
Storage Store in well-closed containers.
Comments A hard, pale amber-colored resin obtained by glycerol esterification of refined gum rosin.[14] The Joint FAO/WHO Expert Committee on Food Additives (JECFA) established the group average daily intake 0–25 mg/kg body weight for glycerol ester of gum rosin and glycerol ester of wood rosin.[12]

A specification for glycerol ester of gum rosin is included in the *Food Chemicals Codex* (FCC)[14]; *see* Table II.

Table II: FCC specification for glycerol ester of gum rosin.

Test	FCC 7
Description	+
Identification	+
Lead	≤1.0 mg/kg
Acid number	3–9
Softening point	82°C

The EINECS number for glycerol ester of gum rosin is 232-482-5.

Hydrogenated rosin

Molecular weight ≈400 g/mol
CAS number [65997-06-0]
Synonyms *FloraRez HR*; *Hydrogral*
Softening point 74–84°C
Solubility Soluble in alcohols, esters, ketones, hydrocarbons, chlorinated solvents, and mineral oils; insoluble in water.
Comments A thermoplastic acidic resin obtained by hydrogenation of rosin.

The EINECS number for hydrogenated rosin is 266-041-3.

Pentaerythritol ester of gum rosin

Molecular weight ≈1 200 g/mol
CAS number [8050-26-8]
Synonyms *FloraRez PE100*; *Granolite*; pentaerythritol ester of rosin; pentaerythritol rosinate; pentaerythritol resin
Softening point 95–105°C
Solubility Soluble in aromatic, aliphatic and chlorinated solvents; insoluble in water and alcohols.
Comments Obtained by pentaerythritol esterification of gum rosin.

The EINECS number for pentaerythritol ester of gum rosin is 232-479-9.

Polymerized rosin

Molecular weight ≈680 g/mol
CAS number [65997-05-9]
Synonyms *FloraRez DR105, 115,140, 95*; *Dertopol*; *Polyrosin;* rosin oligomer
Softening point 90–145°C
Solubility Slightly soluble in alcohols and aliphatic hydrocarbons. Soluble in aromatic hydrocarbons. Insoluble in water.
Comments Obtained by polymerization of abietic acid, which is initially changed to a dimer by linking two of its molecules.[15,16] Polymerized rosin is pale in color and has a lower acid number and higher melting point than rosin.[17]

The EINECS number for polymerized rosin is 500-163-2.

18 Comments

Therapeutically, colophony has been used as an ingredient of ointments and dressings for wounds and minor skin disorders.[18]

The reactive centers of the rosin acid molecule are often used for chemical modifications to synthesize rosin polymers with customized properties. Rosin derivatives show film-forming and coating properties,[19] and may be used for tablet film- and enteric-coating purposes. Polymerized rosins show superior film-forming and sustained-release properties.[20] Plasticized film formulations of rosins may be used to sustain the drug release from drug-layered nonpareil seeds.[21] Rosin and polymerized rosin have also been evaluated for the development of transdermal delivery systems.[22,23]

Rosin-glycerol ester, pentaester, propylene glycol ester, and polymerized colophony derivatives also exhibit wall-forming and encapsulating properties.[24–26] They have been used to formulate microcapsules[27] and nanoparticles.[28] PEGylated rosin derivatives have recently been investigated as controlled-release film formers.[29–31]

Modified abietic acids, such as maleoabietic acid and diabietic acid, are used as hydrophobic matrix materials for controlled release of drugs from tablets.[32,33] Glycerol, sorbitol, and mannitol esters of rosin are used as chewing gum bases for medicinal applications.[34,35] Custom synthesized rosin derivatives can also be used as emulsifying and surface-active agents in formulating creams and lotions.[36]

The degradation and biocompatibility of colophony and rosin acid-based biomaterials (glycerol ester of maleic rosin and pentaerythritol ester of maleic rosin) have been examined *in vitro* and *in vivo*, and they are reported to have good biocompatibility.[6,37] These polymers have been used as implant materials for controlled antibiotic release.[38] Rosin esters are used in fragrances (perfumes) to retard rapid evaporation.

The glycerol ester of hydrogenated rosin [8050-29-1] is included in the FDA Inactive Ingredients Database (nasal ointment).

Monographs of other colophony derivatives such as glycerol ester of partially dimerized rosin, glycerol ester of partially hydrogenated gum rosin, glycerol ester of polymerized rosin, glycerol ester of tall oil rosin, pentaerythritol ester of partially hydrogenated wood rosin, and pentaerythritol ester of wood rosin are included in the *Food Chemicals Codex*.

The EINECS number for colophony is 232-475-7. The PubChem Compound ID (CID) for colophony is 5280656.

19 Specific References

1 Pathak YV. Study of compressibility of rosin coated aspirin microcapsules. *Eastern Pharmacist (India)* 1989; **32**: 121–123.
2 Narender Reddy M, Shirwaikar AA. Rosin, a polymer for microencapsulation of diltiazem hydrochloride for sustained release by an emulsion-solvent evaporation technique. *Indian J Pharm Sci* 2000; **62**: 308–310.
3 Uehara K, *et al.* Dermatological external agent. United States Patent 5,292,531; 1994.
4 Armstrong DP, *et al.* Detergent composition. United States Patent 4,265, 782; 1981.
5 Corzani I, *et al.* Polymeric compositions for sustained release of volatile materials. United States Patent 7,833,515; 2010.
6 Satturwar PM, *et al.* Biodegradation and *in vivo* biocompatibility of rosin: a natural film-forming polymer. *AAPS PharmSciTech* 2003; **4**(4): E55.
7 Silvestre A, Gandini A. Rosin: Major sources, properties and applications. In: Belgacem MN, Gandini A, eds. *Monomers, Polymers and Composites from Renewable Resources*. Kidlington, Oxford: Elsevier, 2008: 67-88.
8 Palkin S, *et al.* A new non-crystallizing gum rosin. *J Am Oil Chem Soc* 1938; **15**(5): 120–122.
9 Brady GS, *et al. Materials Handbook*. 15th edn. New York: McGraw-Hill Inc., 2002; 803.
10 Kanvera L *et al. Handbook of Occupational Dermatitis*. New York: Springer Verlag, 2000: 509.
11 Downs AM, Sansom JE. Colophony allergy: a review. *Contact Dermatitis* 1999; **41**: 305–310.
12 FAO/WHO. Safety evaluation of certain food additives Seventy-first meeting of the Joint FAO/WHO expert committee on food additives. *WHO Food Additive Series* 2010; **62**: 119–132.
13 Lewis RJ, ed. *Sax's Dangerous Properties of Industrial Materials*. 11th edn. New York: Wiley, 2004; 3167.
14 *Food Chemicals Codex*. 7th edn. Bethesda, MD: United States Pharmacopoeia, 2010: 199.
15 Gigante B, *et al.* The structure of an abietic acid dimer. *J Chem Soc Chem Commun* 1986; **13**: 1038–1039.
16 Parkin BA *et al.* Polymerized rosin product and method of preparation of the same. United States Patent 3,891,612; 1975.
17 Brady GS *et al. Materials Handbook*. 15th edn. New York: McGraw-Hill Inc., 2002: 805.
18 Sweetman S, ed. *Martindale: The Complete Drug Reference*. 37th edn. London: Pharmaceutical Press, 2011: 2505.
19 Satturwar PM, *et al.* Rosin derivatives: novel film forming materials for controlled drug delivery. *React Funct Polym* 2002; **50**: 233–242.
20 Fulzele SV, *et al.* Polymerized rosin: novel film forming polymer for drug delivery. *Int J Pharm* 2002; **249**(1–2): 175–184.
21 Satturwar PM, *et al.* Evaluation of new rosin derivatives for pharmaceutical coating. *Int J Pharm* 2004; **270**: 27–36.
22 Horstmann M *et al.* Principles of skin adhesion and methods for measuring adhesion of transdermal systems. In: Mathiowitz E, Chickering DE, Lehr CM, eds. *Bioadhesive Drug Delivery Systems: Fundamentals, Novel Approaches and Development*. New York: Marcel Dekker, 1999: 183.
23 Satturwar PM, *et al.* Evaluation of polymerized rosin for the formulation and development of transdermal drug delivery system: a technical note. *AAPS PharmSciTech* 2005; **6**(4): E649–E654.
24 Pathak YV, Dorle AK. Study of rosin glycerol esters as microencapsulating materials. *J Microencapsul* 1985; **2**(2): 137–140.
25 Pathak YV, *et al. In vivo* performance of pentaester gum-coated aspirin microcapsules. *J Microencapsul* 1987; **4**: 107–110.
26 Sheorey DS, Dorle AK. Preparation and *in vitro* evaluation of rosin microcapsules: solvent evaporation technique. *J Microencapsul* 1990; **7**(2): 261–264.
27 Fulzele SV, *et al.* Preparation and evaluation of microcapsules using polymerized rosin as a novel wall forming material. *J Microencapsul* 2004; **21**(1): 83–89.
28 Lee CM, *et al.* Rosin nanoparticles as a drug delivery carrier for the controlled release of hydrocortisone. *Biotechnol Lett* 2005; **27**(19): 1487–1490.
29 Nande VS. Investigation of PEGylated derivatives of rosin as sustained release film formers. *AAPS PharmSciTech* 2008; **9**(1): 105–111.
30 Nande VS, *et al.* Sustained release microspheres of diclofenac sodium using PEGylated rosin derivatives. *Drug Dev Ind Pharm* 2007; **33**(10): 1090–1100.
31 Morkhade DM, *et al.* A comparative study of aqueous and organic-based films and coatings of PEGylated rosin derivative. *Drug Dev Ind Pharm* 2008; **34**(1): 24–32.
32 Ramani CC, *et al.* Study of diabietic acid as matrix forming material. *Int J Pharm* 1996; **137**: 11–19.
33 Ramani CC, *et al.* Study of rosin–maleoabietic acid as a matrix forming material. *Int J Pharm Adv* 1996; **1**: 344.
34 Urzenis PW *et al.* Chewing gum product including a hydrophilic gum base and method of producing. United States Patent 6,350,480; 2002.
35 Sozzi G *et al.* Non-sticky gum base for chewing gum. United States Patent 7,817,650; 2011.
36 Dhanorkar VT, *et al.* Development and characterization of rosin-based polymer and its application as a cream base. *J Cosmet Sci* 2002; **53**(4): 199–208.
37 Fulzele SV, *et al.* Studies on biodegradation and *in vivo* biocompatibility of novel biomaterials. *Eur J Pharm Sci* 2003; **20**: 53–61.
38 Fulzele SV, *et al.* Novel biopolymers as implant matrix for the delivery of ciprofloxacin: biocompatibility, degradation and *in vitro* antibiotic release. *J Pharm Sci* 2007; **96**(1): 132–144.

20 General References

Americanized Encyclopedia Britannica. Chicago, IL: Belford Clarke Co, 1890; 5132.

ASTM International. ASTM D509 – 05(2011)e1: Standard test methods of sampling and grading rosin. http://www.astm.org/Standards/D509.htm.(accessed 3 August 2011).

Fischer Scientific. Material safety data sheet: *Rosin*, 2007.

FAO/WHO. Compendium of food additive specifications. Seventy-first meeting of Joint FAO/WHO expert committee on food additives. *World Health Organ Tech Rep Ser* 2009; No. 29.

Product data sheets: *Rosin acid derivatives*, 2008. http://www.drtresines.fr/en/products/recherche-de-produits.html (accessed 3 August 2011).

Silicastar Industries. Material safety data sheet: *Gum Rosin*, 2004.

21 Authors

S Fulzele, P Satturwar.

22 Date of Revision

1 March 2012.

Coloring Agents

1 Nonproprietary Names

See Section 17 and Tables I, II, III, and IV.

2 Synonyms

See Section 17 for specific, selected coloring agents.

3 Chemical Name and CAS Registry Number

See Tables I, II, III, and IV.

4 Empirical Formula and Molecular Weight

See Section 17 for specific selected coloring agents.

5 Structural Formula

See Section 17 for specific selected coloring agents.

6 Functional Category

Colorant; opacifier.

7 Applications in Pharmaceutical Formulation or Technology

Coloring agents are used mainly to impart a distinctive appearance to a pharmaceutical dosage form. The main categories of dosage form that are colored are:

- Tablets: either the core itself or the coating.
- Hard or soft gelatin capsules: the capsule shell or coated beads.
- Oral liquids.
- Topical creams and ointments.

Color is a useful tool to help identify a product in its manufacturing and distribution stages. Patients, especially those using multiple products, often rely on color to be able to recognize the prescribed medication.[1] The use of different colors for different strengths of the same drug can also help eliminate errors.

Many drug products look similar; hence color in combination with shape and/or an embossed or printed logo can help with identification. Also, this combination can assist in the prevention of counterfeiting.

Unattractive medication can be made more acceptable to the patient by the use of color, and color can also be used to make a preparation more uniform when an ingredient in the formulation has itself a variable appearance from batch to batch.[2]

Some of the insoluble colors or pigments have the additional benefit when used in tablet coatings or gelatin shells of providing useful opacity, which can contribute to the stability of light-sensitive active materials in the tablet or capsule formulation. Pigments such as the iron oxides, titanium dioxide, and some of the aluminum lakes are especially useful for this purpose.[3]

Of the many classifications possible for pharmaceutical coloring agents, one of the most useful is to simply divide the colors into those that are soluble in water (dyes) and those that are insoluble in water (pigments).

Colors for clear liquid preparations are limited to the dyes;[4] e.g. *see* Section 17.

For surface coloration, which includes coated tablets, the choice of color is usually restricted to insoluble pigments. The reasons for this include their lack of color migration, greater opacity, and enhanced color stability over water-soluble colors.[5]

Lakes are largely water-insoluble forms of the common synthetic water-soluble dyes. They are prepared by adsorbing a sodium or potassium salt of a dye onto a very fine substrate of hydrated alumina, followed by treatment with a further soluble aluminum salt. The lake is then purified and dried.[6]

Lakes are frequently used in coloring tablet coatings since, for this purpose, they have the general advantages of pigments over water-soluble colors. *See* Table V.

8 Description

The physical appearances of coloring agents vary widely. *See* Section 17 for specific selected coloring agents.

9 Pharmacopeial Specifications

Some materials used as pharmaceutical coloring agents are included in various pharmacopeias; for example, titanium dioxide is included in the PhEur 7.4. However, if titanium dioxide is being used exclusively as a colorant, then the specific purity criteria from Directive 2008/128/EC apply.[7]

10 Typical Properties

Typical properties of specific selected coloring agents are shown in Section 17. Selected properties are shown in Tables V, VI, and VII.

11 Stability and Storage Conditions

Pharmaceutical coloring agents form a chemically diverse group of materials that have widely varying stability properties. Specific information for selected colors is shown in Table VII and can be found in Woznicki and Schoneker.[4] *See also* Section 17.

While some colors, notably the inorganic pigments, show excellent stability, other coloring agents, such as some organic colors, have poor stability properties but are used in formulations because of their low toxicity.[8]

Some natural and synthetic organic colors are particularly unstable in light. However, with appropriate manufacturing procedures, combined with effective product packaging, these colors may be used successfully in formulations, thus making a wide choice of colors practically available.

Lakes, inorganic dyes, and synthetic dyes should be stored in well-closed, light-resistant containers at a temperature below 30°C.

For most natural and nature-identical colors, the storage conditions are more stringent and a manufacturer's recommendations for a particular coloring agent should be followed.

To extend their shelf-life, some natural colors are supplied as gelatin-encapsulated or similarly encapsulated powders and may be sealed in containers under nitrogen.

12 Incompatibilities

See Section 17 for incompatibilities of specific selected coloring agents; see also Woznicki and Schoneker,[4] and Walford.[9,10]

13 Method of Manufacture

See Section 17 and Walford[9,10] for information on specific selected coloring agents.

14 Safety

Coloring agents are used in a variety of oral and topical pharmaceutical formulations, in addition to their extensive use in foodstuffs and cosmetic products.

Table I: European Union list of coloring materials authorized for coloring medicinal products up to January 2008. *See also* Section 16.

E number	Common name	CAS number	Alternate name
E100	Curcumin	[458-37-7]	Turmeric
E101	Riboflavin	[83-88-5]	Lactoflavin
E102	Tartrazine	[1934-21-0]	
E104	Quinoline yellow	[8004-92-0]	
E110	Sunset yellow FCF	[2783-94-0]	
E120	Carmine	[1260-17-9]	Cochineal, carminic acid
E122	Carmoisine	[3567-69-9]	
E123	Amaranth	[915-67-3]	
E124	Ponceau 4R	[2611-82-7]	
E127	Erythrosine	[16423-68-0]	
E129	Allura red AC	[25956-17-6]	
E131	Patent blue V	[3536-49-0]	
E132	Indigo carmine	[860-22-0]	Indigotine
E133	Brilliant blue FCF	[2650-18-2]	
E140	Chlorophylls	[479-61-8] for (i) [519-62-0] for (ii)	Magnesium chlorophyll
E141	Copper complexes of chlorophylls and chlorophyllins	—	
E142	Green S	[3087-16-9]	Brilliant green BS
E150	Caramel	[8028-89-5]	
E151	Brilliant black BN	[2519-30-4]	Black PN
E153	Vegetable carbon	[7440-44-0]	Carbo medicinalis vegetabilis
E160	Carotenoids		
	(a) Alpha-, beta-, gamma-carotene	[7235-40-7]	
	(b) Capsanthin	[465-42-9]	Paprika oleoresin
	(c) Capsorubin	[470-38-2]	Paprika oleoresin
	(d) Lycopene	[502-65-8]	
	(e) Beta-apo-8′ carotenal	[1107-26-2]	
	(f) Ethyl ester of beta-apo-8′ carotenoic acid	—	
E161	Xanthophylls		
	(b) Lutein	[127-40-2]	
	(g) Canthaxanthin	[514-78-3]	
E162	Beetroot red	[7659-95-2]	Betanin
E163	Anthocyanins		
	Cyanidin	[528-58-5]	
	Delphidin	[528-53-0]	
	Malvidin	[643-84-5]	
	Pelargonidin	[134-04-3]	
	Peonidin	[134-01-0]	
	Petunidin	[1429-30-7]	
E170[a]	Calcium carbonate	[471-34-1]	
E171	Titanium dioxide	[13463-67-7]	
E172	Iron oxides and hydroxides	[977053-38-5]	
E173	Aluminum	[7429-90-5]	

(a) For surface coloring only.
Note: List of colors taken from Directive 94/36/EC, Annex I and IV. (*Official Journal EC* 1994; L237/13).

Toxicology studies are routinely conducted on an ongoing basis by organizations such as the World Health Organization (WHO), the US Food and Drug Administration (FDA), and the European Commission (EC). The outcome of this continuous review is that the various regulatory bodies around the world have developed lists of permitted colors that are generally regarded as being free from serious adverse toxicological effects. However, owing to the widespread and relatively large use of colors in food, a number of coloring agents in current use have been associated with adverse effects, although in a relatively small number of people.[11,12] Restrictions or bans on the use of some coloring agents have been imposed in some countries, while the same colors may be permitted for use in a different country. As a result the same color may have a different regulatory status in different territories of the world.

In 2007, a study was published linking the use of six colors, tartrazine (E102), quinoline yellow (E104), sunset yellow (E110), carmoisine (E122), ponceau 4R (E124) and allura red (E129)[13] with behavior issues in childen. However, after reviewing the results of the study, the European Food Standards Agency concluded that no change in legislation was needed.

The lake of erythrosine (FD&C red #3), for example, has been delisted (*see* Section 16) in the USA since 1990, following studies in rats that suggested it was carcinogenic. This delisting was as a result of the Delaney Clause, which restricts the use of any color shown to induce cancer in humans or animals in any amount. However, erythrosine was not regarded as being an immediate hazard to health and products containing it were permitted to be used until supplies were exhausted.[14]

Tartrazine (FD&C yellow #5) has also been the subject of controversy over its safety, and restrictions are imposed on its use in some countries; *see* Section 17.

In general, concerns over the safety of coloring agents in pharmaceuticals and foods are associated with reports of hypersensitivity[15–17] and hyperkinetic activity, especially among children.[18]

In the USA, specific labeling requirements are in place for prescription drugs that contain tartrazine (*see* Section 18) as this color was found to be the potential cause of hives in fewer than one in 10 000 people. In the EU, medicinal products containing tartrazine, sunset yellow, carmoisine, amaranth, ponceau 4R or

Table II: Permanently listed color additives subject to US certification in 2008, excluding those approved exclusively for use in medical devices.

Color	Common name	CAS number	21 CFR references to drug use
FD&C blue #1	Brilliant blue FCF	[2650-18-2]	74.1101
FD&C blue #2	Indigotine	[860-22-0]	74.1102
D&C blue #4	Alphazurine FG	[6371-85-3]	74.1104
D&C blue #9	Indanthrene blue	[130-20-1]	74.1109
FD&C green #3	Fast green FCF	[2353-45-9]	74.1203
D&C green #5	Alizarin cyanine green F	[4403-90-1]	74.1205
D&C green #6	Quinizarine green SS	[128-80-3]	74.1206
D&C green #8	Pyranine concentrated	[6358-69-6]	74.1208
D&C orange #4	Orange II	[633-96-5]	74.1254
D&C orange #5	Dibromofluorescein	[596-03-2]	74.1255
D&C orange #10	Diiodofluorescein	[38577-97-8]	74.1260
D&C orange #11	Erythrosine yellowish Na	[38577-97-8]	74.1261
FD&C red #3[a]	Erythrosine	[16423-68-0]	74.1303
FD&C red #4	Ponceau SX	[4548-53-2]	74.1304
D&C red #6	Lithol rubin B	[5858-81-1]	74.1306
D&C red #7	Lithol rubin B Ca	[5281-04-9]	74.1307
D&C red #17	Toney red	[85-86-9]	74.1317
D&C red #21	Tetrabromofluorescein	[15086-94-9]	74.1321
D&C red #22	Eosine	[17372-87-1]	74.1322
D&C red #27	Tetrachlorotetrabromofluorescein	[13473-26-2]	74.1327
D&C red #28	Phloxine B	[18472-87-2]	74.1328
D&C red #30	Helindone pink CN	[2379-74-0]	74.1330
D&C red #31	Brilliant lake red R	[6371-76-2]	74.1331
D&C red #33	Acid fuchsine	[3567-66-6]	74.1333
D&C red #34	Lake bordeaux B	[6417-83-0]	74.1334
D&C red #36	Flaming red	[2814-77-9]	74.1336
D&C red #39	Alba red	[6371-55-7]	74.1339
FD&C red #40	Allura red AC	[25956-17-6]	74.1340
FD&C red #40 lake	Allura Red AC	[68583-95-9]	74.1340
D&C violet #2	Alizurol purple SS	[81-48-1]	74.1602
FD&C yellow #5	Tartrazine	[1934-21-0]	74.1705
FD&C yellow #6	Sunset yellow FCF	[2783-94-0]	74.1706
D&C yellow #7	Fluorescein	[2321-07-5]	74.1707
Ext. D&C yellow #7	Naphthol yellow S	[846-70-8]	74.1707[a]
D&C yellow #8	Uranine	[518-47-8]	74.1708
D&C yellow #10	Quinoline yellow WS	[8004-92-0]	74.1710
D&C yellow #11	Quinoline yellow SS	[8003-22-3]	74.1711

(a) Dye is permanently listed. The lake is not permitted in medicinal products (*see* Table III).

brilliant black BN must carry a warning on the label concerning possible allergic reactions.

15 Handling Precautions

Pharmaceutical coloring agents form a diverse group of materials and manufacturers' data sheets should be consulted for safety and handling data for specific colors.

In general, inorganic pigments and lakes are of low hazard and standard chemical handling precautions should be observed depending upon the circumstances and quantity of material handled. Special care should be taken to prevent excessive dust generation and inhalation of dust.

The organic dyes, natural colors, and nature-identical colors present a greater hazard and appropriate precautions should accordingly be taken.

16 Regulatory Status

Coloring agents have an almost unique status as pharmaceutical excipients in that most regulatory agencies of the world hold positive lists of colors that may be used in medicinal products. Only colors on these lists may be used and some colors may be restricted quantitatively. The legislation also defines purity criteria for the individual coloring agents. In many regions around the world there is a distinction between colors that may be used in drugs and those for food use.

European Union legislation The primary legislation that governs coloring matters that may be added to medicinal products is Council Directive 2009/35/EC of 23 April 2009.[19] This Directive links the pharmaceutical requirements with those for foods in the EU by direct reference to those substances listed in Annex I to Council Directive 94/36/EC.[20] In addition, the Scientific Committee on Medicinal Products and Medical Devices has delivered opinions on the suitability and safety of amaranth,[21] erythrosine,[22] canthaxanthin,[23] aluminum,[24] and silver[25] as colors for medicines. Silver was considered unsuitable. Table I gives the current position taking the above information into account. Directive 2008/128/EC[7] lays down specific purity criteria for food colors and essentially replaces the provisions of the 1962 Directive. EU legislation relating to colors in medicines is clarified by the Committee for Medicinal Products for Human Use note for guidance on excipients in the dossier for application for marketing authorization of a medicinal product, EMEA/CHMP/QWP/396951/2006.[26]

United States legislation The 1960 Color Additive Amendment to the Food Drug and Cosmetic Act defines the responsibility of the Food and Drug Administration in the area of pharmaceutical colorants. Tables II, III, and IV provide lists of permitted colors.[27] The list is superficially long, but many of the coloring agents have restricted use. For the so-called certified colors, the FDA operates a scheme whereby each batch of color produced is certified as analytically correct by the FDA prior to the issuing of a certification number and document that will permit sale of the batch in question. Colors requiring certification are described as FD&C (Food Drug and Cosmetic); D&C (Drug and Cosmetic) or External D&C. The remaining colors are described as

Table III: Provisionally listed color additives subject to US certification in 2008.

Color	Common name	CAS number	21 CFR references to drug use
FD&C lakes	General	*See* individual color	82.51
D&C lakes	General	*See* individual color	82.1051
Ext. D&C lakes	General	*See* individual color	82.2051
FD&C blue #1 lake	Brilliant blue FCF	[53026-57-6]	82.101
FD&C blue #2 lake	Indigotine	[16521-38-3]	82.102
D&C blue #4 lake	Alphazurine FG	[6371-85-3]	82.1104
FD&C green #3 lake	Fast green FCF	[2353-45-9]	82.1203
D&C green #5 lake	Alizarin cyanine green F	[4403-90-1]	82.1205
D&C green #6 lake	Quinizarine green SS	[128-80-3]	82.1206
D&C orange #4 lake	Orange II	[633-56-5]	82.1254
D&C orange #5 lake	Dibromofluorescein	[596-03-2]	74.1255
D&C orange #10 lake	Diiodofluorescein	[38577-97-8]	82.1260
D&C orange #11 lake	Erythosine yellowish Na	[38577-97-8]	82.1261
FD&C red #4 lake	Ponceau SX	[4548-53-2]	82.1304
D&C red #6 lake	Lithol rubin B	[17852-98-1]	82.1306
D&C red #7 lake	Lithol rubin B Ca	[5281-04-9]	82.1307
D&C red #17 lake	Toney red	[85-86-9]	82.1317
D&C red #21 lake	Tetrabromofluorescein	[15086-94-9]	82.1321
D&C red #22 lake	Eosine	[17372-87-1]	82.1322
D&C red #27 lake	Tetrachlorotetrabromofluorescein	[13473-26-2]	82.1327
D&C red #28 lake	Phloxine B	[18472-87-2]	82.1328
D&C red #30 lake	Helindone pink CN	[2379-74-0]	82.1330
D&C red #31 lake	Brilliant lake red R	[6371-76-2]	82.1331
D&C red #33 lake	Acid fuchsine	[3567-66-6]	82.1333
D&C red #34 lake	Lake bordeaux B	[6417-83-0]	82.1334
D&C red #36 lake	Flaming red	[2814-77-9]	82.1336
D&C violet #2 lake	Alizurol purple SS	[81-48-1]	82.1602
FD&C yellow #5 lake	Tartrazine	[12225-21-7]	82.1705
FD&C yellow #6 lake	Sunset yellow FCF	[15790-07-5]	82.1706
D&C yellow #7 lake	Fluorescein	[2321-07-5]	82.1707
Ext. D&C yellow #7 lake	Naphthol yellow S	[846-70-8]	82.2707
D&C yellow #8 lake	Uranine	[518-47-8]	82.1708
D&C yellow #10 lake	Quinoline yellow WS	[68814-04-0]	82.1710

Table IV: List of color additives exempt from certification permitted for drug use in the USA in 2008.

Color	CAS number	21 CFR references to drug use
Alumina	[1332-73-6]	73.1010
Aluminum powder	[7429-90-5]	73.1645
Annatto extract	[8015-67-6]	73.1030
Beta-carotene	[7235-40-7]	73.1095
Bismuth oxychloride	[7787-59-9]	73.1162
Bronze powder	[7440-66-6]	73.1646
Calcium carbonate	[471-34-1]	73.1070
Canthaxanthin	[514-78-3]	73.1075
Caramel	[8028-89-5]	73.1085
Chromium–cobalt–aluminum oxide	[68187-11-1]	73.1015
Chromium hydroxide green	[12182-82-0]	73.1326
Chromium oxide green	[1308-38-9]	73.1327
Cochineal extract; carmine	[1260-17-9] [1390-65-4]	73.1100
Copper powder	[7440-50-6]	73.1647
Dihydroxyacetone	[62147-49-3]	73.1150
Ferric ammonium citrate	[1185-57-5]	73.1025
Ferric ammonium ferrocyanide	[25869-00-5]	73.1298
Ferric ferrocyanide	[14038-43-8]	73.1299
Guanine	[68-94-0] [73-40-5]	73.1329
Iron oxides synthetic	[977053-38-5]	73.1200
Logwood extract	[8005-33-2]	73.1410
Mica	[12001-26-2]	73.1496
Mica-based pearlescent pigments	—	73.1350
Potassium sodium copper chlorophyllin	—	73.1125
Pyrogallol	[87-66-1]	73.1375
Pyrophyllite	[8047-76-5]	73.1400
Talc	[14807-96-6]	73.1550
Titanium dioxide	[13463-67-7]	73.1575
Zinc oxide	[1314-13-2]	73.1991

uncertified colors and are mainly of natural origin. The USA also operates a system of division of certified colors into permanently and provisionally listed colors. Provisionally listed colors require the regular intervention of the FDA Commissioner to provide continued listing of these colors. Should the need arise, the legislative process for removal of these colors from use is comparatively easy.

Licensing authority approval In addition to national approvals and lists, a pharmaceutical licensing authority can impose additional restrictions at the time of application review. Within

Table V: Typical characteristic properties of aluminum lakes.

Average particle size	5–10 μm
Moisture content	12–15%
Oil absorption	40–45[a]
Specific gravity	1.7–2.0 g/cm^3
pH stability range	4.0–8.0

(a) ASTM D281-31, expressed as grams of oil per 100 g of color.

Table VI: Approximate solubilities for selected colors at 25°C (g/100 mL)[a]

Color	Water	Glycerin	Propylene glycol	Ethanol (95%)	Ethanol (50%)
Brilliant blue FCF	18	20	20	1.5	20
Indigo carmine	1.5	1	0.1	Trace	0.2
FD&C green #3	17	15	15	0.2	7
Erythrosine	12	22	22	2	4
Allura red AC	20	3	1.5	Trace	1
Tartrazine	15	18	8	Trace	4
Sunset yellow	18	15	2	Trace	2

(a) The solubility of individual batches of commercial product will differ widely depending on the amounts of salt, pure dye, moisture and subsidiary dyes present.

the EU this generally takes the form of restricting colors, such as tartrazine and other azo colors, in medicinal products for chronic administration, and especially in medicines for allergic conditions.

17 Related Substances

Beta-carotene; indigo carmine; iron oxides; sunset yellow FCF; tartrazine; titanium dioxide.

Beta-carotene

Empirical formula $C_{40}H_{56}$
Molecular weight 536.85
CAS number [7235-40-7]
Synonyms Betacarotene; β-carotene; β,β-carotene; E160a.
Structure

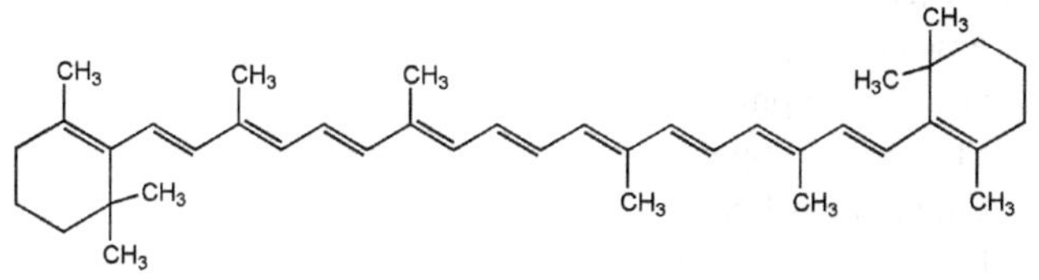

Appearance Occurs in the pure state as red crystals when recrystallized from light petroleum.
Color Index No.
CI 75130 (natural)
CI 40800 (synthetic)
Melting point 183°C
Purity (EU)
Arsenic: ≤3 ppm
Lead: ≤10 ppm
Mercury: ≤1 ppm
Cadmium: ≤1 ppm
Heavy metals: ≤40 ppm
Assay: ≥96% total coloring matters expressed as beta-carotene
Identification: maximum in cyclohexane at 453–456 nm
Sulfated ash: ≤0.2%
Subsidiary coloring matters: carotenoids other than beta-carotene, ≤3.0% of total coloring matters.
Purity (US)
Arsenic: ≤3 ppm
Assay: 96–101%
Lead: ≤10 ppm
Residue on ignition: ≤0.2%
Loss on drying: ≤0.2%
Solubility Soluble 1 in 30 parts of chloroform; practically insoluble in ethanol, glycerin, and water.
Incompatibilities Generally incompatible with oxidizing agents; decolorization will take place.
Stability Beta-carotene is very susceptible to oxidation and antioxidants such as ascorbic acid, sodium ascorbate, or tocopherols should be added. Store protected from light at a low temperature (–20°C) in containers sealed under nitrogen.
Method of manufacture All industrial processes for preparing carotenoids are based on β-ionone. This material can be obtained by total synthesis from acetone and acetylene via dehydrolinalol. The commercially available material is usually 'extended' on a matrix such as acacia or maltodextrin. These extended forms of beta-carotene are dispersible in aqueous systems. Beta-carotene is also available as micronized crystals suspended in an edible oil such as peanut oil.
Comments
Beta-carotene is capable of producing colors varying from pale yellow to dark orange. It can be used as a color for sugar-coated tablets prepared by the ladle process. However, beta-carotene is very unstable to light and air, and products containing this material should be securely packaged to minimize degradation. Beta-carotene is particularly unstable when used in spray-coating processes, probably owing to atmospheric oxygen attacking the finely dispersed spray droplets.
Because of its poor water solubility, beta-carotene cannot be used to color clear aqueous systems, and cosolvents such as ethanol must be used.
Suppositories have been successfully colored with beta-carotene in approximately 0.1% concentration.
The EINECS number for beta-carotene is 230-636-6.

Indigo carmine

Empirical formula $C_{16}H_8N_2Na_2O_8S_2$
Molecular weight 466.37
CAS number [860-22-0]
Synonyms 2-(1,3-Dihydro-3-oxo-5-sulfo-2*H*-indol-2-ylidene)-2,3-dihydro-3-oxo-1*H*-indole-5-sulfonic acid disodium salt; disodium 5,5′-indigotin disulfonate; E132; FD&C blue #2; indigotine; sodium indigotin disulfonate; soluble indigo blue.
Structure

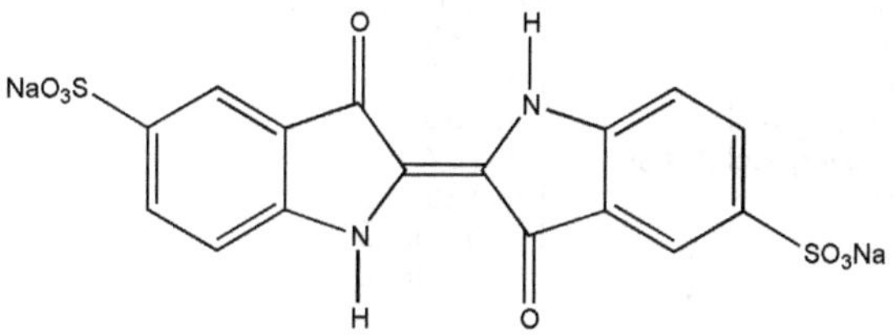

Appearance Dark blue powder. Aqueous solutions are blue or bluish-purple.
Absorption maximum 604 nm
Color Index No. CI 73015
Purity (EU)
Arsenic: ≤3 ppm
Lead: ≤10 ppm
Mercury: ≤1 ppm
Cadmium: ≤1 ppm
Heavy metals: ≤40 ppm
Ether-extractable matter: ≤0.2% under neutral conditions
Accessory colorings: ≤1.0%
Isatin-5-sulfonic acid: ≤1.0%
Water-insoluble matter: ≤0.2%
Assay: ≥85% total coloring matters, calculated as the sodium salt
Disodium 3,3′-dioxo-2,2′-biindoylidene-5,7′-disulfonate: ≤18%.
Water-insoluble matter: ≤0.2%.
Subsidiary coloring matters: excluding provision above, ≤1.0%
Organic compounds other than coloring matters: ≤0.5%
Unsulfonated primary aromatic amines: ≤0.01%, as aniline
Purity (US)
Arsenic: ≤3 ppm
2-(1,3-Dihydro-3-oxo-2*H*-indol-2-ylidene)-2,3-dihydro-3-oxo-1*H*-indole-5-sulfonic acid sodium salt: ≤2%
2-(1,3-Dihydro-3-oxo-7-sulfo-2*H*-indol-2-ylidene)-2,3-dihydro-3-oxo-1*H*-indole-5-sulfonic acid disodium salt: ≤18%
Isatin-5-sulfonic acid: ≤0.4%
Lead: ≤10 ppm
Mercury: ≤1 ppm
5-Sulfoanthranilic acid: ≤0.2%

Table VII: Stability properties of selected colors.

Color	Heat	Light	Acid	Base	Oxidizing agents	Reducing agents
Brilliant blue FCF	Good	Moderate	Very good	Moderate	Moderate	Poor
Indigo carmine	Good	Very poor	Moderate	Poor	Poor	Good
FD&C green #3	Good	Fair	Good	Poor	Poor	Very poor
Erythrosine	Good	Poor	Insoluble	Good	Fair	Very poor
Allura red AC	Good	Moderate	Good	Moderate	Fair	Fair
Tartrazine	Good	Good	Good	Moderate	Fair	Fair
Sunset yellow	Good	Moderate	Good	Moderate	Fair	Fair
D&C yellow #10	Good	Fair	Good	Moderate	Poor	Good

Total color: ≥85%

Volatile matter, chlorides and sulfates (calculated as the sodium salts): ≤15.0% at 135°C

Water-insoluble matter: ≤0.4%

Solubility *see* Table VIII.

Table VIII: Solubility of indigo carmine.

Solvent	Solubility at 20°C unless otherwise stated
Acetone	Practically insoluble
Ethanol (75%)	1 in 1430
Glycerin	1 in 100
Propylene glycol	1 in 1000
Propylene glycol (50%)	1 in 167
Water	1 in 125 at 2°C
	1 in 63 at 25°C
	1 in 45 at 60°C

Incompatibilities Poorly compatible with citric acid and saccharose solutions. Incompatible with ascorbic acid, gelatin, glucose, lactose, oxidizing agents, and saturated sodium bicarbonate solution.

Stability Sensitive to light.

Method of manufacture Indigo is sulfonated with concentrated or fuming sulfuric acid.

Safety

LD_{50} (rat, IV): 93 mg/kg

Comments Indigo carmine is an indigoid dye used to color oral and topical pharmaceutical preparations. It is used with yellow colors to produce green colors. Indigo carmine is also used to color nylon surgical sutures and is used diagnostically as a 0.8% w/v injection.

Sunset yellow FCF

Empirical formula $C_{16}H_{10}N_2Na_2O_7S_2$

Molecular weight 452.37

CAS number [2783-94-0]

Synonyms E110; FD&C yellow #6; 6-hydroxy-5-[(4-sulfophenyl)azo]-2-naphthalenesulfonic acid disodium salt; 1-*p*-sulfophenylazo-2-naphthol-6-sulfonic acid disodium salt; yellow orange S.

Structure

Appearance Reddish yellow powder. Aqueous solutions are bright orange colored.

Absorption maximum 482 nm

Color Index No. CI 15985

Purity (EU)

Arsenic: ≤3 ppm

Lead: ≤10 ppm

Mercury: ≤1 ppm

Cadmium: ≤1 ppm

Heavy metals: ≤40 ppm

Ether-extractable matter: ≤0.2% under neutral conditions

Assay: ≥85% total coloring matters as the sodium salt

Subsidiary colors: ≤5%

1-(Phenylazo)-2-napththalenol (Sudan 1): ≤0.5 mg/kg

Water-insoluble matter: ≤0.2%

Organic compounds other than coloring matters: ≤0.5%

Unsulfonated primary aromatic amines: ≤0.01% as aniline

Ether-extractable matter: ≤0.2% under neutral conditions

Purity (US)

Arsenic: ≤3 ppm

Lead: ≤10 ppm

Mercury: ≤1 ppm

4-Aminobenzenesulfonic acid: ≤0.2% as the sodium salt

6-Hydroxy-2-naphthalenesulfonic acid: ≤0.3% as the sodium salt

6,6′-Oxybis[2-naphthalenesulfonic acid]: ≤1% as the disodium salt

4,4′-(1-Triazene-1,3-diyl)bis[benzenesulfonic acid]: ≤0.1% as the disodium salt

4-Aminobenzene: ≤50 ppb

4-Aminobiphenyl: ≤15 ppb

Aniline: ≤250 ppb

Azobenzene: ≤200 ppb

Benzidine: ≤1 ppb

1,3-Diphenyltriazene: ≤40 ppb

1-(Phenylazo)-2-naphthalenol: ≤10 ppm

Total color: ≥87%

Sum of volatile matter at 135°C, chlorides and sulfates: ≤13.0%

Water-insoluble matter: ≤0.2%

Solubility *see* Table IX.

Incompatibilities Poorly compatible with citric acid, saccharose solutions, and saturated sodium bicarbonate solutions. Incompatible with ascorbic acid, gelatin, and glucose.

Method of manufacture Diazotized sulfanilic acid is coupled with Schaeffer's salt (sodium salt of β-naphthol-6-sulfonic acid).

Safety

LD_{50} (mouse, IP): 4.6 g/kg

LD_{50} (mouse, oral): >6 g/kg

LD_{50} (rat, IP): 3.8 g/kg

LD_{50} (rat, oral): >10 g/kg

Comments
Sunset yellow FCF is a monoazo dye.
The EINECS number for sunset yellow FCF is 220-491-7.

Table IX: Solubility of Sunset yellow FCF.

Solvent	Solubility at 20°C unless otherwise stated
Acetone	1 in 38.5
Ethanol (75%)	1 in 333
Glycerin	1 in 5
Propylene glycol	1 in 45.5
Propylene glycol (50%)	1 in 5
Water	1 in 5.3 at 2°C
	1 in 5.3 at 25°C
	1 in 5 at 60°C

Tartrazine
Empirical formula $C_{16}H_9N_4Na_3O_9S_2$
Molecular weight 534.39
CAS number [1934-21-0]
Synonyms 4,5-Dihydro-5-oxo-1-(4-sulfophenyl)-4-[(4-sulfophenyl)azo]-1*H*-pyrazole-3-carboxylic acid trisodium salt; E102; FD&C yellow #5; hydrazine yellow.
Structure

NaOOC N N SO_3Na NaO_3S N=N OH

Appearance Yellow or orange-yellow powder. Aqueous solutions are yellow-colored; the color is retained upon addition of hydrochloric acid solution, but with sodium hydroxide solution a reddish color is formed.
Absorption maximum 425 nm
Color Index No. CI 19140
Purity (EU)
Arsenic: ≤3 ppm
Lead: ≤10 ppm
Mercury: ≤1 ppm
Cadmium: ≤1 ppm
Heavy metals: ≤40 ppm
Assay: ≥85% total coloring matters as the sodium salt
Organic compounds other than coloring matters: ≤0.5%
Unsulfonated primary aromatic amines: ≤0.01% as aniline
Ether-extractable matter: ≤0.2% under neutral conditions
Accessory colorings: ≤1.0%
Water-insoluble matter: ≤0.2%
Purity (US)
Arsenic: ≤3 ppm
Lead: ≤10 ppm
Mercury: ≤1 ppm
Total color: ≥87.0%
Volatile matter, chlorides and sulfates (calculated as the sodium salts): ≤13.0% at 135°C
Water-insoluble matter: ≤0.2%
4,4′-[4,5-Dihydro-5-oxo-4-[(4-sulfophenyl)hydrazono]-1*H*-pyrazol-1,3-diyl]bis[benzenesulfonic acid]: ≤0.1% as the trisodium salt
4-Aminobenzenesulfonic acid: ≤0.2% as the sodium salt
4,5-Dihydro-5-oxo-1-(4-sulfophenyl)-1*H*-pyrazole-3-carboxylic acid: ≤0.2% as the disodium salt
Ethyl or methyl 4,5-dihydro-5-oxo-1-(4-sulfophenyl)-1*H*- pyrazole-3-carboxylate: ≤0.1% as the sodium salt
4,4′-(1-Triazene-1,3-diyl)bis[benzenesulfonic acid]: ≤0.05% as the disodium salt
4-Aminobenzene: ≤75 ppb
4-Aminobiphenyl: ≤5 ppb
Aniline: ≤100 ppb
Azobenzene: ≤40 ppb
Benzidine: ≤1 ppb
1,3-Diphenyltriazene: ≤40 ppb
Solubility *see* Table X.

Table X: Solubility of tartrazine.

Solvent	Solubility at 20°C unless otherwise stated
Acetone	Practically insoluble
Ethanol (75%)	1 in 91
Glycerin	1 in 5.6
Propylene glycol	1 in 14.3
Propylene glycol (50%)	1 in 5
Water	1 in 26 at 2°C
	1 in 5 at 25°C
	1 in 5 at 60°C

Incompatibilities Poorly compatible with citric acid solutions. Incompatible with ascorbic acid, lactose, 10% glucose solution, and saturated aqueous sodium bicarbonate solution. Gelatin accelerates the fading of the color.
Method of manufacture Phenylhydrazine *p*-sulfonic acid is condensed with sodium ethyl oxalacetate; the product obtained from this reaction is then coupled with diazotized sulfanilic acid.
Safety
LD_{50} (mouse, oral): >6 g/kg
LD_{50} (mouse, IP): 4.6 g/kg
LD_{50} (rat, oral): 10 g/kg
LD_{50} (rat, IP): 3.8 g/kg
Comments
Tartrazine is a monoazo, or pyrazolone, dye. It is used to improve the appearance of a product and to impart a distinctive coloring for identification purposes.
US regulations require that prescription drugs for human use containing tartrazine bear the warning statement:
This product contains FD&C yellow #5 (tartrazine) which may cause allergic-type reactions (including bronchial asthma) in certain susceptible persons.
Although the overall incidence of sensitivity to FD&C yellow #5 (tartrazine) in the general population is low, it is frequently seen in patients who are also hypersensitive to aspirin.

18 Comments

Titanium dioxide is used extensively to impart a white color to film-coated tablets, sugar-coated tablets, and gelatin capsules. It is also used in lakes as an opacifier, to 'extend' the color. *See* Titanium dioxide for further information.

In the EU, colors used in pharmaceutical formulations and colors used in cosmetics are controlled by separate regulations. Cosmetic colors are also classified according to their use, e.g. those that may be used in external products that are washed off after use.

19 Specific References

1 Hess H, Schrank J. Coloration of pharmaceuticals: possibilities and technical problems. *Acta Pharm Technol* 1979; **25**(Suppl. 8): 77–87.
2 Aulton ME, *et al.* The mechanical properties of hydroxypropylmethylcellulose films derived from aqueous systems part 1: the influence of solid inclusions. *Drug Dev Ind Pharm* 1984; **10**: 541–561.

3 Rowe RC. The opacity of tablet film coatings. *J Pharm Pharmacol* 1984; **36**: 569–572.
4 Woznicki EJ, Schoneker DR. Coloring agents for use in pharmaceuticals. In: Swarbrick J, Boylan JC, eds. *Encyclopedia of Pharmaceutical Technology*. New York: Marcel Dekker, 1990: 65–100.
5 Porter SC. Tablet coating. *Drug Cosmet Ind* 1981; **128**(5): 46, 48, 50, 53, 86–93.
6 Marmion DM. *Handbook of US Colorants for Foods, Drugs and Cosmetics*, 3rd edn. New York: Wiley-Interscience, 1991.
7 European Commission. *Official Journal EC*. 2009; L60/20.
8 Delonca H, *et al*. [Stability of principal tablet coating colors II: effect of adjuvants on color stability.] *Pharm Acta Helv* 1983; **58**: 332–337[in French].
9 Walford J, ed. *Developments in Food Colors*, 1. New York: Elsevier, 1980.
10 Walford J, ed. *Developments in Food Colors*, 2. New York: Elsevier, 1980.
11 Weiner M, Bernstein IL. *Adverse Reactions to Drug Formulation Agents: a Handbook of Excipients*. New York: Marcel Dekker, 1989: 159–165.
12 Smolinske SC. *Handbook of Food, Drug, and Cosmetic Excipients*. Boca Raton, FL: CRC Press, 1992.
13 McCann D, *et al*. Food additives and hyperactive behaviour in 3 and 8/9 year old children in the community. *Lancet* 2007; **370**(9598): 1560–1567.
14 Blumenthal D. Red No. 3 and other colorful controversies. *FDA Consumer* 1990; **21**: 18.
15 Bell T. Colourants and drug reactions [letter]. *Lancet* 1991; **338**: 55–56.
16 Lévesque H, *et al*. Reporting adverse drug reactions by proprietary name [letter]. *Lancet* 1991; **338**: 393.
17 Dietemann-Molard A, *et al*. Extrinsic allergic alveolitis secondary to carmine [letter]. *Lancet* 1991; **338**: 460.
18 Pollock I, *et al*. Survey of colourings and preservatives in drugs. *Br Med J* 1989; **299**: 649–651.
19 European Commission. *Official Journal EC*. 2009; L109/10.
20 European Commission. *Official Journal EC*. 1994; L237/13.
21 European Commission (1998). *Opinion on toxicological data on colouring agents for medicinal products: amaranth, adopted by the Scientific Committee on Medicinal Products and Medical Devices on 21 October 1998*. http://ec.europa.eu/health/scientific_committees/emerging/opinions/scmpmd/scmp_out09_en.htm (accessed 1 March 2012).
22 European Commission (1998). *Opinion on toxicological data on colouring agents for medicinal products: erythrosin, adopted by the Scientific Committee on Medicinal Products and Medical Devices on 21 October 1998*. http://ec.europa.eu/health/scientific_committees/emerging/opinions/scmpmd/scmp_out08_en.htm/ (accessed 1 March 2012).
23 European Commission (1998). *Opinion on toxicological data colouring agents for medicinal products: canthaxanthine, adopted by the Scientific Committee on Medicinal Products and Medical Devices on 21 October 1998*. http://ec.europa.eu/health/scientific_committees/emerging/opinions/scmpmd/scmp_out10_en.htm (accessed 1 March 2012).
24 European Commission (1999). *Opinion on toxicological data on colouring agents for medicinal products: aluminum, adopted by the Scientific Committee on Medicinal Products and Medical Devices on 14 April 1999*. http://ec.europa.eu/health/scientific_committees/emerging/opinions/scmpmd/scmp_out21_en.htm (accessed 1 March 2012).
25 European Commission (2000). *Opinion on toxicological data on colouring agents for medicinal products: E174 silver, adopted by the Scientific Committee on Medicinal Products and Medical Devices on 27 June 2000*. http://ec.europa.eu/health/archive/ph_risk/committees/scmp/documents/out30_en.pdf (accessed 1 March 2012).
26 European Agency for the Evaluation of Medicinal Products (EMEA). *Guideline on excipients in the dossier for application for marketing authorisation of a medicinal product*. London, 6 November 2006: EMEA/CHMP/QWP/396951/2006.
27 Code of Federal Regulations. Title 21 Parts 74, 81, 82.

20 General References

Jones BE. Colours for pharmaceutical products. *Pharm Technol Int* 1993; 5(4): 14–1618–20.

21 Author

C Mroz.

22 Date of Revision

1 February 2011.

Copovidone

1 Nonproprietary Names

BP: Copovidone

PhEur: Copovidone

USP–NF: Copovidone

2 Synonyms

Acetic acid vinyl ester, polymer with 1-vinyl-2-pyrrolidinone; copolymer of 1-vinyl-2-pyrrolidone and vinyl acetate in a ratio of 3:2 by mass; copolyvidone; copovidonum; *Kollidon VA 64*; *Luviskol VA*; *Plasdone S-630*; poly(1-vinylpyrrolidone-co-vinyl acetate); polyvinylpyrrolidone-vinyl acetate copolymer; PVP/VA; PVP/VA copolymer.

3 Chemical Name and CAS Registry Number

Acetic acid ethenyl ester, polymer with 1-ethenyl-2-pyrrolidinone [25086-89-9]

4 Empirical Formula and Molecular Weight

$(C_6H_9NO)_n \cdot (C_4H_6O_2)_m$ $\quad (111.1)_n + (86.1)_m$

The ratio of n to m is approximately $n = 1.2m$. Molecular weights of 45 000–70 000 have been determined for *Kollidon VA 64*. The average molecular weight of copovidone is usually expressed as a K-value.

The K-value of *Kollidon VA 64* is nominally 28, with a range of 25.2–30.8. The K-value of *Plasdone S 630* is specified between 25.4 and 34.2. K-values are calculated from the kinematic viscosity of a 1% aqueous solution. Molecular weight can be calculated with the formula:

$$M = 22.22\,(K + 0.075K^2)^{1.65}$$

The USP35–NF30 describes copovidone as a copolymer of 1-vinyl-2-pyrrolidinone and vinyl acetate in the mass proportion of 3:2. The PhEur 7.4 describes copovidone as a copolymer of 1-ethenylpyrrolidin-2-one and ethenyl acetate in the mass proportion of 3:2.

5 Structural Formula

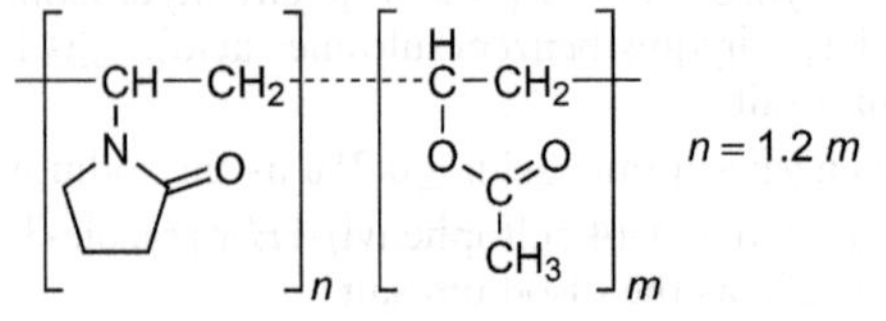

$n = 1.2\,m$

6 Functional Category

Film-forming agent; modified-release agent; tablet and capsule binder.

7 Applications in Pharmaceutical Formulation or Technology

Copovidone is used as a tablet binder, a film-former, and as part of the matrix material used in controlled-release formulations. In tableting, copovidone can be used as a binder for direct compression[1–4] and as a binder in wet granulation.[5,6] Copovidone is often added to coating solutions as a film-forming agent.[7] It provides good adhesion, elasticity, and hardness, and can be used as a moisture barrier.

See Table I.

Table I: Uses of copovidone.

Use	Concentration (%)
Film-forming agent	0.5–5.0[a]
Tablet binder, direct compression	2.0–5.0
Tablet binder, wet granulation	2.0–5.0

(a) This corresponds to the % w/w copovidone in the film-forming solution formulation, before spraying.

8 Description

Copovidone is a white to yellowish-white amorphous powder. It is typically spray-dried with a relatively fine particle size. It has a slight odor and a faint taste.

9 Pharmacopeial Specifications

See Table II. *See also* Section 18.

Table II: Pharmacopeial specifications for copovidone.

Test	PhEur 7.4	USP35–NF30
Aldehydes	≤500 ppm	≤0.05%
Appearance of solution	+	+
Characters	+	—
Ethenyl acetate	35.3–42.0%	35.3–41.4%
Heavy metals	≤20 ppm	—
Hydrazine	≤1 ppm	≤1 ppm
Identification	+	+
K-value	90.0–110.0%	90.0–110.0%
Loss on drying	≤5.0%	≤5.0%
Monomers	≤0.1%	≤0.1%
Nitrogen content	7.0–8.0%	7.0–8.0%
Peroxides	≤400 ppm	≤0.04%
2-Pyrrolidone	≤0.5%	—
Sulfated ash	≤0.1%	—
Residue on ignition	—	≤0.1%
Viscosity, expressed as *K*-value	+	—

10 Typical Properties

Density(bulk) 0.24–0.28 g/cm^3
Density (tapped) 0.35–0.45 g/cm^3
Flash point 215°C
Flowability Relatively free-flowing powder.
Glass transition temperature 106°C for *Plasdone S-630*.[8]
Hygroscopicity At 50% relative humidity, copovidone gains less than 10% weight.
K-value 25.4–34.2 for *Plasdone S-630*.[8]
Melting point 140°C

SEM 1: Excipient: copovidone (*Kollidon VA 64*); manufacturer: BASF; magnification: 400×; voltage: 10 kV.

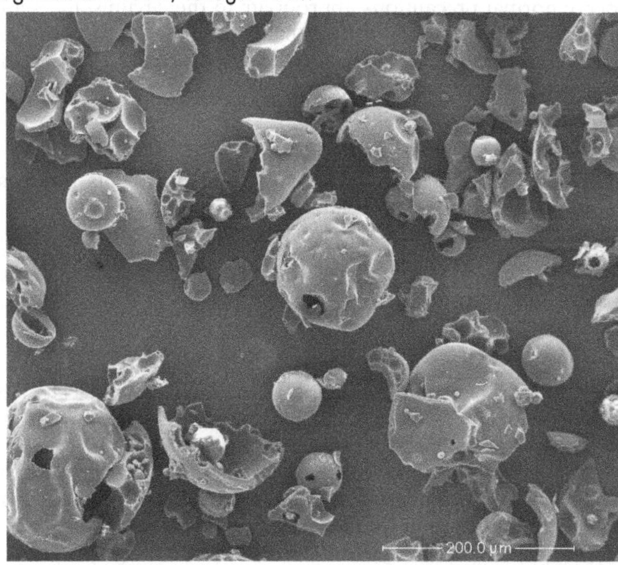

Solubility Greater than 10% solubility in 1,4-butanediol, glycerol, butanol, chloroform, dichloromethane, ethanol (95%), glycerol, methanol, polyethylene glycol 400, propan-2-ol, propanol, propylene glycol, and water. Less than 1% solubility in cyclohexane, diethyl ether, liquid paraffin, and pentane.
Viscosity (dynamic) The viscosity of aqueous solutions depends on the molecular weight and the concentration. At concentrations less than 10%, the viscosity is less than 10 mPa s (25°C).

11 Stability and Storage Conditions

Copovidone is stable and should be stored in a well-closed container in a cool, dry place.

12 Incompatibilities

Copovidone is compatible with most organic and inorganic pharmaceutical ingredients. When exposed to high water levels, copovidone may form molecular adducts with some materials; *see* Crospovidone and Povidone.

13 Method of Manufacture

Copovidone is manufactured by free-radical polymerization of vinylpyrrolidone and vinyl acetate in a ratio of 6 : 4. The synthesis is conducted in an organic solvent owing to the insolubility of vinyl acetate in water.

14 Safety

Copovidone is used widely in pharmaceutical formulations and is generally regarded as nontoxic. However, it is moderately toxic by ingestion, producing gastric disturbances. It has no irritating or sensitizing effects on the skin. A study was conducted to look at the carcinogenicity and chronic toxicity of copovidone (*Kollidon VA 64*) in Wistar rats and Beagle dogs. The results of these studies demonstrated the absence of any significant toxicological findings of high dietary levels of copodivone in rats and dogs, resulting in no-observed-adverse-effect levels of 2800 mg/kg body-weight/day in rats and 2500 mg/kg body-weight/day in dogs, the highest doses tested.[9]

LD_{50} (rat, oral): >0.63 g/kg[10]

15 Handling Precautions

Observe normal precautions appropriate to the circumstances and quantity of material handled. When heated to decomposition, copovidone emits toxic vapors of NO_x. Eye protection, gloves, and a dust mask are recommended.

16 Regulatory Status

Copovidone is included in the FDA Inactive Ingredients Database (oral tablets, oral film-coated tablets, sustained action).

17 Related Substances

Crospovidone; povidone.

18 Comments

Copovidone is one of the materials that have been selected for harmonization by the Pharmacopeial Discussion Group. For further information see the General Information Chapter <1196> in the USP35–NF30, the General Chapter 5.8 in PhEur 7.4, along with the 'State of Work' document on the PhEur EDQM website, and also the General Information Chapter 8 in the JP XV.

Kollidon VA 64 has a spherical structure, with a high proportion of damaged spheres. The shell-like structure reduces flowability, but the damaged spheres cover a greater surface area of the filler particles, increasing the efficacy of its use as a dry binder.[11] Furthermore, when used in transdermal drug delivery systems, copovidone has been shown to significantly alter the melting behavior, by reducing the heat of fusion and the melting point of estradiol and various other sex steroids.[12]

Plasdone S-630 has been used in direct compression experiments with active substances that are difficult to compress, such as acetaminophen (paracetamol); and has been shown to produce harder tablets than those containing the same actives but made with microcrystalline cellulose.[13]

In general, copovidone has better plasticity than povidone as a tablet binder, and is less hygroscopic, more elastic, and less tacky in film-forming applications than povidone.

Up to about 1975, copovidone was marketed by BASF under the name *Luviskol VA 64*. *Luviskol* is currently used only for the technical/cosmetic grade of copovidone.

A specification for copovidone is contained in the *Food Chemicals Codex* (FCC).[14]

19 Specific References

1 Moroni A. A novel copovidone binder for dry granulation and direct-compression tableting. *Pharm Tech* 2001; **25**(Suppl.): 8–24.
2 Selmeczi B. The influence of the compressional force on the physical properties of tablets made by different technological processes. *Arch Pharm (Weinheim)* 1974; **307**(10): 755–760.
3 Stamm A, Mathis C. The liberation of propyromazine from tablets prepared by direct compression. *J Pharm Belg* 1990; **29**(4): 375–389.
4 Herting MG, *et al.* Comparison of different dry binders for roll compaction/dry granulation. *Pharm Dev Tecnol* 2007; **12**(5): 525–532.
5 Vojnovic D, *et al.* Formulation and evaluation of vinylpyrrolidone/vinylacetate copolymer microspheres with griseofulvin. *J Microencapsul* 1993; **10**(1): 89–99.
6 Kristensen HG, *et al.* Granulation in high speed mixers. Part 4: Effect of liquid saturation on the agglomeration. *Pharm Ind* 1984; **46**(7): 763–767.
7 Castellanos G, *et al.* Subcoating with *Kollidon VA 64* as water barrier in a new combined native dextran/HPMC-cetyl alcohol controlled release tablet. *Eur J Pharm Biopharm* 2008; **69**(1): 303–311.
8 International Specialty Products. Technical literature: *Plasdone S-630*, 2002.
9 Mellert W, *et al.* Carcinogenicity and chronic toxicity of copovidone (*Kollidon VA 64*) in Wistar rats and Beagle dogs. *Food Chem Toxicol* 2004; **42**(10): 1573–1587.
10 Lewis RJ, ed. *Sax's Dangerous Properties of Industrial Materials*, 11th edn. New York: Wiley, 2004: 17.
11 Kolter K, Flick D. Structure and dry binding activity of different polymers, including *Kollidon VA 64*. *Drug Dev Ind Pharm* 2000; **26**(11): 1159–1165.
12 Lipp R. Selection and use of crystallization inhibitors for matrix-type transdermal drug-delivery systems containing sex steroids. *J Pharm Pharmacol* 1998; **50**: 1343–1349.
13 International Specialty Products. Technical literature: *Plasdone S0630*: a binder for direct compression and wet/dry granulation, 2002.
14 *Food Chemicals Codex*, 7th edn (Suppl. 1). Bethesda, MD: United States Pharmacopeia, 2010: 1252.

20 General References

BASF. Technical literature: *Kollidon VA 64*, March 2000.
European Directorate for the Quality of Medicines and Healthcare (EDQM). European Pharmacopoeia – State Of Work Of International Harmonisation. *Pharmeuropa* 2010; 22(4): 583–584. www.edqm.eu/en/International-Harmonisation-614.html (accessed 1 March 2012).

21 Author

O AbuBaker.

22 Date of Revision

1 March 2012.

Corn Oil

1 Nonproprietary Names

BP: Refined Maize Oil
JP: Corn Oil
PhEur: Maize Oil, Refined
USP–NF: Corn Oil

2 Synonyms

Maize oil; *Majsao CT*; maydis oleum raffinatum; maydol.

3 Chemical Name and CAS Registry Number

Corn oil [8001-30-7]

4 Empirical Formula and Molecular Weight

Corn oil is composed of fatty acid esters with glycerol, known commonly as triglycerides. Typical corn oil produced in the USA, which is rich in polyunsaturated fatty acids, contains five major fatty acids: linoleic 58.9%; oleic 25.8%; palmitic 11.0%; stearic 1.7%; and linolenic 1.1%. Corn grown outside the USA yields corn oil with lower linoleic, higher oleic, and higher saturated fatty acid levels. Corn oil also contains small quantities of plant sterols.

The USP35–NF30 describes corn oil as the refined fixed oil obtained from the embryo of *Zea mays* Linné (Fam. Gramineae).

5 Structural Formula

See Section 4.

6 Functional Category

Oleaginous vehicle; solvent.

7 Applications in Pharmaceutical Formulation or Technology

Corn oil is used primarily in pharmaceutical formulations as a solvent for intramuscular injections or as a vehicle for topical preparations. Emulsions containing up to 67% corn oil are also used as oral nutritional supplements; *see also* Section 18. When combined with surfactants and gel-forming polymers, it is used to formulate veterinary vaccines.

Corn oil has a long history of use as an edible oil and may be used in tablets or capsules for oral administration.

8 Description

Corn oil occurs as a clear, light yellow-colored, oily liquid with a faint characteristic odor and slightly nutty, sweet taste resembling cooked sweet corn.

9 Pharmacopeial Specifications

See Table I.

Table I: Pharmacopeial specifications for corn oil.

Test	JP XV	PhEur 7.4	USP35–NF30
Identification	–	+	+
Characters	+	+	–
Acid value	≤0.2	≤0.5[a]	≤0.2
Alkaline impurities	–	+	+
Cottonseed oil	–	–	–
Fatty acid composition	–	+	+
Fatty acids <C_{16}	–	≤0.6%	–
Arachidic acid	–	≤0.8%	≤0.8%
Behenic acid	–	≤0.5%	≤0.3%
Oleic acid	–	20.0–42.2%	20.0–42.2%
Eicosenoic acid	–	≤0.5%	≤0.5%
Linoleic acid	–	39.4–65.6%	39.4–62.0%
Linolenic acid	–	0.5–1.5%	0.5–1.5%
Palmitic acid	–	8.6–16.5%	8.6–16.5%
Stearic acid	–	≤3.3%	1.0–3.3%
Other fatty acids	–	≤0.5%	–
Sterols	–	≤0.3%	≤0.3%
Water	–	≤0.1%	≤0.1%
Free fatty acids	–	–	–
Heavy metals	–	–	≤10 ppm
Iodine value	103–130	–	–
Peroxide value	–	≤10.0	≤10.0
Refractive index	–	≈1.474	–
Saponification value	187–195	–	–
Specific gravity	0.915–0.921	≈0.920	–
Unsaponifiable matter	≤1.5%	≤2.8%	≤1.5%

(a) ≤0.3 if intended for parental use.

10 Typical Properties

Acid value 2–6
Autoignition temperature 393°C
Density 0.915–0.918 g/cm^3
Flash point 254°C (closed cup);[1] 321°C
Hydroxyl value 8–12
Iodine value 109–133
Melting point −18 to −10°C
Refractive index

n_D^{25} = 1.470–1.474;

n_D^{40} = 1.464–1.468.

Saponification value 187–196
Solubility Miscible with benzene, chloroform, dichloromethane, ether, hexane, and petroleum ether; practically insoluble in ethanol (95%) and water.
Viscosity (dynamic) 37–39 mPa s (37–39 cP)

11 Stability and Storage Conditions

Corn oil is stable when protected with nitrogen in tightly sealed bottles. Prolonged exposure to air leads to thickening and rancidity. Corn oil may be sterilized by dry heat, maintaining it at 150°C for 1 hour.[2]

Corn oil should be stored in an airtight, light-resistant container in a cool, dry place. Exposure to excessive heat should be avoided. Keep away from oxidizing materials.[1]

12 Incompatibilities

The photooxidation of corn oil is sensitized by cosmetic and drug-grade samples of coated titanium oxide and zinc oxide.[3]

13 Method of Manufacture

Refined corn oil is obtained from the germ or embryo of *Zea mays* Linné (Fam. Gramineae), which contains nearly 50% of the fixed oil compared with 3.0–6.5% in the whole kernel. The oil is obtained from the embryo by expression and/or solvent extraction. Refining involves the removal of free fatty acids, phospholipids, and impurities; decolorizing with solid adsorbents; dewaxing by chilling; and deodorization at high temperature and under vacuum.

14 Safety

Corn oil is generally regarded as a relatively nontoxic and nonirritant material with an extensive history of usage in food preparation. Slightly irritant on contact with skin and eyes, and on inhalation.[1]

15 Handling Precautions

Observe normal precautions appropriate to the circumstances and quantity of material handled. Spillages of this material are very slippery and should be covered with an inert absorbent material prior to disposal. Gloves, eye protection, respirator, and other protective clothing are recommended.

16 Regulatory Status

Included in the FDA Inactive Ingredients Database (IM injections; oral capsules, suspensions, and tablets). Included in the Canadian Natural Health Products Ingredients Database.

17 Related Substances

Almond oil; canola oil; cottonseed oil; peanut oil; sesame oil; soybean oil; sunflower oil.

18 Comments

Corn oil demonstrates good oxidative flavor stability owing to its nonrandom fatty acids distribution on the triglycerides and the presence of antioxidants (tocopherol and ferulic acid) in its composition.

Owing to its high content of unsaturated acids, corn oil has been used as a replacement for fats and oils containing a high content of saturated acids in the diets of patients with hypercholesterolemia.

A specification for corn oil is contained in the *Food Chemicals Codex* (FCC).[4]

The EINECS number for corn oil is 232-281-2.

19 Specific References

1 ScienceLab.com Material safety data sheet: Corn oil, 2010.
2 Pasquale D, *et al.* A study of sterilizing conditions for injectable oils. *Bull Parenter Drug Assoc* 1964; **18**(3): 1–11.
3 Sayre RM, Dowdy JC. Titanium dioxide and zinc oxide induce photooxidation of unsaturated lipids. *Cosmet Toilet* 2000; **115**: 75–8082.
4 *Food Chemicals Codex*, 7th edn. Bethesda, MD: United States Pharmacopeia, 2010: 248.

20 General References

Halbaut L, *et al.* Oxidative stability of semi-solid excipient mixtures with corn oil and its implication in the degradation of vitamin A. *Int J Pharm* 1997; **147**: 31–40.
Mann JI, *et al.* Re-heating corn oil does not saturate its double bonds [letter]. *Lancet* 1977; **ii**: 401.
Watson SA, Ramstead PE, eds. *Corn Chemistry and Technology.* St. Paul, MN: American Association of Cereal Chemists, 1987: 53–78.

21 Author

KS Alexander.

22 Date of Revision

1 March 2012.

Corn Starch and Pregelatinized Starch

1 Nonproprietary Names

None adopted.

2 Synonyms

StarCap 1500.

3 Chemical Name and CAS Registry Number

See Section 8.

4 Empirical Formula and Molecular Weight

See Section 8.

5 Structural Formula

See Section 8.

6 Functional Category

Direct compression excipient; tablet and capsule diluent; tablet and capsule disintegrant.

7 Applications in Pharmaceutical Formulation or Technology

Corn starch and pregelatinized starch can be used in both capsules and tablets to improve flowability, enhance disintegration, and improve hardness.

8 Description

Corn starch and pregelatinized starch occurs as a white free-flowing powder. It is a coprocessed mixture of predominantly corn starch together with pregelatinized starch.

9 Pharmacopeial Specifications

Both corn starch and pregelatinized starch are listed as separate monographs in the JP, PhEur, and USP–NF, but the combination is

SEM 1: Excipient: *StarCap 1500*; manufacturer: Colorcon; magnification: 200×; voltage: 3 kV.

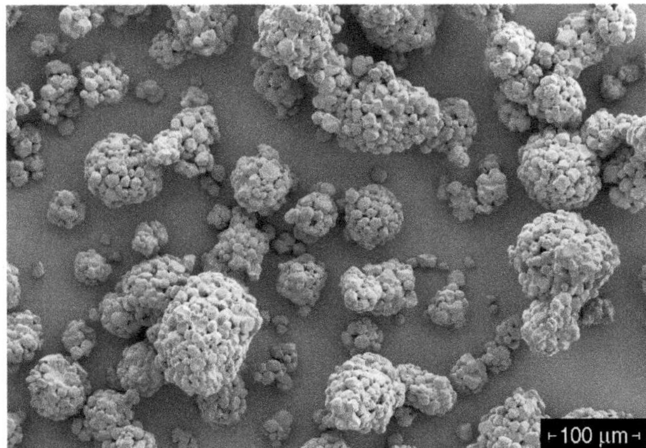

SEM 2: Excipient: *StarCap 1500*; manufacturer: Colorcon; magnification: 500×; voltage: 3 kV.

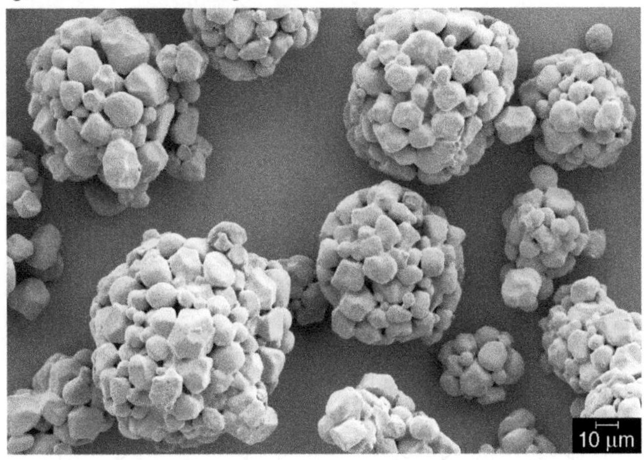

not listed. The pharmacopeial specifications for corn starch have undergone harmonization for many attributes for JP, PhEur, and USP–NF.

See Starch, and Starch, Pregelatinized. *See also* Section 18.

10 Typical Properties

Acidity/alkalinity 4.5–7.0 for *StarCap 1500*
Iron ≤0.001% for *StarCap 1500*
Loss on drying 7–13% for *StarCap 1500*
Microbial content Total aerobes count ≤100 cfu/g; molds and yeasts ≤100 cfu/g (*Escherichia coli*, *Pseudomonas aeruginosa*, and *Salmonella* species absent) for *StarCap 1500*.
Particle size distribution 9–42% retained on #120 mesh (125 μm), 25–50% retained on #200 mesh (74 μm), 20–55% passing #200 mesh (74 μm) for *StarCap 1500*
Solubility Insoluble in water for *StarCap 1500*
Sulfur dioxide ≤0.005% for *StarCap 1500*

11 Stability and Storage Conditions

Store in sealed containers at below 30°C, avoiding high humidity.

12 Incompatibilities

See Starch, and Starch, Pregelatinized.

13 Method of Manufacture

Corn starch and pregelatinized starch is produced by a proprietary spray-drying technique.

14 Safety

See Starch, and Starch, Pregelatinized.

15 Handling Precautions

Observe normal precautions appropriate to the circumstances and quantity of material handled.

16 Regulatory Status

Corn starch and pregelatinized starch is a coprocessed mixture of two materials both of which are regarded as nontoxic:
Starch GRAS listed. Included in the FDA Inactive Ingredients Database (buccal tablets, oral capsules, powders, suspensions and tablets; topical preparations; and vaginal tablets). Included in nonparenteral medicines licensed in the UK. Included in the Canadian Natural Health Products Ingredients Database.
Pregelatinized starch Included in the FDA Inactive Ingredients Database (oral capsules, suspensions, and tablets; vaginal preparations). Included in non-parenteral medicines licensed in the UK.

17 Related Substances

Starch; starch, pregelatinized.

18 Comments

Corn starch has undergone harmonization for many attributes for JP, PhEur, and USP-NF by the Pharmacopeial Discussion Group. For further information see the General Information Chapter <1196> in the USP35–NF30, the General Chapter 5.8 in PhEur 7.4, along with the 'State of Work' document on the PhEur EDQM website, and also the General Information Chapter 8 in the JP XV.

StarCap 1500 is a free-flowing, low-dust excipient with disintegration and dissolution properties independent of medium pH, which help promote deaggregation of the powder mass into primary drug particles and speeds up the dissolution rate of the drug substance, providing rapid disintegration across the pH range present in the human digestive tract.[1–3] *StarCap 1500* has been used in studies to determine the influence of disintegrants on the release rate of theophylline.[4] The coprocessed product has been designed specifically for use in capsules and directly compressed tablets, and has enhanced physical properties that cannot be achieved by single blend. It has been reported as having excellent properties for high-dose, high-solubility capsule formulations, with low weight and good content uniformity.[1] The product acts as a compression aid, diluent, and disintegrant, which allows for robust but simple capsule and directly compressible tablet formulations.

19 Specific References

1 Colorcon. Technical datasheet, version 1: *StarCap 1500*. StarCap 1500 utilized in a direct-fill capsule formulation of a high dose/high solubility active drug – gabapentin capsules 300 mg, August 2007.
2 Colorcon. Product information sheet, version 3: Why *StarCap 1500* in capsules? February 2006.
3 Colorcon. AAPS annual meeting and exposition poster reprint: Evaluation of *StarCap 1500* in a propranolol hydrochloride capsule formulation, November 2005.
4 Slodownik T, *et al*. Influence of disintegrants on theophylline release rate. *Farmacja Polska (Poland)* 2008; **64**: 197–201.

20 General References

Colorcon. Product specification: *StarCap 1500* co-processed starch, January 2007.

Deorkar N. High-functionality excipients: a review. *Tablets and Capsules* 2008: 22–26. www.tabletscapsules.com (accessed 29 November 2011).

European Directorate for the Quality of Medicines and Healthcare (EDQM). European Pharmacopoeia – State Of Work Of International Harmonisation. *Pharmeuropa* 2011; 23(4): 713–714. www.edqm.eu/site/-614.html (accessed 29 November 2011).

Gohel MC, Jogani PD. A review of co-processed directly compressible excipients. *J Pharm Pharm Sci* 2005; 8(1): 76–93.

Nachaegari SK, Bansal AK. Coprocessed excipients for solid dosage forms. *Pharm Tech* 2004; **28**: 52–64.

21 Authors

ME Fenton, RC Rowe.

22 Date of Revision

29 November 2011.

Corn Syrup Solids

1 Nonproprietary Names

PhEur: Glucose Liquid Spray-Dried
USP–NF: Corn Syrup Solids

2 Synonyms

Glucosum liquidum dispersione desiccatum; dehydrated hydrolyzed starch syrup; Maltrin; dried glucose syrup; soluble corn fiber; resistant maltodextrin.

3 Chemical Name and CAS Registry Number

Corn syrup solids [68131-37-3]

4 Empirical Formula and Molecular Weight

$(C_6H_{10}O_5)_n{\cdot}H_2O$ $\leq$4 000

Corn syrup solids are mixtures of amylose (D-glucose units connected by 1→4 glycosidic bonds) and to a lesser extent, amylopectin (D-glucose units connected by 1→4 glycosidic bonds and branched chains connected by 1→6 glycosidic bonds).

The relative amounts of monosaccharides, disaccharides, trisaccharides, and polysaccharides depend on the process used for starch hydrolysis. Dextrose Equivalence (DE) is a quantitative measure of the degree of starch polymer hydrolysis. It is a measure of reducing power compared to a dextrose standard of 100. Products with higher DE have a greater extent of starch hydrolysis. As the product is further hydrolyzed (higher DE), the average molecular weight decreases and the carbohydrate profile changes accordingly.

USP35-NF30 describes corn syrup solids as a dried mixture of saccharides resulting from the partial hydrolysis of edible corn starch by food-grade acids or enzymes. The DE (reducing sugar as D-glucose) is not less than 20.0% on a dried basis.

5 Structural Formula

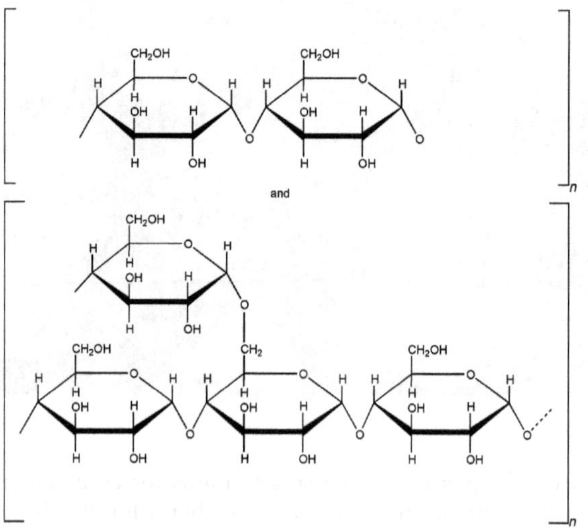

6 Functional Category

Coating agent; humectant; suspending agent; sweetening agent; tablet and capsule binder; tablet and capsule diluent; tonicity agent; viscosity-increasing agent.

7 Applications in Pharmaceutical Formulation or Technology

Corn syrup solids are used in pharmaceutical preparations as sweeteners, viscosity-increasing agents, tablet binders and lubricants.[1–3] They may also be used as carriers for high-potency active ingredients to improve content uniformity for direct compression formulations.[3]

Corn syrup solids are also used in a variety of products including foods, confectionery, and personal care products. They can enhance flavors such as fruit flavors. In some applications, corn syrup solids can be used to control moisture levels.[4]

8 Description

Corn syrup solids occur as bland to slightly sweet, odorless powders or granules. Sweetness increases as DE increases.

9 Pharmacopeial Specifications

See Table I.

Table I: Pharmacopeial specifications for corn syrup solids.

Test	PhEur 7.4	USP 35-NF 30[a]
Identification	+	+
Characters	+	–
pH	4.0–7.0	–
Loss on drying	≤6.0%	–
Microbial limit	–	+
Aerobic	–	≤1000 cfu/g
Mold and yeast	–	≤100 cfu/g
Residue on ignition	≤0.5%	≤0.5%
Heavy metals	≤10 ppm	≤5 μg/g
Starch	–	+
Total solids	–	≥90.0%[b] ≥93.0%[c]
Sulfur dioxide	≤20 ppm	≤40 μg/g
Lead	–	≤0.5 μg/g
Assay (reducing sugar content, dried basis)	≥20.0%	≥20.0%

(a) The USP35-NF30 states that corn syrup solids should be labeled with the nominal DE value and if the amount of sulfur dioxide is greater than 10 μg/g.
(b) When the reducing sugar content is 88.0% or greater
(c) When the reducing sugar content is between 20.0% and 88.0%

10 Typical Properties

Compressibility Compressibility increases as DE increases.
Density (tapped)
0.54–0.70 g/cm^3 for Maltrin QD M200;
0.32–0.45 g/cm^3 for Maltrin QD M600.
Moisture content Hygroscopicity increases as the DE value increases. Generally, the moisture content does not increase substantially at relative humidities less than 50%.
Solubility Freely soluble in water; slightly soluble in ethanol 95%. Solubility increases as the DE value increases.
Viscosity Viscosity decreases as the DE value increases. See Table II.

Table II: Relationship between DE value and viscosity of selected corn syrup solids at 30% and 60% w/w.[5]

DE Value	Viscosity of 30% w/w solution mPa s (cP)	Viscosity of 60 % w/w solution mPa s (cP)
20	~20	~300
25	<20	~200
36	<20	~100

11 Stability and Storage Conditions

Corn syrup solids are stable for at least 2 years when stored at 30°C or below and at a relative humidity less than 50%. Deviations from these conditions will not affect the stability of corn syrups solids except for moisture content which will change depending on the temperature and humidity and should be checked at regular intervals.[6] Store in a well-closed container in a cool, dry place.

12 Incompatibilities

At very low and high pH, corn syrup solids can form degradation products that are colored brown or yellow. In some cases, these degradation products will affect the flavor of the preparation. Corn syrup solids may react with amine compounds (e.g. amino acids) in a Maillard reaction.

13 Method of Manufacture

Corn syrup solids are prepared by heating and treating starch with acid and/or enzymes in the presence of water. This process partially hydrolyzes the starch, to produce a solution of glucose polymers of varying chain length. The solution is filtered, concentrated, and dried to obtain corn syrup solids. The manufacturing process is similar to that of maltodextrin. *See* Maltodextrin.

14 Safety

Corn syrup solids are considered nontoxic and nonirritating materials when used as excipients or as a food ingredient. May cause eye or respiratory tract irritation.

15 Handling Precautions

Observe normal precautions appropriate to the circumstances and quantity of material handled. Corn syrup solids should be handled in a well-ventilated area and excessive dust generation should be avoided. Protective eye goggles and respiratory apparatus are recommended.

16 Regulatory Status

GRAS listed. Listed in the Canadian Natural Health Products Ingredients Database.

17 Related Substances

Dextrates; dextrin; maltodextrin; starch.

18 Comments

Maltodextrins may be used as the term for maltodextrins and corn syrup solids. Corn syrup is used to describe hydrolyzed starches with a DE ranging from 20 to ≈99.

Corn syrup solids are used to control sugar crystallization, with the lower DE values being the most effective. They are used frequently as a secondary sweetener to sucrose because corn syrup solids aid in resisting discoloration caused by heat.

Corn syrup solids are digestible carbohydrates with a nutritional value of ≈17 kJ/g or 4 kcal/g. In oral nutritional supplements, they are used as a carbohydrate source with a lower osmolarity than an isocaloric amount of dextrose. For diabetics, they should be considered as a source of carbohydrate similar to dextrose. Partially hydrolyzed corn syrup solids have been investigated as a replacement for lactose in manufactured liquid diets for neonatal pigs.[7] Corn syrup solids used in the food industry include Clintose CR 24 (Archer Daniels Midland Company, USA), Dry GL (Cargill, USA), Dri-Sweet 36 (Ramsen, USA), Globe (Corn Products, USA), and Star-Dri (Tate & Lyle, USA).

A specification for corn syrup solids (dried glucose syrup) is listed in the *Food Chemicals Codex* (FCC).[8]

19 Specific References

1 Tuerck P, *et al.* A new method of manufacture of tablet granulations II. *J Am Pharm Assoc Am Pharm Assoc* 1960; **49:** 347–349.
2 Allen L, *et al.* Dissolution rates of hydrocortisone and prednisone utilizing sugar solid dispersion systems in tablet form. *J Pharm Sci;* 1978; **67:** 979–981.
3 Kukkar V, *et al.* Mixing and formulation of low dose drugs: underlying problems and solutions. *Thai J Pharm Sci* 2008; **32:** 43–58.
4 Grain Processing Corporation. Product information: *Maltodextrins and corn syrup solids*, 2011. http://www.grainprocessing.com/index.php?option=com_content&Itemid=81&id=56&lang=en&view=article.(accessed 19 August 2011).
5 Bourne M. *Food texture and viscosity: concept and measurement.* 2nd edn. FL, United States. Academic Press, 2002; 80. http://books.google.com/books?id=eSF6zS8n2QoC&=PA80&=viscosity+of+corn+syrup+solids+20+25+36&=en#v=onepage&=viscosity%20of%20corn%20syrup%20solids%2020%2025%2036&=false (accessed 24 February 2012).
6 Product Bulletin for Maltrin QD M600 , Corn Syrup Solids NF, Grain Processing Corporation, Muscatine, IA.
7 Oliver W, *et al.* Efficacy of partially hydrolyzed corn syrup solids as a replacement for lactose in manufactured liquid diets for neonatal pigs. *J Anim Sci* 2002; **80:** 143–153.

8 *Food Chemicals Codex*. 7th edn. Bethesda, MD: United States Pharmacopoeia, 2010; 438.

20 General References

Grain Processing Corporation. Product information: *Maltrin*, 2011. http://www.grainprocessing.com/index.php?option=com_content&Itemid=81&id=56&lang=en&view=article.(accessed 19 August 2011).
Rahman MS, ed. *Handbook of Food Preservation*, 2nd edn. Boca Raton, FL: CRC Press, 2007; 516.
State of Utah Department of Health, WIC Program Resources Product Guide. http://health.utah.gov/wic/pdf/forms_and_modules/Resources/Product%20Guide%2001-2012.pdf (accessed 27 February 2012).

21 Author

JR Creekmore.

22 Date of Revision

1 March 2012.

Cottonseed Oil

1 Nonproprietary Names

USP–NF: Cottonseed Oil

2 Synonyms

Cotton oil; refined cottonseed oil.

3 Chemical Name and CAS Registry Number

Cottonseed oil [8001-29-4]

4 Empirical Formula and Molecular Weight

A typical analysis of refined cottonseed oil indicates the composition of the acids present as glycerides to be as follows: linoleic acid 39.3%; oleic acid 33.1%; palmitic acid 19.1%; stearic acid 1.9%; arachidic acid 0.6%, and myristic acid 0.3%. Also present are small quantities of phospholipid, phytosterols, and pigments. The toxic polyphenolic pigment gossypol is present in raw cottonseed and in the oil cake remaining after expression of oil; it is not found in the refined oil.

5 Structural Formula

See Section 4.

6 Functional Category

Oleaginous vehicle; solvent.

7 Applications in Pharmaceutical Formulation or Technology

Cottonseed oil is used in pharmaceutical formulations primarily as a solvent for intramuscular injections. It has been used in intravenous emulsions as a fat source in parenteral nutrition regimens, although its use for this purpose has been superseded by soybean oil emulsions; *see* Section 14. It is also used as an emollient vehicle for other medications.

Cottonseed oil has been used as a tablet binder for acetaminophen; for characterization of the hot-melt fluid bed coating process;[1] in the manufacturing of stable oral pharmaceutical powders; in encapsulation of enzymes; and as an aqueous dispersion in pharmaceutical coating.

8 Description

Cottonseed oil occurs as a pale yellow or bright golden yellow-colored, clear oily liquid. It is odorless, or nearly so, with a bland, nutty taste. At temperatures below 10°C particles of solid fat may separate from the oil, and at about –5 to 0°C the oil becomes solid or nearly so. If it solidifies, the oil should be remelted and thoroughly mixed before use.

9 Pharmacopeial Specifications

See Table I. *See also* Section 18.

Table I: Pharmacopeial specifications for cottonseed oil.

Test	USP35–NF30
Identification	+
Fatty acid composition	+
Arachidic acid	≤1.0%
Behenic acid	≤0.6%
Erucatic acid	≤0.5%
Lignoceric acid	≤0.5%
Linoleic acid	46.7–58.3%
Linolenic acid	≤1.0%
Myristic acid	0.3–1.0%
Oleic acid	14.0–21.7%
Palmitic acid	18.0–26.4%
Stearic acid	2.1–3.3%
Free fatty acids	+
Heavy metals	≤10 ppm

10 Typical Properties

Autoignition temperature 344°C
Density 0.916 g/cm^3
Flash point 252°C (closed cup);[2] 321°C
Freezing point −5 to 0°C
Heat of combustion 37.1 kJ/g
Refractive index n_D^{40} = 1.4645–1.4655
Solubility Slightly soluble in ethanol (95%); miscible with carbon disulfide, chloroform, ether, hexane, and petroleum ether.
Spectroscopy
IR spectra *see* Figure 1.
Surface tension
35.4 mN/m (35.4 dynes/cm) at 20°C;
31.3 mN/m (31.3 dynes/cm) at 80°C.
Viscosity (dynamic) Up to 70.4 mPa s (70.4 cP) at 20°C.

11 Stability and Storage Conditions

Cottonseed oil is stable if stored in a well-filled, airtight, light-resistant container in a cool, dry place. Avoid exposure to excessive heat. Keep away from oxidizing materials.[2]

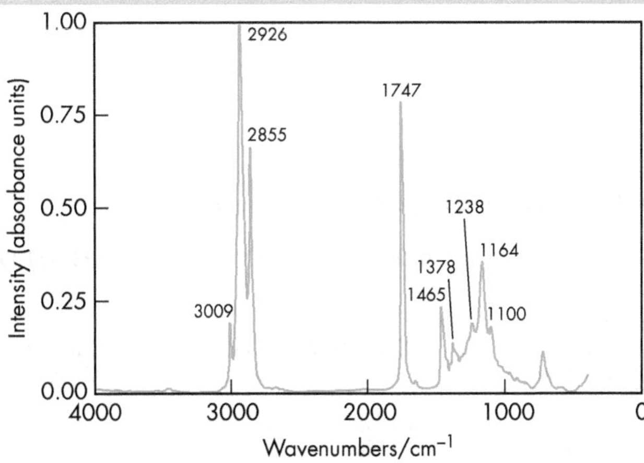

Figure 1: Infrared spectrum of cottonseed oil measured by transmission. Adapted with permission of Informa Healthcare.

12 Incompatibilities

—

13 Method of Manufacture

Cottonseed oil is the refined fixed oil obtained from the seed of cultivated varieties of *Gossypium hirsutum* Linné or of other species of *Gossypium* (Fam. Malvaceae). The seeds contain about 15% oil. The testae of the seeds are first separated and the kernels are then exposed to high pressure in a hydraulic press. The crude oil thus obtained has a bright red or blackish-red color and requires purification before it is suitable for food or pharmaceutical purposes.

Cottonseed oil is refined by treatment with diluted alkali to neutralize acids, decolorized with fuller's earth or activated carbon, deodorized with steam under reduced pressure, and then chilled to separate glycerides and resinous substances of higher melting point to prevent it turning cloudy at 4-10°C or solidifying at about 0°C.

14 Safety

Cottonseed oil emulsions have in the past been used in long-term intravenous nutrition regimens.[3,4] A complex of adverse reactions called the 'overloading syndrome'[5] has been seen with chronic administration of cottonseed oil emulsion. This consisted of anorexia, nausea, abdominal pain, headache, fever, and sore throat. Signs of impaired liver function, anemia, hepatosplenomegaly, thrombocytopenia, and spontaneous hemorrhage due to delayed blood clotting have been reported. For parenteral nutrition purposes, cottonseed oil has been replaced by soybean oil,[3,6,7] especially in pregnant women, where the use of cottonseed lipid emulsion has been associated with adverse effects.[8]

A notable difference between the cottonseed oil emulsion and the soybean oil emulsion is the particle size. The cottonseed oil emulsion has much larger particles than the soybean oil emulsion. These larger particles may have been handled differently by the body, thus perhaps accounting for some of the toxic reactions.

Slightly irritant on contact with skin and eyes, and on inhalation.[2]

15 Handling Precautions

Observe normal precautions appropriate to the circumstances and quantity of material handled. Spillages of this material are very slippery and should be covered with an inert absorbent material prior to disposal.

Cottonseed oil is a combustible liquid when exposed to heat or flame. If it is allowed to impregnate rags or oily waste, there is a risk due to spontaneous heating. Dry chemicals such as carbon dioxide should be used to fight any fires.

Gloves, eye protection, respirator, and other protective clothing are recommended.

16 Regulatory Status

Included in the FDA Inactive Ingredients Database (IM injections; oral, capsule, tablet and sublingual preparations). Included in the Canadian Natural Health Products Ingredients Database.

17 Related Substances

Almond oil; canola oil; corn oil; hydrogenated vegetable oil; peanut oil; sesame oil; soybean oil; sunflower oil.

18 Comments

The USP35–NF30, PhEur 7.4, and BP 2012 also list hydrogenated cottonseed oil.

Cottonseed oil has been used as an adjuvant in cholecystography and as a pediculicide and acaricide. It has the nutritive and emollient properties of fixed vegetable oils. By virtue of its high content of unsaturated acid glycerides (especially linoleic acid), it is used for dietary control of blood cholesterol levels in the prophylaxis and treatment of atherosclerosis. It can also retard gastric secretion and motility, and increase caloric intake.

Cottonseed oil has also been used in the manufacture of soaps, oleomargarine, lard substitutes, glycerin, lubricants, and cosmetics.

Gossypol, a natural toxin in cottonseed, interferes with spermatogenesis and is also a causative factor for paralysis among men. It can increase high-density lipoproteins in the body leading to cardiovascular diseases such as artherosclerosis, heart attack and angina. It may also lead to deficiency of omega-3 essential fatty acid. In a dose of 30 mL or more it is used as a mild cathartic.

A specification for unhydrogenated cottonseed oil is contained in the *Food Chemicals Codex* (FCC)[9] and the *Japanese Pharmaceutical Excipients* (JPE).[10]

The EINECS number for cottonseed oil is 232-280-7.

19 Specific References

1 Jozwiakowski MJ, *et al.* Characterization of a hot-melt fluid bed coating process for fine granules. *Pharm Res* 1990; **7**: 1119–1126.
2 ScienceLab.com Material safety data sheet: Cottonseed oil, 2010.
3 McNiff BL. Clinical use of 10% soybean oil emulsion. *Am J Hosp Pharm* 1977; **34**: 1080–1086.
4 Cole WH. Fat emulsion for intravenous use. *J Am Med Assoc* 1958; **166**: 1042–1043.
5 Goulon M, *et al.* Fat embolism after repeated perfusion of lipid emulsion. *Nouv Presse Med* 1974; **3**: 13–18.
6 Davis SS. Pharmaceutical aspects of intravenous fat emulsions. *J Hosp Pharm* 1974; **32**: 149–160165–171.
7 Singh M, Ravin LJ. Parenteral emulsions as drug carrier systems. *J Parenter Sci Technol* 1986; **41**: 34–41.
8 Amato P. Quercia RA. Historical perspective and review of the safety of lipid emulsion in pregnancy. *Nutr Clin Prac* 1991; **6**(5): 189–192.
9 *Food Chemicals Codex*, 7th edn. Bethesda, MD: United States Pharmacopeia, 2010: 249.
10 Japan Pharmaceutical Excipients Council. *Japanese Pharmaceutical Excipients 2004*. Tokyo: Yakuji Nippo, 2004: 191.

20 General References

—

21 Author

KS Alexander.

22 Date of Revision

1 March 2012.

Cresol

1 Nonproprietary Names

BP: Cresol
JP: Cresol
USP–NF: Cresol

2 Synonyms

Cresylic acid; cresylol; hydroxytoluene; tricresol.

3 Chemical Name and CAS Registry Number

Methylphenol [1319-77-3]

4 Empirical Formula and Molecular Weight

C_7H_8O 108.14

5 Structural Formula

OH

CH_3

m-Cresol

6 Functional Category

Antimicrobial preservative.

7 Applications in Pharmaceutical Formulation or Technology

Cresol is used at 0.15–0.3% concentration as an antimicrobial preservative in intramuscular, intradermal, and subcutaneous injectable pharmaceutical formulations. Cresol is not suitable as a preservative for preparations that are to be freeze-dried.[1]

8 Description

Cresol consists of a mixture of cresol isomers, predominantly *m*-cresol, and other phenols obtained from coal tar or petroleum. It is a colorless, yellowish to pale brownish-yellow, or pink-colored liquid, with a characteristic odor similar to phenol but more tarlike. An aqueous solution has a pungent taste.

9 Pharmacopeial Specifications

See Table I.

Table I: Pharmacopeial specifications for cresol.

Test	BP 2012	JP XV	USP35–NF30
Identification	+	+	+
Characters	+	—	—
Specific gravity	1.029–1.044	1.032–1.041	1.030–1.038
Distilling range	+	196–206°C	195–205°C
Acidity	+	—	—
Hydrocarbons	≤0.15%	+	+
Volatile bases	≤0.15%	—	—
Hydrocarbons and volatile bases combined	≤0.25%	—	—
Phenol	—	—	≤5.0%
Sulfur compounds	+	+	—
Nonvolatile matter	≤0.1%	—	—

10 Typical Properties

Acidity/alkalinity A saturated aqueous solution is neutral or slightly acidic to litmus.

Antimicrobial activity Cresol is similar to phenol but has slightly more antimicrobial activity. It is moderately active against Gram-positive bacteria, less active against Gram-negative bacteria, yeasts, and molds. Cresol is active below pH 9; optimum activity is obtained in acidic conditions. Synergistic effects between cresol and other preservatives have been reported.[2,3] When used as a disinfectant most common pathogens are killed within 10 minutes by 0.3–0.6% solutions. Cresol has no significant activity against bacterial spores.

Solubility *see* Table II.

Table II: Solubility of cresol.

Solvent	Solubility at 20°C
Benzene	Miscible
Chloroform	Freely soluble
Ethanol (95%)	Freely soluble
Ether	Freely soluble
Fixed alkali hydroxides	Freely soluble
Fixed and volatile oils	Freely soluble
Glycerin	Miscible
Water	1 in 50

11 Stability and Storage Conditions

Cresol and aqueous cresol solutions darken in color with age and on exposure to air and light.

Cresol should be stored in a well-closed container, protected from light, in a cool, dry place.

12 Incompatibilities

Cresol has been reported to be incompatible with chlorpromazine.[4] A recent *in-vitro* study in pooled human liver microsomes demonstrated inhibitory effects of *p*-cresol on losartan metabolism.[5] Antimicrobial activity is reduced in the presence of nonionic surfactants.

13 Method of Manufacture

Cresol may be obtained from coal tar or prepared synthetically by either sulfonation or oxidation of toluene.

14 Safety

Reports of adverse reactions to cresol are generally associated with the use of either the bulk material or cresol-based disinfectants, which may contain up to 50% cresol, rather than for its use as a preservative. However, a recent case of cutaneous hypersensitivity reaction to the *m*-cresol component of an insulin formulation detected via intradermal and patch testing has been reported.[6]

Cresol is similar to phenol although it is less caustic and toxic. However, cresol is sufficiently caustic to be unsuitable for skin and wound disinfection. In studies in rabbits, cresol was found to be metabolized and excreted primarily as the glucuronide.[7]

A patient has survived ingestion of 12 g of cresol though with severe adverse effects.[8] Two further cases of nonfatal cresol poisoning have also been reported, although no information on levels of intake were given.[9,10]

LD_{50} (mouse, oral): 0.76 g/kg[11]
LD_{50} (rabbit, skin): 2 g/kg
LD_{50} (rat, oral): 1.45 g/kg

See also Sections 17 and 18.

15 Handling Precautions

Observe normal precautions appropriate to the circumstances and quantity of material handled. Cresol may be irritant to the skin, eyes, and mucous membranes. Eye protection, gloves, and a respirator are recommended. In the US, the permissible and recommended exposure limits are 22 mg/m^3 long-term and 10 mg/m^3 long-term respectively.[12]

16 Regulatory Status

Included in the FDA Inactive Ingredients Database (IM, IV, intradermal, and SC injections). Included in parenteral medicines licensed in the UK. Included in the Canadian Natural Health Products Ingredients Database.

17 Related Substances

Chlorocresol; *m*-cresol; *o*-cresol; *p*-cresol; phenol.

m-Cresol

Empirical formula C_7H_8O
Molecular weight 108.14
CAS number [108-39-4]
Synonyms *m*-Cresylic acid; 3-hydroxytoluene; *meta*-cresol; 3-methylphenol.
Appearance Colorless or yellowish liquid with a characteristic phenolic odor.
Boiling point 202°C
Density 1.034 g/cm^3 at 20°C
Flash point 86°C (closed cup)
Melting point 11–12°C
Refractive index n_D^{20} = 1.5398
Solubility Soluble in organic solvents; soluble 1 in 40 parts of water.
Safety

LD_{50} (cat, SC): 0.15 g/kg[11,13]
LD_{50} (mouse, IP): 0.17 g/kg
LD_{50} (mouse, oral): 0.83 g/kg
LD_{50} (mouse, SC): 0.45 g/kg
LD_{50} (rabbit, IV): 0.28 g/kg
LD_{50} (rabbit, oral): 1.1 g/kg
LD_{50} (rabbit, SC): 0.5 g/kg
LD_{50} (rabbit, skin): 2.05 g/kg
LD_{50} (rat, oral): 2.02 g/kg
LD_{50} (rat, skin): 1.1 g/kg

o-Cresol

Empirical formula C_7H_8O
Molecular weight 108.14
CAS number [95-48-7]
Synonyms *o*-Cresylic acid; 2-hydroxytoluene; 2-methylphenol; *ortho*-cresol.
Appearance Colorless deliquescent solid with a characteristic odor; it becomes yellow on storage.
Boiling point 191–192°C
Density 1.047 g/cm^3 at 20°C
Flash point 81–83°C (closed cup)
Melting point 30°C
Refractive index n_D^{20} = 1.553
Safety

LD_{50} (cat, SC): 0.6 g/kg[11,13]
LD_{50} (mouse, oral): 0.34 g/kg
LD_{50} (mouse, SC): 0.35 g/kg
LD_{50} (mouse, skin): 0.62 g/kg
LD_{50} (rabbit, IV): 0.2 g/kg
LD_{50} (rabbit, oral): 0.8 g/kg
LD_{50} (rabbit, SC): 0.45 g/kg
LD_{50} (rat, oral): 1.35 g/kg
LD_{50} (rat, skin): 0.62 g/kg

p-Cresol

Empirical formula C_7H_8O
Molecular weight 108.14
CAS number [106-44-5]
Synonyms *p*-Cresylic acid; 4-hydroxytoluene; 4-methylphenol; *para*-cresol.
Appearance Crystalline solid.
Boiling point 201.8°C
Density 1.0341 g/cm^3 at 20°C
Flash point 86°C (closed cup)
Melting point 35.5°C
Refractive index n_D^{20} = 1.5395
Solubility Soluble in ethanol (95%) and ether; very slightly soluble in water.
Safety

LD_{50} (cat, SC): 0.08 g/kg[11,13]
LD_{50} (mouse, IP): 0.03 g/kg
LD_{50} (mouse, oral): 0.34 g/kg
LD_{50} (mouse, SC): 0.15 g/kg
LD_{50} (rabbit, IV): 0.16 g/kg
LD_{50} (rabbit, oral): 1.1 g/kg
LD_{50} (rabbit, SC): 0.3 g/kg
LD_{50} (rabbit, skin): 0.3 g/kg
LD_{50} (rat, oral): 1.80 g/kg
LD_{50} (rat, skin): 0.75 g/kg

18 Comments

m-Cresol is generally considered the least toxic of the three cresol isomers.[13] Inhalation of aerosolized *m*-cresol in pulmonary insulin delivery formulations has been shown to be safe in animal models.[14]

Cresol is also used as a preservative in some topical formulations and as a disinfectant.

The PhEur 7.4 contains a specification for cresol, crude.

The EINECS number for cresol is 203-577-9. The PubChem Compound ID (CID) for *m*-cresol is 342.

19 Specific References

1 FAO/WHO. WHO expert committee on biological standardization. Thirty-seventh report. *World Health Organ Tech Rep Ser* 1987; No. 760.

2 Denyer SP, Baird RM, eds. *Guide to Microbiological Control in Pharmaceuticals*. Chichester: Ellis Horwood, 1990: 261.
3 Hugbo PG. Additive and synergistic actions of equipotent admixtures of some antimicrobial agents. *Pharm Acta Helv* 1976; **51**: 284–288.
4 McSherry TJ. Incompatibility between chlorpromazine and metacresol [letter]. *Am J Hosp Pharm* 1987; **44**: 1574.
5 Tsujimoto M, *et al*. Inhibitory effects of uraemic toxins 3-indoxyl sulfate and *p*-cresol on losartan metabolism *in-vitro*. *J Pharm Pharmacol* 2010; **62**: 133–138.
6 Kim D, Baraniuk J. Delayed-type hypersensitivity reaction to the metacresol component of insulin. *Ann Allergy Asthma Immunol* 2007; **99**: 194–195.
7 Cresol. In: *The Pharmaceutical Codex*, 11th edn. London: Pharmaceutical Press, 1979: 232.
8 Côté MA, *et al*. Acute Heinz-body anemia due to severe cresol poisoning: successful treatment with erythrocytapheresis. *Can Med Assoc J* 1984; **130**: 1319–1322.
9 Liu SW, *et al*. A man with black urine. Cresol intoxication. *Ann Emerg Med* 2009; **53**: 836–843.
10 Seak CK, *et al*. A case of black urine and dark skin; cresol poisoning. *Clin Toxicol (Phila)* 2010; **48**(9): 959–960.
11 Lewis RJ, ed. *Sax's Dangerous Properties of Industrial Materials*, 11th edn. New York: Wiley, 2004: 1003.
12 NIOSH. Recommendations for occupational safety and health standard. *MMWR* 1988; **37**(Suppl. S-7): 1–29.
13 Deichmann WB, Keplinger ML. Phenols and phenolic compounds. In: Clayton GD, Clayton FE, eds. *Patty's Industrial Hygiene and Toxicology*, 3rd edn. New York: Wiley, 1981: 2597–2600.
14 Gopalakrishnan V *et al*. Inhalation safety of phenol and *m*-cresol in rodents: a fourteen-day repeat dose study. *Presented at ISAM congress 2001*, Interlaken, Switzerland.

20 General References

Chapman DG *o*-Cresol. In: Board RG *et al*. eds. *Preservatives in the Food, Pharmaceutical and Environmental Industries*. Oxford: Blackwell Scientific, 1987: 184.
Russell AD, *et al*. Reversal of the inhibition of bacterial spore germination and outgrowth by antibacterial agents. *Int J Pharm* 1985; **25**: 105–112.

21 Author

LY Galichet.

22 Date of Revision

1 March 2012.

 Croscarmellose Sodium

1 Nonproprietary Names

BP: Croscarmellose Sodium
JP: Croscarmellose Sodium
PhEur: Croscarmellose Sodium
USP–NF: Croscarmellose Sodium

2 Synonyms

Ac-Di-Sol; carmellosum natricum conexum; crosslinked carboxymethylcellulose sodium; *Explocel*; *Kiccolate*; modified cellulose gum; *Nymcel ZSX*; *Pharmacel XL*; *Primellose*; *Solutab*; *Vivasol*.

3 Chemical Name and CAS Registry Number

Cellulose, carboxymethyl ether, sodium salt, crosslinked [74811-65-7]

4 Empirical Formula and Molecular Weight

Croscarmellose sodium is a crosslinked polymer of carboxymethylcellulose sodium.

See Carboxymethylcellulose sodium.

5 Structural Formula

See Carboxymethylcellulose sodium.

6 Functional Category

Tablet and capsule disintegrant.

7 Applications in Pharmaceutical Formulation or Technology

Croscarmellose sodium is used in oral pharmaceutical formulations as a disintegrant for capsules,[1,2] tablets,[3–13] and granules.

In tablet formulations, croscarmellose sodium may be used in both direct-compression and wet-granulation processes. When used in wet granulations, the croscarmellose sodium should be added in both the wet and dry stages of the process (intra- and extra-granularly) so that the wicking and swelling ability of the disintegrant is best utilized.[11,12] Croscarmellose sodium at concentrations up to 5% w/w may be used as a tablet disintegrant, although normally 2% w/w is used in tablets prepared by direct compression and 3% w/w in tablets prepared by a wet-granulation process.

See Table I.

Table I: Uses of croscarmellose sodium.

Use	Concentration (%)
Disintegrant in capsules	10.0–25.0
Disintegrant in tablets	0.5–5.0

8 Description

Croscarmellose sodium occurs as an odorless, white or grayish-white powder.

9 Pharmacopeial Specifications

The pharmacopeial specifications for croscarmellose sodium have undergone harmonization of many attributes for JP, PhEur, and USP–NF.

See Table II. *See also* Section 18.

10 Typical Properties

Acidity/alkalinity pH = 5.0–7.0 in aqueous dispersions.
Bonding index 0.0456
Brittle fracture index 0.1000

SEM 1: Excipient: croscarmellose sodium (*Ac-Di-Sol*); manufacturer: FMC Biopolymer; magnification: 100×.

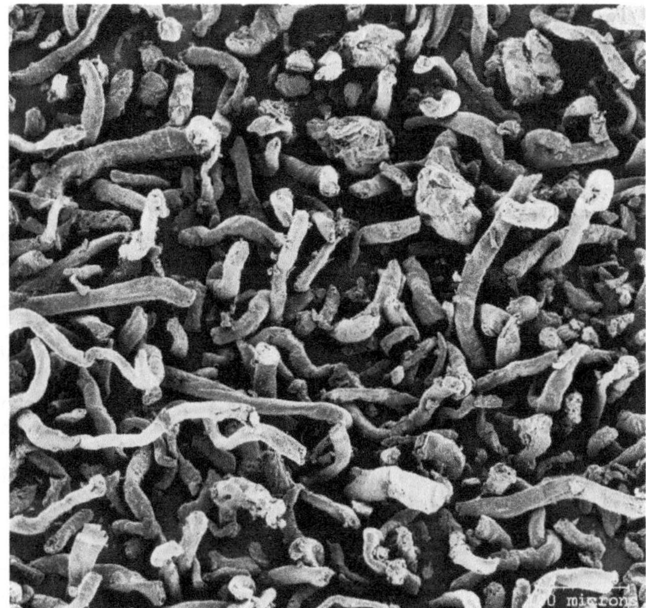

SEM 2: Excipient: croscarmellose sodium (*Ac-Di-Sol*); manufacturer: FMC Biopolymer; magnification: 1000×.

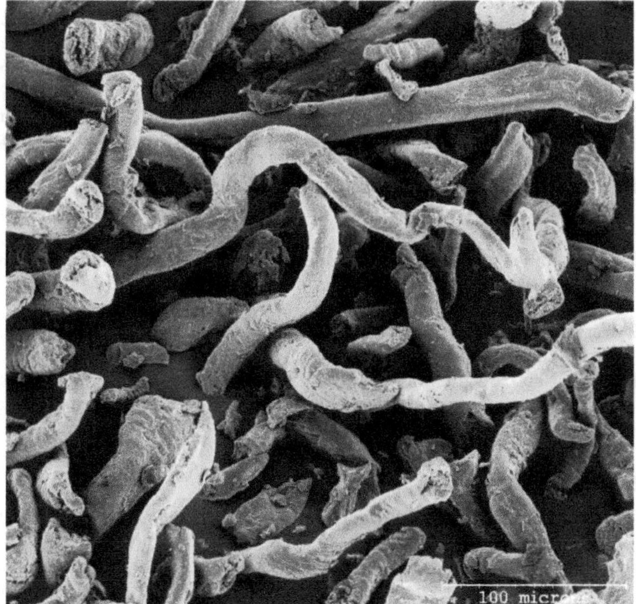

Density (bulk) 0.529 g/cm^3 for *Ac-Di-Sol*
Density (tapped) 0.819 g/cm^3 for *Ac-Di-Sol*
Density (true) 1.543 g/cm^3 for *Ac-Di-Sol*
Particle size distribution

Ac-Di-Sol: not more than 2% retained on a #200 (73.7 μm) mesh and not more than 10% retained on a #325 (44.5 μm) mesh.

Solubility Insoluble in water, although croscarmellose sodium rapidly swells to 4–8 times its original volume on contact with water. Practically insoluble in acetone, ethanol and toluene.

Specific surface area 0.81–0.83 m^2/g

Spectroscopy

IR spectra *see* Figure 1.

NIR spectra *see* Figure 2.

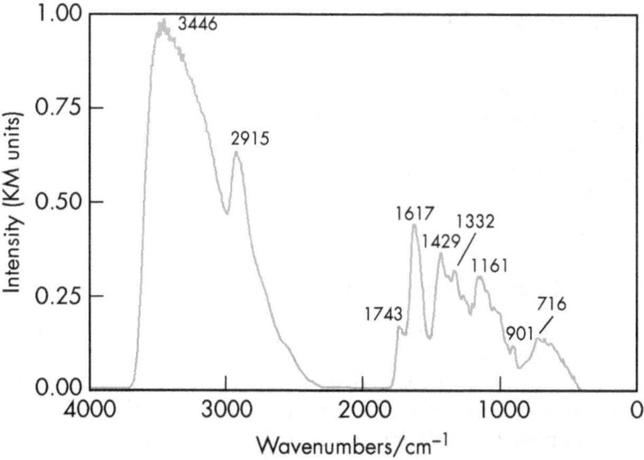

Figure 1: Infrared spectrum of croscarmellose sodium measured by diffuse reflectance. Adapted with permission of Informa Healthcare.

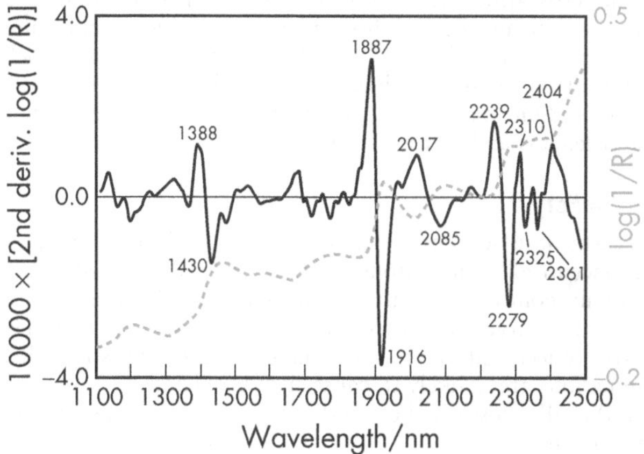

Figure 2: Near-infrared spectrum of croscarmellose sodium measured by reflectance.

Table II: Pharmacopeial specifications for croscarmellose sodium.

Test	JP XV	PhEur 7.4	USP35–NF30
Identification	+	+	+
Characters	+	+	—
pH (1% w/v dispersion)	5.0–7.0	5.0–7.0	5.0–7.0
Loss on drying	≤10.0%	≤10.0%	≤10.0%
Heavy metals[a]	≤10 ppm	≤20 ppm	≤10 ppm
Sodium chloride and sodium glycolate	≤0.5%	≤0.5%	≤0.5%
Sulfated ash	—	14.0–28.0%	—
Residue on ignition	14.0–28.0%	—	14.0–28.0%
Degree of substitution	0.60–0.85	0.60–0.85	0.60–0.85
Content of water-soluble material[a]	1.0–10.0%	≤10.0%	≤10.0%
Settling volume	10.0–30.0 mL	10.0–30.0 mL	10.0–30.0 mL
Microbial contamination	—	+	+
Aerobic	—	10^3 cfu/g	10^3 cfu/g
Fungi	—	10^2 cfu/g	10^2 cfu/g

(a) These tests have not been fully harmonized at the time of publication.

11 Stability and Storage Conditions

Croscarmellose sodium is a stable though hygroscopic material.

A model tablet formulation prepared by direct compression, with croscarmellose sodium as a disintegrant, showed no significant difference in drug dissolution after storage at 30°C for 14 months.[9]

Croscarmellose sodium should be stored in a well-closed container in a cool, dry place.

C

12 Incompatibilities

The efficacy of disintegrants, such as croscarmellose sodium, may be slightly reduced in tablet formulations prepared by either the wet-granulation or direct-compression process that contain hygroscopic excipients such as sorbitol.[10]

Croscarmellose sodium is not compatible with strong acids or with soluble salts of iron and some other metals such as aluminum, mercury, and zinc.

13 Method of Manufacture

Alkali cellulose is prepared by steeping cellulose, obtained from wood pulp or cotton fibers, in sodium hydroxide solution. The alkali cellulose is then reacted with sodium monochloroacetate to obtain carboxymethylcellulose sodium. After the substitution reaction is completed and all of the sodium hydroxide has been used, the excess sodium monochloroacetate slowly hydrolyzes to glycolic acid. The glycolic acid changes a few of the sodium carboxymethyl groups to the free acid and catalyzes the formation of crosslinks to produce croscarmellose sodium. The croscarmellose sodium is then extracted with aqueous alcohol and any remaining sodium chloride or sodium glycolate is removed. After purification, croscarmellose sodium of purity greater than 99.5% is obtained.[4] The croscarmellose sodium may be milled to break the polymer fibers into shorter lengths and hence improve its flow properties.

14 Safety

Croscarmellose sodium is mainly used as a disintegrant in oral pharmaceutical formulations and is generally regarded as an essentially nontoxic and nonirritant material. However, oral consumption of large amounts of croscarmellose sodium may have a laxative effect, although the quantities used in solid dosage formulations are unlikely to cause such problems.

In the UK, croscarmellose sodium is accepted for use in dietary supplements.

The WHO has not specified an acceptable daily intake for the related substance carboxymethylcellulose sodium, used as a food additive, since the levels necessary to achieve a desired effect were not considered sufficient to be a hazard to health.[14]

See also Carboxymethylcellulose Sodium.

15 Handling Precautions

Observe normal precautions appropriate to the circumstances and quantity of material handled. Croscarmellose sodium may be irritant to the eyes; eye protection is recommended.

16 Regulatory Status

Included in the FDA Inactive Ingredients Database (oral capsules, granules, sublingual tablets, and tablets). Included in nonparenteral medicines licensed in the UK. Included in the Canadian Natural Health Products Ingredients Database.

17 Related Substances

Carboxymethylcellulose calcium; carboxymethylcellulose sodium.

18 Comments

Croscarmellose sodium has undergone harmonization of many attributes for JP, PhEur, and USP–NF by the Pharmacopeial Discussion Group. For further information see the General Information Chapter <1196> in the USP35–NF30, the General Chapter 5.8 in PhEur 7.4, along with the 'State of Work' document on the PhEur EDQM website, and also the General Information Chapter 8 in the JP XV.

Typically, the degree of substitution (DS) for croscarmellose sodium is 0.7.

The EINECs number for croscarmellose sodium is 232-674-9.

19 Specific References

1 Botzolakis JE, Augsburger LL. Disintegrating agents in hard gelatin capsules. Part I: mechanism of action. *Drug Dev Ind Pharm* 1988; **14**(1): 29–41.

2 Dahl TC, *et al.* The influence of disintegrant level and capsule size on dissolution of hard gelatin capsules stored in high humidity conditions. *Drug Dev Ind Pharm* 1991; **17**(7): 1001–1016.

3 Gissinger D, Stamm A. A comparative evaluation of the properties of some tablet disintegrants. *Drug Dev Ind Pharm* 1980; **6**(5): 511–536.

4 Shangraw R, *et al.* A new era of tablet disintegrants. *Pharm Technol* 1980; 4(10): 49–57.

5 Zhao N, Augsburger LI. The influence of product brand-to-brand variability on superdisintegrant performance. A case study with croscarmellose sodium. *Pharm Dev Technol* 2006; **11**(2): 179–185.

6 Zhao N, Augsburger LI. The influence of swelling capacity of superdisintegrants in different pH media on the dissolution of hydrochlorthiazide from directly compressed tablets. *AAPS Pharm Sci Tech* 2005; **6**(1): E120–E126.

7 Battu SK, *et al.* Formulation and evaluation of rapidly disintegrating fenoverine tablets: effects of superdisintegrants. *Drug Dev Ind Pharm* 2007; **33**(11): 1225–1232.

8 Gordon MS, Chowhan ZT. Effect of tablet solubility and hygroscopicity on disintegrant efficiency in direct compression tablets in terms of dissolution. *J Pharm Sci* 1987; **76**: 907–909.

9 Gordon MS, Chowhan ZT. The effect of aging on disintegrant efficiency in direct compression tablets with varied solubility and hygroscopicity, in terms of dissolution. *Drug Dev Ind Pharm* 1990; **16**: 437–447.

10 Johnson JR, *et al.* Effect of formulation solubility and hygroscopicity on disintegrant efficiency in tablets prepared by wet granulation, in terms of dissolution. *J Pharm Sci* 1991; **80**: 469–471.

11 Gordon MS, *et al.* Effect of the mode of super disintegrant incorporation on dissolution in wet granulated tablets. *J Pharm Sci* 1993; **82**(2): 220–226.

12 Khattab I, *et al.* Effect of mode of incorporation of disintegrants on the characteristics of fluid-bed wet-granulated tablets. *J Pharm Pharmacol* 1993; **45**(8): 687–691.

13 Ferrero C, *et al.* Disintegrating efficiency of croscarmellose sodium in a direct compression formulation. *Int J Pharm* 1997; **147**: 11–21.

14 FAO/WHO. Evaluation of certain food additives and contaminants. Thirty-fifth report of the joint FAO/WHO expert committee on food additives. *World Health Organ Tech Rep Ser* 1990; No. 789.

20 General References

DMV-Frontera Excipients. Product literature: *Primellose and Primojel*, August 2008.

European Directorate for the Quality of Medicines and Healthcare (EDQM). European Pharmacopoeia – State Of Work Of International Harmonisation. *Pharmeuropa* 2011; 23(4): 713–714. www.edqm.eu/site/-614.html (accessed 5 December 2011).

FMC Biopolymer. Material Safety Datasheet: *Ac-Di-Sol*, May 2008.

JRS Pharma. Product literature: *Vivasol*. www.jrspharma.de/Pharma/wEnglisch/produktinfo/productinfo_vivasol.shtml (accessed 5 December 2011).

21 Author

RT Guest.

22 Date of Revision

5 December 2011.

Crospovidone

1 Nonproprietary Names

BP: Crospovidone
PhEur: Crospovidone
USP–NF: Crospovidone

2 Synonyms

Crospovidonum; *Crospopharm*; crosslinked povidone; E1202; *Kollidon CL*; *Kollidon CL-M*; *Polyplasdone XL*; *Polyplasdone XL-10*; polyvinylpolypyrrolidone; PVPP; 1-vinyl-2-pyrrolidinone homopolymer.

3 Chemical Name and CAS Registry Number

1-Ethenyl-2-pyrrolidinone homopolymer [9003-39-8]

4 Empirical Formula and Molecular Weight

$(C_6H_9NO)_n$ >1 000 000

The USP35–NF30 describes crospovidone as a water-insoluble synthetic crosslinked homopolymer of *N*-vinyl-2-pyrrolidinone. An exact determination of the molecular weight has not been established because of the insolubility of the material.

5 Structural Formula

See Povidone.

6 Functional Category

Tablet and capsule disintegrant.

7 Applications in Pharmaceutical Formulation or Technology

Crospovidone is a water-insoluble tablet disintegrant and dissolution agent used at 2–5% concentration in tablets prepared by direct-compression or wet- and dry-granulation methods.[1–6] It rapidly exhibits high capillary activity and pronounced hydration capacity, with little tendency to form gels. Studies suggest that the particle size of crospovidone strongly influences disintegration of analgesic tablets.[7] Larger particles provide a faster disintegration than smaller particles. Crospovidone has been suggested as an alternative to microcrystalline cellulose as an aid in pelletization.[8]

Crospovidone can also be used as a solubility enhancer. With the technique of co-evaporation, crospovidone can be used to enhance the solubility of poorly soluble drugs. The drug is adsorbed on to crospovidone in the presence of a suitable solvent and the solvent is then evaporated. This technique results in a faster dissolution rate.

8 Description

Crospovidone is a white to creamy-white, finely divided, free-flowing, practically tasteless, odorless or nearly odorless, hygroscopic powder.

9 Pharmacopeial Specifications

The pharmacopeial specifications for crospovidone have undergone harmonization of many attributes for PhEur, and USP–NF.

See Table I. *See also* Section 18.

Table I: Pharmacopeial specifications for crospovidone.

Test	PhEur 7.4	USP35–NF30
Identification	+	+
Characters	+	—
Residue on ignition	≤0.1%	≤0.1%
Water-soluble substances	≤1.0%	≤1.5%
Peroxides	≤400 ppm	—
Heavy metals[a]	≤10 ppm	≤10 ppm
Vinylpyrrolidinone	≤10 ppm	≤10 ppm
Loss on drying	≤5.0%	≤5.0%
Nitrogen content (anhydrous basis)	11.0–12.8%	11.0–12.8%

(a) These tests have not been fully harmonized at the time of publication.

10 Typical Properties

Acidity/alkalinity pH = 5.0–8.0 (1% w/v aqueous slurry)
Density 1.22 g/cm^3
Density (bulk) *see* Table II.
Density (tapped) *see* Table II.

Table II: Density values of commercial grades of crospovidone.

Commercial grade	Density (bulk) (g/cm^3)	Density (tapped) (g/cm^3)
Kollidon CL	0.3–0.4	0.4–0.5
Kollidon CL-M	0.15–0.25	0.3–0.5
Polyplasdone XL	0.213	0.273
Polyplasdone XL-10	0.323	0.461

Moisture content Maximum moisture sorption is approximately 60%.

Particle size distribution Less than 400 μm for *Polyplasdone XL*; less than 74 μm for *Polyplasdone XL-10*. Approximately 50% greater than 50 μm and maximum of 3% greater than 250 μm in size for *Kollidon CL*. Minimum of 90% of particles are below 15 μm for *Kollidon CL-M*. The average particle size for

SEM 1: Excipient: crospovidone (*Polyplasdone XL-10*); manufacturer: ISP Corp.; lot no.: S81031; magnification: 400×; voltage: 10 kV.

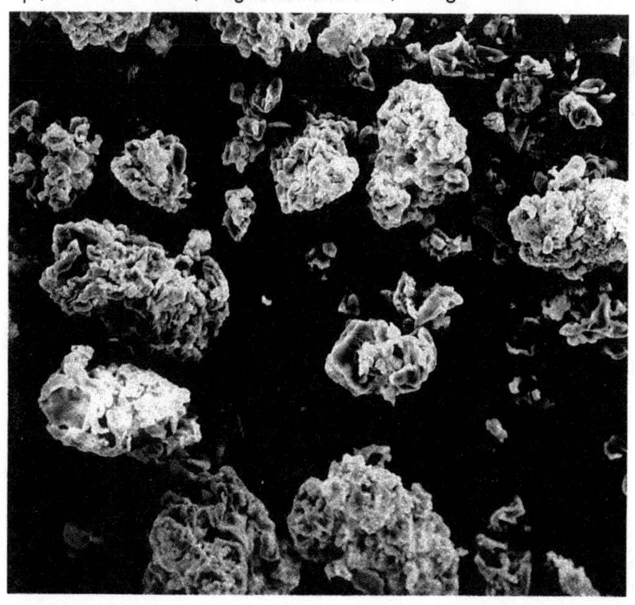

Crospopharm type A is 100 μm and for *Crospopharm* type B it is 30 μm.

Solubility Practically insoluble in water and most common organic solvents.

Specific surface area *see* Table III.

Table III: Specific surface areas for commercial grades of crospovidone.

Commercial grade	Surface area (m^2/g)
Kollidon CL	1.0
Kollidon CL-M	3.0–6.0
Polyplasdone XL	0.6–0.8
Polyplasdone XL-10	1.2–1.4

Spectroscopy

IR spectra *see* Figure 1.

NIR spectra *see* Figure 2.

11 Stability and Storage Conditions

Since crospovidone is hygroscopic, it should be stored in an airtight container in a cool, dry place.

12 Incompatibilities

Crospovidone is compatible with most organic and inorganic pharmaceutical ingredients. When exposed to a high water level, crospovidone may form molecular adducts with some materials; *see* Povidone.

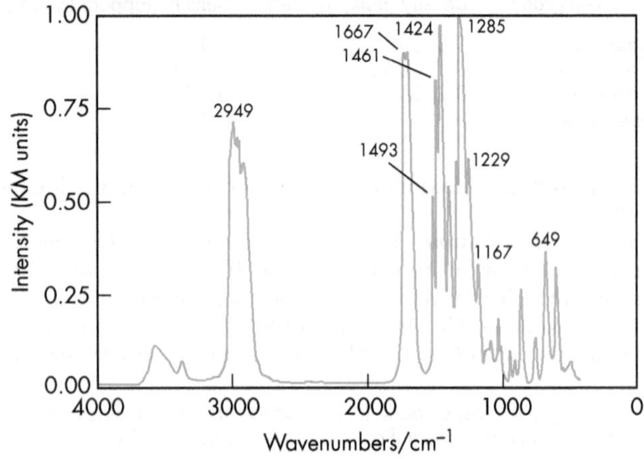

Figure 1: Infrared spectrum of crospovidone measured by diffuse reflectance. Adapted with permission of Informa Healthcare.

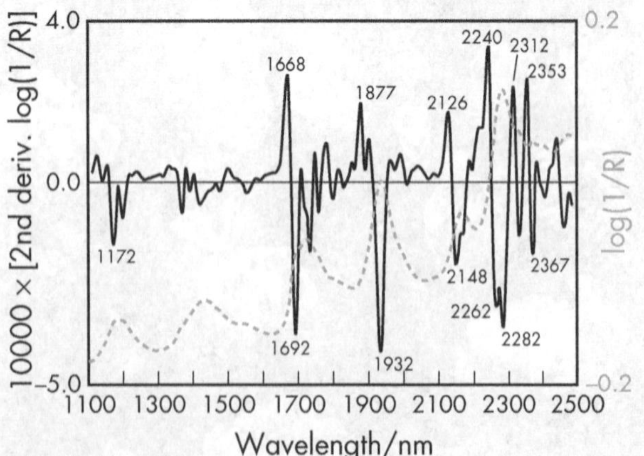

Figure 2: Near-infrared spectrum of crospovidone measured by reflectance.

13 Method of Manufacture

Acetylene and formaldehyde are reacted in the presence of a highly active catalyst to form butynediol, which is hydrogenated to butanediol and then cyclodehydrogenated to form butyrolactone. Pyrrolidone is produced by reacting butyrolactone with ammonia. This is followed by a vinylation reaction in which pyrrolidone and acetylene are reacted under pressure. The monomer vinylpyrrolidone is then polymerized in solution, using a catalyst. Crospovidone is prepared by a 'popcorn polymerization' process.

14 Safety

Crospovidone is used in oral pharmaceutical formulations and is generally regarded as a nontoxic and nonirritant material. Short-term animal toxicity studies have shown no adverse effects associated with crospovidone.[9] However, owing to the lack of available data, an acceptable daily intake in humans has not been specified by the WHO.[9]

LD_{50} (mouse, IP): 12 g/kg

15 Handling Precautions

Observe normal precautions appropriate to the circumstances and quantity of material handled. Eye protection, gloves, and a dust mask are recommended.

16 Regulatory Status

Accepted for use as a food additive in Europe. Included in the FDA Inactive Ingredients Database (IM injections, oral capsules and tablets; topical, transdermal, and vaginal preparations). Included in nonparenteral medicines licensed in the UK. Included in the Canadian Natural Health Products Ingredients Database.

17 Related Substances

Copovidone; povidone.

18 Comments

Crospovidone is one of the materials that have been selected for harmonization by the Pharmacopeial Discussion Group. For further information see the General Information Chapter <1196> in the USP35–NF30, the General Chapter 5.8 in PhEur 7.4, along with the 'State of Work' document on the PhEur EDQM website, and also the General Information Chapter 8 in the JP XV.

Crospovidone has been studied as a superdisintegrant. The ability of the compound to swell has been examined directly using scanning electron microscopy.[10] The impact of crospovidone on percolation has also been examined.[11] The impact of crospovidone on dissolution of poorly soluble drugs in tablets has also been investigated.[12] It has also been evaluated for use in fast disintergrating tablets.[13] Crospovidone has been shown to be effective with highly hygroscopic drugs.[14] It continues to be examined for its uses in a number of tablet formulations.

A specification for crospovidone is contained in the *Food Chemicals Codex* (FCC).[15]

The PubChem Compound ID (CID) for crospovidone is 6917.

19 Specific References

1 Kornblum SS, Stoopak SB. A new tablet disintegrating agent: cross-linked polyvinylpyrrolidone. *J Pharm Sci* 1973; **62**: 43–49.

2 Rudnic EM, *et al*. Studies of the utility of cross linked polyvinylpolypyrrolidine as a tablet disintegrant. *Drug Dev Ind Pharm* 1980; **6**: 291–309.

3 Gordon MS, Chowhan ZT. Effect of tablet solubility and hygroscopicity on disintegrant efficiency in direct compression tablets in terms of dissolution. *J Pharm Sci* 1987; **76**: 907–909.
4 Gordon MS, *et al.* Effect of the mode of super disintegrant incorporation on dissolution in wet granulated tablets. *J Pharm Sci* 1993; **82**: 220–226.
5 Tagawa M, *et al.* Effect of various disintegrants on drug release behavior from tablets. *J Pharm Sci Tech Yakuzaigaku* 2003; **63**(4): 238–248.
6 Hipasawa N, *et al.* Application of nilvadipine solid dispersion to tablet formulation and manufacturing using crospovidone and methylcellulose on dispersion carriers. *Chem Pharm Bull* 2004; **52**(2): 244–247.
7 Schiermeier S, Schmidt PC. Fast dispersible ibuprofen tablets. *Eur J Pharm Sci* 2002; **15**(3): 295–305.
8 Verheyen P, *et al.* Use of crospovidone as pelletization aid as alternative to microcrystalline cellulose: effects on pellet properties. *Drug Dev Ind Pharm* 2009; **35**(11): 1325–1332.
9 FAO/WHO. Evaluation of certain food additives and contaminants. Twenty-seventh report of the joint FAO/WHO expert committee on food additives. *World Health Organ Tech Rep Ser* 1983; No. 696.
10 Thibert R, Hancock BC. Direct visualization of superdisintegrant hydration using environmental scanning electron microscopy. *J Pharm Sci* 1996; **85**: 1255–1258.
11 Caraballo I, *et al.* Influence of disintegrant on the drug percolation threshold in tablets. *Drug Dev Ind Pharm* 1997; **23**(7): 665–669.
12 Yen SY, *et al.* Investigation of dissolution enhancement of nifedipine by deposition on superdisintegrants. *Drug Dev Ind Pharm* 1997; **23**(3): 313–317.
13 Shirsand SB, *et al.* Formulation design and optimization of fast dessolving clonazepam tablets. *Indian J Pharm Sci* 2009; **71**(5): 567–578.
14 Hirai N, *et al.* Improvement of the agitation granulation method to prepare granules containing a high content of a very hygroscopic drug. *J Pharm Pharmacol* 2006; **56**: 1437–1441.
15 *Food Chemicals Codex*, 7th edn (Suppl. 1). Bethesda, MD: United States Pharmacopeia, 2010: 1447.

20 General References

Barabas ES, Adeyeye CM. Crospovidone. In: Brittain HG, ed. *Analytical Profiles of Drug Substances and Excipients*. 24. London: Academic Press, 1996: 87–163.
BASF. Technical literature: *Insoluble Kollidon grades*, 1996.
European Directorate for the Quality of Medicines and Healthcare (EDQM). European Pharmacopoeia – State Of Work Of International Harmonisation. *Pharmeuropa* 2011; 22(4): 583–584. www.edqm.eu/en/International-Harmonisation-614.html (accessed 1 March 2012).
ISP. Technical literature: *Polyplasdone crospovidone NF*, 1999.
NP Pharm. Product data sheet: Crospopharm, 2008.
Wan LSC, Prasad KPP. Uptake of water by excipients in tablets. *Int J Pharm* 1989; **50**: 147–153.

21 Author

AH Kibbe.

22 Date of Revision

1 March 2012.

Cyclodextrins

1 Nonproprietary Names

BP: Alfadex
Betadex
PhEur: Alfadex
Betadex
USP–NF: Alfadex
Betadex
Gamma Cyclodextrin

2 Synonyms

Cyclodextrin *Cavitron*; cyclic oligosaccharide; cycloamylose; cycloglucan; *Encapsin*; Schardinger dextrin.
α-Cyclodextrin alfadexum; alpha-cycloamylose; alpha-cyclodextrin; alpha-dextrin; *Cavamax W6 Pharma*; cyclohexaamylose; cyclomaltohexose.
β-Cyclodextrin beta-cycloamylose; beta-dextrin; betadexum; *Cavamax W7 Pharma*; cycloheptaamylose; cycloheptaglucan; cyclomaltoheptose; E459; *Kleptose*.
γ-Cyclodextrin *Cavamax W8 Pharma*; cyclooctaamylose; cyclomaltooctaose.

3 Chemical Name and CAS Registry Number

α-Cyclodextrin [10016-20-3]
β-Cyclodextrin [7585-39-9]
γ-Cyclodextrin [17465-86-0]

4 Empirical Formula and Molecular Weight

α-Cyclodextrin $C_{36}H_{60}O_{30}$ 972
β-Cyclodextrin $C_{42}H_{70}O_{35}$ 1 135
γ-Cyclodextrin $C_{48}H_{80}O_{40}$ 1 297

5 Structural Formula

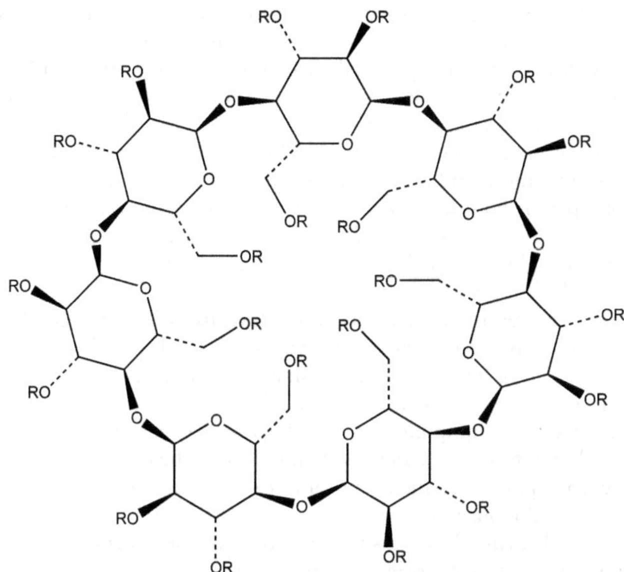

Note: the structure of betadex (β-cyclodextrin) with 7 glucose units is shown.

R = H for 'natural' α, β, and γ-cyclodextrins with 6, 7, and 8 glucose units, respectively

R = H or CH_3 for methyl cyclodextrins

R = H or $CHOHCH_3$ for 2-hydroxyethyl cyclodextrins

R = H or $CH_2CHOHCH_3$ for 2-hydroxypropyl cyclodextrins

6 Functional Category

Complexing agent; solubilizing agent.

7 Applications in Pharmaceutical Formulation or Technology

Cyclodextrins are crystalline, nonhygroscopic, cyclic oligosaccharides derived from starch. Among the most commonly used forms are α-, β-, and γ-cyclodextrin, which have respectively 6, 7, and 8 glucose units; *see* Section 5.

Substituted cyclodextrin derivatives are also available; *see* Section 17.

Cyclodextrins are 'bucketlike' or 'conelike' toroid molecules, with a rigid structure and a central cavity, the size of which varies according to the cyclodextrin type; *see* Section 8. The internal surface of the cavity is hydrophobic and the outside of the torus is hydrophilic; this is due to the arrangement of hydroxyl groups within the molecule. This arrangement permits the cyclodextrin to accommodate a guest molecule within the cavity, forming an inclusion complex.

Cyclodextrins may be used to form inclusion complexes with a variety of drug molecules, resulting primarily in improvements to dissolution and bioavailability owing to enhanced solubility and improved chemical and physical stability; *see* Section 18.

Cyclodextrin inclusion complexes have also been used to mask the unpleasant taste of active materials and to convert a liquid substance into a solid material.

β-Cyclodextrin is the most commonly used cyclodextrin, although it is the least soluble; *see* Section 10. It is the least expensive cyclodextrin; is commercially available from a number of sources; and is able to form inclusion complexes with a number of molecules of pharmaceutical interest. However, β-cyclodextrin is nephrotoxic and should not be used in parenteral formulations; *see* Section 14. β-Cyclodextrin is primarily used in tablet and capsule formulations.

α-Cyclodextrin is used mainly in parenteral formulations. However, as it has the smallest cavity of the cyclodextrins it can form inclusion complexes with only relatively few, small-sized molecules. In contrast, γ-cyclodextrin has the largest cavity and can be used to form inclusion complexes with large molecules; it has low toxicity and enhanced water solubility.

In oral tablet formulations, β-cyclodextrin may be used in both wet-granulation and direct-compression processes. The physical properties of β-cyclodextrin vary depending on the manufacturer. However, β-cyclodextrin tends to possess poor flow properties and requires a lubricant, such as 0.1% w/w magnesium stearate, when it is directly compressed.[1,2]

In parenteral formulations, cyclodextrins have been used to produce stable and soluble preparations of drugs that would otherwise have been formulated using a nonaqueous solvent.

In eye drop formulations, cyclodextrins form water-soluble complexes with lipophilic drugs such as corticosteroids. They have been shown to increase the water solubility of the drug; to enhance drug absorption into the eye; to improve aqueous stability; and to reduce local irritation.[3]

Cyclodextrins have also been used in the formulation of solutions,[4,5] suppositories,[6,7] and cosmetics.[8,9]

8 Description

Cyclodextrins are cyclic oligosaccharides containing at least six D-(+)-glucopyranose units attached by α(1→4) glucoside bonds. The three natural cyclodextrins, α, β, and γ, differ in their ring size and solubility. They contain 6, 7, or 8 glucose units, respectively.

Cyclodextrins occur as white, practically odorless, fine crystalline powders, having a slightly sweet taste. Some cyclodextrin derivatives occur as amorphous powders.

See also Table I.

Table I: Pharmacopeial specifications for α-cyclodextrin (alfadex).

Test	PhEur 7.4	USP35–NF30
Identification	+	+
Characters	+	–
Color and clarity of solution	+	+
pH	5.0–8.0	5.0–8.0
Specific rotation	+147° to +152°	+147° to +152°
Microbial limits	–	≤1000 cfu/g[a]
Sulfated ash	≤0.1%	–
Residue on ignition	–	≤0.1%
Heavy metals	≤10 ppm	≤10 μg/g
Light-absorbing impurities	+	+
Loss on drying	≤11.0%	≤11.0%
Related substances	+	+
Reducing sugars	≤0.2%	≤0.2%
Assay (anhydrous basis)	97.0–102.0%	98.0–101.0%

(a) Tests for *Salmonella* and *Escherichia coli* are negative.

9 Pharmacopeial Specifications

See Tables I, II, and III.

Table II: Pharmacopeial specifications for β-cyclodextrin (betadex).

Test	PhEur 7.4	USP35–NF30
Identification	+	+
Characters	+	–
Color and clarity of solution	+	+
pH	5.0–8.0	5.0–8.0
Specific rotation	+160° to +164°	+160° to +164°
Microbial limits	–	≤1000 cfu/g[a]
Sulfated ash	≤0.1%	–
Residue on ignition	–	≤0.1%
Heavy metals	≤10 ppm	≤5 ppm
Light-absorbing impurities	+	+
Loss on drying	≤16.0%	≤14.0%
Related substances	+	+
Residual solvents	+	–
Reducing sugars	≤0.2%	≤0.2%
Assay (anhydrous basis)	98.0–101.0%	98.0–102.0%

(a) Tests for *Salmonella* and *Escherichia coli* are negative.

Table III: Pharmacopeial specifications for γ-cyclodextrin (gamma cyclodextrin).

Test	USP35–NF30
Identification	+
Color and clarity of solution	+
Specific rotation	+174° to +180°
Microbial limits	≤1000 cfu/g[a]
Residue on ignition	≤0.1%
Heavy metals	≤5 ppm
Loss on drying	≤11.0%
Related substances	+
Reducing sugars	≤0.5%
Assay (anhydrous basis)	98.0–102.0%

(a) Tests for *Salmonella* and *Escherichia coli* are negative.

10 Typical Properties

Compressibility 21.0–44.0% for β-cyclodextrin.

Density (bulk)

α-cyclodextrin: 0.526 g/cm^3;

β-cyclodextrin: 0.523 g/cm^3;

γ-cyclodextrin: 0.568 g/cm^3.

Density (tapped)

α-cyclodextrin: 0.685 g/cm^3;

β-cyclodextrin: 0.754 g/cm^3;

γ-cyclodextrin: 0.684 g/cm^3.

Density (true)

α-cyclodextrin: 1.521 g/cm^3;

γ-cyclodextrin: 1.471 g/cm^3.

Melting point

α-cyclodextrin: 250–260°C;

β-cyclodextrin: 255–265°C;

γ-cyclodextrin: 240–245°C.

Moisture content

α-cyclodextrin: 10.2% w/w;

β-cyclodextrin: 13.0–15.0% w/w;

γ-cyclodextrin: 8–18% w/w.

Particle size distribution β-cyclodextrin: 7.0–45.0 μm

Physical characteristics *see* Table IV.

Table IV: Physical characteristics of cyclodextrins.

Characteristic	Cyclodextrin		
	α	β	γ
Cavity diameter (Å)	4.7–5.3	6.0–6.5	7.5–8.3
Height of torus (Å)	7.9	7.9	7.9
Diameter of periphery (Å)	14.6	15.4	17.5
Approximate volume of cavity (Å^3)	174	262	472
Approximate cavity volume			
Per mol cyclodextrin (mL)	104	157	256
Per g cyclodextrin (mL)	0.1	0.14	0.20

Note: 1 Å = 0.1 nm.

Solubility

α-cyclodextrin: soluble 1 in 7 parts of water at 20°C, 1 in 3 at 50°C.

β-cyclodextrin: soluble 1 in 200 parts of propylene glycol, 1 in 50 of water at 20°C, 1 in 20 at 50°C; practically insoluble in acetone, ethanol (95%), and methylene chloride.

γ-cyclodextrin: soluble 1 in 4.4 parts of water at 20°C, 1 in 2 at 45°C.

Specific rotation

α-cyclodextrin: $[\alpha]_D^{25} = +150.5°$;

β-cyclodextrin: $[\alpha]_D^{25} = +162.0°$;

γ-cyclodextrin: $[\alpha]_D^{25} = +177.4°$.

Spectroscopy

IR spectra *see* Figure 1.

NIR spectra *see* Figures 2, 3 and 4.

Surface tension (at 25°C)

α-cyclodextrin: 71 mN/m (71 dynes/cm);

β-cyclodextrin: 71 mN/m (71 dynes/cm);

γ-cyclodextrin: 71 mN/m (71 dynes/cm).

11 Stability and Storage Conditions

β-Cyclodextrin and other cyclodextrins are stable in the solid state if protected from high humidity.

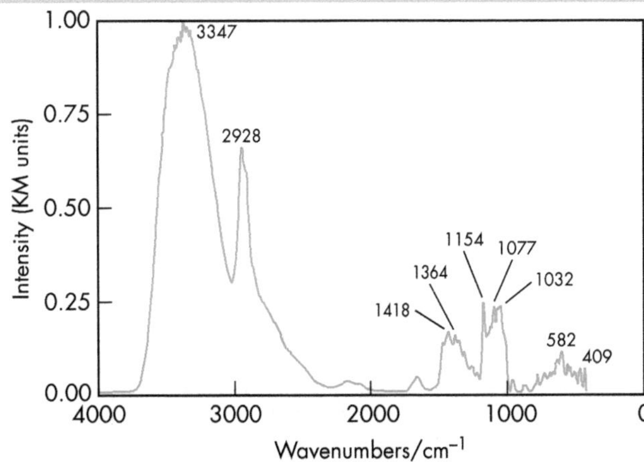

Figure 1: Infrared spectrum of b-cyclodextrin measured by diffuse reflectance. Adapted with permission of Informa Healthcare.

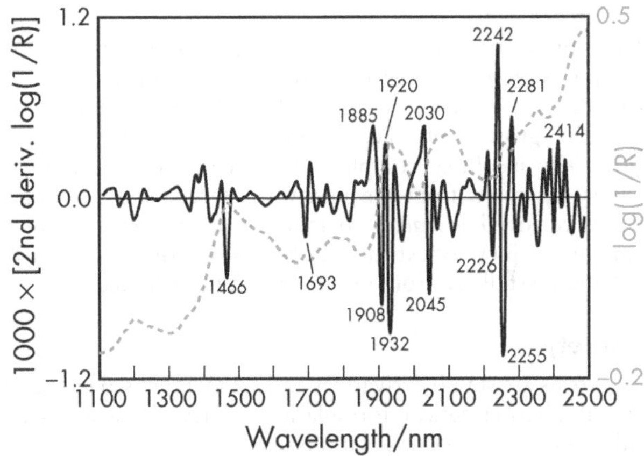

Figure 2: Near-infrared spectrum of α-cyclodextrin measured by reflectance.

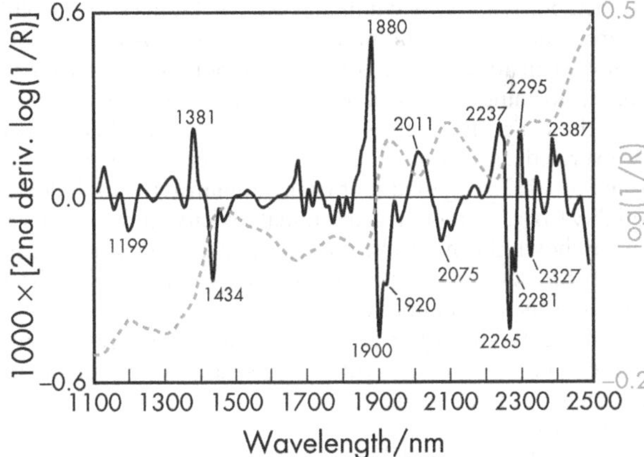

Figure 3: Near-infrared spectrum of β-cyclodextrin measured by reflectance.

Cyclodextrins should be stored in a tightly sealed container, in a cool, dry place.

12 Incompatibilities

—

13 Method of Manufacture

Cyclodextrins are manufactured by the enzymatic degradation of starch using specialized bacteria. For example, β-cyclodextrin is

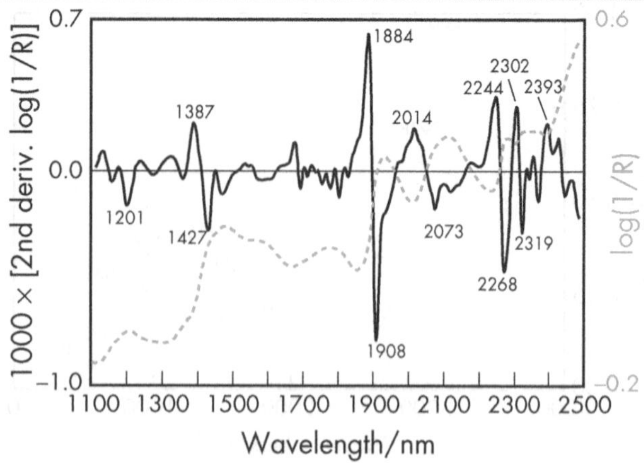

Figure 4: Near-infrared spectrum of γ-cyclodextrin measured by reflectance.

produced by the action of the enzyme cyclodextrin glucosyltransferase upon starch or a starch hydrolysate. An organic solvent is used to direct the reaction that produces β-cyclodextrin, and to prevent the growth of microorganisms during the enzymatic reaction. The insoluble complex of β-cyclodextrin and organic solvent is separated from the noncyclic starch, and the organic solvent is removed *in vacuo* so that less than 1 ppm of solvent remains in the β-cyclodextrin. The β-cyclodextrin is then carbon treated and crystallized from water, dried, and collected.

14 Safety

Cyclodextrins are starch derivatives and are mainly used in oral and parenteral pharmaceutical formulations. They are also used in topical and ophthalmic formulations.(3)

Cyclodextrins are also used in cosmetics and food products, and are generally regarded as essentially nontoxic and nonirritant materials. However, when administered parenterally, β-cyclodextrin is not metabolized but accumulates in the kidneys as insoluble cholesterol complexes, resulting in severe nephrotoxicity.(10)

Cyclodextrin administered orally is metabolized by microflora in the colon, forming the metabolites maltodextrin, maltose, and glucose; these are themselves further metabolized before being finally excreted as carbon dioxide and water. Although a study published in 1957 suggested that orally administered cyclodextrins were highly toxic,(11) more recent animal toxicity studies in rats and dogs have shown this not to be the case, and cyclodextrins are now approved for use in food products and orally administered pharmaceuticals in a number of countries.

Cyclodextrins are not irritant to the skin and eyes, or upon inhalation. There is also no evidence to suggest that cyclodextrins are mutagenic or teratogenic.

α-*Cyclodextrin*

LD_{50} (rat, IP): 1.0 g/kg(12)
LD_{50} (rat, IV): 0.79 g/kg

β-*Cyclodextrin*

LD_{50} (mouse, IP): 0.33 g/kg(13)
LD_{50} (mouse, SC): 0.41 g/kg
LD_{50} (rat, IP): 0.36 g/kg
LD_{50} (rat, IV): 1.0 g/kg
LD_{50} (rat, oral): 18.8 g/kg
LD_{50} (rat, SC): 3.7 g/kg

γ-*Cyclodextrin*

LD_{50} (rat, IP): 4.6 g/kg(12)
LD_{50} (rat ,IV): 4.0 g/kg
LD_{50} (rat, oral): 8.0 g/kg

15 Handling Precautions

Observe normal precautions appropriate to the circumstances and quantity of material handled. Cyclodextrins are fine organic powders and should be handled in a well-ventilated environment. Efforts should be made to limit the generation of dust, which can be explosive.

16 Regulatory Status

Included in the FDA Inactive Ingredients Database: α-cyclodextrin (injection preparations); β-cyclodextrin (oral tablets, topical gels); γ-cyclodextrin (IV injections).

Included in the Canadian Natural Health Products Ingredients Database, and in oral and rectal pharmaceutical formulations licensed in Europe, Japan, and the USA.

17 Related Substances

Dimethyl-β-cyclodextrin; 2-hydroxyethyl-β-cyclodextrin; hydroxypropyl betadex; sulfobutylether β-cyclodextrin; trimethyl-β-cyclodextrin.

Dimethyl-β-cyclodextrin

Molecular weight 1 331
Synonyms DM-β-CD.
Appearance White crystalline powder.
Cavity diameter 6 Å
Melting point 295.0–300.0°C
Moisture content ≤1% w/w
Solubility Soluble 1 in 135 parts of ethanol (95%), and 1 in 1.75 of water at 25°C. Solubility decreases with increasing temperature.
Surface tension 62 mN/m (62 dynes/cm) at 25°C.
Method of manufacture Dimethyl-β-cyclodextrin is prepared from β-cyclodextrin by the selective methylation of all C2 secondary hydroxyl groups and all C6 primary hydroxyl groups (C3 secondary hydroxyl groups remain unsubstituted).
Comments Used in applications similar to those for β-cyclodextrin.

2-Hydroxyethyl-β-cyclodextrin

CAS number [98513-20-3]
Synonyms 2-HE-β-CD.
Appearance White crystalline powder.
Density (bulk) 0.681 g/cm^3
Density (tapped) 0.916 g/cm^3
Density (true) 1.378 g/cm^3
Solubility Greater than 1 in 2 parts of water at 25°C.
Surface tension 68.0–71.0 mN/m (68–71 dynes/cm) at 25°C.
Comments Used in applications similar to those for β-cyclodextrin. The degree of substitution of hydroxyethyl groups can vary.(14)

Trimethyl-β-cyclodextrin

Molecular weight 1 429
Synonyms TM-β-CD.
Appearance White crystalline powder.
Cavity diameter 4.0–7.0 Å
Melting point 157°C
Moisture content ≤1% w/w
Solubility Soluble 1 in 3.2 parts of water at 25°C. Solubility decreases with increasing temperature.
Surface tension 56 mN/m (56 dynes/cm) at 25°C.
Method of manufacture Trimethyl-β-cyclodextrin is prepared from β-cyclodextrin by the complete methylation of all C2 and C3 secondary hydroxyl groups along with all C6 primary hydroxyl groups.
Comments Used in applications similar to those for β-cyclodextrin.

18 Comments

In addition to their use in pharmaceutical formulations, cyclodextrins have also been investigated for use in various industrial applications. Analytically, cyclodextrin polymers are used in chromatographic separations, particularly of chiral materials.

β-Cyclodextrin derivatives are more water-soluble than β-cyclodextrin, and studies have shown that they have greater solubilizing action with some drugs such as ibuproxam, a poorly water-soluble anti-inflammatory agent.[15,16]

Specifications for α-cyclodextrin and β-cyclodextrin are included in the *Japanese Pharmaceutical Excipients* (JPE).[17]

Specifications for α-cyclodextrin, β-cyclodextrin, and γ-cyclodextrin are listed separately in the *Food Chemicals Codex* (FCC).[18]

The EINECS number for cyclodextrin is 231-493-2. The PubChem Compound ID (CID) for cyclodextrins includes 444913 (α-cyclodextrin), 24238 (β-cyclodextrin), and 86575 (γ-cyclodextrin).

19 Specific References

1 El Shaboury MH. Physical properties and dissolution profiles of tablets directly compressed with β-cyclodextrin. *Int J Pharm* 1990; **63**: 95–100.

2 Shangraw RF, *et al.* Characterization of the tableting properties of β-cyclodextrin and the effects of processing variables on inclusion complex formation, compactibility and dissolution. *Drug Dev Ind Pharm* 1992; **18**(17): 1831–1851.

3 Cal K, Centkowska K. Use of cyclodextrins in topical formulations: practical aspects. *Eur J Pharm Biopharm* 2010; **68**: 467–478.

4 Prankerd RJ, *et al.* Degradation of aspartame in acidic aqueous media and its stabilization by complexation with cyclodextrins or modified cyclodextrins. *Int J Pharm* 1992; **88**: 189–199.

5 Palmieri GF, *et al.* Inclusion of vitamin D2 in β-cyclodextrin. Evaluation of different complexation methods. *Drug Dev Ind Pharm* 1993; **19**(8): 875–885.

6 Szente L, *et al.* Suppositories containing β-cyclodextrin complexes, part 1: stability studies. *Pharmazie* 1984; **39**: 697–699.

7 Szente L, *et al.* Suppositories containing β-cyclodextrin complexes, part 2: dissolution and absorption studies. *Pharmazie* 1985; **40**: 406–407.

8 Amann M, Dressnandt G. Solving problems with cyclodextrins in cosmetics. *Cosmet Toilet* 1993; **108**(11): 90–95.

9 Buschmann HJ, Schollmeyer E. Applications of cyclodextrins in cosmetic products: a review. *J Cosmet Sci* 2002; **53**(3): 185–191.

10 Frank DW, *et al.* Cyclodextrin nephrosis in the rat. *Am J Pathol* 1976; **83**: 367–382.

11 French D. The Schardinger dextrins. *Adv Carbohydr Chem* 1957; **12**: 189–260.

12 Sweet DV, ed. *Registry of Toxic Effects of Chemical Substances.* Cincinnati: US Department of Health, 1987: 1721.

13 Lewis RJ, ed. *Sax's Dangerous Properties of Industrial Materials*, 11th edn. New York: Wiley, 2004: 1031.

14 Menard FA, *et al.* Potential pharmaceutical applications of a new beta cyclodextrin derivative. *Drug Dev Ind Pharm* 1988; **14**(11): 1529–1547.

15 Mura P, *et al.* Comparative study of ibuproxam complexation with amorphous beta-cyclodextrin derivatives in solution and in the solid state. *Eur J Pharm Biopharm* 2002; **54**(2): 181.

16 Liu X, *et al.* Biopharmaceuticals of beta-cyclodextrin derivative-based formulations of acitretin in Sprague-Dawley rats. *J Pharm Sci* 2004; **93**(4): 805–815.

17 Japan Pharmaceutical Excipients Council. *Japanese Pharmaceutical Excipients* 2004. Tokyo: Yakuji Nippo, 2004: 200–203.

18 *Food Chemicals Codex*, 7th edn. Bethesda, MD: United States Pharmacopeia, 2010; 259–262.

20 General References

Bekers O, *et al.* Cyclodextrins in the pharmaceutical field. *Drug Dev Ind Pharm* 1991; **17**: 1503–1549.

Bender ML, Komiyama M. *Cyclodextrin Chemistry.* New York: Springer-Verlag, 1978.

Carpenter TO, *et al.* Safety of parenteral hydroxypropyl β-cyclodextrin. *J Pharm Sci* 1995; **84**: 222–225.

Chaubal MV. Drug delivery applications of cyclodextrins Part I. *Drug Dev Technol* 2002; **2**(7): 34–38.

Chaubal MV. Drug delivery applications of cyclodextrins Part II. *Drug Dev Technol* 2003; **3**(2): 34–36.

Darrouzet H. Preparing cyclodextrin inclusion compounds. *Manuf Chem* 1993; **64**(11): 33–34.

Fenyvest É, *et al.* Cyclodextrin polymer, a new tablet disintegrating agent. *Pharmazie* 1984; **39**: 473–475.

Leroy-Lechat F, *et al.* Evaluation of the cytotoxicity of cyclodextrins and hydroxypropylated derivatives. *Int J Pharm* 1994; **101**: 97–103.

Loftsson T. Pharmaceutical applications of β-cyclodextrin. *Pharm Technol* 1999; **23**(12): 40–50.

Loftsson T, Brewster ME. Pharmaceutical applications of cyclodextrins 1: drug solubilization and stabilization. *J Pharm Sci* 1996; **85**(10): 1017–1025.

Loftsson T, Brewster ME. Pharmaceutical applications of cyclodextrins: basic science and product development. *J Pharm Pharmacol* 2010; **62**: 1607–1621.

Loftsson T, Duchêne D. Cyclodextrins and their pharmaceutical applications. *Int J Pharm* 2007; **329**: 1–11.

Loftsson T, Masson M. Cyclodextrins in topical drug formulations: theory and practice. *Int J Pharm* 2001; **225**: 15–30.

Loftsson T, *et al.* Self-association of cyclodextrins and cyclodextrin complexes. *J Pharm Sci* 2004; **93**(4): 1091–1099.

Pande GS, Shangraw RF. Characterization of β-cyclodextrin for direct compression tableting. *Int J Pharm* 1994; **101**: 71–80.

Pande GS, Shangraw RF. Characterization of β-cyclodextrin for direct compression tableting II: the role of moisture in the compactibility of β-cyclodextrin. *Int J Pharm* 1995; **124**: 231–239.

Pitha J *et al.* Molecular encapsulation by cyclodextrin and congeners. In: Bruck SD, ed. *Controlled Drug Delivery.* I. Boca Raton, FL: CRC Press, 1983.

Shao Z, *et al.* Cyclodextrins as nasal absorption promoters of insulin: mechanistic evaluations. *Pharm Res* 1992; **9**: 1157–1163.

Sina VR, *et al.* Cyclodextrins as sustained-release carriers. *Pharm Technol* 2002; **26**(10): 36–46.

Stella VJ, Rajewski RA. Cyclodextrins: their future in drug formulation. *Pharm Res* 1997; **14**(5): 556–567.

Stoddard F, Zarycki R. *Cyclodextrins.* Boca Raton, FL: CRC Press, 1991.

Strattan CE. 2-Hydroxypropyl-β-cyclodextrin, part II: safety and manufacturing issues. *Pharm Technol* 1992; **16**(2): 52, 54, 56, 58.

Szejtli J. Cyclodextrins in drug formulations: part I. *Pharm Technol Int* 1991; 3(2): 15–1820–22.

Szejtli J. Cyclodextrins in drug formulations: part II. *Pharm Technol Int* 1991; 3(3): 16, 18, 20, 22, 24.

Szejtli J. General overview of cyclodextrin. *Chem Rev* 1998; **98**: 1743–2076.

Yamamoto M, *et al.* Some physicochemical properties of branched β-cyclodextrins and their inclusion characteristics. *Int J Pharm* 1989; **49**: 163–171.

21 Author

W Cook.

22 Date of Revision

1 March 2012.

Cyclomethicone

1 Nonproprietary Names

USP–NF: Cyclomethicone

2 Synonyms

Dimethylcyclopolysiloxane; *Dow Corning 245 Fluid*; *Dow Corning 246 Fluid*; *Dow Corning 345 Fluid*.

3 Chemical Name and CAS Registry Number

Cyclopolydimethylsiloxane [69430-24-6]

4 Empirical Formula and Molecular Weight

The USP35–NF30 describes cyclomethicone as a fully methylated cyclic siloxane containing repeating units of the formula $[-(CH_3)_2SiO-]_n$ in which *n* is 4, 5, or 6, or a mixture of them.

5 Structural Formula

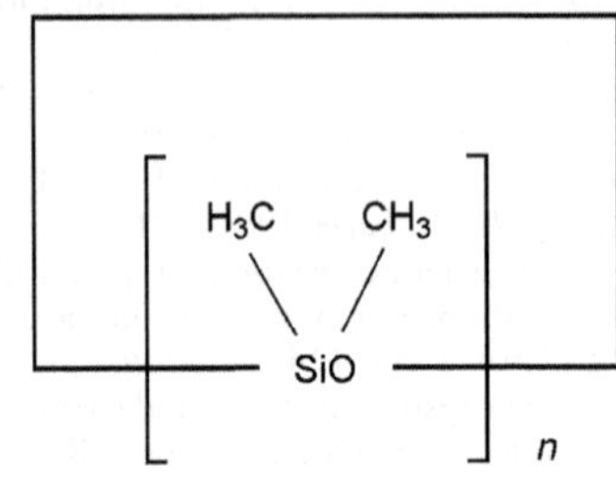

6 Functional Category

Emollient; humectant; viscosity-increasing agent.

7 Applications in Pharmaceutical Formulation or Technology

Cyclomethicone is mainly used in topical pharmaceutical and cosmetic formulations such as water-in-oil creams.[1–3]

Cyclomethicone has been used in cosmetic formulations, at concentrations of 0.1–50%, since the late 1970s and is now the most widely used silicone in the cosmetics industry. Its high volatility and mild solvent properties make it ideal for use in topical formulations. Its low heat of vaporization means that when applied to skin it has a 'dry' feel.

See also Dimethicone.

8 Description

Cyclomethicone occurs as a clear, colorless and tasteless volatile liquid.

9 Pharmacopeial Specifications

See Table I.

Table I: Pharmacopeial specifications for cyclomethicone.

Test	USP35–NF30
Identification	+
Limit of nonvolatile residue	≤0.15%
Assay of $(C_2H_6OSi)_n$ calculated as the sum of cyclomethicone 4, cyclomethicone 5, and cyclomethicone 6	≥98.0%
Assay of individual cyclomethicone components	95.0–105.0%

10 Typical Properties

Solubility Soluble in ethanol (95%), isopropyl myristate, isopropyl palmitate, mineral oil, and petrolatum at 80°C; practially insoluble in glycerin, propylene glycol, and water.

See also Table II.

11 Stability and Storage Conditions

Cyclomethicone should be stored in an airtight container in a cool, dry, place.

12 Incompatibilities

—

13 Method of Manufacture

Cyclomethicone is manufactured by the distillation of crude polydimethylsiloxanes.

14 Safety

Cyclomethicone is generally regarded as a relatively nontoxic and nonirritant material. Although it has been used in oral pharmaceutical applications, cyclomethicone is mainly used in topical pharmaceutical formulations. It is also widely used in cosmetics.[4] Studies of the animal and human toxicology of cyclomethicone suggest that it is nonirritant and not absorbed through the skin. Only small amounts are absorbed orally; an acute oral dose in rats produced no deaths.[5,6]

See also Dimethicone.

15 Handling Precautions

Observe normal precautions appropriate to the circumstances and quantity of material handled.

Table II: Typical physical properties of selected commercially available cyclomethicones.

Grade	Boiling point (°C)	Flash point (°C)	Freezing point (°C)	Refractive index at 25°C	Surface tension (mN/m)	Specific gravity at 25°C	Viscosity (kinematic) (mm^2/s)	Water content (%)
Dow Corning 245 Fluid	205	77	<−50	1.397	18.0	0.95	4.0	0.025
Dow Corning 246 Fluid	245	93	<−40	1.402	18.8	0.96	6.8	0.025
Dow Corning 345 Fluid	217	77	<−50	1.398	20.8	0.957	6.0	0.025

16 Regulatory Status

Included in the FDA Inactive Ingredients Database (oral powder for reconstitution; topical lotion, topical cream, topical emulsion). Included in nonparenteral medicines licensed in the UK. Included in the Canadian Natural Health Products Ingredients Database.

17 Related Substances

Dimethicone; simethicone.

18 Comments

–

19 Specific References

1. Goldenberg RL, *et al.* Silicones in clear formulations. *Drug Cosmet Ind* 1986; **138**(Feb): 34, 38, 40, 44.
2. Chandra D, *et al.* Silicones for cosmetics and toiletries: environmental update. *Cosmet Toilet* 1994; **109**(Mar): 63–66.
3. Forster AH, Herrington TM. Rheology of siloxane-stabilized water in silicone emulsions. *Int J Cosmet Sci* 1997; **19**(4): 173–191.
4. Parente ME, *et al.* Study of sensory properties of emollients used in cosmetics and their correlation with physicochemical properties. *J Cosmet Sci* 2005; **56**(3): 175–182.
5. Anonymous. Final report on the safety assessment of cyclomethicone. *J Am Coll Toxicol* 1991; **10**(1): 9–19.
6. Christopher SM, *et al.* Acute toxicologic evaluation of cyclomethicone. *J Am Coll Toxicol* 1994; **12**(6): 578.

20 General References

—

21 Author

RT Guest.

22 Date of Revision

1 March 2012.

Decyl Oleate

1 Nonproprietary Names

BP: Decyl Oleate
PhEur: Decyl Oleate

2 Synonyms

Ceraphyl 140; *Cetiol V PH*; decyl 9-octadecenoate; decylis oleas; decyloleat; oleic acid, decyl ester; *Tegosoft DO*.

3 Chemical Name and CAS Registry Number

9-Octadecenoic acid (9*Z*)-, decyl ester [3687-46-5]

4 Empirical Formula and Molecular Weight

$C_{28}H_{54}O_2$ 422.74

The BP 2012 defines decyl oleate as a mixture of decyl esters of fatty acids, mainly oleic (*cis*-9-octadecenoic) acid. A suitable antioxidant may be added.

5 Structural Formula

H H
O
H_3C—$(CH_2)_7$ $(CH_3)_7$ —C—O—$(CH_2)_9$—CH_3

6 Functional Category

Emollient.

7 Applications in Pharmaceutical Formulation or Technology

Decyl oleate is used as an emollient in pharmaceutical creams and lotions for skin and personal care products. It has been evaluated as a skin penetration enhancer for flurbiprofen.[1]

Decyl oleate has also been used in combination with carnauba wax as an encapsulation system for inorganic sunscreens.[2–4]

8 Description

Decyl oleate occurs as a clear, colorless, or pale yellow liquid.

9 Pharmacopeial Specifications

See Table I. *See also* Section 18.

Table I: Pharmacopeial specifications for decyl oleate.

Test	PhEur 7.4
Identification	+
Characters	+
Relative density	0.86–0.87
Acid value	≤1.0
Iodine value	55–70
Peroxide value	≤10.0
Saponification value	130–140
Oleic acid	≥60%
Water	≤1.0%
Total ash	≤0.1%

10 Typical Properties

Boiling point 363°C
Flash point 202°C (open cup)
Pour point −6°C for *Tegosoft DO*[5]
Refractive index 1.453–1.456
Solubility Practically insoluble in water; miscible with ethanol (96%), methylene chloride, and light petroleum (40–60°C). Soluble in isopropyl myristate and oleyl alcohol.
Surface tension 31 mN/m at 25°C for *Tegosoft DO*[5]
Viscosity 16.6 mPa s (16.6 cP) at 25°C; approx 14.0 mPa s (14.0 cP) for *Tegosoft DO* at 25°C.[5]

11 Stability and Storage Conditions

Decyl oleate should be stored below 30°C and protected from moisture. Through the addition of mixed tocopherols, the product is protected from auto-oxidation.

Depending on the storage temperature, the peroxide value may increase. The product quality is not impaired as long as the peroxide value remains below 8.0.

12 Incompatibilities

Decyl oleate is incompatible with strong oxidizing agents.

13 Method of Manufacture

Decyl oleate is manufactured from decyl alcohol and oleic acid.

14 Safety

Decyl oleate is used in cosmetics and topical pharmaceutical formulations, and is generally regarded as a nontoxic and nonirritant material.

A 15% solution of decyl oleate in corn oil has been shown to be practically nonirritating in rabbit eye studies and nonsensitizing to guinea pig skin, and a 10% solution has been shown to be nonirritating to rabbit skin.[6] Repeat patch testing in humans with 5.5% decyl oleate produced a low number of reactions with a low total irritation score.[6]

LD_{50} (rat, oral): 34.68 g/kg (40 mL/kg)[6]

15 Handling Precautions

Observe normal precautions appropriate to the circumstances and quantity of the material handled.

16 Regulatory Status

Included in the Canadian Natural Health Products Ingredients Database.

17 Related Substances

Isodecyl oleate.

Isodecyl oleate

Empirical formula $C_{28}H_{54}O_2$
CAS number [59231-34-4]
Synonyms 9-Octadecenoic acid, isodecyl ester; *Schercemol* IDO ester
Appearance A white to straw-colored liquid with characteristic odor.
Acid value ≤5.0 for *Schercemol* IDO ester.[7]
Iodine value 50–65 for *Schercemol* IDO ester.
Refractive index 1.455 for *Schercemol* IDO ester.

Saponification value 130–140 for *Schercemol* IDO ester.
Solubility Soluble in isopropyl myristate, peanut oil, mineral oil, ethanol (95%), and oleyl alcohol; insoluble in water, glycerin, and propylene glycol.
Specific gravity 0.850–0.890 for *Schercemol* IDO ester.
Safety Nonirritating to eyes; mildly irritating to skin.
LD_{50} (rat, oral): >40 mL/kg[6]
Comments Used as an emollient and solvent for pharmaceutical topical formulations. Also used in cosmetic products. The PubChem Compound ID (CID) for isodecyl oleate is 6436737. The EINECS number for isodecyl oleate is 261-673-6.

18 Comments

A specification for decyl oleate is included in the *Japanese Pharmaceutical Excipients* (JPE).[8]

The EINECS number for decyl oleate is 222-981-6. The PubChem Compound ID (CID) for decyl oleate includes 5363234 and 62514.

19 Specific References

1 Cornelio R, Mayorga P. Study of cutaneous penetration of flurbiprofen. *Lat Am J Pharm* 2007; **26**: 883–888.
2 Villalobos-Hernandez JR. Novel nanoparticulate carrier system based on carnauba wax and decyl oleate for the dispersion of inorganic sunscreens in aqueous media. *Eur J Pharm Biopharm* 2005; **60**: 113–122.
3 Villalobos-Hernandez JR. Physical stability, centrifugation tests, and entrapment efficiency studies of carnauba wax–decyl oleate nanoparticles used for the dispersion of inorganic sunscreens in aqueous media. *Eur J Pharm Biopharm* 2006; **63**: 115–127.
4 Villalobos-Hernandez JR. *In vitro* erythemal UV-A protection factors of inorganic sunscreens distributed in aqueous media using carnauba wax–decyl oleate nanoparticles. *Eur J Pharm Biopharm* 2007; **65**: 122–125.
5 Evonik Goldschmidt GmbH. Product literature: *Tegosoft DO*, 2003.
6 Anonymous. Final report on the safety assessment of decyl and isodecyl oleates. *Int J Toxicol* 1982; **1**(2): 85–95.
7 Lubrizol Advanced Materials Inc. Technical data sheet: *Schercemol* IDO Ester, 2007.
8 Japan Pharmaceutical Excipients Council. *Japanese Pharmaceutical Excipients 2004*. Tokyo: Yakuji Nippo, 2004; 211.

20 General References

Chemical Book. Product information: Decyl oleate. http://www.chemicalbook.com/ChemicalProductProperty_EN_CB7458400.htm (accessed 29 February 2012).
Cognis. Product data sheet: *Cetiol V* (version 3.1).
Guillot JP, *et al.* Safety evaluation of cosmetic raw materials. *J Soc Cosmet Chem* 1977; **28**: 377–393.
International Specialty Products. Material safety data sheet: *Ceraphyl 140*, August 2008.

21 Authors

ME Fenton, RC Rowe.

22 Date of Revision

1 March 2012.

Denatonium Benzoate

1 Nonproprietary Names

USP–NF: Denatonium Benzoate

2 Synonyms

Bitrex; *Bitterguard*; denatonii benzoas; *N*-[2-(2,6-dimethylphenyl)amino]-2-oxoethyl]-*N*,*N*-diethylbenzenemethanaminium benzoate monohydrate; lignocaine benzyl benzoate.

3 Chemical Name and CAS Registry Number

Benzyldiethyl[(2,6-xylylcarbamoyl)methyl]ammonium benzoate anhydrous [3734-33-6]

Benzyldiethyl[(2,6-xylylcarbamoyl)methyl]ammonium benzoate monohydrate [86398-53-0]

4 Empirical Formula and Molecular Weight

$C_{28}H_{34}N_2O_3$ 446.59 (for anhydrous)
$C_{28}H_{34}N_2O_3{\cdot}H_2O$ 464.60 (for monohydrate)

5 Structural Formula

CH3 O NH N+ H3C CH3 CH3 O O− · H2O

6 Functional Category

Alcohol denaturant; flavoring agent.

7 Applications in Pharmaceutical Formulation or Technology

Denatonium benzoate is among the most bitter of substances known and is detectable at concentrations of approximately 10 ppb. In pharmaceutical and other industrial applications it is added to some products as a deterrent to accidental ingestion,[1–4] although its effectiveness in reducing the frequency or severity of pediatric antifreeze poisonings has been questioned.[5,6] It is most commonly used at levels of 5–500 ppm. Denatonium benzoate may also be used to replace brucine or quassin as a denaturant for ethanol.

In pharmaceutical formulations, denatonium benzoate has been used as a flavoring agent in placebo tablets.

8 Description

Denatonium benzoate occurs as an odorless, very bitter tasting, white crystalline powder or granules.

D

9 Pharmacopeial Specifications

See Table I.

Table I: Pharmacopeial specifications for denatonium benzoate.

Test	USP35–NF30
Identification	+
Melting range	163–170°C
pH (3% aqueous solution)	6.5–7.5
Loss on drying (monohydrate)	≤3.5–4.5%
Loss on drying (anhydrous)	≤1.0%
Residue on ignition	≤0.1%
Chloride	≤0.2%
Assay (dried basis)	99.5–101.0%

10 Typical Properties

Acidity/alkalinity pH = 6.5–7.5 (3% w/v aqueous solution)
Density (bulk) 0.3–0.6 g/cm^3
Density (tapped) 0.4–0.7 g/cm^3
Partition coefficient Octanol : water = 1.8
Solubility *see* Table II.

Table II: Solubility of denatonium benzoate.

Solvent	Solubility at 20°C
Acetone	Sparingly soluble
Chloroform	1 in 2.9
Ethanol (95%)	1 in 2.4
Ether	1 in 5000
Methanol	Very soluble
Water	1 in 20

Spectroscopy

IR spectra *see* Figure 1.

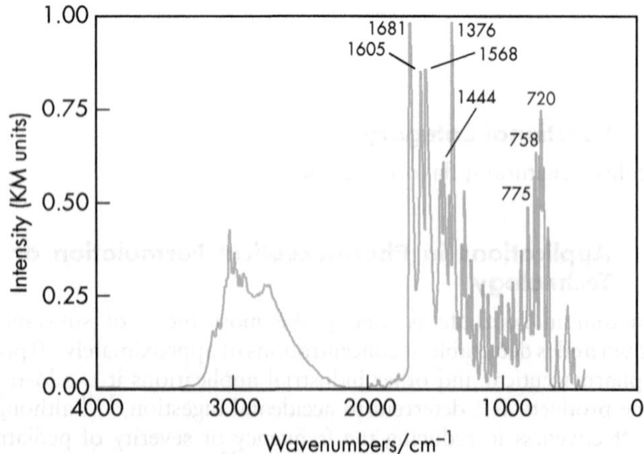

Figure 1: Infrared spectrum of denatonium benzoate measured by diffuse reflectance. Adapted with permission of Informa Healthcare.

11 Stability and Storage Conditions

Denatonium benzoate is stable up to 140°C and over a wide pH range. It should be stored in a well-closed container (such as polythene-lined steel) in a cool, dry place. Aqueous or alcoholic solutions retain their bitterness for several years even when exposed to light.

12 Incompatibilities

Denatonium benzoate is incompatible with strong oxidizing agents.

13 Method of Manufacture

Denatonium benzoate was first synthesized in the 1950s and is usually prepared by reacting denatonium chloride with benzyl benzoate.

14 Safety

Denatonium benzoate is generally regarded as a nonirritant and nonmutagenic substance. However, there has been a single report of contact urticaria attributed to denatonium benzoate occurring in a 30-year-old man who developed asthma and pruritus after using an insecticidal spray denatured with denatonium benzoate.[7]

LD_{50} (rabbit, oral): 0.508 g/kg[8]

LD_{50} (rat, oral): 0.584 g/kg

15 Handling Precautions

Observe normal precautions appropriate to the circumstances and quantity of material handled. Containers should be kept tightly closed and handled in areas with good ventilation. Eye protection, gloves, and a dust mask are recommended. Denatonium benzoate is moderately toxic by ingestion and when heated to decomposition emits toxic vapors of NO_x. Denatonium benzoate may also cause hypersensitization.

16 Regulatory Status

Denatonium benzoate is used worldwide as a denaturant for alcohol. It is included in the FDA Inactive Ingredients Database (topical gel and solution). Included in the Canadian Natural Health Products Ingredients Database.

17 Related Substances

—

18 Comments

In a topical formulation, denatonium benzoate has been used in an anti-nailbiting preparation.[9]

Several HPLC methods of analysis for denatonium benzoate have been reported.[10–12]

The EINECS number for denatonium benzoate is 223-095-2. The PubChem Compound ID (CID) for denatonium benzoate is 19518.

19 Specific References

1 Klein-Schwartz W. Denatonium benzoate: review of efficacy and safety. *Vet Hum Toxicol* 1991; 33(6): 545–547.
2 Sibert JR, Frude N. Bittering agents in the prevention of accidental poisoning: children's reactions to denatonium benzoate (Bitrex). *Arch Emerg Med* 1991; 8(1): 1–7.
3 Hansen SR, *et al*. Denatonium benzoate as a deterrent to ingestion of toxic substances: toxicity and efficacy. *Vet Hum Toxicol* 1993; 35(3): 234–236.
4 Rodgers GC, Tenenbein M. Role of aversive bittering agents in the prevention of pediatric poisonings. *Pediatrics* 1994; 93(Jan): 68–69.
5 White NC, *et al*. The impact of bittering agents on suicidal ingestions of antifreeze. *Clin Toxicol* 2008; 46(6): 507–514.

6 White NC, *et al.* The impact of bittering agents on pediatric ingestions of antifreeze. *Clin Pediatr* 2009; **48**(9): 913–921.
7 Björkner B. Contact urticaria and asthma from denatonium benzoate (Bitrex). *Contact Dermatitis* 1980; **6**(7): 466–471.
8 Lewis RJ, ed. *Sax's Dangerous Properties of Industrial Materials*, 11th edn. New York: Wiley, 2004: 1087.
9 Anonymous. Relief for warts; none for nail biters. *FDA Consum* 1981; **15**(Feb): 13.
10 Sugden K, *et al.* Determination of denaturants in alcoholic toilet preparations 1: denatonium benzoate (Bitrex) by high performance liquid chromatography. *Analyst* 1978; **103**: 653–656.
11 Faulkner A, DeMontigny P. High-performance liquid chromatographic determination of denatonium benzoate in ethanol with 5% polyvinylpyrrolidone. *J Chromatogr-A* 1995; **715**: 189–194.
12 Henderson MC, *et al.* Analysis of denatonium benzoate in Oregon consumer products by HPLC. *Chemosphere* 1998; **36**(1): 203–210.

20 General References

Macfarlan Smith. Bitrex. http://www.bitrex.com (accessed 22 September 2011).
Payne HAS. Bitrex – a bitter solution to safety. *Chem Ind* 1988; **22**: 721–723.
Payne HAS. Bitrex – a bitter solution to product safety. *Drug Cosmet Ind* 1989; **144**(May): 30, 32, 34.

21 Author

PJ Weller.

22 Date of Revision

1 March 2012.

Dextran

1 Nonproprietary Names

BP: Dextran 1 for Injection
Dextran 40 for Injection
Dextran 60 for Injection
Dextran 70 for Injection

JP: Dextran 40
Dextran 70

PhEur: Dextran 1 for Injection
Dextran 40 for Injection
Dextran 60 for Injection
Dextran 70 for Injection

USP–NF: Dextran 1
Dextran 40
Dextran 70

2 Synonyms

Glucose polymer; dextrans.

3 Chemical Name and CAS Registry Number

2,3,4,5-Tetrahydroxy-6-[3,4,5-trihydroxy-6-[[3,4,5-trihydroxy-6-(hydroxymethyl)oxan-2-yl]oxymethyl]oxan-2-yl]oxyhexanal [9004-54-0]

4 Empirical Formula and Molecular Weight

$[C_6H_{10}O_5]_n$

Dextran is a branched high-molecular-weight polymer of α-D-glucose (dextrose). Dextran polymers are available in various average molecular weight fractions from 1000 Da (Dextran 1) to 2 000 000 Da (Dextran 2000), with the most common fractions being 1000 Da, 40 000 Da (Dextran 40), and 70 000 Da (Dextran 70).[1]

The polymer main chain has α(1→6) glycosidic linkages with branching occurring primarily at C3, α(1→3) glycosidic linkages, and occasionally at C4, α(1→4) glycosidic linkages or C2 α(1→2) glycosidic linkages. Commercially available dextrans contain about 95% α(1→6) glycosidic linkages of the main chain and about 5% side branching. About 80% of the side branches are single α-D-glucose subunits with the remaining branches a mixture of side-chains of various lengths.[2]

Dextran is a fermentation product of sucrose by bacteria containing the enzyme dextransucrase capable of synthesizing polysaccharides from sucrose. The native dextran polymer has a broad polydispersity index (range of polymer length and weight). The polydispersity is decreased by purification and separation into molecular weight fractions via partial acid hydrolysis.[1] Polydispersity usually increases with increasing average molecular weight.[3]

5 Structural Formula

See Section 4.

6 Functional Category

Lyophilization aid.

7 Applications in Pharmaceutical Formulation or Technology

Dextran is used in pharmaceutical preparations as a lyophile bulking agent[4] and as a lyoprotectant in lyophilized protein products.[5]

8 Description

Dextran occurs as a white to off-white amorphous hygroscopic powder.

9 Pharmacopeial Specifications

See Table I.

10 Typical Properties

Hygroscopicity Hygroscopic. In lyophiles, dextran is amorphous and hygroscopic before and after lyophilization.[6]

Solubility Readily soluble in water (>50% w/v); soluble in methyl sulfide, formamide, ethylene glycol, and glycerol; insoluble in methanol, ethanol, isopropanol, acetone, and other ketone solvents.[7]

Viscosity Increases linearly with increasing polymer molecular weight.[8]

Table I: Pharmacopeial specifications for dextran.

Test	JP XV S2	PhEur 7.4	USP35–NF30
Identification			
Dextran 1	–	+	+
Dextran 40	+	+	+
Dextran 60	–	+	+
Dextran 70	+	+	+
Characters			
Dextran 1	–	+	–
Dextran 40	+	+	–
Dextran 60	–	+	–
Dextran 70	+	+	–
pH			
Dextran 1	–	–	4.5 to 7.0
Dextran 40	5.0–7.0	–	4.5 to 7.0
Dextran 60	–	–	–
Dextran 70	5.0–7.0	–	4.5 to 7.0
Appearance of solution			
Dextran 1	–	–	–
Dextran 40	+	+	–
Dextran 60	–	+	–
Dextran 70	+	+	–
Absorbance (UV)			
Dextran 1	–	≤0.12 (15% solution)	≤0.12 (15% solution)
Dextran 40	–	–	≤0.20 (1 in 10 solution)
Dextran 60	–	–	–
Dextran 70	–	–	≤0.15 (6 in 100 solution)
Specific rotation			
Dextran 1	–	+148° to +164° (as 3% solution in water)	+148° and +164° (solution in water)
Dextran 40	–	+195° to +201° (as 2% solution in water)	+195° and +203° (as 2% solution in water)
Dextran 60	–	+195° to +201° (as 2% solution in water)	–
Dextran 70	–	+195° to +201° (as 2% solution in water)	+195° and +203° (as 2% solution in water)
Microbial limits			
Dextran 1			
Bacteria (TAMC)	–	≤10^2 cfu/g	≤10^2 cfu/g
Yeasts and molds (TYMC)	–	–	≤10 cfu/g
Dextran 40			
Bacteria (TAMC)	–	≤10^2 cfu/g	–
Yeasts and molds (TYMC)	–	–	–
Dextran 60			
Bacteria (TAMC)	–	≤10^2 cfu/g	–
Yeasts and molds (TYMC)	–	–	–
Dextran 70			
Bacteria (TAMC)	–	≤10^2 cfu/g	–
Yeasts and molds (TYMC)	–	–	–
Bacterial endotoxins			
Dextran 1	–	≤25 EU/g[(a)]	≤25 EU/g
Dextran 40	<2.5 EU/g	≤10 EU/g	≤1.0 EU/mL
Dextran 60	–	≤16 EU/g	–
Dextran 70	–	≤16 EU/g	≤0.5 EU/mL
Pyrogens			
Dextran 1	–	–	–
Dextran 40	–	–	–
Dextran 60	–	–	–
Dextran 70	+	–	–
Chloride			
Dextran 1	–	–	–
Dextran 40	≤0.018%	–	–
Dextran 60	–	–	–
Dextran 70	≤0.018%	–	–
Heavy metals			
Dextran 1	–	≤10 ppm	≤5 μg/g
Dextran 40	≤20 ppm	≤10 ppm	≤5 μg/g
Dextran 60	–	≤10 ppm	–
Dextran 70	≤20 ppm	≤10 ppm	≤5 μg/g
Arsenic			
Dextran 1	–	–	–
Dextran 40	≤1.3 ppm	–	–
Dextran 60	–	–	–
Dextran 70	≤1.3 ppm	–	–
Nitrogen			
Dextran 1	–	≤110 ppm	≤110 ppm
Dextran 40	≤0.010	≤110 ppm	≤0.01%
Dextran 60	–	≤110 ppm	–
Dextran 70	≤0.010%	≤110 ppm	+

Table I (cont)

Test	JP XV S2	PhEur 7.4	USP35–NF30
Reducing substances			
Dextran 1	–	–	–
Dextran 40	+	–	–
Dextran 60	–	–	–
Dextran 70	+	–	–
Loss on drying			
Dextran 1	–	≤5.0%	≤5.0%
Dextran 40	≤5.0%	≤7.0%	≤7.0%
Dextran 60	–	≤7.0%	–
Dextran 70	≤5.0%	≤7.0%	≤7.0%
Residue on ignition			
Dextran 1	–	–	–
Dextran 40	≤0.1%	≤0.3%	–
Dextran 60	–	≤0.3%	–
Dextran 70	≤0.1%	≤0.3%	–
Viscosity (in aqueous solution)			
Dextran 1	–	–	–
Dextran 40	+	–	+
Dextran 60	–	–	–
Dextran 70	+	–	+
Average molecular weight			
Dextran 1	–	850 to 1150	850 to 1150
Dextran 40	≈40 000	35 000 to 45 000	39 000 to 46 000
Dextran 60	–	54 000 to 66 000	–
Dextran 70	≈70 000	64 000 to 76 000	65 000 to 74 000
Molecular weight distribution			
Dextran 1	–	+	–
Dextran 40	–	+	+
Dextran 60	–	+	–
Dextran 70	–	+	+
Antigenicity			
Dextran 1	–	–	–
Dextran 40	+	–	+
Dextran 60	–	–	–
Dextran 70	+	–	+
Sodium chloride			
Dextran 1	–	≤1.5%	≤1.5%
Dextran 40	–	–	–
Dextran 60	–	–	–
Dextran 70	–	–	–
Residual solvents			
Dextran 1	–	–	+
Dextran 40	–	+	+
Dextran 60	–	+	–
Dextran 70	–	+	+
Safety			
Dextran 1	–	–	–
Dextran 40	–	–	+
Dextran 60	–	–	–
Dextran 70	–	–	+
Sulfate			
Dextran 1	–	–	–
Dextran 40	–	–	+
Dextran 60	–	–	–
Dextran 70	–	–	+
Acidity or alkalinity			
Dextran 1	–	+	–
Dextran 40	–	+	–
Dextran 60	–	+	–
Dextran 70	–	+	–
Assay (calculated from optical rotation)			
Dextran 1	–	–	–
Dextran 40	98.0–102.0%	–	–
Dextran 60	–	–	–
Dextran 70	98.0–102.0%	–	–

(a) EU, endotoxin units.

11 Stability and Storage Conditions

Store in well-closed containers to avoid water uptake. The USP35–NF30 states that dextran 1 can be stored between 4°C and 30°C, while dextran 40 and dextran 70 should be stored between 15°C and 30°C. Dextran powder can be stored for up to 5 years at room temperature and aqueous solutions of dextran can be autoclaved.[7] Dextran solutions are stable under mild acidic and basic conditions.[9]

12 Incompatibilities

—

13 Method of Manufacture

Dextran is commercially produced by bacterial enzymatic synthesis with a culture of *Leuconostoc mesenteroides* in presence of sucrose. Ethanol or methanol are used to precipitate the dextran polymer from the culture, and hydrolysis with dilute acid produces the desired molecular weight fraction.[1,10]

14 Safety

Dextran is considered safe for oral administration.[11] There is a risk of severe anaphylactic adverse effects including respiratory distress and death from injection of dextran 40 and dextran 70.[10,12] The anaphylactic response is related to the level of dextran-reactive antibodies present. The risk of anaphylaxis from intravenously administered higher-molecular-weight dextrans (dextran 40) is greatly reduced by pretreating with dextran 1 (15% dextran 1 in normal saline).[13]

LD_{50} (mouse, IV): 12.1 g/kg[14]
LD_{50} (mouse, SC): 13.9 g/kg
LD_{50} (rabbit, IV): 17.4 g/kg

15 Handling Precautions

Observe normal precautions appropriate to the circumstances and quantity of the material handled.

16 Regulatory Status

Included in the FDA Inactive Ingredients Database (IV liquid concentrate injection; IV lyophile powder for injection). Included in the Canadian Natural Health Products Ingredients Database

In the EU, dextran is approved for use in baked products at up to 5% and limited use in clinical nutrition, and as an additive in ice cream and candy.[11]

17 Related Substances

Dextrin; pullulan.

18 Comments

Dextran, dextrin, pullulan, and starch are all polydisperse polysaccharide polymers, with the monomer unit as dextrose (α-D-glucose), but can differ in average polymer length (average molecular weight) and polydispersity, which is a function of the botanical source for plant-based starch, a function of the hydrolysis process used to process starch into dextrin, and a function of the conditions used to fraction native dextran and native pullulan after microbial fermentation.

Dextran and pullulan and are both polysaccharides derived from microbial fermentation of starch or other sugar sources.

Dextran has been evaluated as a tablet binder for immediate-release tablets[15] and as a controlled-release tablet excipient.[15,16] It has also been evaluated for conjugation to proteins for parenteral targeted delivery and protein stabilization,[5,9] for formation of dextran–acid conjugate nanoparticles for parenteral targeted drug delivery to cancer cells,[17,18] and as a mucoadhesive to prolong residence time in the nasal cavity for nasal powder delivery.[19] It has been shown to have limited application as a conjugate for targeted drug delivery by the oral route due to minimal pharmacokinetic absorption.[9]

As an active ingredient, dextran is used for dry eyes in topical ophthalmic artificial tear products and as an intravenous emergency blood plasma expander and anticoagulant.[20]

Dextran was formerly listed as GRAS for food use prior to 1977, but was removed because not enough safety data was available at higher levels, and a survey of food manufacturers indicated that it was no longer used as a food ingredient.

A specification for dextran is included in the *Japanese Pharmaceutical Excipients* (JPE).[21]

The PubChem Compound ID (CID) for dextrans is 4125253.

19 Specific References

1 Amersham Biosciences. Technical literature: *Dextran fractions*, November 2001.
2 Robyt JF. *Essentials of Carbohydrate Chemistry.* New York: Springer Verlag, 1998; 193-221.
3 Ioan C, *et al.* Structure properties of dextran 2. Dilute Solution. *Macromolecules* 2000; **33**(15): 5730–5739.
4 Shalaev E *et al.* Rational choice of excipients for use in lyophilized formulations. In: McNally E, Hastedt J. eds. *Protein Formulation and Delivery.* 2nd edn. New York: Informa Healthcare, 2008; 205-207.
5 Wang W. Lyophilization and development of solid protein pharmaceuticals. *Int J Pharm* 2000; **203**: 1–60.
6 Fakes MG, *et al.* Moisture sorption behavior of selected bulking agents used in lyophilized products. *PDA J Pharm Sci Technol* 2000; **54**(20): 144–149.
7 Pharmacosmos AS. Product information: *Dextran.* http://www.dextran.net/dextran-physical-properties.html (accessed 10 August 2011).
8 Gil EC, *et al.* A sugar cane native dextran as an innovative functional excipient for the development of pharmaceutical tablets. *Eur J Pharm Biopharm* 2008; **68**(2): 319–329.
9 Mehvar R. Dextrans for targeted and sustained delivery of therapeutic and imaging agents. *J Control Release* 2000; **69**: 1–25.
10 Troy D, ed.;1; Plasma expanders and intravenous fluids. In: *Remington: The Science and Practice of Pharmacy,* 21st edn. Philadelphia: Lippincott Williams and Wilkins, 2006: 1321–1323.
11 European Scientific Committee on Food (SCF). Opinion of the SCF on a Dextran preparation on a dextran preparation, produced using *Leuconostoc mesenteroides, Saccharomyces cerevisiae* and *Lactobacillus*, as a novel food ingredient in bakery products, October 2000. http://ec.europa.eu/food/fs/sc/scf/out75_en.pdf (accessed 10 August 2011).
12 World Health Organization (WHO). WHO Model Prescribing Information: *Drugs Used in Anesthesia.* Geneva: World Health Organization Press, 1989. http://archives.who.int/tbs/rational/h2929e.pdf (accessed 10 August 2011).
13 Ljungstrom KG. Pretreatment with dextran 1 makes dextran 40 therapy safer. *J Vasc Surgery* 2006; **43**(5): 1070–1072.
14 Lewis RJ, ed. *Sax's Dangerous Properties of Industrial Materials.* 11th edn. New York: Wiley, 2004; 1100.
15 Castellanos Gil E *et al.* Tablet design. In: SC Gad, ed. *Pharmaceutical Manufacturing Handbook – Production and Processes.* New Jersey: Wiley, 2008; 977-1051.
16 Shrivastava PK, Shrivastava SK. Dextran polysaccharides: successful macromolecular carrier for drug delivery. *Int J Pharm Sci* 2009; **1**(2): 353–368.
17 Tanga M *et al.* One-step synthesis of dextran-based stable nanoparticles assisted by self-assembly *Polymer 2006*; 47: 728-734.
18 Li Y, *et al.* Reversibly stabilized multifunctional dextran nanoparticles efficiently deliver doxorubicin into the nuclei of cancer cells. *Angew Chem Int Ed* 2009; **48**: 9914–9918.
19 Filipović-Gr?ić J, Hafner A. Nasal powder drug delivery. In: SC Gad, ed. *Pharmaceutical Manufacturing Handbook – Production and Processes.* New Jersey: Wiley, 2008; 651-681.

20 de Belder AN. Medical applications of dextran and its derivatives. In: Severian D, ed. *Polysaccharides in Medicinal Applications*. New York: Marcel Dekker, 1996; 505-523.
21 Japan Pharmaceutical Excipients Council. *Japanese Pharmaceutical Excipients* 2004. Tokyo: Yakuji Nippo, 2004; 216.

20 General References

de Belder AN. Dextran. In: Whistler RL, BeMiller JN, eds. *Industrial Gums: Polysaccharides and Their Derivatives*. 3rd edn. San Diego: Academic Press, 1993, 399-426.

21 Author

N Culver.

22 Date of Revision

1 March 2012.

Dextrates

1 Nonproprietary Names

USP–NF: Dextrates

2 Synonyms

Candex; dextratos; *Emdex*.

3 Chemical Name and CAS Registry Number

Dextrates [39404-33-6]

4 Empirical Formula and Molecular Weight

The USP35–NF30 describes dextrates as a purified mixture of saccharides resulting from the controlled enzymatic hydrolysis of starch. It is either hydrated or anhydrous. Its dextrose equivalent is not less than 93.0% and not more than 99.0%, calculated on the dried basis.

5 Structural Formula

See Section 4.

6 Functional Category

Tablet and capsule binder; tablet and capsule diluent.

7 Applications in Pharmaceutical Formulation or Technology

Dextrates is a directly compressible tablet diluent used in chewable, nonchewable, soluble, dispersible, and effervescent tablets.[1–3] It is a free-flowing material and glidants are thus unnecessary. Lubrication with magnesium stearate (0.5–1.0% w/w) is recommended.[4] Dextrates may also be used as a binding agent by the addition of water, no further binder being required.[4]

Tablets made from dextrates increase in crushing strength in the first few hours after manufacture, but no further increase occurs on storage.[5]

8 Description

Dextrates is a purified mixture of saccharides resulting from the controlled enzymatic hydrolysis of starch. It is either anhydrous or hydrated. In addition to dextrose, dextrates contains 3–5% w/w maltose and higher polysaccharides.

Dextrates comprises white spray-crystallized free-flowing porous spheres. It is odorless with a sweet taste (about half as sweet as sucrose).

9 Pharmacopeial Specifications

See Table I.

Table I: Pharmacopeial specifications for dextrates.

Test	USP35–NF30
pH (20% aqueous solution)	3.8–5.8
Loss on drying	
Anhydrous	≤2.0%
Hydrated	7.8–9.2%
Residue on ignition	≤0.1%
Heavy metals	≤5 ppm
Dextrose equivalent (dried basis)	93.0–99.0%

10 Typical Properties

Angle of repose 26.4°[6]
Compressibility *see* Figure 1.[6]
Density (bulk) 0.68 g/cm^3 [6]
Density (tapped) 0.72 g/cm^3 [6]
Density (true) 1.539 g/cm^3
Hausner ratio 1.05
Flowability 9.3 g/s [6]
Heat of combustion 16.8–18.8 J/g (4.0–4.5 cal/g)
Heat of solution −105 J/g (−25 cal/g)
Melting point 141°C
Moisture content 7.8–9.2% w/w (hydrated form). *See also* Figure 2.[7]
Particle size distribution Not more than 3% retained on a 840 μm sieve; not more than 25% passes through a 150 μm sieve. Mean particle size 190–220 μm.
Solubility Soluble 1 in 1 part of water; insoluble in ethanol (95%), propan-2-ol, and common organic solvents.
Specific surface area 0.70 m^2/g
Spectroscopy

IR spectra *see* Figure 3.

NIR spectra *see* Figure 4.

11 Stability and Storage Conditions

Dextrates may be heated to 50°C without any appreciable darkening of color. Dextrates should be stored in a well-closed container in conditions that do not exceed 25°C and 60% relative humidity. When correctly stored in unopened containers, dextrates has a shelf-life of 3 years.

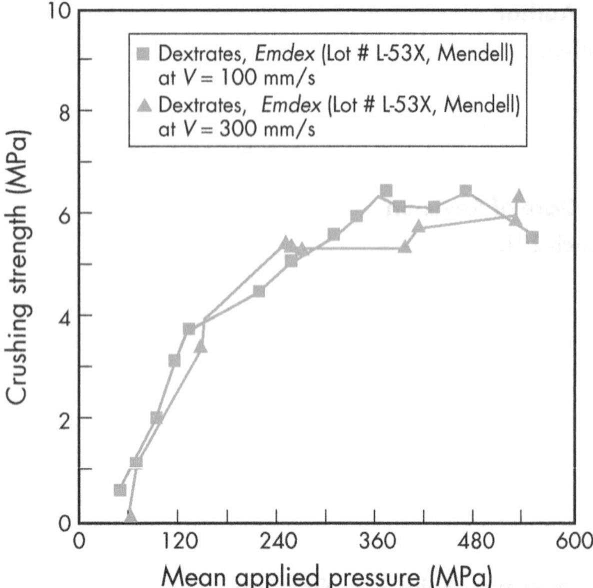

Figure 1: Crushing strength for dextrates.

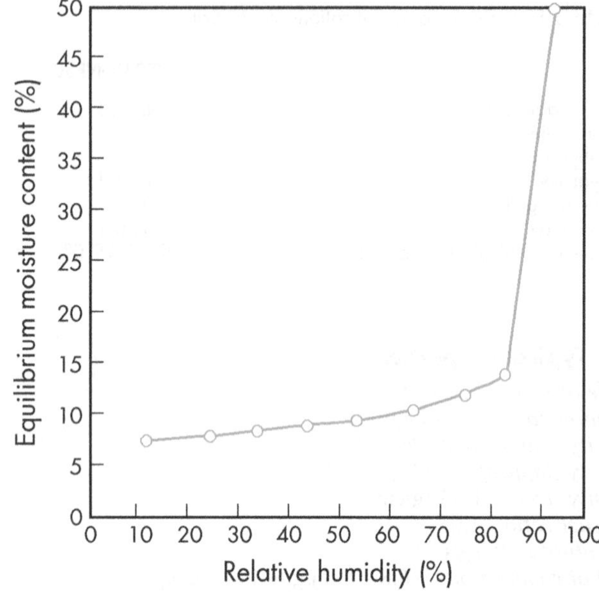

Figure 2: Equilibrium moisture content of dextrates at 25°C.[7]

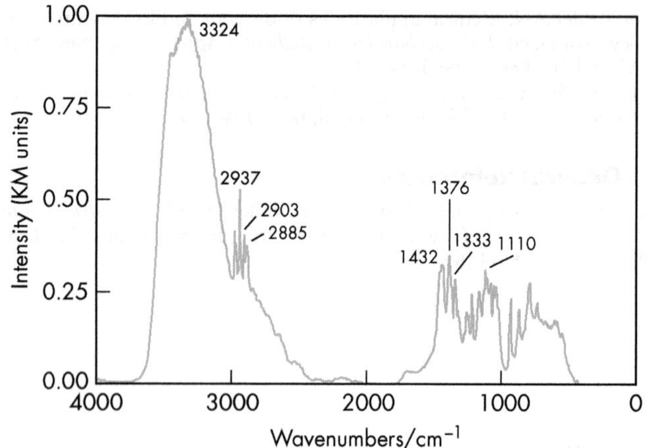

Figure 3: Infrared spectrum of dextrates measured by diffuse reflectance. Adapted with permission of Informa Healthcare.

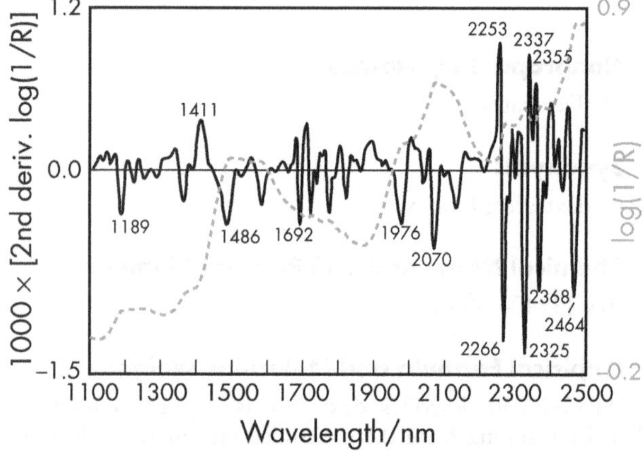

Figure 4: Near-infrared spectrum of dextrates measured by reflectance.

12 Incompatibilities

At high temperatures and humidities, dextrates may react with substances containing a primary amino group (Maillard reaction).[8,9] Also incompatible with oxidizing agents.

13 Method of Manufacture

Dextrates is produced by controlled enzymatic hydrolysis of starch. The product is spray-crystallized, and may be dried to produce an anhydrous form.

14 Safety

Dextrates is used in oral pharmaceutical formulations and is generally regarded as a relatively nontoxic and nonirritant material.

15 Handling Precautions

Observe normal handling precautions appropriate to the circumstances and quantity of material handled. Eye protection, gloves, and a dust mask are recommended.

16 Regulatory Status

GRAS listed. Included in the FDA Inactive Ingredient Database (oral; tablets, chewable and sustained action). Included in nonparenteral medicines licensed in the UK. Included in the Canadian Natural Health Products Ingredients Database.

17 Related Substances

Dextrose.

18 Comments

Only the hydrated form of dextrates is currently commercially available.

19 Specific References

1 Henderson NL, Bruno AJ. Lactose USP (Beadlets) and Dextrose (PAF 2011): two new agents for direct compression. *J Pharm Sci* 1970; 59: 1336–1340.
2 Shukla AJ, Price JC. Effect of moisture content on compression properties of two dextrose-based directly compressible diluents. *Pharm Res* 1991; 8(3): 336–340.
3 Allen LV. Featured excipient: capsule and tablet diluents. *Int J Pharm Compound* 2000; 4(4): 306–310324–325.

4 JRS Pharma. Technical Literature: *Emdex*, 2008.
5 Shangraw RF, *et al.* Morphology and functionality in tablet excipients by direct compression: Part I. *Pharm Technol* 1981; **5**(9): 69–78.
6 Celik M, Okutgen E. A feasibility study for the development of a prospective compaction functionality test and the establishment of a compaction data bank. *Drug Dev Ind Pharm* 1993; **19**: 2309–2334.
7 Callahan JC, *et al.* Equilibrium moisture content of pharmaceutical excipients. *Drug Dev Ind Pharm* 1982; **8**(3): 355–369.
8 Blaug SM, Huang WT. Interaction of dexamphetamine sulphate with dextrates in solution. *J Pharm Sci* 1973; **62**(4): 652–655.
9 Blaug SM, Huang WT. Browning of dextrates in solid-solid mixtures containing dexamphetamine sulfate. *J Pharm Sci* 1974; **63**(9): 1415–1418.

20 General References

Armstrong NA. Tablet manufacture. In: Swarbrick J, Boylan JC, eds. *Encyclopedia of Pharmaceutical Technology*, 2nd edn, 3. New York: Marcel Dekker, 2002: 2713–2732.
Bolhuis GK, Armstrong NA. Excipients for direct compression – an update. *Pharm Dev Technol* 2006; **11**: 111–124.

21 Author

NA Armstrong.

22 Date of Revision

1 March 2012.

Dextrin

1 Nonproprietary Names

BP: Dextrin
JP: Dextrin
PhEur: Dextrin
USP–NF: Dextrin

2 Synonyms

Avedex; British gum; *Caloreen*; canary dextrin; *C*Pharm*; *Crystal Gum*; dextrinum; dextrinum album; *Primogran W*; starch gum; yellow dextrin; white dextrin.

3 Chemical Name and CAS Registry Number

Dextrin [9004-53-9]

4 Empirical Formula and Molecular Weight

$(C_6H_{10}O_5)_n{\cdot}xH_2O$ $(162.14)_n$

Dextrins are a group of low molecular weight carbohydrates that are mixtures of polymers of α-D-glucose subunits. The molecular weight of dextrin is typically 4500–85 000 and depends on the number of $(C_6H_{10}O_5)$ units in the polymer chain.

5 Structural Formula

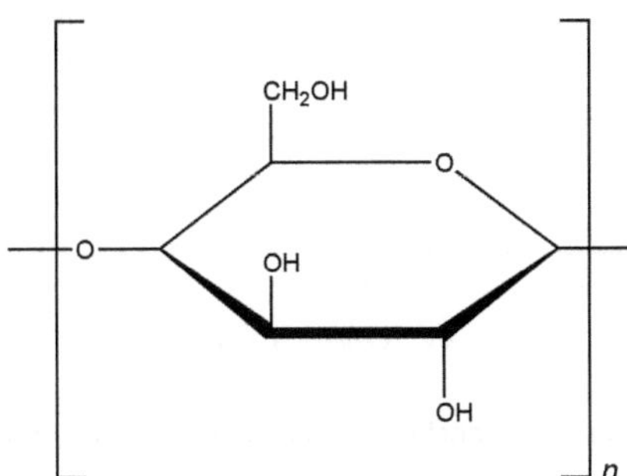

A structure for straight chain dextrin with α-D-(1→4) linkages is shown. Dextrins may be straight chain or contain branches with α-D-(1→6) linkages depending on the method of starch hydrolysis used to produce them.

6 Functional Category

Stiffening agent; suspending agent; tablet and capsule binder; tablet and capsule diluent.

7 Applications in Pharmaceutical Formulation or Technology

Dextrin is a dextrose polymer used as an adhesive and stiffening agent for surgical dressings. It is also used as a tablet and capsule diluent; as a binder for tablet granulation; as a sugar-coating ingredient that serves as a plasticizer and adhesive; and as a thickening agent for suspensions. Dextrin is also used in cosmetics.

8 Description

Dextrin is partially hydrolyzed maize (corn), potato, or cassava starch. It is a white, pale yellow, or brown-colored powder with a slight characteristic odor.

9 Pharmacopeial Specifications

See Table I.

Table I: Pharmacopeial specifications for dextrin.

Test	JP XV	PhEur 7.4	USP35–NF30
Identification	+	+	+
Botanical characteristics	—	—	+
Characters	—	+	—
Appearance of solution	+	—	—
Loss on drying	≤10.0%	≤13.0%	≤13.0%
Acidity	+	—	+
pH	—	2.0–8.0	—
Residue on ignition	≤0.5%	≤0.5%	≤0.5%
Chloride	≤0.013%	≤0.2%	≤0.2%
Sulfate	≤0.019%	—	—
Oxalate	+	—	—
Calcium	+	—	—
Heavy metals	≤50 ppm	≤20 ppm	≤20 ppm
Protein	—	—	≤1.0%
Reducing sugars/substances (calculated as $C_6H_{12}O_6$)	—	≤10.0%	≤10.0%

SEM 1: Excipient: dextrin; manufacturer: Matheson Colleman & Bell; magnification: 600×.

SEM 2: Excipient: dextrin; manufacturer: Matheson Colleman & Bell; magnification: 2400×.

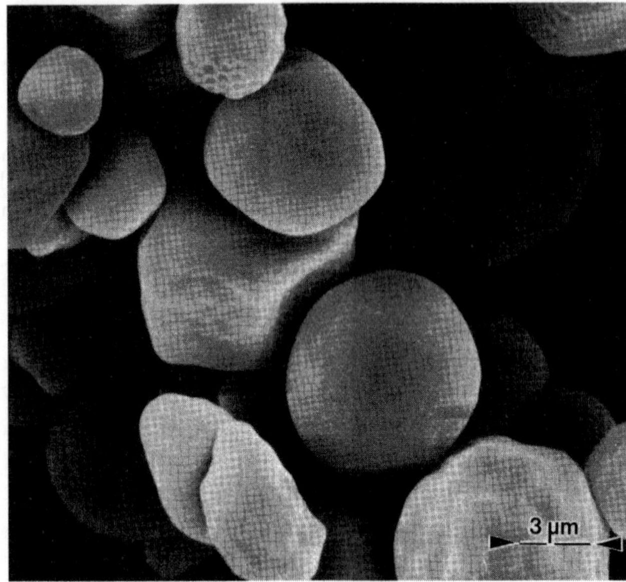

10 Typical Properties

Acidity/alkalinity pH = 2.8–8.0 for a 5% w/v aqueous solution.
Density (bulk) 0.80 g/cm^3
Density (tapped) 0.91 g/cm^3
Density (true) 1.495–1.589 g/cm^3
Melting point 178°C (with decomposition)
Moisture content 5% w/w
Particle size distribution *see* Figure 1.
Solubility Practically insoluble in chloroform, ethanol (95%), ether, and propan-2-ol; slowly soluble in cold water; very soluble in boiling water, forming a mucilaginous solution.
Specific surface area 0.14 m^2/g
Spectroscopy

IR spectra *see* Figure 2.

NIR spectra *see* Figure 3.

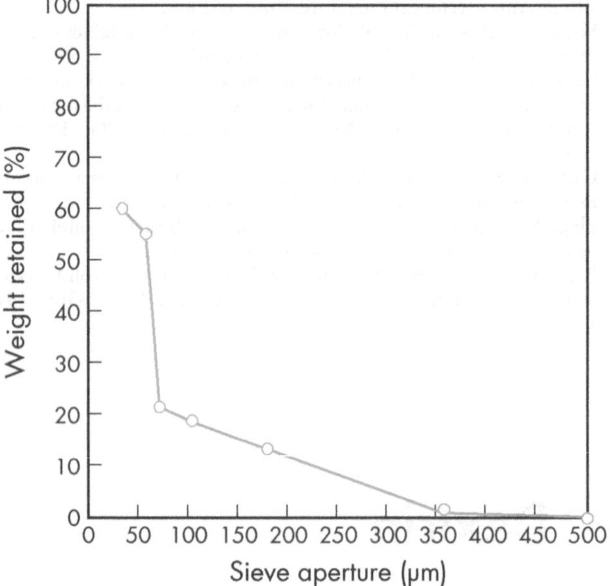

Figure 1: Particle size distribution of dextrin.

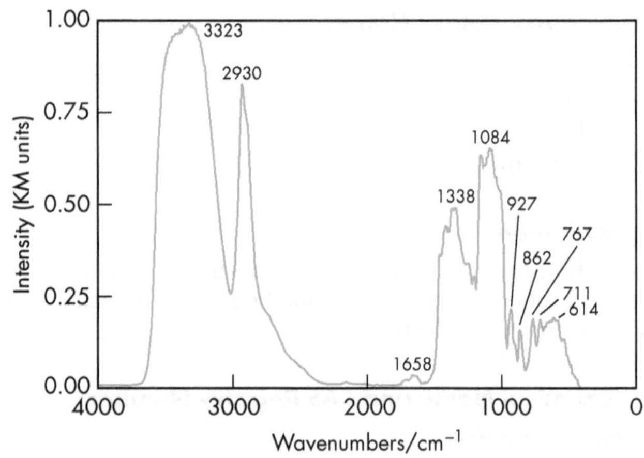

Figure 2: Infrared spectrum of dextrin measured by diffuse reflectance. Adapted with permission of Informa Healthcare.

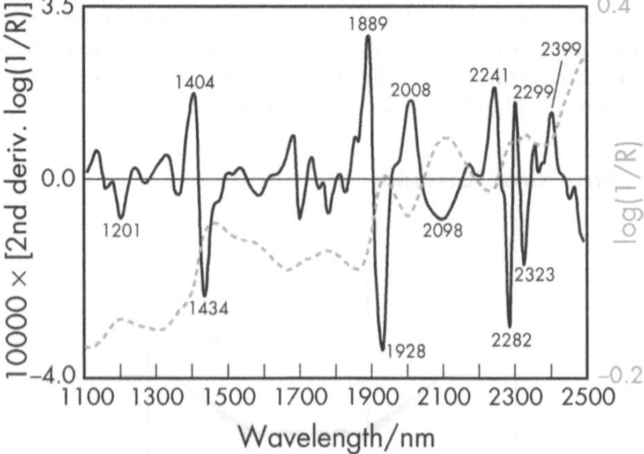

Figure 3: Near-infrared spectrum of dextrin measured by reflectance.

11 Stability and Storage Conditions

Physical characteristics of dextrin may vary slightly depending on the method of manufacture and on the source material. In aqueous solutions, dextrin molecules tend to aggregate as density, temperature, pH, or other characteristics change. An increase in viscosity is caused by gelation or retrogradation as dextrin solutions age, and is particularly noticeable in the less-soluble maize starch dextrins. Dextrin solutions are thixotropic, becoming less viscous when sheared but changing to a soft paste or gel when allowed to stand. However, acids that are present in dextrin as residues from manufacturing can cause further hydrolysis, which results in a gradual thinning of solutions. Residual acid, often found in less-soluble dextrins such as pyrodextrin, will also cause a reduction in viscosity during dry storage. To eliminate these problems, dextrin manufacturers neutralize dextrins of low solubility with ammonia or sodium carbonate in the cooling vessel.

The bulk material should be stored in a well-closed container in a cool, dry place.

12 Incompatibilities

Incompatible with strong oxidizing agents.

13 Method of Manufacture

Dextrin is prepared by the incomplete hydrolysis of starch by heating in the dry state with or without the aid of suitable acids and buffers; moisture may be added during heating. The PhEur 7.4 specifies that dextrin is derived from maize (corn), potato or cassava starch. A specification for cassava is included in the USP35–NF30.

14 Safety

Dextrin is generally regarded as a nontoxic and nonirritant material at the levels employed as an excipient. Larger quantities are used as a dietary supplement without adverse effects, although ingestion of very large quantities may be harmful.

LD_{50} (mouse, IV): 0.35 g/kg[1]

15 Handling Precautions

Observe normal precautions appropriate to the circumstances and quantity of material handled. Dextrin may be irritant to the eyes. Eye protection, gloves, and a dust mask are recommended.

16 Regulatory Status

GRAS listed. Included in the FDA Inactive Ingredients Database (IV injections, oral tablets, and topical preparations). Included in nonparenteral medicines licensed in the UK. Included in the Canadian Natural Health Products Ingredients Database.

17 Related Substances

Dextran; dextrates; dextrose; glucose liquid; maltodextrin; pullulan.

See also Section 18.

18 Comments

Dextrin and starch are both botanical polysaccharides derived from plant sources such as corn, potato, and rice. Starch is purified from the plant source but not further modified, while dextrin is produced by further processing of starch via partial hydrolysis.

Dextrin is available from suppliers in a number of modified forms and mixtures such as dextrimaltose, a mixture of maltose and dextrin obtained by the enzymatic action of barley malt on corn flour. It is a light, amorphous powder, readily soluble in milk or water.

Crystal Gum is a grade of dextrin containing carbohydrate not less than 98% of dry weight. *Caloreen* [2] is a water-soluble mixture of dextrins consisting predominantly of polysaccharides containing an average of 5 dextrose molecules, with a mean molecular weight of 840, which does not change after heating. A 22% w/v solution of *Caloreen* is isoosmotic with serum.

Dextrin has been evaluated as a substrate for fast dissolving buccal delivery (buccal melt wafer),[3] a sustained-release excipient for matrix tablets,[4] a nanogel encapsulation of proteins for protein stabilization,[5] and for conjugation to proteins for protein stabilization and subsequent controlled release.[6]

Dextrin has been used as a source of carbohydrate by people with special dietary requirements because it has a low electrolyte content and is free of lactose and sucrose.[2]

A specification for dextrin is contained in the *Food Chemicals Codex* (FCC).[7]

The EINECS number for dextrin is 232-675-4. The PubChem Compound ID (CID) for dextrin is 62698.

19 Specific References

1. Sweet DV, ed. *Registry of Toxic Effects of Chemical Substances.* Cincinnati: US Department of Health, 1987; 1859.
2. Berlyne GM, *et al.* A soluble glucose polymer for use in renal failure and calorie-deprivation states. *Lancet* 1969; i: 689–692.
3. Testera RF, Qi X. β-Limit dextrin – properties and applications. *Food Hydrocolloids* 2011; **25**(8): 1899–1903.
4. Modi SA, *et al.* Design and evaluation of sustained release drug delivery system of metoclopramide hydrochloride for emesis. *Int J Pharm Technol* 2011; **3**(2): 2590–2598.
5. Carvalho V, *et al.* Self-assembled dextrin nanogel as protein carrier: controlled release and biological activity of IL-10. *Biotechnol Bioeng* 2011; **108**(8): 1977–1986.
6. Hardwicke JT, *et al.* The effect of dextrin-rhEGF on the healing of full-thickness, excisional wounds in the (db/db) diabetic mouse. *J Control Release* 2011; **152**(3): 411–417.
7. *Food Chemicals Codex*, 7th edn. Bethesda, MD: United States Pharmacopeia, 2010.

20 General References

French D. Chemical and physical properties of starch. *J Animal Sci* 1973; **37**: 1048–1061.

Satterthwaite RW, Iwinski DJ. Starch dextrins. In: Whistler RL, Bemiller JN, eds. *Industrial Gums.* New York: Academic Press, 1973; 577–599.

21 Author

N Culver.

22 Date of Revision

1 March 2012.

Dextrose

1 Nonproprietary Names

BP: Glucose
JP: Glucose
PhEur: Glucose Monohydrate
USP: Dextrose

2 Synonyms

Blood sugar; *Caridex*; *Cerelose*; corn sugar; *C*PharmDex*; *Dextrofin*; D-(+)-glucopyranose monohydrate; glucosum monohydricum; grape sugar; *Lycadex PF*; starch sugar; *Tabfine D-100*.

3 Chemical Name and CAS Registry Number

D-(+)-Glucose monohydrate [5996-10-1]
See also Section 17.

4 Empirical Formula and Molecular Weight

$C_6H_{12}O_6{\cdot}H_2O$ 198.17 (for monohydrate)
See also Section 17.

5 Structural Formula

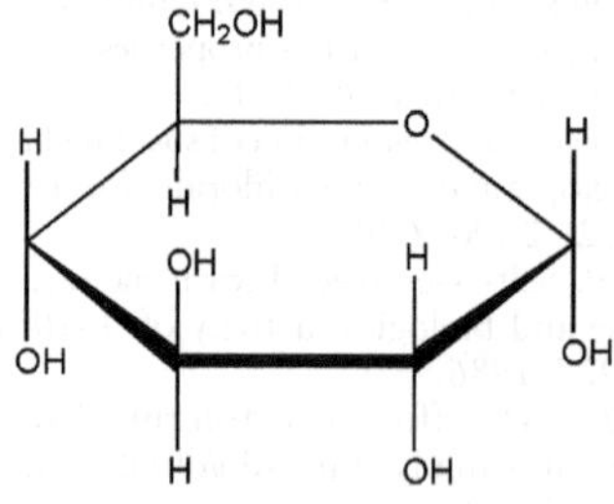

Anhydrous material shown.

6 Functional Category

Tablet and capsule diluent; tonicity agent; sweetening agent.

7 Applications in Pharmaceutical Formulation or Technology

Dextrose is widely used in solutions to adjust tonicity and as a sweetening agent. Dextrose is also used as a wet granulation diluent and binder, and as a direct-compression tablet diluent and binder, primarily in chewable tablets. Although dextrose is comparable as a tablet diluent to lactose, tablets produced with dextrose monohydrate require more lubrication, are less friable, and have a tendency to harden.[1–3] *See also* Section 18.

8 Description

Dextrose occurs as odorless, sweet-tasting, colorless crystals or as a white crystalline or granular powder. The JP XV describes dextrose as dextrose anhydrous; the PhEur 7.4 specifies dextrose as either dextrose anhydrous or dextrose monohydrate; and the USP35–NF30 specifies dextrose as dextrose monohydrate.

9 Pharmacopeial Specifications

See Table I. *See also* Sections 17 and 18.

Table I: Pharmacopeial specifications for dextrose.

Test	JP XV	PhEur 7.4	USP35–NF30
Identification	+	+	+
Characters	+	+	–
Color of solution	+	+	+
Specific optical rotation	–	+52.5° to +53.3°	+52.6° to +53.2°
Acidity or alkalinity	+	+	+
Water			
for monohydrate	–	7.0–9.5%	7.5–9.5%
for anhydrous	≤1.0%	≤1.0%	≤0.5%
Residue on ignition	≤0.1%	≤0.1%	≤0.1%
Chloride	≤0.018%	≤125 ppm	≤0.018%
Sulfate	≤0.024%	≤200 ppm	≤0.025%
Arsenic	≤1.3 ppm	≤1 ppm	≤1 ppm
Barium	–	+	–
Calcium	–	≤200 ppm	–
Heavy metals	≤4 ppm	–	≤5 ppm
Lead	–	≤0.5 ppm	–
Dextrin	+	+	+
Soluble starch, and sulfites	+	+	+
Pyrogens[a]	–	+	–
Assay (dried basis)	≥99.5%	–	–

(a) If intended for large volume parenteral use.

10 Typical Properties

Data are shown for dextrose monohydrate; *see* Section 17 for data for dextrose anhydrous.

Acidity/alkalinity pH = 3.5–5.5 (20% w/v aqueous solution)
Density (bulk) 0.826 g/cm^3
Density (tapped) 1.020 g/cm^3
Density (true) 1.54 g/cm^3
Heat of solution 105.4 J/g (25.2 cal/g)
Melting point 83°C
Moisture content Dextrose anhydrous absorbs significant amounts of moisture at 25°C and a relative humidity of about 85% to form the monohydrate. The monohydrate similarly only

SEM 1: Excipient: dextrose anhydrous (granular); manufacturer: Mallinckrodt Specialty Chemicals Co.; lot no.: KLKZ; magnification: 180×.

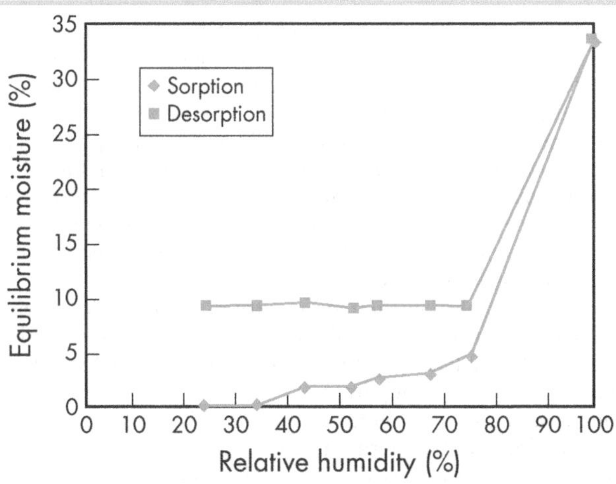

Figure 1: Sorption–desorption isotherm for anhydrous dextrose granules.

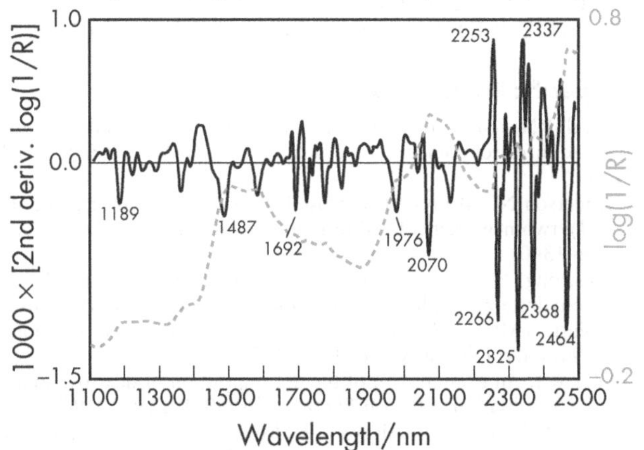

Figure 2: Near-infrared spectrum of dextrose monohydrate measured by reflectance.

Table II: Solubility of dextrose monohydrate.

Solvent	Solubility at 20°C
Chloroform	Practically insoluble
Ethanol (95%)	1 in 60
Ether	Practically insoluble
Glycerin	Soluble
Water	1 in 1

absorbs moisture at around 85% relative humidity and 25°C. *See* Figure 1.

Osmolarity A 5.51% w/v aqueous solution is isoosmotic with serum. However, it is not isotonic since dextrose can pass through the membrane of red cells and cause hemolysis.

Solubility *see* Table II.

Spectroscopy

NIR spectra *see* Figure 2.

11 Stability and Storage Conditions

Dextrose has good stability under dry storage conditions. Aqueous solutions may be sterilized by autoclaving. However, excessive heating can cause a reduction in pH and caramelization of solutions.[(4–7)]

The bulk material should be stored in a well-closed container in a cool, dry place.

12 Incompatibilities

Dextrose solutions are incompatible with a number of drugs such as cyanocobalamin, kanamycin sulfate, novobiocin sodium, and warfarin sodium.[(8)] Erythromycin gluceptate is unstable in dextrose solutions at a pH less than 5.05.[(9)] Decomposition of B-complex vitamins may occur if they are warmed with dextrose.

In the aldehyde form, dextrose can react with amines, amides, amino acids, peptides, and proteins. Brown coloration and decomposition occur with strong alkalis.

Dextrose may cause browning of tablets containing amines (Maillard reaction).

13 Method of Manufacture

Dextrose, a monosaccharide sugar, occurs widely in plants and is manufactured on a large scale by the acid or enzymatic hydrolysis of starch, usually maize (corn) starch. Below 50°C α-D-dextrose monohydrate is the stable crystalline form produced; above 50°C the anhydrous form is obtained; and at still higher temperatures β-D-dextrose is formed, which has a melting point of 148–155°C.

14 Safety

Dextrose is rapidly absorbed from the gastrointestinal tract. It is metabolized to carbon dioxide and water with the release of energy.

Concentrated dextrose solutions given by mouth may cause nausea and vomiting. Dextrose solutions of concentration greater than 5% w/v are hyperosmotic and are liable to cause local vein irritation following intravenous administration. Thrombophlebitis has been observed following the intravenous infusion of isoosmotic dextrose solution with low pH, probably owing to the presence of degradation products formed by overheating during sterilization. The incidence of phlebitis may be reduced by adding sufficient sodium bicarbonate to raise the pH of the infusion above pH 7.

LD_{50} (mouse, IV): 9 g/kg[(10)]
LD_{50} (rat, oral): 25.8 g/kg

15 Handling Precautions

Observe normal precautions appropriate to the circumstances and quantity of material handled. Eye protection and gloves are recommended. Dust generation should be minimized to reduce the risk of explosion.

16 Regulatory Status

Included in the FDA Inactive Ingredients Database (capsules; inhalations; IM, IV, and SC injections; tablets, oral solutions, and syrups). Included in nonparenteral and parenteral medicines licensed in the UK. Included in the Canadian Natural Health Products Ingredients Database.

17 Related Substances

Dextrates; dextrin; dextrose anhydrous; fructose; glucose liquid; polydextrose; sucrose.

Dextrose anhydrous

Empirical formula $C_6H_{12}O_6$

Molecular weight 180.16

CAS number [50-99-7]

Synonyms anhydrous dextrose; anhydrous D-(+)-glucopyranose; anhydrous glucose; dextrosum anhydricum.

Appearance White, odorless, crystalline powder with a sweet taste.

Acidity/alkalinity pH = 5.9 (10% w/v aqueous solution)

Density (bulk) 1.3–1.4 g/cm^3

Density (tapped) 1.1–1.2 g/cm^3

Melting point 146°C

Moisture content *see* Section 10.

NIR spectra *see* Figure 3.

Osmolarity A 5.05% w/v aqueous solution is isoosmotic with serum. *See also* Section 10.

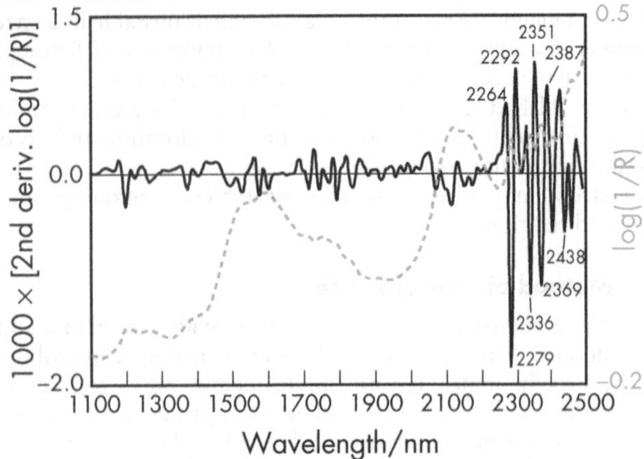

Figure 3: Near-infrared spectrum of dextrose anhydrous measured by reflectance.

Refractive index n_D^{20} = 1.3479 (10% w/v aqueous solution)
Solubility *see* Table III.

Table III: Solubility of dextrose anhydrous.

Solvent	Solubility at 20°C unless otherwise stated
Ethanol (95%)	Sparingly soluble
Ether	Sparingly soluble
Methanol	1 in 120
Water	1 in 1.1 at 25°C
	1 in 0.8 at 30°C
	1 in 0.41 at 50°C
	1 in 0.28 at 70°C
	1 in 0.18 at 90°C

Specific gravity *see* Table IV.

Table IV: Specific gravity of dextrose anhydrous aqueous solutions.

Concentration of aqueous dextrose solution (% w/v)	Specific gravity at 17.5°C
5	1.019
10	1.038
20	1.076
30	1.113
40	1.149

Specific surface area 0.22–0.29 m^2/g

Dextrose anhydrous is listed in the JP XV and PhEur 7.4. Dextrose anhydrous is one of the materials that have been selected for harmonization by the Pharmacopeial Discussion Group. For further information see the General Information Chapter <1196> in the USP35–NF30, the General Chapter 5.8 in PhEur 7.4, along with the 'State of Work' document on the PhEur EDQM website, and also the General Information Chapter 8 in the JP XV.

18 Comments

Dextrose monohydrate is one of the materials that have been selected for harmonization by the Pharmacopeial Discussion Group. For further information see the General Information Chapter <1196> in the USP35–NF30, the General Chapter 5.8 in PhEur 7.4, along with the 'State of Work' document on the PhEur EDQM website, and also the General Information Chapter 8 in the JP XV.

The way in which the strengths of dextrose solutions are expressed varies from country to country. The JP XV requires strengths to be expressed in terms of dextrose monohydrate, while the BP 2012 and USP35–NF30 require strengths to be expressed in terms of anhydrous dextrose. Approximately 1.1 g of dextrose monohydrate is equivalent to 1 g of anhydrous dextrose.

Dextrose has been studied as a diluent in hydrophilic matrix controlled-release tablet formulations, where its high solubility is used to produce an increased initial release rate.[11] It has also been shown that dextrose and other soluble sugars used as diluents in hypromellose matrix tablets can increase the drug release rate in the presence of dissolved sucrose.[12]

Dextrose has been evaluated as a dry powder inhaler carrier. It was reported to produce formulations which were more sensitive to storage at elevated temperature and humidity than formulations containing lactose monohydrate.[13]

Dextrose is also used therapeutically and is the preferred source of carbohydrate in parenteral nutrition regimens.

A specification for dextrose is contained in the *Food Chemicals Codex* (FCC).[14]

The EINECS number for dextrose is 200-075-1. The PubChem Compound ID (CID) for dextrose include 107526 and 66370.

19 Specific References

1 DuVall RN, *et al.* Comparative evaluation of dextrose and spray-dried lactose in direct compression systems. *J Pharm Sci* 1965; **54**: 1196–1200.
2 Henderson NL, Bruno AJ. Lactose USP (beadlets) and dextrose (PAF 2011): two new agents for direct compression. *J Pharm Sci* 1970; **59**: 1336–1340.
3 Armstrong NA *et al.* The compressional properties of dextrose monohydrate and anhydrous dextrose of varying water contents. In: Rubinstein MH, ed. *Pharmaceutical Technology: Tableting Technology.* 1. Chichester: Ellis Horwood, 1987: 127–138.
4 Wing WT. An examination of the decomposition of dextrose solution during sterilisation. *J Pharm Pharmacol* 1960; **12**: 191T–196T.
5 Murty BSR, *et al.* Levels of 5-hydroxymethylfurfural in dextrose injection. *Am J Hosp Pharm* 1977; **34**: 205–206.
6 Sturgeon RJ, *et al.* Degradation of dextrose during heating under simulated sterilization. *J Parenter Drug Assoc* 1980; **34**: 175–182.
7 Durham DG, *et al.* Identification of some acids produced during autoclaving of D-glucose solutions using HPLC. *Int J Pharm* 1982; **12**: 31–40.
8 Patel JA, Phillips GL. A guide to physical compatibility of intravenous drug admixtures. *Am J Hosp Pharm* 1966; **23**: 409–411.
9 Edward M. pH – an important factor in the compatibility of additives in intravenous therapy. *Am J Hosp Pharm* 1967; **24**: 440–449.
10 Lewis RJ, ed. *Sax's Dangerous Properties of Industrial Materials*, 11th edn. New York: Wiley, 2004: 1860–1861.
11 Adhikary A, Vavia PR. Bioadhesive ranitidine hydrochloride for gastroretention with controlled microenvironmental pH. *Drug Dev Ind Pharm* 2008; **34**: 860–869.
12 Williams HD, *et al.* Designing HPMC matrices with improved resistance to dissolved sugar. *Int J Pharm* 2010; **401**: 51–59.
13 Zeng XM, *et al.* Humidity-induced changes of the aerodynamic properties of dry powder aerosol formulations containing different carriers. *Int J Pharm* 2007; **333**: 45–55.
14 *Food Chemicals Codex*, 7th edn. Bethesda, MD: United States Pharmacopeia, 2010: 283.

20 General References

European Directorate for the Quality of Medicines and Healthcare (EDQM). European Pharmacopoeia – State Of Work Of International Harmonisation. *Pharmeuropa* 2011; 23(2): 395–401. www.edqm.eu/en/InternationalHarmonisation-614.html (accessed 26 May 2011).

21 Author

W Cook.

22 Date of Revision

1 March 2012.

Dibutyl Phthalate

1 Nonproprietary Names

BP: Dibutyl Phthalate
PhEur: Dibutyl Phthalate
USP–NF: Dibutyl Phthalate

2 Synonyms

Araldite 502; benzenedicarboxylic acid; benzene-*o*-dicarboxylic acid di-*n*-butyl ester; butyl phthalate; *Celluflex DBP*; DBP; dibutyl 1,2-benzenedicarboxylate; dibutyl benzene 1,2-dicarboxylate; dibutyl ester of 1,2-benzenedicarboxylic acid; dibutylis phthalas; dibutyl-*o*-phthalate; di-*n*-butyl phthalate; *Eastman DBP*; *Elaol*; *Ergoplast FDB*; *Genoplast B*; *Hatcol DBP*; *Hexaplast M/B*; *Kodaflex DBP*; *Monocizer DBP*; *Palatinol C*; phthalic acid dibutyl ester; *Polycizer DBP*; *PX 104*; *RC Plasticizer DBP*; *Staflex DBP*; *Unimoll DB*; *Vestimol C*; *Witcizer 300*.

3 Chemical Name and CAS Registry Number

Dibutyl benzene-1,2-dicarboxylate [84-74-2]

4 Empirical Formula and Molecular Weight

$C_{16}H_{22}O_4$ 278.34

5 Structural Formula

6 Functional Category

Plasticizing agent; solvent.

7 Applications in Pharmaceutical Formulation or Technology

Dibutyl phthalate is used in pharmaceutical formulations as a plasticizing agent in film-coatings. It has been evaluated as a pore-forming agent in novel delivery systems.[3,4] It is also used extensively as a solvent, particularly in cosmetic formulations such as antiperspirants, hair shampoos, and hair sprays.

8 Description

Dibutyl phthalate occurs as an odorless, oily, colorless, or very slightly yellow-colored, viscous liquid.

9 Pharmacopeial Specifications

See Table I.

Table I: Pharmacopeial specifications for dibutyl phthalate.

Test	PhEur 7.4	USP35–NF30
Identification	+	+
Characters	+	—
Appearance	+	+
Relative density	1.043–1.048	1.043–1.048
Refractive index	1.490–1.495	1.490–1.495
Acidity	+	+
Related substances	+	+
Water	≤0.2%	≤0.2%
Sulfated ash	≤0.1%	—
Residue on ignition	—	≤0.1%
Assay	99.0–101.0%	99.0–101.0%

10 Typical Properties

Boiling point 340°C for *Eastman DBP*
Density *see* Table II.
Flash point 171°C (open cup); 190°C for *Eastman DBP*
Melting point −35°C
Partition coefficient
Octanol : water $\log k_{ow} = 4.50$
Refractive index $n_D^{20} = 1.491–1.495$
Solubility Very soluble in acetone, benzene, ethanol (95%), and ether; soluble 1 in 2500 of water at 20°C.
Spectroscopy
IR spectra *see* Figure 1.
Viscosity (dynamic) *see* Table II.

Table II: Density and dynamic viscosity of dibutyl phthalate at specified temperatures.

Temperature (°C)	Density (g/cm³)	Dynamic viscosity (mPa s)
0	1.0627	59
10	1.0546	33
20	1.0465	20
30	1.0384	13
40	1.0303	9
50	1.0222	7

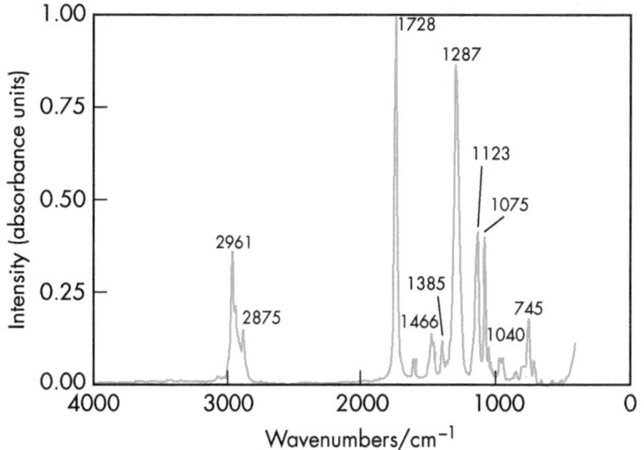

Figure 1: Infrared spectrum of dibutyl phthalate measured by transmission. Adapted with permission of Informa Healthcare.

D

11 Stability and Storage Conditions

Dibutyl phthalate should be stored in a well-closed container in a cool, dry location. Containers may be hazardous when empty since they can contain product residues such as vapors and liquids.

12 Incompatibilities

Dibutyl phthalate reacts violently with chlorine. It also reacts with oxidizing agents, acids, bases, and nitrates.

13 Method of Manufacture

Dibutyl phthalate is produced from *n*-butanol and phthalic anhydride in an ester formation reaction.

14 Safety

Dibutyl phthalate is generally regarded as a relatively nontoxic material, although it has occasionally been reported to cause hypersensitivity reactions. It is widely used in topical cosmetic and some oral pharmaceutical formulations.

LD_{50} (mouse, IV): 0.72 g/kg[1]
LD_{50} (mouse, oral): 5.3 g/kg
LD_{50} (rat, oral): 8.0 g/kg
LD_{50} (rat, IP): 3.05 mL/kg

15 Handling Precautions

Observe normal precautions appropriate to the circumstances and quantity of material handled. Contact with the skin and eyes should be avoided. Decomposition produces toxic fumes, carbon monoxide and carbon dioxide.

In the US, the permitted 8-hour exposure limit for dibutyl phthalate is 5 mg/m^3. In the UK, the long-term (8-hour TWA) workplace exposure limit for dibutyl phthalate is 5 mg/m^3. The short-term (15-minute) workplace exposure limit is 10 mg/m^3.[2]

16 Regulatory Status

Included in the FDA Inactive Ingredients Database (oral capsules; delayed action, enteric coated, and controlled release tablets). Included in nonparenteral medicines licensed in the UK (oral capsules, tablets, granules; topical creams and solutions). Included in the Canadian Natural Health Products Ingredients Database.

17 Related Substances

Diethyl phthalate; dimethyl phthalate; dioctyl phthalate.

Dioctyl phthalate

Empirical formula $C_{24}H_{38}O_4$
Molecular weight 390.55
CAS number Dioctyl phthalate occurs commercially in two isomeric forms: di-*n*-octyl phthalate [117-84-0] and di(2-ethylhexyl) phthalate [117-81-7].
Synonyms 1,2-Benzenedicarboxylic acid bis(2-ethylhexyl) ester; bis(2-ethylhexyl) phthalate; di(2-ethyl-hexyl)phthalate; DEHP; DOP; *Octoil.*
Description Clear, colorless, odorless, and anhydrous liquid.
Boiling point 384°C
Flash point 206°C (closed cup)
Melting point −50°C
Refractive index n_D^{20} = 1.50
Solubility Soluble in conventional organic solvents; practically insoluble in water.
Comments The EINECS number for dioctyl phthalate is 204-214-7.

18 Comments

In addition to a number of industrial applications, dibutyl phthalate is used as an insect repellent, although it is not as effective as dimethyl phthalate.

A specification for dibutyl phthalate is included in the *Japanese Pharmaceutical Excipients* (JPE).[5]

The EINECS number for dibutyl phthalate is 201-557-4. The PubChem Compound ID (CID) for dibutyl phthalate is 3026.

19 Specific References

1 Lewis RJ, ed. *Sax's Dangerous Properties of Industrial Materials*, 11th edn. New York: Wiley, 2004: 1164.
2 Health and Safety Executive. *EH40/2005: Workplace Exposure Limits.* Sudbury: HSE Books, 2011. http://www.hse.gov.uk/pubns/priced/eh40.pdf (accessed 26 May 2011).
3 Choudhury PK, *et al.* Osmotic delivery of flurbiprofen through controlled porosity asymmetric membrane capsule. *Drug Dev Ind Pharm* 2007; 33(10): 1135–1141.
4 He L, *et al.* A novel controlled porosity osmotic pump system for sodium ferulate. *Pharmazie* 2006; **61**(12): 1022–1027.
5 Japan Pharmaceutical Excipients Council. *Japanese Pharmaceutical Excipients* 2004. Tokyo: Yakuji Nippo, 2004: 228–229.

20 General References

Wilson AS. *Plasticisers – Principles and Practice.* London: Institute of Materials, 1995.

21 Author

RT Guest.

22 Date of Revision

27 February 2012.

Dibutyl Sebacate

1 Nonproprietary Names

USP–NF: Dibutyl Sebacate

2 Synonyms

Bis(n-butyl)sebacate; butyl sebacate; DBS; decanedioic acid, dibutyl ester; dibutyl decanedioate; dibutyl 1,8-octanedicarboxylate; *Kodaflex DBS*; *Morflex DBS*.

3 Chemical Name and CAS Registry Number

Decanedioic acid, di-n-butyl ester [109-43-3]

4 Empirical Formula and Molecular Weight

$C_{18}H_{34}O_4$ 314.47

The USP35–NF30 describes dibutyl sebacate as consisting of the esters of n-butyl alcohol and saturated dibasic acids, principally sebacic acid.

5 Structural Formula

$H_3C—(CH_2)_3—O—C(=O)—(CH_2)_8—C(=O)—O—(CH_2)_3—CH_3$

6 Functional Category

Plasticizing agent.

7 Applications in Pharmaceutical Formulation or Technology

Dibutyl sebacate is used in oral pharmaceutical formulations as a plasticizer for film coatings on tablets, beads, and granules, at concentrations of 10–30% by weight of polymer.[1,2] It is also used as a plasticizer in controlled-release tablets and microcapsule preparations.[3,4]

8 Description

Dibutyl sebacate occurs as a clear, colorless, oily liquid with a bland to slight butyl odor.

9 Pharmacopeial Specifications

See Table I.

Table I: Pharmacopeial specifications for dibutyl sebacate.

Test	USP35–NF30
Specific gravity	0.935–0.939
Refractive index	1.429–1.441
Acid value	≤0.1
Saponification value	352–360
Assay (of $C_{18}H_{34}O_4$)	≥92.0%

10 Typical Properties

Acid value 0.02

Boiling point 344–349°C; 180°C at 3 mmHg for *Morflex DBS*.

Flash point 193°C; 178°C (OC) for *Morflex DBS*.

Melting point −11°C

Refractive index n_D^{25} = 1.429–1.441 for *Morflex DBS*.

Solubility Soluble in ethanol (95%), ether, isopropanol, mineral oil, and toluene; practically insoluble in water.

Specific gravity 0.935–0.939 at 20°C

Spectroscopy

IR spectra *see* Figure 1.

Vapor density (relative) 10.8 (air = 1)

Vapor pressure 0.4 kPa (3 mmHg) at 180°C

11 Stability and Storage Conditions

Dibutyl sebacate should be stored in a closed container in a cool, dry location. Dibutyl sebacate is stable under the recommended storage conditions and as used in specified applications under most conditions of use. As an ester, dibutyl sebacate may hydrolyze in the presence of water at high or low pH conditions.

12 Incompatibilities

Dibutyl sebacate is incompatible with strong oxidizing materials and strong alkalis.

13 Method of Manufacture

Dibutyl sebacate is manufactured by the esterification of n-butanol and sebacic acid in the presence of a suitable catalyst, and by the distillation of sebacic acid with n-butanol in the presence of concentrated acid.

14 Safety

Dibutyl sebacate is used in cosmetics, foods, and oral pharmaceutical formulations, and is generally regarded as a nontoxic and nonirritant material. Following oral administration, dibutyl sebacate is metabolized in the same way as fats. In humans, direct eye contact and prolonged or repeated contact with the skin may cause very mild irritation. Acute animal toxicity tests and long-term animal feeding studies have shown no serious adverse effects to be associated with orally administered dibutyl sebacate.

LD_{50} (rat, oral): 16 g/kg[5]

15 Handling Precautions

Observe normal precautions appropriate to the circumstances and quantity of material handled. It is recommended that eye protection

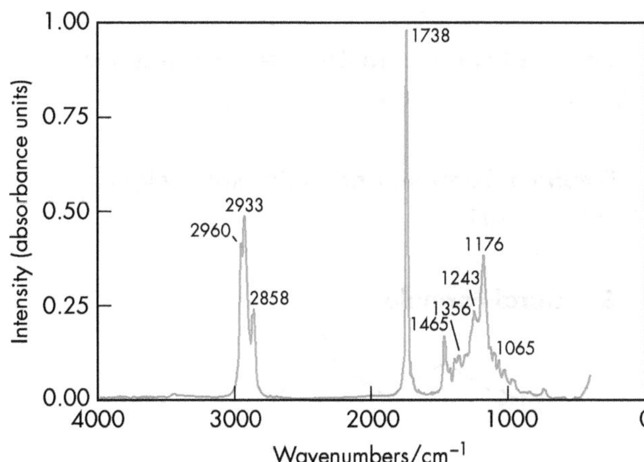

Figure 1: Infrared spectrum of dibutyl sebacate measured by transmission. Adapted with permission of Informa Healthcare.

be used at all times. When heating this product, it is recommended to have a well-ventilated area, and the use of a respirator is advised.

16 Regulatory Status

Included in the FDA Inactive Ingredients Database (oral capsules, granules, film-coated, sustained action, and tablets). Included in the Canadian Natural Health Products Ingredients Database.

17 Related Substances

—

18 Comments

As dibutyl sebacate is an emollient ester, the personal care grade is recommended for use in cosmetics, hair products, lotions, and creams. Dibutyl sebacate is also used as a synthetic flavor and flavor adjuvant in food products;[6] for example, up to 5 ppm is used in ice cream and nonalcoholic beverages.

The EINECS number for dibutyl sebacate is 203-672-5. The PubChem Compound ID (CID) for dibutyl sebacate is 7986.

19 Specific References

1 Goodhart FW, *et al.* An evaluation of aqueous film-forming dispersions for controlled release. *Pharm Technol* 1984; 8(4): 64, 66, 68, 70, 71.
2 Iyer U, *et al.* Comparative evaluation of three organic solvent and dispersion-based ethylcellulose coating formulations. *Pharm Technol* 1990; **14**(9): 68, 70, 72, 74, 76, 78, 80, 82, 84, 86.
3 Lee BJ, *et al.* Controlled release of dual drug loaded hydroxypropyl methylcellulose matrix tablet using drug containing polymeric coatings. *Int J Pharm* 1999; **188**: 71–80.
4 Zhang ZY, *et al.* Microencapsulation and characterization of tramadol-resin complexes. *J Control Release* 2000; **66**: 107–113.
5 Lewis RJ, ed. *Sax's Dangerous Properties of Industrial Materials*, 11th edn. New York: Wiley, 2004; 1165.
6 FAO/WHO. Fifty-ninth report of the Joint FAO/WHO Expert Committee on Food Additives. Dibutyl sebacate, 2002.

20 General References

Appel LE, Zentner GM. Release from osmotic tablets coated with modified Aquacoat lattices. *Proc Int Symp Control Rel Bioact Mater* 1990; **17**: 335–336.
Ozturk AG, *et al.* Mechanism of release from pellets coated with an ethylcellulose-based film. *J Control Release* 1990; **14**: 203–213.
Rowe RC. Materials used in the film coating of oral dosage forms. In: Florence AT, ed. *Materials Used in Pharmaceutical Formulation: Critical Reports on Applied Chemistry.* 6. Oxford: Blackwell Scientific, 1984; 1–36.
Vertellus Specialities Inc. Technical literature: *Morflex DBS*, August 2007.
Wheatley TA, Steurnagel CR. Latex emulsions for controlled drug delivery. In: McGinity JC, ed. *Aqueous Polymeric Coatings for Pharmaceutical Dosage Forms*, 2nd edn. New York: Marcel Dekker, 1996; 13–41.

21 Author

W Cook.

22 Date of Revision

1 March 2012.

Diethanolamine

1 Nonproprietary Names

USP–NF: Diethanolamine

2 Synonyms

Bis(hydroxyethyl)amine; DEA; diethylolamine; 2,2′-dihydroxydiethylamine; diolamine; 2,2′-iminodiethanol.

3 Chemical Name and CAS Registry Number

2,2′-Iminobisethanol [111-42-2]

4 Empirical Formula and Molecular Weight

$C_4H_{11}NO_2$ 105.14

5 Structural Formula

HO–CH2–CH2–NH–CH2–CH2–OH

6 Functional Category

Alkalizing agent; emulsifying agent.

7 Applications in Pharmaceutical Formulation or Technology

Diethanolamine is primarily used in pharmaceutical formulations as a buffering agent, such as in the preparation of emulsions with fatty acids. In cosmetics and pharmaceuticals it is used as a pH adjuster and dispersant.

Diethanolamine has also been used to form the soluble salts of active compounds, such as iodinated organic acids that are used as contrast media. As a stabilizing agent, diethanolamine prevents the discoloration of aqueous formulations containing hexamethylenetetramine-1,3-dichloropropene salts.

Diethanolamine is also used in cosmetics.

8 Description

The USP35–NF30 describes diethanolamine as a mixture of ethanolamines consisting largely of diethanolamine. At about room temperature it is a white, deliquescent solid. Above room temperature diethanolamine is a clear, viscous liquid with a mildly ammoniacal odor.

9 Pharmacopeial Specifications

See Table I.

Table I: Pharmacopeial specifications for diethanolamine.

Test	USP35–NF30
Identification	+
Limit of triethanolamine	≤1.0%
Refractive index at 30°C	1.473–1.476
Water	≤0.15%
Assay (anhydrous basis)	98.5–101.0%

10 Typical Properties

Acidity/alkalinity pH = 11.0 for a 0.1 N aqueous solution.
Autoignition temperature 662°C
Boiling point 268.8°C
Density
1.0881 g/cm^3 at 30°C;
1.0693 g/cm^3 at 60°C.
Dissociation constant pK_a = 8.88
Flash point 138°C (open cup)
Hygroscopicity Very hygroscopic.
Melting point 28°C
Refractive index n_D^{30} = 1.4753
Solubility see Table II.

Table II: Solubility of diethanolamine.

Solvent	Solubility at 20°C
Acetone	Miscible
Benzene	1 in 24
Chloroform	Miscible
Ether	1 in 125
Glycerin	Miscible
Methanol	Miscible
Water	1 in 1

Surface tension 49.0 mN/m (49.0 dynes/cm) at 20°C
Vapor density (relative) 3.65 (air = 1)
Vapor pressure >1 Pa at 20°C
Viscosity (dynamic)
351.9 mPa s (351.9 cP) at 30°C;
53.85 mPa s (53.85 cP) at 60°C.

11 Stability and Storage Conditions

Diethanolamine is hygroscopic and light- and oxygen-sensitive; it should be stored in an airtight container, protected from light, in a cool, dry place.

See Monoethanolamine for further information.

12 Incompatibilities

Diethanolamine is a secondary amine that contains two hydroxy groups. It is capable of undergoing reactions typical of secondary amines and alcohols. The amine group usually exhibits the greater activity whenever it is possible for a reaction to take place at either the amine or a hydroxy group.

Diethanolamine will react with acids, acid anhydrides, acid chlorides, and esters to form amide derivatives, and with propylene carbonate or other cyclic carbonates to give the corresponding carbonates. As a secondary amine, diethanolamine reacts with aldehydes and ketones to yield aldimines and ketimines. Diethanolamine also reacts with copper to form complex salts. Discoloration and precipitation will take place in the presence of salts of heavy metals.

13 Method of Manufacture

Diethanolamine is prepared commercially by the ammonolysis of ethylene oxide. The reaction yields a mixture of monoethanolamine, diethanolamine, and triethanolamine which is separated to obtain the pure products.

14 Safety

Diethanolamine is used in topical and parenteral pharmaceutical formulations, with up to 1.5% w/v being used in intravenous infusions. Experimental studies in dogs have shown that intravenous administration of larger doses of diethanolamine results in sedation, coma, and death.

Animal toxicity studies suggest that diethanolamine is less toxic than monoethanolamine, although in rats the oral acute and subacute toxicity is greater.[(1)] Diethanolamine is said to be heptacarcinogenic in mice and has also been reported to induce hepatic choline deficiency in mice.[(2)]

Diethanolamine is an irritant to the skin, eyes, and mucous membranes when used undiluted or in high concentration. However, in rabbits, aqueous solutions containing 10% w/v diethanolamine produce minor irritation. The lethal human oral dose of diethanolamine is estimated to be 5–15 g/kg body-weight.

The US Cosmetic Ingredient Review Expert Panel evaluated diethanolamine and concluded that it is safe for use in cosmetic formulations designed for discontinuous, brief use followed by thorough rinsing from the surface of the skin. In products intended for prolonged contact with the skin, the concentration of ethanolamines should not exceed 5%. Diethanolamine should not be used in products containing *N*-nitrosating agents.[(1)]

See also Section 18.

LD_{50} (guinea pig, oral): 2.0 g/kg[(3)]
LD_{50} (mouse, IP): 2.3 g/kg
LD_{50} (mouse, oral): 3.3 g/kg
LD_{50} (rabbit, skin): 12.2 g/kg
LD_{50} (rat, IM): 1.5 g/kg
LD_{50} (rat, IP): 0.12 g/kg
LD_{50} (rat, IV): 0.78 g/kg
LD_{50} (rat, oral): 0.71 g/kg
LD_{50} (rat, SC): 2.2 g/kg

15 Handling Precautions

Diethanolamine is irritating to the skin, eyes, and mucous membranes. Protective clothing, gloves, eye protection, and a respirator are recommended. Ideally, diethanolamine should be handled in a fume cupboard.[4] Diethanolamine poses a slight fire hazard when exposed to heat or flame.

16 Regulatory Status

Included in the FDA Inactive Ingredients Database (IV infusions, ophthalmic solutions, and topical preparations). Included in medicines licensed in the UK. Included in the Canadian Natural Health Products Ingredients Database.

17 Related Substances

Monoethanolamine; triethanolamine.

18 Comments

Through a standard battery of rodent studies, diethanolamine has been identified by the US National Toxicology Program as a potential carcinogen following topical administration. Several possible confounding issues have been noted during the review of these studies, which may affect the ultimate conclusion made regarding the carcinogenicity of diethanolamine and the relevance

D

of these findings to humans. Diethanolamine is not permitted for use in cosmetics sold within the EU.

The EINECS number for diethanolamine is 203-868-0. The PubChem Compound ID (CID) for diethanolamine is 8113.

19 Specific References

1 Neudahl GA. Diethanolamine (DEA) and diethanolamides toxicology. *Drug Cosmet Ind* 1998; **162**(4): 26–29.
2 Lehman-McKeeman LD, *et al.* Diethanolamine induces hepatic choline deficiency in mice. *Toxicol Sci* 2002; **67**(1): 38–45.
3 Lewis RJ, ed. *Sax's Dangerous Properties of Industrial Materials*, 11th edn. New York: Wiley, 2004; 1235.

20 General References

—

21 Authors

ME Fenton, PJ Sheskey.

22 Date of Revision

1 March 2012.

Diethylene Glycol Monoethyl Ether

1 Nonproprietary Names

BP: Diethylene Glycol Monoethyl Ether
PhEur: Diethylene Glycol Monoethyl Ether
USP–NF: Diethylene Glycol Monoethyl Ether

2 Synonyms

Carbitol; DEGEE; DGME; diethylenglycoli aether monoethilicus; 3,6-dioxa-1-octanol; ethoxydiglycol; ethyl carbitol; ethyl digol; ethyl dioxitol; 1-hydroxy-3,6-dioxaoctane; *Transcutol HP*; *Transcutol P*.

3 Chemical Name and CAS Registry Number

2-(2-Ethoxyethoxy)ethanol [111-90-0]

4 Empirical Formula and Molecular Weight

$C_6H_{14}O_3$ 134.17

5 Structural Formula

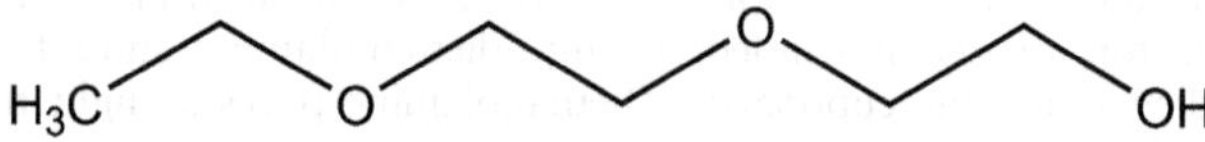

6 Functional Category

Solubilizing agent; solvent.

7 Applications in Pharmaceutical Formulation or Technology

Diethylene glycol monoethyl ether is a solubilizing agent with broad chemical compatibility and widely used in oral, dermal, and parenteral pharmaceutical preparations. It is used as a cosolvent in oral drug delivery systems to enhance drug solubility, with the level of use limited to 10%. Purified diethylene glycol monoethyl ether is also used in some injectable products.[1]

Diethylene glycol monoethyl ether has been reported to enhance skin penetration for a variety of drugs, and is used in creams, lotions, microemulsions, and aqueous gels.[2–14] Diethylene glycol monoethyl ether is also used in cosmetic formulations.

8 Description

Diethylene glycol monoethyl ether is a clear, colorless, hygroscopic liquid with a mild odor.

9 Pharmacopeial Specifications

See Table I.

Table I: Pharmacopeial specifications for diethylene glycol monoethyl ether.

Test	PhEur 7.4	USP35–NF30
Identification	+	+
Characters	+	–
Relative density	≈0.991	–
Refractive index at 20°C	1.426–1.428	1.426–1.428
Acid value	≤0.1	≤0.1
Peroxide value	≤8.0	≤8.0
Limit of free ethylene oxide	≤1 ppm	≤1 μg/g
2-Methoxyethanol	≤50 ppm	≤50 μg/g
2-Ethoxyethanol	≤160 ppm	≤160 μg/g
Ethylene glycol	≤620 ppm	≤620 μg/g
Diethylene glycol	≤250 ppm	≤150 μg/g
Total impurities	≤0.2%	–
Water	≤0.1%	≤0.1%
Assay	–	99–101%

10 Typical Properties

Autoignition temperature

201°C for *Carbitol*;[15]

204°C for *Transcutol*.[16]

Boiling point

201°C for *Carbitol*;

198°C for *Transcutol*.

Density 0.988 g/cm^3

Evaporation rate

0.01 for *Carbitol*;

0.02 for *Transcutol* (*n*-butyl acetate = 1.0).

Explosive limit

1.2% volume at 135°C (lower limit);

23.5% volume at 182°C (upper limit) for *Transcutol*.

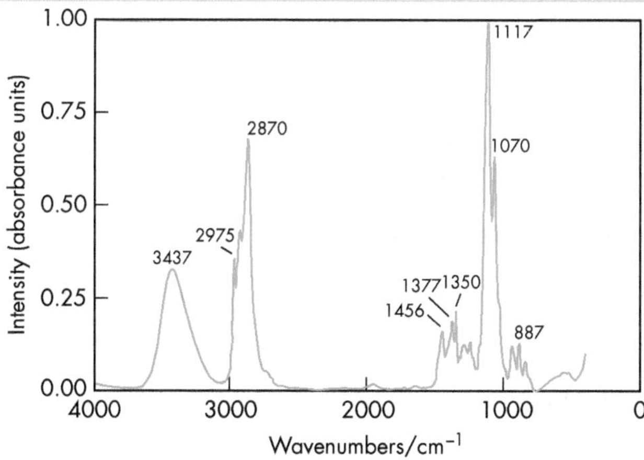

Figure 1: Infrared spectrum of diethylene glycol monoethyl ether measured by transmission. Adapted with permission of Informa Healthcare.

Flash point

90°C for *Carbitol*;

96.1°C for *Transcutol*.

Freezing point

–43°C for *Carbitol*;

–105°C for *Transcutol*.

Hygroscopicity At a relative humidity of 8–50% the equilibrium moisture content at 25°C is approx. 8% w/w, but at a relative humidity of 70% it is approx. 24% (w/w).

Partition coefficient Log *P* (octanol: water) = –0.54

Solubility Soluble in water; miscible in acetone, benzene, chloroform, ethanol (95%), ether, and pyridine; partially soluble in vegetable oils; insoluble in mineral oils.

Specific heat 2.193 J/g/°C (0.52 cal/g/°C) at 25°C for *Carbitol*

Spectroscopy

IR spectra *see* Figure 1.

Vapor density 4.6 (air = 1)

Vapor pressure

9.33 Pa (0.07 mmHg) at 20°C for *Carbitol*;

16.0 Pa (0.12 mmHg) at 20°C for *Transcutol*.

Viscosity 4.5 mPa s (4.5 cP) at 25°C for *Carbitol*.

11 Stability and Storage Conditions

Diethylene glycol monoethyl ether is very stable and inert in its original container. It is hygroscopic and sensitive to water uptake and oxidation. If the drug being formulated is sensitive to oxidation, the addition of a suitable antioxidant prior to processing is required.

Store in a cool, dry, well-ventilated area in a tightly closed container filled with nitrogen. Keep away from sources of ignition.

12 Incompatibilities

—

13 Method of Manufacture

Diethylene glycol monoethyl ether is obtained by O-alkylation of ethanol with two ethylene oxide units.

14 Safety

Diethylene glycol monoethyl ether is used in a wide variety of pharmaceutical formulations for oral, dermal, and parenteral administration, and as a cosmetic ingredient. It is generally regarded as nonirritant and nontoxic.

The Cosmetic Ingredient Review (CIR) Expert Panel has evaluated diethylene glycol monoethyl ether and found it to be safe as presently used in cosmetics at a maximum concentration of 50%.[17]

The acute toxicity of diethylene glycol monoethyl ether after oral, IV, and dermal application as well as inhalation has been studied and can be regarded as very low in all species investigated. The LD_{50} values for acute oral, acute IV, and acute dermal toxicity were found to be generally much higher than 2 g/kg body weight, and the available LC_{50} value for acute inhalation was > 5 mg/L (i.e. 5.24 mg/L) in rat,[18] in each case above the limit doses for classification.

LD_{50} (rat, oral) > 5 g/kg[19]

LD_{50} (mouse, IV): 3.2 g/kg[20]

15 Handling Precautions

Observe normal precautions appropriate to the circumstances and quantities of material handled. Diethylene glycol monoethyl ether should be used with adequate ventilation. Avoid contact with eyes, skin, and clothing. Wear suitable protective clothing.

16 Regulatory Status

Included in the FDA Inactive Ingredients Database (topical and transdermal gels). Approved in the US as an indirect food additive. Included in nonparenteral formulations available in the UK and France. Included in the Canadian Natural Health Products Ingredients Database.

17 Related Substances

—

18 Comments

A number of studies have shown that the inclusion of diethylene glycol monoethyl ether can improve localized drug delivery by increasing drug retention in the skin. This is associated with the swelling of lipid bilayer structures which subsequently act as a depot for drug–solvent complexes, which enables the slow and localized diffusion of the drug over time.[3–5,14]

The EINECS number for diethylene glycol monoethyl ether is 203-919-7. The PubChem Compound ID (CID) for diethylene glycol monoethyl ether is 8146.

19 Specific References

1 Strickley RG. Solubilizing excipients in oral and injectable formulations. *Pharm Res* 2004; **21**(2): 201–230.

2 Puglia C, Bonina F. Effect of polyunsaturated fatty acids and some conventional penetration enhancers on transdermal delivery of atenolol. *Drug Deliv* 2008; **15**(2): 107–112.

3 Ritschel WA, *et al.* Influence of selected solvents on penetration of griseofulvin in rat skin, *in vitro*. 1988; *Pharm Ind*50(4): 483–486.

4 Ritschel WA, *et al.* Development of an intracutaneous depot for drugs binding, drug accumulation and retention studies, and mechanism of depot. *Skin Pharmacol* 1991; **4**: 235–245.

5 Yazdanian M, Chen E. The effect of diethylene glycol monoethyl ether as a vehicle for topical delivery of ivermectin. *Vet Res Commun* 1995; **19**(4): 309–319.

6 Dixit N, *et al.* Nanoemulsion system for the transdermal delivery of a poorly soluble cardiovascular drug. *PDA J Pharm Sci Technol* 2008; **62**(1): 46–55.

7 Gungor S, *et al.* Effect of penetration enhancers on *in vitro* percutaneous penetration of nimesulide through rat skin. *Pharmazie* 2004; **59**(1): 39–41.

8 Papakostantinou E, *et al.* Efficacy of 2 weeks' application of theophylline ointment in psoriasis vulgaris. *J Dermatolog Treat* 2007; **16**(3): 169–170.

9 Touitou E, *et al.* Modulation of caffeine skin delivery by carrier design: liposomes versus permeation enhancers. *Int J Pharm* 1994; **103**(2): 131–136.

10 Touitou E, *et al.* Enhanced permeation of theophylline through the skin and its effect on fibroblast proliferation. *Int J Pharm* 1991; 70(12): 159–166.
11 Harrison JE, *et al.* The relative effect of azone and transcutol on permeant diffusivity and solubility in human stratum corneum. *Pharm Res* 1996; 13(4): 542–546.
12 Watkinson C, *et al.* Aspects of the transdermal delivery of prostaglandins. *Int J Pharm* 1991; 74(23): 229–236.
13 Minghetti P, *et al.* Development of patches for the controlled release of dehydroepiandrosterone. *Drug Dev Ind Pharm* 2001; 27(7): 711–717.
14 Mura P, *et al.* Evaluation of transcutol as a clonazepam transdermal permeation enhancer from hydrophilic gel formulations. *Eur J Pharm Sci* 2000; 9(4): 365–372.
15 Dow Chemical Company. Product information: *Carbitol*, March 2004.
16 Gattefossé. Technical literature: Lipid excipients for oral dosage forms and dermal drug delivery, 2010.
17 Anonymous. Final report on the safety assessment of butylene glycol, hexylene glycol, ethoxydiglycol, and dipropylene glycol. *J Am Coll Toxicol* 1985; 4: 223–232.
18 Palmer AK. Sporadic malformations in laboratory animals and their influence on drug testing. *Adv Exp Med Biol* 1972; 27: 45–60.
19 Gattefossé. *Transcutol HP*: Evaluation of the acute toxicity in mouse after intravenous administration. Report Phycher TAIs-PH-08/0477, 2009.
20 Gattefossé. *Transcutol*: Acute toxicity following a single oral administration (limit test) in the rat. Report No. WPAT94-177/OT2225; Gattefosse No. GAT-94205. New City, NY, USA: GAT, AMA Laboratories, 13 April 1994.

20 General References

European Commission. Scientific Committee on Consumer Products (SCCP): Opinion on diethylene glycol monoethyl ether (DEGEE), 2007. http://ec.europa.eu/health/ph_risk/committees/04_sccp/docs/sccp_o_082.pdf (accessed 29 February 2012).

21 Author

C Subra.

22 Date of Revision

1 March 2012.

Diethyl Phthalate

1 Nonproprietary Names

BP: Diethyl Phthalate

PhEur: Diethyl Phthalate

USP–NF: Diethyl Phthalate

2 Synonyms

DEP; diethyl benzene-1,2-dicarboxylate; diethylis phthalas; *Eastman DEP*; ethyl benzene-1,2-dicarboxylate; ethyl phthalate; *Kodaflex DEP*; *Neantine*; *Palatinol A*; phthalic acid diethyl ester.

3 Chemical Name and CAS Registry Number

1,2-Benzenedicarboxylic acid, diethyl ester [84-66-2]

4 Empirical Formula and Molecular Weight

$C_{12}H_{14}O_4$ 222.24

5 Structural Formula

O, O, CH$_3$, O, CH$_3$, O

6 Functional Category

Plasticizing agent; solvent.

7 Applications in Pharmaceutical Formulation or Technology

Diethyl phthalate is used as a plasticizing agent for film coatings on tablets, beads, and granules at concentrations of 10–30% by weight of polymer.

8 Description

Diethyl phthalate is a clear, colorless, oily liquid. It is practically odorless, or with a very slight aromatic odor and a bitter, disagreeable taste.

9 Pharmacopeial Specifications

See Table I.

Table I: Pharmacopeial specifications for diethyl phthalate.

Test	PhEur 7.4	USP35–NF30
Identification	+	+
Characters	+	—
Specific gravity	1.117–1.121	1.118–1.122
Refractive index	1.500–1.505	1.500–1.505
Appearance	+	—
Acidity	+	+
Related substances	+	—
Water	≤0.2%	≤0.2%
Residue on ignition	—	≤0.02%
Sulfated ash	≤0.1%	—
Assay (anhydrous basis)	99.0–101.0%	98.0–102.0%

10 Typical Properties

Boiling point 295°C; 298°C for *Eastman DEP*

Flash point 160°C (open cup)

Melting point −40°C

Refractive index $n_D^{25} = 1.501$

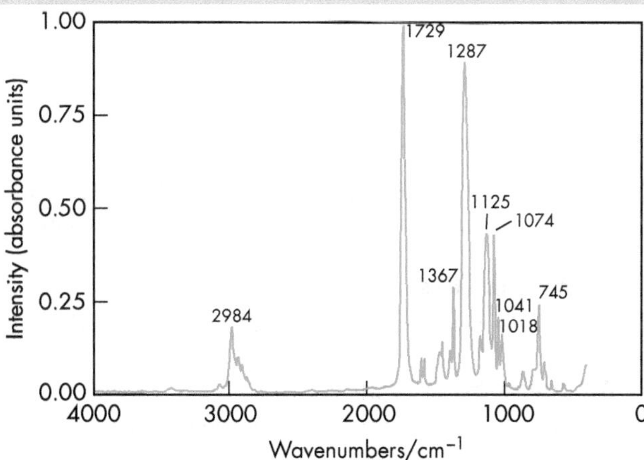

Figure 1: Infrared spectrum of diethyl phthalate measured by transmission. Adapted with permission of Informa Healthcare.

Solubility Miscible with ethanol (95%), ether, and many other organic solvents; practically insoluble in water.
Specific gravity 1.120 at 25°C
Spectroscopy
IR spectra *see* Figure 1.
Vapor density (relative) 7.66 (air = 1)
Vapor pressure 1.87 kPa (14 mmHg) at 163°C

11 Stability and Storage Conditions

Diethyl phthalate is stable when stored in a well-closed container in a cool, dry place.

12 Incompatibilities

Incompatible with strong oxidizing materials, acids, and permanganates.

13 Method of Manufacture

Diethyl phthalate is produced by the reaction of phthalic anhydride with ethanol in the presence of sulfuric acid.

14 Safety

Diethyl phthalate is used in oral pharmaceutical formulations and is generally regarded as a nontoxic and nonirritant material at the levels employed as an excipient. However, if consumed in large quantities it can act as a narcotic and cause paralysis of the central nervous system.

Although some animal studies have suggested that high concentrations of diethyl phthalate may be teratogenic, other studies have shown no adverse effects.[1]

LD_{50} (guinea pig, oral): 8.6 g/kg[2]

LD_{50} (mouse, IP): 2.7 g/kg

LD_{50} (mouse, oral): 6.2 g/kg

LD_{50} (rat, IP): 5.1 g/kg

LD_{50} (rat, oral): 8.6 g/kg

15 Handling Precautions

Observe normal precautions appropriate to the circumstances and quantity of material handled. Diethyl phthalate is irritant to the skin, eyes, and mucous membranes. Protective clothing, eye protection, and nitrile gloves are recommended. Diethyl phthalate should be handled in a fume cupboard or a well-ventilated environment; a respirator is recommended. In the UK, the long-term (8-hour TWA) workplace exposure limit for diethyl phthalate is 5 mg/m^3. The short-term (15-minute) workplace exposure limit is 10 mg/m^3.[3]

16 Regulatory Status

Included in the FDA Inactive Ingredients Database (oral capsules; delayed action, enteric coated, and sustained action tablets). Included in nonparenteral medicines licensed in the UK. Included in the Canadian Natural Health Products Ingredients Database.

D

17 Related Substances

Dibutyl phthalate; dimethyl phthalate.

18 Comments

Diethyl phthalate is used as an alcohol denaturant and as a solvent for cellulose acetate in the manufacture of varnishes and dopes. In perfumery, diethyl phthalate is used as a perfume fixative at a concentration of 0.1–0.5% of the weight of the perfume used.

A specification for diethyl phthalate is included in the *Japanese Pharmaceutical Excipients* (JPE).[4]

The EINECS number for diethyl phthalate is 201-550-6. The PubChem Compound ID (CID) for diethyl phthalate is 6781.

19 Specific References

1 Field EA, *et al.* Developmental toxicity evaluation of diethyl and dimethyl phthalate in rats. *Teratology* 1993; **48**(1): 33–44.
2 Lewis RJ, ed. *Sax's Dangerous Properties of Industrial Materials*, 11th edn. New York: Wiley, 2004: 1284–1285.
3 Health and Safety Executive. *EH40/2005: Workplace Exposure Limits.* Sudbury: HSE Books, 2011. http://www.hse.gov.uk/pubns/priced/eh40.pdf (accessed 27 February 2012).
4 Japan Pharmaceutical Excipients Council. *Japanese Pharmaceutical Excipients* 2004. Tokyo: Yakuji Nippo, 2004: 236.

20 General References

Banker GS. Film coating theory and practice. *J Pharm Sci* 1966; **55**: 81–89.
Berg JA, Mayor GH. Diethyl phthalate not dangerous [letter]. *Am J Hosp Pharm* 1991; **48**: 1448–1449.
Cafmeyer NR, Wolfson BB. Possible leaching of diethyl phthalate into levothyroxine sodium tablets. *Am J Hosp Pharm* 1991; **48**: 735–739.
Cho CW, *et al.* Controlled release of pranoprofen from the ethylene-vinyl acetate matrix using plasticizer. *Drug Dev Ind Pharm* 2007; **33**(7): 747–753.
Chambliss WG. The forgotten dosage form: enteric-coated tablets. *Pharm Technol* 1983; **7**(9): 124, 126, 128, 130, 132, 138.
Eastman. Product data sheet: *Eastman DEP* Plasticizer, 2008.
Health and Safety Executive. Review of the toxicity of the esters of phthalic acid (phthalate esters). *Toxicity Reviews* 14. London: HMSO, 1986.
Kamrin MA, Mayor GH. Diethyl phthalate: a perspective. *J Clin Pharmacol* 1991; **31**: 484–489.
Porter SC, Ridgway K. The permeability of enteric coatings and the dissolution rates of coated tablets. *J Pharm Pharmacol* 1982; **34**: 5–8.
Rowe RC. Materials used in the film coating of oral dosage forms. In: Florence AT, ed. *Materials Used in Pharmaceutical Formulation: Critical Reports on Applied Chemistry.* 6. Oxford: Blackwell Scientific, 1984: 1–36.
Sadeghi F, *et al.* Tableting of Eudragit RS and propranolol hydrochloride solid dispersion: effect of particle size compaction force, and plasticizer addition on drug release. *Drug Dev Ind Pharm* 2004; **30**(7): 759–766.
Wheatley TA, Steurernagel CR. Latex emulsions for controlled drug delivery. In: McGinity JW, ed. *Aqueous Polymeric Coatings for Pharmaceutical Dosage Forms*, 2nd edn. New York: Marcel Dekker, 1996: 41–59.

21 Author

RT Guest.

22 Date of Revision

27 February 2012.

Diethyl Sebacate

1 Nonproprietary Names

None adopted.

2 Synonyms

Bisoflex DES; decanedioic acid, diethyl ester; diethyl 1,10-decanedioate; diethyl 1,8-octanedicarboxylate; ethyl sebacate; *DES-SP*; sebacic acid, diethyl ester.

3 Chemical Name and CAS Registry Number

Diethyl decanedioate [110-40-7]

4 Empirical Formula and Molecular Weight

$C_{14}H_{26}O_4$ 258.35

5 Structural Formula

6 Functional Category

Emollient; penetration enhancer; plasticizing agent.

7 Applications in Pharmaceutical Formulation or Technology

Diethyl sebacate is used as an emollient in topical creams, emulsions, and solutions. It has been shown to aid skin penetration of various anti-inflammatory,[1] antifungal,[2] and local anesthetic[3] drugs, at concentrations of up to 24%.

Diethyl sebacate is also used as a flavoring agent in foods.[4]

8 Description

Diethyl sebacate is a colorless or yellowish liquid with a characteristic fruity odor.

9 Pharmacopeial Specifications

—

10 Typical Properties

Boiling point 307–312°C (with some decomposition)
Density 0.960–0.965 at 20°C
Flash point > 93°C (closed cup); 113°C (open cup)
Melting point 1–3°C
Partition coefficient log *P* (octanol : water) = 3.8–4.3 (calculated)
Refractive index $n_D^{20} = 1.435$
Solubility Miscible with alcohol, ether, hydrocarbons, fixed oils and fats; soluble 1 in 700 cold water, 1 in 50 boiling water.
Surface tension 33.2 mN/m at 20°C
Viscosity (dynamic) 6.1 cP at 20°C

11 Stability and Storage Conditions

Diethyl sebacate is stable to hydrolysis at room temperature and does not tend to go rancid.

Store in tightly closed containers in a cool, well-ventilated area. Protect from light. Keep away from heat and sources of ignition.

12 Incompatibilities

Diethyl sebacate is incompatible with oxidizing agents and alkalis.

13 Method of Manufacture

Diethyl sebacate is prepared by the esterification of sebacic acid and ethanol in the presence of sulfuric acid.

14 Safety

Diethyl sebacate is mildly toxic by ingestion and a skin irritant. It may cause irritation to the eyes and mucous membranes. Contact dermatitis due to diethyl sebacate has been reported.[5]

LD_{50} (guinea pig, oral): 7.28 g/kg[6]
LD_{50} (rat, oral): 14.47 g/kg

15 Handling Precautions

Observe normal precautions appropriate to the circumstances and quantity of the material handled. When heated, diethyl sebacate emits acrid smoke and irritating fumes. Wear suitable protective clothing. Avoid contact with skin and eyes.

16 Regulatory Status

GRAS listed. Included in the FDA Inactive Ingredients Database (topical; emulsions, creams, solutions).

17 Related Substances

Dibutyl sebacate.

18 Comments

A specification for diethyl sebacate is included in the *Food Chemicals Codex* (FCC)[4] and in the *Japanese Pharmaceutical Excipients* (JPE);[7] *see* Table I.

Table I: FCC[4] and JPE[7] specifications for diethyl sebacate.

Test	FCC 7	JPE 2004
Description	+	+
Identification	–	+
Acid value	≤1	≤0.5
Refractive index at 20°C	1.435–1.438	1.435–1.437
Specific gravity at 25°C	0.960–0.965	0.958–0.968
Ester value	–	411–435
Iodine value	–	≤0.5
Heavy metals	–	≤20 ppm
Arsenic	–	≤2 ppm
Residue on ignition	–	≤0.10%
Assay	≥98%	–

The EINECS number for diethyl sebacate is 203-764-5. The PubChem Compound ID (CID) is 8049.

19 Specific References

1 Fukuyasu H, *et al. In vivo* activity and metabolic fate of 2-(2,3,3-triiodoallyl tetrazole (ME1401), a novel antifungal agent. *Drugs Exp Clin Res* 1989; **15**: 335–347.

2 Takahashi K, *et al.* Effect of fatty acid diesters on permeation of anti-inflammatory drugs through rat skin. *Drug Dev Ind Pharm* 2002; **28**: 1285–1294.
3 Matsui R, *et al.* Skin permeation of lidocaine from crystal suspended oily formulations. *Drug Dev Ind Pharm* 2005; **31**: 729–738.
4 *Food Chemicals Codex*, 7th edn. Bethesda, MD: United States Pharmacopeia, 2010; 291.
5 Kimura M, Kawada A. Contact dermatitis due to diethyl sebacate. *Contact Dermatitis* 1999; **40**: 48–49.
6 Lewis RJ, editor. *Sax's Dangerous Properties of Industrial Materials*, 11th edn. New York: John Wiley and Sons Inc, 2004: 1286.
7 Japan Pharmaceutical Excipients Council. *Japanese Pharmaceutical Excipients 2004*. Tokyo: Yakuji Nippo, 2004; 237.

20 General References

ScienceLab.com, Inc. Material safety data sheet: Diethyl sebacate, June 2008.

21 Authors

ME Fenton, RC Rowe.

22 Date of Revision

1 October 2010.

D

Difluoroethane (HFC)

1 Nonproprietary Names

None adopted.

2 Synonyms

Dymel 152a; ethylene fluoride; *Genetron 152a*; halocarbon 152a; HFC 152a; P-152a; propellant 152a; refrigerant 152a; *Solkane 152a*.

3 Chemical Name and CAS Registry Number

1,1-Difluoroethane [75-37-6]

4 Empirical Formula and Molecular Weight

$C_2H_4F_2$ 66.05

5 Structural Formula

```
      F     H
      |     |
H —— C —— C —— H
      |     |
      F     H
```

6 Functional Category

Aerosol propellant.

7 Applications in Pharmaceutical Formulation or Technology

Difluoroethane, a hydrofluorocarbon (HFC), is an aerosol propellant and may be used in topical pharmaceutical formulations.[(1)]

Since difluoroethane does not contain chlorine, there are no environmental controls on the use of this material as a propellant, since it does not deplete the ozone layer and is not a greenhouse gas. However, its use in pharmaceutical aerosols is limited.

8 Description

Difluoroethane is a liquefied gas and exists as a liquid at room temperature when contained under its own vapor pressure, or as a gas when exposed to room temperature and atmospheric pressure. The liquid is practically odorless and colorless. Difluoroethane is noncorrosive and nonirritating.

9 Pharmacopeial Specifications

—

10 Typical Properties

Boiling point −24.7°C
Critical temperature 113.5°C
Density
0.90 g/cm^3 for liquid at 25°C;
0.81 g/cm^3 for liquid at 54.5°C.
Flammability Flammable. Limits of flammability 3.7–18.0% v/v in air.
Melting point −117°C
Solubility Soluble 1 in 357 parts of water at 25°C.
Surface tension 11.25 mN/m (11.25 dynes/cm) for liquid at 20°C.
Vapor density (absolute) 2.949 g/m^3 at standard temperature and pressure.
Vapor density (relative) 2.29 (air = 1)
Vapor pressure
600 kPa (61.7 psig) at 21.1°C;
1317 kPa (191 psia) at 54.5°C.
Viscosity (dynamic) 0.243 mPa s (0.243 cP) for liquid at 20°C.

11 Stability and Storage Conditions

Difluoroethane is a nonreactive and stable material. The liquefied gas is stable when used as a propellant and should be stored in a metal cylinder in a cool, dry place.

12 Incompatibilities

Compatible with the usual ingredients used in the formulation of pharmaceutical aerosols.

13 Method of Manufacture

Difluoroethane is prepared from ethyne by the addition of hydrogen fluoride in the presence of a suitable catalyst. The difluoroethane formed is purified to remove all traces of water, as well as traces of the starting materials.

14 Safety

Difluoroethane is no longer used as an aerosol propellant in topical pharmaceutical formulations.

When propellants are used in topical aerosols they may cause a chilling effect on the skin, although this effect has been somewhat overcome by the use of vapor-tap valves. The propellants quickly vaporize from the skin, and are nonirritating when used as directed.

15 Handling Precautions

Difluoroethane is usually encountered as a liquefied gas and appropriate precautions for handling such materials should be taken. Eye protection, gloves, and protective clothing are recommended. Difluoroethane should be handled in a well-ventilated environment. Fluorocarbon vapors are heavier than air and do not support life; therefore, when cleaning large tanks that have contained these propellants, adequate provision for oxygen supply in the tanks must be made in order to protect workers cleaning the tanks.

Difluoroethane is flammable; *see* Section 10. When it is heated to decomposition, toxic fumes of hydrogen fluoride may be formed.

16 Regulatory Status

Accepted in the US, by the FDA, for use as a topical aerosol propellant. Included in the Canadian Natural Health Products Ingredients Database.

17 Related Substances

Tetrafluoroethane.

18 Comments

Although difluoroethane is no longer available for use in pharmaceutical formulation, the monograph remains in the *Handbook* to assist formulators in their understanding of old products, whose properties they may need to reproduce. It may still be present in some commercial products.

Difluoroethane is useful as an aerosol propellant in that it shows greater miscibility with water than some other fluorocarbons. For a discussion of the numerical nomenclature applied to this aerosol propellant, *see* Chlorofluorocarbons.

The PubChem Compound ID (CID) for difluoroethane is 6368.

19 Specific References

1 Sheridan V. Propelling VOCs down. *Manuf Chem* 1995; **66**(10): 57.

20 General References

Johnson MA. *The Aerosol Handbook*, 2nd edn. Caldwell: WE Dorland, 1982: 305–335.

Johnson MA. Flammability aspects of dimethyl ether, p-22, p-142b, p-152a. *Aerosol Age* 1988; 33(8): 32, 34, 36, 38–39.

Sanders PA. *Handbook of Aerosol Technology*, 2nd edn. New York: Van Nostrand Reinhold, 1979: 19–35.

Sciarra CJ, Sciarra JJ. Aerosols. In: *Remington: The Science and Practice of Pharmacy*, 21st edn. Philadelphia, PA: Lippincott Williams and Wilkins, 2006: 1000–1017.

Sciarra CJ, Sciarra JJ. Pressurized dispensers. In: Schlossman ML. *The Chemistry and Manufacture of Cosmetics*, vol. 1, 4th edn. Carol Stream, IL: Allured Publishing Corporation, 2009: 451–478.

Sciarra JJ Aerosol suspensions and emulsions. In: Lieberman H *et al.* eds. *Pharmaceutical Dosage Forms: Disperse Systems*. 2. 2nd edn. New York: Marcel Dekker, 1996: 319–356.

Sciarra JJ. Pharmaceutical aerosols. In: Banker GS, Rhodes CT, eds. *Modern Pharmaceutics*, 3rd edn. New York: Marcel Dekker, 1996: 547–574.

Sciarra JJ, Stoller L. *The Science and Technology of Aerosol Packaging*. New York: Wiley, 1974: 137–145.

21 Authors

CJ Sciarra, JJ Sciarra.

22 Date of Revision

1 February 2011.

Dimethicone

1 Nonproprietary Names

BP: Dimeticone

PhEur: Dimeticone

USP–NF: Dimethicone

2 Synonyms

ABIL; dimethylpolysiloxane; dimethylsilicone fluid; dimethylsiloxane; dimeticonum; *Dow Corning Q7-9120*; E900; methyl polysiloxane; poly(dimethylsiloxane); *Sentry*.

3 Chemical Name and CAS Registry Number

α-(Trimethylsilyl)-ω-methylpoly[oxy(dimethylsilylene)] [9006-65-9]

4 Empirical Formula and Molecular Weight

The PhEur 7.4 describes dimethicone as a polydimethylsiloxane obtained by hydrolysis and polycondensation of dichlorodimethylsilane and chlorotrimethylsilane. The degree of polymerization (n = 20–400) is such that materials with kinematic viscosities nominally 20–1300 mm^2/s (20–1300 cSt) are produced. Dimethicones with a nominal viscosity of 50 mm^2/s (50 cSt) or lower are intended for external use only.

The USP35–NF30 describes dimethicone as a mixture of fully methylated linear siloxane polymers containing repeating units of the formula $[-(CH_3)_2SiO-]_n$ stabilized with trimethylsiloxy end-blocking units of the formula $[(CH_3)_3SiO-]$, where n has an average value such that the corresponding nominal viscosity is in a discrete range 20–30 000 mm^2/s (20–30 000 cSt).

5 Structural Formula

$H_3C-Si(CH_3)_2-[O-Si(CH_3)_2]_n-O-Si(CH_3)_2-CH_3$

6 Functional Category

Antifoaming agent; emollient; water-repelling agent.

7 Applications in Pharmaceutical Formulation or Technology

Dimethicones of various viscosities are widely used in cosmetic and pharmaceutical formulations. In topical oil-in-water emulsions dimethicone is added to the oil phase as an antifoaming agent. Dimethicone is hydrophobic and is also widely used in topical barrier preparations. *See* Table I.

Table I: Uses of dimethicone.

Use	Concentration (%)
Creams, lotions, and ointments	10.0–30.0
Oil–water emulsions	0.5–5.0

8 Description

Dimethicones are clear, colorless liquids available in various viscosities; *see* Section 4.

9 Pharmacopeial Specifications

See Table II.

Table II: Pharmacopeial specifications for dimethicone.

Test	PhEur 7.4	USP35–NF30
Identification	+	+
Characters	+	—
Acidity	+	+ (a)
Specific gravity	—	+ (a)
Viscosity (kinematic) of the nominal stated value	90–110%	+ (a)
Refractive index	—	+ (a)
Mineral oils	+	—
Phenylated compounds	+	—
Heavy metals	≤5 ppm	≤5 μg/g
Volatile matter (for dimethicones with a viscosity greater than 50 mm^2/s (50 cSt)	≤0.3%	—
Loss on heating	—	+ (a)
Bacterial endotoxins (coating of containers for parenteral use)	—	+
Assay (of polydimethylsiloxane)	—	97.0–103.0%

(a) The USP35–NF30 specifies limits for these tests specific to the nominal viscosity of the dimethicone.

10 Typical Properties

Acid value <0.01

Density 0.94–0.98 g/cm^3 at 25°C

Refractive index $n_D^{25} = 1.401–1.405$

Solubility Miscible with ethyl acetate, methyl ethyl ketone, mineral oil, ether, chloroform, and toluene; soluble in isopropyl myristate; very slightly soluble in ethanol (95%); practically insoluble in glycerin, propylene glycol, and water.

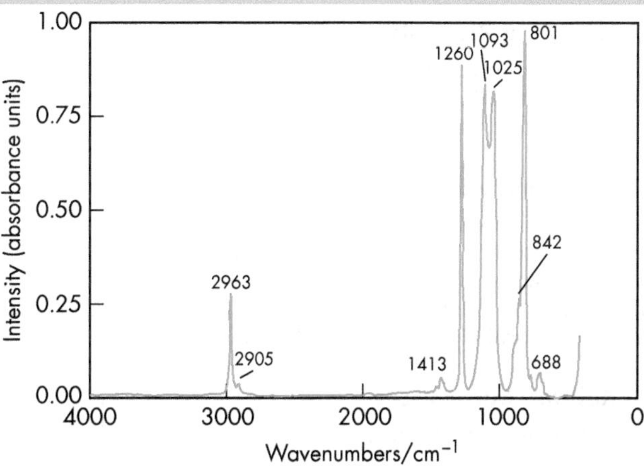

Figure 1: Infrared spectrum of dimethicone measured by transmission. Adapted with permission of Informa Healthcare.

Spectroscopy

IR spectra *see* Figure 1.

Surface tension 20.5–21.2 mN/m at 25°C

11 Stability and Storage Conditions

Dimethicones should be stored in an airtight container in a cool, dry place; they are stable to heat and are resistant to most chemical substances although they are affected by strong acids. Thin films of dimethicone may be sterilized by dry heat for at least 2 hours at 160°C. Sterilization of large quantities of dimethicone by steam autoclaving is not recommended since excess water diffuses into the fluid causing it to become hazy. However, thin films may be sterilized by this method. Gamma irradiation may also be used to sterilize dimethicone. Gamma irradiation can, however, cause crosslinking with a consequent increase in the viscosity of fluids.

12 Incompatibilities

—

13 Method of Manufacture

Dimethicone is a poly(dimethylsiloxane) obtained by hydrolysis and polycondensation of dichlorodimethylsilane and chlorotrimethylsilane. The hydrolysis products contain active silanol groups through which condensation polymerization proceeds. By varying the proportions of chlorotrimethylsilane, which acts as a chain terminator, silicones of varying molecular weight may be prepared. Different grades of dimethicone are produced that may be distinguished by a number placed after the name indicating the nominal viscosity. *See also* Section 4.

14 Safety

Dimethicone is generally regarded as a relatively nontoxic and nonirritant material although it can cause temporary irritation to the eyes. In pharmaceutical formulations it may be used in oral and topical preparations. Dimethicones are also used extensively in cosmetic formulations and in certain food applications.

The WHO has set a tentative estimated acceptable daily intake of dimethicone with a relative molecular mass in the range of 200–300 at up to 1.5 mg/kg body-weight.[1] Health Canada has set a temporary acceptable daily intake of 0.0-0.8 mg/kg body-weight for dimethicone.

Injection of silicones into tissues may cause granulomatous reactions. Accidental intravascular injection has been associated with fatalities.

LD_{50} (mouse, oral): >20 g/kg

15 Handling Precautions

Observe normal precautions appropriate to the circumstances and quantity of material handled. Dimethicone is flammable and should not be exposed to naked flames or heat.

16 Regulatory Status

Accepted for use as a food additive in Europe. Included in the FDA Inactive Ingredients Database (oral capsules and tablets;, topical creams, emulsions, lotions, and transdermal preparations). Included in nonparenteral medicines licensed in the UK. Included in the Canadian Natural Health Products Ingredients Database.

17 Related Substances

Cyclomethicone; simethicone.

18 Comments

Therapeutically, dimethicone may be used with simethicone in oral pharmaceutical formulations used in the treatment of flatulence. Dimethicone is also used to form a water-repellent film on glass containers.

The PubChem Compound ID (CID) for dimethicone includes 24764 and 589984.

19 Specific References

1 FAO/WHO. Evaluation of certain food additives. Twenty-third report of the joint FAO/WHO expert committee on food additives. *World Health Organ Tech Rep Ser* 1980; No. 648.

20 General References

Calogero AV. Regulatory review. *Cosmet Toilet* 2000; **115**(May): 24, 26, 27.

21 Author

RT Guest.

22 Date of Revision

1 March 2012.

Dimethyl Ether

1 Nonproprietary Names

None adopted.

2 Synonyms

Dimethyl oxide; DME; *Dymel A*; methoxymethane; methyl ether; oxybismethane; wood ether.

3 Chemical Name and CAS Registry Number

Methoxymethane [115-10-6]

4 Empirical Formula and Molecular Weight

C_2H_6O 46.07

5 Structural Formula

$$H_3C—O—CH_3$$

6 Functional Category

Aerosol propellant; solvent.

7 Applications in Pharmaceutical Formulation or Technology

Dimethyl ether may be used as an aerosol propellant for topical aerosol formulations in combination with hydrocarbons and other propellants.[1–4] Generally, it cannot be used alone as a propellant owing to its high vapor pressure. Dimethyl ether is a good solvent and has the unique property of high water solubility, compared to other propellants. It has frequently been used with aqueous aerosols. A coarse, wet, spray is formed when dimethyl ether is used as a propellant.

Dimethyl ether is also used as a propellant in cosmetics such as hair sprays, and in other aerosol products such as air fresheners and fly sprays.

8 Description

Dimethyl ether is a liquefied gas and exists as a liquid at room temperature when contained under its own vapor pressure, or as a gas when exposed to room temperature and pressure.

It is a clear, colorless, virtually odorless liquid. In high concentrations, the gas has a faint ether-like odor.

9 Pharmacopeial Specifications

—

10 Typical Properties

Autoignition temperature 350°C
Boiling point −23.6°C
Critical temperature 126.9°C
Density 0.66 g/cm^3 for liquid at 25°C.
Flammability The pure material is flammable; limit of flammability is 3.4–18.2% v/v in air. Aqueous mixtures are nonflammable.
Freezing point −138.5°C
Flash point −41°C
Heat of combustion −28.9 kJ/g (−6900 cal/g)
Kauri-butanol value 60
Solubility Soluble in acetone, chloroform, ethanol (95%), ether, and 1 in 3 parts of water. Dimethyl ether is generally miscible with water, nonpolar materials, and some semipolar materials. For pharmaceutical aerosols, ethanol (95%) is the most useful cosolvent. Glycols, oils, and other similar materials exhibit varying degrees of miscibility with dimethyl ether.
Surface tension 16 mN/m (16 dynes/cm) at −10°C
Vapor density (absolute) 2.058 g/m^3 at standard temperature and pressure.
Vapor density (relative) 1.596 (air = 1)
Vapor pressure
592 kPa at 25°C (63 psig at 21.1°C);
1301 kPa at 54°C.

11 Stability and Storage Conditions

The liquefied gas is stable when used as a propellant. However, exposure to the air for long periods of time may result in explosive peroxides being slowly formed.

Solutions of liquid dimethyl ether should not be concentrated either by distillation or by evaporation. Dimethyl ether should be stored in tightly closed metal cylinders in a cool, dry place.

12 Incompatibilities

Dimethyl ether is an aggressive solvent and may affect the gasket materials used in aerosol packaging. Oxidizing agents, acetic acid, organic acids, and anhydrides should not be used with dimethyl ether. *See also* Section 10.

13 Method of Manufacture

Dimethyl ether is prepared by the reaction of bituminous or lignite coals with steam in the presence of a finely divided nickel catalyst. This reaction produces formaldehyde, which is then reduced to methanol and dimethyl ether. Dimethyl ether may also be prepared by the dehydration of methanol.

14 Safety

Dimethyl ether may be used as a propellant and solvent in topical pharmaceutical aerosols, and is generally regarded as an essentially nontoxic and nonirritant material when used in such applications. However, inhalation of high concentrations of dimethyl ether vapor is harmful. Additionally, skin contact with dimethyl ether liquid may result in freezing of the skin and severe frostbite.

When used in topical formulations, dimethyl ether may exert a chilling effect on the skin, although if it is used as directed the propellant quickly vaporizes and is nonirritating.

LD_{50} (mouse, inhalation): 386 000 ppm/30 min[5]

LD_{50} (rat, inhalation): 308 g/m^3

15 Handling Precautions

Dimethyl ether is usually encountered as a liquefied gas, and appropriate precautions for handling such materials should be taken. Eye protection, gloves, and protective clothing are recommended.

Dimethyl ether should be handled in a well-ventilated environment.

Dimethyl ether vapor is heavier than air and does not support life; therefore, when cleaning large tanks that have contained this material, adequate provisions for oxygen supply in the tanks must be made in order to protect workers cleaning the tanks.

In the UK, the long-term (8-hour TWA) exposure limit for dimethyl ether is 766 mg/m^3 (400 ppm). The short-term (15-minute) exposure limit is 958 mg/m^3 (500 ppm).[6]

Dimethyl ether is flammable; *see* Section 10.

16 Regulatory Status

Included in the FDA Inactive Ingredients Database (topical aerosols). Included in nonparenteral medicines licensed in the UK. Included in the Canadian Natural Health Products Ingredients Database.

17 Related Substances

Hydrocarbons (HC).

18 Comments

Since the solubility of dimethyl ether in water is about 35%, it can be used to good effect in aqueous aerosol products. It also has antimicrobial effects that are organism-dependent.[7]

Dimethyl ether is additionally used as a refrigerant.

The EINECS number for dimethyl ether is 204-065-8. The PubChem Compound ID (CID) for dimethyl ether is 8254.

19 Specific References

1 Bohnenn LJM. DME: an alternative propellant? *Manuf Chem Aerosol News* 1977; 48(9): 40.
2 Bohnenn LJM. DME: further data on this alternative propellant. *Manuf Chem Aerosol News* 1978; 49(8): 39, 63.
3 Bohnenn LJM. 'Alternative' aerosol propellant. *Drug Cosmet Ind* 1979; 125(Nov): 58, 60, 62, 66, 68, 70, 72, 74.
4 Boulden ME. Use of dimethyl ether for reduction of VOC content. *Spray Technol Market* 1992; 2(May): 30, 32, 34, 36.
5 Lewis RJ, ed. *Sax's Dangerous Properties of Industrial Materials*, 11th edn. New York: Wiley, 2004: 2442.
6 Health and Safety Executive. *EH40/2005: Workplace Exposure Limits*. Sudbury: HSE Books, 2011. http://www.hse.gov.uk/pubns/priced/eh40.pdf (accessed 1 March 2012).
7 Ibrahim YK, Sonntag HG. Preservative potentials of some aerosol propellants: effectiveness in some pharmaceutical oils. *Drugs Made Ger* 1995; 38(Apr–Jun): 62–65.

20 General References

Johnson MA. *The Aerosol Handbook*, 2nd edn. Mendham, NJ: WE Dorland, 1982: 305–335.

Johnson MA. Flammability aspects of dimethyl ether, p-22, p-142b, p-152a. *Aerosol Age* 1988; 33(8): 32, 34, 36, 38–39.

Sanders PA. *Handbook of Aerosol Technology*, 2nd edn. New York: Van Nostrand Reinhold, 1979: 44–54.

Sciarra JJ, Stoller L. *The Science and Technology of Aerosol Packaging*. New York: Wiley, 1974: 137–145.

Sciarra JJ. Pharmaceutical aerosols. In: Banker GS, Rhodes CT, eds. *Modern Pharmaceutics*, 3rd edn. New York: Marcel Dekker, 1996: 547–574.

Sciarra CJ, Sciarra JJ. Aerosols. In: Gennaro AR, ed. *Remington: The Science and Practice of Pharmacy*, 21st edn. Philapdelphia, PA: Lippincott, Williams and Wilkins, 2005: 1000–1017.

21 Authors

CJ Sciarra, JJ Sciarra.

22 Date of Revision

1 February 2011.

Dimethyl Phthalate

1 Nonproprietary Names

BP: Dimethyl Phthalate

2 Synonyms

Avolin; 1,2-benzenedicarboxylate; benzenedicarboxylic acid dimethyl ester; dimethyl 1,2-benzenedicarboxylate; dimethyl benzene-*o*-dicarboxylate; dimethyl benzeneorthodicarboxylate; dimethyl *o*-phthalate; *o*-dimethyl phthalate; DMP; *Eastman DMP*; *Fermine*; *Kodaflex DMP*; methyl benzene-1,2-dicarboxylate; *Mipax*; *Palatinol M*; phthalic acid dimethyl ester; phthalic acid methyl ester; *Repeftal*; *Solvanom*; *Solvarone*; *Unimoll DM*.

3 Chemical Name and CAS Registry Number

1,2-Benzenedicarboxylic acid, dimethyl ester [131-11-3]

4 Empirical Formula and Molecular Weight

$C_{10}H_{10}O_4$ 194.19

5 Structural Formula

O
CH3
O
O
CH3
O

6 Functional Category

Plasticizing agent; solvent.

7 Applications in Pharmaceutical Formulation or Technology

Dimethyl phthalate is used in pharmaceutical applications as a solvent and plasticizing agent for film-coatings such as hydroxypropyl methylcellulose, cellulose acetate and cellulose acetate–butyrate mixtures.[1,2]

8 Description

Dimethyl phthalate occurs as a colorless, or faintly colored, odorless, viscous, oily liquid.

9 Pharmacopeial Specifications

See Table I.

Table I: Pharmacopeial specifications for dimethyl phthalate.

Test	BP 2012
Identification	+
Characters	+
Acidity	+
Refractive index	1.515–1.517
Weight per mL	1.186–1.192
Related substances	+
Sulfated ash	≤0.1%
Water	≤0.1%
Assay (dried basis)	99.0–100.5%

10 Typical Properties

Boiling point 280°C, with decomposition; 284°C for *Eastman DMP.*
Density 1.186–1.192 g/cm^3
Flash point 146°C (closed cup); 157°C (open cup) for *Eastman DMP.*
Freezing point The commercial product freezes at 0°C; -1°C for *Eastman DMP.*
Melting point 2.0–5.5°C
Partition coefficient
Octanol : water = 1.56[3]
Refractive index n_D^{20} = 1.515–1.517; n_D^{25}= 1.513 for *Eastman DMP.*
Solubility *see* Table II.

Table II: Solubility of dimethyl phthalate.

Solvent	Solubility at 20°C unless otherwise stated
Benzene	Miscible
Chloroform	Miscible
Ethanol (95%)	Miscible
Ether	Miscible
Mineral oil	1 in 294
Water	1 in 250 at 20°C

Surface tension 41.9 mN/m at 20°C
Vapor density (relative) 6.69 (air = 1)
Vapor pressure 120 Pa at 100°C
Viscosity 17.2 mPa s (17.2 cP) at 25°C.

11 Stability and Storage Conditions

Dimethyl phthalate is sensitive to prolonged exposure to light and it should therefore be stored in a cool, dark, dry, well-ventilated area that is protected from physical damage, and isolated from incompatible substances. Containers of dimethyl phthalate may be hazardous when empty as they may retain product residues such as vapors and liquids. There is a slight fire hazard when exposed to heat, and above the flash point (*see* Section 10) explosive vapor–air mixtures may be formed. Carbon dioxide and carbon monoxide are released when dimethyl phthalate is heated to decomposition. Solutions of dimethyl phthalate in acetone, dimethyl sulfoxide, ethanol (95%), and water are stable for 24 hours under normal laboratory conditions.

12 Incompatibilities

Dimethyl phthalate is incompatible with strong acids or bases, nitrates, and strong oxidizing agents. As with other phthalates, contact with plastics should be avoided.

13 Method of Manufacture

Dimethyl phthalate is produced industrially from phthalic anhydride and methanol.

14 Safety

In pharmaceutical applications, dimethyl phthalate is used in film coating and as a topically applied insect repellent. Acute exposure to the eyes and mucous membranes can cause irritation, although dimethyl phthalate is considered less irritant than diethyl phthalate. Inhalation of dimethyl phthalate can cause irritation of the respiratory tract; oral ingestion can cause a burning sensation in the mouth, vomiting, and diarrhea. Owing to the low water solubility and relatively high lipid solubility, dimethyl phthalate may accumulate in body tissues after chronic exposure, which may cause central nervous system depression.

Although some animal studies have suggested that high concentrations of dimethyl phthalate may be teratogenic or cause mutagenic effects with bacteria,[4,5] other studies have shown no adverse effects.[6] There are no confirmed reports of human reproductive or developmental effects, and the compound is not generally regarded as a carcinogenic material.

LD_{50} (chicken, oral): 8.5 g/kg[7]
LD_{50} (guinea pig, oral): 2.4 g/kg
LD_{50} (mouse, IP): 1.38 g/kg
LD_{50} (mouse, oral): 6.8 g/kg
LD_{50} (rabbit, oral): 4.40 g/kg
LD_{50} (rat, IP): 3.38 g/kg
LD_{50} (rat, oral): 6.80 g/kg

15 Handling Precautions

Observe normal precautions appropriate to the circumstances and quantity of material handled. Skin and eye contact should be avoided; eye goggles or a full face shield should be worn where splashing may occur. Respirators should be used if the compound is heated to decomposition. In the UK, the long-term (8-hour TWA) workplace exposure limit for dimethyl phthalate is 5 mg/m^3. The short-term (15-minute) workplace exposure limit is 10 mg/m^3.[8]

16 Regulatory Status

Dimethyl phthalate is included in a number of topical pharmaceutical formulations. Included in the FDA Inactive Ingredients Database (oral tablets, sustained action). As from 1992, dimethyl phthalate is no longer registered for use as a pesticide in California.

17 Related Substances

Dibutyl phthalate; diethyl phthalate.

18 Comments

In addition to a number of industrial applications, dimethyl phthalate is also widely used as an insect repellent with topical preparations typically applied as a 40% cream or lotion; it has also been applied as a tent fabric treatment.[9]

A specification for dimethyl phthalate is included in the *Japanese Pharmaceutical Excipients* (JPE).[10]

The EINECS number for dimethyl phthalate is 205-011-6. The PubChem Compound ID (CID) for dimethyl phthalate is 8554.

19 Specific References

1 Shah PS, Zatz JL. Plasticization of cellulose esters used in the coating of sustained release solid dosage forms. *Drug Dev Ind Pharm* 1992; **18**: 1759–1772.
2 Wolf B. Bead cellulose products with film formers and solubilisers for controlled drug release. *Int J Pharm* 1997; **156**: 97–107.
3 Ellington JJ, Floyd TL. *EPA/600/5–96: Octanol/water Partition Coefficients for Eight Phthalate Esters.* Athens, GA: US Environmental Protection Agency, 1996.
4 Kozumbo WJ, Rubin RJ. Mutagenicity and metabolism of dimethyl phthalate and its binding to epidermal and hepatic macromolecules. *J Toxicol Environ Health* 1991; **33**(1): 29–46.
5 Niazi JH, *et al.* Initial degradation of dimethyl phthalate by esterases from *Bacillus* species. *FEMS Microbiol Lett* 2001; **196**(2): 201–205.
6 Field EA, *et al.* Developmental toxicity evaluation of diethyl and dimethyl phthalate in rats. *Teratology* 1993; **48**(1): 33–44.
7 Lewis RJ, ed. *Sax's Dangerous Properties of Industrial Materials*, 11th edn. New York: Wiley, 2004: 1460.
8 Health and Safety Executive. *EH40/2005: Workplace Exposure Limits.* Sudbury: HSE Books, 2011. http://www.hse.gov.uk/pubns/priced/eh40.pdf (accessed 27 February 2012).
9 Schreck CE. Permethrin and dimethyl phthalate as tent fabric treatments against *Aedes aegypti*. *J Am Mosq Control Assoc* 1991; 7(4): 533–535.
10 Japan Pharmaceutical Excipients Council. *Japanese Pharmaceutical Excipients* 2004. Tokyo: Yakuji Nippo, 2004: 246–247.

20 General References

Eastman. Product data sheet: *Eastman DMP* Plasticizer, 2008.

21 Author

RT Guest.

22 Date of Revision

27 February 2012.

Dimethyl Sulfoxide

1 Nonproprietary Names

BP: Dimethyl Sulfoxide
PhEur: Dimethyl Sulfoxide
USP: Dimethyl Sulfoxide

2 Synonyms

Deltan; dimexide; dimethylis sulfoxidum; dimethyl sulphoxide; DMSO; *Kemsol*; methylsulfoxide; *Procipient*; *Rimso-50*; sulphinylbismethane

3 Chemical Name and CAS Registry Number

Sulfinylbismethane [67-68-5]

4 Empirical Formula and Molecular Weight

C_2H_6OS 78.13

5 Structural Formula

O
||
H_3C–S–CH_3

6 Functional Category

Penetration enhancer; solvent.

7 Applications in Pharmaceutical Formulation or Technology

Dimethyl sulfoxide is a highly polar substance that is aprotic, therefore lacking acidic and basic properties. It has exceptional solvent properties for both organic and inorganic components, which are derived from its capacity to associate with both ionic species and neutral molecules that are either polar or polarizable. Dimethyl sulfoxide enhances the topical penetration of drugs owing to its ability to displace bound water from the stratum corneum; this is accompanied by the extraction of lipids and configurational changes of proteins.(1) The molecular interactions between dimethyl sulfoxide and the stratum corneum, as a function of depth and time, have been described.(2) Much of the enhancement capacity is lost if the solvent is diluted. Increases in drug penetration have been reported with dimethyl sulfoxide concentrations as low as 15%, but significant increases in permeability generally require concentrations higher than 60–80%. Furthermore, while low molecular weight substances can penetrate quickly into the deep layers of the skin, the appreciable transport of molecules with a molecular weight of more than 3000 is difficult.

Dimethyl sulfoxide is now incorporated into a number of regulated products for healthcare and drug delivery applications, including stabilizing product formulations, sustained-release applications, and for the delivery of medical polymers.(3)

The use of dimethyl sulfoxide to improve transdermal delivery has been reported for diclofenac,(4,5) ciclosporin,(6) timolol,(7) and a wide range of other drugs.(8–10) Dimethyl sulfoxide has also been used in the formulation of an injection containing allopurinol.(11) It has also been investigated for use in an experimental parenteral preparation for the treatment of liver tumors.(12)

In paint formulations of idoxuridine, dimethyl sulfoxide acts both as a solvent to increase drug solubility and a means of enabling penetration of the antiviral agent to the deeper levels of the epidermis. *See* Table I.

Dimethyl sulfoxide has been investigated as a potential cryoprotectant.(13,14)

Table I: Uses of dimethyl sulfoxide.

Use	Concentration (%)
Solvent	≤100
Topical penetration enhancer	≥80

8 Description

Dimethyl sulfoxide occurs as a colorless, viscous liquid, or as colorless crystals that are miscible with water, alcohol, and ether. The material has a slightly bitter taste with a sweet aftertaste, and is odorless, or has a slight odor characteristic of dimethyl sulfoxide. Dimethyl sulfoxide is extremely hygroscopic, absorbing up to 70% of its own weight in water with evolution of heat.

9 Pharmacopeial Specifications

See Table II.

Table II: Pharmacopeial specifications for dimethyl sulfoxide.

Test	PhEur 7.4	USP35–NF30
Characters	+	—
Identification	+	+
Specific gravity	1.100–1.104	1.095–1.101
Freezing point	≥18.3°C	—
Refractive index	1.478–1.479	1.4755–1.4775
Acidity	+	+
Water	≤0.2%	≤0.1%
Ultraviolet absorbance	+	+
Limit of nonvolatile residue	—	≤0.01%
Related substances	+	+
Assay	—	≥99.9%

10 Typical Properties

Acidity/alkalinity pH = 8.5 (for a 50 : 50 mixture with water)
Autoignition temperature 215°C
Boiling point 189°C
Density 1.0955 g/cm^3 at 25°C for *Procipient*
Dielectric constant 48.9 at 20°C
Dipole moment (D) 4.3 at 20°C(15)
Dissociation constant $pK_a = 31.3$(15)
Enthalpy of fusion 3.43 cal/mol(15)
Enthalpy of vaporization 12.64 cal/mol at 25°C(15)
Flash point 95°C (open cup)
Freezing point Markedly reduces the freezing point of water; a 30% aqueous solution in water has a freezing point of −14°C and a 40% aqueous solution has a freezing point of −25°C.
Partition coefficient log (octanol/water) = −2.03
Solubility Miscible with water with evolution of heat; also miscible with ethanol (95%), ether and most organic solvents; immiscible with paraffins, hydrocarbons. Practically insoluble in acetone, chloroform, ethanol (96%), and ether.
Specific heat 0.7 cal/g (liquid)

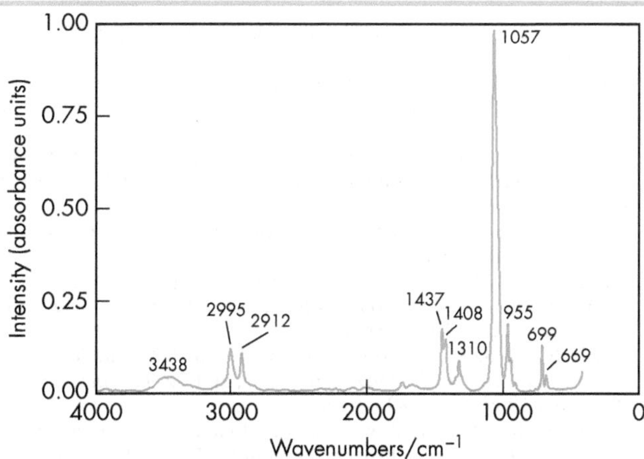

Figure 1: Infrared spectrum of dimethyl sulfoxide measured by transmission. Adapted with permission of Informa Healthcare.

Spectroscopy

IR spectra *see* Figure 1.

Vapor pressure 0.37 mm at 20°C

Viscosity (dynamic)

1.1 mPa s (1.1 cP) at 27°C;

2.0 mPa s (2.0 cP) at 25°C for *Procipient*;

2.47 mPa s (2.47 cP) at 20°C.

11 Stability and Storage Conditions

Dimethyl sulfoxide is reasonably stable to heat, but upon prolonged reflux it decomposes slightly to methyl mercaptan and bis-methylthiomethane. This decomposition is aided by acids, and is retarded by many bases. When heated to decomposition, toxic fumes are emitted.

At temperatures between 40 and 60°C, it has been reported that dimethyl sulfoxide suffers a partial breakdown, which is indicated by changes in physical properties such as refractive index, density, and viscosity.[16]

Dimethyl sulfoxide should be stored in airtight, light-resistant containers. The PhEur 7.4 states that glass containers should be used. Contact with plastics should be avoided.

12 Incompatibilities

Dimethyl sulfoxide can react with oxidizing materials.

13 Method of Manufacture

Dimethyl sulfoxide is prepared by air oxidation of dimethyl sulfide in the presence of nitrogen oxides. It can also be obtained as a by-product of wood pulp manufacture for the paper and allied industries.

14 Safety

Dimethyl sulfoxide has low systemic toxicity but causes local toxic effects.[17–19] It is readily absorbed after injection or after oral or percutaneous administration and is widely distributed throughout the body. Dimethyl sulfoxide acts as a primary irritant on skin, causing redness, burning, itching, and scaling; it also causes urticaria. Systemic symptoms include nausea, vomiting, chills, cramps, and lethargy; dimethyl sulfoxide can also cause increases in intraocular pressure. Administration of dimethyl sulfoxide by any route is followed by a garlic-like odor on the breath.

Intravascular hemolysis and biochemical changes[20] and reversible neurological deterioration[21] have been reported following intravenous administration; however, it has been questioned whether these findings were directly attributable to dimethyl sulfoxide rather than to concomitant drug therapy or contaminants.[22] One report describes massive intracranial hemorrhage associated with ingestion of dimethyl sulfoxide.[23] Recently, a hypersensitivity reaction attributed to dimethyl sulfoxide has been reported.[24] Increased psychomotor impairment has been reported in patients who drank alcohol after topical exposure to dimethyl sulfoxide, including inadvertent occupational exposure.[25]

In 1965, the FDA banned investigation in humans of dimethyl sulfoxide owing to the appearance of changes in the refractive index of the lens of the eye in experimental animals. However, in 1966, the FDA allowed the study of dimethyl sulfoxide in serious conditions such as scleroderma, persistent herpes zoster, and severe rheumatoid arthritis, and in 1968 permitted studies using short-term topical application of the solvent. By 1980, the FDA no longer specifically regulated investigations of dimethyl sulfoxide.[26]

Dimethyl sulfoxide enhances the skin penetration of several drugs, which may result in producing the adverse effects associated with those drugs.

LD_{50} (dog, IV): 2.5 g/kg[27]

LD_{50} (rat, IP): 8.2 g/kg

LD_{50} (rat, IV): 5.3 g/kg

LD_{50} (rat, oral): 14.5 g/kg

LD_{50} (rat, SC): 12 g/kg

LD_{50} (mouse, IP): 2.5 g/kg

LD_{50} (mouse, IV): 3.8 g/kg

LD_{50} (mouse, oral): 7.9 g/kg

15 Handling Precautions

Observe normal precautions appropriate to the circumstances and quantity of material handled. Dimethyl sulfoxide may cause irritation to the skin. Gloves and eye protection are recommended.

16 Regulatory Status

Included in the FDA Inactive Ingredients Database (IV infusions, SC implants, and topical preparations). Available in the US as a 50% solution for irrigation in the treatment of interstitial cystitis. Also available in Canada as a 70% solution for use as a topical antifibrotic, and in Germany as a topical gel containing 10% dimethyl sulfoxide for the treatment of musculoskeletal and joint disorders. Included in topical formulations of idoxuridine licensed in the UK.

17 Related Substances

—

18 Comments

A 2.16% dimethyl sulfoxide solution in water is iso-osmotic with serum. Dimethyl sulfoxide has been used as a 50% aqueous solution for instillation into the bladder in the treatment of interstitial cystitis; it has also been tried clinically for a wide range of indications, including cutaneous and musculoskeletal disorders, but with little evidence of beneficial effects.

Dimethyl sulfoxide has been shown to have bactericidal,[28] bacteriostatic,[28,29] and fungistatic[29] activity, although the concentration required is dependent on the organism present.

Dimethyl sulfoxide has also been investigated as a potential therapeutic agent in conditions such as scleroderma, interstitial cystitis,[30] rheumatoid arthritis, and acute musculoskeletal injuries, and as an analgesic.[26,31–34] It has also been recommended for the treatment of anthracycline extravasation.[35–38]

The EINECS number for dimethyl sulfoxide is 200-664-3. The PubChem Compound ID (CID) for dimethyl sulfoxide is 679.

19 Specific References

1 Anigbogu ANC, *et al.* Fourier transform Raman spectroscopy of interactions between the penetration enhancer dimethyl sulfoxide and human stratum corneum. *Int J Pharm* 1995; **125**: 265–282.

2 Caspers PJ, *et al.* Monitoring the penetration enhancer dimethyl sulfoxide in human stratum corneum *in vivo* by confocal Raman spectroscopy. *Pharm Res* 2002; **19**(10): 1577–1580.

3 McKim AS, Strub R. Dimethyl sulfoxide USP, PhEur in approved pharmaceutical products and medical devices. *Pharm Tech* 2008; **32**(5): 74–85.

4 Roth SH, Shainhouse JZ. Efficacy and safety of a topical diclofenac solution (pennsaid) in the treatment of primary osteoarthritis of the knee – a randomised, double blind, vehicle-controlled clinical trial. *Arch Int Med* 2004; **164**(18): 2017–2023.

5 Bookman AA, *et al.* Effect of topical diclofenac solution for relieving symptoms of primary osteoarthritis of the knee: a randomised controlled trial. *CMAJ* 2004; **171**(4): 333–338.

6 Wang D-P, *et al.* Effect of various physical/chemical properties on the transdermal delivery of ciclosporin through topical application. *Drug Dev Ind Pharm* 1997; **23**(1): 99–106.

7 Soni S, *et al.* Effect of penetration enhancers on transdermal delivery of timolol maleate. *Drug Dev Ind Pharm* 1992; **18**(10): 1127–1135.

8 Barry BW. *Dermatological Formulations*. New York: Marcel Dekker, 1983: 162–167.

9 Motlekar NA, *et al.* Permeation of genistein through human skin. *Pharm Technol* 2003; **27**(3): 140–148.

10 Amrish C, Kumar SP. Transdermal delivery of ketorolac. *Yakugaku Zasshi* 2009; **129**: 373–379.

11 Lee DKT, Wang D-P. Formulation development of allopurinol suppositories and injectables. *Drug Dev Ind Pharm* 1999; **25**(11): 1205–1208.

12 Komemushi A, *et al.* A new liquid embolic material for liver tumors. *Acta Radiol* 2002; **43**(2): 186–191.

13 Higgins J, *et al.* A comparative investigation of glycinebetaine and dimethylsulphoxide as liposome cryoprotectants. *J Pharm Pharmacol* 1987; **39**: 577–582.

14 den Brok MW, *et al.* Pharmaceutical development of a lyophilised dosage form for an investigational anticancer agent Imexon using dimethyl sulfoxide as a solubilising and stabilising agent. *J Pharm Sci* 2005; **94**(5): 1101–1114.

15 MacGregor WS. The chemical and physical properties of DMSO. *Ann NY Acad Sci* 1967; **141**: 3–12.

16 Jacob SW *et al.* eds. *Dimethyl Sulfoxide*. 1. New York: Marcel Dekker, 1971: 81.

17 Brobyn RD. The human toxicology of dimethyl sulfoxide. *Ann NY Acad Sci* 1975; **243**: 497–506.

18 Willhite CC, Katz PI. Toxicology updates: dimethyl sulfoxide. *J Appl Toxicol* 1984; 4: 155–160.

19 Mottu F, *et al.* Organic solvents for pharmaceutical parenterals and embolic liquids: a review of toxicity data. *PDA J Pharm Sci Technol* 2000; 54(6): 456–469.

20 Yellowlees P, *et al.* Dimethylsulphoxide-induced toxicity. *Lancet* 1980; **ii**: 1004–1006.

21 Bond GR, *et al.* Dimethylsulphoxide-induced encephalopathy [letter]. *Lancet* 1989; **i**: 1134–1135.

22 Knott LJ. Safety of intravenous dimethylsulphoxide [letter]. *Lancet* 1980; **ii**: 1299.

23 Topacoglu H, *et al.* Massive intracranial hemorrhage associated with the ingestion of dimethyl sulfoxide. *Vet Hum Toxicol* 2004; **46**(3): 138–140.

24 Creus N, *et al.* Toxicity to topical dimethyl sulfoxide (DMSO) when used as an extravasation antidote. *Pharm World Sci* 2002; **24**(5): 175–176.

25 Sweetman SC, ed. *Martindale: The Complete Drug Reference*, 37th edn. London: Pharmaceutical Press, 2011: 2219.

26 Fischer JM. DMSO: a review. *US Pharm* 1981; **6**(Sept): 25–28.

27 Lewis RJ, ed. *Sax's Dangerous Properties of Industrial Materials*, 11th edn. New York: Wiley, 2004: 1466.

28 Ansel HC, *et al.* Antimicrobial activity of dimethyl sulfoxide against *Escherichia coli*, *Pseudomonas aeruginosa*, and *Bacillus megaterium*. *J Pharm Sci* 1969; **58**(7): 836–839.

29 Placencia AM, *et al.* Sterility testing of fat emulsions using membrane filtration and dimethyl sulfoxide. *J Pharm Sci* 1982; **71**(6): 704–705.

30 Theoharides TC, *et al.* Treatment approaches for painful bladder syndrome/interstitial cystitis. *Drugs* 2007; **67**(2): 215–235.

31 Murdoch L-A. Dimethyl sulfoxide (DMSO): an overview. *Can J Hosp Pharm* 1982; **35**(3): 79–85.

32 Namaka M, Briggs C. DMSO revisited. *Can Pharm J* 1994; **127**(Jun): 248, 249, 255.

33 Parker WA, Bailie GR. Current therapeutic status of DMSO. *Can Pharm J* 1982; **115**(Jul): 247–251.

34 Ely A, Lockwood B. What is the evidence for the safety and efficiency of dimethyl sulfoxide and methylsulfanylmethane in pain relief? *Pharm J* 2002; **269**: 685–687.

35 Bingham JM, Dooley MJ. EXTRA – Extravasation Treatment Record Database: a database to record and review cytotoxic drug extravasation events. *Aust J Hosp Pharm* 1998; **28**(2): 89–93.

36 Bertelli G, *et al.* Dimethylsulphoxide and cooling after extravasation of antitumour agents [letter]. *Lancet* 1993; **341**: 1098–1099.

37 de Lemos ML. Role of dimethyl sulfoxide for management of chemotherapy extravasation. *J Oncol Pharm Pract* 2004; **10**(4): 197–200.

38 Reeves D. Management of anthracycline extravasation injuries. *Ann Pharmacother* 2007; **41**(7–8): 1238–1242.

20 General References

Gaylord Chemical Company. Technical literature: *Procipient*, October 2007.

Mottu F, *et al.* Comparative haemolytic activity of undiluted organic water-miscible solvents for intravenous and intra-arterial injection. *PDA J Pharm Sci Technol* 2001; **55**(1): 16–21.

21 Author

CG Cable.

22 Date of Revision

1 March 2012.

Dimethylacetamide

1 Nonproprietary Names

BP: Dimethylacetamide
PhEur: Dimethylacetamide

2 Synonyms

Acetdimethylamide; acetic acid dimethylamide; acetyldimethylamine; dimethylacetamidum; dimethylacetone amide; dimethylamide acetate; DMA; DMAC.

3 Chemical Name and CAS Registry Number

N,N-Dimethylacetamide [127-19-5]

4 Empirical Formula and Molecular Weight

C_4H_9NO 87.12

5 Structural Formula

O
H3C
N
CH3
CH3

6 Functional Category

Solvent.

7 Applications in Pharmaceutical Formulation or Technology

Dimethylacetamide is used as a solvent in oral and injectable pharmaceutical formulations.(1) It has been used as a cosolvent to solubilize poorly soluble drugs.(2–4) The use of dimethylacetamide has also been investigated as a vehicle for the parenteral delivery of relatively small peptides.(5)

The use of solvents such as dimethylacetamide has been shown to influence the size and rate of release of norfloxacin from nanoparticles.(6)

Dimethylacetamide has also been used in topical formulations and has been evaluated as a permeation enhancer for transdermal drug delivery.(7)

8 Description

Dimethylacetamide occurs as a clear, colorless, slightly hygroscopic liquid. It has a weak ammonia-like or fish-like odor.

9 Pharmacopeial Specifications

See Table I. *See also* Section 18.

Table I: Pharmacopeial specifications for dimethylacetamide.

Test	PhEur 7.4
Identification	+
Characters	+
Appearance	+
Relative density	0.941–0.944
Refractive index	1.435–1.439
Acidity	+
Alkalinity	+
Related substances	+
Heavy metals	≤10 ppm
Nonvolatile matter	≤20 ppm
Water	≤0.1%

10 Typical Properties

Autoignition temperature 490°C
Boiling point 165°C
Dielectric constant $D^{20} = 37.8$
Flash point 70°C
Refractive index $n_D^{22.5} = 1.4371$
Solubility Miscible with ethanol (95%), water, and most common organic solvents.
Specific gravity 0.943
Surface tension 35.7 mN/m (35.7 dyne/cm)
Vapor pressure 0.33 kPa at 20°C
Viscosity (dynamic) 1.02 mPa s (1.02 cP) at 25°C

11 Stability and Storage Conditions

Dimethylacetamide should be stored in an airtight container, protected from light, in a cool, dry place. Dimethylacetamide has an almost unlimited shelf-life when kept in closed containers and under nitrogen. It is combustible.

12 Incompatibilities

Dimethylacetamide is incompatible with carbon tetrachloride, oxidizing agents, halogenated compounds, and iron. It attacks plastic and rubber. Contact with strong oxidizers may cause fire.

13 Method of Manufacture

Dimethylacetamide is manufactured from acetic acid and dimethylamine in a closed system.

14 Safety

Dimethylacetamide is used in pharmaceutical preparations as a solvent in parenteral formulations and is generally regarded as a nontoxic material when used as an excipient. Animal toxicity studies indicate that dimethylacetamide is readily absorbed into the bloodstream following inhalation or topical application. Repeated exposure to dimethylacetamide may be harmful and can result in liver damage. High intravenous doses (>400 mg/kg/day for 3 days) may be hallucinogenic.(8–11)

LD_{50} (rabbit, SC): 9.6 g/kg(12)
LD_{50} (rat, IP): 2.75 g/kg

LD_{50} (rat, IV): 2.64 g/kg
LD_{50} (rat, oral): 4.93 g/kg
LD_{50} (mouse, inhalation): 7.2 g/kg
LD_{50} (mouse, IP): 2.8 g/kg
LD_{50} (mouse, IV): 3.02 g/kg
LD_{50} (mouse, SC): 9.6 g/kg

D

15 Handling Precautions

Observe normal precautions appropriate to the circumstances and quantity of the material handled. Dimethylacetamide can be absorbed into the bloodstream by inhalation and through the skin; it is irritating to the skin and eyes.

16 Regulatory Status

Included in the FDA Inactive Ingredients Database (IM injections, IV injections and infusions). Included in parenteral medicines licensed in the UK.

17 Related Substances

—

18 Comments

A specification for dimethylacetamide is included in the *Japanese Pharmaceutical Excipients* (JPE).[13]

The EINECS number for dimethylacetamide is 204-826-4. The PubChem Compound ID (CID) for dimethylacetamide is 31374.

19 Specific References

1 Strickley RG. Solubilizing excipients in oral and injectable formulations. *Pharm Res* 2004; **21**(2): 201–230.
2 Kawakami K, *et al.* Solubilisation behavior of poorly soluble drugs with combined use of Gelucire 44/14 and cosolvent. *J Pharm Sci* 2004; **93**(6): 1471–1479.
3 Kawakami K, *et al.* Solubilization behaviour of a poorly soluble drug under combined use of surfactants and cosolvents. *Eur J Pharm Sci* 2006; **28**(1–2): 7–14.
4 Taepaiboon P, *et al.* Vitamin-loaded electrospun cellulose acetate nanofiber mats as transdermal and dermal therapeutic agents of vitamin A acid and vitamin E. *Eur J Pharm Biopharm* 2007; **67**(2): 387–397.
5 Larsen SW, *et al.* Kinetics of degradation and oil solubility of ester prodrugs of a model dipeptide (Gly-Phe). *Eur J Pharm Sci* 2004; **22**: 399–408.
6 Jeon HJ, *et al.* Effect of solvent on the preparation of surfactant-free poly (DL-lactide-*co*-glycolide) nanoparticles and norfloxacin release characteristics. *Int J Pharm* 2000; **207**(1–2): 99–108.
7 Sinha VR, Kaur MP. Permeation enhancers for transdermal drug delivery. *Drug Dev Ind Pharm* 2000; **26**(11): 1131–1140.
8 Horn HJ. Toxicology of dimethylacetamide. *Toxicol Appl Pharmacol* 1961; **3**: 12–24.
9 Anschel J. [Solvents and solubilization in injections.] *Pharm Ind* 1965; **27**: 781–787[in German].
10 Kennedy GL, Sherman H. Acute toxicity of dimethylformamide and dimethylacetamide following various routes of administration. *Drug Chem Toxicol* 1986; **9**: 147–170.
11 Kim SN. Preclinical toxicology and pharmacology of dimethylacetamide, with clinical notes. *Drug Metab Rev* 1988; **19**: 345–368.
12 Lewis RJ, ed. *Sax's Dangerous Properties of Industrial Materials*, 11th edn. New York: Wiley, 2004: 1371.
13 Japan Pharmaceutical Excipients Council. *Japanese Pharmaceutical Excipients 2004*. Tokyo: Yakuji Nippo, 2004: 244–245.

20 General References

—

21 Author

RT Guest.

22 Date of Revision

1 March 2012.

Dipropylene Glycol

1 Nonproprietary Names

None adopted.

2 Synonyms

Dipropylene glycol exists as three structural isomers (a, b, c) and a mixture defined by the manufacturing process (d); *see* Sections 3, 5 and 18.

(a) Bis(2-hydroxypropyl) ether; 2,2′-dihydroxydipropyl ether; 4-oxaheptane-2,6-diol; 1,1′-dimethyldiethylene glycol; 1,1′-oxydi-2-propanol; 1,1′-oxydi-2-propanol; 2-propanol, 1,1′-oxybis-

(b) 1,2-Dipropylene glycol; 2,2′-oxydipropanol; 1-propanol, 2,2′-oxybis-

(c) 1-Propanol, 2-(2-hydroxypropoxy)-

(d) 2,2,-Dihydroxyisopropyl ether; DPG; methyl-2(methyl-2)oxybispropanol; 4-oxa-2,6-heptandiol; oxydipropanol; propanol, oxybis-

3 Chemical Name and CAS Registry Number

Dipropylene glycol exists as three structural isomers (a, b, c) and a mixture defined by the manufacturing process (d). *See also* Sections 5 and 18.

(a) 1,1′-Oxybis(2-propanol) [110-98-5]
(b) 2,2′-Oxybis(1-propanol) [108-61-2]
(c) 2-(2-Hydroxypropoxy)propanol [106-62-7]
(d) Oxybispropanol [25265-71-8]

4 Empirical Formula and Molecular Weight

$C_6H_{14}O_3$ 134.2

5 Structural Formula

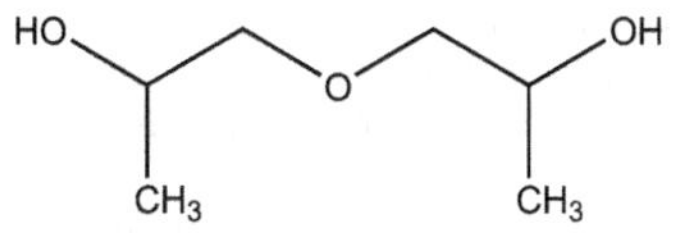

(a) 1,1'-Oxybis(2-propanol)

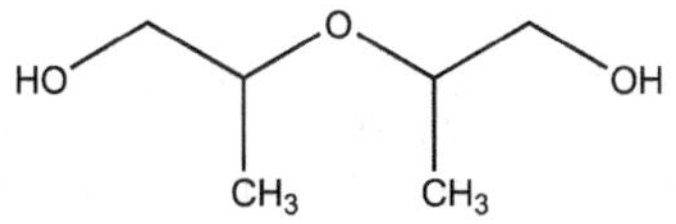

(b) 2,2'-Oxybis(1-propanol)

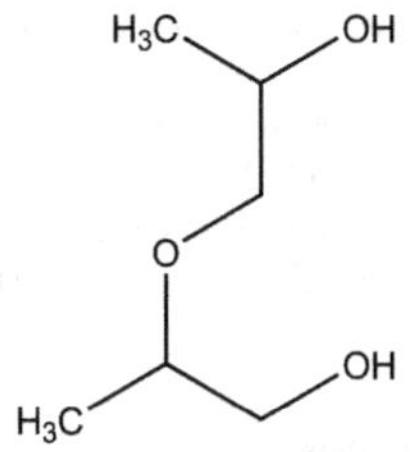

(c) 2-(2-Hydroxypropoxy)propanol

6 Functional Category

Humectant; plasticizing agent; solvent; transdermal delivery component.

7 Applications in Pharmaceutical Formulation or Technology

Dipropylene glycol is used in the preparation of controlled-release buccal and transdermal patches as a component of the adhesive.[1]

Dipropylene glycol is also used in cosmetics, typically skin cleansing preparations, deodorants, lipsticks, and fragrances.

8 Description

Dipropylene glycol occurs as a practically odorless, colorless liquid.

9 Pharmacopeial Specifications

See Section 18.

10 Typical Properties

The properties listed below are for the dominant isomer (a) and mixed isomers (d).[2,3]

Boiling point 228–233°C
Density 1.023 g/cm^3 at 20°C; 0.998 g/cm^3 at 60°C
Flash point 121–124°C (closed cup)
Freezing point Approx. –39°C
Partition coefficient Log *P* (octanol: water) –1.07 to –1.486
Refractive index 1.438–1.442
Solubility Miscible with water, oils, and hydrocarbons.
Specific heat 2.18 J/g/°C (0.52 cal/g/°C) at 25°C
Surface tension 35 mN/m (35 dynes/cm)
Vapor density 4.63 (air = 1)
Vapor pressure 0.0021 kPa (0.016 mmHg) at 25°C
Viscosity 75.0 mPa s (75.0 cP) at 25°C; 10.9 mPa s (10.9 cP) at 60°C.

11 Stability and Storage Conditions

Store in closed containers away from sources of UV light below 40°C.

12 Incompatibilities

Dipropylene glycol is incompatible with strong oxidizing agents.

13 Method of Manufacture

Dipropylene glycol is produced as a byproduct of the manufacture of propylene glycol.

14 Safety

Dipropylene glycol is not acutely toxic by oral, dermal, or inhalation exposure. Ingestion of high doses may cause CNS depression (fatigue, dizziness, possible loss of concentration). Ingestion of dipropylene glycol fog solution has caused acute renal failure, neuropathy, and myopathy in a 32-year old man.[4]

Dipropylene glycol is slightly irritating to skin and eyes. It has a low potential to produce allergic skin reactions.

Dipropylene glycol does not appear to be genotoxic.[5] Carcinogenicity studies in rats and mice have indicated limited carcinogenic potential.[6,7]

LD_{50} (dog, IV): 11.5 g/kg[8]
LD_{50} (guinea pig, oral): 17.6 g/kg
LD_{50} (mouse, IP): 4.5 g/kg
LD_{50} (rabbit, eye): 0.51 g/kg
LD_{50} (rabbit, skin): 0.5 g/kg
LD_{50} (rat, IP): 10.0 g/kg
LD_{50} (rat, IV): 5.8 g/kg
LD_{50} (rat, oral): 14.85 g/kg
LC_{50} (rat, inhalation): 6.0 g/m^3[5]

15 Handling Precautions

Observe normal precautions appropriate to the circumstances and quantity of material handled. Do not heat higher than 28°C unless in an air-free closed system. Wear suitable protective clothing; eye protection is recommended.

16 Regulatory Status

GRAS listed. Dipropylene glycol (a) is included in the FDA Inactive Ingredients Database (topical; emulsions, creams, lotions). Dipropylene glycol (d) is included in the Canadian Natural Health Products Ingredients Database (solvent). Dipropylene glycol (unspecified isomer(s)) is included in nonparenteral medicines licensed in the UK.

17 Related Substances

Polyethylene glycol; propylene glycol.

18 Comments

A specification for dipropylene glycol (unspecified isomer(s)) is contained in the *Japanese Pharmaceutical Excipients* (JPE);[9] see Table I.

Table I: JPE specification for dipropylene glycol.

Test	JPE 2004
Description	+
Identification	+
Specific gravity	1.021–1.027
Acid	+
Chloride	≤0.007%
Heavy metals	≤20 ppm
Water	≤0.5%
Residue on ignition	≤0.05%
Distilling range	220–240°C (≥95.0 vol%)

Dipropylene glycol has not been cleared for use as a direct food additive. In the US it can be used as a component in food contact articles such as adhesives and paper. In the EU it has been cleared for the use of plastic materials for food contact.

Dipropylene glycol is also an ingredient of cutting oils, industrial soaps, and agriculture insecticidal formulations. It is used as an additive for carburettor fuels as a lubricant and antifreezing agent. It is used as a solvent for printing inks, lacquers and coatings.

The EINECS numbers for dipropylene glycol are:

(a) 203-821-4
(b) 203-599-9
(c) 203-416-2
(d) 246-770-3.

The PubChem Compound IDs (CID) for dipropylene glycol are:

(a) 8087
(b) 92739
(c) 32881
(d) 32857.

19 Specific References

1 Marier JF, *et al.* Pharmacokinetics, tolerability, and performance of a novel matrix transdermal delivery system of fentanyl relative to the commercially available reservoir formulation in healthy subjects. *J Clin Pharmacol* 2006; **46**(6): 642–653.
2 Dow Chemical Company. Technical data sheet: Dipropylene glycol, April 2005.
3 ScienceLab.com, Inc. Material safety data sheet: Dipropylene glycol, June 2008.
4 LoVecchio F, *et al.* Acute renal failure, neuropathy, and myopathy after ingestion of dipropylene glycol fog solution. *Am J Emerg Med* 2008; **26**(5): 635.
5 United Nations Environment Programme. OECD SIDS Initial Assessment Report for 11th SIAM: Dipropylene glycol (mixed isomers and dominant isomers), January 2001. http://www.inchem.org/documents/sids/sids/25265-71-8.pdf (accessed 29 February 2012).
6 Hooth MJ, *et al.* Toxicology and carcinogenesis studies of dipropylene glycol in rats and mice. *Toxicology* 2004; **204**(23): 123–140.
7 Anonymous. NTP toxicology and carcinogensis studies of dipropylene glycol (CAS No. 25265-71-8) in F344/N rats and B6C3F1 mice (drinking water studies). *Natl Toxicol Program Tech Rep Ser* 2004; **511**: 6–260.
8 Lewis RJ, ed. *Sax's Dangerous Properties of Industrial Materials*, 11th edn. New York: Wiley, 2004; 2805.
9 Japan Pharmaceutical Excipients Council. *Japanese Pharmaceutical Excipients* 2004. Tokyo: Yakuji Nippo, 2004; 253.

20 General References

—

21 Authors

ME Fenton, RC Rowe.

22 Date of Revision

1 February 2011.

Disodium Edetate

1 Nonproprietary Names

BP: Disodium Edetate
JP: Disodium Edetate Hydrate
PhEur: Disodium Edetate
USP–NF: Edetate Disodium

2 Synonyms

Dinatrii edetas; disodium EDTA; disodium ethylenediaminetetraacetate; edathamil disodium; edetate disodium; edetic acid, disodium salt.

3 Chemical Name and CAS Registry Number

Ethylenediaminetetraacetic acid, disodium salt [139-33-3]
Disodium ethylenediaminetetraacetate dihydrate [6381-92-6]

4 Empirical Formula and Molecular Weight

$C_{10}H_{14}N_2Na_2O_8$ 336.2 (for anhydrous)
$C_{10}H_{18}N_2Na_2O_{10}$ 372.2 (for dihydrate)

5 Structural Formula

6 Functional Category

Complexing agent.

7 Applications in Pharmaceutical Formulation or Technology

Disodium edetate is used as a complexing agent in a wide range of pharmaceutical preparations, including mouthwashes, ophthalmic preparations, and topical preparations,(1–3) typically at concentrations between 0.005 and 0.1% w/v.

Disodium edetate forms stable water-soluble complexes (chelates) with alkaline earth and heavy-metal ions. The chelated form has few of the properties of the free ion, and for this reason chelating agents are often described as 'removing' ions from solution, a process known as sequestering. The stability of the metal–edetate complex is dependent on the metal ion involved and the pH.

See also Edetic acid.

8 Description

Disodium edetate occurs as a white, crystalline, odorless powder with a slightly acidic taste.

9 Pharmacopeial Specifications

See Table I.

Table I: Pharmacopeial specifications for disodium edetate.

Test	JP XV	PhEur 7.4	USP35–NF30
Identification	+	+	+
Characters	+	+	—
Appearance of solution	+	+	—
pH	4.3–4.7	4.0–5.5	4.0–6.0
Iron	—	≤80 ppm	—
Calcium	—	—	+
Heavy metals	≤10 ppm	≤20 ppm	≤0.005%
Cyanide	+	—	—
Arsenic	≤2 ppm	—	—
Limit of nitrilotriacetic acid	—	≤0.1%	≤0.1%
Residue on ignition	37.0–39.0%	—	—
Loss on drying	—	—	8.7–11.4%
Assay	≥99.0%	98.5–101.0%	99.0–101.0%

10 Typical Properties

Acidity/alkalinity pH 4.3–4.7 (1% w/v solution in carbon dioxide-free water)

Freezing point depression 0.14°C (1% w/v aqueous solution)

Melting point Decomposition at 252°C for the dihydrate.

Refractive index 1.33 (1% w/v aqueous solution)

Solubility Practically insoluble in chloroform and ether; slightly soluble in ethanol (95%); soluble 1 part in 11 parts water.

Specific gravity 1.004 (1% w/v aqueous solution)

Spectroscopy

IR spectra *see* Figure 1.

NIR spectra *see* Figure 2.

Viscosity (kinematic) 1.03 mm^2/s (1.03 cSt) (1% w/v aqueous solution).

11 Stability and Storage Conditions

Edetate salts are more stable than edetic acid (*see also* Edetic acid). However, disodium edetate dihydrate loses its water of crystallization when heated to 120°C. Aqueous solutions of disodium edetate may be sterilized by autoclaving, and should be stored in an alkali-free container.

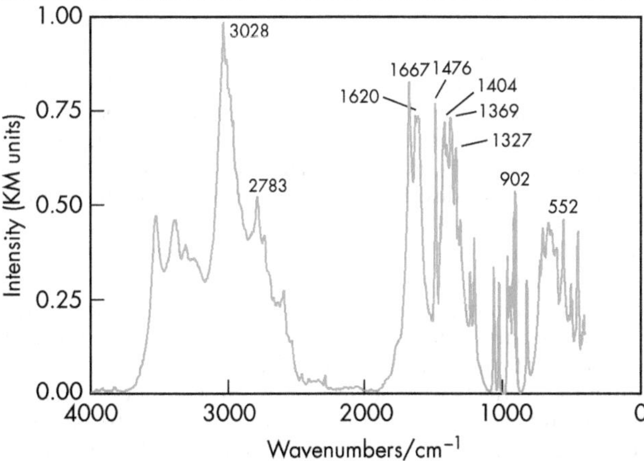

Figure 1: Infrared spectrum of disodium edetate dihydrate measured by diffuse reflectance. Adapted with permission of Informa Healthcare.

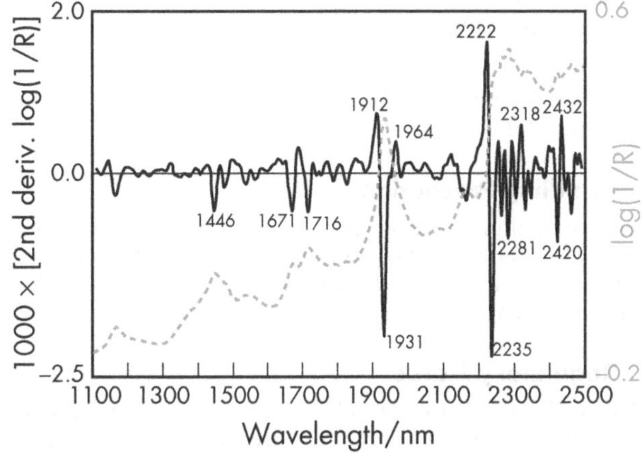

Figure 2: Near-infrared spectrum of disodium edetate dihydrate measured by reflectance.

Disodium edetate is hygroscopic and is unstable when exposed to moisture. It should be stored in a well-closed container in a cool, dry place.

12 Incompatibilities

Disodium edetate behaves as a weak acid, displacing carbon dioxide from carbonates and reacting with metals to form hydrogen. It is incompatible with strong oxidizing agents, strong bases, metal ions, and metal alloys.

See also Edetic acid.

13 Method of Manufacture

Disodium edetate may be prepared by the reaction of edetic acid and sodium hydroxide.

14 Safety

Disodium edetate is used widely in topical, oral, and parenteral pharmaceutical formulations; it is used extensively in cosmetic and food products. Disodium edetate and edetate calcium disodium are used in a greater number and variety of pharmaceutical formulations than is edetic acid. Both disodium edetate and edetate calcium disodium are poorly absorbed from the gastrointestinal tract and are associated with few adverse effects when used as excipients in pharmaceutical formulations.

Disodium edetate, trisodium edetate, and edetic acid readily chelate calcium and can, in large doses, cause calcium depletion (hypocalcemia) if used over an extended period of time, or if administered too rapidly by intravenous infusion. If used in preparations for the mouth, they can also leach calcium from the teeth. However, edetate calcium disodium does not chelate calcium.

Disodium edetate should be used with caution in patients with renal impairment, tuberculosis, and impaired cardiac function.

Although disodium edetate is generally considered safe, there have been reports of disodium edetate toxicity in patients receiving chelation therapy.[4]

Nasal formulations containing benzalkonium chloride and disodium edetate, both known to be local irritants, were shown to produce an inflammatory reaction, and microscopic examination showed an extended infiltration of the mucosa by eosinophils, and pronounced atrophy and disorganization of the epithelium, although these effects were subsequently shown to be reversible.[3]

The WHO has set an estimated acceptable daily intake for disodium EDTA in foodstuffs of up to 2.5 mg/kg body-weight.[5] *See also* Edetic acid.

LD_{50} (mouse, IP): 0.26 g/kg[6]
LD_{50} (mouse, IV): 0.056 g/kg
LD_{50} (mouse, OP): 2.05 g/kg
LD_{50} (rabbit, IV): 0.047 g/kg
LD_{50} (rabbit, OP): 2.3 g/kg
LD_{50} (rat, OP): 2.0 g/kg

15 Handling Precautions

Observe normal precautions appropriate to the circumstances and quantity of material handled. Disodium edetate and its derivatives are mild irritants to the mucous membranes. Eye protection, gloves, and dust masks are recommended.

16 Regulatory Status

GRAS listed. Included in the FDA Inactive Ingredients Database (inhalations; injections; ophthalmic preparations; oral capsules, solutions, suspensions, syrups, and tablets; rectal, topical, and vaginal preparations). Included in nonparenteral and parenteral medicines licensed in the UK. Included in the Canadian Natural Health Products Ingredients Database.

17 Related Substances

Edetic acid.

18 Comments

Disodium edetate has been used experimentally to investigate the stability and skin penetration capacity of captopril gel, in which disodium edetate was shown to exert a potent stabilizing effect, and may be used in the development of a transdermal drug delivery system.[7] A chitosan–EDTA conjugate has been investigated as a novel polymer for use in topical gels. The conjugate was shown to be stable, colorless, and transparent, and it also demonstrated antimicrobial effects.[8] Studies have shown that disodium edetate may be effective in retarding microbial growth in propofol formulations[9] and for the prevention of biofilm formation in catheters.[10]

Therapeutically, disodium edetate is used as an anticoagulant as it will chelate calcium and prevent the coagulation of blood *in vitro*.[11] Concentrations of 0.1% w/v are used in small volumes for hematological testing and 0.3% w/v in transfusions. Studies have shown how disodium edetate is facilitated by methylsulfonylmethane to cross biological membranes for regional chelation therapy.[12]

Disodium edetate is used as a water softener as it will chelate calcium and magnesium ions present in hard water.

A specification for disodium edetate is contained in the *Food Chemicals Codex* (FCC).[13]

The EINECS number for disodium edetate is 205-358-3. The PubChem Compound ID (CID) for disodium edetate includes 8759 and 636371.

19 Specific References

1 Ungphaiboon S, Maitani Y. *In vitro* permeation studies of triamcinolone acetonide mouthwashes. *Int J Pharm* 2001; **220**: 111–117.
2 Kaur IP, *et al.* Formulation and evaluation of ophthalmic preparations of acetazolamide. *Int J Pharm* 2000; **199**: 119–127.
3 Bechgaard E, *et al.* Reversibility and clinical relevance of morphological changes after nasal application of ephedrine nasal drops 1%. *Int J Pharm* 1997; **152**: 67–73.
4 Morgan BW, *et al.* Adverse effects in 5 patients receiving EDTA at an outpatient chelation clinic. *Vet Hum Toxicol* 2002; **44**(5): 274–276.
5 FAO/WHO. Toxicological evaluation of certain food additives with a review of general principles and of specifications. Seventeenth report of the joint FAO/WHO expert committee on food additives. *World Health Organ Tech Rep Ser* 1974; No. 539.
6 Lewis RJ, ed. *Sax's Dangerous Properties of Industrial Materials*, 11th edn. New York: Wiley, 2004: 1660.
7 Huang YB, *et al.* Effect of antioxidants and anti-irritants on the stability, skin irritation and penetration capacity of captopril gel. *Int J Pharm* 2002; **241**: 345–351.
8 Valenta C, *et al.* Chitosan–EDTA conjugate: novel polymer for topical gels. *J Pharm Pharmacol* 1998; **50**: 445–452.
9 Fukada T, Ozaki M. Microbial growth in propofol formulations with disodium edetate and the influence of venous access system dead space. *Anaesthesia* 2007; **62**(6): 575–580.
10 Raad II, *et al.* The role of chelators in preventing biofilm formation and catheter-related bloodstream infections. *Curr Opin Infect Dis* 2008; **21**(4): 385–392.
11 Tilbrook GS, Hider RC. Iron chelators for clinical use. *Met Ions Biol Syst* 1998; **35**: 691–730.
12 Zhang M, *et al.* Assessment of methylsulfonylmethane as a permeability enhancer for regional EDTA chelation therapy. *Drug Deliv* 2009; **16**(5): 243–248.
13 *Food Chemicals Codex*, 7th edn. Bethesda, MD: United States Pharmacopeia 2010: 315.

20 General References

—

21 Authors

SL Cantor, S Shah.

22 Date of Revision

1 March 2012.

Docusate Sodium

1 Nonproprietary Names

BP: Docusate Sodium

PhEur: Docusate Sodium

USP–NF: Docusate Sodium

2 Synonyms

Bis(2-ethylhexyl) sodium sulfosuccinate; dioctyl sodium sulfosuccinate; DSS; natrii docusas; sodium 1,4-bis(2-ethylhexyl) sulfosuccinate; sodium 1,4-bis[(2-ethylhexyl)oxy]-1,4-dioxobutane-2-sulfonate; sodium dioctyl sulfosuccinate; sulfo-butanedioic acid 1,4-bis(2-ethylhexyl) ester, sodium salt; sulfosuccinic acid 1,4-bis(2-ethylhexyl) ester *S*-sodium salt.

3 Chemical Name and CAS Registry Number

Sodium 1,4-bis(2-ethylhexyl) sulfosuccinate [577-11-7]

4 Empirical Formula and Molecular Weight

$C_{20}H_{37}NaO_7S$ 444.56

5 Structural Formula

6 Functional Category

Anionic surfactant.

7 Applications in Pharmaceutical Formulation or Technology

Docusate sodium and docusate salts are widely used as anionic surfactants in pharmaceutical formulations. Docusate sodium is mainly used in capsule, tablet, liquid, and syrup dosage forms, and in direct-compression tablet formulations to assist in wetting and dissolution;[1] *see* Table I.

Table I: Uses of docusate sodium.

Use	Concentration (%)
IM injections	0.015
Surfactant (wetting/dispersing/emulsifying agent)	0.01–1.0
Tablet coating agent	20[a]
Tablet disintegrant	≈0.5

(a) Formulation of a tablet coating solution: 20% docusate sodium; 2–15% sodium benzoate; 0.5% propylene glycol; solution made in ethanol (70%).

8 Description

Docusate sodium is a white or almost white, wax-like, bitter tasting, plastic solid with a characteristic octanol-like odor. It is hygroscopic and usually available in the form of pellets, flakes, or rolls of tissue-thin material.

9 Pharmacopeial Specifications

See Table II.

Table II: Pharmacopeial specifications for docusate sodium.

Test	PhEur 7.4	USP35–NF30
Identification	+	+
Characters	+	—
Alkalinity	+	—
Bis(2-ethylhexyl) maleate	—	≤0.4%
Chlorides	≤350 ppm	—
Clarity of solution	—	+
Heavy metals	≤10 ppm	≤10 ppm
Related nonionic substances	+	—
Residue on ignition	—	15.5–16.5%
Sodium sulfate	≤2.0%	—
Water	≤3.0%	≤2.0%
Assay (dried basis)	98.0–101.0%	99.0–100.5%

10 Typical Properties

Acidity/alkalinity pH = 5.8–6.9 (1% w/v aqueous solution).

Acid value ≤2.5

Critical micelle concentration 0.11% w/v aqueous solution at 25°C.

Density 1.16 g/cm^3

Hydroxyl value 6.0–8.0

Interfacial tension In water versus mineral oil at 25°C, *see* Table III.

Table III: Interfacial tension of docusate sodium.

Concentration (% w/v)	Interfacial tension (mN/m)
0.01	20.7
0.1	5.9
1.0	1.84

Iodine number ≤0.25

Melting point 153–157°C

Moisture content 1.51%

Saponification value 240–253

Solubility *see* Table IV.

Spectroscopy

IR spectra *see* Figure 1.

NIR spectra *see* Figure 2.

Surface tension *see* Table V.

11 Stability and Storage Conditions

Docusate sodium is stable in the solid state when stored at room temperature. Dilute aqueous solutions of docusate sodium between pH 1–10 are stable at room temperature. However, at very low pH (<1) and very high pH (>10) docusate sodium solutions are subject to hydrolysis.

The solid material is hygroscopic and should be stored in an airtight container in a cool, dry place.

12 Incompatibilities

Electrolytes, e.g. 3% sodium chloride, added to aqueous solutions of docusate sodium can cause turbidity.[2,3] However, docusate sodium possesses greater tolerance to calcium, magnesium, and

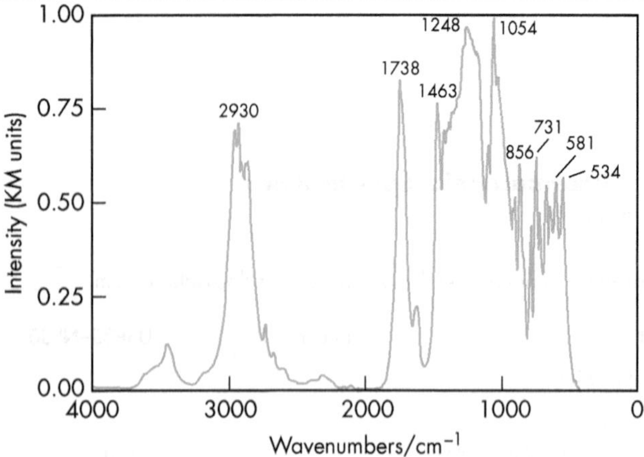

Figure 1: Infrared spectrum of docusate sodium measured by diffuse reflectance. Adapted with permission of Informa Healthcare.

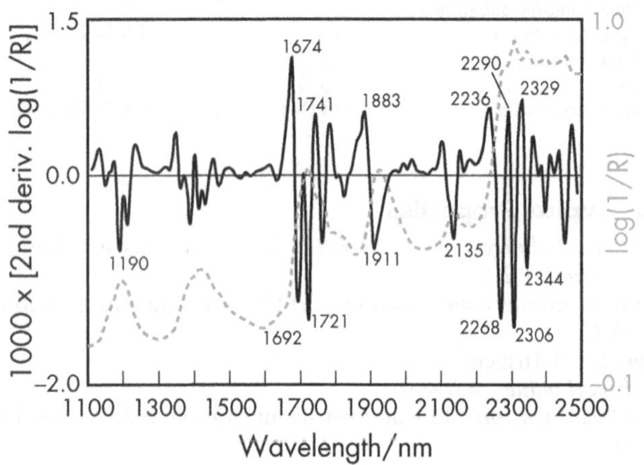

Figure 2: Near-infrared spectrum of docusate sodium measured by reflectance.

Table IV: Solubility of docusate sodium.

Solvent	Solubility at 20°C unless otherwise stated
Acetone	Soluble
Chloroform	1 in 1
Ethanol (95%)	1 in 3
Ether	1 in 1
Glycerin	Freely soluble
Vegetable oils	Soluble
Water	1 in 70 at 25°C[a]
	1 in 56 at 30°C
	1 in 44 at 40°C
	1 in 33 at 50°C
	1 in 25 at 60°C
	1 in 18 at 70°C

(a) In water, higher concentrations form a thick gel.

Table V: Surface tension of docusate sodium.

Concentration in water at 25°C (% w/v)	Surface tension (mN/m)
0.001	62.8
0.1	28.7
1.0	26.0

other polyvalent ions than do some other surfactants. Docusate sodium is incompatible with acids at pH <1 and with alkalis at pH >10.

13 Method of Manufacture

Maleic anhydride is treated with 2-ethylhexanol to produce dioctyl maleate, which is then reacted with sodium bisulfite.

14 Safety

Docusate salts are used in oral formulations as therapeutic agents for their fecal softening and laxative properties. As a laxative in adults, up to 500 mg of docusate sodium is administered daily in divided doses; in children over 6 months old, up to 75 mg in divided doses is used. The quantity of docusate sodium used as an excipient in oral formulations should therefore be controlled to avoid unintended laxative effects.[4] Adverse effects with docusate sodium include diarrhea, nausea, vomiting, abdominal cramps, and skin rashes. As with the chronic use of laxatives, the excessive use of docusate sodium may produce hypomagnesemia.[5]

Docusate salts are absorbed from the gastrointestinal tract and excreted in bile; they may cause alteration of the gastrointestinal epithelium.[6,7] The gastrointestinal or hepatic absorption of other drugs may also be affected by docusate salts, enhancing activity and possibly toxicity. Docusate sodium should not be administered with mineral oil as it may increase the absorption of the oil.

LD_{50} (mouse, IV): 0.06 g/kg[8]
LD_{50} (mouse, oral): 2.64 g/kg
LD_{50} (rat, IP): 0.59 g/kg
LD_{50} (rat, oral): 1.9 g/kg

15 Handling Precautions

Observe normal precautions appropriate to the circumstances and quantity of material handled. Docusate sodium may be irritant to the eyes and skin, and when inhaled. Eye protection, gloves, and a dust mask or respirator are recommended. When heated to decomposition, docusate sodium emits toxic fumes.

16 Regulatory Status

GRAS listed. Included in the FDA Inactive Ingredients Database (IM injections; oral capsules, suspensions, and tablets; topical formulations). Included in nonparenteral medicines licensed in the UK. Included in the Canadian Natural Health Products Ingredients Database.

17 Related Substances

Docusate calcium; docusate potassium.

Docusate calcium

Empirical formula $C_{40}H_{74}CaO_{14}S_2$
Molecular weight 883.23
CAS number [128-49-4]
Synonyms 1,4-Bis(2-ethylhexyl) sulfosuccinate, calcium salt; dioctyl calcium sulfosuccinate.
Appearance White amorphous solid with a characteristic octanol-like odor.
Solubility Soluble 1 in less than 1 of ethanol (95%), chloroform, and ether, and 1 in 3300 of water; very soluble in corn oil and polyethylene glycol 400.

Docusate potassium

Empirical formula $C_{20}H_{37}KO_7S$
Molecular weight 460.67
CAS number [7491-09-0]
Synonyms Dioctyl potassium sulfosuccinate; potassium 1,4-bis(2-ethylhexyl) sulfosuccinate.
Appearance White amorphous solid with a characteristic octanol-like odor.
Solubility Soluble in ethanol (95%) and glycerin; sparingly soluble in water.

18 Comments

A convenient way of making a 1% w/v aqueous solution of docusate sodium is to add 1 g of solid to about 50 mL of water and to apply gentle heat. The docusate sodium dissolves in a short time and the resulting solution can be made up to 100 mL with water. Alternatively, 1 g may be soaked overnight in 50 mL of water and the additional water may then be added with gentle heating and stirring.

Docusate sodium may alter the dissolution characteristics of certain dosage forms and the bioavailability of some drugs.

Therapeutically, docusate salts are used in oral formulations as laxatives and fecal softeners for the short-term relief of constipation.

A specification for docusate sodium is contained in the *Food Chemicals Codex* (FCC).[9]

The EINECS number for docusate sodium is 209-406-4. The PubChem Compound ID (CID) for docusate sodium is 23673837.

19 Specific References

1 Brown S, *et al.* Surface treatment of the hydrophobic drug danazol to improve drug dissolution. *Int J Pharm* 1998; **165**: 227–237.
2 Ahuja S, Cohen J. Dioctyl sodium sulfosuccinate. In: Florey K, ed. *Analytical Profiles of Drug Substances*. 2. New York: Academic Press, 1973: 199–219.
3 Ahuja S, Cohen J. Dioctyl sodium sulfosuccinate. In: Florey K, ed. *Analytical Profiles of Drug Substances*. 12. New York: Academic Press, 1983: 713–720.
4 Guidott JL. Laxative components of a generic drug [letter]. *Lancet* 1996; **347**: 621.
5 Rude RK, Siger FR. Magnesium deficiency and excess. *Annu Rev Med* 1981; **32**: 245–259.
6 Chapman RW, *et al.* Effect of oral dioctyl sodium sulfosuccinate on intake–output studies of human small and large intestine. *Gastroenterology* 1985; **89**: 489–493.
7 Moriarty KJ, *et al.* Studies on the mechanism of action of dioctyl sodium sulfosuccinate in the human jejunum. *Gut* 1985; **26**: 1008–1013.
8 Lewis RJ, ed. *Sax's Dangerous Properties of Industrial Materials*, 11th edn. New York: Wiley, 2004: 1274.
9 *Food Chemicals Codex*, 7th edn. Bethesda, MD: United States Pharmacopeia, 2010: 313.

20 General References

Chambliss WG, *et al.* Effect of docusate sodium on drug release from a controlled release dosage form. *J Pharm Sci* 1981; **70**: 1248–1251.

Hogue DR, *et al.* High-performance liquid chromatographic analysis of docusate sodium in soft gelatin capsules. *J Pharm Sci* 1992; **81**: 359–361.

Shah DN, *et al.* Effect of the pH-zero point of charge relationship on the interaction of ionic compounds and polyols with aluminum hydroxide gel. *J Pharm Sci* 1982; **71**: 266–268.

21 Author

S Murdande.

22 Date of Revision

1 March 2012.

Dodecyl Gallate

1 Non-proprietary Names

BP: Dodecyl Gallate

PhEur: Dodecyl Gallate

2 Synonyms

Benzoic acid, 3,4,5-trihydroxy-, dodecyl ester; dodecylis gallas; E312; gallic acid, dodecyl ester; gallic acid, lauryl ester; lauryl gallate; *Nipagallin LA*; *Progallin LA*.

3 Chemical Name and CAS Registry Number

Dodecyl 3,4,5-trihydroxybenzoate [1166-52-5]

4 Empirical Formula and Molecular Weight

$C_{19}H_{30}O_5$ 338.4

5 Structural Formula

6 Functional Category

Antioxidant.

7 Applications in Pharmaceutical Formulation or Technology

Dodecyl gallate and other alkyl gallates have become widely used as antioxidant preservatives in cosmetics, perfumes, foods, and pharmaceuticals since the use of propyl gallate in preventing autoxidation of oils was first described in 1943.[1–3]

Dodecyl gallate and other alkyl gallates are used as antioxidants in foods, at concentrations of 0.001 to 0.1% w/v, to prevent deterioration and rancidity of fats and oils.[4] They are often used with other antioxidants such as butylated hydroxyanisole or butylated hydroxytoluene, and with sequestrants such as citric acid and zinc salts, to improve acceptability and efficacy.

Dodecyl gallate has been investigated for its effect on the oxidation stability of ergocalciferol in ointments.[5]

8 Description

Dodecyl gallate is a white or almost white, odorless or almost odorless, crystalline powder.

9 Pharmacopeial Specifications

See Table I.

Table I: Pharmacopeial specifications for dodecyl gallate.

Test	PhEur 7.4
Characters	+
Identification	+
Chlorides	≤100 ppm
Heavy metals	≤10 ppm
Loss on drying	≤0.5%
Sulfated ash	≤0.1%
Impurities	+
Assay (dried substance)	97.0–103.0%

D

10 Typical Properties

Boiling point 521.7°C at 760 mmHg
Density 1.112 g/cm^3
Flash point 180.3°C
Melting point 96–97.5°C after drying for 4 hours at 60°C.
Solubility *see* Table II.

Table II: Solubility of dodecyl gallate.

Solvent	Solubility at 20°C
Acetone	1 in 2
Chloroform	1 in 60
Ethanol (95%)	1 in 3.5
Ether	1 in 4
Methanol	1 in 1.5
Peanut oil	1 in 30
Propylene glycol	1 in 60
Water	Practically insoluble

11 Stability and Storage Conditions

Store in a non-metallic container. Protect from light.

12 Incompatibilities

Dodecyl gallate and other alkyl gallates are incompatible with metals, e.g. sodium, potassium and iron, forming intensely colored complexes. Complex formation may be prevented, under some circumstances, by the addition of a sequestering agent, typically citric acid. Dodecyl gallate also reacts with strong oxidizing agents and strong bases.

13 Method of Manufacture

Dodecyl gallate is prepared by the esterification of gallic acid with dodecanol.[6]

14 Safety

Dodecyl gallate is widely used in food products and is generally regarded as a non-toxic and non-irritant material. Studies of experimental sensitization in guinea pigs indicated that, of the alkyl gallates tested, dodecyl gallate was the strongest sensitizer.[7]

LD_{50} (pig, oral): >6 g/kg[8]
LD_{50} (rat, IP): 0.1 g/kg
LD_{50} (rat, oral): 5 g/kg

The WHO has established a temporary estimated acceptable daily intake for dodecyl gallate at up to 0.05 mg/kg body weight.[9]

15 Handling Precautions

Observe normal precautions appropriate to the circumstances and quantity of the material handled. Dodecyl gallate is a skin irritant.[10] When heated to decomposition, dodecyl gallate may emit acrid smoke and irritating fumes. Eye protection and gloves are recommended.

16 Regulatory Status

Accepted for use as a food additive in Europe. Included in the Canadian Natural Health Products Ingredients Database, which has set a limit of up to 0.2 mg/kg body weight daily as the sum of dodecyl, octyl, and propyl gallate.

17 Related Substances

Octyl gallate; propyl gallate.

18 Comments

Dodecyl gallate has been shown to possess antiviral activity as well as antibacterial activity against Gram-positive bacteria, including methicillin-resistant *Staphylococcus aureus* (MRSA) strains.[11,12]

The EINECS number for dodecyl gallate is 214-620-6. The PubChem Compound ID (CID) for dodecyl gallate is 14425.

19 Specific References

1 Boehm E, Williams R. The action of propyl gallate on the autoxidation of oils. *Pharm J* 1943; **151**: 53.
2 Boehm E, Williams R. A study of the inhibiting actions of propyl gallate (normal propyl trihydroxybenzoate) and certain other trihydric phenols on the autoxidation of animal and vegetable oils. *Chemist Drug* 1943; **140**: 146–147.
3 Kubo I, *et al.* Antioxidant activity of dodecyl gallate. *J Agric Food Chem* 2002; **50**(12): 3533–3539.
4 Sweetman SC, ed. *Martindale: The Complete Drug Reference*, 36th edn. London: Pharmaceutical Press, 2009: 1628.
5 Yamaoka K, *et al.* Effect of ointments and ointment constituents on oxidation stability of ergocalciferol. *Jpn J Hosp Pharm* 1983; **9**(1): 57–63.
6 Sas B *et al.* Method of crystallizing and purifying alkyl gallates. United States Patent 6,297,396; 2001.
7 Hausen BM, Beyer W. The sensitizing capacity of the antioxidants propyl, octyl, and dodecyl gallate and some related gallic acid esters. *Contact Dermatitis* 1992; **26**(4): 253–258.
8 Lewis RJ, ed. *Sax's Dangerous Properties of Industrial Materials*, 11th edn. New York: Wiley, 2004; 1547.
9 FAO/WHO. Evaluation of certain food additives and contaminants. Forty-sixth report of the joint FAO/WHO expert committee on food additives. *World Health Organ Tech Rep Ser* 1997; No. 868.
10 Van der Meeren HLM. Dodecyl gallate, permitted in food, is a strong sensitizer. *Contact Dermatitis* 1987; **16**(5): 260–262.
11 Hurtado C, *et al.* Antiviral activity of lauryl gallate against animal viruses. *Antivir Ther* 2008; **13**(7): 909–917.
12 Kubo I, *et al.* Anti-MRSA activity of alkyl gallates. *Bioorg Med Chem Lett* 2002; **12**: 113–116.

20 General References

Changsha Industrial Products & Minerals I/E Co. Ltd. Product data sheet: Dodecyl gallate, October 2008.
Johnson DM, Gu LC. Autoxidation and antioxidants. In: Swarbrick J, Boylan JC, eds. *Encyclopedia of Pharmaceutical Technology*, Vol. 1. New York: Marcel Dekker, 1988; 415–449.

21 Author

RT Guest.

22 Date of Revision

1 March 2012.

Edetic Acid

1 Nonproprietary Names

BP: Edetic Acid
PhEur: Edetic Acid
USP–NF: Edetic Acid

2 Synonyms

Acidum edeticum; *Dissolvine*; edathamil; EDTA; ethylenediaminetetraacetic acid; (ethylenedinitrilo)tetraacetic acid; *Sequestrene AA*; tetracemic acid; *Versene Acid*.

3 Chemical Name and CAS Registry Number

N,N-1,2-Ethanediylbis[*N*-(carboxymethyl)glycine] [60-00-4]

4 Empirical Formula and Molecular Weight

$C_{10}H_{16}N_2O_8$ 292.24

5 Structural Formula

HO—C(=O)—CH₂ and HO—C(=O)—CH₂ on N; N—CH₂—CH₂—N; N bearing CH₂—C(=O)—OH and CH₂—C(=O)—OH

6 Functional Category

Antimicrobial preservative; complexing agent.

7 Applications in Pharmaceutical Formulation or Technology

Edetic acid and edetate salts are used in pharmaceutical formulations, cosmetics, and foods as complexing agents. They form stable water-soluble complexes (chelates) with alkaline earth and heavy metal ions. The complexed form has few of the properties of the free ion, and for this reason complexing agents are often described as 'removing' ions from solution; this process is also called sequestering. The stability of the metal–edetate complex depends on the metal ion involved and also on the pH. The calcium complex is relatively weak and will preferentially exchange calcium for heavy metals, such as iron, copper, and lead, with the release of calcium ions.

Edetic acid and edetates are primarily used as antioxidant synergists, sequestering trace amounts of metal ions, particularly copper, iron, and manganese, that might otherwise catalyze autoxidation reactions. Edetic acid and edetates may be used alone or in combination with true antioxidants, the usual concentration employed being in the range 0.005–0.1% w/v. Edetates have been used to stabilize ascorbic acid; corticosteroids; epinephrine; folic acid; formaldehyde; gums and resins; hyaluronidase; hydrogen peroxide; oxytetracycline; penicillin; salicylic acid; and unsaturated fatty acids. Essential oils may be washed with a 2% w/v solution of edetate to remove trace metal impurities.

Edetic acid and edetates possess some antimicrobial activity but are most frequently used in combination with other antimicrobial preservatives owing to their synergistic effects. Many solutions used for the cleaning, storage, and wetting of contact lenses contain disodium edetate. Typically, edetic acid and edetates are used in concentrations of 0.01–0.1% w/v as antimicrobial preservative synergists; *see* Section 10.

Edetic acid and disodium edetate may also be used as water softeners since they will chelate the calcium and magnesium ions present in hard water; edetate calcium disodium is not effective. Many cosmetic and toiletry products, e.g. soaps, contain edetic acid as a water softener.

E

8 Description

Edetic acid occurs as a white crystalline powder.

9 Pharmacopeial Specifications

See Table I. *See also* Section 17.

Table I: Pharmacopeial specifications for edetic acid.

Test	PhEur 7.4	USP35–NF30
Identification	+	+
Characters	+	—
Appearance of solution	+	—
Residue on ignition	—	≤0.2%
Sulfated ash	≤0.2%	—
Heavy metals	≤20 ppm	≤0.003%
Nitrilotriacetic acid	≤0.1%	≤0.3%
Iron	≤80 ppm	≤0.005%
Chloride	≤200 ppm	—
Assay	98.0–101.0%	98.0–100.5%

10 Typical Properties

Acidity/alkalinity pH = 2.2 for a 0.2% w/v aqueous solution.

Antimicrobial activity Edetic acid has some antimicrobial activity against Gram-negative microorganisms, *Pseudomonas aeruginosa*, some yeasts, and fungi although this activity is insufficient for edetic acid to be used effectively as an antimicrobial preservative on its own.[1,2] However, when used with other antimicrobial preservatives, edetic acid demonstrates a marked synergistic effect in its antimicrobial activity. Edetic acid and edetates are therefore frequently used in combination with such preservatives as benzalkonium chloride; bronopol; cetrimide; imidurea; parabens; and phenols, especially chloroxylenol. Typically, edetic acid is used at a concentration of 0.1–0.15% w/v. In the presence of some divalent metal ions, such as Ca^{2+} or Mg^{2+}, the synergistic effect may be reduced or lost altogether. The addition of disodium edetate to phenylmercuric nitrate[3] and thimerosal[3,4] has also been reported to reduce the antimicrobial efficacy of the preservative. Edetic acid and iodine form a colorless addition compound that is bactericidal.

Dissociation constant

$pK_{a1} = 2.00$;
$pK_{a2} = 2.67$;
$pK_{a3} = 6.16$;
$pK_{a4} = 10.26$.

Melting point Melts above 220°C, with decomposition.

Solubility Soluble in solutions of alkali hydroxides; soluble 1 in 500 of water.

Spectroscopy

NIR spectra *see* Figure 1.

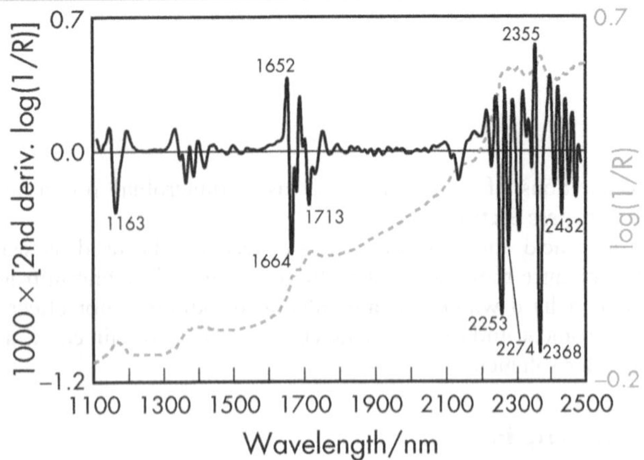

Figure 1: Near-infrared spectrum of edetic acid measured by reflectance.

11 Stability and Storage Conditions

Although edetic acid is fairly stable in the solid state, edetate salts are more stable than the free acid, which decarboxylates if heated above 150°C. Disodium edetate dihydrate loses water of crystallization when heated to 120°C. Edetate calcium disodium is slightly hygroscopic and should be protected from moisture.

Aqueous solutions of edetic acid or edetate salts may be sterilized by autoclaving, and should be stored in an alkali-free container.

Edetic acid and edetates should be stored in well-closed containers in a cool, dry place.

12 Incompatibilities

Edetic acid and edetates are incompatible with strong oxidizing agents, strong bases, and polyvalent metal ions such as copper, nickel, and copper alloy.

Edetic acid and disodium edetate behave as weak acids, displacing carbon dioxide from carbonates and reacting with metals to form hydrogen.

Other incompatibilities include the inactivation of certain types of insulin due to the chelation of zinc, and the chelation of trace metals in total parenteral nutrition (TPN) solutions following the addition of TPN additives stabilized with disodium edetate. Calcium disodium edetate has also been reported to be incompatible with amphotericin and with hydralazine hydrochloride in infusion fluids.

13 Method of Manufacture

Edetic acid may be prepared by the condensation of ethylenediamine with sodium monochloroacetate in the presence of sodium carbonate. An aqueous solution of the reactants is heated to about 90°C for 10 hours, then cooled, and hydrochloric acid is added to precipitate the edetic acid.

Edetic acid may also be prepared by the reaction of ethylenediamine with hydrogen cyanide and formaldehyde with subsequent hydrolysis of the tetranitrile, or under alkaline conditions with continuous extraction of ammonia.

See Section 17 for information on the preparation of edetate salts.

14 Safety

Edetic acid and edetates are widely used in topical, oral, and parenteral pharmaceutical formulations. They are also extensively used in cosmetics and food products.

Edetic acid is generally regarded as an essentially nontoxic and nonirritant material, although it has been associated with dose-related bronchoconstriction when used as a preservative in nebulizer solutions. It has therefore been recommended that nebulizer solutions for bronchodilation should not contain edetic acid.[5]

Edetates, particularly disodium edetate and edetate calcium disodium, are used in a greater number and variety of pharmaceutical formulations than the free acid.

Disodium edetate, trisodium edetate, and edetic acid readily chelate calcium and can, in large doses, cause calcium depletion (hypocalcemia) if used over an extended period or if administered too rapidly by intravenous infusion. If used in preparations for the mouth, they can also leach calcium from the teeth. In contrast, edetate calcium disodium does not chelate calcium.

Edetate calcium disodium is nephrotoxic and should be used with caution in patients with renal impairment.

The WHO has set an estimated acceptable daily intake for disodium edetate in foodstuffs at up to 2.5 mg/kg body-weight.[6]

See also Section 18.

LD_{50} (mouse, IP): 0.25 g/kg[7]

LD_{50} (rat, IP): 0.397 g/kg

15 Handling Precautions

Observe normal precautions appropriate to the circumstances and quantity of material handled. Edetic acid and edetates are mildly irritant to the skin, eyes, and mucous membranes. Ingestion, inhalation, and contact with the skin and eyes should therefore be avoided. Eye protection, gloves, and a dust mask are recommended.

16 Regulatory Status

Included in the FDA Inactive Ingredients Database (oral, otic, rectal, and topical preparations; submucosal injection preparations). Included in nonparenteral medicines licensed in the UK. Included in the Canadian Natural Health Products Ingredients Database.

See also Section 17.

17 Related Substances

Dipotassium edetate; disodium edetate; edetate calcium disodium; sodium edetate; trisodium edetate.

Dipotassium edetate

Empirical formula $C_{10}H_{14}K_2N_2O_8$

Molecular weight 368.46

CAS number [2001-94-7]

Synonyms Dipotassium edathamil; dipotassium ethylenediaminetetraacetate; edathamil dipotassium; edetate dipotassium; edetic acid dipotassium salt; EDTA dipotassium; *N,N'*-1,2-ethanediylbis[*N*-(carboxymethyl)glycine] dipotassium salt; ethylenebis(iminodiacetic acid) dipotassium salt; ethylenediaminetetraacetic acid dipotassium salt; (ethylenedinitrilo)tetraacetic acid dipotassium salt; tetracemate dipotassium.

Appearance White crystalline powder.

Comments The EINECS number for dipotassium edetate is 217-895-0.

Edetate calcium disodium

Empirical formula $C_{10}H_{12}CaN_2Na_2O_8$

Molecular weight 374.28

CAS number [62-33-9] for the anhydrous material and [23411-34-9] for the dihydrate

Synonyms Calcium disodium edetate; calcium disodium ethylenediaminetetraacetate; calcium disodium (ethylenedinitrilo)tetraacetate; E385; edathamil calcium disodium; edetic acid calcium disodium salt; EDTA calcium; ethylenediaminetetraacetic acid calcium disodium chelate; [(ethylenedinitrilo)tetraacetato]calciate(2-) disodium; sodium calciumedetate; *Versene CA.*

Appearance White or creamy-white colored, slightly hygroscopic, crystalline powder or granules; odorless, or with a slight odor; tasteless, or with a faint saline taste.

Acidity/alkalinity pH = 4–5 for a 1% w/v aqueous solution.
Density (bulk) 0.69 g/cm^3
Solubility Practically insoluble in chloroform, ether, and other organic solvents; very slightly soluble in ethanol (95%); soluble 1 in 2 of water.
Method of manufacture Edetate calcium disodium may be prepared by the addition of calcium carbonate to a solution of disodium edetate.
Safety *see also* Section 14.

LD_{50} (mouse, IP): 4.5 g/kg[7]
LD_{50} (rabbit, IP): 6 g/kg
LD_{50} (rabbit, oral): 7 g/kg
LD_{50} (rat, IP): 3.85 g/kg
LD_{50} (rat, IV): 3.0 g/kg
LD_{50} (rat, oral): 10 g/kg

Regulatory status GRAS listed. Accepted for use as a food additive in Europe. Included in the FDA Inactive Ingredients Database (injections; oral capsules, solutions, suspensions, syrups, and tablets).
Comments Used in pharmaceutical formulations as a chelating agent in concentrations between 0.01–0.1% w/v. Usually edetate calcium disodium is used in pharmaceutical formulations in preference to disodium edetate or sodium edetate to prevent calcium depletion occurring in the body. In food products, edetate calcium disodium may also be used in flavors and as a color retention agent. Edetate calcium disodium occurs as the dihydrate, trihydrate, and anhydrous material.

Some pharmacopeias specify that edetate calcium disodium is the dihydrate, others that it is the anhydrous material. The USP35–NF30 specifies that edetate calcium disodium is a mixture of the dihydrate and trihydrate but that the dihydrate predominates. Edetate calcium disodium is one of the materials that have been selected for harmonization by the Pharmacopeial Discussion Group. For further information see the General Information Chapter <1196> in the USP35–NF30, the General Chapter 5.8 in PhEur 7.4, along with the 'State of Work' document on the PhEur EDQM website, and also the General Information Chapter 8 in the JP XV.

The EINECS number for edetate calcium disodium is 200-529-9.

Sodium edetate
Empirical formula $C_{10}H_{12}N_2Na_4O_8$
Molecular weight 380.20
CAS number [64-02-8]
Synonyms Edetate sodium; edetic acid tetrasodium salt; EDTA tetrasodium; *N,N'*-1,2-ethanediylbis[*N*-(carboxymethyl)glycine] tetrasodium salt; ethylenebis(iminodiacetic acid) tetrasodium salt; ethylenediaminetetraacetic acid tetrasodium salt; (ethylenedinitrilo)tetraacetic acid tetrasodium salt; *Sequestrene NA4*; tetracemate tetrasodium; tetracemin; tetrasodium edetate; tetrasodium ethylenebis(iminodiacetate); tetrasodium ethylenediaminetetraacetate; *Versene*.
Appearance White crystalline powder.
Acidity/alkalinity pH = 11.3 for a 1% w/v aqueous solution.
Melting point >300°C
Solubility Soluble 1 in 1 of water.
Safety *see also* Section 14.

LD_{50} (mouse, IP): 0.33 g/kg[7]

Regulatory status Included in the FDA Inactive Ingredients Database (inhalations; injections; ophthalmic preparations; oral capsules and tablets; and topical preparations).
Comments Sodium edetate reacts with most divalent and trivalent metallic ions to form soluble metal chelates and is used in pharmaceutical formulations in concentrations between 0.01–0.1% w/v.

Trisodium edetate
Empirical formula $C_{10}H_{13}N_2Na_3O_8$
Molecular weight 358.20
CAS number [150-38-9]
Synonyms Edetate trisodium; edetic acid trisodium salt; EDTA trisodium; *N,N'*-1,2-ethanediylbis[*N*-(carboxymethyl)glycine] trisodium salt; ethylenediaminetetraacetic acid trisodium salt; (ethylenedinitrilo)tetraacetic acid trisodium salt; *Sequestrene NA3*; trisodium ethylenediaminetetraacetate; *Versene-9*.
Appearance White crystalline powder.
Acidity/alkalinity pH = 9.3 for a 1% w/v aqueous solution.
Melting point >300°C
Method of manufacture Trisodium edetate may be prepared by adding a solution of sodium hydroxide to disodium edetate.
Safety *see also* Section 14.

LD_{50} (mouse, IP): 0.3 g/kg[7]
LD_{50} (mouse, oral): 2.15 g/kg
LD_{50} (rat, oral): 2.15 g/kg

Regulatory status Included in the FDA Inactive Ingredients Database (topical preparations).
Comments More soluble in water than either the disodium salt or the free acid. Trisodium edetate also occurs as the monohydrate and is used in pharmaceutical formulations as a chelating agent. The EINECS number for trisodium edetate is 205-758-8.

18 Comments

Other salts of edetic acid that are commercially available include diammonium, dimagnesium, ferric sodium, and magnesium disodium edetates. Therapeutically, a dose of 50 mg/kg body-weight of disodium edetate, as a slow infusion over a 24-hour period, with a maximum daily dose of 3 g, has been used as a treatment for hypercalcemia. For the treatment of lead poisoning, a dose of 60–80 mg/kg of edetate calcium disodium, as a slow infusion in two daily doses, for 5 days, has been used.

Chelation therapy using edetic acid has been widely used for the treatment of ischemic heart disease. However, it has been suggested that the therapeutic benefits of this treatment may be due to the changes in lifestyle of the patient rather than the administration of edetic acid (40 mg/kg by infusion over a 3-hour period).[8]

The EINECS number for edetic acid is 200-449-4.

19 Specific References

1 Richards RME, Cavill RH. Electron microscope study of effect of benzalkonium chloride and edetate disodium on cell envelope of *Pseudomonas aeruginosa*. *J Pharm Sci* 1976; **65**: 76–80.
2 Whalley G. Preservative properties of EDTA. *Manuf Chem* 1991; **62**(9): 22–23.
3 Richards RME, Reary JME. Changes in antibacterial activity of thiomersal and PMN on autoclaving with certain adjuvants. *J Pharm Pharmacol* 1972; **24**(Suppl.): 84P–89P.
4 Morton DJ. EDTA reduces antimicrobial efficacy of thiomerosal. *Int J Pharm* 1985; **23**: 357–358.
5 Beasley CRW, *et al.* Bronchoconstrictor properties of preservatives in ipratropium bromide (Atrovent) nebuliser solution. *Br Med J* 1987; **294**: 1197–1198.
6 FAO/WHO. Toxicological evaluation of certain food additives with a review of general principles and of specifications. Seventeenth report of the joint FAO/WHO expert committee on food additives. *World Health Organ Tech Rep Ser* 1974; No. 539.
7 Lewis RJ, ed. *Sax's Dangerous Properties of Industrial Materials*, 11th edn. New York: Wiley, 2004; 1660.
8 Knudtson ML, *et al.* Chelation therapy for ischemic heart disease: a randomized controlled trial. *J Am Med Assoc* 2002; **287**(4): 481–486.

20 General References

Chalmers L. The uses of EDTA and other chelates in industry. *Manuf Chem* 1978; **49**(3): 79–8083.
European Directorate for the Quality of Medicines and Healthcare (EDQM). European Pharmacopoeia – State Of Work Of International

Harmonisation. *Pharmeuropa* 2011; 23(4): 713–714 www.edqm.eu/site/-614.html (accessed 30 November 2011).
Hart JR. Chelating agents in cosmetic and toiletry products. *Cosmet Toilet* 1978; 93(12): 28–30.
Hart JR. EDTA-type chelating agents in personal care products. *Cosmet Toilet* 1983; 98(4): 54–58.
Lachman L. Antioxidants and chelating agents as stabilizers in liquid dosage forms. *Drug Cosmet Ind* 1968; 102(2): 43–45146–149.

21 Author

W Cook.

22 Date of Revision

1 March 2012.

Erythorbic Acid

1 Nonproprietary Names

USP–NF: Erythorbic Acid

2 Synonyms

Araboascorbic acid; D-araboascorbic acid; 2,3-D-didehydro-D-*erythro*-hexono-1,4-lactone; E315; erycorbin; D-erythorbic acid; D-*erythro*-ascorbic acid; D-*erythro*-hex-2-enoic acid; D-*erythro*-3-ketohexonic acid lactone; D-*erythro*-3-oxohexonic acid lactone; glucosaccharonic acid; D-isoascorbic acid; isovitamin C; γ-lactone; mercate '5'; neo-cebicure; saccharosonic acid.

3 Chemical Name and CAS Registry Number

D-Isoascorbic acid [89-65-6]

4 Empirical Formula and Molecular Weight

$C_6H_8O_6$ 176.12

5 Structural Formula

6 Functional Category

Antioxidant.

7 Applications in Pharmaceutical Formulation or Technology

Erythorbic acid is used as an antioxidant in pharmaceutical formulations.

8 Description

Erythorbic acid occurs as shiny, granular, white or slightly yellow-colored crystals or powder. It gradually darkens in color upon exposure to light.

9 Pharmacopeial Specifications

See Table I.

Table I: Pharmacopeial specifications for erythorbic acid.

Test	USP35–NF30
Identification	+
Specific rotation	−16.5° to −18.0°
Loss on drying	≤0.4%
Residue on ignition	≤0.3%
Lead	≤10 ppm
Assay (dried basis)	99.0–100.5%

10 Typical Properties

Acidity/alkalinity pH = 2.1 (10% w/v aqueous solution at 25°C)
pH = 5–6 (16% w/v aqueous solution at 25°C)
Density (bulk) 0.704 g/cm^3
Melting point 164–172°C with decomposition at 184°C
Partition coefficient Log P (octanol : water) = −1.880
Solubility see Table II.

Table II: Solubility of erythorbic acid.

Solvent	Solubility at 25°C unless otherwise stated
Acetone	1 in 70
Ethanol (95%)	1 in 20
Ether	Practically insoluble
Glycerol	Slightly soluble
Methanol	1 in 5.5
Propylene glycol	1 in 6.7
Pyridine	Soluble
Water	1 in 2.3
	1 in 1.8 at 38°C
	1 in 1.6 at 50°C

11 Stability and Storage Conditions

Erythorbic acid should be stored in an airtight container, protected from light, in a cool, dry place.

12 Incompatibilities

Erythorbic acid is incompatible with chemically active metals such as aluminum, copper, magnesium, and zinc. It is also incompatible with strong bases and strong oxidizing agents.

13 Method of Manufacture

Erythorbic acid is synthesized by the reaction between methyl 2-keto-D-gluconate and sodium methoxide. It can also be synthesized from sucrose, and produced from *Penicillium* spp.

14 Safety

Erythorbic acid is widely used in oral pharmaceutical formulations and food applications as an antioxidant. Erythorbic acid is generally regarded as nontoxic and nonirritant when used as an excipient. The pure susbstance may cause skin and respiratory tract irritation, and serious eye irritation. It may also be harmful if swallowed or absorbed through the skin.

Erythorbic acid is readily metabolized and does not affect the urinary excretion of ascorbic acid. The WHO has set an acceptable daily intake of erythorbic acid and its sodium salt in foods at up to 5 mg/kg body-weight.[1]

LD_{50} (mouse, oral) = 8.3 g/kg[3]
LD_{50} (rat, oral) = 18.0 g/kg[3]

15 Handling Precautions

Observe normal precautions appropriate to the circumstances and quantity of material handled. When heated to decomposition, erythorbic acid emits acrid smoke and irritating fumes (carbon oxides). Avoid breathing dust or fumes. Wash skin thoroughly after handling. Use outdoors or in a well-ventilated area. Wear suitable protective clothing,

16 Regulatory Status

GRAS listed. Accepted for use as a food additive in Europe. Included in the FDA Inactive Ingredients Database (oral concentrate and tablets).

17 Related Substances

Ascorbic acid; sodium erythorbate

Sodium erythorbate
Empirical formula $C_6H_7NaO_6$
Molecular weight 198.11
CAS number [7378-23-6]
Synonyms E316; D-*erythro*-hex-2-enoic acid sodium salt; erythorbic acid sodium salt.
Acidity/alkalinity pH = 7.2–7.9 for 10% w/v aqueous solution.
Melting point 172°C
Solubility Soluble 1 in 6.5 of water. The sodium salt is less soluble in water than the free acid.
Comments The EINECS number for sodium erythorbate is 228-973-6.

18 Comments

Erythorbic acid is a stereoisomer of L-ascorbic acid, and is used as an antioxidant in oral pharmaceutical formulations. It has approximately 5% of the vitamin C activity of L-ascorbic acid.

Erythorbic acid is also used as an antioxidant in foods. It is used in cured meats to prevent nitrosamine formation from nitrites, and to accelerate color fixing.

A specification for erythorbic acid is included in the *Food Chemicals Codex* (FCC).[2]

The EINECS number for erythorbic acid is 201-928-0. The PubChem Compound ID (CID) for erythorbic acid is 6981.

19 Specific References

1 FAO/WHO. Toxicological evaluation of certain food additives with a review of general principles and of specifications: seventeenth report of the joint FAO/WHO expert committee on food additives. *World Health Organ Tech Rep Ser* 1974; **No. 539**: .
2 *Food Chemicals Codex*, 7th edn. Bethesda, MD: United States Pharmacopeia, 2010.
3 Andersen FA. Final report on the safety assessment of ascorbyl palmitate, ascorbyl dipalmitate, ascorbyl stearate, erythorbic acid, and sodium erythorbate. *Int J Toxicol* 1999; **18**(3): 1–26.

20 General References

Japan Pharmaceutical Excipients Council. *Japanese Pharmaceutical Excipients 2004*. Tokyo: Yakuji Nippo, 2004: 281–282.
Hazardous Substances Data Bank. US National Library of Medicine (2011). http://toxnet.nlm.nih.gov (accessed 23 November 2011).
Hawley GG. *The Condensed Chemical Dictionary*, 9th edn. New York: Van Nostrand Reinhold Co, 1977; 346.

21 Author

DK Chiappetta.

22 Date of Revision

1 March 2012.

Erythritol

1 Nonproprietary Names

BP: Erythritol
PhEur: Erythritol
USP–NF: Erythritol

2 Synonyms

Butane 1,2,3,4-tetrol; 1,2,3,4-butanetetrol; *C*Eridex*; E968; erythrit; erythrite; erythritolum; erythroglucin; erythrol; *meso*-erythritol; L-(-)-threitol;paycite;phycite; phycitol; tetrahydroxybutane; *Zerose*.

3 Chemical Name and CAS Registry Number

(2*R*,3*S*)-Butane 1,2,3,4-tetrol [149-32-6]

4 Empirical Formula and Molecular Weight

$C_4H_{10}O_4$ 122.12

5 Structural Formula

OH
HO OH
OH

6 Functional Category

Sweetening agent; tablet and capsule diluent; taste-masking agent.

7 Applications in Pharmaceutical Formulation or Technology

Erythritol is a naturally occurring noncariogenic excipient used in a variety of pharmaceutical preparations, including in solid dosage forms as a tablet filler,[1] and in coatings.[2,3] It has also been investigated for use in dry powder inhalers.[4,5] It is also used in sugar-free lozenges,[6,7] syrups, and medicated chewing gum.[6]

Erythritol can also be used as a diluent in wet granulation in combination with moisture-sensitive drugs.[8] Erythritol demonstrates superior powder flow behavior and good stability because of its low hygroscopicity, making it a suitable carrier for actives in sachets and capsules, and as a diluent in direct compression tableting.[8] In buccal applications, such as medicated chewing gums, it is used because of its high negative heat of solution which provides a strong cooling effect.[6]

Erythritol is also used as a noncaloric sweetener in syrups;[9] it is used to provide sensorial profile-modifying properties with intense sweeteners; and it is also used to mask unwanted aftertastes.[10]

Erythritol is also used as a noncariogenic sweetener in toothpastes and mouthwash solutions.

See Table I.

Table I: Uses of erythritol.

Use	Concentration (%)
Tablet filler and binder	30.0–90.0
Taste masking in solutions	0.5–3.0
Oral care products	5.0–10.0

8 Description

Erythritol is a sugar alcohol (polyol) that occurs as a white or almost white powder or granular or crystalline substance. It is pleasant tasting with a mild sweetness approximately 60–70% that of sucrose. It also has a high negative heat of solution that provides a strong cooling effect.

9 Pharmacopeial Specifications

See Table II. *See also* Section 18.

Table II: Pharmacopeial specifications for erythritol.

Test	PhEur 7.4	USP35–NF30
Identification	+	+
Melting point	119–122°C	119–123°C
Appearance of solution	+	—
Conductivity	≤ 20 μS/cm	≤ 20 μS/cm
Related substances	≤2.0%	≤2.0%
Lead	≤0.5 ppm	≤0.5 ppm
Water	≤0.5%	≤0.5%
Loss on drying	—	≤0.2%
Residue on ignition	—	≤0.1%
Microbial contamination	+	+
Bacterial endotoxins	+	—
Assay (anhydrous basis)	96.0–102.0%	96.0–102.0%

10 Typical Properties

Acidity/alkalinity pH = 5–7 at 25°C for a 5% w/v aqueous solution.

Boiling point 329–331°C

Caloric value 0.8 kJ/g

Density 1.45 g/cm^3

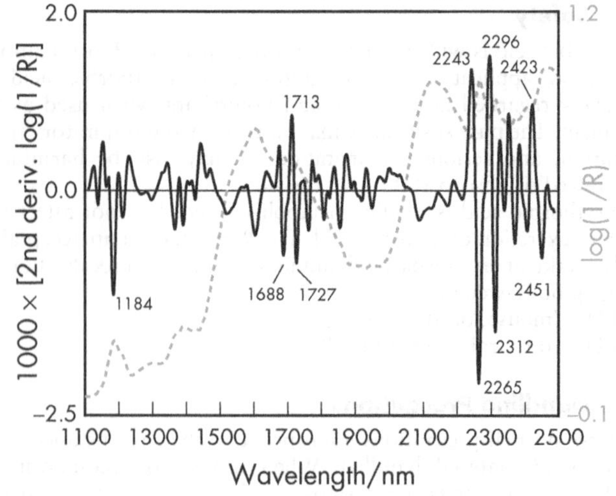

Figure 1: Near-infrared spectrum of erythritol measured by reflectance.

Dissociation constant pK_a = 13.90 at 18°C[11]

Glass transition temperature -42 to -46°C[12]

Heat of solution 23.3 kJ/mol[11]

Hygroscopicity Erythritol is nonhygroscopic; it absorbs approximately 1% w/w of water at 95% relative humidity (RH).[8]

Melting point 121.5°C, with decomposition at 160°C.

Solubility Freely soluble in water; slightly soluble in ethanol (95%); practically insoluble in ether and fats.

Spectroscopy

NIR spectra *see* Figure 1.

Viscosity (dynamic) 3 mPa s (3 cP) at 60°C for a 30% w/w solution.

11 Stability and Storage Conditions

Erythritol has very good thermal and chemical stability. It is nonhygroscopic, and at 25°C does not significantly absorb additional water (<0.5%) up to a relative humidity (RH) of more than 90%.[8] Erythritol resists decomposition both in acidic and alkaline media and remains stable for prolonged periods at pH 2–10.[13]

Store in a cool, dry place, away from sources of ignition. When stored for up to 4 years in ambient conditions (20°C, 50% RH) erythritol has been shown to be stable.[8]

12 Incompatibilities

Erythritol is incompatible with strong oxidizing agents and strong bases.

13 Method of Manufacture

Erythritol is a starch-derived product. The starch is enzymatically hydrolyzed into glucose which is turned into erythritol via a fermentation process, using osmophilic yeasts or fungi (e.g. *Moniliella pollinis*, or *Trichosporonoides megachiliensis*).[14]

14 Safety

Erythritol is used in oral pharmaceutical formulations, confectionery, and food products. It is generally regarded as a nontoxic, nonallergenic, noncarcinogenic, and nonirritant material.[15] However, there has been a case report of urticaria caused by erythritol.[16]

The low molecular weight of erythritol allows more than 90% of the ingested molecules to be rapidly absorbed from the small intestine;[17] it is not metabolized and is excreted unchanged in the urine. Erythritol has a low caloric value (0.8 kJ/g). The WHO has set an acceptable daily intake of 'not specified' for erythritol.[15]

Erythritol is noncariogenic; preliminary studies suggest that it may inhibit the formation of dental plaque.[18]

In general, erythritol is well-tolerated;[19–21] furthermore, excessive consumption does not cause laxative effects. There is no significant increase in the blood glucose level after oral intake, and glycemic response is very low, making erythritol suitable for diabetics.[8]

LD_{50} (dog, IV): 5 g/kg[22]
LD_{50} (mouse, IP): 8–9 g/kg ;[15,22]
LD_{50} (rat, IV): 6.6 g/kg (male)[19]
LD_{50} (rat, IV): 9.6 g/kg (female)[19]
LD_{50} (rat, oral): >13 g/kg[19]

15 Handling Precautions

Observe normal precautions appropriate to the circumstances and quantity of the material handled. Eye protection and a dust mask or respirator are recommended.

16 Regulatory Status

GRAS listed. Accepted for use as a food additive in Europe. Included in the FDA Inactive Ingredients Database (ophthalmic suspensions; oral tablets and sustained-action tablets; topical preparations).

17 Related Substances

Mannitol; sorbitol; xylitol.

18 Comments

Active ingredients can be granulated with erythritol and binders such as maltodextrin or carboxymethylcellulose, resulting in coarser granules with improved flowability.[5] Coprocessing erythritol with a small amount of maltodextrin results in a proprietary compound that may be used in direct compression.[23]

A specification for erythritol is included in the *Japanese Pharmaceutical Excipients* (JPE).[24]

The EINECS number for erythritol is 205-737-3. The PubChem Compound ID (CID) for erythritol is 8998.

19 Specific References

1 Bi YX, *et al.* Evaluation of rapidly disintegrating tablets prepared by a direct compression method. *Drug Dev Ind Pharm* 1999; **25**(5): 571–581.
2 Ohmori S, *et al.* Characteristics of erythritol and formulation of a novel coating with erythritol termed thin-layer sugarless coating. *Int J Pharm* 2004; **278**(2): 447–457.
3 Ohmori S, *et al.* Development and evaluation of the tablets coated with the novel formulation termed thin-layer sugarless coated tablets. *Int J Pharm* 2004; **278**(2): 459–469.
4 Endo K, *et al.* Erythritol-based dry powder of glucagons for pulmonary administration. *Int J Pharm* 2005; **290**: 63–71.
5 Traini D, *et al.* Comparative study of erythritol and lactose monohydrate as carriers for inhalation: atomic force microscopy and *in vitro* correlation. *Eur J Pharm Sci* 2006; **27**(2–3): 243–251.
6 Goossens J, Gonze M. Erythritol. *Manuf Confect* 2000; **80**(1): 71–75.
7 de Cock P. Chewing gum coating with a healthier crunch thanks to erythritol. *Confect Prod* 2003; **6**: 10–11.
8 Michaud J, Haest G. Erythritol: a new multipurpose excipient. *Pharmaceut Technol Eur* 2003; **15**(10): 69–72.
9 de Cock P. Erythritol: a novel noncaloric sweetener ingredient. In: Corti A, ed. *Low-Calorie Sweeteners: Present and Future.* Basel: Karger, 1999: 110–116.
10 de Cock P, Bechert CL. Erythritol. Functionality in noncaloric functional beverages. *Pure Appl Chem* 2002; **74**(7): 1281–1289.
11 Lawson ME. Sugar alcohols. *Kirk-Othmer Encyclopedia of Chemical Technology.* New York: John Wiley & Sons Inc, 2000.
12 Jesus AJL, *et al.* Erythritol – crystal growth from the melt. *Int J Pharm* 2010; **388**: 129–135.
13 Leutner C, ed. *Geigy Scientific Tables.* 1. Basel: Ciba Geigy, 1993; 84–85.
14 Goossens J, Gonze M. Nutritional and application properties of erythritol: a unique combination? Part I: nutritional and functional properties. *Agro Food Ind Hi-tech* 1997; **4**(8): 3–10.
15 FAO/WHO. Evaluation of certain food additives and contaminants. Thirty-fifth report of the joint FAO/WHO expert committee on food additives. *World Health Organ Tech Rep Ser* 2000; No. 896.
16 Hino H, *et al.* A case of allergic urticaria caused by erythritol. *J Dermatol* 2000; **27**(3): 163–165.
17 Bornet FRJ, *et al.* Plasma and urine kinetics of erythritol after oral ingestion by healthy humans. *Regul Toxicol Pharmacol* 1996; **24**: 280–286.
18 Gonze M, Goossens J. Nutritional and application properties of erythritol: a unique combination? Part II: application properties. *Agro Food Ind Hi-tech* 1997; **8**(5): 12–16.
19 Munro IC, *et al.* Erythritol: an interpretive summary of biochemical, metabolic, toxicologic and chemical data. *Food Chem Toxicol* 1998; **36**(12): 1139–1174.
20 Tetzloff W, *et al.* Tolerance to subchronic, high-dose ingestion of erythritol in human volunteers. *Regul Toxicol Pharmacol* 1996; 24(2Pt2): S286–S295.
21 Storey D, *et al.* Gastrointestinal tolerance of erythritol and xylitol ingested in a liquid. *Eur J Clin Nutr* 2007; **61**(3): 349–354.
22 Registry of Toxic Effects of Chemical Substances Database. http://ccinfoweb.ccohs.ca/rtecs/search.html (accessed 23 November 2011).
23 De Sadeleer J, Gonze M. Erythritol compositions. European Patent No. 0497439; 1992.
24 Japan Pharmaceutical Excipients Council. *Japanese Pharmaceutical Excipients 2004.* Tokyo: Yakuji Nippo, 2004; 283–284.

20 General References

Cargill Inc. Zerose. www.zerosesweetener.com (accessed 23 November 2011).

O'Brien Nabors L, Gelardi RC, eds. *Alternative Sweeteners.* New York: Marcel Dekker, 2001.

21 Author

S Mujumdar.

22 Date of Revision

1 March 2012.

Ethyl Acetate

1 Nonproprietary Names

BP: Ethyl Acetate
PhEur: Ethyl Acetate
USP–NF: Ethyl Acetate

2 Synonyms

Acetic acid ethyl ester; acetic ester; acetic ether; acetoxyethane; aethylis acetas; aethylium aceticum; ethyl ethanoate; ethylis acetas; vinegar naphtha.

3 Chemical Name and CAS Registry Number

Ethyl acetate [141-78-6]

4 Empirical Formula and Molecular Weight

$C_4H_8O_2$ 88.1

5 Structural Formula

H_3C–C(=O)–O–CH_2–CH_3

6 Functional Category

Flavoring agent; solvent.

7 Applications in Pharmaceutical Formulation or Technology

In pharmaceutical preparations, ethyl acetate is primarily used as a solvent, although it has also been used as a flavoring agent. As a solvent, it is included in topical solutions and gels, and in edible printing inks used for tablets.

Ethyl acetate has also been shown to increase the solubility of chlortalidone[1] and to modify the polymorphic crystal forms obtained for piroxicam pivalate,[2] mefenamic acid,[3] and fluconazole,[4] and has been used in the formulation of microspheres.[5–8] Ethyl acetate has been used as a solvent in the preparation of a liposomal amphotericin B dry powder inhaler formulation[9] and in the preparation of microparticles.[10] Its use as a chemical enhancer for the transdermal iontophoresis of insulin has been investigated.[11]

In food applications, ethyl acetate is mainly used as a flavoring agent. It is also used in artificial fruit essence and as an extraction solvent in food processing.

8 Description

Ethyl acetate is a clear, colorless, volatile liquid with a pleasant fruity, fragrant, and slightly acetous odor, and has a pleasant taste when diluted. Ethyl acetate is flammable.

9 Pharmacopeial Specifications

See Table I.

Table I: Pharmacopeial specifications for ethyl acetate.

Test	PhEur 7.4	USP35–NF30
Identification	+	+
Characters	+	−
Boiling point	76–78°C	−
Appearance of solution	+	−
Acidity	+	+
Specific gravity	0.898–0.902	0.894–0.898
Refractive index	1.370–1.373	−
Readily carbonizable substances	−	+
Reaction with sulfuric acid	+	−
Chromatographic purity	−[a]	+
Residue on evaporation	≤30 ppm	≤0.02%
Water	≤0.1%	−
Limit of methyl compounds	−	+
Related substances	+	−
Assay	−	99.0–100.5%

(a) The PhEur 7.4 lists impurities in ethyl acetate as methyl acetate, ethanol, and methanol.

10 Typical Properties

Autoignition temperature 486.1°C
Boiling point 77°C
Dielectric constant 6.11
Density 0.902 g/cm^3 at 20°C
Explosive limit 2.2–11.5% (volume in air)
Flash point
+7.2°C (open cup);
−5.0°C (closed cup).
Freezing point −83.6°C
Partition coefficient log P (octanol/water) = 0.7
Refractive index n_D^{20} = 1.3719
Solubility Soluble 1 in 10 of water at 25°C; ethyl acetate is more soluble in water at lower temperatures than at higher temperatures. Soluble in fixed and volatile oils. Miscible with acetone, chloroform, dichloromethane, ethanol (95%), and ether, and with most other organic liquids.
Spectroscopy
IR spectra *see* Figure 1.
NIR spectra *see* Figure 2.
Vapor density 3.04 (air = 1)

11 Stability and Storage Conditions

Ethyl acetate should be stored in an airtight container, protected from light and at a temperature not exceeding 30°C. Ethyl acetate is slowly decomposed by moisture and becomes acidic; the material can absorb up to 3.3% w/w water.

Ethyl acetate decomposes on heating to produce ethanol and acetic acid, and will emit acrid smoke and irritating fumes. It is flammable and its vapor may travel a considerable distance to an ignition source and cause a 'flashback'.

The alkaline hydrolysis of ethyl acetate has been shown to be inhibited by polyethylene glycol and by mixed micelle systems.[12]

12 Incompatibilities

Ethyl acetate can react vigorously with strong oxidizers, strong alkalis, strong acids, and nitrates to cause fires or explosions. It also reacts vigorously with chlorosulfonic acid, lithium aluminum hydride, 2-chloromethylfuran, and potassium *tert*-butoxide.

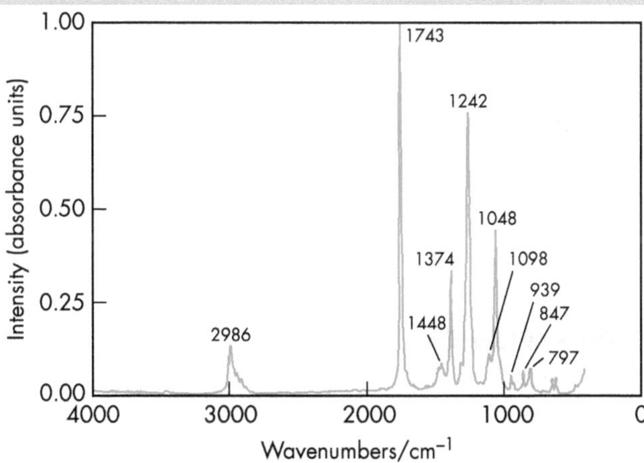

Figure 1: Infrared spectrum of ethyl acetate measured by transmission. Adapted with permission of Informa Healthcare.

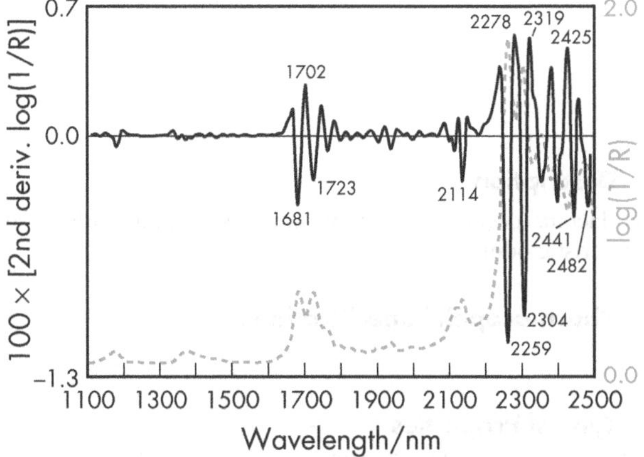

Figure 2: Near-infrared spectrum of ethyl acetate measured by transflectance (1 mm path-length).

13 Method of Manufacture

Ethyl acetate can be manufactured by the slow distillation of a mixture of ethanol and acetic acid in the presence of concentrated sulfuric acid. It has also been prepared from ethylene using an aluminum alkoxide catalyst.

14 Safety

Ethyl acetate is used in foods, and oral and topical pharmaceutical formulations. It is generally regarded as a relatively nontoxic and nonirritant material when used as an excipient.

However, ethyl acetate may be irritant to mucous membranes, and high concentrations may cause central nervous system depression. Potential symptoms of overexposure include irritation of the eyes, nose, and throat, narcosis, and dermatitis.

Ethyl acetate has not been shown to be a human carcinogen or a reproductive or developmental toxin.

The WHO has set an estimated acceptable daily intake of ethyl acetate at up to 25 mg/kg body-weight.[13]

In the UK, it has been recommended that ethyl acetate be temporarily permitted for use as a solvent in food and that the maximum concentration consumed in food should be set at 1000 ppm.[14]

LD_{50} (cat, SC): 3.00 g/kg[15]
LD_{50} (guinea-pig, oral): 5.50 g/kg
LD_{50} (guinea-pig, SC): 3.00 g/kg
LD_{50} (mouse, IP): 0.709 g/kg
LD_{50} (mouse, oral): 4.10 g/kg
LD_{50} (rabbit, oral): 4.935 g/kg
LD_{50} (rat, oral): 5.62 g/kg

15 Handling Precautions

Observe normal precautions appropriate to the circumstances and quantity of material handled. Eye protection and gloves are recommended. In the UK, the workplace exposure limit for ethyl acetate is 400 ppm (short-term) and 200 ppm (long-term).[16]

16 Regulatory Status

Included in the FDA Inactive Ingredients Database (oral tablets and sustained-action tablets, oral sustained-action capsules, and oral suspensions; topical and transdermal preparations; IM injections). Included in nonparenteral medicines licensed in the UK (tablets, topical solutions, and gels). Ethyl acetate is also accepted for use in food applications in a number of countries including the UK. Included in the Canadian Natural Health Products Ingredients Database.

17 Related Substances

—

18 Comments

The following azeotropic mixtures have been reported:

Ethyl acetate (93.9% w/w)–water (6.1% w/w), boiling point 70.4°C

Ethyl acetate (83.2% w/w)–water (7.8% w/w)–ethanol (9.0% w/w), boiling point 70.3°C

Ethyl acetate (69.4%)–ethanol (30.6%), boiling point 71.8°C

Ethyl acetate (77%)–propan-2-ol (23%), boiling point 74.8°C

A specification for ethyl acetate is contained in the *Food Chemicals Codex* (FCC).[17]

The EINECS number for ethyl acetate is 205-500-4. The PubChem Compound ID (CID) for ethyl acetate is 8857.

19 Specific References

1 Lötter J, *et al.* The influence of β-cyclodextrin on the solubility of chlorthalidone and its enantiomers. *Drug Dev Ind Pharm* 1999; **25**(8): 879–884.
2 Giordano F, *et al.* Crystal forms of piroxicam pivalate: preparation and characterization of two polymorphs. *J Pharm Sci* 1998; **87**(3): 333–337.
3 Romero S, *et al.* Solubility behavior of polymorphs I and II of mefenamic acid in solvent mixtures. *Int J Pharm* 1999; **178**: 193–202.
4 Caira MN, *et al.* Preparation and crystal characterisation of a polymorph, a monohydrate, and an ethyl acetate solvate of the antifungal fluconazole. *J Pharm Sci* 2004; **93**(3): 601–611.
5 Abu-Izza K, *et al.* Preparation and evaluation of zidovudine-loaded sustained-release microspheres. 2. Optimization of multiple response variables. *J Pharm Sci* 1996; **85**(6): 572–576.
6 Cleland JL, Jones AJS. Stable formulations of recombinant human growth hormone and interferon-γ for microencapsulation and biodegradable microspheres. *Pharm Res* 1996; **13**(10): 1464–1475.
7 Baccarin MA, *et al.* Ethylcellulose microspheres containing sodium diclofenac: development and characterization. *Acta Farm Bonaerense* 2006; **25**(3): 401–404.
8 Zaghloul AA. Beta-estradiol biodegradable microspheres: effect of formulation parameters on encapsulation efficiency and *in vitro* release. *Pharmazie* 2006; **61**(9): 775–779.
9 Shah SP, Misra A. Development of liposomal amphotericin B dry powder inhaler formulation. *Drug Deliv* 2004; **11**(4): 247–253.
10 Hassan AS, *et al.* Composite microparticles with *in vivo* reduction of the burst release effect. *European J Pharm Biopharm* 2009; **73**: 337–344.
11 Pillai O, *et al.* Transdermal iontophoresis of insulin: IV. Influence of chemical enhancers. *Int J Pharm* 2004; **269**(1): 109–120.

E

12 Xiancheng Z, *et al.* The alkaline hydrolysis of ethyl acetate and ethyl propionate in single and mixed micellar solutions. *J Disper Sci Technol* 1996; **17**(3): 339–348.
13 FAO/WHO. Specifications for the identity and purity of food additives and their toxicological evaluation: some flavouring substances and non-nutritive sweetening agents. Eleventh report of the Joint FAO/WHO Expert Committee on Food Additives. *World Health Organ Tech Rep Ser* 1968; No. 383.
14 Ministry of Agriculture, Fisheries and Food. *Report on the Review of Solvents in Food, FAC/REP/25*. London: HMSO, 1978.
15 Lewis RJ, ed. *Sax's Dangerous Properties of Industrial Materials*, 11th edn. New York: Wiley, 2004: 1625.
16 Health and Safety Executive. *EH40/2005: Workplace Exposure Limits*. Sudbury: HSE Books, 2011. http://www.hse.gov.uk/pubns/priced/eh40.pdf (accessed 1 March 2012).
17 *Food Chemicals Codex*, 7th edn (Suppl. 1). Bethesda, MD: United States Pharmacopeia, 2010: 1465.

20 General References

—

21 Author

CG Cable.

22 Date of Revision

1 March 2012.

Ethyl Lactate

1 Nonproprietary Names

None adopted.

2 Synonyms

Actylol; *Acytol*; *Dermol EL*; ethyl α-hydroxypropionate; ethyl-2-hydroxypropanoate; ethyl-2-hydroxypropionate; ethyl-*S*-(−)-2-hydroxypropionate; 2-hydroxypropanoic acid ethyl ester; lactic acid ethyl ester; propanoic acid 2-hydroxy-ethyl ester; *Purasolv EL*; *Solactol*.

3 Chemical Name and CAS Registry Number

2-Hydroxy-propanoic acid ethyl ester [97-64-3]

4 Empirical Formula and Molecular Weight

$C_5H_{10}O_3$ 118.13

5 Structural Formula

6 Functional Category

Flavoring agent; solvent.

7 Applications in Pharmaceutical Formulation or Technology

Ethyl lactate is used as a solvent or co-solvent in liquid formulations[1,2] and recently as a co-solvent in emulsions and microemulsion technologies. It has also been used as a solvent for nitrocellulose, cellulose acetate, cellulose ethers, polyvinyl and other resins.[3]

Ethyl lactate is also used as a flavoring agent in pharmaceutical preparations, and is found in food products.

8 Description

Ethyl lactate occurs as a clear colorless liquid with a sharp characteristic odor.

9 Pharmacopeial Specifications

—

10 Typical Properties

Acidity/alkalinity pH = 7 (10% w/v aqueous solution)
Boiling point 154–155°C
Density 1.0328 at 20°C
Explosion limits 1.5–11.4%
Flash point 46°C
Heat of combustion 6.5 kcal/g
Melting point −26.0°C
Refractive index n_D^{20} = 1.412–1.414
Solubility Miscible with water (with partial decomposition), ethanol (95%), ether, chloroform, ketones, esters, and hydrocarbons.
Viscosity (dynamic) 0.0261 mPa s (0.0261 cP) at 20°C
Vapor density 4.07 (air = 1)
Vapor pressure 0.732 kPa at 30°C

11 Stability and Storage Conditions

Stable at normal temperature and pressure. Ethyl lactate is a flammable liquid and vapor. Store in a cool, dry, and well-ventilated location away from any fire hazard area, in a tightly closed container.

12 Incompatibilities

Incompatible with bases or strong alkalis and may cause fire or explosion with strong oxidizing agents.

13 Method of Manufacture

Ethyl lactate is produced by the esterification of lactic acid with ethanol in the presence of a little mineral oil, or by combination of acetaldehyde with hydrocyanic acid to form acetaldehyde cyanhydrin. This is followed by treatment with ethanol (95%) and hydrochloric or sulfuric acid. Purification is achieved using fractional distillation. The commercial product is a racemic mixture.

14 Safety

Ethyl lactate is used as a flavoring agent in pharmaceutical preparations, and is found in food products. The estimated acceptable daily intake for lactic acid is 12.5 mg/kg body-weight.

In general, lactate esters have an oral $LD_{50} > 2000$ mg/kg; and the inhalation LC_{50} is generally above 5000 mg/m^3. They have the potential of causing eye and skin irritation (on prolonged contact), but not sensitization.[4] Ethyl lactate is moderately toxic by intraperitoneal, subcutaneous, and intravenous routes. There is low oral and skin contact toxicity; although ingestion may cause nausea, stomach and throat pain, and narcosis. Inhalation of concentrated vapor of ethyl lactate may cause irritation of the mucous membranes, drowsiness, and narcosis.

LD_{50} (rat, oral): >5.0 g/kg[5]
LD_{50} (mouse, oral): 2.5 g/kg
LD_{50} (mouse, SC): 2.5 g/kg
LD_{50} (mouse, IV): 0.6 g/kg
LD_{50} (rabbit, skin): >5.0 g/kg

15 Handling Precautions

Observe normal precautions appropriate to the circumstances and quantity of material handled. Avoid skin and eye contact; eye goggles should be worn, or a full face shield where splashing may occur.

There is a slight explosion hazard in the form of vapor when it is exposed to flame. Avoid ignition sources and use adequate ventilation to keep vapor and mist as low as possible.

When heated to decomposition, it emits acrid smoke and irritating fumes. Facial respirators are recommended when dealing with excessive amounts or with prolonged exposure to the compound.

16 Regulatory Status

GRAS listed. Reported in the EPA TSCA Inventory. Included in the Canadian Natural Health Products Ingredients Database.

17 Related Substances

n-Butyl lactate; methyl lactate.

n-Butyl lactate
Empirical formula $C_7H_{14}O_3$
Molecular weight 146.2
CAS number [138-22-7]
Synonyms Butyl α-hydroxypropionate; propanoic acid 2-hydroxy butyl ester; lactic acid butyl ester; *Purasolv BL*.
Boiling point 188°C
Melting point –43°C
Solubility Partially miscible with water and most organic solvents.
Comments *n*-Butyl lactate is used as a flavoring agent in pharmaceutical preparations.

The EINECS number for *n*-butyl lactate is 205-316-4.

Methyl lactate
Empirical formula $C_4H_8O_3$
Molecular weight 104
CAS number [547-64-8]
Synonyms Methyl hydroxy propionate; *Purasolv ML*.
Appearance Methyl lactate occurs as a clear, colorless liquid.
Boiling point 143.9°C
Comments Methyl lactate is used as a cellulose acetate solvent.

18 Comments

Owing to its biodegradability, ethyl lactate is replacing many solvents in many household products, including packaging, plastics, paints, paint strippers, grease removers, cleansers, aerosols, adhesives, and varnishes.

It has been applied topically in the treatment of acne vulgaris,[6,7] where it accumulates in the sebaceous glands and is hydrolyzed to ethanol and lactic acid, lowering the skin pH and exerting a bactericidal effect.

Ethyl lactate is specified as a flavor chemical in the *Food Chemicals Codex* (FCC).[8]

The EINECS number for ethyl lactate is 202-598-0. The PubChem Compound ID (CID) for ethyl lactate includes 7344 and 92831.

19 Specific References

1 Christensen JM, *et al.* Ethyl lactate–ethanol–water cosolvent for intravenous theophylline. *Res Commun Chem Pathol Pharmacol* 1985; 50(1): 147–150.
2 Mottu F, *et al.* Organic solvents for pharmaceutical parenterals and embolic liquids: a review of toxicity data. *PDA J Pharm Sci Tech* 2000; 54(6): 456–469.
3 Siew LF, *et al.* The properties of amylose–ethylcellulose films cast from organic-based solvents as potential coatings for colonic drug delivery. *Eur J Pharm Sci* 2000; **11**(2): 133–139.
4 Clary JJ, *et al.* Safety assessment of lactate esters. *Regul Toxicol Pharmacol* 1998; **27**(2): 88–97.
5 Lewis RJ, ed. *Sax's Dangerous Properties of Industrial Materials*, 11th edn. New York: Wiley, 2004: 2197.
6 George D, *et al.* Ethyl lactate as a treatment for acne. *Br J Dermatol* 1983; **108**(2): 228–233.
7 Prottey C, *et al.* The mode of action of ethyl lactate as a treatment for acne. *Br J Dermatol* 1984; **110**(4): 475–485.
8 *Food Chemicals Codex*, 7th edn (Suppl. 1). Bethesda, MD: United States Pharmacopeia, 2010: 356.

20 General References

—

21 Author

O AbuBaker.

22 Date of Revision

1 February 2011.

Ethyl Maltol

1 Nonproprietary Names

USP–NF: Ethyl Maltol

2 Synonyms

2-Ethyl pyromeconic acid; 3-hydroxy-2-ethyl-gamma-pyrone; 3-hydroxy-2-ethyl-4-pyrone; *Veltol Plus*.

3 Chemical Name and CAS Registry Number

2-Ethyl-3-hydroxy-4*H*-pyran-4-one [4940-11-8]

4 Empirical Formula and Molecular Weight

$C_7H_8O_3$ 140.14

5 Structural Formula

6 Functional Category

Flavor enhancer; flavoring agent.

7 Applications in Pharmaceutical Formulation or Technology

Ethyl maltol is used in pharmaceutical formulations and food products as a flavoring agent or flavor enhancer in applications similar to maltol. It has a flavor and odor 4–6 times as intense as maltol. Ethyl maltol is used in oral syrups at concentrations of about 0.004% w/v and also at low levels in perfumery.

8 Description

White crystalline solid with characteristic, very sweet, caramel-like odor and taste. In dilute solution it possesses a sweet, fruitlike flavor and odor.

9 Pharmacopeial Specifications

See Table I.

Table I: Pharmacopeial specifications for ethyl maltol.

Test	USP35–NF30
Identification	+
Residue on ignition	≤0.2%
Water	≤0.5%
Assay (anhydrous basis)	≥99.0%

10 Typical Properties

Melting point 85–95°C

Solubility *see* Table II.

Table II: Solubility of ethyl maltol.

Solvent	Solubility at 20°C
Chloroform	1 in 5
Ethanol (95%)	1 in 10
Glycerin	1 in 500
Propylene glycol	1 in 17
Water	1 in 55

Spectroscopy

IR spectra *see* Figure 1.

NIR spectra *see* Figure 2.

11 Stability and Storage Conditions

Solutions may be stored in glass or plastic containers. The bulk material should be stored in a well-closed container, protected from light, in a cool, dry, well-ventilated place, and away from sources of heat or ignition. Store at temperatures 7-35°C and humidity ranges 35-55%. Desiccant bags may be used to reduce moisture build-up.

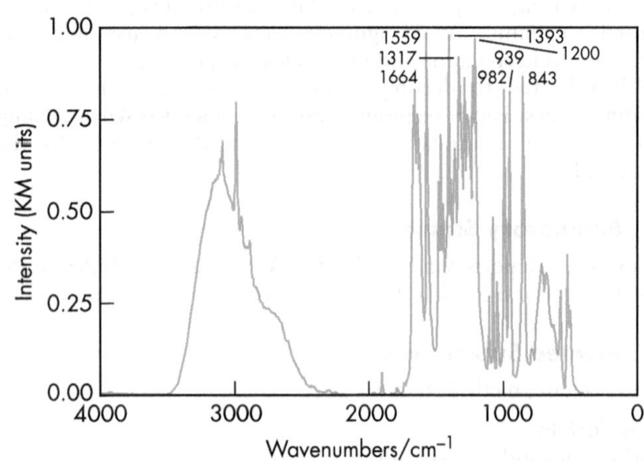

Figure 1: Infrared spectrum of ethyl maltol measured by diffuse reflectance. Adapted with permission of Informa Healthcare.

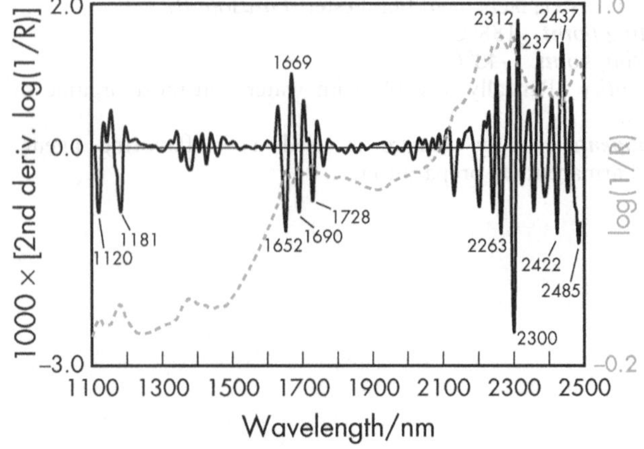

Figure 2: Near-infrared spectrum of ethyl maltol measured by reflectance.

12 Incompatibilities

Ethyl maltol is incompatible with oxidizing agents. It is slightly acidic and forms salts with bases. Ethyl maltol chelates and forms complexes with metals.

13 Method of Manufacture

Unlike maltol, ethyl maltol does not occur naturally. It may be prepared by treating α-ethylfurfuryl alcohol with a halogen to produce 4-halo-6-hydroxy-2-ethyl-2*H*-pyran-3(6*H*)-one, which is converted to ethyl maltol by hydrolysis.

14 Safety

In animal feeding studies, ethyl maltol has been shown to be well tolerated with no adverse toxic, reproductive, or embryogenic effects. It has been reported that while the acute toxicity of ethyl maltol, in animal studies, is slightly greater than maltol, with repeated dosing the opposite is true.[1] The WHO has set an acceptable daily intake for ethyl maltol at up to 2 mg/kg body-weight.[2,3]

LD_{50} (chicken, oral): 1.27 g/kg[4]
LD_{50} (mouse, oral): 0.78 g/kg (male)
LD_{50} (mouse, SC): 0.91 g/kg
LD_{50} (rabbit, skin): >5 g/kg
LD_{50} (rat, oral): 1.15 g/kg (male)
LD_{50} (rat, oral): 1.20 g/kg (female)

15 Handling Precautions

Observe normal precautions appropriate to the circumstances and quantity of material handled. Ethyl maltol should be used in a well-ventilated environment. Dust may be irritant, and eye protection and gloves are recommended.

16 Regulatory Status

GRAS listed. Included in the FDA Inactive Ingredients Database (oral syrup, solutions, elixir). Included in the Canadian Natural Health Products Ingredients Database.

17 Related Substances

Maltol.

18 Comments

See Maltol for further information.

A specification for ethyl maltol is contained in the *Food Chemicals Codex* (FCC).[5]

19 Specific References

1 Gralla EJ, *et al.* Toxicity studies with ethyl maltol. *Toxicol Appl Pharmacol* 1969; **15**: 604–613.
2 FAO/WHO. Evaluation of certain food additives. Eighteenth report of the joint FAO/WHO expert committee on food additives. *World Health Organ Tech Rep Ser* 1974; No. 557.
3 FAO/WHO. Evaluation of certain food additives. Sixty-fifth report of the joint FAO/WHO expert committee on food additives. *World Health Organ Tech Rep Ser* 2006; No. 934.
4 Lewis RJ, ed. *Sax's Dangerous Properties of Industrial Materials*, 11th edn. New York: Wiley, 2004; 1692.
5 *Food Chemicals Codex*, 7th edn. Bethesda, MD: United States Pharmacopeia, 2010; 359–360.

20 General References

Allen LV. Featured excipient: flavor enhancing agents. *Int J Pharm Compound* 2003; **7**(1): 48–50.
LeBlanc DT, Akers HA. Maltol and ethyl maltol: from the larch tree to successful food additives. *Food Technol* 1989; **43**(4): 78–84.
Liu B, *et al.* Solubility of ethyl maltol in aqueous ethanol mixtures. *J Chem Eng Data* 2008; **53**: 2712–2714.

21 Authors

MS Dudhedia, PJ Weller.

22 Date of Revision

1 March 2012.

Ethyl Oleate

1 Nonproprietary Names

BP: Ethyl Oleate
PhEur: Ethyl Oleate
USP–NF: Ethyl Oleate

2 Synonyms

Crodamol EO; ethylis oleas; ethyl 9-octadecenoate; *Kessco EO*; oleic acid, ethyl ester.

3 Chemical Name and CAS Registry Number

(*Z*)-9-Octadecenoic acid, ethyl ester [111-62-6]

4 Empirical Formula and Molecular Weight

$C_{20}H_{38}O_2$ 310.51

5 Structural Formula

H H O
H_3C— $(CH_2)_7$ $(CH_2)_7$ —C—O—C_2H_5

6 Functional Category

Oleaginous vehicle; solvent.

7 Applications in Pharmaceutical Formulation or Technology

Ethyl oleate is primarily used as a vehicle in certain parenteral preparations intended for intramuscular administration. It has also been used as a solvent for drugs formulated as biodegradable capsules for subdermal implantation[1] and in the preparation of microemulsions containing cyclosporin[2] and norcantharidin.[3] Microemulsion formulations containing ethyl oleate have also been proposed for topical[4] and ocular[5] delivery, and for liver targeting following parenteral administration.[6] Ethyl oleate has been used in topical gel formulations,[7] and in self-microemulsifying drug delivery systems for oral administration,[8,9] and intravenous administration.[10]

Ethyl oleate is a suitable solvent for steroids and other lipophilic drugs. Its properties are similar to those of almond oil and peanut oil. However, it has the advantage that it is less viscous than fixed oils and is more rapidly absorbed by body tissues.[11]

Ethyl oleate has also been evaluated as a vehicle for subcutaneous injection.[12]

8 Description

Ethyl oleate occurs as a pale yellow to almost colorless, mobile, oily liquid with a taste resembling that of olive oil and a slight, but not rancid odor.

Ethyl oleate is described in the USP35–NF30 as consisting of esters of ethyl alcohol and high molecular weight fatty acids, principally oleic acid. A suitable antioxidant may be included.

9 Pharmacopeial Specifications

See Table I.

Table I: Pharmacopeial specifications for ethyl oleate.

Test	PhEur 7.4	USP35–NF30
Characters	+	—
Identification	+	—
Specific gravity	0.866–0.874	0.866–0.874
Viscosity		≥5.15 mPa s
Refractive index	—	1.443–1.450
Acid value	≤0.5	≤0.5
Iodine value	75–90	75–85
Saponification value	177–188	177–188
Peroxide value	≤10	—
Oleic acid	≥60.0%	—
Water content	≤1.0%	—
Total ash	≤0.1%	—

10 Typical Properties

Boiling point 205–208°C (some decomposition)
Flash point 175.3°C
Freezing point ≈−32°C
Moisture content At 20°C and 52% relative humidity, the equilibrium moisture content of ethyl oleate is 0.08%.
Refractive index 1.451
Solubility Miscible with chloroform, ethanol (95%), ether, fixed oils, liquid paraffin, and most other organic solvents; practically insoluble in water.
Surface tension 32.3 mN/m (32.3 dynes/cm) at 25°C[11]
Viscosity (dynamic) 3.9 mPa s (3.9 cP) at 25°C[11]
Viscosity (kinematic) 0.046 mm^2/s (4.6 cSt) at 25°C[11]

11 Stability and Storage Conditions

Ethyl oleate should be stored in a cool, dry place in a small, well-filled, well-closed container, protected from light. When a partially filled container is used, the air should be replaced by nitrogen or another inert gas. Ethyl oleate oxidizes on exposure to air, resulting in an increase in the peroxide value. It remains clear at 5°C, but darkens in color on standing. Antioxidants are frequently used to extend the shelf life of ethyl oleate. Protection from oxidation for over 2 years has been achieved by storage in amber glass bottles with the addition of combinations of propyl gallate, butylated hydroxyanisole, butylated hydroxytoluene, and citric or ascorbic acid.[13,14] A concentration of 0.03% w/v of a mixture of propyl gallate (37.5%), butylated hydroxytoluene (37.5%), and butylated hydroxyanisole (25%) was found to be the best antioxidant for ethyl oleate.[14]

Ethyl oleate may be sterilized by heating at 150°C for 1 hour.

12 Incompatibilities

Ethyl oleate dissolves certain types of rubber and causes others to swell.[15,16] It may also react with oxidizing agents.

13 Method of Manufacture

Ethyl oleate is prepared by the reaction of ethanol with oleoyl chloride in the presence of a suitable hydrogen chloride acceptor.

14 Safety

Ethyl oleate is generally considered to be of low toxicity but ingestion should be avoided. Ethyl oleate has been found to cause minimal tissue irritation.[17] No reports of intramuscular irritation during use have been recorded.

15 Handling Precautions

Observe normal precautions appropriate to the circumstances and quantity of material handled. Eye protection and nitrile gloves are recommended. Ethyl oleate is flammable.

16 Regulatory Status

Included in the FDA Inactive Ingredients Database (transdermal preparation). Included in parenteral (IM depot injections) and nonparenteral (transdermal patches) medicines licensed in the UK. Included in the Canadian Natural Health Products Ingredients Database.

17 Related Substances

Methyl oleate; oleic acid.

Methyl oleate

Empirical formula $C_{19}H_{36}O_2$
Molecular weight 296.49
CAS number [112-69-9]
Synonyms Methyl 9-octadecenoate; (Z)-9-octadecenoic acid, methyl ester.
Boiling point 168–170°C
Density 0.879 g/cm^3
Iodine number 85.6
Refractive index n_D^{26} = 1.4510
Solubility Miscible with ethanol (95%) and ether.
Comments Prepared by refluxing oleic acid with *p*-toluenesulfonic acid in methanol.

18 Comments

A specification for ethyl oleate is contained in the *Food Chemicals Codex* (FCC).[18]

The EINECS number for ethyl oleate is 203-889-5. The PubChem Compound ID (CID) for ethyl oleate includes 8123 and 5364430.

19 Specific References

1 Ory SJ, *et al.* Effect of a biodegradable contraceptive capsule (Capronor) containing levonorgestrel on gonadotropin, estrogen and progesterone levels. *Am J Obstet Gynecol* 1983; **145**: 600–605.
2 Kim C-K, *et al.* Preparation and physicochemical characterisation of phase inverted water/oil microemulsion containing cyclosporin A. *Int J Pharm* 1997; **147**: 131–134.

3 Zhang L, *et al.* Formulation and physicochemical characterisation of norcantharidin microemulsion coating lecithin-based surfactants. *STP Pharma Sci* 2004; **14**(6): 461–469.
4 Chen HB, *et al.* Microemulsion-based hydrogel formulation of ibuprofen for topical delivery. *Int J Pharm* 2006; **315**(1–2): 52–58.
5 Chan J, *et al.* Phase transition water-in-oil microemulsions as ocular drug delivery systems: *in vitro* and *in vivo* evaluation. *Int J Pharm* 2007; **328**(1): 65–71.
6 Zhang L, *et al.* An investigation on liver-targeting microemulsions of norcantharidin. *Drug Deliv* 2005; **12**(5): 289–295.
7 El-Megrab NA, *et al.* Formulation and evaluation of meloxicam gels for topical administration. *Saudi Pharm J* 2006; **14**(3–4): 155–162.
8 Cui SM, *et al.* Self-microemulsifying drug delivery systems (SMEDDS) for improving in vitro dissolution and oral absorption of *Pueraria lobata* isoflavone. *Drug Dev Ind Pharm* 2005; **31**(4–5): 349–356.
9 Zhao Y, *et al.* Self-emulsifying drug delivery system (SNEDDS) for oral delvery of Zedoary essential oil: Formulation and bioavailability studies. *Int J Pharm* 2010; **383**: 170–177.
10 Borhade VB, *et al.* Development and characterisation of self-micro-emulsifying drug delivery system of tacrolimus for intravenous administration. *Drug Dev Ind Pharm* 2009; **35**: 619–630.
11 Howard JR, Hadgraft J. The clearance of oily vehicles following intramuscular and subcutaneous injections in rabbits. *Int J Pharm* 1983; **16**: 31–39.
12 Radwan M. *In vivo* screening model for excipients and vehicles used in subcutaneous injections. *Drug Dev Ind Pharm* 1994; **20**: 2753–2762.
13 Alemany P, Del Pozo A. [Autoxidation of ethyl oleate: protection with antioxidants.] *Galenica Acta* 1963; **16**: 335–338[in Spanish].
14 Nikolaeva NM, Gluzman MK. [Conditions for stabilizing ethyl oleate during storage.] *Farmatsiya* 1977; **26**: 25–28[in Russian].
15 Dexter MB, Shott MJ. The evaluation of the force to expel oily injection vehicles from syringes. *J Pharm Pharmacol* 1979; **31**: 497–500.
16 Halsall KG. Calciferol injection and plastic syringes [letter]. *Pharm J* 1985; **235**: 99.
17 Hem SL, *et al.* Tissue irritation evaluation of potential parenteral vehicles. *Drug Dev Commun*1974–75**1**(5): 471–477.
18 *Food Chemicals Codex*, 7th edn (Suppl. 1). Bethesda, MD: United States Pharmacopeia, 2010: 363.

20 General References

Spiegel AJ, Noseworthy MM. Use of nonaqueous solvents in parenteral products. *J Pharm Sci* 1963; **52**: 917–927.

21 Author

CG Cable.

22 Date of Revision

1 March 2012.

Ethyl Vanillin

1 Nonproprietary Names

USP–NF: Ethyl Vanillin

2 Synonyms

Bourbonal; ethavan; ethovan; ethylprotal; ethylprotocatechuic aldehyde; 4-hydroxy-3-ethoxybenzaldehyde; *Rhodiarome*; vanillal.

3 Chemical Name and CAS Registry Number

3-Ethoxy-4-hydroxybenzaldehyde [121-32-4]

4 Empirical Formula and Molecular Weight

$C_9H_{10}O_3$ 166.18

5 Structural Formula

6 Functional Category

Flavoring agent.

7 Applications in Pharmaceutical Formulation or Technology

Ethyl vanillin is used as an alternative to vanillin, i.e. as a flavoring agent in foods, beverages, confectionery, and pharmaceuticals. It is also used in perfumery.

Ethyl vanillin possesses a flavor and odor approximately three times as intense as vanillin; hence the quantity of material necessary to produce an equivalent vanilla flavor may be reduced, causing less discoloration to a formulation and potential savings in material costs. However, exceeding certain concentration limits may impart an unpleasant, slightly bitter taste to a product due to the intensity of the ethyl vanillin flavor. *See* Table I.

Table I: Uses of ethyl vanillin.

Use	Concentration (%)
Foods and confectionery	0.002–0.025
Oral syrups	0.01

8 Description

White or slightly yellowish crystals with a characteristic intense vanilla odor and flavor.

9 Pharmacopeial Specifications

See Table II.

10 Typical Properties

Boiling point 285°C
Density (bulk) 1.05 g/cm^3
Flash point 127°C

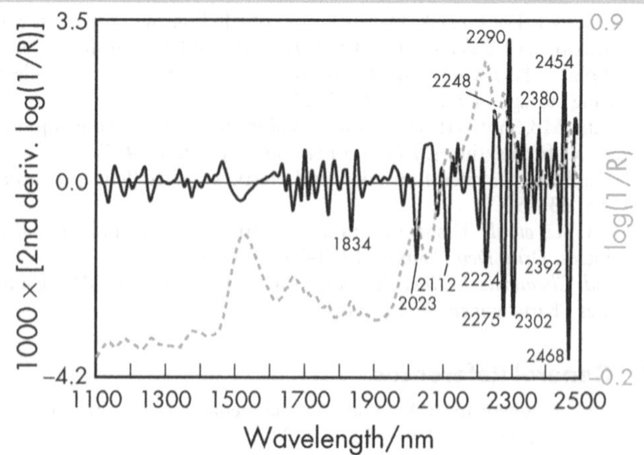

Figure 1: Near-infrared spectrum of ethyl vanillin measured by reflectance.

Table II: Pharmacopeial specifications for ethyl vanillin.

Test	USP35–NF30
Identification	+
Melting range	76–78°C
Loss on drying	≤1.0%
Residue on ignition	≤0.1%
Assay (dried basis)	98.0–101.0%

Table III: Solubility of ethyl vanillin.

Solvent	Solubility at 20°C unless otherwise stated
Alkaline hydroxide solutions	Freely soluble
Chloroform	Freely soluble
Ethanol (95%)	1 in 2
Ether	Freely soluble
Glycerin	Soluble
Propylene glycol	Soluble
Water	1 in 250
	1 in 100 at 50°C

Melting point 76–78°C
Solubility *see* Table III.
Spectroscopy
NIR spectra *see* Figure 1.

11 Stability and Storage Conditions

Store in a well-closed container, protected from light, in a cool, dry place. *See* Vanillin for further information.

12 Incompatibilities

Ethyl vanillin is unstable in contact with iron or steel, forming a red-colored, flavorless compound. In aqueous media with neomycin sulfate or succinylsulfathiazole, tablets of ethyl vanillin produced a yellow color.(1) *See* Vanillin for other potential incompatibilities.

13 Method of Manufacture

Unlike vanillin, ethyl vanillin does not occur naturally. It may be prepared synthetically by the same methods as vanillin, using guethol instead of guaiacol as a starting material; *see* Vanillin.

14 Safety

Ethyl vanillin is generally regarded as an essentially nontoxic and nonirritant material. However, cross-sensitization with other structurally similar molecules may occur; *see* Vanillin.

The WHO has allocated an acceptable daily intake for ethyl vanillin of up to 3 mg/kg body-weight.(2)

LD_{50} (guinea pig, IP): 1.14 g/kg(3,4)
LD_{50} (mouse, IP): 0.75 g/kg
LD_{50} (rabbit, oral): 3 g/kg
LD_{50} (rabbit, SC): 2.5 g/kg
LD_{50} (rat, oral): 1.59 g/kg
LD_{50} (rat, SC): 3.5–4.0 g/kg

15 Handling Precautions

Observe normal precautions appropriate to the circumstances and quantity of material handled. Eye protection is recommended. Heavy airborne concentrations of dust may present an explosion hazard.

16 Regulatory Status

GRAS listed. Included in the FDA Inactive Ingredients Database (oral capsules, suspensions, and syrups). Included in nonparenteral medicines licensed in the UK. Included in the Canadian Natural Health Products Ingredients Database.

17 Related Substances

Vanillin.

18 Comments

Ethyl vanillin can be distinguished analytically from vanillin by the yellow color developed in the presence of concentrated sulfuric acid.

The EINECS number for ethyl vanillin is 204-464-7. The PubChem Compound ID (CID) for ethyl vanillin is 8467.

19 Specific References

1 Onur E, Yalcindag ON. [Double incompatibility of ethyl vanillin (vanillal) in compressed tablets.] *Bull Soc Pharm Bordeaux* 1970; 109(2): 49–51[in French].
2 FAO/WHO. Evaluation of certain food additives and contaminants. Forty-fourth report of the joint FAO/WHO expert committee on food additives. *World Health Organ Tech Rep Ser* 1995; No. 859.
3 Sweet DV, ed. *Registry of Toxic Effects of Chemical Substances*. Cincinnati: US Department of Health, 1987: 721.
4 Lewis RJ, ed. *Sax's Dangerous Properties of Industrial Materials*, 11th edn. New York: Wiley, 2004: 1729.

20 General References

Ali L, *et al.* Rapid method for the determination of coumarin, vanillin, and ethyl vanillin in vanilla extract by reversed-phase liquid chromatography with ultraviolet detection. *J AOAC Int* 2008; **91**(2): 383–386.
Allen LV. Featured excipient: flavor-enhancing agents. *Int J Pharm Compound* 2003; 7(1): 48–50.
Rees DI. Determination of vanillin and ethyl vanillin in food products. *Chem Ind* 1965; **1**: 16–17.

21 Author

PJ Weller.

22 Date of Revision

1 March 2012.

Ethylcellulose

1 Nonproprietary Names

BP: Ethylcellulose
PhEur: Ethylcellulose
USP–NF: Ethylcellulose

2 Synonyms

Aquacoat ECD; *Aqualon*; *Ashacel*; E462; *Ethocel*; ethylcellulosum; *Fetocel*; *Surelease*.

3 Chemical Name and CAS Registry Number

Cellulose ethyl ether [9004-57-3]

4 Empirical Formula and Molecular Weight

Ethylcellulose is partially ethoxylated. Ethylcellulose with complete ethoxyl substitution (DS = 3) is $C_{12}H_{23}O_6(C_{12}H_{22}O_5)_nC_{12}H_{23}O_5$ where *n* can vary to provide a wide variety of molecular weights. Ethylcellulose, an ethyl ether of cellulose, is a long-chain polymer of β-anhydroglucose units joined together by acetal linkages.

5 Structural Formula

$CH_2OC_2H_5$
O
O
OC_2H_5
OC_2H_5
n

6 Functional Category

Coating agent; microencapsulating agent; tablet and capsule binder; tablet and capsule diluent; taste-masking agent; viscosity-increasing agent.

7 Applications in Pharmaceutical Formulation or Technology

Ethylcellulose is widely used in oral and topical pharmaceutical formulations; *see* Table I.

The main use of ethylcellulose in oral formulations is as a hydrophobic coating agent for tablets and granules.[1–8] Ethylcellulose coatings are used to modify the release of a drug,[7–10] to mask an unpleasant taste,[11,12] or to improve the stability of a formulation; for example, where granules are coated with ethylcellulose to inhibit oxidation. Modified-release tablet formulations may also be produced using ethylcellulose as a matrix former.[13–16]

Ethylcellulose, dissolved in an organic solvent or solvent mixture, can be used on its own to produce water-insoluble films. Higher-viscosity ethylcellulose grades tend to produce stronger and more durable films. Ethylcellulose films may be modified to alter their solubility[17] by the addition of hypromellose[18] or a plasticizer;[19–21] *see* Section 18. An aqueous polymer dispersion (or latex) of ethylcellulose such as *Aquacoat ECD* (FMC Biopolymer) or *Surelease* (Colorcon) may also be used to produce ethylcellulose films without the need for organic solvents.

Drug release through ethylcellulose-coated dosage forms can be controlled by diffusion through the film coating. This can be a slow process unless a large surface area (e.g. pellets or granules compared with tablets) is utilized. In those instances, aqueous ethylcellulose dispersions are generally used to coat granules or pellets. Ethylcellulose-coated beads and granules have also demonstrated the ability to absorb pressure and hence protect the coating from fracture during compression.[21]

High-viscosity grades of ethylcellulose are used in drug microencapsulation.[10,22–24]

Release of a drug from an ethylcellulose microcapsule is a function of the microcapsule wall thickness and surface area.

In tablet formulations, ethylcellulose may additionally be employed as a binder, the ethylcellulose being blended dry or wet-granulated with a solvent such as ethanol (95%). Ethylcellulose produces hard tablets with low friability, although they may demonstrate poor dissolution.

Ethylcellulose has also been used as an agent for delivering therapeutic agents from oral (e.g. dental) appliances.[25]

In topical formulations, ethylcellulose is used as a thickening agent in creams, lotions, or gels, provided an appropriate solvent is used.[26] Ethylcellulose has been studied as a stabilizer for emulsions.[27]

Ethylcellulose is additionally used in cosmetics and food products.

Table I: Uses of ethylcellulose.

Use	Concentration (%)
Microencapsulation	10.0–20.0
Sustained-release tablet coating	3.0–20.0
Tablet coating	1.0–3.0
Tablet granulation	1.0–3.0
Taste-masking	5.0-50.0

8 Description

Ethylcellulose is a tasteless, free-flowing, white to light tan-colored powder.

9 Pharmacopeial Specifications

The pharmacopeial specifications for ethylcellulose have undergone harmonization of many attributes for PhEur, and USP–NF.

See Tables II and III. *See also* Section 18.

Table II: Pharmacopeial specifications for ethylcellulose.

Test	PhEur 7.4	USP35–NF30
Identification	+	+
Characters	+	—
Acidity or alkalinity	+	+
Viscosity	*See* Table III	*See* Table III
Loss on drying	≤3.0%	≤3.0%
Residue on ignition	—	≤0.5%
Sulfated ash	≤0.5%	—
Heavy metals[a]	≤20 ppm	≤20 ppm
Acetaldehyde	≤100 ppm	+
Chlorides	≤0.1%	≤0.1%
Assay (of ethoxyl groups)	44.0–51.0%	44.0–51.0%

(a) These tests have not been fully harmonized at the time of publication.

E

SEM 1: Excipient: ethylcellulose; manufacturer: Ashland Aqualon Functional Ingredients; lot no.: 57911; magnification: 60×; voltage: 10 kV.

SEM 2: Excipient: ethylcellulose 10 mPa s (10 cP) fine powder; manufacturer: Dow Chemical Co.; magnification: 600×; voltage: 5 kV.

SEM 3: Excipient: ethylcellulose 100 mPa s (100 cP) fine powder; manufacturer: Dow Chemical Co.; magnification: 600×; voltage: 5 kV.

SEM 4: Excipient: ethylcellulose; manufacturer: Ashland Aqualon Functional Ingredients; lot no.: 57911; magnification: 600×; voltage: 10 kV.

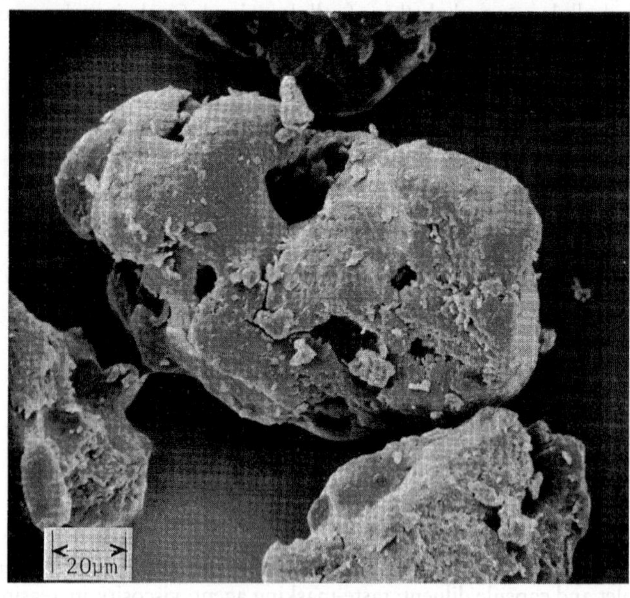

Table III: Pharmacopeial specifications for ethylcellulose viscosity.

Test	PhEur 7.4	USP35–NF30
Nominal viscosity		
≤6 mPa s	75–140% of that stated for its nominal viscosity	75–140% of that stated for its nominal viscosity
>6 mPa s	80–120% of that stated for its nominal viscosity	80–120% of that stated for its nominal viscosity

10 Typical Properties

Density (bulk) 0.4 g/cm^3

Glass transition temperature 129–133°C[28]

Moisture content Ethylcellulose absorbs very little water from humid air or during immersion, and that small amount evaporates readily.[29,30]

See also Figure 1.

Particle size distribution *see* Table IV; *see also* Figures 2 and 3.

Solubility Ethylcellulose is practically insoluble in glycerin, propylene glycol, and water. Ethylcellulose that contains less than 46.5% of ethoxyl groups is freely soluble in chloroform, methyl acetate, and tetrahydrofuran, and in mixtures of aromatic hydrocarbons with ethanol (95%). Ethylcellulose that contains not less than 46.5% of ethoxyl groups is freely soluble in chloroform, ethanol (95%), ethyl acetate, methanol, and toluene.

Specific gravity 1.12–1.15 g/cm^3

Spectroscopy

IR spectra *see* Figure 4.

Viscosity The viscosity of ethylcellulose is measured typically at 25°C using 5% w/v ethylcellulose dissolved in a solvent blend of 80% toluene : 20% ethanol (w/w). Grades of ethylcellulose with various viscosities are commercially available; *see* Table IV. They may be used to produce 5% w/v solutions in organic solvent blends with viscosities nominally ranging from 7 to 100 mPa s (7–100 cP). Specific ethylcellulose grades, or blends of different grades, may be used to obtain solutions of a desired viscosity.

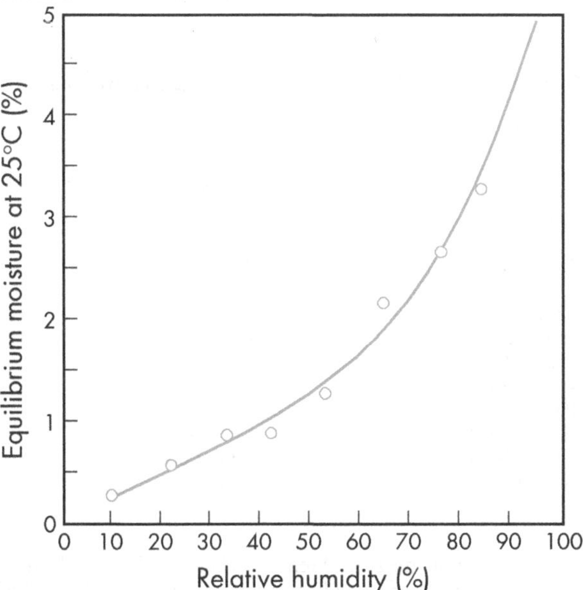

Figure 1: Equilibrium moisture content of ethylcellulose.

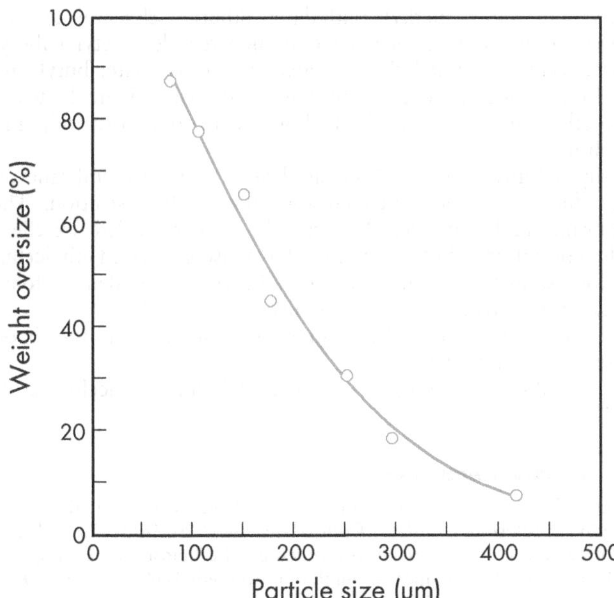

Figure 2: Particle size distribution of ethylcellulose.

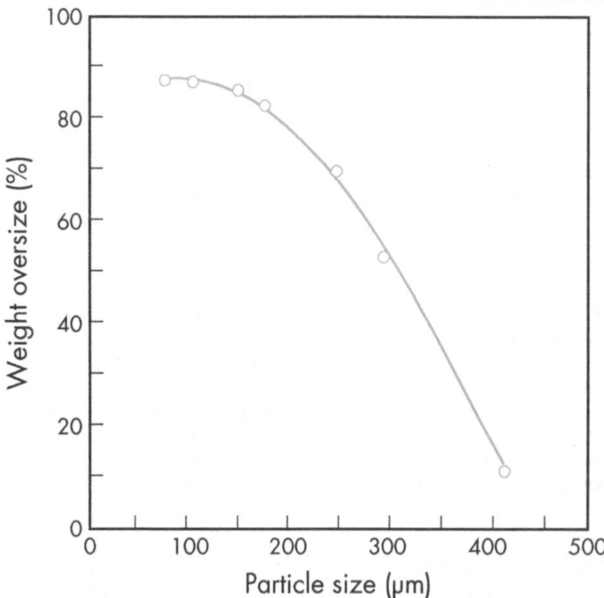

Figure 3: Particle size distribution of ethylcellulose (*Ethocel*, Dow Chemical Co.).

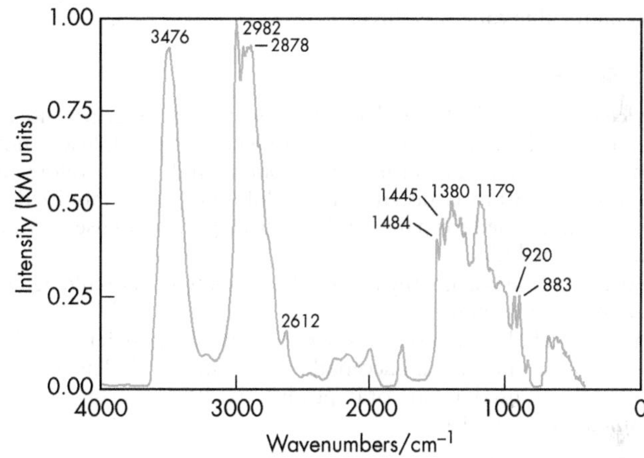

Figure 4: Infrared spectrum of ethylcellulose measured by diffuse reflectance. Adapted with permission of Informa Healthcare.

Solutions of higher viscosity tend to be composed of longer polymer chains and produce strong and durable films.

The viscosity of an ethylcellulose solution increases with an increase in ethylcellulose concentration; e.g. the viscosity of a 5% w/v solution of *Ethocel Standard 4 Premium* is 4 mPa s (4 cP) and of a 25% w/v solution of the same ethylcellulose grade is 850 mPa s (850 cP). Solutions with a lower viscosity may be obtained by incorporating a higher percentage (30–40%) of a low-molecular-weight aliphatic alcohol such as ethanol, butanol, propan-2-ol, or *n*-butanol with toluene. The viscosity of such solutions depends almost entirely on the alcohol content and is independent of toluene.

11 Stability and Storage Conditions

Ethylcellulose is a stable, slightly hygroscopic material. It is chemically resistant to alkalis, both dilute and concentrated, and to salt solutions, although it is more sensitive to acidic materials than are cellulose esters.

Ethylcellulose is subject to oxidative degradation in the presence of sunlight or UV light at elevated temperatures. This may be prevented by the use of antioxidant and chemical additives that absorb light in the 230–340 nm range.

Ethylcellulose should be stored at a temperature not exceeding 32°C (90°F) in a dry area away from all sources of heat. It should not be stored next to peroxides or other oxidizing agents.

12 Incompatibilities

Incompatible with paraffin wax and microcrystalline wax.

13 Method of Manufacture

Ethylcellulose is prepared by treating purified cellulose (sourced from chemical-grade cotton linters and wood pulp) with an alkaline solution, followed by ethylation of the alkali cellulose with chloroethane as shown below, where R represents the cellulose radical:

$$RONa + C_2H_5Cl \rightarrow ROC_2H_5 + NaCl$$

Table IV: Summary of ethylcellulose grades, suppliers, viscosity, and particle size.

Grade	Supplier	Solution viscosity (mPa s)	Mean particle size (μm)
Ethocel Std 4 Premium	Dow Chemical	3.0–5.5	—
N-7	Ashland Specialty Ingredients	5.6–8.0	—
Ethocel Std 7FP Premium	Dow Chemical	6.0–8.0	5.0–15.0
Ethocel Std 7 Premium	Dow Chemical	6.0–8.0	310.0
T-10	Ashland Specialty Ingredients	8.0–11.0	—
N-10	Ashland Specialty Ingredients	8.0–11.0	—
Ethocel Std 10FP Premium	Dow Chemical	9.0–11.0	3.0–15.0
Ethocel Std 10 Premium	Dow Chemical	9.0–11.0	375.0
N-14	Ashland Specialty Ingredients	12.0–16.0	—
Ethocel Std 20 Premium	Dow Chemical	18.0–22.0	—
N-22	Ashland Specialty Ingredients	18.0–24.0	—
Ethocel Std 45 Premium	Dow Chemical	41.0–49.0	—
N-50	Ashland Specialty Ingredients	40.0–52.0	—
N-100	Ashland Specialty Ingredients	80.0–105.0	—
Ethocel Std 100FP Premium	Dow Chemical	90.0–110.0	30.0–60.0
Ethocel Std 100 Premium	Dow Chemical	90.0–110.0	465.0

The manner in which the ethyl group is added to cellulose can be described by the degree of substitution (DS). The DS designates the average number of hydroxyl positions on the anhydroglucose unit that have been reacted with ethyl chloride. Since each anhydroglucose unit of the cellulose molecule has three hydroxyl groups, the maximum value for DS is three.

14 Safety

Ethylcellulose is widely used in oral and topical pharmaceutical formulations. It is also used in food products. Ethylcellulose is not metabolized following oral consumption and is therefore a noncalorific substance. Because ethylcellulose is not metabolized it is not recommended for parenteral products; parenteral use may be harmful to the kidneys.

Ethylcellulose is generally regarded as a nontoxic, nonallergenic, and nonirritating material.

As ethylcellulose is not considered to be a health hazard, the WHO has not specified an acceptable daily intake.[31] The highest reported level used in an oral product is 308.8 mg in an oral sustained release tablet.[32]

LD_{50} (rabbit, skin): >5 g/kg[33]

LD_{50} (rat, oral): >5 g/kg

15 Handling Precautions

It is important to prevent fine dust clouds of ethylcellulose from reaching potentially explosive levels in the air. Ethylcellulose is combustible. Ethylcellulose powder may be an irritant to the eyes and eye protection should be worn.

16 Regulatory Status

GRAS listed. Accepted for use as a food additive in Europe. Included in the FDA Inactive Ingredients Database (oral capsules, suspensions and tablets; topical emulsions and vaginal preparations). Included in nonparenteral medicines licensed in Europe. Included in the Canadian Natural Health Products Ingredients Database.

17 Related Substances

Hydroxyethyl cellulose; hydroxyethylmethyl cellulose; methylcellulose.

18 Comments

Ethylcellulose has undergone harmonization for many attributes for PhEur, and USP–NF by the Pharmacopeial Discussion Group. For further information see the General Information Chapter <1196> in the USP35–NF30 and the General Chapter 5.8 in PhEur 7.4, along with the'State of Work' document on the PhEur EDQM website.

Ethylcellulose is compatible with the following plasticizers: dibutyl phthalate; diethyl phthalate; dibutyl sebacate; triethyl citrate; tributyl citrate; acetylated monoglyceride; acetyl tributyl citrate; triacetin; dimethyl phthalate; benzyl benzoate; butyl and glycol esters of fatty acids; refined mineral oils; oleic acid; stearic acid; ethyl alcohol; stearyl alcohol; castor oil; corn oil; and camphor.

Ethylcellulose has also been used as a backing membrane on mucoadhesive patches intended for buccal administration. The membrane had high tensile strength, and provided excellent unidirectional drug flow.[34] Studies have also suggested ethylcellulose for use in floating microparticles based on low-density foam powder, for gastroretentive drug delivery systems.[35]

A specification for ethylcellulose is contained in the *Food Chemicals Codex* (FCC).[36]

The PubChem Compound ID (CID) for ethylcellulose is 24832091.

19 Specific References

1 Ozturk AG, *et al.* Mechanism of release from pellets coated with an ethyl cellulose-based film. *J Control Release* 1990; **14**(3): 203–213.
2 Narisawa S, *et al.* Porosity-controlled ethyl cellulose film coating. IV. Evaluation of mechanical strength of porous ethyl cellulose film. *Chem Pharm Bull* 1994; **42**(7): 1491–1495.
3 Bodmeier R, Paeratakul O. The effect of curing on drug release and morphological properties of ethylcellulose pseudolatex-coated beads. *Drug Dev Ind Pharm* 1994; **20**(9): 1517–1533.
4 Dressman JB, *et al.* Circumvention of pH-dependent release from ethyl cellulose-coated pellets. *J Control Release* 1995; **36**(3): 251–260.
5 Iyer U, *et al.* Comparative evaluation of three organic solvent and dispersion-based ethyl cellulose coating formulations. *Pharm Technol* 1990; **14**(9): 68–86.
6 Sarisuta N, Sirithunyalug J. Release rate of indomethacin from coated granules. *Drug Dev Ind Pharm* 1988; **14**(5): 683–687.
7 Porter SC. Controlled-release film coatings based on ethylcellulose. *Drug Dev Ind Pharm* 1989; **15**(10): 1495–1521.
8 Sadeghi F, *et al.* Study of drug release from pellets coated with surelease containing hydroxypropylmethylcellulose. *Drug Dev Ind Pharm* 2001; **27**(5): 419–430.
9 Goracinova K, *et al.* Preparation, physical characterization, mechanisms of drug/polymer interactions, and stability studies of controlled-release solid dispersion granules containing weak base as active substance. *Drug Dev Ind Pharm* 1996; **22**(3): 255–262.
10 Lin S. Studies on microencapsulation. 14. Theophylline bioavailability after single oral-administration of sustained-release microcapsules. *Curr Ther Res Clin Exp* 1987; **41**(4): 564–573.
11 Ayenew Z, *et al.* Trends in pharmaceutical taste masking technologies; a patent review. *Recent Pat Drug Deliv Formul* 2009; **3**(1): 26–39.

12 Sohi H, *et al.* Taste masking technologies in oral pharmaceuticals: recent developments and approaches. *Drug Dev Ind Pharm* 2004; **30**(5): 429–448.
13 Pollock D, Sheskey P. Micronized ethylcellulose: opportunities in direct-compression controlled-release tablets. *Pharm Technol* 1996; **20**(9): 120–130.
14 Klinger GH, *et al.* Formulation of controlled release matrices by granulation with a polymer dispersion. *Drug Dev Ind Pharm* 1990; **16**(9): 1473–1490.
15 Katikaneni P, *et al.* Ethyl cellulose matrix controlled-release tablets of a water-soluble drug. *Int J Pharm* 1995; **123**: 119–125.
16 Kulvanich P, *et al.* Release characteristics of the matrices prepared from co-spray-dried powders of theophylline and ethylcellulose. *Drug Dev Ind Pharm* 2002; **28**: 727–739.
17 Kent DJ, Rowe RC. Solubility studies on ethyl cellulose used in film coating. *J Pharm Pharmacol* 1978; **30**: 808–810.
18 Rowe RC. The prediction of compatibility/incompatibility in blends of ethyl cellulose with hydroxypropyl methylcellulose or hydroxypropyl cellulose using 2-dimensional solubility parameter maps. *J Pharm Pharmacol* 1986; **38**: 214–215.
19 Saettone MF, *et al.* Effect of different polymer-plasticizer combinations on '*in vitro*' release of theophylline from coated pellets. *Int J Pharm* 1995; **126**: 83–88.
20 Beck M, Tomka I. On the equation of state of plasticized ethyl cellulose of varying degrees of substitution. *Macromolecules* 1996; **29**(27): 8759–8766.
21 Celik M. Compaction of multiparticulate oral dosage forms. In: Ghebre-Sellassie I, ed. *Multiparticulate Oral Drug Delivery.* New York: Marcel Dekker, 1994: 181–215.
22 Robinson DH. Ethyl cellulose-solvent phase relationships relevant to coacervation microencapsulation processes. *Drug Dev Ind Pharm* 1989; **15**(14–16): 2597–2620.
23 Lavasanifar A, *et al.* Microencapsulation of theophylline using ethyl cellulose: *in vitro* drug release and kinetic modeling. *J Microencapsul* 1997; **14**(1): 91–100.
24 Moldenhauer M, Nairn J. The control of ethyl cellulose microencapsulation using solubility parameters. *J Control Release* 1992; **22**: 205–218.
25 Friedman M, *et al.* Inhibition of plaque formation by a sustained release delivery system for cetylpyridinium chloride. *Int J Pharm* 1988; **44**: 243–247.
26 Ruiz-Martinez A, *et al. In vitro* evaluation of benzylsalicylate polymer interaction in topical formulation. *Pharm Ind* 2001; **63**: 985–988.
27 Melzer E, *et al.* Ethylcellulose: a new type of emulsion stabilizer. *Eur J Pharm Biopharm* 2003; **56**: 23–27.
28 Sakellariou P, *et al.* The thermomechanical properties and glass transition temperatures of some cellulose derivatives used in film coating. *Int J Pharm* 1985; **27**: 267–277.
29 Callahan JC, *et al.* Equilibrium moisture content of pharmaceutical excipients. *Drug Dev Ind Pharm* 1982; **8**(3): 355–369.
30 Velazquez de la Cruz G, *et al.* Temperature effects on the moisture sorption isotherms for methylcellulose and ethylcellulose films. *J Food Engin* 2001; **48**: 91–94.
31 FAO/WHO. Evaluation of certain food additives and contaminants. Thirty-fifth report of the joint FAO/WHO expert committee on food additives. *World Health Organ Tech Rep Ser* 1990; No. 789.
32 Food and Drug Administration, Inactive Ingredient Database, www.accessdata.fda.gov/scripts/cder/iig/index.cfm (accessed 5 December 2011)
33 Lewis RJ, ed. *Sax's Dangerous Properties of Industrial Materials*, 11th edn. New York: Wiley, 2004; 1640.
34 Sharma P, Hamsa V. Formulation and evaluation of buccal mucoadhesive patches of terbutaline sulphate. *STP Pharma Sci* 2001; **11**: 275–281.
35 Streubel A, *et al.* Floating microparticles based on low density foam powder. *Int J Pharm* 2002; **241**: 279–292.
36 *Food Chemicals Codex*, 7th edn. Suppl. 3. Bethesda, MD: United States Pharmacopeia, 2011; 1698.

20 General References

Ashland Specialty Ingredients. Technical literature: *Aqualon* ethylcellulose (EC) physical and chemical properties, 2002.
Ashland Specialty Ingredients. Product literature: Pharmaceutical excipients. www.herc.com/aqualon/product_data/brochures/250_49.pdf#ec (accessed 5 December 2011).
Ashland Specialty Ingredients. Technical literature: Advanced structure-function properties of ethylcellulose: implications for tablet compactibility (PTR-021), 2002.
Ashland Specialty Ingredients. Technical literature: Compaction characteristics of high ethoxyl, low viscosity ethylcellulose (PTR-024), 2003.
Dow Chemical Company. Technical literature: *Ethocel* ethylcellulose polymers technical handbook, 2005.
European Directorate for the Quality of Medicines and Healthcare (EDQM). European Pharmacopoeia – State Of Work Of International Harmonisation. *Pharmeuropa* 2011; 23(4): 713–714. www.edqm.eu/site/-614.html (accessed 5 December 2011).
FMC Biopolymer. Technical literature: *Aquacoat ECD* ethylcellulose aqueous dispersion, 2004.
Majewicz T, Podlas T. Cellulose ethers. In: Kroschwitz J, ed. *Encyclopedia of Chemical Technology.* New York: Wiley, 2000; 445–466.
Morflex Inc. Technical literature: Pharmaceutical Coatings Bulletin, 1995; 102–103.
Morflex Inc. Dibutyl sebacate, NF, 2005.
Rekhi GS, Jambhekar SS. Ethylcellulose – a polymer review. *Drug Dev Ind Pharm* 1995; **21**(1): 61–77.
Vesey CF *et al.* Colorcon. Evaluation of alternative plasticizers for *Surelease*, an aqueous ethylcellulose dispersion for modified release film-coating, Controlled Release Society Annual Meeting, 2005.

21 Author

TC Dahl.

22 Date of Revision

1 March 2012.

Ethylene Glycol and Vinyl Alcohol Grafted Copolymer

1 Nonproprietary Names

PhEur: Macrogol Poly(Vinyl Alcohol) Grafted Copolymer
USP–NF: Ethylene Glycol and Vinyl Alcohol Grafted Copolymer

2 Synonyms

Copolymerum macrogolo et alcoholi poly(vinylico) constatum; ethenol graft polymer with oxirane; *Kollicoat IR*; polyvinyl alcohol-polyethylene glycol graft copolymer; polyvinylalcohol-co-polyethylene glycol; PVA-co-PEG; PVA-PEG graft copolymer.

3 Chemical Name and CAS Registry Number

Polyvinyl alcohol polyethylene glycol copolymer, poly(ethan-1,2-diol-grafted-ethenol) [96734-39-3]

4 Empirical Formula and Molecular Weight

Ethylene glycol and vinyl alcohol grafted copolymer is a synthetic branched copolymer. On average, 2-4 polyvinyl alcohol side chains are grafted onto a polyethylene glycol backbone. The polymer comprises approx. 25% w/w polyethylene glycol and 75% polyvinyl alcohol units. The mean molecular weight of the polymer is 40 000 to 50 000.

Ethylene glycol and vinyl alcohol grafted copolymer may contain a glidant such as colloidal silicon dioxide to improve flowability.

5 Structural Formula

See Section 4.

6 Functional Category

Coating agent; film-forming agent; tablet and capsule binder.

7 Applications in Pharmaceutical Formulation or Technology

Ethylene glycol and vinyl alcohol grafted copolymer is a synthetic polymer that was introduced to the market as an excipient for oral pharmaceutical applications in 2002. It is mainly used as a film-forming agent in instant-release tablet coatings.

The hydrophilic polymer is readily soluble in water, independent of the pH value. The polyvinyl alcohol part of the copolymer provides good film-forming properties, whereas the polyethylene glycol part acts as an internal plasticizer. The grafted copolymer offers functional advantages over the individual components, such as high flexibility of the polymer and low viscosity in aqueous solutions. The low viscosity of aqueous ethylene glycol and vinyl alcohol grafted copolymer solutions allows the processing of highly concentrated coating suspensions.[1,2] Owing to the high flexibility of the polymer, it may be suitable as a carrier in drug-loaded film strips or wafers.[3] It also presents advantages as a hydrophilic pore former in combination with different sustained-release coating agents (e.g. ethylcellulose and polyvinyl acetate), where drug release rates can be adjusted by adding varying amounts of the polymer.[4–8]

Ethylene glycol and vinyl alcohol grafted copolymer has been investigated as a wet binder[2] and as a carrier in solid dispersions.[9–11] It acts as a protective colloid and effectively stabilizes dispersions (e.g. ethylcellulose dispersions) and emulsions (e.g. medium-chain triglycerides/water).

8 Description

Ethylene glycol and vinyl alcohol grafted copolymer occurs as a free-flowing white to slightly yellowish powder. It is an amorphous polymer with partially crystalline domains.[11]

9 Pharmacopeial Specifications

See Table I.

Table I: Pharmacopeial specifications for **ethylene glycol and vinyl alcohol grafted copolymer.**

Test	PhEur 7.4	USP35-NF30
Characters	+	–
Identification	+	+
pH	5.0–8.0	5.0– 8.0
Ester value	10–75	10–75
Ethylene oxide	≤1 ppm	≤1 ppm
Dioxan	≤10 ppm	≤10 ppm
Impurity A (vinyl acetate)	≤100 ppm	≤100 ppm
Impurity B (acetic acid)	≤1.5%	≤1.5%
Sulfated ash	≤3%	–
Residue on ignition	–	≤3%
Loss on drying	≤5%	≤5%
Viscosity	≤250 mPa s[a]	25–250 mPa s

(a) Functionality-related characteristic in PhEur 7.4.

10 Typical Properties

Solubility Dissolves quickly in water and aqueous systems, e.g. weak acids or alkalis up to 40% w/w; solutions of up to 25% (w/w) can be prepared in a 1:1 ethanol–water mixture. Solubility in nonpolar solvents is low. If silicon dioxide is present, solutions are slightly turbid.

11 Stability and Storage Conditions

Ethylene glycol and vinyl alcohol grafted copolymer is slightly hygroscopic. Store in a tightly closed container.

12 Incompatibilities

—

13 Method of Manufacture

Ethylene glycol and vinyl alcohol grafted copolymer is produced using polyethylene glycol and vinyl acetate as starting materials. Polyethylene glycol forms the polymer backbone onto which polyvinyl acetate is grafted. Saponification of the polyvinyl acetate moieties leads to formation of polyvinyl alcohol grafted chains.

14 Safety

Ethylene glycol and vinyl alcohol grafted copolymer is used in oral pharmaceutical preparations, and is generally regarded as a safe material. It has been found to be nonirritating to rabbit eyes and skin, and also nonmutagenic.

LD_{50} (rat, oral) > 2.0 g/kg[12]

15 Handling Precautions

Observe normal precautions appropriate to the circumstances and quantity of the material handled.

16 Regulatory Status

Approved for use in medicinal products in Europe, Japan and the US.

17 Related Substances

Polyethylene glycol; polyvinyl alcohol.

18 Comments

Ethylene glycol and vinyl alcohol grafted copolymer is available as a spray-dried powder (*Kollicoat IR*), in combination with polyvinyl alcohol as a basis for moisture protective coatings (*Kollicoat IR* Protect), or as a ready-to-use colored system (*Kollicoat IR* Coating Systems, *Sepifilm IR*).

19 Specific References

1 BASF SE. Technical information sheet: *Kollicoat IR*, February 2010.
2 BASF SE. Product information: *Kollicoat IR*, 2009. http://www.pharma-ingredients.basf.com/Products (accessed 6 July 2011).
3 Garsuch V, Breitkreutz J. Novel analytical methods for the characterization of wafers. *Eur J Pharm Biopharm* 2009; **73**: 195–201.
4 Siepmann G, *et al.* How to adjust desired drug release patterns from ethylcellulose-coated dosage forms. *J Control Release* 2007; **119**: 182–189.
5 Siepmann F, *et al.* How to improve the storage stability of aqueous polymeric film coatings. *J Control Release* 2008; **126**: 26–33.
6 Ensslin S, *et al.* Modulating pH-independent release from coated pellets: effect of coating composition on solubilization processes and drug release. *Eur J Pharm Biopharm* 2009; **72**: 111–118.
7 Strübing S, *et al.* Mechanistic analysis of drug release from tablets with membrane controlled drug delivery. *Eur J Pharm Biopharm* 2007; **66**: 113–119.
8 Dashevsky A, *et al.* Effect of water-soluble polymers on the physical stability of aqueous polymeric dispersions and the implications on the drug release from coated pellets. *Drug Dev Ind Pharm* 2010; **36**(2): 152–160.
9 Janssens S, *et al.* The use of a new hydrophilic polymer, *Kollicoat IR*, in the formulations of solid dispersions of itraconazole. *Eur J Pharm Sci* 2007; **30**: 288–294.
10 Janssens S, *et al.* Spray drying from complex solvent systems broadens the applicability of *Kollicoat IR* as a carrier in the formulation of solid dispersions. *Eur J Pharm Sci* 2009; **37**: 241–248.
11 Guns S, *et al.* Characterization of the copolymer poly(ethyleneglycol-g-vinylalcohol) as a potential carrier in the formulation of solid dispersions. *Eur J Pharm Biopharm* 2010; **74**: 239–247.
12 Bühler V. *Kollicoat* Grades: Functional polymers for the pharmaceutical industry. BASF SE, January 2007.

20 General References

BASF. Material safety data sheet: *Kollicoat IR*, February 2007.
BASF SE. Excipients. http://www.pharma-ingredients.basf.com/Excipients/Home.aspx (accessed 6 July 2011).

21 Authors

F Guth, T Schmeller.

22 Date of Revision

1 March 2012.

Ethylene Glycol Stearates

1 Nonproprietary Names

BP: Ethylene Glycol Monopalmitostearate
PhEur: Ethylene Glycol Monopalmitostearate
USP–NF: Ethylene Glycol Stearates

2 Synonyms

Ethyleneglycoli monopalmitostearas.

3 Chemical Name and CAS Registry Number

Ethylene glycol palmitostearate
See Sections 8 and 17.

4 Empirical Formula and Molecular Weight

See Section 8.

5 Structural Formula

See Section 8.

6 Functional Category

Emollient; emulsion stabilizing agent.

7 Applications in Pharmaceutical Formulation or Technology

Ethylene glycol stearates are used as stabilizers for water-in-oil emulsions, although they have poor emulsifying properties. They have emollient properties and are also used as opacifying, thickening, and dispersing agents.

In cosmetics, ethylene glycol stearates are used as a ‘fatty body’ for lipsticks, as pearling agents in opalescent and cream shampoos, and as additives for tanning lubricants.

8 Description

The USP35–NF30 and PhEur 7.4 describe ethylene glycol stearates as a mixture of ethylene glycol monoesters and diesters of stearic and palmitic acids, containing not less than 50% of monoesters produced from the condensation of ethylene glycol and stearic acid, of vegetable or animal origin.

Ethylene glycol stearates occur as a white or almost white waxy solid.

9 Pharmacopeial Specifications

See Table I.

Table I: Pharmacopeial specifications for ethylene glycol stearates.

Test	PhEur 7.4	USP35–NF30
Characters	+	–
Identification	+	+
Melting point	54–60°C	54–60°C
Acid value	≤3.0	≤3.0
Iodine value	≤3.0	≤3.0
Saponification value	170–195	170–195
Composition of fatty acids		
Stearic acid	40.0–60.0%	40.0–60.0%
Total of palmitic acid and stearic acid	≥90.0%	≥90.0%
Free ethylene glycol	≤5.0%	≤5.0%
Total ash	≤0.1%	≤0.1%

10 Typical Properties

Melting point 54–60°C

Solubility Soluble in acetone and hot ethanol (95%); practically insoluble in water.

11 Stability and Storage Conditions

Ethylene glycol stearates should be stored in a cool, dark place, protected from light.

12 Incompatibilities

—

13 Method of Manufacture

Ethylene glycol stearates are produced from the condensation of ethylene glycol with stearic acid 50 of vegetable or animal origin.

14 Safety

Ethylene glycol stearates are mainly used in cosmetics and topical pharmaceutical formulations, where they are generally regarded as relatively nontoxic and nonirritant materials.

15 Handling Precautions

Observe normal precautions appropriate to the circumstances and quantity of material handled.

16 Regulatory Status

Included in nonparenteral medicines licensed in Europe.

17 Related Substances

Diethylene glycol monopalmitostearate; ethylene glycol monopalmitate; ethylene glycol monostearate; glyceryl monostearate; glyceryl palmitostearate.

Diethylene glycol monopalmitostearate

Synonyms Diethyleneglycoli monopalmitostearas; diethylene glycol palmitostearate.

Description The PhEur 7.4 describes diethylene glycol monopalmitostearate as a mixture of diethylene glycol monoesters and diesters of stearic and palmitic acids. It contains not less than 45.0% of monoesters produced from the condensation of diethylene glycol and stearic acid 50 of vegetable or animal origin. Diethylene glycol monopalmitostearate occurs as a white or almost white waxy solid.

Acid value ≤4.0

Iodine value ≤3.0

Melting point 43–50°C

Saponification value 150–170

Solubility Soluble in acetone and hot ethanol (95%); practically insoluble in water.

Ethylene glycol monopalmitate

CAS number [4219-49-2]

Ethylene glycol monostearate

Synonyms Ethylene glycol stearate; ethylene glycoli monostearas; ethyleni glycoli stearas; 2-hydroxyethyl ester stearic acid; *Monestriol EN-A*; *Monthyle*.

CAS number [111-60-4]

Empirical formula $C_{20}H_{40}O_3$

Molecular weight 328.60

Description Occurs as pale yellow flakes.

Melting point 57–63°C

Safety

LD_{50} (mouse, IP): 0.20 g/kg[1]

18 Comments

—

19 Specific References

1 Lewis RJ, ed. *Sax's Dangerous Properties of Industrial Materials*, 11th edn. New York: Wiley, 2004; 1669.

20 General References

Sweetman S, ed. *Martindale: The Complete Drug Reference*, 37th edn. London: Pharmaceutical Press, 2011: 2222.

21 Author

PJ Sheskey.

22 Date of Revision

1 March 2012.

Ethylene Vinyl Acetate

1 Nonproprietary Names

None adopted.

2 Synonyms

Acetic acid, ethylene ester polymer with ethane; *CoTran*; ethylene/vinyl acetate copolymer; EVA; EVA copolymer; EVM; poly(ethylene-*co*-vinyl acetate); VA/ethylene copolymer; vinyl acetate/ethylene copolymer.

3 Chemical Name and CAS Registry Number

Ethylene vinyl acetate copolymer [24937-78-8]

4 Empirical Formula and Molecular Weight

$(CH_2CH_2)_x[CH_2CH(CO_2CH_3)]_y$

See Section 5.

5 Structural Formula

Ethylene vinyl acetate copolymer is a random copolymer of ethylene and vinyl acetate.

6 Functional Category

Transdermal delivery component.

7 Applications in Pharmaceutical Formulation or Technology

Ethylene vinyl acetate copolymers are used as membranes and backings in laminated transdermal drug delivery systems. They can also be incorporated as components in backings in transdermal systems. Ethylene vinyl acetate copolymers have been shown to be an effective matrix and membrane for the controlled delivery.[1–11] The system for controlled release of atenolol can be further developed using ethylene vinyl acetate copolymers and plasticizers.[1]

8 Description

Ethylene vinyl acetate copolymers occur as odorless white waxy solids in pellet or powder form. Films are translucent.

9 Pharmacopeial Specifications

—

10 Typical Properties

Density 0.92–0.94 g/cm^3
Flash point 260°C
Melting point 75–102°C depending on polymer ratios.
Moisture vapor transmission rate *see* Table I.
Solubility Insoluble in water
Thickness *see* Table I.
Vinyl acetate content *see* Table I.

11 Stability and Storage Conditions

Ethylene vinyl acetate copolymers are stable under normal conditions and should be stored in a cool, dry, well-ventilated area. Films of ethylene vinyl acetate copolymers should be stored at 0–30°C and less than 75% relative humidity.

12 Incompatibilities

Ethylene vinyl acetate is incompatible with strong oxidizing agents and bases.

13 Method of Manufacture

Various molecular weights of random ethylene vinyl acetate copolymers can be obtained by catalytic copolymerization of ethylene and vinyl acetate, high-pressure radical polymerization, bulk continuous polymerization, or solution polymerization. Film membranes of ethylene vinyl acetate copolymers are manufactured using a melt-cast process.

14 Safety

Ethylene vinyl acetate is mainly used in topical pharmaceutical applications as a membrane or film backing. Generally it is regarded as a relatively nontoxic and nonirritant excipient.

15 Handling Precautions

Observe normal precautions appropriate to the circumstances and quantity of material handled. Ethylene vinyl acetate powder may form an explosive mixture with air.

16 Regulatory Status

Included in the FDA Inactive Ingredients Database (subcutaneous rod; intrauterine suppository; ophthalmic preparations; periodontal film; transdermal film). Included in nonparenteral medicines licensed in the UK.

17 Related Substances

—

18 Comments

Ethylene vinyl acetate copolymers have a wide variety of industrial uses. Properties of ethylene vinyl acetate copolymer films in terms of oxygen and moisture transfer rate are related to the vinyl acetate content and thickness. Higher levels of vinyl acetate result in increased lipophilicity, increased oxygen and moisture vapor permeability, and increased clarity, flexibility, toughness, and solvent solubility.

The PubChem Compound ID (CID) for ethylene vinyl acetate is 32742.

Table I: Characteristics of different *CoTran* (3M Drug Delivery Systems) film grades.

Grade	Thickness (μm)	Vinyl acetate (%)	Moisture vapor transmission rate (g/m^2/24 h)
CoTran 9702	50.8	9	52.8
CoTran 9705	76.2	9	35.2
CoTran 9706	101.6	9	26.4
CoTran 9707	50.8	4.5	15.7
CoTran 9712	50.8	19	97.2
CoTran 9715	76.2	19	64.8
CoTran 9716	101.6	19	48.6
CoTran 9728	50.8	19	97.2

19 Specific References

1 Kim J, Shin SC. Controlled release of atenolol from the ethylene–vinyl acetate matrix. *Int J Pharm* 2004; **273**(1–2): 23–27.
2 Shin SC, Choi JS. Enhanced bioavailability of atenolol by transdermal administration of the ethylene–vinyl acetate matrix in rabbits. *Eur J Pharm Biopharm* 2003; **56**(3): 439–443.
3 Shin SC, Lee HJ. Controlled release of triprolidine using ethylene–vinyl acetate membrane and matrix systems. *Eur J Pharm Biopharm* 2002; **54**(2): 201–206.
4 Shin SC, Lee HJ. Enhanced transdermal delivery of triprolidone from the ethylene–vinyl acetate matrix. *Eur J Pharm Biopharm* 2002; **54**(3): 325–328.
5 Cho CW, *et al.* Controlled release of furosemide from the ethylene-vinyl acetate matrix. *Int J Pharm* 2005; **299**: 127–133.
6 Cho CW, *et al.* Enhanced transdermal absorption and pharmacokinetic evaluation of pranoprofen-ethylene-vinyl acetate matrix containing penetration enhancer in rats. *Arch Pharm Res* 2009; **32**(5): 747–753.
7 Cho CW, *et al.* Enhanced transdermal delivery of loratadine from the EVA matrix. *Drug Deliv* 2009; **16**(4): 230–235.
8 Krishnaiah YS, *et al.* Effect of PEG6000 on the *in vitro* and *in vivo* transdermal permeation of ondansetron hydrochloride from EVA1802 membranes. *Pharm Dev Technol* 2009; **14**(1): 50–61.
9 Cho CW, *et al.* Enhanced transdermal controlled delivery of glimepiride from the ethylene-vinyl acetate matrix. *Drug Deliv* 2009; **16**(6): 320–330.
10 Choi JS, Shin SC. Enhanced bioavailability of ambroxol by transdermal administration of the EVA matrix containing penetration enhancer in rats. *Biomol Ther* 2010; **18**(1): 106–110.
11 Prodduturi S, *et al.* Transdermal delivery of fentanyl from matrix and reservoir systems: effect of heat and compromised skin. *J Pharm Sci* 2010; **99**(5): 2357–2366.

20 General References

3M Drug Delivery Systems. *CoTran*. http://solutions.3m.com/wps/portal/3M/en_WW/3M-DDSD/Drug-Delivery-Systems/Transdermal-Micro-needle-Directory/Componentry/Membranes/ (accessed 1 November 2011).

21 Author

D Traini.

22 Date of Revision

1 November 2011.

Ethylparaben

1 Nonproprietary Names

BP: Ethyl Hydroxybenzoate
JP: Ethyl Parahydroxybenzoate
PhEur: Ethyl Parahydroxybenzoate
USP–NF: Ethylparaben

2 Synonyms

Aethylum hydrobenzoicum; *CoSept E*; E214; ethylis parahydroxybenzoas; ethyl *p*-hydroxybenzoate; *Ethyl parasept*; 4-hydroxybenzoic acid ethyl ester; *Nipagin A*; *Solbrol A*; *Tegosept E*; *Uniphen P-23*.

3 Chemical Name and CAS Registry Number

Ethyl-4-hydroxybenzoate [120-47-8]

4 Empirical Formula and Molecular Weight

$C_9H_{10}O_3$ 166.18

5 Structural Formula

6 Functional Category

Antimicrobial preservative.

7 Applications in Pharmaceutical Formulation or Technology

Ethylparaben is widely used as an antimicrobial preservative in cosmetics,[1] food products, and pharmaceutical formulations.

It may be used either alone or in combination with other paraben esters or with other antimicrobial agents. In cosmetics it is one of the most frequently used preservatives.

The parabens are effective over a wide pH range and have a broad spectrum of antimicrobial activity, although they are most effective against yeasts and molds; *see* Section 10.

Owing to the poor solubility of the parabens, paraben salts, particularly the sodium salt, are frequently used. However, this may cause the pH of poorly buffered formulations to become more alkaline.

See Methylparaben for further information.

8 Description

Ethylparaben occurs as a white, odorless or almost odorless, crystalline powder.

9 Pharmacopeial Specifications

The pharmacopeial specifications for ethylparaben have undergone harmonization of many attributes for JP, PhEur, and USP–NF.

See Table I. *See also* Section 18.

10 Typical Properties

Antimicrobial activity

Ethylparaben exhibits antimicrobial activity from pH 4–8. Preservative efficacy decreases with increasing pH owing to the formation of the phenolate anion. Parabens are more active against yeasts and molds than against bacteria. They are also more active against Gram-positive than against Gram-negative bacteria.

SEM 1: Excipient: ethylparaben; magnification: 600×.

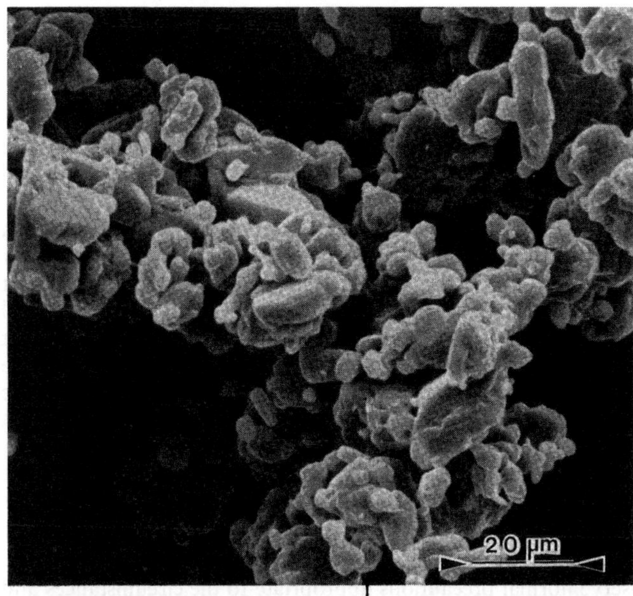

SEM 2: Excipient: ethylparaben; magnification: 3000×.

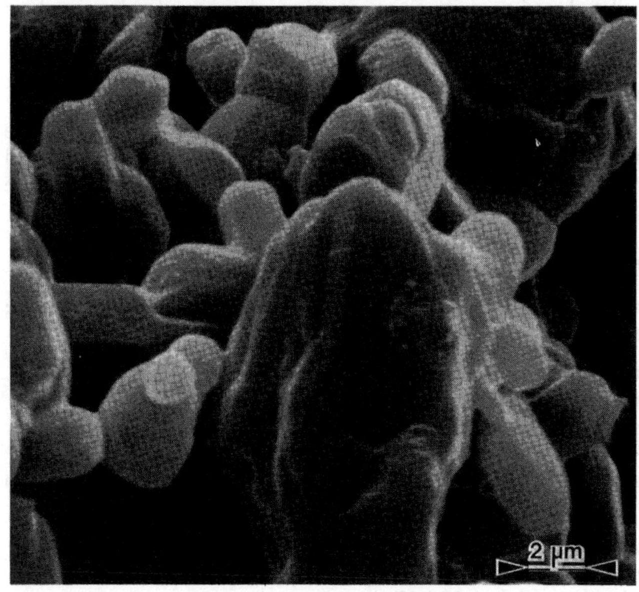

Table I: Pharmacopeial specifications for ethylparaben.

Test	JP XV	PhEur 7.4	USP35–NF30
Identification	+	+	+
Appearance of solution	+	+	+
Characters	—	+	—
Heavy metals[(a)]	≤20 ppm	—	—
Acidity	+	+	+
Melting range	115–118°C	115–118°C	115–118°C
Related substances	+	+	+
Residue on ignition	≤0.1%	≤0.1%	≤0.1%
Assay (dried basis)	98.0–102.0%	98.0–102.0%	98.0–102.0%

(a) These tests have not been fully harmonized at the time of publication.

The activity of the parabens increases with increasing chain length of the alkyl moiety, but solubility decreases. Activity may be improved by using combinations of parabens since synergistic effects occur. Ethylparaben is commonly used with methylparaben and propylparaben in oral and topical formulations (such mixtures are commercially available; for example, *Nipasept* (Nipa Laboratories Inc.). Activity has also been reported to be improved by the addition of other excipients; *see* Methylparaben for further information.

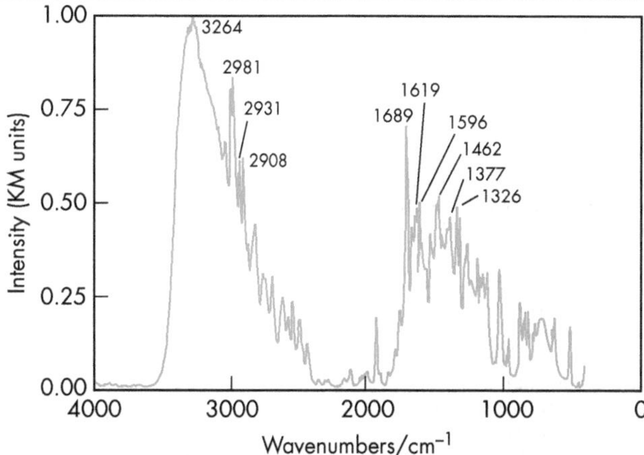

Figure 1: Infrared spectrum of ethylparaben measured by diffuse reflectance. Adapted with permission of Informa Healthcare.

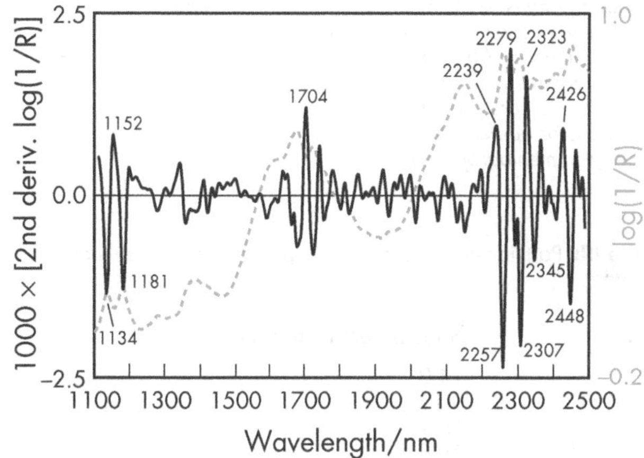

Figure 2: Near-infrared spectrum of ethylparaben measured by reflectance.

See Table II for minimum inhibitory concentrations of ethylparaben.[2]

Boiling point 297–298°C with decomposition.

Melting point 115–118°C

Partition coefficient The values for different vegetable oils vary considerably and are affected by the purity of the oil; *see* Table III.[3]

Solubility *see* Table IV.

Spectroscopy

IR spectra *see* Figure 1.

NIR spectra *see* Figure 2.

11 Stability and Storage Conditions

Aqueous ethylparaben solutions at pH 3–6 can be sterilized by autoclaving, without decomposition.[4] At pH 3–6, aqueous solutions are stable (less than 10% decomposition) for up to about 4 years at room temperature, while solutions at pH 8 or above are subject to rapid hydrolysis (10% or more after about 60 days at room temperature).[5]

Ethylparaben should be stored in a well-closed container in a cool, dry place.

Table II: Minimum inhibitory concentrations (MICs) for ethylparaben in aqueous solution.[2]

Microorganism	MIC (μg/mL)
Aerobacter aerogenes ATCC 8308	1200
Aspergillus niger ATCC 9642	500
Aspergillus niger ATCC 10254	400
Bacillus cereus var. *mycoides* ATCC 6462	1000
Bacillus subtilis ATCC 6633	1000
Candida albicans ATCC 10231	500
Enterobacter cloacae ATCC 23355	1000
Escherichia coli ATCC 8739	1000
Escherichia coli ATCC 9637	1000
Klebsiella pneumoniae ATCC 8308	500
Penicillium chrysogenum ATCC 9480	250
Penicillium digitatum ATCC 10030	250
Proteus vulgaris ATCC 13315	500
Pseudomonas aeruginosa ATCC 9027	>2000
Pseudomonas aeruginosa ATCC 15442	>2000
Pseudomonas stutzeri	1000
Rhizopus nigricans ATCC 6227A	250
Saccharomyces cerevisiae ATCC 9763	500
Salmonella typhosa ATCC 6539	1000
Serratia marcescens ATCC 8100	1000
Staphylococcus aureus ATCC 6538P	1000
Staphylococcus epidermidis ATCC 12228	1000
Trichophyton mentagrophytes	125

Table III: Partition coefficients for ethylparaben in vegetable oil and water.[3]

Solvent	Partition coefficient oil : water
Corn oil	14.0
Mineral oil	0.13
Peanut oil	16.1
Soybean oil	18.8

Table IV: Solubility of ethylparaben in various solvents.

Solvent	Solubility at 20°C unless otherwise stated
Acetone	Freely soluble
Ethanol	1 in 1.4
Ethanol (95%)	1 in 2
Ether	1 in 3.5
Glycerin	1 in 200
Methanol	1 in 0.9
Mineral oil	1 in 4000
Peanut oil	1 in 100
Propylene glycol	1 in 4
Water	1 in 1250 at 15°C
	1 in 910
	1 in 120 at 80°C

12 Incompatibilities

The antimicrobial properties of ethylparaben are considerably reduced in the presence of nonionic surfactants as a result of micellization.[6] Absorption of ethylparaben by plastics has not been reported, although it appears probable given the behavior of other parabens. Ethylparaben is coabsorbed on silica in the presence of ethoxylated phenols.[7] Yellow iron oxide, ultramarine blue, and aluminum silicate extensively absorb ethylparaben in simple aqueous systems, thus reducing preservative efficacy.[8,9]

Ethylparaben is discolored in the presence of iron and is subject to hydrolysis by weak alkalis and strong acids.

See also Methylparaben.

13 Method of Manufacture

Ethylparaben is prepared by the esterification of *p*-hydroxybenzoic acid with ethanol (95%).

14 Safety

Ethylparaben and other parabens are widely used as antimicrobial preservatives in cosmetics, food products, and oral and topical pharmaceutical formulations.

Systemically, no adverse reactions to parabens have been reported, although they have been associated with hypersensitivity reactions. Parabens, *in vivo*, have also been reported to exhibit estrogenic responses in fish.[10] The WHO has set an estimated total acceptable daily intake for methyl-, ethyl-, and propylparabens at up to 10 mg/kg body-weight.[11] A report has been published on the safety assessment of parabens including ethylparaben in cosmetic products.[12]

LD_{50} (mouse, IP): 0.52 g/kg[13]
LD_{50} (mouse, oral): 3.0 g/kg

15 Handling Precautions

Observe normal precautions appropriate to the circumstances and quantity of material handled. Ethylparaben may be irritant to the skin, eyes, and mucous membranes, and should be handled in a well ventilated environment. Eye protection, gloves, and a dust mask or respirator are recommended.

16 Regulatory Status

Accepted as a food additive in Europe. Included in the FDA Inactive Ingredients Database (oral, otic, and topical preparations). Included in nonparenteral medicines licensed in the UK. Included in the Canadian Natural Health Products Ingredients Database.

17 Related Substances

Butylparaben; ethylparaben potassium; ethylparaben sodium; methylparaben; propylparaben.

Ethylparaben potassium
Empirical formula $C_9H_9KO_3$
Molecular weight 204.28
CAS number [36547-19-9]
Synonyms Ethyl 4-hydroxybenzoate potassium salt; potassium ethyl hydroxybenzoate.

Ethylparaben sodium
Empirical formula $C_9H_9NaO_3$
Molecular weight 188.17
CAS number [35285-68-8]
Synonyms E215; ethyl 4-hydroxybenzoate sodium salt; sodium ethyl hydroxybenzoate.

18 Comments

Ethylparaben has undergone harmonization of many attributes for JP, PhEur, and USP–NF by the Pharmacopeial Discussion Group. For further information see the General Information Chapter <1196> in the USP35–NF30, the General Chapter 5.8 in PhEur 7.4, along with the 'State of Work' document on the PhEur EDQM website, and also the General Information Chapter 8 in the JP XV.

See Methylparaben for further information.

The EINECS number for ethylparaben is 204-399-4. The PubChem Compound ID (CID) for ethylparaben is 8434.

19 Specific References

1 Rastogi SC, *et al.* Contents of methyl-, ethyl-, propyl-, butyl- and benzylparaben in cosmetic products. *Contact Dermatitis* 1995; 32(1): 28–30.

2 Haag TE, Loncrini DF. Kabara JJ, ed. *Cosmetic and Drug Preservation.* New York: Marcel Dekker, 1984: 63–77.
3 Wan LSC, *et al.* Partition of preservatives in oil/water systems. *Pharm Acta Helv* 1986; **61**(10–11): 308–313.
4 Aalto TR, *et al.* *p*-Hydroxybenzoic acid esters as preservatives I: uses, antibacterial and antifungal studies, properties and determination. *J Am Pharm Assoc (Sci)* 1953; **42**: 449–457.
5 Kamada A, *et al.* Stability of *p*-hydroxybenzoic acid esters in acidic medium. *Chem Pharm Bull* 1973; **21**: 2073–2076.
6 Aoki M, *et al.* [Application of surface active agents to pharmaceutical preparations I: effect of Tween 20 upon the antifungal activities of *p*-hydroxybenzoic acid esters in solubilized preparations.] *J Pharm Soc Jpn* 1956; **76**: 939–943[in Japanese].
7 Daniels R, Rupprecht H. Effect of coadsorption on sorption and release of surfactant paraben mixtures from silica dispersions. *Acta Pharm Technol* 1985; **31**: 236–242.
8 Sakamoto T, *et al.* Effects of some cosmetic pigments on the bactericidal activities of preservatives. *J Soc Cosmet Chem* 1987; **38**: 83–98.
9 Allwood MC. The adsorption of esters of *p*-hydroxybenzoic acid by magnesium trisilicate. *Int J Pharm* 1982; **11**: 101–107.
10 Pedersen KL, *et al.* The preservatives ethyl-, propyl-, and butylparaben are oestrogenic in an *in vivo* fish assay. *Pharmacol Toxicol* 2000; **86**(3): 110–113.
11 FAO/WHO. Toxicological evaluation of certain food additives with a review of general principles and of specifications. Seventeenth report of the FAO/WHO expert committee on food additives. *World Health Organ Tech Rep Ser* 1974; No. 539.
12 Anonymous. Final amended report on the safety assessment of methylparaben, ethylparaben, propylparaben, isopropylparaben, butylparaben, isobutylparaben, and benzylparaben as used in cosmetic products. *Int J Toxicol* 2008; **27**(Suppl. 4): 1–82.
13 Lewis RJ, ed. *Sax's Dangerous Properties of Industrial Materials*, 11th edn. New York: Wiley, 2004; 2003–2004.

20 General References

European Directorate for the Quality of Medicines and Healthcare (EDQM). European Pharmacopoeia – State Of Work Of International Harmonisation. *Pharmeuropa* 2011; 23(4): 713–714. www.edqm.eu/site/-614.html (accessed 2 December 2011).
Golightly LK, *et al.* Pharmaceutical excipients: adverse effects associated with inactive ingredients in drug products (part I). *Med Toxicol* 1988; **3**: 128–165.
Schnuch A, *et al.* Contact allergy to preservatives. Analysis of IVDK data 1996-2009. *Br J Dermatol* 2011; **164**: 1316–1325.
The HallStar Company. Material safety data sheet: *CoSept E*, 2007.

21 Author

N Sandler.

22 Date of Revision

1 March 2012.

E

Fructose

1 Nonproprietary Names

BP: Fructose
JP: Fructose
PhEur: Fructose
USP–NF: Fructose

2 Synonyms

Advantose FS 95; D-arabino-2-hexulose; D-arabino-hex-2-ulopyranose; *Fructamyl*; *Fructofin*; D-(–)-fructopyranose; β-D-fructose; fructosum; fruit sugar; *Krystar*; laevulose; levulose; nevulose.

3 Chemical Name and CAS Registry Number

D-Fructose [57-48-7]

4 Empirical Formula and Molecular Weight

$C_6H_{12}O_6$ 180.16

5 Structural Formula

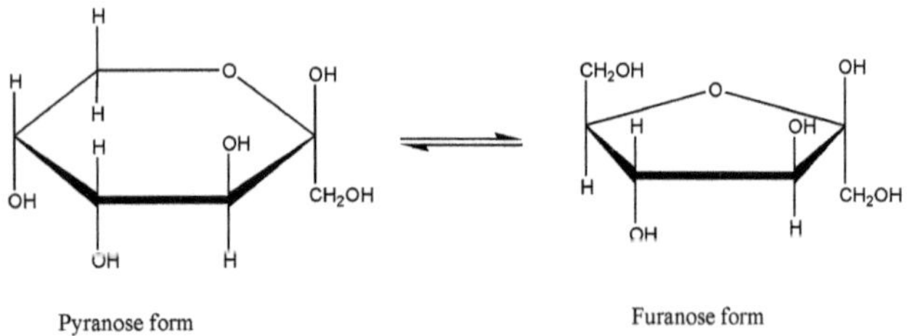

See Section 18.

6 Functional Category

Flavoring agent; sweetening agent; tablet and capsule diluent; taste-masking agent.

7 Applications in Pharmaceutical Formulation or Technology

Fructose is used in tablets, syrups, and solutions as a flavoring and sweetening agent.

When used as a sweetening agent, fructose is perceived as sweeter than mannitol and sorbitol at equivalent concentration. The sweet taste is perceived more rapidly than that of sucrose and dextrose. Fructose has greater solubility in ethanol than dextrose and sucrose and may therefore be a more suitable sweetener for alcohol-based formulations.

Fructose is more water soluble than sucrose and dextrose and can present fewer problems with undesired crystallisation, for example when solutions are refrigerated or deposited on bottle threads (e.g. "cap-locking").

At a given concentration fructose has a higher osmotic potential than sucrose, and therefore lower water activity in solution. This may be beneficial for microbial stability of solutions.

When used as a tablet diluent, fructose produces satisfactory tablet crushing strength only at slow tableting speeds. A fructose : -sorbitol (3 : 1) mixture has better tableting characteristics for direct compression. Pregranulation of fructose with 3.5% povidone also produces a satisfactory tablet excipient.[1]

8 Description

Fructose occurs as odorless, colorless crystals or a white crystalline powder with a very sweet taste.

9 Pharmacopeial Specifications

See Table I.

Table I: Pharmacopeial specifications for fructose.

Test	JP XV	PhEur 7.4	USP35–NF30
Identification	+	+	+
Characters	—	+	—
Color of solution	+	+	+
Acidity	+	+	+
pH	4.0–6.5	—	—
Specific optical rotation	—	–91.0° to –93.5°	—
Foreign sugars	—	+	—
Loss on drying	≤0.5%	—	≤0.5%
Residue on ignition	≤0.1%	≤0.1%	≤0.5%
Chloride	≤0.018%	—	≤0.018%
Sulfate	≤0.024%	—	≤0.025%
Sulfite	+	—	—
Water	—	≤0.5%	—
Arsenic	≤1.3 ppm	—	≤1 ppm
Barium	—	+	—
Calcium and magnesium (as calcium)	+	—	≤0.005%
Lead	—	≤0.5 ppm	—
Heavy metals	≤4 ppm	—	≤5 ppm
Hydroxymethylfurfural	+	+	+
Assay (dried basis)	≥98.0%	—	98.0–102.0%

10 Typical Properties

Acidity/alkalinity pH = 5.35 (9% w/v aqueous solution)
Angle of repose 38.8° for *Advantose FS 95*
Density 1.58 g/cm^3. *See also* Table II.
Heat of combustion 15.3 kJ/g (3.66 kcal/g)
Heat of solution 50.2 kJ/g (12 kcal/g)
Hygroscopicity At 25°C and relative humidities above approximately 60%, fructose absorbs significant amounts of moisture; *see* Figure 1.
Melting point ≈102–105°C (with decomposition); ≈103°C for *Fruitose*.

Osmolarity A 5.05% w/v aqueous solution is isoosmotic with serum.
Particle size distribution The average particle size of standard-grade crystalline fructose is 170–450 μm. The average particle size of powdered fructose is 25–40 μm.
Refractive index *see* Table II.
Solubility *see* Table III.
Specific rotation $[\alpha]_D^{20}$ = −132° to −92° (2% w/v aqueous solution). Note that fructose shows rapid and anomalous mutarotation involving pyranose–furanose interconversion. The final value may be obtained in the presence of hydroxide ions. *See also* Section 18.
Spectroscopy

IR spectra *see* Figure 2.

NIR spectra *see* Figure 3.

Viscosity (dynamic) *see* Table II.

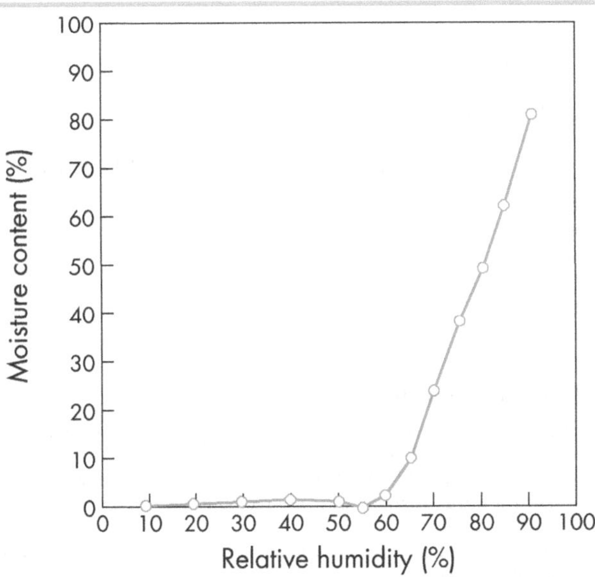

Figure 1: Equilibrium moisture content of fructose at 25°C.

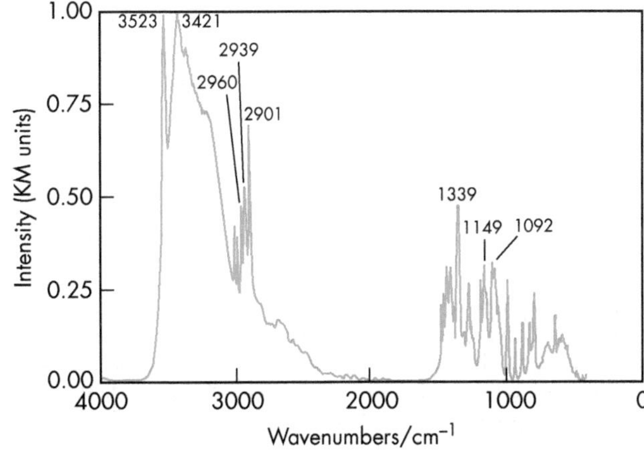

Figure 2: Infrared spectrum of fructose measured by diffuse reflectance. Adapted with permission of Informa Healthcare.

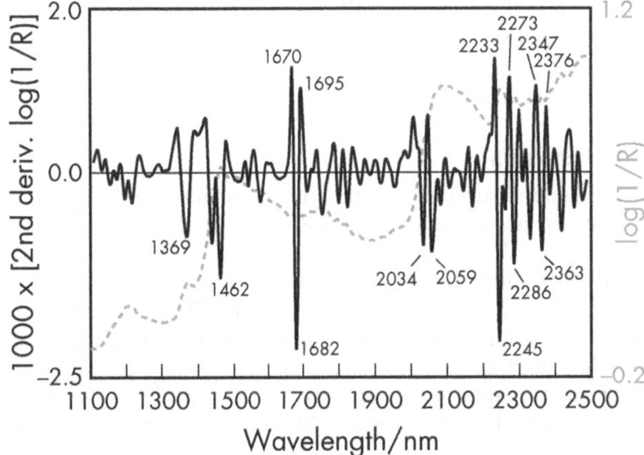

Figure 3: Near-infrared spectrum of fructose measured by reflectance.

Table II: Physical properties of aqueous fructose solutions at 20°C.

Concentration of aqueous fructose solution (% w/w)	Density (g/cm^3)	Refractive index	Viscosity, dynamic (mPa s)
10	1.04	1.3477	1.35
20	1.08	1.3633	1.80
30	1.13	1.3804	2.90
40	1.18	1.3986	5.60
50	1.23	1.4393	34.0
60	1.29	1.4853	309.2

Table III: Solubility of fructose.

Solvent	Solubility at 20°C
Ethanol (95%)	1 in 15
Methanol	1 in 14
Water	1 in 0.3

11 Stability and Storage Conditions

Fructose is hygroscopic and absorbs significant amounts of moisture at relative humidities greater than 60%. If stored in the original sealed packaging at temperatures below 25°C/60% RH it can be expected to retain stability for at least 12 months.

Aqueous solutions are most stable at pH 3–4 and temperatures of 4–70°C; they may be sterilized by autoclaving.

12 Incompatibilities

Incompatible with strong acids or alkalis, forming a brown coloration. In the aldehyde form, fructose can react with amines, amino acids, peptides, and proteins. Fructose may cause browning of tablets containing amines.

13 Method of Manufacture

Fructose, a monosaccharide sugar, occurs naturally in honey and a large number of fruits. It may be prepared from inulin, dextrose, or sucrose by a number of methods. Commercially, fructose is mainly manufactured by crystallization from high-fructose syrup derived from hydrolyzed and isomerized cereal starch or cane and beet sugar.

14 Safety

Although it is absorbed more slowly than dextrose from the gastrointestinal tract, fructose is metabolized more rapidly. Metabolism of fructose occurs mainly in the liver, where it is converted partially to dextrose and the metabolites lactic acid and pyruvic acid. Entry into the liver and subsequent phosphorylation is insulin-independent. Further metabolism occurs by way of a variety of metabolic pathways. In healthy and well regulated diabetics, glycogenesis (glucose stored as glycogen) predominates.

Excessive oral fructose consumption (>75 g daily) in the absence of dietary dextrose in any form (e.g. sucrose, starch, dextrin, etc.) may cause malabsorption in susceptible individuals, which may result in flatulence, abdominal pain, and diarrhea. Except in patients with hereditary fructose intolerance,[3,4] there is no evidence to indicate that oral fructose intake at current levels is a risk factor in any particular disease, other than dental caries.[5]

See also Section 18.

15 Handling Precautions

Observe normal precautions appropriate to the circumstances and quantity of material handled. Fructose may be irritant to the eyes. Eye protection and gloves are recommended.

16 Regulatory Status

Included in the FDA Inactive Ingredients Database (oral solutions, syrup, and suspensions; rectal preparations; intravenous infusions). Included in the Canadian Natural Health Products Ingredients Database.

17 Related Substances

Dextrose; fructose and pregelatinized starch; high-fructose syrup; liquid fructose; powdered fructose; sucrose.

High-fructose syrup

Comments A syrup most commonly containing 42% or 55% fructose, with the remainder consisting of dextrose and small amounts of oligosaccharides. It is a colorless, odorless, highly viscous syrup with a sweet taste.

Liquid fructose

Comments A syrup containing ≥99.5% fructose, made by solubilizing crystalline fructose in water. It is a colorless, odorless, highly viscous syrup with a sweet taste.

Powdered fructose

Comments Finely ground crystalline fructose containing ≤2% silicon dioxide as a glidant.

18 Comments

Co-processed mixtures of starch and fructose have been reported to be suitable for use in direct compression orally disintegrating tablet formulations.[6] A directly compressible grade of fructose, containing a small amount of starch (*Advantose FS 95*, SPI Pharma) is commercially available; *see also* Fructose and Pregelatinized Starch.

The coprecipitation of fructose with hydrophobic drugs such as digoxin has been shown to enhance the dissolution profile of such drugs. Fructose apparently acts as a water-soluble carrier upon coprecipitation, thereby allowing hydrophobic drugs to be more readily wetted.[2]

Fructose can occur in both the furanose and pyranose forms. Fructose present in natural products occurs in the furanose form, while that produced by crystallization occurs in the pyranose form. An aqueous solution at 20°C contains about 20% of the furanose form.

Although fructose has been proposed for use in the diabetic diet, it is not regarded as a suitable source of carbohydrate, although it does have value as a sweetening agent.[7] The British Diabetic Association has recommended that intake of fructose be limited to 25 g daily.[8]

Fructose has been used as an alternative to dextrose in parenteral nutrition, but its use is not recommended by some because of the risk of lactic acidosis. Although popular in many countries, it has therefore been suggested that the use of intravenous infusions containing fructose and sorbitol should be abandoned.[4,9]

Fructose is the sweetest of all sugars; *see* Table IV. A specification for fructose is contained in the *Food Chemicals Codex* (FCC).[10]

The EINECS number for fructose is 200-333-3. The PubChem Compound ID (CID) for fructose is 5984.

Table IV: Relative sweetness of fructose and other sugars.

Sugar	Relative sweetness at 25°C (10% solids)
Fructose	117
Sucrose	100
High fructose syrup-55	99
High fructose syrup-42	92
Dextrose	65

19 Specific References

1 Osberger TF. Tableting characteristics of pure crystalline fructose. *Pharm Technol* 1979; 3(6): 81–86.

2 Ahmed SU, Madan PL. Evaluation of the *in vitro* release profile of digoxin from drug-carbohydrate coprecipitates. *Drug Dev Ind Pharm* 1991; **17**: 831–842.

3 Cox TM. An independent diagnosis: a treatable metabolic disorder diagnosed by molecular analysis of human genes. *Br Med J* 1990; **300**: 1512–1514.

4 Collins J. Metabolic disease. Time for fructose solutions to go. *Lancet* 1993; **341**: 600.

5 Glinsman WH *et al. Evaluation of Health Aspects of Sugars Contained in Carbohydrate Sweeteners: Report of Sugars Task Force*. Washington, DC: Health and Human Services Center for Food Safety and Applied Nutrition, Food and Drug Administration, 1986.

6 Morales JO, *et al.* Orally disintegrating tablets using starch and fructose. *Pharm Technol* 1979; 3(6): 81–86.

7 Anonymous. Has fructose a place in the diabetic diet? *Drug Ther Bull* 1980; **18**(17): 67–68.

8 Clarke BP. Is it harmful to a juvenile diabetic to substitute sorbitol and fructose for ordinary sugar? *Br Med J* 1987; **294**: 422.

9 Sweetman SC, ed. *Martindale: The Complete Drug Reference*, 37th edn. London: Pharmaceutical Press, 2011: 2103.

10 *Food Chemicals Codex*, 7th edn. Bethesda, MD: United States Pharmacopeia, 2010: 413.

20 General References

Cheeseman C. Fructose the odd man out. Why is the genomic control of intestinal GLUT5 expression different? *J Physiol (Oxford, UK)* 2008; 586: 3563.

JRS Pharma. Product information: *ProSolv* ODT, 2010.

Muldering KB. Placebo evaluation of selected sugar-based excipients in pharmaceutical and nutraceutical tableting. *Pharm Technol* 2000; **24**(5): 34, 36, 38, 40, 42, 44.

21 Author

A Balasundaram.

22 Date of Revision

1 March 2012.

Fructose and Pregelatinized Starch

1 Nonproprietary Names

None adopted.

2 Synonyms

Advantose FS 95.

3 Chemical Name and CAS Registry Number

See Section 8.

4 Empirical Formula and Molecular Weight

See Section 8.

5 Structural Formula

See Section 8.

6 Functional Category

Direct compression excipient; taste-masking agent.

7 Applications in Pharmaceutical Formulation or Technology

Fructose and pregelatinized starch is used in the direct compression of tablets to improve flowability and compressibility.[1] Additionally, so-called "chewable" tablets can be prepared using fructose and pregelatinized starch.

Fructose and pregelatinized starch is also used to mask undesirable flavors of components in tablet formulations.

8 Description

Fructose and pregelatinized starch occurs as a white crystalline powder containing 95% fructose and 5% pregelatinized starch (a co-dried system of these two ingredients).[(1)]

9 Pharmacopeial Specifications

Both fructose and pregelatinized starch are listed as separate monographs in the BP 2012, JP XV (only fructose), PhEur 7.4, and USP35–NF30, but the combination is not listed. *See* Fructose and Starch, Pregelatinized.

10 Typical Properties

Angle of repose 12° for *Advantose FS 95*
Density (bulk) 0.55–0.75 g/cm^3 for *Advantose FS 95*
Density (tapped) 0.50–0.75 g/cm^3 for *Advantose FS 95*[2]
Loss on drying $\leq$2.0% for *Advantose FS 95*[(2)]
Moisture content (KF) 1–2% for *Advantose FS 95*
Particle size distribution Over 80% of the crystalline powder is retained on a #60 mesh (250 μm) screen for *Advantose FS 95*.
Solubility Partially soluble in water for *Advantose FS 95*.

See also Fructose and Starch, Pregelatinized.

11 Stability and Storage Conditions

Fructose and pregelatinized starch is a chemically stable mixture. Protect from excessive temperatures and humidity. Optimum storage conditions are 25°C and 50% relative humidity. The container should be kept tightly closed and in a cool, well-ventilated place.[(3)]

12 Incompatibilities

Fructose and pregelatinized starch may react with oxidizing agents.[(3)] Upon decomposition, it emits carbon monoxide, carbon dioxide and/or low molecular weight hydrocarbons. Fructose is a reducing sugar and can react with amines, amino acids, peptides, and proteins.

See Fructose and Starch, Pregelatinized.

13 Method of Manufacture

Fructose and pregelatinized starch is prepared by co-drying a mixture of the two ingredients.

14 Safety

Fructose and pregelatinized starch is a nonflammable white crystalline solid.[3] Eye contact may produce slight irritation. No skin irritation may be expected from single short-term exposure, although prolonged or repeated contact may produce some irritation. Ingestion of large amounts may produce gastrointestinal disturbances. Overexposure to dusts may produce irritation of the respiratory system.

See Fructose and Starch, Pregelatinized.

15 Handling Precautions

Observe normal precautions appropriate to the circumstances and quantity of the material handled. Accumulation of airborne dust may present an explosion hazard.

16 Regulatory Status

Fructose and pregelatinized starch is a mixture of two materials both of which are generally regarded as nontoxic:

Fructose Included in the FDA Inactive Ingredients Database (oral solutions, syrup, and suspensions; rectal preparations; intravenous infusions). Included in the Canadian Natural Health Products Ingredients Database.

Starch, pregelatinized Included in the FDA Inactive Ingredients Database (oral capsules, suspensions, and tablets; vaginal preparations). Included in nonparenteral medicines licensed in the UK.

17 Related Substances

Fructose; starch, pregelatinized.

18 Comments

Fructose and pregelatinized starch (*Advantose FS 95*) has lower hygroscopicity than standard fructose, making it easier to handle.[(1)] The mixture has been used in the pharmaceutical formulation of sublingual or buccal preparations for the administration of active substances with low to poor aqueous solubility.[(4)]

19 Specific References

1 SPI Pharma Inc. Technical Bulletin: *Advantose FS 95* Fructose, June 2008.
2 SPI Pharma Inc. Product Bulletin: *Advantose FS 95* Directly Compressible Fructose, 95% Fructose with 5% Starch, March 2007.
3 SPI Pharma Inc. Material safety data sheet: *Advantose FS 95* Directly Compressible Fructose, July 2007.
4 McCarty JA. Melatonin tablet and methods of preparation and use. United States Patent 2010/0119601 A1; 2010.

20 General References

Bolhuis GK, Armstrong NA. Excipients for direct compaction – an update. *Pharm Dev Tech* 2006; **11**(1): 111–124.
Osberger TF. Tableting characteristics of pure crystalline fructose. *Pharm Technol* 1979; **3**(6): 81–86.
Patel RP, Bhavsar M. Directly compressible materials via co-processing. *Int J PharmTech Res* 2009; **1**(3): 745–753.
Saha S, Shahiwala AF. Multifunctional coprocessed excipients for improved tabletting performance. *Expert Opin Drug Deliv* 2009; **6**(2): 197–208.

21 Author

JT Heinämäki.

22 Date of Revision

1 August 2011.

Fumaric Acid

1 Nonproprietary Names

USP–NF: Fumaric Acid

2 Synonyms

Allomaleic acid; allomalenic acid; boletic acid; butenedioic acid; E297; 1,2-ethenedicarboxylic acid; lichenic acid; *trans*-butenedioic acid; NSC-2752; *trans*-1,2-ethylenedicarboxylic acid; tumaric acid; U-1149; USAF EK-P-583.

3 Chemical Name and CAS Registry Number

(E)-2-Butenedioic acid [110-17-8]

4 Empirical Formula and Molecular Weight

$C_4H_4O_4$ 116.07

Fumaric acid occurs as one of two isomeric unsaturated dicarboxylic acids, the other being maleic acid.

5 Structural Formula

6 Functional Category

Acidulant; antioxidant; complexing agent; flavoring agent; solubilizing agent.

7 Applications in Pharmaceutical Formulation or Technology

Fumaric acid is used primarily in liquid pharmaceutical preparations as an acidulant and flavoring agent, producing an intense acidic sour flavor. Fumaric acid may be included as the acid part of effervescent tablet formulations, although this use is limited as the compound has an extremely low solubility in water. It is also used as a chelating agent which exhibits synergism when used in combination with other true antioxidants. It has been used in film-coated pellet formulations as an acidifying agent and also to increase drug solubility.[1]

8 Description

Fumaric acid occurs as white, odorless or nearly odorless granules or as a crystalline powder that is virtually nonhygroscopic.

9 Pharmacopeial Specifications

See Table I.

Table I: Pharmacopeial specifications for fumaric acid.

Test	USP35–NF30
Identification	+
Water	≤0.5%
Residue on ignition	≤0.1%
Heavy metals	≤10 ppm
Maleic acid	≤0.1%
Assay (anhydrous basis)	99.5–100.5%

10 Typical Properties

Acidity/alkalinity

pH = 2.45 (saturated aqueous solution at 20°C);
pH = 2.58 (0.1% w/v aqueous solution at 25°C);
pH = 2.25 (0.3% w/v aqueous solution at 25°C);
pH = 2.15 (0.5% w/v aqueous solution at 25°C).

Boiling point 290°C (sealed tube)

Density 1.635 g/cm^3 at 20°C

Density (bulk) 0.77 g/cm^3

Density (tapped) 0.93 g/cm^3

Dissociation constant

pK_{a1} = 3.03 at 25°C;
pK_{a2} = 4.54 at 25°C.

Melting point 280–289°C (closed capillary, rapid heating); partial carbonization and formation of maleic anhydride occur at 230°C (open vessel); sublimes at 200°C.

Solubility *see* Table II.

Spectroscopy

IR spectra *see* Figure 1.
NIR spectra *see* Figure 2.

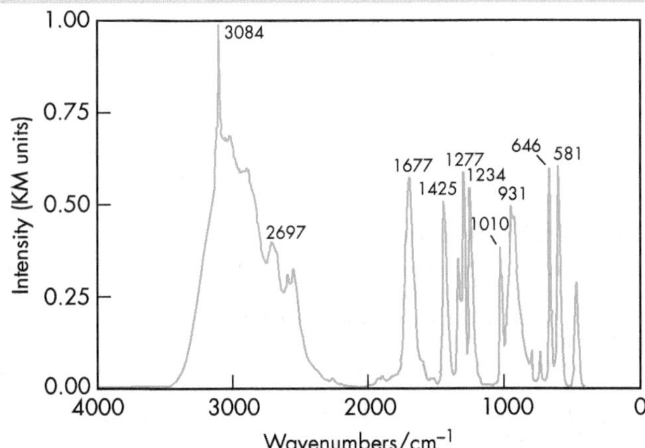

Figure 1: Infrared spectrum of fumaric acid measured by diffuse reflectance. Adapted with permission of Informa Healthcare.

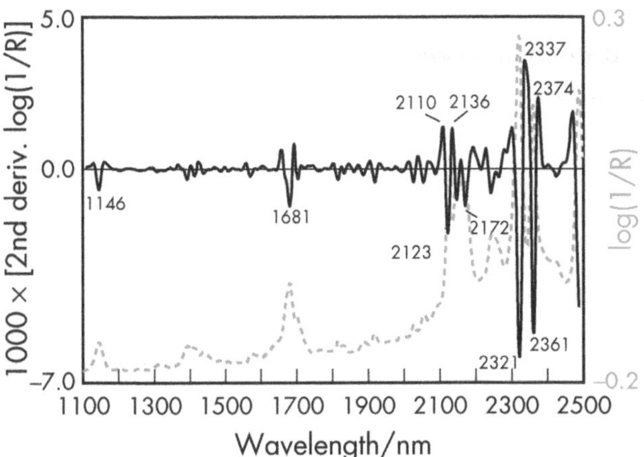

Figure 2: Near-infrared spectrum of fumaric acid measured by reflectance.

Table II: Solubility of fumaric acid.

Solvent	Solubility at 20°C unless otherwise stated
Acetone	1 in 58 at 30°C
Benzene	Very slightly soluble
Carbon tetrachloride	Very slightly soluble
Chloroform	Very slightly soluble
Ethanol	1 in 28
Ethanol (95%)	1 in 17 at 30°C
Ether	Slightly soluble
	1 in 139 at 25°C
Olive oil	Very slightly soluble
Propylene glycol	1 in 33
Water	1 in 200
	1 in 432 at 0°C
	1 in 303 at 10°C
	1 in 159 at 25°C
	1 in 94 at 40°C
	1 in 42 at 60°C
	1 in 10 at 100°C

11 Stability and Storage Conditions

Fumaric acid is stable although it is subject to degradation by both aerobic and anaerobic microorganisms. When heated in sealed vessels with water at 150–170°C it forms DL-malic acid.

The bulk material should be stored in a well-closed container in a cool, dry place.

12 Incompatibilities

Fumaric acid undergoes reactions typical of an organic acid.

13 Method of Manufacture

Commercially, fumaric acid may be prepared from glucose by the action of fungi such as *Rhizopus nigricans*, as a by-product in the manufacture of maleic and phthalic anhydrides, and by the isomerization of maleic acid using heat or a catalyst.

On the laboratory scale, fumaric acid can be prepared by the oxidation of furfural with sodium chlorate in the presence of vanadium pentoxide.

14 Safety

Fumaric acid is used in oral pharmaceutical formulations and food products, and is generally regarded as a relatively nontoxic and nonirritant material. However, acute renal failure and other adverse reactions have occurred following the topical and systemic therapeutic use of fumaric acid and fumaric acid derivatives in the treatment of psoriasis or other skin disorders.[2,3] Other adverse effects of oral therapy have included disturbances of liver function, gastrointestinal effects, and flushing.[4]

The WHO has stated that the establishment of an estimated acceptable daily intake of fumaric acid or its salts was unnecessary since it is a normal constituent of body tissues.[5]

LD_{50} (mouse, IP): 0.1 g/kg[6]
LD_{50} (rabbit skin): 20.0 g/kg
LD_{50} (rat, oral): 9.3 g/kg

15 Handling Precautions

Observe normal precautions appropriate to the circumstances and quantity of material handled. Fumaric acid may be irritating to the skin, eyes, and respiratory system, and should be handled in a well-ventilated environment. Gloves and eye protection are recommended.

16 Regulatory Status

GRAS listed. Accepted for use as a food additive in Europe. Included in the FDA Inactive Ingredients Database (oral capsules, powder for oral solutions, suspensions, syrups, extended release, controlled release, and sustained action chewable tablets). Included in the Canadian Natural Health Products Ingredients Database.

17 Related Substances

Citric acid monohydrate; malic acid; tartaric acid.

18 Comments

In the design of novel pelletized formulations manufactured by extrusion-spheronization, fumaric acid has been used to aid spheronization, favoring the production of fine pellets.[7] It has also been investigated as an alternative filler to lactose in pellets.[2]

Fumaric acid has been investigated as a lubricant for effervescent tablets,[3] and copolymers of fumaric acid and sebacic acid have been investigated as bioadhesive microspheres.[8] Fumaric acid may be capable of creating a microenvironmental pH inside pellets and thus increasing solubility at higher pH.[9]

Fumaric acid is used as a therapeutic agent in the treatment of psoriasis and other skin disorders.[4,10]

Fumaric acid is used as a food additive at concentrations up to 3600 ppm.

A specification for fumaric acid is contained in the *Food Chemical Codex* (FCC)[11] and the *Japanese Pharmaceutical Excipients* (JPE).[12]

The EINECS number for fumaric acid is 203-743-0. The PubChem Compound ID (CID) for fumaric acid is 444972.

19 Specific References

1 Munday DL. Film coated pellets containing verapamil hydrochloride: enhanced dissolution into neutral medium. *Drug Dev Ind Pharm* 2003; 29(5): 575–583.
2 Bianchini R, *et al.* Influence of extrusion-spheronization processing on the physical properties of *d*-indobufen pellets containing pH adjusters. *Drug Dev Ind Pharm* 1992; **18**(14): 1485–1503.
3 Röscheisen G, Schmidt PC. The combination of factorial design and simplex method in the optimization of lubricants for effervescent tablets. *Eur J Pharm Biopharm* 1995; **41**(5): 302–308.
4 Sweetman SC, ed. *Martindale: The Complete Drug Reference*, 37th edn. London: Pharmaceutical Press, 2011: 1739.
5 FAO/WHO. Evaluation of certain food additives and contaminants. Thirty-fifth report of the joint FAO/WHO expert committee on food additives. *World Health Organ Tech Rep Ser* 1990; No. 789.
6 Lewis RJ, ed. *Sax's Dangerous Properties of Industrial Materials*, 11th edn. New York: Wiley, 2004: 1828.
7 Law MFL, Deasy PB. Effect of common classes of excipients on extrusion-spheronization. *J Microencapsul* 1997; 14(5): 647–657.
8 Chickering DE, Mathiowitz E. Bioadhesive microspheres: I. A novel electrobalance-based method to study adhesive interactions between individual microspheres and intestinal mucosa. *J Control Release* 1995; 34: 251–262.
9 Padhy KK, *et al.* Influence of organic acids on drug release pattern of verapamil hydrochloride pellets. *J Adv Pharm Res* 2010; **1**: 65–73.
10 Nieboer C, *et al.* Systemic therapy with fumaric acid derivates: new possibilities in the treatment of psoriasis. *J Am Acad Dermatol* 1989; 20(4): 601–608.
11 *Food Chemicals Codex*, 7th edn. Bethesda, MD: United States Pharmacopeia, 2010: 415.
12 Japan Pharmaceutical Excipients Council. *Japanese Pharmaceutical Excipients 2004*. Tokyo: Yakuji Nippo, 2004: 331–333

20 General References

Allen LV. Featured excipient: flavor-enhancing agents. *Int J Pharm Compound* 2003; 7(1): 48–50.

Anonymous. Malic and fumaric acids. *Manuf Chem Aerosol News* 1964; 35(12): 56–59.

New Zealand Dermatological Society Incorporated (DernNet NZ). Fumaric acid esters, 2011.

Robinson WD, Mount RA. *Kirk-Othmer Encyclopedia of Chemical Technology*, 3rd edn, 14. New York: Wiley-Interscience, 1981: 770–793.

21 Author

BV Kadri.

22 Date of Revision

1 March 2012.

Galactose

1 Nonproprietary Names

BP: Galactose
PhEur: Galactose
USP–NF: Galactose

2 Synonyms

Brain sugar; cerebrose; galactosum; lactoglucose.

3 Chemical Name and CAS Registry Number

α-D-Galactopyranose [3646-73-9]
D-Galactopyranose [59-23-4]

4 Empirical Formula and Molecular Weight

$C_6H_{12}O_6$ 180.16

The USP35–NF30 describes galactose as α-D-galactopyranose, one of the products of the metabolism of lactose, a naturally occurring sugar in dairy products, by the digestive enzyme lactase. The PhEur 7.4 describes galactose as D-galactopyranose.

5 Structural Formula

α-D-Galactopyranose

6 Functional Category

Complexing agent; sweetening agent.

7 Applications in Pharmaceutical Formulation or Technology

Galactose is used as a sweetener in oral pharmaceutical preparations. In recent years, galactose has also been used in nanoparticles and drug–polymer conjugates as a targeting moiety for the liver; *see* Section 18.

8 Description

Galactose occurs as a white or almost white crystalline or finely granulated powder.

9 Pharmacopeial Specifications

See Table I.

Table I: Pharmacopeial specifications for galactose.

Test	PhEur 7.4	USP35–NF30
Characters	+	–
Identification	+	+
Appearance of solution	+	+
Acidity or alkalinity	+	+
Specific optical rotation	+78.0 to +81.5°	+78.0 to +81.5°
Barium	+	+
Lead	≤0.5 ppm	≤0.5 μg/g
Water	≤1.0%	≤1.0%
Sulfated ash	≤0.1%	–
Residue on ignition	–	≤0.1%
Microbial limits	+	+
Aerobic bacteria	≤ 10^2 cfu/g	≤ 10^3 cfu/g[a]
Total combined molds and yeasts	–	≤ 10^2 cfu/g

(a) Tests for *Salmonella*, *Escherichia coli*, *Staphylcoccus aureus*, and *Pseudomonas aeruginosa* are negative.

10 Typical Properties

Density 1.5 g/cm³ for D-galactopyranose.[1]
Dissociation constant pK_a 12.92[2]
Melting point 167°C (α and β forms);[3] 163–170°C for D-galactopyranose.[1]
pH of aqueous solution pH 4.5–6.0 for D-galactopyranose.[1]
Optical rotation $[\alpha]^{20}_D$ +150.7° → 80.2° and +52.8° → 80.2° (water) for the α and β forms, respectively.[3]
Solubility Soluble in 0.5 parts water and in 1.7 parts water for α and β forms, respectively.[3] Soluble in pyridine, alcohol, and ether. Solubility also reported as 47.25 g per 100 g of water (D-galactose);[4] 68.3 g per 100 g of water.[5]
Vapor pressure 1.82×10^{-8} mmHg at 25°C.[2]

11 Stability and Storage Conditions

Keep in a tightly closed container. Store in a cool, dry, ventilated area.

12 Incompatibilities

Galactose is a reducing sugar and may undergo Mailliard type reactions with amines.

13 Method of Manufacture

Galactose is mainly produced by hydrolyzing lactose (a disaccharide consisting of glucose and galactose), which is found in dairy products, such as milk. Non-animal derived D-galactose is manufactured from wood-based or other biomass hydrosylates using an aqueous chromatographic separation process.[6]

14 Safety

Galactose is generally regarded as a safe material, although it has been identified experimentally as having teratogenic or mutagenic effects.[7] Large doses may cause gastrointestinal upset.

Inhalation of dust may cause irritation in the respiratory tract. Contact with eyes and prolonged contact with skin may cause irritation, especially in people with galactosemia, a rare genetic disorder in which the body is unable to convert galactose to dextrose.

15 Handling Precautions

Observe normal precautions appropriate to the circumstances and quantity of material being handled.

When heated to decomposition, galactose emits acrid smoke and irritating fumes.[7] Excessive generation of dust, or inhalation of dust, should be avoided as inhalation of dust may cause irritation to the respiratory tract. Suitable eye, respiratory and skin protection is recommended.

16 Regulatory Status

Included in the FDA Inactive Ingredients Database (oral solutions, tablets; rectal solutions). Included in the Canadian Natural Health Products Ingredients Database.

17 Related Substances

Guar gum; lactose monohydrate; raffinose.

18 Comments

The USP35–NF30 gives the CAS number for α-D-Galactopyranose as 3646-73-9 and PhEur for D-Galactopyranose as 59-23-4.

Galactose is usually found in nature combined with other sugars, for example, in lactose (milk sugar). It is also found in polysaccharides, such as gums, complex carbohydrates and in carbohydrate-containing lipids called glycolipids, which occur in the brain and other nervous tissues of most animals.

Galactose is used as an ultrasound contrast agent, either dissolved in water or as microgranules stabilized with palmitic acid and dispersed in water.

Galactose is also used to test the ability of the liver to remove galactose from the blood and convert it to glycogen in liver function tests, which estimate impaired liver function through measurement of the rate of galactose excretion after ingestion, or injection, of a measured amount of galactose.[8,9] Copolymeric nanoparticles end-capped with galactose, and polymer-drug conjugates containing galactose as a targeting moiety, have been investigated for liver-specific delivery.[10–12]

Galactose is licensed for use in infants and children in echocardiography and for diagnosis of vesicoureteral reflux.[13]

The EINECS number of galactose is 200-416-4. The PubChem ID for α-D-galactose is 439357 and for D-galactose is 6036.

19 Specific References

1 Merck KGaA. Product datasheet: D(+)-Galactose, November 2010.
2 US National Library of Medicine. ChemIDplus Lite: D-Galactose, RN 59-23-4.
3 O'Neil MJ *et al.* eds. *The Merck Index: An Encyclopedia of Chemicals, Drugs, and Biologicals*, 14th edn. Whitehouse Station, NJ: Merck, 2006; 745–746.
4 Gould SP. The final solubility of D-galactose in water. *J Dairy Sci* 1940; 227.
5 Dehn WM. Comparative solubilities in water, pyridine and in aqueous pyridine. *J Am Chem Soc* 1917; **39**: 1399–1404.
6 Danisco A/S. Product information: NAD-D-Galactose, 2011. http://www.danisco.com/products/product_range/rare_sugars/nad_d_galactose/ (accessed 12 July 2011).
7 Lewis RJ, ed. *Sax's Dangerous Properties of Industrial Materials*, 11th edn. New York: Wiley, 2004: 1844.
8 Kuntz HD, Kuntz E. Intravenous loading with galactose as a liver function test. Methods and clinical value. *Fortschr Med* 1983; **101**: 999–1004.
9 Maruyama H, Ebura M. Recent applications of ultrasound: diagnostic and treatment of hepatocellular carcinoma. *Int J Clin Oncol* 2010; **11**: 258–267.
10 Jeong YI, *et al.* Cellular recognition of paclitaxel-loaded polymeric nanoparticles composed of poly(gamma-benzul L-glutamate) and poly(ethylene glycol) diblock copolymer endcapped with galactose moiety. *Int J Pharm* 2005; **296**: 151–161.
11 Kim IS, Kim SH. Development of polymeric nanoparticulate drug delivery systems: evaluation of nanoparticles based on biotinylated poly(ethylene glycol) with sugar moiety. *Int J Pharm* 2003; **257**: 195–203.
12 Duncan R. Development of HPMA copolymer–anticancer conjugates: clinical experience and lessons learnt. *Adv Drug Deliv Rev* 2009; **61**: 1131–1148.
13 Sweetman SC, ed. Galactose. *Martindale: The Complete Drug Reference*, 37th edn. London, Pharmaceutical Press, 2011; 1619.

20 General References

Galactose.org. http://www.galactose.org/ (accessed 12 July 2011).

MeroPharm AG. Product information: D(+)Galactose. http://www.produkte.meropharm-shop.com/ (accessed 12 July 2011).

21 Authors

ME Fenton, RC Rowe.

22 Date of Revision

1 March 2012.

Gelatin

1 Nonproprietary Names

BP: Gelatin
JP: Gelatin
PhEur: Gelatin
USP–NF: Gelatin

2 Synonyms

Byco; *Cryogel*; E441; gelatina; gelatine; *Instagel*; *Kolatin*; *Solugel*; *Vitagel*.

3 Chemical Name and CAS Registry Number

Gelatin [9000-70-8]

4 Empirical Formula and Molecular Weight

Gelatin is a generic term for a mixture of purified protein fractions obtained either by partial acid hydrolysis (type A gelatin) or by partial alkaline hydrolysis (type B gelatin) of animal collagen obtained from cattle and pig bone, cattle skin (hide), pigskin, and fish skin. Gelatin may also be a mixture of both types.

The protein fractions consist almost entirely of amino acids joined together by amide linkages to form linear polymers, varying in molecular weight from 20 000–200 000.

The JP XV also includes a monograph for purified gelatin.

5 Structural Formula

See Section 4.

6 Functional Category

Coating agent; film-forming agent; gelling agent; microencapsulating agent; suspending agent; tablet and capsule binder; viscosity-increasing agent.

7 Applications in Pharmaceutical Formulation or Technology

Gelatin is widely used in a variety of pharmaceutical formulations, including its use as a biodegradable matrix material in an implantable delivery system,[1] although it is most frequently used to form either hard or soft gelatin capsules.[2–4]

Gelatin capsules are unit-dosage forms designed mainly for oral administration. Soft capsules on the market also include those for rectal and vaginal administration. Hard capsules can be filled with solid (powders, granules, pellets, tablets, and mixtures thereof), semisolid or liquid fillings, whereas soft capsules are mainly filled with semisolid or liquid fillings. In hard capsules, the active drug is always incorporated into the filling, while in soft capsules the drug substance can also be incorporated into the thick soft capsule shell. Gelatin is soluble in warm water (>40°C), and a gelatin capsule will initially swell and finally dissolve in gastric fluid to release its contents rapidly.[5]

Hard capsules are manufactured in two pieces by dipping lubricated stainless steel mold pins into a 45–55°C gelatin solution of defined viscosity, which depends on the size of the capsules and whether cap or body are to be formed. The gelatin is taken up by the pins as a result of gelation, and the resulting film thickness is governed by the viscosity of the solution. The capsule shells are passed through a stream of cool air to aid setting of the gelatin, and afterwards they are slowly dried with large volumes of humidity-controlled air heated to a few degrees above ambient temperature and blown directly over the pins. The capsule halves are removed from their pins, trimmed and fitted together. Gelatin that is used to produce hard capsules may contain various coloring agents and antimicrobial preservatives. Surfactants may be present in small quantities in the shells being a residue of the pin lubricant. However, the use of preservatives is no longer encouraged in line with current GMP principles. Capsule shells may be treated with formaldehyde to make them insoluble in gastric fluid. Standard capsules vary in volume from 0.13 to 1.37 mL. For veterinary use, capsules with a volume between 3 and 28 mL are available, and capsules with a capacity of 0.025 mL are available for toxicity studies in rats.

G

In contrast to two-piece hard capsules, soft gelatin capsules are manufactured, filled and sealed in one process. The gelatin used to form the soft shells has a lower gel strength than that used for hard capsules, and the viscosity of the solutions is also lower, which results in more flexible shells. Additives to soft shell formulations are plasticizers such as polyalcohols (glycerin, propylene glycol, polyethylene glycol). Sorbitol can be added as moisturizing agent, whereby the larger amount of water will act as plasticizer. Coloring and opacifying agents are also added. The filling can interact with the gelatin and the plasticizer chemically. There may be migration of filling components into the shell and plasticizer from the shell into the filling. These interactions have to be taken into account during the formulation of the gelatin shell and the filling. The main method to produce soft gelatin capsules is the rotary die method (RP Scherer), and an alternative method for small volumes of round capsules is the Globex system (Industrial Techno-logic Solutions Ltd).[4] *Soflet Gelcaps* (Banner Pharmacaps) are tablets that have been coated with a gelatin film.

Gelatin is also used for the microencapsulation of drugs, where the active drug is sealed inside a microsized capsule or beadlet, which may then be handled as a powder. The first microencapsulated drugs (beadlets) were fish oils and oily vitamins in gelatin beadlets prepared by coacervation.

Low-molecular-weight gelatin has been investigated for its ability to enhance the dissolution of orally ingested drugs.[6] Ibuprofen–gelatin micropellets have been prepared for the controlled release of the drug.[7] Other uses of gelatin include the preparation of pastes, pastilles, pessaries, and suppositories. In addition, it is used as a tablet binder and coating agent, and as a viscosity-increasing agent for solutions and semisolids.

Gelatin is added to vaccines (e.g. MMR, varicella, influenza, yellow fever, and rabies) as a heat stabilizer. Recombinant human gelatins have been developed for use in vaccines and other biological formulations, to reduce the risk of an immune response, and also to eliminate the risk of contamination with pathogens such as viruses and prions.[8,9]

8 Description

Gelatin occurs as a light-amber to faintly yellow-colored, vitreous, brittle solid. It is practically odorless and tasteless, and is available as translucent sheets, flakes, and granules, or as a coarse powder.

9 Pharmacopeial Specifications

See Table I. *See also* Section 18.

Table I: Pharmacopeial specifications for gelatin.

Test	JP XV	PhEur 7.4	USP35–NF30
Identification	+	+	+
Characters	+	+	−
Microbial contamination	−	+	+
Aerobic bacteria	−	≤10³ cfu/g	≤10³ cfu/g
Fungi	−	≤10² cfu/g	−
Residue on ignition	≤2.0%	−	≤2.0%
Loss on drying	≤15.0%	≤15.0%	−
Odor and water-insoluble substances	+	−	+
Isoelectric point	+	+	−
Type A	7.0–9.0	6.0–9.5	−
Type B	4.5–5.0	4.7–5.6	−
Conductivity	−	≤1 mS/cm	−
Sulfur dioxide	−	≤50 ppm	≤0.15%
Sulfite	+	−	−
Arsenic	≤1 ppm	−	≤0.8 ppm
Iron	−	≤30 ppm	−
Chromium	−	≤10 ppm	−
Zinc	−	≤30 ppm	−
Heavy metals	≤50 ppm(a)	−	≤50 ppm
pH	−	3.8–7.6	−
Mercury	≤0.1 ppm	−	−
Peroxides	−	≤10 ppm	−
Gel strength	−	+	−

(a) ≤20 ppm for purified gelatin.

10 Typical Properties

Acidity/alkalinity

For a 1% w/v aqueous solution at 25°C (depending on source and grade):

pH = 3.8–5.5 (type A);

pH = 5.0–7.5 (type B).

Density

1.32 g/cm³ for type A;

1.28 g/cm³ for type B.

Isoelectric point

7.0–9.0 for type A;

4.7–5.4 for type B.

Moisture content 9–11%.[10] *See also* Figures 1 and 2.

Solubility Practically insoluble in acetone, chloroform, ethanol (95%), ether, and methanol. Soluble in glycerin, acids, and alkalis, although strong acids or alkalis cause precipitation. In water, gelatin swells and softens, gradually absorbing between five and 10 times its own weight of water. Gelatin is soluble in water above 40°C, forming a colloidal solution, which gels on cooling to 35–40°C. This gel–sol system is thixotropic and heat-reversible, the melting temperature being slightly higher than the setting point; the melting point can be varied by the addition of glycerin.

Spectroscopy

IR spectra *see* Figure 3.

NIR spectra *see* Figure 4.

Viscosity (dynamic) *see* Table II.[4]

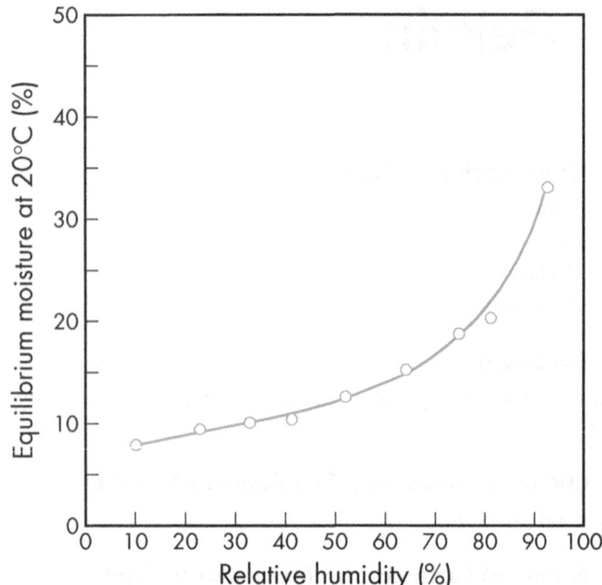

Figure 1: Equilibrium moisture content of gelatin (*Pharmagel A*).

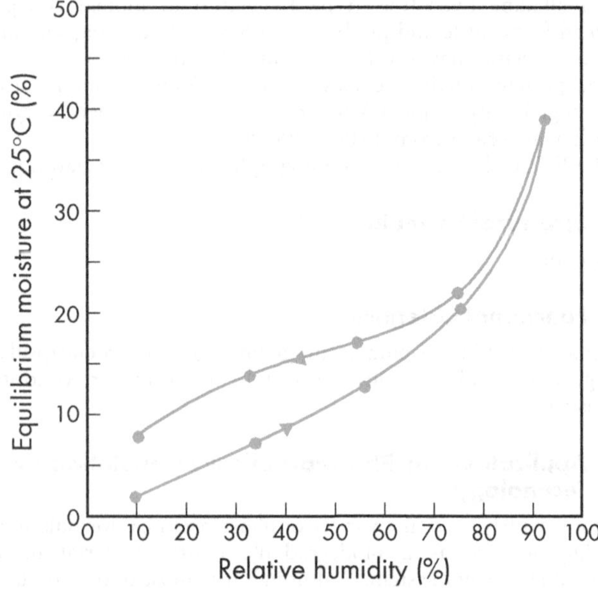

Figure 2: Sorption–desorption isotherm of gelatin.

Table II: Dynamic viscosity of gelatin solutions at 60°C.

Grade	Viscosity (dynamic)/mPa s	
	6.67% w/v aqueous solution	12.5% w/c aqueous solution
Acid ossein	2.7–3.7	12.5–14.5
Acid pigskin	4.2–4.8	19.0–20.5
Fish skin	3.0–4.5	13.0–20.0
Limed ossein/hide	3.6–4.8	19.0–20.5

11 Stability and Storage Conditions

Dry gelatin is stable in air. Aqueous gelatin solutions are also stable for long periods if stored under cool conditions but they are subject to bacterial degradation.[4] At temperatures above about 50°C, aqueous gelatin solutions may undergo slow depolymerization and a reduction in gel strength may occur on resetting. Depolymerization becomes more rapid at temperatures above 65°C, and gel strength may be reduced by half when a solution is heated at 80°C for 1 hour. The rate and extent of depolymerization depends on the molecular weight of the gelatin, with a lower-molecular-weight material decomposing more rapidly.[11]

Gelatin may be sterilized by dry heat.

The bulk material should be stored in an airtight container in a cool, well-ventilated, and dry place.

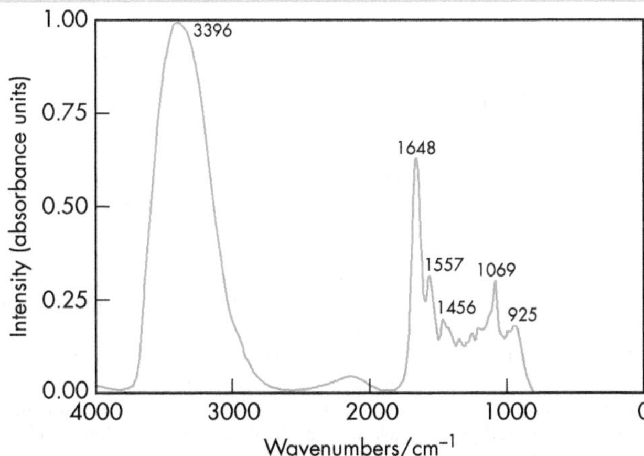

Figure 3: Infrared spectrum of gelatin measured by transmission. Adapted with permission of Informa Healthcare.

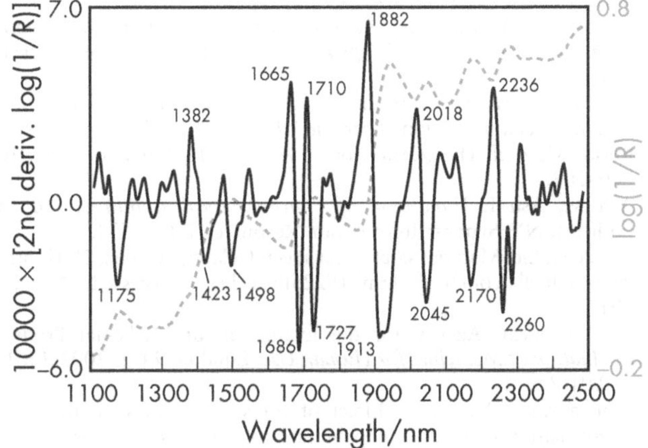

Figure 4: Near-infrared spectrum of gelatin measured by reflectance.

12 Incompatibilities

Gelatin is an amphoteric material and will react with both acids and bases. It is also a protein and thus exhibits chemical properties characteristic of such materials; for example, gelatin may be hydrolyzed by most proteolytic systems to yield its amino acid components.

Gelatin will also react with aldehydes and aldehydic sugars, anionic and cationic polymers, electrolytes, metal ions, plasticizers, preservatives, strong oxidizers, and surfactants. It is precipitated by alcohols, chloroform, ether, mercury salts, and tannic acid. Gels can be liquefied by bacteria unless preserved.

Some of these interactions are exploited to favorably alter the physical properties of gelatin: for example, gelatin is mixed with a plasticizer, such as glycerin, to produce soft gelatin capsules and suppositories; gelatin is treated with formaldehyde to produce gastroresistance; *see* Section 7.

13 Method of Manufacture

Gelatin is extracted from animal tissues rich in collagen such as skin, sinews, and bone. Although it is possible to extract gelatin from these materials using boiling water, it is more practical to first pretreat the animal tissues with either acid or alkali. Gelatin obtained from the acid process is called type A, whereas gelatin obtained from the alkali process is called type B.

The acid-conditioning process (manufacture of type A gelatin) is restricted to soft bone ossein (demineralized bones), sinew, pigskin, calfskin and fish skins for reasons of gaining sufficient yield. The material is cut in pieces and washed in cold water for a few hours to remove superficial fat. It is then treated with mineral acid solutions, mainly HCl or H_2SO_4, at pH 1–3 and 15–20°C until maximum swelling has occurred. This process takes approximately 24 hours. The swollen stock is then washed with water to remove excess acid, and the pH is adjusted to pH 3.5–4.0 (pigskin, fish skin) or 2.0–3.5 (all other tissues) for the conversion to gelatin by hot-water extraction.

The hydrolytic extraction is carried out in a batch-type operation using successive portions of hot water at progressively higher temperatures (50–75°C) until the maximum yield of gelatin is obtained. The gelatin solution is then filtered through previously sterilized cellulose pads, deionized, concentrated to about 20–25% w/v and sterilized by flashing it to 138°C for 4 seconds. The dry gelatin is then formed by chilling the solution to form a gel, which is air-dried in temperature-controlled ovens. The dried gelatin is ground to the desired particle size.

In the alkali process (liming), demineralized bones (ossein) or cattle skins are usually used. The animal tissue is held in a calcium hydroxide (2–5% lime) slurry for a period of 2–4 months at 14–18°C. At the end of the liming, the stock is washed with cold water for about 24 hours to remove as much of the lime as possible. The stock solution is then neutralized with acid (HCl, H_2SO_4, H_3PO_4) and the gelatin is extracted with water in an identical manner to that in the acid process, except that the pH is kept at values between 5.0–6.5 (neutral extraction).

During the preparation of the bovine bones used in the production of gelatin, specified risk materials that could contain transmissible spongiform encephalopathies (TSEs) vectors are removed. TSE infectivity is not present in pharmaceutical grade gelatin.

14 Safety

Gelatin is widely used in a variety of pharmaceutical formulations, including oral and parenteral products.

In general, when used in oral formulations gelatin may be regarded as a nontoxic and nonirritant material. However, there have been rare reports of gelatin capsules adhering to the esophageal lining, which may cause local irritation.[12] Hypersensitivity reactions, including serious anaphylactoid reactions, have been reported following the use of gelatin in parenteral products.[13,14]

There have been concerns over the potential spread of BSE/TSE infections through bovine derived products. However, the risk of such contamination of medicines is extremely low.

LD_{50} (rat, oral): 5 g/kg[15]

TD_{Lo} (mouse, IP): 0.7 g/kg[16]

15 Handling Precautions

Observe normal precautions appropriate to the circumstances and quantity of material handled. Eye protection and gloves are recommended. Gelatin should be handled in a well-ventilated environment and kept away from sources of ignition and heat. Empty containers pose a fire risk, and the gelatin residues should be evaporated under a fume hood.

16 Regulatory Status

GRAS listed. Included in the FDA Inactive Ingredients Database (dental preparations; inhalations; injections; oral capsules, pastilles, solutions, syrups and tablets; topical and vaginal preparations). Included in medicines licensed in the UK, Europe, and Japan. Included in the Canadian Natural Health Products Ingredients Database.

17 Related Substances

—

18 Comments

Gelatin is one of the materials that have been selected for harmonization by the Pharmacopeial Discussion Group. For further information see the General Information Chapter <1196> in the USP35–NF30, the General Chapter 5.8 in PhEur 7.4, along with the 'State of Work' document on the PhEur EDQM website, and also the General Information Chapter 8 in the JP XV.

In the past there has been a significant amount of regulatory activity and legislation due to the attention given to bovine sourced gelatin manufacturing processes and the potential transmission of TSE vectors from raw bovine materials into gelatin.[4] In Europe, the criteria by which the safety is assured involves controlling the geographical sourcing of animals used; the nature of the tissue used (based on scientific data showing where animal BSE infectivity is located); and the method of production.

Gelatin produced with hides as the starting material is considered much safer than using bones, although it is recommended that measures are undertaken to prevent cross-contamination with potentially contaminated materials. When gelatin is produced from bones, the bones should ideally not be sourced from countries classified as Geographical BSE Risk (GBR) I and II, although bones from GBR III countries can be used if the removal of vertebrae from the raw materials is assured (*see* Table III).[17]

Various grades of gelatin are commercially available that differ in particle size, molecular weight, and other properties. Grading is usually by gel strength, expressed as 'Bloom strength', which is the weight in grams that, when applied under controlled conditions to a plunger 12.7 mm in diameter, will produce a depression exactly 4 mm deep in a matured gel containing 6.66% w/w of gelatin in water.

Gelatin–acacia complex coacervation has been used in the preparation of microcapsules of vitamin A.[18] Pindolol-loaded alginate–gelatin beads have been developed for the sustained release of pindolol.[19]

Gelatin-based three-dimensional scaffolds have been used in bone regeneration after surgery,[20] and they have also been investigated in urological procedures.[21] A further potential use of gelatin is in the formulation of drug-eluting stents.[22,23]

Therapeutically, gelatin has been used in the preparation of wound dressings[24] and has been used as a plasma substitute, although anaphylactoid reactions have been reported in the latter application.[13] Absorbable gelatin is available as sterile film, ophthalmic film, sterile sponge, sterile compressed sponge, and sterile powder from sponge. Gelatin sponge has hemostatic properties.

Gelatin is also widely used in food products and photographic emulsions.

A specification for gelatin is contained in the *Food Chemicals Codex* (FCC).[25]

The EINECS number for gelatin is 232-554-6.

Table III: The European Scientific Steering Committee classification of geographical BSE risk (GBR).

GBR level	Presence of one or more cattle clinically or pre-clinically infected with BSE in a geographical region/country
I	Highly unlikely
II	Unlikely but not excluded
III	Likely but not confirmed or confirmed at a lower level
IV	Confirmed at a higher level

19 Specific References

1 Fan H, Dash AK. Effect of cross-linking on the *in vitro* release kinetics of doxorubicin from gelatin implants. *Int J Pharm* 2001; **213**: 103–116.
2 Armstrong NA, *et al.* Drug migration in soft gelatin capsules. *J Pharm Pharmacol* 1982; 34(Suppl.): 5P.
3 Tu J, *et al.* Formulation and pharmacokinetics studies of acyclovir controlled-release capsules. *Drug Dev Ind Pharm* 2001; **27**: 687–692.
4 Podczeck F, Jones BE, eds. *Pharmaceutical Capsules*, 2nd edn. London: Pharmaceutical Press, 2004.
5 Chiwele I, *et al.* The shell dissolution of various empty hard capsules. *Chem Pharm Bull* 2000; **48**: 951–956.
6 Kimura S, *et al.* Evaluation of low-molecular gelatin as a pharmaceutical additive for rapidly absorbed oral dosage formulations. *Chem Pharm Bull* 1991; **39**: 1328–1329.
7 Tayade PT, Kale RD. Encapsulation of water insoluble drug by a cross-linking technique: effect of process and formulation variables on encapsulation efficiency, particle size, and *in vitro* dissolution rate. *AAPS Pharm Sci* 2004; **6**(1): E12.
8 Thyagarajapuram N, *et al.* Stabilization of proteins by recombinant human gelatins. *J Pharm Sci* 2007; **96**(12): 3304–3315.
9 Liska V, *et al.* Evaluation of a recombinant human gelatin as a substitute for a hydrolyzed porcine gelatin in a refrigerator-stable Oka/Merck live varicella vaccine. *J Immune Based Ther Vaccines* 2007; **5**: 4.
10 Callahan JC, *et al.* Equilibrium moisture content of pharmaceutical excipients. *Drug Dev Ind Pharm* 1982; **8**: 355–369.
11 Ling WC. Thermal degradation of gelatin as applied to processing of gel mass. *J Pharm Sci* 1978; **67**: 218–223.
12 Weiner M, Bernstein IL. *Adverse Reactions to Drug Formulation Agents: A Handbook of Excipients*. New York: Marcel Dekker, 1989: 121–123.
13 Blanloeil Y, *et al.* Severe anaphylactoid reactions after infusion of modified gelatin solution. *Therapie* 1983; **38**: 539–546.
14 Kelso JM, *et al.* The gelatin story. *J Allergy Clin Immunol* 1999; **103**: 200–202.
15 Ash M, Ash I. *Handbook of Pharmaceutical Additives*, 3rd edn. Endicott, NY: Synapse Information Resources, 2007: 633.
16 Ted Pella, Inc. Material safety data sheet: Gelatin, purified, 2007. http://www.tedpella.com/msds_html/19225msds.htm (accessed 13 June 2011).
17 The European Agency for the Evaluation of Medicinal Products. *Evaluation of Medicines for Human Use*. London, 9 Dec 2002: EMEA/410/01 Rev. 2.
18 Junyaprasert VB, *et al.* Effect of process variables on the microencapsulation of vitamin A palmitate by gelatin-acacia coacervation. *Drug Dev Ind Pharm* 2001; **27**: 561–566.
19 Almeida PF, Almeida AJ. Cross-linked alginate–gelatin beads: a new matrix for controlled release of pindolol. *J Control Release* 2004; **97**: 431–439.
20 Anders JO, *et al.* Gelatin-based haemostyptic *Spongostan* as a possible three-dimensional scaffold for a chondrocyte matrix?: an experimental study with bovine chondrocytes. *J Bone Joint Surg Br* 2009; **91**(3): 409–416.
21 Singh I, *et al.* Does sealing of the tract with absorbable gelatin (*Spongostan*) facilitate tubeless PCNL? A prospective study. *J Endourol* 2008; **22**: 2485–2493.
22 Takemoto Y, *et al.* Human placental ectonucleoside triphosphate diphosphohydrolase gene transfer via gelatin-coated stents prevents in-stent thrombosis. *Arterioscler Thromb Vasc Biol* 2009; **29**: 857–862.
23 Huang L-Y, Yang M-C. Behaviours of controlled drug release of magnetic-gelatin hydrogel coated stainless steel for drug-eluting-stents application. *J Magn Magn Mater* 2007; **310**: 2874–2876.
24 Thomas S. *Wound Management and Dressings*. London: Pharmaceutical Press, 1990.
25 *Food Chemicals Codex*, 7th edn. Bethesda, MD: United States Pharmacopeia, 2010: 423.

20 General References

European Directorate for the Quality of Medicines and Healthcare (EDQM). European Pharmacopoeia – State Of Work Of International Harmonisation. *Pharmeuropa* 2011; 23(2): 395–401. http://www.edqm.eu/en/InternationalHarmonisation-614.html (accessed 13 June 2011).

Fassihi AR, Parker MS. Influence of gamma radiation on the gel rigidity index and binding capability of gelatin. *J Pharm Sci* 1988; **77**: 876–879.

Hawley AR, *et al.* Physical and chemical characterization of thermosoftened bases for molten filled hard gelatin capsule formulations. *Drug Dev Ind Pharm* 1992; **18**: 1719–1739.

Jones BE. Two-piece gelatin capsules: excipients for powder products, European practice. *Pharm Technol Eur* 1995; **7**(10): 25, 28, 29, 30, 34.

Jones RT. The role of gelatin in pharmaceuticals. *Manuf Chem Aerosol News* 1977; **48**(7): 23–24.

Matthews B. BSE/TSE risks associated with active pharmaceuticals ingredients and starting materials: situation in Europe and the global implications for healthcare manufacturers. *PDA J Pharm Sci Technol* 2001; **55**: 295–329.
Nadkarni SR, Yalkowsky SH. Controlled delivery of pilocarpine 1: *in vitro* characterization of gelfoam matrices. *Pharm Res* 1993; **10**: 109–112.
Ofner CM, Schott H. Swelling studies of gelatin II: effect of additives. *J Pharm Sci* 1987; **76**: 715–723.
Ramsay Olocco K, *et al.* Pre-clinical and clinical evaluation of solution and soft gelatin capsule formulations for a BCS class 3 compound with atypical physicochemical properties. *J Pharm Sci* 2004; **93**: 2214–2221.
Ray-Johnson ML, Jackson IM. Temperature-related incompatibility between gelatin and calcium carbonate in sugar-coated tablets. *J Pharm Pharmacol* 1976; **28**: 309–310.
Schrieber R, Gareis H. *Gelatine Handbook: Theory and Industrial Practice.* Weinheim: Wiley-VCH Verlag GmbH & Co. KGaA, 2007.
Singh S, *et al.* Alteration in dissolution characteristics of gelatin-containing formulations: a review of the problem, test methods, and solutions. *Pharm Technol* 2002; **26**(4): 36–58.
Voigt R, Werchan D. [Radioinduced changes of the properties of gelatin.] *J Pharmazie* 1986; **41**: 120–123[in German].
Ward AG, Courts A, eds. *The Science and Technology of Gelatin.* London: Academic Press, 1977.

21 Author

F Podczeck.

22 Date of Revision

1 March 2012.

Glucose, Liquid

1 Nonproprietary Names

BP: Liquid Glucose
PhEur: Glucose, Liquid
USP–NF: Liquid Glucose

2 Synonyms

Corn syrup; *C*PharmSweet*; *Flolys*; *Glucomalt*; glucose syrup; glucosum liquidum; *Glucosweet*; *Mylose*; *Roclys*; starch syrup.

3 Chemical Name and CAS Registry Number

Liquid glucose [8027-56-3]

4 Empirical Formula and Molecular Weight

See Section 8.

5 Structural Formula

See Section 8.

6 Functional Category

Coating agent; sweetening agent; tablet and capsule binder.

7 Applications in Pharmaceutical Formulation or Technology

Liquid glucose is used as a base in oral solutions and syrups and also as a granulating and coating agent in tablet manufacture. In sugar solutions for tablet coating, liquid glucose is used to retard the crystallization of the sucrose. *See* Table I.

Table I: Uses of liquid glucose.

Use	Concentration (%)
Granulating agent	5–10
Oral syrup vehicle	20–60
Tablet coating	10–20

8 Description

Liquid glucose is an aqueous solution of several compounds, principally dextrose, dextrin, fructose, and maltose, with other oligosaccharides and polysaccharides. It is a colorless, odorless, and viscous sweet-tasting liquid, ranging in color from colorless to straw-colored.

Liquid glucose is classified into four categories according to its degree of hydrolysis, expressed as dextrose equivalent (DE):

Type I: 20–38 DE;
Type II: 38–58 DE;
Type III: 58–73 DE;
Type IV: >73 DE.

9 Pharmacopeial Specifications

See Table II.

Table II: Pharmacopeial specifications for liquid glucose.

Test	PhEur 7.4	USP35–NF30
Identification	+	+
Characters	+	−
Acidity	−	+
pH	4.0–6.0	−
Water	≤30.0%	≤21.0%
Residue on ignition	≤0.5%	≤0.5%
Sulfur dioxide	≤20 ppm[(a)]	−
Sulfite	−	+
Heavy metals	≤10 ppm	≤0.001%
Starch	−	+
Assay for reducing sugars (dextrose equivalent)	within 10% of nominal value	90.0–110.0%
Assay (of dried matter)	≥70.0%	−
Sulfated ash	≤0.5%	−

(a) Or ≤400 ppm if intended for the production of hard boiled candies, provided the final product contains ≤50 ppm.

10 Typical Properties

Density 1.43 g/cm^3 at 20°C

Solubility Miscible with water; partially miscible with ethanol (90%).
Viscosity (dynamic) 13.0–14.5 mPa s (13.0–14.5 cP) at 21°C.

11 Stability and Storage Conditions

Liquid glucose should be stored in a well-closed container in a cool, dry place. Elevated temperatures will cause discoloration.

12 Incompatibilities

Incompatible with strong oxidizing agents.

13 Method of Manufacture

Liquid glucose is prepared by the incomplete acidic or enzymatic hydrolysis of starch.

14 Safety

Liquid glucose is used in oral pharmaceutical formulations and confectionery products and is generally regarded as a nontoxic and nonirritant material.

See also Dextrose.

LD_{50} (mouse, IV): 9 g/kg[1]

15 Handling Precautions

Observe normal precautions appropriate to the circumstances and quantity of material handled.

16 Regulatory Status

GRAS listed. Included in the FDA Inactive Ingredients Database (oral solutions, syrups, and tablets; topical emulsions and gels). Included in nonparenteral medicines licensed in the UK. Included in the Canadian Natural Health Products Ingredients Database.

17 Related Substances

Dextrin; dextrose; maltose.

18 Comments

Liquid glucose is also used in confectionary products at a concentration of 20–60%.

A specification for glucose syrup is contained in the *Food Chemicals Codex* (FCC)[2] and the *Japanese Pharmaceutical Excipients* (JPE).[3] The PhEur 7.4 also includes a specification for glucose, liquid, spray-dried.

The EINECS number for glucose is 200-075-1.

19 Specific References

1 Lewis RJ, ed. *Sax's Dangerous Properties of Industrial Materials*, 11th edn. New York: Wiley, 2004: 1860–1861.
2 *Food Chemicals Codex*, 6th edn. Bethesda, MD: United States Pharmacopeia, 2008: 403.
3 Japan Pharmaceutical Excipients Council. *Japanese Pharmaceutical Excipients 2004*. Tokyo: Yakuji Nippo, 2004: 383-386.

20 General References

Dziedzic SZ, Kearsley MW, eds. *Glucose Syrups: Science and Technology.* New York: Elsevier Applied Science, 1984.
Hoynak RX, Bolcenback GN. *This is Liquid Sugar*, 2nd edn. Yonkers, NY: Refined Syrup and Sugars Inc., 1966: 205, 226.
Inglett GE, ed. *Symposium on Sweeteners*. New York: AVI, 1974.

21 Author

W Yu.

22 Date of Revision

1 March 2012.

Glycerin

1 Nonproprietary Names

BP: Glycerol
JP: Concentrated Glycerin
PhEur: Glycerol
USP: Glycerin

2 Synonyms

Croderol; E422; glicerol; glycerine; glycerolum; *Glycon G-100*; *Kemstrene*; *Optim*; *Pricerine*; 1,2,3-propanetriol; *Speziol G*; trihydroxypropane glycerol.

3 Chemical Name and CAS Registry Number

Propane-1,2,3-triol [56-81-5]

4 Empirical Formula and Molecular Weight

$C_3H_8O_3$ 92.09

5 Structural Formula

HO–CH₂–CH(OH)–CH₂–OH

6 Functional Category

Antimicrobial preservative; emollient; humectant; plasticizing agent; solvent; sweetening agent; tonicity agent.

7 Applications in Pharmaceutical Formulation or Technology

Glycerin is used in a wide variety of pharmaceutical formulations including oral, otic, ophthalmic, topical, and parenteral preparations; *see* Table I.

In topical pharmaceutical formulations and cosmetics, glycerin is used primarily for its humectant and emollient properties. Glycerin is used as a solvent or cosolvent in creams and emulsions.[1–3]

Glycerin is additionally used in aqueous and nonaqueous gels and also as an additive in patch applications.[4–6] In parenteral formulations, glycerin is used mainly as a solvent and cosolvent.[7–10]

In oral solutions, glycerin is used as a solvent,[10] sweetening agent, antimicrobial preservative, and viscosity-increasing agent. It is also used as a plasticizer and in film coatings.[11–14]

Glycerin is used as a plasticizer of gelatin in the production of soft-gelatin capsules and gelatin suppositories.

Glycerin is also used as a food additive.

Table I: Uses of glycerin.

Use	Concentration (%)
Antimicrobial preservative	>20
Emollient	≤30
Gel vehicle, aqueous	5.0–15.0
Gel vehicle, nonaqueous	50.0–80.0
Humectant	≤30
Ophthalmic formulations	0.5–3.0
Patch additive	Variable
Plasticizer in tablet film coating	Variable
Solvent for parenteral formulations	≤50
Sweetening agent in alcoholic elixirs	≤20

8 Description

Glycerin is a clear, colorless, odorless, viscous, hygroscopic liquid; it has a sweet taste, approximately 0.6 times as sweet as sucrose.

9 Pharmacopeial Specifications

See Table II. *See also* Section 18.

Table II: Pharmacopeial specifications for glycerin.

Test	JP XV	PhEur 7.4	USP35–NF30
Identification	+	+	+
Characters	−	+	−
Appearance of solution	+	+	+
Acidity or alkalinity	+	+	−
Refractive index	≥1.470	1.470–1.475	−
Aldehydes	−	≤10 ppm	−
Related substances	−	+	+
Halogenated compounds	−	≤35 ppm	−
Limit of chlorinated compounds	−	−	+
Sugars	−	+	−
Chloride	≤0.001%	≤10 ppm	≤10 ppm
Heavy metals	≤5 ppm	≤5 ppm	≤5 ppm
Water	≤2.0%	≤2.0%	≤5.0%
Sulfated ash	≤0.01%	≤0.01%	≤0.01%
Specific gravity	≥1.258	−	≥1.249
Sulfate	≤0.002%	−	≤20 ppm
Esters	−	+	−
Ammonium	+	−	−
Calcium	+	−	−
Arsenic	≤2 ppm	−	−
Acrolein, glucose or other reducing substances	+	−	−
Fatty acids and esters	+	−	+
Diethylene glycol and ethylene glycol impurities	−	−	+
Readily carbonizable substances	+	−	−
Assay	98.0–101.0%	98.0–101.0%	99.0–101.0%

10 Typical Properties

Boiling point 290°C (with decomposition)

Density

1.2656 g/cm^3 at 15°C;

1.2636 g/cm^3 at 20°C;

1.2620 g/cm^3 at 25°C.

Flash point 176°C (open cup)

Freezing point *see* Table III.

Hygroscopicity Hygroscopic.

Melting point 17.8°C

Osmolarity A 2.6% v/v aqueous solution is isoosmotic with serum.

Refractive index

n_D^{15} = 1.4758;

n_D^{20} = 1.4746;

n_D^{25} = 1.4730.

Solubility *see* Table IV.

Specific gravity *see* Table V.

Spectroscopy

IR spectra *see* Figure 1.

NIR spectra *see* Figure 2.

Surface tension 63.4 mN/m (63.4 dynes/cm) at 20°C.

Vapor density (relative) 3.17 (air = 1)

Viscosity (dynamic) *see* Table VI.

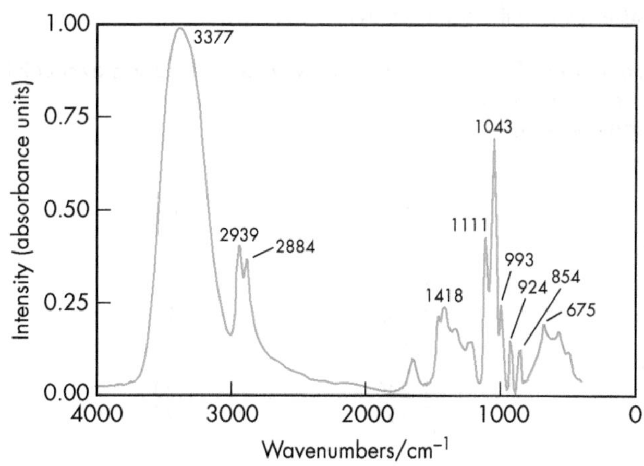

Figure 1: Infrared spectrum of glycerin measured by transmission. Adapted with permission of Informa Healthcare.

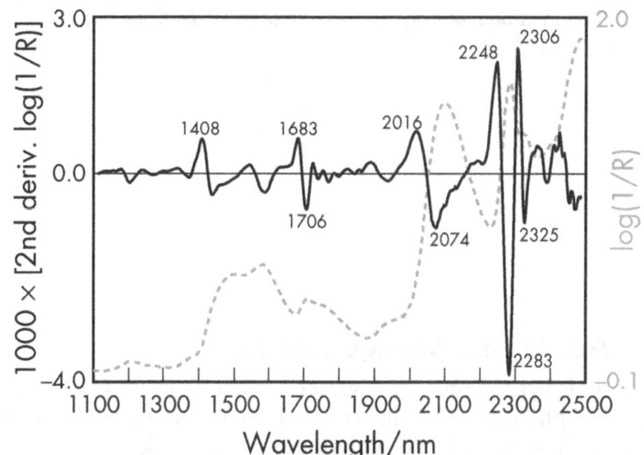

Figure 2: Near-infrared spectrum of glycerin measured by transflectance (1 mm path-length). The small peak at approx. 1950 nm is due to a trace of water (<0.5% m/m).

Table III: Freezing points of aqueous glycerin solutions.

Concentration of aqueous glycerin solution (% w/w)	Freezing point (°C)
10.0	−1.6
20.0	−4.8
30.0	−9.5
40.0	−15.4
50.0	−23
60.0	−34.7
66.7	−46.5
80.0	−20.3
90.0	−1.6

Table IV: Solubility of glycerin.

Solvent	Solubility at 20°C
Acetone	Slightly soluble
Benzene	Practically insoluble
Chloroform	Practically insoluble
Ethanol (95%)	Soluble
Ether	1 in 500
Ethyl acetate	1 in 11
Methanol	Soluble
Oils	Practically insoluble
Water	Soluble

Table V: Specific gravity of glycerin.

Concentration of aqueous glycerin solution (% w/w)	Specific gravity at 15°C	Specific gravity at 20°C
5	1.01	–
10	–	1.024
20	1.049	1.049
30	–	1.075
40	–	1.101
50	1.129	1.128
60	1.157	1.156
70	1.185	–
80	1.213	–
90	1.240	1.238
95	1.253	1.251

Table VI: Viscosity (dynamic) of aqueous glycerin solutions.

Concentration of aqueous glycerin solution (% w/w)	Viscosity at 20°C (mPa s)
5	1.143
10	1.311
25	2.095
50	6.05
60	10.96
70	22.94
83	111.0

11 Stability and Storage Conditions

Glycerin is hygroscopic. Pure glycerin is not prone to oxidation by the atmosphere under ordinary storage conditions, but it decomposes on heating with the evolution of toxic acrolein. Mixtures of glycerin with water, ethanol (95%), and propylene glycol are chemically stable.

Glycerin may crystallize if stored at low temperatures; the crystals do not melt until warmed to 20°C.

Glycerin should be stored in an airtight container, in a cool, dry place.

12 Incompatibilities

Glycerin may explode if mixed with strong oxidizing agents such as chromium trioxide, potassium chlorate, or potassium permanganate. In dilute solution, the reaction proceeds at a slower rate with several oxidation products being formed. Black discoloration of glycerin occurs in the presence of light, or on contact with zinc oxide or basic bismuth nitrate.

An iron contaminant in glycerin is responsible for the darkening in color of mixtures containing phenols, salicylates, and tannin.

Glycerin forms a boric acid complex, glyceroboric acid, that is a stronger acid than boric acid.

13 Method of Manufacture

Glycerin is mainly obtained from oils and fats as a by-product in the manufacture of soaps and fatty acids. It may also be obtained from natural sources by fermentation of, for example, sugar beet molasses in the presence of large quantities of sodium sulfite. Synthetically, glycerin may be prepared by the chlorination and saponification of propylene.

14 Safety

Glycerin occurs naturally in animal and vegetable fats and oils that are consumed as part of a normal diet. Glycerin is readily absorbed from the intestine and is either metabolized to carbon dioxide and glycogen or used in the synthesis of body fats.

Glycerin is used in a wide variety of pharmaceutical formulations including oral, ophthalmic, parenteral, and topical preparations. Adverse effects are mainly due to the dehydrating properties of glycerin.[15]

Oral doses are demulcent and mildly laxative in action. Large doses may produce headache, thirst, nausea, and hyperglycemia. The therapeutic parenteral administration of very large glycerin doses, 70–80 g over 30–60 minutes in adults to reduce cranial pressure, may induce hemolysis, hemoglobinuria, and renal failure.[16] Slower administration has no deleterious effects.[17]

Glycerin may also be used orally in doses of 1.0–1.5 g/kg bodyweight to reduce intraocular pressure.

When used as an excipient or food additive, glycerin is not usually associated with any adverse effects and is generally regarded as a nontoxic and nonirritant material.

LD_{50} (guinea pig, oral): 7.75 g/kg[18]
LD_{50} (mouse, IP): 8.70 g/kg
LD_{50} (mouse, IV): 4.25 g/kg
LD_{50} (mouse, oral): 4.1 g/kg
LD_{50} (mouse, SC): 0.09 g/kg
LD_{50} (rabbit, IV): 0.05 g/kg
LD_{50} (rabbit, oral): 27 g/kg[19]
LD_{50} (rat, IP): 4.42 g/kg
LD_{50} (rat, oral): 5.57 g/kg[19]
LD_{50} (rat, oral): 12.6 g/kg
LD_{50} (rat, SC): 0.1 g/kg

15 Handling Precautions

Observe normal precautions appropriate to the circumstances and quantity of material handled. Eye protection and gloves are recommended. In the UK, the recommended long-term (8-hour TWA) workplace exposure limit for glycerin mist is 10 mg/m^3.[20] Glycerin is combustible and may react explosively with strong oxidizing agents; *see* Section 12.

16 Regulatory Status

GRAS listed. Accepted for use as a food additive in Europe. Included in the FDA Inactive Ingredients Database (dental pastes; buccal preparations; inhalations; injections; nasal and ophthalmic preparations; oral capsules, solutions, suspensions and tablets; otic, rectal, topical, transdermal, and vaginal preparations). Included in nonparenteral and parenteral medicines licensed in the UK. Included in the Canadian Natural Health Products Ingredients Database.

17 Related Substances

—

18 Comments

Glycerin is one of the materials that have been selected for harmonization by the Pharmacopeial Discussion Group. For further information see the General Information Chapter <1196> in the USP35–NF30, the General Chapter 5.8 in PhEur 7.4, along with the 'State of Work' document on the PhEur EDQM website, and also the General Information Chapter 8 in the JP XV.

Some pharmacopeias also contain specifications for diluted glycerin solutions. The JP XV contains a monograph for 'glycerin' that contains 84–87% of propane-1,2,3-triol ($C_3H_8O_3$). The PhEur 7.4 contains a monograph for 'glycerol 85 per cent' that contains 83.5–88.5% of propane-1,2,3-triol ($C_3H_8O_3$).

Glycerin is employed as a therapeutic agent in a variety of clinical applications.[15]

A specification for glycerin is contained in the *Food Chemicals Codex* (FCC).[21]

The EINECS number for glycerin is 200-289-5.

19 Specific References

1 Viegas TX, *et al.* Evaluation of creams and ointments as suitable formulations for peldesine. *Int J Pharm* 2001; **219**(1–2): 73–80.
2 Sheu MT, *et al.* Simultaneous optimization of percutaneous delivery and adhesion for ketoprofen poultice. *Int J Pharm* 2002; **233**: 257–262.
3 Barichello JM, *et al.* Combined effect of liposomalization and addition of glycerol on the transdermal delivery of isosorbide 5-nitrate in rat skin. *Int J Pharm* 2008; **357**(1–2): 199–205.
4 A-Sasutjarit R, *et al.* Viscoelastic properties of carbopol 940 gels and their relationships to piroxicam diffusion coefficients in gel bases. *Pharm Res* 2005; **22**(12): 2134–2140.
5 Chow TC, *et al.* Formulation of hydrophilic non-aqueous gel: drug stability in different solvents and rheological behaviour of gel matrices. *Pharm Res* 2008; **25**(1): 207–217.
6 Abu-Huwaij R, *et al.* Potential mucoadhesive dosage form of lidocaine hydrochloride: II in vitro and in vivo evaluation. *Drug Dev Ind Pharm* 2007; **33**(4): 437–448.
7 Spiegel AJ, Noseworthy MM. Use of nonaqueous solvents in parenteral products. *J Pharm Sci* 1963; **52**(10): 917–927.
8 Strickley RG. Solubilizing excipients in oral and injectable formulations. *Pharm Res* 2004; **21**(2): 201–230.
9 Powell MF, *et al.* Compendium of excipients for parenteral formulations. *PDA J Pharm Sci Technol* 1998; **52**(5): 238–311.
10 Millard JW, *et al.* Solubilization by cosolvents establishing useful constants for the log-linear model. *Int J Pharm* 2002; **245**(1–2): 153–166.
11 Kumar V, *et al.* Preparation and characterization of spray-dried oxidized cellulose particles. *Pharm Dev Technol* 2001; **6**(3): 449–458.
12 Palviainen P, *et al.* Corn starches as film formers in aqueous-based film coating. *Pharm Dev Technol* 2001; **6**(3): 353–361.
13 Pongjanyakul T, Puttipipatkhachorn S. Alginate-magnesium aluminium silicate films: effect on plasticizer on film properties, drug permeation and drug release from coated tablets. *Int J Pharm* 2007; **333**(1–2): 34–44.
14 Bonacucina G, *et al.* Effect of plasticizers on properties of pregelatinised starch acetate (Amprac 01) free films. *Int J Pharm* 2006; **313**(1–2): 72–77.
15 Sweetman SC, ed. *Martindale: The Complete Drug Reference*, 36th edn. London: Pharmaceutical Press, 2009: 2314–2315.
16 Hägnevik K, *et al.* Glycerol-induced haemolysis with haemoglobinuria and acute renal failure. Report of three cases. *Lancet* 1974; **303**(7847): 75–77.
17 Welch KMA, *et al.* Glycerol-induced haemolysis [letter]. *Lancet* 1974; **303**(7847): 416–417.
18 Lewis RJ, ed. *Sax's Dangerous Properties of Industrial Materials*, 11th edn. New York: Wiley, 2004: 1865.
19 CDC National Institute for Occupational Safety and Health. *Registry of toxic effects of chemical substances (RTECS)*, 1993–2008.
20 Health and Safety Executive. *EH40/2005: Workplace Exposure Limits*. Sudbury: HSE Books, 2011. http://www.hse.gov.uk/pubns/priced/eh40.pdf (accessed 1 March 2012).
21 *Food Chemicals Codex*, 7th edn. Bethesda, MD: United States Pharmacopeia, 2010: 442.

20 General References

European Directorate for the Quality of Medicines and Healthcare (EDQM). European Pharmacopoeia – State Of Work Of International Harmonisation. *Pharmeuropa* 2010; 22(4): 583–584. www.edqm.eu/site/-614.html (accessed 1 December 2010).

Grissom CB, *et al.* Use of viscosigens to stabilize vitamin B_{12} solutions against photolysis. *J Pharm Sci* 1993; **82**(6): 641–643.

Jungermann E, Sonntag NOV, eds. *Glycerine: A Key Cosmetic Ingredient*. New York: Marcel Dekker, 1991.

Smolinske SC. *Handbook of Food, Drug, and Cosmetic Excipients*. Boca Raton, FL: CRC Press, 1992: 199–204.

Staples R, *et al.* Gastrointestinal irritant effect of glycerin as compared with sorbitol and propylene glycol in rats and dogs. *J Pharm Sci* 1967; **56**(3): 398–400.

The HallStar Company. Product data sheet: *RTD Glycerine 99.5%*, 2008.

21 Authors

FA Alvarez-Núnez, C Medina.

22 Date of Revision

1 March 2012.

G

Glyceryl Behenate

1 Nonproprietary Names

BP: Glycerol Dibehenate
PhEur: Glycerol Dibehenate
USP–NF: Glyceryl Behenate

2 Synonyms

Behenin; *Compritol 888 ATO*; 1,3-di(docosanoyloxy)propane-2–yl docosanoate; 2,3-dihydroxypropyl docosanoate; docosanoic acid, 2,3-dihydroxypropyl ester; docosanoin; E471; glycerol behenate; glyceroli dibehenas; glyceryl monobehenate; glyceryl tridocosanoate; 1,2,3-propanetriol docosanoate.

Note that tribehenin is used as a synonym for glyceryl tribehenate.

3 Chemical Name and CAS Registry Number

Docosanoic acid, monoester with glycerin [30233-64-8] (glyceryl behenate)
Docosanoic acid, diester with glycerin [94201-62-4] (glyceryl dibehenate)
Docosanoic acid, triester with glycerin [18641-57-1] (glyceryl tribehenate)

4 Empirical Formula and Molecular Weight

Glyceryl dibehenate is a mixture of glycerol esters. The PhEur 7.4 describes glyceryl dibehenate as a mixture of diacylglycerols, mainly dibehenoylglycerol, together with variable quantities of mono- and triacylglycerols obtained by esterification of glycerin with behenic acid (*see* Section 9). The USP35–NF30 describes glyceryl behenate as a mixture of glycerides of fatty acids, mainly behenic acid. It specifies that the content of 1-monoglycerides should be 12.0–18.0%.

5 Structural Formula

See Section 4.

6 Functional Category

Coating agent; modified-release agent; tablet and capsule binder; tablet and capsule lubricant; taste-masking agent; viscosity-increasing agent.

7 Applications in Pharmaceutical Formulation or Technology

Glyceryl behenate is used in cosmetics, foods, and oral and topical pharmaceutical formulations; *see* Table I.

In oral pharmaceutical formulations, glyceryl behenate powder is used at low concentrations as a lubricant in the preparation of oral tablets and capsules.[1–5] It is also used in oral formulations at higher concentrations to produce lipid matrix tablets for sustained drug release. In contact with aqueous fluid, a lipid matrix does not swell or erode, and therefore diffusion is the main drug release mechanism.

Glyceryl behenate decreases ejection force and improves compressibility in pharmaceutical formulations. The well-defined and controlled particle size is associated with reduced friction and improved powder flowability, which aids homogeneity and content uniformity. It is not sensitive to over-blending and does not interfere with tablet disintegration time, dissolution rate, or drug release. It is particularly effective in anhydrous products to formulate drugs that are unstable in aqueous media.

Basic processing techniques such as direct compression or wet granulation can be used to produce tablet matrices using glyceryl behenate.[6,7] Additional processing methods that fortify the sustained-release properties of the matrix may be used in particular cases, for example, in formulations with highly water-soluble drugs.[8-10] Such methods include solid dispersion[11–20] or tablet treatments.[16–18] Glyceryl behenate may also be incorporated via extrusion/spheronization into pellets that can be further compressed into tablets.[19,20] Glyceryl behenate can also be used as a hot-melt coating agent sprayed onto a powder or granules.[21–24]

Along with acid-soluble or swellable polymers glyceryl behenate is also used to mask the bitter or unpleasant taste of the medicament with improved palatability.[25]

Glyceryl behenate has been used for the preparation of ophthalmic inserts.[26,27]

Glyceryl behenate has been investigated for the encapsulation of various drugs in solid lipid nanoparticles (SLN); *see* Section 18.

In cosmetics, glyceryl behenate is used as an emollient and viscosity-increasing agent in emulsions. It also improves the heat stability of emulsions and is a gelifying agent for various oils. For topical formulations, it is used as a thickening agent for oily phases in emulsions and ointments.

See also Section 18.

Table I: Uses of glyceryl behenate.

Use	Concentration (%)
Lipophilic matrix or coating for sustained-released tablets and capsules	>10.0
Tablet and capsule lubricant	1.0–3.0
Taste-masking agent	5.0-15
Viscosity-increasing agent in silicon gels (cosmetics)	1.0–15.0
Viscosity-increasing agent in w/o or o/w emulsions (cosmetics)	1.0–5.0

8 Description

Glyceryl behenate occurs as a fine, off-white or white-yellow powder, as a hard waxy mass or pellet, or as white or almost white unctuous flakes. It has a faint odor and is tasteless.

9 Pharmacopeial Specifications

See Table II.

10 Typical Properties

Flash point 150°C
HLB value 1
Melting point 65–77°C
Solubility Soluble in methylene chloride, and, when heated, it is soluble in chloroform and dichloromethane and in many organic solvents; partly soluble in hot ethanol (96%); practically insoluble in cold ethanol (95%), hexane, mineral oil, and water.
Spectroscopy
IR spectra *see* Figure 1.

11 Stability and Storage Conditions

Glyceryl behenate should be stored in a dry, airtight container, at a temperature less than 35°C. It should be stored away from extreme heat and strong oxidizing agents in a well-ventilated place.

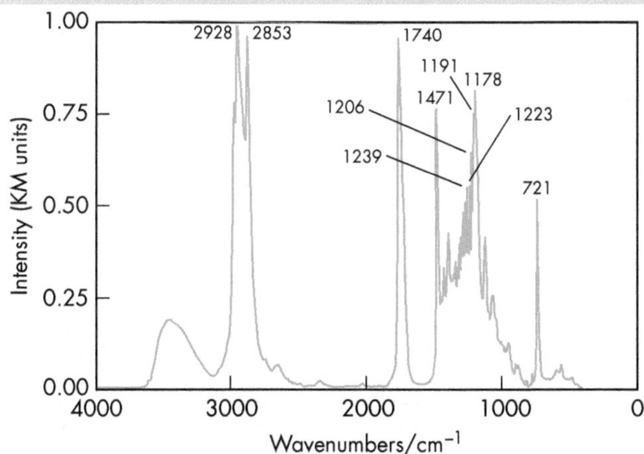

Figure 1: Infrared spectrum of glyceryl behenate measured by diffuse reflectance. Adapted with permission of Informa Healthcare.

Table II: Pharmacopeial specifications for glyceryl behenate.

Test	PhEur 7.4	USP35–NF30
Identification	+	+
Characters	+	—
Acid value	≤4.0	≤4
Iodine value	≤3.0	≤3
Saponification value	145–165	145–165
Residue on ignition	≤0.1%	≤0.1%
Nickel	≤1 ppm	—
Water	≤1.0%	—
Heavy metals	—	≤10 ppm
Melting point	65–77°C	—
Content of 1-monoglycerides	—	12.0–18.0%
Content of acylglycerols (glycerides)	+	—
Monoacylglycerols	15.0–23.0%	—
Diacylglycerols	40–60%	—
Triacylglycerols	21–35%	—
Free glycerin	≤1.0%	≤1.0%
Composition of fatty acids	+	—
Arachidic acid	≤10.0%	—
Behenic acid	≥83.0%	—
Erucic acid	≤3.0%	—
Lignoceric acid	≤3.0%	—
Palmitic acid	≤3.0%	—
Stearic acid	≤5.0%	—

12 Incompatibilities

Glyceryl behenate is incompatible with strong oxidizing agents.

13 Method of Manufacture

Glyceryl behenate is prepared by the esterification of glycerin by behenic acid (C_{22} fatty acid) without the use of catalysts. In the case of *Compritol 888 ATO* (Gattefossé), raw materials used are of vegetable origin, and the esterified material is atomized by spray-cooling.

14 Safety

Glyceryl behenate is used in cosmetics, foods and oral pharmaceutical formulations, and is generally regarded as a relatively nonirritant and nontoxic material. The US Cosmetic Ingredients Review Expert Panel evaluated glyceryl behenate and concluded that it is safe for use in cosmetic formulations in present practices of use and concentration.

LD_{50} (mouse, oral): 5 g/kg[(28)]

15 Handling Precautions

Observe normal precautions appropriate to the circumstances and quantities of material handled. Glyceryl behenate emits acrid smoke and irritating fumes when heated to decomposition.

16 Regulatory Status

GRAS listed. Accepted for use as a food additive in Europe. Included in the FDA Inactive Ingredients Database (oral capsules, tablets, and suspensions). Included in the Canadian Natural Health Products Ingredients Database.

17 Related Substances

Glyceryl palmitostearate.

18 Comments

Glyceryl behenate has been investigated in the preparation of aqueous colloidal dispersions such as solid lipid microparticles (SLM), nanoparticles (SLN), nanostructured lipid carriers (NLC), and lipidic nanospheres[(29)] for the entrapment of lipophilic drugs.[(30)] For example, SLM of ibuprofen have been prepared with a high drug-loading capacity.[(31-33)] Glyceryl behenate SLM for the sunscreen agent octyl-dimethylaminobenzoate (ODAB) showed enhanced sunscreen photostability, and SLM loaded with the polar adenosine A1 receptor agonist N^6-cyclopentyladenosine (CPA) have shown improved drug stability.[(34,35)] SLM have been prepared using glyceryl behenate as carriers for pulmonary delivery of indomethacin with prolonged drug release.[(36, 37)] SLN prepared from glyceryl behenate have been investigated for oral and mucosal delivery with enhanced bioavailability;[(38,39)] for topical and transdermal delivery of vitamin A,[(40)] retinoic acid,[(41)] ketoconazole,[(42)] ketorolac,[(43)] miconazole nitrate;[(44)] for ocular delivery of cyclosporine A;[(45)] for rectal administration of a water-insoluble drug (diazepam) with combined advantages of rapid onset and prolonged drug release;[(46)] and for parenteral drug administration of drugs such as tetracaine, etomidate and prednisolone.[(47–50)] NLC have been prepared by replacing part of solid lipid by oil for improved encapsulation of drugs.[(51)]

A specification for glyceryl behenate is included in the *Food Chemicals Codex* (FCC).[(52)]

The EINECS numbers are: 250-097-0 for glyceryl behenate; 303-650-6 for glyceryl dibehenate; 242-471-7 for glyceryl tribehenate. The PubChem Compound ID (CID) for glyceryl behenate includes 62726 and 121658.

19 Specific References

1 Shah NH, *et al.* Evaluation of two new tablet lubricants – sodium stearyl fumarate and glyceryl behenate. Measurement of physical parameters (compaction, ejection and residual forces) in the tabletting process and the effect on the dissolution rate. *Drug Dev Ind Pharm* 1986; 12: 1329–1346.

2 Baichwal AR, Augsburger LL. Variations in the friction coefficients of tablet lubricants and relationship to their physicochemical properties. *J Pharm Pharmacol* 1988; 40: 569–571.

3 Brossard C, *et al.* Modelling of theophylline compound release from hard gelatin capsules containing *Gelucire* matrix granules. *Drug Dev Ind Pharm* 1991; 17: 1267–1277.

4 N'Diaye A, *et al.* Comparative study of the lubricant performance of Compritol® HD5 ATO and Compritol® 888 ATO: effect of polyethylene glycol behenate on lubricant capacity. *Int J Pharm* 2003; 254(2): 263–269.

5 Lapeyre F, *et al.* Quantitative evaluation of some tablet lubricants. Pratical involvements in tablet formulation. *STP Pharma* 1988; 4(3): 209–214.

6 Barakat NS, *et al.* Formulation release characteristics and bioavailability study of oral monolithic matrix tablets containing carbamazepine. *AAPS PharmSciTech* 2008; 9: 931–938.

7 El-Sayed GM, *et al.* Kinetics of theophylline release from different tablet matrices. *STP Pharma Sci* 1996; 6: 390–397.

8 Gohel M, Nagori SA. Fabrication and evaluation of captopril modified-release oral formulation. *Pharm Dev Technol* 2009; **14**(6): 679–686.
9 Chi N, *et al.* An oral controlled release system for amroxol hydrochloride containing a wax and a water insoluble polymer. *Pharm Dev Technol* 2010; **15**(1): 97–104.
10 Barakat NS, *et al.* Controlled-release carbamazepine matrix granules and tablets comprising lipophilic and hydrophilic components. *Drug Deliv* 2009; **16**(1): 57–65.
11 Prinderre P, *et al.* Evaluation of some protective agents on stability and controlled release of oral pharmaceutical forms by fluid bed technique. *Drug Dev Ind Pharm* 1997; **23**: 817–826.
12 Obaidat AA, Obaidat RM. Controlled release of tramadol hydrochloride from matrices prepared using glyceryl behenate. *Eur J Pharm Biopharm* 2001; **52**(2): 231–235.
13 Jannin V, *et al.* Comparative study of the lubricant performance of *Compritol (R) 888 ATO* either used by blending or by hot melt coating. *Int J Pharm* 2003; **262**(1–2): 39–45.
14 Barthelemy P, *et al. Compritol 888 ATO*: an innovative hot-melt coating agent for prolonged-release drug formulations. *Eur J Pharm Biopharm* 1999; **471**: 87–90.
15 Li FQ, *et al. In vitro* controlled release of sodium ferulate from Compritol® 888 ATO-based matrix tablets. *Int J Pharm* 2006; **324**(2): 152–157.
16 Ozyazici M, *et al.* Release and diffusional modeling of metronidazole lipid matrices. *Eur J Pharm Biopharm* 2006; **63**(3): 331–339.
17 Patel VF, Patel NM. Controlled release of dipyridamole from floating matrices prepared using glyceryl behenate. *Drug Deliv Technol* 2008; **8**(7): 54–59.
18 Perez MA, *et al.* Sustained release phenylpropanolamine hydrochloride from Compritol® 888 ATO matrix. *P R Health Sci J* 1993; **12**: 263–267.
19 Zhang YE, *et al.* Effect of diluents on tablet integrity and controlled drug release. *Pharm Dev Technol* 2000; **26**(7): 761–765.
20 Zhang YE, *et al.* Effect of processing methods and heat treatment on the formation of wax matrix tablets for sustained drug release. *Pharm Dev Technol* 2001; **6**(2): 131–144.
21 Zhang YE, *et al.* Melt granulation and heat treatment for wax matrix-controlled drug release. *Drug Dev Ind Pharm* 2003; **29**(2): 131–138.
22 Iloañusi NO, *et al.* The effect of wax and water on extrusion forces using an instrumented miniextruder. *Drug Dev Ind Pharm* 1996; **22**(7): 667–671.
23 Iloañusi NO, *et al.* The effect of wax on compaction of microcrystalline cellulose beads made by extrusion and spheronization. *Drug Dev Ind Pharm* 1998; **24**(1): 37–44.
24 Faham A, *et al.* Hot-melt coating technology. I: Influence of Compritol 888 ato and granule size on theophylline release. *Drug Dev Ind Pharm* 2000; **26**(2): 167–176.
25 Menjoge A, Kulkarni M. Pharmaceutical composition for improving palatability of drugs and process for preparation thereof. United States Patent No. 7,378,109, 2008.
26 Saettone M, *et al.* Controlled release of pilocarpine from coated polymeric ophthalmic inserts prepared by extrusion. *Int J Pharm* 1992; **86**(2–3): 159–166.
27 Aiache J, Serpin G. Pharmaceutical dosage form for ocular administration and preparation process. United States Patent No. 5,766,619 1998.
28 Sweet DV, ed. *Registry of Toxic Effects of Chemical Substances.* Cincinatti: US Department of Health, 1987.
29 Quintanar-Guerrero D, *et al.* Adaptation and optimization of the emulsification-diffusion technique to prepare lipidic nanospheres. *Eur J Pharm Sci* 2005; **26**(2): 211–218.
30 Souto E, *et al.* Polymorphic behaviour of *Compritol 888 ATO* as bulk lipid and as SLN and NLC. *J Microencapsul* 2006; **23**(4): 417–433.
31 Long C, *et al.* Preparation and crystal modification of ibuprofen-loaded solid lipid microparticles. *Chin J Chem Eng* 2006; **14**(4): 518–525.
32 Long C, *et al.* Dissipative particle dynamics simulation of ibuprofen molecules distribution in the matrix of solid lipid microparticles (SLM). *Computer Aided Chemical Engineering* 2006; **21**(2): 1649–1654.
33 Long C, *et al.* Mesoscale simulation of drug molecules distribution in the matrix of solid lipid microparticles (SLM). *Chem Eng J* 2006; **119**(2–3): 99–106.
34 Tursilli R, *et al.* Solid lipid microparticles containing the sunscreen agent, octyl-dimethylaminobenzoate: effect of the vehicle. *Eur J Pharm Biopharm* 2007; **66**(3): 483–487.
35 Dalpiaz A, *et al.* Solid lipid microparticles for the stability enhancement of the polar drug N^6-cyclopentyladenosine. *Int J Pharm* 2008; **355**(1–2): 81–86.
36 Sanna V, *et al.* Preparation and *in vivo* toxicity study of solid lipid microparticles as carrier for pulmonary administration. *AAPS Pharm Sci Tech* 2003; **5**(2): Article 27.
37 Sedef Erdal M, *et al.* Preparation and in vitro evaluation of indomethacin loaded solid lipid microparticles. *Acta Pharm Sci* 2009; **51**: 203–210.
38 He J, *et al.* Effect of particle size on oral absorption of silymarin-loaded solid lipid nanoparticles. *Zhongguo Zhong Yao Za Zhi* 2005; **30**(21): 1651–1653.
39 Rao K. Polymerized solid lipid nanoparticles for oral or mucosal delivery of therapeutic proteins and peptides. International Patent WO 113665, 2007.
40 Jenning V, *et al.* Vitamin A-loaded solid lipid nanoparticles for topical use: drug release properties. *J Control Release* 2000; **66**(2–3): 115–126.
41 Castro A, *et al.* Development of a new solid lipid nanoparticle formulation containing retinoic acid for topical treatment of acne. *J Microencapsul* 2007; **24**(5): 395–407.
42 Souto E, Müller R. SLN and NLC for topical delivery of ketoconazole. *J Microencapsul* 2005; **22**(5): 501–510.
43 Puglia C, *et al.* Evaluation of alternative strategies to optimize ketorolac transdermal delivery. *AAPS Pharm Sci Tech* 2006; **7**(3): Article 64.
44 Bhalekar MR, *et al.* Preparation and evaluation of miconazole nitrate-loaded solid lipid nanoparticles for topical delivery. *AAPS PharmSci-Tech* 2009; **10**(1): 289–296.
45 Evren HG, *et al.* Cyclosporine A-loaded solid lipid nanoparticles: ocular tolerance and in vivo drug release in rabbit eyes. *Curr Eye Res* 2009; **34**(11): 996–1003.
46 Abdelbary G, Fahmy RH. Diazepam-loaded solid lipid nanoparticles: design and characterization. *AAPS PharmSciTech* 2009; **10**(1): 211–219.
47 Mühlen A, *et al.* Solid lipid nanoparticles (SLN) for controlled drug delivery – drug release and release mechanism. *Eur J Pharm Biopharm* 1998; **45**(2): 149–155.
48 Schwarz C, Mehnert W. Freeze-drying of drug-free and drug-loaded solid lipid nanoparticles (SLN). *Int J Pharm* 1997; **157**(2): 171–179.
49 Weyhers H, *et al.* Solid lipid nanoparticles (SLN): effects of lipid composition on *in vitro* degradation and *in vivo* toxicity. *Pharmazie* 2006; **61**(6): 539–544.
50 Müller R, *et al.* Solid lipid nanoparticles (SLN) for controlled drug delivery – a review of the state of the art. *Eur J Pharm Biopharm* 2000; **50**(1): 161–177.
51 Jores K, *et al.* Investigations on the structure of solid lipid nanoparticles (SLN) and oil-loaded solid lipid nanoparticles by photon correlation spectroscopy, field-flow fractionation and transmission electron microscopy. *J Control Release* 2004; **95**: 217–227.
52 *Food Chemicals Codex*, 7th edn. Bethesda, MD: United States Pharmacopeia, 2010; 450.

20 General References

CosmeticsInfo.org. www.cosmeticsinfo.org/index.php (accessed 23 December 2011).

Gattefossé. Product literature: *Compritol 888 ATO*, 2009.

Hamdani J, *et al.* Physical and thermal characterization of *Precirol* and *Compritol* as lipophilic glycerides used for the preparation of controlled-release matrix pellets. *Int J Pharm* 2003; **260**(1): 47–57.

21 Authors

P Pople, KK Singh, C Subra.

22 Date of Revision

1 March 2012.

 Glyceryl Monooleate

1 Nonproprietary Names

BP: Glycerol Mono-oleate
PhEur: Glycerol Mono-oleate
USP–NF: Glyceryl Monooleate

2 Synonyms

Aldo MO; *Atlas G-695*; *Capmul GMO*; glycerol-1-oleate; glyceroli mono-oleas; glyceryl mono-oleate; glyceryl oleate; *HallStar GMO*; *Imwitor 948*; *Kessco GMO*; *Ligalub*; monolein; *Monomuls 90-O18*; mono-olein; α-mono-olein glycerol; *Peceol*; *Priolube 1408*; *Stepan GMO*; *Tegin*.

3 Chemical Name and CAS Registry Number

9-Octadecenoic acid (*Z*), monoester with 1,2,3-propanetriol [25496-72-4]

4 Empirical Formula and Molecular Weight

$C_{21}H_{40}O_4$ 356.55 (for pure material)

Glyceryl monooleate is a mixture of the glycerides of oleic acid and other fatty acids, consisting mainly of the monooleate; *see* Section 8.

5 Structural Formula

```
      H           O
      |           ||
H — C — O — C — C17H33
      |
H — C — OH
      |
H — C — OH
      |
      H
```

6 Functional Category

Emollient; emulsifying agent; emulsion stabilizing agent; gelling agent; modified-release agent; mucoadhesive; nonionic surfactant.

7 Applications in Pharmaceutical Formulation or Technology

Glyceryl monooleate is a polar lipid that swells in water to give several phases with different rheological properties.[1] It is available in both nonemulsifying (n/e) and self-emulsifying (s/e) grades, the self-emulsifying grade containing about 5% of an anionic surfactant.

The nonemulsifying grade is used in topical formulations as an emollient and as an emulsifying agent for water-in-oil emulsions. It is also a stabilizer for oil-in-water emulsions. The self-emulsifying grade is used as a primary emulsifier for oil-in-water systems.[2]

Glyceryl monooleate gels in excess water, forming a highly ordered cubic phase that can be used to sustain the release of various water-soluble drugs.[3–6] It is also the basis of mucoadhesive drug delivery systems.[7,8]

Glyceryl monooleate is reported to enhance transdermal[9] and buccal penetration.[10]

8 Description

The PhEur 7.4 describes glyceryl monooleate as being a mixture of monoacylglycerols, mainly monooleoylglycerol, together with variable quantities of di- and triacylglycerols. They are defined by the nominal content of monoacylglycerols (*see* Table I) and obtained by partial glycerolysis of vegetable oils mainly containing triacylglycerols of oleic acid or by esterification of glycerol by oleic acid, this fatty acid being of vegetable or animal origin. A suitable antioxidant may be added.

Glyceryl monooleates occur as amber oily liquids, which may be partially solidified at room temperature and have a characteristic odor.

G

Table I: Nominal content of acylglycerols in glycerol monooleate defined in the PhEur 7.4.

	Nominal content of acylglycerol (%)		
	40	**60**	**90**
Monoacylglycerols	32.0–52.0	55.0–65.0	90.0–101.0
Diacylglycerols	30.0–50.0	15.0–35.0	<10.0
Triacylglycerols	5.0–20.0	2.0–10.0	<2.0

9 Pharmacopeial Specifications

See Table II.

Table II: Pharmacopeial specifications for glyceryl monooleate.

Test	PhEur 7.4	USP35–NF30
Identification	+	+
Characters	+	—
Acid value	≤6.0	≤6.0
Iodine value	65.0–95.0	65.0–95.0
Peroxide value	≤12.0	≤12.0
Saponification value	150–175	150–175
Free glycerol	≤6.0%	≤6.0%
Composition of fatty acids		
Palmitic acid	≤12.0%	≤12.0%
Stearic acid	≤6.0%	≤6.0%
Oleic acid	≥60.0%	≥60.0%
Linoleic acid	≤35.0%	≤35.0%
Linolenic acid	≤2.0%	≤2.0%
Arachidic acid	≤2.0%	≤2.0%
Eicosenoic acid	≤2.0%	≤2.0%
Content of acylglycerol	*see* Table I	—
Water	≤1.0%	≤1.0%
Total ash	≤0.1%	≤0.1%

10 Typical Properties

Boiling point 238–240°C
Density 0.942 g/cm^3
Flash point 216°C
HLB value 3.3 (n/e); 4.1 (s/e).
Melting point 35°C (*see also* Section 13)
Refractive index 1.4626
Solubility Soluble in chloroform, ethanol (95%), ether, mineral oil, and vegetable oils; practically insoluble in water. The self-emulsifying grade is dispersible in water.
Spectroscopy
IR spectra *see* Figure 1.
Viscosity (kinematic) 100 m^2/s (100 cSt) at 40°C

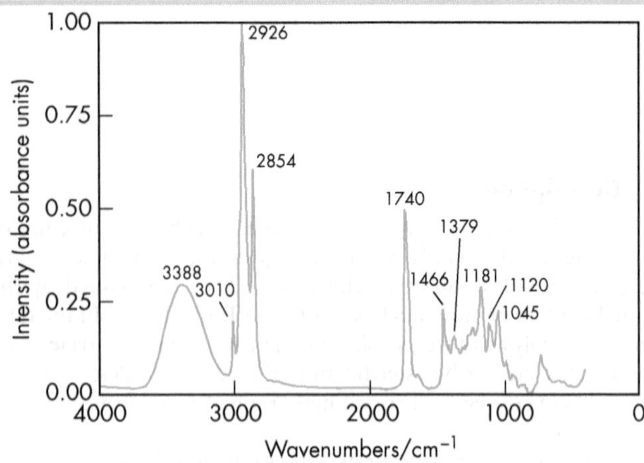

Figure 1: Infrared spectrum of glyceryl monooleate measured by transmission. Adapted with permission of Informa Healthcare.

11 Stability and Storage Conditions

Glyceryl monooleate should be stored in an airtight container, protected from light in a cool, dry place.

12 Incompatibilities

Glyceryl monooleate is incompatible with strong oxidizing agents. The self-emulsifying grade is incompatible with cationic surfactants.

13 Method of Manufacture

Glyceryl monooleate is prepared by the esterification of glycerol with fatty acids, chiefly oleic acid. As the fatty acids are not pure substances, but rather a mixture of fatty acids, the product obtained from the esterification will contain a mixture of esters, including stearic and palmitic. Di- and triesters may also be present. The composition and, therefore, the physical properties of glyceryl monooleate may thus vary considerably from manufacturer to manufacturer; e.g. the melting point may vary from 10–35°C.

14 Safety

Glyceryl monooleate is used in oral and topical pharmaceutical formulations and is generally regarded as a relatively nonirritant and nontoxic excipient.

15 Handling Precautions

Observe normal precautions appropriate to the circumstances and quantity of material handled.

16 Regulatory Status

GRAS listed. Included in the FDA Inactive Ingredients Database (oral capsules, oral powder, oral tablets; creams, controlled-release transdermal films). Included in nonparenteral medicines licensed in the UK. Included in the Canadian Natural Health Products Ingredients Database.

17 Related Substances

Glyceryl monostearate.

18 Comments

A specification for glyceryl monooleate is included in the *Food Chemicals Codex* (FCC).[11]

A specification for glyceryl monooleate is included in the *Japanese Pharmaceutical Excipients* (JPE).[12]

The EINECS number for glyceryl monooleate is 247-038-6. The PubChem Compound ID (CID) for glyceryl monooleate includes 5283468 and 33022.

19 Specific References

1 Engstrom S, *et al.* A study of polar lipid drug carrier systems undergoing a thermoreversible lamellar-to-cubic phase transition. *Int J Pharm* 1992; **86**: 137–145.
2 Ganem-Quintanar A, *et al.* Mono-olein: a review of the pharmaceutical applications. *Drug Dev Ind Pharm* 2000; **26**(8): 809–820.
3 Wyatt DM, Dorschel D. Cubic-phase delivery system composed of glyceryl monooleate and water for sustained release of water-soluble drugs. *Pharm Technol* 1992; **16**: 116–130.
4 Wörle G, *et al.* Influence of composition and preparation parameters on the properties of aqueous monoolein dispersions. *Int J Pharm* 2007; **329**: 150–157.
5 Costa-Balogh FO, *et al.* Drug release from lipid liquid crystalline phases: relation with phase behavior. *Drug Dev Ind Pharm* 2010; **36**(4): 470–481.
6 Shah MH, Paradka A. Effect of HLB of additives on the properties and drug release from glyceryl monooleate matrices. *Eur J Pharm Biopharm* 2007; **67**: 166–174.
7 Neilson LS, *et al.* Bioadhesive drug delivery systems. 1. Characterization of mucoadhesive properties of systems based on glyceryl monooleate and glycerol monolinoleate. *Eur J Pharm Sci* 1998; **6**(9): 231–239.
8 Lee J, *et al.* Water quantitatively induces the mucoadhesion of liquid crystalline phases of glyceryl monooleate. *J Pharm Pharmacol* 2001; **53**(5): 629–636.
9 Ogiso T, *et al.* Effect of various enhancers on transdermal penetration of indomethacin and urea, and relationship between penetration parameters and enhancement factors. *J Pharm Sci* 1995; **84**: 482–488.
10 Lee J, Kellaway IW. Peptide washout and permeability from glyceryl monooleate buccal delivery systems. *Drug Dev Ind Pharm* 2002; **28**: 1158–1162.
11 *Food Chemicals Codex*, 7th edn. (Suppl. 1). Bethesda, MD: United States Pharmacopeia, 2010: 1469.
12 Japan Pharmaceutical Excipients Council. *Japanese Pharmaceutical Excipients* 2004. Tokyo; Yakuji Nippo, 2004; 360.

20 General References

Eccleston GM. Emulsions and microemulsions. In: Swarbrick J, Boylan JC, eds. *Encyclopaedia of Pharmaceutical Technology*, 2nd edn, 2. New York: Marcel Dekker, 2002: 1066–1085.

Weiner AL. Lipid excipients in pharmaceutical dosage forms. In: Swarbrick J, Boylan JC, eds. *Encyclopaedia of Pharmaceutical Technology*, 2nd edn, 2. New York: Marcel Dekker, 2002: 1659–1673.

21 Author

NA Armstrong.

22 Date of Revision

1 March 2012.

Glyceryl Monostearate

1 Nonproprietary Names

BP: Glyceryl Monostearate 40–55
JP: Glyceryl Monostearate
PhEur: Glycerol Monostearate 40–55
USP–NF: Glyceryl Monostearate

Note that the USP35–NF30 also includes a specification for mono- and di-glycerides that corresponds to glyceryl monostearate 40–55 in the PhEur 7.4.

2 Synonyms

Capmul GMS-50; *Cutina GMS*; *Dermowax GMS*; 2,3-dihydroxypropyl octadecanoate; *Geleol*; glycerine monostearate; glycerin monostearate; glycerol monostearate; glyceroli monostearas; glycerol stearate; glyceryl stearate; GMS; *Imwitor 191*; *Imwitor 491*; *Imwitor 900*; *Imwitor 900K*; *Kessco GMS*; *Lonzest GMS*; monoester with 1,2,3-propanetriol; monostearin; *Myvaplex 600P*; *Myvatex*; 1,2,3-propanetriol octadecanoate; *Protachem GMS-450*; *Rheodol MS-165V*; stearic acid, monoester with glycerol; stearic monoglyceride; *Stepan GMS*; *Tegin*; *Tegin 503*; *Tegin 515*; *Tegin 4100*; *Tegin M*; *Unimate GMS*.

3 Chemical Name and CAS Registry Number

Octadecanoic acid, monoester with 1,2,3-propanetriol [31566-31-1]

4 Empirical Formula and Molecular Weight

$C_{21}H_{42}O_4$ 358.6

5 Structural Formula

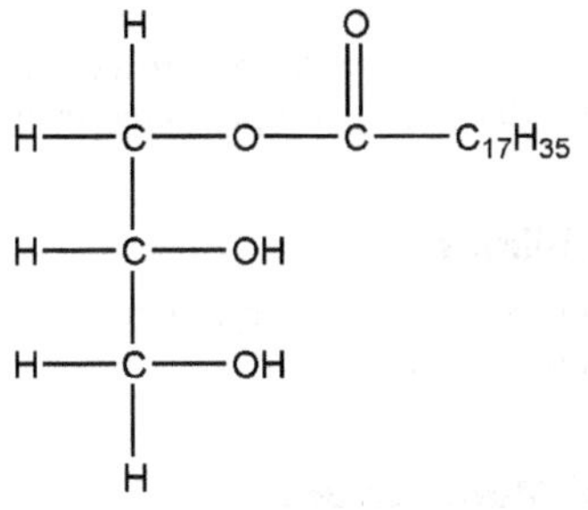

6 Functional Category

Emollient; emulsifying agent; emulsion stabilizing agent; modified-release agent; solubilizing agent; solvent; tablet and capsule lubricant.

7 Applications in Pharmaceutical Formulation or Technology

The many varieties of glyceryl monostearate are used as nonionic emulsifiers, stabilizers, emollients, and plasticizers in a variety of food, pharmaceutical, and cosmetic applications. It acts as an effective stabilizer, that is, as a mutual solvent for polar and nonpolar compounds that may form water-in-oil or oil-in-water emulsions.[1,2] These properties also make it useful as a dispersing agent for pigments in oils or solids in fats, or as a solvent for phospholipids, such as lecithin.

Glyceryl monostearate has also been used in a novel fluidized hot-melt granulation technique for the production of granules and tablets.[3]

Glyceryl monostearate is a lubricant for tablet manufacturing and may be used to form sustained-release matrices for solid dosage forms.[4–6] Sustained-release applications include the formulation of pellets for tablets[7] or suppositories,[8] and the preparation of a veterinary bolus.[9] Glyceryl monostearate has also been used as a matrix ingredient for a biodegradable, implantable, controlled-release dosage form.[10]

When using glyceryl monostearate in a formulation, the possibility of polymorph formation should be considered. The α-form is dispersible and foamy, useful as an emulsifying agent or preservative. The denser, more stable, β-form is suitable for wax matrices. This application has been used to mask the flavor of clarithromycin in a pediatric formulation.[11]

8 Description

While the names glyceryl monostearate and mono- and di-glycerides are used for a variety of esters of long-chain fatty acids, the esters fall into two distinct grades:

40–55 percent monoglycerides The PhEur 7.4 describes glyceryl monostearate 40–55 as a mixture of monoacylglycerols, mostly monostearoylglycerol, together with quantities of di- and triacylglycerols. It contains 40–55% of monoacylglycerols, 30–45% of diacylglycerols, and 5–15% of triacylglycerols. This PhEur grade corresponds to mono- and di-glycerides USP–NF, which has similar specifications (not less than 40% monoglycerides).

90 percent monoglycerides The USP35–NF30 describes glyceryl monostearate as consisting of not less than 90% of mono-glycerides of saturated fatty acids, chiefly glyceryl monostearate ($C_{21}H_{42}O_4$) and glyceryl monopalmitate ($C_{19}H_{38}O_4$).

The commercial products are mixtures of variable proportions of glyceryl monostearate and glyceryl monopalmitate.

Glyceryl monostearate occurs as a white to cream-colored, wax-like solid in the form of beads, flakes, or powder. It is waxy to the touch and has a slight fatty odor and taste.

9 Pharmacopeial Specifications

Table I compares the specifications for the 40–55% grades, glyceryl monostearate PhEur and mono- and di-glycerides USP–NF. PhEur divides glyceryl monostearate 40–55 into three types according to the proportion of stearic acid ester in the mixture, and those specifications are presented in Table II. Table III presents the specifications for glyceryl monostearate USP–NF (90% monoglycerides). Since the JP specifications are broad enough to encompass both grades, JP is included in both Table I and Table III.

See also Section 18.

10 Typical Properties

A wide variety of glyceryl monostearate grades are commercially available, including self-emulsifying grades that contain small amounts of soap or other surfactants. Most grades are tailored for specific applications or made to user specifications and therefore have varied physical properties.

HLB value 3.8
Flash point ≈240°C
Melting point 55–60°C
Polymorphs The α-form is converted to the β-form when heated at 50°C.[12]

Table I: Pharmacopeial specifications for glyceryl monostearate (40–55%).

Test	JP XV	PhEur 7.4	USP35–NF30[a]
Identification	+	+	–
Characters	–	+	–
Acid value	≤15.0	≤3.0	≤4.0
Iodine value	≤3.0	≤3.0	90.0–110.0%[b]
Hydroxyl value	–	–	90.0–110.0%[b]
Saponification value	157–170	158–177	90.0–110.0%[b]
Melting point	≥55°C	54–66°C	–
Residue on ignition	≤0.1%	≤0.1%	≤0.1%
Acidity or alkalinity	+	–	–
Free glycerin	–	≤6.0%	≤7.0%
Composition of fatty acids	–	*see* Table II	–
Heavy metals	–	–	≤0.001%
Arsenic	–	–	≤3 ppm
Nickel	–	≤1 ppm	–
Water	–	≤1.0%	–
Assay (monoglycerides)	–	40.0–55.0%	≤40.0%[c]

(a) Mono- and di-glycerides.
(b) Of the value indicated in the labeling.
(c) 90.0–110.0% of labeled amount.

Table II: Specifications for the composition of fatty acids in glyceryl monostearate 40–55.

Glyceryl monostearate	Fatty acid used in manufacturing	Composition of fatty acids	
		Stearic acid	Sum of palmitic and stearic acids
Type I	Stearic acid 50	40.0–60.0%	≤90.0%
Type II	Stearic acid 70	60.0–80.0%	≤90.0%
Type III	Stearic acid 95	80.0–99.0%	≤96.0%

Table III: Pharmacopeial specifications for glyceryl monostearate (90%).

Test	JP XV	USP35–NF30
Identification	+	–
Acid value	≤15.0	≤6.0
Iodine value	≤3.0	≤3.0
Hydroxyl value	–	290–330
Saponification value	157–170	150–165
Melting point	≥55°C	≥55°C
Residue on ignition	≤0.1%	≤0.5%
Acidity or alkalinity	+	–
Limit of free glycerin	–	≤1.2%
Heavy metals	–	≤10 ppm
Assay (monoglycerides)	–	≥90.0%

Solubility Soluble in hot ethanol, ether, chloroform, hot acetone, mineral oil, and fixed oils. Practically insoluble in water, but may be dispersed in water with the aid of a small amount of soap or other surfactant.

Specific gravity 0.92

Spectroscopy

IR spectra *see* Figure 1.

NIR spectra *see* Figure 2.

11 Stability and Storage Conditions

If stored at warm temperatures, glyceryl monostearate increases in acid value upon aging owing to the saponification of the ester with trace amounts of water. Effective antioxidants may be added, such as butylated hydroxytoluene and propyl gallate.

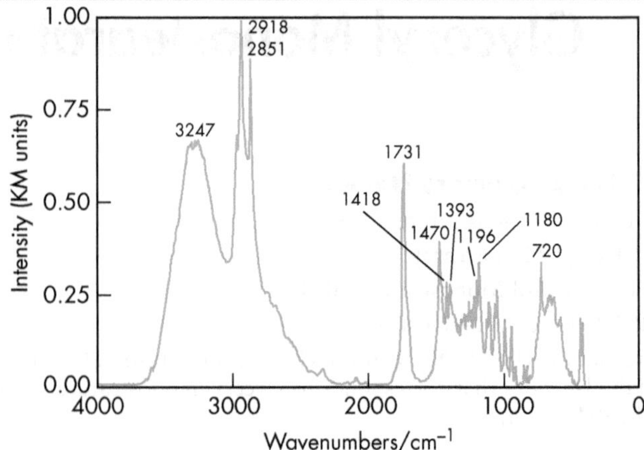

Figure 1: Infrared spectrum of glyceryl monostearate measured by diffuse reflectance. Adapted with permission of Informa Healthcare.

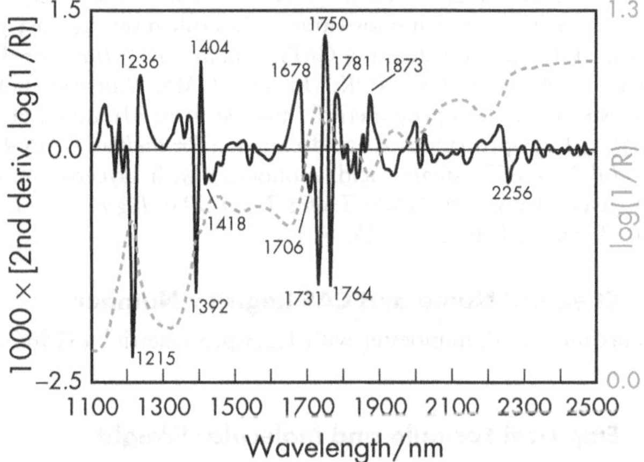

Figure 2: Near-infrared spectrum of glyceryl monostearate measured by reflectance.

Glyceryl monostearate should be stored in a tightly closed container in a cool, dry place, and protected from light.

12 Incompatibilities

The self-emulsifying grades of glyceryl monostearate are incompatible with acidic substances.

13 Method of Manufacture

Glyceryl monostearate is prepared by the reaction of glycerin with triglycerides from animal or vegetable sources, producing a mixture of monoglycerides and diglycerides. The diglycerides may be further reacted to produce the 90% monoglyceride grade. Another process involves reaction of glycerol with stearoyl chloride.

The starting materials are not pure substances and therefore the products obtained from the processes contain a mixture of esters, including palmitate and oleate. Consequently, the composition, and therefore the physical properties, of glyceryl monostearate may vary considerably depending on the manufacturer.

14 Safety

Glyceryl monostearate is widely used in cosmetics, foods, and oral and topical pharmaceutical formulations, and is generally regarded as a nontoxic and nonirritant material.

LD_{50} (mouse, IP): 0.2 g/kg[13]

15 Handling Precautions

Observe normal precautions appropriate to the circumstances and quantity of material handled.

16 Regulatory Status

GRAS listed. Included in the FDA Inactive Ingredients Database (oral capsules and tablets; ophthalmic, otic, rectal, topical, transdermal, and vaginal preparations). Included in nonparenteral medicines licensed in the UK. Included in the Canadian Natural Health Products Ingredients Database.

If glyceryl monostearate is produced from animal fats (tallow), there may be additional regulatory requirements that the source be free of contamination from bovine spongiform encephalopathy.

17 Related Substances

Glyceryl monooleate; glyceryl palmitostearate; self-emulsifying glyceryl monostearate.

Self-emulsifying glyceryl monostearate

Comments A specification for self-emulsifying glyceryl monostearate was previously included in the PhEur. Self-emulsifying glyceryl monostearate is a grade of glyceryl monostearate to which an emusifying agent has been added. The emulsifier may be a soluble soap, a salt of a sulfated alcohol, a nonionic surfactant, or a quaternary compound. It is used primarily as an emulsifying agent for oils, fats, solvents, and waxes. Aqueous preparations should contain an antimicrobial preservative.

18 Comments

Glyceryl monostearate is one of the materials that have been selected for harmonization by the Pharmacopeial Discussion Group. For further information see the General Information Chapter <1196> in the USP35–NF30, the General Chapter 5.8 in PhEur 7.4, along with the 'State of Work' document on the PhEur EDQM website, and also the General Information Chapter 8 in the JP XV.

Glyceryl monostearate and other fatty acid monoesters are not efficient emulsifiers. However, they are useful emollients that are readily emulsified by common emulsifying agents and by incorporation of other fatty materials into the formulation. Addition of the monoester materials provides the creams with smoothness, fine texture, and improved stability.

In topical applications, glyceryl monostearate is less drying than straight stearate creams, and is not drying when used in protective applications. Glyceryl monostearate can form solid lipid nanoparticles, a colloidal carrier system for controlled drug delivery.[14,15] Organogels made with glyceryl monostearate have been shown to improve topical absorption of piroxicam.[16]

A specification for glyceryl monostearate is contained in the *Food Chemicals Codex* (FCC).[17]

A specification for glyceryl monostearate, self-emulsifying type is included in the *Japanese Pharmaceutical Excipients* (JPE).[18]

The EINECS number for glyceryl monostearate is 250-705-4. The PubChem Compound ID (CID) for glyceryl monostearate includes 24699 and 15560611.

19 Specific References

1 O'Laughlin R, *et al.* Effects of variations in physicochemical properties of glyceryl monostearate on the stability of an oil-in-water cream. *J Soc Cosmet Chem* 1989; 40: 215–229.

2 Rafiee-Tehrani M, Mehramizi A. *In vitro* release studies of piroxicam from oil-in-water creams and hydroalcoholic gel topical formulations. *Drug Dev Ind Pharm* 2000; **26**(4): 409–414.

3 Kidokoro M, *et al.* Application of fluidized hot-melt granulation (FHMG) for the preparation of granules for tableting; properties of granules and tablets prepared by FHMG. *Drug Dev Ind Pharm* 2002; **28**(1): 67–76.

4 Peh KK, Yuen KH. Development and *in vitro* evaluation of a novel multiparticulate matrix controlled release formulation of theophylline. *Drug Dev Ind Pharm* 1995; **21**(13): 1545–1555.

5 Peh KK, Yuen KH. *In vivo* perfomance of a multiparticulate matrix, controlled release theophylline preparation. *Drug Dev Ind Pharm* 1995; **22**(4): 349–355.

6 Peh KK, *et al.* Possible mechanism for drug retardation from glyceryl monostearate matrix system. *Drug Dev Ind Pharm* 2000; **26**: 447–450.

7 Thomsen LJ, *et al.* Prolonged release matrix pellets prepared by melt pelletization. I. Process variables. *Drug Dev Ind Pharm* 1993; **19**(15): 1867–1887.

8 Adeyeye CM, Price J. Development and evaluation of sustained-release ibuprofen-wax microspheres. II *In vitro* dissolution studies. *Pharm Res* 1994; **11**(4): 575–579.

9 Evrard B, Delattre L. *In vitro* evaluation of lipid matrices for the development of a sustained-release sulfamethazine bolus for lambs. *Drug Dev Ind Pharm* 1996; **22**(2): 111–118.

10 Peri D, *et al.* Development of an implantable, biodegradable, controlled drug delivery system for local antibiotic therapy. *Drug Dev Ind Pharm* 1994; **20**(8): 1341–1352.

11 Yajima T, *et al.* Optimum heat treatment conditions for masking the bitterness of clarithromycin wax matrix. *Chem Pharm Bull* 2003; **51**(11): 1223–1226.

12 Yajima T, *et al.* Determination of optimum processing temperature for transformation of glyceryl monostearate. *Chem Pharm Bull* 2002; **50**(11): 1430–1433.

13 Lewis RJ, ed. *Sax's Dangerous Properties of Industrial Materials*, 11th edn. New York: Wiley, 2004: 2757–2758.

14 Hu FQ, *et al.* Preparation and characterization of solid lipid nanoparticles containing peptide. *Int J Pharm* 2004; **273**(1–2): 29–35.

15 Gallarate M, *et al.* Preparation of solid lipid nanoparticles from W/O/W emulsions: preliminary studies on insulin encapsulation. *J Microencapsul* 2009; **26**(5): 394–402.

16 Penzes T, *et al.* Topical absorption of piroxicam from organogels – *in vitro* and *in vivo* correlations. *Int J Pharm* 2005; **298**(1): 47–54.

17 *Food Chemicals Codex*, 7th edn. Bethesda, MD: United States Pharmacopeia, 2010: 453.

18 Japan Pharmaceutical Excipients Council. *Japanese Pharmaceutical Excipients* 2004. Tokyo: Yakuji Nippo, 2004: 361–362.

20 General References

Eccleston GM. Emulsions. In: Swarbrick J, Boylan JC, eds. *Encyclopedia of Pharmaceutical Technology.* 5. New York: Marcel Dekker, 1992: 137–188.

European Directorate for the Quality of Medicines and Healthcare (EDQM). European Pharmacopoeia – State Of Work Of International Harmonisation. *Pharmeuropa* 2011; 23(2): 395–401. http://www.edqm.eu/en/InternationalHarmonisation-614.html (accessed 4 July 2011).

Rieger MM. Glyceryl stearate: chemistry and use. *Cosmet Toilet* 1990; **105**(Nov): 51–5456–57.

Schumacher GE. Glyceryl monostearate in some pharmaceuticals. *Am J Hosp Pharm* 1967; **24**: 290–291.

Wisniewski W, Golucki Z. Stability of glycerylmonostearate. *Acta Pol Pharm* 1965; **22**: 296–298.

21 Author

AK Taylor.

22 Date of Revision

1 August 2011.

Glyceryl Palmitostearate

1 Nonproprietary Names

None adopted.

2 Synonyms

Glycerin palmitostearate; glycerol palmitostearate; 2-[(1-oxohexadecyl)-oxy]-1,3-propanediyl dioctadecanoate and 1,2,3-propane triol; *Precirol ATO 5*.

3 Chemical Name and CAS Registry Number

Octadecanoic acid, 2,3-dihydroxypropyl ester mixed with 3-hydroxy-2-[(1-oxohexadecyl)-oxy] propyl octadecanoate [8067-32-1]

4 Empirical Formula and Molecular Weight

Glyceryl palmitostearate is a mixture of mono-, di-, and triglycerides of C_{16} and C_{18} fatty acids.

5 Structural Formula

See Sections 3 and 4.

6 Functional Category

Biodegradable material; coating agent; gelling agent; modified-release agent; tablet and capsule diluent; tablet and capsule lubricant; taste-masking agent.

7 Applications in Pharmaceutical Formulation or Technology

Glyceryl palmitostearate is used in oral solid-dosage pharmaceutical formulations as a lubricant.(1,2) Disintegration times increase(3) and tablet strength decreases(4) with increase in mixing time.

It is used as a lipophilic matrix for sustained-release tablet and capsule formulations.(5,6) Tablet formulations may be prepared by either granulation or a hot-melt technique,(7,8) the former producing tablets that have the faster release profile. Release rate decreases with increased glyceryl palmitostearate content.(5)

Glyceryl palmitostearate is used to form microspheres, which may be used in capsules or compressed to form tablets,(9,10) pellets,(11) coated beads,(12) and biodegradable gels.(13) It is also used for taste-masking.(14)

See Table I.

Table I: Uses of glyceryl palmitostearate.(14)

Use	Concentration (%)
Matrix for sustained release	10.0–20.0
Taste masking	2.0–6.0
Tablet lubricant	1.0–4.0

8 Description

Glyceryl palmitostearate occurs as a fine white powder with a faint odor.

9 Pharmacopeial Specifications

—

10 Typical Properties

Acid value <6.0
Boiling point 200°C
Color <3 (Gardner scale)
Free glycerin content <1.0%
Heavy metals <10 ppm
HLB value 2
Hydroxyl value 60–115
Iodine value <3
Melting point 52–56°C
1-Monoglycerides content 8.0–17.0%
Peroxide value <3.0
Saponification value 175–195
Solubility Freely soluble in chloroform and dichloromethane; practically insoluble in ethanol (95%), mineral oil, and water.
Spectroscopy
IR spectra *see* Figure 1.
NIR spectra *see* Figure 2.
Sulfated ash <0.1%
Unsaponifiable matter <1.0%
Water content <1.0%

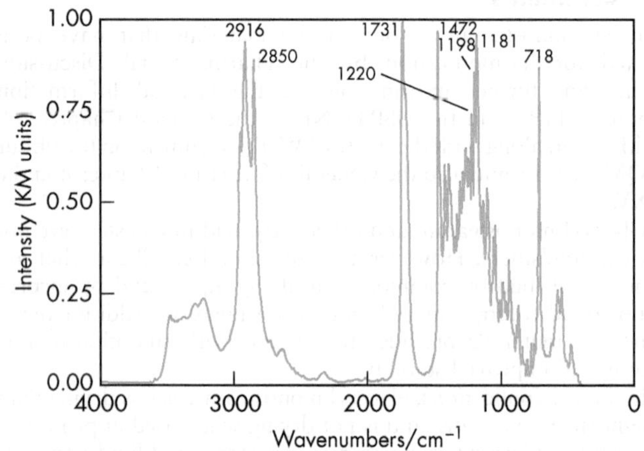

Figure 1: Infrared spectrum of glyceryl palmitostearate measured by diffuse reflectance. Adapted with permission of Informa Healthcare.

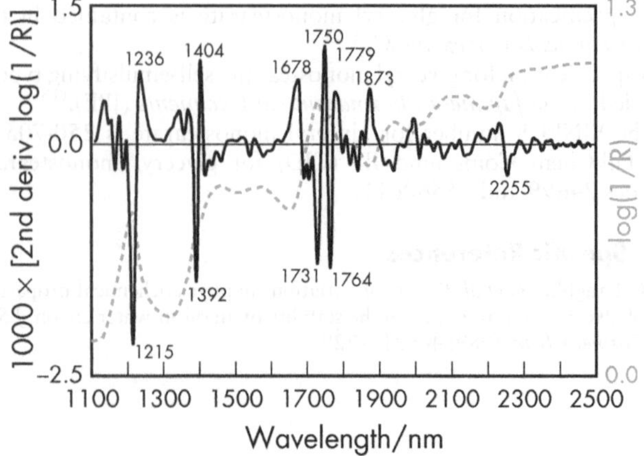

Figure 2: Near-infrared spectrum of glyceryl palmitostearate measured by reflectance.

11 Stability and Storage Conditions

Glyceryl palmitostearate should not be stored at temperatures above 35°C. For storage for periods over 1 month, glyceryl palmitostearate should be stored at a temperature of 5–15°C in an airtight container, protected from light and moisture.

12 Incompatibilities

—

13 Method of Manufacture

Glyceryl palmitostearate is manufactured, without a catalyst, by the direct esterification of palmitic and stearic acids with glycerol.

14 Safety

Glyceryl palmitostearate is used in oral pharmaceutical formulations and is generally regarded as an essentially nontoxic and nonirritant material.

LD_{50} (rat, oral): >6 g/kg[14]

15 Handling Precautions

Observe normal handling precautions appropriate to the circumstances and quantity of material handled.

16 Regulatory Status

GRAS listed. Included in the FDA Inactive Ingredients Database (oral suspension, oral tablet). Included in nonparenteral preparations licensed in Europe. Included in the Canadian Natural Health Products Ingredients Database.

17 Related Substances

Glyceryl behenate; glyceryl monostearate.

18 Comments

The EINECS number for glyceryl palmitostearate is 232-514-8. The PubChem Compound ID (CID) for glyceryl palmitostearate is 114690.

19 Specific References

1 Holzer AW, Sjogren J. Evaluation of some lubricants by the comparison of friction coefficients and tablet properties. *Acta Pharm Suec* 1981; **18**: 139–148.
2 Allen LV. Featured excipient: capsule and tablet lubricants. *Int J Pharm Compound* 2000; **4**(5): 390–392.
3 Sekulovic D. Effect of *Precirol ATO 5* on the properties of tablets. *Pharmazie* 1987; **42**(1): 61–62.
4 Velasco V, *et al.* Force–displacement parameters of maltodextrins after the addition of lubricants. *Int J Pharm* 1997; **152**: 111–120.
5 Reitz C, Kleinebudde P. Solid lipid extrusion of sustained-release dosage forms. *Eur J Pharm Biopharm* 2007; **67**(2): 440–448.
6 Jannin V, *et al.* Influence of poloxamers on the dissolution performance and stability of controlled-release formulations containing *Precirol ATO 5*. *Int J Pharm* 2006; **309**: 6–15.
7 Malamataris S, *et al.* Controlled release from glycerol palmito-stearate matrices prepared by dry-heat granulation and compression at elevated temperature. *Drug Dev Ind Pharm* 1991; **17**(13): 1765–1777.
8 Evrard B, *et al.* Influence of melting and rheological properties of fatty binders in the melt granulation process in a high sheer mixer. *Drug Dev Ind Pharm* 1999; **25**(11): 1177–1184.
9 Shaikh NH, *et al.* Effect of different binders on release characteristics of theophylline from compressed microspheres. *Drug Dev Ind Pharm* 1991; **17**: 793–804.
10 Edimo A, *et al.* Capacity of lipophilic auxiliary substances to give spheres by extrusion-spheronisation. *Drug Dev Ind Pharm* 1993; **19**: 827–842.
11 Hamdani J, *et al.* Physical and thermal characterisation of *Precirol* and *Compritrol* as lipophilic glycerides used for the preparation of controlled release matrix pellets. *Int J Pharm* 2003; **260**: 47–57.
12 Pongjanyakul T, *et al.* Modulation of drug release for glyceryl palmitostearate–alginate beads by heat treatment. *Int J Pharm* 2006; **319**: 20–28.
13 Gao ZH, *et al.* Controlled release of contraceptive steroids from biodegradable and injectable gel: *in vivo* evaluation. *Pharm Res* 1995; **12**: 864–868.
14 Gattefossé. Technical literature: *Precirol ATO 5*, 2008.

20 General References

Armstrong NA. Tablet manufacture. In: Swarbrick J, Boylan JC, eds. *Encyclopedia of Pharmaceutical Technology*, 2nd edn, 3. New York: Marcel Dekker, 2002; 2713–2732.

Armstrong NA. Lubricants, glidants and antiadherents. In: Augsburger LL, Hoag SW, eds. *Pharmaceutical Dosage Forms: Tablets*, 3rd edn, 2. New York: Taylor and Francis, 2008; 1132–1142.

Chan HK, Chew NYK. Excipients-powder and solid dosage forms. In: Swarbrick J, Boylan JC, eds. *Encyclopedia of Pharmaceutical Technology*, 2nd edn, 2. New York: Marcel Dekker, 2002; 1132–1142.

21 Author

NA Armstrong.

22 Date of Revision

21 November 2011.

G

Glycine

1 Nonproprietary Names

BP: Glycine
JP: Glycine
PhEur: Glycine
USP–NF: Glycine

2 Synonyms

Aminoacetic acid; 2-aminoacetic acid; E640; G; Gly; glycinum; glycoamin; glycocoll; glycoll; glycolixir; glycinium; Hampshire glycine; padil; sucre de gélatine.

3 Chemical Name and CAS Registry Number

Aminoethanoic acid [56-40-6]

4 Empirical Formula and Molecular Weight

$C_2H_5NO_2$ 75.07

5 Structural Formula

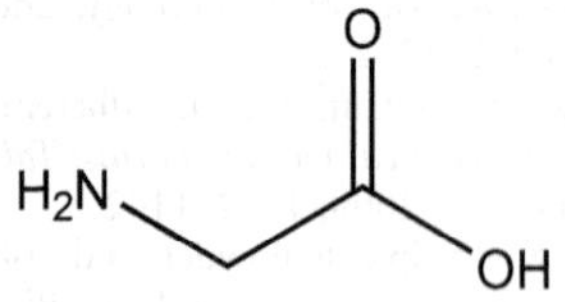

6 Functional Category

Buffering agent; lyophilization aid; tablet and capsule disintegrant.

7 Applications in Pharmaceutical Formulation or Technology

Glycine is routinely used as a cofreeze-dried excipient in protein formulations owing to its ability to form a strong, porous, and elegant cake structure in the final lyophilized product.[1–3] It is one of the most frequently utilized excipients in freeze-dried injectable formulations[4] owing to its advantageous freeze-drying properties.

Glycine has been investigated as a disintegration accelerant in fast-disintegrating formulations owing to its rapid dissolution in aqueous systems.[5,6]

8 Description

Glycine occurs as a white, odorless, crystalline powder, and has a sweet taste.

9 Pharmacopeial Specifications

See Table I.

Table I: Pharmacopeial specifications for glycine.

Test	JP XV	PhEur 7.4	USP35–NF30
Identification	+	+	+
Characters	—	+	—
Loss on drying	≤0.3%	≤0.5%	≤0.2%
Residue on ignition	≤0.1%	—	≤0.1%
pH	5.6–6.6	5.9–6.4	—
Appearance of solution	+	+	—
Chlorides	≤0.021%	≤75 ppm	≤0.007%
Sulfates	≤0.028%	—	≤0.0065%
Ammonium	≤0.02%	—	—
Hydrolyzable substances	—	—	+
Ninhydrin-positive substances	—	+	—
Sulfated ash	—	≤0.1%	—
Heavy metals	≤20 ppm	≤10 ppm	≤20 ppm
Arsenic	≤2 ppm	—	—
Related substances	+	—	—
Assay (dried basis)	≥98.5%	98.5–101.0%	98.5–101.5%

10 Typical Properties

Acidity/alkalinity pH = 4 (0.2 M solution in water)
Density 1.1607 g/cm^3
Melting point 232–236°C
Solubility *see* Table II.

Table II: Solubility of glycine.

Solvent	Solubility at 20°C unless otherwise stated
Ethanol (95%)	1 in 1254
Ether	Very slightly soluble or practically insoluble
Pyridine	1 in 164
Water	1 in 4 at 25°C
	1 in 2.6 at 50°C
	1 in 1.9 at 75°C
	1 in 1.5 at 100°C

11 Stability and Storage Conditions

Glycine starts to decompose at 233°C. Store in well-closed containers. Glycine irrigation solutions (95–105% glycine) should be stored in single dose containers, preferably type I or type II glass.

12 Incompatibilities

Glycine may undergo Maillard reactions with amino acids to produce yellowing or browning. Reducing sugars will also interact with secondary amines to form an imine, but without any accompanying yellow-brown discoloration.

13 Method of Manufacture

Chemical synthesis is the most suitable method of preparation of glycine. Amination of chloroacetic acid and the hydrolysis of aminoacetonitrile are the favored methods of production.[8]

14 Safety

Glycine is used as a sweetener, buffering agent, and dietary supplement. The pure form of glycine is moderately toxic by the IV route and mildly toxic by ingestion.

Systemic absorption of glycine irrigation solutions can lead to disturbances of fluid and electrolyte balance and cardiovascular and pulmonary disorders.[7]

LD_{50} (mouse, IP): 4.45 g/kg[8]
LD_{50} (mouse, IV): 2.37 g/kg
LD_{50} (mouse, oral): 4.92 g/kg
LD_{50} (mouse, SC): 5.06 g/kg
LD_{50} (rat, IV): 2.6 g/kg
LD_{50} (rat, oral): 7.93 g/kg
LD_{50} (rat, SC): 5.2 g/kg

15 Handling Precautions

Observe normal precautions appropriate to the circumstances and quantity of the material handled. When heated to decomposition, glycine emits toxic fumes of nitrogen oxides.

16 Regulatory Status

GRAS listed. Accepted for use as a food additive in Europe. Included in the FDA Inactive Ingredients Database (IM, IV, SC injections; oral; rectal) and approved for irrigant solutions. Included in parenteral (powders for injection; solutions for injection; vaccines; kits for implant) and nonparenteral (orodispersible tablets/oral lyophilizate; powders for inhalation; powders for oral solution; tablets) formulations licensed in the UK. Included in the Canadian Natural Health Products Ingredients Database.

17 Related Substances

Glycine hydrochloride; sodium glycinate.

Glycine hydrochloride
Empirical formula $C_2H_5NO_2 \cdot HCl$
Molecular weight 111.5
CAS number [6000-43-7]
Melting point 182°C
Comments Hygroscopic prisms from hydrochloride. The EINECS number for glycine hydrochloride is 227-841-8.

Sodium glycinate
Empirical formula $C_2H_4NO_2 \cdot Na$
Molecular weight 97.1
CAS number [6000-44-8]
Melting point 197–201°C
Comments The EINECS number for sodium glycinate is 227-842-3.

18 Comments

Anhydrous glycine exists in three polymorphic forms α, β, and γ,[9] which can have implications for product stability in lyophilized systems. The β form is generally produced during freeze-drying. However, the presence of moisture at room temperature can induce a polymorphic conversion to a mixture of the α and γ forms.[10]

Glycine at low levels ($\leq$50 mol/L) has been shown to notably minimize the expected pH reduction in freeze-dried sodium phosphate buffer,[10] which can be detrimental to protein stability during the lyophilization process.

Glycine Irrigation 1.5% USP is a sterile nonpyrogenic, nonhemolytic, nonelectrolytic solution in single-dose containers for use as a urological irrigation solution.

Glycine may be used along with antacids in the treatment of gastric hyperacidity, and it may also be included in aspirin preparations to aid the reduction of gastric irritation.[11] It is also used as a buffering agent and conditioner in cosmetics.

A specification for glycine is contained in the *Food Chemicals Codex* (FCC).[12]

The EINECS number for glycine is 200-272-2. The PubChem Compound ID (CID) for glycine is 750.

19 Specific References

1 Wang W. Lyophilization and development of solid protein pharmaceuticals. *Int J Pharm* 2000; **203**: 1–60.
2 Akers MJ, *et al.* Glycine crystallization during freezing: the effects of salt form, pH and ionic strength. *Pharm Res* 1995; **12**: 1457–1461.
3 Kasraian K, *et al.* Characterisation of the sucrose/glycine/water system by differential scanning calorimetry and freeze-drying microscopy. *Pharm Dev Technol* 1998; 3: 233–239.
4 Chongprasert S, *et al.* Characterisation of frozen solutions of glycine. *J Pharm Sci* 2001; **90**: 1720–1728.
5 Fukami J, *et al.* Evaluation of rapidly disintegrating tablets containing glycine and carboxymethylcellulose. *Int J Pharm* 2006; **310**: 101–109.
6 Fukami J, *et al.* Development of fast disintegrating compressed tablets using amino acid as disintegration accelerator: evaluation of wetting and disintegration of tablet on the basis of surface free energy. *Chem Pharm Bull* 2005; **53**: 1536–1539.
7 Gerhartz W, ed. *Ullman's Encyclopedia of Industrial Chemistry*, 5th edn, A2. Weinheim, Germany: VCH, 66–70.
8 Lewis RJ, ed. *Sax's Dangerous Properties of Industrial Materials*, 11th edn. New York: Wiley, 2004: 1871.
9 Bai SJ, *et al.* Quantification of glycine crystallinity by near-infrared spectroscopy. *J Pharm Sci* 2004; **93**: 2439–2447.
10 Pikal-Cleland KA, *et al.* Effect of glycine on pH changes and protein stability during freeze-thawing in phosphate buffer systems. *J Pharm Sci* 2002; **91**: 1969–1979.
11 Sweetman S, ed. *Martindale: The Complete Drug Reference*, 37th edn. London: Pharmaceutical Press, 2011: 2106.
12 *Food Chemicals Codex*, 7th edn. Bethesda, MD: United States Pharmacopeia, 2010.

20 General References

—

21 Authors

RT Forbes, WL Hulse.

22 Date of Revision

1 March 2012.

G

Glycofurol

1 Nonproprietary Names

None adopted.

2 Synonyms

Glicofurol; glycofural; *Glycofurol 75*; tetraglycol; α-(tetrahydrofuranyl)-ω-hydroxy-poly(oxyethylene); tetrahydrofurfuryl alcohol polyethylene glycol ether; THFP.

Note: tetraglycol is also used as a synonym for tetrahydrofurfuryl alcohol.

3 Chemical Name and CAS Registry Number

α-[(Tetrahydro-2-furanyl)methyl]-ω-hydroxy-poly(oxy-1,2-ethanediyl) [31692-85-0]

4 Empirical Formula and Molecular Weight

$C_9H_{18}O_4$ (average) 190.24 (average)

5 Structural Formula

$CH_2(OCH_2CH_2)_nOH$

Glycofurol 75: n = 1–2

6 Functional Category

Penetration enhancer; solvent.

7 Applications in Pharmaceutical Formulation or Technology

Glycofurol is used as a solvent in parenteral products for intravenous or intramuscular injection in concentrations up to 50% v/v.[1–5] It has also been investigated, mainly in animal studies, for use as a penetration enhancer and solvent in topical[6,7] and intranasal formulations.[8–12] Glycofurol has also been used at 20% v/v concentration in a rectal formulation.[13]

8 Description

Glycofurol is a clear, colorless, almost odorless liquid, with a bitter taste; it produces a warm sensation on the tongue.

9 Pharmacopeial Specifications

—

10 Typical Properties

Boiling point 80–100°C for *Glycofurol 75*
Density 1.070–1.090 g/cm^3 at 20°C
Hydroxyl value 300–400
Moisture content 0.2–5% at ambient temperature and 30% relative humidity.
Refractive index n_D^{40} = 1.4545
Solubility see Table I.

Table I: Solubility of glycofurol.

Solvent	Solubility at 20°C
Arachis oil	Immiscible
Castor oil	Miscible[a]
Ethanol (95%)	Miscible in all proportions
Glycerin	Miscible in all proportions
Isopropyl ether	Immiscible
Petroleum ether	Immiscible
Polyethylene glycol 400	Miscible in all proportions
Propan-2-ol	Miscible in all proportions
Propylene glycol	Miscible in all proportions
Water	Miscible in all proportions[a]

(a) Cloudiness may occur.

Viscosity (dynamic) 8–18 mPa s (8–18 cP) at 20°C for *Glycofurol 75*.

11 Stability and Storage Conditions

Stable if stored under nitrogen in a well-closed container protected from light, in a cool, dry place.

12 Incompatibilities

Glycofurol is incompatible with oxidizing agents.

13 Method of Manufacture

Glycofurol is prepared by the reaction of tetrahydrofurfuryl alcohol with ethylene oxide (followed by a special purification process in the case of *Glycofurol 75*).

14 Safety

Glycofurol is mainly used as a solvent in parenteral pharmaceutical formulations and is generally regarded as a relatively nontoxic and nonirritant material at the levels used as a pharmaceutical excipient. Glycofurol can be irritant when used undiluted; its tolerability is approximately the same as propylene glycol.[1,2]

Glycofurol may have an effect on liver function and may have a low potential for interaction with hepatoxins or those materials undergong extensive hepatic metabolism.[4]

LD_{50} (mouse, IV): 3.5 mL/kg[1,2]

15 Handling Precautions

Observe normal precautions appropriate to the circumstances and quantity of material handled.

16 Regulatory Status

Included in parenteral medicines licensed in Europe.

17 Related Substances

—

18 Comments

Grades other than *Glycofurol 75* may contain significant amounts of tetrahydrofurfuryl alcohol and other impurities. *Glycofurol 75* meets an analytical specification which includes a requirement that the fraction in which n = 1 or 2 amounts to a minimum of 95%; *see* Section 5.

The EINECS number for glycofurol is 227-407-8. The PubChem Compound ID (CID) for glycofurol is 110717.

19 Specific References

1 Spiegelberg H, *et al.* [A new injectable solvent (glycofurol).] *Arzneimittelforschung* 1956; **6**: 75–77[in German].
2 Spiegel AJ, Noseworthy MM. Use of non-aqueous solvents in parenteral products. *J Pharm Sci* 1963; **52**: 917–927.
3 Anschel J. Solvents and solubilisers in injections. *Pharm Ind* 1965; **27**: 781–787.
4 Bury RW, *et al.* Disposition of intravenous glycofurol: effect of hepatic cirrhosis. *Clin Pharmacol Ther* 1984; **36**(1): 82–84.
5 Taub§ll E, *et al.* A new injectable carbamazepine solution: anti-epileptic effects and pharmaceutical properties. *Epilepsy Res* 1990; **7**(1): 59–64.
6 Lashmar UT, *et al.* Topical application of penetration enhancers to the skin of nude mice: a histopathological study. *J Pharm Pharmacol* 1989; **41**(2): 118–122.
7 Barakat NS. Evaluation of glycofurol-based gel as a new vehicle for topical application of naproxen. *AAPS Pharm Sci Tech* 2010; **11**(3): 1138–1146.
8 Bindseil E, *et al.* Morphological examination of rabbit nasal mucosa after exposure to acetylsalicylic acid, glycofurol 75 and ephedrine. *Int J Pharm* 1995; **119**(1): 37–46.
9 Bechgaard E, *et al.* Pharmacokinetic and pharmacodynamic response after intranasal administration of diazepam to rabbits. *J Pharm Pharmacol* 1997; **49**(8): 747–750.
10 Nielson HW, *et al.* Intranasal administration of different liquid formulations of bumetanide to rabbits. *Int J Pharm* 2000; **204**: 35–41.
11 Bagger MA, *et al.* Nasal bioavailability of peptide T in rabbits: absorption enhancement by sodium glycocholate and glycofurol. *Eur J Pharm Sci* 2001; **14**(1): 69–74.
12 Hou H, Siegel RA. Enhanced permeation of diazepam through artificial membranes from supersaturated solutions. *J Pharm Sci* 2006; **95**: 896–905.
13 Dale O, *et al.* Bioavailabilities of rectal and oral methadone in healthy subjects. *Br J Clin Pharmacol* 2004; **58**(2): 156–162.

20 General References

Mottu F, *et al.* Organic solvents for pharmaceutical parenterals and embolic liquids: a review of toxicity data. *PDA J Pharm Sci Technol* 2000; **54**(6): 456–469.

21 Author

PJ Weller.

22 Date of Revision

1 November 2011.

Guar Gum

1 Nonproprietary Names

BP: Guar Galactomannan

PhEur: Guar Galactomannan

USP–NF: Guar Gum

2 Synonyms

E412; *Galactosol*; guar flour; guar galactomannanum; jaguar gum; *Meyprogat*; *Meyprodor*; *Meyprofin*.

3 Chemical Name and CAS Registry Number

Galactomannan polysaccharide [9000-30-0]

4 Empirical Formula and Molecular Weight

$(C_6H_{12}O_6)_n$ ≈220 000

See Section 5.

5 Structural Formula

Guar gum consists of linear chains of (1→4)-β-D-mannopyranosyl units with α-D-galactopyranosyl units attached by (1→6) linkages. The ratio of D-galactose to D-mannose is between 1 : 1.4 and 1 : 2. *See also* Section 8.

6 Functional Category

Bioadhesive material; suspending agent; tablet and capsule binder; tablet and capsule disintegrant; viscosity-increasing agent.

7 Applications in Pharmaceutical Formulation or Technology

Guar gum is a galactomannan, commonly used in cosmetics, food products, and pharmaceutical formulations. It has also been investigated in the preparation of sustained-release matrix tablets in the place of cellulose derivatives such as methylcellulose.[1]

In pharmaceuticals, guar gum is used in solid-dosage forms as a binder and disintegrant,[2–4] *see* Table I; in oral and topical products as a suspending, thickening, and stabilizing agent; and also as a controlled-release carrier. Guar gum has also been examined for use in colonic drug delivery.[5–9] Guar-gum-based three-layer matrix tablets have been used experimentally in oral controlled-release formulations.[10]

Table I: Uses of guar gum.

Use	Concentration (%)
Emulsion stabilizer	1
Tablet binder	Up to 10
Thickener for lotions and creams	Up to 2.5

8 Description

The USP35–NF30 describes guar gum as a gum obtained from the ground endosperms of *Cyamopsis tetragonolobus* (L.) Taub. (Fam. Leguminosae). It consists chiefly of a high-molecular-weight hydrocolloidal polysaccharide, composed of galactan and mannan units combined through glycoside linkages, which may be described chemically as a galactomannan. The PhEur 7.4 similarly describes guar galactomannan as being obtained from the seeds of *Cyamopsis*

G

tetragonolobus (L.) Taub. by grinding the endosperms and subsequent partial hydrolysis.

The main components are polysaccharides composed of D-galactose and D-mannose in molecular ratios of 1 : 1.4 to 1 : 2. The molecule consists of a linear chain of β-(1→4)-glycosidically linked manno-pyranoses and single α-(1→6)-glycosidically linked galacto-pyranoses. *See also* Section 18.

Guar gum occurs as an odorless or nearly odorless, white to yellowish-white powder with a bland taste.

9 Pharmacopeial Specifications

See Table II.

Table II: Pharmacopeial specifications for guar gum.

Test	PhEur 7.4	USP35–NF30
Identification	+	+
Characters	+	−
pH (1% w/w solution)	5.5–7.5	−
Apparent viscosity	+	−
Microbial contamination	≤10^3 cfu/g	−
Loss on drying	≤15.0%	≤15.0%
Ash	≤1.8%	≤1.5%
Acid-insoluble matter	≤7.0%	≤7.0%
Arsenic	−	≤3 ppm
Lead	−	≤10 ppm
Heavy metals	−	≤20 ppm
Protein	≤5.0%	≤10.0%
Starch	−	+
Galactomannans	−	≥66.0%
Tragacanth, sterculia gum, agar, alginates, and carrageenan	+	−

10 Typical Properties

Acidity/alkalinity pH = 5.0–7.0 (1% w/v aqueous dispersion)

Density 1.492 g/cm^3

Solubility Practically insoluble in organic solvents. In cold or hot water, guar gum disperses and swells almost immediately to form a highly viscous, thixotropic sol. The optimum rate of hydration occurs at pH 7.5–9.0. Finely milled powders swell more rapidly and are more difficult to disperse. Two to four hours in water at room temperature are required to develop maximum viscosity.

Spectroscopy

IR spectra *see* Figure 1.

NIR spectra *see* Figure 2.

Viscosity (dynamic) 4.86 Pa s (4860 cP) for a 1% w/v dispersion. Viscosity is dependent upon temperature, time, concentration, pH, rate of agitation, and particle size of the guar gum powder. Synergistic rheological effects may occur with other suspending agents such as xanthan gum; *see* Xanthan Gum.

11 Stability and Storage Conditions

Aqueous guar gum dispersions have a buffering action and are stable at pH 4.0–10.5. However, prolonged heating reduces the viscosity of dispersions.

The bacteriological stability of guar gum dispersions may be improved by the addition of a mixture of 0.15% methylparaben and 0.02% propylparaben as a preservative. In food applications, benzoic acid, citric acid, sodium benzoate, or sorbic acid may be used.

Guar gum powder should be stored in a well-closed container in a cool, dry place.

12 Incompatibilities

Guar gum is compatible with most other plant hydrocolloids such as tragacanth. It is incompatible with acetone, ethanol (95%), tannins, strong acids, and alkalis. Borate ions, if present in the dispersing water, will prevent the hydration of guar gum. However, the addition of borate ions to hydrated guar gum produces cohesive structural gels and further hydration is then prevented. The gel formed can be liquefied by reducing the pH to below 7, or by heating.

Guar gum may reduce the absorption of penicillin V from some formulations by a quarter.[11]

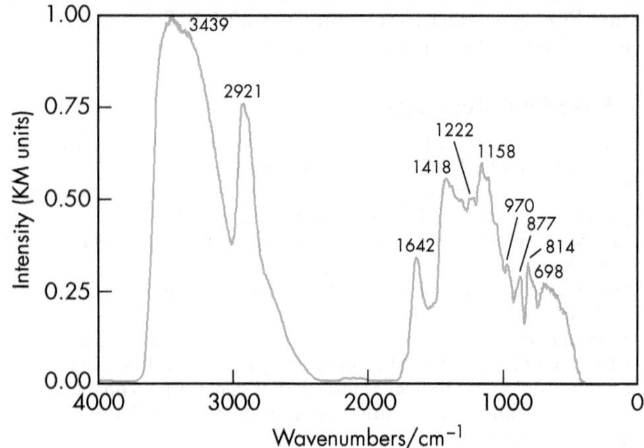

Figure 1: Infrared spectrum of guar gum measured by diffuse reflectance. Adapted with permission of Informa Healthcare.

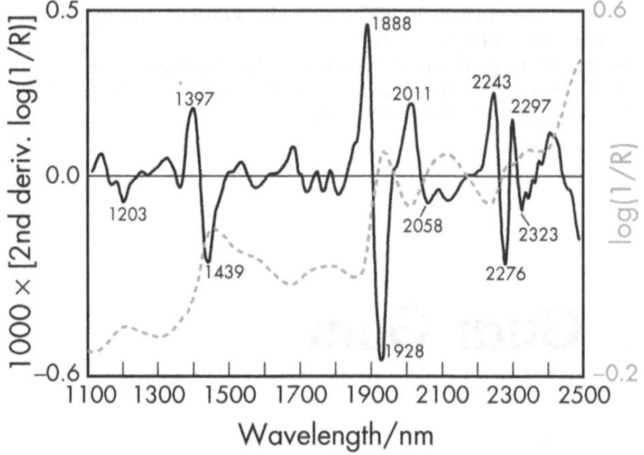

Figure 2: Near-infrared spectrum of guar gum measured by reflectance.

13 Method of Manufacture

Guar gum is obtained from the ground endosperm of the guar plant, *Cyamopsis tetragonolobus* (L.) Taub. (Fam. Leguminosae), which is grown in India, Pakistan, and the semiarid southwestern region of the US.

The seed hull can be removed by grinding, after soaking in sulfuric acid or water, or by charring. The embryo (germ) is removed by differential grinding, since each component possesses a different hardness. The separated endosperm, containing 80% galactomannan is then ground to different particle sizes depending upon final application.

14 Safety

Guar gum is widely used in foods, and oral and topical pharmaceutical formulations. Excessive consumption may cause gastrointestinal disturbance such as flatulence, diarrhea, or nausea. Therapeutically, daily oral doses of up to 25 g of guar gum have been administered to patients with diabetes mellitus.[12]

Although it is generally regarded as a nontoxic and nonirritant material, the safety of guar gum when used as an appetite suppressant has been questioned. When consumed, the gum swells in the stomach to promote a feeling of fullness. However, it is claimed that premature swelling of guar gum tablets may occur and cause obstruction of, or damage to, the esophagus. Consequently, appetite suppressants containing guar gum in tablet form have been banned in the UK.[13,14] However, appetite suppressants containing microgranules of guar gum are claimed to be safe.[15] The use of guar gum for pharmaceutical purposes is unaffected by the ban.

In food applications, an acceptable daily intake of guar gum has not been specified by the WHO.[16]

LD_{50} (hamster, oral): 6.0 g/kg[17]
LD_{50} (mouse, oral): 8.1 g/kg
LD_{50} (rabbit, oral): 7.0 g/kg
LD_{50} (rat, oral): 6.77 g/kg

15 Handling Precautions

Observe normal precautions appropriate to the circumstances and quantity of material handled. Guar gum may be irritating to the eyes. Eye protection, gloves, and a dust mask or respirator are recommended.

16 Regulatory Status

GRAS listed. Accepted for use as a food additive in Europe. Included in the FDA Inactive Ingredients Database (oral suspensions, syrups, and tablets; topical preparations; vaginal tablets). Also included in nonparenteral medicines licensed in the UK. Included in the Canadian Natural Health Products Ingredients Database.

17 Related Substances

Acacia; tragacanth; xanthan gum.

18 Comments

Synthetic derivatives of guar gum, such as guar acetate, guar phthalate, guar acetate phthalate, oxidized guar gum, and sodium carboxymethyl guar, have also been investigated for their pharmaceutical applications. In particular, sodium carboxymethyl guar gives a transparent gel and, when poured over a pool of mercury, produces a flexible, clear, transparent film. Sodium carboxymethyl guar has been used as a polymer matrix in transdermal patches.[18]

Guar gum has been investigated as a matrix gel-forming sustained-release excipient in solid oral dosage forms.[19] It has also been evaluated in buccoadhesive tablets for sustained release along with other gums including xanthan gum and karaya gum, in combination with chitosan.[20] Studies have shown that films composed of guar gum and chitosan appear to be tougher and more bioadhesive.[21]

A specification for guar gum is contained in the *Food Chemicals Codex* (FCC).[22]

The EINECS number for guar gum is 232-536-8.

19 Specific References

1 Khullar R, *et al.* Guar gum as a hydrophilic matrix for preparation of theophylline controlled-release dosage form. *Indian J Pharm Sci* 1999; **61**(6): 342–345.
2 Feinstein W, Bartilucci AJ. Comparative study of selected disintegrating agents. *J Pharm Sci* 1966; **55**: 332–334.
3 Sakr AM, Elsabbagh HM. Evaluation of guar gum as a tablet additive: a preliminary report. *Pharm Ind* 1977; **39**(4): 399–403.
4 Duru C, *et al.* A comparative study of the disintegrating efficiency of polysaccharides in a directly-tabletable formulation. *Pharm Technol Int* 1992; 4(5): 15–23.
5 Adkin DA, *et al.* The use of scintigraphy to provide 'proof of concept' for novel polysaccharide preparations designed for colonic drug delivery. *Pharm Res* 1997; **14**(1): 103–107.
6 Wong D, *et al.* USP Dissolution Apparatus II (reciprocating cylinder) for screening of guar based colonic delivery formulations. *J Control Release* 1997; **47**: 173–179.
7 Sinha VR, *et al.* Colonic drug delivery of 5-fluoracil: an *in vitro* evaluation. *Int J Pharm* 2004; **269**(1): 101–108.
8 Toti US, Aminabhavi TM. Modified guar gum matrix tablet for controlled release of diltriazem hydrochloride. *J Control Release* 2004; **95**(3): 567–577.
9 Tugcu Demiroez F, *et al. In vitro* and *in vivo* evaluation of mesalazine-guar gum matrix tablets for colonic drug delivery. *J Drug Target* 2004; **12**(2): 105–112.
10 Al-Saiden SM, *et al.* Pharmacokinetic evaluation of guar gum-based three-layer matrix tablets for oral controlled delivery of highly soluble metoprolol tartrate as a model drug. *Eur J Pharm Biopharm* 2004; **58**(3): 697–703.
11 Anonymous. Does guar reduce penicillin V absorption? *Pharm J* 1987; **239**: 123.
12 Jenkins DJ, *et al.* Treatment of diabetes with guar gum: reduction of urinary glucose loss in diabetics. *Lancet* 1977; **ii**: 779–780.
13 Uusitupa MIJ. Fibre in the management of diabetes [letter]. *Br Med J* 1990; **301**: 122.
14 Anonymous. Guar slimming tablets ban. *Pharm J* 1989; **242**: 611.
15 Levin R. Guar gum [letter]. *Pharm J* 1989; **242**: 153.
16 WHO. Toxicological evaluation of some food additives including anticaking agents, antimicrobials, antioxidants, emulsifiers and thickening agents. *WHO Food Addit Ser* 1974; No. 5: 321–323.
17 Lewis RJ, ed. *Sax's Dangerous Properties of Industrial Materials*, 11th edn. New York: Wiley, 2004: 1890.
18 Paranjothy KLK, Thampi PP. Development of transdermal patches of verapamil hydrochloride using sodium carboxymethyl guar as a monolithic polymeric matrix and their *in vitro* release studies. *Indian J Pharm Sci* 1997; **59**(2): 49–54.
19 Varshosaz J, *et al.* Use of natural gums and cellulose derivatives in production of sustained release metoprolol tablets. *Drug Deliv* 2006; **13**: 113–119.
20 Park CR, Munday DL. Evaluation of selected polysaccharide excipients in buccoadhesive tablets for sustained release of nicotine. *Drug Dev Ind Pharm* 2004; **30**(6): 609–617.
21 Tiwari S, *et al.* L(9) orthogonal design assisted formulation and evaluation of chitosan-based buccoadhesive films of miconazole nitrate. *Curr Drug Deliv* 2009; **6**(3): 305–316.
22 *Food Chemicals Codex*, 7th edn. Bethesda, MD: United States Pharmacopeia, 2010: 459.

20 General References

Ben-Kerrour L, *et al.* Temperature- and concentration-dependence in pseudoplastic rheological equations for gum guar solutions. *Int J Pharm* 1980; **5**: 59–65.

Bhardwaj TR, *et al.* Natural gums and modified natural gums as sustained-release carriers. *Drug Dev Ind Pharm* 2000; **26**(10): 1025–1038.

Goldstein AM *et al.*Guar gum. In: Whistler RL, ed. *Industrial Gums*, 2nd edn. New York: Academic Press, 1973: 303–321.

Tantry JS, Nagarsenker MS. Rheological study of guar gum. *Indian J Pharm Sci* 2001; **63**(1): 74–76.

Vemuri S. Flow and consistency index dependence of pseudoplastic guar gum solutions. *Drug Dev Ind Pharm* 1988; **14**: 905–914.

21 Author

AH Kibbe.

22 Date of Revision

1 March 2012.

H

Hectorite

1 Nonproprietary Names

None adopted.

2 Synonyms

Accofloc HCX; *Astratone 40*; *Bentone CT*; *Bentone HC*; *Coagu-loid*; *DPI-AW*; *EA 3300*; *Fluorohectorite*; *Ghassoulite*; Hector clay; Hectabrite 200;*Hectabrite AW*; *Hectabrite DP*; *Hectabrite LT*; ; *Laponite*; *Macaloid*; *Optigel SH*; *Rheo-VIS clay*; *SHCa-1*; *Strese & Hofmann's Hectorite*; *Sumecton HE*; *SWN 2739*; *Veegum T.*

3 Chemical Name and CAS Registry Number

Hectorite [12173-47-6]

4 Empirical Formula and Molecular Weight

$\approx Na_{0.3}(Mg,Li)_3Si_4O_{10}(F,OH)_2$ $\approx$383

Hectorite is a mineral with an approximate empirical formula owing to the variability in cation substitution; *see* Table I.

Table I: Approximate composition of hectorite based on chemical analysis.[1]

Component	Wt % range
SiO_2	53.6–55.9
Al_2O_3	0.1–1.1
MgO	24.9–25.4
Fe_2O_3	0–0.05
FeO	0–0.7
CaO	0–0.5
Li_2O	0.4–1.2
Na_2O	0.9–3.0
K_2O	0.05–0.4
TiO_2	0–0.4
F	3.2–6.0
H_2O+ (structural water OH)	5.6–8.3
H_2O- (hydration water)	7.2–9.9

5 Structural Formula

Hectorite is a natural mineral clay, obtained from altered volcanic ash with a high silica content. It is composed of two tetrahedral layers formed by phyllosilicate sheets and one octahedral layer. The apical oxygens of the two tetrahedral sheets project into the octahedral sheet. It is structurally similar to talc but differs by substitution, mainly in the octahedral layer. Common impurities include aluminum, calcium, chlorine, iron, potassium, and titanium.

See Section 4.

6 Functional Category

Adsorbent; emulsifying agent; viscosity-increasing agent.

7 Applications in Pharmaceutical Formulation or Technology

Hectorite is used widely in pharmaceutical preparations as an absorbent, emulsifier, stabilizer, suspending agent, thickener, and viscosity-controlling agent.[2]

Hectorite is a component of other naturally occurring clays and hence may be suitable for use in similar pharmaceutical formulation applications as an adsorbent, oil-in-water emulsifying agent, suspending agent, or viscosity-increasing agent.[3] It is also available as a synthetic material. Hectorite is used to modify the thixotropic behavior of pharmaceutical dispersions[4] and for stabilizing oil-in-water emulsion bases.[5,6]

8 Description

Hectorite is a naturally occurring 2:1 phyllosilicate clay of the smectite (montmorillonite) group and is a principal component of bentonite clay. Hectorite occurs as an odorless, white to cream-colored, waxy, dull powder composed of aggregates of colloidal-sized lath-shaped crystals.

9 Pharmacopeial Specifications

See Section 18.

10 Typical Properties

Cation exchange capacity 43.9 meq/100 g[8]

Crystal data Space group C2/*m*, $a = 5.2$, $b = 9.16$, $c = 16.0$, $\beta \approx 99°$.[9]

Density (true) $\approx$2.3 g/cm^3 (measured)[9]

SEM 1: Excipient: hectorite (*Hectabrite DP*); manufacturer: American Colloid Co.; lot no.: 58905 NFT 288; magnification: 500×.

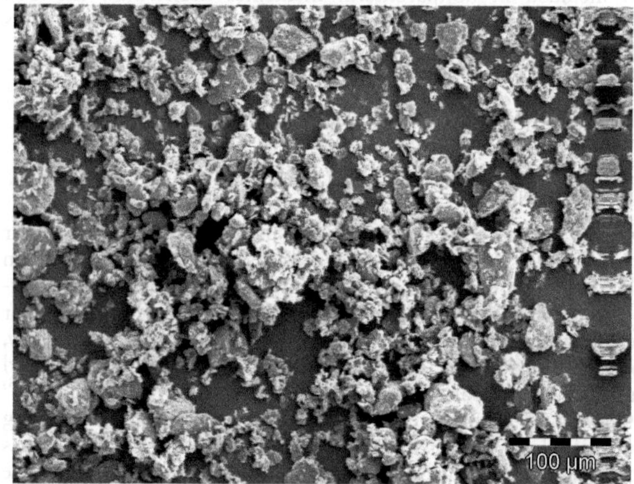

SEM 2: Excipient: hectorite (*Hectabrite DP*); manufacturer: American Colloid Co.; lot no.: 58905 NFT 288; magnification: 1000×.

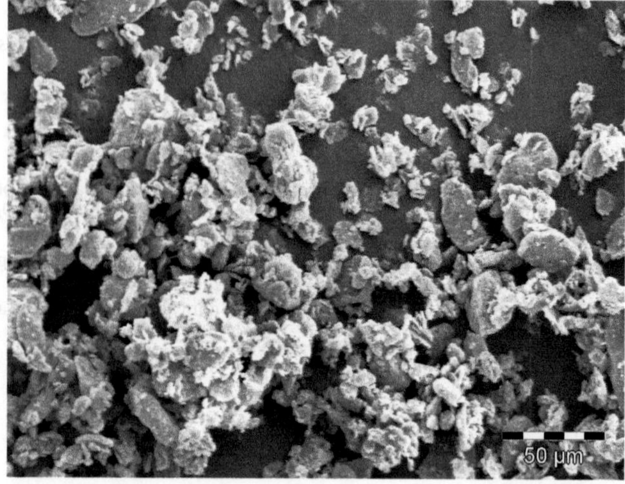

Hardness (Mohs) 1–2[9]
Moisture content Hectorite loses ≈10% of water up to 150°C; ≈2% above 150°C.[10]
Refractive index n_α= 1.49; n_β= 1.50; n_γ= 1.52 (biaxial –).[9]
Specific surface area 63.2 m^2/g. Hectorite swells on the addition of water.[8]

11 Stability and Storage Conditions

Hectorite is a stable material and should be stored in a cool, dry place.

12 Incompatibilities

Contact between hectorite and hydrofluoric acid may generate heat.

13 Method of Manufacture

Naturally occurring hectorite is mined from weathered bentonite deposits. It is further processed to remove grit and impurities so that it is suitable for pharmaceutical and cosmetic applications.

14 Safety

Hectorite is a natural clay mineral that is not considered acutely toxic; therefore no toxicity values have been established. However, hectorite may contain small amounts of crystalline silica in the form of quartz.

Dust can be irritating to the respiratory tract and eyes,[11] and contact with this material may cause drying of the skin. Chronic exposure to crystalline silica may have adverse effects on the respiratory system. EU labeling states that the material is not classified as dangerous.

LD_{50} (rat, oral): >5.0 g/kg[12]

15 Handling Precautions

Observe normal precautions appropriate to the circumstances and quantity of material being handled. Avoid generating and breathing dust, and use eye protection. For dusty conditions, eye protection, gloves, and a dust mask are recommended. The occupational exposure limits for hectorite are 5 mg/m^3 (respirable) PEL-TWA, 3 mg/m^3 (respirable) TLV-TWA, and 10 mg/m^3 (inhalable dust) TLV-TWA.

16 Regulatory Status

The components of hectorite are individually reported in the EPA TSCA Inventory. Included in the FDA Inactive Ingredients Database (transdermal; film and controlled release).

17 Related Substances

Attapulgite; bentonite; kaolin; magnesium aluminum silicate; quaternium 18-hectorite; saponite; stearalkonium hectorite; talc.

Quaternium 18-hectorite

CAS numbers [71011-27-3]; [12001-31-9].
Synonyms *Bentone 38*.
Comments Quaternium 18-hectorite is used in cosmetics as a viscosity-controlling agent. It does not contain crystalline silica. The EINECS numbers for quaternium 18-hectorite are 234-406-6, and 234-406-6.

Stearalkonium hectorite

CAS numbers [94891-33-5]; [71011-26-2].
Synonyms *Bentone 27*.
Comments Steralkonium hectorite is used in cosmetics as a viscosity-controlling agent. Reported in the EPA TSCA Inventory. The EINECS numbers for stearalkonium hectorite are 305-633-9, and 275-126-4.

18 Comments

Hectorite is listed as a component of magnesium aluminum silicate in the USP35-NF30 and as a component of bentonite in the *Food Chemicals Codex* (FCC).[13]

Polyethylene glycols 400, 1500, and 4000 have been shown to increase the consistency of hectorite dispersions.[14] Synthetic hectorite has been conjugated with block copolymers of polyethylene glycol to form hybrid nanocrystal drug carriers.[15] The role of hectorite in the induction of oxidative stress as determined by lipid peroxidation in biological matrices has been examined.[16]

When combined with an appropriate cation, hectorite exhibits properties suitable for use as a contrast agent.[7]

The EINECS number for hectorite is 235-340-0.

19 Specific References

1 Lopez-Galindo A, *et al.* Compositional, technical and safety specifications of clays to be used as pharmaceutical and cosmetic products. *Appl Clay Sci* 2007; **36**(1-3): 51–63.
2 Ash M, Ash I. *Handbook of Pharmaceutical Additives*, 3rd edn. Endicott, NY: Synapse Information Resources, 2008: 649.
3 Viseras C, *et al.* Uses of clay minerals in semisolid health care and therapeutic products. *Appl Clay Sci* 2007; **36**(1-3): 37–50.
4 Plaizier-Vercammen JA. [Viscous behaviour of laponite XLG, a synthetic hectorite and its use in pharmaceutical dispersions.] *Farmaceutisch Tijdschr Belg* 1994; **71**(4–5): 2–9[in Dutch].
5 Plaizier-Vercammen JA. Rheological properties of laponite XLG, a synthetic purified hectorite. *Pharmazie* 1992; **47**(11): 856–861.
6 Burdeska M, Asche H. [Heat sterilization of O/W emulsions using nonionic cream bases as examples: formulation of heat stable cream bases.] *Pharm Ind* 1986; **48**(10): 1171–1177[in German].
7 Balkus KJ, Shi J. A study of suspending agents for gadolinium(III)-exchanged hectorite. An oral magnetic resonance imaging contrast agent. *Langmuir* 1996; **12**(26): 6277–6281.
8 Clay Minerals Society. Source clay minerals: Hectorite *SHCa-1*. http://www.clays.org/SOURCE%20CLAYS/SCdata.html (accessed 23 November 2011).
9 Mineral Data Publishing. Hectorite. In: *Handbook of Minerology*, 2001. http://www.handbookofmineralogy.com/pdfs/hectorite.pdf (accessed 23 November 2011).
10 Dombrowski T. Clays (Survey). In: *Encyclopedia of Chemical Technology*, 4th edn, vol. 6, 1994: 381-405.
11 Elmore AR. Cosmetic Ingredient Review Panel. Final report on the safety assessment of aluminum silicate, calcium silicate, magnesium silicate, magnesium trisilicate, sodium magnesium silicate, zirconium silicate, attapulgite, bentonite, Fuller's earth, hectorite, kaolin, lithium magnesium silicate, lithium magnes sodium silicate, montmorillonite, pyrophyllite, and zeolite. *Int J Toxicol* 2003; **22**(Suppl. 1): 37–102.
12 International Uniform Chemical Information Database (IUCLID). IUCLID dataset: Hectorite clay mineral, 2000. http://esis.jrc.ec.europa.eu/doc/existing-chemicals/IUCLID/data_sheets/12173476.pdf (accessed 23 November 2011).
13 *Food Chemicals Codex*, 7th edn. Bethesda, Md: United States Pharmacopeia, 2010.
14 Omar SM, *et al.* Effect of polyethylene glycols on the rheological characteristics of Macaloid dispersions. *J Drug Res* 1994; **21**(1–2): 91–103.
15 Takahashi T, *et al.* Preparation of a novel peg-clay hybrid as a DDS material: dispersion stability and sustained release profiles. *J. Control. Release* 2005; **107**(3): 408–416.
16 Daria K, *et al.* Determination of lipid peroxidation and cytotoxicity in calcium, magnesium, titanium and hectorite (SHCa-1) suspensions. *Chemosphere* 2011; **82**(3): 418–423.

20 General References

Aguzzi C, *et al.* Use of clays as drug delivery systems: possibilities and limitations. *Appl Clay Sci* 2007; **36**(1–3): 22–36.
Alexander P. Rheological additives. *Manuf Chem* 1986; **57**(Jun): 49–51.
Browne JE, *et al.* Characterization and adsorptive properties of pharmaceutical grade clays. *J Pharm Sci* 1980; **69**(7): 816–823.
Carretero M, *et al.* Clay and non-clay minerals in the pharmaceutical industry. *Appl Clay Sci* 2009; **46**(1): 73–80.
Gormley IP, Addison J. The *in vitro* cytotoxicity of some standard clay mineral dusts of respirable size. *Clay Miner* 1983; **18**(2): 153–163.

H

Earnest CE. Thermal analysis of hectorite. Part I. Thermogravimetry. *Thermochim Acta* 1983; **63:** 277–289.
Earnest CE. Thermal analysis of hectorite. Part II. Differential thermal analysis. *Thermochim Acta* 1983; **63:** 291–306.
Foshaq WR, Woodford AO. Bentonite magnesium clay mineral from California. *Am Mineral* 1936; **21:** 238–244.
Komadel PJ, *et al.* Dissolution of Hectorite in inorganic acids. *Clays Clay Miner* 1996; **44:** 228–236.
Lopez-Galindo A, Viseras C. Pharmaceutical and cosmetic applications of clays. In: Wypych F, Satyanarayana KG, eds. *Clay Surfaces: Fundamentals and Applications.* Amsterdam: Elsevier, 2004; 267–289.
Mineralogy Database. Hectorite mineral data. http://webmineral.com/data/Hectorite.shtml (accessed 23 November 2011).
The Euromin project. Hectorite. http://euromin.w3sites.net//mineraux/hectorite.html (accessed 23 November 2011).
Trottonhorst R, Roberson HE. X-ray diffraction aspects of montmorillonites. *Am Mineral* 1973; **58:** 73–80.
Viseras C, *et al.* Current challenges in clay minerals for drug delivery. *Applied Clay Science* 2010; **48**(3): 291–295.
Viseras C, Lopez-Galindo A. Characteristics of pharmaceutical grade phyllosilicate powders. *Pharm Dev Technol* 2000; **5**(1): 47–52.

21 Author

PE Luner.

22 Date of Revision

1 March 2012.

Heptafluoropropane (HFC)

1 Nonproprietary Names

None adopted.

2 Synonyms

Dymel 227 ea/P; HFA227; HFC227; 2-hydroperfluoropropane; P-227; propellant 227; R-227; *Solkane 227*; *Zephex 227 ea.*

3 Chemical Name and CAS Registry Number

1,1,1,2,3,3,3-Heptafluoropropane [431-89-0]

4 Empirical Formula and Molecular Weight

C_3HF_7 170.0

5 Structural Formula

```
    F   F   F
    |   |   |
F---C---C---C---F
    |   |   |
    F   H   F
```

6 Functional Category

Aerosol propellant.

7 Applications in Pharmaceutical Formulation or Technology

Heptafluoropropane (P-227) is classified as a hydrofluorocarbon (HFC) aerosol propellant since the molecule consists only of carbon, fluorine, and hydrogen atoms. It does not contain any chlorine and consequently does not affect the ozone layer, nor does it have an effect upon global warming. It is therefore considered as an alternative propellant to CFCs for metered-dose inhalers (MDIs). While some of its physical and chemical properties are known, little has been published in regard to its use as a replacement for CFCs in MDIs.

The vapor pressure of heptafluoropropane (P-227) is somewhat lower than that of tetrafluoroethane and dichlorodifluoromethane but considerably higher than the vapor pressure used to formulate most MDIs.

When heptafluoropropane (P-227) is used for pharmaceutical aerosols and MDIs, the pharmaceutical grade must be specified. Industrial grades may not be suitable due to their impurity profile.

Similarly to tetrafluoroethane, heptafluoropropane is not a good solvent for medicinal agents or for the commonly used surfactants and dispersing agents used in the formulation of MDIs.

Heptafluoropropane is used in MDIs as a propellant.

8 Description

Heptafluoropropane is a liquefied gas and exists as a liquid at room temperature when contained under its own vapor pressure, or as a gas when exposed to room temperature and atmospheric pressure. The liquid is practically odorless and colorless. The gas in high concentration has a faint etherlike odor. Heptafluoropropane is noncorrosive, nonirritating, and nonflammable.

9 Pharmacopeial Specifications

—

10 Typical Properties

Boiling point −16.5°C
Density 1.386 g/cm^3 for liquid at 25°C
Flammability Nonflammable.
Freezing point −131°C
Solubility Soluble 1 in 1725 parts of water at 20°C.
Specific gravity 1.41 at 25°C
Vapor pressure 459.81 kPa (66.69 psia) at 25°C

11 Stability and Storage Conditions

Heptafluoropropane is a nonreactive and stable material. The liquefied gas is stable when used as a propellant and should be stored in a metal cylinder in a cool, dry place.

12 Incompatibilities

—

13 Method of Manufacture

—

14 Safety

Heptafluoropropane is used as a fire extinguisher and is applicable as a non-CFC propellant in various MDIs and some topical pharmaceutical preparations. Heptafluoropropane is regarded as nontoxic and nonirritating when used as directed. No acute or chronic hazard is present when it is used normally. Inhaling high concentrations of heptafluoropropane vapors can be harmful and is similar to inhaling vapors of other propellants. Deliberate inhalation of vapors of heptafluoropropane can be dangerous and may cause death. The same labeling required of CFC aerosols would be required for those containing heptafluoropropane as a propellant (except for the EPA requirement). (*See* Chlorofluorocarbons (CFC), Section 14.)

15 Handling Precautions

Heptafluoropropane is usually encountered as a liquefied gas and appropriate precautions for handling such materials should be taken. Eye protection, gloves, and protective clothing are recommended. Heptafluoropropane should be handled in a well-ventilated environment. The vapors are heavier than air and do not support life; therefore, when cleaning large tanks that have contained this propellant, adequate provisions for oxygen supply in the tanks must be made in order to protect workers cleaning the tanks. Although nonflammable, when heated to decomposition heptafluoropropane will emit hydrogen fluoride and carbon monoxide.

16 Regulatory Status

Included in the Canadian Natural Health Products Ingredients Database.

17 Related Substances

Difluoroethane; tetrafluoroethane.

18 Comments

The main disadvantage of using heptafluoropropane is its lack of miscibility with water and its poor solubility characteristics when used with medicinal agents and the commonly used MDI surfactants.

The use of heptafluoropropane as a propellant for MDIs has been the subject of many patents throughout the world. These patents cover the formulation of MDIs, the use of specific surfactants and cosolvents, etc., and the formulator is referred to the patent literature prior to formulating an MDI with any HFC as the propellant. The formulation of MDIs with tetrafluoroethane and heptafluoropropane propellant is complicated since they serve as a replacement for dichlorodifluoromethane or dichlorotetrafluoroethane in MDIs. The use of an HFC as the propellant also requires a change in manufacturing procedure, which necessitates a redesign of the filling and packaging machinery for an MDI.

The PubChem Compound ID (CID) for heptafluoropropane is 62442.

19 Specific References

—

20 General References

Pischtiak AH. Characteristics, supply and use of the hydrofluorocarbons HFA 227 and HFA 134 for medical aerosols in the past, present and future. Manufacturer's perspectives. *Chim Oggi* 2002; 20(3–4): 14–1517–19.

Sciarra CJ, Sciarra JJ. Aerosols. In: *Remington: The Science and Practice of Pharmacy*, 21st edn. Philapdelphia, PA: Lippincott, Williams and Wilkins, 2005: 1000–1017.

21 Authors

CJ Sciarra, JJ Sciarra.

22 Date of Revision

1 August 2011.

Hexetidine

1 Nonproprietary Names

BP: Hexetidine

PhEur: Hexetidine

2 Synonyms

5-Amino-1,3-bis(2-ethylhexyl)hexahydro-5-methylpyrimidine; 5-amino-1,3-di(β-ethylhexyl)hexahydro-5-methylpyrimidine; 1,3-bis (2-ethylhexyl)-5-methylhexahydropyrimidin-5-ylamine; 1,3-bis (β-ethylhexyl)-5-methyl-5-aminohexahydropyrimidine; *Glypesin*; hexetidinum; *Hexigel*; *Hexocil*; *Hexoral*; *Hextril*; *Oraldene*; *Sterisil*; *Steri/Sol*.

3 Chemical Name and CAS Registry Number

1,3-bis(2-Ethylhexyl)-5-methylhexahydro-5-pyrimidinamine [141-94-6]

4 Empirical Formula and Molecular Weight

$C_{21}H_{45}N_3$ 339.61

5 Structural Formula

H_3C NH_2 H_3C CH_3 H_3C N N CH_3 * *

Mixture of stereoisomers

6 Functional Category

Antimicrobial preservative.

7 Applications in Pharmaceutical Formulation or Technology

Hexetidine is used as an antimicrobial preservative in cosmetics and nonparenteral pharmaceutical formulations.

8 Description

Hexetidine is a colorless or faint yellow-colored oily liquid with a characteristic amine odor.

9 Pharmacopeial Specifications

See Table I.

Table I: Pharmacopeial specifications for hexetidine.

Test	PhEur 7.4
Identification	+
Characters	+
Relative density	0.864–0.870
Refractive index	1.461–1.467
Optical rotation	−0.10° to +0.10°
Absorbance	+
Related substances	+
Sulfated ash	≤0.1%
Heavy metals	≤10 ppm
Assay	98.0–102.0%

10 Typical Properties

Antimicrobial activity Hexetidine is a nonantibiotic antimicrobial agent that possesses broad-spectrum antimicrobial activity against Gram-positive and Gram-negative bacteria and fungi such as *Candida albicans.*[1–4] Several studies have identified the antiplaque activity of hexetidine.[3–8] Hexetidine has been shown to be effective against isolates of *Staphylococcus aureus* and *Pseudomonas aeruginosa* in planktonic form and against biofilms of the same microorganisms on PVC.[1] Hexetidine has also been reported to reduce the adherence of *Candida albicans* to human buccal epithelial cells *in vitro.*[9] Hexetidine has been shown to be a promising candidate antimalarial agent, with IC_{50} values being comparable with those of quinine chlorohydrate and chloroquine sulfate.[10] *See also* Table II.

Boiling point 172–176°C

Dissociation constant $pK_a = 8.3$

Density 0.864–0.870 at 20°C

Refractive index $n_D^{20} = 1.463–1.467$

Solubility Soluble in acetone, benzene, chloroform, dichloromethane, ethanol (95%), *n*-hexane, methanol, mineral acids, petroleum ether, and propylene glycol; very slightly soluble in water.

Table II: Minimum inhibitory concentrations (MICs) for hexetidine.

Microorganism	MIC (μg/mL)
Aspergillus niger	<25
Bacillus subtilis	<25
Candida albicans	250–500
Escherichia coli	>500
Pseudomonas aeruginosa	>500
Staphylococcus aureus	>25
Staphylococcus epidermitis	>6

11 Stability and Storage Conditions

Hexetidine is stable and should be stored in a well-closed container in a cool, dry place. Brass and copper equipment should not be used for the handling or storage of hexetidine.

12 Incompatibilities

Hexetidine is incompatible with strong oxidizing agents. Salts are formed with mineral and organic acids; strong acids cause opening of the hexahydropyrimidine ring, releasing formaldehyde.

13 Method of Manufacture

Hexetidine is prepared by hydrogenation under pressure of 1,3-bis(2-ethylhexyl)-5-methyl-4-nitrohexahydropyriminine at 100°C using Raney nickel as a catalyst.

14 Safety

Hexetidine is mainly used in mouthwashes as a bactericidal and fungicidal antiseptic. It is also used as an antimicrobial preservative and is generally regarded as a relatively nontoxic and nonirritant material at concentrations up to 0.1% w/v. Allergic contact dermatitis and altered olfactory and taste perception have occasionally been reported. Hexetidine is toxic when administered intravenously.

Solutions of hexetidine in oil at concentrations of 5–10% w/v cause strong primary irritations without sensitization in humans. Long-term toxicological studies of up to 0.1% w/w of hexetidine in food for 1 year do not show any toxic effect. Fetotoxicity, embryotoxicity, and teratogenicity studies in rats of doses up to 50 mg/kg/day exhibit no sign of toxicity.

LD_{100} (cat, IV): 5–20 mg/kg
LD_{50} (dog, oral): 1.60 g/kg
LD_{50} (mouse, IP): 0.142 g/kg
LD_{50} (mouse, oral): 1.52 g/kg
LD_{50} (rat, oral): 0.61–1.43 g/kg

15 Handling Precautions

Observe normal precautions appropriate to the circumstances and quantity of material handled. Hexetidine may be harmful upon inhalation or on contact with the skin or eyes. Eye protection and gloves are recommended. When significant quantities are being handled, the use of a respirator with an appropriate gas filter is recommended.

16 Regulatory Status

Included in nonparenteral formulations licensed in Europe. Included in the Canadian Natural Health Products Ingredients Database.

17 Related Substances

—

18 Comments

Hexetidine has been quantitatively determined in both commercial formulations and saliva using a reversed-phase HPLC method,[11] with determination being possible at concentrations below the published minimum inhibitory concentrations for a selection of microorganisms.

Therapeutically, hexetidine is mainly used as a 0.1% w/v solution in mouthwash formulations for the prevention and treatment of minor local infections, gingivitis, and mouth ulcers.

The EINECS number for hexetidine is 205-513-5.

19 Specific References

1 Gorman SP, *et al.* The concomitant development of poly(vinyl chloride)-related biofilm and antimicrobial resistance in relation to ventilator-associated pneumonia. *Biomaterials* 2001; **22**(20): 2741–2747.
2 Guiliana G, *et al. In vitro* activities of antimicrobial agents against *Candida* species. *Oral Surg Oral Med Oral Pathol Oral Radiol Endod* 1999; **87**(1): 44–49.
3 Williams MJR, *et al.* The effect of hexetidine 0.1% in the control of dental plaque. *Br Dent J* 1987; **163**(9): 300–302.
4 Wile DB, *et al.* Hexetidine (Oraldene) – a report on its antibacterial and antifungal properties on the oral flora in healthy subjects. *Curr Med Res Opin* 1986; **10**(2): 82–88.
5 Bokor M. The effect of hexetidine spray on dental plaque following periodontal surgery. *J Clin Periodontol* 1996; **23**(12): 1080–1083.
6 Roberts WR, Addy M. Comparison of the *in vivo* and *in vitro* antibacterial properties of antiseptic mouthrinses containing chlorhexidine, alexidine, cetylpyridinium chloride and hexetidine – relevance to mode of action. *J Clin Periodontol* 1981; 8(4): 295–310.
7 Pilloni AP, *et al.* Antimicrobial action of Nitens mouthwash (cetylpyridinium naproxenate) on multiple isolates of pharyngeal microbes: a controlled study against chlorhexidine, benzydamine, hexetidine, amoxicillin clavulanate, clarithromycin and cefaclor. *Chemotherapy* 2002; 48(4): 168–173.
8 Sharma NC, *et al.* Antiplaque and antigingivitis effectiveness of a hexetidine mouthwash. *J Clin Periodontol* 2003; 30(7): 590–594.
9 Jones DS, *et al.* The effects of hexetidine (Oraldene) on the adherence of *Candida albicans* to human buccal epithelial cells *in vitro* and *ex vivo* and on *in vitro* morphogenesis. *Pharm Res* 1997; **14**(12): 1765–1771.
10 Gozalbes R, *et al.* Discovery of new antimalarial compounds by use of molecular connectivity techniques. *J Pharm Pharmacol* 1999; **51**(2): 111–117.
11 McCoy CP, *et al.* Determination of the salivary retention of hexetidine in-vivo by high-performance liquid chromatography. *J Pharm Pharmacol* 2000; 52(11): 1355–1359.

20 General References

Eley BM. Antibacterial agents in the control of supragingival plaque – a review. *Br Dent J* 1999; **186**(6): 286–296.
Jones DS, *et al.* Physicochemical characterization of hexetidine-impregnated endotracheal tube poly(vinyl chloride) and resistance to adherence of respiratory bacterial pathogens. *Pharm Res* 2002; **19**(6): 818–824.

21 Authors

DS Jones, CP McCoy.

22 Date of Revision

1 March 2012.

H

Hydrocarbons (HC)

1 Nonproprietary Names

(a) USP–NF: Butane
(b) USP–NF: Isobutane
(c) USP–NF: Propane

2 Synonyms

(a) A-17; *Aeropres 17*; *n*-butane; E943a
(b) A-31; *Aeropres 31*; E943b; 2-methylpropane
(c) A-108; *Aeropres 108*; dimethylmethane; E944; propyl hydride

3 Chemical Name and CAS Registry Number

(a) Butane [106-97-8]
(b) 2-Methylpropane [75-28-5]
(c) Propane [74-98-6]

4 Empirical Formula and Molecular Weight

(a) C_4H_{10} 58.12
(b) C_4H_{10} 58.12
(c) C_3H_8 44.10

5 Structural Formula

(a) Butane: $CH_3CH_2CH_2CH_3$ (all H shown)

(b) Isobutane

(c) Propane

6 Functional Category

Aerosol propellant.

7 Applications in Pharmaceutical Formulation or Technology

Propane, butane, and isobutane are hydrocarbons (HC). They are used as aerosol propellants: alone, in combination with each other, and in combination with a hydrofluoroalkane propellant. They are used primarily in topical pharmaceutical aerosols (particularly aqueous foam and some spray products).

Depending upon the application, the concentration of hydrocarbon propellant range is 5–95% w/w. Foam aerosols generally use about 4–5% w/w of a hydrocarbon propellant consisting of isobutane (84.1%) and propane (15.9%), or isobutane alone. Spray-type aerosols utilize propellant concentrations of 50% w/w and higher.[(1)]

Hydrocarbon propellants are also used in cosmetics and food products as aerosol propellants.

Only highly purified hydrocarbon grades can be used for pharmaceutical formulations since they may contain traces of unsaturated compounds that not only contribute a slight odor to a product but may also react with other ingredients resulting in decreased stability.

H

8 Description

Hydrocarbon propellants are liquefied gases and exist as liquids at room temperature when contained under their own vapor pressure, or as gases when exposed to room temperature and atmospheric pressure. They are essentially clear, colorless, odorless liquids but may have a slight etherlike odor.

9 Pharmacopeial Specifications

See Table I.

Table I: Pharmacopeial specifications for hydrocarbons from the USP35–NF30.

Test	Butane	Isobutane	Propane
Identification	+	+	+
Water	≤0.001%	≤0.001%	≤0.001%
High-boiling residues	≤5 μg/mL	≤5 μg/mL	≤5 μg/mL
Acidity of residue	+	+	+
Sulfur compounds	+	–	+
Assay	≥97.0%	≥95.0%	≥98.0%

10 Typical Properties

See Table II for selected typical properties.

11 Stability and Storage Conditions

Butane and the other hydrocarbons used as aerosol propellants are stable compounds and are chemically nonreactive when used as propellants. They are, however, highly flammable and explosive when mixed with certain concentrations of air; *see* Section 10.[(2)] They should be stored in a well-ventilated area, in a tightly sealed cylinder. Exposure to excessive heat should be avoided.

12 Incompatibilities

Other than their lack of miscibility with water, butane and the other hydrocarbon propellants do not have any practical incompatibilities with the ingredients commonly used in pharmaceutical aerosol formulations. Hydrocarbon propellants are generally miscible with nonpolar materials and some semipolar compounds such as ethanol.

13 Method of Manufacture

Butane and isobutane are obtained by the fractional distillation, under pressure, of crude petroleum and natural gas. They may be purified by passing through a molecular sieve to remove any unsaturated compounds that are present.

Propane is prepared by the same method. It may also be prepared by a variety of synthetic methods.

14 Safety

The hydrocarbons are generally regarded as nontoxic materials when used as aerosol propellants. However, deliberate inhalation of aerosol products containing hydrocarbon propellants can be fatal as they will deplete oxygen in the lungs when inhaled.

15 Handling Precautions

Butane and the other hydrocarbon propellants are liquefied gases and should be handled with appropriate caution. Direct contact of liquefied gas with the skin is hazardous and may result in serious cold burn injuries. Protective clothing, rubber gloves, and eye protection are recommended.

Butane, isobutane, and propane are asphyxiants and should be handled in a well-ventilated environment; it is recommended that environmental oxygen levels are monitored and not permitted to fall below a concentration of 18% v/v. These vapors do not support life; therefore when cleaning large tanks, adequate provisions for oxygen supply must be provided for personnel cleaning the tanks. Butane is highly flammable and explosive and must only be handled in an explosion-proof room that is equipped with adequate safety warning devices and explosion-proof equipment.

To fight fires, the flow of gas should be stopped and dry powder extinguishers should be used.

Table II: Selected typical properties for hydrocarbon propellants.

	Butane	Isobutane	Propane
Autoignition temperature	405°C	420°C	468°C
Boiling point	−0.5°C	−11.7°C	-42.1°C
Density: liquid at 20°C	0.58 g/cm^3	0.56 g/cm^3	0.50 g/cm^3
Explosive limits			
Lower limit	1.9% v/v	1.8% v/v	2.2% v/v
Upper limit	8.5% v/v	8.4% v/v	9.5% v/v
Flash point	−62°C	−83°C	−104.5°C
Freezing point	−138.3°C	−159.7°C	−187.7°C
Kauri-butanol value	19.5	17.5	15.2
Vapor density			
Absolute	2.595 g/m^3	2.595 g/m^3	1.969 g/m^3
Relative	2.046 (air = 1)	2.01 (air = 1)	1.53 (air = 1)
Vapor pressure at 21°C	113.8 kPa (16.5 psig)	209.6 kPa (30.4 psig)	758.4 kPa (110.0 psig)
Vapor pressure at 54.5°C	—	661.9 kPa (96.0 psig)	1765.1 kPa (256 psig)

16 Regulatory Status

GRAS listed. Butane, isobutane, and propane are accepted for use as food additives in Europe. Included in the FDA Inactive Ingredients Database (aerosol formulations for topical application). Included in nonparental medicines licensed in the UK. Included in the Canadian Natural Health Products Ingredients Database.

17 Related Substances

Dimethyl ether.

18 Comments

Although hydrocarbon aerosol propellants are relatively inexpensive, nontoxic, and environmentally friendly (since they are not damaging to the ozone layer and are not greenhouse gases), their use is limited by their flammability. While hydrocarbon propellants are primarily used in topical aerosol formulations, it is possible that butane may also be useful in metered-dose inhalers as a replacement for chlorofluorocarbons.

Various blends of hydrocarbon propellants that have a range of physical properties suitable for different applications are commercially available, e.g. A-46 (*Aeropres*) is a commonly used mixture for aerosol foams and consists of about 85% isobutane and 15% propane. The number following the letter denotes the approximate vapor pressure of the blend or mixture.

Specifications for butane, isobutane, and propane are contained in the *Food Chemicals Codex* (FCC).[3]

The PubChem Compound IDs (CIDs) for butane, isobutane, and propane are 7843, 6360, and 6334 respectively.

19 Specific References

1 Sciarra JJ. Pharmaceutical aerosols. In: Banker GS, Rhodes CT, eds. *Modern Pharmaceutics*, 3rd edn. New York: Marcel Dekker, 1996: 547–574.
2 Dalby RN. Prediction and assessment of flammability hazards associated with metered-dose inhalers containing flammable propellants. *Pharm Res* 1992; 9: 636–642.
3 *Food Chemicals Codex*, 7th edn. Bethesda, MD: United States Pharmacopeia, 2010: 115, 529, 865.

20 General References

Johnson MA. *The Aerosol Handbook*, 2nd edn. Caldwell: WE Dorland, 1982: 199–255, 335–361.
Randall DS. Solving the problems of hydrocarbon propellants. *Manuf Chem Aerosol News* 1979; 50(4): 43, 44, 47.
Sanders PA. *Handbook of Aerosol Technology*, 2nd edn. New York: Van Nostrand Reinhold, 1979: 36–44.
Sciarra JJ Pharmaceutical aerosols. In: Lackman L *et al.* eds. *The Theory and Practice of Industrial Pharmacy*, 3rd edn. Philadelphia: Lea and Febiger, 1986: 589–618.
Sciarra CJ, Sciarra JJ. Aerosols. In: *Remington: The Science and Practice of Pharmacy*, 21st edn. Philadelphia, PA: Lippincott Williams and Wilkins, 2005: 1000–1017.
Sciarra JJ Aerosol suspensions and emulsions. In: Lieberman H *et al.* eds. *Pharmaceutical Dosage Forms: Disperse Systems*, 2nd edn, 2. New York: Marcel Dekker, 1996: 319–356.

21 Authors

CJ Sciarra, JJ Sciarra.

22 Date of Revision

1 March 2012.

H

Hydrochloric Acid

1 Nonproprietary Names

BP: Hydrochloric Acid
JP: Hydrochloric Acid
PhEur: Hydrochloric Acid, Concentrated
USP–NF: Hydrochloric Acid

2 Synonyms

Acidum hydrochloridum concentratum; chlorohydric acid; concentrated hydrochloric acid; E507.

3 Chemical Name and CAS Registry Number

Hydrochloric acid [7647-01-0]

4 Empirical Formula and Molecular Weight

HCl 36.46

5 Structural Formula

See Section 4.

6 Functional Category

Acidulant.

7 Applications in Pharmaceutical Formulation or Technology

Hydrochloric acid is widely used as an acidulant, in a variety of pharmaceutical and food preparations (*see* Section 16). It may also be used to prepare dilute hydrochloric acid; *see* Section 17.

8 Description

Hydrochloric acid occurs as a clear, colorless, fuming aqueous solution of hydrogen chloride, with a pungent odor.

The JP XV specifies that hydrochloric acid contains 35.0–38.0% w/w of HCl; the PhEur 7.4 specifies that hydrochloric acid contains 35.0–39.0% w/w of HCl; and the USP35–NF30 specifies that hydrochloric acid contains 36.5–38.0% w/w of HCl.

See also Section 9.

9 Pharmacopeial Specifications

See Table I.

10 Typical Properties

Acidity/alkalinity pH = 0.1 (10% v/v aqueous solution)
Boiling point 110°C (constant boiling mixture of 20.24% w/w HCl)
Density ≈1.18 g/cm^3 at 20°C

Table I: Pharmacopeial specifications for hydrochloric acid.

Test	JP XV	PhEur 7.4	USP35–NF30
Identification	+	+	+
Characters	+	+	−
Appearance of solution	−	+	−
Residue on ignition	≤1.0 mg	−	≤0.008%
Residue on evaporation	−	≤0.01%	−
Bromide or iodide	+	−	+
Free bromine	+	−	+
Free chlorine	+	≤4 ppm	+
Sulfate	+	≤20 ppm	+
Sulfite	+	−	+
Arsenic	≤1 ppm	−	−
Heavy metals	≤5 ppm	≤2 ppm	≤5 ppm
Mercury	≤0.04 ppm	−	−
Assay (of HCl)	35.0–38.0%	35.0–39.0%	36.5–38.0%

Freezing point ≈−24°C
Refractive index n_D^{20} = 1.342 (10% v/v aqueous solution)
Solubility Miscible with water; soluble in diethyl ether, ethanol (95%), and methanol.

11 Stability and Storage Conditions

Hydrochloric acid should be stored in a well-closed, glass or other inert container at a temperature below 30°C. Storage in close proximity to concentrated alkalis, metals, and cyanides should be avoided.

12 Incompatibilities

Hydrochloric acid reacts violently with alkalis, with the evolution of a large amount of heat. Hydrochloric acid also reacts with many metals, liberating hydrogen.

13 Method of Manufacture

Hydrochloric acid is an aqueous solution of hydrogen chloride gas produced by a number of methods including: the reaction of sodium chloride and sulfuric acid; the constituent elements; as a by-product from the electrolysis of sodium hydroxide; and as a by-product during the chlorination of hydrocarbons.

14 Safety

When used diluted, at low concentration, hydrochloric acid is not usually associated with any adverse effects. However, the concentrated solution is corrosive and can cause severe damage on contact with the eyes and skin, or if ingested.

LD_{50} (mouse, IP): 1.4 g/kg[1]
LD_{50} (rabbit, oral): 0.9 g/kg

15 Handling Precautions

Caution should be exercised when handling hydrochloric acid, and suitable protection against inhalation and spillage should be taken. Eye protection, gloves, face mask, apron, and respirator are recommended, depending on the circumstances and quantity of hydrochloric acid handled. Spillages should be diluted with copious amounts of water and run to waste. Splashes on the skin and eyes should be treated by immediate and prolonged washing with large amounts of water and medical attention should be sought. Fumes can cause irritation to the eyes, nose, and respiratory system; prolonged exposure to fumes may damage the lungs. In the UK, the recommended short-term workplace exposure limit for hydrogen chloride gas and aerosol mists is 8 mg/m^3 (5 ppm). The long-term exposure limit (8-hour TWA) is 2 mg/m^3 (1 ppm).[2]

16 Regulatory Status

GRAS listed. Accepted for use as a food additive in Europe. Included in the FDA Inactive Ingredients Database (dental solutions; epidural injections; IM, IV, and SC injections; inhalations; ophthalmic preparations; oral solutions; nasal, otic, rectal, and topical preparations). Included in parenteral and nonparenteral medicines licensed in the UK. Included in the Canadian Natural Health Products Ingredients Database.

17 Related Substances

Dilute hydrochloric acid.

Dilute hydrochloric acid
Synonyms Acidum hydrochloridum dilutum; diluted hydrochloric acid.
Density ≈1.05 g/cm^3 at 20°C
Comments The JP XV and PhEur 7.4 specify that dilute hydrochloric acid contains 9.5–10.5% w/w of HCl and is prepared by mixing 274 g of hydrochloric acid with 726 g of water. The USP35–NF30 specifies 9.5–10.5% w/v of HCl, prepared by mixing 226 mL of hydrochloric acid with sufficient water to make 1000 mL.

18 Comments

In pharmaceutical formulations, dilute hydrochloric acid is usually used as an acidulant in preference to hydrochloric acid. It is used intravenously in the management of metabolic alkalosis, and orally for the treatment of achlorhydria. Hydrochloric acid is also used therapeutically as an escharotic.[3]

The PhEur 7.4 also contains a specification for hydrochloric acid, dilute; *see* Section 17. A specification for hydrochloric acid is contained in the *Food Chemicals Codex* (FCC).[4]

The EINECS number for hydrochloric acid is 231-595-7. The PubChem Compound ID (CID) for hydrochloric acid is 313.

19 Specific References

1 Lewis RJ, ed. *Sax's Dangerous Properties of Industrial Materials*, 11th edn. New York: Wiley, 2004; 1980.
2 Health and Safety Executive. *EH40/2005: Workplace Exposure Limits.* Sudbury: HSE Books, 2011. www.hse.gov.uk/pubns/priced/eh40.pdf (accessed 12 April 2012).
3 Sweetman S, ed. *Martindale: The Complete Drug Reference*, 37th edn. London: Pharmaceutical Press, 2009; 2322.
4 *Food Chemicals Codex*, 7th edn. Bethesda, MD: United States Pharmacopeia, 2010.

20 General References

Japan Pharmaceutical Excipients Council. *Japanese Pharmaceutical Excipients Directory 1996*. Tokyo: Yakuji Nippo, 1996: 228.

21 Authors

ME Fenton, PJ Sheskey.

22 Date of Revision

12 April 2012.

Hydrophobic Colloidal Silica

1 Nonproprietary Names

BP: Hydrophobic Colloidal Anhydrous Silica
PhEur: Silica, Hydrophobic Colloidal
USP–NF: Hydrophobic Colloidal Silica

2 Synonyms

Aerosil R972; *HDK*; silica dimethyl silylate; silica hydrophobica colloidalis; silicic acid, silylated; silicon dioxide, silanated.

3 Chemical Name and CAS Registry Number

Silane, dichloro-dimethyl-, reaction products with silica [68611-44-9]

4 Empirical Formula and Molecular Weight

SiO_2 (partly alkylated for hydrophobation) 60.08

5 Structural Formula

See Section 4.

6 Functional Category

Anticaking agent; emulsion stabilizing agent; glidant; suspending agent; viscosity-increasing agent.

7 Applications in Pharmaceutical Formulation or Technology

Hydrophobic colloidal silica has nano-sized primary particles and a large specific surface area,(1) which provide desirable flow characteristics in dry powders used in tableting(2–4) and capsule filling.(3) The hydrophobic grades absorb less moisture(5) and may offer an advantage in moisture-sensitive formulations. Generally, the uses of hydrophobic colloidal silica and the concentrations used are similar to those of the standard hydrophilic colloidal silicon dioxide.

Hydrophobic colloidal silica is also used to thixotropically control viscosity, to thicken and stabilize emulsions, or as a suspending agent in gels and semisolid preparations. Hydrophobic colloidal silica has a less pronounced effect on solution viscosity but can thicken and stabilize the oil phase of a water–oil emulsion.(3,5) With other ingredients of similar refractive index, transparent gels may be formed.(5) The highly hydrophobic particles can be used to form a powder shell that encapsulates water droplets forming "dry water"(6,7) (also known as "dry emulsion" or "liquid marbles"),(6) which can be used to topically deliver water-soluble active agents that are released by rubbing the "dry" powder onto the skin.

8 Description

Hydrophobic colloidal silica occurs as a light, fine, white or almost white amorphous powder, not wettable by water.

9 Pharmacopeial Specifications

See Table I.

Table I: Pharmacopeial specifications for hydrophobic colloidal silica.

Test	PhEur 7.4	USP35–NF30
Identification	+	+
Characters	+	—
Chloride	≤250 ppm	≤0.025%
Water-dispersible fraction	≤3.0%	≤3.0%
Heavy metals	≤25 ppm	—
Lead	—	≤0.0025%
Loss on ignition	≤6.0%	≤6.0%
Assay (on ignited sample)	99.0–101.0%	99.0–101.0%

10 Typical Properties

Density (bulk) 0.094 g/cm3(4)
Density (tapped) see Table II.
Moisture content Less than 0.5% w/w at room temperature between 0 and 100% relative humidity.(5)
Particle size distribution Primary particle size is 16 nm for *Aerosil R972*. Forms loose agglomerates of 10–200 μm.(5)
Acidity/alkalinity pH = 6.5–8.0 for *HDK* H2000;(8)
pH = 3.8–4.8 for *HDK* H13L(9) (4% dispersion in 1:1 mixture of water–methanol).
Refractive index 1.46(5)
Solubility Solubility is 1 in 6.7 parts of water (pH 7, 25°C).(5) Practically insoluble in organic solvents(5) and acids, except hydrofluoric acid; soluble in hot solutions of alkali hydroxide.(10)
Specific gravity 2.0–2.2(5)
Specific surface area see Table II.

Table II: Physical properties of commercial grades of hydrophobic colloidal silica

Grade	Sepcific surface area[(a)] (m^2/g)	Density (tapped) (g/cm^3)
Aerosil R972	110 ± 20	0.05
HDK H13L	125 ± 15	0.07
HDK H2000	200 ± 30	0.2

(a) BET method

11 Stability and Storage Conditions

Hydrophobic colloidal silica should be stored in a well-closed container. It will not absorb moisture but may still absorb volatile substances owing to its high surface area.

12 Incompatibilities

Use of hydrophobic colloidal silica has been shown to reduce the strength of starch-based tablets.(4)

13 Method of Manufacture

Hydrophobic colloidal silica is prepared by the flame hydrolysis of chlorosilanes, such as silicon tetrachloride, at 1800°C using a hydrogen–oxygen flame. It is rapidly cooled to create an amorphous product(11) and immediately treated with dichlorodimethyl silane in a fluid bed reactor.(5) The resulting surface is covered with dimethylsilyl groups.(1) See also Section 18.

14 Safety

The safety profile of hydrophobic colloidal silica is the same as for the hydrophilic silica types, as the modified silica surface does not significantly alter the toxicological properties.(12)

LD_{50} (rat, IV): 0.015 g/kg(13)

15 Handling Precautions

Observe normal precautions appropriate to the circumstances and quantity of the material handled. Eye protection and gloves are recommended. Considered to be a nuisance dust.[11] Inhalation of amorphous hydrophobic colloidal silica dust may cause irritation to the respiratory tract but it is not associated with fibrosis of the lungs (silicosis), which can occur upon exposure to crystalline silica.[11,13]

Precautions should be taken to avoid inhalation of colloidal silicon dioxide. In the absence of suitable containment facilities, a dust mask should be worn when handling small quantities of material. For larger quantities, a dust respirator is recommended.[14] Colloidal silica may build up static charge due to friction during pneumatic conveying and some engineering risk controls may be required.[14,15]

16 Regulatory Status

Approved for use in pharmaceuticals in Europe. Approved by FDA and Europe for food contact articles. Included in nonparenteral medicines (oral tablets; rectal suppositories) licensed in the UK.

H

17 Related Substances

Colloidal silicon dioxide.

18 Comments

Hydrophobic grades have not been considered for use in food intended for human consumption but are permitted in food contact articles in Europe (listed as silicon dioxide, silanated) and as an anticaking agent in animal feed mineral premixes in Europe and the US.

Aerosil R972 is manufactured by modifying the surface of *Aerosil* 130[5] and thus has the same primary particle size and surface area, but the density of silanol groups is reduced from approximately 2.0 $SiOH/nm^2$ to 0.75 $SiOH/nm^2$.[4]

The EINECS number for hydrophobic colloidal silica is 271-893-4.

19 Specific References

1 Matthias J, Wannemaker G. Basic characteristics and applications of Aerosil 30. The chemistry and physics of the Aerosil surface. *J Colloid Interface Sci* 1998; **125**: 61–68.

2 Zimmermann I, *et al*. Nanomaterials as flow regulators in dry powders. *Z Phys Chem* 2004; **218**: 51–102.

3 Jonat S, *et al*. Investigation of compacted hydrophilic and hydrophobic colloidal silicon dioxides as glidants for pharmaceutical excipients. *Powder Technol* 2004; **141**: 31–43.

4 Jonat S, *et al*. Influence of compacted hydrophobic and hydrophilic colloidal silicon dioxide on tabletting properties of pharmaceutical excipients. *Drug Dev Ind Pharm* 2005; **31**: 687–696.

5 Evonik Industries. Technical bulletin fine particles No. 11: Basic characteristics of *Aerosil* fumed silica TB0011-1, 2006.

6 Forny L, *et al*. Storing water in powder form by self-assembling hydrophobic silica nanoparticles. *Powder Technol* 2007; **171**(1): 15–24.

7 S Hasenzahl, *et al*. Dry water for the skin. *SÖFW-J* 2005; **131**(3): 3–8.

8 Wacker-Chemie AG. *Technical data sheet: HDK* H2000, version 1.2, February 2011.

9 Wacker-Chemie AG. *Technical data sheet: HDK* H13L, version 1.1, February 2011.

10 Evonik Industries. Technical literature: *Aerosil* colloidal silicon dioxide for pharmaceuticals TI1281-1, 2006.

11 Waddell WW, ed. Silica amorphous. In: *Kirk-Othmer Encycopedia of Chemical Technology*, 5th edn, 22. New York: Wiley, 2001; 380–406.

12 Lewinson J, *et al*. Characterization and toxicological behavior of synthetic amorphous hydrophobic silica. *Regul Toxicol Pharmacol* 1994; 20(1): 37–57.

13 Lewis RJ, ed. *Sax's Dangerous Properties of Industrial Materials*, 11th edn. New York: Wiley, 2004; 3205.

14 Evonik Industries. Technical bulletin: Fine Particles T28 - Handling of synthetic silica and silicate, May 2011.

15 Evonik Industries. Technical bulletin: Fine Particles T62 - Synthetic silica and electrostatic charges, June 2011.

20 General References

Evonik Degussa Corporation. Technical literature TI 1281: *Aerosil* colloidal silicon dioxide for pharmaceuticals, 2010.

Evonik Industries. Material safety data sheet No. EC 1907/2006: *Aerosil* R972, August 2007.

Evonik Industries. Product data sheet: *Aerosil* R972 Hydrophobic fumed silica, February 2008.

21 Author

KP Hapgood.

22 Date of Revision

1 March 2012.

Hydroxyethyl Cellulose

1 Nonproprietary Names

BP: Hydroxyethylcellulose
PhEur: Hydroxyethylcellulose
USP–NF: Hydroxyethyl Cellulose

2 Synonyms

Cellosize HEC; cellulose hydroxyethyl ether; cellulose 2-hydroxyethyl ether; cellulose hydroxyethylate; ethylhydroxy cellulose; ethylose; HEC; HE cellulose; hetastarch; 2-hydroxyethyl cellulose ether; hydroxyethylcellulosum; hydroxyethyl ether cellulose; hydroxyethyl starch; hyetellose; *Natrosol*; oxycellulose; *Tylose H*; *Tylose PHA*.

3 Chemical Name and CAS Registry Number

Cellulose, 2-hydroxyethyl ether [9004-62-0]

4 Empirical Formula and Molecular Weight

Hydroxyethyl cellulose is a partially substituted poly(hydroxyethyl) ether of cellulose.[1] It is available in several grades that vary in viscosity and degree of substitution; some grades are modified to improve their dispersion in water.

See Section 10.

5 Structural Formula

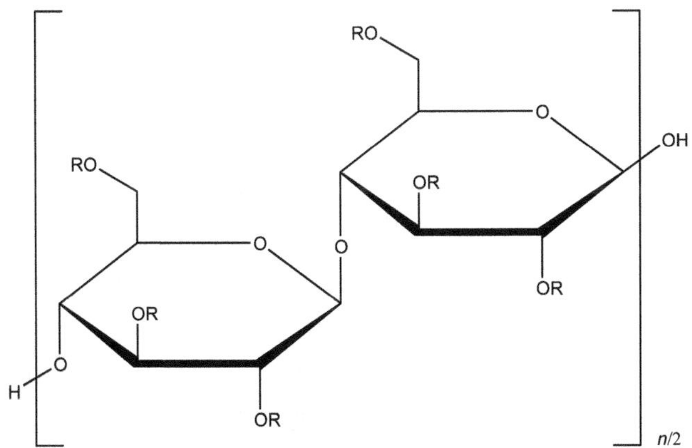

R is H or $[-CH_2CH_2O-]_mH$ where m is a common integral number of cellulose derivatives.

6 Functional Category

Coating agent; suspending agent; tablet and capsule binder; viscosity-increasing agent.

7 Applications in Pharmaceutical Formulation or Technology

Hydroxyethyl cellulose is a nonionic, water-soluble polymer widely used in pharmaceutical formulations. It is primarily used as a thickening agent in ophthalmic[2] and topical formulations.[3] It is also used as a binder[4] and film-coating agent for tablets[4] and capsules, and in oral solutions and suspensions. The concentration of hydroxyethyl cellulose used in a formulation is dependent upon the solvent and the molecular weight of the grade. Hydroxyethyl cellulose hydrogels may be used in various delivery systems.[6]

Hydroxyethyl cellulose is also widely used in cosmetics, hair care and personal care products, including toothpastes, body lotions, and deodorants.[1,7]

8 Description

Hydroxyethyl cellulose occurs as a white, yellowish-white or grayish-white, odorless and tasteless, hygroscopic powder.

9 Pharmacopeial Specifications

See Table I. *See also* Section 18.

Table I: Pharmacopeial specifications for hydroxyethyl cellulose.

Test	PhEur 7.4	USP35–NF30
Identification	+	+
Characters	+	—
Appearance of solution	+	—
Viscosity	75.0–140.0%	+
pH	5.5–8.5	6.0–8.5
Loss on drying	≤10.0%	≤10.0%
Lead	—	≤0.001%
Residue on ignition	—	≤5.0%
Sulfated ash	≤4.0%	—
Chlorides	≤1.0%	—
Heavy metals	≤20 ppm	≤20 μg/g
Nitrates	+	—
Glyoxal	≤20 ppm	—
Ethylene oxide	≤1 ppm	—
2-Chloroethanol	≤10 ppm	—

SEM 1: Excipient: hydroxyethyl cellulose (*Natrosol*); manufacturer: Ashland Aqualon Functional Ingredients; magnification: 120×.

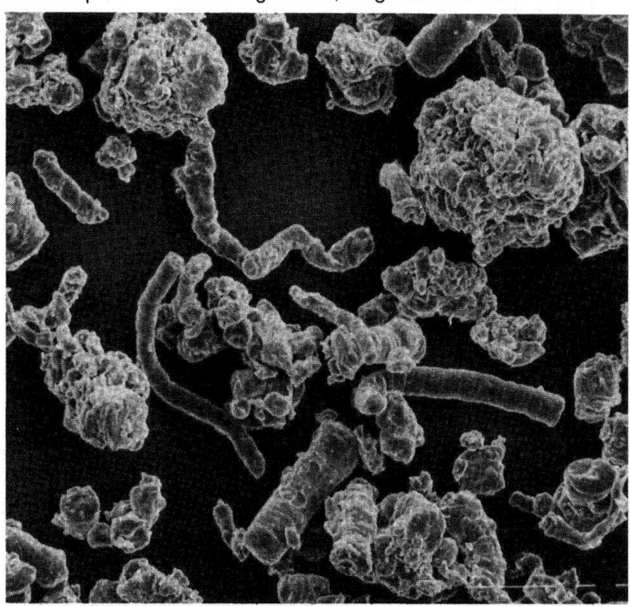

SEM 2: Excipient: hydroxyethyl cellulose (*Natrosol*); manufacturer: Ashland Aqualon Functional Ingredients; magnification: 600×.

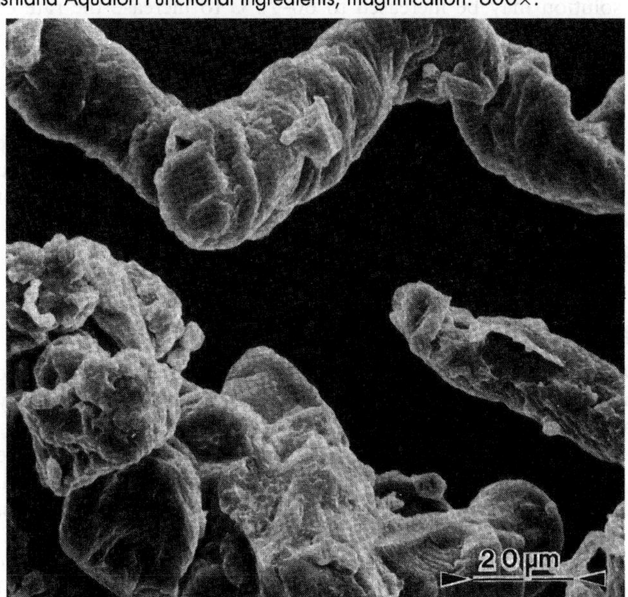

10 Typical Properties

Acidity/alkalinity pH = 6–7 for a 2% w/v aqueous solution.[8]

Density (bulk)

0.35–0.61 g/cm^3 for *Cellosize*;[8]

0.60 g/cm^3 for *Natrosol*.[1]

Melting point Softens at 135–140°C;[1] browns at 205–210°C,[1] decomposes at about 280°C.[8]

Moisture content Commercially available grades of hydroxyethylcellulose contain less than 5% w/w of water.[1] However, as hydroxyethyl cellulose is hygroscopic, the amount of water absorbed depends upon the initial moisture content and the relative humidity of the surrounding air. Typical equilibrium moisture values for *Natrosol* 250 at 25°C are: 6% w/w at 50% relative humidity and 29% w/w at 84% relative humidity.[1]

Particle size distribution

Cellosize: 100% through a US #20 mesh (840 μm);[8]

Natrosol (regular grind): ≤10% retained on a US #40 mesh (420 μm);[1]

Natrosol (X-grind): ≤0.5% retained on a US #60 mesh (250 μm).[1]

Refractive index $n_D^{20} = 1.336$ for a 2% w/v aqueous solution.[1,8]

Solubility Hydroxyethyl cellulose is nonionic and soluble in either hot or cold water, forming clear, smooth, uniform solutions. It does not form a 'cloud point' or precipitate from hot solutions.[1] Practically insoluble in acetone, ethanol (95%), ether, toluene, and most other organic solvents.[1] In some polar organic solvents, such as the glycols, hydroxyethyl cellulose either swells or is partially soluble.[1]

Specific gravity 1.38–1.40 for *Cellosize*; 1.0033 for a 2% w/v aqueous hydroxyethyl cellulose solution.[1]

Spectroscopy

IR spectra *see* Figure 1.

NIR spectra *see* Figure 2.

Surface tension 66.8 mN/m for Natrosol 250LR at 0.1%.[1] *See also* Table II.

Viscosity (dynamic) Hydroxyethyl cellulose is available in a wide range of viscosity types; see Tables III and IV. *See also* Section 18. Aqueous solutions made using a rapidly dispersing material may be prepared by dispersing the hydroxyethyl cellulose in mildly agitated water at 20–25°C. When the hydroxyethyl cellulose has been thoroughly wetted, the temperature of the solution may be increased to 60–70°C to increase the rate of dispersion. Making the solution slightly alkaline also increases the dispersion process. Typically, complete dispersion may be achieved in approximately an hour by controlling the temperature, pH, and rate of stirring.

Normally dispersing grades of hydroxyethyl cellulose require more careful handling to avoid agglomeration during dispersion; the water should be stirred vigorously. Alternatively, a slurry of hydroxyethyl cellulose may be prepared in a nonaqueous solvent, such as ethanol, prior to dispersion in water.

See also Section 11 for information on solution stability.

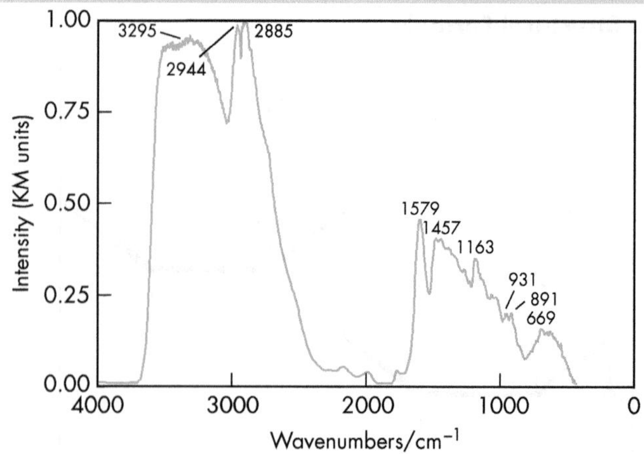

Figure 1: Infrared spectrum of hydroxyethyl cellulose measured by diffuse reflectance. Adapted with permission of Informa Healthcare.

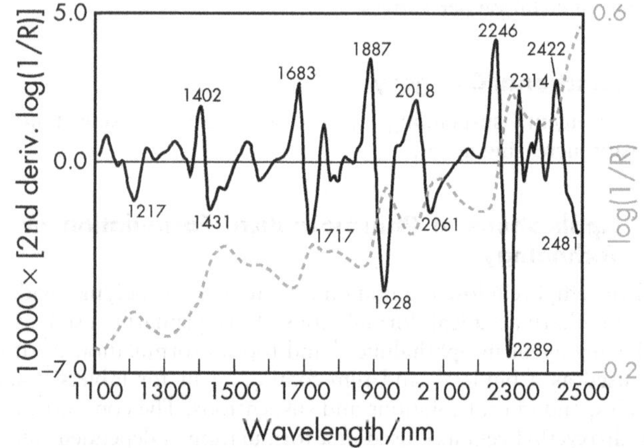

Figure 2: Near-infrared spectrum of hydroxyethyl cellulose measured by reflectance.

Table II: Surface tension (mN/m) of different *Cellosize* (Amerchol Corp.) grades at 25°C.[8]

Concentration of aqueous solution (%w/v)	WP-09	WP-300	QP-4400	QP-52000	QP-100M
0.01	65.7	66.4	66.3	65.9	66.1
0.1	65.4	65.8	65.3	65.4	65.4
1.0	65.1	65.5	65.8	66.1	66.3
2.0	65.0	66.3	67.3	–	–
5.0	64.7	–	–	–	–
10.0	65.9	–	–	–	–

11 Stability and Storage Conditions

Hydroxyethyl cellulose powder is a stable though hygroscopic material. It should be stored in a well-closed container, in a cool, dry place.

Aqueous solutions of hydroxyethyl cellulose are relatively stable at pH 2–12 with the viscosity of solutions being largely unaffected. However, solutions are less stable below pH 5 owing to hydrolysis. At high pH, oxidation may occur.

Increasing the temperature reduces the viscosity of aqueous hydroxyethyl cellulose solutions. However, on cooling, the original viscosity is restored. Solutions may be subjected to freeze–thawing, high-temperature storage, or boiling without precipitation or gelation occurring.

Hydroxyethyl cellulose is subject to enzymatic degradation, with consequent loss in viscosity of its solutions.[10] Enzymes that catalyze this degradation are produced by many bacteria and fungi present in the environment. For prolonged storage, an antimicrobial preservative should therefore be added to aqueous solutions. Aqueous solutions of hydroxyethyl cellulose may also be sterilized by autoclaving.

12 Incompatibilities

Hydroxyethyl cellulose is incompatible with zein and partially compatible with the following water-soluble compounds: casein, gelatin, methylcellulose, polyvinyl alcohol, and starch.[8]

Hydroxyethyl cellulose can be used with a wide variety of water-soluble antimicrobial preservatives. However, sodium pentachlorophenate produces an immediate increase in viscosity when added to hydroxyethyl cellulose solutions.

Hydroxyethyl cellulose has good tolerance for dissolved electrolytes, although it may be salted out of solution when mixed with certain salt solutions. *See also* Section 18.

Hydroxyethyl cellulose is also incompatible with certain fluorescent dyes or optical brighteners, and certain quaternary disinfectants that will increase the viscosity of aqueous solutions.

13 Method of Manufacture

A purified form of cellulose is reacted with sodium hydroxide to produce a swollen alkali cellulose, which is chemically more reactive than untreated cellulose. The alkali cellulose is then reacted with ethylene oxide to produce a series of hydroxyethyl cellulose ethers.[8]

The manner in which ethylene oxide is added to cellulose can be described by two terms, the degree of substitution (DS) and the

Table III: Approximate viscosities of various grades of aqueous *Cellosize* (Amerchol Corp.) solutions at 25°C.[8]

Grade	Concentration (%)	Viscosity[a] (mPa s)
WP/QP 09L	5	75–112
WP-QP 09H	5	113–150
QP 3L	5	215–282
QP 40	2	80–125
WP/QP 300	2	300–400
QP 4400H	2	4800–6000
QP100MH	1	4400–6 000

(a) LVF Brookfield at 25°C.[8]

Table IV: Approximate viscosities of various grades of aqueous *Natrosol 250* (Ashland Aqualon Functional Ingredients) solutions at 25°C.[1]

Type	Viscosity (mPa s) for varying concentrations (% w/v)		
	1%	2%	5%
HHR	3400–5000	–	–
H4R	2600–3300	–	–
HR	1500–2500	–	–
MHR	800–1500	–	–
MR	–	4500–6500	–
KR	–	1500–2500	–
GR	–	150–400	–
ER	–	25–105	–
JR	–	–	150–400
LR	–	–	75–150

molar substitution (MS).[8] The DS designates the average number of hydroxyl positions on the anhydroglucose unit that have been reacted with ethylene oxide. Since each anhydroglucose unit of the cellulose molecule has three hydroxyl groups, the maximum value for DS is 3. MS is defined as the average number of ethylene oxide molecules that have reacted with each anhydroglucose unit. Once a hydroxyethyl group is attached to each unit, it can further react with additional groups in an end-to-end formation. This reaction can continue and there is no theoretical limit for MS.[8]

14 Safety

Hydroxyethyl cellulose is primarily used in ophthalmic and topical pharmaceutical formulations. It is generally regarded as an essentially nontoxic and nonirritant material[11–13] although it is reported in the EPA TSCA Inventory.[14]

Acute and subacute oral toxicity studies in rats have shown no toxic effects attributable to hydroxyethyl cellulose consumption, the hydroxyethyl cellulose being neither absorbed nor hydrolyzed in the rat gastrointestinal tract.[12] However, although used in oral pharmaceutical formulations, hydroxyethyl cellulose has not been approved for direct use in food products;[8] *see* Section 16.

Glyoxal-treated hydroxyethyl cellulose is not recommended for use in oral pharmaceutical formulations or topical preparations that may be used on mucous membranes. Hydroxyethyl cellulose is also not recommended for use in parenteral products.

15 Handling Precautions

Observe normal precautions appropriate to the circumstances and quantity of material handled. Hydroxyethyl cellulose dust may be irritant to the eyes, and eye protection is recommended.

When heated to decomposition, hydroxyethyl cellulose emits acrid smoke and irritating vapors.[14] Hydroxyethyl cellulose powder and dust is combustible and precautions should be taken to minimize the risks of combustion or dust explosion.[15]x

Surfaces covered with a hydroxyethyl cellulose solution can become extremely slippery. Good housekeeping and prompt cleanup of any spills or dust is recommended.[1]

16 Regulatory Status

Included in the FDA Inactive Ingredients Database (buccal films; ophthalmic preparations; oral syrups, capsules, and tablets; otic, topical, and transdermal preparations). Included in nonparenteral medicines licensed in the UK. Included in the Canadian Natural Health Products Ingredients Database.

Hydroxyethyl cellulose is not currently approved for use in food products in Europe or the US, although it is permitted for use in indirect applications such as packaging. This restriction is due to the high levels of ethylene glycol residues that are formed during the manufacturing process.

17 Related Substances

Hydroxyethylmethyl cellulose; hydroxypropyl cellulose; hydroxypropyl cellulose, low-substituted; hypromellose; methylcellulose.

H

18 Comments

Hydroxyethyl cellulose is one of the materials that have been selected for harmonization by the Pharmacopeial Discussion Group. For further information see the General Information Chapter <1196> in the USP35–NF30, the General Chapter 5.8 in PhEur 7.4, along with the 'State of Work' document on the PhEur EDQM website, and also the General Information Chapter 8 in the JP XV.

The following salt solutions will precipitate a 10% w/v solution of *Cellosize* WP-09 and a 2% w/v solution of *Cellosize* WP-440:[8] sodium carbonate 50% and saturated solutions of aluminum sulfate; ammonium sulfate; chromic sulfate; disodium phosphate; magnesium sulfate; potassium ferrocyanide; sodium sulfate; sodium sulfite; sodium thiosulfate; and zinc sulfate.

Natrosol is soluble in most 10% salt solutions, excluding sodium carbonate and sodium sulfate, and many 50% salt solutions with the exception of the following: aluminum sulfate; ammonium sulfate; diammonium phosphate; disodium phosphate; ferric chloride; magnesium sulfate; potassium ferrocyanide; sodium metaborate; sodium nitrate; sodium sulfite; trisodium phosphate; and zinc sulfate. *Natrosol* 150 is generally more tolerant of dissolved salts than is *Natrosol* 250.

Cellosize is manufactured in 11 regular viscosity grades. Hydroxyethyl cellulose grades differ principally in their aqueous solution viscosities which range from 2–20 000 mPa s for a 2% w/v aqueous solution. Two types of *Cellosize* are produced, a WP-type, which is a normal-dissolving material, and a QP-type, which is a rapid-dispersing material. The lowest viscosity grade (02) is available only in the WP-type. Five viscosity grades (09, 3, 40, 300, and 4400) are produced in both WP- and QP-types. Five high-viscosity grades (10000, 15000, 30000, 52000, and 100 M) are produced only in the QP-type. For the standard *Cellosize* grades and types available and their respective viscosity ranges in aqueous solution, see Table III.

Natrosol 250 has a degree of substitution of 2.5 and is produced in 10 viscosity types. The suffix 'R' denotes that *Natrosol* has been surface-treated with glyoxal to aid in solution preparation; *see* Table IV.

19 Specific References

1 Ashland Aqualon Functional Ingredients. Technical literature: *Natrosol* hydroxyethylcellulose – a nonionic water-soluble polymer: Physical and chemical properties, 1999.

2 Grove J, *et al.* The effect of vehicle viscosity on the ocular bioavailability of L-653328. *Int J Pharm* 1990; **66**: 23–28.

3 Gauger LJ. Hydroxyethylcellulose gel as a dinoprostone vehicle. *Am J Hosp Pharm* 1984; **41**: 1761–1762.

4 Delonca H, *et al.* [Influence of temperature on disintegration and dissolution time of tablets with a cellulose component as binder.] *J Pharm Belg* 1978; **33**: 171–178[in French].
5 Kovács B, Merényi G. Evaluation of tack behavior of coating solutions. *Drug Dev Ind Pharm* 1990; **16**(15): 2303–2323.
6 Li J, Xu Z. Physical characterization of a chitosan-based hydrogel delivery system. *J Pharm Sci* 2002; **91**(7): 1669–1677.
7 ShinEtsu SE Tylose GmbH & Co. *Tylose*: Cellulose ethers for personal care products, October 2010.
8 The Dow Chemical Company. *Cellosize*: Hydroxyethyl cellulose, August 2005.
9 Wirick MG. Study of the substitution pattern of hydroxyethyl cellulose and its relationship to enzymic degradation. *J Polym Sci* 1968; 6(Part A-1): 1705–1718.
10 Anonymous. Final report on the safety assessment of hydroxyethylcellulose, hydroxypropylcellulose, methylcellulose, hydroxypropyl methylcellulose and cellulose gum. *J Am Coll Toxicol* 1986; 5(3): 1–60.
11 Durand-Cavagna G, *et al.* Corneal toxicity studies in rabbits and dogs with hydroxyethyl cellulose and benzalkonium chloride. *Fundam Appl Toxicol* 1989; **13**: 500–508.
12 Ashland Aqualon Functional Ingredients. *Natrosol* 250: Hydroxyethylcellulose - Summary of toxicological investigations, 2009.
13 Lewis RJ, ed. *Sax's Dangerous Properties of Industrial Materials*, 11th edn. New York: Wiley, 2004.
14 The Dow Chemical Company. *Cellosize*: Hydroxyethyl cellulose - Explosibility data, March 2005.

20 General References

Chauveau C, *et al.* [*Natrosol 250* part 1: characterization and modeling of rheological behavior.] *Pharm Acta Helv* 1986; **61**: 292–297[in French].
Doelker E. Cellulose derivatives. *Adv Polym Sci* 1993; **107**: 199–265.
European Directorate for the Quality of Medicines and Healthcare (EDQM). European Pharmacopoeia – State Of Work Of International Harmonisation. *Pharmeuropa* 2011; 23(4): 713–714. http://www.edqm.eu/site/-614.html (accessed 28 September 2011).
Haugen P, *et al.* Steady shear flow properties, rheological reproducibility and stability of aqueous hydroxyethylcellulose dispersions. *Can J Pharm Sci* 1978; **13**: 4–7.
Klug ED. Some properties of water-soluble hydroxyalkyl celluloses and their derivatives. *J Polym Sci* 1971; **36**(Part C): 491–508.
Rufe RG. Cellulose polymers in cosmetics and toiletries. *Cosmet Perfum* 1975; 90(3): 93–9499–100.

21 Author

KP Hapgood.

22 Date of Revision

1 March 2012.

Hydroxyethylmethyl Cellulose

1 Nonproprietary Names

BP: Hydroxyethylmethylcellulose

PhEur: Methylhydroxyethylcellulose

2 Synonyms

Cellulose, 2-hydroxyethyl methyl ester; *Culminal MHEC*; HEMC; hydroxyethyl methylcellulose; hymetellose; MHEC; methylhydroxyethylcellulosum; *Tylopur MH*; *Tylopur MHB*; *Tylose MB*; *Tylose MH*; *Tylose MHB*.

3 Chemical Name and CAS Registry Number

Hydroxyethylmethylcellulose [9032-42-2]

4 Empirical Formula and Molecular Weight

The PhEur 7.4 describes hydroxyethylmethyl cellulose as a partly O-methylated and O-(2-hydroxyethylated) cellulose. Various different grades are available, which are distinguished by appending a number indicative of the apparent viscosity in millipascal seconds (mPa s) of a 2% w/v solution measured at 20°C.

5 Structural Formula

See Section 4.

6 Functional Category

Coating agent; suspending agent; tablet and capsule binder; viscosity-increasing agent.

7 Applications in Pharmaceutical Formulation or Technology

Hydroxyethylmethyl cellulose is used as an excipient in a wide range of pharmaceutical products, including oral tablets and suspensions, and topical gel preparations.[1] It has similar properties to methylcellulose, but the hydroxyethyl groups make it more readily soluble in water and solutions are more tolerant of salts and have a higher coagulation temperature.

8 Description

Hydroxyethylmethyl cellulose occurs as a white, yellowish-white or grayish-white powder or granules, hygroscopic after drying.

9 Pharmacopeial Specifications

See Table I.

Table I: Pharmacopeial specifications for hydroxyethylmethyl cellulose.

Test	PhEur 7.4
Identification	+
Characters	+
Appearance of solution	+
pH	5.5–8.0
Apparent viscosity	75–140% of value stated on label
Chlorides	≤0.5%
Heavy metals	0.002%
Loss on drying	≤10.0%
Sulfated ash	≤1.0%

10 Typical Properties

Acidity/alkalinity pH = 5.5–8.0 (2% w/v aqueous solution)
Moisture content ≤10%
Solubility Hydroxyethylmethyl cellulose is practically insoluble in hot water (above 60°C), acetone, ethanol (95%), ether, and toluene. It dissolves in cold water to form a colloidal solution.
Viscosity (dynamic) 8.5–11.5 mPa s (8.5–11.5 cP) for *Culminal MHEC 8000* 2% w/v aqueous solution at 20°C.

11 Stability and Storage Conditions

Hydroxyethylmethyl cellulose is hygroscopic and should therefore be stored under dry conditions away from heat.

12 Incompatibilities

—

13 Method of Manufacture

—

14 Safety

Hydroxyethylmethyl cellulose is used as an excipient in various oral and topical pharmaceutical preparations, and is generally regarded as an essentially nontoxic and nonirritant material.

See Hypromellose for further information.

15 Handling Precautions

Observe normal precautions appropriate to the circumstances and quantity of the material handled. Eye protection and gloves are recommended.

16 Regulatory Status

GRAS listed. Included in nonparenteral medicines licensed in Europe (oral suspensions, tablets, and topical preparations).

17 Related Substances

Ethylcellulose; hydroxyethyl cellulose; hypromellose; methylcellulose.

18 Comments

—

19 Specific References

1 Bogdanova S. Model suspensions of indomethacin 'solvent deposited' on cellulose polymers. *Pharmazie* 2000; 55(11): 829–832.

20 General References

Adden R, *et al.* Comprehensive analysis of the substituent distribution in the glucosyl units and along the polymer chain of hydroxyethylmethyl celluloses and statistical evaluation. *Anal Chem* 2006; 78(4): 1146–1157.

Ashland Aqualon Functional Ingredients. Product data sheet: *Culminal*, April 2004.

21 Author

PJ Sheskey.

22 Date of Revision

1 March 2012.

Hydroxyethylpiperazine Ethane Sulfonic Acid

1 Nonproprietary Names

None adopted.

2 Synonyms

HEPES; 1-[4-(2-hydroxyethyl)-1-piperazinyl]ethane-2-sulfonic acid; 4-(2-hydroxyethyl)-1-piperazineethanesulfonic acid; *N*-(2-hydroxyethyl)piperazine-*N*'-2-ethanesulfonic acid.

3 Chemical Name and CAS Registry Number

2-[4-(2-Hydroxyethyl)piperazin-1-yl]ethanesulfonic acid [7365-45-9]

4 Empirical Formula and Molecular Weight

$C_8H_{18}N_2O_4S$ 238.30

5 Structural Formula

SO_3^- N NH^+ HO

6 Functional Category

Buffering agent.

7 Applications in Pharmaceutical Formulation or Technology

Hydroxyethylpiperazine ethane sulfonic acid (HEPES) is used as a buffering agent in injections.[1–3]

8 Description

HEPES occurs as an odorless white crystalline powder.

9 Pharmacopeial Specifications

–

10 Typical Properties

Acidity/alkalinity pH 5.0–6.5 (1 M aqueous solution)
Dissociation constants

$pK_{a1} \sim 3$;

$pK_{a2} = 7.55$ (20°C);

$\Delta pK_a/°C = -0.014$.

Melting point Approx. 236°C, with decomposition.
Solubility Soluble in water; saturated solution is 2.25 M at 0°C.

11 Stability and Storage Conditions

HEPES is hygroscopic. Store in the original container in a cool, dry place, protected from direct sunlight. Keep the container tightly closed and sealed until ready for use.

H

12 Incompatibilities

HEPES is incompatible with strong oxidizing agents.

13 Method of Manufacture

HEPES is synthetically manufactured via a sulfonation process from sodium bromoethanesulfonate and *N*-2–hydroxyethylpiperazine.

14 Safety

HEPES can cause irritation to the eyes, respiratory system, and skin if improperly handled. May be harmful if swallowed. Ingestion may cause gastrointestinal irritation, nausea, vomiting, and diarrhea. Hazardous thermal decomposition products may include carbon dioxide, carbon monoxide, nitrogen oxides, and sulfur oxides. Toxicological properties have not been fully investigated.

LD_{50} (oral, quail): > 0.316 g/kg

15 Handling Precautions

Observe normal precautions appropriate to the circumstances and quantity of the material handled. Avoid skin and eye contact; eye goggles should be worn, or a full face shield where splashing may occur. There are no specific fire or explosion hazards for HEPES.

16 Regulatory Status

Included in the FDA Inactive Ingredients Database (IV, injection). Included in BP 2012 as a reagent.

17 Related Substances

–

18 Comments

HEPES is widely used in cell culture as a buffering agent where maintenance of pH near or at physiological pH is desired.[1–3] HEPES has been shown to be biologically active *in vitro* as a result of the inhibition of taurine uptake by cultured cells.[4] In addition, cell death has been shown to occur faster upon light exposure in the presence of HEPES.[5,6] HEPES has also been shown to adversely impact *in vitro* pregnancy and implantation rates due to a higher rate of triploid and degenerated oocytes after fertilization.[7]

The EINECS number for HEPES is 230-907-9. The PubChem Compound ID (CID) for HEPES is 23830.

19 Specific References

1 Luo S, *et al.* Effect of HEPES buffer on the uptake and transport of P-glycoprotein substrates and large neutral amino acids. *Mol Pharm* 2010; 7(2): 412–420.
2 Baicu SC, Taylor MJ. Acid–base buffering in organ preservation solutions as a function of temperature: new parameters for comparing buffer capacity and efficiency. *Cryobiology* 2002; **45**(1): 33–48.
3 Swain JE, Pool TB. New pH-buffering system for media utilized during gamete and embryo manipulations for assisted reproduction. *Reprod Biomed Online* 2009; **18**(6): 799–810.
4 Petegnief V, *et al.* Taurine analog modulation of taurine uptake by two different mechanisms in cultured glial cells. *Biochem Pharmacol* 1995; 49(3): 399–410.
5 Lepe-Zuniga JL, *et al.* Toxicity of light-exposed HEPES media. *J Immunol Methods* 1987; **103**(1): 145.
6 Zigler JS, *et al.* Analysis of the cytotoxic effects of light-exposed HEPES-containing culture medium. *In Vitro Cell Dev Biol* 1985; **21**(5): 282–287.
7 Morgia F, *et al.* Use of a medium buffered with *N*-hydroxyethylpiperazine-*N*-ethanesulfonate (HEPES) in intracytoplasmic sperm injection procedures is detrimental to the outcome of *in vitro* fertilization. *Fertil Steril* 2006; **85**(5): 1415–1419.

20 General References

Acros Organics (part of Thermo Fisher Scientific). Material safety data sheet: HEPES, 2010.

21 Author

J Rexroad.

22 Date of Revision

1 February 2011.

 Hydroxypropyl Betadex

1 Nonproprietary Names

BP: Hydroxypropylbetadex
PhEur: Hydroxypropylbetadex
USP–NF: Hydroxypropyl Betadex

2 Synonyms

Cavasol W7; 2-hydroxypropyl-β-cyclodextrin; 2-hydroxypropyl cyclomaltoheptaose; hidroksipropilbetadeksas; hydoxipropylbetadex; hydroksipropylbetadeksi; hydroxypropylbetadeksum; hydroxypropylbetadexum; *Kleptose HPB*.

3 Chemical Name and CAS Registry Number

β-Cyclodextrin, 2-hydroxypropyl ether [94035-02-6] and [128446-35-5]

4 Empirical Formula and Molecular Weight

$C_{42}H_{70}O_{35}(C_3H_6O)_x$ (where x = 7 molar substitution)

The molecular weight depends on the degree of substitution. The molecular weight of unsubstituted β-cyclodextrin is 1134.98.

5 Structural Formula

Hydroxpropyl betadex is a partially substituted ether of β-cyclodextrin. USP35–NF30 requires that the molar substitution is between 0.4 and 1.5 hydroxypropyl groups per anhydroglucose unit.

R = H or $CH_2CH(CH_3)OH$

6 Functional Category

Complexing agent; modified-release agent; solubilizing agent; tonicity agent.

7 Applications in Pharmaceutical Formulation or Technology

Hydroxypropyl betadex has been widely investigated in pharmaceutics and has principally been used as a solubilizer for hydrophobic molecules in oral liquids,[1,2] oral solids,[3] parenterals,[4,5] pressurized metered dose inhalers,[6] dry powder inhalers,[7] and topical formulations.[8] It has also been shown to act as a stabilizer during processing[9] and storage of formulations.[10]

Hydroxypropyl betadex inclusion complexes have been reported to show mechanical properties distinct from the pure materials.[11] The reported advantage of hydroxypropyl betadex over unsubstituted β-cyclodextrin is its greater water solubility.[3]

See also Section 18.

8 Description

Hydroxypropyl betadex occurs as a white or almost white, amorphous or crystalline powder.

9 Pharmacopeial Specifications

See Table I.

Table I: Pharmacopeial specifications for hydroxypropyl betadex.

Test	PhEur 7.4	USP35–NF30
Identification	+	+
Characters	+	–
Clarity of solution	–	+
Appearance of solution	+	–
Microbial limits		
Aerobic microbial count	$\leq 10^3$ cfu/g	≤1000 cfu/g
Yeasts and molds	$\leq 10^2$ cfu/g	≤100 cfu/g
Heavy metals	≤20 ppm	≤20 μg/g
Loss on drying	≤10.0%	≤10.0%
Conductivity	$\leq 200\,\mu S{\cdot}cm^{-1}$	$\leq 200\,\mu S{\cdot}cm^{-1}$
Related substances	+	+
Sterility	–	+
Bacterial endotoxins	<10 IU/g[a]	+
Molar substitution	+	+
Propylene oxide	–	≤0.0001%

(a) If intended for parenteral use.

10 Typical Properties

Acidity/alkalinity pH = 5–8 of a 20 g/L solution at 20°C for *Cavasol W7 HP Pharma*

Density (bulk)

~0.4 g/cm³ for *Cavasol W7 HP*;

0.2–0.3 g/cm³ for *Cavasol W7 HP Pharma.*

Ignition temperature

420°C for *Cavasol W7 HP*;

>400°C for *Cavasol W7 HP Pharma.*

Melting point 278°C; 120–160°C for *Cavasol W7 HP*

Specific rotation $[\alpha]_D^{25}$ = +140° to +145°

Solubility Freely soluble in water and propylene glycol. Soluble in ethanol, methanol, dimethyl sulfoxide and dimethylformamide.

2300 g/L water solubility at 24°C for *Cavasol W7 HP*;

2300 g/L water solubility at 25°C for *Cavasol W7 HP Pharma.*

Water content Typically <3.0%.

11 Stability and Storage Conditions

Store in well-closed containers.

12 Incompatibilities

The activity of some antimicrobial preservatives in aqueous solution can be reduced in the presence of hydroxypropyl betadex.[12–14]

13 Method of Manufacture

Hydroxypropyl betadex is prepared by the treatment of an alkaline solution of β-cyclodextrin with propylene oxide. The substitution pattern can be influenced by varying the pH. Formation of O-6 and O-2 substituted products is favored by high and low alkali concentration, respectively. The mixture of products produced may be refined by preparative chromatography.[15]

14 Safety

The pharmaceutical toxicology of hydroxypropyl betadex has been reviewed,[16] and in general, the material was found to be of low toxicity. It has been suggested that hydroxypropyl betadex may have a synergistic toxic effect with, for example, carcinogens, by increasing their solubility and thus bioavailability.[17]

15 Handling Precautions

Observe normal precautions appropriate to the circumstances and quantity of the material handled.

16 Regulatory Status

Included in oral and parenteral medicinal products. Included in an injectable preparation licensed in the UK for intramuscular or intravenous administration.

17 Related Substances

Cyclodextrins; 3-hydroxypropyl-β-cyclodextrin; sulfobutylether β-cyclodextrin

3-Hydroxypropyl-β-cyclodextrin

Synonyms 3-HP-β-CD.

Appearance White crystalline powder.

Solubility Greater than 1 in 2 parts of water at 25°C.

Surface tension 70.0–71.0 mN/m (70–71 dynes/cm) at 25°C.

Comments Used in applications similar to those for β-cyclodextrin. However, as it is not nephrotoxic it has been suggested for use in parenteral formulations. The degree of substitution of hydroxypropyl groups can vary.

18 Comments

Hydroxypropyl betadex has been investigated as an absorption (permeation) enhancer in oral,[18] transdermal,[19] and nasal[20] systems. It was found to be effective in increasing penetration in some studies, although the mechanism of action may be compound specific.

19 Specific References

1 Pitha J, *et al.* Hydroxypropyl-β-cyclodextrin: preparation and characterization; effects on solubility of drugs. *Int J Pharm* 1986; **29**: 73–82.
2 Strickley RG. Solubilizing excipients in oral and injectable formulations. *Pharm Res* 2004; **21**: 201–230.
3 Albers E, Müller BW. Cyclodextrin derivatives in pharmaceutics. *Crit Rev Ther Drug Carrier Syst* 1995; **12**: 311–337.
4 Buchanan CM, *et al.* Pharmacokinetics of itraconazole after intravenous and oral dosing of itraconazole–cyclodextrin formulations. *J Pharm Sci* 2007; **96**: 3100–3116.
5 Holvoet C, *et al.* Development of an omeprazole parenteral formulation with hydroxypropyl-beta-cyclodextrin. *Pharm Dev Technol* 2007; **12**: 327–336.
6 Williams RO III, Liu J. Influence of formulation technique for hydroxypropyl-beta-cyclodextrin on the stability of aspirin in HFA 134a. *Eur J Pharm Biopharm* 1999; **47**: 145–152.
7 Ungaro F, *et al.* Cyclodextrins in the production of large porous particles: development of dry powders for the sustained release of insulin to the lungs. *Eur J Pharm Sci* 2006; **28**: 423–432.
8 Godwin DA, *et al.* Using cyclodextrin complexation to enhance secondary photoprotection of topically applied ibuprofen. *Eur J Pharm Biopharm* 2006; **62**: 85–93.
9 Branchu S, *et al.* Hydroxypropyl-beta-cyclodextrin inhibits spray-drying-induced inactivation of beta-galactosidase. *J Pharm Sci* 1999; **88**: 905–911.
10 Scalia S, *et al.* Inclusion complexation of the sunscreen agent 2-ethylhexyl-*p*-dimethylaminobenzoate with hydroxypropyl-beta-cyclodextrin: effect on photostability. *J Pharm Pharmacol* 1999; **51**: 1367–1374.
11 Suihko E, *et al.* Deformation behaviors of tolbutamide, hydroxypropyl-beta-cyclodextrin, and their dispersions. *Pharm Res* 2000; **17**: 942–948.
12 Loftsson T, *et al.* Interactions between preservatives and 2-hydroxypropyl-β-cyclodextrin. *Drug Dev Ind Pharm* 1992; **18**(13): 1477–1484.
13 Lehner SJ, *et al.* Interactions between *p*-hydroxybenzoic acid esters and hydroxypropyl-β-cyclodextrin and their antimicrobial effect against *Candida albicans*. *Int J Pharm* 1993; **93**: 201–208.
14 Lehner SJ, *et al.* Effect of hydroxypropyl-β-cyclodextrin on the antimicrobial action of preservatives. *J Pharm Pharmacol* 1994; **46**: 186–191.
15 Pitha J, *et al.* Distribution of substituents in 2-hydroxypropyl ethers of cyclomaltoheptaose. *Carbohydr Res* 1990; **200**: 429–435.
16 Gould S, Scott RC. 2-Hydroxypropyl-beta-cyclodextrin (HP-beta-CD): a toxicology review. *Food Chem Toxicol* 2005; **43**: 1451–1459.
17 Horský J, Pitha J. Hydroxypropyl cyclodextrins: potential synergism with carcinogens. *J Pharm Sci* 1996; **85**: 96–100.
18 Maestrelli F, *et al.* Microspheres for colonic delivery of ketoprofen-hydroxypropyl-beta-cyclodextrin complex. *Eur J Pharm Sci* 2008; **34**: 1–11.
19 Tanaka M. Effect of 2-hydroxypropyl-beta-cyclodextrin on percutaneous absorption of methyl paraben. *J Pharm Pharmacol* 1995; **47**: 897–900.
20 Chavanpatil MD, Vavia PR. The influence of absorption enhancers on nasal absorption of acyclovir. *Eur J Pharm Biopharm* 2004; **57**: 483–487.

20 General References

Loftsson T, Brewster ME. Pharmaceutical applications of cyclodextrins: basic science and product development. *J Pharm Pharmacol* 2010; **62**: 1607–1621.

Wacker. Material safety data sheet No. 60007000: *Cavasol W7 HP*, 2010.

Wacker. Material safety data sheet No. 60015004: *Cavasol W7 HP Pharma*, 2010.

21 Author

W Cook.

22 Date of Revision

1 March 2012.

Hydroxypropyl Cellulose

1 Nonproprietary Names

BP: Hydroxypropylcellulose
JP: Hydroxypropylcellulose
PhEur: Hydroxypropylcellulose
USP–NF: Hydroxypropyl Cellulose

2 Synonyms

AeroWhip; cellulose, hydroxypropyl ether; *Coatcel*; E463; hydroxypropylcellulosum; hyprolose; *Klucel*; *Nisso HPC*; oxypropylated cellulose.

3 Chemical Name and CAS Registry Number

Cellulose, 2-hydroxypropyl ether [9004-64-2]

4 Empirical Formula and Molecular Weight

The PhEur 7.4 and USP35–NF30 describe hydroxypropyl cellulose as a partially substituted poly(hydroxypropyl) ether of cellulose. It may contain not more than 0.6% of silica (SiO_2) or another suitable anticaking agent. Hydroxypropyl cellulose is commercially available in a number of different grades that have various solution viscosities and molecular weights; *see also* Section 10.

5 Structural Formula

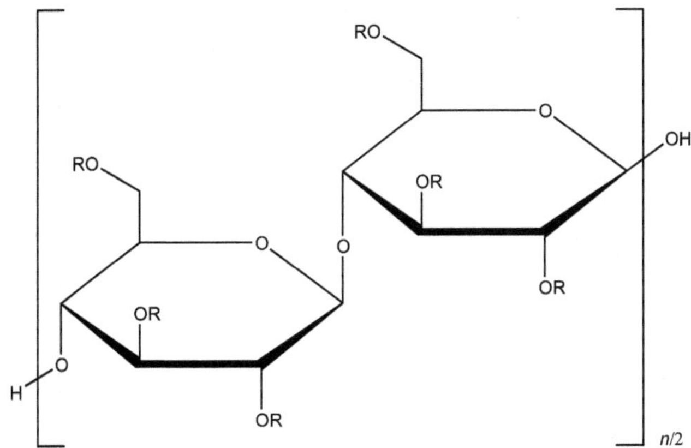

R is H or $[CH_2CH(CH_3)O]_mH$ where m is a common integral number of cellulose derivatives.

Hydroxypropyl cellulose is an ether of cellulose where some of the hydroxyl groups of the cellulose have been hydroxypropylated forming $-OCH_2CH(OH)CH_3$ groups. The average number of hydroxyl groups in the glucose ring substituted is referred to as the degree of substitution (DS). Complete substitution would provide a DS of 3.0. Because the hydroxypropyl group added contains a hydroxyl group, this can also be etherified during preparation of hydroxypropyl cellulose. When this occurs, the number of moles of hydroxypropyl groups per glucose ring, or moles of substitution (MS), can be higher than 3. Hydroxypropyl cellulose must have an MS value of approximately 4 in order to have good solubility in water.

6 Functional Category

Coating agent; emulsifying agent; film-forming agent; modified-release agent; suspending agent; tablet and capsule binder; viscosity-increasing agent.

7 Applications in Pharmaceutical Formulation or Technology

Hydroxypropyl cellulose is widely used in oral and topical pharmaceutical formulations; *see* Table I.

In oral products, hydroxypropyl cellulose is primarily used in tableting as a binder,[1] film-coating,[2] and extended-release-matrix former.[3–5] Concentrations of hydroxypropyl cellulose of 2–6% w/w may be used as a binder in either wet-granulation or dry, direct-compression tableting processes.[6–10] Concentrations of 15–35% w/w of hydroxypropyl cellulose may be used to produce tablets with an extended drug release.[11] The release rate of a drug increases with decreasing viscosity of hydroxypropyl cellulose. Blends of hydroxypropyl cellulose and other cellulosic polymers have been used to improve wet granulation characteristics and tableting characteristics, as well as to achieve better control and manipulation of the rate of drug release.[12–15] As an alternative technology to wet granulation, dry granulation and direct compression of hydroxypropyl cellulose formulations have been reported to exhibit acceptable tableting and flow characteristics for application in extended-release matrix tablets.[16,17]

Solubility of hydroxypropyl cellulose in water as well as in polar organic solvents offers a wide choice in preparing solutions for casting films or for coating purposes. Typically, a 5% w/w solution of hydroxypropyl cellulose may be used to film-coat tablets. Aqueous solutions containing hydroxypropyl cellulose together with an amount of methyl cellulose or ethanolic solutions have been used.[18–22] Hydroxypropyl cellulose cast-films show excellent flexibility and heat sealability, lack tackiness, and provide barrier to oils and fats. It is not necessary to add plasticizers to cast films. In extruded films, plasticizers provide desirable die lubrication, and a reduction in melt viscosity. Plasticizer levels of 5% or less are recommended. Hydroxypropyl cellulose shows good compatibility with several plasticizers including propylene glycol, glycerine, triethylolpropane, and polyethylene glycols. Stearic acid or palmitic acids have also been used as plasticizers for hydroxypropyl cellulose especially in ethanolic solutions. Low-substituted hydroxypropyl cellulose is used as a tablet disintegrant; *see* Hydroxypropyl Cellulose, Low-substituted.

Hydroxypropyl cellulose is also used in microencapsulation processes,[23] as a thickening agent. In topical formulations, hydroxypropyl cellulose is used in transdermal patches and also ophthalmic preparations[24–27] to aid moisture retention, stabilize the tear film, and lubricate the eye.[27]

Hydroxypropyl cellulose is also used in cosmetics and in food products as an emulsifier and stabilizer.[28]

Table I: Typical uses of hydroxypropyl cellulose.

Use	Concentration (%)
Extended release-matrix former	15–35
Tablet binder	2–6
Tablet film coating	5

8 Description

Hydroxypropyl cellulose occurs as a white to off-white or slightly yellow-colored, odorless and tasteless powder.

SEM 1: Excipient: hydroxypropyl cellulose (*Klucel*); manufacturer: Ashland Aqualon Functional Ingredients; magnification: 60×.

SEM 2: Excipient: hydroxypropyl cellulose (*Klucel*); manufacturer: Ashland Aqualon Functional Ingredients; magnification: 600×.

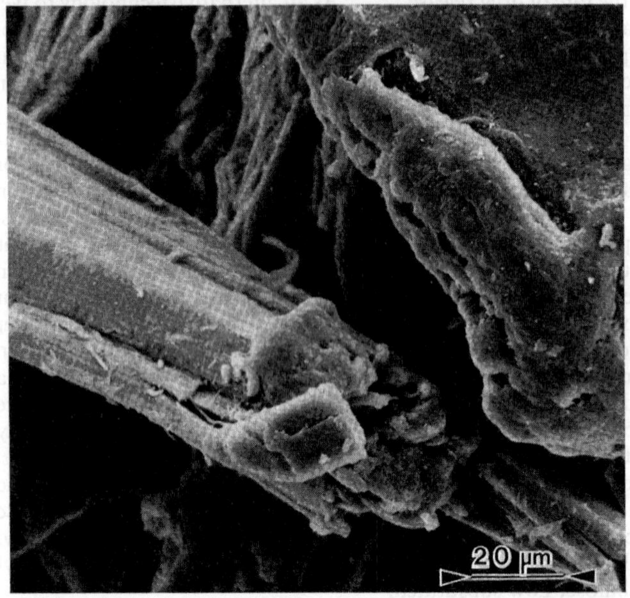

9 Pharmacopeial Specifications

See Table II. *See also* Section 18.

10 Typical Properties

Angle of repose ≈50°
Density (bulk) ≈0.5 g/cm^3
Interfacial tension 12.5 mN/m for a 0.1% w/w aqueous solution compared with mineral oil.
Melting point Softens at 130°C; chars at 260–275°C.
Moisture content Hydroxypropyl cellulose absorbs moisture from the atmosphere; the amount of water absorbed depends upon the initial moisture content and the temperature and relative humidity of the surrounding air. Typical equilibrium moisture content values at 25°C are 4% w/w at 50% relative humidity and 12% w/w at 84% relative humidity. *See* Table III. *See also* Figure 1.

Table II: Pharmacopeial specifications for hydroxypropyl cellulose.

Test	JP XV	PhEur 7.4	USP35–NF30
Identification	+	+	+
Characters	—	+	—
Apparent viscosity	—	+	+
Appearance of solution	+	+	—
pH[a]	5.0–7.5	5.0–8.5	5.0–8.0
Loss on drying	≤5.0%	≤7.0%	≤5.0%
Residue on ignition	≤0.5%	—	≤0.2%
Sulfated ash	—	≤1.6%	—
Arsenic	≤2 ppm	—	—
Chlorides	≤0.142%	≤0.5%	—
Lead	—	—	≤10 ppm
Heavy metals	≤20 ppm	≤20 ppm	≤20 ppm
Silica	—	≤0.6%	—
Sulfate	≤0.048%	—	—
Assay of hydroxypropoxy groups ($-OC_3H_6OH$)	53.4–77.5%	—	≤80.5%

(a) pH: 1 g in 50 mL for JPXV; 1 g in 100 g for PhEur 7.4; 1 g in 100 mL for USP35–NF30.

Table III: Moisture content of *Klucel* (Aqualon).

Grade	Molecular weight[a]	Moisture (%)
Klucel EF	≈80 000	0.59
Klucel LF	≈95 000	2.21
Klucel JF	≈140 000	1.44
Klucel GF	≈370 000	1.67
Klucel MF	≈850 000	1.52
Klucel HF	≈1 150 000	4.27

(a) Weight average molecular weight determined by size exclusion chromatography.

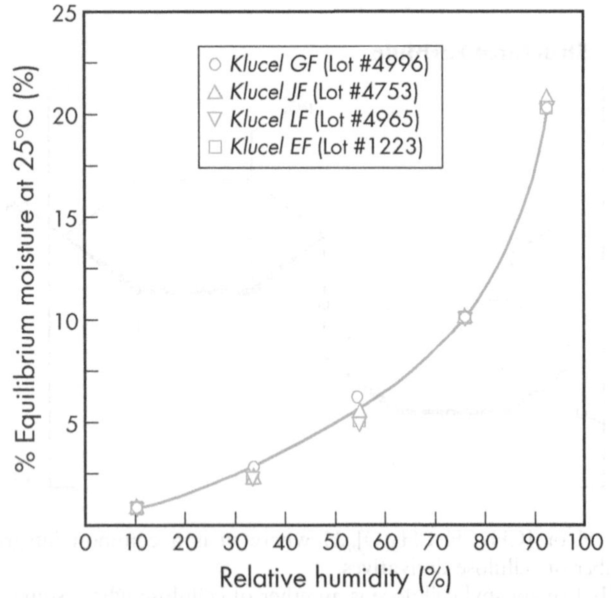

Figure 1: Equilibrium moisture content of various grades of hydroxypropyl cellulose (*Klucel*, Ashland Specialty Ingredients).

Particle size distribution

Klucel (regular grind), minimum 85% (minimun 80% for *Klucel H* grades) through a US #30 mesh (590 μm), and minimum 99% through a US #20 mesh (840 μm);

Klucel (fine-grind), minimum 99.9% through a US #60 mesh (250 μm), minimum 90% through a US #80 mesh (177 μm), and minimum 80% through a US #100 mesh (149 μm);

Nisso HPC-L (regular type): 99% through a US #40 mesh sieve (350 μm);

Nisso HPC-L (fine powder type): 99% through a US #100 mesh sieve (150 μm).

Refractive index n_D^{20} = 1.3353 for a 2% w/v aqueous solution.

Solubility

Hydroxypropyl cellulose is freely soluble in water below 38°C, forming a smooth, clear, colloidal solution. In hot water, it is insoluble and is precipitated as a highly swollen floc at a temperature between 40°C and 45°C; this precipitation is reversible. The polymer dissolves upon cooling the system below 40°C with restoration of original viscosity.

Hydroxypropyl cellulose is soluble in many cold or hot polar organic solvents such as dimethyl formamide; dimethyl sulfoxide; dioxane; ethanol (95%); methanol; propan-2-ol (95%); and propylene glycol. There is no tendency for precipitation in hot organic solvents. As a general rule, hydroxypropyl cellulose should be added to the solution before adding other soluble ingredients. The dissolved materials tend to compete for the solvent and slow down the dissolution rate of hydroxypropyl cellulose. However, the grade of hydroxypropyl cellulose can have a marked effect upon solution quality in some organic liquids that are borderline solvents, such as acetone; butyl acetate; cyclohexanol; dichloromethane; lactic acid; methyl acetate; methyl ethyl ketone; propan-2-ol (99%); and *tert*-butanol. The higher-viscosity grades of hydroxypropyl cellulose tend to produce slightly inferior solutions. However, the solution quality in borderline solvents can often be greatly improved by the use of small quantities (5–15%) of a cosolvent. For example, dichloromethane is a borderline solvent for *Klucel HF* and solutions have a granular texture, but a smooth solution may be produced by adding 10% methanol.

Hydroxypropyl cellulose is compatible with a number of high-molecular-weight, high-boiling waxes and oils, and can be used to modify certain properties of these materials. Examples of materials that are good solvents for hydroxypropyl cellulose at an elevated temperature are acetylated monoglycerides, glycerides, pine oil, polyethylene glycol, and polypropylene glycol.

Hydroxypropyl cellulose is practically insoluble in aliphatic hydrocarbons, aromatic hydrocarbons, carbon tetrachloride, petroleum distillates, glycerin, and oils.

Specific gravity 1.2224 for particles; 1.0064 for a 2% w/v aqueous solution at 20°C; 1.010 for a 2% aqueous solution at 30°C.

Spectroscopy

IR spectra *see* Figure 2.

NIR spectra *see* Figure 3.

Surface tension *see* Table IV. Similar to other polymers, hydroxypropyl cellulose reduces the surface tension of water.

Table IV: Surface tension (mN/m) of aqueous solutions of *Nisso HPC* (Nippon Soda Co. Ltd.) at 20°C.

Grade	Surface tension (mN/m) at 20°C for aqueous solutions of stated concentration			
	0.01%	0.1%	1.0%	10.0%
Nisso HPC-L	51.0	49.1	46.3	45.8
Nisso HPC-M	54.8	49.7	46.3	—

Viscosity (dynamic) A wide range of viscosity types are commercially available; *see* Table V and VI. Solutions should be prepared by gradually adding the hydroxypropyl cellulose to a vigorously stirred solvent. Care should be taken not to create an excessive amount of foam; antifoaming agents may be used to reduce the amount of foam, or a vacuum may be used to dissipate the foam in a timely manner. Increasing concentration produces solutions of increased viscosity.

See also Section 11 for information on solution stability.

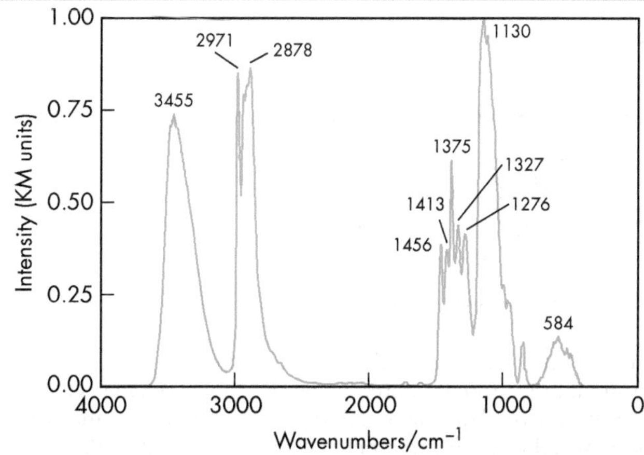

Figure 2: Infrared spectrum of hydroxypropyl cellulose measured by diffuse reflectance. Adapted with permission of Informa Healthcare.

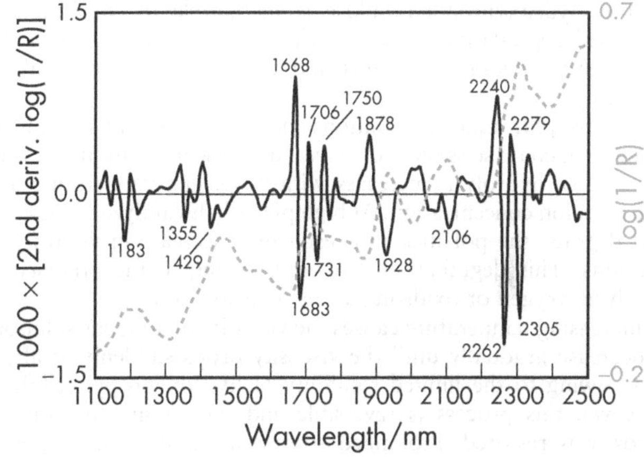

Figure 3: Near-infrared spectrum of hydroxypropyl cellulose measured by reflectance.

Table V: Viscosity of aqueous solutions of *Klucel* (Ashland Aqualon Functional Ingredients) at 25°C.

Grade	Viscosity (mPa s) of various aqueous solutions of stated concentration			
	1%	2%	5%	10%
Klucel HF	1500–3000	—	—	—
Klucel MF	—	4000–6500	—	—
Klucel GF	—	150–400	—	—
Klucel JF	—	—	150–400	—
Klucel LF	—	—	75–150	—
Klucel EF	—	—	—	300–600

11 Stability and Storage Conditions

Hydroxypropyl cellulose powder stability is affected by the initial molecular weight of the polymer and storage conditions. High viscosity grades are more susceptible to viscosity loss over time. Testing after one year and quarterly thereafter is recommended. Hydroxypropyl cellulose also tends to absorb moisture depending

Table VI: Viscosity of aqueous solutions of *Nisso HPC* (Nippon Soda Co. Ltd.) at 20°C.

Grades[a]	Molecular weight	Viscosity (mPa s) of 2% aqueous solution
SSL	40 000	2.0–2.9
SL	100 000	3.0–5.9
L	140 000	6.0–10.0
M	620 000	150–400
H	910 000	1000–4000

(a) Regular and fine powder grades.

on the relative humidity and the temperature of the environment, and should therefore be stored in tightly closed containers and in a dry atmosphere. Aqueous hydroxypropyl cellulose solutions have optimum stability when the pH is maintained at 6.0–8.0, and also when the solution is protected from light, heat, and the action of microorganisms. The viscosity of aqueous solutions, however, remains unchanged over a wide pH range (2–11). Being water-soluble, hydroxypropyl cellulose is susceptible to chemical and biological degradation, which results in a reduction in molecular weight of the dissolved polymer and a decrease in viscosity of the solution.

At low pH, aqueous solutions may undergo acid hydrolysis, resulting in chain scission and hence a decrease in solution viscosity. The rate of hydrolysis increases with increasing temperature and hydrogen ion concentration. At high pH, alkali-catalyzed oxidation may degrade the polymer and result in a decrease in viscosity of solutions. This degradation can occur owing to the presence of dissolved oxygen or oxidizing agents in a solution.

Increasing temperature causes the viscosity of aqueous solutions to decrease gradually until the viscosity drops suddenly at about 45°C owing to the limited solubility of hydroxypropyl cellulose. However, this process is reversible and on cooling the original viscosity is restored. Prolonged storage at room temperature or lower (-20°C) in a 2% aqueous solution does not result in a change in appearance, color, or viscosity.

The high level of substitution of hydroxypropyl cellulose improves the resistance of the polymer to degradation by molds and bacteria.[20] However, aqueous solutions are susceptible to degradation under severe conditions and a viscosity decrease may occur. Certain enzymes produced by microbial action will degrade hydroxypropyl cellulose in solution.[29] Therefore, for prolonged storage, an antimicrobial preservative should be added to aqueous solutions. Solutions of hydroxypropyl cellulose in organic solvents do not generally require preservatives.

Ultraviolet light will also degrade hydroxypropyl cellulose and aqueous solutions may therefore decrease slightly in viscosity if exposed to light for several months.

Hydroxypropyl cellulose powder should be stored in an airtight container in a cool, dry place.

12 Incompatibilities

Hydroxypropyl cellulose in solution demonstrates some incompatibility with substituted phenol derivatives, such as methylparaben and propylparaben. The presence of anionic polymers may increase the viscosity of hydroxypropyl cellulose solutions.

The compatibility of hydroxypropyl cellulose with inorganic salts varies depending upon the salt and its concentration; *see* Table VII. Hydroxypropyl cellulose may not tolerate high concentrations of other dissolved materials.

The balance of the hydrophilic–lipophilic properties of the polymer, which are required for dual solubility, reduces its ability to hydrate with water and it therefore tends to be salted out in the presence of high concentrations of other dissolved materials.

The precipitation temperature of hydroxypropyl cellulose is lower in the presence of relatively high concentrations of other dissolved materials that compete for the water in the system; *see* Table VIII.

Table VII: Compatibility of hydroxypropyl cellulose (*Nisso HPC-L*) with inorganic salts in aqueous solutions.[a]

Salt	Concentration of salt (% w/w)						
	2	3	5	7	10	30	50
Aluminum sulfate	S	S	I	I	I	I	I
Ammonium nitrate	S	S	S	S	S	I	I
Ammonium sulfate	S	S	I	I	I	I	I
Calcium chloride	S	S	S	S	S	T	I
Dichromic acid	S	S	S	S	S	S	S
Disodium hydrogenphosphate	S	S	I	I	I	I	I
Ferric chloride	S	S	S	S	S	I	I
Potassium ferrocyanide	S	S	S	I	I	I	I
Silver nitrate	S	S	S	S	S	S	T
Sodium acetate	S	S	S	S	I	I	I
Sodium carbonate	S	S	I	I	I	I	I
Sodium chloride	S	S	S	S	I	I	I
Sodium nitrate	S	S	S	S	S	I	I
Sodium sulfate	S	S	I	I	I	I	I
Sodium sulfite	S	S	I	I	I	I	I
Sodium thiosulfate	T	T	T	I	I	I	I

(a) S, completely soluble; T, turbid white; I, insoluble.

Table VIII: Variation in precipitation temperature of hydroxypropyl cellulose (*Klucel H*) in the presence of other materials.

Ingredients and concentrations	Precipitation temperature (°C)
1% *Klucel H*	41
1% *Klucel H* + 1.0% sodium chloride	38
1% *Klucel H* + 5.0% sodium chloride	30
0.5% *Klucel H* + 10% sucrose	41
0.5% *Klucel H* + 20% sucrose	36
0.5% *Klucel H* + 30% sucrose	32
0.5% *Klucel H* + 40% sucrose	20
0.5% *Klucel H* + 50% sucrose	7

13 Method of Manufacture

A purified form of cellulose is reacted with sodium hydroxide to produce a swollen alkali cellulose that is chemically more reactive than untreated cellulose. The alkali cellulose is then reacted with propylene oxide at elevated temperature and pressure. The propylene oxide can be substituted on the cellulose through an ether linkage at the three reactive hydroxyls present on each anhydroglucose monomer unit of the cellulose chain. Etherification takes place in such a way that hydroxypropyl substituent groups contain almost entirely secondary hydroxyls. The secondary hydroxyl present in the side chain is available for further reaction with the propylene oxide, and 'chaining-out' may take place. This results in the formation of side chains containing more than 1 mole of combined propylene oxide.

14 Safety

Hydroxypropyl cellulose is widely used as an excipient in oral and topical pharmaceutical formulations. It is also used extensively in cosmetics and food products.

Hydroxypropyl cellulose is generally regarded as an essentially nontoxic and nonirritant material.[30,31] It is not absorbed from the gastrointestinal tract and is fully recovered in feces after oral administration in rats. It does not exhibit skin irritation or skin sensitization. However, the use of hydroxypropyl cellulose as a solid ocular insert has been associated with rare reports of discomfort or irritation, including hypersensitivity and edema of the eyelids.

Adverse reactions to hydroxypropyl cellulose are rare. However, it has been reported that a single patient developed contact dermatitis due to hydroxypropyl cellulose in a transdermal estradiol patch.[32]

The WHO has specified an acceptable daily intake for hydroxypropyl cellulose of up to 1500 mg/kg body-weight.[33] Excessive consumption of hydroxypropyl cellulose may have a laxative effect.

LD_{50} (rat, IV): 0.25 g/kg[34]

LD_{50} (rat, oral): 10.2 g/kg

15 Handling Precautions

Observe normal precautions appropriate to the circumstances and quantity of material handled. Hydroxypropyl cellulose dust may be irritant to the eyes; eye protection is recommended. Excessive dust generation should be avoided to minimize the risk of explosions.

16 Regulatory Status

GRAS listed. Accepted for use as a food additive in Europe. Included in the FDA Inactive Ingredients Database (oral capsules and tablets; buccal films and gums; topical and transdermal preparations). Included in nonparenteral medicines licensed in the UK. Included in the Canadian Natural Health Products Ingredients Database.

17 Related Substances

Hydroxyethyl cellulose; hydroxypropyl cellulose, low-substituted; hypromellose.

18 Comments

Hydroxypropyl cellulose is one of the materials that have been selected for harmonization by the Pharmacopeial Discussion Group. For further information see the General Information Chapter <1196> in the USP35–NF30, the General Chapter 5.8 in PhEur 7.4, along with the 'State of Work' document on the PhEur EDQM website, and also the General Information Chapter 8 in the JP XV.

Hydroxypropyl cellulose is a thermoplastic polymer that can be processed by virtually all fabrication methods used for plastics.

It is also used in hot-melt extruded films for topical use. When it is produced with chlorpheniramine maleate, the matrix is stabilized, allowing film processing at lower temperatures.[35] Mucoadhesive hydroxypropyl cellulose microspheres have been prepared for powder inhalation preparations.[36]

A specification for hydroxypropyl cellulose is included in the *Food Chemicals Codex* (FCC).[37]

The EINECS number for hydroxypropyl cellulose is 220-971-6.

19 Specific References

1 Picker-Freyer KM, Dürig T. Physical mechanical and tablet formation properties of hydroxypropyl cellulose: in pure forms and in mixtures. *AAPS PharmSciTech* 2007; 8(4): 92.

2 Ashland Specialty Ingredients. Technical literature: Pharmaceutical Excipients and Coating Systems, 2009.

3 Divi MK et al. Ashland Specialty Ingredients. Pharmaceutical Technology Report PTR-068: Drug release from hydroxypropylcellulose sustained release matrix tablets: Implications of tablet surface area/volume ratio, molecular weight and drug solubility, 2008.

4 Jeon I, *et al.* Evaluation of roll compaction as a preparation method for hydroxypropyl cellulose-based matrix tablets. *J Pharm Bioallied Sci* 2011; **3**(2): 213–220.

5 Lee DY, Chen CM. Delayed pulse release hydrogel matrix tablet. United States Patent No. 6,103,263; 2000.

6 Machida Y, Nagai T. Directly compressed tablets containing hydroxypropyl cellulose in addition to starch or lactose. *Chem Pharm Bull* 1974; **22**: 2346–2351.

7 Delonca H, *et al.* Binding activity of hydroxypropyl cellulose (200 000 and 1 000 000 mol. wt.) and its effect on the physical characteristics of granules and tablets. *Farmaco (Prat)* 1977; **32**: 157–171.

8 Dürig T et al. Ashland Specialty Ingredients Pharmaceutical Technology Report PTR-030: Use of hydroxypropyl cellulose to enhance roller-compacted granulation compactibility, 2004.

9 Stafford JW, *et al.* Temperature dependence of the disintegration times of compressed tablets containing hydroxypropyl cellulose as binder. *J Pharm Pharmacol* 1978; **30**: 1–5.

10 Desai D, *et al.* Effect of hydroxypropyl cellulose on dissolution rate of hydrochlorothiazide tablets. *Int J Pharm* 2006; **308**: 40–45.

11 Johnson JL, *et al.* Influence of ionic strength on matrix integrity and drug release from hydroxypropyl cellulose compacts. *Int J Pharm* 1993; **90**: 151–159.

12 Skinner GW. Sustained release polymer blend for pharmaceutical applications. United States Patent No. 6,210,710 B1; 2001.

13 Guo JH, Skinner GW. Sustained release polymer blend for pharmaceutical applications. United States Patent No. 6,358,525 B1; 2002.

14 Dürig T. Advances in cellulose ether-based modified-release technologies. In: Rathbone MJ *et al.* eds. *Modified Release Drug Delivery Technology*, 2nd edn. 2008; 143–152.

15 Harcum WW *et al.*. Ashland Specialty Ingredients Pharmaceutical Technology Report, PTR-028: Stability of drug dissolution profiles from hydroxypropyl cellulose (HPC) and hydroxyethyl cellulose (HEC) matrix tablets, 2004.

16 Harcum WW, *et al*Ashland Specialty Ingredients Pharmaceutical Technology Report PTR-018: *Klucel* hydroxypropylcellulose controlled release matrix tablets prepared by roll compaction – effect of polymer, formulation and processing variables, 2002.

17 Dürig T *et al.* Ashland Specialty Ingredients Pharmaceutical Technology Report PTR-019: Compression and drug release characteristics of directly compressible *Klucel* hydroxypropylcellulose controlled release matrix systems, 2002.

18 Lindberg NO. Water vapour transmission through free films of hydroxypropyl cellulose. *Acta Pharm Suec* 1971; **8**: 541–548.

19 Banker G, *et al.* Evaluation of hydroxypropylcellulose and hydroxypropylmethylcellulose as aqueous based film coatings. *Drug Dev Ind Pharm* 1981; **7**: 693–716.

20 Bajdik J, *et al.* The effect of the solvent on film forming parameters of hydroxypropyl cellulose. *Int J Pharm* 2005; **301**: 192–198.

21 Larsson M, *et al.* Effect of ethanol on water permeability of controlled release films of ethyl cellulose and hydroxypropyl cellulose. *Eur J Pharm Biopharm* 2010; **76**: 428–432.

22 Marucci M, *et al.* Osmotic pumping release from ethyl-hydroxypropyl-cellulose coated pellets: a new mechanistic model. *J Control Rel* 2010; **142**: 53–60.

23 Rokhade AP, *et al.* Preparation and characterization of semi interpenetrating polymer network hydrogel microspheres of chitosan and hydroxypropyl cellulose for controlled release of chlorthiazide. *J Microencapsul* 2009; **26**(1): 27–36.

24 Cohen EM *et al.* Solid state ophthalmic medication. United States Patent No. 4,179,497; 1979.

25 Harwood RJ, Schwartz JB. Drug release from compression molded films: preliminary studies with pilocarpine. *Drug Dev Ind Pharm* 1982; **8**: 663–682.

26 Nguyen T, Latkany R. Review of hydroxypropyl cellulose ophthalmic inserts for treatment of dry eye. *Clin Ophthalmol* 2011; **5**: 587–591.

27 Luchs JI, *et al.* Efficacy of hydroxypropyl cellulose ophthalmic inserts (LACRISERT) in subsets of patients with dry eye syndrome: findings from a patient registry. *Cornea* 2010; **29**: 1417–1427.

28 Ashland Specialty Ingredients. Technical literature: *Klucel* hydroxypropyl cellulose: Physical and chemical properties. 2001.

29 Wirick MG. Study of the enzymic degradation of CMC and other cellulose ethers. *J Polym Sci* 1968; 6(Part A-1): 1965–1974.

30 Anonymous. Final report on the safety assessment of hydroxyethylcellulose, hydroxypropylcellulose, methylcellulose, hydroxypropyl methylcellulose and cellulose gum. *J Am Coll Toxicol* 1986; **5**(3): 1–60.

31 Ashland Specialty Ingredients. Technical literature: Bulletin T122E – *Klucel* hydroxypropylcellulose, summary of toxicological investigations, 2004.

32 Schwartz BK, Clendenning WE. Allergic contact dermatitis from hydroxypropyl cellulose in a transdermal estradiol patch. *Contact Dermatitis* 1988; **18**(2): 106–107.

33 FAO/WHO. Evaluation of certain food additives and contaminants. Thirty-fifth report of the joint FAO/WHO expert committee on food additives. *World Health Organ Tech Rep Ser* 1990; No. 789.

34 Lewis RJ, ed. *Sax's Dangerous Properties of Industrial Materials*, 11th edn. New York: Wiley, 2004: 2053.
35 Repka MA, McGinty JW. Influence of chlorpheniramine maleate on topical hydroxypropylcellulose films produced by hot melt extrusion. *Pharm Dev Technol* 2001; **6**(3): 297–304.
36 Sakagami M, *et al*. Enhanced pulmonary absorption following aerosol administration of mucoadhesive powder microspheres. *J Control Release* 2001; 77(1–2): 117–129.
37 *Food Chemicals Codex*, 7th edn. Bethesda, MD: United States Pharmacopeia, 2010; 148.

20 General References

Ashland Specialty Ingredients. Technical literature: *Klucel*, hydroxypropylcellulose modified release formulations, 2008.
Ashland Specialty Ingredients. Technical literature: *Klucel*, hydroxypropylcellulose physical and chemical properties, 2010.
Ashland Specialty Ingredients. Technical literature: *Klucel*, hydroxypropylcellulose. http://www.ashland.com/products/klucel-hydroxypropylcellulose (accessed 6 December 2011).
Ashland Specialty Ingredients. Product data sheet: *Klucel* Pharma hydroxypropyl cellulose, 2011.
Doelker E. Cellulose derivatives. *Adv Polym Sci* 1993; **107**: 199–265.
European Directorate for the Quality of Medicines and Healthcare (EDQM). European Pharmacopoeia – State Of Work Of International Harmonisation. *Pharmeuropa* 2011; 23(4): 713–714. www.edqm.eu/site/-614.html (accessed 6 December 2011).
Ganz AJ. Thermoplastic food production. United States Patent No. 3,769,029; 1973.
Karewicz A, *et al*. Smart alginate-hydroxypropyl cellulose microbeads for controlled release of heparin. *Int J Pharm* 2010; **385**: 163–169.
Klug ED. Some properties of water-soluble hydroxyalkyl celluloses and their derivatives. *J Polym Sci* 1971; **36**(Part C): 491–508.
Nippon Soda Co. Ltd. Product data sheet: *Nisso HPC*. http://www.nissoamerica.com/hpc/datasheet.pdf (accessed 6 December 2011).
Nippon Soda Co. Ltd. Technical data sheet: *Nisso HPC*, TDS-01 http://www.nissoexcipients.com/PDF/TDS-01.pdf (accessed 6 December 2011).
Opota O, *et al*. [Rheological behavior of aqueous solutions of hydroxypropylcellulose: influence of concentration and molecular mass.] *Pharm Acta Helv* 1988; **63**: 26–32[in French].
Paradkar A, *et al*. Shear and extensional rheology of hydroxypropyl cellulose melt using capillary rheometry. *J Pharm Biomed Anal* 2009; **49**: 304–310.
Sudo S. Dielectric properties of the free water in hydroxypropyl cellulose. *J Phys Chem B* 2011; **115**: 2–6.
Yang Q, *et al*. Functionalization of multiwalled carbon nanotubes by pyrene-labeled hydroxypropyl cellulose. *J Phys Chem B* 2008; **112**: 12934–12939.

21 Authors

S Fulzele, E Hamed.

22 Date of Revision

1 March 2012.

Hydroxypropyl Cellulose, Low-substituted

1 Nonproprietary Names

JP: Low Substituted Hydroxypropylcellulose
USP–NF: Low-Substituted Hydroxypropyl Cellulose

2 Synonyms

Cellulose, 2-hydroxypropyl ether; 2-hydroxypropyl ether (low-substituted) cellulose; hyprolose, low-substituted; *L-HPC*; oxypropylated cellulose.

3 Chemical Name and CAS Registry Number

Cellulose, 2-hydroxypropyl ether (low-substituted) [9004-64-2]

4 Empirical Formula and Molecular Weight

The USP35–NF30 describes low-substituted hydroxypropyl cellulose as a low-substituted hydroxypropyl ether of cellulose. Compared to hydroxypropyl cellulose, low-substituted hydroxypropyl cellulose has only a small proportion of the three free hydroxyl groups per glucose subunit converted to a hydroxypropyl ether.(1) When dried at 105°C for 1 hour, it contains not less than 5.0% and not more than 16.0% of hydroxypropoxy groups ($—OCH_2CHOHCH_3$). Low-substituted hydroxypropyl cellulose is commercially available in a number of different grades that have different particle sizes, shapes and substitution levels.

5 Structural Formula

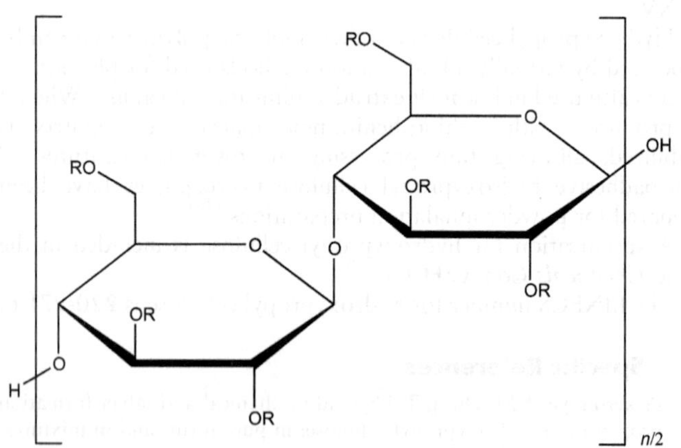

R is H or $[CH_2CH(CH_3)O]_mH$

6 Functional Category

Tablet and capsule binder; tablet and capsule diluent; tablet and capsule disintegrant.

7 Applications in Pharmaceutical Formulation or Technology

Low-substituted hydroxypropyl cellulose (L-HPC) is widely used in oral solid-dosage forms. It is primarily used as a disintegrant, and as a binder for tablets and granules in wet or dry granulation.(1) It has been used in the preparation of rapidly disintegrating tablets produced by direct compression methods.[2–4] In addition, L-HPC has been used as a binder/disintegrant included in the powder layering process on spherical cores and to prepare pellets by extrusion/spheronization.[1,5,6] A low particle size and high hydroxypropyl content is recommended to produce round spheres and rapid dissolution.[1,7]

The typical content of L-HPC in a formulation is approximately 5–50%.

8 Description

Low-substituted hydroxypropyl cellulose occurs as a white to yellowish white powder or granules. It is odorless or has a slight, characteristic odor, and it is tasteless.

9 Pharmacopeial Specifications

See Table I. *See also* Section 18.

10 Typical Properties

Acidity/alkalinity pH = 5.0–7.5 for 1% w/v aqueous suspension.
Angle of repose *see* Table II.
Ash 0.5%

SEM 1: Excipient: low-substituted hydroxypropyl cellulose, type *LH-11*; manufacturer: Shin-Etsu Chemical Co. Ltd.; magnification: 350×; voltage: 3.0 kV.

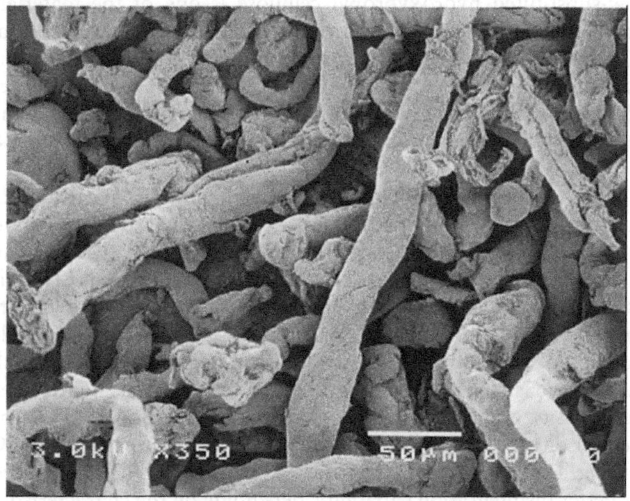

SEM 2: Excipient: low-substituted hydroxypropyl cellulose, type *LH-21*; manufacturer: Shin-Etsu Chemical Co. Ltd.; magnification: 350×; voltage: 3.0 kV.

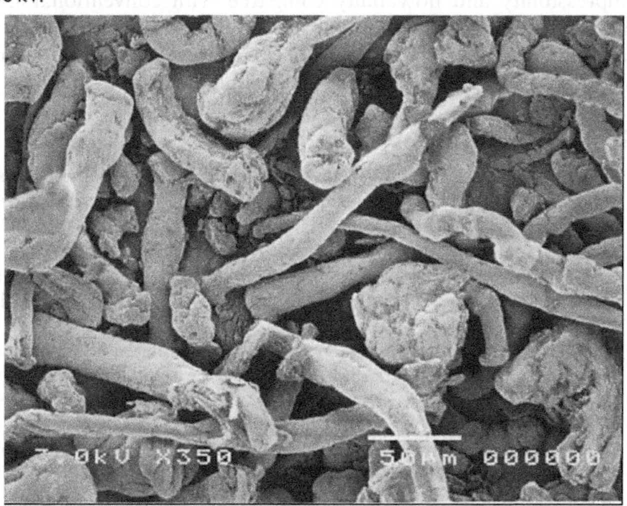

SEM 3: Excipient: low-substituted hydroxypropyl cellulose, type *LH-31*; manufacturer: Shin-Etsu Chemical Co. Ltd.; magnification: 350×; voltage: 3.0 kV.

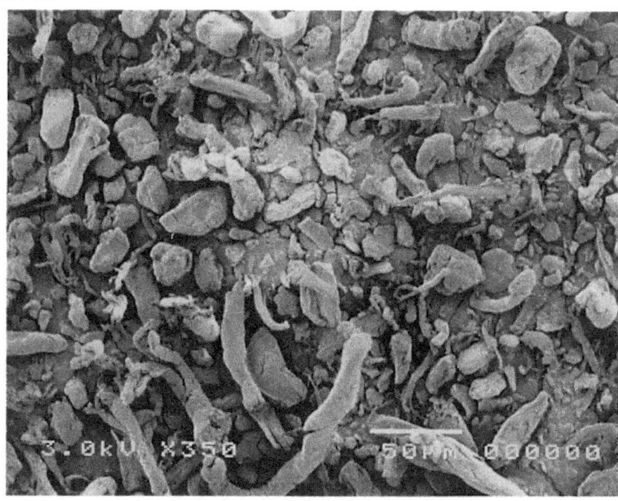

SEM 4: Excipient: low-substituted hydroxypropyl cellulose, type *LH-B1*; manufacturer: Shin-Etsu Chemical Co. Ltd.; magnification: 350×; voltage: 3.0 kV.

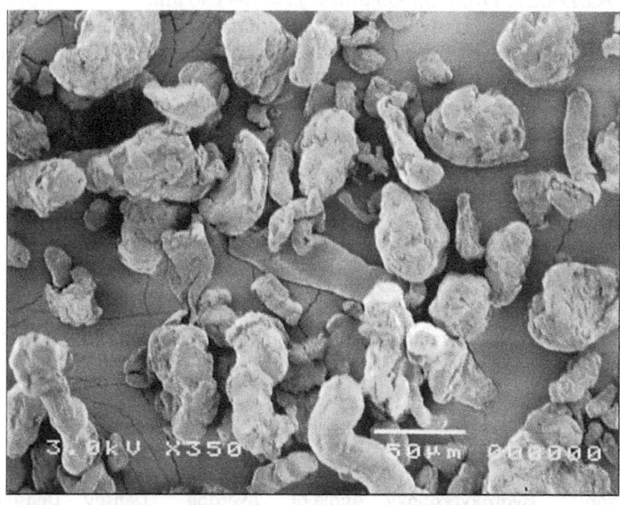

SEM 5: Excipient: low-substituted hydroxypropyl cellulose, type *NBD-021*; manufacturer: Shin-Etsu Chemical Co. Ltd.; Magnification: 350×; voltage: 5.0 kV

Table I: Pharmacopeial specifications for hydroxypropyl cellulose, low-substituted.

Test	JP XV	USP35–NF30
Identification	+	+
Chloride	≤0.335%	≤0.36%
Heavy metals	≤10 ppm	≤0.001%
Arsenic	≤2 ppm	—
pH	5.0–7.5	—
Loss on drying	≤6.0%	≤5.0%
Residue on ignition	≤1.0%	≤0.5%
Assay (of hydroxypropoxy groups)	5.0–16.0%	5.0–16.0%

Density (bulk) *see* Table II.
Density (tapped) *see* Table II.
Density (true) 1.3 g/cm^3
Melting point Decomposition at 290°C.
Moisture content <5%
Particle size distribution

LH-11: D_{50} = 45–65 μm; D_{90} = 150–200 μm;
LH-21: D_{50} = 35–55 μm; D_{90} = 100–150 μm;
LH-31: D_{50} = 17–23 μm; D_{90} = 40–100 μm;
LH-B1: D_{50} = 45–65 μm; D_{90} = 100–150 μm;
NBD-021: D_{50} = 35–55 μm; D_{90} = 70–130 μm.

Solubility Practically insoluble in ethanol (95%) and in ether. Dissolves in a solution of sodium hydroxide (1 in 10) and produces a viscous solution. Insoluble, but swells in water.

Specific surface area

0.8 m^2/g for *LH-21*;
0.5 m^2/g for *LH-B1*;
1.0 m^2/g for *NBD-021*.

11 Stability and Storage Conditions

Low-substituted hydroxypropyl cellulose is a stable, though hygroscopic, material. The powder should be stored in a well-closed container.

Table II: Typical properties of hydroxypropyl cellulose, low-substituted, for selected grades.

Grade	Hydroxypropoxy content (%)	Angle of repose (°)	Average particle size[a] (μm)	Density (bulk) (g/cm^3)	Density (tapped) (g/cm^3)
LH-11	11	54	55	0.33	0.56
LH-21	11	50	45	0.38	0.63
LH-B1	11	40	55	0.48	0.69
LH-31	11	55	20	0.28	0.59
LH-22	8	48	45	0.36	0.63
LH-32	8	53	20	0.21	0.55
NBD-020	14	43	45	0.33	0.53
NBD-021	11	43	45	0.33	0.53
NBD-022	8	43	45	0.33	0.53

(a) By laser diffraction.

12 Incompatibilities

—

13 Method of Manufacture

Low-substituted hydroxypropyl cellulose is manufactured by reacting alkaline cellulose with propylene oxide at elevated temperature. Following the reaction, the product is recrystallized by neutralization, washed, and milled.

14 Safety

Low-substituted hydroxypropyl cellulose is generally regarded as a nontoxic and nonirritant material.

Animal toxicity studies showed no adverse effects in rats fed orally 6 g/kg/day over 6 months. No teratogenic effects were noted in rabbits and rats fed 5 g/kg/day.[8–11]

LD_{50} (rat, oral): >15 g/kg[8]

15 Handling Precautions

Observe normal precautions appropriate to the circumstances and quantity of material handled. Excessive dust generation should be avoided to minimize the risk of explosions.

16 Regulatory Status

Included in the FDA Inactive Ingredients Database (oral capsules, tablets, pellets). Approved for use in pharmaceuticals in Europe, Japan, US, and other countries. Included in the Canadian Natural Health Products Ingredients Database.

17 Related Substances

Hydroxyethylmethyl cellulose; hydroxypropyl cellulose; methylcellulose.

18 Comments

Low-substituted hydroxypropyl cellulose is one of the materials that have been selected for harmonization by the Pharmacopeial Discussion Group. For further information see the General Information Chapter <1196> in the USP35–NF30, the General Chapter 5.8 in PhEur 7.4, along with the "State of Work" document on the PhEur EDQM website, and also the General Information Chapter 8 in the JP XV.

There are a number of grades that have different particle sizes and substitution levels. *LH-11* (Shin-Etsu Chemical Co. Ltd, Japan) has the longest fibrous particles, and is typically used as an anticapping agent and disintegrant for direct compression. *LH-21* is less fibrous and is used as a binder and disintegrant for tablets through the wet-granulation process. *LH-31* is a small-particle grade used especially for extrusion to produce granules, as its small particle size is better for passing a screen. *LH-B1* is the nonfibrous, high-density grade designed for fluid-bed granulation, and can be used for direct compression and/or formulations with a high L-HPC loading. Lower substitution grades *LH-22* and *LH-32* can be used for better disintegration capability, depending on the characteristics of the active ingredients. More recently, *NBD* grades (Shin-Etsu Chemical Co. Ltd, Japan) have become available with better compressibility and flowability compared with conventional *LH* grades. Three grades are available in the NBD series: *NBD-022* for orally disintegrating tablets; *NBD-021* for tablets from direct compression; and *NBD-020* for wet granulation.

19 Specific References

1 Kleinbudde P. Application of low substituted hydroxypropylcellulose (L-HPC) in the production of pellets using extrusion/spheronization. *Int J Pharm* 1993; **96**: 119–128.
2 Kawashima Y, *et al.* Low-substituted hydroxypropylcellulose as a sustained-drug release matrix base or disintegrant depending on its particle size and loading in formulation. *Pharm Res* 1993; **10**(3): 351–355.
3 Ishikawa T, *et al.* Preparation of rapidly disintegrating tablet using new types of microcrystalline cellulose (PH-M series) and low-substituted hydroxypropylcellulose or spherical sugar granules by direct compression method. *Chem Pharm Bull* 2001; **49**(2): 134–139.
4 Alvarez-Lorenzo C, *et al.* Evaluation of low-substituted hydroxypropylcelluloses (L-HPCs) as filler-binders for direct compression. *Int J Pharm* 2000; **197**: 107–116.

5 Kleinbudde P. Shrinking and swelling properties of pellets containing microcrystalline cellulose and low substituted hydroxypropylcellulose: I. Shrinking properties. *Int J Pharm* 1994; **109**: 209–219.
6 Kleinbudde P. Shrinking and swelling properties of pellets containing microcrystalline cellulose and low substituted hydroxypropylcellulose: II. Swelling properties. *Int J Pharm* 1994; **109**: 221–227.
7 Tabata T, *et al*. Manufacturing method of stable enteric granules of a new antiulcer drug (lansoprazole). *Drug Dev Ind Pharm* 1994; **209**: 1661–1672.
8 Kitagawa H, *et al*. Acute, subacute and chronic toxicities of hydroxypropylcellulose of low-substitution in rats. *Pharmacometrics* 1976; **12**: 41–66.
9 Kitagawa H, *et al*. Absorption, distribution, excretion and metabolism of ^{14}C-hydroxypropylcellulose of low-substitution. *Pharmacometrics* 1976; **12**: 33–39.
10 Kitagawa H, *et al*. Teratological study of hydroxypropylcellulose of low substitution (L-HPC) in rabbits. *Pharmacometrics* 1978; **16**: 259–269.
11 Kitagawa H, Saito H. General pharmacology of hydroxypropylcellulose of low substitution (L-HPC). *Pharmacometrics* 1978; **16**: 299–302.

20 General References

European Directorate for the Quality of Medicines and Healthcare (EDQM). European Pharmacopoeia – State Of Work Of International Harmonisation. *Pharmeuropa* 2009; 21(1): 142–143. www.edqm.eu/site/-614.html (accessed 2 December 2011).

Shin-Etsu Chemical Co. Ltd. Technical literature: *L-HPC*, low-substituted hydroxypropyl cellulose, 2008.

21 Author

S Obara.

22 Date of Revision

1 March 2012.

Hydroxypropyl Starch

1 Nonproprietary Names

PhEur: Starch, Hydroxypropyl
USP-NF: Hydroxypropyl Corn Starch
Hydroxypropyl Pea Starch
Hydroxypropyl Potato Starch

2 Synonyms

Amylum hydroxypropylum; E1440.

3 Chemical Name and CAS Registry Number

Hydroxypropyl starch [113894-92-1]

4 Empirical Formula and Molecular Weight

Hydroxypropyl starch is a derivative of natural starch. It is described in the USP35-NF30 and PhEur 7.4 as a partially substituted 2-hydroxypropylether of corn starch, potato starch, tapioca starch, rice starch, or pea starch chemically modified by etherification with the reagent propylene oxide. It may be partially hydrolyzed using acids or enzymes to obtain 'thinned starch' with reduced viscosity

5 Structural Formula

See Section 4.

6 Functional Category

Emulsifying agent; tablet and capsule disintegrant; viscosity-increasing agent.

7 Applications in Pharmaceutical Formulation or Technology

Hydroxypropyl starch is a modified starch and has been used in combination with carrageenan in the production of soft capsules.[1,2] It is also used widely in cosmetics. *See also* Section 18.

8 Description

Hydroxypropyl starch occurs as a free-flowing white to off-white coarse powder.

9 Pharmacopeial Specifications

See Table I. *See also* Section 18.

Table I: Pharmacopeial specifications for hydroxypropyl starch.

Test	PhEur 7.4	USP35-NF30[a]
Identification	+	+
Characters	+	−
pH	4.5-8.0	4.5-8.0
Foreign matter	+	+
Oxidizing substances	≤20 ppm	≤20 μg/g
Sulfur dioxide	≤50 ppm	≤50 ppm
Iron		
Corn, tapioca, potato, or rice starches	≤20 ppm	≤20 μg/g
Pea starch	≤50 ppm	≤50 μg/g
Loss on drying		
Corn, tapioca, pea, or rice starches	≤15.0%	≤15.0%
Potato starch	≤20.0%	≤20.0%
Sulfated ash	≤0.6%	≤0.6%
Microbial contamination		
Bacteria	10^3 cfu/g	10^3 cfu/g
Yeasts and molds	10^2 cfu/g	10^2 cfu/g
Assay (dried basis)	≤7.0%	2.0-7.0%

(a) USP35-NF30 is for hydroxypropyl corn, pea, and potato starches only.

10 Typical Properties

Acidity/alkalinity pH = 4.5–7.0 (10% w/v aqueous dispersion)
Solubility Practically insoluble in water, ethanol (95%), and ether.

11 Stability and Storage Conditions

Hydroxypropyl starch is stable at high humidity and is considered to be inert under normal conditions. It is stable in emulsion systems at pH 3–9.

12 Incompatibilities

See Section 18.

13 Method of Manufacture

Hydroxypropyl starch is produced industrially from natural starch, using propylene oxide as the modifying reagent in the presence of alkali, adding hydroxypropyl ($CH(OH)CH_2CH_3$) groups at the OH positions by an ether linkage.

14 Safety

Hydroxypropyl starch is widely used in cosmetics and food products. It is also used in oral pharmaceutical formulations. The WHO has set an acceptable daily intake for hydroxypropyl starch at 'not limited' since it was well tolerated on oral consumption.[3]

LD_{50} (rat, oral): 0.218 g/kg[4]

H

15 Handling Precautions

Observe normal precautions appropriate to the circumstances and quantity of material handled.

16 Regulatory Status

GRAS listed. Accepted for use as a food additive in Europe.

17 Related Substances

Starch; starch, modified.

18 Comments

Hydroxypropyl starch has been used experimentally in hydrophilic matrices, where it was shown to be an effective matrix for tablets designed for controlled-release drug delivery systems.[5,6]It has also been investigated for the production of hydrophilic matrices by direct compression.[7]

Hydroxypropyl starch–methyl methacrylate (HS-MMA) has also been used experimentally in hydrophilic matrices produced by direct compression.[8] Pregelatinized hydroxypropyl starch has been shown to exhibit good disintegrating properties, and can be used as a binder in wet granulation.[9]

Hydroxypropyl starch is compatible with cationic ingredients (monovalent, divalent), oils, emollients, and silicone. It is used analytically as a bioseparation aqueous-phase-forming polymer.[8]

Aspecification for hydroxypropyl starch is included in the *Japanese Pharmaceutical Excipients* (JPE); *see* Table II.[10]

The EINECS number for hydroxypropyl starch is 232-679-6. The PubChem Compound ID (CID) for hydroxypropyl starch is 24847857.

Table II: JPE 2004 specification for hydroxypropyl starch.[10]

Test	JPE 2004
Description	+
Identification	+
pH	5.0–7.5
Chloride	≤0.142%
Heavy metals	≤20 ppm
Arsenic	≤5 ppm
Loss on drying	≤15.0%
Residue on ignition	≤0.5%
Content of hydroxypropyl group after drying	2.0–7.0%

19 Specific References

1 Draper PR *et al.* Film forming compositions comprising modified starches and iota-carrageenan and methods for manufacturing soft capsules using the same. International Patent WO 013677; 1999.
2 Rinker RA *et al.* Rapidly disintegrating gelatinous coated tablets. United States Patent 7,879,354; 2011.
3 FAO/WHO. Fifteenth report of the Joint FAO/WHO Expert Committee on Food Additives. *World Health Organ Tech Rep Ser* 1972; No. 488.
4 Lewis RJ, ed. *Sax's Dangerous Properties of Industrial Materials*, 11th edn. New York: Wiley, 2004; 2054.
5 Goni I, *et al.* Synthesis of hydroxypropyl methacrylate/polysaccharide graft copolymers as matrices for controlled release tablets. *Drug Dev Ind Pharm* 2002; 28(9): 1101–1115.
6 Echeverria I, *et al.* Ethyl methacrylate grafted on two starches as polymeric matrices for drug delivery. *J App Polymer Sci* 2005; **96**(2): 523–536.
7 Ferrero MC, *et al.* Drug release from a family of graft copolymers of methyl methacrylate. I. *Int J Pharm* 1997; **149**: 233–240.
8 Venacio A, *et al.* Evaluation of crude hydroxypropyl starch as a bioseparation aqueous-phase-forming polymer. *Biotechnol Prog* 1993; 9(6): 635–639.
9 Visavarungroj N, Remon JP. An evaluation of hydroxypropyl starch as disintegrant and binder in tablet formulation. *Drug Dev Ind Pharm* 1991; **17**(10): 1389, 1396.
10 Japan Pharmaceutical Excipients Council. *Japanese Pharmaceutical Excipients 2004*. Tokyo: Yakuji Nippo, 2004: 425–427.

20 General References

Vorwerg W, *et al.* Film properties of hydroxypropyl starch. *Starch-Stärke* 2004; 56: 297–306.

21 Authors

SL Cantor, SA Shah.

22 Date of Revision

2 March 2012.

Hypromellose

1 Nonproprietary Names

BP: Hypromellose
JP: Hypromellose
PhEur: Hypromellose
USP–NF: Hypromellose

2 Synonyms

Anycoat C; *Benecel hypromellose*; E464; *Headcel Cellulose*; HPMC; hydroxypropyl methylcellulose; hypromellosum; *Mecellose*; *Methocel*; methylcellulose propylene glycol ether; methyl hydroxypropylcellulose; *Metolose*; MHPC; *Pharmacoat*; *Rutocel*; *Vivapharm HPMC*.

3 Chemical Name and CAS Registry Number

Cellulose, 2–hydroxypropyl methyl ether [9004-65-3]

4 Empirical Formula and Molecular Weight

The PhEur 7.4 describes hypromellose as a partly O-methylated and O-(2-hydroxypropylated) cellulose. It is available in several grades that vary in viscosity and extent of substitution. Grades may be distinguished by appending a number indicative of the apparent viscosity, in mPa s, of a 2% w/w aqueous solution at 20°C. Hypromellose defined in the USP35-NF30 specifies the substitution type by appending a four-digit number to the nonproprietary name: e.g. hypromellose 1828. The first two digits refer to the nominal content (w/w %) of the methoxy group (OCH_3), calculated on a dried basis. The second two digits refer to the nominal content of the hydroxypropoxy group ($OCH_2CH(OH)CH_3$). Hypromellose contains methoxy and hydroxypropoxy contents conforming to the pharmacopeial limits for the various chemistries; *see* Section 9. Molecular weight ranges from approximately 10 000–1 500 000 Da.

5 Structural Formula

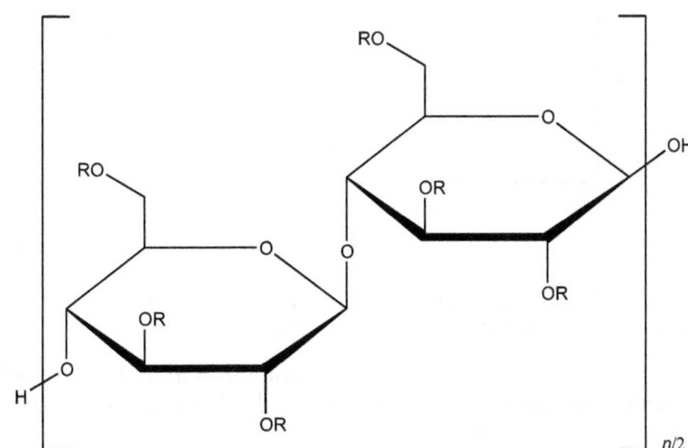

where R is H, CH_3, or $CH_2CH(OH)CH_3$

6 Functional Category

Coating agent; dispersing agent; emulsifying agent; film-forming agent; modified-release agent; solubilizing agent; suspending agent; tablet and capsule binder; viscosity-increasing agent.

7 Applications in Pharmaceutical Formulation or Technology

Hypromellose is widely used in oral, ophthalmic, nasal, and topical formulations.

In oral products, hypromellose is primarily used as a tablet binder,(1) in film-coating,(2–7) and as a matrix for use in extended-release tablet formulations.(8–12) Concentrations between 2% and 5% w/w may be used as a binder in either wet- or dry-granulation processes. Hypromellose is also used in liquid oral dosage forms as a suspending and/or thickening agent at concentrations ranging from 0.25–5.0%.(13) Hypromellose 2910 and hypromellose 2208 (more commonly) are used to retard the release of drugs from monolithic matrices at levels of 10–80% w/w.

Depending upon the viscosity grade, concentrations ranging from 2% to 25% (w/w) are used for film-forming solutions to coat tablets. Lower-viscosity grades are used in aqueous film-coating solutions, while higher-viscosity grades are used with organic solvents.

Hypromellose is also used as a suspending and/or thickening agent in topical formulations. Compared with methylcellulose, hypromellose produces aqueous solutions of greater clarity and with fewer undissolved fibers present, so it is preferentially used in ophthalmic formulations. Hypromellose at concentrations between 0.45% and 1.0% (w/w) may be added as a thickening agent in eye drop and artificial tear formulations. It is also used commercially in liquid nasal formulations at a concentration of 0.1%.(13)

Hypromellose is used as an emulsifier, suspending agent, and stabilizer in topical gels and ointments. As a protective colloid, it can minimize or prevent coalescence or agglomeration of droplets or particles.

Hypromellose is used as film-forming agent in the manufacture of hard-shell capsules, as an adhesive in plastic bandages, and as a wetting agent for hard contact lenses. It is also commonly used in cosmetics and food products.

8 Description

Hypromellose is an odorless and tasteless, white or creamy-white fibrous or granular powder.

See also Section 10.

9 Pharmacopeial Specifications

The pharmacopeial specifications for hypromellose have undergone harmonization of many attributes for JP, PhEur, and USP–NF.

See Table I.

10 Typical Properties

Acidity/alkalinity pH = 5.0–8.0 for a 2% w/w aqueous solution.
Ash $\leq$1.5%
Autoignition temperature 360°C
Density (bulk) 0.341 g/cm^3
Density (tapped) 0.557 g/cm^3
Density (true) 1.326 g/cm^3
Melting point None. Browns at 190–200°C; chars at 225–230°C. Glass transition temperature at 170–180°C.
Moisture content Hypromellose absorbs moisture from the atmosphere; the amount of water absorbed depends upon initial moisture content and the temperature and relative humidity of the surrounding air.
See Figure 1.
Solubility Soluble in cold water, forming a viscous colloidal solution; practically insoluble in hot water, chloroform, ethanol

H

SEM 1: Excipient: *Methocel E5*; manufacturer: Dow Wolff Cellulosics; magnification: 200×; voltage: 3 kV.

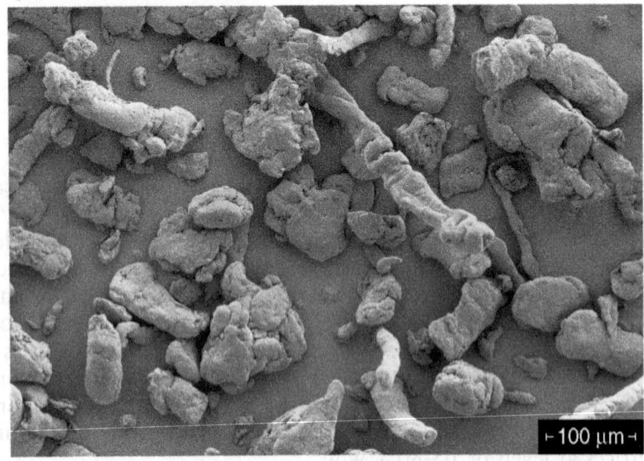

SEM 2: Excipient: *Methocel K4M*; manufacturer: Dow Wolff Cellulosics; magnification: 500×; voltage: 3 kV.

Table I: Pharmacopeial specifications for hypromellose.

Test	JP XV	PhEur 7.4	USP35–NF30
Identification	+	+	+
Characters	−	+	−
Appearance of solution	−	+	−
pH (2% w/w solution)	5.0–8.0	5.0–8.0	5.0–8.0
Apparent viscosity	+	+[a]	+
<600 mPa s	80–120%	80–120%	80–120%
≥600 mPa s	75–140%	75–140%	75–140%
Loss on drying	≤5.0%	≤5.0%	≤5.0%
Residue on ignition	≤1.5%	≤1.5%	≤1.5%
Heavy metals[b]	≤20 ppm	≤20 ppm	≤20 ppm
Methoxy content	+	+[a]	+
Type 1828	16.5–20.0%	16.5–20.0%	16.5–20.0%
Type 2208	19.0–24.0%	19.0–24.0%	19.0–24.0%
Type 2906	27.0–30.0%	27.0–30.0%	27.0–30.0%
Type 2910	28.0–30.0%	28.0–30.0%	28.0–30.0%
Hydroxypropoxy content	+	+[a]	+
Type 1828	23.0–32.0%	23.0–32.0%	23.0–32.0%
Type 2208	4.0–12.0%	4.0–12.0%	4.0–12.0%
Type 2906	4.0–7.5%	4.0–7.5%	4.0–7.5%
Type 2910	7.0–12.0%	7.0–12.0%	7.0–12.0%

(a) May be a functionality related characteristic.
(b) This test has not been fully harmonized at the time of publication.

(95%), and ether, but soluble in mixtures of ethanol and dichloromethane, mixtures of methanol and dichloromethane, and mixtures of water and alcohol. Certain grades of hypromellose are soluble in aqueous acetone solutions, mixtures of dichloromethane and propan-2-ol, and other organic solvents. Some grades are swellable in ethanol.[14]

See also Section 11.

Specific gravity 1.26

Spectroscopy

IR spectra *see* Figures 2 and 3.

NIR spectra *see* Figure 4.

Viscosity (dynamic)

A wide range of viscosity grades are commercially available; see Table II. Aqueous solutions are most commonly prepared,

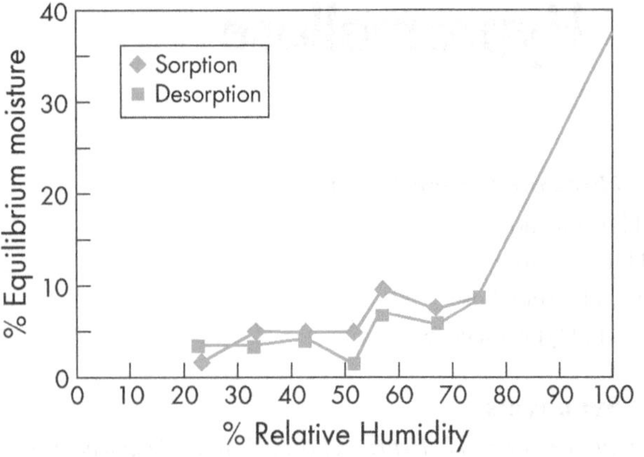

Figure 1: Absorption–desorption isotherm for hypromellose.

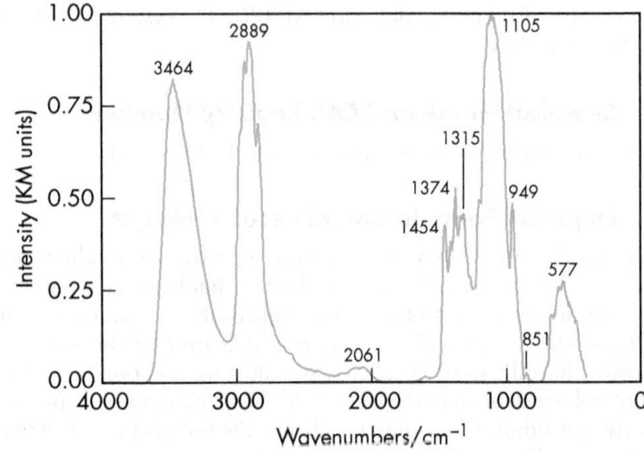

Figure 2: Infrared spectrum of hypromellose (K15M) measured by diffuse reflectance. Adapted with permission of Informa Healthcare.

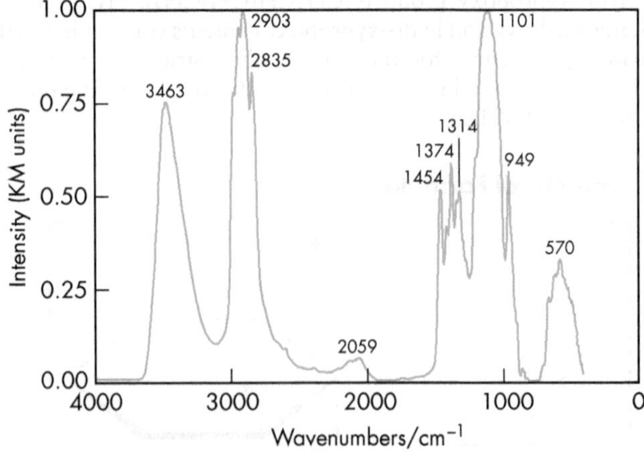

Figure 3: Infrared spectrum of hypromellose (E6) measured by diffuse reflectance. Adapted with permission of Informa Healthcare.

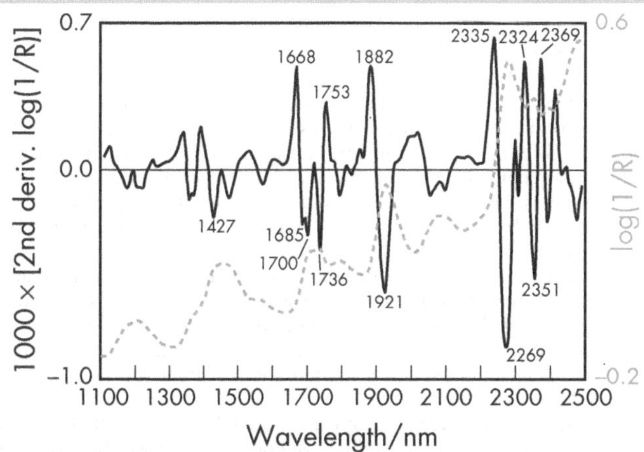

Figure 4: Near-infrared spectrum of hypromellose measured by reflectance.

Table II: Typical viscosity values for 2% (w/v) aqueous solutions of *Methocel* (Dow Wolff Cellulosics) and *Metolose* (Shin-Etsu Chemical Co. Ltd.). Viscosities measured at 20°C.

Methocel and *Metolose* products	JP/PhEur/ USP designation	Nominal viscosity (mPa s)
Methocel K3 Premium LV	2208	3
Methocel K100 Premium LV	2208	100
Methocel K4M Premium	2208	3550
Methocel K15M Premium	2208	17700
Methocel K100M Premium	2208	100000
Methocel E3 Premium LV	2910	3
Methocel E5 Premium LV	2910	5
Methocel E6 Premium LV	2910	6
Methocel E15 Premium LV	2910	15
Methocel E50 Premium LV	2910	50
Methocel E4M Premium	2910	3550
Methocel E10M Premium CR	2910	12700
Methocel F50 Premium LV	2906	50
Methocel F4M Premium	2906	3550
Metolose 60SH	2910	50, 4000, 10000
Metolose 65SH	2906	50, 400, 1500, 4000
Metolose 90SH	2208	100, 400, 4000, 15000, 100000

although hypromellose may also be dissolved in aqueous alcohols such as water/ethanol or water/propan-2-ol provided the alcohol content is less than 50% w/w. Dichloromethane/ethanol mixtures may also be used to prepare hypromellose solutions. Solutions prepared using organic solvents tend to be more viscous. Increasing polymer concentration also produces more viscous solutions.

To prepare an aqueous solution, it is recommended that hypromellose is dispersed and thoroughly hydrated in about 20–30% of the required amount of water. The water should be vigorously stirred and heated to 80–90°C, and then the hypromellose should be added. The heat source can be removed once the hypromellose has been thoroughly dispersed into the hot water. Sufficient cold water should then be added to produce the required volume while continuing to stir.

When aqueous/organic cosolvent mixtures are used for solution preparation, hypromellose should first be dispersed into the organic solvent at a ratio of 5–8 parts of solvent to 1 part of hypromellose. Cold water is then added to produce the final volume. Examples of suitable water-miscible organic solvents include ethanol and glycols. A similar preparation procedure should be used when ethanol or methanol and dichloromethane constitute a completely organic cosolvent mixture.

11 Stability and Storage Conditions

Hypromellose powder is a stable material, although it is hygroscopic after drying. It should be stored in a well-closed container, in a cool, dry place.

Solutions are stable at pH 3–11. Hypromellose undergoes a reversible sol–gel transformation upon heating and cooling, respectively. The gelation temperature ranges from 50°C to 90°C, depending upon hypromellose grade and concentration. For temperatures below the gelation temperature, viscosity of the solution decreases as temperature is increased. Beyond the gelation temperature, viscosity increases as temperature is increased.

Aqueous solutions are relatively enzyme-resistant, providing suitable viscosity stability during long-term storage.(15) Aqueous solutions are, however, susceptible to microbial spoilage and should be preserved with an antimicrobial preservative. When hypromellose is used as a viscosity-increasing agent in ophthalmic solutions, benzalkonium chloride is commonly used as the preservative. Aqueous solutions may also be sterilized by autoclaving. Once cooled, the coagulated polymer can be redispersed by agitation.

12 Incompatibilities

Hypromellose is incompatible with some oxidizing agents. Since it is nonionic, hypromellose will not complex with metallic salts or ionic organics to form insoluble precipitates.

13 Method of Manufacture

A purified form of cellulose, obtained from cotton linters or wood pulp, is reacted with sodium hydroxide solution to produce swollen alkali cellulose that is chemically more reactive than untreated cellulose. The alkali cellulose is then treated with chloromethane and propylene oxide to produce methyl hydroxypropyl ethers of cellulose. The fibrous reaction product is then purified and ground to a fine, uniform powder or granules. Hypromellose can then be exposed to anhydrous hydrogen chloride to induce depolymerization, thus producing low viscosity grades.

14 Safety

Hypromellose is widely used as an excipient in oral, ophthalmic, nasal, and topical pharmaceutical formulations. It is also used extensively in cosmetics and food products.

Hypromellose is generally regarded as a nontoxic and nonirritating material, although excessive oral consumption may have a laxative effect.(16) The WHO has not specified an acceptable daily intake for hypromellose since the levels consumed were not considered to represent a hazard to health.(17) In fact, high dosages of hypromellose are being investigated to treat various metabolic syndromes.(18,19)

LD_{50} (mouse, IP): 5 g/kg(20)
LD_{50} (rat, IP): 5.2 g/kg

15 Handling Precautions

Observe normal precautions appropriate to the circumstances and quantity of material handled. Hypromellose dust may be irritating to the eyes, so eye protection is recommended. Hypromellose is combustible. Excessive dust generation should be avoided to minimize risk of explosion.

16 Regulatory Status

GRAS listed. Accepted for use as a food additive in Europe. Included in the FDA Inactive Ingredients Database (ophthalmic and nasal preparations; oral capsules, suspensions, syrups, and tablets; topical and vaginal preparations). Included in nonparenteral medicines licensed in the UK. Included in the Canadian Natural Health Products Ingredients Database.

17 Related Substances

Ethylcellulose; hydroxyethyl cellulose; hydroxyethylmethyl cellulose; hydroxypropyl cellulose; hypromellose acetate succinate; hypromellose phthalate; methylcellulose.

18 Comments

Hypromellose has undergone harmonization of many attributes for JP, PhEur, and USP–NF by the Pharmacopeial Discussion Group. For further information see the General Information Chapter <1196> in the USP35–NF30, the General Chapter 5.8 in PhEur 7.4, along with the 'State of Work' document on the PhEur EDQM website, and also the General Information Chapter 8 in the JP XV.

Hypromellose has been used in pharmaceutical dosage forms produced via hot-melt extrusion.[21] Premix coating formulations that contain hypromellose as film-forming agent include *Opadry* (Colorcon), *Advantia* Prime Coating Systems (ISP), *Aquapolish* (Biogrund) and *Aquarius* (Ashland Specialty Ingredients). *Methocel* K4M Premium DC and *Methocel* K100M Premium DC (Dow Wolff Cellulosics) have been developed to facilitate direct compression of extended-release tablets.

Powdered or granular, surface-treated grades of hypromellose are also available that are dispersible in cold water. These are not recommended for oral use.

A specification for hypromellose is contained in the *Food Chemicals Codex* (FCC).[22]

The PubChem Compound ID (CID) for hypromellose is 24832095.

19 Specific References

1 Chowhan ZT. Role of binders in moisture-induced hardness increase in compressed tablets and its effect on *in vitro* disintegration and dissolution. *J Pharm Sci* 1980; **69**: 1–4.
2 Rowe RC. The adhesion of film coatings to tablet surfaces – the effect of some direct compression excipients and lubricants. *J Pharm Pharmacol* 1977; **29**: 723–726.
3 Rowe RC. The molecular weight and molecular weight distribution of hydroxypropyl methylcellulose used in the film coating of tablets. *J Pharm Pharmacol* 1980; **32**: 116–119.
4 Banker G, *et al.* Evaluation of hydroxypropyl cellulose and hydroxypropyl methyl cellulose as aqueous based film coatings. *Drug Dev Ind Pharm* 1981; **7**: 693–716.
5 Okhamafe AO, York P. Moisture permeation mechanism of some aqueous-based film coats. *J Pharm Pharmacol* 1982; **34**(Suppl.): 53P.
6 Alderman DA, Schulz GJ. Method of making a granular, cold water dispersible coating composition for tablets. United States Patent No. 4,816,298; 1989.
7 Patell MK. Taste masking pharmaceutical agents. United States Patent No. 4,916,161; 1990.
8 Hardy JG, *et al.* Release rates from sustained-release buccal tablets in man. *J Pharm Pharmacol* 1982; **34**(Suppl.): 91P.
9 Hogan JE. Hydroxypropylmethylcellulose sustained release technology. *Drug Dev Ind Pharm* 1989; **15**: 975–999.
10 Shah AC, *et al.* Gel-matrix systems exhibiting bimodal controlled release for oral delivery. *J Control Release* 1989; **9**: 169–175.
11 Wilson HC, Cuff GW. Sustained release of isomazole from matrix tablets administered to dogs. *J Pharm Sci* 1989; **78**: 582–584.
12 Dahl TC, *et al.* Influence of physicochemical properties of hydroxypropyl methylcellulose on naproxen release from sustained release matrix tablets. *J Control Release* 1990; **14**: 1–10.
13 Food and Drug Administration – Center for Drug Evaluation and Research. www.accessdata.fda.gov/scripts/cder/iig/index.cfm (accessed 6 December 2011).
14 Yamashita K, *et al.* Establishment of new preparation method for solid dispersion formulation of tacrolimus. *Int J Pharm* 2003; **267**: 79–91.
15 Banker G, *et al.* Microbiological considerations of polymer solutions used in aqueous film coating. *Drug Dev Ind Pharm* 1982; **8**: 41–51.
16 Anonymous. Final report on the safety assessment of hydroxyethylcellulose, hydroxypropylcellulose, methylcellulose, hydroxypropyl methylcellulose and cellulose gum. *J Am Coll Toxicol* 1986; **5**(3): 1–60.
17 FAO/WHO. Evaluation of certain food additives and contaminants. Thirty-fifth report of the joint FAO/WHO expert committee on food additives. *World Health Organ Tech Rep Ser* 1990; No. 789.
18 Lynch SK *et al.* Use of water-soluble cellulose derivatives for preventing or treating metabolic syndrome. International Patent WO 2008051794 A2; 2007.
19 Maki KC, *et al.* Hydroxypropylmethylcellulose and methylcellulose consumption reduce postprandial insulinemia in overweight and obese men and women. *J Nutr* 2008; **138**: 292–296.
20 Lewis RJ, ed. *Sax's Dangerous Properties of Industrial Materials*, 11th edn. New York: Wiley, 2004; 2054.
21 Coppens KA, *et al.* Hypromellose, ethylcellulose, and polyethylene oxide use in hot melt extrusion. *Pharm Technol* 2005; **30**(1): 62–70.
22 *Food Chemicals Codex*, 7th edn. Bethesda, MD: United States Pharmacopeia, 2011.

20 General References

Doelker E. Cellulose derivatives. *Adv Polym Sci* 1993; **107**: 199–265.

Dow Chemical Company. *Methocel* Products. www.dow.com/dowexcipients/products/methocel.htm (accessed 6 December 2011).

European Directorate for the Quality of Medicines and Healthcare (EDQM). European Pharmacopoeia – State Of Work Of International Harmonisation. *Pharmeuropa* 2011; 23(4): 713–714. http://www.edqm.eu/site/-614.html (accessed 5 December 2011).

Li CL, *et al.* The use of hypromellose in oral drug delivery. *J Pharm Pharmacol* 2005; **57**: 533–546.

Malamataris S, *et al.* Effect of particle size and sorbed moisture on the compression behavior of some hydroxypropyl methylcellulose (HPMC) polymers. *Int J Pharm* 1994; **103**: 205–215.

Papadimitriou E, *et al.* Probing the mechanisms of swelling of hydroxypropylmethylcellulose matrices. *Int J Pharm* 1993; **98**: 57–62.

Parab PV, *et al.* Influence of hydroxypropyl methylcellulose and of manufacturing technique on *in vitro* performance of selected antacids. *Drug Dev Ind Pharm* 1985; **11**: 169–185.

Radebaugh GW, *et al.* Methods for evaluating the puncture and shear properties of pharmaceutical polymeric films. *Int J Pharm* 1988; **45**: 39–46.

Rowe RC. Materials used in the film coating of oral dosage forms. In: Florence AT, ed. *Critical Reports on Applied Chemistry.* Oxford: Blackwell Scientific, 1984: 1–36.

Sako K, *et al.* Influence of water soluble fillers in hydroxypropylmethylcellulose matrices on *in vitro* and *in vivo* drug release. *J Control Release* 2002; **81**: 165–172.

Sebert P, *et al.* Effect of gamma irradiation on hydroxypropylmethylcellulose powders: consequences on physical, rheological and pharmacotechnical properties. *Int J Pharm* 1993; **99**: 37–42.

Shin-Etsu Chemical Co. Ltd. Technical literature: *Metolose*, 2008.

Shin-Etsu Chemical Co. Ltd. Technical literature: *Pharmacoat*, 2007.

Wan LSC, *et al.* The effect of hydroxypropylmethylcellulose on water penetration into a matrix system. *Int J Pharm* 1991; **73**: 111–116.

21 Author

TL Rogers.

22 Date of Revision

2 March 2012.

Hypromellose Acetate Succinate

1 Nonproprietary Names

USP–NF: Hypromellose Acetate Succinate

2 Synonyms

Aqoat; *Aqoat AS-HF/HG*; *Aqoat AS-LF/LG*; *Aqoat AS-MF/MG*; cellulose, 2-hydroxypropyl methyl ether, acetate succinate; HPMCAS.

3 Chemical Name and CAS Registry Number

Cellulose, 2-hydroxypropylmethyl ether, acetate hydrogen butanedioate [71138-97-1]

4 Empirical Formula and Molecular Weight

The USP35–NF30 describes hypromellose acetate succinate as a mixture of acetic acid and monosuccinic acid esters of hydroxypropylmethyl cellulose. It is available in several subclasses, which vary in extent of substitution, mainly of acetyl and succinoyl groups, and in particle size (fine or granular). When dried at 105°C for one hour, it contains 12.0–28.0% of methoxy groups; 4.0–23.0% of hydroxypropoxy groups; 2.0–16.0% of acetyl groups; and 4.0–28.0% of succinoyl groups.

The molecular weight of hypromellose acetate succinate cannot be characterized by the external calibration or universal calibration methods. It can only be characterized using a mass sensitive detector such as triple detection or multiangle laser light scattering.[1]

5 Structural Formula

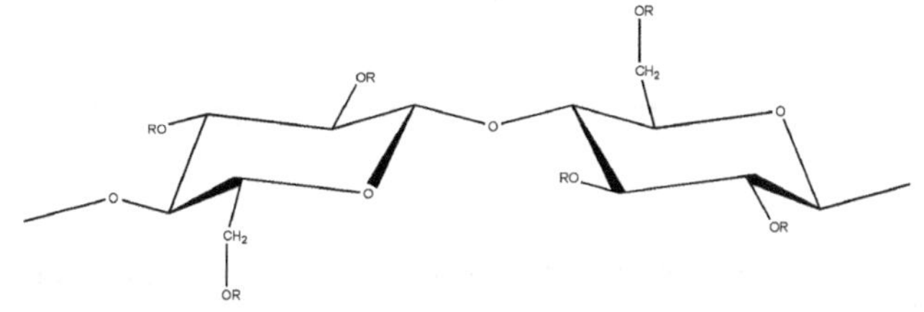

Where —OR represents one of the following functional groups: hydroxyl, methoxyl, 2-hydroxypropoxyl, acetyl, or succinoyl.

6 Functional Category

Coating agent; solubilizing agentmodified-release agent; film-forming agent.

7 Applications in Pharmaceutical Formulation or Technology

Hypromellose acetate succinate is commonly used in oral pharmaceutical formulations as a film coating, as well as an enteric coating material for tablets or granules.[2–4] It is a solubility enhancing agent via solid dispersion. Hypromellose acetate succinate is insoluble in gastric fluid but will swell and dissolve rapidly in the upper intestine. For aqueous film-coating purposes, a dispersion of hypromellose acetate succinate fine powder and triethyl citrate (as a plasticizer) in water is commonly utilized.[5,6] Organic solvents can also be used as vehicles for applying this polymer as a film coating.

Hypromellose acetate succinate may be used alone or in combination with other soluble or insoluble binders in the preparation of granules with sustained drug-release properties; the release rate is pH-dependent.

Dispersions of poorly soluble drugs with hypromellose acetate succinate are prepared using techniques such as mechanical grinding, solvent evaporation, and melt extrusion.[7–15]

8 Description

Hypromellose acetate succinate is a white to off-white powder or granules. It has a faint acetic acid-like odor and a barely detectable taste. Hypromellose acetate succinate is available in several subclasses, according to the pH at which the polymer dissolves (low, L; medium, M; and high, H) and its predominant particle size (cohesive fine powder, F; or free-flowing granules, G).

9 Pharmacopeial Specifications

See Table I. *See also* Section 18.

SEM 1: Excipient: *Aqoat MF*; manufacturer: Shin Etsu Chemical Co. Ltd; magnification: 1000×.

SEM 2: Excipient: *Aqoat MG*; manufacturer: Shin Etsu Chemical Co. Ltd; magnification: 50×.

H

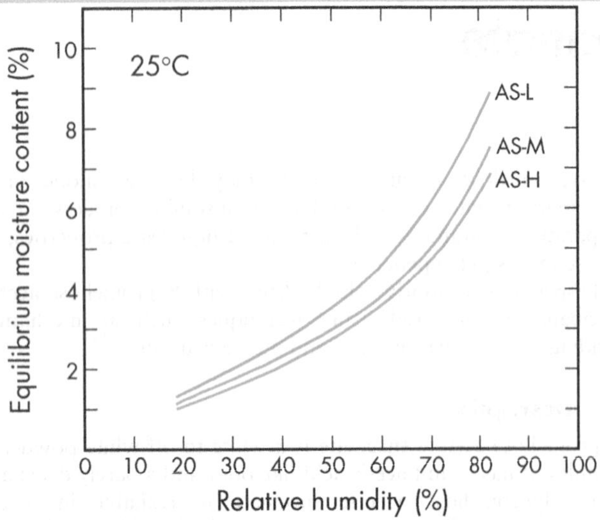

Figure 1: Equilibrium moisture content of *Aqoat* (Shin-Etsu Chemical Co. Ltd) at different relative humidities.[4]

Table I: Pharmacopeial specifications for hypromellose acetate succinate.

Test	USP35–NF30
Identification	+
Viscosity	+
Loss on drying	≤0.5%
Residue on ignition	≤0.2%
Heavy metals	≤0.001%
Limit of free acetic and succinic acids	+
Content of acetyl and succinyl groups	+
Content of methoxy and 2-hydroxypropoxy groups	+

10 Typical Properties

Density (bulk)

0.2–0.3 g/cm^3 for *Aqoat MF*;

0.2–0.5 g/cm^3 for *Aqoat MG*.

Density (tapped)

0.3–0.5 g/cm^3 for *Aqoat MF*;

0.3–0.6 g/cm^3 for *Aqoat MG*.

Density (true) 1.27–1.30 g/cm^3 for *Aqoat*.

Equilibrium moisture content 2–3% w/w at ambient temperature and humidity (≈25°C, 40% RH). *See also* Figure 1.

Glass transition temperature 113 ± 2°C (differential scanning calorimetry; dried sample)

Particle size distribution

10% <1 μm; 50% <5 μm; 90% <10 μm for *Aqoat MF*.

10% <200 μm; 50% <800 μm; 90% <1000 μm for *Aqoat MG*.

Solubility Practically insoluble in ethanol (95%), hexane, unbuffered water, and xylene. On the addition of acetone, or a mixture of ethanol (95%) and dichloromethane (1 : 1), a clear or turbid viscous solution is produced. Hypromellose acetate succinate also forms a clear or turbid solution in buffers of pH greater than 4.5 with the rank order of solubility for the various grades (*see* Section 8) increasing with the ratio of acetyl over succinoyl substitution. The exact pH value at which the polymer starts to swell and dissolve depends on the buffer type and ionic strength, although the rank order for the different subclasses is independent of the buffer used. No solvent has been found to completely dissolve hypromellose acetate succinate polymer; the resulting clear or turbid solutions significantly scatter light in both static and dynamic light scattering experiments, indicating that the polymer may exist as colloids or aggregates. The data is collated from the Shin-Etsu Chemical Co. Ltd and the research work of the authors (R Chen, BC Hancock, and RM Shanker).

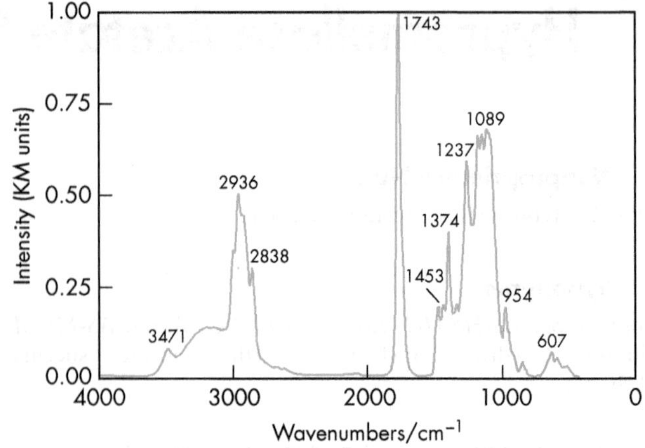

Figure 2: Infrared spectrum of hypromellose acetate succinate (AS-LF) measured by diffuse reflectance. Adapted with permission of Informa Healthcare.

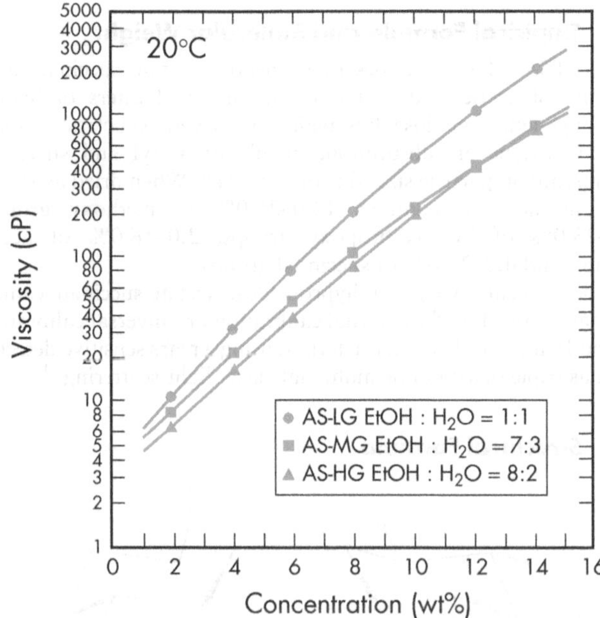

Figure 3: Viscosity of different grades of *Aqoat* (Shin-Etsu Chemical Co. Ltd).

Spectroscopy

IR spectra *see* Figure 2.

Viscosity (dynamic) *see* Figure 3.

11 Stability and Storage Conditions

Hypromellose acetate succinate should be stored in a well-closed container, in a cool, dry place. In such storage conditions, hypromellose acetate succinate is a stable material. Hypromellose acetate succinate is hygroscopic. It is hydrolyzed to acetic acid and succinic acid, and the hypromellose polymer starts to form if dissolved in 1 mol/L sodium hydroxide for more than two hours.[16] The hydrolysis is the main degradation pathway that is responsible for increasing amounts of free acids in storage, especially upon exposure to moisture. Free acids may cause incompatibility with certain active moieties upon storage of drug product formulations.[17]

12 Incompatibilities

Hypromellose acetate succinate is incompatible with strong acids or bases, oxidizing agents, and sustained levels of elevated humidity. It may interact with certain categories of active pharmceutical ingredients.(18)

13 Method of Manufacture

Hypromellose acetate succinate is produced by the esterification of hypromellose with acetic anhydride and succinic anhydride, in a reaction medium of a carboxylic acid, such as acetic acid, and using an alkali carboxylate, such as sodium acetate, as catalyst.(19) The fibrous reaction product is precipitated out by adding a large volume of water to the reaction medium. Purification is achieved by thorough washing with water. The granular grade of hypromellose acetate succinate that is so obtained can be pulverized to a fine powder if required.

14 Safety

The safety and pharmacological profiles of hypromellose acetate succinate are similar to those of other ether and ester derivatives of cellulose.(20–24) All nonclinical studies reported in the literature identify no target organs for toxicity by hypromellose acetate succinate.(25,26) It has also been reported that hypromellose acetate succinate does not alter fertility in rats, does not produce any developmental anomalies in rats and rabbits, and does not alter perinatal and postnatal development in rats when assessed up to 2500 mg/kg body-weight.(27–30) No developmental and reproductive toxicity was observed at doses up to 2500 mg/kg maternal exposure.(32) Additionally a Type V DMF-containing supplemental pre-clinical safety and ADME information to enable chronic use of high doses of polymer in drug products has been filed with the FDA.

15 Handling Precautions

Observe normal precautions appropriate to the circumstances and quantity of material handled. Hypromellose acetate succinate dust may be irritant to the eyes. Excessive dust generation should be avoided to minimize the risks of explosions. Avoid contact with open flame, heat, or sparks. Avoid contact with acids, peroxides, and other oxidizing materials. Eye protection is recommended.

16 Regulatory Status

Included in the FDA Inactive Ingredients Database for use in oral preparations (capsules and delayed-action preparations). Hypromellose acetate succinate has been approved for use in commercial pharmaceutical products in the US and in Japan.

17 Related Substances

Carboxymethyl cellulose; cellulose acetate; cellulose acetate phthalate; cellulose, microcrystalline; ethylcellulose; hypromellose; hypromellose phthalate; hydroxyethyl cellulose; hydroxypropyl cellulose; methylcellulose.

18 Comments

A specification for hypromellose acetate succinate is included in the *Japanese Pharmaceutical Excipients* (JPE).(31)

19 Specific References

1 Chen R. Characterization of hypromellose acetate succinate by size exclusion chromatography (SEC) using Viscotek Triple Detector. *Intern J Polym Anal Charac* 2009; **14**(7): 617–630.

2 Hilton AK, Deasy PB. Use of hydroxypropyl methylcellulose acetate succinate in an enteric polymer matrix to design controlled-release tablets of amoxycillin trihydrate. *J Pharm Sci* 1993; **82**: 737–743.

3 Streubel A, *et al.* Bimodal drug release achieved with multi-layer tablets: transport mechanisms and device design. *J Control Release* 2000; **69**: 455–468.

4 Tezuka Y, *et al.* ^{13}C-NMR structural study on an enteric pharmaceutical coating cellulose derivative having ether and ester substituents. *Carbohyd Res* 1991; **222**: 255–259.

5 Anderson NR *et al.* United States Patent No. 5,508,276; 1996.

6 Nagai T *et al.* Application of HPMC and HPMCAS to aqueous film coating of pharmaceutical dosage forms. In: McGinity JW, ed. *Aqueous Polymeric Coatings for Pharmaceutical Dosage Forms*, 2nd edn. New York: Marcel Dekker, 1997; 177–225.

7 Jeong YI, *et al.* Evaluation of an intestinal pressure-controlled colon delivery capsule prepared by dipping method. *J Control Release* 2001; **71**: 175–182.

8 Nakamichi K. Method of manufacturing solid dispersion. United States Patent No. 5,456,923; 1994.

9 Miyajima M *et al.* Pharmaceutical composition of dihydropyridine compound. United States Patent No. 4,983,593; 1989.

10 Takeichi Y, *et al.* Combinative improving effect of increased solubility and the use of absorption enhancers on the rectal absorption of uracil in beagle dogs. *Chem Pharm Bull* 1990; **38**: 2547–2551.

11 Baba K, *et al.* Molecular behavior and dissolution characteristics of uracil in ground mixtures. *Chem Pharm Bull* 1990; **38**: 2542–2546.

12 Roberts M, *et al.* The effect of spray drying on the compaction properties of hypromellose acetate succinate. *Drug Ind Pharm* 2011; **37**: 268–273.

13 Friesen D, *et al.* Hydroxypropyl methylcellulose acetate succinate based spray-dried dispersions – an overview. *Mol Pharm* 2008; **5**: 1003–1019.

14 Kennedy M, *et al.* Bioavailability of a poorly soluble VR1 antagonist using solid dispersion approach: a case study. *Mol Pharm* 2008; **5**: 981–993.

15 Moser JD, *et al.* Enhancing solubility of poorly soluble drugs using spray dried solid dispersions Part 1 and Part 2. *Am Pharm Rev* 2008.; http://americanpharmaceuticalreview.com.

16 Chen R, *et al.* Development and validation of a cost-effective, efficient, and robust liquid chromatographic method for the simultaneous determination of the acetyl and succinoyl content in hydroxypropyl methylcellulose acetate succinate polymer. *J AOAC Int* 2002; **85**(4): 824–831.(Suppl. Correction 85(6), 125A): .

17 Jansen PJ, *et al.* Characterization of impurities formed by interaction of duloxetine HCl with enteric polymers hydroxypropyl methylcellulose acetate succinate and hydroxypropyl methylcellulose phthalate. *J Pharm Sci* 1998; **87**: 81–85.

18 Dong Z, Choi DS. Hydroxypropyl methylcellulose acetate succinate: potential drug–excipient incompatibility. *AAPS PharmSciTech* 2008; **9**(3): 991–997.

19 Onda Y. Ether-ester derivatives of cellulose and their applications. United States Patent No. 4,226,981; 1980.

20 Final report on the safety assessment of hydroxyethylcellulose, methylcellulose, hydroxypropyl methylcellulose and cellulose gum. *J Am Coll Toxicol* 1986; **5**: 1–59.

21 Informatics: *GRAS (Generally Recognized as Safe) Food ingredients—cellulose and derivatives. For the FDA National Technical Information Service (NTIS)*. 1972, PB No. 22128.

22 Obara S, *et al.* A three month repeated oral administration study of a low viscosity grade of hydroxypropyl methylcellulose in rats. *J Toxicol Sci* 1999; **24**: 33–43.

23 Frawley JP. Studies on the gastro-intestinal absorption of purified sodium carboxymethylcellulose. *Food Cosmet Toxicol* 1964; **2**: 539–543.

24 Kitagawa H, *et al.* Absorption, distribution and excretion of hydroxypropyl methylcellulose phthalate in the rat. *Pharmacometrics* 1971; **5**(1): 1–4.

25 Hoshi N, *et al.* General pharmacological studies of hydroxypropylmethyl cellulose acetate succinate in experimental animals. *J Toxicol Sci* 1985; **10**: 129–146.

26 Hoshi N, *et al.* Toxicological studies of hydroxypropylmethyl cellulose acetate succinate—Acute toxicity in rats and rabbits, and subchronic and chronic toxicities in rats. *J Toxicol Sci* 1985; **10**: 147–185.

27 Hoshi N, *et al.* Studies of hydroxypropylmethyl cellulose acetate succinate on fertility in rats. *J Toxicol Sci* 1985; **10**: 187–201.

H

28 Hoshi N, *et al.* Teratological studies of hydroxypropylmethyl cellulose acetate succinate in rats. *J Toxicol Sci* 1985; **10**: 203–226.
29 Hoshi N, *et al.* Teratological study of hydroxypropylmethyl cellulose acetate succinate in rabbits. *J Toxicol Sci* 1985; **10**: 227–234.
30 Hoshi N, *et al.* Effects of offspring induced by oral administration of hydroxypropylmethyl cellulose acetate succinate to the female rats in peri and post natal periods. *J Toxicol Sci* 1985; **10**: 235–255.
31 Japan Pharmaceutical Excipients Council. *Japanese Pharmaceutical Excipients 2004*. Tokyo: Yakuji Nippo, 2004: 415–420.
32 Cappon GD, *et al.* Embryo/fetal development studies with hydroxypropyl methylcellulose acetate succinate (HPMCAS) in rats and rabbits. Birth Defects Part B. *Dev Repro Toxicol* 2003; **68**: 421–427.

20 General References

Doelker E. Cellulose derivatives. *Adv Polym Sci* 1993; **107**: 199–265.
Tanno F, *et al.* Evaluation of hypromellose acetate succinate (HPMCAS) as a carrier in solid dispersions. *Drug Dev Ind Pharm* 2004; **30(1)**: 9–17.
Shin-Etsu Chemical Co. Ltd. Product data sheet: Aqoat – Enteric coating agent. http://www.harke.com/fileadmin/images/chemicals/ShinEtsu_A-qoat.pdf(accessed 31 October 2011)
Curatolo W, *et al.* Utility of hydroxypropylmethylcellulose acetate succinate (HPMCAS) for initiation and maintenance of drug supersaturation in the GI milieu. *Pharm Res* 2009; **26**: 1419.
Tajarobi F, *et al.* The influence of crystallization inhibition of HPMC and HPMCAS on model substance dissolution and release in swellable matrix tablets. *Eur J Pharm Biopharm* 2011; **78**: 125–133.
Kestur US, Taylor LS. Role of polymer chemistry in influencing crystal growth rates from amorphous felodipine. *Cryst Eng Comm* 2010; **12**: 2390–2397.
Rumondor A, *et al.* Effects of moisture on the growth rate of felodipine crystals in the presence and absence of polymers. *Cryst Growth Des* 2010; **10**: 747–753.

21 Authors

R Chen, BC Hancock, RM Shanker.

22 Date of Revision

1 March 2012.

Hypromellose Phthalate

1 Nonproprietary Names

BP: Hypromellose Phthalate
JP: Hypromellose Phthalate
PhEur: Hypromellose Phthalate
USP–NF: Hypromellose Phthalate

2 Synonyms

Cellulose phthalate hydroxypropyl methyl ether; *Deepcoat*; *HPMCP*; hydroxypropyl methylcellulose benzene-1,2-dicarboxylate; 2-hydroxypropyl methylcellulose phthalate; hypromellosi phthalas; *Mantrocel HP-55*; methylhydroxypropylcellulose phthalate.

3 Chemical Name and CAS Registry Number

Cellulose, hydrogen 1,2-benzenedicarboxylate, 2-hydroxypropyl methyl ether [9050-31-1]

4 Empirical Formula and Molecular Weight

Hypromellose phthalate is a cellulose in which some of the hydroxyl groups are replaced with methyl ethers, 2-hydroxypropyl ethers, or phthalyl esters. Several different types of hypromellose phthalate are commercially available with molecular weights in the range 20 000–200 000. Typical average values are 80 000–130 000.[1]

5 Structural Formula

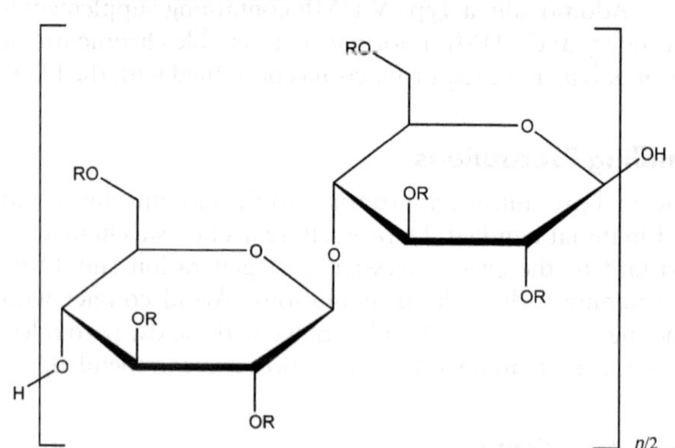

R = H, CH_3, $CH_2CH(OH)CH_3$,

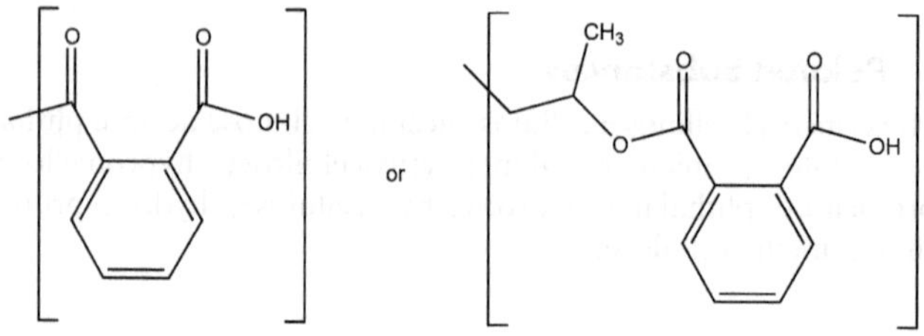

6 Functional Category

Coating agent.

7 Applications in Pharmaceutical Formulation or Technology

Hypromellose phthalate is widely used in oral pharmaceutical formulations as an enteric coating material for tablets or granules.[2–8] Hypromellose phthalate is insoluble in gastric fluid but will swell and dissolve rapidly in the upper intestine. Generally, concentrations of 5–10% of hypromellose phthalate are employed with the material being dissolved in either a dichloromethane-ethanol (50 : 50) or an ethanol-water (80 : 20) solvent mixture. Hypromellose phthalate can normally be applied to tablets and granules without the addition of a plasticizer or other film formers, using established coating techniques. However, the addition of a small amount of plasticizer or water can avoid film cracking problems; many commonly used plasticizers, such as diacetin, triacetin, diethyl and dibutyl phthalate, castor oil, acetyl monoglyceride, and polyethylene glycols, are compatible with hypromellose phthalate. Tablets coated with hypromellose phthalate disintegrate more rapidly than tablets coated with cellulose acetate phthalate.

Hypromellose phthalate can be applied to tablet surfaces using a dispersion of the micronized hypromellose phthalate powder in an aqueous dispersion of a suitable plasticizer such as triacetin, triethyl citrate, or diethyl tartrate together with a wetting agent.[9]

Hypromellose phthalate may be used alone or in combination with other soluble or insoluble binders in the preparation of granules with sustained drug-release properties; the release rate is pH-dependent. Since hypromellose phthalate is tasteless and insoluble in saliva, it can also be used as a coating to mask the unpleasant taste of some tablet formulations. Hypromellose phthalate has also been co-precipitated with a poorly soluble drug to improve dissolution characteristics.[10]

8 Description

Hypromellose phthalate occurs as white to slightly off-white, free-flowing flakes or as a granular powder. It is odorless or with a slightly acidic odor and has a barely detectable taste.

9 Pharmacopeial Specifications

The pharmacopeial specifications for hypromellose phthalate have undergone harmonization of many attributes for JP, PhEur, and USP–NF.

See Table I. *See also* Section 18.

Table I: Pharmacopeial specifications for hypromellose phthalate.

Test	JP XV	PhEur 7.4	USP35–NF30
Identification	+	+	+
Characters	—	+	—
Water	≤5.0%	≤5.0%	≤5.0%
Viscosity (20°C)	+	+	+
Residue on ignition	≤0.2%	≤0.2%	≤0.2%
Chloride	≤0.07%	≤0.07%	≤0.07%
Heavy metals[(a)]	≤10 ppm	≤10 ppm	≤0.001%
Free phthalic acid	≤1.0%	≤1.0%	≤1.0%
Phthalyl content	+	21.0-35.0%	21.0–35.0%
Type 200731	27.0–35.0%	—	—
Type 220824	21.0–27.0%	—	—

(a) This test has not been fully harmonized at the time of publication.

10 Typical Properties

Angle of repose

37° for *HP-50*;

39° for *HP-55*;

38° for *HP-55S*.[11]

Density

1.82 g/cm^3 for *HP-50*;

1.65 g/cm^3 for *HP-55*.[11]

Density (bulk)

0.278 g/cm^3 for *HP-50*;

SEM 1: Excipient: hypromellose phthalate (*HP-55*); manufacturer: Shin-Etsu Chemical Co. Ltd; magnification: 60×.

SEM 2: Excipient: hypromellose phthalate (*HP-55*); manufacturer: Shin-Etsu Chemical Co. Ltd; magnification: 600×.

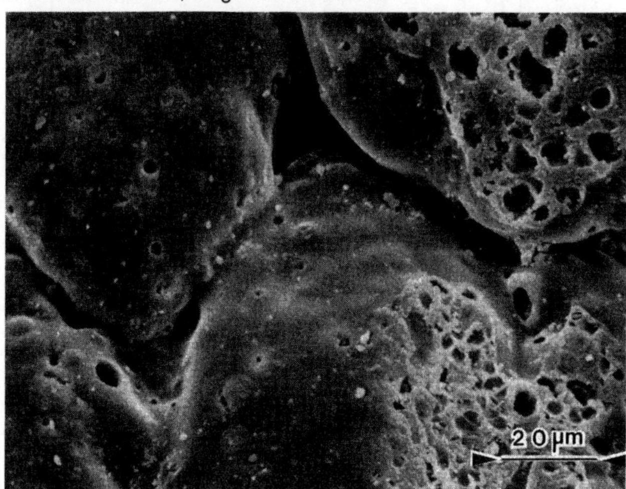

0.275 g/cm^3 for *HP-55*;

0.239 g/cm^3 for *HP-55S*.[11]

Density (tapped)

0.343 g/cm^3 for *HP-50*;

0.306 g/cm^3 for *HP-55*;

0.288 g/cm^3 for *HP-55S*.[11]

Melting point 150°C. Glass transition temperature is 137°C for *HP-50* and 133°C for *HP-55*.[12]

Moisture content Hypromellose phthalate is hygroscopic; it takes up 2–5% of moisture at ambient temperature and humidity conditions. For the moisture sorption isotherm of *HP-50* measured at 25°C, *see* Figure 1.

Particle size distribution *see* Figure 2.

Solubility Readily soluble in a mixture of acetone and methyl or ethyl alcohol (1 : 1), in a mixture of methyl alcohol and dichloromethane (1 : 1), and in aqueous alkali. Practically insoluble in water and dehydrated alcohol and very slightly soluble in acetone. The solubilities of the *HP-50* and *HP-55* grades, in various solvents and solvent mixtures, are shown in Table II.[11]

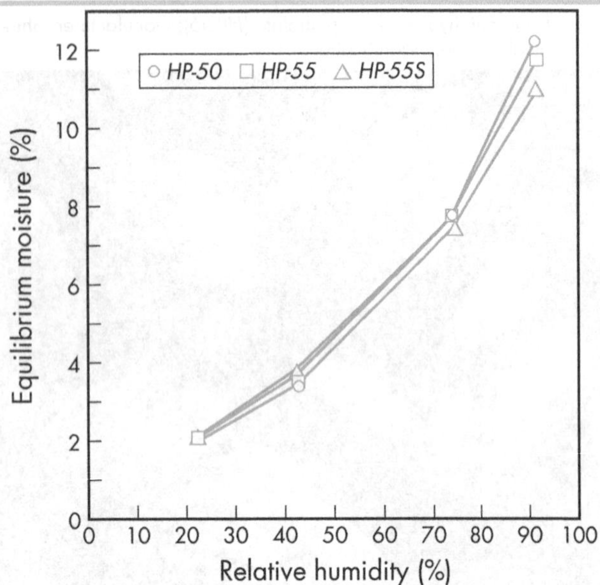

Figure 1: Equilibrium moisture content of hypromellose phthalate (Shin-Etsu Chemical Co. Ltd) at 25°C.[11]

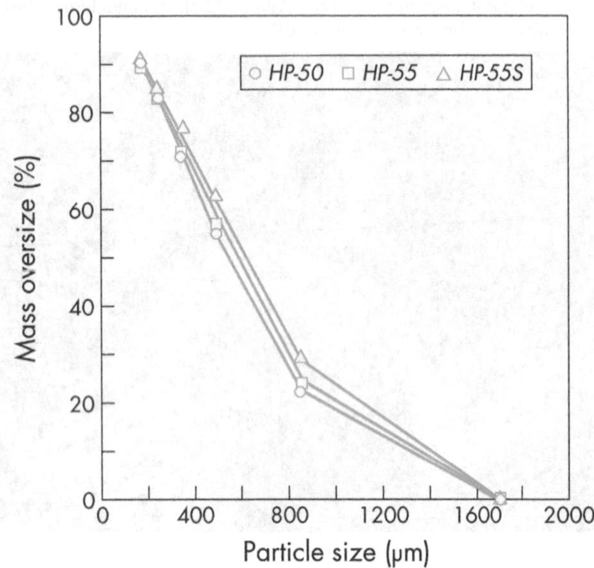

Figure 2: Particle size distribution of hypromellose phthalate (Shin-Etsu Chemical Co. Ltd).[11]

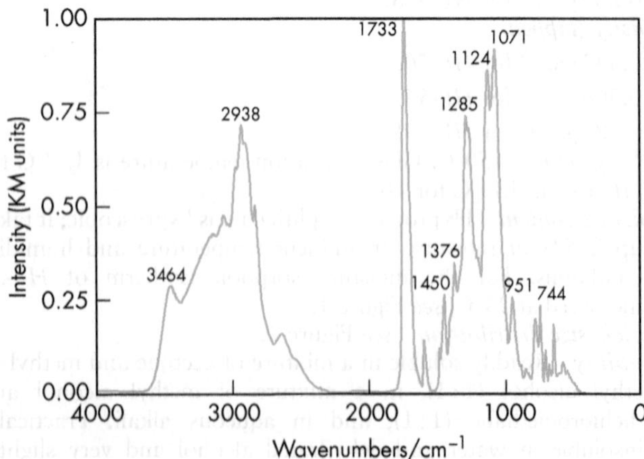

Figure 3: Infrared spectrum of hypromellose phthalate measured by diffuse reflectance. Adapted with permission of Informa Healthcare.

Table II: Solubility of hypromellose phthalate (*HP-50* and *HP-55*, Shin-Etsu Chemical Co. Ltd).

Solvent	Solubility	
	HP-50	*HP-55*
Acetone	S/I	S
Acetone : dichloromethane	S/I	S
Acetone : ethanol	S/S	S
Acetone : methanol	S	S
Acetone : 2-propanol	S/S	S
Acetone : water (95 : 5)	S	S
Benzene : methanol	S	S
Dichloromethane	S/I	S/I
Dichloromethane : ethanol	S	S
Dichloromethane : methanol	S	S
Dichloromethane : 2-propanol	S/S	S
Dioxane	S	S
Ethanol (95%)	S/I	S/I
Ethyl acetate	X	S/I
Ethyl acetate : ethanol	S/S	S
Ethyl acetate : methanol	S	S
Ethyl acetate : 2-propanol	S/I	S
Methanol	S/I	S/I
Methyl ethyl ketone	S/I	S
Propan-2-ol	X	S/I

Note: solubilities are for the pure solvent, or a (1 : 1) solvent mixture, unless otherwise indicated.
S = soluble, clear solution.
S/S = slightly soluble, cloudy solution.
S/I = swells but insoluble.
X = insoluble.

Spectroscopy

IR spectra *see* Figure 3.

Viscosity *see* Figures 4 and 5.

11 Stability and Storage Conditions

Hypromellose phthalate is chemically and physically stable at ambient temperature for at least 3–4 years and for 2–3 months at 40°C and 75% relative humidity.[11] It is stable on exposure to UV light for up to 3 months at 25°C and 70% relative humidity. Drums stored in a cool, dry place should be brought to room temperature before opening to prevent condensation of moisture on inside surfaces. After 10 days at 60°C and 100% relative humidity, 8–9% of carbyoxybenzoyl group were hydrolyzed. In general, hypromellose phthalate is more stable than cellulose acetate phthalate. At ambient storage conditions, hypromellose phthalate is not susceptible to microbial attack.

12 Incompatibilities

Incompatible with strong oxidizing agents.

Splitting of film coatings has been reported rarely, most notably with coated tablets that contain microcrystalline cellulose and calcium carboxymethylcellulose. Film splitting has also occurred when a mixture of acetone-propan-2-ol or dichloromethane-propan-2-ol has been used as the coating solvent, or when coatings have been applied in conditions of low temperature and humidity. However, film splitting may be avoided by careful selection of formulation composition, including solvent, by use of a higher molecular weight grade of polymer, or by suitable selection of plasticizer.

The addition of more than about 10% titanium dioxide to a coating solution of hypromellose phthalate, which is used to produce a colored film coating, may result in coating with decreased elasticity and resistance to gastric fluid.[11]

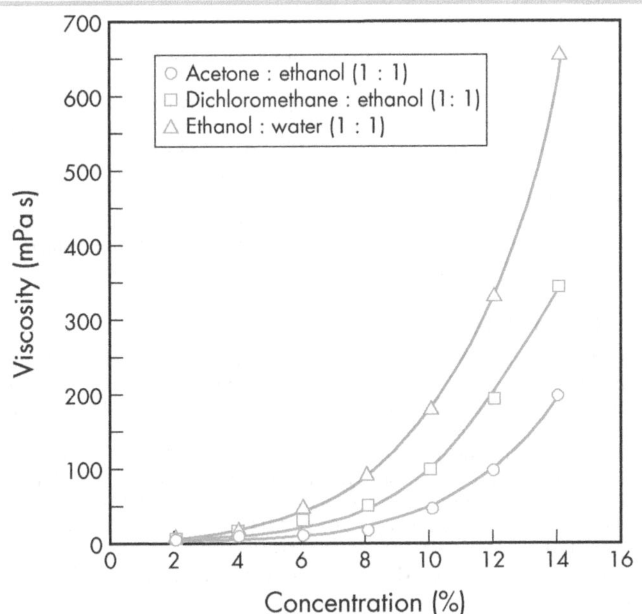

Figure 4: Dynamic viscosity of hypromellose phthalate (*HP-50*) (Shin-Etsu Chemical Co. Ltd) in various solvent mixtures at 20°C.[11]

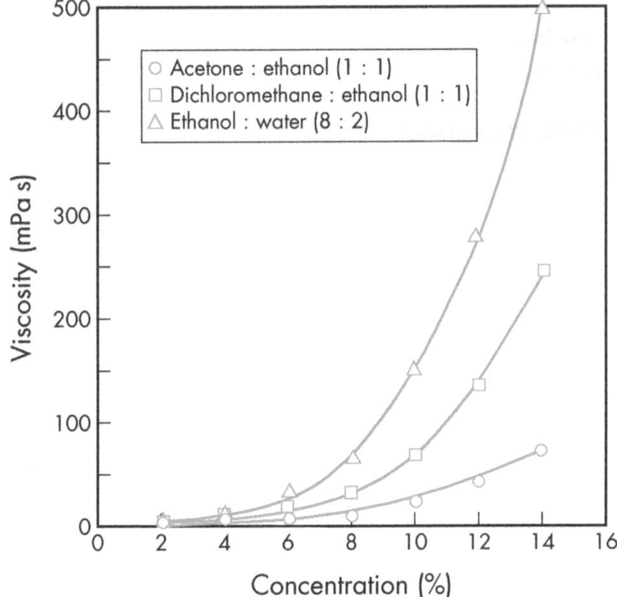

Figure 5: Dynamic viscosity of hypromellose phthalate (*HP-55*) (Shin-Etsu Chemical Co. Ltd) in various solvent mixtures at 20°C.[11]

13 Method of Manufacture

Hypromellose phthalate is prepared by the esterification of hypromellose with phthalic anhydride. The degree of alkyloxy and carboxybenzoyl substitution determines the properties of the polymer and in particular the pH at which it dissolves in aqueous media.

14 Safety

Hypromellose phthalate is widely used, primarily as an enteric coating agent, in oral pharmaceutical formulations. Chronic and acute animal feeding studies on several different species have shown no evidence of teratogenicity or toxicity associated with hypromellose phthalate.[13–17] Hypromellose phthalate is generally regarded as a nonirritant and nontoxic material.

LD_{50} (rat, oral): >15 g/kg[13]

15 Handling Precautions

Observe normal precautions appropriate to the circumstances and quantity of material handled. Eye protection and gloves are recommended. Although no threshold limit value has been set for hypromellose phthalate, it should be handled in a well-ventilated environment and the generation of dust should be minimized.

16 Regulatory Status

Included in the FDA Inactive Ingredients Database (oral capsules and tablets). Included in nonparenteral medicines licensed in the UK. Included in the Canadian Natural Health Products Ingredients Database.

17 Related Substances

Cellulose acetate phthalate; hypromellose.

18 Comments

Hypromellose phthalate has undergone harmonization of many attributes for JP, PhEur, and USP–NF by the Pharmacopeial Discussion Group. For further information see the General Information Chapter <1196> in the USP35–NF30, the General Chapter 5.8 in PhEur 7.4, along with the 'State of Work' document on the PhEur EDQM website, and also the General Information Chapter 8 in the JP XV.

Various grades of hypromellose phthalate are available with differing degrees of substitution and physical properties, e.g. grades *HP-50*, *HP-55*, and *HP-55S* (Shin-Etsu Chemical Co. Ltd). *See* Table III.

The number following 'HP' in each grade designation refers to the pH value (×10) at which the polymer dissolves in aqueous buffer solutions. The designation 'S' in *HP-55S* indicates a higher molecular weight grade, which produces films with a greater resistance to cracking.

Table III: Types of hypromellose phthalate available from Shin-Etsu Chemical Co. Ltd.

Property	Grade of hypromellose phthalate		
	HP-50	*HP-55*	*HP-55S*
Substitution type	220824	200731	200731
Hydroxypropoxy content	6–10%	5–9%	5–9%
Methoxy content	20–24%	18–22%	18–22%
Phthalyl content	21–27%	27–35%	27–35%
Molecular weight	84 000	78 000	132 000

In the US, the substitution type is indicated by a six digit number: the first two digits represent the approximate percentage content of methoxy groups; the next two digits represent the approximate percentage content of hydroxypropoxy groups; and the final two digits represent the approximate percentage content of phthalyl groups.

To dissolve hypromellose phthalate in acetone-ethanol (95%) or dichloromethane-alcohol solvent systems, the hypromellose phthalate should first be well dispersed in alcohol before adding acetone or dichloromethane. When using acetone-dichloromethane, hypromellose phthalate should be first dispersed in the dichloromethane and then the acetone added to the system.

19 Specific References

1 Rowe RC. Molecular weight studies on hydroxypropyl methylcellulose phthalate (HP55). *Acta Pharm Technol* 1982; 28(2): 127–130.
2 Ehrhardt L, *et al.* [Optimization of film coating systems.] *Pharm Ind* 1973; **35**: 719–722[in German].
3 Delporte JP, Jaminet F. [Influence of formulation of enteric coated tablets on the bioavailability of the drug.] *J Pharm Belg* 1976; **31**: 263–276[in French].
4 Patt L, Hartmann V. [Solvent residues in film forming agents.] *Pharm Ind* 1976; **38**: 902–906[in German].
5 Stafford JW. Enteric film coating using completely aqueous dissolved hydroxypropyl methyl cellulose phthalate spray solutions. *Drug Dev Ind Pharm* 1982; **8**: 513–530.
6 Thoma K, *et al.* [Resistance and disintegration behaviour of gastric juice resistant drugs.] *Pharmazie* 1987; **42**: 832–836[in German].
7 Thoma K, Heckenmüller H. [Impact of film formers and plasticizers on stability of resistance and disintegration behaviour.] *Pharmazie* 1987; **42**: 837–841[in German].
8 Takada K, *et al.* Enteric solid dispersion of ciclosporin A (CiA) having potential to deliver CiA into lymphatics. *Chem Pharm Bull* 1989; **37**: 471–474.
9 Muhammad NA, *et al.* Evaluation of hydroxypropyl methylcellulose phthalate 50 as film forming polymer from aqueous dispersion systems. *Drug Dev Ind Pharm* 1992; **18**: 1787–1797.
10 Sertsou G, *et al.* Solvent change co-precipitation with hydroxypropyl methylcellulose phthalate to improve dissolution characteristics of a poorly water-soluble drug. *J Pharm Pharmacol* 2002; **54**(8): 1041–1047.
11 Shin-Etsu Chemical Co. Ltd. Technical literature: Hydroxypropyl methylcellulose phthalate, 1993.
12 Sakellariou P, *et al.* The thermomechanical properties and glass transition temperature of some cellulose derivatives used in film coating. *Int J Pharm* 1985; **27**: 267–277.
13 Kitagawa H, *et al.* Acute and subacute toxicities of hydroxypropyl methylcellulose phthalate. *Pharmacometrics* 1970; 4(6): 1017–1025.
14 Kitagawa H, *et al.* Absorption, distribution and excretion of hydroxypropyl methylcellulose phthalate in the rat. *Pharmacometrics* 1971; 5(1): 1–4.
15 Ito R, Toida S. Studies on the teratogenicity of a new enteric coating material, hydroxypropyl methylcellulose phthalate (HPMCP) in rats and mice. *J Med Soc Toho-Univ* 1972; **19**(5): 453–461.
16 Kitagawa H, *et al.* Chronic toxicity of hydroxypropylmethylcellulose phthalate in rats. *Pharmacometrics* 1973; 7(5): 689–701.
17 Kitagawa H, *et al.* Absorption, distribution, excretion and metabolism of ^{14}C-hydroxypropyl methylcellulose phthalate. *Pharmacometrics* 1974; 8(8): 1123–1132.

20 General References

Deasy PB, O'Connell MJM. Correlation of surface characteristics with ease of production and *in vitro* release of sodium salicylate from various enteric coated microcapsules prepared by pan coating. *J Micoencapsul* 1984; **1**(3): 217–227.
Doelker E. Cellulose derivatives. *Adv Polym Sci* 1993; **107**: 199–265.
European Directorate for the Quality of Medicines and Healthcare (EDQM). European Pharmacopoeia – State Of Work Of International Harmonisation. *Pharmeuropa* 2011; 23(4): 713–714. www.edqm.eu/site/-614.html (accessed 5 December 2011).
Mantrose-Haeser Co. Inc. Material safety data sheet: *Mantrocel HP-55*, 2007.
Rowe RC. Materials used in the film coating of oral dosage forms. In: Florence AT, ed. *Critical Reports on Applied Chemistry.* Oxford: Blackwell Scientific, 1984; 1–36.
Shin-Etsu Chemical Co. Ltd. Product data sheet: *HPMPC*, 2002.

21 Author

SR Goskonda.

22 Date of Revision

2 March 2012.

Imidurea

1 Nonproprietary Names

USP–NF: Imidurea

2 Synonyms

Biopure 100; *Germall 115*; imidazolidinyl urea; methanebis[*N*,*N*′(5-ureido-2,4-diketotetrahydroimidazole)-*N*,*N*-dimethylol]; 1,1′-methylenebis{3-[3-(hydroxymethyl)-2,5-dioxo-4-imidazolidinyl] urea}; *Tri-Stat IU*.

3 Chemical Name and CAS Registry Number

N,*N*″-Methylenebis{*N*′-[3-(hydroxymethyl)-2,5-dioxo-4-imidazolidinyl]urea} [39236-46-9]

4 Empirical Formula and Molecular Weight

$C_{11}H_{16}N_8O_8$ 388.29 (for anhydrous)
$C_{11}H_{16}N_8O_8.H_2O$ 406.33 (for monohydrate)

5 Structural Formula

HO OH H H H H N N N N N N O O HN O O NH O O

6 Functional Category

Antimicrobial preservative.

7 Applications in Pharmaceutical Formulation or Technology

Imidurea is a broad-spectrum antimicrobial preservative used in cosmetics and topical pharmaceutical formulations; typical concentrations used are 0.03–0.5% w/w. It is effective at pH 3–9 and is reported to have synergistic effects when used with parabens; *see* Section 10.

8 Description

Imidurea occurs as a white, free-flowing, odorless powder.

9 Pharmacopeial Specifications

See Table I.

Table I: Pharmacopeial specifications for imidurea.

Test	USP35–NF30
Identification	+
Color and clarity of solution	+
pH (1% w/v solution)	6.0–7.5
Loss on drying	≤3.0%
Residue on ignition	≤3.0%
Heavy metals	≤0.001%
Nitrogen content (dried basis)	26.0–28.0%

10 Typical Properties

Acidity/alkalinity pH = 6.0–7.5 (1% w/v aqueous solution)

Antimicrobial activity Predominantly an antibacterial preservative, imidurea also has some selective antifungal properties. Used at concentrations between 0.03–0.5% w/w it is effective at pH 3–9, although preservative efficacy is best seen in slightly acidic solutions. Synergistic effects have been reported, and preservative activity is considerably enhanced, particularly against fungi, when used in combination with parabens.[1,2] A cosmetic formulation containing 0.5% imidurea, 0.2% methylparaben, and 0.1% propylparaben was effectively preserved against various *Pseudomonas* species.[3] For reported minimum inhibitory concentrations (MICs), *see* Table II.[4]

Table II: Minimum inhibitory concentrations (MICs) for imidurea.

Microorganism	MIC (μg/mL)
Aspergillus niger	8000
Candida albicans	8000
Escherichia coli	2000
Klebsiella pneumoniae	2000
Penicillium notatum	8000
Pseudomonas aeruginosa	2000
Pseudomonas cepacia	2000
Pseudomonas fluorescens	2000
Staphylococcus aureus	1000

Solubility Soluble in water and in glycerol, but insoluble in almost all organic solvents.[4] *See also* Table III.

Table III: Solubility of imidurea.

Solvent	Solubility at 20°C
Ethanol	Very slightly soluble
Ethanol (90%)	Very slightly soluble
Ethanol (70%)	1 in 330
Ethanol (60%)	1 in 25
Ethanol (50%)	1 in 2.5
Ethanol (30%)	1 in 0.8
Ethylene glycol[a]	1 in 0.7
Glycerin[a]	1 in 1
Methanol	Very slightly soluble
Mineral oil	Practically insoluble
Propan-2-ol	Practically insoluble
Propylene glycol[a]	1 in 0.8
Sesame oil	Very slightly soluble
Water	1 in 0.5

(a) Slow to dissolve and requires heating and stirring.

Spectroscopy

NIR spectra *see* Figure 1.

11 Stability and Storage Conditions

Imidurea is hygroscopic and should be stored in a well-closed container in a cool, dry place.

12 Incompatibilities

Imidurea is incompatible with strong oxidants. It is compatible with other preservatives including sorbic acid and quaternary ammonium compounds.[5] It is also compatible with other pharmaceutical

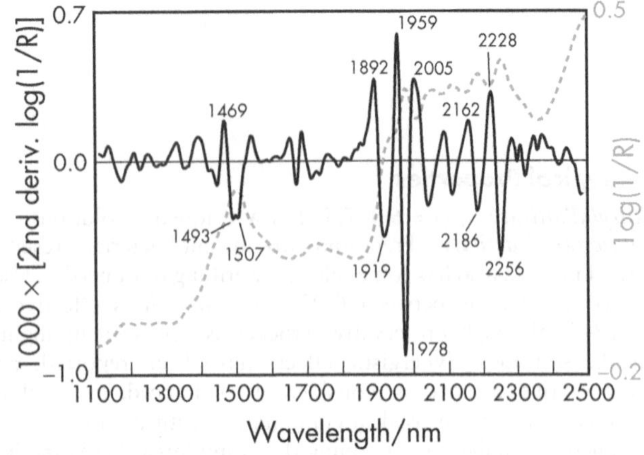

Figure 1: Near-infrared spectrum of imidurea measured by reflectance.

and cosmetic excipients including proteins, nonionic surfactants, and lecithin.[6]

13 Method of Manufacture

Imidurea is produced by the condensation of allantoin with formaldehyde.

14 Safety

Imidurea is widely used in cosmetics and topical pharmaceutical formulations, and is generally regarded as a nontoxic and nonirritant material.[5] However, there have been some reports of contact dermatitis associated with imidurea, although these are relatively few considering its widespread use in cosmetics.[7–13]

Although imidurea releases formaldehyde, it does not appear to be associated with cross-sensitization with formaldehyde or other formaldehyde-releasing compounds.

LD_{50} (mouse, oral): 7.2 g/kg[14,15]
LD_{50} (rabbit, skin): >8 g/kg
LD_{50} (rat, oral): 11.3 g/kg

15 Handling Precautions

Observe normal precautions appropriate to the circumstances and quantity of material handled. Imidurea may be irritant to the eyes. Eye protection and gloves are recommended.

16 Regulatory Status

Included in the FDA Inactive Ingredients Database (topical preparations). Accepted for use in cosmetics in Europe and the US. Included in the Canadian Natural Health Products Ingredients Database.

17 Related Substances

Diazolidinyl urea.

Diazolidinyl urea
Empirical formula $C_8H_{14}N_4O_7$
Molecular weight 278.23
CAS number [78491-02-8]
Synonyms *Germall II*; *N*-(hydroxymethyl)-*N*-(1,3-dihydroxymethyl-2,5-dioxo-4-imidazolidinyl)-*N*′-(hydroxymethyl)urea.
Appearance White, free-flowing hygroscopic powder, with a faint characteristic odor.
Antimicrobial activity Similar to imidurea.[16,17] Diazolidinyl urea is the most active of the imidazolidinyl family of preservatives. Used in concentrations of 0.1–0.5% w/w, at pH 3–9, it has predominantly antibacterial properties. Typical MICs are: *Aspergillus niger* 4000 μg/mL; *Candida albicans* 8000 μg/mL; *Escherichia coli* 1000 μg/mL; *Pseudomonas aeruginosa* 1000 μg/mL; *Staphylococcus aureus* 250 μg/mL.
Solubility Very soluble in water; insoluble in fats.
Safety

LD_{50} (mouse, oral): 3.7 g/kg[18]
LD_{50} (rat, oral): 2.6 g/kg

Comments The EINECS number for diazolidinyl urea is 278-928-2.

18 Comments

Imidurea is the best known of a family of heterocyclic urea derivatives that are effective antimicrobial preservatives. Diazolidinyl urea has the greatest antimicrobial activity.

The EINECS number for imidurea is 254-372-6. The PubChem Compound ID (CID) for imidurea is 38258.

19 Specific References

1 Jacobs G, *et al.* The influence of pH, emulsifier, and accelerated ageing upon preservative requirements of o/w emulsions. *J Soc Cosmet Chem* 1975; **26**: 105–117.
2 Rosen WE, *et al.* Preservation of cosmetic lotions with imidazolidinyl urea plus parabens. *J Soc Cosmet Chem* 1977; **28**: 83–87.
3 Berke PA, Rosen WE. Imidazolidinyl urea activity against *Pseudomonas*. *J Soc Cosmet Chem* 1978; **29**: 757–766.
4 Wallhäusser KH. Imidazolidinyl urea. In: Kabara JJ, ed. *Cosmetic and Drug Preservation Principles and Practice*. New York: Marcel Dekker, 1984; 655–657.
5 Rosen WE, Berke PA. Germall 115: a safe and effective modern preservative. *Cosmet Toilet* 1977; **92**(3): 88–89.
6 Rosen WE, Berke PA. Germall 115 and nonionic emulsifiers. *Cosmet Toilet* 1979; **94**(12): 47–48.
7 Fisher AA. Cosmetic dermatitis: part II. Reactions to some commonly used preservatives. *Cutis* 1980; **26**: 136, 137, 141, 142, 147–148.
8 Dooms-Goossens A, *et al.* Imidazolidinyl urea dermatitis. *Contact Dermatitis* 1986; **14**(5): 322–324.
9 O'Brien TJ. Imidazolidinyl urea (Germall 115) causing cosmetic dermatitis. *Aust J Dermatol* 1987; **28**(1): 36–37.
10 Ziegler V, *et al.* Dose-response sensitization experiments with imidazolidinyl urea. *Contact Dermatitis* 1988; **19**(3): 236–237.
11 Dastychová E, *et al.* Contact sensitization to auxiliary substances in dermatological external and cosmetic preparations. *Ceska Slov Farm* 2004; **53**(3): 151–156.
12 Jong CT, *et al.* Contact sensitivity to preservatives in the UK, 2004-2005: results of multicentre study. *Contact Dermatitis* 2007; **57**(3): 165–168.
13 Garcia-Gavin J, *et al.* Allergic contact dermatitis in a girl due to several cosmetics containing diazolidinyl-urea or imidazolidinyl-urea. *Contact Dermatitis* 2010; **63**(1): 49–50.
14 Elder RL. Final report of the safety assessment for imidazolidinyl urea. *J Environ Pathol Toxicol* 1980; 4(4): 133–146.
15 Sweet DV, ed. *Registry of Toxic Effects of Chemical Substances*. Cincinnati: US Department of Health, 1987: 5023.
16 Berke PA, Rosen WE. Germall II: a new broad-spectrum cosmetic preservative. *Cosmet Toilet* 1982; **97**(6): 49–53.
17 Wallhäusser KH. Germall II. In: Kabara JJ, ed. *Cosmetic and Drug Preservation Principles and Practice*. New York: Marcel Dekker, 1984: 657–659.
18 Lewis RJ, ed. *Sax's Dangerous Properties of Industrial Materials*, 11th edn. New York: Wiley, 2004; 2072.

20 General References

Berke PA, Rosen WE. Germall, a new family of antimicrobial preservatives for cosmetics. *Am Perfum Cosmet* 1970; **85**(3): 55–59.
Croshaw B. Preservatives for cosmetics and toiletries. *J Soc Cosmet Chem* 1977; **28**: 3–16.

Decker RL, Wenninger JA. Frequency of preservative use in cosmetic formulas as disclosed to FDA-1987. *Cosmet Toilet* 1987; **102**(12): 21–24.

Rosen WE, Berke PA. Germall 115: a safe and effective preservative. In: Kabara JJ, ed. *Cosmetic and Drug Preservation Principles and Practice.* New York: Marcel Dekker, 1984; 191–205.

Schnuch A, *et al.* Contact allergy to preservatives. Analysis of IVDK data 1996-2009. *Br J Dermatol* 2011; **164**: 1316–1325.

21 Author

RT Guest.

22 Date of Revision

1 March 2012.

Inulin

1 Nonproprietary Names

BP: Inulin

USP-NF: Inulin

2 Synonyms

Alant starch; alantin; *Beneo*; dahlin; *Frutafit*; oligofructose; *Orafti*; polyfructose; *Raftiline*.

3 Chemical Name and CAS Registry Number

Inulin [9005-80-5]

4 Empirical Formula and Molecular Weight

$C_6H_{11}O_5(C_6H_{11}O_5)_nOH$ ≈5 000

5 Structural Formula

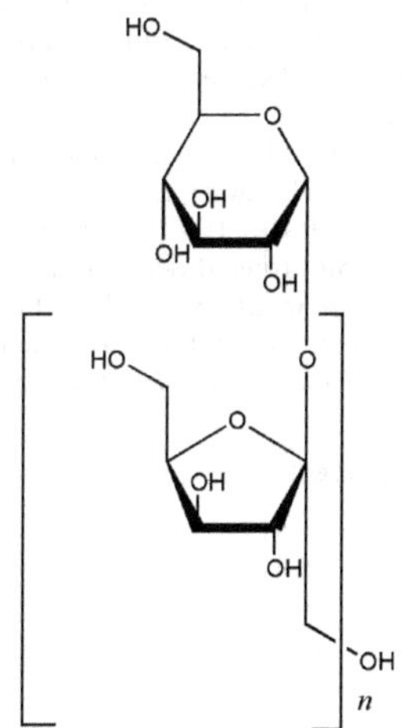

Inulin is a naturally occurring polysaccharide consisting of a linear chain of linked D-fructose molecules, having one terminal glucose molecule.

6 Functional Category

Lyophilization aid; sweetening agent; tablet and capsule diluent.

7 Applications in Pharmaceutical Formulation or Technology

Inulin has many potential uses in pharmaceutical applications, as a filler–binder in tablet formulations,[1,2] to stabilize therapeutic proteins,[3,4] vaccines,[5] and PEGylated lipoplexes[6] during freeze-drying, or to enhance the dissolution of lipophilic drugs.[7–9] Inulin has also been used as a stabilizer for recombinant human deoxyribonuclease when spray-dried to produce a powder for inhalation.[10] *See also* Section 18.

Inulin is used in the food industry as a sweetener and stabilizer.

8 Description

Inulin occurs as an odorless, friable, white amorphous powder with a neutral to slightly sweet taste.

9 Pharmacopeial Specifications

See Table I.

Table I: Pharmacopeial specifications for inulin.

Test	BP 2010	USP35–NF30
Identification	+	—
Characters	+	—
Acidity	+	4.5–7.0
Clarity and color of solution	+	+
Microbial limit	—	≤1000/g
Loss on drying	≤10.0%	≤10.0%
Specific rotation	–36.5° to –40.5°	–32.0° to –40.0°
Residue on ignition	≤0.1%	≤0.05%
Sulfate	≤200 ppm	≤0.05%
Calcium	≤270 ppm	≤0.1%
Chloride	≤170 ppm	≤0.014%
Iron	—	+
Heavy metals	—	≤5 ppm
Arsenic	≤1 ppm	—
Lead	≤2 ppm	—
Oxalate	+	—
Reducing sugars	+	+
Free fructose	—	+
Content of combined glucose	—	+
Assay (dried basis)	—	94.0–102.0%

10 Typical Properties

Acidity/alkalinity pH = 4.5–7.0 (10% w/v aqueous solution)

Density 1.35 g/cm^3

Hygroscopicity Hygroscopic in moist air.

Melting point 178°C

Solubility Soluble in hot water and solutions of dilute acids and alkalis; slightly soluble in cold water and organic solvents.

Specific gravity 1.35

Specific rotation -36.5° to -40.5° (determined on a 2% solution prepared with the aid of heat)

11 Stability and Storage Conditions

Inulin is slightly hygroscopic and should be stored at cool to normal temperatures, in air-tight and water-tight containers. Hydrolysis of inulin yields mainly D-fructose. Crystals of inulin may be deposited on storage of injections; the injection should be heated for not more than 15 minutes to dissolve the crystals and then cooled to a suitable temperature before use. Solutions of inulin for injection are sterilized by filtration.

12 Incompatibilities

Inulin is incompatible with strong oxidizing agents.

13 Method of Manufacture

Inulin is extracted from the tubers of *Dahlia variabilis*, Helianthus, in a procedure similar to the extraction of sugar from sugar beet.

14 Safety

Inulin is a naturally occurring plant polysaccharide and is one of the major constituents of the Compositae family. Inulin is recommended to diabetics, as it has a mild sweet taste, but is not absorbed and does not affect blood sugar levels. It is used widely in the food industry as a sweetener and stabilizer.

15 Handling Precautions

Observe normal precautions appropriate to the circumstances and quantity of material handled. Inulin may cause mild irritation to the skin and the eyes. Eye protection and gloves are recommended.

16 Regulatory Status

GRAS listed.

17 Related Substances

—

18 Comments

Hollow spheres of inulin have been found to have both brittle and ductile properties. On compression, these spheres will undergo fragmentation followed by plastic deformation, resulting in better compressibility over solid inulin spheres. In its amorphous state, inulin has a high glass transition temperature, slow crystallization, and low hygroscopicity. As a binder in solid dosage forms, inulin can increase the dissolution rate of drugs such as diazepam and can enhance the stability of other lipophilic drug molecules.(2,9)

Experimentally, methacrylated inulin hydrogels have been investigated for the development of colon-specific drug delivery systems,(11–13) and coatings composed of inulin and shellac have been shown to release drug in simulated colonic fluid.(14) An inulin-iron complex has been proposed as a potential treatment for iron deficiency anemia.(15)

Inulin is used intravenously as a diagnostic agent to measure the glomerular filtration rate.(16) Inulin has also entered the food supplement market as a prebiotic, where it has been shown to provide protection against inflammatory and malignant colonic diseases in animals,(17,18) and it is used as a noncaloric dietary fiber supplement. Radio-labelled forms of inulin are available as radiochemicals for research.

19 Specific References

1 Eissens AC, *et al.* Inulin as filler-binder for tablets prepared by direct compaction. *Eur J Pharm Sci* 2002; **15**(1): 31–38.
2 Bolhuis GK, *et al.* Hollow filler-binders as excipients for direct compaction. *Pharm Res* 2003; **20**(3): 515–518.
3 Eriksson HJ, *et al.* Investigations into the stabilization of drugs by sugar glasses: I. Tablets prepared from stabilized alkaline phosphate. *Int J Pharm* 2002; **249**(1–2): 59–70.
4 Hinrichs WL, *et al.* Inulin glasses for the stabilisation of therapeutic proteins. *Int J Pharm* 2001; **215**: 163–174.
5 De Jonge J, *et al.* Inulin sugar glasses preserve the structural integrity and biological activity of influenza virosomes during freeze-drying and storage. *Eur J Pharm Sci* 2007; **32**: 33–44.
6 Hinrichs, *et al.* Inulin is a promising cryo- and lyoprotectant for PEGylated lipoplexes. *J Control Release* 2005; **103**: 465–479.
7 Visser MR, *et al.* Inulin solid dispersion technology to improve the absorption of the BCS Class IV drug TMC240. *Eur J Pharm Biopharm* 2010; **74**: 233–238.
8 Fares MM, *et al.* Inulin and poly(acrylic acid) grafted inulin for dissolution enhancement and preliminary controlled release of poorly soluble irbesartan drug. *Int J Pharm* 2011; **410**: 206–211.
9 International Pharmaceutical Excipients Council Europe. *IPEC Europe News*, Jan 2003.
10 Zijlstra GS, *et al.* Formulation and process development of (recombinant human) deoxyribonuclease I as powder for inhalation. *Pharm Dev Technol* 2009; **14**: 358–368.
11 Van den Mooter G, *et al.* Characterization of methacrylated inulin hydrogels designed for colon targeting: *in vitro* release of BSA. *Pharm Res* 2003; **20**(2): 303–307.
12 Windfeld S, *et al.* [^{3}H]Inulin as a marker for glomerular filtration rate. *Am J Physiol Renal Physiol* 2003; **285**(3): 575–576.
13 Reddy BS, *et al.* Effect of dietary oligofructose and inulin on colonic preneoplastic abberant crypt foci inhibition. *Carcinogenesis* 1997; **18**(7): 1371–1374.
14 Maris B, *et al.* Synthesis and characterization of inulin-azo hydrogels designed for colon targeting. *Int J Pharm* 2001; **213**: 143–152.
15 Delzenne N, *et al.* Prebiotics: actual and potential effects in inflammatory and malignant colonic diseases. *Curr Opin Clin Nutr Metab Care* 2003; **6**(5): 581–586.
16 Vervoort L, *et al.* Inulin hydrogels as carriers for colonic drug targetting: I. Synthesis and characterization of methacrylated inulin and hydrogel formation. *Pharm Res* 1997; **14**(12): 1730–1737.
17 Ravi V, *et al.* Novel colon targeted drug delivery system using natural polymers. *Indian J Pharm Sci* 2008; **70**: 111–113.
18 Pitarresi G, *et al.* Inulin-iron complexes: A potential treatment of iron deficiency anemia. *Eur J Pharm Biopharm* 2008; **68**: 267–276.

20 General References

—

21 Author

CG Cable.

22 Date of Revision

1 March 2012.

Iron Oxides

1 Nonproprietary Names

USP–NF: Ferric Oxide

2 Synonyms

(a) Iron oxide black: *Bayferrox 306*; black magnetic oxide; black oxide, precipitated; black rouge; CI 77499; E172; ethiops iron; ferric ferrous oxide; ferrosoferric oxide; *Ferroxide 78P*; *Ferroxide 88P*; iron oxide; iron (II, III) oxide; iron oxides (FeO); magnetite; *Mapico Black EC*; pigment black 11; *Sicovit B80*; *Sicovit B84*; *Sicovit B85*; triiron tetraoxide.

(b) Iron oxide red: anhydrous ferric oxide; anhydrous iron (III) oxide; *Bayferrox 105M*; CI 77491; diiron trioxide; E172; *Ferroxide 212P*; *Ferroxide 226P*; hematite; iron sesquioxide; jeweler's rouge; pigment red 101; red ferric oxide; *Sicovit R30*.

(c) Iron oxide yellow monohydrate: E172; hydrated ferric oxide; iron (III) oxide monohydrate, yellow; pigment yellow 42; yellow ferric oxide. Iron (III) oxide hydrated: *Bayferrox 920Z*; CI 77492; ferric hydroxide; ferric hydroxide oxide; ferric hydrate; ferric oxide hydrated; *Ferroxide 505P*; *Ferroxide 510P*; iron hydrate; iron hydroxide; iron hydroxide oxide; *Mapico Yellow EC*; *Sicovit Y10*; yellow ochre; yellow iron oxide.

3 Chemical Name and CAS Registry Number

Iron oxides [1332-37-2]

(a) Iron oxide black [1317-61-9] and [12227–89–3]

(b) Iron oxide red [1309-37-1]

(c) Iron oxide yellow [51274-00-1] (monohydrate); [20344-49-4] (ferric hydroxide oxide); [12259–21–1] (hydrated ferric oxide)

4 Empirical Formula and Molecular Weight

(a) $FeO{\cdot}Fe_2O_3$	231.54
(b) Fe_2O_3	159.69
(c) $Fe_2O_3{\cdot}H_2O$	177.7 (monohydrate);
$FeO(OH)$	88.85 (hydroxide oxide)

5 Structural Formula

Iron oxides are defined as inorganic compounds consisting of any one of or combinations of synthetically prepared iron oxides, including the hydrated forms.

6 Functional Category

Colorant.

7 Applications in Pharmaceutical Formulation or Technology

Iron oxides are widely used in cosmetics, foods, and pharmaceutical applications as colorants and UV absorbers.[1–3] As as a result of the regulations affecting some synthetic organic dyestuffs, the use of iron oxides (inorganic colorants) is becoming increasingly important. However, iron oxides also have restrictions in some countries on the quantities that may be consumed, and technically their use is restricted because of their limited color range and their abrasiveness.

8 Description

Iron oxides occur as yellow, red, black, or brown powder. The color depends on the particle size and shape, and crystal structure.

9 Pharmacopeial Specifications

See Table I. *See also* Section 18.

Table I: Pharmacopeial specifications for iron oxide (red and yellow).

Test	USP35–NF30
Identification	+
Water-soluble substances	≤1.0%
Acid-insoluble substances	≤0.3%
Organic colors and lakes	+
Mercury	≤3 μg/g
Arsenic	≤3 μg/g
Lead	0.001%
Assay (on ignited basis)	97.0–100.5%

10 Typical Properties

Density

5.1 g/cm^3 for iron oxide black ($FeO{\cdot}Fe_2O_3$);
5.2 g/cm^3 for iron oxide red (Fe_2O_3);
4.1 g/cm^3 for iron oxide yellow ($Fe_2O_3{\cdot}H_2O$).

Melting point

1565°C for iron oxide red (Fe_2O_3). Iron oxide yellow ($Fe_2O_3{\cdot}H_2O$) and iron oxide black ($FeO{\cdot}Fe_2O_3$) are not stable above 80-120°C and undergo dehydration (iron oxide yellow) or oxidation (iron oxide black) transforming into iron oxide red.

Solubility Soluble in mineral acids; insoluble in water.

11 Stability and Storage Conditions

Iron oxides should be stored in well-closed containers in a cool, dry place.

12 Incompatibilities

Iron oxides have been reported to make hard gelatin capsules brittle at higher temperatures when the residual moisture is 11–12%. This factor affects the use of iron oxides for coloring hard gelatin capsules, and will limit the amount that can be incorporated into the gelatin material.

13 Method of Manufacture

Three main manufacturing processes are currently applied for iron oxide pigments:[4]

(a) Solid-state reactions (red, black and brown): calcination of black or yellow iron oxides to red iron oxide; thermal decomposition of ferrous sulfate.

(b) Precipitation process (red, orange, yellow and black): treatment of ferrous sulfate solutions with alkali and oxidation. The Penniman–Zoph process uses ferrous sulfate, alkali, iron powder and air or oxygen.

(c) Laux process or aniline process (red, yellow, and black): reduction of nitrobenzene to aniline with iron.

14 Safety

Iron oxides are widely used in cosmetics, foods, and oral and topical pharmaceutical applications. They are generally regarded as nontoxic and nonirritant excipients. The use of iron oxide colorants is limited in some countries, such as the US, to a maximum ingestion of 5 mg of elemental iron per day.

LD_{50} (mouse, IP): 5.4 g/kg[5]
LD_{50} (rat, IP): 5.5 g/kg

15 Handling Precautions

Observe normal precautions appropriate to the circumstances and quantity of the material handled. In the UK, the workplace exposure limits for iron oxide fumes (as Fe) are 5 mg/m^3 long-term (8-hour TWA) and 10 mg/m^3 short-term.[6]

16 Regulatory Status

Accepted for use as a food additive in Europe. Included in the FDA Inactive Ingredients Database (oral capsules, tablets, and suspensions); iron oxide red is also listed for use in oral drops and buccal films, and iron oxide yellow is listed for use in buccal tablets. Included in nonparenteral medicines licensed in many countries including Japan, UK, and US.

17 Related Substances

—

18 Comments

Although iron oxides are not included in any pharmacopeias, the Joint FAO/WHO Expert Committee on Food Additives has issued specifications for iron oxide; *see* Table II.[7] Specifications for iron oxide black,[8] iron oxide red,[9] and iron oxide yellow monohydrate[10] are included in the *Japanese Pharmaceutical Excipients* (JPE); *see* Table III.

The EINECS number for iron oxide black (Fe_3O_4) is 215-277-5. The EINECS number for iron oxide red (Fe_2O_3) is 215-168-2. The EINECS number for iron oxide yellow ($Fe_2O_3 \cdot H_2O$) is 257-098-5.

Table II: Joint FAO/WHO Expert Committee on Food Additive specifications for iron oxides.

Test	FAO/WHO
Water-soluble matter	≤1.0%
Solubility	+
Loss on drying (iron oxide red)	≤1.0%
Cadmium	≤1 mg/kg
Mercury	≤1 mg/kg
Arsenic	≤3 mg/kg
Lead	≤10 mg/kg
Assay	+

Table III: Specifications for iron oxide black, iron oxide red, and iron oxide yellow monohydrate from JPE 2004.

Test	JPE 2004		
	Iron oxide black (a)	Iron oxide red (b)	Iron oxide yellow monohydrate (c)
Description	+	+	+
Identification	+	+	+
Purity	+	+	+
Heavy metals	≤30 ppm	≤30 ppm	≤30 ppm
Arsenic	≤10 ppm	≤2 ppm	≤2 ppm
Loss on ignition	—	≤2.0%	10.0–13.0%
Water-soluble substances	≤0.75%	≤0.75%	≤0.75%
Loss on drying	≤1.0%	—	—
Assay	≥90.0% (dried basis)	≥98.0% (ignited basis)	≥98.0% (ignited basis)

19 Specific References

1 Rowe RC. Opacity of tablet film coatings. *J Pharm Pharmacol* 1984; 36: 569–572.
2 Rowe RC. Synthetic iron oxides: ideal for pharmaceutical colorants. *Pharm Int* 1984; 5: 221–224.
3 Ceschel GC, Gibellini M. Use of iron oxides in the film coating of tablets. *Farmaco Ed Prat* 1980; 35: 553–563.
4 Buxbaum G. *Industrial Inorganic Pigments*, 2nd edn. Weinheim. Wiley-VCH, 1998: 89–91.
5 Lewis RJ, ed. *Sax's Dangerous Properties of Industrial Materials*, 11th edn. New York: Wiley, 2004: 2111–2112.
6 Health and Safety Executive. *EH40/2005. Workplace Exposure Limits*. Sudbury: HSE Books, 2011. http://www.hse.gov.uk/pubns/priced/eh40.pdf (accessed 21 February 2012).
7 Joint FAO/WHO Expert committee on Food Additives (2008). Iron oxides. www.fao.org/ag/agn/jecfa-additives/details.html?id=893 (accessed 9 December 2011).
8 Japan Pharmaceutical Excipients Council. *Japanese Pharmaceutical Excipients 2004*. Tokyo: Yakuji Nippo, 2006; (Suppl): 34–35.
9 Japan Pharmaceutical Excipients Council. *Japanese Pharmaceutical Excipients 2004*. Tokyo: Yakuji Nippo, 2006; (Suppl): 36–37.
10 Japan Pharmaceutical Excipients Council. *Japanese Pharmaceutical Excipients 2004*. Tokyo: Yakuji Nippo, 2006; (Suppl): 25.

20 General References

—

21 Authors

C Egger, D Schoneker, S Tiwari.

22 Date of Revision

1 March 2012.

Isomalt

1 Nonproprietary Names

BP: Isomalt
PhEur: Isomalt
USP–NF: Isomalt

2 Synonyms

*C*PharmIsoMaltidex*; E953; *galenIQ*; hydrogenated isomaltulose; hydrogenated palatinose; *IsoMaltidex 16500*; isomaltum; *Palatinit*.

3 Chemical Name and CAS Registry Number

Isomalt [64519-82-0]

Isomalt is a mixture of two stereoisomers:

6-*O*-α-D-glucopyranosyl-D-sorbitol (1,6-GPS) [534-73-6]

1-*O*-α-D-glucopyranosyl-D-mannitol dihydrate (1,1-GPM) [20942-99-8]

4 Empirical Formula and Molecular Weight

$C_{12}H_{24}O_{11}$ 344.32 (for anhydrous)
$C_{12}H_{24}O_{11}\cdot 2H_2O$ 380.32 (for dihydrate)

5 Structural Formula

$C_{12}H_{24}O_{11}$
(1,6-GPS)

$C_{12}H_{24}O_{11} \cdot 2H_2O$
(1,1-GPM)

Generally, isomalt comprises a mixture of 1,6-GPS and 1,1-GPM. 1,6-GPS crystallizes without water and is more soluble than 1,1-GPM. By shifting the ratio of the two components, the solubility and crystal water content can be adjusted; *see* Section 10. *galenIQ 720* has a GPM : GPS ratio of 1 : 1; *galenIQ 721* has a GPM : GPS ratio of 1 : 3.

6 Functional Category

Coating agent; direct compression excipient; suspending agent; sweetening agent; tablet and capsule binder; tablet and capsule diluent.

7 Applications in Pharmaceutical Formulation or Technology

Isomalt is a noncariogenic excipient used in a variety of pharmaceutical preparations including tablets or capsules, in coatings as a sugar-free coating agent, sachets, dry powder for suspension, and in effervescent tablets. It can also be used in direct compression, roller compaction, and wet granulation.(1)

In buccal applications such as chewable tablets it is commonly used because of its negligible negative heat of solution, mild sweetness, and 'mouth feel'.(2,3) It is also used widely in medicated high-boiled lozenges such as cough drops, as a sweetening agent in medicated confectionery for diabetics, and in sugar-free chewing gum.

See also Section 18.

8 Description

Isomalt is a sugar alcohol (polyol) that occurs as a white or almost white powder or granular or crystalline substance. It has a pleasant sugar-like taste with a mild sweetness approximately 50–60% of that of sucrose.(2–4)

9 Pharmacopeial Specifications

See Table I.

10 Typical Properties

Angle of repose *see* Table II.
Compressibility Compression characteristics may vary, depending on the grade of isomalt used; *see* Figure 1.
Density (bulk) *see* Table II.
Density (tapped) *see* Table II.
Density (true)
$1.52\,g/cm^3$ for 1,6-GPS;
$1.47\,g/cm^3$ for 1,1-GPM.
Flowability Powder is cohesive; granules are free flowing.(2)

SEM 1: Excipient: *galenIQ 720*; manufacturer: BENEO-Palatinit GmbH; magnification: 400×; voltage: 5 kV.

SEM 2: Excipient: *galenIQ 721*; manufacturer: BENEO-Palatinit GmbH; magnification: 400×; voltage: 5 kV.

SEM 3: Excipient: *galenIQ 810*; manufacturer: BENEO-Palatinit GmbH; magnification: 65×; voltage: 10 kV.

SEM 4: Excipient: *galenIQ 981*; manufacturer: BENEO-Palatinit GmbH; magnification: 90×; voltage: 5 kV.

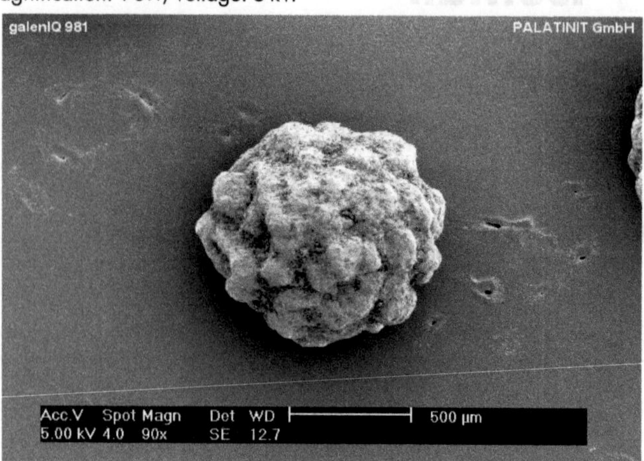

SEM 5: Excipient: *galenIQ 990*; manufacturer: BENEO-Palatinit GmbH; magnification: 130×; voltage: 10 kV.

Table I: Pharmacopoeial specifications for isomalt.

Test	PhEur 7.4	USP35–NF30
Identification	+	+
Characters	+	−
Related products	+	+
Conductivity	≤20 µS cm^{-1}	≤20 µS cm^{-1}
Reducing sugars	≤0.3%	≤0.3%
Lead	≤0.5 ppm	−
Heavy metals	−	≤10 µg/g
Nickel	≤1 ppm	≤1 µg/g
Water	≤7.0%	≤7.0%
Assay	98.0–102.0%	98.0–102.0%

Glass transition temperature

63°C for a 1 : 3 mixture of 1,1-GPM and 1,6-GPS;

68°C for 1,1-GPM;

59°C for 1,6-GPS.[2]

Heat of combustion 0.017 kJ/kg[5]

Heat of solution +14.6 kJ/mol for an equimolar mixture of 1,1-GPM and 1,6-GPS.[2]

Hygroscopicity Not hygroscopic until 85% RH, at 25°C.[2] *See also* Figure 2.

Melting point

141–161°C for a 1 : 3 mixture of 1,1-GPM and 1,6-GPS;

166–168°C for 1,6-GPS;

168–171°C for 1,1-GPM.[2]

Minimum ignition temperature >460°C

Moisture content *see* Figure 2.

Particle size distribution

Approximately 90% >100 µm for *galenIQ 720*;

approximately 58% >20 µm for *galenIQ 800*;

approximately 99% >200 µm for *galenIQ 960*.

pH 3–10[3]

Solubility *see* Figure 3.

11 Stability and Storage Conditions

Isomalt has very good thermal and chemical stability. When it is melted, no changes in the molecular structure are observed. It exhibits considerable resistance to acids and microbial influences.[1] Isomalt is non-hygroscopic, and at 25°C does not significantly absorb additional water up to a relative humidity (RH) of 85%; acetaminophen (paracetamol) tablets based on isomalt were stored for 6 months at 85% RH at 20°C and retained their physical aspect.[1]

Figure 1: Tablet crushing strength of isomalt (*galenIQ 720*, BENEO-Palatinit GmbH).
Formulation: 99.5% isomalt, 0.5% magnesium stearate
Tablet weight: 240 mg
Diameter: 8 mm
Press: Fette P1200
Punch: concave

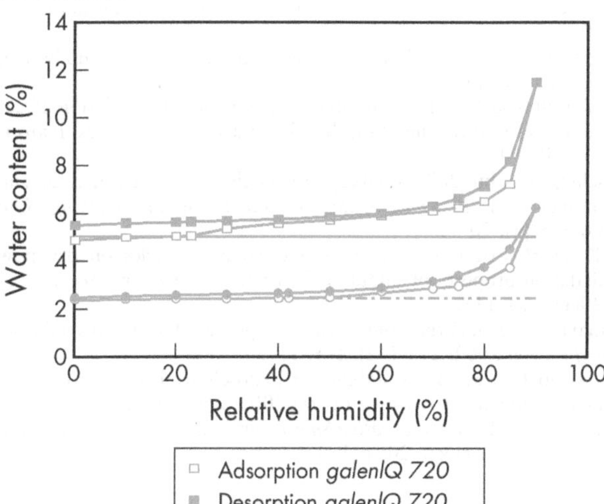

Figure 2: Sorption isotherms of isomalt DC types (*galenIQ*, BENEO-Palatinit GmbH).
Measured using Dynamic Vapor Sorption, Südzucker AG.
1,6-GPS occurs without crystal water and 1,1-GPM crystallizes with 2 mol crystal water (the initial water content in commercial forms, *see* Section 18. The starting point of the curves depends on the water content. The content of free water in the product is typically 0.1–0.5%.

If stored under normal ambient conditions, isomalt is chemically stable for many years. When it is stored in an unopened container at 20°C and 60% RH, a re-evaluation after 3 years is recommended.

Isomalt does not undergo browning reactions; it has no reducing groups, and therefore it does not react with other ingredients in a formulation (e.g. with amines in Maillard reactions).

12 Incompatibilities

—

13 Method of Manufacture

Isomalt is produced from food-grade sucrose in a two-stage process. Beet sugar is converted by enzymatic transglucosidation into the reducing disaccharide isomaltulose. This undergoes catalytical hydrogenation to produce isomalt.

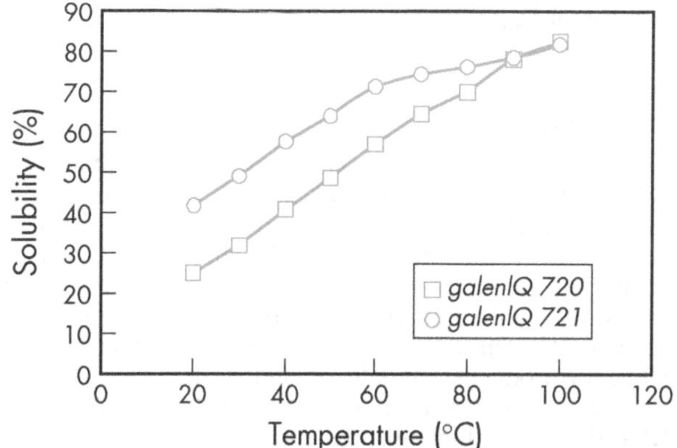

Figure 3: Solubility of isomalt types in water (*galenIQ*, BENEO-Palatinit GmbH).[2]

Table II: Typical physical properties of selected commercially available isomalt grades, *galenIQ* (BENEO-Palatinit GmbH).

Grade	Angle of repose (°)	Density (bulk) (g/cm³)	Density (tapped) (g/cm³)
galenIQ 720	38	0.43	0.48
galenIQ 721	37	0.42	0.45
galenIQ 800	—	0.50	0.65
galenIQ 810	—	0.59	0.70
galenIQ 960	33	0.82	—
galenIQ 980	—	0.82	—
galenIQ 981	—	0.78	—
galenIQ 990	—	0.85	—

14 Safety

Isomalt is used in oral pharmaceutical formulations, confectionery, and food products. It is generally regarded as a nontoxic, nonallergenic, and nonirritant material.

Toxicological and metabolic studies on isomalt[5–10] have been summarized in a WHO report prepared by the FAO/WHO Expert Committee (JECFA), resulting in an acceptable daily intake of 'not specified'.[11]

The glycosidic linkage between the mannitol or sorbitol moiety and the glucose moiety is very stable, limiting the hydrolysis and absorption of isomalt in the small intestine. There is no significant increase in the blood glucose level after oral intake, and glycemic response is very low, making isomalt suitable for diabetics. The majority of isomalt is fermented in the large intestine. In general, isomalt is tolerated very well, although excessive consumption may result in laxative effects.[12–14]

Isomalt is not fermented by bacteria present in the mouth; therefore no significant amount of organic acid is produced that attacks tooth enamel.[15–17]

15 Handling Precautions

Observe normal precautions appropriate to the circumstances and quantity of material handled. Eye protection, gloves, and a dust mask or respirator are recommended.

16 Regulatory Status

GRAS listed. Included in the FDA Inactive Ingredients Database (oral tablets). Accepted as a food additive in Europe. Included in the Canadian Natural Health Products Ingredients Database.

17 Related Substances

—

18 Comments

Compression of isomalt without lubrication is difficult, and problems such as die wall sticking, capping, and lamination have been observed. The addition of a lubricant such as magnesium stearate will reduce die wall adhesion. Co-extrusion of isomalt with acetaminophen (paracetamol) significantly improved the tableting properties of the mixtures, compared to physical mixtures of drug and isomalt.[18] Direct molding is also a potentially suitable technique for producing isomalt-based tablets.[18]

A variety of different grades of isomalt are commercially available that have different applications, e.g. *galenIQ 720* and *721* are used in direct compression and in orally dispersible powders, *galenIQ 801* is used in wet granulation, *galenIQ 981* is used in sugar-free coatings, and *galenIQ 990* is used in high-boiled lozenges.

A specification for isomalt is contained in the *Food Chemicals Codex* (FCC).[19]

19 Specific References

1 Ndindayino F, *et al.* Characterization and evaluation of isomalt performance in direct compression. *Int J Pharm* 1999; **189**: 113–124.
2 BENEO-Palatinit GmbH. Technical literature: Isomalt, *galenIQ*, 2010.
3 Cargill. Technical literature: *C*PharmIsoMaltidex*, 2007.
4 Schiweck H. *Palatinit*—Production, technological characteristics and analytical study of foods containing *Palatinit* [translated from German]. *Alimenta* 1980; **19**: 5–16.
5 Livesey G. The energy values of dietary fibre and sugar alcohols for man. *Nutr Res Rev* 1992; **5**(1): 61–84.
6 Waalkens-Berendsen DH, *et al.* Embryotoxicity/teratogenicity of isomalt in rats and rabbits. *Food Chem Toxicol* 1990; **28**(1): 1–9.
7 Smits-Van Prooije AE, *et al.* Chronic toxicity and carcinogenicity study of isomalt in rats and mice. *Food Chem Toxicol* 1990; **28**(4): 243–251.
8 Waalkens-Berendsen DH, *et al.* Multigeneration reproduction study of isomalt in rats. *Food Chem Toxicol* 1990; **28**(1): 11–19.
9 Waalkens-Berendsen DH, *et al.* Developmental toxicity of isomalt in rats. *Food Chem Toxicol* 1989; **27**(10): 631–637.
10 Pometta D, *et al.* Effects of a 12 week administration of isomalt on metabolic control in type-II-diabetics. *Akt Ernährung* 1985; **10**: 174–177.
11 FAO/WHO. Toxicological evaluation of certain food additives and contaminants. Twentieth report of the joint FAO/WHO expert committee on food additives. *World Health Organ Tech Rep Ser* 1987; No. 539.
12 Livesey G. Tolerance of low-digestible carbohydrates: a general view. *Br J Nutr* 2001; **85**(Suppl. 1): S7–S16.
13 Paige DM, *et al.* Palatinit digestibility in children. *Nutr Res* 1992; **12**: 27–37.
14 Storey DM, *et al.* The comparative gastrointestinal response of young children to the ingestion of 25 g sweets containing sucrose or isomalt. *Br J Nutr* 2002; **87**(4): 291–297.
15 Featherstone DB. Effect of isomalt sweetener on the caries process: A review. *J Clin Dent* 1995; **5**: 82–85.
16 Van der Hoeven JS. Influence of disaccharide alcohols on the oral microflora. *Caries Res* 1979; **13**(6): 301–306.
17 Gehring F, Karle EJ. The sugar substitute *Palatinit* with special emphasis on microbial and caries-preventing aspects. *Z Ernärung* 1981; **20**: 96–106.
18 Ndindayino F. Direct compression and moulding properties of co-extruded isomalt/drug mixtures. *Int J Pharm* 2002; **235**(1–2): 159–168.
19 *Food Chemicals Codex*, 7th edn. Bethesda, MD: United States Pharmacopeia, 2010: 545.

20 General References

Bauer KH *et al. Coated Pharmaceutical Dosage Forms: Fundamentals, Manufacturing Techniques, Biopharmaceutical Aspects, Test Methods and Raw Materials.* Stuttgart: Medpharm Scientific Publications, 1998: 280.

BENEO-Palatinit GmbH www.beneo-palatinit.com/en/Homepage/ (accessed 13 June 2011).

Dörr T, Willibald-Ettle I. Evaluation of the kinetics of dissolution of tablets and lozenges consisting of saccharides and sugar substitutes. *Pharm Ind* 1996; **58**: 947–952.

Fritzsching B, Schmidt T. A survey of isomalt as a sugar free excipient for nutraceuticals. *Pharmaceutical Manufacturing and Packing Sourcer* 2000; (Sept): 70–72.

Iida K, *et al.* Effect of mixing of fine carrier particles on dry powder inhalation property of salbutamol sulfate (SS). *J Pharm Soc Jpn* 2000; **120**(1): 113–119.

Ndindayino F, *et al.* Direct compression properties of melt-extruded isomalt. *Int J Pharm* 2002; **235**(1–2): 149–157.

Ndindayino F, *et al.* Bioavailability of hydrochlorothiazide from isomalt-based moulded tablets. *Int J Pharm* 2002; **246**: 199–202.

O'Brien Nabors L, ed. *Alternative Sweeteners: An Overview*, 3rd edn. New York: Marcel Dekker, 2001: 553.

21 Authors

B Fritzsching, O Luhn, A Schoch.

22 Date of Revision

1 March 2012.

Isopropyl Alcohol

1 Nonproprietary Names

BP: Isopropyl Alcohol
JP: Isopropanol
PhEur: Isopropyl Alcohol
USP-NF: Isopropyl Alcohol

2 Synonyms

Alcohol isopropylicus; dimethyl carbinol; IPA; isopropanol; petrohol; 2-propanol; *sec*-propyl alcohol; rubbing alcohol.

3 Chemical Name and CAS Registry Number

Propan-2-ol [67-63-0]

4 Empirical Formula and Molecular Weight

C_3H_8O 60.1

5 Structural Formula

H_3C, OH, CH_3

6 Functional Category

Antimicrobial preservative; solvent.

7 Applications in Pharmaceutical Formulation or Technology

Isopropyl alcohol (propan-2-ol) is used in cosmetics and pharmaceutical formulations, primarily as a solvent in topical formulations.[1] It is not recommended for oral use owing to its toxicity; *see* Section 14.

Although it is used in lotions, the marked degreasing properties of isopropyl alcohol may limit its usefulness in preparations used repeatedly. Isopropyl alcohol is also used as a solvent both for tablet film-coating and for tablet granulation,[2] where the isopropyl alcohol is subsequently removed by evaporation. It has also been shown to significantly increase the skin permeability of nimesulide from carbomer 934.[3]

8 Description

Isopropyl alcohol occurs as a clear, colorless, mobile, volatile, flammable liquid with a characteristic, spirituous odor resembling that of a mixture of ethanol and acetone; it has a slightly bitter taste.

9 Pharmacopeial Specifications

See Table I.

Table I: Pharmacopeial specifications for isopropyl alcohol.

Test	JP XV	PhEur 7.4	USP35–NF30
Identification	+	+	+
Appearance of solution	+	+	−
Absorbance	−	+	−
Characters	−	+	−
Specific gravity	0.785–0.788	0.785–0.789	0.783–0.787
Refractive index	−	1.376–1.379	1.376–1.378
Acidity or alkalinity	+	+	+
Water	≤0.75%	≤0.5%	−
Nonvolatile residue	≤1.0 mg	≤20 ppm	≤0.005%
Distillation range	81–83°C	−	−
Benzene	−	+	−
Peroxides	−	+	−
Assay	−	−	≥99.0%

10 Typical Properties

Antimicrobial activity Isopropyl alcohol is bactericidal; at concentrations greater than 70% v/v it is a more effective antibacterial preservative than ethanol (95%). The bactericidal effect of aqueous solutions increases steadily as the concentration approaches 100% v/v. Isopropyl alcohol is ineffective against bacterial spores.
Autoignition temperature 425°C
Boiling point 82.4°C
Dielectric constant $D^{20} = 18.62$
Explosive limits 2.5–12.0% v/v in air
Flammability Flammable.
Flash point 11.7°C (closed cup); 13°C (open cup). The water azeotrope has a flash point of 16°C.
Freezing point −89.5°C
Melting point −88.5°C
Moisture content 0.1–13% w/w for commercial grades (13% w/w corresponds to the water azeotrope).
Refractive index
$n_D^{20} = 1.3776$;
$n_D^{25} = 1.3749$.
Solubility Miscible with benzene, chloroform, ethanol (95%), ether, glycerin, and water. Soluble in acetone; insoluble in salt solutions. Forms an azeotrope with water, containing 87.4% w/w isopropyl alcohol (boiling point 80.37°C).
Specific gravity 0.786
Spectroscopy
IR spectra *see* Figure 1.
Vapor density (relative) 2.07 (air = 1)
Vapor pressure
133.3 Pa (1 mmHg) at −26.1°C;
4.32 kPa (32.4 mmHg) at 20°C;
5.33 kPa (40 mmHg) at 23.8°C;
13.33 kPa (100 mmHg) at 39.5°C.
Viscosity (dynamic) 2.43 mPa s (2.43 cP) at 20°C

11 Stability and Storage Conditions

Isopropyl alcohol should be stored in an airtight container in a cool, dry place.

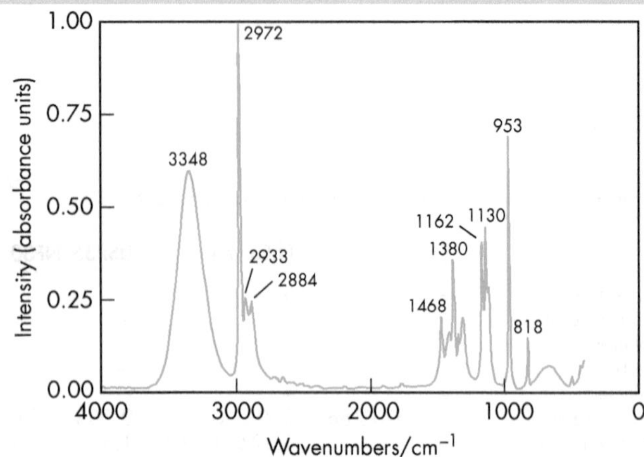

Figure 1: Infrared spectrum of isopropyl alcohol measured by transmission. Adapted with permission of Informa Healthcare.

12 Incompatibilities

Incompatible with oxidizing agents such as hydrogen peroxide and nitric acid, which cause decomposition. Isopropyl alcohol may be salted out from aqueous mixtures by the addition of sodium chloride, sodium sulfate, and other salts, or by the addition of sodium hydroxide.

13 Method of Manufacture

Isopropyl alcohol may be prepared from propylene; by the catalytic reduction of acetone; or by fermentation of certain carbohydrates.

14 Safety

Isopropyl alcohol is widely used in cosmetics and topical pharmaceutical formulations. It is readily absorbed from the gastrointestinal tract and may be slowly absorbed through intact skin. Prolonged direct exposure of isopropyl alcohol to the skin may result in cardiac and neurological deficits.[4] In neonates, isopropyl alcohol has been reported to cause chemical burns following topical application.[5,6]

Isopropyl alcohol is metabolized more slowly than ethanol, primarily to acetone. Metabolites and unchanged isopropyl alcohol are mainly excreted in the urine.

Isopropyl alcohol is about twice as toxic as ethanol and should therefore not be administered orally; isopropyl alcohol also has an unpleasant taste. Symptoms of isopropyl alcohol toxicity are similar to those for ethanol except that isopropyl alcohol has no initial euphoric action, and gastritis and vomiting are more prominent; *see* Alcohol. Delta osmolality may be useful as rapid screen test to identify patients at risk of complications from ingestion of isopropyl alcohol.[7] The lethal oral dose is estimated to be about 120–250 mL although toxic symptoms may be produced by 20 mL.

Adverse effects following parenteral administration of up to 20 mL of isopropyl alcohol diluted with water have included only a sensation of heat and a slight lowering of blood pressure. However, isopropyl alcohol is not commonly used in parenteral products.

Although inhalation can cause irritation and coma, the inhalation of isopropyl alcohol has been investigated in therapeutic applications.[3]

Isopropyl alcohol is most frequently used in topical pharmaceutical formulations where it may act as a local irritant.[8] When applied to the eye it can cause corneal burns and eye damage.

LD_{50} (dog, oral): 4.80 g/kg[8]
LD_{50} (mouse, IP): 4.48 g/kg
LD_{50} (mouse, IV): 1.51 g/kg
LD_{50} (mouse, oral): 3.6 g/kg
LD_{50} (rabbit, oral): 6.41 g/kg
LD_{50} (rabbit, skin): 12.8 g/kg
LD_{50} (rat, IP): 2.74 g/kg
LD_{50} (rat, IV): 1.09 g/kg
LD_{50} (rat, oral): 5.05 g/kg

15 Handling Precautions

Observe normal precautions appropriate to the circumstances and quantity of material handled. Isopropyl alcohol may be irritant to the skin, eyes, and mucous membranes upon inhalation. Eye protection and gloves are recommended. Isopropyl alcohol should be handled in a well-ventilated environment. In the UK, the long-term (8-hour TWA) workplace exposure limit for isopropyl alcohol is 999 mg/m^3 (400 ppm); the short-term (15-minute) workplace exposure limit is 1250 mg/m^3 (500 ppm).[9] OSHA standards state that IPA 8-hour time weighted average airborne level in the workplace cannot exceed 400 ppm. Isopropyl alcohol is flammable and produces toxic fumes on combustion.

16 Regulatory Status

Included in the FDA Inactive Ingredients Database (oral capsules, tablets, and topical preparations). Included in nonparenteral medicines licensed in the UK. Included in the Canadian Natural Health Products Ingredients Database.

17 Related Substances

Propan-1-ol.

Propan-1-ol

Empirical formula C_3H_8O
Molecular weight 60.1
CAS number [71-23-8]
Synonyms Propanol; *n*-propanol; propyl alcohol; propylic alcohol.
Autoignition temperature 540°C
Boiling point 97.2°C
Dielectric constant $D^{25} = 22.20$
Explosive limits 2.15–13.15% v/v in air
Flash point 15°C (closed cup)
Melting point −127°C
Refractive index $n_D^{20} = 1.3862$
Solubility Miscible with ethanol (95%), ether, and water.
Specific gravity 0.8053 at 20°C
Viscosity (dynamic) 2.3 mPa s (2.3 cP) at 20°C
Comments Propan-1-ol is more toxic than isopropyl alcohol. In the UK, the long-term (8-hour TWA) exposure limit for propan-1-ol is 500 mg/m^3 (200 ppm); the short-term (15-minute) exposure limit is 625 mg/m^3 (250 ppm).[9]

18 Comments

Therapeutically, isopropyl alcohol has been investigated for the treatment of postoperative nausea or vomiting.[10]

Isopropyl alcohol has some antimicrobial activity (*see* Section 10) and a 70% v/v aqueous solution is used as a topical disinfectant.

A specification for isopropyl alcohol is contained in the *Food Chemicals Codex* (FCC).[11]

The EINECS number for isopropyl alcohol is 200-661-7. The PubChem Compound ID (CID) for isopropyl alcohol is 3776.

19 Specific References

1 Rafiee Tehrani H, Mehramizi A. *In vitro* release studies of piroxicam from oil-in-water creams and hydroalcoholic gel topical formulations. *Drug Dev Ind Pharm* 2000; **26**(4): 409–414.

2 Ruckmani K, *et al.* Eudragit matrices for sustained release of ketorolac tromethamine: formulation and kinetics of release. *Boll Chim Form* 2000; **139**: 205–208.

3 Guengoer S, Bergisadi N. Effect of penetration enhancers on *in vitro* percutaneous penetration of nimesulide through rat skin. *Pharmazie* 2004; **59**: 39–41.
4 Leeper SC, *et al*. Topical absorption of isopropyl alcohol induced cardiac neurological deficits in an adult female with intact skin. *Vet Hum Toxicol* 2000; **42**: 15–17.
5 Schick JB, Milstein JM. Burn hazard of isopropyl alcohol in the neonate. *Pediatrics* 1981; **68**: 587–588.
6 Weintraub Z, Iancu TC. Isopropyl alcohol burns. *Pediatrics* 1982; **69**: 506.
7 Monaghan MS, *et al*. Use of delta osmolality to predict serum isopropanol and acetone concentrations. *Pharmacotherapy* 1993; **13**(1): 60–63.
8 Lewis RJ, ed. *Sax's Dangerous Properties of Industrial Materials*, 11th edn. New York: Wiley, 2004; 2148–2149.
9 Health and Safety Executive. *EH40/2005: Workplace Exposure Limits*. Sudbury: HSE Books, 2011. www.hse.gov.uk/pubns/priced/eh40.pdf (accessed 12 April 2012).
10 Merritt BA, *et al*. Isopropyl alcohol inhalation: alternative treatment of postoperative nausea and vomiting. *Nurs Res* 2002; **51**(2): 125–128.
11 *Food Chemicals Codex*, 7th edn. Suppl. 3. Bethesda, MD: United States Pharmacopeia, 2011; 1701.

20 General References

—

21 Author

CP McCoy.

22 Date of Revision

12 April 2012.

Isopropyl Isostearate

1 Nonproprietary Names

None adopted.

2 Synonyms

Crodamol IPIS; isostearic acid, isopropyl ester; 1-methylethyl isooctadecanoate; *Nikkol IPIS*; 2-propyl isooctadecanoate; *Schercemol 318 Ester*; *Wickenol 131*.

3 Chemical Name and CAS Registry Number

Propan-2-yl 16-methylheptadecanoate [68171-33-5]

4 Empirical Formula and Molecular Weight

$C_{21}H_{42}O_2$ 326.56

5 Structural Formula

H_3C CH_3 O O CH_3 CH_3

6 Functional Category

Emollient.

7 Applications in Pharmaceutical Formulation or Technology

Isopropyl isostearate is used as an emollient in topical creams and emulsions. It is also used in cosmetics in bath oils, creams, lotions, and lipsticks.

8 Description

Isopropyl isostearate occurs as a clear liquid with a bland odor.

9 Pharmacopeial Specifications

—

10 Typical Properties

Acid value ≤1 for *Schercemol 318 Ester*[1]
Boiling point 360.7°C at 760 mmHg[2]
Contact angle 19.1 for *Schercemol 318 Ester*[1]
Density 0.86 g/cm^3 [2]
Flash point 183.6°C[2]
Freezing point −28°C for *Schercemol 318 Ester*[1]
Interfacial tension 25.8 mN/m for *Schercemol 318 Ester*[1]
Iodine value ≤3 for *Schercemol 318 Ester*[1]
Refractive index 1.445; 1.443 for *Schercemol 318 Ester*[1]
Saponification value 160–180 for *Schercemol 318 Ester*[1]
Solubility Soluble in acetone, ethanol (95%), ethyl acetate, isopropyl alcohol, and mineral oil. Insoluble in water, glycerol, and propylene glycol.
Surface tension 29.8 dyne/cm[2]
Vapor pressure 2.18×10^{-5} mmHg at 25°C[2]
Viscosity 15 mPa s

11 Stability and Storage Conditions

Isopropyl isostearate is resistant to oxidation and hydrolysis, and does not become rancid. It should be stored in a well-closed container in a cool, dry place and protected from light.

12 Incompatibilities

Isopropyl isostearate is incompatible with strong oxidizing agents. Prolonged contact with plastics can result in softening.

13 Method of Manufacture

Isopropyl isostearate may be prepared by the esterification of isostearic acid with propan-2-ol in the presence of a catalyst.

14 Safety

The Cosmetic Ingredient Review (CIR) Expert Panel has found isopropyl isostearate to be safe as presently used in cosmetics at a maximum concentration of 50%.[3] Undiluted isopropyl isostearate has been classified as a slight ocular irritant.[3,4] Repeated applications of a 10.0% aqueous suspension of isopropyl isostearate to the skin of albino rabbits was well-tolerated.[3]

LD_{50} (rat, oral): > 64 g/kg[1]

15 Handling Precautions

Observe normal precautions appropriate to the circumstances and quantity of material handled. When heated to decomposition, isopropyl isostearate emits acrid smoke and irritating vapors.[4]

16 Regulatory Status

Included in the FDA Inactive Ingredients Database (topical creams and sustained-release emulsions). Included in the Canadian Natural Health Products Ingredients Database.

17 Related Substances

Isopropyl myristate; isopropyl palmitate; isopropyl stearate.

Isopropyl stearate

Empirical formula $C_{21}H_{42}O_2$
Molecular weight 326.32
CAS number [112-10-7]
Synonyms Octadecanoic acid; 1-methylethyl ester; *Wickenol 127*; *DUB SIP*.
Boiling point 368.2°C at 760 mmHg
Density 0.861 g/cm^3
Flash point 179.3°C
Melting point 28°C
Partition coefficient Log P (octanol : water) = 9.14
Refractive index 1.445
Vapor pressure 1.3×10^{-5} mmHg at 25°C
Surface tension 30.3 dyne/cm
Solubility Soluble in acetone, chloroform, ethanol (95%), ethyl acetate, fats, fatty alcohols, fixed oils, liquid hydrocarbons, toluene, and waxes. Practically insoluble in glycerin, glycols, and water.
Safety
LD_{Lo} (rat, oral): 8 g/kg[3]
Comments Used as an emollient. The EINECS number for isopropyl stearate is 203-934-9.

18 Comments

Isopropyl isostearate has a low freezing point and exhibits lubricity without oiliness. It imparts good spreadability and slip to topical products.[1]

The EINECS number for isopropyl isostearate is 269-023-3.

19 Specific References

1 Lubrizol. Technical data sheet: *Schercemol* 318 Ester, 2006.
2 Lookchem.com. Isopropyl isostearate. http://www.lookchem.com/Isopropyl-isostearate/(accessed 4 July 2011).
3 Anonymous. Final Report on the safety assessment of isopropyl isostearate. *Int J Toxicol* 1992; **11**(1): 43–49.
4 Lewis RJ, ed. *Sax's Dangerous Properties of Industrial Materials*, 11th edn. New York: John Wiley, 2004: 2161.

20 General References

Alzo International. Product information: *Wickenol 131*. http://www.alzointernational.com/esters.htm (accessed 4 July 2011).
Croda Inc. Product information: *Crodamol IPIS*. http://www.croda.com (accessed 4 July 2011).
Nikko Chemicals Co Ltd. Product information: *Nikkol IPIS*. http://www.nikkol.co.jp/en/ (accessed 4 July 2011).

21 Authors

ME Fenton, RC Rowe.

22 Date of Revision

1 August 2011.

Isopropyl Myristate

1 Nonproprietary Names

BP: Isopropyl Myristate
PhEur: Isopropyl Myristate
USP–NF: Isopropyl Myristate

2 Synonyms

Crodamol IPM; *Dermol IPM*; *Estol IPM*; *Exceparl IPM*; isopropyl ester of myristic acid; isopropylis myristas; *Kessco IPM 95*; *Lexol IPM-NF*; myristic acid isopropyl ester; *Rita IPM*; *Stepan IPM*; *Tegosoft M*; tetradecanoic acid, 1-methylethyl ester; *Waglinol 6014*; *Wickenol 101*.

3 Chemical Name and CAS Registry Number

1-Methylethyl tetradecanoate [110-27-0]

4 Empirical Formula and Molecular Weight

$C_{17}H_{34}O_2$ 270.5

5 Structural Formula

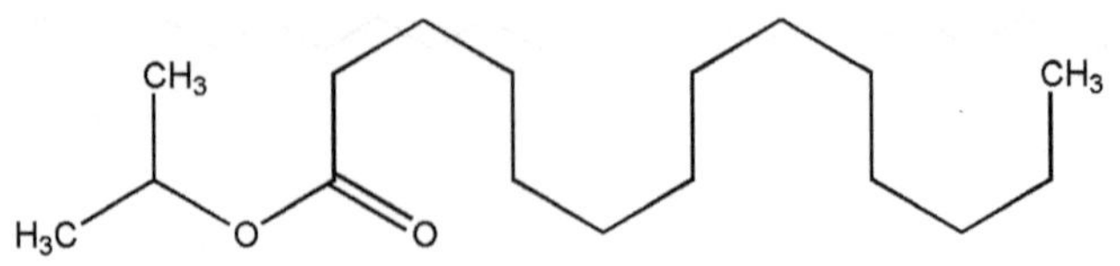

6 Functional Category

Emollient; oleaginous vehicle; penetration enhancer; solvent; transdermal delivery component.

7 Applications in Pharmaceutical Formulation or Technology

Isopropyl myristate is a nongreasy emollient that is absorbed readily by the skin. It is used as a component of semisolid bases and as a solvent for many substances applied topically. Applications in topical pharmaceutical and cosmetic formulations include bath oils, make-up, hair and nail care products, creams, lotions, lip products, shaving products, skin lubricants, deodorants, otic suspensions, and vaginal creams; *see* Table I. For example, isopropyl myristate is a self-emulsifying component of a proposed cold cream formula,[1] which is suitable for use as a vehicle for drugs or dermatological actives; it is also used cosmetically in stable mixtures of water and glycerol.[2]

Isopropyl myristate is used as a penetration enhancer for transdermal formulations, and has been used in conjunction with therapeutic ultrasound and iontophoresis.[3] It has also been used in a water-oil gel prolonged-release emulsion and in various microemulsions. Such microemulsions may increase bioavailability in topical and transdermal applications.[4]

Isopropyl myristate is used in soft adhesives for pressure-sensitive adhesive tapes.[5]

Table I: Uses of isopropyl myristate.

Use	Concentration (%)
Detergent	0.003–0.03
Otic suspension	0.024
Perfumes	0.5–2.0
Microemulsions	<50
Soap	0.03–0.3
Topical aerosols	2.0–98.0
Topical creams and lotions	1.0–10.0

8 Description

Isopropyl myristate occurs as a clear, colorless, practically odorless liquid of low viscosity that congeals at about 5°C. It consists of esters of propan-2-ol and saturated high molecular weight fatty acids, principally myristic acid.

9 Pharmacopeial Specifications

See Table II.

Table II: Pharmacopeial specifications for isopropyl myristate.

Test	PhEur 7.4	USP35–NF30
Identification	+	+
Characters	+	—
Appearance of solution	+	—
Relative density	≈0.853	0.846–0.854
Refractive index	1.434–1.437	1.432–1.436
Residue on ignition	≤0.1%	≤0.1%
Acid value	≤1.0	≤1.0
Saponification value	202–212	202–212
Iodine value	≤1.0	≤1.0
Viscosity	5–6 mPa s	—
Water	≤0.1%	—
Assay (as $C_{17}H_{34}O_2$)	≥90.0%	≥90.0%

10 Typical Properties

Boiling point 140.2°C at 266 Pa (2 mmHg)
Flash point 153.5°C (closed cup)
Freezing point ≈5°C
Solubility Soluble in acetone, chloroform, ethanol (95%), ethyl acetate, fats, fatty alcohols, fixed oils, liquid hydrocarbons, toluene, and waxes. Dissolves many waxes, cholesterol, or lanolin. Practically insoluble in glycerin, glycols, and water.
Spectroscopy
IR spectra *see* Figure 1.
Viscosity (dynamic) 5–7 mPa s (5–7 cP) at 25°C

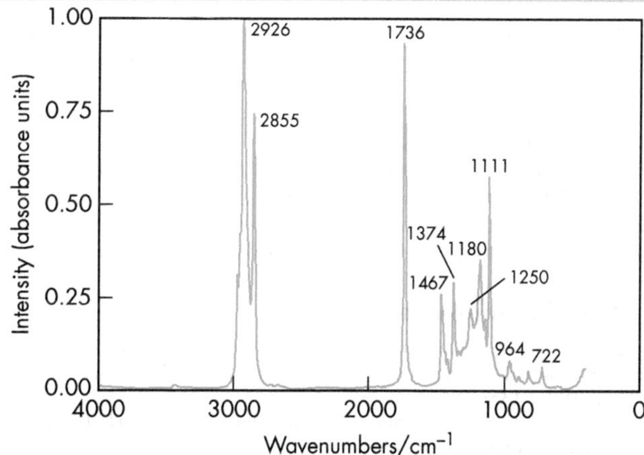

Figure 1: Infrared spectrum of isopropyl myristate measured by transmission. Adapted with permission of Informa Healthcare.

11 Stability and Storage Conditions

Isopropyl myristate is resistant to oxidation and hydrolysis and does not become rancid. It should be stored in a well-closed container in a cool, dry place and protected from light.

12 Incompatibilities

When isopropyl myristate comes into contact with rubber, there is a drop in viscosity with concomitant swelling and partial dissolution of the rubber; contact with plastics, e.g. nylon and polyethylene, results in swelling. Isopropyl myristate is incompatible with hard paraffin, producing a granular mixture. It is also incompatible with strong oxidizing agents.

13 Method of Manufacture

Isopropyl myristate may be prepared either by the esterification of myristic acid with propan-2-ol or by the reaction of myristoyl chloride and propan-2-ol with the aid of a suitable dehydrochlorinating agent. A high-purity material is also commercially available, produced by enzymatic esterification at low temperature.

14 Safety

Isopropyl myristate is widely used in cosmetics and topical pharmaceutical formulations, and is generally regarded as a nontoxic and nonirritant material.[6–8]

LD_{50} (mouse, oral): 49.7 g/kg[9]
LD_{50} (rabbit, skin): 5 g/kg

15 Handling Precautions

Observe normal precautions appropriate to the circumstances and quantity of material handled.

16 Regulatory Status

Included in the FDA Inactive Ingredients Database (otic, topical, transdermal, and vaginal preparations). Used in nonparenteral medicines licensed in the UK. Included in the Canadian Natural Health Products Ingredients Database.

17 Related Substances

Isopropyl palmitate.

18 Comments

Isopropyl myristate has been investigated in the production of microspheres, and has been found to significantly increase the release of drug from etopside-loaded microspheres.[10] It has also been investigated in microemulsion templates to produce nanoparticles as potential drug delivery vehicles for proteins and peptides.[11,12]

Isopropyl myristate 50% has been shown to be an effective pediculicide for the control of head lice.[13]

A specification for isopropyl myristate is included in the *Japanese Pharmaceutical Excipients* (JPE).[14]

The EINECS number for isopropyl myristate is 203-751-4. The PubChem Compound ID (CID) for isopropyl myristate is 8042.

19 Specific References

1 Jimenez SMM, *et al.* Proposal and pharmacotechnical study of a modern dermo-pharmaceutical formulation for cold cream. *Boll Chim Farm* 1996; **135**: 364–373.
2 Ayannides CA, Ktistis G. Stability estimation of emulsions of isopropyl myristate in mixtures of water and glycerol. *J Cosmet Sci* 2002; **53**(3): 165–173.
3 Fang JY, *et al.* Transdermal iontopheresis of sodium nonivaride acetate III: combined effect of pretreatment by penetration enhancers. *Int J Pharm* 1997; **149**: 183–195.
4 Kogan A, Garti N. Microemulsions as transdermal delivery vehicles. *Adv Colloid Interface Sci* 2006; **123–126**: 369–385.
5 Tokumura F, *et al.* Properties of pressure-sensitive adhesive tapes with soft adhesives to human skin and their mechanism. *Skin Res Technol* 2007; **13**(2): 211–216.
6 Stenbäck F, Shubik P. Lack of toxicity and carcinogenicity of some commonly used cutaneous agents. *Toxicol Appl Pharmacol* 1974; **30**: 7–13.
7 Opdyke DL. Monographs on fragrance raw materials. *Food Cosmet Toxicol* 1976; **14**(4): 307–338.
8 Guillot JP, *et al.* Safety evaluation of cosmetic raw materials. *J Soc Cosmet Chem* 1977; **28**: 377–393.
9 Lewis RJ, ed. *Sax's Dangerous Properties of Industrial Materials*, 11th edn. New York: Wiley, 2004: 2164.
10 Schaefer MJ, Singh J. Effect of isopropyl myristic acid ester on the physical characteristics and *in vitro* release of etoposide from PLGA microspheres. *AAPS Pharm Sci Tech* 2000; **1**(4): E32.
11 Graf A, *et al.* Protein delivery using nanoparticles based on microemulsions with different structure types. *Eur J Pharm Sci* 2008; **33**(4–5): 434–444.
12 Graf A, *et al.* Microemulsions containing lecithin and sugar-based surfactant: nonparticle templates for delivery of proteins and peptides. *Int J Pharm* 2008; **350**(1–2): 351–360.
13 Kaul N, *et al.* North American efficacy and safety of a novel pediculicide rinse, isopropyl myristate 50% (results). *J Cutan Med Surg* 2007; **11**(5): 161–167.
14 Japan Pharmaceutical Excipients Council. *Japanese Pharmaceutical Excipients* 2004. Tokyo: Yakuji Nippo, 2004: 443–444.

20 General References

Fitzgerald JE, *et al.* Cutaneous and parenteral studies with vehicles containing isopropyl myristate and peanut oil. *Toxicol Appl Pharmacol* 1968; **13**: 448–453.

Nakhare S, Vyas SP. Prolonged release of rifampicin from internal phase of multiple w/o/w emulsion systems. *IndianJ Pharm Sci* 1995; **57**: 71–77.

21 Author

AK Taylor.

22 Date of Revision

1 March 2012.

Isopropyl Palmitate

1 Nonproprietary Names

BP: Isopropyl Palmitate
PhEur: Isopropyl Palmitate
USP–NF: Isopropyl Palmitate

2 Synonyms

Crodamol IPP; *Dermol IPP*; *Emerest 2316*; *Excepарl IPP*; hexadecanoic acid isopropyl ester; hexadecanoic acid 1-methylethyl ester; isopropyl hexadecanoate; isopropylis palmitas; *Kessco IPP*; *Lexol IPP-NF*; *Liponate IPP*; palmitic acid isopropyl ester; *Propal*; *Protachem IPP*; *Rita IPP*; *Stepan IPP*; *Tegosoft P*; *Unimate IPP*; *Waglinol 6016*; *Wickenol 111*.

3 Chemical Name and CAS Registry Number

1-Methylethyl hexadecanoate [142-91-6]

4 Empirical Formula and Molecular Weight

$C_{19}H_{38}O_2$ 298.51

5 Structural Formula

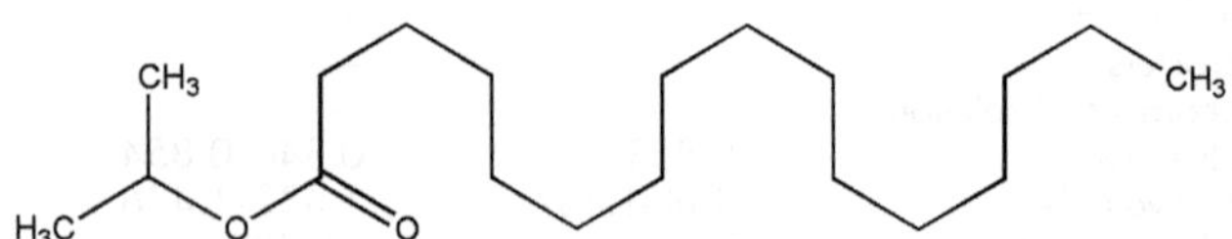

6 Functional Category

Emollient; oleaginous vehicle; penetration enhancer; solvent; transdermal delivery component.

7 Applications in Pharmaceutical Formulation or Technology

Isopropyl palmitate is a nongreasy emollient with good spreading characteristics used in topical pharmaceutical formulations and cosmetics such as bath oils, creams, lotions, make-up, hair care products, deodorants, lip products, suntan preparations, and pressed powders; *see* Table I.

Isopropyl palmitate is an established penetration enhancer for transdermal systems. It has also been used in controlled-release percutaneous films.

Table I: Uses of isopropyl palmitate.

Use	Concentration (%)
Detergent	0.005–0.02
Perfume	0.2–0.8
Soap	0.05–0.2
Topical aerosol spray	3.36
Topical creams and lotions	0.05–5.5

8 Description

Isopropyl palmitate occurs as a clear, colorless to pale yellow-colored, practically odorless viscous liquid that solidifies at less than 16°C.

9 Pharmacopeial Specifications

See Table II.

Table II: Pharmacopeial specifications for isopropyl palmitate.

Test	PhEur 7.4	USP35–NF30
Identification	+	+
Characters	+	—
Acid value	≤1.0	≤1.0
Appearance of solution	+	—
Iodine value	≤1.0	≤1.0
Relative density	≈0.854	0.850–0.855
Residue on ignition	≤0.1%	≤0.1%
Refractive index	1.436–1.440	1.435–1.438
Saponification value	183–193	183–193
Viscosity	5–10 mPa s	—
Water	≤0.1%	—
Assay (of $C_{19}H_{38}O_2$)	≥90.0%	≥90.0%

10 Typical Properties

Boiling point 160°C at 266 Pa (2 mmHg)
Flash Point >100°C (open cup) for *Crodamol IPP.*
Freezing point ≈13–15°C
Solubility Soluble in acetone, chloroform, ethanol (95%), ethyl acetate, mineral oil, propan-2-ol, silicone oils, vegetable oils, and aliphatic and aromatic hydrocarbons; practically insoluble in glycerin, glycols, and water.
Spectroscopy
IR spectra *see* Figure 1.
Surface tension ≈29 mN/m for *Tegosoft P* at 25°C
Viscosity (dynamic) 5–10 mPa s (5–10 cP) at 25°C; 5–6 mPa s (5–6 cP) at 20°C for *Tegosoft P.*

11 Stability and Storage Conditions

Isopropyl palmitate is resistant to oxidation and hydrolysis, and does not become rancid. It should be stored in a well-closed container, above 16°C, and protected from light.

12 Incompatibilities

See Isopropyl Myristate.

13 Method of Manufacture

Isopropyl palmitate is prepared by the reaction of palmitic acid with propan-2-ol in the presence of an acid catalyst. A high-purity material is also commercially available, which is produced by enzymatic esterification at low temperatures.

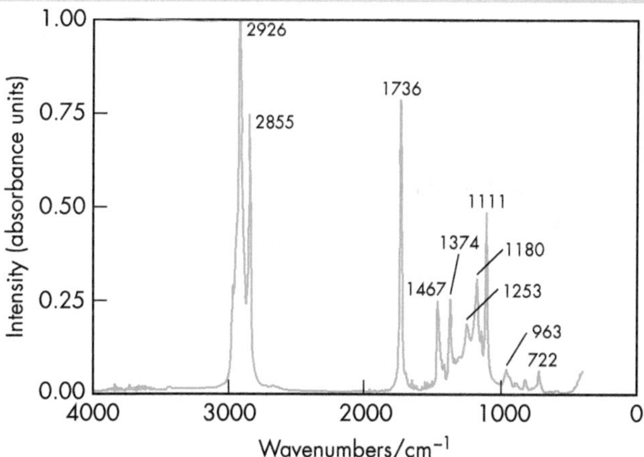

Figure 1: Infrared spectrum of isopropyl palmitate measured by transmission. Adapted with permission of Informa Healthcare.

14 Safety

Isopropyl palmitate is widely used in cosmetics and topical pharmaceutical formulations, and is generally regarded as a relatively nontoxic and nonirritant material.[1–3]

LD_{50} (mouse, IP): 0.1 g/kg[4]

15 Handling Precautions

Observe normal precautions appropriate to the circumstances and quantity of material handled.

16 Regulatory Status

Included in the FDA Inactive Ingredients Database (topical and transdermal preparations). Used in nonparenteral medicines licensed in the UK. Included in the Canadian Natural Health Products Ingredients Database.

17 Related Substances

Isopropyl myristate.

18 Comments

Isopropyl palmitate has been investigated in the production of reversed sucrose ester vehicles.[5] It has also been investigated in microemulsion systems[6] for transdermal delivery, and as the organic phase for formation of lecithin organogels,[7] which may also be used in transdermal delivery systems.

A specification for isopropyl palmitate is included in the *Japanese Pharmaceutical Excipients* (JPE).[8]

The EINECS number for isopropyl palmitate is 205-571-1. The PubChem Compound ID (CID) for isopropyl palmitate is 8907.

19 Specific References

1 Frosch PJ, Kligman AM. The chamber-scarification test for irritancy. *Contact Dermatitis* 1976; **2**: 314–324.
2 Guillot JP, *et al.* Safety evaluation of cosmetic raw materials. *J Soc Cosmet Chem* 1977; **28**: 377–393.
3 Opdyke DL, Letizia C. Monographs on fragrance raw materials. *Food Cosmet Toxicol* 1982; **20**(Suppl.): 633–852.
4 Lewis RJ, ed. *Sax's Dangerous Properties of Industrial Materials*, 11th edn. New York: Wiley, 2004: 2165.
5 Mollee H, *et al.* Stable reversed vesicles in oil: characterization studies and encapsulation of model compounds. *J Pharm Sci* 2000; **89**(7): 930–939.
6 Sintov AC, Shapiro L. New microemulsion vehicle facilitates percutaneous pentration *in vitro* and cutaneous drug bioavailability *in vivo*. *J Control Release* 2004; **95**(2): 173–183.

7 Avramiotis S, *et al*. Lecithin organogels used as bioactive compounds carriers. A microdomain properties investigation. *Langmuir* 2007; 23(8): 4438–4447.
8 Japanese Pharmaceutical Excipients Council. *Japanese Pharmaceutical Excipients* 2004. Tokyo: Yakuji Nippo, 2004: 445–446.

20 General References

Croda Inc. Material safety data sheet: *Crodamol IPP*, 2009.
Evonik Industries. Material safety data sheet: *Tegosoft P*, 2007.

21 Author

AK Taylor.

22 Date of Revision

1 March 2012.

Kaolin

1 Nonproprietary Names

BP: Heavy Kaolin

JP: Kaolin

PhEur: Kaolin, Heavy

USP-NF: Kaolin

Note that the PhEur 7.4 contains a monograph on heavy kaolin. The BP 2012, in addition to the monograph for heavy kaolin, also contains monographs for light kaolin (natural) and light kaolin.

See also Sections 4 and 9.

2 Synonyms

Argilla; bolus alba; China clay; E559; kaolinite; kaolinum ponderosum; *Lion*; porcelain clay; *Sim 90*; weisserton; white bole.

3 Chemical Name and CAS Registry Number

Hydrated aluminum silicate [1332-58-7]

4 Empirical Formula and Molecular Weight

$Al_2H_4O_9Si_2$ 258.16

The USP35–NF30 describes kaolin as a native hydrated aluminum silicate, powdered and freed from gritty particles by elutriation. The BP 2012 similarly describes light kaolin but additionally states that it contains a suitable dispersing agent. Light kaolin (natural) BP contains no dispersing agent. Heavy kaolin is described in the BP 2012 and PhEur 7.4 as a purified, natural hydrated aluminum silicate of variable composition. The JP XV describes kaolin as a native hydrous aluminum silicate.

5 Structural Formula

See Section 4.

6 Functional Category

Adsorbent; suspending agent; tablet and capsule diluent.

7 Applications in Pharmaceutical Formulation or Technology

Kaolin is a naturally occurring mineral used in oral and topical pharmaceutical formulations.

In oral medicines, kaolin has been used as a diluent in tablet and capsule formulations; it has also been used as a suspending vehicle. In topical preparations, sterilized kaolin has been used in poultices and as a dusting powder.[1,2]

8 Description

Kaolin occurs as a white to grayish-white colored, unctuous powder free from gritty particles. It has a characteristic earthy or claylike taste, and when moistened with water it becomes darker in color and develops a claylike odor.

9 Pharmacopeial Specifications

See Table I.

10 Typical Properties

Acidity/alkalinity pH = 4.0–7.5 for a 20% w/v aqueous slurry

Hardness (Mohs) 2.0, very low

Hygroscopicity At relative humidities between about 15–65%, the equilibrium moisture content at 25°C is about 1% w/w, but at relative humidities above about 75%, kaolin absorbs small amounts of moisture.

Particle size distribution Median size = 0.6–0.8 μm

Refractive index 1.56

Solubility Practically insoluble in diethyl ether, ethanol (95%), water, other organic solvents, cold dilute acids, and solutions of alkali hydroxides.

Specific gravity 2.6

SEM 1: Excipient: Kaolin USP; manufacturer: Georgia Kaolin Co.; lot no.: 1672; magnification: 60×; voltage: 10 kV.

SEM 2: Excipient: Kaolin USP; manufacturer: Georgia Kaolin Co.; lot no.: 1672; magnification: 600×; voltage: 10 kV.

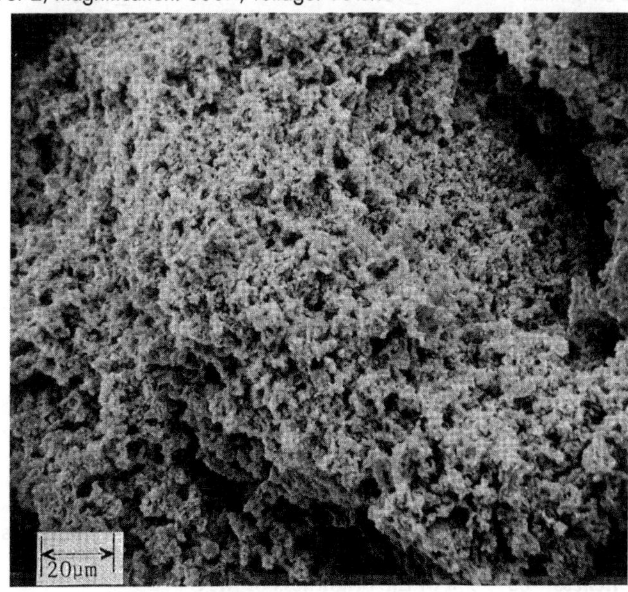

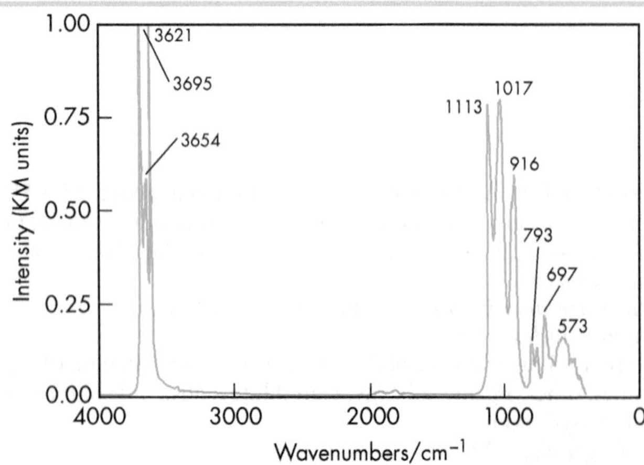

Figure 1: Infrared spectrum of kaolin measured by diffuse reflectance. Adapted with permission of Informa Healthcare.

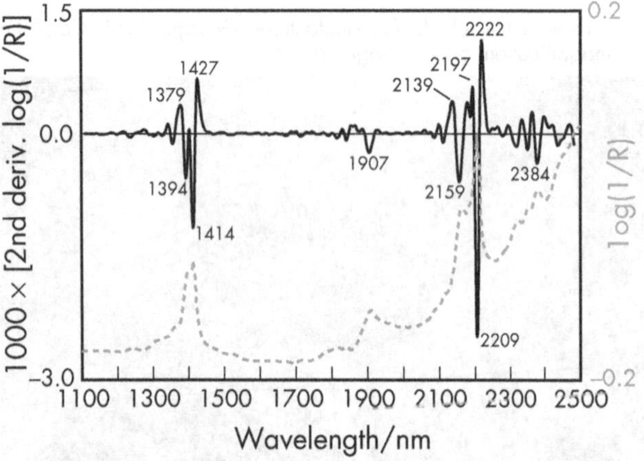

Figure 2: Near-infrared spectrum of kaolin measured by reflectance.

Table I: Pharmacopeial specifications for kaolin.

Test	JP XV	PhEur 7.4	USP 35–NF30
Identification	+	+	+
Characters	–	+	–
Acidity or alkalinity	+	+	–
Microbial limit	–	≤10^3 cfu/g	+
Loss on ignition	≤15.0%	–	≤15.0%
Acid-soluble substances	+	≤1.0%	≤2.0%
Organic impurities	–	+	–
Foreign matter	+	–	–
Adsorption power	–	+	–
Swelling power	–	+	–
Plasticity	+	–	–
Arsenic	≤2 ppm	–	–
Calcium	–	≤250 ppm	–
Carbonate	+	–	+
Chloride	–	≤250 ppm	–
Heavy metals	≤50 ppm	≤50 ppm[(a)]	–
Iron	≤500 ppm	–	+
Lead	–	–	≤0.001%
Sulfate	–	≤0.1%	–

(a) When intended for internal use, the limit is set at ≤25 ppm.

Spectroscopy

IR spectra *see* Figure 1.

NIR spectra *see* Figure 2.

Viscosity (dynamic) 300 mPa s (300 cP) for a 70% w/v aqueous suspension.

Whiteness 85–90% of the brightness of MgO

11 Stability and Storage Conditions

Kaolin is a stable material. Since it is a naturally occurring material, kaolin is commonly contaminated with microorganisms such as *Bacillus anthracis*, *Clostridium tetani*, and *Clostridium welchii*. However, kaolin may be sterilized by heating at a temperature greater than 160°C for not less than 1 hour. When moistened with water, kaolin darkens and becomes plastic.

Kaolin should be stored in a well-closed container in a cool, dry place.

12 Incompatibilities

The adsorbent properties of kaolin may influence the absorption of other orally administered drugs. Drugs reportedly affected by kaolin include amoxicillin;[(3)] ampicillin;[(3)] cimetidine;[(4)] digoxin;[(5)] lincomycin; phenytoin;[(6)] and tetracycline. Warfarin absorption by rat intestine *in vitro* was reported not to be affected by kaolin.[(7)] With clindamycin, the rate (but not the amount) of absorption was affected by kaolin.[(8)]

13 Method of Manufacture

Kaolin is a hydrated aluminum silicate obtained by mining naturally occurring mineral deposits. Large deposits are found in Georgia, US and in Cornwall, England.

Mined kaolin is powdered and freed of coarse, gritty particles either by elutriation or by screening. Impurities such as ferric oxide, calcium carbonate, and magnesium carbonate are removed with an electromagnet and by treatment with hydrochloric acid and/or sulfuric acids.

14 Safety

Kaolin is used in oral and topical pharmaceutical formulations and is generally regarded as an essentially nontoxic and nonirritant material.

Oral doses of about 2–6 g of kaolin every 4 hours have been administered in the treatment of diarrhea.[(1,2)]

15 Handling Precautions

Observe normal precautions appropriate to the circumstances and quantity of material handled. The chronic inhalation of kaolin dust can cause diseases of the lung (silicosis or kaolinosis).[(9)] Eye protection and a dust mask are recommended. In the UK, the long-term (8-hour TWA) workplace exposure limit for kaolin respirable dust is 2 mg/m^3.[(10)]

16 Regulatory Status

Accepted in Europe as a food additive in certain applications. Included in the FDA Inactive Ingredients Database (oral capsules, powders, syrups, and tablets; topical preparations). Included in nonparenteral medicines licensed in the UK.

17 Related Substances

Bentonite; magnesium aluminum silicate.

18 Comments

Kaolin is considered in most countries to be an archaic diluent.

Therapeutically, kaolin has been used in oral antidiarrheal preparations.[(2)]

The name kaolinite was historically used to describe the processed mineral, while the name kaolin was used for the unprocessed clay. However, the two names have effectively become synonymous and kaolin is now generally the only name used. A specification for kaolin is contained in the *Food Chemicals Codex* (FCC).[(11)]

The EINECS number for kaolin is 310-127-6.

19 Specific References

1 Bergman HD. Diarrhea and its treatment. *Commun Pharm* 1999; **91**(3): 31–35.
2 Sweetman SC, ed. *Martindale: The Complete Drug Reference*, 37th edn. London: Pharmaceutical Press, 2011; 1889.
3 Khalil SAH, *et al.* Decreased bioavailability of ampicillin and amoxicillin in presence of kaolin. *Int J Pharm* 1984; **19**: 233–238.
4 Ganjian F, *et al. In vitro* adsorption studies of cimetidine. *J Pharm Sci* 1980; **69**: 352–353.
5 Albert KS, *et al.* Influence of kaolin-pectin suspension on digoxin bioavailability. *J Pharm Sci* 1978; **67**: 1582–1586.
6 McElnay JC, *et al.* Effect of antacid constituents, kaolin and calcium citrate on phenytoin absorption. *Int J Pharm* 1980; **7**: 83–88.
7 McElnay JC, *et al.* The interaction of warfarin with antacid constituents in the gut. *Experientia* 1979; **35**: 1359–1360.
8 Albert KS, *et al.* Pharmacokinetic evaluation of a drug interaction between kaolin-pectin and clindamycin. *J Pharm Sci* 1978; **67**: 1579–1582.
9 Lesser M, *et al.* Silicosis in kaolin workers and firebrick makers. *South Med J* 1978; **71**: 1242–1246.
10 Health and Safety Executive. *EH40/2005: Workplace Exposure Limits.* Sudbury: HSE Books, 2011. http://www.hse.gov.uk/pubns/priced/eh40.pdf (accessed 22 February 2012).
11 *Food Chemicals Codex*, 7th edn. Bethesda, MD: United States Pharmacopeia, 2010.

20 General References

Allen LV. Featured excipient: capsule and tablet diluents. *Int J Pharm Compound* 2000; 4(4): 306–310324–325.

Allwood MC. The adsorption of esters of *p*-hydroxybenzoic acid by magnesium trisilicate. *Int J Pharm* 1982; **11**: 101–107.

Onyekweli AO, *et al.* Adsorptive property of kaolin in some drug formulations. *Trop J Pharm Res* 2003; **2**: 155–159.

21 Author

A Palmieri.

22 Date of Revision

1 March 2012.

 Lactic Acid

1 Nonproprietary Names

BP: Lactic Acid
JP: Lactic Acid
PhEur: Lactic Acid
USP: Lactic Acid

2 Synonyms

Acidum lacticum; E270; *Eco-Lac*; 2-hydroxypropanoic acid; α-hydroxypropionic acid; DL-lactic acid; *Lexalt L*; milk acid; *Patlac LA*; *Purac 88 PH*; *Purac PF 90*; racemic lactic acid.

3 Chemical Name and CAS Registry Number

2-Hydroxypropionic acid [50-21-5]
(*R*)-(−)-2-Hydroxypropionic acid [10326-41-7]
(*S*)-(+)-2-Hydroxypropionic acid [79-33-44]
(*RS*)-(±)-2-Hydroxypropionic acid [598-82-3]
See also Section 8.

4 Empirical Formula and Molecular Weight

$C_3H_6O_3$ 90.08

5 Structural Formula

H
H_3C OH
OH O

6 Functional Category

Acidulant.

7 Applications in Pharmaceutical Formulation or Technology

Lactic acid is used in beverages, foods, cosmetics, and pharmaceuticals (*see* Table I) as an acidulant.

In topical formulations, particularly cosmetics, it is used for its softening and conditioning effect on the skin. Lactic acid may also be used in the production of biodegradable polymers and microspheres, such as poly(D-lactic acid), used in drug delivery systems.[1,2] *See also* Aliphatic Polyesters.

Lactic acid is also used as a food preservative.

Table I: Uses of lactic acid.

Use	Concentration (%)
Injections	0.012–1.16
Topical preparations	0.015–6.6

8 Description

Lactic acid consists of a mixture of 2-hydroxypropionic acid, its condensation products, such as lactoyllactic acid and other polylactic acids, and water. It is usually in the form of the racemate, (*RS*)-lactic acid, but in some cases the (*S*)-(+)-isomer is predominant.

Lactic acid is a practically odorless, colorless or slightly yellow-colored, viscous, hygroscopic, nonvolatile liquid.

9 Pharmacopeial Specifications

See Table II.

Table II: Pharmacopeial specifications for lactic acid.

Test	JP XV	PhEur 7.4	USP35–NF
Identification	+	+	+
Characters	—	+	—
Appearance of solution	—	+	—
Specific rotation	—	—	−0.05° to +0.05°
Calcium	—	≤200 ppm	—
Heavy metals	≤10 ppm	≤10 ppm	≤0.001%
Iron	≤5 ppm	—	—
Sulfate	≤0.01%	≤200 ppm	+
Chloride	≤0.036%	—	+
Citric, oxalic, phosphoric, and tartaric acids	+	+	+
Ether-insoluble substances	—	+	—
Cyanide	+	—	—
Sugars and other reducing substances	+	+	+
Glycerin and mannitol	+	—	—
Methanol and methyl esters	—	≤50 ppm	—
Readily carbonizable substances	+	—	+
Bacterial endotoxins	—	≤5 IU/g	—
Volatile fatty acids	+	—	—
Residue on ignition	≤0.1%	—	≤3.0 mg
Sulfated ash	—	≤0.1%	—
Assay	85.0–92.0%	88.0–92.0%	88.0–92.0%

10 Typical Properties

Boiling point 122°C at 2 kPa (15 mmHg)
Dissociation constant pK_a = 4.14 at 22.5°C
Flash point >110°C
Heat of combustion 15.13 kJ/kg (3615 cal/kg)
Melting point 17°C
Osmolarity A 2.3% w/v aqueous solution is isoosmotic with serum.
Refractive index n_D^{20} = 1.4251
Solubility Miscible with ethanol (95%), ether, and water; practically insoluble in chloroform.
Specific heat 2.11 J/g (0.505 cal/g) at 20°C
Specific gravity 1.21
Specific rotation $[\alpha]_D^{21}$ = −2.6° (8% w/v aqueous solution) for (*R*)-form; +2.6° (2.5% w/v aqueous solution) for (*S*)-form.
Spectroscopy
IR spectra *see* Figure 1.
Viscosity (dynamic) 28.5 mPa s (28.5 cP) for 85% aqueous solution at 25°C.

11 Stability and Storage Conditions

Lactic acid is hygroscopic and will form condensation products such as polylactic acids on contact with water. The equilibrium

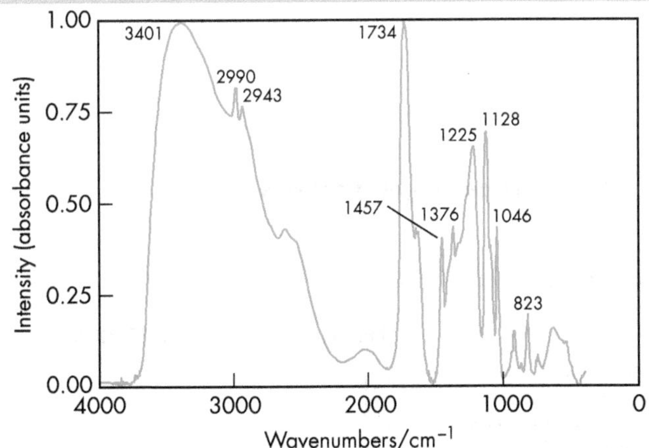

Figure 1: Infrared spectrum of lactic acid measured by transmission. Adapted with permission of Informa Healthcare.

between the polylactic acids and lactic acid is dependent on concentration and temperature. At elevated temperatures lactic acid will form lactide, which is readily hydrolyzed back to lactic acid.

Lactic acid should be stored in a well-closed container in a cool, dry place.

12 Incompatibilities

Incompatible with oxidizing agents, iodides, and albumin. Reacts violently with hydrofluoric acid and nitric acid.

13 Method of Manufacture

Lactic acid is prepared by the fermentation of carbohydrates, such as glucose, sucrose, and lactose, with *Bacillus acidi lacti* or related microorganisms. On a commercial scale, whey, corn starch, potatoes, or molasses are used as a source of carbohydrate. Lactic acid may also be prepared synthetically by the reaction between acetaldehyde and carbon monoxide at 130–200°C under high pressure, or by the hydrolysis of hexoses with sodium hydroxide.(3)

Lactic acid prepared by the fermentation of sugars is levorotatory; lactic acid prepared synthetically is racemic. However, lactic acid prepared by fermentation becomes dextrorotatory on dilution with water owing to the hydrolysis of (*R*)-lactic acid lactate to (*S*)-lactic acid.

14 Safety

Lactic acid occurs in appreciable quantities in the body as an end product of the anaerobic metabolism of carbohydrates and, while harmful in the concentrated form (*see* Section 15), can be considered nontoxic at the levels at which it is used as an excipient. A 1% v/v solution, for example, is harmless when applied to the skin.

There is evidence that neonates have difficulty in metabolizing (*R*)-lactic acid, and this isomer and the racemate should therefore not be used in foods intended for infants aged less than 3 months old.(4)

There is no evidence that lactic acid is carcinogenic, teratogenic, or mutagenic.

LD_{50} (guinea pig, oral): 1.81 g/kg(5)
LD_{50} (mouse, oral): 4.88 g/kg
LD_{50} (mouse, SC): 4.5 g/kg
LD_{50} (rat, oral): 3.73 g/kg

15 Handling Precautions

Lactic acid is caustic in concentrated form and can cause burns on contact with the skin and eyes. It is harmful if swallowed, inhaled, or absorbed through the skin. Observe precautions appropriate to the circumstances and quantity of material handled. Eye protection, rubber gloves, and respirator are recommended. It is advisable to handle the compound in a chemical fume hood and to avoid repeated or prolonged exposure. Spillages should be diluted with copious quantities of water. In case of excessive inhalation, remove the patient to a well-ventilated environment and seek medical attention. Lactic acid presents no fire or explosion hazard but emits acrid smoke and fumes when heated to decomposition.

16 Regulatory Status

GRAS listed. Accepted for use as a food additive in Europe. Included in the FDA Inactive Ingredients Database (IM, IV, and SC injections; oral syrups and tablets; topical and vaginal preparations). Included in medicines licensed in the UK. Included in the Canadian Natural Health Products Ingredients Database.

17 Related Substances

Aliphatic polyesters; sodium lactate.

18 Comments

Therapeutically, lactic acid is used in injections, in the form of lactate, as a source of bicarbonate for the treatment of metabolic acidosis; as a spermicidal agent; in pessaries for the treatment of leukorrhea; in infant feeds; and in topical formulations for the treatment of warts.

A specification for lactic acid is contained in the *Food Chemicals Codex* (FCC).(6)

The EINECS number for lactic acid is 200-018-0. The PubChem Compound ID (CID) for lactic acid includes 612, 107689 and 61503.

L

19 Specific References

1 Brophy MR, Deasy P. Biodegradable polyester polymers as drug carriers. In: Swarbrick J, Boylan JC, eds. *Encyclopedia of Pharmaceutical Technology*. 2. New York: Marcel Dekker, 1990: 1–25.
2 Kim IS, *et al*. Core–shell type polymeric nanoparticles composed of poly(L-lactic acid) and poly(*N*-isopropylacrylamide). *Int J Pharm* 2000; **211**: 1–8.
3 Datta R, Henry M. Lactic acid: recent advances in products, processes and technologies - a review. *J Chem Technol Biotechnol* 2006; **81**: 1119–1129.
4 FAO/WHO. Toxicological evaluation of certain food additives with a review of general principles and specifications. Seventeenth report of the FAO/WHO expert committee on food additives. *World Health Organ Tech Rep Ser* 1974; No. 539.
5 Lewis RJ, ed. *Sax's Dangerous Properties of Industrial Materials*, 11th edn. New York: Wiley, 2004: 2196.
6 *Food Chemicals Codex*, 7th edn. (Suppl. 1), Bethesda, MD: United States Pharmacopeia, 2010: 556.

20 General References

Al-Shammary FJ *et al.Analytical Profiles of Drug Substances and Excipients*. 22. San Diego: Academic Press, 1993: 263–316.

21 Author

MG Lee.

22 Date of Revision

1 March 2012.

Lactitol

1 Nonproprietary Names

BP: Lactitol Monohydrate
PhEur: Lactitol Monohydrate
USP–NF: Lactitol

2 Synonyms

E966; β-galactosido-sorbitol; *Finlac DC.*; lactil; lactite; lactitolum monohydricum; lactobiosit; lactosit; *Lacty.*

3 Chemical Name and CAS Registry Number

4-O-(β-D-Galactopyranosyl)-D-glucitol [585-86-4]
4-O-(β-D-Galactopyranosyl)-D-glucitol monohydrate [81025-04-9]
4-O-(β-D-Galactopyranosyl)-D-glucitol dihydrate [81025-03-8]

4 Empirical Formula and Molecular Weight

$C_{12}H_{24}O_{11}$	344.32 (anhydrous)
$C_{12}H_{24}O_{11}{\cdot}H_2O$	362.34 (monohydrate)
$C_{12}H_{24}O_{11}{\cdot}2H_2O$	380.35 (dihydrate)

5 Structural Formula

6 Functional Category

Direct compression excipient; sweetening agent; tablet and capsule diluent.

7 Applications in Pharmaceutical Formulation or Technology

Lactitol is used as a noncariogenic replacement for sucrose. It is also used as a diluent in solid dosage forms.(1) A direct-compression form is available,(2,3) as is a direct-compression blend of lactose and lactitol.

8 Description

Lactitol occurs as white orthorhombic crystals. It is odorless with a sweet taste that imparts a cooling sensation. It is available in powdered form and in a range of crystal sizes. The directly compressible form is a water-granulated product of microcrystalline aggregates.

9 Pharmacopeial Specifications

See Table I.

Table I: Pharmacopeial specifications for lactitol.

Test	PhEur 7.4	USP35–NF30
Identification	+	+
Characters	+	—
Appearance of solution	+	—
Acidity or alkalinity	+	—
Specific optical rotation	+13.5° to +15.5°	—
Related substances	≤1.0%	≤1.5%
Reducing sugars	≤0.2%	≤0.2% as dextrose
Lead	≤0.5 ppm	—
Nickel	≤1 ppm	—
Water		
monohydrate	4.5–5.5%	4.5–5.5%
dihydrate	—	9.5–10.5%
anhydrous	—	≤0.5%
Microbial contamination	≤10^3 cfu/g	—
Residue on ignition	—	≤0.5%
Sulfated ash	≤0.1%	—
Heavy metals	—	≤5 ppm
Assay	96.5–102.0%	98.0–101.0%

10 Typical Properties

Acidity–alkalinity pH = 4.5–7.0 (10% w/v solution)
Density 1.54 g/cm^3
Heat of solution −54 J/g
Loss of water of crystallization 145–185°C
Moisture content 4.5–5.5% for the monohydrate; ≤0.5% for the anhydrous.
Osmolarity A 7% w/v aqueous solution is isoosmotic with serum.
Refractive index
n_D^{20} = 1.3485 (10% solution);
n_D^{20} = 1.3650 (20% solution);
n_D^{20} = 1.3827 (30% solution);
n_D^{20} = 1.4018 (40% solution);
n_D^{20} = 1.4228 (50% solution);
n_D^{20} = 1.4466 (60% solution).
Solubility Slightly soluble in ethanol (95%) and ether. Soluble 1 in 1.75 of water at 20°C; 1 in 1.61 at 30°C; 1 in 1.49 at 40°C; 1 in 1.39 at 50°C.
Specific rotation $[\alpha]_D^{20}$ = +14.5° to +15°
Spectroscopy
NIR spectra *see* Figure 1.
Viscosity (dynamic)
1.3 mPa s (1.3 cP) for 10% solution at 20°C;
1.9 mPa s (1.9 cP) for 20% solution at 20°C;
3.4 mPa s (3.4 cP) for 30% solution at 20°C;
6.9 mPa s (6.9 cP) for 40% solution at 20°C;
18.9 mPa s (18.9 cP) for 50% solution at 20°C;
80.0 mPa s (80.0 cP) for 60% solution at 20°C.

11 Stability and Storage Conditions

Lactitol as the monohydrate is nonhygroscopic and is stable under humid conditions. It is stable to heat and does not take part in the Maillard reaction. In acidic solution, lactitol slowly hydrolyzes to sorbitol and galactose. Lactitol is very resistant to microbiological

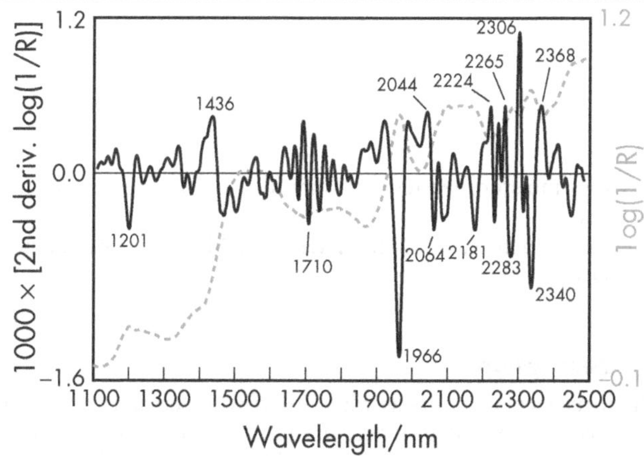

Figure 1: Near-infrared spectrum of lactitol monohydrate measured by reflectance.

breakdown and fermentation. Store in a well-closed container. When the compound is stored in an unopened container at 25°C and 60% relative humidity, a shelf-life in excess of 3 years is appropriate.

12 Incompatibilities

—

13 Method of Manufacture

Lactitol is produced by the catalytic hydrogenation of lactose.

14 Safety

Lactitol is regarded as a nontoxic and nonirritant substance. It is not fermented significantly in the mouth, and is not cariogenic.[4] It is not absorbed in the small intestine, but is broken down by microflora in the large intestine,[5] and is metabolized independently of insulin. In large doses it has a laxative effect; therapeutically, 10–20 g daily in a single oral dose is administered for this purpose.

LD_{50} (mouse, oral): >23 g/kg[6]

LD_{50} (rat, oral): 30 g/kg

15 Handling Precautions

Observe normal precautions appropriate to the circumstances and quantity of material handled. Eye protection is recommended.

16 Regulatory Status

GRAS listed. Included in the FDA Inactive Ingredients Database as lactitol monohydrate (oral and capsule). Accepted for use as a food additive in Europe. Included in the Canadian Natural Health Products Ingredients Database.

17 Related Substances

—

18 Comments

Lactitol has a sweetening power about one-third that of sucrose. It does not promote dental caries and has a caloric value of 9.9 J/g (2.4 cal/g). *Finlac DC* is a commercially available water-granulated directly compressible lactitol.[2]

Lactitol is used therapeutically in the treatment of hepatic encephalopathy and as a laxative;[7] *see also* Section 14.

The EINECS number for lactitol is 209-566-5.

19 Specific References

1 Allen LV. Featured excipient: capsule and tablet diluents. *Int J Pharm Compound* 2000; 4(4): 306–310324–325.
2 Armstrong NA. Direct compression characteristics of lactitol. *Pharm Technol Eur* 1998; 10(2): 42–46.
3 Muzikova J, Vaiglova J. A study of the properties of tablets from the mixture of directly compressible starch and directly compressible lactitol. *Ceska Slov Form* 2007; 5(6): 183–189.
4 Grenby TH, *et al.* Studies on the dental properties of lactitol compared with five other bulk sweeteners *in vitro*. *Caries Res* 1989; **23**: 315–319.
5 Grimble GK, *et al.* Assimilation of lactitol, an unabsorbed disaccharide in the normal human colon. *Gut* 1988; **29**: 1666–1671.
6 Lewis RJ, ed. *Sax's Dangerous Properties of Industrial Materials*, 11th edn. New York: Wiley, 2004; 2198.
7 van Schalk BA. Lactitol versus lactulose in constipation. *Pharm Weekblad* 1991; **126**: 1133–1137.

20 General References

Armstrong NA. Tablet manufacture. In: Swarbrick J, Boylan JC, eds. *Encyclopedia of Pharmaceutical Technology*, 2nd edn, 3. New York: Marcel Dekker, 2002: 2713–2732.
Bolhuis GK, Armstrong NA. Excipients for direct compression – an update. *Pharm Dev Technol* 2006; **11**: 111–124.
van Uyl CH. Technical and commercial aspects of the use of lactitol in foods as a reduced-calorie bulk sweetener. *Dev Sweeteners* 1987; **3**: 65–81.
van Velthuijsen JA. Food additives derived from lactose: lactitol and lactitol palmitate. *J Agric Food Chem* 1979; **27**: 680–686.

21 Author

NA Armstrong.

22 Date of Revision

1 March 2012.

Lactose, Anhydrous

1 Nonproprietary Names

BP: Anhydrous Lactose
JP: Anhydrous Lactose
PhEur: Lactose, Anhydrous
USP–NF: Anhydrous Lactose

2 Synonyms

Anhydrous 60M; *Anhydrous 120M*; *Anhydrous DT*; *Anhydrous DT High Velocity*; *Anhydrous Impalpable*; *Lactopress Anhydrous*; *Lactopress Anhydrous 250*; *Lactopress Anhydrous Crystals*; *Lactopress Anhydrous Microfine*; *Lactopress Anhydrous Powder*; *Lactopress Anhydrous Fine Powder*; lactosum anhydricum; lattosio; milk sugar; *SuperTab 21AN*; *SuperTab 22AN*; *SuperTab 24AN*; saccharum lactis.

3 Chemical Name and CAS Registry Number

O-β-D-Galactopyranosyl-(1→4)-β-D-glucopyranose [63-42-3]

4 Empirical Formula and Molecular Weight

$C_{12}H_{22}O_{11}$ 342.30

5 Structural Formula

Anhydrous α-lactose

Anhydrous β-lactose

The PhEur 7.4 and USP35–NF30 describe anhydrous lactose as O-β-D-galactopyranosyl-(1→4)-β-D-glucopyranose; or a mixture of O-β-D-galactopyranosyl-(1→4)-β-D-glucopyranose and O-β-D-galactopyranosyl-(1→4)-α-D-glucopyranose. The JP XV describes anhydrous lactose as β-lactose or a mixture of β-lactose and α-lactose, and defines these as per the PhEur and USP–NF.

6 Functional Category

Direct compression excipient; dry powder inhaler carrier; lyophilization aid; tablet and capsule diluent.

7 Applications in Pharmaceutical Formulation or Technology

Anhydrous lactose is widely used in direct compression tableting applications, and as a tablet and capsule filler and binder. Anhydrous lactose can be used with moisture-sensitive drugs due to its low moisture content. It may also be used in intravenous injections.

See also Lactose, Inhalation; Lactose, Monohydrate; Lactose, Spray-Dried.

8 Description

Anhydrous lactose occurs as white to off-white crystalline particles or a powder. Several different brands of anhydrous lactose are commercially available which contain anhydrous β-lactose and anhydrous α-lactose. Anhydrous lactose typically contains 70–80% anhydrous β-lactose and 20–30% anhydrous α-lactose.

9 Pharmacopeial Specifications

The pharmacopeial specifications for anhydrous lactose have undergone harmonization for many attributes for JP, PhEur, and USP-NF. *See* Table I. *See also* Section 18.

SEM 1: Excipient: *SuperTab 21AN*; manufacturer: DFE Pharma; magnification: 200×; voltage: 1.5 kV.

SEM 2: Excipient: *SuperTab 22AN*; manufacturer: DFE Pharma; magnification: 55×; voltage: 1.5 kV.

Table I: Pharmacopeial specifications for lactose anhydrous.

Test	JP XV	PhEur 7.4	USP35–NF30
Definition	+	+	+
Identification	+	+	+
Characters	−	+	−
Appearance/color of solution	+	+	+
Optical rotation	+54.4° to +55.9°	+54.4° to +55.9°	+54.4° to +55.9°
Acidity or alkalinity	+	+	+
Heavy metals	≤5 ppm	≤5 ppm	≤5 μg/g
Protein and light absorbing impurities/substances	+	−	+
Absorbance			
210–220 nm	≤0.25	≤0.25	≤0.25
270–300 nm	≤0.07	≤0.07	≤0.07
400 nm	≤0.04	≤0.04	≤0.04
Loss on drying	≤0.5%	+	≤0.5%
Water	≤1.0%	≤1.0%	≤1.0%
Residue on ignition	≤0.1%	−	≤0.1%
Sulfated ash	−	≤1.0%	−
Microbial limit			
Aerobic bacteria	≤100 cfu/g	$\leq 10^2$ cfu/g	≤100 cfu/g
Fungi and yeast	≤50 cfu/g	−	≤50 cfu/g
Absence of *Escherichia coli*	+	+	+
Absence of *Salmonella*	+	−	−
Isomer ratio	+	+[a]	+

(a) Not a mandatory test.

10 Typical Properties

Density (true) 1.589 g/cm^3 for anhydrous β-lactose
Density (bulk)
0.71 g/cm^3 for *SuperTab 21AN*;
0.66 g/cm^3 for *Super-Tab 22AN.*
Density (tapped)
0.88 g/cm^3 for *SuperTab 21AN*;
0.78 g/cm^3 for *Super-Tab 22AN.*
Melting point
223.0°C for anhydrous α-lactose;
252.2°C for anhydrous β-lactose;
232.0°C (typical) for commercial anhydrous lactose.
Particle size distribution see Table II.
Solubility Soluble in water; sparingly soluble in ethanol (95%) and ether; 20 g/100 mL at 25°C for *Lactose Anhydrous DT.*
Spectroscopy
IR spectra *see* Figure 1.
NIR spectra *see* Figure 2.

11 Stability and Storage Conditions

Mold growth may occur under humid conditions (80% RH and above). Lactose may develop a brown coloration on storage, the reaction being accelerated by warm, damp conditions; *see* Section 12. At 80°C and 80% RH, tablets containing anhydrous lactose have been shown to expand 1.2 times after one day.[1]

Lactose anhydrous should be stored in a well-closed container in a cool, dry place.

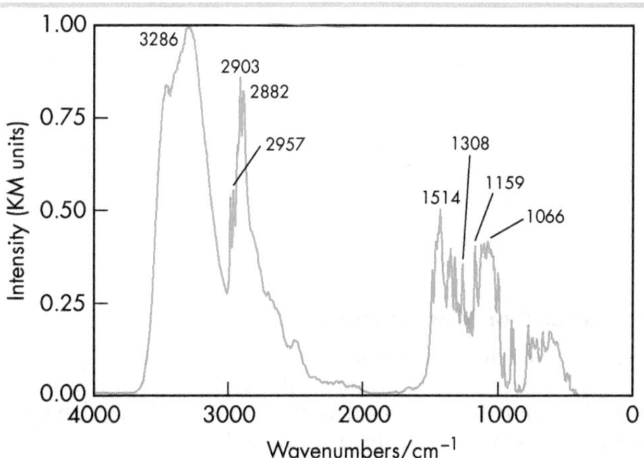

Figure 1: Infrared spectrum of lactose anhydrous measured by reflectance.

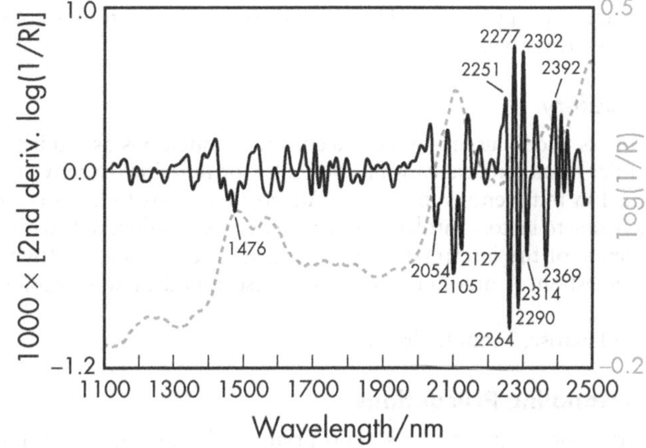

Figure 2: Near-infrared spectrum of lactose anhydrous measured by reflectance.

12 Incompatibilities

Lactose anhydrous is incompatible with strong oxidizers. When mixtures containing a hydrophobic leukotriene antagonist and anhydrous lactose or lactose monohydrate were stored for six weeks at 40°C and 75% RH, the mixture containing anhydrous lactose showed greater moisture uptake and drug degradation.[2]

Table II: Particle size distribution of selected commercially available anhydrous lactose.

Supplier/grade	Percentage less than stated size			
	<45 μm	<53 μm	<150 μm	<250 μm
DFE Pharma				
Lactopress Anhydrous	−	≤30	−	≥80
Lactopress Anhydrous 250	≤20	−	40–65	≥80
SuperTab 21AN	≤20	−	40–65	≥80
SuperTab 22AN	≤7	−	25–50	≥65
Sheffield Bio-Science				
Lactose Anhydrous DT	−	−	35-68	80-90
Lactose Anhydrous DT High Velocity	≤15	−	≤55	60-90

Studies have also shown that in blends of roxifiban acetate (DMP-754) and lactose anhydrous, the presence of lactose anhydrous accelerated the hydrolysis of the ester and amidine groups.[3]

Lactose anhydrous is a reducing sugar with the potential to interact with primary[4] and secondary amines[5] (Maillard reaction) when stored under conditions of high humidity for extended periods.

See Lactose, Monohydrate.

13 Method of Manufacture

There are two anhydrous forms of lactose: α-lactose and β-lactose. The temperature of crystallization influences the ratio of α- and β-lactose. The anhydrous forms that are commercially available may exhibit hygroscopicity at high relative humidities. Anhydrous lactose is produced by roller drying a solution of lactose above 93.5°C. The resulting product is then milled and sieved. Two anhydrous α-lactoses can be prepared using special drying techniques: one is unstable and hygroscopic; the other exhibits good compaction properties.[6,7] However, these materials are not commercially available.

14 Safety

Lactose is widely used in pharmaceutical formulations as a diluent and filler-binder in oral capsule and tablet formulations. It may also be used in intravenous injections. Adverse reactions to lactose are largely due to lactose intolerance, which occurs in individuals with a deficiency of the intestinal enzyme lactase, and is associated with oral ingestion of amounts well over those found in solid dosage forms.

See Lactose, Monohydrate.

15 Handling Precautions

Observe normal precautions appropriate to the circumstances and quantity of materials handled. Excessive generation of dust, or inhalation of dust, should be avoided.

16 Regulatory Status

GRAS listed. Included in the FDA Inactive Ingredients Database (IM, IV: powder for injection solution; sublingual; vaginal; oral: solution, capsules and tablets; powder for inhalation). Included in nonparenteral and parenteral medicines licensed in the UK. Included in the Canadian Natural Health Products Ingredients Database.

17 Related Substances

Lactose, inhalation; lactose, monohydrate; lactose, spray-dried.

18 Comments

Lactose anhydrous has undergone harmonization for many attributes for JP, PhEur, and USP-NF by the Pharmacopeial Discussion Group. For further information see the General Information Chapter <1196> in the USP35–NF30, the General Chapter 5.8 in PhEur 7.4, along with the 'State of Work' document on the PhEur EDQM website, and also the General Information Chapter 8 in the JP XV.

Lactose anhydrous has been used experimentally in hydrophilic matrix tablet formulations[8] and evaluated for dry powder inhalation applications.[9,10] Hydration of anhydrous lactose increases the specific surface area and reduces the flow properties of powders but has no effect on compactibility.[11,12] A specification for lactose is included in the *Food Chemicals Codex* (FCC);[13] *see* Lactose, Monohydrate.

The EINECS number for lactose anhydrous is 200-559-2. The PubChem Compound ID (CID) for lactose anhydrous includes 6134 and 84571.

19 Specific References

1 Du J, Hoag SW. The influence of excipients on the moisture sensitive drugs aspirin and niacinamide: comparison of tablets containing lactose monohydrate with tablets containing anhydrous lactose. *Pharm Dev Technol* 2001; 6(2): 159–166.
2 Jain R, *et al.* Stability of a hydrophobic drug in presence of hydrous and anhydrous lactose. *Eur J Pharm Biopharm* 1998; **46**: 177–182.
3 Badawy SI, *et al.* Effect of different acids on solid state stability of an ester prodrug of a IIb/IIIa glycoprotein receptor antagonist. *Pharm Dev Technol* 1999; 4: 325–331.
4 Castello RA, Mattocks AM. Discoloration of tablets containing amines and lactose. *J Pharm Sci* 1962; **51**: 106–108.
5 Wirth DD, *et al.* Maillard reaction of lactose and fluoxetine hydrochloride, a secondary amine. *J Pharm Sci* 1998; **87**: 31–39.
6 Lerk CF, *et al.* Increased binding capacity and flowability of alpha-lactose monohydrate after dehydration. *J Pharm Pharmacol* 1983; **35**(11): 747–748.
7 Ziffels S, Steckel H. Influence of amorphous content on compaction behaviour of anhydrous alpha-lactose. *Int J Pharm* 2010; **387**: 71–78.
8 Heng PW, *et al.* Investigation of the influence of mean HPMC particle size and number of polymer particles on the release of aspirin from swellable hydrophilic matrix tablets. *J Control Release* 2001; **76**(1–2): 39–49.
9 Larhrib H, *et al.* The use of different grades of lactose as a carrier for aerosolized salbutamol sulphate. *Int J Pharm* 1999; **191**(1): 1–14.
10 Vanderbist F, *et al.* Optimization of a dry powder inhaler formulation of nacystelyn, a new mucoactive agent. *J Pharm Pharmacol* 1999; **51**(11): 1229–1234.
11 Cal S, *et al.* Effects of hydration on the properties of a roller-dried β-lactose for direct compression. *Int J Pharm* 1996; **129**(1–2): 253–261.
12 Shah KR, *et al.* Form conversion of anhydrous lactose during wet granulation and its effect on compactibility. *Int J Pharm* 2008; **357**: 228–234.
13 *Food Chemicals Codex*, 7th edn. Bethesda, MD: United States Pharmacopeia, 2011; 519.

20 General References

Bolhuis GK, Armstrong NA. Excipients for direct compaction: an update. *Pharm Dev Technol* 2006; **11**: 111–124.

Bolhuis GK, Chowhan ZT. Materials for direct compaction. In: Alderborn G, Nystrom C, eds. *Pharmaceutical Powder Compaction Technology.* New York: Marcel Dekker, 1996: 469–473.

DFE Pharma. Technical literature: *Lactopress Anhydrous, Lactopress Anhydrous 250*, 2012. www.dfepharma.com (accessed 16 February 2012).

DFE Pharma. Technical literature: *SuperTab 21AN, SuperTab 22AN, SuperTab 24AN*, 2012. www.dfepharma.com (accessed 16 February 2012).

European Directorate for the Quality of Medicines and Healthcare (EDQM). European Pharmacopoeia – State Of Work Of International Harmonisation. *Pharmeuropa* 2011; 23(4): 713–714. www.edqm.eu/site/-614.html (accessed 16 February 2012).

Kirk JH, *et al.* Lactose: a definite guide to polymorph determination. *Int J Pharm* 2007; **334**: 103–107.

Sheffield Bio-Science. Technical literature: *Anhydrous Lactose*, 2012. www.sheffield-products.com (accessed 16 February 2012).

21 Authors

S Edge, AH Kibbe, J Shur.

22 Date of Revision

1 March 2012.

Lactose, Inhalation

1 Nonproprietary Names

None adopted.

2 Synonyms

InhaLac; inhalation lactose; *Lactohale*; *Respitose*.

For grades, *see* Tables I and II.

3 Chemical Name and CAS Registry Number

Inhalation lactose is lactose monohydrate, *O*-β-D-galactopyranosyl-(1→4)-α-D-glucopyranose monohydrate [5989-81-1]; [10039-26-6]; [64044-51-5] (*see* Lactose, Monohydrate), or anhydrous lactose, *O*-β-D-galactopyranosyl-(1→4)-β-D-glucopyranose [63-42-3], or a mixture of *O*-β-D-galactopyranosyl-(1→4)-β-D-glucopyranose and *O*-β-D-galactopyranosyl-(1→4)-α-D-glucopyranose (*see* Lactose, Anhydrous).

CAS numbers for lactose monohydrate are [5989-81-1] (lactose monohydrate); [10039-26-6] (lactose monohydrate, cyclic); [64044-51-5] (lactose monohydrate, open form).

4 Empirical Formula and Molecular Weight

$C_{12}H_{22}O_{11}$	342.30 (for anhydrous)
$C_{12}H_{22}O_{11}\cdot H_2O$	360.31 (for monohydrate)

5 Structural Formula

See Lactose, Anhydrous; Lactose, Monohydrate.

6 Functional Category

Dry powder inhaler carrier.

7 Applications in Pharmaceutical Formulation or Technology

Inhalation lactose is widely used as a carrier, diluent, and flow aid in dry powder inhalation formulations. Inhalation lactose of suitable particle size can also be used to prepare soft pellets of dry powder inhaler formulations.

See also Lactose, Anhydrous; Lactose, Monohydrate.

8 Description

Lactose occurs as white to off-white crystalline particles or powder. It is odorless and slightly sweet-tasting.

9 Pharmacopeial Specifications

See Lactose, Anhydrous; Lactose, Monohydrate.

10 Typical Properties

Density (bulk) *see* Table I.
Density (tapped) *see* Table I.
Loss on drying *see* Table I.
Particle size distribution *see* Table II.
Surface area *see* Table I.

11 Stability and Storage Conditions

Inhalation lactose should be stored in a well-closed container in a cool, dry place.

SEM 1: Excipient: *Respitose SV003*; manufacturer: DFE Pharma; magnification: 200×; voltage: 5 kV.

SEM 2: Excipient: *Respitose ML001*; manufacturer: DFE Pharma; magnification: 1000×; voltage: 5 kV.

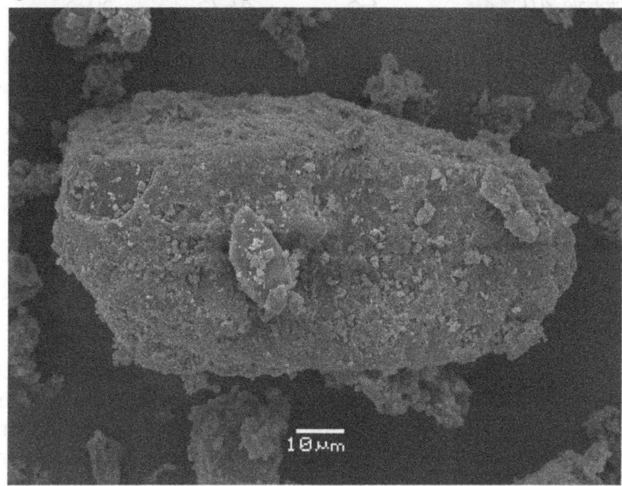

12 Incompatibilities

Lactose is a reducing sugar. Typical reactions include the Maillard reaction with either primary[(1)] or secondary amines.[(2)]

See also Lactose, Anhydrous; Lactose, Monohydrate.

13 Method of Manufacture

Inhalation lactose is manufactured by milling, sieving, air classifying, micronizing and/or blending pharmaceutical grade lactose, typically in dedicated facilities. Although off-the-shelf grades are available, the manufacturing processes can be tailored to produce lactose with properties for a specific application.

14 Safety

Lactose is widely used in pharmaceutical formulations as a diluent in oral capsule and tablet formulations, and has a history of being used in dry powder inhaler formulations.

Table I: Typical physical properties of selected commercially available inhalation lactose.

Supplier/grade	Surface area (m^2/g)	Density (bulk) (g/cm^3)	Density (tapped) (g/cm^3)	Loss on drying (%)
DFE Pharma				
Lactohale 100	–	–	–	≤0.2
Lactohale 200	–	–	–	≤0.2
Lactohale 300	–	–	–	≤0.5
Respitose ML001	0.9	0.57	0.88	–
Respitose ML006	1.6	0.43	0.75	–
Respitose SV003	0.4	0.63	0.78	–
Respitose SV010	0.2	0.69	0.83	–
Meggle GmbH				
InhaLac 70	0.1	0.59	0.69	–
InhaLac 120	0.2	0.69	0.80	–
InhaLac 230	0.2	0.69	0.83	–
InhaLac 250	0.3	0.65	0.85	–

Table II: Typical particle size distribution of selected commercially available inhalation lactose.

Supplier/grade	d_{10} (μm)	d_{50} (μm)	d_{90} (μm)
DFE Pharma			
Lactohale 100[a]	45–65	125–145	200–250
Lactohale 200[a]	5–15	50–100	120–160
Lactohale 201[a]	3–6	20–25	50–60
Lactohale 300[b]	–	<5	≤10
Respitose ML001[b]	4	55	170
Respitose ML006[a]	2	17	45
Respitose SV003[b]	30	60	100
Respitose SV010[a]	50	105	175
Meggle GmbH			
InhaLac 70	120	210	300
InhaLac 120	90	130	180
InhaLac 230	45	100	140
InhaLac 250	20	60	100

(a) Sympatec laser diffraction.
(b) Malvern (wet) laser diffraction.

Adverse reactions to lactose are largely due to lactose intolerance, which occurs in individuals with a deficiency of the enzyme lactase. The presence of milk proteins in lactose-containing dry powder inhalers, which can cause anaphylaxis in cases of severe allergy to cow's milk, has been reported.[3,4] However, a more recent study concluded that milk protein reactions in patients are rare, using the dry powder inhaler products studied.[5]

In view of the route of administration, inhalation lactose should be tested to additional microbiological specifications, for example, endotoxins, as requested by the regulatory authorities. Inhalation lactose is typically supplied with an increased range of microbiological tests.

See also Lactose, Anhydrous; Lactose, Monohydrate.

15 Handling Precautions

Observe normal precautions appropriate to the circumstances and quantity of the material being handled. Excessive generation of dust, or inhalation of dust, should be avoided.

16 Regulatory Status

GRAS listed. Lactose monohydrate and lactose anhydrous are included in the FDA Inactive Ingredients Database (powder for inhalation). Lactose monohydrate and lactose anhydrous are included in inhaled products licensed in the UK.

See also Lactose, Anhydrous; Lactose, Monohydrate.

17 Related Substances

Lactose, Anhydrous; Lactose, Monohydrate.

18 Comments

Lactose is one of a very small number of excipients that are used in marketed dry powder inhaler products. Specific grades of inhalation lactose can be produced from the readily available wide range of pharmaceutical lactose grades using standard pharmaceutical manufacturing processes. Lactose is found in capsule, blister, and reservoir-based dry powder inhaler products. The relatively low mass per dose of lactose in dry powder inhaler products results in a relatively low level of lactose being ingested by the patient compared with conventional oral solid dosage forms.

In view of the importance of particle characteristics for the processing of lactose (micronisation, milling, etc), powder blending and drug product performance,[6–15] it has been suggested that pharmacopeial monograph acceptance criteria are not adequate for controlling key physicochemical characteristics for inhalation applications of this excipient. Accordingly, further material controls may be required to ensure consistent drug product pharmaceutical performance, for example the control of surface properties.

The effect of modifying the surfaces of lactose particles by particle smoothing, crystallization, and co-processing with other excipients on the aerosolization performance has been reported.[16–22]

See also Lactose, Anhydrous; Lactose, Monohydrate.

19 Specific References

1 Castello RA, Mattocks AM. Discoloration of tablets containing amines and lactose. *J Pharm Sci* 1962; **51**: 106–108.

2 Wirth DD, *et al.* Maillard reaction of lactose and fluoxetine hydrochloride, a secondary amine. *J Pharm Sci* 1998; **87**: 31–39.

3 Nowak-Wegrzyn A, *et al.* Contamination of dry powder inhalers for asthma with milk proteins containing lactose. *J Allergy Clin Immunol* 2004; **113**: 558–560.

4 Morisset M, *et al.* Allergy to cow milk proteins contaminating lactose, common excipient of dry powder inhalers for asthma. *J Allergy Clin Immunol* 2006; **117**(Suppl. 1): 95.

5 Spiegel WA, Anolik R. Lack of milk protein allergic reactions in patients using lactose containing dry powder inhalers (DPIs). *J Allergy Clin Immunol* 2010; **125**(Issue 2 (Suppl.1)): AB69.

6 Kawashima Y, *et al.* Effect of surface morphology of carrier lactose on dry powder inhalation property of pralukast hydrate. *Int J Pharm* 1998; **172**: 179–188.

7 Steckel H, *et al.* Functionality testing of inhalation grade lactose. *Eur J Pharm Biopharm* 2004; **57**: 495–505.

8 Telko MJ, Hickey AJ. Dry powder inhaler formulation. *Respir Care* 2005; **50**: 1209–1227.

9 Jones MD, Price R. The influence of fine excipient particles on the performance of carrier-based dry powder inhalation formulations. *Pharm Res* 2006; **23**: 1665–1674.

10 Hickey AJ, *et al.* Physical characterization of component particles included in dry powder inhalers. I. Strategy review and static characteristics. *J Pharm Sci* 2007; **96**: 1282–1301.

11 Hickey AJ, *et al.* Physical characterization of component particles included in dry powder inhalers. II. Dynamic characteristics. *J Pharm Sci* 2007; **96**: 1302–1319.

12 Shur J, *et al.* The role of fines in the modification of the fluidization and dispersion mechanism within dry powder inhaler formulations. *Pharm Res* 2008; **25**: 1931–1940.

13 Larhrib H, *et al.* The use of different grades of lactose as a carrier for aerosolised salbutamol sulfate. *Int J Pharm* 1999; **191**: 1–14.

14 Marek SR. Effects of mild processing pressures on the performance of dry powder inhaler formulations for inhalation therapy (1): Budesonide and lactose. *Eur J Pharm Biopharm* 2011; **78**: 97–106.

15 Shariare MH. The impact of material attributes and process parameters on the micronisation of lactose monohydrate. *Int J Pharm* 2011; **408**: 58–66.

16 Zeng XM, *et al.* The influence of carrier morphology on drug delivery by dry powder inhalers. *Int J Pharm* 2000; **200**: 93–106.

17 Zeng XM, *et al.* The influence of crystallization conditions on the morphology of lactose intended for use as a carrier for dry powder aerosols. *J Pharm Pharmacol* 2004; **52**: 633–643.

18 Islam N, *et al.* Lactose surface modification by decantation: Are drug-fine lactose ratios the key to better dispersion of salmeterol xinofate from lactose interactive mixtures? *Pharm Res* 2004; **21**: 492–499.

19 Young PM, *et al.* Characterisation of a surface modified dry powder inhalation carrier prepared by 'particle smoothing'. *J Pharm Pharmacol* 2002; **54**: 1339–1344.

20 El-Sabawi D, *et al.* Novel temperature controlled surface dissolution of excipient particles for carrier based dry powder inhaler formulations. *Drug Dev Ind Pharm* 2006; **32**: 243–251.

21 Kumon M, *et al.* Application and mechanism of inhalation profile improvement of DPI formulations by mechanofusion with magnesium stearate. *Chem Pharm Bull (Tokyo)* 2008; **56**: 617–625.

22 Guchardi R, *et al.* Influence of fine lactose and magnesium stearate on low dose dry powder inhaler formulations. *Int J Pharm* 2008; **348**: 10–17.

20 General References

CDER. Guidance for industry: Metered dose inhaler (MDI) and dry powder inhaler (DPI) drug products (draft). Rockville, MD: United States Department of Health and Human Services, Food and Drug Administration, Center for Drug Evaluation and Research; 1998.

DFE Pharma. Technical literature: *Respitose*, 2012. http://www.dfepharma.com/ (accessed 17 February 2012).

DFE Pharma. Technical literature: *Lactohale*, 2012. http://www.dfepharma.com/ (accessed 17 February 2012).

EMEA. Committee for Medicinal Products for Human Use. Guidance on the pharmaceutical quality of inhalation and nasal products. EMEA/CHMP/QWP/49313/2005 Corr. London; 21 June 2006.

Kaerger JS, *et al.* Carriers for DPIs: formulation and regulatory challenges. *Pharm Tech Eur* 2006; **18**(10): 25–30.

Kussendrager K *et al.* A new DSC method for the detection of very low amorphous contents in lactose carriers for dry powder inhalers. *Proc AAPS Annual Meeting*, 2005. http://www.aapsj.org/abstracts/AM_2005/AAPS2005-001716.pdf (accessed 17 February 2012).

Meggle GmbH. Technical literature: *InhaLac*, 2012. www.meggle-pharma.de (accessed 17 February 2012).

Sheffield Bio-Science. Technical literature: *Inhalation lactose*, 2012. www.sheffield-products.com (accessed 17 February 2012).

21 Authors

S Edge, JS Kaerger, J Shur.

22 Date of Revision

1 March 2012.

Lactose, Monohydrate

1 Nonproprietary Names

BP: Lactose
JP: Lactose Hydrate
PhEur: Lactose Monohydrate
USP–NF: Lactose Monohydrate

2 Synonyms

CapsuLac; *Foremost*; *GranuLac*; *Lactochem*; lactosum monohydricum; *Monohydrate*; *Pharmatose*; *Lactopress Granulated*; *PrismaLac*; *SacheLac*; *SorboLac*; *SpheroLac*; *SuperTab 30GR*; *Tablettose*.

For grades, *see* Section 10, Tables II and III.

3 Chemical Name and CAS Registry Number

O-β-D-Galactopyranosyl-(1→4)-α-D-glucopyranose monohydrate [5989-81-1]; [10039-26-6]; [64044-51-5]

CAS Registry numbers for lactose monohydrate are [5989-81-1] (lactose monohydrate), [10039-26-6] (lactose monohydrate, cyclic), and [64044-51-5] (lactose monohydrate, open form).

4 Empirical Formula and Molecular Weight

$C_{12}H_{22}O_{11}\cdot H_2O$ 360.31

5 Structural Formula

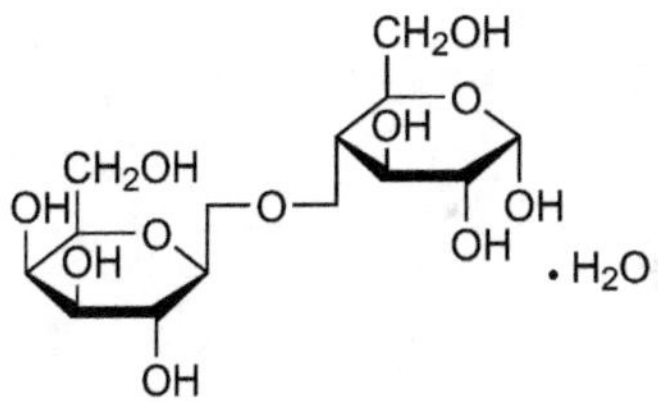

α-Lactose monohydrate

The USP35–NF30 describes lactose monohydrate as a natural disaccharide, obtained from milk, which consists of one galactose and one glucose moiety. The PhEur 7.4 and JP XV describe lactose monohydrate as the monohydrate of O-β-D-galactopyranosyl-(1→4)-α-D-glucopyranose. It is stated in the USP35–NF30 that lactose monohydrate may be modified as to its physical characteristics, and may contain varying proportions of amorphous lactose.

6 Functional Category

Dry powder inhaler carrier; lyophilization aid; tablet and capsule binder; tablet and capsule diluent.

7 Applications in Pharmaceutical Formulation or Technology

Lactose is widely used as a filler and diluent in tablets and capsules, and to a more limited extent in lyophilized products and infant formulas.[1–9] Lactose is also used as a diluent in dry-powder inhalation; *see* Lactose, Inhalation. Various lactose grades are commercially available that have different physical properties such as particle size distribution and flow characteristics. This permits the selection of the most suitable material for a particular application; for example, the particle size range selected for capsules is often dependent on the type of encapsulating machine used. Usually, fine grades of lactose are used in the preparation of tablets by the wet-granulation method or when milling during processing is carried out, since the fine size allows better mixing with other formulation ingredients and utilizes the binder more efficiently.

Other applications of lactose include use in lyophilized products, where lactose is added to freeze-dried solutions to increase plug size and aid cohesion. Lactose is also used in combination with sucrose (approximately 1 : 3) to prepare sugar-coating solutions. It may also be used in intravenous injections. Lactose is also used in the manufacture of dry powder formulations and is used in aqueous film-coating solutions or suspensions.

Direct-compression grades of lactose monohydrate are available as granulated/agglomerated α-lactose monohydrate, containing small amounts of anhydrous lactose.

Direct-compression grades are often used to carry lower quantities of drug and this permits tablets to be made without granulation.

Other directly compressible lactoses are spray-dried lactose and anhydrous lactose; *see* Lactose, Spray-Dried and Lactose, Anhydrous.

8 Description

In the solid state, lactose appears as various isomeric forms, depending on the crystallization and drying conditions, i.e. α-lactose monohydrate, β-lactose anhydrous, and α-lactose anhydrous. The stable crystalline forms of lactose are α-lactose monohydrate, β-lactose anhydrous, and stable α-lactose anhydrous.

Lactose occurs as white to off-white crystalline particles or powder. Lactose is odorless and slightly sweet-tasting; α-lactose is approximately 20% as sweet as sucrose, while β-lactose is 40% as sweet.

SEM 1: Excipient: *Pharmatose 125M*; manufacturer: DFE Pharma; magnification: 100×; voltage: 1.5 kV.

SEM 2: Excipient: *SuperTab 30GR*; manufacturer: DFE Pharma.

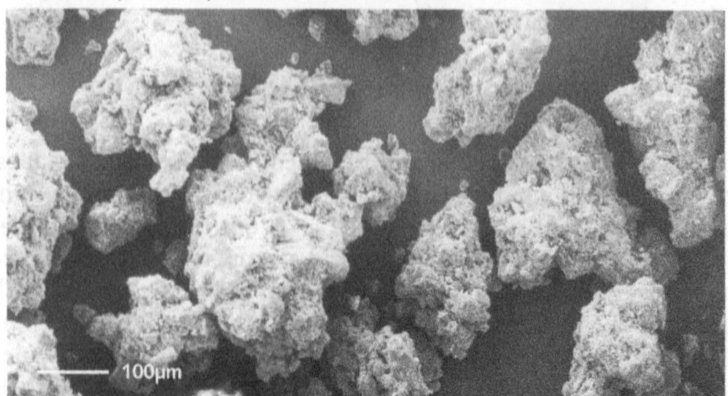

9 Pharmacopeial Specifications

The pharmacopeial specifications for lactose monohydrate have undergone harmonization of many attributes for JP, PhEur, and USP-NF. *See* Table I. *See also* Section 18.

10 Typical Properties

Density (true) 1.545 g/cm^3 (α-lactose monohydrate)

Density (bulk) *see* Table II.

Density (tapped) *see* Table II.

Melting point 201–202°C (for dehydrated α-lactose monohydrate)

Moisture content Lactose monohydrate contains approximately 5% w/w water of crystallization and normally has a range of 4.5–5.5% w/w water content.

Particle size distribution *see* Table III.

Solubility *see* Table IV.

Specific rotation $[\alpha]_D^{20}$ = +54.4° to +55.9° as a 10% w/v solution. Lactose exhibits mutarotation, and an equilibrium mixture containing 62% β-lactose and 38% α-lactose is obtained instantly on the addition of a trace of ammonia.

Spectroscopy

IR spectra *see* Figure 1.

NIR spectra *see* Figure 2.

SEM 1: Excipient: *Lactochem Crystals*; manufacturer: DFE Pharma; magnification: 200×; voltage: 10 kV.

SEM 2: Excipient: *Lactochem Crystals*; manufacturer: DFE Pharma; magnification: 700×; voltage: 10 kV.

Table I: Pharmacopeial specifications for lactose monohydrate.

Test	JP XV	PhEur 7.4	USP35-NF30[a]
Definition	+	+	+
Identification	+	+	+
Characters	+	+	—
Appearance/color of solution	+	+	+
Acidity or alkalinity	+	+	+
Optical rotation	+54.4° to +55.9°	+54.4° to +55.9°	+54.4° to +55.9°
Protein and light absorbing impurities	+	—	+
Absorbance			
210–220 nm	≤0.25	≤0.25	≤0.25
270–300 nm	≤0.07	≤0.07	≤0.07
400 nm	≤0.04	≤0.04	≤0.04
Heavy metals	≤5 ppm	≤5 ppm	≤5 ppm
Water	4.5–5.5%[b]	4.5–5.5%	4.5–5.5%
Sulfated ash	—	≤0.1%	—
Residue on ignition	≤0.1%	—	≤0.1%
Loss on drying	≤0.5%[c]	—	≤0.5%[d]
Microbial limit			
Aerobic bacteria	≤100 cfu/g	≤10^2 cfu/g	≤100 cfu/g
Fungi and yeast	≤50 cfu/g	—	≤50 cfu/g
Absence of *Escherichia coli*	+	+	+
Absence of *Salmonella*	+	—	—

(a) For USP-NF, lactose monohydrate may be modified as to its physical characteristics. It may contain varying proportions of amorphous lactose.
(b) The definition of JP also covers granulated lactose, with a lower limit of 4.0% of water, i.e. 4.0–5.5%.
(c) For granulated powder, ≤1.0%.
(d) For modified monohydrate form, ≤1.0%.

Table II: Typical physical properties of selected commercially available lactose, monohydrate.

Supplier/grade	Density (bulk) (g/cm^3)	Density (tapped) (g/cm^3)
DFE Pharma		
Lactochem Coarse Crystals	0.75	0.88
Lactochem Crystals	0.74	0.86
Lactochem Fine Crystals	0.73	0.85
Lactochem Extra Fine Crystals	0.73	0.86
Lactochem Coarse Powder	0.71	0.95
Lactochem Regular Powder	0.62	0.92
Lactochem Powder	0.64	0.89
Lactochem Fine Powder	0.61	0.84
Lactochem Extra Fine Powder	0.45	0.74
Lactochem Super Fine Powder	0.47	0.74
Pharmatose 50M	0.70	0.82
Pharmatose 60M	0.80	0.98
Pharmatose 70M	0.81	1.02
Pharmatose 80M	0.75	0.92
Pharmatose 90M	0.72	0.90
Pharmatose 100M	0.73	0.88
Pharmatose 110M	0.72	0.88
Pharmatose 125M	0.68	0.85
Pharmatose 130M	0.65	0.96
Pharmatose 150M	0.62	0.90
Pharmatose 200M	0.57	0.84
Pharmatose 350M	0.54	0.80
Pharmatose 450M	0.48	0.74
SuperTab 30GR	0.53	0.66
Meggle GmbH		
CapsuLac 60	0.59	0.70
GranuLac 70	0.72	0.90
GranuLac 140	0.66	0.89
GranuLac 200	0.54	0.80
GranuLac 230	0.47	0.76
PrismaLac 40	0.47	0.54
SacheLac 80	0.60	0.71
SorboLac 400	0.39	0.63
SpheroLac 100	0.69	0.84
Tablettose 70	0.55	0.67
Tablettose 80	0.61	0.74
Tablettose 100	0.57	0.69
Sheffield Bio-Science		
Foremost Lactose 310	0.66	0.92
Foremost Lactose 312	0.53	0.81
Foremost Lactose 313	0.44	0.72

Table IV: Solubility of lactose.

Solvent	Solubility at 20°C unless otherwise stated
Chloroform	Practically insoluble
Ethanol	Practically insoluble
Ether	Practically insoluble
Water	1 in 5.24
	1 in 3.05 at 40°C
	1 in 2.30 at 50°C
	1 in 1.71 at 60°C
	1 in 0.96 at 80°C

11 Stability and Storage Conditions

Mold growth may occur under humid conditions (80% relative humidity and above). Lactose may develop a brown coloration on storage, the reaction being accelerated by warm, damp conditions; *see* Section 12. The purities of different lactoses can vary and color evaluation may be important, particularly if white tablets are being formulated. The color stabilities of various lactoses also differ.

Solutions show mutarotation; *see* Section 10.

Lactose should be stored in a well-closed container in a cool, dry place.

12 Incompatibilities

A Maillard-type condensation reaction is likely to occur between lactose and compounds with a primary amine group to form brown, or yellow-brown-colored products.[10] The Maillard interaction has also been shown to occur between lactose and secondary amine. However, the reaction sequence stops with the formation of the imine, and no yellow-brown coloration develops.[11] It has been reported that minor components such as lactose phosphates may enhance the degradation of steroids.[12]

Lactose is also incompatible with amino acids, amfetamines,[13] and lisinopril.[14]

13 Method of Manufacture

Lactose is a natural disaccharide consisting of galactose and glucose, and is present in the milk of most mammals. Commercially, lactose is produced from the whey of cows' milk; whey being the residual liquid of the milk following cheese and casein production. Cows' milk contains 4.4–5.2% lactose; lactose constitutes 38% of the total solid content of milk.

α-Lactose monohydrate is prepared by crystallization from supersaturated solutions below 93.5°C. Various crystalline shapes are prism, pyramidal, and tomahawk; these are dependent on the method of precipitation and crystallization. Direct compression grades of α-lactose monohydrate are prepared by granulation/agglomeration and spray-drying.

14 Safety

Lactose is widely used in pharmaceutical formulations as a filler and filler-binder in oral capsule and tablet formulations. It may also be

Table III: Particle size distribution of selected commercially available lactose, monohydrate.

Supplier/grade	Typical particle size distribution (%)												
	<10 μm	<32 μm	<45 μm	<63 μm	<75 μm	<100 μm	<150 μm	<200 μm	<250 μm	<315 μm	<400 μm	<600 μm	<800 μm
DFE Pharma													
Lactochem Coarse Crystals	—	—	—	—	—	—	30–80	—	≥65	—	≥90 (a)	—	—
Lactochem Crystals	—	—	—	—	5–30	—	55–95	—	≥90	—	—	—	—
Lactochem Fine Crystals	—	—	—	—	≤30	—	—	—	≥90	—	—	—	—
Lactochem Extra Fine Crystals	—	—	—	—	10–45	—	—	—	≥99	—	—	—	—
Lactochem Coarse Powder	—	—	—	—	15–50	—	≥75	—	≥98	—	—	—	—
Lactochem Regular Powder	—	—	20–42 (b)	—	—	—	—	≥95 (c)	—	—	—	—	—
Lactochem Powder	—	—	—	—	65–80	≥85 (d)	≥95	—	—	—	—	—	—
Lactochem Fine Powder	—	—	55–80 (b)	—	≥80	—	≥98	—	—	—	—	—	—
Lactochem Extra Fine Powder	—	—	≥90 (b)	—	—	—	≥99 (e)	—	—	—	—	—	—
Lactochem Super Fine Powder	—	—	≥95	—	—	—	—	—	—	—	—	—	—
Lactochem Microfine	≤90	—	—	—	—	—	—	—	—	—	—	—	—
Lactopress Granulated	—	—	—	≤20	—	—	—	40-75 (f)	—	—	≥85	≥97 (g)	—
Pharmatose 50M	—	—	—	—	—	—	—	≤20	—	—	≥80	—	—
Pharmatose 60M	—	—	—	—	≤5	—	10–20	—	40–65	—	—	≥98	—
Pharmatose 70M	—	—	—	—	5–13	—	25–45	—	60–80	—	—	≥98	—
Pharmatose 80M	—	—	—	—	—	≤20	—	—	70–90	≥95	—	—	—
Pharmatose 90M	—	—	—	≤15	—	15–30	55–90	—	—	100	—	—	—
Pharmatose 100M	—	—	—	≤15	—	—	60–80	—	≥99	—	—	—	—
Pharmatose 110M	—	—	—	≤20	—	30–60	75–90	—	—	100	—	—	—
Pharmatose 125M	—	—	≤40	40–70	—	≥90	≥97	—	—	—	—	—	—
Pharmatose 130M	—	—	—	—	≤50	—	—	80–90 (f)	—	—	—	—	—
Pharmatose 150M	—	—	≤50	—	—	≥70	≥85	—	—	100	—	—	—
Pharmatose 200M	—	—	50–65	—	—	≥90	≥96	—	≥99	—	—	—	—
Pharmatose 350M	—	—	≥60	—	—	≥96	—	—	100	—	—	—	—
Pharmatose 450M	—	—	≥90	≥98	—	—	100	—	—	—	—	—	—
SuperTab 30GR	—	—	—	—	≤30	—	40–70	—	—	≥90 (h)	—	100 (i)	—
Meggle GmbH													
CapsuLac 60	—	—	—	—	—	5	15	—	60	—	99	100 (g)	—
GranuLac 70	—	—	—	—	—	50	—	—	—	—	99.5	100 (g)	—
GranuLac 140	—	30	—	—	—	90	—	—	—	—	—	—	—
GranuLac 200	—	55	—	—	—	98.4	—	—	—	—	—	—	—
GranuLac 230	—	75	—	96	—	99.5	—	—	—	—	—	—	—
PrismaLac 40	—	—	—	—	—	—	—	4	—	—	65 (i)	90.5 (g)	100
SacheLac 80	—	—	—	—	—	5	—	—	63	—	100	—	—
SorboLac 400	—	97.5	—	99.5	—	—	—	—	—	—	—	—	—
SpheroLac 100	—	—	—	9	—	—	78	98	99.5	—	—	—	—
Tablettose 70	—	—	—	3	—	—	25	56	—	—	93	100 (i)	—
Tablettose 80	—	—	—	13	—	—	—	53 (f)	—	—	93	100 (g)	—
Tablettose 100	—	—	—	12	—	—	42	—	77	—	—	100 (i)	—
Sheffield Bio-Science													
Foremost Lactose 310	—	—	—	—	24–50	69–35 (d)	—	≥98 (f)	—	—	—	—	—
Foremost Lactose 312	—	—	—	64–80	≥94	≥99 (d)	—	—	—	—	—	—	—
Foremost Lactose 313	—	66–83 (j)	91–98	—	≥99	—	—	—	—	—	—	—	—
Foremost 314WG	—	—	—	—	≥94	≥99 (d)	—	100 (f)	—	—	—	—	—
Lactose Monohydrate 80M	—	—	—	—	65–90	—	93–99.5	≥98 (f)	≥99.5	—	—	—	—
Lactose Monohydrate 180M*	—	—	—	—	—	—	—	—	—	—	—	—	—
Lactose Monohydrate 120MS*	—	—	—	—	—	—	—	—	—	—	—	—	—
Lactose Monohydrate 220MS*	—	—	—	—	—	—	—	—	—	—	—	—	—
Lactose Monohydrate Capsulating	—	—	—	—	58–70	—	≥92	≥98 (f)	—	—	—	—	—
Lactose Monohydrate Impalpable	—	—	—	—	>90	—	≥98.5	—	—	—	—	—	—
Sheffield Brand 200M	—	—	—	—	55–90	—	≥85	≥95 (f)	—	—	—	—	—

(a) <425 μm.
(b) <53 μm.
(c) <212 μm.
(d) <106 μm.
(e) <125 μm.
(f) <180 μm.
(g) <630 μm.
(h) <355 μm.
(i) <500 μm.
(j) <37 μm.
*Full laser diffraction ranges available.

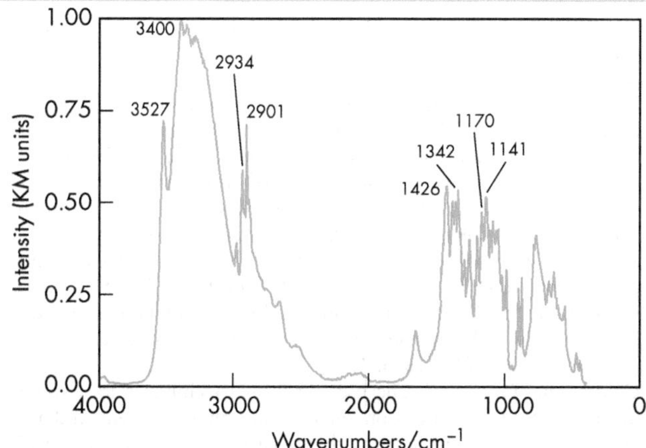

Figure 1: Infrared spectrum of lactose monohydrate measured by diffuse reflectance. Adapted with permission of Informa Healthcare.

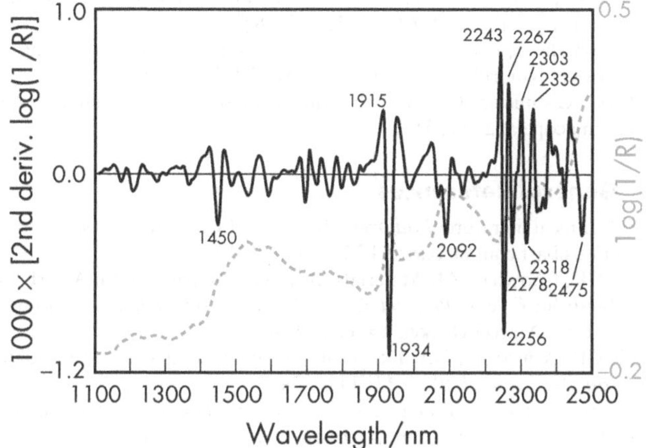

Figure 2: Near-infrared spectrum of lactose monohydrate measured by reflectance.

used in intravenous injections. Adverse reactions to lactose are largely attributed to lactose intolerance, which occurs in individuals with a deficiency of the intestinal enzyme lactase.[15–18] This results in lactose being undigested and may lead to cramps, diarrhea, distension, and flatulence. In lactose-tolerant individuals, lactase hydrolyzes lactose in the small intestine to glucose and galactose, which are then absorbed. Lactase levels are normally high at birth, and levels decline rapidly in early childhood. Malabsorption of lactose (hypolactasia) may occur at an early age (4–8 years) and varies among different ethnic groups. Lactose is excreted unchanged when administered intravenously.

The symptoms of lactose intolerance are caused by the osmotic effect of the unabsorbed lactose, which increases water and sodium levels in the lumen. Unabsorbed lactose, upon reaching the colon, can be fermented by colonic flora, which produces gas, causing abdominal distension and discomfort. A lactose tolerance test has been developed based on the measurement of blood glucose level and the hydrogen level in the breath. However, its usefulness has been questioned as the test is based on a 50 g dose of lactose.

Approximately 10–20% of lactose-intolerant individuals, in two studies, showed clinical symptoms of intolerance after ingestion of 3–5 g of lactose.[15,16] In one of the studies,[15] 75% of the subjects had symptoms with 12 g of lactose (equivalent to 250 mL of milk). In another,[16] eight out of 13 individuals developed diarrhea after the administration of 20 g of lactose, and nine out of 13 after the administration of 25 g.

Lower doses of lactose produce fewer adverse effects, and lactose is better tolerated if taken with other foods. As a result, there is a significant population with lactose malabsorption who are still able to ingest normal amounts of lactose, such as that in milk, without the development of adverse side effects.[17]

Most adults consume about 25 g of lactose per day (500 mL of milk) without symptoms.[18,19] When symptoms appear, they are usually mild and dose-related. The dose of lactose in most pharmaceuticals seldom exceeds 2 g per day. It is unlikely that severe gastrointestinal symptoms can be attributed to the lactose in a conventional oral solid-dosage form, especially in adults who have not previously been diagnosed as severely lactose-intolerant. However, anecdotal reports of drug-induced diarrhea due to lactose intolerance have been made following administration of pharmaceutical preparations containing lactose.

It has also been suggested that lactose intolerance may have a role in irritable bowel syndrome, but this role is currently unclear.[20]

In the past, there have been concerns over the transmissible spongiform encephalopathies (TSE) contamination of animal-derived products. However, in the light of current scientific knowledge, and irrespective of geographical origin, milk and milk derivatives are reported as unlikely to present any risk of TSE contamination; TSE risk is negligible if the calf rennet is produced in accordance with regulations.[21]

LD_{50} (rat, IP): >10 g/kg
LD_{50} (rat, oral): >10 g/kg
LD_{50} (rat, SC): >5 g/kg

15 Handling Precautions

Observe normal precautions appropriate to the circumstances and quantity of material handled. Excessive generation of dust, or inhalation of dust, should be avoided.

16 Regulatory Status

GRAS listed. Included in the FDA Inactive Ingredients Database (IM, IV, and SC: powder for injection; oral: capsules and tablets; inhalation preparations; vaginal preparations). Included in nonparenteral and parenteral medicines licensed in the UK. Included in the Canadian Natural Health Products Ingredients Database.

17 Related Substances

Lactose, anhydrous; lactose, inhalation; lactose, monohydrate and corn starch; lactose, monohydrate and microcrystalline cellulose; lactose, monohydrate and povidone; lactose, monohydrate and powdered cellulose; lactose, spray-dried.

18 Comments

Lactose monohydrate has undergone harmonization for JP, PhEur, and USP-NF by the Pharmacopeial Discussion Group. For further information see the General Information Chapter 1196 in the USP35–NF30, the General Chapter 5.8 in PhEur 7.4, along with the 'State of Work' document on the PhEur EDQM website, and also the General Information Chapter 8 in the JP XV.

A number of different grades of lactose are commercially available that vary in their physical properties, and many studies have been reported in the literature comparing the behavior of these various materials in different formulations.[5,8,9] A number of co-processed excipients that contain lactose are available for direct-compression applications: co-processed lactose and starch (*StarLac*, Meggle/Roquette Fréres),[22] lactose and microcrystalline cellulose (*Microcelac 100*, Meggle);[23] lactose and cellulose powder (*Cellactose 80*, Meggle);[24,25] lactose and hypromellose (*Retalac*, Meggle); lactose, povidone, and crospovidone (*Ludipress*, *Ludipress LCE*, BASF).[26]

Lactose may exhibit complex thermoanalytical transitions because of its several crystalline, as well as amorphous, forms. Differential scanning calorimetry (DSC) can be used effectively to characterize the composition.[27–29] For example, α-lactose becomes anhydrous at approximately 120°C. α-Lactose monohydrate may also contain a small quantity of the β-form. Additionally, lactose monohydrate has been reported to be the least prone to caking compared to anhydrous and spray-dried lactose.[30]

A specification for lactose is included in the *Food Chemicals Codex* (FCC).[31]

The EINECS number for lactose is 200-559-2. The PubChem Compound ID (CID) for lactose monohydrate includes 62223 and 104938.

19 Specific References

1 Alpar O, *et al.* The compression properties of lactose. *J Pharm Pharmacol* 1970; Dec(Suppl.): 1S–7S.
2 Vromans H *et al.* Studies on the tableting properties of lactose: the effect of initial particle size on binding properties and dehydration characteristics of α-lactose monohydrate. In: Rubinstein MH, ed. *Pharmaceutical Technology:Tableting Technology.* 1. Chichester: Ellis Horwood, 1987; 31–42.
3 Thwaites PM, *et al.* An investigation of the effect of high speed mixing on the mechanical and physical properties of direct compression lactose. *Drug Dev Ind Pharm* 1991; **17**: 503–517.
4 Riepma KA, *et al.* The effect of moisture sorption on the strength and internal surface area of lactose tablets. *Int J Pharm* 1992; **87**: 149–159.
5 Çelik M, Okutgen E. A feasibility study for the development of a prospective compaction functionality test and the establishment of a compaction data bank. *Drug Dev Ind Pharm* 1993; **19**: 2309–2334.
6 Lerk CF. Consolidation and compaction of lactose. *Drug Dev Ind Pharm* 1993; **19**: 2359–2398.
7 Otsuka M, *et al.* Effect of humidity on solid-state isomerization of various kinds of lactose during grinding. *J Pharm Pharmacol* 1993; **45**: 2–5.
8 Paronen P. Behaviour of some direct compression adjuvants during the tabletting process. *STP Pharma* 1986; **2**: 682–688.
9 Zuurman K, *et al.* The relationship between bulk density and compactibility of lactose granulations. *Int J Pharm* 1994; **102**: 1–9.
10 Castello RA, Mattocks AM. Discoloration of tablets containing amines and lactose. *J Pharm Sci* 1962; **51**: 106–108.
11 Wirth DD. Maillard reaction of lactose and fluoxetine hydrochloride, a secondary amine. *J Pharm Sci* 1998; **87**: 31–39.
12 Nieuwmeyer F, *et al.* Lactose contaminant as steroid degradation enhancer. *Pharm Res* 2008; **25**(11): 2666–2673.
13 Blaug SM, Huang W. Interaction of dextroamphetamine sulfate with spray-dried lactose. *J Pharm Sci* 1972; **61**: 1770–1775.
14 Eyjolfsson R. Lisinopril–lactose incompatibility. *Drug Dev Ind Pharm* 1998; **24**: 797–798.
15 Bedine MS, Bayless TM. Intolerance of small amounts of lactose by individuals with low lactase levels. *Gastroenterology* 1973; **65**: 735–743.
16 Gudmand-Hoyer E, Simony K. Individual sensitivity to lactose in lactose malabsorption. *Am J Dig Dis* 1977; **22**(3): 177–181.
17 Lorner MC, *et al.* Review article: lactose intolerance in clinical practice – myths and realities. *Aliment Pharmacol Ther* 2008; **27**(2): 93–103.
18 Suarez FL, Savaiano Dennis A. Diet, genetics, and lactose intolerance. *Food Technol* 1997; **51**(3): 74–76.
19 Suarez FL, *et al.* A comparison of symptoms after the consumption of milk or lactose-hydrolyzed milk by people with self-reported lactose intolerance. *N Engl J Med* 1995; **333**: 1–4.
20 Spanier JA. A systemic review of alternative therapies in the irritable bowel syndrome. *Arch Intern Med* 2003; **163**(3): 265–274.
21 The European Agency for the Evaluation of Medicinal Products. *Evaluation of Medicines for Human Use. London, 9 Dec 2002: EMEA/410/01 Rev. 2.*
22 Hauschild K, Picker-Freyer KM. Evaluation of a new coprocessed compound based on lactose and maize starch for tablet formulation. *AAPS Pharm Sci Tech* 2004; **6**(2): e16.
23 Michoel A, *et al.* Comparative evaluation of co-processed lactose and microcrystalline cellulose with their physical mixtures in the formulation of folic acid tablets. *Pharm Dev Technol* 2002; **7**(1): 79–87.
24 Casalderrey M, *et al.* A comparison of drug loading capacity of cellactose with two ad hoc processed lactose-cellulose direct compression excipients. *Chem Pharm Bull (Tokyo)* 2004; **52**(4): 398–401.
25 Arida AI, Al-Tabakha MM. Cellactose a co-processed excipient: a comparison study. *Pharm Dev Technol* 2008; **13**: 165–175.
26 Heinz R, *et al.* Formulation and development of tablets based on Ludipress and scale-up from laboratory to production scale. *Drug Dev Ind Pharm* 2000; **26**: 513–521.
27 Chidavaenzi OC, *et al.* The use of thermal techniques to assess the impact of feed concentration on the amorphous content and polymorphic forms present in spray dried lactose. *Int J Pharm* 1997; **159**: 67–74.
28 Hill VL, *et al.* Characterisation of spray-dried lactose using modulated differential scanning calorimetry. *Int J Pharm* 1998; **161**: 95–107.
29 Lerk CF, *et al.* Alterations of α-lactose during differential scanning calorimetry. *J Pharm Sci* 1984; **73**: 856–857.
30 Listiohadia Y, *et al.* Moisture sorption, compressibility and caking of lactose polymorphs. *Int J Pharm* 2008; **359**: 123–134.
31 *Food Chemicals Codex*, 7th edn. Bethesda, MD: United States Pharmacopeia, 2011; 559.

20 General References

BASF. Technical literature: *Ludipress, Ludipress LCE*, 2012. www.pharma-solutions.basf.com (accessed 17 February 2012).

Bolhuis GK, Chowhan ZT. Materials for direct compaction. In: Alderborn G, Nyström C, eds. *Pharmaceutical Powder Compaction Technology.* New York: Marcel Dekker, 1996; 459–469.

Bolhuis GK, Armstrong NA. Excipients for direct compaction: an update. *Pharm Dev Technol* 2006; **11**: 111–124.

DFE Pharma. Technical literature: *Lactochem, Lactopress Granulated*, 2012. www.dfepharma.com (accessed 17 February 2012).

DFE Pharma. Technical literature: *Pharmatose, SuperTab 30GR*, 2012. www.dfepharma.com (accessed 17 February 2012).

European Directorate for the Quality of Medicines and Healthcare (EDQM). European Pharmacopoeia – State Of Work Of International Harmonisation. Pharmeuropa 2011; 23(4): 713–714.www.edqm.eu/site/-614.html (accessed 17 February 2012).

Kirk JH, *et al.* Lactose: a definite guide to polymorph determination. *Int J Pharm* 2007; **334**: 103–107.

Meggle GmbH. Technical literature: *Lactose excipients*, 2012. www.meggle-pharma.de (accessed 17 February 2012).

Rajah KK, Blenford DE, eds. *The ALM Guide to Lactose Properties and Uses.* The Hague: Association of Lactose Manufacturers, 1998.

Roquette Fréres. Technical literature: *StarLac*, 2012. www.roquette-pharma.com (accessed 17 February 2012).

Sheffield Bio-Science. Technical literature: *Crystalline lactose monohydrate*, 2012. www.sheffield-products.com (accessed 17 February 2012).

21 Authors

S Edge, AH Kibbe, J Shur.

22 Date of Revision

2 March 2012.

Lactose, Monohydrate and Corn Starch

1 Nonproprietary Names
None adopted.

2 Synonyms
StarLac.

3 Chemical Name and CAS Registry Number
See Section 8.

4 Empirical Formula and Molecular Weight
See Section 8.

5 Structural Formula
See Section 8.

6 Functional Category
Direct compression excipient; tablet and capsule diluent; tablet and capsule disintegrant.

7 Applications in Pharmaceutical Formulation or Technology
Lactose monohydrate and corn starch can be used in tablets to improve compressibility, flowability and disintegration properties.[1] It is used in homeopathic and low-dose to mid-dose formulations.

8 Description
α-Lactose monohydrate and corn starch occurs as a white or almost white odorless powder containing 82–88% of lactose monohydrate and 12–18% of corn (maize) starch. It is a free-flowing powder owing to its spherical structure.

SEM 1: Excipient: *StarLac;* manufacturer: Roquette/Meggle; magnification: 200×; voltage: 2 kV.

9 Pharmacopeial Specifications
Both lactose monohydrate and corn (maize) starch are listed as separate monographs in the JP, PhEur, and USP–NF, but the combination is not listed. The pharmacopeial specifications for lactose monohydrate and corn starch have been harmonized for JP, PhEur, and USP–NF.

See Lactose, Monohydrate, and Starch. *See also* Section 18.

10 Typical Properties
Angle of repose ≤29° for *StarLac*
Density (bulk) 0.57 g/cm^3 for *StarLac*
Density (tapped) 0.68 g/cm^3 for *StarLac*
Hausner ratio 1.19 for *StarLac*
Heavy metals 5 ppm for *StarLac*
Loss on drying ≤3.0% for *StarLac*
Microbial content Total viable aerobic count ≤100 cfu/g, molds <10 cfu/g, yeasts <10 cfu/g (*Escherichia coli* and *Salmonella* species absent) for *StarLac.*
Particle size distribution ≤15% <32 μm, 35–65% <160 μm, ≥80% <250 μm for *StarLac.*
Sulfated ash ≤0.25% for *StarLac*
Solubility Partially soluble in cold water for *StarLac.*

11 Stability and Storage Conditions
Store in well-closed containers under dry and odor-free conditions.

12 Incompatibilities
See Lactose, Monohydrate, and Starch.

13 Method of Manufacture
Lactose monohydrate and corn starch is prepared by spray-drying a mixture of the two ingredients.

14 Safety
See Lactose, Monohydrate, and Starch.

15 Handling Precautions
Observe normal precautions appropriate to the circumstances and quantity of material handled.

16 Regulatory Status
Lactose monohydrate and corn starch is a mixture of two materials both of which are generally regarded as nontoxic:

Lactose monohydrate GRAS listed. Included in the FDA Inactive Ingredients Database (IM, IV, and SC: powder for injection; oral: capsules and tablets; inhalation preparations; vaginal preparations). Included in nonparenteral and parenteral medicines licensed in the UK. Included in the Canadian List of Acceptable Non-medicinal Ingredients.

Starch GRAS listed. Included in the FDA Inactive Ingredients Database (buccal tablets, oral capsules, powders, suspensions and tablets; topical preparations; and vaginal tablets). Included in nonparenteral medicines licensed in the UK. Included in the Canadian List of Acceptable Non-medicinal Ingredients.

17 Related Substances
Lactose, monohydrate; starch.

18 Comments

StarLac has been designed for direct compression, combining good flowability and compressibility with fast disintegration properties. Excipients or formulations containing a variety of drugs, namely ascorbic acid, acetaminophen [paracetamol] and theophylline monohydrate show it to be superior to a simple mixture of its components in terms of flowability, tablet strength, friability and disintegration time.[1,2] Starch particles are embedded in a matrix mainly consisting of crystalline lactose monohydrate, and very low quantities of amorphous lactose are detectable. Its balanced elastic and brittle properties make it suitable for roller compaction. Specific quantitative, analytical methods for the assay of starch and lactose in *StarLac* have been developed and validated.[3]

19 Specific References

1 Wagner KG, Dressler JA. A corn starch/alpha-lactose monohydrate compound as a new directly compressible excipient. *Pharm Ind* 2002; 64(9): 992–999.
2 Hauschild K, Picker KM. Evaluation of a new coprocessed compound based on lactose and maize starch for tablet formulation. *AAPS J* 2004; 6(2): 27–38.
3 Roquette/Meggle. Technical data sheet. *StarLac* additional information, August 2007.

20 General References

Barreto LC, da Cunha MS. Co-processed excipients for direct compression tablets. *Lat Am J Pharm* 2009; **28**: 304–312.
Deorkar N. High-functionality excipients: a review. *Tablets and Capsules* 2008: 22–26. www.tabletscapsules.com (accessed 3 March 2009).
European Directorate for the Quality of Medicines and Healthcare (EDQM). European Pharmacopoeia – State Of Work Of International Harmonisation. *Pharmeuropa* 2011; 23(4): 713–714. www.edqm.eu/site/-614.html (accessed 29 November 2011).
Gohel MC, Jogani PD. A review of co-processed directly compressible excipients. *J Pharm Pharm Sci* 2005; **8**(1): 76–93.
Haeusler O. A star is born. *Pharmaceutical Formulation and Quality* 2002; 54–57.
Meggle. Product literature: *Lactose Excipients*, September 2008.
Nachaegari SK, Bansal AK. Coprocessed excipients for solid dosage forms. *Pharm Tech* 2004; (Jan): 52–64.
Roquette. Technical datasheet: *StarLac* – lactose-starch compound for direct compression, October 2005.

21 Authors

ME Fenton, RC Rowe.

22 Date of Revision

29 November 2011.

Lactose, Monohydrate and Microcrystalline Cellulose

1 Nonproprietary Names

None adopted.

2 Synonyms

MicroceLac 100.

3 Chemical Name and CAS Registry Number

See Section 8.

4 Empirical Formula and Molecular Weight

See Section 8.

5 Structural Formula

See Section 8.

6 Functional Category

Tablet and capsule diluent; direct compression excipient.

7 Applications in Pharmaceutical Formulation or Technology

Lactose monohydrate and microcrystalline cellulose can be used in tablets for direct compression.

8 Description

Lactose monohydrate and microcrystalline cellulose occurs as a white or almost white odorless powder containing 73–77% of lactose monohydrate and 23–27% of microcrystalline cellulose.

9 Pharmacopeial Specifications

Both lactose monohydrate and microcrystalline cellulose are listed as separate monographs in the JP, PhEur, and USP–NF, but the combination is not listed. The pharmacopeial specifications for both lactose monohydrate and microcrystalline cellulose have undergone harmonization for many attributes for JP, PhEur, and USP-NF.

See Lactose, Monohydrate, and Cellulose, Microcrystalline. *See also* Section 18.

10 Typical Properties

Acidity/alkalinity pH = 4.0–7.0 for *MicroceLac 100*
Angle of repose 34° for *MicroceLac 100*
Density (bulk) 0.5 g/cm^3 for *MicroceLac 100*
Density (tapped) 0.61 g/cm^3 for *MicroceLac 100*
Hausner ratio 1.16 for *MicroceLac 100*
Heavy metals ≤5 ppm for *MicroceLac 100*
Loss on drying ≤1.5% for *MicroceLac 100*
Microbial content Total viable aerobic count ≤100 cfu/g, molds ≤10 cfu/g, yeasts ≤10 cfu/g (*Escherichia coli* and *Salmonella* species absent) for *MicroceLac 100*
Particle size distribution ≤15% <32 μm, 45–70% <160 μm, ≥90% <250 μm for *MicroceLac 100*
Solubility Partially soluble in water for *MicroceLac 100*
Sulfated ash ≤0.1% for *MicroceLac 100*
Water content 4–6% for *MicroceLac 100*

11 Stability and Storage Conditions

Store at room temperature in well-closed containers under dry and odor-free conditions.

SEM 1: Excipient: *MicroceLac 100*; manufacturer: Meggle; magnification: 200×; voltage: 3 kV.

SEM 2: Excipient: *MicroceLac 100*; manufacturer: Meggle; magnification: 500×; voltage: 3 kV.

12 Incompatibilities

See Lactose, Monohydrate, and Cellulose, Microcrystalline.

13 Method of Manufacture

Lactose monohydrate and microcrystalline cellulose is prepared by spray-drying a mixture of the two ingredients.

14 Safety

See Lactose, Monohydrate, and Cellulose, Microcrystalline.

15 Handling Precautions

Observe normal precautions appropriate to the circumstances and quantity of material handled.

16 Regulatory Status

Lactose monohydrate and microcrystalline cellulose is a mixture of two materials both of which are generally regarded as nontoxic:

Lactose monohydrate GRAS listed. Included in the FDA Inactive Ingredients Database (IM, IV, and SC: powder for injection; oral: capsules and tablets; inhalation preparations; vaginal preparations). Included in nonparenteral and parenteral medicines licensed in the UK. Included in the Canadian Natural Health Products Ingredients Database.

Microcrystalline cellulose GRAS listed. Accepted for use as a food additive in Europe. Included in the FDA Inactive Ingredients Database (inhalations; oral capsules, powders, suspensions, syrups, and tablets; topical and vaginal preparations). Included in nonparenteral medicines licensed in the UK. Included in the Canadian Natural Health Products Ingredients Database.

17 Related Substances

Cellulose, microcrystalline; lactose, monohydrate.

18 Comments

MicroceLac 100 has been designed for formulating high-dose small tablets with a poorly flowable active ingredient. It showed superior flow and binding properties compared to simple mixtures of its components.[1,2] Differences between *MicroceLac 100* and *Cellactose 80* have recently been evaluated.[3]

A specification for lactose and microcrystalline cellulose spheres is contained in the *Japanese Pharmaceutical Excipients* (JPE);[4] *see* Table I.

Table I: JPE specification for lactose and microcrystalline cellulose spheres.

Test	JPE 2004
Description	+
Identification	+
Heavy metals	≤5 ppm
Arsenic	≤2 ppm
Loss on drying	≤5.0%
Water	≤9.0%
Residue on ignition	≤0.1%
Assay (dried basis)	
Lactose	60-80%
Microcrystalline cellulose	20-40%

19 Specific References

1 Michoel A, *et al.* Comparative evaluation of co-processed lactose and microcrystalline cellulose with their physical mixtures in the formulation of folic acid tablets. *Pharm Dev Technol* 2002; 7(1): 79–87.

2 Goto K, *et al.* Pharmaceutical evaluation of multipurpose excipients for direct compressed tablet manufacture: comparisons of the capabilities of multipurpose excipients with those in general use. *Drug Dev Ind Pharm* 1999; 25(8): 869–878.

3 Muzíkova J, Zvolánková J. A study of the properties of tablets from coprocessed dry binders composed of alpha-lactose monohydrate and different types of cellulose. *Ceska Slov Farm* 2007; 56(6): 269–275.

4 Japan Pharmaceutical Excipients Council. *Japanese Pharmaceutical Excipients 2004*. Tokyo: Yakuji Nippo, 2004: 457–459.

20 General References

Barreto LC, da Cunha MS. Co-processed excipients for direct compression tablets. *Lat Am J Pharm* 2009; 28: 304–312.

Deorkar N. High-functionality excipients: a review. *Tablets and Capsules* 2008: 22–26. www.tabletscapsules.com (accessed 29 November 2011).

European Directorate for the Quality of Medicines and Healthcare (EDQM). European Pharmacopoeia – State Of Work Of International Harmonisation. *Pharmeuropa* 2011; 23(4): 713–714. www.edqm.eu/site/-614.html (accessed 29 November 2011).

Gohel MC, Jogani PD. A review of co-processed directly compressible excipients. *J Pharm Pharm Sci* 2005; 8(1): 76–93.

Meggle GmbH. Product literature: Lactose excipients, September 2008.

Meggle GmbH. Product literature: *MicroceLac 100*, March 2006.

Nachaegari SK, Bansal AK. Coprocessed excipients for solid dosage forms. *Pharm Tech* 2004; 28: 52–64.

21 Authors

ME Fenton, RC Rowe.

22 Date of Revision

29 November 2011.

L

Lactose, Monohydrate and Povidone

1 Nonproprietary Names

None adopted.

2 Synonyms

Ludipress LCE.

3 Chemical Name and CAS Registry Number

See Section 8.

4 Empirical Formula and Molecular Weight

See Section 8.

5 Structural Formula

See Section 8.

6 Functional Category

Tablet and capsule diluent; direct compression excipient.

7 Applications in Pharmaceutical Formulation or Technology

Lactose monohydrate and povidone can be used to formulate chewable tablets, lozenges, effervescent tablets, and controlled-release tablets by direct compression.[1,2] It is suitable for low-dose drugs.[3]

8 Description

Lactose monohydrate and povidone occurs as white free-flowing granules, odorless with a neutral taste, containing 96.5% ± 1.8% of lactose monohydrate and 3.5% ± 0.5% of povidone K30.

9 Pharmacopeial Specifications

Both lactose monohydrate and povidone are listed as separate monographs in the JP, PhEur, and USP–NF, but the combination is not listed. The pharmacopeial specifications for both lactose monohydrate and povidone have undergone harmonization for many attributes for JP, PhEur, and USP-NF.

See Lactose, Monohydrate, and Povidone. *See also* Section 18.

10 Typical Properties

Angle of repose 29.5° for *Ludipress LCE*
Density (bulk) 0.56 ± 0.6 g/cm^3 for *Ludipress LCE*
Hausner ratio 1.20 ± 0.10 for *Ludipress LCE*
Heavy metals ≤10 ppm for *Ludipress LCE*
Loss on drying 5.75% for *Ludipress LCE*
Microbial content Mesophilic aerobes ≤1000 cfu/g, yeasts and fungi ≤100 cfu/g (*Escherichia coli*, *Pseudomonas aeruginosa*, *Staphylococcus aureus*, and *Salmonella* species absent), other Enterobacteriaceae ≤100 cfu/g) for *Ludipress LCE*
Particle size distribution ≤20% <63 μm, 40–65% <200 μm, ≤20% <400 μm for *Ludipress LCE*
Solubility Soluble in water for *Ludipress LCE*
Spectroscopy
IR spectra *see* Figure 1.
Water content ≤6.0% for *Ludipress LCE*

11 Stability and Storage Conditions

Store at room temperature in tightly closed containers.

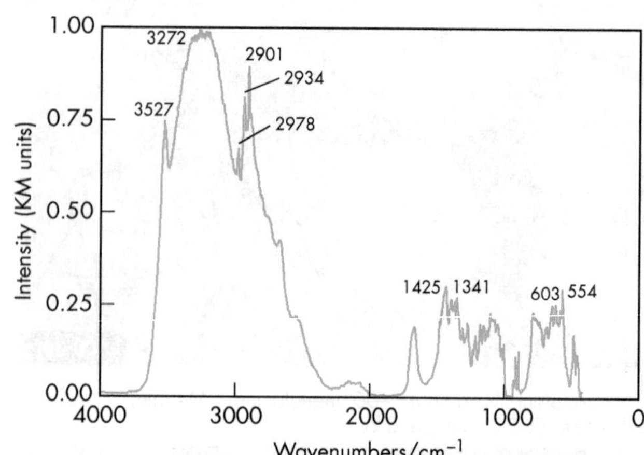

Figure 1: Infrared spectrum of lactose monohydrate and povidone measured by diffuse reflectance. Adapted with permission of Informa Healthcare.

12 Incompatibilities

See Lactose, Monohydrate, and Povidone.

13 Method of Manufacture

Lactose monohydrate and povidone is manufactured by a proprietary agglomeration process.

14 Safety

See Lactose, Monohydrate, and Povidone.

15 Handling Precautions

Observe normal precautions appropriate to the circumstances and quantity of material handled.

16 Regulatory Status

Lactose monohydrate and povidone is a mixture of two materials both of which are generally regarded as nontoxic:

Lactose monohydrate GRAS listed. Included in the FDA Inactive Ingredients Database (IM, IV, and SC: powder for injection; oral: capsules and tablets; inhalation preparations; vaginal preparations). Included in nonparenteral and parenteral medicines licensed in the UK. Included in the Canadian Natural Health Products Ingredients Database.

Povidone Accepted for use in Europe as a food additive. Included in the FDA Inactive Ingredients Database (IM and IV injections; ophthalmic preparations; oral capsules, drops, granules, suspensions, and tablets; sublingual tablets; topical and vaginal preparations). Included in nonparenteral medicines licensed in the UK. Included in the Canadian Natural Health Products Ingredients Database.

17 Related Substances

Lactose, monohydrate; povidone.

18 Comments

Lactose monohydrate and povidone have undergone harmonization for many attributes for JP, PhEur, and USP-NF by the

Pharmacopeial Discussion Group. For further information see the General Information Chapter <1196> in the USP35–NF30, the General Chapter 5.8 in PhEur 7.4, along with the 'State of Work' document on the PhEur EDQM website, and also the General Information Chapter 8 in the JP XV.

Ludipress LCE has been shown to have compression characteristics superior to a simple physical mixture of its constituents.[4] [5]Tablet strength has been shown to be independent of machine speed[5] and tablet geometry,[6]and does not increase on storage.[7] Disintegration time has been shown not to increase at high compression forces.[8]*Ludipress* has also been used in the development of hydrophilic matrices for extended-release tablets,[9] and in solid dispersions to achieve high dissolution.[10]

19 Specific References

1 Kolter K, Fussnegger B. Development of tablet formulations using *Ludipress LCE* as a direct compression excipient. BASF Product Literature, No. 1, November 1998.
2 Sehic S, *et al.* Investigation of intrinsic dissolution behavior of different carbamazepine samples. *Int J Pharm* 2010; **386**: 77–90.
3 Kolter K *et al. Ludipress LCE.* A new direct compression excipient. BASF Product Literature, No. 10, May 2003.
4 Schmidt PC, Rubensdorfer CJ. Evaluation of *Ludipress* as a multipurpose excipient for direct compression. Part 1. Powder characteristics and tableting properties. *Drug Dev Ind Pharm* 1994; **20**(18): 2899–2925.
5 Goto K, *et al.* Pharmaceutical evaluation of multipurpose excipients for direct compressed tablet manufacture: comparisons of the capabilities of multipurpose excipients with those in general use. *Drug Dev Ind Pharm* 1999; **25**(8): 869–878.
6 Heinz R, *et al.* Formulation and development of tablets based on *Ludipress* and scale-up for laboratory to production scale. *Drug Dev Ind Pharm* 2000; **26**(5): 513–521.
7 Baykara T, *et al.* Comparing the compressibility of *Ludipress* with the other direct tableting agents by using acetaminophen as an active ingredient. *Drug Dev Ind Pharm* 1991; **17**(17): 2359–2371.
8 Schmidt PC, Rubensdorfer CJ. Evaluation of *Ludipress* as a multipurpose excipients for direct compression. Part 2. Interactive blending and tableting with micronized glibenclamide. *Drug Dev Ind Pharm* 1994; **20**(18): 2927–2952.
9 Ochiuz L, *et al.* Development of hydrophilic swellable carbopol matrices for extended-release tablets. *Farmacia (Romania)* 2008; **56**: 521–531.
10 Fu JJ, *et al.* Nimodipine (NM) tablets with high dissolution containing NM solid dispersions prepared by hot-melt extrusion. *Drug Dev Ind Pharm* 2011; **37**: 934–944.

20 General References

Barreto LC, da Cunha MS. Co-processed excipients for direct compression tablets. *Lat Am J Pharm* 2009; **28**: 304–312.
Deorkar N. High-functionality excipients: a review. *Tablets and Capsules* 2008: 22–26. www.tabletscapsules.com (accessed 29 November 2011).
European Directorate for the Quality of Medicines and Healthcare (EDQM). European Pharmacopoeia – State Of Work Of International Harmonisation. *Pharmeuropa* 2011; 23(4): 713–714. http://www.edqm.eu/site/-614.html (accessed 29 November 2011).
Gohel MC, Jogani PD. A review of co-processed directly compressible excipients. *J Pharm Pharm Sci* 2005; 8(1): 76–93.
Nachaegari SK, Bansal AK. Coprocessed excipients for solid dosage forms. *Pharm Tech* 2004; **28**: 52–64.
BASF. Technical literature: *Ludipress LCE*, May 1999.

21 Authors

ME Fenton, RC Rowe.

22 Date of Revision

29 November 2011.

L

Lactose, Monohydrate and Powdered Cellulose

1 Nonproprietary Names

None adopted.

2 Synonyms

Cellactose 80.

3 Chemical Name and CAS Registry Number

See Section 8.

4 Empirical Formula and Molecular Weight

See Section 8.

5 Structural Formula

See Section 8.

6 Functional Category

Tablet and capsule diluent.

7 Applications in Pharmaceutical Formulation or Technology

Lactose monohydrate and powdered cellulose can be used in tablets for direct compression to improve compressibility and mouthfeel.[1]

8 Description

Lactose monohydrate and powdered cellulose occurs as a white or almost white odorless powder containing 73–77% of lactose monohydrate and 23–27% of cellulose powder.

9 Pharmacopeial Specifications

Both lactose monohydrate and powdered cellulose are listed as separate monographs in the JP, PhEur, and USP–NF, but the combination is not listed. The pharmacopeial specifications for both lactose monohydrate and powdered cellulose have undergone harmonization for many attributes for JP, PhEur, and USP-NF.

See Lactose, Monohydrate, and Cellulose, Powdered. *See also* Section 18.

10 Typical Properties

Acidity/alkalinity pH = 4.0–7.0 for *Cellactose 80*

Angle of repose 32–35° for *Cellactose 80*
Density (bulk) 0.38 g/cm^3 for *Cellactose 80*
Density (tapped) 0.5 g/cm^3 for *Cellactose 80*
Hausner ratio 1.24 for *Cellactose 80*
Heavy metals ≤5 ppm for *Cellactose 80*
Loss on drying ≤3.5% for *Cellactose 80*
Microbial content Total viable aerobic count ≤100 cfu/g, molds ≤10 cfu/g, yeast ≤10 cfu/g (*Escherichia coli* and *Salmonella* species absent) for *Cellactose 80*
Particle size distribution ≤20% <32 μm, 35–65% <160 μm, ≥80% <250 μm for *Cellactose 80*
Sulfated ash ≤0.2% for *Cellactose 80*
Solubility Partially soluble in water for *Cellactose 80*
Water content 4–7% for *Cellactose 80*

11 Stability and Storage Conditions

Store at room temperature in well-closed containers under dry and odor-free conditions.

12 Incompatibilities

See Lactose, Monohydrate, and Cellulose, Powdered.

13 Method of Manufacture

Lactose monohydrate and powdered cellulose is prepared by spray-drying a mixture of the two ingredients.

14 Safety

See Lactose, Monohydrate, and Cellulose, Powdered.

15 Handling Precautions

Observe normal precautions appropriate to the circumstances and quantity of material handled.

16 Regulatory Status

Lactose monohydrate and powdered cellulose is a mixture of two materials both of which are generally regarded as nontoxic:

Lactose monohydrate GRAS listed. Included in the FDA Inactive Ingredients Database (IM, IV, and SC: powder for injection; oral: capsules and tablets; inhalation preparations; vaginal preparations). Included in nonparenteral and parenteral medicines licensed in the UK. Included in the Canadian Natural Health Products Ingredients Database.

Powdered cellulose GRAS listed. Accepted for use as a food additive in Europe (except for infant food in the UK). Included in nonparenteral medicines licensed in the UK. Included in the Canadian List of Acceptable Natural Health Products Ingredients Database.

17 Related Substances

Lactose, monohydrate; cellulose, powdered.

18 Comments

Lactose monohydrate and powdered cellulose have undergone harmonization for many attributes for JP, PhEur, and USP—NF by the Pharmacopeial Discussion Group. For further information see the General Information Chapter <1196> in the USP35–NF30, the General Chapter 5.8 in PhEur 7.4, along with the 'State of Work' document on the PhEur EDQM website, and also the General Information Chapter 8 in the JP XV.

Cellactose 80 has been designed especially for direct compression. It has been shown to be superior to a simple mixture of its components in terms of dilution potential,[2] compressibility,[3] tensile strength,[4,5] lubricant susceptibility,[6,7] and subsequent tablet properties for a range of drugs.[8] *Cellactose* has been investigated for use in capsules[9] and modified-release tablets,[10,11] and also as a tablet disintegrant.[12]

19 Specific References

1 Nachaegari SK, Bansal AK. Coprocessed excipients for solid dosage forms. *Pharm Tech* 2004; **28**: 52–64.
2 Flores LE, *et al*. Study of load capacity of *Avicel PH-200* and *Cellactose*, two direct-compression excipients, using experimental design. *Drug Dev Ind Pharm* 2000; **26**(4): 465–469.
3 Belda PM, Mielck JB. The tableting behavior of *Cellactose* compared with mixtures of celluloses with lactoses. *Eur J Pharm Biopharm* 1996; **42**(5): 325–330.
4 Muzíkova J, Zvolánková J. A study of the properties of tablets from coprocessed dry binders composed of alpha-lactose monohydrate and different types of cellulose. *Ceska Slov Farm* 2007; **56**(6): 269–275.
5 Arida AI, Al-Tabakha MM. *Cellactose* a co-processed excipient: a comparison study. *Pharm Dev Technol* 2008; **13**(2): 165–175.
6 Konkel P, Mielck JB. Associations of parameters characterizing the time course of the tabletting process on a reciprocating and on a rotary tabletting machine for high-speed production. *Eur J Pharm Biopharm* 1998; **45**(2): 137–148.
7 Flores LE, *et al*. Lubricant susceptibility of *Cellactose* and *Avicel PH-200*: a quantitative relationship. *Drug Dev Ind Pharm* 2000; **26**(3): 297–305.
8 Casalderrey M, *et al*. A comparison of drug loading capacity of cellactose with two ad hoc processed lactose-cellulose direct compression excipients. *Chem Pharm Bull (Tokyo)* 2004; **52**(4): 398–401.
9 Abou-Taleb AE. Formulation and evaluation of rofecoxib capsules. *Saudi Pharm J* 2009; **17**: 40–50.
10 Nalawade P. Formulation and optimization of directly compressible floating tablets of famotidine using 2^3 factorial design. *Lat Am J Pharm* 2010; **29**: 991–999.
11 Mirelabodea. Identification of critical formulation variables for obtaining metoprolol tartrate mini-tablets. *Farmacia (Romania)* 2010; **58**: 719–727.
12 Emeje M. Preparation and standardization of a herbal agent for the therapeutic management of asthma. *Pharm Dev Technol* 2011; **16**: 170–178.

20 General References

Barreto LC, da Cunha MS. Co-processed excipients for direct compression tablets. *Lat Am J Pharm* 2009; **28**: 304–312.
Deorkar N. High-functionality excipients: a review. *Tablets and Capsules* 2008: 22–26. www.tabletscapsules.com (accessed 30 November 2011).
European Directorate for the Quality of Medicines and Healthcare (EDQM). European Pharmacopoeia – State Of Work Of International Harmonisation. *Pharmeuropa* 2011; 23(4): 713–714. http://www.edqm.eu/site/-614.html (accessed 30 November 2011).
Gohel MC, Jogani PD. A review of co-processed directly compressible excipients. *J Pharm Pharm Sci* 2005; **8**(1): 76–93.
Meggle. Product literature: Lactose excipients, September 2008.
Meggle. Technical literature: *Cellactose 80*, August 2006.

21 Authors

ME Fenton, RC Rowe.

22 Date of Revision

30 November 2011.

Lactose, Spray-Dried

1 Nonproprietary Names

None adopted.

2 Synonyms

FlowLac 90; *FlowLac 100*; *Lactopress Spray-Dried*; *Lactopress Spray-Dried 250*; *Lactopress Spray-Dried 260*; *Foremost Lactose 315*; *Foremost Lactose 316 Fast Flo*; *SuperTab 11SD*; *SuperTab 14SD*.

3 Chemical Name and CAS Registry Number

Spray-dried lactose is a mixture of amorphous lactose, which is a 1:1 mixture of α-and-β-lactose, and O-β-D-galactopyranosyl-(1→4)-α-D-glucopyranose monohydrate [5989-81-1]; [10039-26-6]; [64044-51-5].

CAS numbers for lactose monohydrate are [5989-81-1] (lactose monohydrate); [10039-26-6] (lactose monohydrate, cyclic); [64044-51-5] (lactose monohydrate, open form).

4 Empirical Formula and Molecular Weight

$C_{12}H_{22}O_{11}$ 342.30 (for amorphous)
$C_{12}H_{22}O_{11}\cdot H_2O$ 360.31 (for monohydrate)

5 Structural Formula

See Lactose, Anhydrous and Lactose, Monohydrate.

6 Functional Category

Direct compression excipient; tablet and capsule diluent.

7 Applications in Pharmaceutical Formulation or Technology

Spray-dried lactose is widely used as a binder, filler-binder, and flow aid in direct compression tableting.

See also Lactose, Monohydrate; Lactose, Anhydrous.

8 Description

Lactose occurs as white to off-white crystalline particles or powder. It is odorless and slightly sweet-tasting. Spray-dried direct-compression grades of lactose are generally composed of 80–90% specially prepared pure α-lactose monohydrate along with 10–20% of amorphous lactose.

9 Pharmacopeial Specifications

See Section 18. *See also* Lactose, Monohydrate.

10 Typical Properties

Angle of repose 29° for *FlowLac 90.*
Density bulk *see* Table I.
Loss on drying ≤1.0% for *FlowLac 90*
Particle size distribution *see* Table II.
Spectroscopy
IR spectra *see* Figure 1.
Water content *see* Table I.

SEM 1: Excipient: *SuperTab 11SD*; manufacturer: DFE Pharma; magnification: 300×; voltage: 5 kV.

SEM 2: Excipient: *Lactopress Spray-Dried*; manufacturer: DFE Pharma; magnification: 500×; voltage: 3 kV.

Table I: Typical physical properties of selected commercially available spray-dried lactose.

Supplier/grade	Density (bulk) (g/cm^3)	Density (tapped) (g/cm^3)
DFE Pharma		
SuperTab 11SD	0.60	0.71
SuperTab 14SD	0.62	0.72
Meggle GmbH		
FlowLac 90	0.57	0.67
FlowLac 100	0.62	0.71

11 Stability and Storage Conditions

Spray-dried lactose should be stored in a well-closed container in a cool, dry place.

12 Incompatibilities

Lactose is a reducing sugar. The amorphous lactose, which is the most reactive form of lactose present in spray-dried lactose, will

SEM 3: Excipient: *Foremost Lactose 316 Fast Flo*; manufacturer: Sheffield Bio-Science; magnification: 500×; voltage: 3 kV.

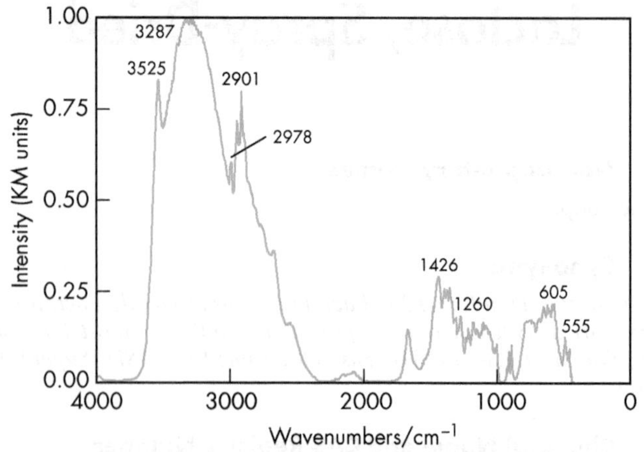

Figure 1: Infrared spectrum of spray-dried lactose measured by diffuse reflectance. Adapted with permission of Informa Healthcare.

interact more readily than conventional crystalline grades.[1,2] Typical reactions include the Maillard reaction with either primary[3] or secondary[4] amines.

See Lactose, Anhydrous and Lactose, Monohydrate.

13 Method of Manufacture

A suspension of α-lactose monohydrate crystals in a lactose solution is atomized and dried in a spray drier.[5,6] Approximately 10–20% of the total amount of lactose is in solution and the remaining 80–90% is present in the crystalline form. The spray-drying process predominantly produces spherical particles. The compactibility of the material and its flow characteristics are a function of the primary particle size of the lactose monohydrate and the amount of amorphous lactose.[7]

14 Safety

Lactose is widely used in pharmaceutical formulations as a diluent in oral capsule and tablet formulations. It may also be used in intravenous injections.

Adverse reactions to lactose are largely due to lactose intolerance, which occurs in individuals with a deficiency of the enzyme lactase.

See Lactose, Monohydrate.

15 Handling Precautions

Observe normal precautions appropriate to the circumstances and quantity of material being handled. Excessive generation of dust, or inhalation of dust, should be avoided.

16 Regulatory Status

See Lactose, Monohydrate.

17 Related Substances

Lactose, anhydrous; lactose, inhalation; lactose, monohydrate.

18 Comments

Spray-dried lactose was one of the first direct-compression excipients. Spray-dried lactose typically comprises lactose monohydrate and amorphous lactose (*see* Section 8); *see* Lactose, Monohydrate for the relevant pharmacopeial information.

It has been shown that during the spray-drying process the effects of nozzle orifice diameter and atomization air flow control the droplet size during atomization; however, it has also been demonstrated that increasing feed concentration results in increased shell thickness of hollow particles that are formed.[8] The physical properties of spray-dried lactose produced from alcoholic media are directly affected by the ethanol-to-water ratio in the feed solution. Lactose spray-dried from pure ethanol was shown to be 100% crystalline, whereas lactose spray-dried from pure water was 100% amorphous. Furthermore, the surface area of the spray-dried lactose increased as a function of amorphous content.[9] Spray-dried lactoses exhibit good flow properties.[10]

Polyethylene glycol (PEG) 4000, when spray-dried with lactose, has been shown to accelerate the rate and extent of crystallization of lactose.[11] It has also been shown that spray-dried lactose composite particles containing an ion complex of chitosan are suitable for the dry-coating of tablets.[12] Spray-dried lactose and

Table II: Particle size distribution of selected commercially available spray-dried lactose.

Supplier/grade	Percentage less than stated size					
	<32 μm	<45 μm	<75 μm	<100 μm	<200 μm	<250 μm
DFE Pharma						
Lactopress Spray-Dried	—	≤25	—	30–60[(a)]	—	≥65
Lactopress Spray-Dried 250	—	≤15	≤50	30–60	—	≥98
SuperTab 11SD	—	≤15	—	30–60	—	≥98
SuperTab 14SD	—	≤15	—	30–60	—	≥98
Meggle GmbH						
FlowLac 90	≤5	—	—	25–40	≥85	—
FlowLac 100	≤10	—	—	20–45	≥80	—
Sheffield Bio-Science						
Foremost Lactose 315	—	—	25–45	40–70[(a)]	—	—
Foremost Lactose 316 Fast Flo	—	—	20–40	44.5–70[(a)]	—	99.5–100

(a) <106 μm.

crystallized spray-dried lactose have been evaluated for dry powder inhalation application.[13,14] Amorphous spray-dried lactose has also been studied in composites with PVP.[15]

Spray-dried lactose has been reported to be more prone to caking than lactose monohydrate.[16]

See also Lactose, Anhydrous, Lactose, Inhalation and Lactose, Monohydrate.

19 Specific References

1 Blaug SM, Huang W. Interaction of dextro amphetamine sulfate with spray-dried lactose. *J Pharm Sci* 1972; **61**: 1770–1775.

2 Flemming A, *et al.* Compaction of lactose drug mixtures: Quantification of the extent of incompatibility by FT-Raman spectroscopy. *Eur J Pharm Biopharm* 2008; **68**: 802–810.

3 Castello RA, Mattocks AM. Discoloration of tablets containing amines and lactose. *J Pharm Sci* 1962; **51**: 106–108.

4 Wirth DD, *et al.* Maillard reaction of lactose and fluoxetine hydrochloride: a secondary amine. *J Pharm Sci* 1998; **87**: 31–39.

5 Hutton JT *et al.* Lactose product and method. United States Patent No. 3,639,170; 1972.

6 Vromans H *et al.* Spray-dried lactose and process for preparing the same. United States Patent No. 4,802,926; 1989.

7 Vromans H, *et al.* Studies on tableting properties of lactose. VII. The effect of variations in primary particle size and percentage of amorphous lactose in spray dried lactose products. *Int J Pharm* 1987; 35(1–2): 29–37.

8 Elversson J, *et al.* Droplet and particle size relationship and shell thickness of inhalable lactose particles during spray drying. *J Pharm Sci* 2003; **92**(4): 900–910.

9 Harjunen PI, *et al.* Effects of ethanol to water ratio in feed solution on the crystallinity of spray dried lactose. *Drug Dev Ind Pharm* 2002; **28**(8): 949–955.

10 Bhattachar SN, *et al.* Evaluation of the vibratory feeder method for assessment of powder flow properties. *Int J Pharm* 2004; **269**: 385–392.

11 Corrigan DO, *et al.* The effects of spray drying solutions of polyethylene glycol (PEG) and lactose/PEG on their physicochemical properties. *Int J Pharm* 2002; **235**(1–2): 193–205.

12 Takeuchi H, *et al.* Spray dried lactose composite particles containing an ion complex of alginate–chitosan for designing a dry coated tablet having a time controlled releasing function. *Pharm Res* 2000; **17**: 94–99.

13 Kawashima Y, *et al.* Effect of surface morphology of carrier lactose on dry powder inhalation property of pranlukast hydrate. *Int J Pharm* 1998; **172**: 179–188.

14 Harjunen P, *et al.* Lactose modifications enhance its drug performance in the novel multiple dose Taifun (R) DPI. *Eur J Pharm Sci* 2002; **16**(4–5): 313–321.

15 Berggren J, *et al.* Compression behaviour and tablet-forming ability of spray-dried amorphous composite particles. *Eur J Pharm Sci* 2004; **22**: 191–200.

16 Listiohadia Y, *et al.* Moisture sorption, compressibility and caking of lactose polymorphs. *Int J Pharm* 2008; **359**: 123–134.

20 General References

Bolhuis GK, Armstrong NA. Excipients for direct compaction: an update. *Pharm Dev Technol* 2006; **11**: 111–124.

Bolhuis GK, Chowhan ZT. Materials for direct compaction. In: Alderborn G, Nyström C, eds. *Pharmaceutical Powder Compaction Technology.* New York: Marcel Dekker, 1996: 473–476.

Bolhuis G *et al.* New developments in spray-dried lactose. *Pharm Tech* 2004; June: 26–31.

DFE Pharma. Technical literature: *Lactopress Spray-Dried, Lactopress Spray-Dried* 250, 2011. http://www.dfepharma.com/ (accessed 20 February 2012).

DFE Pharma. Technical literature: *SuperTab 11SD, SuperTab 14SD*, 2011. http://www.dfepharma.com/ (accessed 20 February 2012).

Fell JT, Newton JM. The characterization of the form of lactose in spray-dried lactose. *Pharm Acta Helv* 1970; **45**: 520–522.

Fell JT, Newton JM. The production and properties of spray-dried lactose, part 1: the construction of an experimental spray drier and the production of spray-dried lactose under various conditions of operation. *Pharm Acta Helv* 1971; **46**: 226–235.

Fell JT, Newton JM. The production and properties of spray-dried lactose, part 2: the physical properties of samples of spray-dried lactose produced on an experimental drier. *Pharm Acta Helv* 1971; **46**: 425–430.

Meggle GmbH. Technical literature: *FlowLac 90, FlowLac 100*, 2012. www.meggle-pharma.de (accessed 20 February 2012).

Price R, Young PM. Visualisation of the crystallisation of lactose from the amorphous state. *J Pharm Sci* 2004; **93**: 155–164.

Sheffield Bio-Science. Technical literature: Foremost *Lactose 315, Foremost Lactose 316 Fast Flo*, 2012. www.sheffield-products.com (accessed 20 February 2012).

21 Authors

S Edge, AH Kibbe, J Shur.

22 Date of Revision

1 March 2012.

Lanolin

1 Nonproprietary Names

BP: Wool Fat

JP: Purified Lanolin

PhEur: Wool Fat

USP-NF: Lanolin

2 Synonyms

Adeps lanae; adeps lanae rurificatus; cera lanae; *Coronet*; E913; *Lanis*; lanolina; lanolin anhydrous; *Lantrol 1650*; *Pharmalan*; *Protalan anhydrous*; purified lanolin; refined wool fat.

3 Chemical Name and CAS Registry Number

Anhydrous lanolin [8006-54-0]

4 Empirical Formula and Molecular Weight

The USP35-NF30 describes lanolin as the purified wax-like substance obtained from the wool of the sheep, *Ovis aries* Linné (Fam. Bovidae), that has been cleaned, decolorized, and deodorized. Lanolin is a complex mixture of naturally occurring esters and polyesters of high molecular weight alcohols (principally sterols) and fatty acids, and is chemically classified as a wax. It contains a minimum 98% esters, of which fatty alcohols and fatty acids comprise approximately a 50:50 ratio. It also contains not more than 0.25% w/w of water and may contain up to 0.02% w/w of a suitable antioxidant; the PhEur 7.4 specifies up to 200 ppm of butylated hydroxytoluene as an antioxidant.

See also Section 18.

5 Structural Formula

See Section 4.

6 Functional Category

Emollient; emulsifying agent; ointment base.

7 Applications in Pharmaceutical Formulation or Technology

Lanolin is widely used in topical pharmaceutical formulations and cosmetics as an emulsifying agent and emollient.

Lanolin may be used as a hydrophobic vehicle and in the preparation of water-in-oil creams and ointments. When mixed with suitable vegetable oils or with soft paraffin, it produces emollient creams that penetrate the skin and hence facilitate the absorption of drugs. Lanolin mixes with about twice its own weight of water, without separation, to produce stable emulsions that do not readily become rancid on storage. Lanolin-based topical products owe their excellent emollient properties to the unique chemical and physical resemblance of lanolin to human skin lipids. In addition to its semiocclusive capabilities, it is also capable of allowing the bidirectional transportation of water.

Lanolin has also been used in pigmented medications (e.g. zinc oxide) as a dispersing agent; in topical products for cutaneous infections (e.g. acne); and in deodorizing toiletries as an antimicrobial; in ophthalmic ointments as an emollient;[1] in suppositories as a carrier for active ingredients; and in surgical adhesive tapes as an impregnating agent and plasticizer.[2]

Lanolin and its many derivatives are used extensively in personal care products, such as facial cosmetics and lip products. It is also used in protective skin treatments for infants.

See also Section 18.

8 Description

Lanolin occurs as a pale yellow-colored, unctuous, waxy substance with a faint, characteristic odor. Melted lanolin is a clear or almost clear, yellow liquid.

9 Pharmacopeial Specifications

See Table I.

Table I: Pharmacopeial specifications for lanolin.

Test	JP XV	PhEur 7.4	USP35–NF30
Identification	+	+	—
Characters	+	+	—
Melting range	37–43°C	38–44°C	38–44°C
Acidity and alkalinity	+	—	+
Loss on drying	≤0.5%	≤0.5%	≤0.25%
Residue on ignition	≤0.1%	—	≤0.1%
Sulfated ash	—	≤0.15%	—
Water-soluble acids and alkalis	—	+	+
Water-soluble oxidizable substances	+	+	+
Chloride	≤0.036%	≤150 ppm	≤0.035%
Ammonia	+	—	+
Acid value	≤1.0	≤1.0	—
Iodine value	18–36	—	18–36
Peroxide value	—	≤20	—
Saponification value	—	90–105	—
Water absorption capacity	—	+	—
Paraffins	—	≤1.0%	—
Petrolatum	+	—	+
Foreign substances (pesticide residues)	—	+	+

10 Typical Properties

Autoignition temperature 445°C

Density 0.95 g/cm^3 at 20°C

Flash point 238°C

Refractive index n_D^{40} = 1.478–1.482

Solubility Freely soluble in chloroform, ether, tetrahydrofuran, and toluene; soluble in diethyl ether and cyclohexane; sparingly soluble in cold ethanol (95%), more soluble in boiling ethanol (95%); practically insoluble in water.

11 Stability and Storage Conditions

Lanolin may gradually undergo autoxidation during storage. To inhibit this process, the inclusion of butylated hydroxytoluene is permitted as an antioxidant. Exposure to excessive or prolonged heating may cause anhydrous lanolin to darken in color and develop a strong rancid like odor. However, lanolin may be sterilized by dry heat at 150°C. Ophthalmic ointments containing lanolin may be sterilized by filtration or by exposure to gamma irradiation.[3]

Lanolin should be stored in a well-filled, well-closed container protected from light, in a cool, dry place, with temperatures not exceeding 25°C. Normal storage life is 2 years.

12 Incompatibilities

Lanolins are incompatible with strong oxidizing agents. Lanolin may contain prooxidants, which may affect the stability of certain active drugs.

13 Method of Manufacture

Lanolin is a naturally occurring wax-like material obtained from the wool of sheep, *Ovis aries* Linné (Fam. Bovidae), and is extracted by washing the wool in hot water with a special wool scouring detergent to remove dirt, crude lanolin, and suint (sweat salts). The crude lanolin is continuously removed during the washing process by centrifugal separators, which concentrate it into a wax-like substance melting at approximately 38°C. The crude lanolin is saponified with a weak alkali and the resultant saponified fat emulsion is centrifuged to remove the aqueous phase. The aqueous phase contains a soap solution from which, on standing, a layer of partially purified lanolin separates. This material is then further refined by treatment with calcium chloride, followed by fusion with unslaked lime to dehydrate the lanolin. The lanolin is finally extracted with acetone and the solvent is removed by distillation.

14 Safety

Lanolin is widely used in cosmetics and a variety of topical pharmaceutical formulations.

Although generally regarded as a nontoxic and nonirritant material, lanolin and lanolin derivatives are associated with skin hypersensitivity reactions, and the use of lanolin in subjects with known sensitivity should be avoided.[4,5] Lanolin can sensitize patients with eczema and ulcerated skin more easily than those with normal skin.[6–9] Other reports suggest that 'sensitivity' arises from false positives in patch testing.[10] However, skin hypersensitivity is relatively uncommon;[11] the incidence of hypersensitivity to lanolin in the general population is estimated to be around 5 per million.[12]

Sensitivity is thought to be associated with the content of free fatty alcohols present in lanolin products rather than the total alcohol content.[13–15] The safety of pesticide residues in lanolin products has also been of concern.[16,17] However, highly refined'-hypoallergenic' grades of lanolin and grades with low pesticide residues are commercially available.[18]

See also Section 18.

LD_{50} (rabbit, skin) >10 ml/kg.

LD_{50} (rat, oral) >16 g/kg;

LD_{50} (rat, oral) >20 g/kg (25% in corn oil);

15 Handling Precautions

Observe normal precautions appropriate to the circumstances and quantity of material handled.

16 Regulatory Status

Included in the FDA Inactive Ingredients Database (ophthalmic, topical, transdermal, and vaginal preparations). Included in nonparenteral medicines licensed in the UK. Included in the Canadian Natural Health Products Ingredients Database.

17 Related Substances

Cholesterol; hydrogenated lanolin; lanolin, hydrous; lanolin alcohols; modified lanolin.

See also Section 18.

Hydrogenated lanolin

Synonyms Adeps lanae hydrogenatus; hydrogenated wool fat.
Acid value $\leq$1.0
Hydroxyl value 140–180
Melting point 45–55°C
Saponification value $\leq$8.0
Water $\leq$3.0%
Comments Some pharmacopeias, such as the PhEur 7.4, contain a monograph for hydrogenated lanolin. This material is a mixture of higher aliphatic alcohols and sterols obtained from the direct, high-pressure, high-temperature hydrogenation of lanolin during which the esters and acids present are reduced to the corresponding alcohols. Hydrogenated lanolin may contain a suitable antioxidant; the PhEur 7.4 specifies not more than 200 ppm of butylated hydroxytoluene.

Modified lanolin

Comments Some pharmacopeias, such as the USP35–NF30, contain a monograph for modified lanolin. This material is lanolin that has been processed to reduce the contents of free lanolin alcohols and detergent and pesticide residues. It contains not more than 0.25% w/w of water. The USP35–NF30 specifies that it may contain not more than 0.02% w/w of a suitable antioxidant.

18 Comments

Lanolin (the anhydrous material) may be confused in some instances with hydrous lanolin since the USP formerly contained monographs for'lanolin' and 'anhydrous lanolin' in which the name 'lanolin' referred to the material containing 25–30% w/w of purified water. However, in the USP35–NF30 the former lanolin monograph (hydrous lanolin) is deleted and the monograph for anhydrous lanolin is renamed 'lanolin'.

Since lanolin is a natural product obtained from various geographical sources, its physical characteristics such as color, consistency, iodine value, saponification value, and hydroxyl value may vary for the products from different sources. Consequently, formulations containing lanolin from different sources may also have different physical properties.

A wide range of grades of lanolin are commercially available that have been refined to different extents in order to produce hypoallergenic grades or grades with low pesticide contents.

Many lanolin derivatives are also commercially available that have properties similar to those of the parent material and include: acetylated lanolin; ethoxylated or polyoxyl lanolin (water-soluble); hydrogenated lanolin; isopropyl lanolate; lanolin oil; lanolin wax; liquid lanolin; and water-soluble lanolin.

Highly purified lanolin is very effective in reducing nipple pain and promoting healing of nipple trauma for breastfeeding/lactating mothers.[19] It is used in burns dressings and wound sprays to support the wound healing process. Lanolin topical ointment has also been evaluated for the treatment of acute anal fissure in children.[20]

A specification for anhydrous lanolin is contained in the *Food Chemicals Codex* (FCC),[21] where it is described as being used as a masticatory substance in chewing gum base. The EINECS number for lanolin is 232-348-6.

19 Specific References

1 So HM, *et al.* Comparing the effectiveness of polyethylene covers (Gladwrap) with lanolin (Duratears) eye ointment to prevent corneal abrasions in critically ill patients: a randomized controlled study. *Int J Nurs Stud* 2008; **45**: 1565–1571.

2 Yixin Chemical Company Ltd. Product information: Lanolin.http://www.xinyi-lanolin.com/Products/Lanolin-Pharmaceutical-Grade.htm (accessed 15 November 2011).

3 Smith GG, *et al.* New process for the manufacture of sterile ophthalmic ointments. *Bull Parenter Drug Assoc* 1975; **29**: 18–25.

4 Anonymous. Lanolin allergy. *Br Med J* 1973; **2**: 379–380.

5 Breit J, Bandmann H-J. Dermatitis from lanolin. *Br J Dermatol* 1973; **88**: 414–416.

6 Lee B, Warshaw E. Lanolin allergy: history, epidemiology, responsible allergens, and management. *Dermatitis* 2008; **19**: 63–72.

7 Uter W, *et al.* The European baseline series in 10 European countries, 2005/2006 – results of the European Surveillance System on Contact Allergies (ESSCA). *Contact Dermatitis* 2009; **61**: 31–38.

8 White-Chu EF, Reddy M. Dry skin in the elderly: complexities of a common problem. *Clin Dermatol* 2011; **29**(1): 37–42.

9 Warshaw EM, *et al.* Positive patch test reactions to lanolin: cross-sectional data from the North American contact dermatitis group, 1994 to 2006. *Dermatitis* 2009; **20**(2): 79–88.

10 Kligman AM. The myth of lanolin allergy. *Contact Dermatitis* 1998; **39**: 103–107.

11 Wakelin SH, *et al.* A retrospective analysis of contact allergy to lanolin. *Br J Dermatol* 2001; **145**(1): 28–31.

12 Clark EW. Estimation of the general incidence of specific lanolin allergy. *J Soc Cosmet Chem* 1975; **26**: 323–335.

13 Clark EW, *et al.* Lanolin with reduced sensitizing potential: a preliminary note. *Contact Dermatitis* 1977; **3**(2): 69–74.

14 Fowler J, *et al.* Lanolin allergy: A 500 patient patch test assessment of six pharmaceutical grade lanolins. *J Am Acad Dermatol* 2009; **60**(3): AB72.

15 Nguyen JC, *et al.* Allergic contact dermatitis caused by lanolin (wool) alcohol contained in an emollient in three postsurgical patients. *J Am Acad Dermatol* 2010; **62**(6): 1064–1065.

16 Copeland CA, *et al.* Pesticide residue in lanolin [letter]. *J Am Med Assoc* 1989; **261**: 242.

17 Cade PH. Pesticide residue in lanolin [letter]. *J Am Med Assoc* 1989; **262**: 613.

18 Steel I. Pure lanolin in treating compromised skin. *Manuf Chem* 1999; **70**(9): 16–17.

19 Abou-Dakn M, *et al.* Positive effect of HPA lanolin versus expressed breastmilk on painful and damaged nipples during lactation. *Skin Pharmacol Physiol* 2011; **24**: 27–35.

20 Büyükyavuz BI, *et al.* Efficacy of lanolin and bovine type I collagen in the treatment of childhood anal fissures: a prospective, randomized, controlled clinical trial. *Surg Today* 2010; **40**: 752–756.

21 *Food Chemicals Codex*, 7th edn. Bethesda, MD: United States Pharmacopeia, 2011.

20 General References

Barnett G. Lanolin and derivatives. *Cosmet Toilet* 1986; **101**(3): 23–44.

Imperial-Oel-Import. http://imperial-oel-import.de/products.html (accessed 15 November 2011).

Osborne DW. Phase behavior characterization of ointments containing lanolin or a lanolin substitute. *Drug Dev Ind Pharm* 1993; **19**: 1283–1302.

Smolinske SC. *Handbook of Food, Drug, and Cosmetic Excipients.* Boca Raton, FL: CRC Press, 1992: 225–229.

21 Authors

HC Shah, KK Singh.

22 Date of Revision

2 March 2012.

L

Lanolin, Hydrous

1 Nonproprietary Names

BP: Hydrous Wool Fat
JP: Hydrous Lanolin
PhEur: Wool Fat, Hydrous

2 Synonyms

Adeps lanae cum aqua; adeps lanae hydrosus; *Lipolan*.

3 Chemical Name and CAS Registry Number

Hydrous lanolin [8020-84-6]

4 Empirical Formula and Molecular Weight

The JP XV describes hydrous lanolin as a mixture of purified lanolin and 25–30% w/w purified water. The PhEur 7.4 describes hydrous lanolin as a mixture of 75% w/w lanolin and 25% w/w purified water; *see also* Section 18. The PhEur 7.4 additionally permits the inclusion of up to 150 ppm of butylated hydroxytoluene as an antioxidant.

See also Lanolin.

5 Structural Formula

See Section 4.

6 Functional Category

Emollient; emulsifying agent; ointment base.

7 Applications in Pharmaceutical Formulation or Technology

Hydrous lanolin is widely used in topical pharmaceutical formulations and cosmetics in applications similar to those for lanolin.

See also Section 18.

8 Description

Hydrous lanolin is a pale yellow-colored, unctuous substance with a faint characteristic odor. When melted by heating on a water bath, hydrous lanolin separates into a clear oily layer and a clear water layer.

9 Pharmacopeial Specifications

See Table I.

Table I: Pharmacopeial specifications for hydrous lanolin.

Test	JP XV	PhEur 7.4
Identification	+	+
Characters	+	+
Melting point	39°C	38–44°C
Acidity/alkalinity	+	—
Water absorption capacity	—	+
Water-soluble acids and alkalis	—	+
Water-soluble oxidizable substances	+	+
Chloride	≤0.036%	≤115 ppm
Ammonia	+	—
Paraffins	—	≤1.0%
Petrolatum	+	—
Acid value	≤1.0	≤0.8
Peroxide value	—	≤15
Iodine value	18–36	—
Saponification value	—	67–79
Residue on evaporation (wool fat content)	70–75%	72.5–77.5%
Sulfated ash	—	≤0.1%

10 Typical Properties

Solubility Soluble in diethyl ether and in cyclohexane, with the separation of water; practically insoluble in chloroform, ether, and water.

11 Stability and Storage Conditions

Hydrous lanolin should be stored in a well-filled, well-closed container protected from light, in a cool, dry place. Normal storage life is 2 years. It is a stable material and does not become rancid.

See also Lanolin.

12 Incompatibilities

See Lanolin.

13 Method of Manufacture

Lanolin is melted, and sufficient purified water is gradually added with constant stirring.

14 Safety

Hydrous lanolin is used in cosmetics and a number of topical pharmaceutical formulations and is generally regarded as a nontoxic and nonirritant material, although it has been associated with hypersensitivity reactions. *See* Lanolin for further information.

15 Handling Precautions

Observe normal precautions appropriate to the circumstances and quantity of material handled.

16 Regulatory Status

Included in nonparenteral medicines licensed in the UK. Included in the Canadian Natural Health Products Ingredients Database.

17 Related Substances

Lanolin; lanolin alcohols.

18 Comments

Lanolin (the anhydrous material) may be confused in some instances with hydrous lanolin since the USP formerly contained monographs for'lanolin' and 'anhydrous lanolin' in which the name 'lanolin' referred to the material containing 25–30% w/w of purified water.

Therapeutically, hydrous lanolin is used in burns dressings to support the wound healing process and enhance dermal repair. It has also been used for treatment of chapped lips, cracked hands, sore nipples, and diaper rash.

19 Specific References

—

20 General References

Barnett G. Lanolin and derivatives. *Cosmet Toilet* 1986; **101**(3): 23–44.
Osborne DW. Phase behavior characterization of ointments containing lanolin or a lanolin substitute. *Drug Dev Ind Pharm* 1993; **19**: 1283–1302.
Smolinske SC. *Handbook of Food, Drug, and Cosmetic Excipients*. Boca Raton, FL: CRC Press, 1992; 225–229.

21 Authors

HC Shah, KK Singh.

22 Date of Revision

2 March 2012.

Lanolin Alcohols

1 Nonproprietary Names

BP: Wool Alcohols
PhEur: Wool Alcohols
USP–NF: Lanolin Alcohols

2 Synonyms

Alcoholes adipis lanae; alcoholia lanae; alcolanum; *Argowax*; lanalcolum; *Lanis AL*; *Lantrol 1780*; *Ritawax*; *Super Hartolan*; wool wax alcohols.

3 Chemical Name and CAS Registry Number

Lanolin alcohols [8027-33-6]

4 Empirical Formula and Molecular Weight

Lanolin alcohols is a mixture of aliphatic alcohols, triterpenoid alcohols, and sterols obtained by hydrolysis of lanolin, including not less than 30% cholesterol. The cholesterol found in lanolin is a mixture of different stereochemical isomers.

The USP35–NF30 permits the inclusion of up to 0.1% w/w of a suitable antioxidant, while the PhEur 7.4 specifies that lanolin alcohols may contain up to 200 ppm of butylated hydroxytoluene as an antioxidant.

5 Structural Formula

See Section 4.

6 Functional Category

Emollient; emulsifying agent; ointment base; solubilizing agent.

7 Applications in Pharmaceutical Formulation or Technology

Lanolin alcohols is used in topical pharmaceutical formulations as an absorption base with emollient properties. It is also used in the preparation of water-in-oil creams and ointments at concentrations as low as 2% w/w. Lanolin alcohols have been used as carrier systems to deliver pharmacologically active substances through the skin.

The proportion of water that can be incorporated into petrolatum is increased threefold by the addition of 5% lanolin alcohols. It is stable in a wide pH range and can be used in acidic and alkaline water-in-oil emulsions, where it acts as a viscosity-increasing agent. Such emulsions do not crack upon the addition of citric, lactic, or tartaric acids.

Lanolin alcohols is also used as a nongelling thickener and emollient in cosmetic creams, lotions, lip balms, lipsticks, and other stick cosmetics.

8 Description

Lanolin alcohols occurs as a pale yellow to golden-brown-colored solid that is plastic when warm but brittle when cold. It has a faint characteristic odor. *See also* Section 4.

9 Pharmacopeial Specifications

See Table I.

Table I: Pharmacopeial specifications for lanolin alcohols.

Test	PhEur 7.4	USP35–NF30
Identification	+	+
Characters	+	—
Melting range	≥56°C	≥56°C
Acidity/alkalinity	+	+
Clarity of solution	+	—
Loss on drying	≤0.5%	≤0.5%
Residue on ignition	≤0.1%	≤0.15%
Copper	—	≤5 ppm
Acid value	≤2.0	≤2.0
Hydroxyl value	120–180	120–180
Peroxide value	≤15	≤15
Saponification value	≤12	≤12
Water absorption capacity	+	—
Content of sterols (as cholesterol)	≥30.0%	≥30.0%

10 Typical Properties

Density 0.98 g/cm^3

Solubility Freely soluble in chloroform, dichloromethane, ether, and light petroleum; soluble 1 in 25 parts of boiling ethanol (95%); slightly soluble in ethanol (90%); practically insoluble in water.

11 Stability and Storage Conditions

Lanolin alcohols may gradually undergo autoxidation during storage. Store in a well-closed, well-filled container, protected from light, in a cool, dry place. Normal storage life is approximately 2 years.

12 Incompatibilities

Incompatible with coal tar, ichthammol, phenol, and resorcinol.

13 Method of Manufacture

Lanolin alcohols is prepared by the saponification of lanolin followed by separation of the fraction containing cholesterol and other alcohols. It is further refined during a multistage molecular distillation process, which improves color and odor, producing a purified, semicrystalline wax.

14 Safety

Lanolin alcohols is widely used in cosmetics and topical pharmaceutical formulations and is generally regarded as a nontoxic material. However, lanolin alcohols may be irritant to the skin and hypersensitivity can occur in some individuals.[1]

LD_{50} (rat, oral) >21 g/kg[2]
LD_{50} (rat, oral) >42.7 g/kg (66% in corn oil)[1]
See also Lanolin.

15 Handling Precautions

Observe normal precautions appropriate to the circumstances and quantity of material handled.

16 Regulatory Status

Included in the FDA Inactive Ingredients Database (ophthalmic and topical preparations). Included in nonparenteral medicines licensed in the UK. Included in the Canadian Natural Health Products Ingredients Database.

17 Related Substances

Cholesterol; lanolin; lanolin, hydrous; petrolatum and lanolin alcohols; mineral oils.

18 Comments

Water-in-oil emulsions prepared with lanolin alcohols, unlike those made with lanolin, do not show surface darkening, nor do they develop an objectionable odor in hot weather.

Therapeutically, lanolin alcohols may help in the wound-healing process.

The EINECS number for lanolin alcohols is 232-430-1.

19 Specific References

1 Nguyen JC, *et al.* Allergic contact dermatitis caused by lanolin (wool) alcohol contained in an emollient in three postsurgical patients. *J Am Acad Dermatol* 2010; **62**: 1064–1065.
2 Making Cosmetics.com. MSDS Lanolin Alcohol. http://www.making-cosmetics.com/msds1/msds-lanolin-alcohol.pdf (accessed 23 February 2012).

20 General References

Barnett G. Lanolin and derivatives. *Cosmet Toilet* 1986; **101**(3): 23–44.
Imperial-Oel-Import. http://imperial-oel-import.de/products.html (accessed 23 February 2012).
Khan AR, *et al.* *In vitro* release of salicylic acid from lanolin alcohols–ethylcellulose films. *J Pharm Sci* 1984; **73**: 302–305.
Osborne DW. Phase behavior characterization of ointments containing lanolin or a lanolin substitute. *Drug Dev Ind Pharm* 1993; **19**: 1283–1302.
Smolinske SC. *Handbook of Food, Drug, and Cosmetic Excipients.* Boca Raton, FL: CRC Press, 1992; 225–229.

21 Authors

HC Shah, KK Singh.

22 Date of Revision

2 March 2012.

Lauric Acid

1 Nonproprietary Names

None adopted.

2 Synonyms

C-1297; dodecanoic acid; dodecoic acid; dodecylic acid; duodecylic acid; *n*-dodecanoic acid; *Hydrofol acid 1255*; *Hydrofol acid 1295*; *Hystrene 9512*; *Kortacid 1299*; laurostearic acid; *Lunac L70*;*Neo-fat 12*; *Neo-fat 12–43*; *Ninol AA62 Extra*; *Prifac 2920*; 1-undecanecarboxylic acid; *Univol U314*; vulvic acid; *Wecoline 1295*.

3 Chemical Name and CAS Registry Number

Dodecanoic acid [143-07-7]

4 Empirical Formula and Molecular Weight

$C_{12}H_{24}O_2$ 200.32

5 Structural Formula

6 Functional Category

Emulsifying agent; penetration enhancer.

7 Applications in Pharmaceutical Formulation or Technology

Lauric acid has been used as an enhancer for topical penetration and transdermal absorption,[1–11] rectal absorption,[12,13] buccal delivery,[14] and intestinal absorption.[15,16] It is also useful for stabilizing oil-in-water emulsions.[17]

8 Description

Lauric acid occurs as a white crystalline powder with a slight odor of bay oil.

9 Pharmacopeial Specifications

See Section 18.

10 Typical Properties

Boiling point 298.9°C at 760 mmHg;[18] 225°C at 100 mmHg.[19]

Density
0.883 g/cm^3 at 20°C;[18]
0.8679 g/cm^3 at 50°C.[20]

Dissociation constant pK_a= 5.3 at 20°C.[21]

Enthalpy of fusion 36.6 kJ mol^{-1};[18] also reported as 36.03 kJ mol^{-1}.[22]

Melting point 44.2°C[18]; also reported as 43.8°C[20] and 48°C.[23]

Partition coefficient Log *P* (octanol : water) = 4.6[24]

Refractive index
n_D^{82} = 1.418;[20]
n_D^{70} = 1.423.[18]

Solubility 4.81 mg/mL in water at 25°C.[24] Very soluble in ether, ethanol (95%), and methanol; soluble in acetone; slightly soluble in chloroform; miscible with benzene.

Surface tension 26.6 mN/m at 70°C[18]

Vapor pressure
10 Pa at 100°C;
100 Pa at 128°C.

Viscosity (dynamic) 7.3 mPa s at 50°C[18]

Viscosity (kinematic) 8.41 mPa s at 50°C[25]

11 Stability and Storage Conditions

Lauric acid is stable at normal temperatures and should be stored in a cool, dry place. Avoid sources of ignition and contact with incompatible materials.

12 Incompatibilities

Lauric acid is incompatible with strong bases, reducing agents, and oxidizing agents.

13 Method of Manufacture

Lauric acid is a fatty carboxylic acid isolated from vegetable and animal fats or oils. For example, coconut oil and palm kernel oil both contain high proportions of lauric acid. Isolation from natural fats and oils involves hydrolysis, separation of the fatty acids, hydrogenation to convert unsaturated fatty acids to saturated acids, and finally distillation of the specific fatty acid of interest.

14 Safety

Lauric acid is widely used in cosmetic preparations, in the manufacture of food-grade additives, and in pharmaceutical formulations. General exposure to lauric acid occurs through the consumption of food and through dermal contact with cosmetics, soaps, and detergent products. Lauric acid is toxic when administered intravenously.

Occupational exposure may cause local irritation of eyes, nose, throat, and respiratory tract,[26] although lauric acid is considered safe and nonirritating for use in cosmetics.[27] No toxicological effects were observed when lauric acid was administered to rats at 35% of the diet for 2 years.[28] Acute exposure tests in rabbits indicate mild irritation.[27] After subcutaneous injection into mice, lauric acid was shown to be noncarcinogenic.[29]

LD_{50} (mouse, IV): 0.13 g/kg[30,31]
LD_{50} (rat, oral): 12 g/kg

15 Handling Precautions

Observe normal precautions appropriate to the circumstances and quantity of material handled. No occupational exposure limits have been established. Under conditions of frequent use or heavy exposure, respiratory protection may be required. When heated, lauric acid emits an acrid smoke and irritating fumes; therefore, use in a well-ventilated area is recommended.

16 Regulatory Status

GRAS listed. Lauric acid is listed as a food additive in the EAFUS list compiled by the FDA. Reported in the EPA TSCA Inventory.

17 Related Substances

Capric acid; myristic acid; palmitic acid; sodium laurate; stearic acid.

Capric acid

Empirical formula $C_{10}H_{20}O_2$
Molecular weight 172.2
CAS number [334-48-5]
Synonyms *n*-Capric acid; caprinic acid; caprynic acid; carboxylic acid C; decanoic acid; *n*-decanoic acid; decoic acid; decyclic acid; *n*-decylic acid; 1-nonanecarboxylic acid.
Appearance White to pale yellow crystals with an unpleasant odor.
Acid value 320–330
Boiling point 270°C
Dissociation constant pK_a = 4.9
Melting point 31.5°C
Partition coefficient Log *P* (octanol : water) = 4.09
Refractive index n_D^{40} = 1.4288
Comments Capric acid is used as a flavoring agent in pharmaceutical preparations, providing a citrus-like flavor. It is used in cosmetics as an emulsifying agent. A specification for capric acid is included in the *Food Chemicals Codex* (FCC).[32] The EINECS number for capric acid is 206-376-4.

L

Sodium laurate

Empirical formula $C_{12}H_{23}O_2Na$
Molecular weight 222.34
CAS number [629-25-4]
Comments Sodium laurate is used as an emulsifying agent and surfactant in cosmetics. The EINECS number for sodium laurate is 211-082-4.

18 Comments

Lauric acid is listed in the BP 2012 appendices as a general reagent when used in the assay of total fatty acids in saw paletto fruit. The content of lauric acid complies with the requirement of being not less than 98% by gas chromatography.

Lauric acid has been evaluated for use in aerosol formulations.[33] It has also been studied for use in sustained-release capsule formulations in combination with drugs and water-soluble polymers.[34] Solid core delivery platforms using lauric acid as a shell material have been prepared using supercritical fluid technology.[35]

A specification for lauric acid is contained in the *Food Chemicals Codex* (FCC);[36] *see* Table I.

Lauric acid is widely used in cosmetics and food products.

The EINECS number for lauric acid is 205-582-1.

Table I: FCC specification for lauric acid.[36]

Test	FCC 7
Acid value	252–287
Lead	≤0.1 mg/kg
Iodine value	≤3
Residue on ignition	≤0.1%
Saponification value	253–287
Solidification point	26–44°C
Unsaponifiable matter	≤0.3%
Water	≤0.2%

19 Specific References

1 Kravchenko IA, *et al*. Effect of lauric acid on transdermal penetration of phenazepam *in vivo*. *Bull Exp Biol Med* 2003; **136**(6): 579–581.

2 Chisty MNA, *et al. In vitro* evaluation of the release of albuterol sulfate from polymer gels: effect of fatty acids on drug transport across biological membranes. *Drug Dev Ind Pharm* 2002; **28**(10): 1221–1229.
3 Stott PW, *et al.* Mechanistic study into the enhanced transdermal permeation of a model beta-blocker, propranolol, by fatty acids: a melting point depression effect. *Int J Pharm* 2001; **219**(1–2): 161–176.
4 Morimoto K, *et al.* Enhancing mechanisms of saturated fatty acids on the permeations of indomethacin and 6-carboxyfluorescein through rat skins. *Drug Dev Ind Pharm* 1995; **21**(17): 1999–2012.
5 Ogiso T, *et al.* Comparison of the *in vitro* penetration of propiverine with that of terodiline. *Biol Pharm Bull* 1995; **18**(7): 968–975.
6 Aungst BJ, *et al.* Contributions of drug solubilization, partitioning, barrier disruption, and solvent permeation to the enhancement of skin permeation of various compounds with fatty acids and amines. *Pharm Res* 1990; **7**(7): 712–718.
7 Ogiso T, Shintani M. Mechanism for the enhancement effect of fatty acids on the percutaneous absorption of propranolol. *J Pharm Sci* 1990; **79**(12): 1065–1071.
8 Pfister WR, Hsieh DST. Permeation enhancers compatible with transdermal drug delivery systems. Part I: Selection and formulation considerations. *Pharm Technol* 1990; **14**(9): 132–140.
9 Green PG, *et al.* Physicochemical aspects of the transdermal delivery of bupranolol. *Int J Pharm* 1989; **55**(2–3): 265–269.
10 Green PG, Guy RH, Hadgraft J. *In vitro* and *in vivo* enhancement of skin permeation with oleic and lauric acids. *Int J Pharm* 1988; **48**(1–3): 103–111.
11 Green PG, Hadgraft J. Facilitated transfer of cationic drugs across a lipoidal membrane by oleic acid and lauric acid. *Int J Pharm* 1987; **37**(3): 251–255.
12 Ogiso T, *et al.* Enhancement effect of lauric acid on the rectal absorption of propranolol from suppository in rats. *Chem Pharm Bull* 1991; **39**(10): 2657–2661.
13 Muranishi S. Characteristics of drug absorption via the rectal route. *Methods Find Exp Clin Pharmacol* 1984; **12**: 763–772.
14 Shojaei AH, Chang RK, *et al.* Systemic drug delivery via the buccal mucosal route. *Pharm Technol* 2001; **25**(6): 70–81.
15 Constantinides PP, *et al.* Water-in-oil microemulsions containing medium-chain fatty acids/salts: formulation and intestinal absorption enhancement evaluation. *Pharm Res* 1996; **13**(2): 210–215.
16 Yamada K, *et al.* Improvement of intestinal absorption of thyrotropin-releasing hormone by chemical modification with lauric acid. *J Pharm Pharmacol* 1992; **44**(9): 717–721.
17 Buszello K. The influence of alkali fatty acids on the properties and the stability of parenteral O/W emulsions modified with Solutol HS 15. *Eur J Pharm Biopharm* 2000; **49**(2): 143–149.
18 *Kirk-Othmer Encyclopedia of Chemical Technology* Online. New York: Wiley, 2011.
19 National Library of Medicine. Toxnet: Lauric acid, 2005.
20 Haynes WM, ed. *Handbook of Chemistry and Physics*, 91st edn. New York: CRC Press, 2010; 3-224.
21 Serjeant EP, Dempsey B. Ionisation constants of organic acids in aqueous solution. International Union of Pure and Applied Chemistry (IUPAC) Chemical Data Series No.23. New York: Pergamon Press, Inc, 1979.
22 David J. Anneken DJ *et al.* Fatty acids. In: *Ullman's Encyclopedia of Industrial Chemistry.* Weinheim, Germany: VCH, 2006.
23 O'Neil MJ *et al* eds. *The Merck Index: an Encyclopedia of Chemicals, Drugs, and Biologicals*, 14th edn. Whitehouse Station, NJ: Merck, 2006: 5382.
24 National Library of Medicine. HSDB Database: ChemIDplus; Lauric acid. Bethesda, MD, 2011. http://chem2.sis.nlm.nih.gov/chemidplus (accessed 24 November 2011).
25 Canadian Centre for Occupational Health and Safety (CCOHS). ChemInfo Search: Lauric acid. http://ccinfoweb.ccohs.ca/cheminfo/search.html (accessed 24 November 2011).
26 Health Evaluation Report on Lauric Acid Exposure during Flaking and Bagging Operations at Emery Industries, Los Angeles, CA. National Institute for Occupational Safety and Health, HHE 80-160-897, NTIS Doc. No. PB 82-25694-2, 1981.
27 Anonymous. Final report on the safety assessment of oleic acid, lauric acid, palmitic acid, myristic acid, and stearic acid. *J Am Coll Toxicol* 1987; **6**(3): 321–401.
28 Verschueren K. *Handbook of Environmental Data of Organic Chemicals*, 2nd edn. New York: Van Nostrand Reinhold, 1983; 793.
29 Swern D, *et al.* Investigation of fatty acids and derivatives for carcinogenic activity. *Cancer Res* 1970; **30**: 1037.
30 Lewis RJ, ed. *Sax's Dangerous Properties of Industrial Materials*, 11th edn. New York: Wiley, 2004: 2204.
31 Oro R, Wretlind A. Pharmacological effects of fatty acids, triolein, and cottonseed oil. *Acta Pharmacol Toxicol* 1961; **18**: 141.
32 *Food Chemicals Codex*, 7th edn. Bethesda, MD: United States Pharmacopeia, 2010: 275.
33 Gupta PK, Hickey AJ. Contemporary approaches in aerosolized drug delivery to the lung. *J Control Release* 1991; **17**(2): 127–147.
34 Divyakant D, *et al.* Fatty acid and water-soluble polymer-based controlled release drug delivery system. *J Pharm Sci* 2011; **100**(5): 1900–1912.
35 Mitchell J, Trivedi V. Pharmaceutical nanomaterials: The preparation of solid core drug delivery systems. *J Pharm Pharmacol* 2010; **62**(10): 1457–1458.
36 *Food Chemicals Codex*, 7th edn. Bethesda, MD: United States Pharmacopeia, 2011: 566.

20 General References

Babu RJ *et al.* Fatty alcohols and fatty acids. In: Smith EW, Maibach HI, eds. *Percutaneous Penetration Enhancers*. Boca Raton, FL: CRC Press, 2006; 137–158.

Cosmetic Ingredient Review. Final report on the safety assessment of oleic acid, lauric acid, palmitic acid, myristic acid, and stearic acid, 2005; 53.

21 Author

PE Luner.

22 Date of Revision

23 November 2011.

Lecithin

1 Nonproprietary Names

USP–NF: Lecithin

See also Section 4.

2 Synonyms

Coatsome NC; E322; egg lecithin; *Epikuron*; *Lipoid*; mixed soybean phosphatides; ovolecithin; *Phosal 53 MCT*; *Phospholipon 100 H*; *ProKote LSC*; soybean lecithin; soybean phospholipids; *Sternfine*; *Sternpur*; vegetable lecithin; *Yelkin*.

3 Chemical Name and CAS Registry Number

Lecithin [8002-43-5]

The chemical nomenclature and CAS Registry numbering of lecithin is complex. The commercially available lecithin, used in cosmetics, pharmaceuticals, and food products, is a complex mixture of phospholipids and other materials. However, it may be referred to in some literature sources as 1,2-diacyl-*sn*-glycero-3-phosphocholine (trivial chemical name, phosphatidylcholine). This material is the principal constituent of egg lecithin and has the same CAS Registry Number. The name lecithin and the CAS Registry Number above are thus used to refer to both lecithin and phosphatidylcholine in some literature sources.

Another principal source of lecithin is from an extract of soybeans (CAS [8030-76-0]). Egg yolk lecithin (CAS [93685-90-6]) is also listed in *Chemical Abstracts*.

See also Section 4.

4 Empirical Formula and Molecular Weight

The USP35–NF30 describes lecithin as a complex mixture of acetone-insoluble phosphatides that consists chiefly of phosphatidylcholine, phosphatidylethanolamine, phosphatidylserine, and phosphatidylinositol, combined with various amounts of other substances such as triglycerides, fatty acids, and carbohydrates as separated from a crude vegetable oil source.

The composition of lecithin (and hence also its physical properties) varies enormously depending upon the source of the lecithin and the degree of purification. Egg lecithin, for example, contains 80.5% phosphatidylcholine and 11.7% phosphatidylethanolamine, along with lysophosphatidylcholine, sphingomyelin, and neutral lipids in minor quantities.[1] Soybean lecithin contains 21% phosphatidylcholine, 22% phosphatidylethanolamine, and 19% phosphatidylinositol, along with other components.[2]

5 Structural Formula

$$\begin{array}{l} CH_2{-}O{-}C({=}O){-}R^1 \\ | \\ CH{-}O{-}C({=}O){-}R^2 \\ | \\ CH_2{-}O{-}P(O^-)({=}O){-}OCH_2CH_2N^+(CH_3)_3 \end{array}$$

α-Phosphatidylcholine

R^1 and R^2 are fatty acids, which may be different or identical.

Lecithin is a complex mixture of materials; *see* Section 4. The structure above shows phosphatidylcholine, the principal component of egg lecithin, in its α-form. In the β-form, the phosphorus-containing group and the R^2 group exchange positions.

6 Functional Category

Emollient; emulsifying agent; solubilizing agent.

7 Applications in Pharmaceutical Formulation or Technology

Lecithins are used in a wide variety of pharmaceutical applications; *see* Table I. They are also used in cosmetics[3] and food products.

Lecithins are mainly used in pharmaceutical products as dispersing, emulsifying, and stabilizing agents, and are included in intramuscular and intravenous injections, parenteral nutrition formulations, and topical products such as emulsions, creams, and ointments.

Lecithins are also used in suppository bases,[4] to reduce the brittleness of suppositories, and have been investigated for their absorption-enhancing properties in an intranasal insulin formulation.[5] Lecithins are also commonly used as a component of enteral and parenteral nutrition formulations.

Table I: Uses of lecithin.

Use	Concentration (%)
Aerosol inhalation	0.1
Biorelevant dissolution media	0.059–0.295
IM injection	0.3–2.3
Oral suspensions	0.25–10.0

8 Description

Lecithins vary greatly in their physical form, from viscous semiliquids to powders, depending upon the free fatty acid content. They may also vary in color from brown to light yellow, depending upon whether they are bleached or unbleached or on the degree of purity. When they are exposed to air, rapid oxidation occurs, also resulting in a dark yellow or brown color.

Lecithins have practically no odor. Those derived from vegetable sources have a bland or nutlike taste, similar to that of soybean oil.

9 Pharmacopeial Specifications

See Table II. *See also* Section 18.

Table II: Pharmacopeial specifications for lecithin.

Test	USP35–NF30
Identification	+
Water	≤1.5%
Lead	≤10 ppm
Heavy metals	≤20 ppm
Acid value	+
Peroxide value	≤10
Hexane-insoluble matter	≤0.3%
Acetone-insoluble matter	+

10 Typical Properties

Density

0.97 g/cm^3 for liquid lecithin;

0.5 g/cm^3 for powdered lecithin.

Iodine number

95–100 for liquid lecithin;

82–88 for powdered lecithin.

Isoelectric point ≈3.5

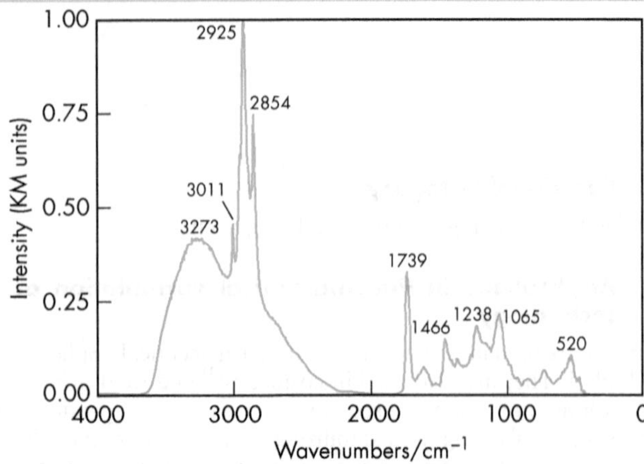

Figure 1: Infrared spectrum of lecithin measured by diffuse reflectance. Adapted with permission of Informa Healthcare.

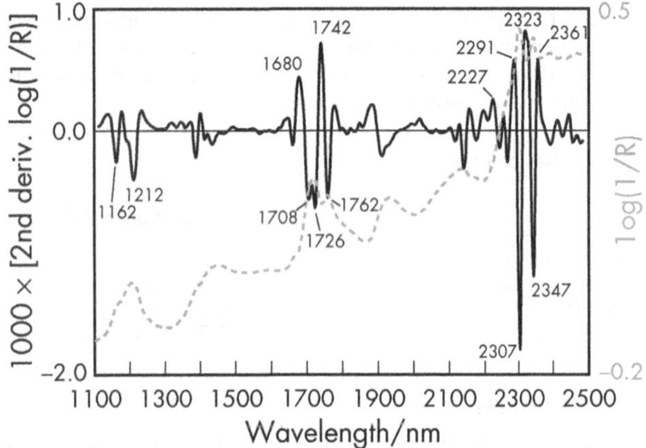

Figure 2: Near-infrared spectrum of lecithin measured by reflectance.

Saponification value 196

Solubility Lecithins are soluble in diethyl ether, chloroform, petroleum ether, mineral oil, and fatty acids. Sparingly soluble in benzene. Practically insoluble in cold vegetable and animal oils, polar solvents, and water. When mixed with water, however, lecithins hydrate to form emulsions.

Spectroscopy

IR spectra *see* Figure 1.

NIR spectra *see* Figure 2.

11 Stability and Storage Conditions

Lecithins decompose at extreme pH. They are also hygroscopic and subject to microbial degradation. When heated, lecithins oxidize, darken, and decompose. Temperatures of 160–180°C will cause degradation within 24 hours.

Store below 25°C, preferably at 2–6°C.

All lecithin grades should be stored in well-closed containers protected from light and oxidation. Purified solid lecithins should be stored in tightly closed containers at subfreezing temperatures.

12 Incompatibilities

Lecithin is incompatible with esterases owing to hydrolysis. It is also incompatible with chlorine, fluorine, and other oxidizing agents.

13 Method of Manufacture

Lecithins are essential components of cell membranes and, in principle, may be obtained from a wide variety of living matter. In practice, however, lecithins are usually obtained from vegetable products such as soybean, peanut, cottonseed, sunflower, rapeseed, corn, or groundnut oils. Soybean lecithin is the most commercially important vegetable lecithin. Lecithin obtained from eggs is also commercially important and was the first lecithin to be discovered.

Vegetable lecithins are obtained as a by-product in the vegetable oil refining process. During the degumming process the crude oil is heated to about 70°C, mixed with 2% water and subjected to thorough stirring for about half an hour to an hour. The addition of water hydrates the polar lipids in the oil, making them insoluble. The resulting lecithin sludge is then separated by centrifugation. This sludge is made up of water, phospholipids and glycolipids, some triglycerides, carbohydrates, traces of sterols, free fatty acids and carotenoids. Polar lipids are extracted with hexane and, after removal of the solvent, a crude vegetable oil is obtained. The crude plant lecithin is obtained by careful drying. The composition and quality of the crude lecithin product are considerably influenced by the quality and origin of the oilseeds as well as the conditions during the degumming process. Following drying, the lecithin may be further purified.[2]

With egg lecithin, a different manufacturing process must be used since the lecithin in egg yolks is more tightly bound to proteins than in vegetable sources. Egg lecithin is thus obtained by solvent extraction from liquid egg yolks using acetone or from freeze-dried egg yolks using ethanol (95%).[2]

Synthetic lecithins may also be produced.

14 Safety

Lecithin is a component of cell membranes and is therefore consumed as a normal part of the diet. Although excessive consumption may be harmful, it is highly biocompatible and oral doses of up to 80 g daily have been used therapeutically in the treatment of tardive dyskinesia.[6] When used in topical formulations, lecithin is generally regarded as a nonirritant and nonsensitizing material.[3] The Cosmetic Ingredients Review Expert Panel (CIR) has reviewed lecithin and issued a tentative report revising the safe concentration of the material from 1.95% to 15.0% in rinse-off and leave-in products. They note, however, that there are insufficient data to rule on products that are likely to be inhaled.[7]

15 Handling Precautions

Observe normal precautions appropriate to the circumstances and quantity of material handled. Lecithins may be irritant to the eyes; eye protection and gloves are recommended. Avoid inhalation.

16 Regulatory Status

GRAS listed. Accepted for use as a food additive in Europe. Included in the FDA Inactive Ingredients Database (inhalations; IM and IV injections; otic preparations; oral capsules, suspensions and tablets; rectal, topical, transdermal, and vaginal preparations). Included in nonparenteral and parenteral medicines licensed in the UK. Included in the Canadian Natural Health Products Ingredients Database.

17 Related Substances

Phospholipids

18 Comments

Poloxamer lecithin organogels have been used in topical formulations for the delivery of non-steroidal anti-inflammatory drugs.[8]

Lecithins contain a variety of unspecified materials; care should therefore be exercised in the use of unpurified lecithin in injectable or topical dosage forms, as interactions with the active substance or other excipients may occur. Unpurified lecithins may also have a greater potential for irritancy in formulations.

Lecithin has been used in various novel drug delivery systems such as formulation of organogels,[9] liposomes,[10] elastic vesicles,[11] self-emulsifying pellets,[12] nanoemulsions,[13] silica-coated nanoemulsions,[14] microemulsions,[15] microcrystal suspensions,[16] nanosuspensions,[17,18] silica-lipid hybrid microcapsules,[19] and mucoadhesive patches.[20] Lecithins have also been employed in stabilizing various types of nanoparticles including solid lipid,[21–23] PLGA/human albumin,[24] chitosan/PLGA,[25] gold,[26] and silver[27] nanoparticles. Lecithin can self-organize into lamellar mesophase when small amounts of water are admixed in its solution in a nonpolar solvent. This property has been utilized in the preparation of crystalline titania nanoparticles[28] and self-assembled phospholipid-based cationic nanocarriers.[29]

There is evidence that phosphatidylcholine (a major component of lecithin) is important as a nutritional supplement to fetal and infant development. Furthermore, choline is a required component of FDA-approved infant formulas.[30] Other studies have indicated that lecithin can protect against alcohol cirrhosis of the liver, lower serum cholesterol levels, and improve mental and physical performance.[31]

Therapeutically, lecithin and derivatives have been used as a pulmonary surfactant in the treatment of neonatal respiratory distress syndrome.

A specification for soybean lecithin is contained in the *Japanese Pharmaceutical Excipients* (JPE).[32] Suppliers' literature should be consulted for information on the different grades of lecithin available and their applications in formulations.

A specification for lecithin is contained in the *Food Chemicals Codex* (FCC).[33]

The EINECS number for lecithin is 232-307-2. The PubChem Compound ID (CID) for lecithin is 24798685.

19 Specific References

1 Palacios LE, Wang T. Egg-yolk lipid fractionation and lecithin characterization. *JAOCS* 2005; **82**(8): 571–578.
2 Schneider M. Achieving purer lecithin. *Drug Cosmet Ind* 1992; **150**(2): 54, 56, 62, 64, 66, 101–103.
3 Anonymous. Lecithin: its composition, properties and use in cosmetic formulations. *Cosmet Perfum* 1974; **89**(7): 31–35.
4 Novak E, *et al.* Evaluation of cefmetazole rectal suppository formulation. *Drug Dev Ind Pharm* 1991; **17**(3): 373–389.
5 Anonymous. Intranasal insulin formulation reported to be promising. *Pharm J* 1991; **247**: 17.
6 Growdon JH, *et al.* Lecithin can suppress tardive dyskinesia [letter]. *N Engl J Med* 1978; **298**: 1029–1030.
7 Anonymous. 'The Rose Sheet'. *FDC Reports* 1997; **18**(39): 8.
8 Franckum J, *et al.* Pluronic lecithin organogel for local delivery of anti-inflammatory drugs. *Int J Pharm Compound* 2004; **8**(2): 101–105.
9 Shchipunov YA. Lecithin organogel A micellar system with unique properties. *Colloids Surf A Physiochem Eng Asp* 2001; **183–185**: 541–554.
10 Rane S, Prabhakar B. Influence of liposome composition on paclitaxel entrapment and pH sensitivity of liposomes. *Int J PharmTech Res* 2009; **1**(3): 914–917.
11 Mura S, *et al.* Transcutol containing vesicles for topical delivery of minoxidil. *J Drug Target* 2011; **19**(3): 189–196.
12 Iosio T , *et al.* Oral bioavailability of silymarin phytocomplex formulated as self-emulsifying pellets. *Phytomedicine* 2011; **18**(6): 505–512.
13 Bruxel F, *et al.* Cationic nanoemulsion as a delivery system for oligonucleotides targeting malarial topoisomerase II. *Int J Pharm* 2011; **416**(2): 402–409.
14 Eskandar NG, *et al.* Interactions of hydrophilic silica nanoparticles and classical surfactants at non-polar oil-water interface. *J Colloid Interface Sci* 2011; **358**(1): 217–225.
15 Graf A, *et al.* Microemulsions containing lecithin and sugar-based surfactants: Nanoparticle templates for delivery of proteins and peptides. *Int J Pharm* 2008; **350**: 351–360.
16 Nippe S, General S. Parenteral oil-based drospirenone microcrystal suspensions – Evaluation of physicochemical stability and influence of stabilising agents. *Int J Pharm* 2011; **416**(1): 181–188.
17 Jain V. Development and characterization of mucoadhesive nanosuspension of ciprofloxacin. *Acta Pol Pharm* 2011; **68**(2): 273–278.
18 Zakir F, *et al.* Nanocrystallization of poorly water soluble drugs for parenteral administration. *J Biomed Nanotechnol* 2011; **7**(1): 127–129.
19 Lim LH, *et al.* Silica-lipid hybrid microcapsules: influence of lipid and emulsifier type on *in vitro* performance. *Int J Pharm* 2011; **409**(1–2): 297–306.
20 Desai KG, *et al.* Development and *in vitro-in vivo* evaluation of fenretinide-loaded oral mucoadhesive patches for site-specific chemoprevention of oral cancer. *Pharm Res* 2011; **28**(10): 2599–2609.
21 Fangueiro JF *et al.* A novel lipid nanocarrier for insulin delivery: production, characterization and toxicity testing. *Pharm Dev Technol* 2011; http://informahealthcare.com/doi/abs/10.3109/10837450.2011.591804 (accessed 8 December 2011)
22 Pandita D , *et al.* Development, characterization and *in vitro* assessment of stearylamine-based lipid nanoparticles of paclitaxel. *Pharmazie* 2011; **66**(3): 171–177.
23 Pandita D , *et al.* Development of lipid-based nanoparticles for enhancing the oral bioavailability of Paclitaxel. *AAPS PharmSciTech* 2011; **12**(2): 712–722.
24 Wohlfart S, *et al.* Efficient chemotherapy of rat glioblastoma using doxorubicin-loaded PLGA nanoparticles with different stabilizers. *PLoS One* 2011; **6**(5): e19121.
25 Murugeshu A, *et al.* Chitosan/PLGA particles for controlled release of α-tocopherol in the GI tract via oral administartion. *Nanomedicine* 2011; **6**(9): 1513–1528.
26 Sharma D. A biologically friendly single step method for gold nanoparticle formation. *Colloids Surf B Biointerfaces* 2011; **85**(2): 330–337.
27 Barani H, *et al.* Nano silver entrapped in phospholipids membrane: synthesis, characteristics and antibacterial kinetics. *Mol Membr Biol* 2011; **28**(4): 206–215.
28 Shchipunov Y, Krekoten A. Crystalline titania nanoparticles synthesized in nonpolar Lα lecithin liquid-crystalline media in one stage at ambient conditions. *Colloids Surf B Biointerfaces* 2011; **87**(2): 203–208.
29 Date AA, *et al.* Lecithin-based novel cationic nanocarriers (Leciplex) II: Improving therapeutic efficacy of quercetin on oral administration. *Mol Pharm* 2011; 8(3): 716–726.
30 US Congress. Infant Formula Act of 1980. Public Law 96-359, 1980.
31 Canty D *et al. Lecithin and Choline Research Update on Health and Nutrition.* Fort Wayne, IN: Central Soya Company, 1996.
32 Japan Pharmaceutical Excipients Council. *Japanese Pharmaceutical Excipients*, 2004. Tokyo: Yakuji Nippo, 2004; 843–844.
33 *Food Chemicals Codex*, 7th edn. Bethesda, MD: United States Pharmacopeia, 2010.

20 General References

Cargill. Product information: Lecithins, 2011. http://www.cargill.com/food/na/en/products/lecithins/manufacturing-process/index.jsp (accessed 8 December 2011).
Guan T , *et al.* Injectable nimodipine-loaded nanoliposomes: preparation, lyophilization and characteristics. *Int J Pharm* 2011; **410**(1–2): 180–187.
Hanin I, Pepeu G, eds. *Phospholipids: Biochemical, Pharmaceutical and Analytical Considerations.* New York: Plenum Press, 1990.
Nieuwenhuyzen WV. Lecithin production and properties. *JAOCS* 1976; **53**(6): 425–427.
Sznitowska M. Lecithin — pharmaceutical applications expanded beyond liposomes. *Cell Mol Biol Lett* 2005; **10**: 52.
Szuhaj BF. Lecithin production and utilization. *JAOCS* 1983; **60**(2): 306–309.
Yurii A, Shchipunov YA. Self-organising structures of lecithin. *Russian Chemical Reviews* 1997; **66**(4): 301.

21 Authors

HC Shah, KK Singh.

22 Date of Revision

2 March 2012.

Leucine

1 Nonproprietary Names

BP: Leucine
JP: L-Leucine
PhEur: Leucine
USP-NF: Leucine

2 Synonyms

α-Aminoisocaproic acid; L-α-aminoisocaproic acid; 2-amino-4-methylpentanoic acid; 2-amino-4-methylvaleric acid; α-amino-γ-methylvaleric acid; 1,2-amino-4-methylvaleric acid; D L-leucine; L-leucine; leu; leucinum; 4-methylnorvaline.

3 Chemical Name and CAS Registry Number

L-Leucine [61-90-5]

4 Empirical Formula and Molecular Weight

$C_6H_{13}NO_2$ 131.17

5 Structural Formula

6 Functional Category

Antiadherent; flavoring agent; tablet and capsule lubricant.

7 Applications in Pharmaceutical Formulation or Technology

Leucine is used in pharmaceutical formulations as a flavoring agent.(1) It has been used experimentally as an antiadherent to improve the deagglomeration of disodium cromoglycate microparticles and other compounds in inhalation preparations;(2) and as a tablet lubricant.(3) Leucine copolymers have been shown to successfully produce stable drug nanocrystals in water.(4)

8 Description

Leucine occurs as a white or almost off-white crystalline powder or shiny flakes.

9 Pharmacopeial Specifications

See Table I.

Table I: Pharmacopeial specifications for leucine.

Test	JP XV	PhEur 7.4	USP35–NF30
Identification	+	+	+
Characters	+	+	–
Optical rotation	+14.5° to +16.0°	+14.5° to +16.5°	+14.9° to +17.3°
pH	5.5–6.5	–	5.5–7.0
Appearance of solution	+	+	–
Chloride	≤0.021%	≤200 ppm	≤0.05%
Sulfate	≤0.028%	≤300 ppm	≤0.03%
Ammonium	≤0.02%	≤200 ppm	–
Ninhydrin-positive substances	–	+	–
Iron	–	≤10 ppm	≤30 ppm
Heavy metals	≤20 ppm	≤10 ppm	≤15 ppm
Arsenic	≤2 ppm	–	–
Related substances	+	–	–
Loss on drying	≤0.30%	≤0.5%	≤0.2%
Residue on ignition	≤0.10%	–	≤0.4%
Sulfated ash	–	≤0.1%	–
Chromatographic purity	–	–	+
Assay	≥98.5%	98.5–101.5%	98.5–101.5%

10 Typical Properties

Density 1.293 g/cm³
Dissociation constant
pK_{a1} = 2.35 at 13°C;
pK_{a2} = 9.60.
Isoelectric point 6.04
Melting point Decomposes at 293–295°C; sublimes at 145–148°C.
Solubility Soluble in acetic acid, ethanol (99%), and water. Practically insoluble in ether.

11 Stability and Storage Conditions

Leucine is sensitive to light and moisture, and should be stored in an airtight container in a cool, dark, dry place.

12 Incompatibilities

Leucine is incompatible with strong oxidizing agents.

13 Method of Manufacture

Leucine is produced microbially by incubating an amino-acid-producing microorganism including but not exclusive to *Pseudomonas*, *Escherichia*, *Bacillus*, or *Staphylococcus* in the presence of oxygen and a hydrocarbon. The nutrient medium should contain an inhibitory amount of a growth inhibitor that is a chemically similar derivative of leucine (e.g. methylallylglycine, α-hydrozinoisocaproic acid, or β-cyclopentanealanine) to inhibit the growth of the organism except for at least one mutant that is resistant to the inhibitory effect. The resistant mutant is then isolated and grown in the presence of oxygen and the hydrocarbon in the absence of the inhibitor. The mutant cells are then harvested and a nutrient medium is formed that includes a hydrocarbon as the sole source of carbon. Finally, the harvested cells are incubated in the medium in the presence of oxygen.(5)

14 Safety

Leucine is an essential amino acid and is consumed as part of a normal diet. It is generally regarded as a nontoxic and nonirritant material. It is moderately toxic by the subcutaneous route.

LD_{50} (rat, IP): 5.379 g/kg[6]

15 Handling Precautions

Observe normal precautions appropriate to the circumstances and quantity of the material handled.

16 Regulatory Status

Included in the FDA Inactive Ingredients Database (IV infusion; oral tablets). Included in nonparenteral medicines licensed in the UK.

17 Related Substances

DL-Leucine

DL-Leucine

Empirical formula $C_6H_{13}NO_2$
Molecular weight 131.20
Appearance White leaflets.
Dissociation constant

pK_{a1} = 2.36;

pK_{a2} = 9.60.

Solubility Soluble in ethanol (90%) and water. Practically insoluble in ether.

18 Comments

A specification for leucine is included in the *Food Chemicals Codex* (FCC).[7] The EINECS number for leucine is 200-522-0.

19 Specific References

1 Ash M, Ash I. *Handbook of Pharmaceutical Additives*, 3rd edn. Endicott, NY: Synapse Information Resources, 2007; 709.
2 Abdolhossien RN, *et al.* The effect of vehicle on physical properties and aerosolisation behaviour of disodium cromoglycate microparticles spray dried alone or with L-leucine. *Int J Pharm* 2004; **285**: 97–108.
3 Gusman S, Gregoriades D. Effervescent potassium chloride tablet. United States Patent No. 3,903,255; 1975.
4 Lee J, *et al.* Amphiphilic amino acid copolymers as stabilizers for the preparation of nanocrystal dispersion. *Eur J Pharm Sci* 2005; **24**: 441–449.
5 Mobil Oil Corp. Synthesis of amino acids. UK Patent No. 1 071 935; 1967.
6 Lewis RJ, ed. *Sax's Dangerous Properties of Industrial Materials*, 11th edn. New York: Wiley, 2004; 2224.
7 *Food Chemicals Codex*, 7th edn. Bethesda, MD: United States Pharmacopeia, 2011.

20 General References

—

21 Author

GE Amidon.

22 Date of Revision

2 March 2012.

Linoleic Acid

1 Nonproprietary Names

None adopted.

2 Synonyms

Emersol 310; *Emersol 315*; leinoleic acid; 9-*cis*,12-*cis*-linoleic acid; 9,12-linoleic acid; linolic acid; *cis*,*cis*-9,12-octadecadienoic acid; *Pamolyn*; *Polylin No. 515*; telfairic acid.

3 Chemical Name and CAS Registry Number

(Z,Z)-9,12-Octadecadienoic acid [60-33-3]

4 Empirical Formula and Molecular Weight

$C_{18}H_{32}O_2$ 280.45

5 Structural Formula

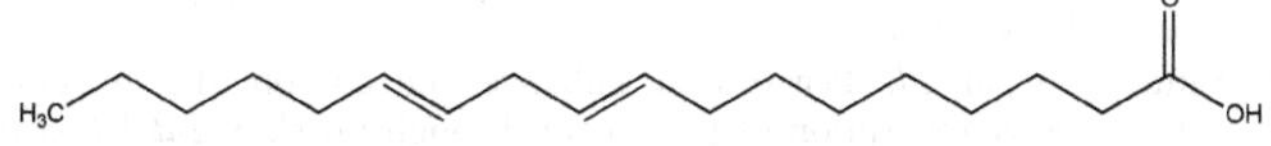

6 Functional Category

Emulsifying agent; penetration enhancer.

7 Applications in Pharmaceutical Formulation or Technology

Linoleic acid is used in topical transdermal formulations,[1–15] in oral formulations as an absorption enhancer,[9,10] in topical cosmetic formulations as an emulsifying agent,[15] and in aqueous microemulsions.[16] It has also been used as a component of nanoparticulate drug carriers.[17,18]

8 Description

Linoleic acid occurs as a colorless to light-yellow-colored oil.

9 Pharmacopeial Specifications

See Section 18.

10 Typical Properties

Boiling point 230°C at 16 mmHg
Density 0.9007 g/cm^3
Iodine value 181.1
Melting point −5°C
Refractive index n_D^{20} = 1.4699
Solubility Freely soluble in ether; soluble in ethanol (95%); miscible with dimethylformamide, fat solvents, and oils.

11 Stability and Storage Conditions

Linoleic acid is sensitive to air, light, moisture, and heat. It should be stored in a tightly sealed container under an inert atmosphere and refrigerated.

12 Incompatibilities

Linoleic acid is incompatible with bases, strong oxidizing agents, and reducing agents.

13 Method of Manufacture

Linoleic acid is obtained by extraction from various vegetable oils such as safflower oil.

14 Safety

Linoleic acid is widely used in cosmetics and topical pharmaceutical formulations, and is generally regarded as a nontoxic material. On exposure to the eyes, skin, and mucous membranes, linoleic acid can cause mild irritation.

15 Handling Precautions

Observe normal precautions appropriate to the circumstances and quantity of material handled. Gloves and eye protection are recommended.

L

16 Regulatory Status

GRAS listed. Approved for use in foods in Europe and the US. Included in the Canadian Natural Health Products Ingredients Database.

17 Related Substances

Ethyl linoleate; methyl linoleate.

Ethyl linoleate

Empirical formula $C_{20}H_{36}O_2$
CAS number [544-35-4]
Synonyms Linoleic acid ethyl ester; 9,12-octadecadienoic acid ethyl ester; vitamin F.
Comments Ethyl linoleate is used in pharmaceutical formulations as an emollient and humectant. It is also used as a solvent for fats. The EINECS number for ethyl linoleate is 208-868-4.

Methyl linoleate

Empirical formula $C_{19}H_{34}O_2$
CAS number [112-63-0]
Synonyms 9,12-Octadecadienoic acid, methyl ester.
Comments Methyl linoleate is used in cosmetics as an emollient. The EINECS number for methyl linoleate is 203-993-0.

18 Comments

Conjugated linoleic acids are a family of linoleic acid isomers in which the double bonds are conjugated. Antioxidant and anticancer properties have been attributed to conjugated linoleic acids, although the different isomers can have markedly different effects. Studies have shown that conjugated linoleic acid increases paracellular permeability across human intestinal-like Caco-2 cell monolayers, and consequently may also, as a dietary supplement, increase calcium absorption *in vivo*.[19]

Linoleic acid is used in parenteral emulsions for total parenteral nutrition and in nonprescription oral dietary supplements.

Linoleic acid has been shown to reduce skin irritation following acute perturbations, exhibiting clinical effects that are comparable to glucocorticoids.[20]

A pre-emulsified linoleic acid system has been used to investigate the protective actions of phenolic compounds against lipid peroxidation.[21]

Linoleic acid was found to have a toxic effect on melanoma cells, causing loss of membrane integrity and/or DNA fragmentation.[22]

Linoleic acid was also found to promote apoptosis and necrosis of human lymphocytes[23] and Jurkat cell death[24] (Jurkat cells are a T-lymphocyte cell line) by mitochondrial depolarization. As a result, it has been suggested that oleic acid may offer a less immunologically problematic alternative to linoleic acid in parenteral nutritional emulsions. However, linoleic acid also reduced genetic damage to normal human lymphocytes caused by benzo(a)-pyrene.[25]

Evidence suggests that linoleate-enriched oil such as sunflower seed oil may enhance skin barrier function,[26] and it has also been shown that linoleic acid is capable of improving epithelial integrity following mucosal injury.[27]

Linoleic acid has been shown to act as a comedolytic agent in acne-prone patients[28] and may have a possible use as a cosmeceutical in acne juvenilis therapy.[29]

Linoleic acid was found to selectively modulate vascular cytotoxicity caused by TNF-alpha and persistent organic pollutants such as polychlorinated biphenyls.[30]

Although not included in any pharmacopeias, a specification for linoleic acid is contained in the *Food Chemicals Codex* (FCC);[31] *see* Table I.

The EINECS number for linoleic acid is 200-470-9.

Table I: FCC specification for linoleic acid.[31]

Test	FCC 7
Identification	+
Acid value	196–202
Lead	≤2 mg/kg
Iodine value	145–160
Residue on ignition	≤0.01%
Unsaponifiable matter	≤2.0%
Water	≤0.5%
Assay (anhydrous basis)	≥60%

19 Specific References

1 Gwak HS, Chun IK. Effect of vehicles and penetration enhancers on the *in vitro* percutaneous absorption of tenoxicam through hairless mouse skin. *Int J Pharm* 2002; **236**(1–2): 57–64.
2 Bhattachrya A, Ghosal SK. Effect of hydrophobic permeation enhancers on the release and skin permeation kinetics from matrix type transdermal drug delivery system of ketotifen fumarate. *Acta Pol Pharm* 2001; **58**(2): 101–105.
3 Gwack HS, Chun IK. Effect of vehicles and enhancers on the *in vitro* skin permeation of aspalatone and its enzymatic degradation across rat skins. *Arch Pharm Res* 2001; **24**(6): 572–577.
4 Shin SC, *et al.* Enhancing effects of fatty acids on piroxicam permeation through rat skins. *Drug Dev Ind Pharm* 2000; **26**(5): 563–566.
5 Meaney CM, O'Driscoll CM. Comparison of the permeation enhancement potential of simple bile salt and mixed bile salt: fatty acid micellar systems using the Caco-2 cell culture model. *Int J Pharm* 2000; **207**(10): 21–30.
6 Bhattacharya A, Ghosal SK. Effect of hydrophobic permeation enhancers on the release and skin permeation kinetics from matrix type transdermal drug delivery system of ketotifen fumarate. *Eastern Pharmacist* 2000; **43**: 109–112.
7 Tanojo H, Junginger HE. Skin permeation enhancement by fatty acids. *J Dispers Sci Technol* 1999; **20**(1–2): 127–138.
8 Bhatia KS, Singh J. Synergistic effect of iontophoresis and a series of fatty acids on LHRH permeability through porcine skin. *J Pharm Sci* 1998; **87**: 462–469.
9 Santoyo S, *et al.* Penetration enhancer effects on the *in vitro* percutaneous absorption of piroxicam through rat skin. *Int J Pharm* 1995; **117**(18): 219–224.
10 Carelli V, *et al.* Enhancement effects in the permeation of alprazolam through hairless mouse skin. *Int J Pharm* 1992; **88**(8): 89–97.
11 Ibrahim SA, *et al.* Formulation and evaluation of some topical antimycotics. Part 3. Effect of promotors on the *in vitro* and *in vivo*

efficacy of clotrimazole ointment. *Bull Pharm Sci Assiut Univ* 1991; **14**(1–2): 82–94.
12 Swafford SK, *et al.* Characterization of swollen micelles containing linoleic acid in a microemulsion system. *J Soc Cosmet Chem* 1991; **42**: 235–247.
13 Mahjour M, *et al.* Skin permeation enhancement effects of linoleic acid and Azone on narcotic analgesics. *Int J Pharm* 1989; **56**(1): 1–11.
14 Gwak HS, *et al.* Transdermal delivery of ondansetron hydrochloride: effects of vehicles and penetration enhancers. *Drug Dev Ind Pharm* 2004; **30**(2): 187–194.
15 Muranushi N, *et al.* Mechanism for the inducement of the intestinal absorption of poorly absorbed drugs by mixed micelles. Part 2. Effect of the incorporation of various lipids on the permeability of liposomal membranes. *Int J Pharm* 1980; **4**: 281–290.
16 Swafford SK, *et al.* Characterization of swollen micelles containing linoleic acid in a microemulsion system. *J Soc Cosmet Chem* 1991; **42**(4): 235–248.
17 Zhao Z, *et al.* Biodegradable nanoparticles based on linoleic acid and poly(β-malic acid) double grafted chitosan derivatives as carriers of anticancer drugs. *Biomacromolecules* 2009; **10**(3): 565–572.
18 Lee CM, *et al.* SPION-loaded chitosan-linoleic acid nanoparticles to target hepatocytes. *Int J Pharm* 2009; **371**((1-2)): 163–169.
19 Jewell C, Cashmen KD. The effect of conjugated linoleic acid and medium-chain fatty acids on transepithelial calcium transport in human intestine-like Caco-2 cells. *Br J Nutr* 2003; **89**(5): 639–647.
20 Schurer NY. Implementation of fatty acid carriers to skin irritation and the epidermal barrier. *Contact Dermatitis* 2002; **47**(4): 199–205.
21 Cheng Z, *et al.* Establishment of a quantitative structure–activity relationship model for evaluating and predicting the protective potentials of phenolic antioxidants on lipid peroxidation. *J Pharm Sci* 2003; **92**(3): 475–484.
22 Andrade LN, *et al.* Toxicity of fatty acids on murine and human melanoma cell lines. *Toxicol In Vitro* 2005; **19**(4): 553–560.
23 Cury-Boaventura MF, *et al.* Comparative toxicity of oleic and linoleic acid on human lymphocytes. *Life Sci* 2006; **78**(13): 1448–1456.
24 Cury-Boaventura MF, *et al.* Comparative toxicity of oleic acid and linoleic acid on Jurkat cells. *Clin Nutr* 2004; **23**(4): 721–732.
25 Das U, *et al.* Precursors of prostaglandins and other N-6 essential fatty acids can modify benzo-a-pyrene-induced chromosomal damage to human lymphocytes *in-vitro*. *Nut Rep Int* 1987; **36**(6): 1267–1272.
26 Ruthig DJ, Meckling-Gill KA. Both (n-3) and (n-6) fatty acids stimulate wound healing in the rat intestinal epithelial cell line, IEC-6. *J Nutr* 1999; **129**(10): 1791–1798.
27 Darmstadt GL, *et al.* Impact of topical oils on the skin barrier: possible implications for neonatal health in developing countries. *Acta Paediatr* 2002; **91**(5): 546–554.
28 Letawe C, *et al.* Digital image analysis of the effect of topically applied linoleic acid on acne microcomedones. *Clin Exp Dermatol* 1998; **23**(2): 56–58.
29 Morganti P, *et al.* Role of topical glycolic acid and phosphatidylcholine linoleic acid-rich in the pathogenesis of acne. Linoleic acid versus squalene. *J Appl Cosmetol* 1997; **15**(1): 33–41.
30 Wang L, *et al.* Different ratios of omega-6 to omega-3 fatty acids can modulate proinflammatory events in endothelial cells induced by TNF-alpha and environmental contaminants. *FASEB J* 2005; **19**(4): (Suppl. S, Part 1): A442.
31 *Food Chemicals Codex*, 7th edn. Bethesda, MD: United States Pharmacopeia, 2010: 592.

20 General References

Gunstone F *et al.* eds. *The Lipid Handbook with CD ROM*. 3rd edn. Boca Raton. FL: CRC Press, 2007.
Wasan KM, ed. *Role of Lipid Excipients in Modifying Oral and Parenteral Drug Delivery*. Hoboken, NJ: John Wiley & Sons, Inc, 2007.

21 Author

MS Tesconi.

22 Date of Revision

1 February 2011.

Lysine Acetate

1 Nonproprietary Names

BP: Lysine Acetate
PhEur: Lysine Acetate
USP–NF: Lysine Acetate

2 Synonyms

(2*S*)-2,6-Diaminohexanoic acid acetate; lysini acetas; L-lysine acetate; L-lysine monoacetate.

3 Chemical Name and CAS Registry Number

L-Lysine monoacetate [57282-49-2]

4 Empirical Formula and Molecular Weight

$C_8H_{18}N_2O_4$ 206.24

5 Structural Formula

6 Functional Category

Buffering agent.

7 Applications in Pharmaceutical Formulation or Technology

Lysine acetate is used as a buffering agent and neutralizing agent for antipyretics and analgesics such as acetylsalicylic acid and ibuprofen. It has been investigated as a component of vaginal sponge-like dressings for treatment of vaginal infections and delivery of peptidic drugs.[1]

8 Description

Lysine acetate occurs as a white or almost white, crystalline powder or colorless crystals.

9 Pharmacopeial Specifications

See Table I.

Table I: Pharmacopeial specifications for lysine acetate.

Test	PhEur 7.4	USP35–NF30
Identification	+	+
Characters	+	–
Appearance of solution	+	–
Specific rotation	+8.5° to +10.0°	+8.4° to +9.9°
Ninhydrin-positive substances	≤0.5%	–
Loss on drying	≤0.5%	≤0.2%
Residue on ignition	≤0.1%	≤0.4%
Chloride	≤200 ppm	≤0.05%
Sulfate	≤300 ppm	≤0.03%
Ammonium	≤200 ppm	–
Iron	≤30 ppm	≤0.003%
Heavy metals	≤10 ppm	≤0.0015%
Chromatographic purity	–	+
Assay	98.5–101.0%	98.0–102.0%

10 Typical Properties

Melting point 281°C[2]

Solubility Freely soluble in water; very slightly soluble in ethanol (96%).

11 Stability and Storage Conditions

Store in well-closed containers protected from light.

12 Incompatibilities

—

13 Method of Manufacture

L-Lysine is manufactured by fermentation from carbohydrate sources. The solution is mixed with acetic acid to produce L-lysine acetate.[3]

14 Safety

Lysine acetate in pure form is moderately toxic by IP, IV, and SC routes, and mildly toxic by ingestion.[4]

LD_{50} (mouse, oral): 14.4 g/kg[4]
LD_{50} (mouse, IP): 5.1 g/kg
LD_{50} (mouse, IV): 3.7 g/kg
LD_{50} (mouse, SC): 5.8 g/kg
LD_{50} (rat, oral): 11.4 g/kg
LD_{50} (rat, IP): 3.7 g/kg
LD_{50} (rat, IV): 2.85 g/kg
LD_{50} (rat, SC): 4.0 g/kg

15 Handling Precautions

When heated to decomposition lysine acetate emits toxic fumes of NO_x.[4]

16 Regulatory Status

Included in the FDA Inactive Ingredient Database (IV injections).

17 Related Substances

Lysine hydrochloride.

18 Comments

The PhEur 7.4 states that lysine acetate exhibits polymorphism.

Therapeutically, lysine acetate is an essential amino acid and is used as a dietary supplement.[5] It is also used for the treatment of mouth and genital lesions caused by herpes zoster viruses.[3]

The EINECS number for lysine acetate is 260-664-4.

19 Specific References

1 Rossi S, *et al.* Development of sponge-like dressings for mucosal/transmucosal drug delivery into vaginal cavity. *Pharm Dev Technol* 2010; **November 15 [Epub ahead of print]**: .
2 ScienceLab.com. Material data safety sheet: *L-Lysine acetate*, January 2010.
3 Ajinomoto Co, Inc. Product information: *L-Lysine acetate*. http://www.ajiaminoscience.com/products/manufactured_products/l-amino_acids/L-Lysine-Acetate.aspx (accessed 20 July 2011).
4 Lewis RJ, ed. *Sax's Dangerous Properties of Industrial Materials*, 11th edn. New York: Wiley, 2004; 2260.
5 Sweetman SC, ed. *Martindale: The Complete Drug Reference*, 37th edn. London: Pharmaceutical Press, 2011; 2114.

20 General References

—

21 Authors

ME Fenton, RC Rowe.

22 Date of Revision

1 March 2012.

Lysine Hydrochloride

1 Nonproprietary Names

BP: Lysine Hydrochloride
JP: L-Lysine Hydrochloride
PhEur: Lysine Hydrochloride
USP–NF: Lysine Hydrochloride

2 Synonyms

Darvyl; (2*S*)-2,6-diaminohexanoic acid hydrochloride; E642; *Enisyl*; L-lysine monohydrochloride; lysine monohydrochloride; lysini hydrochloridum.

3 Chemical Name and CAS Registry Number

L-Lysine, hydrochloride (1 : 1) [657-27-2]

4 Empirical Formula and Molecular Weight

$C_6H_{14}N_2O_2HCl$ 182.65

5 Structural Formula

6 Functional Category

Buffering agent.

7 Applications in Pharmaceutical Formulation or Technology

Lysine hydrochloride is used as a buffering agent in pharmaceutical formulations. It is widely used as a food additive and as a dietary supplement.

8 Description

Lysine hydrochloride occurs as an odorless, white or almost white crystalline powder.

9 Pharmacopeial Specifications

See Table I.

Table I: Pharmacopeial specifications for lysine hydrochloride.

Test	JP XV	PhEur 7.4	USP35–NF30
Identification	+	+	+
Characters	+	+	–
Appearance of solution	+	+	–
Ninhydrin-positive substances	–	+	–
Chlorides	+	+	19.0–19.6%
Residue on ignition	≤0.1%	≤0.1%	≤0.1%
Sulfates	≤0.028%	≤300 ppm	≤0.03%
Ammonium	≤0.02%	≤200 ppm	–
Iron	–	≤30 ppm	≤30 ppm
Arsenic	≤2 ppm	–	–
Heavy metals	≤10 ppm	≤10 ppm	≤15 ppm
Chromatographic purity	+	+	+
Optical rotation	+19.0° to +21.5°	+21.0° to +22.5°	+20.4° to +21.4°
Loss on drying	≤1.0%	≤0.5%	≤0.4%
pH	5.0–6.0	–	–
Related substances	+	–	–
Assay (dried basis)	≥ 98.5%	98.5–101.0%	98.5–101.5%

10 Typical Properties

Melting point 263–264°C
Dissociation constant

$pK_{a1} = 2.18$;
$pK_{a2} = 8.95$;
$pK_{a3} = 10.53$ at 25.9°C.

Solubility Freely soluble 1 in 11 parts in water; slightly soluble in ethanol.

11 Stability and Storage Conditions

Store in well-closed containers protected from light.

12 Incompatibilities

Lysine hydrochloride is incompatible with strong oxidizing agents. Lysine can undergo Maillard reactions with reducing sugars to produce browning.[1] Milling and compression of lactose-containing formulations have been shown to increase the rate of such reactions.[2]

13 Method of Manufacture

Lysine hydrochloride is prepared by titration of an ethanol solution of lysine dihydrochloride with pyridine, and subsequent crystallization from an aqueous solution with the addition of ethanol.[3]

14 Safety

Lysine hydrochloride is commonly used as a dietary supplement. The required dietary allowance in adults is 12 mg/kg/day.

LD_{50} (rat, IP): 10.0 g/kg[4]
LD_{50} (rat, oral): 40.1 g/kg

15 Handling Precautions

Observe normal precautions appropriate to the circumstances and quantity of the material handled.

16 Regulatory Status

Accepted for use as a food additive in Europe.

17 Related Substances

Lysine; lysine acetate.

Lysine

Empirical formula $C_6H_{14}N_2O_2$
Molecular weight 146.19
CAS number [56-87-1]
Synonyms K; lys
Melting point 210–224°C (with decomposition)
Comments Included in the FDA Inactive Ingredient Database (IM and IV injections). Included in the German Pharmacopeia (Deutsches Arzeibuch) as the monohydrate. Lysine is an essential amino acid and is commonly used as a dietary supplement. L-Lysine can be isolated from acid-hydrolyzed proteins, produced by fermentation, or made synthetically.[5] Lysine is considered a Class I counterion for producing salts of drugs.[6]

18 Comments

A specification for lysine hydrochloride is contained in the *Food Chemicals Codex*.[7]

The EINECS for lysine hydrochloride is 211-519-9. The PubChem Compound ID (CID) for lysine hydrochloride includes 69568, 12047, and 517015.

L

19 Specific References

1 Murata M, *et al.* Browning of furfural and amino acids, and a novel yellow compound, furpipate, formed from lysine and furfural. *Biosci Biotechnol Biochem* 2007; **71**: 1717–1723.
2 Qiu Z, *et al.* Effect of milling and compression on the solid-state Maillard reaction. *J Pharm Sci* 2005; **94**: 2568–2580.
3 Rice E.E. A simplified procedure for the isolation of lysine from protein hydrolystates. *J Biol Chem* 1939; **131**: 1–4.
4 Lewis RJ, ed. *Sax's Dangerous Properties of Industrial Materials*, 11th edn. New York: Wiley, 2004; 2260.
5 Warning K, *et al.* A facile synthesis of α, ω-diamino carboxylic acids (lysine, ornithine) [in German]. *Justus Liebigs Ann Chem* 1978; **11**: 1707–1712.
6 Stahl PH, Wermuth CG, eds. *Handbook of Pharmaceutical Salts*. Zurich: Verlag HCA, 2002; 345.
7 *Food Chemicals Codex*, 7th edn. Bethesda, MD: United States Pharmacopeia, 2010; 598.

20 General References

Science Lab. Material safety data sheet: Lysine hydrochloride. http://www.sciencelab.com (accessed 29 February 2012).
Sigma-Aldrich. Material safety data sheet: Lysine hydrochloride. http://www.sigmaaldrich.com (accessed 29 February 2012).
Spectrum Chemical. Material safety data sheet: Lysine hydrochloride. http://www.spectrumchemical.com (accessed 29 February 2012).

21 Author

WJ Lambert.

22 Date of Revision

1 March 2012.

Magnesium Aluminum Silicate

1 Nonproprietary Names

BP: Aluminium Magnesium Silicate

PhEur: Aluminium Magnesium Silicate

USP–NF: Magnesium Aluminum Silicate

2 Synonyms

Aluminii magnesii silicas; aluminosilicic acid, magnesium salt; aluminum magnesium silicate; *Carrisorb*; *Gelsorb*; *Magnabrite*; magnesium aluminosilicate; magnesium aluminum silicate, colloidal; magnesium aluminum silicate, complex colloidal; *Neusilin*; *Pharmasorb*; silicic acid, aluminum magnesium salt; *Veegum*.

3 Chemical Name and CAS Registry Number

Aluminum magnesium silicate [12511-31-8]

Magnesium aluminum silicate [1327-43-1]

4 Empirical Formula and Molecular Weight

Magnesium aluminum silicate is a polymeric complex of magnesium, aluminum, silicon, oxygen, and water. The average chemical analysis is conventionally expressed as oxides:

Silicon dioxide 61.1%
Magnesium oxide 13.7%
Aluminum oxide 9.3%
Titanium dioxide 0.1%
Ferric oxide 0.9%
Calcium oxide 2.7%
Sodium oxide 2.9%
Potassium oxide 0.3%
Carbon dioxide 1.8%
Water of combination 7.2%

5 Structural Formula

The complex is composed of a three-lattice layer of octahedral alumina and two tetrahedral silica sheets. The aluminum is substituted to varying degrees by magnesium (with sodium or potassium for balance of electrical charge). Additional elements present in small amounts include iron, lithium, titanium, calcium, and carbon.

6 Functional Category

Adsorbent; suspending agent; tablet and capsule binder; tablet and capsule disintegrant; viscosity-increasing agent.

7 Applications in Pharmaceutical Formulation or Technology

Magnesium aluminum silicate has been used for many years in the formulation of tablets, ointments, and creams. It is used in oral and topical formulations as a suspending and stabilizing agent either alone or in combination with other suspending agents.[1–3] The viscosity of aqueous dispersions may be greatly increased by combination with other suspending agents, such as xanthan gum, owing to synergistic effects; *see* Xanthan Gum. In tablets, magnesium aluminum silicate is used as a binder and disintegrant in conventional or slow-release formulations.[4,5] *See* Table I.

Magnesium aluminum silicate may cause bioavailability problems with certain drugs; *see* Section 12.

Table I: Uses of magnesium aluminum silicate.

Use	Concentration (%)
Adsorbent	10–50
Binding agent	2–10
Disintegrating agent	2–10
Emulsion stabilizer (oral)	1–5
Emulsion stabilizer (topical)	2–5
Suspending agent (oral)	0.5–2.5
Suspending agent (topical)	1–10
Stabilizing agent	0.5–2.5
Viscosity modifier	2–10

8 Description

The USP35–NF30 describes magnesium aluminum silicate as a blend of colloidal montmorillonite and saponite that has been processed to remove grit and nonswellable ore components. Four types of magnesium aluminum silicate are defined: types IA, IB, IC, and IIA. These types differ according to their viscosity and ratio of aluminum and magnesium content; *see* Table II.

The PhEur 7.4 describes magnesium aluminum silicate (aluminium magnesium silicate) as a mixture of particles with colloidal particle size of montmorillonite and saponite, free from grit and nonswellable ore.

Magnesium aluminum silicate occurs as off-white to creamy white, odorless, tasteless, soft, slippery small flakes, or as a fine, micronized powder. Flakes vary in shape and size from about 0.3 × 0.4 mm to 1.0 × 2.0 mm and about 25–240 μm thick. Many flakes are perforated by scattered circular holes 20–120 μm in diameter. Under dark-field polarized light, innumerable bright specks are observed scattered over the flakes. The powder varies from 45 to 297 μm in size.

Table II: Magnesium aluminum silicate types defined in the USP34–NF29 S1.

Type	Viscosity (mPa s)	Al content/Mg content
IA	225–600	0.5–1.2
IB	150–450	0.5–1.2
IC	800–2200	0.5–1.2
IIA	100–300	1.4–2.8

9 Pharmacopeial Specifications

See Table III.

Table III: Pharmacopeial specifications for magnesium aluminum silicate.

Test	PhEur 7.4	USP35–NF30
Identification	+	+
Characters	+	—
Viscosity (5% w/v suspension)	—	*See* Table II
Microbial limits	$\leq 10^3$ cfu/g	$\leq 10^3$ cfu/g
pH (5% w/v suspension)	9.0–10.0	9.0–10.0
Acid demand	—	+
Loss on drying	≤8.0%	≤8.0%
Arsenic	≤3 ppm	≤3 ppm
Lead	≤15 ppm	≤0.0015%
Assay for Al and Mg content	95.0–105.0	+

SEM 1: Excipient: magnesium aluminum silicate (*Veegum*); manufacturer: RT Vanderbilt Co., Inc.; lot no.: 61A-1; magnification: 600×; voltage: 10 kV.

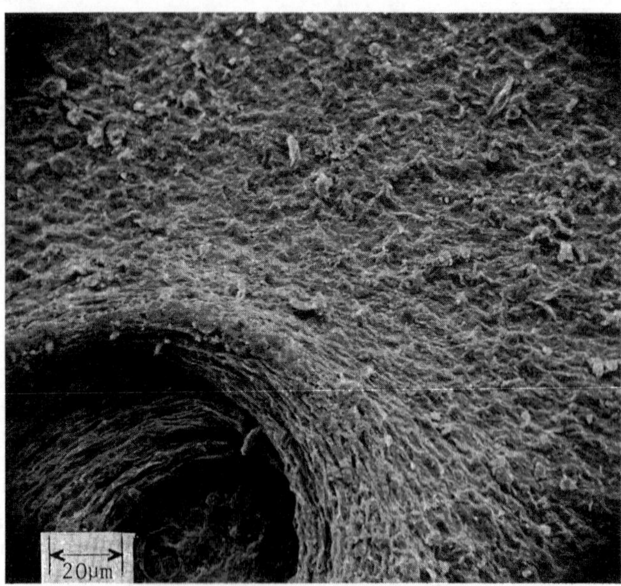

SEM 2: Excipient: magnesium aluminum silicate (*Veegum*); manufacturer: RT Vanderbilt Co., Inc.; lot no.: 61A-1; magnification: 2400×; voltage: 10 kV.

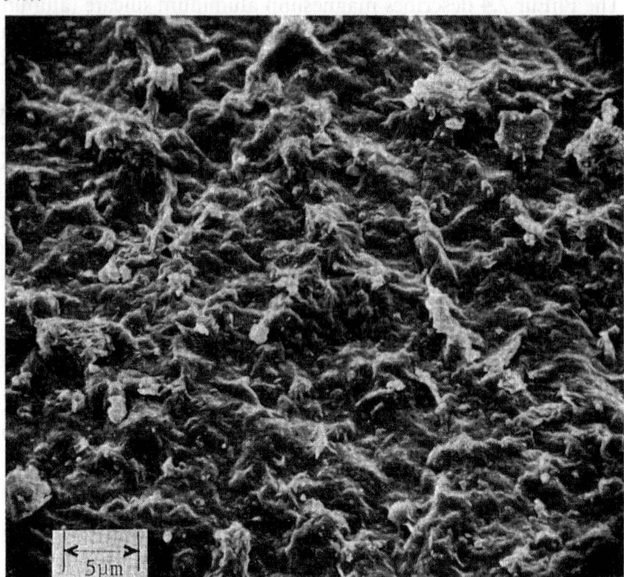

SEM 3: Excipient: magnesium aluminum silicate (*Veegum F*); manufacturer: RT Vanderbilt Co., Inc.; lot no: 61A-2; magnification: 600×; voltage: 10 kV.

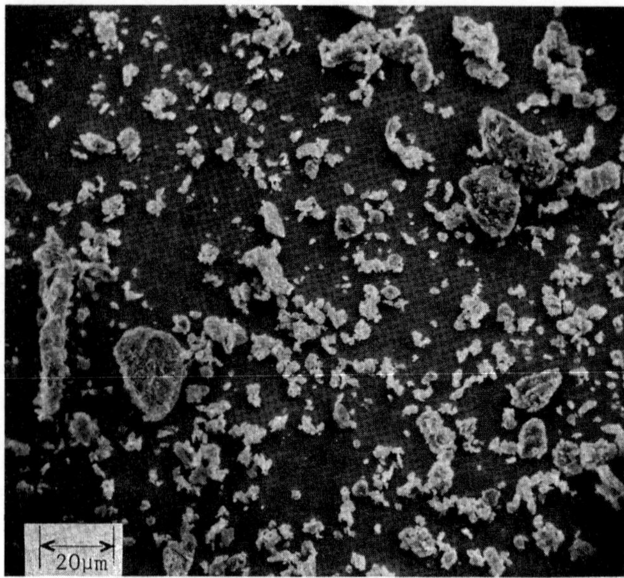

SEM 4: Excipient: magnesium aluminum silicate (*Veegum F*); manufacturer: RT Vanderbilt Co., Inc.; lot no.: 61A-2; magnification: 2400×; voltage: 10 kV.

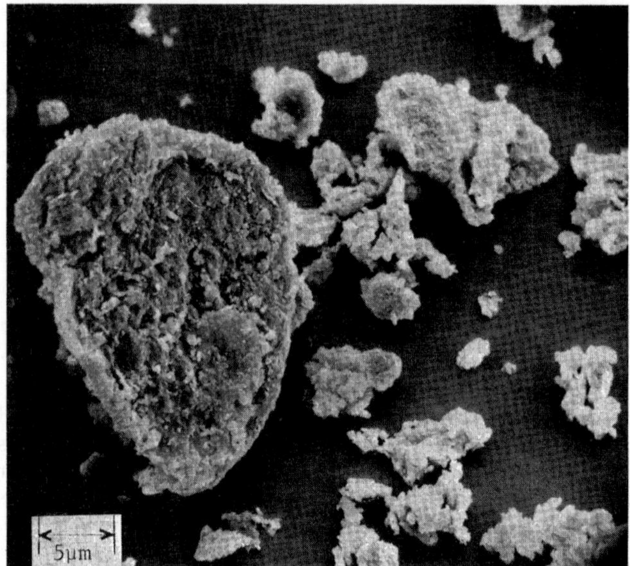

10 Typical Properties

Acid demand 6–8 mL of 0.1 N HCl is required to reduce the pH of 1 g to pH 4.

Density 2.418 g/cm^3

Moisture content 6.0–9.98%.[6] *See also* Figures 1, 2, and 3.[6]

Particle size distribution *see* Section 8.

Solubility Practically insoluble in alcohols, water, and organic solvents.

Swelling capacity Swelling properties are reversible. Magnesium aluminum silicate swells to many times its original volume in water to form colloidal dispersions, and may be dried and rehydrated any number of times.

Viscosity (dynamic) Dispersions in water at the 1–2% w/v level are thin colloidal suspensions. At 3% w/v and above, dispersions are opaque. As the concentration is increased above 3% w/v, the viscosity of aqueous dispersions increases rapidly; at 4–5% w/v, dispersions are thick, white colloidal sols, while at 10% w/v firm gels are formed. Dispersions are thixotropic at concentrations greater than 3% w/v. The viscosity of the suspension increases with heating or addition of electrolytes, and at higher concentrations with aging.

11 Stability and Storage Conditions

Magnesium aluminum silicate is stable indefinitely when stored under dry conditions. It is stable over a wide pH range, has base-exchange capacity, absorbs some organic substances, and is compatible with organic solvents.

Magnesium aluminum silicate should be stored in a well-closed container, in a cool, dry place.

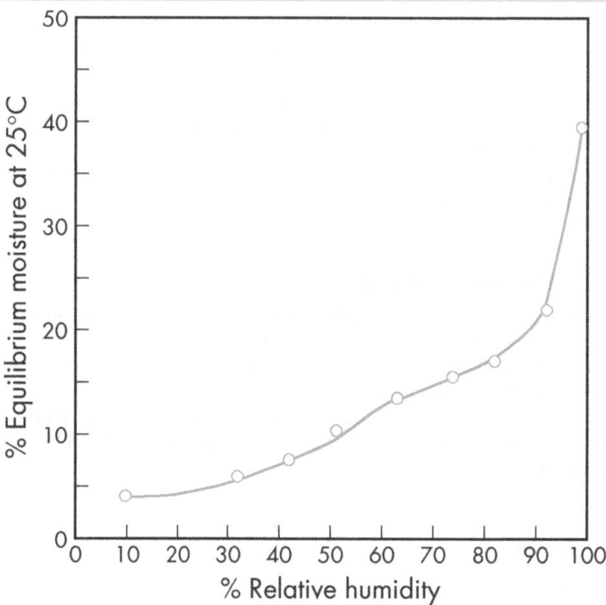

Figure 1: Equilibrium moisture content of magnesium aluminum silicate (*Veegum HV*, RT Vanderbilt Co., Inc.).

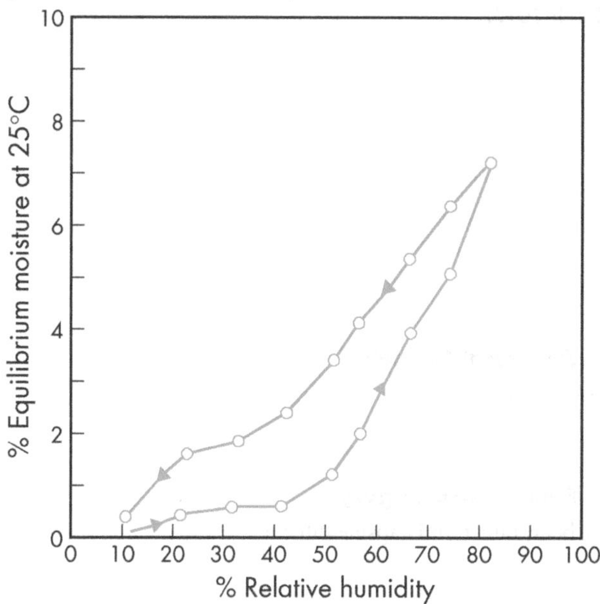

Figure 2: Sorption–desorption isotherm of magnesium aluminum silicate (*Pharmasorb*, BASF).

12 Incompatibilities

Owing to its inert nature, magnesium aluminum silicate has few incompatibilities but is generally unsuitable for acidic solutions below pH 3.5. Magnesium aluminum silicate, as with other clays, may adsorb some drugs.[7,8] This can result in low bioavailability if the drug is tightly bound or slowly desorbed, e.g. amfetamine sulfate,[4] tolbutamide,[9] warfarin sodium,[10] diazepam,[11] and diclofenac sodium.[12]

13 Method of Manufacture

Magnesium aluminum silicate is obtained from silicate ores of the montmorillonite group, which show high magnesium content. The ore is blended with water to form a slurry to remove impurities and separate out the colloidal fraction. The refined colloidal dispersion

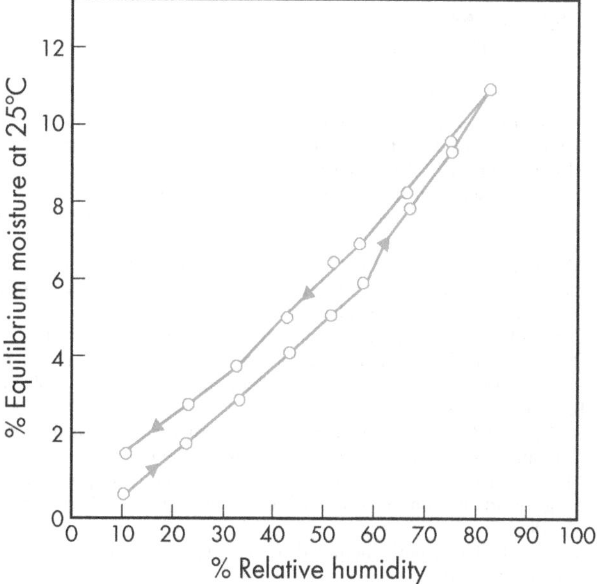

Figure 3: Sorption–desorption isotherm of magnesium aluminum silicate (*Pharmasorb Colloidal*, BASF).

is drum-dried to form a small flake, which is then micro-atomized to form various powder grades.

14 Safety

Magnesium aluminum silicate is generally regarded as nontoxic and nonirritating at the levels employed as a pharmaceutical excipient. Subacute animal feeding studies in rats and dogs fed magnesium aluminum silicate at 10% of the diet, for 90 days, were negative, including autopsy and histopathological examinations.[13]

LD_{50} (rat, oral): > 16 g/kg[14]

15 Handling Precautions

Observe normal precautions appropriate to the circumstances and quantity of material handled. Eye protection and gloves are recommended. Adequate ventilation should be provided and dust generation minimized.

16 Regulatory Status

Included in the FDA Inactive Ingredients Database (oral granules, solutions, suspensions and tablets; rectal preparations; topical preparations; vaginal preparations). Included in nonparenteral medicines licensed in the UK. Included in the Canadian Natural Health Products Ingredients Database.

17 Related Substances

Attapulgite; bentonite; kaolin; magnesium silicate; magnesium trisilicate; montmorillonite; saponite; talc.

Montmorillonite
Empirical formula $Al_2O_5{\cdot}4SiO_2{\cdot}4H_2O$
CAS number [1318-93-0]
Comments A naturally occurring silicate clay.

18 Comments

The EINECS number for magnesium aluminum silicate is 215-478-8. The PubChem Compound ID (CID) for magnesium aluminum silicate is 3084116.

19 Specific References

1 Polon JA. The mechanisms of thickening by inorganic agents. *J Soc Cosmet Chem* 1970; **21**: 347–363.
2 Farley CA, Lund W. Suspending agents for extemporaneous dispensing: evaluation of alternatives to tragacanth. *Pharm J* 1976; **216**: 562–566.
3 Attama AA, *et al.* Effect of Veegum on the suspending properties of Mucuna gum. *Boll Chem Farm* 1997; **136**: 549–553.
4 McGinity JW, Lach JL. Sustained-release applications of montmorillonite interaction with amphetamine sulfate. *J Pharm Sci* 1977; **66**: 63–66.
5 McGinity JW, Harris MR. Optimization of slow-release tablet formulations containing montmorillonite I: properties of tablets. *Drug Dev Ind Pharm* 1980; **6**: 399–410.
6 Grab FL *et al.* Magnesium aluminum silicate. In: *Handbook of Pharmaceutical Excipients*. Washington, DC and London: American Pharmaceutical Association and Pharmaceutical Society of Great Britain, 1986: 166–169.
7 McGinity JW, Lach JL. *In vitro* adsorption of various pharmaceuticals to montmorillonite. *J Pharm Sci* 1976; **65**: 896–902.
8 McGinity JW, Harris MR. Increasing dissolution rates of poorly-soluble drugs by adsorption to montmorillonite. *Drug Dev Ind Pharm* 1980; **6**: 35–48.
9 Varley AB. The generic inequivalence of drugs. *J Am Med Assoc* 1968; **206**: 1745–1748.
10 Wagner JG, *et al. In vivo* and *in vitro* availability of commercial warfarin tablets. *J Pharm Sci* 1971; **60**: 666–677.
11 Munzel K. The desorption of medicinal substances from adsorbents in oral pharmaceutical suspensions. *Acta Pharmacol Toxicol* 1971; **29**(Suppl. 3): 81–87.
12 Pongjanyakul T, *et al.* Influence of magnesium aluminium silicate on rheological, release and permeation characteristics of diclofenac sodium aqueous gels *in vitro*. *J Pharm Pharmacol* 2005; **57**: 429–434.
13 Sakai K, Moriguchi K. Effect of magnesium aluminosilicate administered to pregnant mice on pre- and postnatal development of offsprings. *Oyo Yakri* 1975; **9**: 703.
14 Sweet DV, ed. *Registry of Toxic Effects of Chemical Substances*. Cincinnati: US Department of Health, 1987.

20 General References

RT Vanderbilt Co., Inc. Technical literature: *Veegum*, the versatile ingredient for pharmaceutical formulations, 1992.
Wai K, *et al.* Applications of the montmorillonites in tablet making. *J Pharm Sci* 1966; **55**: 1244–1248.
Yokoi H, *et al.* [Effect of magnesium aluminosilicate on fluidity of pharmaceutical powders.] *J Pharm Soc Jpn* 1978; **98**: 418–425[in Japanese].

21 Author

A Palmieri.

22 Date of Revision

1 March 2012.

Magnesium Carbonate

1 Nonproprietary Names

BP: Heavy Magnesium Carbonate
Light Magnesium Carbonate
JP: Magnesium Carbonate
PhEur: Magnesium Carbonate, Heavy
Magnesium Carbonate, Light
USP: Magnesium Carbonate

2 Synonyms

Carbonic acid, magnesium salt (1:1); carbonate magnesium; *Destab*; E504; hydromagnesite; magnesii subcarbonas levis; magnesii subcarbonas ponderosus. *See* Sections 4 and 17.

3 Chemical Name and CAS Registry Number

Magnesium carbonate anhydrous [546-93-0]
See also Sections 4 and 17.

4 Empirical Formula and Molecular Weight

Magnesium carbonate is not a homogeneous material but may consist of the normal hydrate, the basic hydrate, and the anhydrous material $MgCO_3$, which is rarely encountered. Basic magnesium carbonate is probably the most common form, and may vary in formula between light magnesium carbonate, $(MgCO_3)_3{\cdot}Mg(OH)_2{\cdot}3H_2O$, and magnesium carbonate hydroxide, $(MgCO_3)_4{\cdot}Mg(OH)_2{\cdot}5H_2O$. Normal magnesium carbonate is a hydrous magnesium carbonate with a varying amount of water, $MgCO_3{\cdot}xH_2O$.

See also Sections 8, 13, and 17.

5 Structural Formula

See Section 4.

6 Functional Category

Adsorbent; tablet and capsule diluent.

7 Applications in Pharmaceutical Formulation or Technology

As an excipient, magnesium carbonate is mainly used as a directly compressible tablet diluent in concentrations up to 45% w/w. Heavy magnesium carbonate produces tablets with high crushing strength, low friability, and good disintegration properties.[1–4] However, magnesium carbonate can have varying effects on dissolution and stability.[5,6] *See also* Section 12. Magnesium carbonate has been incorporated in microsphere formulations for the purpose of stabilizing encapsulated proteins.[7] It has also been coencapsulated in poly(lactide-*co*-glycolide) microsphere formulations to neutralize acidity and enhance the immunogenicity of a contraceptive peptide vaccine.[8] Magnesium carbonate has been incorporated as a pore-forming agent in injectable poly(DL-lactide-*co*-glycolide) implants to improve the release rate of a hydrophobic drug.[9] It is also used to absorb liquids, such as flavors, in tableting processes.

Magnesium carbonate is additionally used as a food additive. *See* Table I.

Table I: Uses of magnesium carbonate.

Use	Concentration (%)
Absorbent of liquid, in tableting	0.5–1.0
Tablet excipient (direct compression)	≤45

8 Description

Magnesium carbonate occurs as light, white-colored friable masses or as a bulky, white-colored powder. It has a slightly earthy taste and is odorless but, since it has a high absorptive ability, magnesium carbonate can absorb odors.

The USP35–NF30 describes magnesium carbonate as either a basic hydrated magnesium carbonate or a normal hydrated magnesium carbonate. However, the PhEur 7.4 describes magnesium carbonate as being a hydrated basic magnesium carbonate in two separate monographs: heavy magnesium carbonate and light magnesium carbonate. The molecular formulas for heavy magnesium carbonate and light magnesium carbonate vary, but heavy magnesium carbonate may generally be regarded as the tetrahydrate $[(MgCO_3)_3 \cdot Mg(OH)_2 \cdot 4H_2O]$, while light magnesium carbonate may be regarded as the trihydrate $[(MgCO_3)_3 \cdot Mg(OH)_2 \cdot 3H_2O]$.

The molecular weights of the heavy and light forms of magnesium carbonate are 383.32 and 365.30, respectively.

9 Pharmacopeial Specifications

See Table II.

Table II: Pharmacopeial specifications for magnesium carbonate.

Test	JP XV	PhEur 7.4[(a)]	USP35–NF30
Identification	+	+	+
Characters	–	+	–
Microbial limits	–	–	+
Color of solution	–	+	–
Soluble salts	≤10.0 mg	≤1.0%	≤1.0%
Acid-insoluble substances	≤2.5 mg	≤0.05%	≤0.05%
Arsenic	≤5 ppm	≤2 ppm	≤4 ppm
Calcium	≤0.6%	≤0.75%	≤0.45%
Heavy metals	≤30 ppm	≤20 ppm	≤30 ppm
Iron	≤200 ppm	≤400 ppm	≤200 ppm
Sulfates			
Heavy form	–	≤0.6%	–
Light form	–	≤0.3%	–
Chloride	–	≤700 ppm	–
Precipitation	+	–	–
Assay (as MgO)	40.0–44.0%	40.0–45.0%	40.0–43.5%

(a) Note that except where indicated all of the PhEur test limits apply to both the heavy and light forms of magnesium carbonate.

10 Typical Properties

Angle of repose

42–50° for granular heavy magnesium carbonate;

56–60° for spray-dried heavy magnesium carbonate.[(3)]

Density (bulk)

Heavy magnesium carbonate: 0.207–0.56 g/cm^3;[(10)]

Light magnesium carbonate: ≈0.12 g/cm^3.

Density (tapped)

Heavy magnesium carbonate: 0.314–0.783 g/cm^3;[(10)]

Light magnesium carbonate: ≈0.21 g/cm^3.

Density (true) Heavy magnesium carbonate: 1.966–2.261 g/cm^3 [(10)]

Moisture content At relative humidities between 15% and 65% the equilibrium moisture content of heavy magnesium carbonate at 25°C is about 1% w/w; at relative humidities above 75% the equilibrium moisture content at 25°C is about 5% w/w.[(3)]

Particle size distribution

Heavy magnesium carbonate: 7–43 μm median particle size;[(10)]

Light magnesium carbonate: 99.95% through a 44.5 μm (#350 mesh) sieve for light magnesium carbonate.

Solubility Practically insoluble in water but soluble in water containing carbon dioxide. Insoluble in ethanol (95%) and other solvents. Magnesium carbonate dissolves and effervesces on contact with dilute acids.

Specific surface area

7.8–18.2 m^2/g for granular heavy magnesium carbonate;

4.4–15.5 m^2/g for spray-dried heavy magnesium carbonate;[(3)]

14.64–14.78 m^2/g for basic heavy magnesium carbonate.

Spectroscopy

NIR spectra *see* Figures 1 and 2.

SEM 1: Excipient: magnesium carbonate USP; manufacturer: Mallinckrodt Chemicals Co.; lot no.: KJGJ; magnification: 60×; voltage: 20 kV.

SEM 2: Excipient: magnesium carbonate USP; manufacturer: Mallinckrodt Chemicals Co.; lot no.: KJGJ; magnification: 600×; voltage: 20 kV.

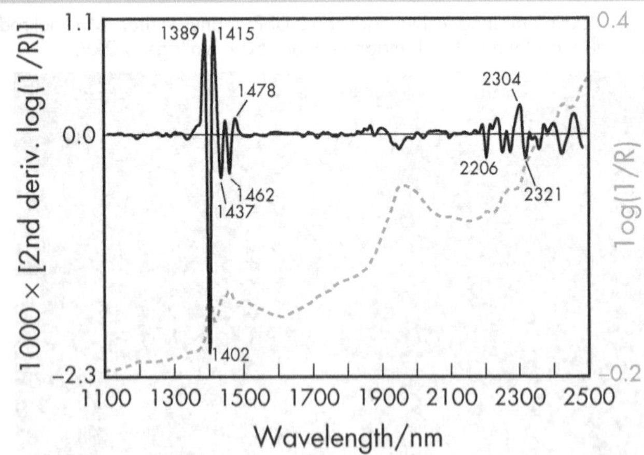

Figure 1: Near-infrared spectrum of heavy magnesium carbonate measured by reflectance.

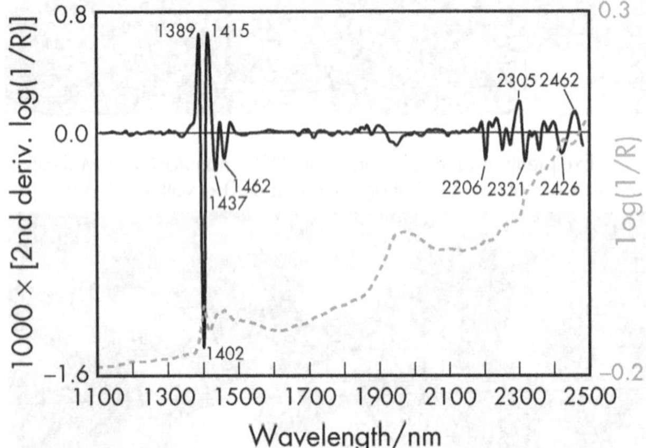

Figure 2: Near-infrared spectrum of light magnesium carbonate measured by reflectance.

11 Stability and Storage Conditions

Magnesium carbonate is stable in dry air and on exposure to light. The bulk material should be stored in a well-closed container in a cool, dry place.

12 Incompatibilities

Incompatible with phenobarbital sodium,(4,11) diazepam solution at a pH ≥ 5,(12) some binary powder mixtures,(13) lansoprazole,(5) and formaldehyde.(14) Acids will dissolve magnesium carbonate, with the liberation of carbon dioxide. Slight alkalinity is imparted to water. Magnesium carbonate was also found to increase the dissolution of acetazolamide formulations at a pH of 1.12; however, dissolution was retarded at a pH of 7.4.(6)

13 Method of Manufacture

Depending upon the manufacturing process used, the composition of the magnesium carbonate obtained may vary from normal hydrated magnesium carbonate to basic hydrated magnesium carbonate.

Light magnesium carbonate may be manufactured by saturating an aqueous suspension of dolomite, $CaMg(CO_3)_2$, with carbon dioxide under pressure. On increase of the temperature, calcium carbonate precipitates almost entirely. The filtered solution is then heated to boiling; the magnesium bicarbonate in the solution loses carbon dioxide and water, and light magnesium carbonate precipitates.

Heavy magnesium carbonate may be manufactured by mixing a hot concentrated solution of magnesium chloride or magnesium sulfate with a solution of sodium carbonate. The heavy magnesium carbonate may be either precipitated to produce a granular material or spray-dried. Varying the temperature of the reaction solutions produces heavy magnesium carbonate with differing physical properties: e.g. material with a higher specific surface area is produced at a lower reaction temperature. Low processing temperature provided the largest surface area, which produced optimum granules or spray-dried powder.(3) If dilute magnesium chloride or magnesium sulfate solutions are used for the reaction, a less dense material is produced.

Magnesium carbonates in varying states of hydration are also found as minerals in nature.

14 Safety

Magnesium carbonate is used as an excipient in oral solid-dosage pharmaceutical formulations and is generally regarded as an essentially nontoxic and nonirritant material. However, the use of magnesium salts, such as magnesium carbonate, is contraindicated in patients with renal impairment. In certain studies, magnesium carbonate has been shown to be an effective phosphate binder in short-term use for patients with chronic kidney disease, but the effects of long-term use require further study.(15) The probable oral lethal dose in humans has been estimated at 0.5–5.0 g/kg bodyweight.(14)

On contact with gastric acid, magnesium carbonate reacts in the stomach to form soluble magnesium chloride and carbon dioxide. Magnesium carbonate should therefore not be used as an antacid by those individuals whose stomachs cannot tolerate the evolution of carbon dioxide. Some magnesium is absorbed but is usually excreted in the urine. As with other magnesium salts, magnesium carbonate has a laxative effect and may cause diarrhea.

Therapeutically, the usual dose of magnesium carbonate as an antacid is 250–500 mg, and 2.0–5.0 g as a laxative.

15 Handling Precautions

Observe normal precautions appropriate to the circumstances and quantity of material handled. Magnesium carbonate may be irritant to the eyes; eye protection is recommended. OSHA standards state that IPA 8-hour time weighted airborne average is 10 mg/m^3.(14)

16 Regulatory Acceptance

GRAS listed. Accepted as a food additive in Europe. Included in the FDA Inactive Ingredients Database (oral capsules and tablets). Included in nonparenteral medicines licensed in the UK. Included in the Canadian Natural Health Products Ingredients Database.

17 Related Substances

Magnesium carbonate anhydrous; magnesium carbonate hydroxide; normal magnesium carbonate.

Magnesium carbonate anhydrous

Empirical formula $MgCO_3$
Molecular weight 84.31
CAS number [546-93-0]
Synonyms Carbonic acid, magnesium salt anhydrous (1 : 1); E504; magnesite.
Appearance Odorless, white-colored bulky powder or light, friable masses.
Melting point Decomposes at 350°C.

Magnesium carbonate hydroxide

Empirical formula $(MgCO_3)_4 \cdot Mg(OH)_2 \cdot 5H_2O$
Molecular weight 485.65
CAS number [39409-82-0]
Synonyms Carbonic acid, magnesium salt (1 : 1), mixture with magnesium hydroxide and magnesium hydrate; dypingite; E504.

Appearance Odorless, white-colored bulky powder or light, friable masses.
Melting point On heating at 700°C it is converted into magnesium oxide.
Specific gravity 1.45
Comments The EINECS number for magnesium carbonate hydroxide is 235-192-7.

Normal magnesium carbonate
Empirical formula $MgCO_3 \cdot xH_2O$
CAS number [23389-33-5]
Synonyms Carbonic acid, magnesium salt (1 : 1), hydrate; magnesium carbonate, normal hydrate; E504.
Appearance Odorless, white-colored bulky powder or light, friable masses.

18 Comments

Therapeutically, magnesium carbonate is used as an antacid.

Magnesium carbonate has been found to increase the dissolution of acetazolamide formulations at a pH of 1.12; however, dissolution was retarded at a pH of 7.4.[6] It has also been found to retard the dissolution of ciprofloxacin, sparfloxacin, and cephradine.[16–18] In addition, magnesium carbonate has been shown to alter the pharmacokinetics of halofantrine, increasing the time to reach maximum plasma concentration and reducing maximum plasma concentrations.[19] Because drug interactions can occur with a variety of antacids,[16–20] the potential for these effects should be considered when designing pharmaceutical formulations containing magnesium carbonate.

A specification for magnesium carbonate is contained in the *Food Chemicals Codex* (FCC).[21]

The EINECS number for magnesium carbonate is 208-915-9.

19 Specific References

1 Haines-Nutt RF. The compression properties of magnesium and calcium carbonates. *J Pharm Pharmacol* 1976; **28**: 468–470.
2 Armstrong NA, Cham T-M. Changes in the particle size and size distribution during compaction of two pharmaceutical powders with dissimilar consolidation mechanisms. *Drug Dev Ind Pharm* 1986; **12**: 2043–2059.
3 Cham T-M. The effect of the specific surface area of heavy magnesium carbonate on its tableting properties. *Drug Dev Ind Pharm* 1987; **13**(9–11): 1989–2015.
4 Peterson CL, *et al.* Characterization of antacid compounds containing both aluminum and magnesium. II: Codried powders. *Pharm Res* 1993; **10**(7): 1005–1007.
5 Tabata T, *et al.* Manufacturing method of stable enteric granules of a new antiulcer drug (lansoprazole). *Drug Dev Ind Pharm* 1994; **20**(9): 1661–1672.
6 Hashim F, El-Din EZ. Effect of some excipients on the dissolution of phenytoin and acetazolamide from capsule formulations. *Acta Pharm Fenn* 1989; **98**: 197–204.
7 Sandor M, *et al.* Effect of lecithin and $MgCO_3$ as additives on the enzymatic activity of carbonic anhydrase encapsulated in poly(lactide-*co*-glycolide) (PLGA) microspheres. *Biochim Biophys Acta* 2002; **1570**(1): 63–74.
8 Cui C, *et al.* Injectable polymer microspheres enhance immunogenicity of a contraceptive peptide vaccine. *Vaccine* 2007; **25**(3): 500–509.
9 Desai KG, *et al.* Effect of formulation parameters on 2-methoxyestradiol release from injectable cylindrical poly(DL-lactide-*co*-glycolide) implants. *Eur J Pharm Biopharm* 2008; **70**(1): 187–198.
10 Freitag F, Kleinebudde P. How do roll compaction/dry granulation affect the tabletting behaviour of inorganic materials? Comparison of four magnesium carbonates. *Eur J Pharm Sci* 2003; **19**: 281–289.
11 Nagavi BG, *et al.* Solid phase interaction of phenobarbitone sodium with some adjuvants. *Indian J Pharm Sci* 1983; **45**(Jul–Aug): 175–177.
12 Jain GK, Kakkar AP. Interaction study of diazepam with excipients in liquid and solid state. *Indian Drugs* 1992; **29**(Jul): 545–546.
13 Jain GK, Kakkar AP. Interaction study of diazepam with excipients in binary powder form. *Indian Drugs* 1992; **29**(Jul): 453–454.
14 Hazardous Substances Data Bank (2010). *Magnesium carbonate.* toxnet.nlm.nih.gov/cgi-bin/sis/htmlgen?HSDB (accessed 17 December 2010).
15 Spiegel DM. The role of magnesium binders in chronic kidney disease. *Semin Dial* 2007; **20**(Jul–Aug): 333–336.
16 Arayne MS, *et al.* Interactions between ciprofloxacin and antacids – dissolution and adsorption studies. *Drug Metabol Drug Interact* 2005; **21**(2): 117–129.
17 Hussain F, *et al.* Interactions between sparfloxacin and antacids – dissolution and adsorption studies. *Pak J Pharm Sci* 2006; **19**(Jan): 16–21.
18 Arayne MS, *et al.* Cephradine antacids interaction studies. *Pak J Pharm Sci* 2007; **20**(3): 179–184.
19 Aideloje SO, *et al.* Altered pharmacokinetics of halofantrine by an antacid, magnesium carbonate. *Eur J Pharm Biopharm* 1998; **463**: 299–303.
20 Sadowski DC, *et al.* Drug interactions with antacids: mechanisms and clinical significance. *Drug Safety* 1994; **116**: 395–407.
21 *Food Chemicals Codex*, 7th edn. Bethesda, MD: United States Pharmacopeia, 2010: 599.

M

20 General References

Freitag F, Kleinebudde P. How do roll compaction/dry granulation affect the tableting behaviour of inorganic materials? Microhardness of ribbons and mercury porosimetry measurements of tablets. *Eur J Pharm Sci* 2004; **22**: 325–333.
Freitag F, *et al.* Coprocessing of powdered cellulose and magnesium carbonate: direct tableting versus tableting after roll compaction/dry granulation. *Pharm Dev Technol* 2005; **10**(3): 353–362.
Jaiyeoba KT, Spring MS. The granulation of ternary mixtures: the effect of solubility of the excipients. *J Pharm Pharmacol* 1980; **32**: 1–5.
Khaled KA. Formulation and evaluation of hydrochlorothiazide liquisolid tablets. *Saudi Pharm J* 1998; **6**(Jan): 39–46.
Law MFL, Deasy PB. Effect of common classes of excipients on extrusion-spheronization. *J Microencapsul* 1997; **14**(May): 647–657.

21 Author

BF Truitt.

22 Date of Revision

1 March 2012.

Magnesium Oxide

1 Nonproprietary Names

BP: Heavy Magnesium Oxide
Light Magnesium Oxide

JP: Magnesium Oxide

PhEur: Magnesium Oxide, Heavy
Magnesium Oxide, Light

USP-NF: Magnesium Oxide

See Section 8.

2 Synonyms

Calcined magnesia; calcinated magnesite; *Descote*; E530; *Magcal*; *Magchem 100*; *Maglite*; magnesia; magnesia monoxide; magnesia usta; magnesii oxidum leve; magnesii oxidum ponderosum; *Magnyox*; *Marmag*; *Oxymag*; periclase.

3 Chemical Name and CAS Registry Number

Magnesium oxide [1309-48-4]

4 Empirical Formula and Molecular Weight

MgO 40.30

5 Structural Formula

See Section 4.

6 Functional Category

Anticaking agent; emulsifying agent; glidant; osmotic agent; tablet and capsule diluent.

7 Applications in Pharmaceutical Formulation or Technology

Magnesium oxide is used as an alkaline diluent in solid-dosage forms to modify the pH of tablets.[1] It can be added to solid-dosage forms to bind excess water and keep the granulation dry. In combination with silica, magnesium oxide can be used as an auxiliary glidant.[2] It is also used as a food additive.

8 Description

Two forms of magnesium oxide exist: a bulky form termed light magnesium oxide and a dense form termed heavy magnesium oxide. The USP35–NF30 and JP XV define both forms in a single monograph, while the BP 2012 and PhEur 7.0 have separate monographs for each form. The JP XV states that when 5 g of magnesium oxide has a volume not more than 30 mL, it may be labelled heavy magnesium oxide. For the heavy variety, 15 g has an apparent volume before settling of not more than 60 mL; for the light variety, 15 g has an apparent volume before settling of not more than 100 mL as defined by the BP 2012 and PhEur 7.4.

Both forms of magnesium oxide occur as fine, white, odorless powders. Magnesium oxide possesses a cubic crystal structure, though the BP 2012 and PhEur 7.4 describe the appearance of light magnesium oxide as an amorphous powder.

9 Pharmacopeial Specifications

See Table I.

Table I: Pharmacopeial specifications for magnesium oxide.

Test	JP XV	PhEur 7.4	USP 35–NF30
Identification	+	+	+
Characters	—	+	—
Loss on ignition	≤10.0%	≤8.0%	≤10.0%
Color of solution	—	+	—
Free alkali and soluble salts	≤0.5%	—	≤2.0%
Soluble substances	—	≤2.0%	—
Acid-insoluble substances	≤0.1%	≤0.1%	≤0.1%
Arsenic	≤10 ppm	≤4 ppm	—
Calcium	—	≤1.5%	≤1.1%
Calcium oxide	≤1.5%	—	—
Carbonate	+	—	—
Heavy metals	≤40 ppm	≤30 ppm	≤20 ppm
Iron	≤500 ppm	+	≤0.05%
Heavy magnesium oxide	—	≤0.07%	—
Light magnesium oxide	—	≤0.1%	—
Chloride	—	+	—
Heavy magnesium oxide	—	≤0.1%	—
Light magnesium oxide	—	≤0.15%	—
Fluoride	≤0.08%	—	—
Sulfate	—	≤1.0%	—
Bulk density	—	—	+
Heavy magnesium oxide	—	≥0.25g/ml	—
Light magnesium oxide	—	≤0.15g/ml	—
Assay	≥96.0%	98.0–100.5%	96.0–100.5%

10 Typical Properties

Acidity/alkalinity pH = 10.3 (saturated aqueous solution)

Boiling point 3600°C

Melting point 2800°C

Particle size distribution 99.98% less than 45 μm in size (light magnesium oxide).

Refractive index 1.735

Solubility Soluble in dilute acids and ammonium salt solutions; very slightly soluble in pure water (≈0.0086 g/100 mL at 30°C; solubility is increased by carbon dioxide); practically insoluble in ethanol (95%).

Specific gravity 3.58 g/cm^3 at 25°C (heavy magnesium oxide).

Spectroscopy

NIR spectra *see* Figure 1.

11 Stability and Storage Conditions

Magnesium oxide is stable at normal temperatures and pressures. However, it forms magnesium hydroxide in the presence of water. Magnesium oxide is hygroscopic and rapidly absorbs water and carbon dioxide on exposure to the air, the light form more readily than the heavy form.

The bulk material should be stored in an airtight container in a cool, dry place.

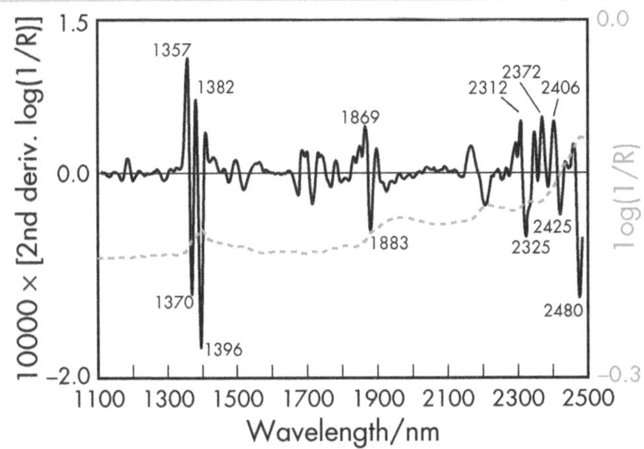

Figure 1: Near-infrared spectrum of magnesium oxide measured by reflectance.

12 Incompatibilities

Magnesium oxide is a basic compound and as such can react with acidic compounds in the solid state to form salts such as Mg(ibuprofen)$_2$ or degrade alkaline-labile drugs.[3] Adsorption of various drugs onto magnesium oxide has been reported, such as antihistamines,[4] antibiotics (especially tetracyclines),[5] salicylates,[6] atropine sulfate,[7] hyoscyamine hydrobromide,[7] acetaminophen, chloroquine;[8] and anthranilic acid derivatives have been reported to adsorb onto the surface of magnesium oxide.[9] Magnesium oxide can also complex with polymers, e.g. *Eudragit RS*, to retard drug release[10–12] and can interact in the solid state with phenobarbital sodium.[13] Magnesium oxide can also reduce the bioavailability of phenytoin,[14] trichlormethiazide,[15] and antiarrhythmics.[16] The presence of magnesium oxide can also have a negative impact on the solid-state chemical stability of drugs, such as diazepam.[17] Magnesium oxide has been used as a stabilizer for omeprazole due to its strong waterproofing effect.[18]

13 Method of Manufacture

Magnesium oxide occurs naturally as the mineral periclase. It can be manufactured by many processes. Limestone containing the mineral dolomite is calcinated at high temperatures to produce dolime, which then reacts with magnesium chloride-rich sea water to produce magnesium hydroxide and calcium chloride.[19] The magnesium hydroxide is then calcinated to produce magnesium oxide and water. In another process, mined magnesite ($MgCO_3$) is calcinated to produce magnesium oxide and carbon dioxide.[19] Purification methods include crushing and size separation, heavy-media separation, and froth flotation. Producing magnesium oxide from sea water is a process that involves heating magnesium chloride concentrated brine from the Dead Sea. The magnesium chloride decomposes into magnesium oxide and hydrochloric acid.[19] Magnesium oxide may also be produced by the thermal decomposition of magnesium chloride, magnesium sulfate, magnesium sulfite, nesquehonite, and the basic carbonate $5MgO{\cdot}4CO_2{\cdot}5H_2O$. Purification of the magnesium oxide produced through thermal degradation is carried out by filtration or sedimentation.

14 Safety

Magnesium oxide is widely used in oral formulations as an excipient and as a therapeutic agent. Therapeutically, 250–500 mg is administered orally as an antacid and 2–5 g as an osmotic laxative. Magnesium oxide is generally regarded as a nontoxic material when employed as an excipient, although adverse effects, due to its laxative action, may occur if high doses are ingested orally.

15 Handling Precautions

Observe normal precautions appropriate to the circumstances and quantity of material handled. Magnesium oxide may be harmful if inhaled, ingested, or absorbed through the skin in quantity, and is irritating to the eyes and respiratory system. Gloves, eye protection, and a dust mask or respirator are recommended. In the US and UK, the long-term (8-hour TWA) workplace exposure limits for magnesium oxide, calculated as magnesium, are 10 mg/m^3 for inhalable dust fume and 4 mg/m^3 for respirable dust.[19,20] The short-term (15-minute) limit for respirable dust is 10 mg/m^3.[19]

16 Regulatory Status

GRAS listed. Accepted for use as a food additive in Europe. Included in the FDA Inactive Ingredients Database (oral capsules, tablets, and buccal). Included in nonparenteral medicines licensed in the UK. Included in the Canadian Natural Health Products Ingredients Database.

17 Related Substances

—

18 Comments

Magnesium oxide is used therapeutically as an osmotic laxative and a magnesium supplement to treat deficiency states. It is also used as an antacid, either alone or in conjunction with aluminum hydroxide. Studies have indicated better bioavailability of magnesium from effervescent tablets than from capsules.[21]

A specification for magnesium oxide is contained in the *Food Chemicals Codex* (FCC).[22]

The EINECS number for magnesium oxide is 215-171-9. The PubChem Compound ID (CID) for magnesium oxide is 14792.

19 Specific References

1 Patel H, *et al.* The effect of excipients on the stability of levothroxine sodium pentahydrate tablets. *Int J Pharm* 2003; **264**: 35–43.
2 Kirk-Othmar. *Encyclopedia of Chemical Technology*, 5th edn, 1. New York: Wiley, 2001.
3 Tugrul TK, *et al.* Solid-state interaction of magnesium oxide and ibuprofen to form a salt. *Pharm Res* 1989; **6**(9): 804–808.
4 Nada AH, *et al. In vitro* adsorption of mepyramine maleate onto some adsorbents and antacids. *Int J Pharm* 1989; **53**: 175–179.
5 Khalil SA, *et al.* The *in vitro* adsorption of some antibiotics on antacids. *Pharmazie* 1976; **31**: 105–109.
6 Naggar VF, *et al.* The *in-vitro* adsorption of some antirheumatics on antacids. *Pharmazie* 1976; **31**: 461–465.
7 Singh A, Mital H. Adsorption of atropine sulfate and hyoscyamine hydrobromide by various antacids. *Acta Pharm Technol* 1979; **25**(3): 217–224.
8 Iwuagwu MA, Aloko KS. Adsorption of paracetamol and chloroquine phosphate by some antacids. *J Pharm Pharmacol* 1992; **44**: 655–658.
9 Monkhouse DC, Lach JL. Drug–Excipient Interactions. *Can J Pharm Sci* 1972; **7**: 29–46.
10 Shanghavi NM, *et al.* Matrix tablets of salbutamol sulfate. *Drug Dev Ind Pharm* 1990; **16**: 1955–1961.
11 Racz I, *et al.* Formulation of controlled release drug preparations with antacid effect. *Pharmazie* 1996; **51**(May): 323–327.
12 Racz I, *et al.* Effect of eudragit type polymers on the drug release from magnesium oxide granules produced by laboratory fluidization. *Drug Dev Ind Pharm* 1995; **21**(18): 2085–2096.
13 Nagavi BG, *et al.* Solid phase interaction of phenobarbitone sodium with some adjuvants. *Indian J Pharm Sci* 1983; **45**(Jul–Aug): 175–177.
14 D'Arcy PF, McElnay JC. Drug–antacid interactions: assessment of clinical importance. *Drug Intell Clin Pharm* 1987; **21**: 607–617.
15 Takahashi H, *et al.* Effect of magnesium oxide on trichlormethiazide bioavailability. *J Pharm Sci* 1985; **74**: 862–865.
16 Remon JP, *et al.* Interaction of antacids with anti-arrhythmics. Part 5. Effect of aluminum hydroxide and magnesium oxide on the bioavailability of quinidine, procainamide, and propranolol in dogs. *Arzneimittel Forschung* 1983; **33**(1): 117–120.

17 Jain G, Kakkar A. Interaction of diazepam with excipients in binary powder form. *Indian Drugs* 1992; 29(Jul): 453–454.
18 Yong CS, *et al.* Physicochemical characterization and evaluation of buccal adhesive tablets containing omeprazole. *Drug Dev Ind Pharm* 2001; 27(5): 447–455.
19 Kirk-Othmer. *Encyclopedia of Chemical Technology*, 5th edn, 15. New York: Wiley, 2007.
20 Health and Safety Executive. *EH40/2005: Workplace Exposure Limits*. Sudbury: HSE Books, 2011. http://www.hse.gov.uk/pubns/priced/eh40.pdf (accessed 16 February 2012).
21 Siener R, *et al.* Bioavailability of magnesium from different pharmaceutical formulations. *Urol Res* 2011; 39(2): 123–127.
22 *Food Chemicals Codex*, 7th edn. Bethesda, MD: United States Pharmacopeia, 2010: 603.

20 General References

—

21 Author

KE Boxell.

22 Date of Revision

16 February 2012.

Magnesium Silicate

1 Nonproprietary Names

JP: Magnesium Silicate
USP–NF: Magnesium Silicate

2 Synonyms

E553a; synthetic magnesium silicate.

3 Chemical Name and CAS Registry Number

Silicic acid, magnesium salt [1343-88-0]

4 Empirical Formula and Molecular Weight

$MgO \cdot SiO_2 \cdot xH_2O$

See also Sections 5 and 17.

5 Structural Formula

Magnesium silicate is a compound of magnesium oxide and silicon dioxide. *See also* Section 17.

The JP XV states that magnesium silicate contains not less than 45.0% of silicon dioxide (SiO_2: molecular weight 60.08) and not less than 20.0% of magnesium oxide (MgO: 40.30), and the ratio of percentage (%) of magnesium oxide to silicon dioxide is not less than 2.2 and not more than 2.5.

The USP35–NF30 describes magnesium silicate as a compound of magnesium oxide (MgO) and silicon dioxide (SiO_2) that contains not less than 15.0% of MgO and not less than 67.0% of SiO_2 calculated on the ignited basis.

6 Functional Category

Anticaking agent; glidant.

7 Applications in Pharmaceutical Formulation or Technology

Magnesium silicate is used in oral pharmaceutical formulations and food products as a glidant and an anticaking agent.

8 Description

Magnesium silicate occurs as an odorless and tasteless, fine, white-colored powder that is free from grittiness.

9 Pharmacopeial Specifications

See Table I.

Table I: Pharmacopeial specifications for magnesium silicate.

Test	JP XV	USP35–NF30
Identification	+	+
pH (10% aqueous suspension)	—	7.0–10.8
Loss on drying	—	≤15%
Soluble salts	≤0.02 g	≤3.0%
Chloride	≤0.053%	—
Free alkali	+	+
Heavy metals	≤30 ppm	≤20 µg/g
Arsenic	≤5 ppm	—
Sulfate	≤0.48%	—
Loss on ignition	≤34%	≤15%
Fluoride	—	≤10 ppm
Lead	—	≤0.001%
Acid-consuming capacity	+	—
Ratio of SiO_2 to MgO	2.2–2.5	2.5–4.5
Assay for MgO	≥20.0%	≥15%
Assay for SiO_2	≥45.0%	≥67%

10 Typical Properties

Moisture content Magnesium silicate is slightly hygroscopic.
Solubility Practically insoluble in ethanol (95%), ether, and water.

11 Stability and Storage Conditions

Magnesium silicate should be stored in a well-closed container in a cool, dry place.

12 Incompatibilities

Magnesium silicate may decrease the oral bioavailability of drugs such as mebeverine hydrochloride,[1] sucralfate, and tetracycline, via chelation or binding, when they are taken together. The dissolution rate of folic acid,[2] erythromycin stearate,[3] paracetamol,[4] and chloroquine phosphate[4] may be retarded by adsorption onto magnesium silicate. Antimicrobial preservatives, such as parabens, may be inactivated by the addition of magnesium silicate.[5]

Magnesium silicate is readily decomposed by mineral acids.

13 Method of Manufacture

Magnesium silicate may be prepared from sodium silicate and magnesium sulfate. The silicate also occurs in nature as the minerals meerschaum, parasepiolite, and sepiolite.

14 Safety

Magnesium silicate is used in oral pharmaceutical formulations and is generally regarded as an essentially nontoxic and nonirritant material.

Orally administered magnesium silicate is neutralized in the stomach to form magnesium chloride and silicon dioxide; some magnesium is absorbed. Caution should be used when greater than 50 mEq of magnesium is given daily to persons with impaired renal function, owing to the risk of hypermagnesemia.

Reported adverse effects include the formation of bladder and renal calculi following the regular use, for many years, of magnesium silicate as an antacid.[6,7]

15 Handling Precautions

Observe normal precautions appropriate to the circumstances and quantity of material handled. Eye protection is recommended.

16 Regulatory Acceptance

GRAS listed. Accepted for use as a food additive in Europe. Included in the FDA Inactive Ingredients Database (oral tablets). Included in the Canadian Natural Health Products Ingredients Database.

17 Related Substances

Magnesium aluminum silicate; magnesium metasilicate; magnesium orthosilicate; magnesium trisilicate; talc.

Magnesium metasilicate

Comments Magnesium metasilicate ($MgSiO_3$) occurs in nature as the minerals clinoenstatite, enstatite, and protoenstatite.

Magnesium orthosilicate

Comments Magnesium orthosilicate (Mg_2SiO_4) occurs in nature as the mineral forsterite.

18 Comments

A specification for magnesium silicate is contained in the *Food Chemicals Codex* (FCC).[8]

The EINECS number for magnesium silicate is 215-681-1. The PubChem Compound ID (CID) for magnesium silicate includes 518821 and 14936.

19 Specific References

1 Al-Gohary OMN. An *in vitro* study of the interaction between mebeverine hydrochloride and magnesium trisilicate powder. *Int J Pharm* 1991; **67**: 89–95.
2 Iwuagwu MA, Jideonwo A. Preliminary investigations into the in-vitro interaction of folic acid with magnesium trisilicate and edible clay. *Int J Pharm* 1990; **65**: 63–67.
3 Arayne MS, Sultana N. Erythromycin–antacid interaction. *Pharmazie* 1993; **48**: 599–602.
4 Iwuagwu MA, Aloko KS. Adsorption of paracetamol and chloroquine phosphate by some antacids. *J Pharm Pharmacol* 1992; **44**: 655–658.
5 Allwood MC. The adsorption of esters of *p*-hydroxybenzoic acid by magnesium trisilicate. *Int J Pharm* 1982; **11**: 101–107.
6 Joekes AM, *et al.* Multiple renal silica calculi. *Br Med J* 1973; **1**: 146–147.
7 Levison DA, *et al.* Silica stones in the urinary bladder. *Lancet* 1982; i: 704–705.
8 *Food Chemicals Codex*, 7th edn. Bethesda, MD: United States Pharmacopeia, 2010: 607.

20 General References

Anonymous. The silicates: attapulgite, kaolin, kieselguhr, magnesium trisilicate, pumice, talc. *Int J Pharmaceut Compound* 1998; **2**(2): 162–163.

21 Author

A Palmieri.

22 Date of Revision

1 March 2012.

M

Magnesium Stearate

1 Nonproprietary Names

BP: Magnesium Stearate

JP: Magnesium Stearate

PhEur: Magnesium Stearate

USP–NF: Magnesium Stearate

2 Synonyms

Cecavon MG 51; Dibasic magnesium stearate; *Kemilub EM-F*; magnesium distearate; magnesii stearas; magnesium octadecanoate; octadecanoic acid, magnesium salt; stearic acid, magnesium salt; *Synpro 90*.

3 Chemical Name and CAS Registry Number

Octadecanoic acid magnesium salt [557-04-0]

4 Empirical Formula and Molecular Weight

$C_{36}H_{70}MgO_4$ 591.24

The USP35–NF30 describes magnesium stearate as a compound of magnesium with a mixture of solid organic acids that consists chiefly of variable proportions of magnesium stearate and magnesium palmitate ($C_{32}H_{62}MgO_4$). The fatty acids are derived from edible sources. It contains not less than 4.0% and not more than 5.0% of Mg, calculated on the dried basis.

The PhEur 7.4 describes magnesium stearate as a mixture of magnesium salts of different fatty acids consisting mainly of stearic (octadecanoic) acid [$(C_{17}H_{35}COO)_2Mg$; M_r 591.3] and palmitic (hexadecanoic) acid [$C_{15}H_{31}COO)_2Mg$; M_r 535.1] with minor proportions of other fatty acids. It contains not less than 4.0% and not more than 5.0% of Mg (A_r 24.30), calculated with reference to the dried substance. The fatty acid fraction contains not less

40.0% of stearic acid and the sum of stearic and palmitic acid is not less than 90.0%.

5 Structural Formula

$[CH_3(CH_2)_{16}COO]_2Mg$

6 Functional Category

Tablet and capsule lubricant.

7 Applications in Pharmaceutical Formulation or Technology

Magnesium stearate is widely used in cosmetics, foods, and pharmaceutical formulations. It is primarily used as a lubricant in capsule and tablet manufacture at concentrations between 0.25% and 5.0% w/w. It is also used in barrier creams.

See also Section 18.

8 Description

Magnesium stearate is a very fine, light white, precipitated or milled, impalpable powder of low bulk density, having a faint odor of stearic acid and a characteristic taste. The powder is greasy to the touch and readily adheres to the skin.

9 Pharmacopeial Specifications

The pharmacopeial specifications for magnesium stearate have undergone harmonization of many attributes for JP, PhEur, and USP–NF.

See Table I. *See also* Section 18.

Table I: Pharmacopeial specifications for magnesium stearate.

Test	JP XV	PhEur 7.4	USP35–NF30
Identification	+	+	+
Characters	—	+	—
Microbial limits	+	+	+
Aerobic microbes	≤1000 cfu/g	≤10^3 cfu/g	≤1000 cfu/g
Fungi and yeasts	≤500 cfu/g	≤10^2 cfu/g	≤500 cfu/g
Acidity or alkalinity	+	+	+
Acid value of the fatty acid	—	195–210	—
Freezing point	—	≥53°C	—
Nickel	—	≤5 ppm	≤5 ppm
Cadmium	—	≤3 ppm	≤3 ppm
Specific surface area	—	—	+
Loss on drying	≤6.0%	≤6.0%	≤6.0%
Chloride	≤0.1%	≤0.1%	≤0.1%
Sulfate	≤1.0%	≤1.0%	≤1.0%
Lead	—	≤10 ppm	≤10 ppm
Heavy metals[(a)]	≤20 ppm	—	—
Relative stearic/palmitic content	+	+	+
Assay (dried, as Mg)	4.0–5.0%	4.0–5.0%	4.0–5.0%

(a) This test has not been fully harmonized at the time of publication.

10 Typical Properties

Crystalline forms High-purity magnesium stearate has been isolated as a trihydrate, a dihydrate, and an anhydrate.

Density (bulk) 0.159 g/cm^3

0.286 g/cm^3

092 g/cm^3

°C

rly flowing, cohesive powder.

to

than

mmercial samples);

gh purity magnesium stearate).

SEM 1: Excipient: magnesium stearate; magnification: 600×.

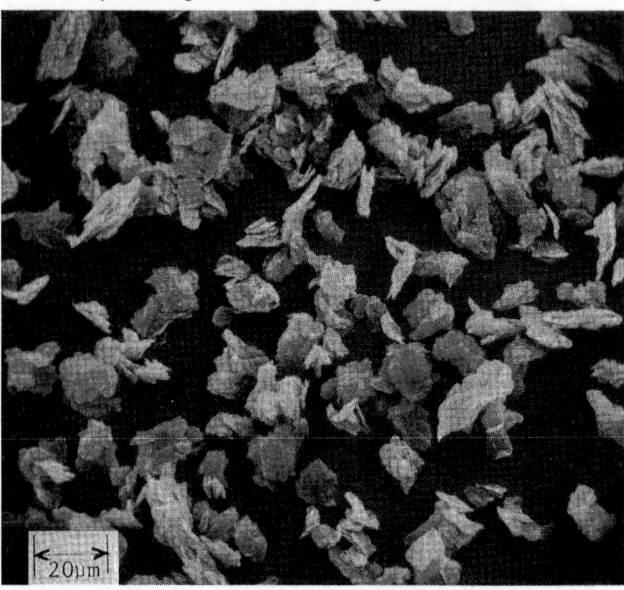

SEM 2: Excipient: magnesium stearate; magnification: 2400×.

Solubility Practically insoluble in ethanol, ethanol (95%), ether and water; slightly soluble in warm benzene and warm ethanol (95%).

Specific surface area 1.6–14.8 m^2/g

Spectroscopy

NIR spectra *see* Figure 1.

11 Stability and Storage Conditions

Magnesium stearate is stable and should be stored in an airtight container in a cool, dry place.

12 Incompatibilities

Incompatible with strong acids, alkalis, and iron salts. Avoid mixing with strong oxidizing materials. Magnesium stearate cannot be used in products containing aspirin, some vitamins, and most alkaloidal salts.

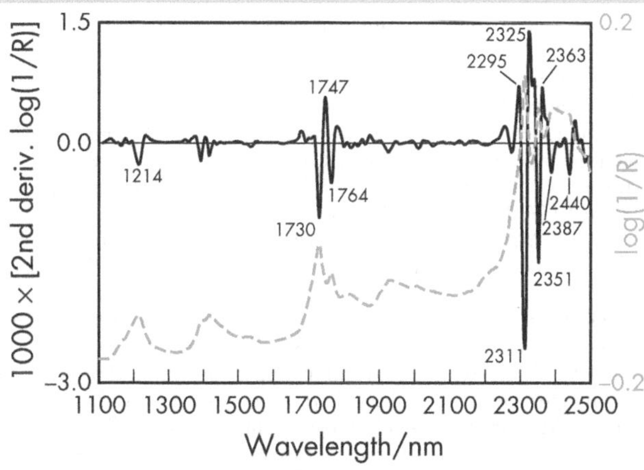

Figure 1: Near-infrared spectrum of magnesium stearate measured by reflectance.

13 Method of Manufacture

Magnesium stearate is prepared either by the interaction of aqueous solutions of magnesium chloride with sodium stearate or by the interaction of magnesium oxide, hydroxide, or carbonate with stearic acid at elevated temperatures.

14 Safety

Magnesium stearate is widely used as a pharmaceutical excipient and is generally regarded as being nontoxic following oral administration. However, oral consumption of large quantities may produce a laxative effect or mucosal irritation.

No toxicity information is available relating to normal routes of occupational exposure. Limits for heavy metals in magnesium stearate have been evaluated in terms of magnesium stearate worst-case daily intake and heavy metal composition.(1)

Toxicity assessments of magnesium stearate in rats have indicated that it is not irritating to the skin, and is nontoxic when administered orally or inhaled.(2,3)

Magnesium stearate has not been shown to be carcinogenic when implanted into the bladder of mice.(4)

LD_{50} (rat, inhalation): >2 mg/L(2)
LD_{50} (rat, oral): >10 g/kg

15 Handling Precautions

Observe normal precautions appropriate to the circumstances and quantity of material handled. Eye protection and gloves are recommended. Excessive inhalation of magnesium stearate dust may cause upper respiratory tract discomfort, coughing, and choking. Finely dispersed particles can form explosive mixtures in air. Magnesium stearate should be handled in a well-ventilated environment; a respirator is recommended. In the US, the OSHA limit is 10 mg/m^3 TWA for magnesium stearate.

16 Regulatory Acceptance

GRAS listed. Accepted as a food additive in the US and UK. Included in the FDA Inactive Ingredients Database (oral capsules, powders, and tablets; buccal and vaginal tablets; topical preparations; intravitreal implants and injections; suspensions; powder for inhalation). Included in nonparenteral medicines licensed in the UK. Included in the Canadian Natural Health Products Ingredients Database. Listed on the US TSCA inventory.

17 Related Substances

Calcium stearate; stearic acid; zinc stearate.

18 Comments

Magnesium stearate has undergone harmonization for many attributes for JP, PhEur, and USP–NF by the Pharmacopeial Discussion Group. For further information see the General Information Chapter <1196> in the USP35–NF30, the General Chapter 5.8 in PhEur 7.4, along with the 'State of Work' document on the PhEur EDQM website, and also the General Information Chapter 8 in the JP XV.

Crystalline form The existence of various crystalline forms of magnesium stearate has been established.(5–9) A trihydrate, a dihydrate, and an anhydrate have been isolated,(7,8,10,11) and an amorphous form has been observed.(12) While the hydrated forms are stable in the presence of moisture, the anhydrous form adsorbs moisture at relative humidity up to 50%, and at higher humidities rehydrates to form the trihydrate. The anhydrate can be formed by drying either of the hydrates at 105°C.(8) Spectroscopic methods have been utilized to assess the polymorphic composition of mixtures of various forms of magnesium stearate.(13)

General physical properties Physical properties of magnesium stearate can vary among batches from different manufacturers(14) because the solid-state characteristics of the powder are influenced by manufacturing variables.(6) Variations in the physical properties of different lots of magnesium stearate from the same vendor have also been observed.(14) Presumably because of these variations, it has not been possible to conclusively correlate the dissolution rate retardation with observed lubricity.(15)

However, various physical properties of different batches of magnesium stearate, such as specific surface area, particle size, crystalline structure, moisture content, and fatty acid composition, have been correlated with lubricant efficacy.(7,12,14,16–22) Due to variations in the specific surface area, the labeling states that specific surface area and the method specified for its determination should be listed on the label. Reduction in dissolution caused by the effects of magnesium stearate in some cases can be overcome by including a highly swelling disintegrant in the formulation.(23)

The functionality of a vegetable and bovine grade of magnesium stearate was compared. The dry granulated formulation containing vegetable-based magnesium stearate showed a lower ejection force than the formulation containing bovine-based magnesium stearate. There was no difference between the dissolution profiles of the tablets using the different sources of magnesium stearate.(24)

Hydrophobicity Magnesium stearate is hydrophobic and may retard the dissolution of a drug from a solid dosage form; the lowest possible concentration is therefore used in such formulations.(10,25–29) Capsule dissolution is also sensitive to both the amount of magnesium stearate in the formulation and the mixing time; higher levels of magnesium stearate and long mixing times can result in the formation of hydrophobic powder beds that do not disperse after the capsule shell dissolves.(30,31)

There is evidence to suggest that the hydrophobic nature of magnesium stearate can vary from batch to batch owing to the presence of water-soluble, surface-active impurities such as sodium stearate. Batches containing very low concentrations of these impurities have been shown to retard the dissolution of a drug to a greater extent than when using batches that contain higher levels of impurities.(15) One study related lubricity to the fatty acid composition (stearate : palmitate) of lubricant lots for tablet formulations based on compaction data and tablet material properties.(21) However, other studies have indicated that fatty acid composition has no influence on lubricant activity(7) and high-purity magnesium stearate was as effective a lubricant as the commercial material.(28) Moisture sorption at different relative humidities can result in morphological changes in the magnesium stearate.(32,33)

Lubricant properties It has not been conclusively established which form of pure magnesium stearate possesses the best lubricating properties.[6,7,12,34,35] Commercial lots of magnesium stearate generally consist of mixtures of crystalline forms.[7,9,12,14,16,34,36] Because of the possibility of conversion of crystalline forms during heating, consideration should be given to the pretreatment conditions employed when determining physical properties of magnesium stearate powders such as surface area.[37,38]

The effects of the concentration of magnesium stearate and the time of lubrication of mixtures with magnesium stearate on the content uniformity of the active ingredient in mixtures have been studied in a model mixture containing lactose and aspirin. This model indicated the presence of an interaction between magnesium stearate concentration and lubrication time. For a given magnesium stearate concentration, there was a significant reduction in the content uniformity of aspirin as the time of lubrication of the mixture with magnesium stearate was increased.[39] The extent to which magnesium stearate improves blend flowability shows a dependency on the specific grade of the filler excipient used[40] and lubrication sensitivity as assessed by tablet tensile strength was dependent on excipient particle size.[41] Magnesium stearate has been shown to be effective for alleviating formulation sticking to rolls during dry granulation by roller compaction but its presence can affect ribbon mass output compared to unlubricated blends for certain roll types[42] as well as affecting the density distribution of the ribbons.[43] In multi-component systems, mixing order has been shown to have an impact blend and tablet performance characteristics.[44] The mechanistics of magnesium stearate lubrication during tablet ejection have been examined using atomic force microscopy.[45]

M

Other An increase in the coefficient of variation of mixing and a decrease in the dissolution rate have been observed following blending of magnesium stearate with a tablet granulation. Tablet dissolution rate and crushing strength decreased as the time of blending increased; and magnesium stearate may also increase tablet friability. Blending times with magnesium stearate should therefore be carefully controlled[46–62] and lubrication kinetics and sensitivity have been shown to vary depending on the filler composition.[63] A variety of online analytical techniques have been investigated to monitor magnesium stearate in powder blends and tablets.[64–66] Inverse gas chromatography has been used to examine the surface coverage of magnesium stearate on powder blends[67] and also to correlate surface energy distribution of magnesium stearate treated powders with flow and cohesive properties.[68,69] Additionally, other advanced surface characterization techniques (XPS, ToF-SIMS) have been used to assess the coating of excipient substrates by magnesium stearate.[70] Magnesium stearate also affects the flow properties of blends.[71]

The impact of magnesium stearate levels on tablet compaction properties and performance of roller compacted granulations has been examined.[72–74] Variation in magnesium stearate physical properties can have a small but observable impact on tablet performance properties in dry granulation processing.[75] In other compaction studies performed with granules, magnesium stearate has been shown to exert an influence on granule relaxation and may help to prevent capping.[76]

Magnesium stearate has been investigated for use in inhalation powders to control their performance.[77]

A specification for magnesium stearate is included in the *Food Chemicals Codex* (FCC).[78]

The EINECS number for magnesium stearate is 209-150-3.

19 Specific References

1 Chowhan ZT. Harmonization of excipient standards. In: Weiner ML, Kotkoskie LA, eds. *Excipient Toxicity and Safety.* New York: Marcel Dekker, 2000; 321–354.
2 Anonymous. Final report of the safety assessment of lithium stearate, aluminum distearate, aluminum stearate, aluminum tristearate, ammonium stearate, calcium stearate, magnesium stearate, potassium stearate, sodium stearate, and zinc stearate. *J Am Coll Toxicol* 1982; **1**: 143–177.
3 Sondergaard D, *et al.* Magnesium stearate given perorally to rats: a short term study. *Toxicology* 1980; **17**: 51–55.
4 Boyland E, *et al.* Further experiments on implantation of materials into the urinary bladder of mice. *Br J Cancer* 1964; **18**: 575–581.
5 Muller BW. The pseudo-polymorphism of magnesium stearate. *Zbl Pharm* 1977; **116**(12): 1261–1266.
6 Miller TA, York P. Physical and chemical characteristics of some high purity magnesium stearate and palmitate powders. *Int J Pharm* 1985; **23**: 55–67.
7 Ertel KD, Carstensen JT. Chemical, physical, and lubricant properties of magnesium stearate. *J Pharm Sci* 1988; **77**: 625–629.
8 Ertel KD, Carstensen JT. An examination of the physical properties of pure magnesium stearate. *Int J Pharm* 1988; **42**: 171–180.
9 Wada Y, Matsubara T. Pseudo-polymorphism and crystalline transition of magnesium stearate. *Thermochim Acta* 1992; **196**: 63–84.
10 Levy G, Gumtow RH. Effect of certain formulation factors on dissolution rate of the active ingredient III: tablet lubricants. *J Pharm Sci* 1963; **52**: 1139–1144.
11 Sharpe SA, *et al.* Physical characterization of the polymorphic variations of magensium stearate and magnesium palmitate hydrate species. *Struct Chem* 1997; **8**(1): 73–84.
12 Leinonen UI, *et al.* Physical and lubrication properties of magnesium stearate. *J Pharm Sci* 1992; **81**(12): 1194–1198.
13 Kauffman JF, *et al.* Near infrared spectroscopy of magnesium stearate hydrates and multivariate calibration of pseudopolymorph composition. *J Pharm Sci* 2008; **97**(7): 2757–2767.
14 Barra J, Somma R. Influence of the physicochemical variability of magnesium stearate on its lubricant properties: possible solutions. *Drug Dev Ind Pharm* 1996; **22**(11): 1105–1120.
15 Billany MR, Richards JH. Batch variation of magnesium stearate and its effect on the dissolution rate of salicylic acid from solid dosage forms. *Drug Dev Ind Pharm* 1982; **8**: 497–511.
16 Dansereau R, Peck GE. The effect of the variability in the physical and chemical properties of magnesium stearate on the properties of compressed tablets. *Drug Dev Ind Pharm* 1987; **13**: 975–999.
17 Frattini C, Simioni L. Should magnesium stearate be assessed in the formulation of solid dosage forms by weight or by surface area? *Drug Dev Ind Pharm* 1984; **10**: 1117–1130.
18 Bos CE, *et al.* Lubricant sensitivity in relation to bulk density for granulations based on starch or cellulose. *Int J Pharm* 1991; **67**: 39–49.
19 Phadke DS, Eichorst JL. Evaluation of particle size distribution and specific surface area of magnesium stearate. *Drug Dev Ind Pharm* 1991; **17**: 901–906.
20 Steffens KJ, Koglin J. The magnesium stearate problem. *Manuf Chem* 1993; **64**(12): 16–19.
21 Marwaha SB, Rubinstein MH. Structure-lubricity evaluation of magnesium stearate. *Int J Pharm* 1988; **43**(3): 249–255.
22 Rao KP, *et al.* Impact of solid-state properties on lubrication efficacy of magnesium stearate. *Pharm Dev Technol* 2005; **10**(3): 423–437.
23 Desai DS, *et al.* Physical interactions of magnesium stearate with starch-derived disintegrants and their effects on capsule and tablet dissolution. *Int J Pharm* 1993; **91**(2–3): 217–226.
24 Hamad ML, *et al.* Functionality of magnesium stearate derived from bovine and vegetable sources: dry granulated tablets. *J Pharm Sci* 2008; **97**(12): 5328–5340.
25 Ganderton D. The effect of distribution of magnesium stearate on the penetration of a tablet by water. *J Pharm Pharmacol* 1969; **21**(Suppl.): 9S–18S.
26 Caldwell HC. Dissolution of lithium and magnesium from lithium carbonate capsules containing magnesium stearate. *J Pharm Sci* 1974; **63**: 770–773.
27 Chowhan ZT, *et al.* Tablet-to-tablet dissolution variability and its relationship to the homogeneity of a water-soluble drug. *Drug Dev Ind Pharm* 1982; **8**: 145–168.
28 Lerk CF, *et al.* Interaction of tablet disintegrants and magnesium stearate during mixing II: effect on dissolution rate. *Pharm Acta Helv* 1982; **57**: 282–286.
29 Hussain MSH, *et al.* Effect of commercial and high purity magnesium stearates on in-vitro dissolution of paracetamol DC tablets. *Int J Pharm* 1992; **78**: 203–207.
30 Samyn JC, Jung WY. *In vitro* dissolution from several experimental capsule formulations. *J Pharm Sci* 1970; **59**: 169–175.

31 Murthy KS, Samyn JC. Effect of shear mixing on *in vitro* drug release of capsule formulations containing lubricants. *J Pharm Sci* 1977; **66**: 1215–1219.

32 Swaminathan V, Kildisig DO. An examination of the moisture sorption characteristics of commercial magnesium stearate. *AAPS PharmSciTech* 2001; **2**(4): 73–79.

33 Bracconi P, *et al.* Structural properties of magnesium stearate pseudopolymorphs: effect of temperature. *Int J Pharm* 2003; **262**(1–2): 109–124.

34 Muller BW. Polymorphism of magnesium stearate and the influence of the crystal structure on the lubricating behavior of excipients. *Acta Pharm Suec* 1981; **18**: 74–75.

35 Okoye P, Wu SH. Lubrication of direct-compressible blends with magnesium stearate monohydrate and dihydrate. *Pharm Technol* 2007; **31**(9): 116–129.

36 Brittain HG. Raw materials. *Drug Dev Ind Pharm* 1989; **15**(13): 2083–2103.

37 Phadke DS, Collier JL. Effect of degassing temperature on the specific surface area and other physical properties of magnesium stearate. *Drug Dev Ind Pharm* 1994; **20**(5): 853–858.

38 Koivisto M, *et al.* Effect of temperature and humidity on vegetable grade magnesium stearate. *Powder Technol* 2004; **147**(1–3): 79–85.

39 Swaminathan V, Kildsig DO. Effect of magnesium stearate on the content uniformity of active ingredient in pharmaceutical powder mixtures. *AAPS PharmSciTech* 2002; 3(3): 27–31.

40 Soppela I, *et al.* Investigation of the powder flow behavior of binary mixtures of microcrystalline celluloses and paracetamol. *Journal of Excipients and Food Chemicals* 2010; **1**(1): 55–67.

41 Almaya A, Aburub A. Effect of particle size on compaction of materials with different deformation mechanisms with and without lubricants. *AAPS PharmSciTech* 2008; **9**(2): 414–418.

42 Dawes J, *et al.* An investigation into the impact of magnesium stearate on powder feeding during roller compaction. *Drug Dev Ind Pharm* 2012; **38**(1): 111–122.

43 Miguelez-Moran AM, *et al.* The effect of lubrication on density distributions of roller compacted ribbons. *Int J Pharm* 2008; **362**(1–2): 52–59.

44 Pingali K, *et al.* Mixing order of glidant and lubricant - Influence on powder and tablet properties. *Int J Pharm* 2011; **409**(1–2): 269–277.

45 Weber D, *et al.* Quantification of Lubricant Activity of Magnesium Stearate by Atomic Force Microscopy. *Drug Dev Ind Pharm* 2008; **34**(10): 1097–1099.

46 Ragnarsson G, *et al.* The influence of mixing time and colloidal silica on the lubricating properties of magnesium stearate. *Int J Pharm* 1979; **3**: 127–131.

47 Bolhuis GK, *et al.* Mixing action and evaluation of tablet lubricants in direct compression. *Drug Dev Ind Pharm* 1980; **6**: 573–589.

48 Bossert J, Stamm A. Effect of mixing on the lubrication of crystalline lactose by magnesium stearate. *Drug Dev Ind Pharm* 1980; **6**: 573–589.

49 Bolhuis GK, *et al.* Interaction of tablet disintegrants and magnesium stearate during mixing I: effect on tablet disintegration. *J Pharm Sci* 1981; **70**: 1328–1330.

50 Sheikh-Salem M, Fell JT. The influence of magnesium stearate on time dependent strength changes in tablets. *Drug Dev Ind Pharm* 1981; **7**: 669–674.

51 Stewart PJ. Influence of magnesium stearate on the homogeneity of a prednisone granule ordered mix. *Drug Dev Ind Pharm* 1981; **7**: 485–495.

52 Jarosz PJ, Parrott EL. Effect of tablet lubricants on axial and radial work of failure. *Drug Dev Ind Pharm* 1982; **8**: 445–453.

53 Mitrevej KT, Augsburger LL. Adhesion of tablets in a rotary tablet press II: effects of blending time, running time, and lubricant concentration. *Drug Dev Ind Pharm* 1982; **8**: 237–282.

54 Khan KA, *et al.* The effect of mixing time of magnesium stearate on the tableting properties of dried microcrystalline cellulose. *Pharm Acta Helv* 1983; **58**: 109–111.

55 Johansson ME. Investigations of the mixing time dependence of the lubricating properties of granular and powdered magnesium stearate. *Acta Pharm Suec* 1985; **22**: 343–350.

56 Johansson ME. Influence of the granulation technique and starting material properties on the lubricating effect of granular magnesium stearate. *J Pharm Pharmacol* 1985; **37**: 681–685.

57 Chowhan ZT, Chi LH. Drug–excipient interactions resulting from powder mixing III: solid state properties and their effect on drug dissolution. *J Pharm Sci* 1986; **75**: 534–541.

58 Chowhan ZT, Chi LH. Drug–excipient interactions resulting from powder mixing IV: role of lubricants and their effect on *in vitro* dissolution. *J Pharm Sci* 1986; **75**: 542–545.

59 Johansson ME, Nicklasson M. Influence of mixing time, particle size and colloidal silica on the surface coverage and lubrication of magnesium stearate. In: Rubinstein MH, ed. *Pharmaceutical Technology: Tableting Technology.* Chichester: Ellis Horwood, 1987: 43–50.

60 Wang LH, Chowhan ZT. Drug–excipient interactions resulting from powder mixing V: role of sodium lauryl sulfate. *Int J Pharm* 1990; **60**: 61–78.

61 Muzikova J, Horacek J. The dry binders, *Vivapur 102*, *Vivapur 12* and the effect of magnesium stearate on the strength of tablets containing these substances. *Ceske Slov Farm* 2003; **52**(4): 176–180.

62 Muzikova J. Effect of magnesium stearate on the tensile strength of tablets made with the binder *Prosolv SMCC 90*. *Ceska Slow Farm* 2002; **51**(1): 41–43.

63 Kushner J IV, Moore F. Scale-up model describing the impact of lubrication on tablet tensile strength. *Int J Pharm* 2010; **399**(1–2): 19–30.

64 Aguirre-Mendez C, Romanach RJ. A Raman spectroscopic method to monitor magnesium stearate in blends and tablets. *Pharmaceut Tech Eur* 2007; **19**(9): 53–61.

65 St-Onge L, *et al.* Rapid quantitative analysis of magnesium stearate in tablets using laser-induced breakdown spectroscopy. *J Pharm Pharmaceut Sci* 2005; 8(2): 272–288.

66 Duong N-H, *et al.* A homogeneity study using NIR spectroscopy: tracking magnesium stearate in Bohle bin-blender. *Drug Dev Ind Pharm* 2003; **29**(6): 679–687.

67 Swaminathan V, *et al.* Measurement of the surface energy of lubricated pharmaceutical powders by inverse gas chromatography. *Int. J. Pharm* 2006; **312**(1–2): 158–165.

68 Das SC, *et al.* Use of surface energy distributions by inverse gas chromatography to understand mechanofusion processing and functionality of lactose coated with magnesium stearate. *Eur J Pharm Sci* 2011; **43**(4): 325–333.

69 Zhou Q, *et al.* Characterization of the surface properties of a model pharmaceutical fine powder modified with a pharmaceutical lubricant to improve flow via a mechanical dry coating approach. *J Pharm Sci* 2011; **100**(8): 3421–3430.

70 Zhou Q, *et al.* Investigation of the extent of surface coating via mechanofusion with varying additive levels and the influences on bulk powder flow properties. *Int J Pharm* 2011; **413**(1–2): 36–43.

71 Faqih AM, *et al.* Effect of moisture and magnesium stearate concentration on flow properties of cohesive granular materials. *Int J Pharm* 2007; **336**(2): 338–345.

72 Wurster DE, *et al.* The influence of magnesium stearate on the Hiestand tableting indices and other related mechanical properties of maltodextrins. *Pharm Dev Technol* 2005; **10**(4): 461–466.

73 Likitlersuang S, *et al.* The effect of binary mixture composition and magnesium stearate concentration on the Hiestand tableting indices and other related mechanical properties. *Pharm Dev Technol* 2007; **12**(5): 533–541.

74 He X, *et al.* Mechanistic study of the effect of roller compaction and lubricant on tablet mechanical strength. *J Pharm Sci* 2007; **96**(5): 1342–1355.

75 Kushner J IV, *et al.* Examining the impact of excipient material property variation on drug product quality attributes: A quality-by-design study for a roller compacted, immediate release tablet. *J Pharm Sci* 2011; **100**(6): 2222–2239.

76 Ebba F, *et al.* Stress relaxation studies of granules as a function of different lubricants. *Eur J Pharm Biopharm* 2001; **52**(2): 211–220.

77 Guchardi R, *et al.* Influence of fine lactose and magnesium stearate on low dose dry powder inhaler formulations. *Int J Pharm* 2008; **348**(1–2): 10–17.

78 *Food Chemicals Codex*, 7th edn. Bethesda, MD: United States Pharmacopeia, 2011.

20 General References

Bohidar NR, *et al.* Selecting key pharmaceutical formulation factors by regression anaysis. *Drug Dev Ind Pharm* 1979; **5**: 175–216.

Butcher AE, Jones TM. Some physical characteristics of magnesium stearate. *J Pharm Pharmacol* 1972; **24**: 1P–9P.

European Directorate for the Quality of Medicines and Healthcare (EDQM). European Pharmacopoeia – State Of Work Of International

M

Harmonisation. *Pharmeuropa* 2011; 23(4): 713–714. www.edqm.eu/site/-614.html (accessed 2 December 2011).
Ford JL, Rubinstein MH. An investigation into some pharmaceutical interactions by differential scanning calorimetry. *Drug Dev Ind Pharm* 1981; 7: 675–682.
Johansson ME. Granular magnesium stearate as a lubricant in tablet formulations. *Int J Pharm* 1984; **21**: 307–315.
Jones TM. The effect of glidant addition on the flowability of bulk particulate solids. *J Soc Cosmet Chem* 1970; **21**: 483–500.
Pilpel N. Metal stearates in pharmaceuticals and cosmetics. *Manuf Chem Aerosol News* 1971; 42(10): 37–40.
York P. Tablet lubricants. In: Florence AT, ed. *Materials Used in Pharmaceutical Formulation.* London: Society of Chemical Industry, 1984: 37–70.
Zanowiak P. Lubrication in solid dosage form design and manufacture. In: Swarbick J, Boylan JC, eds. *Encyclopedia of Pharmaceutical Technology.* 9. New York: Marcel Dekker, 1990: 87–112.

21 Authors

LV Allen Jr, PE Luner.

22 Date of Revision

23 February 2012.

Magnesium Trisilicate

1 Nonproprietary Names

BP: Magnesium Trisilicate
PhEur: Magnesium Trisilicate
USP–NF: Magnesium Trisilicate

2 Synonyms

E553a; magnesii trisilicas; magnesium mesotrisilicate; silicic acid, magnesium salt (1 : 2), hydrate.

3 Chemical Name and CAS Registry Number

Magnesium trisilicate anhydrous [14987-04-3]
Magnesium trisilicate hydrate [39365-87-2]

4 Empirical Formula and Molecular Weight

$Mg_2Si_3O_8 \cdot xH_2O$ 260.86 (anhydrous)

The USP35–NF30 describes magnesium trisilicate as a compound of magnesium oxide (MgO) and silicon dioxide (SiO_2) with varying proportions of water. It contains not less than 20.0% of magnesium oxide and not less than 45.0% of silicon dioxide.

The PhEur 7.4 similarly describes magnesium trisilicate as having a variable composition corresponding to the approximate formula $Mg_2Si_3O_8 \cdot xH_2O$. It contains not less than 29.0% of magnesium oxide and not less than the equivalent of 65.0% of silicon dioxide, both calculated with reference to the ignited substance.

5 Structural Formula

See Section 4.

6 Functional Category

Glidant.

7 Applications in Pharmaceutical Formulation or Technology

Magnesium trisilicate is used in oral pharmaceutical formulations and food products as a glidant.

8 Description

Magnesium trisilicate occurs as an odorless and tasteless, fine, white-colored powder that is free from grittiness.

9 Pharmacopeial Specifications

See Table I.

Table I: Pharmacopeial specifications for magnesium trisilicate.

Test	PhEur 7.4	USP35–NF30
Identification	+	+
Characters	+	—
Ratio of SiO_2 to MgO	—	2.10–2.37
Loss on ignition	17.0–34.0%	17.0–34.0%
Water-soluble salts	≤1.5%	≤1.5%
Chloride	≤500 ppm	≤0.055%
Sulfates	≤0.5%	≤0.5%
Alkalinity	+	+
Arsenic	≤4 ppm	≤8 ppm
Heavy metals	≤40 ppm	≤0.003%
Acid-absorbing capacity	≥100.0 mL[a]	140–160 mL[a]
Assay of MgO	≥29.0%[b]	≥20.0%
Assay of SiO_2	≥65.0%[b]	≥45.0%

(a) Of 0.1 N hydrochloric acid per gram.
(b) With reference to the ignited substance.

10 Typical Properties

Moisture content Magnesium trisilicate is slightly hygroscopic. At relative humidities of 15–65%, the equilibrium moisture content at 25°C is 17–23% w/w; at relative humidities of 75–95%, the equilibrium moisture content is 24–30% w/w.

Solubility Practically insoluble in diethyl ether, ethanol (95%), and water.

Spectroscopy

NIR spectra *see* Figure 1.

11 Stability and Storage Conditions

Magnesium trisilicate is stable if stored in a well-closed container in a cool, dry place.

12 Incompatibilities

Magnesium trisilicate, when taken with drugs such as mebeverine hydrochloride,[2] proguanil,[3] norfloxacin,[4] sucralfate, and tetracycline, may cause a reduction in bioavailability via binding or chelation. The dissolution rate of folic acid,[5] erythromycin stearate,[6] acetaminophen (paracetamol), and chloroquine phos-

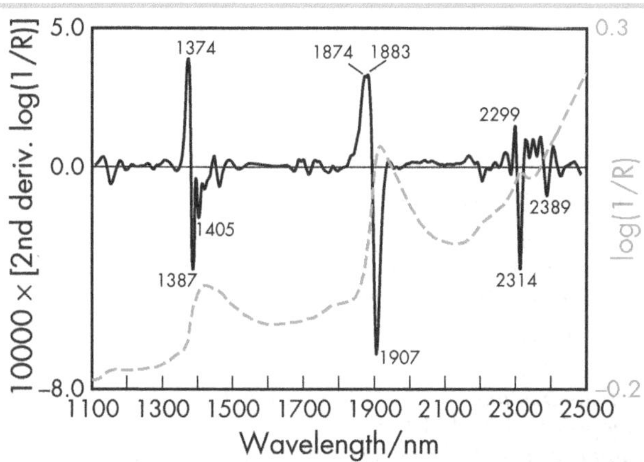

Figure 1: Near-infrared spectrum of magnesium trisilicate measured by reflectance.

phate[7] may be retarded by adsorption onto magnesium trisilicate. Antimicrobial preservatives, such as the parabens, may be inactivated by the addition of magnesium trisilicate.[8]

Magnesium trisilicate is also readily decomposed by mineral acids.

13 Method of Manufacture

Magnesium trisilicate may be prepared from sodium silicate and magnesium sulfate. It also occurs in nature as the minerals meerschaum, parasepiolite, and sepiolite.

14 Safety

Magnesium trisilicate is used in oral pharmaceutical formulations and is generally regarded as an essentially nontoxic and nonirritant material.

When administered orally, magnesium trisilicate is neutralized in the stomach to form magnesium chloride and silicon dioxide; some magnesium may be absorbed. Caution should be used when concentrations greater than 50 mEq of magnesium are given daily to persons with impaired renal function, owing to the risk of hypermagnesemia.

Reported adverse effects include the potential for osmotic diarrhea in the elderly using antacids containing magnesium trisilicate;[9] and the potential for the formation of bladder and renal calculi following the long-term use of magnesium trisilicate as an antacid.[10,11]

15 Handling Precautions

Observe normal precautions appropriate to the circumstances and quantity of material handled. Eye protection is recommended.

16 Regulatory Status

GRAS listed. Accepted for use as a food additive in Europe. Included in the FDA Inactive Ingredients Database (oral tablets). Included in nonparenteral medicines licensed in the UK. Included in the Canadian Natural Health Products Ingredients Database.

17 Related Substances

Calcium silicate; magnesium aluminum silicate; magnesium silicate; magnesium trisilicate anhydrous; talc.

Magnesium trisilicate anhydrous

Empirical formula $Mg_2Si_3O_8$
Molecular weight 260.86
CAS number [14987-04-3]

18 Comments

Magnesium trisilicate is regarded as a type of magnesium silicate. The European food additive code E553a has been applied to both.

Therapeutically, up to about 2 g of magnesium trisilicate may be taken daily as an antacid and it has also been studied for the treatment of ciprofloxacin overdose or toxicity.[11]

The EINECS number for magnesium trisilicate is 239-076-7. The PubChem Compound ID (CID) for magnesium trisilicate is 5311266.

19 Specific References

1 Al-Gohary OMN. An *in vitro* study of the interaction between mebeverine hydrochloride and magnesium trisilicate powder. *Int J Pharm* 1991; **67**: 89–95.
2 Onyeji CO, Babalola CP. The effect of magnesium trisilicate on proguanil absorption. *Int J Pharm* 1993; **100**: 249–252.
3 Okhamafe AO, *et al.* Pharmacokinetic interactions of norfloxacin with some metallic medicinal agents. *Int J Pharm* 1991; **68**: 11–18.
4 Iwuagwu MA, Jideonwo A. Preliminary investigations into the in-vitro interaction of folic acid with magnesium trisilicate and edible clay. *Int J Pharm* 1990; **65**: 63–67.
5 Arayne MS, Sultana N. Erythromycin–antacid interaction. *Pharmazie* 1993; **48**: 599–602.
6 Iwuagwu MA, Aloko KS. Adsorption of paracetamol and chloroquine phosphate by some antacids. *J Pharm Pharmacol* 1992; **44**(8): 655–658.
7 Allwood MC. The adsorption of esters of *p*-hydroxybenzoic acid by magnesium trisilicate. *Int J Pharm* 1982; **11**: 101–107.
8 Ratnaike RN, Jones TE. Mechanisms of drug-induced diarrhoea in the elderly. *Drugs & Aging* 1998; **13**: 245–253.
9 Joekes AM, *et al.* Multiple renal silica calculi. *Br Med J* 1973; **1**: 146–147.
10 Levison DA, *et al.* Silica stones in the urinary bladder. *Lancet* 1982; **i**: 704–705.
11 Ofoefule SI, Okonta M. Adsorption studies of ciprofloxacin: evaluation of magnesium trisilicate, kaolin and starch as alternatives for the management of ciprofloxacin poisoning. *Boll Chim Farm* 1999; **138**: 239–242.

20 General References

Anonymous. The silicates: attapulgite, kaolin, kieselguhr, magnesium trisilicate, pumice, talc. *Int J Pharm Compound* 1998; 2(2): 162–163.

21 Author

AS Kearney.

22 Date of Revision

1 March 2012.

M

Maleic Acid

1 Nonproprietary Names

BP: Maleic Acid
PhEur: Maleic Acid
USP–NF: Maleic Acid

2 Synonyms

Acidum maleicum; *cis*-butenedioic acid; *cis*-2-butenedioic acid; (*Z*)-2-butenedioic acid; *cis*-ethene-1,2-dicarboxylic acid; *cis*-1,2-ethylenedicarboxylic acid; *cis*-maleic acid; maleinic acid; toxilic acid.

3 Chemical Name and CAS Registry Number

Z-But-2-enedioic acid [110-16-7]

4 Empirical Formula and Molecular Weight

$C_4H_4O_4$ 116.07

5 Structural Formula

6 Functional Category

Acidulant; buffering agent.

7 Applications in Pharmaceutical Formulation or Technology

Maleic acid is used in the pharmaceutical industry as a pH modifier and a buffering agent.[1–3] It is used in oral, topical, and parenteral pharmaceutical formulations in addition to food products. Maleic acid is also used to prevent rancidity of oils and fats; a ratio of 1 : 10 000 is usually sufficient to retard rancidity. *See also* Section 18.

8 Description

Maleic acid occurs as a white crystalline (monoclinic) powder and possesses a faint acidulous odor and an astringent taste.

9 Pharmacopeial Specifications

See Table I.

Table I: Pharmacopeial specifications for maleic acid.

Test	PhEur 7.4	USP35–NF30
Identification	+	+
Characters	+	–
Appearance of solution	+	+
Heavy metals	≤10 ppm	≤10 ppm
Fumaric acid	≤1.5%	≤1.5%
Iron	≤5 ppm	≤5 ppm
Residue on ignition	–	≤0.1%
Water	≤2.0%	≤2.0%
Assay (anhydrous basis)	99.0–101.0%	99.0–101.0%

10 Typical Properties

Acidity/alkalinity pH 2 (5% w/v aqueous solution at 25°C)
Boiling point 135°C (with decomposition)
Dissociation constants

$pK_{a1} = 1.91$;

$pK_{a2} = 6.33$.

Heat of combustion 1356.3 kJ/mol (324.2 kcal/mol)
Melting point 130–134°C
Partition coefficient $\log K_{ow} = -0.48$ (octanol/water)
Solubility see Table II.[4]
Specific gravity 1.590 (20°C)

Table II: Solubility of maleic acid.

Solvent	Solubility at 20°C
Benzene	1 in 4167
Carbon tetrachloride	1 in 50 000
Chloroform	1 in 909
Diethyl ether	1 in 13.2
Water	1 in 2.05

11 Stability and Storage Conditions

Maleic acid converts into the much higher-melting fumaric acid (mp: 287°C) when heated to a temperature slightly above its melting point.[5]

Maleic acid is combustible when exposed to heat or flame. The bulk material should be stored in airtight glass containers and protected from light. It is recommended not to store it above 25°C.

12 Incompatibilities

Maleic acid can react with oxidizing materials. Aqueous solutions are corrosive to carbon steels.

13 Method of Manufacture

Maleic anhydride is the main source of maleic acid produced by hydration. Maleic anhydride is prepared commercially by the oxidation of benzene or by the reaction of butane with oxygen in the presence of a vanadium catalyst.

14 Safety

Maleic acid is generally regarded as a nontoxic and nonirritant material when used at low levels as an excipient. However, in pure form it is considered very hazardous in the case of eye contact, which can result in corneal damage. It is also hazardous with respect to skin contact and inhalation. Skin contact can produce inflammation and blistering, with the amount of tissue damage dependent on the length of contact.

LD_{50} (mouse, oral): 2.40 g/kg[6]
LD_{50} (rabbit, skin): 1.56 g/kg
LD_{50} (rat, oral): 0.708 g/kg

15 Handling Precautions

Observe normal precautions appropriate to the circumstances and quantity of the material handled. Gloves, eye protection, and approved or certified respirators should be employed.

16 Regulatory Status

Included in the FDA Inactive Ingredients Database (IM and IV injections; oral tablets and capsules; topical applications). Included in nonparenteral and parenteral medicines licensed in the UK.

17 Related Substances

Fumaric acid.

18 Comments

Maleic acid and fumaric acid are the simplest unsaturated carboxylic diacids. These acids experience two-step dissociation in aqueous solutions.(7) They have the same structural formula but different spatial configurations. Fumaric acid is the *trans* and maleic acid the *cis* isomer. The physical properties of maleic acid and fumaric acid are very different. The *cis* isomer is less stable. Maleic acid is used in the preparation of fumaric acid by catalytic isomerization.

The antimicrobial activity of maleic acid has been evaluated for use in dental irrigants.(8) Novel applications of maleic acid include its use as an intranasal delivery system(9) and the use of styrene-maleic acid copolymer micelles for tumor targeting.(10) Copolymers of maleic acid (butyl monoester of poly[methylvinyl ether/maleic acid]) have found topical applications in medicated nail lacquer and mosquito repellent as film-forming agents. Compositions of certain maleic acid copolymers have been employed for preventing the attachment of dental plaque to the surface of teeth.

Maleic acid is commonly used as a pharmaceutical intermediate to form the maleate salts of several categories of therapeutic agents, such as salts of antihistamines and other drug substances.

The EINECS number for maleic acid is 203-742-5. The PubChem Compound ID (CID) for maleic acid is 444266.

19 Specific References

1 McCarron P, *et al.* Stability of 5-aminolevulinic acid in non-aqueous gel and patch-type systems intended for topical application. *J Pharm Sci* 2005; **94**(8): 1756–1771.
2 Ment W, Naviasky H. Effect of maleic acid in compendial UV absorption assays for antihistamine maleate salts. *J Pharm Sci* 1974; **63**(10): 1604–1609.
3 Zoglio M, *et al.* Pharmaceutical heterogeneous systems III. Inhibition of stearate lubricant induced degradation of aspirin by the use of certain organic acids. *J Pharm Sci* 1968; **57**(11): 1877–1180.
4 Yalkowsky S, ed. *Solubility and Solubilization in Aqueous Media*. New York: Oxford University Press, 1999; 136.
5 O'Neil MJ, ed. *The Merck Index: An Encyclopedia of Chemicals, Drugs and Biologicals*, 14th edn. Whitehouse Station, NJ: Merck, 2006; 986.
6 Lewis RJ, ed. *Sax's Dangerous Properties of Industrial Materials*, 11th edn. New York: Wiley, 2004; 2271–2272.
7 Orlova T, Bychkova S. The heat effects of dissociation of maleic and fumaric acids. *Russian J Phys Chem* 2007; **81**(5): 693–695.
8 Ferrer-Luque CM, *et al.* Antimicrobial activity of maleic acid and combinations of cetrimide with chelating agents against *Enterococcus faecalis* biofilm. *J Endod* 2010; **36**(10): 1673–1675.
9 Kim TK, *et al.* Pharmacokinetic evaluation and modelling of formulated levodopa intranasal delivery systems. *Eur J Pharm Sci* 2009; **38**(5): 525–532.
10 Daruwalla J, *et al.* *In vitro* and *in vivo* evaluation of tumor targeting styrene-maleic acid copolymer-pirarubicin micelles: survival improvement and inhibition of liver metastases. *Cancer Sci* 2010; **101**(8): 1866–1874.

20 General References

Barillaro V, *et al.* Theoretical and experimental investigations of organic acids/cyclodextrin complexes and their consequences upon the formation of miconazole/cyclodextrin/acid ternary inclusion complexes. *Int J Pharm* 2008; **347**(1–2): 62–70.
Barillaro V, *et al.* Theoretical and experimental investigations on miconazole/cyclodextrin/acid complexes: molecular modeling studies. *Int J Pharm* 2007; **342**(1–2): 152–160.
Murat S, *et al.* Controlled release of terbinafine hydrochloride from pH sensitive poly(acrylamide/maleic acid) hydrogels. *Int J Pharm* 2000; **203**(1–2): 149–157.
Wong J, *et al.* Major degradation product identified in several pharmaceutical formulations against the common cold. *Anal Chem* 2006; **78**(22): 7891–7895.
Wong TW, *et al.* Effects of microwave on drug release property of poly(methyl vinyl ether-co-maleic acid) matrix. *Drug Dev Ind Pharm* 2007; **33**(7): 737–746.

21 Author

RT Guest.

22 Date of Revision

1 March 2012.

M

Malic Acid

1 Nonproprietary Names

BP: Malic Acid
PhEur: Malic Acid
USP–NF: Malic Acid

2 Synonyms

Acidum malicum; apple acid; E296; 2-hydroxy-1,4-butanedioic acid; hydroxybutanedioic acid; 1-hydroxy-1,2-ethanedicarboxylic acid; hydroxysuccinic acid; 2-hydroxysuccinic acid; DL-malic acid.

3 Chemical Name and CAS Registry Number

Hydroxybutanedioic acid [6915-15-7]
(*RS*)-(±)-Hydroxybutanedioic acid [617-48-1]

4 Empirical Formula and Molecular Weight

$C_4H_6O_5$ 134.09

5 Structural Formula

O OH
HO OH
O

6 Functional Category

Acidulant; antioxidant; buffering agent; complexing agent; flavoring agent.

7 Applications in Pharmaceutical Formulation or Technology

Malic acid is used in pharmaceutical formulations as a general-purpose acidulant. It possesses a slight apple flavor and is used as a flavoring agent to mask bitter tastes and provide tartness. Malic acid is also used as an alternative to citric acid in effervescent powders, mouthwashes, and tooth-cleaning tablets.

In addition, malic acid has chelating and antioxidant properties.[1] It may be used with butylated hydroxytoluene as a synergist in order to retard oxidation in vegetable oils.

8 Description

White or nearly white, crystalline powder or granules having a slight odor and a strongly acidic taste. It is hygroscopic. The synthetic material produced commercially in Europe and the US is a racemic mixture, whereas the naturally occurring material found in apples and many other fruits and plants is levorotatory.

9 Pharmacopeial Specifications

See Table I.

Table I: Pharmacopeial specifications for malic acid.

Test	PhEur 7.4	USP35–NF30
Identification	+	+
Characters	+	—
Melting point	128–132°C	—
Residue on ignition	≤0.1%	≤0.1%
Appearance of solution	+	—
Water-insoluble substances	≤0.1%	≤0.1%
Heavy metals	≤20 ppm	≤20 ppm
Fumaric acid	—	≤1.0%
Maleic acid	—	≤0.05%
Optical rotation	−0.10° to +0.10°	—
Related substances	+	—
Water	≤2.0%	—
Assay	99.0–101.0%	99.0–100.5%

10 Typical Properties

Data shown below are for the racemate. *See* Section 17 for other data for the D and L forms.

Acidity/alkalinity pH = 2.35 (1% w/v aqueous solution at 25°C)
Boiling point 150°C (with decomposition)
Density (bulk) 0.81 g/cm^3
Density (tapped) 0.92 g/cm^3
Dissociation constant

pK_{a1} = 3.40 at 25°C;

pK_{a2} = 5.05 at 25°C.

Melting point 131–132°C
Solubility Freely soluble in ethanol (95%) and water but practically insoluble in benzene. A saturated aqueous solution contains about 56% malic acid at 20°C. *See* Table II.
Specific gravity

1.601 at 20°C;

1.250 (saturated aqueous solution at 25°C).

Spectroscopy

IR Spectra *see* Figure 1.

NIR spectra *see* Figure 2.

Viscosity (dynamic) 6.5 mPa s (6.5 cP) for a 50% w/v aqueous solution at 25°C.

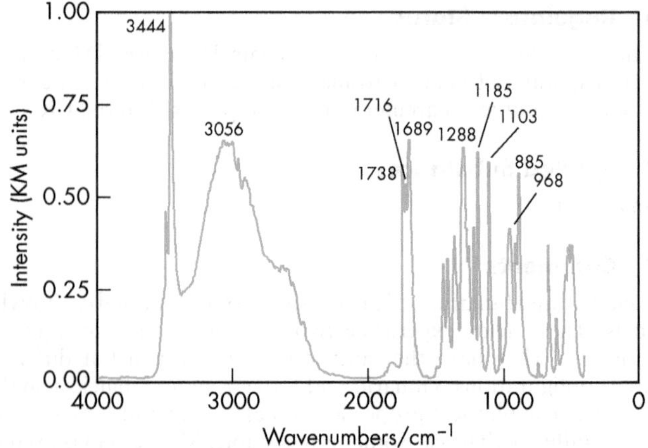

Figure 1: Infrared spectrum of malic acid measured by diffuse reflectance. Adapted with permission of Informa Healthcare.

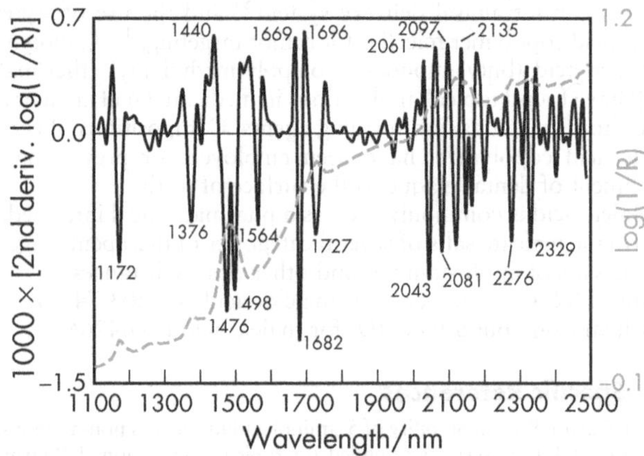

Figure 2: Near-infrared spectrum of malic acid measured by reflectance.

Table II: Solubility of malic acid.

Solvent	Solubility at 20°C
Acetone	1 in 5.6
Diethyl ether	1 in 119
Ethanol (95%)	1 in 2.6
Methanol	1 in 1.2
Propylene glycol	1 in 1.9
Water	1 in 1.5–2.0

11 Stability and Storage Conditions

Malic acid is stable at temperatures up to 150°C. At temperatures above 150°C it begins to lose water very slowly to yield fumaric acid; complete decomposition occurs at about 180°C to give fumaric acid and maleic anhydride. Conditions of high humidity and elevated temperatures should be avoided to prevent caking. The effects of grinding and humidity on malic acid have been investigated.[2] Malic acid is readily degraded by many aerobic and anaerobic microorganisms.

The bulk material should be stored in a well-closed container in a cool, dry place.

12 Incompatibilities

Malic acid can react with oxidizing materials. Aqueous solutions are mildly corrosive to carbon steels.

13 Method of Manufacture

Malic acid is manufactured by hydrating maleic and fumaric acids in the presence of suitable catalysts. The malic acid formed is then separated from the equilibrium product mixture. Malic acid can also be produced by fermentation following an enzymatic conversion of fumaric acid.[3–6]

14 Safety

Malic acid is used in oral, topical, and parenteral pharmaceutical formulations in addition to food products, and is generally regarded as a relatively nontoxic and nonirritant material. However, concentrated solutions may be irritant.

LD_{50} (rat, oral): 1.6 g/kg[7]
LD_{50} (rat, IP): 0.1 g/kg

15 Handling Precautions

Observe normal precautions appropriate to the circumstances and quantity of material handled. Malic acid, and concentrated malic acid solutions may be irritant to the skin, eyes, and mucous membranes. Gloves and eye protection are recommended.

16 Regulatory Status

GRAS listed. Both the racemic mixture and the levorotatory isomer are accepted as food additives in Europe. The DL and L forms are included in the FDA Inactive Ingredients Database (oral preparations). Included in nonparenteral and parenteral medicines licensed in the UK. Included in the Canadian Natural Health Products Ingredients Database.

17 Related Substances

Citric acid; fumaric acid; D-malic acid; L-malic acid; tartaric acid.

D-Malic acid

Empirical formula $C_4H_6O_5$
Molecular weight 134.09
CAS number [636-61-3]
Synonyms (*R*)-(+)-Hydroxybutanedioic acid; D-(+)-malic acid.
Melting point 99–101°C
Specific rotation $[\alpha]_D^{20}$ = +5.2° (in acetone at 18°C).

L-Malic acid

Empirical formula $C_4H_6O_5$
Molecular weight 134.09
CAS number [97-67-6]
Synonyms Apple acid; (*S*)-(−)-hydroxybutanedioic acid; L-(−)-malic acid.
Boiling point ≈140°C (with decomposition)
Melting point 99–100°C
Solubility Practically insoluble in benzene. *See also* Table III.
Specific gravity 1.595 at 20°C
Specific rotation $[\alpha]_D^{20}$ = −5.7° (in acetone at 18°C)

Table III: Solubility of L-malic acid.

Solvent	Solubility at 20°C
Acetone	1 in 1.6
Diethyl ether	1 in 37
Dioxane	1 in 1.3
Ethanol (95%)	1 in 1.2
Methanol	1 in 0.51
Water	1 in 2.8

18 Comments

Therapeutically, malic acid is well-known for its positive effect on skin[8,9] and has been used topically in combination with benzoic acid and salicylic acid to treat burns, ulcers, and wounds. It has also been used orally and parenterally, either intravenously or intramuscularly, in the treatment of liver disorders, and as a sialagogue.[10]

Malic acid can be mineralized by photocatalytic degradation when in the presence of an illuminated source, and therefore has been investigated in order to understand the degradation steps of organic pollutants contained in water.[11–13]

In food products, malic acid may be used in concentrations up to 420 ppm.

A specification for malic acid is contained in the *Food Chemical Codex* (FCC)[14] and the *Japanese Pharmaceutical Excipients* (JPE).[15]

The EINECS number for malic acid is 202-601-5. The PubChem Compound ID (CID) for malic acid is 525.

19 Specific References

1 Troy DB, ed. Remington: The Science and Practice of Pharmacy, 21st edn. Baltimore: Lippincott Williams & Wilkins, 2005: 2260.
2 Piyarom S, *et al.* Effects of grinding and humidification on the transformation of conglomerate to racemic compound in optically active drugs. *J Pharm Pharmacol* 1997; **49**: 384–389.
3 Khachatourians GG, Arora DK, eds. Applied Mycology and Biotechnology, 1st edn. Amsterdam: Elsevier, 2001: 435.
4 Moon SY, *et al.* Metabolic engineering of *Escherichia coli* for the production of malic acid. *Biochem Eng J* 2008; **40**: 312–320.
5 Presečki AV, *et al.* Comparison of the L-malic acid production by isolated fumarase and fumarase in permeabilized baker's yeast cells. *Enzyme Microb Technol* 2007; **41**: 605–612.
6 Nussinovitch A, ed. *Polymer macro- and micro-gel beads: fundamentals and applications.* London: Springer, 2010: 114.
7 Lewis RJ, ed. *Sax's Dangerous Properties of Industrial Materials*, 11th edn. New York: Wiley, 2004: 2273.
8 Al Bawab A, *et al.* Some non–equilibrium phenomena in the malic acid/water/Polysorbate 81 system. *Int J Pharm* 2007; **332**(1–2): 140–146.
9 Baumann L, ed. *Cosmetic Dermatology: Principles and Practice*, 2nd edn. The McGraw-Hill Companies Inc, 2009: 366.
10 Sweetman SC, ed. *Martindale: The Complete Drug Reference*, 37th edn. London: Pharmaceutical Press, 2011: 2563.
11 Herrmann JM, *et al.* Photocatalytic degradation of aqueous hydroxy-butandioic acid (malic acid) in contact with powdered and supported titania in water. *Catal Today* 1999; **54**: 131–141.
12 Sánchez L, *et al.* Solar activated ozonation of phenol and malic acid. *Chemosphere* 2003; **50**(8): 1085–1093.
13 Irawaty W, *et al.* Relationship between mineralization kinetics and mechanistic pathway during malic acid photodegradation. *J Mol Catal A Chem* 2011; **335**: 151–157.
14 *Food Chemicals Codex*, 7th edn. Bethesda, MD: United States Pharmacopeia, 2010: 610.
15 Japan Pharmaceutical Excipients Council. *Japanese Pharmaceutical Excipients 2004*. Tokyo: Yakuji Nippo, 2004: 505-507.

20 General References

Allen LV. Featured excipient: flavor enhancing agents. *Int J Pharm Compound* 2003; 7(1): 48–50.
Anonymous. Malic and fumaric acids. *Manuf Chem Aerosol News* 1964; 35(12): 56–59.
Berger SE. *Kirk-Othmer Encyclopedia of Chemical Technology.* 13. 3rd edn. New York: Wiley-Interscience, 1981: 103.
Brittain HG, ed. Malic acid. In: *Analytical Profiles of Drug Substances and Excipients* 2001; 28: 153-195

21 Author

M Grachet.

22 Date of Revision

1 March 2012.

M

Maltitol

1 Nonproprietary Names

BP: Maltitol
PhEur: Maltitol
USP–NF: Maltitol

2 Synonyms

Amalty; *C*PharmMaltidex*; E965; hydrogenated maltose; *Lycasin*; *Malbit*; *Maltisorb*; *Maltit*; D-maltitol; maltitolum; *SweetPearl*.

3 Chemical Name and CAS Registry Number

4-O-α-D-Glucopyranosyl-D-glucitol [585-88-6]

4 Empirical Formula and Molecular Weight

$C_{12}H_{24}O_{11}$ 344.32

5 Structural Formula

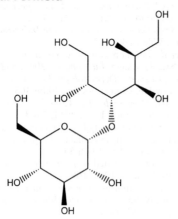

6 Functional Category

Coating agent; tablet and capsule binder; tablet and capsule diluent; sweetening agent.

7 Applications in Pharmaceutical Formulation or Technology

Maltitol is widely used in the pharmaceutical industry in the formulation of oral dosage forms. It is a noncariogenic bulk sweetener, approximately as sweet as sucrose, well adapted as a diluent for different oral dosage forms, wet granulation, and sugar-free hard coating.

8 Description

Maltitol occurs as a white, odorless, sweet, anhydrous crystalline powder. It is a disaccharide consisting of one glucose unit linked with one sorbitol unit via an α-(1→4) bond. The crystal structure is orthorhombic.

SEM 1: Excipient: *SweetPearl P200*; manufacturer: Roquette Frères.

9 Pharmacopeial Specifications

See Table I.

Table I: Pharmacopeial specifications for maltitol.

Test	PhEur 7.4	USP35–NF30
Identification	+	+
Characters	+	−
Appearance of solution	+	−
Conductivity	≤20 µS cm^{-1}	≤20 µS cm^{-1}
Reducing sugars	≤0.2%	≤0.3%
Related substances	+	−
Lead	≤0.5 ppm	−
Nickel	≤1 ppm	≤1.0 ppm
Water	≤1.0%	≤1.0%
Microbial contamination		
Aerobic bacteria	≤1000 cfu/g	≤1000 cfu/g
Fungi	≤100 cfu/g	≤100 cfu/g
Bacterial endotoxins	+	−
Assay (dried basis)	98.0–102.0%	92.0–100.5%

10 Typical Properties

Compressibility 9.5%
Density (bulk) 0.79 g/cm^3 (1)
Density (crystal) 1.6238 (calculated from crystallographic data).(2)
Density (tapped) 0.95 g/cm^3 (1)
Flowability 5 seconds(1)
Melting point 148–151°C
Particle size distribution 95% ≤ 500 µm, 40% ≥ 100 µm in size for *SweetPearl P200* (Roquette); 95% ≤ 200 µm, 50% ≥ 100 µm in size for *SweetPearl P90* (Roquette).
Solubility Freely soluble in water. *See also* Table II.
Spectroscopy
IR spectra *see* Figure 1.
NIR spectra *see* Figure 2.
Viscosity (dynamic) *see* Table III.

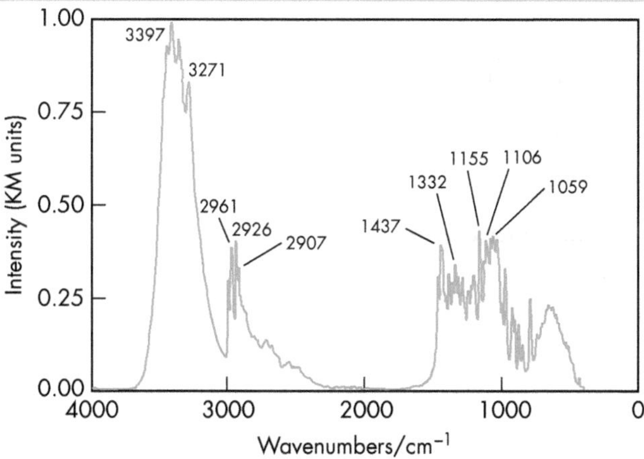

Figure 1: Infrared spectrum of maltitol measured by diffuse reflectance. Adapted with permission of Informa Healthcare.

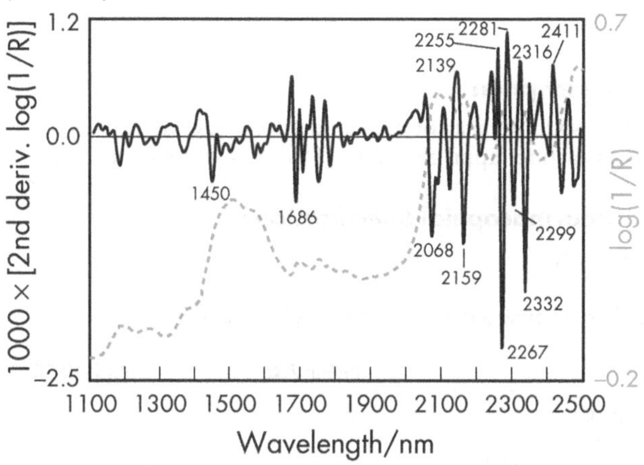

Figure 2: Near-infrared spectrum of maltitol measured by reflectance.

Table II: Solubility of maltitol (*SweetPearl*).[1]

Solvent	Solubility at 20°C unless otherwise stated
Water	1 in 0.67
	1 in 0.48 at 40°C
	1 in 0.33 at 60°C
	1 in 0.22 at 80°C
	1 in 0.18 at 90°C

Table III: Viscosity (dynamic) of aqueous maltitol (*SweetPearl*) solutions at 20°C.[1]

Concentration of aqueous maltitol solution (% w/v)	Viscosity (mPa s)
10	8
20	10
30	11
40	15
50	24
60	70

11 Stability and Storage Conditions

Maltitol has good thermal and chemical stability. When it is heated at temperatures above 200°C, decomposition begins (depending on time, temperature, and other prevailing conditions). Maltitol does not undergo browning reactions with amino acids, and absorbs atmospheric moisture only at relative humidities of 89% and above, at 20°C.

12 Incompatibilities

—

13 Method of Manufacture

Maltitol is obtained from hydrogenated maltose syrup. Starch is hydrolyzed to yield a high-concentration maltose syrup, which is hydrogenated with a catalyst. After purification and concentration, the syrup is crystallized.

14 Safety

Maltitol is used in oral pharmaceutical formulations, confectionery, and food products, and is considered to be noncariogenic. It is generally regarded as a nontoxic, nonallergenic, and nonirritant material.

Digestion of maltitol follows two different metabolic pathways: absorption in the small intestine and fermentation in the large intestine (colon). These two metabolic pathways must thus be considered when evaluating the energy value.

The hydrolysis of maltitol in the small intestine releases sorbitol and glucose. Glucose is actively transported and rapidly absorbed, whereas sorbitol absorption is passive. The nonabsorbed sorbitol and nonhydrolyzed maltitol are fermented by the microflora in the colon. The relative importance of the two absorption pathways depends on numerous individual factors and is related to the quantity of maltitol ingested. Excessive oral consumption (>50 g daily) may cause flatulence and diarrhea.

Maltitol exhibits a low glycemic index and can therefore, under medical supervision, have a place in the diet of diabetic patients. The intake of maltitol must be taken into account for the calculation of the daily glucidic allowance.

The WHO, in considering the safety of maltitol, did not set a value for the acceptable daily intake since the levels used in food to achieve a desired effect were not considered a hazard to health.[3,4]

15 Handling Precautions

Observe normal precautions appropriate to circumstances and quantity of material handled. Eye protection and gloves are recommended.

16 Regulatory Status

GRAS listed. Accepted for use as a food additive in Europe. Included in oral pharmaceutical formulations. Included in the Canadian Natural Health Products Ingredients Database.

17 Related Substances

Sorbitol.

18 Comments

Maltitol is not fermented by oral bacteria and is neither acidogenic nor cariogenic. A specification for maltitol syrup is contained in the *Food Chemicals Codex* (FCC).[5]

The EINECS number for maltitol is 209-567-0. The PubChem Compound ID (CID) for maltitol includes 3871 and 493591.

19 Specific References

1 Roquette Frères. Product information: *SweetPearl*, 2009.
2 Schouten A, *et al*. A redetermination of the crystal and molecular structure of maltitol (4-O-α-d-glucopyranosyl-d-glucitol). *Carbohydr Res* 1999; 322(3): 298–302.
3 FAO/WHO. Evaluation of certain food additives and contaminants. Thirty-third report of the Joint FAO/WHO Expert Committee on Food Additives. *World Health Organ Tech Rep Ser* 1989; No. 776.

4 FAO/WHO. Evaluation of certain food additives and contaminants. Forty-sixth report of the Joint FAO/WHO Expert Committee on Food Additives. *World Health Organ Tech Rep Ser* 1997; No. 868.
5 *Food Chemicals Codex*, 7th edn. (Suppl. 1), Bethesda, MD: United States Pharmacopeia, 2010: 1477.

20 General References

Moskowitz AH. Maltitol and hydrogenated starch hydrolysate. In: Nabors LO, Gelardi RC, eds. *Alternative Sweeteners*, 2nd edn. New York: Marcel Dekker, 1991: 259–282.

Portman MO, Kilcast D. Psycho-physical characterization of new sweeteners of commercial importance for the EC food industry. *Food Chem* 1996; 56(3): 291–302.

21 Author

D Simon.

22 Date of Revision

1 March 2012.

 Maltitol Solution

1 Nonproprietary Names

BP: Liquid Maltitol
PhEur: Maltitol, Liquid
USP–NF: Maltitol Solution

2 Synonyms

E965; hydrogenated glucose syrup; *Finmalt L*; *Lycasin HBC*; *Lycasin 75/75*; *Lycasin 80/55*; *Lycasin 85/55*; *Maltisorb 75/75*; *Maltisweet 3145*; maltitol syrup; maltitolum liquidum.

3 Chemical Name and CAS Registry Number

Maltitol solution [9053-46-7]

4 Empirical Formula and Molecular Weight

The PhEur 7.4 describes liquid maltitol as an aqueous solution of a hydrogenated, partly hydrolyzed starch, with not less than 68% w/w of solid matter and not more than 85% w/w. This is composed of a mixture of mainly D-maltitol (≥50% w/w), D-sorbitol (≤8% w/w), and hydrogenated oligo- and polysaccharides, all quoted on an anhydrous basis.

The USP35–NF30 describes maltitol solution as a water solution containing, on the anhydrous basis, not less than 50% w/w of D-maltitol ($C_{12}H_{24}O_{11}$) and not more than 8.0% w/w of D-sorbitol ($C_6H_{14}O_6$). *See also* Section 18.

5 Structural Formula

See Section 4.

6 Functional Category

Suspending agent; sweetening agent.

7 Applications in Pharmaceutical Formulation or Technology

Maltitol solution is used in oral pharmaceutical formulations as a bulk sweetening agent, either alone or in combination with other excipients, such as sorbitol. Maltitol solution is also used as a suspending agent in oral suspensions as an alternative to sucrose syrup since it is viscous, noncariogenic, and has a low calorific value. It is also noncrystallizing and therefore prevents 'cap-locking' in syrups and elixirs.

Maltitol solution is additionally used in the preparation of pharmaceutical lozenges,(1) and is also used in confectionery and food products.

8 Description

Maltitol solution is a colorless and odorless, clear viscous liquid. It is sweet-tasting (approximately 75% the sweetness of sucrose).

9 Pharmacopeial Specifications

See Table I.

Table I: Pharmacopeial specifications for maltitol solution.

Test	PhEur 7.4	USP35–NF30
Identification	+	+
Characters	+	—
Appearance of solution	+	—
Conductivity	≤10 μS·cm^{-1}	—
pH	—	5.0–7.5
Reducing sugars	≤0.2%	≤0.3%
Lead	≤0.5 ppm	—
Nickel	≤1 ppm	≤1 ppm
Water	15.0–32.0%	≤31.5%
Residue on ignition	—	≤0.1%
Maltitol (dried basis)	≥50.0%	≥50.0%
Sorbitol (dried basis)	≤8.0%	≤8.0%
Microbial contamination		
Aerobic bacteria	—	≤1000 cfu/g
Fungi	—	≤100 cfu/g

10 Typical Properties

Boiling point 105°C
Flash point >150°C
Density 1.36 g/cm^3 at 20°C
Heat of combustion 10.0 kJ/g (2.4 kcal/g)
Osmolarity The osmolarity of an aqueous maltitol solution is similar to that of a sucrose solution of the same concentration. A 10% v/v aqueous solution of *Lycasin 80/55* (Roquette) is iso-osmotic with serum.
Refractive index n_D^{20} = 1.478
Solubility Miscible with ethanol (provided the ethanol concentration is less than 55%), glycerin, propylene glycol, and water. Insoluble in mineral and vegetable oils.
Viscosity (dynamic) Maltitol solution is a viscous, syrupy, liquid. At 20°C, a solution of *Lycasin 80/55* (Roquette) containing 75% of dry substances has a viscosity of approximately 2000 mPa s (2000 cP). With increasing temperature, the viscosity of a maltitol solution is reduced; *see* Figure 1. The viscosity of maltitol solutions also decreases with decreasing concentration

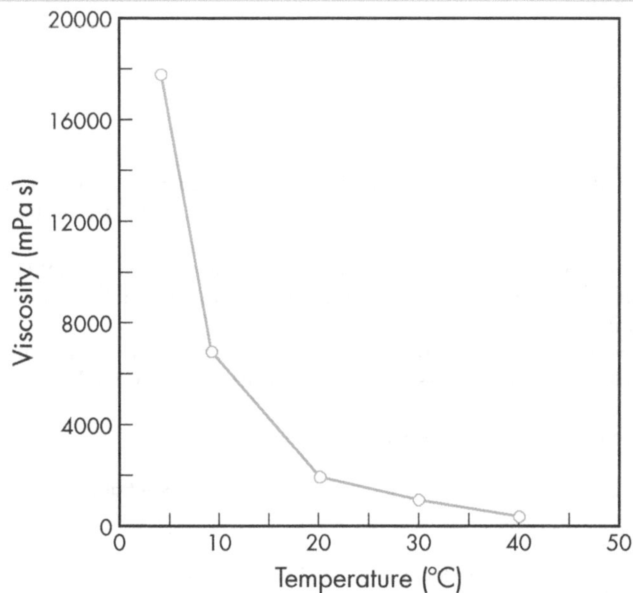

Figure 1: Viscosity of maltitol solution (*Lycasin 80/55*, Roquette Frères), containing 75% of dry substances, at different temperatures.

of dry solids, at a constant temperature. Maltitol solution may also be mixed with sorbitol solution to obtain blends of a desired viscosity.

11 Stability and Storage Conditions

Maltitol solution is stable for at least 2 years at room temperature and pH 3–9. Following storage for 3 months at 50°C, maltitol solution at pH 2 underwent slight hydrolysis (1.2%) and became yellow colored. At pH 3, and the same storage conditions, no color change was apparent although very slight hydrolysis occurred (0.2%). At pH 4–9, no hydrolysis occurred although a very slight yellow color was formed under alkaline conditions.[2]

Formulations containing maltitol solution should be preserved with an antimicrobial preservative such as sodium benzoate or a mixture of parabens. Maltitol solution is noncrystallizing.

Maltitol solution should be stored in a well-closed container, in a cool, dry place.

12 Incompatibilities

—

13 Method of Manufacture

Maltitol solution is prepared by the hydrogenation of a high-maltose syrup that is obtained from starch by enzymatic hydrolysis. The maltitol solution produced from this process consists of the hydrogenated homologs of the oligosaccharides contained in the original syrup.

14 Safety

Maltitol solution is used in oral pharmaceutical formulations, confectionery, and food products, and is considered to be less cariogenic than sucrose.[3–6] It is generally regarded as a nontoxic, nonallergenic, and nonirritant material. However, excessive oral consumption (more than 50 g daily) may cause flatulence and diarrhea.

The WHO, in considering the safety of maltitol solution, did not set a value for the acceptable daily intake since the levels used in food to achieve a desired effect were not considered a hazard to health.[7,8]

LD_{50} (rat, IP): 20 g/kg[9]

15 Handling Precautions

Observe normal precautions appropriate to the circumstances and quantity of material handled.

16 Regulatory Status

Included in the FDA Inactive Ingredients Database (oral solutions). Accepted for use in confectionery, foods, and nonparenteral pharmaceutical formulations in Europe and the USA. Included in the Canadian Natural Health Products Ingredients Database.

17 Related Substances

Maltitol; sorbitol.

18 Comments

Hydrogenated glucose syrup is a generic term used to describe aqueous mixtures containing mainly D-maltitol, along with D-sorbitol and hydrogenated oligosaccharides and polysaccharides. Such mixtures can vary widely in their composition and hence physical and chemical properties. Products containing up to 90% of maltitol are usually known as maltitol syrup or maltitol solution. Preparations containing a minimum of 98% of maltitol are designated maltitol.

M

19 Specific References

1 Grenby TH. Dental properties of antiseptic throat lozenges formulated with sugars or Lycasin. *J Clin Pharm Ther* 1995; **20**: 235–241.
2 Roquette. Technical literature: *Lycasin* the sweetener for sugarless pharmaceuticals, 1993.
3 Frostell G, Birkhed D. Acid production from Swedish *Lycasin* (candy quality) and French *Lycasin* (80/55) in human dental plaques. *Caries Res* 1978; **12**: 256–263.
4 Grenby TH. Dental and nutritional effects of Lycasins replacing sucrose in the diet of laboratory rats. *J Dent Res* 1982; **61**: 557.
5 Würsch P, Koellreutter B. Maltitol and maltotriitol as inhibitors of acid production in human dental plaque. *Caries Res* 1982; **16**: 90–95.
6 Havenaar R, *et al.* Potential cariogenicity of *Lycasin 80/55* in comparison to starch, sucrose, xylitol, sorbitol and L-sorbose in rats. *Caries Res* 1984; **18**: 375–384.
7 FAO/WHO. Evaluation of certain food additives and contaminants: Thirty-third report of the Joint FAO/WHO Expert Committee on Food Additives. *World Health Organ Tech Rep Ser* 1989; No. 776.
8 FAO/WHO. Evaluation of certain food additives and contaminants: Forty-sixth report of the Joint FAO/WHO Expert Committee on Food Additives. *World Health Organ Tech Rep Ser* 1997; No. 868.
9 Sweet DV. ed. *Registry of Toxic Effects of Chemical Substances.* Cincinnati: US Department of Health, 1987.

20 General References

Le Bot Y. *Lycasin* for confections. *Manuf Confect* 1983; (Dec): 69–74.

21 Author

D Simon.

22 Date of Revision

1 March 2012.

Maltodextrin

1 Nonproprietary Names

BP: Maltodextrin
PhEur: Maltodextrin
USP–NF: Maltodextrin

2 Synonyms

Amidex; *Cargill Dry*; *C*Dry MD*; *C*PharmDry*; *Clintose*; *Globe*; *Globe Plus*; *Glucidex*; *Glucodry*; *Lycatab DSH*; *Maldex*; *Maldex G*; *Malta*Gran*; maltodextrinum; *Maltosweet*; *Maltrin*; *Maltrin QD*; *Paselli MD10 PH*; *Rice*Trin*; *Star-Dri*; *Tapi*.

3 Chemical Name and CAS Registry Number

Maltodextrin [9050-36-6]

4 Empirical Formula and Molecular Weight

$(C_6H_{10}O_5)_n{\cdot}H_2O$ 900–9 000

The USP35–NF30 describes maltodextrin as a nonsweet, nutritive saccharide mixture of polymers that consist of D-glucose units, with a dextrose equivalent (DE) less than 20; *see also* Section 18. The D-glucose units are linked primarily by α-(1→4) bonds but there are branched segments linked by α-(1→6) bonds. It is prepared by the partial hydrolysis of a food-grade starch with suitable acids and/or enzymes.

5 Structural Formula

[Structural formula: α-(1→4)-linked D-glucose chain units, n, and α-(1→6)-branched units, n]

6 Functional Category

Biodegradable material; coating agent; direct compression excipient; osmotic agent; tablet and capsule binder; tablet and capsule diluent; viscosity-increasing agent.

7 Applications in Pharmaceutical Formulation or Technology

Maltodextrin is used in tablet formulations as a binder and diluent in both direct-compression and wet-granulation or agglomeration processes.(1–7) Maltodextrin appears to have no adverse effect on the rate of dissolution of tablet and capsule formulations; magnesium stearate 0.5–1.0% may be used as a lubricant. It has been used as a carrier in a spray-dried redispersible oil-in-water emulsion to improve the bioavailability of poorly soluble drugs.(8) Maltodextrin may also be used as a tablet film former in aqueous film-coating processes. Maltodextrin grades with a high DE value are particularly useful in chewable tablet formulations.

Maltodextrin may also be used in pharmaceutical formulations to increase the viscosity of solutions and to prevent the crystallization of syrups.

Maltodextrin is also widely used in confectionery and food products, as well as personal care applications. *See* Table I.

Table I: Uses of maltodextrin.

Use	Concentration (%)
Aqueous film-coating	2–10
Carrier	10–99
Crystallization inhibitor for lozenges and syrups	5–20
Osmolarity regulator for solutions	10–50
Spray-drying aid	20–80
Tablet binder (direct compression)	2–40
Tablet binder (wet granulation)	3–10

8 Description

Maltodextrin occurs as a nonsweet, odorless, white powder or granules. The solubility, hygroscopicity, sweetness, and compressibility of maltodextrin increase as the DE increases. The USP35–NF30 states that it may be physically modified to improve its physical and functional characteristics.

9 Pharmacopeial Specifications

See Table II.

10 Typical Properties

Angle of repose

35.2° for *Maltrin QD M500*;[5]

28.4° for *Maltrin M510*.[5]

SEM 1: Excipient: maltodextrin (*Maltrin M100*); manufacturer: Grain Processing Corp.; magnification: 100×.

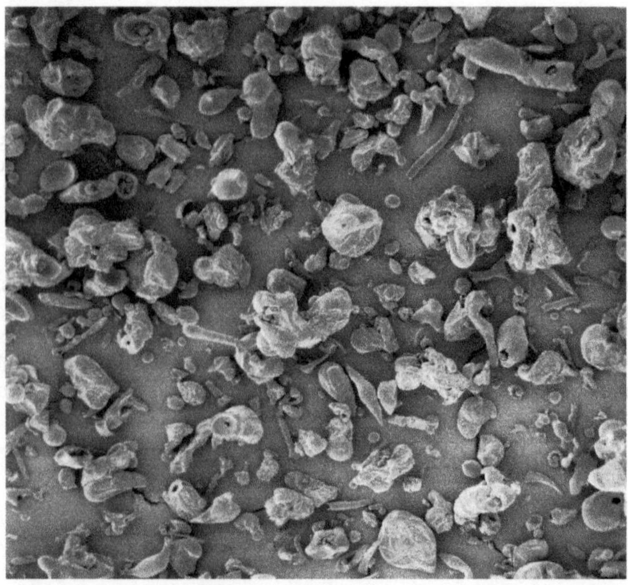

SEM 2: Excipient: maltodextrin (*Maltrin QD M500*); manufacturer: Grain Processing Corp.; magnification: 100×.

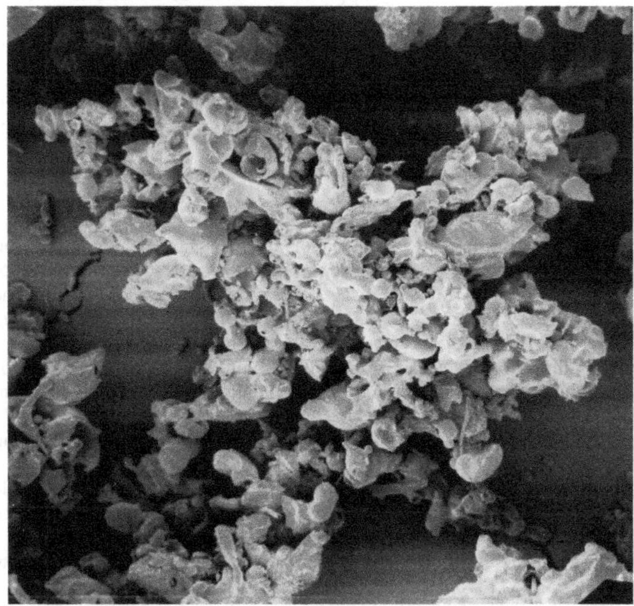

Table II: Pharmacopeial specifications for maltodextrin.

Test	PhEur 7.4	USP35–NF30
Identification	+	–
Characters	+	–
Microbial limits	+	+
pH (20% w/v solution)	4.0–7.0	4.0–7.0
Loss on drying	≤6.0%	≤6.0%
Residue on ignition	≤0.5%	≤0.5%
Heavy metals	≤10 ppm	≤5 ppm
Protein	–	≤0.1%
Sulfur dioxide	≤20 ppm	≤40 ppm
Dextrose equivalent	+	<20

Density (bulk)

0.43 g/cm^3 for *Lycatab DSH*;

0.26 g/cm^3 for *Maltrin QD M500*;

0.51 g/cm^3 for *Maltrin M040*;

0.54 g/cm^3 for *Maltrin M100*;

0.57 g/cm^3 for *Maltrin M150*;

0.61 g/cm^3 for *Maltrin M180*;

0.30 g/cm^3 for *Maltrin QD M440*;

0.56 g/cm^3 for *Maltrin M510*;

0.37 g/cm^3 for *Maltrin QD M550*;

0.40 g/cm^3 for *Maltrin QD M580*;

0.13 g/cm^3 for *Maltrin M700*.

Density (tapped)

0.63 g/cm^3 for *Lycatab DSH*;

0.32 g/cm^3 for *Maltrin QD M500*;

0.54 g/cm^3 for *Maltrin M510*.[(5)]

Density (true)

1.419 g/cm^3;

1.334 g/cm^3 for Maltodextrin FCC;

1.410 g/cm^3 for *Maltrin M500*;

1.425 g/cm^3 for *Maltrin M510*.

Moisture content Hygroscopicity increases as DE increases. Maltodextrin is slightly hygroscopic at relative humidities less than 50%. At relative humidities greater than 50%, the hygroscopicity of maltodextrin increases nonlinearly.

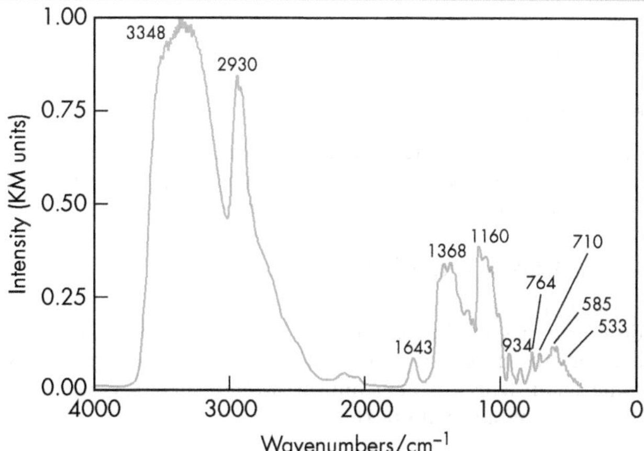

Figure 1: Infrared spectrum of maltodextrin measured by diffuse reflectance. Adapted with permission of Informa Healthcare.

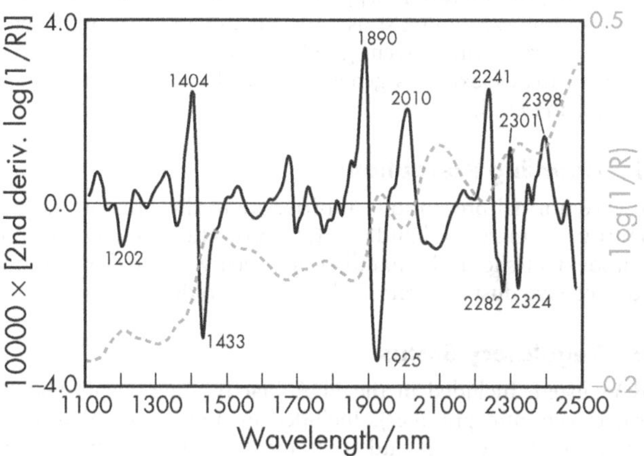

Figure 2: Near-infrared spectrum of maltodextrin measured by reflectance.

Particle size distribution

Maltrin is available in various grades with different particle size distributions.

For *Lycatab DSH*: maximum of 15% greater than 200 μm, and minimum of 80% greater than 50 μm in size.

Solubility Freely soluble in water; slightly soluble in ethanol (95%). Solubility increases as DE increases.

Specific surface area

0.54 m^2/g for *Maltrin QD M500*;

0.31 m^2/g for *Maltrin M510*.[(5)]

Spectroscopy

IR spectra *see* Figure 1.

NIR spectra *see* Figure 2.

Viscosity (dynamic)

Less than 20 mPa s (20 cP) for a 20% w/v aqueous solution of *Lycatab DSH*. The viscosity of maltodextrin solutions decreases as the DE increases.

Viscosity is 3.45 mPa s for a 20% w/v aqueous dispersion of *Star-Dri* (Tate & Lyle).

11 Stability and Storage Conditions

Maltodextrin is stable for at least 1 year when stored at a cool temperature (<30°C) and less than 50% relative humidity. Maltodextrin solutions may require the addition of an antimicrobial preservative.

M

Maltodextrin should be stored in a well-closed container in a cool, dry place.

12 Incompatibilities

Under certain pH and temperature conditions maltodextrin may undergo Maillard reactions with amino acids to produce yellowing or browning. Incompatible with strong oxidizing agents.

13 Method of Manufacture

Maltodextrin is prepared by heating and treating starch with acid and/or enzymes in the presence of water. This process partially hydrolyzes the starch, to produce a solution of glucose polymers of varying chain length. This solution is then filtered, concentrated, and dried to obtain maltodextrin.

14 Safety

Maltodextrin is a readily digestible carbohydrate with a nutritional value of approximately 17 kJ/g (4 kcal/g). In the US, it is generally recognized as safe (GRAS) as a direct human food ingredient at levels consistent with current good manufacturing practices. As an excipient, maltodextrin is generally regarded as a nonirritant and nontoxic material.

15 Handling Precautions

Observe normal precautions appropriate to the circumstances and quantity of material handled. Eye protection is recommended. Maltodextrin should be handled in a well-ventilated environment and excessive dust generation should be avoided.

M

16 Regulatory Status

GRAS listed. Included in the FDA Inactive Ingredients Database (oral tablets and granules). Included in nonparenteral medicines licensed in the UK. Included in the Canadian Natural Health Products Ingredients Database.

17 Related Substances

Corn syrup solids; dextrates; dextrin; starch.

Corn syrup solids

Comments Corn syrup solids are glucose polymers with a DE $\geq$20 and are prepared, in a similar manner to maltodextrin, by the partial hydrolysis of starch.

18 Comments

Various different grades of maltodextrin are commercially available for food and pharmaceutical applications from a number of suppliers: e.g. *Lycatab DS* (Roquette Frères), *Maltrin* (Grain Processing Corp.) and *Star-Dri* (Tate & Lyle). The grades have different physical properties such as solubility and viscosity, depending upon their DE value. The dextrose equivalent (DE) value is a measure of the extent of starch-polymer hydrolysis and is defined as the reducing power of a substance expressed in grams of D-glucose per 100 g of the dry substance.

Therapeutically, maltodextrin is often used as a carbohydrate source in oral nutritional supplements because solutions with a lower osmolarity than isocaloric dextrose solutions can be prepared. At body osmolarity, maltodextrin solutions provide a higher caloric density than sugars.

A specification for maltodextrin is contained in the *Food Chemicals Codex* (FCC).[9] The EINECS number for maltodextrin is 232-940-4.

19 Specific References

1 Li LC, Peck GE. The effect of moisture content on the compression properties of maltodextrins. *J Pharm Pharmacol* 1990; **42**(4): 272–275.
2 Li LC, Peck GE. The effect of agglomeration methods on the micrometric properties of a maltodextrin product *Maltrin 150*. *Drug Dev Ind Pharm* 1990; **16**: 1491–1503.
3 Papadimitriou E, *et al.* Evaluation of maltodextrins as excipients for direct compression tablets and their influence on the rate of dissolution. *Int J Pharm* 1992; **86**: 131–136.
4 Visavarungroj N, Remon JP. Evaluation of maltodextrin as binding agent. *Drug Dev Ind Pharm* 1992; **18**: 1691–1700.
5 Mollan MJ, Çelik M. Characterization of directly compressible maltodextrins manufactured by three different processes. *Drug Dev Ind Pharm* 1993; **19**: 2335–2358.
6 Muñoz-Ruiz A, *et al.* Physical and rheological properties of raw materials. *STP Pharma Sci* 1993; **3**: 307–312.
7 Symecko CW, *et al.* Comparative evaluation of two pharmaceutical binders in the wet granulation of hydrochlorothiazide: Lycatab DSH vs. Kollidon 30. *Drug Dev Ind Pharm* 1993; **19**: 1131–1141.
8 Dollo G, *et al.* Spray-dried redispersible oil-in-water emulsion to improve oral bioavailability of poorly soluble drugs. *Eur J Pharm Sci* 2003; **19**(4): 273–280.
9 *Food Chemicals Codex*, 6th edn. Bethesda, MD: United States Pharmacopeia, 2008: 576.

20 General References

Archer Daniels Midland Company. www.adm.com (accessed 1 November 2011).
Cargill. www.cargill.com (accessed 1 November 2011).
Grain Processing Corporation. Technical literature: Ingredients for the Pharmaceutical and Nutraceutical Industries, 2010.
Primera Foods. Maltodextrins. www.primerafoods.com/Specialty.asp (accessed 1 November 2011).
Grain Processing Corporation. Technical literature: *Maltrin* maltodextrins and corn syrup solids, 2010.
Roquette Frères. www.roquette.com (accessed 1 November 2011).
Roquette Frères. Technical literature: *Lycatab DSH* excipient for wet granulation, 1992.
Shah A, *et al.* Characterisation of maltodextrins using isothermal microcalorimetry. *J Pharm Pharmacol* 2000; **52**(Suppl.): 183.
Tate & Lyle. www.tateandlyle.com (accessed 1 November 2011).
Corn Products. http://www.cornproducts.com (accessed 1 November 2011).

21 Author

SO Freers.

22 Date of Revision

1 March 2012.

Maltol

1 Nonproprietary Names

USP–NF: Maltol

2 Synonyms

Corps praline; 3-hydroxy-2-methyl-(1,4-pyran); 3-hydroxy-2-methyl-4-pyrone; larixinic acid; larixix acid; 2-methyl-3-hydroxy-4-pyrone; 2-methyl pyromeconic acid; *Palatone*; *Talmon*; *Veltol.*

3 Chemical Name and CAS Registry Number

3-Hydroxy-2-methyl-4 *H* -pyran-4-one [118-71-8]

4 Empirical Formula and Molecular Weight

$C_6H_6O_3$ 126.11

5 Structural Formula

6 Functional Category

Flavor enhancer; flavoring agent.

7 Applications in Pharmaceutical Formulation or Technology

Maltol is used in pharmaceutical formulations and food products as a flavoring agent or flavor enhancer. In foods, it is used at concentrations up to 30 ppm, particularly with fruit flavorings, although it is also used to impart a freshly baked odor and flavor to bread and cakes. When used at concentrations of 5–75 ppm, maltol potentiates the sweetness of a food product, permitting a reduction in sugar content of up to 15% while maintaining the same level of sweetness. Maltol is also used at low levels in perfumery.

8 Description

White crystalline solid with a characteristic, caramel-like odor and taste. In dilute solution it possesses a sweet, strawberry-like or pineapple-like flavor and odor.

9 Pharmacopeial Specifications

See Table I.

Table I: Pharmacopeial specifications for maltol.

Test	USP35–NF30
Identification	+
Melting range	160–164°C
Water	≤0.5%
Residue on ignition	≤0.2%
Lead	≤10 ppm
Heavy metals	≤20 ppm
Assay (anhydrous basis)	≥99.0%

10 Typical Properties

Acidity/alkalinity pH = 5.3 (0.5% w/v aqueous solution)
Melting point 162–164°C (begins to sublime at 93°C)
Solubility *see* Table II.

Table II: Solubility of maltol.

Solvent	Solubility at 20°C
Chloroform	Freely soluble
Diethyl ether	Sparingly soluble
Ethanol (95%)	1 in 21
Glycerin	1 in 80
Propan-2-ol	1 in 53
Propylene glycol	1 in 28
Water	1 in 82

Spectroscopy

IR spectra *see* Figure 1.

NIR spectra *see* Figure 2.

11 Stability and Storage Conditions

Maltol solutions may be stored in glass or plastic containers. The bulk material should be stored in a well-closed container, protected from light, in a cool, dry place. *See also* Section 12.

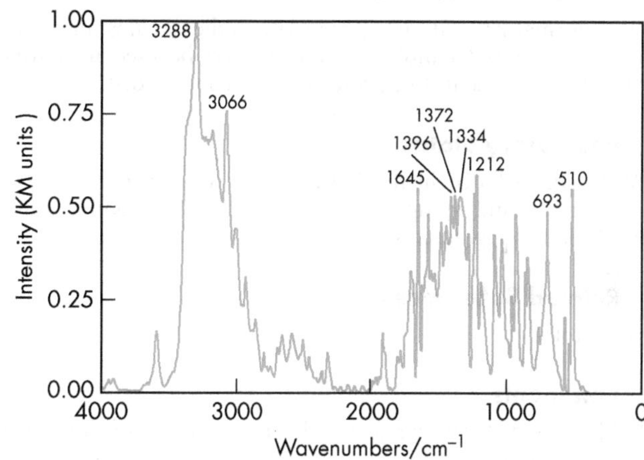

Figure 1: Infrared spectrum of maltol measured by diffuse reflectance. Adapted with permission of Informa Healthcare.

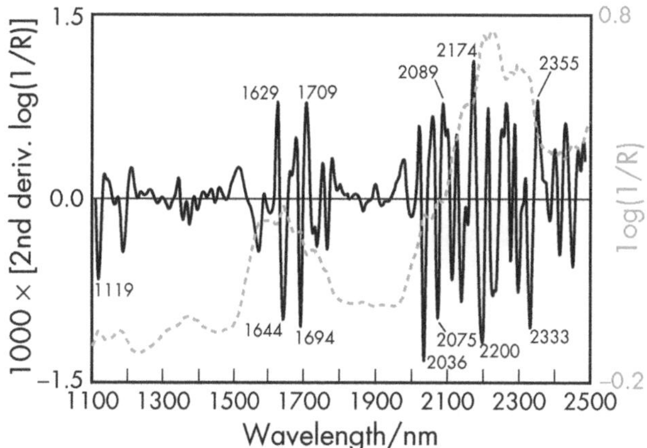

Figure 2: Near-infrared spectrum of maltol measured by reflectance.

12 Incompatibilities

Concentrated solutions in metal containers, including some grades of stainless steel, may discolor on storage.

13 Method of Manufacture

Maltol is mainly isolated from naturally occurring sources such as beechwood and other wood tars; pine needles; chicory; and the bark of young larch trees. It may also be synthesized by the alkaline hydrolysis of streptomycin salts or by a number of other synthetic methods.

14 Safety

Maltol is generally regarded as an essentially nontoxic and nonirritant material. In animal feeding studies, it has been shown to be well tolerated with no adverse toxic, reproductive, or embryogenic effects observed in rats and dogs fed daily intakes of up to 200 mg/kg body-weight of maltol, for 2 years.(1) The WHO has set an acceptable daily intake for maltol at up to 1 mg/kg body-weight.(2,3) A case of allergic contact dermatitis, attributed to the use of maltol in a lip ointment, has been reported.(4)

LD_{50} (chicken, oral): 3.72 g/kg(5)
LD_{50} (guinea pig, oral): 1.41 g/kg
LD_{50} (mouse, oral): 0.85 g/kg
LD_{50} (mouse, SC): 0.82 g/kg
LD_{50} (rabbit, oral): 1.62 g/kg
LD_{50} (rat, oral): 1.41 g/kg

15 Handling Precautions

Observe normal precautions appropriate to the circumstances and quantity of material handled. Maltol should be used in a well-ventilated environment. Eye protection is recommended.

16 Regulatory Status

GRAS listed. Included in the FDA Inactive Ingredients Database (oral solutions and syrups). Included in the Canadian Natural Health Products Ingredients Database.

17 Related Substances

Ethyl maltol.

18 Comments

Maltol is a good chelating agent and various metal complexes, e.g. aluminum maltol and ferric maltol, have been investigated as potentially useful therapeutic or experimental agents.(6–9)

Maltol is a constituent of Korean red ginseng.(10)

A specification for maltol is included in the *Food Chemicals Codex*.(11)

The EINECS number for maltol is 204-271-8. The PubChem Compound ID (CID) for maltol is 8369.

19 Specific References

1 Gralla EJ, *et al.* Toxicity studies with ethyl maltol. *Toxicol Appl Pharmacol* 1969; **15**: 604–613.
2 FAO/WHO. Evaluation of certain food additives. Twenty-fifth report of the joint FAO/WHO expert committee on food additives. *World Health Organ Tech Rep Ser* 1981; No. 669.
3 FAO/WHO: Evaluation of certain food additives. Sixty-fifth report of the joint FAO/WHO expert committee on food additives. *World Health Organ Tech Rep Ser* 2006; No. 934.
4 Taylor AE, *et al.* Allergic contact dermatitis from strawberry lipsalve. *Contact Dermatitis* 1996; **34**(2): 142–143.
5 Lewis RJ, ed. *Sax's Dangerous Properties of Industrial Materials*, 11th edn. New York: Wiley, 2004; 2275.
6 Finnegan MM, *et al.* A neutral water-soluble aluminum complex of neurological interest. *J Am Chem Soc* 1986; **108**: 5033–5035.
7 Barrand MA, *et al.* Effects of the pyrones, maltol and ethyl maltol, on iron absorption from the rat small intestine. *J Pharm Pharmacol* 1987; **39**: 203–211.
8 Singh RK, Barrand MA. Lipid peroxidation effects of a novel iron compound, ferric maltol. A comparison with ferrous sulfate. *J Pharm Pharmacol* 1990; **42**: 276–279.
9 Kelsey SM, *et al.* Absorption of low and therapeutic doses of ferric maltol, a novel ferric iron compound, in iron deficient subjects using a single dose iron absorption test. *J Clin Pharm Ther* 1991; **16**: 117–122.
10 Wei J. [Studies on the constituents of Korean red ginseng – the isolation and identification of 3-hydroxy-2-methyl-4-pyrone.] *Acta Pharmaceutica Sinica* 1982; **17**: 549–550[in Chinese].
11 *Food Chemicals Codex*, 7th edn. Bethesda, MD: United States Pharmacopeia, 2010.

20 General References

Allen LV. Featured excipient: flavor enhancing agents. *Int J Pharm Compound* 2003; **7**(1): 48–50.

LeBlanc DT, Akers HA. Maltol and ethyl maltol: from the larch tree to successful food additives. *Food Technol* 1989; **43**(4): 78–84.

Registry of Toxic Effects of Chemical Substances Database. http://ccinfoweb.ccohs.ca/rtecs/search.html (accessed 24 November 2011).

21 Author

MF Tanenbaum.

22 Date of Revision

2 March 2012.

Maltose

1 Nonproprietary Names

JP: Maltose Hydrate
USP–NF: Maltose

2 Synonyms

Advantose 100; cextromaltose; *Finetose*; *Finetose F*; 4-O-α-D-glucopyranosyl-β-D-glucose; 4-(α-D-glucosido)-D-glucose; malt sugar; maltobiose; *Maltodiose*; *Maltose HH*; *Maltose HHH*; *Maltose PH*; *Sunmalt*; *Sunmalt S*.

3 Chemical Name and CAS Registry Number

4-O-α-D-Glucopyranosyl-β-D-glucopyranose anhydrous [69-79-4]
4-O-α-D-Glucopyranosyl-β-D-glucopyranose monohydrate [6363-53-7]

4 Empirical Formula and Molecular Weight

$C_{12}H_{22}O_{11}$	342.30 (anhydrous)
$C_{12}H_{22}O_{11}\cdot H_2O$	360.31 (monohydrate)

5 Structural Formula

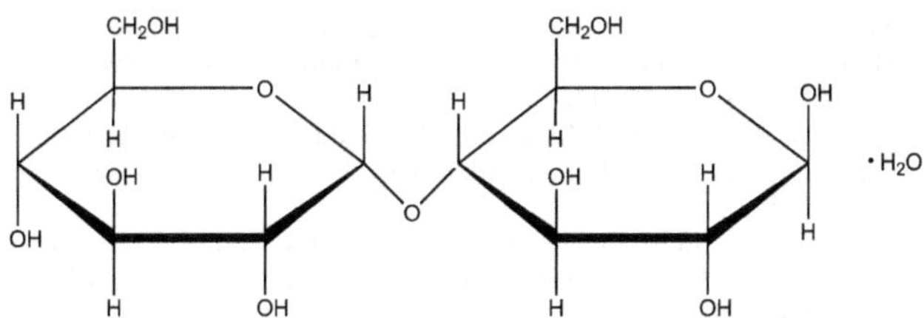

6 Functional Category

Sweetening agent; tablet and capsule diluent.

7 Applications in Pharmaceutical Formulation or Technology

Maltose is a disaccharide carbohydrate widely used in foods and pharmaceuticals. In parenteral products, maltose may be used as a source of sugar, particularly for diabetic patients.

Crystalline maltose is used as a direct-compression tablet excipient in chewable and nonchewable tablets.(1–3)

8 Description

Maltose occurs as white crystals or as a crystalline powder. It is odorless and has a sweet taste approximately 30% that of sucrose.

SEM 1: Excipient: crystalline maltose; manufacturer: SPI Pharma Group; lot no.: 8K110947; magnification: 100×; voltage 10 kV.

9 Pharmacopeial Specifications

See Table I. *See also* Section 18.

Table I: Pharmacopeial specifications for maltose.

Test	JP XV	USP35–NF30
Identification	+	+
Specific rotation	+126° to +131°	—
pH	4.5–6.5	—
for anhydrous	—	3.7–4.7
for monohydrate	—	4.0–5.5
Clarity and color of solution	+	—
Chloride	≤0.018%	—
Sulfate	≤0.024%	—
Heavy metals	≤4 ppm	≤5 μg/g
Arsenic	≤1.3 ppm	—
Dextrin, soluble starch and sulfite	+	+
Nitrogen	≤0.01%	—
Related substances	+	—
Loss on drying	≤0.5%	—
Water	—	—
for anhydrous	—	≤1.5%
for monohydrate	—	4.5–6.5%
Residue on ignition	≤0.1%	≤0.05%
Assay	+	+

10 Typical Properties

Acidity/alkalinity pH = 4.5–6.5 for a 10% w/v aqueous solution.
Angle of repose 25° for *Advantose 100*.(1)
Density (bulk) 0.67–0.72 g/cm^3 for *Advantose 100*.(1)
Density (tapped) 0.73–0.81 g/cm^3 for *Advantose 100*.(1)
Dissociation constant pK_a = 12.05 at 21°C
Flowability 18% (Carr compressibility index) for *Advantose 100*.(3)
Melting point 102–125°C.(1–4)

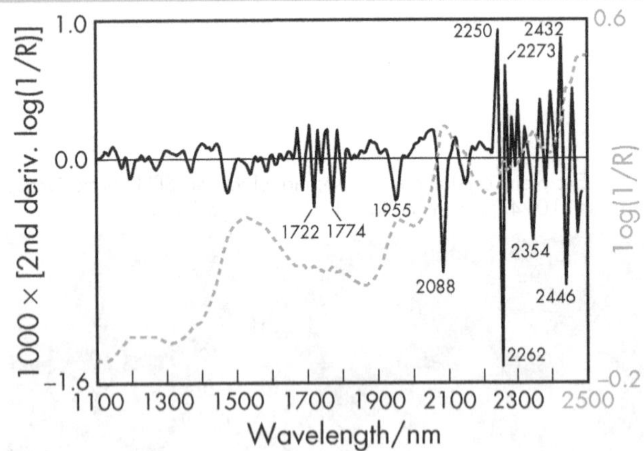

Figure 1: Near-infrared spectrum of maltose monohydrate measured by reflectance.

Particle size distribution Not more than 15% greater than 210 μm, and not more than 15% smaller than 74 μm in size for *Advantose 100*.[1]

Solubility Very soluble in water; very slightly soluble in cold ethanol (95%); practically insoluble in ether.

Specific surface area 0.08 m^2/g for *Advantose 100*.[1]

Spectroscopy

NIR spectra *see* Figure 1.

11 Stability and Storage Conditions

Maltose should be stored in a well-closed container in a cool, dry place.

12 Incompatibilities

Maltose may react with oxidizing agents. A Maillard-type reaction may occur between maltose and compounds with a primary amine group, e.g. glycine, to form brown-colored products.[6]

13 Method of Manufacture

Maltose monohydrate is prepared by the enzymatic degradation of starch.

14 Safety

Maltose is used in oral and parenteral pharmaceutical formulations and is generally regarded as an essentially nontoxic and nonirritant material. However, there has been a single report of a liver transplantation patient with renal failure who developed hyponatremia following intravenous infusion of normal immunoglobulin in 10% maltose. The effect, which recurred on each of four successive infusions, resembled that of hyperglycemia and was thought to be due to accumulation of maltose and other osmotically active metabolites in the extracellular fluid.[5]

LD_{50} (mouse, IV): 26.8 g/kg[7]
LD_{50} (mouse, SC): 38.6 g/kg
LD_{50} (rabbit, IV): 25.2 g/kg
LD_{50} (rat, IP): 30.6 g/kg
LD_{50} (rat, IV): 15.3 g/kg
LD_{50} (rat, oral): 34.8 g/kg

15 Handling Precautions

Observe normal precautions appropriate to the circumstances and quantity of material handled. Eye protection, rubber or plastic gloves, and a dust respirator are recommended. When heated to decomposition, maltose emits acrid smoke and irritating fumes.

16 Regulatory Status

In the US, maltose is considered as a food by the FDA and is therefore not subject to food additive and GRAS regulations. Included in the FDA Inactive Ingredients Database (oral solutions). Included in the Canadian Natural Health Products Ingredients Database. Included in parenteral products available in a number of countries worldwide.

17 Related Substances

Glucose, liquid.

18 Comments

Crystalline maltose, e.g. *Advantose 100* (SPI Pharma, Inc.), is spray-dried to produce spherical particles with good flow properties. The material is also nonhygroscopic and is highly compressible.

A specification for maltose syrup powder is contained in the *Japanese Pharmaceutical Excipients* (JPE).[8]

The EINECS number for maltose is 200-716-5. The PubChem Compound ID (CID) for maltose includes 6255 and 23724983.

19 Specific References

1 SPI Pharma, Inc. Technical literature: *Advantose 100* Maltose powder for direct compression, 2008.
2 Bowe KE, *et al.* Crystalline maltose: a direct compression pharmaceutical excipient. *Pharm Technol Eur* 1998; **10**(5): 40.
3 Mulderrig KB. Placebo evaluation of selected sugar-based excipients in pharmaceutical and nutraceutical tableting. *Pharm Technol* 2000; 24(5): 34, 36, 38, 40, 42, 44.
4 Mallinckrodt Chemical. Material safety data sheet: Maltose, 2008.
5 Palevsky PM, *et al.* Maltose-induced hyponatremia. *Ann Intern Med* 1993; **118**(7): 526–528.
6 Mundt S, Wedzicha BL. Role of glucose in the Maillard browning of maltose and glycine: a radiochemical approach. *J Agric Food Chem* 2005; 53: 6798–6803.
7 Lewis RJ, ed. *Sax's Dangerous Properties of Industrial Materials*, 11th edn. New York: Wiley, 2004: 2275.
8 Japan Pharmaceutical Excipients Council. *Japan Pharmaceutical Excipients*, 2004. Tokyo: Yakuji Nippo, 2004: 516–518.

20 General References

—

21 Author

CK Tye.

22 Date of Revision

1 March 2012.

Mannitol

1 Nonproprietary Names

BP: Mannitol
JP: D-Mannitol
PhEur: Mannitol
USP–NF: Mannitol

2 Synonyms

Compressol; Cordycepic acid; *C*PharmMannidex*; E421; *Emprove*; *Ludiflash*; manita; manitol; manna sugar; mannit; D-mannite; mannite; mannitolum; *Mannogem*; *Pearlitol.*

3 Chemical Name and CAS Registry Number

D-Mannitol [69-65-8]

4 Empirical Formula and Molecular Weight

$C_6H_{14}O_6$ 182.17

5 Structural Formula

6 Functional Category

Lyophilization aidplasticizing agent; sweetening agent; tablet and capsule diluent; tonicity agent.

7 Applications in Pharmaceutical Formulation or Technology

Mannitol is widely used in pharmaceutical formulations and food products. In pharmaceutical preparations it is primarily used as a diluent (10–90% w/w) in tablet formulations, where it is of particular value since it is not hygroscopic and may thus be used with moisture-sensitive active ingredients.[1,2]

Mannitol may be used in direct-compression tablet applications,[3,4] for which the granular and spray-dried forms are available,[5] or in wet granulations.[6,7] Granulations containing mannitol have the advantage of being easily dried. Specific tablet applications include antacid preparations, glyceryl trinitrate tablets, and vitamin preparations. Mannitol is commonly used as an excipient in the manufacture of chewable tablet formulations because of its negative heat of solution, sweetness, and 'mouth feel'.[7,8] It is also used as a diluent in rapidly dispersing oral dosage forms.[9,10]

Mannitol has been used to prevent thickening in aqueous antacid suspensions of aluminum hydroxide (<7% w/v). It has also been used as a plasticizer in soft-gelatin capsules, as a component of sustained-release tablet formulations,[11] as a carrier in dry powder inhalers,[12,13] and as a diluent in rapidly dispersing oral dosage forms.[14,15]

In lyophilized preparations, mannitol (20–90% w/w) has been used as a carrier to produce a stiff, homogeneous cake that improves the appearance of the lyophilized plug in a vial.[16–18] A pyrogen-free form is available specifically for this use.

Mannitol is used in food applications as a bulking agent.

8 Description

Mannitol is D-mannitol. It is a hexahydric alcohol related to mannose and is isomeric with sorbitol.

Mannitol occurs as a white, odorless, crystalline powder, or free-flowing granules. It has a sweet taste, approximately as sweet as glucose and half as sweet as sucrose, and imparts a cooling sensation in the mouth. Microscopically, it appears as orthorhombic needles when crystallized from alcohol. Mannitol shows polymorphism.[19]

9 Pharmacopeial Specifications

See Table I. *See also* Section 18.

10 Typical Properties

Compressibility *see* Figure 1.

SEM 1: Excipient: mannitol; manufacturer: Merck; magnification: 50×; voltage: 3.5 kV.

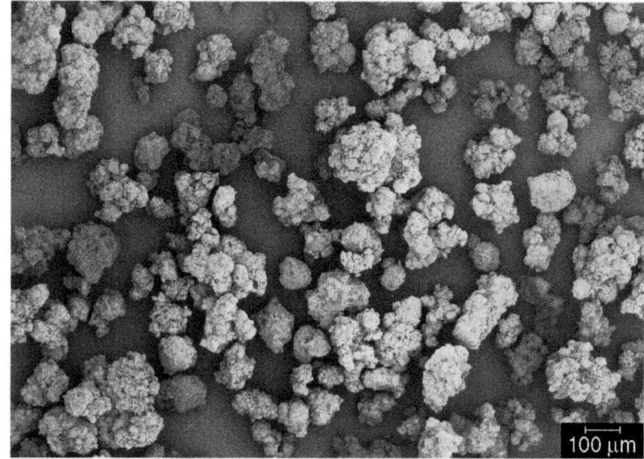

SEM 2: Excipient: mannitol; manufacturer: Merck; magnification: 500×; voltage: 3.5 kV.

SEM 3: Excipient: mannitol powder; manufacturer: SPI Polyols Inc.; lot no: 3140G8; magnification: 100×.

SEM 4: Excipient: mannitol granular; manufacturer: SPI Polyols Inc.; lot no: 2034F8; magnification: 100×.

Density (bulk)

0.430 g/cm^3 for powder;

0.7 g/cm^3 for granules.

Density (tapped)

0.734 g/cm^3 for powder;

0.8 g/cm^3 for granules.

Density (true) 1.514 g/cm^3

Dissociation constant pK_a = 13.5 at 18°C

Flash point <150°C

Flowability Powder is cohesive, granules are free flowing.

Heat of combustion 16.57 kJ/g (3.96 kcal/g)

Heat of solution −120.9 J/g (−28.9 cal/g) at 25°C

Melting point 166–168°C

Moisture content *see* Figure 2.

Osmolarity A 5.07% w/v aqueous solution is isoosmotic with serum.

Particle size distribution

Pearlitol 300 DC: maximum of 0.1% greater than 500 μm and minimum of 90% greater than 200 μm in size;

Pearlitol 400 DC: maximum of 20% greater than 500 μm and minimum of 85% greater than 100 μm in size;

Pearlitol 500 DC: maximum of 0.5% greater than 841 μm and minimum of 90% greater than 150 μm in size.

Table I: Pharmacopeial specifications for mannitol.

Test	JP XV	PhEur 7.4	USP35–NF30
Identification	+	+	+
Characters	−	+	−
Appearance of solution	+	+	−
Melting range	166–169°C	165–170°C	164–169°C
Specific rotation	+137° to +145°	+23° to +25°	+137° to +145°
Conductivity	–	≤20 μS·cm^{-1}	−
Acidity	+	−	+
Loss on drying	≤0.3%	≤0.5%	≤0.3%
Chloride	≤0.007%	−	≤0.007%
Sulfate	≤0.01%	−	≤0.01%
Arsenic	≤1.3 ppm	−	≤1 ppm
Lead	–	≤0.5 ppm	−
Nickel	+	≤1 ppm	−
Heavy metals	≤5 ppm	−	−
Reducing sugars	+	≤0.2%	+
Residue on ignition	≤0.10%	−	−
Related substances	−	+	−
Bacterial endotoxins[(a)]	−	≤4 IU/g[(b)] ≤2.5 IU/g[(c)]	−
Microbial contamination	−	≤100 cfu/g	−
Assay (dried basis)	≥ 98.0%	98.0–102.0%	96.0–101.5%

(a) Test applied only if the mannitol is to be used in the manufacture of parenteral dosage forms.
(b) For parenteral preparations having a concentration of 100 g/L or less of mannitol.
(c) For parental preparations having a concentration of more than 100 g/L of mannitol.

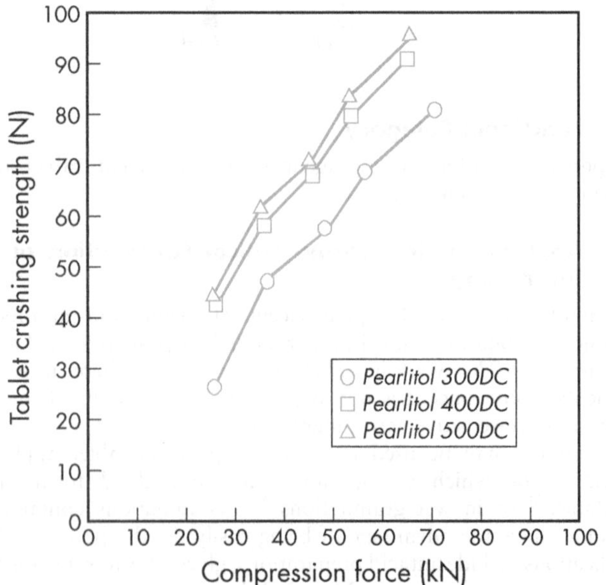

Figure 1: Compression characteristics of granular mannitol (*Pearlitol*, Roquette Frères).
Tablet diameter: 20 mm. Lubricant: magnesium stearate 0.7% w/w for *Pearlitol 400 DC* and *Pearlitol 500 DC*; magnesium stearate 1% w/w for *Pearlitol 300 DC*.

Average particle diameter is 250 μm for *Pearlitol 300 DC*, 360 μm for *Pearlitol 400 DC* and 520 μm for *Pearlitol 500 DC*.[20] *See also* Figure 3.

Refractive index n_D^{20} = 1.333

Solubility *see* Table II.

Specific surface area 0.37–0.39 m^2/g

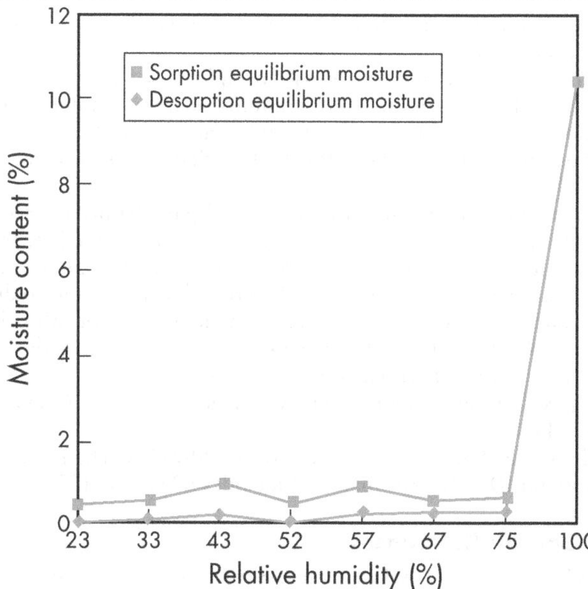

Figure 2: Sorption–desorption isotherm for mannitol.

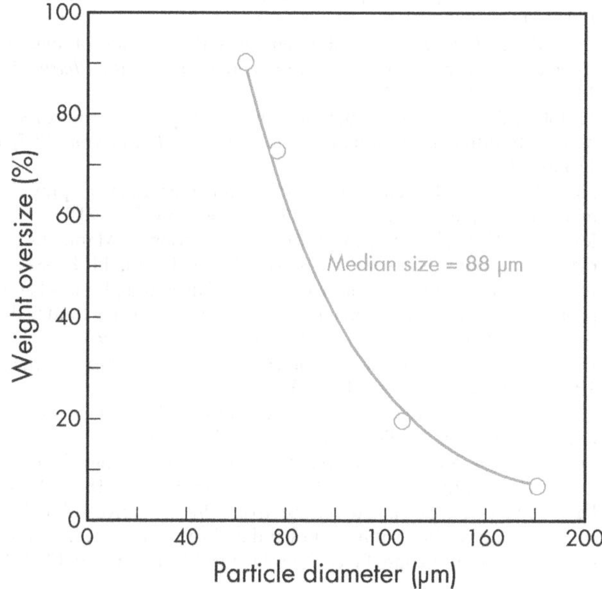

Figure 3: Particle size distribution of mannitol powder.

Table II: Solubility of mannitol.

Solvent	Solubility at 20°C
Alkalis	Soluble
Ethanol (95%)	1 in 83
Ether	Practically insoluble
Glycerin	1 in 18
Propan-2-ol	1 in 100
Water	1 in 5.5

Spectroscopy

IR spectra *see* Figure 4.

NIR spectra *see* Figure 5.

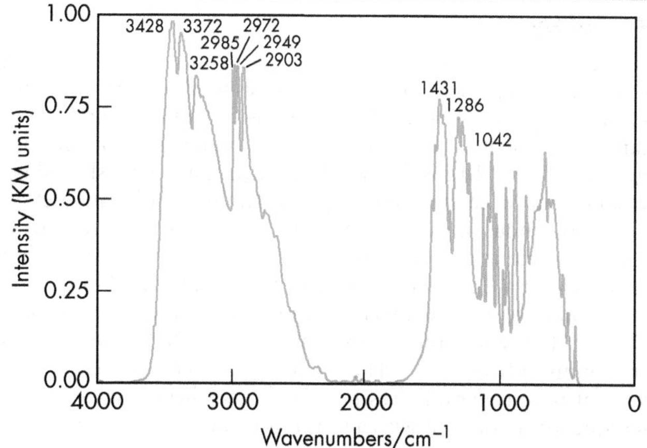

Figure 4: Infrared spectrum of mannitol measured by diffuse reflectance. Adapted with permission of Informa Healthcare.

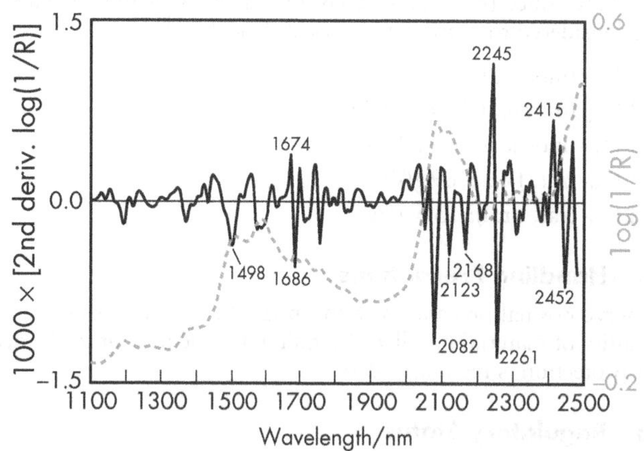

Figure 5: Near-infrared spectrum of mannitol measured by reflectance.

11 Stability and Storage Conditions

Mannitol is stable in the dry state and in aqueous solutions. Solutions may be sterilized by filtration or by autoclaving and if necessary may be autoclaved repeatedly with no adverse physical or chemical effects.[22] In solution, mannitol is not attacked by cold, dilute acids or alkalis, nor by atmospheric oxygen in the absence of catalysts. Mannitol does not undergo Maillard reactions.

The bulk material should be stored in a well-closed container in a cool, dry place.

12 Incompatibilities

Mannitol solutions, 20% w/v or stronger, may be salted out by potassium chloride or sodium chloride.[23] Precipitation has been reported to occur when a 25% w/v mannitol solution was allowed to contact plastic.[24] Sodium cephapirin at 2 mg/mL and 30 mg/mL concentration is incompatible with 20% w/v aqueous mannitol solution. Mannitol is incompatible with xylitol infusion and may form complexes with some metals such as aluminum, copper, and iron. Reducing sugar impurities in mannitol have been implicated in the oxidative degradation of a peptide in a lyophilized formation.[25] Mannitol was found to reduce the oral bioavailability of cimetidine compared to sucrose.[26]

13 Method of Manufacture

Mannitol may be extracted from the dried sap of manna and other natural sources by means of hot alcohol or other selective solvents. It is commercially produced by the catalytic or electrolytic reduction of monosaccharides such as mannose and glucose.

14 Safety

Mannitol is a naturally occurring sugar alcohol found in animals and plants; it is present in small quantities in almost all vegetables. Laxative effects may occur if mannitol is consumed orally in large quantities.[27] If it is used in foods as a bodying agent and daily ingestion of over 20 g is foreseeable, the product label should bear the statement 'excessive consumption may have a laxative effect'. After intravenous injection, mannitol is not metabolized to any appreciable extent and is minimally reabsorbed by the renal tubule, about 80% of a dose being excreted in the urine in 3 hours.[28]

A number of adverse reactions to mannitol have been reported, primarily following the therapeutic use of 20% w/v aqueous intravenous infusions.[29] The quantity of mannitol used as an excipient is considerably less than that used therapeutically and is consequently associated with a lower incidence of adverse reactions. However, allergic, hypersensitive-type reactions may occur when mannitol is used as an excipient.

An acceptable daily intake of mannitol has not been specified by the WHO since the amount consumed as a sweetening agent was not considered to represent a hazard to health.[30]

LD_{50} (mouse, IP): 14 g/kg[31]
LD_{50} (mouse, IV): 7.47 g/kg
LD_{50} (mouse, oral): 22 g/kg
LD_{50} (rat, IV): 9.69 g/kg
LD_{50} (rat, oral): 13.5 g/kg

15 Handling Precautions

Observe normal precautions appropriate to the circumstances and quantity of material handled. Mannitol may be irritant to the eyes; eye protection is recommended.

16 Regulatory Status

GRAS listed. Accepted for use as a food additive in Europe. Included in the FDA Inactive Ingredients Database (IP, IM, IV, and SC injections; infusions; buccal, oral and sublingual tablets, powders and capsules; ophthalmic preparations; topical solutions). Included in nonparenteral and parenteral medicines licensed in the UK. Included in the Canadian Natural Health Products Ingredients Database.

17 Related Substances

Sorbitol.

18 Comments

Mannitol is one of the materials that have been selected for harmonization by the Pharmacopeial Discussion Group. For further information see the General Information Chapter <1196> in the USP35–NF30, the General Chapter 5.8 in PhEur 7.4, along with the 'State of Work' document on the PhEur EDQM website, and also the General Information Chapter 8 in the JP XV.

Mannitol is an isomer of sorbitol, the difference between the two polyols occurring in the planar orientation of the OH group on the second carbon atom. Each isomer is characterized by its own individual set of properties, the most important difference being the response to moisture. Sorbitol is hygroscopic, while mannitol resists moisture sorption, even at high relative humidities.

Granular mannitol flows well and imparts improved flow properties to other materials. However, it usually cannot be used with concentrations of other materials exceeding 25% by weight. Recommended levels of lubricant are 1% w/w calcium stearate or 1–2% w/w magnesium stearate. Suitable binders for preparing granulations of powdered mannitol are gelatin, methylcellulose 400, starch paste, povidone, and sorbitol. Usually, 3–6 times as much magnesium stearate or 1.5–3 times as much calcium stearate is needed for lubrication of mannitol granulations than is needed for other excipients.

Ludiflash (BASF) is a coprocessed excipient used as a tablet filler, binder, and disintegrant, and contains mainly mannitol with povidone, crospovidone and polyvinyl acetate. It is used for direct compression, and especially for tablets that disintegrate in the mouth.[32]

Therapeutically, mannitol administered parenterally is used as an osmotic diuretic, as a diagnostic agent for kidney function, as an adjunct in the treatment of acute renal failure, and as an agent to reduce intracranial pressure, treat cerebral edema, and reduce intraocular pressure. Given orally, mannitol is not absorbed significantly from the gastrointestinal tract, but in large doses it can cause osmotic diarrhea; *see* Section 14.

A specification for mannitol is contained in the *Food Chemicals Codex* (FCC).[33]

The EINECS number for mannitol is 200-711-8. The PubChem Compound ID (CID) for mannitol includes 6251 and 453.

19 Specific References

1 Allen LV. Featured excipient: capsule and tablet diluents. *Int J Pharm Compound* 2000; **4**(4): 306–310324–325.
2 Yoshinari T, *et al.* Improved compaction properties of mannitol after a moisture induced polymorphic transition. *Int J Pharm* 2003; **258**(1–2): 121–131.
3 Debord B, *et al.* Study of different crystalline forms of mannitol: comparative behaviour under compression. *Drug Dev Ind Pharm* 1987; **13**: 1533–1546.
4 Molokhia AM, *et al.* Aging of tablets prepared by direct compression of bases with different moisture content. *Drug Dev Ind Pharm* 1987; **13**: 1933–1946.
5 Hulse WL, *et al.* The characterisation and compression of spray-dried mannitol samples. *Drug Dev Ind Pharm* 2009; **35**: 712–718.
6 Mendes RW, *et al.* Wet granulation: a comparison of Manni-Tab and mannitol. *Drug Cosmet Ind* 1978; **122**(3): 36, 38, 40, 44, 87–88.
7 Bouffard J, *et al.* Influence of processing variables and physicochemical properties on the granulation mechanisms of mannitol in a fluid bed top spray granulator. *Drug Dev Ind Pharm* 2005; **31**: 923–933.
8 Daoust RG, Lynch MJ. Mannitol in chewable tablets. *Drug Cosmet Ind* 1963; **93**(1): 26–2888, 92, 128–129.
9 Seager H. Drug development products and the Zydis fast dissolving dosage form. *J Pharm Pharmacol* 1998; **50**: 375–382.
10 Okuda Y, *et al.* A new formulation for orally disintegrating tablets using a suspension spray-coating method. *Int J Pharm* 2009; **382**: 80–87.
11 Parab PV, *et al.* Sustained release from *Precirol* (glycerol palmitostearate) matrix. Effect of mannitol and hydroxypropyl methylcellulose on the release of theophylline. *Drug Dev Ind Pharm* 1986; **12**: 1309–1327.
12 Steckel H, Bolzen N. Alternative sugars as potential carriers for dry powder inhalers. *Int J Pharm* 2004; **270**(1–2): 297–306.
13 Kataly W, *et al.* The enhanced aerosol performance of salbutamol from dry powders containing engineered mannitol as excipients. *Int J Pharm* 2010; **392**: 178–188.
14 Glover W, *et al.* Effect of particle size of dry powder mannitol on the lung deposition in healthy volunteers. *Int J Pharm* 2008; **349**: 314–322.
15 Cavatur RK, *et al.* Crystallization behavior of mannitol in frozen aqueous solutions. *Pharm Res* 2002; **19**: 894–900.
16 Liao XM, *et al.* Influence of processing conditions on the physical state of mannitol – implications in freeze drying. *Pharm Res* 2007; **24**: 370–376.
17 Pyne A, *et al.* Crystallization of mannitol below T_g' during freeze-drying in binary and ternary aqueous systems. *Pharm Res* 2002; **19**: 901–908.
18 Schneid S, *et al.* Influence of common excipients on the crystalline modification of freeze-dried mannitol. *Pharm Technol* 2008; **32**: 178–184.
19 Bauer H, *et al.* Investigations on polymorphism of mannitol/sorbitol mixtures after spray drying using differential scanning calorimetry, x-ray diffraction and near infrared spectroscopy. *Pharm Ind* 2000; **62**(3): 231–235.
20 Roquette Frères. Technical literature: *Pearlitol*, 2004.
21 Weast RC, ed. *Handbook of Chemistry and Physics*, 60th edn. Boca Raton: CRC Press, 1979; c–369.

22 Murty BSR, Kapoor JN. Properties of mannitol injection (25%) after repeated autoclavings. *Am J Hosp Pharm* 1975; **32**: 826–827.
23 Jacobs J. Factors influencing drug stability in intravenous infusions. *J Hosp Pharm* 1969; **27**: 341–347.
24 Epperson E. Mannitol crystallization in plastic containers [letter]. *Am J Hosp Pharm* 1978; **35**: 1337.
25 Dubost DC, *et al.* Characterization of a solid state reaction product from a lyophilized formulation of a cyclic heptapeptide. A novel example of an excipient-induced oxidation. *Pharm Res* 1996; **13**: 1811–1814.
26 Adkin DA, *et al.* The effect of mannitol on the oral bioavailability of cimetidine. *J Pharm Sci* 1995; **84**: 1405–1409.
27 Anonymous. Flatulence, diarrhoea, and polyol sweeteners. *Lancet* 1983; **ii**: 1321.
28 Porter GA, *et al.* Mannitol hemodilution–perfusion: the kinetics of mannitol distribution and excretion during cardiopulmonary bypass. *J Surg Res* 1967; **7**: 447–456.
29 McNeill IY. Hypersensitivity reaction to mannitol. *Drug Intell Clin Pharm* 1985; **19**: 552–553.
30 FAO/WHO. Evaluation of certain food additives and contaminants. Thirtieth report of the joint FAO/WHO expert committee on food additives. *World Health Organ Tech Rep Ser* 1987; No. 751.
31 Lewis RJ, ed. *Sax's Dangerous Properties of Industrial Materials*, 11th edn. New York: Wiley, 2004: 1944–1945.
32 BASF. Product literature: *Ludiflash*, 2004. http://www.pharma-solutions.basf.com (accessed 2 December 2011).
33 *Food Chemicals Codex*, 7th edn. Bethesda, MD: United States Pharmacopeia, 2010.

20 General References

Armstrong NA. Tablet manufacture. Diluents. In: Swarbrick J, Boylan JC, eds. *Encyclopedia of Pharmaceutical Technology*, 2nd edn, 3. New York: Marcel Dekker, 2002: 2713–2732.

Bolhuis GK, Armstrong NA. Excipients for direct compression – an update. *Pharm Tech Technol* 2006; **11**: 111–124.

European Directorate for the Quality of Medicines and Healthcare (EDQM). European Pharmacopoeia – State Of Work Of International Harmonisation. *Pharmeuropa* 2011; 23(4): 713–714. www.edqm.eu/site/-614.html (accessed 2 December 2011).

Pikal MJ. Freeze drying. In: Swarbrick J, Boylan JC, eds. *Encyclopedia of Pharmaceutical Technology*, 2nd edn, 2. New York: Marcel Dekker, 2002: 1299–1326.

21 Author

NA Armstrong.

22 Date of Revision

2 March 2012.

Mannitol and Sorbitol

M

1 Nonproprietary Names

None adopted.

2 Synonyms

Compressol S; *Compressol SM*.

3 Chemical Name and CAS Registry Number

See Section 8.

4 Empirical Formula and Molecular Weight

See Section 8.

5 Structural Formula

See Section 8.

6 Functional Category

Direct compression excipient.

7 Applications in Pharmaceutical Formulation or Technology

Mannitol and sorbitol is used in direct compression of tablets with high active loading and/or with actives that are difficult to compress as it has good flow and compressibility properties.(1,2) The coprocessed mixture with its improved functionality has been used in tableting of moisture-sensitive active substances as it has a lower hygroscopicity than standard sorbitol.(1–7)

Mannitol and sorbitol coprocessed mixture has a pleasant taste and favorable mouthfeel, making it especially suitable for chewable tablets.(1,2)

8 Description

Mannitol and sorbitol occurs as a white to off-white fine powder with a smooth mouthfeel.

9 Pharmacopeial Specifications

Both mannitol and sorbitol are listed as separate monographs in the BP, JP, PhEur, and USP–NF, but the combination is not listed. *See* Mannitol and Sorbitol.

10 Typical Properties

Density (tapped) 0.5–0.65 g/cm^3 for *Compressol SM* (8)
Loss on drying ≤1% for *Compressol SM* (8)
Particle size distribution

≤10% of the powder is retained on the 60-mesh screen

≤18% goes through the 325-mesh screen (for *Compressol SM*)(8)

Solubility Soluble in water (for *Compressol SM*)(9)
Total polyols 96–101% (for *Compressol SM*)(8)

11 Stability and Storage Conditions

Protect from excessive temperatures and humidity. Store in a tightly closed container.

12 Incompatibilities

See Mannitol and Sorbitol. The coprocessed mixture may react with strong acids or oxidizing agents.(4) Upon decomposition, this product emits carbon monoxide, carbon dioxide, and/or low-molecular-weight hydrocarbons.

13 Method of Manufacture

Mannitol and sorbitol coprocessed mixture is manufactured by a proprietary process.

14 Safety

Mannitol and sorbitol is a nonflammable white powder.[4] Eye contact may produce slight irritation. No skin irritation can be expected from single short-term exposure to this product. Prolonged or repeated contact may produce some irritation. Ingestion of large amounts may produce gastrointestinal disturbances including a laxative action. Overexposure to dusts may produce irritation of the respiratory system.

See also Mannitol and Sorbitol.

15 Handling Precautions

Observe normal precautions appropriate to the circumstances and quantity of the material handled.

Accumulation of airborne dust may present an explosion hazard in the presence of an ignition source. Avoid breathing dusts from this material.[4] Avoid contact with the eyes, and wash thoroughly after handling.

16 Regulatory Status

Mannitol and sorbitol is a coprocessed mixture of two materials both of which are GRAS listed and accepted for use as a food additive in Europe. *See* Mannitol and Sorbitol.

Mannitol GRAS listed. Accepted for use as a food additive in Europe. Included in the FDA Inactive Ingredients Database (IP, IM, IV, and SC injections; infusions; buccal, oral and sublingual tablets, powders and capsules; ophthalmic preparations; topical solutions). Included in nonparenteral and parenteral medicines licensed in the UK. Included in the Canadian Natural Health Products Ingredients Database.

Sorbitol GRAS listed. Accepted for use as a food additive in Europe. Included in the FDA Inactive Ingredients Database (intra-articular and IM injections; nasal; oral capsules, solutions, suspensions, syrups and tablets; rectal, topical, and vaginal preparations). Included in parenteral and nonparenteral medicines licensed in the UK. Included in the Canadian Natural Health Products Ingredients Database.

17 Related Substances

Mannitol; sorbitol.

18 Comments

—

19 Specific References

1 SPI Pharma Inc. Technical bulletin: *Compressol SM, Co-processed polyol*, January 2009.
2 SPI Pharma Inc. Technical bulletin: *Compressol S, Co-processed polyol*, September 2007.
3 Bolhuis GK, *et al.* Polyols as filler-binders for disintegrating tablets prepared by direct compaction. *Drug Dev Ind Pharm* 2009; **35**(6): 671–677.
4 Patel RP, Bhavsar M. Directly compressible materials via co-processing. *Int J PharmTech Res* 2009; **1**(3): 745–753.
5 Saha S, Shahiwala AF. Multifunctional coprocessed excipients for improved tabletting performance. *Expert Opin Drug Deliv* 2009; **6**(2): 197–208.
6 Bolhuis GK, Armstrong NA. Excipients for direct compaction – an update. *Pharm Dev Tech* 2006; **11**(1): 111–124.
7 Jivraj M, *et al.* An overview of the different excipients useful for the direct compression of tablets. *PSTT* 2000; **3**(2): 58–63.
8 SPI Pharma Inc. Product bulletin: *Compressol SM, Co-processed polyol*, November 2010.
9 SPI Pharma Inc. Material safety data sheet: *Compressol SM*, November 2009.

20 General References

—

21 Author

JT Heinämäki.

22 Date of Revision

1 May 2012.

D-Mannose

1 Nonproprietary Names

None adopted.

2 Synonyms

Carubinose; dextra mannose; D-mannopyranose; *MannoTab*; seminose.

3 Chemical Name and CAS Registry Number

(3*S*,4*S*,5*S*,6*R*)-6-(Hydroxymethyl)oxane-2,3,4,5-tetrol [3458-28-4]

4 Empirical Formula and Molecular Weight

$C_6H_{12}O_6$ 180.16

5 Structural Formula

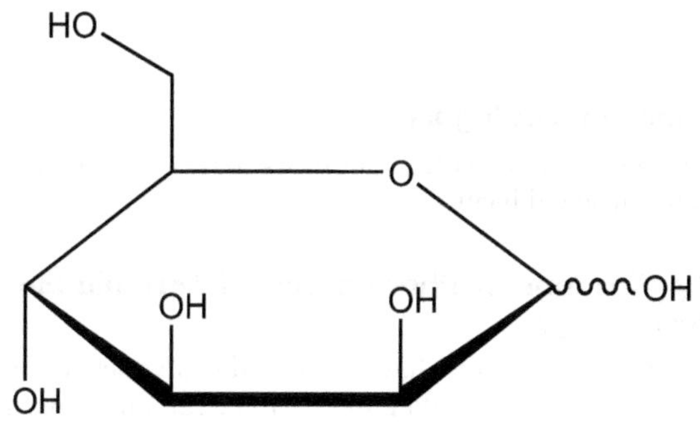

6 Functional Category

Antioxidant; sweetening agent.

7 Applications in Pharmaceutical Formulation or Technology

D-Mannose is used as a sweetening agent in oral pharmaceutical products. A directly compressible grade of D-mannose is used as an antioxidant in dietary applications.[1]

8 Description

D-Mannose is a colorless or white crystalline powder with a sweet taste but bitter aftertaste. The ambiguity of taste perception of D-mannose has been traced to actual differences in taste between the α and β forms.[2]

9 Pharmacopeial Specifications

–

10 Typical Properties

Density 1.54 g/cm^3
Dissociation constant $pK_a = 11.98$
Melting point 132–140°C (with decomposition); 133°C (α-form)
Solubility *see* Table I.

Table I: Solubility of D-mannose.

Solvent	Solubility at 20°C unless otherwise stated
Ethanol	1 in 250
Methanol	1 in 120
Pyridine	1 in 3.5
Water	1 in 0.4

Specific rotation $[\alpha]_D^{20}$ +13.7 to +14.2° (20% w/v in water containing approx. 0.05% w/v NH_3).

11 Stability and Storage Conditions

D-Mannose is stable in the dry state and in aqueous solutions. It can be fermented by yeast. Store in tightly closed containers in a cool, dry place.

12 Incompatibilities

—

13 Method of Manufacture

D-Mannose is prepared from glucose in the Lobry-de Bruyn-van Ekenstein transformation. It can also be prepared by the oxidation of mannitol.

D-Mannose can also be manufactured from wood-based or other biomass hydrolysates using an aqueous chromatographic separation process.[1]

14 Safety

D-Mannose is used in oral preparations. The majority of ingested D-mannose is excreted unconverted into the urine within 30–60 minutes, with no significant increase in blood-glucose levels during this time.[3]

15 Handling Precautions

Observe normal precautions appropriate to the circumstances and quantity of the material handled.

16 Regulatory Status

Included in the FDA Inactive Ingredients Database (oral, tablet). Included in the Canadian Natural Health Products Ingredients Database.

17 Related Substances

Dextrose; mannitol.

18 Comments

Included in the BP 2012 as a general reagent. D-Mannose is a naturopathic remedy for urinary tract infections.

D-Mannose has recently been investigated as a surface modifier for nanoparticles and liposomes to enhance oral and alveolar delivery by macrophage targeting.[4–7] Research has also shown that D-mannose is useful in freeze-dried suspensions to prevent aggregation.[8]

The EINECS number for D-mannose is 222-392-4. The PubChem Compound ID (CID) for D-mannose is 18950.

19 Specific References

1 Danisco. Product information: D-mannose. http://www.danisco.com/wps/wcm/connect/www/corporate/products/product_range/rare_sugars/d_mannose (accessed 29 February 2012).
2 Steinhardt RG, *et al.* Taste-structure correlation with α-D-Mannose and β-D-Mannose. *Science* 1962; **135**: 367–368.
3 Alton G, *et al.* Direct utilization of mannose for mammalian glycoprotein biosynthesis. *Glycobiology* 1998; **8**(3): 285–295.
4 Nimje N, *et al.* Mannosylated nanoparticulate carriers of rifabutin for alveolar targeting. *J Drug Target* 2009; **17**: 777–787.
5 Fievez V, *et al.* Targeting nanoparticles to M cells with non-peptidic ligands for oral vaccination. *Eur J Pharm Biopharm* 2009; **73**: 16–24.
6 Chono S. Effect of surface-mannose modification on aerosolized liposomal delivery to alveolar macrophages. *Drug Dev Ind Pharm* 2010; **36**: 102–107.
7 Nahar M, *et al.* *In vitro* evaluation of surface functionalized gelatin nanoparticles for macrophage targeting in the therapy of visceral leishmaniasis. *J Drug Target* 2010; **18**: 93–105.
8 Kamiya S. Physical properties of griseofulvin-lipid nanoparticles in suspension and their novel interaction mechanism with saccharide during freeze-drying. *Eur J Pharm Biopharm* 2010; **74**: 461–466.

20 General References

—

21 Authors

ME Fenton, RC Rowe.

22 Date of Revision

1 February 2011.

Medium-chain Triglycerides

M

1 Nonproprietary Names

BP: Medium-chain Triglycerides
PhEur: Triglycerides, Medium-Chain
USP–NF: Medium-Chain Triglycerides

2 Synonyms

Bergabest; caprylic/capric triglyceride; *Captex 300*; *Captex 355*; *Coconad*; *Crodamol GTCC*; *Delios VF*; glyceryl tricaprylate/caprate; *Labrafac CC*; *Labrafac Lipo*; *Labrafac WL1349*; MCT oil; *Miglyol 810*; *Miglyol 812*; *Myritol*; *Neobee M5*; *Nesatol*; oleum neutrale; oleum vegetable tenue; *ProKote 2855*; thin vegetable oil; triglycerida saturata media; *Waglinol 3/9280*.

3 Chemical Name and CAS Registry Number

Medium-chain triglycerides [438544-49-1]

4 Empirical Formula and Molecular Weight

≈500 (average)

The PhEur 7.4 describes medium-chain triglycerides as the fixed oil extracted from the hard, dried fraction of the endosperm of *Cocos nucifera* L. or from the dried endosperm of *Elaeis guineensis* Jacq. They consist of a mixture of triglycerides of saturated fatty acids, mainly of caprylic acid and of capric acid. They contain not less than 95% of saturated fatty acids.

5 Structural Formula

H–C(H)–O–R^1 / H–C–O–R^2 / H–C(H)–O–R^3

where R^1, R^2 and R^3 = $-C(=O)-(CH_2)_nCH_3$

n = 6-8

See also Section 4.

6 Functional Category

Emulsifying agent; solvent; suppository base; suspending agent; tablet and capsule diluent.

7 Applications in Pharmaceutical Formulation or Technology

Medium-chain triglycerides have been used in a variety of pharmaceutical formulations including oral, parenteral, and topical preparations.

In oral formulations, medium-chain triglycerides are used as the base for the preparation of oral emulsions, microemulsions, self-emulsifying systems, solutions, or suspensions of drugs that are unstable or insoluble in aqueous media, e.g. calciferol. Medium-chain triglycerides have also been investigated as intestinal-absorption enhancers[1,2] and have additionally been used as a filler in capsules and sugar-coated tablets, and as a lubricant or antiadhesion agent in tablets.

In parenteral formulations, medium-chain triglycerides have similarly been used in the production of emulsions, solutions, or suspensions intended for intravenous administration.[3–9] In rectal formulations, medium-chain triglycerides have been used in the preparation of suppositories containing labile materials. In cosmetics and topical pharmaceutical preparations, medium-chain triglycerides are used as a component of ointments, creams, and liquid emulsions.[5]

Although similar to long-chain triglycerides, medium-chain triglycerides have a number of advantages in pharmaceutical formulations, which include better spreading properties on the skin; no impedance of skin respiration; good penetration properties; good emollient and cosmetic properties; no visible film on the skin surface; good compatibility; good solvent properties; and good stability against oxidation.

8 Description

A colorless to slightly yellowish oily liquid that is practically odorless and tasteless. It solidifies at about 0°C. The oil is free from catalytic residues or the products of cracking.

9 Pharmacopeial Specifications

See Table I.

Table I: Pharmacopeial specifications for medium-chain triglycerides.

Test	PhEur 7.4	USP35–NF30
Identification	+	+
Characters	+	−
Appearance	+	+
Alkaline impurities	+	+
Relative density	0.93–0.96	0.93–0.96
Refractive index	1.440–1.452	1.440–1.452
Viscosity	25–33 mPa s	25–33 mPa s
Acid value	≤0.2	≤0.2
Hydroxyl value	≤10	≤10
Iodine value	≤1.0	≤1.0
Peroxide value	≤1.0	≤1.0
Saponification value	310–360	310–360
Unsaponifiable matter	≤0.5%	≤0.5%
Composition of fatty acids		
Caproic acid	≤2.0%	≤2.0%
Caprylic acid	50.0–80.0%	50.0–80.0%
Capric acid	20.0–50.0%	20.0–50.0%
Lauric acid	≤3.0%	≤3.0%
Myristic acid	≤1.0%	≤1.0%
Heavy metals[a]	≤10 ppm	10 µg/g
Water	≤0.2%	≤0.2%
Total ash	≤0.1%	≤0.1%
Chromium	≤0.05 ppm	≤0.05 µg/g
Copper[a]	≤0.1 ppm	≤0.1 µg/g
Lead[a]	≤0.1 ppm	≤0.1 µg/g
Nickel[a]	≤0.2 ppm	≤0.1 µg/g
Tin[a]	≤0.1 ppm	≤0.1 µg/g

(a) For medium-chain triglycerides intended for use in parenteral nutrition, the test for heavy metals is replaced by the tests for chromium, copper, lead, nickel, and tin.

10 Typical Properties

Acid value

≤0.1 for *Crodamol GTCC*;
≤0.1 for *Miglyol 810*;
≤0.1 for *Miglyol 812*;
≤0.05 for *Neobee M5*.

Cloud point

≤5°C for *Crodamol GTCC*;
≈10°C for *Miglyol 810*;
≈10°C for *Miglyol 812*.

Color

≤60 (Hazen color index) for *Crodamol GTCC*;
≤90 (Hazen color index) for *Miglyol 810*;
≤60 (Hazen color index) for *Miglyol 812*;
≤100 (Hazen color index) for *Neobee M5*.

Density

0.94–0.96 g/cm^3 for *Crodamol GTCC* at 20°C;
0.94–0.95 g/cm^3 for *Miglyol 810* at 20°C;
0.94–0.95 g/cm^3 for *Miglyol 812* at 20°C;
0.94 g/cm^3 for *Neobee M5* at 20°C.

Freezing point −5°C for *Neobee M5*

Hydroxyl value ≤8 for *Neobee M5*

Iodine number

≤1.0 for *Crodamol GTCC*;
≤0.5 for *Miglyol 810*;
≤0.5 for *Miglyol 812*;
≤0.5 for *Neobee M5*.

Moisture content

≤0.15% w/w for *Crodamol GTCC*;
≤0.10% w/w for *Miglyol 810*;
≤0.10% w/w for *Miglyol 812*;
≤0.15% w/w for *Neobee M5*.

Peroxide value

≤1.0 for *Miglyol 810*;
≤1.0 for *Miglyol 812*;
≤0.5 for *Neobee M5*.

Refractive index

1.4485–1.4500 for *Crodamol GTCC* at 20°C;
1.4485–1.4505 for *Miglyol 810* at 20°C;
1.4490–1.4510 for *Miglyol 812* at 20°C;
1.4480–1.4510 for *Neobee M5* at 20°C.

Saponification value

325–345 for *Crodamol GTCC*;
335–355 for *Miglyol 810*;
325–345 for *Miglyol 812*;
335–360 for *Neobee M5*.

Solubility Soluble in all proportions at 20°C in acetone, benzene, 2-butanone, carbon tetrachloride, chloroform, dichloromethane, ethanol, ethanol (95%), ether, ethyl acetate, petroleum ether, special petroleum spirit (boiling range 80–110°C), propan-2-ol, toluene, and xylene. Miscible with long-chain hydrocarbons and triglycerides; practically insoluble in water.

Surface tension

32.2 mN/m for *Crodamol GTCC* at 25°C;
31.0 mN/m for *Miglyol 810* at 20°C;
31.1 mN/m for *Miglyol 812* at 20°C;
32.3 mN/m for *Neobee M5* at 25°C.

Viscosity (dynamic)

27–30 mPa s (27–30 cP) for *Miglyol 810* at 20°C;
28–32 mPa s (28–32 cP) for *Miglyol 812* at 20°C;
23 mPa s (23 cP) for *Neobee M5* at 25°C.

11 Stability and Storage Conditions

Medium-chain triglycerides are stable over the wide range of storage temperatures that can be experienced in tropical and temperate climates. Ideally, however, they should be stored at temperatures not exceeding 25°C and not exposed to temperatures above 40°C for long periods. At low temperatures, samples of medium-chain triglycerides may become viscous or solidify. Samples should therefore be well melted and mixed before use, although overheating should be avoided.

In the preparation of microemulsions and self-emulsifying systems, emulsions, or aqueous suspensions of medium-chain triglycerides, care should be taken to avoid microbiological contamination of the preparation, since lipase-producing micro-organisms, which become active in the presence of moisture, can cause hydrolysis of the triglycerides. Hydrolysis of the triglycerides is revealed by the characteristic unpleasant odor of free medium-chain fatty acids.

Medium-chain triglycerides may be sterilized by maintaining at 170°C for 1 hour.

Medium-chain triglycerides should be stored protected from light in a well-filled and well-closed container. When stored dry, in sealed containers, medium-chain triglycerides remain stable for many years.

12 Incompatibilities

Preparations containing medium-chain triglycerides should not come into contact with polystyrene containers or packaging components since the plastic rapidly becomes brittle upon contact. Low-density polyethylene should also not be used as a packaging material as the medium-chain triglycerides readily penetrate the plastic, especially at high temperatures, forming an oily film on the outside. High-density polyethylene is a suitable packaging material. Closures based on phenol resins should be tested before use for compatibility with medium-chain triglycerides. Polyvinyl chloride packaging should also be tested for compatibility since medium-chain triglycerides can dissolve some plasticizers, such as phthalates, out of the plastic.

Materials recommended as safe for packaging medium-chain triglycerides are low-density polyethylene, polypropylene, glass, and metal.

13 Method of Manufacture

Medium-chain triglycerides are obtained from the fixed oil extracted from the hard, dried fraction of the endosperm of *Cocos nucifera* L. Hydrolysis of the fixed oil followed by distillation yields the required fatty acids, which are then re-esterified to produce the medium-chain triglycerides.

Although the PhEur 7.4 specifies that medium-chain fatty acids are obtained from coconut oil, medium-chain triglycerides are also to be found in substantial amounts in the kernel oils of certain other types of palm-tree, e.g. palm kernel oil and babassu oil. Some animal products, such as milk-fat, also contain small amounts (up to 4%) of the medium-chain fatty acid esters.

14 Safety

Medium-chain triglycerides are used in a variety of pharmaceutical formulations including oral, parenteral, and topical products, and are generally regarded as essentially nontoxic and nonirritant materials.

In acute toxicology studies in animals and humans, no irritant or other adverse reactions have been observed; for example, when they were patch-tested on more than 100 individuals, no irritation was produced on either healthy or eczematous skin. Medium-chain triglycerides are not irritating to the eyes.

Similarly, chronic toxicology studies in animals have shown no harmful adverse effects associated with medium-chain triglycerides following inhalation or intraperitoneal, oral, and parenteral administration.

In humans, administration of 0.5 g/kg body-weight medium-chain triglycerides to healthy individuals produced no change in blood or serum triglycerides compared to subjects receiving the same dose of the long-chain triglyceride triolein.

In patients consuming diets based on medium-chain triglycerides, adverse effects reported include abdominal pain and diarrhea.

LD_{50} (mouse, IV): 3.7 g/kg
LD_{50} (mouse, oral): 29.6 g/kg
LD_{50} (rat, oral): 33.3 g/kg

15 Handling Precautions

Observe normal precautions appropriate to the circumstances and quantity of material handled.

16 Regulatory Status

GRAS listed. Included in the FDA Inactive Ingredients Database (topical preparations). Included in nonparenteral and parenteral medicines licensed in Europe. Included in the Canadian Natural Health Products Ingredients Database.

17 Related Substances

Coconut oil; suppository bases, hard fat; vegetable oil, hydrogenated.

18 Comments

Medium-chain triglycerides may also be known as fractionated coconut oil, which contains three saturated lipid chains bound to a glycerin backbone, and are distinguished from other triglycerides by the length of the carbon chains, normally between 6 and 10.

Therapeutically, medium-chain triglycerides have been used as nutritional agents.[10] Diets containing medium-chain triglycerides are used in conditions associated with the malabsorption of fat, such as cystic fibrosis, since medium-chain triglycerides are more readily digested than long-chain triglycerides. Medium-chain triglycerides have been particularly investigated for their use in total parenteral nutrition (TPN) regimens in combination with long-chain triglycerides.[4,11]

A specification for medium-chain fatty acid triglycerides is included in the *Japanese Pharmaceutical Excipients* (JPE).[12]

19 Specific References

1 Swenson ES, Curatolo WJ. Intestinal permeability enhancement for proteins, peptides and other drugs: mechanisms and potential toxicity. *Adv Drug Delivery Rev* 1992; **8**: 39–92.
2 Spencer SA, *et al.* Evaluation of a special low birth weight formula, with and without the use of medium chain triglycerides. *Early Hum Dev* 1986; **13**: 87–95.
3 Bach A, *et al.* Metabolic effects following a short and medium-chain triglycerides load in dogs I: infusion of an emulsion of short and medium-chain triglycerides. *Arch Sci Physiol* 1972; **26**: 121–129.
4 Hatton J, *et al.* Safety and efficacy of a lipid emulsion containing medium-chain triglycerides. *Clin Pharm* 1990; **9**: 366–371.
5 Adams U, Neuwald F. Comparative studies of the release of salicylic acid from medium-chain triglyceride gel and paraffin ointment bases: *in vitro* and *in vivo*. *Pharm Ind* 1982; **44**: 625–629.
6 Pietkiewicz J, Sznitowska M. The choice of lipids and surfactants for injectable extravenous microspheres. *Pharmazie* 2004; **59**: 325–326.
7 Schaub E, *et al.* Pain on injection: a double-blind comparison of propofol with lidocaine pretreatment versus propofol formulated with long- and medium-chain triglycerides. *Anaesth Analg* 2004; **99**: 1699–1702.
8 Cournarie F, *et al.* Insulin-loaded w/o/w multiple emulsions: comparison of the performances of systems prepared with medium-chain triglycerides and fish oil. *Eur J Pharm Biopharm* 2004; **58**: 477–482.
9 Holmberg I, *et al.* Absorption of a pharmacological dose of vitamin D3 from two different lipid vehicles in man: comparison of peanut oil and a medium chain triaglyceride. *Biopharm Drug Dispos* 1990; **11**: 807–815.
10 Ruppin DC, Middleton WRJ. Clinical use of medium-chain triglycerides. *Drugs* 1980; **20**: 216–224.
11 Wolfram G. Medium-chain triglycerides (MCT) for total parenteral nutrition. *World J Surgery* 1986; **10**: 33–37.
12 Japan Pharmaceutical Excipients Council. *Japanese Pharmaceutical Excipients* 2004. Tokyo: Yakuji Nippo, 2004: 519.

20 General References

Akkar A, *et al.* Solubilizing poorly soluble antimycotic agents by emulsification via a solvent-free process. *AAPS Pharm Sci Tech* 2004; 5: E24.

21 Author

G Moss.

22 Date of Revision

1 March 2012.

M

Meglumine

1 Nonproprietary Names

BP: Meglumine
JP: Meglumine
PhEur: Meglumine
USP–NF: Meglumine

2 Synonyms

Meglumin; meglumina; megluminum; 1-methylamino-1-deoxy-D-glucitol; *N*-methylglucamine; *N*-methyl-D-glucamine.

3 Chemical Name and CAS Registry Number

1-Deoxy-1-(methylamino)-D-glucitol [6284-40-8]

4 Empirical Formula and Molecular Weight

$C_7H_{17}NO_5$ 195.21

5 Structural Formula

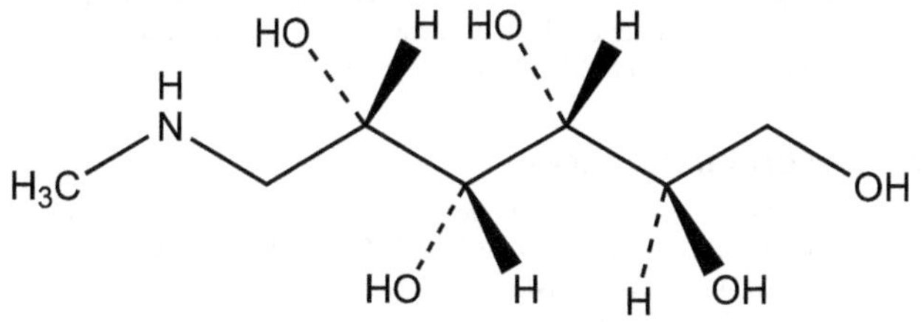

6 Functional Category

Buffering agent.

7 Applications in Pharmaceutical Formulation or Technology

Meglumine is an organic base used as a pH-adjusting agent and solubilizing agent, primarily in the preparation of soluble salts of iodinated organic acids used as X-ray contrast media.

8 Description

Meglumine occurs as a white to slightly yellow-colored crystalline powder; it is odorless or with a slight odor.

9 Pharmacopeial Specifications

See Table I.

Table I: Pharmacopeial specifications for meglumine.

Test	JP XV	PhEur 7.4	USP35–NF30
Identification	+	+	+
Appearance of solution	+	+	+
Melting range	128–131°C	128°C	128–132°C
Specific optical rotation	−16.0° to −17.0°	−16.0° to −17.0°	−15.7° to −17.3°
Reducing substances	—	≤0.2%	—
Loss on drying	≤0.5%	≤0.5%	≤1.0%
Residue on ignition	≤0.1%	≤0.1%	≤0.1%
Absence of reducing substances	+	—	+
Bacterial endotoxins	—	≤1.5 IU/g	—
Heavy metals	≤10 ppm	≤10 ppm	≤0.002%
Iron	—	≤10 ppm	—
Arsenic	≤1 ppm	—	—
Chloride	≤0.009%	≤100 ppm	—
Sulfate	≤0.019%	≤150 ppm	—
Nickel	—	≤5 ppm	—
Aluminum	—	≤5 ppm	—
Assay (dried basis)	≥99.0%	99.0–101.0%	99.0–100.5%

10 Typical Properties

Acidity/alkalinity pH = 10.5 (1% w/v aqueous solution).
Dissociation constant pK_a = 9.5 at 20°C
Melting point 128–132°C
Osmolarity A 5.02% w/v aqueous solution is iso-osmotic with serum.
Solubility *see* Table II.

Table II: Solubility of meglumine.

Solvent	Solubility at 20°C unless otherwise stated
Chloroform	Practically insoluble
Ethanol (95%)	1 in 80
	1 in 4.8 at 70°C
Ether	Practically insoluble
Water	1 in 1

Specific rotation $[\alpha]_D^{20}$ = −16.5° (10% w/v aqueous solution)
Spectroscopy

IR spectra *see* Figure 1.

NIR spectra *see* Figure 2.

11 Stability and Storage Conditions

Meglumine does not polymerize or dehydrate unless heated above 150°C for prolonged periods.

The bulk material should be stored in a well-closed container in a cool, dry place. Meglumine should not be stored in aluminum containers since it reacts to evolve hydrogen gas; it discolors if stored in containers made from copper or copper alloys. Stainless steel containers are recommended.

12 Incompatibilities

Incompatible with aluminum, copper, mineral acids, and oxidizing materials. Differential scanning calorimetry studies suggest meglumine is incompatible with glipizide.[1]

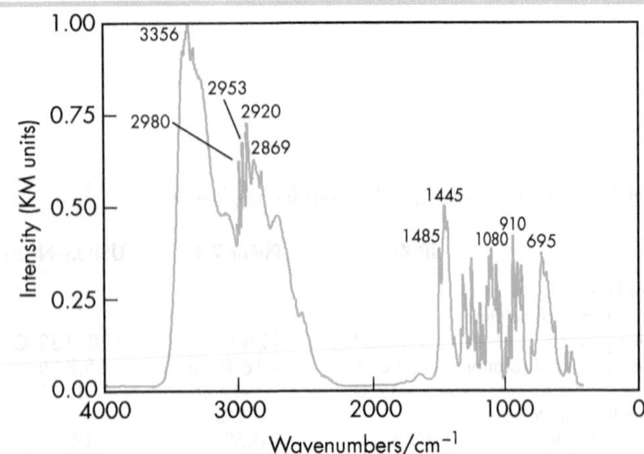

Figure 1: Infrared spectrum of meglumine measured by diffuse reflectance. Adapted with permission of Informa Healthcare.

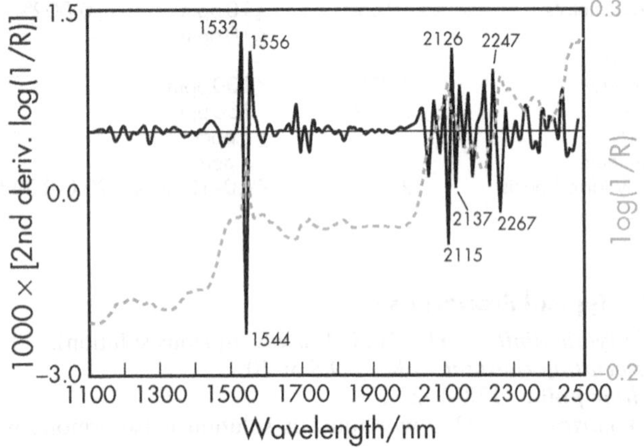

Figure 2: Near-infrared spectrum of meglumine measured by reflectance.

13 Method of Manufacture

Meglumine is prepared by the imination of glucose and monomethylamine, in an alcoholic solution, followed by catalytic hydrogenation.

14 Safety

Meglumine is widely used in parenteral pharmaceutical formulations and is generally regarded as a nontoxic material at the levels usually employed as an excipient.

LD_{50} (mouse, IP): 1.68 g/kg

15 Handling Precautions

Observe normal precautions appropriate to the circumstances and quantity of material handled. Meglumine should be handled in a well-ventilated environment, and eye protection, gloves, and a respirator are recommended. Exposure to meglumine dust should be kept below 10 mg/m^3 for total inhalable dust (8-hour TWA) or 5 mg/m^3 for respirable dust (8-hour TWA). There is a risk of explosion when meglumine dust is mixed with air.

16 Regulatory Status

Included in the FDA Inactive Ingredients Database (injections; oral tablets, capsules). Included in parenteral medicines licensed in the UK. Included in the Canadian Natural Health Products Ingredients Database.

17 Related Substances

Eglumine.

Eglumine

Empirical formula $C_8H_{19}NO_5$
Molecular weight 209.24
CAS number [14216-22-9]
Synonyms 1-Deoxy-1-(ethylamino)-D-glucitol; *N*-ethylglucamine.
Melting point ≈138°C
Comments Eglumine is prepared similarly to meglumine except that monoethylamine is used as the precursor, instead of monomethylamine.

18 Comments

The EINECS number for meglumine is 228-506-9. The PubChem Compound ID (CID) for meglumine is 8567.

19 Specific References

1 Verma RK, Garg S. Selection of excipients for extended release formulations of glipizide through drug-excipient compatibility testing. *J Pharm Biomed Anal* 2005; **38**: 633–644.

20 General References

Bremecker KD, *et al.* [Polyacrylate gels: use of new bases in drug formulation.] *Dtsch Apoth Ztg* 1990; **130**(8): 401–403[in German].

Chromy V, *et al.* D-(−)-*N*-Methylglucamine buffer for pH 8.5 to 10.5. *Clin Chem* 1978; **24**(2): 379–381.

Chromy V, *et al.* Use of *N*-methyl-D-glucamine as buffer in the determination of serum alkaline phosphatase activity. *Clin Chem* 1981; **27**(10): 1729–1732.

Japan Pharmaceutical Excipients Council. *Japanese Pharmaceutical Excipients Directory 1996*. Tokyo: Yakuji Nippon, 1996: 305.

21 Author

W Yang.

22 Date of Revision

1 March 2012.

Menthol

1 Nonproprietary Names

BP: Racementhol
JP: *dl*-Menthol
PhEur: Menthol, Racemic
USP: Menthol

2 Synonyms

Hexahydrothymol; 2-isopropyl-5-methylcyclohexanol; 4-isopropyl-1-methylcyclohexan-3-ol; 3-*p*-menthanol; *p*-menthan-3-ol; *dl*-menthol; mentholum racemicum; menthomenthol; mentoli; mentolis; peppermint camphor; racemic menthol.

3 Chemical Name and CAS Registry Number

(1*RS*,2*RS*,5*RS*)-(±)-5-Methyl-2-(1-methylethyl)cyclohexanol [15356-70-4]

Note that the following CAS numbers have also been used: [1490-04-6] and [89-78-1].

4 Empirical Formula and Molecular Weight

$C_{10}H_{20}O$ 156.27

5 Structural Formula

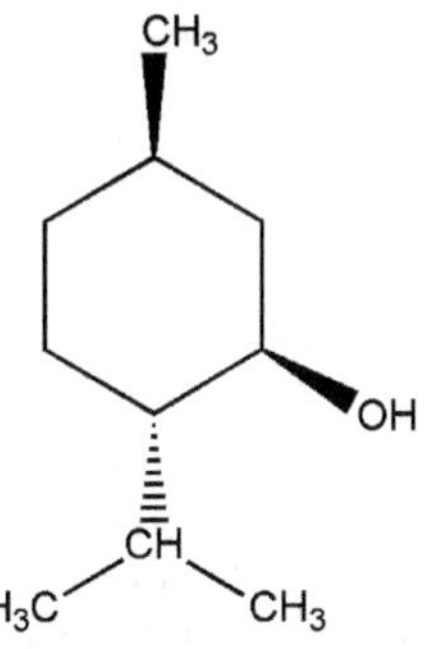

6 Functional Category

Flavoring agent.

7 Applications in Pharmaceutical Formulation or Technology

Menthol is widely used in pharmaceuticals, confectionery, and toiletry products as a flavoring agent or odor enhancer. In addition to its characteristic peppermint flavor, *l*-menthol, which occurs naturally, also exerts a cooling or refreshing sensation that is exploited in many topical preparations; *see* Section 18. Unlike mannitol, which exerts a similar effect due to a negative heat of solution, *l*-menthol interacts directly with the body's coldness receptors. *d*-Menthol has no cooling effect, while racemic menthol exerts an effect approximately half that of *l*-menthol. The propensity of menthol to sublime has been exploited to prepare formulated granules with increased porosity, which can increase dissolution and disintegration rate.[1]

When used to flavor tablets, menthol is generally dissolved in ethanol (95%) and sprayed onto tablet granules and not used as a solid excipient.

Menthol has been investigated as a skin-penetration enhancer and is also used in perfumery, tobacco products and chewing gum.

See Table I.

Figure 1: Photomicrograph of large DL-menthol crystals; magnification 7×. Manufacturer: Charkit Chemical Corp., USA.

Table I: Uses of menthol.

Use	Concentration (%)
Pharmaceutical products	
Inhalation	0.02–0.05
Oral suspension	0.003
Oral syrup	0.005–0.015
Tablets	0.2–0.4
Topical formulations	0.05–10.0
Cosmetic products	
Toothpaste	0.4
Mouthwash	0.1–2.0
Oral spray	0.3

8 Description

Racemic menthol is a mixture of equal parts of the (1*R*,2*S*,5*R*)- and (1*S*,2*R*,5*S*)-isomers of menthol. It is a free-flowing or agglomerated crystalline powder, or colorless, prismatic, or acicular shiny crystals, or hexagonal or fused masses with a strong characteristic odor and taste. The crystalline form may change with time owing to sublimation within a closed vessel. The USP35–NF30 specifies that menthol may be either naturally occurring *l*-menthol or synthetically prepared racemic or *dl*-menthol. However, the JP XV and PhEur 7.4, along with other pharmacopeias, include two separate monographs for racemic and *l*-menthol.

See Figure 1.

9 Pharmacopeial Specifications

See Table II.

10 Typical Properties

Boiling point 212°C
Flash point 91°C
Melting point 34°C
Refractive index n_D^{20} = 1.4615
Solubility Very soluble in ethanol (95%), chloroform, ether, fatty oils and liquid paraffin; freely soluble in glacial acetic acid;

M

Table II: Pharmacopeial specifications for menthol.

Test	JP XV	PhEur 7.4	USP35–NF30
Identification	+	+	+
Characters	−	+	−
Acidity or alkalinity	−	+	−
Congealing range	27–28°C	−	27–28°C
Melting point			
dl-menthol	−	≈34°C	−
l-menthol	42–44°C	≈43°C	41–44°C
Specific optical rotation			
dl-menthol	−2° to +2°	−0.2° to +0.2°	−2° to +2°
l-menthol	−45° to −51°	−48° to −51°	−45° to −51°
Readily oxidizable substances	−	−	+
Chromatographic purity	−	−	+
Related substances	−	+	−
Appearance of solution	−	+	−
Nonvolatile residue	+	−	≤0.05%
Residue on evaporation	−	≤0.05%	−
Thymol	+	−	−
Nitromethane or nitroethane	+	−	−
Assay	≥98.0%	−	−

soluble in acetone and benzene; very slightly soluble in glycerin; practically insoluble in water.

Specific gravity 0.904 at 15°C

Specific rotation

$[\alpha]_D^{20}$ = −2° to +2° (10% w/v alcoholic solution)

See also Section 17.

Spectroscopy

IR spectra *see* Figure 2.

NIR spectra *see* Figure 3.

11 Stability and Storage Conditions

A formulation containing menthol 1% w/w in aqueous cream has been reported to be stable for up to 18 months when stored at room temperature.[2]

Menthol should be stored in a well-closed container at a temperature not exceeding 25°C, since it sublimes readily.

12 Incompatibilities

Incompatible with: butylchloral hydrate; camphor; chloral hydrate; chromium trioxide; β-naphthol; phenol; potassium permanganate; pyrogallol; resorcinol; and thymol.

13 Method of Manufacture

Menthol occurs widely in nature as *l*-menthol and is the principal component of peppermint and cornmint oils obtained from the *Mentha piperita* and *Mentha arvensis* species. Commercially, *l*-menthol is mainly produced by extraction from these volatile oils. It may also be prepared by partial or total synthetic methods.

Racemic menthol is prepared synthetically via a number of routes, e.g. by hydrogenation of thymol.

14 Safety

Almost all toxicological data for menthol relate to its use as a therapeutic agent rather than as an excipient. Inhalation or ingestion of large quantities can result in serious adverse reactions such as ataxia[3] and CNS depression,[4] hypersensitivity reactions, severe abdominal pain, nausea, vomiting, vertigo, drowsiness, and

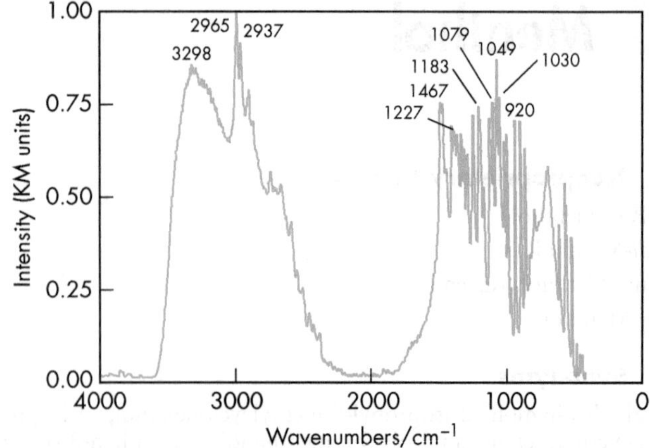

Figure 2: Infrared spectrum of menthol measured by diffuse reflectance. Adapted with permission of Informa Healthcare.

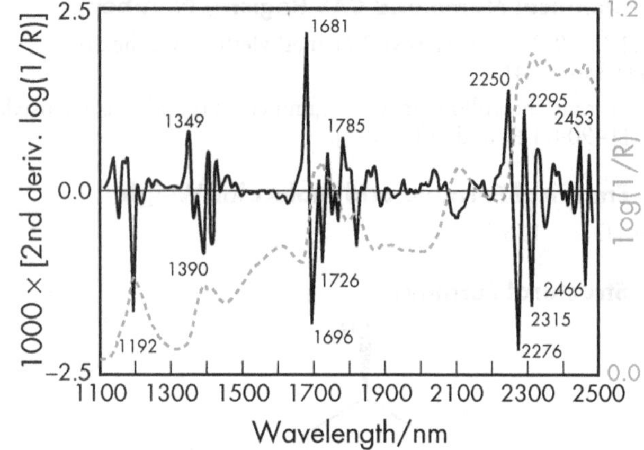

Figure 3: Near-infrared spectrum of menthol measured by reflectance.

coma.[5] Although menthol is essentially nonirritant there have been some reports of hypersensitivity following topical application.[6,7] In a Polish study approximately 1% of individuals were determined as being sensitive to menthol.[8] There have been reports of apnea and instant collapse in infants after the local application of menthol to their nostrils.[5]

The WHO has set an acceptable daily intake of menthol at up to 0.4 mg/kg body-weight.[9]

LD_{50} (rat, IM): 10.0 g/kg[10]

LD_{50} (rat, oral): 3.18 g/kg

15 Handling Precautions

May be harmful by inhalation or ingestion in large quantities; may be irritant to the skin, eyes, and mucous membranes. Observe normal precautions appropriate to the circumstances and quantity of material handled. Eye protection, chemical resistant gloves, and respirators are recommended.

Avoid prolonged or repeated exposure.

16 Regulatory Status

Included in the FDA Inactive Ingredients Database (dental preparations, inhalations, oral aerosols, capsules, solutions, suspensions, syrups, and tablets; also topical preparations). Menthol is listed as a food additive in the EAFUS list compiled by the FDA. Included in nonparenteral medicines licensed in the UK. Accepted for use in foods and confectionery as a flavoring agent of natural origin. Included in the Canadian Natural Health Products Ingredients Database.

17 Related Substances

d-Menthol; *l*-menthol; thymol.

d-Menthol

Empirical formula $C_{10}H_{20}O$
Molecular weight 156.27
CAS number [15356-60-2]
Synonyms (1*S*,2*R*,5*S*)-(+)-5-Methyl-2-(1-methylethyl)cyclohexanol.
Appearance Colorless, prismatic or acicular, shiny crystals, without the characteristic odor, taste, and cooling effect of *l*-menthol. The crystalline form may change with time owing to sublimation within a closed vessel.
Flash point 91°C
Melting point 43–44°C
Specific rotation $[\alpha]_D^{23} = +48°$ (10% w/v alcoholic solution)

l-Menthol

Empirical formula $C_{10}H_{20}O$
Molecular weight 156.27
CAS number [2216-51-5]
Synonyms Levomenthol; levomentholum; (−)menthol; (1*R*,2*S*,5*R*)-(−)-5-methyl-2-(1-methylethyl)cyclohexanol.
Appearance Colorless, prismatic, or acicular, shiny crystals, with a strong, characteristic odor, taste, and cooling effect. The crystalline form may change with time owing to sublimation within a closed vessel.
Flash point >100°C
Melting point 41–44°C
Refractive index $n_D^{20} = 1.4600$
Specific rotation $[\alpha]_D^{20} = -50°$ (10% w/v alcoholic solution)
Safety

LD_{50} (mouse, IP): 6.6 g/kg[10]
LD_{50} (mouse, oral): 3.4 g/kg
LD_{50} (rat, IP): 0.7 g/kg
LD_{50} (rat, oral): 3.3 g/kg

18 Comments

Therapeutically, when applied to the skin, menthol dilates the blood vessels, causing a sensation of coldness followed by an analgesic effect. It relieves itching and is used in creams, lotions, and ointments. When administered orally in small doses menthol has a carminative action.

It should be noted that considerable variation in the chemical composition of natural menthol oils can occur depending upon their country of origin.

A specification for menthol is contained in the *Food Chemicals Codex* (FCC).[11]

The EINECS number for menthol is 201-939-0. The PubChem Compound ID (CID) for menthol is 1254.

19 Specific References

1 Kummar R, *et al.* Formulation evaluation of mouth dissolving tablets of fenofibrate using sublimation technique. *Int J ChemTech Res* 2009; 1(4): 840–850.
2 Gallagher P, Jones S. A stability and validation study of 1% w/w menthol in aqueous cream. *Int J Pharm Pract* 1997; **5**: 101–104.
3 Luke E. Addiction to mentholated cigarettes [letter]. *Lancet* 1962; **i**: 110–111.
4 O'Mullane NM, *et al.* Adverse CNS effects of menthol-containing olabas oil [letter]. *Lancet* 1982; **i**: 1121.
5 Sweetman S, ed. *Martindale: The Complete Drug Reference*, 37th edn. London: Pharmaceutical Press, 2009: 2567.
6 Papa CM, Shelley WB. Menthol hypersensitivity. *J Am Med Assoc* 1964; **189**: 546–548.
7 Hayakawa R, *et al.* Contact dermatitis from *l*-menthol. *Cosmet Toilet* 1996; **111**(7): 28–29.
8 Rudzki E, Kleniewska D. The epidemiology of contact dermatitis in Poland. *Br J Dermatol* 1970; **83**: 543–545.
9 FAO/WHO. Evaluation of certain food additives: Fifty-first report of the joint FAO/WHO expert committee on food additives. *World Health Organ Tech Rep Ser* 2000; No. 891.
10 Lewis RJ, ed. *Sax's Dangerous Properties of Industrial Materials*, 11th edn. New York: Wiley, 2004: 2297.
11 *Food Chemical Codex*, 7th edn (Suppl. 1). Bethesda, MD: United States Pharmacopeia, 2010: 638.

M

20 General References

Bauer K *et al.Common Fragrance and Flavor Materials*. Weinheim: VCH, 1990: 43–46.
Eccles R. Menthol and related cooling compounds. *J Pharm Pharmacol* 1994; **46**: 618–630.
Walker T. Menthol. Properties, uses and some methods of manufacture. *Manuf Chem Aerosol News* 1967; 53.

21 Authors

BA Langdon, MP Mullarney.

22 Date of Revision

1 March 2012.

Methionine

1 Nonproprietary Names

BP: Methionine
JP: L-Methionine
PhEur: Methionine
USP–NF: Methionine

2 Synonyms

α-Amino-γ-methylmercaptobutyric acid; (*S*)-2-amino-4-(methylthio)butanoic acid; 2-amino-4-(methylthio)butyric acid; L-methionine; methioninum; γ-methylthio-α-aminobutyric acid.

3 Chemical Name and CAS Registry Number

(2*S*)-2-Amino-4-methylsulfanylbutanoic acid [63-68-3]

4 Empirical Formula and Molecular Weight

$C_5H_{11}NO_2S$ 149.21

5 Structural Formula

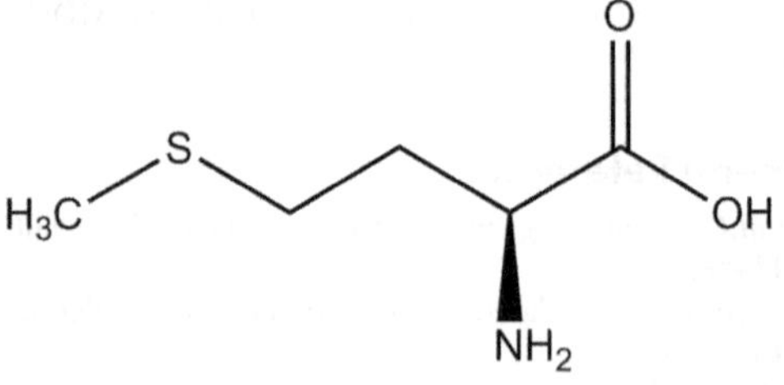

6 Functional Category

Buffering agent; flavoring agent.

7 Applications in Pharmaceutical Formulation or Technology

Methionine is used in oral pharmaceutical formulations as a flavoring agent.[1] It has been included in parenteral formulations as a pH controlling agent.[2]

8 Description

Methionine occurs as a white or almost white, crystalline powder or colorless crystals.

9 Pharmacopeial Specifications

See Table I.

10 Typical Properties

Acidity/alkalinity pH = 5.6–6.1 (1% w/v aqueous solution)[1]
Density 1.34 g/cm^3[1]
Melting point 280–282°C[3]
Solubility Soluble in water, dilute acids, and alkalis. Insoluble in absolute ethanol, ethanol (95%), benzene, acetone, and ether.

11 Stability and Storage Conditions

Methionine is sensitive to light and should be stored in a cool, dark place.

SEM 1: Excipient: methionine; manufacturer: Sigma-Aldrich; magnification: 60×; voltage: 10 kV.

Table I: Pharmacopeial specifications for methionine.

Test	JP XV	PhEur 7.4	USP35–NF30
Identification	+	+	+
Specific rotation	+21.0° to 25.0°	+22.5° to 24.0°	+22.4° to 24.7°
pH	5.2–6.2	5.5–6.5	5.6–6.1
Appearance of solution	+	+	—
Chloride	≤0.021%	≤200 ppm	≤0.05%
Sulfate	≤0.028%	≤300 ppm	≤0.03%
Ammonium	≤0.02%	≤200 ppm	—
Heavy metals	≤20 ppm	≤10 ppm	≤0.0015%
Arsenic	≤2 ppm	—	—
Related substances	+	—	—
Loss on drying	≤0.3%	≤0.5%	≤0.3%
Residue on ignition	≤0.1%	—	≤0.4%
Chromatographic purity	—	—	+
Ninhydrin-positive substances	—	+	—
Iron	—	≤10ppm	≤0.003%
Sulfated ash	—	≤0.1%	—
Assay (dried basis)	≥98.5%	99.0–101.0%	98.5–101.5%

12 Incompatibilities

Methionine is incompatible with strong oxidizing agents.

13 Method of Manufacture

Numerous methods have been described for manufacture of methionine, including hydrolysis of methionine amide[4] and 5-(β-methylmercaptoethyl)-hydantoin.[5]

14 Safety

Methionine is used in oral pharmaceutical formulations. The pure form of methionine is mildly toxic by ingestion and by the IP route.

LD_{50} (rat, IP): 4.328 g/kg[6]
LD_{50} (rat, oral): 36 g/kg

15 Handling Precautions

Observe normal precautions appropriate to the circumstances and quantity of the material handled.

16 Regulatory Status

Included in the FDA Inactive Ingredients Database (oral tablets). Included in parenteral preparations (injection solutions; powders for reconstitution) licensed in the UK. Included in the Canadian Natural Health Products Ingredients Database.

17 Related Substances

D-Methionine; DL-methionine.

D-Methionine
CAS number [348-67-4]
Comments The EINECS number for D-methionine is 206-483-6.

DL-Methionine
CAS number [59-51-8]
Acidity/alkalinity pH = 5.6–6.1 (1% w/v aqueous solution)
Density 1.34 g/cm^3
Dissociation constant pK_{a1} = 2.28 at 25°C; pK_{a2} = 9.21 at 25°C
Melting point 281°C
Solubility Soluble in dilute acids and alkalis. *See also* Table II.
Comments The EINECS number for DL-methionine is 200-432-1.

Table II: Solubility of DL-methionine.

Solvent	Solubility at 20°C unless otherwise stated
Ethanol (95%)	Very slightly soluble
Ether	Insoluble
Water	1 in 55 at 0°C
	1 in 30 at 25°C
	1 in 16.5 at 50°C
	1 in 9.5 at 75°C
	1 in 5.7 at 100°C

18 Comments

Methionine has been used experimentally as an antioxidant with antibodies,[7] and in lyophilized lipid/DNA complexes.[8] It has also been studied in a topical protein formulation to prevent viscosity loss.[9]

Methionine is used in paracetamol poisoning to prevent hepatotoxicity, and is frequently included in paracetamol formulations for this purpose.[10,11] L-Methionine is an essential amino acid and is included in amino acid solutions for parenteral nutrition.[10] It has been used to treat liver disorders and also to lower urinary pH.[10]

A specification for L-methionine is contained in the *Food Chemicals Codex* (FCC).[12]

The EINECS number for L-methionine is 200-562-9. The PubChem Compound ID (CID) for L-methionine is 6137.

19 Specific References

1 Ash M, Ash I. *Handbook of Pharmaceutical Additives*, 3rd edn. Endicott, NY: Synapse Information Resources, 2007: 731.
2 Pharmacia & Upjohn. Stabilized aqueous suspensions for parenteral use. WO Patent WO0187266; 2001.
3 O'Neil MJ, ed. *The Merck Index: An Encyclopedia of Chemicals, Drugs, and Biologicals*, 14th edn. Whitehouse Station, NJ: Merck, 2006: 1032–1033.
4 Rhone Poulenc Animal Nutrition. Process for the production of methionine. WP Patent 0160788; 2001.
5 Degussa. Continuous process for the manufacture of methionine. US Patent No. 4,069,251; 1978.
6 Lewis RJ, ed. *Sax's Dangerous Properties of Industrial Materials*, 11th edn. New York: Wiley, 2004: 2334.
7 Lam XM, *et al.* Antioxidants for prevention of methionine oxidation in recombinant monoclonal antibody HER2. *J Pharm Sci* 1997; **86**(11): 1250–1255.
8 Molina MD, Anchordoquy TJ. Formulation strategies to minimize oxidative damage in lyophilized lipid/DNA complexes during storage. *J Pharm Sci* 2008; **97**(12): 5089–5105.
9 Ji JA. *et al.*Effect of EDTA and methionine on preventing loss of viscosity of cellulose-based topical gel. *AAPS PharmSciTech* 2009; **10**(2): 678–683.
10 Sweetman SC, ed. *Martindale: The Complete Drug Reference*, 37th edn. London, UK: Pharmaceutical Press, 2011: 1591.
11 Neuvonen PJ, *et al.* Methionine in paracetamol tablets, a tool to reduce paracetamol toxicity. *Int J Clin Pharmacol Ther Toxicol* 1985; **23**(9): 497–500.
12 *Food Chemicals Codex*, 7th edn. Bethesda, MD: United States Pharmacopeia, 2010: 642.

20 General References

—

21 Authors

JC Hooton, N Sandler.

22 Date of Revision

1 March 2012.

M

Methylcellulose

1 Nonproprietary Names

BP: Methylcellulose
JP: Methylcellulose
PhEur: Methylcellulose
USP–NF: Methylcellulose

2 Synonyms

Benecel; *Cellacol*; *Culminal MC*; E461; *Mapolose*; *Methocel*; methylcellulosum; *Metolose*; *Rutocel A 55 RT*; *Tylose*; *Viscol.*

3 Chemical Name and CAS Registry Number

Cellulose methyl ether [9004-67-5]

4 Empirical Formula and Molecular Weight

Methylcellulose is a long-chain substituted cellulose in which approximately 27–32% of the hydroxyl groups are in the form of the methyl ether. The various grades of methylcellulose have degrees of polymerization in the range 50–1000, with molecular weights (number average) in the range 10 000–220 000 Da. The degree of substitution of methylcellulose is defined as the average number of methoxyl (CH_3O) groups attached to each of the anhydroglucose units along the chain. The degree of substitution also affects the physical properties of methylcellulose, such as its solubility.

5 Structural Formula

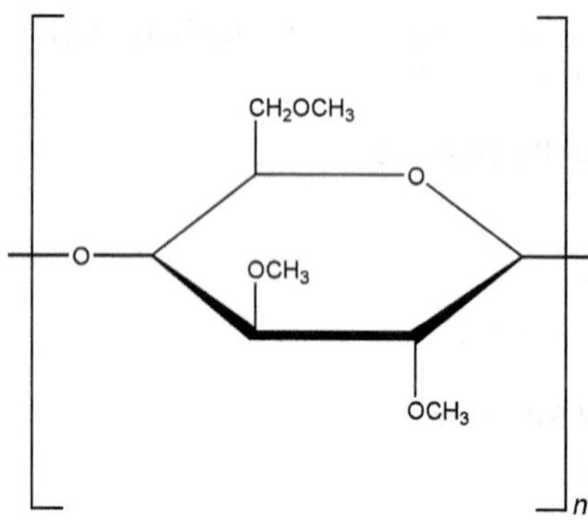

The structure shown is with complete substitution of the available hydroxyl units of methoxyl substitution. Note that methoxyl substitution can occur at any combination of the hydroxyl groups of the anhydroglucose ring of cellulose at positions 2, 3, and 6.

See Section 4.

6 Functional Category

Coating agent; emulsifying agent; modified-release agent; suspending agent; tablet and capsule binder; tablet and capsule disintegrant; viscosity-increasing agent.

7 Applications in Pharmaceutical Formulation or Technology

Methylcellulose is widely used in oral and topical pharmaceutical formulations; *see* Table I.

In tablet formulations, low- or medium-viscosity grades of methylcellulose are used as binding agents, the methylcellulose being added either as a dry powder or in solution.[1–3] High-viscosity grades of methylcellulose may also be incorporated in tablet formulations as a disintegrant.[4] Methylcellulose may be added to a tablet formulation to produce sustained-release preparations.[5]

Tablet cores may also be spray-coated with either aqueous or organic solutions of highly substituted low-viscosity grades of methylcellulose to mask an unpleasant taste or to modify the release of a drug by controlling the physical nature of the granules.[6] Methylcellulose coats are also used for sealing tablet cores prior to sugar coating.

Low-viscosity grades of methylcellulose are used to emulsify olive, peanut, and mineral oils.[7] They are also used as suspending or thickening agents for orally administered liquids, methylcellulose commonly being used in place of sugar-based syrups or other suspension bases.[8] Methylcellulose delays the settling of suspensions and increases the contact time of drugs, such as antacids, in the stomach.

High-viscosity grades of methylcellulose are used to thicken topically applied products such as creams and gels.

In ophthalmic preparations, a 0.5–1.0% w/v solution of a highly substituted, high-viscosity grade of methylcellulose has been used as a vehicle for eye drops.[9] However, hypromellose-based formulations are now preferred for ophthalmic preparations. Methylcellulose is also used in injectable formulations.

Table I: Uses of methylcellulose.

Use	Concentration (%)
Bulk laxative	5.0–30.0
Creams, gels, and ointments	1.0–5.0
Emulsifying agent	1.0–5.0
Ophthalmic preparations	0.5–1.0
Suspensions	1.0–2.0
Sustained-release tablet matrix	5.0–75.0
Tablet binder	1.0–5.0
Tablet coating	0.5–5.0
Tablet disintegrant	2.0–10.0

8 Description

Methylcellulose occurs as a white, fibrous powder or granules. It is practically odorless and tasteless. It should be labeled to indicate its viscosity type (viscosity of a 1 in 50 solution).

9 Pharmacopeial Specifications

The pharmacopeial specifications for methylcellulose have undergone harmonization of many attributes for JP, PhEur, and USP–NF.

See Table II. *See also* Section 18.

10 Typical Properties

Acidity/alkalinity pH = 5.0–8.0 for a 1% w/v aqueous suspension.
Autoignition temperature >350°C for *Methocel A4M.*
Degree of substitution 1.64–1.92
Density (bulk) 0.276 g/cm^3
Density (tapped) 0.464 g/cm^3
Density (true) 1.341 g/cm^3
Glass transition temperature (T_g) 196°C for *Methocel A4M.*
Melting point Begins to brown at 190–200°C; begins to char at 225–230°C.

SEM 1: Excipient: methylcellulose; manufacturer: Dow Chemical Co.; lot no.: KC16012N21; magnification: 60×; voltage: 5 kV.

SEM 2: Excipient: methylcellulose; manufacturer: Dow Chemical Co.; lot no.: KC16012N21; magnification: 600×; voltage: 5 kV.

Table II: Pharmacopeial specifications for methylcellulose.

Test	JP XV	PhEur 7.4	USP35–NF30
Identification	+	+	+
Characters	—	+	—
Appearance of solution	—	+	—
pH	5.0–8.0	5.0–8.0	5.0–8.0
Apparent viscosity	+	+	+
Loss on drying	≤5.0%	≤5.0%	≤5.0%
Residue on ignition	≤1.5%	≤1.5%	≤1.5%
Heavy metals[a]	≤20 ppm	≤20 ppm	≤20 ppm
Assay (of methoxyl groups)	26.0–33.0%	26.0–33.0%	26.0–33.0%

(a) These tests have not been fully harmonized at the time of publication.

Refractive index of solution n_D^{20} = 1.336 (2% aqueous solution).

Solubility Practically insoluble in acetone, methanol, chloroform, ethanol (95%), ether, saturated salt solutions, toluene, and hot water. Soluble in glacial acetic acid and in a mixture of equal volumes of ethanol and chloroform. In cold water, methylcellulose swells and disperses slowly to form a clear to opalescent, viscous, colloidal dispersion.

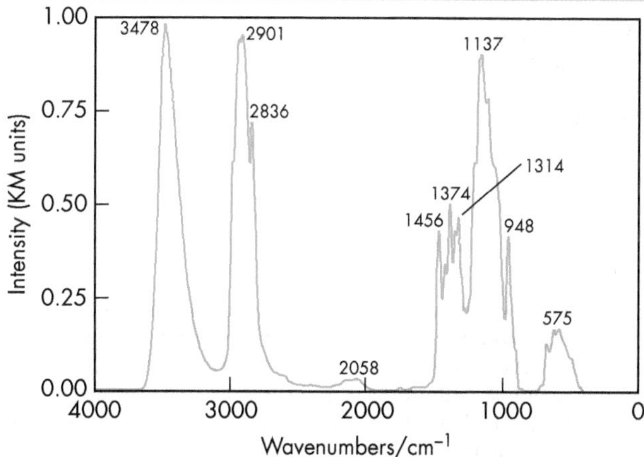

Figure 1: Infrared spectrum of methylcellulose measured by diffuse reflectance. Adapted with permission of Informa Healthcare.

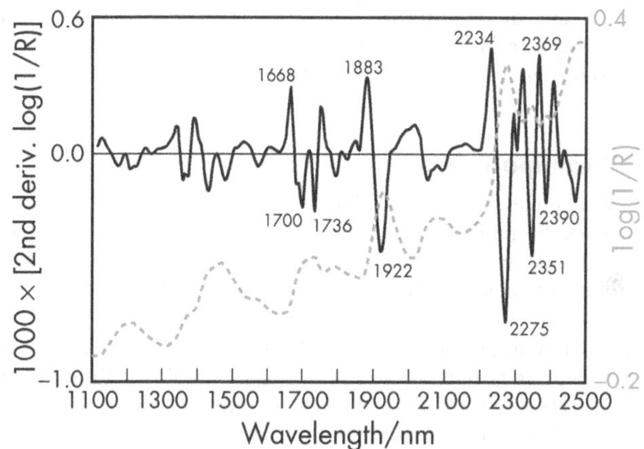

Figure 2: Near-infrared spectrum of methylcellulose measured by reflectance.

Spectroscopy

IR spectra *see* Figure 1.

NIR spectra *see* Figure 2.

Surface tension

53–59 mN/m (53–59 dynes/cm) for a 0.05% w/v solution at 25°C;

45–55 mN/m for 0.1% at 20°C.

Interfacial tension of solution versus paraffin oil is 19–23 mN/m for 0.1% w/v solution at 20°C.

Viscosity (dynamic) Various grades of methylcellulose are commercially available that vary in their degree of polymerization. Aqueous solutions at concentrations of 2% w/v will produce viscosities between 5 and 75 000 mPa s. Individual grades of methylcellulose have a stated, narrowly defined viscosity range measured for a 2% w/v solution. The viscosity of solutions may be increased by increasing the concentration of methylcellulose. Increased temperatures reduce the viscosity of solutions until gel formation occurs at 50–60°C. The process of thermogelation is reversible, with a viscous solution being reformed on cooling. *See also* Table III.

11 Stability and Storage Conditions

Methylcellulose powder is stable, although slightly hygroscopic. The bulk material should be stored in an airtight container in a cool, dry place.

Table III: Typical viscosity values for 2% w/v aqueous solutions of *Methocel* (Dow Chemical Co.) at 20°C.

Methocel grade	Viscosity (mPa s)
A4MP	4000
A15-LV	15
A15CP	1500
A4CP	400

Solutions of methylcellulose are stable to alkalis and dilute acids at pH 3–11, at room temperature. At pH less than 3, acid-catalyzed hydrolysis of the glucose–glucose linkages occurs and the viscosity of methylcellulose solutions is reduced.[10] On heating, solution viscosity is reduced until gel formation occurs at approximately 50°C; *see* Section 10.

Methylcellulose solutions are liable to microbial spoilage and antimicrobial preservatives should therefore be used. Solutions may also be sterilized by autoclaving, although this process can decrease the viscosity of a solution.[11,12] The change in viscosity after autoclaving is related to solution pH. Solutions at pH less than 4 had viscosities reduced by more than 20% subsequent to autoclaving.[11]

12 Incompatibilities

Methylcellulose is incompatible with aminacrine hydrochloride; chlorocresol; mercuric chloride; phenol; resorcinol; tannic acid; silver nitrate; cetylpyridinium chloride; *p*-hydroxybenzoic acid; *p*-aminobenzoic acid; methylparaben; propylparaben; and butylparaben.

Salts of mineral acids (particularly polybasic acids), phenols, and tannins will coagulate solutions of methylcellulose, although this can be prevented by the addition of ethanol (95%) or glycol diacetate. Complexation of methylcellulose occurs with highly surface-active compounds such as tetracaine and dibutoline sulfate.

High concentrations of electrolytes increase the viscosity of methylcellulose mucilages owing to the 'salting out' of methylcellulose. With very high concentrations of electrolytes, the methylcellulose may be completely precipitated in the form of a discrete or continuous gel. Methylcellulose is incompatible with strong oxidizing agents such as bleach, perchloric acid, nitric acid, perchlorates, alkali nitrates, alkali nitrites, and calcium oxide.

13 Method of Manufacture

Methylcellulose is prepared from wood pulp (cellulose) by treatment with alkali followed by methylation of the alkali cellulose with methyl chloride. The product is then purified and ground to powder form.

14 Safety

Methylcellulose is widely used in a variety of oral and topical pharmaceutical formulations. It is also extensively used in cosmetics and food products, and is generally regarded as a nontoxic, nonallergenic, and nonirritant material.[13]

Following oral consumption, methylcellulose is not digested or absorbed and is therefore a noncaloric material. Ingestion of excessive amounts of methylcellulose may temporarily increase flatulence and gastrointestinal distention.

In the normal individual, oral consumption of large amounts of methylcellulose has a laxative action and medium- or high-viscosity grades are therefore used as bulk laxatives.

Esophageal obstruction may occur if methylcellulose is swallowed with an insufficient quantity of liquid. Consumption of large quantities of methylcellulose may additionally interfere with the normal absorption of some minerals. However, this and the other adverse effects discussed above relate mainly to the use of methylcellulose as a bulk laxative and are not significant factors when methylcellulose is used as an excipient in oral preparations.

Methylcellulose is not commonly used in parenteral products, although it has been used in intra-articular and intramuscular injections. Studies in rats have suggested that parenterally administered methylcellulose may cause glomerulonephritis and hypertension.[13] Methylcellulose is considered to be toxic by the intraperitoneal route of administration.

The WHO has not specified an acceptable daily intake of methylcellulose since the level of use in foods was not considered to be a hazard to health.[14]

LD_{50} (mouse, IP): 275 g/kg[15]

15 Handling Precautions

Observe normal precautions appropriate to the circumstances and quantity of material handled. Dust may be irritant to the eyes and eye protection should be worn. Use in a well-ventilated area. Excessive dust generation should be avoided to minimize the risk of explosion. Methylcellulose is combustible. Spills of the dry powder or solution should be cleaned up immediately, as the slippery film that forms can be dangerous.

16 Regulatory Status

GRAS listed. Accepted as a food additive in the US, Europe and Japan. Included in the FDA Inactive Ingredients Database (sublingual tablets; IM injections; intrasynovial injections; nasal preparations; ophthalmic preparations; oral capsules, suspensions, and tablets; topical and vaginal preparations). Included in nonparenteral medicines licensed in the UK. Included in the Canadian Natural Health Products Ingredients Database. Reported in the EPA TSCA inventory.

17 Related Substances

Ethylcellulose; hydroxyethyl cellulose; hydroxyethylmethyl cellulose; hypromellose.

18 Comments

Methylcellulose has undergone harmonization of many attributes for JP, PhEur, and USP–NF by the Pharmacopeial Discussion Group. For further information see the General Information Chapter <1196> in the USP35–NF30, the General Chapter 5.8 in PhEur 7.4, along with the 'State of Work' document on the PhEur EDQM website, and also the General Information Chapter 8 in the JP XV.

The thermal gelation temperature for methylcellulose decreases as a function of concentration. The presence of additives can increase or decrease the thermal gelation temperature. The presence of drugs can influence the properties of methylcellulose gels.[16] In addition, the viscosity of methylcellulose solutions can be modified by the presence of drugs or other additives.[17] Aqueous solutions of methylcellulose can be frozen and do not undergo phase separation upon freezing.

Methylcellulose is best dissolved in water by one of three methods, the most suitable being chosen for a particular application.

The most commonly used method is to add methylcellulose initially to hot water. The appropriate quantity of methylcellulose required to produce a solution of specified viscosity is mixed with water at 70°C; about half the desired final volume of water is used. Cold water or ice is then added to the hot methylcellulose slurry in order to reduce the temperature to below 20°C. A clear, aqueous methylcellulose solution is obtained.[18]

Alternatively, either methylcellulose powder may be dry-blended with another powder prior to mixing with cold water, or methylcellulose powder may be moistened with an organic solvent such as ethanol (95%) prior to the addition of water.

In general, methylcellulose solutions exhibit pseudoplastic flow and there is no yield point. Nonthixotropic flow properties are observed below the gelation temperature.

Note that some cellulose ether products possess hydroxypropyl substitutions in addition to methyl substitutions but are designated with the same trade name in a product line, differing only by a unique identifier code. These products should not be confused with the products that contain only methyl substitutions. Methylcellulose has been investigated as a stabilizer for liposome dispersions.[19]

Therapeutically, methylcellulose is used as a bulk laxative. It has also been used to aid appetite control in the management of obesity, but there is little evidence supporting its efficacy.

A specification for methylcellulose is contained in the *Food Chemicals Codex* (FCC).[20]

19 Specific References

1 Wan LSC, Prasad KPP. Uptake of water by excipients in tablets. *Int J Pharm* 1989; **50**: 147–153.

2 Pharmpedia.com. Tablet: formulation of tablets/binders. http://www.pharmpedia.com (accessed 5 December 2011).

3 Itiola OA, Pilpel N. Formulation effects on the mechanical properties of metronidazole tablets. *J Pharm Pharmacol* 1991; **43**: 145–147.

4 Esezobo S. Disintegrants: effects of interacting variables on the tensile strengths and dissolution times of sulfaguanidine tablets. *Int J Pharm* 1989; **56**: 207–211.

5 Sanghavi NM, *et al.* Sustained release tablets of theophylline. *Drug Dev Ind Pharm* 1990; **16**: 1843–1848.

6 Wan LSC, Lai WF. Factors affecting drug release from drug-coated granules prepared by fluidized-bed coating. *Int J Pharm* 1991; **72**: 163–174.

7 Hong ST. Changes in the stability properties of methylcellulose emulsions as affected by competitive adsorption between methylcellulose and Tween 20. *J Korean Soc Food Sci Nutr* 2008; **37**(10): 1278–1286.

8 Dalal PS, Narurkar MM. *In vitro* and *in vivo* evaluation of sustained release suspensions of ibuprofen. *Int J Pharm* 1991; **73**: 157–162.

9 Gerbino PP. Topical drugs. In: *Remington: The Science and Practice of Pharmacy*, 21st edn. Philadelphia, PA: Lippincott Williams and Wilkins, 2005; 1280.

10 Huikari A, Karlsson A. Viscosity stability of methylcellulose solutions at different pH and temperature. *Acta Pharm Fenn* 1989; **98**(4): 231–238.

11 Huikari A. Effect of heat sterilization on the viscosity of methylcellulose solutions. *Acta Pharm Fenn* 1986; **95**(1): 9–17.

12 Huikari A, *et al.* Effect of heat sterilization on the molecular weight of methylcellulose determined using high pressure gel filtration chromatography and viscometry. *Acta Pharm Fenn* 1986; **95**(3): 105–111.

13 Anonymous. Final report on the safety assessment of hydroxyethylcellulose, hydroxypropylcellulose, methylcellulose, hydroxypropyl methylcellulose and cellulose gum. *J Am Coll Toxicol* 1986; **5**(3): 1–60.

14 FAO/WHO. Evaluation of certain food additives and contaminants. Thirty-fifth report of the joint FAO/WHO expert committee on food additives. *World Health Organ Tech Rep Ser* 1990; No. 789.

15 Lewis RJ, ed. *Sax's Dangerous Properties of Industrial Materials*, 11th edn. New York: Wiley, 2004; 2408.

16 Mitchell K, *et al.* Influence of drugs on the properties of gels and swelling characteristics of matrices containing methylcellulose or hydroxypropylmethylcellulose. *Int J Pharm* 1993; **100**(1–3): 165–173.

17 Huikari A, Kristoffersson E. Rheological properties of methylcellulose solutions: general flow properties and effects of added substances. *Acta Pharm Fenn* 1985; **94**(4): 143–154.

18 Allen Jr LV *et al*, eds. *Ansel's Pharmaceutical Dosage Forms and Drug Delivery Systems*, 9th edn. Baltimore MD: Wolters Kluwer-LWW, 2011; 410.

19 Csempesz F, Puskas I. Controlling the physical stability of liposomal colloids. In: Tadros TF, ed. *Colloid Stability and Application in Pharmacy.* Weinheim, Germany: Wiley-VCH, 2007; 79–89.

20 *Food Chemicals Codex*, 7th edn. Bethesda, MD: United States Pharmacopeia, 2010; 683.

20 General References

Dow Chemical Company. Technical literature: *Methocel* cellulose ethers, September 2002. http://www.dow.com/dowwolff/en/pdfs/192-01062.pdf (accessed 5 December 2011).

Dow Chemical Company. Material safety data sheet: *Methocel A4M*, 2004.

European Directorate for the Quality of Medicines and Healthcare (EDQM). European Pharmacopoeia – State Of Work Of International Harmonisation. *Pharmeuropa* 2011; 23(4): 713–714. www.edqm.eu/site/-614.html (accessed 5 December 2011).

Kamide K. *Cellulose and Cellulose Derivatives*, Amsterdam: Elsevier Science, 2005.

Li J, Mei X. Applications of cellulose and cellulose derivatives in immediate release solid dosage. In: ACS Symposium Series, 934 , *Polysaccharides for Drug Delivery and Pharmaceutical Applications*. 2006: 19–55.

Muratora SA, *et al.* Studying interactions in microcrystalline cellulose-drug systems. *Pharm Chem J* 2002; **36**(11): 619–622.

Oxford University. Material safety data sheet: Methyl Cellulose, 2006. msds.chem.ox.ac.uk/ME.methyl_cellulose.html (accessed 5 December 2011).

Rowe RC. The molecular weight of methyl cellulose used in pharmaceutical formulation. *Int J Pharm* 1982; **11**: 175–179.

Tapia Villanueva C, Sapag Hagar J. Methylcellulose: its pharmaceutical applications. *Acta Farm Bonaerense* 1995; **14**(Jan–Mar): 41–47.

Viriden A, *et al.* Influence of substitution pattern on solution behaviour of hydroxypropyl methylcellulose. *Biomacromolecules* 2009; **10**(3): 522–529.

Wan LS, Prasad KP. Influence of quantity of granulating liquid on water uptake and disintegration of tablets with methylcellulose. *Pharm Ind* 1989; **51**(1): 105–109.

Wan LS, Prasad KP. Studies on the swelling of composite disintegrant–methylcellulose films. *Drug Dev Ind Pharm* 1990; **16**(2): 191–200.

21 Author

LV Allen Jr.

22 Date of Revision

2 March 2012.

Methylparaben

1 Nonproprietary Names

BP: Methyl Hydroxybenzoate
JP: Methyl Parahydroxybenzoate
PhEur: Methyl Parahydroxybenzoate
USP–NF: Methylparaben

2 Synonyms

Aseptoform M; *CoSept M*; E218; 4-hydroxybenzoic acid methyl ester; metagin; *Methyl Chemosept*; methylis parahydroxybenzoas; methyl *p*-hydroxybenzoate; *Methyl Parasept*; *Nipagin M*; *Solbrol M*; *Tegosept M*; *Uniphen P-23*.

3 Chemical Name and CAS Registry Number

Methyl-4-hydroxybenzoate [99-76-3]

4 Empirical Formula and Molecular Weight

$C_8H_8O_3$ 152.15

5 Structural Formula

O OCH3

OH

6 Functional Category

Antimicrobial preservative.

7 Applications in Pharmaceutical Formulation or Technology

Methylparaben is widely used as an antimicrobial preservative in cosmetics, food products, and pharmaceutical formulations; *see* Table I. It may be used either alone or in combination with other parabens or with other antimicrobial agents. In cosmetics, methylparaben is the most frequently used antimicrobial preservative.(1)

The parabens are effective over a wide pH range and have a broad spectrum of antimicrobial activity, although they are most effective against yeasts and molds. Antimicrobial activity increases as the chain length of the alkyl moiety is increased, but aqueous solubility decreases; therefore a mixture of parabens is frequently used to provide effective preservation. Preservative efficacy is also improved by the addition of propylene glycol (2–5%), or by using parabens in combination with other antimicrobial agents such as imidurea; *see* Section 10.

Owing to the poor solubility of the parabens, paraben salts (particularly the sodium salt) are more frequently used in formulations. However, this raises the pH of poorly buffered formulations.

Methylparaben (0.18%) together with propylparaben (0.02%) has been used for the preservation of various parenteral pharmaceutical formulations; *see* Section 14.

Table I: Uses of methylparaben.

Use	Concentration (%)
IM, IV, SC injections(a)	0.065–0.25
Inhalation solutions	0.025–0.07
Intradermal injections	0.10
Nasal solutions	0.033
Ophthalmic preparations(a)	0.015–0.2
Oral solutions and suspensions	0.015–0.2
Rectal preparations	0.1–0.18
Topical preparations	0.02–0.3
Vaginal preparations	0.1–0.18

(a) *See* Section 14.

8 Description

Methylparaben occurs as colorless crystals or a white crystalline powder. It is odorless or almost odorless and has a slight burning taste.

9 Pharmacopeial Specifications

The pharmacopeial specifications for methylparaben have undergone harmonization of many attributes for JP, PhEur, and USP–NF.

See Table II. *See also* Section 18.

10 Typical Properties

Antimicrobial activity *See* Table III. Methylparaben exhibits antimicrobial activity of pH 4–8. Preservative efficacy decreases with increasing pH owing to the formation of the phenolate anion. Parabens are more active against yeasts and molds than against bacteria. They are also more active against Gram-positive bacteria than against Gram-negative bacteria.

SEM 1: Excipient: methylparaben; supplier: Bate Chemical Co. Ltd; magnification: 600×.

Table II: Pharmacopeial specifications for methylparaben.

Test	JP XV	PhEur 7.4	USP35–NF30
Identification	+	+	+
Characters	—	+	—
Appearance of solution	+	+	+
Acidity	+	+	+
Heavy metals[a]	≤20 ppm	—	—
Impurities	—	+	+
Melting range	—	125–128°C	125–128°C
Related substances	+	+	+
Sulfated ash	—	≤0.1%	—
Residue on ignition	≤0.1%	—	≤0.1%
Assay (dried basis)	98.0–102.0%	98.0–102.0%	98.0–102.0%

(a) This test has not been fully harmonized at the time of publication.

Methylparaben is the least active of the parabens; antimicrobial activity increases with increasing chain length of the alkyl moiety. Activity may be improved by using combinations of parabens as synergistic effects occur. Therefore, combinations of methyl-, ethyl-, propyl-, and butylparaben are often used together. Activity has also been reported to be enhanced by the addition of other excipients such as: propylene glycol (2–5%);[2] phenylethyl alcohol;[3] and edetic acid.[4] Activity may also be enhanced owing to synergistic effects by using combinations of parabens with other antimicrobial preservatives such as imidurea.[5]

The hydrolysis product *p*-hydroxybenzoic acid has practically no antimicrobial activity.

See also Section 12.

Table III: Minimum inhibitory concentrations (MICs) of methylparaben in aqueous solution.[4]

Microorganism	MIC (μg/mL)
Aerobacter aerogenes ATCC 8308	2000
Aspergillus oryzae	600
Aspergillus niger ATCC 9642	1000
Aspergillus niger ATCC 10254	1000
Bacillus cereus var. *mycoides* ATCC 6462	2000
Bacillus subtilis ATCC 6633	2000
Candida albicans ATCC 10231	2000
Enterobacter cloacae ATCC 23355	1000
Escherichia coli ATCC 8739	1000
Escherichia coli ATCC 9637	1000
Klebsiella pneumoniae ATCC 8308	1000
Penicillium chrysogenum ATCC 9480	500
Penicillium digitatum ATCC 10030	500
Proteus vulgaris ATCC 8427	2000
Proteus vulgaris ATCC 13315	1000
Pseudomonas aeruginosa ATCC 9027	4000
Pseudomonas aeruginosa ATCC 15442	4000
Pseudomonas stutzeri	2000
Rhizopus nigricans ATCC 6227A	500
Saccharomyces cerevisiae ATCC 9763	1000
Salmonella typhosa ATCC 6539	1000
Sarcina lutea	4000
Serratia marcescens ATCC 8100	1000
Staphylococcus aureus ATCC 6538P	2000
Staphylococcus epidermidis ATCC 12228	2000
Trichoderma lignorum ATCC 8678	250
Trichoderma mentagrophytes	250

Density (true) 1.352 g/cm^3
Dissociation constant pK_a = 8.4 at 22°C
Melting point 125–128°C
Partition coefficients Values for different vegetable oils vary considerably and are affected by the purity of the oil; *see* Table IV.
Solubility *see* Table V.

Table IV: Partition coefficients of methylparaben in vegetable oil and water.[6,7]

Solvent	Partition coefficient oil : water
Almond oil	7.5
Castor oil	6.0
Corn oil	4.1
Diethyl adipate	200
Isopropyl myristate	18.0
Lanolin	7.0
Mineral oil	0.1
Peanut oil	4.2
Soybean oil	6.1

Table V: Solubility of methylparaben in various solvents.[4]

Solvent	Solubility at 25°C unless otherwise stated
Ethanol	1 in 2
Ethanol (95%)	1 in 3
Ethanol (50%)	1 in 6
Ether	1 in 10
Glycerin	1 in 60
Mineral oil	Practically insoluble
Peanut oil	1 in 200
Propylene glycol	1 in 5
Water	1 in 400
	1 in 50 at 50°C
	1 in 30 at 80°C

Spectroscopy

IR spectra *see* Figure 1.

NIR spectra *see* Figure 2.

11 Stability and Storage Conditions

Aqueous solutions of methylparaben at pH 3–6 may be sterilized by autoclaving at 120°C for 20 minutes, without decomposition.[8] Aqueous solutions at pH 3–6 are stable (less than 10% decomposition) for up to about 4 years at room temperature, while aqueous solutions at pH 8 or above are subject to rapid hydrolysis (10% or more after about 60 days storage at room temperature); *see* Tables VI and VII.[9]

Methylparaben should be stored in a well-closed container in a cool, dry place.

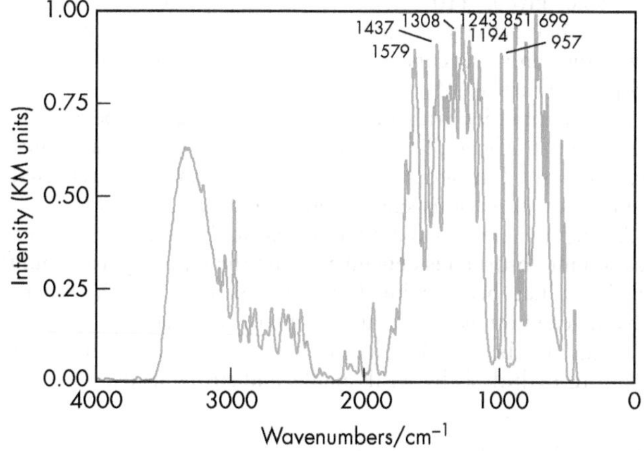

Figure 1: Infrared spectrum of methylparaben measured by diffuse reflectance. Adapted with permission of Informa Healthcare.

M

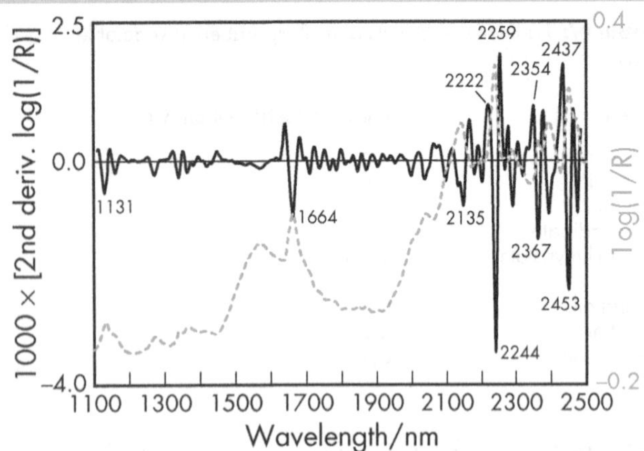

Figure 2: Near-infrared spectrum of methylparaben measured by reflectance.

Table VI: Predicted rate constants and half-lives for methylparaben dissolved in dilute hydrochloric acid solution, at 25°C.

Initial pH of solution	Rate constant $k \pm \sigma^{(a)}$ (hour^{-1})	Half-life $t_{1/2} \pm \sigma^{(a)}$ (day)
1	$(1.086 \pm 0.005) \times 10^{-4}$	266 ± 13
2	$(1.16 \pm 0.12) \times 10^{-5}$	2490 ± 260
3	$(6.1 \pm 1.5) \times 10^{-7}$	47000 ± 12000
4	$(3.27 \pm 0.64) \times 10^{-7}$	88000 ± 17000

(a) Indicates the standard error.

M

Table VII: Predicted remaining amount of methylparaben dissolved in dilute hydrochloric acid solution, after autoclaving.

Initial pH of solution	Rate constant $k \pm \sigma^{(a)}$ (hour^{-1})	Predicted residual amount after autoclaving (%)
1	$(4.96 \pm 0.16) \times 10^{-1}$	84.77 ± 0.46
2	$(4.49 \pm 0.37) \times 10^{-2}$	98.51 ± 0.12
3	$(2.79 \pm 0.57) \times 10^{-3}$	99.91 ± 0.02
4	$(1.49 \pm 0.22) \times 10^{-3}$	99.95 ± 0.01

(a) Indicates the standard error.

12 Incompatibilities

The antimicrobial activity of methylparaben and other parabens is considerably reduced in the presence of nonionic surfactants, such as polysorbate 80, as a result of micellization.(10,11) However, propylene glycol (10%) has been shown to potentiate the antimicrobial activity of the parabens in the presence of nonionic surfactants and prevents the interaction between methylparaben and polysorbate 80.(12)

Incompatibilities with other substances, such as bentonite,(13) magnesium trisilicate,(14) talc, tragacanth,(15) sodium alginate,(16) essential oils,(17) sorbitol,(18) and atropine,(19) have been reported. It also reacts with various sugars and related sugar alcohols.(20)

Absorption of methylparaben by plastics has also been reported; the amount absorbed is dependent upon the type of plastic and the vehicle. It has been claimed that low-density and high-density polyethylene bottles do not absorb methylparaben.(21)

Methylparaben is discolored in the presence of iron and is subject to hydrolysis by weak alkalis and strong acids.

13 Method of Manufacture

Methylparaben is prepared by the esterification of *p*-hydroxybenzoic acid with methanol.

14 Safety

Methylparaben and other parabens are widely used as antimicrobial preservatives in cosmetics and oral and topical pharmaceutical formulations. Although parabens have also been used as preservatives in injections and ophthalmic preparations, they are now generally regarded as being unsuitable for these types of formulations owing to the irritant potential of the parabens. These experiences may depend on immune responses to enzymatically formed metabolites of the parabens in the skin. A comprehensive review of the evaluation of the health aspects of methylparaben based on existing literature has been published.(22)

Parabens are nonmutagenic, nonteratogenic, and noncarcinogenic. Sensitization to the parabens is rare, and these compounds do not exhibit significant levels of photocontact sensitization or phototoxicity.

Hypersensitivity reactions to parabens, generally of the delayed type and appearing as contact dermatitis, have been reported. However, given the widespread use of parabens as preservatives, such reactions are relatively uncommon; the classification of parabens in some sources as high-rate sensitizers may be overstated.(23)

Immediate hypersensitivity reactions following injection of preparations containing parabens have also been reported.(24–26) Delayed-contact dermatitis occurs more frequently when parabens are used topically, but has also been reported to occur after oral administration.(27–29)

Unexpectedly, preparations containing parabens may be used by patients who have reacted previously with contact dermatitis provided they are applied to another, unaffected, site. This has been termed the paraben paradox.(30)

Concern has been expressed over the use of methylparaben in infant parenteral products because bilirubin binding may be affected, which is potentially hazardous in hyperbilirubinemic neonates.(31)

Although methylparaben has been considered a safe preservative in cosmetics, a report has indicated that it may have harmful effects on human skin when exposed to sunlight.(32) Another study has concluded that methylparaben may cause skin damage involving carcinogenesis through the combined activation of sunlight irradiation and skin esterases.(33)

The WHO has set an estimated total acceptable daily intake for methyl-, ethyl-, and propylparabens at up to 10 mg/kg bodyweight.(34)

LD_{50} (dog, oral): 3.0 g/kg(35)
LD_{50} (mouse, IP): 0.96 g/kg
LD_{50} (mouse, SC): 1.20 g/kg

15 Handling Precautions

Observe normal precautions appropriate to the circumstances and quantity of material handled. Methylparaben may be irritant to the skin, eyes, and mucous membranes, and should be handled in a well-ventilated environment. Eye protection, gloves, and a dust mask or respirator are recommended.

16 Regulatory Status

Methylparaben and propylparaben are affirmed GRAS Direct Food Substances in the US at levels up to 0.1%. All esters except the benzyl ester are allowed for injection in Japan. In cosmetics, the EU and Brazil allow use of each paraben at 0.4%, but the total of all parabens may not exceed 0.8%. The upper limit in Japan is 1.0%.

Accepted for use as a food additive in Europe. Included in the FDA Inactive Ingredients Database (IM, IV, and SC injections; inhalation preparations; ophthalmic preparations; oral capsules, tablets, solutions and suspensions; otic, rectal, topical, and vaginal preparations). Included in medicines licensed in the UK. Included in the Canadian Natural Health Products Ingredients Database.

17 Related Substances

Butylparaben; ethylparaben; methylparaben potassium; methylparaben sodium; propylparaben.

Methylparaben potassium

Empirical formula $C_8H_7KO_3$
Molecular weight 190.25
CAS number [26112-07-2]
Synonyms Methyl 4-hydroxybenzoate potassium salt; potassium methyl hydroxybenzoate.
Comments Methylparaben potassium may be used instead of methylparaben because of its greater aqueous solubility.

Methylparaben sodium

Empirical formula $C_8H_7NaO_3$
Molecular weight 174.14
CAS number [5026-62-0]
Synonyms E219; methyl 4-hydroxybenzoate sodium salt; sodium methyl hydroxybenzoate; soluble methyl hydroxybenzoate.
Appearance A white, odorless or almost odorless, hygroscopic crystalline powder.
Acidity/alkalinity pH = 9.5–10.5 (0.1% w/v aqueous solution)
Solubility 1 in 50 of ethanol (95%); 1 in 2 of water; practically insoluble in fixed oils.
Comments Methylparaben sodium may be used instead of methylparaben because of its greater aqueous solubility. However, it may cause the pH of a formulation to become more alkaline.

18 Comments

Methylparaben has undergone harmonization of many attributes for JP, PhEur, and USP–NF by the Pharmacopeial Discussion Group. For further information see the General Information Chapter <1196> in the USP35–NF30, the General Chapter 5.8 in PhEur 7.4, along with the 'State of Work' document on the PhEur EDQM website, and also the General Information Chapter 8 in the JP XV.

The BP 2012, PhEur 7.4 and USP35–NF30 also list Methylparaben Sodium as a separate monograph.

In addition to the most commonly used paraben esters, some other less-common esters have also been used; *see* Table VIII. A specification for methylparaben is contained in the *Food Chemicals Codex* (FCC).[36]

The EINECS number for methylparaben is 202-785-7.

The PubChem Compound ID (CID) for methylparaben is 7456.

Table VIII: CAS numbers of less common paraben esters.

Name	CAS Number
Benzylparaben	94-18-8
Isobutylparaben	4247-02-3
Isopropylparaben	4191-73-5

19 Specific References

1 Decker RL, Wenninger JA. Frequency of preservative use in cosmetic formulas as disclosed to FDA—1987. *Cosmet Toilet* 1987; **102**(12): 21–24.
2 Prickett PS, *et al.* Potentiation of preservatives (parabens) in pharmaceutical formulations by low concentrations of propylene glycol. *J Pharm Sci* 1961; **50**: 316–320.
3 Richards RME, McBride RJ. Phenylethanol enhancement of preservatives used in ophthalmic preparations. *J Pharm Pharmacol* 1971; **23**: 141S–146S.
4 Haag TE, Loncrini DF. Esters of para-hydroxybenzoic acid. In: Kabara JJ, ed. *Cosmetic and Drug Preservation*. New York: Marcel Dekker, 1984: 63–77.
5 Rosen WE, *et al.* Preservation of cosmetic lotions with imidazolidinyl urea plus parabens. *J Soc Cosmet Chem* 1977; **28**: 83–87.
6 Hibbott HW, Monks J. Preservation of emulsions—*p*-hydroxybenzoic ester partition coefficient. *J Soc Cosmet Chem* 1961; **12**: 2–10.
7 Wan LSC, *et al.* Partition of preservatives in oil/water systems. *Pharm Acta Helv* 1986; **61**: 308–313.
8 Aalto TR, *et al.* *p*-Hydroxybenzoic acid esters as preservatives I: uses, antibacterial and antifungal studies, properties and determination. *J Am Pharm Assoc Sci* 1953; **42**: 449–457.
9 Kamada A, *et al.* Stability of *p*-hydroxybenzoic acid esters in acidic medium. *Chem Pharm Bull* 1973; **21**: 2073–2076.
10 Aoki M, *et al.* [Application of surface active agents to pharmaceutical preparations I: effect of Tween 20 upon the antifungal activities of *p*-hydroxybenzoic acid esters in solubilized preparations.] *J Pharm Soc Jpn* 1956; **76**: 939–943[in Japanese].
11 Patel N, Kostenbauder HB. Interaction of preservatives with macromolecules I: binding of parahydroxybenzoic acid esters by polyoxyethylene 20 sorbitan monooleate (Tween 80). *J Am Pharm Assoc Sci* 1958; **47**: 289–293.
12 Poprzan J, deNavarre MG. The interference of nonionic emulsifiers with preservatives VIII. *J Soc Cosmet Chem* 1959; **10**: 81–87.
13 Yousef RT, *et al.* Effect of some pharmaceutical materials on the bactericidal activities of preservatives. *Can J Pharm Sci* 1973; **8**: 54–56.
14 Allwood MC. The adsorption of esters of *p*-hydroxybenzoic acid by magnesium trisilicate. *Int J Pharm* 1982; **11**: 101–107.
15 Eisman PC, *et al.* Influence of gum tragacanth on the bactericidal activity of preservatives. *J Am Pharm Assoc Sci* 1957; **46**: 144–147.
16 Myburgh JA, McCarthy TJ. The influence of suspending agents on preservative activity in aqueous solid/liquid dispersions. *Pharm Weekbl Sci* 1980; **2**: 143–148.
17 Chemburkar PB, Joslin RS. Effect of flavoring oils on preservative concentrations in oral liquid dosage forms. *J Pharm Sci* 1975; **64**: 414–417.
18 Runesson B, Gustavii K. Stability of parabens in the presence of polyols. *Acta Pharm Suec* 1986; **23**: 151–162.
19 Deeks T. Oral atropine sulfate mixtures. *Pharm J* 1983; **230**: 481.
20 Ma M, *et al.* Interaction of methylparaben preservative with selected sugars and sugar alcohols. *J Pharm Sci* 2002; **91**(7): 1715–1723.
21 Kakemi K, *et al.* Interactions of parabens and other pharmaceutical adjuvants with plastic containers. *Chem Pharm Bull* 1971; **19**: 2523–2529.
22 Soni MG, *et al.* Evaluation of the health aspects of methyl paraben: a review of the published literature. *Food Chem Toxicol* 2002; **40**: 1335–1373.
23 Weiner M, Bernstein IL. *Adverse Reactions to Drug Formulation Agents: A Handbook of Excipients*. New York: Marcel Dekker, 1989; 298–300.
24 Aldrete JA, Johnson DA. Allergy to local anesthetics. *J Am Med Assoc* 1969; **207**: 356–357.
25 Latronica RJ, *et al.* Local anesthetic sensitivity: report of a case. *Oral Surg* 1969; **28**: 439–441.
26 Nagel JE, *et al.* Paraben allergy. *J Am Med Assoc* 1977; **237**: 1594–1595.
27 Michäelsson G, Juhlin L. Urticaria induced by preservatives and dye additives in food and drugs. *Br J Dermatol* 1973; **88**: 525–532.
28 Warin RP, Smith RJ. Challenge test battery in chronic urticaria. *Br J Dermatol* 1976; **94**: 401–406.
29 Kaminer Y, *et al.* Delayed hypersensitivity reaction to orally administered methylparaben. *Clin Pharm* 1982; **1**(5): 469–470.
30 Fisher AA. Cortaid cream dermatitis and the 'paraben paradox' [letter]. *J Am Acad Dermatol* 1982; **6**: 116–117.
31 Loria CJ, *et al.* Effect of antibiotic formulations in serum protein: bilirubin interaction of newborn infants. *J Pediatr* 1976; 89(3): 479–482.
32 Handa O, *et al.* Methylparaben potentiates UV-induced damage of skin keratinocytes. *Toxicology* 2006; **227**((1-2)): 62–72.
33 Okamoto Y, *et al.* Combined activation of methyl paraben by light irradiation and esterase metabolism toward oxidative DNA damage. *Chem Res Toxicol* 2008; **21**(8): 1594–1599.
34 FAO/WHO. Toxicological evaluation of certain food additives with a review of general principles and of specifications. Seventeenth report of the joint FAO/WHO expert committee on food additives. *World Health Organ Tech Rep Ser* 1974; No. 539.
35 Lewis RJ, ed. *Sax's Dangerous Properties of Industrial Materials*, 11th edn. New York: Wiley, 2004; 2004.
36 *Food Chemicals Codex*, 7th edn. Suppl. 3. Bethesda, MD: United States Pharmacopeia, 2011; 1703.

20 General References

Bando H, *et al.* Effects of skin metabolism on percutaneous penetration of lipophilic drugs. *J Pharm Sci* 1997; **86**(6): 759–761.
European Directorate for the Quality of Medicines and Healthcare (EDQM). European Pharmacopoeia – State Of Work Of International Harmonisation. *Pharmeuropa* 2011; 23(4): 713–714. www.edqm.eu/site/-614.html (accessed 3 December 2011).
Forster S, *et al.* The importance of chain length on the wettability and solubility of organic homologs. *Int J Pharm* 1991; **72**: 29–34.
Golightly LK, *et al.* Pharmaceutical excipients: adverse effects associated with inactive ingredients in drug products (part I). *Med Toxicol* 1988; **3**: 128–165.
Grant DJW, *et al.* Non-linear van't Hoff solubility–temperature plots and their pharmaceutical interpretation. *Int J Pharm* 1984; **18**: 25–38.
Jian L, Li Wan Po A. Ciliotoxicity of methyl- and propyl-*p*-hydroxybenzoates: a dose-response and surface-response study. *J Pharm Pharmacol* 1993; **45**: 925–927.
Jones PS, *et al.* *p*-Hydroxybenzoic acid esters as preservatives III: the physiological disposition of *p*-hydroxybenzoic acid and its esters. *J Am Pharm AssocSci* 1956; **45**: 268–273.
Kostenbauder HB. Physical chemical aspects of preservative selection for pharmaceutical and cosmetic emulsions. *Dev Ind Microbiol* 1962; **1**: 286–296.
Marouchoc SR. Cosmetic preservation. *Cosmet Technol* 1980; **2**(10): 38–44.
Matthews C, *et al.* *p*-Hydroxybenzoic acid esters as preservatives II: acute and chronic toxicity in dogs, rats and mice. *J Am Pharm AssocSci* 1956; **45**: 260–267.
Sakamoto T, *et al.* Effects of some cosmetic pigments on the bactericidal activities of preservatives. *J Soc Cosmet Chem* 1987; **38**: 83–98.
Schnuch A, *et al.* Contact allergy to preservatives. Analysis of IVDK data 1996-2009. *Br J Dermatol* 2011; **164**: 1316–1325.
Sokol H. Recent developments in the preservation of pharmaceuticals. *Drug Standards* 1952; **20**: 89–106.

21 Author

N Sandler.

22 Date of Revision

2 March 2012.

Mineral Oil

1 Nonproprietary Names

BP: Liquid Paraffin
JP: Liquid Paraffin
PhEur: Paraffin, Liquid
USP-NF: Mineral Oil

2 Synonyms

Avatech; *Drakeol*; heavy mineral oil; heavy liquid petrolatum; liquid petrolatum; paraffin oil; paraffinum liquidum; *Sirius*; white mineral oil.

3 Chemical Name and CAS Registry Number

Mineral oil [8012-95-1]

4 Empirical Formula and Molecular Weight

Mineral oil is a mixture of refined liquid saturated aliphatic (C_{14}–C_{18}) and cyclic hydrocarbons obtained from petroleum.

5 Structural Formula

See Section 4.

6 Functional Category

Emollient; lubricant; oleaginous vehicle; solvent; vaccine adjuvant.

7 Applications in Pharmaceutical Formulation or Technology

Mineral oil is used primarily as an excipient in topical pharmaceutical formulations, where its emollient properties are exploited as an ingredient in ointment bases; *see* Table I. It is additionally used in oil-in-water emulsions,[1–5] as a solvent, and as a lubricant in capsule and tablet formulations, and to a limited extent as a mold-release agent for cocoa butter suppositories. It has also been used in the preparation of microspheres and as a vaccine adjunct.[6–10]

Table I: Uses of mineral oil.

Use	Concentration (%)
Ophthalmic ointments	3.0–60.0
Otic preparations	0.5–3.0
Topical emulsions	1.0–32.0
Topical lotions	1.0–20.0
Topical ointments	0.1–95.0

8 Description

Mineral oil is a transparent, colorless, viscous oily liquid, without fluorescence in daylight. It is practically tasteless and odorless when cold, and has a faint odor of petroleum when heated.

9 Pharmacopeial Specifications

See Table II.

10 Typical Properties

Boiling point >360°C
Flash point 210–224°C
Pour point −12.2 to −9.4°C
Refractive index n_D^{20} = 1.4756–1.4800
Surface tension ≈35 mN/m at 25°C
Solubility Practically insoluble in ethanol (95%), glycerin, and water; soluble in acetone, benzene, chloroform, carbon disulfide, ether, and petroleum ether. Miscible with volatile oils and fixed oils, with the exception of castor oil.
Spectroscopy
IR spectra *see* Figure 1.
Viscosity (dynamic) 110–230 mPa s (110–230 cP) at 20°C

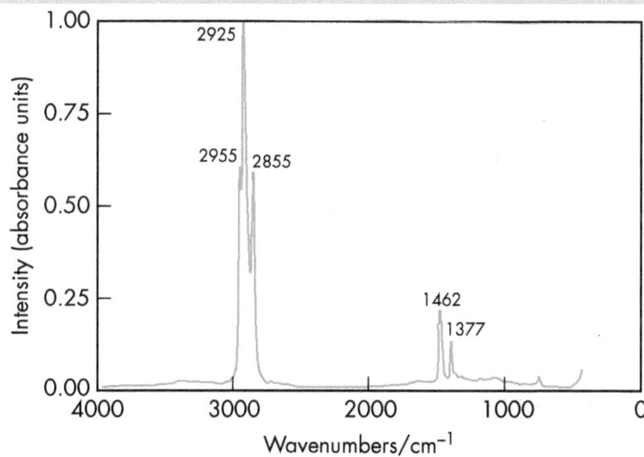

Figure 1: Infrared spectrum of mineral oil measured by transmission. Adapted with permission of Informa Healthcare.

Table II: Pharmacopeial specifications for mineral oil.

Test	JP XV	PhEur 7.4	USP35–NF30
Identification	+	+	+
Characters	–	+	–
Specific gravity	0.860–0.890	0.827–0.890	0.845–0.905
Viscosity	≥37 mm^2/s[a]	110–230 mPa s	≥34.5 mm^2/s[b]
Odor	+	–	–
Acidity or alkalinity	+	+	+
Heavy metals	≤10 ppm	–	–
Arsenic	≤2 ppm	–	–
Solid paraffin	+	+	+
Sulfur compounds	+	–	+
Polycyclic aromatic compounds	+	+	+
Readily carbonizable substances	+	+	+

(a) At 37.8°C.
(b) At 40.0 ± 0.1°C.

11 Stability and Storage Conditions

Mineral oil undergoes oxidation when exposed to heat and light. Oxidation begins with the formation of peroxides, exhibiting an 'induction period'. Under ordinary conditions, the induction period may take months or years. However, once a trace of peroxide is formed, further oxidation is autocatalytic and proceeds very rapidly. Oxidation results in the formation of aldehydes and organic acids, which impart taste and odor. Stabilizers may be added to retard oxidation; butylated hydroxyanisole, butylated hydroxytoluene, and alpha tocopherol are the most commonly used antioxidants.

Sterilization of petrolatum eye ointment base (containing mineral oil) by gamma irradiation exposures of 15–50 kGy has been shown to produce radiolytic volatile degradation products in proportion to the radiation dose. The principal radiolysis products from mineral oil were reported to be short-chain hydrocarbons, and as such was shown to be a suitable sterilization method.[11]

Mineral oil may be sterilized by dry heat.

Mineral oil should be stored in an airtight container, protected from light, in a cool, dry place.

12 Incompatibilities

Mineral oil is incompatible with strong oxidizing agents.

13 Method of Manufacture

Mineral oil is obtained by distillation of petroleum. The lighter hydrocarbons are first removed by distillation and the residue is then redistilled between 330–390°C. The distillate is chilled and the solid fractions are removed by filtration. The filtrate is then further purified and decolorized by high-pressure hydrogenation or sulfuric acid treatment; the purified filtrate is then filtered through adsorbents. The liquid portion obtained is distilled and the portion boiling below 360°C is discarded. A suitable stabilizer may be added to the mineral oil; *see* Section 11.

14 Safety

Mineral oil is used as an excipient in a wide variety of pharmaceutical formulations; *see* Section 16. It is also used in cosmetics and in some food products.

Therapeutically, mineral oil has been used in the treatment of constipation, as it acts as a lubricant and stool softener when taken orally. Daily doses of up to 45 mL have been administered orally, while doses of up to 120 mL have been used as an enema. However, excessive dosage of mineral oil, either orally or rectally, can result in anal seepage and irritation, and its oral use as a laxative is not considered desirable.

Chronic oral consumption of mineral oil may impair the appetite and interfere with the absorption of fat-soluble vitamins. Prolonged use should be avoided. Mineral oil is absorbed to some extent when emulsified and can lead to granulomatous reactions. Similar reactions also occur upon injection of the oil;[12] injection may also cause vasospasm.

The most serious adverse reaction to mineral oil is lipoid pneumonia caused by aspiration of the oil.[13,14] Mineral oil can enter the bronchial tree without eliciting the cough reflex.[15] With the reduction in the use of mineral oil in nasal formulations, the incidence of lipoid pneumonia has been greatly reduced. However, lipoid pneumonia has also been associated with the use of mineral oil-containing cosmetics[16] and ophthalmic preparations.[17] It is recommended that products containing mineral oil not be used in very young children, the elderly, or persons with debilitating illnesses.

Given its widespread use in many topical products, mineral oil has been associated with few instances of allergic reactions.

The WHO has not specified an acceptable daily intake of mineral oil given the low concentration consumed in foods.[18]

LD_{50} (mouse, oral): 22 g/kg[19]

15 Handling Precautions

Observe precautions appropriate to the circumstances and quantity of material handled. Avoid inhalation of vapors and wear protective clothing to prevent skin contact. Mineral oil is combustible.

16 Regulatory Status

GRAS listed. Accepted in the UK for use in certain food applications. Included in the FDA Inactive Ingredients Database (dental preparations; IV injections; ophthalmic preparations; oral capsules and tablets; otic, topical, transdermal, and vaginal preparations). Included in nonparenteral medicines licensed in the UK. Included in the Canadian List of Natural Health Products Ingredients Database.

17 Related Substances

Mineral oil and lanolin alcohols; light mineral oil; paraffin; petrolatum.

18 Comments

Therapeutically, mineral oil has been used as a laxative; *see* Section 14. It is indigestible and thus has limited absorption. Mineral oil is used in ophthalmic formulations for its lubricant properties. It is also used in cosmetics, and was historically used in some food products.[20]

M

19 Specific References

1 Zatz JL. Effect of formulation additives on flocculation of dispersions stabilized by a non-ionic surfactant. *Int J Pharm* 1979; **4**: 83–86.
2 Wepierre J, *et al.* Factors in the occlusivity of aqueous emulsions. *J Soc Cosmet Chem* 1982; **33**: 157–167.
3 Fong-Spaven F, Hollenbeck RG. Thermal rheological analysis of triethanolamine-stearate stabilized mineral oil in water emulsions. *Drug Dev Ind Pharm* 1986; **12**: 289–302.
4 Abd Elbary A, *et al.* Physical stability and rheological properties of w/o/w emulsions as a function of electrolytes. *Pharm Ind* 1990; **52**: 357–363.
5 Jayaraman SC, *et al.* Topical delivery of erythromycin from various formulations: an in-vivo hairless mouse study. *J Pharm Sci* 1996; **85**: 1082–1084.
6 Zinotti C, *et al.* Preparation and characterization of ethyl cellulose microspheres containing 5-fluorouracil. *J Microencapsul* 1994; **11**: 555–563.
7 O'Donnell PB, *et al.* Properties of multiphase microspheres of poly(D, 2-lactic-co-glycolic acid) prepared by a potentiometric dispersion technique. *J Microencapsul* 1995; **12**: 155–163.
8 Bachtsi AR, Kiparissides C. An experimental investigation of enzyme release from poly(vinyl alcohol) crosslinked microspheres. *J Microencapsul* 1995; **12**: 23–35.
9 Xiao C, *et al.* Improvement of a commercial foot-and-mouth disease vaccine by supplement of Quil A. *Vaccine* 2007; **25**(25): 4795–4800.
10 Jansen T, *et al.* Structure- and oil type-based efficacy of emulsion adjuvants. *Vaccine* 2006; **24**(26): 5400–5405.
11 Hong L, Altorfer A. Characterization of gamma irradiated petrolatum eye ointment base by headspace-gas chromatography-mass spectrometry. *J Pharm Biomed Anal* 2002; **29**(1-2): 263–275.
12 Bloem JJ, van der Waal I. Paraffinoma of the face: a diagnostic and therapeutic problem. *Oral Surg* 1974; **38**: 675–680.
13 Volk BW, *et al.* Incidence of lipoid pneumonia in a survey of 389 chronically ill patients. *Am J Med* 1951; **10**: 316–324.
14 Smolinske SC. *Handbook of Food, Drug, and Cosmetic Excipients.* Boca Raton, FL: CRC Press, 1992; 231–234.
15 Bennet JC, Plum F, eds. *Textbook of Medicine*. Philadelphia: WB Saunders, 1996; 407–408, 1016.
16 Becton DL, *et al.* Lipoid pneumonia in an adolescent girl secondary to use of lip gloss. *J Pediatr* 1984; **105**: 421–423.
17 Prakash UBS, Rosenow EC. Pulmonary complications from ophthalmic preparations. *Mayo Clin Proc* 1990; **65**: 521.
18 FAO/WHO. Evaluation of certain food additives and contaminants. Thirty-seventh report of the Joint FAO/WHO Expert Committee on Food Additives. *World Health Organ Tech Rep Ser* 1991; No. 806.
19 Lewis RJ, ed. *Sax's Dangerous Properties of Industrial Materials*, 11th edn. New York: Wiley, 2004; 2554–2555.
20 Anonymous. Mineral hydrocarbons to be banned from foods. *Pharm J* 1989; **242**: 187.

20 General References

Davis SS, Khanderia MS. Rheological characterization of Plastibases and the effect of formulation variables on the consistency of these vehicles part 3: oscillatory testing. *Int J PharmTechnol Prod Manuf* 1981; **2**(Apr): 13–18.

Deasy PB, Gouldson MP. *In-vitro* evaluation of pellets containing enteric coprecipitates of nifedipine formed by non-aqueous spheronization. *Int J Pharm* 1996; **132**: 131–141.

Gosselin RE *et al.* eds. *Clinical Toxicology of Commercial Products*, 5th edn. Baltimore: Williams & Wilkins, 1984; II-156–II-157.

Rhodes RK. Highly refined petroleum products in skin lotions. *Cosmet Perfum* 1974; **89**(3): 53–56.

21 Author

W Cook.

22 Date of Revision

2 March 2012.

Mineral Oil, Light

1 Nonproprietary Names

BP: Light Liquid Paraffin
JP: Light Liquid Paraffin
PhEur: Paraffin, Light Liquid
USP–NF: Light Mineral Oil

2 Synonyms

905 (mineral hydrocarbons); *Citation*; light liquid petrolatum; light white mineral oil; paraffinum perliquidum.

3 Chemical Name and CAS Registry Number

Light mineral oil [8012-95-1]

4 Empirical Formula and Molecular Weight

Light mineral oil is a mixture of refined liquid saturated hydrocarbons obtained from petroleum. It is less viscous and has a lower specific gravity than mineral oil. The USP35–NF30 specifies that light mineral oil may contain a suitable stabilizer.

5 Structural Formula

A mixture of refined liquid hydrocarbons, essentially paraffins and naphthenic in nature, obtained from petroleum.

6 Functional Category

Emollient; oleaginous vehicle; solvent; tablet and capsule lubricant.

7 Applications in Pharmaceutical Formulation or Technology

Light mineral oil is used in applications similar to those of mineral oil. It is used primarily as an excipient in topical pharmaceutical formulations where its emollient properties are exploited in ointment bases;[1–3] *see* Table I. It is also used in ophthalmic formulations.[4,5] Light mineral oil is additionally used in oil-in-water and polyethlylene glycol/glycerol emulsions;[6–9] as a solvent and lubricant in capsules and tablets; as a solvent and penetration enhancer in transdermal preparations;[10] and as the oily medium used in the microencapsulation of many drugs.[11–20]

Light mineral oil is also used in cosmetics and certain food products.

Table I: Uses of light mineral oil.

Use	Concentration (%)
Ophthalmic ointments	≤15.0
Otic preparations	≤50.0
Topical emulsions	1.0–20.0
Topical lotions	7.0–16.0
Topical ointments	0.2–23.0

8 Description

Light mineral oil occurs as a transparent, colorless liquid, without fluorescence in daylight. It is practically tasteless and odorless when cold, and has a faint odor when heated.

9 Pharmacopeial Specifications

See Table II.

Table II: Pharmacopeial specifications for light mineral oil.

Test	JP XV	PhEur 7.4	USP35–NF30
Identification	+	+	+
Characters	—	+	—
Specific gravity	0.830–0.870	0.810–0.875	0.818–0.880
Viscosity	<37 mm^2/s[(a)]	25–80 mPa s	3.0–34.0 mm^2/s[(b)]
Acidity or alkalinity	+	+	+
Heavy metals	≤10 ppm	—	—
Arsenic	≤2 ppm	—	—
Sulfur compounds	+	—	+
Readily carbonizable substances	+	+	+
Polycyclic aromatic compounds	+	+	+
Odor	+	—	—
Solid paraffin	+	+	+

(a) At 37.8°C.
(b) At 40.0 ± 0.1°C.

10 Typical Properties

Solubility Soluble in chloroform, ether, and hydrocarbons; sparingly soluble in ethanol (95%); practically insoluble in water.

11 Stability and Storage Conditions

Light mineral oil undergoes oxidation when exposed to heat and light. Oxidation begins with the formation of peroxides, exhibiting an 'induction period'. Under typical storage conditions, the induction period may take months or years. However, once a trace of peroxide is formed, further oxidation is autocatalytic and proceeds very rapidly. Oxidation results in the formation of aldehydes and organic acids, which impart taste and odor. The USP35–NF30 permits the addition of suitable stabilizers to retard oxidation, butylated hydroxyanisole, butylated hydroxytoluene, and alpha tocopherol being the most commonly used antioxidants.

Light mineral oil may be sterilized by dry heat.

Light mineral oil should be stored in an airtight container in a cool, dry place and protected from light.

12 Incompatibilities

Light mineral oil is incompatible with strong oxidizing agents.

13 Method of Manufacture

Light mineral oil is obtained by the distillation of petroleum. A suitable stabilizer may be added to the oil; *see* Section 11.

See also Mineral Oil for further information.

14 Safety

Light mineral oil is used in applications similar to those of mineral oil. Mineral oil is considered safe by the FDA for direct use in foods. However, oral ingestion of large doses of light mineral oil or chronic consumption may be harmful. Chronic use may impair appetite and interfere with the absorption of fat-soluble vitamins. It is absorbed to some extent when emulsified, leading to granulomatous reactions. Oral and intranasal use of mineral oil or products containing mineral oil by infants or children is not recommended because of the possible danger of causing lipoid pneumonia.

See Mineral Oil for further information.

15 Handling Precautions

Observe normal precautions appropriate to the circumstances and quantity of material handled. Since light mineral oil is combustible, it should not be handled or stored near heat, sparks, or flame. Light mineral oil should not be mixed with or stored with strong oxidants. Inhalation of mineral oil vapors may be harmful.

16 Regulatory Status

GRAS listed. Accepted in the UK for use in certain food applications. Light mineral oil is included in the FDA Inactive Ingredients Database (ophthalmic preparations; oral capsules and tablets; otic, rectal, topical, and transdermal preparations). Included in the Canadian Natural Health Products Ingredients Database.

M

17 Related Substances

Mineral oil; mineral oil and lanolin alcohols; paraffin; petrolatum.

18 Comments

—

19 Specific References

1 Jolly ER. Clinical evaluation of baby oil as a dermal moisturizer. *Cosmet Toilet* 1976; **91**: 51–52.
2 Magdassi S, *et al.* Correlation between nature of emulsifier and multiple emulsion stability. *Drug Dev Ind Pharm* 1985; **11**: 791–798.
3 Tanaka S, *et al.* Solubility and distribution of dexamethasone acetate in oil-in-water creams and its release from the creams. *Chem Pharm Bull* 1985; **33**: 3929–3934.
4 Merritt JC, *et al.* Topical Δ9-tetrahydrocannabinol and aqueous dynamics in glaucoma. *J Clin Pharmacol* 1981; **21**: 467S–471S.
5 Jay WM, Green K. Multiple-drop study of topically applied 1% delta 9-tetrahydrocannabinol in human eyes. *Arch Ophthalmol* 1983; **101**: 591–593.
6 Hallworth GW, Carless JE. Stablization of oil-in-water emulsions by alkyl sulfates: influence of the nature of the oil on stability. *J Pharm Pharmacol* 1972; **24**: 71–83.
7 Magdassi S. Formation of oil-in-polyethylene glycol/water emulsions. *J Disper Sci Technol* 1988; **9**: 391–399.
8 Magdassi S, Frank SG. Formation of oil in glycerol/water emulsions: effect of surfactant ethylene oxide content. *J Disper Sci Technol* 1990; **11**: 519–528.
9 Moaddel T, Frierg SE. Phase equilibria and evaporation rates in a four component emulsion. *J Disper Sci Technol* 1995; **16**: 69–97.
10 Pfister WR, Hsieh DST. Permeation enhancers compatible with transdermal drug delivery systems part II: system design considerations. *Pharm Technol* 1990; **14**(10): 54, 56–5860.
11 Beyger JW, Nairn JG. Some factors affecting the microencapsulation of pharmaceuticals with cellulose acetate phthalate. *J Pharm Sci* 1986; **75**: 573–578.

12 Pongpaibul Y, Whitworth CW. Preparation and *in vitro* dissolution characteristics of propranolol microcapsules. *Int J Pharm* 1986; 33: 243–248.
13 Sheu M-T, Sokoloski TD. Entrapment of bioactive compounds within native albumin beads III: evaluation of parameters affecting drug release. *J Parenter Sci Technol* 1986; 40: 259–265.
14 D'Onofrio GP, *et al.* Encapsulated microcapsules. *Int J Pharm* 1979; 2: 91–99.
15 Huang HP, Ghebre Sellassie I. Preparation of microspheres of water-soluble pharmaceuticals. *J Microencapsul* 1989; 6(2): 219–225.
16 Ghorab MM, *et al.* Preparation of controlled release anticancer agents I: 5-fluorouracil–ethyl cellulose microspheres. *J Microencapsul* 1990; 7(4): 447–454.
17 Ruiz R, *et al.* A study on the manufacture and *in vitro* dissolution of terbutaline sulfate microcapsules and their tablets. *Drug Dev Ind Pharm* 1990; **16**: 1829–1842.
18 Sanghvi SP, Nairn JG. Phase diagram studies for microencapsulation of pharmaceuticals using cellulose acetate trimellitate. *J Pharm Sci* 1991; 80: 394–398.
19 Iwata M, McGinity JW. Preparation of multi-phase microspheres of poly(D,L-lactic acid) and poly(D,L-lactic co-glycolic acid) containing a w/o emulsion by a multiple emulsion solvent evaporation technique. *J Microencapsul* 1992; 9(2): 201–214.
20 Sanghvi SP, Nairn JG. Effect of viscosity and interfacial tension on particle size of cellulose acetate trimellitate microspheres. *J Microencapsul* 1992; 9(2): 215–227.

20 General References

Allen LV. Featured excipient: capsule and tablet lubricants. *Int J Pharm Compound* 2000; 4(5): 390–392.

Allen LV. Featured excipient: oleaginous vehicles. *Int J Pharm Compound* 2000; 4(6): 470–473484–485.

See also Mineral Oil.

21 Author

W Cook.

22 Date of Revision

2 March 2012.

 Mineral Oil and Lanolin Alcohols

1 Nonproprietary Names

None adopted.

2 Synonyms

Amerchol L-101; liquid paraffin and lanolin alcohols; *Protalan M-16*; *Protalan M-26*.

3 Chemical Name and CAS Registry Number

Mineral oil [8012-95-1]

Lanolin alcohols [8027-33-6]

4 Empirical Formula and Molecular Weight

A mixture of mineral oil and lanolin alcohols.

5 Structural Formula

See Section 4.

6 Functional Category

Emollient; emulsifying agent.

7 Applications in Pharmaceutical Formulation or Technology

Mineral oil and lanolin alcohols is an oily liquid used in topical pharmaceutical formulations and cosmetics as an emulsifying agent with emollient properties; *see* Table I. It is used as a primary emulsifier in the preparation of water-in-oil creams and lotions and as an auxiliary emulsifier and stabilizing agent in oil-in-water creams and lotions.

Table I: Uses of mineral oil and lanolin alcohols.

Use	Concentration (%)
Emollient	3.0–6.0
Emulsifier in w/o creams and lotions	5.0–15.0
Emulsifier in o/w creams and lotions	0.5–6.0

8 Description

A pale yellow-colored, oily liquid with a faint characteristic sterol odor.

9 Pharmacopeial Specifications

Lanolin alcohols and mineral oil are listed as separate monographs in BP, JP, PhEur and USP–NF but the combination is not listed; *see* Lanolin Alcohols and Mineral Oil.

10 Typical Properties

Acid value ≤ 1

Arsenic ≤ 2 ppm

Ash $\leq 0.2\%$

Heavy metals ≤ 20 ppm

HLB value ≈ 8

Hydroxyl value 10–15

Iodine number ≤ 12

Microbiological count The total bacterial count, when packaged, is less than 10 per gram of sample.

Moisture content $\leq 0.2\%$

Saponification value ≤ 2

Solubility Soluble 1 in 2 parts of chloroform, 1 in 4 parts of castor oil, and 1 in 4 parts of corn oil. Practically insoluble in ethanol (95%) and water. Precipitation occurs in hexane.

Specific gravity 0.840–0.860 at 25°C

11 Stability and Storage Conditions

Mineral oil and lanolin alcohols is stable and should be stored in a well-closed container in a cool, dry place.

12 Incompatibilities

Lanolin alcohols are incompatible with coal tar, ichthammol, phenol, and resorcinol.

13 Method of Manufacture

Lanolin alcohols are dissolved in mineral oil.

14 Safety

Mineral oil and lanolin alcohols is generally regarded as an essentially nontoxic and nonirritant material. However, lanolin alcohols may be irritant to the skin and cause hypersensitivity in some individuals.[1,2]

15 Handling Precautions

Observe normal precautions appropriate to the circumstances and quantity of material handled.

16 Regulatory Status

Accepted for use in topical pharmaceutical formulations and cosmetics. Included in the Canadian Natural Health Products Ingredients Database.

17 Related Substances

Lanolin alcohols; mineral oil; petrolatum and lanolin alcohols.

18 Comments

See Lanolin Alcohols and Mineral Oil for further information.

Studies have been carried out into the use of synthetic blends containing lanolin alcohols and mineral oils that may avoid an allergic response whilst still producing an ointment base with the appropriate physical characteristics.[3]

19 Specific References

1 Wakelin SH, *et al.* A retrospective analysis of contact allergy to lanolin. *Br J Dermatol* 2001; **145**(1): 28–31.
2 Nguyen JC, *et al.* Allergic contact dermatitis caused by lanolin (wool) alcohol contained in an emollient in three postsurgical patients. *J Am Acad Dermatol* 2010; **62**(6): 1064–1065.
3 Savić S, *et al.* An alkylpolyglucoside surfactant as a prospective pharmaceutical excipient for topical formulations: the influence of oil polarity on the colloidal structure and hydrocortisone *in vitro/in vivo* permeation. *Eur J Pharm Sci* 2007; **30**(5): 441–450.

20 General References

Davis SS. Viscoelastic properties of pharmaceutical semisolids I: ointment bases. *J Pharm Sci* 1969; **58**: 412–418.
Prosperio G, *et al.* Lanolin and its derivatives for cosmetic creams and lotions. *Cosmet Toilet* 1980; **95**(4): 81–85.

21 Author

AH Kibbe.

22 Date of Revision

1 August 2011.

M

Monoethanolamine

1 Nonproprietary Names

BP: Ethanolamine

USP–NF: Monoethanolamine

2 Synonyms

β-Aminoethyl alcohol; colamine; ethylolamine; β-hydroxyethylamine; 2-hydroxyethylamine.

3 Chemical Name and CAS Registry Number

2-Aminoethanol [141-43-5]

4 Empirical Formula and Molecular Weight

C_2H_7NO 61.08

5 Structural Formula

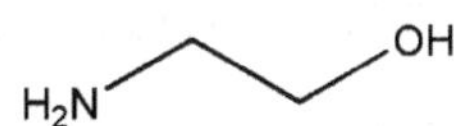

6 Functional Category

Alkalizing agent; emulsifying agent; solvent.

7 Applications in Pharmaceutical Formulation or Technology

Monoethanolamine is used primarily in pharmaceutical formulations for pH adjustment and in the preparation of emulsions. Other uses include as a solvent for fats and oils and as a stabilizing agent in an injectable dextrose solution of phenytoin sodium.

8 Description

Monoethanolamine occurs as a clear, colorless or pale yellow-colored, moderately viscous liquid with a mild, ammoniacal odor.

9 Pharmacopeial Specifications

See Table I.

10 Typical Properties

Acidity/alkalinity pH = 12.1 for a 0.1 N aqueous solution.

Boiling point 170.8°C

Critical temperature 341°C

Density

1.0117 g/cm^3 at 25°C;

0.9998 g/cm^3 at 40°C;

0.9844 g/cm^3 at 60°C.

Dissociation constant $pK_a = 9.4$ at 25°C

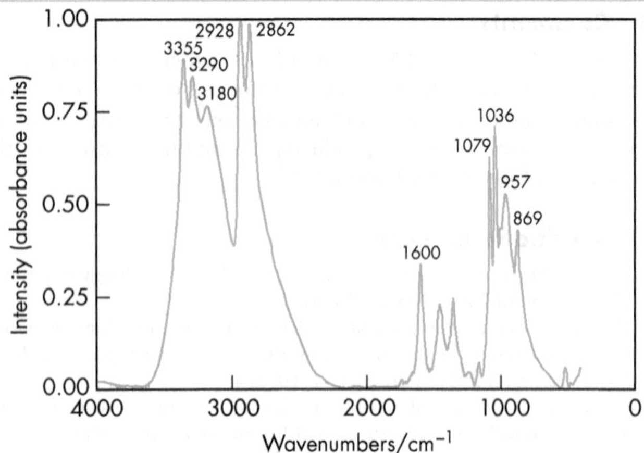

Figure 1: Infrared spectrum of monoethanolamine measured by transmission. Adapted with permission of Informa Healthcare.

Table I: Pharmacopeial specifications for monoethanolamine.

Test	BP 2012	USP35–NF30
Identification	+	+
Characters	+	—
Specific gravity	1.014–1.023	1.013–1.016
Refractive index	1.453–1.459	—
Related substances	≤2.0%	—
Distilling range	—	167–173°C
Residue on ignition	—	≤0.1%
Assay	98.0–100.5%	98.0–100.5%

Flash point (open cup) 93°C
Hygroscopicity Very hygroscopic.
Melting point 10.3°C
Refractive index $n_D^{20} = 1.4539$
Solubility *see* Table II.

Table II: Solubility of monoethanolamine.

Solvent	Solubility at 20°C
Acetone	Miscible
Benzene	1 in 72
Chloroform	Miscible
Ethanol (95%)	Miscible
Ethyl ether	1 in 48
Glycerol	Miscible
Methanol	Miscible
Water	Miscible

Spectroscopy
IR spectra *see* Figure 1.
Surface tension 48.8 mN/m at 20°C
Vapor density (relative) 2.1 (air = 1)
Vapor pressure 53.3 Pa (0.4 mmHg) at 20°C
Viscosity (dynamic)
18.95 mPa s (18.95 cP) at 25°C;
5.03 mPa s (5.03 cP) at 60°C.

11 Stability and Storage Conditions

Monoethanolamine is very hygroscopic and is unstable when exposed to light. Aqueous monoethanolamine solutions may be sterilized by autoclaving.

When monoethanolamine is stored in large quantities, stainless steel is preferable for long-term storage. Copper, copper alloys, zinc, and galvanized iron are corroded by amines and should not be used for construction of storage containers. Ethanolamines readily absorb moisture and carbon dioxide from the air; they also react with carbon dioxide. This can be prevented by sealing the monoethanolamine under an inert gas. Smaller quantities of monoethanolamine should be stored in an airtight container, protected from light, in a cool, dry place.

12 Incompatibilities

Monoethanolamine contains both a hydroxy group and a primary amine group and will thus undergo reactions characteristic of both alcohols and amines. Ethanolamines will react with acids to form salts and esters. Discoloration and precipitation will take place in the presence of salts of heavy metals. Monoethanolamine reacts with acids, acid anhydrides, acid chlorides, and esters to form amide derivatives, and with propylene carbonate or other cyclic carbonates to give the corresponding carbonates.

As a primary amine, monoethanolamine will react with aldehydes and ketones to yield aldimines and ketimines. Additionally, monoethanolamine will react with aluminum, copper, and copper alloys to form complex salts. A violent reaction will occur with acrolein, acrylonitrile, epichlorohydrin, propiolactone, and vinyl acetate.

13 Method of Manufacture

Monoethanolamine is prepared commercially by the ammonolysis of ethylene oxide. The reaction yields a mixture of monoethanolamine, diethanolamine, and triethanolamine, which is separated to obtain the pure products. Monoethanolamine is also produced from the reaction between nitromethane and formaldehyde.

14 Safety

Monoethanolamine is an irritant, caustic material, but when it is used in neutralized parenteral and topical pharmaceutical formulations it is not usually associated with adverse effects, although hypersensitivity reactions have been reported. Monoethanolamine salts are generally regarded as being less toxic than monoethanolamine.

LD_{50} (mouse, IP): 0.05 g/kg[1]
LD_{50} (mouse, oral): 0.7 g/kg
LD_{50} (rabbit, skin): 1.0 g/kg
LD_{50} (rat, IM): 1.75 g/kg
LD_{50} (rat, IP): 0.07 g/kg
LD_{50} (rat, IV): 0.23 g/kg
LD_{50} (rat, oral): 1.72 g/kg
LD_{50} (rat, SC): 1.5 g/kg

15 Handling Precautions

When handling concentrated solutions of monoethanolamine, personal protective equipment such as an appropriate respirator, chemically resistant gloves, safety goggles, and other protective clothing should be worn. Transfer or prepare monoethanolamine solutions only in a chemical fume hood.

Vapors may flow along surfaces to distant ignition sources and flash back. Closed containers exposed to heat may explode. Contact with strong oxidizers may cause fire.

In the UK, the short-term (15-minute) workplace exposure limit for monoethanolamine is 7.6 mg/m^3 (3 ppm) and the long-term exposure limit (8-hour TWA) is 2.5 mg/m^3 (1 ppm).[2]

16 Regulatory Status

Included in the FDA Inactive Ingredients Database (oral tablets; delayed action). Included in parenteral and nonparenteral medicines licensed in the UK and US. Included in the Canadian Natural Health Products Ingredients Database.

17 Related Substances

Diethanolamine; triethanolamine.

18 Comments

Monoethanolamine is used to produce a variety of salts with therapeutic uses. For example, a salt of monoethanolamine with vitamin C is used for intramuscular injection, while the salicylate and undecenoate monoethanolamine salts are utilized respectively in the treatment of rheumatism and as an antifungal agent. However, the most common therapeutic use of monoethanolamine is in the production of ethanolamine oleate injection, which is used as a sclerosing agent.[3]

A specification for monoethanolamine is included in the *Japanese Pharmaceutical Excipients* (JPE).[4]

The EINECS number for monoethanolamine is 205-483-3.

19 Specific References

1 Lewis RJ, ed. *Sax's Dangerous Properties of Industrial Materials*, 11th edn. New York: Wiley, 2004: 1607–1608.

2 Health and Safety Executive. *EH40/2005: Workplace Exposure Limits*. Sudbury: HSE Books, 2011. http://www.hse.gov.uk/pubns/priced/eh40.pdf (accessed 28 February 2012).

3 Crotty B, *et al*. The management of acutely bleeding varices by injection sclerotherapy. *Med J Aust* 1986; **145**: 130–133.

4 Japan Pharmaceutical Excipients Council. *Japanese Pharmaceutical Excipients* 2004. Tokyo: Yakuji Nippo, 2004: 562–563.

20 General References

Kubis A, *et al*. Studies on the release of solubilized drugs from ointment bases. *Pharmazie* 1984; **39**: 168–170.

21 Author

SR Goskonda.

22 Date of Revision

28 February 2012.

Monosodium Glutamate

1 Nonproprietary Names

USP–NF: Monosodium Glutamate

2 Synonyms

Chinese seasoning; E621; glutamic acid monosodium salt; glutamic acid, sodium salt; MSG; monosodium L-glutamate monohydrate; natrii glutamas; sodium L-glutamate; sodium glutamate monohydrate; sodium hydrogen L-(+)-2-aminoglutarate monohydrate.

3 Chemical Name and CAS Registry Number

Glutamic acid monosodium salt monohydrate [142-47-2]

4 Empirical Formula and Molecular Weight

$C_5H_8NO_4Na$ 169.13 (anhydrous)
$C_5H_8NO_4Na{\cdot}H_2O$ 187.13 (monohydrate)

5 Structural Formula

O O H HO O^-Na^+ NH_2

6 Functional Category

Buffering agent; flavoring agent.

7 Applications in Pharmaceutical Formulation or Technology

Monosodium glutamate is used in oral pharmaceutical formulations as a buffer and a flavor enhancer. For example, it is used with sugar to improve the palatability of bitter-tasting drugs and can reduce the metallic taste of iron-containing liquids. It has also been used in subcutaneous live vaccine injections such as measles, mumps, rubella and varicella-zoster live vaccine (*ProQuad*).

8 Description

Monosodium glutamate occurs as white free-flowing crystals or a crystalline powder. It is practically odorless and has a meat-like taste.

9 Pharmacopeial Specifications

See Table I.

Table I: Pharmacopeial specifications for monosodium glutamate.

Test	USP35–NF30
Identification	+
Clarity and color of solution	+
Specific rotation	+24.8° to +25.3°
pH (5% solution)	6.7–7.2
Loss on drying	≤0.5%
Chloride	≤0.25%
Lead	≤10 ppm
Heavy metals	≤20 ppm
Assay	99.0–100.5%

10 Typical Properties

Acidity/alkalinity pH = 7.0 (0.2% w/v aqueous solution)
Melting point 232°C
Solubility Soluble in water; sparingly soluble in ethanol (95%).

Specific rotation $[\alpha]_D^{25}$ = +24.2° to +25.5° at 25°C (8.0% w/v in 1.0 N HCl).

11 Stability and Storage Conditions

Aqueous solutions of monosodium glutamate may be sterilized by autoclaving. Monosodium glutamate should be stored in a tight container in a cool, dry place.

12 Incompatibilities

—

13 Method of Manufacture

Monosodium glutamate is the monosodium salt of the naturally occurring L-form of glutamic acid. It is commonly manufactured by fermentation of carbohydrate sources such as sugar beet molasses. In general, sugar beet products are used in Europe and the USA. Other carbohydrate sources such as sugar cane and tapioca are used in Asia.

14 Safety

Monosodium glutamate is widely used in foods and oral pharmaceutical formulations. It is generally regarded as moderately toxic on ingestion or intravenous administration. Adverse effects include somnolence, hallucinations and distorted perceptions, headache, dyspnea, nausea or vomiting, and dermatitis. The lowest lethal oral dose in humans is reported to be 43 mg/kg.[1] The use of monosodium glutamate in foods has been controversial due to the so-called 'Chinese Restaurant Syndrome' (*see* Section 18), although it is generally regarded as safe at intake levels of up to 6 mg/kg body-weight.[2] In Europe, total glutamate intake from food ranges from 5–12 g/day.[2]

There has been a report of a foreign body granuloma caused by monosodium glutamate after a BCG vaccination.[3]

15 Handling Precautions

Observe normal precautions appropriate to the circumstances and quantity of material handled. When heated to decomposition, monosodium glutamate emits toxic fumes of NO_x and Na_2O.

16 Regulatory Status

GRAS listed. Accepted in Europe for use as a food additive in certain applications. Included in the FDA Inactive Ingredients Database (oral syrup). Included in nonparenteral medicines licensed in the UK. Included in subcutaneous vaccine injections. Included in the Canadian Natural Health Products Ingredients Database.

17 Related Substances

—

18 Comments

The most widespread use of monosodium glutamate is as a flavor enhancer in food products. Typically, 0.2–0.9% is used in normally salted foods, although products such as soy protein can contain 10–30%.

Monosodium glutamate has been associated with reports of adverse reactions termed 'Chinese Restaurant Syndrome' after it was first self-reported by a physician who regularly experienced numbness and palpitations after consuming Chinese food.[4]

Subsequent to this first report, numerous other anecdotal reports of adverse reactions to monosodium glutamate were made, with symptoms occurring at doses of 1.5–12 g. Reactions include paresthesias or a skin burning sensation, facial pressure or tightness sensation, and substernal chest pressure. Severity of reaction corresponded with increased dose. Reports of 'Chinese Restaurant Syndrome' in children are rare. A variety of other adverse reactions to monosodium glutamate have also been reported including flushing, asthma, headache, behavioral abnormalities, and ventricular tachycardia.[5–7]

Placebo-controlled, blinded, trials of monosodium glutamate consumption have, however, largely failed to reproduce the full effects of 'Chinese Restaurant Syndrome' as it was originally described, and symptoms may be simply due to dyspepsia.[8] Some dose-dependent adverse reactions may be attributed to monosodium glutamate, with doses of 5 g producing reactions in 30% of individuals tested.[9] In the US, the FDA has stated that monosodium glutamate and related substances are safe food ingredients for most people when used at 'customary' levels.[10–12]

Monosodium glutamate monohydrate 32 g is approximately equivalent to anhydrous monosodium glutamate 29 g or glutamic acid 25 g. Each gram of monosodium glutamate monohydrate represents 5.3 mmol (5.3 mEq) of sodium.

A specification for monosodium glutamate is contained in the *Food Chemicals Codex* (FCC).[13]

The EINECS number for monosodium glutamate is 205-538-1. The PubChem Compound ID (CID) for monosodium glutamate is 23689119.

19 Specific References

1 Lewis RJ, ed. *Sax's Dangerous Properties of Industrial Materials*, 11th edn. New York: Wiley, 2004: 2573.
2 Beyreuther K, *et al.* Consensus meeting: monosodium glutamate – an update. *Eur J Clin Nutr* 2007; **61**(3): 304–313.
3 Chin YK, *et al.* Foreign body granuloma caused by monosodium glutamate after BCG vaccination. *J Am Acad Dermatol* 2006; **55**(2 Suppl.): S1–S5.
4 Kwok HM. Chinese restaurant syndrome. *N Engl J Med* 1968; **278**: 796.
5 Alston RM. Chinese restaurant syndrome. *N Engl J Med* 1976; **294**: 225.
6 Allen DH, Baker GH. Chinese restaurant asthma. *N Engl J Med* 1981; **305**: 1154–1155.
7 Smolinske SC. *Handbook of Food, Drug and Cosmetic Excipients*. Boca Raton, FL: CRC Press, 1992: 235–241.
8 Kenney RA. Chinese restaurant syndrome. *Lancet* 1980; **i**: 311–312.
9 Kenney RA. The Chinese restaurant syndrome: an anecdote revisited. *Food Chem Toxicol* 1986; **24**: 351–354.
10 Anonymous. Monosodium glutamate safe for most people, says FDA. *Pharm J* 1996; **256**: 83.
11 Meadows M. MSG: a common flavor enhancer. *FDA Consumer* 2003; **37**(1): 34–35.
12 Freeman M. Reconsidering the effects of monosodium glutamate: a literature review. *J Am Acad Nurse Pract* 2006; **18**(10): 482–486.
13 *Food Chemicals Codex*, 7th edn. Bethesda, MD: United States Pharmacopeia, 2010: 698.

20 General References

Chevassus H, *et al.* Effects of oral monosodium L-glutamate on insulin secretion and glucose tolerance in healthy volunteers. *Br J Clin Pharmacol* 2002; **53**(6): 641–643.

Japan Pharmaceutical Excipients Council. *Japanese Pharmaceutical Excipients Directory 1996*. Tokyo: Yakuji Nippo, 1996: 335.

Rhys Williams AT, Winfield SA. Determination of monosodium glutamate in food using high-performance liquid chromatography and fluorescence detection. *Analyst* 1982; **107**(1278): 1092–1094.

Walker R. The significance of excursions above the ADI. Case study: monosodium glutamate. *Reg Toxicol Pharmacol* 1999; **30**: S119–S121.

21 Author

PJ Weller.

22 Date of Revision

1 March 2012.

Monothioglycerol

1 Nonproprietary Names

USP–NF: Monothioglycerol

2 Synonyms

1-Mercaptoglycerol; 1-mercapto-2,3-propanediol; monothioglycerin; α-monothioglycerol; thioglycerin; 1-thioglycerol.

3 Chemical Name and CAS Registry Number

3-Mercapto-1,2-propanediol [96-27-5]

4 Empirical Formula and Molecular Weight

$C_3H_8O_2S$ 108.16

5 Structural Formula

HS–CH2–CH(OH)–CH2–OH

6 Functional Category

Antimicrobial preservative; antioxidant.

7 Applications in Pharmaceutical Formulation or Technology

Monothioglycerol is used as an antioxidant in pharmaceutical formulations, mainly in parenteral preparations.(1) Monothioglycerol is reported to have some antimicrobial activity.(2–4) It is also widely used in cosmetic formulations such as depilating agents.

8 Description

Monothioglycerol occurs as a colorless or pale-yellow colored, viscous, hygroscopic liquid with a slight odor of sulfide.

9 Pharmacopeial Specifications

See Table I.

Table I: Pharmacopeial specifications for monothioglycerol.

Test	USP35–NF30
Specific gravity	1.241–1.250
Refractive index	1.521–1.526
pH (10% aqueous solution)	3.5–7.0
Water	≤5.0%
Residue on ignition	≤0.1%
Selenium	≤0.003%
Heavy metals	≤0.002%
Assay (anhydrous basis)	97.0–101.0%

10 Typical Properties

Acidity/alkalinity pH = 3.5–7.0 (10% w/v aqueous solution)
Boiling point 118°C
Flash point 110°C
Refractive index n_D^{25} = 1.521–1.526
Solubility Miscible with ethanol (95%); freely soluble in water; practically insoluble in ether.

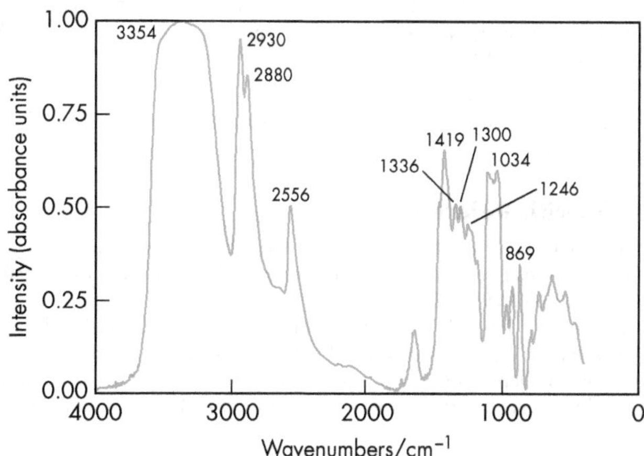

Figure 1: Infrared spectrum of monothioglycerol measured by transmission. Adapted with permission of Informa Healthcare.

Specific gravity 1.241–1.250
Spectroscopy
IR spectra *see* Figure 1.

11 Stability and Storage Conditions

Monothioglycerol is unstable in alkaline solutions. Monothioglycerol should be stored in a well-closed container in a cool, dry place.

12 Incompatibilities

Monothioglycerol can react with oxidizing materials.

13 Method of Manufacture

Monothioglycerol is prepared by heating an ethanolic solution of 3-chloro-1,2-propanediol with potassium bisulfide.

14 Safety

Monothioglycerol is generally regarded as a relatively nontoxic and nonirritant material at the concentrations used as a pharmaceutical excipient. It is used in topical and injectable preparations.

Undiluted monothioglycerol is considered a poison by the IP and IV routes; it has also been reported to be mutagenic.(5)

LD_{50} (cat, IV): 0.22 g/kg(5)
LD_{50} (mouse, IP): 0.34 g/kg
LD_{50} (rabbit, IV): 0.25 g/kg
LD_{50} (rat, IP): 0.39 g/kg

15 Handling Precautions

Observe normal precautions appropriate to the circumstances and quantity of material handled. Monothioglycerol is flammable when exposed to heat or flame; when heated to decomposition it emits toxic fumes of SO_x.

16 Regulatory Status

Included in the FDA Inactive Ingredients Database (IM, IV and other injections). Included in the Canadian Natural Health Products Ingredients Database.

17 Related Substances

—

18 Comments

Therapeutically, monothioglycerol has been used in a 0.02% w/w aqueous solution to stimulate wound healing, and as a 0.1% w/w jelly in atrophic rhinitis.

The EINECS number for monothioglycerol is 202-495-0. The PubChem Compound ID (CID) for monothioglycerol includes 7291 and 447638.

19 Specific References

1 Kasraian K, *et al.* Developing an injectable formula containing an oxygen sensitive drug: case study of danofloxacin injectable. *Pharm Dev Technol* 1999; 4(4): 475–480.
2 Jensen KK, Javor GT. Inhibition of *Escherichia coli* by thioglycerol. *Antimicrob Agents Chemother* 1981; 19: 556–561.
3 Javor GT. Depression of adenosylmethionine content of *Escherichia coli* by thioglycerol. *Antimicrob Agents Chemother* 1983; 24: 860–867.
4 Javor GT. Inhibition of respiration of *Escherichia coli* by thioglycerol. *Antimicrob Agents Chemother* 1983; 24: 868–870.
5 Lewis RJ, ed. *Sax's Dangerous Properties of Industrial Materials*, 11th edn. New York: Wiley, 2004; 2574.

20 General References

Modi S, *et al.* Determination of thio-based additives for biopharmaceuticals by pulsed electrochemical detection following HPLC. *J Pharm Biomed Anal* 2005; 37(1): 19–25.

Nealon DA, *et al.* Diluent pH and the stability of the thiol group in monothioglycerol, *N*-acetyl-L-cysteine, and 2-mercaptoethanol. *Clin Chem* 1981; 27(3): 505–506.

Sherman F, Kuselman I. Water determination in drugs containing thiols. *Int J Pharm* 1999; 190(2): 193–196.

21 Author

PJ Sheskey.

22 Date of Revision

1 March 2012.

Myristic Acid

M

1 Nonproprietary Names

None adopted.

2 Synonyms

Edenor C14 98-100; *n*-tetradecanoic acid; 1-tridecanecarboxylic acid.

3 Chemical Name and CAS Registry Number

3-Mercapto-1,2-propanediol [544-63-8]

4 Empirical Formula and Molecular Weight

$C_{14}H_{28}O_2$ 228.37

5 Structural Formula

H_3C–(zigzag chain)–COOH

6 Functional Category

Emulsifying agent; penetration enhancer; tablet and capsule lubricant.

7 Applications in Pharmaceutical Formulation or Technology

Myristic acid is used in oral and topical pharmaceutical formulations. *See also* Section 18.

8 Description

Myristic acid occurs as an oily white crystalline solid with a faint odor.

9 Pharmacopeial Specifications

See Section 18.

10 Typical Properties

Boiling point 326.2°C
Flash point >110°C
Melting point 54.5°C
Solubility Soluble in acetone, benzene, chloroform, ethanol (95%), ether, and aromatic and chlorinated solvents; practically insoluble in water.
Specific gravity 0.860–0.870

11 Stability and Storage Conditions

The bulk material should be stored in a well-closed container in a cool, dry place.

12 Incompatibilities

Myristic acid is incompatible with strong oxidizing agents and bases.

13 Method of Manufacture

Myristic acid occurs naturally in nutmeg butter and in most animal and vegetables fats. Synthetically, it may be prepared by electrolysis of methyl hydrogen adipate and decanoic acid or by Maurer oxidation of myristyl alcohol.

14 Safety

Myristic acid is used in oral and topical pharmaceutical formulations and is generally regarded as nontoxic and nonirritant at the levels employed as an excipient. However, myristic acid is reported to be an eye and skin irritant at high levels and is poisonous by intravenous administration. Mutation data have also been reported.[1]

LD_{50} (mouse, IV): 0.043 g/kg[1]
LD_{50} (rat, oral): >10 g/kg

15 Handling Precautions

Observe normal precautions appropriate to the circumstances and quantity of the material handled. Acrid smoke and irritating fumes are emitted when myristic acid is heated to decomposition.

16 Regulatory Status

GRAS listed. Included in the FDA Inactive Ingredients Database (oral capsules). Included in nonparenteral medicines licensed in the UK.

17 Related Substances

Lauric acid; myristyl alcohol; palmitic acid; potassium myristate; sodium myristate; stearic acid.

Potassium myristate
Empirical formula $C_{14}H_{28}O_2K$
Molecular weight 267.52
CAS number [13429-27-1]
Comments Potassium myristate is used as a surfactant and emulsifying agent in pharmaceutical formulations. The EINECS number for potassium myristate is 236-550-5.

Sodium myristate
Empirical formula $C_{14}H_{28}O_2Na$
Molecular weight 251.41
CAS number [822-12-8]
Comments Sodium myristate is used as an emulsifying agent in pharmaceutical formulations. The EINECS number for sodium myristate is 212-487-9.

18 Comments

Myristic acid has been evaluated as a penetration enhancer in melatonin transdermal patches in rats[2] and bupropion formulations on human cadaver skin.[3] Further studies have assessed the suitability of myristic acid in oxymorphone formulations,[4] clobetasol 17-propionate topical applications,[5] and diltiazem hydrochloride hydroxypropyl methylcellulose gel formulations.[6] Furthermore, polyvinyl alcohol substituted with myristic acid (as well as other fatty acids) at different substitution degrees has been studied for the preparation of biodegradable microspheres containing progesterone or indomethacin.[7] Myristic acid has also been investigated for use in solid lipid nanoparticles for enhanced oral bioavailability of low molecular weight heparins,[8] and for colon-specific delivery of peptide drugs.[9]

Therapeutically, myristic acid has been investigated for use in gene therapy[10,11] and chemotherapy.[12]

Although not included in any pharmacopeias, a specification for myristic acid is contained in the *Food Chemicals Codex* (FCC)[13] and in the *Japanese Pharmaceutical Excipients* (JPE);[14] *see* Table I.

The EINECS number for myristic acid is 208-875-2. The PubChem Compound ID (CID) for myristic acid is 11005.

Table I: *Food Chemicals Codex*[13] and *Japanese Pharmaceutical Excipients*[14] specifications for myristic acid.

Test	FCC 7	JPE 2008
Identification	—	+
Acid value	242–249	240–250
Heavy metals	—	+
Lead	≤2 mg/kg	—
Iodine value	≤1.0	≤1.0
Residue on ignition	≤0.1%	≤0.1%
Saponification value	242–251	—
Melting point	48–55.5°C	—
Unsaponifiable matter	≤1%	—
Water	≤0.2%	—
Ester value	—	≤3

19 Specific References

1 Lewis RJ. *Sax's Dangerous Properties of Industrial Materials*, 11th edn. New York: Wiley, 2004; 2586.
2 Kanikkannan N, *et al.* Formulation and *in vitro* evaluation of transdermal patches of melatonin. *Drug Dev Ind Pharm* 2004; **30**: 205–212.
3 Gondaliya D, Pundarikakshudu K. Studies in formulation and pharmacotechnical evaluation of controlled release transdermal delivery system of bupropion. *AAPS Pharm Sci Tech* 2003; **4**: E3.
4 Aungst BJ, *et al.* Transdermal oxymorphone formulation development and methods for evaluating flux and lag times for two skin permeation-enhancing vehicles. *J Pharm Sci* 1990; **79**: 1072–1076.
5 Fang JY, *et al.* Evaluation of topical application of clobetasol 17-propionate from various cream bases. *Drug Dev Ind Pharm* 1999; **25**: 7–14.
6 Karakatsani M, *et al.* The effect of permeation enhancers on the viscosity and the release profile of transdermal hydroxypropyl methylcellulose gel formulations containing diltiazem HCl. *Drug Dev Ind Pharm* 2010; **36**(10): 1195–1206.
7 Orienti I, *et al.* Fatty acid substituted polyvinyl alcohol as a supporting material for microsphere preparation. *J Microencapsul* 2001; **18**: 77–87.
8 Paliwal R, *et al.* Biomimetic solid lipid nanoparticles for oral bioavailability enhancement of low molecular weight heparin and its lipid conjugates: *in vitro* and *in vivo* evaluation. *Mol Pharm* 2011; **8**(4): 1314–1321.
9 Luppi B , *et al.* New environmental sensitive system for colon-specific delivery of peptidic drugs. *Int J Pharm* 2008; **358**(1-2): 44–49.
10 Li J, *et al.* The use of myristic acid as a ligand of polyethylenimine/DNA nanoparticles for targeted gene therapy of glioblastoma. *Nanotechnology* 2011; **22**(43): 435101.
11 Meng Q, *et al.* Myristic acid-conjugated polyethylenimine for brain-targeting delivery: *in-vivo* and *ex-vivo* imaging evaluation. *J Drug Target* 2010; **18**(6): 438–446.
12 Chhikara BS, *et al.* Synthesis and evaluation of fatty acyl ester derivatives of cytarabine as anti-leukemia agents. *Eur J Med Chem* 2010; **45**(10): 4601–4608.
13 *Food Chemicals Codex*, 7th edn. Bethesda, MD: United States Pharmacopoeia, 2010; 702–703.
14 Japan Pharmaceutical Excipients Council. *Japanese Pharmaceutical Excipients 2004*. Tokyo: Yakuji Nippo, 2004: 572.

20 General References

—

21 Authors

S Cantor, S Shah.

22 Date of Revision

1 November 2011.

M

Myristyl Alcohol

1 Nonproprietary Names

USP–NF: Myristyl Alcohol

2 Synonyms

Alcohol miristilo; *Dytol R-52*; *Lanette Wax KS*; *Lorol C14-95*; *Loxanol V*; myristic alcohol; *Nacol 14-95*; *Nacol 14-98*; 1-tetradecanol; *n*-tetradecanol-1; *n*-tetradecyl alcohol; tetradecyl alcohol; *Unihydag WAX-14*.

3 Chemical Name and CAS Registry Number

Tetradecan-1-ol [112-72-1]

4 Empirical Formula and Molecular Weight

$C_{14}H_{30}O$ 214.4

5 Structural Formula

6 Functional Category

Emollient; emulsion stabilizing agent; nonionic surfactant; oleaginous vehicle; viscosity-increasing agent.

7 Applications in Pharmaceutical Formulation or Technology

Myristyl alcohol is used in oral, parenteral, and topical pharmaceutical formulations. *See also* Section 18.

8 Description

Myristyl alcohol occurs as a white crystalline solid with a waxy odor. Also reported as opaque leaflets or crystals from ethanol.(1)

9 Pharmacopeial Specifications

See Table I.

Table I: Pharmacopeial specifications for myristyl alcohol.

Test	USP35–NF30
Identification	+
Melting range	36–42°C
Acid value	≤2
Iodine value	≤1
Hydroxyl value	250–267
Assay	≥90.0%

10 Typical Properties

Boiling point 167°C at 1.5 mPa (15 atm)

Density 0.8355 g/cm^3 at 20°C for solid; 0.8236 g/cm^3 at 38°C for liquid(1)

Flash point 140°C (open cup)(1)

Melting point 38°C; also reported as 37.6°C(1)

Solubility Practically insoluble in water; soluble in ether, slightly soluble in ethanol (95%).

Specific gravity 0.824

Vapor pressure 1.33 Pa (0.01 mmHg) at 20°C(1)

11 Stability and Storage Conditions

The bulk material should be stored in a well-closed container in a cool, dry place.

12 Incompatibilities

Myristyl alcohol is combustible when exposed to heat or flame. It can react with oxidizing materials. When heated to decomposition, it emits acrid smoke and irritating fumes.(1)

13 Method of Manufacture

Myristyl alcohol is found in spermaceti wax and sperm oil, and may be synthesized by sodium reduction of fatty acid esters or the reduction of fatty acids by lithium aluminum hydride. It can also be formed from acetaldehyde and dimethylamine.(2)

14 Safety

Myristyl alcohol is used in oral, parenteral, and topical pharmaceutical formulations. The pure form of myristyl alcohol is mildly toxic by ingestion and may be carcinogenic; experimental tumorigenic data are available.(4) It is also a human skin irritant. In animal studies of the skin permeation enhancement effect of saturated fatty alcohols, myristyl alcohol exhibited a lower effect when compared with decanol, undecanol, or lauryl alcohol but caused greater skin irritation.(3)A study investigating contact sensitization to myristyl alcohol revealed that patch testing of myristyl alcohol 10% petrolatum should not be carried out owing to observed irritant effects; thus the use of a lower concentration of myristyl alcohol for such tests (5% petrolatum) was recommended.(4) Myristyl alcohol has been associated with some reports of contact allergy.(5,6)A moderate-to-severe erythema and moderate edema are seen when 75 mg is applied to human skin intermittently in three doses over 72 hours.(1)

LD_{50}(rabbit, skin): 7.1 g/kg(1)
LD_{50}(rat, oral): 33.0 g/kg(1)

15 Handling Precautions

Observe normal precautions appropriate to the circumstances and quantity of the material handled. The use of gloves is recommended.

16 Regulatory Status

Included in the FDA Inactive Ingredients Database (oral tablet: sustained-release; and topical formulations: cream, lotion, suspension). Included in nonparenteral (topical cream) formulations licensed in the UK.

17 Related Substances

Lauric acid; myristic acid; palmitic acid; potassium myristate; sodium myristate; stearic acid.

18 Comments

Myristyl alcohol has been evaluated as a penetration enhancer in melatonin transdermal patches in rats.(7) The steady-state flux value of melatonin across human skin using myristyl alcohol as a permeation enhancer was reported as 18.2 μg/(cm^3 h).(8) It has also been investigated for use in microemulsions as transdermal drug delivery vehicles.(9)

Myristyl alcohol has been studied in controlled-release lipid particle systems.(10–12) It has also been tested as a bilayer stabilizer in niosome formulations containing ketorolac tromethamine,[13] and zidovudine.[14] Niosomes containing myristyl alcohol showed a considerably slower release rate of ketorolac tromethamine than those containing cholesterol.[13] This was also observed with the zidovudine formulation.[14]

A specification for myristic acid is contained in the *Food Chemicals Codex* (FCC).[15]

The EINECS number for myristyl alcohol is 204-000-3. The PubChem Compound ID (CID) for myristyl alcohol is 8209.

19 Specific References

1 Lewis RJ, ed. *Sax's Dangerous Properties of Industrial Materials*, 11th edn. New York: Wiley, 2004; 3384.
2 Opdyke DLJ. Alcohol C-14 myristic. *Food Cosmet Toxicol* 1975; **13**: 699–700.
3 Kanikkannan N, Singh M. Skin permeation enhancement effect and skin irritation of saturated fatty alcohols. *Int J Pharm* 2002; **248**: 219–228.
4 Geier J, *et al.* Patch testing with myristyl alcohol. *Contact Dermatitis* 2006; **55**: 366–367.
5 De Groot AC, *et al.* Cosmetic allergy from myristyl alcohol. *Contact Dermatitis* 1988; **19**: 76–77.
6 Pecegueiro M, *et al.* Contact dermatitis to *Hirudoid* cream. *Contact Dermatitis* 1987; **17**: 290–293.
7 Kanikkannan N, *et al.* Formulation and *in vitro* evaluation of transdermal patches of melatonin. *Drug Dev Ind Pharm* 2004; **30**: 205–212.
8 Andega S, *et al.* Comparison of the effect of fatty alcohols on the permeation of melatonin between porcine and human skin. *J Control Release* 2001; **77**: 17–25.
9 Kogan A, Garti N. Microemulsions as transdermal drug delivery vehicles. *Adv Colloid Interface Sci* 2006; **123–126**: 369–385..
10 Lee SJ, *et al.* Characterization of nano oxaliplatin prepared by novel Fat Employing Supercritical Nano System, the *FESNS*. *Pharm Dev Technol* 2010; Nov 15 (Epub).
11 Yi D, *et al.* Magnetic activated release of umbelliferone from lipid matrices. *Int J Pharm* 2010; **394**(1-2): 143–146.
12 Zeng P, *et al.* Collisional solute release from thermally activated lipid particles. *Drug Dev Ind Pharm* 2009; **35**(1): 12–18.
13 Devaraj GN, *et al.* Release studies on niosomes containing fatty alcohols as bilayer stabilizers instead of cholesterol. *J Colloid Interface Sci* 2002; **251**: 360–365.
14 Gopinath D, *et al.* Pharmacokinetics of zidovudine following intravenous bolus administration of a novel niosome preparation devoid of cholesterol. *Arzneimittelforschung* 2001; **51**: 924–930.
15 *Food Chemicals Codex*, 7th edn. Bethesda, MD: United States Pharmacopoeia, 2010; 703.

20 General References

—

21 Authors

S Cantor, S Shah.

22 Date of Revision

1 March 2012.

M

N

Neohesperidin Dihydrochalcone

1 Nonproprietary Names

BP: Neohesperidin Dihydrochalcone

PhEur: Neohesperidin Dihydrochalcone

2 Synonyms

Citrosa; 1-[4-[[2-O-6-deoxy-α-L-mannopyranosyl)-β-D-glycopyranosyl]oxy]-2,6-dihydroxyphenyl]-3-(3-hydroxy-4-methoxyphenyl); 3,5-dihydroxy-4-(3-hydroxy-4-methoxyhydrocinnamoyl)phenyl-2-O-(6-deoxy-α-L-mannopyranosyl)-β-D-glucopyranoside; 3,5-dihydroxy-4-[3-(3-hydroxy-4-methoxyphenyl)propionyl]phenyl-2-O-(6-deoxy-α-L-mannopyranosyl)-β-D-glucopyranoside; E959; neohesperidin DC; neohesperidin DHC; neohesperidin dihydrochalconum; neohesperidine dihydrochalcone; NHDC; 1-propanone;*Sukor.*

3 Chemical Name and CAS Registry Number

1-[4-[[2-O-(6-Deoxy-α-L-mannopyranosyl)-β-D-glucopyranosyl]oxy]-2,6-dihydroxyphenyl]-3-(3-hydroxy-4-methoxyphenyl)propan-1-one [20702-77-6]

4 Empirical Formula and Molecular Weight

$C_{28}H_{36}O_{15}$ 612.58

5 Structural Formula

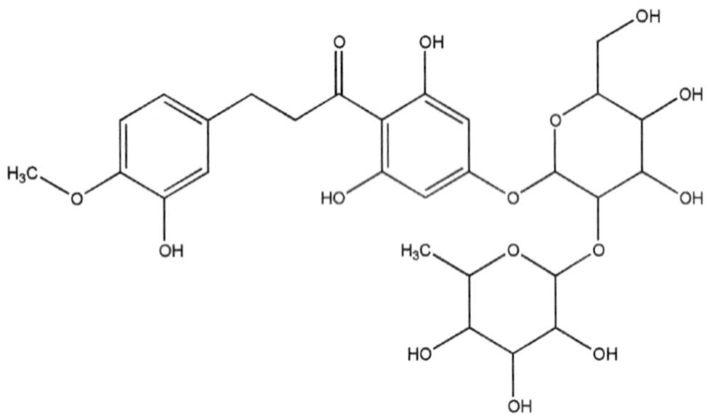

6 Functional Category

Flavor enhancer; sweetening agent.

7 Applications in Pharmaceutical Formulation or Technology

Neohesperidin dihydrochalcone is a synthetic intense sweetening agent approximately 1500–1800 times sweeter than sucrose and 20 times sweeter than saccharin. Structurally it is an analogue of neohesperidin, a flavanone that occurs naturally in Seville oranges (*Citrus aurantium*). Neohesperidin dihydrochalcone is used in pharmaceutical and food applications as a sweetening agent and flavor enhancer. The sweetness profile is characterized by a lingering sweet/menthol-like aftertaste.[1] The typical level used in foods is 1–5 ppm although much higher levels may be used in certain applications such as chewing gum. Synergistic effects occur with other intense and bulk sweeteners such as acesulfame K, aspartame, polyols, and saccharin.[2]

In pharmaceutical applications, neohesperidin dihydrochalcone is useful in masking the unpleasant bitter taste of a number of drugs such as antacids, antibiotics, and vitamins. In antacid preparations, levels of 10–30 ppm result in improved palatability.

8 Description

Neohesperidin dihydrochalcone occurs as a white or yellowish-white powder with an intensely sweet taste.

9 Pharmacopeial Specifications

See Table I.

Table I: Pharmacopeial specifications for neohesperidin dihydrochalcone.

Test	PhEur 7.4
Identification	+
Characters	+
Appearance of solution	+
Related substances	+
Heavy metals	≤10 ppm
Water	≤12.0%
Sulfated ash	≤0.2%
Assay (anhydrous substance)	96.0–101.0%

10 Typical Properties

Hygroscopicity Slightly hygroscopic; absorbs up to 15% of water.

Melting point 156–158°C

Solubility *see* Table II.

Table II: Solubility of neohesperidin dihydrochalcone.

Solvent	Solubility at 25°C unless otherwise stated
Dichloromethane	Practically insoluble
Dimethyl sulfoxide	Freely soluble
Methanol	Soluble
Water	1 in 2000 at 22°C
	1 in 1.54 at 80°C

11 Stability and Storage Conditions

Neohesperidin dihydrochalcone is stable for over three years when stored at room temperature.[1]

Accelerated stability studies on aqueous solutions stored at 30–60°C and pH 1–7 for 140 days indicate that neohesperidin dihydrochalcone solutions are likely to be stable for 12 months at room temperature and pH 2–6.[3] Solutions formulated with some or all of the water replaced by solvents with a lower dielectric constant are reported to have longer shelf-lives.[4]

The bulk material should be stored in a cool, dry place protected from light.

12 Incompatibilities

—

13 Method of Manufacture

Neohesperidin dihydrochalcone is synthesized commercially from either of the bitter-flavanones neohesperidin or naringin by catalytic hydrogenation under alkaline conditions in a process first described in the 1960s, in which neohesperidin is purified by recrystallization from water solutions.[5] Neohesperidin dihydrochalcone is obtained by the alkaline hydrogenation of neohesperidin.[6]

14 Safety

Neohesperidin dihydrochalcone is accepted for use in food products either as a sweetener or flavor modifier in a number of areas including Europe, US, Australia, New Zealand, and several countries in Africa and Asia. It is also used in a number of oral pharmaceutical formulations.

Animal toxicity studies suggest that neohesperidin dihydrochalcone is a nontoxic, nonteratogenic, and noncarcinogenic material at the levels used in foods and pharmaceuticals.[7,8] In Europe, an acceptable daily intake of 0–5 mg/kg body-weight has been established.[9,10]

15 Handling Precautions

Observe normal precautions appropriate to the circumstances and quantity of material handled.

16 Regulatory Status

GRAS listed. Accepted for use as a food additive in Europe. Included in the Canadian Natural Health Products Ingredients Database.

17 Related Substances

Hesperidin.

Hesperidin

Empirical formula $C_{28}H_{34}O_{15}$
Molecular weight 610.56
CAS number [520-26-3]
Synonyms (2*S*)-7-[[6-*O*-(6-Deoxy-α-L-mannopyranosyl)-β-D-glucopyranosyl]oxy]-2,3-dihydro-5-hydroxy-2-(3-hydroxy-4-methoxyphenyl)-4*H*-1-benzopyran-4-one; hesperitin 7-rhamnoglucoside; hesperetin-7-rutinoside.
Melting point 258–262°C
Solubility Freely soluble in diluted alkalis and pyridines; soluble in formamide; slightly soluble in methanol and hot glacial acetic acid.
Comments Hesperedin is the predominant flavonoid in lemons and sweet oranges (*Citrus sinensis*).

18 Comments

Neohesperidin dihydrochalcone is sufficiently soluble in aqueous solutions for most pharmaceutical and food applications; however, solubility may be improved by dissolving in ethanol, glycerin, propylene glycol, or aqueous mixtures of these solvents.[10] Solubility may also be improved by mixing with other intense or bulk sweeteners.[2]

Neohesperidin dihydrochalcone in weak concentrations has been shown not to enhance the taste of aqueous sucrose solutions.[6]

The EINECS number for neohesperidin dihydrochalcone is 243-978-6. The PubChem Compound ID (CID) for neohesperidin dihydrochalcone is 30231.

19 Specific References

1 Cano J, *et al*. Masking the bitter taste of pharmaceuticals. *Manuf Chem* 2000; **71**(7): 16–17.
2 Benavente-Garcia O, *et al*. Improved water solubility of neohesperidin dihydrochalcone sweetener blends. *J Agric Food Chem* 2001; **49**(1): 189–191.
3 Canales I, *et al*. Neohesperidin dihydrochalcone stability in aqueous buffer solutions. *J Food Sci* 1993; **58**: 589–591643.
4 Montijano H, Borrego F. Hydrolysis of the intense sweetener neohesperidine dihydrochalcone in water–organic solvent mixtures. *Int J Food Sci Technol* 1999; **34**: 291–294.
5 Horowitz RM, Gentili B. Dihydrochalcone derivatives and their use as sweetening agents. US Patent No. 3,087,821; 1963.
6 Kroeze JH. Neohesperidine dihydrochalcone is not a taste enhancer in aqueous solutions. *Chem Senses* 2000; **25**(5): 555–559.
7 Lina BAR, *et al*. Subchronic (13-week) oral toxicity of neohesperidin dihydrochalcone in rats. *Food Chem Toxicol* 1990; **28**(7): 507–513.
8 Waalkens-Berendsen DH, *et al*. Embryotoxicity and teratogenicity study with neohesperidin dihydrochalcone in rats. *Regul Toxicol Pharmacol* 2004; **40**(1): 74–79.
9 Horowitz RM, Gentili B. Dihydrochalcone sweeteners from citrus flavanones. In: O'Brien Nabors L, Gelardi RC, eds. *Alternative Sweeteners*, 2nd edn. New York: Marcel Dekker, 1991: 97–115.
10 Borrego F, Montijano H. Neohesperidin dihydrochalcone. In: O'Brien Nabors L, ed. *Alternative Sweeteners*, 3rd edn. New York: Marcel Dekker, 2001: 87–104.

20 General References

Borrego F, Montijano H. [Potential applications of the sweetener neohesperidin dihydrochalcone in drugs.] *Pharm Ind* 1995; **57**: 880–882[in German].
Borrego F. Neohesperidine DC. In: Birch G, ed. *Ingredients Handbook: Sweeteners*, 2nd edn. Leatherhead: Leatherhead Publishing, 2000: 205–220.
Colaizzi JL. Synthetic sweeteners—toxicity problems and current status. *J Am Pharm Assoc* 1971; **11**(3): 135–138.
DuBois GE, *et al*. Non-nutritive sweeteners: taste–structure relationships for some new simple dihydrochalcones. *Science* 1977; **195**: 397–399.
Lautenbacher L. Neohesperidin DC (PhEur): an exceptional sweetener from Spanish bitter oranges—application and approval in finished drugs. *Pharm Ind* 2003; **65**: 82–83.
Lindley MG. Neohesperidine dihydrochalcone: recent findings and technical advances. In: Grenby TH, ed. *Advances in Sweeteners*. Glasgow: Blackie Academic and Professional, 1996: 240–252.
Nakazato M, *et al*. Determination of neohesperidin dihydrochalcone in foods. *Shokuhin Eiseigaku Zasshi* 2001; **42**(1): 40–44.
Uchiyama N, *et al*. HPLC separation of naringin, neohesperidin and their C-2 epimers in commercial samples and herbal medicines. *J Pharm Biomed Anal* 2008; **46**(5): 864–869.

21 Author

PJ Weller.

22 Date of Revision

1 March 2012.

Neotame

1 Nonproprietary Names

USP–NF: Neotame

2 Synonyms

3-(3,3-Dimethylbutylamino)-*N*-(α-carboxyphenethyl)succinamic acid methyl ester; *N*-[*N*-(3,3-dimethylbutyl)-L-α-aspartyl]-L-phenylalanine 1-methyl ester; L-phenylalanine, *N*-[*N*-(3,3-dimethylbutyl)-L-α-aspartyl]-1-methyl ester.

3 Chemical Name and CAS Registry Number

(3*R*)-3-(3,3-Dimethylbutylamino)-4-[[(2*R*)-1-methoxy-1-oxo-3-phenylpropan-2-yl]amino]-4-oxobutanoic acid [165450-17-9]

4 Empirical Formula and Molecular Weight

$C_{20}H_{30}N_2O_5$ 378.47

5 Structural Formula

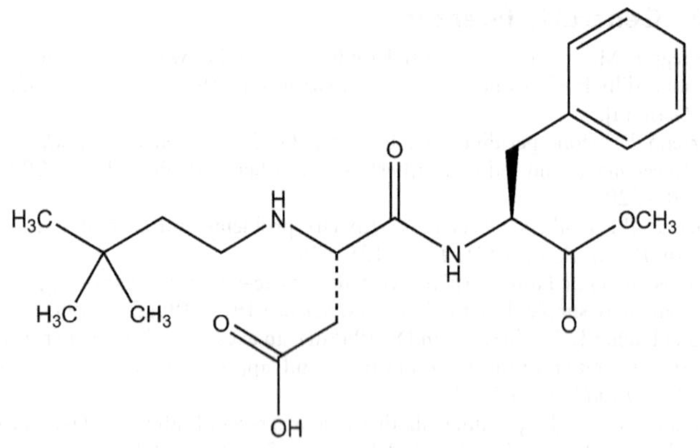

H3C H3C CH3 H N O O OH N H OCH3 O

6 Functional Category

Flavor enhancer; sweetening agent.

7 Applications in Pharmaceutical Formulation or Technology

Neotame is a water-soluble, nonnutritive, intense sweetening agent used in beverages and foods. It is structurally related to aspartame and is about 7000–13 000 times sweeter than sucrose, and about 30–60 times sweeter than aspartame, making it the sweetest artificial sweetener available. Neotame is said to have a 'clean' sweet taste in contrast to the bitter, metallic aftertaste associated with saccharin. Although neotame has approximately the same caloric value as sucrose (1.2 kJ/g) the small quantities used to achieve a desired level of sweetness in a formulation mean that it is essentially nonnutritive.

Neotame may be used in sub-sweetening quantities as a flavor enhancer, e.g. with mint or strawberry flavor.

8 Description

Neotame occurs as an odorless, white to off-white powder. It has an intense sweet taste 7000–13 000 times sweeter than sucrose depending on the matrix.

9 Pharmacopeial Specifications

See Table I.

Table I: Pharmacopeial specifications for neotame.

Test	USP35–NF30
Identification	+
Specific optical rotation	−40.0° to −43.4°
Water	≤5.0%
Residue on ignition	≤0.2%
Lead	≤2 ppm
Related compounds	+
Assay (anhydrous basis)	97.0–102.0%

10 Typical Properties

Acidity/alkalinity pH = 5.0–7.0 (0.5% w/v aqueous solution)

Dissociation constant

pK_{a1} = 3.01;

pK_{a2} = 8.02.

Melting point 80–83°C

Solubility *see* Table II.

Table II: Solubility of neotame.

Solvent	Solubility at 25°C unless otherwise stated
Ethanol	1 in 1.05
Ethyl acetate	1 in 23 at 15°C
	1 in 13
	1 in 1 at 60°C
Water	1 in 94 at 15°C
	1 in 79
	1 in 21 at 60°C

11 Stability and Storage Conditions

Neotame stability is affected by moisture, pH, and temperature. Neotame is stable in bakery products and pasteurized dairy products.

The bulk material should be stored in a well-closed container, in a cool, dry place; it is stable for up to 5 years at room temperature.

12 Incompatibilities

—

13 Method of Manufacture

Neotame is manufactured by the reaction of aspartame and 3,3-dimethylbutyraldehyde, followed by purification, drying, and milling.[1–3]

14 Safety

Neotame is a nonnutritive intense sweetening agent used in beverages and foods. Studies in animals and humans have shown that neotame is a relatively nontoxic, nonteratogenic, and noncarcinogenic substance. It is reported as safe for use during pregnancy and lactation, and by children and persons with diabetes.

At least 30% of ingested neotame is rapidly absorbed. Neotame is metabolized to de-esterified neotame and methanol, with practically all neotame being eliminated from the body in the urine and feces. Peak plasma concentrations of neotame are observed at approximately 30–60 minutes after ingestion. Human studies in healthy and diabetic patients suggest that neotame is well-tolerated at doses up to 1.5 mg/kg body-weight daily (the highest dose studied). Following reviews of over 100 animal and human toxicity studies the European Food Safety Authority and WHO have established an acceptable daily intake for neotame at up to 2 mg/kg body-weight.[4,5]

15 Handling Precautions

Observe normal precautions appropriate to the circumstances and quantity of the material handled. Eye protection is recommended.

16 Regulatory Status

Accepted for use as a food additive in several countries including the US, Mexico, Australia, and New Zealand. Included in the FDA Inactive Ingredients Database (oral solutions). Approved for use in India in pharmaceutical preparations.[6] Included in the Canadian Natural Health Products Ingredients Database.

17 Related Substances

Aspartame.

18 Comments

Neotame does not degrade to diketopiperazine and does not require special labeling for phenylketonuria.

The PubChem Compound ID (CID) for neotame is 3081923.

19 Specific References

1 Nofre C, Tinti J-M. *N*-Substituted derivatives of aspartame useful as sweetening agents. United States Patent 5,480,668; 1996.
2 Prakash I. Method for preparing and purifying an *N*-alkylated aspartame derivative. United States Patent 5,728,862; 1998.
3 Witt J. Discovery and development of neotame. *World Rev Nutr Diet* 1999; **85**: 52–57.
4 FAO/WHO. Evaluation of certain food additives and contaminants. *Tech Rep Ser Wld Hlth Org* 2004; **922**: 1–176.
5 European Food Safety Authority. Scientific opinion of the panel on food additives, flavourings, processing aids and materials in contact with food on a request from European Commission on neotame as a sweetener and flavour enhancer. *The EFSA Journal* 2007; **581**: 1–3.
6 In-Pharmatechnologist.com. India clears Neotame for use in drugs. http://www.in-pharmatechnologist.com (accessed 22 September 2011).

20 General References

American Dietic Association. Position of the American Dietic Association: use of nutritive and nonnutritive sweeteners. *J Am Diet Assoc* 2004; **104**(2): 255–275.
Anonymous. Neotame – a new artificial sweetener. *Med Lett Drugs Ther* 2002; **44**(1137): 73–74.
Anonymous. Artificial sweeteners: no calories – sweet. *FDA Consum* 2006; **40**(4): 27–28.
Burdock GA. *Fenaroli's Handbook of Flavor Ingredients*, 4th edn. Boca Raton, FL: CRC Press, 2002: 1284–1285.
Stargel WW *et al*.Neotame. In: O'Brien Nabors L, ed. *Alternative Sweeteners*, 3rd edn. New York: Marcel Dekker, 2001: 129–145.
The NutraSweet Company. Neotame: a scientific overview. http://www.neotame.com (accessed 22 September 2011).
Weihrauch MR, Diehl V. Artificial sweeteners – do they bear a carcinogenic risk? *Ann Oncol* 2004; **15**(10): 1460–1465.

21 Author

PJ Weller.

22 Date of Revision

1 March 2012.

N

Nitrogen

1 Nonproprietary Names

BP: Nitrogen
JP: Nitrogen
PhEur: Nitrogen
USP–NF: Nitrogen

2 Synonyms

Azote; E941; nitrogenium.

3 Chemical Name and CAS Registry Number

Nitrogen [7727-37-9]

4 Empirical Formula and Molecular Weight

N_2 28.01

5 Structural Formula

See Section 4.

6 Functional Category

Aerosol propellant; air displacement.

7 Applications in Pharmaceutical Formulation or Technology

Nitrogen and other compressed gases such as carbon dioxide and nitrous oxide are used as propellants for topical pharmaceutical aerosols. They are also used in other aerosol products that work satisfactorily with the coarse aerosol spray produced with compressed gases, e.g. furniture polish and window cleaner. Nitrogen is insoluble in water and other solvents, and therefore remains separated from the actual pharmaceutical formulation.

Advantages of compressed gases as aerosol propellants are that they are less expensive; of low toxicity; and practically odorless and tasteless. In contrast to liquefied gases, their pressures change relatively little with temperature. However, there is no reservoir of propellant in the aerosol and as a result the pressure decreases as the product is used, changing the spray characteristics.

Misuse of a product by the consumer, such as using a product inverted, results in the discharge of the vapor phase instead of the liquid phase. Most of the propellant is contained in the vapor phase and therefore some of the propellant will be lost and the spray characteristics will be altered. Additionally, the sprays produced using compressed gases are very wet. However, developments in valve technology have reduced the risk of misuse by making available valves which will spray only the product (not propellant) regardless of the position of the container. Additionally, barrier systems will also prevent loss of propellant, and have been used for pharmaceuticals and cosmetic aerosol sprays and foams utilizing nitrogen as the propellant.

Nitrogen is also used to displace air from solutions subject to oxidation, by sparging, and to replace air in the headspace above products in their final packaging, e.g. in parenteral products packaged in glass ampoules. Nitrogen is also used for the same purpose in many food products.

8 Description

Nitrogen occurs naturally as approximately 78% v/v of the atmosphere. It is a nonreactive, noncombustible, colorless, tasteless, and odorless gas. It is usually handled as a compressed gas, stored in metal cylinders.

9 Pharmacopeial Specifications

See Table I.

Table I: Pharmacopeial specifications for nitrogen.

Test	JP XV	PhEur 7.4	USP35–NF30
Identification	+	+	+
Characters	–	+	–
Production	–	+	–
Odor	–	–	+
Carbon monoxide	–	≤5 ppm	≤0.001%
Carbon dioxide	+	≤300 ppm	–
Water	–	≤67 ppm	–
Oxygen	–	≤50 ppm	≤1.0%
Assay	≥99.5%	≥99.5%	≥99.0%

10 Typical Properties

Density 0.967 g/cm^3 for vapor at 21°C.

Flammability Nonflammable

Solubility Practically insoluble in water and most solvents; soluble in water under pressure.

Vapor density (absolute) 1.25 g/cm^3 at standard temperature and pressure.

Vapor density (relative) 0.97 (air = 1)

11 Stability and Storage Conditions

Nitrogen is stable and chemically unreactive. It should be stored in tightly sealed metal cylinders in a cool, dry place.

12 Incompatibilities

Generally compatible with most materials encountered in pharmaceutical formulations and food products.

13 Method of Manufacture

Nitrogen is obtained commercially, in large quantities, by the fractional distillation of liquefied air.

14 Safety

Nitrogen is generally regarded as a nontoxic and nonirritant material. However, it is an asphyxiant and inhalation of large quantities is therefore hazardous. *See also* Section 18.

15 Handling Precautions

Handle in accordance with procedures for handling metal cylinders containing liquefied or compressed gases. Eye protection, gloves, and protective clothing are recommended. Nitrogen is an asphyxiant and should be handled in a well-ventilated environment.

16 Regulatory Status

GRAS listed. Included in the FDA Inactive Ingredients Database (injections; dental preparations; nasal sprays; oral solutions; rectal gels). Accepted for use as a food additive in Europe. Included in parenteral and nonparenteral medicines licensed in the UK and US. Included in the Canadian Natural Health Products Ingredients Database.

17 Related Substances

Carbon dioxide; nitrous oxide.

18 Comments

Different grades of nitrogen are commercially available that have, for example, especially low moisture levels.

Nitrogen is commonly used as a component of the gas mixtures breathed by divers. Under high pressure, such as when diving at great depths, nitrogen will dissolve in blood and lipid. If decompression is too rapid, decompression sickness may occur when the nitrogen effervesces from body stores to form gas emboli.

A specification for nitrogen is contained in the *Food Chemicals Codex* (FCC).[1]

The EINECS number for nitrogen is 231-783-9. The PubChem Compound ID (CID) for nitrogen is 947.

19 Specific References

1 *Food Chemicals Codex*, 7th edn. Bethesda, MD: United States Pharmacopeia, 2010: 717.

20 General References

Johnson MA. *The Aerosol Handbook*, 2nd edn. New Jersey: WE Dorland, 1982: 361–372.

Sanders PA. *Handbook of Aerosol Technology*, 2nd edn. New York: Van Nostrand Reinhold, 1979: 44–54.

Sciarra JJ. Pharmaceutical aerosols. In: Banker GS, Rhodes CT, eds. *Modern Pharmaceutics*, 3rd edn. New York: Marcel Dekker, 1996: 547–574.

Sciarra CJ, Sciarra JJ. Aerosols. In: *Remington: The Science and Practice of Pharmacy*, 21st edn. Philadelphia, PA: Lippincott Williams and Wilkins, 2005: 1000–1017.

Sciarra JJ, Stoller L. *The Science and Technology of Aerosol Packaging*. New York: Wiley, 1974: 137–145.

21 Authors

CJ Sciarra, JJ Sciarra.

22 Date of Revision

1 March 2012.

N

Nitrous Oxide

1 Nonproprietary Names

BP: Nitrous Oxide
JP: Nitrous Oxide
PhEur: Nitrous Oxide
USP–NF: Nitrous Oxide

2 Synonyms

Dinitrogenii oxidum; dinitrogen monoxide; E942; laughing gas; nitrogen monoxide.

3 Chemical Name and CAS Registry Number

Dinitrogen oxide [10024-97-2]

4 Empirical Formula and Molecular Weight

N_2O 44.01

5 Structural Formula

See Section 4.

6 Functional Category

Aerosol propellant.

7 Applications in Pharmaceutical Formulation or Technology

Nitrous oxide and other compressed gases such as carbon dioxide and nitrogen are used as propellants for topical pharmaceutical aerosols. They are also used in other aerosol products that work satisfactorily with the coarse aerosol spray that is produced with compressed gases, e.g. furniture polish and window cleaner.

The advantages of compressed gases as aerosol propellants are that they are less expensive, of low toxicity, and practically odorless and tasteless. In contrast to liquefied gases, their pressures change relatively little with temperature. However, there is no reservoir of propellant in the aerosol, and as a result the pressure decreases as the product is used, changing the spray characteristics.

Misuse of a product by the consumer, such as using a product inverted, results in the discharge of the vapor phase instead of the liquid phase. Since most of the propellant is contained in the vapor phase, some of the propellant will be lost and the spray characteristics will be altered. Additionally, the sprays produced using compressed gases are very wet. However, recent developments in valve technology have reduced the risk of misuse by making available valves which will spray only the product (not propellant) regardless of the position of the container. Additionally, barrier systems will also prevent loss of propellant, and have found increased use with this propellant.

8 Description

Nitrous oxide is a nonflammable, colorless and odorless, sweet-tasting gas. It is usually handled as a compressed gas, stored in metal cylinders.

9 Pharmacopeial Specifications

See Table I.

Table I: Pharmacopeial specifications for nitrous oxide.

Test	JP XV	PhEur 7.4	USP 35–NF30
Identification	+	+	+
Characters	–	+	–
Production	–	+	–
Acidity or alkalinity	+	–	–
Carbon dioxide	+	≤300 ppm	≤0.03%
Carbon monoxide	+	≤5 ppm	≤0.001%
Nitric oxide	–	–	≤1 ppm
Nitrogen dioxide	–	–	≤1 ppm
Nitric monoxide and nitrogen dioxide	–	≤2 ppm	–
Halogens	–	–	≤1 ppm
Oxidizing substances	+	–	–
Potassium permanganate-reducing substances	+	–	–
Ammonia	–	–	≤0.0025%
Chloride	+	–	–
Air	–	–	≤1.0%
Water	–	≤67 ppm	≤150 mg/m^3
Assay	≥97.0%	≥98.0%	≥99.0%

10 Typical Properties

Density 1.53 g/cm^3
Flammability Nonflammable, but supports combustion.
Solubility Freely soluble in chloroform, ethanol (95%), ether, and oils; soluble 1 in 1.5 volumes of water at 20°C and 101.3 kPa pressure.
Vapor density (absolute) 1.97 g/cm^3 at standard temperature and pressure.
Vapor density (relative) 1.52 (air = 1)

11 Stability and Storage Conditions

Nitrous oxide is essentially nonreactive and stable except at high temperatures; at a temperature greater than 500°C nitrous oxide decomposes to nitrogen and oxygen. Explosive mixtures may be formed with other gases such as ammonia, hydrogen, and other fuels. Nitrous oxide should be stored in a tightly sealed metal cylinder in a cool, dry place.

12 Incompatibilities

Nitrous oxide is generally compatible with most materials encountered in pharmaceutical formulations, although it may react as a mild oxidizing agent.

13 Method of Manufacture

Nitrous oxide is prepared by heating ammonium nitrate to about 170°C. This reaction also forms water.

14 Safety

Nitrous oxide is most commonly used therapeutically as an anesthetic and analgesic. Reports of adverse reactions to nitrous oxide therefore generally concern its therapeutic use, where relatively large quantities of the gas may be inhaled, rather than its use as an excipient.

The main complications associated with nitrous oxide inhalation occur as a result of hypoxia. Prolonged administration may also be harmful. Nitrous oxide is rapidly absorbed on inhalation.

15 Handling Precautions

Handle in accordance with procedures for handling metal cylinders containing liquefied or compressed gases. Eye protection, gloves, and protective clothing are recommended. Nitrous oxide is an anesthetic gas and should be handled in a well-ventilated environment. In the UK, the recommended long-term (8-hour TWA) workplace exposure limit for nitrous oxide is 183 mg/m^3 (100 ppm).[1]

16 Regulatory Status

GRAS listed. Accepted for use as a food additive in Europe. Included in nonparenteral medicines licensed in the UK and US. Included in the Canadian Natural Health Products Ingredients Database.

17 Related Substances

Carbon dioxide; nitrogen.

18 Comments

A mixture of 50% nitrous oxide and 50% oxygen (*Entonox*, BOC) is commonly used as an analgesic administered by inhalation.

Therapeutically, nitrous oxide is best known as an anesthetic administered by inhalation. When used as an anesthetic it has strong analgesic properties but produces little muscle relaxation. Nitrous oxide is always administered in conjunction with oxygen since on its own it is hypoxic.

A specification for nitrous oxide is contained in the *Food Chemicals Codex* (FCC).[2]

The EINECS number for nitrous oxide is 233-032-0. The PubChem Compound ID (CID) for nitrous oxide is 948.

N

19 Specific References

1 Health and Safety Executive. *EH40/2005: Workplace Exposure Limits.* Sudbury: HSE Books, 2011. http://www.hse.gov.uk/pubns/priced/eh40.pdf (accessed 28 February 2012).

2 *Food Chemicals Codex*, 7th edn. Bethesda, MD: United States Pharmacopeia, 2010: 719.

20 General References

Johnson MA. *The Aerosol Handbook*, 2nd edn. New Jersey: WE Dorland, 1982: 361–372.

Sanders PA. *Handbook of Aerosol Technology*, 2nd edn. New York: Van Nostrand Reinhold, 1979: 44–54.

Sciarra JJ. Aerosol suspensions and emulsions. In: *Pharmaceutical Dosage Forms; Disperse Systems*, 2nd edn. New York: Marcel Dekker, 1996: 319–356.

Sciarra JJ. Pharmaceutical aerosols. In: Banker GS, Rhodes CT, eds. *Modern Pharmaceutics*, 3rd edn. New York: Marcel Dekker, 1996: 547–574.

Sciarra CJ, Sciarra JJ. Aerosols. In: *Remington: The Science and Practice of Pharmacy*, 21st edn. Philadelphia, PA: Lippincott Williams and Wilkins, 2005: 1000–1017.

Sciarra JJ, Stoller L. *The Science and Technology of Aerosol Packaging*. New York: Wiley, 1974: 137–145.

21 Authors

CJ Sciarra, JJ Sciarra.

22 Date of Revision

1 March 2012.

Octyldodecanol

1 Nonproprietary Names

BP: Octyldodecanol
PhEur: Octyldodecanol
USP–NF: Octyldodecanol

2 Synonyms

Eutanol G PH; isoarachidyl alcohol; *Jarcol 1-20*; *Jeecol ODD*; octildodecanol; octyldodecanolum; 2-octyl-1-dodecanol; 2-octyl-dodecanol; *Standamul G*; 2-octyldodecyl alcohol.

3 Chemical Name and CAS Registry Number

Octyldodecanol [5333-42-6]

4 Empirical Formula and Molecular Weight

$C_{20}H_{42}O$ 298.62

5 Structural Formula

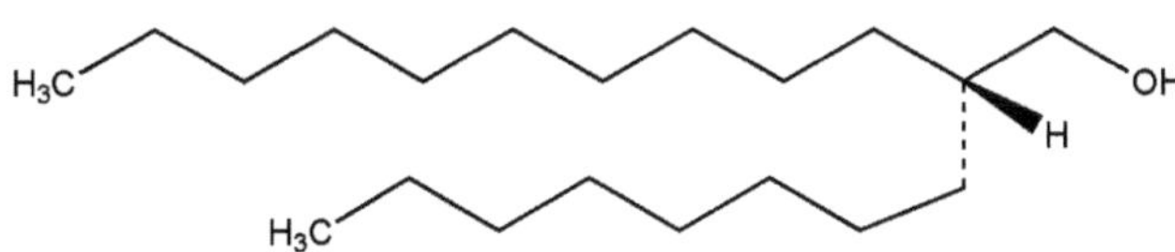

6 Functional Category

Emollient; emulsifying agent; lubricant; solvent; viscosity-increasing agent.

7 Applications in Pharmaceutical Formulation or Technology

Octyldodecanol is widely used in cosmetics and pharmaceutical applications as an emulsifying and opacifying agent. It is primarily used in topical applications because of its lubricating and emollient properties.

Octyldodecanol has been used in the preparation of oil/water microemulsions and nanoemulsions investigated as vehicles for the dermal administration of drugs having no or low skin penetration.[1–3] Octyldodecanol has also been evaluated as a solvent for naproxen when applied topically.[4] Studies of estimated permeability coefficient suggest that octyldodecanol could be a potential dermal permeation enhancer.[5]

8 Description

Octyldodecanol occurs as a clear, colorless, or yellowish, oily liquid.

9 Pharmacopeial Specifications

See Table I. *See also* Section 18.

Table I: Pharmacopeial specifications for octyldodecanol.

Test	PhEur 7.4	USP35–NF30
Identification	+	+
Characters	+	—
Acidity or alkalinity	+	—
Relative density	≈0.840	—
Refractive index	≈1.455	—
Optical rotation	−0.10° to +0.10°	—
Hydroxyl value	175–190	175–190
Iodine value	≤8.0	≤8
Saponification value	≤5.0	≤5
Acid value	—	≤0.5
Peroxide value	≤5.0	—
Heavy metals	≤10 ppm	—
Water	≤0.5%	—
Sulfated ash	≤0.1%	—
Assay	≥90.0%	≥90.0%

10 Typical Properties

Flash point 180°C–200°C
Melting point <−20°C
Refractive index n_D^{20} = 1.45–1.46
Solubility Miscible with ethanol (95%); practically insoluble in water.
Specific gravity 0.83–0.85 at 20°C
Viscosity (dynamic) 58–64 mPa s (58–64 cP) at 20°C

11 Stability and Storage Conditions

The bulk material should be stored in a well-closed container in a cool, dry place, protected from light. In the original unopened container, octyldodecanol can be stored for 2 years protected from moisture at below 30°C.

12 Incompatibilities

Octyldodecanol is generally compatible with most materials encountered in cosmetic and pharmaceutical formulations.

13 Method of Manufacture

Octyldodecanol is produced by the condensation of two molecules of decyl alcohol. It also occurs naturally in small quantities in plants.

14 Safety

Octyldodecanol is widely used in cosmetics and topical pharmaceutical formulations, and is generally regarded as nontoxic and nonirritant at the levels employed as an excipient.

In acute oral toxicity studies in rats fed 5 g/kg of undiluted octyldodecanol, no deaths were observed.[6] In an acute dermal toxicity study, intact and abraded skin sites of guinea pigs were treated with 3 g/kg of undiluted octyldodecanol under occlusive patches; no deaths occurred and no gross skin lesions were observed.[6] Octyldodecanol caused either no ocular irritation or minimal, transient irritation in the eyes of rabbits.[6] However, some sources describe undiluted octyldodecanol as an eye and severe skin irritant.

15 Handling Precautions

Observe normal precautions appropriate to the circumstances and quantity of material handled. When heated to decomposition, octyldodecanol emits acrid smoke and irritating fumes.

16 Regulatory Status

Included in the FDA Inactive Ingredients Database (topical, transdermal, and vaginal preparations). Included in nonparenteral medicines licensed in the UK. Included in the Canadian Natural Health Products Ingredients Database.

17 Related Substances

—

18 Comments

A specification for octyldodecanol is included in *Japanese Pharmaceutical Excipients* (JPE).[7]

The EINECS number for octyldodecanol is 226-242-9. The PubChem Compound ID (CID) for octyldodecanol includes 21414 and 11983377.

19 Specific References

1 Shukla A, *et al*. Investigation of pharmaceutical oil/water microemulsions by small-angle scattering. *Pharm Res* 2002; **19**(6): 881–886.

2 Fasolo D, *et al*. Development of topical nanoemulsions containing quercetin and 3-*O*-methylquercetin. *Pharmazie* 2009; **64**(11): 726–739.

3 Silva AP, *et al*. Development of topical nanoemulsions containing the isoflavone genistein. *Pharmazie* 2009; **64**(1): 32–35.

4 Contreras Claramonte MD, *et al*. An application of regular solution theory in the study of the solubility of naproxen in some solvents used in topical preparations. *Int J Pharm* 1993; **94**: 23–30.

5 Mbah CJ. Studies on the lipophilicity of vehicles (or covehicles) and botanical oils used in cosmetic products. *Pharmazie* 2007; **62**(5): 351–353.

6 Elder RL. Final report on the safety assessment of stearyl alcohol, oleyl alcohol and octyl dodecanol. *J Am Coll Toxicol* 1985; **4**: 1–29.

7 Japan Pharmaceutical Excipients Council. *Japanese Pharmaceutical Excipients 2004*. Tokyo: Yakuji Nippo, 2004; 583–585.

20 General References

Allen LV. Featured excipient: oligeanous vehicles. *Int J Pharm Compound* 2000; **4**(6): 470–473484–485.

Filippi U, *et al*. Proposal for the pharmacopeia; octyl dodecanol. *Bell Clin Form* 1982; **121**: 425–427.

21 Author

RT Guest.

22 Date of Revision

2 March 2012.

Octyl Gallate

1 Nonproprietary Names

BP: Octyl Gallate

PhEur: Octyl Gallate

2 Synonyms

Benzoic acid, 3,4,5-trihydroxy-octyl ester; E311; gallic acid, octyl ester; octylis gallas.

3 Chemical Name and CAS Registry Number

Octyl 3,4,5-trihydroxybenzoate [1034-01-1]

4 Empirical Formula and Molecular Weight

$C_{15}H_{22}O_5$ 282.3

5 Structural Formula

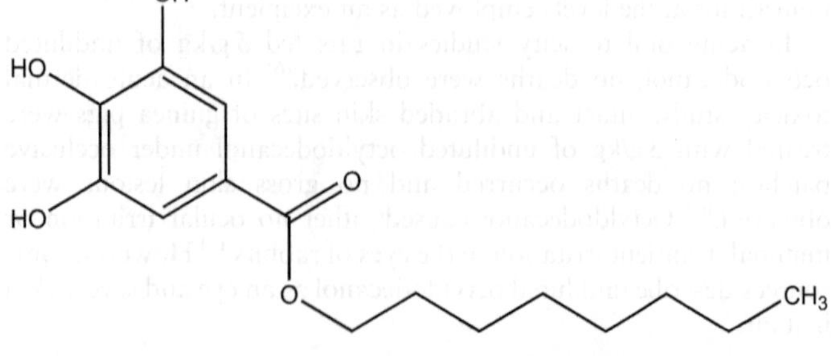

6 Functional Category

Antioxidant.

7 Applications in Pharmaceutical Formulation or Technology

Octyl gallate and other alkyl gallates have become widely used as antioxidant preservatives in cosmetics, perfumes, foods, and pharmaceuticals since the use of propyl gallate in preventing autoxidation of oils was first described in 1943.[1,2]

Octyl gallate and other alkyl gallates are used as antioxidants in foods, at concentrations of 0.001 to 0.1% w/v, to prevent deterioration and rancidity of fats and oils.[3] They are often used with other antioxidants such as butylated hydroxyanisole or butylated hydroxytoluene, and with sequestrants such as citric acid and zinc salts, to improve acceptability and efficacy.

Octyl gallate has been investigated as a means of increasing the bioavailability of orally administered veterinary pharmaceutical compounds.[4]

8 Description

Octyl gallate is a white or almost white, odorless or almost odorless, crystalline powder.

9 Pharmacopeial Specifications

See Table I.

Table I: Pharmacopeial specifications for octyl gallate.

Test	PhEur 7.4
Characters	+
Identification	+
Chlorides	≤100 ppm
Heavy metals	≤10 ppm
Loss on drying	≤0.5%
Sulfated ash	≤0.1%
Impurities	+
Assay (dried substance)	97.0–103.0%

10 Typical Properties

Melting point 94–95°C; 100–102°C after 6 hours drying at 90°C.
Solubility *See* Table II.

Table II: Solubility of octyl gallate.

Solvent	Solubility at 20°C
Acetone	1 in 1
Chloroform	1 in 30
Ethanol (95%)	1 in 2.5
Ether	1 in 3
Methanol	1 in 0.7
Peanut oil	1 in 33
Propylene glycol	1 in 7
Water	Practically insoluble

11 Stability and Storage Conditions

Store in non-metallic airtight containers. Protect from light.

12 Incompatibilities

The alkyl gallates are incompatible with metals, e.g. sodium, potassium, and iron, forming intensely colored complexes. Complex formation may be prevented, under some circumstances, by the addition of a sequestering agent, typically citric acid. Octyl gallate also reacts with strong oxidizing agents and strong bases.

13 Method of Manufacture

Octyl gallate is prepared by the esterification of gallic acid with octanol.[5]

14 Safety

Octyl gallate is widely used in food products and is generally regarded as a non-toxic and non-irritant material. It can cause allergic contact dermatitis.[6,7]

LD_{50} (IP, rat): 0.06 g/kg[8]

LD_{50} (oral, rat): 1.96 g/kg

The WHO has established a temporary estimated acceptable daily intake for octyl gallate at up to 0.1 mg/kg body weight.[9]

15 Handling Precautions

Observe normal precautions appropriate to the circumstances and quantity of the material handled. When heated to decomposition, octyl gallate emits acrid and irritating fumes. Eye protection and gloves are recommended.

16 Regulatory Status

Accepted for use as a food additive in Europe. Included in the Canadian Natural Health Products Ingredients Database, which has set a limit of up to 0.1 mg/kg body weight daily for octyl gallate.

17 Related Substances

Dodecyl gallate; propyl gallate.

18 Comments

Octyl gallate has been investigated for its antifungal activity,[10–12] and for its antiviral effect against influenza and other viruses.[13,14]

The EINECS number for octyl gallate is 252-073-5. The PubChem Compound ID (CID) for octyl gallate is 61253.

19 Specific References

1 Boehm E, Williams R. The action of propyl gallate on the autoxidation of oils. *Pharm J* 1943; **151**: 53.
2 Boehm E, Williams R. A study of the inhibiting actions of propyl gallate (normal propyl trihydroxybenzoate) and certain other trihydric phenols on the autoxidation of animal and vegetable oils. *Chemist Drug* 1943; **140**: 146–147.
3 Sweetman SC, ed. *Martindale: The Complete Drug Reference*, 36th edn. London: Pharmaceutical Press, 1999; 1628.
4 Wacher V, Benet LZ. Use of gallic acid esters to increase bioavailability of orally administered pharmaceutical compounds. WO Patent WO0051643; 2000.
5 Sas B *et al.* Method of crystallizing and purifying alkyl gallates. United States Patent 6,297,396; 2001.
6 Linares L G-M, *et al.* Allergic contact dermatitis from gallates. *J Am Acad Dermatol* 2007; **56**(2): AB75.
7 Garcia-Melgares ML, *et al.* Sensitization to gallates: review of 46 cases. *Actas Dermosifiliogr* 2007; **98**: 688–693.
8 Lewis RJ, ed. *Sax's Dangerous Properties of Industrial Materials*, 11th edn. New York: Wiley, 2004; 2769.
9 FAO/WHO. Evaluation of certain food additives and contaminants. Forty-sixth report of the joint FAO/WHO expert committee on food additives. *World Health Organ Tech Rep Ser* 1997; No. 868.
10 Fujita KI, Kubo I. Antifungal activity of octyl gallate. *Int J Food Micro* 2002; **79**(3): 193–201.
11 Kubo I, *et al.* Antifungal activity of octyl gallate: structural criteria and mode of action. *Bioorg Med Chem Lett* 2001; **11**(3): 347–350.
12 Hsu FL, *et al.* Evaluation of antifungal properties of octyl gallate and its synergy with cinnamaldehyde. *Bioresour Tech* 2007; **98**: 734–738.
13 Yamasaki H, *et al.* Antiviral effect of octyl gallate against influenza and other RNA viruses. *Int J Mol Med* 2007; **19**(4): 685–688.
14 Uozaki M, *et al.* Antiviral effect of octyl gallate against DNA and RNA viruses. *Antiviral Res* 2007; **73**(2): 85–91.

20 General References

Johnson DM, Gu LC. Autoxidation and antioxidants. In: Swarbrick J, Boylan JC, eds. *Encyclopedia of Pharmaceutical Technology*, Vol. 1. New York: Marcel Dekker, 1988; 415–449.

21 Author

RT Guest.

22 Date of Revision

2 March 2012.

O

Oleic Acid

1 Nonproprietary Names

BP: Oleic Acid
PhEur: Oleic Acid
USP–NF: Oleic Acid

2 Synonyms

Acidum oleicum; *Crodolene*; *Crossential 094*; elaic acid; *Emersol*; *Glycon*; *Groco*; *Hy-Phi*; *Industrene*; *Metaupon*; *Neo-Fat*; *cis*-9-octadecenoic acid; 9,10-octadecenoic acid; oleinic acid; *Priolene*.

3 Chemical Name and CAS Registry Number

(Z)-9-Octadecenoic acid [112-80-1]

4 Empirical Formula and Molecular Weight

$C_{18}H_{34}O_2$ 282.47

5 Structural Formula

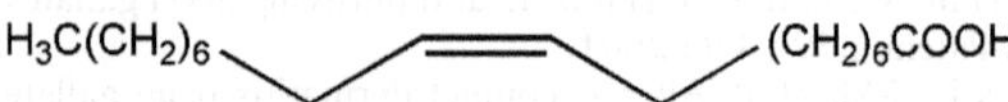

6 Functional Category

Emulsifying agent; penetration enhancer.

7 Applications in Pharmaceutical Formulation or Technology

Oleic acid is used as an emulsifying agent in foods and topical pharmaceutical formulations. It has been investigated as a penetration enhancer in transdermal formulations,[1–14] to improve the bioavailability of poorly water-soluble drugs in tablet formulations,[15] and as part of a vehicle in soft gelatin capsules, in topical nanoemulsion[16,17] and microemulsion[18–22] formulations, in oral self-emulsifying drug delivery systems,[23,24] in oral mucoadhesive patches,[25] and in a metered-dose inhaler.[26]

The phase behavior of sonicated dispersions of oleic acid has been described,[27] and mechanisms for the topical penetration-enhancing actions of oleic acid have been presented.[28] Oleic acid has also been studied as a plasticizer in drug-loaded casein beads.[29]

8 Description

A yellowish to pale brown, oily liquid with a characteristic lard-like odor and taste.

Oleic acid consists chiefly of (Z)-9-octadecenoic acid together with varying amounts of saturated and other unsaturated acids. It may contain a suitable antioxidant.

9 Pharmacopeial Specifications

See Table I.

Table I: Pharmacopeial specifications for oleic acid.

Test	PhEur 7.4	USP35–NF30
Identification	+	+
Characters	+	—
Specific gravity	≈0.892	0.889–0.895
Residue on ignition	—	≤1 mg
Total ash	≤0.1%	—
Mineral acids	—	+
Neutral fat or mineral oil	—	+
Fatty acid composition	+	—
Myristic acid	≤5.0%	—
Palmitic acid	≤16.0%	—
Palmitoleic acid	≤8.0%	—
Stearic acid	≤6.0%	—
Oleic acid	65.0–88.0%	—
Linoleic acid	≤18.0%	—
Linolenic acid	≤4.0%	—
Fatty acids of chain length greater than C_{18}	≤4.0%	—
Acid value	195–204	196–204
Iodine value	89–105	85–95
Peroxide value	≤10.0	—
Congealing temperature	—	+
From animal sources	—	3–10°C
From vegetable sources	—	10–16°C
Margaric acid	+	—
From animal sources	≤4.0%	—
From vegetable sources	≤0.2%	—
Color of solution	+	—
Assay	65–88%	—

10 Typical Properties

Acidity/alkalinity pH = 4.4 (saturated aqueous solution)
Autoignition temperature 363°C
Boiling point 286°C at 13.3 kPa (100 mmHg) (decomposition at 80–100°C)
Density 0.895 g/cm^3
Flash point 189°C
Melting point 6–12°C; pure oleic acid solidifies at 4°C[30]
Refractive index n_D^{26} = 1.4585
Solubility Miscible with benzene, chloroform, ethanol (95%), ether, hexane, and fixed and volatile oils; practically insoluble in water.
Vapor pressure 133 Pa (1 mmHg) at 176.5°C
Viscosity (dynamic) 26 mPa s (26 cP) at 25°C

11 Stability and Storage Conditions

On exposure to air, oleic acid gradually absorbs oxygen, darkens in color, and develops a more pronounced odor. At atmospheric pressure, it decomposes when heated at 80–100°C.

Oleic acid should be stored in a well-filled, well-closed container, protected from light, in a cool, dry place.

12 Incompatibilities

Incompatible with aluminum, calcium, heavy metals, iodine solutions, perchloric acid, and oxidizing agents. Oleic acid reacts with alkalis to form soaps.

13 Method of Manufacture

Oleic acid is obtained by the hydrolysis of various animal and vegetable fats or oils, such as olive oil, followed by separation of the liquid acids. It consists chiefly of (Z)-9-octadecenoic acid. Oleic acid that is to be used systemically should be prepared from edible sources.

14 Safety

Oleic acid is used in oral and topical pharmaceutical formulations.

In vitro tests have shown that oleic acid causes rupture of red blood cells (hemolysis), and intravenous injection or ingestion of a large quantity of oleic acid can therefore be harmful. The effects of oleic acid on alveolar[(31)] and buccal[(32)] epithelial cells *in vitro* have also been studied; the *in vitro* and *in vivo* effects of oleic acid on rat skin have been reported.[(33)] Oleic acid is a moderate skin irritant; it should not be used in eye preparations.

An acceptable daily intake for the calcium, sodium, and potassium salts of oleic acid was not specified by the WHO since the total daily intake of these materials in foods was such that they did not pose a hazard to health.[(34)]

LD_{50} (mouse, IV): 0.23 g/kg[(35)]

LD_{50} (rat, IV): 2.4 mg/kg

LD_{50} (rat, oral): 74 g/kg

15 Handling Precautions

Observe normal precautions appropriate to the circumstances and quantity of material handled. Gloves and eye protection are recommended.

16 Regulatory Status

GRAS listed. Included in the FDA Inactive Ingredients Database (inhalation, nasal and oral aerosols; oral tablets; topical emulsions and solutions; transdermal films and patches). Included in nonparenteral medicines (metered-dose inhalers; oral prolonged-release capsules; oral prolonged-release granules) licensed in the UK. Included in the Canadian Natural Health Products Ingredients Database.

17 Related Substances

Ethyl oleate.

18 Comments

The USP35–NF30 requires the label for oleic acid to indicate if it is for external use only, whether it is derived from animal or vegetable sources, and the name and quantity of any added substances. The PhEur 7.4 requires the label to state whether the oleic acid is obtained from animal or vegetable sources.

Several grades of oleic acid are commercially available, ranging in color from pale yellow to reddish brown. Different grades become turbid at varying temperatures depending upon the amount of saturated acid present. Usually, oleic acid contains 7–12% saturated acids, such as stearic and palmitic acid, together with other unsaturated acids, such as linoleic acid.

Oleic acid has been shown to be an important factor in the hypoglycemic effect produced by multiple emulsions containing insulin intended for intestinal delivery of insulin.[(36)]

Oleic acid has been reported to act as an ileal 'brake' that slows down the transit of luminal contents through the distal portion of the small bowel.[(37)]

Oleic acid labeled with ^{131}I and ^{3}H is used in medical imaging.

A specification for oleic acid is contained in the *Food Chemicals Codex* (FCC)[(38)] and in the *Japanese Pharmaceutical Excipients* (JPE).[(39)]

The EINECS number for oleic acid is 204-007-1. The PubChem Compound ID (CID) for oleic acid includes 965 and 445639.

19 Specific References

1 Schroeder IZ, *et al.* Delivery of ethinylestradiol from film forming polymeric solutions across human epidermis *in vitro* and *in vivo* in pigs. *J Control Release* 2007; **118**(2): 196–203.

2 Francoeur ML, *et al.* Oleic acid: its effects on stratum corneum in relation to (trans)dermal drug delivery. *Pharm Res* 1990; **7**: 621–627.

3 El-Gendy NA, *et al.* Transdermal delivery of salbutamol sulphate: formulation and evaluation. *Pharm Dev Technol* 2009; **14**(2): 216–225.

4 Abd-ElGawad AH, *et al.* Effect of penetration enhancers on the permeability of ketoconazole gels through rabbit skin. *Bull Pharm Sci* 2006; **29 (part 1)**: 67–77.

5 Swain K, *et al.* Drug in adhesive type transdermal matrix systems of ondansetron hydrochloride: optimization of permeation pattern using response surface methodology. *J Drug Target* 2010; **18**(2): 106–114.

6 Mehdizadeh A, *et al.* Effects of pressure sensitive adhesives and chemical permeation enhancers on the permeability of fentanyl through excised rat skin. *Acta Pharm* 2006; **56**(2): 219–229.

7 Rastogi R, *et al.* Investigation on the synergistic effect of a combination of chemical enhancers and modulated iontophoresis for transdermal delivery of insulin. *Drug Dev Ind Pharm* 2010; **36**(8): 993–1004.

8 Jain AK, Panchagnula R. Combination of penetration enhancers for transdermal drug delivery – studies with imipramine hydrochloride. *Pharmazeutische Industrie* 2004; **66**(4): 478–482.

9 Narishetty ST, Panchagnula R. Transdermal delivery system for zidovudine: *in vitro*, *ex vivo* and *in vivo* evaluation. *Biopharm Drug Dispos* 2004; **25**(1): 9–20.

10 Casiraghi A, *et al.* The effects of excipients for topical preparations on the human skin permeability of terpinen-4-ol contained in Tea tree oil: infrared spectroscopic investigations. *Pharm Dev Technol* 2010; **15**(5): 545–552.

11 Zakir F, *et al.* Development and characterization of oleic acid vesicles for the topical delivery of fluconazole. *Drug Deliv* 2010; **17**(4): 238–248.

12 Kim MJ, *et al.* Skin permeation enhancement of diclofenac by fatty acids. *Drug Deliv* 2008; **15**(6): 373–379.

13 Wang Y, *et al.* Effects of fatty acids and iontophoresis on the delivery of midodrine hydrochloride and the structure of human skin. *Pharm Res* 2003; **20**(10): 1612–1618.

14 Gwak HS, *et al.* Transdermal delivery of ondansetron hydrochloride: effects of vehicles and penetration enhancers. *Drug Dev Ind Pharm* 2004; **30**(2): 187–194.

15 Tokumura T, *et al.* Enhancement of the oral bioavailability of cinnarizine in oleic acid in beagle dogs. *J Pharm Sci* 1987; **76**: 286–288.

16 Talegaonkar S, *et al.* Design and development of oral oil-in-water nanoemulsion formulation bearing atorvastatin: *in vitro* assessment. *J Disp Sci Technol* 2010; **31**: 690–701.

17 Kumar D, *et al.* Investigation of a nanoemulsion as vehicle for transdermal delivery of amlodipine. *Pharmazie* 2009; **64**(2): 80–85.

18 Suppasansatorn P, *et al.* Microemulsions as topical delivery vehicles for the anti-melanoma prodrug, temozolomide hexyl ester (TMZA-HE). *J Pharm Pharmacol* 2007; **59**(6): 787–794.

19 El-Megrab NA, *et al.* Formulation and evaluation of meloxicam gels for topical administration. *Saudi Pharm J* 2006; **14**(3–4): 155–162.

20 Zhao X, *et al.* Enhancement of transdermal delivery of theophylline using microemulsion vehicle. *Int J Pharm* 2006; **327**(1–2): 58–64.

21 Park ES, *et al.* Transdermal delivery of piroxicam using microemulsions. *Arch Pharm Res* 2005; **28**(2): 243–248.

22 Hathout RM, *et al.* Microemulsion formulations for the transdermal delivery of testosterone. *Eur J Pharm Sci* 2010; **40**(3): 188–196.

23 Quan DQ, *et al.* Studies on preparation and absolute bioavailability of a self-emulsifying system containing puerarin. *Chem Pharm Bull (Tokyo)* 2007; **55**(5): 800–803.

24 Park MJ, *et al.* *In vitro* and *in vivo* comparative study of itiraconazole bioavailability when formulated in highly soluble self-emulsifying system and in solid dispersion. *Biopharm Drug Dispos* 2007; **28**(4): 199–207.

25 Onishi H, *et al.* Novel mucoadhesive oral patch containing diazepam. *Drug Dev Ind Pharm* 2005; **31**(7): 607–613.

26 Saso Y, *et al.* Formulation design and pharmaceutical evaluation of an HFA 227-based furosemide metered dose inhaler. *STP Pharma Sci* 2004; **14**(2): 135–140.

27 Ferreira DA, *et al.* Cryo-TEM investigation of phase behaviour and aggregate structure in dilute dispersions of monoolein and oleic acid. *Int J Pharm* 2006; **310**(1–2): 203–212.

28 Rowat AC, *et al.* Interactions of oleic acid and model stratum corneum membranes as seen by ^{2}HNMR. *Int J Pharm* 2006; **307**(2): 225–231.

29 Bani-Jaber A, *et al.* Drug-loaded casein beads: influence of different metal-types as cross-linkers and oleic acid as a plasticizer on some properties of the beads. *J Drug Deliv Sci Technol* 2009; **19**: 125–131.

O

30 Troy DB, ed. *Remington: The Science and Practice of Pharmacy*, 21st edn. Philadelphia: Lippincott, Wiliams and Wilkins, 2006: 1077.
31 Wang LY, *et al.* Alveolar permeability enhancement by oleic acid and related fatty acids: evidence for a calcium-dependent mechanism. *Pharm Res* 1994; **11**: 513–517.
32 Turunen TM, *et al.* Effect of some penetration enhancers on epithelial membrane lipid domains: evidence from fluorescence spectroscopy studies. *Pharm Res* 1994; **11**: 288–294.
33 Fang JY, *et al. In vitro* and *in vivo* evaluations of the efficacy and safety of skin permeation enhancers using flurbiprofen as a model. *Int J Pharm* 2003; **255**(1–2): 153–166.
34 FAO/WHO. Evaluation of certain food additives and contaminants. Thirty-third report of the joint FAO/WHO expert committee on food additives. *World Health Organ Tech Rep Ser* 1989: No. 776.
35 Lewis RJ, ed. *Sax's Dangerous Properties of Industrial Materials*, 11th edn. New York: Wiley, 2004: 2778.
36 Onuki Y, *et al.* Formulation optimization of water-in-oil-in-water multiple emulsion for intestinal insulin delivery. *J Control Release* 2004; **97**(1): 91–99.
37 Dobson CL, *et al.* The effects of ileal brake activators on the oral bioavailability of atenolol in man. *Int J Pharm* 2002; **248**(1–2): 61–70.
38 *Food Chemical Codex*, 7th edn. Bethesda, MD: United States Pharmacopeia, 2010: 743.
39 Japan Pharmaceutical Excipients Council. *Japanese Pharmaceutical Excipients* 2004, Tokyo: Yakuji Nippo, 2004: 589–590.

20 General References

—

21 Author

CG Cable.

22 Date of Revision

2 March 2012.

Oleyl Alcohol

1 Nonproprietary Names

BP: Oleyl Alcohol
PhEur: Oleyl Alcohol
USP–NF: Oleyl Alcohol

2 Synonyms

Alcohol oleicus; *HD-Eutanol V PH*; *Novol*; *Ocenol*; *cis*-9-octadecen-1-ol; oleic alcohol; oleo alcohol; oleol.

3 Chemical Name and CAS Registry Number

(Z)-9-Octadecen-1-ol [143-28-2]

4 Empirical Formula and Molecular Weight

$C_{18}H_{36}O$ 268.48

The USP35–NF30 describes oleyl alcohol as a mixture of unsaturated and saturated high molecular weight fatty alcohols consisting chiefly of oleyl alcohol.

5 Structural Formula

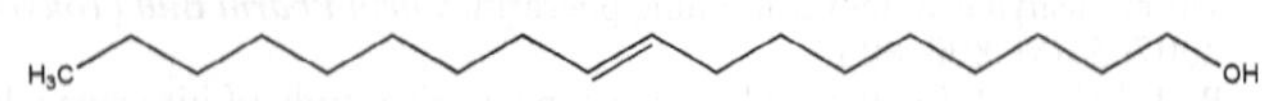

6 Functional Category

Emollient; emulsifying agent; penetration enhancer; solubilizing agent.

7 Applications in Pharmaceutical Formulation or Technology

Oleyl alcohol is mainly used as an emollient, solubilizer, and emulsifying agent in topical pharmaceutical formulations, and it has also been used in transdermal delivery formulations.[1–8] It has been utilized in the development of biodegradable injectable thermoplastic oligomers,[9] pH-sensitive liposome formulations,[10,11] and in aerosol formulations of insulin[12] and albuterol.[13]

8 Description

Oleyl alcohol occurs as a colorless or light yellow oily liquid with a faint characteristic odor and bland taste.

9 Pharmacopeial Specifications

See Table I. *See also* Section 18.

Table I: Pharmacopeial specifications for oleyl alcohol.

Test	PhEur 7.4	USP35–NF30
Appearance	+	—
Cloud point	<10°C	<10°C
Refractive index	1.458–1.461	1.458–1.460
Acid value	≤1.0	≤1
Hydroxyl value	205–215	205–215
Iodine value	—	85–95
Saponification value	≤2.0	—
Composition of fatty alcohols	+	—
Palmityl alcohol	≤8%	—
Stearyl alcohol	≤5%	—
Oleyl alcohol (sum of oleyl and elaidyl alcohols)	≥80%	—
Linoleyl alcohol	≤3%	—
Linolenyl alcohol	≤0.5%	—
Arachidyl alcohol	≤0.3%	—

10 Typical Properties

Boiling point 330–360°C
Density 0.847–0.850 g/cm^3 at 40°C
Flash point 189–191°C
Melting point 13–19°C
Partition coefficient Log *P* (octanol : water) = 7.50.
Solubility Soluble in ethanol (95%), ether, isopropyl alcohol, and light mineral oil; insoluble in water.
Spectroscopy
IR spectra *see* Figure 1.

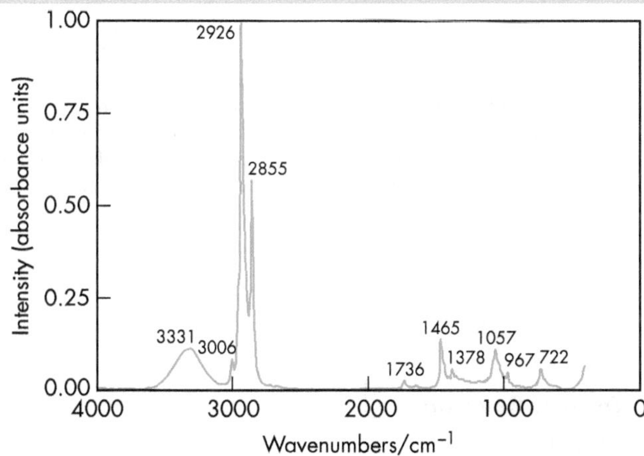

Figure 1: Infrared spectrum of oleyl alcohol measured by transmission. Adapted with permission of Informa Healthcare.

11 Stability and Storage Conditions

The bulk material should be stored in a well-closed container in a cool, dry, place.

12 Incompatibilities

—

13 Method of Manufacture

Oleyl alcohol is produced by catalytic, high-pressure hydrogenation of oleic acid followed by filtration and distillation. Synthetically, it can be prepared from butyl oleate by a Bouveault–Blanc reduction with sodium and butyl alcohol. An alternative method of manufacture is by the hydrogenation of triolein in the presence of zinc chromite.

14 Safety

Oleyl alcohol is mainly used in topical pharmaceutical formulations and is generally regarded as a nontoxic and nonirritant material at the levels employed as an excipient. However, contact dermatitis due to oleyl alcohol has been reported.[14–18]

The results of acute oral toxicity and percutaneous studies in animals with products containing 8% oleyl alcohol indicate a very low toxicity.[19] Formulations containing 8% or 20% oleyl alcohol administered by gastric intubation, at doses up to 10 g/kg bodyweight, caused no deaths and no toxic effects in rats.[19]

LD_{50} (rat, oral): >5 g/kg

15 Handling Precautions

Observe normal precautions appropriate to the circumstances and quantity of material handled.

16 Regulatory Status

Included in the FDA Inactive Ingredients Database (topical emulsions, creams, and ointments; transdermal controlled-release films). Included in nonparenteral medicines licensed in the UK. Included in the Canadian Natural Health Products Ingredients Database.

17 Related Substances

Oleic acid; oleyl oleate.

Oleyl oleate

Empirical formula $C_{36}H_{68}O_2$
Molecular weight 532.9
CAS number [3687-45-4]
Refractive index n_D^{25} = 1.464–1.466
Specific gravity 0.861–0.882
Solubility Miscible with chloroform and with diethyl ether; slightly soluble in ethanol.

18 Comments

Oleyl alcohol is a mixture of unsaturated and staturated long-chain fatty alcohols consisting mainly of octadec-9-enol (oleyl alcohol and elaidyl alcohol). It may be of animal or vegetable origin.

Oleyl alcohol occurs naturally in fish oils. Therapeutically, it has been suggested that oleyl alcohol may exhibit antitumor properties via transmembrane permeation.[20,21]

A specification for oleyl alcohol is included in the *Japanese Pharmaceutical Excipients* (JPE).[22]

The EINECS number for oleyl alcohol is 205-597-3. The PubChem Compound ID (CID) for oleyl alcohol is 5284499.

19 Specific References

1 Agyralides GG, *et al.* Development and *in vitro* evaluation of furosemide transdermal formulations using experimental design techniques. *Int J Pharm* 2004; **281**: 35–43.
2 Gwak HS, *et al.* Transdermal delivery of ondansetron hydrochloride: effects of vehicles and penetration enhancers. *Drug Dev Ind Pharm* 2004; **30**: 187–194.
3 Lee PJ, *et al.* Novel microemulsion enhancer formulation for simultaneous transdermal delivery of hydrophilic and hydrophobic drugs. *Pharm Res* 2003; **20**: 264–269.
4 Kim MJ, *et al.* Skin permeation enhancement of diclofenac by fatty acids. *Drug Deliv* 2008; **15**: 373–379.
5 Joo HH, *et al. In vitro* permeation study of hinokitiol: effects of vehicles and enhancers. *Drug Deliv* 2008; **15**: 19–22.
6 Mbah CJ. Studies on the lipophilicity of vehicles (or co-vehicles) and botanical oils used in cosmetic products. *Pharmazie* 2007; **62**: 351–353.
7 Lee PJ, *et al.* Evaluation of chemical enhancers in the transdermal delivery of lidocaine. *Int J Pharm* 2006; **308**: 33–39.
8 Fetih G, *et al.* Design and characterization of transdermal films containing ketorolac tromethamine. *Int J PharmTech Res* 2011; **3**: 449–458.
9 Amsden B, *et al.* Development of biodegradable injectable thermoplastic oligomers. *Biomacromolecules* 2004; **5**: 637–642.
10 Sudimack JJ, *et al.* A novel pH-sensitive liposome formulation containing oleyl alcohol. *Biochim Biophys Acta* 2002; **1564**: 31–37.
11 Vyas SP, *et al.* pH sensitive liposomes enhances immunogenicity of 19 kDa carboxylterminal fragment of *Plasmodium falciparum. Int J Pharm Sci Nanotechnol* 2008; **1**: 78–86.
12 Lee SW, Sciarra JJ. Development of an aerosol dosage form containing insulin. *J Pharm Sci* 1976; **65**: 567–572.
13 Tiwari D, *et al.* Formulation and evaluation of albuterol metered dose inhalers containing tetrafluoroethane (P132a), a non-CFC propellant. *Pharm Dev Technol* 1998; **3**: 163–174.
14 Guidetti MS, *et al.* Contact dermatitis due to oleyl alcohol. *Contact Dermatitis* 1994; **31**: 260–261.
15 Koch P. Occupational allergic contact dermatitis from oleyl alcohol and monoethanolamine in a metalworking fluid. *Contact Dermatitis* 1995; 33(4): 273.
16 Anderson KE, Broesby-Olsen S. Allergic contact dermatitis from oleyl alcohol in Elidel cream. *Contact Dermatitis* 2006; **55**(6): 354–356.
17 Inui S, *et al.* Recurrent contact cheilitis because of glyceryl isostearate, diisostearyl maleate, oleyl alcohol, and *Lithol Rubine BCA* in lipsticks. *Contact Dermatitis* 2009; **60**: 231–232.
18 Geier J, *et al.* Patch testing with components of water-based metalworking fluids: results of a multicentre study with a second series. *Contact Dermatitis* 2006; **55**: 322–329.
19 CFTA. Final report on the safety assessment of stearyl alcohol, oleyl alcohol and octyl dodecanol. *The Cosmetic Ingredient Review Program* 1985; No. 4.
20 Takada Y, *et al.* Correlation of DNA synthesis-inhibiting activity and the extent of transmembrane permeation into tumor cells by unsaturated or saturated fatty alcohols of graded chain-length upon hyperthermia. *Oncol Rep* 2001; **8**: 547–551.
21 Orienti I, *et al.* Enhancement of oleyl alcohol anti tumor activity through complexation in polyvinylacohol amphiphilic derivatives. *Drug Deliv* 2007; **14**: 209–217.

O

22 Japan Pharmaceutical Excipients Council. *Japanese Pharmaceutical Excipients 2004*. Tokyo: Yakuji Nippo, 2004: 593–595.

20 General References

CosmeticsInfo.org. http://www.cosmeticsinfo.org/ingredient_details.php?ingredient_id=1426 (accessed 8 December 2011).

Jiang N, *et al*. Perforated vesicles as intermediate structures in the transition from vesicles to micelles in dilute aqueous systems containing long chain alcohols and ionic surfactants. *J Disper Sci Technol* 2009; 30(6): 802–808.

Konno Y, *et al*. Phase behavior and hydrated solid structure in lysophospholipid/long-chain alcohol/water system and effect of cholesterol addition. *J Oleo Sci* 2010; 59: 581–587.

Malcolm RK, *et al*. A dynamic mechanical method for determining the silicone elastomer solubility of drugs and pharmaceutical excipients in silicone intravaginal drug delivery rings. *Biomaterials* 2002; 23: 3589–3594.

Murakami R, *et al*. Aggregate formation in oil and adsorption at oil/water interface: thermodynamics and its application to the oleyl alcohol system. *J Colloid Interface Sci* 2004; **270**: 262–269.

Murota K, *et al*. Oleyl alcohol inhibits intestinal long-chain fatty acid absorption in rats. *J Nutr Sci Vitaminol (Tokyo)* 2000; **46**: 302–308.

Tokuyama H, Kato Y. Preparation of poly(*N*-isopropylacrylamide) emulsion gels and their drug release behaviors. *Colloids Surf B Biointerfaces* 2008; 67: 92–98.

21 Authors

HC Shah, KK Singh.

22 Date of Revision

2 March 2012.

Olive Oil

1 Nonproprietary Names

BP: Refined Olive Oil
JP: Olive Oil
PhEur: Olive Oil, Refined
USP–NF: Olive Oil

2 Synonyms

Gomenoleo oil; olea europaea oil; oleum olivae; olivae oleum raffinatum; pure olive oil.

3 Chemical Name and CAS Registry Number

Olive oil [8001-25-00]

4 Empirical Formula and Molecular Weight

Olive oil is a mixture of fatty acid glycerides. Analysis of olive oil shows a high proportion of unsaturated fatty acids, and a typical analysis shows that the composition of the fatty acids is as follows:

Myristic acid (14 : 0), ≤0.5%
Palmitic acid (16 : 0), 7.5–20.0%
Palmitoleic acid (16 : 1), 0.3–5.0%
Hepatodecenoic acid (17 : 1), ≤0.3%
Stearic acid (18 : 0), 0.5–5.0%
Oleic acid (18 : 1), 55.0–83.0%
Linoleic acid (18 : 2), 3.5–21.0%
Linolenic acid (18 : 3), ≤0.9%
Arachidic acid (20 : 0), ≤0.6%
Eicosenoic acid (20 : 1), ≤0.4%
Behenic acid (22 : 0), ≤0.2%
Lignoceric acid (24 : 0), ≤1.0%
Sterols are also present.

5 Structural Formula

See Section 4.

6 Functional Category

Oleaginous vehicle.

7 Applications in Pharmaceutical Formulation or Technology

Olive oil has been used in enemas, liniments, ointments, plasters, and soap. It has also been used in oral capsules and solutions, and as a vehicle for oily injections including targeted delivery systems.(1) Olive oil has been investigated for use in self-emulsifying and self-microemulsifying formulations for a variety of drug molecules. It has been used in topically applied lipogels of methyl nicotinate.(2) Olive oil has been used in combination with soybean oil to prepare lipid emulsion for use in pre-term infants.(3)

8 Description

Olive oil is the fixed oil obtained by cold expression or other suitable mechanical means from the ripe drupes of *Olea europaea*. It occurs as a clear, colorless or yellow, transparent oily liquid. It may contain suitable antioxidants.

Refined olive oil is obtained by refining crude olive oil such that the glyceride content of the oil is unchanged. A suitable antioxidant may be added.

9 Pharmacopeial Specifications

The regulation of olive oil is different in different countries. The pharmacopeial specifications are also different, and may refer to different materials. In JP XV the monograph covers either mixtures of refined olive oil and virgin olive oil or just refined olive oil. In the USP35–NF30 and PhEur 7.4, the monographs are specific for refined olive oil.

See Table I.

10 Typical Properties

Flash point 225°C
Refractive index n_D^{25} = 1.4657–1.4893
Smoke point 160–188°C
Solubility Slightly soluble in ethanol (95%); miscible with ether, chloroform, light petroleum (50–70°C), and carbon disulfide.

Table I: Pharmacopeial specifications for olive oil.

Test	JP XV	PhEur 7.4[a]	USP35–NF30[b]
Identification	–	+	–
Characters	+	+	–
Acid value	≤1.0	≤0.3	≤0.3
Peroxide value	–	≤10.0[c]	≤10.0
Saponification value	186–194	–	–
Unsaponifiable matter	≤1.5%	≤1.5%	≤1.5%
Iodine value	79–88	–	–
Specific gravity	0.908–0.914	–	–
Alkaline impurities	–	+	+
Absorbance at 270 nm	–	≤1.20	≤1.20
Composition of fatty acids	–	+	+
Saturated fatty acids of chain length less than C_{16}	–	≤0.1%	≤0.1%
Palmitic acid	–	7.5–20.0%	7.5–20.0%
Palmitoleic acid	–	≤3.5%	≤3.5%
Stearic acid	–	0.5–5.0%	0.5–5.0%
Oleic acid	–	56.0–85.0%	56.0–85.0%
Linoleic acid	–	3.5–20.0%	3.5–20.0%
Linolenic acid	–	≤1.2%	≤1.2%
Arachidic acid	–	≤0.7%	≤0.7%
Eicosenoic acid	–	≤0.4%	≤0.4%
Behenic acid	–	≤0.2%	≤0.2%
Lignoceric acid	–	≤0.2%	≤0.2%
Sterols	–	+	+
Sum of contents of β-sitostanol, Δ^5,24-stigmastadienol, clerosterol, sitostanol, Δ^5-avenasterol, and Δ^5,23-stigmastadienol	–	≥93.0%	≥93.0%
Cholesterol	–	≤0.5%	≤0.5%
Δ^7-Stigmasterol	–	≤0.5%	≤0.5%
Campesterol	–	≤4.0%	≤4.0%
Stigmasterol	–	Not greater than that of campesterol	Not greater than that of campesterol
Sesame oil	–	+	–
Absence of sesame oil	–	–	+
Water	–	≤0.1%	≤0.1%
Drying oil	+	–	–
Peanut oil	+	–	–
Heavy metals	–	–	≤10 ppm

(a) The PhEur 7.4 monograph refers to refined olive oil.
(b) The USP35–NF30 monograph refers to refined olive oil.
(c) If intended for use in the manufacture of parenteral preparations, the PhEur imposes a limit of ≤5.0

11 Stability and Storage Conditions

When cooled, olive oil becomes cloudy at approximately 10°C, and becomes a butter-like mass at 0°C.

Olive oil should be stored in a cool, dry place in a tight, well-filled container, protected from light.

For refined oil intended for use in the manufacture of parenteral dosage forms, the PhEur 7.4 requires that the bulk oil be stored under an inert gas.

12 Incompatibilities

Olive oil may be saponified by alkali hydroxides. As it contains a high proportion of unsaturated fatty acids, olive oil is prone to oxidation and is incompatible with oxidizing agents.

13 Method of Manufacture

Virgin olive oil is produced by crushing olives (the fruit of *Olea europaea*), typically using an edge runner mill. The oil is then expressed from the crushed mass solely by mechanical or other physical methods under conditions that do not cause deterioration of the oil. Any further treatment that the oil undergoes is limited to washing, decantation, centrifugation, and filtration.

Refined olive oil is obtained from virgin olive oil by refining methods that do not alter the initial glyceride content of the oil.

14 Safety

Olive oil is used widely as an edible oil and in food preparations and products such as cooking oils and salad dressings. It is used in cosmetics and topical pharmaceutical formulations. Olive oil is generally regarded as a relatively nonirritant and nontoxic material when used as an excipient.

Olive oil is a demulcent and has mild laxative properties when taken orally. It has been used in topical formulations as an emollient and to sooth inflamed skin; to soften the skin and crusts in eczema; in massage oils; and to soften earwax.[4]

There have been isolated reports that olive oil may cause a reaction in hypersensitive individuals. However, this is relatively uncommon.[5–7] Olive oil is an infrequent sensitizer and does not appear to be a significant allergen in the US, possibly due to the development of oral tolerance.

15 Handling Precautions

Observe normal precautions appropriate to the circumstances and quantity of material handled. Olive oil spills are slippery and an inert oil absorbent should be used to cover the oil, which can then be disposed of according to the appropriate legal regulations.

16 Regulatory Status

Olive oil is an edible oil. Included in the FDA Inactive Ingredients Database (oral capsules and solution; topical solutions). Included in nonparenteral medicines licensed in Europe. Included in the Canadian Natural Health Products Ingredients Database. For nontopical uses, refined olive oil is generally preferred.

17 Related Substances

Crude olive-pomace oil; extra virgin olive oil; fine virgin olive oil; lampante virgin olive oil; olive-pomace oil; refined olive-pomace oil; virgin olive oil.

Crude olive-pomace oil

Comments Crude olive-pomace oil is olive-pomace oil that is intended for refining prior to its use in food for human consumption, or that is intended for technical purposes.

Extra virgin olive oil

Comments Extra virgin oil is a virgin oil that has an organoleptic rating of not less than 6.5, and a free acidity (as oleic acid) of not more than 1.0 g per 100 g.

Fine virgin olive oil

Comments Fine virgin oil has an organoleptic rating of not less than 5.5, and a free acidity (as oleic acid) of not more than 1.5 g per 100 g.

Lampante virgin olive oil

Comments Lampante virgin olive oil is virgin olive oil that is not fit for consumption unless it is further processed. This grade of oil is intended for refining or technical purposes.

Olive-pomace oil

Comments Olive-pomace oil is the oil obtained from the solvent extraction of olive pomace, but does not include oils obtained by reesterification processes or any mixture with oils of any kind. Olive-pomace oil of commerce is a blend of refined olive-pomace oil and virgin olive oil that is fit for human consumption. *See also* Section 18.

Refined olive-pomace oil

Comments Refined olive-pomace oil is obtained from crude olive-pomace oil by refining methods that do not alter the initial glyceride structure. It is intended for consumption, or blended with virgin olive oil.

Virgin olive oil

Comments Virgin olive oil has an organoleptic rating of not less than 3.5, and a free acidity (as oleic acid) of not more than 3.3 g per 100 g. The PhEur 7.4 contains a monograph on virgin olive oil as well as refined olive oil.

18 Comments

Olive oil is available in a variety of different grades; *see* Section 17. All olive oils are graded according to the degree of acidity.

Olive oil is used widely in the food industry as a cooking oil and for preparing salad dressings. The flavor, color, and fragrance of olive oils may vary, depending on the region where the olives are grown, the condition of the crops, and the type of olive used. Therapeutically, it has been used to soften ear wax.[4]

In cosmetics, olive oil is used as a solvent, and also as a skin and hair conditioner. Types of products containing olive oil include shampoos and hair conditioners, cleansing products, topical creams and lotions, and sun-tan products.

Olive-pomace oil is obtained from the olive pomace by solvent extraction. The use of solvent extraction causes small changes in the typical fatty acid composition of the oil, and changes in organoleptic properties and impurities. Other oils can be prepared by reesterification of the appropriate combination of fatty acids with glycerol. Olive-pomace oils or reesterified oils cannot be called olive oil.

19 Specific References

1 Jakate AS, *et al.* Preparation, characterization, and preliminary application of fibrinogen-coated olive oil droplets for the targeted delivery of docetaxol to solid malignancies. *Cancer Res* 2003; **63**: 7314–7320.

2 Realdon N, *et al.* Effect of gelling conditions and mechanical treatment on drug availability from a lipogel. *Drug Dev Ind Pharm* 2001; **27**(2): 165–170.

3 Koletzko B, *et al.* Parenteral fat emulsions based on olive and soybean oils: a randomized clinical trial in preterm infants. *J Paed Gastroenterology Nutr* 2003; **37**(2): 161–167.

4 Smythe O. Ear care. *NZ Pharm* 1998; **18**(12): 25–2628.

5 Kranke B, *et al.* Olive oil – contact sensitizer or irritant. *Contact Dermatitis* 1997; **36**(1): 5–10.

6 Jung HD, Holzegel K. Contact allergy to olive oil. *Derm Beruf Umwelt* 1987; **35**(4): 131–133.

7 Van Joost T, *et al.* Sensitization to olive oil (*Olea europeae*). *Contact Dermatitis* 1981; **7**(6): 309–310.

20 General References

Allen LV. Featured excipient: oleaginous vehicles. *Int J Pharm Compound* 2000; **4**(6): 470–473484–485.

Croucher P. Olive oil as a functional food. *NZ Pharm* 2002; **22**(8): 40–42.

Garcia Del Pozo JA, Alvarez Martinez MO. Olive oil: attainment, composition and properties. *Farm (El Farmaceutico)* 2000; **241**: 94, 96, 98–100102, 104–105.

21 Author

RC Moreton.

22 Date of Revision

2 March 2012.

O

Palmitic Acid

1 Nonproprietary Names

BP: Palmitic Acid
PhEur: Palmitic Acid
USP–NF: Palmitic Acid

2 Synonyms

Acidum palmiticum; cetylic acid; *Edenor C16 98-100*; *Emersol 140*; *Emersol 143*; *n*-hexadecoic acid; hexadecylic acid; *Hydrofol*; *Hystrene 9016*; *Industrene 4516*; *Lunac P-95*; *NAA-160*; 1-pentadecanecarboxylic acid.

3 Chemical Name and CAS Registry Number

Hexadecanoic acid [57-10-3]

4 Empirical Formula and Molecular Weight

$C_{16}H_{32}O_2$ 256.42

5 Structural Formula

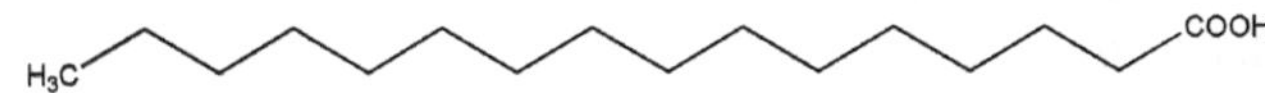

6 Functional Category

Emulsifying agent; penetration enhancer; tablet and capsule lubricant.

7 Applications in Pharmaceutical Formulation or Technology

Palmitic acid is used in oral and topical pharmaceutical formulations as an emulsifying agent, skin penetration enhancer, and tablet and capsule lubricant. *See also* Section 18.

8 Description

Palmitic acid occurs as white crystalline scales with a slight characteristic odor and taste.

9 Pharmacopeial Specifications

See Table I. *See also* Section 18.

Table I: Pharmacopeial specifications for palmitic acid.

Test	PhEur 7.4	USP35–NF30
Identification	+	+
Characters	+	−
Appearance	+	−
Acidity	+	−
Acid value	216–220	216–220
Color	−	+
Freezing point	60–66°C	60–66°C
Iodine value	<1	<1
Stearic acid	<6%	≤6%
Mineral acid	−	+
Heavy metals	−	≤10 ppm
Nickel	<1 ppm	−
Assay	>92.0%	≥92.0%

10 Typical Properties

Boiling point 351–352°C; 271.5°C at 100 mmHg
Flash point >110°C
Melting point 63–64°C
Solubility Soluble in ethanol (95%); practically insoluble in water.
Specific gravity 0.849–0.851

11 Stability and Storage Conditions

The bulk material should be stored in a well-closed container in a cool, dry place.

12 Incompatibilities

Palmitic acid is incompatible with strong oxidizing agents and bases.

13 Method of Manufacture

Palmitic acid occurs naturally in all animal fats as the glyceride, palmitin, and in palm oil partly as the glyceride and partly uncombined. Palmitic acid is most conveniently obtained from olive oil after removal of oleic acid, or from Japanese beeswax. Synthetically, palmitic acid may be prepared by heating cetyl alcohol with soda lime to 270°C or by fusing oleic acid with potassium hydrate.

14 Safety

Palmitic acid is used in oral and topical pharmaceutical formulations and is generally regarded as nontoxic and nonirritant at the levels employed as an excipient. However, palmitic acid is reported to be an eye and skin irritant at high levels and is poisonous by intravenous administration.

LD_{50} (mouse, IV): 57 mg/kg[(1)]

15 Handling Precautions

Observe normal precautions appropriate to the circumstances and quantity of material handled. When palmitic acid is heated to decomposition, carbon dioxide and carbon monoxide are formed.

16 Regulatory Status

GRAS listed. Included in the FDA Inactive Ingredients Database (oral tablets). Included in nonparenteral medicines licensed in the UK. Included in the Canadian Natural Health Ingredients Database.

17 Related Substances

Lauric acid; myristic acid; palmitin; sodium palmitate; stearic acid.

Palmitin

Empirical formula $C_{51}H_{98}O_6$
Molecular weight 807.29
CAS number [555-44-2]
Refractive index n_D^{25} = 1.4381
Specific gravity 0.886
Solubility Soluble in benzene, chloroform, and ether; practically insoluble in ethanol (95%) and in water.

Sodium palmitate

Synonyms Hexadecanoic acid sodium salt; palmitic acid sodium salt; sodium hexadecanoate.
Empirical formula $C_{16}H_{31}O_2Na$
Molecular weight 278.47

CAS number [408-35-5]
Melting point 283–290°C
Comments Sodium palmitate is used as a surfactant and emulsifying agent in pharmaceutical formulations. The EINECS number for sodium palmitate is 206-988-1.

18 Comments

Palmitic acid has been investigated for use in implants for sustained release of insulin in rats.[2,3]

Therapeutically, palmitic acid has been studied as an inhibitor of HIV infection.[4-6]

A specification for palmitic acid is included in the *Food Chemicals Codex*[9] and in the *Japanese Pharmaceutical Excipients* 2004 (JPE).[10]

The EINECS number for palmitic acid is 200-312-9. The PubChem Compound ID (CID) for palmitic acid is 985.

19 Specific References

1 Lewis RJ, ed. *Sax's Dangerous Properties of Industrial Materials*, 11th edn. New York: Wiley, 2004; 2813.
2 Wang PY. Palmitic acid as an excipient in implants for sustained release of insulin. *Biomaterials* 1991; **12**: 57–62.
3 Hashizume M, *et al.* Improvement of large intestinal absorption of insulin by chemical modification with palmitic acid in rats. *J Pharm Pharmacol* 1992; **44**: 555–559.
4 Wang PY. Palmitic acid as an excipient in implants for sustained release of insulin. *Biomaterials* 1991; **12**: 57–62.
5 Hashizume M, *et al.* Improvement of large intestinal absorption of insulin by chemical modification with palmitic acid in rats. *J Pharm Pharmacol* 1992; **44**: 555–559.
6 Lin X, *et al.* Inhibition of HIV-1 infection in *ex-vivo* cervical tissue model of human vagina by palmitic acid; implications for a microbicide development. *PLoS One* 2011; **6**(9): e24803.
7 Lee DY, *et al.* Palmitic acid is a novel CD4 fusion inhibitor that blocks HIV entry and infection. *AIDS Res Hum Retroviruses* 2009; **25**(12): 1231–1241.
8 Paskaleva EE, *et al.* Palmitic acid analogs exhibit nanomolar binding affinity for the HIV-1 CD4 receptor and nanomolar inhibition of gp120-to-CD4 fusion. *PLoS One* 2010; **5**(8): 312168.
9 *Food Chemicals Codex*, 7th edn. Bethesda, MD: United States Pharmacopeia, 2010.
10 Japan Pharmaceutical Excipients Council. *Japanese Pharmaceutical Excipients 2004*. Tokyo: Yakuji Nippo, 2004: 601.

20 General References

Bhattacharya A, Ghosal SK. Permeation kinetics of ketotifen fumarate alone and in combination with hydrophobic permeation enhancers through human cadaver epidermis. *Boll Chim Farm* 2000; **139**: 177–181.
International Labour Organization. *International Chemical Safety Card (ICSC)* 0530: Palmitic acid, 1997.
Yagi S, *et al.* Factors determining drug residence in skin during transdermal absorption: studies on beta-blocking agents. *Biol Pharm Bull* 1998; **21**: 1195–1201.

21 Authors

S Cantor, S Shah.

22 Date of Revision

2 March 2012.

P

 Palm Kernel Oil

1 Nonproprietary Names

BP: Fractionated Palm Kernel Oil
USP–NF: Palm Kernel Oil

2 Synonyms

Elaeis guineensis seed oil; oleum palmae nuclei; palm-nut oil

3 Chemical Name and CAS Registry Number

Palm kernel oil [8023-79-8]

4 Empirical Formula and Molecular Weight

Palm kernel oil consists of glyceride esters of fatty acids, primarily lauric (48%), myristic (16%), and oleic (15%) acids.

5 Structural Formula

See Section 4.

6 Functional Category

Emollient; suppository base.

7 Applications in Pharmaceutical Formulation or Technology

Palm kernel oil is widely used in cosmetics, food products, and pharmaceuticals. Fractionated palm kernel oil is used as a basis for suppositories, whereas the unfractionated oil has been used as an emollient and ointment base.

8 Description

Palm kernel oil occurs as a white to yellowish, odorless or almost odorless, solid, brittle fat.

9 Pharmacopeial Specifications

See Table I.

10 Typical Properties

Density 0.925–0.935
Refractive index n_D^{40} = 1.449–1.452
Solubility Practically insoluble in water; miscible with chloroform, ether, and petroleum spirit (boiling range, 40–60°C); practically insoluble in ethanol (96%).

11 Stability and Storage Conditions

Preserve in well-closed containers. Store at a temperature not exceeding 25°C.

Table I: Pharmacopeial specifications for palm kernel oil.[a]

Test	BP 2012	USP35–NF30
Identification	–	+
Characters	+	–
Acid value	≤0.2	≤2.0
Iodine value	≤6.0	–
Melting point	31–36°C	27–29°C
Refractive index	1.445–1.447	–
Saponification value	246–250	–
Peroxide value	–	≤10.0
Peroxides	+	–
Unsaponifiable matter	–	≤1.5%
Water	–	≤0.1%
Limit of lead	–	≤0.1 μg/g
Fatty acid composition	–	+

(a) BP is fractionated palm kernel oil and USP–NF is palm kernel oil.

12 Incompatibilities

—

13 Method of Manufacture

Palm kernel oil is obtained from the kernels of the oil palm *Elaeis guineensis* by either mechanical expression or solvent extraction.

14 Safety

Palm kernel oil is used in cosmetics and foods as well as topically in pharmaceutical formulations. It is considered safe to use in cosmetic formulations.[1]

15 Handling Precautions

Observe normal precautions appropriate to the circumstances and quantity of the material handled. When heated to decomposition, unhydrogenated palm kernel oil emits acrid smoke and irritating fumes.[2]

16 Regulatory Status

Included in the FDA Inactive Ingredients Database (rectal; suppository). Included in the Canadian Natural Health Products Ingredients Database.

17 Related Substances

Palm oil.

18 Comments

Palm kernel oil is used in soap production and the oleochemical industry.

A specification for palm kernel oil (unhydrogenated) is contained in the *Food Chemicals Codex*.[3]

The EINECS number for palm kernel oil is 232-425-4.

19 Specific References

1 Andersen AF. Final report on the safety assessment of *Elaeis guineensis* (palm) oil, *Elaeis guineensis* (palm) kernel oil, hydrogenated palm oil, and hydrogenated palm kernel oil. *Int J Toxicol* 2000; **19**: 7–28.

2 Lewis RJ, ed. *Sax's Dangerous Properties of Industrial Materials*, 11th edn. New York: Wiley, 2004; 2813.

3 *Food Chemicals Codex*, 7th edn. Bethesda, MD: United States Pharmacopeia, 2010; 755.

20 General References

Pantzaris TP, Ahmed MJ (eds). Malaysian Palm Oil Board. Properties and utilization of palm kernel oil. *Palm Oil Devel* 2001; **35**: 11–23.

Cornelius JA. Palm oil and palm kernel oil. *Prog Chem Fats Other Lipids* 1977; **15**: 5–27.

21 Author

JC Hooton.

22 Date of Revision

2 March 2012.

Palm Oil

1 Nonproprietary Names

USP–NF: Palm Oil

2 Synonyms

Cegesoft; *Elaeis guineensis* oil; palm butter; palm fat; palm grease; palm tallow.

3 Chemical Name and CAS Registry Number

Palm oil [8002-75-3]

4 Empirical Formula and Molecular Weight

The USP35–NF30 states that palm oil is the refined fixed oil obtained from the pulp of the fruit of the oil palm *Elaeis guineensis* (Fam. Aracaceae). It may contain suitable antioxidants.

5 Structural Formula

See Section 4.

6 Functional Category

Emollient.

7 Applications in Pharmaceutical Formulation or Technology

Palm oil is used as an emollient in pharmaceutical creams and ointments. It is also used as a constituent for suppositories.[1]

Palm oil is widely used in the food processing industry.

See also Section 18.

8 Description

Palm oil occurs as a white to yellowish fatty solid with a faint odor of violets.

9 Pharmacopeial Specifications

See Table I.

Table I: Pharmacopeial specifications for palm oil.

Test	USP35–NF30
Identification	+
Melting range	30–40°C
Acid value	≤2.0
Peroxide value	≤5.0
Unsaponifiable matter	≤1.0%
Fatty acid composition	+
Residue on ignition	≤0.1%
Water	≤0.1%
Heavy metals	≤0.001%
Alkaline impurities	+

10 Typical Properties

Density 0.952 g/cm^3

Refractive index n_D^{20} = 1.453–1.459

Solubility Soluble in ethanol (96%), ether, chloroform, and carbon disulfide. Insoluble in water.

11 Stability and Storage Conditions

Store in well-closed containers. Do not store above 55°C.

12 Incompatibilities

—

13 Method of Manufacture

Palm oil is obtained by boiling, centrifugation, and mechanical expression. It is refined, bleached, and deodorized to substantially remove free fatty acids, phospholipids, color, odor and flavor components, and other miscellaneous non-oil materials.

14 Safety

Palm oil is widely used in foods as well as topically in pharmaceutical formulations. It is also considered safe for use in cosmetic formulations.[2]

TD_{LO} (oral, rat): 55 g/kg

15 Handling Precautions

Observe normal precautions appropriate to the circumstances and quantity of the material handled. When heated to decomposition, palm oil emits acrid smoke and irritating fumes.[3]

16 Regulatory Status

Included in the Canadian Natural Health Products Ingredients Database.

17 Related Substances

Palm oil, hydrogenated; palm kernel oil.

18 Comments

Palm oil mixed with polysorbate 80 has been shown to enhance the oral bioavailability of the antioxidant astaxanthin.[4]

Palm oil is widely used in the manufacture of soaps, detergents, and other surfactants. Unrefined palm oil is rich in carotenoids, which are largely or wholly removed during standard refinement and processing.[5] The crude oil is deep orange-red with a characteristic odor, and is semisolid at 21–27°C.[5,6]

Palm oil is used for liver disorders in the Philippines (*Livermin*, Korea Ginseng Research).[5]

A specification for palm oil (unhydrogenated) is contained in the *Food Chemicals Codex*.[6]

The EINECS number for palm oil is 232-316-1.

19 Specific References

1 Allen LV, ed. *Suppositories*. London: Pharmaceutical Press, 2008; 28.

2 Andersen AF. Final report on the safety assessment of *Elaeis guineensis* (palm) oil, *Elaeis guineensis* (palm) kernel oil, hydrogenated palm oil, and hydrogenated palm kernel oil. *Int J Toxicol* 2000; **19**: 7–28.

3 Lewis RJ, ed. *Sax's Dangerous Properties of Industrial Materials*. 11th edn. New York: Wiley, 2004; 2813.

4 Odeberg JM, *et al*. Oral bioavailability of the antioxidant astaxanthin in humans is enhanced by incorporation of lipid based formulations. *Eur J Pharm Sci* 2003; **19**: 299–304.

5 Sweetman SC, ed. *Martindale: The Complete Drug Reference*. 37th edn. London: Pharmaceutical Press, 2011; 2235.

6 *Food Chemicals Codex*. 7th edn. Bethesda, MD: United States Pharmacopeia, 2010; 755.

20 General References

Cognis. and Product data sheet, Cognis. Product data sheet: *Cegesoft GPO*. http://www.cospha.ro/dbimg/Cegesoft%20GPO.pdf (accessed 25 July 2011).

Cornelius JA. Palm oil and palm kernel oil. *Prog Chem Fats Other Lipids* 1977; **15**: 5–27.

Edem DO. Palm oil: biochemical, physiological, nutritional, hematological, and toxicological aspects: a review. *Plant Foods Hum Nutr* 2002; **57**: 319–341.

Sundram K, *et al*. Palm fruit chemistry and nutrition. *Asia Pac J Clin Nutr* 2003; **12**: 355–362.

21 Authors

ME Fenton, RC Rowe.

22 Date of Revision

2 March 2012.

P

Palm Oil, Hydrogenated

1 Nonproprietary Names

USP–NF: Hydrogenated Palm Oil

2 Synonyms

Cegesoft; *Dynasan P60*.

3 Chemical Name and CAS Registry Number

Hydrogenated palm oil [68514-74-9]

4 Empirical Formula and Molecular Weight

Hydrogenated palm oil consists mainly of triglycerides of palmitic and stearic acids.

5 Structural Formula

See Section 4.

6 Functional Category

Emulsifying agent; tablet and capsule lubricant; viscosity-increasing agent.

7 Applications in Pharmaceutical Formulation or Technology

Hydrogenated palm oil is used in pharmaceutical formulations as a constituent of bases for suppositories, as an emulsifying agent and viscosity-increasing agent for creams, gels, and ointments, and as a lubricant and binder in tablet and capsule formulation.[1,2] It is also widely used in the cosmetics industry.

8 Description

Hydrogenated palm oil occurs as a white to yellowish, fatty solid to semi-solid.

9 Pharmacopeial Specifications

See Table I.

Table I: Pharmacopeial specifications for hydrogenated palm oil.

Test	USP35-NF30
Identification	+
Melting range	58–62°C
Acid value	≤2.0
Peroxide value	≤5.0
Unsaponifiable matter	≤0.8%
Fatty acid composition	+
Loss on drying	≤1.0%
Residue on ignition	≤0.1%
Heavy metals	≤0.001%
Alkaline impurities	+
Nickel	≤1.0 μg/g

10 Typical Properties

Solubility Practically insoluble in water; very slightly soluble in alcohol; freely soluble in ether.

11 Stability and Storage Conditions

Store in airtight containers. Protect from light.

12 Incompatibilities

—

13 Method of Manufacture

Hydrogenated palm oil is obtained by refining and hydrogenating the oil obtained from the pulp of the fruit of the oil palm *Elaeis guineensis*, and consists mainly of triglycerides of palmitic and stearic acids.

14 Safety

Hydrogenated palm oil is used in oral pharmaceutical formulations and is regarded as a safe material.

15 Handling Precautions

Observe normal precautions appropriate to the circumstances and quantity of the material handled.

16 Regulatory Status

Included in the FDA Inactive Ingredient Database (rectal suppositories; vaginal gels). Included in the Canadian Natural Health Products Ingredients Database.

17 Related Substances

Palm oil; palm kernel oil.

18 Comments

The EINECS number for hydrogenated palm oil 271-0563.

19 Specific References

1 Dash AK, Cudworth GC 2nd. Evaluation of an acetic acid ester of monoglyceride as a suppository base with unique properties. *AAPS PharmSciTech* 2001; 2(3): E13.

2 Sasol. Product information: Excipients for pharmaceuticals. http://www.sasoltechdata.com/MarketingBrochures/Excipients_Pharmaceuticals.pdf. (accessed 25 July 2011).

20 General References

Cornelius JA. Palm oil and palm kernel oil. *Prog Chem Fats Other Lipids* 1977; **15**: 5–27.

21 Author

RC Rowe.

22 Date of Revision

2 March 2012.

P

Paraffin

1 Nonproprietary Names

BP: Hard Paraffin
JP: Paraffin
PhEur: Paraffin, Hard
USP–NF: Paraffin

2 Synonyms

Hard wax; paraffinum durum; paraffinum solidum; paraffin wax.

3 Chemical Name and CAS Registry Number

Paraffin [8002-74-2]

4 Empirical Formula and Molecular Weight

Paraffin is a purified mixture of solid saturated hydrocarbons having the general formula C_nH_{2n+2}, and is obtained from petroleum or shale oil.

5 Structural Formula

See Section 4.

6 Functional Category

Coating agent; ointment base; stiffening agent.

7 Applications in Pharmaceutical Formulation or Technology

Paraffin is mainly used in topical pharmaceutical formulations as a component of creams and ointments. In ointments, it may be used to increase the melting point of a formulation or to add stiffness. Paraffin is additionally used as a coating agent for capsules and tablets, and is used in some food applications. Paraffin coatings can also be used to affect the release of drug from ion-exchange resin beads.[1]

8 Description

Paraffin is an odorless and tasteless, translucent, colorless, or white solid. It feels slightly greasy to the touch and may show a brittle fracture. Microscopically, it is a mixture of bundles of microcrystals. Paraffin burns with a luminous, sooty flame. When melted, paraffin is essentially without fluorescence in daylight; a slight odor may be apparent.

9 Pharmacopeial Specifications

See Table I.

10 Typical Properties

Density ≈0.84–0.89 g/cm³ at 20°C
Melting point Various grades with different specified melting ranges are commercially available.
Solubility Soluble in chloroform, ether, volatile oils, and most warm fixed oils; slightly soluble in ethanol; practically insoluble in acetone, ethanol (95%), and water. Paraffin can be mixed with most waxes if melted and cooled.
Spectroscopy
NIR spectra *see* Figure 1.

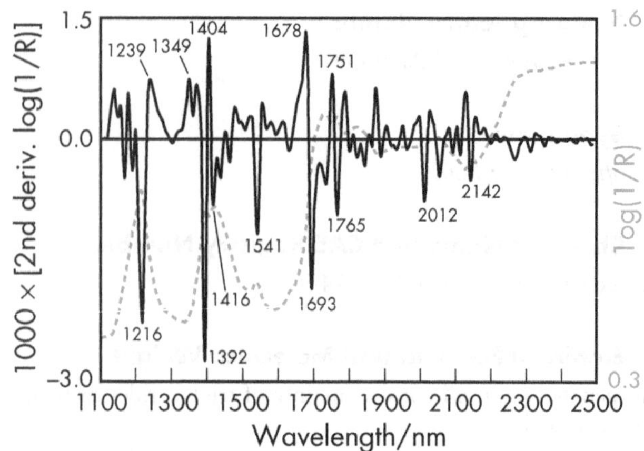

Figure 1: Near-infrared spectrum of paraffin measured by reflectance.

Table I: Pharmacopeial specifications for paraffin.

Test	JP XV	PhEur 7.4	USP35–NF30
Identification	+	+	+
Characters	—	+	—
Congealing range	50–75°C	—	47–65°C
Melting point	—	50–61°C	—
Heavy metals	≤10 ppm	—	—
Arsenic	≤2 ppm	—	—
Sulfates	+	≤150 ppm	+
Polycyclic aromatic hydrocarbons	—	+	+
Readily carbonizable substances	+	—	+
Acidity or alkalinity	+	+	+

11 Stability and Storage Conditions

Paraffin is stable, although repeated melting and congealing may alter its physical properties. Paraffin should be stored at a temperature not exceeding 40°C in a well-closed container.

12 Incompatibilities

—

13 Method of Manufacture

Paraffin is manufactured by the distillation of crude petroleum or shale oil, followed by purification by acid treatment and filtration. Paraffins with different properties may be produced by controlling the distillation and subsequent congealing conditions.

Synthetic paraffin, synthesized from carbon monoxide and hydrogen is also available; *see* Section 17.

14 Safety

Paraffin is generally regarded as an essentially nontoxic and nonirritant material when used in topical ointments and as a coating agent for tablets and capsules. However, granulomatous reactions (paraffinomas) may occur following injection of paraffin into tissue for cosmetic purposes or to relieve pain. Long-term inhalation of aerosolized paraffin may lead to interstitial pulmonary disease. Ingestion of a substantial amount of white soft paraffin has led to intestinal obstruction in one instance.[2–6]

See also Mineral Oil for further information.

15 Handling Precautions

Observe normal precautions appropriate to the circumstances and quantity of material handled. In the UK, the recommended workplace exposure limits for paraffin wax fumes are 2 mg/m^3 long-term (8-hour TWA) and 6 mg/m^3 short-term.[7]

16 Regulatory Status

Accepted in the UK for use in certain food applications. Included in the FDA Inactive Ingredients Database (oral capsules and tablets; topical emulsions; ointments). Included in nonparenteral medicines licensed in the UK. Included in the Canadian Natural Health Products Ingredients Database.

17 Related Substances

Light mineral oil; microcrystalline wax; petrolatum; synthetic paraffin.

Synthetic paraffin

Molecular weight 400–1400

Description A hard, odorless, white wax consisting of a mixture of mostly long-chain, unbranched, saturated hydrocarbons along with a small amount of branched hydrocarbons.

Melting point 96–105°C

Viscosity (dynamic) 5–15 mPa s (5–15 cP) at 135°C.

Comments The USP35–NF30 states that synthetic paraffin is synthesized by the Fischer–Tropsch process from carbon monoxide and hydrogen, which are catalytically converted to a mixture of paraffin hydrocarbons. The lower molecular weight fractions are removed by distillation and the residue is hydrogenated and further treated by percolation through activated charcoal. This mixture may be fractionated into its components by a solvent-separation method. Synthetic paraffin may contain not more than 0.005% w/w of a suitable antioxidant.

18 Comments

The more highly purified waxes are used in preference to paraffin in many applications because of their specifically controlled physical properties such as hardness, malleability, and melting range.

A specification for synthetic paraffin is contained in the *Food Chemicals Codex* (FCC).[8]

The EINECS numbers for paraffin are 232-315-6 and 265-154-5.

19 Specific References

1 Motyckas S, Nairn J. Influence of wax coatings on release rate of anions from ion-exchange resin beads. *J Pharm Sci* 1978; **67**: 500–503.
2 Crosbie RB, Kaufman HD. Self-inflicted oleogranuloma of breast. *Br Med J* 1967; **3**: 840–841.
3 Bloem JJ, van der Waal I. Paraffinoma of the face: a diagnostic and therapeutic problem. *Oral Surg* 1974; **38**: 675–680.
4 Greaney MG, Jackson PR. Oleogranuloma of the rectum produced by Lasonil ointment. *Br Med J* 1977; **2**: 997–998.
5 Pujol J, *et al.* Interstitial pulmonary disease induced by occupation exposure to paraffin. *Chest* 1990; **97**: 234–236.
6 Goh D, Buick R. Intestinal obstruction due to ingested Vaseline. *Arch Dis Child* 1987; **62**: 1167–1168.
7 Health and Safety Executive. *EH40/2005: Workplace Exposure Limits*. Sudbury: HSE Books, 2011. http://www.hse.gov.uk/pubns/priced/eh40.pdf (accessed 28 February 2012).
8 *Food Chemicals Codex*, 7th edn. Bethesda, MD: United States Pharmacopeia, 2010: 758.

20 General References

—

21 Author

AH Kibbe.

22 Date of Revision

2 March 2012.

Peanut Oil

1 Nonproprietary Names

BP: Arachis Oil

JP: Peanut Oil

PhEur: Arachis Oil, Refined

USP–NF: Peanut Oil

2 Synonyms

Aextreff CT; arachidis oleum raffinatum; earthnut oil; groundnut oil; katchung oil; nut oil.

3 Chemical Name and CAS Registry Number

Peanut oil [8002-03-7]

4 Empirical Formula and Molecular Weight

A typical analysis of refined peanut oil indicates the composition of the acids present as glycerides to be: arachidic acid 2.4%; behenic acid 3.1%; palmitic acid 8.3%; stearic acid 3.1%; lignoceric acid 1.1%; linoleic acid 26.0%, and oleic acid 56.0%.[1]

5 Structural Formula

See Section 4.

6 Functional Category

Oleaginous vehicle; solvent.

7 Applications in Pharmaceutical Formulation or Technology

Peanut oil is used as an excipient in pharmaceutical formulations primarily as a solvent for sustained-release intramuscular injections. It is also used as a vehicle for topical preparations and as a solvent for vitamins and hormones. In addition, it has been studied as part of sustained-release bead formulations,[2] nasal drug delivery systems,[3] controlled-release injectables,[4] and slow-release intramuscular injections.[5]

8 Description

Peanut oil is a colorless or pale yellow-colored liquid that has a faint nutty odor and a bland, nutty taste. At about 3°C it becomes cloudy, and at lower temperatures it partially solidifies.

9 Pharmacopeial Specifications

See Table I.

Table I: Pharmacopeial specifications for peanut oil.

Test	JP XV	PhEur 7.4	USP35–NF30
Identification	+	+	+
Characters	–	+	–
Solidification range	22–33°C	≈2°C	–
Acid value	≤0.2	≤0.5	≤0.2
Peroxide value	–	≤5.0	≤5.0
Unsaponifiable matter	≤1.5%	≤1.0%	≤1.5%
Specific gravity	0.909–0.916	≈0.915	–
Alkaline impurities	–	+	+
Rancidity	–	–	+
Iodine value	84–103	–	–
Saponification value	188–196	–	–
Heavy metals	–	–	≤10 ppm
Water	–	≤0.1%	≤0.1%
Free fatty acids	–	–	+
Composition of fatty acids	–	+	+
Saturated fatty acids $<C_{14}$	–	–	≤0.1%
Saturated fatty acids $<C_{16}$	–	≤0.4%	–
Myristic acid	–	–	≤0.2%
Palmitic acid	–	5.0–14.0%	7.0–16.0%
Palmitoleic acid	–	–	≤1.0%
Stearic acid	–	1.3–6.5%	1.3–6.5%
Oleic acid	–	35.0–72.0%	35.0–72.0%
Linoleic acid	–	12.0–43.0%	13.0–43.0%
Linolenic acid	–	≤0.6%	≤0.6%
Lignoceric acid	–	0.5–3.0%	0.5–3.0%
Arachidic acid	–	0.5–3.0%	0.5–3.0%
Eicosenoic acid	–	≤0.5–3.0%	≤0.5–2.1%
Behenic acid	–	1.0–5.0%	1.0–5.0%
Erucic acid	–	≤0.5%	≤0.5%

10 Typical Properties

Autoignition temperature 443°C
Density 0.915 g/cm^3 at 25°C
Flash point 283°C
Freezing point −5°C
Hydroxyl value 2.5–9.5
Interfacial tension 19.9 mN/m at 25°C[6]
Refractive index n_D^{25} = 1.466–1.470
Solubility Very slightly soluble in ethanol (95%); soluble in benzene, carbon tetrachloride, and oils; miscible with carbon disulfide, chloroform, ether, and hexane.
Spectroscopy
IR spectra *see* Figure 1.
Surface tension 37.5 mN/m at 25°C[6]
Viscosity (dynamic) 35.2 mPa s (35.2 cP) at 37°C[6]
Viscosity (kinematic) 39.0 mm^2/s (39.0 cSt) at 37°C[6]

11 Stability and Storage Conditions

Peanut oil is an essentially stable material.[7] However, on exposure to air it can slowly thicken and may become rancid. Solidified peanut oil should be completely melted and mixed before use. Peanut oil may be sterilized by aseptic filtration or by dry heat, for example, by maintaining it at 150°C for 1 hour.[8]

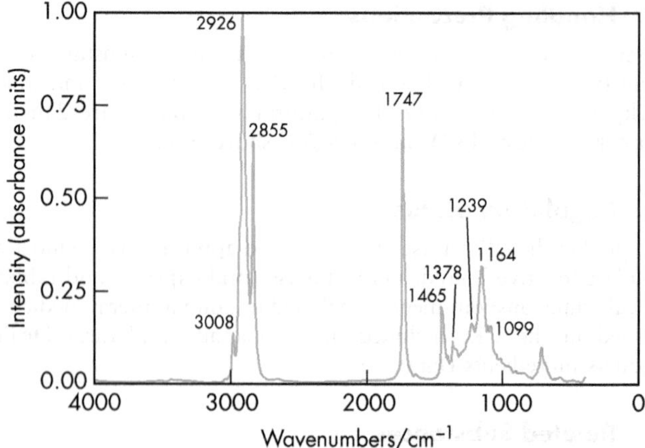

Figure 1: Infrared spectrum of peanut oil measured by transmission. Adapted with permission of Informa Healthcare.

Peanut oil should be stored in a well-filled, airtight, light-resistant container, at a temperature not exceeding 40°C. Material intended for use in parenteral dosage forms should be stored in a glass container.

12 Incompatibilities

Peanut oil may be saponified by alkali hydroxides.

13 Method of Manufacture

Refined peanut oil is obtained from the seeds of *Arachis hypogaea* Linné (Fam. Leguminosae). The seeds are separated from the peanut shells and are expressed in a powerful hydraulic press. The crude oil has a light yellow to light brown color, and is then purified to make it suitable for food or pharmaceutical purposes. A suitable antioxidant may be added.

14 Safety

Peanut oil is mildly laxative at a dosage of 15–60 mL orally or of 100–500 mL rectally as an enema.

Adverse reactions to peanut oil in foods and pharmaceutical formulations have been reported extensively.[9–19] These include severe allergic skin rashes[9,10] and anaphylactic shock following consumption of peanut butter.[11] Some workers have suggested that the use in infancy of preparations containing peanut oil, including infant formula and topical preparations, is associated with sensitization to peanut, with a subsequent risk of hypersensitivity reactions, and that such products should therefore be avoided or banned.[9–13] However, the role of pharmaceutical preparations in later development of hypersensitivity is disputed since such preparations contain highly refined peanut oil that should not contain the proteins associated with allergic reactions in susceptible individuals.[14–16]

Peanut oil is harmful if administered intravenously and it should not be used in such formulations.[17]

See also Section 18.

15 Handling Precautions

Observe normal handling precautions appropriate to the circumstances and quantity of material handled. Spillages of peanut oil are slippery and should be covered with an inert absorbent material prior to disposal.

16 Regulatory Status

Included in the FDA Inactive Ingredients Database (IM injections; topical preparations; oral capsules; vaginal emulsions). Included in parenteral and nonparenteral medicines licensed in the UK.

Included in the Canadian Natural Health Products Ingredients Database.

17 Related Substances

Almond oil; canola oil; corn oil; cottonseed oil; sesame oil; soybean oil; sunflower oil.

18 Comments

Therapeutically, emulsions containing peanut oil have been used in nutrition regimens, in enemas as a fecal softener, and in otic drops to soften ear wax. It is also administered orally, usually with sorbitol, as a gall bladder evacuant prior to cholecystography.

Peanut oil is also widely used as an edible oil. As a result of the potentially fatal reactions noted in Section 14, certain food products are now commonly labeled with a statement that they contain peanut oil.

A specification for unhydrogenated peanut oil is contained in the *Food Chemicals Codex* (FCC).[20]

The EINECS number for peanut oil is 232-296-4.

19 Specific References

1 Allen A, *et al.* Fatty acid composition of some soapmaking fats and oils. Part 4: Groundnut (peanut oil). *Soap Perfum Cosmet* 1969; **42**: 725–726.
2 Santucci E, *et al.* Gellan for the formulation of sustained delivery beads. *J Control Release* 1996; **42**: 157–164.
3 Maitani Y, *et al.* Modelling analysis of drug absorption and administration from ocular, naso-lacrimal duct, and nasal routes in rabbits. *Int J Pharm* 1995; **126**: 89–94.
4 Matsubara K, *et al.* Controlled release of the LHRH agonist buserelin acetate from injectable suspensions containing triacetylated cyclodextrins in an oil vehicle. *J Control Release* 1994; **31**: 173–180.
5 Liu KS, *et al.* Novel depots of buprenorphine have a long-acting effect for the management of physical dependence to morphine. *J Pharm Pharmacol* 2006; **58**(3): 337–344.
6 Howard JR, Hadgraft J. The clearance of oily vehicles following intramuscular and subcutaneous injections in rabbits. *Int J Pharm* 1983; **16**: 31–39.
7 Selles E, Ruiz A. [Study of the stability of peanut oil.] *Ars Pharm* 1981; **22**: 421–427[in Spanish].
8 Pasquale D, *et al.* A study of sterilizing conditions for injectable oils. *Bull Parenter Drug Assoc* 1964; **18**(3): 1–11.
9 Moneret-Vautrin DA, *et al.* Allergenic peanut oil in milk formulas [letter]. *Lancet* 1991; **338**: 1149.
10 Brown HM. Allergenic peanut oil in milk formulas [letter]. *Lancet* 1991; **338**: 1523.
11 De Montis G, *et al.* Sensitization to peanut and vitamin D oily preparations [letter]. *Lancet* 1993; **341**: 1411.
12 Lever LR. Peanut and nut allergy: creams and ointments containing peanut oil may lead to sensitisation. *Br Med J* 1996; **313**: 299.
13 Wistow S, Bassan S. Peanut allergy. *Pharm J* 1999; **262**: 709–710.
14 Hourihane JO, *et al.* Randomized, double blind, crossover challenge study of allergenicity of peanut oils in subjects allergic to peanuts. *Br Med J* 1997; **314**: 1084–1088.
15 Committee on Toxicity of Chemicals in Food. *Consumer Products and the Environment: Peanut Allergy.* London: Department of Health, 1998.
16 Anonymous. Questions raised over new advice following research into peanut oil. *Pharm J* 2001; **266**: 773.
17 Lynn KL. Acute rhabdomyolysis and acute renal failure after intravenous self-administration of peanut oil. *Br Med J* 1975; **4**: 385–386.
18 Ewan PW. Clinical study of peanut and nut allergy in 62 consecutive patients: new features and associations. *Br Med J* 1996; **312**: 1074–1078.
19 Tariq SM, *et al.* Cohort study of peanut and tree nut sensitisation by age of 4 years. *Br Med J* 1996; **313**: 514–517.
20 *Food Chemicals Codex*, 7th edn. Bethesda, MD: United States Pharmacopeia, 2010: 763.

20 General References

Strickley RG. Solubilizing excipients in oral and injectable formulations. *Pharm Res* 2004; **21**(2): 201–230.

21 Author

AH Kibbe.

22 Date of Revision

2 March 2012.

P

Pectin

1 Nonproprietary Names

USP–NF: Pectin

2 Synonyms

Citrus pectin; E440; *Genu*; methopectin; methyl pectin; methyl pectinate; mexpectin; pectina; pectinic acid.

3 Chemical Name and CAS Registry Number

Pectin [9000-69-5]

4 Empirical Formula and Molecular Weight

Pectin is a high-molecular-weight, carbohydrate-like plant constituent consisting primarily of chains of galacturonic acid units linked as 1,4-α-glucosides, with a molecular weight of 30 000–100 000.

5 Structural Formula

HO_2C O OH HO O H_3CO_2C O OH HO O n

Pectin is a complex polysaccharide comprising mainly esterified D-galacturonic acid residues in an α-(1–4) chain. The acid groups along the chain are largely esterified with methoxy groups in the natural product. The hydroxyl groups may also be acetylated.

Pectin gelation characteristics can be divided into two types: high-methoxy and low-methoxy gelation, and sometimes the low-methoxy pectins may contain amine groups. Gelation of high-methoxy pectin usually occurs at pH <3.5. Low-methoxy pectin is gelled with calcium ions and is not dependent on the presence of acid or high solids content. Amidation may interfere with gelation, causing the process to be delayed. However, gels from amidated pectins have the ability to re-heal after shearing.[1]

The USP35–NF30 describes pectin as a purified carbohydrate product obtained from the dilute acid extract of the inner portion of the rind of citrus fruits or from apple pomace. It consists chiefly of partially methoxylated polygalacturonic acids.

6 Functional Category

Emulsifying agent; gelling agent; viscosity-increasing agent.

7 Applications in Pharmaceutical Formulation or Technology

Pectin is used as a gelling agent in intranasal formulations to modify drug absorption characteristics.[2] It has also been used as an emulsion stabilizer.[3]

Experimentally, pectin has been used in gel formulations for the oral sustained delivery of ambroxol.[4] Pectin gel beads have been shown to be an effective medium for controlling the release of a drug within the gastrointestinal (GI) tract.[5] It has also been used in a colon-biodegradable pectin matrix with a pH-sensitive polymeric coating, which retards the onset of drug release, overcoming the problems of pectin solubility in the upper GI tract.[6–9] Amidated pectin matrix patches have been investigated for the transdermal delivery of chloroquine,[10] and gelling pectin formulations for the oral sustained delivery of acetaminophen (paracetamol) have been investigated *in situ*.[11] Pectin-based matrices with varying degrees of esterification have been evaluated as oral controlled-release tablets. Low-methoxy pectins were shown to have a release rate more sensitive to the calcium content of the formulation.[12] Pectins have been used as a component in the preparation of mixed polymer microsphere systems with the intention of producing controlled drug release.[13]

8 Description

Pectin occurs as a coarse or fine, yellowish-white, odorless powder that has a mucilaginous taste.

9 Pharmacopeial Specifications

See Table I.

Table I: Pharmacopeial specifications for pectin.

Test	USP35–NF30
Identification	+
Loss on drying	≤10.0%
Arsenic	≤3 ppm
Lead	≤50 ppm
Sugars and organic acids	+
Microbial limits	+
Assay	
Galacturonic acid	≥74.0%

10 Typical Properties

Acidity/alkalinity pH = 6.0–7.2

Solubility Soluble in water; insoluble in ethanol (95%) and other organic solvents.

Spectroscopy

NIR spectra *see* Figure 1.

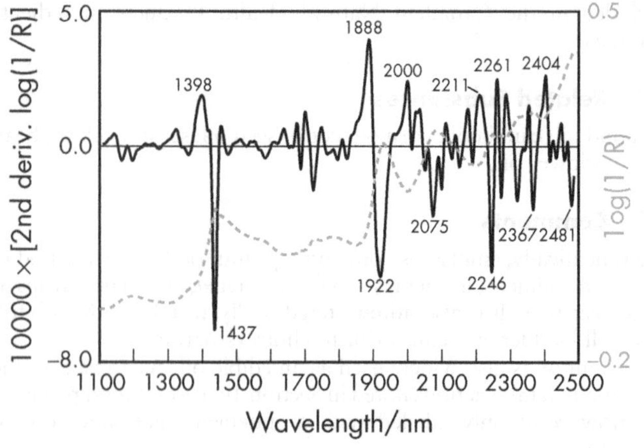

Figure 1: Near-infrared spectrum of pectin measured by reflectance.

11 Stability and Storage Conditions

Pectin is a nonreactive and stable material; it should be stored in a cool, dry place.

12 Incompatibilities

—

13 Method of Manufacture

Pectin is obtained from the diluted acid extract from the inner portion of the rind of citrus fruits or from apple pomace.

14 Safety

Pectin is used in oral pharmaceutical formulations and food products, and is generally regarded as an essentially nontoxic and nonirritant material.

Low toxicity by the subcutaneous route has been reported.[14]

LD_{50} (mouse, SC): 6.4 g/kg[14]

15 Handling Precautions

Observe normal precautions appropriate to the circumstances and quantity of material handled. When pectin is heated to decomposition, acrid smoke and irritating fumes are emitted.

16 Regulatory Status

GRAS listed. Accepted for use as a food additive in Europe. Included in the FDA Inactive Ingredients Database (dental paste; oral powders; topical pastes). Included in nonparenteral medicines licensed in the UK. Included in the Canadian Natural Health Products Ingredients Database.

17 Related Substances

—

18 Comments

Pectin has been used in film-coating formulations containing chitosan and hydroxypropylmethyl cellulose in the investigation of the biphasic drug-release properties of film-coated acetaminophen tablets, both *in vitro*,[15,16] and *in vivo*.[17] It has been shown that chitosan acts as a crosslinking agent for concentrated pectin solutions.[18]

Pectin gel systems have been used to show the partition and release of aroma compounds in foods during storage.[19]

A specification for pectins is included in the *Food Chemical Codex* (FCC).[20] In the food industry it is used as an emulsifying agent, gelling agent, thickener, and stabilizer. Cosmetically, it is used as a binder, emulsifying agent and viscosity-controlling agent.

Pectin has been used therapeutically as an adsorbent and bulk-forming agent, and is present in multi-ingredient preparations for the management of diarrhea, constipation, and obesity.[21]

The EINECS number for pectin is 232-553-0.

19 Specific References

1 Cybercolloids Ltd. *Introduction to pectins: properties*. www.cybercolloids.net/library/pectin/properties.php (accessed 30 November 2011).
2 Watts P, Smith A. PecSyst: *in situ* gelling system for optimised nasal drug delivery. *Expert Opin Drug Deliv* 2009; **6**(5): 543–552.
3 Lund W, ed. *The Pharmaceutical Codex: Principles and Practice of Pharmaceutics*, 12th edn. London: Pharmaceutical Press, 1994; 88.
4 Kubo W, *et al.* Oral sustained delivery of ambroxol from in-situ gelling pectin formulations. *Int J Pharm* 2004; **271**(1–2): 233–240.
5 Murata Y, *et al.* Drug release properties of a gel bead prepared with pectin and hydrolysate. *J Control Release* 2004; **95**(1): 61–66.
6 Sriamornsak P, *et al.* Composite film-coated tablets intended for colon-specific delivery of 5-aminosalicylic acid: using deesterified pectin. *Pharm Dev Technol* 2003; 8(3): 311–318.
7 Liu L, *et al.* Pectin-based systems for colon-specific drug delivery via oral route. *Biomaterials* 2003; **24**(19): 3333–3343.
8 Tho I, *et al.* Disintegrating pellets from a water-insoluble pectin derivative produced by extrusion/spheronisation. *Eur J Pharm Biopharm* 2003; **56**(3): 371–380.
9 Chourasia MK, Jain SK. Pharmaceutical approaches to colon targeted drug delivery systems. *J Pharm Pharm Sci* 2003; **6**(1): 33–66.
10 Musabayane CT, *et al.* Transdermal delivery of chloroquine by amidated pectin hydrogel matrix patch in the rat. *Ren Fail* 2003; 25(4): 525–534.
11 Kubo W, *et al. In situ* gelling pectin formulations for oral sustained delivery of paracetamol. *Drug Dev Ind Pharm* 2004; **30**(6): 593–599.
12 Sungthongjeen S, *et al.* Effect of degree of esterification of pectin and calcium amount on drug release from pectin-based matrix tablets. *AAPS Pharm Sci Tech* 2004; 5(1): E9.
13 Pillay V, *et al.* A crosslinked calcium-alginate–pectinate–cellulose acetophthalate gelisphere system for linear drug release. *Drug Delivery* 2002; **9**(2): 77–86.
14 Lewis RJ, ed. *Sax's Dangerous Properties of Industrial Materials*, 11th edn. New York: Wiley, 2004; 2825–2826.
15 Ofori-Kwakye K, Fell JT. Biphasic drug release from film-coated tablets. *Int J Pharm* 2003; **250**(2): 431–440.
16 Ofori-Kwakye K, Fell JT. Leaching of pectin from mixed films containing pectin, chitosan and HPMC intended for biphasic drug delivery. *Int J Pharm* 2003; **250**(1): 251–257.
17 Ofori-Kwake K, *et al.* Gamma scintigraphic evaluation of film-coated tablets intended for colonic or biphasic release. *Int J Pharm* 2004; **270**(1–2): 307–313.
18 Marudova M, *et al.* Pectin–chitosan interactions and gel formation. *Carbohydr Res* 2004; **339**(11): 1933–1939.
19 Hansson A, *et al.* Partition and release of 21 aroma compounds during storage of a pectin gel system. *J Agric Food Chem* 2003; **51**(7): 2000–2005.
20 *Food Chemicals Codex*, 7th edn. Bethesda, MD: United States Pharmacopeia, 2010.
21 Sweetman SC, ed. *Martindale: The Complete Drug Reference*, 37th edn. London: Pharmaceutical Press, 2011: 2236.

20 General References

Lofgren C, *et al.* Microstructure and rheological behavior of pure and mixed pectin gels. *Biomacromolecules* 2002; **3**(6): 1144–1153.

21 Author

W Cook.

22 Date of Revision

2 March 2012.

P

Pentetic Acid

1 Nonproprietary Names

USP–NF: Pentetic Acid

2 Synonyms

Acidicum penteticum; *N,N*-bis[2-[bis(carboxymethyl)amino]ethyl]-glycine; [[(carboxymethyl)imino]bis(ethylenenitrilo)]tetraacetic acid; diethylenetriamine pentaacetic acid; diethylenetriamine-*N,N,N′,N′,N″*-pentaacetic acid; (diethylenetrinitrilo)pentaacetic acid; DTPA; glycine, *N,N*-bis[2-[bis(carboxymethyl)amino]ethyl]; pentacarboxymethyl diethylenetriamine; *Versenex*; ZK-43649.

3 Chemical Name and CAS Registry Number

2-[Bis[2-(bis(carboxymethyl)amino)ethyl]amino]acetic acid [67-43-6]

4 Empirical Formula and Molecular Weight

$C_{14}H_{23}N_3O_{10}$ 393.35

5 Structural Formula

O OH HO N O OH N N O O HO OH O

6 Functional Category

Antimicrobial preservative; complexing agent.

7 Applications in Pharmaceutical Formulation or Technology

Pentetic acid is mainly used as a complexing agent in the preparation of imaging and contrast agents for radionuclide and magnetic resonance imaging.[1,2] It is also used as a carrier excipient for neutron-capture isotopes in, for example, radiotherapy.[3] Pentetic acid–isotope complexes have also been considered as surrogate active substances in scintigraphic imaging studies.[4] Pentetic acid has been used to produce antioxidant effects by forming complexes with metal ions to reduce formation of reactive oxygen species during lyophilization.[5]

See also Table I.

Table I: Uses of pentetic acid.[6]

Use	Concentration (%)
Antioxidant	0.1–0.3
Copreservative	0.05

8 Description

Pentetic acid occurs as a white crystalline solid and is almost odorless.

9 Pharmacopeial Specifications

See Table II.

Table II: Pharmacopeial specifications for pentetic acid.

Test	USP35–NF30
Identification	+
Residue on ignition	≤0.2%
Iron	≤0.01%
Heavy metals	≤0.005%
Nitrilotriacetic acid	≤0.1%
Assay	98.0–100.5%

10 Typical Properties

Acidity/alkalinity pH = 2.1–2.5 (1% w/v aqueous solution) for *Versenex*[7]

Melting point 220–222°C

Solubility 1 in 200 parts of water at 25°C for *Versenex*[7]

11 Stability and Storage Conditions

Pentetic acid should be stored in well-closed containers in a cool, dry place.

12 Incompatibilities

The activity of pentetic acid as a complexing agent may cause unwanted effects in formulations containing metal ions. The desired chelate may be displaced by other ions from the formulation.

13 Method of Manufacture

Pentetic acid is a pentaacetic acid triamine formed during the preparation of the amino carboxylic acid and its salt.[8]

14 Safety

Pentetic acid is used in intrathecal and intravenous injection preparations. The pure form of pentetic acid is moderately toxic by the intraperitoneal route.

LD_{50} (mouse, IP): 0.54 g/kg[9]
LD_{50} (mouse, oral): 4.84 g/kg[10]

15 Handling Precautions

Observe normal precautions appropriate to the circumstances and quantity of the material handled. When heated to decomposition pentetic acid emits acrid smoke and irritating fumes.

16 Regulatory Status

Included in the FDA Inactive Ingredients Database (intrathecal and intravenous injections). Included in intravenous and intra-articular injections licensed in the UK.

17 Related Substances

Calcium trisodium pentetate; pentasodium pentetate; zinc trisodium pentetate.

Calcium trisodium pentetate

Empirical formula $C_{14}H_{18}CaN_3Na_3O_{10}$

Molecular weight 497.35

CAS number [12111-24-9]

Synonyms Calcium DTPA; calcium pentetate; calcium trisodium DTPA; pentacin; pentetate calcium trisodium; trisodium calcium diethylenetriaminepentaacetate.

Regulatory status Included in the FDA Inactive Ingredients Database (intrathecal and intravenous injections).

Safety Calcium trisodium pentetate is moderately toxic by IV and IP routes. When heated to decomposition, it emits toxic fumes.

Solubility Soluble in water; practically insoluble in ethanol.

Comments The EINECS number for calcium trisodium pentetate is 235-169-1.

Pentasodium pentetate

Empirical formula $C_{14}H_{18}N_3Na_5O_{10}$

Molecular weight 503.25

CAS number [140-01-2]

Synonyms DTPAN; pentasodium DTPA; pentetate pentasodium; sodium DTPA.

Regulatory status Included in the FDA Inactive Ingredients Database (intravenous injections).

Safety Pentasodium pentetate is moderately irritating to the skin and mucous membranes.

Comments The EINECS number for pentasodium pentetate is 205-391-3.

Zinc trisodium pentetate

Synonyms Pentetate zinc trisodium; trisodium zinc diethylenetriaminepentaacetate; zinc DTPA (zinc pentetate or zinc trisodium pentetate).

Empirical formula $C_{14}H_{18}N_3Na_3O_{10}Zn$

Molecular weight 522.7

CAS number [65229-17-6] (zinc pentetate); [125833-02-5] (zinc trisodium pentetate)

18 Comments

The EINECS number for pentetic acid is 200-652-8. The PubChem Compound ID (CID) for pentetic acid is 3053.

Calcium and zinc pentetates have been evaluated as therapeutic agents for the removal of radioisotopes following occupational exposure.[11]

19 Specific References

1 Laniado M, *et al.* MR imaging of the gastrointestinal tract: value of Gd-DTPA. *Am J Roentgenol* 1988; **150**: 817–821.
2 Sinha VR, *et al. In vivo* evaluation of time and site of disintegration of polysaccharide tablet prepared for colon-specific drug delivery. *Int J Pharm* 2005; **289**: 79–85.
3 Le Uyen M, Cui Z. Long-circulating gadolinium-encapsulated liposomes for potential application in tumor neutron capture therapy. *Int J Pharm* 2006; **312**: 105–112.

4 Teran M, *et al.* Hydrophilic and lipophilic radiopharmaceuticals as tracers in pharmaceutical development: *in vitro–in vivo* studies. *BMC Nucl Med* 2005; 5: 5.
5 Molina MC, Anchordoquy TJ. Metal contaminants promote degradation of lipid/DNA complexes during lyophilisation. *Biochim Biophys Acta* 2007; **1768**: 669–677.
6 Dow Chemical Company. Product data sheet No. 113-01505-1207AMS: *Versene Chelating Agents*, 2007.
7 Dow Chemical Company. Technical data sheet No. 113-01346: *Versenex DTPA Acid*, June 2002.
8 Curme GO, Chitwood HC, Clark JW. Preparation of amino carboxylic acids and their salts. United States Patent 2,384,816; 1945.
9 Lewis RJ, ed. *Sax's Dangerous Properties of Industrial Materials*, 11th edn. New York: Wiley, 2004; 1271.
10 Mallinckrodt Baker. Safety data sheet: Diethylene triamine pentaacetic acid, 2006.
11 Bertelli L, *et al.* Three plutonium chelation cases at Los Alamos National Laboratory. *Health Phys* 2010; **99**(4): 532–538.

20 General References

—

21 Author

W Cook.

22 Date of Revision

2 March 2012.

Petrolatum

1 Nonproprietary Names

BP: Yellow Soft Paraffin

JP: Yellow Petrolatum

PhEur: Paraffin, Yellow Soft

USP–NF: Petrolatum

2 Synonyms

Merkur; mineral jelly; petroleum jelly; *Pinnacle*; *Silk*; *Silkolene*; *Snow White*; *Soft White*; vaselinum flavum; yellow petrolatum; yellow petroleum jelly.

3 Chemical Name and CAS Registry Number

2-[Bis[2-(bis(carboxymethyl)amino)ethyl]amino]acetic acid [8009-03-8]

4 Empirical Formula and Molecular Weight

Petrolatum is a purified mixture of semisolid saturated hydrocarbons having the general formula C_nH_{2n+2}, and is obtained from petroleum. The hydrocarbons consist mainly of branched and unbranched chains although some cyclic alkanes and aromatic molecules with paraffin side chains may also be present. The USP35–NF30 and PhEur 7.4 material may contain a suitable stabilizer (antioxidant) that must be stated on the label. The inclusion of a stabilizer is not discussed in the JP XV monograph.

5 Structural Formula

See Section 4.

6 Functional Category

Emollient; ointment base.

7 Applications in Pharmaceutical Formulation or Technology

Petrolatum is mainly used in topical pharmaceutical formulations as an emollient-ointment base; it is poorly absorbed by the skin. Petrolatum is also used in creams and transdermal formulations and as an ingredient in lubricant formulations for medicated confectionery together with mineral oil.

Petrolatum is additionally widely used in cosmetics and in some food applications. *See* Table I.

Table I: Uses of petrolatum.

Use	Concentration (%)
Emollient topical creams	10–30
Topical emulsions	4–25
Topical ointments	Up to 100

8 Description

Petrolatum is a pale yellow to yellow-colored, translucent, soft unctuous mass. It is odorless, tasteless, and not more than slightly fluorescent by daylight, even when melted.

9 Pharmacopeial Specifications

See Table II. *See also* Sections 17 and 18.

10 Typical Properties

Refractive index n_D^{60} = 1.460–1.474

Solubility Practically insoluble in acetone, ethanol, hot or cold ethanol (95%), glycerin, and water; soluble in benzene, carbon disulfide, chloroform, ether, hexane, and most fixed and volatile oils.

Spectroscopy

IR spectra *see* Figure 1.

NIR spectra *see* Figure 2.

Viscosity (dynamic) The rheological properties of petrolatum are determined by the ratio of the unbranched chains to the branched chains and cyclic components of the mixture. Petrolatum contains relatively high amounts of branched and cyclic hydrocarbons, in contrast to paraffin, which accounts for its softer character and makes it an ideal ointment base.[1–4]

11 Stability and Storage Conditions

Petrolatum is an inherently stable material owing to the unreactive nature of its hydrocarbon components; most stability problems occur because of the presence of small quantities of impurities. On

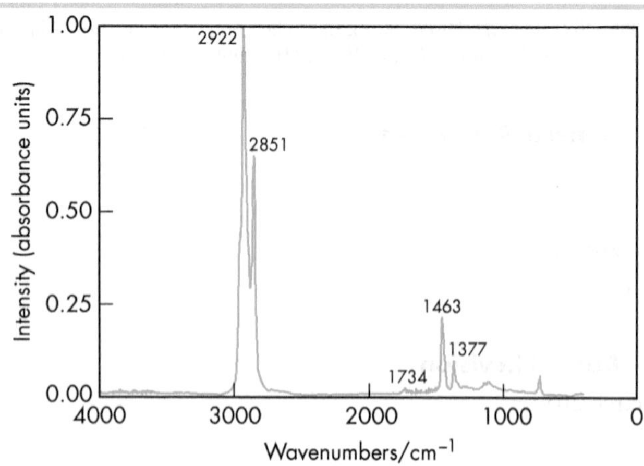

Figure 1: Infrared spectrum of petrolatum measured by transmission. Adapted with permission of Informa Healthcare.

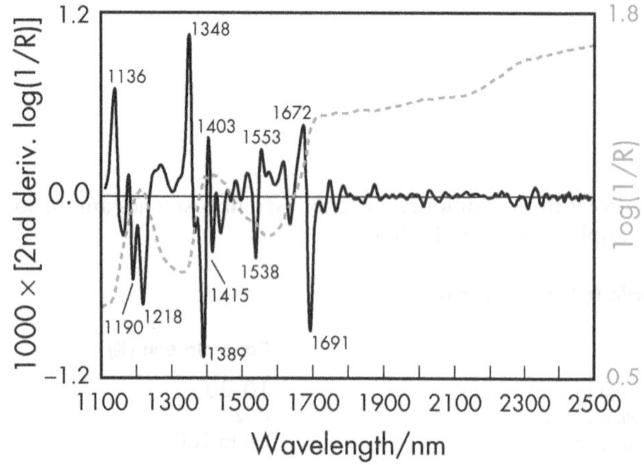

Figure 2: Near-infrared spectrum of petrolatum measured by reflectance.

Table II: Pharmacopeial specifications for petrolatum.

Test	JP XV	PhEur 7.4	USP35–NF30
Identification	–	+	–
Characters	–	+	–
Specific gravity at 60°C	–	–	0.815–0.880
Melting range	38–60°C	–	38–60°C
Drop point	–	40–60°C	–
Consistency	–	100–300	100–300
Alkalinity	+	+	+
Acidity	+	+	+
Residue on ignition	≤0.05%	–	≤0.1%
Sulfated ash	–	≤0.05%	–
Organic acids	+	–	+
Polycyclic aromatic hydrocarbons	–	+	–
Fixed oils, fats and resins	+	–	+
Color/appearance	+	+	+
Light absorption	–	+	–
Heavy metals	≤30 ppm	–	–
Arsenic	≤2 ppm	–	–
Sulfur compounds	+	–	–

exposure to light, these impurities may be oxidized to discolor the petrolatum and produce an undesirable odor. The extent of the oxidation varies depending upon the source of the petrolatum and the degree of refinement. Oxidation may be inhibited by the inclusion of a suitable antioxidant such as butylated hydroxyanisole, butylated hydroxytoluene, or alpha tocopherol.

Petrolatum should not be heated for extended periods above the temperature necessary to achieve complete fluidity (approximately 70°C). *See also* Section 18.

Petrolatum may be sterilized by dry heat. Although petrolatum may also be sterilized by gamma irradiation, this process affects the physical properties of the petrolatum such as swelling, discoloration, odor, and rheological behavior.[5,6]

Petrolatum should be stored in a well-closed container, protected from light, in a cool, dry place.

12 Incompatibilities

Petrolatum is an inert material with few incompatibilities.

13 Method of Manufacture

Petrolatum is manufactured from the semisolid residue that remains after the steam or vacuum distillation of petroleum.[7] This residue is dewaxed and/or blended with stock from other sources, along with lighter fractions, to give a product with the desired consistency. Final purification is performed by a combination of high-pressure hydrogenation or sulfuric acid treatment followed by filtration through adsorbents. A suitable antioxidant may be added.

14 Safety

Petrolatum is mainly used in topical pharmaceutical formulations and is generally considered to be a nonirritant and nontoxic material.

Animal studies, in mice, have shown petrolatum to be nontoxic and noncarcinogenic following administration of a single subcutaneous 100 mg dose. Similarly, no adverse effects were observed in a 2-year feeding study with rats fed a diet containing 5% of petrolatum blends.[8]

Although petrolatum is generally nonirritant in humans following topical application, rare instances of allergic hypersensitivity reactions have been reported,[9–11] as have cases of acne, in susceptible individuals following repeated use on facial skin.[12] However, given the widespread use of petrolatum in topical products, there are few reports of irritant reactions. The allergic components of petrolatum appear to be polycyclic aromatic hydrocarbons present as impurities. The quantities of these materials found in petrolatum vary depending upon the source and degree of refining. Hypersensitivity appears to occur less with white petrolatum and it is therefore the preferred material for use in cosmetics and pharmaceuticals.

Petrolatum has also been tentatively implicated in the formation of spherulosis of the upper respiratory tract following use of a petrolatum-based ointment packing after surgery,[13] and lipoid pneumonia following excessive use in the perinasal area.[14] Other adverse reactions to petrolatum include granulomas (paraffinomas) following injection into soft tissue.[15] Also, when taken orally, petrolatum acts as a mild laxative and may inhibit the absorption of lipids and lipid-soluble nutrients.

Petrolatum is widely used in direct and indirect food applications. In the US, the daily dietary exposure to petrolatum is estimated to be 0.404 mg/kg body-weight.[16]

For further information *see* Mineral Oil and Paraffin.

15 Handling Precautions

Observe normal precautions appropriate to the circumstances and quantity of material handled. For recommended workplace exposure limits *see* Mineral Oil and Paraffin.

16 Regulatory Status

GRAS listed. Accepted for use in certain food applications in many countries worldwide. Included in the FDA Inactive Ingredients Database (ophthalmic preparations; oral capsules and tablets; otic, topical, and transdermal preparations). Included in nonparenteral medicines licensed in the UK. Included in the Canadian Natural Health Products Ingredients Database.

17 Related Substances

Mineral oil; mineral oil light; paraffin; petrolatum and lanolin alcohols; white petrolatum.

White petrolatum

Synonyms Vaselinum album; white petroleum jelly; white soft paraffin.

Appearance White petrolatum is a white to pale yellow-colored, translucent, soft unctuous mass. It is odorless and tasteless, and not more than slightly fluorescent by daylight, even when melted.

Method of manufacture White petrolatum is petrolatum that has been highly refined so that it is wholly or nearly decolorized. It may contain a stabilizer.

Comments White petrolatum is listed in the JP XV, PhEur 7.4, and USP35–NF30. White petrolatum is one of the materials that have been selected for harmonization by the Pharmacopeial Discussion Group. For further information see the General Information Chapter <1196> in the USP35–NF30, the General Chapter 5.8 in PhEur 7.4, along with the 'State of Work' document on the PhEur EDQM website, and also the General Information Chapter 8 in the JP XV. White petrolatum is associated with fewer instances of hypersensitivity reactions and is the preferred petrolatum for use in cosmetics and pharmaceuticals, *see* Section 14.

18 Comments

Petrolatum is one of the materials that have been selected for harmonization by the Pharmacopeial Discussion Group. For further information see the General Information Chapter <1196> in the USP35–NF30, the General Chapter 5.8 in PhEur 7.4, along with the "State of Work" document on the PhEur EDQM website, and also the General Information Chapter 8 in the JP XV.

Various grades of petrolatum are commercially available, which vary in their physical properties depending upon their source and refining process. Petrolatum obtained from different sources may therefore behave differently in a formulation.(17)

Care is required in heating petrolatum because of its large coefficient of thermal expansion. It has been shown by both rheological and spectrophotometric methods that petrolatum undergoes phase transition at temperatures between 30–40°C.

Therapeutically, sterile gauze dressings containing petrolatum may be used for nonadherent wound dressings or as a packing material.(18)

Additives, such as microcrystalline wax, may be used to add body to petrolatum. A specification for petrolatum is contained in the *Food Chemicals Codex* (FCC).(19)

The EINECS number for petrolatum is 232-373-2.

19 Specific References

1 Boylan JC. Rheological estimation of the spreading characteristics of pharmaceutical semisolids. *J Pharm Sci* 1967; **56**: 1164–1169.
2 Longworth AR, French JD. Quality control of white soft paraffin. *J Pharm Pharmacol* 1969; **21**(Suppl.): 1S–5S.
3 Barry BW, Grace AJ. Grade variation in the rheology of white soft paraffin BP. *J Pharm Pharmacol* 1970; **22**(Suppl.): 147S–156S.
4 Barry BW, Grace AJ. Structural, rheological and textural properties of soft paraffins. *J Texture Studies* 1971; **2**: 259–279.
5 Jacob BP, Leupin K. [Sterilization of eye–nose ointments by gamma radiation]. *Pharm Acta Helv* 1974; **49**: 12–20 [in German].
6 Davis SS, *et al.* Effect of gamma radiation on rheological properties of pharmaceutical semisolids. *J Texture Studies* 1977; **8**: 61–80.
7 Schindler H. Petrolatum for drugs and cosmetics. *Drug Cosmet Ind* 1961; **89**(1): 36, 37, 76, 78–8082.
8 Oser BL, *et al.* Toxicologic studies of petrolatum in mice and rats. *Toxicol Appl Pharmacol* 1965; **7**: 382–401.
9 Dooms-Goossens A, Degreef H. Contact allergy to petrolatums I: sensitivity capacity of different brands of yellow and white petrolatums. *Contact Dermatitis* 1983; **9**: 175–185.
10 Dooms-Goossens A, Degreef H. Contact allergy to petrolatums II: attempts to identify the nature of the allergens. *Contact Dermatitis* 1983; **9**: 247–256.
11 Dooms-Goossens A, Dooms M. Contact allergy to petrolatums III: allergenicity prediction and pharmacopeial requirements. *Contact Dermatitis* 1983; **9**: 352–359.
12 Verhagen AR. Pomade acne in black skin [letter]. *Arch Dermatol* 1974; **110**: 465.
13 Rosai J. The nature of myospherulosis of the upper respiratory tract. *Am J Clin Pathol* 1978; **69**: 475–481.
14 Cohen MA, *et al.* Exogenous lipoid pneumonia caused by facial application of petrolatum. *JAMA* 2003; **49**: 1128–1130.
15 Crosbie RB, Kaufman HD. Self-inflicted oleogranuloma of breast. *Br Med J* 1967; **3**: 840–841.
16 Heimbach JT, *et al.* Dietary exposure to mineral hydrocarbons from food-use applications in the United States. *Food Chem Toxicol* 2002; **40**: 555–571.
17 Kneczke M, *et al.* *In vitro* release of salicylic acid from two different qualities of white petrolatum. *Acta Pharm Suec* 1986; **23**: 193–204.
18 Smack DP, *et al.* Infection and allergy incidence in ambulatory surgery patients using white petrolatum vs bacitracin ointment: randomized controlled trial. *JAMA* 1996; **276**: 972–977.
19 *Food Chemicals Codex*, 7th edn. Bethesda, MD: United States Pharmacopeia, 2010: 775.

20 General References

Bandelin FJ, Sheth BB. Semisolid preparations. In: Swarbrick J, Boylan JC, eds. *Encyclopedia of Pharmaceutical Technology*. vol. 14. New York: Marcel Dekker, 1996: 31–61.

Barker G. New trends in formulating with mineral oil and petrolatum. *Cosmet Toilet* 1977; **92**(1): 43–46.

Davis SS. Viscoelastic properties of pharmaceutical semisolids I: ointment bases. *J Pharm Sci* 1969; **58**: 412–418.

De Muynck C, *et al.* Chemical and physicochemical characterization of petrolatums used in eye ointment formulations. *J Pharm Pharmacol* 1993; **45**: 500–503.

De Rudder D, *et al.* Structural stability of ophthalmic ointments containing soft paraffin. *Drug Dev Ind Pharm* 1987; **13**: 1799–1806.

European Directorate for the Quality of Medicines and Healthcare (EDQM). European Pharmacopoeia – State Of Work Of International Harmonisation. *Pharmeuropa* 2010; 22(4): 583–584. www.edqm.eu/en/International-Harmonisation-614.html (accessed 1 December 2010).

Morrison DS. Petrolatum: a useful classic. *Cosmet Toilet* 1996; **111**(1): 59–6669.

Smolinske SC. *Handbook of Food, Drug, and Cosmetic Excipients*. Boca Raton, FL: CRC Press, 1992: 265–269.

Sucker H. Petrolatums: technological properties and quality assessment. *Cosmet Perfum* 1974; **89**(2): 37–43.

21 Author

WJ Lambert.

22 Date of Revision

2 March 2012.

Petrolatum and Lanolin Alcohols

1 Nonproprietary Names

None adopted.

2 Synonyms

Amerchol CAB; *Forlan 500*; petrolatum and wool alcohols; *Vilvanolin CAB*; white soft paraffin and lanolin alcohols; yellow soft paraffin and lanolin alcohols.

3 Chemical Name and CAS Registry Number

Petrolatum [8009-03-8] and
Lanolin alcohols [8027-33-6]

4 Empirical Formula and Molecular Weight

A mixture of petrolatum and lanolin alcohols.

5 Structural Formula

See Section 4.

6 Functional Category

Emollient; ointment base; plasticizing agent.

7 Applications in Pharmaceutical Formulation or Technology

Petrolatum and lanolin alcohols is a soft solid used in topical pharmaceutical formulations and cosmetics as an ointment base with emollient properties. It is also used in the preparation of creams and lotions.

See Table I.

Table I: Uses of petrolatum and lanolin alcohols.

Use	Concentration (%)
Absorption base component	10.0–50.0
Emollient and plasticizer in ointments	5.0–50.0

8 Description

Petrolatum and lanolin alcohols occurs as a pale ivory-colored, soft solid with a faint, characteristic sterol odor.

9 Pharmacopeial Specifications

Petrolatum and lanolin alcohols are listed as separate monographs in the BP, JP, PhEur, and also USP–NF but the combination is not listed; *see* individual monographs on Petrolatum and Lanolin Alcohols.

See also Section 18.

10 Typical Properties

Acid value ≤1 for *Vilvanolin CAB*
Arsenic ≤2 ppm for *Vilvanolin CAB*
Ash ≤0.2% for *Vilvanolin CAB*
Heavy metals ≤20 ppm for *Vilvanolin CAB*
Hydroxyl value 11–15 for *Vilvanolin CAB*
Iodine value 6–13 for *Forlan 500*
Melting range 46–53°C for *Forlan 500*; 40–46°C for *Vilvanolin CAB*.
Microbiological count The total bacterial count, when packaged, is less than 10 per gram of sample for *Vilvanolin CAB*.
Moisture content ≤0.25% for *Forlan 500*; ≤0.2% for *Vilvanolin CAB*.
Saponification value ≤2 for *Vilvanolin CAB*
Solubility Soluble 1 in 20 parts of chloroform, and 1 in 100 parts of mineral oil; precipitates at higher concentrations. Precipitation occurs in ethanol (95%), hexane, and water. May be dispersed in isopropyl palmitate. Forms a gel in castor oil and corn oil.
Specific gravity 0.96 at 25°C for *Forlan 500*

11 Stability and Storage Conditions

Petrolatum and lanolin alcohols is stable and should be stored in a well-closed container in a cool, dry place.

12 Incompatibilities

Petrolatum and lanolin alcohols is incompatible with coal tar, ichthammol, phenol, and resorcinol.

13 Method of Manufacture

Lanolin alcohols is blended with petrolatum.

14 Safety

Petrolatum and lanolin alcohols is generally regarded as an essentially nontoxic and nonirritant material. However, lanolin alcohols may be irritant to the skin and cause hypersensitivity in some individuals.

15 Handling Precautions

Observe normal precautions appropriate to the circumstances and quantity of material handled.

16 Regulatory Status

Accepted for use in topical pharmaceutical formulations and cosmetics. Included in the Canadian Natural Health Products Ingredients Database.

17 Related Substances

Lanolin alcohols; lanolin alcohols ointment; mineral oil and lanolin alcohols; petrolatum.

Lanolin alcohols ointment

Synonyms *Argobase EU*; wool alcohols ointment.
Appearance White-colored ointment if prepared using white petrolatum; a yellow-colored ointment if yellow petrolatum is used in its preparation.
Comments

The BP 2012 describes lanolin alcohols ointment (wool alcohols ointment BP) as a mixture consisting of:

Lanolin alcohols 60 g
Paraffin 240 g
Yellow or white petrolatum 100 g
Mineral oil 600 g

However, the proportions of paraffin, petrolatum, and mineral oil may be varied to produce an ointment of the desired physical properties.

18 Comments

Petrolatum is one of the materials that have been selected for harmonization by the Pharmacopeial Discussion Group. For further information see the General Information Chapter <1196> in the USP35–NF30, the General Chapter 5.8 in PhEur 7.4, along with the 'State of Work' document on the PhEur EDQM website, and also the General Information Chapter 8 in the JP XV.

Therapeutically, petrolatum and lanolin alcohols can be used to absorb wound exudates.

See individual monographs on Lanolin Alcohols and Petrolatum for further information.

19 Specific References

—

20 General References

Davis SS. Viscoelastic properties of pharmaceutical semisolids I: ointment bases. *J Pharm Sci* 1969; 58: 412–418.
European Directorate for the Quality of Medicines and Healthcare (EDQM). European Pharmacopoeia – State Of Work Of International Harmonisation. *Pharmeuropa* 2011; 23(4): 713–714. http://www.edqm.eu/site/-614.html (accessed 29 November 2011).
Lubrizol Corporation. Product literature: *Vivanolin CAB* Lanolin Product, 1 March 2006.
Rita Corporation. Product literature: *Forlan 500*, 31 May 2002.
Rita Corporation. Material safety data sheet: *Forlan 500*, 6 May 2004.

21 Author

W Cook.

22 Date of Revision

2 March 2012.

Phenol

1 Nonproprietary Names

BP: Phenol
JP: Phenol
PhEur: Phenol
USP–NF: Phenol

2 Synonyms

Carbolic acid; hydroxybenzene; oxybenzene; phenic acid; phenolum; phenyl hydrate; phenyl hydroxide; phenylic acid; phenylic alcohol.

3 Chemical Name and CAS Registry Number

Phenol [108-95-2]

4 Empirical Formula and Molecular Weight

C_6H_6O 94.11

5 Structural Formula

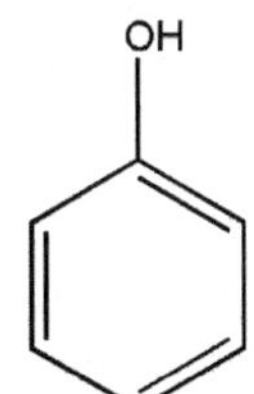

6 Functional Category

Antimicrobial preservative.

7 Applications in Pharmaceutical Formulation or Technology

Phenol is used mainly as an antimicrobial preservative in parenteral pharmaceutical products. It has also been used in topical pharmaceutical formulations and cosmetics; *see* Table I. Phenol should not be used to preserve preparations that are to be freeze-dried.[1]

Table I: Uses of phenol.

Use	Concentration (%)
Disinfectant	5.0
Injections (preservative)	0.5
Local anesthetic	0.5–1.0
Mouthwash	≤1.4

8 Description

Phenol occurs as colorless to light pink, caustic, deliquescent needle-shaped crystals or crystalline masses with a characteristic odor. When heated gently phenol melts to form a highly refractive liquid. The USP35–NF30 permits the addition of a suitable stabilizer; the name and amount of substance used for this purpose must be clearly stated on the label.

9 Pharmacopeial Specifications

See Table II.

Table II: Pharmacopeial specifications for phenol.

Test	JP XV	PhEur 7.4	USP35–NF30
Identification	+	+	+
Characters	–	+	–
Clarity of solution	+	+	+
Acidity	+	+	+
Congealing temperature	≈40°C	≥39.5°C	≥39°C
Water	–	–	≤0.5%
Nonvolatile residue	≤0.05%	≤0.05%	≤0.05%
Assay	≥98.0%	99.0–100.5%	99.0–100.5%

10 Typical Properties

Acidity/alkalinity pH = 6.0 (saturated aqueous solution)

Antimicrobial activity Phenol exhibits antimicrobial activity against a wide range of microorganisms such as Gram-negative and Gram-positive bacteria, mycobacteria and some fungi, and viruses; it is only very slowly effective against spores. Aqueous solutions of 1% w/v concentration are bacteriostatic, while stronger solutions are bactericidal. Phenol shows most activity in acidic solutions; increasing temperature also increases the antimicrobial activity. Phenol is inactivated by the presence of organic matter.

Autoignition temperature 715°C

Boiling point 181.8°C

Density 1.071 g/cm^3

Dissociation constant pK_a = 10 at 25°C

Flash point 79°C (closed cup)

Explosive limits 2% lower limit; 9% upper limit.

Freezing point 40.9°C

Melting point 43°C

Osmolarity A 2.8% w/v solution is iso-osmotic with serum.

Partition coefficient Octanol : water = 1.46

Refractive index n_D^{41} = 1.5425

Solubility *see* Table III.

Spectroscopy

IR spectra *see* Figure 1.

NIR spectra *see* Figure 2.

Vapor density (relative) 3.24 (air = 1)

Vapor pressure 133 Pa (1 mmHg) at 40°C

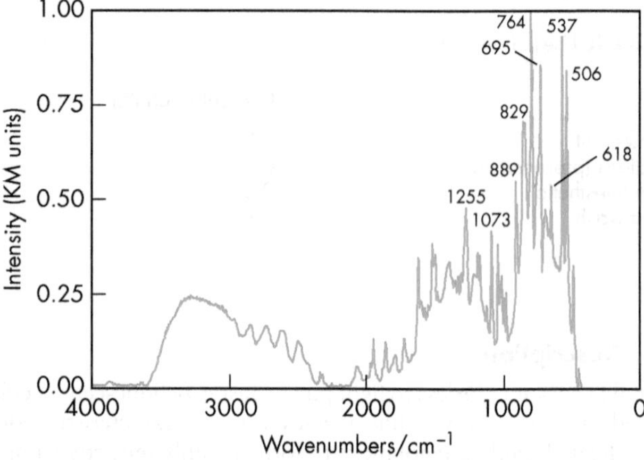

Figure 1: Infrared spectrum of phenol measured by diffuse reflectance. Adapted with permission of Informa Healthcare.

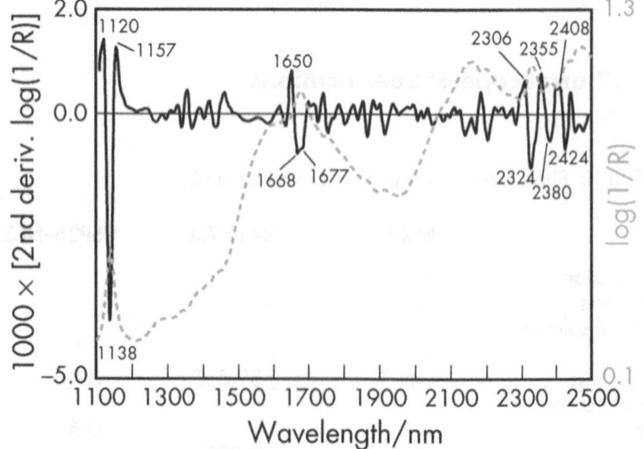

Figure 2: Near-infrared spectrum of phenol measured by reflectance.

11 Stability and Storage Conditions

When exposed to air and light, phenol turns a red or brown color, the color being influenced by the presence of metallic impurities. Oxidizing agents also hasten the color change. Aqueous solutions of phenol are stable. Oily solutions for injection may be sterilized in hermetically sealed containers by dry heat. The bulk material should be stored in a well-closed, light-resistant container at a temperature not exceeding 15°C.

Table III: Solubility of phenol.

Solvent	Solubility at 20°C
Carbon disulfide	Very soluble
Chloroform	Very soluble
Ethanol (95%)	Very soluble
Ether	Very soluble
Fixed oils	Very soluble
Glycerin	Very soluble
Mineral oil	1 in 70
Volatile oils	Very soluble
Water	1 in 15

12 Incompatibilities

Phenol undergoes a number of chemical reactions characteristic of alcohols; however, it possesses a tautomeric enol structure that is weakly acidic. It will form salts with sodium hydroxide or potassium hydroxide, but not with their carbonates or bicarbonates.

Phenol is a reducing agent and is capable of reacting with ferric salts in neutral to acidic solutions to form a greenish-colored complex. Phenol decolorizes dilute iodine solutions, forming hydrogen iodide and iodophenol; stronger solutions of iodine react with phenol to form the insoluble 2,4,6-triiodophenol.

Phenol is incompatible with albumin and gelatin as they are precipitated. It forms a liquid or soft mass when triturated with compounds such as camphor, menthol, thymol, acetaminophen, phenacetin, chloral hydrate, phenazone, ethyl aminobenzoate, methenamine, phenyl salicylate, resorcinol, terpin hydrate, sodium phosphate, or other eutectic formers. Phenol also softens cocoa butter in suppository mixtures.

13 Method of Manufacture

Historically, phenol was produced by the distillation of coal tar. Today, phenol is prepared by one of several synthetic methods, such as the fusion of sodium benzenesulfonate with sodium hydroxide followed by acidification; the hydrolysis of chlorobenzene by dilute sodium hydroxide at high temperature and pressure to give sodium phenate, which on acidification liberates phenol (Dow process); or the catalytic vapor-phase reaction of steam and chlorobenzene at 500°C (Raschig process).

14 Safety

Phenol is highly corrosive and toxic, the main effects being on the central nervous system. The lethal human oral dose is estimated to be 1 g for an adult.

Phenol is absorbed from the gastrointestinal tract, skin, and mucous membranes, and is metabolized to phenylglucuronide and phenyl sulfate, which are excreted in the urine.

Although there are a number of reports describing the toxic effects of phenol, these largely concern instances of accidental poisoning[2,3] or adverse reactions during its use as a therapeutic agent.[4,5] Adverse reactions associated with phenol used as a preservative are less likely owing to the smaller quantities that are used; however, it has been suggested that the body burden of phenol should not exceed 50 mg in a 10-hour period.[6] This amount could be exceeded following administration of large volumes of phenol-preserved medicines.

LD_{50} (mouse, IV): 0.11 g/kg[7]
LD_{50} (mouse, oral): 0.3 g/kg
LD_{50} (rabbit, skin): 0.85 g/kg
LD_{50} (rat, skin): 0.67 g/kg
LD_{50} (rat, oral): 0.32 g/kg
LD_{50} (rat, SC): 0.46 g/kg

15 Handling Precautions

Phenol is toxic on contact with the skin or if swallowed or inhaled. Phenol is strongly corrosive, producing possibly irreversible damage to the cornea and severe skin burns, although the skin burns are painless owing to the anesthetic effects of phenol.

Phenol should be handled with caution, particularly when hot, owing to the release of corrosive and toxic fumes. The use of fume cupboards, enclosed plants, or other environmental containment is recommended. Protective polyvinyl chloride or rubber clothing is recommended, together with gloves, eye protection, and respirators. Spillages on the skin or eyes should be washed with copious amounts of water. Affected areas of the skin should be washed with water followed by application of a vegetable oil. Medical attention should be sought.

Phenol poses a slight fire hazard when cold and a moderate hazard when hot and exposed to heat or flame.

In the UK, the workplace exposure limits for phenol are 2 ppm long-term (8-hour TWA).[8] In the US, the permissible exposure limit is 19 mg/m^3 long-term and the recommended exposure limits are 20 mg/m^3 long-term, and a maximum of 60 mg/m^3 short-term.

16 Regulatory Status

Included in the FDA Inactive Ingredients Database (IM, IV, and SC injections). Included in medicines licensed in the UK. Included in the Canadian Natural Health Products Ingredients Database.

17 Related Substances

Liquefied phenol.

Liquefied phenol

Appearance Liquefied phenol is phenol maintained as a liquid by the presence of approximately 10% water. It is a colorless liquid, with a characteristic aromatic odor, which may develop a red coloration on exposure to air and light. It contains less not less than 88% of phenol.
Specific gravity 1.065 at 25°C
Comments Liquefied phenol is often more convenient to use in a formulation than the crystalline form. However, liquefied phenol should not be used with fixed or mineral oils, although the crystalline solid may be used. Caution should be observed when handling liquified phenol to avoid contact with skin, as this could cause serious burns.

18 Comments

Although phenol is soluble in approximately 12 parts of water at ambient temperatures, larger amounts of phenol in water produce a two-phase system of phenol solution floating on a lower layer of wet phenol. At 20°C, 100 parts of phenol may be liquefied by the addition of 10 parts of water. At 84°C phenol is miscible with water in all proportions.

Therapeutically, phenol is widely used as an antiseptic and disinfectant.

The EINECS number for phenol is 203-632-7. The PubChem Compound ID (CID) for phenol is 996.

19 Specific References

1 FAO/WHO. WHO expert committee on biological standardization. Thirty-seventh report. *World Health Organ Tech Rep Ser* 1987; No. 760.
2 Foxall PJD, *et al.* Acute renal failure following accidental cutaneous absorption of phenol: application of NMR urinalysis to monitor the disease process. *Hum Toxicol* 1989; **9**: 491–496.
3 Christiansen RG, Klaman JS. Successful treatment of phenol poisoning with charcoal hemoperfusion. *Vet Hum Toxicol* 1996; **38**: 27–28.
4 Warner MA, Harper JV. Cardiac dysrhythmias associated with chemical peeling with phenol. *Anesthesiology* 1985; **62**: 366–367.
5 Ho SL, Hollinrake K. Acute epiglottitis and Chloraseptic. *BMJ* 1989; **298**: 1584.
6 Brancato DJ. Recognizing potential toxicity of phenol. *Vet Hum Toxicol* 1982; **24**: 29–30.
7 Lewis RJ, ed. *Sax's Dangerous Properties of Industrial Materials*, 11th edn. New York: Wiley, 2004; 2885.
8 Health and Safety Executive. *EH40/2005: Workplace Exposure Limits*. Sudbury: HSE Books, 2011. http://www.hse.gov.uk/pubns/priced/eh40.pdf (accessed 16 February 2012).

20 General References

Karabit MS. Studies on the evaluation of preservative efficacy V. Effect of concentration of micro-organisms on the antimicrobial activity of phenol. *Int J Pharm* 1990; **60**: 147–150.

21 Author

RT Guest.

22 Date of Revision

2 March 2012.

P

Phenoxyethanol

1 Nonproprietary Names

BP: Phenoxyethanol
PhEur: Phenoxyethanol
USP–NF: Phenoxyethanol

2 Synonyms

Arosol; *Dowanol EPh*; *Emeressence 1160*; ethyleneglycol monophenyl ether; β-hydroxyethyl phenyl ether; 1-hydroxy-2-phenoxyethane; *Phenoxen*; *Phenoxetol*; phenoxyethanolum; β-phenoxyethyl alcohol; *Phenyl Cellosolve*; *Protectol*.

3 Chemical Name and CAS Registry Number

2-Phenoxyethanol [122-99-6]

4 Empirical Formula and Molecular Weight

$C_8H_{10}O_2$ 138.16

5 Structural Formula

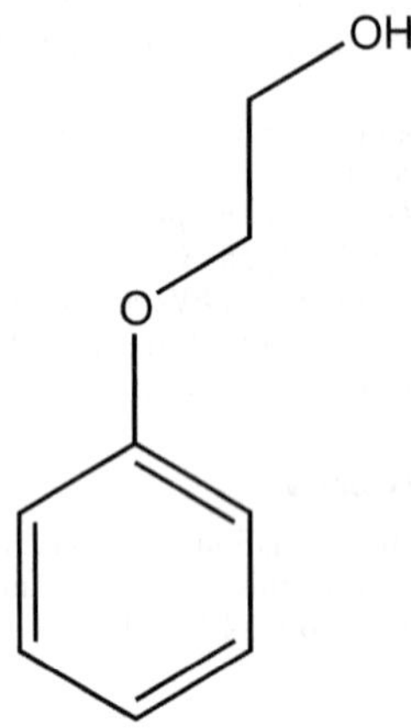

6 Functional Category

Antimicrobial preservative.

7 Applications in Pharmaceutical Formulation or Technology

Phenoxyethanol is an antimicrobial preservative used in cosmetics and topical pharmaceutical formulations at a concentration of 0.5–1.0%; it may also be used as a preservative and antimicrobial agent for vaccines.[1,2]

Phenoxyethanol has a narrow spectrum of activity and is thus frequently used in combination with other preservatives; *see* Section 10.

8 Description

Phenoxyethanol is a colorless, slightly viscous liquid with a faint pleasant odor and burning taste.

9 Pharmacopeial Specifications

See Table I.

Table I: Pharmacopeial specifications for phenoxyethanol.

Test	PhEur 7.4	USP35–NF30
Identification	+	+
Characters	+	–
Refractive index	1.537–1.539	–
Relative density	1.105–1.110	1.105–1.110
Phenol	≤0.1%	≤0.1%
Chromatographic purity	–	+
Related substances	+	–
Assay	99.0–100.5%	98.0–102.0%

10 Typical Properties

Acidity/alkalinity pH = 6.0 for a 1% v/v aqueous solution.

Antimicrobial activity Phenoxyethanol is an antibacterial preservative effective over a wide pH range against strains of *Pseudomonas aeruginosa* and to a lesser extent against *Proteus vulgaris* and other Gram-negative organisms. It is most frequently used in combination with other preservatives, such as parabens, to obtain a wider spectrum of antimicrobial activity.[3–5] *See also* Section 12. For reported minimum inhibitory concentrations (MICs) *see* Table II.[6,7]

Table II: Minimum inhibitory concentrations (MICs) of phenoxyethanol.

Microorganism	MIC (mg/mL)
Aspergillus niger ATCC 16404	2.5
Candida albicans ATCC 10231	3.2
Escherichia coli NCIB 9517	3.2
Pseudomonas aeruginosa NCTC 6750	10
Staphylococcus aureus ATCC 6538	7.5

Autoignition temperature 135°C
Boiling point 245.2°C
Dissociation constant $pK_a = 15.1$[8]
Flash point 121°C (open cup)
Melting point 14°C
Partition coefficients
Octanol : water = 1.16;[9]
Isopropyl palmitate : water = 2.9;[10]
Mineral oil : water = 0.3;[10]
Peanut oil : water = 2.6.[10]
Refractive index $n_D^{20} = 1.537–1.539$
Solubility *see* Table III.

Table III: Solubility of phenoxyethanol.

Solvent	Solubility at 20°C
Acetone	Miscible
Ethanol (95%)	Miscible
Glycerin	Miscible
Isopropyl palmitate	1 in 26
Mineral oil	1 in 143
Olive oil	1 in 50
Peanut oil	1 in 50
Water	1 in 43

Specific gravity 1.11 at 20°C

11 Stability and Storage Conditions

Aqueous phenoxyethanol solutions are stable and may be sterilized by autoclaving. The bulk material is also stable and should be stored in a well-closed container in a cool, dry place.

12 Incompatibilities

The antimicrobial activity of phenoxyethanol may be reduced by interaction with nonionic surfactants and possibly by absorption by polyvinyl chloride.[11] The antimicrobial activity of phenoxyethanol against *Pseudomonas aeruginosa* may be reduced in the presence of cellulose derivatives (methylcellulose, sodium carboxymethylcellulose, and hypromellose (hydroxypropylmethylcellulose)).[12]

13 Method of Manufacture

Phenoxyethanol is prepared by treating phenol with ethylene oxide in an alkaline medium.

14 Safety

Phenoxyethanol produces a local anesthetic effect on the lips, tongue, and other mucous membranes. The pure material is a moderate irritant to the skin and eyes. In animal studies, a 10% v/v solution was not irritant to rabbit skin and a 2% v/v solution was not irritant to the rabbit eye.[7] Long-term exposure to phenoxyethanol may result in CNS toxic effects similar to other organic solvents.[13] Safety issues related to preservatives used in vaccines, including 2-phenoxyethanol have been reviewed.[14] Contact urticaria has been reported upon exposure to 2-phenoxyethanol-containing cosmetics.[15]

The US FDA has recommended avoiding at least one topical product containing phenoxyethanol due to concerns over inadvertant exposure to nursing infants.[16]

LD_{50} (rabbit, skin): 5 g/kg[17]
LD_{50} (rat, oral): 1.26 g/kg[9]

15 Handling Precautions

Observe normal precautions appropriate to the circumstances and quantity of material handled. Phenoxyethanol may be irritant to the skin and eyes; eye protection and gloves are recommended.

16 Regulatory Status

Included in the FDA Inactive Ingredients Database (topical preparations). Included in nonparenteral medicines licensed in the UK. Included in the Canadian Natural Health Products Ingredients Database.

Under European regulations for cosmetics (76/768/EEC), the maximum authorized concentration (MAC) of 2-phenoxyethanol is 1.0%.[18]

17 Related Substances

Chlorobutanol; chlorophenoxyethanol; phenoxypropanol.

Chlorophenoxyethanol
Empirical formula $C_8H_9ClO_2$
Molecular weight 172.60
CAS number [29533-21-9]

Phenoxypropanol
Empirical formula $C_9H_{12}O_2$
Molecular weight 152.18
CAS number [4169-04-4]
Synonyms 1-Phenoxypropan-2-ol.

18 Comments

Aqueous solutions are best prepared by shaking phenoxyethanol with hot water until dissolved, followed by cooling and adjusting the volume to the required concentration.

Therapeutically, a 2.2% solution or 2.0% cream has been used as a disinfectant for superficial wounds, burns, and minor infections of the skin and mucous membranes.[19–21]

The EINECS number for phenoxyethanol is 204-589-7. The PubChem Compound ID (CID) for phenoxyethanol is 31236.

19 Specific References

1 Meyer BKH, *et al.* Antimicrobial preservative use in parenteral products. *J Pharm Sci* 2007; **96**: 3155–3167.
2 Lowe I, Southern J. The antimicrobial activity of phenoxyethanol in vaccines. *Lett Appl Microbiol* 1994; **18**(2): 115–116.
3 Abdelaziz AA, El-Nakeeb MA. Sporicidal activity of local anaesthetics and their binary combinations with preservatives. *J Clin Pharm Ther* 1988; **13**: 249–256.
4 Denyer SP, *et al.* Synergy in preservative combinations. *Int J Pharm* 1985; **25**: 245–253.
5 Onawunmi GO. *In vitro* studies on the antibacterial activity of phenoxyethanol in combination with lemon grass oil. *Pharmazie* 1988; **43**: 42–44.
6 Hall AL. Cosmetically acceptable phenoxyethanol. In: Kabara JJ, ed. *Cosmetic and Drug Preservation Principles and Practice.* New York: Marcel Dekker, 1984: 79–108.
7 BASF AG. Technical information: *Protectol S, Protectol PE S.* http://www.basf.co.kr/02_products/03_chemicals/specialty_chemicals/data/TI_Protectol%20PE_ES1494e_Jul2004.pdf (accessed 1 December 2011).
8 Serjeant EP, Dempsey B. *IUPAC Chemical Data Series*, No. 23: Ionization Constants of Organic Acids in Aqueous Solution, 1979.
9 *The Hazardous Substances Data Bank.* US National Library of Medicine (2008). http://toxnet.nlm.nih.gov (accessed 4 February 2009)
10 Anonymous. Final Report on the Safety Assessment of Phenoxyethanol. *J Amer Col Toxicol* 1990; **9**: 259–277.
11 Lee MG. Phenoxyethanol absorption by polyvinyl chloride. *J Clin Hosp Pharm* 1984; **9**: 353–355.
12 Kurup TRR, *et al.* Interaction of preservatives with macromolecules. Part II: cellulose derivatives. *Pharm Acta Helv* 1995; **70**(2): 187–193.
13 Morton WE. Occupational phenoxyethanol neurotoxicity: a report of three cases. *J Occup Med* 1990; **32**(1): 42–45.
14 Heidary N, Cohen DE. Hypsersensitivity reactions to vaccine components. *Dermatitis* 2005; **16**: 115–120.
15 Birnie AJ, English JS. 2-phenoxyethanol induced contact urticaria. *Contact Dermatitis* 2006; **54**: 349.
16 US FDA. FDA warns consumers against using Mommy's Bliss nipple cream (*FDA News* online, 2008). www.fda.gov/ (accessed 1 December 2011).
17 Lewis RJ, ed. *Sax's Dangerous Properties of Industrial Materials*, 11th edn. New York: Wiley, 2004; 2904.
18 European Commission (2008). Consolidated version of Cosmetics Directive 76/768/EECEU: Annex VI. *List of preservatives which cosmetics products may contain*, 2008: 112. ec.europa.eu/enterprise/cosmetics/html/consolidated_dir.htm (accessed 1 December 2011).
19 Thomas B. *et al*Sensitivity of urine-grown cells of *Providencia stuartii* to antiseptics. *J Clin Pathol* 1978; **31**: 929–932.
20 Lawrence JC. *et al*Evaluation of phenoxetol-chlorhexidine cream as a prophylactic antibacterial agent in burns. *Lancet* 1982; **1**(8280): 1037–1040.
21 Bollag U. Phenoxetol–chlorhexidine cream as a prophylactic antibacterial agent in burns [letter]. *Lancet* 1982; **2**(8289): 106.

20 General References

Baird RM. Proposed alternative to calamine cream BPC. *Pharm J* 1974; **213**(5781): 153–154.
Denyer SP, Baird RM, eds. *Guide to Microbiological Control in Pharmaceuticals.* Chichester: Ellis Horwood, 1990.
Fitzgerald KA, *et al.* Effect of chlorhexidine and phenoxyethanol, alone and in combination, on leakage from Gram-negative bacteria. *J Pharm Pharmacol* 1990; **42**(Suppl.): 104P.

P

Gilbert P, *et al.* The action of phenoxyethanol upon respiration and dehydrogenase enzyme systems in *Escherichia coli*. *J Pharm Pharmacol* 1976; **28**(Suppl.): 51P.
Hall AL. Phenoxyethanol: a cosmetically acceptable preservative. *Cosmet Toilet* 1981; **96**(3): 83–85.
Schnuch A, *et al.* Contact allergy to preservatives. Analysis of IVDK data 1996-2009 . *Br J Dermatol* 2011; **164**: 1316–1325.

21 Author

W Cook.

22 Date of Revision

2 March 2012.

Phenylethyl Alcohol

1 Nonproprietary Names

USP–NF: Phenylethyl Alcohol

2 Synonyms

Benzeneethanol; benzyl carbinol; benzylmethanol; β-fenylethanol; β-fenethylalkohol;β-hydroxyethyl benzene; PEA; phenethanol; β-phenylethyl alcohol; 2-phenylethyl alcohol; phenylethanol.

3 Chemical Name and CAS Registry Number

2-Phenylethanol [60-12-8]

4 Empirical Formula and Molecular Weight

$C_8H_{10}O$ 122.17

5 Structural Formula

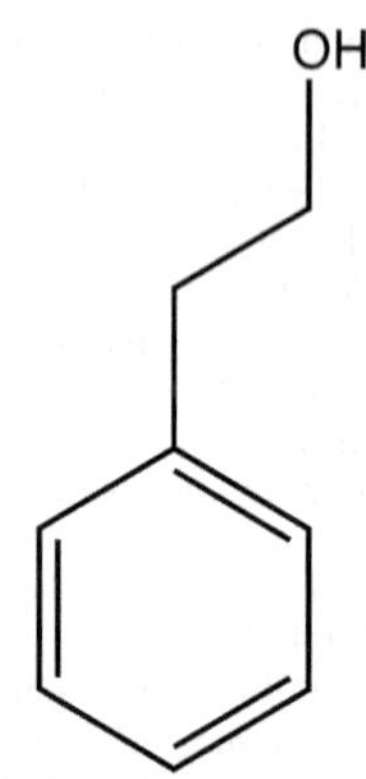

6 Functional Category

Antimicrobial preservative; flavoring agent.

7 Applications in Pharmaceutical Formulation or Technology

Phenylethyl alcohol is used as an antimicrobial preservative in nasal, ophthalmic, and otic formulations at 0.25–0.5% v/v concentration; it is generally used in combination with other preservatives.[1–3] Phenylethyl alcohol has also been used on its own as an antimicrobial preservative at concentrations up to 1% v/v in topical preparations. At this concentration, mycoplasmas are inactivated within 20 minutes, although enveloped viruses are resistant.[4] Phenylethyl alcohol is also used in flavors.

8 Description

Phenylethyl alcohol occurs as a clear, colorless liquid with an odor of rose oil. It has a burning taste that irritates and then anesthetizes mucous membranes.

9 Pharmacopeial Specifications

See Table I.

Table I: Pharmacopeial specifications for phenylethyl alcohol.

Test	USP35–NF30
Identification	+
Specific gravity	1.017–1.020
Refractive index	1.531–1.534
Residue on ignition	≤0.005%
Chlorinated compounds	+
Aldehyde	+

10 Typical Properties

Antimicrobial activity Phenylethyl alcohol has moderate antimicrobial activity although it is relatively slow acting; it is not sufficiently active to be used alone.[5] Greatest activity occurs at less than pH 5; it is inactive above pH 8. Synergistic effects have been reported when combined with benzalkonium chloride, chlorhexidine gluconate or diacetate, polymyxin B sulfate, and phenylmercuric nitrate.[6–10] With either benzalkonium chloride or chlorhexidine, synergistic effects were observed against *Pseudomonas aeruginosa* and apparently additive effects against Gram-positive organisms. With phenylmercuric nitrate, the effect was additive against *Pseudomonas aeruginosa*. Additive effects against *Pseudomonas cepacia* in combination with either benzalkonium chloride or chlorhexidine have also been reported.[11]
See also Section 12.
Bacteria Fair activity against Gram-positive bacteria; for *Staphylococcus aureus*, the minimum inhibitory concentration (MIC) may be more than 5 mg/mL. Greater activity is shown against Gram-negative organisms.[12] Typical MIC values are: *Salmonella typhi* 1.25 mg/mL; *Pseudomonas aeruginosa* 2.5 mg/mL; *Escherichia coli* 5.0 mg/mL.
Fungi Poor activity against molds and fungi.
Spores Inactive, e.g. at 0.6% v/v concentration, reported to be ineffective against spores of *Bacillus stearothermophilus* at 100°C for 30 minutes.

Boiling point 219–221°C
Flash point 102°C (open cup)
Melting point −27°C

Partition coefficients

Chloroform : water = 15.2;

Heptane : water = 0.58;

Octanol : water = 21.5.

Solubility *see* Table II.

Table II: Solubilty of phenylethyl alcohol.

Solvent	Solubility at 20°C
Benzyl benzoate	Very soluble
Chloroform	Very soluble
Diethyl phthalate	Very soluble
Ethanol (95%)	Very soluble
Ether	Very soluble
Fixed oils	Very soluble
Glycerin	Very soluble
Mineral oil	Slightly soluble
Propylene glycol	Very soluble
Water	1 in 60

11 Stability and Storage Conditions

Phenylethyl alcohol is stable in bulk, but is volatile and sensitive to light and oxidizing agents. It is reasonably stable in both acidic and alkaline solutions. Aqueous solutions may be sterilized by autoclaving. If stored in low-density polyethylene containers, phenylethyl alcohol may be absorbed by the containers. Losses to polypropylene containers have been reported to be insignificant over 12 weeks at 30°C. Sorption to rubber closures is generally small.

The bulk material should be stored in a well-closed container, protected from light, in a cool, dry place.

12 Incompatibilities

Incompatible with oxidizing agents and protein, e.g. serum. Phenylethyl alcohol is partially inactivated by polysorbates, although this is not as great as the reduction in antimicrobial activity that occurs with parabens and polysorbates.[13]

13 Method of Manufacture

Phenylethyl alcohol is prepared by reduction of ethyl phenylacetate with sodium in absolute alcohol; by hydrogenation of phenylacetaldehyde in the presence of a nickel catalyst; or by addition of ethylene oxide or ethylene chlorohydrin to phenylmagnesium bromide, followed by hydrolysis. Phenylethyl alcohol also occurs naturally in a number of essential oils, especially rose oil.

14 Safety

Phenylethyl alcohol is generally regarded as a nontoxic and nonirritant material. However, at the concentration used to preserve eye-drops (about 0.5% v/v) or above, eye irritation may occur.[14]

LD_{50} (rabbit, skin): 0.79 g/kg[15]

LD_{50} (rat, oral): 1.79 g/kg

15 Handling Precautions

Observe normal precautions appropriate to the circumstances and quantity of material handled. Phenylethyl alcohol is combustible when exposed to heat or flame, and emits acrid smoke when heated to decomposition. Eye protection and gloves are recommended.

16 Regulatory Status

Included in the FDA Inactive Ingredients Database (nasal, ophthalmic, and otic preparations). Included in nonparenteral medicines licensed in the UK. Included in the Canadian Natural Health Products Ingredients Database.

17 Related Substances

Chlorobutanol.

18 Comments

Phenylethyl alcohol is used as a perfumery component, especially in rose perfumes.

The EINECS number for phenylethyl alcohol is 200-456-2. The PubChem Compound ID (CID) for phenylethyl alcohol is 6054.

19 Specific References

1 Goldstein SW. Antibacterial agents in compounded ophthalmic solutions. *J Am Pharm Assoc (*Pract Pharm*)* 1953; **14**: 498–524.

2 Heller WM, *et al.* Preservatives in solutions. *J Am Pharm Assoc (Pract Pharm)* 1955; **16**: 29–36.

3 Hodges NA, *et al.* Preservative efficacy tests on formulated nasal products: reproducibility and factors affecting preservative activity. *J Pharm Pharmacol* 1996; **48**: 1237–1242.

4 Staal SP, Rowe WP. Differential effect of phenylethyl alcohol on mycoplasmas and enveloped viruses. *J Virol* 1974; **14**: 1620–1622.

5 Kohn SR, *et al.* Effectiveness of antibacterial agents presently employed in ophthalmic preparations as preservatives against *Pseudomonas aeruginosa*. *J Pharm Sci* 1963; **52**: 967–974.

6 Richards RME, McBride RJ. Cross-resistance in *Pseudomonas aeruginosa* resistant to phenylethanol. *J Pharm Sci* 1972; **61**: 1075–1077.

7 Richards RME, McBride RJ. The preservation of ophthalmic solutions with antibacterial combinations. *J Pharm Pharmacol* 1972; **24**: 145–148.

8 Richards RME, McBride RJ. Effect of 3-phenylpropan-1-ol, 2-phenylethanol, and benzyl alcohol on *Pseudomonas aeruginosa*. *J Pharm Sci* 1973; **62**: 585–587.

9 Richards RME, McBride RJ. Enhancement of benzalkonium chloride and chlorhexidine acetate activity against *Pseudomonas aeruginosa* by aromatic alcohols. *J Pharm Sci* 1973; **62**: 2035–2037.

10 Richards RME, McBride RJ. Antipseudomonal effect of polymyxin and phenylethanol. *J Pharm Sci* 1974; **63**: 54–56.

11 Richards RME, Richards JM. *Pseudomonas cepacia* resistance to antibacterials. *J Pharm Sci* 1979; **68**: 1436–1438.

12 Lilley BD, Brewer JH. The selective antibacterial action of phenylethyl alcohol. *J Am Pharm Assoc (Sci)* 1953; **42**: 6–8.

13 Bahal CK, Kostenbauder HB. Interaction of preservatives with macromolecules V: binding of chlorobutanol, benzyl alcohol, and phenylethyl alcohol by nonionic agents. *J Pharm Sci* 1964; **53**: 1027–1029.

14 Boer Y. Irritation by eyedrops containing 2-phenylethanol. *Pharm Weekbl (*Sci*)* 1981; **3**: 826–827.

15 Lewis RJ, ed. *Sax's Dangerous Properties of Industrial Materials*, 11th edn. New York: Wiley, 2004; 2879.

20 General References

Silver S, Wendt L. Mechanism of action of phenylethyl alcohol: breakdown of the cellular permeability barrier. *J Bacteriol* 1967; **93**: 560–566.

Sklubalova Z. Antimicrobial substances in ophthalmic drugs. *Ceska Slov Farm* 2004; 53(3): 107–116.

21 Author

W Cook.

22 Date of Revision

2 March 2012.

P

Phenylmercuric Acetate

1 Nonproprietary Names

BP: Phenylmercuric Acetate
PhEur: Phenylmercuric Acetate
USP–NF: Phenylmercuric Acetate

2 Synonyms

(Acetato-*O*)phenylmercury; acetoxyphenylmercury; phenylhydrargyri acetas; phenylmercury acetate; PMA; PMAC; PMAS.

3 Chemical Name and CAS Registry Number

(Acetato)phenylmercury [62-38-4]

4 Empirical Formula and Molecular Weight

$C_8H_8HgO_2$ 336.74

5 Structural Formula

HgO–C(=O)–CH_3 attached to benzene ring

6 Functional Category

Antimicrobial preservative.

7 Applications in Pharmaceutical Formulation or Technology

Phenylmercuric acetate is used as an alternative antimicrobial preservative to phenylmercuric borate or phenylmercuric nitrate in a limited range of cosmetics (in concentrations not exceeding 0.007% of mercury calculated as the metal) and pharmaceuticals. It may be used in preference to phenylmercuric nitrate owing to its greater solubility.

Phenylmercuric acetate is also used as a spermicide; *see* Table I. *See also* Phenylmercuric Nitrate.

Table I: Uses of phenylmercuric acetate.

Use	Concentration (%)
Bactericide in parenterals and eye-drops	0.001–0.002
Spermicide in vaginal suppositories and jellies (active ingredient)	0.02

8 Description

Phenylmercuric acetate occurs as a white to creamy white, odorless or almost odorless, crystalline powder or as small white prisms or leaflets.

9 Pharmacopeial Specifications

See Table II.

10 Typical Properties

Acidity/alkalinity pH ≈ 4 for a saturated aqueous solution at 20°C.

SEM 1: Excipient: phenylmercuric acetate; manufacturer: Eastman Fine Chemicals; magnification: 600×.

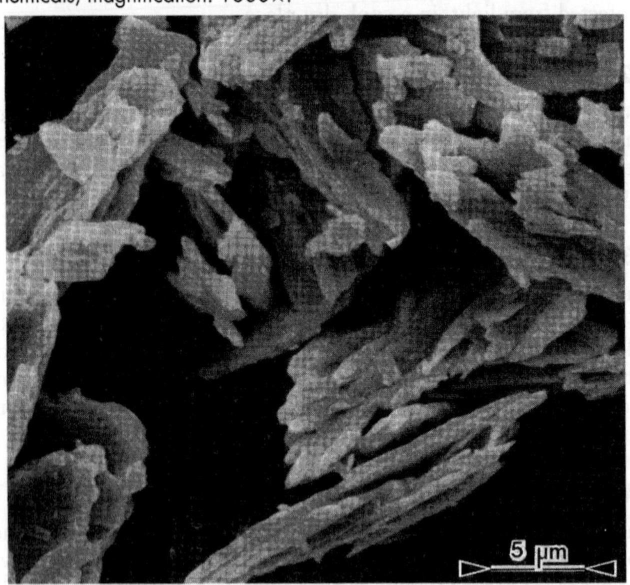

SEM 2: Excipient: phenylmercuric acetate; manufacturer: Eastman Fine Chemicals; magnification: 1800×.

Antimicrobial activity Phenylmercuric acetate is a broad-spectrum antimicrobial preservative with slow bactericidal and fungicidal activity similar to phenylmercuric nitrate; *see* Phenylmercuric Nitrate.

Dissociation constant $pK_a = 3.3$

Melting point 149°C

Partition coefficients Mineral oil : water = 0.1

Solubility *see* Table III.

Spectroscopy

NIR spectra *see* Figure 1.

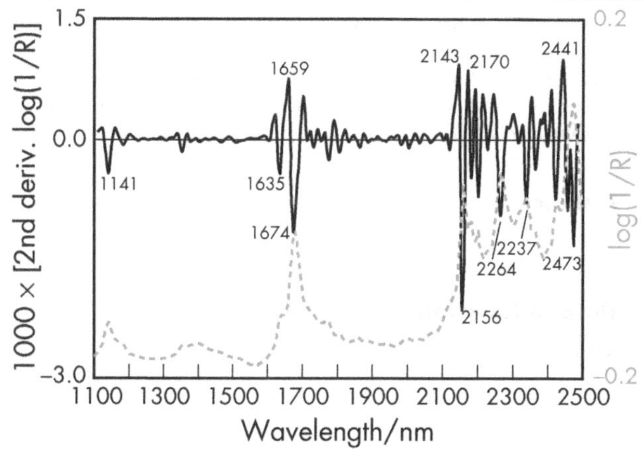

Figure 1: Near-infrared spectrum of phenylmercuric acetate measured by reflectance.

Table II: Pharmacopeial specifications for phenylmercuric acetate.

Test	PhEur 7.4	USP35–NF30
Identification	+	+
Characters	+	−
Appearance of solution	+	−
Ionized mercury	≤0.2%	−
Loss on drying	≤0.5%	−
Polymercurated benzene compounds	≤1.5%	≤1.5%
Melting range	−	149–153°C
Residue on ignition	−	≤0.2%
Mercuric salts and heavy metals	−	+
Assay	98.0–100.5%	98.0–100.5%

Table III: Solubility of phenylmercuric acetate.

Solvent	Solubility at 20°C[a]
Acetone	1 in 19
Chloroform	1 in 6.8
Ethanol (95%)	1 in 225
Ether	1 in 200
Water	1 in 180

(a) Compendial values for solubility vary considerably and in most instances do not show close agreement with laboratory-determined values, which also vary.

11 Stability and Storage Conditions

As for other phenylmercuric salts; *see* Phenylmercuric Nitrate.

Phenylmercuric acetate should be stored in a well-closed container, protected from light, in a cool, dry place.

12 Incompatibilities

As for other phenylmercuric salts; *see* Phenylmercuric Nitrate.

Incompatible with: halides; anionic emulsifying agents and suspending agents; tragacanth; starch; talc; sodium metabisulfite; sodium thiosulfate; disodium edetate; silicates; aluminum and other metals; amino acids; ammonia and ammonium salts; sulfur compounds; rubber; and some plastics.

Phenylmercuric acetate is reported to be incompatible with cefuroxime and ceftazidime.[1]

13 Method of Manufacture

Phenylmercuric acetate is readily formed by heating benzene with mercuric acetate.

14 Safety

Phenylmercuric acetate is mainly used as an antimicrobial preservative in topical pharmaceutical formulations. A number of adverse reactions to mercury-containing preservatives have been reported; *see* Phenylmercuric Nitrate.

LD_{50} (chicken, oral): 60 mg/kg[2]
LD_{50} (mouse, IP): 13 mg/kg
LD_{50} (mouse, IV): 18 mg/kg
LD_{50} (mouse, oral): 13 mg/kg
LD_{50} (mouse, SC): 12 mg/kg
LD_{50} (rat, oral): 41 mg/kg

15 Handling Precautions

Observe normal precautions appropriate to the circumstances and quantity of material handled. Phenylmercuric acetate may be irritant to the skin, eyes, and mucous membranes. Eye protection, gloves, and a respirator are recommended. Chronic exposure via any route can lead to central nervous system damage.

16 Regulatory Status

Included in the FDA Inactive Ingredients Database (ophthalmic ointments; topical emulsions/creams; vaginal emulsions/creams). Included in the Canadian Natural Health Products Ingredients Database (ophthalmic, nasal and otic preparations up to 0.004%; however there must be no other suitable alternative preservative available).

Phenylmercuric acetate is no longer permitted to be used as a pesticide in the US. Its use in cosmetic products in the US is limited to eye area cosmetics at not more than 0.0065% provided that there is no other suitable available preservative. It is specifically prohibited in vaginal contraceptive drug products and antimicrobial diaper rash drug products in the US. Phenylmercuric compounds are prohibited from use in cosmetic products in Canada.

In Europe, use in cosmetic products is limited to eye makeup and eye makeup remover at concentrations not exceeding 0.007% mercury alone or in combination with other permitted mercurial compounds.[3] In France, a maximum concentration of 0.01% is permitted for use in pharmaceuticals. The use of mercurial compounds in cosmetics in Japan is limited to concentrated shampoo or cream at not more than 0.003% Hg and eye makeup at not more than 0.0065% Hg.

P

17 Related Substances

Phenylmercuric borate; phenylmercuric nitrate; thimerosal.

18 Comments

The EINECS number for phenylmercuric acetate is 200-532-5.

19 Specific References

1 Hill DB, Barnes AR. Compatibility of phenylmercuric acetate with cefuroxime and ceftazidime eye drops. *Int J Pharm* 1997; **147**: 127–129.
2 Lewis RJ, ed. *Sax's Dangerous Properties of Industrial Materials*, 11th edn. New York: Wiley, 2004: 33–34.
3 Statutory Instruments (SI) 2004: No.2152. Consumer Protection: The Consumer Products (Safety) Regulations 2004. London: HMSO, 2004.

20 General References

Abdelaziz AA, El-Nakeeb MA. Sporicidal activity of local anaesthetics and their binary combinations with preservatives. *J Clin Pharm Ther* 1988; **13**: 249–256.
Barkman R, *et al.* Preservatives in eye drops. *Acta Ophthalmol* 1969; **47**: 461–475.

Grier N. Mercurials inorganic and organic. In: Block SS, ed. *Disinfection, Sterilization and Preservation*, 3rd edn. Philadelphia: Lea and Febiger, 1983: 346–374.

Hecht G. Ophthalmic preparations. In: Gennaro AR, ed. *Remington: The Science and Practice of Pharmacy*, 20th edn. Baltimore: Lippincott Williams and Wilkins, 2000: 821–835.

Parkin JE. The decomposition of phenylmercuric nitrate in sulphacetamide drops during heat sterilization. *J Pharm Pharmacol* 1993; **45**: 1024–1027.

Parkin JE, *et al*. The decomposition of phenylmercuric nitrate caused by disodium edetate in neomycin eye drops during the process of heat sterilization. *J Clin Pharm Ther* 1992; **17**: 191–196.

Parkin JE, *et al*. The chemical degradation of phenylmercuric nitrate by disodium edetate during heat sterilization at pH values commonly encountered in ophthalmic products. *J Clin Pharm Ther* 1992; **17**: 307–314.

Kodym A, *et al*. Influence of additives and storage temperature on physiological and microbiological peroperties of eye drops containing ceftazidime. *Acta Pol Pharm* 2006; **63**(6): 507–513.

21 Author

BR Matthews.

22 Date of Revision

2 March 2012.

Phenylmercuric Borate

1 Nonproprietary Names

BP: Phenylmercuric Borate
PhEur: Phenylmercuric Borate

2 Synonyms

(Dihydrogen borato)phenylmercury; phenylhydrargyri boras; phenylmercuriborate; phenylmercury borate; PMB.

3 Chemical Name and CAS Registry Number

[Orthoborato(3-)-O]-phenylmercurate(2-)dihydrogen [102-98-7]

The CAS Registry Number, chemical name and synonyms all refer to phenylmercuric borate alone, rather than the compound. The name phenylmercuric borate and the synonyms may, however, be applied to the PhEur 7.4 material, which is a compound or a mixture of compounds; *see* Section 4. Unique CAS Registry Numbers for phenylmercuric borate and the compounds are as follows:

$C_6H_7BHgO_3$ [102-98-7]
$C_{12}H_{13}BHg_2O_4$ [8017-88-7]
$C_{12}H_{11}BHg_2O_3$ [6273-99-0]

4 Empirical Formula and Molecular Weight

The PhEur 7.4 material is a compound consisting of equimolecular proportions of phenylmercuric hydroxide and phenylmercuric orthoborate ($C_{12}H_{13}BHg_2O_4$) or of the dehydrated form (metaborate, $C_{12}H_{11}BHg_2O_3$), or a mixture of the two compounds.

Phenylmercuric hydroxide and phenylmercuric orthoborate: $C_{12}H_{13}BHg_2O_4$ 633.2

Phenylmercuric hydroxide and phenylmercuric metaborate: $C_{12}H_{11}BHg_2O_3$ 615.2

5 Structural Formula

HgOB(OH)$_2$ HOHg

Phenylmercuric orthoborate and phenylmercuric hydroxide

HgOBO HOHg

Phenylmercuric metaborate and phenylmercuric hydroxide

6 Functional Category

Antimicrobial preservative.

7 Applications in Pharmaceutical Formulation or Technology

Phenylmercuric borate is used as an alternative antimicrobial preservative to phenylmercuric acetate or phenylmercuric nitrate. It is more soluble than phenylmercuric nitrate and has also been reported to be less irritant than either phenylmercuric acetate or phenylmercuric nitrate.[1] *See* Table I. *See also* Phenylmercuric Nitrate.

Table I: Uses of phenylmercuric borate.

Use	Concentration (%)
Antimicrobial agent in ophthalmics	0.002–0.004
Antimicrobial agent in parenterals	0.002

8 Description

Phenylmercuric borate occurs as colorless, shiny flakes or as a white or slightly yellow, odorless, crystalline powder.

9 Pharmacopeial Specifications

See Table II.

Table II: Pharmacopeial specifications for phenylmercuric borate.

Test	PhEur 7.4
Identification	+
Characters	+
Appearance of solution	+
Ionized mercury (as heavy metals)	≤0.01%
Loss on drying (at 45°C)	≤3.5%
Assay (dried basis) of	
Mercury	64.5–66.0%
Borates (as H_3BO_3)	9.8–10.3%

10 Typical Properties

Acidity/alkalinity pH = 5.0–7.0 for 0.6% w/v aqueous solution at 20°C.

Antimicrobial activity Phenylmercuric borate is a broad-spectrum antimicrobial preservative with slow bactericidal and fungicidal activity similar to that of phenylmercuric nitrate; *see* Phenylmercuric Nitrate.

Dissociation constant $pK_a = 3.3$

Melting point 112–113°C

Solubility *see* Table III.

Table III: Solubility of phenylmercuric borate.

Solvent	Solubility at 20°C[a] unless otherwise stated
Ethanol (95%)	1 in 150
Glycerin	Soluble
Propylene glycol	Soluble
Water	1 in 125
	1 in 100 at 100°C

(a) Compendial values for solubility vary considerably.

11 Stability and Storage Conditions

As for other phenylmercuric salts; *see* Phenylmercuric Nitrate. Solutions may be sterilized by autoclaving.

Phenylmercuric borate should be stored in a well-closed container, protected from light, in a cool, dry place.

12 Incompatibilities

As for other phenylmercuric salts; *see* Phenylmercuric Nitrate.

Incompatible with: halides; anionic emulsifying agents and suspending agents; tragacanth; starch; talc; sodium metabisulfite; sodium thiosulfate; disodium edetate; silicates; aluminum and other metals; amino acids; ammonia and ammonium salts; sulfur compounds; rubber; and some plastics.

13 Method of Manufacture

Phenylmercuric borate may be prepared by heating mercuric borate with benzene or by evaporating to dryness, under vacuum, an alcoholic solution containing equimolar proportions of phenylmercuric hydroxide and boric acid.

14 Safety

Phenylmercuric borate is mainly used as an antimicrobial preservative in topical pharmaceutical formulations. A number of adverse reactions to mercury-containing preservatives have been reported; *see* Phenylmercuric Nitrate.

Although phenylmercuric borate is an irritant, it has been reported to be less so than either phenylmercuric acetate or phenylmercuric nitrate.[1] There is, however, some cross-sensitization potential with other mercurial preservatives.

Systemic absorption has been reported following regular use of a hand disinfectant soap containing 0.04% phenylmercuric borate, resulting in an increase in the estimated total daily body load of mercury from 30–100 μg per 24 hours.[2]

15 Handling Precautions

Observe normal precautions appropriate to the circumstances and quantity of material handled. Phenylmercuric borate may be irritant to the skin, eyes, and mucous membranes. Eye protection, gloves, and a respirator are recommended.

16 Regulatory Status

Included in nonparenteral medicines licensed in Europe. Included in the Canadian Natural Health Products Ingredients Database (ophthalmic, nasal and otic preparations up to 0.004%; however there must be no other suitable alternative preservative).

In Europe, use is limited to eye makeup and eye makeup remover at not more than 0.007% mercury alone or in combination with other permitted mercurial substances.[3] In France, a maximum concentration of up to 0.01% is permitted for use in pharmaceutical formulations. Prohibited from use in cosmetic products in Canada. Prohibited in antimicrobial diaper rash drug products in the US. Limited use in Japan (*see* Phenylmercuric Nitrate).

17 Related Substances

Phenylmercuric acetate; phenylmercuric nitrate; thimerosal.

18 Comments

The EINECS number for phenylmercuric borate is 203-068-1.

19 Specific References

1 Marzulli FN, Maibach HI. Antimicrobials: experimental contact sensitization in man. *J Soc Cosmet Chem* 1973; 24: 399–421.

2 Peters-Haefeli L, *et al*. Urinary excretion of mercury after the use of an antiseptic soap containing 0.04% of phenylmercuric borate [in French]. *Schweiz Med Wochenschr* 1976; **106**(6): 171–178.

3 Statutory Instrument (SI) 2004: No. 2152. Consumer Protection: The Consumer Products (Safety) Regulations 2004. London: HMSO, 2004.

20 General References

Abdelaziz AA, El-Nakeeb MA. Sporicidal activity of local anaesthetics and their binary combinations with preservatives. *J Clin Pharm Ther* 1988; 13: 249–256.

Barkman R, *et al*. Preservatives in eye drops. *Acta Ophthalmol* 1969; 47: 461–475.

Grier N. Mercurials inorganic and organic. In: Block SS, ed. *Disinfection, Sterilization and Preservation*, 3rd edn. Philadelphia: Lea and Febiger, 1983: 346–374.

Hecht G. Ophthalmic preparations. In: Gennaro AR, ed. *Remington: The Science and Practice of Pharmacy*, 20th edn. Baltimore: Lippincott Williams and Wilkins, 2000: 821–835.

Parkin JE. The decomposition of phenylmercuric nitrate in sulphacetamide drops during heat sterilization. *J Pharm Pharmacol* 1993; 45: 1024–1027.

Parkin JE, *et al*. The decomposition of phenylmercuric nitrate caused by disodium edetate in neomycin eye drops during the process of heat sterilization. *J Clin Pharm Ther* 1992; 17: 191–196.

Parkin JE, *et al*. The chemical degradation of phenylmercuric nitrate by disodium edetate during heat sterilization at pH values commonly encountered in ophthalmic products. *J Clin Pharm Ther* 1992; 17: 307–314.

21 Author

BR Matthews.

22 Date of Revision

2 March 2012.

P

Phenylmercuric Nitrate

1 Nonproprietary Names

BP: Phenylmercuric Nitrate
PhEur: Phenylmercuric Nitrate
USP–NF: Phenylmercuric Nitrate

2 Synonyms

Basic phenylmercury nitrate; mercuriphenyl nitrate; merphenyl nitrate; nitratophenylmercury; phenylhydrargyri nitras; phenylmercury nitrate; *Phe-Mer-Nite*; PMN.

Note that the synonyms above are usually used to refer to phenylmercuric nitrate alone. However, confusion with nomenclature and CAS Registry Number has led to these synonyms also being applied to the PhEur 7.4 and USP35–NF30 material, which is a compound of phenylmercuric nitrate and phenylmercuric hydroxide.

3 Chemical Name and CAS Registry Number

There are two CAS Registry Numbers associated with phenylmercuric nitrate. One refers to the mixture of phenylmercuric nitrate and phenylmercuric hydroxide ($C_{12}H_{11}Hg_2NO_4$) while the other refers to phenylmercuric nitrate alone ($C_6H_5HgNO_3$). The PhEur 7.4, and USP35–NF30 use the name phenylmercuric nitrate to describe the mixture and use the CAS Registry Number [55-68-5].

Hydroxyphenylmercury mixture with (nitrato-O)phenylmercury:
$C_{12}H_{11}Hg_2NO_4$ [8003-05-2]
(Nitrato-O)phenylmercury:
$C_6H_5HgNO_3$ [55-68-5]

4 Empirical Formula and Molecular Weight

$C_{12}H_{11}Hg_2NO_4$ 634.45

5 Structural Formula

C_6H_5–$HgNO_3$ HOHg–C_6H_5

6 Functional Category

Antimicrobial preservative.

7 Applications in Pharmaceutical Formulation or Technology

Phenylmercuric salts are used as antimicrobial preservatives mainly in ophthalmic preparations, but are also used in cosmetics (*see* Section 16), parenteral, and topical pharmaceutical formulations; *see* Table I.

Phenylmercuric salts are active over a wide pH range against bacteria and fungi and are usually used in neutral to alkaline solutions, although they have also been used effectively at slightly acid pH; *see* Section 10. In acidic formulations, phenylmercuric nitrate may be preferred to phenylmercuric acetate or phenylmercuric borate as it does not precipitate.

Phenylmercuric nitrate is also an effective spermicide, although its use in vaginal contraceptives is no longer recommended; *see* Section 14.

A number of adverse reactions to phenylmercuric salts have been reported, and concern at the toxicity of mercury compounds may preclude the use of phenylmercuric salts under certain circumstances; *see* Section 14.

Table I: Uses of phenylmercuric nitrate.

Use	Concentration (%)
Bactericide in parenterals and eye drops	0.001–0.002
Bactericide/spermacide in vaginal suppositories and jellies	0.02

8 Description

Phenylmercuric nitrate PhEur 7.4 and USP35–NF30 is an equimolecular compound of phenylmercuric hydroxide and phenylmercuric nitrate; it occurs as a white, crystalline powder with a slight aromatic odor.

9 Pharmacopeial Specifications

See Table II.

Table II: Pharmcopeial specifications for phenylmercuric nitrate.

Test	PhEur 7.4	USP35–NF30
Identification	+	+
Characters	+	—
Appearance of solution	+	—
Loss on drying	≤1.0%	—
Residue on ignition	—	≤0.1%
Mercury ions	—	+
Inorganic mercuric compounds	+	—
Assay (dried basis) of:		
Mercury	62.5–64.0%	62.75–63.50%
Phenylmercuric ion	—	87.0–87.9%

10 Typical Properties

Acidity/alkalinity A saturated aqueous solution is acidic to litmus.

Antimicrobial activity Phenylmercuric salts are broad-spectrum, growth-inhibiting agents at the concentrations normally used for the preservation of pharmaceuticals. They possess slow bactericidal and fungicidal activity. Antimicrobial activity tends to increase with increasing pH, although in solutions of pH 6 and below, activity against *Pseudomonas aeruginosa* has been demonstrated. Phenylmercuric salts are included in several compendial eye drop formulations of acid pH.

Activity is also increased in the presence of phenylethyl alcohol, and in the presence of sodium metabisulfite at acid pH. Activity is decreased in the presence of sodium metabisulfite at alkaline pH.(1–3) When used as preservatives in topical creams, phenylmercuric salts are active at pH 5–8.(4)

Bacteria (Gram-positive): ***Staphylococcus aureus*** Good inhibition, more moderate cidal activity. Minimum inhibitory concentration (MIC) against *Staphylococcus aureus* is 0.5 μg/mL.

Bacteria (Gram-negative): ***Pseudomonas aeruginosa*** Inhibitory activity for most Gram-negative bacteria is similar to that for Gram-positive bacteria (MIC is approximately 0.3–0.5 μg/mL). Phenylmercuric salts are less active against

SEM 1: Excipient: phenylmercuric nitrate; manufacturer: Eastman Fine Chemicals; magnification: 180×.

SEM 2: Excipient: phenylmercuric nitrate; manufacturer: Eastman Fine Chemicals; magnification: 1800×.

some *Pseudomonas* species, and particularly *Pseudomonas aeruginosa* (MIC is approximately 12 µg/mL).

Fungi: *Candida albicans* and *Aspergillus niger* Most fungi are inhibited by 0.3–1 µg/mL; phenylmercuric salts exhibit both inhibitory and fungicidal activity; e.g. for phenylmercuric acetate against *Candida albicans*, MIC is 0.8 µg/mL; for phenylmercuric acetate against *Aspergillus niger*, MIC is approximately 10 µg/mL.

Spores Phenylmercuric salts may be active in conjunction with heat. The BP 1980 included heating at 100°C for 30 minutes in the presence of 0.002% w/v phenylmercuric acetate or phenylmercuric nitrate as a sterilization method. However, in practice this may not be sufficient to kill spores and heating with a bactericide no longer appears as a sterilization method in the BP 2012.

Dissociation constant $pK_a = 3.3$

Melting point 187–190°C with decomposition.

Partition coefficients

Mineral oil : water = 0.58;

Peanut oil : water = 0.4.

Solubility More soluble in the presence of either nitric acid or alkali hydroxides. *See* Table III.

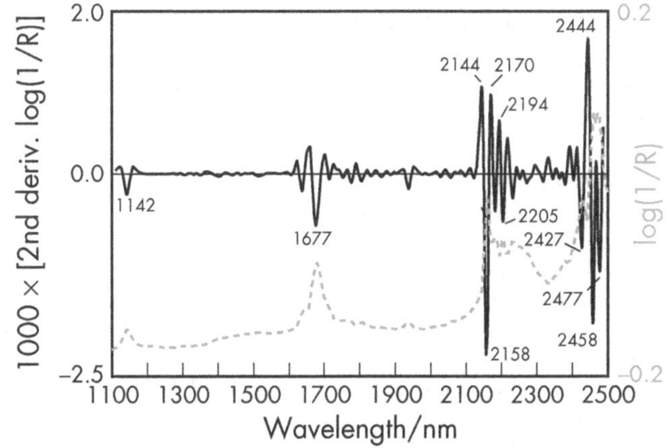

Figure 1: Near-infrared spectrum of phenylmercuric nitrate measured by reflectance.

Table III: Solubility of phenylmercuric nitrate.

Solvent	Solubility at 20°C[(a)] unless otherwise stated
Ethanol (95%)	1 in 1000
Fixed oils	Soluble
Glycerin	Slightly soluble
Water	1 in 600–1500
	1 in 160 at 100°C

(a) Compendial values for solubility vary considerably.

Spectroscopy

NIR spectra *see* Figure 1.

11 Stability and Storage Conditions

All phenylmercuric compound solutions form a black residue of metallic mercury when exposed to light or after prolonged storage. Solutions may be sterilized by autoclaving, although significant amounts of phenylmercuric salts may be lost, hence reducing preservative efficacy, owing to incompatibilities with packaging components or other excipients, e.g. sodium metabisulfite.[(5–7)] *See* Section 12.

Phenylmercuric nitrate should be stored in a well-closed container, protected from light, in a cool, dry place.

12 Incompatibilities

The antimicrobial activity of phenylmercuric salts may be reduced in the presence of anionic emulsifying agents and suspending agents, tragacanth, starch, talc, sodium metabisulfite,[(8)] sodium thiosulfate,[(2)] disodium edetate,[(2)] and silicates (bentonite, aluminum magnesium silicate, magnesium trisilicate, and kaolin).[(9,10)]

Phenylmercuric salts are incompatible with halides, particularly bromides and iodides, as they form less-soluble halogen compounds. At concentrations of 0.002% w/v precipitation may not occur in the presence of chlorides. Phenylmercuric salts are also incompatible with aluminum and other metals, ammonia and ammonium salts, amino acids, and with some sulfur compounds, e.g. in rubber.

Phenylmercuric salts are absorbed by rubber stoppers and some types of plastic packaging components; uptake is usually greatest to natural rubbers and polyethylene, and least to polypropylene.[(11–16)]

Incompatibilities with some types of filter membranes may also result in loss of phenylmercuric salts following sterilization by filtration.[(17)]

P

13 Method of Manufacture

Phenylmercuric nitrate is readily formed by heating benzene with mercuric acetate, and treating the resulting acetate with an alkali nitrate.[18]

14 Safety

Phenylmercuric nitrate and other phenylmercuric salts have been widely used as antimicrobial preservatives in parenteral and topical pharmaceutical formulations. However, concern over the use of phenylmercuric salts in pharmaceuticals has increased as a result of greater awareness of the toxicity of mercury and other mercury compounds. This concern must, however, be balanced by the effectiveness of these materials as antimicrobial preservatives and the low concentrations in which they are employed.

Phenylmercuric salts are irritant to the skin at 0.1% w/w concentration in petrolatum.[19] In solution, they may give rise to erythema and blistering 6–12 hours after administration. In a modified repeated insult patch test, a 2% w/v solution was found to produce extreme sensitization of the skin.[20,21]

Eye drops containing phenylmercuric nitrate as a preservative should not be used continuously for prolonged periods as mercurialentis, a brown pigmentation of the anterior capsule of the lens may occur. Incidence is 6% in patients using eye drops for greater than 6 years; however, the condition is not associated with visual impairment.[22,23] Cases of atypical band keratopathy have also been attributed to phenylmercuric nitrate preservative in eye drops.[24]

Concern that the absorption of mercury from the vagina may be harmful has led to the recommendation that phenylmercuric nitrate should not be used in intravaginal formulations.[25]

LD_{50} (mouse, IV): 27 mg/kg[26]
LD_{50} (mouse, oral): 50 mg/kg
LD_{50} (rat, SC): 63 mg/kg

15 Handling Precautions

Observe normal precautions appropriate to the circumstances and quantity of material handled. Phenylmercuric nitrate may be irritant to the skin, eyes, and mucous membranes. Eye protection, gloves, and a respirator are recommended.

16 Regulatory Status

Included in the FDA Inactive Ingredients Database (parenteral and ophthalmic preparations). Included in parenteral products and eye drops in Europe. Included in the Canadian Natural Health Products Ingredients Database (ophthalmic, nasal and otic preparations only up to 0.002%; however there must be no other suitable alternative preservative).

Prohibited in first aid antiseptic drug products, antimicrobial diaper rash drug products, and vaginal contraceptive drug products in the US. Limited uses permitted in Japan and Europe for cosmetics (*see* Phenylmercuric Acetate).

17 Related Substances

Phenylmercuric acetate; phenylmercuric borate; thimerosal.

18 Comments

Phenylmercuric salts should be used in preference to benzalkonium chloride as a preservative for salicylates and nitrates and in solutions of salts of physostigmine and epinephrine that contain 0.1% sodium sulfite.

19 Specific References

1 Buckles J, *et al.* The inactivation of phenylmercuric nitrate by sodium metabisulphite. *J Pharm Pharmacol* 1971; 23(Suppl.): 237S–238S.

2 Richards RME, Reary JME. Changes in antibacterial activity of thiomersal and PMN on autoclaving with certain adjuvants. *J Pharm Pharmacol* 1972; 24(Suppl.): 84P–89P.

3 Richards RME, *et al.* Interaction between sodium metabisulphite and PMN. *J Pharm Pharmacol* 1972; 24: 999–1000.

4 Parker MS. The preservation of pharmaceuticals and cosmetic products. In: Russell AD *et al.* eds. *Principles and Practice of Disinfection, Preservation and Sterilization.* Oxford: Blackwell Scientific, 1982: 287–305.

5 Hart A. Antibacterial activity of phenylmercuric nitrate in zinc sulphate and adrenaline eye drops BPC 1968. *J Pharm Pharmacol* 1973; 25: 507–508.

6 Miezitis EO, *et al.* Concentration changes during autoclaving of aqueous solutions in polyethylene containers: an examination of some methods for reduction of solute loss. *Aust J Pharm Sci* 1979; 8(3): 72–76.

7 Parkin JE, Marshall CA. The instability of phenylmercuric nitrate in APF ophthalmic products containing sodium metabisulfite. *Aust J Hosp Pharm* 1991; 20: 434–436.

8 Collins AJ, *et al.* Incompatibility of phenylmercuric acetate with sodium metabisulphite in eye drop formulations. *J Pharm Pharmacol* 1985; 37(Suppl.): 123P.

9 Yousef RT, *et al.* Effect of some pharmaceutical materials on the bactericidal activities of preservatives. *Can J Pharm Sci* 1973; 8: 54–56.

10 Horn NR, *et al.* Interactions between powder suspensions and selected quaternary ammonium and organomercurial preservatives. *Cosmet Toilet* 1980; 95(2): 69–73.

11 Ingversen J, Andersen VS. Transfer of phenylmercuric compounds from dilute aqueous solutions to vials and rubber closures. *Dansk Tidsskr Farm* 1968; 42: 264–271.

12 Eriksson K. Loss of organomercurial preservatives from medicaments in different kinds of containers. *Acta Pharm Suec* 1967; 4: 261–264.

13 Christensen K, Dauv E. Absorption of preservatives by drip attachments in eye drop packages. *J Mond Pharm* 1969; 12(1): 5–11.

14 Aspinall JE, *et al.* The effect of low density polyethylene containers on some hospital-manufactured eye drop formulations I: sorption of phenylmercuric acetate. *J Clin Hosp Pharm* 1980; 5: 21–29.

15 McCarthy TJ. Interaction between aqueous preservative solutions and their plastic containers, III. *Pharm Weekbl* 1972; 107: 1–7.

16 Aspinall JE, *et al.* The effect of low density polyethylene containers on some hospital-manufactured eye drop formulations II: inhibition of the sorption of phenylmercuric acetate. *J Clin Hosp Pharm* 1983; 8: 233–240.

17 Naido NT, *et al.* Preservative loss from ophthalmic solutions during filtration sterilization. *Aust J Pharm Sci* 1972; 1(1): 16–18.

18 Pyman FL, Stevenson HA. Phenylmercuric nitrate. *Pharm J* 1934; 133: 269.

19 Koby GA, Fisher AA. Phenylmercuric acetate as primary irritant. *Arch Dermatol* 1972; 106: 129.

20 Kligman AM. The identification of contact allergens by human assay, III. The maximization test: a procedure for screening and rating contact sensitizers. *J Invest Dermatol* 1966; 47: 393–409.

21 Galindo PA, *et al.* Mercurochrome allergy: immediate and delayed hypersensitivity. *Allergy* 1997; 52(11): 1138–1141.

22 Garron LK, *et al.* A clinical and pathologic study of mercurialentis medicamentosus. *Trans Am Ophthalmol Soc* 1976; 74: 295–320.

23 Winder AF, *et al.* Penetration of mercury from ophthalmic preservatives into the human eye. *Lancet* 1980; ii: 237–239.

24 Brazier DJ, Hitchings RA. Atypical band keratopathy following long-term pilocarpine treatment. *Br J Ophthalmol* 1989; 73: 294–296.

25 Lohr L. Mercury controversy heats up. *Am Pharm* 1978; 18(9): 23.

26 Sweet DV, ed. *Registry of Toxic Effects of Chemical Substances.* Cincinnati: US Department of Health, 1987: 3060–3093.

20 General References

Abdelaziz AA, El-Nakeeb MA. Sporicidal activity of local anaesthetics and their binary combinations with preservatives. *J Clin Pharm Ther* 1988; 13: 249–256.

Barkman R, *et al.* Preservatives in eye drops. *Acta Ophthalmol* 1969; 47: 461–475.

Grier N. Mercurials inorganic and organic. In: Block SS, ed. *Disinfection, Sterilization and Preservation*, 3rd edn. Philadelphia: Lea and Febiger, 1983: 346–374.

Hecht G. Ophthalmic preparations. In: Gennaro AR, ed. *Remington: The Science and Practice of Pharmacy*, 20th edn. Baltimore: Lippincott Williams and Wilkins, 2000: 821–835.

Mehta AC, *et al.* High pressure liquid chromatographic determination of phenylmercuric nitrate in eye drops. *J Clin Pharm Therap* 1976; **1**(4): 177–180.

Parkin JE. A high performance liquid chromatographic assay for phenylmercuric nitrate in the presence of zinc and its application to an assessment of the stability of phenylmercuric nitrate in zinc sulphate and zinc sulphate and adrenaline eye drops. *J Clin Pharm Therap* 1991; **16**(3): 197–201.

Parkin JE. The decomposition of phenylmercuric nitrate in sulphacetamide drops during heat sterilization. *J Pharm Pharmacol* 1993; **45**: 1024–1027.

Parkin JE, *et al.* The decomposition of phenylmercuric nitrate caused by disodium edetate in neomycin eye drops during the process of heat sterilization. *J Clin Pharm Ther* 1992; **17**: 191–196.

Parkin JE, *et al.* The chemical degradation of phenylmercuric nitrate by disodium edetate during heat sterilization at pH values commonly encountered in ophthalmic products. *J Clin Pharm Ther* 1992; **17**: 307–314.

Wood RW, Welles HL. Determination of phenylmercuric nitrate by potentiometric titration. *J Pharm Sci* 2006; **68**(10): 1272–1274.

21 Author

BR Matthews.

22 Date of Revision

2 March 2012.

Phospholipids

1 Nonproprietary Names

See Section 9.

2 Synonyms

Coatsome; glycerol phosphatides; *Lipoid*; phosphatides; phosphatidic acid; phosphatidylcholine; phosphatidylethanolamine; phosphatidylglycerol; phosphatidylinositol; phosphatidylserine; phosphoglycerides; *PhosphoLipid*; purified egg yolk PC; sphingomyelin.

See also Table I.

3 Chemical Name and CAS Registry Number

See Table II.

4 Empirical Formula and Molecular Weight

Phospholipids are formed from two combinations of apolar and 'backbone' moieties: a glycerol (or other polyol) moiety substituted with one or two acyl or alkyl chains; or an *N*-acylated sphingoid base (a ceramide).[1] Typically, the molecular weights range from 600 to 5000.

See also Table III.

5 Structural Formula

See Table IV.

6 Functional Category

Anionic surfactant; biodegradable material; cationic surfactant; dispersing agent; emulsifying agent; emulsion stabilizing agent; membrane-forming agent; nonionic surfactant; solubilizing agent; suspending agent.

7 Applications in Pharmaceutical Formulation or Technology

Phospholipids are amphiphilic molecules and are the major component of most cell membranes.[2] They are able to self-associate and form a variety of structures, including micelles and liposomes.[3]

Numerous pharmaceutical formulations use phospholipids to form various types of liposomes, including unilamellar (one bilayer membrane surrounding an aqueous chamber), multilamellar (two or more concentric membranes, each surrounding an aqueous chamber), and multivesicular (numerous aqueous chambers joined in a honeycomb-like arrangement) liposomes.[4] Modified phospholipids have been used to enhance the properties of the resulting liposomes. The covalent attachment of polyethylene glycol (PEG) to the phospholipid, or PEGylation, provides steric hindrance to the surface of the liposomes, resulting in decreased uptake by the reticuloendothelial system (RES), also known as the mononuclear phagocyte system, and a prolonged circulation half-life following intravenous administration; the so-called "stealth liposomes".[5] Conjugation with antibodies produces immunoliposomes, which are able to target specific cell types and deliver a payload of encapsulated drug.[6]

Phospholipids can be anionic, cationic, or neutral in charge. Because of their amphiphilic nature, phospholipids will associate at hydrophobic/hydrophilic interfaces. The charged lipids can be used to provide electrostatic repulsion and physical stability to suspended particles. Thus, they have been used to physically stabilize emulsions and suspensions.[7,8] Phospholipids have also been used in formulations administered as lung surfactants, intravenous fat emulsions, topical and transdermal formulations, and oral solutions.

8 Description

Phospholipids occur as white powders. They are sometimes supplied as clear, nearly colorless chloroform or methylene chloride solutions. Phosphatidylglycerols, phosphatidic acids, and phosphatidylserines are available as sodium or ammonium salts.

9 Pharmacopeial Specifications

Egg Phospholipids USP–NF is defined as a mixture of naturally occurring phospholipids obtained from the yolk of hens' eggs that is suitable for use as an emulsifying agent in injectable emulsions. The content of phosphatidylcholine, phosphatidylethanolamine, lysophosphatidylcholine, and other related phospholipids is to be reported in the certificate of analysis. It may also contain a suitable stabilizer. *See* Table V. *See also* Lecithin.

Table I: Specific synonyms of selected phospholipids.

Common name	Trade name	Manufacturer	Synonym
Dilauroyl phosphatidylcholine	*Coatsome MC-2020*	NOF	DLPC
	PhosphoLipid-DLAPC	Nippon	
Dimyristoyl phosphatidylcholine	*Coatsome MC-4040*	NOF	DMPC
	Lipoid PC 14:0/14:0 (DMPC)	Lipoid	
	PhosphoLipid-DMPC	Nippon	
Dipalmitoyl phosphatidylcholine	*Coatsome MC-6060*	NOF	DPPC
	Lipoid PC 16:0/16:0 (DPPC)	Lipoid	
	PhosphoLipid-DPPC	Nippon	
Distearoyl phosphatidylcholine	*Coatsome MC-8080*	NOF	DSPC
	Lipoid PC 18:0/18:0 (DSPC)	Lipoid	
	PhosphoLipid-DSPC	Nippon	
Dioleoyl phosphatidylcholine	*Coatsome MC-8181*	NOF	DOPC
	Lipoid PC 18:1/18:1 (DOPC)	Lipoid	
	PhosphoLipid-DOPC	Nippon	
Dierucoyl phosphatidylcholine	*PhosphoLipid-DERPC*	Nippon	DEPC
Palmitoyloleoyl phosphatidylcholine	*Coatsome MC-6081*	NOF	POPC
	PhosphoLipid-POPC	Nippon	
Dimyristoyl phosphatidylglycerol, sodium salt	*Coatsome MG-4040LS*	NOF	DMPG
	Lipoid PG 14:0/14:0 (DMPG)	Lipoid	
	PhosphoLipid-DMPG	Nippon	
Dipalmitoyl phosphatidylglycerol, sodium salt	*Coatsome MG-6060LS*	NOF	DPPG
	Lipoid PG 16:0/16:0 (DPPG)	Lipoid	
	PhosphoLipid-DPPG	Nippon	
Distearoyl phosphatidylglycerol, sodium salt	*Coatsome MG-8080LS*	NOF	DSPG
	Lipoid PG 18:0/18:0 (DSPG)	Lipoid	
	PhosphoLipid-DSPG	Nippon	
Dioleoyl phosphatidylglycerol, sodium salt	*Lipoid PG 18:1/18:1 (DOPG)*	Lipoid	DOPG
	PhosphoLipid-DOPG	Nippon	
Palmitoyloleoyl phosphatidylglycerol, sodium salt	*Lipoid PG 16:0/18:1 (POPG)*	Lipoid	POPG
	PhophoLipid-POPG	Nippon	
Dimyristoyl phosphatidylethanolamine	*Coatsome ME-4040*	NOF	DMPE
	Lipoid PE 14:0/14:0 (DMPE)	Lipoid	
Dipalmitoyl phosphatidylethanolamine	*Coatsome ME-6060*	NOF	DPPE
	Lipoid PE 16:0/16:0 (DPPE)	Lipoid	
Distearoyl phosphatidylethanolamine	*Coatsome ME-8080*	NOF	DSPE
	Lipoid PE 18:0/18:0 (DSPE)	Lipoid	
Dioleoyl phosphatidylethanolamine	*Coatsome ME-8181*	NOF	DOPE
	Lipoid PE 18:1/18:1 (DOPE)	Lipoid	
Dimyristoyl phosphatidic acid, sodium salt	*Coatsome MA-4040LS*	NOF	DMPA
Dipalmitoyl phosphatidic acid, sodium salt	*Coatsome MA-6060LS*	NOF	DPPA
	Lipoid PA 16:0/16:0 (DPPA)	Lipoid	
Distearoyl phosphatidic acid, sodium salt	*Coatsome MA-8080LS*	NOF	DSPA
	Lipoid PA 18:0/18:0 (DSPA)	Lipoid	
Dioleoyl phosphatidylserine, sodium salt	*Coatsome MS-8181LS*	NOF	DOPS

Table II: Chemical name and CAS registry number of selected phospholipids.

Name	IUPAC Name	CAS number
Dilauroyl phosphatidylcholine	1,2-Didodecanoyl-*sn*-glycero-3-phosphocholine	18194-25-7
Dimyristoyl phosphatidylcholine	1,2-Ditetradecanoyl-*sn*-glycero-3-phosphocholine	18194-24-6
Dipalmitoyl phosphatidylcholine	1,2-Dihexadecanoyl-*sn*-glycero-3-phosphocholine	63-89-8
Distearoyl phosphatidylcholine	1,2-Dioctadecanoyl-*sn*-glycero-3-phosphocholine	816-94-4
Dioleoyl phosphatidylcholine	1,2-Dioctadecenoyl-*sn*-glycero-3-phosphocholine	4235-95-4
Dierucoyl phosphatidylcholine	1,2-Didocosenoyl-*sn*-glycero-3-phosphocholine	51779-95-4
Palmitoyloleoyl phosphatidylcholine	1-Hexadecanoyl-2-octadecenoyl-*sn*-glycero-3-phosphocholine	26853-31-6
Dimyristoyl phosphatidylglycerol, sodium salt	1,2-Ditetradecanoyl-*sn*-glycero-3-[phospho-*rac*-(1-glycerol)]	67232-80-8
Dipalmitoyl phosphatidylglycerol, sodium salt	1,2-Dihexadecanoyl-*sn*-glycero-3-[phospho-*rac*-(1-glycerol)]	67232-81-9
Distearoyl phosphatidylglycerol, sodium salt	1,2-Dioctadecanoyl-*sn*-glycero-3-[phospho-rac-(1-glycerol)]	67232-82-0
Dioleoyl phosphatidylglycerol, sodium salt	1,2-Dioctadecenoyl-*sn*-glycero-3-[phospho-*rac*-(1-glycerol)]	62700-69-0
Palmitoyloleoyl phosphatidylglycerol, sodium salt	1-Hexadecanoyl-2-octadecenoyl-*sn*-glycero-3-[phospho-*rac*-(1-glycerol)]	81490-05-3
Dimyristoyl phosphatidylethanolamine	1,2-Ditetradecanoyl-*sn*-glycero-3-phosphoethanolamine	998-07-2
Dipalmitoyl phosphatidylethanolamine	1,2-Dihexadecanoyl-*sn*-glycero-3-phosphoethanolamine	923-61-5
Distearoyl phosphatidylethanolamine	1,2-Dioctadecanoyl-*sn*-glycero-3-phosphoethanolamine	1069-79-0
Dioleoyl phosphatidylethanolamine	1,2-Dioctadecenoyl-*sn*-glycero-3-phosphoethanolamine	4004-05-1
Dimyristoyl phosphatidic acid, sodium salt	1,2-Ditetradecanoyl-*sn*-glycero-3-phosphatidic acid	80724-31-8
Dipalmitoyl phosphatidic acid, sodium salt	1,2-Dihexadecanoyl-*sn*-glycero-3-phosphatidic acid	74427-52-4
Distearoyl phosphatidic acid, sodium salt	1,2-Dioctadecanoyl-*sn*-glycero-3-phosphatidic acid	108321-18-2
Dioleoyl phosphatidylserine, sodium salt	1,2-Dioctadecenoyl-*sn*-glycero-3-phosphoserine	70614-14-1

Table III: Empirical formula and molecular weight of selected phospholipids.

Name	Empirical formula	Molecular weight
Dilauroyl phosphatidylcholine	$C_{32}H_{64}O_8NP$	621.8
Dimyristoyl phosphatidylcholine	$C_{36}H_{72}O_8NP$	677.9
Dipalmitoyl phosphatidylcholine	$C_{40}H_{80}O_8NP$	734.0
Distearoyl phosphatidylcholine	$C_{44}H_{88}O_8NP$	790.2
Dioleoyl phosphatidylcholine	$C_{44}H_{84}O_8NP$	786.1
Dierucoyl phosphatidylcholine	$C_{52}H_{100}O_8NP$	898.4
Palmitoyloleoyl phosphatidylcholine	$C_{34}H_{82}O_8NP$	760.1
Dimyristoyl phosphatidylglycerol, sodium salt	$C_{34}H_{66}O_{10}PNa$	688.9
Dipalmitoyl phosphatidylglycerol, sodium salt	$C_{38}H_{74}O_{10}PNa$	745.0
Distearoyl phosphatidylglycerol, sodium salt	$C_{42}H_{82}O_{10}PNa$	801.1
Dioleoyl phosphatidylglycerol, sodium salt	$C_{42}H_{78}O_{10}PNa$	797.0
Palmitoyloleoyl phosphatidylglycerol, sodium salt	$C_{40}H_{76}O_{10}PNa$	771.0
Dimyristoyl phosphatidylethanolamine	$C_{33}H_{66}O_8NP$	653.9
Dipalmitoyl phosphatidylethanolamine	$C_{37}H_{74}O_8NP$	692.0
Distearoyl phosphatidylethanolamine	$C_{41}H_{82}O_8NP$	748.1
Dioleoyl phosphatidylethanolamine	$C_{41}H_{78}O_8NP$	744.0
Dimyristoyl phosphatidic acid, sodium salt	$C_{31}H_{60}O_8PNa$	614.8
Dipalmitoyl phosphatidic acid, sodium salt	$C_{35}H_{68}O_8PNa$	670.9
Distearoyl phosphatidic acid, sodium salt	$C_{39}H_{76}O_8PNa$	727.0
Dioleoyl phosphatidylserine, sodium salt	$C_{42}H_{77}O_{10}NPNa$	810.0

10 Typical Properties

Phospholipids are amphiphilic, surface-active molecules with a high tendency to form aggregates (phases) both in the dry state and when fully hydrated. The temperature of transition from crystalline to mesomorphic (liquid crystalline) state is the transition temperature (T_m).[9] *See* Table VI. Most phospholipids are freely soluble in organic solvents.

11 Stability and Storage Conditions

Phospholipids are stable in the solid state if protected from oxygen, heat, and light. Chloroform or dichloromethane solutions are also stable. Both the solid-state and solution forms should be stored at −20°C. Liposomal phospholipids are known to degrade via oxidation and hydrolysis. To minimize oxidation, liposomes can be prepared under oxygen-free environments and antioxidants, such as butylated hydroxytoluene (BHT), can be added.[11,12] To minimize hydrolysis, water can be removed from liposomes by lyophilization. In cases where liposomes are unstable to lyophilization, long-term storage at 2–8°C is recommended. The ester hydrolysis of phospholipids in liposomes typically follows a V-shaped curve, with the minimum at around pH 6.5.[13,14]

12 Incompatibilities

—

13 Method of Manufacture

Phospholipids can be manufactured from naturally occurring materials, especially soybean and egg. The manufacturing process typically involves extraction, fractionation, and purification. They can also be synthesized chemically by reacting glycerol phosphocholine (PC), glycerol phosphoglycerol (PG), glycerol phosphoserine (PS), glycerol phosphoethanolamine (PE), or glycerol phosphoinositol (PI) with purified fatty acids.[15]

14 Safety

Generally, phospholipids have little or no acute toxicity (i.e. they are well tolerated even when administered at doses in the g/kg range).[16] While naturally occurring antibodies to phospholipids are observed, the immunology and safety of the antibodies are not considered to be clinically relevant.[17] The clearance of most phospholipids occurs by well-known metabolic pathways.[18]

Liposomes containing stearylamines (cationic liposomes) have been found to induce cytotoxicity through apoptosis in the macrophage-like cell line RA W2647[19] and inhibit the growth of cells *in vitro*.[20,21] In nine cancer-derived cell lines and one normal

Table IV: Structural formulas of selected phospholipids.

Series name	Phosphatidyl moiety	Headgroup
Phosphatidylcholines	CH_2OCOR_1 – $CHOCOR_2$ – CH_2O—P(=O)(O⁻)— ; R^1, R^2 = acyl chains	$OCH_2CH_2N^+(CH_3)_3$
Phosphatidylglycerol sodium salts	CH_2OCOR_1 – $CHOCOR_2$ – CH_2O—P(=O)(O-)— Na^+ ; R^1, R^2 = acyl chains	$OCH_2CH(OH)CH_2OH$
Phosphatidylethanolamines	CH_2OCOR_1 – $CHOCOR_2$ – CH_2O—P(=O)(O⁻)— ; R^1, R^2 = acyl chains	$OCH_2CH_2N^+H_3$
Phosphatidic acid sodium salts	CH_2OCOR_1 – $CHOCOR_2$ – CH_2O—P(=O)(O-)— Na^+ ; R^1, R^2 = acyl chains	
Phosphatidylserines	CH_2OCOR_1 – $CHOCOR_2$ – CH_2O—P(=O)(O-)— Na^+ ; R^1, R^2 = acyl chains	$OCH(N^+H_3)CO_2^-$

Table V: Pharmacopeial specifications for egg phospholipids.

Test	USP35–NF30
Acid value	≤20.0
Peroxide value	≤3.0
Bacterial endotoxins	≤6 IU/g
Microbial limit	≤100 cfu/g
Water	≤6.0%
Heavy metals	≤10 ppm
Nonphosphatidyl lipids	≤7.0%
Assay (lysophosphatidylcholine)	≤3.0%

Table VI: Properties of common phospholipids.[10]

Name	Transition temperature T_m(°C)	Net charge at pH 7.4
Dilauroyl phosphatidylcholine	−1	0
Dimyristoyl phosphatidylcholine	23	0
Dipalmitoyl phosphatidylcholine	41	0
Distearoyl phosphatidylcholine	55	0
Dioleoyl phosphatidylcholine	−20	0
Dimyristoyl phosphatidylethanolamine	50	0
Dipalmitoyl phosphatidylethanolamine	63	0
Dioleoyl phosphatidylethanolamine	−16	0
Dimyristoyl phosphatidic acid, sodium salt	50	−1.3
Dipalmitoyl phosphatidic acid, sodium salt	67	−1.3
Dioleoyl phosphatidic acid, sodium salt	−8	−1.3
Dimyristoyl phosphatidylglycerol, sodium salt	23	−1
Dipalmitoyl phosphatidylglycerol, sodium salt	41	−1
Dioleoyl phosphatidylglycerol, sodium salt	−18	−1
Dimyristoyl phosphatidylserine, sodium salt	35	−1
Dipalmitoyl phosphatidylserine, sodium salt	54	−1
Dioleoyl phosphatidylserine, sodium salt	−11	−1

cultured human cell line, stearylamine- and cardiolipin-containing liposomes were toxic (LD_{50}) at 200 μM liposomal lipid concentration or less, whereas PG- and PS-containing liposomes were toxic in the range 130–3000 μM.[22] Positively charged lipids such as stearylamine can increase the toxicity of liposomes.[23] These studies reported an LD_{50} (IV) of 1.1 g/kg and 7.5 g/kg with and without stearylamine, respectively.

The safety of phospholipids delivered by the intravenous route is complicated by their tendency to form particles that are recognized by macrophages of the RES.[24] Uptake by the RES is dependent on particle size and composition.

15 Handling Precautions

Observe normal precautions appropriate to the circumstances and quantity of the material handled. Eye protection and gloves are recommended.

16 Regulatory Status

Included in the FDA Inactive Ingredient Database (oral, otic, buccal, vaginal, topical, epidural, intravenous, intramuscular, and inhalation aerosol). A number of phospholipids such as DPPG and DOPC are present in approved products in Europe and the US. Included in the Canadian Natural Health Products Ingredients Database.

17 Related Substances

Lecithin.

18 Comments

Therapeutically, oral formulations of phosphatidylcholine with nonsteroidal anti-inflammatory drugs (NSAIDS) have been shown to reduce gastrointestinal bleeding and ulceration associated with NSAIDS.[25]

Phospholipids can be purified from natural sources, such as eggs or soybeans, or can be chemically synthesized. Egg or soy phospholipids are composed of mixtures of individual phospholipids including phosphatidylcholine, phosphatidylethanolamine and lysophosphatidylcholine. The fatty acid content of egg phospholipids depends on the types of fat in the diets of the hen as well as the strain of hen.[26] Likewise, the composition of soy is known to vary.[27] To ensure batch-to-batch reproducibility, phospholipids from synthetic sources are commonly used. Lecithins are partially purified mixtures of naturally occurring phospholipids and other material.

19 Specific References

1 Silvius JR. Structure and nomenclature. In: Cevc G, ed. *Phospholipids Handbook*. New York: Marcel Dekker, 1993: 1–22.
2 Yorek MA. Biological distribution. In: Cevc G, ed. *Phospholipids Handbook*. New York: Marcel Dekker, 1993: 745–776.
3 Seddon JM, Cevc G. Lipid polymorphism: structure and stability of lyotrophic mesophases of phospholipids. In: Cevc G, ed. *Phospholipids Handbook*. New York: Marcel Dekker, 1993: 403–454.
4 Mantripragada S. A lipid based depot (*DepoFoam* technology) for sustained-release drug delivery. *Prog Lipid Res* 2002; **41**(15): 392–406.
5 Needham D *et al.* Surface chemistry of the sterically stablilized PEG-liposome: general principles. In: Janoff AS, ed. *Liposomes: Rational Design*. New York: Marcel Dekker, 1999: 13–62.
6 Allen TM, Stuart DD. Liposome pharmacokinetics: classical, sterically stabilized, cationic liposomes and immunoliposomes. In: Janoff AS, ed. *Liposomes: Rational Design*. New York: Marcel Dekker, 1999: 63–68.
7 Young TJ, *et al.* Phospholipid-stabilized nanoparticles of cyclosporine A by rapid expansion from supercritical to aqueous solution. *AAPS Pharm Sci Tech* 2004; **5**(1): E11.
8 Han J, *et al.* Design and evaluation of an emulsion vehicle for paclitaxel. I. Physicochemical properties and plasma stability. *Pharm Res* 2004; **21**(9): 1573–1580.
9 Hauser H. Phospholipids vesicles. In: Cevc G, ed. *Phospholipids Handbook*. New York: Marcel Dekker, 1993: 603–638.
10 Silvius JR. Thermotropic phase transitions of pure lipids in model membranes and their modifications by membrane proteins. In: Jost PC, Griffith OH, eds. *Lipid-Protein Interactions*. 2. New York: Wiley, 1982: 239–281.
11 Rabinovich-Guilatt L, *et al.* Phospholipid hydrolysis in a pharmaceutical emulsion assessed by physicochemical parameters and a new analytical method. *Eur J Pharm Biopharm* 2005; **61**: 69–76.
12 Zuidam NJ, Crommelin DJ. Chemical hydrolysis of phospholipids. *J Pharm Sci* 1995; **84**(9): 1113–1119.
13 Grit M, Crommelin DJ. Chemical stability of liposomes: implications for their physical stability. *Chem Phys Lipids* 1993; **64**: 3–18.
14 Grit M, *et al.* Hydrolysis of partially saturated egg phosphatidylcholine in aqueous liposome dispersions and the effect of cholesterol incorporation on hydrolysis kinetics. *J Pharm Pharmacol* 1993; **45**: 490–495.
15 Brittman R. Chemical preparation of glycerolipids: a review of recent syntheses. In: Cevc G, ed. *Phospholipids Handbook*. New York: Marcel Dekker, 1993: 141–232.
16 Hart IR, *et al.* Toxicity studies of liposome-encapsulated immunomodulators administered intravenously to dogs and mice. *Cancer Immunol Immunother* 1981; **10**: 157–166.
17 CR Alving. Antibodies to lipids and liposomes: immunology and safety. *J Lipsome Res* 2006; **16**(3): 157–166.
18 Scherphof GL. Phospholipid metabolism in animal cells. In: Cevc G, ed. *Phospholipids Handbook*. New York: Marcel Dekker, 1993: 777–800.
19 Takano S, *et al.* Physicochemical properties of liposomes affecting apoptosis induced by cationic liposomes in macrophages. *Pharm Res* 2003; **20**(7): 962–968.

20 Allen TM. A study of phospholipid interactions between high-density lipoproteins and small unilamellar vesicles. *Biochim Biophys Acta* 1981; **640**: 385–397.

21 Heath TD, *et al.* The effects of liposome size and surface charge on liposome-mediated delivery of MTX-γ-asp to cells *in vitro*. *Biochim Biophys Acta* 1985; **820**: 74–84.

22 Mayhew E, *et al.* Toxicity of non-drug-containing liposomes for cultured human cells. *Exp Cell Res* 1987; **171**: 195–202.

23 Olson F, *et al.* Characterization, toxicity and therapeutic efficacy of adriamycin encapsulated in liposomes. *Eur J Cancer Clin Oncol* 1982; **18**: 167–176.

24 Allen TM. Toxicity and systematic effects of phospholipids. In: Cevc G, ed. *Phospholipids Handbook*. New York: Marcel Dekker, 1993: 801–816.

25 Lichtenberger LM, *et al.* Association of phosphatidylcholine and NSAIDs as a novel strategy to reduce gastrointestinal toxicity. *Drugs Today (Barc)* 2009; **45**(12): 877–890.

26 Stadelman WJ, Cotterill OJ. The chemistry of eggs and egg products. In: *Egg Science and Technology*, 4th edn. Philadelphia: Haworth Press, 1995: 113–114.

27 Porter MA, Jones AM. Variability in soy flour composition. *J Am Oil Chem Soc* 2003; **80**(6): 557–562.

20 General References

Betageri GV *et al.* eds. *Liposome Drug Delivery Systems*. London: Informa Healthcare, 1993.

Duzgunes N. Liposomes. In: *Methods in Enzymology Part A*. 367. New York: Academic Press, 2003: 3–128.

Florence AT, ed. *Drug Targeting and Delivery*. 2. Boca Raton, FL: CRC Press, 1993.

Gregoriadis G, ed. *Liposome Technology*, 3rd edn. London: Informa Healthcare, 2006.

Janoff AS, ed. *Liposomes: Rational Design*. London: Informa Healthcare, 1998.

Lasic DD, Martin FJ. *Handbooks in Pharmacology and Toxicology*. Boca Raton, FL: CRC Press, 1995.

Lieberman HA *et al.* eds. *Pharmaceutical Dosage Forms – Disperse Systems*. London: Informa Healthcare, 1996.

Ostro MJ, ed. *Liposomes*. London: Informa Healthcare, 1987.

Torchilin V, Weissig V. *Liposomes: A Practical Approach*. The Practical Approach Series , 264. New York: Oxford University Press, 2003.

21 Authors

W Lambert, B Richard, J Schrier, P Ying.

22 Date of Revision

2 March 2012.

Phosphoric Acid

1 Nonproprietary Names

BP: Phosphoric Acid

PhEur: Phosphoric Acid, Concentrated

USP–NF: Phosphoric Acid

See also Section 17.

2 Synonyms

Acid fosforico; acide phosphorique; acidum phosphoricum concentratum; E338; hydrogen phosphate; syrupy phosphoric acid.

3 Chemical Name and CAS Registry Number

Orthophosphoric acid [7664-38-2]

4 Empirical Formula and Molecular Weight

H_3PO_4 98.00

5 Structural Formula

See Section 4.

6 Functional Category

Acidulant; buffering agent.

7 Applications in Pharmaceutical Formulation or Technology

Phosphoric acid is widely used as an acidulant in a variety of pharmaceutical formulations. It is used in pharmaceutical products as part of a buffer system when combined with a phosphate salt such as sodium phosphate, monobasic or dibasic. It is also widely used in food preparations as an acidulant, flavor, and synergistic antioxidant (0.001–0.005%) and sequestrant.

Phosphoric acid 35% gel has also been used to etch tooth enamel and to enhance delivery of drugs through the nail.[1]

8 Description

Concentrated phosphoric acid occurs as a colorless, odorless, syrupy liquid.

9 Pharmacopeial Specifications

See Table I.

Table I: Pharmacopeial specifications for phosphoric acid.

	PhEur 7.4	USP35–NF30
Identification	+	+
Characters	+	−
Appearance of solution	+	−
Relative density	≈1.7	−
Sulfate	≤100 ppm	+
Chloride	≤50 ppm	−
Heavy metals	≤10 ppm	≤10 ppm
Substances precipitated with ammonia	+	−
Arsenic	≤2 ppm	−
Iron	≤50 ppm	−
Alkali phosphates	−	+
Limit of nitrate	−	+
Phosphorous or hypophosphorous acid	+	+
Assay (of H_3PO_4)	84.0–90.0%	85.0–88.0%

P

10 Typical Properties

Acidity/alkalinity pH = 1.6 (1% w/w aqueous solution)
Boiling point 117.87°C
Dissociation constant
$pK_{a1} = 2.15$;
$pK_{a2} = 7.09$;
$pK_{a3} = 12.32$.
Melting point 42.35°C
Refractive index
$n_D^{17.5}$ = 1.35846 (30% w/w aqueous solution);
$n_D^{17.5}$ = 1.35032 (20% w/w aqueous solution);
$n_D^{17.5}$ = 1.3423 (10% w/w aqueous solution).
Solubility Miscible with ethanol (95%) and water with the evolution of heat.
Specific gravity
1.874 (100% w/w) at 25°C;
1.6850 (85% w/w aqueous solution) at 25°C;
1.3334 (50% w/w aqueous solution) at 25°C;
1.0523 (10% w/w aqueous solution) at 25°C.

11 Stability and Storage Conditions

When stored at a low temperature, phosphoric acid may solidify, forming a mass of colorless crystals, comprising the hemihydrate, which melts at 28°C. Phosphoric acid should be stored in an airtight container in a cool, dry place. Stainless steel containers may be used.

12 Incompatibilities

Phosphoric acid is a strong acid and reacts with alkaline substances. Mixtures with nitromethane are explosive.

13 Method of Manufacture

The majority of phosphoric acid is made by digesting phosphate rock (essentially tricalcium phosphate) with sulfuric acid; the phosphoric acid is then separated by slurry filtration. Purification is achieved via chemical precipitation, solvent extraction, crystallization, or ion exchange.

P

14 Safety

In the concentrated form, phosphoric acid is an extremely corrosive and harmful acid. However, when used in pharmaceutical formulations it is usually very diluted and is generally regarded as an essentially nontoxic and nonirritant material.

The lowest lethal oral dose of concentrated phosphoric acid in humans is reported to be 1286 μL/kg.[2]

LD_{50} (rabbit, skin): 2.74 g/kg[2]
LD_{50} (rat, oral): 1.53 g/kg

15 Handling Precautions

Observe normal precautions appropriate to the circumstances and quantity of material handled. Phosphoric acid is corrosive and can cause burns on contact with the skin, eyes and mucous membranes; contact should be avoided. Splashes should be washed with copious quantities of water. Protective clothing, gloves and eye protection are recommended.

Phosphoric acid is also irritant on inhalation. In the UK, the workplace exposure limit for phosphoric acid is 1 mg/m^3 long-term (8-hour TWA) and 2 mg/m^3 short-term (15-minutes).[3]

Phosphoric acid emits toxic fumes on heating.

16 Regulatory Status

GRAS listed. Accepted as a food additive in Europe. Included in the FDA Inactive Ingredients Database (infusions, injections, oral solutions, topical creams, lotions, ointments and solutions, and vaginal preparations). Included in nonparenteral and parenteral medicines licensed in the UK. Included in the Canadian Natural Health Products Ingredients Database.

17 Related Substances

Dilute phosphoric acid.

Dilute phosphoric acid

Synonyms acidum phosphoricum dilutum; diluted phosphoric acid.

Comments The PhEur 7.4 states that dilute phosphoric acid contains 9.5–10.5% w/w H_3PO_4 and may be prepared by mixing phosphoric acid 115 g with 885 g of water. The USP35–NF30 contains a monograph for diluted phosphoric acid and states that it contains 9.5–10.5% w/v H_3PO_4 and may be prepared by mixing phosphoric acid 69 mL with water to 1000 mL.

18 Comments

Therapeutically, dilute phosphoric acid has been used well-diluted in preparations used in the treatment of nausea and vomiting.

Nanosized hydroxyapatite powder was made by combining phosphoric acid with egg shells.[4]

In the UK, a 1 in 330 aqueous solution of phosphoric acid is approved as a disinfectant for foot-and-mouth disease. A specification for phosphoric acid is contained in the *Food Chemicals Codex* (FCC).[5]

The EINECS number for phosphoric acid is 231-633-2. The PubChem Compound ID (CID) for phosphoric acid is 1004.

19 Specific References

1 Repka MA. *Delivery of medicaments to the nail.* United States Patent No.20040197280, 2004.
2 Lewis RJ, ed. *Sax's Dangerous Properties of Industrial Materials*, 11th edn. New York: Wiley, 2004: 2948–2949.
3 Health and Safety Executive. *EH40/2005: Workplace Exposure Limits.* Sudbury: HSE Books, 2011. http://www.hse.gov.uk/pubns/priced/eh40.pdf (accessed 1 March 2012).
4 Lee SJ, *et al.* Nanosized hydroxyapatite powder synthesized from eggshell and phosphoric acid. *Nanotechnol* 2007; 7(11): 4061–4064.
5 *Food Chemicals Codex*, 7th edn. (Suppl. 1). Bethesda, MD: United States Pharmacopeia, 2010: 1482.

20 General References

—

21 Author

WG Chambliss.

22 Date of Revision

1 March 2012.

Polacrilin Potassium

1 Nonproprietary Names

USP–NF: Polacrilin Potassium

2 Synonyms

Amberlite IRP-88; methacrylic acid polymer with divinylbenzene, potassium salt; polacrilinum kalii.

3 Chemical Name and CAS Registry Number

2-Methyl-2-propenoic acid polymer with divinylbenzene, potassium salt [39394-76-5]

4 Empirical Formula and Molecular Weight

See Sections 5, 13 and 18.

5 Structural Formula

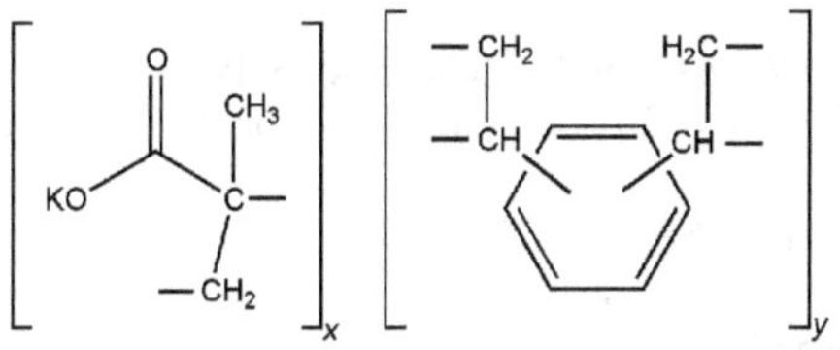

6 Functional Category

Modified-release agent; tablet and capsule disintegrant.

7 Applications in Pharmaceutical Formulation or Technology

Polacrilin potassium is a cation-exchange resin used in oral pharmaceutical formulations as a tablet disintegrant.[1–3] Concentrations of 2–10% w/w have been used for this purpose, although 2% w/w of polacrilin potassium is usually sufficient. Other polacrilin ion-exchange resins have been used as excipients to stabilize drugs, to mask or modify the taste of drugs, and in the preparation of sustained-release dosage forms[4] and drug carriers.

Polacrilin resins are also used in the analysis and manufacture of pharmaceuticals and food products.

8 Description

Polacrilin potassium occurs as a cream-colored, odorless and tasteless, free-flowing powder. Aqueous dispersions have a bitter taste.

9 Pharmacopeial Specifications

See Table I.

Table I: Pharmacopeial specifications for polacrilin potassium.

Test	USP35–NF30
Identification	+
Loss on drying	≤10.0%
Powder fineness	≤1.0% on a #100 mesh
	≤30.0% on a #200 mesh
Iron	≤0.01%
Sodium	≤0.20%
Heavy metals	≤0.002%
Assay of potassium (dried basis)	20.6%–25.1%

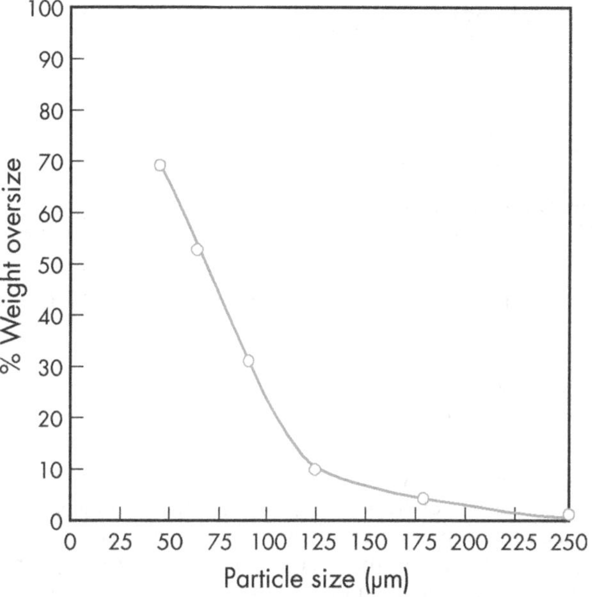

Figure 1: Particle size distribution of polacrilin potassium (*Amberlite IRP-88*, Rohm and Haas Co.).

10 Typical Properties

Density (bulk) 0.48 g/cm^3 for *Amberlite IRP-88*.[3]
Density (tapped) 0.62 g/cm^3 for *Amberlite IRP-88*.[3]
Particle size distribution *see* Figure 1.[3]
Solubility Practically insoluble in water and most other liquids, although polacrilin resins swell rapidly when wetted.

11 Stability and Storage Conditions

Polacrilin potassium and other polacrilin resins are stable to light, air, and heat up to their maximum operation temperature; *see* Table II. Excessive heating can cause thermal decomposition of the resins and may yield one or more oxides of carbon, nitrogen, sulfur, and/or amines.

Polacrilin resins should be stored in well-closed containers in a cool, dry place.

12 Incompatibilities

Incompatible with strong oxidizing agents, amines, particularly tertiary amines, and some other substances that interact with polacrilin resins.[5]

13 Method of Manufacture

Polacrilin resin (*Amberlite IRP-64*) is prepared by the copolymerization of methacrylic acid with divinylbenzene (DVB). Polacrilin potassium (*Amberlite IRP-88*) is then produced by neutralizing this resin with potassium hydroxide.

Other resins are similarly produced by copolymerization between styrene and divinylbenzene (*Amberlite IRP-69*, *Amberlite IRP-67*, *Amberlite IR-120*, and *Amberlite IRA-400*). Phenolic-based polyamine condensates (*Amberlite IRP-58*) may also be produced.

The homogeneity of the resin structure depends on the purity, nature, and properties of the copolymers used as well as the controls

Table II: Summary of physicochemical properties of pharmaceutical grade *Amberlite* resins.

Amberlite grade	Copolymer	Type	Functional structure	Ionic form	Particle size (mesh)	Parent resin	Maximum moisture (%)	pH range	Maximum temperature (°C)	Application
Cation-exchange resins										
IRP-69	Styrene and DVB[a]	Strongly acidic	$SO_3^-Na^+$	Na^+	100–500	*IR-120*	10	0–14	120	Carrier for cationic drugs that are bases or salts
IRP-64	Methacrylic acid and DVB	Weakly acidic	COO^-H^+	H^+	100–500	*IRC-50*	10	5–14	120	Carrier for cationic drugs
IRP-88	Methacrylic acid and DVB	Weakly acidic	COO^-K^+	K^+	100–500	*IRC-50*	10	5–14	120	Tablet disintegrant
Anion-exchange resins										
IRP-58	Phenolic polyamine	Weakly basic	NH_2NH_2	Free base	100–500	*IR-4B*	10	0–7	60	Carrier for anionic drugs that are acids
IRP-67	Styrene and DVB	Strongly basic	$N(CH_3)_3{}^+Cl^-$	Cl^-	100–500	*IRA-400*	10	0–12	60	Carrier for anionic drugs that are acids or salts

Note that all of the above grades, with the exception of *Amberlite IRP-88*, are available in particle-size grades <325 mesh.
(a) DVB: divinylbenzene.

and conditions employed during the polymerization reaction. The nature and degree of crosslinking have significant influence on the physicochemical properties of the resin matrix. The functional groups introduced on the matrix confer the property of ion exchange. Depending upon the acidity or basicity of the functional groups, strongly acidic to strongly basic types of ion-exchange resins may be produced.

14 Safety

Polacrilin potassium and other polacrilin resins are used in oral pharmaceutical formulations and are generally regarded as nontoxic and nonirritant materials. However, excessive ingestion of polacrilin resins may disturb the electrolyte balance of the body.

15 Handling Precautions

Observe normal precautions appropriate to the circumstances and quantity of material handled. Polacrilin potassium may be irritating to the eyes; eye protection and gloves are recommended.

16 Regulatory Status

Included in the FDA Inactive Ingredients Database (oral capsules and tablets). Included in non-parenteral medicines licensed in the UK. Included in the Canadian Natural Health Products Ingredients Database.

17 Related Substances

Polacrilin.

Polacrilin

CAS number [54182-62-6]

Synonyms

Amberlite IRP-64; methacrylic acid polymer with divinylbenzene; 2-methyl-2-propenoic acid polymer with divinylbenzene.

See also Section 18.

18 Comments

A number of other polacrilin (*Amberlite*) resins are commercially available that have a variety of industrial and pharmaceutical applications; *see* Table II.

19 Specific References

1 Van Abbé NJ, Rees JT. Amberlite resin XE-88 as a tablet disintegrant. *J Am Pharm Assoc (Sci)* 1958; 47: 487–489.
2 Khan KA, Rhodes CT. Effect of disintegrant concentration on disintegration and compression characteristics of two insoluble direct compression systems. *Can J Pharm Sci* 1973; 8: 77–80.
3 Rudnic EM, *et al.* Evaluation of the mechanism of disintegrant action. *Drug Dev Ind Pharm* 1982; 8: 87–109.
4 Smith HA, *et al.* The development of a liquid antihistaminic preparation with sustained release properties. *J Am Pharm Assoc (Sci)* 1960; 49: 94–97.
5 Borodkin S, Yunker MH. Interaction of amine drugs with a polycarboxylic acid ion-exchange resin. *J Pharm Sci* 1970; 59: 481–486.

20 General References

—

21 Author

A Palmieri.

22 Date of Revision

2 March 2012.

Poloxamer

1 Nonproprietary Names

BP: Poloxamers

PhEur: Poloxamers

USP–NF: Poloxamer

2 Synonyms

Lutrol; *Monolan*; *Pluracare*; *Pluronic*; poloxalkol; poloxamera; polyethylene–propylene glycol copolymer; polyoxyethylene–polyoxypropylene copolymer; *Supronic*; *Surfonic*; *Synperonic*.

3 Chemical Name and CAS Registry Number

α-Hydro-ω-hydroxypoly(oxyethylene)poly(oxypropylene) poly(oxyethylene) block copolymer [9003-11-6]

4 Empirical Formula and Molecular Weight

The poloxamer polyols are a series of closely related block copolymers of ethylene oxide and propylene oxide conforming to the general formula $HO(C_2H_4O)_a(C_3H_6O)_b(C_2H_4O)_aH$. The grades included in the PhEur 7.4 and USP35–NF30 are shown in Table I. The PhEur 7.4 and the USP35–NF30 both states that a suitable antioxidant may be added.

Table I: Typical poloxamer grades.

Poloxamer	Physical form	*a*	*b*	Average molecular weight
124	Liquid	12	20	2090–2360
188	Solid	80	27	7680–9510
237	Solid	64	37	6840–8830
338	Solid	141	44	12 700–17 400
407	Solid	101	56	9840–14 600

5 Structural Formula

6 Functional Category

Dispersing agent; emulsifying agent; nonionic surfactant; solubilizing agent; tablet and capsule lubricant.

7 Applications in Pharmaceutical Formulation or Technology

Poloxamers are nonionic polyoxyethylene–polyoxypropylene copolymer surfactants used primarily in pharmaceutical formulations as emulsifying or solubilizing agents.[1-3] The polyoxyethylene segment is hydrophilic while the polyoxypropylene segment is hydrophobic. All of the poloxamers are chemically similar in composition, differing only in the relative amounts of propylene and ethylene oxides added during manufacture. Their physical and surface-active properties vary over a wide range and a number of different types are commercially available; *see* Sections 4, 9, 10, and 18.

Poloxamers are used as emulsifying agents in intravenous fat emulsions, and as solubilizing and stabilizing agents to maintain the clarity of elixirs and syrups. Poloxamers may also be used as wetting agents in ointments, suppository bases, and gels, and as tablet binders and coatings. Solid poloxamers often include butylated hydroxytoluene (50–125 ppm) as an antioxidant.

Poloxamer 338 and 407 are used in solutions for contact lens care. *See* Table II. Poloxamers are also used in the cosmetics industry. *See also* Section 18.

Table II: Uses of poloxamer.

Use	Concentration (%)
Fat emulsifier	0.3
Flavor solubilizer	0.3
Fluorocarbon emulsifier	2.5
Gelling agent	15–50
Spreading agent	1
Stabilizing agent	1–5
Suppository base	4–6 or 90
Tablet coating	10
Tablet excipient	5–10
Wetting agent	0.01–5

8 Description

Poloxamers 188, 237, 338, and 407 occur as white or almost white, waxy powders, microbeads or flakes. Poloxamer 124 is a colorless or almost colorless liquid. Poloxamers are practically odorless and tasteless.

9 Pharmacopeial Specifications

See Table III.

10 Typical Properties

Acidity/alkalinity pH = 5.0–7.5 for a 2.5% w/v aqueous solution.

Cloud point

>100°C for a 1% w/v aqueous solution, and a 10% w/v aqueous solution of poloxamer 188, 237, 338, and 407;

71–75°C for a 10% solution, and 65°C for a 1% solution of poloxamer 124;

15–19°C for a 10% solution, and 24°C for a 1% solution of poloxamer 181;

22–26°C for a 10% solution, and 32°C for a 1% solution of poloxamer 182;

9–13°C for a 10% solution, and 15°C for a 1% solution of poloxamer 331.

Density 1.06 g/cm^3 at 25°C

Flash point

≈240°C for poloxamer 124;

≈235°C for poloxamer 181 and poloxamer 182;

≈260°C for poloxamer 188;

≈244°C for poloxamer 237;

≈257°C for poloxamer 338;

>204°C for poloxamer 331;

>150°C for poloxamer 407.

Flowability Solid poloxamers are free flowing.

Table III: Pharmacopeial specifications for poloxamer.

Test	PhEur 7.4	USP35–NF30
Identification	+	+
Characters	+	−
Appearance of solution	+	−
Average molecular weight		
Poloxamer 124	2090–2360	2090–2360
Poloxamer 188	7680–9510	7680–9510
Poloxamer 237	6840–8830	6840–8830
Poloxamer 338	12700–17400	12700–17400
Poloxamer 407	9840–14600	9840–14600
Weight percent oxyethylene		
Poloxamer 124	44.8–48.6	46.7 ± 1.9
Poloxamer 188	79.9–83.7	81.8 ± 1.9
Poloxamer 237	70.5–74.3	72.4 ± 1.9
Poloxamer 338	81.4–84.9	83.1 ± 1.7
Poloxamer 407	71.5–74.9	73.2 ± 1.7
pH (aqueous solution)	5.0–7.5	5.0–7.5
Unsaturation (mEq/g)		
Poloxamer 124	−	0.020 ± 0.008
Poloxamer 188	−	0.026 ± 0.008
Poloxamer 237	−	0.034 ± 0.008
Poloxamer 338	−	0.031 ± 0.008
Poloxamer 407	−	0.048 ± 0.017
Oxypropylene : oxyethylene ratio	+	−
Total ash	≤0.4%	−
Heavy metals	−	≤20 ppm
Water	≤1.0%	−
Free ethylene oxide, propylene oxide and 1,4-dioxane	+	+
Ethylene oxide	≤1 ppm	≤1 µg/g
Propylene oxide	≤5 ppm	≤5 µg/g
1,4-Dioxane	≤10 ppm	≤5 µg/g

HLB value
12–18 for poloxamer 124;
1–7 for poloxamer 181;
1–7 for poloxamer 182;
>24 for poloxamer 188, 237, and 338;
1–7 for poloxamer 331;
18–23 for poloxamer 407.

Melting point
16°C for poloxamer 124;
-29°C for poloxamer 181;
-4°C for poloxamer 182;
52°C for poloxamer 188;
49°C for poloxamer 237;
-23°C for poloxamer 331;
57°C for poloxamer 338;
56°C for poloxamer 407.

Moisture content Poloxamers generally contain less than 0.5% w/w water and are hygroscopic only at relative humidity greater than 80%. *See also* Figure 1.

Solubility Solubility varies according to the poloxamer type; *see also* Table IV.

The aqueous solubility of poloxamers decreases as the temperature increases due to reduced hydration of the polymer caused by the breaking of hydrogen bonds at higher temperatures. Grades of poloxamer are available that range from hydrophobic liquids that are practically insoluble in water, to solids that are very soluble in water; aqueous solubility increases as the polyoxyethylene content of the molecule increases.

Spectroscopy
IR spectra *see* Figure 2.

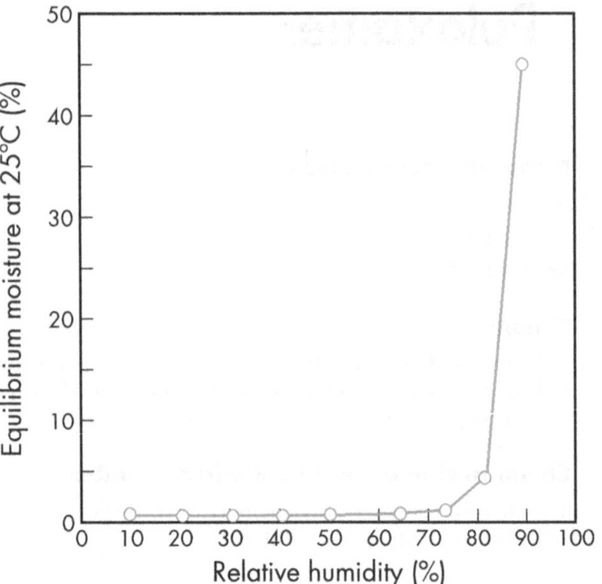

Figure 1: Equilibrium moisture content of poloxamer 188 (*Pluronic F-68*, BASF Corp.).

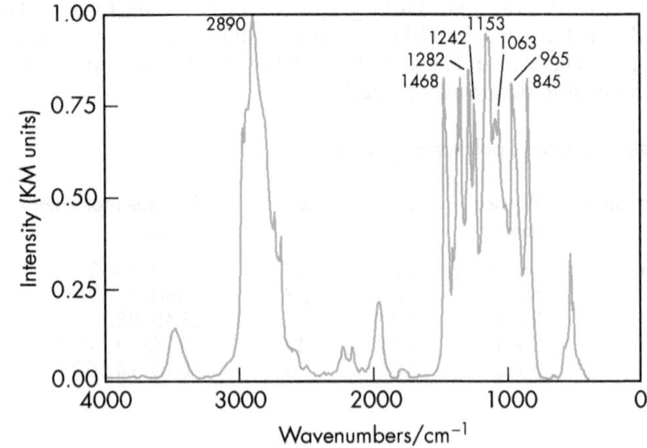

Figure 2: Infrared spectrum of poloxamer measured by diffuse reflectance. Adapted with permission of Informa Healthcare.

Surface tension
45 mN/m (45 dynes/cm) for a 0.1% aqueous solution of poloxamer 124 at 25°C;
43 mN/m (43 dynes/cm) for a 0.1% aqueous solution of poloxamer 182 at 25°C;
50 mN/m (50 dynes/cm) for a 0.1% aqueous solution of poloxamer 188 at 25°C;
44 mN/m (44 dynes/cm) for a 0.1% aqueous solution of poloxamer 237 at 25°C;
41 mN/m (41 dynes/cm) for a 0.1% aqueous solution of poloxamer 338 at 25°C;
41 mN/m (41 dynes/cm) for a 0.1% aqueous solution of poloxamer 407 at 25°C.

Viscosity (dynamic)
440 mPa s (440 cP) as a liquid at 25°C for poloxamer 124;
325 mPa s (325 cP) as a liquid at 25°C for poloxamer 181;
450 mPa s (450 cP) as a liquid at 25°C for poloxamer 182;
1000 mPa s (1000 cP) as a melt at 70°C for poloxamer 188;
700 mPa s (700 cP) as a melt at 70°C for poloxamer 237;
800 mPa s (800 cP) as a liquid at 25°C for poloxamer 331;

Table IV: Solubility at 20°C for various types of poloxamer in different solvents.

Type	Solvent				
	Ethanol (95%)	Propan-2-ol	Propylene glycol	Water	Xylene
Poloxamer 124	Freely soluble	Freely soluble	Freely soluble	Freely soluble	Freely soluble
Poloxamer 188	Freely soluble	—	—	Freely soluble	—
Poloxamer 237	Freely soluble	Sparingly soluble	—	Freely soluble	Sparingly soluble
Poloxamer 338	Freely soluble	—	Sparingly soluble	Freely soluble	—
Poloxamer 407	Freely soluble	Freely soluble	—	Freely soluble	—

2800 mPa s (2800 cP) as a melt at 70°C for poloxamer 338;
3100 mPa s (3100 cP) as a melt at 70°C for poloxamer 407.

11 Stability and Storage Conditions

Poloxamers are stable materials that are not liable to polymerization. Aqueous solutions are stable in the presence of acids, alkalis, and metal ions. However, aqueous solutions support mold growth.

Poloxamer solutions can be sterilized by autoclaving. At high temperatures, poloxamers may be combustible, producing carbon oxides and phosphates.

Poloxamer 335 micelles stabilized with a cross-linked network of diethylacrylamide were exposed to ultrasound frequencies of 70 kHz and 476 kHz; it was shown that ultrasound did not disrupt the covalent network stabilizing the micelles.[4] Perflunafene emulsions stabilized with poloxamer 188 were more stable than emulsions prepared with lecithin, but required the addition of soybean oil to stabilize the emulsions to enable them to withstand the temperatures required for sterilization by autoclaving.[5]

The material should be stored in an airtight container in a cool, dry place.

12 Incompatibilities

Depending on the relative concentrations, poloxamer 188 is incompatible with phenols and parabens. Poloxamer 331 may react violently with isocyanates. Dimethyl sulfoxide has been shown to induce reductions in the critical micellization and the gelation temperatures of poloxamer systems, as well as influencing the rate of drug release from these systems.[6]

13 Method of Manufacture

Poloxamer polymers are prepared by reacting propylene oxide with propylene glycol to form a polyoxypropylene chain. Polyoxyethylene chains are then added to each end of the polyoxypropylene chain to form the block copolymer.

14 Safety

Poloxamers are used in a variety of oral, parenteral, and topical pharmaceutical formulations, and are generally regarded as nontoxic and nonirritant materials. Poloxamers are not absorbed from the gastrointestinal tract, and may increase the absorption of liquid paraffin and other fat-soluble substances. Poloxamers are not metabolized in the body. Poloxamer 407 has been reported to cause alterations to lipid profiles and possible renal toxicity, which have compromised its use in parenteral applications.[7]

Animal toxicity studies, with dogs and rabbits, have shown poloxamers to be nonirritating and nonsensitizing when applied in 5% w/v and 10% w/v concentration to the eyes, gums, and skin.

In a 14-day study of intravenous administration at concentrations up to 0.5 g/kg/day to rabbits, no overt adverse effects were noted. A similar study with dogs also showed no adverse effects at dosage levels up to 0.5 g/kg/day. In a longer-term study, rats fed 3% w/w or 5% w/w of poloxamer in food for up to 2 years did not exhibit any significant symptoms of toxicity. However, rats receiving 7.5% w/w of poloxamer in their diet showed some decrease in growth rate.

No hemolysis of human blood cells was observed over 18 hours at 25°C, with 0.001–10% w/v poloxamer solutions.

Acute animal toxicity data for poloxamer 188:[8]

LD_{50} (mouse, IV): 1 g/kg
LD_{50} (mouse, oral): 15 g/kg
LD_{50} (mouse, SC): 5.5 g/kg
LD_{50} (rat, IV): 7.5 g/kg
LD_{50} (rat, oral): 9.4 g/kg

15 Handling Precautions

Observe normal precautions appropriate to the circumstances and quantity of material handled. Eye protection and gloves are recommended.

16 Regulatory Status

Included in the FDA Inactive Ingredients Database. Included in nonparenteral medicines licensed in the UK. Included in the Canadian Natural Health Products Ingredients Database. *See also* Table V.

Table V: Poloxamers listed in the US FDA Inactive Ingredients Database and included in licensed medicines in the UK.

Type	US FDA	UK
Poloxamer 124	Oral suspensions Topical gels and lotions	—
Poloxamer 181	Topical lotions	—
Poloxamer 182	Topical gels	Topical gels
Poloxamer 188	Oral capsules, concentrates, extended release tablets, granules, powders for oral suspensions, powders for oral solution, solutions, suspensions, syrups, and tablets Topical creams Intravenous injections, powders for intravenous injections, subcutaneous injections Ophthalmic solutions	Oral capsules, granules for suspensions, orodispersible tablets, prolonged-release tablets, solutions, tablets, and suspensions Topical gels Subcutaneous injections, and powders for reconstitution for intravenous infusions
Poloxamer 237	Topical sponges	—
Poloxamer 331	Oral suspensions, powders for oral suspensions	—
Poloxamer 338	Enteric-coated pellets for oral capsules	—
Poloxamer 407	Oral enteric-coated capsule pellets, solutions, suspensions, and tablets Topical creams, gels, solutions and suspensions Ophthalmic solutions	Oral tablets Oromucosal spray Topical gels and creams

17 Related Substances

—

18 Comments

Although the USP35–NF30 and the PhEur 7.4 contain specifications for five poloxamer grades, many more different poloxamers are commercially available that vary in their molecular weight and the proportion of oxyethylene present in the polymer. A series of poloxamers with greatly varying physical properties are thus available.

The nonproprietary name 'poloxamer' is followed by a number, the first two digits of which, when multiplied by 100, correspond to the approximate average molecular weight of the polyoxypropylene portion of the copolymer and the third digit, when multiplied by 10, corresponds to the percentage by weight of the polyoxyethylene portion. Similarly, with many of the trade names used for poloxamers, e.g. *Pluronic F-68* (BASF Corp.), the first digit arbitrarily represents the molecular weight of the polyoxypropylene portion and the second digit represents the weight percent of the oxyethylene portion. The letters 'L', 'P', and 'F', stand for the physical form of the poloxamer: liquid, paste, or flakes; *see also* Table VI.

Table VI: Nonproprietary name and corresponding commercial grade.

Nonproprietary name	Commercial grade
Poloxamer 124	L-44
Poloxamer 188	F-68
Poloxamer 237	F-87
Poloxamer 338	F-108
Poloxamer 407	F-127

Note that in the US the trade name *Pluronic* is used by BASF Corp. for pharmaceutical-grade and industrial-grade poloxamers, while in Europe the trade name *Lutrol* is used by BASF Corp. for the pharmaceutical-grade material.

Poloxamer 188 has been used as an emulsifying agent for fluorocarbons used as artificial blood substitutes, in the preparation of solid-dispersion systems, as a tablet lubricant,[9] and in liquid suppositories.[10],[11] More recently, poloxamers have found use in drug-delivery systems.[12]–[15]

The thermoreversible, rheological, and solubilizing properties of poloxamer 407 have been reviewed, and immunomodulation and cytotoxicity-promoting properties demonstrated.[7]

The thermoresponsive properties of poloxamers, particularly poloxamer 407, have been exploited in a range of gel formulations for buccal,[16]–[18] ophthalmic,[19]–[24] rectal,[25],[26] vaginal,[27,28] nasal,[29] and oral[30] drug delivery. Poloxamers have also been investigated for use in the preparation of nanoparticles,[()]–[33] subcutaneous implants,[34] intravenous emulsions,[35] topical foams,[36] organogels,[37] and self-emulsifying drug delivery systems.[38]

Poloxamer has been used in a poly(lactic-*co*-glycolic acid) (PLGA) : poloxamer and PLGA : poloxamine blend nanoparticle composition as novel carriers for gene delivery.(39)

Poloxamers for use in the cosmetic industry as oil-in-water emulsifiers, cleansers for mild facial products, and dispersing agents are marketed by BASF Corp. as *Pluracare*; the grades available are listed in Table VII.

Therapeutically, poloxamer 188 is administered orally as a wetting agent and stool lubricant in the treatment of constipation; it is usually used in combination with a laxative such as danthron. Poloxamers may also be used therapeutically as wetting agents in eye-drop formulations, in the treatment of kidney stones, and as skin-wound cleansers.

Specifications for poloxamer 331 and poloxamer 407 are contained in the *Food Chemicals Codex* (FCC).[40]

The PubChem Compound ID (CID) for poloxamer is 24751.

Table VII: Nonproprietary name and corresponding *Pluracare* grade (BASF Corp.).

Nonproprietary name	Commercial grade	HLB value	pH of 2.5% w/v aqueous solution
Poloxamer 184	L-64	12–18	5–7.5
Poloxamer 185	P-65	12–18	6–7.4
Poloxamer 407	F-127	18–23	6–7.4

19 Specific References

1 Lee K, *et al.* Fluorescence spectroscopy studies on micellization of poloxamer 407 solution. *Arch Pharm Res* 2003; **26:** 653–658.

2 Mata JP, *et al.* Concentration, temperature and salt induced micellization of a triblock copolymer *Pluronic L64* in aqueous media. *J Colloid Interface Sci* 2005; **292:** 548–556.

3 Jebari MM, *et al.* Aggregation behaviour of *Pluronic L64* surfactant at various temperatures and concentrations examined by dynamic light scattering and viscosity measurements. *Polym Int* 2006; **55:** 176–183.

4 Husseini GA, *et al.* Degradation kinetics of stabilised *Pluronic* micelles under the action of ultrasound. *J Control Rel* 2009; **138:** 45–48.

5 Johnson OL, *et al.* Long-term stability studies of fluorocarbon oxygen transport emulsions. *Int J Pharm* 1990; **63:** 65–72.

6 Ur-Rehman T, *et al.* Effect of DMSO on micellisation, gelation and drug release profile of poloxamer 407. *Int J Pharm* 2010; **394:** 92–98.

7 Dumortier G, *et al.* A review of poloxamer 407 pharmaceutical and pharmacological characteristics. *Pharm Res* 2006; **23**(12): 2709–2728.

8 Sweet DV, ed. *Registry of Toxic Effects of Chemical Substances.* Cincinnati: US Department of Health, 1987.

9 Desai D, *et al.* Evaluation of selected micronized poloxamers as tablet lubricants. *Drug Deliv* 2007; **14:** 413–426.

10 Yong CS, *et al.* Preparation of ibuprofen-loaded liquid suppository using eutectic mixture system with menthol. *Eur J Pharm Sci* 2004; **23:** 347–353.

11 El-Kamel A, El-Khatib M. Thermally reversible *in situ* gelling carbamazepine liquid suppository. *Drug Deliv* 2006; **13:** 143–148.

12 Lu G, Jun HW. Diffusion studies of methotrexate in carbopol and poloxamer gels. *Int J Pharm* 1998; **160**(1): 1–9.

13 Oh T, *et al.* Micellar formulations for drug delivery based on mixtures of hydrophobic and hydrophilic *Pluronic (R)* block copolymers. *J Control Release* 2004; **94**(10): 411–422.

14 Anderson BC, *et al.* Understanding drug release from poly(ethylene oxide)-*b*-(propylene oxide)-*b*-poly(ethlene oxide) gels. *J Control Release* 2001; **70:** 157–167.

15 Moore T, *et al.* Experimental investigation and mathematical modelling of *Pluromic F127* gel dissolution: drug release in stirred systems. *J Control Release* 2000; **67:** 191–202.

16 Monti D. *et al.*Poloxamer 407 microspheres for orotransmucosal drug delivery. Part II: *In vitro/in vivo* evaluation. *Int J Pharm* 2010; **400:** 32–36.

17 Albertini B, *et al.* Poloxamer 407 microspheres for orotransmucosal drug delivery. Part I: Formulation, manufacturing and characterisation. *Int J Pharm* 2010; **399:** 71–79.

18 Shidhaye SS, *et al.* Buccal delivery of pravastatin sodium. *AAPS PharmSciTech* 2010; **11:** 416–424.

19 Qian Y, *et al.* Preparation and evaluation of *in situ* gelling ophthalmic drug delivery system for methazolamide. *Drug Dev Ind Pharm* 2010; **36:** 1340–1347.

20 Gratieri T, *et al.* A poloxamer/chitosan *in situ* forming gel with prolonged retention time for ocular delivery. *Eur J Pharm Biopharm* 2010; **75:** 186–193.

21 Dumortier G, *et al.* Development of a thermogelling ophthalmic formulation of cysteine. *Drug Dev Ind Pharm* 2006; **32:** 72.

22 Qi H, *et al.* Development of poloxamer analogs/carbopol-based in situ gelling and mucoadhesive ophthalmic delivery system for puerarin. *Int J Pharm* 2007; **337:** 178–187.

23 Bochot A, *et al.* Liposomes dispersed within a thermosensitive gel: a new dosage form for ocular delivery. *Pharm Res* 1998; **15:** 1364–1369.

24 Kim EK, *et al.* rhEGF/HP-β-CD complex in poloxamer gel for ophthalmic delivery. *Int J Pharm* 2002; **233:** 159–167.

25 Barakat NS. *In vitro* and *in vivo* characteristics of a thermogelling rectal delivery system of etodolac. *AAPS PharmSciTech* 2009; **10:** 724–731.

26 Fawaz F, *et al.* Comparative *in vitro-in vivo* study of two quinine rectal gel formulations. *Int J Pharm* 2004; **280**: 151–162.
27 Liu Y, *et al.* Effect of carrageenan on poloxamer-based *in situ* gel for vaginal use: Improved *in vitro* and *in vivo* sustained-release properties. *Eur J Pharm Sci* 2009; **37**: 306–312.
28 Darwish AM, *et al.* Evaluation of a novel vaginal bromocriptine mesylate formulation. *Fertil Steril* 2005; **83**(4): 1053–1055.
29 Bhalerao AV, *et al.* Nasal mucoadhesive *in situ* gel of ondansetron hydrochloride. *Indian J Pharm Sci* 2009; **71**: 711–713.
30 Jones DS, *et al.* Rheological, mechanical and mucoadhesive properties of thermoresponsive, bioadhesive binary mixtures composed of poloxamer 407 and *Carbopol 974P* designed as platforms for implantable drug delivery systems for use in the oral cavity. *Int J Pharm* 2009; **372**: 49–58.
31 Oh KS, *et al.* Temperature-induced gel formation of core/shell nanoparticles for the regeneration of ischemic heart. *J Control Rel* 2010; **146**: 207–211.
32 Kalam MA, *et al.* Preparation, characterisation, and evaluation of gatifloxacin loaded solid lipid nanoparticles as colloidal ocular drug delivery system. *J Drug Target* 2010; **18**: 191–204.
33 Kim JY, *et al. In vivo* tumour targeting of pluronic-based nano-carriers. *J Control Rel* 2010; **147**: 109–117.
34 Mei L, *et al.* A novel mifepristone-loaded implant for long-term treatment of endometriosis: *In vitro* and *in vivo* studies. *Eur J Pharm Sci* 2010; **39**: 421–427.
35 Xia XJ, *et al.* Formulation, characterisation and hypersensitivity evaluation of an intravenous emulsion loaded with a paclitaxel-cholesterol complex. *Chemical and Pharmaceutical Bulletin* 2011; **59**: 321–326.
36 Zhao YJ, *et al.* Engineering novel topical foams using hydrofluoroalkane emulsions stabilised with pluronic surfactants. *Eur J Pharm Sci* 2009; **37**: 370–377.
37 Pandey M, *et al.* Pluronic lecithin organogel as a topical drug delivery system. *Drug Deliv* 2010; **17**: 38–47.
38 Hong JY, *et al.* A new self-emulsifying formulation of itraconazole with improved dissolution and oral absorption. *J Control Rel* 2006; **110**: 332–338.
39 Csaba N, *et al.* PLGA:poloxamer and PLGA:poloxamine blend nanoparticles: new carriers for gene delivery. *Biomacromolecules* 2005; **6**(1): 271–278.
40 *Food Chemicals Codex*, 7th edn. Bethesda, MD: United States Pharmacopeia, 2010; 807–811.

20 General References

BASF. Technical literature: Pluronics. www.worldaccount.basf.com/wa/NAFTA/Catalog/ChemicalsNAFTA/pi/BASF/Brand/pluronic (accessed 28 October 2011).
Science Lab.com. Technical literature: Poloxamers. www.sciencelab.com (accessed 28 October 2011).
Spectrum. Material safety data sheet: Poloxamers. www.spectrumchemical.com/OA_HTML/ibeCCtpSctDspRte.jsp?section=10568&minisite=10020&respid=50577 (accessed 28 October 2011)..

21 Author

CG Cable.

22 Date of Revision

2 March 2012.

Polycarbophil

1 Nonproprietary Names

USP–NF: Polycarbophil

2 Synonyms

Noveon AA-1.

3 Chemical Name and CAS Registry Number

Polycarbophil [9003-97-8]

4 Empirical Formula and Molecular Weight

Polycarbophil is a high molecular weight acrylic acid polymer crosslinked with divinyl glycol. The molecular weight of this polymer is theoretically estimated to range from 700 000 to 3–4 billion. However, there are no methods currently available to measure the actual molecular weight of a crosslinked (i.e. three-dimensional) polymer of this type.

5 Structural Formula

See Section 4.

6 Functional Category

Adsorbent; bioadhesive material; emulsifying agent; gelling agent; modified-release agent; suspending agent; tablet and capsule binder; viscosity-increasing agent.

7 Applications in Pharmaceutical Formulation or Technology

Conventionally, polycarbophil is used as a viscosity-increasing agent at very low concentrations (less than 1%) to produce a wide range of viscosities and flow properties in topical lotions, creams, and gels, in oral suspensions, and in transdermal gel reservoirs. It is also used as an emulsifying agent in topical oil-in-water systems.

Polycarbophil is an excellent bioadhesive in various applications.[1–10] Buccal tablets prepared using polycarbophil have shown high bioadhesive force and prolonged residence time, and proved to be non-irritant in *in vivo* trials with human buccal mucosa.[11] Polycarbophil has been used in combination with hydroxypropyl methylcellulose to develop a bilayered buccal bioadhesive film formulation of nicotine hydrogen tartrate for smoking cessation therapy.[12] Thyrotropin loaded bioadhesive buccal patches prepared with polycarbophil and sodium alginate have shown a sustained release profile, maximum adhesion force with highest water uptake and swelling capacity compared to other polymers.[13] Polycarbophil is also useful in designing controlled-release formulations[14] and for drugs that undergo first-pass metabolism.[15] Sublingual tablets of buprenorphine formulated using polycarbophil have shown superior mucoadhesive strength when compared to those using carbomer.[16]

Polycarbophil has also been used in ocular drug delivery systems.[17–20]

Polycarbophil with carboxymethylcellulose sodium are the polymers of choice for the formulation of an acid-buffering bioadhesive vaginal tablet of clotrimazole and metronidazole.[21] Mucoadhesive vaginal vaccine delivery systems using polycarbophil have proved to be effective in the induction of mucosal and systemic immune responses.[22]

Polycarbophil gels have been used to deliver granulocyte-macrophage colony-stimulating factor (GM-CSF) effectively to genital preneoplastic lesions.[23] Polycarbophil microspheres have been formulated for drug delivery to oral[24,25] and nasal[26] cavities. Tablet matrices prepared using a combination of polycarbophil with hydrophobic polymer ethyl cellulose showed a sigmodial release pattern that is ideal for colonic drug delivery.[27,28] Floating-bioadhesive microspheres coated with polycarbophil have been found to be a useful gastroretentive drug delivery system for the treatment of *Helicobacter pylori.*[29]

Conjugation with L-cysteine (thiolated polycarbophil) greatly enhances the mucoadhesive properties of polycarbophil[30] and can be used as a platform for oral[31] and nasal[32] polypeptide delivery (e.g. heparin,[33–35] human growth hormone,[36] insulin[37,38]). These compounds have shown higher stability and more controlled drug release, and have also been reported to act as a permeation enhancer.[39,40] Due to its likelihood for inhibiting P-glycoprotein, thiolated polycarbophil has demonstrated improved bioavailability of an oral paclitaxel formulation.[41]

8 Description

Polycarbophil occurs as fluffy, white to off-white, mildly acidic polymer powder with slightly ester-like odor.

9 Pharmacopeial Specifications

See Table I.

Table I: Pharmacopeial specifications for polycarbophil.

Test	USP35–NF30
Identification	+
pH (1% dispersion)	≤4.0
Loss on drying	≤1.5%
Absorbing power	≥62 g/g
Limit of acrylic acid	≤0.3%
Limit of ethyl acetate	≤0.45%
Residue on ignition	≤4.0%

10 Typical Properties

Density (bulk) ≤0.24 g/cm^3

Dissociation constant pK_a = 6.0 ± 0.5

Equilibrium moisture content 8–10% (at 50% relative humidity)

Glass transition temperature 100–105°C

Particle size distribution Polycarbophils are produced from primary polymer particles of an average diameter of about 0.2 μm. These polymers are then flocculated, resulting in powders averaging 2–7 μm in diameter. Once formed, the flocculated agglomerates cannot be broken down into their primary particles.

Residual solvents

Benzene 0.50 ppm;

Ethyl acetate 0.45%.

Solubility Polycarbophil polymers do not dissolve in water but can swell in water to around 1000 times their original volume (and ten times their original diameter) to form gels when exposed to a pH environment above 4–6. Since the pK_a of these polymers is 6.0 ± 0.5, the carboxylate groups on the polymer backbone ionize, resulting in electrostatic repulsion between the negative particles, which extends the molecule, adding to the swelling of the polymer.

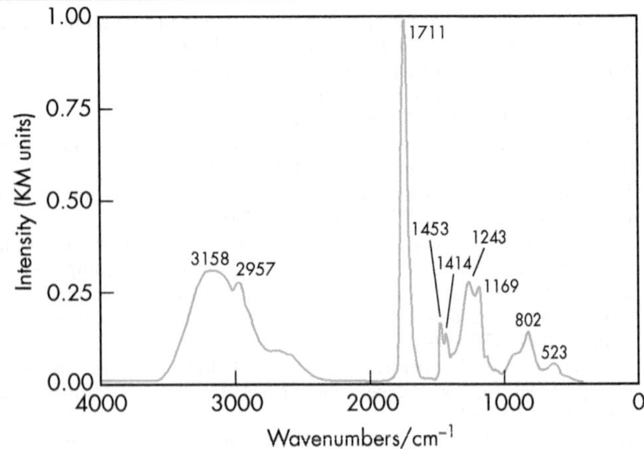

Figure 1: Infrared spectrum of polycarbophil measured by diffuse reflectance. Adapted with permission of Informa Healthcare.

Specific gravity 1.41 at 20°C

Spectroscopy

IR spectra *see* Figure 1.

11 Stability and Storage Conditions

Polycarbophil polymers are stable, hygroscopic materials. They do not undergo hydrolysis or oxidation under normal conditions. Heating at temperatures below 104°C and for up to 2 hours does not affect the efficiency of the dry polymer. However, prolonged exposure to excessive temperatures can result in sintering of the polymer, which alters the drug release rate from tablets, and slows dispersion and gel formation in liquid formulations. Discoloration may occur depending on temperature and exposure time. Complete decomposition occurs with heating for 30 minutes at 260°C.

Polycarbophil polymers do not support bacteria, mold, or fungal growth in dry powder form. Microbial growth may occur in mucilages of the polymer solution. Although the gel properties are not affected by such growth, this phenomenon is usually unacceptable. The addition of appropriate preservatives prevents mold and bacterial growth in these mucilages. Mucilages and emulsions containing these polymers are stable under freeze–thaw conditions but exposure to high temperatures results in a drop in viscosity.

Polycarbophil polymers are very hygroscopic and should be packed in airtight, corrosion-resistant containers. They should be stored in a cool, dry place, and the container should be kept closed when not in use. Moisture pickup does not affect the efficiency of the resins, but resin containing high levels of moisture is more difficult to disperse and weigh accurately. Glass, plastic, or resin-lined containers are recommended for products containing polycarbophil. Packaging in aluminum tubes usually requires formulations to have a pH of 6.5 or less, and for packaging in other metallic tubes or containers a pH of 7.7 or greater to is preferred.

12 Incompatibilities

Heat may be generated if polycarbophil comes into contact with strong basic materials such as ammonia, sodium hydroxide, potassium hydroxide, or strongly basic amines. Polycarbophil polymers are not compatible with cationic polymers, strong acids, and high levels of electrolytes, as electrolytes tend to reduce the viscosity of polycarbophil-based gels.

13 Method of Manufacture

Polycarbophils are synthetic, high-molecular-weight, crosslinked polymers of acrylic acid. These poly(acrylic acid) polymers are crosslinked with divinyl glycol. They are synthesized via precipitation polymerization in ethyl acetate and then dried.

14 Safety

Polycarbophil polymers have a long history of safe and effective use in topical gels, creams, lotions, and ointments. They have been shown to have extremely low irritancy properties and are nonsensitizing with repeated usage.

Polycarbophil dust can be an irritant to eyes, mucous membranes, and the respiratory tract. Dust inhalation may cause coughing, mucus production, and shortness of breath. Contact dermatitis may occur in individuals under extreme conditions of prolonged and repeated contact, high exposure, high temperature, and occlusion (being held onto the skin) by clothing.

The use of these polymers is supported by extensive toxicological studies.[42]

LD_{50} (guinea pig, oral): 2.0 g/kg
LD_{50} (mouse, IP): 0.039 g/kg
LD_{50} (mouse, IV): 0.070 g/kg
LD_{50} (mouse, oral): 4.6 g/kg
LD_{50} (rat, oral): >2.5 g/kg
LD_{50} (rabbit, skin): >3.0 g/kg

15 Handling Precautions

Observe normal precautions appropriate to the circumstances and quantity of material handled. Excessive dust generation should be minimized to avoid the risk of explosion (lowest explosive concentration is 130 g/m^3). Gloves, eye protection, and a dust respirator are recommended during handling. Polycarbophil should be used in well-ventilated conditions.

16 Regulatory Status

GRAS listed. Included in the FDA Inactive Ingredients Database (buccal tablets and films; ophthalmic preparations; topical patches; vaginal gel). Included in nonparenteral medicines licensed in the UK.

17 Related Substances

Calcium polycarbophil; carbomer.

Calcium polycarbophil

Empirical formula Calcium polycarbophil is the calcium salt of polyacrylic acid crosslinked with divinyl glycol.

Molecular weight The molecular weight of these polymers is theoretically estimated to range from 700 000 to 3–4 billion. There are, however, no methods currently available to measure the actual molecular weight of a crosslinked (i.e. three-dimensional) polymer of this type.

CAS number [9003-97-8]

Synonyms *Noveon CA-1*; *Noveon CA-2*.

Appearance White powder with slightly acetic odor.

Acidity/alkalinity pH = 6.0–8.0 (1% w/v aqueous dispersion)

Density (bulk) 0.86 g/cm^3 (*Noveon CA-1*); 0.55 g/cm^3 (*Noveon CA-2*).

Moisture content <10%

Particle size distribution

75 μm (*Noveon CA-1*);
25 μm (*Noveon CA-2*).

Pharmacopeial specifications *see* Table II.

Table II: Pharmacopeial specifications for calcium polycarbophil.

Test	USP35–NF30
Identification	+
Loss on drying	≤10%
Absorbing power	≥35 g/g
Calcium content (on dried basis)	18–22%

Safety

LD_{50}: (rat, oral): >2.5 g/kg
LD_{50}: (rabbit, skin): >3.0 g/kg

Regulatory status GRAS listed. Included in the FDA Inactive Ingredients Database (oral, troche).

Comments *Noveon CA-1* is a coarsely ground grade of calcium polycarbophil and is ideally suited for formulating swallowable bulk laxative tablets, while *Noveon CA-2* is a finely ground grade and is designed for formulating chewable or swallowable bulk laxative tablets. Both grades swell in the intestinal tract, taking advantage of the natural water absorbency of polycarbophil. The swollen polycarbophil gel then acts as a bulk laxative as it moves through the gastrointestinal tract. Calcium polycarbophil is useful in improving colonic transit, bowel movements, stool form and abdominal pain in patients with irritable bowel syndrome.[43,44]

18 Comments

A novel interpolyelectrolyte complex of chitosan–polycarbophil has demonstrated a high potential as an excipient for intranasal,[45] topical, and transdermal drug delivery, and for the production of monolithic swellable matrix systems with controlled drug release properties.[46–48]

19 Specific References

1 Sangeetha S, *et al.* Mucosa as a route for systemic drug delivery. *RJPBCS* 2010; **1**(3): 178–187.
2 Jones DS, *et al.* Physicochemical characterization and preliminary *in vivo* efficacy of bioadhesive, semisolid formulations containing flurbiprofen for the treatment of gingivitis. *J Pharm Sci* 1999; **88**(6): 592–598.
3 Jones DS, *et al.* Viscoelastic properties of bioadhesive, chlorhexidine-containing semi-solids for topical application to the oropharynx. *Pharm Res* 1998; **15**(7): 1131–1136.
4 Jones DS, *et al.* Development and mechanical characterization of bioadhesive semi-solid, polymeric systems containing tetracycline for the treatment of periodontal diseases. *Pharm Res* 1996; **13**(11): 1734–1738.
5 Jones DS, *et al.* Design, characterisation and preliminary clinical evaluation of a novel mucoadhesive topical formulation containing tetracycline for the treatment of periodontal disease. *J Control Release* 2000; **67**(2–3): 357–368.
6 Park JS, *et al. In situ* gelling and mucoadhesive polymer vehicles for controlled intranasal delivery of plasmid DNA. *J Biomed Mater Res* 2002; **59**(1): 144–151.
7 Bregni C, *et al.* Release study of diclofenac from new carbomer gels. *Pak J Pharm Sci* 2008; **21**(1): 12–16.
8 Acartürk F. Mucoadhesive vaginal drug delivery systems. *Recent Pat Drug Deliv Formul* 2009; **3**(3): 193–205.
9 Gupta S, *et al.* Dual-drug delivery system based on *in situ* gel-forming nanosuspension of forskolin to enhance antiglaucoma efficacy. *AAPS PharmSciTech* 2010; **11**(1): 322–335.
10 Hosny EA. Relative hypoglycemia of rectal insulin suppositories containing deoxycholic acid, sodium taurocholate, polycarbophil, and their combinations in diabetic rabbits. *Drug Dev Ind Pharm* 1999; **25**(6): 745–752.
11 Nafee NA, *et al.* Mucoadhesive delivery systems. I. Evaluation of mucoadhesive polymers for buccal tablet formulation. *Drug Dev Ind Pharm* 2004; **30**(9): 985–993.
12 Garg S, Kumar G. Development and evaluation of a buccal bioadhesive system for smoking cessation therapy. *Pharmazie* 2007; **624**: 266–272.
13 Chinwala MG, Lin S. Application of hydrogel polymers for development of thyrotropin releasing hormone-loaded adhesive buccal patches. *Pharm Dev Technol* 2010; **15**(3): 311–327.
14 Jain AC, *et al.* Development and *in vivo* evaluation of buccal tablets prepared using danazol–sulfobutylether 7 beta-cyclodextrin (SBE 7) complexes. *J Pharm Sci* 2002; **91**(7): 1659–1668.
15 Akbari J, *et al.* Development and evaluation of buccoadhesive propranolol hydrochloride tablet formulations: effect of fillers. *Farmaco* 2004; **59**(2): 155–161.

P

16 Das NG, Das SK. Development of mucoadhesive dosage forms of buprenorphine for sublingual drug delivery. *Drug Deliv* 2004; **11**(2): 89–95.
17 Nagarsenker MS, *et al.* Preparation and evaluation of liposomal formulations of tropicamide for ocular delivery. *Int J Pharm* 1999; **190**(1): 63–71.
18 Akpek EK, *et al.* Ocular surface distribution and pharmacokinetics of a novel ophthalmic 1% azithromycin formulation. *J Ocul Pharmcol Ther* 2009; **25**(5): 433–439.
19 Bowman LM, *et al.* Development of a topical polymeric mucoadhesive ocular delivery system for azithromycin. *J Ocul Pharmacol Ther* 2009; **25**(2): 133–139.
20 Sensoy D, *et al.* Bioadhesive sulfacetamide sodium microspheres: evaluation of their effectiveness in the treatment of bacterial keratitis caused by *Staphylococcus aureus* and *Pseudomonas aeruginosa* in a rabbit model. *Eur J Pharm Biopharm* 2009; **72**(3): 487–495.
21 Alam MA, *et al.* Development and evaluation of acid-buffering bioadhesive vaginal tablet for mixed vaginal infections. *AAPS Pharm Sci Tech* 2007; **84**: E109.
22 Oh YK, *et al.* Enhanced mucosal and systemic immune responses to a vaginal vaccine coadministered with RANTES-expressing plasmid DNA using in situ-gelling mucoadhesive delivery system. *Vaccine* 2003; **21**(17–18): 1980–1988.
23 Hubert P, *et al.* Delivery of granulocyte-macrophage colony-stimulating factor in bioadhesive hydrogel stimulates migration of dendritic cells in models of human papillomavirus-associated (pre)neoplastic epithelial lesions. *Antimicrob Agents Chemother* 2004; **48**(11): 4342–4348.
24 Kockisch S, *et al.* Polymeric microspheres for drug delivery to the oral cavity: an *in vitro* evaluation of mucoadhesive potential. *J Pharm Sci* 2003; **92**(8): 1614–1623.
25 Kockisch S, *et al.* In situ evaluation of drug-loaded microspheres on a mucosal surface under dynamic test conditions. *Int J Pharm* 2004; **276**(1–2): 51–58.
26 Leitner VM, *et al.* Nasal delivery of human growth hormone: *in vitro* and *in vivo* evaluation of a thiomer/glutathione microparticulate delivery system. *J Control Release* 2004; **100**(1): 87–95.
27 Ali Asghar LF, *et al.* Design and *in vitro* evaluation of formulations with pH and transit time controlled sigmoidal release profile for colon-specific delivery. *Drug Deliv* 2009; **16**(6): 295–303.
28 Hamman JH. Chitosan based polyelectrolyte complexes as potential carrier materials in drug delivery systems. *Mar Drugs* 2010; **8**: 1305–1322.
29 Umamaheswari RB, *et al.* Floating-bioadhesive microspheres containing acetohydroxamic acid for clearance of *Helicobacter pylori*. *Drug Deliv* 2002; **9**(4): 223–231.
30 Langoth N, *et al.* Development of buccal drug delivery systems based on a thiolated polymer. *Int J Pharm* 2003; **252**(1–2): 141–148.
31 Bernkop-Schnürch A, Thaler SC. Polycarbophil–cysteine conjugates as platforms for oral polypeptide delivery systems. *J Pharm Sci* 2000; **89**(7): 901–909.
32 Bernkop-Schnürch A, *et al.* *In vitro*evaluation of the potential of thiomers for the nasal administration of Leu-enkephalin. *Amino Acids* 2006; **304**: 417–423.
33 Kast CE, *et al.* Development and *in vivo* evaluation of an oral delivery system for low molecular weight heparin based on thiolated polycarbophil. *Pharm Res* 2003; **20**(6): 931–936.
34 Schmitz T, *et al.* Oral heparin delivery: design and *in vitro* evaluation of a stomach-targeted mucoadhesive delivery system. *J Pharm Sci* 2005; **945**: 966–973.
35 Bernkop-Schnürch A, *et al.* Thiomers for oral delivery of hydrophilic macromolecular drugs. *Expert Opin Drug Deliv* 2004; **11**: 87–98.
36 Leitner VM, *et al.* Thiomers in noninvasive polypeptide delivery: *in vitro* and *in vitro* characterization of a polycarbophil-cysteine/glutathione gel formulation for human growth hormone. *J Pharm Sci* 2004; **937**: 1682–1691.
37 Marschutz MK, *et al.* Design and *in vivo* evaluation of an oral delivery system for insulin. *Pharm Res* 2000; **17**(12): 1468–1474.
38 Grabovac V, *et al.* Design and *in vivo* evaluation of a patch delivery system for insulin based on thiolated polymers. *Int J Pharm* 2008; **348**(1–2): 169–174.
39 Di Colo G, *et al.* Polymeric enhancers of mucosal epithelia permeability: synthesis, transepithelial penetration-enhancing properties, mechanism of action, safety issues. *J Pharm Sci* 2007; **975**: 1652–1680.
40 Ross BP, Toth I. Gastrointestinal absorption of heparin by lipidization or coadministration with penetration enhancers. *Curr Drug Deliv* 2005; **23**: 277–287.
41 Föger F, *et al.* Effect of a thiolated polymer on oral paclitaxel absorption and tumor growth in rats. *J Drug Target* 2008; **162**: 149–155.
42 *The Registry of Toxic Effects of Chemical Substances*. Atlanta: National Institute for Occupational Safety and Health, 2004.
43 Chiba T, *et al.* Colonic transit, bowel movements, stool form, and abdominal pain in irritable bowel syndrome by treatments with calcium polycarbophil. *Hepatogastroenterology* 2005; **5265**: 1416–1420.
44 Shibata C, *et al.* Effect of calcium polycarbophil on bowel function after restorative proctocolectomy for ulcerative colitis: a randomized controlled trial. *Dig Dis Sci* 2007; **526**: 1423–1426.
45 Kumar M, *et al.* Formulation and characterization of nanoemulsion of olanzapine for intranasal delivery. *PDA J Pharm Sci Technol* 2009; **63**(6): 501–511.
46 Lu Z, *et al.* Chitosan-polycarbophil interpolyelectrolyte complex as an excipient for bioadhesive matrix systems to control macromolecular drug delivery. *Pharm Dev Technol* 2008; **131**: 37–47.
47 Lu Z, *et al.* Matrix polymeric excipients: comparing a novel interpolyelectrolyte complex with hydroxypropylmethylcellulose. *Drug Deliv* 2008; **152**: 87–96.
48 Lu Z, *et al.* Chitosan-polycarbophil complexes in swellable matrix systems for controlled drug release. *Curr Drug Deliv* 2007; **44**: 257–263.

20 General References

Lubrizol Advanced Materials Inc. *Polycarbophil*. www.pharma.lubrizol.com/products/polycarbophils_overview.asp (accessed 12 December 2011).

21 Author

KK Singh.

22 Date of Revision

2 March 2012.

Polydextrose

1 Nonproprietary Names

USP–NF: Polydextrose

2 Synonyms

E1200; *Litesse*; polydextrose A; polydextrose K; *STA-Lite*.

3 Chemical Name and CAS Registry Number

Polydextrose [68424-04-4]

4 Empirical Formula and Molecular Weight

$(C_6H_{12}O_6)_x$ 1 200–2 000 (average)

5 Structural Formula

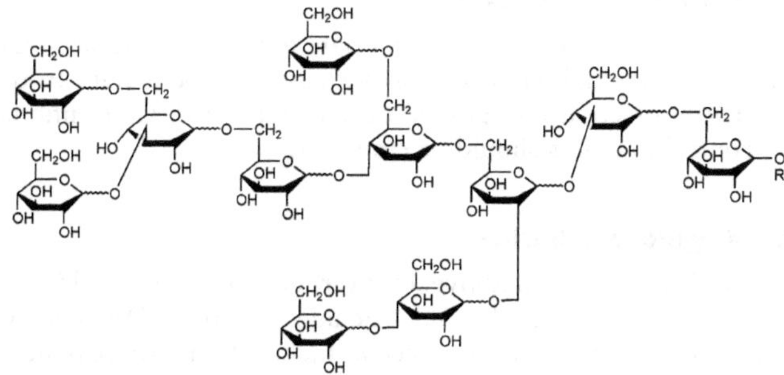

See Section 18.

6 Functional Category

Coating agent; flavor enhancer; humectant; tablet and capsule binder; tablet and capsule diluent; taste-masking agent; viscosity-increasing agent.

7 Applications in Pharmaceutical Formulation or Technology

Polydextrose is widely used in pharmaceutical formulations and food products. In food products it is predominantly used as a low-calorie speciality carbohydrate with texturizing and humectant properties, whilst also facilitating a number of nutrition claims.

Although polydextrose can be used in a wide range of pharmaceutical formulations, its primary use is in solid-dosage forms.

In tableting, polydextrose solutions are used as binders in wet-granulation processes. Polydextrose is also used in the manufacture of directly compressible tableting excipients. Polydextrose solutions may also be used, in conjunction with other materials, as a film and tablet coating agent.

Polydextrose acts as a bulking agent in the formulation of 'sugar-free' confectionery-type dosage forms. In conjunction with isomalt, lactitol, or maltitol, polydextrose can be used in the manufacture of 'sugar-free' hard-boiled candies and acacia lozenges or pastilles as a base for medicated confectionery.

The combination of high water solubility and high viscosity of polydextrose facilitates the processing of sugar-free candies of excellent quality. Polydextrose is amorphous and does not crystallize at low temperatures or high concentrations, so it can be used to control the crystallization of polyols and sugars and therefore the structure and texture of the final product.

8 Description

Polydextrose occurs as an odorless, off-white to light tan powder with a bland, slightly sweet to slightly tart taste, dependent upon grade. Polydextrose is also available as a clear, light yellow to colorless liquid (70% dry substance), which is odorless with a slightly sweet taste.

9 Pharmacopeial Specifications

See Table 1.

Table I: Pharmacopeial specifications for polydextrose.

Test	USP35–NF30
Identification	+
Water	≤4.0%
pH (10% solution)	2.5–5.0
Residue on ignition	≤0.3%
5-Hydroxymethylfurfural and related compounds	≤0.1%
Molecular weight limit	≤22 000
Limit of monomers	
Glucose and sorbitol	≤6.0%
1,6 Anhydrous-D-glucose (levoglucosan)	≤4.0%
Lead	≤0.5 μg/g
Dextrose polymer assay (anhydrous basis)	≥90%

10 Typical Properties

Acidity/alkalinity pH = 2.5–7.0 (10% aqueous solution)

Density (bulk) 0.7–0.8 g/cm^3 (dependent upon grade)

Heat of solution 8 kcal/g

Melting point Polydextrose is an amorphous polymer that does not have a melting range. However, it can undergo a viscosity transition at a temperature as low as 150–160°C.

Moisture content ≤4.0%. At relative humidities above approximately 60%, polydextrose powders absorb significant amounts of moisture; *see* Section 11. *See also* Figure 1.

Refractive index n_D^{20} = 1.3477 (10% w/v aqueous solution)

Solubility Completely miscible in water. Sparingly soluble to insoluble in most organic solvents. Polydextrose has a higher water solubility than most carbohydrates and polyols, allowing the preparation of 80% w/w solutions at 20°C. Polydextrose is soluble in ethanol and only partially soluble in glycerin and propylene glycol.

Viscosity (dynamic) Polydextrose solutions behave as Newtonian fluids. Polydextrose has a higher viscosity than sucrose or sorbitol at equivalent temperatures. This characteristic enables polydextrose to provide the desirable mouthfeel and textural qualities that are important when formulating syrups and viscous solutions. *See* Figure 2.

11 Stability and Storage Conditions

Polydextrose powder is hygroscopic and absorbs significant amounts of moisture at relative humidities greater than 60%. Under dry storage conditions, and in original sealed packaging, polydextrose powders can be expected to retain stability for at least 3 years. Solution grades have a shorter shelf-life of 3 to 6 months (dependent upon grade) at an ambient temperature of 25°C, although this can be extended to 12 months through the use of refrigeration.

Bulk material should be stored in a cool, dry place in well-closed containers.

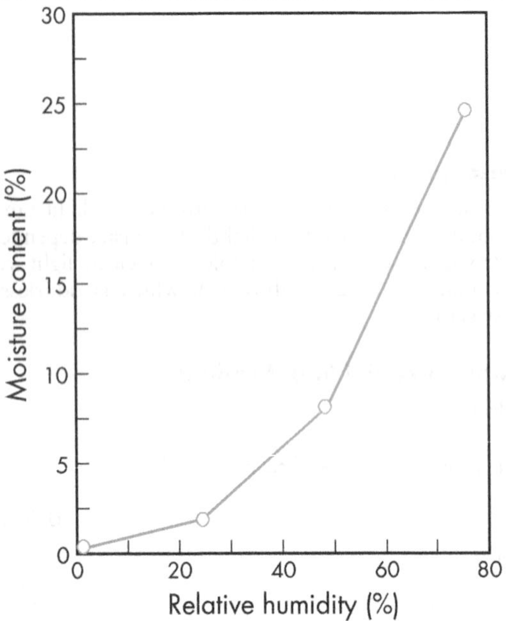

Figure 1: Moisture content of polydextrose at 20°C.

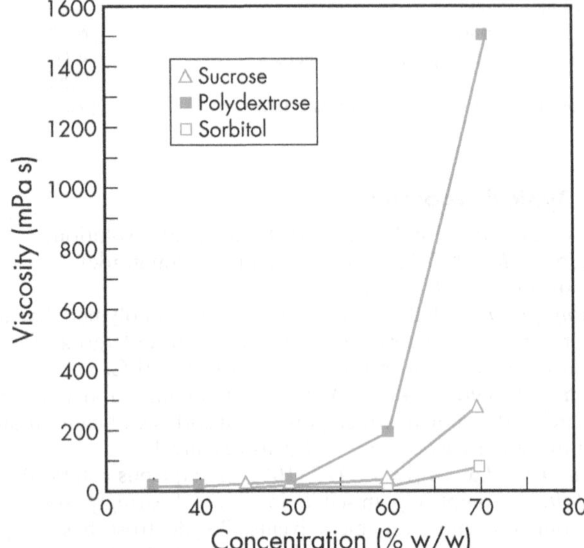

Figure 2: Viscosity of polydextrose solutions at 25°C at various concentrations.

12 Incompatibilities

Incompatible with oxidizing agents, strong acids, and alkalis, forming a brown coloration and depolymerizing.

13 Method of Manufacture

Polydextrose is prepared by the bulk melt polycondensation of glucose and sorbitol in conjunction with small amounts of food-grade acid *in vacuo*. Further purification steps are then involved to generate a range of products with improved organoleptic properties by the removal of acidity and flavor notes generated during the condensation reaction. A partially hydrogenated version of polydextrose, which is suited for high inclusion rates, for sugar-free applications, and where Maillard reactions are not required, is also available.[1]

14 Safety

Polydextrose is used in oral pharmaceutical applications, food products, and confectionery, and is generally regarded as a relatively nontoxic and nonirritant material.[2,3]

However, excessive consumption of non-digestible carbohydrates, such as polydextrose, can lead to gastrointestinal distress. After evaluating a series of clinical studies, the Joint FAO/WHO Expert Committee on Food Additives (JECFA) and the European Commission Scientific Committee for Food (EC/SCF) concluded that polydextrose was better tolerated than other nondigestible carbohydrates such as polyols. The committee concluded that polydextrose has a mean laxative threshold of approximately 90 g/day (1.3 g/kg body-weight) or 50 g as a single dose.[4] [5]*See also* Section 18.

LD_{50} (dog, oral): >20 g/kg[3]
LD_{50} (dog, IV): >2 g/kg[3]
LD_{50} (mouse, oral): >30 g/kg[3]
LD_{50} (rat, oral): >19 g/kg[3]

15 Handling Precautions

Observe normal precautions appropriate to the circumstances and quantity of material handled. Polydextrose may be irritant to the eyes. Eye protection and gloves are recommended. Conventional dust-control practices should be employed.

16 Regulatory Status

Approved as a food additive in over 60 countries worldwide, including Europe and the US. Included in the FDA Inactive Ingredients Database (oral tablets). Included in non-parenteral medicines licensed in the UK. Included in the Canadian Natural Health Products Ingredients Database.

17 Related Substances

Dextrose.

18 Comments

Polydextrose is a randomly bonded polymer prepared by the condensation of a melt that consists of approximately 90% w/w D-glucose, 10% w/w sorbitol, and 1% w/w citric acid or 0.1% w/w phosphoric acid.

The 1,6 glycosidic linkage predominates in the polymer, but all other possible bonds are present. The product contains small quantities of free glucose, sorbitol, and D-anhydroglucoses (levoglucosan), with traces of citric or phosphoric acid.

Polydextrose may be partially reduced by transition-metal catalytic hydrogenation in aqueous solution. It may be neutralized with any food-grade base and/or decolorized and deionized for further purification.

Polydextrose is not absorbed in the small intestine, but passes intact to the large intestine where approximately 50% of the ingested dose is fermented by intestinal microorganisms to produce volatile fatty acids, whilst the remaining 50% of the dose is excreted. The volatile fatty acids are subsequently absorbed in the large intestine. Because of the inefficient way the human body derives energy from volatile fatty acids, and the amount of the ingested dose subsequently excreted, polydextrose contributes only one-quarter of the energy of the equivalent weight of sugar, i.e. ≈4 kJ/g (1 kcal/g).[6–9]

When consumed, polydextrose has a negligible effect on blood glucose levels and is metabolized independently of insulin, making it suitable for diabetics and those on carbohydrate-controlled diets.

A specification for polydextrose is contained in the *Food Chemicals Codex* (FCC).[10]

19 Specific References

1 Mitchell H *et al.* Bulking agents: Multi-functional ingredients. In: Mitchell H, ed. *Sweeteners and Sugar Alternatives in Food Technology.* Oxford: Blackwell Publishing, 2006: 367–380.
2 Flood MT, *et al.* A review of the clinical toleration studies of polydextrose in food. *Food Chem Toxicol* 2004; **42**(9): 1531–1542.
3 Burdock GA, Flamm WG. A review of the studies of the safety of polydextrose in food. *Food Chem Toxicol* 1999; **37**(2–3): 233–264.
4 FAO/WHO. Evaluation of certain food additives and contaminants. Thirty-first report of the Joint FAO/WHO Expert Committee on Food Additives (JECFA). *World Health Organization Technical Report Series* 1987: No. 759.
5 European Commission Scientific Committee for Food (EC/SCF). Excerpt from the minutes of the 71st meeting of the Scientific Committee for Foods: 4.2 Polydextrose. 25–26 January 1990.
6 Figdor SK, Rennhard HH. Caloric utilization and disposition of [^{14}C]polydextrose in the rat. *J Agric Food Chem* 1981; **29**: 1181–1189.
7 Juhr N, Franke J. A method for estimating the available energy of incompletely digested carbohydrates in rats. *J Nutr* 1992; **122**: 1425–1433.
8 Achour L, *et al.* Gastrointestinal effects and energy value of polydextrose in healthy non-obese men. *Am J Clin Nutr* 1994; **59**: 1362–1368.
9 Auerbach MH, *et al.* Caloric availability of polydextrose. *Nutr Rev* 2007; **12**(1): 544–549.
10 *Food Chemicals Codex*, 7th edn. Bethesda, MD: United States Pharmacopeia, 2010: 811.

20 General References

Allingham RP. *Chemistry of Foods and Beverages: Recent Developments.* New York: Academic Press, 1982: 293–303.

Murphy O. Non-polyol low-digestible carbohydrates: food applications and functional benefits. *Br J Nutr* 2001; **85**(Suppl. 1): S47–S53.

Slade L, Levine H. Glass transitions and water–food interaction. In: *Advances in Food and Nutrition Research.* San Diego: Academic Press, 1994.

21 Author

M Bond.

22 Date of Revision

2 March 2012.

Poly(DL-Lactic Acid)

1 Nonproprietary Names

None adopted.

2 Synonyms

DL-Dilactide homopolymer; DL-dilactide polymer; DL-PLA; DL-3,6-dimethyl-1,4-dioxane-2,5-dione homopolymer; DL-PLA; lactic acid homopolymer; D,L-lactic acid homopolymer; D,L-lactic acid polymer; D-lactic acid-L copolymer; DL-lactide polymer; D-lactide-L-lactide copolymer; PDLLA; poly (*RS*)-2-hydroxypropanoic acid; D,L-polylactic acid; poly(*dl*-lactic acid); polylactide; poly(DL-lactide); poly-DL-lactide; *RS*-propanoic acid, 2-hydroxy-, homopolymer.

3 Chemical Name and CAS Registry Number

Poly[oxy(1-methyl-2-oxo-1,2-ethanediyl)] [26023-30-3] or [26680-10-4]

4 Empirical Formula and Molecular Weight

$(C_3H_4O_2)_n$

The molecular weight of this polymer varies according to the intended application. The end group of the polymer chains may be altered to become ester, hydroxyl, or carboxylic acid.

5 Structural Formula

6 Functional Category

Biodegradable material; coating agent; modified-release agent.

7 Applications in Pharmaceutical Formulation or Technology

Poly(DL-lactic acid) is used in drug delivery systems in implants, injections, and oral solid dispersions. It is also used as a coating agent.

8 Description

Poly(DL-lactic acid) is a glassy material, occurring as white to golden-yellow pellets or granules.

9 Pharmacopeial Specifications

—

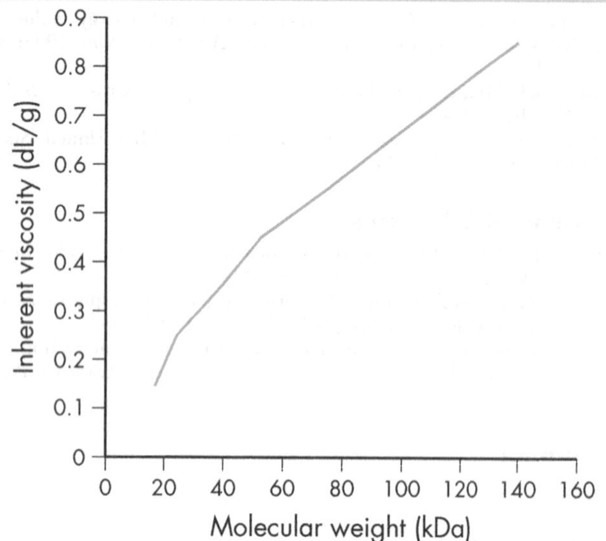

Figure 1: Relationship between the inherent viscosity (IV) and molecular weight (MW) for poly(DL-lactic acid). IV determined in chloroform at 30°C; MW determined by gel permeation chromatography in chloroform. The graph represents best fitted line. (Adapted with permission from the Durect Corporation).

10 Typical Properties

Thermal and mechanical properties of poly(DL-lactic acid) are directly affected by the molecular weight and the composition of the polymer.[1–3]

Density 1.21–1.28 g/cm^3

Elongation (%) 2.5–7.0 (according to molecular weight)

Glass transition temperature 40–69°C (according to the molecular weight and the percentage of DL-lactic acid monomers)

Inherent viscosity *see* Figure 1.

Melting point Amorphous (some sources quote a melting point in the range 165–180°C)

Solubility Soluble in dichloromethane, tetrahydrofuran, ethyl acetate, chloroform, hexafluoroisopropanol, and acetone. Insoluble in water.

Tensile strength 35–85 MPa (according to molecular weight)

11 Stability and Storage Conditions

Poly(DL-lactic acid) is stable under dry conditions. However, it typically biodegrades over a period of 10–15 months according to the molecular weight. Increasing moisture and temperature enhances biodegradation; the onset of degradation in water at 25°C is 6 months.[4] In contrast to many other biodegradable polymers, poly(DL-lactic acid) degrades through a two-step mechanism. The primary degradation step involves the hydrolysis of the ester bonds independently of microbial activity to produce a low-molecular-weight polymer. When the molecular weight drops below 10 000, microorganisms digest the polymer into carbon dioxide and water.[4] The nature of the functional group at the end of the chains has a significant impact on the water uptake of the polymer. Chains with carboxylic end groups are more hydrophilic and therefore exhibit a higher hydrolysis rate and drug release profile than those with hydroxyl or ester end groups.[5] In general, poly(DL-lactic acid) is more stable than poly(L-lactic acid) or poly(D-lactic acid) alone.[6]

Poly(DL-lactic acid) should be stored in a dry inert environment at a temperature of −15°C to −20°C.

12 Incompatibilities

Poly(DL-lactic acid) is incompatible with strong acids or alkaline materials.

13 Method of Manufacture

Lactic acid is a chiral molecule and has two optically active forms: L-lactic acid and D-lactic acid. Poly(DL-lactic acid) is produced from the racemic mixture of lactic acid. Lactic acid is produced either from ethylene (petrochemical pathway) or by bacterial fermentation of D-glucose derived from food stocks. The former pathway involves an oxidation step followed by treatment with hydrogen cyanide and produces only racemic DL-lactic acid. In contrast, lactic acid produced by fermentation occurs mainly as L-lactic acid. Low-molecular-weight poly-DL-(lactic acid) (500–10 000 Da) is produced directly from lactic acid by condensation. Higher-molecular-weight product is produced by one of two major pathways. The first involves a depolymerization of low-molecular-weight polymer into the cyclic dimer form (lactide) followed by ring-opening polymerization. Alternatively, it can be produced by a direct condensation using azeotropic distillation.

14 Safety

Poly(DL-lactic acid) degrades to produce lactic acid, which is considered a well-tolerated nontoxic material. Several *in vitro* and *in vivo* studies demonstrated that poly(lactic acid) in general (including poly(DL-lactic acid)) is well tolerated and does not induce a significant immune response.[7–11] However, some studies have illustrated signs of a mild immune response.[12,13] The FDA has also reported some rare cases of inflammatory responses in patients treated with cosmetic poly(DL-lactic acid) injections.

15 Handling Precautions

Observe normal precautions appropriate to the circumstances and quantity of the material handled. Eye and skin protection are recommended. Handle under dry, inert conditions.

16 Regulatory Status

Included in the FDA Inactive Ingredients Database (IM, powder for injection, lyophilized suspension; periodontal drug delivery system). Poly(DL-lactic acid) is considered as 'not hazardous' according to the European Directive 67/548/EEC. Included in parenteral preparations (prolonged-release powder for suspension for subcutaneous or intramuscular injection) licensed in the UK.

17 Related Substances

Aliphatic polyesters; lactic acid.

18 Comments

Poly(DL-lactic acid) has various IUPAC names, CAS registry numbers, empirical and structural formulae, which is due to some sources quoting the reactants (either lactic acid or lactide) as the repeating unit.

Owing to its high brittleness, poly(DL-lactic acid) is rarely used alone. It is often mixed and copolymerized with other polymers (poly(L-lactide-co-glycolide) (PLGA),[14] poly(ethylene oxide) (PEO),[15] poly(ethylene glycol) (PEG),[16,17] poly(vinylpyrrolidone) (PVP),[16] and poly(vinyl alcohol) (PVA).[18,19] The method of sterilization can affect the mechanical properties of the polymer.

Poly(DL-lactic acid) is a biodegradable thermoplastic polymer. It is used as a component of medical devices such as surgical dressings, sutures, stents, scaffolds for tissue engineering, and dental and bone fixation.

19 Specific References

1 Bio Invigor Corporation. Material safety data sheet: Poly-DL-lactic acid, March 2005.
2 Durect Corporation. Material safety data sheet: *Lactel*, 2005.
3 Purac Biomaterials. Material safety data sheet: *Purasorb PDL 05*, 2007.

4 Lunt J. Large-scale production, properties and commercial applications of polylactic acid polymers. *Polym Degrad Stab* 1998; **59**(1–3): 145–152.
5 Wigginsa JS, *et al.* Hydrolytic degradation of poly(DL-lactide) as a function of end group: carboxylic acid vs. hydroxyl. *Polymer* 2006; **47**(6): 1960–1969.
6 Tsuji H, Fukui I. Enhanced thermal stability of poly(lactide)s in the melt by enantiomeric polymer blending. *Polymer* 2003; **44**(10): 2891–2896.
7 Athanasiou KA, *et al.* Sterilization, toxicity, biocompatibility and clinical applications of polylactic acid polyglycolic acid copolymers. *Biomaterials* 1996; **17**(2): 93–102.
8 Yamada A, *et al.* A three-dimensional microfabrication system for biodegradable polymers with high resolution and biocompatibility. *J Micromech Microeng* 2008; **18**(2): Article 025035.
9 Morgan SM, *et al.* Expansion of human bone marrow stromal cells on poly-(DL-lactide-co-glycolide) (P(DL)LGA) hollow fibres designed for use in skeletal tissue engineering. *Biomaterials* 2007; **28**(35): 5332–5343.
10 Zhang LF, *et al.* An ionically crosslinked hydrogel containing vancomycin coating on a porous scaffold for drug delivery and cell culture. *Int J Pharm* 2008; **353**(1–2): 74–87.
11 Majola A, *et al.* Absorption, biocompatibility, and fixation properties of polylactic acid in bone tissue – an experimental study in rats. *Clin Orthop Relat Res* 1991; **268**: 260–269.
12 Schakenraad JM, *et al. In-vivo* and *in-vitro* degradation of glycine DL-lactic acid copolymers. *J Biomed Mater Res* 1989; **23**(11): 1271–1288.
13 Ren J, *et al.* Poly (D,L-lactide)/nano-hydroxyapatite composite scaffolds for bone tissue engineering and biocompatibility evaluation. *J Mater Sci Mater Med* 2008; **19**(3): 1075–1082.
14 Wu L, *et al.* 'Wet-state' mechanical properties of three-dimensional polyester porous scaffolds. *J Biomed Mater Res A* 2006; **76**(2): 264–271.
15 Bacakova L, *et al.* Adhesion and growth of vascular smooth muscle cells in cultures on bioactive RGD peptide-carrying polylactides. *J Mater Sci Mater Med* 2007; **18**(7): 1317–1323.
16 Gaucher G, *et al.* Pharmaceutical nanotechnology – poly(*N*-vinylpyrrolidone)-block-poly(D,L-lactide) as polymeric emulsifier for the preparation of biodegradable nanoparticles. *J Pharm Sci* 2007; **96**(7): 1763–1775.
17 Joo MK, *et al.* Stereoisomeric effect on reverse thermal gelation of poly(ethylene glycol)/poly(lactide) multiblock copolymer. *Macromolecules* 2007; **40**(14): 5111–5115.
18 Jovanovic I, *et al.* Preparation of smallest microparticles of poly-D,L-lactide by modified precipitation method: influence of the process parameters. *Microsc Res Tech* 2008; **71**(2): 86–92.
19 Cui F, *et al.* Preparation and characterization of melittin-loaded poly (DL-lactic acid) or poly (DL-lactic-co-glycolic acid) microspheres made by the double emulsion method. *J Control Rel* 2005; **107**(2): 310–319.

20 General References

Durect Corporation. *Lactel* Absorbable Polymers. www.durect.com/wt/durect/page_name/bp (accessed 29 September 2011).
Purac Biomaterials. *Purasorb*. www.puracbiomaterials.com (accessed 29 September 2011).

21 Authors

RT Forbes, M Isreb.

22 Date of Revision

1 November 2011.

Polyethylene Glycol

1 Nonproprietary Names

BP: Macrogols
JP: Macrogol 400
Macrogol 1500
Macrogol 4000
Macrogol 6000
Macrogol 20000
PhEur: Macrogols
USP–NF: Polyethylene Glycol

2 Synonyms

Carbowax; *Carbowax Sentry*; *Lipoxol*; *Lutrol E*; macrogola; PEG; *Pluriol E*; polyoxyethylene glycol.

3 Chemical Name and CAS Registry Number

α-Hydro-ω-hydroxypoly(oxy-1,2-ethanediyl) [25322-68-3]

4 Empirical Formula and Molecular Weight

$HOCH_2(CH_2OCH_2)_mCH_2OH$ where *m* represents the average number of oxyethylene groups.

Alternatively, the general formula $H(OCH_2CH_2)_nOH$ may be used to represent polyethylene glycol, where *n* is a number *m* in the previous formula + 1.

See Table I for the average molecular weights of typical polyethylene glycols. Note that the number that follows PEG indicates the average molecular weight of the polymer.

Table I: Structural formula and molecular weight of typical polyethylene glycol polymers.

Grade	*m*	Average molecular weight
PEG 200	4.2	190–210
PEG 300	6.4	285–315
PEG 400	8.7	380–420
PEG 540 (blend of PEG 300 and 1450)	–	500–600
PEG 600	13.2	570–613
PEG 900	15.3	855–900
PEG 1000	22.3	950–1 050
PEG 1450	32.5	1 300–1 600
PEG 1540	28.0–36.0	1 300–1 600
PEG 2000	40.0–50.0	1 800–2 200
PEG 3000	60.0–75.0	2 700–3 300
PEG 3350	75.7	3 000–3 700
PEG 4000	69.0–84.0	3 000–4 800
PEG 4600	104.1	4 400–4 800
PEG 8000	181.4	7 000–9 000

5 Structural Formula

$$HO-\underset{H}{\overset{H}{C}}-(CH_2-O-CH_2)_m-\underset{H}{\overset{H}{C}}-OH$$

6 Functional Category

Coating agent; ointment base; plasticizing agent; solvent; suppository base; tablet and capsule diluent; tablet and capsule lubricant.

7 Applications in Pharmaceutical Formulation or Technology

Polyethylene glycols (PEGs) are widely used in a variety of pharmaceutical formulations, including parenteral, topical, ophthalmic, oral, and rectal preparations. Polyethylene glycol has been used experimentally in biodegradable polymeric matrices used in controlled-release systems.[1]

Polyethylene glycols are stable, hydrophilic substances that are essentially nonirritant to the skin; *see* Section 14. They do not readily penetrate the skin, although the polyethylene glycols are water-soluble and are easily removed from the skin by washing, making them useful as ointment bases.[2] Solid grades are generally employed in topical ointments, with the consistency of the base being adjusted by the addition of liquid grades of polyethylene glycol.

Mixtures of polyethylene glycols can be used as suppository bases,[3] for which they have many advantages over fats. For example, the melting point of the suppository can be modified by choice of polyethylene glycol of an appropriate molecular weight to withstand exposure to warmer climates; release of the drug is not dependent upon melting point; the physical stability on storage is better; and suppositories are readily miscible with rectal fluids. Polyethylene glycols have the following disadvantages: they are chemically more reactive than fats; greater care is needed in processing to avoid inelegant contraction holes in the suppositories; the rate of release of water-soluble medications decreases with the increasing molecular weight of the polyethylene glycol; and polyethylene glycols tend to be more irritating to mucous membranes than fats.

Aqueous polyethylene glycol solutions can be used either as suspending agents or to adjust the viscosity and consistency of other suspending vehicles. When used in conjunction with other emulsifiers, polyethylene glycols can act as emulsion stabilizers.

Liquid polyethylene glycols are used as water-miscible solvents for the contents of soft gelatin capsules. However, if not controlled in the formulation, polyethylene glycols may cause hardening of the capsule shell by preferential absorption of moisture from gelatin in the shell. Gelatin capsule hardening can also be accelerated by inherent trace carbonyl impurities found in polyethylene glycols.

In concentrations up to approximately 30% v/v, PEG 300 and PEG 400 have been used as the vehicle for parenteral dosage forms.

In solid-dosage formulations, higher-molecular-weight polyethylene glycols can enhance the effectiveness of tablet binders and impart plasticity to granules.[4] However, they have only limited binding action when used alone, and can prolong disintegration if present in concentrations greater than 5% w/w. When used for thermoplastic granulations,[5–7] a mixture of the powdered constituents with 10–15% w/w PEG 6000 is heated to 70–75°C. The mass becomes pastelike and forms granules if stirred while cooling. This technique is useful for the preparation of dosage forms such as lozenges when prolonged disintegration is required.

Polyethylene glycols can also be used to enhance the aqueous solubility or dissolution characteristics of poorly soluble compounds by making solid dispersions with an appropriate polyethylene glycol.[8] Animal studies have been performed using polyethylene glycols as solvents for steroids in osmotic pumps.

In film coatings, solid grades of polyethylene glycol can be used alone (6 000-8 000 molecular weight is common) for the film-coating of tablets or can be useful as hydrophilic polishing materials. Solid grades are also widely used as plasticizers in conjunction with film-forming polymers.[9] The presence of polyethylene glycols in film coats, especially of lower molecular weight liquid grades, tends to increase their water permeability and may reduce protection against low pH in enteric-coating films. Polyethylene glycols are useful as plasticizers in microencapsulated products to avoid rupture of the coating film when the microcapsules are compressed into tablets.

Polyethylene glycol grades with molecular weights of 6 000 and above are useful as lubricants, particularly for soluble tablets. The lubricant action is not as effective as that of magnesium stearate, and stickiness may develop if the material becomes too warm during compression. An antiadherent effect is also exerted, again subject to the avoidance of overheating.

Polyethylene glycols have been used in the preparation of urethane hydrogels, which are used as controlled-release agents. Polyethylene glycol has also been used in insulin-loaded microparticles for the oral delivery of insulin;[10,11] it has been used in inhalation preparations to improve aerosolization;[12] polyethylene glycol nanoparticles have been used to improve the oral bioavailability of cyclosporine;[13] it has been used in self-assembled polymeric nanoparticles as a drug carrier;[14] and copolymer networks of polyethylene glycol grafted with poly(methacrylic acid) have been used as bioadhesive controlled drug delivery formulations.[15]

8 Description

The USP35–NF30 describes polyethylene glycol as being an addition polymer of ethylene oxide and water. Polyethylene glycol grades 200–600 are liquids under common ambient temperature conditions; grades 900 and above are solids at ambient temperatures.

Liquid grades (PEG 200–600) occur as clear, colorless or slightly yellow-colored, viscous liquids. They have a slight but characteristic odor and a bitter, slightly burning taste. PEG 600 can occur as a solid at ambient temperatures (melting point 15–25°C).

Solid grades (PEG >900) are white or off-white in color, and range in consistency from pastes to hard waxes. They have a faint, sweet odor. Grades of PEG 3000 and above are available as free-flowing powders.

9 Pharmacopeial Specifications

See Table II. *See also* Section 18.

10 Typical Properties

Density

1.11–1.14 g/cm^3 at 25°C for liquid PEGs;
1.15–1.21 g/cm^3 at 25°C for solid PEGs.

Flash point

182°C for PEG 200;
213°C for PEG 300;
238°C for PEG 400;
250°C for PEG 600.

Freezing point

<−65°C PEG 200 sets to a glass;
−15 to −8°C for PEG 300;
4–8°C for PEG 400;
15–25°C for PEG 600.

Melting point

37–40°C for PEG 1000;
44–48°C for PEG 1500;

Table II: Pharmacopeial specifications for polyethylene glycol.

Test	JP XV	PhEur 7.4	USP35–NF30
Identification	+	+	—
Characters	—	+	—
Acidity or alkalinity	+	+	—
Appearance of solution	+[(a)]	+	+
Density	1.110–1.140[(b)]	*See* Table IV	—
Freezing point	*See* Table III	*See* Table IV	—
Viscosity	—	*See* Table IV	*See* Table V
Average molecular weight	*See* Table III	—	*See* Table V
pH (5% w/v solution)	*See* Table III	—	4.5–7.5
Hydroxyl value	—	*See* Table IV	—
Reducing substances	—	+	—
Residue on ignition	*See* Table III	—	≤0.1%
Sulfated ash	—	≤0.2%	—
Limit of ethylene glycol and diethylene glycol	≤0.25%	≤0.4%	≤0.25%
Ethylene oxide	—	≤1 ppm	≤10 μg/g
1,4-Dioxane	—	≤10 ppm	≤10 μg/g
Heavy metals	—	≤20 ppm	≤5 ppm
Water	≤1.0%	≤2.0%	—
Formaldehyde	—	≤30 ppm	—

(a) For PEG 1500, 4000, 6000, 20000.
(b) For PEG 400.

Table III: Specifications from JP XV.

Type of PEG	Average molecular weight	Freezing point (°C)	pH (5% w/v solution)	Residue on ignition
400	380–420	4–8	4.0–7.0	≤0.1%
1500	—	37–41	4.0–7.0	≤0.1%
4000	2 600–3 800	53–57	4.0–7.5	≤0.2%
6000	7 300–9 300	56–61	4.5–7.5	≤0.2%
20000	15 000–25 000	56–64	4.5–7.5	≤0.2%

40–48°C for PEG 1540;
45–50°C for PEG 2000;
48–54°C for PEG 3000;
50–58°C for PEG 4000;
55–63°C for PEG 6000;
60–63°C for PEG 8000;
60–63°C for PEG 20000.

Moisture content Liquid polyethylene glycols are very hygroscopic, although hygroscopicity decreases with increasing molecular weight. Solid grades, e.g. PEG 4000 and above, are not hygroscopic. *See* Figures 1, 2, and 3.

Particle size distribution *see* Figures 4 and 5.

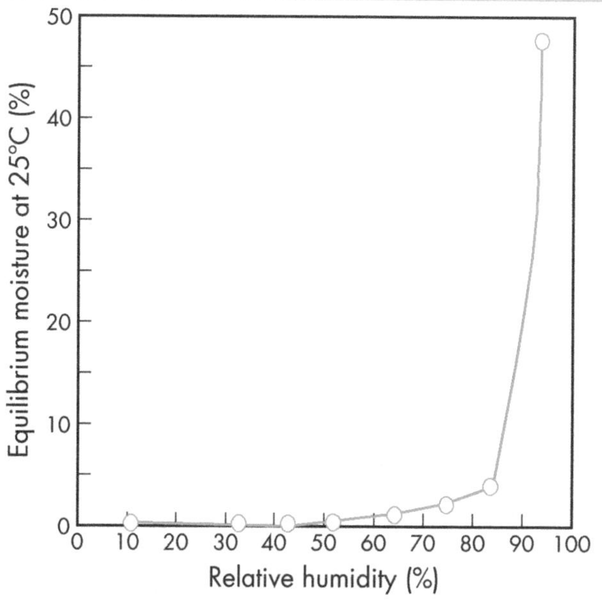

Figure 1: Equilibrium moisture content of PEG 4000 (McKesson, Lot No. B192–8209) at 25°C.

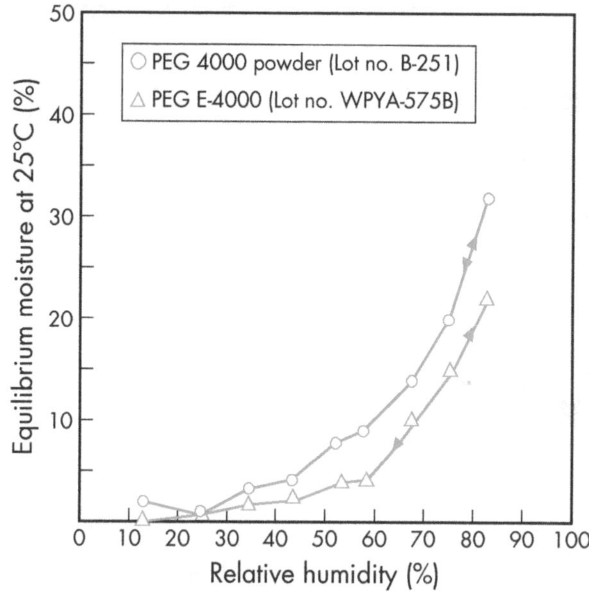

Figure 2: Equilibrium moisture content of PEG 4000 (Dow Chemical Company) and PEG E-4000 (BASF) at 25°C.

Table IV: Specifications from PhEur 7.4.

Type of PEG	Density (g/cm^3)	Freezing point (°C)	Hydroxyl value	Viscosity (dynamic) [mPa s (cP)]	Viscosity (kinematic) [mm^2/s (cSt)]
300	1.120	—	340–394	80–105	71–94
400	1.120	—	264–300	105–130	94–116
600	1.080	15–25	178–197	15–20	13.9–18.5
1000	1.080	35–40	107–118	22–30	20.4–27.7
1500	1.080	42–48	70–80	34–50	31–46
3000	1.080	50–56	34–42	75–100	69–93
3350	1.080	53–57	30–38	83–120	76–110
4000	1.080	53–59	25–32	110–170	102–158
6000	1.080	55–61	16–22	200–270	185–250
8000	1.080	55–62	12–16	260–510	240–472
20000	1.080	≥57	—	2 700–3 500	2 500–3 200
35000	1.080	≥57	—	11 000–14 000	10 000–13 000

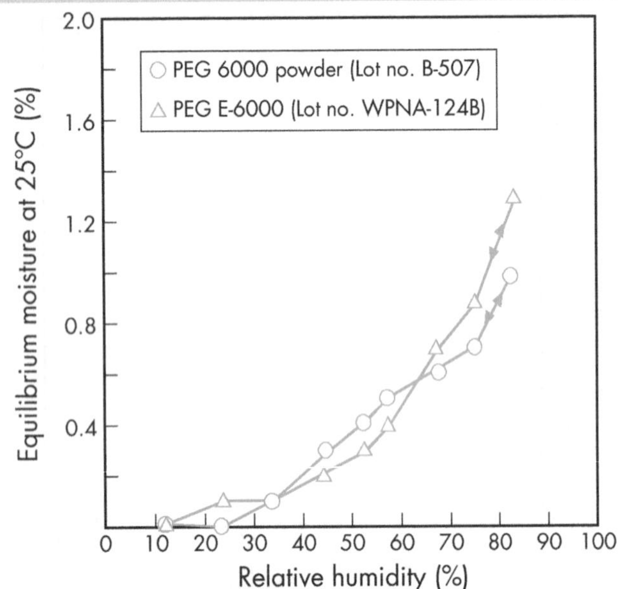

Figure 3: Equilibrium moisture content of PEG 6000 (Dow Chemical Company) and PEG E-6000 (BASF) at 25°C.

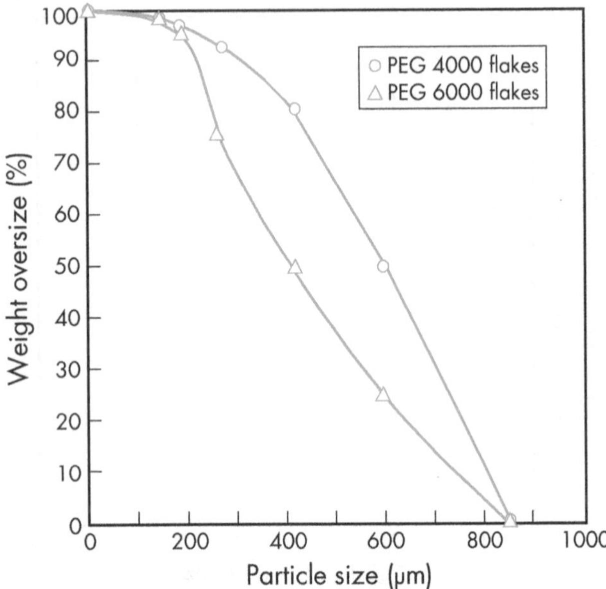

Figure 4: Particle size distribution of PEG 4000 and PEG 6000 flakes.

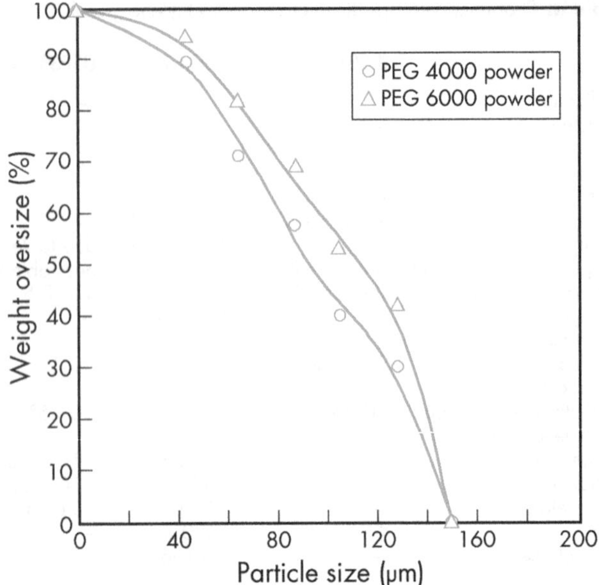

Figure 5: Particle size distribution of PEG 4000 and PEG 6000 powder.

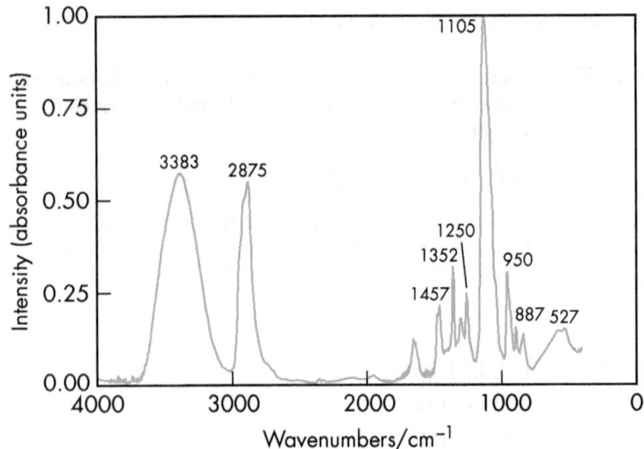

Figure 6: Infrared spectrum of polyethylene glycol (300 NF) measured by transmission. Adapted with permission of Informa Healthcare.

Refractive index

n_D^{25} = 1.459 for PEG 200;
n_D^{25} = 1.463 for PEG 300;
n_D^{25} = 1.465 for PEG 400;
n_D^{25} = 1.467 for PEG 600.

Solubility All grades of polyethylene glycol are soluble in water and miscible in all proportions with other polyethylene glycols (after melting, if necessary). Aqueous solutions of higher-molecular-weight grades may form gels. Liquid polyethylene glycols are soluble in acetone, alcohols, benzene, glycerin, and glycols. Solid polyethylene glycols are soluble in acetone, dichloromethane, ethanol (95%), and methanol; they are slightly soluble in aliphatic hydrocarbons and ether, but insoluble in fats, fixed oils, and mineral oil.

Spectroscopy

IR spectra *see* Figure 6.

Surface tension Approximately 44 mN/m (44 dynes/cm) for liquid polyethylene glycols; approximately 55 mN/m (55 dynes/cm) for 10% w/v aqueous solution of solid polyethylene glycol.

Viscosity (kinematic) *see* Tables IV, V, and VI.

11 Stability and Storage Conditions

Polyethylene glycols exhibit good chemical stability in air and in solution, although grades with a molecular weight less than 2000 are hygroscopic. Polyethylene glycols do not support microbial growth, and they do not become rancid.

Polyethylene glycols and aqueous polyethylene glycol solutions can be sterilized by autoclaving, filtration, or gamma irradiation.[16] Sterilization of solid grades by dry heat at 150°C for 1 hour may induce oxidation, darkening, and the formation of acidic degradation products. Ideally, sterilization should be carried out in an inert atmosphere. Oxidation of polyethylene glycols may also be inhibited by the inclusion of a suitable antioxidant such as butylated hydroxytoluene.

If heated tanks are used to maintain normally solid polyethylene glycols in a molten state for extended periods of time, care must be

Table V: Specification for viscosity of polyethylene glycol of the given nominal molecular weight at 98.9°C ± 0.3°C from the USP35–NF30.

Type of PEG (nominal average molecular weight)	Viscosity (kinematic) [mm²/s (cSt)]
200	3.9–4.8
300	5.4–6.4
400	6.8–8.0
500	8.3–9.6
600	9.9–11.3
700	11.5–13.0
800	12.5–14.5
900	15.0–17.0
1 000	16.0–19.0
1 100	18.0–22.0
1 200	20.0–24.5
1 300	22.0–27.5
1 400	24–30
1 450	25–32
1 500	26–33
1 600	28–36
1 700	31–39
1 800	33–42
1 900	35–45
2 000	38–49
2 100	40–53
2 200	43–56
2 300	46–60
2 400	49–65
2 500	51–70
2 600	54–74
2 700	57–78
2 800	60–83
2 900	64–88
3 000	67–93
3 250	73–105
3 350	76–110
3 500	87–123
3 750	99–140
4 000	110–158
4 250	123–177
4 500	140–200
4 750	155–228
5 000	170–250
5 500	206–315
6 000	250–390
6 500	295–480
7 000	350–590
7 500	405–735
8 000	470–900

Table VI: Viscosity of selected polyethylene glycols at 25°C and 99°C.

Type of PEG	Viscosity [mm²/s (cSt)]	
	25°C	99°C
PEG 200	39.9	4.4
PEG 300	68.8	5.9
PEG 400	90.0	7.4
PEG 600	131	11.0
PEG 1000 solid	19.5	—
PEG 2000 solid	47	—
PEG 4000 solid	180	—
PEG 6000 solid	580	—
PEG 20000 solid	6 900	—

taken to avoid contamination with iron, which can lead to discoloration. The temperature must be kept to the minimum necessary to ensure fluidity; oxidation may occur if polyethylene glycols are exposed for long periods to temperatures exceeding 50°C. However, storage under nitrogen reduces the possibility of oxidation.

Polyethylene glycols should be stored in well-closed containers in a cool, dry place. Stainless steel, aluminum, glass, or lined steel containers are preferred for the storage of liquid grades.

12 Incompatibilities

The chemical reactivity of polyethylene glycols is mainly confined to the two terminal hydroxyl groups, which can be either esterified or etherified. However, all grades can exhibit some oxidizing activity owing to the presence of trace peroxide impurities and secondary products formed by autoxidation.

Liquid and solid polyethylene glycol grades may be incompatible with some coloring agents.

The antibacterial activity of certain antibiotics is reduced in polyethylene glycol bases, particularly that of penicillin and bacitracin. The preservative efficacy of the parabens may also be impaired owing to binding with polyethylene glycols.

Physical effects caused by polyethylene glycol bases include softening and liquefaction in mixtures with phenol, tannic acid, and salicylic acid. Discoloration of sulfonamides and dithranol can also occur, and sorbitol may be precipitated from mixtures. Migration of polyethylene glycol can occur from tablet film coatings, leading to interaction with core components.

13 Method of Manufacture

Polyethylene glycol polymers are formed by the reaction of ethylene oxide and ethylene glycols under pressure in the presence of a catalyst.

14 Safety

Polyethylene glycols are widely used in a variety of pharmaceutical formulations. Generally, they are regarded as having a very low order of toxicity and as nonirritant materials.[17–19]

Adverse reactions to polyethylene glycols have been reported, the greatest toxicity being with glycols of low molecular weight. However, the toxicity of glycols is relatively low.

Polyethylene glycols administered topically may cause stinging, especially when applied to mucous membranes. Hypersensitivity reactions to polyethylene glycols applied topically have also been reported, including urticaria and delayed allergic reactions.[20]

The most serious adverse effects associated with polyethylene glycols are hyperosmolarity, metabolic acidosis, and renal failure following the topical use of polyethylene glycols in burn patients.[21] Topical preparations containing polyethylene glycols should therefore be used cautiously in patients with renal failure, extensive burns, or open wounds.

Liquid polyethylene glycols may be absorbed when taken orally, but the higher-molecular-weight polyethylene glycols are not significantly absorbed from the gastrointestinal tract. Absorbed polyethylene glycol is excreted largely unchanged in the urine, although polyethylene glycols of low molecular weight may be partially metabolized.

The WHO has set an estimated acceptable daily intake of polyethylene glycols at up to 10 mg/kg body-weight.[22]

In parenteral products, the maximum recommended concentration of PEG 300 is approximately 30% v/v as hemolytic effects have been observed at concentrations greater than about 40% v/v.

For animal toxicity data, *see* Table VII.[23]

P

Table VII: Animal toxicity data (LD_{50}) for various grades of polyethylene glycol.[23]

PEG grade	LD_{50} (g/kg)								
	Guinea pig (oral)	**Mouse (IP)**	**Mouse (IV)**	**Mouse (oral)**	**Rabbit (oral)**	**Rabbit (IV)**	**Rat (IP)**	**Rat (IV)**	**Rat (oral)**
PEG 200	—	7.5	—	34	19.9	—	—	—	28.0
PEG 300	19.6	—	—	—	17.3	—	—	—	27.5
PEG 400	15.7	10.0	8.6	28.9	26.8	—	9.7	7.3	—
PEG 600	—	—	—	47	—	—	—	—	38.1
PEG 1000	—	20	—	—	—	—	15.6	—	32
PEG 1500	28.9	—	—	—	28.9	8	17.7	—	44.2
PEG 4000	50.9	—	16	—	76	—	11.6	—	50
PEG 6000	50	—	—	—	—	—	6.8	—	—

15 Handling Precautions

Observe normal precautions appropriate to the circumstances and quantity of material handled. Eye protection is recommended.

16 Regulatory Status

Included in the FDA Inactive Ingredients Database (dental preparations; IM and IV injections; ophthalmic preparations; oral capsules, solutions, syrups, and tablets; rectal, topical, and vaginal preparations). Included in nonparenteral medicines licensed in the UK. Included in the Canadian Natural Health Products Ingredients Database.

17 Related Substances

Polyoxyethylene alkyl ethers; polyethylene oxide; polyoxyethylene sorbitan fatty acid esters; polyoxyethylene stearates; suppository bases.

18 Comments

Polyethylene glycol is one of the materials that have been selected for harmonization by the Pharmacopeial Discussion Group. For further information see the General Information Chapter <1196> in the USP35–NF30, the General Chapter 5.8 in PhEur 7.4, along with the 'State of Work' document on the PhEur EDQM website, and also the General Information Chapter 8 in the JP XV.

Oral administration of large quantities of polyethylene glycols can have a laxative effect and are now commonly available for the relief of occasional constipation. Therapeutically, up to 4 L of an aqueous mixture of electrolytes and high-molecular-weight polyethylene glycol is consumed by patients undergoing bowel cleansing.[24]

A specification for polyethylene glycol is contained in the *Food Chemicals Codex* (FCC).[25]

Specifications for PEG 200, PEG 300, PEG 600, PEG 1000, and PEG 1540 are included in the *Japanese Pharmaceutical Excipients* (JPE).[26]

19 Specific References

1 Mohl S, Winter G. Continuous release of rh-interferon alpha-2a from triglyceride matrices. *J Control Release* 2004; **97**(1): 67–78.
2 Hadia IA, *et al.* Formulation of polyethylene glycol ointment bases suitable for tropical and subtropical climates I. *Acta Pharm Hung* 1989; **59**: 137–142.
3 Kellaway IW, Marriott C. Correlations between physical and drug release characteristics of polyethylene glycol suppositories. *J Pharm Sci* 1975; **64**: 1162–1166.
4 Wells JI, *et al.* Improved wet massed tableting using plasticized binder. *J Pharm Pharmacol* 1982; **34**(Suppl.): 46P.
5 Chiou WL, Riegelman S. Pharmaceutical applications of solid dispersion systems. *J Pharm Sci* 1971; **60**: 1281–1302.
6 Ford JL, Rubinstein MH. Formulation and ageing of tablets prepared from indomethacin–polyethylene glycol 6000 solid dispersions. *Pharm Acta Helv* 1980; **55**: 1–7.
7 Vila-Jato JL, *et al.* The effect of the molecular weight of polyethylene glycol on the bioavailability of paracetamol–polyethylene glycol solid dispersions. *J Pharm Pharmacol* 1986; **38**: 126–128.
8 Miralles MJ, *et al.* Combined water-soluble carriers for coprecipitates of tolbutamide. *J Pharm Sci* 1982; **71**: 302–304.
9 Okhamafe AO, York P. Moisture permeation mechanism of some aqueous-based film coats. *J Pharm Pharmacol* 1982; **34**(Suppl.): 53P.
10 Marishita M, *et al.* Mucosal insulin delivery systems based on complexation polymer hydrogels: effect of particle size on insulin enteral absorption. *J Control Release* 2004; **97**(1): 67–78.
11 Marcel T, *et al.* Calcium phosphate-PEG-insulin-casein (CAPIC) particles as oral delivery systems for insulin. *Int J Pharm* 2004; **277**(1–2): 91–97.
12 Fiegel J, *et al.* Poly(ether-anhydride) dry powder aerosols for sustained drug delivery in the lungs. *J Control Release* 2004; **96**(3): 411–423.
13 Jaiswal J, *et al.* Preparation of biodegradable cyclosporine nanoparticles by high-pressure emulsion-solvent evaporation process. *J Control Release* 2004; **96**(1): 169–178.
14 Jung SW, *et al.* Self-assembled polymeric nanoparticles of poly(ethylene glycol) grafted pullulan acetate as a novel drug carrier. *Arch Pharmacal Res* 2004; **27**(5): 562–569.
15 Peppas NA. Devices based on intelligent biopolymers for oral protein delivery. *Int J Pharm* 2004; **277**(1–2): 11–17.
16 Bhalla HL, *et al.* Radiation sterilization of polyethylene glycols. *Int J Pharm* 1983; **17**: 351–355.
17 Smyth HF, *et al.* The toxicology of the polyethylene glycols. *J Am Pharm Assoc (Sci)* 1950; **39**: 349–354.
18 Tusing TW, *et al.* The chronic dermal toxicity of a series of polyethylene glycols. *J Am Pharm Assoc (Sci)* 1954; **43**: 489–490.
19 Smyth HF, *et al.* The chronic oral toxicology of the polyethylene glycols. *J Am Pharm Assoc (Sci)* 1955; **44**: 27–30.
20 Fisher AA. Immediate and delayed allergic contact reactions to polyethylene glycol. *Contact Dermatitis* 1978; **4**: 135–138.
21 Anonymous. Topical PEG in burn ointments. *FDA Drug Bull* 1982; **12**: 25–26.
22 FAO/WHO. Evaluation of certain food additives. Twenty-third report of the Joint FAO/WHO Expert Committee on Food Additives. *World Health Organ Tech Rep Ser* 1980: No. 648.
23 Lewis RJ, ed. *Sax's Dangerous Properties of Industrial Materials*, 11th edn. New York: Wiley, 2004: 3001.
24 Sweetman SC, ed. *Martindale: The Complete Drug Reference*, 37th edn. London: Pharmaceutical Press, 2011; 2562.
25 *Food Chemicals Codex*, 7th edn. Bethesda, MD: United States Pharmacopeia, 2010: 818.
26 Japan Pharmaceutical Excipients Council. *Japanese Pharmaceutical Excipients* 2004. Tokyo: Yakuji Nippo, 2004: 486, 488, 490, 492, 494.

20 General References

Buggins T, *et al.* The effects of pharmaceutical excipients on drug disposition. *Adv Drug Del Rev* 2007; **59**: 1482–1503.

Donovan MD, *et al.* Absorption of polyethylene glycols 600 through 2000: molecular weight dependence of gastrointestinal and nasal absorption. *Pharm Res* 1990; **7**: 863–867.

European Directorate for the Quality of Medicines and Healthcare (EDQM). European Pharmacopoeia – State Of Work Of International

Harmonisation. *Pharmeuropa* 2011; 23(2): 395–401. http://www.edqm.eu/en/International-Harmonisation-614.html (accessed 13 July 2011).
Mi YI, Wood J. The application and mechanisms of polyethylene glycol 8000 on stabilizing lactate dehydrogenase during lyophilization. *PDA J Pharm Sci Technol* 2004; 58(4): 192–202.
Dow Chemical Company. Technical literature: *Carbowax* and *Carbowax Sentry*, March 2006.
Van Dam J, Daenens P. Molecular weight identification of polyethylene glycols in pharmaceutical preparations by gel permeation chromatography. *J Pharm Sci* 1993; 82: 938–941.
Yamaoka T, *et al*. Distribution and tissue uptake of poly(ethylene glycol) with different molecular weights after intravenous administration to mice. *J Pharm Sci* 1994; 83: 601–606.

21 Authors

B Barner, D Wallick.

22 Date of Revision

2 March 2012.

Polyethylene Oxide

1 Nonproprietary Names

USP–NF: Polyethylene Oxide

2 Synonyms

Polyox; polyoxiante; polyoxirane; polyoxyethylene.

3 Chemical Name and CAS Registry Number

Polyethylene oxide [25322-68-3]

4 Empirical Formula and Molecular Weight

See Table I.

5 Structural Formula

The USP35–NF30 describes polyethylene oxide as a nonionic homopolymer of ethylene oxide, represented by the formula $(CH_2CH_2O)_n$, where n represents the average number of oxyethylene groups. It may contain up to 3% of silicon dioxide or suitable antioxidant.

6 Functional Category

Mucoadhesive; coating agent; tablet and capsule binder; viscosity-increasing agent.

7 Applications in Pharmaceutical Formulation or Technology

Polyethylene oxide can be used as a tablet binder at concentrations of 5–85%. The higher molecular weight grades provide delayed drug release via the hydrophilic matrix approach;[1,2] *see* Table I. Polyethylene oxide has also been shown to facilitate coarse extrusion for tableting[3] as well as being an aid in hot-melt extrusion.[4,5]

The relationship between swelling capacity and molecular weight is a good guide when selecting products for use in immediate- or sustained-release matrix formulations; *see* Figure 1.

Polyethylene oxide has been shown to be an excellent mucoadhesive polymer.[6] Low levels of polyethylene oxide are effective thickeners, although alcohol is usually added to water-based formulations to provide improved viscosity stability; *see* Table II. Polyethylene oxide films demonstrate good lubricity when wet. This property has been utilized in the development of coatings for medical devices. Polyethylene oxide can be radiation crosslinked in solution to produce a hydrogel that can be used in wound care applications.

Table I: Number of repeat units and molecular weight as a function of polymer grade for polyethylene oxide.

Polyox grade	Approximate number of repeating units	Approximate molecular weight
WSR N-10	2 275	100 000
WSR N-80	4 500	200 000
WSR N-750	6 800	300 000
WSR N-3000	9 100	400 000
WSR 205	14 000	600 000
WSR 1105	20 000	900 000
WSR N-12K	23 000	1 000 000
WSR N-60K	45 000	2 000 000
WSR 301	90 000	4 000 000
WSR Coagulant	114 000	5 000 000
WSR 303	159 000	7 000 000

Note: molecular weight based on dilute viscosity measurements.

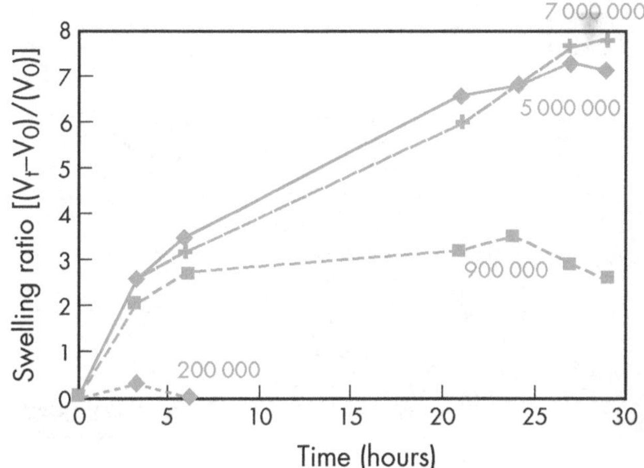

Figure 1: Swelling capacity of polyethylene oxide (*Polyox* WSR, Dow Chemical Company). Measured for four molecular weight grades; 28 mm tablets in 300 mL of water.

8 Description

White to off-white, free-flowing powder. Slight ammoniacal odor.

9 Pharmacopeial Specifications

See Table III.

Table II: Polyethylene oxide viscosity at 25°C (mPa s).

Polyox grade	5% solution	2% solution	1% solution
WSR N-10	30–50	–	–
WSR N-80	55–90	–	–
WSR N-750	600–1 200	–	–
WSR N-3000	2 250–4 500	–	–
WSR 205	4 500–8 800	–	–
WSR 1105	8 800–17 600	–	–
WSR N-12K	–	400–800	–
WSR N-60K	–	2 000–4 000	–
WSR 301	–	–	1 650–5 500
WSR Coagulant	–	–	5 500–7 500
WSR 303	–	–	7 500–10 000

Note: all solution concentrations are based on the water content of the hydro-alcoholic solutions.

SEM 1: Excipient: *Polyox* Coagulant LEO NF; manufacturer: Dow Chemical Company; magnification; 100×

SEM 2: Excipient: *Polyox* WSR N-80; manufacturer: Dow Chemical Company; magnification; 200×

Table III: Pharmacopeial specifications for polyethylene oxide.

Test	USP35–NF30
Identification	+
Loss on drying	≤1.0%
Silicon dioxide and nonsilicon dioxide residue on ignition	≤2.0%
Silicon dioxide	≤3.0%
Heavy metals	≤10 ppm
Free ethylene oxide	≤0.001%

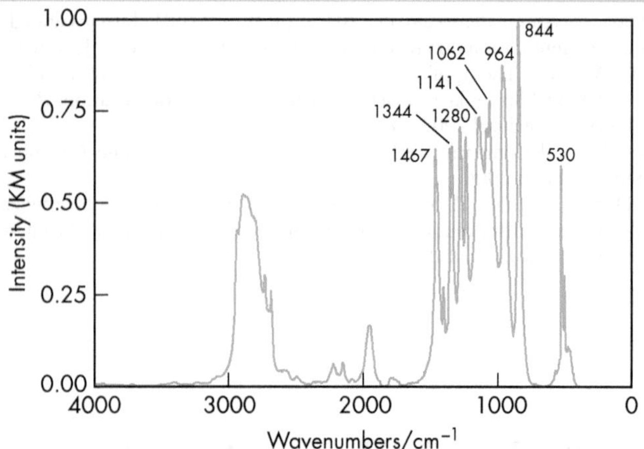

Figure 2: Infrared spectrum of polyethylene oxide measured by diffuse reflectance. Adapted with permission of Informa Healthcare.

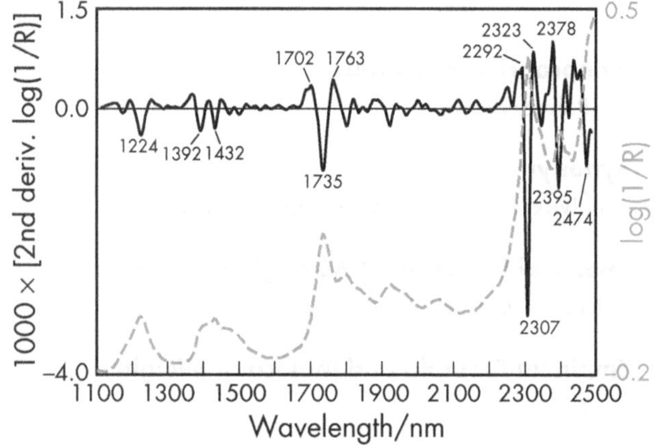

Figure 3: Near-infrared spectrum of polyethylene oxide measured by reflectance.

10 Typical Properties

Angle of repose 34°
Density (true) 1.3 g/cm^3
Melting point 65–70°C
Moisture content <1%
Solubility Polyethylene oxide is soluble in water and a number of common organic solvents such as acetonitrile, chloroform, and methylene chloride. It is insoluble in aliphatic hydrocarbons, ethylene glycol, and most alcohols.[7]
Spectroscopy
IR spectra *see* Figure 2.
NIR spectra *see* Figure 3.
Viscosity (dynamic) *see* Table II.

11 Stability and Storage Conditions

Store in tightly sealed containers in a cool, dry place. Avoid exposure to high temperatures since this can result in reduction in viscosity.

12 Incompatibilities

Polyethylene oxide is incompatible with strong oxidizing agents.

13 Method of Manufacture

Polyethylene oxide is prepared by the polymerization of ethylene oxide using a suitable catalyst.[6]

14 Safety

Animal studies suggest that polyethylene oxide has a low level of toxicity regardless of the route of administration. It is poorly absorbed from the gastrointestinal tract but appears to be completely and rapidly eliminated. The resins are neither skin irritants nor sensitizers, and they do not cause eye irritation.

15 Handling Precautions

Observe normal precautions appropriate to the circumstances and quantity of material handled. Polyethylene oxide may form an explosive dust–air mixture. Gloves, eye protection, a respirator, and other protective clothing should be worn.

16 Regulatory Status

Included in the FDA Inactive Ingredients Database (sustained-release tablets). Included in the Canadian Natural Health Products Ingredients Database.

17 Related Substances

Polyethylene glycol.

18 Comments

Polyethylene oxide and polyethylene glycol have the same CAS Registry Number [25322-68-3].

19 Specific References

1 Dhawan S, *et al.* High molecular weight poly(ethylene oxide)-based drug delivery systems. Part I: hydrogels and hydrophilic matrix systems. *Pharm Technol* 2005; **29**(5): 72–7476–80.
2 Dhawan S, *et al.* Applications of poly(ethylene oxide) in drug delivery systems. Part II. *Pharm Technol* 2005; **29**(9): 82–96.
3 Pinto JF, *et al.* Evaluation of the potential use of poly(ethylene oxide) as tablet-and-extrudate forming material. *AAPS PharmSci* 2004; **6**(2): Article 15.
4 Repka MA, McGinity JW. Influence of vitamin E TPGS on the properties of hydrophilic films provided by hot-melt extrusion. *Int J Pharm* 2000; **202**: 63–70.
5 Coppens KA, *et al.* Hypromellose, ethylcellulose, and polyethylene oxide use in hot melt extrusion. *Pharm Technol* 2005; **30**(1): 62–70.
6 Bottenberg P, *et al.* Development and testing of bioadhesive, fluoride-containing slow-release tablets for oral use. *J Pharm Pharmacol* 1991; 43: 457–464.
7 Bailey FE, Kolesky JV. *Poly(ethylene oxide)*. London: Academic Press, 1976.

20 General References

Dow Chemical Company. Technical literature: *Polyox* water-soluble resins, 2002.
Yu DM, *et al.* Viscoelastic properties of poly(ethylene oxide) solution. *J Pharm Sci* 1994; 83: 1443–1449.

21 Author

JS Maximilien.

22 Date of Revision

2 March 2012.

Polymethacrylates

1 Nonproprietary Names

BP:
Ammonio Methacrylate Copolymer (Type A)
Ammonio Methacrylate Copolymer (Type B)
Basic Butylated Methacrylate Copolymer
Methacrylic Acid–Ethyl Acrylate Copolymer (1 : 1)
Methacrylic Acid–Ethyl Acrylate Copolymer (1 : 1) Dispersion 30 per cent
Methacrylic Acid–Methyl Methacrylate Copolymer (1 : 1)
Methacrylic Acid–Methyl Methacrylate Copolymer (1 : 2)
Polyacrylate Dispersion (30 per cent)

PhEur:
Ammonio Methacrylate Copolymer (Type A)
Ammonio Methacrylate Copolymer (Type B)
Basic Butylated Methacrylate Copolymer
Methacrylic Acid–Ethyl Acrylate Copolymer (1 : 1)
Methacrylic Acid–Ethyl Acrylate Copolymer (1 : 1) Dispersion 30 per cent
Methacrylic Acid–Methyl Methacrylate Copolymer (1 : 1)
Methacrylic Acid–Methyl Methacrylate Copolymer (1 : 2)
Polyacrylate Dispersion 30 per cent

USP–NF:
Amino Methacrylate Copolymer
Ammonio Methacrylate Copolymer
Ammonio Methacrylate Copolymer Dispersion
Ethyl Acrylate and Methyl Methacrylate Copolymer Dispersion
Methacrylic Acid Copolymer
Methacrylic Acid Copolymer Dispersion
Partially Neutralized Methacrylic Acid and Ethyl Acrylate Copolymer
Methacrylic Acid and Ethyl Acrylate Copolymer
Methacrylic Acid and Ethyl Acrylate Copolymer Dispersion

Note that nine separate monographs applicable to polymethacrylates are contained in the USP35–NF30. Several different types of material are defined in the same monographs. The PhEur 7.4 contains eight separate monographs applicable to polymethacrylates. *See also* Section 9.

2 Synonyms

Acidi methacrylici et ethylis acrylatis polymerisatum; acidi methacrylici et methylis methacrylatis polymerisatum; *Acryl-EZE*ammonio methacrylatis copolymerum; copolymerum methacrylatis butylati basicum; *Eastacryl*; *Eudragit*; *Kollicoat MAE*; polyacrylatis dispersio 30 per centum; polymeric methacrylates. *See also* Table I.

3 Chemical Name and CAS Registry Number

See Table I.

Table I: Chemical name and CAS Registry Number of polymethacrylates.

Chemical name	Trade name	Company name	CAS number
Poly(butyl methacrylate, (2-dimethylaminoethyl) methacrylate, methyl methacrylate) 1 : 2 : 1	*Eudragit E 100*	Evonik Industries	[24938-16-7]
	Eudragit E 12.5	Evonik Industries	
	Eudragit E PO	Evonik Industries	
Poly(ethyl acrylate, methyl methacrylate) 2 : 1	*Eudragit NE 30 D*	Evonik Industries	[9010-88-2]
	Eudragit NE 40 D	Evonik Industries	
	Eudragit NM 30 D	Evonik Industries	
Poly(methacrylic acid, methyl methacrylate) 1 : 1	*Eudragit L 100*	Evonik Industries	[25806-15-1]
	Eudragit L 12.5	Evonik Industries	
Poly(methacrylic acid, ethyl acrylate) 1 : 1	*Acryl-EZE 93A*	Colorcon	[25212-88-8]
	Acryl-EZE MP	Colorcon	
	Eudragit L 30 D-55	Evonik Industries	
	Eudragit L 100-55	Evonik Industries	
	Eastacryl 30D	Eastman Chemical	
	Kollicoat MAE 30 DP	BASF Fine Chemicals	
	Kollicoat MAE 100 P	BASF Fine Chemicals	
Poly(methacrylic acid, methyl methacrylate) 1 : 2	*Eudragit S 100*	Evonik Industries	[25086-15-1]
	Eudragit S 12.5	Evonik Industries	
	Eudragit FS 30D	Evonik Industries	
Poly(ethyl acrylate, methyl methacrylate, methacrylic acid) 7 : 3 : 1	*Eudragit RL 100*	Evonik Industries	[33434-24-1]
	Eudragit RL PO	Evonik Industries	
	Eudragit RL 30 D	Evonik Industries	
	Eudragit RL 12.5	Evonik Industries	
	Eudragit RS 100	Evonik Industries	
	Eudragit RS PO	Evonik Industries	
	Eudragit RS 30 D	Evonik Industries	
	Eudragit RS 12.5	Evonik Industries	

4 Empirical Formula and Molecular Weight

The PhEur 7.4 describes methacrylic acid–ethyl acrylate copolymer (1 : 1) as a copolymer of methacrylic acid and ethyl acrylate having a mean relative molecular mass of about 250 000. The ratio of carboxylic groups to ester groups is about 1 : 1. The substance is in the acid form (type A) or partially neutralized using sodium hydroxide (type B). It may contain suitable surfactants such as sodium dodecyl sulfate or polysorbate 80. An aqueous 30% w/v dispersion of this material is also defined in a separate monograph. Methacrylic acid–methyl methacrylate copolymer (1 : 1) is described in the PhEur 7.4 as a copolymer of methacrylic acid and methyl methacrylate having a mean relative molecular mass of about 135 000. The ratio of carboxylic acid to ester groups is about 1 : 1. A further monograph in the PhEur 7.4 describes methacrylic acid–methyl methacrylate copolymer (1 : 2), where the ratio of carboxylic acid to ester groups is about 1 : 2. The PhEur 7.4 describes basic butylated methyacrylate copolymer as a copolymer of (2-dimethylaminoethyl) methacrylate, butyl methyacrylate, and methyl methacrylate having a mean relative molecular mass of about 150 000. The ratio of (2-dimethylaminoethyl) methacrylate groups to butyl methyacrylate and methyl methacrylate groups is about 2 : 1 : 1. The PhEur 7.4 describes ammonio methyacrylate copolymer as a poly(ethyl propenoate-*co*-methyl 2-methylpropenoate-*co*-2-(trimethylammonio)ethyl 2-methylpropenoate) chloride having a mean relative molecular mass of about 150 000. The ratio of ethyl propenoate to methyl 2-methylpropenoate to 2-(trimethylammonio)ethyl 2-methylpropenoate is about 1 : 2 : 0.2 for Type A and 1 : 2 : 0.1 for Type B. Polyacrylate dispersion (30 per cent) is described in the PhEur 7.4 as a dispersion in water of a copolymer of ethyl acrylate and methyl methacrylate having a mean relative molecular mass of about 800 000. It may contain a suitable emulsifier.

The USP35–NF30 describes methacrylic acid copolymer as a fully polymerized copolymer of methacrylic acid and an acrylic or methacrylic ester. Three types of copolymers, namely Type A, Type B, and Type C, are defined in the monograph. They vary in their methacrylic acid content and solution viscosity. All types A, B, and C may contain suitable surface-active agents. Ammonio methacrylate copolymers Type A and Type B, consisting of fully polymerized copolymers of acrylic and methacrylic acid esters with a low content of quaternary ammonium groups, are also described in the USP35–NF30. They vary in their ammonio methacrylate units. The USP35–NF30 also describes amino methacrylate copolymer as a fully polymerized copolymer of 2-dimethylaminoethyl methacrylate, butyl methacrylate and methyl methacrylate. See Sections 9 and 18. Further monographs for aqueous dispersions of Type C methacrylic acid copolymer, ammonio methacrylate copolymer, and also ethyl acrylate and methyl methacrylate copolymer are also defined; *see* Section 9.

The molecular weight information currently reported in the pharmacopeias is ≥100 00. Using size exclusion chromatography (SEC), the average molecular weights are typically ≥30 000.[1,2]

5 Structural Formula

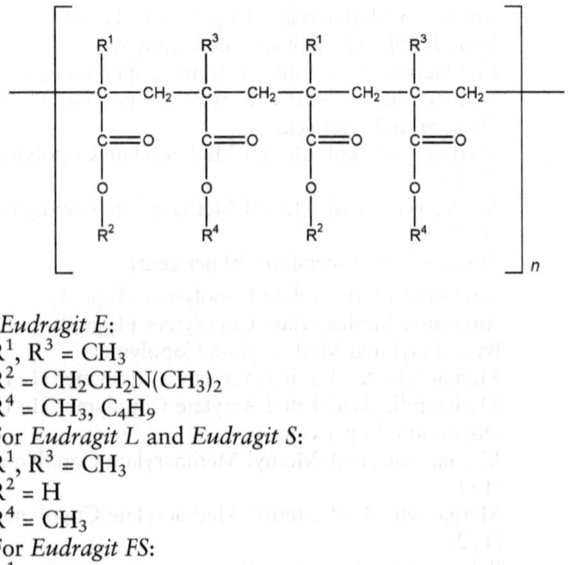

For *Eudragit E*:
$R^1, R^3 = CH_3$
$R^2 = CH_2CH_2N(CH_3)_2$
$R^4 = CH_3, C_4H_9$
For *Eudragit L* and *Eudragit S*:
$R^1, R^3 = CH_3$
$R^2 = H$
$R^4 = CH_3$
For *Eudragit FS*:
$R^1 = H$
$R^2 = H, CH_3$
$R^3 = CH_3$

Table II: Summary of properties and uses of commercially available polymethacrylates.

Grade	Supply form	Polymer dry weight content	Recommended solvents or diluents	Solubility/permeability	Applications
Eudragit E 12.5	Organic solution	12.5%	Acetone, alcohols	Soluble in gastric fluid to pH 5	Film coating
Eudragit E 100	Granules	98%	Acetone, alcohols	Soluble in gastric fluid to pH 5	Film coating
Eudragit E PO	Powder	98%	Acetone, alcohols	Soluble in gastric fluid to pH 5	Film coating
Eudragit L 12.5	Organic solution	12.5%	Acetone, alcohols	Soluble in intestinal fluid from pH 6	Enteric coatings
Eudragit L 100	Powder	95%	Acetone, alcohols	Soluble in intestinal fluid from pH 6	Enteric coatings
Eudragit L 100–55	Powder	30%	Water	Soluble in intestinal fluid from pH 5.5	Enteric coatings
Eudragit S 12.5	Organic solution	30%	Water	Soluble in intestinal fluid from pH 7	Enteric coatings
Eudragit S 100	Powder	95%	Acetone, alcohols	Soluble in intestinal fluid from pH 7	Enteric coatings
Eudragit FS 30 D	Organic solution	30%	Water	Soluble in intestinal fluid from pH 7	Enteric coatings
Eudragit RL 12.5	Powder	95%	Acetone, alcohols	High permeability	Enteric coatings
Eudragit RL 100	Granules	97%	Water	High permeability	Enteric coatings
Eudragit RL PO	Powder	97%	Acetone, alcohols	High permeability	Sustained release
Eudragit RL 30 D	Aqueous dispersion	30%	Water	High permeability	Sustained release
Eudragit RS 12.5	Powder	97%	Acetone, alcohols	Low permeability	Sustained release
Eudragit RS 100	Granules	97%	Water	Low permeability	Sustained release
Eudragit RS PO	Powder	97%	Acetone, alcohols	Low permeability	Sustained release
Eudragit RS 30 D	Granules	30%	Water	Low permeability	Sustained release
Eudragit NE 30 D	Powder	30%	Water	Low permeability	Sustained release
Eudragit NE 40 D	Aqueous dispersion	40%	Water	Low permeability	Sustained release
Eudragit NM 30 D	Aqueous dispersion	30%	Water	Swellable, permeable	Sustained release, tablet matrix
Eastacryl 30 D	Aqueous dispersion	30%	Water	Soluble in intestinal fluid from pH 5.5	Enteric coatings
Kollicoat MAE 30 DP	Aqueous dispersion	30%	Water	Soluble in intestinal fluid from pH 5.5	Enteric coatings
Kollicoat MAE 100 P	Powder	95%	Acetone, alcohols	Soluble in intestinal fluid from pH 5.5	Enteric coatings
Acryl-EZE	Powder	40–60%	Water	Soluble in intestinal fluid from pH 5.5	Enteric coatings

Note: Recommended plasticizers for the above polymers include dibutyl phthalate, polyethylene glycols, triethyl citrate, triacetin, and 1,2-propylene glycol. The recommended concentration of the plasticizer is approximately 10–25% of the dry polymer weight. A plasticizer is not necessary with *Eudragit E 12.5, Eudragit E PO, Eudragit E 100* and *Eudragit NE 30 D.*

$R^4 = CH_3$
For *Eudragit RL* and *Eudragit RS*:
$R^1 = H, CH_3$
$R^2 = CH_3, C_2H_5$
$R^3 = CH_3$
$R^4 = CH_2CH_2N(CH_3)_3\,^+Cl^-$
For *Eudragit NE 30 D* and *Eudragit NE 40 D*:
$R^1, R^3 = H, CH_3$
$R^2, R^4 = CH_3, C_2H_5$
For *Acryl-EZE 93A* and *Acryl-EZE MP; Eudragit L 30 D-55* and *Eudragit L 100-55, Eastacryl 30 D, Kollicoat MAE 100 P,* and *Kollicoat MAE 30 DP*:
$R^1, R^3 = H, CH_3$
$R^2 = H$
$R^4 = CH_3, C_2H_5$

6 Functional Category

Film-forming agent; modified-release agent; solubilizing agent; tablet and capsule binder.

7 Applications in Pharmaceutical Formulation or Technology

Polymethacrylates are primarily used in oral capsule and tablet formulations as film-coating agents.[3–23] Depending on the type of polymer used, films of different solubility characteristics can be produced; *see* Table II.

Eudragit E is used as a plain or insulating film former. It is soluble in gastric fluid below pH 5 and is used effectively for moisture protection and taste-masking, as well as solubility enhancement. In contrast, *Eudragit L, S* and *FS* types are used as enteric coating agents because they are resistant to gastric fluid. Different types of enteric coatings are soluble at different pH values: e.g. *Eudragit L* is soluble at pH > 6 whereas *Eudragit S* and *FS* are soluble at pH > 7 and suitable for colonic delivery applications. The S grade is generally used for coating tablets, while the flexible *FS 30 D* dispersion is preferred for coating particles for further compression.

Eudragit RL, RS, NE 30 D, NE 40 D, and *NM 30 D* are used to form water-insoluble film coats for sustained-release products. *Eudragit RL* films are more permeable than those of *Eudragit RS,* and films of varying permeability can be obtained by mixing the two types together. The neutral Eudragit *NE/NM* grades do not have functional ionic groups. They swell in aqueous media independently of pH without dissolving.

Eudragit L 30 D-55 is used as an enteric coating film former for solid-dosage forms. The coating is resistant to gastric juice but dissolves readily at above pH 5.5.

Eudragit L 100-55 is the spray-dried version of *Eudragit L 30 D-55.* It is commercially available as a redispersible powder.

Table III: Solubility of commercially available polymethacrylates in various solvents.

Grade	Solvent						
	Acetone and alcohols[(a)]	Dichloromethane	Ethyl acetate	1 N HCl	1 N NaOH	Petroleum ether	Water
Eudragit E 12.5	M	M	M	M	—	M	—
Eudragit E 100	S	S	S	—	—	I	I
Eudragit L 12.5	M	M	M	—	M	P	P
Eudragit L 100-55	S	I	I	—	S	I	I
Eudragit L 100	S	I	I	—	S	I	I
Eudragit L 30 D-55[(b)]	M[(c)]	—	—	M	—	M	—
Eudragit S 12.5	M	M	M	—	M	P	P
Eudragit S 100	S	I	I	—	S	I	I
Eudragit RL 12.5	M	M	M	—	—	P	M
Eudragit RL 100	S	S	S	—	—	I	I
Eudragit RL PO	S	S	S	—	I	I	I
Eudragit RL 30 D[(b)]	M[(c)]	M	M	—	I	I	M
Eudragit RS 12.5	M	M	M	—	—	P	M
Eudragit RS 100	S	S	S	—	—	I	I
Eudragit RS PO	S	S	S	—	I	I	I
Eudragit RS 30 D[(b)]	M[(c)]	M	M	—	I	I	M
Eastacryl 30 D[(b)]	M[(c)]	—	—	—	M	—	M
Eudragit RS 30 D[(b)]	M	—	—	—	M	—	M
Kollicoat MAE 100 P	S[(c)]	—	—	—	M	—	M
Acryl-EZE	S	I	I	—	S	I	I

(a) Alcohols including ethanol (95%), methanol, and propan-2-ol.
(b) Supplied as a milky-white aqueous dispersion.
(c) A 1 : 5 mixture forms a clear, viscous, solution.
S = soluble; M = miscible; I = insoluble; P = precipitates.
1 part of *Eudragit RL 30 D* or of *Eudragit RS 30 D* dissolves completely in 5 parts acetone, ethanol (95%), or propan-2-ol to form a clear or slightly turbid solution. However, when mixed in a ratio of 1 : 5 with methanol, *Eudragit RL 30 D* dissolves completely, whereas *Eudragit RS 30 D* dissolves only partially.

Kollicoat MAE 100 P, *Acryl-EZE 93A* and *Acryl-EZE MP* are also commercially available as redispersible powder forms, which are designed for enteric coating of tablets or beads.

Eastacryl 30 D and *Kollicoat MAE 30 DP* are aqueous dispersions of methacrylic acid–ethyl acrylate copolymers. They are also used as enteric coatings for solid-dosage forms.

Polymethacrylates are also used as binders in both aqueous and organic wet-granulation processes. Larger quantities (5–20%) of dry polymer are used to control the release of an active substance from a tablet matrix. Solid polymers may be used in direct-compression processes in quantities of 10–50%.

Polymethacrylate polymers may additionally be used to form the matrix layers of transdermal delivery systems and have also been used to prepare novel gel formulations for rectal administration.[(24)]

See also Section 18.

8 Description

Polymethacrylates are synthetic cationic and anionic polymers of dimethylaminoethyl methacrylates, methacrylic acid, and methacrylic acid esters in varying ratios. Several different types are commercially available and may be obtained as the dry powder, as an aqueous dispersion, or as an organic solution. A (60 : 40) mixture of acetone and propan-2-ol is most commonly used as the organic solvent. *See* Tables I and III.

Eudragit E is a cationic polymer based on dimethylaminoethyl methacrylate and other neutral methacrylic acid esters. It is soluble in gastric fluid as well as in weakly acidic buffer solutions (up to pH ≈ 5). *Eudragit E* is available as a 12.5% ready-to-use solution in propan-2-ol–acetone (60 : 40). It is light yellow in color with the characteristic odor of the solvents. Solvent-free granules (*Eudragit E 100*) contain ≈98% dried weight content of *Eudragit E*. *Eudragit E PO* is a white free-flowing powder with at least 98% of dry polymer.

Eudragit L and *S*, also referred to as methacrylic acid copolymers in the USP35–NF30 monograph, are anionic copolymerization products of methacrylic acid and methyl methacrylate. The ratio of free carboxyl groups to the ester is approximately 1 : 1 in *Eudragit L* (Type A) and approximately 1 : 2 in *Eudragit S* (Type B). Both polymers are readily soluble in neutral to weakly alkaline conditions (pH 6–7) and form salts with alkalis, thus affording film coats that are resistant to gastric media but soluble in intestinal fluid. They are available as a 12.5% solution in propan-2-ol without plasticizer (*Eudragit L 12.5* and *S 12.5*). Solutions are colorless, with the characteristic odor of the solvent. *Eudragit L 100* and *Eudragit S-100* are white free-flowing powders with at least 95% of dry polymers.

Eudragit FS 30 D is the aqueous dispersion of an anionic copolymer based on methyl acrylate, methyl methacrylate, and methacrylic acid. The ratio of free carboxyl groups to ester groups is approximately 1 : 10. The dispersion also contains 0.3% sodium lauryl sulfate and 1.2% polysorbate 80 as emulsifiers. It is a highly flexible polymer, designed for use in enteric-coated solid-dosage forms, and dissolves in aqueous systems at pH >7.

Eudragit RL and *Eudragit RS*, also referred to as ammonio methacrylate copolymers in the USP35–NF30 monograph, are copolymers synthesized from acrylic acid and methacrylic acid esters, with *Eudragit RL* (Type A) having 10% of functional quaternary ammonium groups and *Eudragit RS* (Type B) having 5% of functional quaternary ammonium groups. The ammonium groups are present as salts and give rise to pH-independent permeability of the polymers. Both polymers are water-insoluble, and films prepared from *Eudragit RL* are freely permeable to water, whereas films prepared from *Eudragit RS* are only slightly permeable to water. They are available as 12.5% ready-to-use solutions in propan-2-ol–acetone (60 : 40). Solutions are colorless or slightly yellow in color, and may be clear or slightly turbid; they have an odor characteristic of the solvents. Solvent-free granules (*Eudragit RL 100* and *Eudragit RS 100*) contain ≥97% of the dried weight content of the polymer.

Eudragit RL PO and *Eudragit RS PO* are fine, white powders with a slight amine-like odor. They are characteristically the same polymers as *Eudragit RL* and *RS*. They contain ≥97% of dry polymer.

Eudragit RL 30 D and *Eudragit RS 30 D* are aqueous dispersions of copolymers of acrylic acid and methacrylic acid esters with a low content of quaternary ammonium groups. The dispersions contain 30% polymer. The quaternary groups occur as salts and are responsible for the permeability of films made from these polymers. Films prepared from *Eudragit RL 30 D* are readily permeable to water, whereas films prepared from *Eudragit RS 30 D* are less permeable to water. Film coatings prepared from both polymers give pH-independent release of active substance. Plasticizers are usually added to improve film properties.

Eudragit NE 30 D and *Eudragit NE 40 D* are aqueous dispersions of a neutral copolymer based on ethyl acrylate and methyl methacrylate. The dispersions are milky-white liquids of low viscosity and have a weak aromatic odor. The *Eudragit NE 30 D* and *40 D* dispersions contain 30% or 40% dry substance. It also contains 1.5% or 2.0% nonoxynol 100 as an emulsifier, respectively. Films prepared from the lacquer swell in water, to which they become permeable. Thus, films produced are insoluble in water, but give pH-independent drug release.

Eudragit NM 30 D is an aqueous dispersion of a neutral copolymer based on ethyl acrylate and methyl methacrylate. The dispersion also contains 0.7% macrogol stearyl ether (20) as an emulsifier. The product complies with the USP–NF monograph 'Ethyl Acrylate Methyl Methacrylate Copolymer Dispersion' except for the slightly lower molecular weight.

Eudragit L 30 D-55 is an aqueous dispersion of an anionic copolymer based on methacrylic acid and ethyl acrylate. The copolymer corresponds to USP35–NF30 methacrylic acid copolymer, Type C. The ratio of free-carboxyl groups to ester groups is 1 : 1. Films prepared from the copolymers dissolve above pH 5.5, forming salts with alkalis, thus affording coatings that are insoluble in gastric media but soluble in the small intestine.

Eastacryl 30D and *Kollicoat MAE 30 DP* are also aqueous dispersions of the anionic copolymer based on methacrylic acid and ethyl acrylate. The copolymer also corresponds to USP35–NF30 methacrylic acid copolymer, Type C. The ratio of free-carboxyl groups to ester groups is 1 : 1. Films prepared from the copolymers dissolve above pH 5.5, forming salts with alkalis, thus affording coatings that are insoluble in gastric media but soluble in the small intestine.

Eudragit L 100-55 (prepared by spray-drying *Eudragit L 30 D-55*) is a white, free-flowing powder that is redispersible in water to form a latex that has properties similar to those of *Eudragit L 30 D-55*.

Acryl-EZE 93A and *Acryl-EZE MP* are commercially available as redispersible powder forms, which are designed for enteric coating of tablets and beads, respectively.

9 Pharmacopeial Specifications

Specifications for polymethacrylates from PhEur 7.4 are shown in Table IV, and those from the USP35–NF30 in Table V. *See also* Section 18.

10 Typical Properties

Acid value

300–330 for *Eudragit L 12.5, L 100, L 30 D-55, L 100-55, Eastacryl 30 D, Kollicoat MAE 100 P*, and *Kollicoat MAE 30 DP*;

180–200 for *Eudragit S 12.5* and *S 100*.

Alkali value

162–198 for *Eudragit E 12.5* and *E 100*;

23.9–32.3 for *Eudragit RL 12.5, RL 100*, and *RL PO*;

27.5–31.7 for *Eudragit RL 30 D*;

12.1–18.3 for *Eudragit RS 12.5, RS 100*, and *RS PO*;

16.5–22.3 for *Eudragit RS 30 D*.

Density (bulk) 0.390 g/cm³

Density (tapped) 0.424 g/cm³

Density (true)

0.811–0.821 g/cm³ for *Eudragit E*;

0.83–0.85 g/cm³ for *Eudragit L, S 12.5*;

1.058–1.068 g/cm³ for *Eudragit FS 30 D*;

0.831–0.852 g/cm³ for *Eudragit L, S 100*;

1.062–1.072 g/cm³ for *Eudragit L 30 D-55*;

0.821–0.841 g/cm³ for *Eudragit L 100-55*;

0.816–0.836 g/cm³ for *Eudragit RL* and *RS 12.5*;

0.816–0.836 g/cm³ for *Eudragit RL* and *RS PO*;

1.047–1.057 g/cm³ for *Eudragit RL* and *RS 30 D*;

1.037–1.047 g/cm³ for *Eudragit NE 30 D*;

1.062–1.072 g/cm³ for *Eastacryl 30 D*;

1.062–1.072 g/cm³ for *Kollicoat MAE 100 P* and *Kollicoat MAE 30 DP*.

Refractive index

n_D^{20} = 1.38–1.385 for *Eudragit E*;

n_D^{20} = 1.39–1.395 for *Eudragit L* and *S*;

n_D^{20} = 1.387–1.392 for *Eudragit L 100-55*;

n_D^{20} = 1.38–1.385 for *Eudragit RL* and *RS*.

Solubility *see* Table III.

Spectroscopy

NIR spectra see Figures 1, 2, 3, 4, 5, 6, and 7.

Viscosity (dynamic)

3–12 mPa s for *Eudragit E*;

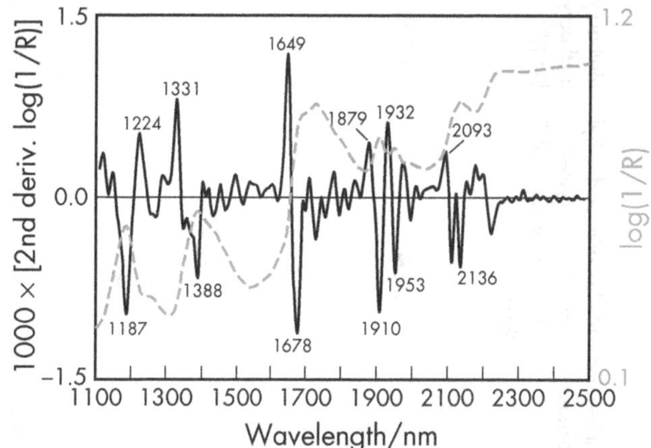

Figure 1: Near-infrared spectrum of polymethacrylates (*Eudragit E 100*) measured by reflectance.

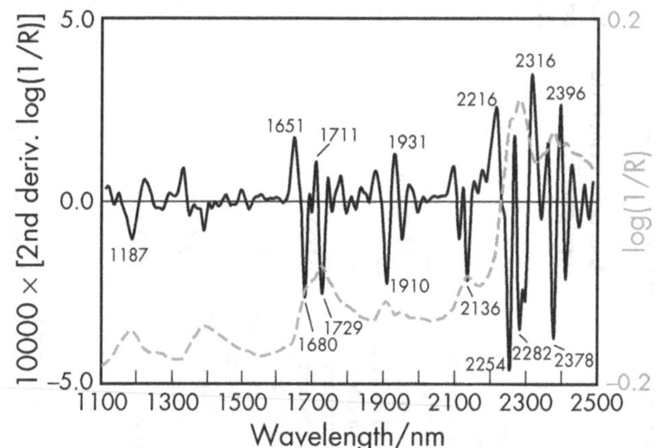

Figure 2: Near-infrared spectrum of polymethacrylates (*Eudragit E PO*) measured by reflectance.

Table IV: Specifications from PhEur 7.4.

Test	Ammonio methacrylate copolymer[(a)]	Methacrylic acid–ethyl acrylate copolymer (1 : 1)[(b)]	Methacrylic acid–ethyl acrylate copolymer (1 : 1) dispersion 30%[(c)]	Methacrylic acid–methyl methacrylate copolymer (1 : 1)[(d)]	Methacrylic acid–methyl methacrylate copolymer (1 : 2)[(e)]	Basic butylated methacrylate copolymer[(f)]	Polyacrylate dispersion 30%[(g)]
Identification	+	+	+	+	+	+	+
Characters	+	+	+	+	+	+	+
Appearance of a film	+	—	+	+	+	+	+
Relative density	—	—	—	—	—	—	1.037–1.047
Apparent viscosity	≤15 mPa s	—	≤15 mPa s	50–200 mPa s	50–200 mPa s	3–6 mPa s	≤50 mPa s
Absorbance at 420 nm	—	—	—	—	—	≤0.3	—
Particulate matter	—	—	≤1.0%	—	—	—	≤0.5%
Limit of monomers	—	—	—	—	—	≤0.3%	≤100 ppm
Ethyl acrylate and methacrylic acid	≤100 ppm	≤0.1%	≤0.1%	—	—	—	—
Methyl methacrylate and methacrylic acid	≤50 ppm	—	—	≤0.1%	≤0.1%	—	—
Residue on evaporation	—	—	28.5–31.5%	—	—	—	28.5–31.5%
Loss on drying	≤3.0%	≤5.0%	—	≤5.0%	≤5.0%	≤2.0%	—
Methanol	≤1.5%	—	—	—	—	—	—
Heavy metals	≤20 ppm	—	—	—	—	≤20 ppm	≤20 ppm
Sulfated ash	—	+	≤0.2%	≤0.1%	≤0.1%	≤0.1%	≤0.4%
Type A	—	0.4%	—	—	—	—	—
Type B	—	0.5–3.0%	—	—	—	—	—
Microbial contamination	—	—	≤10^3 cfu/g	—	—	—	≤10^3 cfu/g
Assay	Ammonio methacrylate units	Methacrylic acid units	Methacrylic acid units	Methacrylic acid units	Methacrylic acid units	Dimethylaminoethyl units	Residue on evaporation
	+	+	46.0–50.6%	46.0–50.6%	27.6–30.7%	20.8–25.5%	28.5%–31.5%
Type A	8.9–12.3%	46.0–50.6%	—	—	—	—	—
Type B	4.5–7.0%	43.0–48.0%	—	—	—	—	—

(a) Corresponds to *Eudragit RL/RS*.
(b) Corresponds to *Eudragit L100-55*.
(c) Corresponds to *Eudragit L 30 D-55*.
(d) Corresponds to *Eudragit L*.
(e) Corresponds to *Eudragit S*.
(f) Corresponds to *Eudragit E*.
(g) Corresponds to *Eudragit NE 30 D* and *NMD 30*.

Table V: Specifications from USP35–NF30.

Test	Amino methacrylate copolymer[(a)]	Ammonio methacrylate copolymer[(b)]	Ammonio methacrylate copolymer dispersion[(c)]	Ethyl acrylate and methyl methacrylate copolymer dispersion[(d)]	Methacrylic acid copolymer[(e)]	Methacrylic acid copolymer dispersion[(f)]
Identification	+	+	+	+	+	+
Color of solution	+	—	—	—	—	—
Viscosity	3–6 mPa s	—	—	2–20 mPa s	—	—
Type A	—	≤15 mPa s	≤100 mPa s	—	60–120 mPa s	—
Type B	—	≤15 mPa s	≤100 mPa s	—	50–200 mPa s	—
Type C	—	—	—	—	100–200 mPa s	2–15 mPa s
Loss on drying	≤2.0%	—	—	68.5–71.5%[(g)]	—	—
Type A	—	≤3.0%	68.5–71.5%[(g)]	—	≤5.0%	—
Type B	—	≤3.0%	68.5–71.5%[(g)]	—	≤5.0%	—
Type C	—	—	—	—	≤5.0%	68.5–71.5%[(g)]
Residue on ignition	≤0.1%	—	—	≤0.4%	—	—
Type A	—	≤0.1%	≤0.5%	—	≤0.1%	—
Type B	—	≤0.1%	≤0.5%	—	≤0.1%	—
Type C	—	—	—	—	≤0.4%	≤0.2%[(g)]
Heavy metals	—	≤20 ppm	—	—	≤20 ppm	≤20 ppm[(g)]
Limit of total monomers	—	+	—	≤0.01%	≤0.05%	≤0.01%
Limit of methyl methacrylate	≤0.1%	≤50 ppm	≤0.002%	—	—	—
Limit of butyl methacrylate	≤0.1%	—	—	—	—	—
Limit of 2-dimethylaminoethyl methacrylate	≤0.1%	—	—	—	—	—
Limit of ethyl acrylate	—	≤100 ppm	≤0.008%	—	—	—
Coagulum content	—	—	≤1.0%[(g)]	≤1.0%[(g)]	—	≤1%[(g)]
Microbial contamination	—	—	—	—	—	—
Aerobic bacteria	—	—	—	10^3 cfu/g	—	—
Yeast and mold	—	—	—	10^2 cfu/g	—	—
pH	—	—	—	5.5–8.6	—	2.0–3.0
Assay (dried basis)	Dimethylaminoethyl units20.8–25.5%	Ammonio methacrylate units	Ammonio methacrylate units	—	Methacrylic acid units	Methacrylic acid units
Type A	—	8.85–11.96%	10.18–13.73%	—	46.0–50.6%	—
Type B	—	4.48–6.77%	6.11–8.26%	—	27.6–30.7%	—
Type C	—	—	—	—	46.0–50.6%	46.0–50.6%

(a) Corresponds to *Eudragit E*.
(b) Corresponds to *Eudragit RL* and *RS*.
(c) Corresponds to *Eudragit RL 30 D* and *RS 30 D*.
(d) Corresponds to *Eudragit NE 30 D*.
(e) Corresponds to *Eudragit L*, *S* and *L100–55*.
(f) Corresponds to *Eudragit L 30 D–55*.
(g) Calculated based on undried dispersion basis.

P

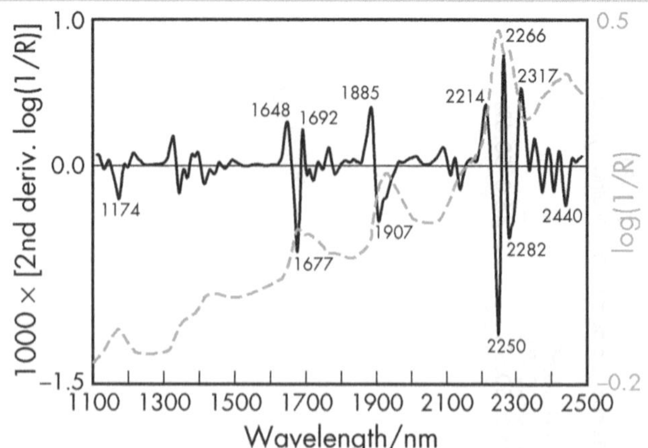

Figure 3: Near-infrared spectrum of polymethacrylates (*Eudragit L 100*) measured by reflectance.

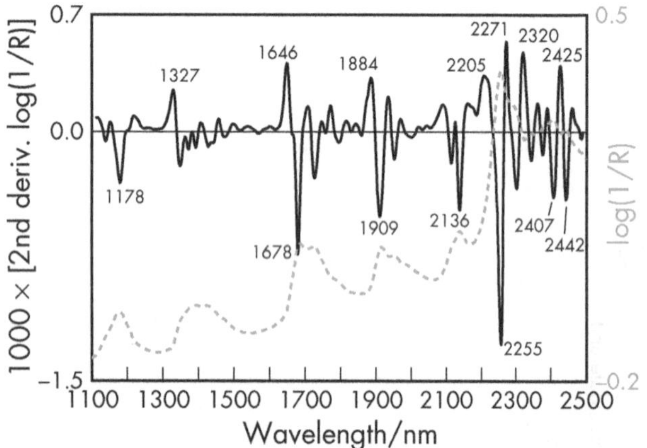

Figure 4: Near-infrared spectrum of polymethacrylates (*Eudragit RL PO*) measured by reflectance.

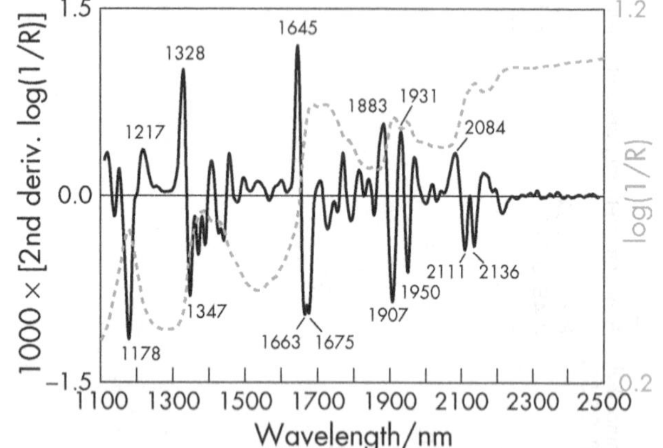

Figure 5: Near-infrared spectrum of polymethacrylates (*Eudragit RS 100*) measured by reflectance.

≤50 mPa s for *Eudragit NE 30 D*;
50–200 mPa s for *Eudragit L* and *S*;
≤20 mPa s for *Eudragit FS 30 D*;
≤15 mPa s for *Eudragit L 30 D-55*;
100–200 mPa s for *Eudragit L 100-55*;
≤15 mPa s for *Eudragit RL* and *RS*;
≤200 mPa s for *Eudragit RL* and *RS 30 D*;
≤15 mPa s for *Kollicoat MAE 100 P* and *Kollicoat MAE 30 DP*;
145 mPa s for *Eastacryl 30D*.

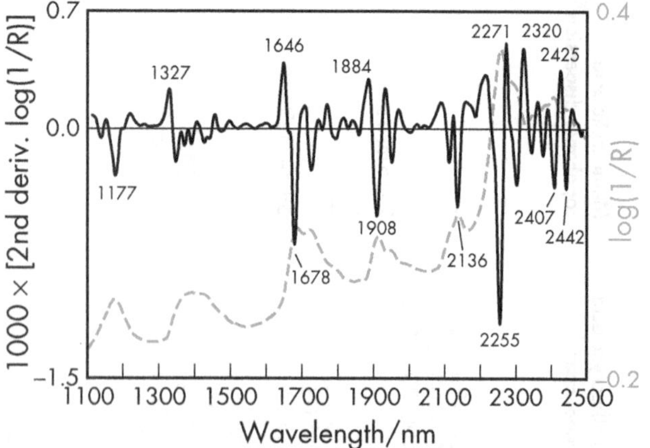

Figure 6: Near-infrared spectrum of polymethacrylates (*Eudragit RS PO*) measured by reflectance.

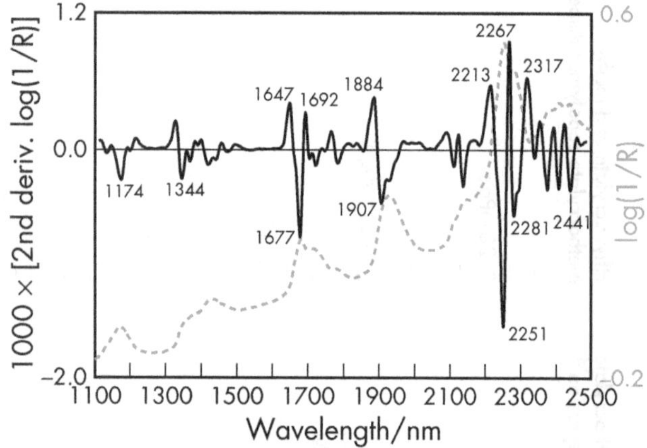

Figure 7: Near-infrared spectrum of polymethacrylates (*Eudragit S 100*) measured by reflectance.

11 Stability and Storage Conditions

Dry powder polymer forms are stable at temperatures less than 30°C. Above this temperature, powders tend to form clumps, although this does not affect the quality of the substance and the clumps can be readily broken up. Dry powders are stable for at least 3 years if stored in a tightly closed container at less than 30°C.

Dispersions are sensitive to extreme temperatures and phase separation occurs below 0°C. Dispersions should therefore be stored at temperatures between 5 and 25°C and are stable for at least 18 months after shipping from the manufacturer's warehouse if stored in a tightly closed container at the above conditions.

12 Incompatibilities

Incompatibilities occur with certain polymethacrylate dispersions depending upon the ionic and physical properties of the polymer and solvent. For example, coagulation may be caused by soluble electrolytes, pH changes, some organic solvents, and extremes of temperature; *see* Table II. For example, dispersions of *Eudragit L 30 D*, *RL 30 D*, *L 100-55*, and *RS 30 D* are incompatible with magnesium stearate. *Eastacryl 30 D*, *Kollicoat MAE 100 P*, and *Kollicoat MAE 30 DP* are also incompatible with magnesium stearate.

Interactions between polymethacrylates and some drugs can occur, although solid polymethacrylates and organic solutions are generally more compatible than aqueous dispersions.

13 Method of Manufacture

Prepared by the polymerization of acrylic and methacrylic acids or their esters, e.g. butyl ester or dimethylaminoethyl ester.

14 Safety

Polymethacrylate copolymers are widely used as film-coating materials in oral pharmaceutical formulations. They are also used in topical formulations and are generally regarded as nontoxic and nonirritant materials.

Acute and chronic toxicity studies have shown that polymethacrylate copolymers may be regarded as essentially safe in humans.[25,26,27,28]

See also Section 15.

15 Handling Precautions

Observe normal precautions appropriate to the circumstances and quantity of material handled. Additional measures should be taken when handling organic solutions of polymethacrylates. Eye protection, gloves, and a dust mask or respirator are recommended. Polymethacrylates should be handled in a well-ventilated environment and measures should be taken to prevent dust formation.

Acute and chronic adverse effects have been observed in workers handling the related substances methyl methacrylate and poly(methyl methacrylate) (PMMA).[29] In the UK, the workplace exposure limit for methyl methacrylate has been set at 208 mg/m^3 (50 ppm) long-term (8-hour TWA), and 416 mg/m^3 (100 ppm) short-term.[30]

See also Section 17.

16 Regulatory Status

Included in the FDA Inactive Ingredients Database (oral capsules and tablets). Included in nonparenteral medicines licensed in the UK. Included in the Canadian Natural Health Products Ingredients Database.

17 Related Substances

Methyl methacrylate; poly(methyl methacrylate).

Methyl methacrylate

Empirical formula $C_5H_8O_2$
Molecular weight 100.13
CAS number [80-62-6]
Synonyms Methacrylic acid, methyl ester; methyl 2-methacrylate; methyl 2-methylpropenoate; MME.
Safety

LD_{50} (dog, SC): 4.5 g/kg
LD_{50} (mouse, IP): 1 g/kg
LD_{50} (mouse, oral): 5.2 g/kg
LD_{50} (mouse, SC): 6.3 g/kg
LD_{50} (rat, IP): 1.33 g/kg
LD_{50} (rat, SC): 7.5 g/kg

Comments Methyl methacrylate forms the basis of acrylic bone cements used in orthopedic surgery.

Poly(methyl methacrylate)

Empirical formula $(C_5H_8O_2)_n$
Synonyms Methyl methacrylate polymer; PMMA.
Comments Poly(methyl methacrylate) has been used as a material for intraocular lenses, for denture bases, and as a cement for dental prostheses.

18 Comments

A number of different polymethacrylates are commercially available that have different applications and properties; *see* Table II.

For spray coating, polymer solutions and dispersions should be diluted with suitable solvents. Some products need the addition of a plasticizer such as dibutyl sebacate, dibutyl phthalate, glyceryl triacetate, or polyethylene glycol. Different types of plasticizer may be mixed to optimize the polymer properties for special requirements.

The *Japanese Pharmaceutical Excipients* (JPE) 2004 includes specifications for aminoalkyl methacrylate copolymer RS, aminoalkyl methacrylate copolymer E, dried methacrylic acid copolymer LD, ethyl acrylate and methyl methacrylate copolymer dispersion, methacrylic acid copolymer L, methacrylic acid copolymer S, and methacrylic acid copolymer LD.

19 Specific References

1 Adler M, *et al.* Molar mass characterization of hydrophilic copolymers, 2. Size exclusion chromatography of cationic (meth)acrylate copolymers. *e-Polymers* 2005; **57**: 1–11.
2 Adler M, *et al.* Molar mass characterization of hydrophilic copolymers, 1. Size exclusion chromatography of neutral and anionic (meth)acrylate copolymers. *e-Polymers* 2004; **55**: 1–16.
3 Akhgari A, *et al.* Combination of time-dependent and pH-dependent polymethacrylates as a single coating formulation for colonic delivery of indomethacin pellets. *Int J Pharm* 2006; **320**(1–2): 137–142.
4 Arno EA, *et al. Eudragit NE30D* based metformin/gliclazide extended release tablets: formulation, characterisation and *in vitro* release studies. *Chem Pharm Bull (Tokyo)* 2002; **50**(11): 1495–1498.
5 Ceballos A, *et al.* Influence of formulation and process variables on in vitro release of theophylline from directly-compressed *Eudragit* matrix tablets. *Farmaco* 2005; **60**(11-12): 913–8.
6 Cerea M, *et al.* A novel powder coating process for attaining taste masking and moisture protective films applied to tablets. *Int J Pharm* 2004; **279**(1–2): 127–139.
7 Cheng G, *et al.* Time- and pH-dependent colon-specific drug delivery for orally administered diclofenac sodium and 5-aminosalicylic acid. *World J Gastroenterol* 2004; **10**(12): 1769–1774.
8 Cole ET, *et al.* Enteric coated HPMC capsules designed to achieve intestinal targeting. *Int J Pharm* 2002; **231**(1): 83–95.
9 Dittgen M, *et al.* Acrylic polymers: a review of pharmaceutical applications. *STP Pharma Sciences* 1997; **7**(6): 403–437.
10 Felton L, McGinity J. Enteric film coating of soft gelatin capsules. *Drug Deliv Technol* 2003; **3**(6): 46–51.
11 Gao Y, *et al.* Preparation of roxithromycin-polymeric microspheres by the emulsion solvent diffusion method for taste masking. *Int J Pharm* 2006; **318**(1–2): 62–69.
12 Hamed E, Sakr A. Effect of curing conditions and plasticizer level on the release of highly lipophilic drug from coated multiparticulate drug delivery system. *Pharm Dev Technol* 2003; **8**(4): 397–407.
13 Ibekwe VC, *et al.* A comparative *in vitro* assessment of the drug release performance of pH-responsive polymers for ileo-colonic delivery. *Int J Pharm* 2006; **308**(1–2): 52–60.
14 Kidokoro M, *et al.* Properties of tablets containing granulations of ibuprofen and an acrylic copolymer prepared by thermal processes. *Pharm Dev Technol* 2001; **6**(2): 263–275.
15 Lorenzo-Lamosa M, *et al.* Development of a microencapsulated form of cefuroxime axetil using pH-sensitive acrylic polymers. *J Microencapsul* 1997; **14**(5): 607–616.
16 Nisar ur R, *et al.* Drug-polymer mixed coating: a new approach for controlling drug release rates in pellets. *Pharm Dev Technol* 2006; **11**(1): 71–77.
17 Rahman N, Yuen K. *Eudragit NE40*-drug mixed coating system for controlling drug release of core pellets. *Drug Dev Ind Pharm* 2005; **31**(4–5): 339–347.
18 Rao VM, *et al.* Design of pH-independent controlled release matrix tablets for acidic drugs. *Int J Pharm* 2003; **252**(1–2): 81–86.
19 Saravanan M, *et al.* The effect of tablet formulation and hardness on *in vitro* release of cephalexin from *Eudragit L100* based extended release tablets. *Biol Pharm Bull* 2002; **25**(4): 541–545.
20 Signorino C, *et al.* The use of acrylic resins for improved aqueous enteric coating. *Pharm Technol Excipients Solid Dosage Forms* 2004; (Suppl.): 32–39.
21 Sohi H, *et al.* Taste masking technologies in oral pharmaceuticals: recent developments and approaches. *Drug Dev Ind Pharm* 2004; **30**(5): 429–448.
22 Zheng W, *et al.* Influence of hydroxyethylcellulose on the drug release properties of theophylline pellets coated with *Eudragit RS 30 D*. *Eur J Pharm Biopharm* 2005; **59**(1): 147–154.

P

23 Zhu Y, *et al.* Influence of thermal processing on the properties of chlorpheniramine maleate tablets containing an acrylic polymer. *Pharm Dev Technol* 2002; 7(4): 481–489.
24 Umejima H, *et al.* Preparation and evaluation of *Eudragit* gels VI: *in vivo* evaluation of Eudispert rectal hydrogel and xerogel containing salicylamide. *J Pharm Sci* 1993; 82: 195–199.
25 European Food Safety Agency. EFSA Panel on Food Additives and Nutrient Sources added to Food (ANS): Scientific opinion on the use of basic methacrylate copolymer as a food additive on request from the European Commission. *EFSA Journal* 2010; 8(2) :1513.
26 European Food Safety Agency. EFSA Panel on Food Additives and Nutrient Sources added to Food (ANS): Scientific opinion on the safety of anionic methacrylate copolymer for the proposed uses as a food additive. *EFSA Journal* 2010; 8(7): 1656.
27 European Food Safety Agency. EFSA Panel on Food Additives and Nutrient Sources added to Food (ANS): Scientific opinion on the safety of neutral methacrylate copolymer for the proposed uses as a food additive. *EFSA Journal* 2010; 8(7): 1655.
28 Evonik Industries. Eudragit. http://eudragit.evonik.com/product/eudragit (accessed 31 October 2011).
29 Routledge R. Possible hazard of contact lens manufacture [letter]. *Br Med J* 1973; 1: 487–488.
30 Health and Safety Executive. *EH40/2005: Workplace Exposure Limits.* Sudbury: HSE Books, 2011. http://www.hse.gov.uk/pubns/priced/eh40.pdf (accessed 16 February 2012).

20 General References

Japan Pharmaceutical Excipients Council. *Japanese Pharmaceutical Excipients 2004*. Tokyo: Yakuji Nippo, 2004.

McGinity JW, Felton LA. *Aqueous Polymeric Coatings for Pharmaceutical Dosage Forms*, 3rd edn. New York: Informa Healthcare, 2008.

Petereit HU *et al. Eudragit Application Guidelines*, 11th edn. Darmstadt, Germany: Evonik Industries AG, 2009.

21 Authors

RK Chang, Y Peng, N Trivedi, JR Johnson.

22 Date of Revision

2 March 2012.

Poly(methyl vinyl ether/maleic anhydride)

1 Nonproprietary Names

None adopted.

2 Synonyms

Butyl ester of poly(methylvinyl ether-co-maleic anhydride); calcium and sodium salts of poly(methylvinyl ether-co-maleic anhydride); *Gantrez*.

3 Chemical Name and CAS Registry Number

See Table I.

4 Empirical Formula and Molecular Weight

$(C_4H_2O_3 \cdot C_3H_6O)_x$ *See* Table II.

5 Structural Formula

See Section 4.

6 Functional Category

Bioadhesive material; complexing agent; emulsion stabilizing agent; film-forming agent; transdermal delivery component; viscosity-increasing agent.

7 Applications in Pharmaceutical Formulation or Technology

Poly(methylvinyl ether/maleic anhydride) copolymers and derivatives are used in denture adhesive bases,[1] controlled-release coatings, enteric coatings, ostomy adhesives,[2] transdermal patches,[3] toothpastes,[4] mouthwashes,[5] and transdermal gels.[6,7]

Table I: Chemical name and CAS registry number for poly(methylvinyl ether/maleic anhydride) copolymers and derivatives.

Chemical name	Trade name	CAS number
Poly(methylvinyl ether/maleic anhydride)	*Gantrez AN-119* *Gantrez AN-903* *Gantrez AN-139* *Gantrez AN-149* *Gantrez AN-169* *Gantrez AN-179*	[9011-16-9]
Poly(methylvinyl ether/maleic acid)	*Gantrez S-95* *Gantrez S-96* *Gantrez S-97*	[25153-40-6]
Monoethyl ester of poly(methylvinyl ether/maleic acid) (48–52%) in ethanol (48–52%)	*Gantrez ES-225 50% Alcoholic Solution*	[25087-06-3][64-17-5]
Mixture of monoethyl ester of poly(methylvinyl ether/maleic acid) and monobutyl ester of poly(methylvinyl ether/maleic acid) (48–52%) in ethanol (43–47%) and *n*-butyl alcohol (≈5%)	*Gantrez ES-425 50% Alcoholic Solution*	[25087-06-3][25119-68-0] [64-17-5] [200-751-6]
Mixed sodium/calcium salts of poly(methylvinyl ether/maleic anhydride)	*Gantrez MS-955*	[62386-95-2]

Table II: Molecular weights of selected commercially available copolymers of poly(methylvinyl ether/maleic anhydride).

Grade	Approximate molecular weight
Gantrez AN-119	200 000
Gantrez AN-903	800 000
Gantrez AN-139	1 000 000
Gantrez AN-169	2 000 000
Gantrez S-96	700 000
Gantrez S-97 (powder)	1 200 000
Gantrez S-97 (solution)	1 500 000
Gantrez MS-995	1 000 000
Gantrez ES-225	100 000–150 000
Gantrez ES-425	90 000–150 000

8 Description

In the solid state, poly(methylvinyl ether/maleic anhydride) copolymers are a white to off-white free flowing, odorless, hygroscopic powders. In solution, poly(methylvinyl ether/maleic anhydride) is a slightly hazy, odorless, viscous liquid.

9 Pharmacopeial Specifications

—

10 Typical Properties

See Table III.

11 Stability and Storage Conditions

Poly(methylvinyl ether/maleic anhydride) and related free acids are hygroscopic powders, and therefore excessive exposure to moisture should be avoided. Aqueous solutions exhibit decreases in viscosity upon exposure to UV light. Poly(methylvinyl ether/maleic anhydride) should be stored in a cool, dry place out of direct sunlight.

12 Incompatibilities

Poly(methylvinyl ether/maleic anhydride) and copolymers are incompatible with strong oxidizing agents and reducing agents, concentrated nitric acid, sulfuric acid, nitrofoam, oleum, potassium *t*-butoxide, aluminum, aluminum triisopropoxide, and crotonaldehyde. In addition, the anhydride will hydrolyze in water to form a water-soluble free acid that can subsequently be ionized to form salts in the presence of cations (Na^+, Zn^{2+}, Ca^{2+}, and Al^{3+}). Excessive addition of bivalent and trivalent metal ions to aqueous solution will result in precipitation, particularly in solutions containing high polymer concentrations.

13 Method of Manufacture

Poly(methylvinyl ether/maleic anhydride) and copolymers are manufactured from methylvinyl ether and maleic anhydride. *See also* Section 18.

14 Safety

Poly(methylvinyl ether/maleic anhydride) and copolymers are widely used in a diverse range of topical and oral pharmaceutical formulations.[8] These copolymers are generally regarded as nontoxic and nonirritant, although there are exceptions; *see* Section 18.

LD_{50} (rat, oral): 8 g/kg (*Gantrez AN-130 Powder*)[9]
LD_{50} (rat, oral): 40 ml/kg (*Gantrez AN-139* 20% w/w aqueous solution)
LD_{50} (rat, oral): >25.6 g/kg (*Gantrez ES-225*)
LD_{50} (rat, oral): 25.6 g/kg (*Gantrez ES-425* 40% w/v corn oil solution)
LD_{50} (rat, oral): >5.0 g/kg (*Gantrez MS-955* 20% aqueous solution)

15 Handling Precautions

Observe normal precautions appropriate to the circumstances and quantity of material handled. Excessive dust generation should be avoided when using powders, and an appropriate ventilation area and dust mask are recommended. Hand and eye protection is also recommended. *See also* Section 18.

16 Regulatory Status

GRAS listed. Included in nonparenteral medicines licensed in the UK.

Table III: Typical physical properties of selected commercially available copolymers of poly(methylvinyl ether/maleic anhydride).

Grade	Specific viscosity (1% in MEK)	T_g (°C)	Specific gravity (25°C, 5% solids)	Bulk density (g/cm³)	Polydispersity (M_n/M_w)	Moisture content (% w/w)	Viscosity (mPa s) of 5% w/w solution at 25°C	Dissociation constant
***Gantrez AN* copolymers**								
AN-119	0.1–0.5	152	1.018	0.34	2.74	<1	15	—
AN-903	0.8–1.2	156	1.017	0.33	—	—	30	—
AN-139	1.0–1.5	151	1.016	0.33	3.47	<1	40	—
AN-149	1.5–2.5	153	1.017	0.35	2.58	<1	45	—
AN-169	2.5–3.5	154	1.017	0.32	2.06	<1	85	—
AN-179	3.5–5.0	154	1.017	0.33	2.12	<1	135	—
***Gantrez S* copolymers**								
S-95	1.0–2.0	139	1.015	—	2.71	≤17	20	3.51–6.41
S-96 Solution	≈4.0	—	—	—	—	86–88	150	3.51–6.41
S-97	4.0–10.0	143	1.015	—	2.06	≤6	70	3.47–6.47
S-97 Solution	4.0–10.0	—	—	—	—	86–88	1000	3.50–6.50
***Gantrez ES* and *MS* copolymers**								
ES-225	0.36–0.45	102	0.983	—	2.5–3.0	≤0.5	18,800	5.33
ES-425	0.37–0.45	96	0.977	—	2.5–3.4	≤0.5	14,400	5.28
MS-955	—	—	1.061[a]	—	2.3	≤15	700–3000[b]	—

(a) 13% solids at 30°C.
(b) Viscosity of 11.1% solids aqueous solution.

17 Related Substances

—

18 Comments

Gantrez AN-119 has been used to manufacture specific bioadhesive ligand-nanoparticle conjugates[10] to aid gastrointestinal retention for oral drug delivery applications. *Gantrez* has also been used to develop novel polyethylene surface-modified medical devices with enhanced hydrophilicity and wettability.[11]

The *S*, *ES*, and *MS* grades of *Gantrez* are manufactured by dispersing *AN* copolymers in a number of different solvents or salt solutions.[9]

The *A*, *ES*, and *MS* copolymers of *Gantrez* are extremely irritating to the eyes and a NIOSH-approved respirator and suitable eye protection are recommended when using *Gantrez ES-435*, *Gantrez ES-225*, and *Gantrez A-425*.

19 Specific References

1 Shay K. The retention of complete dentures. In: Zarb GA *et al.* eds. *Boucher's Prosthodontic Treatment for Edentulous Patients*. Toronto, Ontario: Mosby, 1997: 400–411.
2 Scalf BS, Fowler JF. Peristomal allergic contact dermatitis due to *Gantrez* in stomadhesive paste. *J Am Acad Dermatol* 2000; **42**: 355–356.
3 Woolfson AD, *et al.* Development and characterization of a moisture-activated bioadhesive drug delivery system for percutaneous local anesthesia. *Int J Pharm* 1998; **169**: 83–94.
4 Busscher HJ, *et al.* A surface physicochemical rationale for calculus formation in the oral cavity. *J Cryst Growth* 2004; **261**: 87–92.
5 Kockisch S, *et al.* A direct-staining method to evaluate the mucoadhesion of polymers from aqueous dispersion. *J Control Release* 2001; **77**: 1–6.
6 Jones DS, *et al.* Examination of the flow rheological and textural properties of polymer gels composed of poly(methylvinylether-co-maleic anhydride) and poly(vinylpyrrolidone): rheological and mathematical interpretation of textural parameters. *J Pharm Sci* 2002; **91**(9): 2090–2101.
7 Jones DS, *et al.* Rheological and mucoadhesive characterization of polymeric systems composed of poly(methylvinylether-co-maleic anhydride) and poly(vinylpyrrolidone) designed as platforms for topical drug delivery. *J Pharm Sci* 2003; **92**(5): 995–1007.
8 Sharma NC, *et al.* The clinical effectiveness of a dentrifice containing triclosan and a copolymer for controlling breath odor measured organoleptically twelve hours after toothbrushing. *J Clin Dent* 1999; **10**: 131–134.
9 ISP. Technical literature: *Gantrez Copolymers*, 2005.
10 Arbos P, *et al. Gantrez* AN as a new polymer for the preparation of ligand-nanoparticle conjugates. *J Control Release* 2002; **83**: 321–330.
11 Kuzuya M, *et al.* Plasma technique for the fabrication of a durable functional surface on organic polymers. *Surf Coat Tech* 2003; **169**: 587–591.

20 General References

—

21 Authors

GP Andrews, DS Jones.

22 Date of Revision

1 August 2011.

P

Polyoxyethylene Alkyl Ethers

1 Nonproprietary Names

The polyoxyethylene alkyl ethers are a series of polyoxyethylene glycol ethers of *n*-alcohols (lauryl, oleyl, myristyl, cetyl, and stearyl alcohol). Of the large number of different materials commercially available, four types are listed in the USP35–NF30, one type in the JP XV, and four types in the PhEur 7.4.

BP: Macrogol Cetostearyl Ether
Macrogol Lauryl Ether
Macrogol Oleyl Ether
Macrogol Stearyl Ether
Lauromacrogol

PhEur: Macrogol Cetostearyl Ether
Macrogol Lauryl Ether
Macrogol Oleyl Ether
Macrogol Stearyl Ether

USP–NF: Polyoxyl 20 Cetostearyl Ether
Polyoxyl 10 Oleyl Ether
Polyoxyl Lauryl Ether
Polyoxyl Stearyl Ether

Polyoxyethylene alkyl ethers are employed extensively in cosmetics, where the CTFA names laureth-*N*, myreth-*N*, ceteth-*N*, and steareth-*N* are commonly used. In this nomenclature, *N* is the number of ethylene oxide groups, e.g. steareth-20.

See also Sections 2–5.

2 Synonyms

Polyoxyethylene alkyl ethers are nonionic surfactants produced by the polyethoxylation of linear fatty alcohols. Products tend to be mixtures of polymers of slightly varying molecular weights, and the numbers used to describe polymer lengths are average values.

Two systems of nomenclature are used to describe these materials. The number '10' in the name *Brij C10* refers to the approximate polymer length in oxyethylene units (i.e. *y*; *see* Section 5). The number '1000' in the name 'cetomacrogol 1000' refers to the average molecular weight of the polymer chain.

Synonyms applicable to polyoxyethylene alkyl ethers are shown below.

Brij; *Cremophor A*; *Cyclogol 1000*; *Emalex*; *Emulgen*; *Ethosperse*; *Genapol*; *Hetoxol*; *Hostacerin*; *Jeecol*; *Lipocol*; *Lumulse*; *Nikkol*; macrogol ethers; macrogoli aether cetostearylicus; macrogoli aether laurilicus; macrogoli aether oleicus; macrogoli aether stearylicus; polyoxyethylene lauryl alcohol ether; *Procol*; *Ritoleth*; *Ritox*.

Table I shows synonyms for specific materials.

3 Chemical Name and CAS Registry Number

Polyethylene glycol monocetyl ether [9004-95-9]
Polyethylene glycol monolauryl ether [9002-92-0]

Table I: Synonyms of selected polyoxyethylene alkyl ethers.

Name	Synonym
Cetomacrogol 1000	Polyethylene glycol 1000; macrocetyl ether; polyoxyethylene glycol 1000 monocetyl ether; *Cresmer 1000*.
Polyoxyl 6 cetostearyl ether	*Ceteareth 6; Cremophor A6*.
Polyoxyl 20 cetostearyl ether	*Brij CS-20; Ceteareth 20; Cremophor A 20 polyether; Genapol T200; Hetoxol CS-20; Jeecol CS-20; Lipocol SC-20; Lumulse CS-20; Ritacet 20*.
Polyoxyl 25 cetostearyl ether	*Brij CS25; Ceteareth 25; Cremophor A25; Hetoxol CS-25*.
Polyoxyl 2 cetyl ether	*Brij C2*; ceteth-2; *Hetoxol CA-2; Jeecol CA-10; Lipocol C-2; Nikkol BC-2; Procol CA-2*.
Polyoxyl 10 cetyl ether	*Brij C10*; ceteth-10; *Jeecol CA-10; Lipocol C-10; Nikkol BC-10TX; Procol CA-10*.
Polyoxyl 20 cetyl ether	*Brij C20*; ceteth-20; *Hetoxol CA-20; Jeecol CA-20; Lipocol C-20; Nikkol BC-20TX*.
Polyoxyl 26 glyceryl ether	*Ethosperse G-26; Genapol G-260*; Glycereth-26; *Hetoxide G-26; Jeechem GL-26*.
Polyoxyl 4 lauryl ether	*Brij L4; Ethosperse LA-4; Genapol LA 040; Hetoxol LA-4; Jeecol LA-4*; laureth-4; *Lipocol L-4; Lumulse L-4; Nikkol BL-4.2; Procol LA-4*.
Polyoxyl 9 lauryl ether	*Brij L9; Hetoxol LA-9; Jeecol LA-9*; laureth-9; lauromacrogol 400; *Nikkol BL-9EX*; polidocanol.
Polyoxyl 12 lauryl ether	*Hetoxol LA-12; Jeecol LA-12*; laureth-12; *Lipocol L-12; Lumulse L-12; Procol LA-12*.
Polyoxyl 23 lauryl ether	*Brij L23; Ethosperse LA-23; Genapol LA 230; Hetoxol LA-23; Jeecol LA-23*; laureth-23; *Lipocol L-23; Lumulse L-23; Procol LA-23; Ritox 35*.
Polyoxyl 2 oleyl ether	*Brij O2; Genapol O 020; Jeecol OA-2; Lipocol O-2; Nikkol BO-2V*; oleth-2; *Procol OA-2; Ritoleth 2*.
Polyoxyl 10 oleyl ether	*Brij O10; Genapol O 100; Hetoxol OA-10; Jeecol OA-10; Lipocol O-10; Nikkol BO-10V*; oleth-10; polyethylene glycol monooleyl ether; *Procol OA-10; Ritoleth 10*.
Polyoxyl 20 oleyl ether	*Brij O20; Genapol O 200; Jeecol OA-20; Lipocol O-20; Nikkol BO-20V*; oleth-20; *Procol OA-20; Ritoleth 20*.
Polyoxyl 2 stearyl ether	*Brij S2; Genapol HS 020; Hetoxol STA-2; Jeecol SA-2; Lipocol S-2; Nikkol BS-2; Procol SA-2*; steareth-2.
Polyoxyl 10 stearyl ether	*Brij S10; Hetoxol STA-10; Jeecol SA-10; Lipocol S-10; Procol SA-10*; steareth-10.
Polyoxyl 21 stearyl ether	*Brij S721; Jeecol SA-21; Lipocol S-21; Ritox 721*; steareth-21.
Polyoxyl 100 stearyl ether	*Brij S100; Hetoxol STA-100; Jeecol SA-100*; steareth-100.

Polyethylene glycol monooleyl ether [9004-98-2]

Polyethylene glycol monostearyl ether [9005-00-9]

4 Empirical Formula and Molecular Weight

See Sections 1, 2, and 5.

5 Structural Formula

$CH_3(CH_2)_x(OCH_2CH_2)_yOH$

In the formula, $(x + 1)$ is the number of carbon atoms in the alkyl chain, typically:

12 lauryl (dodecyl)

14 myristyl (tetradecyl)

16 cetyl (hexadecyl)

18 stearyl (octadecyl)

and y is the number of ethylene oxide groups in the hydrophilic chain, typically 10–60.

The polyoxyethylene alkyl ethers tend to be mixtures of polymers of slightly varying molecular weights, and the numbers quoted are average values. In cetomacrogol 1000, for example, x is 15 or 17, and y is 20–24.

6 Functional Category

Dispersing agent; emulsifying agent; gelling agent; nonionic surfactant; penetration enhancer; solubilizing agent; viscosity-increasing agent.

7 Applications in Pharmaceutical Formulation or Technology

Polyoxyethylene alkyl ethers are nonionic surfactants widely used in topical pharmaceutical formulations and cosmetics, primarily as emulsifying agents for water-in-oil and oil-in-water emulsions, and the stabilization of microemulsions and multiple emulsions.

Polyoxyethylene alkyl ethers are used as solubilizing agents for essential oils, perfumery chemicals, vitamin oils, and drugs of low-water solubility such as cortisone acetate, griseofulvin, menadione,(1) chlordiazepoxide(2) and cholesterol.(3) They have applications as antidusting agents for powders; wetting and dispersing agents for coarse-particle liquid dispersions; and detergents, especially in shampoos, face washes and similar cosmetic cleaning preparations. They are used as gelling and foaming agents.

Polyoxyethylene alkyl ethers have also been used in suppository formulations to increase the drug release from the suppository bases.(4–6)

Polyoxyethylene alkyl ethers (especially laureth-23) have been used as solubilizers and coating agents to provide hydrophilicity to polymeric nanoparticles.(7–9)

Polyoxyethylene alkyl ethers such as polidocanol are suitable for use as solubilizers or dispersants.(10)

8 Description

Polyoxyethylene alkyl ethers vary considerably in their physical appearance from liquids, to pastes, to solid waxy substances. They occur as colorless, white, cream-colored or pale yellow materials with a slight odor.

9 Pharmacopeial Specifications

See Table II.

10 Typical Properties

See Tables III and IV.

11 Stability and Storage Conditions

Polyoxyethylene alkyl ethers are chemically stable in strongly acidic or alkaline conditions. The presence of strong electrolytes may, however, adversely affect the physical stability of emulsions containing polyoxyethylene alkyl ethers.

On storage, polyoxyethylene alkyl ethers can undergo autoxidation, resulting in the formation of peroxides with an increase in acidity. Many commercially available grades are thus supplied with added antioxidants. Typically, a mixture of 0.01% butylated hydroxyanisole and 0.005% citric acid is used for this purpose.

Polyoxyethylene alkyl ethers should be stored in an airtight container, in a cool, dry place.

12 Incompatibilities

Discoloration or precipitation may occur with iodides, mercury salts, phenolic substances, salicylates, sulfonamides, and tannins.

Table II: Pharmacopeial specifications for polyoxyethylene alkyl ethers.

Test	JP XV	PhEur 7.4				USP35–NF30			
	Lauromacrogol	Macrogol cetostearyl ether	Macrogol lauryl ether	Macrogol oleyl ether	Macrogol stearyl ether	Polyoxyl 20 cetostearyl ether	Polyoxyl 10 oleyl ether	Polyoxyl lauryl ether	Polyoxyl stearyl ether
Characters	–	+	+	+	+	–	–	–	–
Identification	+	+	+	+	+	+	+	+	+
Appearance of solution	–	+	+	+	+	–	–	+	+
Water	–	≤3.0%	≤3.0%	≤3.0%	≤3.0%	≤1.0%	≤3.0%	≤3.0%	≤3.0%
pH (1 in 10 solution)	–	–	–	–	–	4.5–7.5	–	–	–
Alkalinity	–	+	+	+	+	–	–	+	+
Acidity	+	–	–	–	–	–	–	–	–
Residue on ignition	≤0.2%	–	–	–	–	≤0.4%	≤0.4%	–	–
Heavy metals	–	–	–	–	–	≤0.002%	≤0.002%	–	–
Acid value	–	≤1.0	≤1.0	≤1.0	≤1.0	≤0.5	≤1.0	≤1.0	≤1.0
Hydroxyl value	–	+	+	+	+	42–60	75–95	+	+
Iodine value	–	≤2.0	≤2.0	+	≤2.0	–	23–40	≤2.0	≤2.0
Saponification value	–	≤3.0	≤3.0	≤3.0	≤3.0	≤2.0	≤3.0	≤3.0	≤3.0
Free polyethylene glycols	–	–	–	–	–	≤7.5%	≤7.5%	–	–
Free ethylene oxide	–	≤1 ppm	≤1 ppm	≤1 ppm	≤1 ppm	≤0.01%	≤0.01%	≤1 μg/g	≤1 μg/g
Dioxan	–	≤10 ppm	≤10 ppm	≤10 ppm	≤10 ppm	–	–	≤10 μg/g	≤10 μg/g
Peroxide value	–	–	–	≤10.0	–	–	–	–	–
Average polymer length	–	–	–	–	–	17.2–25.0	9.1–10.9	≈3.0–23.0	≈2.0–20.0
Total ash	–	≤0.2%	≤0.2%	≤0.2%	–	–	–	≤0.2%	–
Unsaturated compound	+	–	–	–	–	–	–	–	–

Polyoxyethylene alkyl ethers are also incompatible with benzocaine, tretinoin[11] and oxidizable drugs.[12]

The antimicrobial efficacy of some phenolic preservatives, such as the parabens, is reduced owing to hydrogen bonding. Cloud points are similarly depressed by phenols owing to hydrogen bonding between ether oxygen atoms and phenolic hydroxyl groups. Salts, other than nitrates, iodides, and thiocyanates (which cause an increase) can also depress cloud points.[13]

13 Method of Manufacture

Polyoxyethylene alkyl ethers are prepared by the condensation of linear fatty alcohols with ethylene oxide. The reaction is controlled so that the required ether is formed with the polyethylene glycol of the desired molecular weight.

14 Safety

Polyoxyethylene alkyl ethers are used as nonionic surfactants in a variety of topical pharmaceutical formulations and cosmetics. The polyoxyethylene alkyl ethers form a series of materials with varying physical properties; manufacturers' literature should be consulted for information on the applications and safety of specific materials.

Although generally regarded as essentially nontoxic and non-irritant materials, some polyoxyethylene alkyl ethers, particularly when used in high concentration (>20%), appear to have a greater irritant potential than others. It has been demonstrated that the toxicity exerted by these surfactants to erythrocytes can be mitigated when combined with egg phosphatidylcholine.[14]

Animal toxicity studies suggest that polyoxyethylene alkyl ethers have a similar oral toxicity to other surfactants and can be regarded as being moderately toxic. An acute toxicity and genotoxicity study in aquatic organisms has shown that although not genotoxic (1000 mg/L), polyoxyethylene alkyl ethers are toxic (LC_{50} between 1 and 10 mg/L) or even highly toxic (LC_{50} below 1 mg/L) with higher molecular weight compounds showing higher toxicity.[15]

Polyoxyethylene cetyl ether:[16]

LD_{50} (mouse, oral): 2.60 g/kg
LD_{50} (rabbit, skin): 40 g/kg/4 week intermittent
LD_{50} (rat, oral): 2.50 g/kg

Polyoxyethylene lauryl ether:[16]

LD_{50} (mouse, IP): 0.16 g/kg
LD_{50} (mouse, IV): 0.10 g/kg
LD_{50} (mouse, oral): 4.94 g/kg
LD_{50} (mouse, SC): 0.79 g/kg
LD_{50} (rat, IV): 0.027 g/kg
LD_{50} (rat, oral): 8.60 g/kg
LD_{50} (rat, SC): 0.95 g/kg

Polyoxyethylene oleyl ether:[16] LD_{50} (rat, oral): 25.8 g/kg

15 Handling Precautions

Observe normal precautions appropriate to the circumstances and quantity of material handled. Eye protection and gloves are recommended.

16 Regulatory Status

Included in nonparenteral medicines licensed in the US and UK. Included in the Canadian Natural Health Products Ingredients Database.

17 Related Substances

Nonionic emulsifying wax.

18 Comments

Many other polyoxyethylene ethers are commercially available and are also used as surfactants. In addition to their surfactant properties, the series of polyoxyethylene ethers with lauryl side chains, e.g. nonoxynol 10, are also widely used as spermicides. *Brij* (Croda Chemicals) surfactants show synergistic effects when used in combination and readily produce stable difficult-to-formulate emulsions, such as clear gel microemulsions and alkali-based hair straighteners, e.g. *Brij S2* and *S721* work well in combination giving rise to oleosomes, thereby providing flexibility to emulsify a wide range of oils and can be used in formulating emulsions from water-thin milks to thicker medicated creams.

Polyoxyethylene alkyl ethers have been studied in drug delivery systems containing hydrogels,[17] oleosomes, hydrosomes, phosphosomes, vesicles[18] and niosomes.[19–22] An increased flux of

Table III: Typical properties of selected commercially available grades of polyoxyethylene alkyl ethers.

Name	Physical form	Acid value	HLB value	Hydroxyl value	Iodine number	Saponification value	Density (g/cm³) at 20°C unless otherwise stated	Water content (%)	Boiling point (°C)	Melting point or pour point (°C)	Cloud point (°C) for 1% aqueous solution	pH aqueous solution
Brij C2	White waxy solid	≤1	5.3	160–180	—	—	≈0.95	≤1.0	—	33	—	5–8 (10% in 1 : 1 IPA : water)[a]
Brij C10	White waxy solid	≤1	12.9	75–90	—	—	≈1.06 at 25°C	≤3.0	—	31	—	—
Brij C20	White waxy solid	≤1	15.7	45–60	—	—	1.02 at 25°C	≤3.0	—	38	—	—
Brij CS20	Solid	—	15.7	—	—	—	—	—	—	—	—	—
Brij L4	Clear, colorless liquid	≤2	9.7	145–165	—	—	≈0.95 at 25°C	≤1.0	>100	≈2	—	—
Brij L9	Liquid precipitates on standing	—	13.3	—	—	—	—	—	—	≈19	—	—
Brij L23	White waxy solid	≤5	16.9	40–60	—	—	≈1.05 at 25°C	≤3.0	>100	≈33	—	—
Brij O10	Pale yellow liquid	≤1	12.4	80–95	—	—	≈1.0 at 25°C	≤3.0	>100	16	>100	>100
Brij O20	Pale yellow liquid	≤1	15.5	50–65	—	—	≈1.07 at 25°C	≤3.0	>100	33	—	5–8 (10% in 1 : 4 IPA : water)
Brij S2	White waxy solid or semi-solid or pastille form	≤1	4.9	150–170	—	—	≈0.97 at 25°C	≤1.0	—	43	—	—
Brij S10	White waxy solid	≤1	12.4	75–90	—	—	≈1.05 at 25°C	≤3.0	>100	38	—	5–8 (10% in 1 : 1 IPA : water)
Brij S20	White waxy solid	≤1	15.3	45–60	—	—	≈1.09 at 25°C	≤3.0	—	38	—	5–8 (10% in 1 : 4 IPA : water)
Brij S721	White waxy solid or white powder or pastille form	<2	15.5	44–61	—	—	≈1.0 at 25°C	≤2.0	—	45	—	—
Brij S100	White to pale yellow waxy solid	—	18.8	—	—	—	—	—	—	—	—	—
Cremophor A6	Whitish waxy substance	≤1	10–12	115–134	≤1	≤3	0.896–0.906 at 60°C	≤1.0	—	41–45	—	—
Cremophor A 20 polyether	White waxy solid	≤5	15	45–60	—	—	0.98 at 70°C	<1.0	>149	56	—	—
Cremophor A25	Whitish microbeads	≤1	15–17	36–45	≤1	≤3	1.020–1.028 at 60°C	≤1.0	—	44–48	—	5–7 (10%)
Emulgen 104P	Clear liquid	—	9.6	—	—	—	—	—	—	—	—	—
Emulgen 109P	Light yellow liquid	—	13.6	—	—	—	—	—	—	—	83	—
Emulgen 123P	White solid	—	16.9	—	—	—	—	—	—	—	>100	—
Emulgen 210P	Light yellow solid	—	10.7	—	—	—	—	—	—	—	—	—
Emulgen 220	Light yellow solid	—	14.2	—	—	—	—	—	—	—	98	—
Emulgen 306P	Light yellow wax	—	9.4	—	—	—	—	—	—	—	—	—
Emulgen 320P	White solid	—	13.9	—	—	—	—	—	—	—	91	—
Emulgen 404	Light yellow liquid	—	8.8	—	—	—	—	—	—	—	—	—
Emulgen 409P	Light yellow liquid	—	12.0	—	—	—	—	—	—	—	55	—
Ethosperse LA-4	Water white liquid at 25°C	≤0.3	10.0	145–160	—	—	—	≤0.5	—	—	—	—
Ethosperse LA-23	White, waxy solid at 25°C	≤0.3	17.0	45–52	—	—	—	≤3.0	—	30–45	—	—
Ethosperse G26	Clear liquid at 25°C	≤0.5	18.0	123–138	—	—	—	≤0.6	—	—	—	6.0 (5%)
Genapol HS 020	Wax at 44°C	—	5.0	—	—	—	—	—	—	—	—	—
Genapol HS 200	Wax	—	15.0	—	—	—	—	—	—	—	—	—
Genapol G-260	Clear to slightly turbid liquid	—	18.0	—	—	—	—	—	—	—	—	—
Genapol LA 040	Liquid	—	9.0	—	—	—	—	—	—	—	—	—
Genapol LA 230	Wax	—	17.0	—	—	—	—	—	—	—	—	—
Genapol O 020	Liquid	—	5.0	—	—	—	—	—	—	—	—	—
Genapol O 100	Soft wax at 25°C	—	12.0	—	—	—	—	—	—	—	—	—
Genapol O 200	Wax at 34°C	—	15.0	—	—	—	—	—	—	—	—	—
Genapol T 150	Wax at 44°C	—	14.0	—	—	—	—	—	—	—	—	—
Genapol T 200	Wax at 48°C	—	15.0	—	—	—	—	—	—	—	—	—
Genapol T 250P	Powder at 50°C	—	16.0	—	—	—	—	—	—	—	—	—
Genapol T500P	Powder at 56°C	—	18.0	—	—	—	—	—	—	—	—	—
Hetoxol CA-2	Solid	—	5.1	—	—	—	—	—	—	—	—	—

cont.

P

P

Table cont.

Name	Physical form	Acid value	HLB value	Hydroxyl value	Iodine number	Saponification value	Density (g/cm³) at 20°C unless otherwise stated	Water content (%)	Boiling point (°C)	Melting point or pour point (°C)	Cloud point (°C) for 1% aqueous s olution	pH aqueous solution
Hetoxol CA-20	Solid	—	15.5	—	—	—	—	—	—	—	—	—
Hetoxol CS-20	Solid	—	15.5	—	—	—	—	—	—	—	—	—
Hetoxol CS-25	Solid	—	16.3	—	—	—	—	—	—	—	—	—
Hetoxol G-26	Liquid	—	18.4	—	—	—	—	—	—	—	—	—
Hetoxol LA-4	Liquid	—	9.4	—	—	—	—	—	—	—	—	—
Hetoxol LA-9	Soft solid	—	13.3	—	—	—	—	—	—	—	—	—
Hetoxol LA-12	Semi-solid	—	14.6	—	—	—	—	—	—	—	—	—
Hetoxol LA-23	Solid	—	16.7	—	—	—	—	—	—	—	—	—
Hetoxol OA-10	Semi-solid	—	12.4	—	—	—	—	—	—	—	—	—
Hetoxol OA-35	Solid	—	16.9	—	—	—	—	—	—	—	—	—
Hetoxol STA-2	Solid	—	4.9	—	—	—	—	—	—	—	—	—
Hetoxol STA-10	Solid	—	12.4	—	—	—	—	—	—	—	—	—
Hetoxol STA-100	Solid	—	18.8	—	—	—	—	—	—	—	—	—
Hostacerin T 3	Soft wax at 40°C	—	6.0	—	—	—	—	—	—	—	—	—
Jeechem GL-26	Clear to hazy straw liquid at 25°C	≤1	—	128–138	—	—	1.13 at 15.6°C	≤0.5	—	—	—	5.5–7.0 (5%)
Jeecol CA-2	White waxy solid at 25°C	≤1	—	160–180	—	≤2	0.8456	≤1	110	—	—	—
Jeecol CA-10	White waxy solid at 25°C	≤1	—	75–90	—	—	1.003 at 25°C	≤3	—	—	—	—
Jeecol CA-20	White waxy solid at 25°C	≤1	—	45–60	—	—	1.02	≤3	—	—	—	5.0–6.5 (5%)
Jeecol CS-10	Off-white solid at 25°C	≤1	—	70–84	—	—	—	≤0.5	—	38	77–87	5.0–6.5 (5%)
Jeecol CS-20	White waxy solid at 25°C	≤1	—	45–55	—	—	1.02 at 45°C	—	—	—	—	5.0–6.5 (5%)
Jeecol LA-4	Water white liquid at 25°C	≤1	—	150–165	—	—	—	≤1.0	—	—	—	5.0–6.5 (5%)
Jeecol LA-9	Clear to hazy liquid precipitates on standing at 25°C	≤1	—	90–110	—	—	—	≤1.0	—	—	73–77	5.5–7.0 (5%)
Jeecol LA-12	Soft white solid at 25°C	≤1	—	73–83	—	—	—	≤1.0	—	—	94–100	5.0–6.5 (5%)
Jeecol LA-23	White waxy solid at 25°C	≤1	—	40–60	—	—	—	≤1.0	—	—	—	5.0–7.5 (5%)
Jeecol OA-2	Yellow liquid at 25°C	≤1	—	150–175	50–60	—	—	≤1.0	—	—	—	5.5–7.0 (5% 50:50 IPA:water)
Jeecol OA-10	Yellow soft solid at 25°C	≤1	—	83–95	—	—	—	≤3.0	—	—	56–62	5.0–6.5 (5%)
Jeecol OA-20	White solid at 25°C	≤1	—	56–70	—	—	—	≤1.0	—	—	—	5.0–6.5 (5%)
Jeecol SA-2	White waxy solid at 25°C	≤1	—	148–165	—	≤2.0	—	≤1.0	—	42–46	—	5.0–6.5 (5%)
Jeecol SA-10	White waxy solid at 25°C	≤1	—	75–85	≤2.0	≤2.0	—	≤1.0	—	—	—	5.0–7.0 (3%)
Jeecol SA-20	White waxy solid at 25°C	≤1	—	45–60	—	≤2.0	—	≤2.0	—	—	—	—
Lipocol C-2	Off-white waxy solid	—	5.3	—	—	—	1.02	—	—	27–35	—	—

Table cont.

Name	Physical form	Acid value	HLB value	Hydroxyl value	Iodine number	Saponification value	Density (g/cm^3) at 20°C unless otherwise stated	Water content (%)	Boiling point (°C)	Melting point or pour point (°C)	Cloud point (°C) for 1% aqueous s olution	pH aqueous solution
Lipocol C-10	White waxy solid	—	12.9	—	—	—	1.00	—	>100	—	—	—
Lipocol C-20	White waxy solid	—	15.7	—	—	—	1.02	—	—	—	—	—
Lipocol L-4	Colorless to slightly yellow liquid	—	9.7	—	—	—	0.94	—	>199	—	—	—
Lipocol L-12	White waxy solid	—	14.5	—	—	—	0.94	—	>199	—	—	—
Lipocol L-23	—	—	16.9	—	—	—	—	—	—	—	—	—
Lipocol O-3	Clear to slightly hazy liquid	—	6.6	—	—	—	0.9	—	>100	—	—	—
Lipocol O-10	White to yellow turbid liquid to soft solid	—	12.4	—	—	—	—	—	—	—	—	—
Lipocol O-20	Off-white waxy solid	—	15.3	—	—	—	1.0	—	>100	—	—	—
Lipocol S-2	White to off-white waxy solid	—	4.9	—	—	—	—	—	—	—	—	6.5
Lipocol S-20	White waxy solid	—	15.3	—	—	—	1.0	—	>100	—	—	—
Lipocol S–21	White waxy solid	—	15.5	—	—	—	1.02	—	>100	—	—	—
Lipocol SC-20	White waxy solid	—	15.5	—	—	—	1.0	—	>100	—	—	—
Lumulse CS-20	Solid at 25°C	≤0.5	15.2	45–60	0.5	—	—	<0.5	—	40	—	7.0 (5%)
Lumulse L-4	Liquid at 25°C	≤1	9.5	145–160	0.1	—	0.95 at 25°C	<0.5	—	12	—	7.0 (5%)
Lumulse L-12	Solid at 25°C	≤1	14.5	70–80	0.1	—	0.97 at 25°C	<0.5	—	30	—	6.0 (5%)
Lumulse L-23	Solid at 25°C	≤1	16.4	40–55	0.1	—	0.99 at 25°C	<1.0	—	40	—	6.0 (5%)
Nikkol BC-2	White solid	—	8.0	—	—	—	—	—	—	—	—	—
Nikkol BC-10	White solid	—	13.5	—	—	—	—	—	—	—	—	—
Nikkol BC-20	White solid	—	17.0	—	—	—	—	—	—	—	—	—
Nikkol BL-4.2	Colorless liquid	—	11.5	—	—	—	—	—	—	—	—	—
Nikkol BL-9EX	Colorless liquid	—	14.5	—	—	—	—	—	—	—	—	—
Nikkol BO-2V	Light yellow liquid	—	7.5	—	—	—	—	—	—	—	—	—
Nikkol BO-10V	Pale yellow liquid containing waxy substances	—	14.5	—	—	—	—	—	—	—	—	—
Nikkol BO-20V	Light yellow solid	—	17.0	—	—	—	—	—	—	—	—	—
Nikkol BS-2	White solid	—	8.0	—	—	—	—	—	—	—	—	—
Nikkol BS-20	White solid	—	18.0	—	—	—	—	—	—	—	—	—
Ritacet 20	Light yellow solid	—	—	—	—	—	—	—	—	—	—	—
Ritoleth 2	Clear to slightly yellow liquid	<1.0	4.9	150–180	—	—	0.92 at 25°C	<1.0	149	—	—	—
Ritoleth 5	Clear to slightly yellow liquid	<2.0	8.8	120–133	—	—	0.94 at 25°C	<3.0	149	—	—	—
Ritoleth 10	White paste	<10.0	12.3	80–90	32–40	<2.0	0.94 at 25°C	<3.0	149	—	47–55	4.5–7.5 (10%)
Ritoleth 20	White to yellow solid	<0.5	15.4	45–70	—	<2.0	1.01 at 25°C	<3.0	149	—	—	—
Ritox 35	White waxy solid	<5.0	16.7	40–60	—	—	1.05 at 25°C	<3.0	—	—	—	—
Ritox 721	—	—	—	—	—	—	—	—	—	—	—	—

(a) IPA: isopropyl alcohol.

P

Table IV: Typical properties of selected commercially available grades of polyoxyethylene alkyl ethers.

Name	Critical micelle concentration (%)	Surface tension of aqueous solution (mN/m)	Dynamic viscosity at 25°C or pour point (mPas)	Refractive index at 60°C	Solubility					Flash point (°C)
					Ethanol	Fixed oils	Mineral oil	Propylene glycol	Water	
Brij L4	—	—	≈30	—	S	D	D	S	I	>149
Brij L9	—	—	≈24 at 50°C	—	S	I	S	S	S	>149
Brij L23	0.013	—	—	—	S	I	I	S	S	>149
Brij C2	—	—	—	—	S	D	H	D	I	>149
Brij C10	—	—	—	—	H	D	I	D	H	>149
Brij C20	—	—	—	—	S	D	I	I	S	>149
Brij S2	—	—	—	—	S	—	S	I	I	>149
Brij S10	—	—	—	—	S	D	I	D	S	>149
Brij S20	—	—	—	—	S	D	I	I	D	>149
Brij S721	—	—	—	—	S (warm)	I	D	I	D	>110
Brij O10	—	—	100	—	S	D	H	S	S	—
Brij O20	—	—	—	—	S	D	I	D	S	>149
Brij CS20	—	—	—	—	S	—	—	—	S	>100
Cremophor A6	—	33.4 (0.5%) at 23°C	13.5 at 60°C	1.4420–1.4424	S	S	S	—	S	190
Cremophor A20 polyether	—	—	—	—	—	—	—	—	S	>149
Cremophor A25	—	42.0 (0.5%) at 23°C	—	1.4512–1.4520	S	S	S	—	S	—
Jeechem GL-26	—	—	—	—	—	—	—	—	S	>100
Jeecol CA-2	—	—	—	—	—	—	—	—	PS	—
Jeecol CA-10	—	—	—	—	—	—	—	—	S	>149
Jeecol CA-20	—	—	—	—	—	—	—	—	D	>149
Jeecol CS-10	—	—	—	—	—	—	—	—	S	>200
Jeecol CS-20	—	—	—	—	—	—	—	—	S at 25°C	—
Jeecol LA-4	—	—	—	1.4450–1.4550 at 25°C	—	—	—	—	—	—
Jeecol LA-9	—	—	—	—	—	—	—	—	S	—
Jeecol LA-12	—	—	—	—	—	—	—	—	S	—
Jeecol LA-23	—	—	—	—	—	—	—	—	S	—
Jeecol OA-2	—	—	—	—	—	—	—	—	D	—
Jeecol OA-10	—	—	—	—	—	—	—	—	S	—
Jeecol OA-20	—	—	—	—	—	—	—	—	S	—
Jeecol SA-2	—	—	—	—	—	—	—	—	<1%	—
Jeecol SA-10	—	—	—	—	—	—	—	—	PS	—
Jeecol SA-20	—	—	—	—	—	—	—	—	D	—
Lipocol C-2	—	—	—	—	S	S	S	I	I	>149
Lipocol C-10	—	—	—	—	S	D (Low concentration)	D	—	D	>149
Lipocol C-20	—	—	—	—	S	I	I	—	S	>149
Lipocol L-4	—	—	—	—	S	PS	SH	S	D	>149
Lipocol L-12	—	—	—	—	S	I	I	—	S	>149
Lipocol O-3	—	—	—	—	S	—	S	S	I	>149
Lipocol O-10	—	—	—	—	S	S	I	—	S	>250
Lipocol O-20	—	—	—	—	S	S	I	—	S	>149
Lipocol S-2	—	—	—	—	S	SH	SH	—	I	>149
Lipocol S-20	—	—	—	—	S	SH	SH	—	I	>149
Lipocol S-21	—	—	—	—	S	SH	SH	—	I	>149
Lipocol SC-20	—	—	—	—	—	—	I	—	S	>149
Lumulse CS-20	—	—	—	—	—	—	—	—	S	>175
Lumulse L-4	—	—	—	—	—	—	—	—	I	>177
Lumulse L-12	—	—	—	—	—	—	—	—	S	>177
Lumulse L-23	—	—	—	—	—	—	—	—	S	>177

S = soluble; H = soluble with haze; I = insoluble; D = dispersible; PS = partially soluble; SH = soluble on heating.
Suppliers: BASF Corporation (*Cremophor*); Clariant (*Genapol/Hostacerin*); Croda Chemicals (*Brij*); Global Seven (*Hetoxol*); Jeen International Corporation (*Jeecol*); Kao Global Chemicals (*Emulgen*); Lambent Technologies (*Lumulse*); Lipo Chemicals (*Lipocol*); Lonza Group Ltd (*Ethosperse*); Nikko Chemicals Co. Ltd (*Nikkol*); Rita Corporation (*Ritoleth, Ritox*).

estradiol niosomes through human stratum corneum *in vitro* has been demonstrated.[23] Polyoxyethylene alkyl ether niosomes encapsulating insulin have been investigated for oral drug delivery.[24]

Polyoxyethylene alkyl ethers have been found to have an enhancing effect on the skin permeation of drugs such as ibuprofen,[25] methyl nicotinate,[26] and clotrimazole.[27] Enhanced ocular absorption of insulin from eye drops,[28] and an ocular insert device,[29] have been observed using polyoxyethylene alkyl ethers in the formulation systems. Increased buccal absorption of verapamil through porcine esophageal mucosa has also been reported.[30] A combination of a mucolytic agent possessing a free thiol group and a polyoxyethylene alkyl ether surfactant of similar polyoxyethylene and alkyl chain length have been reported to show effective enhancement in the intestinal absorption of poorly absorbed hydrophilic compounds.[31]

Long chain polyoxyethylene alkyl ethers have proved useful in the soluble expression of membrane proteins up to a certain threshold concentration,[32] and in sieving matrices for the separation of DNA fragments by capillary electrophoresis.[33] It has been demonstrated that the antibacterial activity of butyl *p*-hydroxybenzoate is enhanced by the presence of polyoxyethylene alkyl ethers, particularly those with 12 carbons in the hydrophobic group.[34]

19 Specific References

1 Elworthy PH, Patel MS. Demonstration of maximum solubilization in a polyoxyethylene alkyl ether series of non-ionic surfactants. *J Pharm Pharmacol* 1982; **34**: 543–546.
2 Abdel Rahman AA, *et al.* Factors affecting chlordiazepoxide solubilization by non-ionic surfactants. *Bull Pharm Sci Assiut Univ* 1991; **14**(1–2): 35–45.
3 Mueller-Goymann CC, Usselmann BS. Solubilization of cholesterol in liquid crystals of aqueous systems of polyoxyethylene cetyl ethers. *Acta Pharm Jugosl* 1988; **38**(4): 327–329.
4 Al Gohary OM, Foda NH. Pharmaceutical and microbiological aspects of nalidixic acid suppositories. *Egypt J Pharm Sci* 1996; **37**(1–6): 273–284.
5 El Assasy AH, *et al.* Formulation of flurbiprofen suppositories. *Egypt J Pharm Sci* 1995; **36**(1–6): 31–53.
6 El Assasy AH, *et al.* Release characteristics and bioavailability of pirprofen from suppository bases. *Egypt J Pharm Sci* 1995; **36**(1–6): 15–29.
7 Harmia-Pulkkinen T, Ojantakanen S. *In vitro* release kinetics of timolol and timol oleate from polyethylcyanoacrylate nanoparticles. Part 2. Nanoparticles manufacture with timolol maleate using different surfactants and organic solvents. *Acta Pharm Fenn* 1992; **101**(2): 57–63.
8 Muller RH, *et al. In vitro* characterization of poly(methyl-methacrylate) nanoparticles and correlation to their *in vivo* fate. *J Control Release* 1992; **20**: 237–246.
9 Troster SD, *et al.* Modification of the body distribution of poly(-methylmethacrylate) nanoparticles in rats by coating with surfactants. *Int J Pharm* 1990; **61**: 85–100.
10 Cabrera J, *et al.* Ultrasound-guided injection of polidocanol microfoam in the management of venous leg ulcers. *Arch Dermatol* 2004; **140**(6): 667–673.
11 Brisaert MG, *et al.* Chemical stability of tretinoin in dermatological preparations. *Pharm Acta Helv* 1995; **70**(2): 161–166.
12 Azaz E, *et al.* Incompatibility of non-ionic surfactants with oxidizable drugs. *Pharm J* 1975; **211**: 15.
13 McDonald C, Richardson C. The effect of added salts on solubilization by a non-ionic surfactant. *J Pharm Pharmacol* 1981; **33**: 38–39.
14 Gould LA, *et al.* Mitigation of surfactant erythrocyte toxicity by egg phosphatidylcholine. *J Pharm Pharmacol* 2000; **52**(10): 1203–1209.
15 Liwarska-Bizukojc E, *et al.* Acute toxicity and genotoxicity of five selected anionic and non-ionic surfactants. *Chemosphere* 2005; **58**(9): 1249–1253.
16 The Registry of Toxic Effects of Chemical Substances. Atlanta, GA: National Institute for Occupational Safety and Health, 2000.
17 Park SK, *et al.* Reusable ultrasonic tissue mimicking hydrogels containing non-ionic surface active agents for visualizing thermal lesions. *IEEE Trans Biomed Eng* 2010; **57**(1): 194–202.
18 Friberg SE, *et al.* Preparation of vesicles from hydrotope solutions. *J Dispersion Sci Technol* 1998; **19**(1): 19–30.
19 Arunothayanun P, *et al. In vitro/in vivo* characterization of polyhedral niosomes. *Int J Pharm* 1999; **183**(1): 57–61.
20 Parthasarathi G, *et al.* Formulation and *in vitro* evaluation of vincristine encapsulated niosomes. *Indian J Pharm Sci* 1994; **56**(3): 90–94.
21 Tabbakhian M, *et al.* Enhancement of follicular delivery of finasteride by liposomes and niosomes – 1. *In vitro* permeation and *in vivo* deposition studies using hamster flank and ear models. *Int J Pharm* 2006; **323**(1–2): 1–10.
22 Balakrishnan P, *et al.* Formulation and *in vitro* assessement of minoxidil niosomes for enhanced skin delivery. *Int J Pharm* 2009; **377**(1–2): 1–8.
23 Van Hal D, *et al.* Diffusion of estradiol from non-ionic surfactant vesicles through human stratum corneum *in vitro*. *STP Pharm Sci* 1996; **6**(1): 72–78.
24 Pardakhty A, *et al. In vitro* study of polyoxyethylene alkyl ether niosomes for delivery of insulin. *Int J Pharm* 2007; **328**(2): 130–141.
25 Park ES, *et al.* Enhancing effect of polyoxyethylene alkyl ethers on the skin permeation of ibuprofen. *Int J Pharm* 2000; **209**(1–2): 109–119.
26 Ashton P, *et al.* Surfactant effects in percutaneous absorption. I. Effects on the transdermal flux of methyl nicotinate. *Int J Pharm* 1992; **87**(1–3): 261–264.
27 Ibrahim SA, *et al.* Formulation and evaluation of some topical anti mycotics. Part 3. Effect of promoters on the *in vitro* and *in vivo* efficacy of clotrimazole ointment. *Bull Pharm Sci* 1991; **14**(1–2): 82–94.
28 Zhang WY, Zhang LH. Study of absorption enhancers of insulin eye drops. *J China Pharm Univ* 1997; **28**(5): 275–277.
29 Lee YC, *et al.* Effect of Brij-78 on systemic delivery of insulin from an ocular device. *J Pharm Sci* 1997; **86**(4): 430–433.
30 Sawicki W, Janicki S. Influence of polyoxyethylene-10-oleylether on *in vitro* verapamil hydrochloride penetration through mucous membrane from model buccal drug formulation. *STP Pharma Sci* 1998; **8**(2): 107–111.
31 Takatsuka S, *et al.* Influence of various combinations of mucolytic agent and non-ionic surfactant on intestinal absorption of poorly absorbed hydrophilic compounds. *Int J Pharm* 2008; **349**(1–2): 94–100.
32 Klammt C, *et al.* Evaluation of detergents for the soluble expression of alpha-helical and beta-barrel-type integral membrane proteins by a preparative scale individual cell-free expression system. *FEBS J* 2005; **272**(23): 6024–6038.
33 Wei W, Yeung ES. DNA capillary electrophoresis in entangled dynamic polymers of surfactant molecules. *Anal Chem* 2001; **73**(8): 1776–1783.
34 Fukahori M, *et al.* Effects of the structures of polyoxyethylene alkyl ethers on uptake of butyl *p*-hydroxybenzoate by *Escherichia coli* and its antibacterial activity. *Chem Pharm Bull* 1996; **44**(12): 2335–2337.

20 General References

Ammar HO, Khali RM. Solubilization of certain analgesics by Cetomacrogol 1000. *Egypt J Pharm Sci* 1996; **37**: 261–271.
Croda Europe Ltd. Technical literature (DH027/2 and DH081/1): *Brij*, *Brij S2*, and *Brij S721*, 2009.
Elworthy PH, Guthrie WG. Adsorption of non-ionic surfactants at the griseofulvin-solution interface. *J Pharm Pharmacol* 1970; **22**(Suppl.): 114S–120S.
Guveli D, *et al.* Viscometric studies on surface agent solutions and the examination of hydrophobic interactions. *J Pharm Pharmacol* 1974; **26**(Suppl.): 127P–128P.
Malcolmson C, *et al.* Effect of oil on the level of solubilization of testosterone propionate into non-ionic oil-in-water microemulsions. *J Pharm Sci* 1998; **87**: 109–116.
Vasiljevic D, *et al.* Influence of emulsifier concentration on the rheological behavior of w/o/w multiple emulsions. *Pharmazie* 1994; **49**: 933–934.
Walters KA, *et al.* Non-ionic surfactants and gastric mucosal transport of paraquat. *J Pharm Pharmacol* 1981; **33**: 207–213.

21 Authors

RT Gupta, KK Singh.

22 Date of Revision

2 March 2012.

Polyoxyethylene Castor Oil Derivatives

1 Nonproprietary Names

BP: Polyoxyl Castor Oil
Hydrogenated Polyoxyl Castor Oil

PhEur: Macrogolglycerol Ricinoleate
Macrogolglycerol Hydroxystearate

USP–NF: Polyoxyl 35 Castor Oil
Polyoxyl 40 Hydrogenated Castor Oil

Polyoxyethylene castor oil derivatives are a series of materials obtained by reacting varying amounts of ethylene oxide with either castor oil or hydrogenated castor oil. Several different types of material are commercially available, the best-known being the *Cremophor* series. Of these, two castor oil derivatives are listed in the PhEur 7.4 and USP35–NF30.

See also Sections 2, 3, and 4.

2 Synonyms

Synonyms applicable to polyoxyethylene castor oil derivatives are shown below. *See* Table I for information on specific materials.

Acconon; *Cremophor*; *Etocas*; *Eumulgin*; *Jeechem*; *Lipocol*; *Lumulse*; macrogolglyceroli hydroxystearas; macrogolglyceroli ricinoleas; *Nikkol*; *Protachem*; *Simulsol*.

3 Chemical Name and CAS Registry Number

Polyethoxylated castor oil [61791-12-6]

4 Empirical Formula and Molecular Weight

Polyoxyethylene castor oil derivatives are complex mixtures of various hydrophobic and hydrophilic components. Members within each range have different degrees of ethoxylation (moles)/PEG units as indicated by their numerical suffix (n). The chemical structures of the polyethoxylated hydrogenated castor oils are analogous to polyethoxylated castor oils with the exception that the double bond in the fatty chain has been saturated by hydrogenation.

The PhEur 7.4 states that polyoxyl castor oil contains mainly ricinoleyl glycerol ethoxylated with 30–50 molecules of ethylene oxide (nominal value), with small amounts of macrogol ricinoleate, and of the corresponding free glycols. The PhEur 7.4 also states that polyoxyl hydrogenated castor oil contains mainly trihydroxystearyl glycerol ethoxylated with 7–60 molecules of ethylene oxide (nominal value), with small amounts of macrogol hydroxystearate, and of the corresponding free glycols.

In polyoxyl 35 castor oil, the relatively hydrophobic constituents comprise about 83% of the total mixture, the main component being glycerol polyethylene glycol ricinoleate. Other hydrophobic constituents include fatty acid esters of polyethylene glycol along with some unchanged castor oil. The hydrophilic part (17%) consists of free polyethylene glycols and glycerol ethoxylates. *Cremophor ELP*, a 'purified' grade of *Cremophor EL* is also a polyoxyl 35 castor oil; it has a lower content of water (<0.5%), potassium ions (<15 ppm), and free fatty acids (C_{12}–C_{18} <1%), particularly ricinoleic (<0.2%), oleic (<0.1%) and palmitic (<0.1%) acids, and hence is claimed to contribute to improved stability of some sensitive active ingredients.

In polyoxyl 40 hydrogenated castor oil and polyoxyl 60 hydrogenated castor oil the main constituent is glycerol polyethylene glycol oxystearate, which together with fatty acid glycerol polyglycol esters, forms the hydrophobic constituent. The hydrophilic portion consists of polyethylene glycols and glycerol ethoxylate. *Cremophor RH 410* (90% *Cremophor RH 40* + 10% water), *Cremophor CO 40*, *Cremophor 410* (90% *Cremophor CO 40* + 10% water), and *Cremophor CO 455* (90% *Cremophor CO 40* + 5% water + 5% propylene glycol) are cosmetic grades of polyoxyl hydrogenated castor oils.

Table I: Synonyms of selected polyoxyethylene castor oil derivatives.

Name	Synonym
Polyoxyl 5 castor oil	Castor oil POE-5; *Etocas 5*; *Hetoxide C-5*; *Jeechem CA-5*; *Lumulse CO-5*; PEG-5 castor oil; polyoxyethylene 5 castor oil.
Polyoxyl 9 castor oil	Castor oil POE-9; *Jeechem CA-9*; PEG-9 castor oil; polyoxyethylene 9 castor oil; *Protachem CA-9*.
Polyoxyl 15 castor oil	Castor oil POE-15; *Jeechem CA-15*; PEG-15 castor oil; polyoxyethylene 15 castor oil; *Protachem CA-15*.
Polyoxyl 35 castor oil	Castor oil POE-35; *Cremophor EL*; *Cremophor ELP*; *Etocas 35*; *Emulgin RO 35 PH*; glycerol polyethyleneglycol ricinoleate; PEG-35 castor oil; polyethoxylated castor oil; polyoxyethylene 35 castor oil.
Polyoxyl 40 castor oil	*Acconon CA-40*; Castor oil POE-40; *Croduret 40*; *Etocas 40*; *Emulgin RO 40*; *Hetoxide C40*; *Jeechem CA-40*; *Lumulse CO-40*; *Marlowet R40*; *Nikkol CO 40TX*; PEG-40 castor oil; polyoxyethylene 40 castor oil; *Protachem CA-40*.
Polyoxyl 40 hydrogenated castor oil	*Cremophor RH 40*; *Croduret 40*; *Eumulgin HRE 40PH*; glycerol polyethyleneglycol oxystearate; *Hetoxide HC40*; hydrogenated castor oil POE-40; *Jeechem CAH-40*; PEG-40 hydrogenated castor oil; polyethoxylated hydrogenated castor oil; polyoxyethylene 40 hydrogenated castor oil; *Lipocol HCO-40*; *Lipocol LAV HCO 40*; *Lumulse HCO 40*; *Nikkol HCO 40 Pharma*; *Protachem CAH-40*.
Polyoxyl 60 castor oil	Castor oil POE-60;*Jeechem CA-60*; *Nikkol CO 60TX*; PEG-60 castor oil; polyoxyethylene 60 castor oil.
Polyoxyl 60 hydrogenated castor oil	*Cremophor RH 60*; *Croduret 60*; *Eumulgin HRE 60*; *Hetoxide HC60*; hydrogenated castor oil POE-60; *Jeechem CAH-60*; PEG-60 hydrogenated castor oil; polyoxyethylene 60 hydrogenated castor oil; *Lipocol HCO-60*; *Nikkol HCO 60 Pharma*; *Protachem CAH-60*.
Polyoxyl 100 castor oil	Hydrogenated castor oil POE-100; *Jeechem CA-100*; PEG-100 hydrogenated castor oil; polyoxyethylene 100 hydrogenated castor oil.
Polyoxyl 100 hydrogenated castor oil	*Jeechem CA-100*; *Nikkol HCO 100*; polyoxyethylene 100 hydrogenated castor oil.
Polyoxyl 200 castor oil	*Hetoxide C200*; *Jeechem CA-200*; polyoxyethylene 200 castor oil; PEG-200 castor oil; castor oil POE-200.
Polyoxyl 200 hydrogenated castor oil	Hydrogenated castor oil POE-200; *Jeechem CAH-200*; PEG-200 hydrogenated castor oil; polyoxyethylene 200 hydrogenated castor oil.

5 Structural Formula

See Section 4.

6 Functional Category

Emulsifying agent; nonionic surfactant; solubilizing agent.

Table II: Pharmacopeial specifications for polyoxyethylene castor oil derivatives.

Test	PhEur 7.4		USP35–NF30	
	Macrogolglycerol ricinoleate	Macrogolglycerol hydroxystearate	Polyoxyl 35 castor oil	Polyoxyl 40 hydrogenated castor oil
Identification	+	+	+	+
Characters	+	+	—	—
Appearance of solution	+	+	—	—
Alkalinity	+	+	—	—
Relative density	≈1.05	—	—	—
Specific gravity	—	—	1.05–1.06	—
Congealing temperature	—	—	—	16–26°C
Viscosity at 25°C	500–800 mPa s	—	600–850 cP s	—
Water	≤3.0%	≤3.0%	≤3.0%	≤3.0%
Total ash	≤0.3%	≤0.3%	—	—
Residue on ignition	—	—	≤0.3%	≤0.3%
Heavy metals	≤10 ppm	≤10 ppm	≤0.001%	≤0.001%
Acid value	≤2.0	≤2.0	≤2.0	≤2.0
Hydroxyl value	+	+	65–80	60–80
Iodine value	25–35	≤5.0	25–35	≤2.0
Saponification value	+	+	60–75	45–69
1,4-Dioxan	≤10 ppm	≤10 ppm	—	—
Free ethylene oxide	≤1 ppm	≤1 ppm	—	—

7 Applications in Pharmaceutical Formulation or Technology

Polyoxyethylene castor oil derivatives are nonionic solubilizers and emulsifying agents used in oral, topical, and parenteral pharmaceutical formulations.

Polyoxyl 35 castor oil is mainly used as an emulsifying and solubilizing agent, and is particularly suitable for the production of aqueous liquid preparations containing volatile oils, fat-soluble vitamins, and other hydrophobic substances, including aqueous solutions of hydrophobic drugs (e.g. miconazole, hexetidine, clotrimazole, benzocaine). In oral formulations, the taste of polyoxyl 35 castor oil can be masked by a banana flavor. Polyoxyl 35 castor oil has been used as a buffering agent for aqueous tropicamide eyedrops.[1] It has also been used in an aqueous mixture together with caprylic/capric glyceride for mucosal vaccination, providing a potential alternative to parenteral vaccination.[2] Polyoxyl 35 castor oil has been used to enhance the permeability of peptides across monolayers of Caco-2 cells by inhibiting the apically polarized efflux system, enhancing intestinal absorption of some drugs.[3] Polyoxyl 35 castor oil is also used in the production of glycerin suppositories to enhance drug release.[4,5] In cosmetics, polyoxyl 35 castor oil is mainly used as a solubilizing agent for perfume bases and volatile oils in vehicles containing 30–50% v/v alcohol (ethanol or propan-2-ol). In hand lotions, it can be used to replace castor oil.

Polyoxyl 40 hydrogenated castor oil may be used in preference to polyoxyl 35 castor oil in oral formulations as a solubilizer for fat soluble vitamins, essential oils and other hydrophobic pharmaceuticals. It has very little odor and it is almost tasteless. In aqueous alcoholic or completely aqueous solutions, polyoxyl 40 hydrogenated castor oil can be used to solubilize vitamins, essential oils, and certain drugs. In aerosol vehicles that include water, the addition of polyoxyl 40 hydrogenated castor oil improves the solubility of the propellant in the aqueous phase. Foam formation in aqueous ethanol solutions containing polyoxyl 40 hydrogenated castor oil can be suppressed by the addition of small amounts of polypropylene glycol 2000.

Polyoxyl 60 hydrogenated castor oil derivatives have been reported to provide a self-microemulsifying system with enhanced oral absorption,[6] and a drastic reduction in plasma clearance of lipid emulsions.[7] They have been used in the formulation of liposomes,[8] and it has been suggested that more than 60% aids in the targeting of liposomes to the liver.[9] Polyoxyl 60 hydrogenated castor oil micellar solutions of cyclosporine A were found to deliver the drug via the gastrointestinal tract to the lymphatics with an extremely high selectivity.[10,11]

8 Description

Polyoxyl 35 castor oil occurs as a pale yellow, oily liquid that is clear at temperatures above 26°C. It has a faint but characteristic odor and can be completely liquefied by heating to 26°C.

Polyoxyl 40 hydrogenated castor oil occurs as a white to yellowish, semisolid paste at 20°C that liquefies at 30°C. It has a very faint characteristic odor and is almost tasteless in aqueous solution.

Polyoxyl 60 hydrogenated castor oil occurs as a white paste at room temperature. It has little taste or odor in aqueous solution.

See also Table III.

9 Pharmacopeial Specifications

See Table II.

10 Typical Properties

See Tables III, IV, and V.

11 Stability and Storage Conditions

Polyoxyl 35 castor oil forms stable solutions in many organic solvents such as chloroform, ethanol, and propan-2-ol; it also forms clear, stable, aqueous solutions. Polyoxyl 35 castor oil is miscible with other polyoxyethylene castor oil derivatives and on heating with fatty acids, fatty alcohols, and certain animal and vegetable oils. Solutions of polyoxyl 40 hydrogenated castor oil in aqueous alcohols and purely aqueous solutions are also stable. Solutions become cloudy as temperature increases.

On heating of an aqueous solution, the solubility of polyoxyl 35 castor oil is reduced and the solution becomes turbid. Aqueous solutions of polyoxyl hydrogenated castor oil heated for prolonged periods may separate into solid and liquid phases on cooling. However, the product can be restored to its original form by homogenization.

Aqueous solutions of polyoxyl 35 castor oil are stable in the presence of low concentrations of electrolytes such as acids or salts, with the exception of mercuric chloride; *see* Section 12. Heating together with very acidic or basic substances results in saponification.

Table III: Typical physical properties of selected commercially available polyoxyethylene castor oil derivatives.

Name	Description	Acid value	HLB value	Hydroxyl value	Iodine number	Saponification value	Water content (%)	Melting point (°C)	Solidification point (°C)
Castor oil derivatives									
Polyoxyl 35 castor oil (*Cremophor EL*)	Pale yellow oily liquid, clear above 26°C with faint characteristic odor	≤2.0	12–14	65–78	25–35	65–70	2.8	—	—
Polyoxyl 35 castor oil, purified (*Cremophor ELP*)	White to slightly yellowish waxy paste or cloudy liquid with weak characteristic odor	≤2.0	12–14	65–78	25–35	65–70	≤0.5	—	—
Etocas 5	Yellow liquid with characteristic odor at 25°C	—	3.8	—	—	—	—	—	—
Etocas 29	Clear yellow liquid at 25°C	—	11.7	—	—	—	—	—	—
Etocas 30	Pale yellow liquid at 25°C	—	12	—	—	—	—	—	—
Etocas 35	Clear yellow liquid at 25°C	—	12.7	—	—	—	—	—	—
Etocas 40	Clear pale yellow liquid at 25°C	—	13	—	—	—	—	—	—
Etocas 200	Yellow solid at 25°C	—	18.1	—	—	—	—	—	—
Hetoxide C-5	Liquid	—	4	—	—	—	—	—	—
Hetoxide C-16	Liquid	—	8.6	—	—	—	—	—	—
Hetoxide C-25	Liquid	—	10.8	—	—	—	—	—	—
Hetoxide C-30	Liquid	—	11.8	—	—	—	—	—	—
Hetoxide C-36	Liquid	—	12.6	—	—	—	—	—	—
Hetoxide C-40	Liquid	—	13	—	—	—	—	—	—
Hetoxide C-200	Solid	—	18.1	—	—	—	—	—	—
Jeechem CA-5	Yellow liquid	≤1.5	—	128–140	63–73	138–153	≤1.0	—	—
Jeechem CA-9	Clear viscous liquid	≤2.0	—	—	—	120–136	≤3.0	—	—
Jeechem CA-15	Clear amber liquid	≤1.0	—	—	—	95–105	≤1.0	—	—
Jeechem CA-25	Yellow liquid	—	—	75–85	—	77–85	≤1.0	—	—
Jeechem CA-30	Clear yellow liquid	≤2.0	—	77–90	—	65–78	≤2.0	—	—
Jeechem CA-40	Yellow to amber liquid	≤2.0	—	77–89	24–30	57–64	≤3.0	—	—
Jeechem CA-60	Viscous clear yellow liquid	≤2.0	—	42–55	—	28–38	≤12.0	—	—
Jeechem CA-100	Tan solid	≤2.0	—	—	—	27–37	≤1.0	—	—
Jeechem CA-200	Solid	≤2.0	—	20–34	—	14–20	≤1.0	125	—
Lumulse CO-5	Liquid	≤2.0	4	—	—	138–153	≤1.0	–26	—
Lumulse CO-25	Clear liquid	≤2.0	10.8	—	—	75–88	≤1.0	5	—
Lumulse CO-40	Clear yellow liquid	≤2.0	13	—	—	58–64	≤0.5	14	—
Hydrogenated castor oil derivatives									
Polyoxyl 40 hydrogenated castor oil (*Cremophor RH 40*)	White to yellowish paste at 20°C with very little odor in aqueous solutions, almost tasteless	≤0.8	14–16	60–75	≤1	50–60	≤2.0	≈30	16–26
Polyoxyl 60 hydrogenated castor oil	White to yellowish soft or flowing paste with faint odor or taste in aqueous solutions	≤1.0	15–17	50–70	≤1	40–50	≤2.0	≈40	—
Croduret 7	Pale yellow liquid at 25°C	—	5	—	—	—	—	—	—
Croduret 40	Off white paste at 25°C	—	13	—	—	—	—	—	—
Croduret 50	Off white solid at 25°C	—	14.1	—	—	—	—	—	—
Croduret 60	Off white solid at 25°C	—	14.7	—	—	—	—	—	—
Eumulgin HRE 40	White to slightly yellow lard-like fat mass with mild odor	≤1.0	—	60–75	≤2	50–60	≤1.0	—	—
Eumulgin HRE 40 PH	White to slightly yellow lard-like fat mass with mild odor	≤2.0	—	60–80	≤5	45–69	≤3.0	—	—
Eumulgin HRE 60	White lard-like fat mass with mild odor	≤1.0	—	50–67	—	40–50	≤1.0	—	≤22
Hetoxide HC-16	Liquid	—	8.6	—	—	—	—	—	—
Hetoxide HC-40	Semi-solid	—	13.1	—	—	—	—	—	—
Hetoxide HC-60	Solid	—	14.8	—	—	—	—	—	—
Jeechem CAH-16	Yellow liquid clear to slightly hazy	≤1.5	—	85–102	≤1.5	95–105	≤1.0	—	—
Jeechem CAH-25	Viscous yellow liquid	≤2.0	—	73–84	≤1.0	77–87	≤2.0	—	—

Table cont.

Name	Description	Acid value	HLB value	Hydroxyl value	Iodine number	Saponification value	Water content (%)	Melting point (°C)	Solidification point (°C)
Jeechem CAH-40	Yellow liquid to semi-solid	≤3.0	—	59–68	≤2.0	50–65	≤1.0	—	—
Jeechem CAH-60	Off-white waxy solid	≤1.5	—	39–49	—	41–51	≤1.0	—	—
Jeechem CAH-100	Creamy waxy solid	≤1.5	—	39–49	—	41–51	≤1.0	—	—
Jeechem CAH-200	White to light-yellow waxy solid	≤2.0	—	20–33	—	14–22	≤1.0	125	—
Lipocol HCO-40	Slightly yellow viscous soft pasty solid with characteristic fatty odor	≤3.0	15.0	—	≤2.0	60–67	—	—	—
Lipocol HCO-60	Off-white solid with mild characteristic odor	≤1.0	16.0	50–70	≤1.0	40–50	≤2.0	—	—
Lumulse HCO-25	Yellow liquid	≤3.0	10.8	75–95	—	77–87	—	5	—
Lumulse HCO-40	Light yellow liquid	≤2.0	—	60–68	≤2.0	60–67	≤1.0	—	—
Lumulse HCO-50	Soft paste	≤3.0	14.1	59–80	—	50–67	≤2.0	14	—
Nikkol HCO-40 Pharm	White to pale yellow liquid with faint characteristic odor	—	12.5	—	—	—	—	—	—
Nikkol HCO-60 Pharm	White to pale yellow solid with faint characteristic odor	—	14	—	—	—	—	—	—
Nikkol HCO-80	White to pale yellow waxy solid	—	15	—	—	—	—	—	—
Nikkol HCO-100	White to pale yellow waxy solid	—	16.5	—	—	—	—	—	—

P

Table IV: Typical physical properties of selected commercially available polyoxyethylene castor oil derivatives.

Name	Density (g/cm³) at 25°C	pH	Refractive index at 20°C	Surface tension at 23°C (5 g/L) (mN/m)	Viscosity at 25°C (mPa s)	Critical micelle concentration (%)
Castor oil derivatives						
Polyoxyl 35 castor oil (*Cremophor EL*)	1.05–1.06	6–8[(a)]	1.471	40.9	—	≈0.02
Poloxyl 35 castor oil, purified (*Cremophor ELP*)	1.05–1.06	5–7[(a)]	—	—	600–750	≈0.02
Jeechem CA-5	1.0	6–8[(c)]	—	—	—	—
Jeechem CA-9	1.02	5.5–7.5[(d)]	—	—	—	—
Jeechem CA-15	1.021	6.0–7.5[(e)]	—	—	—	—
Jeechem CA-25	1.04	6.0–7.5[(c)]	—	—	—	—
Jeechem CA-30	1.01	6.0–7.5[(c)]	—	—	—	—
Jeechem CA-40	1.1	5.0–8.0[(c)]	—	—	—	—
Jeechem CA-60	1.068	5.0–7.0[(c)]	—	—	—	—
Jeechem CA-100	—	5.5–7.0[(e)]	—	—	—	—
Jeechem CA-200	1.08	5.0–7.0[(c)]	—	—	—	—
Lumulse CO-5	0.99	—	—	—	—	—
Lumulse CO-25	1.04	—	—	—	—	—
Lumulse CO-40	1.06	—	—	—	—	—
Hydrogenated castor oil derivatives						
Polyoxyl 40 hydrogenated castor oil (*Cremophor RH 40*)	—	5–7[(a)]	1.453–1.457	41.9	20–40[(b)]	0.039
Polyoxyl 60 hydrogenated castor oil	—	6–7	—	40.4	—	—
Eumulgin HRE 40	1.022–1.026 at 70°C	6–7[(a)]	—	—	—	—
Eumulgin HRE 60	1.034–1.038 at 70°C	6–7[(a)]	—	—	—	—
Jeechem CAH-16	1.02	6.0–7.5[(c)]	1.4665–1.4685	—	—	—
Jeechem CAH-25	1.03	5.0–7.5[(c)]	—	—	—	—
Jeechem CAH-40	1.1	5.5–7.0[(e)]	—	—	—	—
Jeechem CAH-60	—	3.5–6.1[(c)]	—	—	—	—
Jeechem CAH-100	1.1	3.5–6.1[(c)]	—	—	—	—
Jeechem CAH-200	1.1	—	—	—	—	—
Lipocol HCO-40	1.0	—	—	—	—	—
Lipocol HCO-60	1.05	—	—	—	—	—
Lumulse HCO-25	1.03	—	—	—	—	—
Lumulse HCO-50	1.03	—	—	—	—	—
Nikkol HCO 40 Pharm	—	5–7[(c)]	—	—	—	—
Nikkol HCO 60 Pharm	—	3.6–6[(c)]	—	—	—	—
Nikkol HCO 100	—	4.5–6.5[(c)]	—	—	—	—

(a) 10% in water.
(b) 30% w/v aqueous solution.
(c) 5% in water.
(d) 1% in water.
(e) 3% in water.

Table V: Solubility of selected commercially available polyoxyethylene castor oil derivatives.

Name	Solubility							
	Castor oil	Chloroform	Ethanol	Fatty acids	Fatty alcohols	Olive oil	Mineral oil	Water
Castor oil derivatives								
Polyoxyl 35 castor oil (*Cremophor EL*)	S	S	S	S	S	S	—	S
Poloxyl 35 castor oil, purified (*Cremophor ELP*)	S	S	S	S	S	S	—	S
Etocas 5	S	—	S	—	S	—	I	I
Etocas 29	S	—	S	—	S	—	I	S
Etocas 35	S	—	S	—	S	—	I	S
Etocas 40	S	—	S	—	PS	—	I	S
Jeechem CA-5	—	—	—	—	—	—	—	D
Jeechem CA-9	—	—	—	—	—	—	—	D
Jeechem CA-15	—	—	—	—	—	—	—	PS
Jeechem CA-25	—	—	—	—	—	—	—	S
Jeechem CA-30	—	—	—	—	—	—	—	S
Jeechem CA-40	—	—	—	—	—	—	—	S
Jeechem CA-60	—	—	—	—	—	—	—	PS
Jeechem CA-200	—	—	—	—	—	—	—	S
Lumulse CO-5	—	—	—	—	—	—	—	I
Lumulse CO-25	—	—	—	—	—	—	—	D
Lumulse CO-40	—	—	—	—	—	—	—	S
Hydrogenated castor oil derivatives								
Polyoxyl 40 hydrogenated castor oil (*Cremophor RH 40*)	S	S	S	S[(a)]	S[(a)]	S	—	S
Polyoxyl 60 hydrogenated castor oil	S	—	S[(b)]	S	S	S	—	S
Croduret 7	S	—	PS	—	S	—	I	I
Croduret 40	D	—	S	—	D	—	I	S
Croduret 50	D	—	S	—	I	—	I	S
Croduret 60	D	—	S	—	D	—	I	S
Jeechem CAH-16	—	—	—	—	—	—	—	D
Jeechem CAH-25	—	—	—	—	—	—	—	D
Jeechem CAH-40	—	—	—	—	—	—	—	S
Jeechem CAH-60	—	—	—	—	—	—	—	S
Jeechem CAH-100	—	—	—	—	—	—	—	S
Jeechem CAH-200	—	—	—	—	—	—	—	S
Lipocol HCO-40	—	—	—	—	—	—	—	S
Lipocol HCO-60	—	—	—	—	—	—	—	S
Lumulse HCO-25	—	—	—	—	—	—	—	S
Lumulse HCO-50	—	—	—	—	—	—	—	S
Nikkol HCO 40 Pharm	—	—	—	—	—	—	—	S
Nikkol HCO 60 Pharm	—	—	—	—	—	—	—	S
Nikkol HCO-80	—	—	—	—	—	—	—	S
Nikkol HCO-100	—	—	—	—	—	—	—	S

(a) At elevated temperatures only.
(b) Need to add 0.5–1.0% water to maintain a clear solution.
S = soluble, PS = partially soluble, I = insoluble, D = dispersible.

P

Aqueous solutions of polyoxyl 35 castor oil can be sterilized by autoclaving for 30 minutes at 120°C. In this process, a product may acquire a deeper color but this has no significance for product stability. Aqueous solutions of polyoxyl hydrogenated castor oil can similarly be sterilized by autoclaving at 120°C, but this may cause a slight decrease in the pH value. Phase separation may also be observed during sterilization, but can be remedied by agitating the solution while it is still hot.

Although the method of manufacture used for polyoxyethylene castor oil derivatives ensures that they are near-sterile, microbial contamination can occur on storage.

Polyoxyethylene castor oil derivatives should be stored in a well-filled, airtight container, protected from light, in a cool, dry place. They are stable for at least 2 years if stored in the unopened original containers at room temperature (maximum 25°C).

12 Incompatibilities

In strongly acidic or alkaline solutions, the ester components of polyoxyethylene hydrogenated castor oil are liable to saponify.

In aqueous solution, polyoxyl 35 castor oil is stable toward most electrolytes in the concentrations normally employed. However, it is incompatible with mercuric chloride since precipitation occurs.

Some organic substances may cause precipitation at certain concentrations, especially compounds containing phenolic hydroxyl groups, e.g. phenol, resorcinol, and tannins.

Polyoxyl 40 hydrogenated castor oil and polyoxyl 60 hydrogenated castor oil are largely unaffected by the salts that cause hardness in water.

13 Method of Manufacture

Polyoxyethylene castor oil derivatives are prepared by reacting varying amounts of ethylene oxide with either castor oil or hydrogenated castor oil under controlled conditions.

Polyoxyl 35 castor oil is produced in this way by reacting 1 mole of castor oil with 35 moles of ethylene oxide, which may be followed by purification.

Polyoxyl 40 hydrogenated castor oil is produced by reacting 1 mole of hydrogenated castor oil with 40–45 moles of ethylene oxide. Polyoxyl 60 hydrogenated castor oil is similarly produced by reacting 1 mole of hydrogenated castor oil with 60 moles of ethylene oxide.

14 Safety

Polyoxyethylene castor oil derivatives are used in a variety of oral, topical, and parenteral pharmaceutical formulations.

Acute and chronic toxicity tests in animals have shown polyoxyethylene castor oil derivatives to be essentially nontoxic and nonirritant materials; *see* Table VI.[12,13] However, there are reports of cardiovascular changes and nephrotoxicity in various species of animals.[14] Several serious anaphylactic reactions,[15–26] cardiotoxicity,[27–29] nephrotoxicity,[30,31] neurotoxicity,[32] and pulmonary toxicity[33] have also been observed in humans and animals following parenteral administration of formulations containing polyoxyethylene castor oil derivatives. The precise mechanism of the reaction is not known.

Table VI: LD_{50} values of selected polyoxyethylene castor oil derivatives.

Name	Animal and route	LD_{50} (g/kg body-weight)
Polyoxyl 35 castor oil (*Cremophor EL*)	Rat (oral)	>10
	Dog (IV)	0.64[28]
	Mouse (IV)	6.5[28]
Polyoxyl 40 hydrogenated castor oil (*Cremophor RH 40*) and polyoxyl 60 hydrogenated castor oil	Rat (oral)	>20.1

15 Handling Precautions

Observe normal precautions appropriate to the circumstances and quantity of material handled. Eye protection and gloves are recommended.

16 Regulatory Status

Included in the FDA Inactive Ingredients Database (IV injections and ophthalmic solutions). Included in parenteral medicines licensed in the UK. Included in the Canadian Natural Health Products Ingredients Database.

17 Related Substances

Polyoxyethylene alkyl ethers; polyoxyethylene stearates.

18 Comments

Note that the trade name *Cremophor* (BASF Corp) is also used for other polyoxyethylene derivatives e.g., the *Cremophor A*: series are polyoxyethylene alkyl ethers of cetostearyl alcohol.

Polyoxyl 60 hydrogenated castor oil derivative has been investigated as an absorption enhancer in the absorption of erythropoietin from rat small intestine using gastrointestinal patches.[34] In another study, lipiodol and polyoxyl 60 hydrogenated castor oil derivative have been found to play an important role in the prolongation and selective retention of w/o emulsion or w/o/w multiple emulsion of doxorubicin hydrochloride *in vitro* and *in vivo*.[35] Polyoxyethylene castor oil derivatives have also been used experimentally as surfactants for a controlled-release matrix pellet formulation containing nanocrystalline ketoprofen,[36] and for the transdermal delivery of vinpocetin.[37] Itraconazole has been incorporated in an aqueous parenteral formulation in an o/w microemulsion system containing polyoxyl 40 hydrogenated castor oil[38] and polyoxyl 50 hydrogenated castor oil.[39] A novel o/w microemulsion containing various emulsifiers including polyoxyl 40 hydrogenated castor oil was found to increase the solubility from 60 to 20 000 times of as many as nine poorly water soluble compounds, as well as to enhance the oral bioavailability of these compounds.[40] Hydrogenated castor oil derivatives containing more than 20 oxyethylene units were found to prolong the plasma circulation times of menatetrenone incorporated in lipid emulsions.[41,42]

Cremophor EL (BASF) has been used as a solubilizing agent for drugs like cyclosporine A,[43] paclitaxel,[44] and cisplatin.[45] A self-microemulsifying drug delivery system (SMEDDS) containing *Cremophor EL* has been reported to enhance the oral bioavailability of halofantrin,[46] and simvastatin.[47] *Cremophor EL* has been found to improve solubilization of volatile oil[48] and the dissolution profile of naproxen in solid dosage form.[49]

Cremophor ELP (BASF) and *Super Refined Etocas 35* (Croda) are purity grades of polyoxyl 35 castor oil for topical, oral and parenteral use, designed specially to provide the highest level of purity in formulations containing extremely sensitive APIs, and improve the solubilization of vitamins and drugs (e.g. paclitaxel).

Cremophor RH 40 and *RH 60* (BASF) have been used as additives to enhance drug release from suppository formulations.[4,5] *Cremophor RH 40* was found to prolong the dissolution time of digoxin tablets.[50]

Cremophor has been used as a vehicle for boron neutron-capture therapy in mice, which is a form of radiation therapy used in the treatment of glioblastoma multiforme.[51]

In veterinary practice, polyoxyl 35 castor oil can be used to emulsify cod liver oil, and oils and fats incorporated into animal feeding stuffs. *Cremophor EL* can enhance the bioavailability of substances such as vitamins in feed and veterinary medicines, improving their efficacy.

19 Specific References

1 Carmignani C, *et al.* Ophthalmic vehicles containing polymer-solubilized tropicamide: *in vitro–in vivo* evaluation. *Drug Dev Ind Pharm* 2002; **28**: 101–105.
2 Gizurarson S, *et al.* Intranasal vaccination: pharmaceutical evaluation of the vaccine delivery system and immunokinetic characteristics of the immune response. *Pharm Dev Technol* 1998; **3**: 385–394.
3 Nerurkar MM, *et al.* The use of surfactants to enhance the permeability of peptides through Caco-2 cells by inhibition of an apically polarized efflux system. *Pharm Res* 1996; **13**: 528–534.
4 Berko S, *et al. Solutol* and *Cremophor* products as new additives in suppository formulation. *Drug Dev Ind Pharm* 2002; **28**: 203–206.
5 Berko S, *et al. In vitro* and *in vivo* study in rats of rectal suppositories containing furosemide. *Eur J Pharm Biopharm* 2002; **53**: 311–315.
6 Itoh K, *et al.* Improvement of physicochemical properties of N-4472. Part II: characterization of N-4472 microemulsion and the enhanced oral absorption. *Int J Pharm* 2002; **246**: 75–83.
7 Sakeada T, Hirano K. Effect of composition on biological fate of oil particles after intravenous injection of O/W lipid emulsions. *J Drug Target* 1998; **6**: 273–284.
8 Kato Y, *et al.* Modification of liposomes by addition of HCO60. II. Encapsulation of doxorubicin into liposomes containing HCO60. *Biol Pharm Bull* 1993; **16**: 965–969.
9 Kato Y, *et al.* Modification of liposomes by addition of HCO60. I. Targeting of liposomes to liver by addition of HCO60 to liposomes. *Biol Pharm Bull* 1993; **16**: 960–964.
10 Takada K, *et al.* Biological and pharmaceutical factors affecting the absorption and lymphatic delivery of cyclosporin A from gastrointestinal tract. *J Pharmacobiodyn* 1988; **11**: 80–87.
11 Takada K, *et al.* Promotion of the selective lymphatic delivery of cyclosporin A by lipid-surfactant mixed micelles. *J Pharmacobiodyn* 1985; **8**: 320–323.
12 Lewis RJ, ed. *Sax's Dangerous Properties of Industrial Materials*, 11th edn. New York: John Wiley, 2004; 727.
13 BASF Corporation. Material safety data sheet: *Cremophor EL*, 2004; *Cremophor RH40*,2002; *Cremophor RH 60*, 2005.
14 Final report on the safety assessment of PEG-30, 33, 35, 36, and 40 castor oil and PEG-30 and 40 hydrogenated castor oil. *Int J Toxicol* 1997; **16**(3): 269–306.
15 Forrest ARW, *et al.* Long-term althesin infusion and hyperlipidaemia. *Br Med J* 1977; **2**: 1357–1358.
16 Dye D, Watkins J. Suspected anaphylactic reaction to *Cremophor EL*. *Br Med J* 1980; **280**: 1353.
17 Knell AJ, *et al.* Potential hazard of steroid anaesthesia for prolonged sedation [letter]. *Lancet* 1983; **i**: 526.
18 Lawler PGP, *et al.* Potential hazards of prolonged steroid anaesthesia [letter]. *Lancet* 1983; **i**: 1270–1271.
19 Moneret-Vautrin DA, *et al.* Anaphylaxis caused by anti-*Cremophor EL* IgG STS antibodies in a case of reaction to althesin. *Br J Anaesth* 1983; **55**: 469–471.
20 Chapuis B, *et al.* Anaphylactic reaction to intravenous cyclosporine. *N Engl J Med* 1985; **312**: 1259.
21 Howrie DL, *et al.* Anaphylactoid reactions associated with parenteral cyclosporine use: possible role of *Cremophor EL*. *Drug Intell Clin Pharm* 1985; **19**: 425–427.
22 van Hooff JP, *et al.* Absence of allergic reaction to cyclosporin capsules in patient allergic to standard oral and intravenous solution of cyclosporin [letter]. *Lancet* 1987; **ii**: 1456.
23 Siddall SJ, *et al.* Anaphylactic reactions to teniposide. *Lancet* 1989; **i**: 394.
24 McCormick PA, *et al.* Reformulation of injectable vitamin A: potential problems. *Br Med J* 1990; **301**: 924.
25 Fjällskog M-L, *et al.* Is *Cremophor EL*, solvent for paclitaxel, cytotoxic? *Lancet* 1993; **342**: 873.
26 Liebmann J, *et al. Cremophor EL*, solvent for paclitaxel, and toxicity. *Lancet* 1993; **342**: 1428.
27 Badary OA, *et al.* Effect of *Cremophor EL* on the pharmacokinetics, antitumor activity and toxicity of doxorubicin in mice. *Anticancer Drugs* 1998; **9**: 809–815.
28 Sanchez H, *et al.* Immunosuppressive treatment affects cardiac and skeletal muscle mitochondria by the toxic effect of vehicle. *J Mol Cell Cardiol* 2000; **32**: 323–331.
29 Bowers VD, *et al.* The hemodynamic effects of *Cremophor-EL*. *Transplantation* 1991; **51**: 847–850.
30 Verani R. Cyclosporin nephrotoxicity in the Fischer rat. *Clin Nephrol* 1986; **25**(Suppl 1): S9–S13.
31 Thiel G, *et al.* Acutely impaired renal function during the intravenous administration of cyclosporin A: a cremophore side-effect. *Clin Nephrol* 1986; **25**(Suppl 1): S40–S42.
32 Windebank AJ, *et al.* Potential neurotoxicity of the solvent vehicle for cyclosporin. *J Pharmacol Exp Ther* 1994; **268**: 1051–1056.
33 Kiorpes AL, *et al.* Pulmonary changes in rats following the administration of 3-methylindole in *Cremophor EL*. *Histol Histopathol* 1988; **3**: 125–132.
34 Venkatesan N, *et al.* Gastro-intestinal patch system for the delivery of erythropoietin. *J Control Rel* 2006; **111**: 19–26.
35 Lin SY, *et al. In vitro* release, pharmacokinetic and tissue distribution studies of doxorubicin hydrochloride (adriamycin HCl (R)) encapsulated in lipiodolized w/o emulsions and w/o/w multiple emulsions. *Pharmazie* 1992; **47**: 439–443.
36 Vergote GJ, *et al.* An oral controlled release matrix pellet formulation containing nanocrystalline ketoprofen. *Int J Pharm* 2001; **219**: 81–87.
37 Hua L, *et al.* Preparation and evaluation of microemulsion of vinpocetin for transdermal delivery. *Pharmazie* 2004; **59**: 274–278.
38 Spernath A, *et al.* Fully dilutable microemulsions embedded with phospholipids and stabilized by short chain organic acids and polyols. *J Colloid Interface Sci* 2006; **299**: 900–909.
39 Rhee YS, *et al.* Formulation of parenteral microemulsion containing itraconazole. *Arch Pharm Res* 2007; **30**: 114–123.
40 Araya H, *et al.* The novel formulation design of o/w microemulsion for improving the gastrointestinal absorption of poorly water soluble compounds. *Int J Pharm* 2005; **305**: 61–74.
41 Ueda K, *et al.* Prolonged circulation of menatetrenone by emulsions with hydrogenated castor oils in rats. *J Control Rel* 2004; **95**: 93–100.
42 Ueda K, *et al.* Effect of oxyethylene moieties in hydrogenated castor oil on the pharmacokinetics of menatetrenone incorporated in O/W lipid emulsions prepared with hydrogenated castor oil and soybean oil in rats. *J Drug Target* 2003; **11**: 37–43.
43 Ran Y, *et al.* Solubilization of cyclosporin A. *AAPS Pharm Sci Tech* 2001; **2**(1): Article 2.
44 Gelderblom H, *et al. Cremophor EL*: the drawbacks and advances of vehicle selection for drug formulation. *Eur J Cancer* 2001; **37**: 1590–1598.
45 Gelderblom H, *et al.* Modulation of cisplatin pharmacodynamics by *Cremophor EL*: experimental and clinical studies. *Eur J Cancer* 2002; **38**: 205–213.
46 Holm R, *et al.* Examination of oral absorption and lymphatic transport of halofantrine in a triple-cannulated canine model after administration in self-microemulsifying drug delivery systems (SMEDDS) containing structured triglycerides. *Eur J Pharm Sci* 2003; **20**: 91–97.
47 Kang BK, *et al.* Development of self-microemulsifying drug delivery systems (SMEDDS) for oral bioavailability enhancement of simvastatin in beagle dogs. *Int J Pharm* 2004; **274**: 65–73.
48 Hong Y, *et al.* Solubilization of o/w microemulsion for volatile oil from *Houttuynia cordra*. *Zhongguo Zhong Yao Za Zhi* 2010; **35**: 49–52.
49 Ngiik T, *et al.* Effects of liquisolid formulations on dissolution of naproxen. *Eur J Pharm Biopharm* 2009; **73**: 373–384.
50 Tayrouz Y, *et al.* Pharmacokinetic and pharmaceutic interaction between digoxin and *Cremophor RH40*. *Clin Pharm Ther* 2003; **73**: 397–405.
51 Miura M, *et al.* Synthesis of a nickel tetracarbonylphenylporphyrin for boron neutron-capture therapy: biodistribution and toxicity in tumor-bearing mice. *Int J Cancer* 1996; **68**: 114–119.

20 General References

BASF Corporation. Technical literature: *Cremophor Grades*, January 2008.
BASF Corporation. Technical literature: *Cremophor EL*, Castor Oil, June 2008.
BASF Corporation. Technical literature: *Cremophor ELP*, July 2008.
BASF Corporation. Technical literature: *Cremophor RH 40*, November 2008.
Rischin D, *et al. Cremophor* pharmacokinetics in patients receiving 3, 6, and 24 hour infusions of paclitaxel. *J Natl Cancer Inst* 1996; **88**: 1297–1301.

21 Authors

RT Gupta, KK Singh.

22 Date of Revision

2 March 2012.

P

Polyoxyethylene Sorbitan Fatty Acid Esters

1 Nonproprietary Names

BP: Polysorbate 20
Polysorbate 40
Polysorbate 60
Polysorbate 80

JP: Polysorbate 80

PhEur: Polysorbate 20
Polysorbate 40
Polysorbate 60
Polysorbate 80

USP–NF: Polysorbate 20
Polysorbate 40
Polysorbate 60
Polysorbate 80

2 Synonyms

For synonyms of selected polysorbates, *see* Table I; *see also* Section 3.

3 Chemical Names and CAS Registry Numbers

See Table II.

4 Empirical Formula and Molecular Weight

Approximate molecular weights for selected polysorbates are shown in Table III.

5 Structural Formula

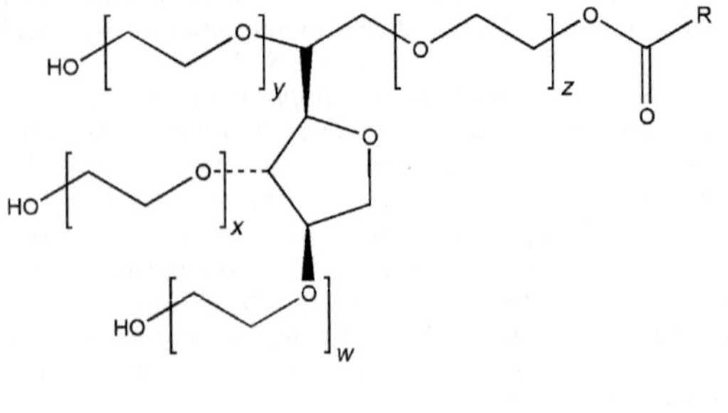

Polyoxyethylene sorbitan monoester

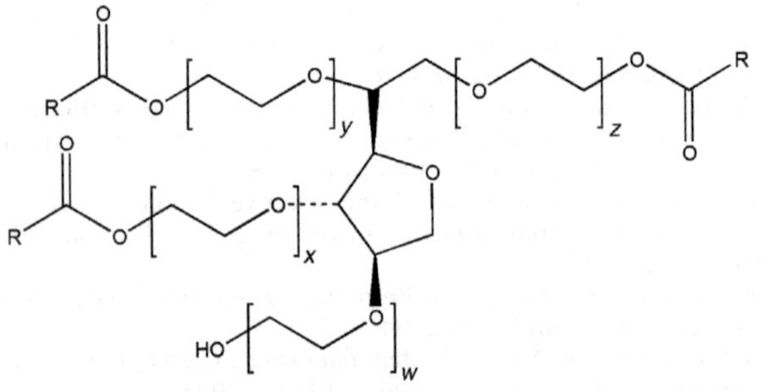

Polyoxyethylene sorbitan triester

$w + x + y + z = 20$ (Polysorbates 20, 40, 60, 65, 80, and 85)
$w + x + y + z = 5$ (Polysorbates 81)
$w + x + y + z = 4$ (Polysorbates 21 and 61)
R = fatty acid

Table I: Synonyms of selected polysorbates.

Polysorbate	Synonym
Polysorbate 20	*Armotan PML 20; Crillet 1; Drewmulse;* E432; *Durfax 20;* E432; *Eumulgin SML; Glycosperse L-20; Hodag PSML-20; Lamesorb SML-20; Liposorb L-20; Liposorb L-20K; Montanox 20; Nissan Nonion LT-221; Norfox Sorbo T-20; POE-SML;* polysorbatum 20; *Ritabate 20; Sorbax PML-20;* sorbitan monododecanoate; *Sorgen TW-20; T-Maz 20; T-Maz 20K;* poly(oxy-1,2-ethanediyl) derivatives; polyoxyethylene 20 laurate; *Protasorb L-20; Tego SML 20; Tween 20.*
Polysorbate 21	*Crillet 11; Hodag PSML-4; Protasorb L-5; Tween 21.*
Polysorbate 40	*Crillet 2;* E434; *Eumulgin SMP; Glycosperse S-20; Hodag PSMP-20; Lamesorb SMP-20; Liposorb P-20; Lonzest SMP-20; Montanox 40;* poly(oxy-1,2-ethanediyl) derivatives; polysorbatum 40; *Protasorb P-20; Ritabate 40;* sorbitan monohexadecanoate; *Sorbax PMP-20; Tween 40.*
Polysorbate 60	*Atlas 70K; Atlas Armotan PMS 20; Cremophor PS 60; Crillet 3; Drewpone 60K; Durfax 60; Durfax 60K;* E435; *Emrite 6125; Eumulgin SMS; Glycosperse S-20; Glycosperse S-20FG; Glycosperse S-20FKG; Hodag PSMS-20; Hodag SVS-18; Lamsorb SMS-20; Liposorb S-20; Liposorb S-20K; Lonzest SMS-20; Montanox 60; Nikkol TS-10; Norfox SorboT-60; Polycon T 60 K;* polyoxyethylene 20 stearate; polysorbatum 60; *Protasorb S-20; Ritabate 60; Sorbax PMS-20;* sorbitan monooctadecanoate poly(oxy-1,2-ethanediyl) derivatives; *T-Maz 60; T-Max 60KHS; Tween 60; Tween 60K; Tween 60 VS.*
Polysorbate 61	*Crillet 31; Hodag PSMS-4; Liposorb S-4; Protasorb S-4; Tween 61.*
Polysorbate 65	*Alkamuls PSTS-20; Crillet 35;* E436; *Glycosperse TS-20; Glycosperse TS-20 FG; Glycosperse TS-20 KFG; Hodag PSTS-20; Lamesorb STS-20; Lanzet STS-20; Liposorb TS-20; Liposorb TS-20A; Liposorb TS-20K; Montanox 65; Protasorb STS-20; Sorbax PTS-20;* sorbitan trioctadecanoate poly(oxy-1,2-ethanediyl) derivatives; *T-Maz 65K; Tween 65; Tween 65K; Tween 65V.*
Polysorbate 80	*Atlas E; Armotan PMO 20; Cremophor PS 80; Crillet 4; Crillet 50; Drewmulse POE-SMO; Drewpone 80K; Durfax 80; Durfax 80K;* E433; *Emrite 6120; Eumulgin SMO; Glycosperse O-20; Hodag PSMO-20; Liposorb O-20; Liposorb O-20K; Montanox 80;* polyoxyethylene 20 oleate; polysorbatum 80; *Protasorb O-20; Ritabate 80;* (*Z*)-sorbitan mono-9-octadecenoate poly(oxy1,2-ethanediyl) derivatives; *Tego SMO 80; Tego SMO 80V; Tween 80.*
Polysorbate 81	*Crillet 41; Hetsorb O-5; Hodag PSMO-5; Protasorb O-5; Sorbax PMO-5;* sorbitan mono-9-octadecenoate poly(oxy-1,2-ethanediyl) derivatives; *T-Maz 81; Tego SMO 81; Tween 81.*
Polysorbate 85	*Alkamuls PSTO-20; Crillet 45; Glycosperse TO-20; Hodag PSTO-20; Liposorb TO-20; Lonzest STO-20; Montanox 85; Protasorb TO-20; Sorbax PTO-20;* sorbitan tri-9-octadecenoate poly(oxy1,2-ethanediyl) derivatives; *Tego STO 85; Tween 85.*
Polysorbate 120	*Crillet 6.*

Table II: Chemical names and CAS Registry Numbers of selected polysorbates.

Polysorbate	Chemical name	CAS number
Polysorbate 20	Polyoxyethylene 20 sorbitan monolaurate	[9005-64-5]
Polysorbate 21	Polyoxyethylene (4) sorbitan monolaurate	[9005-64-5]
Polysorbate 40	Polyoxyethylene 20 sorbitan monopalmitate	[9005-66-7]
Polysorbate 60	Polyoxyethylene 20 sorbitan monostearate	[9005-67-8]
Polysorbate 61	Polyoxyethylene (4) sorbitan monostearate	[9005-67-8]
Polysorbate 65	Polyoxyethylene 20 sorbitan tristearate	[9005-71-4]
Polysorbate 80	Polyoxyethylene 20 sorbitan monooleate	[9005-65-6]
Polysorbate 81	Polyoxyethylene (5) sorbitan monooleate	[9005-65-6]
Polysorbate 85	Polyoxyethylene 20 sorbitan trioleate	[9005-70-3]
Polysorbate 120	Polyoxyethylene 20 sorbitan monoisostearate	[66794-58-9]

Table III: Empirical formula and molecular weight of selected polysorbates.

Polysorbate	Formula	Molecular weight
Polysorbate 20	$C_{58}H_{114}O_{26}$	1128
Polysorbate 21	$C_{26}H_{50}O_{10}$	523
Polysorbate 40	$C_{62}H_{122}O_{26}$	1284
Polysorbate 60	$C_{64}H_{126}O_{26}$	1312
Polysorbate 61	$C_{32}H_{62}O_{10}$	607
Polysorbate 65	$C_{100}H_{194}O_{28}$	1845
Polysorbate 80	$C_{64}H_{124}O_{26}$	1310
Polysorbate 81	$C_{34}H_{64}O_{11}$	649
Polysorbate 85	$C_{100}H_{188}O_{28}$	1839
Polysorbate 120	$C_{64}H_{126}O_{26}$	1312

6 Functional Category

Dispersing agent; emollient; emulsifying agent; nonionic surfactant; plasticizing agent; solubilizing agent; suspending agent.

7 Applications in Pharmaceutical Formulation or Technology

Polyoxyethylene sorbitan fatty acid esters (polysorbates) are a series of partial fatty acid esters of sorbitol and its anhydrides copolymerized with approximately 20, 5, or 4 moles of ethylene oxide for each mole of sorbitol and its anhydrides. The resulting product is therefore a mixture of molecules of varying sizes rather than a single uniform compound. Polysorbates are used in a wide variety of pharmaceutical formulations including oral, otic, ophthalmic, topical, and parenteral preparations.

Polysorbates containing 20 units of oxyethylene are hydrophilic nonionic surfactants that are widely used as emulsifying agents in the preparation of stable oil-in-water pharmaceutical emulsions. In pharmaceutical applications, polysorbate 80 is the most commonly used surfactant in FDA-approved parenteral products.(1) Polysorbates may also be used as solubilizing agents for a variety of substances including essential oils and oil-soluble vitamins, and as nonionic surfactants in the formulation of oral, otic, topical, and parenteral suspensions. Polysorbates have also been used to improve chemical stability of solution formulations.(2) It has also been reported that polysorbates can be used as plasticizing agents to reduce the melting temperature of drug molecules and the glass transition temperature of solid dispersion formulations.(3) In topical applications, polysorbates are also used as emollients to increase the water holding capacity of the formulation. Polysorbates have been found to be useful in improving the oral bioavailability of drug molecules that are substrates for P-glycoprotein.(4,5)

Polysorbates are also widely used in cosmetics and food products. *See* Table IV.

Table IV: Uses of polysorbates.

Use	Concentration (%)
Emollient in topical formulations	1–15
Emulsifying agent in oil-in-water emulsions	1–15
Nonionic surfactant in ophthalmic preparations	≤1
Nonionic surfactant in lipophilic bases	≤5
Solubilizing agent in lipophilic bases	≤15
Used as or in combination with plasticizing agents	≤10

8 Description

Polysorbates have a characteristic odor and a bitter taste. The appearances of polysorbates at 25°C are shown in Table V, although it should be noted that the absolute color intensity of the products may vary from batch to batch and from manufacturer to manufacturer.

Table V: Appearance of selected polysorbates at 25°C.

Polysorbate	Color and form at 25°C
Polysorbate 20	Yellow oily liquid
Polysorbate 21	Yellow oily liquid
Polysorbate 40	Yellow oily liquid
Polysorbate 60	Yellow oily liquid
Polysorbate 61	Tan solid
Polysorbate 65	Tan solid
Polysorbate 80	Yellow oily liquid
Polysorbate 81	Amber liquid
Polysorbate 85	Amber liquid
Polysorbate 120	Yellow liquid

9 Pharmacopeial Specifications

The pharmacopeial specifications for polysorbate 80 have undergone harmonization of many attributes for JP, PhEur, and USP–NF.

See Tables VI and VII. *See also* Section 18.

Table VII: Pharmacopeial specifications for the fatty acid composition of polysorbate 20, 40, 60, and 80.

Fatty acid	Polysorbate 20 (PhEur 7.4)	Polysorbate 40 (PhEur 7.4)	Polysorbate 60 (PhEur 7.4)	Polysorbate 80 (PhEur 7.4 and USP35–NF30)
Caproic acid	≤1.0%	—	—	—
Caprylic acid	≤10.0%	—	—	—
Capric acid	≤10.0%	—	—	—
Lauric acid	40.0–60.0%	—	—	—
Myristic acid	14.0–25.0%	—	—	≤5.0%
Palmitic acid	7.0–15.0%	≥92.0%	+(a)	≤16.0%
Palmitoleic acid	—	—	—	≤8.0%
Stearic acid	≤7.0%	—	40.0–60.0%	≤6.0%
Oleic acid	≤11.0%	—	—	≥58.0%
Linolenic acid	—	—	—	≤4.0%
Linoleic acid	≤3.0%	—	—	≤18.0%

(a) Sum of the contents of palmitic and stearic acids ≥90.0%.

P

Table VI: Pharmacopeial specifications for polysorbates.

Test	JP XV	PhEur 7.4	USP35–NF30
Identification			
Polysorbate 20	—	+	+
Polysorbate 40	—	+	+
Polysorbate 60	—	+	+
Polysorbate 80	+	+	+
Characters	—	+	—
Saponification value			
Polysorbate 20	—	40–50	40–50
Polysorbate 40	—	41–52	41–52
Polysorbate 60	—	45–55	45–55
Polysorbate 80	45–55	45–55	45–55
Composition of fatty acids			
Polysorbate 20	—	*see* Table VII	—
Polysorbate 40	—	*see* Table VII	—
Polysorbate 60	—	*see* Table VII	—
Polysorbate 80	—	*see* Table VII	*see* Table VII
Hydroxyl value			
Polysorbate 20	—	96–108	96–108
Polysorbate 40	—	89–105	89–105
Polysorbate 60	—	81–96	81–96
Polysorbate 80	—	65–80	65–80
Water			
Polysorbate 20	—	≤3.0%	≤3.0%
Polysorbate 40	—	≤3.0%	≤3.0%
Polysorbate 60	—	≤3.0%	≤3.0%
Polysorbate 80	≤3.0%	≤3.0%	≤3.0%
Residue on ignition			
Polysorbate 20	—	≤0.25%	≤0.25%
Polysorbate 40	—	≤0.25%	≤0.25%
Polysorbate 60	—	≤0.25%	≤0.25%
Polysorbate 80	≤0.1%	≤0.25%	≤0.25%
Arsenic			
Polysorbate 80	≤2 ppm	—	—
Heavy metals[a]			
Polysorbate 20	—	≤10 ppm	≤10 ppm
Polysorbate 40	—	≤10 ppm	≤0.001%
Polysorbate 60	—	≤10 ppm	≤10 ppm
Polysorbate 80	≤20 ppm	≤10 ppm	≤10 ppm
Acid value			
Polysorbate 20	—	≤2.0	≤2.2
Polysorbate 40	—	≤2.0	≤2.2
Polysorbate 60	—	≤2.0	≤2.2
Polysorbate 80	≤2.0	≤2.0	≤2.0
Iodine value			
Polysorbate 80	19–24	—	—
Specific gravity			
Polysorbate 20	—	≈1.10	—
Polysorbate 40	—	≈1.10	—
Polysorbate 60	—	≈1.10	—
Polysorbate 80	1.065–1.095	≈1.10	1.06–1.09
Viscosity at 25°C			
Polysorbate 20	—	≈400 mPa s	—
Polysorbate 80	345–445 mm^2/s	≈400 mPa s	300–500 cSt
Viscosity at 30°C			
Polysorbate 40	—	≈400 mPa s	—
Polysorbate 60	—	≈400 mPa s	—
Peroxide value			
Polysorbate 20	—	≤10	—
Polysorbate 40	—	≤10	—
Polysorbate 60	—	≤10	—
Polysorbate 80	—	≤10	≤10
Residual ethylene oxide			
Polysorbate 20	—	≤1 ppm	—
Polysorbate 40	—	≤1 ppm	—
Polysorbate 60	—	≤1 ppm	—
Polysorbate 80	—	≤1 ppm	≤1 ppm
Residual dioxan			
Polysorbate 20	—	≤10 ppm	—
Polysorbate 40	—	≤10 ppm	—
Polysorbate 60	—	≤10 ppm	—
Polysorbate 80	—	≤10 ppm	≤10 ppm

(a) This test has not been fully harmonized at the time of publication.

10 Typical Properties

Acid value *see* Table VIII.
Acidity/alkalinity pH = 6.0–8.0 for a 5% w/v aqueous solution.
Critical micelle concentration (CMC) *see* Table IX.[6]
Flash point 149°C
HLB value *see* Table IX.
Hydroxyl value *see* Table VIII.
Moisture content *see* Table VIII.
Saponification value *see* Table VIII.
Solubility *see* Table X.
Specific gravity *see* Table IX.
Surface tension For 0.1% w/v solutions, *see* Table XI.
Viscosity (dynamic) *see* Table IX.

Table VIII: Typical properties of selected polysorbates.

Polysorbate	Acid value (%)	Hydroxyl value	Moisture content	Saponification value
Polysorbate 20	2.0	96–108	3.0	40–50
Polysorbate 21	3.0	225–255	3.0	100–115
Polysorbate 40	2.0	90–105	3.0	41–52
Polysorbate 60	2.0	81–96	3.0	45–55
Polysorbate 61	2.0	170–200	3.0	95–115
Polysorbate 65	2.0	44–60	3.0	88–98
Polysorbate 80	2.0	65–80	3.0	45–55
Polysorbate 81	2.0	134–150	3.0	96–104
Polysorbate 85	2.0	39–52	3.0	80–95
Polysorbate 120	2.0	65–85	5.0	40–50

Table IX: Typical properties of selected polysorbates.

Polysorbate	CMC at 25°C (g/100 mL)	HLB value	Specific gravity at 25°C	Viscosity (mPa s)
Polysorbate 20	≈0.0060	16.7	1.1	400
Polysorbate 21		13.3	1.1	500
Polysorbate 40	≈0.0031	15.6	1.08	500
Polysorbate 60	≈0.0028	14.9	1.1	600
Polysorbate 61		9.6	1.06	Solid
Polysorbate 65	≈0.0040-0.0060	10.5	1.05	Solid
Polysorbate 80	≈0.0014	15.0	1.08	425
Polysorbate 81		10.0	–	450
Polysorbate 85	≈0.0023	11.0	1.00	300
Polysorbate 120		14.9	–	–

Table X: Solubilities of selected polysorbates in various solvents.

Polysorbate	Solvent			
	Ethanol	Mineral oil	Vegetable oil	Water
Polysorbate 20	S	I	I	S
Polysorbate 21	S	I	I	D
Polysorbate 40	S	I	I	S
Polysorbate 60	S	I	I	S
Polysorbate 61	SW	SW	SWT	D
Polysorbate 65	SW	SW	DW	D
Polysorbate 80	S	I	I	S
Polysorbate 81	S	S	ST	D
Polysorbate 85	S	I	ST	D
Polysorbate 120	S	I	I	S

D = dispersible; I = insoluble; S = soluble; T = turbid; W = on warming.

Table XI: Surface tension of related polysorbates.

Polysorbate	Surface tension at 20°C (mN/m)
Polysorbate 21	34.7
Polysorbate 40	41.5
Polysorbate 60	42.5
Polysorbate 61	41.5
Polysorbate 80	42.5
Polysorbate 85	41.0

11 Stability and Storage Conditions

Polysorbates are stable in the presence of electrolytes and weak acids and bases; gradual saponification occurs with strong acids and bases. Polysorbates are hygroscopic and should be examined for water content prior to use and dried if necessary. Upon storage, polysorbates are prone to oxidation and formation of peroxides.

Polysorbates should be stored in a well-closed container, protected from light, in a cool, dry place.

12 Incompatibilities

Discoloration and/or precipitation occur with various substances, especially phenols, tannins, tars, and tarlike materials. The antimicrobial activity of paraben preservatives is reduced in the presence of polysorbates.[7] *See* Methylparaben.

13 Method of Manufacture

Polysorbates are prepared from sorbitol in a three-step process. Water is initially removed from the sorbitol to form a sorbitan (a cyclic sorbitol anhydride). The sorbitan is then partially esterified with a fatty acid, such as oleic or stearic acid, to yield a hexitan ester. Finally, ethylene oxide is chemically added in the presence of a catalyst to yield the polysorbate.

14 Safety

Polysorbates are widely used in cosmetics, food products, and oral, parenteral and topical pharmaceutical formulations, and are generally regarded as nontoxic and nonirritant materials. There have, however, been occasional reports of hypersensitivity to polysorbates following their topical and intramuscular use.[8] Polysorbates have also been associated with serious adverse effects, including some deaths, in low-birthweight infants following intravenous administration of a vitamin E preparation containing a mixture of polysorbates 20 and 80.[9,10] When heated to decomposition, the polysorbates emit acrid smoke and irritating fumes.

The WHO has set an estimated acceptable daily intake for polysorbates 20, 40, 60, 65, and 80, calculated as total polysorbate esters, at up to 25 mg/kg body-weight.[11]

Polysorbate 20 Moderate toxicity by IP and IV routes. Moderately toxic by ingestion. Human skin irritant.

LD_{50} (hamster, oral): 18 g/kg[12]

LD_{50} (mouse, IV): 1.42 g/kg

LD_{50} (rat, oral): 37 g/kg

Polysorbate 21 Moderately toxic by IV route.

Polysorbate 40 Moderately toxic by IV route.

LD_{50} (rat, IV): 1.58 g/kg[12]

Polysorbate 60 Moderately toxic by IV route. Experimental tumorigen; reproductive effects.

LD_{50} (rat, IV): 1.22 g/kg[12]

Polysorbate 61 Moderately toxic by IV route.

Polysorbate 80 Moderately toxic by IV route. Mildly toxic by ingestion. Eye irritation. Experimental tumorigen, reproductive effects. Mutagenic data.

LD_{50} (mouse, IP): 7.6 g/kg[12]

LD_{50} (mouse, IV): 4.5 g/kg

LD_{50} (mouse, oral): 25 g/kg

LD_{50} (rat, IP): 6.8 g/kg

LD_{50} (rat, IV): 1.8 g/kg

Polysorbate 85 Skin irritant.

P

15 Handling Precautions

Observe normal precautions appropriate to the circumstances and quantity of material handled. Eye protection and gloves are recommended.

16 Regulatory Status

Polysorbates 60, 65, and 80 are GRAS listed. Polysorbates 20, 40, 60, 65, and 80 are accepted as food additives in Europe. Polysorbates 20, 40, 60, and 80 are included in the FDA Inactive Ingredients Database (IM, IV, oral, ophthalmic, otic, rectal, topical, and vaginal preparations). Polysorbates are included in parenteral and nonparenteral medicines licensed in the UK. Polysorbates 20, 21, 40, 60, 61, 65, 80, 81, 85, and 120 are included in the Canadian Natural Health Products Ingredients Database.

17 Related Substances

Polyethylene glycol; sorbitan esters (sorbitan fatty acid esters).

18 Comments

Polysorbate 80 has undergone harmonization of many attributes for JP, PhEur, and USP–NF by the Pharmacopeial Discussion Group. For further information see the General Information Chapter <1196> in the USP35–NF30, the General Chapter 5.8 in PhEur 7.4, along with the 'State of Work' document on the PhEur EDQM website, and also the General Information Chapter 8 in the JP XV.

The PubChem Compound ID (CID) for polysorbates includes 443314 and 5281955.

19 Specific References

1 Nema S, *et al.* Excipients and their use in injectable products. *PDA J Pharm Sci Tech* 1997; **51**: 166–171.
2 Yalkowski SH. *Solubility and Solubilization in Aqueous Media.* New York, NY: Oxford University Press, 1999; 310–312.
3 Ghebremeskel AN, *et al.* Use of surfactants as plasticizers in preparing solid dispersions of poorly soluble API: selection of polymer-surfactant combinations using solubility parameters and testing the processability. *Int J Pharm* 2007; **328**: 119–129.
4 Nerurkar MM, *et al.* The use of surfactants to enhance the permeability of peptides through Caco-2 cells by inhibition of an apically polarized efflux system. *Pharm Res* 1996; **13**(4): 528–534.
5 Zhang H, *et al.* Commonly used surfactant, Tween 80, improves absorption of P-glycoprotein substrate, digoxin, in rats. *Arch Pharm Res* 2003; **26**: 768–772.
6 Wan LSC, *et al.* CMC of polysorbates. *J Pharm Sci* 1974; **63**: 136–137.
7 Blanchard J. Effect of polyols on interaction of paraben preservatives with polysorbate 80. *J Pharm Sci* 1980; **69**: 169–173.
8 Shelley WB, *et al.* Polysorbate 80 hypersensitivity [letter]. *Lancet* 1995; **345**: 1312–1313.
9 Alade SL, *et al.* Polysorbate 80 and E-Ferol toxicity. *Pediatrics* 1986; **77**: 593–597.
10 Balistreri WF, *et al.* Lessons from the E-Ferol tragedy. *Pediatrics* 1986; **78**: 503–506.
11 FAO/WHO. Toxicological evaluation of certain food additives with a review of general principles and of specifications, Seventeenth report of the joint FAO/WHO expert committee on food additives. *World Health Organ Tech Rep Ser* 1974; No. 539.
12 Lewis RJ, ed. *Sax's Dangerous Properties of Industrial Materials*, 11th edn. New York: Wiley, 2004; 3013.

20 General References

Allen LV, *et al.* Effect of surfactant on tetracycline absorption across everted rat intestine. *J Pharm Sci* 1981; **70**: 269–271.

Chowhan ZT, Pritchard R. Effect of surfactants on percutaneous absorption of naproxen I: comparisons of rabbit, rat, and human excised skin. *J Pharm Sci* 1978; **67**: 1272–1274.

Donbrow M, *et al.* Autoxidation of polysorbates. *J Pharm Sci* 1978; **67**: 1676–1681.

European Directorate for the Quality of Medicines and Healthcare (EDQM). European Pharmacopoeia – State Of Work Of International Harmonisation. *Pharmeuropa* 2011; 23(4): 713–714. www.edqm.eu/site/-614.html (accessed 13 December 2011).

Khossravi M, *et al.* Analysis methods of polysorbate 20: a new method to assess the stability of polysorbate 20 and established methods that may overlook degraded polysorbate 20. *Pharm Res* 2002; **19**(5): 634–639.

Smolinske SC. *Handbook of Food, Drug, and Cosmetic Excipients.* Boca Raton, FL: CRC Press, 1992: 295–301.

21 Authors

C Medina, FA Alvarez-Nunez.

22 Date of Revision

2 March 2012.

Polyoxyethylene Stearates

1 Nonproprietary Names

The polyoxyethylene stearates are a series of polyethoxylated derivatives of stearic acid. Of the large number of different materials commercially available, one type is listed in the USP35–NF30.

JP: Polyoxyl 40 Stearate

USP–NF: Polyoxyl 40 Stearate

See also Sections 2, 3, 4, and 5.

2 Synonyms

Ethoxylated fatty acid esters; macrogol stearates; *Marlosol*; PEG fatty acid esters; PEG stearates; polyethylene glycol stearates; poly(oxy-1,2-ethanediyl) α-hydro-ω-hydroxyoctadecanoate; polyoxyethylene glycol stearates.

Polyoxyethylene stearates are nonionic surfactants produced by polyethoxylation of stearic acid. Two systems of nomenclature are used for these materials. The number '8' in the names 'poloxyl 8 stearate' or 'polyoxyethylene 8 stearate' refers to the approximate polymer length in oxyethylene units. The same material may also be designated 'polyoxyethylene glycol 400 stearate' or 'macrogol stearate 400' in which case, the number '400' refers to the average molecular weight of the polymer chain.

For synonyms applicable to specific polyoxyethylene stearates, *see* Table I.

3 Chemical Name and CAS Registry Number

Polyethylene glycol stearate [9004-99-3]

Polyethylene glycol distearate [9005-08-7]

4 Empirical Formula and Molecular Weight

See Table II.

Table II: Empirical formulas and molecular weights of selected polyoxyethylene stearates.

Name	Empirical formula	Molecular weight
Polyoxyl 6 stearate	$C_{30}H_{60}O_8$	548.80
Polyoxyl 8 stearate	$C_{34}H_{68}O_{10}$	636.91
Polyoxyl 12 stearate	$C_{42}H_{84}O_{14}$	813.12
Polyoxyl 20 stearate	$C_{58}H_{116}O_{22}$	1165.55
Polyoxyl 40 stearate	$C_{98}H_{196}O_{42}$	2046.61
Polyoxyl 50 stearate	$C_{118}H_{236}O_{52}$	2487.15
Polyoxyl 100 stearate	$C_{218}H_{436}O_{102}$	4689.80

5 Structural Formula

R–C(=O)–$(OCH_2CH_2)_n$–OH

Structure A

R–C(=O)–$(OCH_2CH_2)_n$–O–C(=O)–R

Structure B

Structure A applies to the monostearate; where the average value of *n* is 6 for polyoxyl 6 stearate, 8 for polyoxyl 8 stearate, and so on.

Structure B applies to the distearate; where the average value of *n* is 12 for polyoxyl 12 distearate, 32 for polyoxyl 32 distearate, and so on.

Table I: Synonyms of selected polyoxyethylene stearates and distearates.

Name	Synonym
Polyoxyl 2 stearate	*Hodag DGS*; *Lipo DGS*; *Lipopeg 2-DEGS*; PEG-2 stearate.
Polyoxyl 4 stearate	*Acconon 200-MS*; *Hodag 20-S*; *Lipopeg 2-DEGS*; PEG-4 stearate; polyethylene glycol 200 monostearate; polyoxyethylene (4) monostearate; *Protamate 200-DPS*.
Polyoxyl 6 stearate	*Cerasynt 616*; *Kessco PEG 300 Monostearate*; *Lipal 300S*; *Lipopeg 3-S*; PEG-6 stearate; polyethylene glycol 300 monostearate; polyoxyethylene (6) monostearate; *Polystate C*; *Protamate 300-DPS*.
Polyoxyl 8 stearate	*Acconon 400-MS*; *Cerasynt 660*; *Cithrol 4MS*; *Crodet S8*; *Emerest 2640*; *Grocor 400*; *Hodag 40-S*; *Kessco PEG-400 Monostearate*; *Lipopeg 4-S*; macrogol stearate 400; *Myrj 45*; PEG-8 stearate; *Pegosperse 400 MS*; polyethylene glycol 400 monostearate; polyoxyethylene (8) monostearate; *Protamate 400-DPS*; *Ritapeg 400 MS*.
Polyoxyl 12 stearate	*Hodag 60-S*; *Kessco PEG 600 Monostearate*; *Lipopeg 6-S*; PEG-12 stearate; *Pegosperse 600 MS*; polyethylene glycol 600 monostearate; polyoxyethylene (12) monostearate; *Protamate 600-DPS*.
Polyoxyl 20 stearate	*Cerasynt 840*; *Hodag 100-S*; *Kessco PEG 1000 Monostearate*; *Lipopeg 10-S*; *Myrj 49*; *Pegosperse 1000 MS*; PEG-20 stearate; polyethylene glycol 1000 monostearate; polyoxyethylene (20) monostearate; *Protamate 1000-DPS*.
Polyoxyl 30 stearate	*Myrj 51*; PEG-30 stearate; polyoxyethylene (30) stearate.
Polyoxyl 40 stearate	*Crodet S40*; E431; *Emerest 2672*; *Hodag POE (40) MS*; *Lipal 395*; *Lipopeg 39-S*; macrogol stearate 2000; *Myrj 52*; PEG-40 stearate; polyoxyethylene glycol 2000 monostearate; polyoxyethylene (40) monostearate; *Protamate 2000-DPS*; *Ritox 52*.
Polyoxyl 50 stearate	*Atlas G-2153*; *Crodet S50*; *Lipal 505*; *Myrj 53*; PEG-50 stearate; polyoxyethylene (50) monostearate.
Polyoxyl 100 stearate	*Lipopeg 100-S*; *Myrj 59*; PEG-100 stearate; polyethylene glycol 4400 monostearate; polyoxyethylene (100) monostearate; *Protamate 4400-DPS*; *Ritox 53*.
Polyoxyl 150 stearate	*Hodag 600-S*; PEG-150 stearate; *Ritox 59*.
Polyoxyl 4 distearate	*Hodag 22-S*; PEG-4 distearate.
Polyoxyl 8 distearate	*Hodag 42-S*; *Kessco PEG 400 DS*; PEG-8 distearate; polyethylene glycol 400 distearate; *Protamate 400-DS*.
Polyoxyl 12 distearate	*Hodag 62-S*; *Kessco PEG 600 Distearate*; PEG-12 distearate; polyethylene (12) distearate; polyethylene glycol 600 distearate; *Protamate 600-DS*.
Polyoxyl 32 distearate	*Hodag 154-S*; *Kessco PEG 1540 Distearate*; PEG-32 distearate; polyethylene glycol 1540 distearate; polyoxyethylene (32) distearate.
Polyoxyl 150 distearate	*Hodag 602-S*; *Kessco PEG 6000 DS*; *Lipopeg 6000-DS*; PEG-150 distearate; polyethylene glycol 6000 distearate; polyoxyethylene (150) distearate; *Protamate 6000-DS*.

P

In both structures, R represents the alkyl group of the parent fatty acid. With stearic acid, R is $CH_3(CH_2)_{16}$. However, it should be noted that stearic acid usually contains other fatty acids, primarily palmitic acid, and consequently a polyoxyethylene stearate may also contain varying amounts of other fatty acid derivatives such as palmitates.

6 Functional Category

Emulsifying agent; nonionic surfactant.

7 Applications in Pharmaceutical Formulation or Technology

Polyoxyethylene stearates are generally used as emulsifiers in oil-in-water-type creams and lotions. Their hydrophilicity or lipophilicity depends on the number of ethylene oxide units present: the larger the number, the greater the hydrophilic properties. Polyoxyl 40 stearate has been used as an emulsifying agent in intravenous infusions.[1]

Polyoxyethylene stearates are particularly useful as emulsifying agents when astringent salts or other strong electrolytes are present. They can also be blended with other surfactants to obtain any hydrophilic–lipophilic balance for lotions or ointment formulations. *See* Table III.

Table III: Uses of polyoxyethylene stearates.

Use	Concentration (%)
Auxiliary emulsifier for o/w intravenous fat emulsion	0.5–5
Emulsifier for o/w creams or lotions	0.5–10
Ophthalmic ointment	7
Suppository component	1–10
Tablet lubricant	1–2

8 Description

See Table IV.

Table IV: Description of various polyoxyethylene stearates.

Name	Description
Polyoxyl 6 stearate	Soft solid
Polyoxyl 8 stearate	Waxy cream
Polyoxyl 12 stearate	Pasty solid
Polyoxyl 20 stearate	Waxy solid
Polyoxyl 40 stearate	Waxy solid, with a faint, bland, fat-like odor, off-white to light tan in color
Polyoxyl 50 stearate	Solid, with a bland, fat-like odor or odorless
Polyoxyl 100 stearate	Solid
Polyoxyl 12 distearate	Paste
Polyoxyl 32 distearate	Solid
Polyoxyl 150 distearate	Solid

9 Pharmacopeial Specifications

See Table V.

10 Typical Properties

Flash point >149°C for poloxyl 8 stearate (*Myrj 45*).
Solubility *see* Table VI.
See also Table VII.

Table V: Pharmacopeial specifications for polyoxyethylene stearates.

Test	JP XV Polyoxyl 40 stearate	USP35–NF30 Polyoxyl 40 stearate
Identification	–	+
Clarity and color of solution	+	–
Congealing range	39–44°C	37–47°C
Congealing point of the fatty acid	≥53°C	–
Residue on ignition	≤0.1%	–
Water	–	≤3.0%
Arsenic	≤3 ppm	–
Heavy metals	≤10 ppm	≤0.001%
Acid value	≤1	≤2
Hydroxyl value	–	25–40
Saponification value	25–35	25–35
Free polyethylene glycols	–	17–27%

Table VI: Solubility of polyoxyethylene stearates.

Name	Solvent		
	Ethanol (95%)	Mineral oil	Water
Polyoxyl 6 stearate	S	S	DH
Polyoxyl 8 stearate	S	I	D
Polyoxyl 12 stearate	S	I	S
Polyoxyl 20 stearate	S	I	S
Polyoxyl 40 stearate	S	I	S
Polyoxyl 50 stearate	S	I	S
Polyoxyl 100 stearate	S	I	S
Polyoxyl 12 distearate	S	–	DH
Polyoxyl 32 distearate	S	–	S
Polyoxyl 150 distearate	I	–	S

D = dispersible; I = insoluble; S = soluble; DH = dispersible (with heat).

11 Stability and Storage Conditions

Polyoxyethylene stearates are generally stable in the presence of electrolytes and weak acids or bases. Strong acids and bases can cause gradual hydrolysis and saponification.

The bulk material should be stored in a well-closed container, in a dry place, at room temperature.

12 Incompatibilities

Polyoxyethylene stearates are unstable in hot alkaline solutions owing to hydrolysis, and will also saponify with strong acids or bases. Discoloration or precipitation can occur with salicylates, phenolic substances, iodine salts, and salts of bismuth, silver, and tannins.[2–4] Complex formation with preservatives may also occur.[5] The antimicrobial activity of some materials such as bacitracin, chloramphenicol, phenoxymethylpenicillin, sodium penicillin, and tetracycline may be reduced in the presence of polyoxyethylene stearate concentrations greater than 5% w/w.[6,7]

13 Method of Manufacture

Polyoxyethylene stearates are prepared by the direct reaction of fatty acids, particularly stearic acid, with ethylene oxide.

14 Safety

Although polyoxyethylene stearates are primarily used as emulsifying agents in topical pharmaceutical formulations, certain materials, particularly polyoxyl 40 stearate, have also been used in intravenous injections and oral preparations.[1,4]

Polyoxyethylene stearates have been tested extensively for toxicity in animals[8–13] and are widely used in pharmaceutical

Table VII: Typical properties of polyoxyethylene stearates.

Name	Acid value	Free ethylene oxide	HLB value	Hydroxyl value	Iodine number	Melting point (°C)	Saponification value	Water content (%)
Polyoxyl 6 stearate	≤5.0	≤100 ppm	9.7	—	≤0.5	28–32	95–110	—
Polyoxyl 8 stearate	≤2.0	≤100 ppm	11.1	87–105	≤1.0	28–33	82–95	≤3.0
Polyoxyl 12 stearate	≤8.5	≤100 ppm	13.6	55–75	≤1.0	≈37	62–78	≤1.0
Polyoxyl 20 stearate	≤1.0	≤100 ppm	14	50–62	≤1.0	≈28	46–56	≤1.0
Polyoxyl 30 stearate	≤2.0	—	16	35–50	—	—	30–45	≤3.0
Polyoxyl 40 stearate	≤1.0	—	16.9	27–40	—	≈38	25–35	≤3.0
Polyoxyl 50 stearate	≤2.0	—	17.9	23–35	—	≈42	20–28	≤3.0
Polyoxyl 100 stearate	≤1.0	≤100 ppm	18.8	15–30	—	≈46	9–20	≤3.0
Polyoxyl 8 distearate	≤10.0	—	—	≤15	≤0.5	≈36	115–124	—
Polyoxyl 12 distearate	≤10.0	≤100 ppm	10.6	≤20	≤1.0	≈39	93–102	≤1.0
Polyoxyl 32 distearate	≤10.0	≤100 ppm	14.8	≤20	≤0.25	≈45	50–62	≤1.0
Polyoxyl 150 distearate	7–9	≤100 ppm	18.4	≤15	≤0.1	53–57	14–20	≤1.0

formulations and cosmetics. They are generally regarded as essentially nontoxic and nonirritant materials.

Polyoxyl 8 stearate LD_{50} (hamster, oral): 27 g/kg

LD_{50} (rat, oral): 64 g/kg

Polyoxyl 20 stearate LD_{50} (mouse, IP): 0.2 g/kg

LD_{50} (mouse, IV): 0.87 g/kg

15 Handling Precautions

Observe normal precautions appropriate to the circumstances and quantity of material handled.

Polyoxyethylene stearates that contain greater than 100 ppm of free ethylene oxide may present an explosion hazard when stored in a closed container. This is due to the release of ethylene oxide into the container headspace, where it can accumulate and so exceed the explosion limit.

16 Regulatory Status

Included in the FDA Inactive Ingredients Database (dental solutions; IV injections; ophthalmic preparations; oral capsules and tablets; otic suspensions; topical creams, emulsions, lotions, ointments, and solutions; and vaginal preparations). Included in nonparenteral medicines licensed in the UK. Included in the Canadian Natural Health Products Ingredients Database.

17 Related Substances

Polyethylene glycol; stearic acid.

18 Comments

It has been reported that polyoxyl 40 stearate may also enhance the activity of chemotherapeutic agents and reverse multidrug resistance of tumor cells.[14]

19 Specific References

1 Cohn I, *et al.* New intravenous fat emulsion. *J Am Med Assoc* 1963; **183**: 755–757.

2 Thoma K, *et al.* [The antibacterial activity of phenols in the presence of polyoxyethylene stearates and polyethylene glycols.] *Arch Pharm* 1970; **303**: 289–296[in German].

3 Thoma K, *et al.* [Dimensions and cause of the reaction between phenols and polyoxyethylene stearates.] *Arch Pharm* 1970; **303**: 297–304[in German].

4 Duchene D, *et al.* [Tablet study III: influence of nonionic surfactants with ester linkage on the quality of sulfanilamide grains and tablets.] *Ann Pharm Fr* 1970; **28**: 289–298[in French].

5 Chakravarty D, *et al.* Study of complex formation between polyoxyl 40 stearate and some pharmaceuticals. *Drug Standards* 1957; **25**: 137–140.

6 Ullmann E, Moser B. [Effect of polyoxyethylene stearates on the antibacterial activity of antibiotics.] *Arch Pharm* 1962; **295**: 136–143[in German].

7 Thoma K, *et al.* [Investigation of the stability of penicillin G sodium in the presence of nonionic surface active agents (polyethylene glycol derivatives).] *Arch Pharm* 1962; **295**: 670–678[in German].

8 Culver PJ, *et al.* Intermediary metabolism of certain polyoxyethylene derivatives in man I: recovery of the polyoxyethylene moiety from urine and feces following ingestion of polyoxyethylene (20) sorbitan monooleate and of polyoxyethylene (40) mono-stearate. *J Pharmacol Exp Ther* 1951; **103**: 377–381.

9 Oser BL, Oser M. Nutritional studies on rats on diets containing high levels of partial ester emulsifiers I: general plan and procedures; growth and food utilization. *J Nutr* 1956; **60**: 367–390.

10 Oser BL, Oser M. Nutritional studies on rats on diets containing high levels of partial ester emulsifiers II: reproduction and lactation. *J Nutr* 1956; **60**: 489–505.

11 Oser BL, Oser M. Nutritional studies on rats on diets containing high levels of partial ester emulsifiers III: clinical and metabolic observations. *J Nutr* 1957; **61**: 149–166.

12 Oser BL, Oser M. Nutritional studies on rats on diets containing high levels of partial ester emulsifiers IV: mortality and postmortem pathology; general conclusions. *J Nutr* 1957; **61**: 235–252.

13 Fitzhugh OG, *et al.* Chronic oral toxicities of four stearic acid emulsifiers. *Toxicol Appl Pharmacol* 1959; **1**: 315–331.

14 Luo L, *et al.* Polyoxyethylene 40 stearate modulates multi-drug resistance and enhances antitumor activity of vinblastine sulphate. *AAPS J* 2007; **9**: E329–E335.

20 General References

Satkowski WB *et al.* Polyoxyethylene esters of fatty acids. In: Schick MJ, ed. *Nonionic Surfactants*. New York: Marcel Dekker, 1967: 142–174.

21 Author

J Shur.

22 Date of Revision

2 March 2012.

Polyoxyl 15 Hydroxystearate

1 Nonproprietary Names

BP: Macrogol 15 Hydroxystearate
PhEur: Macrogol 15 Hydroxystearate
USP–NF: Polyoxyl 15 Hydroxystearate

2 Synonyms

12-Hydroxyoctadecanoic acid polymer with α-hydro-ω-hydroxypoly(oxy-1,2-ethanediyl); 12-hydroxystearic acid polyethylene glycol copolymer; macrogol 15 hydroxistearate; macrogoli 15 hydroxystearas; polyethylene glycol-15-hydroxystearate; polyethylene glycol 660 12-hydroxystearate; *Solutol HS 15*.

3 Chemical Name and CAS Registry Number

2-Hydroxyethyl-12-hydroxyoctadecanoate [70142-34-6]

4 Empirical Formula and Molecular Weight

The PhEur 7.4 describes polyoxyl 15 hydroxystearate as a mixture of mainly monoesters and diesters of 12-hydroxystearic acid (lypophilic) and macrogols (hydrophilic) obtained by the ethoxylation of 12-hydroxystearic acid. The number of moles of ethylene oxide reacted per mole of 12-hydroxystearic acid is 15 (nominal value). It contains about 30% free macrogols.

$C_{20}H_{40}O_4$ 344.53

5 Structural Formula

See Section 4.

6 Functional Category

Emulsifying agent; nonionic surfactant; solubilizing agent.

7 Applications in Pharmaceutical Formulation or Technology

Polyoxyl 15 hydroxystearate is mainly used as a nonionic solubulizing agent for aqueous parenteral preparations, and is frequently used in preclinical testing of drugs.[1–4] The solubilizing capacity for some tested drugs (clotrimazole, carbamazepine, 17β-estradiol, sulfathiazole, and piroxicam) increases almost linearly with increasing concentration of solubilizing agent; *see* Figure 1. This is due to the formation of spherical micelles even at high concentrations of polyoxyl 15 hydroxystearate. Similarly, tests have revealed that viscosity increases with increasing amount of solubilizer, but the amount of solubilized drugs does not have any additional influence on the kinematic viscosity; *see* Figure 2. Lipid nanocapsules comprising polyoxyl 15 hydroxystearate and soybean phosphatidylcholine containing 3% docetaxel have been successfully prepared by a solvent-free inversion process.

Polyoxyl 15 hydroxystearate has been used in the manufacture of aqueous parenteral preparations with vitamin A, D, E and K, and a number of other lipophilic pharmaceutical active agents, such as propanidid, miconazole, alfadolone, alfaxalone, nifedipine, carbamazepine, and piroxicam. It is very efficient at solubilizing substances like fat-soluble vitamins and active ingredients of hydrophobic nature. It is an excellent solubilizer for parenteral use, at a concentration of 20%, and the water solubility of different drugs may be enhanced by a factor of 10–100, depending on the structure of the drug molecule.

Polyoxyl 15 hydroxystearate is not restricted solely to parenteral use, but is also suitable for oral applications.

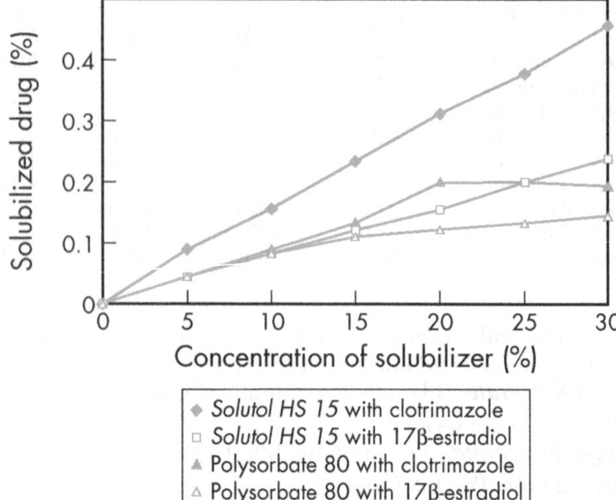

Figure 1: Solubilizing capacity of polyoxyl 15 hydroxystearate (*Solutol HS 15*, BASF plc).

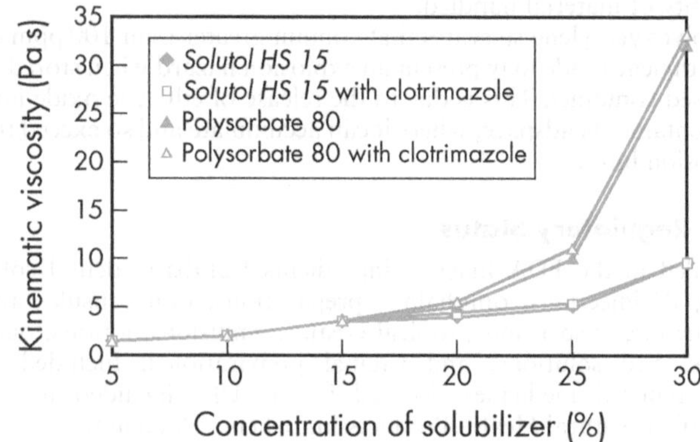

Figure 2: Kinematic viscosity of polyoxyl 15 hydroxystearate (*Solutol HS 15*, BASF plc).

8 Description

Polyoxyl 15 hydroxystearate occurs as a yellowish-white, almost odorless waxy mass or paste at room temperature, which becomes liquid at approximately 30°C.

9 Pharmacopeial Specifications

See Table I.

10 Typical Properties

Acidity/alkalinity pH = 6–7 (10% w/v aqueous solution at 20°C)
Critical micelle concentration 0.005–0.02%
Density 1.03 g/cm³
Flash point 272°C
Heavy metals ≤10 ppm
HLB value 14–16
Ignition temperature 360°C
Solidification temperature 25–30°C
Solubility Soluble in organic solvents such as ethanol (96%), propan-2-ol, and very soluble in water to form clear solutions.

Table I: Pharmacopeial specifications for polyoxyl 15 hydroxystearate.

Test	PhEur 7.4	USP35–NF30
Identification	+	+
Characters	+	−
Appearance of solution	+	−
Acid value	≤1.0	≤1.0
Hydroxyl value	90–110	90–110
Iodine value	≤2.0	≤2.0
Peroxide value	≤5.0	≤5.0
Saponification value	53–63	53–63
Free macrogols	27.0–39.0%	27.0–39.0%
Ethylene oxide	≤1 ppm	≤1 ppm
Dioxane	≤50 ppm	≤50 ppm
Nickel	≤1 ppm	1 μg/g
Water	≤1.0%	≤1.0%
Total ash	≤0.3%	≤0.3%

The solubility in water is >200 g/L at 30°C and decreases with increasing temperature. It is insoluble in liquid paraffin.

Viscosity (dynamic) 12 mPa s (12 cP) for a 30% w/v aqueous solution at 25°C; 73 mPa s (73 cP) for a 30% w/v aqueous solution at 60°C.

11 Stability and Storage Conditions

Polyoxyl 15 hydroxystearate has a high chemical stability. The prolonged action of heat may induce physical separation into a liquid and a solid phase after cooling, which can be reversed by subsequent homogenization. Polyoxyl 15 hydroxystearate is stable for at least 24 months if stored in unopened airtight containers at room temperature (maximum 25°C). Aqueous solutions of polyoxyl 15 hydroxystearate can be heat-sterilized (121°C, 0.21 MPa). The pH may drop slightly during heating, which should be taken into account. Separation into phases may also occur, but agitating the hot solution can reverse this. Aqueous solutions can be stabilized with the standard preservatives used in pharmaceuticals.

Polyoxyl 15 hydroxystearate should be stored in airtight containers in a dry place and protected from light.

12 Incompatibilities

—

13 Method of Manufacture

Polyoxyl 15 hydroxystearate is produced by reacting 15 moles of ethylene oxide with 1 mole of 12-hydroxystearic acid.

14 Safety

Polyoxyl 15 hydroxystearate is used in parenteral pharmaceutical preparations in concentrations up to 50% to solubilize diclofenac, propanidid, and vitamin K1. It has also been used in preclinical formulations in preparing supersaturated injectable formulations of water-insoluble molecules. It is generally regarded as a relatively nontoxic and nonirritant excipient. However, although nonirritating to skin, it may cause sensitization on skin contact in guinea pigs.

Polyoxyl 15 hydroxystearate is reported not to be mutagenic in bacteria, mammalian cell cultures and mammals.

LD_{50} (dog, IV): >3.10 g/kg[5]

LD_{50} (mouse, IP): >0.0085 g/kg

LD_{50} (mouse, IV): >3.16 g/kg

LD_{50} (mouse, oral): >20 g/kg

LD_{50} (rabbit, IV): 1.0–1.4 g/kg

LD_{50} (rat, oral): >20 g/kg

LD_{50} (rat, IV): 1.0–1.47 g/kg

15 Handling Precautions

Observe normal precautions appropriate to the circumstances and quantity of material handled. Polyoxyl 15 hydroxystearate undergoes thermal decomposition at >300°C. Ensure adequate ventilation and wear suitable respiratory protection.

16 Regulatory Status

Included in the Canadian Natural Health Products Ingredients Database.

17 Related Substances

Polyethylene glycol.

18 Comments

Polyoxyl 15 hydroxystearate has been investigated as a nonionic surfactant and coemulsifier in the preparation of parenteral emulsions,[6] microemulsions,[1,7–10] nanoemulsions,[11] self-emulsifying drug delivery systems (SEDDS),[12] self-nanoemulsifying drug delivery systems (SNEDDS),[13] nanostructured lipid carriers (NLC),[14] and lipid nanoparticles.[15,16]

Oral bioavailability of the highly lipophilic and poorly water-soluble immunosuppressive agent, cyclosporine A, showed twofold higher bioavailability with a polyoxyl 15 hydroxystearate-based formulation compared to a microsuspension.[17] It has also been studied along with microcrystalline cellulose to prepare self-emulsifying pellets using an extrusion/spheronization technique to increase the bioavailability of lipophilic drugs.[18]

Polyoxyl 15 hydroxystearate has been incorporated as a solubility-increasing additive in a rectal suppository dosage form to study the increase in bioavailability of poorly water-soluble drugs.[19]

Polyoxyl 15 hydroxystearate has been investigated as a therapeutic agent in the preparation of lipid nanoparticles of an anticancer drug,[20] and has also been shown to be effective for reversing multidrug resistance, with low toxicity *in vivo*.[21] It has also been investigated as a weak inhibitor of cytochrome P450 3A activity on the metabolism of colchicine and midazolum.[22–25]

Polyoxyl 15 hydroxystearate is also used in veterinary injections.

The PubChem Compound ID (CID) for *Solutol HS 15* is 124898.

P

19 Specific References

1 Ryoo HK, *et al.* Development of propofol-loaded microemulsion systems for parenteral delivery. *Arch Pharm Res* 2005; **28**(12): 1400–1404.

2 Buszello K, *et al.* The influence of alkali fatty acids on the properties and the stability of parenteral O/W emulsions modified with *Solutol HS 15*. *Eur J Pharm Biopharm* 2000; **49**: 143–149.

3 Bittner B, Mountfield RJ. Formulations and related activities for the oral administration of poorly water-soluble compounds in early discovery animal studies. *Pharm Ind* 2002; **64**: 800.

4 Strickley R. Solubilizing excipients in oral and injectable formulations. *Pharm Res* 2004; **21**: 201–230.

5 *BASF. Solutol HS 15*. http://www.pharma-ingredients.basf.com/Statements/Technical%20Informations/EN/Pharma%20Solutions/03_030748e_Solutol%20HS%2015.pdf (accessed 6 December 2011).

6 Sun H, *et al.* Formulation of a stable and high-loaded quercetin injectable emulsion. *Pharm Dev Technol* 2010; **Aug 19**: [Epub ahead of pub].

7 Date A, Nagarsenker MS. Parenteral microemulsions: an overview. *Int J Pharm* 2008; **355**(1–2): 19–30.

8 Zhao X, *et al.* Synthesis of ibuprofen eugenol ester and its microemulsion formulation for parenteral delivery. *Chem Pharm Bull (Tokyo)* 2005; **53**(10): 1246–1250.

9 Lazzari P, *et al.* Antinociceptive activity of delta9–tetrahydrocannabinol non-ionic microemulsions. *Int J Pharm* 2010; **393**(1–2): 238–243.

10 Gan L, *et al.* Novel microemulsion in situ electrolyte-triggered gelling system for ophthalmic delivery of lipophilic cyclosporine A: *in vitro* and *in vivo* results. *Int J Pharm* 2009; **365**(1–2): 143–149.

11 Kim B, *et al. In vitro* permeation studies of nanoemulsions containing ketoprofen as a model drug. *Drug Deliv* 2008; **15**(7): 465–469.

12 Abdalla A, *et al.* A new self-emulsifying drug delivery system (SEDDS) for poorly soluble drugs: Characterization, dissolution, *in vitro* digestion and incorporation into solid pettets. *Eur J Pharm Sci* 2008; **35**(5): 475–464.
13 Nepel P, *et al.* Preparation and *in vitro-in vivo* evaulation of *Witepsol H35* based self-nanoemulsifying drug delivery systems (SNEDDS) of coenzyme Q(10). *Eur J Pharm Sci* 2010; **39**(4): 224–232.
14 Yang C, *et al.* Preparation, optimization and characteristic of huperzine a loaded nanostructured lipid carriers. *Chem Pharm Bull (Tokyo)* 2010; **58**(5): 565–661.
15 Malzert-Freon A, *et al.* Partial least square analysis and mixture design for the study of the influence of composition variables on lipidic nanoparticle characterisitics. *J Pharm Sci* 2010; **99**(11): 4603–4615.
16 Subedi R, *et al.* Preparation and characterization of solid lipid nanoparticles loaded with doxorubicin. *Eur J Pharm Sci* 2009; **37**(3–4): 508–513.
17 Roberto CB, *et al.* Improved oral bioavailability of cyclosporin A in male Wistar rats. Comparison of a *Solutol HS 15* containing self-dispersing formulation and a microsuspension. *Int J Pharm* 2002; **245**(1–2): 143–151.
18 Abdalla A, Mäder K. Preparation and characterization of a self-emulsifying pellet formulation. *Eur J Pharm Biopharm* 2007; **66**(2): 220–226.
19 Berkó S, *et al.* Solutol and cremophor products as new additives in suppository formulation. *Drug Dev Ind Pharm* 2002; **28**(2): 203–206.
20 Malzert-Fréon A, *et al.* Formulation of sustained release nanoparticles loaded with a tripentone, a new anticancer agent. *Int J Pharm* 2006; **320**(1–2): 157–164.
21 Garcion E , *et al.* A new generation of anticancer, drug-loaded, colloidal vectors reverses multidrug resistance in glioma and reduces tumor progression in rats. *Mol Cancer Ther* 2006; **5**(7): 1710–1722.
22 Bravo González RC, *et al.* *In vitro* investigation on the impact of the surface-active excipients *Cremophor EL*, *Tween 80* and *Solutol HS 15* on the metabolism of midazolam. *Biopharm Drug Dispos* 2004; **25**(1): 37–49.
23 Bittner B, *et al.* Impact of *Solutol HS 15* on the pharmacokinetic behavior of midazolam upon intravenous administration to male Wistar rats. *Eur J Pharm Biopharm* 2003; **56**(1): 143–146.
24 Bittner B, *et al.* Impact of *Solutol HS 15* on the pharmacokinetic behaviour of colchicine upon intravenous administration to male Wistar rats. *Biopharm Drug Dispos* 2003; **24**(4): 173–181.
25 Roberto CB, *et al.* *In vitro* investigation on the impact of *Solutol HS 15* on the uptake of colchicine into rat hepatocytes. *Int J Pharm* 2004; **279**(1–2): 27–31.

20 General References

BASF. Materials safety data sheet: *Solutol HS 15*, March 2006.
BASF. Technical information sheet: *Solutol HS 15*, August 2010.
BASF. Product data sheet: *Solutol HS 15*, September 2010.
Bley H, *et al.* Characterization and stability of solid dispersions based on PEG/polymer blends. *Int J Pharm* 2010; **390**(2): 165–173.
Didriksen EJ, Gerlach AL. Solubilized pharmaceutical composition for parenteral administration. European Patent 1171100; 2003.
Frömming K-H, *et al.* Physico-chemical properties of the mixed micellar system of *Solutol HS 15* and sodium deoxycholate. *Acta Pharm Technol* 1990; **36**: 214–220.
Kim K, *et al.* Pharmacokinetics and pharmacodynamics of propofol microemulsion and lipid emulsion after an intravenous bolus and variable rate infusion. *Anesthesiology* 2007; **106**(5): 924–934.
OECD Guideline 406. http://iccvam.niehs.nih.gov/methods/immunotox/llnadocs/OECDtg406.pdf (accessed 6 December 2011).

21 Author

KK Singh.

22 Date of Revision

1 March 2012.

Polyoxylglycerides

1 Nonproprietary Names

BP: Caprylocaproyl Macrogolglycerides
Lauroyl Macrogolglycerides
Linoleoyl Macrogolglycerides
Oleoyl Macrogolglycerides
Stearoyl Macrogolglycerides

PhEur: Caprylocaproyl Macrogolglycerides
Lauroyl Macrogolglycerides
Linoleoyl Macrogolglycerides
Oleoyl Macrogolglycerides
Stearoyl Macrogolglycerides

USP–NF: Caprylocaproyl Polyoxylglycerides
CLauroyl Polyoxylglycerides
Linoleoyl Polyoxylglycerides
Oleoyl Polyoxylglycerides
Stearoyl Polyoxylglycerides

2 Synonyms

Polyoxylglycerides are referred to as macrogolglycerides in Europe; *see* Table I.

Table I: Synonyms of polyoxylglycerides (macrogolglycerides).

Name	Synonyms
Caprylocaproyl polyoxylglycerides	*Labrasol*; macrogolglyceridorum caprylocaprates; PEG 400 caprylic/capric glycerides
Lauroyl polyoxylglycerides	*Gelucire 44/14*; hydrogenated coconut oil PEG 1500 esters; hydrogenated palm/palm kernel oil PEG 300 esters; *Labrafil M2130CS* ; macrogolglyceridorum laurates
Linoleoyl polyoxylglycerides	Corn oil PEG 300 esters; *Labrafil M2125CS*; macrogolglyceridorum linoleates
Oleoyl polyoxylglycerides	Apricot kernel oil PEG 300 esters; *Labrafil M1944CS*; macrogolglyceridorum oleates; peglicol-5-oleate
Stearoyl polyoxylglycerides	*Gelucire 50/13*; hydrogenated palm oil PEG 1500 esters; macrogolglyceridorum stearates

3 Chemical Name and CAS Registry Number

See Table II.

Table II: Chemical names and CAS registry numbers of polyoxylglycerides.

Name	CAS number		Chemical name
Caprylocaproyl polyoxylglycerides	[73398-61-5]	[223129-75-7]	Decanoic acid, mixed monoesters with glycerol and octanoic acid; poly(oxy-1,2-ethanediyl), α-hydro-ω-hydroxy-, mixed decanoate and octanoate
Lauroyl polyoxylglycerides	[57107-95-6]	[27194-74-7]	Lauric acid, diester with glycerol; poly(oxy-1,2-ethanediyl), α-(1-oxododecyl)-ω-[(1-oxododecyl)oxy]-
Linoleoyl polyoxylglycerides	[61789-25-1]		Corn oil, ethoxylated; 9,12-octadecadienoic acid (9*E*,12*E*)-monoester with 1,2,3-propanetriol
Oleoyl polyoxylglycerides	[68424-61-3]	[9004-96-0]	9-Octadecenoic acid (9*Z*)-, monoester with 1,2,3-propanetriol; poly(oxy-1,2-ethanediyl), α-[(9*Z*)-1-oxo-9-octadecenyl]-ω-hydroxy-
Stearoyl polyoxylglycerides	[1323-83-7]	[9005-08-7]	Distearic acid, diester with glycerol; poly(oxy-1,2-ethanediyl), α-(1-oxooctadecyl)-ω-[(1-oxooctadecyl)-oxy]

4 Empirical Formula and Molecular Weight

Polyoxylglycerides are mixtures of monoesters, diesters, and triesters of glycerol, and monoesters and diesters of polyethylene glycols (PEG).

Caprylocaproyl polyoxylglycerides Mixtures of monoesters, diesters, and triesters of glycerol and monoesters and diesters of polyethylene glycols with mean relative molecular mass between 200 and 400. They are obtained by partial alcoholysis of medium-chain triglycerides using polyethylene glycol or by esterification of glycerin and polyethylene glycol with caprylic (octanoic) acid and capric (decanoic) acid or a mixture of glycerin esters and condensates of ethylene oxide with caprylic acid and capric acid. They may contain free polyethylene glycols.

Lauroyl polyoxylglycerides Mixtures of monoesters, diesters, and triesters of glycerol and monoesters and diesters of polyethylene glycols with mean relative molecular mass between 300 and 1500. They are obtained by partial alcoholysis of saturated oils mainly containing triglycerides of lauric (dodecanoic) acid, using polyethylene glycol, or by esterification of glycerol and polyethylene glycol with saturated fatty acids, or by mixing glycerol esters and condensates of ethylene oxide with the fatty acids of these hydrogenated oils.

Linoleoyl polyoxylglycerides Mixtures of monoesters, diesters, and triesters of glycerol and monoesters and diesters of polyethylene glycols. They are obtained by partial alcoholysis of an unsaturated oil mainly containing triglycerides of linoleic (*cis*,*cis*-9,12-octadecadienoic) acid, using polyethylene glycol with mean relative molecular mass between 300 and 400, or by esterification of glycerol and polyethylene glycol with unsaturated fatty acids, or by mixing glycerol esters and condensates of ethylene oxide with the fatty acids of this unsaturated oil.

Oleoyl polyoxylglycerides Mixtures of monoesters, diesters, and triesters of glycerol and monoesters and diesters of polyethylene glycols. They are obtained by partial alcoholysis of an unsaturated oil mainly containing triglycerides of oleic (*cis*-9-octadecenoic) acid, using polyethylene glycol with mean relative molecular mass between 300 and 400, or by esterification of glycerol and polyethylene glycol with unsaturated fatty acids, or by mixing glycerol esters and condensates of ethylene oxide with the fatty acids of this unsaturated oil.

Stearoyl polyoxylglycerides Mixtures of monoesters, diesters, and triesters of glycerol and monoesters and diesters of polyethylene glycols with mean relative molecular mass between 300 and 4000. They are obtained by partial alcoholysis of saturated oils containing mainly triglycerides of stearic (octadecanoic) acid, using polyethylene glycol, or by esterification of glycerol and polyethylene glycol with saturated fatty acids, or by mixture of glycerol esters and condensates of ethylene oxide with the fatty acids of these hydrogenated oils.

5 Structural Formula

See Section 4.

6 Functional Category

Emulsifying agent; modified-release agent; nonionic surfactant; penetration enhancer; solubilizing agent.

7 Applications in Pharmaceutical Formulation or Technology

Polyoxylglycerides are used as self-emulsifying and solubilizing agents in oral and topical pharmaceutical formulations. They are also used in cosmetic and food products.

See also Tables III, IV, V, VI, and VII.

Table III: Uses of caprylocaproyl polyoxylglycerides.

Use	Concentration	Reference
Dermal route	10–55%	1–11
Nasal route	2–22%	12, 13
Oral route		
Capsule	10–99%	14–32
Sublingual route	10–35%	33

Table IV: Uses of lauroyl polyoxylglycerides.

Use	Concentration	Reference
Oral route		
Adsorption (tablet)	<80%	34, 35
Capsule	60–99%	14, 29, 31, 32, 34, 36–40, 41–44
Melt granulation	15–50%	34, 44, 45
Spray drying	<60%	30, 34, 35

Table V: Uses of linoleoyl polyoxylglycerides.

Use	Concentration	Reference
Dermal route	5–20%	23
Oral route		
Capsule	10–90%	16, 18, 46, 47

Table VI: Uses of oleoyl polyoxylglycerides.

Use	Concentration	Reference
Dermal route	5–20%	2
Nasal route	8%	13
Oral route		
Capsule	10–90%	16, 18, 26, 31, 46, 48, 49

Table VII: Uses of stearoyl polyoxylglycerides.

Use	Concentration	Reference
Oral route		
Adsorption (tablet)	<80%	34, 35, 50
Capsule	60–99%	34, 39, 51–54
Melt granulation	15–50%	34, 55–57
Spray congealing	95%	58
Spray drying	<60%	34, 54

8 Description

Polyoxylglycerides are inert liquid or semi-solid waxy materials and are amphiphilic in character. Caprylocaproyl polyoxylglycerides are pale-yellow oily liquids. Lauroyl polyoxylglycerides and stearoyl polyoxylglycerides occur as pale-yellow waxy solids. Oleoyl polyoxylglycerides and linoleoyl polyoxylglycerides occur as amber oily liquids, which may give rise to a deposit after prolonged periods at 20°C.

9 Pharmacopeial Specifications

See Tables VIII and IX.

10 Typical Properties

Solubility

Caprylocaproyl and lauroyl polyoxylglycerides: dispersible in hot water; freely soluble in methylene chloride.

Linoleoyl and oleoyl polyoxylglycerides: practically insoluble but dispersible in water; freely soluble in methylene chloride.

Stearoyl polyoxylglycerides: dispersible in warm water and warm liquid paraffin; soluble in warm ethanol; freely soluble in methylene chloride.

Viscosity

Linoleoyl polyoxylglycerides: 70–90 mPa s at 20°C, ≈35 mPa s at 40°C for PEG 300.

Oleoyl polyoxylglycerides: 75–95 mPa s at 20°C, ≈35 mPa s at 40°C for PEG 300.

See also Section 9.

See also Table X.

11 Stability and Storage Conditions

Polyoxylglycerides are very stable and inert. However, preventive measures against the risk of oxidation or hydrolysis may be taken to ensure stability during handling. *See* Section 15.

Polyoxylglycerides should be preserved in their original containers, and exposure to air, light, heat, and moisture should be prevented.

12 Incompatibilities

—

13 Method of Manufacture

Polyoxylglycerides are obtained by partial alcoholysis of vegetable oils using macrogols, by esterification of glycerol and macrogols with unsaturated fatty acids, or by mixing glycerol esters and condensates of ethylene oxide with the fatty acids of the vegetable oil.

Table XI: EINECS numbers for polyoxylglycerides.

Name	EINECS number
Caprylocaproyl polyoxylglycerides	277-452-2
Lauroyl polyoxylglycerides	248-315-4
Oleoyl polyoxylglycerides	270-312-1
Stearoyl polyoxylglycerides	215-359-0

14 Safety

Polyoxylglycerides are used in oral and topical pharmaceutical formulations, and also in cosmetics and food products. They are generally regarded as relatively nonirritant and nontoxic materials.

Caprylocaproyl polyoxylglycerides:
LD_{50} (rat, oral): >22 ml/(kg day).[59]
Lauroyl polyoxylglycerides:
LD_{50} (rat, oral): >2004 mg/(kg day).[60]

15 Handling Precautions

Observe normal precautions appropriate to the circumstances and quantities of the material handled (refer to manufacturers' safety information).

Polyoxylglycerides are heterogeneous. Owing to their composition and physical characteristics, semisolid polyoxylglycerides can segregate by molecular weight over time during storage in containers, resulting in a nonhomogenous distribution. In addition, semisolid polyoxylglycerides must be heated to at least 20°C above melting point in order to ensure that all crystallization clusters are fully melted. Therefore, it is essential that the entire contents of each container are melted to facilitate sample withdrawal or transfer, ensuring sample homogeneity.

For liquid polyoxylglycerides, owing to their composition and physical characteristics, partial crystallization of saturated glycerides may be observed after long-term storage. In case of crystallization, heat to 60–70°C before use.

Polyoxylglycerides are hygroscopic. Only heat in a water bath if the materials are contained in a sealed glass container or are for immediate use. Otherwise, heat in a microwave or convention oven. Avoid exposure to excessive and repeated high temperatures (i.e. above 100°C) and cooling cycles.

To ensure stability during handling, and avoid the risk of oxidation or hydrolysis, the following measures should be taken:

Risk of oxidation:
- minimize aeration of the mixture (avoid use of high-speed homogenizers);
- minimize and control the degree of exposure to heat and light;
- use a nitrogen blanket.

Risk of hydrolysis:
- minimize and control relative humidity;
- do not heat near a source of humidity (e.g. water bath).

Table X: Typical properties of polyoxylglycerides.

Property	Caprylocaproyl polyoxylglycerides	Lauroyl polyoxylglycerides	Linoleoyl polyoxylglycerides	Oleoyl polyoxylglycerides	Stearoyl polyoxylglycerides
HLB value					
PEG 300	—	4	4	4	—
PEG 400	14	—	—	—	—
PEG 1500	—	14	—	—	13
Relative density (at 20°C)	1.0	—	0.95	0.95	—
Refractive index (at 20°C)	1.450–1.470	—	1.465–1.475	1.465–1.475	—

Table VIII: Pharmacopeial specifications for polyoxylglycerides.

Test	Caprylocaproyl polyoxylglycerides		Lauroyl polyoxylglycerides		Linoleoyl polyoxylglycerides		Oleoyl polyoxylglycerides		Stearoyl polyoxylglycerides	
	PhEur 7.4	USP35–NF30	PhEur 7.4	USP35–NF30	PhEur 7.4	USP35–NF30	PhEur 7.4	USP35–NF30	PhEur 7.4	USP35–NF30
Identification	+	+	+	+	+	+	+	+	+	+
Characters	+	—	+	—	+	—	+	—	+	—
Drop point (°C)										
PEG 300	—	—	33–38	—	—	—	—	—	—	—
PEG 400	—	—	36–41	—	—	—	—	—	—	—
PEG 600	—	—	38–43	—	—	—	—	—	—	—
PEG 1500	—	—	42.5–47.5	—	—	—	—	—	—	—
Viscosity at 20°C ± 5°C										
PEG 200	30–50	—	—	—	—	—	—	—	—	—
PEG 300	60–80	—	—	—	—	—	—	—	—	—
PEG 400	80–110	—	—	—	—	—	—	—	—	—
Acid value	≤2.0	≤2.0	≤2.0	≤2.0	≤2.0	≤2.0	≤2.0	≤2.0	≤2.0	≤2.0
Hydroxyl value	+	+	+	+	+	45–65	+	45–65	+	25–56
PEG 200	80–120	80–120	—	—	—	—	—	—	25–56	—
PEG 300	140–180	140–180	65–85	65–85	45–65	—	45–65	—	25–56	—
PEG 400	170–205	170–205	60–80	60–80	45–65	—	45–65	—	25–56	—
PEG 600	—	—	50–70	50–70	—	—	—	—	25–56	—
PEG 1500	—	—	36–56	36–56	—	—	—	—	25–56	—
Iodine value	≤2.0	≤2.0	≤2.0	≤2.0	90–110	90–110	75–95	75–95	≤2.0	≤2.0
Peroxide value	≤6.0	≤6.0	≤6.0	≤6.0	≤12.0	≤12.0	≤12.0	≤12.0	≤6.0	≤6.0
Saponification value	+	+	+	+	+	150–170	+	150–170	+	67–112
PEG 200	265–285	265–285	—	—	150–170	—	150–170	—	67–112	—
PEG 300	170–190	170–190	190–204	190–204	150–170	150–170	150–170	—	67–112	—
PEG 400	85–105	85–105	170–190	170–190	—	—	—	—	67–112	—
PEG 600	—	—	150–170	150–170	—	—	—	—	67–112	—
PEG 1500	—	—	79–93	79–93	—	—	—	—	67–112	—
Alkaline impurities	+	+	+	+	+	+	+	+	+	+
Free glycerol	≤5.0%	≤5.0%	≤3.0%	≤5.0%	≤3.0%	≤5.0%	≤3.0%	≤5.0%	≤3.0%	≤5.0%
Ethylene oxide	≤1 ppm	≤1 µg/g	≤1 ppm	≤1 µg/g	≤1 ppm	≤1 µg/g	≤1 ppm	≤1 µg/g	≤1 ppm	≤1 µg/g
Dioxane	≤10 ppm	≤10 µg/g	≤10 ppm	≤10 µg/g	≤10 ppm	≤10 µg/g	≤10 ppm	≤10 µg/g	≤10 ppm	≤10 µg/g
Heavy metals	≤10 ppm	≤0.001%	≤10 ppm	≤0.001%	≤10 ppm	≤0.001%	≤10 ppm	≤0.001%	≤10 ppm	≤0.001%
Water	≤1.0%	≤1.0%	≤1.0%	≤1.0%	≤1.0%	≤1.0%	≤1.0%	≤1.0%	≤1.0%	≤1.0%
Total ash	≤0.1%	≤0.1%	≤0.1%	≤0.1%	≤0.1%	≤0.1%	≤0.1%	≤0.1%	≤0.2%	≤0.2%

Table IX: Pharmacopeial specifications for polyoxylglycerides (fatty acids composition).

Fatty acid	Caprylocaproyl polyoxylglycerides		Lauroyl polyoxylglycerides		Linoleoyl polyoxylglycerides		Oleoyl polyoxylglycerides		Stearoyl polyoxylglycerides	
	PhEur 7.4	USP35–NF30	PhEur 7.4	USP35–NF30	PhEur 7.4	USP35–NF30	PhEur 7.4	USP35–NF30	PhEur 7.4	USP35–NF30
C_6 = Caproic acid	<2%	≤2%	—	—	—	—	—	—	—	—
C_8 = Caprylic acid	50–80%	50–80%	<15%	≤15%	—	—	—	—	—	—
C_{10} = Capric acid	20–50%	20–50%	<12%	≤12%	—	—	—	—	—	—
C_{12} = Lauric acid	<3%	≤3%	30–50%	30–50%	—	—	—	—	<5%	≤5%
C_{14} = Myristic acid	<1%	≤1%	5–25%	5–25%	—	—	—	—	<5%	≤5%
C_{16} = Palmitic acid	—	—	4–25%	4–25%	4–20%	4–20%	4–9%	4–9%	>90%	40–50%
C_{18} = Stearic acid	—	—	5–35%	5–35%	≤6%	≤6%	≤6%	≤6%	>90%	48–58%
$C_{18:1}$ = Oleic acid	—	—	—	—	20–35%	20–35%	58–80%	58–80%	—	—
$C_{18:2}$ = Linoleic acid	—	—	—	—	50–65%	50–65%	15–35%	15–35%	—	—
$C_{18:3}$ = Linolenic acid	—	—	—	—	≤2%	≤2%	≤2%	≤2%	—	—
C_{20} = Arachidic acid	—	—	—	—	≤1%	≤1%	≤2%	≤2%	—	—
$C_{20:1}$ = Eicosenoic acid	—	—	—	—	≤1%	≤1%	≤2%	≤2%	—	—

16 Regulatory Status

Lauroyl polyoxylglycerides and stearoyl polyoxylglycerides are approved as food additives in the US. Included in the FDA Inactive Ingredients Database (oral route: capsules, tablets, solutions; topical route: emulsions, creams, lotions; vaginal route: emulsions, creams). Oleyl polyoxylglycerides are included in a topical cream formulation licensed in the UK.

17 Related Substances

—

18 Comments

See Table XI for EINECS numbers for polyoxylglycerides.

19 Specific References

1 Bugaj A, *et al.* The effect of skin permeation enhancers on the formation of porphyrins in mouse skin during topical application of the methyl ester of 5-aminolevulinic acid. *J Photochem Photobiol B* 2006; **83**(2): 94–97.

2 Ceschel G, *et al.* Solubility and transdermal permeation properties of a dehydroepiandrosterone cyclodextrin complex from hydrophilic and lipophilic vehicles. *Drug Deliv* 2005; **12**(5): 275–280.

3 Cheong HA, Choi HK. Effect of ethanolamine salts and enhancers on the percutaneous absorption of piroxicam from a pressure sensitive adhesive matrix. *Eur J Pharm Sci* 2003; **18**(2): 149–153.

4 Jurkovic P, *et al.* Skin protection against ultraviolet induced free radicals with ascorbyl palmitate in microemulsions. *Eur J Pharm Biopharm* 2003; **56**(1): 59–66.

5 Kikwai L, *et al.* Effect of vehicles on the transdermal delivery of melatonin across porcine skin *in vitro*. *J Control Release* 2002; **83**(2): 307–311.

6 Kim J, *et al.* Effect of vehicles and pressure sensitive adhesives on the permeation of tacrine across hairless mouse skin. *Int J Pharm* 2000; **196**(1): 105–113.

7 Kreilgaard M. Influence of microemulsions on cutaneous drug delivery. *Bulletin technique Gattefossé* 2002; **95**: 79–100.

8 Kreilgaard M, *et al.* Influence of a microemulsion vehicle on cutaneous bioequivalence of a lipophilic model drug assessed by microdialysis and pharmacodynamics. *Pharm Res* 2001; **18**(5): 593–599.

9 Minghetti P, *et al.* Evaluation of *ex vivo* human skin permeation of genistein and daidzein. *Drug Deliv* 2006; **13**(6): 411–415.

10 Rhee YS, *et al.* Transdermal delivery of ketoprofen using microemulsions. *Int J Pharm* 2001; **228**(1–2): 161–170.

11 Zhao X, *et al.* Enhancement of transdermal delivery of theophylline using microemulsion vehicle. *Int J Pharm* 2006; **327**(1–2): 58–64.

12 Dingemanse J, *et al.* Pronounced effect of caprylocaproyl macrogolglycerides on nasal absorption of IS-159, a peptide serotonin 1B/1D-receptor agonist. *Clin Pharmacol Ther* 2000; **68**(2): 114–121.

13 Zhang Q, *et al.* Preparation of nimodipine-loaded microemulsion for intranasal delivery and evaluation on the targeting efficiency to the brain. *Int J Pharm* 2004; **275**(1–2): 85–96.

14 Aungst BJ, *et al.* Improved oral bioavailability of an HIV protease inhibitor using gelucire 44/14 and labrasol vehicles. *Bulletin technique Gattefossé* 1994; **87**: 49–54.

15 Chang RK, Shojaei AH. Effect of a lipoidic excipient on the absorption profile of compound UK 81252 in dogs after oral administration. *J Pharm Pharm Sci* 2004; **7**(1): 8–12.

16 Cirri M, *et al.* Liquid spray formulations of xibornol by using self-microemulsifying drug delivery systems. *Int J Pharm* 2007; **340**(1–2): 84–91.

17 Demirel M, *et al.* Formulation and *in vitro-in vivo* evaluation of piribedil solid lipid micro- and nanoparticles. *J Microencapsul* 2001; **18**(3): 359–371.

18 Devani M, *et al.* The emulsification and solubilisation properties of polyglycolysed oils in self-emulsifying formulations. *J Pharm Pharmacol* 2004; **56**(3): 307–316.

19 Djordjevic L, *et al.* Characterization of caprylocaproyl macrogolglycerides based microemulsion drug delivery vehicles for an amphiphilic drug. *Int J Pharm* 2004; **271**(1–2): 11–19.

20 Esposito E, *et al.* Amphiphilic association systems for amphotericin B delivery. *Int J Pharm* 2003; **260**(2): 249–260.

21 Hu Z, *et al.* A novel emulsifier, *Labrasol*, enhances gastrointestinal absorption of gentamicin. *Life Sci* 2001; **69**(24): 2899–2910.

22 Ito Y, *et al.* Oral solid gentamicin preparation using emulsifier and adsorbent. *J Control Release* 2005; **105**(1–2): 23–31.

23 Kim HJ, *et al.* Preparation and *in vitro* evaluation of self-microemulsifying drug delivery systems containing idebenone. *Drug Dev Ind Pharm* 2000; **26**(5): 523–529.

24 Kommuru TR, *et al.* Self-emulsifying drug delivery systems (SEDDS) of coenzyme Q10: formulation development and bioavailability assessment. *Int J Pharm* 2001; **212**(2): 233–246.

25 Rama Prasad YV, *et al.* Evaluation of oral formulations of gentamicin containing labrasol in beagle dogs. *Int J Pharm* 2003; **268**(1–2): 13–21.

26 Shen H, Zhong M. Preparation and evaluation of self-microemulsifying drug delivery systems (SMEDDS) containing atorvastatin. *J Pharm Pharmacol* 2006; **58**(9): 1183–1191.

27 Shibata N, *et al.* Application of pressure-controlled colon delivery capsule to oral administration of glycyrrhizin in dogs. *J Pharm Pharmacol* 2001; **53**(4): 441–447.

28 Subramanian N, *et al.* Formulation design of self-microemulsifying drug delivery systems for improved oral bioavailability of celecoxib. *Biol Pharm Bull* 2004; **27**(12): 1993–1999.

29 Venkatesan N, *et al. Gelucire* 44/14 and *Labrasol* in enhancing oral absorption of poorly absorbable drugs. *Bulletin technique Gattefossé* 2006; **99**: 79–88.

30 Venkatesan N, *et al.* Liquid filled nanoparticles as a drug delivery tool for protein therapeutics. *Biomaterials* 2005; **26**(34): 7154–7163.

31 Wei L, *et al.* Preparation and evaluation of SEDDS and SMEDDS containing carvedilol. *Drug Dev Ind Pharm* 2005; **31**(8): 775–784.

32 Yüksel N, *et al.* Enhanced bioavailability of piroxicam using *Gelucire 44/14* and *Labrasol*: *in vitro* and *in vivo* evaluation. *Eur J Pharm Biopharm* 2003; **56**(3): 453–459.

33 Shephard SE, *et al.* Pharmacokinetic behaviour of sublingually administered 8-methoxypsoralen for PUVA therapy. *Photodermatol Photoimmunol Photomed* 2001; **17**(1): 11–21.

34 Jannin V, *et al.* Approaches for the development of solid and semi-solid lipid-based formulations. *Adv Drug Deliv Rev* 2008; **60**(6): 734–746.

35 Chauhan B, *et al.* Preparation and evaluation of glibenclamide-polyglycolized glycerides solid dispersions with silicon dioxide by spray drying technique. *Eur J Pharm Sci* 2005; **26**(2): 219–230.

36 Aungst BJ, *et al.* Amphiphilic vehicles improve the oral bioavailability of a poorly soluble HIV protease inhibitor at high doses. *Int J Pharm* 1997; **156**(1): 79–88.

37 Barakat NS. Etodolac-liquid-filled dispersion into hard gelatin capsules: an approach to improve dissolution and stability of etodolac formulation. *Drug Dev Ind Pharm* 2006; **32**(7): 865–876.

38 Barker SA, *et al.* An investigation into the structure and bioavailability of α-tocopherol dispersions in *Gelucire 44/14*. *J Control Release* 2003; **91**(3): 477–488.

39 Bowtle W. Lipid formulations for oral drug delivery. *Pharm Technol Eur* 2000; **12**(9): 20–30.

40 Iwanaga K, *et al.* Disposition of lipid-based formulation in the intestinal tract affects the absorption of poorly water-soluble drugs. *Biol Pharm Bull* 2006; **29**(3): 508–512.

41 Kane A, *et al.* A statistical mixture design approach for formulating poorly soluble compounds in liquid filled hard shell capsules. *Bulletin technique Gattefossé* 2006; **99**: 43–49.

42 Khoo SM, *et al.* The formulation of Halofantrine as either non-solubilising PEG 6000 or solubilising lipid based solid dispersions: physical stability and absolute bioavailability assessment. *Int J Pharm* 2000; **205**(1–2): 65–78.

43 Schamp K, *et al.* Development of an *in vitro/in vivo* correlation for lipid formulations of EMD 50733, a poorly soluble, lipophilic drug substance. *Eur J Pharm Biopharm* 2006; **62**(3): 227–234.

44 Chambin O, Jannin V. Interest of multifunctional lipid excipients: case of *Gelucire 44/14*. *Drug Dev Ind Pharm* 2005; **31**(6): 527–534.

45 Yang D, *et al.* Effect of the melt granulation technique on the dissolution characteristics of griseofulvin. *Int J Pharm* 2007; **329**(1–2): 72–80.

46 Dordunoo SK. Sustained release liquid filled hard gelatin capsules in drug discovery and development: a small pharmaceutical company's perspectives. *Bulletin technique Gattefossé* 2004; **97**: 29–39.

47 Bravo González RC, *et al.* Improved oral bioavailability of cyclosporin a in male Wistar rats. Comparison of a *Solutol HS 15* containing self-dispersing formulation and a microsuspension. *Int J Pharm* 2002; **245**(1–2): 143–151.

48 Fernández-Carballido A, *et al.* Biodegradable ibuprofen-loaded PLGA microspheres for intraarticular administration. Effect of *Labrafil* addition on release *in vitro*. *Int J Pharm* 2004; **279**: 33–41.

P

49 Kang BK, *et al.* Controlled release of paclitaxel from microemulsion containing PLGA and evaluation of anti-tumor activity *in vitro* and *in vivo*. *Int J Pharm* 2004; **286**(1–2): 147–156.
50 Gupta MK, *et al.* Enhanced drug dissolution and bulk properties of solid dispersions granulated with a surface adsorbent. *Pharm Dev Technol* 2001; **6**(4): 563–572.
51 Craig DQM. Lipid matrices for sustained release – an academic overview. *Bulletin technique Gattefossé* 2004; **97**: 9–19.
52 Galal S, *et al.* Study of *in-vitro* release characteristics of carbamazepine extended release semisolid matrix filled capsules based on Gelucires. *Drug Dev Ind Pharm* 2004; **30**(8): 817–829.
53 Khan N, Craig DQ. The influence of drug incorporation on the structure and release properties of solid dispersions in lipid matrices. *J Control Release* 2003; **93**(3): 355–368.
54 Shimpi SL, *et al.* Stabilization and improved *in vivo* performance of amorphous etoricoxib using *Gelucire 50/13*. *Pharm Res* 2005; **22**(10): 1727–1734.
55 Ochoa L, *et al.* Preparation of sustained release hydrophilic matrices by melt granulation in a high-shear mixer. *J Pharm Pharm Sci* 2005; **8**(2): 132–140.
56 Schaefer T. Pelletisation with meltable binders. *Bulletin technique Gattefossé* 2004; **97**: 113–124.
57 Seo A, *et al.* The preparation of agglomerates containing solid dispersions of diazepam by melt agglomeration in a high shear mixer. *Int J Pharm* 2003; **259**(1–2): 161–171.
58 Passerini N, *et al.* Evaluation of melt granulation and ultrasonic spray congealing as techniques to enhance the dissolution of praziquantel. *Int J Pharm* 2006; **318**(1–2): 92–102.
59 Gattefossé. *Labrasol: GAT-89107*. Test to evaluate the acute oral toxicity following a single oral administration (LD_{50}) in rats, 1989.
60 Gattefossé. *Labrafil M2130CS: GAT-8815*. Test to evaluate oral toxicity following a single oral administration in the rat (limit test), 1988.

20 General References

Gattefossé. Technical literature: *Gelucire 44/14*, 2007.
Gattefossé. Technical literature: *Gelucire 50/13*, 2005.
Gattefossé. Technical literature: Oral route excipients, 2007.

21 Author

M Julien.

22 Date of Revision

2 March 2012.

Polyvinyl Acetate Dispersion

1 Nonproprietary Names

PhEur: Poly(Vinyl Acetate) Dispersion 30 per cent
USP–NF: Polyvinyl Acetate Dispersion

2 Synonyms

Kollicoat SR 30 D; poly(vinylis acetas) dispersio 30 per centum.

3 Chemical Name and CAS Registry Number

Aqueous dispersion of vinyl acetate homopolymer [900-20-7]

USP35–NF30 states that polyvinyl acetate dispersion contains 25–30% polyvinyl acetate. It may contain suitable surface active agents and stabilizers. Suitable such agents are sodium monododecyl sulfate and 1-vinyl-2-pyrrolidinone polymer.

4 Empirical Formula and Molecular Weight

Polyvinyl acetate$[C_4H_6O_2]_n$where n is between approx. 100 to 17 000 (molecular mass approx. 450 000).

5 Structural Formula

O
CH3
O
*
*
n

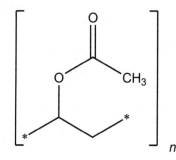

6 Functional Category

Coating agent; film-forming agent; modified-release agent.

7 Applications in Pharmaceutical Formulation or Technology

Aqueous polyvinyl acetate dispersion was developed as a film-coating agent for oral sustained-release applications. Most formulations require a plasticizer such as triethyl citrate, triacetin or propylene glycol, usually at a concentration of 5–10% based on the polymer weight.[(1)] Polyvinyl acetate films are insoluble in water (independent of the pH value) and act as an effective diffusion barrier. The permeability of the film can be adjusted by the addition of hydrophilic pore formers, such as povidone K-30 or polyethyleneglycol–polyyvinyl alcohol graft copolymer.[(2–7)] The films have a good mechanical stability and their self-healing properties help to reduce the risk of dose dumping.[(8,9)] Owing to these characteristics, polyvinyl acetate dispersion is highly suitable for manufacturing multiple unit particulate systems (MUPS).[(10)]

Polyvinyl acetate dispersion is also used for sustained-release matrix tablets by granulation and subsequent compression. The release profile depends on the solubility and particle size of the active ingredient, the amount of polyvinyl acetate, and the use of other excipients (hydrophilic fillers or pore formers) in the formulation.[(10–12)]

8 Description

Polyvinyl acetate dispersion occurs as an opaque, white or off-white, slightly viscous liquid. It is an aqueous colloidal dispersion of polyvinyl acetate, where the mean size of the polymer particles is approx. 170 nm.

9 Pharmacopeial Specifications

See Table I.

Table I: Pharmacopeial specifications for polyvinyl acetate dispersion.

Test	PhEur 7.4	USP35–NF30
Identification	+	+
Characters	+	–
Agglomerates	≤0.5 g	–
Coagulum content	–	≤0.5 g
Vinyl acetate	≤100 ppm	≤100 ppm
Povidone	≤4.0%	≤4.0%
Acetic acid	≤1.5%	≤1.5%
pH	–	3.0–5.5
Residue on evaporation	28.5–31.5%	–
Loss on drying	–	28.5–31.5%
Sulfated ash	≤0.5%	–
Residue on ignition	–	≤0.5%
Microbial contamination		
Aerobic bacteria	≤10^3 cfu/g	≤10^3 cfu/g
Fungi and yeast	≤10^2 cfu/g	≤10^2 cfu/g
Apparent viscosity	≤100 mPa s	–
Solubility of a film	+	–
Assay (polyvinyl acetate)	25.0–30.0%	25.0–30.0%

10 Typical Properties

Density 1.045–1.065 g/mL

Minimum film-forming temperature 18°C

Solubility Miscible with water, ethanol and acetone; the polymer first precipitates on adding less hydrophilic solvents, but dissolves again on adding further solvent.

Viscosity <100 mPa s at 20°C; low viscosity very similar to water, and is stable against sedimentation.[13,14]

11 Stability and Storage Conditions

Polyvinyl acetate dispersion should be stored between 5°C and 25°C in a tightly closed container. Exposure to frost or high shear forces will cause an irreversible phase separation.

12 Incompatibilities

The addition of methacrylic acid–ethyl acrylate (1 : 1) copolymer dispersion will cause an irreversible coagulation. Alkaline substances (i.e. strong bases) favor the hydrolysis of the polymer.

As polyvinyl acetate films are soluble in aqueous sodium lauryl sulfate solutions,[15] this surfactant cannot be used as a solubilizer in the dissolution medium for poorly soluble drugs.

13 Method of Manufacture

Polyvinyl acetate dispersion is manufactured by emulsion polymerization of the monomer vinyl acetate.

14 Safety

Polyvinyl acetate dispersion is used in oral pharmaceutical preparations and is generally regarded as safe, being based on polyvinyl acetate, the safety of which is documented by a variety of studies.[16–22]

15 Handling Precautions

Observe normal precautions appropriate to the circumstances and quantity of the material handled.

16 Regulatory Status

Included in the FDA Inactive Ingredients Database (sustained-release tablets, transdermal films and patches).

17 Related Substances

Polyvinyl acetate.

Polyvinyl acetate

Comments Listed in PhEur 7.4 and USP35–NF30. Included in the Japanese Standard for Food Additives. Included in the Canadian Natural Health Products Ingredients Database.

18 Comments

Polyvinyl acetate dispersion is used as a base for chewing gum; it has also been used to coat fruits and vegetables in Asian countries.

Aqueous sodium lauryl sulfate solutions may be used to clean coating equipment.

A specification for polyvinyl acetate is contained in *Japanese Pharmaceutical Excipients* (JPE).[23]

19 Specific References

1 BASF SE. Technical information: *Kollicoat SR 30 D*, 2009.

2 Ensslin S, *et al.* New insight into modified release pellets – Internal structure and drug release mechanism. *J Control Release* 2008; **128**: 149–156.

3 Ensslin S, *et al.* Modulating pH-independent release from coated pellets: effect of coating composition on solubilization processes and drug release. *Eur J Pharm Biopharm* 2009; **72**: 111–118.

4 Strübing S, *et al.* Mechanistic analysis of drug release from tablets with membrane controlled drug delivery. *Eur J Pharm Biopharm* 2007; **66**: 113–119.

5 Strübing S, *et al.* Monitoring of dissolution induced changes in film coat composition by 1 H NMR spectroscopy and SEM. *J Control Release* 2007; **119**: 190–196.

6 Dashevsky A, *et al.* Effect of water-soluble polymers on the physical stability of aqueous polymeric dispersions and the implications on the drug release from coated pellets. *Drug Dev Ind Pharm* 2010; **36**(2): 152–160.

7 Ho L, *et al.* Effects of film coating thickness and drug layer uniformity on *in vitro* release from sustained-release coated pellets: a case study using terahertz pulsed imaging. *Int J Pharm* 2009; **382**: 151–159.

8 Kolter K, Gerbert S. Coated drug delivery systems based on Kollicoat SR 30 D. APV/APGI *4th World Meeting on Pharmaceutics, Biopharmaceutics and Pharmaceutical Technology*, Italy 2002.

9 Ensslin S, *et al.* Safety and robustness of coated pellets: self-healing film properties and storage stability. *Pharm Res* 2009; **26**(6): 1534–1543.

10 Dashevsky A, *et al.* Compression of pellets coated with various aqueous polymer dispersions. *Int J Pharm* 2004; **279**: 19–26.

11 Bordaweka MS, *et al.* Evaluation of polyvinyl acetate dispersion as a sustained release polymer for tablets. *Drug Deliv* 2006; **13**: 121–131.

12 Al-Zoubi N, *et al.* Sustained-release of buspirone HCl by co spray-drying with aqueous polymeric dispersions. *Eur J Pharm Biopharm* 2008; **69**: 735–742.

13 Dashevsky A, *et al.* Physicochemical and release properties of pellets coated with *Kollicoat SR* 30, a new aqueous polyvinyl acetate dispersion for extended release. *Int J Pharm* 2005; **290**: 15–23.

14 Dashevsky A, *et al.* pH-independent release of a basic drug from pellets coated with the extended release polymer dispersion *Kollicoat SR 30 D* and the enteric polymer dispersion *Kollicoat MAE 30 DP*. *Eur J Pharm Biopharm* 2004; **58**: 45–49.

15 Myers D, ed. Polymeric surfactants and surfactant–polymer interactions. In: *Surfactant Science and Technology*, 3rd edn. Hoboken, New Jersey: John Wiley & Sons, 2006: 242.

16 Scherbak BI *et al. Uch Zap-Mosk Nauch Issled Inst Gig* 1975; 22: 74-80.(Cited in IARC *Monographs on the Evaluation of the Carcinogenic Risk of Chemicals to Humans* 1979; 19: 351).

17 Weeks MH, Pope CR. *Toxicological Evaluation of Polyvinyl Acetate (Pva) Emulsion Dust Control Material*, May 1973–March 1974 (AD 784603). US Army Environmental Hygiene Agency, Aberdeen, March 1974.

18 Hood DB. In: Clayton GD, Clayton FE, eds. *Patty's Industrial Hygiene and Toxicology*, 4th edn, vol 2, Part E. John Wiley & Sons, 1994; 3799.

19 Carpenter WM, *et al.* Histocompatibility of polyvinyl acetate, an ingredient of chewing gum. *Oral Surg Oral Med Oral Pathol* 1976; **42**: 461–469.

20 Ishidate MJJr, *et al.* Primary mutagenicity screening of food additives currently used in Japan. *Food Chem Toxicol* 1984; **22**: 623–636.

21 Nothdurft H. Experimental development of sarcoma by means of enclosed foreign bodies. *Strahlentherapie* 1956; **100**: 192–210(Suppl. [in German]): .

22 Miyasaki K. Experimental polymer storage disease in rabbits. An approach to the histogenesis of sphingolipidoses. *Virchows Arch A Pathol Anat Histol* 1975; **365**: 351–365.

23 Japan Pharmaceutical Excipients Council. *Japanese Pharmaceutical Excipients 2004*. Tokyo: Yakuji Nippo, 2004: 706.

20 General References

BASF SE. Excipients. http://www.pharma-ingredients.basf.com/excipients.aspx (accessed 6 July 2011).

Bühler V. *Kollicoat* Grades: Functional polymers for the pharmaceutical industry. BASF SE, January 2007.

McGinity JW. Aqueous Polymeric Coatings for Pharmaceutical Dosage Forms, 2nd edn. New York: Marcel Dekker, 1997.

21 Authors

F Guth, T Schmeller.

22 Date of Revision

1 March 2012.

Polyvinyl Acetate Phthalate

1 Nonproprietary Names

USP–NF: Polyvinyl Acetate Phthalate

2 Synonyms

Phthalavin; PVAP; *Opaseal*; *Sureteric*.

3 Chemical Name and CAS Registry Number

Polyvinyl acetate phthalate [34481-48-6]

4 Empirical Formula and Molecular Weight

The USP35–NF30 describes polyvinyl acetate phthalate as a reaction product of phthalic anhydride and a partially hydrolyzed polyvinyl acetate. It contains not less than 55.0% and not more than 62.0% of phthalyl (*o*-carboxybenzoyl, $C_8H_5O_3$) groups, calculated on an anhydrous acid-free basis.

It has been reported that the free phthalic acid content is dependent on the source of the material.[1]

5 Structural Formula

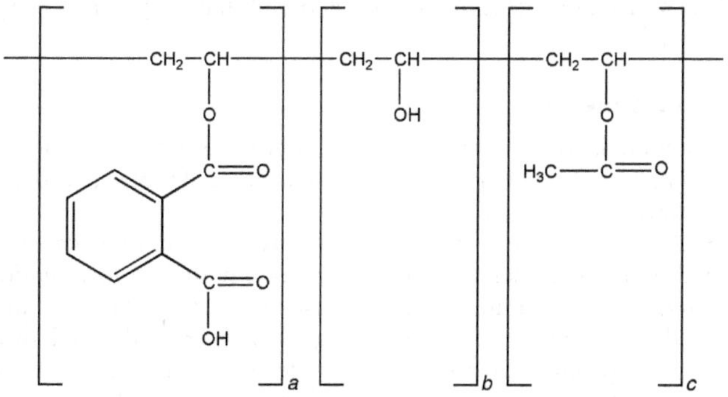

Depending on the phthalyl content, *a* will vary with *b* in mole percent. The acetyl content *c* remains constant depending on the starting material.

6 Functional Category

Coating agent; film-forming agent.

7 Applications in Pharmaceutical Formulation or Technology

Polyvinyl acetate phthalate is a viscosity-modifying agent that is used in pharmaceutical formulations to produce enteric coatings for products and for the core sealing of tablets prior to a sugar-coating process. Polyvinyl acetate phthalate does not exhibit tackiness during coating and produces strong robust films.

Plasticizers are often included in polyvinyl acetate phthalate coating formulations to enable a continuous, homogeneous, noncracking film to be produced. Polyvinyl acetate phthalate has been shown to be compatible with several plasticizers such as glyceryl triacetate, triethyl citrate, acetyl triethylcitrate, diethyl phthalate and polyethylene glycol 400.

For enteric coating applications, polyvinyl acetate phthalate is dissolved in a solvent system together with other additives such as diethyl phthalate and stearic acid. Methanol may be used as the solvent if a colorless film is required; for a colored film, methanol or ethanol/water may be used depending on the amount of pigment to be incorporated. A weight increase of up to 8% is necessary for nonpigmented systems, whereas for pigmented systems a weight increase of 6% is usually required. Hot-melt extrusion of coating polymers, such as polyvinyl acetate phthalate, has been described for the enteric coating of capsules.[2]

Polyvinyl acetate phthalate has superseded materials such as shellac in producing the initial layers of coating (the sealing coat) in the sugar coating process for tablets. The sealing coating should be kept as thin as possible while providing an adequate barrier to moisture, a balance that is often difficult to achieve in practice. A solvent system containing a high proportion of industrial methylated spirits and other additives can be used. Two coats are usually sufficient to seal most tablets, although up to five may be necessary for tablets containing alkaline ingredients. If an enteric coating is also required, between six and 12 coats may be necessary; *see* Table I.

The properties of polyvinyl acetate phthalate enteric coating have been compared with those of other enteric polymers such as cellulose acetate phthalate.[3,4] The factors that affect the release kinetics from polyvinyl acetate phthalate enteric-coated tablets have also been described.[5] A method for enteric coating hypromellose capsules that avoids the sealing step prior to coating has been developed. The properties of several enteric coating polymers, including polyvinyl acetate phthalate, were assessed.[6]

8 Description

Polyvinyl acetate phthalate occurs as a free-flowing white to off-white powder and may have a slight odor of acetic acid. The material is essentially amorphous.[7]

Table I: Uses of polyvinyl acetate phthalate.

Use	Concentration (%)
Tablet enteric film coating	9–10
Tablet sealant (sugar-coating)	28–29

Table II: Pharmacopeial specifications for polyvinyl acetate phthalate.

Test	USP35–NF30
Identification	+
Apparent viscosity at 25°C	7–11 cP
Water	≤5.0%
Residue on ignition	≤1.0%
Free phthalic acid	≤0.6%
Free acid other than phthalic	≤0.6%
Phthalyl content	55.0–62.0%

9 Pharmacopeial Specifications

See Table II.

10 Typical Properties

The characteristics of polyvinyl acetate phthalate from two sources have been compared; values for molecular weight (60 700; 47 000), moisture content (3.74%; 2.20%) and density (1.31 g/cm^3; 1.37 g/cm^3) have been reported. The solubility of each polyvinyl acetate phthalate in a range of different solvents was described and scanning electron photomicrographs were produced to give evidence of the different polymer morphology.[8]

Glass transition temperature A glass transition temperature of 42.5°C has been reported for polyvinyl acetate phthalate; the glass transition temperature was shown to fall with the addition of increasing amounts of the plasticizer diethyl phthalate.[7]

Solubility Soluble in ethanol and methanol; sparingly soluble in acetone and propan-2-ol; practically insoluble in chloroform, dichloromethane, and water. In buffer solutions, polyvinyl acetate phthalate (200 mg/L) is insoluble below pH 5 and becomes soluble at pH values above 5. Polyvinyl acetate pththalate shows a sharp solubility response with pH; this occurs at pH 4.5–5.0, which is lower than for most other polymers used for enteric coatings. Solubility is also influenced by ionic strength. *See* Table III.

Table III: Solubility of polyvinyl acetate phthalate.

Solvent	Solubility at 25°C
Acetone/ethanol (1 : 1 w/w)	1 in 3
Acetone/methanol (1 : 1 w/w)	1 in 4
Ethanol (95%)	1 in 4
Methanol	1 in 2
Methanol/dichloromethane (1 : 1 w/w)	1 in 3

Viscosity (dynamic) The viscosity of a solution of polyvinyl acetate phthalate/methanol (1 : 1) is 5000 mPa s. In methanol/dichloromethane systems, viscosity increases as the concentration of methanol in the system increases.

11 Stability and Storage Conditions

Polyvinyl acetate phthalate should be stored in airtight containers. It is relatively stable to temperature and humidity, and does not age, giving predictable release profiles even after prolonged storage.

At high temperature and humidity, polyvinyl acetate phthalate undergoes less hydrolysis than other commonly used enteric coating polymers. In aqueous colloidal dispersions of polyvinyl acetate phthalate, the formation of free phthalic acid through hydrolysis was found to adversely affect physical stability.[1] *See also* Section 18.

12 Incompatibilities

Polyvinyl acetate phthalate reacts with povidone to form an insoluble complex that precipitates out of solution;[9] benzocaine is also incompatible with polyvinyl acetate phthalate.[10] Erythromycin disperses in polyvinyl acetate phthalate and has been shown to be physically stable[11] while omeprazole exists in the amorphous form in polyvinyl acetate phthalate coatings with no evidence of interaction.[12]

13 Method of Manufacture

Polyvinyl acetate phthalate is a reaction product of phthalic anhydride, sodium acetate, and a partially hydrolyzed polyvinyl alcohol. The polyvinyl alcohol is a low molecular weight grade, and 87–89 mole percent is hydrolyzed. Therefore, the polyvinyl acetate phthalate polymer is a partial esterification of a partially hydrolyzed polyvinyl acetate.

See also Section 4.

14 Safety

Polyvinyl acetate phthalate is used in oral pharmaceutical formulations and is generally regarded as an essentially nonirritant and nontoxic material when used as an excipient.

15 Handling Precautions

Observe normal precautions appropriate to the circumstances and quantity of material handled. Gloves and eye protection are recommended.

16 Regulatory Status

Included in the FDA Inactive Ingredients Database (enteric-coated and sustained-action oral tablets). Included in nonparenteral medicines (enteric coated tablets; in printing ink formulations used for oral tablets and capsules) licensed in Europe. Included in the Canadian Natural Health Products Ingredients Database.

17 Related Substances

Cellulose acetate phthalate; hypromellose phthalate; polymethacrylates; shellac.

18 Comments

A formulated, aqueous-based coating solution (*Sureteric*, Colorcon) is available commercially for the enteric coating of tablets, hard and soft gelatin capsules and granules.

Following storage at room temperature for 9 months, capsules coated with a commercial polyvinyl acetate phthalate formulation (*Coateric*) were found to retain gastroresistant properties and showed no apparent physical change; however, a delayed drug dissolution profile was observed after storage. Storage at 37°C, or 37°C and 80% relative humidity for 3 months resulted in capsules having an unsatisfactory appearance.[4]

Polyvinyl acetate phthalate has been investigated as an enteric concentration enhancing polymer and has been found to improve the oral bioavailability of itraconazole from solid dispersions.[13]

19 Specific References

1 Davis MB. Preparation and stability of aqueous-based enteric polymer dispersions. *Drug Dev Ind Pharm* 1986; **12**(10): 1419–1448.
2 Mehuys E, *et al*. Production of enteric capsules by means of hot-melt extrusion. *Eur J Pharm Sci* 2007; **24**(2–3): 207–212.
3 Porter SC, Ridgway K. The permeability of enteric coatings and the dissolution rates of coated tablets. *J Pharm Pharmacol* 1982; **34**: 5–8.

4 Murthy KS, *et al.* A comparative evaluation of aqueous enteric polymers in capsule coatings. *Pharm Technol* 1986; **10**(10): 36–44.
5 Ozturk SS, *et al.* Kinetics of release from enteric-coated tablets. *Pharm Res* 1988; **5**(9): 550–565.
6 Huyghebaert N, *et al.* Alternative method for enteric coating of HPMC capsules resulting in ready-to-use enteric-coated capsules. *Eur J Pharm Sci* 2004; **21**(5): 617–623.
7 Porter SC, Ridgway K. An evaluation of the properties of enteric coating polymers: measurement of glass transition temperature. *J Pharm Pharmacol* 1983; **35**: 341–344.
8 Nesbitt RU, *et al.* Evaluation of polyvinyl acetate phthalate as an enteric coating material. *Int J Pharm* 1985; **26**: 215–226.
9 Kumar V, *et al.* Interpolymer complexation I: preparation and characterization of a polyvinyl acetate phthalate–polyvinylpyrrolidone (PVAP-PVP) complex. *Int J Pharm* 1999; **188**: 221–232.
10 Kumar V, Banker GS. Incompatibility of polyvinyl acetate phthalate with benzocaine: isolation and characterization of 4-phthalimidobenzoic acid ethyl ester. *Int J Pharm* 1992; **79**: 61–65.
11 Sarisuta N, *et al.* Physico-chemical characterization of interactions between erythromycin and various film polymers. *Int J Pharm* 1999; **186**: 109–118.
12 Sarisuta N, Kumpugdee M. Crystallinity of omeprazole in various film polymers. *Pharm Pharmacol Commun* 2000; **6**: 7–11.
13 DiNunzio JC, *et al.* Amorphous compositions using concentration enhancing polymers for improved bioavailability of itraconazole. *Mol Pharm* 2008; **5**(6): 968–980.

20 General References

—

21 Author

CG Cable.

22 Date of Revision

1 March 2012.

Polyvinyl Alcohol

1 Nonproprietary Names

PhEur: Poly(Vinyl Alcohol)
USP–NF: Polyvinyl Alcohol

2 Synonyms

Airvol; *Alcotex*; *Celvol*; *Elvanol*; *Gelvatol*; *Gohsenol*; *Kollicoat*; *Lemol*; *Mowiol*; poly(alcohol vinylicus); *Polyvinol*; PVA; vinyl alcohol polymer.

3 Chemical Name and CAS Registry Number

Ethenol, homopolymer [9002-89-5]

4 Empirical Formula and Molecular Weight

$(C_2H_4O)_n$ 20 000–200 000

Polyvinyl alcohol is a water-soluble synthetic polymer represented by the formula $(C_2H_4O)_n$. The value of n for commercially available materials lies between 500 and 5000, equivalent to a molecular weight range of approximately 20 000–200 000; *see* Table I.

Table I: Commercially available grades of polyvinyl alcohol.

Grade	Molecular weight
High viscosity	≈200 000
Medium viscosity	≈130 000
Low viscosity	≈20 000

5 Structural Formula

$$\left[-CH_2-\underset{\displaystyle OH}{\underset{|}{CH}}- \right]_n$$

6 Functional Category

Coating agent; emulsion stabilizing agent; film-forming agent; lubricant; viscosity-increasing agent.

7 Applications in Pharmaceutical Formulation or Technology

Polyvinyl alcohol is used primarily in topical pharmaceutical and ophthalmic formulations; *see* Table II.[1–3] It is used as a stabilizing agent for emulsions (0.25–3.0% w/v). Polyvinyl alcohol is also used as a viscosity-increasing agent for viscous formulations such as ophthalmic products. It is used in artificial tears and contact lens solutions for lubrication purposes, in sustained-release formulations for oral administration,[4] and in transdermal patches.[5] Polyvinyl alcohol may be made into microspheres when mixed with a glutaraldehyde solution[6] and is a commonly used emulsifier in the formulation of polylactide and poly(DL-lactide-*co*-glycolide) (PLGA) polymeric nanoparticles.[7–9]

Polyvinyl alcohol is also used in coating formulations for tablets as the film forming polymer.[10]

Table II: Uses of polyvinyl alcohol.

Use	Concentration (%)
Emulsions	0.5
Ophthalmic formulations	0.25–3.00
Topical lotions	2.5

8 Description

Polyvinyl alcohol occurs as an odorless, white to cream-colored granular powder.

9 Pharmacopeial Specifications

See Table III.

Table III: Pharmacopeial specifications for polyvinyl alcohol.

Test	PhEur 7.4	USP35–NF30
Identification	+	+
Characters	+	—
Appearance of solution	+	—
Viscosity	85.0–115.0%	85.0–115.0%
pH	4.5–6.5	5.0–8.0
Acid value	≤3.0	≤3.0
Ester value	90.0–110.0%	—
Heavy metals	≤10 ppm	≤10 μg/g
Loss on drying	≤5.0%	≤5.0%
Residue on ignition	—	≤1.0%
Sulfated ash	≤1.0%	—
Water-insoluble substances	—	≤0.1%
Degree of hydrolysis	—	+
Methanol	—	≤1.0%
Methyl acetate	—	≤1.0%

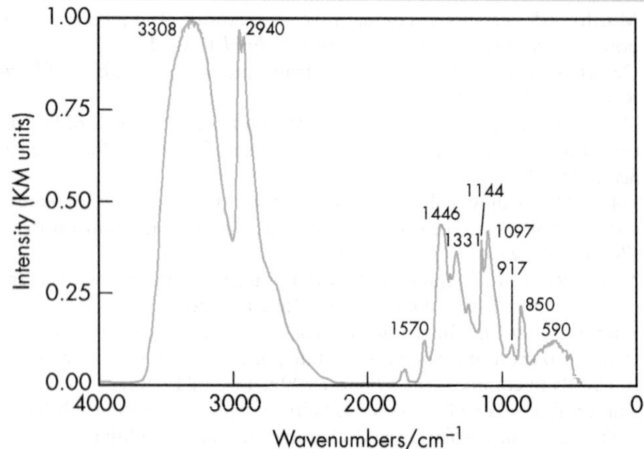

Figure 1: Infrared spectrum of polyvinyl alcohol measured by diffuse reflectance. Adapted with permission of Informa Healthcare.

10 Typical Properties

Melting point

228°C for fully hydrolyzed grades;

180–190°C for partially hydrolyzed grades.

Refractive index n_D^{25} = 1.49–1.53

Solubility Soluble in water; slightly soluble in ethanol (95%); insoluble in organic solvents. Dissolution requires dispersion (wetting) of the solid in water at room temperature followed by heating the mixture to about 90°C for approximately 5 minutes. Mixing should be continued while the heated solution is cooled to room temperature.

Specific gravity

1.19–1.31 for solid at 25°C;

1.02 for 10% w/v aqueous solution at 25°C.

Specific heat 1.67 J/g (0.4 cal/g)

Spectroscopy

IR spectra *see* Figure 1.

Viscosity (dynamic) *see* Table IV.

Table IV: Viscosity of commercial grades of polyvinyl alcohol.

Grade	Dynamic viscosity of 4% w/v aqueous solution at 20°C (mPa s)
High viscosity	40.0–65.0
Medium viscosity	21.0–33.0
Low viscosity	4.0–7.0

11 Stability and Storage Conditions

Polyvinyl alcohol is stable when stored in a tightly sealed container in a cool, dry place. Aqueous solutions are stable in corrosion-resistant sealed containers. Preservatives may be added to the solution if extended storage is required. Polyvinyl alcohol undergoes slow degradation at 100°C and rapid degradation at 200°C; it is stable on exposure to light.

12 Incompatibilities

Polyvinyl alcohol undergoes reactions typical of a compound with secondary hydroxy groups, such as esterification. It decomposes in strong acids, and softens or dissolves in weak acids and alkalis. It is incompatible at high concentration with inorganic salts, especially sulfates and phosphates; precipitation of polyvinyl alcohol 5% w/v can be caused by phosphates. Gelling of polyvinyl alcohol solution may occur if borax is present.

13 Method of Manufacture

Polyvinyl alcohol is produced through the hydrolysis of polyvinyl acetate. The repeating unit of vinyl alcohol is not used as the starting material because it cannot be obtained in the quantities and purity required for polymerization purposes. The hydrolysis proceeds rapidly in methanol, ethanol, or a mixture of alcohol and methyl acetate, using alkalis or mineral acids as catalysts.

14 Safety

Polyvinyl alcohol is generally considered a nontoxic material. It is nonirritant to the skin and eyes at concentrations up to 10%; concentrations up to 7% are used in cosmetics.

Studies in rats have shown that polyvinyl alcohol 5% w/v aqueous solution injected subcutaneously can cause anemia and infiltrate various organs and tissues.[11]

LD_{50} (mouse, oral): 14.7 g/kg

LD_{50} (rat, oral): >20 g/kg

15 Handling Precautions

Observe normal precautions appropriate to the circumstances and quantity of material handled. Eye protection and gloves are recommended. Polyvinyl alcohol dust may be an irritant on inhalation. Handle in a well-ventilated environment.

16 Regulatory Status

Included in the FDA Inactive Ingredients Database (ophthalmic preparations and oral tablets). Included in nonparenteral medicines licensed in the UK. Included in the Canadian Natural Health Products Ingredients Database.

17 Related Substances

—

18 Comments

Various grades of polyvinyl alcohol are commercially available. The degree of polymerization and the degree of hydrolysis are the two determinants of their physical properties. Pharmaceutical grades are partially hydrolyzed materials and are named according to a coding system. The first number following a trade name refers to the degree of hydrolysis and the second set of numbers indicates the approximate viscosity (dynamic), in mPa s, of a 4% w/v aqueous solution at 20°C.

A specification for polyvinyl alcohol is contained in the *Food Chemicals Codex* (FCC).[12]

19 Specific References

1 Krishna N, Brow F. Polyvinyl alcohol as an ophthalmic vehicle: effect on regeneration of corneal epithelium. *Am J Ophthalmol* 1964; **57**: 99–106.
2 Patton TF, Robinson JR. Ocular evaluation of polyvinyl alcohol vehicle in rabbits. *J Pharm Sci* 1975; **64**: 1312–1316.
3 Anonymous. New method of ocular drug delivery launched. *Pharm J* 1993; **250**: 174.
4 Carstensen JT, *et al.* Bonding mechanisms and hysteresis areas in compression cycle plots. *J Pharm Sci* 1981; **70**: 222–223.
5 Wan LSC, Lim LY. Drug release from heat-treated polyvinyl alcohol films. *Drug Dev Ind Pharm* 1992; **18**: 1895–1906.
6 Thanoo BC, *et al.* Controlled release of oral drugs from crosslinked polyvinyl alcohol microspheres. *J Pharm Pharmacol* 1993; **45**: 16–20.
7 Sahoo SK, *et al.* Residual polyvinyl alcohol associated with poly(D,L-lactide-*co*-glycolide) nanoparticles affects their physical properties and cellular uptake. *J Control Rel* 2002; **82**(1): 105–114.
8 Westedt U, *et al.* Poly(vinyl alcohol)-graft-poly(lactide-*co*-glycolide) nanoparticles for local delivery of paclitaxel for restenosis treatment. *J Control Rel* 2007; **119**: 41–51.
9 Lee SC, *et al.* Quantitative analysis of polyvinyl alcohol on the surface of poly(D,L-lactide-*co*-glycolide) microparticles prepared by solvent evaporation method: effect of particle size and PVA concentration. *J Control Rel* 1999; **59**(2): 123–132.
10 Rajabi-Siahboomi AR, Farrell TP. *Drugs and the Pharmaceutical Sciences* (Aqueous ploymeric coatings for pharmaceutical dosage forms (3rd edition)). Wiley 2008.
11 Hall CE, Hall O. Polyvinyl alcohol: relationship of physicochemical properties to hypertension and other pathophysiologic sequelae. *Lab Invest* 1963; **12**: 721–736.
12 *Food Chemicals Codex*, 7th edn. Bethesda, MD: United States Pharmacopeia, 2010: 832.

20 General References

Chudzikowski R. Polyvinyl alcohol. *Manuf Chem Aerosol News* 1970; **41**(7): 31–37.
Finch CA, ed. *Polyvinyl Alcohol Developments*. Chichester: Wiley, 1992.

21 Author

O AbuBaker.

22 Date of Revision

1 March 2012.

Potassium Alginate

1 Nonproprietary Names

USP–NF: Potassium Alginate

2 Synonyms

Alginic acid, potassium salt; E402; *Improved Kelmar*; potassium polymannuronate.

3 Chemical Name and CAS Registry Number

Potassium alginate [9005-36-1]

4 Empirical Formula and Molecular Weight

$(C_6H_7O_6K)_n$

Potassium alginate is the potassium salt of alginic acid, a polyuronide made up of a sequence of two hexuronic acid residues, namely D-mannuronic acid and L-guluronic acid. The two sugars form blocks of up to 20 units along the chain, with the proportion of the blocks dependent on the species of seaweed and also the part of the seaweed used. The number and length of the blocks is important in determining the physical properties of the alginate produced; the number and sequence of the mannuronate and guluronate residues varies in the naturally occurring alginate.

The USP35–NF30 describes potassium alginate as consisting chiefly of the potassium salt of alginic acid, a linear glycuronoglycan consisting of β-1,4 linked D-mannuronic acid and L-guluronic acid units in the pyranose form.

5 Structural Formula

See Section 4.

6 Functional Category

Emulsifying agent; gelling agent; suspending agent; viscosity-increasing agent.

7 Applications in Pharmaceutical Formulation or Technology

Potassium alginate is widely used in foods as a stabilizer, thickener, and emulsifier; however, its use as a pharmaceutical excipient is currently limited to experimental hydrogel systems. The viscosity, adhesiveness, elasticity, stiffness, and cohesiveness of potassium alginate hydrogels have been determined and compared with values from a range of other hydrogel-forming materials.(1) The effect of calcium ions on the rheological properties of procyanidin hydrogels containing potassium alginate and intended for oral administration has also been investigated.(2)

8 Description

Potassium alginate occurs as a white to yellowish, fibrous or granular powder; it is almost odorless and tasteless.

9 Pharmacopeial Specifications

See Table I.

Table I: Pharmacopeial specifications for potassium alginate.

Test	USP35–NF30
Identification	+
Microbial limits	
Total aerobic count	$\leq 10^3$ cfu/g
Total combined molds and yeasts	$\leq 10^2$ cfu/g
Loss on drying	≤15%
Total ash	24–32%
Arsenic	1.5 ppm
Lead	≤10 ppm
Heavy metals	≤40 ppm
Assay	89.2–105.5%

P

10 Typical Properties

Particle size distribution Average particle size ≈150 μm (*Improved Kelmar*).

Solubility Potassium alginate is soluble in water, dissolving to form a viscous hydrophilic colloidal solution. It is insoluble in ethanol (95%) and in hydroalcoholic solutions in which the alcohol content is greater than 30% by weight; also insoluble in chloroform, ether, and acids having a pH lower than about 3. When preparing solutions of potassium alginate it is important to ensure proper dispersion of the particles, as poor dispersion will lead to the formation of large lumps of unhydrated powder and significantly extended hydration times.

Viscosity (dynamic)

400 mPa s (for a 1% dispersion of *Improved Kelmar*). Vicosities of 4.32×10^3 mPa s (2.5% dispersion) and 31.1×10^3 mPa s (4% dispersion) have been reported.[1]

Potassium alginate hydrates readily in hot or cold water; in solution, the acid groups of the alginate become ionized and a viscous solution is obtained. The viscosity is proportional to the concentration and molecular weight of the material used. As the temperature rises, a reversible decrease in viscosity occurs. The addition of calcium ions to potassium alginate solutions results in crosslinking and in the formation of gels; where the crosslinks formed are strong and numerous, the gel becomes thermally irreversible.

11 Stability and Storage Conditions

In the solid state, potassium alginate is a stable material that is not prone to microbial spoilage. Over time, a slow reduction in the degree of polymerization can occur, which may be reflected in a reduction in the viscosity of solutions. As both temperature and moisture can impair the performance of potassium alginate, storage below 25°C is recommended.

Potassium alginate solutions are stable at pH 4–10; long-term storage outside this range can result in depolymerization of the polymer through hydrolysis. Gelation or precipitation of the alginate can occur at pH values less than 4. Liquid or semisolid alginate formulations should be preserved: suitable preservatives are sodium benzoate, potassium sorbate, or parabens.

Potassium alginate should be stored under cool, dry conditions in a well-closed container.

12 Incompatibilities

Incompatible with strong oxidizers.

13 Method of Manufacture

Alginate obtained from brown seaweed is subjected to demineralization, extraction, and precipitation of alginic acid. Following neutralization, the potassium alginate obtained is dried and milled.

14 Safety

Potassium alginate is widely used in food products. It is currently used as an excipient only in experimental pharmaceutical formulations.

15 Handling Precautions

Observe normal precautions appropriate to the circumstances and quantity of material handled. When heated to decomposition, potassium alginate emits acrid smoke and irritating fumes. Potassium alginate may be irritant to the skin, eyes and lungs. Gloves, eye protection, suitable protective clothing, and respiratory equipment should be worn.

16 Regulatory Status

GRAS listed. Accepted for use in foods in the US and Europe. Included in the Canadian Natural Health Products Ingredients Database.

17 Related Substances

Alginic acid; ammonium alginate; calcium alginate; propylene glycol alginate; sodium alginate.

18 Comments

Studies on elastic alginates, such as potassium alginate and sodium alginate, have shown that they may be more appropriate for tableting pressure-sensitive materials than microcrystalline cellulose.[3]

A specification for potassium alginate is contained in the *Food Chemicals Codex* (FCC).[3]

19 Specific References

1 Vennat B, *et al.* Comparative texturometric analysis of hydrogels based on cellulose derivatives, carraghenates and alginates. Evaluation of adhesiveness. *Drug Dev Ind Pharm* 1998; **24**(1): 27–35.

2 Vennat B, *et al.* Procyanidin hydrogels. Influence of calcium on the gelling of alginate solutions. *Drug Dev Ind Pharm* 2003; **20**(17): 2707–2714.

3 Schmid W, Picker-Freyer KM. Tableting and tablet properties of alginates: characterisation and potential for Soft Tableting. *Eur J Pharm Biopharm* 2009; 72(1): 165–172.

4 *Food Chemicals Codex*, 7th edn. Bethesda, MD: United States Pharmacopeia, 2010: 840.

20 General References

—

21 Author

CG Cable.

22 Date of Revision

1 March 2012.

Potassium Alum

1 Nonproprietary Names

BP: Alum
JP: Aluminum Potassium Sulfate Hydrate
PhEur: Alum
USP–NF: Potassium Alum

2 Synonyms

Alumen; alum flour; alum meal; alum potassium; aluminum potassium alum; aluminum potassium disulfate; aluminum potassium sulfate; dialuminum dipotassium sulfate; kalinite; potash alum; potassium aluminum sulfate (1 : 1 : 2); potassium aluminum sulfate-12-hydrate; rock alum; sulfuric acid aluminum potassium salt (2 : 1 : 1) dodecahydrate.

3 Chemical Name and CAS Registry Number

Aluminum potassium sulfate anhydrous [10043-67-1]
Aluminum potassium sulfate dodecahydrate [7784-24-9]

4 Empirical Formula and Molecular Weight

$AlK(SO_4)_2$ 258.21 (for anhydrous)
$AlK(SO_4)_2 \cdot 12H_2O$ 474.39 (for dodecahydrate)

5 Structural Formula

See Section 4.

6 Functional Category

Vaccine adjuvant.

7 Applications in Pharmaceutical Formulation or Technology

Potassium alum has the ability to precipitate proteins and is utilized in the manufacture of vaccines, where purified proteins are coprecipitated with and adsorbed onto potassium alum.[1,2]

Potassium alum is often included in preparations used as mouthwashes or gargles and in dermatological preparations.

8 Description

The PhEur 7.4 describes potassium alum as a granular powder, or colorless, transparent, crystalline masses. The JP XV describes it as colorless or white crystals or powder. Potassium alum is odorless and has a slightly sweet, strongly astringent taste.

9 Pharmacopeial Specifications

See Table I.

Table I: Pharmacopeial specifications for potassium alum.

Test	JP XV	PhEur 7.4	USP35–NF30
Identification	+	+	+
Characters	—	+	—
Appearance of solution	—	+	—
Loss on drying	—	—	43.0–46.0%
pH	—	3.0–3.5	—
Ammonium	—	≤0.2%	—
Iron	≤20 ppm	≤100 ppm	+
Arsenic	≤3.3 ppm	—	—
Heavy metals	≤20 ppm	≤20 ppm	≤0.002%
Assay	≥99.5%	99.0–100.5%	99.0–100.5%

10 Typical Properties

Acidity/Alkalinity pH = 3.0–3.5 (10% w/v aqueous solutions at 20°C).
Density (bulk) 1 g/cm^3[3]
Density (true) 1.725 g/cm^3
Melting point 92.5°C
Solubility Freely soluble in water, very soluble in boiling water; soluble in glycerol; practically insoluble in ethanol (96%).
Vapor density (relative) 16.4 (air = 1)[4]

11 Stability and Storage Conditions

Store in a cool, dry place in tightly closed containers. Stable under normal temperatures and pressures. When kept for a long time at 60–65°C (or over sulfuric acid) potassium alum dodecahydrate loses water, which is reabsorbed on exposure to air. It becomes anhydrous at about 200°C.

12 Incompatibilities

Potassium alum is incompatible with strong oxidizing agents, aluminum, copper, steel, and zinc. When it is dispersed in powders with phenol, salicylates, or tannic acid, gray or green colors may be developed owing to traces of iron in the alum.

13 Method of Manufacture

Potassium alum is manufactured by treating bauxite with sulfuric acid and then potassium sulfate. Alternatively, aluminum sulfate is reacted with potassium sulfate.[5]

14 Safety

Potassium alum is often included in preparations used as mouthwashes or gargles and in dermatological preparations.

Large doses of potassium alum act as an irritant and may be corrosive; gum necrosis and gastrointestinal hemorrhage have occurred. Acute encephalopathy has been reported[6,7] following bladder irrigation with alum solutions in the treatment of bladder hemorrhage. Anecdotal evidence suggests that this practice should be avoided in patients with renal insufficiency.[6]

15 Handling Precautions

Observe normal precautions appropriate to the circumstances and quantity of the material handled. It causes eye and skin irritation and may cause respiratory tract irritation.

During a fire, irritating and highly toxic gases may be generated by thermal decomposition or combustion of potassium alum. Hazardous decomposition products include oxides of sulfur, aluminum oxide, and oxides of potassium.

The American Conference of Governmental Industrial Hygienists (ACGIH) value for potassium alum is 2 mg/m^3 TWA (as aluminum).

16 Regulatory Status

GRAS listed. Included in the FDA Inactive Ingredients Database (vaginal; suppository). Included in medicines licensed in the UK. Included in the Canadian Natural Health Products Ingredients Database.

17 Related Substances

—

18 Comments

Potassium alum is a powerful astringent, and may also be used therapeutically as a topical hemostatic, either as a solid or as a solution. Intravesical instillation of potassium alum, typically as a 1% solution, has been used for hemorrhagic cystitis.

The JP XV has separate monographs for aluminum potassium sulfate and dried aluminum potassium sulfate. A specification for potassium alum is contained in the *Food Chemicals Codex* (FCC).[8]

The EINECS number for potassium alum (anhydrous) is 233-141-3. The PubChem Compound ID (CID) for potassium alum dodecahydrate is 62667.

19 Specific References

1 Merck & Co, Inc. Product literature: *Recombivax HB* Hepatitis B Vaccine (Recombinant). December 2007.
2 Aventis Pasteur Inc. Product literature: *Tripedia* Diphtheria and Tetanus Toxoids and Acellular Pertussis Vaccine, Adsorbed. December 2003.
3 Parchem. Datasheet: Potassium Aluminum Sulfate. www.parchem.com (accessed 11 October 2011).
4 ScienceLab.com, Inc. Material safety data sheet: Aluminum potassium sulfate, 9 October 2005.
5 Darragh KV, Ertell CA, eds. Aluminum sulfate and alums. In: *Kirk-Othmer Encyclopedia of Chemical Technology*, 5th edn. New York: Wiley, 2003.
6 Phelps KR, *et al.* Encephalopathy after bladder irrigation with alum: case report and literature review. *Am J Med Sci* 1991; **318:** 181–185.
7 Nakamura H, *et al.* Acute encephalopathy due to aluminium toxicity successfully treated by combined intravenous deferoxamine and hemodialysis. *J Clin Pharmacol* 2000; **40:** 296–300.
8 *Food Chemicals Codex*, 7th edn. Bethesda, MD: United States Pharmacopeia, 2010.

20 General References

—

21 Author

RT Guest.

22 Date of Revision

1 March 2012.

Potassium Benzoate

1 Nonproprietary Names

USP–NF: Potassium Benzoate

2 Synonyms

Benzoate of potash; benzoic acid potassium salt; E212; kalium benzoat; potassium salt trihydrate; *ProBenz PG*.

3 Chemical Name and CAS Registry Number

Potassium benzoate [582-25-2]

4 Empirical Formula and Molecular Weight

$C_7H_5KO_2$ 160.21

5 Structural Formula

6 Functional Category

Antimicrobial preservative; tablet and capsule lubricant.

7 Applications in Pharmaceutical Formulation or Technology

Potassium benzoate is predominantly used as an antimicrobial preservative in a wide range of beverages, foods and some pharmaceutical formulations. Preservative efficacy increases with decreasing pH; it is most effective at pH 4.5 or below. However, at low pH undissociated benzoic acid may produce a slight though discernible taste in food products.

Increasingly, potassium benzoate is used as an alternative to sodium benzoate in applications where a low sodium content is desirable. *See also* Table I.

Table I: Uses of potassium benzoate.

Use	Concentration (%)
Carbonated beverages	0.03–0.08
Food products	≤0.1

8 Description

Potassium benzoate occurs as a slightly hygroscopic, white, odorless or nearly odorless crystalline powder or granules. Aqueous solutions are slightly alkaline and have a sweetish astringent taste.

9 Pharmacopeial Specifications

See Table II.

10 Typical Properties

Acidity/alkalinity Aqueous solutions are slightly alkaline.
Melting point >300°C
Solubility *see* Table III.
Specific gravity 1.5

Table II: Pharmacopeial specifications for potassium benzoate.

Test	USP35–NF30
Identification	+
Alkalinity	+
Water	≤1.5%
Heavy metals	≤10 ppm
Assay (anhydrous basis)	99.0–100.5%

Table III: Solubility of potassium benzoate.

Solvent	Solubility at 20°C unless otherwise stated
Ethanol (95%)	1 in 75
Ethanol (90%)	1 in 50
Ether	Practically insoluble
Methanol	Very slightly soluble
Water	1 in 2.46 at 13°C
	1 in 2.43 at 17.5°C
	1 in 2.36
	1 in 2.27 at 33.3°C
	1 in 2.23 at 41°C
	1 in 2.15 at 50°C

11 Stability and Storage Conditions

Potassium benzoate is stable at room temperature under normal storage conditions. Since it is slightly hygroscopic, potassium benzoate should be stored in sealed containers. Exposure to conditions of high humidity and elevated temperatures should be avoided.

12 Incompatibilities

Potassium benzoate is incompatible with strong acids and strong oxidizing agents.

13 Method of Manufacture

Potassium benzoate is prepared from the acid–base reaction between benzoic acid and potassium hydroxide.

14 Safety

Potassium benzoate is widely used in food products and is generally regarded as a nontoxic and nonirritant material. However, people with a history of allergies may show allergic reactions when exposed to potassium benzoate. Ingestion is inadvisable for asthmatics. Higher concentrations of potassium benzoate have been reported to cause irritation to mucous membranes.

The WHO acceptable daily intake of total benzoates including potassium benzoate, calculated as benzoic acid, has been estimated at up to 5 mg/kg of body-weight.[1,2]

15 Handling Precautions

Observe normal precautions appropriate to the circumstances and quantity of material handled. Potassium benzoate may be irritant to the eyes and skin. Eye protection and gloves are recommended. When exposed to heat, and when heated to decomposition, potassium benzoate emits acrid smoke and irritating fumes.

16 Regulatory Status

GRAS listed. Accepted as a food additive in Europe. Included in the Canadian Natural Health Products Ingredients Database.

17 Related Substances

Benzoic acid; sodium benzoate.

18 Comments

Therapeutically, potassium benzoate has been used in the management of hypokalemia.

The EINECS number for potassium benzoate is 209-481-3. The PubChem Compound ID (CID) for potassium benzoate is 23661960.

19 Specific References

1 FAO/WHO. Toxicological evaluation of certain food additives with a review of general principles and of specifications. Seventeenth report of the joint FAO/WHO expert committee on food additives. *World Health Organ Tech Rep Ser* 1974; No. 539.

2 FAO/WHO. Evaluation of certain food additives and contaminants. Twenty-seventh report of the joint FAO/WHO expert committee on food additives. *World Health Organ Tech Rep Ser* 1983; No. 696.

20 General References

—

21 Author

CP McCoy.

22 Date of Revision

2 March 2012.

P

Potassium Bicarbonate

1 Nonproprietary Names

BP: Potassium Bicarbonate
PhEur: Potassium Hydrogen Carbonate
USP–NF: Potassium Bicarbonate

2 Synonyms

Carbonic acid monopotassium salt; E501; kalii hydrogenocarbonas; monopotassium carbonate; potassium acid carbonate; potassium hydrogen carbonate.

3 Chemical Name and CAS Registry Number

Potassium bicarbonate [298-14-6]

4 Empirical Formula and Molecular Weight

$KHCO_3$ 100.11

5 Structural Formula

See Section 4.

6 Functional Category

Alkalizing agent.

7 Applications in Pharmaceutical Formulation or Technology

As an excipient, potassium bicarbonate is generally used in formulations as a source of carbon dioxide in effervescent preparations, at concentrations of 25–50% w/w. It is of particular use in formulations where sodium bicarbonate is unsuitable, for example, when the presence of sodium ions in a formulation needs to be limited or is undesirable. Potassium bicarbonate is often formulated with citric acid or tartaric acid in effervescent tablets or granules; on contact with water, carbon dioxide is released through chemical reaction, and the product disintegrates. On occasion, the presence of potassium bicarbonate alone may be sufficient in tablet formulations, as reaction with gastric acid can be sufficient to cause effervescence and product disintegration.

Potassium bicarbonate has also been investigated as a gas-forming agent in alginate raft systems.[1,2] The effects of potassium bicarbonate on the stability and dissolution of paracetamol and ibuprofen have been described.[3]

Potassium bicarbonate is also used in food applications as an alkali and a leavening agent, and is a component of baking powder.

8 Description

Potassium bicarbonate occurs as colorless, transparent crystals or as a white granular or crystalline powder. It is odorless, with a saline or weakly alkaline taste.

9 Pharmacopeial Specifications

See Table I.

Table I: Pharmacopeial specifications for potassium bicarbonate.

Test	PhEur 7.4	USP35–NF30
Identification	+	+
Characters	+	—
Appearance	+	—
Carbonates	+	—
Normal carbonates	—	≤2.5%
Chloride	≤150 ppm	—
Sulfate	≤150 ppm	—
Ammonium	≤20 ppm	—
Calcium	≤100 ppm	—
Heavy metals	≤10 ppm	≤10 ppm
Iron	≤20 ppm	—
Sodium	≤0.5%	—
Loss on drying	—	≤0.3%
Assay	99.0–101.0%	99.5–101.5%

10 Typical Properties

Acidity/alkalinity pH = 8.2 (for a 0.1 M aqueous solution); a 5% solution in water has a pH of ≤8.6.

Solubility Soluble 1 in 4.5 of water at 0°C, 1 in 2.8 of water at 20°C, 1 in 2 of water at 50°C; practically insoluble in ethanol (95%).

Specific gravity 2.17

11 Stability and Storage Conditions

Potassium bicarbonate should be stored in a well-closed container in a cool, dry location. Potassium bicarbonate is stable in air at normal temperatures, but when heated to 100–200°C in the dry state, or in solution, it is gradually converted to potassium carbonate.

12 Incompatibilities

Potassium bicarbonate reacts with acids and acidic salts with the evolution of carbon dioxide.

13 Method of Manufacture

Potassium bicarbonate can be made by passing carbon dioxide into a concentrated solution of potassium carbonate, or by exposing moist potassium carbonate to carbon dioxide, preferably under moderate pressure.

Potassium bicarbonate also occurs naturally in the mineral calcinite.

14 Safety

Potassium bicarbonate is used in cosmetics, foods, and oral pharmaceutical formulations, where it is generally regarded as a relatively nontoxic and nonirritant material when used as an excipient. However, excessive consumption of potassium bicarbonate or other potassium salts may produce toxic manifestations of hyperkalemia. Ingestion of potassium salts can cause gastrointestinal adverse effects, and tablet formulations may cause contact irritation due to high local concentrations of potassium.

15 Handling Precautions

Observe normal precautions appropriate to the circumstances and quantity of material handled. Eye protection and gloves are recommended.

16 Regulatory Status

GRAS listed. Accepted as a food additive in Europe (the E number E501 refers to potassium carbonates). Included in the FDA Inactive Ingredients Database (oral formulations). Included in nonparenteral medicines licensed in the UK and US (chewable tablets; effervescent granules; effervescent tablets; lozenges; oral granules; oral suspensions; powder for oral solutions). Included in the Canadian Natural Health Products Ingredients Database.

17 Related Substances

Sodium bicarbonate.

18 Comments

One gram of potassium bicarbonate represents approximately 10 mmol of potassium and of bicarbonate; 2.56 g of potassium bicarbonate is approximately equivalent to 1 g of potassium.

Therapeutically, potassium bicarbonate is used as an alternative to sodium bicarbonate in the treatment of certain types of metabolic acidosis. It is also used as an antacid to neutralize acid secretions in the gastrointestinal tract and as a potassium supplement.

A specification for potassium bicarbonate is contained in the *Food Chemicals Codex* (FCC).[4]

The EINECS number for potassium bicarbonate is 206-059-0. The PubChem Compound ID (CID) for potassium bicarbonate is 516893.

19 Specific References

1 Johnson FA, *et al.* The effects of alginate molecular structure and formulation variables on the physical characteristics of alginate raft systems. *Int J Pharm* 1997; **159**: 35–42.
2 Johnson FA, *et al.* The use of image analysis as a means of monitoring bubble formation in alginate rafts. *Int J Pharm* 1998; **170**: 179–185.
3 Shaw LR, *et al.* The effect of selected water-soluble excipients on the dissolution of paracetamol and ibuprofen. *Drug Dev Ind Pharm* 2005; **31**(6): 515–525.
4 *Food Chemicals Codex*, 7th edn. Bethesda, MD: United States Pharmacopeia, 2011: 841.

20 General References

—

21 Author

CG Cable.

22 Date of Revision

1 March 2012.

 Potassium Chloride

1 Nonproprietary Names

BP: Potassium Chloride
JP: Potassium Chloride
PhEur: Potassium Chloride
USP–NF: Potassium Chloride

2 Synonyms

Chloride of potash; chloropotassuril; dipotassium dichloride; E508; kalii chloridum; potassium monochloride; potassium muriate; tripotassium trichloride.

3 Chemical Name and CAS Registry Number

Potassium chloride [7447-40-7]

4 Empirical Formula and Molecular Weight

KCl 74.55

5 Structural Formula

See Section 4.

6 Functional Category

Tonicity agent.

7 Applications in Pharmaceutical Formulation or Technology

Potassium chloride is widely used in a variety of parenteral and nonparenteral pharmaceutical formulations. Its primary use, in parenteral and ophthalmic preparations, is to produce isotonic solutions.

Many solid dosage forms of potassium chloride exist including: tablets prepared by direct compression[1–4] or granulation;[5,6] effervescent tablets; coated, sustained-release tablets;[7–10] sustained-release wax matrix tablets;[11] microcapsules;[12] pellets; and osmotic pump formulations.[13,14] *See also* Section 18.

Potassium chloride is also used widely in the food industry as a dietary supplement, salt substitute, pH control agent, stabilizer, thickener, and gelling agent. It can also be used in infant formulations.

8 Description

Potassium chloride occurs as odorless, colorless crystals or a white crystalline powder, with an unpleasant, saline taste. The crystal lattice is a face-centered cubic structure.

9 Pharmacopeial Specifications

See Table I.

10 Typical Properties

Acidity/alkalinity pH ≈ 7 for a saturated aqueous solution at 15°C.
Boiling point Sublimes at 1500°C.
Compressibility *see* Figure 1.[3,4]
Density 1.99 g/cm^3; 1.17 g/cm^3 for a saturated aqueous solution at 15°C.
Dielectric constant 4.68 at 10^6 Hz
Heat of fusion 337.7 kJ/kg
Melting point 770°C; also reported as 773°C

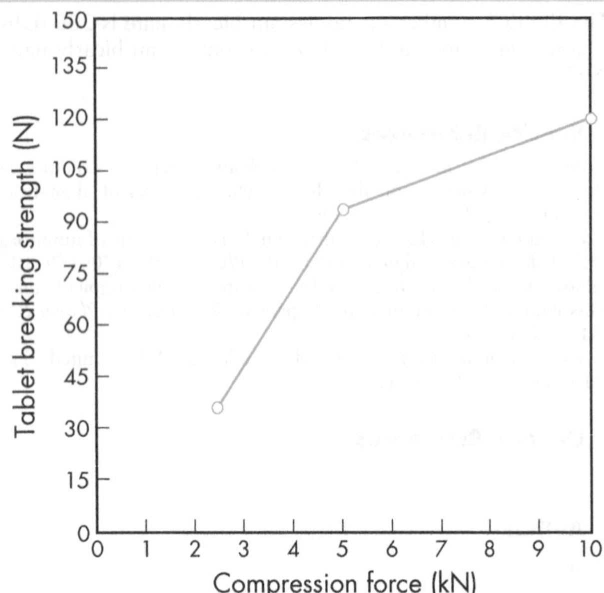

Figure 1: Compression characteristics of potassium chloride.[3] Tablet diameter = 10 mm.

Table I: Pharmacopeial specifications for potassium chloride.

Test	JP XV	PhEur 7.4	USP35–NF30
Identification	+	+	+
Characters	–	+	–
Acidity or alkalinity	+	+	+
Appearance of solution	+	+	–
Loss on drying	≤0.5%	≤1.0%	≤1.0%
Iodide or bromide	+	+	+
Aluminum	–	≤1 ppm	≤1 ppm
Arsenic	≤2 ppm	–	–
Barium	–	+	–
Calcium and magnesium	+	≤200 ppm	+
Heavy metals	≤5 ppm	≤10 ppm	≤10 ppm
Iron	–	≤20 ppm	–
Sodium	+	≤0.1%	+
Sulfates	–	≤300 ppm	–
Assay (dried basis)	≥99.0%	99.0–100.5%	99.0–100.5%

Table II: Solubility of potassium chloride.

Solvent	Solubility at 20°C unless otherwise stated
Acetone	Practically insoluble
Ethanol (95%)	1 in 250
Ether	Practically insoluble
Glycerin	1 in 14
Water	1 in 2.8
	1 in 1.8 at 100°C

Osmolarity A 1.19% w/v solution is isoosmotic with serum.
Refractive index 1.4903
Solubility *see* Table II.
Specific heat 0.694 J/g/K
Spectroscopy

NIR spectra *see* Figure 2.

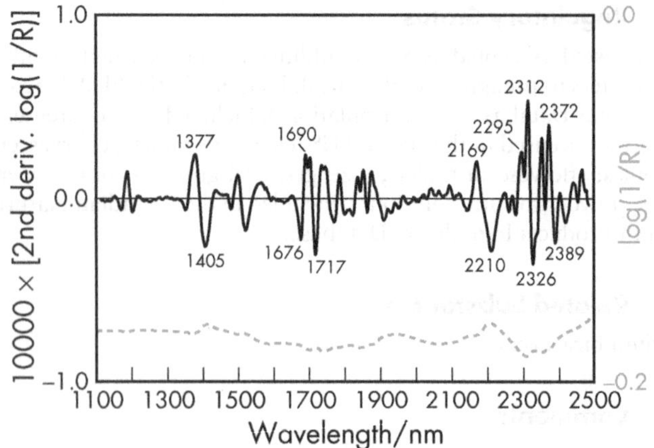

Figure 2: Near-infrared spectrum of potassium chloride measured by reflectance. Potassium chloride does not absorb in the near-infrared region; however, it will generally show some peaks due to traces of moisture (approx. 1450 nm and 1950 nm).

11 Stability and Storage Conditions

Potassium chloride tablets become increasingly hard on storage at low humidities due to hygroscopicity. However, tablets stored at 76% relative humidity showed no increase or only a slight increase in hardness.[2] The addition of lubricants, such as 2% w/w magnesium stearate,[1] reduces tablet hardness and hardness on aging.[2] Aqueous potassium chloride solutions may be sterilized by autoclaving or by filtration.

Potassium chloride is stable and should be stored in a well-closed container in a cool, dry place.

12 Incompatibilities

Potassium chloride reacts violently with bromine trifluoride and with a mixture of sulfuric acid and potassium permanganate. The presence of hydrochloric acid, sodium chloride, or magnesium chloride decreases the solubility of potassium chloride in water. Aqueous solutions of potassium chloride form precipitates with lead and silver salts. Intravenous aqueous potassium chloride solutions are incompatible with protein hydrolysate.

13 Method of Manufacture

Potassium chloride occurs naturally as the mineral sylvite or sylvine; it also occurs in other minerals such as sylvinite, carnallite, and kainite. Commercially, potassium chloride is obtained by the solar evaporation of brine or by the mining of mineral deposits followed by milling, washing, screening, flotation, crystallization, refining, and drying.

14 Safety

Potassium chloride is used in a large number of pharmaceutical formulations, including oral, parenteral, and topical preparations, both as an excipient and as a therapeutic agent.

Potassium ions play an important role in cellular metabolism, such as enzymatic reactions, nerve conduction, muscle contraction, and carbohydrate metabolism, and imbalances can result in serious clinical effects. Orally ingested potassium chloride is rapidly absorbed from the gastrointestinal tract and excreted by the kidneys, with more side effects observed in patients with renal insufficiency. Potassium chloride is more irritant than sodium chloride when adminstered orally, and ingestion of large quantities of potassium chloride can cause effects such as gastrointestinal irritation, nausea, vomiting, and diarrhea.

High localized concentrations of potassium chloride in the gastrointestinal tract can cause ulceration, hence the development of the many enteric-coated and wax matrix sustained-release preparations that are available.[15] However, many enteric-coated tablets of potassium chloride are no longer recommended because of the high incidence of severe injury to gastrointestinal tissue during tablet dissolution.[16] Although it is claimed that some formulations cause less ulceration than others, it is often preferred to administer potassium chloride as an aqueous solution. However, solutions have also been associated with problems, mainly due to their unpleasant taste.

Parenterally, rapid injection of strong potassium chloride solutions can cause cardiac arrest; in the adult, solutions should be infused at a rate not greater than 750 mg/hour.

Therapeutically, in adults, up to 10 g orally, in divided doses, has been administered daily, while intravenously up to 6 g daily has been used.

LD_{50} (guinea pig, oral): 2.5 g/kg[17]
LD_{50} (mouse, IP): 1.18 g/kg
LD_{50} (mouse, IV): 0.12 g/kg
LD_{50} (mouse, oral): 0.38 g/kg
LD_{50} (rat, IP): 0.66 g/kg
LD_{50} (rat, IV): 0.14 g/kg
LD_{50} (rat, oral): 2.6 g/kg

15 Handling Precautions

Observe normal precautions appropriate to the circumstances and quantity of material handled.

16 Regulatory Status

GRAS listed. Accepted as a food additive in Europe. Included in the FDA Inactive Ingredients Database (injections, ophthalmic preparations, oral capsules, and tablets). Included in nonparenteral and parenteral medicines licensed in the UK. Included in the Canadian Natural Health Products Ingredients Database.

17 Related Substances

Sodium chloride.

18 Comments

Each gram of potassium chloride represents approximately 13.4 mmol of potassium; 1.91 g of potassium chloride is approximately equivalent to 1 g of potassium.

Experimentally, potassium chloride is frequently used as a model drug in the development of new solid-dosage forms, particularly for sustained-release or modified-release products.

Potassium chloride is used therapeutically in the treatment of hypokalemia, and as an electrolyte replenisher. For diets where the intake of sodium chloride is restricted, salt substitutes for use in cooking or as table salt are available and contain mainly potassium chloride, e.g. *LoSalt* (Klinge Chemicals Ltd) is a blend of 2/3 potassium chloride and 1/3 sodium chloride with magnesium carbonate added as a flow-promoting agent.

A specification for potassium chloride is contained in the *Food Chemicals Codex* (FCC).[18]

The EINECS number for potassium chloride is 231-211-8. The PubChem Compound ID (CID) for potassium chloride is 4873.

19 Specific References

1 Hirai Y, Okada J. Calculated stress and strain conditions of lubricated potassium chloride powders during die-compression. *Chem Pharm Bull* 1982; 30: 2202–2207.

2 Lordi N, Shiromani P. Mechanism of hardness of aged compacts. *Drug Dev Ind Pharm* 1984; **10**: 729–752.

3 Pintye-Hodi K, Sohajda-Szücs E. [Study on the compressibility of potassium chloride part 1: direct pressing without auxiliary products.] *Pharm Ind* 1984; **46**: 767–769[in German].

4 Pintye-Hodi K, Sohajda-Szücs E. [Study on the compressibility of potassium chloride part 2: direct compressing with microgranulous celluloses.] *Pharm Ind* 1984; **46**: 1080–1083[in German].

5 Niskanen T, *et al.* Granulation of potassium chloride in instrumental fluidized bed granulator part 1: effect of flow rate. *Acta Pharm Fenn* 1990; **99**: 13–22.

6 Niskanen T, *et al.* Granulation of potassium chloride in instrumental fluidized bed granulator part 2: evaluation of the effects of two independent process variables using 3^2-factorial design. *Acta Pharm Fenn* 1990; **99**: 23–30.

7 Fee JV, *et al.* The effect of surface coatings on the dissolution rate of a non-disintegrating solid (potassium chloride). *J Pharm Pharmacol* 1973; **25**(Suppl.): 149P–150P.

8 Thomas WH. Measurement of dissolution rates of potassium chloride from various slow release potassium chloride tablets using a specific ion electrode. *J Pharm Pharmacol* 1973; **25**: 27–34.

9 Cartwright AC, Shah C. An *in vitro* dissolution test for slow release potassium chloride tablets. *J Pharm Pharmacol* 1977; **29**: 367–369.

10 Beckett AH, Samaan SS. Sustained release potassium chloride products *in vitro–in vivo* correlations. *J Pharm Pharmacol* 1978; **30**(Suppl.): 69P.

11 Flanders P, *et al.* The control of drug release from conventional melt granulation matrices. *Drug Dev Ind Pharm* 1987; **13**: 1001–1022.

12 Harris MS. Preparation and release characteristics of potassium chloride microcapsules. *J Pharm Sci* 1981; **70**: 391–394.

13 Ramadan MA, Tawashi R. The effect of hydrodynamic conditions and delivery orifice size on the rate of drug release from the elementary osmotic pump system (EOP). *Drug Dev Ind Pharm* 1987; **13**: 235–248.

14 Lindstedt B, *et al.* Osmotic pumping release from KCl tablets coated with porous and non-porous ethylcellulose. *Int J Pharm* 1991; **67**: 21–27.

15 McMahon FG, *et al.* Effect of potassium chloride supplements on upper gastrointestinal mucosa. *Clin Pharmacol Ther* 1984; **35**: 852–855.

16 United States Pharmacopeial Convention, Inc. USP Dispensing Information: Drug Information for the Health Care Professional, vol. 1. Greenwood Village, CO: Thomson/Micromedex, 2006; 2483.

17 Lewis RJ, ed. *Sax's Dangerous Properties of Industrial Materials*, 11th edn. New York: Wiley, 2004; 3025–3026.

18 *Food Chemicals Codex*, 7th edn. Bethesda, MD: United States Pharmacopeia, 2010.

20 General References

Love DW, *et al.* Comparison of the taste and acceptance of three potassium chloride preparations. *Am J Hosp Pharm* 1978; **35**(5): 586–588.

ScienceLab.com. Material safety data sheet: Potassium chloride, 2010.

Staniforth JN, Rees JE. Segregation of vibrated powder mixes containing different concentrations of fine potassium chloride and tablet excipients. *J Pharm Pharmacol* 1983; **35**: 549–554.

Ullmann's Encyclopedia of Industrial Chemistry, 7th edn. Weinheim Germany: Wiley-VCH, 2011.

21 Author

C Bhugra.

22 Date of Revision

2 March 2012.

P

Potassium Citrate

1 Nonproprietary Names

BP: Potassium Citrate
PhEur: Potassium Citrate
USP–NF: Potassium Citrate

2 Synonyms

Citrate of potash; citric acid potassium salt; E332; kalii citras; tripotassium citrate.

3 Chemical Name and CAS Registry Number

2-Hydroxy-1,2,3-propanetricarboxylic acid tripotassium salt monohydrate [6100-05-6]

2-Hydroxy-1,2,3-propanetricarboxylic acid tripotassium salt anhydrous [866-84-2]

4 Empirical Formula and Molecular Weight

$C_6H_5K_3O_7{\cdot}H_2O$ 324.41 (for monohydrate)
$C_6H_5K_3O_7$ 306.40 (for anhydrous)

5 Structural Formula

```
     H    O
     |    ||
 H — C —— C — OK
     |
     |    O
     |    ||
HO — C —— C — OK     • H2O
     |
     |    O
     |    ||
 H — C —— C — OK
     |
     H
```

6 Functional Category

Alkalizing agent; buffering agent; complexing agent.

7 Applications in Pharmaceutical Formulation or Technology

Potassium citrate is used in beverages, foods, and oral pharmaceutical formulations as a buffering and alkalizing agent. It is also used as a sequestering agent. *See* Table I.

Table I: Uses of potassium citrate.

Use	Concentration (%)
Buffer for solutions	0.3–2.0
Complexing agent	0.3–2.0

8 Description

Potassium citrate occurs as transparent prismatic crystals or as a white, granular powder. It is hygroscopic and odorless, and has a cooling, saline taste.

9 Pharmacopeial Specifications

See Table II.

Table II: Pharmacopeial specifications for potassium citrate.

Test	PhEur 7.4	USP35–NF30
Identification	+	+
Characters	+	—
Acidity or alkalinity	+	+
Loss on drying	4.0–7.0%	3.0–6.0%
Appearance of solution	+	—
Tartrate	—	+
Heavy metals	≤10 ppm	≤10 ppm
Sodium	≤0.3%	—
Chlorides	≤50 ppm	—
Oxalates	≤300 ppm	—
Sulfates	≤150 ppm	—
Readily carbonizable substances	+	—
Assay (dried basis)	99.0–101.0%	99.0–100.5%

10 Typical Properties

Acidity/alkalinity pH = 8.5 (saturated aqueous solution).
Density 1.98 g/cm^3
Melting point 230°C (loses water of crystallization at 180°C).
Solubility *see* Table III.

Table III: Solubility of potassium citrate.

Solvent	Solubility at 20°C
Ethanol (95%)	Practically insoluble
Glycerin	1 in 2.5
Water	1 in 0.65

Spectroscopy
IR spectra *see* Figure 1.
NIR spectra *see* Figure 2.

11 Stability and Storage Conditions

Potassium citrate is a stable, though hygroscopic material, and should be stored in an airtight container in a cool, dry place. It is deliquescent in moist air.

12 Incompatibilities

Aqueous potassium citrate solutions are slightly alkaline and will react with acidic substances. Potassium citrate may also precipitate alkaloidal salts from their aqueous or alcoholic solutions. Calcium and strontium salts will cause precipitation of the corresponding citrates. Potassium citrate is incompatible with strong oxidizing agents.

13 Method of Manufacture

Potassium citrate is prepared by adding either potassium bicarbonate or potassium carbonate to a solution of citric acid until

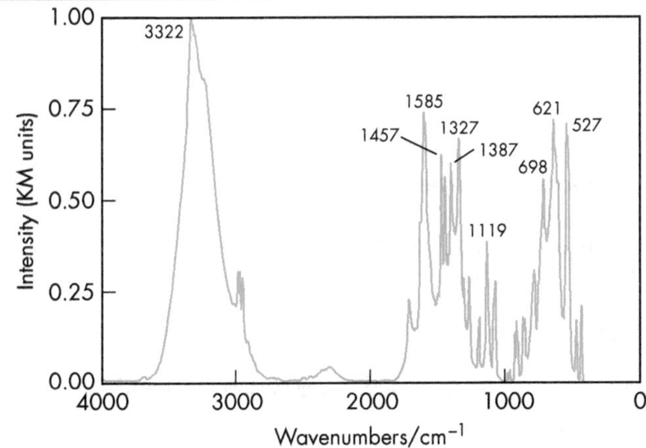

Figure 1: Infrared spectrum of potassium citrate measured by diffuse reflectance. Adapted with permission of Informa Healthcare.

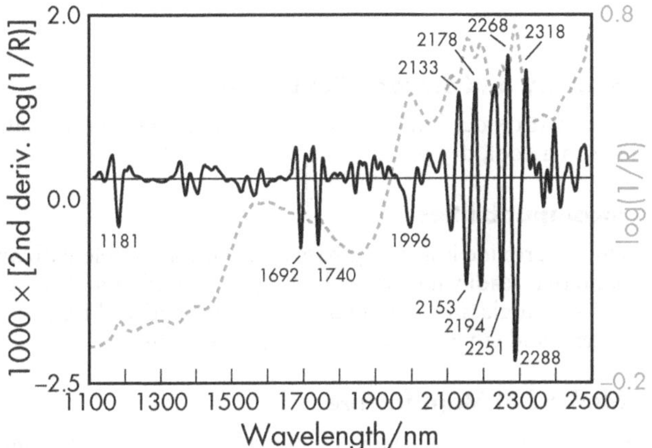

Figure 2: Near-infrared spectrum of potassium citrate, monohydrate measured by reflectance.

effervescence ceases. The resulting solution is then filtered and evaporated to dryness to obtain potassium citrate.

14 Safety

Potassium citrate is used in oral pharmaceutical formulations and is generally regarded as a nontoxic and nonirritant material by this route of administration.

Most potassium citrate safety data relate to its use as a therapeutic agent, for which up to 10 g may be administered daily, in divided doses, as a treatment for cystitis. Although there are adverse effects associated with excessive ingestion of potassium salts, the quantities of potassium citrate used as a pharmaceutical excipient are insignificant in comparison to those used therapeutically.

LD_{50} (dog, IV): 0.17 g/kg[1]

15 Handling Precautions

Observe normal precautions appropriate to the circumstances and quantity of material handled. Potassium citrate may be irritant to the skin and eyes and should be handled in a well-ventilated environment. Eye protection and gloves are recommended. When heated to decomposition, potassium citrate emits toxic fumes of potassium oxide.[6]

16 Regulatory Status

GRAS listed. Accepted as a food additive in Europe. Included in the FDA Inactive Ingredients Database (oral solutions and suspensions; topical emulsions and aerosol foams). Included in nonparenteral medicines (cutaneous foams and emulsions; oral liquids, granules, mixtures and soluble tablets; topical liquids, emulsions and mousses) licensed in the UK. Included in the Canadian Natural Health Products Ingredients Database.

17 Related Substances

—

18 Comments

Each gram of potassium citrate monohydrate represents approximately 9.3 mmol of potassium and 3.08 mmol of citrate. Potassium citrate monohydrate 2.77 g is equivalent to about 1 g of potassium. Each gram of potassium citrate anhydrous represents approximately 9.8 mmol of potassium and 3.26 mmol of citrate.

The flow properties of a range of powder size fractions of potassium citrate have been investigated using a method involving the indentation of a sphere.[2] The influence of the powder particle size of potassium citrate on the drained angle and the correlation between the drained angle and the mass flow rate have been reported.[3]

Therapeutically, potassium citrate is used to alkalinize the urine and to relieve the painful irritation caused by cystitis.[4–8]

A specification for potassium citrate is contained in the *Food Chemicals Codex* (FCC).[7]

The EINECS number for potassium citrate is 212-755-5. The PubChem Compound ID (CID) for potassium citrate monohydrate includes 22470 and 2735208.

19 Specific References

1 Lewis RJ, ed. *Sax's Dangerous Properties of Industrial Materials*, 11th edn. New York: Wiley, 2004: 3026.
2 Zatloukal Z, Sklubalová Z. Indentation test for free-flowable powder excipients. *Pharm Dev Technol* 2008; **13**(1): 85–92.
3 Sklubalová Z, Zatloukal Z. The relationship between drained angle and flow rate of size fractions of powder excipients. *Pharmazie* 2009; **64**(12): 846–847.
4 Elizabeth JE, Carter NJ. Potassium citrate mixture: soothing but not harmless? *Br Med J* 1987; **295**: 993.
5 Gabriel R. Potassium sorbate mixture: soothing but not harmless? [letter]. *Br Med J* 1987; **295**: 1487.
6 Liak TL, *et al.* The effects of drug therapy on urinary pH: excipient effects and bioactivation of methenamine. *Int J Pharm* 1987; **36**: 233–242.
7 Fjellstedt E, *et al.* A comparison of the effects of potassium citrate and sodium bicarbonate in the alkalinization of urine in homozygous cystinuria. *Urol Res* 2001; **29**(5): 295–302.
8 Domrongkitchaiporn S, *et al.* Dosage of potassium citrate in the correction of urinary abnormalities in pediatric distal renal tubular acidosis patients. *Am J Kidney Dis* 2002; **39**(2): 383–391.
9 *Food Chemicals Codex*, 7th edn. Bethesda, MD: United States Pharmacopeia, 2010: 846.

20 General References

Cole ET, *et al.* Relations between compaction data for some crystalline pharmaceutical materials. *Pharm Acta Helv* 1975; **50**: 28–32.

21 Author

CG Cable.

22 Date of Revision

1 March 2012.

Potassium Hydroxide

1 Nonproprietary Names

BP: Potassium Hydroxide
JP: Potassium Hydroxide
PhEur: Potassium Hydroxide
USP–NF: Potassium Hydroxide

2 Synonyms

Caustic potash; E525; kalii hydroxidum; kalium hydroxydatum; potash lye; potassium hydrate.

3 Chemical Name and CAS Registry Number

Potassium hydroxide [1310-58-3]

4 Empirical Formula and Molecular Weight

KOH 56.11

5 Structural Formula

See Section 4.

6 Functional Category

Alkalizing agent.

7 Applications in Pharmaceutical Formulation or Technology

Potassium hydroxide is widely used in pharmaceutical formulations to adjust the pH of solutions. It can also be used to react with weak acids to form salts.

8 Description

Potassium hydroxide occurs as a white or nearly white fused mass. It is available in small pellets, flakes, sticks and other shapes or forms. It is hard and brittle and shows a crystalline fracture. Potassium hydroxide is hygroscopic and deliquescent; on exposure to air, it rapidly absorbs carbon dioxide and water with the formation of potassium carbonate.

9 Pharmacopeial Specifications

See Table I.

Table I: Pharmacopeial specifications for potassium hydroxide.

Test	JP XV	PhEur 7.4	USP35–NF30
Identification	+	+	+
Characters	—	+	—
Appearance of solution	+	+	—
Aluminum	—	≤0.2 ppm[(a)]	—
Chloride	≤0.05%	≤50 ppm	—
Heavy metals	≤30 ppm	≤10 ppm	≤30 ppm
Insoluble substances	—	—	+
Iron	—	≤10 ppm	—
Phosphates	—	≤20 ppm	—
Potassium carbonate	≤2.0%	≤2.0%	—
Sodium	+	≤1.0%	—
Sulfates	—	≤50 ppm	—
Assay	≥85.0%	85.0–100.5%	≥85.0%

(a) If intended for use in the manufacture of hemodialysis solutions.

10 Typical Properties

Acidity/alkalinity pH = 13.5 (0.1 M aqueous solution)
Melting point 360°C; 380°C when anhydrous
Solubility *see* Table II.

Table II: Solubility of potassium hydroxide.

Solvent	Solubility at 20°C unless otherwise stated
Ethanol (95%)	1 in 3
Ether	Practically insoluble
Glycerin	1 in 2.5
Water	1 in 0.9
	1 in 0.6 at 100°C

11 Stability and Storage Conditions

Potassium hydroxide should be stored in an airtight, nonmetallic container in a cool, dry place.

12 Incompatibilities

Potassium hydroxide is a strong base and is incompatible with any compound that readily undergoes hydrolysis or oxidation. It should not be stored in glass or aluminum containers, and will react with acids, esters, and ethers, especially in aqueous solution.

13 Method of Manufacture

Potassium hydroxide is made by the electrolysis of potassium chloride. Commercial grades may contain chlorides as well as other impurities.

14 Safety

Potassium hydroxide is widely used in the pharmaceutical and food industries and is generally regarded as a nontoxic material at low concentrations. At high concentrations it is a corrosive irritant to the skin, eyes, and mucous membranes.

LD_{50} (rat, oral): 0.273 g/kg[(1)]

15 Handling Precautions

Potassium hydroxide is a corrosive irritant to the skin, eyes, and mucous membranes. The solid and solutions cause burns, often with deep ulceration. It is very toxic on ingestion and harmful on inhalation. Observe normal handling precautions appropriate to the quantity and concentration of material handled. Gloves, eye protection, respirator, and other protective clothing should be worn.

Potassium hydroxide is strongly exothermic when dissolved in ethanol (95%) or water, and considerable heat is generated. The reaction between potassium hydroxide solutions and acids is also strongly exothermic.

In the UK, the workplace exposure limit for potassium hydroxide has been set at 2 mg/m^3 short-term.[(2)]

16 Regulatory Status

GRAS listed. Accepted for use in Europe in certain food applications. Included in the FDA Inactive Ingredients Database (injections; infusions; oral capsules and solutions). Included in nonparenteral and parenteral medicines licensed in the UK. Included in the Canadian Natural Health Products Ingredients Database.

17 Related Substances

Sodium hydroxide.

18 Comments

Therapeutically, potassium hydroxide is used in various dermatological applications.

A specification for potassium hydroxide is contained in the *Food Chemicals Codex* (FCC).[3]

The EINECS number for potassium hydroxide is 215-181-3.

19 Specific References

1 Lewis RJ, ed. *Sax's Dangerous Properties of Industrial Materials*, 11th edn. New York: Wiley, 2004: 3033–3034.

2 Health and Safety Executive. *EH40/2005: Workplace Exposure Limits*. Sudbury: HSE Books, 2011. http://www.hse.gov.uk/pubns/priced/eh40.pdf (accessed 29 February 2012).

3 *Food Chemicals Codex*, 7th edn. Bethesda, MD: United States Pharmacopeia, 2010: 849.

20 General References

—

21 Author

AH Kibbe.

22 Date of Revision

29 February 2012.

Potassium Metabisulfite

1 Nonproprietary Names

BP: Potassium Metabisulphite
PhEur: Potassium Metabisulphite
USP–NF: Potassium Metabisulfite

2 Synonyms

Disulfurous acid; dipotassium pyrosulfite; dipotassium salt; E224; kali disulfis; potassium pyrosulfite.

3 Chemical Name and CAS Registry Number

Dipotassium pyrosulfite [16731-55-8]

4 Empirical Formula and Molecular Weight

$K_2S_2O_5$ 222.32

5 Structural Formula

See Section 4.

6 Functional Category

Antimicrobial preservative; antioxidant.

7 Applications in Pharmaceutical Formulation or Technology

Potassium metabisulfite is used in pharmaceutical applications similar to those of sodium metabisulfite. It is used as an antioxidant, antimicrobial preservative, and sterilizing agent.

8 Description

Potassium metabisulfite occurs as white or colorless free-flowing crystals, crystalline powder, or granules, usually with an odor of sulfur dioxide.

9 Pharmacopeial Specifications

See Table I.

10 Typical Properties

Acidity/alkalinity pH = 3.5–4.5 (5% w/v aqueous solution)
Density (bulk) 1.1–1.3 g/cm^3
Density (tapped) 1.2–1.5 g/cm^3
Melting point 190°C although potassium metabisulfite decomposes at temperatures above 150°C.
Solubility Soluble 1 in 2.2 of water; practically insoluble in ethanol (96%).
Spectroscopy
IR spectra *see* Figure 1.

11 Stability and Storage Conditions

Potassium metabisulfite should be stored in a cool, dark place. When stored at a maximum temperature of 25°C and maximum relative humidity of 45%, the shelf-life is 6 months. Potassium metabisulfite decomposes at temperatures above 150°C. In the air, it oxidizes to the sulfate, more readily in the presence of moisture.

In aqueous solution, potassium metabisulfite forms potassium bisulfite ($KHSO_3$) which exerts a strong reducing effect.

12 Incompatibilities

Potassium metabisulfite is incompatible with strong acids, water, and most common metals. It reacts with nitrites and sodium nitrate at room temperature, which occasionally results in the formation of flame. The reaction may be explosive if water is present. Potassium metabisulfite liberates SO_2 with acids.

Sulfites, including potassium metabisulfite, can react with various pharmaceutical compounds including sympathomimetics such as epinephrine (adrenaline),[1] chloramphenicol,[1] cisplatin,[2]

Table I: Pharmacopeial specifications for potassium metabisulfite.

Test	PhEur 7.4	USP35–NF30
Identification	+	+
Characters	+	—
Appearance of solution	+	—
pH	3.0–4.5	—
Thiosulfates	+	—
Selenium	≤10 ppm	—
Zinc	≤25 ppm	—
Iron	≤10 ppm	≤10 ppm
Heavy metals	≤10 ppm	≤10 ppm
Assay (as SO_2)	—	51.8–57.6%
Assay (dried substance)	95.0–101.0%	—

P

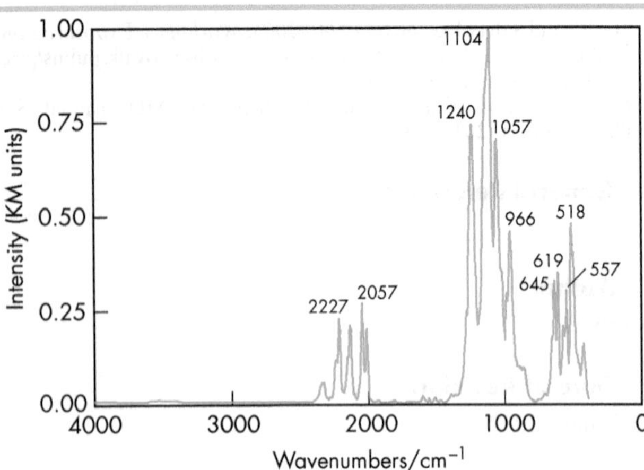

Figure 1: Infrared spectrum of potassium metabisulfite measured by diffuse reflectance. Adapted with permission of Informa Healthcare.

and amino acids,[3] which can result in their pharmacological inactivation. Sulfites are also reported to react with phenylmercuric nitrate,[4,5] and may adsorb onto rubber closures.

See also Section 18.

13 Method of Manufacture

—

14 Safety

Potassium metabisulfite is used in a variety of foods and pharmaceutical preparations, including oral, otic, rectal, and parenteral preparations. Potassium metabisulfite is considered a very irritating material, and may cause dermatitis on exposed skin.[6,7]

Hypersensitivity reactions to potassium metabisulfite and other sulfites, mainly used as preservatives in food products, have been reported. Reactions include bronchospasm and anaphylaxis; some deaths have also been reported, especially in those with a history of asthma or atopic allergy.[8–12] These reactions have led to restrictions by the FDA on the use of sulfites in food applications.[13] However, this restriction has not been extended to their use in pharmaceutical applications. Indeed, epinephrine (adrenaline) injections used to treat severe allergic reactions may contain sulfites.[12,13]

The WHO has set an acceptable daily intake of sulfites, as SO_2, at up to 0.35 mg/kg body-weight.[14]

15 Handling Precautions

Observe normal precautions appropriate to the circumstances and quantity of material handled. Protective gloves and safety goggles are recommended, and precautions should be taken to minimize exposure to the mucous membranes and respiratory tract. When heated to decomposition, it emits toxic fumes of SO_2.

See also Section 12.

16 Regulatory Status

GRAS listed. Accepted in Europe for use as a food additive in certain applications. Included in the FDA Inactive Ingredients Database (IM and IV injection; otic and rectal solutions and suspensions). Included in the Canadian Natural Health Products Ingredients Database.

17 Related Substances

Potassium bisulfite; sodium metabisulfite.

Potassium bisulfite

Empirical formula $KHSO_3$

Molecular weight 120.2

CAS number [7773-03-7]

Synonyms E228; potassium acid sulfite; potassium bisulphite; potassium hydrogen sulfite.

Comments Accepted in Europe as a food additive in certain applications. Included in food and pharmaceutical applications similarly to potassium metabisulfite.

18 Comments

Potassium metabisulfite is used in applications similar to those of sodium metabisulfite in the food, brewing, and wine making industries. Like all sulfites, potassium metabisulfite is not recommended for use in foods that are a source of thiamin, owing to the instability of the vitamin in their presence. Such foods include meat, raw fruits and vegetables, fresh potatoes, and foods that are a source of vitamin B_{12}.

A specification for potassium metabisulfite is contained in the *Food Chemicals Codex* (FCC).[15]

The EINECS number for potassium metabisulfite is 240-795-3. The PubChem Compound ID (CID) for potassium metabisulfite includes 28019 and 516928.

19 Specific References

1 Higuchi T, Schroeter LC. Reactivity of bisulfite with a number of pharmaceuticals. *J Am Pharm Assoc (Sci)* 1959; **48**: 535–540.
2 Garren KW, Repta AJ. Incompatibility of cisplatin and Reglan Injectable. *Int J Pharm* 1985; **24**: 91–99.
3 Brawley V, *et al.* Effect of sodium metabisulphite on hydrogen peroxide production in light-exposed pediatric parenteral amino acid solutions. *Am J Health Syst Pharm* 1998; **55**: 1288–1292.
4 Richards RME, Reary JME. Changes in antibacterial activity of thiomersal and PMN on autoclaving with certain adjuvants. *J Pharm Pharmacol* 1972; **24**(Suppl.): 84P–89P.
5 Collins AJ, *et al.* Incompatibility of phenylmercuric acetate with sodium metabisulfite in eye drop formulations. *J Pharm Pharmacol* 1985; **37**(Suppl.): 123P.
6 Nater JP. Allergic contact dermatitis caused by potassium metabisulfite. *Dermatologica* 1968; **136**(6): 477–478.
7 Vena GA, *et al.* Sulfite contact allergy. *Contact Dermatitis* 1994; **31**(3): 172–175.
8 Twarog FJ. Metabisulfite sensitivity in asthma: a review. *N Engl Reg Allergy Proc* 1983; **4**(2): 100–103.
9 Mathison DA, *et al.* Precipitating factors in asthma: aspirin, sulfites, and other drugs and chemicals. *Chest* 1985; **87**(Suppl.): 50S–54S.
10 Anonymous. Sulfites in drugs and food. *Med Lett Drugs Ther* 1986; **28**: 74–75.
11 Belchi-Hernandez J, *et al.* Sulfite-induced urticaria. *Ann Allergy* 1993; **71**(3): 230–232.
12 Sweetman SC, ed. *Martindale: The Complete Drug Reference*, 37th edn. London: Pharmaceutical Press, 2011: 1807.
13 Anonymous. Warning for prescription drugs containing sulfites. *FDA Drug Bull* 1987; **17**: 2–3.
14 FAO/WHO. Evaluation of the toxicity of a number of antimicrobials and antioxidants. Sixth report of the joint FAO/WHO expert committee on food additives. *World Health Organ Tech Rep Ser* 1962; No. 228.
15 *Food Chemicals Codex*, 7th edn. Bethesda, MD: United States Pharmacopeia, 2010.

20 General References

Smolinske SC. *Handbook of Food, Drug and Cosmetic Excipients*. Boca Raton, FL: CRC Press, 1992; 393–406.

Valade J-P, Le Bras G. Sulfur dioxide release from effervescent tablets. *Rev Fr Oenol* 1998; **171**: 22–25.

21 Author

PJ Sheskey.

22 Date of Revision

1 March 2012.

P

Potassium Metaphosphate

1 Nonproprietary Names

USP–NF: Potassium Metaphosphate

2 Synonyms

E452; *Europhos KMP FG*; potassium Kurrol's salt; potassium polymetaphosphate; potassium polyphosphate.

3 Chemical Name and CAS Registry Number

Potassium metaphosphate [7790-53-6]

4 Empirical Formula and Molecular Weight

KPO_3 118.07

5 Structural Formula

K^+

$O{=}P({=}O){-}O^-$

6 Functional Category

Buffering agent.

7 Applications in Pharmaceutical Formulation or Technology

Potassium metaphosphate is used in pharmaceutical formulations as a buffering agent for oral suspensions.[1]

8 Description

Potassium metaphosphate occurs as a white, odorless powder.

9 Pharmacopeial Specifications

See Table I.

Table I: Pharmacopeial specifications for potassium metaphosphate.

Test	USP35-NF30
Identification	+
Viscosity	6.5–15 cps
Lead	≤5 ppb
Heavy metals	≤0.002%
Fluoride	≤0.001%
Assay (P_2O_5)	>59.0 to ≤61.0%

10 Typical Properties

Density 2.45 g/cm^3

Melting point 807°C

Solubility Insoluble in water; soluble in dilute solutions of sodium salts.

11 Stability and Storage Conditions

Stable under normal conditions of handling and storage. Store in well-closed containers.

12 Incompatibilities

—

13 Method of Manufacture

Potassium metaphosphate is prepared by the thermal dehydration of potassium dihydrogen phosphate.

14 Safety

LD_{50} (mouse, oral): 1.7 g/kg [2]
LD_{50} (rat, dermal): >7.94 g/kg
LD_{50} (rat, oral): 7.1 g/kg

15 Handling Precautions

Observe normal precautions appropriate to the circumstances and quantity of the material handled.

16 Regulatory Status

GRAS listed. Included in the FDA Inactive Ingredients Database (oral suspensions). Included in the Canadian Natural Health Products Ingredients Database.

17 Related Substances

—

18 Comments

Potassium metaphosphate has been investigated for use in a powder for suspension in the treatment of gastric-related disorders, comprising a therapeutically effective amount of an acid-labile, substituted benzimidazol proton pump inhibitor with a buffering agent in approx. 0.1–2.5 mEq per mg of the proton pump inhibitor. [1]

The low water solubility of potassium metaphosphate has led to it being evaluated as a polishing material in dental formulations, such as toothpaste and tooth powder. [3]

The EINECS number for potassium metaphosphate is 232-212-6. The PubChem Compound ID (CID) for potassium metaphosphate is 16133895.

19 Specific References

1 Phillips JO. Substituted benzimidazole dosage forms and method of using same. United States Patent 6,780,882; 2004.
2 Prayon. Material safety data sheet: *Europhos* KMP FG, 2003. http://www.prayon.com/media/pdf/gamme/ENGLISH/MSDS/food%20-grade/KMP_FG_EN1.pdf (accessed 1 July 2011).
3 Gaffar A *et al.* Antibacterial antiplaque, anticalculus oral composition. United States Patent 5,686,064; 1997.

20 General References

—

21 Author

RT Guest.

22 Date of Revision

1 March 2012.

Potassium Nitrate

1 Nonproprietary Names

BP: Potassium Nitrate
PhEur: Potassium Nitrate
USP–NF: Potassium Nitrate

2 Synonyms

E252; kalii nitras; kalium nitricum; niter; nitrate of potash; nitric acid, potassium salt; saltpeter.

3 Chemical Name and CAS Registry Number

Potassium nitrate [7757-79-1]

4 Empirical Formula and Molecular Weight

KNO_3 101.1

5 Structural Formula

6 Functional Category

Antimicrobial preservative.

7 Applications in Pharmaceutical Formulation or Technology

Potassium nitrate is used as a preservative in eye lotions and eye drops. It is widely used in foods and oral care products; *see* Section 18.

8 Description

Potassium nitrate occurs as a white or almost white crystalline powder or colorless crystals, with a cooling, saline, pungent taste.

9 Pharmacopeial Specifications

See Table I.

Table I: Pharmacopeial specifications for potassium nitrate.

Test	PhEur 7.4	USP35-NF30
Characters	+	–
Identification	+	+
Appearance of solution	+	–
Acidity or alkalinity	+	–
Reducible substances	+	–
Chlorides	≤20 ppm[a]	≤0.03%
Sulfates	≤150 ppm	≤0.1%
Ammonium	≤100 ppm[b]	–
Calcium	≤100 ppm[b]	–
Iron	≤20 ppm[c]	≤10 ppm
Sodium	≤0.1%	≤0.1%
Lead	–	≤10 ppm
Heavy metals	≤10 ppm	≤20 ppm
Loss on drying	≤0.5%	–
Limit of nitrite	–	≤5 μg/g
Assay (dried substance)	99.0–101.0%	99.0–100.5%

(a) If for ophthalmic use.
(b) ≤50 ppm if for ophthalmic use.
(c) ≤10 ppm if for ophthalmic use.

10 Typical Properties

Boiling point 400°C
Density 2.11 g/cm^3
Melting point 333°C
Solubility Freely soluble in water; very soluble in boiling water; practically insoluble in alcohol; soluble in glycerol.

11 Stability and Storage Conditions

Store in airtight containers in a cool, well-ventilated place.

12 Incompatibilities

Potassium nitrate is incompatible with strong oxidizers. An explosive reaction occurs with potassium chlorate or bromine trifluoride.

13 Method of Manufacture

Potassium nitrate is produced by double decomposition of sodium nitrate with potassium chloride.

14 Safety

Potassium nitrate in its pure form is irritating to the respiratory tract, and may cause coughing and shortness of breath. It is also irritating to the skin and eyes, causing redness, itching, and pain. Ingestion of large quantities may cause violent gastroenteritis. Prolonged exposure to small amounts may produce anemia, methemoglobinemia, and nephritis. The acceptable daily intake up to 3.7 mg/kg body weight daily is expressed as the nitrate ion.

LD_{50} (rabbit, oral): 1.9 g/kg[1]
LD_{50} (rat, oral): 3.75 g/kg

15 Handling Precautions

Observe normal precautions appropriate to the circumstances and quantity of the material handled. Contact with other materials may cause fire. Thermal decomposition produces highly toxic fumes. Avoid creating and inhaling dust. Protective eye goggles, clothing, and respiratory equipment should be worn.

16 Regulatory Status

Included in the Canadian Natural Health Products Ingredients Database.

17 Related Substances

Phenylmercuric nitrate.

18 Comments

Potassium nitrate is included in dentrifices to reduce tooth hypersensitivity.[2] When taken by mouth in dilute solution, potassium nitrate acts as a diuretic, and was formerly used for this purpose.[3]

Potassium nitrate has been used experimentally to induce corneal neovascularization in animal models.[4]

A specification for potassium nitrate is included in the *Food Chemicals Codex* (FCC).[5]

The EINECS number for potassium nitrate is 231-818-8.

19 Specific References

1 Lewis RJ, ed. *Sax's Dangerous Properties of Industrial Materials*. 11th edn. New York: Wiley, 2004; 3036.
2 Salian S, *et al*. A randomized controlled clinical study evaluating the efficacy of two desensitizing dentrifices. *J Clin Dent* 2010; **21**(3): 82–87.
3 Sweetman SC, ed. *Martindale: The Complete Drug Reference*. 37th edn. London: Pharmaceutical Press, 2011; 1803.
4 Avisar I, *et al*. Effect of subconjunctival and intraocular bevacizumab injections on corneal neovascularization in a mouse model. *Curr Eye Res* 2010; **35**(2): 108–115.
5 *Food Chemicals Codex*. 7th edn. Suppl. 3. Bethesda, MD: United States Pharmacopoeia, 2011; 1705.

20 General References

Jost Chemical and Company. Material safety data sheet: *Potassium nitrate*, January 2011. http://www.jostchemical.com/chemicals/3665.html (accessed 24 February 2012).
ScienceLab.com. Material safety data sheet: *Potassium nitrate*, June 2010.

21 Author

RT Guest.

22 Date of Revision

1 March 2012.

Potassium Phosphate, Dibasic

1 Nonproprietary Names

BP: Dipotassium Hydrogen Phosphate
PhEur: Dipotassium Phosphate
USP–NF: Dibasic Potassium Phosphate

2 Synonyms

Dikalii phosphas; dipotassium hydrogen phosphate; dipotassium phosphate; E340; phosphoric acid, dipotassium salt; potassium phosphate.

3 Chemical Name and CAS Registry Number

Dipotassium hydrogen orthophosphate [7758-11-4]
Dipotassium hydrogen orthophosphate [16788-57-1]

4 Empirical Formula and Molecular Weight

K_2HPO_4	174.2
$K_2HPO_4\ 3H_2O$	228.22

5 Structural Formula

See Section 4.

6 Functional Category

Buffering agent.

7 Applications in Pharmaceutical Formulation or Technology

Dibasic potassium phosphate is mainly used in pharmaceutical formulations as a buffering agent in oral and parenteral products.

8 Description

Dibasic potassium phosphate occurs as a very hygroscopic white powder or colorless crystals.

9 Pharmacopeial Specifications

See Table I.

Table I: Pharmacopeial specifications for dibasic potassium phosphate.

Test	PhEur 7.4[a]	USP35–NF30
Identification	+	+
Characters	+	–
Insoluble substances	–	≤0.2%
Chloride	≤200 ppm	≤0.03%
Sulfate	≤0.1%	≤0.1%
Arsenic	≤2 ppm	≤3 ppm
Iron	≤10 ppm	≤30 ppm
Heavy metals	≤10 ppm	≤10 ppm
Sodium	≤0.1%	+
Fluoride	–	≤0.001%
pH	–	8.5–9.6
Loss on drying	≤2.0%	≤1.0%
Carbonate	–	+
Limit of monobasic or tribasic salt	–	+
Limit of monobasic salt	≤2.5%	–
Reducing substances	+	–
Assay (dried basis)	98.0–101.0%	98.0–100.5%

(a) The PhEur 7.4 includes a limit for endotoxins for material intended for parenteral use.

10 Typical Properties

Acidity/alkalinity pH 8.5–9.6 for a 5% w/v solution in water at 25°C.
Density 2.44 g/cm^3
Melting point 465°C with decomposition.
Solubility Freely soluble in water (167 g/100 mL at 20 °C); very slightly soluble in ethanol (95%).

11 Stability and Storage Conditions

Dibasic potassium phosphate is very hygroscopic. Store in tightly closed containers.

12 Incompatibilities

Soluble phosphates are incompatible in solution with alkaloids, antipyrene, chloral hydrate, pyrogallol and resorcinol.[1] There is the potential for the formation of insoluble salts with divalent metal ions such as calcium; this should be considered when multiple injections and/or infusion fluids are to be administered to a patient.

13 Method of Manufacture

Dibasic potassium phosphate is produced when two moles of potassium hydroxide are added stoichiometrically to one mole of phosphoric acid (this is a very exothermic reaction). The solution is taken to a pH endpoint, filtered, evaporated and cooled. The crystallized product is centrifuged and dried.

14 Safety

Dibasic potassium phosphate powder may cause irritation of the gastrointestinal tract if ingested, and the respiratory tract if inhaled. It may cause irritation to skin, eyes, lungs, mucous membranes and the gastrointestinal tract.

15 Handling Precautions

Observe normal precautions appropriate to the circumstances and quantity of the material handled. Gloves, eye protection, a respirator, and other protective clothing should be worn.

16 Regulatory Status

GRAS listed. Accepted for use as a food additive in Europe. Included in the FDA Inactive Ingredients Database (oral capsules, powders, suspensions, syrups; IM, IV, SC injections). Included in the Canadian Natural Health Products Ingredients Database.

Dibasic potassium phosphate may be a component of oral rehydration mixtures. It is the active ingredient of Dipotassium Hydrogen Phosphate Injection BP.

17 Related Substances

Sodium phosphate, dibasic; monobasic potassium phosphate; tripotassium phosphate.

Monobasic potassium phosphate

Empirical formula KH_2PO_4
Molecular weight 136.1
CAS number [7778-77-0]
Synonyms Monopotassium phosphate; potassium acid phosphate; potassium biphosphate; potassium dihydrogen orthophosphate; potassium dihydrogen phosphate.
Appearance A white, crystalline powder or colorless crystals.
Solubility Freely soluble in water; practically insoluble in alcohol.
Comments The USP35-NF30 lists Monobasic Potassium Phosphate and the PhEur 7.4 lists Potassium Dihydrogen Phosphate. A specification for monobasic potassium phosphate is included in the *Food Chemicals Codex*.[2]

Tripotassium phosphate

Empirical formula K_3PO_4
Molecular weight 212.27
CAS number [7778-53-2]
Synonyms Tribasic potassium phosphate; tripotassium orthophosphate.
Appearance A white, crystalline powder or colorless crystals.
Solubility Freely soluble in water; practically insoluble in alcohol.
Comments Deliquescent. A food additive used in the preparation of whips and emulsions, and as a foam stabilizer. Aqueous solutions are highly alkaline. A specification for tribasic potassium phosphate is included in the *Food Chemicals Codex*.[2]

18 Comments

One gram of dibasic potassium phosphate contains approximately 11.5 mmol potassium and 5.7 mmol phosphate.

Dibasic potassium phosphate is used in fermentation and cell culture. A 2.08% w/v aqueous solution of dibasic potassium phosphate is isoosmotic with serum.

A specification for dibasic potassium phosphate is included in the *Food Chemicals Codex*.[2]

19 Specific References

1 Discher CA, Medwick T, Bailey LC. *Modern Inorganic Pharmaceutical Chemistry*, 2nd edn. Prospect Heights, IL: Waveland Press, 1985: 577.
2 *Food Chemicals Codex*, 7th edn. (Suppl. 2). Bethesda, MD: United States Pharmacopoeia, 2011; 1483–1484.

20 General References

—

21 Author

RC Moreton.

22 Date of Revision

1 March 2012.

P

Potassium Sorbate

1 Nonproprietary Names

BP: Potassium Sorbate
PhEur: Potassium Sorbate
USP–NF: Potassium Sorbate

2 Synonyms

E202; 2,4-hexadienoic acid (*E*,*E*)-potassium salt; kalii sorbas; potassium (*E*,*E*)-hexa-2,4-dienoate; potassium (*E*,*E*)-sorbate; sorbic acid potassium salt.

3 Chemical Name and CAS Registry Number

2,4-Hexadienoic acid potassium salt [24634-61-5]

4 Empirical Formula and Molecular Weight

$C_6H_7O_2K$ 150.22

5 Structural Formula

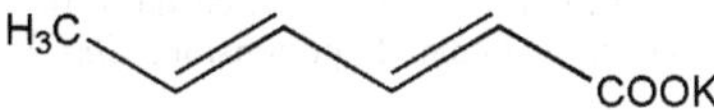

6 Functional Category

Antimicrobial preservative.

7 Applications in Pharmaceutical Formulation or Technology

Potassium sorbate is an antimicrobial preservative, with antibacterial and antifungal properties used in pharmaceuticals, foods, enteral preparations, and cosmetics. Generally, it is used at concentrations of 0.1–0.2% in oral and topical formulations, especially those containing nonionic surfactants. Potassium sorbate has been used to enhance the ocular bioavailability of timolol.(1)

Potassium sorbate is used in approximately twice as many pharmaceutical formulations as is sorbic acid owing to its greater solubility and stability in water. Like sorbic acid, potassium sorbate has minimal antibacterial properties in formulations above pH 6.

8 Description

Potassium sorbate occurs as a white crystalline powder with a faint, characteristic odor.

9 Pharmacopeial Specifications

See Table I.

Table I: Pharmacopeial specifications for potassium sorbate.

Test	PhEur 7.4	USP35–NF30
Identification	+	+
Characters	+	−
Appearance of solution	+	−
Acidity or alkalinity	+	+
Loss on drying	≤1.0%	≤1.0%
Heavy metals	≤10 ppm	≤10 ppm
Aldehydes (as C_2H_4O)	≤0.15%	−
Assay (dried basis)	99.0–101.0%	98.0–101.0%

10 Typical Properties

Antimicrobial activity Potassium sorbate is predominantly used as an antifungal preservative, although it also has antibacterial properties. Similarly to sorbic acid, the antimicrobial activity is dependent on the degree of dissociation; there is practically no antibacterial activity above pH 6. Preservative efficacy is increased with increasing temperature,(2) and increasing concentration of potassium sorbate.(2) The efficacy of potassium sorbate is also increased when used in combination with other antimicrobial preservatives or glycols since synergistic effects occur.(3) Reported minimum inhibitory concentrations (MICs) at the pH values indicated are shown in Table II.(3)

Table II: Minimum inhibitory concentrations (MIC) of potassium sorbate.

Microorganism	MIC (μg/mL) at the stated pH		
	5.5	6.0	7.0
Escherichia coli	1400	1500	3800
Pseudomonas aeruginosa	1600–2300	1900–2500	5600–9000
Staphylococcus aureus	1200	1000	3800

Density 1.363 g/cm^3
Melting point 270°C with decomposition.
Solubility *see* Table III.

Table III: Solubility of potassium sorbate.

Solvent	Solubility at 20°C unless otherwise stated
Acetone	1 in 1000
Benzene	Practically insoluble
Chloroform	Very slightly soluble
Corn oil	Very slightly soluble
Ethanol	1 in 50
Ethanol (95%)	1 in 35
Ethanol (5%)	1 in 1.7
Ether	Very slightly soluble
Propylene glycol	1 in 1.8
	1 in 2.1 at 50°C
	1 in 5 at 100°C
Water	1 in 1.72
	1 in 1.64 at 50°C
	1 in 1.56 at 100°C

Spectroscopy
IR spectra *see* Figure 1.
NIR spectra *see* Figure 2.

11 Stability and Storage Conditions

Potassium sorbate is more stable in aqueous solution than sorbic acid; aqueous solutions may be sterilized by autoclaving.

The bulk material should be stored in a well-closed container, protected from light, at a temperature not exceeding 40°C.

12 Incompatibilities

Some loss of antimicrobial activity occurs in the presence of nonionic surfactants and some plastics. *See also* Sorbic Acid.

13 Method of Manufacture

Potassium sorbate is prepared from sorbic acid and potassium hydroxide.

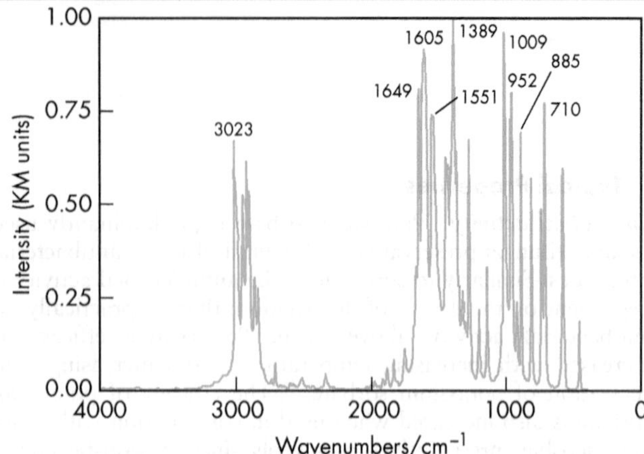

Figure 1: Infrared spectrum of potassium sorbate measured by diffuse reflectance. Adapted with permission of Informa Healthcare.

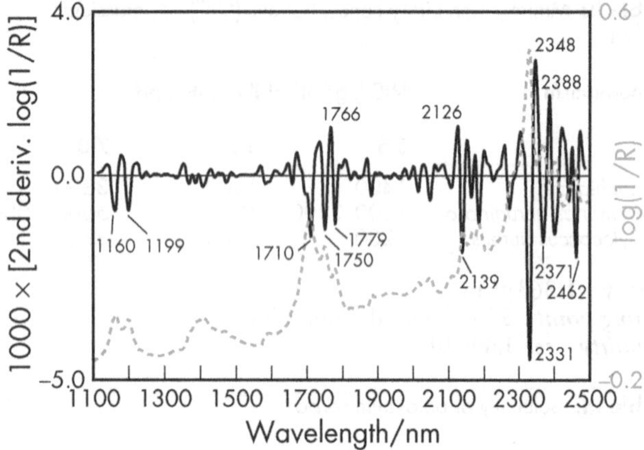

Figure 2: Near-infrared spectrum of potassium sorbate measured by reflectance.

14 Safety

Potassium sorbate is used as an antimicrobial preservative in oral and topical pharmaceutical formulations and is generally regarded as a relatively nontoxic material. However, some adverse reactions to potassium sorbate have been reported, including irritant skin reactions which may be of the allergic, hypersensitive type. There have been no reports of adverse systemic reactions following oral consumption of potassium sorbate.

The WHO has set an estimated total acceptable daily intake for sorbic acid, calcium sorbate, potassium sorbate, and sodium sorbate expressed as sorbic acid at up to 25 mg/kg body-weight.[4,5]

LD_{50} (mouse, IP): 1.3 g/kg[6]
LD_{50} (rat, oral): 4.92 g/kg
See also Sorbic Acid.

15 Handling Precautions

Observe normal precautions appropriate to the circumstances and quantity of material handled. Potassium sorbate is irritant to the skin, eyes, and mucous membranes; eye protection and gloves are recommended. In areas of limited ventilation, a respirator is also recommended.

16 Regulatory Status

GRAS listed. Accepted for use as a food additive in Europe. Included in the FDA Inactive Ingredients Database (nasal sprays; oral capsules, solutions, suspensions, syrups, tablets; topical creams and lotions). Included in nonparenteral medicines licensed in the UK. Included in the Canadian Natural Health Products Ingredients Database.

17 Related Substances

Sorbic acid.

18 Comments

Much of the information contained in the sorbic acid monograph on safety, incompatibilities, and references also applies to potassium, calcium, and sodium sorbates. *See* Sorbic Acid for further information.

Potassium sorbate has less antimicrobial activity than sorbic acid, but is more water soluble. Most potassium sorbate compounds will contain sorbic acid.

A specification for potassium sorbate is contained in the *Food Chemicals Codex* (FCC)[7] and the *Japanese Pharmaceutical Excipients* (JPE).[8]

The EINECS number for potassium sorbate is 246-376-1. The PubChem Compound ID (CID) for potassium sorbate includes 23676745 and 24184641.

19 Specific References

1 Mandorf TK, *et al.* A 12 month, multicentre, randomized, double-masked, parallel group comparison of timolol-LA once daily and timolol maleate ophthalmic solution twice daily in the treatment of adults with glaucoma or ocular hypertension. *Clin Ther* 2004; **26**(4): 541–551.
2 Lusher P, *et al.* A note on the effect of dilution and temperature on the bactericidal activity of potassium sorbate. *J Appl Bacteriol* 1984; **57**: 179–181.
3 Woodford R, Adams E. Sorbic acid. *Am Perfum Cosmet* 1970; **85**(3): 25–30.
4 FAO/WHO. Toxicological evaluation of certain food additives with a review of general principles and of specifications. Seventeenth report of the joint FAO/WHO expert committee on food additives. *World Health Organ Tech Rep Ser* 1974; No. 539.
5 FAO/WHO. Evaluation of certain food additives and contaminants. Twenty-ninth report of the joint FAO/WHO expert committee on food additives. *World Health Organ Tech Rep Ser* 1986; No. 733.
6 Lewis RJ, ed. *Sax's Dangerous Properties of Industrial Materials*, 11th edn. New York: Wiley, 2004: 3043.
7 *Food Chemicals Codex*, 7th edn. Bethesda, MD: United States Pharmacopeia, 2010: 859.
8 Japan Pharmaceutical Excipients Council. *Japanese Pharmaceutical Excipients 2004*. Tokyo: Yakuji Nippo, 2004: 722-723.

20 General References

Stopforth JD *et al.* Sorbic acid and sorbates. In: Davidson PM *et al*, eds. *Antimicrobials in Foods*, 3rd edn. Boca Raton, FL: CRC Press, 2005: 50-54.
Ho CY, *et al. In vitro* effects of preservatives in nasal sprays on human nasal epithelial cells. *Am J Rhinol* 2008; **22**(2): 125–129.
Scheler S, *et al.* Preservation of liquid drug preparations for oral administration. *J Pharm Sci* 2010; **99**(1): 357–367.
Smith J, Hong-Shum L. *Food Additives Databook*. New York, Wiley, 2007: chapter 267.
Smolinske SC, ed. *Handbook of Food, Drug, and Cosmetic Excipients*. Boca Raton, FL: CRC Press, 1992: 363–367.
Surhone LM *et al. Potassium sorbate*. Betascript Publishing, 2010.
Walker R. Toxicology of sorbic acid and sorbates. *Food Add Contam* 1990; **7**(5): 671–676.

21 Author

KE Boxell.

22 Date of Revision

1 March 2012.

Povidone

1 Nonproprietary Names

BP: Povidone
JP: Povidone
PhEur: Povidone
USP: Povidone

2 Synonyms

E1201; *Kollidon*; *Plasdone*; poly[1-(2-oxo-1-pyrrolidinyl)ethylene]; polyvidone; polyvinylpyrrolidone; povidonum; *Povipharm*; PVP; 1-vinyl-2-pyrrolidinone polymer.

3 Chemical Name and CAS Registry Number

1-Ethenyl-2-pyrrolidinone homopolymer [9003-39-8]

4 Empirical Formula and Molecular Weight

$(C_6H_9NO)_n$ 2 500–3 000 000

The USP35–NF30 describes povidone as a synthetic polymer consisting essentially of linear 1-vinyl-2-pyrrolidinone groups, the differing degree of polymerization of which results in polymers of various molecular weights. It is characterized by its viscosity in aqueous solution, relative to that of water, expressed as a *K*-value, in the range 10–120. The *K*-value is calculated using Fikentscher's equation:[1]

$$\log z = c\left[\frac{75k^2}{1+1.5kc}\right]+k$$

where z is the relative viscosity of the solution of concentration c (in % w/v), and k is the *K*-value × 10^{-3}.

Alternatively, the *K*-value may be determined from the following equation:

$$K\text{-value} = \sqrt{\frac{300c \log z\,(c+1.5c\log z)^2+1.5}{0.15c+0.003c^2}}$$

where z is the relative viscosity of the solution of concentration c (in % w/v).

Approximate molecular weights for different povidone grades are shown in Table I.

Table I: Approximate molecular weights for different grades of povidone.

K-value	Approximate molecular weight
12	2 500
15	8 000
17	10 000
25	30 000
30	50 000
60	400 000
90	1 000 000
120	3 000 000

See also Section 8.

5 Structural Formula

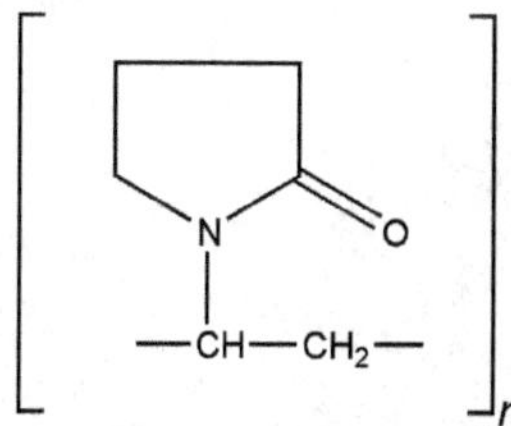

6 Functional Category

Coating agent; film-forming agent; solubilizing agent; suspending agent; tablet and capsule binder; tablet and capsule disintegrant.

7 Applications in Pharmaceutical Formulation or Technology

Although povidone is used in a variety of pharmaceutical formulations, it is primarily used in solid-dosage forms. In tableting, povidone solutions are used as binders in wet-granulation processes.[2,3] Povidone is also added to powder blends in the dry form and granulated *in situ* by the addition of water, alcohol, or hydroalcoholic solutions. Povidone is used as a solubilizer in oral and parenteral formulations, and has been shown to enhance dissolution of poorly soluble drugs from solid-dosage forms.[4–6] Povidone solutions may also be used as coating agents or as binders when coating active pharmaceutical ingredients on a support such as sugar beads.

Povidone is additionally used as a suspending, stabilizing, or viscosity-increasing agent in a number of topical and oral suspensions and solutions. The solubility of a number of poorly soluble active drugs may be increased by mixing with povidone. *See* Table II.

Special grades of pyrogen-free povidone are available and have been used in parenteral formulations; *see* Section 14.

Table II: Uses of povidone.

Use	Concentration (%)
Carrier for drugs	10–25
Dispersing agent	Up to 5
Eye drops	2–10
Suspending agent	Up to 5
Tablet binder, tablet diluent, or coating agent	0.5–5

8 Description

Povidone occurs as a fine, white to creamy-white colored, odorless or almost odorless, hygroscopic powder. Povidones with *K*-values equal to or lower than 30 are manufactured by spray-drying and occur as spheres. Povidone K-90 and higher *K*-value povidones are manufactured by drum drying and occur as plates.

9 Pharmacopeial Specifications

The pharmacopeial specifications for povidone have undergone harmonization of many attributes for JP, PhEur, and USP–NF.

See Table III. *See also* Section 18.

P

SEM 1: Excipient: povidone K-15 (*Plasdone K-15*); manufacturer: ISP; lot no.: 82A-1; magnification: 60×; voltage: 5 kV.

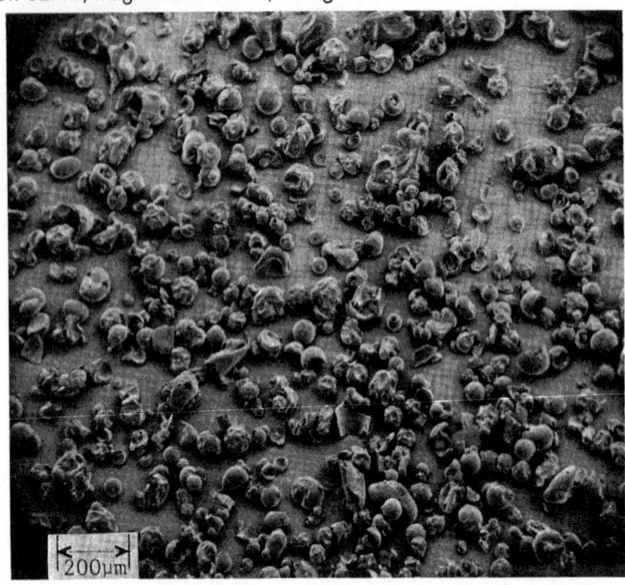

SEM 2: Excipient: povidone K-15 (*Plasdone K-15*); manufacturer: ISP; lot no.: 82A-1; magnification: 600×; voltage: 5 kV.

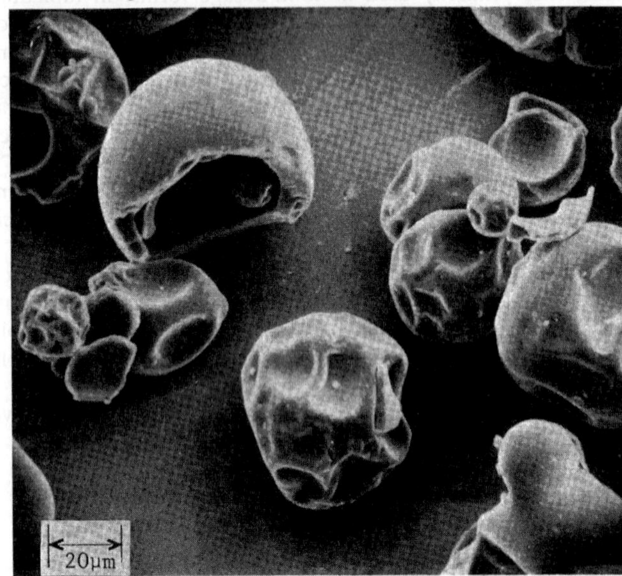

SEM 3: Excipient: povidone K-26/28 (*Plasdone K-26/28*); manufacturer: ISP; lot no.: 82A-2; magnification: 60×; voltage: 5 kV.

SEM 4: Excipient: povidone K-26/28 (*Plasdone K-26/28*); manufacturer: ISP; lot no.: 82A-2; magnification: 600×; voltage: 10 kV.

10 Typical Properties

Acidity/alkalinity pH = 3.0–7.0 (5% w/v aqueous solution); pH = 4.0–7.0 (5% w/v aqueous solution) for *Povipharm K-90*.

Density (bulk) 0.29–0.39 g/cm^3 for *Plasdone*.

Density (tapped) 0.39–0.54 g/cm^3 for *Plasdone*.

Density (true) 1.180 g/cm^3

Flowability

20 g/s for povidone K-15;

16 g/s for povidone K-29/32.

Melting point Softens at 150°C.

Moisture content Povidone is very hygroscopic, significant amounts of moisture being absorbed at low relative humidities. *See* Figures 1 and 2.

Particle size distribution

Kollidon 25/30: 90% >50 μm, 50% >100 μm, 5% >200 μm;

Kollidon 90: 90% >200 μm, 95% >250 μm.[7]

Solubility Freely soluble in acids, chloroform, ethanol (95%), ketones, methanol, and water; practically insoluble in ether, hydrocarbons, and mineral oil. In water, the concentration of a solution is limited only by the viscosity of the resulting solution, which is a function of the *K*-value.

Spectroscopy

IR spectra *see* Figure 3.

NIR spectra *see* Figure 4.

Surface tension Water-soluble polymers such as povidone will have an impact on the surface tension of water. They do not have the classic impact associated with surfactants where the drop in surface tension occurs at low concentration and plateaus at the critical micelle concentration (CMC), but instead continue to have an impact on the surface tension until the concentration reaches the gel point; *see* Figures 5 and 6.[8]

Viscosity (dynamic) The viscosity of aqueous povidone solutions depends on both the concentration and the molecular weight of the polymer employed. *See* Tables IV and V.[7]

SEM 5: Excipient: povidone K-30 (*Plasdone K-30*); manufacturer: ISP; lot no.: 82A-4; magnification: 60×; voltage: 10 kV.

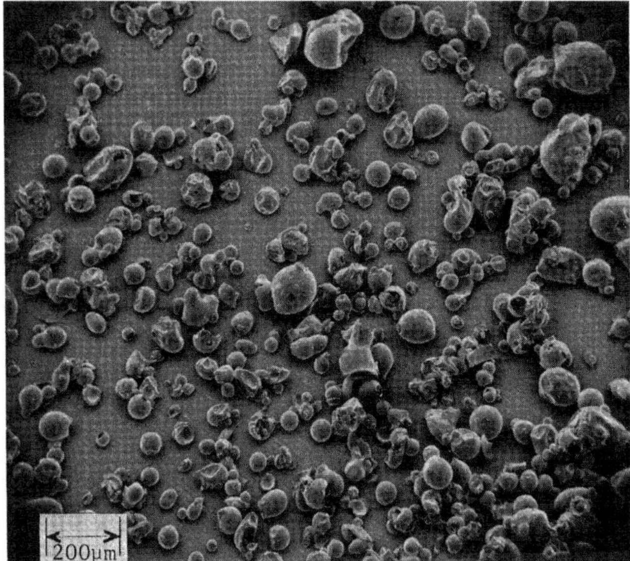

SEM 6: Excipient: povidone K-30 (*Plasdone K-30*); manufacturer: ISP; lot no.: 82A-4; magnification: 600×; voltage: 10 kV.

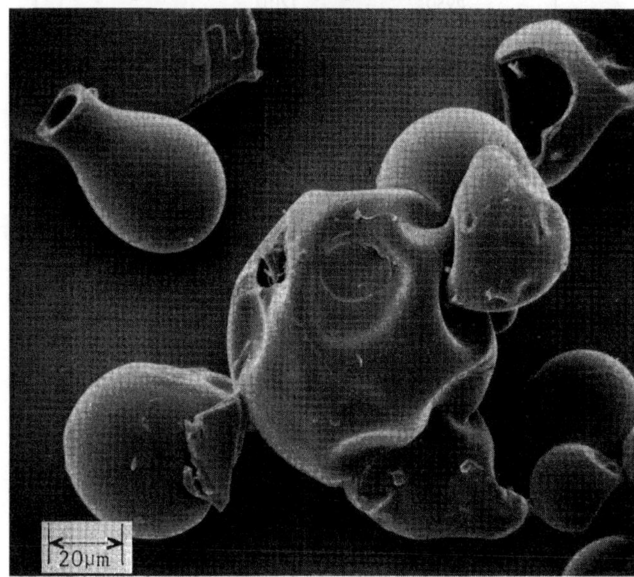

SEM 7: Excipient: povidone K-29/32 (*Plasdone K-29/32*); manufacturer: ISP; lot no.: 82A-3; magnification: 60×; voltage: 5 kV.

SEM 8: Excipient: povidone K-29/32 (*Plasdone K-29/32*); manufacturer: ISP; lot no.: 82A-3; magnification: 600×; voltage: 10 kV.

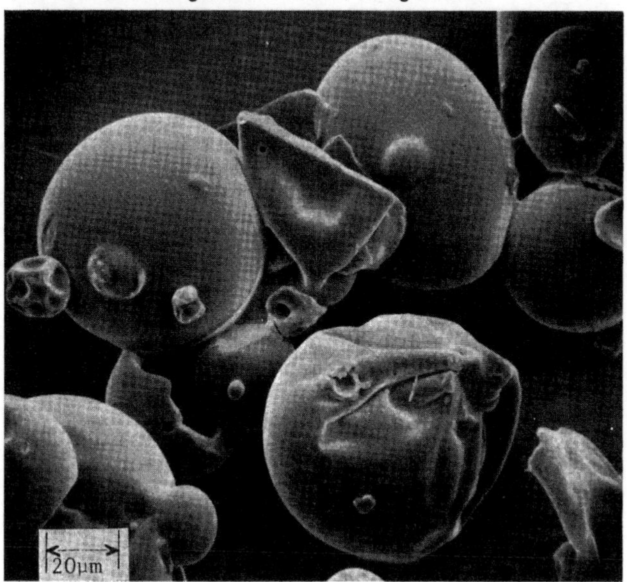

11 Stability and Storage Conditions

Povidone darkens to some extent on heating at 150°C, with a reduction in aqueous solubility. It is stable to a short cycle of heat exposure around 110–130°C; steam sterilization of an aqueous solution does not alter its properties. Aqueous solutions are susceptible to mold growth and consequently require the addition of suitable preservatives.

Povidone may be stored under ordinary conditions without undergoing decomposition or degradation. However, since the powder is hygroscopic, it should be stored in an airtight container in a cool, dry place.

12 Incompatibilities

Povidone is compatible in solution with a wide range of inorganic salts, natural and synthetic resins, and other chemicals. It forms molecular adducts in solution with sulfathiazole, sodium salicylate, salicylic acid, phenobarbital, tannin, and other compounds; *see* Section 18. The efficacy of some preservatives, e.g. thimerosal, may be adversely affected by the formation of complexes with povidone.

13 Method of Manufacture

Povidone is manufactured by the Reppe process. Acetylene and formaldehyde are reacted in the presence of a highly active copper acetylide catalyst to form butynediol, which is hydrogenated to butanediol and then cyclodehydrogenated to form butyrolactone. Pyrrolidone is produced by reacting butyrolactone with ammonia. This is followed by a vinylation reaction in which pyrrolidone and acetylene are reacted under pressure. The monomer, vinylpyrrolidone, is then polymerized in the presence of a combination of catalysts to produce povidone.

Table III: Pharmacopeial specifications for povidone.

Test	JP XV	PhEur 7.4	USP35–NF30
Identification	+	+	+
Characters	—	+	—
pH	+	+	+
$K \leq 30$	3.0–5.0	3.0–5.0	3.0–5.0
$K > 30$	4.0–7.0	4.0–7.0	4.0–7.0
Appearance of solution	+	+	—
Viscosity	—	+	+
Water	≤5.0%	≤5.0%	≤5.0%
Residue on ignition	≤0.1%	≤0.1%	≤0.1%
Lead	—	—	≤10 ppm
Aldehydes	≤500 ppm[(a)]	≤500 ppm[(a)]	≤0.05%
Formic acid	—	≤0.5%	≤0.5%
Hydrazine	≤1 ppm	≤1 ppm	≤1 ppm
Vinylpyrrolidinone	≤10 ppm	—	≤0.001%
Pyrrolidone	—	≤3.0%	≤3.0%
Peroxides	≤400 ppm[(b)]	≤400 ppm[(b)]	≤400 ppm[(b)]
K-value	25–90	—	—
≤15	90.0–108.0%	85.0–115.0%	85.0–115.0%
>15	90.0–108.0%	90.0–108.0%	90.0–108.0%
Heavy metals[(c)]	≤10 ppm	≤10 ppm	—
Assay (nitrogen content)	11.5–12.8%	11.5–12.8%	11.5–12.8%

(a) Expressed as acetaldehyde.
(b) Expressed as hydrogen peroxide.
(c) These tests have not been fully harmonized at the time of publication.

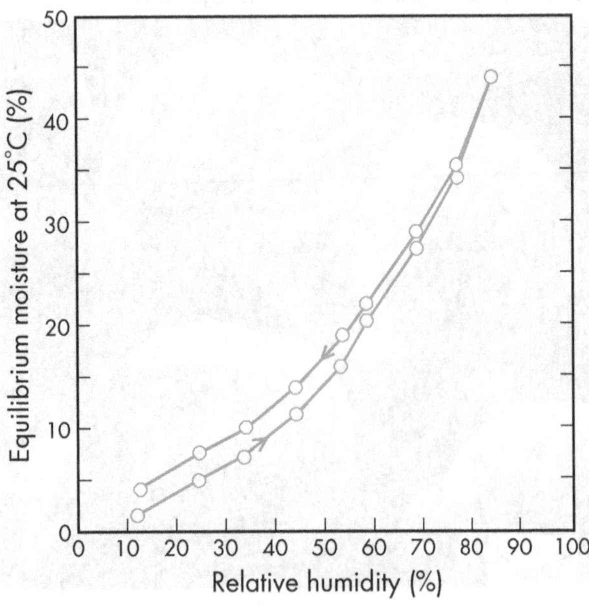

Figure 1: Sorption–desorption isotherm of povidone K-15 (*Plasdone K-15*, ISP).

14 Safety

Povidone has been used in pharmaceutical formulations for many years, being first used in the 1940s as a plasma expander, although it has now been superseded for this purpose by dextran.[(9)]

Povidone is widely used as an excipient, particularly in oral tablets and solutions. When consumed orally, povidone may be regarded as essentially nontoxic since it is not absorbed from the gastrointestinal tract or mucous membranes.[(9)] Povidone additionally has no irritant effect on the skin and causes no sensitization.

Reports of adverse reactions to povidone primarily concern the formation of subcutaneous granulomas at the injection site of intramuscular injections formulated with povidone.[(10)] Evidence also exists that povidone may accumulate in the organs of the body following intramuscular injection.[(11)]

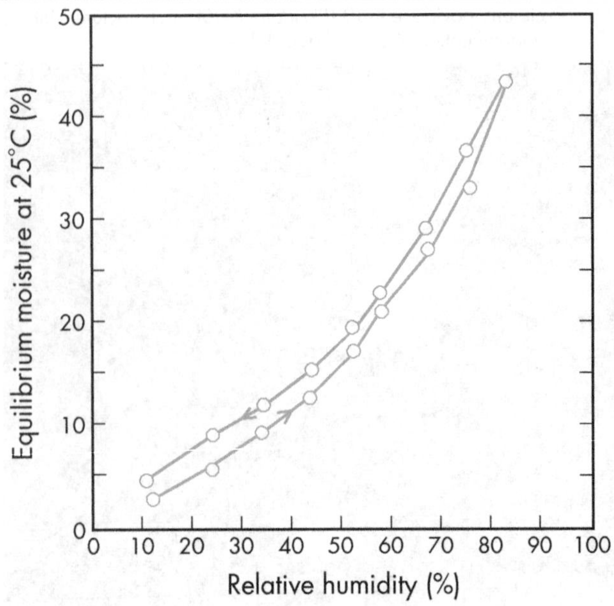

Figure 2: Sorption–desorption isotherm of povidone K-29/32 (*Plasdone K-29/32*, ISP).

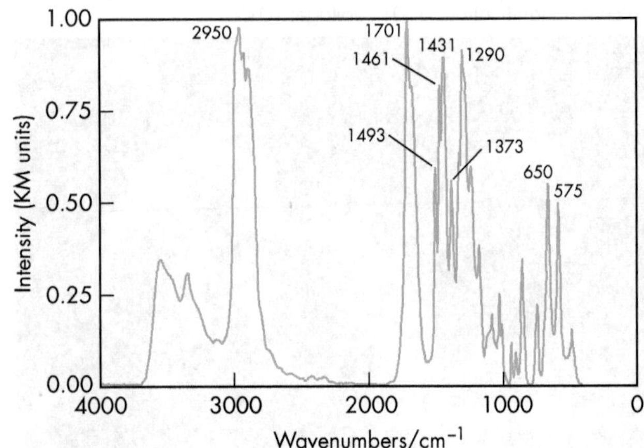

Figure 3: Infrared spectrum of povidone measured by diffuse reflectance. Adapted with permission of Informa Healthcare.

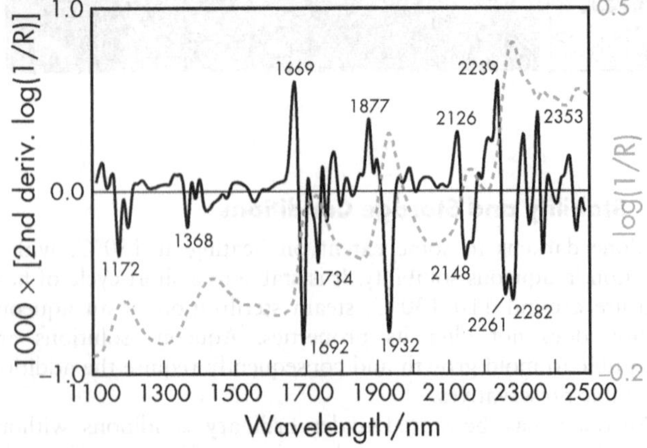

Figure 4: Near-infrared spectrum of povidone measured by reflectance.

A temporary acceptable daily intake for povidone has been set by the WHO at up to 25 mg/kg body-weight.[(12)]

LD_{50} (mouse, IP): 12 g/kg[(13)]

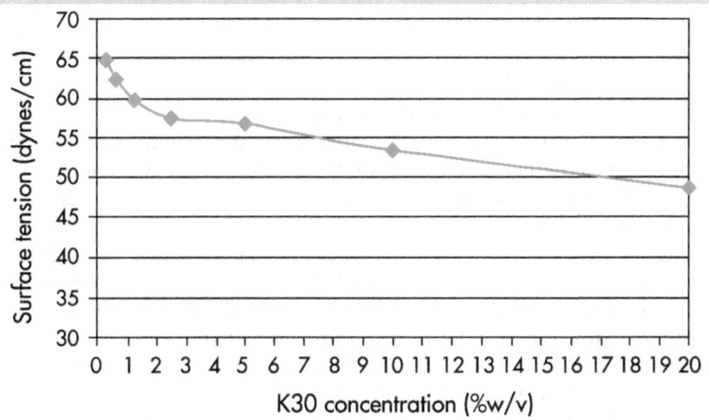

Figure 5: Surface tension of povidone K-30 solutions.

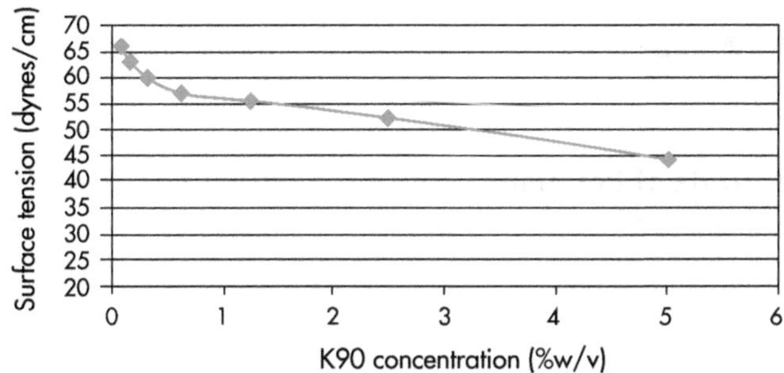

Figure 6: Surface tension of povidone K-90 solutions.

Table IV: Dynamic viscosity of 10% w/v aqueous povidone (*Kollidon*) solutions at 20°C.[7]

Grade	Dynamic viscosity (mPa s)
K-11/14	1.3–2.3
K-16/18	1.5–3.5
K-24/27	3.5–5.5
K-28/32	5.5–8.5
K-85/95	300–700

Table V: Dynamic viscosity of 5% w/v povidone (*Kollidon*) solutions in ethanol (95%) and propan-2-ol at 25°C.[7]

Grade	Dynamic viscosity (mPa s)	
	Ethanol (95%)	Propan-2-ol
K-12PF	1.4	2.7
K-17PF	1.9	3.1
K-25	2.7	4.7
K-30	3.4	5.8
K-90	53.0	90.0

15 Handling Precautions

Observe normal precautions appropriate to the circumstances and quantity of material handled. Eye protection, gloves, and a dust mask are recommended.

16 Regulatory Status

Accepted for use in Europe as a food additive. Included in the FDA Inactive Ingredients Database (IM and IV injections; ophthalmic preparations; oral capsules, drops, granules, suspensions, and tablets; sublingual tablets; topical and vaginal preparations). Included in nonparenteral medicines licensed in the UK. Included in the Canadian Natural Health Products Ingredients Database.

17 Related Substances

Crospovidone.

18 Comments

Povidone is one of the materials that have been selected for harmonization by the Pharmacopeial Discussion Group. For further information see the General Information Chapter <1196> in the USP35–NF30, the General Chapter 5.8 in PhEur 7.4, along with the 'State of Work' document on the PhEur EDQM website, and also the General Information Chapter 8 in the JP XV.

Povidone has been extensively studied for its impact on dissolution and stability in various formulations, such as gelleable drug forms,[14] spray-dried dispersions,[15] nasal delivery systems,[16] and drug nanoparticles.[17]

The molecular adduct formation properties of povidone may be used advantageously in solutions, slow-release solid-dosage forms, and parenteral formulations. Perhaps the best-known example of povidone complex formation is povidone–iodine, which is used as a topical disinfectant.

For accurate standardization of solutions, the water content of the solid povidone must be determined before use and taken into account for any calculations. Many excipients such as povidone may contain peroxides as trace contaminants. These can lead to degradation of an active pharmaceutical ingredient that is sensitive to oxidation.

A specification for povidone is contained in the *Food Chemicals Codex* (FCC).[18]

19 Specific References

1 Fikentscher H, Herrle K. Polyvinylpyrrolidone. *Modern Plastics* 1945; 23(3): 157–161212, 214, 216, 218.
2 Becker D, *et al*. Effectiveness of binders in wet granulation: comparison using model formulations of different tabletability. *Drug Dev Ind Pharm* 1997; 23(8): 791–808.
3 Stubberud L, *et al*. Water–solid interactions. Part 3. Effect of glass transition temperature, T_g and processing on tensile strength of compacts of lactose and lactose/polyvinyl pyrrolidone. *Pharm Dev Technol* 1996; 1(2): 195–204.
4 Iwata M, Ueda H. Dissolution properties of glibenclamide in combinations with polyvinylpyrrolidone. *Drug Dev Ind Pharm* 1996; **22**: 1161–1165.
5 Lu WG, *et al*. Development of nifedipine (NE) pellets with a high bioavailability. *Chin Pharm J Zhongguo Yaoxue Zazhi* 1995; **30**: 24–26.
6 Chowdary KP, Ramesh KV. Microencapsulation of solid dispersions of nifedipine – novel approach for controlling drug release. *Indian Drugs* 1995; 32(Oct): 477–483.
7 BASF Corporation. Technical literature: Soluble Kollidon Grades, Soluble Polyvinylpyrrolidone for the Pharmaceutical Industry, 1997.
8 Kibbe AH, Jadeja VJ. Characterization of compounds: Effect of surfactants, polymers, and nonionic compounds on surface tension at liquid-air interphase. AAPS Meeting Report, November 2009.
9 Wessel W, *et al*. Polyvinylpyrrolidone (PVP), its diagnostic, therapeutic and technical application and consequences thereof. *Arzneimittelforschung* 1971; **21**: 1468–1482.
10 Hizawa K, *et al*. Subcutaneous pseudosarcomatous polyvinylpyrrolidone granuloma. *Am J Surg Pathol* 1984; **8**: 393–398.
11 Christensen M, *et al*. Storage of polyvinylpyrrolidone (PVP) in tissues following long-term treatment with a PVP containing vasopressin preparation. *Acta Med Scand* 1978; **204**: 295–298.
12 FAO/WHO. Evaluation of certain food additives and contaminants. Twenty-seventh report of the joint FAO/WHO expert committee on food additives. *World Health Organ Tech Rep Ser* 1983; No. 696.
13 Lewis RJ, ed. *Sax's Dangerous Properties of Industrial Materials*, 11th edn. New York: Wiley, 2004: 3016–3017.
14 Fan C, *et al*. Impact of polymers on dissolution performance of an amorphous gelleable drug from surface-coated beads. *Eur J Pharm Sci* 2009; 37(1): 1–10.

15 Thybo P, *et al.* Characterization and physical stability of spray-dried solid dispersions of probucol and PVP-K30. *Pharm Dev Technol* 2008; **13**(5): 375–386.
16 Alsarra IA, *et al.* Mucoadhesive polymeric hydrogels for nasal delivery of acyclovir. *Drug Dev Ind Pharm* 2009; **35**(3): 352–362.
17 Sepassi S, *et al.* Effect of polymer molecular weight on the production of drug nanoparticles. *J Pharm Sci* 2007; **96**(10): 2655–2666.
18 *Food Chemicals Codex*, 7th edn. Bethesda, MD: United States Pharmacopeia, 2010: 862.

20 General References

Adeyeye CM, Barabas E. Povidone. In: Brittain HG, ed. *Analytical Profiles of Drug Substances and Excipients*. 22. London: Academic Press, 1993: 555–685.
European Directorate for the Quality of Medicines and Healthcare (EDQM). European Pharmacopoeia – State Of Work Of International Harmonisation. *Pharmeuropa* 2011; 22(4): 583–584. www.edqm.eu/en/International-Harmonisation-614.html (accessed 4 December 2011).
Genovesi A, *et al.* Binder evaluation in tabletting. *Manuf Chem* 2004; **175**(6): 29–30.
Horn D, Ditter W. Chromatographic study of interactions between polyvinylpyrrolidone and drugs. *J Pharm Sci* 1982; **71**: 1021–1026.
Hsiao CH, *et al.* Fluorescent probe study of sulfonamide binding to povidone. *J Pharm Sci* 1977; **66**: 1157–1159.
ISP. Technical literature: *Plasdone* povidone USP, 1999.
Jager KF, Bauer KH. Polymer blends from PVP as a means to optimize properties of fluidized bed granulates and tablets. *Acta Pharm Technol* 1984; **30**(1): 85–92.
NP Pharm. Product data sheet: Povipharm, 2008.
Plaizier-Vercammen JA, DeNève RE. Interaction of povidone with aromatic compounds III: thermodynamics of the binding equilibria and interaction forces in buffer solutions at varying pH values and varying dielectric constant. *J Pharm Sci* 1982; **71**: 552–556.
Robinson BV *et al. PVP: A Critical Review of the Kinetics and Toxicology of Polyvinylpyrrolidone (Povidone)*. Chelsea, MI: Lewis Publishers, 1990.
Shefter E, Cheng KC. Drug–polyvinylpyrrolidone (PVP) dispersions. A differential scanning calorimetric study. *Int J Pharm* 1980; **6**: 179–182.
Smolinske SC. *Handbook of Food, Drug, and Cosmetic Excipients*. Boca Raton, FL: CRC Press, 1992: 303–305.
Wasylaschuk WR, *et al.* Evaluation of hydroperoxides in common pharmaceutical excipients. *J Pharm Sci* 2007; **96**: 106–116.

21 Author

AH Kibbe.

22 Date of Revision

1 March 2012.

Propionic Acid

1 Nonproprietary Names

USP–NF: Propionic Acid

2 Synonyms

Carboxyethane; ethanecarboxylic acid; E280; ethylformic acid; metacetonic acid; methylacetic acid; propanoic acid; pseudoacetic acid.

3 Chemical Name and CAS Registry Number

Propionic acid [79-09-4]

4 Empirical Formula and Molecular Weight

$C_3H_6O_2$ 74.08

5 Structural Formula

OH
H_3C
O

6 Functional Category

Acidulant; antimicrobial preservative; antioxidant.

7 Applications in Pharmaceutical Formulation or Technology

Propionic acid is primarily used as an antioxidant and antimicrobial preservative in foods, and in oral and topical pharmaceutical applications.

8 Description

Propionic acid occurs as a corrosive, oily liquid having a slightly pungent, disagreeable, rancid odor. It is flammable.

9 Pharmacopeial Specifications

See Table I.

Table I: Pharmacopeial specifications for propionic acid.

Test	USP35–NF30
Specific gravity	0.988–0.993
Distilling range	138.5–142.5°C
Heavy metals	≤10 ppm
Limit of nonvolatile residue	≤0.01%
Readily oxidizable substances	+
Limit of aldehydes	+
Assay	99.5–100.5%

10 Typical Properties

Antimicrobial activity *see* Table II.
Autoignition temperature 955°C

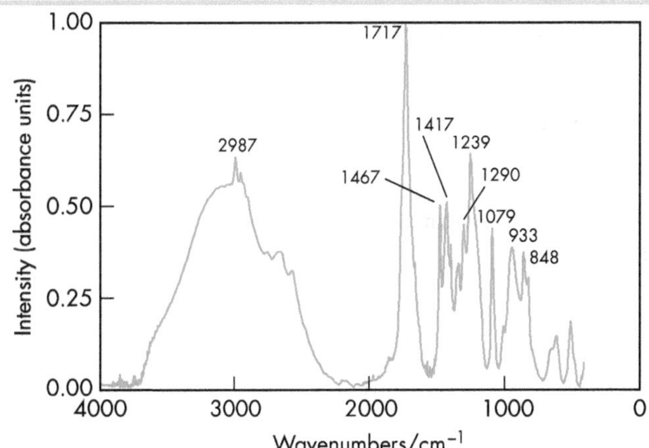

Figure 1: Infrared spectrum of propionic acid measured by transmission. Adapted with permission of Informa Healthcare.

Table II: Typical minimum inhibitory concentrations (MICs) for propionic acid at pH 3.9.[1]

Microorganism	MIC (μg/mL)
Aspergillus niger	2000
Candida albicans	2000
Escherichia coli	2000
Klebsiella pneumoniae	1250
Penicillium notatum	2000
Pseudomonas aeruginosa	3000
Pseudomonas cepacia	3000
Pseudomonas fluorescens	1250
Staphylococcus aureus	2000

Boiling point 141.1°C
Dissociation constant $pK_a = 4.874$
Flash point 52–58°C (open cup)
Melting point −21.5°C
Partition coefficients Octanol : water = 0.33.
Refractive index $n_D^{25} = 1.3848$
Solubility Miscible with chloroform, ethanol (95%), ether, and water.
Specific gravity 0.9934
Spectroscopy
IR spectra *see* Figure 1.
Surface tension 27.21 mN/m (27.21 dynes/cm) at 15°C
Vapor density (relative) 2.56 (air = 1)
Vapor pressure 320 Pa (2.4 mmHg) at 20°C
Viscosity (dynamic) *see* Table III.

Table III: Dynamic viscosity of propionic acid.

Viscosity (dynamic) (mPa s)	Temperature
1.175	15°C
1.02	25°C
0.956	30°C
0.668	60°C
0.495	90°C

11 Stability and Storage Conditions

Although stable, propionic acid is flammable. It should be stored in an airtight container away from heat and flames.

12 Incompatibilities

Propionic acid is incompatible with alkalis, ammonia, amines, and halogens. It can be salted out of aqueous solutions by the addition of calcium chloride or other salts.

13 Method of Manufacture

Propionic acid can be obtained from wood pulp waste liquor by fermentation. It can also be prepared from ethylene, carbon monoxide and steam; from ethanol and carbon monoxide using boron trifluoride catalyst; from natural gas; or as a by-product in the pyrolysis of wood. Very pure propionic acid can be obtained from propionitrile. Propionic acid can be found in dairy products in small amounts.

14 Safety

Propionic acid is generally regarded as a nontoxic and nonirritant material when used in low levels as an excipient. Up to 1% may be used in food applications (up to 0.3% in flour and cheese products). Propionic acid is readily metabolized.

The pure form of propionic acid is corrosive and will cause burns to any area of contact. Both liquid and vapor forms are flammable. Concentrated propionic acid is harmful if swallowed, inhaled or absorbed through the skin. *See also* Sodium Propionate.

LD_{50} (mouse, IV): 0.63 g/kg[2]
LD_{50} (rabbit, skin): 0.5 g/kg
LD_{50} (rat, oral): 2.6 g/kg

15 Handling Precautions

Propionic acid is corrosive and can cause eye and skin burns. It may be harmful if swallowed, inhaled or absorbed through the skin as a result of prolonged or widespread contact. Eye protection, PVC gloves, and suitable protective clothing should be worn. Propionic acid should be handled in a well-ventilated environment away from heat and flames. In the UK, the workplace exposure limits for propionic acid are 31 mg/m^3 (10 ppm) long-term (8-hour TWA) and 46 mg/m^3 (15 ppm) short-term.[3]

16 Regulatory Status

GRAS listed. Accepted for use in Europe as a food additive. In Japan, propionic acid is restricted to use as a flavoring agent. Included in the Canadian Natural Health Products Ingredients Database.

17 Related Substances

Calcium propionate; sodium propionate.

18 Comments

A specification for propionic acid is contained in the *Food Chemicals Codex* (FCC)[4] and the *Japanese Pharmaceutical Excipients* (JPE).[5]

The EINECS number for propionic acid is 201-176-3. The PubChem Compound ID (CID) for propionic acid is 1032.

19 Specific References

1 Wallhäusser KH. Propionic acid. In: Kabara JJ, ed. *Cosmetic and Drug Preservation: Principles and Practice.* New York: Marcel Dekker, 1984: 665–666.

2 Lewis RJ, ed. *Sax's Dangerous Properties of Industrial Materials*, 11th edn. New York: Wiley, 2004: 3069–3070.
3 Health and Safety Executive. *EH40/2005 Workplace Exposure Limits*. Sudbury: HSE Books, 2011. http://www.hse.gov.uk/pubns/priced/eh40.pdf (accessed 16 February 2012).
4 *Food Chemicals Codex*, 6th edn. Bethesda, MD: United States Pharmacopeia, 2008: 825.
5 Japan Pharmaceutical Excipients Council. *Japanese Pharmaceutical Excipients 2004*. Tokyo: Yakuji Nippo, 2004: 732-733.

20 General References

—

21 Author

GE Amidon.

22 Date of Revision

16 February 2012.

Propyl Gallate

1 Nonproprietary Names

BP: Propyl Gallate
PhEur: Propyl Gallate
USP–NF: Propyl Gallate

2 Synonyms

E310; gallic acid propyl ester; *n*-propyl gallate; *Progallin P*; propyl 3,4,5-trihydroxybenzoate; propylis gallas; *Tenox PG*.

3 Chemical Name and CAS Registry Number

3,4,5-Trihydroxybenzoic acid propyl ester [121-79-9]

4 Empirical Formula and Molecular Weight

$C_{10}H_{12}O_5$ 212.20

5 Structural Formula

HO, HO, OH, O, O, CH_3

6 Functional Category

Antioxidant; antimicrobial preservative.

7 Applications in Pharmaceutical Formulation or Technology

Propyl gallate has become widely used as an antioxidant in cosmetics, perfumes, foods, and pharmaceuticals since its use in preventing autoxidation of oils was first described in 1943.[1,2] It is primarily used, in concentrations up to 0.1% w/v, to prevent the rancidity of oils and fats;[3] it may also be used at concentrations of 0.002% w/v to prevent peroxide formation in ether, and at 0.01% w/v to prevent the oxidation of paraldehyde. Synergistic effects with other antioxidants such as butylated hydroxyanisole and butylated hydroxytoluene have been reported. Propyl gallate is also said to possess some antimicrobial properties; *see* Section 10.

Studies have shown that, when added to powder blends containing ketorolac, propyl gallate significantly increases the drug stability in the preparation.[4]

Propyl gallate has approximately equivalent antioxidant properties to other alkyl gallates when used in equimolar concentration; however, solubilities vary; *see* Section 17.

8 Description

Propyl gallate is a white, odorless or almost odorless crystalline powder, with a bitter astringent taste that is not normally noticeable at the concentrations employed as an antioxidant.

9 Pharmacopeial Specifications

See Table I. *See also* Section 18.

Table I: Pharmacopeial specifications for propyl gallate.

Test	PhEur 7.4	USP35–NF30
Identification	+	+
Characters	+	–
Melting range	148–151°C	146–150°C
Appearance of solution	+	–
Gallic acid	+	–
Loss on drying	≤0.5%	≤0.5%
Residue on ignition	–	≤0.1%
Sulfated ash	≤0.1%	–
Total chlorine	≤200 ppm	–
Chloride	≤100 ppm	–
Heavy metals	≤10 ppm	≤10 ppm
Zinc	≤25 ppm	–
Assay (dried basis)	97.0–103.0%	98.0–102.0%

10 Typical Properties

Acidity/alkalinity pH = 5.9 (0.1% w/v aqueous solution)

Antimicrobial activity Propyl gallate has been reported to possess some antimicrobial activity against Gram-negative, Gram-positive, and fungal species.[5] Its effectiveness as a preservative may be improved when used in combination with zinc salts, such as zinc sulfate, owing to synergistic effects.[6] For reported minimum inhibitory concentrations (MICs) for aqueous solutions containing 4% v/v ethanol as cosolvent, *see* Table II.[5]

Dissociation constant $pK_a = 8.11$

Melting point 150°C

Partition coefficients

Octanol : water = 32;

Oleyl alcohol : water = 17.

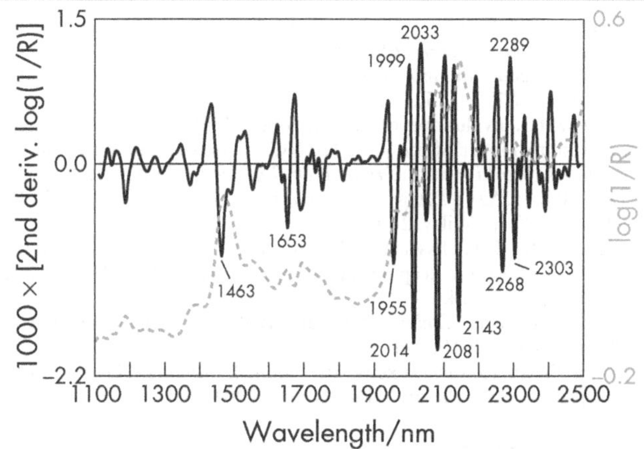

Figure 1: Near-infrared spectrum of propyl gallate measured by reflectance.

Table II: Minimum inhibitory concentrations (MICs) for aqueous solutions containing propyl gallate and 4% v/v ethanol.

Microorganism	MIC (mg/mL)
Candida albicans	1500
Escherichia coli	330
Staphylococcus aureus	600

Table III: Solubility of propyl gallate.

Solvent	Solubility at 20°C unless otherwise stated
Almond oil	1 in 44
Castor oil	1 in 4.5
Cottonseed oil	1 in 81 at 30°C
Ethanol (95%)	1 in 3
	1 in 0.98 at 25°C
Ether	1 in 3
	1 in 1.2 at 25°C
Lanolin	1 in 16.7 at 25°C
Lard	1 in 88 at 45°C
Mineral oil	1 in 200
Peanut oil	1 in 2000
Propylene glycol	1 in 2.5 at 25°C
Soybean oil	1 in 100 at 25°C
Water	1 in 1000
	1 in 286 at 25°C

Solubility *see* Table III.
Spectroscopy
NIR spectra *see* Figure 1.

11 Stability and Storage Conditions

Propyl gallate is unstable at high temperatures and is rapidly destroyed in oils that are used for frying purposes.

The bulk material should be stored in a well-closed, nonmetallic container, protected from light, in a cool, dry place.

12 Incompatibilities

The alkyl gallates are incompatible with metals, e.g. sodium, potassium, and iron, forming intensely colored complexes. Complex formation may be prevented, under some circumstances, by the addition of a sequestering agent, typically citric acid. Propyl gallate may also react with oxidizing materials.

13 Method of Manufacture

Propyl gallate is prepared by the esterification of 3,4,5-trihydroxybenzoic acid (gallic acid) with *n*-propanol. Other alkyl gallates are prepared similarly using an appropriate alcohol of the desired alkyl chain length.

14 Safety

It has been reported, following animal studies, that propyl gallate has a strong contact sensitization potential.[7] Propyl gallate has also produced cytogenic effects in CHO-K1 cells.[8] However, despite this, there have been few reports of adverse reactions to propyl gallate.[9] Those that have been described include contact dermatitis, allergic contact dermatitis,[9–13] and methemoglobinemia in neonates.[14]

The WHO has set an estimated acceptable daily intake for propyl gallate at up to 1.4 mg/kg body-weight.[15]

LD_{50} (cat, oral): 0.4 g/kg[16]
LD_{50} (mouse, oral): 1.7 g/kg
LD_{50} (rat, oral): 2.1 g/kg
LD_{50} (rat, IP): 0.38 g/kg

15 Handling Precautions

Observe normal precautions appropriate to the circumstances and quantity of material handled. Eye protection and gloves are recommended. When heated to decomposition, propyl gallate may emit toxic fumes and smoke.

16 Regulatory Status

GRAS listed. Accepted for use as a food additive in Europe. Included in the FDA Inactive Ingredients Database (IM injections; oral, and topical preparations). Included in nonparenteral medicines licensed in the UK. Included in the Canadian Natural Health Products Ingredients Database.

17 Related Substances

Dodecyl gallate; ethyl gallate; octyl gallate.

Ethyl gallate
Empirical formula $C_9H_{10}O_5$
Molecular weight 198.17
CAS number [831-61-8]
Synonym Ethyl 3,4,5-trihydroxybenzoate
Appearance White, odorless or almost odorless, crystalline powder.
Melting point 151–154°C
Solubility *see* Table IV.

Table IV: Solubility of ethyl gallate.

Solvent	Solubility at 20°C
Ethanol (95%)	1 in 3
Ether	1 in 3
Peanut oil	Practically insoluble
Water	Slightly soluble

18 Comments

Propyl gallate has been reported to impart an 'off' flavor to corn and cottonseed oils when used as an antioxidant.[17]

A specification for propyl gallate is contained in the *Food Chemicals Codex* (FCC) and the *Japanese Pharmaceutical Excipients* (JPE).[18,19]

The EINECS number for propyl gallate is 204-498-2. The PubChem Compound ID (CID) for propyl gallate is 4947.

19 Specific References

1 Boehm E, Williams R. The action of propyl gallate on the autoxidation of oils. *Pharm J* 1943; **151**: 53.
2 Boehm E, Williams R. A study of the inhibiting actions of propyl gallate (normal propyl trihydroxy benzoate) and certain other trihydric phenols on the autoxidation of animal and vegetable oils. *Chemist Drug* 1943; **140**: 146–147.
3 Okide GB, Adikwu MU. Kinetic study of the auto-oxidation of arachis oil. *Boll Chim Farm* 1998; **137**: 277–280.
4 Brandl M, *et al*. Approaches for improving the stability of ketorolac in powder blends. *J Pharm Sci* 1995; **84**: 1151–1153.
5 Zeelie JJ, McCarthy TJ. The potential antimicrobial properties of antioxidants in pharmaceutical systems. *S Afr Pharm J* 1982; **49**: 552–554.
6 McCarthy TJ, *et al*. The antimicrobial action of zinc ion/antioxidant combinations. *J Clin Pharm Ther* 1992; **17**: 51–54.
7 Kahn G, *et al*. Propyl gallate contact sensitization and orally induced tolerance. *Arch Dermatol* 1974; **109**: 506–509.
8 Tayama S, Nakagawa Y. Cytogenetic effects of propyl gallate in CHO-K1 cells. *Mutat Res* 2001; **498**(1–2): 117–127.
9 Golightly LK, *et al*. Pharmaceutical excipients: adverse effects associated with 'inactive' ingredients in drug products (part II). *Med Toxicol* 1988; **3**: 209–240.
10 Cusano F, *et al*. Safety of propyl gallate in topical products. *J Am Acad Dermatol* 1987; **17**: 308–309.
11 Bojs G, *et al*. Allergic contact dermatitis to propyl gallate. *Contact Dermatitis* 1987; **17**: 294–298.
12 Anonymous. Final report on the amended safety assessment of propyl gallate. *Int J Toxicol* 2007; **26**(Suppl. 3): 89–118.
13 Perez A, *et al*. Positive rates to propyl gallate on patch testing: a change in trend. *Contact Dermatitis* 2008; **58**(1): 47–48.
14 Nitzan M, *et al*. Infantile methemoglobinemia caused by food additives. *Clin Toxicol* 1979; **15**(3): 273–280.
15 FAO/WHO. Evaluation of certain food additives and contaminants. Forty-sixth report of the joint FAO/WHO expert committee on food additives. *World Health Organ Tech Rep Ser* 1997; No. 868.
16 Lewis RJ, ed. *Sax's Dangerous Properties of Industrial Materials*, 11th edn. New York: Wiley, 2004; 3084.
17 McConnell JEW, Esselen WB. Effect of storage conditions and antioxidants on the keeping quality of packaged oils. *J Am Oil Chem Soc* 1947; **24**: 6–14.
18 *Food Chemicals Codex*, 7th edn. Bethesda, MD: United States Pharmacopeia, 2010; 872.
19 Japan Pharmaceutical Excipients Council. *Japanese Pharmaceutical Excipients 2004*. Tokyo: Yakuji Nippo, 2004; 741.

20 General References

Johnson DM, Gu LC. Autoxidation and antioxidants. In: Swarbrick J, Boylan JC, eds. *Encyclopedia of Pharmaceutical Technology*. 1: New York: Marcel Dekker, 1988; 415–449.

21 Author

ME Fenton.

22 Date of Revision

1 March 2012.

Propylene Carbonate

P

1 Nonproprietary Names

USP–NF: Propylene Carbonate

2 Synonyms

Carbonic acid;cyclic methylethylene carbonate;cyclic propylene ester; cyclic propylene carbonate; 4-methyl-2-oxo-1,3-dioxolane; 1,2-propanediol cyclic carbonate; 1,2-propylene carbonate.

3 Chemical Name and CAS Registry Number

±-4-Methyl-1,3-dioxolan-2-one [108-32-7]

4 Empirical Formula and Molecular Weight

$C_4H_6O_3$ 102.09

5 Structural Formula

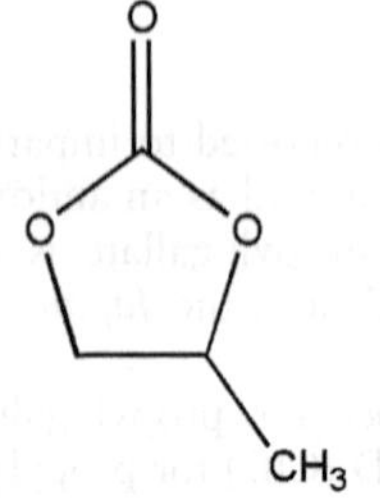

6 Functional Category

Solvent.

7 Applications in Pharmaceutical Formulation or Technology

Propylene carbonate is used mainly as a solvent in oral and topical pharmaceutical formulations.

In topical applications, propylene carbonate has been used in combination with propylene glycol as a solvent for corticosteroids. The corticosteroid is dissolved in the solvent mixture to yield microdroplets that can then be dispersed in petrolatum.[1] Propylene carbonate has been used as a dispensing solvent in topical preparations.[2]

Propylene carbonate has also been used in hard gelatin capsules as a nonvolatile, stabilizing, liquid carrier. For formulations with a low dosage of active drug, a uniform drug content may be obtained by dissolving the drug in propylene carbonate and then spraying this solution on to a solid carrier such as compressible sugar; the sugar may then be filled into hard gelatin capsules.[3]

Propylene carbonate may additionally be used as a solvent, at room and elevated temperatures, for many cellulose-based polymers and plasticizers. Propylene carbonate is also used in cosmetics.

8 Description

Propylene carbonate is a clear, colorless, odorless, mobile liquid.

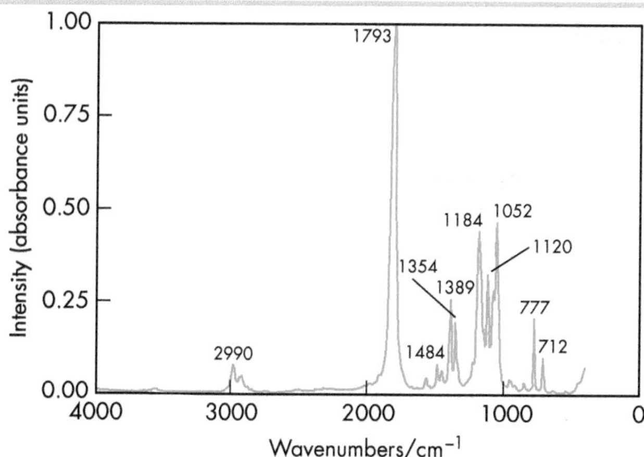

Figure 1: Infrared spectrum of propylene carbonate measured by transmission. Adapted with permission of Informa Healthcare.

9 Pharmacopeial Specifications

See Table I.

Table I: Pharmacopeial specifications for propylene carbonate.

Test	USP35–NF30
Identification	+
Specific gravity at 20°C	1.203–1.210
pH[a]	6.0–7.5
Residue on ignition	≤0.01%
Assay	99.0–100.5%

(a) 10% v/v aqueous solution with 0.3% saturated KCl.

10 Typical Properties

Boiling point 242°C
Dielectric constant 66.14 (15–30°C)
Flash point 135°C
Freezing point −48.8°C
Heat of combustion 14.21 kJ/mol (3.40 kcal/mol)
Heat of vaporization 55.2 kJ/mol (13.2 kcal/mol) at 150°C.
Refractive index $n_D^{20} = 1.4189$
Solubility Practically insoluble in hexane; freely soluble in water. Miscible with acetone, benzene, chloroform, ethanol, ethanol (95%), and ether.
Specific heat 2.57 J/g/°C (0.62 cal/g/°C) at 20°C.
Spectroscopy
IR spectra *see* Figure 1.
Vapor pressure 4 Pa (0.03 mmHg) at 20°C.
Viscosity (dynamic) 2.5 mPa s (2.5 cP) at 25°C.

11 Stability and Storage Conditions

Propylene carbonate and its aqueous solutions are stable but may degrade in the presence of acids or bases, or upon heating; *see also* Section 12.

Store in a well-closed container in a cool, dry place.

12 Incompatibilities

Propylene carbonate hydrolyzes rapidly in the presence of strong acids and bases, forming mainly propylene oxide and carbon dioxide. Propylene carbonate can also react with primary and secondary amines to yield carbamates.

13 Method of Manufacture

Propylene oxide reacts with carbon dioxide to yield propylene carbonate.[4] Propylene carbonate may also be prepared by the reaction of sodium bicarbonate with propylene chlorohydrin.[5]

14 Safety

Propylene carbonate is used as a solvent in oral and topical pharmaceutical formulations, and is generally regarded as an essentially nontoxic and nonirritant material.

In animal studies, propylene carbonate was found to cause tissue necrosis after parenteral administration.[6]

LD_{50} (mouse, oral): 20.7 g/kg [7]
LD_{50} (mouse, SC): 15.8 g/kg [8]
LD_{50} (rat, oral): 29.1 mL/kg [9]
LD_{50} (rat, SC): 11.1 g/kg [8]

15 Handling Precautions

Observe normal precautions appropriate to the circumstances and quantity of material handled. Propylene carbonate may be irritant to the eyes and mucous membranes. When heated to decomposition, propylene carbonate produces carbon monoxide and carbon dioxide.[9] Eye protection and gloves are recommended.

16 Regulatory Status

Included in the FDA Inactive Ingredients Database (topical ointments). Included in the Canadian Natural Health Products Ingredient Database.

17 Related Substances

(*S*)-Propylene carbonate.

(*S*)-Propylene carbonate
Empirical formula $C_4H_6O_3$
Molecular weight 102.09
CAS number [51260-39-0]
Specific rotation $[\alpha]_D^{25} = -1.7°$ (0.92% v/v solution in ethanol)
Comments The (*S*)-enantiomer of ±-propylene carbonate.[10]

18 Comments

The EINECS number for propylene carbonate is 203-572-1. The PubChem Compound ID (CID) for propylene carbonate is 7924.

19 Specific References

1 Burdick KH, *et al.* Corticosteroid ointments: comparison by two human bioassays. *Curr Ther Res* 1973; **15**: 233–242.
2 Yoshida H, *et al. In vitro* release of tacrolimus from tacrolimus ointment and its speculated mechanism. *Int J Pharm* 2004; **270**(1–2): 55–64.
3 Dahl TC, Burke G. Feasibility of manufacturing a solid dosage form using a liquid nonvolatile drug carrier: a physicochemical characterization. *Drug Dev Ind Pharm* 1990; **16**: 1881–1891.
4 Peppel WJ. Preparation and properties of the alkylene carbonates. *Ind Eng Chem* 1958; **50**(5): 767–770.
5 Najer H, *et al.* [Study of organic cyclic carbonates and their derivatives.] *Bull Soc Chim Fr* 1954; 1142–1148[in French].
6 Hem SL, *et al.* Tissue irritation evaluation of potential parenteral vehicles. *Drug Dev Commun*1974–751: 471–477.
7 Anonymous. Final report on the assessment of propylene carbonate. *J Am Coll Toxicol* 1987; **6**(1): 23–51.
8 Shikoku Igaku Zasshi. *Shikoku Medical Journal* 1972; 28: 276.
9 Burdick & Jackson. Material safety data sheet: Propylene carbonate, 2002.
10 Usieli V, *et al.* Chiroptical properties of cyclic esters and ketals derived from (*S*)-1,2-propylene glycol and (*S,S*)- and (*R,R*)-2,3-butylene glycol. *J Org Chem* 1974; **39**: 2073–2079.

20 General References

Cheng H, Gadde RR. Determination of propylene carbonate in pharmaceutical formulations using liquid chromatography. *J Pharm Sci* 1985; 74: 695–696.
Haynes WM, ed. *Handbook of Chemistry and Physics*, 91st edn. New York: CRC Press, 2010.
Ursin C, *et al*. Permeability of commercial solvents through living human skin. *Am Ind Hyg J* 1995; 56: 651–660.

21 Author

J Gunn.

22 Date of Revision

2 March 2012.

Propylene Glycol

1 Nonproprietary Names

BP: Propylene Glycol
JP: Propylene Glycol
PhEur: Propylene Glycol
USP-NF: Propylene Glycol

2 Synonyms

1,2-Dihydroxypropane; E1520; 2-hydroxypropanol; methyl ethylene glycol; methyl glycol; propane-1,2-diol; propylenglycolum.

3 Chemical Name and CAS Registry Number

1,2-Propanediol [57-55-6]
(−)-1,2-Propanediol [4254-14-2]
(+)-1,2-Propanediol [4254-15-3]
(±)-Propane-1,2-diol [4254-16-4]

4 Empirical Formula and Molecular Weight

$C_3H_8O_2$ 76.09

5 Structural Formula

OH
OH
H_3C

6 Functional Category

Antimicrobial preservative; humectant; plasticizing agent; solvent.

7 Applications in Pharmaceutical Formulation or Technology

Propylene glycol has become widely used as a solvent, extractant, and preservative in a variety of parenteral and nonparenteral pharmaceutical formulations. It is a better general solvent than glycerin and dissolves a wide variety of materials, such as corticosteroids, phenols, sulfa drugs, barbiturates, vitamins (A and D), most alkaloids, and many local anesthetics.

Propylene glycol is commonly used as a plasticizer in aqueous film-coating formulations.

Propylene glycol is also used in cosmetics and in the food industry as a carrier for emulsifiers and as a vehicle for flavors in preference to ethanol, since its lack of volatility provides a more uniform flavor. See Table I.

Table I: Uses of propylene glycol.

Use	Dosage form	Concentration (%)
Humectant	Topicals	≈15
Preservative	Solutions, semisolids	15–30
Solvent or cosolvent	Aerosol solutions	10–30
	Oral solutions	10–25
	Parenterals	10–60
	Topicals	5–80

8 Description

Propylene glycol is a clear, colorless, viscous, practically odorless liquid, with a sweet, slightly acrid taste resembling that of glycerin.

9 Pharmacopeial Specifications

See Table II. *See also* Section 18.

Table II: Pharmacopeial specifications for propylene glycol.

Test	JP XV	PhEur 7.4	USP35-NF30
Identification	+	+	+
Appearance	—	+	—
Specific gravity	1.035–1.040	1.035–1.040	1.035–1.037
Acidity	+	+	+
Water	≤0.5%	≤0.2%	≤0.2%
Residue on ignition	≤0.005%	—	≤3.5 mg/50 g
Sulfated ash	—	≤0.01%	—
Chloride	≤0.007%	—	≤70 ppm
Sulfate	≤0.002%	—	≤60 ppm
Heavy metals	≤5 ppm	≤5 ppm	≤5 ppm
Refractive index	—	1.431–1.433	—
Oxidizing substances	—	+	—
Reducing substances	—	+	—
Arsenic	≤2 ppm	—	—
Glycerin	+	—	—
Boiling point	184–189°C	184–189°C	—
Assay	—	—	≥99.5%

10 Typical Properties

Autoignition temperature 371°C
Boiling point 187–188°C
Density 1.0361 g/cm^3 at 20°C[1]
Dissociation constant pKa = 14.8 at 25°C[2,3]

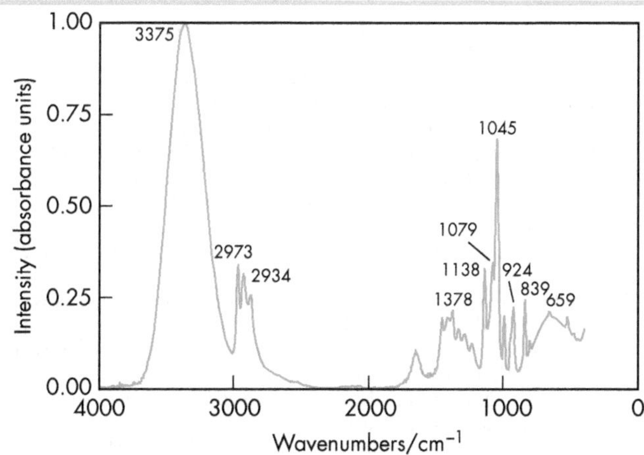

Figure 1: Infrared spectrum of propylene glycol measured by transmission. Adapted with permission of Informa Healthcare.

Flammability

Upper limit, 12.5% v/v in air;

lower limit, 2.6% v/v in air.

Flash point

99°C (open cup);[4]

104°C (closed cup).[4]

Heat of combustion 1803.3 kJ/mol (431.0 kcal/mol)

Heat of vaporization 705.4 J/g (168.6 cal/g) at b.p.

Melting point −59°C

Osmolarity A 2.0% v/v aqueous solution is iso-osmotic with serum.

Partition coefficient Log P (octanol:water) = -0.92 [3,4]

Refractive index n_D^{20} = 1.4324

Specific rotation

$[\alpha]_D^{20}$ = −15.0° (neat) for (R)-form;

$[\alpha]_D^{20}$ = +15.8° (neat) for (S)-form.

Solubility Miscible with acetone, chloroform, ethanol (95%), methanol; glycerin, and water; soluble at 1 in 6 parts of ether; not miscible with light mineral oil or fixed oils, but will dissolve some essential oils.

Specific heat 2.47 J/g (0.590 cal/g) at 20°C.

Spectroscopy

IR spectra *see* Figure 1.

Surface tension 40.1 mN/m (40.1 dynes/cm) at 25°C.

Vapor density (relative) 2.62 (air = 1)

Vapor pressure 9.33 Pa (0.07 mmHg) at 20°C.

Viscosity (dynamic) 58.1 mPa s (58.1 cP) at 20°C.

11 Stability and Storage Conditions

At cool temperatures, propylene glycol is stable in a well-closed container, but at high temperatures, in the open, it tends to oxidize, giving rise to products such as propionaldehyde, lactic acid, pyruvic acid, and acetic acid. Propylene glycol is chemically stable when mixed with ethanol (95%), glycerin, or water; aqueous solutions may be sterilized by autoclaving.

Propylene glycol is hygroscopic and should be stored in a well-closed container, protected from light, in a cool, dry place.

12 Incompatibilities

Propylene glycol is incompatible with oxidizing reagents such as potassium permanganate.

13 Method of Manufacture

Propylene is converted to chlorohydrin by chlorine water and hydrolyzed to 1,2-propylene oxide. With further hydrolysis, 1,2-propylene oxide is converted to propylene glycol.

14 Safety

Propylene glycol is used in a wide variety of pharmaceutical formulations and is generally regarded as a relatively nontoxic material. It is also used extensively in foods and cosmetics. Probably as a consequence of its metabolism and excretion, propylene glycol is less toxic than other glycols. Propylene glycol is rapidly absorbed from the gastrointestinal tract; there is also evidence that it is absorbed topically when applied to damaged skin. It is extensively metabolized in the liver, mainly to lactic and pyruvic acids, and is also excreted unchanged in the urine.[5,6]

In topical preparations, propylene glycol is regarded as minimally irritant,[7] although it is more irritant than glycerin. There have been some reports of contact dermatitis associated with propylene glycol.[8,9] Some local irritation is produced upon application to mucous membranes or when it is used under occlusive conditions.[10] Parenteral administration may cause pain or irritation when propylene glycol is used in high concentration.

Propylene glycol is estimated to be one-third as intoxicating as ethanol, with administration of large volumes being associated with adverse effects most commonly on the central nervous system, especially in neonates and children.[11–13] Other adverse reactions reported, though generally isolated, include: ototoxicity;[14] cardiovascular effects; seizures; and hyperosmolarity[15] and lactic acidosis, both of which occur most frequently in patients with renal impairment. Adverse effects are more likely to occur following consumption of large quantities of propylene glycol or on adminstration to neonates, children under 4 years of age, pregnant women, and patients with hepatic or renal failure. Adverse events may also occur in patients treated with disulfiram or metronidazole.[16]

On the basis of metabolic and toxicological data, the WHO has set an acceptable daily intake of propylene glycol at up to 25 mg/kg body-weight.[17] Formulations containing 35% propylene glycol can cause hemolysis in humans.

In animal studies, there has been no evidence that propylene glycol is teratogenic or mutagenic. Rats can tolerate a repeated oral daily dose of up to 30 mL/kg body-weight in the diet over 6 months, while the dog is unaffected by a repeated oral daily dose of 2 g/kg in the diet for 2 years.[18]

LD_{50} (mouse, IP): 9.72 g/kg[19]

LD_{50} (mouse, IV): 6.63 g/kg

LD_{50} (mouse, oral): 22.0 g/kg

LD_{50} (mouse, SC): 17.37 g/kg

LD_{50} (rat, IM): 14.0 g/kg

LD_{50} (rat, IP): 6.66 g/kg

LD_{50} (rat, IV): 6.42 g/kg

LD_{50} (rat, oral): 20.0 g/kg

LD_{50} (rat, SC): 22.5 g/kg

15 Handling Precautions

Observe normal precautions appropriate to the circumstances and quantity of material handled. Propylene glycol should be handled in a well-ventilated environment; eye protection is recommended. In the UK, the long-term (8-hour TWA) workplace exposure limit for propylene glycol vapor and particulates is 474 mg/m^3 (150 ppm) and 10 mg/m^3 for particulates.[20]

16 Regulatory Status

GRAS listed. Accepted for use as a food additive in Europe. Included in the FDA Inactive Ingredients Database (dental preparations; IM and IV injections; inhalations; ophthalmic, oral, otic, percutaneous, rectal, topical, and vaginal preparations). Included in nonparenteral and parenteral medicines licensed in the UK. Included in the Canadian Natural Health Products Ingredients Database.

17 Related Substances

Propylene glycol alginate.

18 Comments

Propylene glycol is one of the materials that have been selected for harmonization by the Pharmacopeial Discussion Group. For further information see the General Information Chapter <1196> in the USP35–NF30, the General Chapter 5.8 in PhEur 7.4, along with the 'State of Work' document on the PhEur EDQM website, and also the General Information Chapter 8 in the JP XV.

As an antiseptic it is similar to ethanol, and against molds it is similar to glycerin and only slightly less effective than ethanol. In addition to its uses as an excipient, propylene glycol is used in veterinary medicine as an oral glucogenic in ruminants.[21]

A specification for potassium glycol is contained in the *Food Chemicals Codex* (FCC).[22]

The EINECS number for propylene glycol is 200-338-0. The PubChem Compound ID (CID) for propylene glycol is 1030.

19 Specific References

1 *Ullman's Encyclopedia of Industrial Chemistry.* Propanediols. Weinheim, Germany: VCH, 2006.
2 Riddick JA et al. Organic Solvents. In: *Techniques of Chemistry*, 4th edn, vol. II. New York: John Wiley and Sons, 1985; 266.
3 Hansch C et al. Exploring QSAR - Hydrophobic, electronic, and steric constants. Washington, DC: American Chemical Society, 1995; 7.
4 *Kirk-Othmer Encyclopedia of Chemical Technology* Glycols. Online. New York: Wiley, 2011.
5 Yu DK, *et al.* Pharmacokinetics of propylene glycol in humans during multiple dosing regimens. *J Pharm Sci* 1985; 74: 876–879.
6 Speth PAJ, *et al.* Propylene glycol pharmacokinetics and effects after intravenous infusion in humans. *Ther Drug Monit* 1987; 9: 255–258.
7 Lessmann H, *et al.* Skin-sensitizing and irritant properties of propylene glycol. *Contact Dermatitis* 2005; 53(5): 247–259.
8 Kuznetsov AV, *et al.* Contact allergy to propylene glycol and dodecyl gallate mimicking seborrheic dermatitis. *Contact Dermatitis* 2006; 55(5): 307–308.
9 Lowther A, *et al.* Systemic contact dermatitis from propylene glycol. *Dermatitis* 2008; **19**(2): 105–108.
10 Motoyoshi K, *et al.* The safety of propylene glycol and other humectants. *Cosmet Toilet* 1984; **99**(10): 83–91.
11 Arulanantham K, Genel M. Central nervous system toxicity associated with ingestion of propylene glycol. *J Pediatr* 1978; **93**: 515–516.
12 MacDonald MG, *et al.* Propylene glycol: increased incidence of seizures in low birth weight infants. *Pediatrics* 1987; **79**: 622–625.
13 Martin G, Finberg L. Propylene glycol: a potentially toxic vehicle in liquid dosage form. *J Pediatr* 1970; **77**: 877–878.
14 Morizono T, Johnstone BM. Ototoxicity of chloramphenicol ear drops with propylene glycol as solvent. *Med J Aust* 1975; **2**: 634–638.
15 Fligner CL, *et al.* Hyperosmolality induced by propylene glycol: a complication of silver sulfadiazine therapy. *J Am Med Assoc* 1985; **253**: 1606–1609.
16 Anonymous. US warning on HIV drug excipient. *Pharm J* 2000; **264**: 685.
17 FAO/WHO. Toxicological evaluation of certain food additives with a review of general principles and of specifications. Seventeenth report of the FAO/WHO expert committee on food additives. *World Health Organ Tech Rep Ser* 1974: No. 539.
18 Clayton GD, Clayton FE, eds. *Patty's Industrial Hygiene and Toxicology*, 3rd edn. Chichester: Wiley, 1987.
19 Lewis RJ, ed. *Sax's Dangerous Properties of Industrial Materials*, 11th edn. New York: Wiley, 2004: 3061.
20 Health and Safety Executive. *EH40/2005: Workplace Exposure Limits.* Sudbury: HSE Books, 2011. www.hse.gov.uk/pubns/priced/eh40.pdf (accessed 12 April 2012).
21 Bishop Y, ed. *The Veterinary Formulary*, 6th edn. London: Pharmaceutical Press, 2005: 420.
22 *Food Chemicals Codex*, 7th edn. Bethesda, MD: United States Pharmacopeia, 2010.

20 General References

Doenicke A, *et al.* Osmolalities of propylene glycol-containing drug formulations for parenteral use: should propylene glycol be used as a solvent? *Anesth Analg* 1992; **75**(3): 431–435.

European Directorate for the Quality of Medicines and Healthcare (EDQM). European Pharmacopoeia – State Of Work Of International Harmonisation. *Pharmeuropa* 2011; 23(4): 714–715. www.edqm.eu/site/-614.html (accessed 24 November 2011).

Krzyzaniak JF, *et al.* Lysis of human red blood cells 2: effect of contact time on cosolvent induced hemolysis. *Int J Pharm* 1997; **152**: 193–200.

Strickley RG. Solubilizing excipients in oral and injectable formulations. *Pharm Res* 2004; **21**(2): 201–230.

Wells JI, *et al.* Improved wet massed tableting using plasticized binder. *J Pharm Pharmacol* 1982; **34**(Suppl.): 46P.

Williams AC, Barry BW. Penetration enhancers. *Adv Drug Delivery Rev* 2004; **56**(5): 603–618.

Yu CD, Kent JS. Effect of propylene glycol on subcutaneous absorption of a benzimidazole hydrochloride. *J Pharm Sci* 1982; **71**: 476–478.

21 Author

NS Ladyzhynsky.

22 Date of Revision

12 April 2012.

P

Propylene Glycol Alginate

1 Nonproprietary Names

USP–NF: Propylene Glycol Alginate

2 Synonyms

Alginic acid, propylene glycol ester; E405; hydroxypropyl alginate; *Kelcoloid*; *Kimiloid*; *Manucol Ester*; *Profoam*; *Pronova*; propane-1,2-diol alginate; *Protanal*; *TIC Pretested*.

3 Chemical Name and CAS Registry Number

Propylene glycol alginate [9005-37-2]

4 Empirical Formula and Molecular Weight

Propylene glycol alginate is a propylene glycol ester of alginic acid, a linear glycuronan polymer consisting of a mixture of β-(1→4)-D-mannosyluronic acid and α-(1→4)-L-gulosyluronic acid residues.

5 Structural Formula

See Section 4.

6 Functional Category

Emulsifying agent; foam stabilizing agent; suspending agent; viscosity-increasing agent.

7 Applications in Pharmaceutical Formulation or Technology

Propylene glycol alginate is used as a stabilizing, suspending, gelling, and emulsifying agent in oral and topical pharmaceutical formulations. Typically, a concentration of 0.3–5.0% w/v is used, although this may vary depending upon the specific application and the grade of propylene glycol alginate used.

Propylene glycol alginate is also used in cosmetics and food products.

8 Description

Propylene glycol alginate occurs as a white to yellowish colored, practically odorless and tasteless, fibrous or granular powder.

9 Pharmacopeial Specifications

See Table I. *See also* Section 18.

Table I: Pharmacopeial specifications for propylene glycol alginate.

Test	USP35–NF30
Identification	+
Microbial limits	≤200 cfu/g
Loss on drying	≤20.0%
Ash	≤10.0%
Arsenic	≤3 ppm
Lead	≤10 ppm
Heavy metals	≤40 ppm
Free carboxyl groups	+
Esterified carboxyl groups	+
Assay (of alginates)	+

10 Typical Properties

Solubility Soluble in dilute organic acids and water, forming stable, viscous, colloidal solutions at pH 3. Depending upon the degree of esterification, propylene glycol alginate is also soluble in aqueous ethanol/water mixtures containing up to 60% w/w of ethanol (95%). The interfacial and foaming properties of propylene glycol alginate aqueous solutions have been studied, and were found to be affected by the degree of esterification and molecular weight.(1)

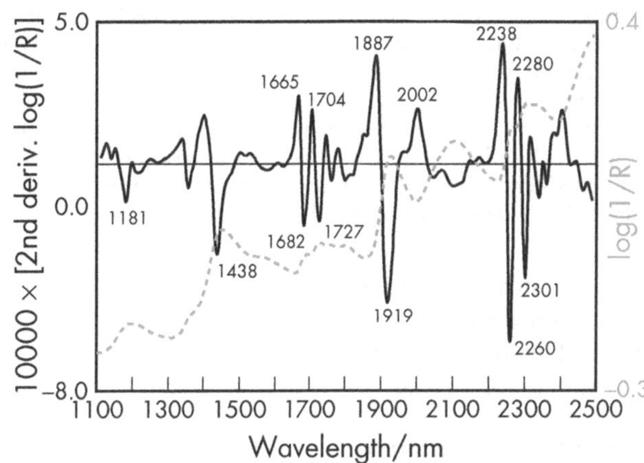

Figure 1: Near-infrared spectrum of propylene glycol alginate measured by reflectance.

Spectroscopy

NIR spectra *see* Figure 1.

Viscosity (dynamic) The viscosity of aqueous solutions depends upon the grade of material used. Typically, a 1% w/v aqueous solution has a viscosity of 20–400 mPa s (20–400 cP). Viscosity may vary depending upon concentration, pH, temperature, or the presence of metal ions. *See also* Sodium Alginate.

11 Stability and Storage Conditions

Propylene glycol alginate is a stable material, although it will gradually become less soluble if stored at elevated temperatures for extended periods.

Propylene glycol alginate solutions are most stable at pH 3–6. In alkaline solutions, propylene glycol alginate is rapidly saponified. Alginate solutions are susceptible to microbial spoilage and should be sterilized or preserved with an antimicrobial preservative. However, sterilization processes may adversely affect the viscosity of propylene glycol alginate solutions; *see* Sodium Alginate.

The bulk material should be stored in an airtight container in a cool, dry place.

12 Incompatibilities

—

13 Method of Manufacture

Alginic acid, extracted from brown seaweed, is reacted with propylene oxide to form propylene glycol alginate. Various grades may be obtained that differ in composition according to the degree of esterification and the percentage of free and neutralized carboxyl groups present in the molecule; complete esterification of alginic acid is impractical.

14 Safety

Propylene glycol alginate is used in oral and topical pharmaceutical formulations, cosmetics, and food products. It is generally regarded as a nontoxic and nonirritant material, although excessive oral consumption may be harmful. A study in five healthy male volunteers fed a daily intake of 175 mg/kg body-weight of propylene glycol alginate for 7 days, followed by a daily intake of 200 mg/kg body-weight of propylene glycol alginate for a further 16 days, showed no significant adverse effects.[2]

Inhalation of alginate dust may be irritant and has been associated with industrially related asthma in workers involved in alginate production. However, it appears that the cases of asthma were linked to exposure to seaweed dust rather than pure alginate dust.[3]

LD_{50} (hamster, oral): 7.0 g/kg[4]
LD_{50} (mouse, oral): 7.8 g/kg
LD_{50} (rabbit, oral): 7.6 g/kg
LD_{50} (rat, oral): 7.2 g/kg

15 Handling Precautions

Observe normal precautions appropriate to the circumstances and quantity of material handled. Propylene glycol alginate may be irritant to the eyes or respiratory system if inhaled as dust; *see* Section 14. Eye protection, gloves, and a dust respirator are recommended. Propylene glycol alginate should be handled in a well-ventilated environment.

16 Regulatory Status

GRAS listed. Accepted in Europe for use as a food additive. Included in the FDA Inactive Ingredients Database (oral preparations). Included in nonparenteral medicines licensed in the UK. Included in the Canadian Natural Health Products Ingredients Database.

17 Related Substances

Alginic acid; propylene glycol; sodium alginate.

18 Comments

A specification for propylene glycol alginate is contained in the *Food Chemicals Codex* (FCC).[5] A specification for propylene glycol alginate is also contained in the *Japanese Pharmaceutical Excipients* (JPE).[6] *See* Alginic Acid and Sodium Alginate for further information.

19 Specific References

1 Baeza R, *et al.* Interfacial and foaming properties of propylenglycol alginates. Effect of degree of esterification and molecular weight. *Colloids Surf B Biointerfaces* 2004; **36**(3–4): 139–145.
2 Anderson DM, *et al.* Dietary effects of propylene glycol alginate in humans. *Food Addit Contam* 1991; **8**(3): 225–236.
3 Henderson AK, *et al.* Pulmonary hypersensitivity in the alginate industry. *Scott Med J* 1984; **29**(2): 90–95.
4 Lewis RJ, ed. *Sax's Dangerous Properties of Industrial Materials*, 11th edn. New York: Wiley, 2004: 3080–3081.
5 *Food Chemicals Codex*, 7th edn. Bethesda, MD: United States Pharmacopeia, 2010: 875.
6 Japan Pharmaceutical Excipients Council. *Japan Pharmaceutical Excipients* 2004. Tokyo: Yakuji Nippo, 2004: 736–737.

20 General References

McDowell RH. New reactions of propylene glycol alginate. *J Soc Cosmet Chem* 1970; **21**: 441–457.

21 Authors

RD Reddy, JLP Soh.

22 Date of Revision

1 March 2012.

P

Propylene Glycol Dilaurate

1 Nonproprietary Names

BP: Propylene Glycol Dilaurate
PhEur: Propylene Glycol Dilaurate
USP–NF: Propylene Glycol Dilaurate

2 Synonyms

Dodecanoic acid, 1-methyl-1,2-ethanediyl ester; E477; *Emalex PG di-L*; propane-1,2-diyl didodecanoate; propylene dilaurate; propylenglycoli dilauras.

3 Chemical Name and CAS Registry Number

2-Dodecanoyloxypropyl dodecanoate [22788-19-8]

4 Empirical Formula and Molecular Weight

$C_{27}H_{52}O_4$ 440.7

The USP35–NF30 and PhEur 7.4 describe propylene glycol dilaurate as a mixture of the propylene glycol mono- and di-esters of lauric acid, containing not less than 70% of diesters and not more than 30% of monoesters.

5 Structural Formula

O
O
CH3
H3C
O
CH3
O

6 Functional Category

Emollient; emulsifying agent; nonionic surfactant.

7 Applications in Pharmaceutical Formulation or Technology

Propylene glycol dilaurate is used in the formulation of emulsions, lotions, and creams for topical use.

8 Description

Propylene glycol dilaurate occurs as a colorless or slightly yellow, clear oily liquid.

9 Pharmacopeial Specifications

See Table I.

Table I: Pharmacopeial specifications for propylene glycol dilaurate.

Test	PhEur 7.4	USP35–NF30
Identification	+	+
Characters	+	–
Acid value	≤4	≤4
Iodine value	≤1	≤1
Saponification value	230–250	230–250
Free propylene glycol	≤2%	≤2%
Water	≤1.0%	≤1.0%
Ash	≤0.1%	≤0.1%
Fatty acid composition		
Caprylic acid C8	≤0.5%	≤0.5%
Capric acid C10	≤2.0%	≤2.0%
Lauric acid C12	≥95.0%	≥95.0%
Myristic acid C14	≤3.0%	≤3.0%
Palmitic acid C16	≤1.0%	≤1.0%
Assay		
Diesters	≥70%	≥70%
Monoesters	≤30%	≤30%

10 Typical Properties

Boiling point 504.7°C at 760 mmHg.[1]
Density 0.915 g/cm^3 [1]
Flash point 235.8°C[1]
HLB value Approx. 2.2 for *Emalex PG-di-L*
Refractive index n^{20}_{D} 1.4545[2]
Solubility Practically insoluble in water; very soluble in alcohol, methanol, and methylene chloride.

11 Stability and Storage Conditions

Store in tightly closed containers protected from light and moisture.

12 Incompatibilities

—

13 Method of Manufacture

Propylene glycol dilaurate is produced from a reaction between propylene glycol and lauric acid.

14 Safety

Propylene glycol dilaurate is generally regarded as safe for use in cosmetic formulations in the present practice of use.[3]

15 Handling Precautions

Observe normal precautions appropriate to the circumstances and quantity of the material handled.

16 Regulatory Status

Accepted for use as a food additive in Europe.

17 Related Substances

Propylene glycol monolaurate.

18 Comments

The EINECS number for propylene glycol dilaurate is 245-217-3. The PubChem Compound (CID) number for propylene glycol dilaurate is 90838.

19 Specific References

1 Lookchem.com. Propylene glycol dilaurate. http://www.lookchem.com/cas-227/22788-19-8.html (accessed 7 July 2011).
2 Pfaltz & Bauer Inc. Material safety data sheet: Propylene glycol dilaurate, 2007.
3 Johnson W Snr. Final report on the safety assessment of propylene glycol (PG) dicaprylate, PG dicaprylate/dicaprate, PG dicocoate, PG dipelargonate, PG isostearate, PG laurate, PG myristate, PG oleate, PG oleate SE, PG dioleate, PG dicaprate, PG diisostearate, and PG dilaurate. *Int J Toxicol* 1999; **18**(2): 35–52.

20 Authors

ME Fenton, RC Rowe.

21 Date of Revision

1 March 2012.

22 General References

Nihon-Emulsion. Product information: *Emalex PG-di-L*. http://www.nihon-emulsion.co.jp/english/products/hlblist.html (accessed 7 July 2011).

Propylene Glycol Monolaurate

1 Nonproprietary Names

BP: Propylene Glycol Monolaurate
PhEur: Propylene Glycol Monolaurate
USP–NF: Propylene Glycol Monolaurate

2 Synonyms

Capmul PG-12; *Cithrol PGML*; dodecanoic acid, monoester with 1,2 propanediol; E477; *Emalex PGML*; *Imwitor 412*; lauric acid, monoester with 1,2 propanediol; *Lauroglycol 90*; propylene glycol monododecanoate; propylenglycoli monolauras; *Schercemol PGML*.

3 Chemical Name and CAS Registry Number

1,2 Propanediol monolaurate [27194-74-7]

4 Empirical Formula and Molecular Weight

$C_{15}H_{30}O_3$ 258.4

The USP35–NF30 describes propylene glycol monolaurate as a mixture of the propylene glycol mono- and di-esters of lauric acid. The requirements for monoester and diester content differ for the two types (Type I and II).

5 Structural Formula

6 Functional Category

Emollient; emulsifying agent; nonionic surfactant; penetration enhancer.

7 Applications in Pharmaceutical Formulation or Technology

Propylene glycol monolaurate is used as a diluent in soft and hard gelatin capsules. It is also used as a penetration enhancer in transdermal drug delivery systems,[1–8] and as a secondary emulsifier in microemulsions. Propylene glycol monolaurate is used as a solvent for lipophilic actives.

Propylene glycol monolaurate is also used in skin lotions and lipsticks. In skin lotions it is used to increase the resistance to freezing and decrease the tendency to produce surface crusts.[9]

8 Description

Propylene glycol monolaurate occurs as a colorless or slightly yellow oily liquid.

9 Pharmacopeial Specifications

See Table I.

Table I: Pharmacopeial specifications for propylene glycol monolaurate.

Test	PhEur 7.4	USP35–NF30
Identification	+	+
Characters	+	–
Acid value	≤4	≤4
Iodine value	≤1	≤1
Saponification value		
Type I	210–245	210–245
Type II	200–230	200–230
Free propylene glycol		
Type I	≤5%	≤5%
Type II	≤1%	≤1%
Water	≤1.0%	≤1.0%
Ash	≤0.1%	≤0.1%
Fatty acid composition		
Caprylic acid C8	≤0.5%	≤0.5%
Capric acid C10	≤2.0%	≤2.0%
Lauric acid C12	≥95.0%	≥95.0%
Myristic acid C14	≤3.0%	≤3.0%
Palmitic acid C16	≤1.0%	≤1.0%
Assay		
Type I		
Monoesters	45–70%	45–70%
Diesters	30–55%	30–55%
Type II		
Monoesters	≥90%	≥90%
Diesters	≤10%	≤10%

10 Typical Properties

Boiling point 362.5°C at 760 mmHg.[10]
Density 0.91–0.95 g/cm^3
Flash point 188°C (open cup); also reported as 139.3°C.[10]
HLB value 4.5; 5 for *Lauroglycol 90*.
Hydroxyl value 110–140 for Imwitor 412.
Melting point 15°C; 10°C for *Schercemol PGML*.
Monoglyceride content >50% for *Imwitor 412*.
Refractive index $n^{60}{}_{D}$1.438
Solubility Practically insoluble in water; very soluble in alcohol, acetone, *n*-hexane, methanol, and methylene chloride. Soluble in fats and oils.

11 Stability and Storage Conditions

Store in tightly closed containers protected from light and moisture.

12 Incompatibilities

—

13 Method of Manufacture

Propylene glycol monolaurate is manufactured by the esterification of propylene glycol and lauric acid.

14 Safety

Propylene glycol monolaurate is generally regarded as safe for use in cosmetic formulations in the present practice of use.[11] Prolonged or frequently repeated skin contact may cause allergic reactions in hypersensitive individuals exposed to this product.

15 Handling Precautions

Observe normal precautions appropriate to the circumstances and quantity of the material handled.

16 Regulatory Status

GRAS listed. Accepted for use as a food additive in Europe. Included in the FDA Inactive Ingredients Database (transdermal; film, controlled release).

17 Related Substances

1,2-Propanediol, 1-laurate; propylene glycol dilaurate.

1,2-Propanediol, 1-laurate
Molecular formula $C_{15}H_{30}O_3$
CAS number [142-55-2]
Synonyms Lauric acid, 2-hydroxypropyl ester; 2-hydroxy
Safety
LD_{50} (rat, oral): 34.6 g/kg
Comments Included in the Canadian Natural Health Products Ingredients Database. The EINECS number for 1,2-propanediol, 1-laurate is 205-542-3.

18 Comments

Propylene glycol monolaurate has been investigated as a cosolvent in a self-emulsifying oral formulation of coenzyme Q10.(12)

The EINECS number for propylene glycol monolaurate is 248-315-4. The PubChem Compound (CID) number for propylene glycol monolaurate is 14870.

19 Specific References

1 Gwak HS, Chun IK. Effect of vehicles and penetration enhancers on the *in vitro* percutaneous absorption of tenoxicam through hairless mouse skin. *Int J Pharm* 2002; **236**: 57–64.
2 Gwak HS, *et al.* Effect of vehicles and enhancers on the *in vitro* permeation of melatonin through hairless mouse skin. *Arch Pharm Res* 2002; **25**: 392–396.
3 Cho YA, Gwak HS. Transdermal delivery of ketorolac tromethamine: effects of vehicles and penetration enhancers. *Drug Dev Ind Pharm* 2004; 557–564.
4 Kim HW, *et al.* The effect of vehicles and pressure sensitive adhesives on the percutaneous absorption of quercetin through the hairless mouse skin. *Arch Pharm Res* 2004; **27**: 763–768.
5 Shin SC, Cho CW. Enhanced transdermal delivery of pranoprofen from the bioadhesive gels. *Arch Pharm Res* 2006; **29**: 928–933.
6 Choi JS, *et al.* Formulation and evaluation of ketorolac transdermal systems. *Drug Deliv* 2007; **14**: 69–74.
7 Kim T, *et al.* Pharmacokinetics of formulated tenoxicam transdermal delivery systems. *J Pharm Pharmacol* 2008; **60**: 135–138.
8 Jung SY, *et al.* Formulation and evaluation of ubidecarenone transdermal delivery systems. *Drug Dev Ind Pharm* 2009; **35**: 1029–1034.
9 Idson B. Nonionic skin lotions. *Cosmet Perfum* 1974; **89**: 59–61.
10 Lookchem.com. Propylene glycol monolaurate. http://www.lookchem.com/cas-271/27194-74-7.html (accessed 7 July 2011).
11 Johnson WSnr. Final report on the safety assessment of propylene glycol (PG) dicaprylate, PG dicaprylate/dicaprate, PG dicocoate, PG dipelargonate, PG isostearate, PG laurate, PG myristate, PG oleate, PG oleate SE, PG dioleate, PG dicaprate, PG diisostearate, and PG dilaurate. *Int J Toxicol* 1999; **18**(2): 35–52.
12 Seo DW, *et al.* Self-microemulsifying formulation-based oral solution of coenzyme Q10. *Yakugaku Zasshi* 2009; **129**: 1559–1563.

20 General References

Abitec. Product data sheet: *Capmul*, 2007.
Gattefosse. Product information: *Lauroglycol 90*, 2007.
Lubrizol. Product data sheet: *Schercemol*, 2007.
Nihon-Emulsion. Product information: *Emalex PGML*. http://www.nihon-emulsion.co.jp/english/products/list/E-PG-M-Lf.htm (accessed 7 July 2011).
Sasol. Material safety data sheet: *Imwitor 412*, 2001.

21 Authors

ME Fenton, RC Rowe.

22 Date of Revision

1 March 2012.

P

Propylparaben

1 Nonproprietary Names

BP: Propyl Hydroxybenzoate
JP: Propyl Parahydroxybenzoate
PhEur: Propyl Parahydroxybenzoate
USP–NF: Propylparaben

2 Synonyms

Aseptoform P; *CoSept P*; E216; 4-hydroxybenzoic acid propyl ester; *Nipagin P*; *Nipasol M*; propagin; *Propyl Aseptoform*; propyl butex; *Propyl Chemosept*; propylis parahydroxybenzoas; propyl *p*-hydroxybenzoate; *Propyl Parasept*; *Solbrol P*; *Tegosept P*; *Uniphen P-23*.

3 Chemical Name and CAS Registry Number

Propyl 4-hydroxybenzoate [94-13-3]

4 Empirical Formula and Molecular Weight

$C_{10}H_{12}O_3$ 180.20

5 Structural Formula

O
O
CH_3
HO

6 Functional Category

Antimicrobial preservative.

7 Applications in Pharmaceutical Formulation or Technology

Propylparaben is widely used as an antimicrobial preservative in cosmetics, food products, and pharmaceutical formulations; *see* Table I.

It may be used alone, in combination with other paraben esters, or with other antimicrobial agents. It is one of the most frequently used preservatives in cosmetics.[1]

The parabens are effective over a wide pH range and have a broad spectrum of antimicrobial activity, although they are most effective against yeasts and molds; *see* Section 10.

Owing to the poor solubility of the parabens, the paraben salts, particularly the sodium salt, are frequently used in formulations. This may cause the pH of poorly buffered formulations to become more alkaline.

Propylparaben (0.02% w/v) together with methylparaben (0.18% w/v) has been used for the preservation of various parenteral pharmaceutical formulations; *see* Section 14.

See Methylparaben for further information.

Table I: Uses of propylparaben in pharmaceutical preparations.

Use	Concentration (%)
IM, IV, SC injections	0.005–0.2
Inhalation solutions	0.015
Intradermal injections	0.02–0.26
Nasal solutions	0.017
Ophthalmic preparations	0.005–0.01
Oral solutions and suspensions	0.01–0.02
Rectal preparations	0.02–0.1
Topical preparations	0.01–0.6
Vaginal preparations	0.02–0.1

8 Description

Propylparaben occurs as a white, crystalline, odorless, and tasteless powder.

9 Pharmacopeial Specifications

The pharmacopeial specifications for propylparaben have undergone harmonization of many attributes for JP, PhEur, and USP–NF. *See* Table II. *See also* Section 18.

Table II: Pharmacopeial specifications for propylparaben.

Test	JP XV	PhEur 7.4	USP35–NF30
Identification	+	+	+
Characters	—	+	—
Melting range	96.0–99.0°C	96.0–99.0°C	96.0–99.0°C
Acidity	+	+	+
Residue on ignition	≤0.1%	—	≤0.1%
Sulfated ash	—	≤0.1%	—
Appearance of solution	+	+	+
Heavy metals[a]	≤20 ppm	—	—
Related substances	+	+	+
Assay	98.0–102.0%	98.0–102.0%	98.0–102.0%

(a) This test has not been fully harmonized at the time of publication.

10 Typical Properties

Antimicrobial activity

Propylparaben exhibits antimicrobial activity between pH 4–8. Preservative efficacy decreases with increasing pH owing to the formation of the phenolate anion. Parabens are more active against yeasts and molds than against bacteria. They are also more active against Gram-positive than against Gram-negative bacteria. The activity of the parabens increases with increasing chain length of the alkyl moiety; however, solubility decreases.

Activity may be improved by using combinations of parabens, as additive effects occur. Propylparaben has been used with methylparaben in parenteral preparations, and is used in combination with other parabens in topical and oral formulations. Activity has also been reported to be improved by the addition of other excipients; *see* Methylparaben.

Reported minimum inhibitory concentrations (MICs) for propylparaben are provided in Table III.[2]

Table III: Minimum inhibitory concentrations (MICs) for propylparaben in aqueous solution.[2]

Microorganism	MIC (μg/mL)
Aerobacter aerogenes ATCC 8308	1000
Aspergillus niger ATCC 9642	500
Aspergillus niger ATCC 10254	200
Bacillus cereus var. *mycoides* ATCC 6462	125
Bacillus subtilis ATCC 6633	500
Candida albicans ATCC 10231	250
Enterobacter cloacae ATCC 23355	1000
Escherichia coli ATCC 8739	500
Escherichia coli ATCC 9637	100
Klebsiella pneumoniae ATCC 8308	500
Penicillium chrysogenum ATCC 9480	125
Penicillium digitatum ATCC 10030	63
Proteus vulgaris ATCC 13315	250
Pseudomonas aeruginosa ATCC 9027	>1000
Pseudomonas aeruginosa ATCC 15442	>1000
Pseudomonas stutzeri	500
Rhizopus nigricans ATCC 6227A	125
Saccharomyces cerevisiae ATCC 9763	125
Salmonella typhosa ATCC 6539	500
Serratia marcescens ATCC 8100	500
Staphylococcus aureus ATCC 6538P	500
Staphylococcus epidermidis ATCC 12228	500
Trichophyton mentagrophytes	65

Boiling point 295°C
Density (bulk) 0.426 g/cm^3
Density (tapped) 0.706 g/cm^3
Density(true) 1.288 g/cm^3
Dissociation constant pK_a = 8.4 at 22°C
Flash point 140°C
Partition coefficients Values for different vegetable oils vary considerably and are affected by the purity of the oil; *see* Table IV.

Table IV: Partition coefficients for propylparaben in vegetable oil and water.[3]

Solvent	Partition coefficient oil : water
Corn oil	58.0
Mineral oil	0.5
Peanut oil	51.8
Soybean oil	65.9

Refractive index n_D^{14} = 1.5049
Solubility *see* Table V.
Spectroscopy

IR spectra *see* Figure 1.

NIR spectra *see* Figure 2.

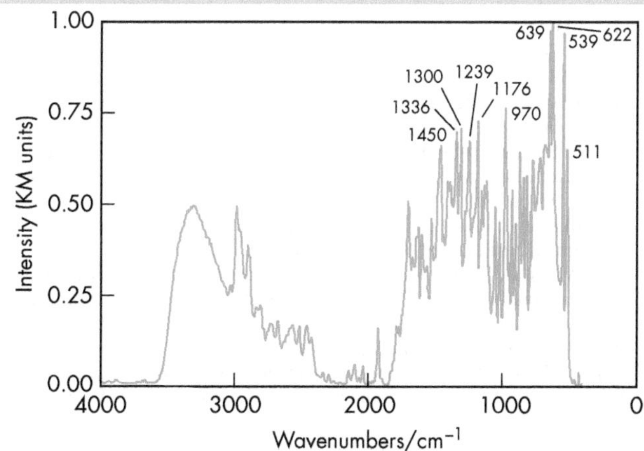

Figure 1: Infrared spectrum of propylparaben measured by diffuse reflectance. Adapted with permission of Informa Healthcare.

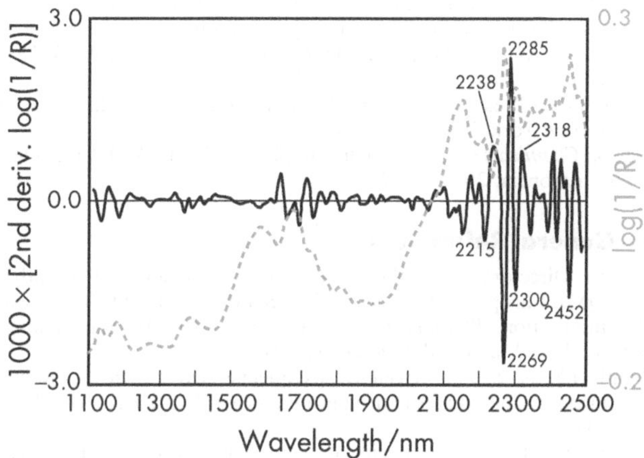

Figure 2: Near-infrared spectrum of propylparaben measured by reflectance.

Table V: Solubility of propylparaben in various solvents.[2]

Solvent	Solubility at 20°C unless otherwise stated
Acetone	Freely soluble
Ethanol (95%)	1 in 1.1
Ethanol (50%)	1 in 5.6
Ether	Freely soluble
Glycerin	1 in 250
Mineral oil	1 in 3330
Peanut oil	1 in 70
Propylene glycol	1 in 3.9
Propylene glycol (50%)	1 in 110
Water	1 in 4350 at 15°C
	1 in 2500
	1 in 225 at 80°C

11 Stability and Storage Conditions

Aqueous propylparaben solutions at pH 3–6 can be sterilized by autoclaving, without decomposition.[4] At pH 3–6, aqueous solutions are stable (less than 10% decomposition) for up to about 4 years at room temperature, while solutions at pH 8 or above are subject to rapid hydrolysis (10% or more after about 60 days at room temperature).[5]

See Table VI, for the predicted rate constants and half-lives at 25°C for propylparaben.[5]

Propylparaben should be stored in a well-closed container in a cool, dry place.

Table VI: Predicted rate constants and half-lives at 25°C for propylparaben dissolved in hydrochloric acid solution.

Initial pH of solution	Rate constant $k \pm \sigma^{(a)}$ (h^{-1})	Half-life $t_{1/2} \pm \sigma^{(a)}$ (day)
1	$(1.255 \pm 0.042) \times 10^{-4}$	230 ± 7.6
2	$(1.083 \pm 0.081) \times 10^{-5}$	2670 ± 200
3	$(8.41 \pm 0.96) \times 10^{-7}$	$34\,300 \pm 3900$
4	$(2.23 \pm 0.37) \times 10^{-7}$	$130\,000 \pm 22\,000$

(a) σ indicates the standard error.

The predicted amount of propylparaben remaining after autoclaving is given in Table VII.[5]

Table VII: Predicted amount of propylparaben dissolved in hydrochloric acid, after autoclaving.

Initial pH of solution	Rate constant $k \pm \sigma^{(a)}$ (h^{-1})	Predicted residual amount after sterilization (%)
1	$(4.42 \pm 0.10) \times 10^{-1}$	86.30 ± 0.30
2	$(4.67 \pm 0.19) \times 10^{-2}$	98.46 ± 0.06
3	$(2.96 \pm 0.24) \times 10^{-3}$	99.90 ± 0.01
4	$(7.8 \pm 1.1) \times 10^{-4}$	99.97 ± 0.004

(a) σ indicates the standard error.

12 Incompatibilities

The antimicrobial activity of propylparaben is reduced considerably in the presence of nonionic surfactants as a result of micellization.[6] Absorption of propylparaben by plastics has been reported, with the amount absorbed dependent upon the type of plastic and the vehicle.[7] Magnesium aluminum silicate, magnesium trisilicate, yellow iron oxide, and ultramarine blue have also been reported to absorb propylparaben, thereby reducing preservative efficacy.[8,9]

Propylparaben is discolored in the presence of iron and is subject to hydrolysis by weak alkalis and strong acids.

See also Methylparaben.

13 Method of Manufacture

Propylparaben is prepared by the esterification of *p*-hydroxybenzoic acid with *n*-propanol.

14 Safety

Propylparaben and other parabens are widely used as antimicrobial preservatives in cosmetics, food products, and oral and topical pharmaceutical formulations.

Propylparaben and methylparaben have been used as preservatives in injections and ophthalmic preparations; however, they are now generally regarded as being unsuitable for these types of formulations owing to the irritant potential of the parabens.

Systemically, no adverse reactions to parabens have been reported, although they have been associated with hypersensitivity reactions. The WHO has set an estimated acceptable total daily intake for methyl, ethyl, and propyl parabens at up to 10 mg/kg body-weight.[10]

LD_{50} (mouse, IP): 0.2 g/kg[11]

LD_{50} (mouse, oral): 6.33 g/kg

LD_{50} (mouse, SC): 1.65 g/kg

15 Handling Precautions

Observe normal precautions appropriate to the circumstances and quantity of material handled. Propylparaben may be irritant to the skin, eyes, and mucous membranes, and should be handled in a well-ventilated environment. Eye protection, gloves, and a dust mask or respirator are recommended.

16 Regulatory Status

Propylparaben and methylparaben are affirmed GRAS direct food substances in the US at levels up to 0.1%. All esters except the benzyl ester are allowed for injection in Japan.

In cosmetics, the EU and Brazil allow use of each paraben at 0.4%, but the total of all parabens may not exceed 0.8%. The upper limit in Japan is 1.0%.

Accepted as a food additive in Europe. Included in the FDA Inactive Ingredients Database (IM, IV, and SC injections; inhalations; ophthalmic preparations; oral capsules, solutions, suspensions, and tablets; otic, rectal, topical, and vaginal preparations). Included in parenteral and nonparenteral medicines licensed in the UK. Included in the Canadian Natural Health Products Ingredients Database.

17 Related Substances

Butylparaben; ethylparaben; methylparaben; propylparaben potassium; propylparaben sodium.

Propylparaben potassium
Empirical formula $C_{10}H_{11}KO_3$
Molecular weight 218.30
CAS number [84930-16-5]
Synonyms Potassium propyl hydroxybenzoate; propyl 4-hydroxybenzoate potassium salt.

18 Comments

Propylparaben has undergone harmonization for many attributes for JP, PhEur, and USP–NF by the Pharmacopeial Discussion Group. For further information see the General Information Chapter <1196> in the USP35-NF30, the General Chapter 5.8 in PhEur 7.4, along with the 'State of Work' document on the PhEur EDQM website, and also the General Information Chapter 8 in the JP XV.

A specification for propylparaben is contained in the *Food Chemicals Codex* (FCC).[12]

The EINECS number for propylparaben is 202-307-7. The PubChem Compound ID (CID) for propylparaben is 7175.

See Methylparaben for further information and references.

19 Specific References

1 Decker RL, Wenninger JA. Frequency of preservative use in cosmetic formulas as disclosed to FDA—1987. *Cosmet Toilet* 1987; **102**(12): 21–24.
2 Haag TE, Loncrini DF. Esters of *para*-hydroxybenzoic acid. In: Kabara JJ, ed. *Cosmetic and Drug Preservation.* New York: Marcel Dekker, 1984; 63–77.
3 Wan LSC, *et al.* Partition of preservatives in oil/water systems. *Pharm Acta Helv* 1986; **61**: 308–313.
4 Aalto TR, *et al. p*-Hydroxybenzoic acid esters as preservatives I: uses, antibacterial and antifungal studies, properties and determination. *J Am Pharm Assoc (Sci)* 1953; **42**: 449–457.
5 Kamada A, *et al.* Stability of *p*-hydroxybenzoic acid esters in acidic medium. *Chem Pharm Bull* 1973; **21**: 2073–2076.
6 Aoki M, *et al.* [Application of surface active agents to pharmaceutical preparations I: effect of Tween 20 upon the antifungal activities of *p*-hydroxybenzoic acid esters in solubilized preparations.] *J Pharm Soc Jpn* 1956; **76**: 939–943[in Japanese].
7 Kakemi K, *et al.* Interactions of parabens and other pharmaceutical adjuvants with plastic containers. *Chem Pharm Bull* 1971; **19**: 2523–2529.
8 Allwood MC. The adsorption of esters of *p*-hydroxybenzoic acid by magnesium trisilicate. *Int J Pharm* 1982; **11**: 101–107.
9 Sakamoto T, *et al.* Effects of some cosmetic pigments on the bactericidal activities of preservatives. *J Soc Cosmet Chem* 1987; **38**: 83–98.
10 FAO/WHO. Toxicological evaluation of certain food additives with a review of general principles and of specifications. Seventeenth report of the joint FAO/WHO expert committee on food additives. *World Health Organ Tech Rep Ser* 1974; No. 539.
11 Lewis RJ, ed. *Sax's Dangerous Properties of Industrial Materials*, 11th edn. New York: Wiley, 2004; 2053.
12 *Food Chemicals Codex*, 7th edn. Suppl. 3. Bethesda, MD: United States Pharmacopeia, 2011; 1707.

20 General References

European Directorate for the Quality of Medicines and Healthcare (EDQM). European Pharmacopoeia – State Of Work Of International Harmonisation. *Pharmeuropa* 2011; 23(4): 713–714. www.edqm.eu/site/-614.html (accessed 2 December 2011).

Golightly LK, *et al.* Pharmaceutical excipients: adverse effects associated with inactive ingredients in drug products (part I). *Med Toxicol* 1988; **3**: 128–165.

Jian L, Li Wan Po A. Ciliotoxicity of methyl- and propyl-*p*-hydroxybenzoates: a dose-response and surface-response study. *J Pharm Pharmacol* 1993; **45**: 925–927.

Šafra J, Pospíšilová M. Separation and determination of ketoprofen, methylparaben and propylparaben in pharmaceutical preparation by micellar electrokinetic chromatography. *J Biomed Anal* 2008; **48**(2): 452–455.

Schnuch A, *et al.* Contact allergy to preservatives. Analysis of IVDK data 1996-2009. *Br J Dermatol* 2011; **164**: 1316–1325.

21 Author

N Sandler.

22 Date of Revision

2 December 2011.

P

Propylparaben Sodium

1 Nonproprietary Names

BP: Sodium Propyl Hydroxybenzoate
PhEur: Sodium Propyl Parahydroxybenzoate
USP–NF: Propylparaben Sodium

2 Synonyms

E217; 4-hydroxybenzoic acid propyl ester, sodium salt; *Nipasol M Sodium*; parasept; propyl 4-hydroxybenzoate, sodium salt; propyl *p*-hydroxybenzoate, sodium salt; propylis parahydroxybenzoas natricus; sodium 4-propoxycarbonylphenolate; sodium propyl *p*-hydroxybenzoate; soluble propyl hydroxybenzoate.

3 Chemical Name and CAS Registry Number

Sodium 4-propoxycarbonylphenolate [35285-69-9]

4 Empirical Formula and Molecular Weight

$C_{10}H_{11}NaO_3$ 202.2

5 Structural Formula

6 Functional Category

Antimicrobial preservative.

7 Applications in Pharmaceutical Formulation or Technology

Propylparaben sodium is used as an antibacterial or antifungal preservative in oral pharmaceuticals and in many water-based cosmetics. It is generally used in combination with other parabens esters.

8 Description

Propylparaben sodium occurs as a white, crystalline, odorless or almost odorless powder.

9 Pharmacopeial Specifications

See Table I.

Table I: Pharmacopeial specifications for propylparaben sodium.

Test	PhEur 7.4	USP35–NF30
Identification	+	+
Characters	+	—
Completeness of solution	—	+
Appearance of solution	+	—
pH	9.5–10.5	9.5–10.5
Related substances	+	—
Water	≤5.0%	≤5.0%
Chloride	350 ppm	≤0.035%
Sulfate	300 ppm	≤0.12%
Heavy metals	10 ppm	—
Assay (anhydrous basis)	99.0–104.0	98.5–101.5%

10 Typical Properties

Acidity/alkalinity pH = 9.5–10.5 (0.1% w/v aqueous solution)
Dissociation constant pK_a = 8.4 at 22°C
Partition coefficient log *P* (octanol : water) = 3.0
Solubility *see* Table II.

Table II: Solubility of propylparaben sodium.

Solvent	Solubility at 20°C
Ethanol (50%)	1 in 2
Ethanol (95%)	1 in 50
Fixed oils	Practically insoluble
Methylene chloride	Practically insoluble
Water	1 in 1

11 Stability and Storage Conditions

Propylparaben sodium is stable under normal conditions. It decomposes on heating. Store in a tightly closed container.

12 Incompatibilities

The activity of propylparaben sodium can be adversely affected by the presence of other excipients or active ingredients, such as atropine, essential oils, iron, magnesium trisilicate, talc, polysorbate 80 and other nonionic surfactants, sorbitol, weak alkalis, and strong acids.[1]

13 Method of Manufacture

Propylparaben sodium is produced from benzoic acid.

14 Safety

Propylparaben sodium is used in oral pharmaceuticals and cosmetics. The pure form is toxic by the IV route and moderately toxic by ingestion and the IP route. Propylparaben sodium may cause asthma, rashes, and hyperactivity.

LD_{50} (mouse, IP): 0.49 g/kg
LD_{50} (mouse, IV): 0.18 g/kg
LD_{50} (mouse, oral): 3.7 g/kg[2]

15 Handling Precautions

Observe normal precautions appropriate to the circumstances and quantity of the material handled. Avoid inhalation or contact with eyes, skin, and clothing. Avoid prolonged or repeated exposure. Wear suitable protective clothing, gloves, eye/face protection, and a respirator.

16 Regulatory Status

Included in the FDA Inactive Ingredients Database (oral capsules, tablets, suspensions). Accepted for use as a food additive in Europe. Included in nonparenteral medicines (oral capsules, mixtures, orodispersible tablets, solutions and suspensions; cutaneous emulsions) licensed in the UK.

17 Related Substances

Butylparaben; ethylparaben; methylparaben; propylparaben.

18 Comments

Propylparaben sodium may be used instead of propylparaben because of its greater aqueous solubility. However, it may cause the pH of a formulation to become more alkaline.

The EINECS number for propylparaben sodium is 252-488-1. The PubChem Compound ID (CID) for propylparaben sodium is 23679044.

19 Specific References

1 Sweetman SC, ed. *Martindale: The Complete Drug Reference*, 37th edn. London: Pharmaceutical Press, 2011: 1793.
2 Lewis RJ, ed. *Sax's Dangerous Properties of Industrial Materials*, 11th edn. New York: Wiley, 2004; 3085.

20 General References

Schnuch A, *et al.* Contact allergy to preservatives. Analysis of IVDK data 1996-2009. *Br J Dermatol* 2011; **164**: 1316–1325.
ScienceLab.com, Inc. Material safety data sheet: Propyl paraben sodium, 12 July 2005.
Sigma Aldrich. Material safety data sheet: Propyl 4 hydroxybenzoate sodium salt, 13 February 2006.

21 Authors

RT Forbes, WL Hulse.

22 Date of Revision

2 March 2012.

Pullulan

1 Nonproprietary Names

JP: Pullulan
USP–NF: Pullulan

2 Synonyms

E1204; 1,4-1,6-α-D-glucan; 1,6-α-D-glucan; 1,6-α-linked maltotriose; *Pullulan PI-10*; *Pullulan PI-20*.

3 Chemical Name and CAS Registry Number

Poly[6-α-D-glucopyranosyl-(1→4)-α-D-glucopyranosyl-(1→4)-α-D-glucopyranosyl-(1→] [9057-02-07]

4 Empirical Formula and Molecular Weight

$(C_{36}H_{60}O_{30})_n$ $n = 1250$ approx. (for *Pullulan PI-20*)

The molecular weight may range from 8000 to more than 2 000 000 Da, depending upon conditions under which the producing organism is grown.

The USP35–NF 30 describes pullulan as a neutral, simple polysaccharide produced by the growth of the polymorphic fungus *Aureobasidium pullulans* with a chain structure of repeated α-1,6 bonds of maltotriose composed of three glucoses in α-1,4 bonds. It may contain some maltotetraosyl units.

5 Structural Formula

6 Functional Category

Coating agent; film-forming agent; tablet and capsule diluent; tablet and capsule binder; viscosity-increasing agent.

7 Applications in Pharmaceutical Formulation or Technology

Pullulan has been used in pharmaceutical formulation as a polymer carrier for producing quick-dissolving edible films by a solvent casting method.(1–3) It can also be formed into capsule shells that are water-soluble with relatively low oxygen permeability, which may be useful for ingredients sensitive to oxidation.(4)

Granular pullulan can be used in tablets to improve the tableting performance of both drugs and other excipients.(4) Pullulan can also be used in the formulation of tablet coatings. When used as a base coat it is claimed to provide improved adhesion; and when used as a topcoat it increases the strength and resilience of the coating.(4)

The films formed by drying pullulan solutions are highly oxygen impermeable and have excellent mechanical properties.(5)

Pullulan is widely used in cosmetics for the preparation of shampoos, skin care, and facial cleansing products. It has also been used in Japan for many years as a food additive and ingredient for coatings on food.(4)

8 Description

Pullulan occurs as a white to off-white odorless powder.

9 Pharmacopeial Specifications

See Table I.

Table I: Pharmacopeial specifications for pullulan.

Test	JP XV	USP35-NF30
Identification	+	+
Viscosity (kinematic)	100–180 $mm^2 s^{-1}$	100–180 $mm^2 s^{-1}$
Microbial limits		
Bacteria	–	<100 cfu/g
Yeasts and molds	–	<100 cfu/g
pH	4.5–6.5	4.5–6.5
Loss on drying	≤6.0%	≤6.0%
Residue on ignition	≤0.3%	≤0.3%
Heavy metals	≤5 ppm	≤5 μg/g
Content of monosaccharide, disaccharide, and oligosaccharides	≤10.0%	≤10.0%
Nitrogen content	≤0.05%	≤0.05%

10 Typical Properties

Ash 0–0.16 for *Pullulan PI-20*[6]

Combined mono-, di-, and oligosaccharides 5.0–8.7% for *Pullulan PI-20*[6]

Hygroscopicity Pullulan is not hygroscopic. Moisture content is 10–15% at relative humidity of less than 70%.

Lead <1 mg/kg for *Pullulan PI-20*[(6)]

Nitrogen 0.002–0.004 for *Pullulan PI-20*[6]

pH 5-7 for *Pullulan PI-20* (10% w/w aqueous solution)[6]

Purity 91.2–95.0% dry weight for *Pullulan PI-20*[6]

Refractive index Significant positive linear correlation of concentration and refractive index at 20 and 45°C.

Solubility Freely soluble in cold and hot water. Soluble in dimethylformamide or dimethyl sulfoxide. Practically insoluble in ethanol. Insoluble in organic solvents such as methylene chloride.

Viscosity 132–179 $mm^2 s^{-1}$ for *Pullulan PI-20* 10% w/w aqueous solution at 30°C.[6]

11 Stability and Storage Conditions

Pullulan is nonreducing and relatively stable. It decomposes and carbonizes at 250–280°C. It is stable in aqueous solution over a wide pH range and in the presence of salts and most metal ions; only prolonged heating at pH <3 leads to a decrease in viscosity, which is indicative of hydrolytic depolymerization.[6]

Store in well-closed containers at room temperature.

12 Incompatibilities

—

13 Method of Manufacture

Pullulan is produced by mesophilic (22–30°C) fermentation of starch syrup with the black yeast *Aureobasidium pullulans*. The culture is filtered, sterilized, decolorized, and further filtered before removal of salts and protein contaminants with an ion-exchange resin. The liquid is concentrated, decolorized, and filtered before being evaporated to dryness. The solid is then pulverized.

14 Safety

Studies have indicated that pullulan is hydrolyzed only very slowly by gastrointestinal enzymes but is fermented to short-chain fatty acids by intestinal microorganisms.[7] A dietary level of 10% pullulan equal to 7.9 g/kg per day was tolerated by male and female rats without toxicological effects.[7] The acceptable daily intake for humans is <10 g.

There was no indication of genotoxic potential in a mutagenicity study in *Salmonella typhimurium*.[8]

LD_{50} (mice, oral): >14 g/kg[7]
LD_{50} (rat, oral): >24 g/kg

15 Handling Precautions

Observe normal precautions appropriate to the circumstances and quantity of the material handled.

16 Regulatory Status

GRAS listed. Accepted as a food additive in Europe. Approved as a food ingredient in Japan. Included in the Canadian Natural Health Products Ingredients Database. Included in nonparenteral medicines licensed in the UK.

17 Related Substances

Dextran; dextrin.

18 Comments

In the brand name *Pullulan PI-20*, P stands for pullulan, I stands for deionized, and 20 designates the average molecular weight of approx. 200 000 Da.

Pullulan has been found to stabilize a moisture-sensitive drug in a lyophilized formulation owing to the restricted water mobility in the polysaccharide.[9] Studies have indicated that nanoparticles of pullulan have merit as possible carriers for intracellular delivery of genes and nucleotide drugs.[10]

Chemically substituted pullulans capable of forming hydrogels have been described.[11] Pullulan-stearic acid conjugates have been used to prepare nanoparticles for drug delivery.[12] Pullulan acetate phthalate has been shown to improve the dissolution rate of diazepam from co-ground mixtures.[13]

Pullulan and dextran are both polysaccharides derived from microbial fermentation of starch or other sugar sources.

A specification for pullulan is included in the *Food Chemicals Codex* (FCC)[14] and the *Japanese Pharmaceutical Excipients* (JPE).[15]

The EINECS number for pullulan is 232-945-1.

P

19 Specific References

1 Gounga ME, *et al.* Film forming mechanism and mechanical and thermal properties of whey protein isolate-based edible films as affected by protein concentration, glycerol ratio and pullulan content. *J Food Biochem*34(3): 501–519.

2 Smutzer G, *et al.* A test for measuring gustatory function. *Laryngoscope* 2008; **118**(8): 1411–1416.

3 Kulkarni AS, *et al.* Exploration of different polymers for use in the formulation of oral fast dissolving strips. *J Curr Pharm Res* 2010; **2**: 3–35.

4 Hayashibara International. Product information: *Pullulan*. www.hayashibara-intl.com (accessed 1 December 2010).

5 Sakata Y, Otsuka M. Evaluation of relationship between molecular behaviour and mechanical strength of pullulan films. *Int J Pharm* 2009; **374**(1–2): 33–38.

6 FAO/WHO. Joint FAO/WHO Expert Committee on Food Additives (JECFA). *Chemical and Technical Assessment 65th JECFA: Pullulan*. Geneva: WHO; 2006.

7 Dixon B *et al.* WHO publications: *Pullulan*, 2006. http://whqlibdoc.who.int/publications/2006/9241660562_part1_d_eng.pdf (accessed 1 December 2010).

8 Kimoto T. Repeated toxicity studies and genotoxicity testing of pullulan. *Food Chem Toxicol* 1997; **35**(8): 855.

9 Moribe K, *et al.* Stabilization mechanism of limaprost in solid dosage form. *Int J Pharm* 2007; **338**(1–2): 1–6.

10 Gupta M, Gupta AK. In vitro cytotoxicity studies of hydrogel pullulan nanoparticles prepared by AOT/*n*-hexane micellar system. *J Pharm Pharmaceut Sci* 2004; **7**(1): 38–46.

11 Coviello T, *et al.* Polysaccharide hydrogels for modified release formulations. *J Control Release* 2007; **119**(1): 5–24.

12 Kim IS, Oh IJ. Preparation and characterization of stearic acid-pullulan nanoparticles. *Arch Pharm Res (Korea)* 2010; **33**: 761–767.

13 Choudhari KB, Sanghavi NM. Dissolution behavior and characterization of diazepam–pullulan coground mixtures. *Int J Pharm* 1993; **89**: 207–211.
14 *Food Chemicals Codex*. 7th edn. Bethesda, MD: United States Pharmacopoeia, 2010; 878.
15 Japan Pharmaceutical Excipients Council. *Japanese Pharmaceutical Excipients 2004*. Tokyo: Yakuji Nippo, 2004; 743.

20 General References

Kato T, *et al.* Solution properties and chain flexibility of pullulan in aqueous solution. *Biopolymers* 1982; **21**(8): 1623–1633.

Tsujisaka Y, Mitsuhashi M. Pullulan. In: BeMiller JN, Whistler RL eds. *Industrial Gums: Polysaccharides and Their Derivatives*. 3rd edn. New York: Academic Press, 1993; 447–460.

21 Authors

SW Hoag, H-P Lim.

22 Date of Revision

1 March 2012.

Pyroxylin

1 Nonproprietary Names

BP: Pyroxylin
JP: Pyroxylin
USP–NF: Pyroxylin

2 Synonyms

Celloidin; cellulose nitrate; cellulose tetranitrate; collodion (see Section 7); collodion wool; colloxylin; pyroxylin solution; nitrocellulose; *Parlodion*; soluble guncotton; xyloidin.

3 Chemical Name and CAS Registry Number

Pyroxylin [9004-70-0]

4 Empirical Formula and Molecular Weight

$(C_{12}H_{16}N_4O_{18})_n$ $(504.3)_n$

5 Structural Formula

Pyroxylin is a mixture of nitrates of cellulose, corresponding to the introduction of 2 to 2.5 nitrate groups per sugar unit.

Collodion USP35–NF30 contains pyroxylin 40 g, ether 750 mL, and alcohol 250 mL and is prepared by adding the alcohol and the ether to the pyroxylin and shaking occasionally until the pyroxylin is dissolved.

BP 2012 states that collodion for the preparation of flexible collodion (*see* Section 17) is prepared by adding 100 g of pyroxylin to 900 mL of a mixture of 3 volumes of solvent ether and 1 volume of ethanol (90 %) and agitating until dissolved.

6 Functional Category

Film-forming agent.

7 Applications in Pharmaceutical Formulation or Technology

Pyroxylin is used in the preparation of collodions, which are applied to the skin for the protection of small cuts and abrasions. In addition, pyroxylin has been investigated in the preparation of microcapsules,[1] and in the preparation of thermo-responsive controlled-release systems.[2–4]

Collodion is a liquid preparation intended for local external application to the skin; it includes pyroxylin as the film-forming agent. Application of collodion to the skin results in the evaporation of the solvent, and a transparent, occlusive, flexible film forms on the skin. Flexibility of the film is conferred by the inclusion of a plasticizing agent such as castor oil, while the presence of camphor in the formulation makes the films waterproof. Collodion can be used to seal off minor cuts or wounds that have partially healed, or as a means of holding a dissolved medicament, such as salicylic acid, in contact with the skin for long periods.

Collodion has been used as a vehicle for fluorouracil in an attempt to reduce or eliminate the side effects of fluorouracil treatment of solar keratoses;[5] fluorouracil in collodion has also been used to treat common warts on the hands.[6]

8 Description

Pyroxylin occurs as white or almost white cuboid granules, white flakes, or fibrous material. The fibrous material resembles absorbent cotton but is harsher to the touch and more powdery. Both the granules and the fibrous material appear moist and are highly flammable.

Collodion is a colorless or pale yellow syrupy liquid with an odor of ether.

9 Pharmacopeial Specifications

See Tables I and II. *See also* Section 17 and 18.

Table I: Pharmacopeial specifications for pyroxylin.

Test	BP 2012	JP XV	USP35–NF30
Identification	+	+	–
Characteristics	+	+	–
Clarity and color of solution	+	+	–
Viscosity (kinematic)	1160–2900 mm^2/s	–	–
Nitrogen	11.7–12.2%	–	–
Acidity	–	+	+
Water-soluble substances	–	≤ 1.5 mg	≤ 1.5 mg
Residue on ignition	–	≤ 0.3%	≤ 0.3%
Viscosity (dynamic)	–	–	110–147 mPa s

10 Typical Properties

The properties of nitrated celluloses largely depend upon the number of nitrate groups introduced into the cellulose molecule.

Autoignition temperature 180–190°C (collodion)
Boiling point 38°C (collodion)
Explosion limits 2–36% (collodion)

Table II: Pharmacopeial specifications for collodion.

Test	BP 2012	USP35–NF30
Identification	–	+
Characteristics	+	–
Specific gravity	0.785–0.795	0.765–0.775
Acidity	–	+
Alcohol content	–	22.0–26.0%
Viscosity (kinematic)	405–700 mm^2/s	–
Assay	–	≥5.0% w/w

Flash point –45°C (closed cup) (collodion)
Melting point –123°C (collodion)
Solubility Pyroxylin is freely soluble in acetone, and very slightly soluble in diethyl ether. It is soluble in methanol, amyl acetate, and glacial acetic acid but is insoluble in water and ether–alcohol mixtures. Collodion is insoluble in water; on the addition of water to collodion, pyroxylin may precipitate.

11 Stability and Storage Conditions

Pyroxylin should be stored loosely packed, protected from light, and stored at a temperature not exceeding 15°C, remote from fire. The containers should be suitably designed to disrupt should the internal pressure reach or exceed 1400 kPa. The amount of damping or moistening fluid must not be allowed to fall below 25% w/w; should this happen, the material should either be rewetted or used immediately for the preparation of collodion.

Upon heating or exposure to light, pyroxylin is decomposed with the evolution of nitrous acid vapors, leaving a carbonaceous residue.

Collodion USP should be stored in airtight containers at a temperature not exceeding 30°C and remote from fire. The label of collodion USP should include a caution statement to the effect that collodion is highly flammable.

When heated to decomposition, collodion emits toxic fumes including oxides of carbon and oxides of nitrogen.

12 Incompatibilities

Pyroxylin is incompatible with oxidizing agents.

Collodion is incompatible with acetyl peroxide, bromoazide, chlorine, strong oxidizing agents, strong acids, strong bases and amines. It readily forms explosive mixtures with air.

13 Method of Manufacture

Pyroxylin is a nitrated cellulose obtained by the action of a mixture of nitric and sulfuric acids on wood pulp or cotton linters that have been freed from fatty matter. It must be damped or moistened with not less than 25% by weight of isopropyl alcohol or of an alcohol such as industrial methylated spirit.

14 Safety

Pyroxylin may cause allergic skin reactions.

Collodion causes irritation to the skin, eyes, and respiratory tract. Collodion should not be considered as a tablet or sugar coating as it is toxic and remains insoluble in the gastrointestinal tract.

15 Handling Precautions

Observe normal precautions appropriate to the circumstances and quantity of the material handled.

Pyroxylin is highly flammable. Particular care should be exercised when drying pyroxylin as the material obtained is explosive and sensitive to ignition by impact or friction; it should be handled as carefully as possible.

Collodion as liquids and vapors is highly flammable and it is explosive. Keep away from heat, sparks, and flames and use only with adequate ventilation.

16 Regulatory Status

Pyroxylin is included in gels, solutions, and paints for topical administration in the UK. Included in the Canadian Natural Health Products Ingredients Database.

17 Related Substances

Flexible collodion.

Flexible collodion

Nonproprietary names BP: Flexible Collodion
USP: Flexible Collodion

Description Flexible collodion USP contains camphor 20 g, castor oil 30 g, and collodion to 1000 g; the ingredients are weighed into a bottle and the mixture shaken until the camphor is dissolved.

Flexible collodion BP is a solution of colophony in a mixture of virgin castor oil and collodion. If prepared extemporaneously, it contains colophony 25 g, virgin castor oil 25 g, and collodion to 1000 mL. The ingredients are mixed and stirred until the colophony has dissolved, allowing any deposit to settle and decanting the clear liquid.

Pharmacopeial specifications *See* Table III.

Table III: Pharmacopeial specifications for flexible collodion.

Test	BP 2012	USP35–NF30
Identification	+	+
Specific gravity	–	0.770–0.790
Alcohol content	20–23% v/v	21–25%

pH Neutral (for a 1% solution of flexible collodion in water).
Solubility Flexible collodion is very slightly soluble in cold water but insoluble in hot water; it is soluble in methanol, diethyl ether and acetone.
Stability and storage conditions USP35–NF30 states that flexible collodion should be stored in airtight containers at a temperature not exceeding 30°C and remote from fire. The label should include a caution statement to the effect that flexible collodion is highly flammable. Flexible collodion is stable under normal conditions of use, but is liable to form explosive peroxides on exposure to light and air.
Safety Flexible collodion vapors may cause drowsiness and dizziness. Allergic contact dermatitis to colophony included in the formulation of flexible collodion BP has been reported.[7]
Comments Flexible collodion is included in salicylic acid collodion BP, which is used in the treatment of warts and corns. The uptake of salicylic acid into the stratum corneum of the forearm of human volunteers was found to be greater from flexible collodion than an ointment formulation; the partitioning of the salicylic acid from the collodion formulation was concentration independent.[8]

Cantharidin in flexible collodion has been used in the treatment of molluscum contagiosum in children,[9] while triamcinolone acetonide in flexible collodion has been investigated as a treatment for a range of dermatological conditions.[10]

The BP 2012 states that when collodion is prescribed or demanded, flexible collodion shall be dispensed or supplied. Flexible collodion is included in solutions for topical administration in the UK. It is also included in the Canadian Natural Health Products Ingredients Database as a film-former for topical use only.

P

18 Comments

Collodion has also been used in the preparation of model membranes,[11,12] and as dressings for central venous catheters.[13]

A monograph for salicylic acid collodion is included in the BP 2012 and USP35–NF30.

19 Specific References

1 Habibi-Moini S, D'mello AP. Evaluation of possible reasons for the low phenylalanine ammonia lyase activity in cellulose nitrate membrane microcapsules. *Int J Pharm* 2001; **215**: 185–196.
2 Haupenthal H, Moll F. Thermoactive liberation of ibuprofen from TDO-embedded membranes. *Pharmazie* 1997; **52**: 532–535.
3 Lin YY, *et al.* Development and investigation of a thermo-responsive cholesteryl oleyl carbonate-embedded membrane. *J Control Release* 1996; **41**: 163–170.
4 Lin SY, *et al.* Manufacturing factors affecting the drug delivery function of thermoresponsive membrane prepared by adsorption of binary liquid crystals. *Eur J Pharm Sci* 2002; **17**: 153–160.
5 Goncalves JC. Treatment of solar keratoses with a 5-fluorouracil and salicylic acid varnish. *Br J Dermatol* 1975; **92**: 85–88.
6 Goncalves JC. 5-Fluorouracil in the treatment of common warts of the hands. *Br J Dermatol* 1975; **92**: 89–91.
7 Lachapelle JM, Leroy B. Allergic contact dermatitis to colophony included in the formulation of flexible collodion BP, the vehicle of a salicylic and lactic acid wart paint. *Dermatol Clin* 1990; **8**: 143–146.
8 Tsai J-C, *et al.* Distribution of salicylic acid in human stratum corneum following topical application *in vivo*: a comparison of six different formulations. *Int J Pharm* 1999; **188**: 145–153.
9 Silverberg NB, *et al.* Childhood molluscum contagiosum: experience with cantharidin therapy in 300 patients. *J Am Acad Dermatol* 2000; 43: 503–507.
10 Brock W, Cullen SI. Triamcinolone acetonide in flexible collodion for dermatologic therapy. *Arch Dermatol* 1967; **96**: 193–194.
11 Alkrad JA, *et al.* The release profiles of intact and enzymatically digested hyaluronic acid from semisolid formulations using multi-layer membrane system. *Eur J Pharm Biopharm* 2003; **56**: 37–41.
12 Yamauchi A, *et al.* Membrane characteristics of composite collodion membrane IV. Transport properties across blended collodion/Nafion membrane. *J Memb Sci* 2000; **170**: 1–7.
13 Babycos CR, *et al.* Collodion as a safe, cost-effective dressing for central venous catheters. *South Med J* 1990; **83**: 1286–1288.

20 General References

—

21 Author

C Cable.

22 Date of Revision

1 March 2012.

Pyrrolidone

1 Nonproprietary Names

BP: Pyrrolidone
PhEur: Pyrrolidone

2 Synonyms

γ-Aminobutyric acid lactam; 4-aminobutyric acid lactam; γ-aminobutyric lactam; γ-aminobutyrolactam; butyrolactam; γ-butyrolactam; 2-ketopyrrolidine; 2-oxopyrrolidine; *2-Pyrol*; α-pyrrolidinone; pyrrolidin-2-one; 2-pyrrolidone; α-pyrrolidone; pyrrolidonum; *Soluphor P.*

3 Chemical Name and CAS Registry Number

2-Pyrrolidinone [616-45-5]

4 Empirical Formula and Molecular Weight

C_4H_7NO 85.11

5 Structural Formula

H
N
O

6 Functional Category

Penetration enhancer; plasticizing agent; solubilizing agent; solvent.

7 Applications in Pharmaceutical Formulation or Technology

Pyrrolidone and *N*-methylpyrrolidone (*see* Section 17) are mainly used as solvents in veterinary injections. Pyrrolidone has been shown to be a better solubilizer than glycerin, propylene glycol, or ethanol.[1] They have also been suggested for use in human pharmaceutical formulations as solvents in parenteral, oral, and topical applications. In topical applications, pyrrolidones appear to be effective penetration enhancers.[2–6] Pyrrolidones have also been investigated for their application in controlled-release depot formulations.[4,7]

8 Description

Pyrrolidone occurs as a clear, colorless or slightly grayish liquid, as white or almost white crystals, or colorless crystal needles. It has a characteristic odor.

9 Pharmacopeial Specifications

See Table I.

10 Typical Properties

Acidity/alkalinity pH = 8.2–10.8 for a 10% v/v aqueous solution.

Antimicrobial activity Pyrrolidone has also been shown to possess strong antimicrobial activity against Gram-positive and Gram-negative bacteria, and mold.

Boiling point 245°C

Dipole moment 2.3 Debye at 25°C

Enthalpy of vaporization 48.21 ± 3.0 kJ/mol

Flash point 129°C (open cup)

Table I: Pharmacopeial specifications for pyrrolidone.

Test	PhEur 7.4
Identification	+
Characters	+
Appearance	+
Alkalinity	+
Related substances	+
Heavy metals	≤10 ppm
Water	≤0.1%
Sulfated ash	≤0.1%
Melting point	≈25°C
Boiling point	≈245°C
Relative density	1.112–1.115
Refractive index	1.487–1.490

Refractive index n_D^{25} = 1.480–1.490
Solubility Miscible with ethanol (95%), propan-2-ol, and water. Also miscible with other organic solvents such as aromatic hydrocarbons including benzene, carbon disulfide, chloroform, ether, and ethyl acetate.
Specific gravity 1.11 at 25°C
Viscosity (dynamic) 13.3 mPa s (13.3 cP) at 25°C

11 Stability and Storage Conditions

Pyrrolidone is chemically stable and, if it is kept in unopened original containers, the shelf-life is approximately one year. Pyrrolidone should be stored in a well-closed container protected from light and oxidation, at temperatures below 20°C.

12 Incompatibilities

Pyrrolidone is incompatible with oxidizing agents and strong acids.

13 Method of Manufacture

Pyrrolidone is prepared from butyrolactone by a Reppe process, in which acetylene is reacted with formaldehyde.

14 Safety

Pyrrolidones are mainly used in veterinary injections and have also been suggested for use in human oral, topical, and parenteral pharmaceutical formulations. In mammalian species, pyrrolidones are biotransformed to polar metabolites that are excreted via the urine.(8,9) Pyrrolidone is mildly toxic by ingestion and subcutaneous routes; mutagenicity data have been reported.(10)

LD_{50} (guinea pig, oral): 6.5 g/kg(10)
LD_{50} (rat, oral): 6.5 g/kg

15 Handling Precautions

Observe normal precautions appropriate to the circumstances and quantity of material handled. Some pyrrolidones in their pure state are considered toxic, corrosive, and flammable; contact with skin and eyes should be avoided. Vapors or sprays should not be inhaled. Suitable eye and skin protection and a respirator are recommended. When heated to decomposition, pyrrolidone emits toxic fumes of NO_x.

16 Regulatory Status

—

17 Related Substances

N-Methylpyrrolidone.

N-Methylpyrrolidone

Synonyms 1-Methyl-2-pyrrolidinone; *N*-methyl-α-pyrrolidinone; *N*-methyl-γ-butyrolactam; *N*-methyl-2-pyrrolidinone; 1-methylazacyclopentan-2-one; *N*-methylpyrrolidonum; MP; NMP; *Pharmasolve*; m-*Pyrol*.
Empirical formula C_5H_9NO
Molecular weight 99.14
CAS number [872-50-4]
Description *N*-Methylpyrrolidone occurs as a clear, hygroscopic liquid with a mild amine odor.
Typical Properties **Boiling point**202°C
Dielectric constant 32.2 at 25°C
Dipole moment 4.09 Debye at 25°C
Enthalpy of evaporation 43.82 ± 3.0 kJ/mol
Flash point (closed cup) 93°C
Flash point (open cup) 96°C
Freezing point/melting point −24°C
Heat of combustion 719 kcal/mol
Refractive index n_D^{25} = 1.4690
Solubility Miscible with ethanol (95%), water, and most other organic solvents.
Specific gravity 1.028 at 25°C
Surface tension 40.7 mN/m (40.7 dyne/cm) at 25°C
Vapor pressure 0.33 mmHg at 23.2°C; 5.00 mmHg at 65.0°C.
Viscosity 1.65 mPa s (1.65 cP) at 25°C

Safety
N-Methylpyrrolidone is considered a poison when injected via the intravenous route. It is moderately toxic by ingestion, skin contact, and intraperitoneal routes. It is an experimental teratogen; mutagenicity data have been reported.(11)
LD_{50} (mouse, IP): 3.05 g/kg(11)
LD_{50} (mouse, IV): 0.155 g/kg
LD_{50} (mouse, oral): 5.13 g/kg
LD_{50} (rabbit, SC): 8.0 g/kg
LD_{50} (rat, IP): 2.472 g/kg
LD_{50} (rat, IV): 0.0805 g/kg
LD_{50} (rat, oral): 3.914 g/kg

Handling precautions In the UK, the workplace exposure limits for *N*-methylpyrrolidone are 103 mg/m^3 (25 ppm) long-term (8-hour TWA) and 309 mg/m^3 (75 ppm) short-term (15 minutes).(12)
Comments *N*-Methylpyrrolidone is produced by the condensation of butyrolactone with methylamine. The EINECS number for *N*-methylpyrrolidone is 212-828-1. A specification for *N*-methylpyrrolidone is included in the USP35–NF30, PhEur 7.4 and *Japanese Pharmaceutical Excipients* (JPE).(13)

18 Comments

The EINECS number for pyrrolidone is 204-648-7. The PubChem Compound ID (CID) for pyrrolidone is 12025.

19 Specific References

1. Jain P, Yalkowsky SH. Solubilization of poorly soluble compounds using 2-pyrrolidone. *Int J Pharm* 2007; **342**: 1–5.
2. Babu R, Pandit J. Effect of penetration enhancers on the transdermal delivery of bupranolol through rat skin. *Drug Deliv* 2005; **12**: 165–169.
3. Babu RJ, *et al.* Cardiovascular effects of transdermally delivered bupranolol in rabbits: effect of chemical penetration enhancers. *Life Sci* 2008; **82**(5–6): 273–278.
4. Kranz H, Bodmeier R. Structure formation and characterization of injectable drug loaded biodegradable devices: in situ implants versus in situ microparticles. *Eur J Pharm Sci* 2008; **34**(2–3): 164–172.
5. Medi B *et al.* Assessing efficacy of penetration enhancers. In: Riviere JE, ed. *Dermal Absorption Models in Toxicology and Pharmacology*, 2006. 213–249.
6. Shishu, Aggarwal N. Preparation of hydrogels of griseofulvin for dermal application. *Int J Pharm* 2006; **326**(1–2): 20–24.

P

7 Lu Y, *et al.* Sucrose acetate isobutyrate as an in situ forming system for sustained risperidone release. *J Pharm Sci* 2007; **96**(12): 3252–3262.
8 Akesson B, Jonsson BA. Major metabolic pathway for *N*-methyl-2-pyrrolidone in humans. *Drug Metab Dispos* 1997; **25**: 267–269.
9 Bader M, *et al.* Human experimental exposure study on the uptake and urinary elimination of *N*-methyl-2-pyrrolidone (NMP) during simulated workplace conditions. *Arch Toxicol* 2007; **81**(5): 335–346.
10 Lewis RJ, ed. *Sax's Dangerous Properties of Industrial Materials*, 11th edn. New York: Wiley, 2004: 3122.
11 Lewis RJ, ed. *Sax's Dangerous Properties of Industrial Materials*, 11th edn. New York: Wiley, 2004: 2523.
12 Health and Safety Executive. *EH40/2005: Workplace Exposure Limits.* Sudbury: HSE Books, 2011. http://www.hse.gov.uk/pubns/priced/eh40.pdf (accessed 29 February 2012).
13 Japan Pharmaceutical Excipients Council. *Japanese Pharmaceutical Excipients 2004*. Tokyo: Yakuji Nippo, 2004: 547–548.

20 General References

BASF. Technical Information: *Soluphor P*, July 2008. http://www.pharma-ingredients.basf.com/Statements/Technical%20Informations/EN/Pharma%20Solutions/EMP%20030747e_Soluphor%20P.pdf (accessed 13 July 2011).

International Specialty Products. Material data safety sheet: *2-Pyrol*, November 2010. http://online1.ispcorp.com/en-us/pages/msds.aspx (accessed 13 July 2011).

21 Authors

RK Chang, JR Johnson, N Trivedi, W Qu.

22 Date of Revision

29 February 2012.

Raffinose

1 Nonproprietary Names

None adopted.

2 Synonyms

Gossypose; melitose; melitriose; D-raffinose; D-(+)-raffinose.

3 Chemical Name and CAS Registry Number

β-D-Fructofuranosyl-O-α-D-galactopyranosyl-(1→6)-α-D-glucopyranoside, anhydrous [512-69-6]

β-D-Fructofuranosyl-O-α-D-galactopyranosyl-(1→6)-α-D-glucopyranoside pentahydrate [17629-30-0]

4 Empirical Formula and Molecular Weight

$C_{18}H_{32}O_{16}$	504.44 (for anhydrous)
$C_{18}H_{32}O_{16}\cdot 5H_2O$	594.52 (for pentahydrate)

5 Structural Formula

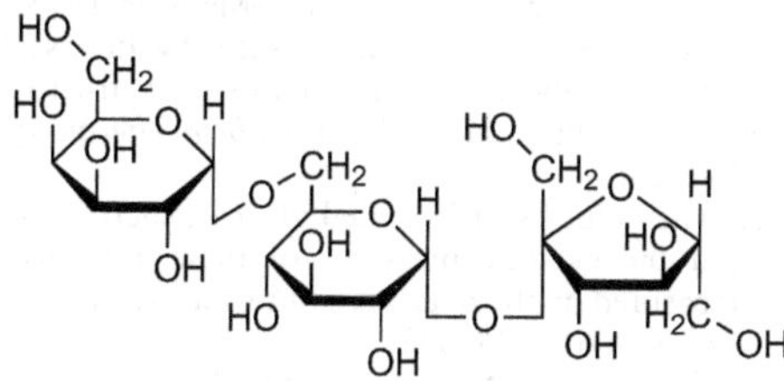

D-Raffinose anhydrous

6 Functional Category

Lyophilization aid.

7 Applications in Pharmaceutical Formulation or Technology

Raffinose is a trisaccharide carbohydrate that is used as a bulking agent, stabilizing agent, and water scavenger in freeze-drying where it acts as a stabilizer for freeze-dried formulations.[1,2] It is also used as a crystallization inhibitor in sucrose solutions.[3–5]

8 Description

Raffinose is a white crystalline powder. It is odorless and has a sweet taste approximately 10% that of sucrose.[6]

9 Pharmacopeial Specifications

—

10 Typical Properties

Collapse temperature −26°C[2]
Decomposition temperature 118–119°C (anhydrous); 130°C (pentahydrate).[7]
Density (bulk) 0.67 g/cm^3 (pentahydrate)
Density (tapped) 0.98 g/cm^3 (pentahydrate)
Density (true) 1.465 g/cm^3 (anhydrous)
Diffusion coefficient (infinite dilution) 0.33×10^{-5} cm^2/s (water at 15°C)[8]
Glass transition temperature 114°C (amorphous)[9]
Heat of solution at infinite dilution (25°C) 52 kJ/mol (crystalline pentahydrate); −38 kJ/mol (amorphous).[1]

SEM 1: Excipient: D-(+)-raffinose pentahydrate; manufacturer: Sigma-Aldrich (Lot No. 092K01211); magnification: 100×.

SEM 2: Excipient: D-(+)-raffinose pentahydrate; manufacturer: Sigma-Aldrich (Lot No. 092K01211); magnification: 500×.

Melting point 80°C (pentahydrate);[7] 118°C (anhydrous).[10]
Optical rotation 105° (pentahydrate); 123° (anhydrous).[11]
Specific gravity 1.465 (pentahydrate)[7]
Solubility in methanol 0.10 g/mL[11]
Solubility in water 0.14 g/mL[7]
Solubility Soluble 1 in 10 of methanol, in pyridine and 1 in 7.1 of water; slightly soluble in ethanol (95%); insoluble in diethyl ether.

The data for the crystal structure,[12,13] NMR structure,[14] powder x-ray diffraction pattern,[15] water vapor sorption isotherms,[15,16] glass transition temperature as a function of water,[15] heat capacity,[1] heat of solution properties,[1] vapor pressure,[17] and osmotic pressure[18] are described in the literature.

11 Stability and Storage Conditions

Raffinose is stable under ordinary conditions of use and storage. Excessive heat should be avoided to prevent degradation. Thermal decomposition products are carbon monoxide and carbon dioxide.[19,20]

12 Incompatibilities

Raffinose is incompatible with strong oxidizers.[21]

13 Method of Manufacture

Raffinose occurs naturally in Australian manna, cottonseed meal, and seeds of various food legumes. It can be isolated from beet sugar molasses through sucrose separation, seed-crystallization, and filtration.[13,22]

14 Safety

Raffinose is a naturally occurring trisaccharide investigated for use in freeze-dried pharmaceutical formulations. It occurs in a number of plants that are consumed widely (*see* Section 13).

15 Handling Precautions

Observe normal precautions appropriate to the circumstances and quantity of material handled. Gloves and safety glasses are recommended. Dust generation should be kept to reasonable levels to avoid ignition or explosion. Short-term exposure has caused respiratory and eye irritation. Long-term exposure has shown adverse reproductive effects in animals. No occupational exposure limits have been established. Dust or air mixtures may ignite or explode.[19,20]

16 Regulatory Status

Raffinose is a naturally occurring trisaccharide and is consumed as part of a normal diet.

17 Related Substances

Raffinose is composed of three monosaccharides: galactose, glucose, and fructose. It shares related structures with sucrose and melibiose. It is also related to stachyose, which possesses an additional (1→6)-linked α-D-galactopyranosyl unit.

Two solvated forms[22] and an amorphous form[14,23,24] of raffinose can be synthesized.

18 Comments

Raffinose has been shown to accumulate in organisms that can survive extreme desiccation, and has therefore been examined as an excipient in stabilizing co-lyophilized protein and labile preparations during storage at elevated temperatures.[25,26]

When exposed to elevated relative humidity (RH) of 75% at 25°C, raffinose has been shown to form different hydrate levels.[27]

Raffinose is indigestible by humans because of a lack of an α-galactosidase and undergoes fermentation in the colon, causing production of carbon dioxide, hydrogen, and methane gases.[10]

Raffinose is used as a blood substitute.

The PubChem Compound ID (CID) for raffinose includes 10542, 219993, and 439242.

19 Specific References

1 Miller DP, de Pablo JJ. Calorimetric solution properties of simple saccharides and their significance for the stabilization of biological structure and function. *J Phys Chem* 2000; **B104**: 8876–8883.
2 Mackenzie AP. Basic principles of freeze-drying for pharmaceuticals. *Bull Parenter Drug Assoc* 1966; **20**(4): 101–129.
3 Caffrey M, *et al.* Lipid–sugar interactions: relevance to anhydrous biology. *Plant Physiol* 1988; **86**: 754–758.
4 Liang B, *et al.* Effects of raffinose to anhydrous biology. *AIChE J* 1989; 35(12): 2053–2057.
5 Van Scoik KG, Carstensen JT. Nucleation phenomena in amorphous sucrose systems. *Int J Pharm* 1990; **vol. 58**: 185–196.
6 Halsam E, ed. *Comprehensive Organic Chemistry: The Synthesis and Reactions of Organic Compounds*. 5. Oxford: Pergamon Press, 1979: 749.
7 Perry RH, Green DW. *Perry's Chemical Engineer's Handbook*, 7th edn. New York: McGraw Hill, 1997.
8 Lide DR. *Handbook of Chemistry and Physics*, 83rd edn. Boca Raton, FL: CRC Press, 2002.
9 Taylor LS, Zografi G. Sugar–polymer hydrogen bond interactions in lyophilized amorphous mixtures. *J Pharm Sci* 1998; **87**(12): 1615–1621.
10 *Kirk-Othmer Encyclopedia of Chemical Technology*, 4th edn. New York: Wiley, 1992: 903.
11 O'Neil MJ, ed. *Merck Index*, 14th edn. Whitehouse Station, NJ: Merck, 2006: 1394.
12 Van Alsenoy C, *et al. Ab initio*-MIA and molecular mechanics studies of the distorted sucrose linkage of raffinose. *J Am Chem Soc* 1994; **116**: 9590–9595.
13 Berman, HM. The crystal structure of a trisaccharaide, raffinose pentahydrate. *Acta Crystallogr* 1970; **B26**: 290–299.
14 Neubauer H, *et al.* NMR structure determination of saccharose and raffinose by means of homo- and heteronuclear dipolar couplings. *Helv Chim Acta* 2001; **84**(1): 243–258.
15 Saleki-Gerhardt A, *et al.* Hydration and dehydration of crystalline and amorphous forms of raffinose. *J Pharm Sci* 1995; **84**(3): 318–323.
16 Saleki-Gerhardt A. Role of water in the solid state properties of crystalline and amorphous form of sugars. *Doctor of Philosophy Thesis*. Wisconsin: University of Wisconsin-Madison 1993; 104–108.
17 Cooke SA, Jonsdottir SO. The vapour pressure of water as a function of solute concentration above aqeous solutions of fructose, sucrose, raffinose, erythitol, xylitol, and sorbitol. *J Chem Thermodynam* 2002; **34**(10): 1545–1555.
18 Kiyosawa K. The volumes of hydrated glucose, sucrose and raffinose molecules, and the osmotic pressures of these aqueous saccharide solutions as measured by the freezing-point-of-depression method. *Bull Chem Soc Jpn* 1988; **61**: 633–642.
19 Mallinckrodt Baker, Inc. Material safety data sheet No. R0300: Raffinose, 5-hydrate, 29 October 2001.
20 Acros Organics N.V. Material safety data sheet No. 93702: D -Raffinose pentahydrate , 2 August 2000.
21 MDL Information Systems, Inc. Material safety data sheet: D-Raffinose pentahydrate, 22 March 2001.
22 Hungerford EH, Nees AR. Raffinose preparation and properties. *Ind Eng Chem* 1934; **26**(4): 462–464.
23 Collins PM, ed. *Carbohydrates*. London: Chapman and Hall, 1997: 431.
24 Jeffrey GA, Huang D. The hydrogen bonding in the crystal structure of raffinose pentahydrate. *Carbohydr Res* 1990; **206**: 173–182.
25 Davidson P, Sun QW. Effect of sucrose/raffinose mass ratios on the stability of co-lyophilized protein during storage above the T_g. *Pharm Res* 2001; **18**(4): 474–479.
26 Kazuhito K, *et al.* Structural and dynamic properties of crystalline and amorphous phases in raffinose–water mixtures. *Pharm Res* 1999; **16**(9): 1441–1448.
27 Hogan SE, Buckton G. Water sorption/desorption—near IR and calorimetric study of crystalline and amorphous raffinose. *Int J Pharm* 2001; **227**: 57–69.

20 General References

—

21 Authors

BC Hancock, MP Mullarney.

22 Date of Revision

1 February 2011.

Saccharin

1 Nonproprietary Names

BP: Saccharin
JP: Saccharin
PhEur: Saccharin
USP–NF: Saccharin

2 Synonyms

1,2-Benzisothiazolin-3-one 1,1-dioxide; benzoic acid sulfimide; benzoic sulfimide; benzosulfimide; 1,2-dihydro-2-ketobenzisosulfonazole; 2,3-dihydro-3-oxobenzisosulfonazole; E954; *Garantose*; gluside; *Hermesetas*; sacarina; saccarina; saccharin insoluble; saccharinum; *o*-sulfobenzimide; *o*-sulfobenzoic acid imide; *Syncal.*

3 Chemical Name and CAS Registry Number

1,2-Benzisothiazol-3(2*H*)-one 1,1-dioxide [81-07-2]

4 Empirical Formula and Molecular Weight

$C_7H_5NO_3S$ 183.18

5 Structural Formula

O O S NH O

6 Functional Category

Sweetening agent.

7 Applications in Pharmaceutical Formulation or Technology

Saccharin is an intense sweetening agent used in beverages, food products, table-top sweeteners, and oral hygiene products such as toothpastes and mouthwashes. In oral pharmaceutical formulations, it is used at a concentration of 0.02–0.5% w/w. It has been used in chewable tablet formulations as a sweetening agent.[1,2] Saccharin has been used to form various pharmaceutical cocrystals.[3] Saccharin can be used to mask some unpleasant taste characteristics or to enhance flavor systems. Its sweetening power is approximately 300–600 times that of sucrose.

8 Description

Saccharin occurs as odorless white crystals or a white crystalline powder. It has an intensely sweet taste, with a metallic or bitter aftertaste that at normal levels of use can be detected by approximately 25% of the population. The aftertaste can be masked by blending saccharin with other sweeteners.

SEM 1: Excipient: saccharin; magnification: 600×.

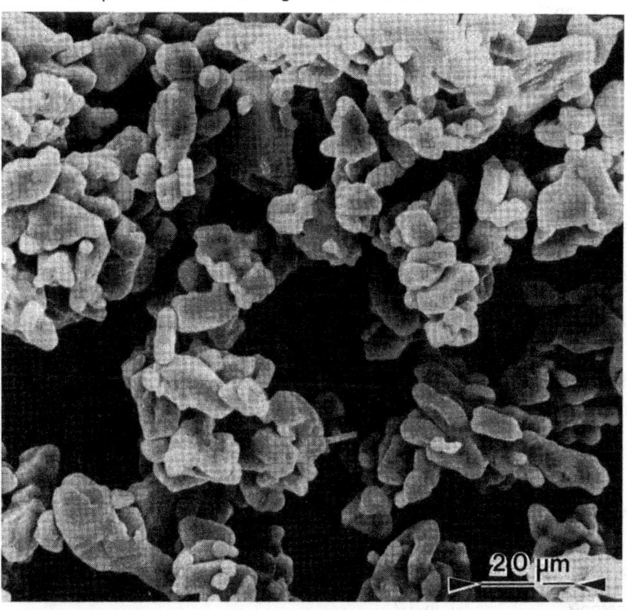

SEM 2: Excipient: saccharin; magnification: 2400×.

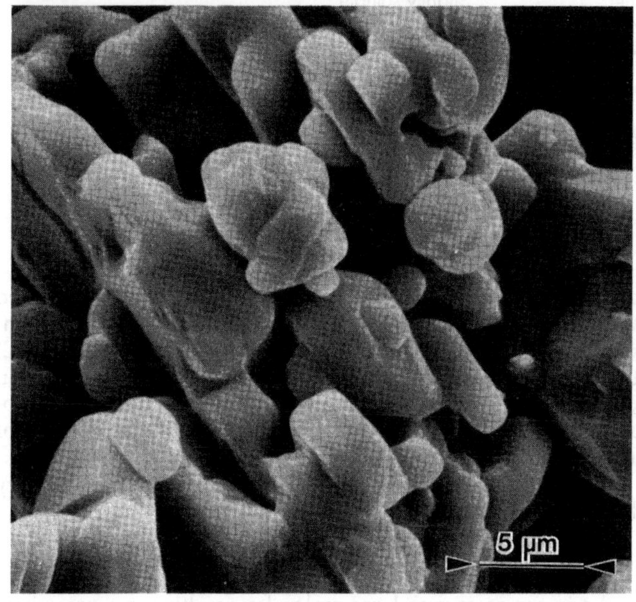

9 Pharmacopeial Specifications

The pharmacopeial specifications for saccharin have undergone harmonization of many attributes for JP, PhEur, and USP–NF.

See Table I. *See also* Section 18.

10 Typical Properties

Acidity/alkalinity pH = 2.0 (0.35% w/v aqueous solution)
Density (bulk) 0.7–1.0 g/cm^3
Density (tapped) 0.9–1.2 g/cm^3
Dissociation constant pK_a = 1.6 at 25°C
Heat of combustion 3644.3 kJ/mol (871 kcal/mol)

Table I: Pharmacopeial specifications for saccharin.

Test	JP XV	PhEur 7.4	USP35–NF30
Identification	+	+	+
Characters	+	+	–
Clarity and color of solution[a]	+	+	+
Melting range	226–230°C	226–230°C	226–230°C
Loss on drying	≤1.0%	≤1.0%	≤1.0%
Residue on ignition	≤0.2%	–	≤0.2%
Sulfated ash	–	≤0.2%	–
Toluenesulfonamides	+	+	+
Heavy metals[a]	≤10 ppm	≤20 ppm	≤10 ppm
Readily carbonizable substances	+	+	+
Benzoic and salicylic acids	+	–	+
Assay (dried basis)	99.0–101.0%	99.0–101.0%	99.0–101.0%

(a) These tests have not been fully harmonized at the time of publication.

Moisture content 0.1%

Solubility Readily dissolved by dilute ammonia solutions, alkali hydroxide solutions, or alkali carbonate solutions (with the evolution of carbon dioxide). *See* Table II.

Table II: Solubility of saccharin.

Solvent	Solubility at 20°C unless otherwise stated
Acetone	1 in 12
Chloroform	Slightly soluble
Ethanol (95%)	1 in 31
Ether	Slightly soluble
Glycerin	1 in 50
Water	1 in 290
	1 in 25 at 100°C

Spectroscopy

IR spectra *see* Figure 1.

NIR spectra *see* Figure 2.

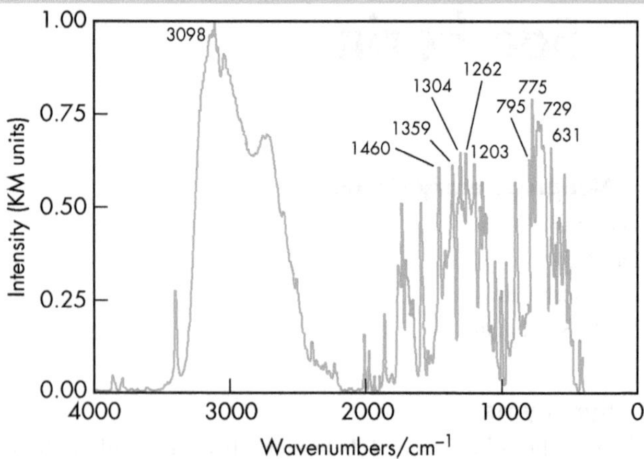

Figure 1: Infrared spectrum of saccharin measured by diffuse reflectance. Adapted with permission of Informa Healthcare.

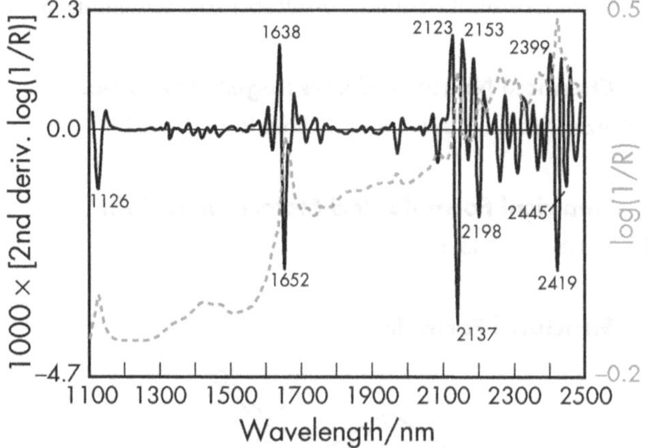

Figure 2: Near-infrared spectrum of saccharin measured by reflectance.

11 Stability and Storage Conditions

Saccharin is stable under the normal range of conditions employed in formulations. In the bulk form it shows no detectable decomposition and only when it is exposed to a high temperature (125°C) at a low pH (pH 2) for over 1 hour does significant decomposition occur. The decomposition product formed is (ammonium-o-sulfo)benzoic acid, which is not sweet.[4] The aqueous stability of saccharin is excellent.

Saccharin should be stored in a well-closed container in a dry place.

12 Incompatibilities

Saccharin can react with large molecules, resulting in a precipitate being formed. It does not undergo Maillard browning.

13 Method of Manufacture

Saccharin is prepared from toluene by a series of reactions known as the Remsen–Fahlberg method. Toluene is first reacted with chlorosulfonic acid to form *o*-toluenesulfonyl chloride, which is reacted with ammonia to form the sulfonamide. The methyl group is then oxidized with dichromate, yielding *o*-sulfamoylbenzoic acid, which forms the cyclic imide saccharin when heated.

An alternative method involves a refined version of the Maumee process. Methyl anthranilate is initially diazotized to form 2-carbomethoxybenzenediazonium chloride; sulfonation followed by oxidation then yields 2-carbomethoxybenzenesulfonyl chloride. Amidation of this material, followed by acidification, forms insoluble acid saccharin.

14 Safety

There has been considerable controversy concerning the safety of saccharin, which has led to extensive studies since the mid-1970s.

Two-generation studies in rats exposed to diets containing 5.0–7.5% total saccharin (equivalent to 175 g daily in humans) suggested that the incidence of bladder tumors was significantly greater in saccharin-treated males of the second generation than in controls.[5,6] Further experiments in rats suggested that a contaminant of commercial saccharin, *o*-toluene sulfonamide, might also account for carcinogenic effects. In view of these studies, a ban on the use of saccharin was proposed in several countries. However, in 1977 a ban by the FDA led to a Congressional moratorium that permitted the continued use of saccharin in the US.

From the available data it now appears that the development of tumors is a sex-, species-, and organ-specific phenomenon, and extensive epidemiological studies have shown that saccharin intake is not related to bladder cancer in humans.[7,8]

The WHO has set a temporary acceptable daily intake for saccharin, including its calcium, potassium, and sodium salts, at up to 2.5 mg/kg body-weight.[9] In the UK, the Committee on Toxicity of Chemicals in Food, Consumer Products, and the Environment (COT) has set an acceptable daily intake for saccharin and its calcium, potassium, and sodium salts (expressed as saccharin sodium) at up to 5 mg/kg body-weight.[10]

Adverse reactions to saccharin, although relatively few in relation to its widespread use, include: urticaria with pruritus following ingestion of saccharin-sweetened beverages[11] and photosensitization reactions.[12]

LD_{50} (mouse, oral): 17.5 g/kg[13]
LD_{50} (rat, IP): 7.10 g/kg
LD_{50} (rat, oral): 14.2 g/kg

15 Handling Precautions

Observe normal precautions appropriate to the circumstances and quantity of material handled. When heated to decomposition, saccharin releases toxic gases, including NO_x and SO_x. Eye protection and a dust mask are recommended.

16 Regulatory Status

Accepted for use as a food additive in Europe. Note that the EU number 'E954' is applied to both saccharin and saccharin salts. Included in the FDA Inactive Ingredients Database (oral solutions, syrups, tablets; topical preparations; inhalation, aerosols). Included in nonparenteral medicines licensed in the UK. Included in the Canadian Natural Health Products Ingredients Database.

17 Related Substances

Acesulfame potassium; alitame; aspartame; isomalt; lactitol; maltitol; mannitol; neotame; saccharin ammonium; saccharin calcium; saccharin sodium; sodium cyclamate; sorbitol; sucralose; tagatose; thaumatin; xylitol.

Saccharin ammonium
Empirical formula $C_7H_8N_2O_3S$
Molecular weight 200.2
CAS number [6381-61-9]

Saccharin calcium
Empirical formula $C_{14}H_8CaN_2O_6S_2 \cdot 3H_2O$
Molecular weight 467.48
CAS number
[6381-91-5] for the hydrated form
[6485-34-3] for the anhydrous form
Synonyms *Syncal CAS*.
Appearance White, odorless crystals or crystalline powder with an intensely sweet taste.
Solubility 1 in 4.7 ethanol (95%); 1 in 2.6 of water.

18 Comments

Saccharin has undergone harmonization of many attributes for JP, PhEur, and USP–NF by the Pharmacopeial Discussion Group. For further information see the General Information Chapter <1196> in the USP35–NF30, the General Chapter 5.8 in PhEur 7.4, along with the 'State of Work' document on the PhEur EDQM website, and also the General Information Chapter 8 in the JP XV.

The perceived intensity of sweeteners relative to sucrose depends upon their concentration, temperature of tasting, and pH, and on the flavor and texture of the product concerned.

Intense sweetening agents will not replace bulk, textural, or preservative characteristics of sucrose if sucrose is removed from a formulation.

Synergistic effects for combinations of sweeteners have been reported. Saccharin is often used in combination with cyclamates and aspartame since the saccharin content may be reduced to minimize any aftertaste.

A specification for saccharin is contained in the *Food Chemicals Codex* (FCC).[14]

The EINECS number for saccharin is 201-321-0. The PubChem Compound ID (CID) for saccharin is 5143.

19 Specific References

1 Suzuki H, *et al.* Acetaminophen-containing chewable tablets with suppressed bitterness and improved oral feeling. *Int J Pharm* 2004; **278**(1): 57–61.
2 Mullarney MP, *et al.* The powder flow and compact mechanical properties of sucrose and three high-density sweetners used in chewable tablets. *Int J Pharm* 2003; **257**(1–2): 227–236.
3 Banerjee R, *et al.* Saccharin salts of active pharmaceutical ingredients, their crystal structures, and increased water solubilities. *Cryst Growth Des* 2005; **5**: 2299–2309.
4 DeGarmo O, *et al.* Hydrolytic stability of saccharin. *J Am Pharm Assoc* 1952; **41**: 17–18.
5 Arnold DL, *et al.* Long-term toxicity of *ortho*-toluenesulfonamide and sodium saccharin in the rat. *Toxicol Appl Pharmacol* 1980; **52**: 113–152.
6 Arnold DL. Two-generation saccharin bioassays. *Environ Health Perspect* 1983; **50**: 27–36.
7 Council on Scientific Affairs. Saccharin: review of safety issues. *J Am Med Assoc* 1985; **254**: 2622–2624.
8 Morgan RW, Wong O. A review of epidemiological studies on artificial sweeteners and bladder cancer. *Food Chem Toxicol* 1985; **23**: 529–533.
9 FAO/WHO. Evaluation of certain food additives and contaminants. Twenty-eighth report of the FAO/WHO expert committee on food additives. *World Health Organ Tech Rep Ser* 1984; No. 710.
10 Food Advisory Committee. FAC further advice on saccharin. FdAC/REP/9. London: MAFF, 1990.
11 Miller R, *et al.* A case of episodic urticaria due to saccharin ingestion. *J Allergy Clin Immunol* 1974; **53**: 240–242.
12 Gordon HH. Photosensitivity to saccharin. *J Am Acad Dermatol* 1983; **8**: 565.
13 Lewis RJ, ed. *Sax's Dangerous Properties of Industrial Materials*, 11th edn. New York: Wiley, 2004; 3277.
14 *Food Chemicals Codex*, 7th edn. Bethesda, MD: United States Pharmacopeia, 2010.

20 General References

Anonymous. Saccharin is safe. *Chem Br* 2001; **37**(4): 18.
European Directorate for the Quality of Medicines and Healthcare (EDQM). European Pharmacopoeia – State Of Work Of International Harmonisation. *Pharmeuropa* 2011; 23(4): 713–714. www.edqm.eu/site/-614.html (accessed 5 December 2011).
Lindley MG. Sweetener markets, marketing and product development. In: Marie S, Piggott JR, eds. *Handbook of Sweeteners*. Glasgow: Blackie, 1991; 186.
Nelson AL. *Sweeteners: Alternative*. St Paul, MN: Eagan Press, 2000: 1–99.
Zubair MU, Hassan MMA. Saccharin. In: Florey K, ed. *Analytical Profiles of Drug Substances*. 13. Orlando, FL: Academic Press, 1984; 487–519.

21 Author

P Heljo.

22 Date of Revision

1 March 2012.

Saccharin Sodium

1 Nonproprietary Names

BP: Saccharin Sodium
JP: Saccharin Sodium Hydrate
PhEur: Saccharin Sodium
USP–NF: Saccharin Sodium

2 Synonyms

1,2-Benzisothiazolin-3-one 1,1-dioxide, sodium salt; *Crystallose*; E954; gendorf 450; saccharinum natricum; sodium *o*-benzosulfimide; soluble gluside; soluble saccharin; sucaryl sodium; *Syncal*.

3 Chemical Name and CAS Registry Number

1,2-Benzisothiazol-3(2*H*)-one 1,1-dioxide, sodium salt
[6155-57-3] for the dihydrate
[128-44-9] for the anhydrous material
See also Section 8.

4 Empirical Formula and Molecular Weight

$C_7H_4NNaO_3S$	205.16
$C_7H_4NNaO_3S \cdot {}^2\!/_3H_2O$ (84%)	217.24
$C_7H_4NNaO_3S \cdot 2H_2O$ (76%)	241.19

5 Structural Formula

•$2H_2O$ 76% saccharin sodium (dihydrate)

•${}^2\!/_3H_2O$ 84% saccharin sodium

6 Functional Category

Sweetening agent.

7 Applications in Pharmaceutical Formulation or Technology

Saccharin sodium is an intense sweetening agent used in beverages, food products, table-top sweeteners,(1) and pharmaceutical formulations such as tablets, powders, medicated confectionery, gels, suspensions, liquids, and mouthwashes;(2) *see* Table I. It is also used in vitamin preparations.

Saccharin sodium is considerably more soluble in water than saccharin, and is more frequently used in pharmaceutical formulations. Its sweetening power is approximately 300–600 times that of sucrose. Saccharin sodium enhances flavor systems and may be used to mask some unpleasant taste characteristics.

Injection of saccharin sodium has been used to measure the arm-to-tongue circulation time.

Table I: Uses of saccharin sodium.

Use	Concentration (%)
Dental paste/gel	0.12–0.3
IM/IV injections	0.9
Oral solution	0.075–0.6
Oral syrup	0.04–0.25

8 Description

Saccharin sodium occurs as a white, odorless or faintly aromatic, efflorescent, crystalline powder. It has an intensely sweet taste, with a metallic or bitter aftertaste that at normal levels of use can be detected by approximately 25% of the population. The aftertaste can be masked by blending saccharin sodium with other sweeteners. Saccharin sodium can contain variable amounts of water.

9 Pharmacopeial Specifications

The pharmacopeial specifications for saccharin sodium have undergone harmonization of many attributes for JP, PhEur, and USP–NF.

See Table II. *See also* Section 18.

10 Typical Properties

Unless stated, data refer to either 76% or 84% saccharin sodium.
Acidity/alkalinity pH = 6.6 (10% w/v aqueous solution)
Density (bulk)
0.8–1.1 g/cm^3 (76% saccharin sodium);
0.86 g/cm^3 (84% saccharin sodium).
Density (particle) 1.70 g/cm^3 (84% saccharin sodium)
Density (tapped)
0.9–1.2 g/cm^3 (76% saccharin sodium);
0.96 g/cm^3 (84% saccharin sodium).
Melting point Decomposes upon heating.

SEM 1: Excipient: saccharin sodium; magnification: 35×; voltage: 5 kV.

Table II: Pharmacopeial specifications for saccharin sodium.

Test	JP XV	PhEur 7.4	USP35–NF30
Identification	+	+	+
Characters	+	+	−
Clarity and color of solution[(a)]	+	+	+
Acidity or alkalinity	+	+	+
Water	≤15.0%	≤15.0%	≤15.0%
Benzoate and salicylate	+	−	+
Toluenesulfonamides	+	+	+
Heavy metals[(a)]	≤10 ppm	≤20 ppm	≤10 ppm
Readily carbonizable substances	+	+	+
Assay (anhydrous basis)	99.0–101.0%	99.0–101.0%	99.0–101.0%

(a) These tests have not been fully harmonized at the time of publication.

Moisture content Saccharin sodium 76% contains 14.5% w/w water; saccharin sodium 84% contains 5.5% w/w water. During drying, water evolution occurs in two distinct phases. The 76% material dries under ambient conditions to approximately 5.5% moisture (84% saccharin sodium); the remaining moisture is then removed only by heating.
Solubility *see* Table III.

Table III: Solubility of saccharin sodium.

Solvent	Solubility at 20°C unless otherwise stated
Buffer solutions:	
pH 2.2 (phthalate)	1 in 1.15 1 in 0.66 at 60°C
pH 4.0 (citrate–phosphate)	1 in 1.21 1 in 0.69 at 60°C
pH 7.0 (citrate–phosphate)	1 in 1.21 1 in 0.66 at 60°C
pH 9.0 (borate)	1 in 1.21 1 in 0.69 at 60°C
Ethanol	1 in 102
Ethanol (95%)	1 in 50
Ether	Practically insoluble
Propylene glycol	1 in 3.5
Propan-2-ol	Practically insoluble
Water	1 in 1.2

Specific surface area 0.25 m^2/g
Spectroscopy

IR spectra *see* Figure 1.

NIR spectra *see* Figure 2.

11 Stability and Storage Conditions

Saccharin sodium is stable under the normal range of conditions employed in formulations. Only when it is exposed to a high temperature (125°C) at a low pH (pH 2) for over 1 hour does significant decomposition occur. The 84% grade is the most stable form of saccharin sodium since the 76% form will dry further under ambient conditions. Solutions for injection can be sterilized by autoclave.

Saccharin sodium should be stored in a well-closed container in a dry place.

12 Incompatibilities

Saccharin sodium does not undergo Maillard browning.

13 Method of Manufacture

Saccharin is produced by the oxidation of *o*-toluene sulfonamide by potassium permanganate in a solution of sodium hydroxide.

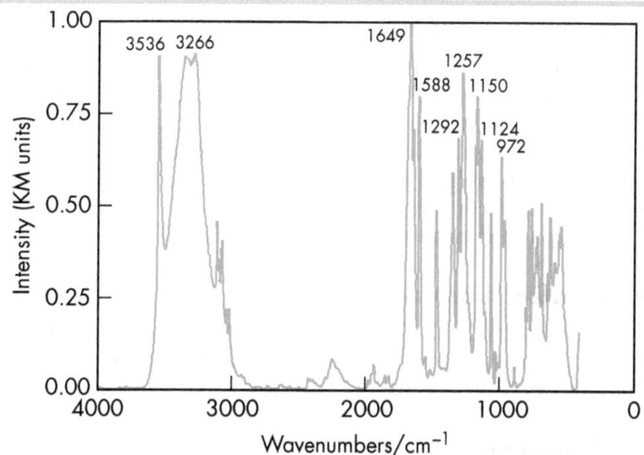

Figure 1: Infrared spectrum of saccharin sodium dihydrate measured by diffuse reflectance. Adapted with permission of Informa Healthcare.

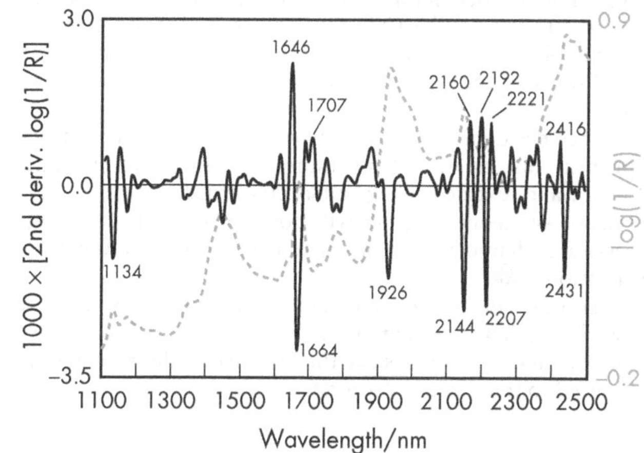

Figure 2: Near-infrared spectrum of saccharin sodium measured by reflectance.

Acidification of the solution precipitates saccharin, which is then dissolved in water at 50°C and neutralized by addition of sodium hydroxide. Rapid cooling of the solution initiates crystallization of saccharin sodium from the liquors.

14 Safety

There has been considerable controversy concerning the safety of saccharin and saccharin sodium in recent years; however, it is now generally regarded as a safe, intense sweetener. *See* Saccharin for further information.

The WHO has set a temporary acceptable daily intake of up to 2.5 mg/kg body-weight for saccharin, including its salts.[(3)] In the UK, the Committee on Toxicity of Chemicals in Food, Consumer Products, and the Environment (COT) has set an acceptable daily intake for saccharin and its salts (expressed as saccharin sodium) at up to 5 mg/kg body-weight.[(4)]

LD_{50} (mouse, oral): 17.5 g/kg[(5)]

LD_{50} (rat, IP): 7.1 g/kg

LD_{50} (rat, oral): 14.2 g/kg

15 Handling Precautions

Observe normal precautions appropriate to the circumstances and quantity of material handled. When heated to decomposition, saccharin sodium releases toxic gases, including NO_x, Na_2O, and SO_x. Eye protection and a dust mask are recommended.

S

16 Regulatory Status

Accepted for use as a food additive in Europe; 'E954' is applied to both saccharin and saccharin salts. Included in the FDA Inactive Ingredients Database (buccal and dental preparations; IM and IV injections; oral preparations; topical preparations; inhalation preparations; rectal tablets, solutions, and suspensions). Included in nonparenteral medicines licensed in the UK. Included in the Canadian Natural Health Products Ingredients Database.

17 Related Substances

Acesulfame potassium; alitame; aspartame; isomalt; lactitol; maltitol; mannitol; neotame; saccharin; sorbitol; sucralose; tagatose; thaumatin; xylitol.

18 Comments

Saccharin sodium has undergone harmonization of many attributes for JP, PhEur, and USP–NF by the Pharmacopeial Discussion Group. For further information see the General Information Chapter <1196> in the USP35–NF30, the General Chapter 5.8 in PhEur 7.4, along with the 'State of Work' document on the PhEur EDQM website, and also the General Information Chapter 8 in the JP XV.

The perceived intensity of sweeteners relative to sucrose depends upon their concentration, temperature of tasting, and pH, and on the flavor and texture of the product concerned.

Intense sweetening agents will not replace bulk, textural, or preservative characteristics of sugar if sugar is removed from a formulation.

Synergistic effects for combinations of sweeteners have been reported. Saccharin sodium is often used in combination with cyclamates and aspartame since the saccharin sodium content may be reduced to minimize any aftertaste.

The PubChem Compound ID (CID) for saccharin sodium includes 656582 and 23691045.

19 Specific References

1 Kloesel L. Sugar substitutes. *Int J Pharm Compound* 2000; 4(2): 86–87.
2 Ungphaiboon S, Maitani Y. *In vitro* permeation studies of triamcinolone acetonide mouthwashes. *Int J Pharm* 2001; 220(1–2): 111–117.
3 FAO/WHO. Evaluation of certain food additives and contaminants. Twenty-eighth report of the FAO/WHO expert committee on food additives. *World Health Organ Tech Rep Ser* 1984; No. 710.
4 Food Advisory Committee. FAC further advice on saccharin. FdAC/REP/9. London: MAFF, 1990.
5 Lewis RJ, ed. *Sax's Dangerous Properties of Industrial Materials*, 11th edn. New York: Wiley, 2004; 3277.

See Saccharin for further references.

20 General References

Anonymous. Saccharin is safe. *Chem Br* 2001; 37(4): 18.

European Directorate for the Quality of Medicines and Healthcare (EDQM). European Pharmacopoeia – State Of Work Of International Harmonisation. *Pharmeuropa* 2011; 23(4): 713–714. www.edqm.eu/site/-614.html (accessed 5 December 2011).

Lindley MG. Sweetener markets, marketing and product developments. In: Marie S, Piggott JR, eds. *Handbook of Sweeteners*. Glasgow: Blackie, 1991: 186.

Nelson AL. *Sweeteners: Alternative*. St Paul, MN: Eagan Press, 2000; 1–99.

US Food and Drug Administration. Artificial sweeteners: no calories...sweet! *FDA Consumer Magazine* 2006; 40(4): 27–28.

21 Author

P Heljo.

22 Date of Revision

1 March 2012.

Safflower Oil

S

1 Nonproprietary Names

BP: Refined Safflower Oil
PhEur: Safflower Oil, Refined
USP–NF: Safflower Oil

2 Synonyms

Aceite de alazor; aceite de cartamo; carthami oleum raffinatum; dygminu aliejus, rafinuotas; huile de carthame; safflorolja, raffinerad; safflower oil (unhydrogenated); saflonoljy puhdistettu.

3 Chemical Name and CAS Registry Number

Safflower oil [8001-23-8]

4 Empirical Formula and Molecular Weight

The PhEur 7.4 defines the composition of the fatty acid fraction of two types of refined safflower oil (type I and type II); *see* Section 9.

5 Structural Formula

See Section 4.

6 Functional Category

Emollient; oleaginous vehicle; solvent.

7 Applications in Pharmaceutical Formulation or Technology

Safflower oil is mainly used as an oleaginous vehicle in oral and topical formulations.

Safflower oil has been used as a vehicle in the development of an oral dosage form containing a novel viral-specific inhibitor of the replication of human rhinoviruses.[1] It has also been used as a solvent for a capsule formulation containing a new antilipemic agent; formulations containing safflower oil were found to have the greatest bioavailability in dogs compared with formulations containing PEG 300 or water.[2]

Safflower oil is used in cosmetics products such as soaps, lotions, creams, and hair-care preparations.

8 Description

Refined safflower oil is a clear, viscous, yellow to pale-yellow liquid, with a slight vegetable odor.

9 Pharmacopeial Specifications

See Table I. *See also* Section 18.

The PhEur 7.4 defines the composition of the fatty acid fraction, refractive index, and relative density values of two types of refined safflower oil (type I and type II); *see* Tables II and III.

Table I: Pharmacopeial specifications for safflower oil.

Test	PhEur 7.4	USP35–NF30
Identification	+	—
Characters	+	—
Acid value	≤0.5	—
Peroxide value	≤10.0[a]	≤10
Unsaponifiable matter	≤1.5%	≤1.5%
Alkaline impurities	+	—
Composition of fatty acids	+[b]	+
Brassicasterol	≤0.3% in sterol fraction of oil	—
Water	≤0.1%	—
Free fatty acids	—	+
Iodine value	—	135–150
Heavy metals	—	≤10 ppm

(a) ≤5.0 if for parenteral use.
(b) Compositions of the fatty acid fraction of type I and type II refined safflower oil are defined.

Table II: Composition of the fatty acids for safflower oil.

Fatty acid	PhEur 7.4		USP35–NF30
	Type I	Type II	
Saturated fatty acids of chain length less than C_{14}	≤0.2%	≤0.2%	—
Myristic acid	≤0.2%	≤0.2%	—
Palmitic acid	4.0–10.0%	3.6–6.0%	2–10%
Stearic acid	1.0–5.0%	1.0–5.0%	1–10%
Oleic acid	8.0–21.0%		70.0–84.0%
7–42%			
Linoleic acid			68.0–83.0%
7.0–23.0%	72–84%		
Linolenic acid	≤0.5%	≤0.5%	—
Arachidic acid	≤0.5%	≤1.0%	—
Eicosenoic acid	≤0.5%	≤1.0%	—
Behenic acid	≤1.0%	≤1.2%	—

Table III: Refractive index and relative density values for type I and type II refined safflower oils.

	Refined safflower oil	
	Type I	Type II
Refractive index	≈1.476	≈1.472
Relative density	≈0.922	≈0.914

10 Typical Properties

Acid value 1.0–9.7

Flash point >287.8°C (closed cup)

Hydroxyl value 2.9–6.0

Iodine value 140–150

Refractive index

n_D^{25} = 1.472–1.475;

n_D^{40} = 1.4690–1.4692.

See also Table III.

Relative density *see* Table III.

Saponification value 188–194

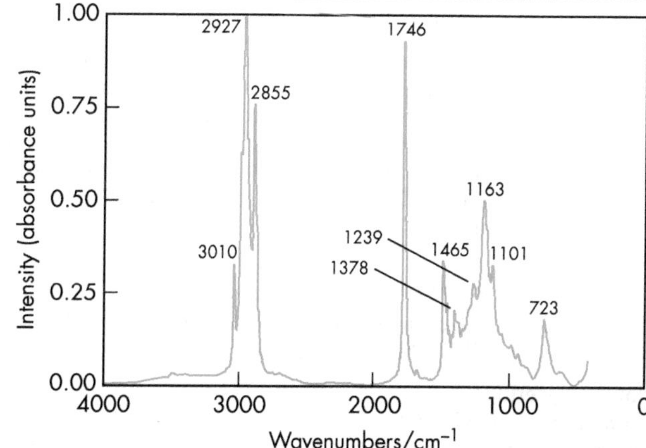

Figure 1: Infrared spectrum of safflower oil measured by transmission. Adapted with permission of Informa Healthcare.

Solubility Soluble in organic solvents. Refined safflower oil is miscible with ether, chloroform, light petroleum (bp 40–60°C); practically insoluble in alcohol. Solubility in water is <0.1%.

Spectroscopy

IR spectra *see* Figure 1.

11 Stability and Storage Conditions

Safflower oil thickens and becomes rancid on prolonged exposure to air. It is also sensitive to light. Safflower oil should be preserved in tight, light-resistant containers. Refined safflower oil should be stored in a well-filled, airtight container, protected from light.

Parenteral fat emulsions containing safflower oil are destabilized by electrolytes; severe droplet coalescence in the emulsion occurs 3–5 days after the addition of 10% v/v dimethyl sulfoxide, and after 10 days if 5% v/v is added.[1] Parenteral fat emulsions are prone to bacterial and fungal growth. Generally, fat emulsions containing safflower oil or soybean oil show similar growth patterns,[2,3] although growth of *Candida albicans* has been reported to be higher in safflower oil containing fat emulsions than in other types of emulsion.[4]

12 Incompatibilities

Safflower oil is incompatible with strong oxidizing agents.

13 Method of Manufacture

Refined safflower oil is the fatty oil obtained from the seeds of *Carthamus tinctorius* L. (type I) or from seeds of hybrids of *Carthamus tinctorius* L. (type II) by expression and/or extraction followed by refining. Type II refined safflower oil is rich in oleic (*cis*-9-octadecenoic) acid. It may contain a suitable antioxidant.

Safflower oil USP35–NF30 is the refined fixed oil yielded by the seed of *Carthamus tinctorius* Linné (Fam. Compositae).

14 Safety

Safflower oil is an edible oil and generally presents no significant health hazards following eye contact, skin contact, oral ingestion, or inhalation. Skin irritation or allergic reactions, or eye irritation may occur. Ingestion of large doses can cause vomiting. Safflower oil may cause diarrhea.

LD_{50} (mouse, IP): >50 g/kg[5]

15 Handling Precautions

Observe normal precautions appropriate to the circumstances and quantity of the material handled. When heated to decomposition, safflower oil emits acrid smoke and irritating fumes.

16 Regulatory Status

Included in the FDA Inactive Ingredients Database (topical lotion). Included in the Canadian Natural Health Products Ingredients Database. Included in an intravenous fat emulsion (*Liposyn II*) available in the US. Included in a capsule formulation available in Canada and in a non-medicinal capsule formulation previously available in the UK. It is also a component of a Canadian enteral nutrition preparation.

17 Related Substances

Almond oil; canola oil; corn oil; cottonseed oil; peanut oil; safflower glycerides; sesame oil; soybean oil; sunflower oil.

Safflower glycerides

CAS number [79982-97-1]

Comments Safflower glycerides (safflower oil monoglycerides) are used in cosmetics as emollients and emulsifying agents. The EINECS number for safflower glycerides is 279-360-8.

18 Comments

The PhEur 7.4 requires the label for refined safflower oil to state, where applicable, that the substance is suitable for use in the manufacture of parenteral dosage forms, and the type of oil (type I or type II). The PhEur 7.4 also lists a monograph for safflower flower, while JP XV includes an unofficial monograph for safflower.

Safflower oil is used as a component of parenteral fat emulsions for the preparation of parenteral nutrition solutions. A topical lotion containing 3% safflower oil is commercially available, and parenteral fat emulsions containing a mixture of safflower oil 5% and soya oil 5%, or 10% and 10%, respectively, have been administered as part of total parenteral nutrition regimes.

Safflower oil is used as a food, being consumed in the form of soft margarine, salad oils, and cooking oils.

Safflower oil has been investigated as a vehicle in the development of an oral dosage form containing a novel viral-specific inhibitor of the replication of human rhinoviruses.[6] It has also been investigated as a solvent for a capsule formulation containing a new antilipemic agent; formulations containing safflower oil were found to have the greatest bioavailability in dogs compared with formulations containing PEG 300 or water.[7] Safflower oil has been studied as a vehicle in an *in situ*-forming oleogel implant[8] as well as a vehicle for a subcutaneous injection of high-dose lysozyme microparticles.[9]

A specification for safflower oil is listed in *Japanese Pharmaceutical Excipients* (JPE).[10] A monograph for safflower oil (unhydrogenated) is contained in the *Food Chemicals Codex* (FCC).[11]

The EINECS number for safflower oil is 232-276-5.

19 Specific References

1 Li J, *et al.* Method for the early evaluation of the effects of storage and additives on the stability of parenteral fat emulsions. *Pharm Res* 1993; 10: 535–541.
2 Crocker KS, *et al.* Microbial growth comparisons of five commercial parenteral lipid emulsions. *J Parenter Enteral Nutr* 1984; 8: 391–395.
3 Kim CH, *et al.* Bacterial and fungal growth in intravenous fat emulsion. *Am J Hosp Pharm* 1983; 40: 2159–2161.
4 Keammerer D, *et al.* Microbial growth patterns in intravenous fat emulsions. *Am J Hosp Pharm* 1983; 40: 1650–1653.
5 Lewis RJ, ed. *Sax's Dangerous Properties of Industrial Materials*, 11th edn. New York: Wiley, 2004: 3176.
6 Simmons DM, *et al.* Oral dosage development of a human rhinovirus and non-polio enterovirus inhibitor. *Drug Dev Ind Pharm* 1997; **23**(8): 783–789.
7 Burcham DL, *et al.* Improved bioavailability of the hypocholesterolemic DMP565 in dogs following oral dosing in oil and glycol solutions. *Biopharm Drug Dispos* 1997; **18**(8): 737–742.
8 Vintiloiu A, *et al. In situ*-forming oleogel implant for rivastigmine delivery. *Pharm Res* 2008; **25**(4): 845–852.
9 Miller MA, *et al.* Low viscosity highly concentrated injectable nonaqueous suspensions of lysozyme microparticles. *Langmuir* 2010; **26**(2): 1067–1074.
10 Japan Pharmaceutical Excipients Council. *Japan Pharmaceutical Excipients* 2004. Tokyo: Yakuji Nippo, 2004: 752.
11 *Food Chemicals Codex*, 7th edn. Bethesda, MD: United States Pharmacopeia, 2010: 899.

20 General References

Parchem Trading Ltd. Product specification sheet: Safflower oil. http://www.parchem.com/Safflower-Oil-001444.aspx (accessed 13 July 2011).

21 Author

CG Cable.

22 Date of Revision

1 March 2012.

Saponite

1 Nonproprietary Names

None adopted.

2 Synonyms

Afrodit; aluminum-saponite; auxite; cathkinite; ferroan saponite; griffithite; *Imvite 1016*; *Ionite P*; *Laponite*; licianite; lucianite; piotine; *SapCa-1*; *Smectiton SA*; *SMI 200H*; *Stevensonite*; *Sumecton 5A*; *SY 5*; *Veegum S6198*; zebedassite.

3 Chemical Name and CAS Registry Number

Saponite [1319-41-1]

4 Empirical Formula and Molecular Weight

$(Ca_{0.5}Na)_{0.3}(Mg,Fe^{2+})_3(Si,Al)_4O_{10}(OH)_2 \cdot 4H_2O$ ≈480

Saponite is a mineral with an approximate empirical formula owing to the variability in cation substitution; *see* Table I.

Table I: Approximate composition of saponite based on chemical analysis.[1]

Component	Wt % range
SiO_2	39.6–54.7
Al_2O_3	3.9–10.2
MgO	15.8–33.3
Fe_2O_3	0.2–12.0
FeO	0–7.8
CaO	0–2.9
Na_2O	0–0.7
K_2O	0–0.3
TiO_2	0–0.4
MnO	0–0.3
H_2O+ (structural water OH)	4.2–12.0
H_2O- (hydration water)	7.2–17.4

5 Structural Formula

Saponite is composed of two tetrahedral layers formed by phylosilicate sheets and one octahedral layer. Common impurities include manganese, nickel, phosphorus, potassium, and titanium.

See Section 4.

6 Functional Category

Adsorbent; emulsifying agent; viscosity-increasing agent.

7 Applications in Pharmaceutical Formulation or Technology

Saponite is a colloidal material present in various naturally occurring clays such as magnesium aluminum silicates[2] and is therefore suitable for use in pharmaceutical formulation applications as an adsorbent, viscosity-increasing agent, suspending agent, or as an oil-in-water emulsifying agent.

Saponite, as a component of magnesium aluminium silicates, is useful as a formulation component in semisolid cosmetic and health care products.[3]

8 Description

Saponite is a naturally occurring 2:1 phyllosilicate clay of the smectite (montmorillonite) group. It is a magnesium-rich hydrated aluminum silicate and is present as a component of some commercial magnesium aluminum silicate clays. It occurs in soft, amorphous masses in the cavities of certain rocks.

Saponite occurs as a white to off-white, dull powder composed of fine-grained crystals of colloidal size. The material is greasy or soapy to the touch and swells on the addition of water.

9 Pharmacopeial Specifications

See Section 18.

10 Typical Properties

Density (true) ~2.24-2.3 g/cm^3 (measured; may vary with clay source)

Crystal data Monoclinic: $a = 5.3$, $b = 9.14$, $c = 16.9$, $\beta \approx 97°$.

Hardness (Mohs) 1–2

Moisture content

~13.7% water loss up to 150°C;

~6.9% water loss above 150°C.

11 Stability and Storage Conditions

Saponite is a stable material and should be stored in a cool, dry place.

12 Incompatibilities

Saponite may generate heat in contact with hydrofluoric acid.

13 Method of Manufacture

Naturally occurring saponite is mined from deposits in various localities around the world.

14 Safety

Saponite is a natural clay mineral that is not acutely toxic; therefore, no toxicity values have been established. However, it may contain small amounts of crystalline silica in the form of quartz. Chronic exposure to crystalline silica can have adverse effects on the respiratory system. EU labeling states the material is not classified as dangerous.

Saponite dust can be irritating to the respiratory tract and eyes. Contact with this material may cause drying of the skin.

15 Handling Precautions

Observe normal precautions appropriate to the circumstances and quantity of material being handled. Avoid generating and breathing dust, and use eye protection. For dusty conditions, eye protection, gloves, and a dust mask are recommended. The occupational exposure limits for saponite are 5 mg/m^3 (respirable) PEL-TWA, 3 mg/m^3 (respirable) TLV-TWA, and 10 mg/m^3 (inhalable) dust TLV-TWA.

16 Regulatory Status

The components of saponite are individually reported in the EPA TSCA Inventory.

17 Related Substances

Attapulgite; bentonite; kaolin; hectorite; magnesium aluminum silicate; talc.

18 Comments

Saponite is listed as a component of aluminum magnesium silicate in the PhEur 7.4, BP 2012, and USP35–NF30.

Saponite is a swelling clay with a low cation exchange capacity, and when mixed with water it displays thixotropic properties. Saponite is similar to bentonite, and has the capacity to adsorb drugs through cationic exchange.[4] Drug–saponite adsorbates show a slight reduction in dissolution rate[5] and the mechanistics of adsorption of drug molecules to saponite have been examined.[6] Saponite is useful in the formulation of gastrointestinal X-ray contrast agents[7] and formulations designed for sustained drug delivery to the gastrointestinal tract.[8] The adsorption mechanism of an amine drug to saponite in mixtures with montmorillonite as well as the drug's subsequent release kinetics from these complexes has been studied by a variety of physical characterization techniques.[9] The intercalation tendency of donepezil into saponite and other smectites has been examined in relation to their drug release kinetics.[10]

The EINECS number for saponite is 215-289-0.

19 Specific References

1 Lopez-Galindo A, *et al.* Compositional, technical and safety specifications of clays to be used as pharmaceutical and cosmetic products. *Appl Clay Sci* 2007; **36**(1-3): 51–63.
2 Browne JE, *et al.* Characterization and adsorptive properties of pharmaceutical grade clays. *J Pharm Sci* 1980; **69**(7): 816–823.
3 Viseras C, *et al.* Uses of clay minerals in semisolid health care and therapeutic products. *Appl Clay Sci* 2007; **36**(1-3): 37–50.
4 Vico LI, Acebal SG. Some aspects about the adsorption of quinoline on fibrous silicates and patagonian saponite. *Appl Clay Sci* 2006; **33**(2): 142–148.
5 El-Gindy GA, *et al.* Preparation and formulation of sustained-release terbutaline sulphate microcapsules. *Bull Pharm Sci Assiut Univ* 2000; **23**(1): 55–63.
6 Akalin E, *et al.* Adsorption and interaction of 5-fluorouracil with montmorillonite and saponite by FT-IR spectroscopy. *J. Mol. Struct* 2007; **834–836**: 477–481.
7 Ruddy SB *et al.* Formulations of oral gastrointestinal therapeutic agents in combination with pharmaceutically acceptable clays. International Patent WO96/2096; 1996.
8 Ruddy SB *et al.* X-ray contrast compositions containing iodoaniline derivatives and pharmaceutically acceptable clays. United States Patent No. 5,424,056; 1995.
9 Pongjanyakul T. *et al*Physicochemical characterizations and release studies of nicotine-magnesium aluminum silicate complexes. *Appl Clay Sci* 2009; **44**(3-4): 242–250.
10 Park JK. *et al*Controlled release of donepezil intercalated in smectite clays. *Int J Pharm* 2008; **359**(1-2): 198–204.

20 General References

Aguzzi C, *et al.* Use of clays as drug delivery systems: possibilities and limitations. *Appl Clay Sci* 2007; **36**(1–3): 22–36.

Carretero M, *et al.* Clay and non-clay minerals in the pharmaceutical industry. *Appl Clay Sci* 2009; **46**(1): 73–80.

Gormley IP, Addison J. The *in vitro* cytotoxicity of some standard clay mineral dusts of respirable size. *Clay Miner* 1983; **18**(2): 153–163.

Kirk-Othmer Encyclopedia of Chemical Technology. 6: 4th edn. New York: Wiley, 1994; 381–405.

Lopez-Galindo A, Viseras C. Pharmaceutical and cosmetic applications of clay. In: Wypych F, Satyanarayana, KG, eds. *Clay Surfaces: Fundamentals and Applications.* Elsevier, 2004; 267–289.

Polon JA. Mechanisms of thickening by inorganic agents. *J Soc Cosmet Chem* 1970; **21**: 347–363.

Post JL. Saponite from near Ballarat, California. *Clays Clay Miner* 1984; **32**: 147–153.

The Source Clays Repository. Material safety data sheet: *SapCa-2*, 2004.

Viseras C, *et al.* Current challenges in clay minerals for drug delivery. *Appl Clay Sci* 2010; **48**(3): 291–295.

Viseras C, Lopez-Galindo A. Characteristics of pharmaceutical grade phylosilicate powders. *Pharm Dev Technol* 2000; **5**(1): 47–52.

21 Author

PE Luner.

22 Date of Revision

1 March 2012.

S

Sesame Oil

1 Nonproprietary Names

BP: Refined Sesame Oil
JP: Sesame Oil
PhEur: Sesame Oil, Refined
USP–NF: Sesame Oil

2 Synonyms

Benne oil; gingelly oil; gingili oil; jinjili oil; *Lipovol SES*; sesami oleum raffinatum; teel oil.

3 Chemical Name and CAS Registry Number

Sesame oil [8008-74-0]

4 Empirical Formula and Molecular Weight

A typical analysis of refined sesame oil indicates the composition of the acids, present as glycerides, to be: arachidic acid 0.8%; linoleic acid 40.4%; oleic acid 45.4%; palmitic acid 9.1%; and stearic acid 4.3%. Sesamin, a complex cyclic ether, and sesamolin, a glycoside, are also present in small amounts.

Note that other reported analyses may vary slightly from that above.[1]

The monographs for Sesame Oil in the USP35–NF30 and Refined Sesame Oil in the PhEur 7.4 specify the acceptable range of eight triglycerides found in sesame oil.

5 Structural Formula

See Section 4.

6 Functional Category

Oleaginous vehicle; solvent.

7 Applications in Pharmaceutical Formulation or Technology

The major use of sesame oil in pharmaceutical formulations is as a solvent in the preparation of sustained-release intramuscular injections of steroids, such as estradiol valerate, hydroxyprogesterone caproate, testosterone enanthate, and nandrolone decanoate,[2] or other oil-soluble drug substances, such as the decanoate or enanthate esters of fluphenazine, haloperidol, and ergocalciferol. The disappearance of sesame oil from the injection site, following subcutaneous or intramuscular administration to pigs, has been reported to have a half-life of about 23 days.[3] The *in vitro* drug release rates from oily depot formulations containing sesame oil intended for intra-articular administration have been reported,[4] as has the *in vitro* release of local anesthetics from sesame oil suspensions.[5]

Sesame oil may be used as a solvent in the preparation of subcutaneous injections,[6] oral capsules,[7,8] rectal suppositories,[9] and ophthalmic preparations;[10] it may also be used in the formulation of suspensions[11] and emulsions.[11–13]

8 Description

Refined sesame oil is a clear, pale-yellow colored liquid with a slight, pleasant odor and a bland taste. It solidifies to a soft mass at about −4°C.

9 Pharmacopeial Specifications

See Table I.

Table I: Pharmacopeial specifications for sesame oil.

Test	JP XV	PhEur 7.4	USP35–NF30
Identification	+	+	+
Characters	—	+	—
Specific gravity	0.914–0.921	≈0.919	0.916–0.921
Refractive index at 20°C	—	≈1.473	—
Heavy metals	—	—	≤0.001%
Cottonseed oil	—	+	+
Solidification range of fatty acids	20–25°C	—	20–25°C
Free fatty acids	—	—	+
Acid value	≤0.2	≤0.5	—
	—	≤0.3[a]	—
Iodine value	103–118	—	103–116
Peroxide value	—	≤10.0	—
	—	≤5.0[a]	—
Saponification value	187–194	—	188–195
Unsaponifiable matter	≤2.0%	≤2.0%	≤1.5%
Composition of triglycerides	—	+	+
Alkaline impurities	—	+	—
Water	—	≤0.1%	—

(a) In sesame oil intended for parenteral use.

10 Typical Properties

Density 0.916–0.920 g/cm^3

Flash point 338°C (open cup)

Freezing point −5°C

Refractive index n_D^{40} = 1.4650–1.4665

Solubility Insoluble in water; practically insoluble in ethanol (95%); miscible with diethyl ether, carbon disulfide, chloroform, ether, hexane, and light petroleum.

Specific rotation $[\alpha]_D^{25}$ = +1° to +9°

Viscosity (dynamic) 43 mPa s (43 cP)

11 Stability and Storage Conditions

Sesame oil is more stable than most other fixed oils and does not readily become rancid; this has been attributed to the antioxidant effect of some of its characteristic constituents. The PhEur 7.4 permits the addition of a suitable antioxidant to sesame oil.

Sesame oil may be sterilized by aseptic filtration or dry heat. It has been reported that suitable conditions for the sterilization of injections containing sesame oil are a temperature of 170°C for 2 hours; it has been suggested that 150°C for 1 hour is inadequate.[14] However, it has been demonstrated that dry heat sterilization of sesame oil at 150°C for 1 hour was sufficient to kill all added *Bacillus subtilis* spores.[15]

Sesame oil should be stored in a well-filled, airtight, light-resistant container, at a temperature not exceeding 40°C. Sesame oil intended for use in the manufacture of parenteral dosage forms should be stored under an inert gas in an airtight glass container.

The PhEur 7.4 states that when the container is opened, its contents are to be used as soon as possible; any part of the contents not used at once is protected by an atmosphere of an inert gas.

12 Incompatibilities

Sesame oil may be saponified by alkali hydroxides.

13 Method of Manufacture

Sesame oil is obtained from the ripe seeds of one or more cultivated varieties of *Sesamum indicum* Linné (Fam. Pedaliaceae) by expression in a hydraulic press or by solvent extraction. The crude oil thus obtained is refined to obtain an oil suitable for food or pharmaceutical use. Improved color and odor may be obtained by further refining.

14 Safety

Sesame oil is mainly used in intramuscular and subcutaneous injections; it should not be administered intravenously. It is also used in topical pharmaceutical formulations and consumed as an edible oil.

Although it is generally regarded as an essentially nontoxic and nonirritant material,[16] there have been rare reports of hypersensitivity to sesame oil, with sesamin suspected as being the primary allergen.[17–20] Anaphylactic reactions to sesame seeds have also been reported. However, it is thought that the allergens in the seeds may be inactivated or destroyed by heating as heat-extracted sesame seed oil or baked sesame seeds do not cause anaphylactic reactions in sesame seed-allergic individuals.[21]

Subcutaneous nodules have been reported after the subcutaneous administration of sesame oil.[22]

LD_{50} (rabbit, IV): 678 μg/kg[23]

15 Handling Precautions

Observe normal precautions appropriate to the circumstances and quantity of material handled. Spillages of sesame oil are slippery and should be covered with an inert absorbent material prior to disposal.

16 Regulatory Status

Included in the FDA Inactive Ingredients Database (IM and SC injections; oral capsules, concentrates, emulsions, and tablets; also topical preparations). Included in parenteral (IM injections) and nonparenteral (oral capsules) medicines licensed in the UK. Included in the Canadian Natural Health Products Ingredients Database.

17 Related Substances

Almond oil; canola oil; corn oil; cottonseed oil; peanut oil; soybean oil; sunflower oil.

18 Comments

Multiple-emulsion formulations, in which sesame oil was one of the oil phases incorporated, have been investigated as a prolonged-release system for rifampicin;[24] microemulsions containing sesame oil have been prepared for the transdermal delivery of ketoprofen.[25] Sesame oil has also been included in self-microemulsifying drug delivery systems,[26] and fast-disintegrating lyophilized dry emulsion tablets[27] for oral administration. It has also been used in the preparation of liniments, pastes, ointments, and soaps. A sesame paste (tahini), composed of crushed sesame seeds in sesame oil, has been investigated as a novel suspending agent.[28]

Sesame oil is additionally used as an edible oil and in the preparation of oleomargarine.

19 Specific References

1 British Standards Institute. *Specification for Crude Vegetable Fats, BS 7207*. London: BSI, 1990.
2 Williams JS, *et al.* Nandrolone decanoate therapy for patients receiving hemodialysis. *Arch Intern Med* 1974; **134**: 289–292.
3 Larsen SW, *et al.* Determination of the disappearance rate of iodine-125 labelled oils from the injection site after intramuscular and subcutaneous administration to pigs. *Int J Pharm* 2001; **230**(1–2): 67–75.
4 Larsen SW, *et al. In vitro* assessment of drug release from oil depot formulations intended for intra-articular administration. *Eur J Pharm Sci* 2006; **29**(5): 348–354.
5 Larsen SW, *et al.* On the mechanism of drug release from oil suspensions *in vitro* using local anesthetics as model drug compounds. *Eur J Pharm Sci* 2008; **34**(1): 37–44.
6 Hirano K, *et al.* Studies on the absorption of practically water-insoluble drugs following injection V: subcutaneous absorption in rats from solutions in water immiscible oils. *J Pharm Sci* 1982; **71**: 495–500.
7 Perez-Reyes M, *et al.* Pharmacology of orally administered Δ^9-tetrahydrocannabinol. *Clin Pharmacol Ther* 1973; **14**: 48–55.
8 Sallan SE, *et al.* Antiemetic effect of delta-9-tetrahydrocannabinol in patients receiving cancer chemotherapy. *N Engl J Med* 1975; **293**: 795–797.
9 Tanabe K, *et al.* [Effect of different suppository bases on release of indomethacin.] *Yakuzaigaku* 1984; **44**: 115–120[in Japanese].
10 Ahuja M, *et al.* Stability studies on aqueous and oily ophthalmic solutions of diclofenac. *Yakugaku Zasshi* 2009; **129**(4): 495–502.
11 Shinkuma D, *et al.* Bioavailability of phenytoin from oil suspension and emulsion in dogs. *Int J Pharm* 1981; **9**: 17–28.
12 Rosenkrantz H, *et al.* Oral and parenteral formulations of marijuana constituents. *J Pharm Sci* 1972; **61**: 1106–1112.
13 Unno K, *et al.* [Preparation and tissue distribution of 5-fluorouracil emulsion.] *J Nippon Hosp Pharm Assoc* 1980; **6**(1): 14–20[in Japanese].
14 Pasquale D, *et al.* A study of sterilizing conditions for injectable oils. *Bull Parenter Drug Assoc* 1964; **18**(3): 1–11.
15 Kupiec TC, *et al.* Dry-heat sterilisation of parenteral oil vehicles. *Int J Pharm Compound* 2000; **4**(3): 223–224.
16 Hem SL, *et al.* Tissue irritation evaluation of potential parenteral vehicles. *Drug Dev Commun* 1974–75; **1**: 471–477.
17 Neering H, *et al.* Allergens in sesame oil contact dermatitis. *Acta Dermatol Venerol* 1975; **55**: 31–34.
18 Weiner M, Bernstein IL. *Adverse Reactions to Drug Formulation Agents: A Handbook of Excipients*. New York: Marcel Dekker, 1989: 212–213.
19 Perkins MS. Sesame allergy is also a problem [letter]. *Br Med J* 1996; **313**: 300.
20 Perkins MS. Raising awareness of sesame allergy. *Pharm J* 2001; **267**: 757–758.
21 Kägi MK, Wüthrich B. Falafel-burger anaphylaxis due to sesame seed allergy [letter]. *Lancet* 1991; **338**: 582.
22 Darsow U, *et al.* Subcutaneous oleomas induced by self-injection of sesame seed oil for muscle augmentation. *J Am Acad Dermatol* 2000; **42**(2 Pt 1): 292–294.
23 Lewis RJ, ed. *Sax's Dangerous Properties of Industrial Materials*, 11th edn. New York: Wiley, 2004: 3203.
24 Nakhare S, Vyas SP. Prolonged release of rifampicin from internal phase of multiple w/o/w emulsion systems. *Indian J Pharm Sci* 1995; **57**(2): 71–77.
25 Rhee Y-S, *et al.* Transdermal delivery of ketoprofen using microemulsions. *Int J Pharm* 2001; **228**(1–2): 161–170.
26 Grove B, *et al.* Bioavailability of seocalcitol II: Development and characterisation of self-microemulsifying drug delivery systems (SMEDDS) for oral administration containing medium and long triglycerides. *Eur J Pharm Sci* 2006; **28**(3): 233–242.
27 Ahmed IS, Aboul-Einien MH. *In vitro* and *in vivo* evaluation of a fast-disintegrating lyophilised dry emulsion tablet containing griseofulvin. *Eur J Pharm Sci* 2007; **32**(1): 58–68.
28 Al-Achi A, *et al.* Calamine lotion: experimenting with a new suspending agent. *Int J Pharm Compound* 1999; **3**(6): 490–492.

20 General References

—

21 Author

CG Cable.

22 Date of Revision

1 March 2012.

S

Shellac

1 Nonproprietary Names

BP: Shellac
JP: Purified Shellac
White Shellac
PhEur: Shellac
USP–NF: Shellac

2 Synonyms

AT 10-1010; *Blonde*; *Bulls Eye Shellac*; *CertiSeal FC 300A*; *Crystalac*; E904; *Excelacs 3-Circles*; *Excelacs 3-Stars*; *Gifu Shellac GBN-PH*; *Gifu Shellac Pearl-811*; lac; lacca; *Mantrolac R-49*; *Mantrolac R-52*; *Marcoat 125*; *Opaglos R*; *Sepifilm SN*; *SSB Aquagold*; *SSB 55 Pharma*; *SSB 56 Pharma*; *SSB 57 Pharma*; *Swanlac*.

3 Chemical Name and CAS Registry Number

Shellac [9000-59-3]

4 Empirical Formula and Molecular Weight

Shellac is the general term for the refined form of lac, a natural polyester resin secreted by insects.

PhEur 7.4 and USP35–NF30 define four types of shellac depending on the refining method, and JP XV mentions only two types; *see* Section 13.

Elementary analysis reveals that shellac contains carbon, hydrogen, oxygen, and a negligible amount of ash. Orange shellac contains approx. 68% carbon, 9% hydrogen and 23% oxygen, and with a molecular weight of 1006 (bleached shellac is 949) the empirical formula for the average shellac molecule is $C_{60}H_{90}O_{15}$.

Even with this relatively low molecular weight, shellac has excellent film-forming properties.

Lac is a complex mixture of aliphatic and alicyclic acids. The major components are aleuritic, jalaric and shellolic acids, as well as butolic and kerrolic acids. Seed lac and orange shellac contain approximately 5–6% wax and two coloring components, the water soluble laccaic acid and the water insoluble erythrolaccin.

5 Structural Formula

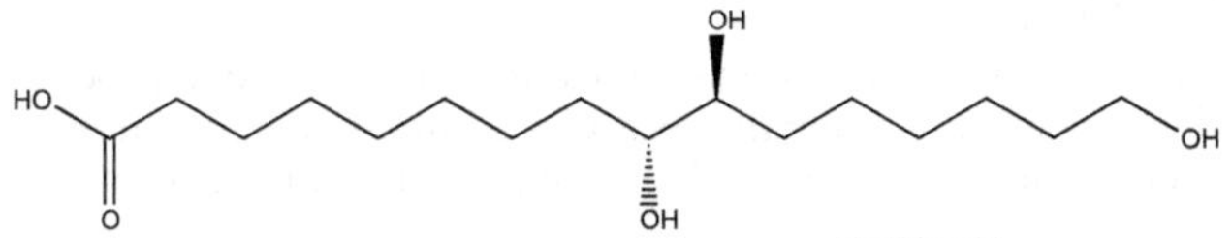

Aleuritic acid

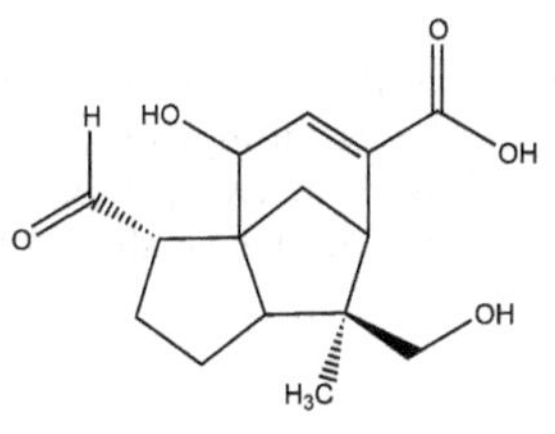

Jalaric acid

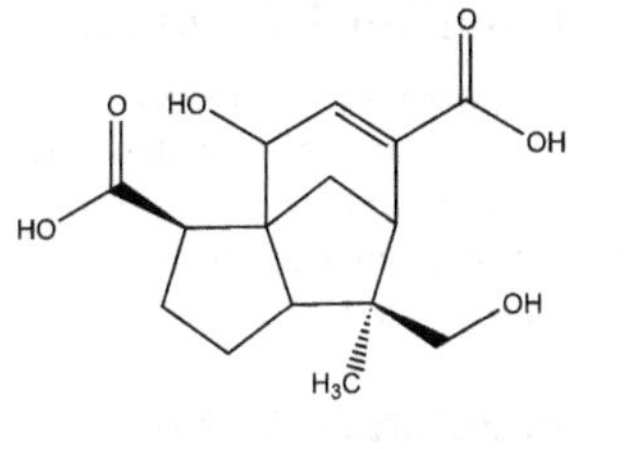

Shellolic acid

6 Functional Category

Coating agent; film-forming agent; microencapsulating agent; modified-release agent.

7 Applications in Pharmaceutical Formulation or Technology

Shellac is widely used as a moisture barrier coating for tablets and pellets due to its low water vapor and oxygen permeability. It has usually been applied in the form of alcoholic or aqueous solutions (pharmaceutical glazes); *see also* Section 18.

Shellac, particularly novel aqueous shellac solutions, is mainly used in food products and nutritional supplements.(1,2) Aqueous ammonium shellac solutions, based on dewaxed orange shellac, do not show the problems exhibited by alcoholic shellac solutions and are used as enteric coatings for granules, pellets, tablets, soft and hard gelatine capsules, primarily in nutritional supplements.(3)

Shellac is a primary ingredient of pharmaceutical printing inks for capsules and tablets, and can be applied as a 40% w/v alcoholic solution. It has also been used to apply one or two sealing coats to tablet cores to protect them from moisture before being film- or sugar-coated.

Other applications of shellac are the coating or encapsulation of powders or granules, e.g. in probiotics.(4) Prior to the introduction of film coating, a combination of shellac, cetostearyl alcohol and stearic acid was used as an enteric coating. In cosmetics, shellac is used in hairsprays, mascara and lipstick formulations.(5) Aqueous shellac solutions are also used for colonic drug delivery.(6)

8 Description

Shellac is tasteless and may have a faint odor. The typical odor of shellac is the result of a complex fragrance system.(7)

Shellac is a natural resin that may be obtained in a variety of colors ranging from light yellow to dark red in the form of hard, brittle flakes with or without wax, depending on the refining process; *see* Sections 4 and 13. The different types of shellac include bleached shellac, bleached dewaxed shellac, dewaxed and decolorized shellac, dewaxed flake shellac, dewaxed orange shellac, dewaxed shellac, orange shellac, purified shellac, refined bleached shellac, regular bleached shellac, regular waxy shellac, wax-containing shellac, and white shellac. The flakes may be crushed or milled to a coarse or fine powder. Bleached shellac is supplied as a coarse off-white powder.

9 Pharmacopeial Specifications

See Table I.

Table I: Pharmacopeial specifications for shellac.

Test	JP XV	PhEur 7.4	USP35–NF30
Identification	—	+	+
Characters	—	+	—
Chloride	≤0.14%(a)	—	—
Sulfate	≤0.11%(a)	—	—
Heavy metals	≤10 ppm	≤10 ppm	≤0.001%
Arsenic	≤5 ppm	≤3 ppm	—
Ethanol-insoluble substances	≤2.0%	—	—
Rosin	+	+	+
Total ash	≤1.0%	—	—
Acid value (on dried basis)	+	+	+
Dewaxed orange shellac	60–80	65–95	71–79
Orange shellac	—	65–95	68–76
Refined bleached shellac	65–90	65–95	75–91
Regular bleached shellac	—	65–95	73–89
Loss on drying	+	+	+
Dewaxed orange shellac	≤2.0%	≤2.0%	≤2.0%
Orange shellac	—	≤2.0%	≤2.0%
Refined bleached shellac	≤6.0%	≤6.0%	≤6.0%
Regular bleached shellac	—	≤6.0%	≤6.0%
Wax	+	+(b)	+
Dewaxed orange shellac	≤20 mg	—	≤0.2%
Orange shellac	—	—	≤5.5%
Refined bleached shellac	≤20 mg	—	≤0.2%
Regular bleached shellac	—	—	≤5.5%

(a) For white shellac.
(b) Refer to Identification test B.

10 Typical Properties

The properties of shellac depend on the insect strain and host tree as well as the method used for refining the crude lac (seed lac).

Density 1.035–1.140 g/cm^3

Dissociation constant pK_a value = 5.60–6.59(8)

Glass transition temperature 33–52°C;(8) the wide range in temperature is a result of the process used in refining and the type of seed lac used as a starting material.

Hydroxyl value 230–280

Iodine number 10–18

Melting point 77–90°C

Refractive index n_D^{20} = 1.514–1.524

Saponification value 185–260

Solubility *see* Table II.

Table II: Solubility of shellac.

Solvent	Solubility at 20°C
Alkalis	Soluble
Benzene	1 in 10
Ethanol	1 in 2
Ethanol (95%)	1 in 1.2 (very slowly soluble)
Ether	1 in 8
Hexane	Practically insoluble
Propylene glycol	1 in 10
Water	Practically insoluble

11 Stability and Storage Conditions

After long periods of storage, shellac becomes less readily soluble in alcohol, less fluid on heating, and darker in color; *see also* Section 18.

Shellac should be stored in a well-closed container at temperatures below 15°C. Wax-containing grades should be mixed before use to ensure uniform distribution of the wax. Orange and dewaxed orange shellac have a shelf-life of 1 to 2 years. The shelf-life of bleached shellac is approximately 6 months.

12 Incompatibilities

Shellac is chemically reactive with aqueous alkalis, organic bases, alcohols, and agents that esterify carboxyl groups. Therefore, shellac should be used with caution in the presence of such compounds.

13 Method of Manufacture

Shellac or lac is cultivated and refined from lacca, a resinous secretion produced by the tiny insect *Kerria lacca* (Kerr) Lindinger (Coccideae), formerly *Laccifer lacca* (Kerr). The insects are parasitic on certain trees, mainly in India. In Thailand and South China, the resin is secreted by another species, *Laccifer chinensis* (Madihassan) on different trees.[8,9] The insects pierce through the bark of the tree and transform the sap into a natural polyester resin, called stick lac, which is secreted through the surface of their body. The resin forms thick encrustations on the smaller branches and twigs, which are then scraped off the twigs and further processed to produce seed lac, as it is known at this stage. Seed lac is then refined to become shellac.

The chemical composition, properties and the color of shellac depend on the insect or insect strain, and thus the host tree, as well as the process used for refining.[8] Three very different processes are used for refining the seed lac to shellac (bleaching, melting, and solvent extraction),[8–10] resulting in products with different characteristics and properties.

Bleaching process Refined bleached or white shellac is obtained by dissolving seed lac in an aqueous alkaline solution, which is then filtered, dewaxed, and bleached with sodium hypochlorite to completely remove the color. However, changes in the molecular structure and the addition of chlorine substituents may lead to self-crosslinking and polymerization.

Melting process After melting the seed lac, the highly viscous molten lac is pressed through a filter and drawn to a thin film. Once cooled, the film breaks into thin flakes. The shellac wax is not removed by this process and the color depends on the type of seed lac used.

Solvent extraction process Solvent extraction is a very gentle process for refining shellac. The seed lac is dissolved in ethanol, and wax and impurities are removed by filtration. Activated carbon is used to produce light-colored grades. After a further filtration step and the removal of ethanol, the resin is drawn to a thin film, which breaks into flakes after cooling. The properties of the final product depend on the type of seed lac used and are influenced by the processing parameters and the grade of activated carbon.

PhEur 7.4 and USP35–NF30 define four types of shellac depending on the refining method, and the JP XV mentions two types; *see* Table III.

Table III: Types of shellac according to the refining process.

Refining process	JP XV	PhEur 7.4	USP35–NF30
Bleaching	—	Bleached shellac	Regular bleached shellac
Bleaching and dewaxing	White shellac	Bleached dewaxed shellac	Refined bleached shellac
Melting	—	Wax-containing shellac	Orange shellac
Solvent extraction	Purified shellac	Dewaxed shellac	Dewaxed orange shellac

The use of the term 'pharmaceutical grade' as well as the quality of the shellac depends on the manufacturer.

Seed lac is mainly produced in India, Thailand and China. Orange shellac, refined by the melting process, is manufactured by several companies in India, Thailand and South-East Asia. Bleached shellac is produced in the US, Canada, Japan, India, Thailand and South China. Dewaxed orange shellac is refined by the solvent extraction process in Germany, Japan and India.

14 Safety

Shellac is used in oral pharmaceutical formulations, food products, and cosmetics. It is generally regarded as an essentially nonirritant and nontoxic material at the levels employed as an excipient.

15 Handling Precautions

Observe normal precautions appropriate to the circumstances and quantity of material handled. Shellac can be irritating to the eyes, and to the respiratory system if inhaled as dust. Eye protection, gloves, and a dust respirator are recommended. Shellac should be handled in a well-ventilated environment.

16 Regulatory Status

Accepted as a food additive in the US, Europe, and Japan. Included in the FDA Inactive Ingredients Database (oral capsules and tablets). Included in nonparenteral medicines (oral tablets and capsules, often in printing ink formulations) licensed in the UK. Included in the Canadian Natural Health Products Ingredients Database.

17 Related Substances

Aleuritic acid; aqueous shellac solution; laccaic acid B; pharmaceutical glaze.

Aleuritic acid

Empirical formula $C_{16}H_{32}O_5$
Molecular weight 304.42
CAS number [533-87-9]
Synonyms DL-*erythro*-9,10,16,-Trihydroxyhexadecanoic acid; 9,10,16-trihydroxypalmitic acid; 8,9,15-trihydroxypentadecane-1-carboxylic acid.
Melting point 100–101°C
Solubility Soluble in hot water and lower alcohols.
Comments Main component of shellac. Isolated by saponification, and the starting material for the synthesis of macrocyclic musk compounds for fragrances and pheromones.[11]
The EINECS number for aleuritic acid is 208-578-8.

Aqueous shellac solution

CAS number [68308-35-0]
Synonyms Shellac ammonium salt
Comments Aqueous solution of shellac with 20-25 % solids at a pH of 7–7.5. The EINECS number is 269-647-6. Used as a coating material for granules, pellets, tablets, hard and soft gelatine capsules. Suitable for sustained-release and enteric coatings, as moisture barrier and as encapsulation agent.[3,9,10]

Laccaic acid B

Empirical formula $C_{24}H_{16}O_{12}$
Molecular weight 496
Synonyms Lac dye; natural red 25; 9,10-dihydro-3,5,6,8-tetrahydroxy-7-[2-hydroxy-5-(2-hydroxyethyl)phenyl]-9,10-dioxo-1,2-anthracenedicarboxylic acid.
CAS number [17249-00-2]
Solubility Water soluble
Comments Laccaic acid consists of 5 compounds, laccaic acid A, B, C, D and E. Laccaic acid B is used as a food color in Japan. The color is pH dependent: orange at pH 3 to reddish purple at above pH 7.

Pharmaceutical glaze

Synonyms Confectionary glaze, alcoholic shellac solution

Comments Pharmaceutical glaze is a specially denatured alcoholic solution of shellac containing between 20% and 57% of shellac. It may be prepared using either ethanol or ethanol (95%), and may contain waxes and titanium dioxide as an opacifying agent.

18 Comments

Under the general term shellac, many grades are available. The pharmacopeial specifications are very narrow with regard to purity; however, they allow a wide range for the acid value. The variability of the physicochemical properties of different shellac types has to be considered, especially for the manufacture of controlled-release formulations.[8,12]

Shellac in the form of its alcoholic solution (pharmaceutical glaze) has been used for many years as an enteric coating for pharmaceutical applications. However, due to significant problems with delayed disintegration and changes in release profiles of the coated dosage forms after storage, alcoholic shellac solutions have limited use as enteric coatings in the pharmaceutical industry today.[3,13,14] Problems are due to an esterification of the carboxyl groups of shellac with alcohol and a polymerization due to trapped alcohol residues in the dry film. The use of bleached shellac, where the molecular structure is partly changed by the treatment with sodium hypochlorite, increases the polymerization problems.

Shellac films from ammoniated aqueous shellac solutions using dewaxed orange shellac do not have these problems and have very stable release characteristics even after extended storage times. Furthermore, they can be formulated in combination with other polymers such as HPMC, CMC, sodium alginates, or modified starch together with plasticizers to meet the disintegration requirements of the USP, PhEur and JP.[2,3,9,10] To avoid cracking of the coating film, a higher glass transition temperature of the ammonium salt form[15] has to be considered for the manufacture of formulations coated with aqueous ammoniacal solutions.[16]

A specification for bleached shellac is contained in the *Food Chemicals Codex* (FCC).[17]

The EINECS number for shellac is 232-549-9.

19 Specific References

1 Freed P, *et al.* Moisture vapor transmission comparison between acrylic, aqueous shellac, and HPMC coating systems. *Proc AAPS Annual Meeting* 2007.
2 Pearnchop N, *et al.* Shellac used as coating material for solid pharmaceutical dosage forms. *STP Pharma Sci* 2003; **13**: 387–396.
3 Signorino C. Shellac film coatings providing release at selected pH and method. US Patent 6,620,431; 2003.
4 Stummer S, *et al.* Application of shellac for the development of probiotic formulations. *Food Research Int* 2010; **43**(5): 1312–1320.
5 Tannert U. Shellac – a natural polymer for hair care products. *Seifen Öle Fette Wachse* 1992; **17**: 1079–1083.
6 Roda A, *et al.* A new oral formulation for the release of sodium butyrate in the ileo-cecal region and colon. *World J Gastroenterol* 2007; **13**: 1079–1084.
7 Buchbauer G, *et al.* Headspace constituents of shellac. *Zeitschr Naturforsch* 1993; **48b**: 247–248.
8 Buch K, *et al.* Investigations of various shellac grades: additional analysis for identity. *Drug Dev Ind Pharm* 2009; **35**(6): 639–648.
9 Penning M. Aqueous shellac solutions for controlled release coatings. In: Karsa DR, Stephenson RA, eds. *Chemical Aspects of Drug Delivery Systems.* London: Royal Society of Chemistry, 1996: 146–154.
10 Specht F, *et al.* The application of shellac acidic polymer for enteric coating. *Pharm Technol Eur* 1998; **10**(9): 20–28.
11 Subramanian GB, Bushan KH. Aleuritic acid in perfumery and pheromones. *Perfumer Flavorist* 1993; **18**: 41–44.
12 Farag Y, Leopold CS. Investigation of drug release from pellets coated with different shellac types. *Drug Dev Ind Pharm* 2010; Aug 12 (Epub).
13 Luce GT. Disintegration of tablets enteric coated with CAP. *Manuf Chem Aerosol News* 1978; **49**(7): 50–67.
14 Chambliss WG. The forgotten dosage form: enteric-coated tablets. *Pharm Technol* 1983; **7**: 124–140.
15 Farag Y, Leopold CS. Physicochemical properties of various shellac types. *Dissolution Technol* 2009; **16**: 33–39.
16 Farag Y, Leopold CS. Influence of the inlet air temperature in a fluid bed coating process on drug release from shellac-coated pellets. *Drug Dev Ind Pharm* 2010; Sep 6 (Epub).
17 *Food Chemicals Codex*, 7th edn. Bethesda, MD: United States Pharmacopeia, 2010: 916.

20 General References

Bose PK *et al. Chemistry of Lac.* Ranchi, India: Indian Lac Research Institute, 1962.

Cockeram HS, Levine SA. The physical and chemical properties of shellac. *J Soc Cosmet Chem* 1961; **12**: 316–323.

Martin J. Shellac. In: Kirk-Othmer RE, ed. *Encyclopedia of Chemical Technology*, 3rd edn. NJ: John Wiley, 1982: 20. 737–747.

Misra GS, Sengupta SC. Shellac. In: *Encyclopedia of Polymer Science and Technology.* NJ: John Wiley, 1970: 12. 419–440.

Mukhopadhyay B, Muthana MS. *A Monograph on Lac.* Ranchi, India: Indian Lac Research Institute, 1962.

Stroever Schellack Bremen. Product data sheet: *SSB* (Shellac). www.stroever.de/applications-pharma-24.html (accessed 4 January 2011).

21 Authors

BR Jasti, CS Leopold, X Li, M Penning.

22 Date of Revision

1 March 2012.

Simethicone

1 Nonproprietary Names

BP: Simeticone
PhEur: Simeticone
USP–NF: Simethicone

2 Synonyms

Dow Corning Q7-2243 LVA; *Dow Corning Q7-2587*; polydimethylsiloxane–silicon dioxide mixture; *Sentry Simethicone*; simeticonum.

3 Chemical Name and CAS Registry Number

α-(Trimethysilyl-ω-methylpoly[oxy(dimethylsilylene)], mixture with silicon dioxide [8050-81-5]

4 Empirical Formula and Molecular Weight

See Section 8.

5 Structural Formula

$$H_3C-\left[Si(CH_3)_2-OSi(CH_3)_2\right]_n-CH_3 \quad + SiO_2$$

where n = 200–350

6 Functional Category

Antifoaming agent; tablet and capsule diluent; water-repelling agent.

7 Applications in Pharmaceutical Formulation or Technology

The main use of simethicone as an excipient is as an antifoaming agent in pharmaceutical manufacturing processes, for which 1–50 ppm is used. It is also included in antacid products such as tablets or capsules.[1–5]

When simethicone is used in aqueous formulations, it should be emulsified to ensure compatibility with the aqueous system and components.

In the US, up to 10 ppm of simethicone may be used in food products.

8 Description

The PhEur 7.4 and USP35–NF30 describe simethicone as a mixture of fully methylated linear siloxane polymers containing repeating units of the formula $[-(CH_3)_2SiO-]_n$, stabilized with trimethylsiloxy end-blocking units of the formula $[(CH_3)_3\ SiO-]$, and silicon dioxide. It contains not less than 90.5% and not more than 99.0% of the polydimethylsiloxane $[-(CH_3)_2SiO-]_n$, and not less than 4.0% and not more than 7.0% of silicon dioxide. The PhEur 7.4 additionally states that the degree of polymerization is between 20–400.

Simethicone occurs as a translucent, gray-colored, viscous fluid. It has a molecular weight of 14 000–21 000.

9 Pharmacopeial Specifications

See Table I.

Table I: Pharmacopeial specifications for simethicone.

Test	PhEur 7.4	USP 35–NF30
Identification	+	+
Characters	+	—
Production	+	—
Acidity	+	—
Defoaming activity	≤15 seconds	≤15 seconds
Loss on heating	—	≤18.0%
Volatile matter	≤1.0%	—
Heavy metals	≤5 ppm	≤5 μg/g
Mineral oils	+	—
Phenylated compounds	+	—
Assay (silicon dioxide)	—	4.0–7.0%
Assay (silica)	4.0–7.0%	—
Assay (polydimethylsiloxane)	90.5–99.0%	90.5–99.0%

10 Typical Properties

Boiling point 35°C
Refractive index n_D^{20} = 0.965–0.970
Solubility Practically insoluble in ethanol (95%) and water. The liquid phase is soluble in benzene, chloroform, and ether, but silicon dioxide remains as a residue in these solvents.
Specific gravity 0.95–0.98 at 25°C
Spectroscopy
IR spectra *see* Figure 1.
Viscosity (kinematic) 370 mm²/s (370 cSt) at 25°C for *Dow Corning Q7-2243 LVA.*

11 Stability and Storage Conditions

Simethicone is generally regarded as a stable material when stored in the original unopened container. A shelf-life of 18 months from the date of manufacture is typical. However, some simethicone products have a tendency for the silicon dioxide to settle slightly and containers of simethicone should therefore be shaken thoroughly to ensure uniformity of contents before sampling or use. Simethicone

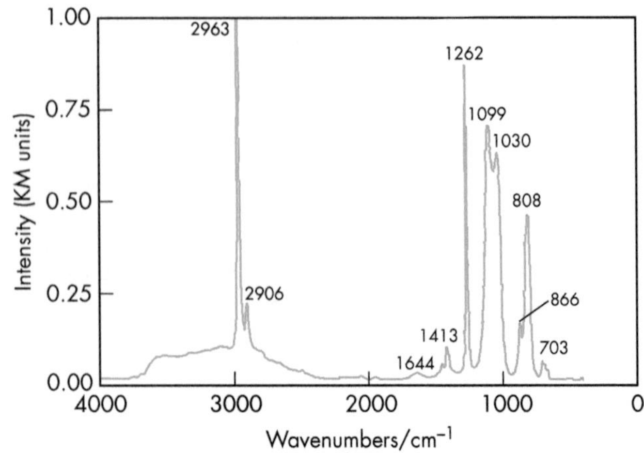

Figure 1: Infrared spectrum of simethicone measured by diffuse reflectance. Adapted with permission of Informa Healthcare.

should be stored in a cool, dry location away from oxidizing materials.

Simethicone can be sterilized by dry heating or autoclaving. With dry heating, a minimum of 4 hours at 160°C is required.

12 Incompatibilities

Simethicone as supplied is not generally compatible with aqueous systems and will float like an oil on a formulation unless it is first emulsified. It should not be used in formulations or processing conditions that are very acidic (below pH 3) or highly alkaline (above pH 10), since these conditions may have some tendency to break the polydimethylsiloxane polymer. Simethicone cannot normally be mixed with polar solvents of any kind because it is very minimally soluble. Simethicone is incompatible with oxidizing agents.

13 Method of Manufacture

Silicon dioxide is initially rendered hydrophobic in one of a variety of proprietary processes specific to a particular manufacturer. It is then slowly mixed with the silicone fluids in a formulation. After mixing, the simethicone is milled to ensure uniformity.

14 Safety

Simethicone is used in cosmetics, foods, and oral and topical pharmaceutical formulations, and is generally regarded as a relatively nontoxic and nonirritant material when used as an excipient. Direct contact with the eye may cause irritation.

Therapeutically, oral doses of 125–250 mg of simethicone, three or four times daily, have been given as an antiflatulent. Doses of 20–40 mg of simethicone have been given with feeds to relieve colic in infants.[6]

LD_{50} (dog, IV): 0.9 g/kg

15 Handling Precautions

Observe normal precautions appropriate to the circumstances and quantity of material handled. Eye protection and gloves are recommended. Simethicone should be handled in areas with adequate ventilation.

16 Regulatory Status

GRAS listed. Included in the FDA Inactive Ingredients Database (oral emulsions, powders, solutions, suspensions, tablets; IM-IV powder for injection solutions; rectal and topical preparations). Included in nonparenteral medicines licensed in the UK. Included in the Canadian Natural Health Products Ingredients Database.

17 Related Substances

Cyclomethicone; dimethicone.

18 Comments

Therapeutically, simethicone is included in a number of oral pharmaceutical formulations as an antiflatulent, although its therapeutic benefit is questionable.[7,8] In some types of surgical or gastroscopic procedures where gas is used to inflate the body cavity, a defoaming preparation containing simethicone may be used in the area to control foaming of the fluids.

The PubChem Compound ID (CID) for simethicone includes 6433516 and 9794495.

19 Specific References

1 Sox T. Simethicone and sulfasalazine for treatment of ulcerative colitis. United States Patent 6,100,245; 1999.

2 Holtman G. *et al*Randomized double-blind comparison of simethicone with cisapride in functional dyspepsia. *Aliment Pharmacol Ther* 1999; **13**(11): 1459–1465.

3 Tiongson A. Process of making an aqueous calcium carbonate suspension. International Patent WO 9945937; 1999.

4 Luber J *et al.* Antifoam oral solid dosage forms comprising simethicone and anhydrous calcium phosphate. European Patent 891776; 1999.

5 Devlin BT, Hoy MR. Semisolid composition containing an antiflatulent agent. European Patent 815864; 1998.

6 Metcalf TJ, *et al.* Simethicone in the treatment of infant colic: randomized, placebo-controlled, multicenter trial. *Pediatrics* 1994; **94**: 29–34.

7 Anonymous. Simethicone for gastrointestinal gas. *Med Lett Drugs Ther* 1996; **38**: 57–58.

8 Azpiroz F, Serra J. Treatment of excessive intestinal gas. *Curr Treat Options Gastroenterol* 2004; 7(4): 299–305.

20 General References

Daher L. Lubricants for use in tabletting. United States Patent 5,922,351; 1999.

Rider JA, *et al.* Further analysis of standards for antacid simethicone defoaming properties. *Curr Ther Res* 1997; **58**(12): 955–963.

21 Author

RT Guest.

22 Date of Revision

1 March 2012.

Sodium Acetate

1 Nonproprietary Names

BP: Sodium Acetate Trihydrate
JP: Sodium Acetate Hydrate
PhEur: Sodium Acetate Trihydrate
USP–NF: Sodium Acetate

2 Synonyms

Acetic acid, sodium salt; E262; natrii acetas trihydricus; sodium ethanoate.

3 Chemical Name and CAS Registry Number

Sodium acetate anhydrous [127-09-3]
Sodium acetate trihydrate [6131-90-4]

4 Empirical Formula and Molecular Weight

$C_2H_3NaO_2$ 82.0 (for anhydrous)
$C_2H_3NaO_2{\cdot}3H_2O$ 136.1 (for trihydrate)

Note that the trihydrate is the material described in the JP XV, PhEur 7.4 and USP35–NF30, although the PhEur 7.4 is the only pharmacopeia that makes this explicit with the title of the monograph.

5 Structural Formula

O

H_3C ONa

6 Functional Category

Antimicrobial preservative; buffering agent; flavoring agent.

7 Applications in Pharmaceutical Formulation or Technology

Sodium acetate is used as part of a buffer system when combined with acetic acid in various intramuscular, intravenous, topical, ophthalmic, nasal, oral, otic, and subcutaneous formulations. It may be used to reduce the bitterness of oral pharmaceuticals.[1] It can be used to enhance the antimicrobial properties of formulations; it has been shown to inhibit the growth of *S. aureus* and *E. coli*, but not *C. albicans* in protein hydrolysate solutions.[2] It is widely used in the food industry as a preservative.[3]

8 Description

Sodium acetate occurs as colorless, transparent crystals or a granular crystalline powder with a slight acetic acid odor.

9 Pharmacopeial Specifications

See Table I.

Table I: Pharmacopeial specifications for sodium acetate.

Test	JP XV	PhEur 7.4	USP35–NF30
Identification	+	+	+
Description	+	—	—
Characters	—	+	—
Appearance of solution	+	+	—
Acid or alkali	+	—	—
pH	—	7.5–9.0	7.5–9.2
Insoluble matter	—	—	≤0.05%
Chloride	≤0.011%	≤200 ppm	≤350 ppm
Sulfate	≤0.017%	≤200 ppm	≤50 ppm
Heavy metals	≤10 ppm	≤10 ppm	≤10 ppm
Calcium and magnesium	+	≤50 ppm	+
Potassium	—	—	+
Arsenic	≤2 ppm	≤2 ppm	—
Iron	—	≤10 ppm	—
Reducing substances	+	+	—
Aluminum	—	≤0.2 ppm	≤0.2 ppm
Loss on drying			
Anhydrous	—	—	≤1.0%
Trihydrate	39.0–40.5%	39.0–40.5%	38.0–41.0%
Assay (dried basis)	≥99.5%	99.0–101.0%	99.0–101.0%

10 Typical Properties

Acidity/alkalinity pH = 7.5–9.0 (5% w/v aqueous solution)
Hygroscopicity The anhydrous and trihydrate sodium acetate are hygroscopic.
Melting point 58°C for trihydrate; 324°C for anhydrous.[4]
Solubility Soluble 1 in 0.8 in water, 1 in 20 in ethanol (95%).
Specific gravity 1.53
Spectroscopy
IR spectra *see* Figure 1.

11 Stability and Storage Conditions

Sodium acetate should be stored in airtight containers.

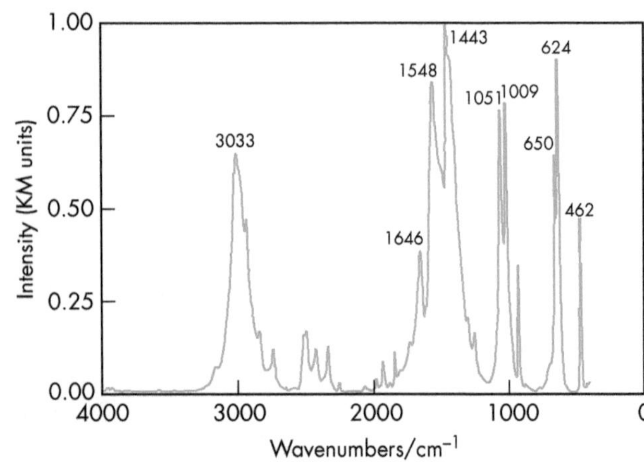

Figure 1: Infrared spectrum of sodium acetate (anhydrous) measured by diffuse reflectance. Adapted with permission of Informa Healthcare.

12 Incompatibilities

Sodium acetate reacts with acidic and basic components. It will react violently with fluorine, potassium nitrate, and diketene.

13 Method of Manufacture

Sodium acetate is prepared by neutralization of acetic acid with sodium carbonate.

14 Safety

Sodium acetate is widely used in cosmetics, foods, and pharmaceutical formulations (*see* Section 18), and is generally regarded as a nontoxic and nonirritant material.

A short-term feeding study in chickens with a diet supplemented with 5.44% sodium acetate showed reduced growth rates that were attributed to the sodium content.[5] Sodium acetate is poisonous if injected intravenously, is moderately toxic by ingestion, and is an irritant to the skin and eyes.[6]

LD_{50} (rat, oral): 3.53 g/kg[6]
LD_{50} (mouse, IV): 0.38 g/kg[7]
LD_{50} (mouse, SC): 8.0 g/kg[6]

15 Handling Precautions

Observe normal precautions appropriate to the circumstances and quantity of material handled. Sodium acetate is a mild skin and eye irritant; gloves and eye protection are recommended. On exposure, wash eyes and skin with large amounts of water. Inhalation of dust may cause pulmonary tract problems. When heated to decomposition, sodium acetate emits toxic fumes of NaO_2.[6]

16 Regulatory Status

GRAS listed. Accepted as a food additive in Europe. Included in the FDA Inactive Ingredients Database (injections, nasal, otic, ophthalmic, and oral preparations). Included in the Canadian Natural Health Products Ingredients Database.

17 Related Substances

—

18 Comments

Sodium acetate was shown to enhance aqueous humor to plasma concentration ratio of timolol by about 20-fold in an ophthalmic monoisopropyl PVM-MA matrix system, presumably by decreasing systemic absorption.[8]

Sodium acetate has also been used experimentally in matrix tablet formulations, where it increased the effect of carbomer as a sustained release matrix.[9]

Sodium acetate has also been used therapeutically for the treatment of metabolic acidosis in premature infants,[10,11] and in hemodialysis solutions.[12,13]

A specification for sodium acetate is contained within the *Food Chemicals Codex* (FCC).[14] The PhEur 7.4 also contains a monograph on sodium acetate [1-^{11}C] injection under Radiopharmaceutical Preparations.

The EINECS number for sodium acetate is 204-823-8. The PubChem Compound ID (CID) for sodium acetate trihydrate is 23665404.

19 Specific References

1 Keast RS, Breslin PA. Modifying the bitterness of selected oral pharmaceuticals with cation and anion series of salts. *Pharm Res* 2002; **19**(7): 1019–1026.
2 Frech G, Allen LV. Sodium acetate as a preservative in protein hydrolysate solutions. *Am J Hosp Pharm* 1979; **36**: 1672–1675.
3 Bedie GK, *et al*. Antimicrobials in the formulation to control *Listeria monocytogenes* postprocessing contamination on frankfurters stored at 4°C in vacuum packages. *J Food Prot* 2001; **64**(12): 1949–1955.
4 Ash M, Ash I. *Handbook of Pharmaceutical Additives*, 3rd edn. Endicott, NY: Synapse Information Resources, 2007: 893.
5 Waterhouse HN, Scott HM. Effect of sex, feathering, rate of growth and acetates on chicks need for glycine. *Poultry Sci* 1962; **41**: 1957–1962.
6 Lewis RJ, ed. *Sax's Dangerous Properties of Industrial Materials*, 11th edn. New York: Wiley, 2004: 3225.
7 Spector WS. *Handbook of Toxicology*. Philadelphia: WB Saunders, 1956: 268.
8 Finne U, *et al*. Sodium acetate improves the ocular/systemic absorption ratio of timolol applied ocularly in monoisopropyl PVM-MA matrices. *Int J Pharm* 1991; **75**: R1–R4.
9 Meshali MM, *et al*. Effect of added substances on theophylline release from carbopol 934P matrix. *STP Pharma Sci* 1997; 7(3): 195–198.
10 Ekblad H, *et al*. Slow sodium acetate infusion in the correction of metabolic acidosis in premature infants. *Am J Dis Child* 1985; **139**(7): 708–710.
11 Kasik JW, *et al*. Sodium acetate infusion to correct acidosis in premature infants. *Am J Dis Child* 1986; **140**(1): 9–10.
12 Katiuchi T, Mabuchi H, *et al*. Hemodynamic change during hemodialysis, especially on cardiovascular effects of sodium acetate. *Jpn J Artif Organs* 1982; **11**(2): 456–459.
13 Jackson JK, Derleth DP. Effects of various arterial infusion solutions on red blood cells in the newborn. *Arch Dis Child Fetal Neonatal Ed* 2000; **83**(2): F130–F134.
14 *Food Chemicals Codex*, 7th edn. Bethesda, MD: United States Pharmacopeia, 2010: 919.

20 General References

—

21 Author

WG Chambliss.

22 Date of Revision

1 March 2012.

S

Sodium Alginate

1 Nonproprietary Names

BP: Sodium Alginate
PhEur: Sodium Alginate
USP–NF: Sodium Alginate

2 Synonyms

Alginato sodico; algin; alginic acid, sodium salt; E401; *Kelcosol*; *Keltone*; natrii alginas; *Pronova;Protanal*; sodium polymannuronate.

3 Chemical Name and CAS Registry Number

Sodium alginate [9005-38-3]

4 Empirical Formula and Molecular Weight

Sodium alginate consists chiefly of the sodium salt of alginic acid, which is a mixture of polyuronic acids composed of residues of D-mannuronic acid and L-guluronic acid.

The block structure and molecular weight of sodium alginate samples have been investigated.[1]

5 Structural Formula

See Section 4.

6 Functional Category

Modified-release agentsuspending agent; tablet and capsule binder; tablet and capsule disintegrant; viscosity-increasing agent.

7 Applications in Pharmaceutical Formulation or Technology

Sodium alginate is used in a variety of oral and topical pharmaceutical formulations.[2] In tablet formulations, sodium alginate may be used as both a binder and disintegrant;[3] it has been used as a diluent in capsule formulations.[4] Sodium alginate has also been used in the preparation of sustained-release oral formulations since it can delay the dissolution of a drug from tablets,[5–7] capsules,[8] and aqueous suspensions.[9] It has also been used in oral formulations of enteric capsules,[10] multiparticulate floating systems,[11] sustained-release liquids,[12] and soft tablets.[13] The effects of particle size, viscosity and chemical composition of sodium alginate on drug release from matrix tablets have been described.[14]

In topical formulations, sodium alginate is widely used as a thickening and suspending agent in a variety of pastes, creams, and gels, and as a stabilizing agent for oil-in-water emulsions.

Recently, sodium alginate has been used for the aqueous microencapsulation of drugs,[15] in contrast with the more conventional microencapsulation techniques which use organic-solvent systems. It has also been used in the formation of nanoparticles,[16,17] microcapsules,[18] and beads.[19]

The adhesiveness of hydrogels prepared from sodium alginate has been investigated,[20] and drug release from oral mucosal adhesive tablets,[21,22] buccal gels,[23–27] fast-disintegrating tablets,[28,29] films,[30,31] and vaginal tablets[32] based on sodium alginate have been reported. The esophageal bioadhesion of sodium alginate suspensions may provide a barrier against gastric reflux or site-specific delivery of therapeutic agents.[33,34] Other novel delivery systems containing sodium alginate include ophthalmic solutions that form a gel *in situ* when administered to the eye;[35–37] ophthalmic mucoadhesive systems[38] and intraocular implants;[39] *in situ* forming gels containing paracetamol,[40] and baclofen[41] for oral administration; nasal delivery systems based on mucoadhesive microspheres;[42] and a freeze-dried device intended for the delivery of bone-growth factors.[43]

Hydrogel systems containing alginates have also been investigated for delivery of proteins and peptides.[44] In addition, sodium alginate microspheres have been used in the preparation of a foot-mouth disease DNA vaccine,[45] and in an oral vaccine for *Helicobacter pylori*;[46] chitosan nanoparticles coated with sodium alginate may have applications in mucosal vaccine delivery systems.[47]

Sodium alginate is also used in cosmetics and food products; *see* Table I.

Table I: Uses of sodium alginate.

Use	Concentration (%)
Pastes and creams	5–10
Stabilizer in emulsions	1–3
Suspending agent	1–5
Tablet binder	1–3
Tablet disintegrant	2.5–10

8 Description

Sodium alginate occurs as an odorless and tasteless, white to pale yellowish-brown colored powder.

9 Pharmacopeial Specifications

See Table II.

Table II: Pharmacopeial specifications for sodium alginate.

Test	PhEur 7.4	USP35–NF30
Characters	+	—
Identification	+	+
Appearance of solution	+	—
Microbial limits	≤1000 cfu/g	≤200 cfu/g
Loss on drying	≤15.0%	≤15.0%
Ash	—	18.0–27.0%
Sulfated ash	30.0–36.0%	—
Arsenic	—	≤1.5 ppm
Calcium	≤1.5%	—
Chlorides	≤1.0%	—
Lead	—	≤10 ppm
Heavy metals	≤20 ppm	≤40 ppm
Assay (dried basis)	—	90.8–106.0%

10 Typical Properties

Acidity/alkalinity pH ≈ 7.2 (1% w/v aqueous solution)

Solubility Practically insoluble in ethanol (95%), ether, chloroform, and ethanol/water mixtures in which the ethanol content is greater than 30% w/w. Also, practically insoluble in other organic solvents and aqueous acidic solutions in which the pH is less than 3. Slowly soluble in water, forming a viscous colloidal solution.

Spectroscopy

IR spectra *see* Figure 1.

NIR spectra *see* Figure 2.

Viscosity (dynamic) Various grades of sodium alginate are commercially available that yield aqueous solutions of varying

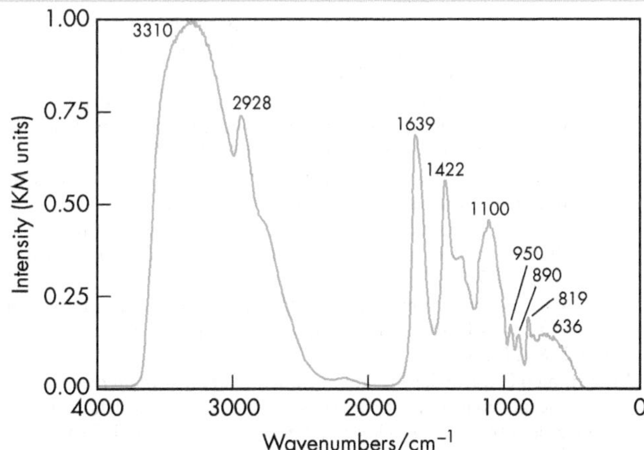

Figure 1: Infrared spectrum of sodium alginate measured by diffuse reflectance. Adapted with permission of Informa Healthcare.

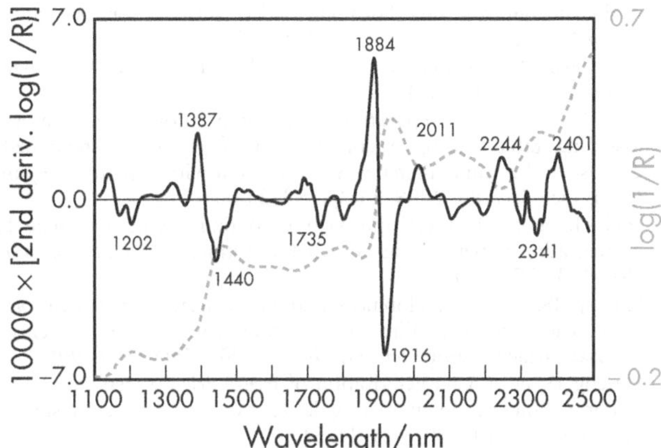

Figure 2: Near-infrared spectrum of sodium alginate measured by reflectance.

viscosity. Typically, a 1% w/v aqueous solution, at 20°C, will have a viscosity of 20–400 mPa s (20–400 cP). Viscosity may vary depending upon concentration, pH, temperature, or the presence of metal ions.[55–57] Above pH 10, viscosity decreases; *see also* Alginic Acid and Section 11.

11 Stability and Storage Conditions

Sodium alginate is a hygroscopic material, although it is stable if stored at low relative humidities and a cool temperature.

Aqueous solutions of sodium alginate are most stable at pH 4–10. Below pH 3, alginic acid is precipitated. A 1% w/v aqueous solution of sodium alginate exposed to differing temperatures had a viscosity 60–80% of its original value after storage for 2 years.[58] Solutions should not be stored in metal containers.

Sodium alginate solutions are susceptible on storage to microbial spoilage, which may affect solution viscosity. Solutions are ideally sterilized using ethylene oxide, although filtration using a 0.45 μm filter also has only a slight adverse effect on solution viscosity.[59] Heating sodium alginate solutions to temperatures above 70°C causes depolymerization with a subsequent loss of viscosity. Autoclaving of solutions can cause a decrease in viscosity, which may vary depending upon the nature of any other substances present.[59,60] Gamma irradiation should not be used to sterilize sodium alginate solutions since this process severely reduces solution viscosity.[59,61]

Preparations for external use may be preserved by the addition of 0.1% chlorocresol, 0.1% chloroxylenol, or parabens. If the medium is acidic, benzoic acid may also be used.

The bulk material should be stored in an airtight container in a cool, dry place.

12 Incompatibilities

Sodium alginate is incompatible with acridine derivatives, crystal violet, phenylmercuric acetate and nitrate, calcium salts, heavy metals, and ethanol in concentrations greater than 5%. Low concentrations of electrolytes cause an increase in viscosity but high electrolyte concentrations cause salting-out of sodium alginate; salting-out occurs if more than 4% of sodium chloride is present.

13 Method of Manufacture

Alginic acid is extracted from brown seaweed and is neutralized with sodium bicarbonate to form sodium alginate.

14 Safety

Sodium alginate is widely used in cosmetics, food products, and pharmaceutical formulations, such as tablets and topical products, including wound dressings. It is generally regarded as a nontoxic and nonirritant material, although excessive oral consumption may be harmful. A study in five healthy male volunteers fed a daily intake of 175 mg/kg body-weight of sodium alginate for 7 days, followed by a daily intake of 200 mg/kg body-weight of sodium alginate for a further 16 days, showed no significant adverse effects.[62]

The WHO has not specified an acceptable daily intake for alginic acid and alginate salts as the levels used in food do not represent a hazard to health.[63]

Inhalation of alginate dust may be irritant and has been associated with industrial-related asthma in workers involved in alginate production. However, it appears that the cases of asthma were linked to exposure to seaweed dust rather than pure alginate dust.[64]

LD_{50} (cat, IP): 0.25 g/kg[65]
LD_{50} (mouse, IV): 0.2 g/kg
LD_{50} (rabbit, IV): 0.1 g/kg
LD_{50} (rat, IV): 1 g/kg
LD_{50} (rat, oral): >5 g/kg

15 Handling Precautions

Observe normal precautions appropriate to the circumstances and quantity of material handled. Sodium alginate may be irritant to the eyes or respiratory system if inhaled as dust; *see* Section 14. Eye protection, gloves, and a dust respirator are recommended. Sodium alginate should be handled in a well-ventilated environment.

16 Regulatory Status

GRAS listed. Accepted in Europe for use as a food additive. Included in the FDA Inactive Ingredients Database (oral capsules, suspensions, syrups, tablets, and troches; oral sustained-action tablets). Included as an excipient in nonparenteral medicines (oral capsules, chewable tablets, modified-release tablets, enteric-coated tablets, suspensions, powders, and lozenges) licensed in the UK. Included in the Canadian Natural Health Products Ingredients Database.

17 Related Substances

Alginic acid; calcium alginate; potassium alginate; propylene glycol alginate.

18 Comments

A number of different grades of sodium alginate, which have different solution viscosities, are commercially available. Many different alginate salts and derivatives are also commercially available including ammonium alginate; calcium alginate; magnesium alginate, and potassium alginate.

To assist in the preparation of dispersions of sodium alginate, the material may be mixed with a dispersing agent such as sucrose, ethanol, glycerol, or propylene glycol.

Therapeutically, sodium alginate has been used in combination with an H_2-receptor antagonist in the management of gastroesophageal reflux,[48] and as a hemostatic agent in surgical dressings.[49,50] Alginate dressings, used to treat exuding wounds, often contain significant amounts of sodium alginate as this improves the gelling properties,[51] and wound dressing systems containing nitrofurazone have been reported.[52] Sponges composed of sodium alginate and chitosan produce a sustained drug release and may be useful as wound dressings or as tissue engineering matrices.[53] Lyophilized wound healing wafers composed of sodium alginate have been found to exhibit large reductions in viscosity following gamma irradiation.[54]

A specification for sodium alginate is contained in the *Food Chemicals Codex* (FCC)[66] and the *Japanese Pharmaceutical Excipients* (JPE).[67]

The PubChem Compound ID (CID) for sodium alginate is 6850754.

See also Alginic acid for further information.

19 Specific References

1 Johnson FA, *et al.* Characterization of the block structure and molecular weight of sodium alginates. *J Pharm Pharmacol* 1997; **49**: 639–643.

2 Tonnesen HH, Karlsen J. Alginate in drug delivery systems. *Drug Dev Ind Pharm* 2002; **28**(6): 621–630.

3 Sakr AM, *et al.* Effect of the technique of incorporating sodium alginate on its binding and/or disintegrating effectiveness in sulfathiazole tablets. *Pharm Ind* 1978; **40**(10): 1080–1086.

4 Veski P, Marvola M. Sodium alginates as diluents in hard gelatin capsules containing ibuprofen as a model drug. *Pharmazie* 1993; **48**(10): 757–760.

5 Al Zoubi N, *et al.* Optimisation of extended-release hydrophilic matrix tablets by support vector regression. *Drug Dev Ind Pharm* 2011; **37**: 80–87.

6 Holte O, *et al.* Sustained release of water-soluble drug from directly compressed alginate tablets. *Eur J Pharm Sci* 2003; **20**(4–5): 403–407.

7 Azarmi S, *et al.* '*In situ*' cross-linking of polyanionic polymers to sustain the drug-release of acetazolamide tablets. *Pharm Ind* 2003; **63**(9): 877–881.

8 Veski P, *et al.* Biopharmaceutical evaluation of pseudoephedrine hydrochloride capsules containing different grades of sodium alginate. *Int J Pharm* 1994; **111**: 171–179.

9 Zatz JL, Woodford DW. Prolonged release of theophylline from aqueous suspensions. *Drug Dev Ind Pharm* 1987; **13**: 2159–2178.

10 Smith AM, *et al.* Polymer film formulations for the preparation of enteric pharmaceutical capsules. *J Pharm Pharmacol* 2010; **62**(2): 167–172.

11 Gaikwad M, *et al.* Formulation and evaluation of floating, pulsatile, multiparticulates using pH-dependent swellable polymers. *Pharm Dev Technol* 2010; **15**(2): 209–216.

12 Itoh K, *et al. In situ* gelling xyloglucan/alginate liquid formulation for oral sustained drug delivery to dysphagic patients. *Drug Dev Ind Pharm* 2010; **36**(4): 449–455.

13 Schmid W, Picker-Freyer KM. Tableting and tablet properties of alginates: Characterisation and potential for soft tableting. *Eur J Pharm Biopharm* 2009; **72**(1): 165–172.

14 Liew CV, *et al.* Evaluation of sodium alginate as drug release modifier in matrix tablets. *Int J Pharm* 2006; **309**(1–2): 25–37.

15 Bodmeier R, Wang J. Microencapsulation of drugs with aqueous colloidal polymer dispersions. *J Pharm Sci* 1993; **82**: 191–194.

16 Nanjwade BK, *et al.* Preparation and evaluation of carboplatin biodegradable polymeric nanoparticles. *Int J Pharm* 2010; **385**(1–2): 176–180.

17 Motwani SK, *et al.* Chitosan-sodium alginate nanoparticles as submicroscopic reservoirs for ocular delivery: Formulation, optimisation and *in vitro* characterisation. *Eur J Pharm Biopharm* 2008; **68**(3): 513–525.

18 Sarfaraz M, *et al.* Formulation and characterisation of rifampicin microcapsules. *Indian J Pharm Sci* 2010; **72**: 101–105.

19 Takka S, Gürel A. Evaluation of chitosan/alginate beads using experimental design: formulation and *in vitro* characterization. *AAPS PharmSciTech* 2010; **11**(1): 460–466.

20 Vennat B, *et al.* Comparative texturometric analysis of hydrogels based on cellulose derivatives, carraghenates, and alginates: evaluation of adhesiveness. *Drug Dev Ind Pharm* 1998; **24**(1): 27–35.

21 Miyazaki S, *et al.* Drug release from oral mucosal adhesive tablets of chitosan and sodium alginate. *Int J Pharm* 1995; **118**: 257–263.

22 El-Gindy GA. Formulation development and *in-vivo* evaluation of buccoadhesive tablets of verapamil hydrochloride. *Bull Pharm Sci* 2004; **27**: 293–306.

23 Attia MA, *et al.* Transbuccal permeation, anti-inflammatory and clinical efficacy of piroxicam formulated in different gels. *Int J Pharm* 2004; **276**: 11–28.

24 Mohammed FA, Kheder H. Preparation and *in vitro/in vivo* evaluations of the buccal bioadhesive properties of slow-release tablets containing miconazole nitrate. *Drug Dev Ind Pharm* 2003; **29**(3): 321–337.

25 Kotagale NR, *et al.* Carbopol 934-sodium alginate-gelatin mucoadhesive ondansetron tablets for buccal delivery: Effect of pH modifiers. *Indian J Pharm Sci* 2010; **72**(4): 471–479.

26 Chinwala MG Lin S. Application of hydrogel polymers for development of thyrotropin releasing hormone-loaded adhesive buccal patches. *Pharm Dev Technol* 2010; **15**(3): 311–327.

27 Skulason S, *et al.* Evaluation of polymeric films for buccal drug delivery. *Pharmazie* 2009; **64**(3): 197–201.

28 Vora N, Rana V. Preparation and optimization of mouth/orally dissolving tablets using a combination of glycine, carboxymethyl cellulose and sodium alginate: A comparison with superdisintegrants. *Pharm Dev Technol* 2008; **13**(3): 233–243.

29 Goel H, *et al.* A novel approach to optimize and formulate fast disintegrating tablets for nausea and vomiting. *AAPS PharmSciTech* 2008; **9**(3): 774–781.

30 Boateng JS, *et al.* Development and mechanical characterization of solvent-cast polymeric films as potential drug delivery systems to mucosal surfaces. *Drug Dev Ind Pharm* 2009; **35**(8): 986–996.

31 Boateng JS, *et al.* Characterisation of freeze-dried wafers and solvent evaporated films as potential drug delivery systems to mucosal surfaces. *Int J Pharm* 2010; **389**(1–2): 24–31.

32 Sharma G, *et al.* Once daily bioadhesive vaginal clotrimazole tablets: design and evaluation. *Acta Pharm* 2006; **56**(3): 337–345.

33 Richardson JC, *et al.* Oesophageal bioadhesion of sodium alginate suspensions 2. Suspension behaviour on oesophageal mucosa. *Eur J Pharm Sci* 2005; **24**(1): 107–114.

34 Richardson JC, *et al.* Oesophageal bioadhesion of sodium alginate suspensions: particle swelling and mucosal retention. *Eur J Pharm Sci* 2005; **24**(1): 49–56.

35 Cohen S, *et al.* A novel *in situ*-forming ophthalmic drug delivery system from alginates undergoing gelation in the eye. *J Control Release* 1997; **44**: 201–208.

36 Balasubramaniam J, Pandit JK. Ion-activated *in situ* gelling systems for sustained ophthalmic delivery of ciprofloxacin hydrochloride. *Drug Delivery* 2003; **10**(3): 185–191.

37 Mali MN, Hajare AA. Ion activated *in situ* gel system for ophthalmic delivery of moxifloxacin hydrochloride. *Lat Am J Pharm* 2010; **29**: 876–882.

38 Kesavan K, *et al.* Sodium alginate based mucoadhesive system for gatifloxacin and its *in vitro* antibacterial activity. *Sci Pharm* 2010; **78**(4): 941–957.

39 Balasubramaniam J, *et al.* Studies on indomethacin intraocular implants using different *in vitro* release methods. *Indian J Pharm Sci* 2008; **70**(2): 216–221.

40 Kubo W, *et al.* Oral sustained delivery of paracetamol from *in-situ* gelling gellan and sodium alginate formulations. *Int J Pharm* 2003; **258**(1–2): 55–64.

41 Jivani RR, *et al.* The influence of variation of gastric pH on gelation and release characteristics of *in situ* gelling sodium alginate formulations. *Acta Pharm Sci* 2010; **52**: 365–369.

42 Gavini E, *et al.* Mucoadhesive microspheres for nasal administration of an antiemetic drug, metoclopramide: *in vitro/ex-vivo* studies. *J Pharm Pharmacol* 2005; **57**(3): 287–294.

43 Duggirala S, DeLuca PP. Buffer uptake and mass loss characteristics of freeze-dried cellulosic and alginate devices. *PDA J Pharm Sci Technol* 1996; **50**(5): 297–305.

44 Gombotz WR, Pettit DK. Biodegradable polymers for protein and peptide drug delivery. *Bioconjug Chem* 1995; **6**: 332–351.

S

45 Liu SK, *et al.* Preparation and *in vitro* release of foot-mouth-disease vaccine-loaded sodium alginate microspheres. *Pharm Care Resch (Yaoxue Fuwu Yu Yanjiu)* 2004; 4(2): 107–110.
46 Wang YC, *et al.* A preliminary non-clinical study of *Helicobacter pylori* microsphere vaccine. *Chin J New Drugs* 2007; **16**(7): 539–543.
47 Borges O, *et al.* Preparation of coated nanoparticles for a new mucosal vaccine delivery system. *Int J Pharm* 2005; **299**(1-2): 155–166.
48 Stanciu C, Bennett JR. Alginate/antacid in the reduction of gastro-oesophageal reflux. *Lancet* 1974; **i**: 109–111.
49 Thomas S. *Wound Management and Dressings.* London: Pharmaceutical Press, 1990: 43–49.
50 Qin Y, Gilding DK. Alginate fibres and wound dressings. *Med Device Technol* 1996; **Nov**: 32–41.
51 Thomas S. Alginate dressings in surgery and wound management—Part 1. *J Wound Care* 2000; **9**(2): 56–60.
52 Kim JO, *et al.* Development of polyvinyl alcohol-sodium alginate gel-matrix-based wound dressing system containing nitrofurazone. *Int J Pharm* 2008; **359**(1–2): 79–86.
53 Lai HL, *et al.* The preparation and characteristics of drug-loaded alginate and chitosan sponges. *Int J Pharm* 2003; **251**: 175–181.
54 Matthews KH, *et al.* Gamma-irradiation of lyophilised wound healing wafers. *Int J Pharm* 2006; **313**(1–2): 78–86.
55 Bugaj J, Górecki M. Kinetics of dynamic viscosity changes of aqueous sodium carboxymethylcellulose and sodium alginate solutions. *Pharmazie* 1995; **50**(11): 750–752.
56 Duggirala S, DeLuca PP. Rheological characterization of cellulosic and alginate polymers. *PDA J Pharm Sci Technol* 1996; **50**(5): 290–296.
57 Bugaj J, Górecki M. Rheometrical estimation of physical properties of sodium alginate and sodium carboxymethylcellulose aqueous solutions. *Acta Pol Pharm Drug Res* 1996; **53**(2): 141–146.
58 Pávics L. [Comparison of rheological properties of mucilages.] *Acta Pharm Hung* 1970; **40**: 52–59[in Hungarian].
59 Coates D, Richardson G. A note on the production of sterile solutions of sodium alginate. *Can J Pharm Sci* 1974; **9**: 60–61.
60 Vandenbossche GMR, Remon J-P. Influence of the sterilization process on alginate dispersions. *J Pharm Pharmacol* 1993; **45**: 484–486.
61 Hartman AW, *et al.* Viscosities of acacia and sodium alginate after sterilization by cobalt-60. *J Pharm Sci* 1975; **64**: 802–805.
62 Anderson DM, *et al.* Dietary effects of sodium alginate in humans. *Food Addit Contam* 1991; 8(3): 237–248.
63 FAO/WHO. Evaluation of certain food additives and naturally occurring toxicants. Thirty-ninth report of the joint FAO/WHO expert committee on food additives. *World Health Organ Tech Rep Ser* 1992; No. 828.
64 Henderson AK, *et al.* Pulmonary hypersensitivity in the alginate industry. *Scott Med J* 1984; **29**(2): 90–95.
65 Lewis RJ, ed. *Sax's Dangerous Properties of Industrial Materials*, 11th edn. New York: Wiley, 2004: 3225–3226.
66 *Food Chemicals Codex*, 6th edn. Bethesda, MD: United States Pharmacopeia, 2008: 872.
67 Japan Pharmaceutical Excipients Council. *Japanese Pharmaceutical Excipients 2004*. Tokyo: Yakuji Nippo, 2004: 765–767.

20 General References

—

21 Author

CG Cable.

22 Date of Revision

1 March 2012.

 Sodium Ascorbate

1 Nonproprietary Names

BP: Sodium Ascorbate

PhEur: Sodium Ascorbate

USP–NF: Sodium Ascorbate

2 Synonyms

L-Ascorbic acid monosodium salt; E301; 3-oxo-L-gulofuranolactone sodium enolate; natrii ascorbas; *SA-99*; vitamin C sodium.

3 Chemical Name and CAS Registry Number

Monosodium L-(+)-ascorbate [134-03-2]

4 Empirical Formula and Molecular Weight

$C_6H_7NaO_6$ 198.11

5 Structural Formula

6 Functional Category

Antioxidant.

7 Applications in Pharmaceutical Formulation or Technology

Sodium ascorbate is used as an antioxidant in pharmaceutical formulations, and also in food products where it increases the effectiveness of sodium nitrite against growth of *Listeria monocytogenes* in cooked meats. It improves gel cohesiveness and sensory firmness of fiberized products regardless of vacuum treatment.

8 Description

Sodium ascorbate occurs as a white or slightly yellow-colored, practically odorless, crystalline powder with a pleasant saline taste.

9 Pharmacopeial Specifications

See Table I.

10 Typical Properties

Acidity/alkalinity pH = 7–8 (10% w/v aqueous solution)

Density (tapped)

0.6–1.1 g/cm^3 for fine powder;

0.8–1.1 g/cm^3 for fine granular grade.

Density (true) 1.826 g/cm^3

SEM 1: Excipient: sodium ascorbate USP-NF; manufacturer: Pfizer Ltd.; lot no: 9B-1 (C92220-C4025); magnification: 120×; voltage: 20 kV.

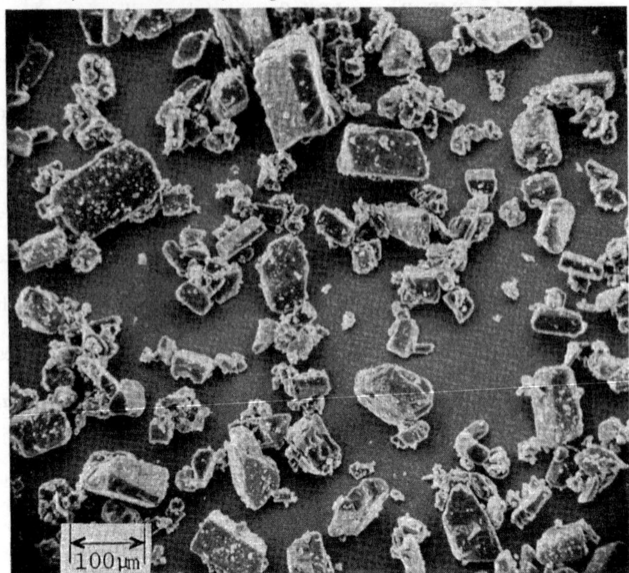

SEM 2: Excipient: sodium ascorbate USP; manufacturer: Pfizer Ltd.; lot no: 9B-1 (C92220-C4025); magnification: 600×; voltage: 20 kV.

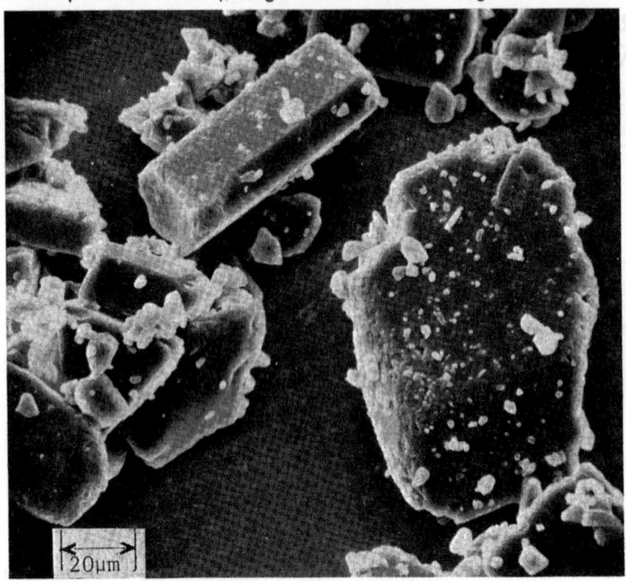

Hygroscopicity Not hygroscopic. Sodium ascorbate adsorbs practically no water up to 80% relative humidity at 20°C and less than 1% w/w of water at 90% relative humidity.

Melting point 218°C (with decomposition)

Particle size distribution Various grades of sodium ascorbate with different particle-size distributions are commercially available, e.g. approximately 98% passes through a 149 μm mesh for a fine powder grade (Takeda), and approximately 95% passes through a 840 μm mesh for a standard grade (Takeda).

Solubility *see* Table II.

Specific gravity

1.782 for powder at 20°C;

1.005 for 1% w/v aqueous solution at 25°C;

1.026 for 5% w/v aqueous solution at 25°C.

Specific rotation $[\alpha]_D^{20} = +104.4°$ (10% w/v aqueous solution)

Spectroscopy

IR spectra *see* Figure 1.

NIR spectra *see* Figure 2.

Table I: Pharmacopeial specifications for sodium ascorbate.

Test	PhEur 7.4	USP35–NF30
Identification	+	+
Characters	+	—
Appearance of solution	+	—
pH	7.0–8.0	7.0–8.0
Specific optical rotation(10% w/v aqueous solution)	+103° to +108°	+103° to +108°
Oxalic acid	≤0.30%	—
Related substances	+	—
Sulfates	≤150 ppm	—
Copper	≤5 ppm	—
Iron	≤2 ppm	—
Nickel	≤1 ppm	—
Heavy metals	≤10 ppm	≤20 ppm
Loss on drying	≤0.25%	≤0.25%
Assay (dried basis)	99.0–101.0%	99.0–101.0%

Table II: Solubility of sodium ascorbate.

Solvent	Solubility at 20°C unless otherwise stated
Chloroform	Practically insoluble
Ethanol (95%)	Very slightly soluble
Ether	Practically insoluble
Water	1 in 1.6
	1 in 1.3 at 75°C

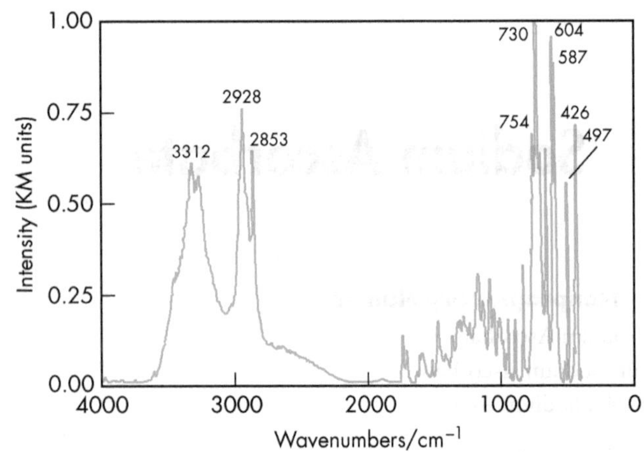

Figure 1: Infrared spectrum of sodium ascorbate measured by diffuse reflectance. Adapted with permission of Informa Healthcare.

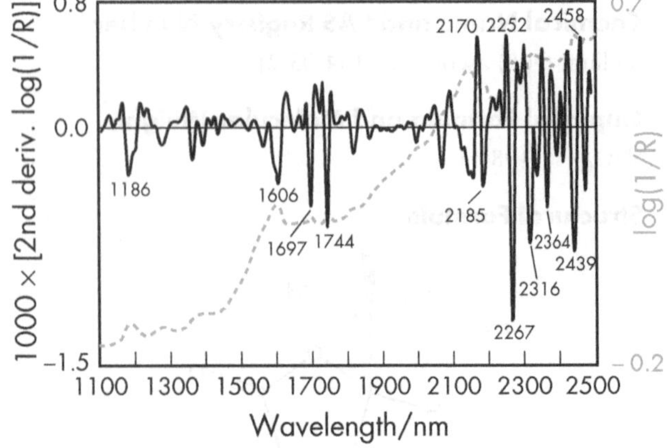

Figure 2: Near-infrared spectrum of sodium ascorbate measured by reflectance.

11 Stability and Storage Conditions

Sodium ascorbate is relatively stable in air, although it gradually darkens on exposure to light. Aqueous solutions are unstable and subject to rapid oxidation in air at pH > 6.0.

The bulk material should be stored in a well-closed nonmetallic container, protected from light, in a cool, dry place.

12 Incompatibilities

Incompatible with oxidizing agents, heavy metal ions, especially copper and iron, methenamine, sodium nitrite, sodium salicylate, and theobromine salicylate. The aqueous solution is reported to be incompatible with stainless steel filters.[1]

13 Method of Manufacture

An equivalent amount of sodium bicarbonate is added to a solution of ascorbic acid in water. Following the cessation of effervescence, the addition of propan-2-ol precipitates sodium ascorbate.

14 Safety

The parenteral administration of 0.25–1.00 g of sodium ascorbate, given daily in divided doses, is recommended in the treatment of vitamin C deficiencies. Various adverse reactions have been reported following the administration of 1 g or more of sodium ascorbate, although ascorbic acid and sodium ascorbate are usually well tolerated; *see* Ascorbic acid. There have been no reports of adverse effects associated with the much lower concentrations of sodium ascorbate and ascorbic acid, which are employed as antioxidants.

The WHO has set an acceptable daily intake of ascorbic acid, potassium ascorbate, and sodium ascorbate, as antioxidants in food, at up to 15 mg/kg body-weight in addition to that naturally present in food.[2]

15 Handling Precautions

Observe normal precautions appropriate to the circumstances and quantity of material handled. Sodium ascorbate may be irritant to the eyes. Eye protection and rubber or plastic gloves are recommended.

16 Regulatory Status

GRAS listed. Accepted for use as a food additive in Europe. Included in the FDA Inactive Ingredients Database (IV preparations; oral tablets). Included in nonparenteral and parenteral medicines licensed in the UK. Included in the Canadian Natural Health Products Ingredients Database.

17 Related Substances

Ascorbic acid; ascorbyl palmitate; calcium ascorbate.

Calcium ascorbate

Empirical formula $C_{12}H_{14}O_{12}Ca$

Molecular weight 390.31

CAS number [5743-27-1]

Synonyms calcium L-(+)-ascorbate; *CCal-97*; E302.

18 Comments

1 mg of sodium ascorbate is equivalent to 0.8890 mg of ascorbic acid (1 mg of ascorbic acid is equivalent to 1.1248 mg of sodium ascorbate); 1 g of sodium ascorbate contains approximately 5 mmol of sodium.

Sodium ascorbate is used therapeutically as a source of vitamin C in tablets and parenteral preparations. Sodium ascorbate is used as an antioxidant in food products where it increases the effectiveness of sodium nitrite against growth of *Listeria monocytogenes* in cooked meats. It improves gel cohesiveness and sensory firmness of fiberized products regardless of vacuum treatment.

A specification for sodium ascorbate is contained in the *Food Chemicals Codex* (FCC).[3]

The EINECS number for sodium ascorbate is 205-126-1. The PubChem Compound ID (CID) for sodium ascorbate is 23666832.

19 Specific References

1 Buck GW, Wolfe KR. Interaction of sodium ascorbate with stainless steel particulate filter needles [letter]. *Am J Hosp Pharm* 1991; **48**: 1191.

2 FAO/WHO. Toxicological evaluation of certain food additives with a review of general principles and of specifications. Seventeenth report of the FAO/WHO expert committee on food additives. *World Health Organ Tech Rep Ser* 1974; No. 539.

3 *Food Chemicals Codex*, 7th edn. Bethesda, MD: United States Pharmacopeia, 2010.

20 General References

Dahl GB, *et al.* Vitamin stability in a TPN mixture stored in an EVA plastic bag. *J Clin Hosp Pharm* 1986; **11**: 271–279.

DeRitter E, *et al.* Effect of silica gel on stability and biological availability of ascorbic acid. *J Pharm Sci* 1970; **59**: 229–232.

Dettman IC. Sterilization of ascorbates by heat and absolute ethanol. United States Patent No. 4,816,223; 1989.

Iida S *et al.* Stable ascorbic acid solutions. Japanese Patent No. 61,130,205; 1986.

Pfeifer HJ, Webb JW. Compatibility of penicillin and ascorbic acid injection. *Am J Hosp Pharm* 1976; **33**: 448–450.

Sekine K *et al.* Powdery pharmaceutical compositions containing ascorbic acids for intranasal administration. Japanese Patent No. 63,115,820; 1988.

Thielemann AM, *et al.* Biopharmaceutical study of a vitamin C controlled-release formulation. *Farmaco (Prat)* 1988; **43**: 387–395.

21 Author

CP McCoy.

22 Date of Revision

1 March 2012.

S

Sodium Benzoate

1 Nonproprietary Names

BP: Sodium Benzoate
JP: Sodium Benzoate
PhEur: Sodium Benzoate
USP–NF: Sodium Benzoate

2 Synonyms

Benzoate of soda;benzoic acid sodium salt; E211; natrii benzoas; natrium benzoicum; sobenate; sodii benzoas; sodium benzoic acid.

3 Chemical Name and CAS Registry Number

Sodium benzoate [532-32-1]

4 Empirical Formula and Molecular Weight

$C_7H_5NaO_2$ 144.11

5 Structural Formula

O
ONa

6 Functional Category

Antimicrobial preservative; tablet and capsule lubricant.

7 Applications in Pharmaceutical Formulation or Technology

Sodium benzoate is used primarily as an antimicrobial preservative in cosmetics, foods, and pharmaceuticals. It is used in concentrations of 0.02–0.5% in oral medicines, 0.5% in parenteral products, and 0.1–0.5% in cosmetics. The usefulness of sodium benzoate as a preservative is limited by its effectiveness over a narrow pH range; *see* Section 10. The inhibitory concentration of sodium benzoate required in emulsions increases with oil content.

Sodium benzoate is used in preference to benzoic acid in some circumstances, owing to its greater solubility. However, in some applications it may impart an unpleasant flavor to a product. Sodium benzoate has also been used as a tablet lubricant[1] at 2–5% w/w concentrations, providing rapid disintegration times.[2]

8 Description

Sodium benzoate occurs as a white granular or crystalline, slightly hygroscopic powder. It is odorless, or with faint odor of benzoin and has an unpleasant sweet and saline taste.

9 Pharmacopeial Specifications

See Table I.

10 Typical Properties

Acidity/alkalinity pH = 8.0 (saturated aqueous solution at 25°C). It is relatively inactive above approximately pH 5.

Antimicrobial activity Sodium benzoate has both bacteriostatic and antifungal properties attributed to undissociated benzoic acid; hence preservative efficacy is best seen in acidic solutions (pH 2–5). In alkaline conditions it is almost without effect.

SEM 1: Excipient: sodium benzoate; manufacturer: Bush Boake Allen Corp.; magnification: 60×.

SEM 2: Excipient: sodium benzoate; manufacturer: Bush Boake Allen Corp.; magnification: 2400×.

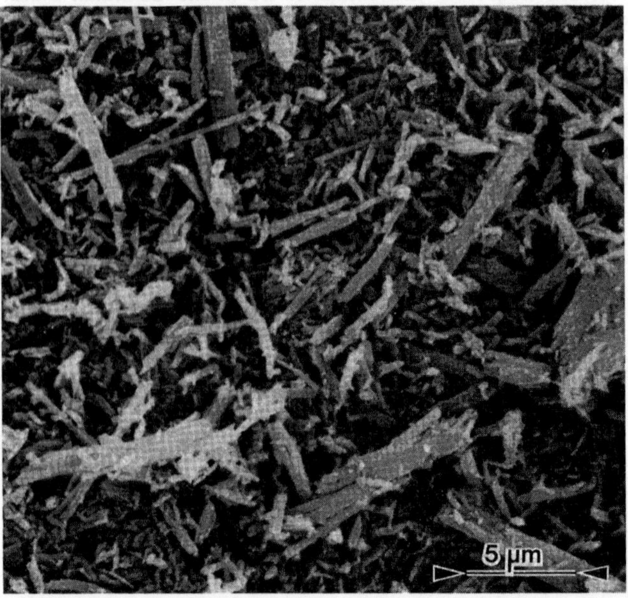

Table I: Pharmacopeial specifications for sodium benzoate.

Test	JP XV	PhEur 7.4	USP35–NF30
Identification	+	+	+
Characters	+	+	—
Acidity or alkalinity	+	+	+
Appearance of solution	+	+	—
Arsenic	≤2 ppm	—	—
Chloride	+	≤200 ppm	—
Heavy metals	≤20 ppm	≤10 ppm	≤10 ppm
Loss on drying	≤1.5%	≤2.0%	≤1.5%
Phthalic acid	+	—	—
Sulfate	≤0.120%	—	—
Total chlorine	—	≤300 ppm	—
Assay (dried basis)	≥99.0%	99.0–100.5%	99.0–100.5%

Table II: Solubility of sodium benzoate.

Solvent	Solubility at 20°C unless otherwise stated
Ethanol (95%)	1 in 75
Ethanol (90%)	1 in 50
Water	1 in 1.8
	1 in 1.4 at 100°C

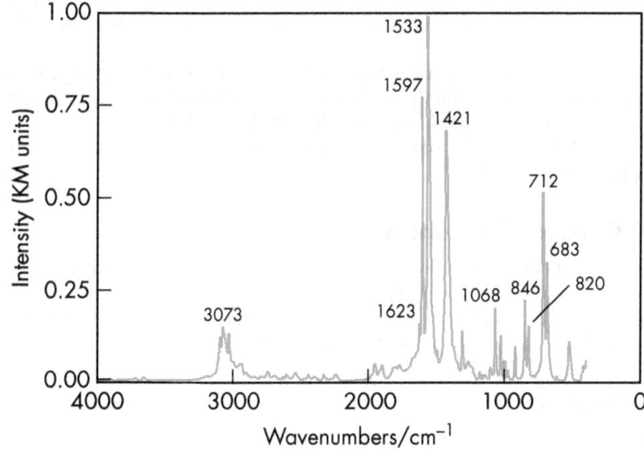

Figure 1: Infrared spectrum of sodium benzoate measured by diffuse reflectance. Adapted with permission of Informa Healthcare.

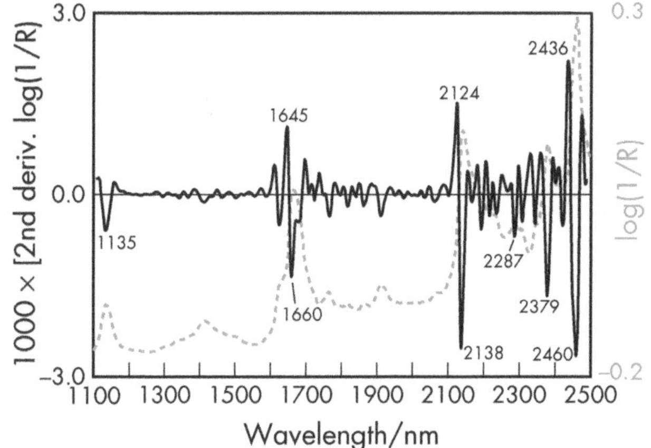

Figure 2: Near-infrared spectrum of sodium benzoate measured by reflectance.

Density 1.497–1.527 g/cm^3 at 24°C
Freezing point depression 0.24°C (1.0% w/v)
Osmolarity A 2.25% w/v aqueous solution is iso-osmotic with serum.
Partition coefficients Vegetable oil : water = 3–6
Solubility *see* Table II.
Spectroscopy

IR spectra *see* Figure 1.

NIR spectra *see* Figure 2.

11 Stability and Storage Conditions

Aqueous solutions may be sterilized by autoclaving or filtration.

The bulk material should be stored in a well-closed container, in a cool, dry place.

12 Incompatibilities

Sodium benzoate is incompatible with quaternary compounds, gelatin, ferric salts, calcium salts, and salts of heavy metals, including silver, lead, and mercury. Preservative activity may be reduced by interactions with kaolin[3] or nonionic surfactants. Sodium benzoate may also react with ascorbic acid to form benzene.[4]

13 Method of Manufacture

Sodium benzoate is prepared by the treatment of benzoic acid with either sodium carbonate or sodium bicarbonate.

14 Safety

Ingested sodium benzoate is conjugated with glycine in the liver to yield hippuric acid, which is excreted in the urine. Symptoms of systemic benzoate toxicity resemble those of salicylates.[5] Whereas oral administration of the free-acid form may cause severe gastric irritation, benzoate salts are well tolerated in large quantities: e.g. 6 g of sodium benzoate in 200 mL of water is administered orally as a liver function test.

Clinical data have indicated that sodium benzoate can produce nonimmunological contact urtcaria and nonimmunological immediate contact reactions.[6] However, it is also recognized that these reactions are strictly cutaneous, and sodium benzoate can therefore be used safely at concentrations up to 5%. Nevertheless, this nonimmunological phenomenon should be considered when designing formulations for infants and children.

Other adverse effects include anaphylaxis[7–9] and urticarial reactions, although a controlled study has shown that the incidence of urticaria in patients given sodium benzoate was very low compared with the placebo-controlled group.[10]

It has been recommended that caffeine and sodium benzoate injection should not be used in neonates;[11] however, sodium benzoate has been used by others in the treatment of some neonatal metabolic disorders.[12] It has been suggested that there is a general adverse effect of benzoate preservatives on the behavior of 3-year-old children, which is detectable by parents, but not by a simple clinical assessment.[13]

The WHO acceptable daily intake of total benzoates, calculated as benzoic acid, has been estimated at up to 5 mg/kg of body-weight.[14,15]

LD_{50} (mouse, IM): 2.3 g/kg[15,16]
LD_{50} (mouse, IV): 1.4 g/kg
LD_{50} (mouse, oral): 1.6 g/kg
LD_{50} (rabbit, oral): 2.0 g/kg
LD_{50} (rat, IV): 1.7 mg/kg
LD_{50} (rat, oral): 4.1 g/kg

See also Benzoic Acid.

S

15 Handling Precautions

Observe normal precautions appropriate to the circumstances and quantity of material handled. Sodium benzoate may be irritant to the eyes and skin. Eye protection and rubber or plastic gloves are recommended.

16 Regulatory Status

GRAS listed. Accepted as a food additive in Europe. Included in the FDA Inactive Ingredients Database (dental preparations; IM and IV injections; oral capsules, solutions and tablets; rectal; and topical preparations). Included in nonparenteral medicines licensed in the UK. Included in the Canadian Natural Health Products Ingredients Database.

17 Related Substances

Benzoic acid; potassium benzoate.

18 Comments

Sodium benzoate has been studied for use in caffeine preparations and anesthetic formulations.[17]

Sodium benzoate has been used as an antimicrobial agent used in polymeric films in food packaging.[18] Solutions of sodium benzoate have also been administered, orally or intravenously, in order to determine liver function.

A specification for sodium benzoate is contained in the *Food Chemicals Codex* (FCC).[19]

The EINECS number for sodium benzoate is 208-534-8. The PubChem Compound ID (CID) for sodium benzoate is 517055.

19 Specific References

1 Saleh SI, *et al.* Improvement of lubrication capacity of sodium benzoate: effects of milling and spray drying. *Int J Pharm* 1988; **48**: 149–157.

2 Lotter AP, *et al.* Identification and prevention of insoluble reaction products forming after dissolution of effervescent multi-vitamin tablets. *Drug Dev Ind Pharm* 1995; **21**(17): 1989–1998.

3 Clarke CD, Armstrong NA. Influence of pH on the adsorption of benzoic acid by kaolin. *Pharm J* 1972; **209**: 44–45.

4 Ju HK, *et al.* Evaluation of headspace-gas chromatography/mass spectrometry for the analysis of benzene in vitamin C drinks; pitfalls of headspace in benzene detection. *Biomed Chromatogr* 2008; **22**(8): 900–905.

5 Michils A, *et al.* Anaphylaxis with sodium benzoate [letter]. *Lancet* 1991; **337**: 1424–1425.

6 Nair B. Final report on the safety assessment of benzyl alcohol, benzoic acid, and sodium benzoate. *Int J Toxicol* 2001; **20**(Suppl. 3): 23–50.

7 Rosenhall L. Evaluation of intolerance to analgesics, preservatives and food colorants with challenge tests. *Eur J Respir Dis* 1982; **63**: 410–419.

8 Michaëlsson G, Juhlin L. Urticaria induced by preservatives and dye additives in food and drugs. *Br J Dermatol* 1973; **88**: 525–532.

9 Warin RP, Smith RJ. Challenge test battery in chronic urticaria. *Br J Dermatol* 1976; **94**: 401–406.

10 Nettis E. *et al*Sodium benzoate-induced repeated episodes of acute urticaria/angioedema: randomized controlled trial. *Br J Dermatol* 2004; **151**(4): 898–902.

11 Edwards RC, Voegeli CJ. Inadvisability of using caffeine and sodium benzoate in neonates. *Am J Hosp Pharm* 1984; **41**: 658.

12 Brusilow SW, *et al.* Treatment of episodic hyperammonemia in children with inborn errors of urea synthesis. *N Engl J Med* 1984; **310**: 1630–1634.

13 Bateman B, *et al.* The effects of a double blind, placebo controlled, artificial food colorings and benzoate preservative challenge on hyperactivity in a general population sample of preschool children. *Arch Dis Child* 2005; **90**(8): 875.

14 FAO/WHO. Toxicological evaluation of certain food additives with a review of general principles and of specifications. Seventeenth report of the joint FAO/WHO expert committee on food additives. *World Health Organ Tech Rep Ser* 1974; No. 539.

15 FAO/WHO. Evaluation of certain food additives and contaminants. Twenty-seventh report of the joint FAO/WHO expert committee on food additives. *World Health Organ Tech Rep Ser* 1983; No. 696.

16 Lewis RJ, ed. *Sax's Dangerous Properties of Industrial Materials*, 11th edn. New York: Wiley, 2004; 3232.

17 Noh ES Chun IK. Formulation of caffeine nasal sprays and its enhanced permeation through rabbit nasal mucosa. *Yakche Hakhoechi* 2004; 34(2): 131–138.

18 Buonocore GG, *et al.* A general approach to describe the antimicrobial agent release from highly swellable films intended for food packaging applications. *J Control Release* 2003; **90**(1): 97–107.

19 *Food Chemicals Codex*, 7th edn. Bethesda, MD: United States Pharmacopeia, 2010.

20 General References

Nishijo J, Yonetani I. Interaction of theobromine with sodium benzoate. *J Pharm Sci* 1982; **71**: 354–356.

Schnuch A, *et al.* Contact allergy to preservatives. Analysis of IVDK data 1996-2009 . *Br J Dermatol* 2011; **164**: 1316–1325.

21 Author

X He.

22 Date of Revision

1 March 2012.

S

Sodium Bicarbonate

1 Nonproprietary Names

BP: Sodium Bicarbonate
JP: Sodium Bicarbonate
PhEur: Sodium Hydrogen Carbonate
USP–NF: Sodium Bicarbonate

2 Synonyms

Baking soda; E500; *Effer-Soda*; monosodium carbonate; natrii hydrogenocarbonas; Sal de Vichy; sodium acid carbonate; sodium hydrogen carbonate.

3 Chemical Name and CAS Registry Number

Carbonic acid monosodium salt [144-55-8]

4 Empirical Formula and Molecular Weight

$NaHCO_3$ 84.01

5 Structural Formula

See Section 4.

6 Functional Category

Alkalizing agent; tablet and capsule diluent.

7 Applications in Pharmaceutical Formulation or Technology

Sodium bicarbonate is generally used in pharmaceutical formulations as a source of carbon dioxide in effervescent tablets and granules. It is also widely used to produce or maintain an alkaline pH in a preparation.

In effervescent tablets and granules, sodium bicarbonate is usually formulated with citric and/or tartaric acid;[1] combinations of citric and tartaric acid are often preferred in formulations as citric acid alone produces a sticky mixture that is difficult to granulate, while if tartaric acid is used alone, granules lose firmness. When the tablets or granules come into contact with water, a chemical reaction occurs, carbon dioxide is evolved, and the product disintegrates.[2,3] Melt granulation in a fluidized bed dryer has been suggested as a one-step method for the manufacture of effervescent granules composed of anhydrous citric acid and sodium bicarbonate, for subsequent compression into tablets.[4]

Tablets may also be prepared with sodium bicarbonate alone since the acid of gastric fluid is sufficient to cause effervescence and disintegration. Sodium bicarbonate is also used in tablet formulations to buffer drug molecules that are weak acids, thereby increasing the rate of tablet dissolution and reducing gastric irritation.[5–7]

The effects of tablet binders, such as polyethylene glycols, microcrystalline cellulose, silicified microcrystalline cellulose, pregelatinized starch, and povidone, on the physical and mechanical properties of sodium bicarbonate tablets have also been investigated.[8,9]

Additionally, sodium bicarbonate is used in solutions as a buffering agent for erythromycin,[10] lidocaine,[11] local anesthetic solutions,[12] and total parenteral nutrition (TPN) solutions.[13] In some parenteral formulations, e.g. niacin, sodium bicarbonate is used to produce a sodium salt of the active ingredient that has enhanced solubility. Sodium bicarbonate has also been used as a freeze-drying stabilizer[14] and in toothpastes.

See Table I.

Table I: Uses of sodium bicarbonate.

Use	Concentration (%)
Buffer in tablets	10–40
Effervescent tablets	25–50
Isotonic injection/infusion	1.39

8 Description

Sodium bicarbonate occurs as an odorless, white, crystalline powder with a saline, slightly alkaline taste. The crystal structure is monoclinic prisms. Grades with different particle sizes, from a fine powder to free-flowing uniform granules, are commercially available.

9 Pharmacopeial Specifications

See Table II.

Table II: Pharmacopeial specifications for sodium bicarbonate.

Test	JP XV	PhEur 7.4	USP35–NF30
Identification	+	+	+
Characters	—	+	—
Loss on drying	—	—	≤0.25%
Insoluble substances	—	—	+
pH (5% w/v aqueous solution)	7.9–8.4	—	—
Appearance	+	+	—
Carbonate	+	+	≤0.23%[a]
Normal carbonate	—	—	+
Chloride	≤0.04%	≤150 ppm	≤0.015%
Sulfate	—	≤150 ppm	≤0.015%
Ammonia	—	—	+
Ammonium	+	≤20 ppm	—
Aluminum	—	—	≤2 μg/g[a]
Arsenic	≤2 ppm	≤2 ppm	≤2 ppm
Calcium	—	≤100 ppm	≤0.01%[a]
Magnesium	—	—	≤0.004%[a]
Copper	—	—	≤1 ppm[a]
Iron	—	≤20 ppm	≤5 ppm[a]
Heavy metals	≤5 ppm	≤10 ppm	≤5 ppm
Limit of organics	—	—	≤0.01%[a]
Assay (dried basis)	≥99.0%	99.0–101.0%	99.0–100.5%

(a) Where it is labeled as intended for use in hemodialysis.

10 Typical Properties

Acidity/alkalinity pH = 8.3 for a freshly prepared 0.1 M aqueous solution at 25°C; alkalinity increases on standing, agitation, or heating.

Density (bulk) 0.869 g/cm^3

Density (tapped) 1.369 g/cm^3

Density(true) 2.173 g/cm^3

Freezing point depression 0.381°C (1% w/v solution)

Melting point 270°C (with decomposition)

Moisture content Below 80% relative humidity, the moisture content is less than 1% w/w. Above 85% relative humidity, sodium bicarbonate rapidly absorbs excessive amounts of water and may start to decompose with loss of carbon dioxide.

Osmolarity A 1.39% w/v aqueous solution is isoosmotic with serum.

SEM 1: Excipient: sodium bicarbonate; manufacturer: Merck Ltd; magnification: 120×.

SEM 2: Excipient: sodium bicarbonate; manufacturer: Merck Ltd; magnification: 600×.

Refractive index n_D^{20} = 1.3344 (1% w/v aqueous solution)
Solubility *see* Table III.
Spectroscopy

IR spectra *see* Figure 1.

NIR spectra *see* Figure 2.

11 Stability and Storage Conditions

When heated to about 50°C, sodium bicarbonate begins to dissociate into carbon dioxide, sodium carbonate, and water; on heating to 250–300°C for a short time, sodium bicarbonate is completely converted into anhydrous sodium carbonate. However, the process is both time- and temperature-dependent, with conversion 90% complete within 75 minutes at 93°C. The reaction proceeds via surface-controlled kinetics; when sodium bicarbonate crystals are heated for a short period of time, very fine needle-shaped crystals of anhydrous sodium carbonate are formed on the sodium bicarbonate surface.[15]

The effects of relative humidity and temperature on the moisture sorption and stability of sodium bicarbonate powder have been investigated. Sodium bicarbonate powder is stable below 76% relative humidity at 25°C and below 48% relative humidity at 40°C.[16] At 54% relative humidity, the degree of pyrolytic decarboxylation of sodium bicarbonate should not exceed 4.5% in order to avoid detrimental effects on stability.[17]

At ambient temperatures, aqueous solutions slowly decompose with partial conversion into the carbonate; the decomposition is accelerated by agitation or heat. Aqueous solutions begin to break up into carbon dioxide and sodium carbonate at about 20°C, and completely on boiling.

Aqueous solutions of sodium bicarbonate may be sterilized by filtration or autoclaving. To minimize decomposition of sodium

Table III: Solubility of sodium bicarbonate.

Solvent	Solubility at 20°C unless otherwise stated
Ethanol (95%)	Practically insoluble
Ether	Practically insoluble
Water	1 in 11
	1 in 4 at 100°C[a]
	1 in 10 at 25°C
	1 in 12 at 18°C

(a) Note that in hot water, sodium bicarbonate is converted to the carbonate.

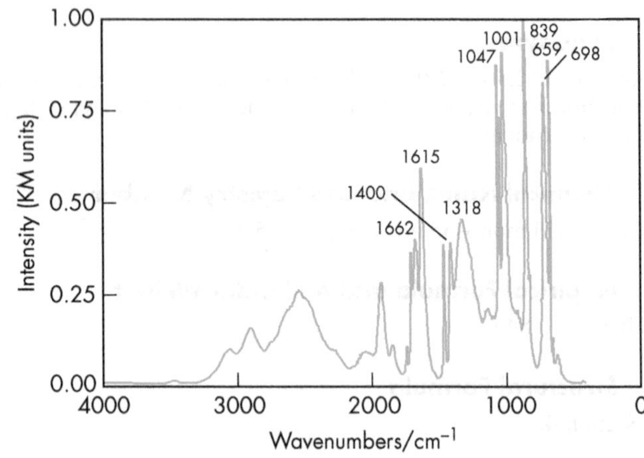

Figure 1: Infrared spectrum of sodium bicarbonate measured by diffuse reflectance. Adapted with permission of Informa Healthcare.

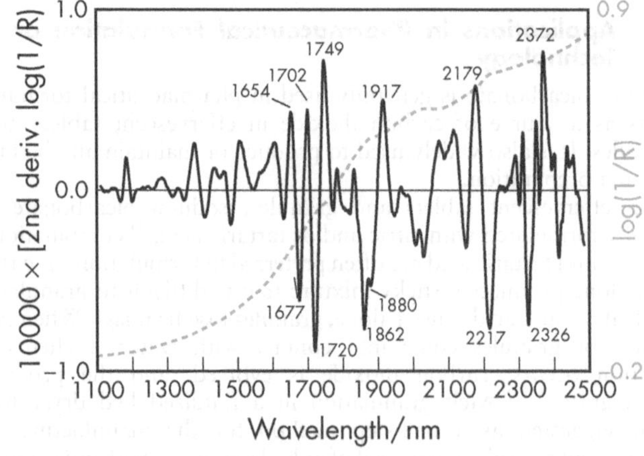

Figure 2: Near-infrared spectrum of sodium bicarbonate measured by reflectance.

bicarbonate by decarboxylation on autoclaving, carbon dioxide is passed through the solution in its final container, which is then hermetically sealed and autoclaved. The sealed container should not be opened for at least 2 hours after it has returned to ambient temperature, to allow time for the complete reformation of the bicarbonate from the carbonate produced during the heating process.

Aqueous solutions of sodium bicarbonate stored in glass containers may develop deposits of small glass particles. Sediments of calcium carbonate with traces of magnesium or other metal carbonates have been found in injections sterilized by autoclaving; these are due to impurities in the bicarbonate or to extraction of calcium and magnesium ions from the glass container. Sedimentation may be retarded by the inclusion of 0.01–0.02% disodium edetate.[18–20]

Sodium bicarbonate is stable in dry air but slowly decomposes in moist air and should therefore be stored in a well-closed container in a cool, dry place.

12 Incompatibilities

Sodium bicarbonate reacts with acids, acidic salts, and many alkaloidal salts, with the evolution of carbon dioxide. Sodium bicarbonate can also intensify the darkening of salicylates.

In powder mixtures, atmospheric moisture or water of crystallization from another ingredient is sufficient for sodium bicarbonate to react with compounds such as boric acid or alum. In liquid mixtures containing bismuth subnitrate, sodium bicarbonate reacts with the acid formed by hydrolysis of the bismuth salt.

In solution, sodium bicarbonate has been reported to be incompatible with many drug substances such as ciprofloxacin,[21,22] amiodarone,[23] nicardipine,[24] and levofloxacin.[25]

13 Method of Manufacture

Sodium bicarbonate is manufactured either by passing carbon dioxide into a cold saturated solution of sodium carbonate, or by the ammonia–soda (Solvay) process, in which first ammonia and then carbon dioxide is passed into a sodium chloride solution to precipitate sodium bicarbonate while the more soluble ammonium chloride remains in solution.

14 Safety

Sodium bicarbonate is used in a number of pharmaceutical formulations including injections and ophthalmic, otic, topical, and oral preparations.

Sodium bicarbonate is metabolized to the sodium cation, which is eliminated from the body by renal excretion, and the bicarbonate anion, which becomes part of the body's bicarbonate store. Any carbon dioxide formed is eliminated via the lungs. Administration of excessive amounts of sodium bicarbonate may thus disturb the body's electrolyte balance, leading to metabolic alkalosis or possibly sodium overload with potentially serious consequences. The amount of sodium present in antacids and effervescent formulations has been sufficient to exacerbate chronic heart failure, especially in elderly patients.[26]

Orally ingested sodium bicarbonate neutralizes gastric acid with the evolution of carbon dioxide and may cause stomach cramps and flatulence.

When used as an excipient, sodium bicarbonate is generally regarded as an essentially nontoxic and nonirritant material.

LD_{50} (mouse, oral): 3.36 g/kg[27]
LD_{50} (rat, oral): 4.22 g/kg

15 Handling Precautions

Observe normal precautions appropriate to the circumstances and quantity of material handled. Eye protection and gloves are recommended.

16 Regulatory Status

GRAS listed. Accepted for use as a food additive in Europe. Included in the FDA Inactive Ingredients Database (buccal chewing gums and tablets; intravenous, intramuscular, intraperitoneal, intrathecal, intravitreal, and subcutaneous injections; ophthalmic preparations; oral capsules, solutions, and tablets). Included in parenteral (intravenous infusions and injections; intramuscular and subcutaneous injections) and nonparenteral medicines (chewing gums; ear drops; eye lotions; mouthwashes; oral capsules, chewable tablets, effervescent powders, effervescent tablets, granules, soluble tablets, orodispersible tablets, suspensions, and tablets; suppositories; throat spray; vaginal insert) licensed in the UK. Included in the Canadian Natural Health Products Ingredients Database.

17 Related Substances

Potassium bicarbonate.

18 Comments

Each gram of sodium bicarbonate represents approximately 11.9 mmol of sodium and of bicarbonate. Each gram of sodium bicarbonate will neutralize 12 mEq of gastric acid in 60 minutes.

The yield of carbon dioxide from sodium bicarbonate is approximately 52% by weight.

Three molecules of sodium bicarbonate are required to neutralize one molecule of citric acid, and two molecules of sodium bicarbonate to neutralize one molecule of tartaric acid.

Therapeutically, sodium bicarbonate may be used as an antacid,[28] and as a source of the bicarbonate anion in the treatment of metabolic acidosis. Sodium bicarbonate may also be used as a component of oral rehydration salts and as a source of bicarbonate in dialysis fluids; it has also been suggested as a means of preventing radiocontrast-induced nephrotoxicity.[29]

Recently, sodium bicarbonate has been used as a gas-forming agent in alginate raft systems[30–32] and in floating, controlled-release oral dosage forms for a range of drugs.[33–42] Formulations containing sodium bicarbonate have also been used for fast-dissolving tablets[43,44] and orodispersible tablets.[45] Tablet formulations containing sodium bicarbonate have been shown to increase the absorption of paracetamol,[46,47] and improve the stability of levothyroxine.[48] Sodium bicarbonate has also been included in formulations of vaginal bioadhesive tablets,[49] sustained-release matrix tablets,[50] and in carbon dioxide-releasing suppositories.[51]

Sodium bicarbonate is used in food products as an alkali or as a leavening agent, e.g. baking soda.

A specification for sodium bicarbonate is contained in the *Food Chemicals Codex* (FCC).[52]

The EINECS number for sodium bicarbonate is 205-633-8. The PubChem Compound ID (CID) for sodium bicarbonate includes 516892 and 24192197.

19 Specific References

1 Usui F, Carstensen JT. Interactions in the solid state I: interactions of sodium bicarbonate and tartaric acid under compressed conditions. *J Pharm Sci* 1985; **74**(12): 1293–1297.

2 Anderson NR, *et al.* Quantitative evaluation of pharmaceutical effervescent systems I: design of testing apparatus. *J Pharm Sci* 1982; **71**(1): 3–6.

3 Anderson NR, *et al.* Quantitative evaluation of pharmaceutical effervescent systems II: stability monitoring by reactivity and porosity measurements. *J Pharm Sci* 1982; **71**(1): 7–13.

4 Yanze FM, *et al.* A process to produce effervescent tablets: fluidised bed dryer melt granulation. *Drug Dev Ind Pharm* 2000; **26**(11): 1167–1176.

5 Javaid KA, Cadwallader DE. Dissolution of aspirin from tablets containing various buffering agents. *J Pharm Sci* 1972; **61**(9): 1370–1373.

6 Rainsford KD. Gastric mucosal ulceration induced in pigs by tablets but not suspensions or solutions of aspirin. *J Pharm Pharmacol* 1978; **30**: 129–131.

7 Mason WD, Winer N. Kinetics of aspirin, salicylic acid and salicyluric acid following oral administration of aspirin as a tablet and two buffered solutions. *J Pharm Sci* 1981; 70(3): 262–265.

8 Olsson H, *et al.* Evaluation of the effects of polyethylene glycols of differing molecular weights on the mechanical strength of sodium chloride and sodium bicarbonate tablets. *Int J Pharm* 1998; **171**(1): 31–44.

9 Mattsson S, Nyström C. Evaluation of critical binder properties affecting the compactibility of binary mixtures. *Drug Dev Ind Pharm* 2001; 27(3): 181–194.

10 Allwood MC. The influence of buffering on the stability of erythromycin injection in small-volume infusions. *Int J Pharm* 1992; 80(Suppl.): R7–R9.

11 Doolan KL. Buffering lidocaine with sodium bicarbonate. *Am J Hosp Pharm* 1994; **51**: 2564–2565.

12 Erramouspe J. Buffering local anesthetic solutions with sodium bicarbonate: literature review and commentary. *Hosp Pharm* 1996; **31**(10): 1275–1282.

13 MacKay MW, *et al.* The solubility of calcium and phosphate in two specialty amino acid solutions. *J Parenter Enteral Nutr* 1996; 20: 63–66.

14 Connolly M, *et al.* Freeze crystallization of imipenem. *J Pharm Sci* 1996; **85**(2): 174–177.

15 Shefter E, *et al.* A kinetic study of the solid state transformation of sodium bicarbonate to sodium carbonate. *Drug Dev Commun* 1974; **1**: 29–38.

16 Kuu WY, *et al.* Effect of humidity and temperature on moisture sorption and stability of sodium bicarbonate powder. *Int J Pharm* 1998; **166**(2): 167–175.

17 Ljunggren L, *et al.* Calorimetry a method to be used to characterise pyrolytically decarboxylated bicarbonate and assess its stability at elevated humidities. *Int J Pharm* 2000; **202**(1–2): 71–77.

18 Hadgraft JW, Hewer BD. Molar injection of sodium bicarbonate [letter]. *Pharm J* 1964; **192**: 544.

19 Hadgraft JW. Unsatisfactory infusions of sodium bicarbonate [letter]. *Lancet* 1966; **i**: 603.

20 Smith G. Unsatisfactory infusions of sodium bicarbonate [letter]. *Lancet* 1966; **i**: 658.

21 Gilbert DL, *et al.* Compatibility of ciprofloxacin lactate with sodium bicarbonate during simulated Y- site administration. *Am J Health Syst Pharm* 1997; **54**: 1193–1195.

22 Trissel LA. Concentration-dependent precipitation of sodium bicarbonate with ciprofloxacin lactate [letter]. *Am J Health Syst Pharm* 1996; **53**: 84–85.

23 Korth-Bradley JM, *et al.* Incompatibility of amiodarone hydrochloride and sodium bicarbonate injections [letter]. *Am J Health Syst Pharm* 1995; **52**: 2340.

24 Baaske DM, *et al.* Stability of nicardipine hydrochloride in intravenous solutions. *Am J Health Syst Pharm* 1996; **53**: 1701–1705.

25 Williams NA, *et al.* Stability of levofloxacin in intravenous solutions in polyvinyl chloride bags. *Am J Health Syst Pharm* 1996; **53**: 2309–2313.

26 Panchmatia K, Jolobe OM. Contra-indications of Solpadol [letter]. *Pharm J* 1993; **251**: 73.

27 Lewis RJ, ed. *Sax's Dangerous Properties of Industrial Materials*, 11th edn. New York: Wiley, 2004: 3233.

28 Zoungas S, *et al.* Systematic review: Sodium bicarbonate treatment regimens for the prevention of contrast-induced nephropathy. *Ann Intern Med* 2009; **151**(9): 631–638.

29 Bajdik J, *et al.* Formulation of intelligent tablets with an antacid effect. *Pharm Dev Technol* 2009; **14**(5): 471–475.

30 Johnson FA, *et al.* The effects of alginate molecular structure and formulation variables on the physical characteristics of alginate raft systems. *Int J Pharm* 1997; **159**(1): 35–42.

31 Johnson FA, *et al.* The use of image analysis as a means of monitoring bubble formation in alginate rafts. *Int J Pharm* 1998; **170**(2): 179–185.

32 Choi BY, *et al.* Preparation of alginate beads for floating drug delivery system: effects of carbon dioxide gas-forming agents. *Int J Pharm* 2002; **239**(1–2): 81–91.

33 Singh AV, *et al.* Development, evaluation and optimisation of baclofen oral floating tablet. *Lat Am J Pharm* 2010; **29**: 562–567.

34 Tadros MI. Controlled-release effervescent floating matrix tablets of ciprofloxacin hydrochloride: Development, optimization and *in vitro-in vivo* evaluation in healthy human volunteers. *Eur J Pharm Biopharm* 2010; 74(2): 332–339.

35 Hassan MA, *et al.* Design and evaluation of a floating ranitidine tablet as a drug delivery system for oral application. *J Drug Deliv Sci Technol* 2007; **17**(2): 125–128.

36 Basak SC, *et al.* Design and *in vitro* testing of a floatable gastroretentive tablet of metformin hydrochloride. *Pharmazie* 2007; **62**(2): 145–148.

37 Sungthongjeen S, *et al.* Preparation and *in vitro* evaluation of a multiple-unit floating drug delivery system based on gas formation technique. *Int J Pharm* 2006; **324**(2): 136–143.

38 Hamdani J, *et al. In vitro* and *in vivo* evaluation of floating riboflavin pellets developed using the melt pelletization process. *Int J Pharm* 2006; **323**(1–2): 86–92.

39 Hamdani J, *et al.* Development and *in vitro* evaluation of a novel floating multiple unit dosage form obtained by melt pelletization. *Int J Pharm* 2006; **322**(1–2): 96–103.

40 Tous SS, *et al.* Formulation and *in vitro* evaluation of nitrofurantoin floating matrix tablets. *J Drug Deliv Sci Technol* 2006; **16**(3): 217–221.

41 Rahman Z, *et al.* Design and evaluation of bilayer floating tablets of captopril. *Acta Pharm* 2006; **56**(1): 49–57.

42 Elkheshen SA, *et al. In vitro* and *in vivo* evaluation of floating controlled release dosage forms of verapamil hydrochloride. *Pharmazeutische Industrie* 2004; **66**(11): 1364–1372.

43 Nagendrakumar D, *et al.* Fast dissolving tablets of fexofenadine HCl by effervescent method. *Indian J Pharm Sci* 2009; **71**(2): 116–119.

44 Jacob S, *et al.* Preparation and evaluation of fast-disintegrating effervescent tablets of glibenclamide. *Drug Dev Ind Pharm* 2009; **35**(3): 321–328.

45 Swamy PV. Preparation and evaluation of orodispersible tablets of pheniramine maleate by effervescent method. *Indian J Pharm Sci* 2009; **71**(2): 151–154.

46 Rostami-Hodjegan A, *et al.* A new rapidly absorbed paracetamol tablet containing sodium bicarbonate. I. A four-way crossover study to compare the concentration–time profile of paracetamol from the new paracetamol/sodium bicarbonate tablet and a conventional paracetamol tablet in fed and fasted volunteers. *Drug Dev Ind Pharm* 2002; *28*(5): 523–531.

47 Rostami-Hodjegan A, *et al.* A new rapidly absorbed paracetamol tablet containing sodium bicarbonate. II. Dissolution studies and *in vitro/in vivo* correlation. *Drug Dev Ind Pharm* 2002; **28**(5): 533–543.

48 Patel H, *et al.* The effect of excipients on the stability of levothyroxine pentahydrate tablets. *Int J Pharm* 2003; **264**(1–2): 35–43.

49 Du LD, *et al.* Preparation of actinomycin D vaginal bioadhesive tablets. *Chin J New Drugs* 2007; **16**(5): 387–390.

50 Hamza YE, Aburahma MH. Design and *in vitro* evaluation of novel sustained-release matrix tablets for lornoxicam based on the combination of hydrophilic matrix formers and basic pH-modifiers. *Pharm Dev Technol* 2010; **15**(2): 139–153.

51 Lazzaroni M, *et al.* Role for carbon dioxide-releasing suppositories in the treatment of chronic functional constipation – a double-blind, randomised, placebo-controlled trial. *Clin Drug Investig* 2005; **25**(8): 499–505.

52 *Food Chemicals Codex*, 7th edn. Bethesda, MD: United States Pharmacopeia, 2010.

20 General References

Hannula A-M, *et al.* Release of ibuprofen from hard gelatin capsule formulations: effect of sodium bicarbonate as a disintegrant. *Acta Pharm Fenn* 1989; **98**: 131–134.

Sendall FEJ, *et al.* Effervescent tablets. *Pharm J* 1983; **230**: 289–294.

Travers DN, White RC. The mixing of micronized sodium bicarbonate with sucrose crystals. *J Pharm Pharmacol* 1971; **23**: 260S–261S.

21 Author

CG Cable.

22 Date of Revision

1 March 2012.

Sodium Borate

1 Nonproprietary Names

BP: Borax
JP: Sodium Borate
PhEur: Borax
USP–NF: Sodium Borate

2 Synonyms

Borax decahydrate; boric acid disodium salt; E285; natrii tetraboras; natrii tetraboras decahydricus;sodium biborate decahydrate; sodium pyroborate decahydrate; sodium tetraborate decahydrate.

3 Chemical Name and CAS Registry Number

Disodium tetraborate decahydrate [1303-96-4]

4 Empirical Formula and Molecular Weight

$Na_2B_4O_7 \cdot 10H_2O$ 381.37

5 Structural Formula

See Section 4.

6 Functional Category

Alkalizing agent; antimicrobial preservative; buffering agent; emulsifying agent.

7 Applications in Pharmaceutical Formulation or Technology

Sodium borate is used in pharmaceutical formulations as an antimicrobial preservative in ointments and topical creams, and as an emulsifying agent in creams.[1] It has also been used in lozenges, mouthwashes, otic preparations (0.3% w/v), and ophthalmic solutions (0.03–1.0% w/v).

Sodium borate is used in cosmetics such as moisturizers, deodorants, and shampoos. It is also used in foods as an antimicrobial preservative.

8 Description

Sodium borate occurs as white, hard crystals, granules, or crystalline powder. It is odorless and efflorescent.

9 Pharmacopeial Specifications

See Table I.

Table I: Pharmacopeial specifications for sodium borate.

Test	JP XV	PhEur 7.4	USP35–NF30
Identification	+	+	+
Characters	—	+	—
Carbonate and bicarbonate	+	—	+
Color of solution	+	+	—
pH	9.1–9.6	9.0–9.6	—
Heavy metals	≤20 ppm	≤25 ppm	≤20 ppm
Arsenic	≤5 ppm	≤5 ppm	—
Calcium	—	≤100 ppm	—
Ammonium	—	≤10 ppm	—
Sulfates	—	≤50 ppm	—
Assay	99.0–103.0%	99.0–103.0%	99.0–105.0%

10 Typical Properties

Acidity/alkalinity pH = 9.0–9.6 (4% w/v aqueous solution)
Density 1.73 g/cm^3
Melting point 75°C when rapidly heated. At 100°C it loses $5H_2O$; at 150°C it loses $9H_2O$; and at 320°C it becomes anhydrous. At about 880°C the substance melts into a glassy state: 'borax beads'.
Solubility 1 in 1 of glycerin; 1 in 1 of boiling water; 1 in 16 of water; practically insoluble in ethanol (95%), ethanol (99.5%), and diethyl ether.

11 Stability and Storage Conditions

Sodium borate should be stored in a well-closed container in a cool, dry place. *See also* Section 18.

12 Incompatibilities

Sodium borate is incompatible with acids, metallic and alkaloidal salts, and strong reducing agents.

13 Method of Manufacture

Sodium borate can be prepared from minerals such as borosodium calcite, pandermite, or tinkal; these are natural sodium or calcium borates. Treatment of the mineral with sodium carbonate and sodium hydrogencarbonate yields the sodium borate decahydrate. In the US, brine from salt lakes is also an important source of sodium borate.[2]

14 Safety

Sodium borate has weak bacteriostatic and astringent properties. Historically, sodium borate has been used as a disinfectant in skin lotions and eye-, nose-, and mouthwashes. However, boric acid is easily absorbed via mucous membranes and damaged skin, and severe toxicity has been observed, especially in babies and children.[3] Consequently, the use of sodium borate as a disinfectant is now considered somewhat obsolete and careful use is recommended. The toxic effects of sodium borate include vomiting, diarrhea, erythema, CNS depression, and kidney damage. The lethal oral intake is approximately 20 g in adults and 5 g in children.[4]

LD_{50} (guinea pig, oral): 5.33 g/kg[4,5]
LD_{50} (mouse, IP): 2.711 g/kg
LD_{50} (mouse, IV): 1.320 g/kg
LD_{50} (mouse, oral): 2.0 g/kg
LD_{50} (rat, oral): 5.66 g/kg

15 Handling Precautions

Observe normal precautions appropriate to the circumstances and the quantity of material handled; do not combine with acids.

16 Regulatory Status

Accepted for use as a food additive in Europe. Included in the FDA Inactive Ingredients Database (otic preparations; ophthalmic solutions and suspensions). Included in nonparenteral medicines licensed in the UK, Italy, France, Germany, and Japan. Included in the Canadian Natural Health Products Ingredients Database.

17 Related Substances

Boric acid; sodium borate anhydrous.

S

Sodium borate anhydrous
Synonyms Borax glass; disodium tetraborate anhydrous; fused borax; fused sodium borate; sodium pyroborate; sodium tetraborate anhydrous.
Empirical formula $Na_2B_4O_7$
Molecular weight 201.2
CAS number [1330-43-4]
Boiling point 1575°C (decomposes)
Melting point 741°C
Solubility Slightly soluble in glycerin and water; practically insoluble in ethanol (95%).
Specific gravity 2.367
Comments The EINECS number for sodium borate anhydrous is 215-540-4.

18 Comments

Commercially available sodium borate decahydrate is usually present as monoclinic prismatic crystals that become opaque on the surface in dry air. In addition to the decahydrate, a pentahydrate exists; this is also known as 'jeweller's borax'. The anhydrous substance is also available and is called 'pyroborax'.

Sodium borate has been investigated in the prevention of crystal formation in freeze-dried solutions.[6]

Therapeutically, sodium borate has been used externally as a mild astringent. Preparations of sodium borate in honey have historically been used as paints for the throat, tongue, and mouth, but such use is now inadvisable because of concerns about toxicity in such applications; *see* Section 14.

The EINECS number for sodium borate is 271-536-2. The PubChem Compound ID (CID) for sodium borate is 11954323.

19 Specific References

1 Prince LM. Beeswax/borax reaction in cold creams. *Cosmet Perfum* 1974; 89(May): 47–49.
2 Lyday PA. Boron. In: *Mineral Yearbook*. 1. Washington, DC: US Department of the Interior US Geological Survey, 1992; 249.
3 Gordon AS, *et al*. Seizure disorders and anemia associated with chronic borax intoxication. *Can Med Assoc J* 1973; 108: 719–721724.
4 Lewis RJ. *Sax's Dangerous Properties of Industrial Materials*, 11th edn. New York: Wiley, 2004; 3234.
5 Smyth HF, *et al*. Range-finding toxicity data: list VII. *Am Ind Hyg Assoc J* 1969; 30(5): 470–476.
6 Izutsu K, *et al*. Effects of sodium tetraborate and boric acid on nonisothermal mannitol crystallization in frozen solutions and freeze-dried solids. *Int J Pharm* 2004; 273(1): 85–93.

20 General References

—

21 Author

H Ehlers.

22 Date of Revision

1 March 2012.

Sodium Carbonate

1 Nonproprietary Names

BP: Anhydrous Sodium Carbonate
JP: Dried Sodium Carbonate
PhEur: Sodium Carbonate, Anhydrous
USP–NF: Sodium Carbonate

2 Synonyms

Bisodium carbonate; calcined soda; carbonic acid disodium salt; Cenzias de Soda; crystol carbonate; disodium carbonate; E500i; *Gran-Plus*; natrii carbonas anhydricus; *Novacarb*; soda ash; soda calcined; *Sodium Carbonate IPH*.

3 Chemical Name and CAS Registry Number

Sodium carbonate anhydrous [497-19-8]
Sodium carbonate monohydrate [5968-11-6]
Sodium carbonate decahydrate [6132-02-1]

4 Empirical Formula and Molecular Weight

Na_2CO_3	105.9
$Na_2CO_3{\cdot}H_2O$	124.0
$Na_2CO_3{\cdot}10H_2O$	286.1

5 Structural Formula

See Section 4.

6 Functional Category

Alkalizing agent; buffering agent; dispersing agent; tablet and capsule diluent.

7 Applications in Pharmaceutical Formulation or Technology

Sodium carbonate is used as an alkalizing agent in injectable, ophthalmic, oral, and rectal formulations.[1]

In effervescent tablets or granules, sodium carbonate is used in combination with an acid, typically citric acid or tartaric acid.[2] When the tablets or granules come into contact with water, an acid–base reaction occurs in which carbon dioxide gas is produced and the product disintegrates.[3] Raw materials with low moisture contents are required to prevent the early triggering of the effervescent reaction.[3]

As an alkalizing agent, concentrations of sodium carbonate between 2% and 5% w/w are used in compressed tablet formulations.[2,4] As an effervescent agent, concentrations of sodium carbonate up to 10% w/w can be used.[3]

8 Description

Sodium carbonate occurs as a white, almost white, or colorless inorganic salt, produced as crystalline powder or granules. It is hygroscopic and odorless.

9 Pharmacopeial Specifications

See Table I.

Table I: Pharmacopeial specifications for sodium carbonate.

Test	JP XV	PhEur 7.4	USP35–NF30
Identification	+	+	+
Characters	—	+	—
Appearance of solution	+	+	—
Alkali hydroxides and bicarbonates	—	+	—
Chlorides	≤0.071%	≤125 ppm	—
Sulfates	—	≤250 ppm	—
Arsenic	≤3.1 ppm	≤5 ppm	—
Iron	—	≤50 ppm	—
Heavy metals	≤20 ppm	≤50 ppm	≤10 ppm
Loss on drying	≤2.0%	≤1.0%	≤0.5%
Assay (dried basis)	>99.0%	99.5–100.5%	99.5–100.5%

10 Typical Properties

Acidity/alkalinity

Strongly alkaline;

pH = 11.37 (1% w/v aqueous solution at 25°C);

pH = 11.58 (5% w/v aqueous solution at 25°C);

pH = 11.70 (10% w/v aqueous solution at 25°C).[5] *See also* Figure 1.[6]

Hygroscopicity One mole of sodium carbonate will gradually absorb 1 mole of water (approximately 15%) on exposure to air.

Melting point 851°C

Refractive index n_D^{20} = 1.3341 at 1.0% w/w solution; 1.3440 at 5.0% w/w solution; 1.3547 at10.0% w/w solution.[7]

Solubility Freely soluble in water, with solubility initially increasing with temperature and then settling at 30.8% w/w above 80°C[6] (*see* Figure 2). Soluble in glycerin; practically insoluble in ethanol (95%).

Specific gravity 2.53

Spectroscopy

IR spectra *see* Figure 3.

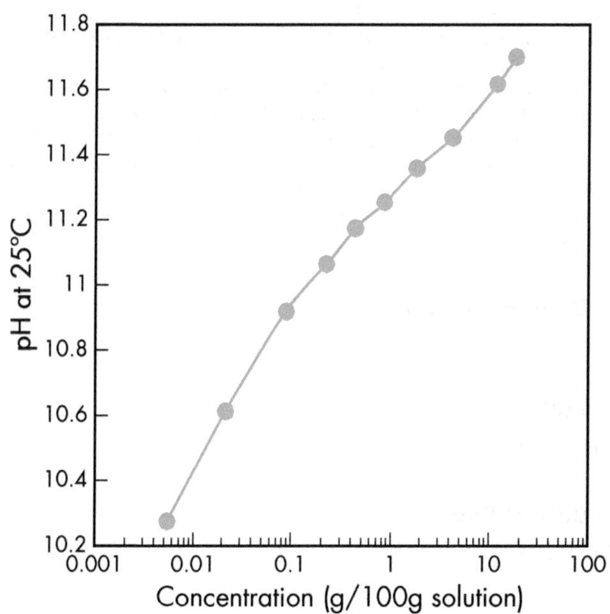

Figure 1: pH of sodium carbonate in water at 25°C.[5] Adapted with permission.

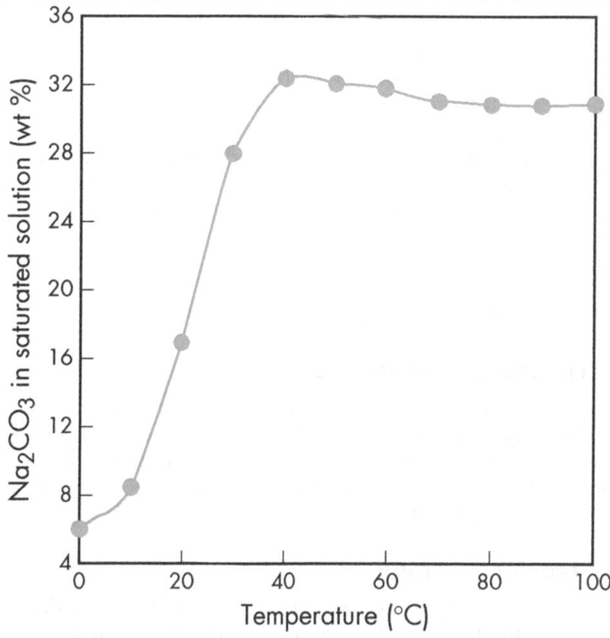

Figure 2: Solubility of sodium carbonate in water.[6] Adapted with permission.

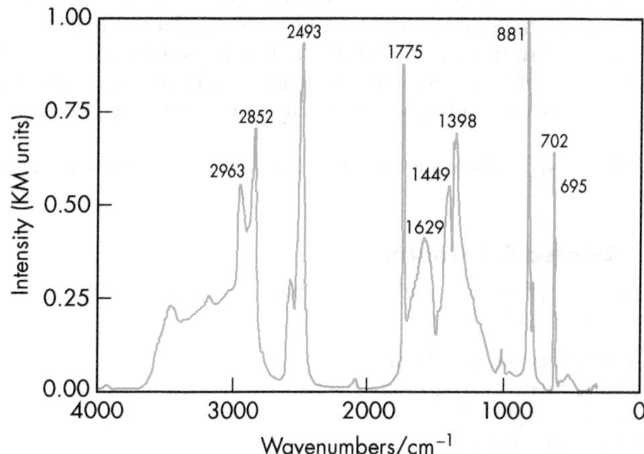

Figure 3: Infrared spectrum of sodium carbonate measured by diffuse reflectance. Adapted with permission of Informa Healthcare.

11 Stability and Storage Conditions

Sodium carbonate converts to the monohydrate form when in contact with water and produces heat. It begins to lose carbon dioxide at temperatures above 400°C[8] and decomposes before boiling. Store in airtight containers.

12 Incompatibilities

Sodium carbonate decomposes when in contact with acids in the presence of water to produce carbon dioxide and effervescence. It may react violently with aluminum, phosphorous pentoxide, sulfuric acid, fluorine, and lithium.

13 Method of Manufacture

Sodium carbonate is produced by the ammonia-soda process, also known as the Solvay process.[6,8]

14 Safety

Sodium carbonate is used in injectable, oral, and rectal pharmaceutical formulations. The pure form of sodium carbonate is mildly toxic by ingestion, moderately toxic by inhalation and SC routes, and very toxic by the IP route. It is irritating to the skin and eyes. Dust and vapors of sodium carbonate may irritate mucous membranes, causing coughing and shortness of breath. It also has experimental reproductive effects.

Sodium carbonate can migrate to food from packaging materials. When used as an excipient or antacid, sodium carbonate is generally regarded as a nontoxic and nonirritating material.

LD_{50} (mouse, IP): 0.12 g/kg[9]
LD_{50} (mouse, SC): 2.21 g/kg
LD_{50} (rat, oral): 4.09 g/kg

15 Handling Precautions

Observe normal precautions appropriate to the circumstances and quantity of the material handled. When heated to decomposition it emits toxic fumes of sodium oxide. Eye protection and gloves are recommended. Respiratory protection is also recommended if inhalable dust is present.

16 Regulatory Status

GRAS listed. Accepted for use as a food additive in Europe. Included in the FDA Inactive Ingredients Database (intra-arterial, intramuscular, intraperitoneal, intrapleural, intratumor, intravascular, intravenal, and nerve block injections; ophthalmic solutions; oral granules, capsules, buccal chewing gum, and tablets; rectal suspensions). Included in the Canadian Natural Health Products Ingredients Database. Included in parenteral (powder for solution for injection) and nonparenteral medicines (oral effervescent tablets, soluble tablets, granules, lozenges, chewing gums) licensed in the UK.

USP35–NF30 allows either the anhydrous or the monohydrate form.

17 Related Substances

Sodium bicarbonate; sodium carbonate decahydrate; sodium carbonate monohydrate.

Sodium carbonate decahydrate
Empirical formula $Na_2CO_3 \cdot 10H_2O$
Molecular weight 286.1
CAS number [6132-02-1]
Description Colorless, transparent, or white crystals or powder.
Solubility Freely soluble in water; practically insoluble in ethanol (95%).
Comments Listed in PhEur 7.4 and JP XV. Used in alkaline baths.[10]

Sodium carbonate monohydrate
Empirical formula $Na_2CO_3 \cdot H_2O$
Molecular weight 124.0
CAS number [5968-11-6]
Description Colorless or white crystals or granules.
Solubility Soluble in 3 parts water, 1.8 parts boiling water, or 7 parts glycerin; practically insoluble in ethanol (95%). Dries out in warm dry air or above 50°C, and converts to anhydrous form above 100°C.
Comments Listed in PhEur 7.4 and USP35–NF30. Commonly used in antacid preparations and as a reagent.[10]

18 Comments

Sodium carbonate is more resistant to effervescent reactions than sodium bicarbonate, and therefore sodium bicarbonate is most commonly used in effervescent formulations.[3] Sodium carbonate can be added to these formulations as a stabilizing agent (up to 10% w/w) as it absorbs moisture, preventing early effervescent reactions.[3] This effect is exploited in *Effer-Soda*, in which a sodium bicarbonate core is protected by a surface layer of sodium carbonate, equivalent to 8–12% w/w.[11]

The technical grade of sodium carbonate anhydrous (approximately 99% purity) is known as soda ash.[8] *Sodium Carbonate IPH* (Solvay Pharmaceuticals) is not authorized for parenteral formulations.

Sodium carbonate has been investigated as part of a controlled-release, taste-masking formulation,[12] and in a novel effervescent dry powder inhaler formulation.[13]

Therapeutically, sodium carbonate is used as an oral antacid.[10]

A specification for sodium carbonate is contained in the *Food Chemicals Codex* (FCC).[14]

The EINECS number for sodium carbonate is 207-838-8. The PubChem Compound ID (CID) for sodium carbonate is 10340.

19 Specific References

1 He W, *et al.* Influences of sodium carbonate on physicochemical properties of lansoprazole in designed multiple coating pellets. *AAPS PharmSciTech* 2010; **11**(3): 1287–1293.
2 Niazi S. Compressed solid dosage formulations. In: Niazi SK, ed. *Handbook of Pharmaceutical Manufacturing Formulations.* 1. Part II. Boca Raton FL: CRC Press, 2004.
3 Bertuzzi D. Effervescent granulation. In: Parikh D, ed. *Handbook of Pharmaceutical Granulation Technology*, 2nd edn. Boca Raton FL: Taylor and Francis, 2005; 365.
4 Badawy S, *et al.* Effect of processing and formulation variables on the stability of a salt of a weakly basic drug candidate. *Pharm Dev Technol* 2004; 9: 239–245.
5 Thieme C. Sodium Carbonates. In: *Ullmann's Encyclopedia of Industrial Chemistry.* Wiley-VCH Verlag GmbH & Co. KGaA, 2000.
6 Eggeman T. Sodium carbonate. In: *Kirk-Othmer Encyclopedia of Chemical Technology*, 5th edn, 22. New York: Wiley, 2001; 787–797.
7 Lide DR, ed. *CRC Handbook of Chemistry and Physics*, 88th edn. Boca Raton FL: CRC Press/Taylor and Francis, 2008; 8–52.
8 O'Neil MJ, ed. *Merck Index: An Encyclopedia of Chemicals, Drugs and Biologicals*, 14th edn. Whitehouse Station NJ: Merck, 2006; 1480–1481.
9 Lewis RJ, ed. *Sax's Dangerous Properties of Industrial Chemicals*, 11th edn. New York: Wiley, 2004; 3236.
10 Sweetman SC, ed. *Martindale: the Complete Drug Reference*, 37th edn. London: Pharmaceutical Press, 2011; 2624–2625.
11 SPI Pharma. Technical Bulletin No. 117/0300: *Effer-Soda*, 2007.
12 Yoshida T, *et al.* Mechanism of controlled drug release from a salting-out taste-masking system. *J Control Release* 2008; **131**(1): 47–53.
13 Ely L, *et al.* Effervescent dry powder for respiratory drug delivery. *Eur J Pharm Biopharm* 2007; **65**(3): 346–353.
14 *Food Chemicals Codex*, 67th edn. Bethesda, MD: United States Pharmacopeia, 2010.

20 General References

—

21 Author

KP Hapgood.

22 Date of Revision

1 March 2012.

Sodium Chloride

1 Nonproprietary Names

BP: Sodium Chloride
JP: Sodium Chloride
PhEur: Sodium Chloride
USP–NF: Sodium Chloride

2 Synonyms

Alberger; chlorure de sodium; common salt; hopper salt; natrii chloridum; natural halite; rock salt; saline; salt; sea salt; table salt.

3 Chemical Name and CAS Registry Number

Sodium chloride [7647-14-5]

4 Empirical Formula and Molecular Weight

NaCl 58.44

5 Structural Formula

See Section 4.

6 Functional Category

Tablet and capsule diluent; tonicity agent.

7 Applications in Pharmaceutical Formulation or Technology

Sodium chloride is widely used in a variety of parenteral and nonparenteral pharmaceutical formulations, where the primary use is to produce isotonic solutions.

Sodium chloride has been used as a lubricant and diluent in capsules and direct-compression tablet formulations in the past,[1–5] although this practice is no longer common. Sodium chloride has also been used as a channeling agent[6,7] and as an osmotic agent[8,9] in the cores of controlled-release tablets. It has been used as a porosity modifier in tablet coatings,[10] and to control drug release from microcapsules.[11,12]

The addition of sodium chloride to aqueous spray-coating solutions containing hydroxypropyl cellulose or hypromellose suppresses the agglomeration of crystalline cellulose particles.[13] Sodium chloride can also be used to modify drug release from gels[14] and from emulsions.[15] It can be used to control micelle size,[16–18] and to adjust the viscosity of polymer dispersions by altering the ionic character of a formulation.[19,20]

See Table I.

Table I: Uses of sodium chloride.

Use	Concentration (%)
Capsule diluent	10–80
Controlled flocculation of suspensions	≤1
Direct compression tablet diluent	10–80
To produce isotonic solutions in intravenous or ophthalmic preparations	≤0.9
Water-soluble tablet lubricant	5–20

8 Description

Sodium chloride occurs as a white crystalline powder or colorless crystals; it has a saline taste. The crystal lattice is a face-centered cubic structure. Solid sodium chloride contains no water of crystallization although, below 0°C, salt may crystallize as a dihydrate.

9 Pharmacopeial Specifications

The pharmacopeial specifications for sodium chloride have undergone harmonization of many attributes for JP, PhEur, and USP–NF.

See Table II. *See also* Section 18.

10 Typical Properties

Acidity/alkalinity pH = 6.7–7.3 (saturated aqueous solution)

Angle of repose 38° for cubic crystals

Boiling point 1413°C

Compressibility With sodium chloride powder of less than 30 μm particle size, tablets are formed by plastic deformation; above this size, both plastic deformation and fracture occur.[1,3,4]

SEM 1: Excipient: sodium chloride, powder; manufacturer: Mallinckrodt Speciality Chemicals Co.; magnification: 600×.

SEM 2: Excipient: sodium chloride, granular; manufacturer: Van Waters & Rogers, Inc.; magnification: 120×.

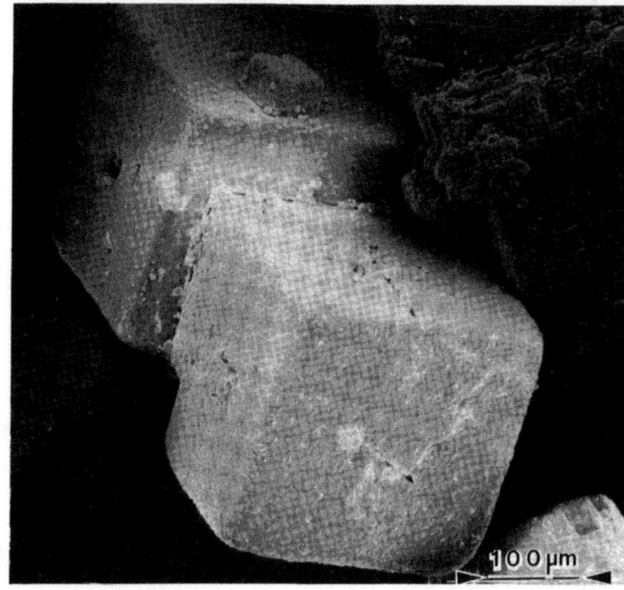

SEM 3: Excipient: sodium chloride, granular; manufacturer: Van Waters & Rogers, Inc.; magnification: 600×.

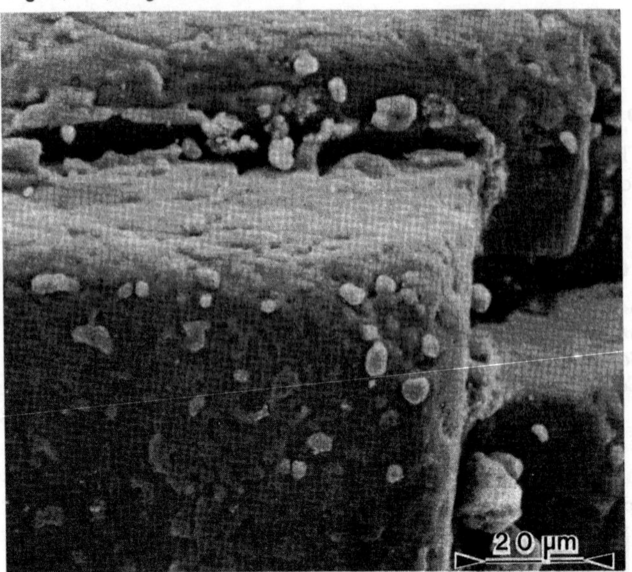

Table II: Pharmacopeial specifications for sodium chloride.

Test	JP XV	PhEur 7.4	USP35–NF30
Identification	+	+	+
Characters	+	+	−
Appearance of solution	+	+	+
Acidity or alkalinity	+	+	+
Loss on drying	≤0.5%	≤0.5%	≤0.5%
Arsenic	≤2 ppm	≤1 ppm	≤1 ppm
Bromides	+	≤100 ppm	≤100 ppm
Chloride	−	−	+
Barium	+	+	+
Nitrites	−	+	+
Aluminum	−	≤0.2 ppm[a]	≤0.2 µg/g[a]
Magnesium and alkaline earth metals	+	≤100 ppm	≤100 ppm
Iodide	+	+	+
Iron	+	≤2 ppm	≤2 ppm
Sulfate	+	≤200 ppm	≤200 ppm
Ferrocyanides	+	+	+
Heavy metals[c]	≤3 ppm	≤5 ppm	≤5 ppm
Phosphate	+	≤25 ppm	≤25 ppm
Potassium	−	≤500 ppm[a] [b]	≤500 ppm[a] [b]
Sterility	−	−	+
Bacterial endotoxins[c]	−	≤5 IU/g[b]	+
Assay (dried basis)	99.0–100.5%	99.0–100.5%	99.0–100.5%

(a) If for use in peritoneal dialysis, hemodialysis or hemofiltration solutions.
(b) If for parenteral use.
(c) These tests have not been fully harmonized at the time of publication.

See also Figure 1.

Density

2.17 g/cm^3;

1.20 g/cm^3 for saturated aqueous solution.

Density (bulk) 0.93 g/cm^3

Density (tapped) 1.09 g/cm^3

Dielectric constant 5.9 at 1 MHz

Freezing point depression *see* Table III.

Hardness (Mohs) 2–2.5

Hygroscopicity Hygroscopic above 75% relative humidity.

Melting point 804°C

Osmolarity A 0.9% w/v aqueous solution is iso-osmotic with serum.

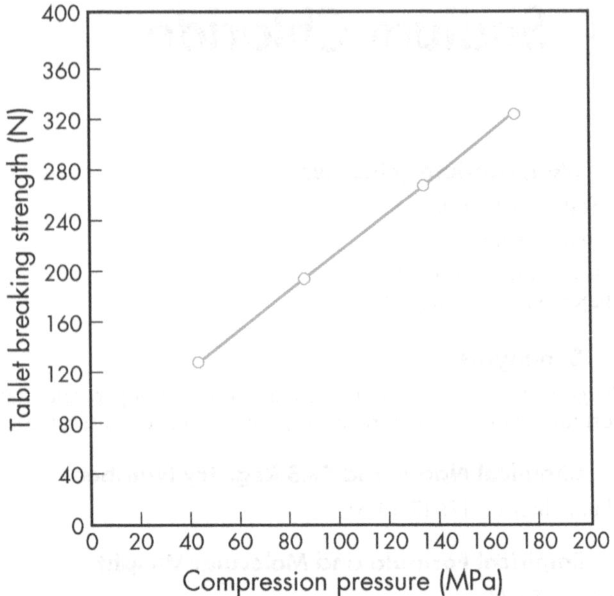

Figure 1: Compression characteristics of sodium chloride (cubic crystals).[3] Tablet diameter = 12 mm.

Table III: Freezing point depression values of aqueous sodium chloride.

Aqueous sodium chloride solution (% w/v)	Freezing point depression (°C)
11.69	6.90
17.53	10.82
23.38	15.14
30.39	21.12

Table IV: Solubility of sodium chloride.

Solvent	Solubility at 20°C unless otherwise stated
Ethanol	Slightly soluble
Ethanol (95%)	1 in 250
Glycerin	1 in 10
Water	1 in 2.8
	1 in 2.6 at 100°C

Refractive index n_D^{20} = 1.343 for a 1 M aqueous solution.

Solubility *see* Table IV.

Thermal conductivity 1.15 Wm/K at 273 K

Specific heat capacity 854 J/kg/K

Spectroscopy

NIR spectra *see* Figure 2.

Vapor pressure

133.3 Pa at 865°C for solid;

1759.6 Pa at 20°C for a saturated aqueous solution (equivalent to 75.3% relative humidity).

Viscosity A 10% w/v solution has a viscosity of 1.19 mPa s (1.19 cP).

11 Stability and Storage Conditions

Aqueous sodium chloride solutions are stable but may cause the separation of glass particles from certain types of glass containers. Aqueous solutions may be sterilized by autoclaving or filtration. The solid material is stable and should be stored in a well-closed container, in a cool, dry place.

It has been shown that the compaction characteristics and the mechanical properties of tablets are influenced by the relative humidity of the storage conditions under which sodium chloride was kept.[21,22]

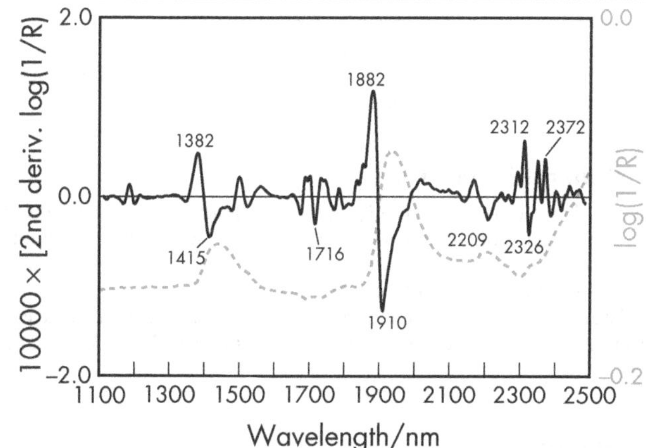

Figure 2: Near-infrared spectrum of sodium chloride measured by reflectance. Sodium chloride does not absorb in the near-infrared region; however, it will generally show some peaks due to traces of moisture (approx. 1450 nm and 1950 nm).

12 Incompatibilities

Aqueous sodium chloride solutions are corrosive to iron. They also react to form precipitates with silver, lead, and mercury salts. Strong oxidizing agents liberate chlorine from acidified solutions of sodium chloride. The solubility of the antimicrobial preservative methylparaben is decreased in aqueous sodium chloride solutions[23] and the viscosity of carbomer gels and solutions of hydroxyethyl cellulose or hydroxypropyl cellulose is reduced by the addition of sodium chloride.

13 Method of Manufacture

Sodium chloride occurs naturally as the mineral halite. Commercially, it is obtained by the solar evaporation of sea water, by mining, or by the evaporation of brine from underground salt deposits.

14 Safety

Sodium chloride is the most important salt in the body for maintaining the osmotic tension of blood and tissues. About 5–12 g of sodium chloride is consumed daily, in the normal adult diet, and a corresponding amount is excreted in the urine. As an excipient, sodium chloride may be regarded as an essentially nontoxic and nonirritant material. However, toxic effects following the oral ingestion of 0.5–1.0 g/kg body-weight in adults may occur. The oral ingestion of larger quantities of sodium chloride, e.g. 1000 g in 600 mL of water,[24] is harmful and can induce irritation of the gastrointestinal tract, vomiting, hypernatremia, respiratory distress, convulsions, or death.

In rats, the minimum lethal intravenous dose is 2.5 g/kg body-weight.

LD_{50} (mouse, IP): 6.61 g/kg[25]
LD_{50} (mouse, IV): 0.65 g/kg
LD_{50} (mouse, oral): 4.0 g/kg
LD_{50} (mouse, SC): 3.0 g/kg
LD_{50} (rat, oral): 3.0 g/kg

15 Handling Precautions

Observe normal precautions appropriate to the circumstances and quantity of material handled. If heated to high temperatures, sodium chloride evolves a vapor irritating to the eyes.

16 Regulatory Status

GRAS listed. Included in the FDA Inactive Ingredients Database (injections; inhalations; nasal, ophthalmic, oral, otic, rectal, and topical preparations). Included in nonparenteral and parenteral medicines licensed in the UK. Included in the Canadian Natural Health Products Ingredients Database.

17 Related Substances

Potassium chloride.

18 Comments

Sodium chloride has undergone harmonization of many attributes for JP, PhEur, and USP–NF by the Pharmacopeial Discussion Group. For further information see the General Information Chapter <1196> in the USP35–NF30, the General Chapter 5.8 in PhEur 7.4, along with the 'State of Work' document on the PhEur EDQM website, and also the General Information Chapter 8 in the JP XV.

Domestic table salt may contain sodium iodide (as a prophylactic substance against goiter) and agents such as magnesium carbonate, calcium phosphate, or starch, which reduce the hygroscopic characteristics of the salt and maintain the powder in a free-flowing state.

Food-grade dendritic salt, which is porous, can be used as an absorbent for liquid medications, and as a tablet diluent in specific formulations.

Each gram of sodium chloride represents approximately 17.1 mmol of sodium and 17.1 mmol of chloride; 2.54 g of sodium chloride is approximately equivalent to 1 g of sodium.

A saturated solution of sodium chloride can be used as a constant-humidity solution; at 25°C, a relative humidity of 75% is produced. A specification for sodium chloride is contained in the *Food Chemicals Codex* (FCC).[26]

The EINECS number for sodium chloride is 231-598-3. The PubChem Compound ID (CID) for sodium chloride is 5234.

19 Specific References

1 Leigh S, *et al.* Compression characteristics of some pharmaceutical materials. *J Pharm Sci* 1967; **56**: 888–892.
2 Rees JE, Shotton E. Some observations on the ageing of sodium chloride compacts. *J Pharm Pharmacol* 1970; **22**: 17S–23S.
3 Shotton E, Obiorah BA. The effect of particle shape and crystal habit on the properties of sodium chloride. *J Pharm Pharmacol* 1973; **25**: 37P–43P.
4 Roberts RJ, *et al.* Brittle-ductile transitions in die compaction of sodium chloride. *Chem Eng Sci* 1989; **44**: 1647–1651.
5 Hammouda Y, *et al.* The use of sodium chloride as a directly compressible filler. Part III: Drug-to-filler ratio. *Pharm Ind* 1978; 40(9): 987–992.
6 González-Rodriguez ML, *et al.* Design and evaluation of a new central core matrix tablet. *Int J Pharm* 1997; **146**: 175–180.
7 Korsatko-Wabnegg B. [Development of press-coated tablets with controlled release effect using poly-D-(–)-3-hydroxybutyric acid.] *Pharmazie* 1990; **45**: 842–844[in German].
8 Moussa IS, Cartilier LH. Evaluation of crosslinked amylose press-coated tablets for sustained drug delivery. *Int J Pharm* 1997; **149**: 139–149.
9 Özdemir N, Sahin J. Design of a controlled release osmotic pump system of ibuprofen. *Int J Pharm* 1997; **158**: 91–97.
10 Shivanand P, Sprockel OL. A controlled porosity drug delivery system. *Int J Pharm* 1998; **167**: 83–96.
11 Tirkkonen S, Paronen P. Enhancement of drug release from ethylcellulose microcapsules using solid sodium chloride in the wall. *Int J Pharm* 1992; **88**: 39–51.
12 Tirkkonen S, Paronen P. Release of indomethacin from tabletted ethylcellulose microcapsules. *Int J Pharm* 1993; **92**: 55–62.
13 Yuasa H, *et al.* Suppression of agglomeration in fluidized bed coating I. Suppression of agglomeration by adding sodium chloride. *Int J Pharm* 1997; **158**: 195–201.
14 Pandit NK, Wang D. Salt effects on the diffusion and release rate of propranolol from poloxamer 407 gels. *Int J Pharm* 1998; **167**: 183–189.
15 Mishra B, Pandit JK. Multiple water-oil-water emulsions as prolonged release formulations of pentazocine. *J Control Release* 1990; **14**: 53–60.

S

16 Shah D, *et al.* Coacervate formation by inorganic salts with benzalkonium chloride. *J Pharm Sci* 1973; **62**: 1741–1742.
17 Richard AJ. Ultracentrifugal study of effect of sodium chloride on micelle size of fusidate sodium. *J Pharm Sci* 1975; **64**: 873–875.
18 McDonald C, Richardson C. The effect of added salts on solubilization by a non-ionic surfactant. *J Pharm Pharmacol* 1981; **33**: 38–39.
19 Mattha AG. Rheological studies on *Plantago albicans* (*Psyllium*) seed gum dispersions II: effect of some pharmaceutical additives. *Pharm Acta Helv* 1977; **52**: 214–217.
20 Okor RS. The effect of phenol on the electrolyte flocculation of certain polymeric dispersions to thixotropic gels. *Pharm Res* 1993; **10**: 220–222.
21 Elamin AA, *et al.* The effect of pre-compaction processing and storage conditions on powder and compaction properties of some crystalline materials. *Int J Pharm* 1994; **108**: 213–224.
22 Ahlneck C, Alderborn G. Moisure adsorption and tabletting. II. The effect on tensile strength and air permeability of the relative humidity during storage of tablets of 3 crystalline materials. *Int J Pharm* 1989; **56**: 143–150.
23 McDonald C, Lindstrom RE. The effect of urea on the solubility of methyl *p*-hydroxybenzoate in aqueous sodium chloride solution. *J Pharm Pharmacol* 1974; **26**: 39–45.
24 Calam J, *et al.* Extensive gastrointestinal damage following a saline emetic. *Dig Dis Sci* 1982; **27**: 936–940.
25 Lewis RJ, ed. *Sax's Dangerous Properties of Industrial Materials*, 11th edn. New York: Wiley, 2004; 3238–3239.
26 *Food Chemicals Codex*, 7th edn. Suppl. 3. Bethesda, MD: United States Pharmacopeia, 2011; 1708.

20 General References

European Directorate for the Quality of Medicines and Healthcare (EDQM). European Pharmacopoeia – State Of Work Of International Harmonisation. *Pharmeuropa* 2011; 23(4): 713–714. www.edqm.eu/site/-614.html (accessed 5 December 2011).

Heng PW, *et al.* Influence of osmotic agents in diffusion layer on drug release from multilayer coated pellets. *Drug Dev Ind Pharm* 2004; 30(2): 213–220.

21 Author

JS Maximilien.

22 Date of Revision

1 March 2012.

Sodium Citrate Dihydrate

1 Nonproprietary Names

BP: Sodium Citrate
JP: Sodium Citrate Hydrate
PhEur: Sodium Citrate
USP-NF: Sodium Citrate

2 Synonyms

Citric acid trisodium salt; E331; natrii citras; sodium citrate tertiary; trisodium citrate.

3 Chemical Name and CAS Registry Number

Trisodium 2-hydroxypropane-1,2,3-tricarboxylate dihydrate [6132-04-3]

4 Empirical Formula and Molecular Weight

$C_6H_5Na_3O_7{\cdot}2H_2O$ 294.10

5 Structural Formula

6 Functional Category

Alkalizing agent; buffering agent; complexing agent; emulsifying agent.

7 Applications in Pharmaceutical Formulation or Technology

Sodium citrate, as either the dihydrate or anhydrous material, is widely used in pharmaceutical formulations; *see* Table I. It is also used as a complexing agent to bind metal ions in solution. The anhydrous material is used in effervescent tablet formulations.(1)

Table I: Uses of sodium citrate dihydrate.

Use	Concentration (%)
Buffering agent	0.3–2.0
Injections	0.02–4.0
Ophthalmic solutions	0.1–2.0
Complexing agent	0.3–2.0

8 Description

Sodium citrate dihydrate consists of odorless, colorless, monoclinic crystals, or a white crystalline powder with a cooling, saline taste. It is slightly deliquescent in moist air, and in warm dry air it is efflorescent. Although most pharmacopeias specify that sodium citrate is the dihydrate, the USP35–NF30 states that sodium citrate may be either the dihydrate or anhydrous material.

9 Pharmacopeial Specifications

See Table II.

10 Typical Properties

Acidity/alkalinity pH = 7.0–9.0 (5% w/v aqueous solution)
Density (bulk) 1.12 g/cm^3
Density (tapped) 0.99 g/cm^3
Density (true) 1.19 g/cm^3
Melting point Converts to the anhydrous form at 150°C.

SEM 1: Excipient: sodium citrate dihydrate (granular); manufacturer: Pfizer Ltd; magnification: 60×.

SEM 2: Excipient: sodium citrate dihydrate (granular); manufacturer: Pfizer Ltd; magnification: 600×.

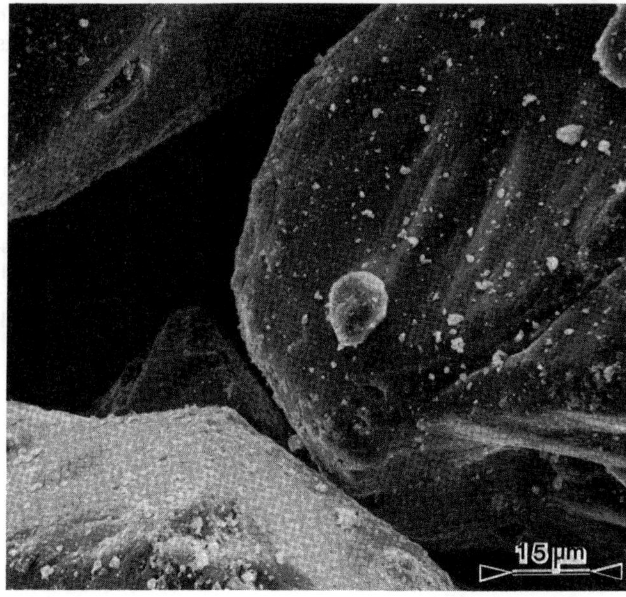

Osmolarity A 3.02% w/v aqueous solution is iso-osmotic with serum.

Particle size distribution Various grades of sodium citrate dihydrate with different particle sizes are commercially available.

Solubility Soluble 1 in 1.5 of water, 1 in 0.6 of boiling water; practically insoluble in ethanol (95%).

Spectroscopy

IR spectra *see* Figure 1.

NIR spectra *see* Figure 2.

Table II: Pharmacopeial specifications for sodium citrate dihydrate.

Test	JP XV	PhEur 7.4	USP35–NF30
Identification	+	+	+
Characters	–	+	–
pH	7.5–8.5	–	–
Appearance of solution	+	+	–
Acidity or alkalinity	–	+	+
Loss on drying	10.0–13.0%	–	–
Water	–	11.0–13.0%	10.0–13.0%
Oxalate	+	≤300 ppm	–
Sulfate	≤0.048%	≤150 ppm	–
Heavy metals	≤10 ppm	≤10 ppm	≤10 ppm
Arsenic	≤2 ppm	–	–
Chloride	≤0.015%	≤50 ppm	–
Tartrate	+	–	+
Readily carbonizable substances	+	+	–
Pyrogens	–	+[a]	–
Assay (anhydrous basis)	99.0–101.0%	99.0–101.0%	99.0–100.5%

(a) If intended for use in large-volume preparations for parenteral use, compliance with a test for pyrogens may be required.

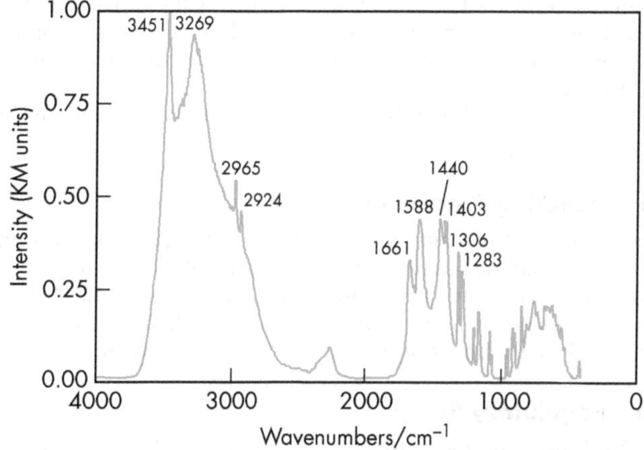

Figure 1: Infrared spectrum of sodium citrate dihydrate measured by diffuse reflectance. Adapted with permission of Informa Healthcare.

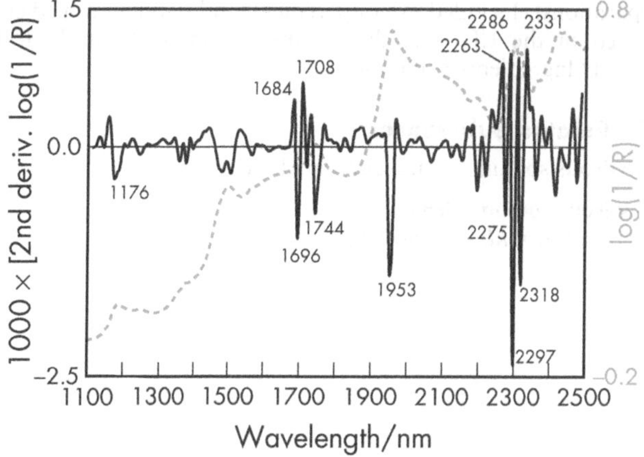

Figure 2: Near-infrared spectrum of sodium citrate dihydrate measured by reflectance.

11 Stability and Storage Conditions

Sodium citrate dihydrate is a stable material. Aqueous solutions may be sterilized by autoclaving. On storage, aqueous solutions may cause the separation of small, solid particles from glass containers.

The bulk material should be stored in an airtight container in a cool, dry place.

S

12 Incompatibilities

Aqueous solutions are slightly alkaline and will react with acidic substances. Alkaloidal salts may be precipitated from their aqueous or hydro-alcohol solutions. Calcium and strontium salts will cause precipitation of the corresponding citrates. Other incompatibilities include bases, reducing agents, and oxidizing agents.

13 Method of Manufacture

Sodium citrate is prepared by adding sodium carbonate to a solution of citric acid until effervescence ceases. The resulting solution is filtered and evaporated to dryness.

14 Safety

After ingestion, sodium citrate is absorbed and metabolized to bicarbonate. Although it is generally regarded as a nontoxic and nonirritant excipient, excessive consumption may cause gastrointestinal discomfort or diarrhea. Therapeutically, in adults, up to 15 g daily of sodium citrate dihydrate may be administered orally, in divided doses, as an aqueous solution to relieve the painful irritation caused by cystitis.

Citrates and citric acid enhance intestinal aluminum absorption in renal patients, which may lead to increased, harmful serum aluminum levels. It has therefore been suggested that patients with renal failure taking aluminum compounds to control phosphate absorption should not be prescribed citrate- or citric acid-containing products.[2]

See Section 17 for anhydrous sodium citrate animal toxicity data.

15 Handling Precautions

Observe normal precautions appropriate to the circumstances and quantity of material handled. Sodium citrate dihydrate dust may be irritant to the eyes and respiratory tract. Eye protection and gloves are recommended. Sodium citrate should be handled in a well-ventilated environment or a dust mask should be worn.

16 Regulatory Status

GRAS listed. Accepted for use as a food additive in Europe. Included in the FDA Inactive Ingredients Database (inhalations; injections; ophthalmic products; oral solutions, suspensions, syrups and tablets; nasal, otic, rectal, topical, transdermal, and vaginal preparations). Included in nonparenteral and parenteral medicines licensed in the UK. Included in the Canadian Natural Health Products Ingredients Database.

17 Related Substances

Anhydrous sodium citrate; citric acid monohydrate.

Anhydrous sodium citrate

Empirical formula $C_6H_5Na_3O_7$

Molecular weight 258.07

CAS number [68-04-2]

Synonyms Anhydrous trisodium citrate; citric acid trisodium salt anhydrous; trisodium 2-hydroxy-1,2,3-propanetricarboxylic acid.

Appearance Colorless crystals or a white crystalline powder.

Safety

LD_{50} (mouse, IP): 1.36 g/kg[3]

LD_{50} (mouse, IV): 0.17 g/kg

LD_{50} (rabbit, IV): 0.45 g/kg

LD_{50} (rat, IP): 1.55 g/kg

18 Comments

Each gram of sodium citrate dihydrate represents approximately 10.2 mmol of sodium and 3.4 mmol of citrate. Each gram of anhydrous sodium citrate represents approximately 11.6 mmol of sodium and 3.9 mmol of citrate.

Therapeutically, sodium citrate is used to relieve the painful irritation caused by cystitis, and also to treat dehydration and acidosis due to diarrhea; *see* Section 14.

Sodium citrate, as either the dihydrate or anhydrous material, is used in food products, primarily to adjust the pH of solutions.

Sodium citrate is additionally used as a blood anticoagulant either alone or in combination with other citrates such as disodium hydrogen citrate.

The EINECS number for sodium citrate is 200-675-3. The PubChem Compound ID (CID) for sodium citrate dihydrate is 71474.

19 Specific References

1 Anderson NR, *et al.* Quantitative evaluation of pharmaceutical effervescent systems II: stability monitoring of reactivity and porosity measurements. *J Pharm Sci* 1982; 71: 7–13.

2 Main J, Ward MK. Potentiation of aluminum absorption by effervescent analgesic tablets in a haemodialysis patient. *Br Med J* 1992; 304: 1686.

3 Lewis RJ, ed. *Sax's Dangerous Properties of Industrial Materials*, 11th edn. New York: Wiley, 2004: 2572.

20 General References

—

21 Author

GE Amidon.

22 Date of Revision

1 March 2012.

S

Sodium Cyclamate

1 Nonproprietary Names

BP: Sodium Cyclamate
PhEur: Sodium Cyclamate

2 Synonyms

Cyclamate sodium; cyclohexylsulfamic acid monosodium salt; E952; natrii cyclamas; sodium cyclohexanesulfamate.

3 Chemical Name and CAS Registry Number

Sodium *N*-cyclohexylsulfamate [139-05-9]

4 Empirical Formula and Molecular Weight

$C_6H_{12}NNaO_3S$ 201.22

5 Structural Formula

(cyclohexyl)—NH—$SO_3^-Na^+$

6 Functional Category

Sweetening agent.

7 Applications in Pharmaceutical Formulation or Technology

Sodium cyclamate is used as an intense sweetening agent in pharmaceutical formulations, foods, beverages, and table-top sweeteners. In dilute solution, up to about 0.17% w/v, the sweetening power is approximately 30 times that of sucrose. However, at higher concentrations this is reduced and at a concentration of 0.5% w/v a bitter taste becomes noticeable. Sodium cyclamate enhances flavor systems and can be used to mask some unpleasant taste characteristics. In most applications, sodium cyclamate is used in combination with saccharin, often in a ratio of 10 : 1.[1]

8 Description

Sodium cyclamate occurs as white, odorless or almost odorless crystals, or as a crystalline powder with an intensely sweet taste.

9 Pharmacopeial Specifications

See Table I.

Table I: Pharmacopeial specifications for sodium cyclamate.

Test	PhEur 7.4
Identification	+
Characters	+
Appearance of solution	+
pH (10% w/v aqueous solution)	5.5–7.5
Absorbance at 270 nm	≤0.10
Sulfamic acid	+
Aniline	≤1 ppm
Cyclohexylamine	≤10 ppm
Dicyclohexylamine	≤1 ppm
Sulfates	≤0.1%
Heavy metals	≤10 ppm
Loss on drying	≤1.0%
Assay (dried basis)	98.5–101.0%

10 Typical Properties

Acidity/alkalinity pH = 5.5–7.5 for a 10% w/v aqueous solution.
Solubility *see* Table II.

Table II: Solubility of sodium cyclamate.

Solvent	Solubility at 20°C unless otherwise stated
Benzene	Practically insoluble
Chloroform	Practically insoluble
Ethanol (95%)	1 in 250
Ether	Practically insoluble
Propylene glycol	1 in 25
Water	1 in 5
	1 in 2 at 45°C

Spectroscopy

IR spectra *see* Figure 1.

NIR spectra *see* Figure 2.

11 Stability and Storage Conditions

Sodium cyclamate is hydrolyzed by sulfuric acid and cyclohexylamine at a very slow rate that is proportional to the hydrogen ion concentration. Therefore, for all practical considerations, it can be regarded as stable. Solutions are also stable to heat, light, and air over a wide pH range.

Samples of tablets containing sodium cyclamate and saccharin have shown no loss in sweetening power following storage for up to 20 years.

The bulk material should be stored in a well-closed container in a cool, dry place.

12 Incompatibilities

—

13 Method of Manufacture

Cyclamates are prepared by the sulfonation of cyclohexylamine in the presence of a base. Commercially, the sulfonation can involve sulfamic acid, a sulfate salt, or sulfur trioxide. Tertiary bases such as triethylamine or trimethylamine may be used as the condensing agent. The amine salts of cyclamate that are produced are converted to the sodium, calcium, potassium, or magnesium salt by treatment with the appropriate metal oxide.

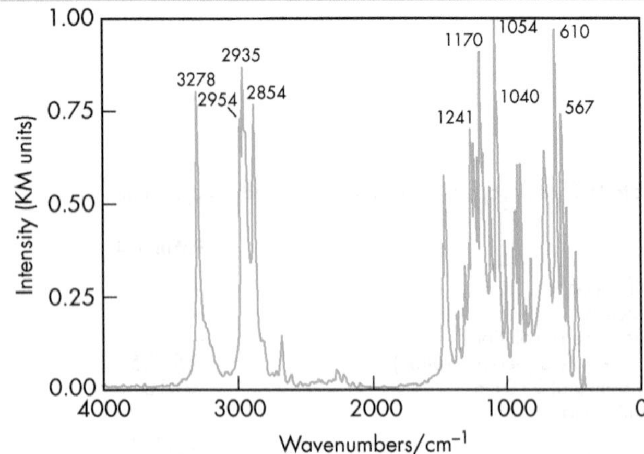

Figure 1: Infrared spectrum of sodium cyclamate measured by diffuse reflectance. Adapted with permission of Informa Healthcare.

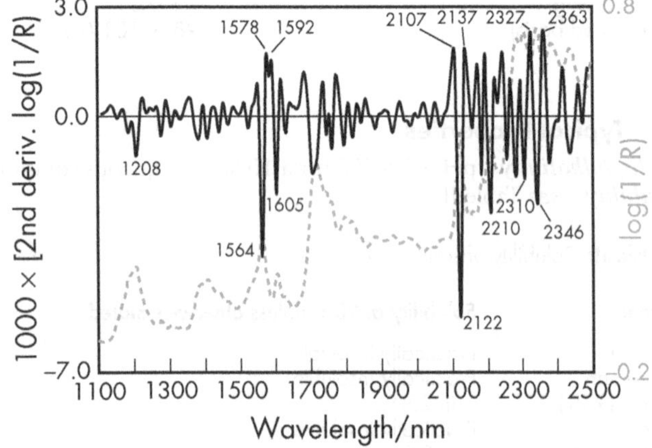

Figure 2: Near-infrared spectrum of sodium cyclamate measured by reflectance.

14 Safety

There has been considerable controversy concerning the safety of cyclamate following the FDA decision in 1970 to ban its use in the US.[2–4] This decision resulted from a feeding study in rats that suggested that cyclamate could cause an unusual form of bladder cancer. However, that study has been criticized because it involved very high doses of cyclamate administered with saccharin, which has itself been the subject of controversy concerning its safety; *see* Saccharin. Although excreted almost entirely unchanged in the urine, a potentially harmful metabolite of sodium cyclamate, cyclohexylamine, has been detected in humans.[5] In addition, there is evidence to suggest cyclamate is metabolized to cyclohexylamine by the microflora in the large intestine of some individuals (approximately 25% of the population with higher precedence in Japanese than Europeans or North Americans). Cyclohexylamine, following absorption, is metabolized to an extent of 1-2% to cyclohexanol and cyclohexane-1,2-diol. Established no-observed-effect level (NOEL) and acceptable daily intake (ADI) values are based on cyclohexylamine levels of high cyclamate converters.[6,7]

Extensive long-term animal feeding studies and epidemiological studies in humans have failed to show any evidence that cyclamate is carcinogenic or mutagenic.[8,9] As a result, sodium cyclamate is now accepted in many countries for use in foods and pharmaceutical formulations. *See also* Section 16.

Few adverse reactions to cyclamate have been reported, although its use has been associated with instances of photosensitive dermatitis.[10]

The WHO has set an estimated acceptable daily intake for sodium and calcium cyclamate, expressed as cyclamic acid, at up to 11 mg/kg body-weight.[11] In Europe, a temporary acceptable daily intake for sodium and calcium cyclamate, expressed as cyclamic acid, has been set at up to 1.5 mg/kg body-weight.

LD_{50} (mouse, IP): 1.15 g/kg[12]
LD_{50} (mouse, IV): 4.8 g/kg
LD_{50} (mouse, oral): 17 g/kg
LD_{50} (rat, IP): 1.35 g/kg
LD_{50} (rat, IV): 3.5 g/kg
LD_{50} (rat, oral): 15.25 g/kg

15 Handling Precautions

Observe normal precautions appropriate to the circumstances and quantity of material handled. Eye protection is recommended.

16 Regulatory Status

The use of cyclamates as artificial sweetners in food, soft drinks, and artificial sweetening tablets was at one time prohibited in the UK and some other countries owing to concern about the metabolite cyclohexylamine. However, this is no longer the case, and cyclamates are now permitted for use as a food additive in Europe.

Included in the FDA Inactive Ingredients Database (oral powder, solutions, chewable tablets, and suspensions). Included in nonparenteral medicines licensed in the UK. Included in the Canadian Natural Health Products Ingredients Database.

17 Related Substances

Alitame; calcium cyclamate; cyclamic acid.

Calcium cyclamate
Empirical formula $C_{12}H_{24}CaN_2O_6S_2 \cdot 2H_2O$
Molecular weight 432.57
CAS number
[5897-16-5] for the dihydrate;
[139-06-0] for the anhydrous form.
Synonyms Calcium *N*-cyclohexylsulfamate dihydrate; *Cyclan*; cyclohexanesulfamic acid calcium salt; cyclohexylsulfamic acid calcium salt; E952; *Sucaryl calcium*.
Appearance White, odorless or almost odorless crystals or a crystalline powder with an intensely sweet taste.
Acidity/alkalinity pH = 5.5–7.5 for a 10% w/v aqueous solution.
Solubility Freely soluble in water; practically insoluble in benzene, chloroform, ethanol (95%), and ether.

Cyclamic acid
Empirical formula $C_6H_{13}NO_3S$
Molecular weight 179.23
CAS number [100-88-9]
Synonyms Cyclamate; cyclohexanesulfamic acid; *N*-cyclohexylsulfamic acid; E952; hexamic acid; *Sucaryl*.
Appearance White, odorless or almost odorless crystals or a crystalline powder with an intensely sweet taste.
Melting point 169–170°C
Solubility Slightly soluble in water.

18 Comments

The perceived intensity of sweeteners relative to sucrose depends upon their concentration, temperature of tasting, and pH, and on the flavor and texture of the product concerned.

Intense sweetening agents will not replace the bulk, textural, or preservative characteristics of sucrose if sucrose is removed from a formulation.

Synergistic effects for combinations of sweeteners have been reported, e.g. sodium cyclamate with saccharin sodium or acesulfame potassium.

Sodium cyclamate has also been used to increase the solubility of neohesperidin dihydrochalcone in sweetener blends.[13]

A specification for sodium cyclamate is contained in the *Food Chemicals Codex* (FCC).[14]

The PubChem Compound ID (CID) for sodium cyclamate is 23665706.

19 Specific References

1 Bernryma GH, *et al.* A case for safety of cyclamate and cyclamate-saccharin combinations. *Am J Clin Nutr* 1968; **21**(6): 673–687.
2 Nabors LO, Miller WT. Cyclamate: a toxicological review. *Commen Toxicol* 1989; 3(4): 307–315.
3 Lecos C. The sweet and sour history of saccharin, cyclamate and aspartame. *FDA Consumer* 1981; **15**(7): 8–11.
4 Anonymous. Cyclamate alone not a carcinogen. *Am Pharm* 1985; **NS25**(9): 11.
5 Kojima S, Ichibagase H. Studies on synthetic sweetening agents VIII. Cyclohexylamine, a metabolite of sodium cyclamate. *Chem Pharm Bull* 1966; **14**: 971–974.
6 Mitchell H, ed. *Sweeteners and Sugar Alternatives in Food Technology.* Oxford: Blackwell Scientific, 2006: 123.
7 Bopp B, Price P. *Alternative Sweeteners*, 3rd edn. New York: Marcel Dekker, Inc., 2001: 63–85.
8 D'Arcy PF. Adverse reactions to excipients in pharmaceutical formulations. In: Florence AT, Salole EG, eds. *Formulation Factors in Adverse Reactions*. London: Wright, 1990: 1–22.
9 Schmähl D, Habs M. Investigations on the carcinogenicity of the artificial sweeteners sodium cyclamate and sodium saccharin in rats in a two-generation experiment. *Arzneimittelforschung* 1984; **34**: 604–606.
10 Yong JM, Sanderson KV. Photosensitive dermatitis and renal tubular acidosis after ingestion of calcium cyclamate. *Lancet* 1969; **ii**: 1273–1274.
11 FAO/WHO. Evaluation of certain food additives and contaminants. Twenty-sixth report of the joint FAO/WHO expert committee on food additives. *World Health Organ Tech Rep Ser* 1982; No. 683.
12 Lewis RJ, ed. *Sax's Dangerous Properties of Industrial Materials*, 11th edn. New York: Wiley, 2004: 3243.
13 Benavente-Garcia O, *et al.* Improved water solubility of neohesperidin dihydrochalcone in sweetener blends. *J Agric Food Chem* 2001; **49**(1): 189–191.
14 *Food Chemicals Codex*, 7th edn. Suppl. 3. Bethesda, MD: United States Pharmacopeia, 2011; 1711.

20 General References

Anonymous. Saccharin is safe. *Chem Br* 2001; **37**(4): 18.

Schiffman SS, *et al.* Effect of temperature, pH, and ions on sweet taste. *Physiol Behav* 2000; **68**(4): 469–481.

21 Author

PL Goggin.

22 Date of Revision

1 March 2012.

Sodium Formaldehyde Sulfoxylate

1 Nonproprietary Names

USP–NF: Sodium Formaldehyde Sulfoxylate

2 Synonyms

Formaldehyde hydrosulfite; formaldehyde sodium sulfoxylate; formaldehydesulfoxylic acid sodium salt; methanesulfinic acid, hydroxy-, monosodium salt; monosodium hydroxymethane sulfinate; *Rongalite*; sodium hydroxymethane sulfinate; sodium hydroxymethylsulfinate; sodium methanalsulfoxylate; sodium sulfinomethanolate.

3 Chemical Name and CAS Registry Number

Sodium formaldehyde sulfoxylate [149-44-0]

Sodium formaldehyde sulfoxylate dihydrate [6035-47-8]

4 Empirical Formula and Molecular Weight

CH_3NaO_3S	118.09
$CH_3NaO_3S{\cdot}2H_2O$	154.11

5 Structural Formula

HO–CH2–S(=O)–ONa

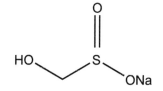

6 Functional Category

Antioxidant.

7 Applications in Pharmaceutical Formulation or Technology

Sodium formaldehyde sulfoxylate is a water-soluble antioxidant and is generally used as the dihydrate. It is used in the formulation of injection products at a level of up to 0.1% w/v in the final preparation administered to the patient.

8 Description

When freshly prepared, sodium formaldehyde sulfoxylate occurs as white, odorless crystals, which quickly develop a characteristic garlic odor on standing.

9 Pharmacopeial Specifications

See Table I.

10 Typical Properties

Acidity/alkalinity pH = 9.5–10.5 (2% w/v aqueous solution)

Melting point 64–68°C (dihydrate)

Solubility Freely soluble in water; slightly soluble in ethanol, chloroform, ether and benzene.

Table I: Pharmacopeial specifications for sodium formaldehyde sulfoxylate.

Test	USP35–NF30
Identification	+
Clarity and color of solution	+
Alkalinity	+
pH (1 in 50 solution)	9.5–10.5
Loss on drying	≤27.0%
Sulfide	+
Iron	+
Sodium sulfite	≤5.0%
Assay (as sulfur dioxide)	45.5–54.5%

Note: USP35–NF30 also states that sodium formaldehyde sulfoxylate may contain a suitable stabilizer such as sodium carbonate.

11 Stability and Storage Conditions

Store in well-closed, light-resistant containers at controlled room temperature (15–30°C).

12 Incompatibilities

Sodium formaldehyde sulfoxylate is incompatible with strong oxidizing agents; it is decomposed by dilute acid.

13 Method of Manufacture

Sodium formaldehyde sulfoxylate is manufactured from sodium dithionate and formaldehyde in water.

14 Safety

The toxicological properties of sodium formaldehyde sulfoxylate have not been fully investigated. However, it is used in the formulation of injection products at a level to 0.1% w/v in the final preparation administered to the patient.

Sodium formaldehyde sulfoxylate is moderately toxic by ingestion, and when heated to decomposition it emits toxic fumes of sulfur dioxide and sodium oxide.[1]

LD_{50} (mouse, oral): 4 g/kg[1,2]
LD_{50} (rat, IP): >2 g/kg[2]
LD_{50} (rat, oral): >2 g/kg[2]

S

15 Handling Precautions

Observe normal precautions appropriate to the circumstances and quantity of the material handled. May cause irritation of the eyes, skin, respiratory tract and digestive tract; the use of eye protection, a respirator and gloves is strongly recommended.

16 Regulatory Status

Included in the FDA Inactive Ingredients Database (parenteral products up to 0.1% via the IM, IV, and SC routes).

17 Related Substances

Zinc formaldehyde sulfoxylate.

Zinc formaldehyde sulfoxylate
Empirical formula $C_2H_6O_6S_2Zn$
Molecular weight 256.5
CAS number [24887-06-7]
Comments Used as an additive in polymers and textiles. The EINECS number is 246-515-6.

18 Comments

Sodium formaldehyde sulfoxylate has been investigated as an antidote to mercury poisoning, but is considered less effective than dimercaprol (British anti-lewisite (BAL)) and other treatments.[3,4] It is also used as an industrial bleach. It is used in chemical synthesis as a nucleophilic agent in the preparation of sulfones. The empirical formula and molecular weight are also given as CH_4O_3SNa and 119.1, respectively.[1]

A specification for sodium formaldehyde sulfoxylate is included in the *Japanese Pharmaceutical Excipients* (JPE).[5]

The EINECS number for sodium formaldehyde sulfoxylate is 205-739-4. The PubChem Compound ID for sodium formaldehyde sulfoxylate is 23725019.

19 Specific References

1 Lewis RJ, ed. *Sax's Dangerous Properties of Industrial Materials*, 11th edn. New York: Wiley, 2004: 1815.
2 Sigma-Aldrich. Material safety data sheet: Sodium formaldehyde sulfoxylate, Australia, 2004.
3 Stocken LA. British anti-lewisite as an antidote for acute mercury poisoning. *Biochem J* 1947; **41**: 358–360.
4 Lehotzky K. Protection by spironolactone and different antidotes against acute organic mercury poisoning of rats. *Int Arch Occup Environ Health* 1974; **33**: 329–334.
5 Japan Pharmaceutical Excipients Council. *Japanese Pharmaceutical Excipients 2004*. Tokyo: Yakuji Nippo, 2004: 796-798.

20 General References

—

21 Author

RC Moreton.

22 Date of Revision

1 March 2012.

Sodium Hyaluronate

1 Nonproprietary Names

BP: Sodium Hyaluronate
PhEur: Sodium Hyaluronate

2 Synonyms

Hyaluronan; hyaluronate sodium; natrii hyaluronas; *RITA HA C-1-C.*

3 Chemical Name and CAS Registry Number

Sodium hyaluronate [9067-32-7]

4 Empirical Formula and Molecular Weight

$(C_{14}H_{20}NO_{11}Na)_n$ $(401.3)_n$

5 Structural Formula

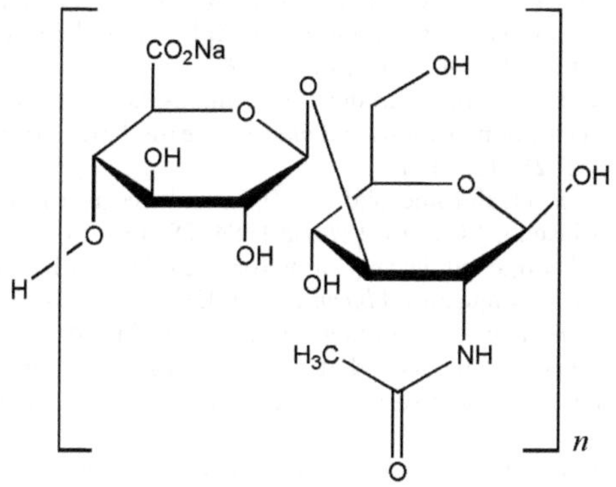

6 Functional Category

Humectant; lubricant; modified-release agent.

7 Applications in Pharmaceutical Formulation or Technology

Sodium hyaluronate is the predominant form of hyaluronic acid at physiological pH. The name hyaluronan is used when the polysaccharide is mentioned in general terms, and in the literature the terms hyaluronic acid and sodium hyaluronate are used interchangeably.

Crosslinked hyaluronan gels are used as drug delivery systems.(1) Hyaluronan is the most common negatively charged glycosaminoglycan in the human vitreous humor, and is known to interact with polymeric and liposomal DNA complexes,(2) where hyaluronan solutions have been shown to decrease the cellular uptake of complexes.(3) This is useful for enhancing the availability and retention time of drugs administered to the eye. It is immunoneutral, which makes it useful for the attachment of biomaterials for use in tissue engineering and drug delivery systems;(4)

See also Section 18.

8 Description

The PhEur 7.4 describes sodium hyaluronate as the sodium salt of hyaluronic acid, a glycosaminoglycan consisting of D-glucuronic acid and *N*-acetyl-D-glucosamine disaccharide units.

Sodium hyaluronate occurs as white to off-white powder or granules. It is very hygroscopic.

9 Pharmacopeial Specifications

See Table I.

Table I: Pharmacopeial specification for sodium hyaluronate.

Test	PhEur 7.4
Characters	+
Identification	+
Appearance of solution	+
pH	5.0–8.5
Intrinsic viscosity	+
Sulfated glycosaminoglycans	≤1%
Nucleic acids	≤0.5
Protein	≤0.3%[a]
Chlorides	≤0.5%
Iron	≤80 ppm
Heavy metals	≤20 ppm[b]
Loss on drying	≤20.0%
Microbial contamination	≤10^2 cfu/g
Bacterial endotoxins	<0.5 IU/mg[c]
Assay	95.0–105.0%

(a) <0.1% for parenteral dosage forms.
(b) ≤10 ppm for parenteral preparations.
(c) ≤0.5 IU/mg for parenteral dosage forms.

10 Typical Properties

Acidity/alkalinity pH = 5.0–8.5 (0.5% w/v aqueous solution)

Solubility Soluble in water, although speed of dissolution depends upon molecular weight (higher molecular weights are slower to dissolve, although this process can be increased by gentle agitation). Slightly soluble in mixtures of organic solvents with water.(5)

11 Stability and Storage Conditions

Sodium hyaluronate should be stored in a cool, dry place in tightly sealed containers. The powder is stable for 3 years if stored in unopened containers.

12 Incompatibilities

—

13 Method of Manufacture

Sodium hyaluronate occurs naturally in vitreous humor, serum, chicken combs, shark skin, and whale cartilage; it is usually extracted and purified from chicken combs. It may also be manufactured by fermentation of selected *Streptococcus zooepidemicus* bacterial strains; sodium hyaluronate is removed from the fermentation medium by filtration and purified by ultrafiltration. It is then precipitated with an organic solvent and dried.

14 Safety

Sodium hyaluronate is used in cosmetics and in topical, parenteral, and ophthalmic pharmaceutical formulations. It is generally regarded as a relatively nontoxic and nonirritant material. Sodium hyaluronate has been reported to be an experimental teratogen.(6)

LD_{50} (mouse, IP): 1.5 g/kg(6)
LD_{50} (rabbit, IP): 1.82 g/kg
LD_{50} (rat, IP): 1.77 g/kg

15 Handling Precautions

Observe normal precautions appropriate to the circumstances and quantity of material handled. When heated to decomposition, sodium hyaluronate emits toxic fumes of Na_2O.

16 Regulatory Status

Included in the FDA Inactive Ingredients Database (topical gel preparation).

17 Related Substances

Hyaluronic acid.

Hyaluronic acid

Molecular weight Hyaluronic acid molecules have a molecular weight of 300–2000 kDa as the number of repeating disaccharide units in each molecule is variable. In its natural form, hyaluronic acid exists as a high-molecular-weight polymer of 10 000 000–100 000 000 Da.

CAS number [9067-32-7]

Appearance Hyaluronic acid appears as a white to off-white powder or granules.

Comments Hyaluronic acid is used as an adjuvant for ophthalmic drug delivery,[7] and has been found to enhance the absorption of drugs and proteins via mucosal tissue.[8] It has also been used experimentally in controlled-release films that are suitable for application to surgical sites for the prevention of adhesion formation,[9] and in matrix formulations used in gene delivery systems.[10] The EINECS number for hyaluronic acid is 232-678-0.

18 Comments

Microspheres prepared from hyaluronan esters have been evaluated for the vaginal administration of calcitonin in the treatment of postmenopausal osteoporosis.[11] Microspheres prepared from hyaluronan esters have also been used experimentally as delivery devices for nerve growth factors,[12] and as a nasal delivery system for insulin.[13]

An *N*-(2-hydroxypropyl)methacrylamide (HPMA)–hyaluronan polymeric drug delivery system has been studied for the targeted delivery of doxorubicin to cancer cells. This copolymer exhibited increased toxicity due to hyaluronan receptor-mediated uptake of the macromolecular drug.[14]

Hyaluronan is used therapeutically to treat osteoarthritis in the knee, and is an effective treatment for arthritic pain.[15] It also has important applications in the fields of vascosurgery and vascosupplementation.[16]

The EINECS number for sodium hyaluronate is 232-678-0. The PubChem Compound ID (CID) for sodium hyaluronate is 3084049.

19 Specific References

1 Dehayza P, Cheng L. Sodium hyaluronate microspheres. US Patent No. 2,004,127,459; 2004.

2 Pitkänen L, *et al.* Vitreous is a barrier in nonviral gene transfer by cationic lipids and polymers. *Pharm Res* 2003; **20**(4): 576–583.

3 Ruponen M, *et al.* Interactions of polymeric and liposomal gene delivery systems with extracellular glycosaminoglycans: physicochemical and transfection studies. *Biochim Biophys Acta* 1999; **1415**: 331–341.

4 Vercruysse KP, Prestwich GD. Hyaluronate derivatives in drug delivery. *Crit Rev Ther Carrier Syst* 1998; **15**: 513–555.

5 Contipro C a.s. *Sodium hyaluronate*. www.contipro.cz (accessed 1 December 2011).

6 Lewis RJ, ed. *Sax's Dangerous Properties of Industrial Materials*, 11th edn. New York: Wiley, 2004; 1970.

7 Saettone MF, *et al.* Mucoadhesive ophthalmic vehicles: evaluation of polymeric low-viscosity formulations. *J Ocul Pharm* 1994; **10**: 83–92.

8 Cho KY, *et al.* Release of ciprofloxacin from polymer-graft-hyaluronic acid hydrogels *in vitro*. *Int J Pharm* 2003; **260**(1): 83–91.

9 Jackson JK, *et al.* Paclitaxel-loaded crosslinked hyaluronic acid films for the prevention of postsurgical adhesions. *Pharm Res* 2002; **19**(4): 411–417.

10 Kim A, *et al.* Characterization of DNA-hyaluronan matrix for sustained gene transfer. *J Control Release* 2003; **90**(1): 81–75.

11 Rochira M, *et al.* Novel vaginal delivery systems for calcitonin II. Preparation and characterisation of HYAFF microspheres containing calcitonin. *Int J Pharm* 1996; **144**: 19–26.

12 Ghezzo E, *et al.* Hyaluronan derivative microspheres as NGF delivery devices: preparation methods and *in vitro* release characterization. *Int J Pharm* 1992; **29**: 133–141.

13 Illum L, *et al.* Hyaluronic acid ester microspheres as a nasal delivery system for insulin. *J Control Release* 1994; **29**: 133–141.

14 Luo Y, *et al.* Targetted delivery of doxorubicin by HPMA copolymer–hyaluronan bioconjugates. *Pharm Res* 2002; **19**(4): 396–402.

15 Castellacci E, Polieri T. Antalgic effect and clinical tolerability of hyaluronic acid in patients with degenerative diseases of knee cartilage: an outpatient treatment survey. *Drugs Exp Clin Res* 2004; **30**(2): 67–73.

16 Balazs EA *et al.* Clinical uses of hyaluronan. In: Evered D, Whelan J, eds. *The Biology of Hyaluronan*. Chichester: Wiley, 1989; 265–280.

20 General References

—

21 Authors

ME Fenton, PJ Sheskey.

22 Date of Revision

2 March 2012.

Sodium Hydroxide

1 Nonproprietary Names

BP: Sodium Hydroxide
JP: Sodium Hydroxide
PhEur: Sodium Hydroxide
USP–NF: Sodium Hydroxide

2 Synonyms

Caustic soda; E524; lye; natrii hydroxidum; soda lye; sodium hydrate.

3 Chemical Name and CAS Registry Number

Sodium hydroxide [1310-73-2]

4 Empirical Formula and Molecular Weight

NaOH 40.00

5 Structural Formula

See Section 4.

6 Functional Category

Alkalizing agent.

7 Applications in Pharmaceutical Formulation or Technology

Sodium hydroxide is widely used in pharmaceutical formulations to adjust the pH of solutions.(1) It can also be used to react with weak acids to form salts.

8 Description

Sodium hydroxide occurs as a white or nearly white fused mass. It is available in small pellets, flakes, sticks, and other shapes or forms. It is hard and brittle and shows a crystalline fracture. Sodium hydroxide is very deliquescent and on exposure to air it rapidly absorbs carbon dioxide and water.

9 Pharmacopeial Specifications

See Table I.

Table I: Pharmacopeial specifications for sodium hydroxide.

Test	JP XV	PhEur 7.4	USP35–NF30
Identification	+	+	+
Characters	−	+	−
Appearance of solution	+	+	−
Insoluble substances and organic matter	−	−	+
Sodium carbonate	≤2.0%	≤2.0%	−
Sulfates	−	≤50 ppm	−
Chlorides	≤0.05%	≤50 ppm	−
Iron	−	≤10 ppm	−
Mercury	+	−	−
Heavy metals	≤30 ppm	≤20 ppm	≤30 ppm
Potassium	+	−	+
Assay (total alkali calculated as NaOH)	≥95.0%	97.0–100.5%	95.0–100.5%

10 Typical Properties

Acidity/alkalinity

pH ≈ 12 (0.05% w/w aqueous solution);
pH ≈ 13 (0.5% w/w aqueous solution);
pH ≈ 14 (5% w/w aqueous solution).

Melting point 318°C
Solubility *see* Table II.

Table II: Solubility of sodium hydroxide.

Solvent	Solubility at 20°C unless otherwise stated
Ethanol	1 in 7.2
Ether	Practically insoluble
Glycerin	Soluble
Methanol	1 in 4.2
Water	1 in 0.9
	1 in 0.3 at 100°C

11 Stability and Storage Conditions

Sodium hydroxide should be stored in an airtight nonmetallic container in a cool, dry place. When exposed to air, sodium hydroxide rapidly absorbs moisture and liquefies, but subsequently becomes solid again owing to absorption of carbon dioxide and formation of sodium carbonate.

12 Incompatibilities

Sodium hydroxide is a strong base and is incompatible with any compound that readily undergoes hydrolysis or oxidation. It will react with acids, esters, and ethers, especially in aqueous solution.

13 Method of Manufacture

Sodium hydroxide is manufactured by electrolysis of brine using inert electrodes. Chlorine is evolved as a gas at the anode and hydrogen is evolved as a gas at the cathode. The removal of chloride and hydrogen ions leaves sodium and hydroxide ions in solution. The solution is dried to produce the solid sodium hydroxide.

A second method uses the Kellner–Solvay cell. Saturated sodium chloride solution is electrolyzed between a carbon anode and a flowing mercury cathode. In this case the sodium is produced at the cathode rather than the hydrogen because of the readiness of sodium to dissolve in the mercury. The sodium–mercury amalgam is then exposed to water and a sodium hydroxide solution is produced.

14 Safety

Sodium hydroxide is widely used in the pharmaceutical and food industries and is generally regarded as a nontoxic material at low concentrations. At high concentrations it is a corrosive irritant to the skin, eyes, and mucous membranes.

LD_{50} (mouse, IP): 0.04 g/kg(2)
LD_{50} (rabbit, oral): 0.5 g/kg

15 Handling Precautions

Observe normal handling precautions appropriate to the quantity and concentration of material handled. Gloves, eye protection, a respirator, and other protective clothing should be worn.

S

Sodium hydroxide is a corrosive irritant to the skin, eyes, and mucous membranes. The solid and solutions cause burns, often with deep ulceration. It is moderately toxic on ingestion and harmful on inhalation.

In the UK, the workplace exposure limit for sodium hydroxide has been set at 2 mg/m^3 short-term.[3]

16 Regulatory Status

GRAS listed. Accepted for use as a food additive in Europe. Included in the FDA Inactive Ingredients Database (dental preparations; injections; inhalations; nasal, ophthalmic, oral, otic, rectal, topical, and vaginal preparations). Included in nonparenteral and parenteral medicines licensed in the UK. Included in the Canadian Natural Health Products Ingredients Database.

17 Related Substances

Potassium hydroxide.

18 Comments

Sodium hydroxide is most commonly used in solutions of fixed concentration. Sodium hydroxide has some antibacterial and antiviral properties and is used as a disinfectant in some applications.[4–6]

A specification for sodium hydroxide is contained in the *Food Chemicals Codex* (FCC).[7]

The EINECS number for sodium hydroxide is 215-185-5. The PubChem Compound ID (CID) for sodium hydroxide is 14798.

19 Specific References

1 Zhan X, *et al.* Improved stability of 25% vitamin C parenteral formulation. *Int J Pharm* 1998; **173**: 43–49.
2 Lewis RJ, ed. *Sax's Dangerous Properties of Industrial Materials*, 11th edn. New York: Wiley, 2004: 3254–3255.
3 Health and Safety Executive. *EH40/2005: Workplace Exposure Limits*. Sudbury: HSE Books, 2011. http://www.hse.gov.uk/pubns/priced/eh40.pdf (accessed 29 February 2012).
4 Brown P, *et al.* Sodium hydroxide decontamination of Creutzfeldt–Jakob disease virus. *N Engl J Med* 1984; **320**: 727.
5 Gasser G. Creutzfeldt–Jakob disease [letter]. *Br Med J* 1990; **300**: 1523.
6 Perkowski CA. Operational aspects of bioreactor contamination control. *J Parenter Sci Technol* 1990; **44**: 113–117.
7 *Food Chemicals Codex*, 7th edn. Bethesda, MD: United States Pharmacopeia, 2010: 941.

20 General References

—

21 Author

AH Kibbe.

22 Date of Revision

2 March 2012.

Sodium Lactate

1 Nonproprietary Names

BP: Sodium Lactate Solution
PhEur: Sodium Lactate Solution
USP–NF: Sodium Lactate Solution

2 Synonyms

E325; 2-hydroxypropanoic acid monosodium salt; *Lacolin*; lactic acid monosodium salt; lactic acid sodium salt; natrii lactatis solutio; *Patlac*; *Purasal*; *Ritalac NAL*; sodium α-hydroxypropionate.

3 Chemical Name and CAS Registry Number

Sodium lactate [72-17-3]

4 Empirical Formula and Molecular Weight

$C_3H_5NaO_3$ 112.06

5 Structural Formula

The PhEur 7.4 and USP35–NF30 describe sodium lactate solution as a mixture of the enantiomers of sodium 2-hydroxypropanoate in approximately equal proportions.

6 Functional Category

Antimicrobial preservative; buffering agent; emulsifying agent; flavoring agent; humectant.

7 Applications in Pharmaceutical Formulation or Technology

Sodium lactate is widely used in cosmetics,[1,2] food products, and pharmaceutical applications including parenteral and topical formulations.

8 Description

Sodium lactate occurs as a clear, colorless, slightly syrupy liquid. It is odorless, or has a slight odor with a characteristic saline taste. It is hygroscopic.

9 Pharmacopeial Specifications

See Table I.

10 Typical Properties

Acidity/alkalinity pH = 7 for an aqueous solution.
Boiling point 112°C
Hygroscopicity Very hygroscopic.
Melting point 17°C with decomposition at 140°C.
Solubility Miscible with ethanol (95%), and with water.
Specific gravity 1.31–1.34

11 Stability and Storage Conditions

Sodium lactate should be stored in a well-closed container in a cool, dry place. Sodium lactate is combustible and decomposes upon heating.

S

Table I: Pharmacopeial specifications for sodium lactate solution.

Test	PhEur 7.4	USP35–NF30
Characters	+	—
Identification	+	+
Appearance of solution	+	—
pH	6.5–9.0	5.0–9.0
Reducing sugars and sucrose	+	+
Methanol	≤50 ppm[a]	—
Methanol and methyl esters	—	≤0.025%
Chlorides	≤50 ppm	≤0.05%
Oxalates and phosphates	+	—
Citrate, oxalate, phosphate or tartrate	—	+
Sulfates	≤100 ppm	+
Aluminum	≤0.1 ppm[a]	—
Barium	+	—
Iron	≤10 ppm	—
Heavy metals	≤10 ppm	≤0.001%
Bacterial endotoxins	≤5 IU/g[b]	—
Assay	96.0–104.0%[c]	98.0–102.0%

(a) If intended for use in the manufacture of parenteral dosage forms, hemodialysis, or hemofiltration solutions.
(b) If intended for use in the manufacture of parenteral dosage forms without a further appropriate procedure for the removal of bacterial endotoxins.
(c) PhEur 7.4 also lists sodium (S) lactate solution, which has an additional requirement for the content of the (S)-enantiomer of ≤95%.

12 Incompatibilities

See Lactic Acid.

13 Method of Manufacture

See Lactic Acid.

14 Safety

Sodium lactate occurs naturally in the body and is involved in physiological processes. It is generally regarded as a relatively nontoxic and nonirritant material when used as an excipient. Low concentrations are well tolerated by skin and eye mucosa, although higher concentrations should be avoided.

LD_{50} (rat, IP): 2 g/kg[3]

15 Handling Precautions

Observe normal precautions appropriate to the circumstances and quantity of material handled. Sodium lactate may cause eye irritation. When heated to decomposition, sodium lactate emits toxic fumes of Na_2O.[3]

16 Regulatory Status

GRAS listed (not for infant formulas). Included in the FDA Inactive Ingredient Database (epidural, IM, IV, and SC injections; oral suspensions; topical gels and solutions). Included in nonparenteral medicines licensed in the UK. Included in the Canadian Natural Health Products Ingredients Database.

17 Related Substances

Lactic acid.

18 Comments

Generally, the commercially available product is a mixture with water containing 70–80% sodium lactate.

Therapeutically, sodium lactate is used in infusions as a component of Ringer-lactate solution; as an alternative for sodium biocarbonate in light acidosis; as a rehydrating agent; and as a carrier for electrolyte concentrates or medicines in perfusion/infusion solutions.

A specification for sodium lactate solution is contained in the *Food Chemicals Codex* (FCC).[4]

The EINECS number for sodium lactate is 200-772-0. The PubChem Compound ID (CID) for sodium lactate is 23666456.

19 Specific References

1 Suomela A, Kristoffersson E. Dry skin and moisturizing agents. *Acta Pharm Fenn* 1983; 92(2): 67–76.
2 Middleton JD. Sodium lactate as a moisturizer. *Cosmet Toiletries* 1978; 93(3): 85–86.
3 Lewis RJ, ed. *Sax's Dangerous Properties of Industrial Materials*, 11th edn. New York: Wiley, 2004: 2197–2198.
4 *Food Chemicals Codex*, 7th edn. Bethesda, MD: United States Pharmacopeia, 2010: 944.

20 General References

—

21 Author

MG Lee.

22 Date of Revision

2 March 2012.

Sodium Lauryl Sulfate

1 Nonproprietary Names

BP: Sodium Lauryl Sulphate
JP: Sodium Lauryl Sulfate
PhEur: Sodium Laurilsulfate
USP–NF: Sodium Lauryl Sulfate

2 Synonyms

Dodecyl alcohol hydrogen sulfate, sodium salt; dodecyl sodium sulfate; dodecylsulfate sodium salt; *Elfan 240*; lauryl sodium sulfate; lauryl sulfate, sodium salt; monododecyl sodium sulfate; natrii laurilsulfas; sodium dodecyl sulfate; sodium *N*-dodecyl sulfate; sodium laurilsulfate; sodium monododecyl sulfate; sodium monolauryl sulfate; SDS; SLS; sulfuric acid monododecyl ester, sodium salt; *Texapon K12P.*

3 Chemical Name and CAS Registry Number

Sulfuric acid monododecyl ester sodium salt (1 : 1) [151-21-3]

4 Empirical Formula and Molecular Weight

$C_{12}H_{25}NaO_4S$ 288.38

The USP35–NF30 describes sodium lauryl sulfate as a mixture of sodium alkyl sulfates consisting mainly of sodium lauryl sulfate $[CH_3(CH_2)_{10}CH_2OSO_3Na]$. The combined content of sodium chloride and sodium sulfate is not more than 8.0%. The PhEur 7.4 states that sodium lauryl sulfate should contain not less than 85% of sodium alkyl sulfates calculated as $C_{12}H_{25}NaO_4S$. The JP XV describes sodium lauryl sulfate as a mixture of sodium alkyl sulfate consisting mainly of sodium lauryl sulfate ($C_{12}H_{25}NaO_4S$).

5 Structural Formula

$$H-CH_2-(CH_2)_{10}-CH_2-O-SO_2-O^- \; Na^+$$

6 Functional Category

Anionic surfactant; emulsifying agent; modified-release agent; penetration enhancer; solubilizing agent; tablet and capsule lubricant.

7 Applications in Pharmaceutical Formulation or Technology

Sodium lauryl sulfate is an anionic surfactant employed in a wide range of nonparenteral pharmaceutical formulations and cosmetics; *see* Table I.

It is a detergent and wetting agent effective in both alkaline and acidic conditions. In recent years it has found application in analytical electrophoretic techniques: SDS (sodium dodecyl sulfate) polyacrylamide gel electrophoresis is one of the more widely used techniques for the analysis of proteins;[1] and sodium lauryl sulfate has been used to enhance the selectivity of micellar electrokinetic chromatography (MEKC).[2]

Table I: Uses of sodium lauryl sulfate.

Use	Concentration (%)
Anionic emulsifier, forms self-emulsifying bases with fatty alcohols	0.5–2.5
Detergent in medicated shampoos	≈10
Skin cleanser in topical applications	1
Solubilizer in concentrations greater than critical micelle concentration	>0.0025
Tablet lubricant	0.5–2.0
Wetting agent in dentrifices	1.0–2.0

8 Description

Sodium lauryl sulfate occurs as white or cream to pale yellow-colored crystals, flakes, or powder having a smooth feel, a soapy, bitter taste, and a faint odor of fatty substances.

9 Pharmacopeial Specifications

See Table II. *See also* Section 18.

Table II: Pharmacopeial specifications for sodium lauryl sulfate.

Test	JP XV	PhEur 7.4	USP35–NF30
Identification	+	+	+
Characters	—	+	—
Alkalinity	+	+	+
Heavy metals	—	—	≤20 ppm
Sodium chloride and sodium sulfate combined content	≤8.0%	≤8.0%	≤8.0%
Unsulfated alcohols	≤4.0%	—	≤4.0%
Nonesterified alcohols	—	≤4.0%	—
Total alcohols	≥59.0%	—	≥59.0%
Water	≤5.0%	—	—
Assay (as $C_{12}H_{25}NaO_4S$)	—	≥85.0%	—

10 Typical Properties

Acidity/alkalinity pH = 7.0–9.5 (1% w/v aqueous solution)

Acid value 0

Antimicrobial activity Sodium lauryl sulfate has some bacteriostatic action against Gram-positive bacteria but is ineffective against many Gram-negative microorganisms. It potentiates the fungicidal activity of certain substances such as sulfanilamide and sulfathiazole. It has also been demonstrated that sodium lauryl sulfate has microbicidal activity against human immunodeficiency virus type I (HIV-1).[3]

Critical micelle concentration

8.2 mmol/L (2.365 g/L) at 20°C;

8.6 mmol/L (2.480 g/L) at 40°C.[4,5]

Table III: CMC values of sodium lauryl sulfate in various media

Method	Condition	CMC (mol/L)
Conductance	No additives	8.0×10^{-3}
Conductance	[TMU] = 1 mol/L	$1.4–1.5 \times 10^{-2}$
Conductance	[HCl] = 6 mol/L	5.5×10^{-5}
BZA-abs	No additives	7.8×10^{-3}
BZA-abs	[HCl] = 33 mol/L	2.8×10^{-3}
Florescence	No additives	7.4×10^{-3}

TMU – Tetramethylurea
BZA – Benzoylacetone

SEM 1: Excipient: sodium lauryl sulfate; manufacturer: Canadian Alcolac Ltd; magnification: 120×.

SEM 2: Excipient: sodium lauryl sulfate; manufacturer: Canadian Alcolac Ltd; magnification: 600×.

The micelle formation and surface activity of sodium lauryl sulfate has also been evaluated at various temperatures.[4,5]

Density 1.07 g/cm^3 at 20°C
Dissociation constant pK_a = 1.9
HLB value ≈40
Interfacial tension 11.8 mN/m (11.8 dynes/cm) for a 0.05% w/v solution (unspecified nonaqueous liquid) at 30°C.
Melting point 204–207°C (for pure substance)
Moisture content ≤5%; sodium lauryl sulfate is not hygroscopic.
Solubility Freely soluble in water, giving an opalescent solution; practically insoluble in chloroform and ether.
Spectroscopy

IR spectra *see* Figure 1.

NIR spectra *see* Figure 2.

Spreading coefficient −7.0 (0.05% w/v aqueous solution) at 30°C

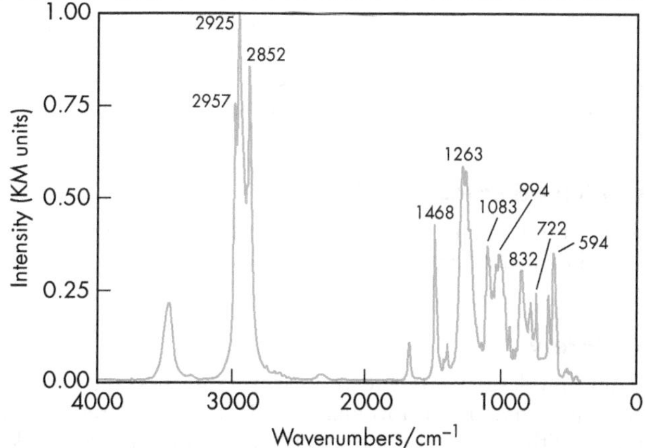

Figure 1: Infrared spectrum of sodium lauryl sulfate measured by diffuse reflectance. Adapted with permission of Informa Healthcare.

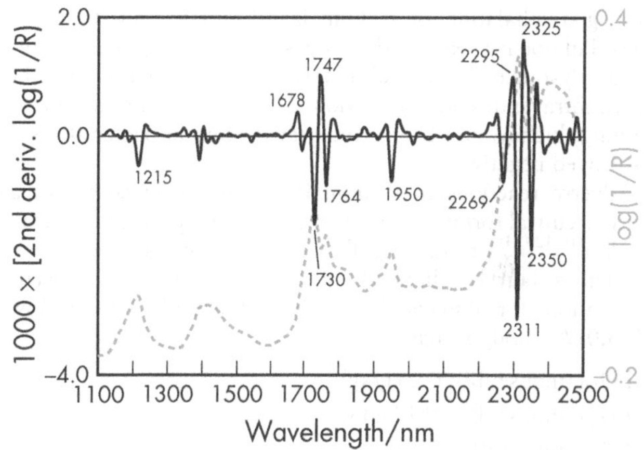

Figure 2: Near-infrared spectrum of sodium lauryl sulfate measured by reflectance.

Surface tension 25.2 mN/m (25.2 dynes/cm) for a 0.05% w/v aqueous solution at 30°C
Wetting time (Draize test) 118 seconds (0.05% w/v aqueous solution) at 30°C

11 Stability and Storage Conditions

Sodium lauryl sulfate is stable under normal storage conditions. However, in solution, under extreme conditions, i.e. pH 2.5 or below, it undergoes hydrolysis to lauryl alcohol and sodium bisulfate.

The bulk material should be stored in a well-closed container away from strong oxidizing agents in a cool, dry place.

12 Incompatibilities

Sodium lauryl sulfate reacts with cationic surfactants, causing loss of activity even in concentrations too low to cause precipitation. Unlike soaps, it is compatible with dilute acids and calcium and magnesium ions.

Sodium lauryl sulfate is incompatible with salts of polyvalent metal ions, such as aluminum, lead, tin or zinc, and precipitates with potassium salts. Solutions of sodium lauryl sulfate (pH 9.5–10.0) are mildly corrosive to mild steel, copper, brass, bronze, and aluminum.

Sodium lauryl sulfate (an anionic surfactant) has been shown to bind to nonionic cellulose ethers to form a strong gel matrix, which can subsequently retard drug release from a matrix system.[6–8] Sodium lauryl sulfate has also been shown to negatively impact dissolution of drugs that have been formulated in gelatin capsules. This phenomenon has been observed when sodium lauryl sulfate is incorporated into the dissolution media at a pH<5 due to the formation of a less soluble precipitate of gelatin.[9]

13 Method of Manufacture

Sodium lauryl sulfate is prepared by sulfation of lauryl alcohol, followed by neutralization with sodium carbonate.

14 Safety

Sodium lauryl sulfate is widely used in cosmetics and oral and topical pharmaceutical formulations. It is a moderately toxic material with acute toxic effects including irritation to the skin, eyes, mucous membranes, upper respiratory tract, and stomach. Repeated, prolonged exposure to dilute solutions may cause drying and cracking of the skin; contact dermatitis may develop.[10] Prolonged inhalation of sodium lauryl sulfate will damage the lungs. Pulmonary sensitization is possible, resulting in hyperactive airway dysfunction and pulmonary allergy. Animal studies have shown intravenous administration to cause marked toxic effects to the lung, kidney, and liver. Mutagenic testing in bacterial systems has proved negative.[11]

Adverse reactions to sodium lauryl sulfate in cosmetics and pharmaceutical formulations mainly concern reports of irritation to the skin[10,12–14] or eyes[15] following topical application.

Sodium lauryl sulfate should not be used in intravenous preparations for humans. The probable human lethal oral dose is 0.5–5.0 g/kg body-weight.

LD_{50} (mouse, IP): 0.25 g/kg[16]
LD_{50} (mouse, IV): 0.12 g/kg
LD_{50} (rat, oral): 1.29 g/kg
LD_{50} (rat, IP): 0.21 g/kg
LD_{50} (rat, IV): 0.12 g/kg

15 Handling Precautions

Observe normal precautions appropriate to the circumstances and quantity of material handled. Inhalation and contact with the skin and eyes should be avoided; eye protection, gloves, and other protective clothing, depending on the circumstances, are recommended. Adequate ventilation should be provided or a dust respirator should be worn. Prolonged or repeated exposure should be avoided. Sodium lauryl sulfate emits toxic fumes on combustion.

S

16 Regulatory Status

GRAS listed. Included in the FDA Inactive Ingredients Database (dental preparations; oral capsules, suspensions, and tablets; topical and vaginal preparations). Included in nonparenteral medicines licensed in the UK. Included in the Canadian Natural Health Products Ingredients Database.

17 Related Substances

Cetostearyl alcohol; cetyl alcohol; magnesium lauryl sulfate; wax, anionic emulsifying.

Magnesium lauryl sulfate

Empirical formula $C_{12}H_{26}O_4S{\cdot}HMg$
CAS number [3097-08-3]
Comments A soluble tablet lubricant.[17] The EINECS number for magnesium lauryl sulfate is 221-450-6.

18 Comments

Sodium lauryl sulfate is one of the materials that have been selected for harmonization by the Pharmacopeial Discussion Group. For further information see the General Information Chapter <1196> in the USP35–NF30, the General Chapter 5.8 in PhEur 7.4, along with the 'State of Work' document on the PhEur EDQM website, and also the General Information Chapter 8 in the JP XV.

A specification for sodium lauryl sulfate is contained in the *Food Chemicals Codex* (FCC).[18]

The EINECS number for sodium lauryl sulfate is 205-788-1. The PubChem Compound ID (CID) for sodium lauryl sulfate is 3423265.

19 Specific References

1 Smith BJ. SDS polyacrylamide gel electrophoresis of proteins. *Methods Mol Biol* 1994; 32: 23–34.
2 Riekkola ML, *et al.* Selectivity in capillary electrophoresis in the presence of micelles, chiral selectors and non-aqueous media. *J Chromatogr* 1997; **792A**: 13–35.
3 Bestman-Smith J, *et al.* Sodium lauryl sulphate abrogates human immunodeficiency virus by affecting viral attachment. *Antimicrob Agents Chemother* 2001; **45**(8): 2229–2237.
4 Bayrak Y. Micelle formation in sodium dodecyl sulfate and dodecyltrimethylammonium bromide at different temperatures. *Turk J Chem* 2003; **27**: 487–492.
5 Tang Y, *et al.* Temperature effects on surface activity and application in oxidation of toluene derivatives of CTAB-SDS with $KMnO_4$. *J Chem Sci* 2006; **118**(3): 281–285.
6 Nokhodchi A, *et al.* The effect of various surfactants on the release rate of propranolol hydrochloride from hydroxypropylmethylcellulose (HPMC)-*Eudragit* matrices. *Eur J Pharm Biopharm* 2002; **54**: 349–356.
7 Walderhaug H. Interactions of ionic surfactants with a nonionic ether in solution in the gel state studied by pulsed field gradient NMR. *J Phys Chem* 1995; **99**: 4672–4678.
8 Nokhodchi A, *et al.* Effect of surfactants and their concentration on the controlled release of captopril from polymeric matrices. *Acta Pharm* 2008; **58**: 151–162.
9 Zhao F, *et al.* Effect of sodium lauryl sulfate in dissolution media of hard gelatin capsules. *Pharm Res* 2004; **21**: 144–148.
10 Wigger-Alberti W, *et al.* Experimental irritant contact dermatitis due to cumulative epicutaneous exposure to sodium lauryl sulphate and toluene: single and concurrent application. *Br J Dermatol* 2000; **143**: 551–556.
11 Mortelmans K, *et al.* Salmonella mutagenicity tests II: results from the testing of 270 chemicals. *Environ Mutagen* 1986; 8(Suppl. 7): 1–119.
12 Blondeel A, *et al.* Contact allergy in 330 dermatological patients. *Contact Dermatitis* 1978; 4(5): 270–276.
13 Bruynzeel DP, *et al.* Delayed time course of irritation by sodium lauryl sulfate: observations on threshold reactions. *Contact Dermatitis* 1982; 8(4): 236–239.
14 Eubanks SW, Patterson JW. Dermatitis from sodium lauryl sulfate in hydrocortisone cream. *Contact Dermatitis* 1984; **11**(4): 250–251.
15 Grant WM. *Toxicology of the Eye*, 2nd edn. Springfield, IL: Charles C Thomas, 1974: 964.
16 Lewis RJ, ed. *Sax's Dangerous Properties of Industrial Materials*, 11th edn. New York: Wiley, 2004; 3258–3259.
17 Caldwell HC, Westlake WJ. Magnesium lauryl sulfate–soluble lubricant [letter]. *J Pharm Sci* 1972; **61**: 984–985.
18 *Food Chemicals Codex*, 7th edn. Bethesda, MD: United States Pharmacopeia, 2010.

20 General References

European Directorate for the Quality of Medicines and Healthcare (EDQM). European Pharmacopoeia – State Of Work Of International Harmonisation. *Pharmeuropa* 2011; 23(4): 713–714. www.edqm.eu/site/-614.html (accessed 6 December 2011).
Hadgraft J, Ashton P. The effect of sodium lauryl sulfate on topical drug bioavailability. *J Pharm Pharmacol* 1985; 37(Suppl.85P).
Nakagaki M, Yokoyama S. Acid-catalyzed hydrolysis of sodium dodecyl sulfate. *J Pharm Sci* 1985; **74**: 1047–1052.

Vold RD, Mittal KL. Determination of sodium dodecyl sulfate in the presence of lauryl alcohol. *Anal Chem* 1972; 44(4): 849–850.
Wan LSC, Poon PKC. The interfacial activity of sodium lauryl sulfate in the presence of alcohols. *Can J Pharm Sci* 1970; 5: 104–107.
Wang L-H, Chowhan ZT. Drug–excipient interactions resulting from powder mixing V: role of sodium lauryl sulfate. *Int J Pharm* 1990; 60: 61–78.

21 Authors

A Pirjanian, F Alvarez-Nunez.

22 Date of Revision

2 March 2012.

Sodium Metabisulfite

1 Nonproprietary Names

BP: Sodium Metabisulphite
JP: Sodium Pyrosulfite
PhEur: Sodium Metabisulphite
USP–NF: Sodium Metabisulfite

2 Synonyms

Disodium disulfite; disodium pyrosulfite; disulfurous acid, disodium salt; E223; natrii disulfis; natrii metabisulfis; sodium acid sulfite.

3 Chemical Name and CAS Registry Number

Sodium pyrosulfite [7681-57-4]

4 Empirical Formula and Molecular Weight

$Na_2S_2O_5$ 190.1

Sodium metabisulfite contains 24.19% sodium, 42.08% oxygen, and 33.73% sulfur.

5 Structural Formula

See Section 4.

6 Functional Category

Antimicrobial preservative; antioxidant.

7 Applications in Pharmaceutical Formulation or Technology

Sodium metabisulfite is used as an antioxidant in oral, parenteral, and topical pharmaceutical formulations, at concentrations of 0.01–1.0% w/v, and at a concentration of approximately 27% w/v in intramuscular injection preparations. Primarily, sodium metabisulfite is used in acidic preparations; for alkaline preparations, sodium sulfite is usually preferred; *see* Section 18. Sodium metabisulfite also has some antimicrobial activity, which is greatest at acid pH, and may be used as a preservative in oral preparations such as syrups.

In the food industry and in wine production, sodium metabisulfite is similarly used as an antioxidant, antimicrobial preservative, and antibrowning agent. However, at concentrations above about 550 ppm it imparts a noticeable flavor to preparations.

Sodium metabisulfite usually contains small amounts of sodium sulfite and sodium sulfate.

8 Description

Sodium metabisulfite occurs as colorless, prismatic crystals or as a white to creamy-white crystalline powder that has the odor of sulfur dioxide and an acidic, saline taste. Sodium metabisulfite crystallizes from cold water as a hydrate containing seven equivalents of water per mole.

9 Pharmacopeial Specifications

See Table I.

Table I: Pharmacopeial specifications for sodium metabisulfite.

Test	JP XV	PhEur 7.4	USP35–NF30
Identification	+	+	+
Characters	–	+	–
Appearance of solution	+	+	–
pH (5% w/v solution)	–	3.5–5.0	–
Chloride	–	–	≤0.05%
Thiosulfate	+	+	≤0.05%
Arsenic	≤4 ppm	≤5 ppm	–
Heavy metals	≤20 ppm	≤20 ppm	≤20 ppm
Iron	≤20 ppm	≤20 ppm	≤20 ppm
Assay (as $Na_2S_2O_5$)	≥95.0%	95.0–100.5%	–
Assay (as SO_2)	–	–	65.0–67.4%

10 Typical Properties

Acidity/alkalinity pH = 3.5–5.0 for a 5% w/v aqueous solution at 20°C.
Melting point Sodium metabisulfite melts with decomposition at less than 150°C.
Osmolarity A 1.38% w/v aqueous solution is isoosmotic with serum.
Solubility *see* Table II.

Table II: Solubility of sodium metabisulfite.

Solvent	Solubility at 20°C unless otherwise stated
Ethanol (95%)	Slightly soluble
Glycerin	Freely soluble
Water	1 in 1.9
	1 in 1.2 at 100°C

Spectroscopy
IR spectra *see* Figure 1.
NIR spectra *see* Figure 2.

11 Stability and Storage Conditions

On exposure to air and moisture, sodium metabisulfite is slowly oxidized to sodium sulfate with disintegration of the crystals.[1] Addition of strong acids to the solid liberates sulfur dioxide.

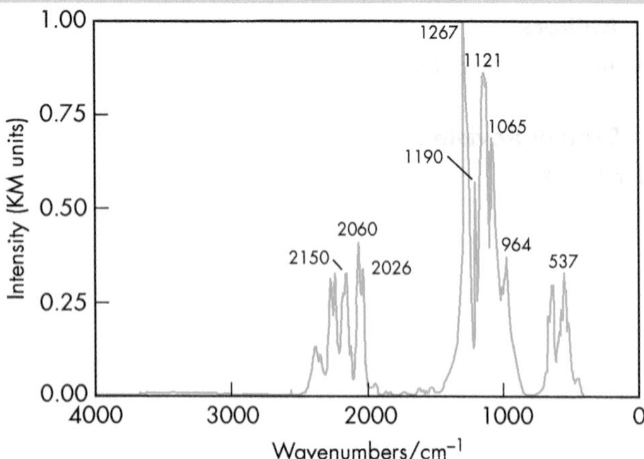

Figure 1: Infrared spectrum of sodium metabisulfite measured by diffuse reflectance. Adapted with permission of Informa Healthcare.

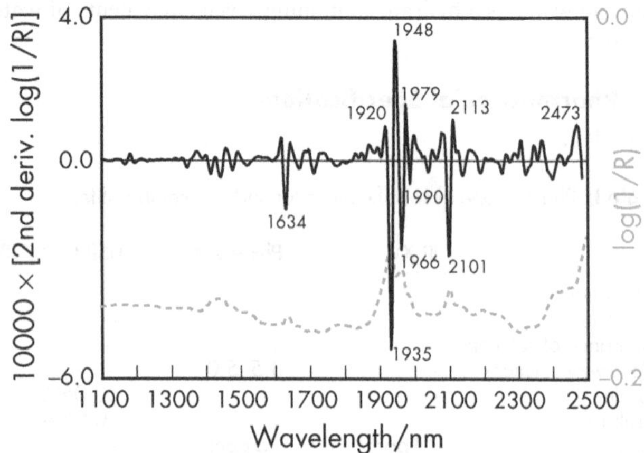

Figure 2: Near-infrared spectrum of sodium metabisulfite measured by reflectance.

In water, sodium metabisulfite is immediately converted to sodium (Na^+) and bisulfite (HSO_3^-) ions. Aqueous sodium metabisulfite solutions also decompose in air, especially on heating. Solutions that are to be sterilized by autoclaving should be filled into containers in which the air has been replaced with an inert gas, such as nitrogen. The addition of dextrose to aqueous sodium metabisulfite solutions results in a decrease in the stability of the metabisulfite.[2]

The bulk material should be stored in a well-closed container, protected from light, in a cool, dry place.

12 Incompatibilities

Sodium metabisulfite reacts with sympathomimetics and other drugs that are *ortho-* or *para-*hydroxybenzyl alcohol derivatives to form sulfonic acid derivatives possessing little or no pharmacological activity. The most important drugs subject to this inactivation are epinephrine (adrenaline) and its derivatives.[3] In addition, sodium metabisulfite is incompatible with chloramphenicol owing to a more complex reaction;[3] it also inactivates cisplatin in solution.[4,5]

It is incompatible with phenylmercuric acetate when autoclaved in eye drop preparations.[6]

Sodium metabisulfite may react with the rubber caps of multidose vials, which should therefore be pretreated with sodium metabisulfite solution.[7]

13 Method of Manufacture

Sodium metabisulfite is prepared by saturating a solution of sodium hydroxide with sulfur dioxide and allowing crystallization to occur; hydrogen is passed through the solution to exclude air. Sodium metabisulfite may also be prepared by saturating a solution of sodium carbonate with sulfur dioxide and allowing crystallization to occur, or by thermally dehydrating sodium bisulfite.

14 Safety

Sodium metabisulfite is widely used as an antioxidant in oral, topical, and parenteral pharmaceutical formulations; it is also widely used in food products.

Although it is extensively used in a variety of preparations, sodium metabisulfite and other sulfites have been associated with a number of severe to fatal adverse reactions.[8–19] These are usually hypersensitivity-type reactions and include bronchospasm and anaphylaxis. Allergy to sulfite antioxidants is estimated to occur in 5–10% of asthmatics, although adverse reactions may also occur in nonasthmatics with no history of allergy.

Following oral ingestion, sodium metabisulfite is oxidized to sulfate and is excreted in urine. Ingestion may result in gastric irritation, owing to the liberation of sulfurous acid, while ingestion of large amounts of sodium metabisulfite can cause colic, diarrhea, circulatory disturbances, CNS depression, and death.

In Europe, the acceptable daily intake of sodium metabisulfite and other sulfites used in foodstuffs has been set at up to 3.5 mg/kg body-weight, calculated as sulfur dioxide (SO_2). The WHO has similarly also set an acceptable daily intake of sodium metabisulfite, and other sulfites, at up to 7.0 mg/kg body-weight, calculated as sulfur dioxide (SO_2).[20]

LD_{50} (rat, IV): 0.12 g/kg[21]

15 Handling Precautions

Observe normal precautions appropriate to the circumstances and quantity of material handled. Sodium metabisulfite may be irritant to the skin and eyes; eye protection and gloves are recommended. In the UK, the long-term (8-hour TWA) workplace exposure limit for sodium metabisulfite is 5 mg/m^3.[22]

16 Regulatory Status

GRAS listed. Accepted for use as a food additive in Europe. Included in the FDA Inactive Ingredients Database (epidural; inhalation; IM and IV injections; ophthalmic solutions; oral preparations; rectal, topical, and vaginal preparations). Included in nonparenteral and parenteral medicines licensed in the UK. Included in the Canadian Natural Health Products Ingredients Database.

17 Related Substances

Potassium metabisulfite; sodium bisulfite; sodium sulfite.

Sodium bisulfite

Empirical formula $NaHSO_3$

Molecular weight 104.07

CAS number [7631-90-5]

Synonyms E222; sodium hydrogen sulfite.

Appearance White crystalline powder.

Density 1.48 g/cm^3

Solubility Soluble 1 in 3.5 parts of water at 20°C; 1 in 2 parts of water at 100°C; and 1 in 70 parts of ethanol (95%). Freely soluble in glycerol. Aqueous solution is acidic.

Comments Most substances sold as sodium bisulfite contain significant, variable amounts of sodium metabisulfite, as the latter is less hygroscopic and more stable during storage and shipment. *See* Section 18.

18 Comments

Sodium metabisulfite is used as an antioxidant at low pH, sodium bisulfite at intermediate pH, and sodium sulfite at higher pH values. A specification for sodium metabisulfite is contained in the *Food Chemicals Codex* (FCC).(23)

The EINECS number for sodium metabisulfite is 231-673-0.

19 Specific References

1 Schroeter LC. Oxidation of sulfurous acid salts in pharmaceutical systems. *J Pharm Sci* 1963; **52**: 888–892.
2 Schumacher GE, Hull RL. Some factors influencing the degradation of sodium bisulfite in dextrose solutions. *Am J Hosp Pharm* 1966; **23**: 245–249.
3 Higuchi T, Schroeter LC. Reactivity of bisulfite with a number of pharmaceuticals. *J Am Pharm Assoc (Sci)* 1959; **48**: 535–540.
4 Hussain AA, *et al.* Reaction of cis-platinum with sodium bisulfite. *J Pharm Sci* 1980; **69**(3): 364–365.
5 Garren KW, Repta AJ. Incompatibility of cisplatin and Reglan injectable. *Int J Pharm* 1985; **24**: 91–99.
6 Collins AJ, *et al.* Incompatibility of phenylmercuric acetate with sodium metabisulphite in eye drop formulations. *J Pharm Pharmacol* 1985; **37S**: 123P.
7 Schroeter LC. Sulfurous acid salts as pharmaceutical antioxidants. *J Pharm Sci* 1961; **50**(11): 891–901.
8 Jamieson DM, *et al.* Metabisulfite sensitivity: case report and literature review. *Ann Allergy* 1985; **54**(4): 115–121.
9 Anonymous. Possible allergic-type reactions. *FDA Drug Bull* 1987; **17**: 2.
10 Tsevat J, *et al.* Fatal asthma after ingestion of sulfite-containing wine [letter]. *Ann Intern Med* 1987; **107**(2): 263.
11 Weiner M, Bernstein IL. *Adverse Reactions to Drug Formulation Agents: a Handbook of Excipients.* New York: Marcel Dekker, 1989; 314–320.
12 Fitzharris P. What advances if any, have been made in treating sulfite allergy? *Br Med J* 1992; **305**: 1478.
13 Smolinske SC. *Handbook of Food, Drug and Cosmetic Excipients.* Boca Raton, FL: CRC Press, 1992; 393–406.
14 Anonymous. Sulfites in drugs and food. *Med Lett Drugs Ther* 1986; **28**: 74–75.
15 Baker GJ. Bronchospasm induced by bisulfite containing food and drugs. *Med J Aust* 1981; **ii**: 614–617.
16 Fwarog FJ, Leung DYM. Anaphylaxis to a component of isoethane. *J Am Med Assoc* 1982; **248**: 2030–2031.
17 Koephe JW. Dose dependent bronchospasm from sulfites in isoethane. *J Am Med Assoc* 1984; **251**: 2982–2983.
18 Mikolich DJ, McCloskey WW. Suspected gentamicin allergy could be sulfite sensitivity. *Clin Pharm* 1988; **7**: 269.
19 Deziel-Evans LM, Hussey WJ. Possible sulfite sensitivity with gentamicin infusion. *DICP Ann Pharmacother* 1989; **23**: 1032–1033.
20 FAO/WHO. Evaluation of certain food additives and contaminants. Thirtieth report of the joint FAO/WHO expert committee on food additives. *World Health Organ Tech Rep Ser* 1987; No. 751.
21 Lewis RJ, ed. *Sax's Dangerous Properties of Industrial Materials*, 11th edn. New York: Wiley, 2004; 3261.
22 Health and Safety Executive. *EH40/2005: Workplace Exposure Limits.* Sudbury: HSE Books, 2011. www.hse.gov.uk/pubns/priced/eh40.pdf (accessed 12 April 2012).
23 *Food Chemicals Codex*, 7th edn. Bethesda, MD: United States Pharmacopeia, 2010.

20 General References

Halsby SF, Mattocks AM. Absorption of sodium bisulfite from peritoneal dialysis solutions. *J Pharm Sci* 1965; **54**: 52–55.
Wilkins JW, *et al.* Toxicity of intraperitoneal bisulfite. *Clin Pharmacol Ther* 1968; **9**: 328–332.

21 Author

W Cook.

22 Date of Revision

12 April 2012.

Sodium Phosphate, Dibasic

S

1 Nonproprietary Names

BP: Anhydrous Disodium Hydrogen Phosphate
Disodium Hydrogen Phosphate Dihydrate
Disodium Hydrogen Phosphate Dodecahydrate

JP: Dibasic Sodium Phosphate Hydrate

PhEur: Disodium Phosphate, Anhydrous
Disodium Phosphate Dihydrate
Disodium Phosphate Dodecahydrate

USP–NF: Dibasic Sodium Phosphate

Note that the BP 2012 and PhEur 7.4 contain three separate monographs for the anhydrous, the dihydrate, and the dodecahydrate; the JP XV contains one monograph for the dodecahydrate; and the USP35–NF30 contains one monograph that includes the anhydrous, the monohydrate, the dihydrate, the heptahydrate, and the dodecahydrate. *See also* Section 8.

2 Synonyms

Dinatrii phosphas anhydricus; dinatrii phosphas dihydricus; dinatrii phosphas dodecahydricus; disodium hydrogen phosphate; disodium phosphate; E339; phosphoric acid, disodium salt; secondary sodium phosphate; sodium orthophosphate.

3 Chemical Name and CAS Registry Number

Anhydrous dibasic sodium phosphate [7558-79-4]
Dibasic sodium phosphate dihydrate [10028-24-7]
Dibasic sodium phosphate dodecahydrate [10039-32-4]
Dibasic sodium phosphate heptahydrate [7782-85-6]
Dibasic sodium phosphate hydrate [10140-65-5]
Dibasic sodium phosphate monohydrate [118830-14-1]

4 Empirical Formula and Molecular Weight

Na_2HPO_4	141.96
$Na_2HPO_4 \cdot H_2O$	159.94
$Na_2HPO_4 \cdot 2H_2O$	177.98
$Na_2HPO_4 \cdot 7H_2O$	268.03
$Na_2HPO_4 \cdot 12H_2O$	358.08

Table I: Pharmacopeial specifications for sodium phosphate, dibasic.

Test	JP XV[a]	PhEur 7.4[b]	USP35–NF30
Identification	+	+	+
Characters	+	+	–
Appearance of solution	+	+	–
pH	9.0–9.4	–	–
Reducing substances	–	+	–
Insoluble substances	–	–	≤0.4%
Monosodium phosphate	–	≤2.5%	–
Carbonate	+	–	–
Chloride	≤0.014%	+	≤0.06%
Anhydrous	–	≤200 ppm	–
Dihydrate	–	≤400 ppm	–
Dodecahydrate	–	≤200 ppm	–
Water	–	+	–
Anhydrous	–	–	–
Dihydrate	–	–	–
Dodecahydrate	–	57.0–61.0%	–
Sulfates	≤0.038%	+	≤0.2%
Anhydrous	–	≤500 ppm	–
Dihydrate	–	≤0.1%	–
Dodecahydrate	–	≤500 ppm	–
Arsenic	≤2 ppm	+	≤16 ppm
Anhydrous	–	≤2 ppm	–
Dihydrate	–	≤4 ppm	–
Dodecahydrate	–	≤2 ppm	–
Heavy metals	≤10 ppm	+	≤0.002%
Anhydrous	–	≤10 ppm	–
Dihydrate	–	≤20 ppm	–
Dodecahydrate	–	≤10 ppm	–
Iron	–	+	–
Anhydrous	–	≤20 ppm	–
Dihydrate	–	≤40 ppm	–
Dodecahydrate	–	≤20 ppm	–
Loss on drying	57.0–61.0%	+	+
Anhydrous	–	≤1.0%	≤5.0%
Monohydrate	–	–	10.3–12.0%
Dihydrate	–	19.5–21.0%	18.5–21.5%
Heptahydrate	–	–	43.0–50.0%
Dodecahydrate	–	–	55.0–64.0%
Assay (dried basis)	≥98.0%	98.0–101.0% 98.5–102.5%[a]	98.0–100.5%

(a) JP XV for the dodecahydrate.
(b) PhEur 7.4 for the anhydrous, dihydrate and dodecahydrate.

5 Structural Formula

See Section 4.

6 Functional Category

Buffering agent; complexing agent.

7 Applications in Pharmaceutical Formulation or Technology

Dibasic sodium phosphate is used in a wide variety of pharmaceutical formulations as a buffering agent and as a complexing agent.

Dibasic sodium phosphate is also used in food products; for example as an emulsifier in processed cheese.

8 Description

The USP35–NF30 states that dibasic sodium phosphate is dried or contains, 1, 2, 7, or 12 molecules of water of hydration.

Anhydrous dibasic sodium phosphate occurs as a white powder. The dihydrate occurs as white or almost white, odorless crystals. The heptahydrate occurs as colorless crystals or as a white granular or caked salt that effloresces in warm, dry air. The dodecahydrate occurs as strongly efflorescent, colorless or transparent crystals.

9 Pharmacopeial Specifications

See Table I.

10 Typical Properties

Acidity/alkalinity pH = 9.1 for a 1% w/v aqueous solution of the anhydrous material at 25°C. A saturated aqueous solution of the dodecahydrate has a pH of about 9.5.

Ionization constants

pK_{a1} = 2.15 at 25°C;[1]

pK_{a2} = 7.20 at 25°C;

pK_{a3} = 12.38 at 25°C.

Moisture content The anhydrous form is hygroscopic and will absorb up to 7 moles of water on exposure to air, whereas the heptahydrate is stable in air.

Osmolarity A 2.23% w/v aqueous solution of the dihydrate is isoosmotic with serum; a 4.45% w/v aqueous solution of the dodecahydrate is isoosmotic with serum.

Solubility Very soluble in water, more so in hot or boiling water; practically insoluble in ethanol (95%). The anhydrous material is soluble 1 in 8 parts of water, the heptahydrate 1 in 4 parts of water, and the dodecahydrate 1 in 3 parts of water.

Spectroscopy

IR spectra *see* Figure 1.

NIR spectra *see* Figure 2.

11 Stability and Storage Conditions

The anhydrous form of dibasic sodium phosphate is hygroscopic. When heated to 40°C, the dodecahydrate fuses; at 100°C it loses its water of crystallization; and at a dull-red heat (about 240°C) it is converted into the pyrophosphate, $Na_4P_2O_7$. Aqueous solutions of dibasic sodium phosphate are stable and may be sterilized by autoclaving.

The bulk material should be stored in an airtight container, in a cool, dry place.

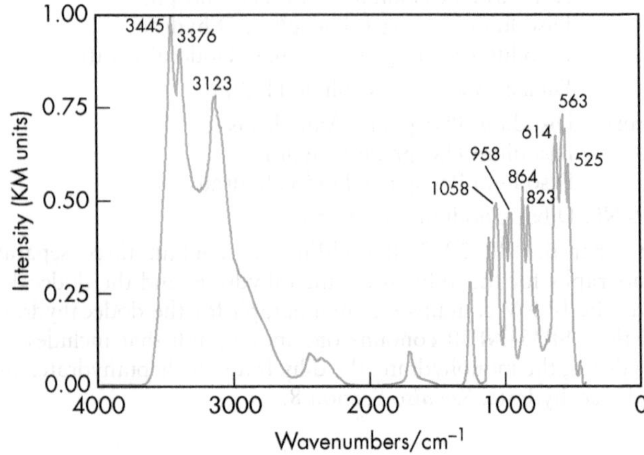

Figure 1: Infrared spectrum of anhydrous dibasic sodium phosphate measured by diffuse reflectance. Adapted with permission of Informa Healthcare.

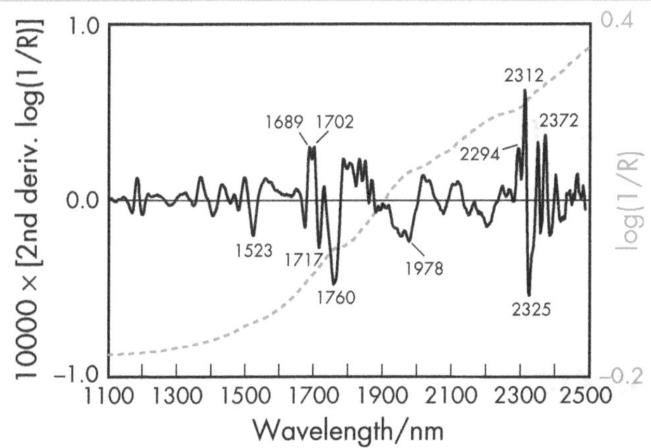

Figure 2: Near-infrared spectrum of dibasic sodium phosphate measured by reflectance.

12 Incompatibilities

Dibasic sodium phosphate is incompatible with alkaloids, antipyrine, chloral hydrate, lead acetate, pyrogallol, resorcinol and calcium gluconate, and ciprofloxacin.(2) Interaction between calcium and phosphate, leading to the formation of insoluble calcium–phosphate precipitates, is possible in parenteral admixtures.

13 Method of Manufacture

Either bone phosphate (bone ash), obtained by heating bones to whiteness, or the mineral phosphorite is used as a source of tribasic calcium phosphate, which is the starting material in the industrial production of dibasic sodium phosphate.

Tribasic calcium phosphate is finely ground and digested with sulfuric acid. This mixture is then leached with hot water and neutralized with sodium carbonate, and dibasic sodium phosphate is crystallized from the filtrate.

14 Safety

Dibasic sodium phosphate is widely used as an excipient in parenteral, oral, and topical pharmaceutical formulations.

Phosphate occurs extensively in the body and is involved in many physiological processes since it is the principal anion of intracellular fluid. Most foods contain adequate amounts of phosphate, making hypophosphatemia (phosphate deficiency)(3) virtually unknown except for certain disease states(4) or in patients receiving total parenteral nutrition. Treatment is usually by the oral administration of up to 100 mmol of phosphate daily.

Approximately two-thirds of ingested phosphate is absorbed from the gastrointestinal tract, virtually all of it being excreted in the urine, and the remainder is excreted in the feces.

Excessive administration of phosphate, particularly intravenously, rectally, or in patients with renal failure, can cause hyperphosphatemia that may lead to hypocalcemia or other severe electrolyte imbalances.(5,6) Adverse effects occur less frequently following oral consumption, although phosphates act as mild saline laxatives when administered orally or rectally. Consequently, gastrointestinal disturbances including diarrhea, nausea, and vomiting may occur following the use of dibasic sodium phosphate as an excipient in oral formulations. However, the level of dibasic sodium phosphate used as an excipient in a pharmaceutical formulation is not usually associated with adverse effects.

LD_{50} (rat, oral): 17 g/kg(7)

15 Handling Precautions

Observe normal precautions appropriate to the circumstances and quantity of material handled. Dibasic sodium phosphate may be irritating to the skin, eyes, and mucous membranes. Eye protection and gloves are recommended.

16 Regulatory Status

GRAS listed. Accepted in Europe for use as a food additive. Included in the FDA Inactive Ingredients Database (injections; infusions; nasal, ophthalmic, oral, otic, topical, and vaginal preparations). Included in nonparenteral and parenteral medicines licensed in the UK. Included in the Canadian Natural Health Products Ingredients Database.

17 Related Substances

Dibasic potassium phosphate; sodium phosphate, monobasic; tribasic sodium phosphate.

Tribasic sodium phosphate

Empirical formula $Na_3PO_4{\cdot}xH_2O$

Molecular weight

163.94 for the anhydrous material

380.06 for the dodecahydrate ($12H_2O$)

CAS number [7601-54-9] for the anhydrous material.

Synonyms E339; trisodium orthophosphate; trisodium phosphate; TSP.

Acidity/alkalinity pH = 12.1 for a 1% w/v aqueous solution of the anhydrous material at 25°C. A 1% w/v aqueous solution of the dodecahydrate at 25°C has a pH of 12.0–12.2.

Density

1.3 g/cm^3 for the anhydrous material;

0.9 g/cm^3 for the dodecahydrate.

Solubility The anhydrous material is soluble 1 in 8 parts of water, while the dodecahydrate is soluble 1 in 5 parts of water at 20°C.

18 Comments

One gram of anhydrous dibasic sodium phosphate represents approximately 14.1 mmol of sodium and 7.0 mmol of phosphate.

One gram of dibasic sodium phosphate dihydrate represents approximately 11.2 mmol of sodium and 5.6 mmol of phosphate.

One gram of dibasic sodium phosphate heptahydrate represents approximately 7.5 mmol of sodium and 3.7 mmol of phosphate.

One gram of dibasic sodium phosphate dodecahydrate represents approximately 5.6 mmol of sodium and 2.8 mmol of phosphate.

Therapeutically, dibasic sodium phosphate is used as a mild laxative and in the treatment of hypophosphatemia.(3,4)

A specification for sodium phosphate, dibasic is contained in the *Food Chemicals Codex* (FCC).(8)

The PubChem Compound ID (CID) for anhydrous dibasic sodium phosphate is 24203, and for dibasic sodium phosphate dodecahydrate is 61456.

19 Specific References

1 Albert A, Serjearnt EP. *Ionization Constants of Acids and Bases*, 2nd edn. Edinburgh: Chapman and Hall, 1971.
2 Benjamin BE. Ciprofloxacin and sodium phosphates not compatible during actual Y-site injection [letter]. *Am J Health Syst Pharm* 1996; **53**: 1850–1851.
3 Lloyd CW, Johnson CE. Management of hypophosphatemia. *Clin Pharm* 1988; **7**: 123–128.
4 Holland PC, *et al.* Prenatal deficiency of phosphate, phosphate supplementation, and rickets in very-low-birthweight infants. *Lancet* 1990; **335**: 697–701.
5 Haskell LP. Hypocalcaemic tetany induced by hypertonic-phosphate enema [letter]. *Lancet* 1985; **ii**: 1433.
6 Martin RR, *et al.* Fatal poisoning from sodium phosphate enema: case report and experimental study. *J Am Med Assoc* 1987; **257**: 2190–2192.
7 Lewis RJ, ed. *Sax's Dangerous Properties of Industrial Materials*, 11th edn. New York: Wiley, 2004: 3273.

S

8 *Food Chemicals Codex*, 7th edn. Bethesda, MD: United States Pharmacopeia, 2010: 954.

20 General References

Sweetman SC, ed. *Martindale: The Complete Drug Reference*, 37th edn. London: Pharmaceutical Press, 2011: 1828.

21 Author

AS Kearney.

22 Date of Revision

2 March 2012.

 Sodium Phosphate, Monobasic

S

1 Nonproprietary Names

BP: Anhydrous Sodium Dihydrogen Phosphate
Sodium Dihydrogen Phosphate Monohydrate
Sodium Dihydrogen Phosphate Dihydrate

PhEur: Sodium Dihydrogen Phosphate Dihydrate

USP–NF: Monobasic Sodium Phosphate

Note that the BP 2012 contains three separate monographs for the anhydrous, the monohydrate, and the dihydrate; the PhEur 7.4 contains a single monograph for the dihydrate; and the USP35–NF30 contains one monograph for the anhydrous, the monohydrate and the dihydrate. *See also* Section 8.

2 Synonyms

Acid sodium phosphate; E339; *Kalipol 32*; monosodium orthophosphate; monosodium phosphate; natrii dihydrogenophosphas dihydricus; phosphoric acid, monosodium salt; primary sodium phosphate; sodium biphosphate; sodium dihydrogen orthophosphate; sodium dihydrogen phosphate.

3 Chemical Name and CAS Registry Number

Anhydrous monobasic sodium phosphate [7558-80-7]
Monobasic sodium phosphate monohydrate [10049-21-5]
Monobasic sodium phosphate dihydrate [13472-35-0]

4 Empirical Formula and Molecular Weight

NaH_2PO_4	119.98
$NaH_2PO_4 \cdot H_2O$	137.99
$NaH_2PO_4 \cdot 2H_2O$	156.01

5 Structural Formula

See Section 4.

6 Functional Category

Buffering agent; complexing agent.

7 Applications in Pharmaceutical Formulation or Technology

Monobasic sodium phosphate is used in a wide variety of pharmaceutical formulations as a buffering agent and as a complexing agent.

Monobasic sodium phosphate is also used in food products, for example, in baking powders, and as a dry acidulant and sequestrant.

8 Description

The USP35–NF30 states that monobasic sodium phosphate contains one or two molecules of water of hydration or is anhydrous.

The hydrated forms of monobasic sodium phosphate occur as odorless, colorless or white, slightly deliquescent crystals. The anhydrous form occurs as a white crystalline powder or granules.

9 Pharmacopeial Specifications

See Table I.

Table I: Pharmacopeial specifications for sodium phosphate, monobasic.

Test	PhEur 7.4	USP35–NF30
Identification	+	+
Characters	+	—
Appearance of solution	+	—
Aluminum, calcium and related elements	—	+
Arsenic	≤2 ppm	≤8 ppm
Chloride	≤200 ppm	≤0.014%
Insoluble substances	—	≤0.2%
Heavy metals	≤10 ppm	≤0.002%
Iron	≤10 ppm	—
pH	4.2–4.5	4.1–4.5
Reducing substances	+	—
Sulfate	≤300 ppm	≤0.15%
Water	+	+
Anhydrous	—	≤2.0%
Monohydrate	—	10.0–15.0%
Dihydrate	21.5–24.0%	18.0–26.5%
Assay (dried basis)	98.0–100.5%	98.0–103.0%

10 Typical Properties

Acidity/alkalinity pH = 4.1–4.5 for a 5% w/v aqueous solution of the monohydrate at 25°C.

Density 1.915 g/cm^3 for the dihydrate.

Dissociation constant pK_a = 2.15 at 25°C

Solubility Soluble 1 in 1 of water; very slightly soluble in ethanol (95%).

Spectroscopy

IR spectra *see* Figure 1.

NIR spectra *see* Figures 2 and 3.

11 Stability and Storage Conditions

Monobasic sodium phosphate is chemically stable, although it is slightly deliquescent. On heating at 100°C, the dihydrate loses all of its water of crystallization. On further heating, it melts with

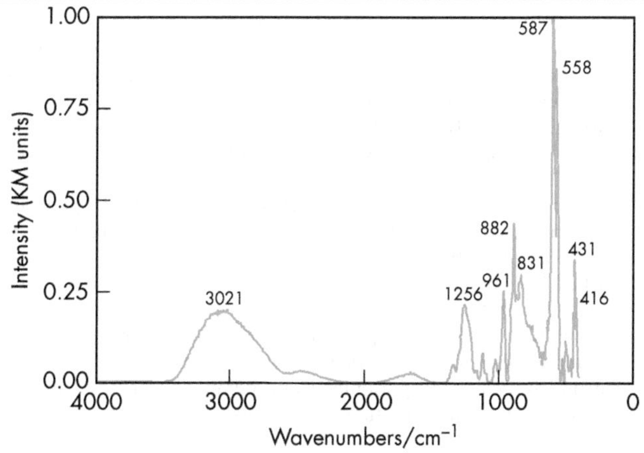

Figure 1: Infrared spectrum of anhydrous monobasic sodium phosphate measured by diffuse reflectance. Adapted with permission of Informa Healthcare.

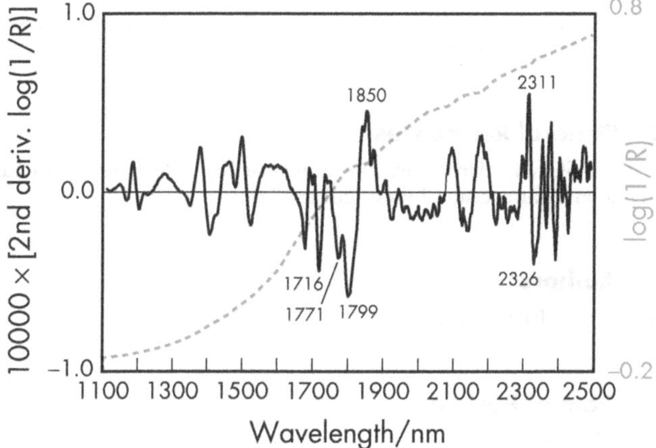

Figure 2: Near-infrared spectrum of anhydrous monobasic sodium phosphate measured by reflectance.

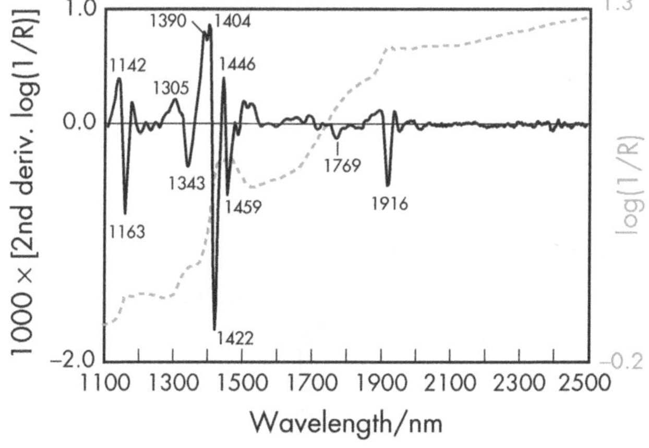

Figure 3: Near-infrared spectrum of monobasic sodium phosphate dihydrate measured by reflectance.

decomposition at 205°C, forming sodium hydrogen pyrophosphate, $Na_2H_2P_2O_7$. At 250°C it leaves a final residue of sodium metaphosphate, $NaPO_3$.

Aqueous solutions are stable and may be sterilized by autoclaving.

Monobasic sodium phosphate should be stored in an airtight container in a cool, dry place.

12 Incompatibilities

Monobasic sodium phosphate is an acid salt and is therefore generally incompatible with alkaline materials and carbonates; aqueous solutions of monobasic sodium phosphate are acidic and will cause carbonates to effervesce.

Monobasic sodium phosphate should not be administered concomitantly with aluminum, calcium, or magnesium salts since they bind phosphate and could impair its absorption from the gastrointestinal tract. Interaction between calcium and phosphate, leading to the formation of insoluble calcium phosphate precipitates, is possible in parenteral admixtures.[1–3]

13 Method of Manufacture

Monobasic sodium phosphate is prepared by adding phosphoric acid to a hot, concentrated solution of disodium phosphate until the liquid ceases to form a precipitate with barium chloride. This solution is then concentrated and the monobasic sodium phosphate is crystallized.

14 Safety

Monobasic sodium phosphate is widely used as an excipient in parenteral, oral, and topical pharmaceutical formulations.

Phosphate occurs extensively in the body and is involved in many physiological processes since it is the principal anion of intracellular fluid. Most foods contain adequate amounts of phosphate, making hypophosphatemia[4] virtually unknown except in certain disease states[5] or in patients receiving total parenteral nutrition. Treatment is usually by the oral administration of up to 100 mmol of phosphate daily.

Approximately two-thirds of ingested phosphate is absorbed from the gastrointestinal tract, virtually all of it being excreted in the urine, and the remainder is excreted in the feces.

Excessive administration of phosphate, particularly intravenously, rectally, or in patients with renal failure, can cause hyperphosphatemia that may lead to hypocalcemia or other severe electrolyte imbalances.[6–8] Adverse effects occur less frequently following oral consumption, although phosphates act as mild saline laxatives when administered orally or rectally (2–4 g of monobasic sodium phosphate in an aqueous solution is used as a laxative). Consequently, gastrointestinal disturbances including diarrhea, nausea, and vomiting may occur following the use of monobasic sodium phosphate as an excipient in oral formulations. However, the level of monobasic sodium phosphate used as an excipient in a pharmaceutical formulation is not usually associated with adverse effects.

LD_{50} (rat, IM): 0.25 g/kg[9]

LD_{50} (rat, oral): 8.29 g/kg

15 Handling Precautions

Observe normal precautions appropriate to the circumstances and quantity of material handled. Monobasic sodium phosphate may be irritant to the skin, eyes, and mucous membranes. Eye protection and gloves are recommended.

16 Regulatory Status

GRAS listed. Accepted for use as a food additive in Europe. Included in the FDA Inactive Ingredients Database (injections; infusions; ophthalmic, oral, topical, and vaginal preparations). Included in nonparenteral and parenteral medicines licensed in the UK. Included in the Canadian Natural Health Products Ingredients Database.

17 Related Substances

Dibasic sodium phosphate; monobasic potassium phosphate.

Monobasic potassium phosphate
Empirical formula KH_2PO_4
Molecular weight 136.09
CAS number [7778-77-0]
Synonyms E340; monopotassium phosphate; potassium acid phosphate; potassium biphosphate; potassium dihydrogen orthophosphate.
Appearance Colorless crystals or a white, odorless, granular or crystalline powder.
Acidity/alkalinity pH $\approx$ 4.5 for a 1% w/v aqueous solution at 25°C.
Solubility Freely soluble in water; practically insoluble in ethanol (95%).
Comments 1 g of monobasic potassium phosphate represents approximately 7.3 mmol of potassium and of phosphate.
The EINECS number for monobasic potassium phosphate is 231-913-4.

18 Comments

One gram of anhydrous monobasic sodium phosphate represents approximately 8.3 mmol of sodium and of phosphate.

One gram of monobasic sodium phosphate monohydrate represents approximately 7.2 mmol of sodium and of phosphate.

One gram of monobasic sodium phosphate dihydrate represents approximately 6.4 mmol of sodium and of phosphate.

Therapeutically, monobasic sodium phosphate is used as a mild saline laxative and in the treatment of hypophosphatemia.[4,5,10]

A specification for sodium phosphate monobasic is contained in the *Food Chemicals Codex* (FCC).[11]

The EINECS number for monobasic sodium phosphate is 231-449-2. The PubChem Compound ID (CID) for monobasic sodium phosphate dihydrate is 23673460.

19 Specific References

1 Eggert LD, *et al.* Calcium and phosphorus compatibility in parenteral nutrition solutions for neonates. *Am J Hosp Pharm* 1982; **39**: 49–53.
2 Niemiec PW, Vanderveen TW. Compatibility considerations in parenteral nutrient solutions. *Am J Hosp Pharm* 1984; **41**: 893–911.
3 Pereira-da-Silva L, *et al.* Compatibility of calcium and phosphate in four parenteral nutrition solutions for preterm neonates. *Am J Health Syst Pharm* 2003; **60**: 1041–1044.
4 Lloyd CW, Johnson CE. Management of hypophosphatemia. *Clin Pharm* 1988; **7**: 123–128.
5 Holland PC, *et al.* Prenatal deficiency of phosphate, phosphate supplementation, and rickets in very-low-birthweight infants. *Lancet* 1990; **335**: 697–701.
6 Haskell LP. Hypocalcaemic tetany induced by hypertonic-phosphate enema [letter]. *Lancet* 1985; **ii**: 1433.
7 Larson JE, *et al.* Laxative phosphate poisoning: pharmacokinetics of serum phosphorus. *Hum Toxicol* 1986; **5**: 45–49.
8 Martin RR, *et al.* Fatal poisoning from sodium phosphate enema: case report and experimental study. *J Am Med Assoc* 1987; **257**: 2190–2192.
9 Lewis RJ, ed. *Sax's Dangerous Properties of Industrial Materials*, 11th edn. New York: Wiley, 2004: 3274.
10 Rosen GH, *et al.* Intravenous phosphate repletion regimen for critically ill patients with moderate hypophosphatemia. *Crit Care Med* 1995; **23**: 1204–1210.
11 *Food Chemicals Codex*, 7th edn. (Suppl. 1). Bethesda, MD: United States Pharmacopeia, 2010: 1491.

20 General References

Sweetman SC, ed. *Martindale: The Complete Drug Reference*, 37th edn. London: Pharmaceutical Press, 2011: 1828.

21 Authors

ME Fenton, PJ Sheskey.

22 Date of Revision

2 March 2012.

Sodium Propionate

1 Nonproprietary Names

BP: Sodium Propionate
PhEur: Sodium Propionate
USP–NF: Sodium Propionate

2 Synonyms

E281; ethylformic acid, sodium salt, hydrate; methylacetic acid, sodium salt, hydrate; natrii propionas; sodium propanoate hydrate; sodium propionate hydrate.

3 Chemical Name and CAS Registry Number

Propionic acid, sodium salt, hydrate [6700-17-0]
Propionic acid, sodium salt, anhydrous [137-40-6]

4 Empirical Formula and Molecular Weight

$C_3H_5NaO_2 \cdot xH_2O$	114.06 (for monohydrate)
$C_3H_5NaO_2$	96.06 (for anhydrous)

5 Structural Formula

H_3C O^-Na^+ O

6 Functional Category

Antimicrobial preservative.

7 Applications in Pharmaceutical Formulation or Technology

As an excipient, sodium propionate is used in oral pharmaceutical formulations as an antimicrobial preservative. Like propionic acid, sodium propionate and other propionic acid salts are fungistatic and bacteriostatic against a number of Gram-positive cocci.

Propionates are more active against molds than is sodium benzoate, but have essentially no activity against yeasts; *see* Section 10.

8 Description

Sodium propionate occurs as colorless transparent crystals or as a granular, free-flowing, crystalline powder. It is odorless, or with a slight characteristic odor, and is deliquescent in moist air. Sodium propionate has a characteristic, slightly cheese like taste, although by itself it is unpalatable.

9 Pharmacopeial Specifications

See Table I.

Table I: Pharmacopeial specifications for sodium propionate.

Test	PhEur 7.4	USP35–NF30
Identification	+	+
Characters	+	–
Appearance of solution	+	–
Alkalinity	–	+
pH	7.8–9.2	–
Water	–	≤1.0%
Heavy metals	≤10 ppm	≤10 ppm
Related substances	+	–
Readily oxidizable substances	+	–
Iron	≤10 ppm	–
Loss on drying	0.5%	–
Assay (dried basis)	99.0–101.0%	99.0–100.5%

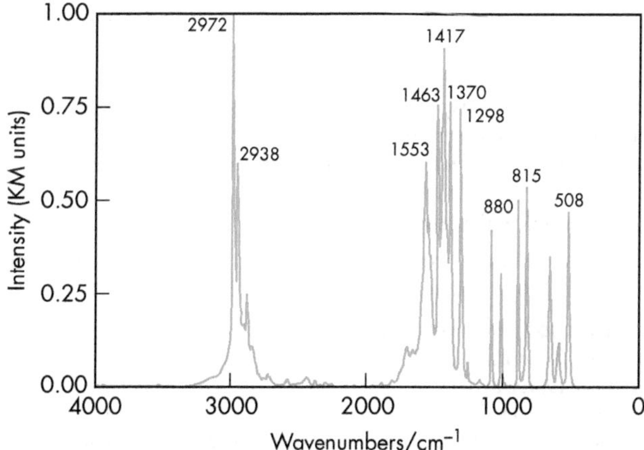

Figure 1: Infrared spectrum of anhydrous sodium propionate measured by diffuse reflectance. Adapted with permission of Informa Healthcare.

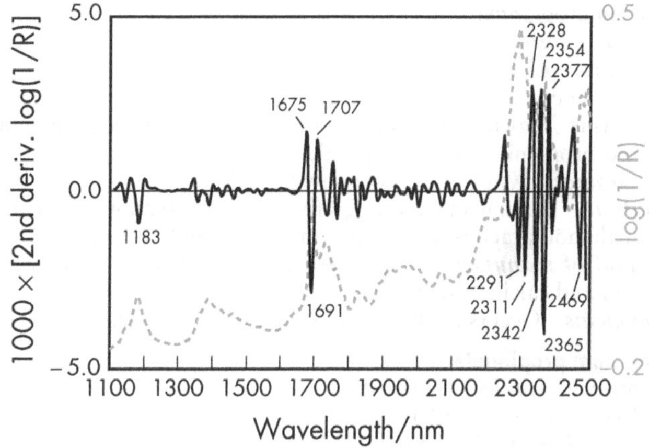

Figure 2: Near-infrared spectrum of sodium propionate measured by reflectance.

10 Typical Properties

Antimicrobial activity Sodium propionate, propionic acid, and other propionates possess mainly antifungal activity and are used as preservatives primarily against molds; they exhibit essentially no activity against yeasts. Although, in general, propionates exhibit little activity against bacteria, sodium propionate is effective against *Bacillus mesenterium*, the organism that causes 'rope' in bread. Antimicrobial activity is largely dependent upon the presence of the free acid and hence propionates exhibit optimum activity at acid pH, notably at less than pH 5. Synergistic effects occur between propionates and carbon dioxide or sorbic acid. *See also* Propionic acid.

Melting point 285°C

Solubility Soluble 1 in 24 of ethanol (95%), 1 in 1 of water, and 1 in 0.65 of boiling water; practically insoluble in chloroform, ether, and methylene chloride.

Spectroscopy

IR spectra *see* Figure 1.

NIR spectra *see* Figure 2.

11 Stability and Storage Conditions

Sodium propionate is deliquescent and should therefore be stored in an airtight container in a cool, dry place.

12 Incompatibilities

Incompatibilities for sodium propionate are similar to those of other weak organic acids.

13 Method of Manufacture

Sodium propionate is prepared by the reaction of propionic acid with sodium carbonate or sodium hydroxide.

14 Safety

Sodium propionate and other propionates are used in oral pharmaceutical formulations, food products, and cosmetics. The free acid, propionic acid, occurs naturally at levels up to 1% w/w in certain cheeses.

Following oral consumption, propionate is metabolized in mammals in a manner similar to that of fatty acids. Toxicity studies in animals have shown sodium propionate and other propionates to be relatively nontoxic materials.[1,2] In veterinary medicine, sodium propionate is used as a therapeutic agent for cattle and sheep.[5]

In humans, 6 g of sodium propionate has been administered daily without harm.[1] However, allergic reactions to propionates can occur.

LD_{50} (mouse, oral): 6.33 g/kg[3]
LD_{50} (mouse, SC): 2.1 g/kg
LD_{50} (rabbit, skin): 1.64 g/kg

15 Handling Precautions

Observe normal precautions appropriate to the circumstances and quantity of material handled. Sodium propionate may be irritant to the eyes and skin. Gloves, eye protection, and a dust-mask are recommended. When heated to decomposition, sodium propionate emits toxic fumes of sodium monoxide, Na_2O.

In the UK, the workplace exposure limits for propionic acid are 31 mg/m^3 (10 ppm) long-term (8-hour TWA) and 46 mg/m^3 (15 ppm) short-term.[4]

16 Regulatory Status

GRAS listed. Accepted for use as a food additive in Europe. In cheese products, propionates are limited to 0.3% w/w concentration; a limit of 0.32% w/w is applied in flour and white bread rolls, while a limit of 0.38% w/w is applied in whole wheat products.

Included in the FDA Inactive Ingredients Database (oral tablets, capsules, powder, suspensions, and syrups). Included in nonparenteral medicines licensed in the UK. Included in the Canadian Natural Health Products Ingredients Database.

17 Related Substances

Anhydrous sodium propionate; calcium propionate; potassium propionate; propionic acid; zinc propionate.

Anhydrous sodium propionate
Empirical formula $C_3H_5O_2Na$
Molecular weight 96.06
CAS number [137-40-6]
Synonyms E281; propanoic acid, sodium salt, anhydrous.
Safety
LD_{50} (mouse, oral): 2.35 g/kg[4]
LD_{50} (rat, oral): 3.92 g/kg

Calcium propionate
Empirical formula $C_6H_{10}O_4Ca$
Molecular weight 186.22
CAS number [4075-81-4]
Synonyms Calcium dipropionate; E282; propanoic acid, calcium salt; propionic acid, calcium salt.
Appearance White crystalline powder.
Solubility Soluble in water; slightly soluble in ethanol (95%) and methanol; practically insoluble in acetone and benzene.
Method of manufacture Prepared by the reaction of propionic acid and calcium hydroxide.
Comments Occurs as the monohydrate or trihydrate.

Potassium propionate
Empirical formula $C_3H_5O_2K$
Molecular weight 112.17
CAS number [327-62-8]
Synonyms E283; propanoic acid, potassium salt; propionic acid, potassium salt.
Appearance White crystalline powder.
Comments Occurs as the anhydrous form and the monohydrate. Decomposes in moist air to give off propionic acid.

Zinc propionate
Empirical formula $C_6H_{10}O_4Zn$
Molecular weight 211.52
CAS number [557-28-8]
Synonyms Propanoic acid, zinc salt; propionic acid, zinc salt.
Appearance White platelets or needlelike crystals (for the monohydrate).
Solubility The anhydrous form is soluble 1 in 36 of ethanol (95%) at 15°C, 1 in 6 of boiling ethanol (95%), and 1 in 3 of water at 15°C.
Method of manufacture Prepared by dissolving zinc oxide in dilute propionic acid solution.
Comments Occurs as the anhydrous form and the monohydrate. Decomposes in moist air to give off propionic acid.

18 Comments

Propionate salts are used as antimicrobial preservatives in preference to propionic acid since they are less corrosive.

Therapeutically, sodium propionate has been used topically in concentrations up to 10% w/w alone or in combination with other propionates, caprylates, or other antifungal agents, in the form of ointments or solutions for the treatment of dermatophyte infections. Eye drops containing 5% w/v sodium propionate have also been used.

In food processes, particularly baking, sodium propionate is used as an antifungal agent; it may also be used as a flavoring agent in food products. In veterinary medicine, sodium propionate is used therapeutically as a glucogenic substance in ruminants.[5]

The therapeutic use of sodium propionate in topical antifungal preparations has largely been superseded by a new generation of antifungal drugs.

A specification for sodium propionate is contained in the *Food Chemicals Codex* (FCC).[6]

The EINECS number for sodium propionate is 205-290-4. The PubChem Compound ID (CID) for sodium propionate is 23663426.

19 Specific References

1 Heseltine WW. A note on sodium propionate. *J Pharm Pharmacol* 1952; 4: 120–122.
2 Graham WD, *et al.* Chronic toxicity of bread additives to rats. *J Pharm Pharmacol* 1954; 6: 534–545.
3 Lewis RJ, ed. *Sax's Dangerous Properties of Industrial Materials*, 11th edn. New York: Wiley, 2004: 3276.
4 Health and Safety Executive. *EH40/2005 Workplace Exposure Limits*. Sudbury: HSE Books, 2011. http://www.hse.gov.uk/pubns/priced/eh40.pdf (accessed 16 February 2012).
5 Bishop Y, ed. *The Veterinary Formulary*, 6th edn. London: Pharmaceutical Press, 2005: 419–420.
6 *Food Chemicals Codex*, 7th edn. Bethesda, MD: United States Pharmacopeia, 2010: 959.

20 General References

Doores S. Organic acids. In: Branen AL, Davidson PM, eds. *Antimicrobials in Foods*. New York: Marcel Dekker, 1983: 85–87.
Furia TE, ed. *CRC Handbook of Food Additives*. Cleveland, OH: CRC Press, 1972: 137–141.

21 Author

BV Kadri.

22 Date of Revision

2 March 2012.

S

Sodium Starch Glycolate

1 Nonproprietary Names

BP: Sodium Starch Glycolate

PhEur: Sodium Starch Glycolate

USP–NF: Sodium Starch Glycolate

2 Synonyms

Carboxymethyl starch, sodium salt; carboxymethylamylum natricum; *Explosol*; *Explotab*; *Glycolys*; *Primojel*; *SSG Sanaq*; starch carboxymethyl ether, sodium salt; *Tablo*; *Vivastar Low pH*; *Vivastar P*.

3 Chemical Name and CAS Registry Number

Sodium carboxymethyl starch [9063-38-1]

4 Empirical Formula and Molecular Weight

The USP35–NF30 describes two types of sodium starch glycolate, Type A and Type B, and states that sodium starch glycolate is the sodium salt of a carboxymethyl ether of starch or of a crosslinked carboxymethyl ether of starch.

The PhEur 7.4 describes three types of material: Type A and Type B are described as the sodium salt of a crosslinked partly *O*-carboxymethylated potato starch. Type C is described as the sodium salt of a partly *O*-carboxymethylated starch, crosslinked by physical dehydration. Types A, B, and C are differentiated by their pH, sodium, and sodium chloride content.

The PhEur and USP–NF monographs have been harmonized for Type A and Type B variants.

Sodium starch glycolate may be characterized by the degree of substitution and crosslinking. The molecular weight is typically 500 000–1 000 000.

5 Structural Formula

CH_2COONa; CH_2OH; CH_2O; O; OH; n

6 Functional Category

Tablet and capsule disintegrant.

7 Applications in Pharmaceutical Formulation or Technology

Sodium starch glycolate is widely used in oral pharmaceuticals as a disintegrant in capsule[1–6] and tablet formulations.[7–10] It is commonly used in tablets prepared by either direct-compression[11–13] or wet-granulation processes.[14–16] The usual concentration employed in a formulation is between 2% and 8%, with the optimum concentration about 4%, although in many cases 2% is sufficient. Disintegration occurs by rapid uptake of water followed by rapid and enormous swelling.[17–20]

Although the effectiveness of many disintegrants is affected by the presence of hydrophobic excipients such as lubricants, the disintegrant efficiency of sodium starch glycolate is unimpaired. Increasing the tablet compression pressure also appears to have no effect on disintegration time.[10–12]

Sodium starch glycolate has also been investigated for use as a suspending vehicle.[21]

8 Description

Sodium starch glycolate occurs as a white or almost white free-flowing very hygroscopic powder. The PhEur 7.4 states that when examined under a microscope it is seen to consist of: granules, irregularly shaped, ovoid or pear-shaped, 30–100 μm in size, or rounded, 10–35 μm in size; compound granules consisting of 2–4 components occur occasionally; the granules have an eccentric hilum and clearly visible concentric striations. Between crossed nicol prisms, the granules show a distinct black cross intersecting at the hilum; small crystals are visible at the surface of the granules. The granules show considerable swelling in contact with water.

9 Pharmacopeial Specifications

The pharmacopeial specifications for sodium starch glycolate have undergone harmonization of many attributes for PhEur, and USP–NF.

See Table I. *See also* Section 18.

10 Typical Properties

Acidity/alkalinity *See* Section 9.

Density (bulk)

0.75 g/cm^3 for *Explotab*;

0.756 g/cm^3 for *Glycolys*;

0.81 g/cm^3 for *Primojel*;

0.67 g/cm^3 for *Tablo*.

Density (tapped)

0.88 g/cm^3 for *Explotab*;

0.945 g/cm^3 for *Glycolys*;

SEM 1: Excipient: sodium starch glycolate (*Explotab*); manufacturer: JRS Pharma; magnification: 300×; voltage: 5 kV.

SEM 2: Excipient: sodium starch glycolate (*Glycolys*); manufacturer: Roquettes Frères.

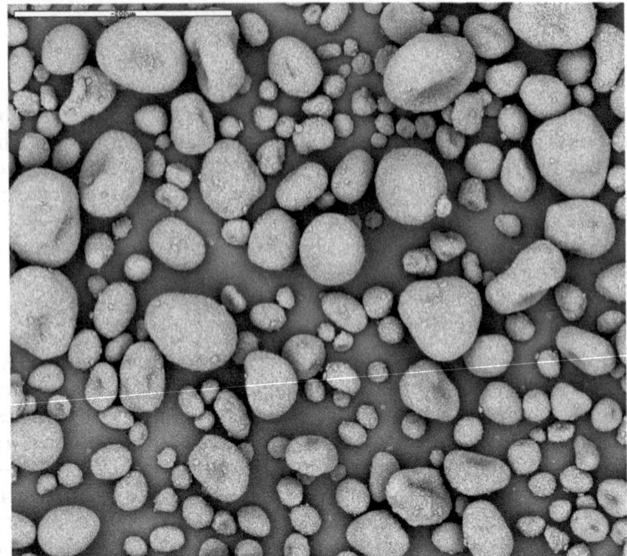

SEM 3: Excipient: sodium starch glycolate (*Primojel*); manufacturer: DMV-Fonterra Excipients; magnification: 200×; voltage: 1.5 kV.

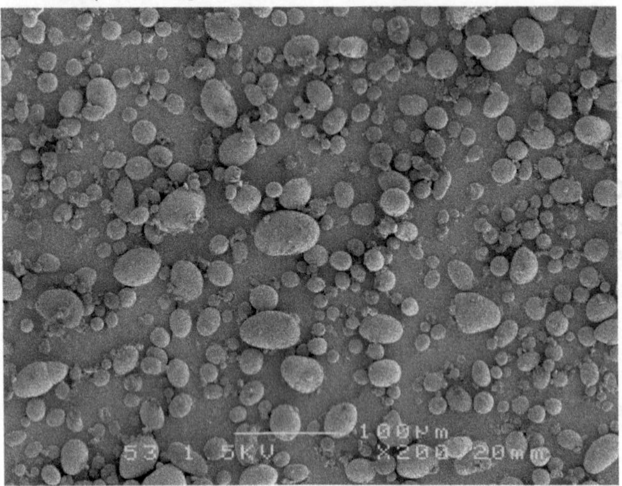

0.98 g/cm^3 for *Primojel*;

0.83 g/cm^3 for *Tablo*.

Density (true)

1.51 g/cm^3 for *Explotab*;

1.56 g/cm^3 for *Primojel*;

1.49 g/cm^3 for *Tablo*.

Melting point Does not melt, but chars at approximately 200°C.

Particle size distribution 100% of particles less than 106 μm in size. Average particle size (d_{50}) is 38 μm and 42 μm for *Primojel* by microscopy and sieving, respectively.

Solubility Practically insoluble in methylene chloride. It gives a translucent suspension in water.

Specific surface area

0.202 m^2/g for *Explotab*;

0.24 m^2/g for *Glycolys*;

0.185 m^2/g for *Primojel*;

0.335 m^2/g for *Tablo*.

Spectroscopy

IR spectra *see* Figure 1.

NIR spectra *see* Figure 2.

SEM 4: Excipient: sodium starch glycolate (*Vivastar P*); manufacturer: JRS Pharma; magnification: 300×; voltage: 5 kV.

Table I: Pharmacopeial specifications for sodium starch glycolate.

Test	PhEur 7.4	USP35–NF30
Identification	+	+
Characters	+	—
Appearance of solution	+	—
pH	+	+
Type A	5.5–7.5	5.5–7.5
Type B	3.0–5.0	3.0–5.0
Type C	5.5–7.5	—
Heavy metals[b]	≤20 ppm	≤20 ppm
Iron	≤20 ppm	≤0.002%
Loss on drying	+	≤10%
Type A	≤10.0%	—
Type B	≤10.0%	—
Type C	≤7.0%	—
Microbial limits	+[a]	+[a]
Sodium chloride	+	≤7.0%
Type A	≤7.0%	—
Type B	≤7.0%	—
Type C	≤1.0%	—
Sodium glycolate	+	≤2.0%
Type A	≤2.0%	—
Type B	≤2.0%	—
Type C	≤2.0%	—
Assay (of Na)	+	+
Type A	2.8–4.2%	2.8–4.2%
Type B	2.0–3.4%	2.0–3.4%
Type C	2.8–5.0%	—

(a) Complies with tests for *Salmonella* and *Escherichia coli*.
(b) This test has not been fully harmonized at the time of publication.

Swelling capacity In water, sodium starch glycolate swells to up to 300 times its initial volume.

Viscosity (dynamic) ≤200 mPa s (200 cP) for a 4% w/v aqueous dispersion; viscosity is 4.26 mPa s for a 2% w/v aqueous dispersion (depending on source and grade).

11 Stability and Storage Conditions

Tablets prepared with sodium starch glycolate have good storage properties.[22–24] Sodium starch glycolate is stable although very hygroscopic, and should be stored in a well-closed container in order to protect it from wide variations of humidity and temperature, which may cause caking.

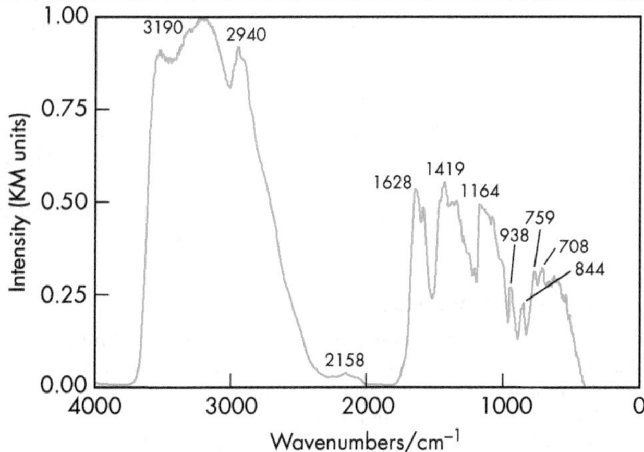

Figure 1: Infrared spectrum of sodium starch glycolate measured by diffuse reflectance. Adapted with permission of Informa Healthcare.

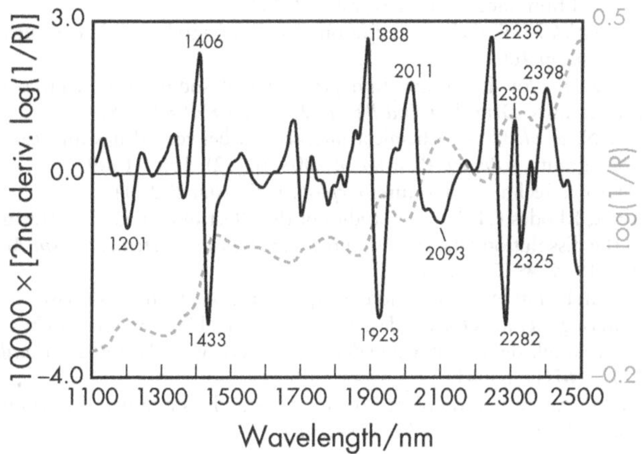

Figure 2: Near-infrared spectrum of sodium starch glycolate measured by reflectance.

The physical properties of sodium starch glycolate remain unchanged for up to 3 years if it is stored at moderate temperatures and humidity.

12 Incompatibilities

Sodium starch glycolate is incompatible with ascorbic acid.[25]

13 Method of Manufacture

Sodium starch glycolate is a substituted derivative of potato starch. Typically, commercial products are also crosslinked using either sodium trimetaphosphate (Types A and B) or dehydration (Type C).[26]

Starch is carboxymethylated by reacting it with sodium chloroacetate in an alkaline, nonaqueous medium, typically denatured ethanol or methanol, followed by neutralization with citric acid, acetic acid, or some other acid. *Vivastar P* is manufactured in methanolic medium, and *Explotab* in ethanolic medium.

14 Safety

Sodium starch glycolate is widely used in oral pharmaceutical formulations and is generally regarded as a nontoxic and nonirritant material. However, oral ingestion of large quantities may be harmful.

15 Handling Precautions

Observe normal precautions appropriate to the circumstances and quantity of material handled. Sodium starch glycolate may be irritant to the eyes; eye protection and gloves are recommended. A dust mask or respirator is recommended for processes that generate a large quantity of dust.

16 Regulatory Status

Included in the FDA Inactive Ingredients Database (oral capsules and tablets). Included in nonparenteral medicines licensed in the UK. Included in the Canadian Natural Health Products Ingredients Database.

17 Related Substances

Pregelatinized starch; starch.

18 Comments

Sodium starch glycolate is has undergone harmonization of many attributes for PhEur, and USP–NF by the Pharmacopeial Discussion Group. For further information see the General Information Chapter <1196> in the USP35–NF30, the General Chapter 5.8 in PhEur 7.4, along with the 'State of Work' document on the PhEur EDQM website, and also the General Information Chapter 8 in the JP XV.

The physical properties of sodium starch glycolate, and hence its effectiveness as a disintegrant, are affected by the degree of crosslinkage, extent of carboxymethylation, and purity.[27,28]

Sodium starch glycolate has been reported to interact with glycopeptide antibiotics,[29,30] basic drugs, and increase the photostability of norfloxacin.[31] The solubility of the formulation matrix and mode of incorporation in wet granulation can affect the disintegration time; disintegration times can be slower in tablets containing high levels of soluble excipients.[32]

Commercially, sodium starch glycolate is available in a number of speciality grades, e.g. low pH (*Explotab Low pH, Glycolys Low pH*); low viscosity (*Explotab CLV, Glycolys LV*); low solvent (*Vivastar PSF*); and low moisture *Glycolys LM*.

A specification for sodium starch glycolate is included in the *Japanese Pharmaceutical Excipients* (JPE).[33]

19 Specific References

1 Newton JM, Razzo FN. Interaction of formulation factors and dissolution fluid and the *in vitro* release of drug from hard gelatin capsules. *J Pharm Pharmacol* 1975; **27**: 78P.

2 Stewart AG, *et al.* The release of a model low-dose drug (riboflavine) from hard gelatin capsule formulations. *J Pharm Pharmacol* 1979; **31**: 1–6.

3 Chowhan ZT, Chi L-H. Drug–excipient interactions resulting from powder mixing III: solid state properties and their effect on drug dissolution. *J Pharm Sci* 1986; **75**: 534–541.

4 Botzolakis JE, Augsburger LL. Disintegrating agents in hard gelatin capsules part 1: mechanism of action. *Drug Dev Ind Pharm* 1988; **14**(1): 29–41.

5 Hannula A-M, *et al.* Release of ibuprofen from hard gelatin capsule formulations: effect of modern disintegrants. *Acta Pharm Fenn* 1989; **98**: 189–196.

6 Marvola M, *et al.* Effect of sodium bicarbonate and sodium starch glycolate on the *in vivo* disintegration of hard gelatin capsules – a radiological study in the dog. *Acta Pharm Nord* 1989; **1**: 355–362.

7 Khan KA, Rooke DJ. Effect of disintegrant type upon the relationship between compressional pressure and dissolution efficiency. *J Pharm Pharmacol* 1976; **28**: 633–636.

8 Rubinstein MH, Price EJ. *In vivo* evaluation of the effect of five disintegrants on the bioavailability of frusemide from 40 mg tablets. *J Pharm Pharmacol* 1977; **29**: 5P.

9 Caramella C, *et al.* The influence of disintegrants on the characteristics of coated acetylsalicylic acid tablets. *Farmaco (Prat)* 1978; **33**: 498–507.

S

10 Gebre Mariam T, *et al*. Evaluation of the disintegration efficiency of a sodium starch glycolate prepared from enset starch in compressed tablets. *Eur J Pharm Biopharm* 1996; **42**(2): 124–132.
11 Cid E, Jaminet F. [Influence of adjuvants on the dissolution rate and stability of acetylsalicylic acid in compressed tablets.] *J Pharm Belg* 1971; **26**: 38–48[in French].
12 Gordon MS, Chowhan ZT. Effect of tablet solubility and hygroscopicity on disintegrant efficiency in direct compression tablets in terms of dissolution. *J Pharm Sci* 1987; **76**: 907–909.
13 Cordoba-Diaz M, *et al*. Influence of pharmacotechnical design on the interaction and availability of norfloxacin in directly compressed tablets with certain antacids. *Drug Dev Ind Pharm* 2000; **26**: 159–166.
14 Sekulović D, *et al*. The investigation of the influence of Explotab on the disintegration of tablets. *Pharmazie* 1986; **41**: 153–154.
15 Bolhius GK, *et al*. Improvement of dissolution of poorly soluble drugs by solid deposition on a super disintegrant. Part 2. Choice of super disintegrants and effect of granulation. *Eur J Pharm Sci* 1997; **5**(2): 63–69.
16 Gordon MS, *et al*. Effect of the mode of super disintegrant incorporation on dissolution in wet granulated tablets. *J Pharm Sci* 1993; **82**: 220–226.
17 Khan KA, Rhodes CT. Disintegration properties of calcium phosphate dibasic dihydrate tablets. *J Pharm Sci* 1975; **64**: 166–168.
18 Khan KA, Rhodes CT. Water-sorption properties of tablet disintegrants. *J Pharm Sci* 1975; **64**: 447–451.
19 Wan LSC, Prasad KPP. Uptake of water by excipients in tablets. *Int J Pharm* 1989; **50**: 147–153.
20 Thibert R, Hancock BC. Direct visualization of superdisintegrant hydration using environmental scanning electron microscopy. *J Pharm Sci* 1996; **85**: 1255–1258.
21 Danckwerts MP, *et al*. Pharmaceutical formulation of a fixed-dose anti-tuberculosis combination. *Int J Tuberc Lung D* 2003; **7**: 289–297.
22 Horhota ST, *et al*. Effect of storage at specified temperature and humidity on properties of three directly compressible tablet formulations. *J Pharm Sci* 1976; **65**: 1746–1749.
23 Sheen P-C, Kim S-I. Comparative study of disintegrating agents in tiaramide hydrochloride tablets. *Drug Dev Ind Pharm* 1989; **15**(3): 401–414.
24 Gordon MS, Chowhan ZT. The effect of aging on disintegrant efficiency in direct compression tablets with varied solubility and hygroscopicity, in terms of dissolution. *Drug Dev Ind Pharm* 1990; **16**(3): 437–447.
25 Botha SA, *et al*. DSC screening for drug–excipient and excipient–excipient interactions in polypharmaceuticals intended for the alleviation of the symptoms of colds and flu. III. *Drug Dev Ind Pharm* 1987; **13**(7): 1197–1215.
26 Bolhuis GK, *et al*. On the similarity of sodium starch glycolate from different sources. *Drug Dev Ind Pharm* 1986; **12**(4): 621–630.
27 Rudnic EM, *et al*. Effect of molecular structure variation on the disintegrant action of sodium starch glycolate. *J Pharm Sci* 1985; **74**: 647–650.
28 Bolhuis GK, *et al*. Effect of variation of degree of substitution, crosslinking and purity on the disintegrant efficiency of sodium starch glycolate. *Acta Pharm Technol* 1984; **30**: 24–32.
29 Claudius JS, Neau SH. Kinetic and equilibrium characterization of interactions between glycopeptide antibiotics and sodium carboxymethyl starch. *Int J Pharm* 1996; **144**: 71–79.
30 Claudius JS, Neau SH. The solution stability of vancomycin in the presence and absence of sodium carboxymethyl starch. *Int J Pharm* 1998; **168**: 41–48.
31 Cordoba-Borrego M, *et al*. Validation of a high performance liquid chromatographic method for the determination of norfloxacin and its application to stability studies (photostability study of norfloxacin). *J Pharm Biomed Anal* 1999; **18**: 919–926.
32 Gordon MS, *et al*. Effect of the mode of super disintegrant incorporation on dissolution in wet granulated tablets. *J Pharm Sci* 1993; **82**: 220–226.
33 Japanese Pharmaceutical Excipients Council. *Japanese Pharmaceutical Excipients 2004*. Tokyo: Yakuji Nippo, 2004; 774.

20 General References

Augsburger LL *et al*. Superdisintegrants: characterisation and function. In: Swarbrick J, ed. *Encyclopedia of Pharmaceutical Technology*, 3rd edn. London: Informa Healthcare, 2007: 2553–2567.
Blanver. Technical literature: *Tablo, Explosol*, 2008.
DMV-Fonterra Excipients. Technical literature: *Primojel*, 2008.
European Directorate for the Quality of Medicines and Healthcare (EDQM). European Pharmacopoeia – State Of Work Of International Harmonisation. *Pharmeuropa* 2011; 23(4): 713–714. www.edqm.eu/site/-614.html (accessed 5 December 2011).
Edge S, *et al*. Chemical characterisation of sodium starch glycolate particles. *Int J Pharm* 2002; **240**: 67–78.
Edge S, *et al*. Powder compaction properties of sodium starch glycolate disintegrants. *Drug Dev Ind Pharm* 2002; **28**(8): 989–999.
Fransén N, *et al*. Physicochemical interactions between drugs and superdisintegrants. *J Pharm Pharmacol* 2008; **60**(12): 1583–1589.
JRS Pharma. Technical literature: *Explotab, Vivastar P*, 2004.
Khan KA, Rhodes CT. Further studies of the effect of compaction pressure on the dissolution efficiency of direct compression systems. *Pharm Acta Helv* 1974; **49**: 258–261.
Mantovani F, *et al*. A combination of vapor sorption and dynamic laser light scattering methods for the determination of the Flory parameter chi and the crosslink density of a powdered polymeric gel. *Fluid Phase Equilib* 2000; **167**(1): 63–81.
Mendell E. An evaluation of carboxymethyl starch as a tablet disintegrant. *Pharm Acta Helv* 1974; **49**: 248–250.
Mittapalli RK, *et al*. Varying efficacy of superdisintegrants in orally disintegrating tablets among different manufacturers. *Pharmazie* 2010; **65**(11): 805–810.
Roquette Frères. Technical literature: *Glycolys*, 2008.
Shah U, Augsburger L. Multiple sources of sodium starch glycolate NF: evaluation of functional equivalence and development of standard performance tests. *Pharm Dev Tech* 2002; **7**(3): 345–359.
Young PM, *et al*. Interaction of moisture with sodium starch glycolate. *Pharm Dev Tech* 2007; **12**: 211–216.
Young PM, *et al*. The effect of moisture on the compressibility and compactability of sodium starch glycolates. *Pharm Dev Tech* 2007; **12**: 217–222.

21 Author

JX Li.

22 Date of Revision

2 March 2012.

S

Sodium Stearate

1 Nonproprietary Names

BP: Sodium Stearate
PhEur: Sodium Stearate
USP–NF: Sodium Stearate

2 Synonyms

Kemilub ES; natrii stearas; octadecanoic acid, sodium salt; *Prodhygine*; stearic acid, sodium salt.

3 Chemical Name and CAS Registry Number

Sodium octadecanoate [822-16-2]

4 Empirical Formula and Molecular Weight

$C_{18}H_{35}NaO_2$ 306.5

The USP35–NF30 describes sodium stearate as a mixture of sodium stearate and sodium palmitate, which together constitute not less than 90% of the total content. Sodium stearate contains small amounts of the sodium salts of other fatty acids.

5 Structural Formula

H_3C O O^- Na^+

6 Functional Category

Emulsifying agent; gelling agent; glidant; modified-release agent; stiffening agent; tablet and capsule lubricant.

7 Applications in Pharmaceutical Formulation or Technology

Sodium stearate is used as an emulsifying and stiffening agent in a variety of topical creams[1] and rectal preparations (Glycerin Suppositories USP). It is used as a tablet and capsule lubricant in immediate-release tablets and gastro-resistant capsules both containing omeprazole. It is also used as a glidant and modified-release agent in tablets. It is used in the preparation of microemulsions.[2]

Medicated solidified sodium stearate sticks containing vitamins, fungicides, and local anesthetics have been extensively studied for their physical[3] and rheological[4] properties, stability,[5] and biological activity *in vitro* and in animals.[6] The stick dosage form proved as effective as the ointment dosage form, and showed good stability over an 18-month period.

Sodium stearate is used as a gelling agent in deodorant sticks,[7,8] and in cosmetics, shampoos, and bubble baths.

8 Description

Sodium stearate occurs as a white or yellowish fine powder, greasy to the touch with a slight, tallow-like odor.

9 Pharmacopeial Specifications

See Table I.

Table I: Pharmacopeial specifications for sodium stearate.

Test	PhEur 7.4	USP35-NF30
Identification	+	+
Characters	+	–
Freezing point/solidification temperature	≥53°C	≥54°C
Acid value	195–210	196–211
Iodine value	–	≤4
Loss on drying	≤5%	≤5%
Acidity	+	+
Chlorides	≤0.2%	–
Sulfates	≤0.3%	–
Nickel	≤5 ppm	–
Microbial contamination		
Bacteria	10^3 cfu/g	–
Yeast	10^2 cfu/g	–
Alcohol-insoluble substances	–	+
Assay		
Sodium	7.4–8.5%	–
Stearic acid	≥40%	–
Sum of stearic and palmitic acids	≥90%	–
Sodium stearate	–	≥40%
Sum of sodium stearate and sodium palmitate	–	≥90%

10 Typical Properties

Boiling point 359.4°C at 760 mmHg.
Flash point 162.4°C
Melting point 205–255°C
pH 10–11 for a 5% aqueous solution.
Solubility Slightly soluble in water, glycols, and ethanol (96%) at room temperature but readily dissolves on heating.
Specific gravity 1.02

11 Stability and Storage Conditions

Store in a well-closed container in a cool, dry place, protected from light. Keep away from sources of ignition.

12 Incompatibilities

Sodium stearate is incompatible with strong acids and oxidizing agents.

13 Method of Manufacture

Sodium stearate is prepared by reacting stearic acid with an equimolar portion of sodium hydroxide.[9]

14 Safety

Sodium stearate is generally nonirritating to skin although irritation has been reported after prolonged contact.[10] Ingestion may cause nausea, vomiting, and diarrhea. Momentary eye irritation is common. Irritant to respiratory tract and occupational asthma has been reported.[10] LD_{50} (rabbit, skin): > 3 g/kg[11]

LD_{50} (rat, oral): > 5 g/kg[11]

LD_{Lo} (dog, IV): 0.01 g/kg[11–13]

S

15 Handling Precautions

Observe normal precautions appropriate to the circumstances and quantity of material handled. When heated to decomposition, sodium stearate emits toxic and flammable fumes.

Eye protection and gloves are recommended. Sodium stearate may be harmful on inhalation and should be used in a well-ventilated environment; a respirator is recommended. In the US, the OSHA limit is 15 mg/m^3 for total dust for sodium stearate.[13] The American Conference of Governmental Industrial Hygienists (ACGIH) TLV is 10 mg/m^3 for total dust for stearates.[11,13]

16 Regulatory Status

Included in the FDA Inactive Ingredient Database (oral, immediate and modified-release tablets). Included in nonparenteral medicines licensed in the UK (immediate-release tablets and gastroresistant capsules). Included in the Canadian Natural Health Products Ingredients Database.

17 Related Substances

Calcium stearate; magnesium stearate; zinc stearate.

18 Comments

Sodium stearate has been investigated as an adjunct to tobramycin microparticulate powders as pulmonary formulations in dry powder inhalers.[14] The presence of sodium stearate had a direct influence on the aerosol performance. In addition, preliminary analysis of the toxic effect of sodium stearate on a cell line showed no effect on cell viability compared to the pure drug. The pulmonary powders showed no overt toxicity to lung cells.

Sodium stearate has also been investigated in semisolid oleaginous ointment bases for ophthalmic use.[15]

Studies have shown that sodium stearate induces platelet aggregation,[16] which is probably the reason why intravenous infusion is known to cause massive thrombosis in dogs.[17]

Therapeutically, sodium stearate has been used in sycosis and other skin diseases.[9]

A specification for sodium stearate is contained in the *Japanese Pharmaceutical Excipients* (JPE).[18]

The EINECS number for sodium stearate is 212-490-5.

S

19 Specific References

1 Schuster G. Experimental tests on the problem of cold emulsification with monodiglyceride dispersions. *Cosmet Toil* 1979; 94: 49–50, 52, 54–56, 61–62, 65, 67.

2 Jayakrishan A. Microemulsions: evolving technology for cosmetic applications. *J Soc Cosmet Chem* 1983; **34**: 335–350.

3 Kassem AA. Influence of panthenol, chlorphenesin and lignocaine on the physical characteristics of solidified sodium stearate-based sticks (SSSS). *Drug Dev Ind Pharm* 1987; **13**: 2277–2299.

4 Mattha AG. Influence of panthenol, chlorphenesin and lignocaine on the rheological properties of solidified sodium stearate based sticks. *Drug Dev Ind Pharm* 1984; **10**: 111–125.

5 Kassem AA. Stability of panthenol, chlorphenesin and lignocaine in solidified sodium stearate based sticks (SSSS). *Drug Dev Ind Pharm* 1984; **10**: 481–490.

6 Abdel-Hamid M. Evaluation of the biological activity of some medicated solidified sodium stearate based sticks (SSSS). *Drug Dev Ind Pharm* 1984; **10**: 685–697.

7 Barker G. Sodium stearate based sticks: proposed structure. *Cosmet Toil* 1987; **102**(7172): 77–80.

8 Barker G. Solidified sodium stearate based sticks. *Cosmet Toil* 1977; **92**: 73–75.

9 Osol A *et al.*, eds. *Remington's Pharmaceutical Sciences*. 15th edn. Easton, Pennsylvania: Mack Publishing, 1975: 1266.

10 National Library of Medicine. Toxnet: Sodium stearate, 2005.

11 Ash M, Ash I. *Handbook of Pharmaceutical Additives*, 3rd edn. Endicott, NY: Synapse Information Resources, 2007: 907–908.

12 Lewis RJ, ed. *Sax's Dangerous Properties of Industrial Materials*, 11th edn. New York: John Wiley, 2004: 3280.

13 Mathe Norac Inc. Material safety data sheet: Sodium stearate, 2006.

14 Parlati C. Pulmonary spray dried powders of tobramycin containing sodium stearate to improve aerosolization efficiency. *Pharm Res* 2009; **26**: 1084–1092.

15 Jurgens RWJr. Semisolid oleaginous ointment bases for ophthalmic use. *J Pharm Sci* 1974; **63**: 443–445.

16 Zentner GM. Free fatty acid induced platelet aggregation: studies with solubilized and nonsolubilized fatty acids. *J Pharm Sci* 1981; 975–981.

17 Connor WE, *et al.* Massive thrombosis produced by fatty acid infusion. *J Clin Invest* 1963; **42**(6): 860–866.

18 Japan Pharmaceutical Excipients Council. *Japanese Pharmaceutical Excipients 2004*. Tokyo: Yakuji Nippo, 2004: 821.

20 General References

Lower ES. Soaps. Part I: Sodium stearate – Sodium palmitate. *Seifen Oele Fette Wachse* 2002; **128**(42-44): 46–48.

Lower ES. Sodium stearate. Part II: Uses. *Soap, Perfumery and Cosmetics*. (England) 1982; 55: 85–89.

21 Authors

ME Fenton, RC Rowe.

22 Date of Revision

2 March 2012.

Sodium Stearyl Fumarate

1 Nonproprietary Names

BP: Sodium Stearyl Fumarate
PhEur: Sodium Stearyl Fumarate
USP–NF: Sodium Stearyl Fumarate

2 Synonyms

Fumaric acid, octadecyl ester, sodium salt; *Lubripharm*; natrii stearylis fumaras; *Pruv*; sodium monostearyl fumarate.

3 Chemical Name and CAS Registry Number

2-Butenedioic acid, monooctadecyl ester, sodium salt [4070-80-8]

4 Empirical Formula and Molecular Weight

$C_{22}H_{39}NaO_4$ 390.5

5 Structural Formula

H_3C–[CH_2CH_2]$_8$–CH_2–O–C(=O)–CH=CH–CO_2Na

6 Functional Category

Tablet and capsule lubricant.

7 Applications in Pharmaceutical Formulation or Technology

Sodium stearyl fumarate is used as a lubricant in capsule and tablet formulations at concentrations of 0.5–2.0% w/w.[1–9] It is also used in certain food applications; *see* Section 16.

8 Description

Sodium stearyl fumarate for use as a pharmaceutical tablet lubricant occurs as a fine, white powder comprising agglomerates of flat, hexagonal-shaped particles about 5–10 μm in diameter.

9 Pharmacopeial Specifications

See Table I.

Table I: Pharmacopeial specifications for sodium stearyl fumarate.

Test	PhEur 7.4	USP35–NF30
Identification	+	+
Characters	+	—
Water	≤5.0%	≤5.0%
Lead	—	≤10 ppm
Heavy metals	—	≤20 ppm
Related substances	≤5.0%	—
Sodium stearyl maleate	—	≤0.25%
Stearyl alcohol	—	≤0.5%
Saponification value (anhydrous basis)	—	142.2–146.0
Assay (anhydrous basis)	99.0–101.5%	99.0–101.5%

10 Typical Properties

Acidity/alkalinity pH = 8.3 for a 5% w/v aqueous solution at 90°C.
Density 1.107 g/cm^3
Density (bulk) 0.2–0.35 g/cm^3
Density (tapped) 0.3–0.5 g/cm^3
Melting point 224–245°C (with decomposition)
Solubility *see* Table II.

SEM 1: Excipient: sodium stearyl fumarate; manufacturer: JRS Pharma LP; lot no.: 255-01; magnification: 300×.

SEM 2: Excipient: sodium stearyl fumarate; manufacturer: JRS Pharma LP; lot no.: 255-01; magnification: 500×.

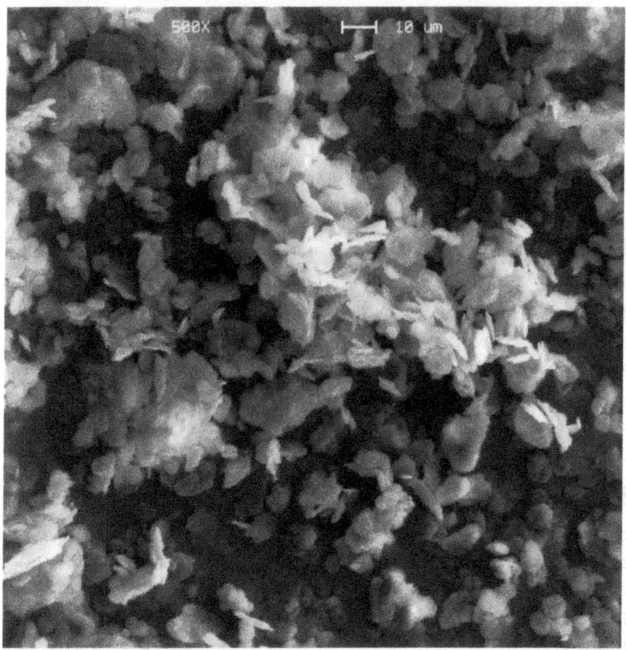

SEM 3: Excipient: sodium stearyl fumarate; manufacturer: JRS Pharma LP; lot no.: 255-01; magnification: 1000×.

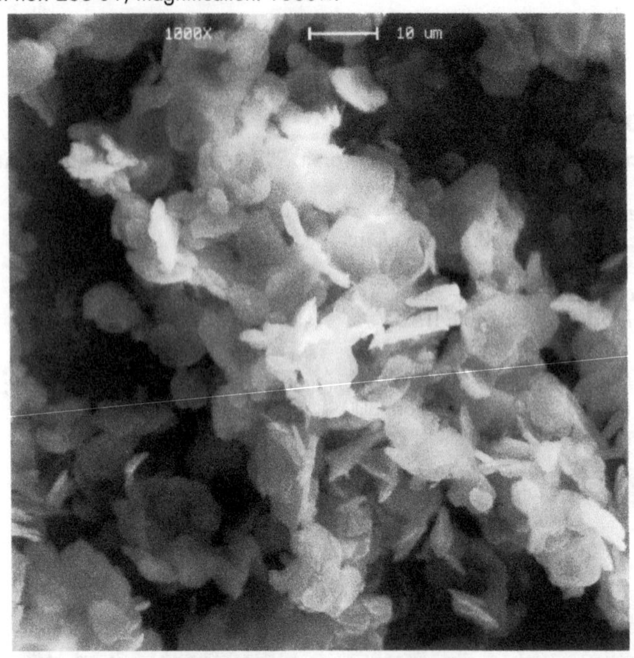

Table II: Solubility of sodium stearyl fumarate.

Solvent	Solubility at 20°C unless otherwise stated
Acetone	Practically insoluble
Chloroform	Practically insoluble
Ethanol	Practically insoluble
Methanol	Slightly soluble
Water	1 in 20 000 at 25°C
	1 in 10 at 80°C
	1 in 5 at 90°C

Specific surface area 1.2–2.0 m^2/g
Spectroscopy
IR spectra *see* Figure 1.

11 Stability and Storage Conditions

At ambient temperature, sodium stearyl fumarate is stable for up to 3 years when stored in amber glass bottles with polyethylene screw caps.

The bulk material should be stored in a well-closed container in a cool, dry place.

12 Incompatibilities

Sodium stearyl fumarate is reported to be incompatible with chlorhexidine acetate.[10] Since it is a sodium salt, it will also undergo the typical incompatibilities of sodium salts especially at higher water activities. Primary amines can interact with the olefinic double bond of the fumarate moiety to form an adduct in a manner analogous to a Michael addition.

13 Method of Manufacture

Stearyl alcohol is reacted with maleic anhydride in stoichiometric ratio to form monostearyl maleate. The product of this reaction then undergoes an isomerization step to form the monostearyl fumarate, followed by salt formation to produce sodium stearyl fumarate.

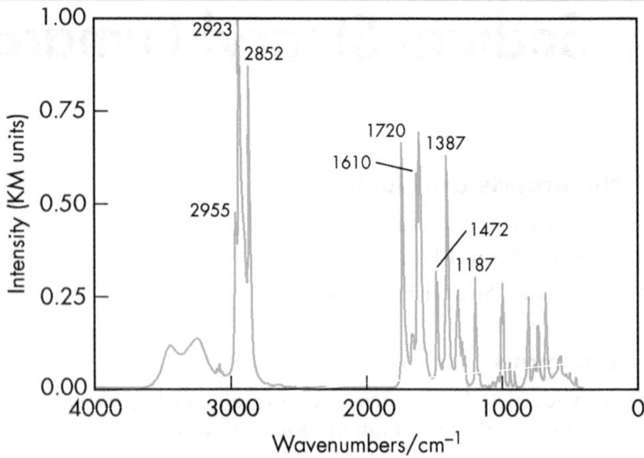

Figure 1: Infrared spectrum of sodium stearyl fumarate measured by diffuse reflectance. Adapted with permission of Informa Healthcare.

14 Safety

Sodium stearyl fumarate is used in oral pharmaceutical formulations and is generally regarded as a nontoxic and nonirritant material.

Metabolic studies of sodium stearyl fumarate in the rat and dog indicated that approximately 80% was absorbed and 35% was rapidly metabolized. The fraction absorbed was hydrolyzed to stearyl alcohol and fumaric acid, with the stearyl alcohol further oxidized to stearic acid. In the dog, sodium stearyl fumarate that was not absorbed was excreted unchanged in the feces within 24 hours.[11]

Stearyl alcohol and stearic acid are naturally occurring constituents in various food products, while fumaric acid is a normal constituent of body tissue. Stearates and stearyl citrate have been reviewed by the WHO and an acceptable daily intake for stearyl citrate has been set at up to 50 mg/kg body-weight.[12] The establishment of an acceptable daily intake for stearates[12] and fumaric acid[13] was thought unnecessary.

Disodium fumarate has been reported to have a toxicity not greatly exceeding that of sodium chloride.[14,15]

See Fumaric Acid, Stearic Acid, and Stearyl Alcohol for further information.

15 Handling Precautions

Observe normal precautions appropriate to the circumstances and quantity of material handled. Sodium stearyl fumarate should be handled in a well-ventilated environment; eye protection is recommended.

16 Regulatory Status

GRAS listed. Permitted by the FDA for direct addition to food for human consumption as a conditioning or stabilizing agent in various bakery products, flour-thickened foods, dehydrated potatoes, and processed cereals up to 0.2–1.0% by weight of the food. Included in nonparenteral medicines licensed in the UK. Included in the FDA Inactive Ingredients Database (oral capsules and tablets). Included in the Canadian Natural Health Products Ingredients Database.

17 Related Substances

—

18 Comments

Sodium stearyl fumarate can exist as two polymorphic forms; as small hexagonal plates or as needle crystals. The plate form is used as a tablet lubricant. Sodium stearyl fumarate is less hydrophobic than magnesium stearate or stearic acid and has less of a retardant effect on drug dissolution than magnesium stearate.

A specification for sodium stearyl fumarate is contained in the *Food Chemicals Codex* (FCC).[16]

A specification for sodium stearyl fumarate is included in the *Japanese Pharmaceutical Excipients* (JPE).[17]

The EINECS number for sodium stearyl fumarate is 223-781-1. The PubChem Compound ID (CID) for sodium stearyl fumarate is 23665634.

19 Specific References

1 Surén G. Evaluation of lubricants in the development of tablet formula. *Dansk Tidsskr Farm* 1971; **45**: 331–338.
2 Hölzer AW, Sjögren J. Evaluation of sodium stearyl fumarate as a tablet lubricant. *Int J Pharm* 1979; **2**: 145–153.
3 Hölzer AW, Sjögren J. Evaluation of some lubricants by the comparison of friction coefficients and tablet properties. *Acta Pharm Suec* 1981; **18**: 139–148.
4 Saleh SI, *et al.* Evaluation of some water soluble lubricants for direct compression. *Lab Pharm Prob Tech* 1984; **32**: 588–591.
5 Chowhan ZT, Chi L-H. Drug–excipient interactions resulting from powder mixing IV: role of lubricants and their effect on in vitro dissolution. *J Pharm Sci* 1986; **75**: 542–545.
6 Shah NH, *et al.* Evaluation of two new tablet lubricants sodium stearyl fumarate and glyceryl behenate. Measurement of physical parameters (compaction, ejection and residual forces) in the tableting process and the effect on the dissolution rate. *Drug Dev Ind Pharm* 1986; **12**: 1329–1346.
7 Davies PN, *et al.* Some pitfalls in accelerated stability testing with tablet and capsule lubricants. *J Pharm Pharmacol* 1987; **39**: 86P.
8 Mu X, *et al.* Investigations into the food effect on a polysaccharide dosage form. *Eur J Pharm Sci* 1996; 4(Suppl. 1): S184.
9 Michoel A, *et al.* Comparative evaluation of co-processed lactose and microcrystalline cellulose with their physical mixtures in the formulation of folic acid tablets. *Pharm Dev Technol* 2002; 7(1): 79–87.
10 Pesonen T, *et al.* Incompatibilities between chlorhexidine diacetate and some tablet excipients. *Drug Dev Ind Pharm* 1995; **21**: 747–752.
11 Figdor SK, Pinson R. The absorption and metabolism of orally administered tritium labelled sodium stearyl fumarate in the rat and dog. *J Agric Food Chem* 1970; **18**(5): 872–877.
12 FAO/WHO. Toxicological evaluation of certain food additives with a review of general principles and of specifications. Seventeenth report of the joint FAO/WHO expert committee on food additives. *World Health Organ Tech Rep Ser* 1974; No. 539.
13 FAO/WHO. Evaluation of certain food additives and contaminants. Thirty-fifth report of the FAO/WHO expert committee on food additives. *World Health Organ Tech Rep Ser* 1990; No. 789.
14 Bodansky O, *et al.* The toxicity and laxative action of sodium fumarate. *J Am Pharm Assoc (Sci)* 1942; **31**: 1–8.
15 Locke A, *et al.* The comparative toxicity and cathartic efficiency of disodium tartrate and fumarate, and magnesium fumarate, for the mouse and rabbit. *J Am Pharm Assoc (Sci)* 1942; **31**: 12–14.
16 *Food Chemicals Codex*, 7th edn. Bethesda, MD: United States Pharmacopeia, 2010: 965.
17 Japan Pharmaceutical Excipients Council. *Japanese Pharmaceutical Excipients 2004*. Tokyo: Yakuji Nippo, 2004: 822-823.

20 General References

Nicklasson M, Brodin A. The coating of disk surfaces by tablet lubricants, determined by an intrinsic rate of dissolution method. *Acta Pharm Suec* 1982; **19**: 99–108.

Zanowiak P. Lubrication in solid dosage form design and manufacture. In: Swarbrick J, Boylan JC, eds. *Encyclopedia of Pharmaceutical Technology*. 9. New York: Marcel Dekker, 1994: 87–111.

21 Author

RC Moreton.

22 Date of Revision

2 March 2012.

Sodium Sulfite

1 Nonproprietary Names

BP: Anhydrous Sodium Sulfite
JP: Dried Sodium Sulfite
PhEur: Sodium Sulfite, Anhydrous
USP–NF: Sodium Sulfite

2 Synonyms

Disodium sulfite; exsiccated sodium sulfite; E221; natrii sulfis anhydricus; sulfurous acid disodium salt.

3 Chemical Name and CAS Registry Number

Sodium sulfite [7757-83-7]

4 Empirical Formula and Molecular Weight

Na_2SO_3 126.04

5 Structural Formula

See Section 4.

6 Functional Category

Antimicrobial preservative; antioxidant.

7 Applications in Pharmaceutical Formulation or Technology

Sodium sulfite is used as an antioxidant in applications similar to those for sodium metabisulfite.[1] It is also an effective antimicrobial preservative, particularly against fungi at low pH (0.1% w/v of sodium sulfite is used). Sodium sulfite is used in cosmetics, food products, and pharmaceutical applications such as parenteral formulations, inhalations, oral formulations, and topical preparations.

See also Sodium Metabisulfite.

8 Description

Sodium sulfite occurs as an odorless white powder or hexagonal prismatic crystals. Note that the commercially available sodium sulfite is often presented as a white to tan- or pink-colored powder that would not conform to the pharmacopeial specification.

Table I: Pharmacopeial specifications for sodium sulfite.

Test	JP XV	PhEur 7.4	USP35–NF30
Characters	+	+	—
Identification	+	+	+
Appearance of solution	—	+	+
Heavy metals	≤20 ppm	≤10 ppm	≤10 ppm
Arsenic	≤4 ppm	—	—
Iron	—	≤10 ppm	≤10 ppm
Selenium	—	≤10 ppm	≤10 ppm
Thiosulfates	+	≤0.1%	≤0.1%
Zinc	—	≤25 ppm	≤25 ppm
Assay	≥97%	95.0–100.5%	95.0–100.5%

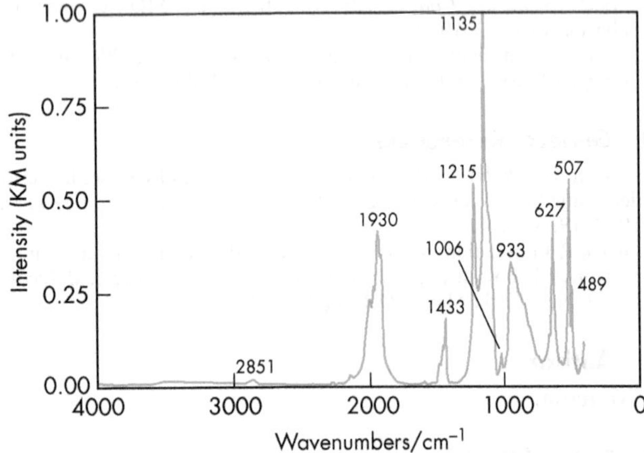

Figure 1: Infrared spectrum of sodium sulfite measured by diffuse reflectance. Adapted with permission of Informa Healthcare.

9 Pharmacopeial Specifications

See Table I.

10 Typical Properties

Acidity/alkalinity pH = 8.5–10.5 in aqueous solution.
Density 2.633 g/cm^3
Hygroscopicity Hygroscopic.
Solubility Soluble 1 in 3.2 parts of water; soluble in glycerin; practically insoluble in ethanol (95%), acetone, and most other organic solvents.
Spectroscopy
IR spectra *see* Figure 1.

11 Stability and Storage Conditions

Sodium sulfite should be stored in a well-closed container in a cool, dry place. It is stable in dry air at ambient temperatures or at 100°C, but undergoes rapid oxidation to sodium sulfate in moist air. In solution, sodium sulfite is slowly oxidized to sulfate by dissolved oxygen; strong acids lead to formation of sulfurous acid/sulfur dioxide. On heating, sodium sulfite decomposes liberating sulfur oxides.

12 Incompatibilities

Sodium sulfite is incompatible with acids, oxidizing agents, many proteins, and vitamin B_1.

See also Sodium Metabisulfite.

13 Method of Manufacture

Sodium bisulfite is prepared by reacting sulfur dioxide gas with aqueous sodium hydroxide or sodium carbonate. The solid material is obtained by evaporation of water. Further neutralization with sodium hydroxide or sodium carbonate while keeping the temperature above 33.6°C leads to crystallization of the anhydrous sodium sulfite (below this temperature the heptahydrate form is obtained).

14 Safety

Sodium sulfite is widely used in food and pharmaceutical applications as an antioxidant and antimicrobial preservative. It is generally regarded as relatively nontoxic and nonirritant when used as an excipient.[2,3] However, contact dermatitis and hypersensitivity reactions have been reported, some of which are occasionally severe.[4–6] The acceptable daily intake for sodium sulfite has been set at up to 700 μg/kg body-weight, expressed as sulfur dioxide.[7]

LD_{50} (mouse, IP): 0.950 g/kg[8]
LD_{50} (mouse, IV): 0.130 g/kg
LD_{50} (mouse, oral): 0.820 g/kg
LD_{50} (rabbit, IV): 0.065 g/kg
LD_{50} (rabbit, oral): 1.181 g/kg
LD_{50} (rat, IV): 0.115 g/kg
LD_{50} (rat, oral): 3.56 g/kg [9]

15 Handling Precautions

Observe normal precautions appropriate to the circumstances and quantity of material handled.

16 Regulatory Status

GRAS listed. Accepted for use as a food additive in Europe. Included in FDA Inactive Ingredients Database (epidural, IM, IV, and SC injections; inhalation solution; ophthalmic solutions; oral syrups and suspensions; otic solutions; topical creams and emulsions). Included in nonparenteral medicines licensed in the UK.

17 Related Substances

Sodium sulfite heptahydrate; sodium metabisulfite.

Sodium sulfite heptahydrate
Synonyms Natrii sulfis heptahydricus
CAS number [7785-83-7]
Molecular weight 252.15
Description Colorless crystals.
Density 1.56 g/cm^3
Solubility 1 in 1.6 of water; 1 in 30 of glycerin; sparingly soluble in ethanol (95%).
Comments Sodium sulfite heptahydrate is included in the PhEur 7.4. The heptahydrate is unstable, oxidizing in the air to the sulfate.

18 Comments

A specification for sodium sulfite is contained in the *Food Chemicals Codex* (FCC).[10]

The EINECS number for sodium sulfite is 231-821-4. The PubChem Compound ID (CID) for sodium sulfite is 24437.

19 Specific References

1 Islam MS, *et al.* Photoprotection of daunorubicin hydrochloride with sodium sulfite. *PDA J Pharm Sci Technol* 1995; **49**: 122–126.
2 Nair B, Elmore AR. Final report on the safety assessment of sodium sulfite, potassium sulfite, ammonium sulfite, sodium bisulfite, ammonium bisulfite, sodium metabisulfite and potassium metabisulfite. *Int J Toxicol* 2003; 22(2): 63–88.
3 Gunnisson AF. Sulphite toxicity: a critical review of *in vitro* and *in vivo* data. *Food Cosmet Toxicol* 1981; **19**: 667–682.
4 Vissers-Croughs KJ, *et al.* Allergic contact dermatitis from sodium sulfite. *Contact Dermatitis* 1988; **18**(4): 252–253.

5 Gunnisson AF, Jacobsen DW. Sulphite hypersensitivity: a critical review. *CRC Crit Review Toxicol* 1987; **17**(3): 185–214.

6 Vally H, *et al.* Clinical effect of sulphite additives. *Clin Exp Allergy* 2009; **39**(11): 1643–1651.

7 FAO/WHO. Safety evaluation of certain food additives. Fifty—first meeting of the joint FAO/WHO Expert Committee on Food Additives. *World Health Organ Tech Rep Ser* 2000: No. 891.

8 Lewis RJ, ed. *Sax's Dangerous Properties of Industrial Materials*, 11th edn. New York: Wiley, 2004; 3281–3282.

9 Parent RA. Acute toxicity data submissions. *Int J Toxicol* 2000; **19**(5): 331–373.

10 *Food Chemicals Codex*, 7th edn. Bethesda, MD: United States Pharmacopeia, 2011; 967.

20 General References

Barbera JJ et al. Sulfites, thiosulfates, and dithionites. In: *Ullmann's Encyclopedia of Industrial Chemistry* 7th edn. Weinheim: Wiley-VCH Verlag GmbH & Co. KgaA, 2010.

Fisher Scientific. Material safety data sheet: Sodium sulfite, January 2010.

Haynes WM, ed. *Handbook of Chemistry and Physics*, 91st edn. New York: CRC Press, 2010.

Lim S, *et al.* Effect of antibrowning agents on browning and intermediate formation in the Glucose-Glutamic Acid model. *J Food Sci* 2010; 8: C678–C683.

Sigma-Aldrich. Material safety data sheet: Sodium sulfite, March 2011.

US Food and Drug Adminstration. Code of Federal Regulations Title 21 (21 CFR 182.3798): Sodium sulfite, April 2011.

Weil ED *et al.* Sulfur compounds. In: *Kirk-Othmer Encyclopedia of Chemical Technology* Wiley, 2000.

21 Author

S Sienkiewicz.

22 Date of Revision

2 March 2012.

Sodium Thiosulfate

1 Nonproprietary Names

BP: Sodium Thiosulphate

JP: Sodium Thiosulfate Hydrate

PhEur: Sodium Thiosulfate

USP–NF: Sodium Thiosulfate

2 Synonyms

Ametox; disodium thiosulfate; disodium thiosulfate pentahydrate; natrii thiosulfas; natrium thiosulfuricum; sodium hyposulfite; sodium subsulfite; *Sodothiol*; *Sulfothiorine*; thiosulfuric acid disodium salt.

3 Chemical Name and CAS Registry Number

Sodium thiosulfate anhydrous [7772-98-7]

Sodium thiosulfate pentahydrate [10102-17-7]

4 Empirical Formula and Molecular Weight

$Na_2S_2O_3$	158.11 (for anhydrous)
$Na_2S_2O_3{\cdot}5H_2O$	248.2 (for pentahydrate)

5 Structural Formula

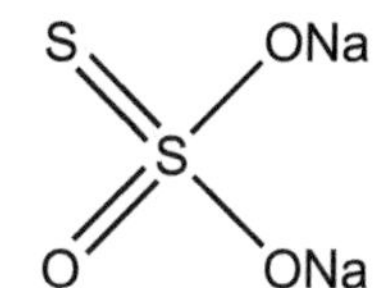

6 Functional Category

Antioxidant.

7 Applications in Pharmaceutical Formulation or Technology

Sodium thiosulfate is used as an antioxidant in pharmaceuticals (ophthalmic, intravenous, and oral preparations).

8 Description

Sodium thiosulfate occurs as odorless and colorless crystals, a crystalline powder or granules. It is efflorescent in dry air and deliquescent in moist air.

9 Pharmacopeial Specifications

See Table I.

10 Typical Properties

Acidity/alkalinity Aqueous solution practically neutral at pH 6.5–8.0 (pentahydrate).

Density 1.69 g/cm^3 (pentahydrate)

Hygroscopicity Slightly deliquesces in moist air (pentahydrate).

Melting point 48°C (pentahydrate)

Solubility Soluble in water; practically insoluble in ethanol (95%).

SEM 1: Excipient: sodium thiosulfate; magnification: 100×; voltage: 10 kV.

S

Table I: Pharmacopeial specifications for sodium thiosulfate.

Test	JP XV	PhEur 7.4	USP35–NF30
Identification	+	+	+
Characters	–	+	–
pH	6.0–8.0	6.0–8.4	–
Appearance of solution	+	+	–
Water	–	–	32.0–37%
Calcium	+	–	+
Heavy metals	≤20 ppm	≤10 ppm	≤0.002%
Arsenic	≤5 ppm	–	–
Loss on drying	+	–	–
Sulfides	–	+	–
Sulfates and sulfites	–	≤0.2%	–
Assay (dried basis)	99.0–101.0%	99.0–101.0%	99.0–100.5%

11 Stability and Storage Conditions

Sodium thiosulfate decomposes on heating. The bulk powder should be stored in a cool place, and the container should be kept tightly closed in a dry and well-ventilated place. It should not be stored near acids.

12 Incompatibilities

Sodium thiosulfate is incompatible with iodine, with acids, and with lead, mercury, and silver salts. It may reduce the activity of some preservatives, including bronopol, phenylmercuric salts, and thimerosal.[1]

13 Method of Manufacture

On an industrial scale, sodium thiosulfate is produced chiefly from liquid waste products of sodium sulfide or sulfur dye manufacture. Small-scale synthesis is done by boiling an aqueous solution of sodium sulfite with sulfur.[2,3]

14 Safety

Sodium thiosulfate is used in ophthalmic, intravenous, and oral pharmaceutical preparations. Apart from osmotic disturbances, sodium thiosulfate is relatively nontoxic. It is moderately toxic by the subcutaneous route and mildly irritating to respiratory tract and skin. Large oral doses have a cathartic action.[1]

LD_{50} (mouse, IP): 5.6 g/kg[4]

LD_{50} (mouse, IV): 2.4 g/kg

15 Handling Precautions

Observe normal precautions appropriate to the circumstances and quantity of the material handled. Protective gloves are recommended for prolonged or repeated contact use. Hazardous products (sulfur oxides) are formed when heated to decomposition.

16 Regulatory Status

GRAS listed. Included in the FDA Inactive Ingredients Database (IV solutions; ophthalmic solutions and suspensions; oral capsules, solutions, and tablets). Included in the Canadian Natural Health Products Ingredients Database.

17 Related Substances

—

18 Comments

Sodium thiosulfate has been used as an antidote to cyanide poisoning.[5,6] Thiosulfate acts as a sulfur donor for the conversion of cyanide to thiocyanate (which can then be safely excreted in the urine), catalyzed by the enzyme rhodanase.

Therapeutically, sodium thiosulfate has been shown to produce clinical improvement of calciphylaxis lesions in adults with chronic kidney disease.[7] It has also been used for its antifungal properties[1] and as a reagent in analytical chemistry.

There is a specification for sodium thiosulfate in the *Food Chemicals Codex* (FCC).[8]

The EINECS number for sodium thiosulfate is 231-867-5. The PubChem Compound ID (CID) for sodium thiosulfate pentahydrate is 516922.

19 Specific References

1 Sweetman SC, ed. *Martindale: The Complete Drug Reference*, 37th edn. London, UK: Pharmaceutical Press, 2011: 1607–1608.
2 Lowenheim FA, Moran MK, eds. *Faith, Keyes & Clarks Industrial Chemicals*, 4th edn. New York: Wiley-Interscience, 1975: 769–773.
3 Holleman AF, Wiberg E. *Inorganic Chemistry.* San Diego: Academic Press, 2001: 1937.
4 Lewis RJ, ed. *Sax's Dangerous Properties of Industrial Materials*, 11th edn. New York: Wiley, 2004: 3284–3285.
5 Frankenberg L, Sörbo B. Effect of cyanide antidotes on the metabolic conversion of cyanide to thiocyanate. *Arch Toxicol* 1975; **14**: 81–89.
6 Sylvester DM, *et al.* Effects of thiosulfate on cyanide pharmacokinetics in dogs. *Toxicol Appl Pharmacol* 1983; **69**: 265–271.
7 Raymond CB, Wazny LD. Sodium thiosulfate, bisphosphonates, and cinacalcet for treatment of calciphylaxis. *Am J Health Syst Pharm* 2008; 65(15): 1419–1429.
8 *Food Chemicals Codex*, 7th edn. Bethesda, MD: United States Pharmacoepeia, 2010: 968.

20 General References

—

21 Authors

JC Hooton, N Sandler.

22 Date of Revision

2 March 2012.

S

Sorbic Acid

1 Nonproprietary Names

BP: Sorbic Acid
PhEur: Sorbic Acid
USP–NF: Sorbic Acid

2 Synonyms

Acidum sorbicum; (2-butenylidene) acetic acid; crotylidene acetic acid; E200; hexadienic acid; hexadienoic acid; 2,4-hexadienoic acid; 1,3-pentadiene-1-carboxylic acid; 2-propenylacrylic acid; (*E*,*E*)-sorbic acid; *Sorbistat*.

3 Chemical Name and CAS Registry Number

(*E*,*E*)-Hexa-2,4-dienoic acid [22500-92-1]

4 Empirical Formula and Molecular Weight

$C_6H_8O_2$ 112.13

5 Structural Formula

H_3C OH O

6 Functional Category

Antimicrobial preservative.

7 Applications in Pharmaceutical Formulation or Technology

Sorbic acid is an antimicrobial preservative[1] with antibacterial and antifungal properties used in pharmaceuticals, foods, enteral preparations, and cosmetics. Generally, it is used at concentrations of 0.05–0.2% in oral and topical pharmaceutical formulations, especially those containing nonionic surfactants. Sorbic acid is also used with proteins, enzymes, gelatin, and vegetable gums.[2] It has been shown to be an effective preservative for promethazine hydrochloride solutions in a concentration of 1 g/L.[3]

Sorbic acid has limited stability and activity against bacteria and is thus frequently used in combination with other antimicrobial preservatives or glycols, when synergistic effects appear to occur; *see* Section 10.

8 Description

Sorbic acid occurs as a tasteless, white to yellow-white crystalline powder with a faint characteristic odor.

9 Pharmacopeial Specifications

See Table I.

SEM 1: Excipient: sorbic acid; manufacturer: Pfizer Ltd.; magnification: 60×.

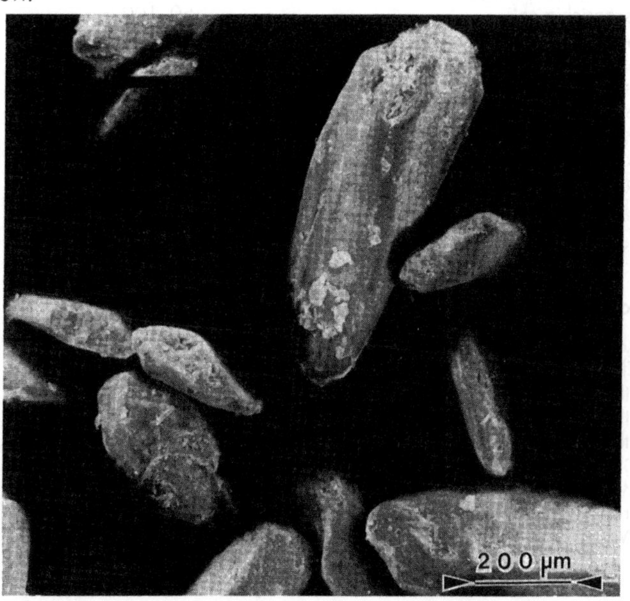

Table I: Pharmacopeial specifications for sorbic acid.

Test	PhEur 7.4	USP35–NF30
Identification	+	+
Characters	+	—
Appearance of solution	+	—
Melting range	132–136°C	132–135°C
Water	≤1.0%	≤0.5%
Residue on ignition	—	≤0.2%
Sulfated ash	≤0.2%	—
Heavy metals	≤10 ppm	≤10 ppm
Aldehyde (as C_2H_4O)	≤0.15%	—
Assay (anhydrous basis)	99.0–101.0%	99.0–101.0%

10 Typical Properties

Antimicrobial activity Sorbic acid is primarily used as an antifungal agent, although it also possesses antibacterial properties. The optimum antibacterial activity is obtained at pH 4.5; and practically no activity is observed above pH 6.[4,5] The efficacy of sorbic acid is enhanced when it is used in combination with other antimicrobial preservatives or glycols since synergistic effects occur.[6] Reported minimum inhibitory concentrations (MICs) at pH 6 are shown in Table II.[7]

Boiling point 228°C with decomposition.

Density 1.20 g/cm^3

Dissociation constant $pK_a = 4.76$

Flash point 127°C

Melting point 134.5°C

Solubility *see* Table III. In syrup, the solubility of sorbic acid decreases with increasing sugar content.

Spectroscopy

IR spectra *see* Figure 1.

NIR spectra *see* Figure 2.

Vapor pressure <1.3 Pa (<0.01 mmHg) at 20°C

Table II: Minimum inhibitory concentrations (MICs) of sorbic acid at pH 6.

Microorganism	MIC (μg/mL)
Aspergillus niger	200–500
Candida albicans	25–50
Clostridium sporogenes	100–500
Escherichia coli	50–100
Klebsiella pneumoniae	50–100
Penicillium notatum	200–300
Pseudomonas aeruginosa	100–300
Pseudomonas cepacia	50–100
Pseudomonas fluorescens	100–300
Saccharomyces cerevisiae	200–500
Staphylococcus aureus	50–100

Table III: Solubility of sorbic acid.

Solvent	Solubility at 20°C unless otherwise stated
Acetone	1 in 11
Chloroform	1 in 15
Ethanol	1 in 8
Ethanol (95%)	1 in 10
Ether	1 in 30
Glycerin	1 in 320
Methanol	1 in 8
Propylene glycol	1 in 19
Water	1 in 400 at 30°C
	1 in 26 at 100°C

11 Stability and Storage Conditions

Sorbic acid is sensitive to oxidation, particularly in the presence of light; oxidation occurs more readily in aqueous solution than in the solid form. Sorbic acid may be stabilized by phenolic antioxidants such as 0.02% propyl gallate.(6)

Sorbic acid is combustible when exposed to heat or flame. When heated to decomposition, it emits acrid smoke and irritating fumes. The bulk material should be stored in a well-closed container, protected from light, at a temperature not exceeding 40°C.

12 Incompatibilities

Sorbic acid is incompatible with bases, oxidizing agents, and reducing agents. Some loss of antimicrobial activity occurs in the presence of nonionic surfactants and plastics. Oxidation is catalyzed by heavy-metal salts. Sorbic acid will also react with sulfur-containing amino acids, although this can be prevented by the addition of ascorbic acid, propyl gallate, or butylhydroxytoluene.

When stored in glass containers, the solution becomes very pH sensitive; therefore, preparations using sorbic acid as a preservative should be tested for their microbial purity after prolonged periods of storage.

Aqueous solutions of sorbic acid without the addition of antioxidants are rapidly decomposed when stored in polypropylene, polyvinylchloride, and polyethylene containers.

13 Method of Manufacture

Naturally occurring sorbic acid may be extracted as the lactone (parasorbic acid) from the berries of the mountain ash *Sorbus aucuparia* L. (Fam. Rosaceae). Synthetically, sorbic acid may be prepared by the condensation of crotonaldehyde and ketene in the presence of boron trifluoride; by the condensation of crotonaldehyde and malonic acid in pyridine solution; or from 1,1,3,5-tetraalkoxyhexane. Fermentation of sorbaldehyde or sorbitol with bacteria in a culture medium has also been used.

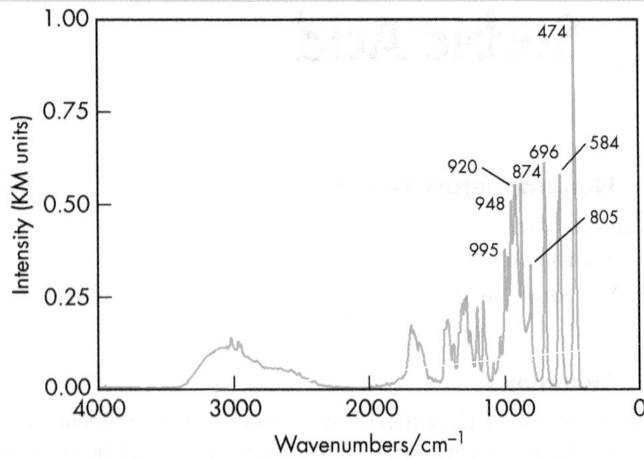

Figure 1: Infrared spectrum of sorbic acid measured by diffuse reflectance. Adapted with permission of Informa Healthcare.

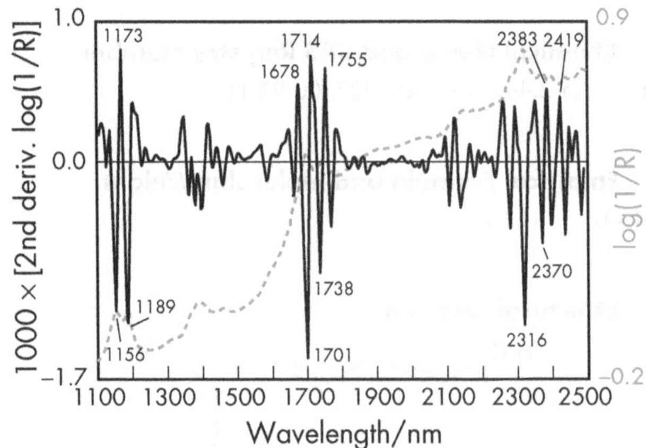

Figure 2: Near-infrared spectrum of sorbic acid measured by reflectance.

14 Safety

Sorbic acid is used as an antimicrobial preservative in oral and topical pharmaceutical formulations and is generally regarded as a nontoxic material. However, adverse reactions to sorbic acid and potassium sorbate, including irritant skin reactions(8–11) and allergic hypersensitivity skin reactions (which are less frequent), have been reported.(12–14)

Other adverse reactions that have been reported include exfoliative dermatitis due to ointments that contain sorbic acid,(15) and allergic conjunctivitis caused by contact lens solutions preserved with sorbic acid.(16)

No adverse reactions have been described after systemic administration of sorbic acid, and it has been reported that it can be ingested safely by patients who are allergic to sorbic acid.(17) However, perioral contact urticaria has been reported.(11)

The WHO has set an estimated total acceptable daily intake for sorbic acid, calcium sorbate, potassium sorbate, and sodium sorbate, expressed as sorbic acid, at up to 25 mg/kg bodyweight.(18,19)

Animal toxicological studies have shown no mammalian carcinogenicity or teratogenicity for sorbic acid consumed at up to 10% of the diet.(20)

LD_{50} (mouse, IP): 2.82 g/kg(21)
LD_{50} (mouse, oral): 3.20 g/kg
LD_{50} (mouse, SC): 2.82 g/kg
LD_{50} (rat, oral): 7.36 g/kg

15 Handling Precautions

Observe normal precautions appropriate to the circumstances and quantity of material handled. Sorbic acid can be irritant to the skin, eyes, and respiratory system. Eye protection, gloves, and a dust mask or respirator are recommended.

16 Regulatory Status

GRAS listed. Accepted as a food additive in Europe. Included in the FDA Inactive Ingredients Database (ophthalmic solutions; oral capsules, solutions, syrups, tablets; topical and vaginal preparations). Included in nonparenteral medicines licensed in the UK. Included in the Canadian Natural Health Products Ingredients Database.

17 Related Substances

Calcium sorbate; potassium sorbate; sodium sorbate.

Calcium sorbate

Empirical formula $C_{12}H_{14}O_4Ca$
Synonyms E203
Molecular weight 262.33
CAS number [7492-55-9]
Appearance White, odorless, tasteless, crystalline powder.
Solubility Soluble 1 in 83 parts of water; practically insoluble in fats.
Comments The EINECS number for calcium sorbate is 231-321-6.

Sodium sorbate

Empirical formula $C_6H_7O_2Na$
Synonyms E201; sodium (*E*,*E*)-hexa-2,4-dienoate.
Molecular weight 134.12
CAS number [42788-83-0]
Appearance Light, white, crystalline powder.
Solubility Soluble 1 in 3 parts of water.
Comments The EINECS number for sodium sorbate is 231-819-3.

18 Comments

The *trans,trans*-isomer of sorbic acid is the commercial product. A specification for sorbic acid is contained in the *Food Chemicals Codex* (FCC).[22]

The EINECS number for sorbic acid is 203-768-7. The PubChem Compound ID (CID) for sorbic acid includes 643460 and 1550734.

19 Specific References

1 Charvalos E, *et al.* Controlled release of water-soluble polymeric complexes of sorbic acid with antifungal activities. *Appl Microbiol Biotechnol* 2001; **57**(5–6): 770–775.
2 Weiner M, Bernstein IL. *Adverse Reactions to Drug Formulation Agents: A Handbook of Excipients.* New York: Marcel Dekker, 1989; 179.
3 Van-Doorne H, Leijen JB. Preservation of some oral liquid preparations: replacement of chloroform by other preservatives. *Pharm World Sci* 1994; **16**(Feb 18): 18–21.
4 Golightly LK, *et al.* Adverse effects associated with inactive ingredients in drug products (part I). *Med Toxicol* 1988; **3**: 128–165.
5 Eklund T. The antimicrobial effect of dissociated and undissociated sorbic acid at different pH levels. *J Appl Bacteriol* 1983; **54**: 383–389.
6 Woodford R, Adams E. Sorbic acid. *Am Perfum Cosmet* 1970; **85**(3): 25–30.
7 Wallhäusser KH. Sorbic acid. In: Kabara JJ, ed. *Cosmetic and Drug Preservation Principles and Practice.* New York: Marcel Dekker, 1984; 668–670.
8 Soschin D, Leyden JJ. Sorbic acid-induced erythema and edema. *J Am Acad Dermatol* 1986; **14**: 234–241.
9 Fisher AA. Erythema limited to the face due to sorbic acid. *Cutis* 1987; **40**: 395–397.
10 Clemmensen OJ, Schiodt M. Patch test reaction of the buccal mucosa to sorbic acid. *Contact Dermatitis* 1982; **8**(5): 341–342.
11 Clemmensen O, Hjorth N. Perioral contact urticaria from sorbic acid and benzoic acid in a salad dressing. *Contact Dermatitis* 1982; **3**: 1–6.
12 Saihan EM, Harman RRM. Contact sensitivity to sorbic acid in 'Unguentum Merck'. *Br J Dermatol* 1978; **99**: 583–584.
13 Fisher AA. Cutaneous reactions to sorbic acid and potassium sorbate. *Cutis* 1980; **25**: 350, 352, 423.
14 Fisher AA. Allergic reactions to the preservatives in over-the-counter hydrocortisone topical creams and lotions. *Cutis* 1983; **32**: 222, 224, 230.
15 Coyle HE, *et al.* Sorbic acid sensitivity from Unguentum Merck. *Contact Dermatitis* 1981; **7**: 56–57.
16 Fisher AA. Allergic reactions to contact lens solutions. *Cutis* 1985; **36**: 209–211.
17 Klaschka F, Beiersdorff HU. Allergic eczematous reaction from sorbic acid used as a preservative in external medicaments. *Munch Med Wschr* 1965; **107**: 185–187.
18 FAO/WHO. Toxicological evaluation of certain food additives with a review of general principles and of specifications. Seventeenth report of the joint FAO/WHO expert committee on food additives. *World Health Organ Tech Rep Ser* 1974; No. 539.
19 FAO/WHO. Evaluation of certain food additives and contaminants. Twenty-ninth report of the joint FAO/WHO expert committee on food additives. *World Health Organ Tech Rep Ser* 1986; No. 733.
20 Walker R. Toxicology of sorbic acid and sorbates. *Food Addit Contam* 1990; **7**(5): 671–676.
21 Lewis RJ, ed. *Sax's Dangerous Properties of Industrial Materials*, 11th edn. New York: Wiley, 2004; 3291.
22 *Food Chemicals Codex*, 7th edn. Bethesda, MD: United States Pharmacopeia, 2010.

20 General References

Radus TP, Gyr G. Determination of antimicrobial preservatives in pharmaceutical formulations using reverse-phase liquid chromatography. *J Pharm Sci* 1983; **72**: 221–224.
Schnuch A, *et al.* Contact allergy to preservatives. Analysis of IVDK data 1996-2009. *Br J Dermatol* 2011; **164**: 1316–1325.
Sofos JN, Busta FF. Sorbates. In: Branen AL, Davidson PM, eds. *Antimicrobials in Foods.* New York: Marcel Dekker, 1983; 141–175.
Warth A. Mechanism of resistance of *Saccharomyces bailii* to benzoic, sorbic and other weak acids used as food preservatives. *J Appl Bacteriol* 1977; **43**: 215–230.

21 Author

W Cook.

22 Date of Revision

2 March 2012.

Sorbitan Esters (Sorbitan Fatty Acid Esters)

1 Nonproprietary Names

BP: Sorbitan Laurate
Sorbitan Oleate
Sorbitan Palmitate
Sorbitan Sesquioleate
Sorbitan Stearate
Sorbitan Trioleate

JP: Sorbitan Sesquioleate

PhEur: Sorbitan Laurate
Sorbitan Oleate
Sorbitan Palmitate
Sorbitan Sesquioleate
Sorbitan Stearate
Sorbitan Trioleate

USP–NF: Sorbitan Monolaurate (sorbitan, esters monodecanoate)
Sorbitan Monooleate
Sorbitan Monopalmitate
Sorbitan Monostearate
Sorbitan Sesquioleate
Sorbitan Trioleate

2 Synonyms

See Table I.

3 Chemical Names and CAS Registry Numbers

See Table II.

4 Empirical Formula and Molecular Weight

See Table III.

5 Structural Formula

$H_2C{-}R^3$ / $HC{-}R^2$ / O / HO / R^1

$R^1 = R^2 = OH$, $R^3 = R$ (see below) for sorbitan monoesters
$R^1 = OH$, $R^2 = R^3 = R$ for sorbitan diesters
$R^1 = R^2 = R^3 = R$ for sorbitan triesters
where R = $(C_{17}H_{35})COO$ for isostearate
$(C_{11}H_{23})COO$ for laurate
$(C_{17}H_{33})COO$ for oleate
$(C_{15}H_{31})COO$ for palmitate
$(C_{17}H_{35})COO$ for stearate
The sesquiesters are equimolar mixtures of monoesters and diesters.

6 Functional Category

Dispersing agent; emulsifying agent; nonionic surfactant; solubilizing agent; suspending agent.

Table III: Empirical formula and molecular weight of selected sorbitan esters.

Name	Formula	Molecular weight
Sorbitan diisostearate	$C_{42}H_{80}O_7$	697
Sorbitan dioleate	$C_{42}H_{76}O_7$	693
Sorbitan monoisostearate	$C_{24}H_{46}O_6$	431
Sorbitan monolaurate	$C_{18}H_{34}O_6$	346
Sorbitan monooleate	$C_{24}H_{44}O_6$	429
Sorbitan monopalmitate	$C_{22}H_{42}O_6$	403
Sorbitan monostearate	$C_{24}H_{46}O_6$	431
Sorbitan sesquiisostearate	$C_{33}H_{63}O_{6.5}$	564
Sorbitan sesquioleate	$C_{33}H_{60}O_{6.5}$	561
Sorbitan sesquistearate	$C_{33}H_{63}O_{6.5}$	564
Sorbitan triisostearate	$C_{60}H_{114}O_8$	964
Sorbitan trioleate	$C_{60}H_{108}O_8$	958
Sorbitan tristearate	$C_{60}H_{114}O_8$	964

7 Applications in Pharmaceutical Formulation or Technology

Sorbitan monoesters are a series of mixtures of partial esters of sorbitol and its mono- and di-anhydrides with fatty acids. Sorbitan diesters are a series of mixtures of partial esters of sorbitol and its monoanhydride with fatty acids.

Sorbitan esters are widely used in cosmetics, food products, and pharmaceutical formulations as lipophilic nonionic surfactants. They are mainly used in pharmaceutical formulations as emulsifying agents in the preparation of creams, emulsions, and ointments for topical application. When used alone, sorbitan esters produce stable water-in-oil emulsions and microemulsions, but are frequently used in combination with varying proportions of a polysorbate to produce water-in-oil or oil-in-water emulsions or creams of varying consistencies, and also in self-emulsifying drug delivery systems for poorly soluble compounds.[1]

Sorbitan monolaurate, sorbitan monopalmitate and sorbitan trioleate have also been used at concentrations of 0.01–0.05% w/v in the preparation of an emulsion for intramuscular administration. *See* Table IV.

8 Description

Sorbitan esters occur as cream- to amber-colored liquids or solids with a distinctive odor and taste; *see* Table V.

9 Pharmacopeial Specifications

See Table VI.

10 Typical Properties

Acid value *see* Table VII.
Density *see* Table VII.
Flash point >149°C
HLB value *see* Table VII.
Hydroxyl value *see* Table VII.
Iodine number *see* Table VII.
Melting point *see* Table VII.
Moisture content *see* Table VIII.
Pour point *see* Table VII.
Saponification value *see* Table VIII.

Table I: Synonyms of selected sorbitan esters.

Name	Synonym
Sorbitan monoisostearate	1,4-Anhydro-D-glucitol, 6-isooctadecanoate; anhydrosorbitol monoisostearate; *Arlacel 987*; *Crill 6*; sorbitan isostearate.
Sorbitan monolaurate	*Arlacel 20*; *Armotan ML*; *Crill 1*; *Dehymuls SML*; E493; *Glycomul L*; *Hodag SML*; *Liposorb L*; *Montane 20*; *Protachem SML*; *Sorbester P12*; *Sorbirol L*; sorbitan laurate; sorbitani lauras; *Span 20*; *Tego SML*.
Sorbitan monooleate	*Ablunol S-80*; *Arlacel 80*; *Armotan MO*; *Capmul O*; *Crill 4*; *Crill 50*; *Dehymuls SMO*; *Drewmulse SMO*; *Drewsorb 80K*; E494; *Glycomul O*; *Hodag SMO*; *Lamesorb SMO*; *Liposorb O*; *Montane 80*; *Nikkol SO-10*; *Nissan Nonion OP-80R*; *Norfox Sorbo S-80*; *Polycon S80 K*; *Proto-sorb SMO*; *Protachem SMO*; *S-Maz 80K*; *Sorbester P17*; *Sorbirol O*; sorbitan oleate; sorbitani oleas; *Sorgen 40*; *Sorgon S-40-H*; *Span 80*; *Tego SMO*.
Sorbitan monopalmitate	1,4-Anhydro-D-glucitol, 6-hexadecanoate; *Ablunol S-40*; *Arlacel 40*; *Armotan MP*; *Crill 2*; *Dehymuls SMP*; E495; *Glycomul P*; *Hodag SMP*; *Lamesorb SMP*; *Liposorb P*; *Montane 40*; *Nikkol SP-10*; *Nissan Nonion PP-40R*; *Protachem SMP*; *Proto-sorb SMP*; *Sorbester P16*; *Sorbirol P*; sorbitan palmitate; sorbitani palmitas; *Span 40*.
Sorbitan monostearate	*Ablunol S-60*; *Alkamuls SMS*; 1,4-Anhydro-D-glucitol, 6-octadecanoate; anhydrosorbitol monostearate; *Arlacel 60*; *Armotan MS*; *Atlas 110K*; *Capmul S*; *Crill 3*; *Dehymuls SMS*; *Drewmulse SMS*; *Drewsorb 60K*; *Durtan 60*; *Durtan 60K*; E491; *Famodan MS Kosher*; *Glycomul S FG*; *Glycomul S KFG*; *Hodag SMS*; *Lamesorb SMS*; *Liposorb S*; *Liposorb SC*; *Liposorb S-K*; *Montane 60*; *Nissan Nonion SP-60R*; *Norfox Sorbo S-60FG*; *Polycon S60K*; *Protachem SMS*; *Prote-sorb SMS*; *S-Maz 60K*; *S-Maz 60KHS*; *Sorbester P18*; *Sorbirol S*; sorbitan stearate; sorbitani stearas; *Sorgen 50*; *Span 60*; *Span 60K*; *Span 60 VS*; *Tego SMS*.
Sorbitan sesquiisostearate	*Protachem SQI*.
Sorbitan sesquioleate	*Arlacel C*; *Arlacel 83*; *Crill 43*; *Glycomul SOC*; *Hodag SSO*; *Liposorb SQO*; *Montane 83*; *Nikkol SO-15*; *Nissan Nonion OP-83RAT*; *Protachem SOC*; sorbitani sesquioleas; *Sorgen 30*; *Sorgen S-30-H*.
Sorbitan trilaurate	*Span 25*.
Sorbitan trioleate	*Ablunol S-85*; *Arlacel 85*; *Crill 45*; *Glycomul TO*; *Hodag STO*; *Liposorb TO*; *Montane 85*; *Nissan Nonion OP-85R*; *Protachem STO*; *Prote- sorb STO*; *S-Maz 85K*; *Sorbester P37*; sorbitani trioleas; *Span 85*; *Tego STO*.
Sorbitan tristearate	*Alkamuls STS*; *Crill 35*; *Crill 41*; *Drewsorb 65K*; E492; *Famodan TS Kosher*; *Glycomul TS KFG*; *Hodag STS*; *Lamesorb STS*; *Liposorb TS*; *Liposorb TS-K*; *Montane 65*; *Protachem STS*; *Proteo-sorb STS*; *Sorbester P38*; *Span 65*; *Span 65K*.

Table II: Chemical name and CAS Registry Number of selected sorbitan esters.

Name	Chemical name	CAS number
Sorbitan diisostearate	Sorbitan diisooctadecanoate	[68238-87-9]
Sorbitan dioleate	(Z,Z)-Sorbitan di-9-octadecanoate	[29116-98-1]
Sorbitan monolaurate	Sorbitan monododecanoate	[1338-39-2]
Sorbitan monoisostearate	Sorbitan monoisooctadecanoate	[71902-01-7]
Sorbitan monooleate	(Z)-Sorbitan mono-9-octadecenoate	[1338-43-8]
Sorbitan monopalmitate	Sorbitan monohexadecanoate	[26266-57-9]
Sorbitan monostearate	Sorbitan mono-octadecanoate	[1338-41-6]
Sorbitan sesquiisostearate	Sorbitan sesquiisooctadecanoate	[71812-38-9]
Sorbitan sesquioleate	(Z)-Sorbitan sesqui-9-octadecenoate	[8007-43-0]
Sorbitan sesquistearate	Sorbitan sesqui-octadecanoate	[51938-44-4]
Sorbitan triisostearate	Sorbitan triisooctadecanoate	[54392-27-7]
Sorbitan trioleate	(Z,Z,Z)-Sorbitan tri-9-octadecenoate	[26266-58-0]
Sorbitan tristearate	Sorbitan tri-octadecanoate	[26658-19-5]

Table IV: Uses of sorbitan esters.

Use	Concentration (%)
Emulsifying agent	
Used alone in water-in-oil emulsions	1–15
Used in combination with hydrophilic emulsifiers in oil-in-water emulsions	1–10
Used to increase the water-holding properties of ointments	1–10
Nonionic surfactant	
For insoluble, active constituents in lipophilic bases	0.1–3
Solubilizing agent	
For poorly soluble, active constituents in lipophilic bases	1–10
For insoluble, active constituents in lipophilic bases	0.1–3

Table V: Appearance of selected sorbitan esters.

Name	Appearance
Sorbitan monoisostearate	Yellow viscous liquid
Sorbitan monolaurate	Yellow viscous liquid
Sorbitan monooleate	Yellow viscous liquid
Sorbitan monopalmitate	Cream solid
Sorbitan monostearate	Cream solid
Sorbitan sesquioleate	Amber viscous liquid
Sorbitan trioleate	Amber viscous liquid
Sorbitan tristearate	Cream/yellow solid

Solubility Sorbitan esters are generally soluble or dispersible in oils; they are also soluble in most organic solvents. In water, although insoluble, they are generally dispersible.

Spectroscopy

IR spectra *see* Figure 1.

NIR spectra *see* Figure 2.

Surface tension *see* Table VIII.

Viscosity (dynamic) *see* Table VIII.

11 Stability and Storage Conditions

Gradual soap formation occurs with strong acids or bases; sorbitan esters are stable in weak acids or bases.

Sorbitan esters should be stored in a well-closed container in a cool, dry place.

12 Incompatibilities

—

13 Method of Manufacture

Sorbitol is dehydrated to form a hexitan (1,4-sorbitan), which is then esterified with the desired fatty acid.

14 Safety

Sorbitan esters are widely used in cosmetics, food products, and oral and topical pharmaceutical formulations, and are generally regarded as nontoxic and nonirritant materials. However, there have been occasional reports of hypersensitive skin reactions

Table VI: Pharmacopeial specifications for sorbitan esters.

Test	JP XV	PhEur 7.4	USP35–NF30
Identification	+	+	+
Characters	–	+	–
Acidity	+	–	–
Acid value			
Sorbitan monolaurate	–	≤7.0	≤8
Sorbitan monooleate	–	≤8.0	≤8
Sorbitan monopalmitate	–	≤8.0	≤8
Sorbitan monostearate	–	≤10.0	≤10
Sorbitan sesquioleate	–	≤16.0	≤14
Sorbitan trioleate	–	≤16.0	≤17
Hydroxyl value			
Sorbitan monolaurate	–	330–358	330–358
Sorbitan monooleate	–	190–210	190–215
Sorbitan monopalmitate	–	270–305	275–305
Sorbitan monostearate	–	235–260	235–260
Sorbitan sesquioleate	–	180–215	182–220
Sorbitan trioleate	–	55–75	50–75
Iodine value			
Sobitan monolaurate	–	≤10.0	–
Sorbitan monooleate	–	62–76	62–76
Sorbitan sesquioleate	–	70–95	65–75
Sorbitan trioleate	–	76–90	77–85
Peroxide value			
Sorbitan monolaurate	–	≤5.0	–
Sorbitan monooleate	–	≤10.0	–
Sorbitan monopalmitate	–	≤5.0	–
Sorbitan monostearate	–	≤5.0	–
Sorbitan sesquioleate	–	≤10.0	–
Sorbitan trioleate	–	≤10.0	–
Saponification value			
Sorbitan monolaurate	–	158–170	158–170
Sorbitan monooleate	–	145–160	145–160
Sorbitan monopalmitate	–	140–155	140–150
Sorbitan monostearate	–	147–157	147–157
Sorbitan sesquioleate	150–168	145–166	143–165
Sorbitan trioleate	–	170–190	169–183
Water			
Sorbitan monolaurate	–	≤1.5%	≤1.5%
Sorbitan monooleate	–	≤1.5%	≤1.0%
Sorbitan monopalmitate	–	≤1.5%	≤1.5%
Sorbitan monostearate	–	≤1.5%	≤1.5%
Sorbitan sesquioleate	≤3.0%	≤1.5%	≤1.0%
Sorbitan trioleate	–	≤1.5%	≤0.7%
Residue on ignition			
Sorbitan monolaurate	–	–	≤0.5%
Sorbitan monooleate	–	–	≤0.5%
Sorbitan monopalmitate	–	–	≤0.5%
Sorbitan monostearate	–	–	≤0.5%
Sorbitan sesquioleate	≤1.0%	–	≤1.4%
Sorbitan trioleate	–	–	≤0.25%
Total ash	–	≤0.5%	–
Heavy metals	≤20 ppm	≤10 ppm	≤0.001%
Arsenic	≤2 ppm	–	
Specific gravity			
Sorbitan monolaurate	–	≈0.98	–
Sorbitan monooleate	–	≈0.99	–
Sorbitan sesquioleate	0.960–1.020	≈0.99	–
Sorbitan trioleate	–	≈0.98	–
Melting point			
Sorbitan palmitate	–	44–51°C	–
Sorbitan monostearate	–	50–60°C	–
Assay for fatty acids			
Sorbitan monolaurate	–	+	55.0–63.0%
Sorbitan monooleate	–	+	72.0–78.0%
Sorbitan monopalmitate	–	+	63.0–71.0%
Sorbitan monostearate	–	+	68.0–76.0%
Sorbitan sesquioleate	–	+	74.0–80.0%
Sorbitan trioleate	–	+	85.5–90.0%
Assay for polyols			
Sorbitan monolaurate	–	–	39.0–45.0%
Sorbitan monooleate	–	–	25.0–31.0%
Sorbitan monopalmitate	–	–	32.0–38.0%
Sorbitan monostearate	–	–	27.0–34.0%
Sorbitan sesquioleate	–	–	22.0–28.0%
Sorbitan trioleate	–	–	13.0–19.0%

table continues

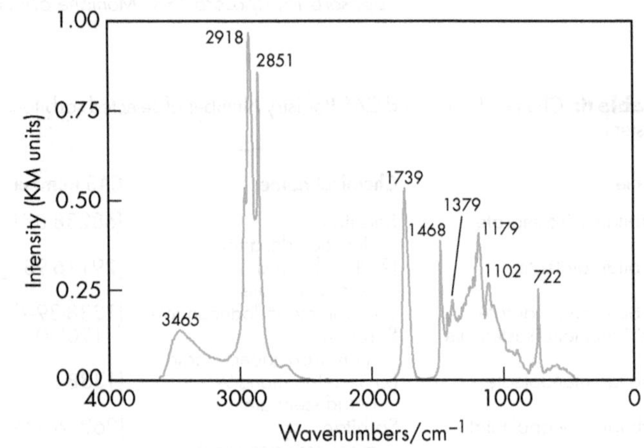

Figure 1: Infrared spectrum of sorbitan tristearate measured by diffuse reflectance. Adapted with permission of Informa Healthcare.

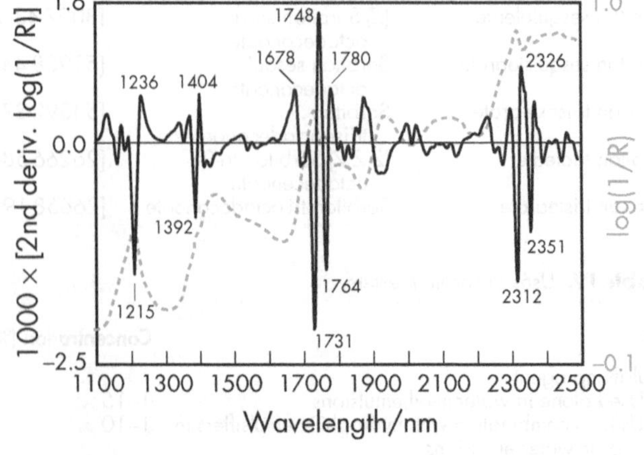

Figure 2: Near-infrared spectrum of sorbitan esters (sorbitan fatty acid esters) measured by reflectance.

following the topical application of products containing sorbitan esters.[2–5] When heated to decomposition, the sorbitan esters emit acrid smoke and irritating fumes.

Table VII: Typical properties of selected sorbitan esters.

Name	Acid value	Density (g/cm³)	HLB value	Hydroxyl value	Iodine number	Melting point (°C)	Pour point (°C)
Sorbitan monoisostearate	≤8	—	4.7	220–250	—	—	—
Sorbitan monolaurate	≤7	1.01	8.6	159–169	≤7	—	16–20
Sorbitan monooleate	≤8	1.01	4.3	193–209	—	—	−12
Sorbitan monopalmitate	3–7	1.0	6.7	270–303	≤1	43–48	—
Sorbitan monostearate	5–10	—	4.7	235–260	≤1	53–57	—
Sorbitan sesquioleate	8.5–13	1.0	3.7	188–210	—	—	—
Sorbitan trioleate	10–14	0.95	1.8	55–70	—	—	—
Sorbitan tristearate	≤7	—	2.1	60–80	—	—	—

Table VIII: Typical properties of selected sorbitan esters.

Name	Saponification value	Surface tension of 1% aqueous solution (mN/m)	Viscosity at 25°C (mPa s)	Water content (%)
Sorbitan monoisostearate	143–153	—	—	≤1.0
Sorbitan monolaurate	159–169	28	3900–4900	≤0.5
Sorbitan monooleate	149–160	30	970–1080	≤0.5
Sorbitan monopalmitate	142–152	36	Solid	≤1.0
Sorbitan monostearate	147–157	46	Solid	≤1.0
Sorbitan sesquioleate	149–160	—	1500	≤1.0
Sorbitan trioleate	170–190	32	200–250	≤1.0
Sorbitan tristearate	172–185	48	Solid	≤1.0

The WHO has set an estimated acceptable daily intake of sorbitan monopalmitate, monostearate, and tristearate,(6) and of sorbitan monolaurate and monooleate(7) at up to 25 mg/kg body-weight calculated as total sorbitan esters.

Sorbitan monolaurate LD_{50} (rat, oral): 33.6 g/kg.(8)

Experimental neoplastigen.

Sorbitan monostearate LD_{50} (rat, oral): 31 g/kg.(8)

Very mildly toxic by ingestion. Experimental reproductive effects.

15 Handling Precautions

Observe normal precautions appropriate to the circumstances and quantity of material handled. Eye protection and gloves are recommended.

16 Regulatory Status

Certain sorbitan esters are accepted as food additives in the UK. Sorbitan esters are included in the FDA Inactive Ingredients Database (inhalations; IM injections; ophthalmic, oral, topical, and vaginal preparations). Sorbitan esters are used in nonparenteral medicines licensed in the UK. Sorbitan esters are included in the Canadian Natural Health Products Ingredients Database.

17 Related Substances

Polyoxyethylene sorbitan fatty acid esters.

18 Comments

EINECS numbers

Sorbitan diisostearate 269-410-7
Sorbitan dioleate 249-448-0
Sorbitan laurate 215-663-3
Sorbitan oleate 215-665-4
Sorbitan palmitate 247-568-8
Sorbitan sesquiolate 232-360-1
Sorbitan sesquistearate 257-529-7
Sorbitan stearate 215-664-9
Sorbitan triisostearate 259-141-3
Sorbitan trioleate 247-569-3
Sorbitan tristearate 247-891-4

19 Specific References

1 Fatouros DG, *et al.* Clinical studies with oral lipid based formulations of poorly soluble compounds. *Clin Risk Manag* 2007; **3**: 591–604.
2 Finn OA, Forsyth A. Contact dermatitis due to sorbitan monolaurate. *Contact Dermatitis* 1975; **1**: 318.
3 Hannuksela M, *et al.* Allergy to ingredients of vehicles. *Contact Dermatitis* 1976; **2**: 105–110.
4 Austad J. Allergic contact dermatitis to sorbitan monooleate (*Span 80*). *Contact Dermatitis* 1982; **8**: 426–427.
5 Boyle J, Kennedy CTC. Contact urticaria and dermatitis to *Alphaderm*. *Contact Dermatitis* 1984; **10**: 178.
6 FAO/WHO. Toxicological evaluations of certain food additives with a review of general principles and of specifications. Seventeenth report of the joint FAO/WHO expert committee on food additives. *World Health Organ Tech Rep Ser* 1974; No. 539.
7 FAO/WHO. Evaluation of certain food additives and contaminants. Twenty-sixth report of the joint FAO/WHO expert committee on food additives. *World Health Organ Tech Rep Ser* 1982; No. 683.
8 Lewis RJ, ed. *Sax's Dangerous Properties of Industrial Materials*, 11th edn. New York: Wiley, 2004; 3291.

20 General References

Konno K, *et al.* Solubility, critical aggregating or micellar concentration and aggregate formation of non-ionic surfactants in non-aqueous solutions. *J Colloid Interface Sci* 1974; **49**: 383.

Mittal KL, ed. *Micellization, Solubilization and Microemulsions*. New York: Plenum Press, 1977.

Smolinske SC. *Handbook of Food, Drug, and Cosmetic Excipients*. Boca Raton, FL: CRC Press, 1992; 369–370.

Suzuki E, *et al.* Studies on methods of particle size reduction of medicinal compounds VIII: size reduction by freeze-drying and the influence of pharmaceutical adjuvants on the micromeritic properties of freeze-dried powders. *Chem Pharm Bull* 1979; **27**: 1214–1222.

Whitworth CW, Pongpaibul Y. The influence of some additives on the stability of aspirin in an oleaginous suppository base. *Can J Pharm Sci* 1979; **14**: 36–38.

21 Author

S Wallace.

22 Date of Revision

2 March 2012.

S

Sorbitol

1 Nonproprietary Names

BP: Sorbitol
JP: D-Sorbitol
PhEur: Sorbitol
USP–NF: Sorbitol

2 Synonyms

*C*PharmSorbidex*; E420; 1,2,3,4,5,6-hexanehexol; *Liponic 70-NC*; *Liponic 76-NC*; *Meritol*; *Neosorb*; *Sorbitab*; *Sorbitab SD 250*; sorbite; D-sorbitol; *Sorbitol Instant*; *Sorbitol Special*; sorbitolum; *Sorbogem*.

3 Chemical Name and CAS Registry Number

D-Glucitol [50-70-4]

4 Empirical Formula and Molecular Weight

$C_6H_{14}O_6$ 182.17

5 Structural Formula

6 Functional Category

Humectant; plasticizing agent; sweetening agent; tablet and capsule diluent.

7 Applications in Pharmaceutical Formulation or Technology

Sorbitol is widely used as an excipient in pharmaceutical formulations. It is also used extensively in cosmetics and food products; *see* Table I.

Sorbitol is used as a diluent in tablet formulations prepared by either wet granulation or direct compression.[1–6] It is particularly useful in chewable tablets owing to its pleasant, sweet taste and cooling sensation. Sorbitol has been studied for use in fast disintegrating tablets and pellets[7–9] and dry powder inhaler formulations.[10] It has been shown to increase the strength of methylcellulose gel formulations.[11] In capsule formulations it is used as a plasticizer for gelatin. Sorbitol has been used as a plasticizer in film formulations.[12,13]

In liquid preparations[14] sorbitol is used as a vehicle in sugar-free formulations, for example in drugs,[15] vitamins,[16,17] and antacid suspensions. Furthermore, sorbitol is used as an excipient in liquid parenteral biologic formulations.[18] It has also been shown to be a suitable carrier to enhance the *in vitro* dissolution rate of indometacin.[19] In syrups it is effective in preventing crystallization around the cap of bottles. Sorbitol is additionally used in injectable[20] and topical preparations.

Table I: Uses of sorbitol.

Use	Concentration (%)
Humectant	3–15
IM injections	10–25
Moisture control agent in tablets	3–10
Oral solutions	20–35
Oral suspensions	70
Plasticizer for gelatin and cellulose	5–20
Prevention of 'cap locking' in syrups and elixirs	15–30
Substitute for glycerin and propylene glycol	25–90
Tablet binder and filler	25–90
Toothpastes	20–60
Topical emulsions	2–18

8 Description

Sorbitol is D-glucitol. It is a hexahydric alcohol related to mannose and is isomeric with mannitol.

Sorbitol occurs as an odorless, white or almost colorless, crystalline, hygroscopic powder. Four crystalline polymorphs and one amorphous form of sorbitol have been identified that have slightly different physical properties, e.g. melting point.[3] Sorbitol is available in a wide range of grades and polymorphic forms, such as granules, flakes, or pellets that tend to cake less than the powdered form and have more desirable compression characteristics. Sorbitol has a pleasant, cooling, sweet taste and has approximately 50–60% of the sweetness of sucrose.

See also Section 18.

9 Pharmacopeial Specifications

See Table II. *See also* Section 18.

10 Typical Properties

Acidity/alkalinity pH = 4.5–7.0 for a 10% w/v aqueous solution.
Compressibility Compression characteristics and the degree of lubrication required vary, depending upon the particle size and

SEM 1: Excipient: sorbitol; manufacturer: SPI Pharma; lot no.: 5224F8; magnification: 100×.

Table II: Pharmacopeial specifications for sorbitol.

Test	JP XV	PhEur 7.4	USP35–NF30
Identification	+	+	+
Characters	–	+	–
Acidity or alkalinity	+	–	–
pH	–	–	3.5–7.0
Appearance of solution	+	+	+
Arsenic	≤1.3 ppm	–	–
Chloride	≤0.005%	–	≤0.005%
Sulfate	≤0.006%	–	≤0.01%
Conductivity	–	≤20 μS cm^{-1}	–
Glucose	+	–	–
Heavy metals	≤5 ppm	–	–
Lead	–	≤0.5 ppm	–
Microbial contamination	–	+	–
Bacterial	–	≤10^3 cfu/g	≤10^3 cfu/g
Fungi	–	≤10^2 cfu/g	≤10^2 cfu/g
Bacterial endotoxins	–	+	+[(a)]
Nickel	+	≤1 ppm	≤1 ppm
Reducing sugars	–	≤0.2%	≤0.3%
Related products	–	≤0.1%	–
Residue on ignition	≤0.02%	–	≤0.1%
Total sugars	+	–	–
Water	≤2.0%	≤1.5%	≤1.5%
Assay (anhydrous basis)	≥97.0%	97.0–102.0%	91.0–100.5%

(a) ≤4 USP–NF Endotoxin Units per g for parenteral dosage forms having a concentration of less than 100 g of sorbitol per L and ≤2.5 USP–NF Endotoxin Units per g for parenteral dosage forms having a concentration of 100 g or more of sorbitol per L.

grade of sorbitol used. The spray-dried forms of sorbitol afford greater compression characteristics than standard grades of sorbitol.

Density 1.49 g/cm^3

Density (bulk)

0.448 g/cm^3;

0.6–0.7 g/cm^3 for *Sorbitab SD 250*;

0.5–0.6 g/cm^3 for *Sorbitab SD 500*.

Density (tapped)

0.400 g/cm^3;

0.7 g/cm^3 for *Sorbitab SD 250*;

0.6 g/cm^3 for *Sorbitab SD 500*.

Density (true) 1.507 g/cm^3

Flowability Flow characteristics vary depending upon the particle size and grade of sorbitol used. Fine powder grades tend to be poorly flowing, while granular grades have good flow properties.

Heat of solution –110.9 J/g (–26.5 cal/g)

Melting point

Anhydrous form: 110–112°C;

Gamma polymorph: 97.7°C;

Metastable form: 93°C.

Moisture content Sorbitol is a very hygroscopic powder and relative humidities greater than 60% at 25°C should be avoided when sorbitol is added to direct-compression tablet formulas. *See also* Figure 1.

Osmolarity A 5.48% w/v aqueous solution of sorbitol hemihydrate is iso-osmotic with serum.

Particle size distribution Particle size distribution varies depending upon the grade of sorbitol; *see* Table III. For fine powder grades, typically 87% <125 μm in size; for granular grades, 22% <125 μm, 45% between 125 and 250 μm, and 33% between 250 and 590 μm. Individual suppliers' literature should be consulted for further information.

Solubility *see* Table IV. *See also* Section 17.

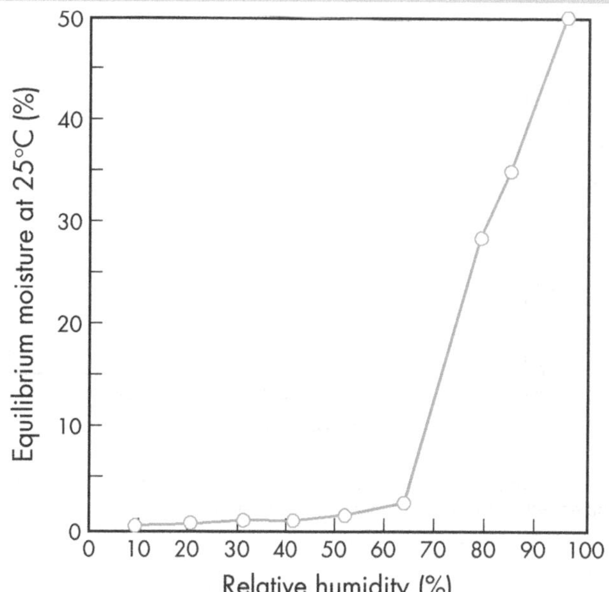

Figure 1: Equilibrium moisture content of sorbitol USP–NF.

Table III: Mean particle sizes for various grades of sorbitol.

Grade	Mean particle size (μm)
Neosorb P100T	140
Neosorb P20/60	650
Neosorb P30/60	480
Neosorb P60	220
Neosorb P60W	260
Sorbitab SD 250	250
Sorbitab SD 500	500

Table IV: Solubility of sorbitol.

Solvent	Solubility at 20°C
Chloroform	Practically insoluble
Ethanol (95%)	1 in 25
Ethanol (82%)	1 in 8.3
Ethanol (62%)	1 in 2.1
Ethanol (41%)	1 in 1.4
Ethanol (20%)	1 in 1.2
Ethanol (11%)	1 in 1.14
Ether	Practically insoluble
Methanol	Slightly soluble
Water	1 in 0.5

Spectroscopy

IR spectra *see* Figure 2.

NIR spectra *see* Figure 3.

11 Stability and Storage Conditions

Sorbitol is chemically relatively inert and is compatible with most excipients. It is stable in air in the absence of catalysts and in cold, dilute acids and alkalis. Sorbitol does not darken or decompose at elevated temperatures or in the presence of amines. It is nonflammable, noncorrosive, and nonvolatile.

Although sorbitol is resistant to fermentation by many microorganisms, a preservative should be added to sorbitol solutions. Solutions may be stored in glass, plastic, aluminum, and stainless steel containers. Solutions for injection may be sterilized by autoclaving.

S

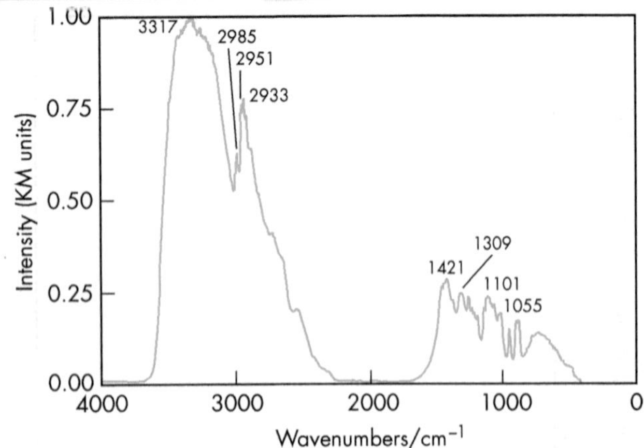

Figure 2: Infrared spectrum of sorbitol measured by diffuse reflectance. Adapted with permission of Informa Healthcare.

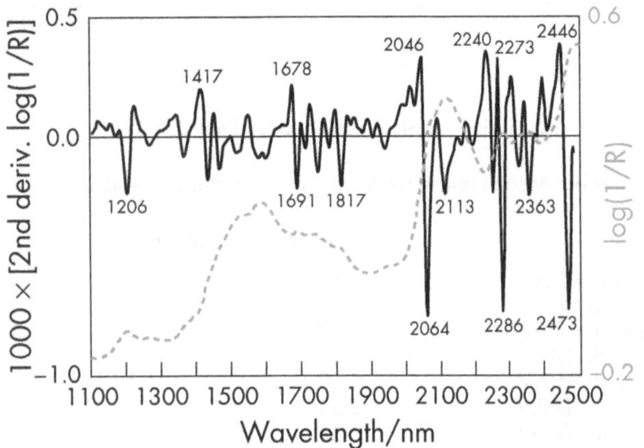

Figure 3: Near-infrared spectrum of sorbitol measured by reflectance.

The bulk material is hygroscopic and should be stored in an airtight container in a cool, dry place.

12 Incompatibilities

Sorbitol will form water-soluble chelates with many divalent and trivalent metal ions in strongly acidic and alkaline conditions. Addition of liquid polyethylene glycols to sorbitol solution, with vigorous agitation, produces a waxy, water-soluble gel with a melting point of 35–40°C. Sorbitol solutions also react with iron oxide to become discolored.

Sorbitol increases the degradation rate of penicillins in neutral and aqueous solutions.[21]

13 Method of Manufacture

Sorbitol occurs naturally in the ripe berries of many trees and plants. It was first isolated in 1872 from the berries of the Mountain Ash (*Sorbus americana*).

Industrially, sorbitol is prepared by high-pressure hydrogenation with a copper–chromium or nickel catalyst, or by electrolytic reduction of glucose and corn syrup. If cane or beet sugars are used as a source, the disaccharide is hydrolyzed to dextrose and fructose prior to hydrogenation.

14 Safety

Sorbitol is widely used in a number of pharmaceutical products and occurs naturally in many edible fruits and berries. It is absorbed more slowly from the gastrointestinal tract than sucrose and is metabolized in the liver to fructose and glucose. Its caloric value is approximately 16.7 J/g (4 cal/g). Sorbitol is better tolerated by diabetics than sucrose and is widely used in many sugar-free liquid vehicles. However, it is not considered to be unconditionally safe for diabetics.

Reports of adverse reactions to sorbitol are largely due to its action as an osmotic laxative when ingested orally,[22–24] which may be exploited therapeutically. Ingestion of large quantities of sorbitol (>20 g/day in adults) should therefore be avoided.

Sorbitol is not readily fermented by oral microorganisms and has little effect on dental plaque pH; hence, it is generally considered to be noncariogenic.[25]

Sorbitol is generally considered to be more irritating than mannitol.

LD_{50} (mouse, IV): 9.48 g/kg[26]
LD_{50} (mouse, oral): 17.8 g/kg
LD_{50} (rat, IV): 7.1 g/kg
LD_{50} (rat, SC): 29.6 g/kg

15 Handling Precautions

Sorbitol may be harmful if ingested in great quantities. It may be irritant to the eyes. Observe normal precautions appropriate to the circumstances and quantity of material handled. Eye protection, gloves, and a dust mask or respirator are recommended.

16 Regulatory Status

GRAS listed. Accepted for use as a food additive in Europe. Included in the FDA Inactive Ingredients Database (intra-articular and IM injections; nasal; oral capsules, solutions, suspensions, syrups and tablets; rectal, topical, and vaginal preparations). Included in parenteral and nonparenteral medicines licensed in the UK. Included in the Canadian Natural Health Products Ingredients Database.

17 Related Substances

Maltitol solution; mannitol; xylitol.

18 Comments

Sorbitol may be substituted for sucrose to prepare 70–90% w/v syrups.

Several different grades of sorbitol, with different polymorphic form, particle size, and other physical characteristics are commercially available, e.g. *Neosorb* (Roquette Frères). Pyrogen-free grades are also available from some suppliers.

Sorbitol is also available in liquid form and occurs as a clear, colorless, syrupy liquid, which is miscible with water (*see* Table V). Liquid sorbitol is an aqueous solution of a hydrogenated, partly hydrolyzed starch. Partially dehydrated sorbitol solutions are also available, which are produced by partial dehydration of liquid sorbitol. *Sorbo* sorbitol solution (Corn Products Specialty Ingredients) is used as a bulking agent, sweetener and humectant. *Sorbitol Special* (SPI Pharma) is a noncrystallizing polyol solution used for soft gelatin capsules. The USP35–NF30 and JP XV list sorbitol solution. The BP 2012 and PhEur 7.4 also include partially dehydrated liquid sorbitol, liquid sorbitol (crystallizing), and liquid sorbitol (non-crystallizing).

A study has shown that sorbitol may affect the bioavailability/bioequivalence of drugs by increasing gastrointestinal fluid influx and motility, which reduces time for drug absorption. It may also be employed as a cathartic in the management of poisoning.[27] Sorbitol may also be used analytically as a marker for assessing liver blood flow,[28] and therapeutically as an osmotic laxative.

A specification for sorbitol is contained in the *Food Chemicals Codex* (FCC).[29]

The EINECS number for sorbitol is 200-061-5. The PubChem Compound ID (CID) for sorbitol includes 5780 and 82170.

Table V: Physical properties of sorbitol in water solutions.

Concentration (% w/w) at 25°C	Density (g/cm^3) at 25°C	Viscosity (mPa s) at 25°C	Refractive index	Freezing point (°C)
10	1.034	1.2	1.348	−1.1
20	1.073	1.7	1.365	−3.8
30	1.114	2.5	1.383	−8.0
40	1.155	4.4	1.400	−13.0
50	1.197	9.1	1.418	−26.0
60	1.240	26.0	1.437	—
70	1.293	110.0	1.458	—
80	1.330	900.0	1.478	—

19 Specific References

1 Molokhia AM, *et al.* Effect of storage conditions on the hardness, disintegration and drug release from some tablet bases. *Drug Dev Ind Pharm* 1982; 8: 283–292.

2 Bolton S, Atluri R. Crystalline sorbitol tablets: effect of mixing time and lubricants on manufacturing. *Drug Cosmet Ind* 1984; **135**(5): 44, 46, 47, 48, 50.

3 DuRoss JW. Modification of the crystalline structure of sorbitol and its effects on tableting characteristics. *Pharm Technol* 1984; 8(9): 42–53.

4 Basedow AM, Möschl GA. Sorbitol instant – an excipient with unique tableting properties. *Drug Dev Ind Pharm* 1986; **12**: 2061–2089.

5 Schmidt PC, Vortisch W. [Influence of manufacturing method of fillers and binders on their tableting properties: comparison of 8 commercially available sorbitols.] *Pharm Ind* 1987; **49**: 495–503[in German].

6 Yamamoto K, *et al.* Preparation and evaluation of medicinal carbon tablets with different saccharides as binders. *Chem Pharm Bull (Tokyo)* 2009; **57**(10): 1058–1060.

7 Bolhuis GK, *et al.* Polyols as filler-binders for disintegrating tablets prepared by direct compaction. *Drug Dev Ind Pharm* 2009; **35**(6): 671–677.

8 Chandrasekhar R, *et al.* The role of formulation excipients in the development of lyophilised fast-disintegrating tablets. *Eur J Pharm Biopharm* 2009; **72**(1): 119–129.

9 Goyanes A, *et al.* Control of drug release by incorporation of sorbitol or mannitol in microcrystalline-cellulose-based pellets prepared by extrusion-spheronization. *Pharm Dev Technol* 2010; **15**(6): 626–635.

10 Hamishehkar H, *et al.* Effect of carrier morphology and surface characteristics on the development of respirable PLGA microcapsules for sustained-release pulmonary delivery of insulin. *Int J Pharm* 2010; **389**(1–2): 74–85.

11 Itoh K, *et al.* Effect of D-sorbitol on the thermal gelation of methylcellulose formulations for drug delivery. *Chem Pharm Bull (Tokyo)* 2010; **58**(2): 247–249.

12 Krogars K, *et al.* Development and characterization of aqueous amylose-rich maize starch dispersion for film formation. *Eur J Pharm Biopharm* 2003; **56**(2): 215–221.

13 Cervera MF, *et al.* Solid state and mechanical properties of aqueous chitosan-amylose starch films plasticized with polyols. *AAPS Pharm Sci Tech* 2004; **5**(1): E15.

14 Daoust RG, Lynch MJ. Sorbitol in pharmaceutical liquids. *Drug Cosmet Ind* 1962; **90**(6): 689–691773, 776, 777, 779, 781–785.

15 Sabatini GR, Gulesich JJ. Formulation of a stable and palatable oral suspension of procaine penicillin G. *J Am Pharm Assoc (Pract Pharm)* 1956; **17**: 806–808.

16 Bandelin FJ, Tuschhoff JV. The stability of ascorbic acid in various liquid media. *J Am Pharm Assoc (Sci)* 1955; **44**: 241–244.

17 Parikh BD, Lofgren FV. A further stability study of an oral multivitamin liquid preparation. *Drug Standards* 1958; **26**: 56–61.

18 Piedmonte DM, *et al.* Sorbitol crystallization can lead to protein aggregation in frozen protein formulations. *Pharm Res* 2007; **24**(1): 136–146.

19 Valizdeh H, *et al.* Physicochemical characterization of solid dispersions of indometacin with PEG 6000, Myri 52, lactose, sorbitol, dextrin, and Eudragit (R) E100. *Drug Dev Ind Pharm* 2004; **30**(3): 303–317.

20 Lindvall S, Andersson NSE. Studies on a new intramuscular haematinic, iron–sorbitol. *Br J Pharmacol* 1961; **17**: 358–371.

21 Bundgaard H. Drug allergy: chemical and pharmaceutical aspects. In: Florence AT, Salole EG, eds. *Formulation Factors in Adverse Reactions.* London: Wright, 1990: 23–55.

22 Jain NK, *et al.* Sorbitol intolerance in adults. *Am J Gastroenterol* 1985; **80**: 678–681.

23 Brown AM, Masson E. 'Hidden' sorbitol in proprietary medicines – a cause for concern? *Pharm J* 1990; **245**: 211.

24 Greaves RRSH, *et al.* An air stewardess with puzzling diarrhoea. *Lancet* 1996; **348**: 1488.

25 Ayers CS, Abrams RA. Noncariogenic sweeteners: sugar substitutes for caries control. *Dental Hygiene* 1987; **61**: 162–167.

26 Lewis RJ, ed. *Sax's Dangerous Properties of Industrial Materials*, 11th edn. New York: Wiley, 2004: 3292.

27 Chen M-L, *et al.* A modern view of excipient effects on bioequivalence: case study of sorbitol. *Pharm Res* 2007; **1**: 73–80.

28 Burggraaf J, *et al.* Sorbitol as a marker for drug-induced decreases of variable duration in liver blood flow in healthy volunteers. *Eur J Pharm Sci* 2000; **12**(2): 133–139.

29 *Food Chemicals Codex*, 7th edn. Bethesda, MD: United States Pharmacopeia, 2010: 977.

20 General References

Barr M, *et al.* The solubility of sorbitol in hydroalcoholic solutions. *Am J Pharm* 1957; **129**: 102–106.

Blanchard J, *et al.* Effect of sorbitol on interaction of phenolic preservatives with polysorbate 80. *J Pharm Sci* 1977; **66**: 1470–1473.

Burgess S. Sorbitol instant: a unique excipient. *Manuf Chem* 1987; **58**(6): 55, 57, 59.

Cargill Pharmaceutical Excipients. Product literature: *C*PharmSorbidex*, 2008.

Collins J. Metabolic disease: time for fructose solutions to go. *Lancet* 1993; **341**: 600.

Rabinowitz MP, *et al.* GLC assay of sorbitol as cyclic *n*-butylboronate. *J Pharm Sci* 1974; **63**: 1601–1604.

Roquette Frères. Technical literature: *Neosorb*, 2004.

Shah DN, *et al.* Mechanism of interaction between polyols and aluminum hydroxide gel. *J Pharm Sci* 1981; **70**: 1101–1104.

SPI Pharma. Technical literature: *Sorbitab SD 250/500*, 2007.

Zatz JL, Lue R-Y. Flocculation of suspensions containing nonionic surfactants by sorbitol. *J Pharm Sci* 1987; **76**: 157–160.

21 Author

J Shur.

22 Date of Revision

2 March 2012.

S

Soybean Oil

1 Nonproprietary Names

BP: Refined Soya Oil
JP: Soybean Oil
PhEur: Soya-Bean Oil, Refined
USP–NF: Soybean Oil

2 Synonyms

Aceite de soja; *Calchem IVO-114*; *Lipex 107*; *Lipex 200*; *Shogun CT*; soiae oleum raffinatum; soja bean oil; soyabean oil; soya bean oil.

3 Chemical Name and CAS Registry Number

Soybean oil [8001-22-7]

4 Empirical Formula and Molecular Weight

A typical analysis of refined soybean oil indicates the composition of the acids, present as glycerides, to be: linoleic acid 50–57%; linolenic acid 5–10%; oleic acid 17–26%; palmitic acid 9–13%; and stearic acid 3–6%. Other acids are present in trace quantities.[1]

5 Structural Formula

See Sections 4 and 8.

6 Functional Category

Oleaginous vehicle; solvent.

7 Applications in Pharmaceutical Formulation or Technology

Emulsions containing soybean oil have also been used as vehicles for the oral and intravenous administration of drugs;[2,3] drug substances that have been incorporated into such emulsions include amphotericin,[4–6] diazepam, retinoids,[7] vitamins,[8] poorly water-soluble steroids,[9,10] fluorocarbons,[11,12] ibuprofen,[13] and insulin.[14]

8 Description

The USP35–NF30 describes soybean oil as the refined fixed oil obtained from the seeds of the soya plant *Glycine max* Merr. (Fabaceae); if an antoxidant is added, the name and quantity must be specified on the label. The PhEur 7.4 defines refined soybean oil as the fatty oil obtained from the seeds of *Glycine max* (L.) Merr. (*G. hispida* (Moench) Maxim.) by extraction and subsequent refining; it may contain a suitable antioxidant. The PhEur 7.4 also includes a monograph for hydrogenated soybean oil. *See* Vegetable Oil, hydrogenated, type 1.

Soybean oil is a clear, pale-yellow colored, odorless or almost odorless liquid, with a bland taste that solidifies between −10 and −16°C.

9 Pharmacopeial Specifications

See Table I.

Table I: Pharmacopeial specifications for soybean oil.

Test	JP XV	PhEur 7.4	USP35–NF30
Identification	—	+	+
Characters	—	+	—
Specific gravity	0.916–0.922	≈0.922	—
Refractive index	—	≈1.475	—
Heavy metals	—	—	≤0.001%
Fatty acid composition	—	+	+
Acid value	≤0.2	≤0.5	≤0.3
Iodine value	126–140	—	—
Saponification value	188–195	—	—
Unsaponifiable matter	≤1.0%	≤1.5%	≤1.5%
Peroxide	—	≤10.0 or ≤5.0[a]	≤10.0
Alkaline impurities	—	+	+
Brassicasterol	—	≤0.3%	≤0.3%
Water	—	≤0.1%	≤0.1%

(a) In soybean oil intended for parenteral use.

10 Typical Properties

Autoignition temperature 445°C
Density 0.916–0.922 g/cm^3 at 25°C
Flash point 282°C
Freezing point −10 to −16°C
Hydroxyl value 4–8
Interfacial tension 50 mN/m (50 dynes/cm) at 20°C.
Refractive index n_D^{25} = 1.471–1.475
Solubility Practically insoluble in ethanol (95%) and water; miscible with ether, carbon disulfide, chloroform, ether, and light petroleum.
Spectroscopy
IR spectra *see* Figure 1.
Surface tension 25 mN/m (25 dynes/cm) at 20°C.
Viscosity (dynamic)
172.9 mPa s (172.9 cP) at 0°C;
99.7 mPa s (99.7 cP) at 10°C;
50.09 mPa s (50.09 cP) at 25°C;
28.86 mPa s (28.86 cP) at 40°C.

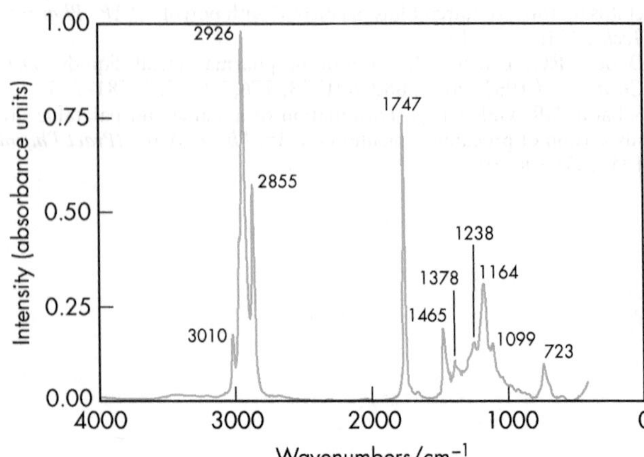

Figure 1: Infrared spectrum of soybean oil measured by transmission. Adapted with permission of Informa Healthcare.

11 Stability and Storage Conditions

Soybean oil is a stable material if protected from atmospheric oxygen.

The formation of undesirable flavors in soybean oil is accelerated by the presence of 0.01 ppm copper and 0.1 ppm iron, which act as catalysts for oxidation; this can be minimized by the addition of chelating agents.

Prolonged storage of soybean oil emulsions, particularly at elevated temperatures, can result in the formation of free fatty acids, with a consequent reduction in the pH of the emulsion; degradation is minimized at pH 6–7. However, soybean oil emulsions are stable at room temperature if stored under nitrogen in a light-resistant glass container. Plastic containers are permeable to oxygen and should not be used for long-term storage since oxidative degradation can occur.

The stability of soybean oil emulsions is considerably influenced by other additives in a formulation.[15–21]

Soybean oil should be stored in a well-filled, airtight, light-resistant container at a temperature not exceeding 25°C.

12 Incompatibilities

Soybean oil emulsions have been reported to be incompatible at 25°C with a number of materials including calcium chloride, calcium gluconate, magnesium chloride, phenytoin sodium, and tetracycline hydrochloride.[22] Lower concentrations of these materials, or lower storage temperatures, may result in improved compatibility. The source of the material may also affect compatibility; for example, while one injection from a particular manufacturer might be incompatible with a fat emulsion, an injection with the same amount of active drug substance from another manufacturer might be compatible.

Amphotericin B has been reported to be incompatible with soybean oil containing fat emulsions under certain conditions.[23]

Soybean oil emulsions are also incompatible with many other drug substances, IV infusion solutions, and ions (above certain concentrations).

When plastic syringes are used to store soybean oil emulsion, silicone oil may be extracted into the emulsion; swelling of the syringe components may also occur, resulting in the necessity for increased forces to maintain the motion of the plunger.[24]

13 Method of Manufacture

Obtained by solvent extraction using petroleum hydrocarbons, or to a lesser extent by expression using continuous screw-press operations, of the seeds of either *Glycine max* (Leguminosae) or *Glycine soja* (Leguminosae). The oil is refined, deodorized, and clarified by filtration at about 0°C. Any phospholipids or sterols present are removed by refining with alkali.

14 Safety

Soybean oil is widely used intramuscularly as a drug vehicle or as a component of emulsions used in parenteral nutrition regimens; it is also consumed as an edible oil. Generally, soybean oil is regarded as an essentially nontoxic and nonirritant material. However, serious adverse reactions to soybean oil emulsions administered parenterally have been reported. These include cases of hypersensitivity,[25] CNS reactions,[26] and fat embolism.[27] Interference with the anticoagulant effect of warfarin has also been reported.[28]

Anaphylactic reactions have also been reported following the consumption of foods derived from, or containing, soybeans. Recently there has been concern at the concentration of phytoestrogens in some soy-derived products. Administration of soy protein to humans has resulted in significantly decreased serum lipid concentrations.[29]

In 1999, the UK Medical Devices Agency announced the voluntary withdrawal of a breast implant that contained soybean oil. The decision was taken because not enough was known at that time about the long-term safety and the rate of breakdown of the soybean oil in the filling and its possible effects on the body.[30]

LD_{50} (mouse, IV): 22.1 g/kg[31]

LD_{50} (rat, IV): 16.5 g/kg

15 Handling Precautions

Observe normal precautions appropriate to the circumstances and quantity of material handled. Spillages of soybean oil are slippery and should be covered with an inert absorbent material prior to disposal.

16 Regulatory Status

Included in the FDA Inactive Ingredients Database (IV injections and emulsion injections; oral capsules, tablets, and chewable tablets; topical lotions and solutions). Included in nonparenteral (oral capsules, tablets, chewable tablets, effervescent tablets and granules, solutions; topical bath additives; transdermal patches) and parenteral (emulsions for IV injection or infusion) medicines licensed in the UK. Included in the Canadian Natural Health Products Ingredients Database.

17 Related Substances

Canola oil; corn oil; cottonseed oil; peanut oil; sesame oil; sunflower oil.

18 Comments

The stability of soybean oil emulsions may be readily disturbed by the addition of other materials, and formulations containing soybean oil should therefore be evaluated carefully for their compatibility and stability.

In pharmaceutical preparations, soybean oil emulsions are primarily used as a fat source in total parenteral nutrition (TPN) regimens.[32] Although other oils, such as peanut oil, have been used for this purpose, soybean oil is now preferred because it is associated with fewer adverse reactions.

In addition, soybean oil has been used in the formulation of many drug delivery systems such as liposomes,[33] microspheres,[34] dry emulsions,[35] self-emulsifying systems,[36,41] microemulsions,[38,39] nanoemulsions[40,41] and nanocapsules,[40] solid-in-oil suspensions,[42] and multiple emulsions.[43] In the formation of organogels, 12-hydroxystearic acid has been suggested as a gelator for soybean oil.[44]

Soybean oil may also be used in cosmetics and is consumed as an edible oil. As soybean oil has emollient properties, it is used as a bath additive in the treatment of dry skin conditions.

A specification for soybean oil (hydrogenated) is contained in the USP35–NF30 and the *Food Chemicals Codex* (FCC).[45]

S

19 Specific References

1 British Standards Institute. *Specification for Crude Vegetable Fats, BS 7207*. London: HMSO, 1990.

2 Jeppsson R. Effects of barbituric acids using an emulsion form intravenously. *Acta Pharm Suec* 1972; **9**: 81–90.

3 Medina J, *et al*. Use of ultrasound to prepare lipid emulsions of lorazepam for intravenous injection. *Int J Pharm* 2001; **216**(1–2): 1–8.

4 Wasan KM. Amphotericin B-intralipid. *Drugs of the Future* 1994; **19**(3): 225–227.

5 Vita E. Intralipid in prophylaxis of amphotericin B nephrotoxicity. *Ann Pharmacother* 1994; **28**: 1182–1183.

6 Pascual B, *et al*. Administration of lipid-emulsion versus conventional amphotericin B in patients with neutropenia. *Ann Pharmacother* 1995; **29**: 1197–1201.

7 Nankevis R, *et al*. Studies on the intravenous pharmacokinetics of three retinoids in the rat. *Int J Pharm* 1994; **101**: 249–256.

8 Dahl GB, *et al*. Stability of vitamins in soybean oil fat emulsion under conditions simulating intravenous feeding of neonates and children. *J Parenter Enteral Nutr* 1994; **18**(3): 2234–2239.

9 Malcolmson C, Lawrence MJ. A comparison of the incorporation of model steroids into non-ionic micellar and microemulsion systems. *J Pharm Pharmacol* 1993; **45**: 141–143.
10 Steroid anaesthetic agents [editorial]. *Lancet* 1992; **340**: 83–84.
11 Johnson OL, *et al.* Thermal stability of fluorocarbon emulsions that transport oxygen. *Int J Pharm* 1990; **59**: 131–135.
12 Johnson OL, *et al.* Long-term stability studies of fluorocarbon oxygen transport emulsions. *Int J Pharm* 1990; **63**: 65–72.
13 Rathi LG, *et al.* Intravenous ibuprofen lipid emulsions – formulation and *in vitro* evaluation. *Eur J Parenter Pharm Sci* 2007; **12**(1): 17–21.
14 Morishita M, *et al.* Improving insulin enteral absorption using water-in-oil emulsion. *Int J Pharm* 1998; **172**(1–2): 189–198.
15 Takamura A, *et al.* Study of intravenous hyperalimentation: effect of selected amino acids on the stability of intravenous fat emulsions. *J Pharm Sci* 1984; **73**: 91–94.
16 Driscoll DF, *et al.* Practical considerations regarding the use of total nutrient admixtures. *Am J Hosp Pharm* 1986; **43**: 416–419.
17 Washington C. The stability of intravenous fat emulsions in total parenteral nutrition mixtures. *Int J Pharm* 1990; **66**: 1–21.
18 Manning RJ, Washington C. Chemical stability of total parenteral nutrition mixtures. *Int J Pharm* 1992; **81**: 1–20.
19 Jumaa M, Müller BW. The effect of oil components and homogenisation conditions on the physicochemical properties and stability of parenteral fat emulsions. *Int J Pharm* 1998; **163**(1–2): 81–89.
20 Jumaa M, Müller BW. The stabilisation of parenteral fat emulsion using non-ionic ABA copolymer surfactant. *Int J Pharm* 1998; **174**(1–2): 29–37.
21 Warisnoicharoen W, *et al.* Non-ionic oil-in-water microemulsions: the effects of oil type on phase behaviour. *Int J Pharm* 2000; **198**(1): 7–27.
22 Trissel LA. *Handbook on Injectable Drugs*, 9th edn. Bethesda, MD: American Society of Hospital Pharmacists, 1996: 435–447.
23 Trissel LA. Amphotericin B does not mix with fat emulsion [letter]. *Am J Health Syst Pharm* 1995; **52**: 1463–1464.
24 Capes DF, *et al.* The effect on syringe performance of fluid storage and repeated use: implications for syringe pumps. *PDA J Pharm Sci Technol* 1996; **50**(Jan–Feb): 40–50.
25 Hiyama DT, *et al.* Hypersensitivity following lipid emulsion infusion in an adult patient. *J Parenter Enteral Nutr* 1989; **13**: 318–320.
26 Jellinek EH. Dangers of intravenous fat infusions [letter]. *Lancet* 1976; **ii**: 967.
27 Estebe JP, Malledant Y. Fat embolism after lipid emulsion infusion [letter]. *Lancet* 1991; **337**: 673.
28 Lutomski DM, *et al.* Warfarin resistance associated with intravenous lipid administration. *J Parent Enteral Nutr* 1987; **11**(3): 316–318.
29 Anderson JW, *et al.* Meta-analysis of the effects of soy protein intake on serum lipids. *N Engl J Med* 1995; **333**(5): 276–282.
30 Bradbury J. Breast implants containing soy-bean oil withdrawn in UK [news]. *Lancet* 1999; **353**: 903.
31 Sweet DV, ed. *Registry of Toxic Effects of Chemical Substances.* Cincinnati: US Department of Health, 1987: 4454.
32 McNiff BL. Clinical use of 10% soybean oil emulsion. *Am J Hosp Pharm* 1977; **34**: 1080–1086.
33 Stricker H, Müller H. [The storage stability of dispersions of soybean-lecithin liposomes.] *Pharm Ind* 1984; **46**: 1175–1183[in German].
34 Salmerón MD, *et al.* Encapsulation study of 6-methylprednisolone in liquid microspheres. *Drug Dev Ind Pharm* 1997; **23**(2): 133–136.
35 Pedersen GP, *et al.* Solid state characterisation of a dry emulsion: a potential drug delivery system. *Int J Pharm* 1998; **171**(2): 257–270.
36 Buyukozturk F, *et al.* Impact of emulsion-based drug delivery systems on intestinal permeability and drug release kinetics. *J Control Release* 2010; **142**: 22–30.
37 Taha EI, *et al.* Response surface methodology for the development of self-nanoemulsified drug delivery system (SNEDDS) of all-trans-retinol acetate. *Pharm Dev Technol* 2005; **10**(3): 363–370.
38 Karasulu HY, *et al.* Controlled release of methotrexate from W/O microemulsion and its *in vitro* antitumour activity. *Drug Deliv* 2007; **14**(4): 225–233.
39 Hwang SR, *et al.* Phospholipid-based microemulsion formulation of all-trans-retinoic acid for parenteral administration. *Int J Pharm* 2004; **276**(1–2): 175–183.
40 Ganta S, *et al.* Pharmacokinetics and pharmacodynamics of chlorambucil delivered in long-circulating nanoemulsion. *J Drug Target* 2010; **18**: 125–133.
41 Fang JY, *et al.* Lipid nano-submicron emulsions as vehicles for topical flurbiprofen delivery. *Drug Deliv* 2004; **11**(2): 97–105.
42 Piao H, *et al.* Oral delivery of diclofenac sodium using a novel solid-in-oil suspension. *Int J Pharm* 2006; **313**: 159–162.
43 Bokir A, Hayta G. Preparation and evaluation of multiple emulsions water-in-oil-in-water (w/o/w) as delivery system for influenza virus antigens. *J Drug Target* 2004; **12**(3): 157–164.
44 Iwanaga K, *et al.* Characterisation of organogel as a novel oral controlled release formulation for lipophilic compounds. *Int J Pharm* 2010; **388**: 123–128.
45 *Food Chemicals Codex*, 7th edn. Bethesda, MD: United States Pharmacopeia, 2010: 983.

20 General References

Benita S, Levy MY. Submicron emulsions as colloidal drug carriers for intravenous administration: comprehensive physicochemical characterization. *J Pharm Sci* 1993; **82**: 1069–1079.
Delaveau P, Hotellier F. [Oils of pharmaceutical, dietetic, and cosmetic interest, part I: maize, soybean, sunflower.] *Ann Pharm Fr* 1971; **29**: 399–412[in French].
Mirtallo JM, Oh T. A key to the literature of total parenteral nutrition: update 1987. *Drug Intell Clin Pharm* 1987; **21**: 594–606.
Smolinske SC. *Handbook of Food, Drug, and Cosmetic Excipients.* Boca Raton: FL: CRC Press, 1992: 383–385.
Wolf WJ. *Kirk-Othmer Encyclopedia of Chemical Technology.* 21. 3rd edn. New York: Wiley-Interscience, 1981: 417–442.

21 Author

CG Cable.

22 Date of Revision

2 March 2012.

S

Squalane

1 Nonproprietary Names

BP: Squalane
PhEur: Squalane
USP–NF: Squalane

2 Synonyms

Cosbiol; dodecahydrosqualene; *Fitoderm*; perhydrosqualene; *Robane*; spinacane; squalanum; *Vitabiosol*; *Wax-O-Sol*.

3 Chemical Name and CAS Registry Number

2,6,10,15,19,23-Hexamethyltetracosane [111-01-3]

4 Empirical Formula and Molecular Weight

$C_{30}H_{62}$ 422.8

Squalane is a saturated derivative of squalene, which is a constituent of human sebum. The USP35–NF30 describes squalane as a saturated hydrocarbon obtained by hydrogenation of squalene, an aliphatic triterpene occurring in some fish oils. The PhEur 7.4 states that squalane may be of vegetable (unsaponifiable matter of olive oil) or animal (shark liver oil) origin.

5 Structural Formula

CH₃ CH₃ CH₃ H₃C CH₃ CH₃ CH₃ CH₃

6 Functional Category

Emollient; ointment base; penetration enhancer.

7 Applications in Pharmaceutical Formulation or Technology

Squalane is included in topical pharmaceutical preparations to increase skin permeability owing to its miscibility with human sebum. It is also used as a carrier of lipid-soluble drugs in suppositories.

Squalane is used as an adjuvant and as a component of nanoemulsions that have been developed for drug and vaccine applications.[1–5]

Squalane is also used as an emollient in cosmetics, hair, and shaving products.

8 Description

Squalane occurs as a colorless, almost odorless, transparent oil.

9 Pharmacopeial Specifications

See Table I.

Table I: Pharmacopeial specifications for squalane.

Test	PhEur 7.4	USP35–NF30
Characters	+	–
Identification	+	+
Appearance	+	–
Specific gravity	–	0.807–0.810
Refractive index	1.450–1.454	1.4510–1.4525
Acid value	≤0.2	≤0.2
Iodine value	≤0.4	≤0.4
Saponification value	≤3.0	≤2.0
Nickel	≤1 ppm	–
Total ash	≤0.5%	≤0.5%
Chromatographic purity	–	≥97%
Assay	96.0–103.0%	–

10 Typical Properties

Boiling point
≈350°C at 760 mmHg;
263°C at 10 mmHg;
248°C at 5 mmHg;
210–215°C at 1 mmHg;
176°C at 0.05 mmHg.

Density ≈ 0.815 g/cm^3

Flash point 218–230°C

Melting point −38°C

Solubility Insoluble in water; very slightly soluble in dehydrated alcohol; miscible with chloroform and with ether; slightly soluble in acetone.

Viscosity 34 mPa s (34 cP) at 20°C

11 Stability and Storage Conditions

Stable under normal temperatures and pressures. Store in a cool, dry, well-ventilated area away from incompatible substances, and in airtight containers.

12 Incompatibilities

Squalane is incompatible with oxidizing agents.

13 Method of Manufacture

Squalane is prepared by complete hydrogenation of squalene. Commercial grades are obtained by direct hydrogenation of shark liver oil and may contain some batyl alcohol. It can also be isolated from the unsaponifiable fraction of olive oil by extraction, followed by hydrogenation.

14 Safety

Squalane is generally considered to be a safe material. The Cosmetic Ingredient Review (CIR) Expert Panel have assessed squalane and concluded that the material is safe as used in cosmetics and personal care products.[6] May cause irritation to the skin, eyes, and respiratory tract. May cause gastrointestinal irritation.

LD_{50} >5.0 g/kg[7]

15 Handling Precautions

Observe normal precautions appropriate to the circumstances and quantity of the material handled. When heated to decomposition, squalane emits acrid smoke and irritating vapors. Avoid contact with eyes, and skin, and avoid ingestion and inhalation. Protective clothing and safety goggles are recommended.

S

16 Regulatory Status

Included in the FDA Inactive Ingredients Database (topical emulsions, creams, and solutions). Included in the Canadian Natural Health Products Ingredients Database. Included in nonparenteral medicines licensed in the UK.

17 Related Substances

—

18 Comments

Squalane has been investigated for inducing bone formation in a protein delivery system and was found to be effective for the slow local release of bone morphogenetic protein from gelatin capsules.[8]

The EINECS number for squalane is 203-825-6. The PubChem Compound ID (CID) is 8089.

19 Specific References

1 Allison AC. Squalene and squalane emulsions as adjuvants. *Methods* 1999; **19**(1): 87–93.
2 Fox CB. Squalene emulsions for parenteral vaccine and drug delivery. *Molecules* 2009; **14**: 3286–3312.
3 Hilgers LA. Sulfolipo-cyclodextrin in squalane-in-water as a novel and safe vaccine adjuvant. *Vaccine* 1999; **17**(3): 219–228.
4 Hilgers LA, Blom AG. Sucrose fatty acid sulphate esters as novel vaccine adjuvant. *Vaccine* 2006; **24**(Suppl2): 81–82.
5 Shahiwala A, Amiji MM. Enhanced mucosal and systemic immune response with squalane oil-containing multiple emulsions upon intranasal and oral administration in mice. *J Drug Target* 2008; **16**(4): 302–310.
6 Cosmetic Ingredient Review Expert Panel. Final report on the safety assessment of squalane and squalene. *Int J Toxicol* 1982; 1(2): 37–56.
7 MakingCosmetics.com, Inc. Material safety data sheet: *Squalane*, February 2008.
8 Kawakami T. Evaluation of heterotopic bone formation induced by squalane and bone morphogenetic protein composite. *Clin Orthop Relat Res* 1997; **337**: 261–266.

20 General References

Cognis Iberia, Cognis Iberia SL. Product data sheet: *Fitoderm*. http://www.cospha.ro/dbimg/Fitoderm.pdf .(accessed 11 August 2011).
Jeen International Corp. Material safety data sheet: *Squalane*, 2000.
Rosenthal, Rosenthal ML. Squalane: the natural moisturizer. In: Schlossman ML, ed. *Chemistry and Manufacture of Cosmetics*. 3rd edn. Vol. 3 (Book 2). Carol Steam, IL: Allured Publishing Corporation, 2002: 869-875.
ScienceLab.com. Material safety data sheet: *Squalane*, 2010.

21 Authors

ME Fenton, RC Rowe.

22 Date of Revision

2 March 2012.

Starch

1 Nonproprietary Names

BP: Maize starch
Potato starch
Pea Starch
Rice Starch
Tapioca Starch
Wheat Starch

JP: Corn Starch
Potato Starch
Rice Starch
Wheat Starch

PhEur: Maize Starch
Pea Starch
Potato Starch
Rice Starch
Wheat Starch

USP–NF: Corn Starch
Pea Starch
Potato Starch
Rice Starch
Tapioca Starch
Wheat Starch

The USP35–NF30 and PhEur 7.4 share a common monograph for Pea Starch (*Pisum sativum* L.). Tapioca Starch (*Manihot utilissima* Pohl) is described in individual monographs in the USP35–NF30 and BP 2012. The USP35–NF30 also includes a monograph for topical starch.

See also Sections 9 and 18.

2 Synonyms

Amido; amidon; amilo; amylum; *C*PharmGel*; *Eurylon*; fecule; *Hylon*; *Maisita*; maydis amylum; *Melojel*; *Meritena*; oryzae amylum; *Pearl*; *Perfectamyl*; pisi amylum; *Pure-Dent*; *Purity 21*; *Purity 826*; *Remy*; solani amylum; *Stärkina*; tritici amylum; *Uni-Pure*.

See also Sections 1 and 18.

3 Chemical Name and CAS Registry Number

Starch [9005-25-8]

4 Empirical Formula and Molecular Weight

$(C_6H_{10}O_5)_n$ where n = 300–1000.

Starch consists of linear amylose and branched amylopectin, two polysaccharides based on α-(D)-glucose. Both polymers are organized in a semicrystalline structure, and in the starch granule, amylopectin forms the crystalline portion. The exact structure of starch is not yet fully understood. There is no specific distribution pattern of amylose and amylopectin molecules in the starch grain. Both molecules are organized in similar structures, probably as clusters according to the most recent scientifically recognized models. The different configurations of these molecules result in different behavior in cold aqueous solutions. Amylose (only linear 1,4 bonds) shows a high tendency for crystallization (retrogradation) resulting in insoluble adducts, whereas amylopectin (branched polymer) shows slow jellification, forming opaque and highly viscous preparations after some days.

See also Sections 5 and 10.

The molecular weight depends on the origin and the nature of the starch. It can range between 50 and 500 million Da, with amylopectin having a higher molecular weight than amylose.

5 Structural Formula

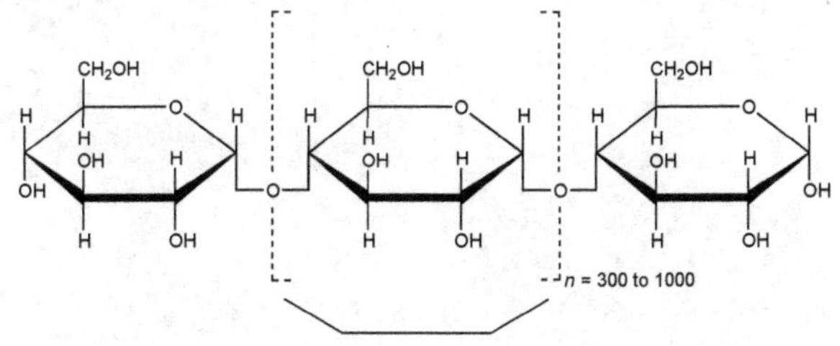

n = 300 to 1000

Amylose

Glucose unit

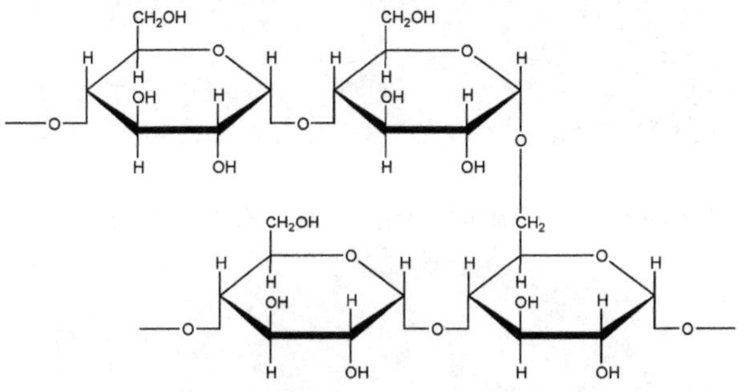

Segment of amylopectin molecule

6 Functional Category

Tablet and capsule binder; tablet and capsule diluent; tablet and capsule disintegrant; viscosity-increasing agent.

7 Applications in Pharmaceutical Formulation or Technology

Starch is a versatile excipient used primarily in oral solid-dosage formulations where it is utilized as a binder, diluent, and disintegrant.

As a diluent, starch is used for the preparation of standardized triturates of colorants, potent drugs, and herbal extracts, facilitating subsequent mixing or blending processes in manufacturing operations. Starch is also used in dry-filled capsule formulations for volume adjustment of the fill matrix,[1] and to improve powder flow, especially when using dried starches. Starch quantities of 3–10% w/w can act as an antiadherent and lubricant in tableting and capsule filling.

In tablet formulations, freshly prepared starch paste is used at a concentration of 3–20% w/w (usually 5–10%, depending on the starch type) as a binder for wet granulation. The required binder ratio should be determined by optimization studies, using parameters such as tablet friability and hardness, disintegration time, and drug dissolution rate.

Starch is one of the most commonly used tablet disintegrants at concentrations of 3–25% w/w;[2–7] a typical concentration is 15%. When using starch, a prior granulation step is required in most cases to avoid problems with insufficient flow and segregation. However, starch that is not pregelatinized does not compress well and tends to increase tablet friability and capping if used in high concentrations.[8] Balancing the elastic properties of starch with adapted excipients has been shown to improve the compaction properties in tableting.[9–11]

Starch, particularly the fine powders of rice and wheat starch, is also used in topical preparations for its absorbency of liquids. Starch paste is used in ointment formulations, usually in the presence of higher ratios of glycerin.

Starch has been investigated as an excipient in novel drug delivery systems for nasal,[12] and other site-specific delivery systems.[13,14] The retrogradation of starch can be used to modify the surface properties of drug particles.[15] Starches are useful carriers for amorphous drug preparations, such as pellets with immediate or delayed drug release obtained, for example, by melt extrusion,[16,17] and they can improve the bioavailability of poorly soluble drugs. Some specific starch varieties, especially waxy corn starch and high amylose corn starch, are valuable candidates to replace MCC in extrusion/spheronization processes,[18–20] and may also be used for colon-targeting applications.[21,22]

Native starches conforming to pharmacopeial specifications are used as the raw materials for the production of starch-based excipients and active pharmaceutical ingredients, frequently covered with their own pharmacopeial monographs.

See also Section 17.

8 Description

Starch occurs as an odorless and tasteless, fine, white to off-white powder. It consists of very small spherical or ovoid granules or grains whose size and shape are characteristic for each botanical variety.

9 Pharmacopeial Specifications

The pharmacopeial specifications for corn (maize) starch (*Zea mays*), potato starch (*Solanum tuberosum* L.), rice starch (*Oryza sativa* L.), and wheat starch (*Triticum aestivum* L.) have undergone harmonization of many attributes for USP–NF, PhEur, and JP XV.

See Table I. *See also* Section 18.

10 Typical Properties

Acidity/alkalinity Aqueous dispersions of starch usually have a pH in the range 4.0–8.0. Starch does not exhibit a significant self-buffering capacity.

Amylose content

24–28% for corn starch;

35–39% for pea starch;

20–23% for potato starch;

17–20% for tapioca starch;

24–28% for wheat starch.

Compactability *see* Figure 1.

Density (bulk) (depending on the industrial process and humidity)

0.45–0.58 g/cm^3 for corn starch;[23]

0.56–0.82 g/cm^3 for potato starch;[23]

≈0.50 g/cm^3 for wheat starch.

Density (tapped) (depending on the industrial process and humidity)

0.69–0.77 g/cm^3 for corn starch;[23]

0.80–0.90 g/cm^3 for potato starch;[23]

≈0.76 g/cm^3 for wheat starch.

Density (true) 1.478 g/cm^3 for corn starch.

Flowability Commercial starch is generally cohesive and has poor flow characteristics. The flow properties depend strictly on the moisture content,[23,24] and drying can result in a free-flowing material.

Gelatinization temperature (measured at 20% w/w in water with differential scanning colorimetry (peak))

71°C for corn starch;

62°C for pea starch,

64°C for potato starch;

S

SEM 1: Excipient: corn starch; manufacturer: Roquette Frères; magnification: 750×; voltage: 5 kV.

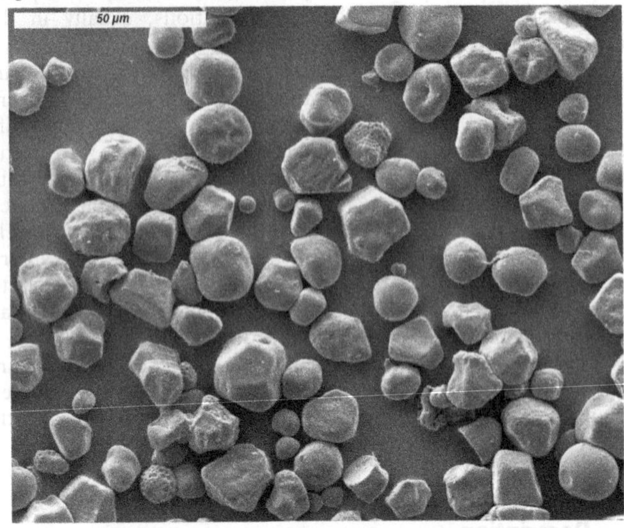

SEM 2: Excipient: pea starch; manufacturer: Roquette Frères; magnification: 750×; voltage: 5 kV.

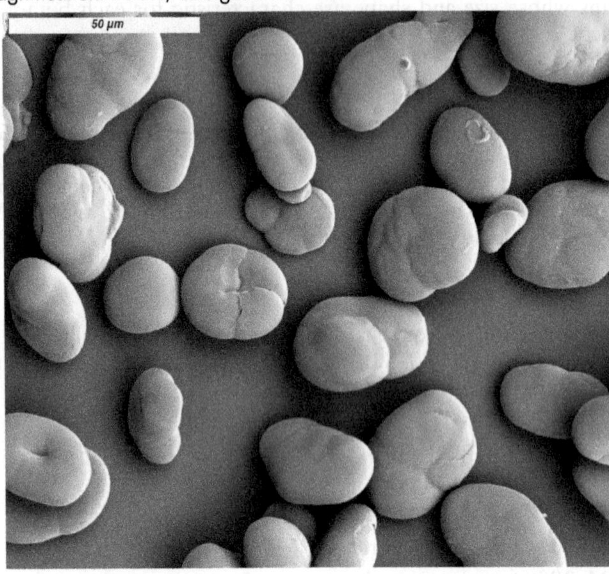

SEM 3: Excipient: potato starch; manufacturer: Roquette Frères; magnification: 750×; voltage: 5 kV.

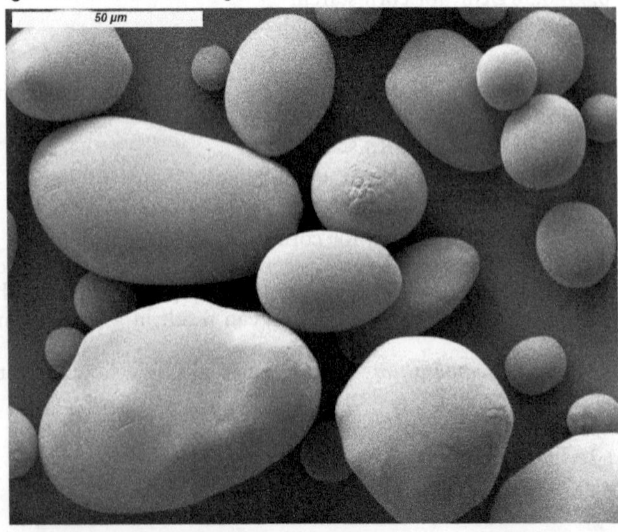

SEM 4: Excipient: rice starch; manufacturer: Remy Industries NV; magnification: 750×; voltage: 5 kV.

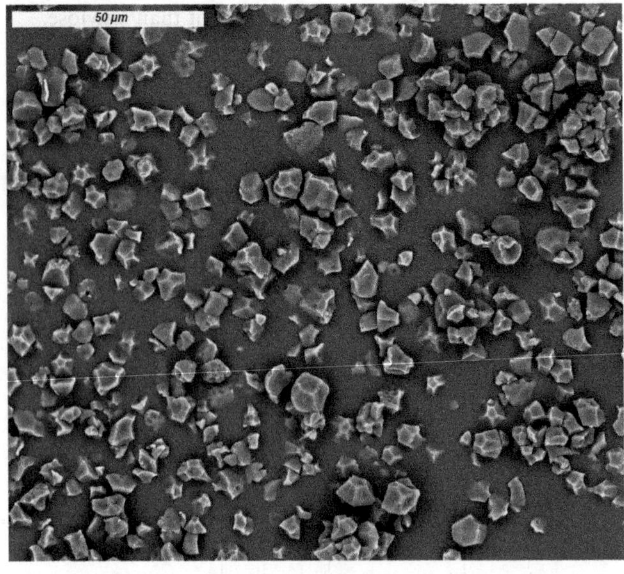

SEM 5: Excipient: tapioca starch; magnification: 750×; voltage: 5 kV.

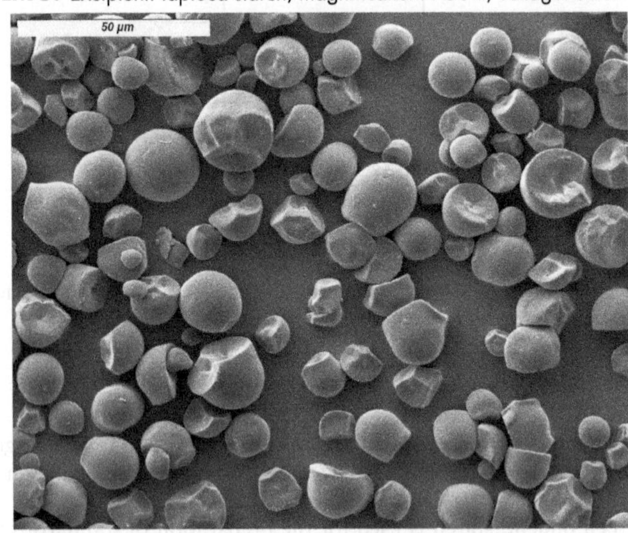

SEM 6: Excipient: wheat starch; manufacturer: Roquette Frères; magnification: 750×; voltage: 5 kV.

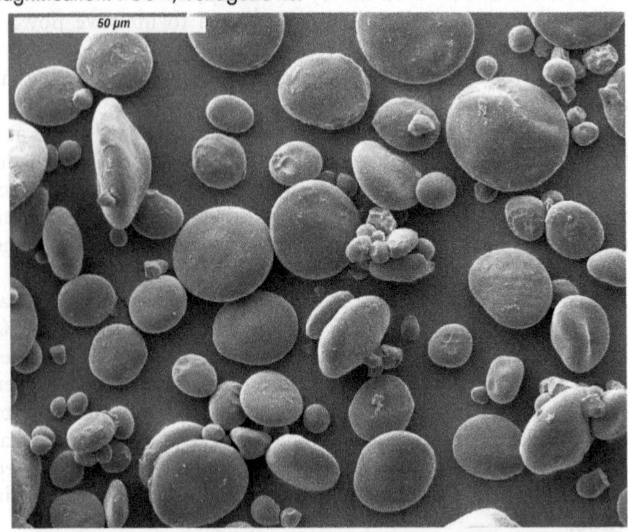

Table I: Pharmacopeial specifications for starch.

Test	JP XV	PhEur 7.4	USP35–NF30
Identification	+	+	+
Characters	+	+	+
Microbial limits	—	+	+
pH			
Corn starch	4.0–7.0	—	4.0–7.0
Pea starch	—	5.0–8.0	5.0-8.0
Potato starch	—	5.0–8.0	5.0-8.0
Rice starch	—	5.0–8.0	5.0-8.0
Tapioca starch	—	—	4.5–7.0
Wheat starch	4.5–7.0	4.5–7.0	4.5–7.0
Loss on drying			
Corn starch	≤15.0%	—	≤15.0%
Pea starch	—	≤16.0%	≤16.0%
Potato starch	≤20.0%	≤20.0%	≤20.0%
Rice starch	≤15.0%	≤15.0%	≤15.0%
Tapioca starch	—	—	≤16.0%
Wheat starch	≤15.0%	≤15.0%	≤15.0%
Residue on ignition			
Corn starch	≤0.6%	—	≤0.6%
Pea starch	—	—	≤0.6%
Potato starch	≤0.6%	—	≤0.6%
Rice starch	≤1.0%	—	≤0.6%
Tapioca starch	—	—	≤0.6%
Wheat starch	≤0.6%	—	≤0.6%
Sulfated ash			
Corn starch	—	—	—
Pea starch	—	≤0.6%	—
Potato starch	—	≤0.6%	—
Rice starch	—	≤0.6%	—
Wheat starch	—	≤0.6%	—
Iron			
Corn starch	≤10 ppm	—	≤10 ppm
Pea starch	—	≤50 ppm	≤50 µg/g
Potato starch	≤10 ppm	≤10 ppm	≤10 ppm
Rice starch	—	≤10 ppm	≤10 ppm
Tapioca starch	—	—	≤0.002%
Wheat starch	≤10 ppm	≤10 ppm	≤10 ppm
Oxidizing substances			
Corn starch	≤20 ppm	—	≤20 ppm
Pea starch	—	≤20 ppm	≤20 µg/g
Potato starch	≤20 ppm	≤20 ppm	≤20 ppm
Rice starch	—	≤0.002%	≤20 ppm
Tapioca starch	—	—	≤0.002%
Wheat starch	≤20 ppm	≤20 ppm	≤20 ppm
Sulfur dioxide			
Corn starch	≤50 ppm	—	≤50 ppm
Pea starch	—	≤50 ppm	≤50 µg/g
Potato starch	≤50 ppm	≤50 ppm	≤50 ppm
Rice starch	—	≤50 ppm	≤50 ppm
Tapioca starch	—	—	≤0.005%
Wheat starch	≤50 ppm	≤50 ppm	≤50 ppm
Total protein			
Wheat starch	—	≤0.3%	≤0.3%
Foreign matter			
Corn starch	—	—	—
Pea starch	—	+	+
Potato starch	—	+	—
Rice starch	+	+	—
Wheat starch	—	+	—

68°C for rice starch;

59°C for wheat starch.

Gelatinization causes the rupture of the starch grains and is an irreversible loss of the structure of the starch particle.(25)

Moisture content All starches are hygroscopic and absorb atmospheric moisture to reach the equilibrium humidity.(26–28) The approximate equilibrium moisture is characteristic for each starch. At 50% relative humidity:

12% for corn starch;

14% for pea starch,

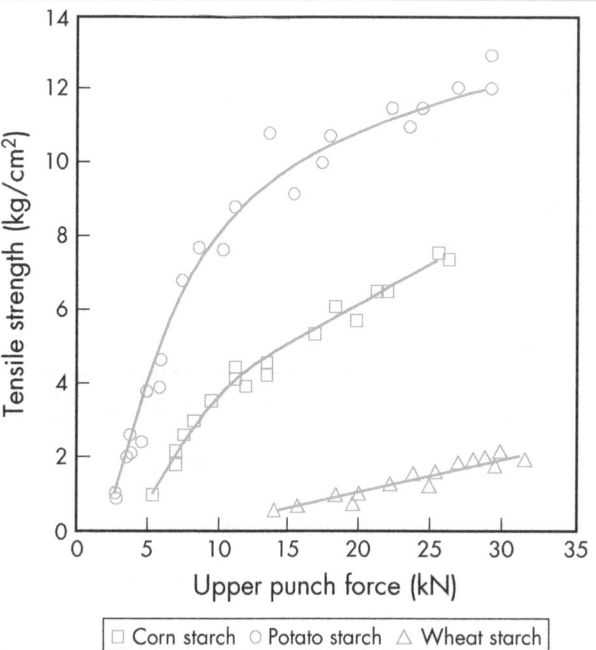

Figure 1: Compaction characteristics of corn, potato and wheat starches. Tablet machine: Manesty F; speed: 50 per min; weight: 490–510 mg. Strength test: Diametral compression between flat-faced rams. Upper ram stationary, lower moving at 66 µm/s.

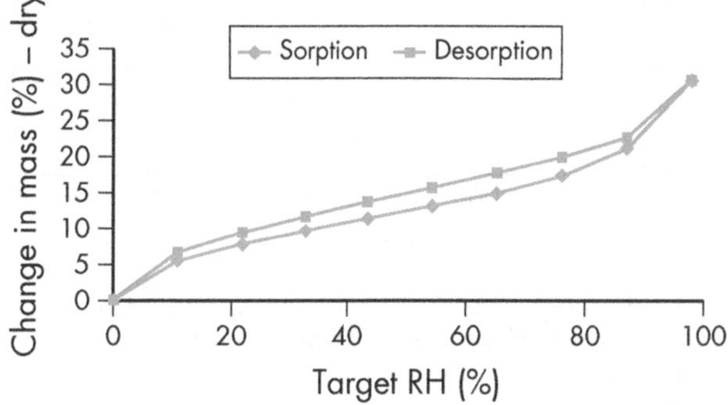

Figure 2: Sorption–desorption isotherm of wheat starch at 20°C.

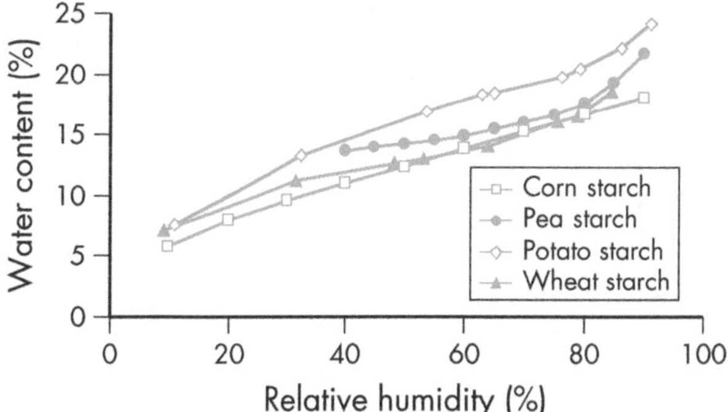

Figure 3: Sorption isotherm of various granular starches at 20°C, measured with dynamic vapor sorption equipment.

18% for potato starch;

14% for rice starch;

13% for wheat starch.

Excessively dried starches with a humidity lower than the equilibrium humidity, are commercially available. These products should be stored in hermetically sealed containers to maintain their low moisture content. *See also* Figures 2 and 3.

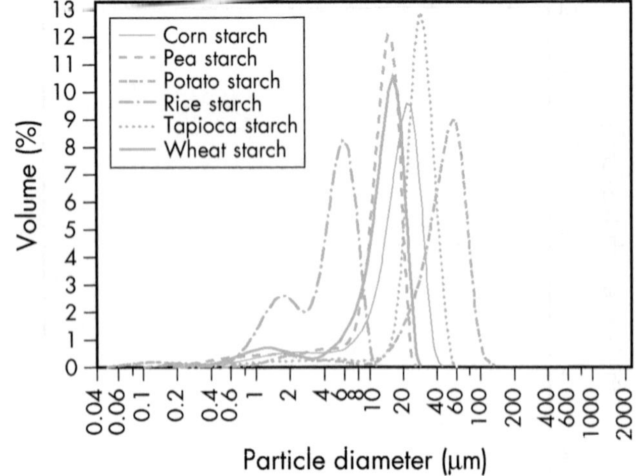

Figure 4: Particle size distribution of commercial starches (laser method, volume distribution).

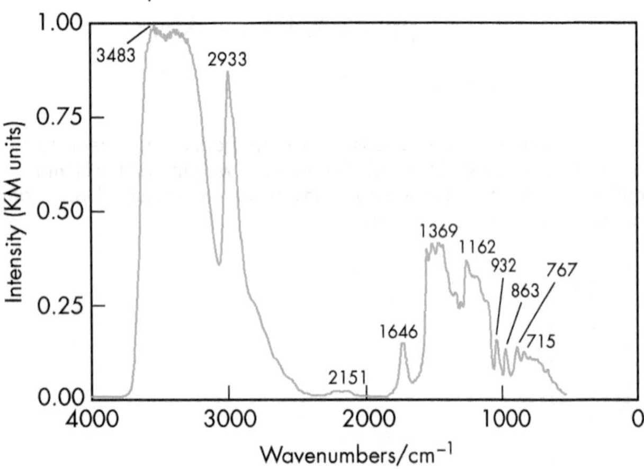

Figure 5: Infrared spectrum of corn starch measured by diffuse reflectance. Adapted with permission of Informa Healthcare.

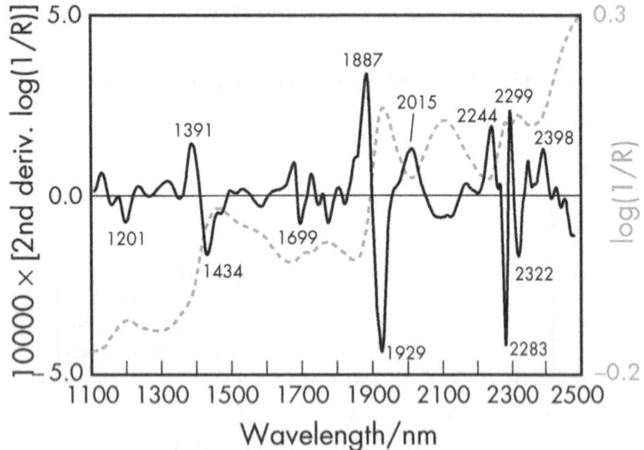

Figure 6: Near-infrared spectrum of starch measured by reflectance.

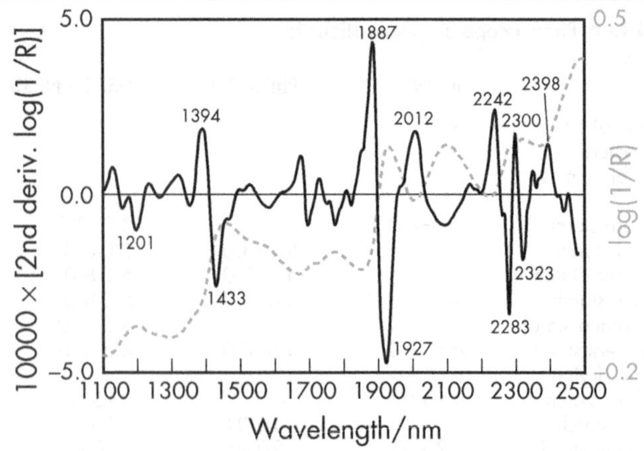

Figure 7: Near-infrared spectrum of corn starch measured by reflectance.

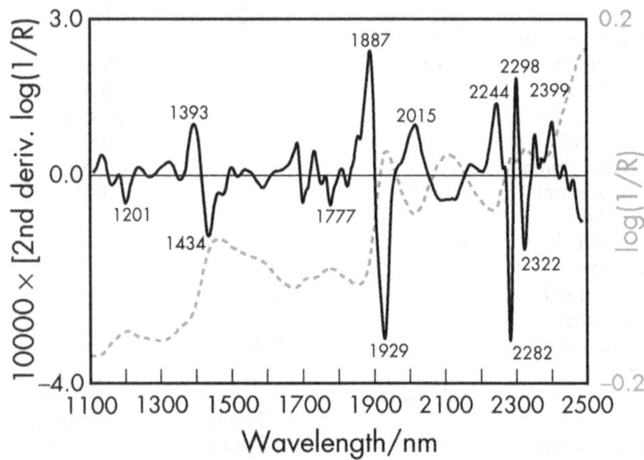

Figure 8: Near-infrared spectrum of rice starch measured by reflectance.

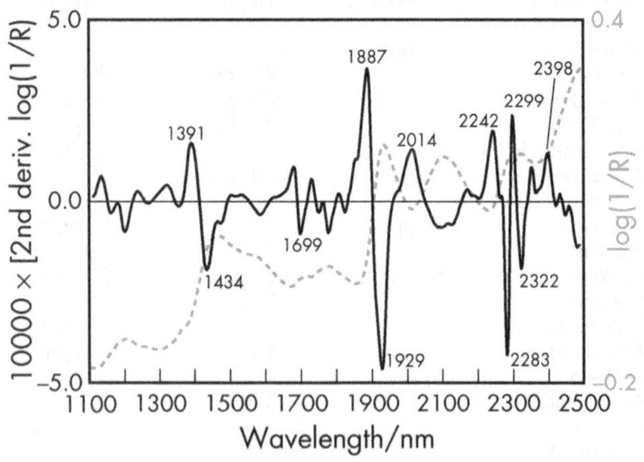

Figure 9: Near-infrared spectrum of wheat starch measured by reflectance.

Particle size distribution

Corn starch: 2–32 μm; average particle diameter 13 μm;
Pea starch: 5–90 μm; average particle diameter 30 μm;
Potato starch: 10–100 μm; average particle diameter 46 μm;
Rice starch: 2–20 μm; average particle diameter 5 μm;
Tapioca starch: 5–35 μm; average particle diameter 13 μm;
Wheat starch: 2–45 μm; bimodal particle size distribution, peak values approx. 2 μm and 20 μm.

See also Figure 4.

Solubility Practically insoluble in cold ethanol (96%) and in cold water. Starch swells instantaneously in water by about 5–10% at 37°C.[3] Starch becomes soluble in hot water at temperatures above the gelatinization temperature. Starches are partially soluble in dimethylsulfoxide and dimethylformamide.

Specific surface area 0.40–0.54 m^2/g for corn starch.[29]

Spectroscopy

IR spectra *see* Figure 5.

NIR spectra *see* Figures 6, 7, 8, and 9.

Swelling temperature Swelling is a reversible process.[25]

64°C for corn starch;
63°C for potato starch;
72°C for rice starch;
55°C for wheat starch.

Viscosity (dynamic) Nonmodified starches are not the preferred polymer for regulating the viscosity of pharmaceutical preparations, except for clinical nutrition products. This is due to the physical and microbial instability of starch paste. In food applications, starch contributes to higher viscosity via the volume effect of the swollen particles and not with its dynamic viscosity. The viscosities of starch paste, obtained and measured under similar conditions may be ranked as follows: potato starch >> tapioca starch > corn starch. Note that aqueous starch dispersions show significant rheopexy, especially at concentrations above 40% w/w.

11 Stability and Storage Conditions

Dry starch is stable if protected from high humidity. Starch is considered to be chemically and microbiologically inert under normal storage conditions. Starch solutions or pastes are physically unstable and are readily metabolized by microorganisms; they should therefore be freshly prepared when used for wet granulation.

Starch should be stored in an airtight container in a cool, dry place.

12 Incompatibilities

Starch is incompatible with strongly oxidizing substances. Colored inclusion compounds are formed with iodine.

13 Method of Manufacture

Starch is extracted from plant sources with specific processes according to the botanical origin. Typical production steps are steeping (corn), wet milling (corn, potato), dry milling (wheat), or sieving and physical separation with hydrocyclones. The last production step is usually a centrifugal separation from the starch slurry followed by drying with hot air. The starch separation process may use sulfur dioxide or peroxides as a processing aid, improving the separation process and the microbial quality of the final product.

14 Safety

Starch is an edible food substance, considered a food ingredient and not a food additive. It is regarded as an essentially nontoxic and nonirritant material.[30] Starch is therefore widely used as an excipient in pharmaceutical formulations.

Both amylose and amylopectin have been evaluated as safe and without limitation for daily intake.[31] Allergic reactions to starch are extremely rare and individuals apparently allergic to one particular starch may not experience adverse effects with a starch from a different botanical source. The wheat proteins (gluten) are problematic for conditions such as celiac disease.

Contamination of surgical wounds with the starch glove powder used by surgeons has resulted in the development of granulomatous lesions.[32]

LD_{50} (mouse, IP): 6.6 g/kg[33]

15 Handling Precautions

Observe normal precautions appropriate to the circumstances and quantity of material handled. Eye protection and a dust mask are recommended. Excessive dust generation should be avoided to minimize the risks of explosion. The minimal explosive concentration of corn starch is 30–60 g/m^3 air.

In the UK, the long-term (8-hour TWA) workplace exposure limits for starch are 10 mg/m^3 for total inhalable dust and 4 mg/m^3 for respirable dust.[34]

16 Regulatory Status

GRAS listed. Included in the FDA Inactive Ingredients Database (buccal tablets, oral capsules, powders, suspensions and tablets; topical preparations; and vaginal tablets). Included in nonparenteral medicines licensed in the UK. Included in the Canadian Natural Health Products Ingredients Database.

17 Related Substances

Dextrin; hydroxypropyl starch; maltodextrin; sodium starch glycolate; starch, modified; starch, pregelatinized; starch, sterilizable maize.

18 Comments

Note that corn starch is also known as maize starch and that tapioca starch is also known as cassava or manioc starch.

Corn starch, potato starch, and wheat starch have undergone harmonization of many attributes for JP, PhEur, and USP–NF by the Pharmacopeial Discussion Group. Rice starch has also been selected for harmonization. For further information see the General Information Chapter <1196> in the USP35–NF30, the General Chapter 5.8 in PhEur 7.4, along with the 'State of Work' document on the PhEur EDQM website, and also the General Information Chapter 8 in the JP XV.

Starch is isolated from vegetable sources. Pure starch should only contain traces of foreign matter (e.g. tissue fragments) and no traces of starches other than from the declared botanical origin. Inside their crystalline structure, starch particles contain smaller quantities of lipids (0–0.8%) and proteins (0–0.5%). The contents are relatively stable and typical for each starch variety. Starches from different plant sources differ in particle size, humidity, and their amylose/amylopectin ratio (*see also* Section 10). Differences in the physical properties of the various starches mean that they are not automatically interchangeable in a given pharmaceutical application.[35]

Corn starch is also available in a naturally white variety (extra white corn starch), containing low levels of carotenoids (especially lutein and zeaxanthin). This starch variety is extracted from specific and nongenetically modified organism hybrids of *Zea mays* L. Bleached starches are considered as modified (oxidized) starches. They are not interchangeable with nontreated starches for regulatory and technical reasons.[36]

The use of granular starch in direct compression applications could also be improved by compounding. Lactose–starch and mannitol–starch compounds have been introduced on the market, using starch for improving the tableting process and the faster disintegration of the tablets produced.[30,37]

The pharmacopeial monographs for starch do not include an assay for starch content. Possible analytical methods for quantification are polarimetric[38] or enzymatic tests.[39]

Starch, particularly rice starch, has been used in the treatment of children's diarrheal diseases. Specific starch varieties with a high amylose content (resistant starches) are used as insoluble fiber in clinical nutrition.

19 Specific References

1 York P. Studies of the effect of powder moisture content on drug release from hard gelatin capsules. *Drug Dev Ind Pharm* 1980; 6: 605–627.
2 Ingram JT, Lowenthal W. Mechanism of action of starch as a tablet disintegrant I: factors that affect the swelling of starch grains at 37°. *J Pharm Sci* 1966; 55: 614–617.
3 Patel NR, Hopponen RE. Mechanism of action of starch as a disintegrating agent in aspirin tablets. *J Pharm Sci* 1966; 55: 1065–1068.
4 Lowenthal W. Mechanism of action of tablet disintegrants. *Pharm Acta Helv* 1973; 48: 589–609.
5 Shangraw RF, *et al.* Morphology and functionality in tablet excipients for direct compression: part II. *Pharm Technol* 1981; 5(10): 44–60.

6 Kitamori N, Makino T. Improvement in pressure-dependent dissolution of trepibutone tablets by using intragranular disintegrants. *Drug Dev Ind Pharm* 1982; **8**: 125–139.
7 Kottke MK, *et al.* Comparison of disintegrant and binder activity of three corn starch products. *Drug Dev Ind Pharm* 1992; **18**: 2207–2223.
8 Bos CE, *et al.* Native starch in tablet formulations: properties on compaction. *Pharm Weekbl Sci* 1987; **9**: 274–282.
9 Shlieout G, *et al.* Evaluating the elastic behaviour of pharmaceutical excipients and binary mixtures using the modified Fraser-Suzuki function. *Pharm Tech Eur* 2002; **14**(6): 24–30.
10 McKenna A, McCafferty DF. Effect of particle size on the compaction mechanism and tensile strength of tablets. *J Pharm Pharmacol* 1982; **34**: 347–351.
11 Berthomieu D *et al.* Tabletting study for ODTs. *Manuf Chem* (online), 2010. http://www.manufacturingchemist.com/technical/article_page/Tabletting_study_for_ODTs/57763 (accessed 1 March 2012).
12 Callens C, *et al.* Rheological study on mucoadhesivity of some nasal powder formulations. *Eur J Pharm Biopharm* 2003; **55**: 323–328.
13 Clausen AE, Bernkop-Schnurch A. Direct compressible polymethacrylic acid-starch compositions for site-specific drug delivery. *J Control Release* 2001; **75**: 93–102.
14 Palviainen P, *et al.* Corn starches as film formers in aqueous-based film coating. *Pharm Dev Technol* 2001; **6**: 353–361.
15 Rein H, Steffens KJ. Surface modification of water insoluble drug particles with starch. *Starch* 1997; **49**(9): 364–371.
16 Henrist D, *et al. In vitro* and *in vivo* evaluation of starch-based hot stage extruded double matrix systems. *J Control Release* 2001; **75**: 391–400.
17 Rein, H. Steffens, KJ. Method for producing a water insoluble amorphous controlled release matrix. International Patent, WO 00644106; 2000.
18 Dukic A, *et al.* Development of starch-based pellets via extrusion/spheronisation. *Eur J Pharm Biopharm* 2007; **66**: 83–94.
19 Junnila J, *et al.* Waxy corn starch: A potent cofiller in pellets produced by extrusion-spheronisation. *Pharm Dev Technol* 2000; **5**(1): 67–76.
20 Otero-Espinar FJ. Non-MCC materials as extrusion-spheronization aids in pellets production. *J Drug Del Sci Tech* 2010; **20**(4): 303–318.
21 Basit A, Bloor J. Perspectives on colonic drug delivery. *Business Briefing, Pharmatech* 2003; 185–190.
22 Karrout Y, *et al.* Peas starch-based film coatings for site-specific drug delivery to the colon. *J Appl Polym Sci* 2011; **119**(2): 1176–1184.
23 Cooke JL, Freeman R. The flowability of powders and the effect of flow additives. *World Congress on Particle Technology 5*, Orlando, 2006.
24 Shimada Y, *et al.* Measurement and evaluation of adhesive force between particles by direct separation method. *J Pharm Sci* 2003; **92**(3): 560–567.
25 Atwell WA, *et al.* The terminology and methodology associated with basic starch phenomena. *Cereal Foods World* 1988; **33**: 306–311.
26 Callahan JC, *et al.* Equilibrium moisture content of pharmaceutical excipients. *Drug Dev Ind Pharm* 1982; **8**: 355–369.
27 Wurster DE, *et al.* A comparison of the moisture adsorption–desorption properties of corn starch, USP, and directly compressible starch. *Drug Dev Ind Pharm* 1982; **8**: 343–354.
28 Faroongsarng D, Peck GE. The swelling and water uptake of tablets III: moisture sorption behavior of tablets disintegrants. *Drug Dev Ind Pharm* 1994; **20**(5): 779–798.
29 Hauschild K, Picker-Freyer KM. Evaluation of a new coprocessed compound based on lactose and maize starch for tablet formulation. *AAPS Pharm Sci* 2004; **6**: Article 16.
30 Weiner M, Bernstein IL. *Adverse Reactions to Drug Formulation Agents: A Handbook of Excipients*. New York: Marcel Dekker, 1989: 91–92.
31 FAO/WHO. Toxicological evaluation of certain food additives with a review of general principles and of specifications. Seventeenth report of the Joint FAO/WHO Expert Committee on Food Additives. *World Health Organ Tech Rep Ser* 1974; No. 539.
32 Michaels L, Shah NS. Dangers of corn starch powder [letter]. *Br Med J* 1973; **2**: 714.
33 Lewis RJ, ed. *Sax's Dangerous Properties of Industrial Materials*, 11th edn. New York: Wiley, 2004: 3299.
34 Health and Safety Executive. *EH40/2005: Workplace Exposure Limits*. Sudbury: HSE Books, 2011. http://www.hse.gov.uk/pubns/priced/eh40.pdf (accessed 1 March 2012).
35 Smallenbroek AJ, *et al.* The effect of particle size of disintegrants on the disintegration of tablets. *Pharm Weekbl* 1981; **116**: 1048–1051.
36 Iwuagu MA, Agidi AA. The effect of bleaching on the disintegrant properties of maize starch. *STP Pharma Sci* 2000; **10**: 143–147.
37 Wagner KJ, Dressler JA. A corn starch/α-lactose monohydrate compound as a new directly compressible excipient. *Pharm Ind* 2002; **64**(9): 992–999.
38 EEC Commission of European Communities. EEC 3rd Commission Directive 72/199: Determination of Starch, 1999.
39 Brunt K, *et al.* The enzymatic determination of starch in food, feed and raw materials of the food industry. *Starch* 1998; **50**(10): 413–419.

20 General References

Center for Research on Macromolecules Végétales. Cyber Starch – Starch Structure and Morphology. http://www.cermav.cnrs.fr/ (accessed 1 March 2012).

European Directorate for the Quality of Medicines and Healthcare (EDQM). European Pharmacopoeia – State Of Work Of International Harmonisation. *Pharmeuropa* 2011; 23(4): 713–714. www.edqm.eu/site/-614.html (accessed 1 March 2012).

21 Author

O Häusler.

22 Date of Revision

1 March 2012.

Starch, Modified

1 Nonproprietary Names

USP–NF (a) Modified Starch
(b) Pregelatinized Modified Starch

Both monographs in the USP35–NF30 describe a vast family of food modified starches, all conforming to the US Code of Federal Regulations 21 CFR 172.892. The two monographs cover either the granular and cold-water-insoluble starches or the cold-water-soluble pendants, obtained via an additional pregelatinization process. Some specific pharmaceutical starches without food conformity are covered in separate monographs (e.g. *see* Sodium Starch Glycolate).

The PhEur 7.4 describes only hydroxypropylated (food modified) starches in a dedicated monograph (*see* Hydroxypropyl Starch).

2 Synonyms

Amprac; amylum modificatum; *Capsul*; *E1401-1452*; *Eratab*; *Hi-Cap*; *Instant Pure-Cote*; *Pure-Gel*; *Lycoat*; *Pure-Coat*; *Purity*; *Uni-Pure*.

According to the International Numbering System (INS) for Food Additives[1] the following starches should comply with food legal regulations:

1401 Acid-treated starch
1402 Alkaline-treated starch
1403 Bleached starch
1404 Oxidized starch
1405 Starches, enzyme-treated
1410 Monostarch phosphate
1412 Distarch phosphate
1413 Phosphated distarch phosphate
1414 Acetylated distarch phosphate
1420 Starch acetate
1422 Acetylated distarch adipate
1440 Hydroxypropyl starch
1442 Hydroxypropyl distarch phosphate
1443 Hydroxypropyl distarch glycerol
1450 Starch sodium octenyl succinate
1451 Acetylated oxidized starch

3 Chemical Name and CAS Registry Number

The chemical name and the corresponding CAS number depend on the botanical origin of the starch and the nature of the chemical modification(s). Some examples are given in this nonexhaustive list:

Acetylated distarch adipate, waxy maize basis [63798-35-6]
Acid-treated maize starch [65996-63-6]
Acid-treated waxy maize starch [68909-37-5]
Distarch phosphate, waxy maize basis [55963-33-2]
Oxidized waxy maize starch [65996-62-5]
Sodium octenyl succinate starch [66829-29-6]

4 Empirical Formula and Molecular Weight

$(C_6H_{10}O_5)_n$ where n = 300–10 000

The USP35–NF30 describes modified starch as starch modified by chemical means. Food starch may be acid-modified, bleached, oxidized, esterified, etherified, or treated enzymatically to change its functional properties (21 CFR 172.892).

The USP35–NF30 describes pregelatinized modified starch as modified starch that has been chemically or mechanically processed, or both, to rupture all or part of the granules to produce a product that swells in cold water.

Moderate chemical modification of food modified starches does not affect the basic structure of starch. Granular products maintain the semicrystalline nature of the starch granule. Crosslinking reactions significantly increase the molecular weight of modified starches. *See also* Section 18.

5 Structural Formula

See Section 4.

6 Functional Category

Coating agent; emulsion stabilizing agent; emulsifying agent; modified-release agent; tablet and capsule diluent; tablet and capsule disintegrant; viscosity-increasing agent.

7 Applications in Pharmaceutical Formulation or Technology

Modified starches are used in pharmaceutical preparations as binders for wet granulation (especially oxidized starches) and as tablet disintegrants. Some types of modified starch are used for their film-forming properties in nonfunctional and functional film coatings[2] or as wall-forming material in capsules and soft capsules.

Modified starches have also been developed as viscosity-increasing agents and texture modifiers in various food applications, and therefore are used as thickening agents and stabilizers in emulsions.[3] They are also valuable additives in the production of jellies with or without gelatin.

Amphiphilic octenyl succinate starches (E1450) are widely used as emulsifying agents for the manufacture of vitamins and herbal extracts, and in flavor encapsulation.[3–5]

Pregelatinized modified starches are soluble in cold water or swell with development of high viscosity, and are occasionally used as viscosity-increasing agents in syrups.

S

8 Description

Modified starch occurs as an odorless and tasteless, fine, white to off-white powder.

9 Pharmacopeial Specifications

See Table I.

10 Typical Properties

Density Depends on the nature of the starch and mainly on the industrial process, especially for pregelatinized products.

Gelatinization temperature Chemical modification has a major influence on the gelatinization of starches. So-called'stabilization' typically decreases the gelatinization temperature owing to the weakening of the crystalline structure of the starch particles, whereas crosslinking causes the opposite effect (*see also* Section 18).

Moisture content All starches (granular or pregelatinized) are hygroscopic and absorb atmospheric moisture to reach the equilibrium humidity.

Particle size distribution Chemically modified granular starches preserve the particle size and microscopic image of the starch raw material (*see also* Starch). The particle size of pregelatinized

Table I: Pharmacopeial specifications for modified starch and pregelatinized starch

Test	USP35–NF30
Identification	
Modified starch	+
Pregelatinized modified starch	+
Microbial limits	
Modified starch	+
Bacteria	≤1000 cfu/g
Molds and yeasts	≤100 cfu/g
pH	
Modified starch	3.0–9.0
Pregelatinized modified starch	3.0–9.0
Loss on drying	
Modified corn starch	≤15.0%
Modified wheat starch	≤15.0%
Modified tapioca starch	≤18.0%
Modified potato starch	≤21.0%
Pregelatinized modified starch	≤15.0%
Residue on ignition	
Modified starch	≤1.5%
Pregelatinized modified starch	≤1.5%
Iron	
Modified starch	≤20 ppm
Oxidizing substances	
Modified starch	≤0.018%
Pregelatinized modified starch	+
Sulfur dioxide	
Modified starch	≤50 ppm
Pregelatinized modified starch	≤0.005%

modified starches is determined by the technical production process.

Solubility Granular modified starches are practically insoluble in cold ethanol (96%) and in cold water. Pregelatinized starches are at least partially soluble in cold water and do typically develop viscosity. All food modified starches preserve the hydrophilic characteristics of native starch.

Viscosity (dynamic) The viscosity of starch is correlated with the botanical origin of the starch and chemical modification(s). Food modified starches for viscosity or texture adjustment are typically crosslinked and stabilized products (*see also* Section 18).

S

11 Stability and Storage Conditions

Modified starch is considered to be chemically and microbiologically inert under normal storage conditions. Store in an airtight container in a cool, dry place.

12 Incompatibilities

Modified starch is incompatible with strongly oxidizing substances. Colored inclusion compounds are formed with iodine.

13 Method of Manufacture

The most common production method is chemical modification in starch slurry. The nature of the reactive(s) and the specific reaction conditions are determined by the product in synthesis. A double modification may be carried out when authorized in the food regulations. An additional reduction of the starch viscosity is possible, e.g. via an acid hydrolysis.

After modification, starch particles are separated from the aqueous reaction medium (e.g. via filtration or centrifugation), washed and dried. Chemical reactions may alternatively be done in dry phase, e.g. in an extruder.

Pregelatinized modified starch is prepared by cooking and drying a starch paste on steam-heated rollers. A starch slurry or paste is dried on the surface of these drums and removed for milling to the desired particle size. An alternative process is so-called 'spray cooking,' which produces instant-swelling starches. These starches keep their initial particle size and form, but swell immediately in contact with cold water.

14 Safety

Modified starch is regarded as an essentially nontoxic and nonirritant material. The Joint FAO/WHO Expert Committee on Food Additives (JECFA),[6] the European Commission Science Committee on Food (EC/SCF),[7] and the Food and Drug Administration (FDA),[8] have evaluated the safety of starch derivatives, resulting in an acceptable daily intake of 'not specified.'

LD_{50} (cat, oral): >9 g/kg[9]
LD_{50} (mouse, oral): >24 g/kg
LD_{50} (rat, oral): >35 g/kg

15 Handling Precautions

Observe normal precautions appropriate to the circumstances and quantity of the material handled. Eye protection and a dust mask are recommended. Excessive dust generation should be avoided to minimize the risks of explosion.

16 Regulatory Status

Included in the FDA Inactive Ingredients Database (oral capsules and tablets; powders for suspension).

In the US, all food modified starches are considered as edible food additives, and are authorized for nearly all applications. The pharmaceutical authorization of modified starch is correlated with FDA regulations.[8] In the EU, some modified starches (E1401, E1402, and E1403) are considered as food ingredients, and others as food additives; the conditions of use are regulated in EU Directive 95/2/EC.[10]

Products must comply with purity criteria as defined by food regulations.[6,11] The labeling of the botanical source of the starch is compulsory if it may contain gluten.

Food products for the EU market containing starch derivatives should be labeled with the group category 'modified starch,' and do not require an E-number.

17 Related Substances

Hydroxypropyl starch; starch; starch, pregelatinized.

18 Comments

The chemical modification of starch, along with its botanical origin, has a major impact on its physicochemical properties.[12,13] Chemical modification may also influence the digestion speed, and therefore insulinemic and satiating properties of the starch.[14] The maximum permitted level of chemical modification depends on the nature of the substitution.[6,11]

Starch chemists differentiate between two basic starch modifications:

Stabilization A monofunctional substitution (e.g., esterification or etherification) on the starch hydroxyl groups with low substitution level. This stabilization improves the physical stability of the starch paste by steric hindrance of the starch retrogradation.

Crosslinking Starch chains are chemically linked with very small amounts of bi- or multifunctional reagents. Crosslinked starches are more resistant to swelling and gelatinization. These starches have better heat stability during the cooking process in food preparation, and they better resist shear forces.

Cold-water-soluble starch derivatives have been investigated as controlled-release agents in hydrophilic matrix systems.[15–20] Amphiphilic octenyl succinate starches (E1450) have also been investigated as wetting agents and alternatives to commonly used solubilizing or emulsifying agents in ophthalmic formulations.[21] Starches and starch derivatives are increasingly being investigated in the development of nanoparticles[22] and in targeted drug preparations.[23]

Dry-Flo (National Starch) is a specially processed modified food starch used as a dusting and lubricating agent in a variety of food and pharmaceutical products.[24]

A specification for modified food starch is included in the *Food Chemicals Codex* (FCC).[25]

19 Specific References

1 FAO/WHO Codex Alimentarius Commission. *General Standard for Food Additives (GFSA)*. http://www.fao.org/ag/agn/jecfa-additives/details.html?id=840 (accessed 8 August 2011).

2 Bonacucina J, *et al.* Effect of plasticisers on properties of pregelatinised starch acetate (Amprac 01) films. *Int J Pharm* 2006; **26**: 72–77.

3 Yusoff A, Murray BS. Modified starch granules as particle stabilizers of oil-in-water emulsions. *Food Hydrocolloids* 2011; **25**: 42–55.

4 Tech S, *et al.* Stabilisation of emulsions by OSA starches. *J Food Eng* 2002; **54**(2): 167–174.

5 Kuentz M, *et al.* A technical feasibility study of surfactant-free drug suspensions using octenyl succinate-modified starches. *Eur J Pharm Biopharm* 2006; **63**: 37–43.

6 FAO/WHO. *Combined Compendium of Food Additive Specifications, Monograph 7*. Seventy-first report of the Joint FAO/WHO Expert Committee on Food Additives (JECFA), 2009. http://www.fao.org/ag/agn/jecfa-additives/details.html?id=840 (accessed 9 August 2011).

7 European Commission. European Science Committee on Food (EU SCF): Reports on modified starches 1976, 1981, 1995 and 1997. http://ec.europa.eu/food/fs/sc/scf/reports_en.html (accessed 9 August 2011).

8 US Food and Drug Administration. Code of Federal Regulations. Title 21 (21 CFR 172.892): Food starch-modified, 2010. http://www.accessdata.fda.gov/scripts/cdrh/cfdocs/cfcfr/CFRSearch.cfm?fr=172.892 (accessed 9 August 2011).

9 Lewis RJ, ed. *Sax's Dangerous Properties of Industrial Materials*. 11th edn. New York: Wiley, 2004: 3299.

10 European Commission. Directive 95/2/EC: Food additives other than colours and sweeteners. *Official Journal* L061, March 1995: 1–40. http://eur-lex.europa.eu/LexUriServ/LexUriServ.do?uri=CELEX:31995L0002:EN:HTML (accessed 9 August 2011).

11 European Commission. Directive 2008/84/EC: Laying down specific purity criteria on food additives other than colours and sweeteners. *Official Journal* L 253, September 2008: 164–170. http://eur-lex.europa.eu/LexUriServ/LexUriServ.do?uri=OJ:L:2008:253:0001:0175:EN:PDF (accessed 9 August 2011).

12 Singh J, *et al.* Factors influencing the physico-chemical, morphological, thermal and rheological properties of some chemically modified starches for food applications – A review. *Food Hydrocolloids* 2006; **21**: 1–22.

13 Wootton M, Bamunuarachchi A. Water binding capacity of commercial produced native and modified starches. *Starch* 1978; **30**(9): 306–309.

14 Singh J, *et al.* Starch digestibility in food matrix: a review. *Trends Food Sci Tech* 2010; **4**: 168–180.

15 Vandenbossche GMR, *et al.* Performance of a modified starch hydrophilic matrix for the sustained release of theophylline in health volunteers. *J Pharm Sci* 1992; **81**(3): 245–248.

16 O'Brien S, *et al.* Starch phosphates prepared by reactive extrusion as a sustained release agent. *Carbohydr Polym* 2009; **76**: 557–566.

17 Sanghvi PP, *et al.* Evaluation of Preflo modified starches as new direct compression excipients. I. Tabletting characteristics. *Pharm Res* 1993; **10**(11): 1597–1603.

18 Visavarungroj N, *et al.* Crosslinked starch as sustained release agent. *Drug Dev Ind Pharm* 1990; **16**(7): 1091–1108.

19 Onofre FO, *et al.* Sustained-release properties of crosslinked corn starches with varying amylose contents in monolithic tablets. *Starch* 2010; **62**: 165–172.

20 van Veen B, *et al.* The effect of powder blend and tablet structure on drug release mechanism of hydrophobic starch acetate matrix tablets. *Eur J Pharm Biopharm* 2005; **61**: 149–157.

21 Baydoun L, *et al.* New surface-active polymers for ophthalmic formulation: evaluation of ocular tolerance. *Eur J Pharm Biopharm* 2004; **58**: 169–175.

22 Preetz C, *et al.* Preparation and characterization of biocompatible oil-loaded polyelectrolyte nanocapsules. *Nanomedicine* 2008; **4**: 106–114.

23 Thiele C, *et al.* Nanoparticles of anionic starch and cationic cyclodextrin derivatives for the targeted delivery of drugs. *Polym Chem* 2011; **2**: 209.

24 National Starch and Chemical Co. Technical Service Bulletin. *Dry-Flo*. http://eu.foodinnovation.com/docs/DRY-FLO.pdf (accessed 24 February 2012)

25 *Food Chemicals Codex*. 7th edn. Bethesda, MD: United States Pharmacopeia, 2010; 407.

20 General References

French AD. In: Whistler RL, BeMiller JN Paschall EF eds. *Starch: Chemistry and Technology*. 2nd edn. New York: Academic Press, 1984; 183–247.

Wurzburg OB. In: Alistair M *et al.*, eds. *Food Polysaccharides and Their Applications*. Chapter 3. CRC Press, 2006.

21 Author

O Häusler.

22 Date of Revision

2 March 2012.

S

Starch, Pregelatinized

1 Nonproprietary Names

BP: Pregelatinised Starch
PhEur: Starch, Pregelatinised
USP–NF: Pregelatinized Starch

2 Synonyms

Amylum pregelificatum; compressible starch; *C*PharmGel*; *Instastarch*; *Lycatab C*; *Lycatab PGS*; *Merigel*; *National 78-1551*; *Pharma-Gel*; *Prejel*; *Sepistab ST200*; *Spress B820*; *Starch 1500 G*; *Tablitz*; *Unipure LD*; *Unipure WG220*.

3 Chemical Name and CAS Registry Number

Pregelatinized starch [9005-25-8]

4 Empirical Formula and Molecular Weight

$(C_6H_{10}O_5)_n$ where n = 300–1000.

Pregelatinized starch is a starch that has been chemically and/or mechanically processed to rupture all or part of the starch granules. Both fully and partially pregelatinized grades are commercially available. Partial pregelatinization renders the starch flowable and directly compressible. Full pregelatinization produces a cold-water soluble starch that can be used as a wet granulation binder. Typically, pregelatinized starch contains 5% of free amylose, 15% of free amylopectin, and 80% unmodified starch.

The USP35–NF30 does not specify the botanical origin of the original starch, but the PhEur 7.4 specifies that pregelatinized starch is obtained from maize (corn), potato, or rice starch. *See also* Starch and Section 13. Normally the fully pregelatinized starch contains 20–30% amylose and the rest amylopectin, which is about the same ratio (1 : 3) as for the partially pregelatinized form. There are ways to increase the amylose portion.(1)

5 Structural Formula

See Starch.

6 Functional Category

Tablet and capsule binder; tablet and capsule diluent; tablet and capsule disintegrant.

7 Applications in Pharmaceutical Formulation or Technology

Partially pregelatinized starch is a modified starch used in oral capsule and tablet formulations as a binder, diluent,(2,3) and disintegrant.(4)

In comparison to starch, partially pregelatinized starch may be produced with enhanced flow and compression characteristics such that the pregelatinized material may be used as a tablet binder in dry-compression or direct compression processes.(5–15) In such processes, pregelatinized starch is self-lubricating. However, when it is used with other excipients it may be necessary to add a lubricant to a formulation. Although magnesium stearate 0.25% w/w is commonly used for this purpose, concentrations greater than this may have adverse effects on tablet strength and dissolution. Therefore, stearic acid is generally the preferred lubricant with pregelatinized starch.(16)

Partially pregelatinized starch is used in oral dry powder hard capsule formulations.

Both partially and fully pregelatinized starch may also be used in wet granulation processes.(17) *See* Table I.

Fully pregelatinized starches can be used to make soft capsules, shells, and coatings as well as binders in tablets.

Table I: Uses of pregelatinized starch.

Use	Concentration (%)
Diluent (hard gelatin capsules)	5–75
Tablet binder (direct compression)	5–20
Tablet binder (wet granulation)	5–10
Tablet disintegrant	5–10

8 Description

Pregelatinized starch occurs as a moderately coarse to fine, white to off-white colored powder. It is odorless and has a slight characteristic taste.

Examination of fully pregelatinized starch as a slurry in cold water, under a polarizing microscope, reveals no significant ungelatinized granules, i.e. no 'maltese crosses' characteristic of the starch birefringence pattern. Examination of samples suspended in glycerin shows characteristic forms depending upon the method of drying used during manufacture: either irregular chunks from drum drying or thin plates. Partially pregelatinized starch (e.g. *Starch 1500G* and *Sepistab ST200*) show retention of birefringence patterns typical of unmodified starch granules.

9 Pharmacopeial Specifications

See Table II. *See also* Section 18.

10 Typical Properties

Acidity/alkalinity pH = 4.5–7.0 for a 10% w/v aqueous dispersion.
Angle of repose 40.7° (6)
Density (bulk) 0.586 g/cm^3
Density (tapped) 0.879 g/cm^3
Density (true) 1.516 g/cm^3
Flowability 18–23% (Carr compressibility index)(18)

SEM 1: Excipient: *Lycatab PGS*; manufacturer: Roquette Frères.

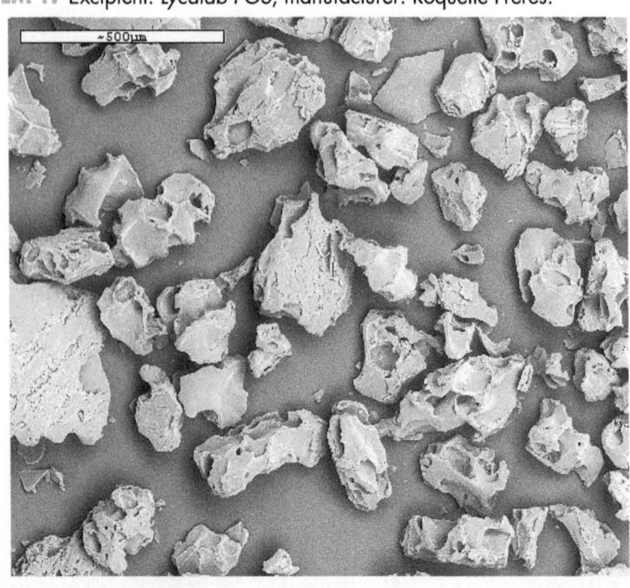

SEM 2: Excipient: pregelatinized starch; manufacturer: Cargill; magnification: 200×; voltage: 3 kV.

SEM 3: Excipient: pregelatinized starch; manufacturer: Cargill; magnification: 200×; voltage: 3 kV.

SEM 4: Excipient: pregelatinized starch; manufacturer: Cargill; magnification: 200×; voltage: 3 kV.

Moisture content Pregelatinized maize starch is hygroscopic.(15,19,20) *See also* Figure 1.

Particle size distribution 30–150 µm, median diameter 52 µm. For partially pregelatinized starch, greater than 90% through a US #100 mesh (149 µm); and less than 0.5% retained on a US #40 mesh (420 µm).

Solubility Practically insoluble in organic solvents. Slightly soluble to soluble in cold water, depending upon the degree of pregelatinization. Pastes can be prepared by sifting the pregelatinized starch into stirred, cold water. Cold-water-soluble matter for partially pregelatinized starch is 10–20%.

SEM 5: Excipient: pregelatinized starch; manufacturer: Cargill; magnification 200×; voltage: 3 kV.

Table II: Pharmacopeial specifications for pregelatinized starch.

Test	PhEur 7.4	USP35–NF30
Identification	+	+
Characters	+	—
pH (10% w/v slurry)	4.5–7.0	4.5–7.0
Iron	≤20 ppm	≤20 ppm
Oxidizing substances	+	+
Sulfur dioxide	≤50 ppm	≤80 ppm
Microbial limits	+	+
Loss on drying	≤15.0%	≤14.0%
Residue on ignition	—	≤0.5%
Foreign matter	+	—
Sulfated ash	≤0.6%	—

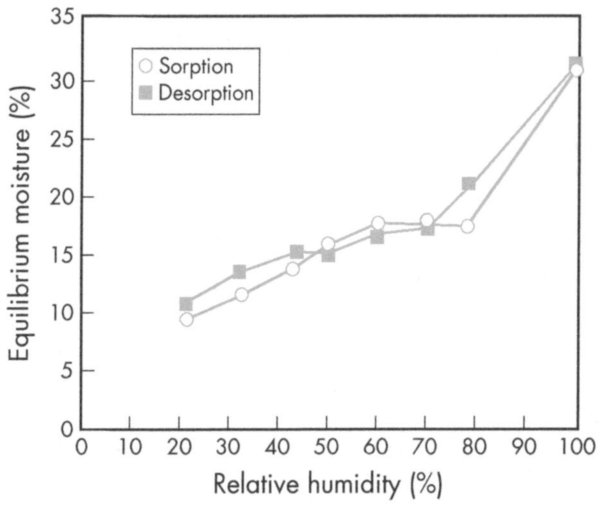

Figure 1: Pregelatinized starch sorption–desorption isotherm.

Specific surface area

$0.26\,m^2/g$ (Colorcon);

$0.18–0.28\,m^2/g$ (Roquette).

Spectroscopy

IR spectra *see* Figure 2.

NIR spectra *see* Figures 3 and 4.

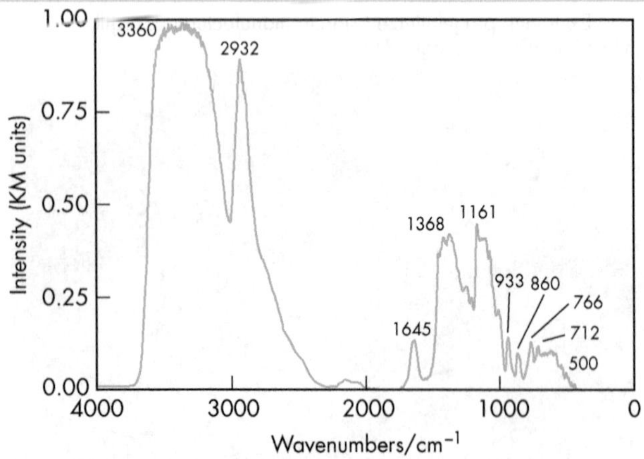

Figure 2: Infrared spectrum of pregelatinized starch measured by diffuse reflectance. Adapted with permission of Informa Healthcare.

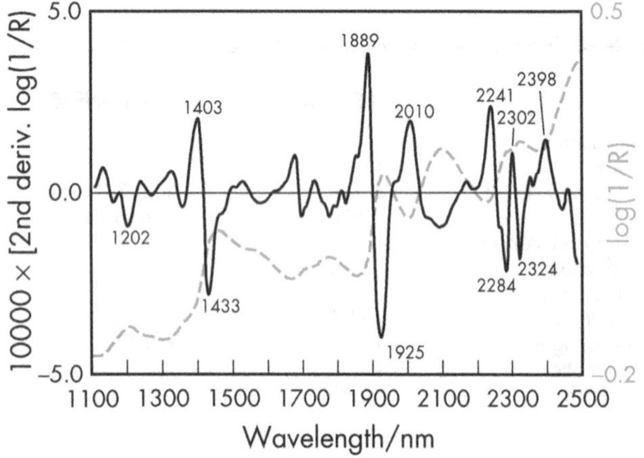

Figure 3: Near-infrared spectrum of pregelatinized maize starch measured by reflectance.

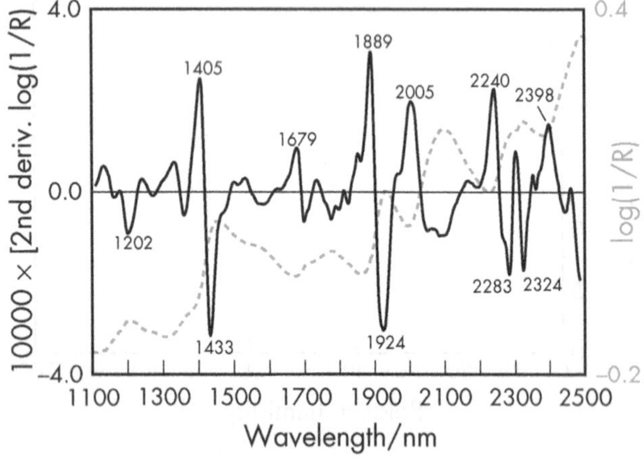

Figure 4: Near-infrared spectrum of pregelatinized rice starch measured by reflectance.

Viscosity (dynamic) 8–10 mPa s (8–10 cP) for a 2% w/v aqueous dispersion at 25°C.

11 Stability and Storage Conditions

Pregelatinized starch is a stable but hygroscopic material, which should be stored in a well-closed container in a cool, dry place.

12 Incompatibilities

—

13 Method of Manufacture

Food-grade pregelatinized starches are prepared by heating an aqueous slurry containing up to 42% w/w of starch at 62–72°C. Chemical additives that may be included in the slurry are gelatinization aids (salts or bases) and surfactants, added to control rehydration or minimize stickiness during drying. After heating, the slurry may be spray-dried, roll-dried, extruded, or drum-dried. In the last case, the dried material may be processed to produce a desired particle size range.

Pharmaceutical grades of fully pregelatinized starch use no additives and are prepared by spreading an aqueous suspension of ungelatinized starch on hot drums where gelatinization and subsequent drying take place. Partially pregelatinized starch is produced by subjecting moistened starch to mechanical pressure. The resultant material is ground and the moisture content is adjusted to specifications.

14 Safety

Pregelatinized starch and starch are widely used in oral solid-dosage formulations. Pregelatinized starch is generally regarded as a nontoxic and nonirritant excipient. However, oral consumption of large amounts of pregelatinized starch may be harmful.

See Starch for further information.

15 Handling Precautions

Observe normal precautions appropriate to the circumstances and quantity of material handled. Eye protection and a dust mask are recommended. Excessive dust generation should be avoided to minimize the risks of explosions.

In the UK, the long-term (8-hour TWA) workplace exposure limits for starch are 10 mg/m^3 for total inhalable dust and 4 mg/m^3 for respirable dust.[21]

16 Regulatory Status

Included in the FDA Inactive Ingredients Database (oral capsules, suspensions, and tablets; vaginal preparations). Included in non-parenteral medicines licensed in the UK.

17 Related Substances

Corn starch and pregelatinized starch; starch; starch, modified; starch, sterilizable maize.

18 Comments

Pregelatinized starch is one of the materials that have been selected for harmonization by the Pharmacopeial Discussion Group. For further information see the General Information Chapter <1196> in the USP35–NF30, the General Chapter 5.8 in PhEur 7.4, along with the 'State of Work' document on the PhEur EDQM website, and also the General Information Chapter 8 in the JP XV.

Pregelatinized starch has been shown to cause a decrease in mRNA production in tissue culture systems, which may hold potential for excipient–drug interactions by repression of CYP3A4 expression.[22]

The USP35–NF30 also lists pregelatinized modified starch. A low-moisture grade of pregelatinized starch, *Starch 1500 LM* (Colorcon), containing less than 7% of water, specifically intended for use as a diluent in capsule formulations is commercially available.[16]

Sepistab ST200 is described as an agglomerate of starch granules consisting of native and pregelatinized corn starch.[23] Compression characteristics of pregelatinized starches from sorghum and plantain have been evaluated against traditional corn-based products.[24]

StarCap 1500 (Colorcon) is a coprocessed mixture of pregelatinized starch and corn starch promoted for use in dry-powder, hard-capsule fillings; *see* Corn Starch and Pregelatinized Starch.

A specification for partly pregelatinized starch is contained in the *Japanese Pharmaceutical Excipients* (JPE).[25]

19 Specific References

1 Carbone D *et al.* Process for cooking/drying high-amylose starches. United States Patent 7,118,764; 2002.
2 Small LE, Augsburger LL. Aspects of the lubrication requirements for an automatic capsule filling machine. *Drug Dev Ind Pharm* 1978; **4**: 345–372.
3 Mattson S, Nyström C. Evaluation of critical binder properties affecting the compactability of binary mixtures. *Drug Dev Ind Pharm* 2001; **27**: 181–194.
4 Rudnic EM, *et al.* Evaluations of the mechanism of disintegrant action. *Drug Dev Ind Pharm* 1982; **8**: 87–109.
5 Manudhane KS, *et al.* Tableting properties of a directly compressible starch. *J Pharm Sci* 1969; **58**: 616–620.
6 Underwood TW, Cadwallader DE. Influence of various starches on dissolution rate of salicylic acid from tablets. *J Pharm Sci* 1972; **61**: 239–243.
7 Bolhuis GK, Lerk CF. Comparative evaluation of excipients for direct compression. *Pharm Weekbl* 1973; **108**: 469–481.
8 Sakr AM, *et al.* Sta-Rx 1500 starch: a new vehicle for the direct compression of tablets. *Arch Pharm Chem (Sci)* 1974; **2**: 14–24.
9 Schwartz JB, *et al.* Intragranular starch: comparison of starch USP and modified cornstarch. *J Pharm Sci* 1975; **64**: 328–332.
10 Rees JE, Rue PJ. Work required to cause failure of tablets in diametral compression. *Drug Dev Ind Pharm* 1978; **4**: 131–156.
11 Shangraw RF, *et al.* Morphology and functionality in tablet excipients for direct compression: part II. *Pharm Technol* 1981; 5(10): 44–60.
12 Chilamkurti RW, *et al.* Some studies on compression properties of tablet matrices using a computerized instrumental press. *Drug Dev Ind Pharm* 1982; **8**: 63–86.
13 Malamataris S, *et al.* Moisture sorption and tensile strength of some tableted direct compression excipients. *Int J Pharm* 1991; **68**: 51–60.
14 Iskandarani B, *et al.* Scale-up feasability in high-shear mixers: determination through statistical procedures. *Drug Dev Ind Pharm* 2001; **27**: 651–657.
15 Shiromani PK, Clair J. Statistical comparison of high-shear versus low-shear granulation using a common formulation. *Drug Dev Ind Pharm* 2000; **26**: 357–364.
16 Colorcon. Technical literature: *Starch 1500*. 1997.
17 Jaiyeoba KT, Spring MS. The granulation of ternary mixtures: the effect of the stability of the excipients. *J Pharm Pharmacol* 1980; **32**: 1–5.
18 Carr RL. Particle behaviour storage and flow. *Br Chem Eng* 1970; **15**: 1541–1549.
19 Callahan JC, *et al.* Equilibrium moisture content of pharmaceutical excipients. *Drug Dev Ind Pharm* 1982; **8**: 355–369.
20 Wurster DE, *et al.* A comparison of the moisture adsorption–desorption properties of corn starch USP, and directly compressible starch. *Drug Dev Ind Pharm* 1982; **8**: 343–354.
21 Health and Safety Executive. *EH40/2005: Workplace Exposure Limits*. Sudbury: HSE Books, 2011. www.hse.gov.uk/pubns/priced/eh40.pdf (accessed 12 April 2012).
22 Tompkins L, *et al.* Effects of commonly used excipients on the expression of CYP3A4 in colon and liver cells. *Pharm Res* 2010; **27**(8): 1703–1712.
23 Seppic. Technical Literature: *Sepistab ST200*. 1997.
24 Alebiowu G, Itiola OA. Compression characteristics of native and pregelatinized forms of sorghum, plantain, and corn starches, and the mechanical properties of their tablets. *Drug Dev Ind Pharm* 2002; **28**(6): 663–672.
25 Japan Pharmaceutical Excipients Council. *Japanese Pharmaceutical Excipients* 2004. Tokyo: Yakuji Nippo, 2004; 856.

20 General References

Alebiowu G, Itiola OA. The influence of pregelatinized starch disintegrants on interacting variables that act on disintegrant properties. *Pharm Tech* 2003; **24**(8): 28–33.

European Directorate for the Quality of Medicines and Healthcare (EDQM). European Pharmacopoeia – State Of Work Of International Harmonisation. *Pharmeuropa* 2011; 23(4): 713–714. www.edqm.eu/site/-614.html (accessed 16 November 2011).

Monedeero Perales MC, *et al.* Comparative tableting and microstructural properties of a new starch for direct compression. *Drug Dev Ind Pharm* 1996; **22**: 689–695.

Rees JH, Tsardaka KD. Some effects of moisture on the viscoelastic behavior of modified starch during powder compaction. *Eur J Pharm Biopharm* 1994; **40**: 193–197.

Roquette Frères. Technical literature: *Lycatab PGS*. 2001.

Sanghvi PP, *et al.* Evaluation of Preflo modified starches as new direct compression excipients I: tabletting characteristics. *Pharm Res* 1993; **10**: 1597–1603.

21 Author

AH Kibbe.

22 Date of Revision

12 April 2012.

Starch, Sterilizable Maize

1 Nonproprietary Names

USP–NF: Absorbable Dusting Powder

2 Synonyms

Bio-sorb; double-dressed, white maize starch; *Fluidamid R444P*; *Keoflo ADP*; *Meritena*; modified starch dusting powder; *Pure-Dent B851*; starch-derivative dusting powder; sterilizable corn starch.

3 Chemical Name and CAS Registry Number

Sterilizable maize starch

4 Empirical Formula and Molecular Weight

$(C_6H_{10}O_5)_n$ where n = 300–1000.

Sterilizable maize starch is a modified corn (maize) starch that may also contain up to 2.0% of magnesium oxide.

See also Starch.

5 Structural Formula

See Starch.

6 Functional Category

Lubricant.

7 Applications in Pharmaceutical Formulation or Technology

Sterilizable maize starch is a chemically or physically modified corn (maize) starch that does not gelatinize on exposure to moisture or steam sterilization. Sterilizable maize starch is primarily used as a lubricant for examination and surgeons' gloves, although because of safety concerns unlubricated gloves are now generally recommended; *see* Section 14. It is also used as a vehicle for medicated dusting powders.

8 Description

Sterilizable maize starch occurs as an odorless, white, free-flowing powder. Particles may be rounded or polyhedral in shape.

9 Pharmacopeial Specifications

See Table I.

Table I: Pharmacopeial specifications for sterilizable maize starch.

Test	USP35–NF30
Identification	+
Stability to autoclaving	+
Sedimentation	+
pH (1 in 10 suspension)	10.0–10.8
Loss on drying	≤12%
Residue on ignition	≤3.0%
Magnesium oxide	≤2.0%
Heavy metals	≤0.001%

10 Typical Properties

Acidity/alkalinity pH = 9.5–10.8 for a 10% w/v suspension at 25°C.

Density 1.48 g/cm^3

Density (bulk) 0.47–0.59 g/cm^3

Density (tapped) 0.64–0.83 g/cm^3

Flowability 24–30% (Carr compressibility index)[1]

Moisture content 10–15%

Particle size distribution 6–25 μm; median diameter is 16 μm.

Solubility Very slightly soluble in chloroform and ethanol (95%); practically insoluble in water.

Specific surface area 0.50–1.15 m^2/g

11 Stability and Storage Conditions

Sterilizable maize starch may be sterilized by autoclaving at 121°C for 20 minutes, by ethylene oxide, or by irradiation.[2]

SEM 1: Excipient: sterilizable maize starch; manufacturer: Corn Products; magnification: 2000×.

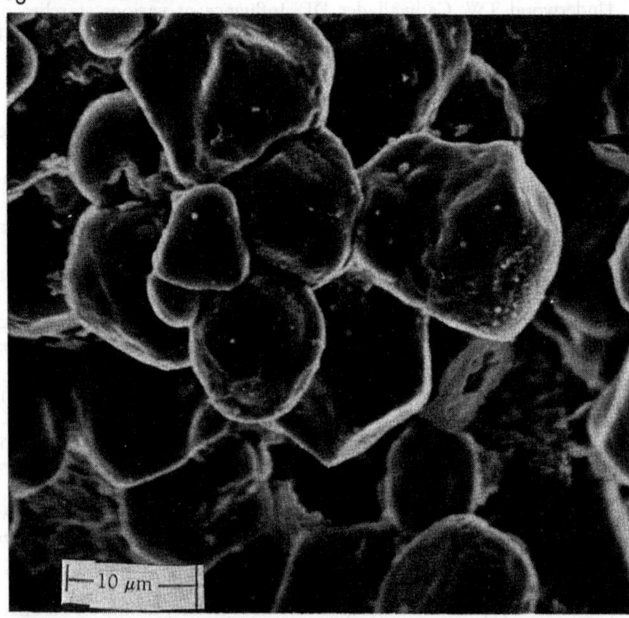

SEM 2: Excipient: sterilizable maize starch; manufacturer: Biosorb; magnification: 2000×.

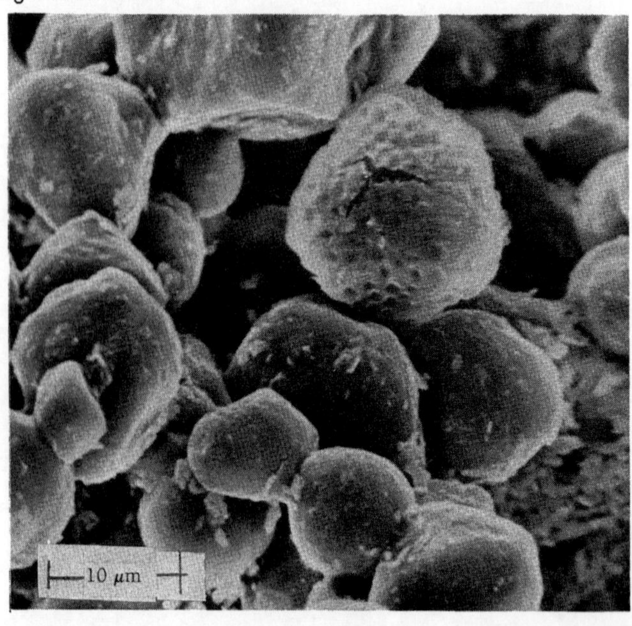

SEM 3: Excipient: sterilizable maize starch; manufacturer: J & W Starches Ltd; magnification: 2000×.

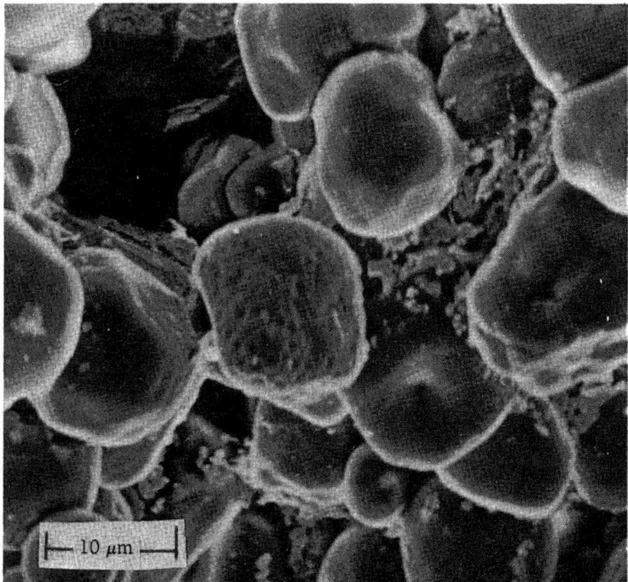

Sterilizable maize starch should be stored in a well-closed container in a cool, dry place.

12 Incompatibilities

—

13 Method of Manufacture

Corn starch (maize starch) is physically or chemically modified by treatment with either phosphorus oxychloride or epichlorhydrin so that the branched-chain and straight-chain starch polymers cross-link. Up to 2.0% of magnesium oxide may also be added to the starch.

See also Starch.

14 Safety

Sterilizable maize starch is primarily used as a lubricant for surgeons' gloves and as a vehicle for topically applied dusting powders.

Granulomatous reactions, peritonitis and inflammation at operation sites have been attributed to contamination with surgical glove powders containing sterilizable maize starch. In addition, glove powder may be a risk factor in the development of latex allergy. As a consequence, it has been suggested that the use of sterilizable maize starch in latex gloves should be prohibited.[3–12]

See also Starch.

15 Handling Precautions

Observe normal precautions appropriate to the circumstances and quantity of material handled. Eye protection and a dust mask are recommended. Excessive dust generation should be avoided to minimize the risks of explosions.

In the UK, the long-term (8-hour TWA) occupational exposure limits for starch are 10 mg/m^3 for total inhalable dust and 4 mg/m^3 for respirable dust.[13]

16 Regulatory Status

Included in the FDA Inactive Ingredients Database (oral tablets and topical preparations). Included in nonparenteral medicines licensed in the UK. Included in the Canadian Natural Health Products Ingredients Database.

17 Related Substances

Starch; starch, pregelatinized.

18 Comments

—

19 Specific References

1 Carr RL. Particle behaviour storage and flow. *Br Chem Eng* 1970; **15**: 1541–1549.

2 Kelsey JC. Sterilization of glove powder by autoclaving. *Mon Bull Minist Health* 1962; **21**: 17–21.

3 Neely J, Davis JD. Starch granulomatosis of the peritoneum. *Br Med J* 1971; **3**: 625–629.

4 Michaels L, Shah NS. Dangers of corn starch powder [letter]. *Br Med J* 1973; **2**: 714.

5 Karcioglu ZA, *et al.* Inflammation due to surgical glove powders in the rabbit eye. *Arch Ophthalmol* 1988; **106**(6): 808–811.

6 Ruhl CM, *et al.* A new hazard of cornstarch, an absorbable dusting powder. *J Emerg Med* 1994; **12**(1): 11–14.

7 Anonymous. AAAAI and ACAAI joint statement concerning the use of powdered and non-powdered natural rubber latex gloves. *Ann Allergy Asthma Immunol* 1997; **79**: 487.

8 Haglund U, Junghanus K. Glove powder – the hazards which demand a ban. *Eur J Surg* 1997; **163**(Suppl 579): 1–55.

9 Cote SJ, *et al.* Ease of donning commercially available latex examination gloves. *J Biomed Mater Res* 1998; **43**(3): 331–337.

10 Dave J, *et al.* Glove powder: implications for infection control. *J Hosp Infect* 1999; **42**: 282–285.

11 Truscott W. Post-surgical complications associated with the use of USP Absorbable Dusting Powder. *Surg Technol Int* 2000; **VIII**: 65–73.

12 Edlich RF, Reddy VR. A call to ban glove cornstarch. *Arch Surg* 2001; **136**: 116.

13 Health and Safety Executive. *EH40/2005: Workplace Exposure Limits.* Sudbury: HSE Books, 2011. www.hse.gov.uk/pubns/priced/eh40.pdf (accessed 12 April 2012).

20 General References

El Saadany RMA, *et al.* Degradation of corn starch under the influence of gamma irradiation. *Staerke* 1976; **28**: 208–211.

Greenwood CT. The thermal degradation of starch. *Adv Carbohydr Chem Biochem* 1967; **22**: 483–515.

Greenwood CT. Starch. *Adv Cereal Sci Technol* 1976; **1**: 119–157.

21 Author

PJ Weller.

22 Date of Revision

12 April 2012.

Stearic Acid

1 Nonproprietary Names

BP: Stearic Acid
JP: Stearic Acid
PhEur: Stearic Acid
USP–NF: Stearic Acid

2 Synonyms

Acidum stearicum; cetylacetic acid; *Crodacid*; *Cristal G*; *Cristal S*; *Dermofat 4919*; *Dervacid*; E570; *Edenor*; *Emersol*; *Extra AS*; *Extra P*; *Extra S*; *Extra ST*; 1-heptadecanecarboxylic acid; *Hystrene*; *Industrene*; *Kortacid 1895*; *Pearl Steric*; *Pristerene*; *Speziol L2SM GF*; stereophanic acid; *Tegostearic*; *TriStar.*

3 Chemical Name and CAS Registry Number

Octadecanoic acid [57-11-4]

4 Empirical Formula and Molecular Weight

$C_{18}H_{36}O_2$ 284.47 (for pure material)

The USP35–NF30 describes stearic acid as a mixture of stearic acid ($C_{18}H_{36}O_2$) and palmitic acid ($C_{16}H_{32}O_2$). In the USP35–NF30, the content of stearic acid is not less than 40.0% and the sum of the two acids is not less than 90.0%. The USP35–NF30 also contains a monograph for purified stearic acid; *see* Section 17. The PhEur 7.4 contains a single monograph for stearic acid but defines stearic acid 50, stearic acid 70, and stearic acid 95 as containing specific amounts of stearic acid ($C_{18}H_{36}O_2$); *see* Section 9.

5 Structural Formula

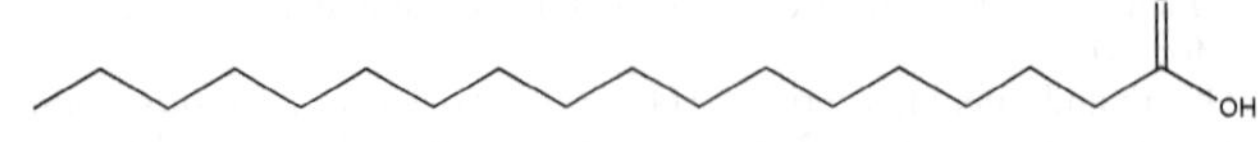

6 Functional Category

Emulsifying agent; solubilizing agent; tablet and capsule lubricant.

7 Applications in Pharmaceutical Formulation or Technology

Stearic acid is widely used in oral and topical pharmaceutical formulations. It is mainly used in oral formulations as a tablet and capsule lubricant;(1–3) *see* Table I, although it may also be used as a binder(4) or in combination with shellac as a tablet coating. It has also been suggested that stearic acid may be used in enteric tablet coatings and as a sustained-release drug carrier.(5)

In topical formulations, stearic acid is used as an emulsifying and solubilizing agent. When partially neutralized with alkalis or triethanolamine, stearic acid is used in the preparation of creams.(6,7) The partially neutralized stearic acid forms a creamy base when mixed with 5–15 times its own weight of aqueous liquid, the appearance and plasticity of the cream being determined by the proportion of alkali used.

Stearic acid is used as the hardening agent in glycerin suppositories.

Stearic acid is also widely used in cosmetics and food products.

Table I: Uses of stearic acid.

Use	Concentration (%)
Ointments and creams	1–20
Tablet lubricant	1–3

8 Description

Stearic acid occurs as a hard, white or faintly yellow-colored, somewhat glossy, crystalline solid or a white or yellowish white powder. It has a slight odor (with an odor threshold of 20 ppm) and taste suggesting tallow.

See also Section 13.

9 Pharmacopeial Specifications

The pharmacopeial specifications for stearic acid have undergone harmonization of many attributes for JP, PhEur, and USP–NF.

See Table II. *See also* Section 18.

10 Typical Properties

Acid value 195–212
Boiling point 383°C
Density (bulk) ≈0.537 g/cm^3
Density (tapped) 0.571 g/cm^3
Density (true)
0.980 g/cm^3;
0.847 g/cm^3 at 70°C
Flash point 113°C (closed cup)
Melting point 69–70°C
Moisture content Contains practically no water.
Partition coefficient Log (oil : water) = 8.2
Refractive index 1.43 at 80°C
Saponification value 200–220

SEM 1: Excipient: stearic acid, 95% (*Emersol 153*); manufacturer: Emery Industries; lot no.: 18895; magnification: 120×; voltage: 10 kV.

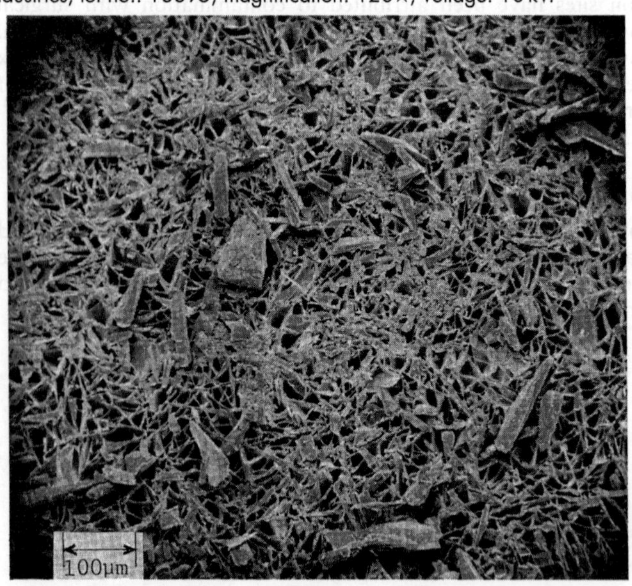

SEM 2: Excipient: stearic acid, food grade (*Emersol 6332*); manufacturer: Emery Industries; lot no.: 18895; magnification: 120×; voltage: 10 kV.

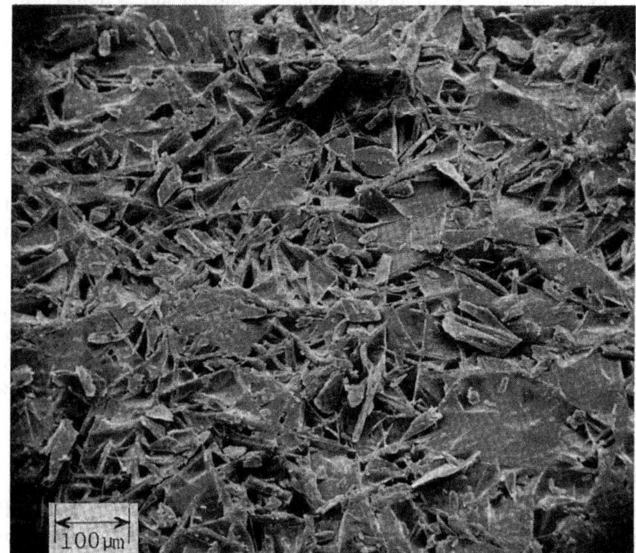

Table II: Pharmacopeial specifications for stearic acid.

Test	JP XV	PhEur 7.4	USP35–NF30
Identification	—	+	+
Characters	—	+	—
Acidity	—	+	+
Acid value	194–210	194–212	194–212
Appearance	—	+	+
Content of stearic acid	—	+	≥40.0%
Stearic acid 50	—	40–60%	40–60%
Stearic acid 70	—	60–80%	60–80%
Stearic acid 95	—	≥90.0%	≥90.0%
Content of stearic and palmitic acids	—	+	+
Stearic acid 50	—	≥90.0%	≥90.0%
Stearic acid 70	—	≥90.0%	≥90.0%
Stearic acid 95	—	≥96.0%	≥96.0%
Congealing temperature	56.0–72.0°C	—	≥54°C
Freezing point	—	+	—
Stearic acid 50	—	53–59°C	53–59°C
Stearic acid 70	—	57–64°C	57–64°C
Stearic acid 95	—	64–69°C	64–69°C
Iodine value	≤4.0	+	≤4.0
Stearic acid 50	—	≤4.0%	≤4.0%
Stearic acid 70	—	≤4.0%	≤4.0%
Stearic acid 95	—	≤1.5%	≤1.5%
Nickel	—	≤1 ppm	—
Residue on ignition	≤0.1%	—	≤0.1%
Heavy metals[a]	≤20 ppm	—	≤10 ppm
Neutral fat or paraffin	+	—	—
Mineral acid	+	—	—

(a) This test has not been fully harmonized at the time of publication.

Solubility Freely soluble in benzene, carbon tetrachloride, chloroform, and ether; soluble in ethanol (95%), hexane, and propylene glycol; practically insoluble in water.[8]

Specific surface area

0.51–0.53 m^2/g

See also Section 17 and Table III.

Spectroscopy

IR spectra *see* Figure 1.

NIR spectra *see* Figure 2.

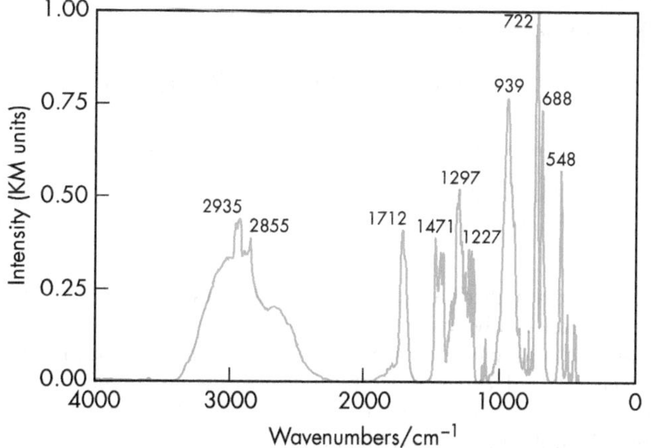

Figure 1: Infrared spectrum of stearic acid measured by diffuse reflectance. Adapted with permission of Informa Healthcare.

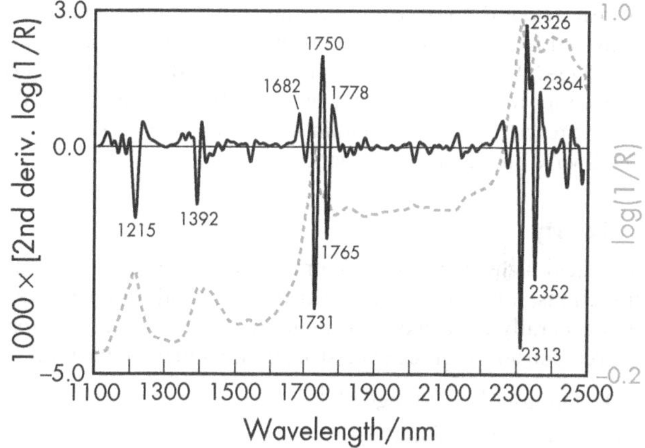

Figure 2: Near-infrared spectrum of stearic acid measured by reflectance.

11 Stability and Storage Conditions

Stearic acid is a stable material; an antioxidant may also be added to it; *see* Section 13. It slowly volatilizes at 90-100°C. The bulk material should be stored in a well-closed container in a cool, dry place.

12 Incompatibilities

Stearic acid is incompatible with most metal hydroxides and may be incompatible with bases, reducing agents, and oxidizing agents.

Ointment bases made with stearic acid may show evidence of drying out or lumpiness due to such a reaction when compounded with zinc or calcium salts.

A number of differential scanning calorimetry studies have investigated the compatibility of stearic acid with drugs. Although such laboratory studies have suggested incompatibilities, e.g. with naproxen,[9] they may not necessarily be applicable to formulated products.

Stearic acid has been reported to cause pitting in the film coating of tablets applied using an aqueous film-coating technique; the pitting was found to be a function of the melting point of the stearic acid.[10]

13 Method of Manufacture

Stearic acid is manufactured by hydrolysis of fat by continuous exposure to a countercurrent stream of high-temperature water and fat in a high-pressure chamber. The resultant mixture is purified by

S

Table III: Specifications of different stearic acid grades.

Product	Stearic acid content (%)	Melting range (°C)	Acid value	Iodine value	Saponification value	Unsaponifiable matter (%)
Hystrene 5016	44	54.5–56.5	206–210	≤0.5	206–211	≤0.2
Hystrene 7018	68.5	61.0–62.5	200–205	≤0.5	200–206	≤0.2
Hystrene 9718	90	66.5–68.0	196–201	≤0.8	196–202	≤0.3
Industrene 7018	65	58.0–62.0	200–207	≤1.5	200–208	≤0.5
Industrene 8718	87	64.5–67.5	196–201	≤2.0	196–202	≤1.5

vacuum steam distillation and the distillates are then separated using selective solvents.

Stearic acid may also be manufactured by the hydrogenation of cottonseed and other vegetable oils; by the hydrogenation and subsequent saponification of olein followed by recrystallization from alcohol; and from edible fats and oils by boiling with sodium hydroxide, separating any glycerin, and decomposing the resulting soap with sulfuric or hydrochloric acid. The stearic acid is then subsequently separated from any oleic acid by cold expression.

Stearic acid is derived from edible fat sources unless it is intended for external use, in which case nonedible fat sources may be used. The USP35–NF30 states that stearic acid labeled solely for external use is exempt from the requirement that it be prepared from edible sources. Stearic acid may contain a suitable antioxidant such as 0.005% w/w butylated hydroxytoluene.

14 Safety

Stearic acid is widely used in oral and topical pharmaceutical formulations; it is also used in cosmetics and food products. Stearic acid is generally regarded as a nontoxic and nonirritant material. However, consumption of excessive amounts may be harmful.

LD_{50} (mouse, IV): 23 mg/kg[11]

LD_{50} (rat, IV): 21.5 mg/kg

15 Handling Precautions

Observe normal precautions appropriate to the circumstances and quantity of material handled. Stearic acid dust may be irritant to the skin, eyes, and mucous membranes. Eye protection, gloves, and a dust respirator are recommended. Stearic acid is combustible.

16 Regulatory Status

GRAS listed. Accepted as a food additive in Europe (fatty acids). Included in the FDA Inactive Ingredients Database (sublingual tablets; oral capsules, solutions, suspensions, and tablets; topical and vaginal preparations). Included in nonparenteral medicines licensed in the UK. Included in the Canadian Natural Health Products Ingredients Database.

17 Related Substances

Calcium stearate; magnesium stearate; polyoxyethylene stearates; purified stearic acid; zinc stearate.

Purified stearic acid

Empirical formula $C_{18}H_{36}O_2$
Molecular weight 284.47
CAS number [57-11-4]
Synonyms Octadecanoic acid
Acid value 195–200
Boiling point 361°C
Density 0.847 g/cm^3 at 70°C
Flash point 196°C
Iodine number ≤1.5
Melting point 66–69°C
Refractive index n_D^{80} = 1.4299
Solubility Soluble 1 in 5 parts benzene, 1 in 6 parts carbon tetrachloride, 1 in 2 parts chloroform, 1 in 15 parts ethanol, 1 in 3 parts ether; practically insoluble in water.
Vapor density (relative) 9.80 (air = 1)
Comments The USP35–NF30 describes purified stearic acid as a mixture of stearic acid ($C_{18}H_{36}O_2$) and palmitic acid ($C_{16}H_{32}O_2$), which together constitute not less than 96.0% of the total content. The content of $C_{18}H_{36}O_2$ is no less than 90.0% of the total.

18 Comments

Stearic acid has undergone harmonization of many attributes for JP, PhEur, and USP–NF by the Pharmacopeial Discussion Group. For further information see the General Information Chapter <1196> in the USP35–NF30, the General Chapter 5.8 in PhEur 7.4, along with the 'State of Work' document on the PhEur EDQM website, and also the General Information Chapter 8 in the JP XV.

A wide range of different grades of stearic acid are commercially available that have varying chemical compositions and hence different physical and chemical properties; *see* Table III.[12] Stearic acid is highly soluble in structurally diverse solvents. Stearic acid/solvent packing within a 24.8 Å^3 cubic volume explains the stoichiometry of stearic acid solubility at multiple temperatures in multiple solvents.[13]

In one study the release of an active drug in a formulation containing stearic acid was independent of compression pressure in the range 1–7 tons; the particle size of the stearic acid did have a significant influence on the drug release.[14]

A potential application of stearic acid is in the preparation of 'cushioning pellets', composed of stearic acid : microcrystalline cellulose (4 : 1 w/w). The use of these pellets may avoid rupture of the coating of pellets during the compression step of manufacturing.[15]

If stearic acid is intended for external use only, the labelling so indicates; *see* Section 13.

A specification for stearic acid is contained in the *Food Chemicals Codex* (FCC).[16]

The EINECS number for stearic acid is 200-313-4.

19 Specific References

1 Roberts M, *et al.* Effect of lubricant type and concentration on the punch tip adherence of model ibuprofen formulations. *J Pharm Pharmacol* 2004; 56(3): 299–305.
2 Jarosz PJ, Parrott EL. Effect of tablet lubricants on axial and radial work of failure. *Drug Dev Ind Pharm* 1982; 8: 445–453.
3 Mitrevej KT, Augsburger LL. Adhesion of tablets in a rotary tablet press II: effects of blending time, running time, and lubricant concentration. *Drug Dev Ind Pharm* 1982; 8: 237–282.
4 Musikabhumma P, *et al.* Evaluation of stearic acid and polyethylene glycol as binders for tabletting potassium phenethicillin. *Drug Dev Ind Pharm* 1982; 8: 169–188.
5 Zhang Q, *et al.* Studies on the cyclosporin A loaded stearic acid nanoparticles. *Int J Pharm* 2000; 200: 153–159.
6 Wisegeek.com. What are the different uses of stearic acid? http://www.wisegeek.com/what-are-the-different-uses-of-stearic-acid.htm (accessed 5 December 2011).
7 Mores LR. Application of stearates in cosmetic creams and lotions. *Cosmet Toilet* 1980; 95(3): 79, 81–84.
8 Yalkowsky SH, He Y, eds. *Handbook of Solubility Data.* Boca Raton, FL: CRC Press, 2003; 1119–1120.

9 Botha SA, Lötter AP. Compatibility study between naproxen and tablet excipients using differential scanning calorimetry. *Drug Dev Ind Pharm* 1990; **16**: 673–683.
10 Rowe RC, Forse SF. Pitting: a defect on film-coated tablets. *Int J Pharm* 1983; **17**: 347–349.
11 Lewis RJ, ed. *Sax's Dangerous Properties of Industrial Materials*, 11th edn. New York: Wiley, 2004; 3229–3300.
12 Phadke DS, *et al.* Evaluation of batch-to-batch and manufacturer-to-manufacturer variability in the physical properties of talc and stearic acid. *Drug Dev Ind Pharm* 1994; **20**: 859–871.
13 Schmidt WF, *et al.* Stearic acid solubility and cubic phase volume. *Chem Phys Lipids* 2006; **142**(1–2): 23–32.
14 Killen BU, Corrigon OI. Factors influencing drug release from stearic acid based compacts. *Int J Pharm* 2001; **228**(1–2): 189–198.
15 Qi XL, *et al.* Preparation of tablets containing enteric-coated diclofenac sodium pellets. *Yao Xue Xue Bao* 2008; **43**(1): 97–101.
16 *Food Chemicals Codex*, 7th edn. Bethesda, MD: United States Pharmacopeia, 2010; 989.

20 General References

Allen LV. Featured excipient: capsule and tablet lubricants. *Int J Pharm Compound* 2000; **4**(5): 390–392404–405.
European Directorate for the Quality of Medicines and Healthcare (EDQM). European Pharmacopoeia – State Of Work Of International Harmonisation. *Pharmeuropa* 2011; 23(4): 713–714. www.edqm.eu/site/-614.html (accessed 5 December 2011).
ScienceLab.com, Inc. Material safety data sheet: Stearic acid, 2005. November 2010.

21 Author

LV Allen Jr.

22 Date of Revision

2 March 2012.

Stearyl Alcohol

1 Nonproprietary Names

BP: Stearyl Alcohol
JP: Stearyl Alcohol
PhEur: Stearyl Alcohol
USP–NF: Stearyl Alcohol

2 Synonyms

Alcohol stearylicus; *Cachalot*; *Crodacol S95*; *Hyfatol 18-95*; *Hyfatol 18-98*; *Lanette 18*; *Lipocol S*; *Lipocol S-DEO*; *Nacol 18-98*; *Nacol 18-98P*; *n*-octadecanol; octadecyl alcohol; *Rita SA*; *Speziol C18 Pharma*; *Stearol*; *Stenol*; *Tego Alkanol 18*; *Vegarol 1898*;*Vegerol 1895*.

3 Chemical Name and CAS Registry Number

1-Octadecanol [112-92-5]

4 Empirical Formula and Molecular Weight

$C_{18}H_{38}O$ 270.48 (for pure material)

The PhEur 7.4 describes stearyl alcohol as a mixture of solid alcohols containing not less than 95% of 1-octadecanol, $C_{18}H_{38}O$ of animal or vegetable origin. The USP35–NF30 states that stearyl alcohol contains not less than 90% of 1-octadecanol, the remainder consisting chiefly of related alcohols.

5 Structural Formula

6 Functional Category

Stiffening agent.

7 Applications in Pharmaceutical Formulation or Technology

Stearyl alcohol is used in topical pharmaceutical creams and ointments as a stiffening agent. By increasing the viscosity of an emulsion, stearyl alcohol increases its stability. Stearyl alcohol also has some emollient and weak emulsifying properties, and is used to increase the water-holding capacity of ointments, e.g. petrolatum. *See also* Section 18.

Stearyl alcohol is also used as a stiffening agent in cosmetics.[1,2]

8 Description

Stearyl alcohol occurs as hard, white, waxy pieces, flakes, or granules with a slight characteristic odor and bland taste.

9 Pharmacopeial Specifications

See Table I.

Table I: Pharmacopeial specifications for stearyl alcohol.

Test	JP XV	PhEur 7.4	USP35–NF30
Identification	—	+	+
Characters	—	+	—
Appearance of solution	+	+	—
Melting range	56–62°C	57–60°C	55–60°C
Acid value	≤1.0	≤1.0	≤2.0
Iodine value	≤2.0	≤2.0	≤2.0
Hydroxyl value	200–220	197–217	195–220
Saponification value	—	≤2.0	—
Ester value	≤3.0	—	—
Residue on ignition	≤0.05%	—	—
Assay (of $C_{18}H_{38}O$)	—	≥95%	≥90.0%

10 Typical Properties

Autoignition temperature 450°C
Boiling point 210.5°C at 2 kPa (15 mmHg)
Flash point 191°C (open cup)
Freezing point 55–57°C
Melting point 59.4–59.8°C for the pure material.
Refractive index n_D^{60} = 1.4388 at 60°C
Solidification point 56–59°C for *Nacol 18-98*; 55–58°C for *Speziol C18 Pharma*.
Solubility Soluble in chloroform, ethanol (95%), ether, hexane, propylene glycol, benzene, acetone, and vegetable oils; practically insoluble in water.

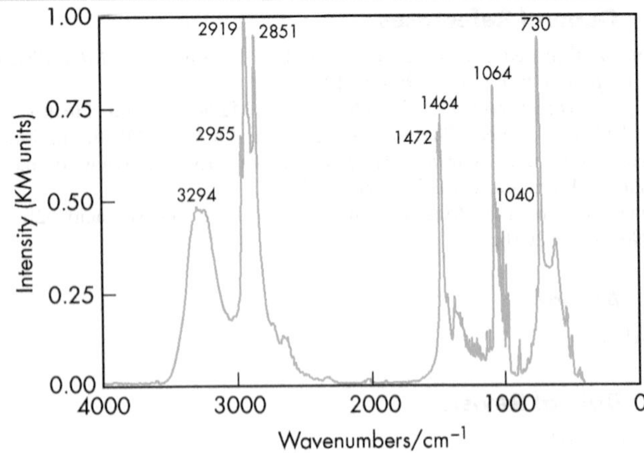

Figure 1: Infrared spectrum of stearyl alcohol measured by diffuse reflectance. Adapted with permission of Informa Healthcare.

Spectroscopy

IR spectra *see* Figure 1.

Vapor pressure 133.3 Pa (1 mmHg) at 150.3°C

11 Stability and Storage Conditions

Stearyl alcohol is stable to acids and alkalis and does not usually become rancid. It should be stored in a well-closed container in a cool, dry place.

12 Incompatibilities

Stearyl alcohol is incompatible with strong oxidizing agents and strong acids.

13 Method of Manufacture

Historically, stearyl alcohol was prepared from sperm whale oil but is now largely prepared synthetically by reduction of ethyl stearate with lithium aluminum hydride.

14 Safety

Stearyl alcohol is generally considered to be an innocuous, nontoxic material. However, adverse reactions to stearyl alcohol present in topical preparations have been reported. These include contact urticaria and hypersensitivity reactions, which are possibly due to impurities contained in stearyl alcohol rather than stearyl alcohol itself.[3–7]

The probable lethal oral human dose is greater than 15 g/kg.

LD_{50} (rat, oral): 20 g/kg[8]

15 Handling Precautions

Observe normal precautions appropriate to the circumstances and quantity of material handled. Eye protection and gloves are recommended. Stearyl alcohol is not a fire hazard, although it will burn and may give off noxious fumes containing carbon monoxide.

16 Regulatory Status

Included in the FDA Inactive Ingredients Database (oral tablets; rectal, topical, and vaginal preparations). Included in nonparenteral medicines licensed in the UK. Included in the Canadian Natural Health Products Ingredients Database.

17 Related Substances

Cetostearyl alcohol; cetyl alcohol.

18 Comments

Stearyl alcohol has been studied for use in controlled-release tablets,[9,10] suppositories,[11,12] and microspheres.[13,14] It has also been investigated for use as a transdermal penetration enhancer[15] and as an alternative to cholesterol as a membrane-stabilizing agent for liposomes.[16]

The EINECS number for stearyl alcohol is 204-017-6. The PubChem Compound ID (CID) for stearyl alcohol is 8221.

19 Specific References

1 Egan RR, Portwood O. Higher alcohols in skin lotions. *Cosmet Perfum* 1974; 89(3): 39–42.
2 Alexander P. Organic rheological additives. *Manuf Chem* 1986; 57(9): 49, 52.
3 Gaul LE. Dermatitis from cetyl and stearyl alcohols. *Arch Dermatol* 1969; 99: 593.
4 Fisher AA. Contact dermatitis from stearyl alcohol and propylene glycol. *Arch Dermatol* 1974; **110**: 636.
5 Black H. Contact dermatitis from stearyl alcohol in Metosyn (flucinonide) cream. *Contact Dermatitis* 1975; **1**: 125.
6 Cronin E. *Contact Dermatitis*. Edinburgh: Churchill Livingstone, 1980; 808.
7 Yesudian PD, King CM. Allergic contact dermatitis from stearyl alcohol in Efudix cream. *Contact Dermatitis* 2001; **45**: 313–314.
8 Lewis RJ, ed. *Sax's Dangerous Properties of Industrial Materials*, 11th edn. New York: Wiley, 2004; 2758.
9 Cao DY, *et al.* Preparation of tetramethylpyrazine phosphate pulsed-release tablets. *Chin J New Drugs* 2005; **14**(6): 723–726.
10 Cao QR, *et al.* Photoimages and the release characteristics of lipophilic matrix tablets containing highly water-soluble potassium citrate with high drug loadings. *Int J Pharm* 2007; 339(1–2): 19–24.
11 Kaiho F, *et al.* Application of fatty alcohols to pharmaceutical dosage forms. *Yakuzaigaku* 1984; **44**: 99–102.
12 Tanabe K, *et al.* Effect of additives on release of ibuprofen from suppositories. *Yakuzaigaku* 1988; **48**: 262–269.
13 Passerini N, *et al.* Controlled release of verapamil hydrochloride from waxy microparticles prepared by spray congealing. *J Control Release* 2003; 88(2): 263–275.
14 Liggins RT, Burt HM. Paclitaxel loaded poly(L-lactic acid) microspheres: properties of microspheres made with low molecular weight polymers. *Int J Pharm* 2001; 222(1): 19–33.
15 Takahashi Y, *et al.* Trial for transdermal administration of sulfonylureas. *J Pharm Soc Japan* 1997; **117**: 1022–1027.
16 Habib Ali M, *et al.* Solubilisation of drugs within liposomal bilayers: alternatives to cholesterol as a membrane stabilising agent. *J Pharm Pharmacol* 2010; **62**(11): 1646–1655.

20 General References

Barry BW. Continuous shear, viscoelastic and spreading properties of a new topical vehicle, FAPG base. *J Pharm Pharmacol* 1973; **25**: 131–137.
Cognis. Product literature: *Speziol C18 Pharma*, 2004.
Cognis. Product literature: *Drug-Delivery Agents – Pharmaline*, 2008.
Japan Pharmaceutical Excipients Council. *Japanese Pharmaceutical Excipients Directory 1996*. Tokyo: Yakuji Nippo, 1996: 527.
Madan PL, *et al.* Microencapsulation of a waxy solid: wall thickness and surface appearance studies. *J Pharm Sci* 1974; **63**: 280–284.
Rowe RC. A quantitative assessment of the reactivity of the fatty alcohols with cetrimide using immersion calorimetry. *J Pharm Pharmacol* 1987; **39**: 50–52.
Sasol. Product literature: Sasol olefins and surfactants, 2005.
Schott H, Han SK. Effect of inorganic additives on solutions of nonionic surfactants II. *J Pharm Sci* 1975; **64**: 658–664.
Wan LSC, Poon PKC. The interfacial activity of sodium lauryl sulfate in the presence of alcohols. *Can J Pharm Sci* 1970; **5**: 104–107.

21 Author

RT Guest.

22 Date of Revision

2 March 2012.

S

Sucralose

1 Nonproprietary Names

PhEur: Sucralose
USP–NF: Sucralose

2 Synonyms

Splenda; sucralosa; sucralosum; *SucraPlus*; TGS; 1′,4′,6′-trichlorogalactosucrose; 4,1′,6′-trichloro-4,1′,6′-trideoxy-*galacto*-sucrose; *Unisweet*.

3 Chemical Name and CAS Registry Number

1,6-Dichloro-1,6-dideoxy-β-D-fructofuranosyl-4-chloro-4-deoxy-α-D-galactopyranoside [56038-13-2]

4 Empirical Formula and Molecular Weight

$C_{12}H_{19}Cl_3O_8$ 397.64

5 Structural Formula

6 Functional Category

Sweetening agent.

7 Applications in Pharmaceutical Formulation or Technology

Sucralose is used as a sweetening agent in beverages, foods, and pharmaceutical formulations. It has a sweetening power approximately 300–1000 times that of sucrose and has no aftertaste. The usual concentration for use in food products is 0.03–0.24%.

8 Description

Sucralose occurs as a white to off-white colored, free-flowing, crystalline powder.

9 Pharmacopeial Specifications

See Table I.

10 Typical Properties

Acidity/alkalinity pH = 5–6 (10% w/v aqueous solution at 20°C)
Density (bulk) 0.35 g/cm^3
Density (tapped) 0.62 g/cm^3
Density (true) 1.63 g/cm^3
Melting point 130°C (for anhydrous crystalline form); 36.5°C (for pentahydrate).
Particle size distribution 90% <12 μm in size for the powder grade (Tate & Lyle).
Partition coefficient $\log_{10} P = -0.51$ (octanol : water)
Refractive index 1.33 to 1.37

Table I: Pharmacopeial specifications for sucralose.

Test	PhEur 7.4	USP35–NF30
Identification	+	+
Specific rotation	+84.0° to +87.5°	+84.0° to +87.5°
Water	≤2.0%	≤2.0%
Residue on ignition	≤0.7%	≤0.7%
Heavy metals	≤10 ppm	≤10 ppm
Limit of hydrolysis products	—	≤0.1%
Limit of methanol	—	≤0.1%
Limit of impurities[(a)]	≤0.1%	—
Related compounds	≤0.5%	≤0.5%
Assay (dried basis)	98.0–102.0%	98.0–102.0%

(a) PhEur 7.4 lists impurities as 1,6-dichloro-1,6-dideoxy-β-D-fructofuranose and 4-chloro-4-deoxy-α-D-galactopyranose.

Solubility Freely soluble in ethanol (95%), methanol, and water; slightly soluble in ethyl acetate.
Specific rotation $[\alpha]_D^{20}$ = +84.0° to +87.5° (1% w/v aqueous solution); +68.2° (1.1% w/v solution in ethanol).
Viscosity 0.6–3.8 mPa s (0.6–3.8 cP) (10–60% w/w in water at 20–60°C).

11 Stability and Storage Conditions

Sucralose is a relatively stable material. In aqueous solution, at highly acidic conditions (pH <3), and at high temperatures (≥35°C), it is hydrolyzed to a limited extent, producing 4-chloro-4-deoxygalactose and 1,6-dichloro-1,6-dideoxyfructose. In food products, sucralose remains stable throughout extended storage periods, even at low pH. However, it is most stable at pH 5–6.

Sucralose should be stored in a well-closed container in a cool, dry place, at a temperature not exceeding 21°C. Sucralose, when heated at elevated temperatures, may break down with the release of carbon dioxide, carbon monoxide, and minor amounts of hydrogen chloride.

12 Incompatibilities

—

SEM 1: Excipient: sucralose; manufacturer: Tate & Lyle; magnification: 1000×; voltage 3.0 kV.

13 Method of Manufacture

Sucralose may be prepared by a variety of methods that involve the selective substitution of three sucrose hydroxyl groups by chlorine. Sucralose can also be synthesized by the reaction of sucrose (or an acetate) with thionyl chloride.

14 Safety

Sucralose is generally regarded as a nontoxic and nonirritant material and is approved, in a number of countries, for use in food products. Following oral consumption, sucralose is mainly unabsorbed and is excreted in the feces.[1–3]

The WHO has set an acceptable daily intake for sucralose of up to 15 mg/kg body-weight.[4]

LD_{50} (mouse, oral): > 16 g/kg
LD_{50} (rat, oral): > 10 g/kg

15 Handling Precautions

Observe normal precautions appropriate to the circumstances and quantity of material handled.

16 Regulatory Status

The FDA, in April 1998, approved sucralose for use as a tabletop sweetener and as an additive in a variety of food products, and so is included in the EAFUS list. In the UK, sucralose was fully authorized for use in food products in 2005.[5] It is also accepted for use in many other countries worldwide. Included in the Canadian Natural Health Products Ingredients Database.

17 Related Substances

Sucrose.

18 Comments

The sweetening effect of sucralose is not reduced by heating, and food products containing sucralose may be subjected to high-temperature processes such as pasteurization, sterilization, UHT processing and baking. Sucralose is often blended with maltodextrin or dextrose as bulking agents in its granular form.

Sucralose has no nutritional value, is noncariogenic, does not promote dental caries, and produces no glycemic response.

A specification for sucralose is contained in the *Food Chemicals Codex* (FCC).[6]

The EINECS number for sucralose is 259-952-2. The PubChem Compound ID (CID) for sucralose includes 71485 and 5066234.

19 Specific References

1 Grice HC, Goldsmith LA. Sucralose – an overview of the toxicity data. *Food Chem Toxicol* 2000; **38**(Suppl. 2): S1–S6.
2 Roberts A, *et al.* Sucralose metabolism and pharmacokinetics in man. *Food Chem Toxicol* 2000; **38**(Suppl. 2): S31–S41.
3 Mclean Baird I, *et al.* Repeated dose study of sucralose tolerance in human subjects. *Food Chem Toxicol* 2000; **38**(Suppl. 2): S123–S129.
4 FAO/WHO. Evaluation of certain food additives and contaminants. Thirty-seventh report of the joint FAO/WHO expert committee on food additives. *World Health Organ Tech Rep Ser* 1991; No. 806: 21–23.
5 Statutory Instrument (SI) 2005: No. 1156. The Sweeteners in Food (Amendment) (Wales) Regulations 2005. London: Stationery Office, 2005.
6 *Food Chemicals Codex*, 7th edn. Bethesda, MD: United States Pharmacopeia, 2010: 993.

20 General References

American Dietetic Association. Position of the American Dietetic Association: use of nutritive and nonnutritive sweeteners. *J Am Diet Assoc* 2004; **104**: 255–275.

Anonymous. Artificial sweeteners. *Can Pharm J* 1996; **129**(Apr): 22.

Anonymous. Sucralose – a new artificial sweetener. *Med Lett Drugs Ther* 1998; 40: 67–68.

Jenner MR, Smithson A. Physicochemical properties of the sweetener sucralose. *J Food Sci* 1989; **54**(6): 1646–1649.

Kloesel L. Sugar substitutes. *Int J Pharm Compound* 2000; 4(2): 86–87.

Knight I. The development and applications of sucralose, a new high-intensity sweetener. *Can J Physiol Pharmacol* 1994; 72(4): 435–439.

Kroschwitz JI, Howe-Grant M, eds. *Kirk-Othmer Encyclopedia of Chemical Technology*, 4th edn, **11**. New York: John Wiley & Sons, 1994: 295.

Marti N, *et al.* An update on alternative sweeteners. *International Sugar Journal* 2008; **110**(1315): 425–429.

McNeil Nutritionals. *Splenda*: the online guide to cooking, eating and living well. http://www.splenda.com (accessed 14 July 2011).

Tate and Lyle. Technical literature: Sucralose, 2001.

21 Authors

BA Langdon, MP Mullarney.

22 Date of Revision

2 March 2012.

Sucrose

1 Nonproprietary Names

BP: Sucrose
JP: Sucrose
PhEur: Sucrose
USP–NF: Sucrose

2 Synonyms

Beet sugar; cane sugar; α-D-glucopyranosyl-β-D-fructofuranoside; refined sugar; saccharose; saccharosum; saccharum; sucrosum; sugar.

3 Chemical Name and CAS Registry Number

β-D-fructofuranosyl-α-D-glucopyranoside [57-50-1]

4 Empirical Formula and Molecular Weight

$C_{12}H_{22}O_{11}$ 342.30

5 Structural Formula

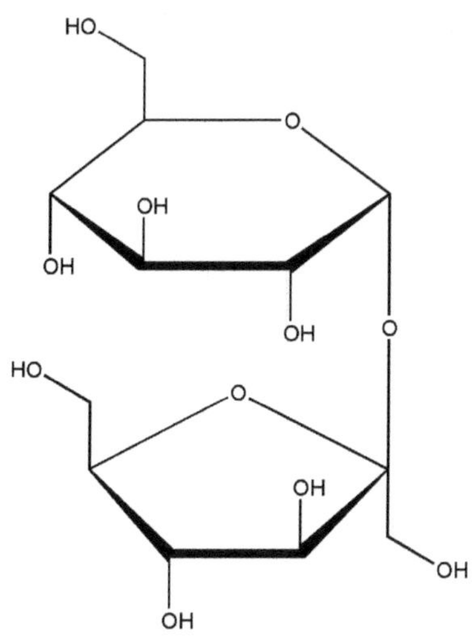

6 Functional Category

Coating agent; suspending agent; sweetening agent; tablet and capsule binder; tablet and capsule diluent; viscosity-increasing agent.

7 Applications in Pharmaceutical Formulation or Technology

Sucrose is widely used in oral pharmaceutical formulations.

Sucrose syrup, containing 50–67% w/w sucrose, is used in tableting as a binding agent for wet granulation. In the powdered form, sucrose serves as a dry binder (2–20% w/w) or as a bulking agent and sweetener in chewable tablets and lozenges.[1–3] Directly compressible sucrose is available in a range of particle sizes.[4] Tablets that contain large amounts of sucrose may harden to give poor disintegration.

Sucrose syrups are used as tablet-coating agents at concentrations between 50% and 67% w/w. With higher concentrations, partial inversion of sucrose occurs, which makes sugar coating difficult.

Sucrose syrups are also widely used as vehicles in oral liquid-dosage forms to enhance palatability or to increase viscosity.[5,6]

Sucrose has been used as a diluent in freeze-dried protein products.[7,8]

Sucrose is also widely used in foods and confectionery.

See Table I.

Table I: Uses of sucrose.

Use	Concentration (% w/w)
Syrup for oral liquid formulations	67
Sweetening agent	67
Tablet binder (dry granulation)	2–20
Tablet binder (wet granulation)	50–67
Tablet coating (syrup)	50–67

8 Description

Sucrose is a sugar obtained from sugar cane (*Saccharum officinarum* Linné (Fam. Gramineae)), sugar beet (*Beta vulgaris* Linné (Fam. Chenopodiaceae)), and other sources. It contains no added substances. Sucrose occurs as colorless crystals, as crystalline masses or blocks, or as a white crystalline powder; it is odorless and has a sweet taste.

9 Pharmacopeial Specifications

The pharmacopeial specifications for sucrose have undergone harmonization of many attributes for JP, PhEur, and USP–NF.

See Table II. *See also* Section 18.

10 Typical Properties

Density (bulk)

0.93 g/cm^3 (crystalline sucrose);

0.60 g/cm^3 (powdered sucrose).

Density (tapped)

SEM 1: Excipient: sucrose; manufacturer: Great Western Sugar Co.; lot no.: 1-2-80; magnification: 60×; voltage: 10 kV.

S

SEM 2: Excipient: sucrose; manufacturer: Great Western Sugar Co.; lot no.: 1-2-80; magnification: 600×; voltage: 10 kV.

Table II: Pharmacopeial specifications for sucrose.

Test	JP XV	PhEur 7.4	USP35–NF30
Identification	+	+	+
Characters	–	+	+
Appearance of solution	+	+	–
Acidity or alkalinity	+	+	+
Specific optical rotation	+66.3° to +67.0°	+66.3° to +67.0°	+66.3° to +67.0°
Color value	–	≤45	≤45
Conductivity	+	35 µS cm^{-1}	35 µS cm^{-1}
Loss on drying	≤0.1%	≤0.1%	≤0.1%
Bacterial endotoxins[(a)]	≤0.25 IU/mg	≤0.25 IU/mg	≤0.25 IU/mg
Dextrins[(a)]	+	+	–
Reducing sugars	–	+	+
Invert sugar	+	–	–
Chloride	–	–	≤35 ppm
Sulfate	–	–	≤60 ppm
Sulfites	≤15 ppm	≤10 ppm	–
Calcium	–	–	–
Lead	≤0.5 ppm	–	–
Residue on ignition	–	–	≤0.05%

(a) If sucrose is to be used in large volume infusions.

1.03 g/cm^3 (crystalline sucrose);

0.82 g/cm^3 (powdered sucrose).

Density (true) 1.6 g/cm^3

Dissociation constant pK_a = 12.62

Flowability Crystalline sucrose is free flowing, whereas powdered sucrose is a cohesive solid.

Melting point 160–186°C (with decomposition)

Moisture content Finely divided sucrose is hygroscopic and absorbs up to 1% water.[(9)]

See Figure 1.

Osmolarity A 9.25% w/v aqueous solution is isoosmotic with serum.

Particle size distribution Powdered sucrose is a white, irregular-sized granular powder. The crystalline material consists of colorless crystalline, roughly cubic granules.

See Figures 2 and 3.

Refractive index n_D^{25} = 1.34783 (10% w/v aqueous solution)

Solubility *see* Table III.

Specific gravity *see* Table IV.

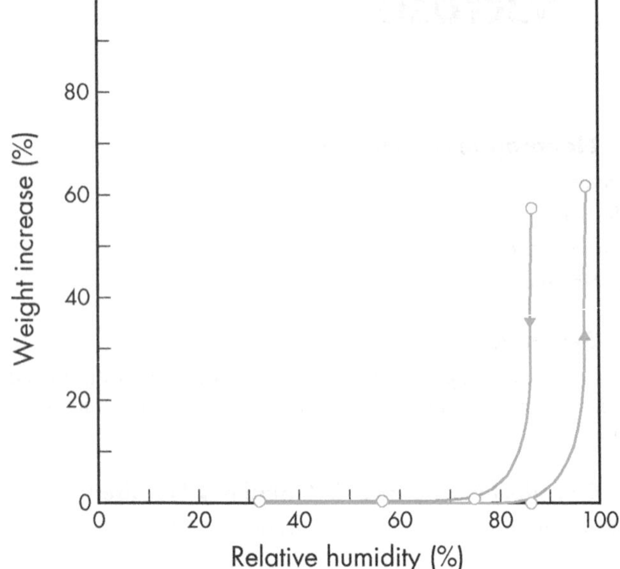

Figure 1: Moisture sorption–desorption isotherm of powdered sucrose. Samples dried initially at 60°C over silica gel for 24 hours. Note: at 90% relative humidity, sufficient water was absorbed to cause dissolution of the solid.

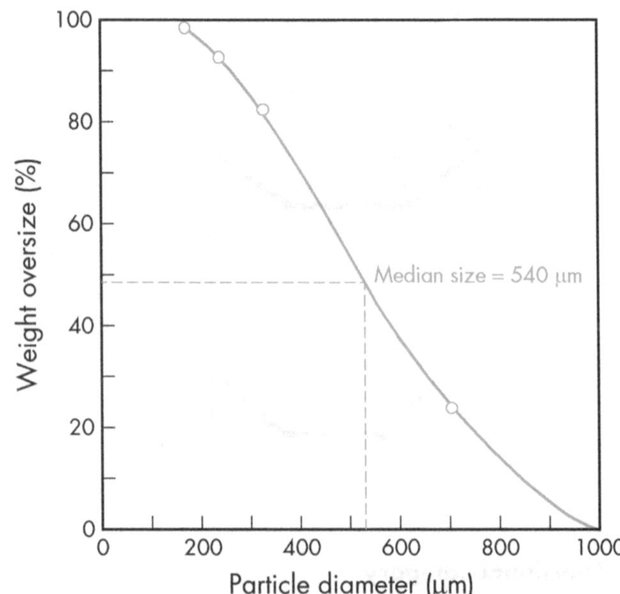

Figure 2: Particle size distribution of crystalline sucrose.

Spectroscopy

IR spectra *see* Figure 4.

NIR spectra *see* Figure 5.

11 Stability and Storage Conditions

Sucrose has good stability at room temperature and at moderate relative humidity. It absorbs up to 1% moisture, which is released upon heating at 90°C. Sucrose caramelizes when heated to temperatures above 160°C. Dilute sucrose solutions are liable to fermentation by microorganisms but resist decomposition at higher concentrations, e.g. above 60% w/w concentration. Aqueous solutions may be sterilized by autoclaving or filtration.

When sucrose is used as a base for medicated confectionery, the cooking process, at temperatures rising from 110 to 145°C, causes some inversion to form dextrose and fructose (invert sugar). The

Table III: Solubility of sucrose.

Solvent	Solubility at 20°C unless otherwise stated
Chloroform	Practically insoluble
Ethanol	1 in 400
Ethanol (95%)	1 in 170
Propan-2-ol	1 in 400
Water	1 in 0.5
	1 in 0.2 at 100°C

Table IV: Specific gravity of aqueous sucrose solutions.

Concentration of aqueous sucrose solution (% w/w)	Specific gravity at 20°C
2	1.0060
6	1.0219
10	1.0381
20	1.0810
30	1.1270
40	1.1764
50	1.2296
60	1.2865
70	1.3471
76	1.3854

fructose imparts stickiness to confectionery but prevents cloudiness due to graining. Inversion is accelerated particularly at temperatures above 130°C and by the presence of acids.

The bulk material should be stored in a well-closed container in a cool, dry place.

12 Incompatibilities

Powdered sucrose may be contaminated with traces of heavy metals, which can lead to incompatibility with active ingredients, e.g. ascorbic acid. Sucrose may also be contaminated with sulfite from the refining process. With high sulfite content, color changes can occur in sugar-coated tablets; for certain colors used in sugar-coating the maximum limit for sulfite content, calculated as sulfur, is 1 ppm. In the presence of dilute or concentrated acids, sucrose is hydrolyzed or inverted to dextrose and fructose (invert sugar). Sucrose may attack aluminum closures.[10]

13 Method of Manufacture

Sucrose is obtained from the sugar cane plant, which contains 15–20% sucrose, and sugar beet, which contains 10–17% sucrose. Juice from these sources is heated to coagulate water-soluble proteins, which are removed by skimming. The resultant solution is then decolorized with an ion-exchange resin or charcoal and concentrated. Upon cooling, sucrose crystallizes out. The remaining solution is concentrated again and yields more sucrose, brown sugar, and molasses.

14 Safety

Sucrose is hydrolyzed in the small intestine by the enzyme sucrase to yield dextrose and fructose, which are then absorbed. When administered intravenously, sucrose is excreted unchanged in the urine.

Although sucrose is very widely used in foods and pharmaceutical formulations, sucrose consumption is a cause of concern and should be monitored in patients with diabetes mellitus or other metabolic sugar intolerance.[11]

Sucrose is also considered to be more cariogenic than other carbohydrates since it is more easily converted to dental plaque. For this reason, its use in oral pharmaceutical formulations is declining.

Although sucrose has been associated with obesity, renal damage, and a number of other diseases, conclusive evidence linking sucrose intake with some diseases could not be established.[12,13] It was, however, recommended that sucrose intake in the diet should be reduced.[13]

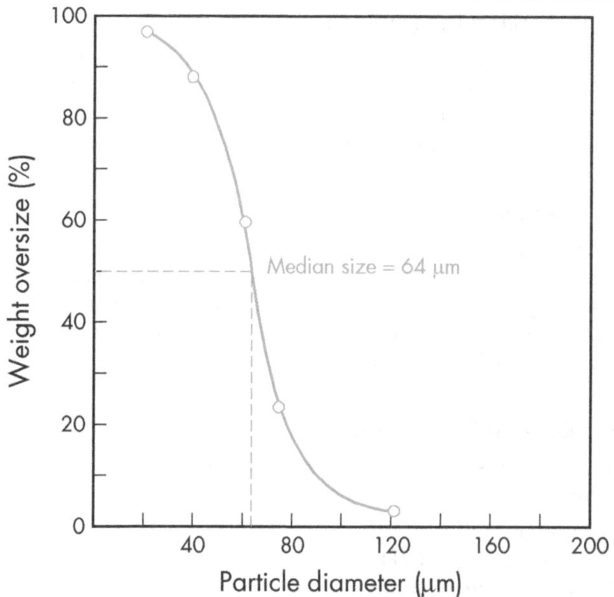

Figure 3: Particle size distribution of powdered sucrose.

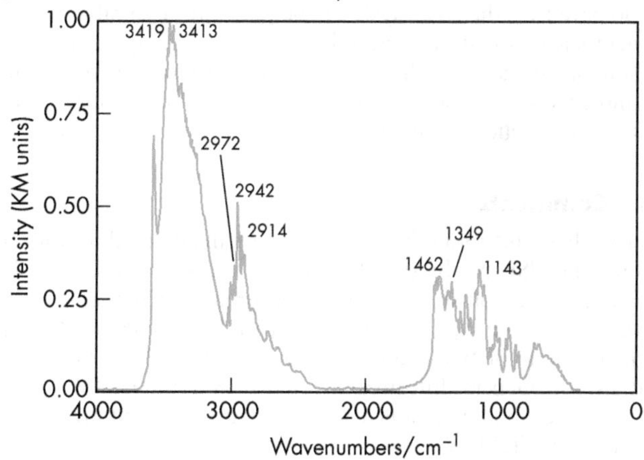

Figure 4: Infrared spectrum of sucrose measured by diffuse reflectance. Adapted with permission of Informa Healthcare.

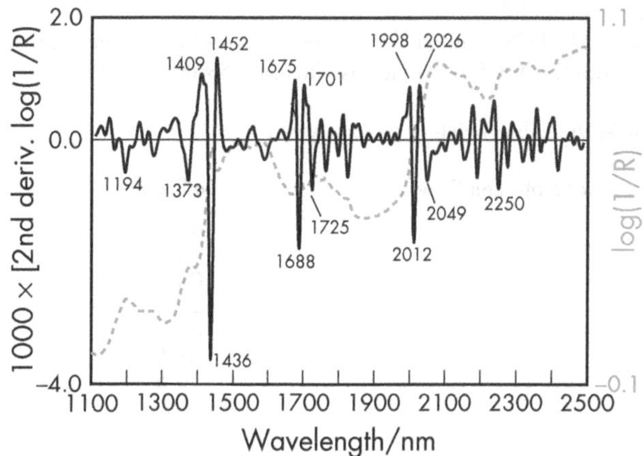

Figure 5: Near-infrared spectrum of sucrose measured by reflectance.

LD_{50} (mouse, IP): 14 g/kg[14]

LD_{50} (rat, oral): 29.7 g/kg

15 Handling Precautions

Observe normal precautions appropriate to the circumstances and quantity of material handled. Eye protection and gloves are recommended. In the UK, the workplace exposure limit for sucrose is 10 mg/m^3 long-term (8-hour TWA) and 20 mg/m^3 short-term.[15]

16 Regulatory Status

GRAS listed. Included in the FDA Inactive Ingredients Database (injections; oral capsules, solutions, syrups, and tablets; topical preparations). Included in nonparenteral and parenteral medicines licensed in the UK. Included in the Canadian Natural Health Products Ingredients Database.

17 Related Substances

Compressible sugar; confectioner's sugar; invert sugar; sugar spheres.

Invert sugar

Empirical formula $C_6H_{12}O_6$

Molecular weight 180.16

CAS number [8013-17-0]

Comments An equimolecular mixture of dextrose and fructose prepared by the hydrolysis of sucrose with a suitable mineral acid such as hydrochloric acid. Invert sugar may be used as a stabilizing agent to help prevent crystallization of sucrose syrups and graining in confectionery. A 10% aqueous solution is also used in parenteral nutrition.

18 Comments

Sucrose has undergone harmonization of many attributes for JP, PhEur, and USP–NF by the Pharmacopeial Discussion Group. For further information see the General Information Chapter <1196> in the USP35–NF30, the General Chapter 5.8 in PhEur 7.4, along with the 'State of Work' document on the PhEur EDQM website, and also the General Information Chapter 8 in the JP XV.

For typical boiling points of sucrose syrups, without inversion of the sugar, *see* Table V.

Therapeutically sugar pastes are used to promote wound healing.[16,17]

A specification for sucrose is contained in the *Food Chemicals Codex* (FCC).[18]

The EINECS number for sucrose is 200-334-9. The PubChem Compound ID (CID) for sucrose includes 5988 and 1115.

Table V: Boiling points of sucrose syrups.

Sucrose concentration (% w/v)	Boiling point (°C)
50	101.5
60	103
64	104
72	105.5
75	107
77.5	108.5
80	110.5

19 Specific References

1 Allen LV. Featured excipient: capsule and tablet diluents. *Int J Pharm Compound* 2000; 4(4): 306–310324–325.
2 Mullarney MP, *et al.* The powder flow and compact mechanical properties of sucrose and three high intensity sweeteners used in chewable tablets. *Int J Pharm* 2003; **257**(1–2): 227–236.
3 Sugimoto M, *et al.* Development of a manufacturing method for rapidly disintegrating oral tablets using the crystalline transition of amorphous sucrose. *Int J Pharm* 2006; **320**: 71–78.
4 British Sugar Plc. Technical literature: Sucrose. http://www.britishsugar.co.uk (accessed 2 December 2011).
5 Salazar DSM, Saavedra C. Application of a sensorial response model to the design of an oral liquid pharmaceutical dosage form. *Drug Dev Ind Pharm* 2000; **26**(1): 55–60.
6 Cooper J. A question of taste: uses of sucrose. *Manuf Chem* 2003; **74**(10): 71–72, 74.
7 Imamura K, *et al.* Imprints of compression on crystallisation behaviour of freeze-dried amorphous sucrose. *J Pharm Sci* 2010; **99**: 1452–1463.
8 Johnson RE, *et al.* Mannitol-sucrose mixtures: versatile formulations for protein lyophilisation. *J Pharm Sci* 2002; **91**(4): 914–922.
9 Hancock BC, Dalton CR. Effect of temperature on water vapour sorption by some amorphous pharmaceutical sugars. *Pharm Dev Technol* 1999; **4**(1): 125–131.
10 Tressler LJ. Medicine bottle caps [letter]. *Pharm J* 1985; **235**: 99.
11 Golightly LK, *et al.* Pharmaceutical excipients: adverse effects associated with 'inactive' ingredients in drug products (part II). *Med Toxicol* 1988; **3**: 209–240.
12 Yudkin J. Sugar and disease. *Nature* 1972; **239**: 197–199.
13 Anon. *Report on Health and Social Subjects* 37. London: HMSO, 1989.
14 Lewis RJ, ed. *Sax's Dangerous Properties of Industrial Materials*, 11th edn. New York: Wiley, 2004: 3318.
15 Health and Safety Executive. *EH40/2005: Workplace Exposure Limits.* Sudbury: HSE Books, 2011. http://www.hse.gov.uk/pubns/priced/eh40.pdf (accessed 23 February 2012).
16 Middleton KR, Seal D. Sugar as an aid to wound healing. *Pharm J* 1985; **235**: 757–758.
17 Thomas S. *Wound Management and Dressings.* London: Pharmaceutical Press, 1990: 62–63.
18 *Food Chemicals Codex*, 7th edn. Bethesda, MD: United States Pharmacopeia, 2010.

20 General References

Armstrong NA. In: Swarbrick J, Boylan JC, eds. *Encyclopedia of Pharmaceutical Technology*, 2nd edn, 3. New York: Marcel Dekker, 2002: 2713–2732.

European Directorate for the Quality of Medicines and Healthcare (EDQM). European Pharmacopoeia – State Of Work Of International Harmonisation. *Pharmeuropa* 2011; 23(4): 713–714. www.edqm.eu/site/-614.html (accessed 2 December 2011).

Jackson EB, ed. *Sugar Confectionery Manufacture.* Glasgow: Blackie, 1990.

Lipari JM, Reiland TL. Flavors and flavor modifiers. In: Swarbrick J, Boylan JC, eds. *Encyclopedia of Pharmaceutical Technology*, 2nd edn, 2. New York: Marcel Dekker, 2002: 1255–1263.

Wolraich ML, *et al.* Effects of diets high in sucrose or aspartame on the behavior and cognitive performance of children. *N Engl J Med* 1994; 330: 301–307.

21 Author

NA Armstrong.

22 Date of Revision

2 March 2012.

Sucrose Octaacetate

1 Nonproprietary Names

USP–NF: Sucrose Octaacetate

2 Synonyms

α-D-Glucopyranoside, 1,3,4,6-tetra-*O*-acetyl-β-D-fructofuranosyl-, tetraacetate; octaacetylsucrose.

3 Chemical Name and CAS Registry Number

Acetic acid[(2*S*,3*S*,4*R*,5*R*)-4-acetoxy-2,5-bis(acetoxymethyl)-2-[[(2*R*,3*R*,4*S*,5*R*,6*R*)-3,4,5-triacetoxy-6-(acetoxymethyl)-2-tetrahydropyranyl]oxy]-3-tetrahydrofuranyl]ester [126-14-7]

4 Empirical Formula and Molecular Weight

$C_{28}H_{38}O_{19}$ 678.59

5 Structural Formula

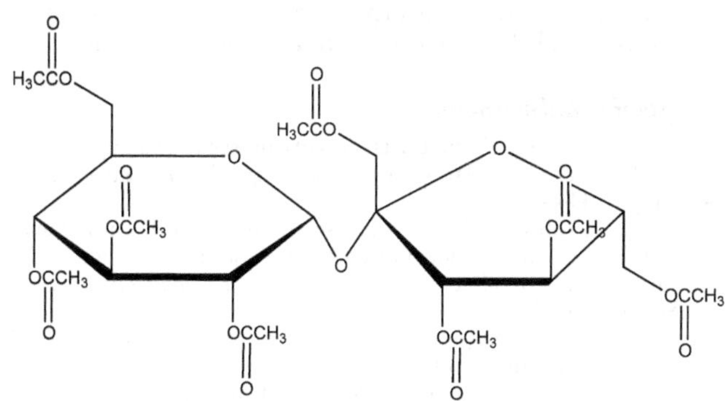

6 Functional Category

Alcohol denaturant; bitter flavoring agent.

7 Applications in Pharmaceutical Formulation or Technology

Sucrose octaacetate is used in pharmaceutical formulations as an alcohol denaturant and as a bittering agent.

8 Description

Sucrose octaacetate occurs as white hygroscopic powder. It is practically odorless with a bitter taste.

9 Pharmacopeial Specifications

See Table I.

Table I: Pharmacopeial specifications for sucrose octaacetate.

Test	USP35–NF30
Water	≤1.0%
Residue on ignition	≤0.1%
Melting temperature	≥78°C
Acidity	+
Assay (anhydrous)	98.0–100.5%

10 Typical Properties

Boiling point 260°C
Flash point 307.3°C
Melting point 89°C (decomposes above 285°C)
Refractive index 1.47
Solubility Very soluble in methanol and chloroform; soluble in ethanol (95%) and ether; very slightly soluble in water; *see also* Table II.
Specific gravity 1.28 at 20°C (water = 1)
Specific rotation $[\alpha]_D^{25.4} = +58.5°$
Spectroscopy

IR spectra *see* Figure 1.

SEM 1: Excipient: sucrose octaacetate; manufacturer: Sigma-Aldrich Inc.; lot no.: RS33841'08; magnification: 100×; voltage: 5 kV.

SEM 2: Excipient: sucrose octaacetate; manufacturer: Sigma-Aldrich Inc.; lot no.: RS33841'08; magnification: 500×; voltage: 5 kV.

Table II: Solubility of sucrose octaacetate.

Solvent	Solubility at 20°C
Acetone	1 in 0.3
Benzene	1 in 0.6
Ethanol (95%)	1 in 11
Glacial acetic acid	1 in 0.7
Toluene	1 in 0.5
Water	1 in 1100

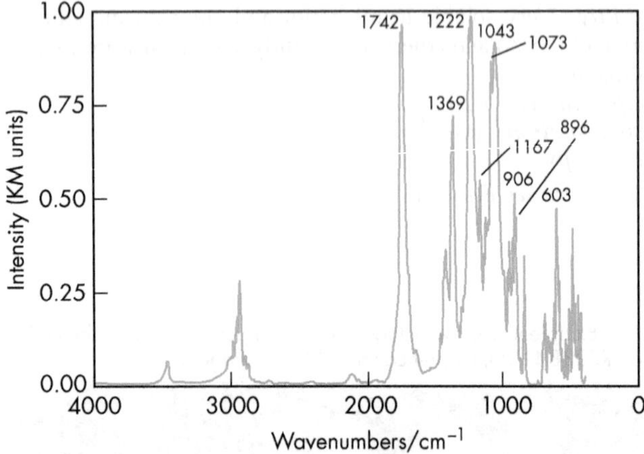

Figure 1: Infrared spectrum of sucrose octaacetate measured by diffuse reflectance. Adapted with permission of Informa Healthcare.

11 Stability and Storage Conditions

Sucrose octaacetate is a stable material and should be stored in a well-closed, airtight container. Store in a cool (2–8°C), dry place; moisture may cause instability.

12 Incompatibilities

Sucrose octaacetate is incompatible with strong oxidizing agents.

13 Method of Manufacture

Sucrose octaacetate is typically produced by chemical synthesis; one reported synthetic method is by pyridine-catalyzed acetylation of sucrose.[1,2]

14 Safety

Sucrose octaacetate is generally regarded as safe. It is considered slightly hazardous in cases of skin contact (irritant), ingestion, or inhalation.

LD_{50} (rabbit, skin): >5 g/kg
LD_{50} (rat, oral): >5 g/kg[3]

15 Handling Precautions

Observe normal precautions appropriate to the circumstances and quantity of the material handled. When heated to decomposition, sucrose octaacetate emits acrid smoke and irritating vapors. Compatible chemical-resistant gloves and eye safety goggles are recommended. Respiratory protection is not required, but dust masks may be used for protection from nuisance levels of dust.

16 Regulatory Status

GRAS listed. Approved by the FDA as both a direct and an indirect food additive, and as a nail-biting deterrent for over-the-counter drug products.[4]

17 Related Substances

Sodium acetate; sucrose.

18 Comments

Sucrose octaacetate is a naturally occurring substance that has been isolated from plant material: the root of *Clematis japonica* contains 0.15% of sucrose octaacetate by dry weight. At a concentration of 0.06% sucrose octaacetate renders sugar too bitter for human consumption.[5]

The EINECS number for sucrose octaacetate is 204-772-1. The PubChem Compound ID (CID) for sucrose octaacetate is 219904.

Sucrose octaacetate is incorporated as an active agent in preparations intended to deter nail-biting or thumb-sucking.

19 Specific References

1 Amagasa M, Yanagita T. Industrial uses of cane sugar I. Catalytic effects of pyridine on the acetylation of sucrose. *Kogyo Kagaku Zasshi* 1940; 43(Suppl.): 444–445.
2 Amagasa M, Tobisima H. Industrial uses of cane sugar II. The influence of temperature and the amount of acetic anhydride upon the acetylation of sucrose with pyridine as catalyst. *Kogyo Kagaku Zasshi* 1941; 44(Suppl.): 40–41.
3 Lewis RJ, ed. *Sax's Dangerous Properties of Industrial Materials*, 11th edn. New York: Wiley, 2004; 2755.
4 Jarvis CM. Reassessment of the exemption from the requirement of a tolerance for sucrose octaacetate (CAS Reg. No. 126-14-7). US EPA Memorandum, December 21, 2005; 1–10.
5 Thierry DM, *et al.* Preparation of sucrose octaacetate – a bitter-tasting compound. *J Chem Educ* 1992; 69(8): 668–669.

20 General References

Sciencelab, Inc. Material safety data sheet: Sucrose octaacetate, 9 October 2005.
Sigma-Aldrich. Material safety data sheet: D-(+)-Sucrose octaacetate, 11 April 2008.

21 Author

C-M Lee.

22 Date of Revision

2 March 2012.

S

Sucrose Palmitate

1 Nonproprietary Names

BP: Sucrose Monopalmitate
PhEur: Sucrose Monopalmitate
USP-NF: Sucrose Palmitate

2 Synonyms

2-[3,4-Dihydroxy-2,5-bis(hydroxymethyl)oxolan-2-yl]oxy-6-(hydroxymethyl)oxane-3,4,5-triol, hexadecanoic acid; *DUB SE*; E-473; α-D-glucopyranoside, β-D-fructofuranosyl, monohexadecanoate; *Ryoto*; sacchari monopalmitas; *Sisterna PS750-C*; sucrose hexadecanoate; sucrose palmitate; *Surfhope SE Cosme*; *Surfhope SE Pharma*.

3 Chemical Name and CAS Registry Number

[6-[3,4-Dihydroxy-2,5-bis(hydroxymethyl)oxolan-2-yl]oxy-3,4,5-trihydroxyoxan-2-yl]methyl hexadecanoate [26446-38-8]

4 Empirical Formula and Molecular Weight

$C_{28}H_{52}O_{12}$ 580.17

The USP35–NF30 and PhEur 7.4 describe sucrose palmitate as a mixture of sucrose monoesters, mainly sucrose monopalmitate, containing variable quantities of mono-, di-, tri-, and polyesters.

5 Structural Formula

R = H or

6 Functional Category

Dispersing agent; emulsifying agent; nonionic surfactant; solubilizing agent.

7 Applications in Pharmaceutical Formulation or Technology

Sucrose palmitate is used for the stabilization of emulsions as a surfactant or cosurfactant.[1–3] It has also been used as a solubilizing agent, and has been reported to improve the solubility of felodipine in a controlled-release composition.[4]

Sucrose palmitate has been used as an absorption enhancer for oral delivery of poorly absorbed drugs through the intestinal mucosa.[5] It has excellent skin compatibility properties and has been shown to impove the skin permeation of acyclovir when applied topically as an o/w cream.[6]

Sucrose palmitate has demonstrated temperature-dependent viscoelastic gel-forming properties, which can be exploited for the formulation of temperature-sensitive controlled drug delivery systems.[7] It has been used as a tablet matrix-forming agent along with microcrystalline cellulose and has shown slight retardation of drug release of two model drugs, theophylline monohydrate and ibuprofen.[8]

Sucrose palmitate is used in many cosmetic formulations such as creams and lotions, lipsticks, antiperspirant, and bath products. It provides good foam quality and excellent skin mildness, emolliency, and moisturizing properties. In general, it is used in cosmetic applications between 0.5% and 5%.

8 Description

Sucrose palmitate occurs as a white or almost white, tasteless, odorless, unctuous powder.

9 Pharmacopeial Specifications

See Table I.

Table I: Pharmacopeial specifications for sucrose palmitate.

Test	PhEur 7.4	USP35–NF30
Identification	+	+
Characteristics	+	–
Acid value (on a 3 g sample)	≤6.0	≤6.0
Composition of fatty acids	+	+
Lauric acid	≤3%	≤3%
Myristic acid	≤3%	≤3%
Palmitic acid	70–85%	70–85%
Stearic acid	10–25%	10–25%
Sum of the contents of palmitic acid and stearic acid	≥90%	≥90%
Free sucrose	≤4%	≤4%
Water	≤4%	≤4%
Total ash	≤1.5%	≤1.5%
Assay		
Monoesters	≥55%	≥55%
Diesters	≤40%	≤40%
Sum of triesters and polyesters	≤20%	≤20%

10 Typical Properties

Decomposition temperature Starts at 130°C, completes at 240°C.
HLB value *see* Table II.

Table II: Typical properties of selected commercially available grades of sucrose palmitate.

Name	HLB value	Monoesters (%)	Main fatty acid (%)
DUB SE 15P	15	70	90
DUB SE 16P	16	80	90
Surfhope SE Pharma D-1615	15	70	80
Surfhope SE Pharma D-1616	16	80	80

Melting point Approx. 60°C
Monoesters and fatty acids *see* Table II.
Solubility Very slightly soluble in water; sparingly soluble in ethanol (96%) and glycerin. Insoluble in oil. *See also* Table III.

11 Stability and Storage Conditions

Sucrose palmitate should be stored at 15–25°C and protected from humidity.

Table III: Solubility of sucrose palmitate.[a]

Solvent	DUB SE	
	15P	16P
Ethanol		
25°C	I	I
75°C	S	S
Glycerin		
25°C	D	D
75°C	D	S
Oil		
25°C	D	D
75°C	I	I
Water		
25°C	D	S
75°C	S	S

(a) I = insoluble; D = dispersible; S = soluble.

12 Incompatibilities

—

13 Method of Manufacture

Sucrose palmitate is obtained by transesterification of palmitic acid methyl esters of edible tallow or hydrogenated edible tallow or edible vegetable oils with sucrose. The manufacture of the fatty acid methyl esters includes a distillation step.

Ethyl acetate, methyl ethyl ketone or dimethyl sulfoxide and isobutyl alcohol (2-methyl-1-propanol) may be used in the preparation of sucrose fatty acid esters.[9]

14 Safety

The Joint FAO/WHO Expert Committee on Food Additives (JECFA) has reviewed the safety of sucrose fatty acid esters and recommended an acceptable daily intake of up to 30 mg/kg. Sucrose fatty acid esters may have a laxative effect at higher doses.[10]

Sucrose palmitate is nonirritant for the skin and eyes. Acute toxicity performed in rats and mice with an oral dose totalling 20 g/kg, administered in 10 equally divided doses at 30 to 60 minute intervals produced no deaths.[11]

LD_{50} (mouse, oral): > 2g/kg

15 Handling Precautions

Observe normal precautions appropriate to the circumstances and quantity of material handled. Dust formation should be avoided. Eye protection and gloves are recommended.

16 Regulatory Status

Accepted for use as a food additive in US, Europe, and Japan. Included in the FDA Inactive Ingredients Database (oral; powder for suspension). Included in nonparenteral medicines licensed in the UK. Included in the Canadian Natural Health Products Ingredients Database.

17 Related Substances

Sucrose monolaurate; sucrose stearate.

Sucrose monolaurate

Empirical formula: $C_{24}H_{44}O_{12}$

Molecular weight: 524.6

CAS number: [25339-99-5]

Synonyms: [(2*S*,3*R*,4*S*,5*S*,6*R*)-2-[(2*S*,3*S*,4*S*,5*R*)-3,4-Dihydroxy-2,5-bis(hydroxymethyl)oxolan-2-yl]-3,4,5-trihydroxy-6-(hydroxymethyl)oxan-2-yl] dodecanoate; α-D-glucopyranoside, β-D-fructofuranosyl, monododecanoate; β-D-fructofuranosyl-α-D-glucopyranoside, monododecanoate.

Appearance: Occurs as a white powder.

Solubility: Soluble in water

Comments: Used as an emulsifier for pharmaceutical emulsions and aids solubilization.

18 Comments

Sucrose palmitate has shown a promising stabilizing effect on the droplet size of a soya bean oil nanoemulsion when used in combination with phosphatidyl choline and sodium palmitate.[12] It has also been investigated for the formulation of niosomes for peroral vaccine delivery.[13]

Ryoto sugar ester is widely used in foods such as wheat products, dairy products, bread and bakery products, confectionery, and packed beverages.

The EINECS number of sucrose palmitate is 247-706-7. The PubChem Compound ID (CID) for sucrose palmitate is 518683.

19 Specific References

1 Chen FJ, Patel MV. Clear-oil-containing pharmaceutical compositions. United States Patent 6,267,985; 2001.
2 Voorspoels JFM. Self-microemulsifying drug delivery systems of a HIV protease inhibitor. United States Patent 0104740A1; 2007.
3 Burton GW, Daroszweski J. Topical formulations for treatment of skin conditions. United States Patent 0025929A1; 2008.
4 Lee KK *et al.* . Controlled release composition comprising felodipine and method of preparation thereof. International Patent WO 105905; 2003.
5 Choi SH, Cho SW. Oral formulation for delivery of poorly absorbed drugs. United States Patent 7,666,466; 2010.
6 Santus G, Marcelloni L, Golzi R. Topical aciclovir formulations. International Patent WO 001390; 2000.
7 Szuts A, *et al.* Study of gel-forming properties of sucrose esters for thermosensitive drug delivery systems. *Int J Pharm* 2010; **383**: 132–137.
8 Ntawukulilyayo JD, *et al.* Microcrystalline cellulose-sucrose esters as tablet matrix forming agents. *Int J Pharm* 1995; **121**: 205–210.
9 US Food and Drug Adminstration. Code of Federal Regulations Title 21 (21 CFR 172.859): Sucrose fatty acid esters, April 2009.
10 FAO/WHO. Sucrose esters of fatty acids and sucroglycerides (WHO Food Additives Series 35). http://www.inchem.org/documents/jecfa/jecmono/v35je06.htm (accessed 29 February 2012).
11 FAO/WHO. Toxicological evaluation of certain food additives (WHO Food Additives Series 10). Twentieth report of the Joint FAO/WHO Expert Committee on Food Additives (JECFA). *World Health Organ Tech Rep Ser* 1976; No. 599. http://www.inchem.org/documents/jecfa/jecmono/v10je11.htm (accessed 29 February 2012).
12 Takegami S, *et al.* Preparation and characterization of a new lipid nano-emulsion containing two cosurfactants, sodium palmitate for droplet size reduction and sucrose palmitate for stability enhancement. *Chem Pharm Bull* 2008; **56**: 1097–1102.
13 Rentel CO, *et al.* Niosomes as a novel peroral vaccine delivery system. *Int J Pharm* 1999; **186**: 161–167.

20 General References

Stearinerie Dubois Fils. Product information: *DUB SE*. http://www.stearinerie-dubois.com/gammes/template_e.php?gamme=4 (accessed 29 February 2012).

Mitsubishi-Kagaku Foods Corporation. Product information: Sugar esters. http://www.mfc.co.jp/english/index.htm (accessed 29 February 2012).

Sisterna BV. Product information: *Sisterna*. http://www.sisterna.com (accessed 29 February 2012).

21 Authors

HC Shah, KK Singh.

22 Date of Revision

2 March 2012.

Sucrose Stearate

1 Nonproprietary Names

BP: Sucrose Stearate
PhEur: Sucrose Stearate
USP–NF: Sucrose Stearate

2 Synonyms

Crodesta F; *DUB SE*; E473; sacchari stearas; saccharose stearate; *Sisterna SP*; *Stelliesters SE 15S*; sucrose monostearate; sucrose monostearic acid ester; sucrose octadecanoate; sucrose stearate ester; *Surfhope SE*; *Tegosoft PSE*.

3 Chemical Name and CAS Registry Number

Sucrose monostearate [25168-73-4]
Sucrose distearate [27195-16-0]
Sucrose tristearate [27923-63-3]

4 Empirical Formula and Molecular Weight

$C_{30}H_{56}O_{12}$ (sucrose monostearate) 608.72
$C_{48}H_{90}O_{13}$ (sucrose distearate) 875.22
$C_{66}H_{124}O_{14}$ (sucrose tristearate) 1 141.68

Sucrose stearate, a sucrose fatty acid ester, is a nonionic surfactant with a sugar substituent, sucrose, as the polar head group and stearic acids as apolar groups. Sucrose has eight free hydroxyl groups and can therefore be esterified with up to eight fatty acids to form esters ranging from monoesters to octaesters.

The USP35–NF30 and PhEur 7.4 describe sucrose stearate as a mixture of sucrose esters, mainly sucrose stearate, obtained by transesterification of stearic acid methyl esters derived from vegetable origin with sucrose. It contains variable quantities of mono-, di-, tri,- and polyesters.

The mono- and diesters requirements differ for Type I and Type II sucrose stearate in the USP–NF, and Type I, Type II, and Type III sucrose stearate in the PhEur.

5 Structural Formula

OR
O
HO
O
O
OR
HO
OH
HO
OH

R = H or

6 Functional Category

Dispersing agent; emulsifying agent; modified-release agent; nonionic surfactant; solubilizing agent; tablet and capsule lubricant.

7 Applications in Pharmaceutical Formulation or Technology

Sucrose stearate is used as an emulsifying and dispersing agent, and as an aid for solubilization in pharmaceutical skin preparations. It coats the interfaces of an emulsion very efficiently, leading to emulsions that are exceptionally heat stable.

Sucrose stearate has been studied as a skin penetration enhancer to provide a useful vehicle for delivery of active ingredients to the skin.[1–6]

Sucrose stearate has been found to be a promising excipient for use in oral drug delivery formulations as a lubricant, disintegrant, and a controlled-release agent, and has been shown to sustain the release of drugs with various solubilities from matrix tablet systems.[7] It has also been shown to improve the plasticity and compressibility of powder mixtures. The sustained-release properties of sucrose stearate are attributed to its swelling characteristics, which could be further related to their chemical composition and hydrophilic–lipophilic properties.

Sucrose stearate, having various HLB values, may also be used to influence the rate of dissolution of drugs, where a low HLB value has been found to reduce the ejection force sufficiently during the tableting process, and may serve as an alternative to magnesium stearate. The addition of sucrose stearate with a high HLB value has been shown to contribute to the fast disintegration of tablets when formulated using a homogenous mixture of drug/sucrose stearate/microcrystalline cellulose and directly compressed with croscarmellose sodium and xylitol.[8,9] Sucrose distearate has been shown to act as an amphiphilic solubilizer that improves the solubilization capacity of discrete o/w or w/o type microemulsions.[10]

Sucrose stearate has demonstrated temperature-dependent viscoelastic gel-forming properties, which may be exploited for the formulation of thermosensitive controlled drug delivery systems.[11,12] The gelling of sucrose stearate is temperature- and concentration-dependent giving gel structures stronger than sucrose palmitate.[11] Sucrose stearate has also been explored as a droplet stabilizer for the preparation of polymeric microspheres.[13,14] Increasing concentrations of sucrose stearate resulted in increased surface porosity of the microspheres.[13]

Sucrose stearate has been used to develop controlled-release proniosome-derived niosomes for the nebulizable delivery of cromolyn sodium. The HLB values had a pronounced effect on the drug release and the nebulization efficiency.[15] *In-vitro* inhalation properties[16] and the fine particle fraction[17] of a dry powder inhalation drug formulation have been found to increase when the lactose carrier particles were surface covered with sucrose tristearate.

Sucrose stearate is used in many cosmetic formulations such as skin care creams, shampoos, shower gels, and toothpaste due to its foam quality, emolliency, and moisturizing properties. It is also used as an emulsifying agent in the food industry.

8 Description

Sucrose stearate occurs as a white or almost white, tasteless, unctuous powder.

9 Pharmacopeial Specifications

See Table I.

10 Typical Properties

Decomposition temperature 130–240°C for *DUB SE*.
HLB value Depends on the monoester content and ranges from 1–16; *see* Table II and Table III.
Melting point Approx. 60°C
Solubility Sparingly soluble in ethanol (96%); very slightly soluble in water. Soluble in tetrahydrofuran. *See also* Table IV.

11 Stability and Storage Conditions

Sucrose stearate should be stored at room temperature and protected from humidity.

S

Table I: Pharmacopeial specifications for sucrose stearate.

Test	PhEur 7.4	USP35–NF30
Identification	+	+
Acid value	≤6.0	≤6.0
Free sucrose	≤4%	≤4%
Water	≤4%	≤4%
Total ash	≤1.5%	≤1.5%
Composition of fatty acids		
Lauric acid	≤3%	≤3%
Myristic acid	≤3%	≤3%
Palmitic acid	25–40%	25–40%
Stearic acid	55–75%	55–75%
Sum of the contents of palmitic acid and stearic acid	≥90%	≥90%
Assay		
Type I		
Monoesters	≥50%	≥50%
Diesters	≤40%	≤40%
Sum of triesters and polyesters	≤25%	≤25%
Type II		
Monoesters	20–45%	20–45%
Diesters	30–40%	30–40%
Sum of triesters and polyesters	≤30%	≤30%
Type III		
Monoesters	15–25%	—
Diesters	30–45%	—
Sum of triesters and polyesters	35–50%	—

Table II: HLB values of commercially available grades of sucrose stearate (*Surfhope SE*).

Grade	Main fatty acid (%)	Ester composition (%)		HLB value
		Monoester	Di, tri, polyester	
Surfhope SE D-1803 F	≈70	≈20	≈80	3
Surfhope SE D-1805	70	30	70	5
Surfhope SE D-1807	70	40	60	7
Surfhope SE D-1809	70	50	50	9
Surfhope SE D-1811	70	55	45	11
Surfhope SE D-1811 F	70	55	45	11
Surfhope SE D-1815	70	70	30	15
Surfhope SE D-1816	70	75	25	16

Table III: HLB values of commercially available grades of sucrose stearate (*Sisterna SP*).

Grade	Monoester composition (%)	HLB value
Sisterna SP01-C	<1	1
Sisterna SP10-C	10	2
Sisterna SP30-C	30	6
Sisterna SP40-C	40	8
Sisterna SP50-C	50	11
Sisterna SP60-C	60	13
Sisterna SP70-C	70	15

12 Incompatibilities

—

13 Method of Manufacture

Sucrose stearate is obtained by transesterification of stearic acid methyl esters derived from vegetable origin with sucrose. The manufacture of the fatty acid methyl esters includes a distillation step.

Ethyl acetate or methyl ethyl ketone or dimethyl sulfoxide and isobutyl alcohol (2-methyl-1-propanol) may be used in the preparation of sucrose fatty acid esters.[18]

Table IV: Solubility of sucrose stearate (*DUB SE*).[(a)]

Solvent	Grade		
	DUB SE 11S	DUB SE 5S	DUB SE 3S
Ethanol			
25°C	I	I	I
75°C	S	S	S
Glycerin			
25°C	D	D	D
75°C	D	D	D
Oil			
25°C	D	D	D
75°C	I	I	S
Water			
25°C	D	I	I
75°C	S	S	D

(a) I= insoluble; D= dispersible; S= soluble.

14 Safety

Sucrose stearate is a nontoxic, nonionic carbohydrate-based surfactant, widely used in the pharmaceutical, food and cosmetic industries.

The Joint FAO/WHO Expert Committee on Food Additives (JECFA) has reviewed the safety of sucrose fatty acid esters and recommended an acceptable daily intake of up to 30 mg/kg. Sucrose fatty acid esters may have a laxative effect at higher doses.[19]

Acute toxicity performed in rats and mice with an oral dose totalling 20 000 mg/kg, administered in 10 equally divided doses at 30 to 60 min intervals produced no deaths.[20] Sucrose stearate is nonirritant to the skin and slightly irritant to the eyes.

LD_{50} (mice, oral): >2.0 g/kg
NOAEL (rat): approx. 2.0 g/kg/day[21]

15 Handling Precautions

Observe normal precautions appropriate to the circumstances and quantity of material handled. Dust formation should be avoided. Eye protection and gloves are recommended.

16 Regulatory Status

Accepted for use as a food additive in the US, Europe, and Japan. Included in the FDA Inactive Ingredients Database (oral capsules, tablets for extended-release or sustained action); sucrose distearate (topical emulsions and creams). Included in nonparenteral medicines licensed in the UK. Sucrose stearate, sucrose distearate, and sucrose tristearate are included in the Canadian Natural Health Products Ingredients Database.

17 Related Substances

Sucrose monolaurate; sucrose palmitate.

18 Comments

Sucrose fatty acid esters, such as sucrose stearate, have been shown to improve the activity of lipase inhibitors.[18]

Only sucrose mono, di and tri- stearates are available commercially. *Ryoto* sugar ester is widely used in foods such as wheat products, dairy products, bread and bakery products, confectionery and packed beverages.

A specification for sucrose esters of fatty acids is contained in the *Japanese Pharmaceutical Excipients* (JPE).[22]

The EINECS numbers for sucrose stearate are 246-705-9 (monostearate), 248-317-5 (distearate), and 248-731-6 (tristearate). The PubChemID numbers for sucrose stearate are 5360773 (monostearate) and 5360827 (distearate).

19 Specific References

1 Csóka G, *et al.* 2007; Application of sucrose fatty acid esters in transdermal therapeutic systems. *Eur J Pharm Biopharm*65(2): 233–237.
2 Klang V, *et al.* 2010; Enhancement of stability and skin permeation by sucrose stearate and cyclodextrins in progesterone nanoemulsions. *Int J Pharm*393: 152–160.
3 Chen FJ, Patel MV. Clear-oil-containing pharmaceutical compositions. United States Patent 09/877541 2004.
4 Burton GW, Daroszweski J. Topical formulations for treatment of skin conditions. United States Patent 12771074 2010.
5 Choi SH, Cho SW. Oral formulation for delivery of poorly absorbed drugs. United States Patent 7666446 2010.
6 Mercier MF *et al.* Multi-lamellar liquid crystal emulsion system. European Patent 20050738399 2007.
7 Chansanroj K, Betz G. 2010; Sucrose esters with various hydrophilic–lipophilic properties: novel controlled release agents for oral drug delivery matrix tablets prepared by direct compaction. *Acta Biomater*6: 3101–3109.
8 Koseki T, *et al.* 2009; Preparation and evaluation of novel directly-compressed fast-disintegrating furosemide tablets with sucrose stearic acid ester. *Biol Pharm Bull*32(6): 1126–1130.
9 Koseki T, *et al.* 2008; Development of novel fast-disintegrating tablets by direct compression using sucrose stearic acid ester as a disintegration-accelerating agent. *Chem Pharm Bull*56(10): 1384–1388.
10 Aramaki K, *et al.* 2001; Effect of adding an amphiphilic solubilization improver, sucrose distearate, on the solubilization capacity of nonionic microemulsions. *J Colloid Interface Sci*236(1): 14–19.
11 Szuts A, *et al.* 2010; Study of gel-forming properties of sucrose esters for thermosensitive drug delivery systems. *Int J Pharm*383(12): 132–137.
12 Ntawukulilyayo JD, *et al.* 1995; Microcrystalline cellulose-sucrose esters as tablet matrix-forming agents. *Int J Pharm*121: 205–210.
13 Yüksel N, Baykara T. 1997; Preparation of polymeric microspheres by the solvent evaporation method using sucrose stearate as a droplet stabilizer. *J Microencapsul*14(6): 725–733.
14 Squillante E, *et al.* 2003; Microencapsulation of beta-galactosidase with Eudragit L-100. *J Microencapsul*20(2): 153–167.
15 Abd-Elbary A, *et al.* 2008; Sucrose stearate-based proniosome-derived niosomes for the nebulisable delivery of cromolyn sodium. *Int J Pharm*357(12): 189–198.
16 Iida K, *et al.* 2003; Effect of surface covering of lactose carrier particles on dry powder inhalation properties of salbutamol sulfate. *Chem Pharm Bull*51(12): 1455–1457.
17 Kumon M, *et al.* 2006; Novel approach to DPI carrier lactose with mechanofusion process with additives and evaluation by IGC. *Chem Pharm Bull*54(11): 1508–1514.
18 US Food and Drug Adminstration. Code of Federal Regulations Title 21 (21 CFR 172.859): Sucrose fatty acid esters, April 2009.
19 FAO/WHO. Sucrose esters of fatty acids and sucroglycerides (WHO Food Additives Series 40). http://www.inchem.org/documents/jecfa/jecmono/v040je04.htm (accessed 7 July 2011).
20 FAO/WHO. Toxicological evaluation of certain food additives (WHO Food additives series 10) http://www.inchem.org/documents/jecfa/jecmono/v10je11.htm (accessed 7 July 2011).
21 FAO/WHO. Sucrose esters of fatty acids and sucroglycerides (WHO Food Additives Series 35). http://www.inchem.org/documents/jecfa/jecmono/v35je06.htm (accessed 7 July 2011).
22 Japan Pharmaceutical Excipients Council. *Japanese Pharmaceutical Excipients 2004*. Tokyo: Yakuji Nippo, 2004: 883–884.

20 General References

Croda Inc. Product information: www.crodausa.com (accessed 7 July 2011).
Evonik Goldschmidt GmbH. Product information: www.evonik.com/personal-care (accessed 7 July 2011).
Mitsubishi-Kagaku Foods Corporation. Product information: Sugar esters. http://www.mfc.co.jp/english/index.htm (accessed 7 July 2011).
Stearinerie Dubois Fils. Product information: *DUB SE*. http://www.stearinerie-dubois.com/gammes/template_e.php?gamme=4 (accessed 7 July 2011).
Sisterna BV. Product information: *Sisterna*. http://www.sisterna.com (accessed 7 July 2011).

21 Authors

HC Shah, KK Singh.

22 Date of Revision

2 March 2012.

 Sugar, Compressible

S

1 Nonproprietary Names

BP: Compressible Sugar

USP–NF: Compressible Sugar

2 Synonyms

Compressuc; *Comprima*; *Di-Pac*; direct compacting sucrose; directly compressible sucrose; *Nu-Tab*.

3 Chemical name and CAS Registry Number

See Sections 4 and 18.

4 Empirical Formula and Molecular Weight

The BP 2012 and USP35–NF30 state that compressible sugar contains not less than 95.0% and not more than 98.0% of sucrose ($C_{12}H_{22}O_{11}$). It may contain starch, maltodextrin, or invert sugar, and may contain a suitable lubricant.

5 Structural Formula

See Section 4.

6 Functional Category

Sweetening agent; tablet and capsule binder; tablet and capsule diluent.

7 Applications in Pharmaceutical Formulation or Technology

Compressible sugar is used primarily in the preparation of direct-compression chewable tablets. Its tableting properties can be influenced by changes in moisture level;[1] *see* Table I for typical uses.

8 Description

Compressible sugar is a free-flowing, sweet-tasting, white powder (or crystalline agglomerates).

SEM 1: Excipient: compressible sugar; manufacturer: SPI Polyols; magnification: 100×. Reproduced from Bowe KE, 1998[2] with permission.

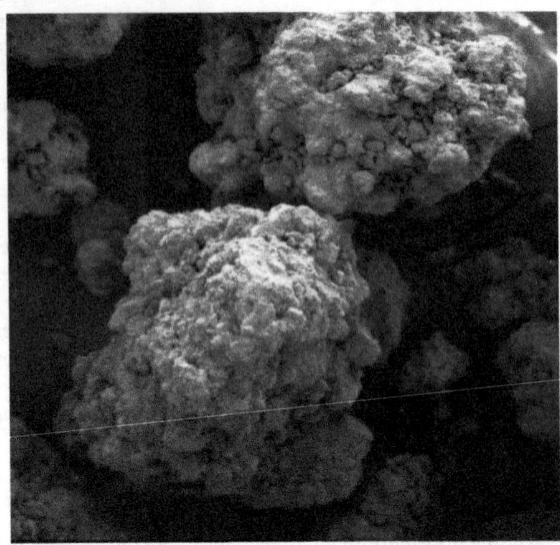

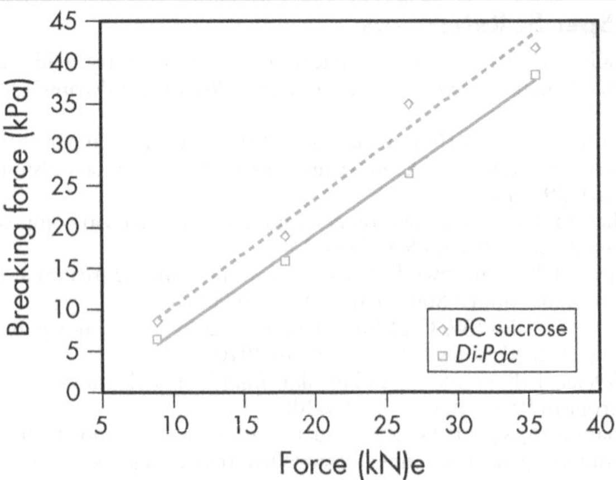

Figure 1: Tablet breaking force as a function of compaction force for *Di-Pac* (SPI Polyols) and directly compressible (DC) sucrose. Compaction conditions: 1.5 g, 5/8 inch flat-faced bevelled edge, 0.5% magnesium stearate. Adapted from Bowe KE, 1998[2] with permission.

Table I: Uses of compressible sugar.

Use	Concentration (wt%)
Dry binder in tablet formulations	5–20
Filler in chewable tablets	20–60
Filler in tablets	20–60
Sweetener in chewable tablets	10–50

Table II: Pharmacopeial specifications for compressible sugar.

Test	BP 2012	USP35–NF30
Identification	+	+
Calcium	+	+
Chloride	≤125 ppm	≤0.014%
Lead	≤0.5 ppm	—
Heavy metals	—	≤5 ppm
Loss on drying	0.25–1.0%	≤1.0%
Residue on ignition	—	≤0.1%
Microbial limits	—	+
Sulfate	≤100 ppm	≤0.010%
Assay	95.0–98.0%	95.0–98.0%

9 Pharmacopeial Specifications

See Table II.

10 Typical Properties

Compaction profile *see* Figure 1.
Flowability Free flowing
Density (bulk) 0.609–0.673 g/cm^3 for *Di-Pac.*
Hygroscopicity Moisture content of *Di-Pac* depends on factors such as temperature, pressure, relative humidity, and time of exposure to a given environment.[3] Up to 80% relative humidity at 20°C, water uptake is typically <1% w/w.
Melting point 186°C for *Di-Pac.*
Particle size distribution For *Di-Pac*, 3% maximum retained on a #40 (425 μm) mesh; 75% minimum through a #100 (150 μm) mesh; 5% maximum through #200 (75 μm) mesh.
Solubility The sucrose portion is soluble in water.
Specific surface area 0.13–0.14 m^2/g for *Di-Pac.*

11 Stability and Storage Conditions

Compressible sugar is physically stable at room temperature and low relative humidity. It deliquesces at above 80% relative humidity at 25°C. The bulk material should be stored in a well-closed container in a cool, dry place.

12 Incompatibilities

Incompatible with dilute acids, which cause hydrolysis of sucrose to invert sugar, and with alkaline earth hydroxides, which react with sucrose to form sucrates.

13 Method of Manufacture

Compressible sugar is prepared by cocrystallization of sucrose with other excipients such as maltodextrin.[1] Compressible sugar may also be prepared using a dry granulation process or fluid bed granulation process.

14 Safety

Compressible sugar is generally regarded as a relatively nontoxic and nonirritant material. *See also* Sucrose.

15 Handling Precautions

Observe normal precautions appropriate to the circumstances and quantity of material handled. *See also* Sucrose.

16 Regulatory Status

Included in the FDA Inactive Ingredients Database (oral capsules and tablets). Included in nonparenteral medicines licensed in the UK. Included in the Canadian Natural Health Products Ingredients Database.

17 Related Substances

Confectioner's sugar; sucrose; sugar spheres; *Sugartab.*

Sugartab

Appearance *Sugartab* (JRS Pharma) is a compressible sugar that does not conform to the USP35–NF30 specification. It is an agglomerated sugar product containing approximately 90–93% sucrose, the balance being invert sugar.
Density (bulk) 0.60 g/cm^3
Density (tapped) 0.69 g/cm^3
Flowability 42.7 g/s
Moisture content 0.20–0.57%.
Particle size distribution 30% through a #20 (850 μm) mesh; 3% through a #30 (600 μm) mesh.

S

18 Comments

—

19 Specific References

1 Rizzuto AB, *et al.* Modification of the sucrose crystal structure to enhance pharmaceutical properties of excipient and drug substances. *Pharm Technol* 1984; 8(9): 32, 34, 36, 38–39.
2 Bowe KE. Recent advances in sugar-based excipients. *Pharm Sci Technol Today* 1998; **1**(4): 166–173.
3 Tabibi SE, Hollenbeck RG. Interaction of water vapor and compressible sugar. *Int J Pharm* 1984; **18**: 169–183.

20 General References

Domino Foods Inc. Dominio Specialty Ingredients. Product literature: *Di-Pac*, July 2008.
JRS Pharma. Technical literature: *Sugartab*, 2003.
Mendes RW, *et al.* Nu-tab as a chewable direct compression carrier. *Drug Cosmet Ind* 1974; **115**(6): 42–46130–133.
Ondari CO, *et al.* Comparative evaluation of several direct compression sugars. *Drug Dev Ind Pharm* 1983; **9**: 1555–1572.
Ondari CO, *et al.* Comparative evaluation of several direct compression sugars. *Drug Dev Ind Pharm* 1988; **14**: 1517–1527.
Shangraw RF, *et al.* Morphology and functionality in tablet excipients for direct compression. *Pharm Technol* 1981; **5**: 69–78.
Chow K, *et al.* Engineering of pharmaceutical materials: an industrial perspective. *J Pharm Sci* 2008; **97**: 2855–2877.
Marwaha M, *et al.* Coprocessing of excipients: a review on excipient development for improved tableting performance. *Int J Appl Pharm* 2010; **2**: 41–47.

21 Author

CC Sun.

22 Date of Revision

2 March 2012.

Sugar, Confectioner's

1 Nonproprietary Names

USP–NF: Confectioner's Sugar

2 Synonyms

Icing sugar; powdered sugar.

3 Chemical Name and CAS Registry Number

See Section 4.

4 Empirical Formula and Molecular Weight

The USP35–NF30 describes confectioner's sugar as a mixture of sucrose ($C_{12}H_{22}O_{11}$) and corn starch that has been ground to a fine powder; it contains not less than 95.0% sucrose calculated on the dried basis.

5 Structural Formula

See Section 4 and Sucrose.

6 Functional Category

Coating agent; sweetening agent; tablet and capsule diluent; viscosity-increasing agent.

7 Applications in Pharmaceutical Formulation or Technology

Confectioner's sugar is used in pharmaceutical formulations when a rapidly dissolving form of sugar is required for flavoring or sweetening. It is used as a diluent in solid-dosage formulations when a small particle size is necessary to achieve content uniformity in blends with finely divided active ingredients. In solutions, at high concentrations (70% w/v), confectioner's sugar provides increased viscosity along with some preservative effects. Confectioner's sugar is also used in the preparation of sugar-coating solutions and in wet granulations as a binder/diluent. *See* Table I.

Table I: Uses of confectioner's sugar.

Use	Concentration (%)
Sweetening agent in tablets	10–20
Tablet diluent	10–50

See also Section 18.

8 Description

Confectioner's sugar occurs as a sweet-tasting, fine, white, odorless powder.

SEM 1: Excipient: confectioner's sugar; manufacturer: Frost; lot no.: 101A-1; magnification: 60×; voltage: 20 kV.

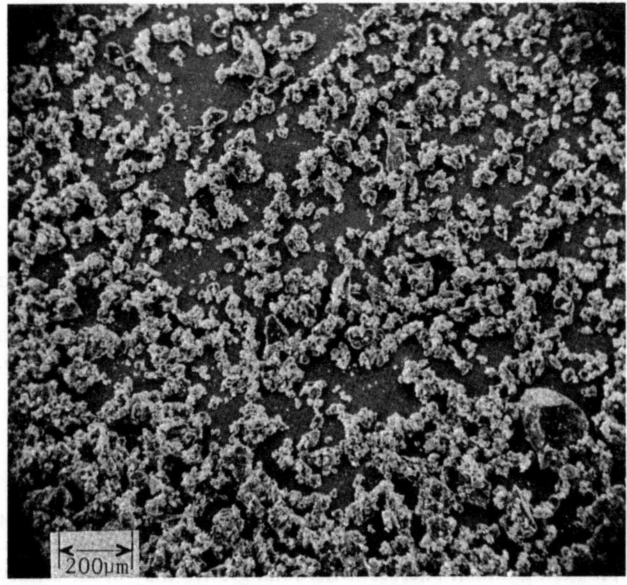

SEM 2: Excipient: confectioner's sugar; manufacturer: Frost; lot no.: 101A-1; magnification: 600×; voltage: 20 kV.

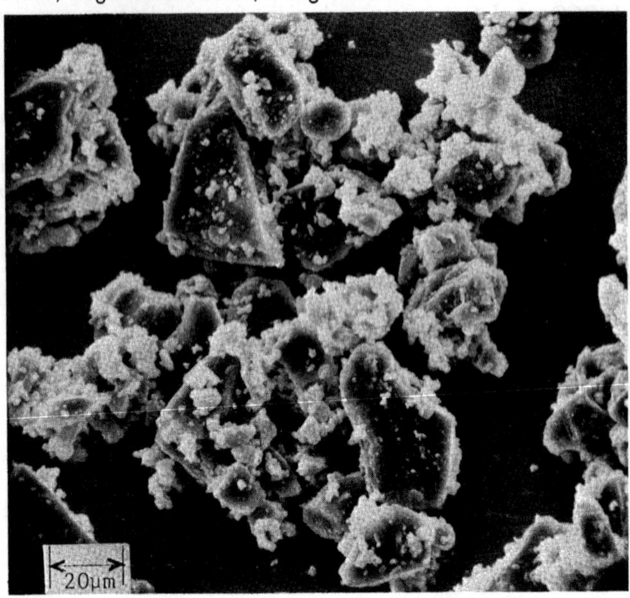

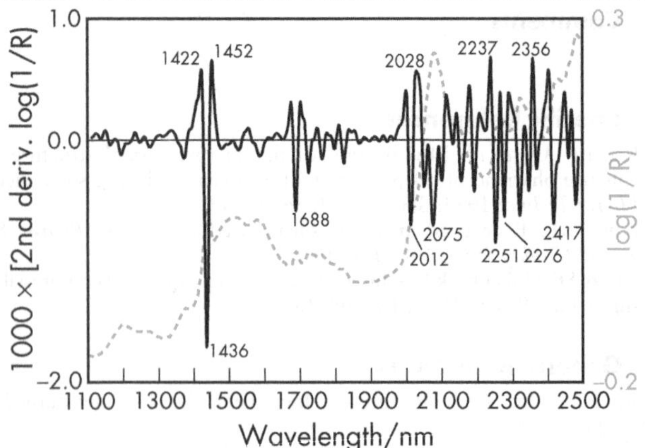

Figure 1: Near-infrared spectrum of confectioner's sugar measured by reflectance.

9 Pharmacopeial Specifications

See Table II.

Table II: Pharmacopeial specifications for confectioner's sugar.

Test	USP35–NF30
Identification	+
Chloride	≤0.014%
Calcium	+
Heavy metals	≤5 ppm
Loss on drying	≤1.0%
Microbial limits	+
Residue on ignition	≤0.08%
Specific rotation	≥+62.6°
Sulfate	≤0.006%
Assay	≤95.0%

S

10 Typical Properties

Density (bulk) 0.465 g/cm³
Density (tapped) 0.824 g/cm³
Moisture content 0.1–0.31%
Particle size distribution Various grades with different particle sizes are commercially available, e.g. 6X, 10X, and 12X grades of confectioner's sugar from the Domino Sugar Corp. Mean particle size is 14.3 μm.

For 6X, 94% through a #200 (75 μm) mesh.

For 10X, 99.9% through a #100 (150 μm) mesh and 97.5% through a #200 (75 μm) mesh.

For 12X, 99% through a #200 (75 μm) mesh and 96% through a #325 (45 μm) mesh.

Solubility The sucrose portion is water-soluble while the starch portion is insoluble in water, although it forms a cloudy solution.

Spectroscopy

NIR spectra *see* Figure 1.

11 Stability and Storage Conditions

Confectioner's sugar is stable in air at moderate temperatures but may caramelize and decompose above 160°C. It is more hygroscopic than granular sucrose. Microbial growth may occur on dry storage if adsorbed moisture is present or in dilute aqueous solutions.

Confectioner's sugar should be stored in a well-closed container in a cool, dry place.

12 Incompatibilities

Confectioner's sugar is incompatible with dilute acids, which cause the hydrolysis of sucrose to invert sugar. It is also incompatible with alkaline earth hydroxides, which react with sucrose to form sucrates.

13 Method of Manufacture

Confectioner's sugar is usually manufactured by grinding refined granulated sucrose with corn starch to produce a fine powder. Other anticaking agents, such as tricalcium phosphate and various silicates, have also been used but are less common.

14 Safety

Confectioner's sugar is used in confectionery and oral pharmaceutical formulations. It is generally regarded as a relatively nontoxic and nonirritant material. *See also* Sucrose.

15 Handling Precautions

Observe normal precautions appropriate to the circumstances and quantity of material handled. *See also* Sucrose.

16 Regulatory Status

Included in the FDA Inactive Ingredients Database (capsules and tablets). Included in the Canadian Natural Health Products Ingredients Database.

17 Related Substances

Compressible sugar; sucrose; sugar spheres.

18 Comments

Confectioner's sugar is not widely used in pharmaceutical formulations because the poor flow characteristics prevent its use in direct-compression blends. However, confectioner's sugar is used when a smooth mouth feel or a rapidly dissolving sweetener is required, and when a milled/micronized active ingredient must be blended with a diluent of similar particle size for powders or wet granulations.

Low-starch grades of confectioner's sugar containing 0.01% w/w starch are also commercially available.

19 Specific References

—

20 General References

Barry RH, *et al.* Stability of phenylpropanolamine hydrochloride in liquid formulations containing sugars. *J Pharm Sci* 1982; **71**: 116–118.

Czeisler JL, Perlman KP. Diluents. In: Swarbrick J, Boylan JC, eds. *Encyclopedia of Pharmaceutical Technology.* 4. New York: Marcel Dekker, 1988; 37–84.

Edwards WP. *The Science of Sugar Confectionery.* Cambridge: Royal Society of Chemistry, 2000.

Jackson EB, ed. *Sugar Confectionery Manufacture.* Glasgow: Blackie, 1990.

Onyekweli AO, Pilpel N. Effect of temperature changes on the densification and compression of griseofulvin and sucrose powders. *J Pharm Pharmacol* 1981; **33**: 377–381.

Wolraich ML, *et al.* Effects of diets high in sucrose or aspartame on the behavior and cognitive performance of children. *N Engl J Med* 1994; **330**: 301–307.

21 Author

AH Kibbe.

22 Date of Revision

2 March 2012.

Sugar Spheres

1 Nonproprietary Names

BP: Sugar Spheres
PhEur: Sugar Spheres
USP–NF: Sugar Spheres

2 Synonyms

Non-pareil; non-pareil seeds; *Nu-Core*; *Nu-Pareil PG*; sacchari sphaerae; sugar seeds; *Suglets*.

3 Chemical Name and CAS Registry Number

—

4 Empirical Formula and Molecular Weight

See Section 8.

5 Structural Formula

See Section 8.

6 Functional Category

Tablet and capsule diluent.

7 Applications in Pharmaceutical Formulation or Technology

Sugar spheres are mainly used as inert cores in capsule and tablet formulations, particularly multiparticulate sustained-release formulations.(1–4) They form the base upon which a drug is coated, usually followed by a release-modifying polymer coating.

Alternatively, a drug and matrix polymer may be coated onto the cores simultaneously. The active drug is released over an extended period either via diffusion through the polymer or through to the controlled erosion of the polymer coating.

Complex drug mixtures contained within a single-dosage form may be prepared by coating the drugs onto different batches of sugar spheres with different protective polymer coatings.

8 Description

The USP35–NF30 describes sugar spheres as approximately spherical granules of a labeled nominal-size range with a uniform diameter and containing not less than 62.5% and not more than 91.5% of sucrose, calculated on the dried basis. The remainder is chiefly starch.

The PhEur 7.4 states that sugar spheres contain not more than 92% of sucrose calculated on the dried basis. The remainder consists of corn (maize) starch and may also contain starch hydrolysates and color additives. The diameter of sugar spheres varies from 200 to 2000 μm, and the upper and lower limits of the size of the sugar spheres are stated on the label.

9 Pharmacopeial Specifications

See Table I.

Table I: Pharmacopeial specifications for sugar spheres.

Test	PhEur 7.4	USP35–NF30
Identification	+	+
Heavy metals	≤5 ppm	≤5 μg/g
Loss on drying	≤5.0%	≤4.0%
Microbial limits	+	+
Particle size distribution	+	+
Residue on ignition	—	≤0.25%
Sulfated ash	≤0.2%	—
Specific rotation	—	+41° to +61°
Sucrose (dried basis)	≤92%	62.5–91.5%

10 Typical Properties

Density

1.57–1.59 g/cm^3 for *Suglets* less than 500 μm in size;

1.55–1.58 g/cm^3 for *Suglets* more than 500 μm in size.

Flowability <10 seconds, free flowing.

Particle size distribution Sugar spheres are of a uniform diameter. The following sizes are commercially available from various suppliers (US standard sieves):

45–60 mesh (250–355 μm)
40–50 mesh (300–425 μm)
35–45 mesh (355–500 μm)
35–40 mesh (420–500 μm)
30–35 mesh (500–600 μm)
25–30 mesh (610–710 μm)
20–25 mesh (710–850 μm)

18–20 mesh (850–1000 μm)
16–20 mesh (850–1180 μm)
14–18 mesh (1000–1400 μm)

Solubility Solubility in water varies according to the sucrose-to-starch ratio. The sucrose component is freely soluble in water, whereas the starch component is practically insoluble in cold water.

Specific surface area
0.1–0.2 m^2/g for *Suglets* less than 500 μm in size;
>0.2 m^2/g for *Suglets* more than 500 μm in size.

11 Stability and Storage Conditions

Sugar spheres are stable when stored in a well-closed container in a cool, dry place.

12 Incompatibilities

See Starch and Sucrose for information concerning the incompatibilities of the component materials of sugar spheres.

13 Method of Manufacture

Sugar spheres are prepared from crystalline sucrose, which is coated using sugar syrup and a starch dusting powder.

14 Safety

Sugar spheres are used in oral pharmaceutical formulations. The sucrose and starch components of sugar spheres are widely used in edible food products and oral pharmaceutical formulations.

The adverse reactions and precautions necessary with the starch and sucrose components should be considered in any product containing sugar spheres. For example, sucrose is generally regarded as more cariogenic than other carbohydrates, and in higher doses is also contraindicated in diabetic patients.

See Starch and Sucrose for further information.

15 Handling Precautions

Observe normal precautions appropriate to the circumstances and quantity of material handled.

16 Regulatory Status

Included in the FDA Inactive Ingredients Database (oral capsules and tablets). Included in nonparenteral medicines licensed in the UK and Europe. The sucrose and starch components of sugar spheres are individually approved for use as food additives in Europe and the USA.

17 Related Substances

Compressible sugar; confectioner's sugar; microcrystalline cellulose spheres; *NPTAB*; starch; sucrose.

Microcrystalline cellulose spheres

Comments Microcrystalline cellulose spheres are prepared from microcrystalline cellulose by spherization. They are available from several manufacturers and in different sizes, and the size grades can vary between the different manufacturers. Typical size grades available are:
120–230 mesh (63–125 μm);
70–140 mesh (106–212 μm);
30–100 mesh (150–300 μm);
45–70 mesh (212–355 μm);
35–45 mesh (355–500 μm);
25–35 mesh (500–710 μm);
18–25 mesh (710–1000 μm);
14–18 mesh (1000–1400 μm).

NPTAB

Appearance *NPTAB* (NP Pharm) is a compressible sugar that does not conform to the USP35–NF30 specification. It is an agglomerated sugar product containing not more than 92% sucrose, the balance being corn (maize) starch.

Density 1.55–1.59 g/cm^3 (varies with particle size)

Flowability <10% (Carr compressibility index)

Particle size distribution
NPTAB 190 (180–212 μm);
NPTAB 220 (212–250 μm);
NPTAB 270 (250–300 μm);
NPTAB 320 (300–350 μm).

18 Comments

In addition, sugar spheres have been used as inert cores in sustained-release formulations, which also contain kaolin and talc.

Sugar spheres are also used in confectionery products.

19 Specific References

1 Narsimhan R, *et al.* Timed-release noscapine microcapsules. *Indian J Pharm Sci* 1988; 50: 120–122.
2 Bansal AK, Kakkar AP. Solvent deposition of diazepam over sucrose pellets. *Indian J Pharm Sci* 1990; 52: 186–187.
3 Ho H-O, *et al.* The preparation and characterization of solid dispersions on pellets using a fluidized-bed system. *Int J Pharm* 1996; 139: 223–229.
4 Miller RA, *et al.* The compression of spheres coated with an aqueous ethylcellulose dispersion. *Drug Dev Ind Pharm* 1999; 25(4): 503–511.

20 General References

Birch GG, Parker KJ, eds. *Sugar: Science and Technology.* London: Applied Science Publications, 1979.
NP Pharm. Product data sheet: *NPTAB*, 2008.

21 Author

RC Moreton.

22 Date of Revision

2 March 2012.

S

Sulfobutylether β-Cyclodextrin

1 Nonproprietary Names

None adopted.

2 Synonyms

Betadex sulfobutyl ether sodium; beta-cyclodextrin sulfobutylethers, sodium salts; β-cyclodextrin sulfobutylethers, sodium salts; *Captisol*; $(SBE)_m$-beta-CD; SBE7-β-CD; SBECD; sulfobutylether-β-cyclodextrin, sodium salt.

3 Chemical Name and CAS Registry Number

β-Cyclodextrin sulfobutylether, sodium salt [182410-00-0]

4 Empirical Formula and Molecular Weight

$C_{42}H_{70-n}O_{35}\cdot(C_4H_8SO_3Na)_n$ 2 163 (where n = approximately 6.5)

β-Cyclodextrin is a cyclic oligosaccharide containing seven D-(+)-glucopyranose units attached by α(1→4) glucoside bonds (*see* Cyclodextrins). Sulfobutylether β-cyclodextrin is an anionic β-cyclodextrin derivative with a sodium sulfonate salt separated from the hydrophobic cavity by a butyl spacer group. The substituent is introduced at positions 2, 3, and 6 in at least one of the glucopyranose units in the cyclodextrin structure. Introducing the sulfobutylether (SBE) into β-cyclodextrin can produce materials with different degrees of substitution, theoretically from 1 to 21; the hepta-substituted preparation (SBE7-β-CD) is the cyclodextrin with the most desirable drug carrier properties.[1]

5 Structural Formula

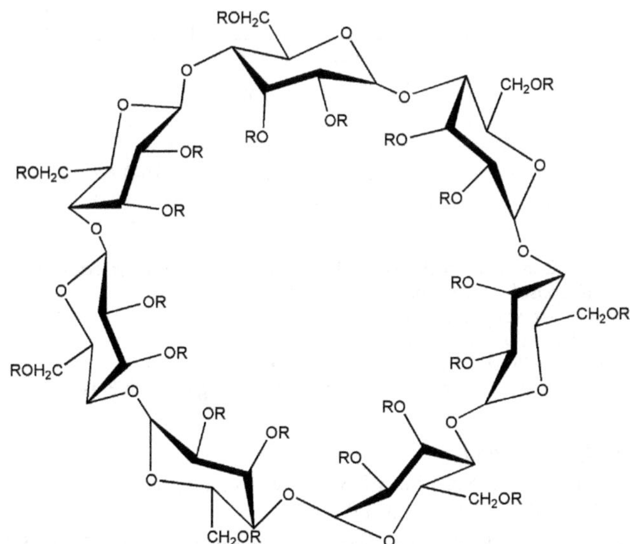

R = H_{21-n} or $(CH_2CH_2CH_2CH_2SO_2ONa)_n$ where n = 6.2–6.9

Note: the substitution pattern is random, yielding a heterogeneous mixture both in terms of the site of substitution as well as degree of substitution. The n value is an average number derived from the average degree of substitution.

6 Functional Category

Complexing agent; osmotic agent; solubilizing agent; stabilizing agent; tablet and capsule diluent; taste-masking agent; viscosity-increasing agent.

7 Applications in Pharmaceutical Formulation or Technology

Cyclodextrins are crystalline, nonhygroscopic, cyclic oligosaccharides derived from starch (*see* Cyclodextrins). Sulfobutylether β-cyclodextrin is an amorphous, anionic substituted β-cyclodextrin derivative (*see* Section 8); other substituted cyclodextrin derivatives are also available (*see* Section 17).

Sulfobutylether β-cyclodextrin can form noncovalent complexes with many types of compounds including small organic molecules, peptides,[2] and proteins.[3] It can also enhance their solubility[4,5] and stability[5–7] in water. Sulfobutylether β-cyclodextrin has primarily been used in parenteral drug products, but it can also be used in oral solid[8,9] and liquid[10] dosage forms, and ophthalmic,[11,12] inhalation, and intranasal formulations[13] for improved bioavailability and efficient drug delivery. Sulfobutylether β-cyclodextrin can also function as an osmotic agent and/or a solubilizer for controlled-release delivery,[9] and has antimicrobial preservative properties when present at sufficient concentrations.[14]

The amount of sulfobutylether β-cyclodextrin that may be used is dependent on the purpose for inclusion in the formulation, the route of administration, and the ability of the cyclodextrin to complex with the drug being delivered. The minimum amount required for solubilization is, in general, a cyclodextrin/drug molar ratio of approximately 1–5 (the exact ratio being experimentally determined from complexation data). The maximum use in a formulation may be limited by physicochemical constraints such as viscosity (e.g. syringeable concentrations may be considered up to 50% w/v), tonicity, or the total weight and size of solid dosage forms (e.g. less than a gram in an individual tablet). It may also be limited by pharmacokinetic/pharmacodynamic (PK/PD) considerations. As dilution of a cyclodextrin formulation leads to an increase in the amount of uncomplexed drug, formulations that are not diluted upon administration, such as ophthalmic formulations, are sensitive to cyclodextrin concentration. In formulations such as these, cyclodextrin concentrations greater than the minimum required for solubilization can reduce the availability of uncomplexed drug and thereby affect PK/PD expectations by producing effects such as slower onset, lower C_{max}, and bioavailability.

8 Description

Sulfobutylether β-cyclodextrin occurs as a white amorphous powder.

9 Pharmacopeial Specifications

—

10 Typical Properties

Acidity/alkalinity pH = 4.0–6.8 (30% w/v aqueous solution)[1]

Angle of repose

20.5° for freeze-dried *Captisol*;

31.6° for spray-dried *Captisol*.

Appearance/clarity of solution A 30% w/v solution in water is clear and essentially free from particles of foreign matter.

Average degree of substitution 6.2–6.9[1]

Compressibility *see* Figure 1.

Density (bulk)

0.446–0.482 g/cm³ for freeze-dried *Captisol*;

0.524 g/cm³ for spray-dried *Captisol*;

0.482 g/cm³ for spray-agglomerated *Captisol*.

SEM 1: Excipient: freeze-dried sulfobutylether β-cyclodextrin sodium (*Captisol*); manufacturer: CyDex Pharmaceuticals; magnification: 150×; voltage 15 kV.

SEM 2: Excipient: spray-dried sulfobutylether β-cyclodextrin sodium (*Captisol*); manufacturer: CyDex Pharmaceuticals; magnification: 150×; voltage 15 kV.

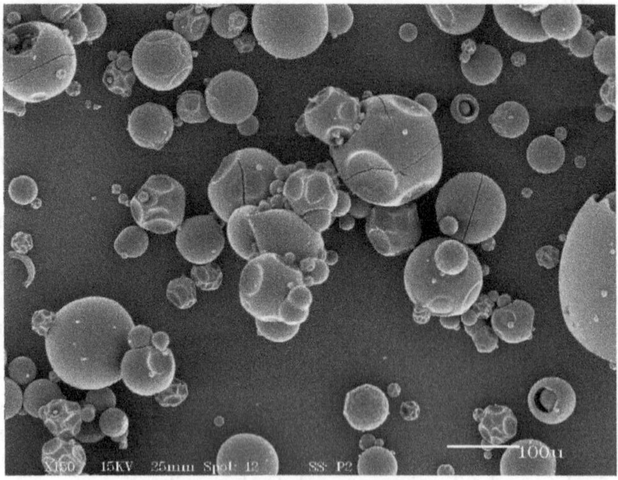

SEM 3: Excipient: spray-agglomerated sulfobutylether β-cyclodextrin sodium (*Captisol*); manufacturer: CyDex Pharmaceuticals; magnification: 150×; voltage 15 kV.

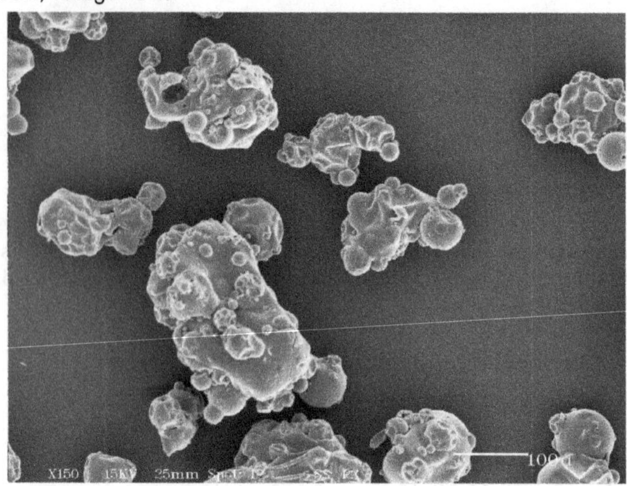

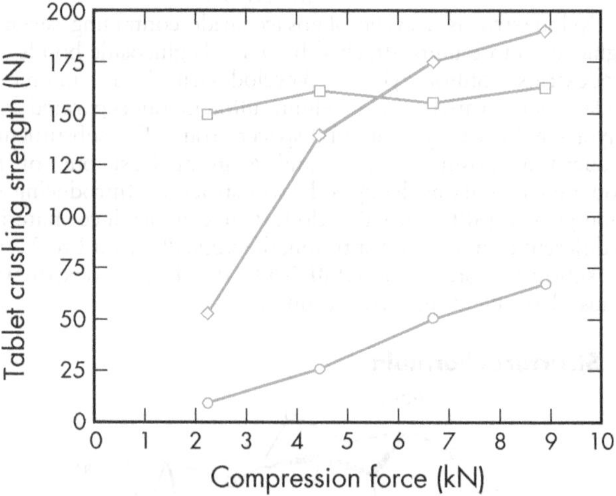

Figure 1: Compression characteristics of sulfobutylether β-cyclodextrin sodium (*Captisol*, CyDex Pharmaceuticals).
Mean tablet weight: 220 mg
Tablet dimensions: 5/16 inch std concave
Lubricated with 0.5% magnesium stearate
Tablet machine: Instrumented Stokes Model F, Single Punch Press

Density (tapped)
0.565–0.597 g/cm^3 for freeze-dried *Captisol*;
0.624 g/cm^3 for spray-dried *Captisol*;
0.595 g/cm^3 for spray-agglomerated *Captisol*.

Flowability 50 g/s for freeze-dried *Captisol*.

Glass transition temperature −25°C[15]

Hygroscopicity Reversibly picks up water at relative humidities (RH) up to 60%. Equilibration at RH equal to or above 60% will result in deliquescence and a water content of approximately 16% w/w.
See Figure 2.

Melting point Decomposition at 275°C.

Moisture content 3–6% typically; maximum 10%.

Osmolarity Solutions of *Captisol* in the range 9.5–11.4% w/v are isoosmotic with serum.[1]

Particle size distribution Typical mean particle size for spray-dried sulfobutylether β-cyclodextrin sodium is 70–120 μm. Various processing and handling methods may result in different nominal mean particle sizes.

Specific rotation $[\alpha]_D^{20}$ = +94° ± 3°

Solubility Soluble 1 in less than 1 of water; 1 in 30–40 of methanol; practically insoluble in ethanol, *n*-hexane, 1-butanol, acetonitrile, 2-propanol, and ethyl acetate.

Viscosity (dynamic)
1.75 mPa s (1.75 cP) for a 8.5% w/w aqueous solution at 25°C, 1.09 mPa s (1.09 cP) at 60°C;
528 mPa s (528 cP) for a 60% w/w aqueous solution at 25°C, 87 mPa s (87 cP) at 60°C.[15]

11 Stability and Storage Conditions

Sulfobutylether β-cyclodextrin is stable in the solid state and should be protected from high humidity. It should be stored in a tightly sealed container in a cool, dry place.

It will reversibly take up moisture without any effect on the appearance of the material at humidities up to 60% RH. Equilibration at RH values above 60% will result in deliquescence. Once in this state, the material can be dried, but will give a glasslike

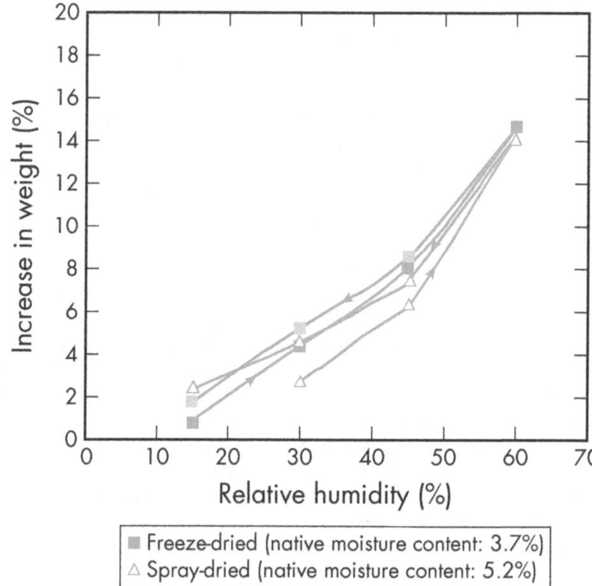

Figure 2: Moisture sorption–desorption isotherm of sulfobutylether β-cyclodextrin sodium at 30°C.

product. This water absorption behavior is typical of amorphous hygroscopic materials.

Sulfobutylether β-cyclodextrin is stable in aqueous solutions at values above about pH 1. It can degrade in highly acidic (pH <1) solutions, only apparent at elevated temperatures, producing the ring-opened form, followed by hydrolysis of the α(1→4) glucoside bonds.

Sulfobutylether β-cyclodextrin solutions may be autoclaved.[1]

12 Incompatibilities

The preservative activity of benzalkonium chloride is reduced in the presence of sulfobutylether β-cyclodextrin.

13 Method of Manufacture

Sulfobutylether β-cyclodextrin is prepared by alkylation of β-cyclodextrin using 1,4-butane sultone under basic conditions. The degree of substitution in β-cyclodextrin is controlled by the stoichiometric ratio of β-cyclodextrin to sultone used in the process.

14 Safety

Sulfobutylether β-cyclodextrin is derived from β-cyclodextrin, which is nephrotoxic when administered parenterally (*see* Cyclodextrins). However, studies have shown that sulfobutylether β-cyclodextrin is well tolerated when administered via intravenous, intramuscular, and subcutaneous injections, orally, and opthalmically.[16–18]

Sulfobutylether β-cyclodextrin has been subjected to an extensive battery of *in vitro* and *in vivo* genotoxicity and pharmacological evaluations. No genotoxic or mutagenic changes were observed with sulfobutylether β-cyclodextrin administration. Sulfobutylether β-cyclodextrin is biocompatible and is rapidly eliminated unmetabolized when administered intravenously.[1]

15 Handling Precautions

Observe normal precautions appropriate to the circumstances and quantity of material handled.

16 Regulatory Status

Sulfobutylether β-cyclodextrin is included in IV and IM injectable products currently approved and marketed in the US, Europe, and Japan. It is included in the FDA Inactive Ingredients Database for IM and IV use. Its use by other routes, including SC, oral, inhalation, nasal and ophthalmic, is being evaluated in clinical studies.

17 Related Substances

α-Cyclodextrin; β-cyclodextrin; γ-cyclodextrin; dimethyl-β-cyclodextrin; 2-hydroxyethyl-β-cyclodextrin; hydroxypropyl betadex; 3-hydroxypropyl-β-cyclodextrin; trimethyl-β-cyclodextrin.

18 Comments

Therapeutically, sulfobutylether β-cyclodextrin has demonstrated a protective effect on the kidney against certain renal toxic drugs and diagnostics.[19]

In addition to its use in pharmaceutical formulations, sulfobutylether β-cyclodextrin is also used in chromatographic separations, particularly in chiral separations by HPLC[20] and capillary electrophoresis,[21–24] and in tissue imaging.[25]

19 Specific References

1 CyDex Pharmaceuticals Inc. Technical information: *Captisol* sulfobutyl ether β-cyclodextrin, http://www.captisol.com/faq/ (accessed 14 December 2011).
2 Johnson MD, *et al.* Solubilization of a tripeptide HIV protease inhibitor using a combination of ionization and complexation with chemically modified cyclodextrins. *J Pharm Sci* 1994; **83**(8): 1142–1146.
3 Tokihiro K, *et al.* Varying effects of cyclodextrin derivatives on aggregation and thermal behavior of insulin in aqueous solution. *Chem Pharm Bull* 1997; **45**(3): 525–531.
4 Zia V, *et al.* Effect of cyclodextrin charge on complexation of neutral and charged substrates: comparison of $(SBE)_{7m}$-Beta-CD to HP-Beta-CD. *Pharm Res* 2001; **18**(5): 667–673.
5 Ueda H, *et al.* Evaluation of a sulfobutyl ether beta-cyclodextrin as a solubilizing/stabilizing agent for several drugs. *Drug Dev Ind Pharm* 1998; **24**(9): 863–867.
6 Uekama K, *et al.* Stabilizing and solubilizing effects of sulfobutyl ether β-cyclodextrin on prostaglandin E_1 analogue. *Pharm Res* 2001; **18**(11): 1578–1585.
7 Seto Y, *et al. In vitro* photobiochemical characterization of sulfobutylether-β-cyclodextrin formulation of bufexamac. *J Pharm Biomed Anal* 2011; **55**(3): 591–596.
8 Lefeuvre C, *et al.* Biopharmaceutics and pharmacokinetics of 5-phenyl-1,2-dithiole-3-thione complexed with sulfobutyl ether-7-beta-cyclodextrin in rabbits. *J Pharm Sci* 1999; **88**(10): 1016–1020.
9 Okimoto K, *et al.* Design and evaluation of an osmotic pump tablet (opt) for prednisolone, a poorly water soluble drug, using (SBE)(7m)-beta-CD. *Pharm Res* 1998; **15**(10): 1562–1568.
10 Kaukonen AM, *et al.* Water-soluble beta-cyclodextrins in paediatric oral solutions of spironolactone: preclinical evaluation of spironolactone bioavailability from solutions of beta-cyclodextrin derivatives in rats. *J Pharm Pharmacol* 1998; **50**(6): 611–619.
11 Jarho P, *et al.* The use of cyclodextrins in ophthalmic formulations of dipivefrin. *Int J Pharm* 1997; **153**: 225–233.
12 Jarho P, *et al.* Modified beta-cyclodextrin (SBE7-β-CyD) with viscous vehicle improves the ocular delivery and tolerability of pilocarpine prodrug in rabbits. *J Pharm Pharmacol* 1996; **48**: 263–269.
13 Salapatek AM, *et al.* Solubilized nasal steroid (CDX-947) when combined in the same solution with an antihistamine (CDX-313) provides improved, fast-acting symptom relief in patients with allergic rhinitis. *Allergy Asthma Proc* 2011; **32**: 1–9.
14 Mosher GL *et al.* Use of sulfoalkyl ether cyclodextrins as a preservative. US Patent No. 20050164986; 2005.
15 Liu J, *et al.* A study of the impact of freezing on the lyophilization of a concentrated formulation with a high fill depth. *Pharm Dev Technol* 2005; **10**(2): 261–272.
16 Luke DR, *et al.* Review of the basic and clinical pharmacology of sulfobutylether-β-cyclodextrin (SBECD). *J Pharm Sci* 2010; **99**(8): 3291–3301.
17 Stella VJ, He Q. Cyclodextrins. *Toxicol Pathol* 2008; **36**(1): 30–42.

S

18 ClinicalTrials.gov. A study of the pharmacokinetics (PK) and safety of IV carbamazepine relative to oral carbamazepine in adults with epilepsy. Identifier No. NCT01079351, April 2009.
19 Rowe V. Compositions useful for reducing nephrotoxicity and methods of use thereof. US Patent No. 20070270380; 2007.
20 Owens PK, *et al.* Method development in liquid chromatography with a charged cyclodextrin additive for chiral resolution of rac-amlodipine utilizing a central composite design. *Chirality* 1996; **8**(7): 466–476.
21 Dolezalova M, Fanali S. Enantiomeric separation of dihydroxyphenylalanine (dopa), methyldihydroxyphenylalanine (Mdopa) and hydrazinomethyldihydroxyphenylalanine (Cdopa) by using capillary electrophoresis with sulfobutyl ether-beta-cyclodextrin as a chiral selector. *Electrophoresis* 2000; **21**(15): 3264–3269.
22 Fanali S, *et al.* Separation of reboxetine enantiomers by means of capillary electrophoresis. *Electrophoresis* 2002; **23**(12): 1870–1877.
23 Aumatell A, Wells RJ. Enantiomeric differentiation of a wide range of pharmacologically active substances by cyclodextrin-modified micellar electrokinetic capillary chromatography using a bile salt. *J Chromatogr A* 1994; **688**(1–2): 329–337.
24 Chankvetadze B, *et al.* About some aspects of the use of charged cyclodextrins for capillary electrophoresis enantio-separation. *Electrophoresis* 1994; **15**(6): 804–807.
25 Kay AR, *et al.* Imaging synaptic activity in intact brain and slices with FM1-43 in *C. elegans*, lamprey, and rat. *Neuron* 1999; **24**(4): 809–817.

20 General References

Brewster ME, Loftsson T. Cyclodextrins as pharmaceutical solubilizers. *Adv Drug Deliv Rev* 2007; **59**(7): 645–666.
Hafner V, *et al.* Pharmacokinetics of sulfobutylether-beta-cyclodextrin and voriconazole in patients with end-stage renal failure during treatment with two hemodialysis systems and hemodiafiltration. *Antimicrob Agents Chemother* 2010; **54**(6): 2596–2602.
Hardeep SS, *et al.* The effects of substituted cyclodextrins on the colloidal and conformational stability of selected proteins. *J Pharm Sci* 2010; **99**(6): 2800–2818.
Irie T, Uekama K. Pharmaceutical applications of cyclodextrins. III. Toxicological issues and safety evaluation. *J Pharm Sci* 1997; **86**(2): 147–162.
Loftsson T, Duchene D. Cyclodextrins and their pharmaceutical applications. *Int J Pharm* 2007; **329**(1–2): 1–11.
Loftsson T, *et al.* Cyclodextrins in drug delivery. *Expert Opin Drug Deliv* 2005; **2**(2): 335–351.
Messner M, *et al.* Self-assembled cyclodextrin aggregates and nanoparticles. *Int J Pharm* 2010; **387**: 199–208.
Mosher G, Thompson DO. Complexation: Cyclodextrins. In: Swarbrick J, ed. *Encyclopedia of Pharmaceutical Technology* 3rd edn, vol. 2. New York: Informa Healthcare USA, Inc, 2006; 671-696.
Rajewski RA, Stella VJ. Pharmaceutical applications of cyclodextrins. II. *In vivo* drug delivery. *J Pharm Sci* 1996; **85**(11): 1142–1169.
Schneiderman E, Stalcup AM. Cyclodextrins: a versatile tool in separation science. *J Chromatogr B* 2000; **745**(1): 83–102.
Souney PF, *et al.* PM101: intravenous amiodarone formulation changes can improve medication safety. *Expert Opin Drug Saf* 2010; **9**(2): 319–333.
Stella V. SBE7-β-CD, a new, novel and safe polyanionic β-cyclodextrin derivative: characterization and biomedical applications. In: Szejtli J, Szente L, eds. *Proceedings 8th International Symposium, Cyclodextrins.* Dordrecht: Kluwer Academic Publishers, 1996: 471–476.
Stella VJ, *et al.* Mechanisms of drug release from cyclodextrin complexes. *Adv Drug Delivery Rev* 1999; **36**(1): 3–16.
Stella VJ, Rajewski RA. Cyclodextrins: their future in drug formulation and delivery. *Pharm Res* 1997; **14**(5): 556–567.
Thompson DO. Cyclodextrins-enabling excipients: their present and future use in pharmaceuticals. *Critl Rev Ther Drug Carrier Syst* 1997; **14**(1): 1–104.
Uekama K. Design and evaluation of cyclodextrin-based drug formulation. *Chem Pharm Bull (Tokyo)* 2004; **52**(8): 900–915.

21 Authors

GL Mosher, VD Antle, JD Pipkin.

22 Date of Revision

14 December 2011.

S

Sulfur Dioxide

1 Nonproprietary Names

USP–NF: Sulfur Dioxide

2 Synonyms

E220; sulfur(IV) oxide; sulfurous anhydride; sulfurous oxide.

3 Chemical Name and CAS Registry Number

Sulfur dioxide [7446-09-5]

4 Empirical Formula and Molecular Weight

SO_2 64.06

5 Structural Formula

See Section 4.

6 Functional Category

Antimicrobial preservative; antioxidant.

7 Applications in Pharmaceutical Formulation or Technology

Sulfur dioxide is used as an antioxidant for pharmaceutical injections.

8 Description

Sulfur dioxide occurs as a colorless gas or liquid at room temperature and pressure, with a strong, suffocating, pungent odor. It is noncombustible and is a strong reducing agent.

9 Pharmacopeial Specifications

See Table I.

Table I: Pharmacopeial specifications for sulfur dioxide.

Test	USP35–NF30
Water	≤2.0%
Nonvolatile residue	≤0.0025%
Sulfuric acid	≤0.002%
Assay	≥97.0%

10 Typical Properties

Boiling point −10.0°C at 760 mmHg
Melting point −75.5°C
Solubility *see* Table II.
Specific gravity 1.436 at 0°C
Vapor density 2.264 at 0°C (air = 1.0)
Vapor pressure 338.4 kPa (2538 mmHg) at 21.1°C

Table II: Solubility of sulfur dioxide.

Solvent	Solubility at 20°C unless otherwise stated
Acetic acid	Soluble
Chloroform	Soluble
Ethanol 95%	1 in 4
Ether	Soluble
Methanol	1 in 3
Water	1 in 5.7 at 0°C
	1 in 8.4 at 15°C
	1 in 11.8 at 25°C
	1 in 15.6 at 35°C

11 Stability and Storage Conditions

Sulfur dioxide is noncorrosive and stable when dry. It is usually stored under pressure in cylinders, and should be kept in a cool, dry, well-ventilated area, away from flammable materials.

12 Incompatibilities

Sulfur dioxide reacts vigorously with strong alkalis and oxidizing agents. The moist gas corrodes most metals. Sulfur dioxide is incompatible with chlorates, fluorine, interhalogens, powdered metals, metal oxides, metal acetylides, sodium hydroxide, and diethyl zinc. It is also incompatible with thiamine and gelatin.

See further details under Sodium Metabisulfite.

13 Method of Manufacture

Sulfur dioxide can be made by burning sulfur, or by roasting sulfide ores such as pyrites, sphalerite, and cinnabar.

14 Safety

Sulfur dioxide is used in food and pharmaceutical products. However, in large amounts, sulfur dioxide gas is highly irritant to the eyes, skin, and mucous membranes. Inhalation can lead to severe irritation of the respiratory tract. Direct contact with the liquid form may cause frostbite. Sulfur dioxide and sulfites may also cause allergic reactions and asthma.[1–3] In rabbits, sulfur dioxide causes mild to moderate irritation to the skin, and mild to severe irritation to the eyes.

LC_{50} (mouse, inhalation): 3000 ppm

LC_{50} (rat, inhalation): 2520 ppm

15 Handling Precautions

Observe normal precautions appropriate to the circumstances and quantity of the material handled. Sulfur dioxide forms sulfurous acid on contact with water. When heated to decomposition, it emits toxic fumes of sulfur oxide gases. Avoid inhalation and contact with eyes and skin.

16 Regulatory Status

GRAS listed. Accepted for use as a food additive in Europe. Included in the FDA Inactive Ingredients Database (IV infusions; injection solutions). Included in the Canadian Natural Health Products Ingredients Database.

17 Related Substances

Potassium metabisulfite; sodium metabisulfite; sodium sulfite.

18 Comments

The US Environmental Protection Agency (EPA) has strengthened the primary National Ambient Air Quality Standard (NAAQS) for sulfur dioxide whereby the revised standard will improve public health protection, especially for children, the elderly, and people with asthma.[4]

Sulfur dioxide is also used as a preservative in the food and cosmetics industries.

Sulfur dioxide is a byproduct of cement manufacture. A specification for sulfur dioxide is contained in the *Food Chemicals Codex* (FCC).[5]

The EINECS number for sulfur dioxide is 231-195-2. The PubChem Compound ID (CID) for sulfur dioxide is 1119.

19 Specific References

1 Smolinske SC, ed. *Handbook of Food, Drug, and Cosmetic Excipients.* Boca Raton FL: CRC Press, 1992: 393–406.
2 Weiner M, Bernstein IL. *Adverse Reactions to Drug Formulation Agents: A Handbook of Excipients.* New York: Marcel Dekker, 1989: 318–319.
3 Freedman BJ. Sulfur dioxide in foods and beverage: its use as a preservative and its effect on asthma. *Br J Dis Chest* 1980; 74: 128–134.
4 US Environmental Protection Agency. Fact sheet – Revisions to the Primary National Ambient Air Quality Standard (NAAQS), monitoring network, and data reporting requirements for sulfur dioxide, 2010. http://www.epa.gov/air/sulfurdioxide/pdfs/20100602fs.pdf (accessed 23 September 2011).
5 *Food Chemicals Codex*, 6th edn. Bethesda, MD: United States Pharmacopeia, 2008: 939.

20 General References

Agency for Toxic Substances and Disease Registry (ATSDR). Toxprofiles: Sulfur dioxide, December 1998.
BOC Gases. Material safety data sheet no. G-79: Sulfur dioxide, July 1996.
Thatcher Company. Material safety data sheet: Sulfur dioxide, September 2005.

21 Author

BV Kadri.

22 Date of Revision

2 March 2012.

Sulfuric Acid

1 Nonproprietary Names

BP: Sulphuric Acid
PhEur: Sulphuric Acid
USP–NF: Sulfuric Acid

2 Synonyms

Acidum sulfuricum; E513; hydrogen sulfate; oil of vitriol.

3 Chemical Name and CAS Registry Number

Sulfuric acid [7664-93-9]

4 Empirical Formula and Molecular Weight

H_2SO_4 98.08

5 Structural Formula

See Section 4.

6 Functional Category

Acidulant.

7 Applications in Pharmaceutical Formulation or Technology

Sulfuric acid is used as an acidulant in a variety of pharmaceutical and food preparations. It may also be used to prepare dilute sulfuric acid for use as an excipient. Sulfuric acid has been used in parenteral, oral, topical, and ophthalmic pharmaceutical formulations.

8 Description

Sulfuric acid occurs as a clear, colorless, odorless, oily liquid. It is very corrosive and has a great affinity for water.

The USP35–NF30 specifies that sulfuric acid contains not less than 95% and not more than 98%, by weight, of H_2SO_4; the remainder is water. *See also* Section 9.

9 Pharmacopeial Specifications

See Table I.

Table I: Pharmacopeial specifications for sulfuric acid.

Test	PhEur 7.4	USP35–NF30
Identification	+	+
Characters	+	—
Appearance of solution	+	—
Residue on ignition	—	≤0.005%
Chloride	≤50 ppm	≤50 ppm
Arsenic	≤1 ppm	≤1 ppm
Heavy metals	≤5 ppm	≤5 ppm
Weight per mL	≈1.84	—
Iron	≤25 ppm	—
Nitrate	+	—
Reducing substances	—	+
Assay (of H_2SO_4)	95.0–100.5%	95.0–98.0%

10 Typical Properties

Boiling point

≈290°C for H_2SO_4 (95%–98% w/w);
330°C for H_2SO_4 (100% w/w).

Density ≈1.84 g/cm^3 at 20°C
Dissociation constant

$pK_{a1} = -3.00$;
$pK_{a2} = 1.99$.

Freezing point

10°C for H_2SO_4 (100% w/w);
3°C for H_2SO_4 (98% w/w);
−32°C for H_2SO_4 (93% w/w).

Solubility Miscible with ethanol and water.
Vapor density 3.4 (air = 1.0)
Vapor pressure <0.3 mmHg at 20°C

11 Stability and Storage Conditions

Sulfuric acid is stable but very corrosive and hygroscopic. It will draw moisture from the atmosphere. Sulfuric acid should be stored in a tightly closed container in an explosion-proof area. Containers should be stored out of direct sunlight and away from heat. Avoid heat and moisture. Isolate from incompatible materials. *See also* Section 12.

12 Incompatibilities

Avoid storage in close proximity to water, most common metals, organic materials, strong reducing agents, combustible materials, strong bases, carbonates, sulfides, cyanides, strong oxidizing agents, and carbides.

Sulfuric acid is a powerful oxidizer and may ignite or explode on contact with many materials.

It can react violently with the evolution of a large amount of heat. Oxides of sulfur and hydrogen can be generated during reactions.

Great care must be exercised when mixing with other liquids. Always add sulfuric acid to the diluent with great caution.

13 Method of Manufacture

Sulfuric acid may be prepared industrially by either the contact process or the chamber process.[1,2]

Contact Process

$2SO_2 + O_2 \rightarrow 2SO_3$
$SO_3 + H_2O \rightarrow H_2SO_4$

Chamber Process

$2NO + O_2 \rightarrow 2NO_2$
$NO_2 + SO_2 + H_2O \rightarrow H_2SO_4 + NO$

14 Safety

Sulfuric acid is widely used in a variety of pharmaceutical formulations. Although concentrated sulfuric acid is very corrosive, it is normally used well diluted in formulations. Concentrated sulfuric acid will react violently with water and much heat is generated. When diluting sulfuric acid, the acid should always be added to the other liquid with great caution.

The concentrated solution is extremely corrosive and can cause severe damage or necrosis on contact with the eyes and skin. Ingestion may cause severe injury or death. Inhalation of concentrated vapors can cause serious lung damage.

LD_{50} (rat, oral): 2.14 g/kg[3]

S

15 Handling Precautions

Caution should be exercised when handling sulfuric acid and suitable protection against inhalation and spillage should be made. Respiratory protection may not be required where adequate ventilation exists. Eye protection (safety goggles and face shield), rubber gloves, and apron are recommended, depending on the circumstances and quantity of sulfuric acid handled. Do not dilute spills of concentrated acid with water since an exothermic reaction will occur. Spills should be neutralized with soda ash or lime. Splashes on the skin and eyes should be treated by immediate and prolonged washing (10–15 minutes) with large amounts of water, and medical attention should be sought. Do not neutralize acid in contact with skin or eyes as the exothermic reaction can increase the severity of the burn. Remove contaminated clothing immediately.

Fumes can cause irritation or permanent damage to the eyes, nose, and respiratory system; prolonged exposure to fumes may damage the lungs.

16 Regulatory Status

GRAS listed. Accepted for use as a food additive in Europe. Included in the FDA Inactive Ingredients Database (IM, IV, and IP injections; inhalation solutions, irrigation solutions, nasal, ophthalmic solutions and suspensions; oral solutions; topical emulsions and creams). Included in nonparenteral and parenteral medicines licensed in Europe. Included in the Canadian Natural Health Products Ingredients Database.

The United Nations Convention Against Illicit Traffic in Narcotic Drugs and Psychotropic Substances (1988) lists sulfuric acid as a chemical frequently used in the illicit manufacture of narcotic drugs or psychotropic substances.[4] In the US, sulfuric acid is included in the list of essential or precursor chemicals established pursuant to the Chemical Diversion and Trafficking Act. Accordingly, transactions of sulfuric acid such as imports, exports, sales, and transfers are subject to regulation and monitoring by the Drug Enforcement Administration.[5]

17 Related Substances

Dilute sulfuric acid; fuming sulfuric acid.

Dilute sulfuric acid

Density 1.062–1.072 g/cm^3

Comments Prepared by adding 104 g of sulfuric acid to 896 g of purified water with constant stirring and cooling. Dilute sulfuric acid contains between 9.5% and 10.5% w/w of H_2SO_4.

Fuming sulfuric acid

Synonyms Oleum.

Comments

Fuming sulfuric acid consists of H_2SO_4 with free sulfur trioxide (SO_3). It is prepared by adding sulfur trioxide to sulfuric acid. Available in grades containing up to about 80% free SO_3.

Fuming sulfuric acid is a colorless or slightly colored, viscous liquid that emits choking fumes of sulfur trioxide. It is extremely corrosive and should be handled with great care and stored in tightly closed glass-stoppered bottles.

18 Comments

Sulfuric acid may be used to prepare dilute sulfuric acid, which, in addition to its use as an excipient, has some therapeutic use for the treatment of gastric hypoacidity, as an astringent in diarrhea, or to stimulate appetite.

A specification for sulfuric acid is contained in the *Food Chemicals Codex* (FCC).[6]

The EINECS number for sulfuric acid is 231-639-5. The PubChem Compound ID (CID) for sulfuric acid is 1118.

19 Specific References

1 Druecker WW, West JR. *The Manufacture of Sulfuric Acid.* New York: Reinhold, 1959; 515.

2 Nickless G, ed. *Inorganic Sulphur Chemistry.* New York: Elsevier, 1968: 535–561.

3 Lewis RJ, ed. *Sax's Dangerous Properties of Industrial Materials*, 11th edn. New York: Wiley, 2004; 3331–3332.

4 United Nations Convention Against Illicit Traffic in Narcotic Drugs and Psychotropic Substances, 1988. List of precursors and chemicals frequently used in the illicit manufacture of narcotid drugs and psychotropic substances under international control. January 2007.

5 US Department of Justice –Drug Enforcement Administration. Code of Federal Regulations Section 1310.02(b), April 2006. www.deadiversion.usdoj.gov/21cfr/cfr/1310/1310_02.htm (accessed 21 November 2011).

6 *Food Chemicals Codex*, 7th edn. Bethesda, MD: United States Pharmacopeia, 2010.

20 General References

—

21 Author

GE Amidon.

22 Date of Revision

2 March 2012.

S

Sunflower Oil

1 Nonproprietary Names

BP: Refined Sunflower Oil
PhEur: Sunflower Oil, Refined
USP–NF: Sunflower Oil

2 Synonyms

Helianthi annui oleum raffinatum; huile de tournesol; oleum helianthi; sunflowerseed oil.

3 Chemical Name and CAS Registry Number

Sunflower oil [8001-21-6]

4 Empirical Formula and Molecular Weight

See Section 5.

5 Structural Formula

Sunflower oil is classified as an oleic–linoleic acid oil. Its composition includes linoleic acid (66%), oleic acid (21.3%), palmitic acid (6.4%), arachidic acid (4.0%), stearic acid (1.3%), and behenic acid (0.8%).

The USP35–NF30 describes sunflower oil as a refined fixed oil obtained from the seeds of *Helianthus annus* Linné (Fam. Asteraceae alt. Compositae).

The PhEur 7.4 describes sunflower oil as the refined fatty oil obtained from the seeds of *Helianthus annus* C. by mechanical expression or by extraction. A suitable antioxidant may be added.

6 Functional Category

Emollient; solvent; tablet and capsule diluent.

7 Applications in Pharmaceutical Formulation or Technology

Sunflower oil is used extensively in cosmetics and pharmaceutical formulations. It is also widely used as an edible oil, primarily in oleomargarine. High oleic acid content sunflower oil with good oxidative stability and emollient properties is commercially available for use in cosmetic formulations.(1) Sunflower oil with marked oxidative stability is particularly suitable for the manufacture of sunscreen formulations.(2)

See also Section 18.

8 Description

Sunflower oil occurs as a clear, light yellow-colored liquid with a bland, agreeable taste.

9 Pharmacopeial Specifications

See Table I.

Table I: Pharmacopeial specifications for sunflower oil.

Test	PhEur 7.4	USP35–NF30
Identification	+	—
Characters	+	—
Acid value	≤0.5	—
Peroxide value	≤10.0	—
Unsaponifiable matter	≤1.5%	≤1.0%
Alkaline impurities	+	—
Specific gravity	—	0.914–0.924
Iodine value	—	+
Saponification value	—	180–200
Refractive index	—	1.472–1.474
Heavy metals	—	≤10 ppm
Limit of peroxide	—	≤10.0 mEq/kg
Composition of fatty acids	+	+
Palmitic acid	4.0–9.0%	—
Stearic acid	1.0–7.0%	—
Oleic acid	14.0–40.0%	—
Linoleic acid	48.0–74.0%	—

10 Typical Properties

Boiling point 40–60°C
Density 0.915–0.919 g/cm^3
Hydroxyl value 14–16
Iodine number 125–140
Melting point −18°C
Refractive index
n_D^{25} = 1.472–1.474;
n_D^{40} = 1.466–1.468.

Solubility Miscible with benzene, chloroform, carbon tetrachloride, diethyl ether, and light petroleum; practically insoluble in ethanol (95%) and water.

11 Stability and Storage Conditions

Sunflower oil should be stored in an airtight, well-filled container, protected from light. Stability may be improved by the addition of an antioxidant such as butylated hydroxytoluene.

12 Incompatibilities

The oxidative stability of sunflower oil is reduced in the presence of iron oxides and zinc oxide.(3)

Sunflower oil forms a 'skin' after being exposed to air for 2–3 weeks.

13 Method of Manufacture

Sunflower oil is obtained from the fruits and seeds (achenes) of the sunflower, *Helianthus annus* (Compositae), by mechanical means or by extraction.

14 Safety

Sunflower oil is widely used in food products and on its own as an edible oil. It is also used extensively in cosmetics and topical pharmaceutical formulations, and is generally regarded as a relatively nontoxic and nonirritant material.

15 Handling Precautions

Observe normal precautions appropriate to the circumstances and quantity of material handled. When heated to decomposition, sunflower oil emits acrid smoke and irritating fumes.

16 Regulatory Status

GRAS listed. Included in nonparenteral medicines licensed in the UK.

17 Related Substances

Corn oil; cottonseed oil; peanut oil; sesame oil; soybean oil.

18 Comments

Therapeutically, sunflower oil is used to provide energy and essential fatty acids for parenteral nutrition. Studies have shown that sunflower oil may be used in intramuscular injections without inducing tissue damage.[4]

Sunflower oil should be labeled to indicate the name and concentration of any antioxidant added, and also whether the oil was obtained by mechanical expression or extraction.

A specification for sunflower oil is contained in the *Food Chemicals Codex* (FCC).[5]

The EINECS number for sunflower oil is 232-273-9.

19 Specific References

1 Arquette DJ, *et al.* A natural oil made to last. *Cosmet Toilet* 1997; **112**(1): 67–72.
2 Arquette DJ, *et al.* Oils and fats: place in the sun. *Soap Perfum Cosmet* 1994; **67**(Nov): 49, 51.
3 Brown JH, *et al.* Oxidative stability of botanical emollients. *Cosmet Toilet* 1997; **112**(7): 87–9092, 94, 96–98.
4 Vinardell MP, Vives MA. Plasma creatine kinase activity after intramuscular injection of oily vehicles in rabbits. *Pharm Pharmacol Lett* 1996; **6**(2): 54–55.
5 *Food Chemicals Codex*, 7th edn. Bethesda, MD: United States Pharmacopeia, 2010.

20 General References

—

21 Author

PJ Sheskey.

22 Date of Revision

2 March 2012.

Suppository Bases, Hard Fat

1 Nonproprietary Names

BP: Hard Fat

PhEur: Hard Fat

USP–NF: Hard Fat

2 Synonyms

Adeps neutralis; adeps solidus; *Akosoft*; *Akosol*; *Cremao CS-34*; *Cremao CS-36*; hydrogenated vegetable glycerides; *Massa Estarinum*; *Massupol*; *Novata*; semisynthetic glycerides; *Suppocire*; *Wecobee*; *Witepsol*.

3 Chemical Name and CAS Registry Number

Hard fat triglyceride esters

4 Empirical Formula and Molecular Weight

Hard fat suppository bases consist mainly of mixtures of the triglyceride esters of the higher saturated fatty acids ($C_8H_{17}COOH$ to $C_{18}H_{37}COOH$) along with varying proportions of mono- and diglycerides. Special grades may contain additives such as beeswax, lecithin, polysorbates, ethoxylated fatty alcohols, and ethoxylated partial fatty glycerides.

5 Structural Formula

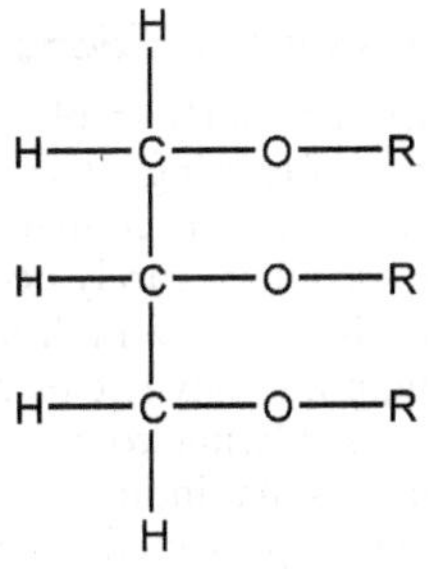

```
        H
        |
   H—C—O—R
        |
   H—C—O—R
        |
   H—C—O—R
        |
        H
```

where R = H or $OC(CH_2)_nCH_3$; n = 7–17

Not all Rs can be H at the same time.

6 Functional Category

Suppository base.

7 Applications in Pharmaceutical Formulation or Technology

The primary application of hard fat suppository bases, or semisynthetic glycerides, is as a vehicle for the rectal or vaginal administration of a variety of drugs, either to exert local effects or to achieve systemic absorption.

Selection of a suppository base cannot usually be made in the absence of knowledge of the physicochemical properties and intrinsic thermodynamic activity of the drug substance. Other drug-related factors that can affect release and absorption and which must therefore be considered are the particle size distribution of insoluble solids, the oil : water partition coefficient, and the dissociation constant. The displacement value should also be known, as well as the ratio of drug to base. Properties of the suppository base that may or may not be modified by the drug, or that can influence drug release, are the melting characteristics, chemical reactivity, and rheology. The presence of additives in the base can also affect performance.

Melting characteristics Fatty-based suppositories intended for systemic use should liquefy at just below body temperature. Softening or dispersion may be adequate for suppositories intended for local action or modified release. High-melting-point bases may be indicated for fat-soluble drugs that tend to depress the melting point of bases or for suppositories used in warm climates. Drugs that dissolve in bases when hot may create problems if they deposit as crystals of different form or increased size on cooling or on storage. Low-melting-point bases, particularly those that melt to liquids of low viscosity, can be of value when large volumes of insoluble substances are to be incorporated; there is a risk of sedimentation in such instances. An important factor during processing is the time required for setting. This is affected by the temperature difference between the melting point and the solidification point.[1,2]

Chemical reactivity Although the use of bases with low hydroxyl values (low partial ester content) is indicated to minimize the risk of interaction with chemically reactive compounds, formulators should be aware that hydroxyl values are also related to hydrophilic properties, which, in turn, can modify both release and absorption rates. Bases with low hydroxyl values tend to be less plastic than those with higher values and, if cooled rapidly, may become excessively brittle. Peroxide values give a measure of the resistance of the base to oxidation and are a guide to the onset of rancidity.

Rheology The viscosity of the melted base can affect the uniformity of distribution of suspended solids during manufacture. It can also influence the release and absorption of the drug in the rectum. Further reduction in the particle size of insoluble solids is the method of choice to minimize the risk of sedimentation. However, the presence of a high content of fine suspended particles is likely to increase viscosity. It may also make pouring difficult, delay melting, and induce brittleness on solidification. Additives are sometimes included to modify rheological properties and to maintain homogeneity, e.g. microcrystalline wax, but the extent of their effect on drug release should first be assessed. Release from a base in which viscosity has been enhanced by an added thickener may vary and be related to the aqueous solubility of the drug itself.

Additives Some grades of commercial bases already contain additives, and these are usually identified by the manufacturers by means of suitable letters and numbers. Additives may also be incorporated by formulators. Properties of suppositories that have been modified and additives or types of additives that have been used are shown in Table I. Water is undesirable as an additive because it enhances hydrolysis and the potential for a chemical reaction between constituents of the suppository. In low concentration, water plays little part in drug release and can serve as a medium for microbial growth.

8 Description

A white or almost white, practically odorless, waxy, brittle mass. When heated to 50°C it melts to give a colorless or slightly yellowish liquid.

9 Pharmacopeial Specifications

See Table II.

10 Typical Properties

Acid value see Table III.

Color number

≤3 for *Massa Estarinum* (iodine color index);

≤3 for *Suppocire* excluding L grades (Gardener scale);

≤5 for *Suppocire* L grades (Gardener scale);

≤3 for *Witepsol* (iodine color index).

Density

0.955–0.975 g/cm^3 for *Massa Estarinum* at 20°C;

0.950–0.960 g/cm^3 for *Suppocire* at 20°C;

0.950–0.980 g/cm^3 for *Witepsol* at 20°C.

Heat of melting (22–40°C)

≈145 J/g/°C for *Massa Estarinum*;

100–130 J/g/°C for *Suppocire*;

≈145 J/g/°C for *Witepsol*.

Hydroxyl value see Table III.

Iodine value see Table III.

Melting point see Table III.

Moisture content

≤0.2% w/w for *Massa Estarinum*;

<0.5% w/w for *Suppocire*;

≤0.2% w/w for *Witepsol*.

Table I: Selected suppository additives.

Property	Additive
Dispersants (release and/or absorption enhancers)	Surfactants
Hygroscopicity (reduced)	Colloidal silicon dioxide
Hardeners (or increasing melting point)	Beeswax
	Cetyl alcohol
	Stearic acid
	Stearyl alcohol
	Aluminum monostearate (or di- and tristearate)
	Bentonite
	Magnesium stearate
	Colloidal silicon dioxide
Plasticizers (or decreasing melting point)	Glyceryl monostearate
	Myristyl alcohol
	Polysorbate 80
	Propylene glycol

Table II: Pharmacopeial specifications for suppository bases.

Test	PhEur 7.4	USP35–NF30
Identification	+	—
Characters	+	—
Melting range	30–45°C	27–44°C
Residue on ignition	—	≤0.05%
Total ash	≤0.05%	—
Acid value	≤0.5	≤1.0
Iodine value	≤3.0	≤7.0
Saponification value	210–260	215–255
Hydroxyl value	≤50	≤70
Peroxide value	≤3.0	—
Unsaponifiable matter	≤0.6%	≤3.0%
Alkaline impurities	+	+
Heavy metals	≤10 ppm	—

Peroxide value

≤3 for *Massa Estarinum*;

≤1.2 for *Suppocire*;

≤3 for *Witepsol*.

Saponification value see Table III.

Solidification point see Table III.

Solubility Freely soluble in carbon tetrachloride, chloroform, ether, toluene, and xylene; slightly soluble in warm ethanol; practically insoluble in water.

Specific heat

≈2.6 J/g/°C for *Massa Estarinum*;

1.7–2.5 J/g/°C for *Suppocire*;

≈2.6 J/g/°C for *Witepsol*.

Unsaponifiable matter see Table III.

11 Stability and Storage Conditions

Hard fat suppository bases are fairly stable toward oxidation and hydrolysis, with the iodine value being a measure of their resistance to oxidation and rancidity. Water content is usually low and deterioration due to hygroscopicity rarely occurs.

Melting characteristics, hardness, and drug-release profiles alter with time, and the melting point may rise by more than 1.0°C after storage for several months. Owing to the complexity of bases, elucidation of the mechanisms that induce these changes on aging is difficult. Evidence has been presented[3] that supports a finite transition from amorphous to crystalline forms in which polymorphism may or may not contribute, whereas other workers have found melting point changes to be closely associated with the

Table III: Typical properties of suppository bases.

Product		Acid value	Hydroxyl value	Iodine value	Melting point (°C)	Saponification value	Solidification point (°C)	Unsaponifiable matter (%)
Cremao	CS-34	<0.3	—	<2	33–35	250	—	—
	CS-36	<0.3	—	<1	34–37	250	—	—
Massa Estarinum	B	≤0.3	20–30	≤3	33–35.5	225–240	31–33	≤0.3
	BC	≤0.3	30–40	≤3	33.5–35.5	225–240	30.5–32.5	≤0.3
	C	≤0.3	20–30	≤3	36–38	225–235	33–35	≤0.3
	299	≤0.3	≤2	≤3	33.5–35.5	240–255	32–34.5	≤0.3
Massupol	—	—	≤2	34–36	240–250	31–32.5	—	—
Massupol 15	—	—	≤3	35–37	220–230	31–33	—	—
Suppocire	A	<0.5	20–30	<2	35–36.5	225–245	—	≤0.5
	AM	<0.2	≤6	<2	35–36.5	225–245	—	≤0.5
	AML	<0.5	≤6	<2	35–36.5	225–245	—	≤0.6
	AIML	<0.5	≤6	<3	33–35	225–245	—	≤0.6
	AS_2	<0.5	15–25	<2	35–36.5	225–245	—	≤0.5
	AS_2X	<0.5	15–25	<2	35–36.5	225–245	—	≤0.6
	AT	<0.5	25–35	<2	35–36.5	225–245	—	≤0.5
	AP	<1.0	30–50	<1	33–35	200–220	—	≤0.5
	AI	<0.5	20–30	<2	33–35	225–245	—	≤0.5
	AIX	<0.5	20–30	<2	33–35	220–240	—	<0.6
	AIM	<0.3	<6	<2	33–35	225–245	—	≤0.5
	AIP	<1.0	30–50	<1	30–33	205–225	—	<0.5
	B	<0.5	20–30	<2	36–37.5	225–245	—	≤0.5
	BM	<0.2	<6	<2	36–37.5	225–245	—	≤0.5
	BML	<0.5	<6	<3	36–37.5	225–245	—	≤0.6
	BS_2	<0.5	15–25	<2	36–37.5	225–245	—	≤0.5
	BS_2X	<0.5	15–25	≤3	36–37.5	220–240	—	≤0.6
	BT	<0.5	25–35	<2	36–37.5	225–245	—	≤0.5
	BP	<1.0	30–50	<1	36–37	200–220	—	≤0.5
	C	<0.5	20–30	<2	38–40	220–240	—	≤0.5
	CM	<0.2	<6	<2	38–40	225–245	—	≤0.5
	CS_2	<0.5	15–25	<2	38–40	220–240	—	≤0.5
	CS_2X	<0.5	15–25	<2	38–40	220–240	—	<0.6
	CT	<0.5	25–35	<2	38–40	220–240	—	≤0.5
	CP	<1.0	≤50	<1	37–39	200–220	—	<0.5
	D	<0.5	20–30	<2	42–45	215–235	—	≤0.5
	DM	<0.2	<6	<2	42–45	215–235	—	≤0.5
	NA	<0.5	<40	<2	35.5–37.5	225–245	—	<0.5
	NB	<0.5	<40	<2	36.5–38.5	215–235	—	<0.5
	NC	<0.5	<40	<2	38.5–40.5	220–240	—	<0.5
	NAI 0	<0.5	≤3	<2	33.5–35.5	220–245	—	<0.5
	NAI 5	<0.5	≤5	<2	33.5–35.5	220–245	—	<0.5
	NAI 10	<0.5	<15	<2	33.5–35.5	220–245	—	<0.5
	NAI	<0.5	<40	<2	33.5–35.5	225–245	—	<0.5
	NAIL	<1.0	<40	<3	33.5–35.5	225–245	—	<0.6
	NAIX	<0.5	<40	<2	33.5–35.5	220–240	—	<0.6
	NA 0	<0.5	≤3	<2	35.5–37.5	225–245	—	<0.5
	NA 5	<0.5	≤5	<2	35.5–37.5	225–245	—	<0.5
	NA 10	<0.5	≤15	<2	35.5–37.5	225–245	—	<0.5
	NAL	<0.5	<40	<2	33.5–35.5	225–245	—	<0.6
	NAX	<0.5	<40	<2	35.5–37.5	220–240	—	<0.6
	NBL	<0.5	<40	<3	36.5–38.5	220–240	—	<0.6
	NBX	<0.5	<40	<2	36.5–38.5	215–235	—	<0.6
	ND	<0.5	<40	<2	42–45	210–230	—	<0.5
Witepsol	H5	≤0.2	≤5	≤2	34–36	235–245	33–35	≤0.3
	H12	≤0.2	5–15	≤3	32–33.5	240–255	29–33	≤0.3
	H15	≤0.2	5–15	≤3	33.5–35.5	230–245	32.5–34.5	≤0.3
	H19[a]	≤0.2	20–30	≤7	33.5–35.5	230–240	—	≤0.3
	H32	≤0.2	≤3	≤3	31–33	240–250	30–32.5	≤0.3
	H35	≤0.2	≤3	≤3	33.5–35.5	240–250	32–35	≤0.3
	H37	≤0.2	≤3	≤3	36–38	225–245	35–37	≤0.3
	H175[a]	≤0.7	5–15	≤3	34.5–36.5	225–245	32–34.5	≤1.0
	H185	≤0.2	5–15	≤3	38–39	220–235	34–37	≤0.3
	W25	≤0.3	20–30	≤3	33.5–35.5	225–240	29–33	≤0.3
	W31	≤0.3	25–35	≤3	35–37	225–240	30–33	≤0.5
	W32	≤0.3	40–50	≤3	32–33.5	225–245	25–30	≤0.3
	W35	≤0.3	40–50	≤3	33.5–35.5	225–235	27–32	≤0.3
	W45	≤0.3	40–50	≤3	33.5–35.5	225–235	29–34	≤0.3
	S51[a]	≤1.0	55–70	≤8	30–32	215–230	25–27	≤2.0
	S52[a]	≤1.0	50–65	≤3	32–33.5	220–230	27–30	≤2.0
	S55[a]	≤1.0	50–65	≤3	33.5–35.5	215–230	28–33	≤2.0
	S58[a]	≤1.0	60–70	≤7	31.5–33	215–225	27–29	≤2.0
	E75[a]	≤1.3	5–15	≤3	37–39	220–230	32–36	≤3.0
	E76	≤0.3	30–40	≤3	37–39	220–230	31–35	≤0.5
	E85	≤0.3	5–15	≤3	42–44	220–230	37–42	≤0.5

(a) Note that these types are mixtures containing hard fat and therefore do not comply with the specifications of the PhEur 7.4 and USP35–NF30.

S

conversion of triglycerides to more stable polymorphic forms.[4] Before melting point determinations are made, bases are 'conditioned' to a stable crystalline form.

Suppository bases should be stored protected from light in an airtight container at a temperature at least 5°C less than their stated melting point. Refrigeration is usually recommended for molded suppositories.

Suppositories that are not effectively packaged may develop a 'bloom' of powdery crystals at the surface. This is usually due to the presence of high-melting-point components in the base and can often be overcome by using a different base. Alternatively, the base can be precrystallized prior to pouring, since the crystals will cause a quick and complete crystallization into its end crystal form. This process is called 'tempering.'

12 Incompatibilities

Incompatibilities with suppository bases are not now extensively reported in the literature. The occurrence of a chemical reaction between a hard fat suppository base and a drug is relatively rare, but any potential for such a reaction may be indicated by the magnitude of the hydroxyl value of the base. The risk of hydrolysis of aspirin, for example, may be reduced by the use of a base with a low hydroxyl value (<5) and, additionally, by minimization of the water content of both the base and the aspirin.

There is evidence that aminophylline reacts with the glycerides in some hard fat bases to form diamides. On aging or exposure to elevated temperatures, degradation is accompanied by hardening and suppositories tend to exhibit a marked increase in melting point. The ethylenediamine content is also reduced.[5,6]

Certain fat-soluble medications, such as chloral hydrate, may depress the melting point when incorporated into a base. Similarly, when large amounts of an active substance, either solid or liquid, have to be dispersed into a base, the rheological characteristics of the resultant suppository may be changed, with concomitant effects on release and absorption. Careful selection of bases or the inclusion of additives may therefore be necessary.

13 Method of Manufacture

The most common method of manufacture involves the hydrolysis of natural vegetable oils such as coconut or palm kernel oil, followed by fractional distillation of the free fatty acids produced. The C_8 to C_{18} fractions are then hydrogenated and reesterified under controlled conditions with glycerin to form a mixture of tri-, di-, and monoglycerides of the required characteristics and hydroxyl value. This process is used for *Witepsol.*

In an alternative procedure, coconut or palm kernel oil is directly hydrogenated and then subjected to an interesterification either with itself or with glycerin to form a mixture of tri-, di-, and monoglycerides of the required characteristics and hydroxyl value, e.g. *Suppocire.*

14 Safety

Suppository bases are generally regarded as nontoxic and nonirritant materials when used in rectal formulations. However, animal studies have suggested that some bases, particularly those types with a high hydroxyl value, may be irritant to the rectal mucosa.[7]

15 Handling Precautions

Observe normal precautions appropriate to the circumstances and quantity of material handled. There is a slight fire hazard on exposure to heat or flame.

16 Regulatory Status

Included in the FDA Inactive Ingredients Database (rectal and vaginal preparations). Included in nonparenteral medicines licensed in the UK. Included in the Canadian Natural Health Products Ingredients Database.

17 Related Substances

Glycerin; medium-chain triglycerides; polyethylene glycol; theobroma oil.

Theobroma oil

CAS number [8002-31-1]

Synonyms Cocoa butter; oleum cacao; oleum theobromatis.

Appearance A yellowish or white, brittle solid with a slight odor of cocoa.

Melting point 31–34°C

Solubility Freely soluble in chloroform, ether, and petroleum spirit; soluble in boiling ethanol; slightly soluble in ethanol (95%).

Stability and storage conditions Heating theobroma oil to more than 36°C during the preparation of suppositories can result in an appreciable lowering of the solidification point owing to the formation of metastable states; this may lead to difficulties in the setting of the suppository. Theobroma oil should be stored at a temperature not exceeding 25°C.

Comments Theobroma oil is a fat of natural origin used as a suppository base. It comprises a mixture of the triglycerides of saturated and unsaturated fatty acids, in which the unsaturated acid is preferentially situated on the 2-position of the glyceride. Theobroma oil is also a major ingredient of chocolate.

18 Comments

A specification for hard fat is included in the *Japanese Pharmaceutical Excipients* (JPE).[8]

19 Specific References

1 Setnikar I, Fantelli S. Softening and liquefaction temperature of suppositories. *J Pharm Sci* 1963; **52**: 38–43.
2 Krówczynski L. [A simple device for testing suppositories.] *Diss Pharm* 1959; **11**: 269–273[in Polish].
3 Coben LJ, Lordi NG. Physical stability of semisynthetic suppository bases. *J Pharm Sci* 1980; **69**: 955–960.
4 Liversidge GG, *et al.* Influence of physicochemical interactions on the properties of suppositories I: interactions between the constituents of fatty suppository bases. *Int J Pharm* 1981; **7**: 211–223.
5 Brower JF, *et al.* Decomposition of aminophylline in suppository formulations. *J Pharm Sci* 1980; **69**: 942–945.
6 Taylor JB, Simpkins DE. Aminophylline suppositories: *in vitro* dissolution and bioavailability in man. *Pharm J* 1981; **227**: 601–603.
7 De Muynck C, *et al.* Rectal mucosa damage in rabbits after subchronical application of suppository bases. *Pharm Res* 1991; **8**: 945–950.
8 Japan Pharmaceutical Excipients Council. *Japanese Pharmaceutical Excipients 2004.* Tokyo: Yakuji Nippo, 2004: 373.

20 General References

Allen LV. Compounding suppositories Part I: Theoretical considerations. *Int J Pharm Compound* 2000; **4**(4): 289–293324–325.

Allen LV. Compounding suppositories Part II: Extemporaneous preparation. *Int J Pharm Compound* 2000; **4**(5): 371–373404–405.

Allen LV. Suppositories. London: Pharmaceutical Press, 2008.

Anschel J, Lieberman HA. Suppositories. In: Lachman L *et al.* eds. *The Theory and Practice of Industrial Pharmacy*, 2nd edn. Philadelphia: Lea and Febiger, 1976: 245–269.

Realdon N, *et al.* Effects of silicon dioxide on drug release from suppositories. *Drug Dev Ind Pharm* 1997; **23**(11): 1025–1041.

Realdon N, *et al.* Layered excipient suppositories: the possibility of modulating drug availability. *Int J Pharm* 1997; **148**: 155–163.

Schoonen AJM, *et al.* Release of drugs from fatty suppository bases I: the release mechanism. *Int J Pharm* 1979; **4**: 141–152.

Senior N. Review of rectal suppositories 1: formulation and manufacture. *Pharm J* 1969; **203**: 703–706.

Senior N. Review of rectal suppositories 2: resorption studies and medical applications. *Pharm J* 1969; **203**: 732–736.

Senior N. Rectal administration of drugs. In: Bean HS *et al.* eds. *Advances in Pharmaceutical Sciences*. London: Academic Press, 1974: 363–435.

Sutananta W, *et al.* An evaluation of the mechanism of drug release from glyceride bases. *J Pharm Pharmacol* 1995; **47**: 182–187.

21 Author

RC Moreton.

22 Date of Revision

2 March 2012.

Tagatose

1 Nonproprietary Names

USP–NF: Tagatose

2 Synonyms

D-*lyxo*-Hexulose; (3*S*,4*S*,5*R*)-2-(hydroxymethyl)oxane-2,3,4,5-tetrol; *Naturlose*; D-tagatose; tagatosum; tagatoza.

3 Chemical Name and CAS Registry Number

(3*S*,4*S*,5*R*)-1,3,4,5,6-pentahydroxyhexan-2-one [87-81-0]

4 Empirical Formula and Molecular Weight

$C_6H_{12}O_6$ 180.16

5 Structural Formula

6 Functional Category

Sweetening agent.

7 Applications in Pharmaceutical Formulation or Technology

Tagatose is used as a sweetening agent in beverages, foods, and pharmaceutical applications. A 10% solution of tagatose is about 92% as sweet as a 10% sucrose solution.(1) It is a low-calorie sugar with approximately 38% of the calories of sucrose per gram. It occurs naturally in low levels in milk products.(1) Like other sugars (fructose, glucose, sucrose), it is also used as a bulk sweetener, humectant, texturizer, and stabilizer, and may be used in dietetic foods with a low glycemic index.(2,3)

8 Description

Tagatose is a white, anhydrous crystalline solid. It is a carbohydrate, a ketohexose, an epimer of D-fructose inverted at C-4. It can exist in several tautomeric forms.(4,5)

9 Pharmacopeial Specifications

See Table I.

Table I: Pharmacopeial specifications for tagatose.

Test	USP35–NF30
Identification	+
Specific rotation	−4° to −7°
Melting range	133–144°C
Microbial limits	
Aerobic bacteria	≤1000 cfu/g
Molds and yeast	≤100 cfu/g
Water	≤0.5%
Total ash	≤0.1%
Lead	≤1 ppm
Assay (dried basis)	≥98.0%

10 Typical Properties

Hygroscopicity Crystalline D-tagatose has low hygroscopicity similar to that of sucrose.
Melting point 132–135°C
Solubility Very soluble in water: 1 in 0.7 parts water (58% w/w) at 21°C. Slightly soluble in ethanol: 1 in 5000 parts ethanol.

11 Stability and Storage Conditions

Tagatose is stable under pH conditions typically encountered in foods (pH >3). It is a reducing sugar and undergoes the Maillard reaction.

Tagatose is stable under typical storage conditions. It caramelizes at elevated temperature.

12 Incompatibilities

A Maillard-type condensation reaction is likely to occur between tagatose and compounds with a primary amine group to form brown or yellow-brown colored Amidori compounds. Reducing sugars will also interact with secondary amines to form an imine, but without any accompanying yellow-brown discoloration.

13 Method of Manufacture

Tagatose is obtained from D-galactose by isomerization under alkaline conditions in the presence of calcium.

14 Safety

Tagatose is safe for use in food and beverages. It has been used in pharmaceutical products.(1)

15 Handling Precautions

Observe normal precautions appropriate to the circumstances and quantity of the material handled. Excessive generation of dust, or inhalation of dust, should be avoided.

16 Regulatory Status

GRAS listed. Included in the FDA Inactive Ingredients Database (oral and rectal solutions).

17 Related Substances

DL-Tagatose; L-tagatose.

DL-Tagatose
Empirical formula $C_6H_{12}O_6$
CAS number [17598-81-1]
Synonyms *lyxo*-2-Hexulose

L-Tagatose
Empirical formula $C_6H_{12}O_6$
CAS number [17598-82-2]
Melting point 134–135°C
Specific rotation α_D^{16} = +1° (2% aqueous solution)

Comments Sweetening agent for pharmaceutical and personal aid products.

18 Comments

The EINECS number for tagatose is 201-772-3. The PubChem Compound ID (CID) for tagatose is 92092.

19 Specific References

1 Levin GV. Tagatose, the new GRAS sweetener and health product. *J Med Food* 2002; 5: 23–36.
2 Lu Y. Humectancies of *d*-tagatose and *d*-sorbitol. *Int J Cosmet Sci* 2001; 23: 175–181.
3 FAO/WHO. Joint FAO/WHO Expert Committee on Food Additives. Chemical and Technical Assessment, 2003; 61.
4 Freimund S, *et al*. Convenient chemo-enzymatic synthesis of D-tagatose. *J Carbohydr Chem* 1996; **15**(1): 115–120.
5 Que L, Gray GR. ^{13}C Nuclear magnetic resonance spectra and the tautomeric equilibria of ketohexoses in solution. *Biochemistry* 1974; **13**(1): 146–153.

20 General References

—

21 Author

GE Amidon.

22 Date of Revision

1 March 2012.

Talc

1 Nonproprietary Names

BP: Purified Talc

JP: Talc

PhEur: Talc

USP: Talc

2 Synonyms

Altalc; E553b; hydrous magnesium calcium silicate; hydrous magnesium silicate; *Imperial*; *Luzenac Pharma*; magnesium hydrogen metasilicate; *Magsil Osmanthus*; *Magsil Diamond*; *Magsil Star*; powdered talc; purified French chalk; *Purtalc*; soapstone; steatite; *Superiore*; talcum.

3 Chemical Name and CAS Registry Number

Talc [14807-96-6]

4 Empirical Formula and Molecular Weight

Talc is a purified, hydrated, magnesium silicate, approximating to the formula $Mg_6(Si_2O_5)_4(OH)_4$. It may contain small, variable amounts of aluminum silicate and iron.

5 Structural Formula

See Section 4.

6 Functional Category

Anticaking agent; glidant; tablet and capsule diluent; tablet and capsule lubricant.

7 Applications in Pharmaceutical Formulation or Technology

Talc was once widely used in oral solid dosage formulations as a lubricant, glidant, and diluent, *see* Table I,[1–3] although today it is less commonly used. However, it is widely used as a dissolution retardant in the development of controlled-release products.[4–6] Talc is also used as a lubricant in tablet formulations,[7] in a novel powder coating for extended-release pellets,[8] and as an adsorbent.[9]

In topical preparations, talc is used as a dusting powder, although it should not be used to dust surgical gloves; *see* Section 14. Talc is a natural material; it may therefore frequently contain microorganisms and should be sterilized when used as a dusting powder; *see* Section 11.

Talc continues to be widely used in coatings, and the ratio of talc to polymer is important in the release rate of a drug from coated beads.[10] Talc has also been shown to be useful in oral disintegrating tablets as it is insensitive to tablet hardness unlike other lubricants.[11]

Talc is additionally used to clarify liquids and is also used in cosmetics and food products, mainly for its lubricant properties.

Table I: Uses of talc.

Use	Concentration (%)
Dusting powder	90.0–99.0
Glidant and tablet lubricant	1.0–10.0
Tablet and capsule diluent	5.0–30.0

T

8 Description

Talc occurs as a very fine, white to grayish-white, odorless, impalpable, unctuous, crystalline powder. It adheres readily to the skin and is soft to the touch and free from grittiness.

9 Pharmacopeial Specifications

The pharmacopeial specifications for talc have undergone harmonization of many attributes for JP, PhEur, and USP–NF.

See Table II. *See also* Section 18.

10 Typical Properties

Acidity/alkalinity pH = 7–10 for a 20% w/v aqueous dispersion.

Hardness (Mohs) 1.0–1.5

Moisture content Talc absorbs insignificant amounts of water at 25°C and relative humidities up to about 90%.

Particle size distribution Varies with the source and grade of material. Two typical grades are ≥99% through a 74 μm (#200 mesh) or ≥99% through a 44 μm (#325 mesh).

Refractive index n_D^{20} = 1.54–1.59

Solubility Practically insoluble in dilute acids and alkalis, organic solvents, and water.

SEM 1: Excipient: talc (*Purtalc*); manufacturer: Charles B Chrystal Co., Inc.; lot no.: 1102A-2; magnification: 1200×; voltage: 10 kV.

SEM 2: Excipient: talc; magnification: 1000×; voltage: 3 kV.

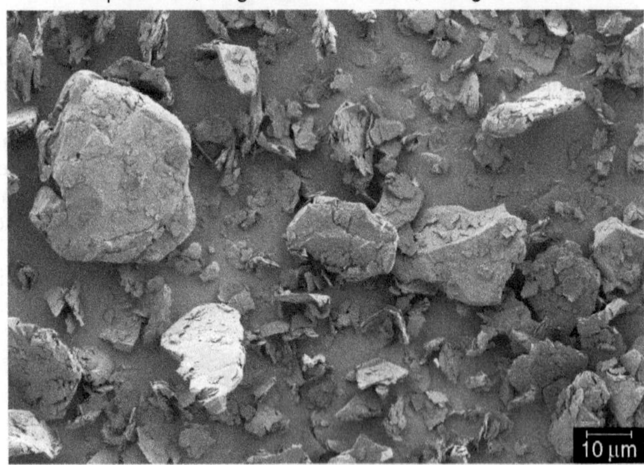

SEM 3: Excipient: talc; magnification: 1000×; voltage: 3 kV.

Specific gravity 2.7–2.8
Specific surface area 2.41–2.42 m^2/g
Spectroscopy

IR spectra *see* Figure 1.

NIR spectra *see* Figure 2.

Table II: Pharmacopeial specifications for talc.

Test	JP XV	PhEur 7.4	USP35–NF30
Identification	+	+	+
Characters	+	+	—
Acid-soluble substances	≤2.0%	—	—
Acidity or alkalinity	—	+	+
Production	—	+	—
pH	—	—	—
Water-soluble substances	—	≤0.2%	≤0.1%
Aluminum	—	≤2.0%	2.0%
Calcium	—	≤0.9%	0.9%
Iron	—	≤0.25%	0.25%
Lead	—	≤10 ppm	≤10 ppm
Magnesium	—	17.0–19.5%	17.0–19.5%
Loss on ignition	≤5.0%	≤7.0%	≤7.0%
Microbial contamination	—	+	+
Aerobic bacteria	—	≤10^2 cfu/g	10^2 cfu/g[a] 10^3 cfu/g[b]
Fungi	—	≤10^2 cfu/g	50 cfu/g[a] 10^2 cfu/g[b]
Acid and alkali-soluble substances	≤4.0 mg	—	≤2.0%
Water-soluble iron	+	—	—
Arsenic	≤4 ppm	—	—
Absence of asbestos	—	—	+

(a) If intended for topical administration.
(b) If intended for oral administration.

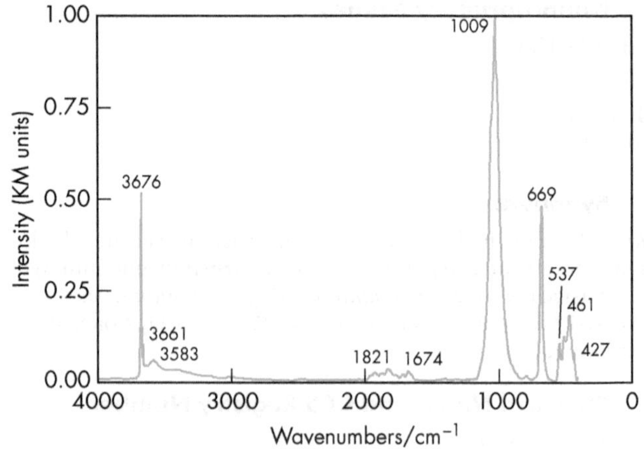

Figure 1: Infrared spectrum of talc measured by diffuse reflectance. Adapted with permission of Informa Healthcare.

11 Stability and Storage Conditions

Talc is a stable material and may be sterilized by heating at 160°C for not less than 1 hour. It may also be sterilized by exposure to ethylene oxide or gamma irradiation.(12)

Talc should be stored in a well-closed container in a cool, dry place.

12 Incompatibilities

Incompatible with quaternary ammonium compounds.

13 Method of Manufacture

Talc is a naturally occurring hydropolysilicate mineral found in many parts of the world including Australia, China, Italy, India, France, and the US.(13)

The purity of talc varies depending on the country of origin. For example, Italian types are reported to contain calcium silicate as the contaminant; Indian types contain aluminum and iron oxides; French types contain aluminum oxide; and American types contain calcium carbonate (California), iron oxide (Montana), aluminum

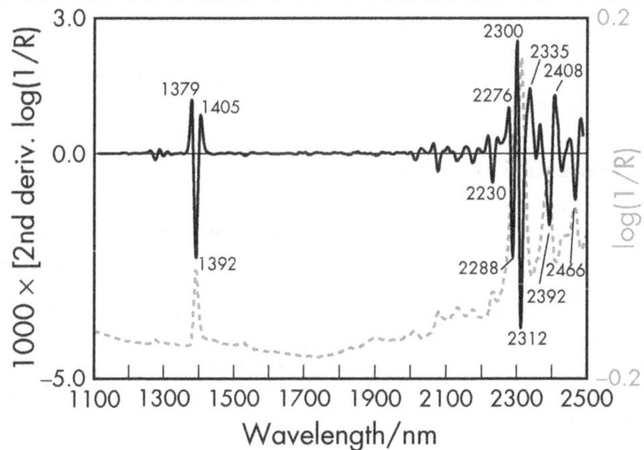

Figure 2: Near-infrared spectrum of talc measured by reflectance.

and iron oxides (North Carolina), or aluminum oxide (Alabama).[14]

Naturally occurring talc is mined and pulverized before being subjected to flotation processes to remove various impurities such as asbestos (tremolite); carbon; dolomite; iron oxide; and various other magnesium and carbonate minerals. Following this process, the talc is finely powdered, treated with dilute hydrochloric acid, washed with water, and then dried. The processing variables of agglomerated talc strongly influence its physical characteristics.[15–17]

14 Safety

Talc is used mainly in tablet and capsule formulations. Talc is not absorbed systemically following oral ingestion and is therefore regarded as an essentially nontoxic material. However, intranasal or intravenous abuse of products containing talc can cause granulomas in body tissues, particularly the lungs.[18–20] Contamination of wounds or body cavities with talc may also cause granulomas; therefore, it should not be used to dust surgical gloves. Inhalation of talc causes irritation and may cause severe respiratory distress in infants;[21] *see also* Section 15.

Although talc has been extensively investigated for its carcinogenic potential, and it has been suggested that there is an increased risk of ovarian cancer in women using talc, the evidence is inconclusive.[22,23] However, talc contaminated with asbestos has been proved to be carcinogenic in humans, and asbestos-free grades should therefore be used in pharmaceutical products.[24]

Also, long-term toxic effects of talc contaminated with large quantities of hexachlorophene caused serious irreversible neurotoxicity in infants accidentally exposed to the substance.[25]

15 Handling Precautions

Observe normal precautions appropriate to the circumstances and quantity of material handled. Talc is irritant if inhaled and prolonged excessive exposure may cause pneumoconiosis.

In the UK, the workplace exposure limit for talc is 1 mg/m^3 of respirable dust long-term (8-hour TWA).[26] Eye protection, gloves, and a respirator are recommended.

16 Regulatory Status

Accepted for use as a food additive in Europe. Included in the FDA Inactive Ingredients Database (buccal tablets; oral capsules and tablets; rectal and topical preparations). Included in nonparenteral medicines licensed in the UK. Included in the Canadian Natural Health Products Ingredients Database.

17 Related Substances

Bentonite; magnesium aluminum silicate; magnesium silicate; magnesium trisilicate.

18 Comments

Talc has undergone harmonization of many attributes for JP, PhEur, and USP–NF by the Pharmacopeial Discussion Group. For further information see the General Information Chapter <1196> in the USP35–NF30, the General Chapter 5.8 in PhEur 7.4, along with the 'State of Work' document on the PhEur EDQM website, and also the General Information Chapter 8 in the JP XV.

Various grades of talc are commercially available that vary in their chemical composition depending upon their source and method of preparation.[13,27,28]

Talc derived from deposits that are known to contain associated asbestos is not suitable for pharmaceutical use. Tests for amphiboles and serpentines should be carried out to ensure that the product is free of asbestos.

A specification for talc is contained in the *Food Chemicals Codex* (FCC).[29]

The EINECS number for talc is 238-877-9. The PubChem Compound ID (CID) for talc includes 26924, 443754 and 16211421.

19 Specific References

1 Dawoodbhai S, Rhodes CT. Pharmaceutical and cosmetic uses of talc. *Drug Dev Ind Pharm* 1990; **16**: 2409–2429.

2 Dawoodbhai S, *et al.* Optimization of tablet formulations containing talc. *Drug Dev Ind Pharm* 1991; **17**: 1343–1371.

3 Wang DP, *et al.* Formulation development of oral controlled release pellets of diclofenac sodium. *Drug Dev Ind Pharm* 1997; **23**: 1013–1017.

4 Fassihi RA, *et al.* Potential use of magnesium stearate and talc as dissolution retardants in the development of controlled release drug delivery systems. *Pharm Ind* 1994; **56**: 579–583.

5 Fassihi R, *et al.* Application of response surface methodology to design optimization in formulation of a typical controlled release system. *Drugs Made Ger* 1996; **39**(Oct–Dec): 122–126.

6 Schultz P, *et al.* New multiparticulate delayed release system. Part 2. Coating formulation and properties of free films. *J Control Release* 1997; **47**: 191–199.

7 Oetari RA, *et al.* Formulation of PGV-O a new antiinflammatory agent as a tablet dosage form. *Indonesian J Pharm* 2003; **14**(4): 160–168.

8 Pearnchob N, Bodmeier R. Dry powder coating of pellets with micronized Eudragil (R) RS for extended drug release. *Pharm Res* 2003; **20**(12): 1970–1976.

9 Mani N, *et al.* Microencapsulation of a hydrophilic drug into a hydrophobic matrix using a salting-out procedure: II. Effects of adsorbents on microsphere properties. *Drug Dev Ind Pharm* 2004; **30**(1): 83–93.

10 El-Malah Y, Nazzal S. Effect of *Eudragit RS 30D* and talc powder on verapamil hydrochloride release from beads coated with drug layer matrices. *AAPS PharmSciTech* 2008; **9**(1): 75–83.

11 Kuno Y, *et al.* Effect of the type of lubricant on characteristics of oral disintegrating tablets manufactured using the phase transition of sugar alcohol. *Eur J Pharm Biopharm* 2008; **69**(3): 986–992.

12 Bubik JS. Preparation of sterile talc for treatment of pleural effusion [letter]. *Am J Hosp Pharm* 1992; **49**: 562–563.

13 Grexa RW, Parmentier CJ. Cosmetic talc properties and specifications. *Cosmet Toilet* 1979; **94**(2): 29–33.

14 Hoepfner EM *et al.* eds. *Fiedler Encyclopedia of Excipients for Pharmaceuticals, Cosmetics and Related Areas*, 5th edn, II. Aulendorf: Editio Cantor Verlag, 2002: 1556–1559.

15 Lin K, Peck GE. Development of agglomerated talc. Part 1. Evaluation of fluidized bed granulation parameters on the physical properties of agglomerated talc. *Drug Dev Ind Pharm* 1995; **21**: 447–460.

16 Lin K, Peck GE. Development of agglomerated talc. Part 2. Optimization of the processing parameters for the preparation of granulated talc. *Drug Dev Ind Pharm* 1995; **21**: 159–173.

17 Lin K, Peck GE. Development of agglomerated talc. Part 3. Comparisons of the physical properties of the agglomerated talc prepared by

three different processing methods. *Drug Dev Ind Pharm* 1996; 22: 383–392.
18 Schwartz IS, Bosken C. Pulmonary vascular talc granulomatosis. *J Am Med Assoc* 1986; 256: 2584.
19 Johnson DC, *et al.* Foreign body pulmonary granulomas in an abuser of nasally inhaled drugs. *Pediatrics* 1991; 88: 159–161.
20 Sparrow SA, Hallam LA. Talc granulomas [letter]. *Br Med J* 1991; 303: 58.
21 Pairaudeau PW, *et al.* Inhalation of baby powder: an unappreciated hazard. *Br Med J* 1991; 302: 1200–1201.
22 Longo DL, Young RC. Cosmetic talc and ovarian cancer. *Lancet* 1979; ii: 349–351.
23 Phillipson IM. Talc quality [letter]. *Lancet* 1980; i: 48.
24 International Agency for Research on Cancer/World Health Organization. *Silica and Some Silicates: IARC Monographs on the Evaluation of the Carcinogenic Risk of Chemicals to Humans.* Geneva: WHO, 1987: 42.
25 Anonymous. Long-term sequelae of hexachlorophene poisoning. *Prescrire Int* 1992; 1: 168.
26 Health and Safety Executive. *EH40/2005: Workplace Exposure Limits.* Sudbury: HSE Books, 2011. http://www.hse.gov.uk/pubns/priced/eh40.pdf (accessed 23 February 2012).
27 Phadke DS, *et al.* Evaluation of batch-to-batch and manufacturer-to-manufacturer variability in the physical properties of talc and stearic acid. *Drug Dev Ind Pharm* 1994; 20: 859–871.
28 Lin K, Peck GE. Characterization of talc samples from different sources. *Drug Dev Ind Pharm* 1994; 20: 2993–3003.
29 *Food Chemicals Codex*, 7th edn (Suppl 1). Bethesda, MD: United States Pharmacopeia, 2010.

20 General References

European Directorate for the Quality of Medicines and Healthcare (EDQM). European Pharmacopoeia – State Of Work Of International Harmonisation. *Pharmeuropa* 2011; 23(4): 713–714. www.edqm.eu/site/-614.html (accessed 23 February 2012).
Gold G, Campbell JA. Effects of selected USP talcs on acetylsalicylic acid stability in tablets. *J Pharm Sci* 1964; 53: 52–54.
Leterne P, *et al.* Influence of the morphogranulometry and hydrophobicity of talc on its antisticking power in the production of tablets. *Int J Pharm* 2005; 289: 109–115.

21 Author

AH Kibbe.

22 Date of Revision

1 March 2012.

Tartaric Acid

1 Nonproprietary Names

BP: Tartaric Acid
JP: Tartaric Acid
PhEur: Tartaric Acid
USP–NF: Tartaric Acid

2 Synonyms

Acidum tartaricum; dextrotartaric acid; L-(+)-2,3-dihydroxybutanedioic acid; (2*R*,3*R*)-2,3-dihydroxybutane-1,4-dioic acid; 1,2-dihydroxyethane-1,2-dicarboxylic acid; 2,3-dihydroxysuccinic acid; *d*-α,β-dihydroxysuccinic acid; E334; succinic acid, 2,3-dihydroxy; *d*-tartaric acid; L-(+)-tartaric acid; threaric acid.

3 Chemical Name and CAS Registry Number

[*R*-(*R**,*R**)]-2,3-Dihydroxybutanedioic acid [87-69-4]

4 Empirical Formula and Molecular Weight

$C_4H_6O_6$ 150.09

5 Structural Formula

O H OH OH HO H OH O

6 Functional Category

Acidulant; complexing agent; flavoring agent.

7 Applications in Pharmaceutical Formulation or Technology

Tartaric acid is used in beverages, confectionery, food products, and pharmaceutical formulations as an acidulant. It may also be used as a sequestering agent and as an antioxidant synergist. In pharmaceutical formulations, it is widely used in combination with bicarbonates, as the acid component of effervescent granules, powders, and tablets.[1]

Tartaric acid is also used to form molecular compounds (salts and cocrystals) with active pharmaceutical ingredients to improve physicochemical properties such as dissolution rate and solubility.[2,3]

8 Description

Tartaric acid occurs as colorless monoclinic crystals, or a white or almost white crystalline powder. It is odorless, with an extremely tart taste.

9 Pharmacopeial Specifications

See Table I.

10 Typical Properties

Acidity/alkalinity pH = 2.2 (0.1 N solution)
Density 1.76 g/cm^3
Dissociation constant
pK_{a1} = 2.98 at 25°C;
pK_{a2} = 4.34 at 25°C.
Heat of combustion 1151 kJ/mol (275.1 kcal/mol)

Table I: Pharmacopeial specifications for tartaric acid.

Test	JP XV	PhEur 7.4	USP35–NF30
Identification	+	+	+
Characters	–	+	–
Appearance of solution	–	+	–
Specific rotation	–	+12.0° to +12.8°	+12.0° to +13.0°
Loss on drying	≤0.5%	≤0.2%	≤0.5%
Sulfated ash	–	≤0.1%	–
Residue on ignition	≤0.05%	–	≤0.1%
Chloride	–	≤100 ppm	–
Oxalic acid	–	≤350 ppm	–
Oxalate	+	–	+
Sulfate	≤0.048%	≤150 ppm	+
Calcium	+	≤200 ppm	–
Heavy metals	≤10 ppm	≤10 ppm	≤10 ppm
Arsenic	≤1 ppm	–	–
Assay (dried basis)	≥99.7%	99.5–101.0%	99.7–100.5%

Table II: Solubility of tartaric acid.

Solvent	Solubility at ≈25°C unless otherwise stated
Chloroform	Practically insoluble
Ethanol	1 in 3
Ether	1 in 250
Glycerin	Soluble
Methanol	1 in 1.7
Propan-1-ol	1 in 10.5
Water	1 in 0.75
	1 in 0.5 at 100°C

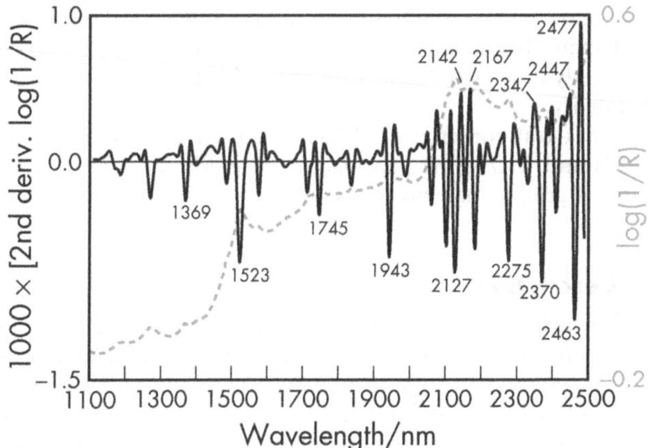

Figure 1: Near-infrared spectrum of tartaric acid measured by reflectance.

Heat of fusion 32.3 kJ/mol[4]
Melting point 168–170°C
Osmolarity A 3.9% w/v aqueous solution is isoosmotic with serum.
Solubility *see* Table II.
Specific heat 1.20 J/g (0.288 cal/g) at 20°C.
Specific rotation $[\alpha]_D^{20}$ = +12.0° (20% w/v aqueous solution).
Spectroscopy
NIR spectra *see* Figure 1.

11 Stability and Storage Conditions

The bulk material is stable and should be stored in a well-closed container in a cool, dry place.

12 Incompatibilities

Tartaric acid is incompatible with silver and reacts with metal carbonates and bicarbonates (a property exploited in effervescent preparations).

13 Method of Manufacture

Tartaric acid occurs naturally in many fruits as the free acid or in combination with calcium, magnesium, and potassium.

Commercially, L-(+)-tartaric acid is manufactured from potassium tartrate (cream of tartar), a by-product of wine making. Potassium tartrate is treated with hydrochloric acid, followed by the addition of a calcium salt to produce insoluble calcium tartrate. This precipitate is then removed by filtration and reacted with 70% sulfuric acid to yield tartaric acid and calcium sulfate.

14 Safety

Tartaric acid is widely used in food products and oral, topical, and parenteral pharmaceutical formulations. It is generally regarded as a nontoxic and nonirritant material; however, strong tartaric acid solutions are mildly irritant and if ingested undiluted may cause gastroenteritis.

An acceptable daily intake for L-(+)-tartaric acid has not been set by the WHO, although an acceptable daily intake of up to 30 mg/kg body-weight for monosodium L-(+)-tartrate has been established.[5]

LD_{50} (mouse, IV): 0.49 g/kg[6]
LD_{Lo} (dog, oral): 5 g/kg [6]
LD_{Lo} (rabbit, oral): 5 g/kg[6]
LD_{Lo} (rat, oral): 7.5 g/kg[7]
TD_{Lo} (monkey, IV): 376 mg/kg (2 day, intermittent)[8]

15 Handling Precautions

Observe normal precautions appropriate to the circumstances and quantity of material handled. Tartaric acid may be irritating to the eyes; eye protection and rubber or plastic gloves are recommended. When heated to decomposition, tartaric acid emits acrid smoke and fumes.

16 Regulatory Status

GRAS listed. Accepted for use as a food additive in Europe. Included in the FDA Inactive Ingredients Database (IM and IV injections; oral solutions, syrups and tablets; sublingual tablets; topical films; rectal and vaginal preparations). Included in the EAFUS database and listed on the TSCA inventory. Included in nonparenteral medicines licensed in the UK. Included in the Canadian Natural Health Products Ingredients Database.

T

17 Related Substances

Citric acid monohydrate; fumaric acid; malic acid.

18 Comments

L-(+)-Tartaric acid, the optical isomer usually encountered, is the naturally occurring form and is specified as tartaric acid in the PhEur 7.4 and USP35–NF30.

Tartaric acid has been studied for use in tablets as a pH modifier to enhance dissolution.[8]

A specification for tartaric acid is contained in the *Food Chemicals Codex* (FCC).[9]

The EINECS number for tartaric acid is 201-766-0.

19 Specific References

1 Shirsand SB, *et al.* Formulation design and optimization of fast disintegrating Lorazepam tablets by effervescent method. *Indian J Pharm Sci* 2010; 72(4): 431–436.

2 Black SN, *et al.* Structure, solubility, screening, and synthesis of molecular salts. *J Pharm Sci* 2007; **96**: 1053–1068.
3 Childs SL, Hardcastle KI. Cocrystals of piroxicam with carboxylic acids. *Crystal Growth Design* 2007; **7**: 1291–1304.
4 Mura P, *et al.* Differential scanning calorimetry in compatibility testing of picotamide with pharmaceutical excipients. *Thermochim Acta* 1998; **321**: 59–65.
5 FAO/WHO. Evaluation of certain food additives. Twenty-first report of the joint FAO/WHO expert committee on food additives. *World Health Organ Tech Rep Ser* 1978; No. 617.
6 Lewis RJ, ed. *Sax's Dangerous Properties of Industrial Materials*, 11th edn. New York: Wiley, 2004; 3349.
7 National Library of Medicine. Toxnet: Registry of Toxic Effects of Chemical Substances: Tartaric acid, 2011.
8 Badawy SI, *et al.* Formulation of solid dosage forms to overcome gastric pH interaction of the factor Xa inhibitor, BMS-561389. *Pharm Res* 2006; **23**(5): 989–996.
9 *Food Chemicals Codex*, 7th edn. Bethesda, MD: United States Pharmacopeia, 2011; 1492.

20 General References

Kassaian J-M. Tartaric Acid. In: *Ullmann's Encyclopedia of Industrial Chemistry*, 7th edn, 2010.

Schmidt PC, Christin I. Effervescent tablets, an almost forgotten pharmaceutical form [in German]. *Pharmazie* 1990; **45**(2): 89–101.

Sendall FEJ, Staniforth JN. A study of powder adhesion to metal surfaces during compression of effervescent pharmaceutical tablets. *J Pharm Pharmacol* 1986; **38**: 489–493.

Usui F, Carstensen JT. Interactions in the solid state I: interactions of sodium bicarbonate and tartaric acid under compressed conditions. *J Pharm Sci* 1985; **74**: 1293–1297.

21 Authors

PE Luner, BJ Murphy.

22 Date of Revision

1 March 2012.

Tetrafluoroethane (HFC)

1 Nonproprietary Names

None adopted.

2 Synonyms

Dymel 134a/P; fluorocarbon 134a; *Frigen 134a*; *Genetron 134a*; HFA 134a; HFC 134a; *Isceon 134a*; *Klea 134a*; propellant 134a; refrigerant 134a; *Solkane 134a*; *Suva 134a*; *Zephex 134a*.

3 Chemical Name and CAS Registry Number

1,1,1,2-Tetrafluoroethane [811-97-2]

4 Empirical Formula and Molecular Weight

$C_2H_2F_4$ 102.0

5 Structural Formula

```
     F     F
     |     |
F————C—————C————H
     |     |
     F     H
```

6 Functional Category

Aerosol propellant.

7 Applications in Pharmaceutical Formulation or Technology

Tetrafluoroethane is a hydrofluorocarbon (HFC) or hydrofluoroalkane (HFA) aerosol propellant (contains hydrogen, fluorine, and carbon) as contrasted to a CFC (chlorine, fluorine, and carbon). The lack of chlorine in the molecule and the presence of hydrogen reduce the ozone depletion activity to practically zero. Hence tetrafluoroethane is an alternative to CFCs in the formulation of metered-dose inhalers (MDIs).[1–9] It has replaced CFC-12 as a refrigerant and propellant since it has essentially the same vapor pressure. Its very low Kauri-butanol value and solubility parameter indicate that it is not a good solvent for the commonly used surfactants for MDIs. Sorbitan trioleate, sorbitan sesquioleate, oleic acid, and soya lecithin show limited solubility in tetrafluoroethane and the amount of surfactant that actually dissolves may not be sufficient to keep a drug readily dispersed. Up to 10% ethanol may be used to increase its solubility.

When tetrafluoroethane (P-134a) is used for pharmaceutical aerosols and MDIs, the pharmaceutical grade must be specified. Industrial grades may not be satisfactory due to their impurity profiles.

8 Description

Tetrafluoroethane is a liquefied gas and exists as a liquid at room temperature when contained under its own vapor pressure, or as a gas when exposed to room temperature and atmospheric pressure. The liquid is practically odorless and colorless. The gas in high concentrations has a slight etherlike odor. Tetrafluoroethane is noncorrosive, nonirritating, and nonflammable.

9 Pharmacopeial Specifications

—

10 Typical Properties

Boiling point −26.5°C
Density 1.21 g/cm^3 for liquid at 25°C
Flammability Nonflammable
Freezing point −108°C
Kauri-butanol value 8
Solubility Soluble in ethanol (95%), ether, and 1 in 1294 parts of water at 20°C.
Specific gravity 1.208 at 25°C
Vapor density (absolute) 4.466 g/L at standard temperature and pressure.
Vapor density (relative) 3.6 (air = 1) at 25°C
Vapor pressure
662 kPa (96 psia) at 25°C

Viscosity (dynamic)

0.222 mPa s (0.222 cP) for liquid at 20°C;

0.210 mPa s (0.210 cP) for liquid at 25°C.

11 Stability and Storage Conditions

Tetrafluoroethane is a nonreactive and stable material. The liquified gas is stable when used as a propellant and should be stored in a metal cylinder in a cool dry place.

12 Incompatibilities

The major incompatibility of tetrafluoroethane is its lack of miscibility with water. Since it has a very low Kauri-butanol value, tetrafluoroethane is considered to be a very poor solvent for most drugs used in MDI formulations. It also shows a low solubility for some of the commonly used MDI surfactants.

13 Method of Manufacture

Tetrafluoroethane can be prepared by several different routes; however, the following routes of preparation illustrate the methods used:

Isomerization/hydrofluorination of 1,1,2-trichloro-1,2,2-trifluoroethane (CFC-113) to 1,1-dichloro-1,2,2,2-tetrafluoroethane (CFC-114a), followed by hydrodechlorination of the latter.

Hydrofluorination of trichloroethylene, via 1-chloro-1,1,1-trifluoroethane (HCFC-133a).

14 Safety

Tetrafluoroethane is used as a refrigerant and as a non-CFC propellant in various aerosols including topical pharmaceuticals and MDIs. Tetrafluoroethane is regarded as nontoxic and nonirritating when used as directed. No acute or chronic hazard is present when exposures to the vapor are below the acceptable exposure limit (AEL) of 1000 ppm, 8-hour and 12-hour time weighed average (TWA).[10] In this regard it has the same value as the threshold limit value (TLV) for CFC-12. Inhaling a high concentration of tetrafluoroethane vapors can be harmful and is similar to inhaling vapors of CFC-12. Intentional inhalation of vapors of tetrafluoroethane can be dangerous and may cause death. The same labeling required on CFC aerosols would be required for those containing tetrafluoroethane as a propellant (except for the EPA requirement). *See* Chlorofluorocarbons, Section 14.

15 Handling Precautions

Tetrafluoroethane is usually encountered as a liquefied gas and appropriate precautions for handling should be taken. Eye protection, gloves, and protective clothing are recommended. Tetrafluoroethane should be handled in a well-ventilated environment. The vapors are heavier than air and do not support life; therefore, when cleaning large tanks that have contained the propellant, adequate provisions for oxygen supply in the tanks must be made in order to protect workers cleaning the tanks.

Although nonflammable, when heated to decomposition tetrafluoroethane emits toxic fumes.

In the UK, the long-term workplace exposure limit (8-hour TWA) for tetrafluoroethane is 4240 mg/m^3 (1000 ppm).[11]

16 Regulatory Status

Included in the FDA Inactive Ingredients Database (aerosol formulations for inhalation and nasal applications). Included in nonparenteral medicines licensed in the UK.

17 Related Substances

Difluoroethane; heptafluoropropane.

18 Comments

The use of tetrafluoroethane as a propellant for MDIs has been the subject of numerous patents throughout the world. These patents cover the formulation of MDIs and use of specific surfactants, cosolvents, etc. A US patent claims a self-propelling aerosol formulation that may be free of CFCs and which comprises a medicament, 1,1,1,2-tetrafluoroethane, a surface-active agent, and at least one compound having a higher polarity than 1,1,1,2-tetrafluoroethane.[12] Another patent has been issued by the European Patent Office and has 14 claims, among them a claim that includes tetrafluoroethane, an alcohol (such as ethanol), surfactant, and medicament.[13] The formulator is referred to the patent literature prior to formulating a MDI with tetrafluoroethane as the propellant. The formulation of MDI with this non-CFC propellant is complicated since tetrafluoroethane serves as a replacement for dichlorodifluoromethane or dichlorotetrafluoroethane. The use of an HFC as the propellant also requires a change in manufacturing procedure, which necessitates a redesign of the filling and packaging machinery for a MDI.[14]

Currently, there are no pharmacopeial specifications for tetrafluoroethane. However, typical specifications are shown in Table I.

Table I: Typical product specifications for tetrafluoroethane.

Test	Value
Appearance	Clear and colorless
High boiling impurities	≤0.01%
Acidity as HCl	≤0.1 ppm
Non-volatile residue	≤5 ppm
Non-absorbable gases	≤1.5%
Water	≤10 ppm
Total unidentified impurities	≤10 ppm
Assay	≥99.99%

19 Specific References

1 Strobach DR. Alternative to CFCs. *Aerosol Age* 1988; **33**(7): 32–3342–43.

2 Daly J. Properties and toxicology of CFC alternatives. *Aerosol Age* 1990; **35**(2): 26–2740.

3 Dalby RN, *et al.* CFC propellant substitution: P-134a as a potential replacement for P-12 in MDIs. *Pharm Technol* 1990; **14**(3): 26–33.

4 Kontny MJ, *et al.* Issues surrounding MDI formulation development with non-CFC propellants. *J Aerosol Med* 1991; **4**(3): 181–187.

5 Anonymous. 3M first with a CFC-free asthma inhaler. *Pharm J* 1995; **254**: 388.

6 Taggart SCO, *et al.* GR106642X: a new, non-ozone depleting propellant for inhalers. *Br Med J* 1995; **310**: 1639–1640.

7 Elvecrog J. Metered dose inhalers in a CFC-free future. *Pharm Technol Eur* 1997; **9**(1): 52–55.

8 Tansey IP. Changing to CFC-free inhalers: the technical and clinical challenges. *Pharm J* 1997; **259**: 896–898.

9 McDonald KJ, Martin GP. Transition to CFC-free metered dose inhalers: into the new millenium. *Int J Pharm* 2000; **201**: 89–107.

10 DuPont. Technical literature: *Dymel 134a/P* pharmaceutical grade HFC-134a propellant, 1996.

11 Health and Safety Executive. *EH40/2005: Workplace Exposure Limits*. Sudbury: HSE Books, 2011. http://www.hse.gov.uk/pubns/priced/eh40.pdf (accessed 12 April 2012).

12 Purewal TS, Greenleaf DJ. *Medicinal aerosol formulations*. United States Patent No. 5,605,674, 1997.

13 Purewal TS, Greenleaf DJ. *Medicinal aerosol formulations*. European Patent 372777B1, 1993.

14 Tzou T, *et al.* Drug form selection in albuterol-containing metered-dose inhaler formulations and its impact on chemical and physical stability. *J Pharm Sci* 1997; **86**: 1352–1357.

T

20 General References

Harrison LI, *et al*. Twenty-eight day double-blind safety study of an HFA 134a inhalation aerosol system in healthy subjects. *J Pharm Pharmacol* 1996; 48: 596–600.

Hoet P, *et al*. Epidemic of liver disease caused by hydrochlorofluorocarbons used as ozone-sparing substitutes of chlorofluorocarbons. *Lancet* 1997; 350: 556–559.

Sawyer E, *et al*. Microorganism survival in non-CFC propellant P134a and a combination of CFC propellants P11 and P12. *Pharm Technol* 2001; 25(3): 90–96.

Sciarra CJ, Sciarra JJ. Aerosols. In: *Remington: The Science and Practice of Pharmacy*, 21st edn. Philapdelphia, PA: Lippincott, Williams and Wilkins, 2005: 1000–1017.

Steed KP, *et al*. The oropharyngeal and lung deposition patterns of a fusafungine MDI spray delivered by HFA 134a propellant or by CFC 12 propellant. *Int J Pharm* 1995; **123**: 291–293.

Tiwari D, *et al*. Compatibility evaluation of metered-dose inhaler valve elastomers with tetrafluoroethane (P134a), a non-CFC propellant. *Drug Dev Ind Pharm* 1998; **24**: 345–352.

21 Authors

CJ Sciarra, JJ Sciarra.

22 Date of Revision

12 April 2012.

Thaumatin

1 Nonproprietary Names

None adopted.

2 Synonyms

E957; katemfe; *Talin*; taumatin; taumatina; thalin; thaumatine; thaumatins; thaumatins protein.

3 Chemical Name and CAS Registry Number

Thaumatin [53850-34-3]

4 Empirical Formula and Molecular Weight

See Section 5.

5 Structural Formula

Thaumatin is a mixture of five thaumatin proteins: thaumatins I, II, III, and a and b, where thaumatins I and II predominate. Thaumatins I and II consist of almost identical sequences of amino acids. There are no unusual side-chains or peptide linkages, and there are no end-group substitutions.

6 Functional Category

Flavor enhancer; sweetening agent.

7 Applications in Pharmaceutical Formulation or Technology

Thaumatin is a naturally occurring intense sweetening agent approximately 2000–3000 times as sweet as sucrose. It has a delayed-onset taste profile and long (up to one hour) licorice-like aftertaste. It is used extensively in food applications as a sweetening agent and flavor enhancer, and has potential for use in pharmaceutical applications such as oral suspensions.[1]

8 Description

Thaumatin occurs as a pale-brown colored, odorless, hygroscopic powder with an intensely sweet taste.

9 Pharmacopeial Specifications

—

10 Typical Properties

Solubility *see* Table I.

Table I: Solubility of thaumatin.

Solvent	Solubility at 25°C unless otherwise stated
Acetone	Practically insoluble
Ethanol (95%)	Soluble
Glycerin	Soluble
Propylene glycol	Soluble
Water	1 in 5 at pH 3

11 Stability and Storage Conditions

Thaumatin is stable in aqueous solutions at pH 2–8. It is also heat-stable at less than pH 5.5 (e.g. during baking, canning, pasteurizing, or UHT processes).

12 Incompatibilities

—

13 Method of Manufacture

Thaumatin is a naturally occurring intense sweetener isolated from the fruit of the African plant *Thaumatococcus daniellii* (Benth).[2] Commercially, thaumatin is produced by aqueous extraction under reduced pH conditions followed by other physical processes such as reverse osmosis.

14 Safety

Thaumatin is accepted for use in food products either as a sweetener or as a flavor modifier in a number of areas including Europe and Australia. It is also used in oral hygiene products such as mouthwashes and toothpastes, and has been proposed for use in oral pharmaceutical formulations. Thaumatin is generally regarded as a relatively nontoxic and nonirritant material when used as an excipient. In Europe, because of its lack of toxicity, an ADI has been set of 'not specified'.[3,4]

LD_{50} (mouse, oral): >20 g/kg[4]

LD_{50} (rat, oral): >20 g/kg

15 Handling Precautions

Observe normal precautions appropriate to the circumstances and quantity of material handled.

16 Regulatory Status

GRAS listed. Accepted for use as a food additive in Europe. Included in nonparenteral medicines licensed in the UK. Included in the Canadian Natural Health Products Ingredients Database.

17 Related Substances

—

18 Comments

The typical thaumatin level used in foods is 0.5–3 ppm, although higher levels are used in certain applications such as chewing gum. Synergistic effects with other intense sweeteners such as acesulfame K and saccharin occur. The extensive disulfide crosslinking within thaumatin maintains the tertiary structure of the polypeptide: cleavage of just one disulfide bridge has been shown to result in the loss of the sweet taste of thaumatin.[5]

As thaumatin is a protein it has some calorific value; however, in food products and pharmaceutical formulations the quantities used are so small that the calorific value is insignificant.

A specification for thaumatin is included in the *Japanese Pharmaceutical Excipients* (JPE).[6]

The EINECS number for thaumatin is 258-822-2.

19 Specific References

1 Odusote MO, Nasipuri RN. Effect of pH and storage conditions on the stability of a novel chloroquine phosphate syrup formulation. *Pharm Ind* 1988; **50**(3): 367–369.
2 Daniell WF. Katemfe, or the miraculous fruit of the Soudan. *Pharm J* 1855; **14**: 158–160.
3 Higginbotham JD. Safety evaluation of thaumatin (Talin protein). *Food Chem Toxicol* 1983; **21**(6): 815–823.
4 FAO/WHO. Toxicological evaluation of certain food additives and contaminants. Twenty-ninth report of the joint FAO/WHO expert committee on food additives. *WHO Food Add Ser* 1985; No. 20.
5 Iyengar RB. The complete amino-acid sequence of the sweet protein thaumatin. *Eur J Biochem* 1979; **96**: 193–204.
6 Japan Pharmaceutical Excipients Council. *Japanese Pharmaceutical Excipients 2004*. Tokyo: Yakuji Nippo, 2004: 900-901.

20 General References

Dodson AG, Wright SJC. New sweeteners: confectioner's viewpoint. *Food Flavour Ingred Packag Process* 1982; **4**(Sep): 29, 31, 32, 59.

Green C. Thaumatin: a natural flavour ingredient. *World Rev Nutr Diet* 1999; **85**: 129–132.

Hart H. Thaumatin. In: Birch G, ed. *Ingredients Handbook: Sweeteners*, 2nd edn. Leatherhead: Leatherhead Publishing, 2000: 255–263.

Higginbotham JD. Talin protein (thaumatin). In: O'Brien Nabors L, Gelardi RC, eds. *Alternative Sweeteners*. New York: Marcel Dekker, 1986: 103–134.

Kinghorn AD, Compadre CM. Naturally occurring intense sweeteners. *Pharm Int* 1985; **6**(Aug): 201–204.

Kinghorn AD, Compadre CM. Less common high-potency sweeteners. In: O'Brien Nabors L, ed. *Alternative Sweeteners*, 3rd edn. New York: Marcel Dekker, 2001: 214–215.

Naturex. *Talin*: a Naturex product. http://www.thaumatinnaturally.com (accessed 23 September 2011).

Sanyude S. Alternative sweeteners. *Can Pharm J* 1990; **123**(Oct): 455–456459–460.

Temussi PA. Natural sweet macromolecules: how sweet proteins work. *Cell Mol Life Sci* 2006; **63**(16): 1876–1888.

Unterhalt B, *et al.* Enzyme immunoassay for the determination of the sweetener thaumatin. *Pharmazie* 1997; **52**: 641–642.

Witty M, Higginbotham JD. *Thaumatin*. Boca Raton, FL: CRC Press, 1994.

21 Author

PJ Weller.

22 Date of Revision

1 November 2011.

Thimerosal

1 Nonproprietary Names

BP: Thiomersal

PhEur: Thiomersal

USP–NF: Thimerosal

2 Synonyms

Benzoic acid; 2-mercapto-, mercury complex; [(*o*-Carboxyphenyl)thio]ethylmercury sodium salt; elcide; ethyl (2-mercaptobenzoato-*S*)-mercury, sodium salt; ethyl (sodium *o*-mercaptobenzoato)mercury; mercurothiolate; merzonin sodium; sodium ethylmercurithiosalicylate; *Thimerosal Sigmaultra*; thiomersalate; thiomersalum.

3 Chemical Name and CAS Registry Number

Ethyl[2-mercaptobenzoato(2–)-*O*,*S*]-mercurate(1–) sodium [54-64-8]

4 Empirical Formula and Molecular Weight

$C_9H_9HgNaO_2S$ 404.81

Thimerosal contains approximately 49% mercury by weight.[1]

5 Structural Formula

6 Functional Category

Antimicrobial preservative.

7 Applications in Pharmaceutical Formulation or Technology

Thimerosal has been used as an antimicrobial preservative in biological and pharmaceutical preparations since the 1930s;[2] *see* Table I.

Table I: Uses of thimerosal.

Use	Concentration (%)
IM, IV, SC injections	0.01
Ophthalmic solutions	0.001–0.15
Ophthalmic suspensions	0.001–0.004
Otic preparations	0.001–0.01
Topical preparations	0.01

8 Description

Thimerosal is a light cream-colored crystalline powder with a slight, characteristic metallic odor and taste.

9 Pharmacopeial Specifications

See Table II.

Table II: Pharmacopeial specifications for thimerosal.

Test	PhEur 7.4	USP35–NF30
Identification	+	+
Characters	+	—
Appearance of solution	+	—
Melting point	103–115°C	—
pH	6.0–8.0	—
Inorganic mercury compounds	≤0.70%	—
Loss on drying	≤0.5%	≤0.5%
Ether-soluble substances	—	≤0.8%
Mercury ions	—	≤0.70%
Readily carbonizable substances	—	+
Assay (dried basis)	97.0–101.0%	97.0–101.0%

10 Typical Properties

Acidity/alkalinity pH = 6.7 for a 1% w/v aqueous solution at 20°C.

Antimicrobial activity Thimerosal is bactericidal at acidic pH, bacteriostatic and fungistatic at alkaline or neutral pH. Thimerosal is not effective against spore-forming organisms. *See also* Section 12. For reported minimum inhibitory concentrations (MICs), *see* Table III.[3]

Table III: Reported minimum inhibitory concentrations (MICs) for thimerosal.[3]

Microorganism	MIC (μg/mL)
Aspergillus niger	128.0
Candida albicans	32.0
Escherichia coli	4.0
Klebsiella pneumoniae	4.0
Penicillium notatum	128.0
Pseudomonas aeruginosa	8.0
Pseudomonas cepacia	8.0
Pseudomonas fluorescens	4.0
Staphylococcus aureus	0.2

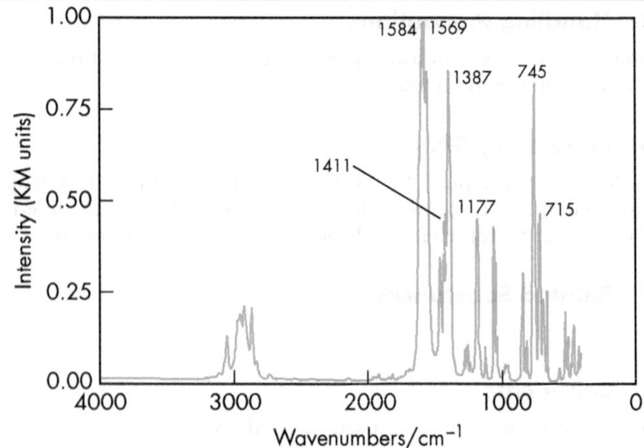

Figure 1: Infrared spectrum of thimerosal measured by diffuse reflectance. Adapted with permission of Informa Healthcare.

Density (bulk) <0.33 g/cm^3

Dissociation constant pK_a = 3.05 at 25°C.

Melting point 232–233°C with decomposition.

Solubility Freely soluble in water; soluble 1 in 8 of ethanol (95%); practically insoluble in benzene, ether, and methylene chloride.

Spectroscopy

IR spectra *see* Figure 1.

11 Stability and Storage Conditions

Thimerosal is stable at normal temperatures and pressures; exposure to light may cause discoloration.

Aqueous solutions may be sterilized by autoclaving but are sensitive to light. The rate of oxidation in solutions is increased by the presence of trace amounts of copper and other metals. Edetic acid or edetates may be used to stabilize solutions but have been reported to reduce the antimicrobial efficacy of thimerosal solutions; *see* Section 12.

The solid material should be stored in a well-closed container, protected from light, in a cool, dry place.

12 Incompatibilities

Incompatible with aluminum and other metals, strong oxidizing agents such as peroxides, permanganates, and nitric acid, strong acids and bases, sodium chloride solutions,[4] lecithin, phenylmercuric compounds, quaternary ammonium compounds, thioglycolate, and proteins. The presence of sodium metabisulfite, edetic acid, and edetates in solutions can reduce the preservative efficacy of thimerosal.[5]

In solution, thimerosal may be adsorbed by plastic packaging materials, particularly polyethylene. It is strongly adsorbed by treated or untreated rubber caps that are in contact with solutions.[6,7]

When it was used with cyclodextrin, the effectiveness of thimerosal was reduced; however, this was related to the lipid nature of the other ingredients in the preparation.[8]

13 Method of Manufacture

Thimerosal is prepared by the interaction of ethylmercuric chloride, or hydroxide, with thiosalicylic acid and sodium hydroxide, in ethanol (95%).

14 Safety

Thimerosal is widely used as an antimicrobial preservative in parenteral and topical pharmaceutical formulations. However, concern over the use of thimerosal in pharmaceuticals has increased as a result of a greater awareness of the toxicity of mercury and other associated mercury compounds.[9,10] The increasing number of reports of adverse reactions, particularly hypersensitivity,[11–13] to thimerosal and doubts as to its effectiveness as a preservative have led to suggestions that it should not be used as a preservative in eye drops[14] or vaccines.[15–17] In both Europe and the US, regulatory bodies have recommended that thimerosal in vaccines be phased out.[18–20]

More recent studies assessing the safety of thimerosal in vaccines have, however, suggested that while the risk of hypersensitivity reactions is present, the relative risk of neurological harm in infants is negligible given the quantities of thimerosal present in vaccines.[21–24] Regulatory bodies in Europe and the US have therefore updated their advice on the use of thimerosal in vaccines by stating that while it would be desirable for thimerosal not to be included in vaccines and other formulations the benefits of vaccines far outweigh any risks of adverse effects associated with their use.[25–28]

The most frequently reported adverse reaction to thimerosal, particularly in vaccines,[15–31] is hypersensitivity, usually with erythema and papular or vesicular eruptions. Although not all thimerosal-sensitive patients develop adverse reactions to vaccines containing thimerosal, there is potential risk. Patch testing in humans and animal experiments have suggested that 0.1% w/v thimerosal can sensitize children.[32] The incidence of sensitivity to thimerosal appears to be increasing; a study of 256 healthy subjects showed approximately 6% with positive sensitivity.[33]

Adverse reactions to thimerosal used to preserve contact lens solutions have also been reported. Reactions include ocular redness, irritation, reduced lens tolerance, and conjunctivitis.[34–36] One estimate suggests that approximately 10% of contact lens wearers may be sensitive to thimerosal.[37]

Thimerosal has also been associated with false positive reactions to old tuberculin,[38] ototoxicity,[39] and an unusual reaction to aluminum[40] in which a patient suffered a burn 5 cm in diameter at the site of an aluminum foil diathermy electrode after preoperative preparation of the skin with a 0.1% w/v thimerosal solution in ethanol (50%). Investigation showed that considerable heat was generated when such a solution came into contact with aluminum.

An interaction between orally administered tetracyclines and thimerosal, which resulted in varying extents of ocular irritation, has been reported in patients using a contact lens solution preserved with thimerosal.[41]

Controversially, some have claimed a connection between the use of thimerosal in vaccines and the apparent rise in the incidence of autism.[42,43] However, recent studies have shown no association between thimerosal exposure and autism.[25–28,44,45]

Serious adverse effects and some fatalities have been reported following the parenteral and topical use of products containing thimerosal. Five fatal poisonings resulted from the use of 1000 times the normal concentration of thimerosal in a chloramphenicol preparation for intramuscular injection.[46]

Ten out of 13 children died as a result of treatment of umbilical hernia (omphalocele) with a topical tincture of thimerosal.[47] It has therefore been recommended that organic mercurial disinfectants should be restricted or withdrawn from use in hospital since absorption occurs readily through intact membranes.

In a case of attempted suicide, a 44-year-old man drank 83 mg/kg of a thimerosal-containing solution. Despite spontaneously vomiting after 15 minutes, gastric lavage and administration of chelating agents on hospital admission, serious symptoms ultimately ending in coma occurred. The patient survived and after 5 months treatment made a full recovery except for sensory defects in two toes.[48]

LD_{50} (mouse, oral): 91 mg/kg[49]
LD_{50} (rat, oral): 75 mg/kg
LD_{50} (rat, SC): 98 mg/kg

15 Handling Precautions

Observe normal precautions appropriate to the circumstances and quantity of material handled. Thimerosal is irritant to the skin and mucous membranes, and may be systemically absorbed through the skin and upper respiratory tract. Thimerosal should be handled in a well-ventilated environment. Eye protection, gloves, and a respirator are recommended.

Chemical decomposition may cause the release of toxic fumes containing oxides of carbon, sulfur, and mercury in addition to mercury vapor.

16 Regulatory Status

Included in the FDA Inactive Ingredients Database (IM, IV, and SC injections [powder, for sustained action]; ophthalmic, otic, and topical preparations). Included in nonparenteral and parenteral medicines licensed in the UK. In the UK, the use of thimerosal in cosmetics is limited to 0.003% w/w (calculated as mercury) as a preservative in shampoos and hair-creams, which contain nonionic emulsifiers that would render other preservatives ineffective. The total permitted concentration (calculated as mercury) when mixed with other mercury compounds is 0.007% w/w.[50] Included in the Canadian Natural Health Products Ingredients Database.

17 Related Substances

Phenylmercuric acetate; phenylmercuric borate; phenylmercuric nitrate.

18 Comments

Some variation between the results obtained when comparing different thimerosal assay methods has been reported.[51,52]

Recent studies have indicated that the removal of preservatives from hepatitis B vaccine does not adversely affect its immunogenicity both in the short and in the longer term.[53]

Therapeutically, thimerosal is occasionally used as a bacteriostatic and fungistatic mercurial antiseptic, which is usually applied topically at a concentration of 0.1% w/w.[54] However, its use is declining owing to its toxicity and effects on the environment.

Thimerosal is also used in cosmetics (*see* Section 16) and to preserve soft contact lens solutions.

The EINECS number for thimerosal is 200-210-4. The PubChem Compound ID (CID) for thiomersal is 16684434.

19 Specific References

1. National Library of Medicine. Toxnet: Thimerosal, 2011.
2. Amieson WA, Powell HM. Merthiolate as a preservative for biological products. *Am J Hyg* 1931; **14**: 218–224.
3. Wallhäusser KH. Thimerosal. In: Kabara JJ, ed. *Cosmetic and Drug Preservation Principles and Practice.* New York: Marcel Dekker, 1984: 735–737.
4. Reader MJ. Influence of isotonic agents on the stability of thimerosal in ophthalmic formulations. *J Pharm Sci* 1984; **73**(6): 840–841.
5. Richards RME, Reary JME. Changes in antibacterial activity of thiomersal and PMN on autoclaving with certain adjuvants. *J Pharm Pharmacol* 1972; **24**(Suppl.): 84P–89P.
6. Wiener S. The interference of rubber with the bacteriostatic action of thiomersalate. *J Pharm Pharmacol* 1955; **7**: 118–125.
7. Birner J, Garnet JR. Thimerosal as a preservative in biological preparations III: factors affecting the concentration of thimerosal in aqueous solutions and in vaccines stored in rubber-capped bottles. *J Pharm Sci* 1964; **53**: 1424–1426.
8. Lehner SJ, *et al.* Effect of hydroxypropyl-beta-cyclodextrin on the antimicrobial action of preservatives. *J Pharm Pharmacol* 1994; **46**(3): 186–191.
9. Van't Veen AJ. Vaccines without thiomersal: why so necessary, why so long in coming? *Drugs* 2001; **61**(5): 565–572.
10. Clements CJ, *et al.* Thimerosal in vaccines: is removal warranted? *Drug Saf* 2001; **24**(8): 567–574.
11. Suneja T, Belsito DV. Thimerosal in the detection of clinically relevant allergic, contact reactions. *J Am Acad Dermatol* 2001; **45**(1): 23–27.

12 Audicana MT, *et al.* Allergic contact dermatitis from mercury antiseptics and derivatives: study protocol of tolerance to intramuscular injections of thimerosal. *Am J Contact Dermat* 2002; **13**(1): 3–9.
13 Freiman A, *et al.* Patch testing with thiomersal in a Canadian center: an 11-year experience. *Am J Contact Dermatol* 2003; **14**(3): 138–143.
14 Ford JL, *et al.* A note on the contamination of eye-drops following use by hospital out-patients. *J Clin Hosp Pharm* 1985; **10**(2): 203–209.
15 Cox NH, Forsyth A. Thiomersal allergy and vaccination reactions. *Contact Dermatitis* 1988; **18**: 229–233.
16 Seal D, *et al.* The case against thiomersal [letter]. *Lancet* 1991; **338**(8762): 315–316.
17 Noel I, *et al.* Hypersensitivity to thiomersal in hepatitis B vaccine [letter]. *Lancet* 1991; **338**: 705.
18 Anonymous. Thiomersal to be removed from vaccines in the US. *Pharm J* 1999; **263**: 112.
19 European Agency for the Evaluation of Medicinal Products (EMEA). EMEA public statement on thiomersal containing medicinal products, 8 July 1999. EMEA publication no. 20962/99.
20 American Academy of Pediatrics, United States Public Health Service. Thimerosal in vaccines: a joint statement of the American Academy of Pediatrics and the Public Health Service. *MMWR* 1999; **48**: 563–565.
21 Clements CJ. The evidence for the safety of thimerosal in newborn and infant vaccines. *Vaccine* 2004; **22**(15–16): 1854–1861.
22 Counter SA, Buchanan LH. Mercury exposure in children: a review. *Toxicol Appl Pharmacol* 2004; **198**(2): 209–230.
23 Bigham M, Copes R. Thimerosal in vaccines: balancing the risks of adverse effects with the risk of vaccine-preventable disease. *Drug Safety* 2005; **28**(2): 89–101.
24 Geier DA, *et al.* A review of Thimerosal (Merthiolate) and its ethylmercury breakdown product: specific historical considerations regarding safety and effectiveness. *J Toxicol Environ Health B Crit Rev* 2007; **10**(8): 575–596.
25 European Medicines Evaluation Agency (EMEA). EMEA public statement on thiomersal in vaccines for human use—recent evidence supports safety of thiomersal-containing vaccines, 2004. www.emea.eu.int/pdfs/human/press/pus/119404eu.pdf (accessed 23 September 2011).
26 Committee on Safety of Medicines. Safety of thiomersal-containing vaccines. *Current Problems* 2003; **29**: 9.
27 FDA. Thiomersal in vaccines (updated 14 January 2009). http://www.fda.gov/BiologicsBloodVaccines/SafetyAvailability/VaccineSafety/UCM096228 (accessed 16 February 2012).
28 WHO. Guidelines on regulatory expectations related to the elimination, reduction or replacement of thiomersal in vaccines. *World Health Organ Tech Rep Ser* 2004; No. 926. www.who.int/biologicals/publications/trs/areas/vaccines/thiomersal/Annex%204%20(95-102)TRS926thiomersal.pdf (accessed 23 September 2011).
29 Rietschel RL, Adams RM. Reactions to thimerosal in hepatitis B vaccines. *Dermatol Clin* 1990; **8**(1): 161–164.
30 Golightly LK, *et al.* Pharmaceutical excipients: adverse effects associated with inactive ingredients in drug products (part I). *Med Toxicol* 1988; **3**: 128–165.
31 Lee-Wong M, *et al.* A generalized reaction to thimerosal from an influenza vaccine. *Ann Allergy Asthma Immunol* 2005; **94**(1): 90–94.
32 Osawa J, *et al.* A probable role for vaccines containing thimerosal in thimerosal hypersensitivity. *Contact Dermatitis* 1991; **24**(3): 178–182.
33 Seidenari S, *et al.* [Sensitization after contact with thimerosal in a healthy population.] *G Ital Dermatol Venereol* 1989; **124**(7-8): 335–339[in Italian].
34 Mondino BJ, Groden LR. Conjunctival hyperemia and corneal infiltrates with chemically disinfected soft contact lenses. *Arch Ophthalmol* 1980; **98**(10): 1767–1770.
35 Sendele DD, *et al.* Superior limbic keratoconjunctivitis in contact lens wearers. *Ophthalmology* 1983; **90**: 616–622.
36 Fisher AA. Allergic reactions to contact lens solutions. *Cutis* 1985; **36**(3): 209–211.
37 Miller JR. Sensitivity to contact lens solutions. *West J Med* 1984; **140**: 791.
38 Hansson H, Möller H. Intracutaneous test reactions to tuberculin containing merthiolate as a preservative. *Scand J Infect Dis* 1971; **3**: 169–172.
39 Honigman JL. Disinfectant ototoxicity. *Pharm J* 1975; **215**: 523.
40 Jones HT. Danger of skin burns from thiomersal. *Br Med J* 1972; **2**: 504–505.
41 Crook TG, Freeman JJ. Reactions induced by the concurrent use of thimerosal and tetracyclines. *Am J Optom Physiol Opt* 1983; **60**: 759–761.
42 Schultz ST. Does thimerosal or other mercury exposure increase the risk for autism? A review of current literature. *Acta Neurobiol Exp (Wars)* 2010; **70**(2): 187–195.
43 Baker PJ. Mercury, vaccines, and autism: one controversy, three histories. *Am J Public Health* 2008; **98**(2): 244–253.
44 Department of Health. Public letter from the Chief Medical Officer: current vaccine and immunisation issues, 15 October 2001, PL/CMO/2001/5.
45 Parker SK, *et al.* Thimerosal-containing vaccines and autistic spectrum disorder: a critical review of published original data. *Pediatrics* 2004; **114**(3): 793–804.
46 Axton JHM. Six cases of poisoning after a parenteral organic mercurial compound (Merthiolate). *Postgrad Med J* 1972; **48**: 417–421.
47 Fagan DG, *et al.* Organ mercury levels in infants with omphaloceles treated with organic mercurial antiseptic. *Arch Dis Child* 1977; **52**(12): 962–964.
48 Pfab R, *et al.* Clinical course of severe poisoning with thiomersal. *J Toxicol Clin Toxicol* 1996; **34**(4): 453–460.
49 Lewis RJ, ed. *Sax's Dangerous Properties of Industrial Materials*, 11th edn. New York: Wiley, 2004: 2321.
50 Statutory Instrument 2233. Consumer protection: the consumer products (safety) regulations 1989. London: HMSO, 1989.
51 Fleitman JS, *et al.* Thimerosal analysis in ketorolac tromethamine ophthalmic solution. *Drug Dev Ind Pharm* 1991; **17**: 519–530.
52 Hu OY-P, *et al.* Simultaneous determination of thiomersal and chlorhexidine in solutions for soft contact lenses and its application in stability studies. *J Chromatogr* 1990; **523**: 321–326.
53 Van Damme P, *et al.* Long-term immunogenicity of preservative-free hepatitis B vaccine formulations in adults. *J Med Virol* 2009; **81**(10): 1710–1715.
54 Sweetman SC, ed. *Martindale: the Complete Drug Reference*, 37th edn. London: Pharmaceutical Press, 2011: 1808.

20 General References

Caraballo I, *et al.* Study of thimerosal degradation mechanism. *Int J Pharm* 1993; **89**: 213–221.
Rabasco AM, *et al.* Formulation factors affecting thimerosal stability. *Drug Dev Ind Pharm* 1993; **19**: 1673–1691.
Tan M, Parkin JE. Route of decomposition of thiomersal (thimerosal). *Int J Pharm* 2000; **208**: 23–34.

21 Author

BV Kadri.

22 Date of Revision

1 March 2012.

T

Thymol

1 Nonproprietary Names

BP: Thymol
JP: Thymol
PhEur: Thymol
USP–NF: Thymol

2 Synonyms

Acido trimico; 3-*p*-cymenol; *p*-cymen-3-ol; *Flavinol*; 3-hydroxy-*p*-cymene; 3-hydroxy-1-methyl-4-isopropylbenzene; *Intrasol*; isopropyl cresol; isopropyl-*m*-cresol; 6-isopropyl-*m*-cresol; isopropyl metacresol; 2-isopropyl-5-methylphenol; 1-methyl-3-hydroxy-4-isopropylbenzene; 5-methyl-2-isopropylphenol; 5-methyl-2-(1-methylethyl) phenol; *Medophyll*; thyme camphor; thymic acid; *m*-thymol; thymolum; timol.

3 Chemical Name and CAS Registry Number

Thymol [89-83-8]

4 Empirical Formula and Molecular Weight

$C_{10}H_{14}O$ 150.24

5 Structural Formula

6 Functional Category

Antioxidant; flavoring agent; penetration enhancer.

7 Applications in Pharmaceutical Formulation or Technology

Thymol is a phenolic antiseptic, which has antibacterial and antifungal activity. However, it is not suitable for use as a preservative in pharmaceutical formulations because of its low aqueous solubility. The antimicrobial activity of thymol against eight oral bacteria has been studied *in vitro*. Inhibitory activity was noted against almost all organisms, and a synergistic effect was observed for combinations of thymol and eugenol, and of thymol and carvacrol.[1] The activity of thymol against bacteria commonly involved in upper respiratory tract infections has also been shown.[2] The continued suppression of bacterial growth following limited exposure to thymol has been described.[3]

Thymol is also a true antioxidant and has been used at concentrations of 0.01% as an antioxidant for halothane, trichloroethylene, and tetrachloroethylene. The antioxidant activity of thymol[4,5] and thymol analogues[4] has been described.

8 Description

Thymol occurs as colorless or often large translucent crystals, or as a white crystalline powder with a herbal odor (aromatic and thyme-like) and a pungent caustic taste.

9 Pharmacopeial Specifications

See Table I.

Table I: Pharmacopeial specifications for thymol.

Test	JP XV	PhEur 7.4	USP35–NF30
Identification	+	+	+
Characters	—	+	—
Melting range	49–51°C	48–52°C	48–51°C
Appearance of solution	—	+	—
Acidity	—	+	—
Related substances	—	+	—
Residue on evaporation	—	≤0.05%	≤0.05%
Other phenols	+	—	—
Assay	≥98.0%	—	99.0–101.0%

10 Typical Properties

Acidity/alkalinity A 4% solution in ethanol (50%) is neutral to litmus; a 1% solution in water has a pH of 7.
Boiling point About 233°C.
Density 0.97 g/cm^3 at 25°C; has a greater density than water, but when liquefied by fusion is less dense than water.
Dissociation constant $pK_a = 10.6$ at 20°C
Melting point 48–51°C, but, once melted, remains liquid at a considerably lower temperature.
Partition coefficient log (octanol/water) = 3.3
Phenol coefficient About 50.
Refractive index
$n_D^{25} = 0.15204$;
$n_D^{20} = 0.15227$.
Solubility Soluble 1 in 0.7–1.0 of chloroform; 1 in 1 of ethanol (95%); 1 in 1.5 of ether, glacial acetic acid; 1 in 1.7–2.0 of olive oil; 1 in 1000 of water. Freely soluble in essential oils, fixed oils, and fats. Sparingly soluble in glycerin. Dissolves in dilute solutions of alkali hydroxides, forming salts that have increased solubility but whose solutions darken on standing.
Spectroscopy
IR spectra *see* Figure 1.
Vapor pressure 0.04 mmHg at 20°C
Volatility Appreciable volatility at 100°C; volatile in water vapor at 25°C.

11 Stability and Storage Conditions

Thymol should be stored in well-closed, light-resistant containers, in a cool, dry, place. Thymol is affected by light.

12 Incompatibilities

Thymol is incompatible with iodine, alkalis, and oxidizing agents. It liquefies, or forms soft masses, on trituration with acetanilide, antipyrine, camphor, monobromated camphor, chloral hydrate, menthol, phenol, or quinine sulfate. The antimicrobial activity of thymol is reduced in the presence of proteins.

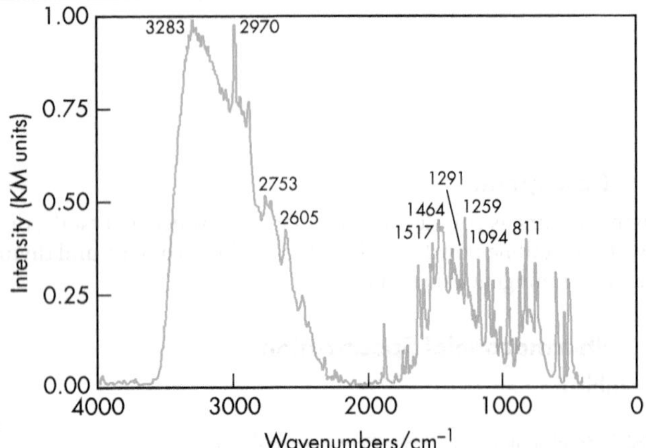

Figure 1: Infrared spectrum of thymol measured by diffuse reflectance. Adapted with permission of Informa Healthcare.

13 Method of Manufacture

Thymol is obtained from the volatile oil of thyme (*Thymus vulgaris* Linné (Fam. Labiatae)) by fractional distillation followed by extraction and recrystallization. Thyme oil yields about 20–30% thymol. Thymol may also be produced synthetically from *p*-cymene, menthone, or piperitone, or by the interaction of *m*-cresol with isopropyl chloride.

14 Safety

Thymol is used in cosmetics, foods, and pharmaceutical applications as an excipient. However, thymol may be irritating when inhaled or following contact with the skin or eyes. It may also cause abdominal pain and vomiting, and sometimes stimulation followed by depression of the central nervous system following oral consumption; fats and alcohol increase absorption and aggravate symptoms.

Respiratory arrest, attributed to acute nasal congestion and edema, has been reported in a 3-week-old patient due to the erroneous intranasal application of *Karvol*, a combination product that includes thymol. The patient recovered, but it was recommended that inhalation decongestants should not be used in children under the age of 5 years.[6]

LD_{50} (guinea pig, oral): 0.88 g/kg[7]
LD_{50} (mouse, IP): 0.11 g/kg
LD_{50} (mouse, IV): 0.1 g/kg
LD_{50} (mouse, oral): 0.64 g/kg
LD_{50} (mouse, SC): 0.243 g/kg
LD_{50} (rat, oral): 0.98 g/kg

15 Handling Precautions

Observe normal precautions appropriate to the circumstances and quantity of material handled. Special precautions should be taken to avoid inhalation, or contact with the skin or eyes. Eye protection and gloves are recommended. When thymol is heated to decomposition, carbon dioxide and carbon monoxide are formed.

16 Regulatory Status

GRAS listed. Included in the FDA Inactive Ingredients Database (inhalation, liquid; oral, powder for solution). Included in nonparenteral medicines (oral lozenges; capsules for inhalation, inhalation drops; topical creams and ointments) licensed in the UK. Included in the Canadian Natural Health Products Ingredients Database.

17 Related Substances

Menthol.

18 Comments

The inhalation of thymol, in combination with other volatile substances, is used to alleviate the symptoms of colds, coughs, and associated respiratory disorders. Externally, thymol has been used in dusting powders for the treatment of fungal skin infections; thymol has been shown to have synergistic antifungal effects when combined with ketoconazole.[8] Thymol was formerly used in the treatment of hookworm infections but has now been superseded by less toxic substances.

In dentistry, thymol has been mixed with phenol and camphor to prepare cavities before filling, and mixed with zinc oxide to form a protective cap for dentine. Thymol has also been used as a pesticide and fungicide.

Thymol is a more powerful disinfectant than phenol, but its low water solubility, its irritancy to tissues, and its inactivation by organic material, such as proteins, limit its use as a disinfectant. Thymol is chiefly used as a deodorant in antiseptic mouthwashes, gargles, and toothpastes, such as in Compound Thymol Glycerin BP, in which it has no antiseptic action.

More recently, thymol has been shown to enhance the *in vitro* percutaneous absorption of a number of drugs, including 5-fluorouracil,[9] piroxicam,[10] propranolol,[11] naproxen,[12] and tamoxifen.[13] Studies have also demonstrated that the melting point of lidocaine is significantly lowered when it is mixed with thymol.[14,15]

Thymol has been included in food, perfume, and cosmetic products.

A specification for thymol is included in the *Food Chemicals Codex* (FCC).[16]

The EINECS number for thymol is 201-944-8. The PubChem Compound ID (CID) for thymol is 6989.

19 Specific References

1 Didry N, *et al.* Activity of thymol, carvacrol, cinnamaldehyde and eugenol on oral bacteria. *Pharm Acta Helv* 1994; **69**(1): 25–28.
2 Didry N, *et al.* Antimicrobial activity of thymol, carvacrol and cinnamaldehyde alone or in combination. *Pharmazie* 1993; **48**: 301–304.
3 Zarrini G, *et al.* Post-antibacterial effect of thymol. *Pharm Biol* 2010; **48**(6): 633–636.
4 Shen AY, *et al.* Thymol analogues with antioxidant and L-type calcium current inhibitory activity. *Drug Dev Res* 2005; **64**(4): 195–202.
5 Yanishlieva NV, Marinova FM. Antioxidant activity of some natural antioxidants in lipids at ambient temperature. *Seifen, Oele, Fette, Wachse* 2006; **132**(6): 30–34.
6 Blake KD. Dangers of common cold treatments in children. *Lancet* 1993; **341**: 640.
7 Lewis RJ. *Sax's Dangerous Properties of Industrial Materials*, 11th edn. New York: Wiley, 2004: 3462–3463.
8 Shin S, Kim JH. Antifungal activities of essential oils from *Thymus quinquecostatus* and *T. magnus*. *Planta Med* 2004; **70**(11): 1090–1092.
9 Gao S, Singh J. Mechanism of transdermal transport of 5-fluorouracil by terpenes: carvone, 1,8-cineole and thymol. *Int J Pharm* 1997; **154**(1): 67–77.
10 Doliwa A, *et al.* Effect of passive and iontophoretic skin pretreatments with terpenes on the *in vitro* skin transport of piroxicam. *Int J Pharm* 2001; **229**(1–2): 37–44.
11 Songkro S, *et al.* The effects of *p*-menthane monoterpenes and related compounds on the percutaneous absorption of propranolol hydrochloride across newborn pig skin I. *In vitro* skin permeation and retention studies. *STP Pharma Sci* 2003; **13**(5): 349–357.
12 Ray S, Ghosal SK. Release and skin permeation studies of naproxen from hydrophilic gels and effect of terpenes as enhancers on its skin permeation. *Boll Chim Farm* 2003; **142**(3): 125–129.

13 Gao S, Singh J. *In vitro* percutaneous absorption enhancement of the lipophilic drug tamoxifen by terpenes. *J Control Release* 1998; **51**: 193–199.
14 Kang L, *et al.* Preparation and characterisation of two-phase melt systems of lignocaine. *Int J Pharm* 2001; **222**(1): 35–44.
15 Kang L, Jun HW. Formulation and efficacy studies of new topical anaesthetic creams. *Drug Dev Ind Pharm* 2003; **29**(5): 505–512.
16 *Food Chemicals Codex*, 7th edn. Bethesda, MD: United States Pharmacopeia, 2010: 1033.

20 General References

—

21 Author

CG Cable.

22 Date of Revision

1 March 2012.

Titanium Dioxide

1 Nonproprietary Names

BP: Titanium Dioxide
JP: Titanium Oxide
PhEur: Titanium Dioxide
USP–NF: Titanium Dioxide

2 Synonyms

Anatase titanium dioxide; brookite titanium dioxide; color index number 77891; E171; *Hombitan AFDC*; *Hombitan FF-Pharma*; *Kronos 1171*; pigment white 6; *Pretiox AV-01-FG*; rutile titanium dioxide; *Tioxide*; *TiPure*; titanic anhydride; titanii dioxidum; *Tronox*.

3 Chemical Name and CAS Registry Number

Dioxotitanium [13463-67-7]

4 Empirical Formula and Molecular Weight

TiO_2 79.88

5 Structural Formula

See Section 4.

6 Functional Category

Coating agent; opacifier; pigment.

7 Applications in Pharmaceutical Formulation or Technology

Titanium dioxide is widely used in confectionery, cosmetics, and foods, in the plastics industry, and in topical and oral pharmaceutical formulations as a white pigment.

Owing to its high refractive index, titanium dioxide has light-scattering properties that may be exploited in its use as a white pigment and opacifier. The range of light that is scattered can be altered by varying the particle size of the titanium dioxide powder. For example, titanium dioxide with an average particle size of 230 nm scatters visible light, while titanium dioxide with an average particle size of 60 nm scatters ultraviolet light and reflects visible light.[1]

In pharmaceutical formulations, titanium dioxide is used as a white pigment in film-coating suspensions,[2,3] sugar-coated tablets, and gelatin capsules. Titanium dioxide may also be admixed with other pigments.

Titanium dioxide is also used in dermatological preparations and cosmetics, such as sunscreens.[1,4]

8 Description

Titanium dioxide occurs as a white, amorphous, odorless, and tasteless nonhygroscopic powder. Although the average particle size of titanium dioxide powder is less than 1 μm, commercial titanium dioxide generally occurs as aggregated particles of approximately 100 μm diameter.

Titanium dioxide may occur in several different crystalline forms: rutile; anatase; and brookite. Of these, rutile and anatase are the only forms of commercial importance. Rutile is the more thermodynamically stable crystalline form, but anatase is the form most commonly used in pharmaceutical applications.

9 Pharmacopeial Specifications

See Table I. The PhEur has introduced a nonmandatory functionality related characteristics section in the monograph for titanium dioxide. This section applies when the material is being used as an opacifier in solid oral dosage forms and in preparations for cutaneous application. The appropriate tests are for particle size distribution and for bulk and tapped density. As these are nonmandatory tests no limits are given.

SEM 1: Excipient: titanium dioxide; magnification: 1200×; voltage: 10 kV.

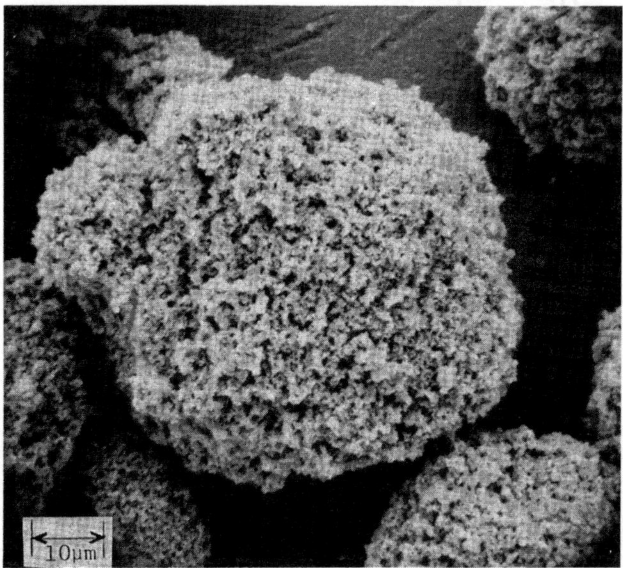

Table I: Pharmacopeial specifications for titanium dioxide.

Test	JP XV	PhEur 7.4	USP35–NF30
Identification	+	+	+
Characters	—	+	—
Appearance of solution	—	+	—
Acidity or alkalinity	—	+	—
Water-soluble substances	≤5.0 mg	≤0.5%	≤0.25%
Antimony	—	≤100 ppm	—
Arsenic	≤10 ppm	≤5 ppm	≤1 ppm
Barium	—	+	—
Heavy metals	—	≤20 ppm	—
Iron	—	≤200 ppm	—
Loss on drying	≤0.5%	—	≤0.5%
Loss on ignition	—	—	≤13%
Acid-soluble substances	—	—	≤0.5%
Lead	≤60 ppm	—	—
Assay	≥98.5%	98.0–100.5%	99.0–100.5%

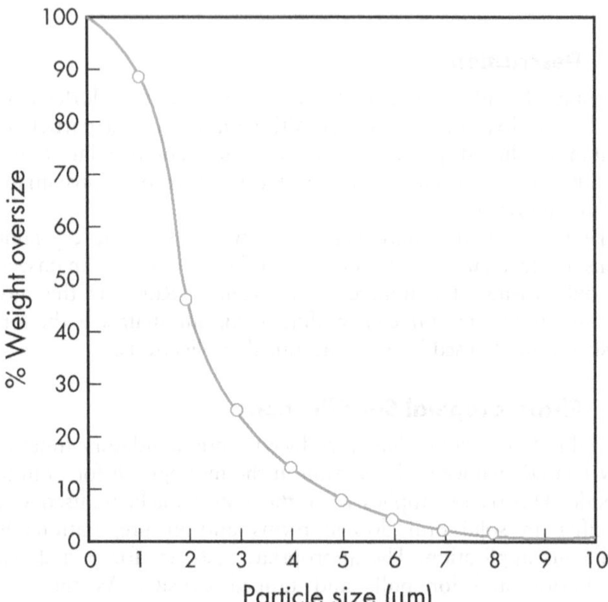

Figure 1: Particle-size distribution of titanium dioxide (fine powder).

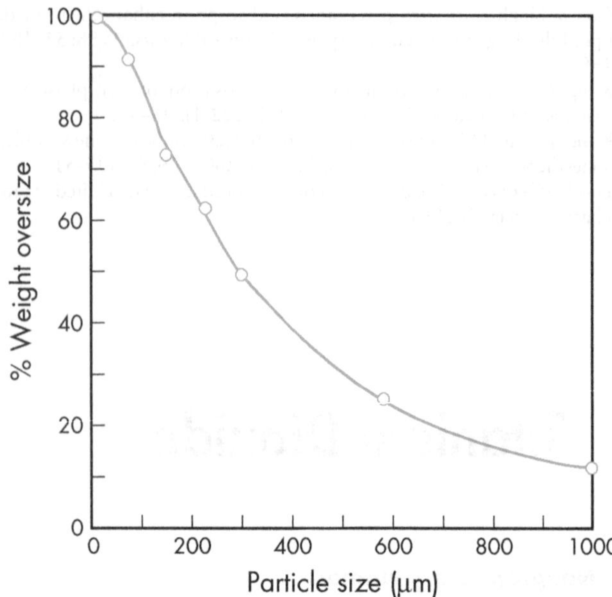

Figure 2: Particle-size distribution of titanium dioxide (agglomerated particles).

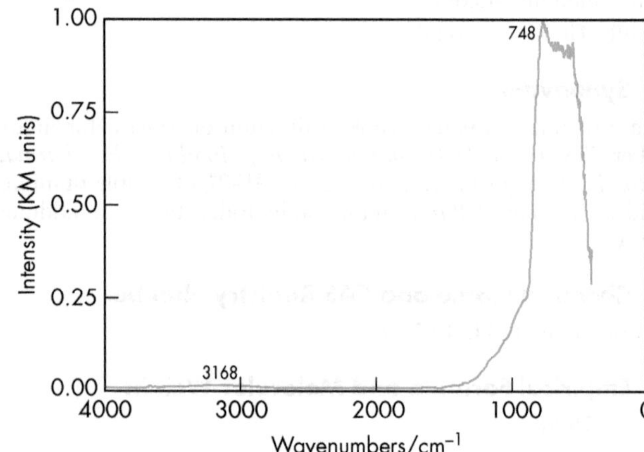

Figure 3: Infrared spectrum of titanium dioxide measured by diffuse reflectance. Adapted with permission of Informa Healthcare.

10 Typical Properties

Density (bulk) 0.4–0.62 g/cm^3 [5]
Density (tapped) 0.625–0.830 g/cm^3 [6]
Density (true)
3.8–4.1 g/cm^3 for anatase;
≈3.9 g/cm^3 for *Hombitan FF-Pharma*;
3.9–4.2 g/cm^3 for rutile.
Dielectric constant
48 for anatase;
114 for rutile.
Hardness (Mohs)
5–6 for anatase;
6–7 for rutile.
See also Section 18.
Melting point 1855°C
Moisture content 0.44%
Particle size distribution Average particle size = 1.05 μm;[5] ≈0.3 μm for *Hombitan FF-Pharma. See also* Figures 1 and 2.
Refractive index
2.55 for anatase;
≈2.5 for *Hombitan FF-Pharma*;
2.76 for rutile.
Solubility Practically insoluble in dilute sulfuric acid, hydrochloric acid, nitric acid, organic solvents, and water. Soluble in hydrofluoric acid and hot concentrated sulfuric acid. Solubility depends on previous heat treatment; prolonged heating produces a less-soluble material.
Specific heat
0.71 J/g (0.17 cal/g) for anatase;
0.71 J/g (0.17 cal/g) for rutile.
Specific surface area
9.90–10.77 m^2/g;
≈10.0 m^2/g for *Hombitan FF-Pharma*.
Spectroscopy
IR spectra *see* Figure 3.
NIR spectra *see* Figure 4.
Tinting strength (Reynolds)
1200–1300 for anatase;
1650–1900 for rutile.

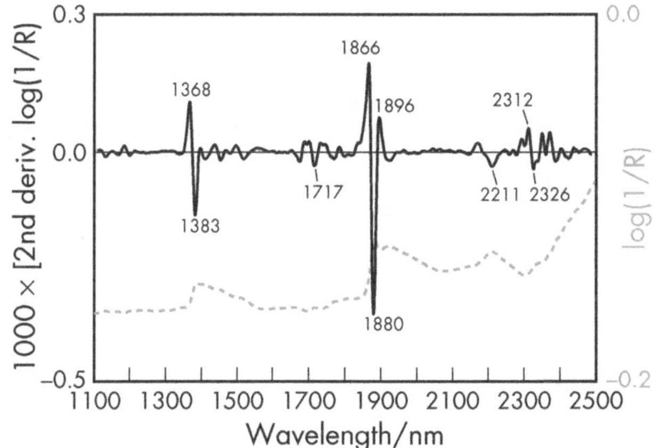

Figure 4: Near-infrared spectrum of titanium dioxide measured by reflectance. Titanium dioxide shows no significant absorption in the near-infrared region; however, it will generally show some peaks due to moisture (approx. 1450 nm and 1950 nm).

11 Stability and Storage Conditions

Titanium dioxide is extremely stable at high temperatures. This is due to the strong bond between the tetravalent titanium ion and the bivalent oxygen ions. However, titanium dioxide can lose small, unweighable amounts of oxygen by interaction with radiant energy. This oxygen can easily recombine again as a part of a reversible photochemical reaction, particularly if there is no oxidizable material available. These small oxygen losses are important because they can cause significant changes in the optical and electrical properties of the pigment.

Titanium dioxide should be stored in a well-closed container, protected from light, in a cool, dry place.

12 Incompatibilities

Owing to a photocatalytic effect, titanium dioxide may interact with certain active substances, e.g. famotidine.[7] Studies have shown that titanium dioxide monatomically degrades film mechanical properties and increases water vapor permeability of polyvinyl alcohol coatings when used as an inert filler and whitener.[6]

Titanium dioxide has also been shown to induce photooxidation of unsaturated lipids.[8]

13 Method of Manufacture

Titanium dioxide occurs naturally as the minerals rutile (tetragonal structure), anatase (tetragonal structure), and brookite (orthorhombic structure).

Titanium dioxide may be prepared commercially by either the sulfate or chloride process. In the sulfate process a titanium containing ore, such as ilmenite, is digested in sulfuric acid. This step is followed by dissolving the sulfates in water, then precipitating the hydrous titanium dioxide using hydrolysis. Finally, the product is calcinated at high temperature. In the chloride process, the dry ore is chlorinated at high temperature to form titanium tetrachloride, which is subsequently oxidized to form titanium dioxide.

14 Safety

Titanium dioxide is widely used in foods and oral and topical pharmaceutical formulations. It is generally regarded as an essentially nonirritant and nontoxic excipient.

15 Handling Precautions

Observe normal precautions appropriate to the circumstances and quantity of material handled. Eye protection, gloves, and a dust mask are recommended. Titanium dioxide is regarded as a relatively innocuous nuisance dust,[9] that may be irritant to the respiratory tract. In the UK, the long-term (8-hour TWA) workplace exposure limit is 10 mg/m^3 for total inhalable dust and 4 mg/m^3 for respirable dust.[10]

Titanium dioxide particles in the 500 nm range have been reported to translocate to all major body organs after oral administration in the rat.[11]

16 Regulatory Status

Accepted as a food additive in Europe. Included in the FDA Inactive Ingredients Database (dental paste; intrauterine suppositories; ophthalmic preparations; oral capsules, suspensions, tablets; topical and transdermal preparations). Included in nonparenteral medicines licensed in the UK. Included in the Canadian Natural Health Products Ingredients Database.

17 Related Substances

Coloring agents.

18 Comments

Titanium dioxide is one of the materials that have been selected for harmonization by the Pharmacopeial Discussion Group. For further information see the General Information Chapter <1196> in the USP35–NF30, the General Chapter 5.8 in PhEur 7.4, along with the 'State of Work' document on the PhEur EDQM website, and also the General Information Chapter 8 in the JP XV.

Titanium dioxide is a hard, abrasive material. Coating suspensions containing titanium dioxide have been reported to cause abrasion and wear of a steel-coated pan surface, which led to white tablets being contaminated with black specks.[12]

If titanium dioxide is used as a pigment in the EU, it should conform to the appropriate food standards specifications, which are more demanding than the pharmacopeial specifications.

When mixed with methylcellulose, titanium dioxide can reduce the elongation and tensile strength of the film but slightly increase the adhesion between pigmented film and the tablet surface.[13]

A specification for titanium dioxide is contained in the *Food Chemicals Codex* (FCC).[14]

The EINECS number for titanium dioxide is 236-675-5. The PubChem Compound ID (CID) for titanium dioxide is 26042.

19 Specific References

1 Hewitt JP. Titanium dioxide: a different kind of sunshield. *Drug Cosmet Ind* 1992; **151**(3): 26, 28, 30, 32.
2 Rowe RC. Quantitative opacity measurements on tablet film coatings containing titanium dioxide. *Int J Pharm* 1984; **22**: 17–23.
3 Béchard SR, *et al*. Film coating: effect of titanium dioxide concentration and film thickness on the photostability of nifedipine. *Int J Pharm* 1992; **87**: 133–139.
4 Alexander P. Ultrafine titanium dioxide makes the grade. *Manuf Chem* 1991; **62**(7): 21, 23.
5 Brittain HG *et al.*Titanium dioxide. In: Brittain HG, ed. *Analytical Profiles of Drug Substances and Excipients*. 21. San Diego: Academic Press, 1992: 659–691.
6 Hsu ER, *et al*. Effects of plasticizers and titanium dioxide on the properties of poly(vinyl alcohol) coatings. *Pharm Dev Technol* 2001; **6**(2): 277–284.
7 Kakinoki K, *et al*. Effect of relative humidity on the photocatalytic activity of titanium dioxide and photostability of famotidine. *J Pharm Sci* 2004; **93**(3): 582–589.
8 Sayre RM, Dowdy JC. Titanium dioxide and zinc oxide induce photooxidation of unsaturated lipids. *Cosmet Toilet* 2000; **115**: 75–8082.
9 Driscoll KE, *et al*. Respiratory tract responses to dust: relationships between dust burden, lung injury, alveolar macrophage fibronectin

T

release, and the development of pulmonary fibrosis. *Toxicol Appl Pharmacol* 1990; **106**: 88–101.
10 Health and Safety Executive. *EH40/2005: Workplace Exposure Limits.* Sudbury: HSE Books, 2011. www.hse.gov.uk/pubns/priced/eh40.pdf (accessed 12 April 2012).
11 Jani PU, *et al.* Titanium dioxide (rutile) particle uptake from the rat GI tract and translocation to systemic organs after oral administration. *Int J Pharm* 1994; **105**(May 2): 157–168.
12 Rosoff M, Sheen P-C. Pan abrasion and polymorphism of titanium dioxide in coating suspensions. *J Pharm Sci* 1983; **72**: 1485.
13 Lehtola VM, *et al.* Effect of titanium dioxide on mechanical, permeability and adhesion properties of aqueous-based hydroxypropyl methylcellulose films. *Boll Chim Farm* 1994; **133**(Dec): 709–714.
14 *Food Chemicals Codex*, 7th edn. Bethesda, MD: United States Pharmacopeia, 2010: 1033.

20 General References

European Directorate for the Quality of Medicines and Healthcare (EDQM). European Pharmacopoeia – State Of Work Of International Harmonisation. *Pharmeuropa* 2011; 23(2): 395–401. www.edqm.eu/site/-614.html (accessed 13 September 2011).

Judin VPS. The lighter side of TiO_2. *Chem Br* 1993; **29**(6): 503–505.

Loden M, *et al.* Novel method for studying photolability of topical formulations: a case study of titanium dioxide stabilization of ketoprofen. *J Pharm Sci* 2005; **94**(4): 781–787.

Ortyl TT, Peck GE. Surface charge of titanium dioxide and its effect on dye adsorption and aqueous suspension stability. *Drug Dev Ind Pharm* 1991; **17**: 2245–2268.

Rowe RC. Materials used in the film coating of oral dosage forms. In: Florence AT, ed. *Critical Reports on Applied Chemistry.* 6. Oxford: Blackwell Scientific, 1984: 1–36.

Sachtleben. Product literature: *Hombitan FF-Pharma,* 2005/08.

21 Author

C Mroz.

22 Date of Revision

1 March 2012.

Tragacanth

1 Nonproprietary Names

BP: Tragacanth
JP: Tragacanth
PhEur: Tragacanth
USP–NF: Tragacanth
See also Section 18.

2 Synonyms

E413; goat's horn; gum benjamin; gum dragon; gum tragacanth; persian tragacanth; trag; tragant; tragacantha.

3 Chemical Name and CAS Registry Number

Tragacanth gum [9000-65-1]

4 Empirical Formula and Molecular Weight

Tragacanth is a natural biopolymer of high molecular weight of about 84 000 Da. Most commercial gum is obtained from *Astragalus gummifer* Labillardière and, to a lesser extent, other species of *Astragalus* grown in Western Asia; *see* Section 13.

The gum consists of a mixture of water-insoluble and water-soluble polysaccharides. Bassorin or tragacanthic acid constitutes 60–70% of the gum and is the main water-insoluble portion, that swells greatly in contact with water. The remainder of the gum consists of the water-soluble material tragacanthin, which is a neutral polysaccharide. On hydrolysis, tragacanthin yields *l*-arabinose, *l*-fucose, *d*-xylose, *d*-galactose, and *d*-galacturonic acid; *see* Section 11. Tragacanth gum also contains small amounts of cellulose, starch, protein, and ash.

5 Structural Formula

See Section 4.

6 Functional Category

Emulsifying agent; suspending agent; viscosity-increasing agent.

7 Applications in Pharmaceutical Formulation or Technology

Tragacanth gum is used as an emulsifying and suspending agent in a variety of pharmaceutical formulations. It is used in creams, gels, and emulsions at various concentrations according to the application of the formulation and the grade of gum used.It has been used as a diluent and, to a lesser extent, as a binder and a sustained-release polymer in tablet formulations.[1–3]

Tragacanth gum is also used similarly in cosmetics and food products.

8 Description

Tragacanth gum occurs as flattened, lamellated, frequently curved fragments, or as straight or spirally twisted linear pieces from 0.5–2.5 mm in thickness; it may also be obtained in a powdered form. White to yellowish in color, tragacanth is a translucent, odorless substance, with an insipid mucilaginous taste.

9 Pharmacopeial Specifications

See Table I.

10 Typical Properties

Acidity/alkalinity pH = 5–6 for a 1% w/v aqueous dispersion.
Acid value 2–5
Moisture content ≤15% w/w
Particle size distribution For powdered grades 50% w/w passes through a 73.7 μm mesh; ≤90% passes through a 150 μm mesh.[2]
Solubility Practically insoluble in water, ethanol (95%), and other organic solvents. Although insoluble in water, tragacanth gum swells rapidly in 10 times its own weight of either hot or cold

Table I: Pharmacopeial specifications for tragacanth.

Test	JP XV	PhEur 7.4	USP35–NF30
Identification	+	+	+
Botanical characteristics	−	−	+
Microbial limits	−	+	+
Flow time	−	≥10 s[a]	−
Lead	−	−	≤10 ppm
Heavy metals	−	−	≤20 ppm
Methylcellulose	−	+	−
Acacia	−	+	−
Foreign matter	−	≤1.0%	−
Karaya gum	+	−	+
Sterculia gum	−	+	−
Ash	≤4.0%	≤4.0%	−

(a) 50 s for emulsions.

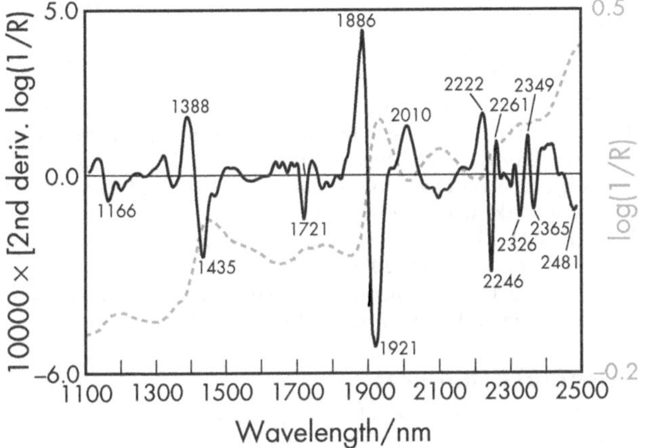

Figure 1: Near-infrared spectrum of tragacanth measured by reflectance.

water to produce viscous colloidal sols or semigels.[2] *See also* Section 18.

Specific gravity 1.250–1.385

Spectroscopy

NIR spectra *see* Figure 1.

Viscosity (dynamic) Viscosity is the most important factor used to evaluate quality and functionality of tragacanth. In general, the ribbon form gives a higher viscosity than the flake form.[4] The viscosity of tragacanth dispersions varies according to the grade and source of the material. Typically, 1% w/v aqueous dispersions may range in viscosity from 100–4000 mPa s (100–4000 cP) at 25°C and 20 rpm. Maximum viscosity is obtained after 24 hours at room temperature or by heating at 50°C for 2 hours. Maximum initial viscosity occurs at pH 8, although the greatest stability of tragacanth dispersions occurs at about pH 5. Solutions show pseudoplastic behavior typical of most gums. *See also* Sections 11 and 12.

11 Stability and Storage Conditions

Both the flaked and powdered forms of tragacanth are stable. Tragacanth gels are liable to exhibit microbial contamination with enterobacterial species, and stock solutions should therefore contain suitable antimicrobial preservatives. In emulsions, glycerin or propylene glycol are used as preservatives; in gel formulations, tragacanth is usually preserved with either 0.1% w/v benzoic acid, sorbic acid, or sodium benzoate below pH 6. A combination of 0.17% w/v methylparaben and 0.03% w/v propylparaben is also an effective preservative for tragacanth gels;[5] *see also* Section 12. Gels may be sterilized by autoclaving. Sterilization by gamma irradiation causes a marked reduction in the viscosity of tragacanth dispersions.[6]

The use of heat on tragacanth results in a reversible viscosity thinning effect. However, prolonged heating can degrade the gum and reduce viscosity permanently. It may be pulverized by heating to 50°C.

A higher tendency of tragacanth gum to hydrolyze compared to other hydrocolloids, such as guar or xanthan gums, has been described and is attributed to its highly branched and complex structure.[7]

Tragacanth dispersions are most stable at pH 4–8, although stability is satisfactory at higher pH or as low as pH 2. It is one of the most acid-resistant gums. However, it is recommended that acids be added only after the gum has fully hydrated.

The bulk material should be stored in an airtight container in a cool, dry place.

12 Incompatibilities

At pH 7, tragacanth has been reported to considerably reduce the efficacy of the antimicrobial preservatives benzalkonium chloride, chlorobutanol, and methylparaben, and to a lesser extent that of phenol and phenylmercuric acetate.[8] However, at pH <5 tragacanth was reported to have no adverse effects on the preservative efficacy of benzoic acid, chlorobutanol, or methylparaben.[5]

The addition of strong mineral and organic acids can reduce the viscosity of tragacanth dispersions. Viscosity may also be reduced by the addition of alkali or sodium chloride, particularly if the dispersion is heated. Tragacanth is compatible with relatively high salt concentrations and most other natural and synthetic suspending agents such as acacia, carboxymethylcellulose, starch, and sucrose. A yellow colored, stringy precipitate is formed with 10% w/v ferric chloride solution.

In emulsions, tragacanth is unstable under alkaline conditions.

13 Method of Manufacture

Tragacanth gum is the air-dried gum obtained from *Astragalus gummifer* Labillardière and other species of *Astragalus* grown principally in Iran, Syria, and Turkey. The primary source of this gum is the large tap roots of the bush. A low-quality gum is obtained by collecting the natural air-dried exudate or by making incisions in the bark of branches. The exudate is left to drain from the incision and dry naturally in the air before being collected. Drying conditions determine the color of the gum. After collection, the tragacanth gum is sorted by hand into various grades, such as ribbons or flakes. The gum is usually ground into powder with varying particle sizes, according to the desired viscosity.

T

14 Safety

Tragacanth has been used for many years in oral pharmaceutical formulations and food products, and is generally regarded as an essentially nontoxic material. It can be ingested in large amounts with little danger except for diarrhea and flatulence.[9] Tragacanth has been shown to be noncarcinogenic in mice.[10] However, hypersensitivity reactions, which are occasionally severe, have been reported following ingestion of products containing tragacanth.[11,12] Contact dermatitis has also been reported following the topical use of tragacanth formulations.[13] It is a skin and eye irritant.

The WHO has not specified an acceptable daily intake for tragacanth gum, as the daily intake necessary to achieve a desired effect, and its background levels in food, were not considered to be a hazard to health.[14]

LD_{50} (hamster, oral): 8.8 g/kg[15]

LD_{50} (mouse, oral): 10 g/kg

LD_{50} (rabbit, oral): 7.2 g/kg

LD_{50} (rat, oral): 16.4 g/kg

15 Handling Precautions

Observe normal precautions appropriate to the circumstances and quantity of material handled. Tragacanth gum may be irritant to the skin and eyes. Eye protection, gloves, and a dust mask are recommended.

16 Regulatory Status

GRAS listed. Accepted for use as a food additive in Europe. Included in the FDA Inactive Ingredients Database (buccal/sublingual tablets, oral powders, suspensions, syrups, and tablets). Included in nonparenteral medicines licensed in the UK. Included in the Canadian Natural Health Products Ingredients Database. Classified as 'acceptable daily intake (ADI) not specified' (the highest category of safety evaluation) by the Joint WHO/FAO Expert Committee on Food Additives.

17 Related Substances

See Section 18.

18 Comments

Tragacanth gum is a naturally occurring material whose physical properties vary greatly according to the grade and source of the material. Samples can contain relatively high levels of bacterial contamination.(16,17)

Powdered tragacanth gum tends to form lumps when added to water, and aqueous dispersions should therefore be agitated vigorously with a high-speed mixer. However, aqueous dispersions are more readily prepared by first prewetting the tragacanth with a small quantity of a wetting agent such as ethanol (95%), glycerin, or propylene glycol. If lumps form, they usually disperse on standing. Dispersion is generally complete after 1 hour. If other powders, such as sucrose, are to be incorporated into a tragacanth formulation the powders are best mixed together in the dry state.

Xanthan gum has been used in industrial applications as a substitute for tragacanth. In combination with gum arabic, an 'unusual' viscosity reduction has been observed with tragacanth.(2)

Some pharmacopeias, such as JP XV, contain a specification for powdered tragacanth.

19 Specific References

1 Whistler RL. Exudate Gums. In: Whistler RL, Bemiller JN, eds. *Industrial gums: polysaccharides and their derivatives,* San Diego, Academic Press, 1993; 318-337.
2 Weiping W. Tragacanth and Karaya. In: Philips GO, Williams PA, eds. *Handbook of hydrocolloids.* Cambridge: Woodhead, 2000: 155–168.
3 Ngwuluka NC, *et al.* Formulation and evaluation of paracetamol tablets manufactured using the dried fruit of *Phoenix dactylifera* Linn as an excipient. *Res Pharm Biotech* 2010; 2(3): 25–32.
4 Stauffer KR. Gum tragacanth. In: Davidson RL, ed. *Handbook of water-soluble gums and resins.* New York: McGraw-Hill, 1980; chapter 11.
5 Taub A, *et al.* Conditions for the preservation of gum tragacanth jellies. *J Am Pharm Assoc (Sci)* 1958; **47**: 235–239.
6 Jacobs GP, Simes R. The gamma irradiation of tragacanth: effect on microbial contamination and rheology. *J Pharm Pharmacol* 1979; **31**: 333–334.
7 ?ahin H, Özdemir F. Effect of some hydrocolloids on the serum separation of different formulated ketchups. *J Food Eng* 2007; **81**(2): 437–446.
8 Eisman PC, *et al.* Influence of gum tragacanth on the bactericidal activity of preservatives. *J Am Pharm Assoc (Sci)* 1957; **46**: 144–147.
9 Gosselin RE *et al. Clinical Toxicology of Commercial Products.* 4th edn. Baltimore: Williams and Wilkins, 1976; 11-52.
10 Hagiwara A, *et al.* Lack of carcinogenicity of tragacanth gum in B6C3F1 mice. *Food Chem Toxicol* 1992; **30**(8): 673–679.
11 Danoff D, *et al.* Big Mac attack [letter]. *N Engl J Med* 1978; **298**: 1095–1096.
12 Rubinger D. Hypersensitivity to tablet additives in transplant recipients on prednisone [letter]. *Lancet* 1978; **ii**: 689.
13 Coskey RJ. Contact dermatitis caused by ECG electrode jelly. *Arch Dermatol* 1977; **113**: 839–840.
14 FAO/WHO. Evaluation of certain food additives and contaminants. Twenty-ninth report of the joint FAO/WHO expert committee on food additives. *World Health Organ Tech Rep Ser* 1986; No. 733.
15 Lewis RJ, ed. *Sax's Dangerous Properties of Industrial Materials*, 11th edn. New York: Wiley, 2004; 3500.
16 Westwood N. Microbial contamination of some pharmaceutical raw materials. *Pharm J* 1971; **207**: 99–102.
17 De La Rosa MC, *et al.* Microbiological quality of pharmaceutical raw materials. *Pharm Acta Helv* 1995; **70**: 227–232.

20 General References

Caballero B, ed. Properties of individual gums. In: *Encyclopedia of Food Science and Nutrition*, 2003; 2992-3001.
Eastwood MA, *et al.* The effects of dietary gum tragacanth in man. *Toxicol Lett* 1984; **21**: 73–81.
Fairbairn JW. The presence of peroxidases in tragacanth [letter]. *J Pharm Pharmacol* 1967; **19**: 191.
Furia TE. Gum tragacanth. In: *Handbook of Food Additives*, 2nd edn. Boca Raton, FL: CRC Press, 1980; 315-316.
Verbeken D, *et al.* Exudate gums: occurence, production, and applications. *Appl Microbiol Biotechnol* 2003; **63**(1): 10–21.

21 Author

C Telang.

22 Date of Revision

1 March 2012.

Trehalose

1 Nonproprietary Names

PhEur: Trehalose Dihydrate
USP–NF: Trehalose

2 Synonyms

α-D-Glucopyranosyl-α-D-glucopyranoside; (α-D-glucosido)-α-D-glucoside; mycose; natural trehalose; α,α-trehalose; *Treha*; trehalose dihydrate.

3 Chemical Name and CAS Registry Number

Tragacanth gum [99-20-7]

(2*R*,3*R*,4*S*,5*R*,6*R*)-2-(Hydroxymethyl)-6-[(2*R*,3*R*,4*S*,5*R*,6*R*)-3,4,5-trihydroxy-6-(hydroxymethyl)oxan-2-yl]oxy-oxane-3,4,5-triol dihydrate [6138-23-4]

See also Section 17.

4 Empirical Formula and Molecular Weight

$C_{12}H_{22}O_{11}$ 342.30 (anhydrous)
$C_{12}H_{22}O_{11}{\cdot}2H_2O$ 378.33 (dihydrate)

5 Structural Formula

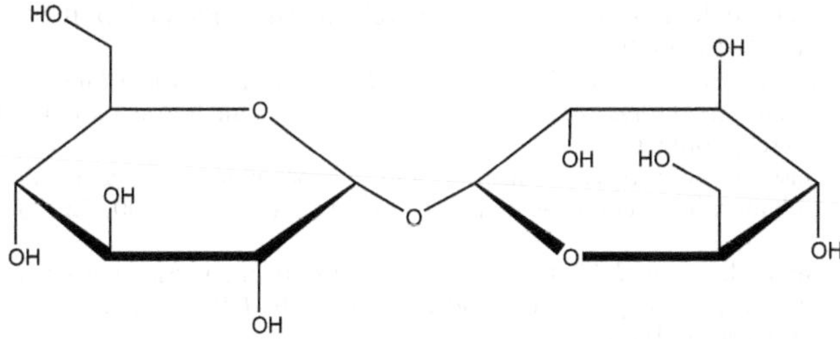

α,α-Trehalose dihydrate

6 Functional Category

Complexing agent; flavor enhancer; humectant; lyophilization aid; sweetening agent; tablet and capsule diluent; viscosity-increasing agent.

7 Applications in Pharmaceutical Formulation or Technology

Trehalose is used for the lyoprotection of therapeutic proteins, particularly for parenteral administration. Other pharmaceutically relevant applications include use as an excipient for diagnostic assay tablets;[1] for stabilization during the freeze–thaw and lyophilization of liposomes;[2,3] and for stabilization of blood cells,[4] cosmetics,[5] and monoclonal antibodies.[6] Trehalose may also be used in formulations for topical application.[7]

8 Description

Trehalose occurs as virtually odorless, white or almost white crystals with a sweet taste (approximately 45% of the sweetness of sucrose).[8]

9 Pharmacopeial Specifications

See Table I.

Table I: Pharmacopeial specifications for trehalose.

Test	PhEur 7.4	USP35–NF30
Identification	+	+
Color and clarity of solution	+	+
Specific rotation	+197–+201°	+197–+201°
Microbial limits	+	+
Bacterial endotoxins	+	+
pH	4.5–6.5	4.5–6.5
Water		
Anhydrous form	–	≤1.0%
Dihydrate form	9.0-11.0%	9.0–11.0%
Residue on ignition	–	≤0.1%
Soluble starch	+	+
Chloride	≤125 ppm	≤0.0125%
Sulfate	≤200 ppm	≤0.0200%
Heavy metals	≤5 ppm	≤5 ppm
Nitrogen content	–	≤0.005%
Related substances	+	+
Assay (anhydrous basis)	97.0–102.0%	97.0–102.0%

10 Typical Properties

Acidity/alkalinity pH = 4.5–6.5 (30% w/v aqueous solution)
Melting point 97°C (for the dihydrate)[8]
Moisture content 9.5% (for the dihydrate)
Solubility Soluble in water; very slightly soluble in ethanol (95%); practically insoluble in ether.

See also Section 18.

11 Stability and Storage Conditions

Trehalose is a relatively stable material. At 60°C for 5 hours it loses not more than 1.5% w/w of water (the dihydrate water of crystallization is retained). Open stored powder may liquefy at high relative humidity (≥90%).

Trehalose should be stored in a cool, dry place in a well-sealed container.

12 Incompatibilities

Trehalose is incompatible with strong oxidizing agents, especially in the presence of heat.

13 Method of Manufacture

Trehalose is prepared from liquefied starch by a multistep enzymatic process.[8] The commercial product is the dihydrate.

14 Safety

Trehalose is used in cosmetics, foods, and parenteral and nonparenteral pharmaceutical formulations. It is generally regarded as a relatively nontoxic and nonirritant material when used as an excipient.

In the gut, trehalose is rapidly metabolized to glucose by the specific enzyme trehalase. A small minority of the population exhibits a primary (hereditary) or secondary (acquired) trehalase deficiency and thus may experience intestinal discomfort after ingestion of excessive amounts of trehalose owing to the osmotic activity of undigested trehalose in the gut. However, smaller amounts of trehalose are tolerated by such individuals without any symptoms.[8]

Trehalose is used as a sweetener and is reported to have substantially less cariogenic potential than sucrose.

LD_{50} (dog, IV): >1 g/kg
LD_{50} (dog, oral): >5 g/kg
LD_{50} (mouse, IV): >1 g/kg
LD_{50} (mouse, oral): >5 g/kg
LD_{50} (rat, IV): >1 g/kg
LD_{50} (rat, oral): >5 g/kg

15 Handling Precautions

Observe normal precautions appropriate to the circumstances and quantity of material handled. Eye protection and gloves are recommended.

16 Regulatory Status

GRAS listed. In Canada it is classified as an NHP under schedule 1, item 2 (an isolate) of the Natural Health Products regulations. In the UK trehalose may be used in certain food applications.[9] Included in parenteral and nonparenteral investigational formulations.

17 Related Substances

Isotrehalose; neotrehalose.

Isotrehalose
CAS number [499-23-0]
Synonyms β,β-Trehalose.

Neotrehalose
CAS number [585-91-1]
Synonyms α,β-Trehalose.

18 Comments

α,α-Trehalose is the only naturally occurring isomer of trehalose and occurs as the dihydrate. However, α,β-trehalose (neotrehalose) and β,β-trehalose (isotrehalose) have been synthesized and are also available commercially. *See also* Section 17.

Trehalose is a nonreducing sugar and therefore does not react with amino acids or proteins as a part of Maillard browning. It is relatively stable under low-pH conditions compared to other disaccharides.

It should be noted that although trehalose dihydrate is quoted to have a melting point of 97°C, the true nature of this melting process has been the subject of debate in the literature,[10–12] including the transformation of the dihydrate into the anhydrous form. Anhydrous crystalline trehalose has been reported to melt at 203°C,[13] although higher values (215°C) have also been quoted in the literature.[14]

The glass transition temperature of trehalose is reported to be approximately 120°C (anhydrous amorphous phase).[15]

A specification for trehalose is contained in the *Food Chemicals Codex* (FCC).[16]

The EINECS number for trehalose is 202-739-6. The PubChem Compound ID (CID) for trehalose is 7427.

19 Specific References

1 Bollin E, Fletcher G. Trehalose as excipient and stabilizer for diagnostic assay tablets. United States Patent No. 4,678,812; 1987.
2 Holovati JL, Acker JP. Spectrophotometric measurement of intraliposomal trehalose. *Cryobiology* 2007; 55(2): 98–107.
3 Ohtake S, *et al.* Phase behavior of freeze-dried phospholipid-cholesterol mixtures stabilized with trehalose. *Biochim Biophys Acta* 2005; **1713**: 57–64.
4 Ligler FS, *et al.* Liposome encapsulated hemoglobin; stabilization, encapsulation and storage. *Prog Clin Biol Res* 1989; **319**: 435–455.
5 Pauly M. Pharmaceuticals and cosmetics containing glucidic compounds as active agents for skin regeneration. French Patent 2 609 397; 1988.
6 Matsuo E, Yamazaki S. Freeze-dried composition containing enzyme-labeled antihuman β-interferon antibody. International Patent 09 402 05; 1989.
7 Giandala G, *et al.* Trehalose-hydroxyethylcellulose microspheres containing vancomycin for topical drug delivery. *Eur J Pharm Biopharm* 2001; **52**(1): 83–89.
8 Richards AB, *et al.* Trehalose: a review of properties, history of use and human tolerance, and results of multiple safety studies. *Food Chem Toxicol* 2002; **40**: 871–898.
9 Advisory Committee On Novel Foods And Processes UK/2000/001. Opinion on an application under the Novel Food Regulation from Bioresco Ltd for clearance of Trehalose produced by a novel enzymatic process. www.food.gov.uk/multimedia/pdfs/trehafin.pdf (accessed 31 October 2011).
10 Sussich F, *et al.* Reversible dehydration of trehalose and anhydrobiosis: from solution state to an exotic crystal? *Carbohydr Res* 2001; **334**: 165–176.
11 Taylor LS, York P. Characterisation of the phase transitions of trehalose dihydrate on heating and subsequent dehydration. *J Pharm Sci* 1998; **87**: 347–355.
12 McGarvey OS, *et al.* An investigation into the crystallization of alpha,alpha-trehalose from the amorphous state. *J Phys Chem B* 2003; **107**: 6614–6620.
13 O'Neil MJ, ed. Trehalose. In: *The Merck Index: an Encyclopedia of Chemicals, Drugs, and Biologicals*, 14th edn. Whitehouse Station, NJ: Merck, 2006; 1647.
14 Sussich F, Cesaro A. Transitions and phenomenology of α,α-trehalose polymorphs inter-conversion. *J Therm Anal Calorim* 2000; **62**: 757–768.
15 Simperler A, *et al.* Glass transition temperature of glucose, sucrose, and trehalose: an experimental and in silico study. *J Phys Chem B* 2006; **110**: 19678–19684.
16 *Food Chemicals Codex*, 7th edn. Suppl. 3. Bethesda, MD: United States Pharmacopeia, 2011; 1718.

20 General References

Cargill. Product literature: *Treha*, 2008. www.cargill.com (accessed 12 August 2010).
Ohtake S, Wang J. Trehalose: Current use and future applications. *J Pharm Sci* 2011; **100**: 2020–2053.
Pikal MJ. Freeze drying. In: Swarbrick J, Boylan JC, eds. *Encyclopedia of Pharmaceutical Technology*, 2nd edn, vol. 2. New York: Marcel Dekker, 2002; 1299–1326.

21 Author

VL Kett.

22 Date of Revision

1 March 2012.

Triacetin

1 Nonproprietary Names

BP: Triacetin
PhEur: Triacetin
USP-NF: Triacetin

2 Synonyms

Captex 500; E1518; glycerol triacetate; glyceryl triacetate; *Speziol GTA*; triacetinum; triacetyl glycerine.

3 Chemical Name and CAS Registry Number

1,2,3-Propanetriol triacetate [102-76-1]

4 Empirical Formula and Molecular Weight

$C_9H_{14}O_6$ 218.21

5 Structural Formula

6 Functional Category

Humectant; plasticizing agent; solvent.

7 Applications in Pharmaceutical Formulation or Technology

Triacetin is mainly used as a hydrophilic plasticizer in both aqueous and solvent-based polymeric coating of capsules, tablets, beads, and granules; typical concentrations used are 10–35% w/w.[1,2]

8 Description

Triacetin is a colorless, viscous liquid with a slightly fatty odor.

9 Pharmacopeial Specifications

See Table I.

Table I: Pharmacopeial specifications for triacetin.

Test	PhEur 7.4	USP35–NF30
Appearance	+	—
Characters	+	—
Identification	+	+
Specific gravity	1.159–1.164	1.152–1.158
Refractive index	1.429–1.432	1.429–1.430
Acidity	+	+
Water	≤0.2%	≤0.2%
Assay (anhydrous basis)	97.0–100.5%	97.0–100.5%

10 Typical Properties

Autoignition temperature 432°C
Boiling point 258°C
Density 1.16 g/cm^3 at 25°C
Explosive limits
1.05% at 189°C lower limit;
7.73% at 215°C upper limit.
Flash point 153°C (open cup)
Freezing point 3.2°C (supercools to about −70°C)
Melting point −78°C
Refractive index n_D^{25} = 1.4296
Solubility *see* Table II.

Table II: Solubility of triacetin.

Solvent	Solubility at 20°C
Carbon disulfide	Miscible
Chloroform	Miscible
Ethanol	Miscible
Ethanol (95%)	Miscible
Ether	Miscible
Toluene	Miscible
Water	1 in 14

Spectroscopy
IR spectra *see* Figure 1.
Vapor density (relative) 7.52 (air = 1)
Vapor pressure 133 Pa (1 mmHg) at 100°C
Viscosity (dynamic)
1111 mPa s (1111 cP) at −17.8°C;
107 mPa s (107 cP) at 0°C;
17.4 mPa s (17.4 cP) at 25°C;
1.8 mPa s (1.8 cP) at 100°C.

11 Stability and Storage Conditions

Triacetin is stable and should be stored in a well-closed, nonmetallic container, in a cool, dry place.

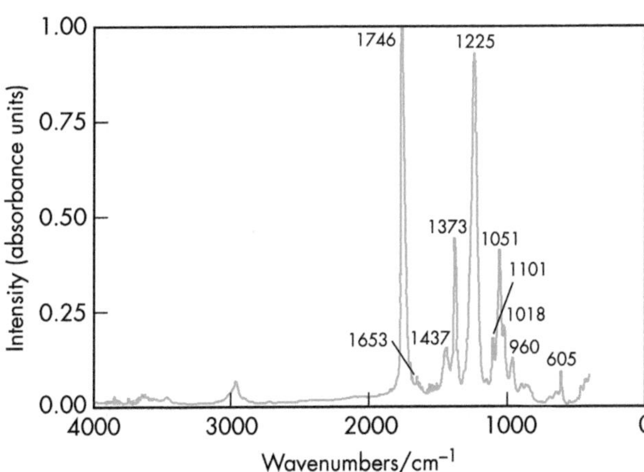

Figure 1: Infrared spectrum of triacetin measured by transmission. Adapted with permission of Informa Healthcare.

12 Incompatibilities

Triacetin is incompatible with metals and may react with oxidizing agents. Triacetin may destroy rayon fabric.

13 Method of Manufacture

Triacetin is prepared by the esterification of glycerin with acetic anhydride.

14 Safety

Triacetin is used in oral pharmaceutical formulations and is generally regarded as a relatively nontoxic and nonirritant material at the levels employed as an excipient.[3]

LD_{50} (dog, IV): 1.5 g/kg[4]
LD_{50} (mouse, IP): 1.4 g/kg
LD_{50} (mouse, IV): 1.6 g/kg
LD_{50} (mouse, oral): 1.1 g/kg
LD_{50} (mouse, SC): 2.3 g/kg
LD_{50} (rabbit, IV): 0.75 g/kg
LD_{50} (rat, IP): 2.1 g/kg
LD_{50} (rat, oral): 3 g/kg
LD_{50} (rat, SC): 2.8 g/kg

15 Handling Precautions

Observe normal precautions appropriate to the circumstances and quantity of material handled. Triacetin may be irritant to the eyes; eye protection and gloves are recommended.

16 Regulatory Status

GRAS listed. Accepted in Europe as a food additive in certain applications. Included in the FDA Inactive Ingredients Database (oral capsules and tablets; gels). Included in nonparenteral medicines licensed in the UK. Included in the Canadian Natural Health Products Ingredients Database.

17 Related Substances

—

18 Comments

Triacetin is used in cosmetics, perfumery, and foods as a solvent and as a fixative in the formulation of perfumes and flavors.

A specification for triacetin is contained in the *Food Chemicals Codex* (FCC).[5]

The EINECS number for triacetin is 203-051-9. The PubChem Compound ID (CID) for triacetin is 5541.

19 Specific References

1 Shah PS, Zatz JL. Plasticization of cellulose esters used in the coating of sustained release solid dosage forms. *Drug Dev Ind Pharm* 1992; **18**: 1759–1772.
2 Williams RO, *et al.* Influence of plasticization and curing conditions on the mechanical properties of aqueous based cellulose acetate films. *STP Pharma Sci* 1999; 9(6): 545–553.
3 Fiume MZ. Final report on the safety assessment of triacetin. *Int J Toxicol* 2003; 22(Suppl 2): 1–10.
4 Lewis RJ, ed. *Sax's Dangerous Properties of Industrial Materials*, 11th edn. New York: Wiley, 2004; 3503.
5 *Food Chemicals Codex*, 7th edn. Bethesda, MD: United States Pharmacopeia, 2010.

20 General References

Gutierrez-Rocca JC, McGinity JW. Influence of aging on the physical-mechanical properties of acrylic resin films cast from aqueous dispersions and organic solutions. *Drug Dev Ind Pharm* 1993; **19**: 315–332.

Johnson K, *et al.* Effect of triacetin and polyethylene glycol 400 on some physical properties of hydroxypropyl methylcellulose free films. *Int J Pharm* 1991; 73: 197–208.

Lehmann KOR. Chemistry and application properties of polymethacrylate coating systems. In: McGinity JW, ed. *Aqueous Polymeric Coatings for Pharmaceutical Dosage Forms*. New York: Marcel Dekker, 1989; 224.

Lin S-Y, *et al.* The effect of plasticizers on compatibility, mechanical properties, and adhesion strength of drug-free Eudragit E films. *Pharm Res* 1991; 8: 1137–1143.

Rowe RC. Materials used in the film coating of oral dosage forms. In: Florence AT, ed. *Critical Reports on Applied Chemistry*. Oxford: Blackwell Scientific, 1984; 1–36.

21 Author

A Palmieri.

22 Date of Revision

1 March 2012.

T

Tributyl Citrate

1 Nonproprietary Names

USP–NF: Tributyl Citrate

2 Synonyms

Citric acid, tributyl ester; *Citroflex 4*; *Citrofol BI*; TBC; tri-*n*-butyl citrate; tributyl 2-hydroxy-1,2,3-propanetricarboxylate.

3 Chemical Name and CAS Registry Number

1,2,3-Propanetricarboxylic acid, 2-hydroxy, tributyl ester [77-94-1]

4 Empirical Formula and Molecular Weight

$C_{18}H_{32}O_7$ 360.5

5 Structural Formula

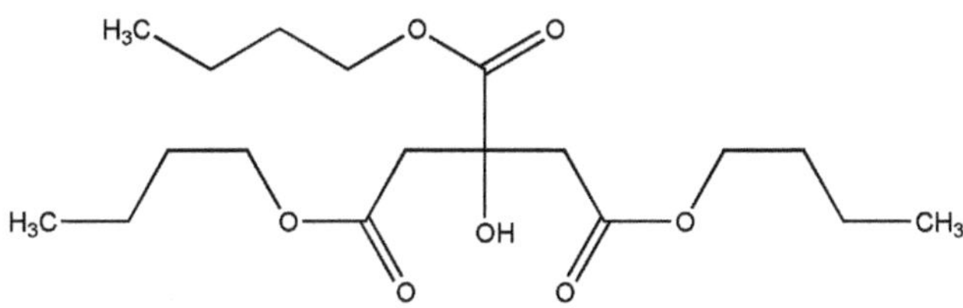

6 Functional Category

Plasticizing agent.

7 Applications in Pharmaceutical Formulation or Technology

Tributyl citrate is used to plasticize polymers in formulated pharmaceutical coatings. The coating applications include capsules, tablets, beads, and granules for taste masking, immediate release, sustained-release, and enteric formulations.[1–6]

8 Description

Tributyl citrate is a clear, odorless, practically colorless, oily liquid.

9 Pharmacopeial Specifications

See Table I.

Table I: Pharmacopeial specifications for tributyl citrate.

Test	USP35–NF30
Identification	+
Specific gravity	1.037–1.045
Refractive index	1.443–1.445
Acidity	+
Water	≤0.2%
Heavy metals	≤0.001%
Assay (anhydrous basis)	≥99.0%

10 Typical Properties

Acid value 0.02
Boiling point 322°C (decomposes)
Flash point 185°C
Pour point −62°C

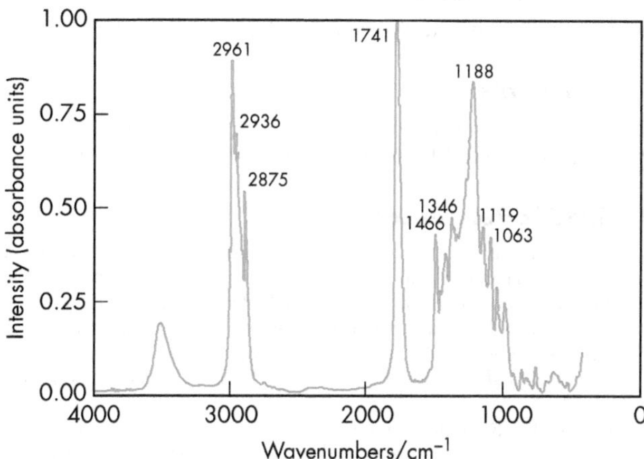

Figure 1: Infrared spectrum of tributyl citrate measured by transmission. Adapted with permission of Informa Healthcare.

Refractive index n_D^{25} = 1.443–1.445
Solubility Miscible with acetone, ethanol, and vegetable oil; practically insoluble in water.
Specific gravity 1.037–1.045 for *Citroflex 4*.
Spectroscopy
IR spectra *see* Figure 1.
Viscosity 32 mPa s (32 cP) at 25°C

11 Stability and Storage Conditions

Tributyl citrate should be stored in well-closed containers in a cool, dry location at temperatures not exceeding 38°C. When stored in accordance with these conditions, tributyl citrate is a stable material.

12 Incompatibilities

Tributyl citrate is incompatible with strong alkalis and oxidizing materials.

13 Method of Manufacture

Tributyl citrate is prepared by the esterification of citric acid with butanol.

14 Safety

Tributyl citrate is used in oral pharmaceutical formulations. It is generally regarded as an essentially nontoxic and nonirritating material. However, ingestion of large quantities may be harmful.

LD_{50} (cat, oral): >50 mL/kg[7]
LD_{50} (mouse, IP): 2.9 g/kg
LD_{50} (rat, oral): >30 mL/kg

15 Handling Precautions

Observe normal precautions appropriate to the circumstances and quantity of material handled. Tributyl citrate may be irritating to the eyes. It may also be irritating to the respiratory system at elevated temperatures.

Gloves and eye protection are recommended for normal handling, and a respirator is recommended for elevated temperatures.

16 Regulatory Status

Approved in the US for indirect food contact in food films. Included in the Canadian Natural Health Products Ingredients Database.

17 Related Substances

Acetyltributyl citrate; acetyltriethyl citrate; triethyl citrate.

18 Comments

The EINECS number for tributyl citrate is 201-071-2. The PubChem Compound ID (CID) for tributyl citrate is 6507.

19 Specific References

1 Gutierrez-Rocca JC, McGinity JW. Influence of water soluble and insoluble plasticizer on the physical and mechanical properties of acrylic resin copolymers. *Int J Pharm* 1994; **103**: 293–301.
2 Lehmann K. Chemistry and application properties of polymethacrylate coating systems. In: McGinity JW, ed. *Aqueous Polymeric Coatings for Pharmaceutical Dosage Forms*. New York: Marcel Dekker, 1989; 153–245.
3 Steurnagel CR. Latex emulsions for controlled drug delivery. In: McGinity JW, ed. *Aqueous Polymeric Coatings for Pharmaceutical Dosage Forms*. New York: Marcel Dekker, 1989; 1–61.
4 Gutierrez-Rocca JC, McGinity JW. Influence of aging on the physical-mechanical properties of acrylic resin films cast from aqueous dispersions and organic solutions. *Drug Dev Ind Pharm* 1993; **19**(3): 315–332.
5 Felton LA, McGinity JW. Influence of plasticisers on the adhesive properties of an acrylic resin copolymer to hydrophilic and hydrophobic tablet compacts. *Int J Pharm* 1997; **154**(2): 167–178.
6 Okarter TU, Singla K. The effects of plasticisers on the release of metoprolol tartrate from granules coated with a polymethacrylate film. *Drug Dev Ind Pharm* 2000; **26**(3): 323–329.
7 Lewis RJ, ed. *Sax's Dangerous Properties of Industrial Materials*, 11th edn. New York: Wiley, 2004; 3513.

20 General References

Morflex Inc. Technical literature: *Citroflex 4* (tri-*n*-butyl citrate), 2005.

21 Authors

ME Fenton, PJ Sheskey.

22 Date of Revision

1 March 2012.

Tricaprylin

1 Nonproprietary Names

None adopted.

2 Synonyms

Caprylic acid, 1,2,3-propanetriyl ester; caprylic acid triglyceride; *Captex 8000*; glycerin tricaprylate; glycerol tricaprylate; glycerol trioctanoate; glyceryl tricaprylate; glyceryl trioctanoate; *Hest TC*; MCT; *Miglyol 808*; *n*-octanoic acid glycerol triester; octanoic acid, 1,2,3-propanetriyl ester; *Panacet 800/875*; *Rofetan GTC*; tricaprilin; tricapryloglycerol; tricaprylylglycerin; trioctanoin; trioctonolglycerol.

3 Chemical Name and CAS Registry Number

1,3-Di(octanoyloxy)propan-2-yl octanoate [538-23-8]

4 Empirical Formula and Molecular Weight

$C_{27}H_{50}O_6$ 470.70

5 Structural Formula

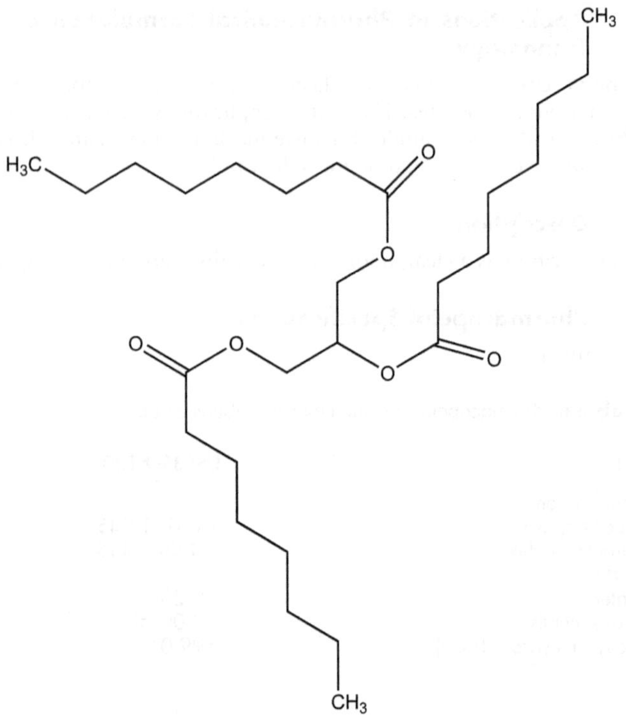

6 Functional Category

Dispersing agent; emollient; lubricant; nonionic surfactant; penetration enhancer; plasticizing agent; solubilizing agent; solvent.

7 Applications in Pharmaceutical Formulation or Technology

Tricaprylin is used in pharmaceutical preparations as a neutral carrier, absorption promoter, and solubilizer for active drugs. It has been used as an oily phase to prepare water-in-oil-in-water multiple emulsions for incorporating water-soluble drugs such as cefadroxil, cephradine, 4-aminoantipyrine, and antipyrine,[1] and also for obtaining stable microcapsules[2] and oral microemulsions.[3]

Tricaprylin acts as a vehicle for topical creams and lotions and cosmetic preparations, and is used as a penetration-enhancing lipid base and emollient. Owing to its nongreasy characteristics and low viscosity, it has very good spreadability. In spite of being skin-permeable, tricaprylin does not obstruct natural skin respiration, and hence it is used in baby oils, massage oils, and face masks. It is an excellent dispersant, and acts as a solubilizer, wetting agent and binder in color cosmetics. Being readily miscible with natural oils and surfactants, tricaprylin is used as the fat component in two-phase foam baths. It is used in sunscreen creams and oils because of its compatibility with organic and inorganic filter agents. It is also used as a fixative for perfumes/fragrances.

8 Description

Tricaprylin occurs as a clear, colorless to pale-yellow liquid. It forms crystals from acetone/ethanol (95%). Tricaprylin is odorless.

See also Table I.

Table I: Description of commercially available grades of tricaprylin.

Grade	Description
Captex 8000	Clear to light-yellow liquid with bland taste and neutral odor
Hest TC	Light liquid
Miglyol 808	Clear, virtually colorless oily liquid with neutral odor and taste
Panacet 800/875	Liquid
Rofetan GTC	Liquid

9 Pharmacopeial Specifications

See Section 18.

10 Typical Properties

Acid value
≤0.1 for *Captex 8000*;
≤0.1 for *Miglyol 808*;
≤0.1 for *Rofetan GTC*.
Ash 0.1% for *Rofetan GTC*
Boiling point 233°C at 133.3 Pa (1 mmHg)
Cloud point −5.0°C for *Rofetan GTC*
Color
≤150.0 (Hazen color index) for *Captex 8000*;
≤50.0 (Hazen color index) for *Miglyol 808*;
≤50.0 (Hazen color index) for *Rofetan GTC*.
Density
0.95 g/cm^3;
0.94–0.96 g/cm^3 for *Rofetan GTC* at 20°C.
Fatty acid distribution *see* Table II.
Flash point
209°C;
246°C (open cup) for *Captex 8000*.
Freezing point ≤5.0 for *Panacet 800/875*
Heavy metals 10 mg/kg for *Rofetan GTC*
HLB value 7.0 for *Hest TC*
Hydroxyl value
≤0.5 for *Miglyol 808*;
≤0.5 for *Rofetan GTC*.
Iodine number
≤0.3 for *Miglyol 808*;
≤0.3 for *Rofetan GTC*.
Melting point 9–10°C
Moisture content
≤0.1% w/w for *Captex 8000*;
≤0.1% w/w for *Miglyol 808*;
≤0.1% w/w for *Rofetan GTC*.
Partition coefficient log P_{ow} (octanol : water) = 9.20[3]
Peroxide value
≤0.1 for *Miglyol 808*;
≤0.1 for *Rofetan GTC*.
Refractive index
1.4484;
1.447 for *Rofetan GTC* at 20°C.
Saponification value
345–360 for *Captex 8000*;
340–370 for *Miglyol 808*;
340–370 for *Rofetan GTC*.
Solubility Miscible with most organic solvents including ethanol (95%). *Captex 8000* is insoluble in water.
Specific gravity 0.94 for *Captex 8000* at 25°C
Specific heat of vaporisation 0.0165 kJ/g
Spreading rate 254 mm^2 for *Hest TC*
Surface tension 0.030 N/m
Vapor pressure <133.3 Pa (<1 mmHg) for *Captex 8000* at 25°C
Viscosity (dynamic)
11.57 mPa s at 39.9°C;
23–29 mPa s (23–29 cP) for *Miglyol 808* at 20°C;
23–29 mPa s (23–29 cP) for *Rofetan GTC* at 20°C.
Viscosity (kinematic) 20.9 mm^2/s (20.9 cSt) for *Captex 8000* at 25°C

Table II: Typical fatty acid distribution of commercially available grades of tricaprylin.

Grade	Fatty acid distribution by gas–liquid chromatography (GLC)
Captex 8000	
Caproic acid (C_6)	≤1.0%
Caprylic acid (C_8)	≥90.0%
Capric acid (C_{10})	≤5.0%
Lauric acid (C_{12})	≤1.0%
Rofetan GTC	
Caproic acid (C_6)	≤0.5%
Caprylic acid (C_8)	≥95.0%
Capric acid (C_{10})	≤5.0%
Lauric acid (C_{12})	≤0.5%

11 Stability and Storage Conditions

Tricaprylin is classified as a stable compound. It has high stability against oxidation and is not heat sensitive. Even in hot climates cooling is not necessary. However, exposure to high temperatures near the flash point (246°C) should be avoided. Owing to its very low water content, it is not sensitive to hydrolytic and microbial splitting. Although polymerization of tricaprylin will not occur, it is reported to decompose into carbon monoxide and carbon dioxide.

Tricaprylin should be stored in well-closed containers, protected from light, in a dry place at ambient temperature. High-density polyethylene, polypropylene, metal (aluminum), and glass are suitable for packaging. Some plastics, especially those containing plasticizers, can become brittle or expand in the presence of tricaprylin. Polystyrene and polyvinyl chloride are not suitable for its storage. Tricaprylin has a high tendency to migrate, and therefore care should be taken when selecting seal-closure elastomer material.

12 Incompatibilities

Tricaprylin is incompatible with strong oxidizing agents.

13 Method of Manufacture

Tricaprylin is a triglyceride manufactured by esterification of caprylic acid and glycerin.

14 Safety

Tricaprylin is used in pharmaceutical and cosmetic formulations.

The Cosmetic Ingredient Review (CIR) Expert Panel found that dermal application of tricaprylin has not been associated with significant irritation in rabbit skin.[5] However, as a penetration enhancer, tricaprylin may allow other chemicals to penetrate deeper into the skin, increasing their concentration so that they may reach the bloodstream. Ocular exposures of tricaprylin were found to be only mildly irritating to rabbit eyes.[5] Little or no acute, subchronic, or chronic oral toxicity was observed in animal studies unless levels approached a significant percentage of caloric intake.[5] Subcutaneous injections of tricaprylin in rats over a period of 5 weeks caused a granulomatous reaction.[5]

Tricaprylin has not been found to be teratogenic in rats, mice, or hamsters, but some reproductive effects have been seen in rabbits.[5] Dose-related central nervous system toxicity in dogs has also been observed.[6]

LD_{50} (mouse, IP): >27.8 g/kg[7]
LD_{50} (mouse, IV): 3.7 g/kg[8]
LD_{50} (mouse, oral): 29.6 g/kg[8]
LD_{50} (mouse, SC): >27.8 g/kg[7]
LD_{50} (rat, IP): 0.05 g/kg[8]
LD_{50} (rat, oral): 33.3 g/kg[8]
LD_{50} (rat, SC): >27.8 g/kg[7]
LD_{Lo} (rat, IV): 4 g/kg[7]

15 Handling Precautions

Observe normal precautions appropriate to the circumstances and quantity of the material handled. Use of a mask and/or respirator is recommended. When heated to decomposition, tricaprylin emits acrid smoke and irritating fumes. Ventilation is recommended to control dust or fumes from the heated material. Chemical splash goggles are recommended for eye protection, and neoprene-type gloves are also recommended.

T

16 Regulatory Status

Included in the FDA Inactive Ingredients Database (epidural injections).

17 Related Substances

Glyceryl triisooctanoate; medium-chain triglycerides.

Glyceryl triisooctanoate
Empirical formula $C_{27}H_{50}O_6$
Molecular weight 470.68
CAS number [7360-38-5]
Synonyms Glyceryl tris-2-ethylhexanoate
Residue on ignition ≤0.5%
Specific gravity 0.945–0.950
Comments A clear, colorless to pale yellow, oily liquid; odorless. Miscible with ethanol, 2-propanol, and diethyl ether; practically insoluble in water. Listed in JPE 2004.[9]

18 Comments

Although it is not currently included in the pharmacopeias, a specification for tricaprylin is included in the *Japanese Pharmaceutical Excipients Directory* (JPED); *see* Table III. It is included in the Cosmetics Ingredient Review (CIR) Category 1 and Category 37 as safe for use in cosmetics.[10]

Tricaprylin has been used as a skin permeation enhancer in studies of transdermal drug delivery systems and it has been shown to improve skin permeability of various drugs with different lipophilicity within a tricaprylin/ethanol (60/40) lipophilic binary vehicle.[11–17] Tricaprylin has also been investigated as the oily phase to fabricate 'hairy' colloidosomes[18] and colloidosome microcapsules[2,19] for drug delivery applications. Tricaprylin/glycerol monostearate/water systems are increasingly being used as mesophases existing in biological systems for the study of drug behavior across membranes as they have structural resemblance to human membranes.[20,21] Reverse hexagonal liquid-crystalline structures composed of monoolein/tricaprylin/water have been demonstrated to solubilize large quantities of cyclosporine A[22] and gabapentin,[23] as well as dermal penetration enhancers.

Tricaprylin is also used as an oily phase in various model emulsions, which are used as substrates in food and agricultural research[24] or for synergistic stabilization of oil-in-water emulsions in the presence of surface active nanoparticles and surfactants.[25] An intramuscular injection of triolein/tricaprylin multivesicular liposome formulation for the sustained delivery of breviscapine has been reported.[26] Tricaprylin has also been studied for use in a novel solid lipid nanoparticle system, in which a superfine nanostructure was formed within the solid lipid, maintaining the nano size range.[27]

The EINECS number of tricaprylin is 208-686-5. The PubChem Compound ID for tricaprylin is 10850.

Table III: JPED specification for tricaprylin.[28]

Test	JPED 1993
Assay	≥98.0%
Refractive index	1.440–1.445
Specific gravity	0.945–0.960
Acid value	≤0.2
Hydroxyl value	≤10
Chloride	≤0.004%
Heavy metals	≤10 ppm
Arsenic	≤1 ppm
Zinc	+
Water	≤0.30%
Residue on ignition	≤0.10%

19 Specific References

1 Zhang W, *et al.* Preparation of stable W/O/W type multiple emulsion containing water-soluble drugs and *in vitro* evaluation of its drug-releasing properties. *Yakugaku Zasshi* 1992; **112**(1): 73–80.
2 Adachi S, *et al.* Preparation of an water-in-oil-in-water (W/O/W) type microcapsules by a single-droplet-drying method and change in encapsulation efficiency of a hydrophilic substance during storage. *Biosci Biotechnol Biochem* 2003; **67**(6): 1376–1381.
3 Kim SK, *et al.* Tricaprylin microemulsion for oral delivery of low molecular weight heparin conjugates. *J Control Release* 2005; **105**(1–2): 32–42.
4 Meylan WM, Howard PH. Atom/fragment contribution method for estimating octanol–water partition coefficients. *J Pharm Sci* 1995; **84**: 83–92.
5 Johnson W Jr. Cosmetic Ingredient Review Expert Panel. Final report on the safety assessment of trilaurin, triarachidin, tribehenin, tricaprin, tricaprylin, trierucin, triheptanoin, triheptylundecanoin, triisononanoin, triisopalmitin, triisostearin, trilinolein, trimyristin, trioctanoin, triolein, tripalmitin, tripalmitolein, triricinolein, tristearin, triundecanoin, glyceryl triacetyl hydroxystearate, glyceryl triacetyl ricinoleate, and glyceryl stearate diacetate. *Int J Toxicol* 2001; **20**(Suppl. 4): 61–94.

6 Miles JM, *et al.* Metabolic and neurologic effects of an intravenous medium-chain triglyceride emulsion. *J Parenter Enteral Nutr* 1991; **15**(1): 37–41.
7 The Registry of Toxic Effects of Chemical Substances, Trioctanoin, RTECS#: YJ7700000, National Institute for Occupational Safety and Health, February 2003. www.cdc.gov/niosh/rtecs/yj757e20.html (accessed 14 December 2011).
8 Lewis RJ, ed. *Sax's Dangerous Properties of Industrial Materials*, 11th edn. New York: Wiley, 2004: 3605.
9 Japan Pharmaceutical Excipients Council. *Japan Pharmaceutical Excipients 2004*. Tokyo: Yakuki Nippo, 2004: 363–364.
10 Cosmetic Ingredient Review. Compendium containing abstracts, discussions, and conclusions of CIR cosmetic ingredient safety assessments. Washington DC: Cosmetic Ingredient Review, 2003.
11 Goto S. Studies on development of pharmaceutical preparation with the purpose of improving controlled-release and bioavailability. *Yakugaku Zasshi* 1995; **115**(11): 871–891.
12 Goto S, *et al.* Effect of various vehicles on ketoprofen permeation across excised hairless mouse skin. *J Pharm Sci* 1993; **82**(9): 959–963.
13 Uchida T, *et al.* Enhancement effect of an ethanol/Panasate 800 binary vehicle on anti-inflammatory drug permeation across excised hairless mouse skin. *Biol Pharm Bull* 1993; **16**(2): 168–171.
14 Lee CK, *et al.* Transdermal delivery of theophylline using an ethanol/ panasate 800-ethylcellulose gel preparation. *Biol Pharm Bull* 1995; **18**(1): 176–180.
15 Lee CK, *et al.* Skin permeability of various drugs with different lipophilicity. *J Pharm Sci* 1994; **83**(4): 562–565.
16 Lee CK, *et al.* Effect of hydrophilic and lipophilic vehicles on skin permeation of tegafur, alclofenac and ibuprofen with or without permeation enhancers. *Biol Pharm Bull* 1993; **16**(12): 1264–1269.
17 Lee CK, *et al.* Skin permeation enhancement of tegafur by ethanol/ panasate 800 or ethanol/water binary vehicle and combined effect of fatty acids and fatty alcohols. *J Pharm Sci* 1993; **82**(11): 1155–1159.
18 Noble PF, *et al.* Fabrication of 'hairy' colloidosomes with shells of polymeric microrods. *J Am Chem Soc* 2004; **126**: 8092–8093.
19 Paunov VN, *et al.* Fabrication of novel types of colloidosome microcapsules for drug delivery applications. *Mater Res Soc Sym Proc* 2005; **845**: AA5.18.1–AA5.18.5.
20 Amar-Yuli I, *et al.* Hexosome and hexagonal phases mediated by hydration and polymer stabilizer. *Langmuir* 2007; **23**: 3637–3645.
21 Garti N, Amar-Yuli I. Transitions induced by solubilized fat into reverse hexagonal mesophases. *Colloids Surf B Biointerfaces* 2005; **43**: 72–82.
22 Libster D, *et al.* An HII liquid crystal-based delivery system for cyclosporin A: physical characterization. *J Colloid Interface Sci* 2007; **308**: 514–524.
23 Achrai B, *et al.* Solubilization of gabapentin into H-II mesophases. *J Physical Chem* 2011; **115**(5): 825–835.
24 van Koningsveld GA, *et al.* Effects of protein composition and enzymatic activity on formation and properties of potato protein stabilized emulsions. *J Agric Food Chem* 2006; **54**: 6419–6427.
25 Binks BP, *et al.* Synergistic stabilization of emulsions by a mixture of surface-active nanoparticles and surfactant. *Langmuir* 2007; **23**: 1098–1106.
26 Zhong H, *et al.* Multivesicular liposome formulation for the sustained delivery of breviscapine. *Int J Pharm* 2005; **301**: 15–24.
27 Hou D-Z, *et al.* [Microstructure of novel solid lipid nanoparticle loaded triptolide.] *Yaoxue Xuebao* 2007; **42**(4): 429–433[in Chinese].
28 Japanese Pharmaceutical Excipients Council. *Japanese Excipients Directory* 1996. Tokyo: Yakuji Nippo, 1996: 560.

20 General References

Abitec Corporation. Technical bulletin, version 6: *Captex 8000*, April 2008.
Abitec Corporation. Material safety data sheet no. 05-8503-00: *Captex 8000*, July 2005.
Ecogreen Oleochemicals GmbH. Product specification: *Rofetan GTC*, December 2002.
Sasol Germany GmbH Oleochemicals. Product information: *Miglyol 808*, February 2005.

21 Authors

RT Gupta, KK Singh.

22 Date of Revision

14 December 2011.

Triethanolamine

1 Nonproprietary Names

BP: Triethanolamine
PhEur: Trolamine
USP–NF: Trolamine

2 Synonyms

TEA; *Tealan*; triethylolamine; trihydroxytriethylamine; tris (hydroxyethyl)amine; trolaminum.

3 Chemical Names and CAS Registry Number

2,2′,2″-Nitrilotriethanol [102-71-6]

4 Empirical Formula and Molecular Weight

$C_6H_{15}NO_3$ 149.19

5 Structural Formula

HO N OH OH

6 Functional Category

Alkalizing agent; emulsifying agent.

7 Applications in Pharmaceutical Formulation or Technology

Triethanolamine is widely used in topical pharmaceutical formulations, primarily in the formation of emulsions.

When mixed in equimolar proportions with a fatty acid, such as stearic acid or oleic acid, triethanolamine forms an anionic soap with a pH of about 8, which may be used as an emulsifying agent to produce fine-grained, stable oil-in-water emulsions. Concentrations that are typically used for emulsification are 2–4% v/v of triethanolamine and 2–5 times that of fatty acids. In the case of mineral oils, 5% v/v of triethanolamine will be needed, with an appropriate increase in the amount of fatty acid used. Preparations that contain triethanolamine soaps tend to darken on storage. However, discoloration may be reduced by avoiding exposure to light and contact with metals and metal ions.

Triethanolamine is also used in salt formation for injectable solutions and in topical analgesic preparations. It is also used in sun screen preparations.[1]

8 Description

Triethanolamine is a clear, colorless to pale yellow-colored viscous liquid having a slight ammoniacal odor. It is a mixture of bases, mainly 2,2′,2″-nitrilotriethanol, although it also contains 2,2′-iminobisethanol (diethanolamine) and smaller amounts of 2-aminoethanol (monoethanolamine).

9 Pharmacopeial Specifications

See Table I.

Table I: Pharmacopeial specifications for triethanolamine.

Test	PhEur 7.4	USP35–NF30
Characters	+	–
Identification	+	+
Appearance of solution	+	–
Related substances	+	–
Heavy metals	≤10 ppm	–
Water	≤1.0%	≤0.5%
Residue on ignition	–	≤0.05%
Sulfated ash	≤0.1%	–
N-Nitrosodiethanolamine	+	–
Specific gravity	–	1.120–1.128
Refractive index	–	1.481–1.486
Assay	99.0–103.0%	99.0–107.4%

10 Typical Properties

Acidity/alkalinity pH = 10.5 (0.1 N solution)
Boiling point 335°C
Flash point 208°C
Freezing point 21.6°C
Hygroscopicity Very hygroscopic.
Melting point 20–21°C
Moisture content 0.09%
Solubility *see* Table II.

Table II: Solubility of triethanolamine.

Solvent	Solubility at 20°C
Acetone	Miscible
Benzene	1 in 24
Carbon tetrachloride	Miscible
Ethyl ether	1 in 63
Methanol	Miscible
Water	Miscible

Spectroscopy

IR spectra *see* Figure 1.

Surface tension 48.9 mN/m (48.9 dynes/cm) at 25°C
Viscosity (dynamic) 590 mPa s (590 cP) at 30°C

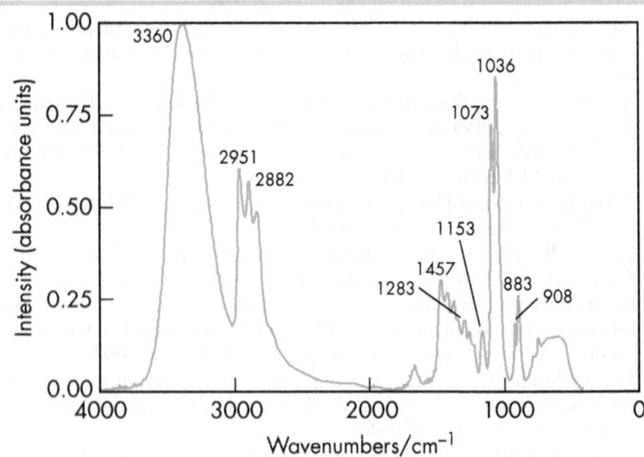

Figure 1: Infrared spectrum of triethanolamine measured by transmission. Adapted with permission of Informa Healthcare.

11 Stability and Storage Conditions

Triethanolamine may turn brown on exposure to air and light.

The 85% grade of triethanolamine tends to stratify below 15°C; homogeneity can be restored by warming and mixing before use.

Triethanolamine should be stored in an airtight container protected from light, in a cool, dry place.

See Monoethanolamine for further information.

12 Incompatibilities

Triethanolamine is a tertiary amine that contains hydroxy groups; it is capable of undergoing reactions typical of tertiary amines and alcohols. Triethanolamine will react with mineral acids to form crystalline salts and esters. With the higher fatty acids, triethanolamine forms salts that are soluble in water and have characteristics of soaps. Triethanolamine will also react with copper to form complex salts. Discoloration and precipitation can take place in the presence of heavy metal salts.

Triethanolamine can react with reagents such as thionyl chloride to replace the hydroxy groups with halogens. The products of these reactions are very toxic, resembling other nitrogen mustards.

13 Method of Manufacture

Triethanolamine is prepared commercially by the ammonolysis of ethylene oxide. The reaction yields a mixture of monoethanolamine, diethanolamine, and triethanolamine, which are separated to obtain the pure products.

14 Safety

Triethanolamine is used primarily as an emulsifying agent in a variety of topical pharmaceutical preparations. Although generally regarded as a nontoxic material,[2] triethanolamine may cause hypersensitivity or be irritant to the skin when present in formulated products. The lethal human oral dose of triethanolamine is estimated to be 5–15 g/kg body-weight.

Following concern about the possible production of nitrosamines in the stomach, the Swiss authorities have restricted the use of triethanolamine to preparations intended for external use.[3]

LD_{50} (guinea pig, oral): 5.3 g/kg[4]
LD_{50} (mouse, IP): 1.45 g/kg
LD_{50} (mouse, oral): 7.4 g/kg
LD_{50} (rat, oral): 8 g/kg

15 Handling Precautions

Triethanolamine may be irritant to the skin, eyes, and mucous membranes. Inhalation of vapor may be harmful. Protective clothing, gloves, eye protection, and a respirator are recommended. Ideally, triethanolamine should be handled in a fume cupboard. On heating, triethanolamine forms highly toxic nitrous fumes. Triethanolamine is combustible.

16 Regulatory Status

Included in the FDA Inactive Ingredients Database (rectal, topical, and vaginal preparations). Included in nonparenteral medicines licensed in the UK. Included in the Canadian Natural Health Products Ingredients Database.

17 Related Substances

Diethanolamine; monoethanolamine.

18 Comments

Various grades of triethanolamine are available. The standard commercial grade contains 85% triethanolamine. The superior grade contains 98–99% triethanolamine.

One part by volume of triethanolamine with 5–7 parts of a mixture of CaO_2 and ZnO_2 is used as a filling material that enhances the restorative process in periodontal tissues.

Triethanolamine is used as an intermediate in the manufacture of surfactants, textile specialties, waxes, polishes, herbicides, petroleum demulsifiers, toilet goods, cement additives, and cutting oils. Triethanolamine is also claimed to be used for the production of lubricants for the rubber gloves and textile industries. Other general uses are as buffers, solvents, and polymer plasticizers, and as a humectant. Triethanolamine is recommended as the preferred stabilizer to be used in latex polymerization because of its weak mutagenic effect in the Ames tests.

The EINECS number for triethanolamine is 203-049-8.

19 Specific References

1 Turkoglu M, Yener S. Design and *in vivo* evaluation of ultrafine inorganic-oxide-containing-sunscreen formulations. *Int J Cosmet Sci* 1997; **19**(4): 193–201.

2 Maekawa A, *et al.* Lack of carcinogenicity of triethanolamine in F344 rats. *J Toxicol Environ Health* 1986; **19**(3): 345–357.

3 Anonymous. Trolamine: concerns regarding potential carcinogenicity. *WHO Drug Inf* 1991; **5**: 9.

4 Lewis RJ, ed. *Sax's Dangerous Properties of Industrial Materials*, 11th edn. New York: Wiley, 2004: 3568.

20 General References

Friberg SE, *et al.* The influence of solvent on nonaqueous lyotropic liquid crystalline phase formed by triethanolamine oleate. *J Pharm Sci* 1985; **74**(7): 771–773.

Ramsay B, *et al.* The effect of triethanolamine application on anthralin-induced inflammation and therapeutic effect in psoriasis. *J Am Acad Dermatol* 1990; **23**: 73–76.

Yano H, Noda A, *et al.* Generation of Maillard-type compounds from triethanolamine alone. *J Am Oil Chem Soc* 1997; **74**(7): 891–893.

21 Author

SR Goskonda.

22 Date of Revision

1 March 2012.

Triethyl Citrate

1 Nonproprietary Names

BP: Triethyl Citrate

PhEur: Triethyl Citrate

USP–NF: Triethyl Citrate

2 Synonyms

Citric acid ethyl ester; citric acid triethyl ester; *Citroflex 2*; *Citrofol AI*; *Crodamol TC*; E1505; ethyl citrate; *Hydagen CAT*; 1,2,3-propanetricarboxylic acid, 2-hydroxy-, triethyl ester (9CI); TEC; triethyl 2-hydroxypropane-1,2,3-tricarboxylate; triethylis citras; *Uniflex TEC*; *Uniplex 80*.

3 Chemical Name and CAS Registry Number

2-Hydroxy-1,2,3-propanetricarboxylic acid triethyl ester [77-93-0]

4 Empirical Formula and Molecular Weight

$C_{12}H_{20}O_7$ 276.29

5 Structural Formula

H_3C O O H_3C CH_3 O OH O O O

6 Functional Category

Plasticizing agent; solvent.

7 Applications in Pharmaceutical Formulation or Technology

Triethyl citrate is used to plasticize polymers in formulated pharmaceutical coatings, particularly aqueous polymeric dispersions used for extended-release and enteric formulations.[1–5]

For extended-release applications, drug release from polymeric films plasticized with triethyl citrate is generally faster compared to films plasticized with water-insoluble plasticizers due to the aqueous solubility of triethyl citrate.[1,4] Triethyl citrate is a suitable plasticizer for polymethacrylate polymeric film coatings where it is typically used at concentrations of 10-20% relative to the polymer dry weight, although higher concentrations may be needed depending on the physicochemical properties of the polymers and the properties of the drug and the drug-containing substrate.[4,6] Triethyl citrate is also reported to be a suitable plasticizer for ethylcellulose aqueous dispersions (10-35%),[1] polyvinyl acetate (5%),[7] hypromellose phthalate (20-40%),[8] and hypromellose acetate succinate (20-40%).[8]

In addition to its use in aqueous polymeric films, triethyl citrate is also used as a plasticizer for polymers used in hot melt extrusion.[9,10] However, care must be exercised in setting up the temperature ranges for these processes to avoid triethyl citrate loss due to evaporation. Triethyl citrate is also used as a plasticizer for dry powder coatings,[11,12] and as a plasticizer and pore-forming agent in microspheres.[13]

8 Description

Triethyl citrate occurs as a clear, viscous, odorless, bitter, and practically colorless, hygroscopic oily liquid.

9 Pharmacopeial Specifications

See Table I.

Table I: Pharmacopeial specifications for triethyl citrate.

Test	PhEur 7.4	USP35–NF30
Identification	+	+
Characters	+	–
Appearance	+	–
Specific gravity	–	1.135–1.139
Refractive index	1.440–1.446	1.439–1.441
Acidity	+	+
Related substances	+	–
Sulfated ash	≤0.1%	–
Heavy metals	≤5 ppm	≤10 ppm
Water	≤0.25%	≤0.25%
Assay (anhydrous basis)	98.5–101.0%	99.0–100.5%

10 Typical Properties

Acid value 0.02
Boiling point 294°C
Density 1.135–1.139 g/cm^3 at 25°C
Flash point 150–155°C
Pour point −45°C
Saponification value 610
Solubility Soluble 1 in 125 of peanut oil, 1 in 15 of water. Miscible with ethanol (95%), ether, acetone, and propan-2-ol.
Spectroscopy
IR spectra *see* Figure 1.
Viscosity (dynamic) 35.2 mPa s (35.2 cP) at 25°C

11 Stability and Storage Conditions

Triethyl citrate should be stored in a closed container in a cool, dry location. When stored in accordance with these conditions, triethyl citrate is a stable product.

12 Incompatibilities

Triethyl citrate is incompatible with strong alkalis and oxidizing materials.

13 Method of Manufacture

Triethyl citrate is prepared by the esterification of citric acid and ethanol in the presence of a catalyst.

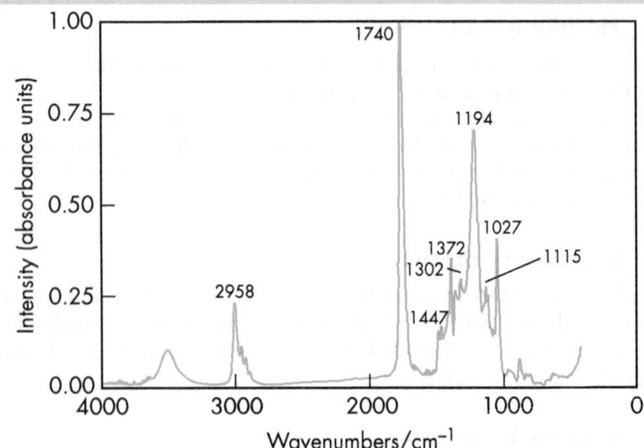

Figure 1: Infrared spectrum of triethyl citrate measured by transmission. Adapted with permission of Informa Healthcare.

14 Safety

Triethyl citrate is used in oral pharmaceutical formulations and as a direct food additive. It is generally regarded as a nontoxic and nonirritant material. However, ingestion of large quantities may be harmful.

LD_{50} (mouse, IP): 1.75 g/kg[14]
LD_{50} (rat, IP): 4 g/kg
LD_{50} (rat, oral): 5.9 g/kg
LD_{50} (rat, SC): 6.6 g/kg

15 Handling Precautions

Observe normal precautions appropriate to the circumstances and quantity of material handled. Triethyl citrate is irritating to the eyes and may irritate the skin. It is irritating to the respiratory system as a mist or at elevated temperatures. Gloves, eye protection, and a respirator are recommended.

16 Regulatory Status

GRAS listed. Accepted for use as a food additive in Europe. Included in the FDA Inactive Ingredients Database (oral capsules and tablets). Included in the Canadian Natural Health Ingredients Database.

17 Related Substances

Acetyltributyl citrate; acetyltriethyl citrate; tributyl citrate.

18 Comments

Triethyl citrate is used as an active agent in deodorants, where it inhibits enzymatic decomposition of sweat components.

Triethyl citrate is also used as a carrier for flavors and processing aids in foods. As a food additive (E1505), triethyl citrate is used to stabilize foams, especially as a whipping aid for egg white. It is also used as a solvent and fixative in perfumes, and as a tackiness reducer for low volatile hair sprays.[15]

A specification for triethyl citrate is contained in the *Food Chemicals Codex* (FCC).[16]

The EINECS number for triethyl citrate is 201-070-7. The PubChem Compound ID (CID) for triethyl citrate is 6506.

19 Specific References

1 Siepmann J *et al.* Process and formulation factors affecting drug release from pellets coated with ethylcellulose pseudolatex *Aquacoat*. In: McGinity JW, Felton LA, eds. *Aqueous Polymeric Coatings for*

Pharmaceutical Dosage Forms*, 3rd edn. New York, London: Informa Healthcare, 2008: 203-236.
2 Hamed E, Sakr A. Application of multiple response optimization technique to extended release formulations design. *J Control Release* 2001; **73**: 329–338.
3 Gutierrez-Rocca JC, McGinity JW. Influence of water soluble and insoluble plasticizers on the physical and mechanical properties of acrylic resin copolymers. *Int J Pharm* 1994; **103**: 293–301.
4 Skalsky B, Petereit H-U. Chemistry and application properties of polymethacrylate systems. In: McGinity JW, Felton LA, eds. *Aqueous Polymeric Coatings for Pharmaceutical Dosage Forms*, 3rd edn. New York, London: Informa Healthcare, 2008: 237-278.
5 Carlin B, Li JX, Felton LA. Pseudolatex dispersions for controlled drug delivery. In: McGinity JW, Felton LA, eds. *Aqueous Polymeric Coatings for Pharmaceutical Dosage Forms*, 3rd edn. New York, London: Informa Healthcare, 2008: 1–46.
6 Hamed E, Sakr A. Effect of curing conditions and plasticizer level on the release of highly lipophilic drug from coated multiparticulate drug delivery system. *Pharm Dev Technol* 2003; **8**(4): 397–407.
7 Kolter K, Ruchatz F. *Kollicoat* SR 30 D, a new sustained-release excipient. http://worldaccount.basf.com/wa/NAFTA/Catalog/Pharma/info/BASF/exact/kollicoat_sr_30_d (accessed 14 December 2011).
8 Obara S, Kokubo H. Application of HPMC and HPMCAS to aqueous film coating of pharmaceutical dosage forms. In: McGinity JW, Felton LA, eds. *Aqueous Polymeric Coatings for Pharmaceutical Dosage Forms*, 3rd edn. New York, London: Informa Healthcare, 2008: 279-322.
9 Schilling SU, *et al.* Influence of plasticizer type and level on the properties of *Eudragit S100* matrix pellets prepared by hot-melt extrusion. *J Microencapsul* 2010; **27**(6): 521–523.
10 Zhu Y, *et al.* Influence of plasticizer level on the drug release from sustained release film coated and hot-melt extruded dosage forms. *Pharm Dev Technol* 2006; **11**(3): 285–294.
11 Sauer D, *et al.* Influence of processing parameters and formulation factors on the drug release from tablets powder-coated with *Eudragit* L 100-55. *Eur J Pharm Biopharm* 2007; **67**(2): 464–475.
12 Kablitz CD, Urbanetz NA. Evaluating the process parameters of the dry coating process using a 2(5–1) factorial design. *Pharm Dev Technol* 2011; Epub 19 August.
13 Sengel-Turk CT, *et al.* Ethyl cellulose-based matrix type microspheres: influence of plasticizer ratio as pore forming agent. *AAPS PharmSciTech* 2011; **12**(4): 1127–1135.
14 Lewis RJ, ed. *Sax's Dangerous Properties of Industrial Materials*, 11th edn. New York: Wiley, 2004: 3546.
15 Jungbunzlauer Inc. Product information: *Citrofol AI* - triethyl citrate. http://www.jungbunzlauer.com/products-applications/products/specialties/citrofolR-a-i/general-information.html (accessed 23 February 2012).
16 *Food Chemicals Codex*, 7th edn. Bethesda, MD: United States Pharmacopeia, 2010.

20 General References

Chemicalbook.com. Product information: Triethyl citrate. http://www.chemicalbook.com (accessed 14 December 2011).
Fadda HM, *et al.* The use of dynamic mechanical analysis (DMA) to evaluate plasticization of acrylic polymer films under simulated gastrointestinal conditions. *Eur J Pharm Biopharm* 2010; **76**(3): 493–497.
FAO/WHO. Evaluation of certain food additives. Seventy-first report of the Joint FAO/WHO Expert Committee on Food Additives. World Health Organ Tech Rep Ser. 2010; 956: 1-80,
Jungbunzlauer Inc. Material safety data sheet: *Citrofol AI*, 2010.
Qiao M, *et al.* Sustained release coating of tablets with *Eudragit* RS/RL using a novel electrostatic dry powder coating process. *Int J Pharm* 2010; **399**(1–2): 37–43.
Sigma-Aldrich. Material safety data sheet: Triethyl citrate, 2011.
Vertellus Specialties Inc. Technical data sheet: *Citroflex* 2, 2010.

21 Authors

E Hamed, S Fulzele.

22 Date of Revision

23 February 2012.

Triolein

1 Nonproprietary Names

None adopted.

2 Synonyms

Captex GTO; glycerol trielaidate; glyceryl trioleate; 9-octadecenoic acid-1,2,3-propanetriyl ester; olein; 1,2,3-propanetriyl tris((*E*)-9-octadecenoate); trielaidin; trielaidoylglycerol; 1,2,3-tri(*cis*-9-octadecenoyl)glycerol.

3 Chemical Name and CAS Registry Number

2,3-bis[[(*Z*)-octadec-9-enoyl]oxy]propyl (*Z*)-octadec-9-enoate [122-32-7]

4 Empirical Formula and Molecular Weight

$C_{57}H_{104}O_6$ 885.43

5 Structural Formula

$CH_2OC(=O)(CH_2)_7CH{=}CH(CH_2)_7CH_3$
$|$
$CHOC(=O)(CH_2)_7CH{=}CH(CH_2)_7CH_3$
$|$
$CH_2OC(=O)(CH_2)_7CH{=}CH(CH_2)_7CH_3$

6 Functional Category

Emollient; penetration enhancer; solubilizing agent; solvent.

7 Applications in Pharmaceutical Formulation or Technology

Triolein is used as a solubilizer and solvent in injectable preparations. It has been used in marketed preparations of sustained-release injections of cytarabine and multivesicular liposomal injections of morphine sulfate. It has also been used in enteric coatings for oral preparations in combination with other enteric coating excipients to protect against degradation by pancreatic lipase.[1]

Triolen is used in personal care products as a skin-conditioning and viscosity-controlling agent.

8 Description

Triolein occurs as a clear, colorless to yellowish oily liquid, and is tasteless and odorless.

9 Pharmacopeial Specifications

—

10 Typical Properties

Boiling point 235–240°C
Density 0.915 g/cm^3
Flash point 330°C
Free fatty acids ≤0.1%
Iodine value 80–100
Melting point −5 to −4°C
Peroxide value ≤2
Refractive index
n_D^{20} = 1.4676;
n_D^{60} = 1.4561.
Solubility Soluble in chloroform, ether, carbon tetrachloride; slightly soluble in ethanol (95%); practically insoluble in water.
Specific gravity 0.9 at 25°C (water = 1)
Vapor density >1 (air = 1)
Vapor pressure <133.3 Pa (<1 mmHg) at 25°C
Viscosity (kinematic) 74 mm^2/s (74 cSt) at 25°C

11 Stability and Storage Conditions

Triolein is classified as a stable compound but is sensitive to air and light. It should be stored in airtight containers in a dry area at 2–8°C. Thermal decomposition of triolein may lead to release of irritating gases and vapors such as carbon oxides. Exposure to air and moisture over prolonged periods should be avoided.

12 Incompatibilities

Triolein is incompatible with strong oxidizing agents and spontaneously flammable products. Being a triglyceride ester, triolein can be hydrolyzed by strong acids, and particularly by strong bases. It is possible for primary amines to form an adduct across the olefinic double bonds (analogous to a Michael addition).

13 Method of Manufacture

Triolein is manufactured by the esterification of fractionated fatty acids, mainly oleic acid and glycerin.

14 Safety

Triolein is used in injectable preparations, in enteric coatings for oral preparations, and in personal care products. Chronic exposure may cause nausea and vomiting, and higher exposures may cause unconsciousness.

The Cosmetic Ingredient Review (CIR) Expert Panel found that dermal application of triolein was not associated with significant irritation, and no evidence of sensitization or photosensitization was observed.[2] Ocular exposures were found to be only mildly irritating to eyes. Triolein has not been found to be genotoxic in a number of *in vitro* and *in vitro* assay systems. Subcutaneous injections of triolein in rats showed no tumors at the injection site. The CIR Expert Panel also noted that metabolism data indicated that glyceryl triesters (including triolein) followed the same metabolic pathways as fats in food. They were split into monoglycerides, free fatty acids, and glycerol, all of which were absorbed into the intestinal mucosa and metabolized further. Therefore, oral exposure to these compounds was not found to be a concern.[2]

A triolein-based amphotericin emulsion showed better safety with a higher LD_{50} in rats as compared with the conventional amphotericin deoxycholate.[3]

15 Handling Precautions

Observe normal precautions appropriate to the circumstances and quantity of the material handled. Use of a mask and/or respirator is recommended in case aerosol/dust is formed. Ventilation is recommended to control dust or fumes from the material. For eye protection, safety glasses with side shields are recommended. For hand protection, PVC or other plastic material gloves are recommended.

16 Regulatory Status

Included in the FDA Inactive Ingredients Database (liposomal suspension for epidural injections). Included in parenteral medicines (suspension for intrathecal injection) licensed in the UK.

Triolein is included in the CIR category as safe for use in cosmetics and personal care products. Its use as an indirect food additive has been approved by the FDA.

17 Related Substances

—

18 Comments

Studies have shown that triolein enhances the transfection efficiency of polycation nanostructured lipid carriers.[4] Transferrin-conjugated solid lipid nanoparticles containing triolein have been found to enhance the delivery of quinine dihydrochloride to the brain for the treatment of cerebral malaria.[5] A polymer-lipid hybrid nanoparticle containing triolein has been reported to have higher transfection efficiency with acceptable cytotoxicity.[6]

A nanoemulsion lipoprotein delivery system, comprising triolein in its oily phase, has been found to show lower cytotoxicity than conventional systems in *in vitro* gene transfection in human glioma cells.[7] Gadolinium-containing lipid nanoemulsions have also been prepared using triolein.[8] A w/o/w insulin emulsion system containing triolein in its oily phase has demonstrated strong hypoglycemic effects.[9] An intra-arterial delivery of triolein emulsion has been reported to increase the vascular permeability in skeletal muscles of rabbits, suggesting its application in drug delivery.[10] Triolein has also been studied in multivesicular liposome formulations for sustained delivery, and binary lipid nanoparticles for controlled release.[11–13]

A paclitaxel prodrug has been incorporated into a lipid nanoparticle formulation comprising triolein in a mixture of lipids and has shown promising results in the treatment of folate receptor tumors.[14] Improved drug distribution to the tumor has been reported with parenteral administration of a submicrometer lipid emulsion of paclitaxel with triolein as the oily core.[15] A synthetic nano-low-density lipoprotein vehicle in lipid emulsion containing triolein has been found to have selective targeting for glioblastoma multiforme cells.[16,17] Folate receptor-targeted solid lipid nanoparticles of hematoporphyrin containing triolein have also shown specific receptor binding and potential as a targeted drug delivery system.[18]

The EINECS number for triolein is 204-534-7. The PubChem Compound ID (CID) for triolein is 5497163.

19 Specific References

1 Yoshitomi H, *et al.* Evaluation of enteric coated tablet sensitive to pancreatic lipase. II. In vivo evaluation. *Biol Pharm Bull* 1993; **16**: 1260–1263.
2 Johnson W Jr. Cosmetic Ingredient Review Expert Panel. Final report on the safety assessment of trilaurin, triarachidin, tribehenin, tricaprin, tricaprylin, trierucin, triheptanoin, triheptylundecanoin, triisonanoin, triisopalmitin, triisostearin, trilinolein, trimyristin, trioctanoin, triolein, tripalmitin, tripalmitolein, triricinolein, tristearin, triundecanoin, glyceryl triacetyl hydroxystearate, glyceryl triacetyl ricinoleate, and glyceryl state diacetate. *Int J Toxicol* 2001; **20**(Suppl. 4): 61–94.
3 Souza LC, Campa A. Pharmacological parameters of intravenously administered amphotericin B in rats: comparison of the conventional formulation with amphotericin B associated with a triglyceride-rich emulsion. *J Antimicrob Chemother* 1999; **44**: 77–84.
4 Zhang Z, *et al.* Polycation nanostructured lipid carrier, a novel nonviral vector constructed with triolein for efficient gene delivery. *Biochem Biophys Res Commun* 2008; **370**: 478–482.
5 Gupta Y, *et al.* Transferrin-conjugated solid lipid nanoparticles for enhanced delivery of quinine dihydrochloride to the brain. *J Pharm Pharmacol* 2007; **59**: 935–940.
6 Li J, *et al.* A novel polymer-lipid hybrid nanoparticle for efficient nonviral gene delivery. *Acta Pharmacologica Sinica* 2010; **31**(4): 509–514.
7 Pan G, *et al. In vitro* gene transfection in human glioma cells using a novel and less cytotoxic artificial lipoprotein delivery system. *Pharm Res* 2003; **20**: 738–744.
8 Ichikawa H, *et al.* Formulation considerations of gadolinium lipid nanoemulsion for intravenous delivery to tumors in neutron-capture therapy. *Curr Drug Deliv* 2007; **4**: 131–140.
9 Morishita M, *et al.* Improving insulin enteral absorption using water-in-oil-in-water emulsion. *Int J Pharm* 1998; **172**(1–2): 189–198.
10 Kim HJ, *et al.* Intra-arterial delivery of triolein emulsion increases vascular permeability in skeletal muscles of rabbits. *Acta Vet Scand* 2009; **51**: article 30.
11 Zhong H, *et al.* Multivesicular liposome formulation for the sustained delivery of breviscapine. *Int J Pharm* 2005; **301**: 15–24.
12 Jiao Y, *et al.* Preparation of sustained release multivesicular liposome for thymopentin and preliminary study on its pharmacokinetics in rats. *Yao Xue Xue Bao* 2008; **43**(7): 756–760.
13 Kim JK, *et al.* Development of a binary lipid nanoparticles formulation of itraconazole for parenteral administration and controlled release. *Int J Pharm* 2010; **383**(1–2): 209–215.
14 Stevens PJ, *et al.* A folate receptor-targeted lipid nanoparticle formulation for a lipophilic paclitaxel prodrug. *Pharm Res* 2004; **21**: 2153–2157.
15 Lundberg BB. A submicron lipid emulsion coated with amphipathic polyethylene glycol for parenteral administration of paclitaxel (*Taxol*). *J Pharm Pharmacol* 1997; **49**: 16–21.
16 Nikanjam M, *et al.* Synthetic nano-LDL with paclitaxel oleate as a targeted drug delivery vehicle for glioblastoma multiforme. *J Control Rel* 2007; **124**(3): 163–171.
17 Nikanjam M, *et al.* Synthetic nano-low density lipoprotein as targeted drug delivery vehicle for glioblastoma multiforme. *Int J Pharm* 2007; **328**(1): 86–94.
18 Stevens PJ, *et al.* Synthesis and evaluation of a hematoporphyrin derivative in a folate receptor-targeted solid-lipid nanoparticle formulation. *Anticancer Res* 2004; **24**: 161–165.

20 General References

Abitec Corporation. Material safety data sheet No. 05-8360-00: *Captex GTO*, 19 July 2005.
Cosmeticsinfo.org. www.cosmeticsinfo.org (accessed 21 December 2011).
MP Biomedicals, LLC, USA. Material safety data sheet: *Triolein*, April 2006.

21 Authors

RT Gupta, KK Singh.

22 Date of Revision

21 December 2011.

Tromethamine

1 Nonproprietary Names

BP: Trometamol
PhEur: Trometamol
USP–NF: Tromethamine

2 Synonyms

Aminotrimethylolmethane; *Pehanorm*; *Talatrol*; TRIS; trisamine; TRIS buffer; tris(hydroxymethyl)aminomethane; tris(hydroxymethyl)methanamine; tris(hydroxymethyl)methylamine; 1,1,1-tris(hydroxymethyl)methylamine; *Trizma*; trometamolum.

3 Chemical Name and CAS Registry Number

2-Amino-2-(hydroxymethyl)propane-1,3-diol [77-86-1]

4 Empirical Formula and Molecular Weight

$C_4H_{11}NO_3$ 121.14

5 Structural Formula

OH
HO OH
NH_2

6 Functional Category

Alkalizing agent; buffering agent; emulsifying agent.

7 Applications in Pharmaceutical Formulation or Technology

Tromethamine is widely used as a buffering agent in parenteral, oral, topical, and ophthalmic preparations. It is used to prepare water-soluble salts of NSAIDs (e.g., ketorolac or ibuprofen), which can then be entrapped in polymeric vehicles for continuous release.[1–6]

Tromethamine is also used as an emulsifying agent for cosmetic creams and lotions, and mineral oil and paraffin wax emulsions.[7]

8 Description

Tromethamine occurs as a white or almost white crystalline powder or colorless crystals.

9 Pharmacopeial Specifications

See Table I.

Table I: Pharmacopeial specifications for tromethamine.

Test	PhEur 7.4	USP35–NF30
Identification	+	+
Characters	+	−
Appearance of solution	+	−
Melting range	168–174°C	168–172°C
pH	10.0–11.5	10.0–11.5
Loss on drying	≤0.5%	≤1.0%
Residue on ignition	≤0.1%	≤0.1%
Heavy metals	≤10 ppm	≤0.001%
Related substances	+	−
Chlorides	≤100 ppm	−
Iron	≤10 ppm	−
Bacterial endotoxins	<0.03 IU/mg	−
Assay (dried basis)	99.0–100.5%	99.0–101.0%

10 Typical Properties

Boiling point 219–220°C (10 mmHg)[7]
Density 1.353 g/cm^3 [8]
Dissociation constants

pK_a = 8.3 at 20°C;
pK_a = 7.82 at 37°C.[7]

Solubility *See* Table II.

Table II: Solubility of tromethamine.

Solvent	Solubility of tromethamine at 20°C unless otherwise stated
Acetone	Soluble
Carbon tetrachloride	Practically insoluble
Chloroform	Practically insoluble
Cyclohexane	Very slightly soluble
Ethanol (96%)	Sparingly soluble
Ethyl acetate	Very slightly soluble
Ethylene glycol	Soluble
Methanol	Soluble
Olive oil	Very slightly soluble
Water	Freely soluble

Vapor pressure ≈2.2×10^{-5} mmHg at 25°C[9]

11 Stability and Storage Conditions

Store in airtight containers in a cool, dry place. The bulk material is hygroscopic.

12 Incompatibilities

Tromethamine is incompatible with strong oxidizing agents, bases, copper, brass, and aluminum.

13 Method of Manufacture

Tromethamine may be prepared by reduction or catalytic hydrogenation of the corresponding nitro compound.

14 Safety

Tromethamine as the pure material is moderately toxic by ingestion and IV routes. It is an irritant to eyes, skin, and the respiratory tract.

LD_{50} (mouse, IP): 3.3 g/kg[10]
LD_{50} (mouse, IV): 3.5 g/kg[9,11]
LD_{50} (mouse, oral): 5.5 g/kg[10]
LD_{50} (rat, IV): 1.8 g/kg[11]
LD_{50} (rat, oral): 5.9 g/kg[11]

15 Handling Precautions

Observe normal precautions appropriate to the circumstances and quantity of the material handled. Tromethamine is combustible. When heated to decomposition, it produces nitrogen oxides, carbon monoxide, and carbon dioxide.[11] Suitable eye, skin, and respiratory protection is recommended.

16 Regulatory Status

Included in the FDA Inactive Ingredients Database (intravenous, intramuscular, subcutaneous, intrathecal, intra-arterial injections and infusions; ophthalmics (solutions and suspensions); oral preparations (tablets and capsules); topical (gels and solutions); inhalation, rectal, and urethral solutions). Included in parenteral and nonparenteral medicines licensed in the UK. Included in the Canadian Natural Health Products Ingredients Database.

17 Related Substances

—

18 Comments

Therapeutically, tromethamine is an organic amine proton acceptor used as an alkalizing agent in the treatment of metabolic acidosis, though it is mainly used during cardiac bypass surgery and during cardiac arrest.[12] THAM (Fisher Scientific) is a sterile, nonpyrogenic 0.3 molar solution of tromethamine, adjusted to a pH of approximately 8.6 with glacial acetic acid, administered by IV injection.

Tromethamine may be restricted in certain sports as it is considered to be a member of a prohibited group (diuretics).[12]

The EINECS number for tromethamine is 201-064-4. The PubChem Compound ID (CID) number for tromethamine is 6503.

19 Specific References

1 Alsarra IA, *et al.* Clinical evaluation of novel buccoadhesive film containing ketorolac in dental and post-oral surgery pain management. *Pharmazie* 2007; **62**: 773–778.
2 Manjappa AS, *et al.* Sustained ophthalmic *in situ* gel of ketorolac tromethamine: rheology and *in vivo* studies. *Drug Dev Res* 2009; **70**: 417–424.
3 Chelladurai S, *et al.* Design and evaluation of bioadhesive in-situ nasal gel of ketorolac tromethamine. *Chem Pharm Bull (Tokyo)* 2008; **56**: 1596–1599.
4 Basu SK, *et al.* Evaluation of ketorolac tromethamine microspheres by chitosan/gelatin B complex coacervation. *Sci Pharm* 2010; **78**: 79–92.
5 Rao MR, *et al.* Effect of processing and sintering on controlled release wax matrix tablets of ketorolac tromethamine. *Indian J Pharm Sci* 2009; **71**: 538–544.
6 Al Omari MM, *et al.* Novel inclusion complex of ibuprofen tromethamine with cyclodextrins: physico-chemical characterization. *J Pharm Biomed Anal* 2009; **50**: 449–458.
7 O'Neil MJ, ed. *The Merck Index: An Encyclopedia of Chemicals, Drugs, and Biologicals.* 14th edn. Whitehouse Station, NJ: Merck, 2006; 1677-1678.
8 ChemicalBook.com. Product information: *Trometamol*, 2007.
9 National Library of Medicine. Toxnet: *Tromethamine*, 2008.
10 *Fiedler – Encyclopedia of Excipients.* Stuttgart: Wissenschaftliche Verlagsgesellschaft, 2011. http://www.justscience.de.(accessed 22 August 2011).

11 Lewis RJ, ed. *Sax's Dangerous Properties of Industrial Materials*. 11th edn. New York: Wiley, 2004; 3436.
12 Sweetman SC, ed. *Martindale: The Complete Drug Reference*. 37th edn. London: Pharmaceutical Press, 2011; 2643.

20 General References

ScienceLab.com. Material safety data sheet: *Tromethamine*, 2008.

21 Authors

ME Fenton, RC Rowe.

22 Date of Revision

1 March 2012.

T

Vanillin

1 Nonproprietary Names

BP: Vanillin
PhEur: Vanillin
USP–NF: Vanillin

2 Synonyms

4-Hydroxy-*m*-anisaldehyde; *p*-hydroxy-*m*-methoxybenzaldehyde; 3-methoxy-4-hydroxybenzaldehyde; methylprotocatechuic aldehyde; *Rhovanil*; vanillic aldehyde; vanillinum.

3 Chemical Name and CAS Registry Number

4-Hydroxy-3-methoxybenzaldehyde [121-33-5]

4 Empirical Formula and Molecular Weight

$C_8H_8O_3$ 152.15

5 Structural Formula

6 Functional Category

Flavoring agent; taste-masking agent.

7 Applications in Pharmaceutical Formulation or Technology

As a pharmaceutical excipient, vanillin is used in tablets, solutions (0.01–0.02% w/v), syrups, and powders to mask the unpleasant taste and odor characteristics of certain formulations, such as caffeine tablets and polythiazide tablets. It is similarly used in film coatings to mask the taste and odor of vitamin tablets.

Vanillin has also been investigated as a photostabilizer in furosemide 1% w/v injection, haloperidol 0.5% w/v injection, and thiothixene 0.2% w/v injection.(1)

8 Description

White or cream, crystalline needles or powder with characteristic vanilla odor and sweet taste.

9 Pharmacopeial Specifications

See Table I.

Table I: Pharmacopeial specifications for vanillin.

Test	PhEur 7.4	USP35–NF30
Identification	+	+
Characters	+	—
Appearance of solution	+	—
Melting range	81–84°C	81–83°C
Loss on drying	≤1.0%	≤1.0%
Sulfated ash	≤0.05%	—
Residue on ignition	—	≤0.05%
Related substances	+	—
Reaction with sulfuric acid	+	—
Assay (dried basis)	99.0–101.0%	97.0–103.0%

10 Typical Properties

Acidity/alkalinity Aqueous solutions are acid to litmus.
Boiling point 284–285°C (with decomposition)
Density (bulk) 0.6 g/cm^3
Flash point 153°C (closed cup)
Melting point 81–83°C
Solubility *see* Table II.

Table II: Solubility of vanillin.

Solvent	Solubility at 20°C unless otherwise stated
Acetone	Soluble
Alkali hydroxide solutions	Soluble
Chloroform	Soluble
Ethanol (95%)	1 in 2
Ethanol (70%)	1 in 3
Ether	Soluble
Glycerin	1 in 20
Methanol	Soluble
Oils	Soluble
Water	1 in 100
	1 in 16 at 80°C

Specific gravity 1.056 (liquid)
Spectroscopy
IR spectra *see* Figure 1.
NIR spectra *see* Figure 2.

11 Stability and Storage Conditions

Vanillin oxidizes slowly in moist air and is affected by light.

Solutions of vanillin in ethanol decompose rapidly in light to give a yellow-colored, slightly bitter tasting solution of 6,6′-dihydroxy-5,5′-dimethoxy-1,1′-biphenyl-3,3′-dicarbaldehyde. Alkaline solutions also decompose rapidly to give a brown-colored solution. However, solutions stable for several months may be produced by adding sodium metabisulfite 0.2% w/v as an antioxidant.(2)

The bulk material should be stored in a well-closed container, protected from light, in a cool, dry place.

12 Incompatibilities

Incompatible with acetone, forming a brightly colored compound.(3) A compound practically insoluble in ethanol is formed with glycerin.

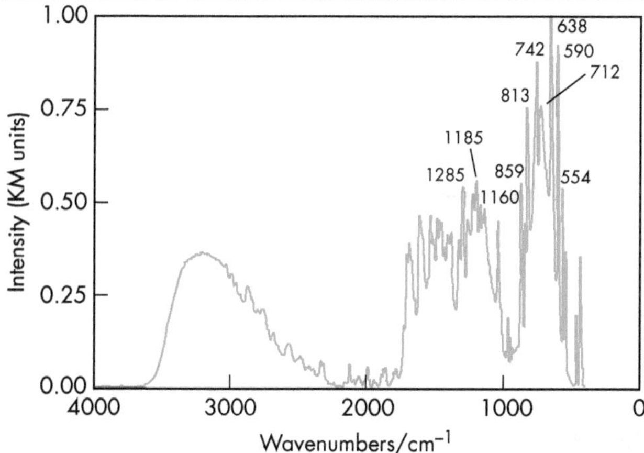

Figure 1: Infrared spectrum of vanillin measured by diffuse reflectance. Adapted with permission of Informa Healthcare.

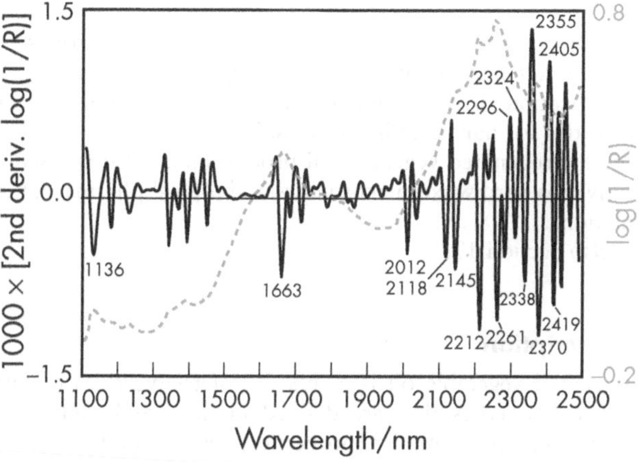

Figure 2: Near-infrared spectrum of vanillin measured by reflectance.

13 Method of Manufacture

Vanillin occurs naturally in many essential oils and particularly in the pods of *Vanilla planifolia* and *Vanilla tahitensis*. Industrially, vanillin is prepared from lignin, which is obtained from the sulfite wastes produced during paper manufacture. Lignin is treated with alkali at elevated temperature and pressure, in the presence of a catalyst, to form a complex mixture of products from which vanillin is isolated. Vanillin is then purified by successive recrystallizations.

Vanillin may also be prepared synthetically by condensation, in weak alkali, of a slight excess of guaiacol with glyoxylic acid at room temperature. The resultant alkaline solution, containing 4-hydroxy-3-methoxymandelic acid is oxidized in air, in the presence of a catalyst, and vanillin is obtained by acidification and simultaneous decarboxylation. Vanillin is then purified by successive recrystallizations.

14 Safety

There have been few reports of adverse reactions to vanillin, although it has been speculated that cross-sensitization with other structurally similar molecules, such as benzoic acid, may occur.[4] Adverse reactions that have been reported include contact dermatitis[5] and bronchospasm caused by hypersensitivity.[6]

The WHO has allocated an estimated acceptable daily intake for vanillin of up to 10 mg/kg body-weight.[7]

LD_{50} (guinea pig, IP): 1.19 g/kg[8]
LD_{50} (guinea pig, oral): 1.4 g/kg
LD_{50} (mouse, IP): 0.48 g/kg
LD_{50} (rat, IP): 1.16 g/kg
LD_{50} (rat, oral): 1.58 g/kg
LD_{50} (rat, SC): 1.5 g/kg

15 Handling Precautions

Observe normal precautions appropriate to the quantity of material handled. Eye protection is recommended. Heavy airborne concentrations of dust may present an explosion hazard.

16 Regulatory Status

GRAS listed. Included in the FDA Inactive Ingredients Database (oral solutions, suspensions, syrups, and tablets). Included in nonparenteral medicines licensed in the UK. Included in the Canadian Natural Health Products Ingredients Database.

17 Related Substances

Ethyl vanillin.

18 Comments

Vanillin is widely used as a flavor in pharmaceuticals, foods, beverages, and confectionery products, to which it imparts a characteristic taste and odor of natural vanilla. It is also used in perfumes, as an analytical reagent and as an intermediate in the synthesis of a number of pharmaceuticals, particularly methyldopa. Additionally, it has been investigated as a potential therapeutic agent in sickle cell anemia[9] and is claimed to have some antifungal properties.[10]

In food applications, vanillin has been investigated as a preservative.[11,12]

One part of synthetic vanillin is equivalent to 400 parts of vanilla pods.

A specification for vanillin is included in the *Japanese Pharmaceutical Excipients* (JPE).[13]

The EINECS number for vanillin is 204-465-2. The PubChem Compound ID (CID) for vanillin is 1183.

19 Specific References

1 Thoma K, Klimek R. Photostabilization of drugs in dosage forms without protection from packaging materials. *Int J Pharm* 1991; **67**: 169–175.
2 Jethwa SA, *et al.* Light stability of vanillin solutions in ethanol. *Drug Dev Ind Pharm* 1979; **5**: 79–85.
3 Thakur AB, Dayal S. Schiff base formation with nitrogen of a sulfonamido group. *J Pharm Sci* 1982; **71**: 1422.
4 Weiner M, Bernstein IL. *Adverse Reactions to Drug Formulation Agents: A Handbook of Excipients*. New York: Marcel Dekker, 1989: 238–239.
5 Wang X-S, *et al.* Occupational contact dermatitis in manufacture of vanillin. *Chin Med J* 1987; **100**: 250–254.
6 Van Assendelft AHW. Bronchospasm induced by vanillin and lactose. *Eur J Respir Dis* 1984; **65**: 468–472.
7 FAO/WHO. Specifications for the identity and purity of food additives and their toxicological evaluation: some flavouring substances and non-nutritive sweetening agents. Eleventh report of the joint FAO/WHO expert committee on food additives. *World Health Organ Tech Rep Ser* 1968; No. 383.
8 Lewis RJ, ed. *Sax's Dangerous Properties of Industrial Materials*, 11th edn. New York: Wiley, 2004: 3661–3662.
9 Abraham DJ, *et al.* Vanillin, a potential agent for the treatment of sickle cell anemia. *Blood* 1991; **77**: 1334–1341.
10 Lisá M, *et al.* [A contribution to the antifungal effect of propolis.] *Folia Pharm* 1989; **13**(1): 29–44[in German].
11 Fitzgerald DJ, *et al.* Analysis of the inhibition of food spoilage yeasts by vanillin. *Int J Food Microbiol* 2003; **86**(1–2): 113–122.
12 Fitzgerald DJ, *et al.* The potential application of vanillin in preventing yeast spoilage of soft drinks and fruit juices. *J Food Prot* 2004; **67**(2): 391–395.
13 Japan Pharmaceutical Excipients Council. *Japanese Pharmaceutical Excipients 2004*. Tokyo: Yakuji Nippo, 2004: 927-928.

20 General References

Ali L, *et al*. Rapid method for the determination of coumarin, vanillin, and ethyl vanillin in vanilla extract by reversed-phase liquid chromatography with ultraviolet detection. *J AOAC Int* 2008; **91**(2): 383–386.

Allen LV. Featured excipient: flavor-enhancing agents. *Int J Pharm Compound* 2003; 7(1): 48–50.

Clark GS. Vanillin. *Perfum Flavor* 1990; **15**(Mar/Apr): 45–54.

Rees DI. Determination of vanillin and ethyl vanillin in food products. *Chem Ind* 1965; **1**: 16–17.

21 Author

PJ Weller.

22 Date of Revision

1 March 2012.

Vegetable Oil, Hydrogenated

1 Nonproprietary Names

BP: Hydrogenated Vegetable Oil
JP: Hydrogenated Oil
USP–NF: Hydrogenated Vegetable Oil
See also Sections 8, 9, and 17.

2 Synonyms

Some trade names for materials derived from stated vegetable oils are shown below:

Hydrogenated cottonseed oil: *Akofine*; *Lubritab*; *Sterotex*.
Hydrogenated palm oil: *Softisan 154*.
Hydrogenated soybean oil: *Lipovol HS-K*; *Sterotex HM*.

3 Chemical Name and CAS Registry Number

Hydrogenated vegetable oil [68334-00-9]
Hydrogenated soybean oil [8016-70-4]

4 Empirical Formula and Molecular Weight

The USP35–NF30 defines two types of hydrogenated vegetable oil, type I and type II, which differ in their physical properties and applications; *see* Sections 9 and 17.

5 Structural Formula

$R^1COOCH_2—CH(OOCR^2)—CH_2OOCR^3$
where R^1, R^2, and R^3 are mainly C_{15} and C_{17}.

6 Functional Category

Modified-release agent; tablet and capsule binder; tablet and capsule lubricant.

7 Applications in Pharmaceutical Formulation or Technology

Hydrogenated vegetable oil type I may be used as a lubricant in tablet and capsule formulations.[1,2] In this application it is used at concentrations of 1–6% w/w, usually in combination with talc, silica, or a silicate to prevent sticking to tablet punch faces. It may also be used as an auxiliary binder in tablet formulations.

Hydrogenated vegetable oil type I is additionally used as the matrix-forming material in lipophilic-based controlled-release formulations;[3–6] it may also be used as a coating aid in controlled-release formulations. It has also been investigated in hydrophobic melt agglomeration.[7]

Other uses of hydrogenated vegetable oil type I include use as a viscosity modifier in the preparation of oil-based liquid and semisolid formulations; in the preparation of suppositories, to reduce the sedimentation of suspended components and to improve the solidification process; and in the formulation of liquid and semisolid fills for hard gelatin capsules.[8]

Fully hydrogenated vegetable oil products may also be used as alternatives to hard waxes in cosmetics and topical pharmaceutical formulations.

See also Section 17.

8 Description

Hydrogenated vegetable oil is a mixture of triglycerides of fatty acids. The two types that are defined in the USP35–NF30 are characterized by their physical properties; *see* Section 9.

Hydrogenated vegetable oil type I occurs in various forms, e.g. fine powder, flakes, or pellets. The color of the material depends on the manufacturing process and the form. In general, the material is white to yellowish-white with the powder grades appearing more white-colored than the coarser grades.

9 Pharmacopeial Specifications

See Table I.

Table I: Pharmacopeial specifications for hydrogenated vegetable oil.

Test	BP 2012	JP XV[a]	USP35–NF30	
			Type I	Type II
Identification	+	–	–	–
Characters	+	+	–	–
Melting range	57–70°C	–	57–85°C	20–50°C
Heavy metals	≤10 ppm	+	≤0.001%	≤0.001%
Moisture and coloration	–	+	–	–
Alkali	–	+	–	–
Chloride	–	+	–	–
Nickel	–	+	–	–
Iodine value	≤5	–	0–5	55–80
Saponification value	175–205	–	175–200	175–200
Loss on drying	≤0.1%	–	≤0.1%	≤0.1%
Acid value	≤4.0	≤2.0	≤4.0	≤4.0
Unsaponifiable matter	≤0.8%	–	≤0.8%	≤0.8%
Residue on ignition	–	≤0.1%	–	–

(a) Note that Hydrogenated Oil JP may be of fish, animal or vegetable origin.

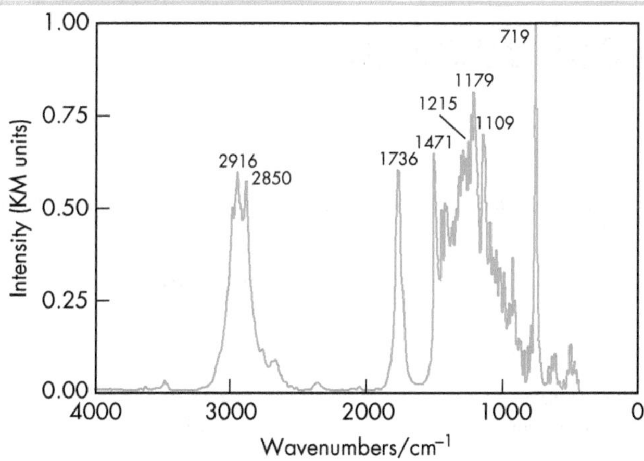

Figure 1: Infrared spectrum of hydrogenated vegetable oil measured by diffuse reflectance. Adapted with permission of Informa Healthcare.

10 Typical Properties

Density (tapped) 0.57 g/cm^3 for *Lubritab*
Melting point 61–66°C for *Lubritab*
Particle size distribution 85% < 177 μm, 25% < 74 μm in size for *Lubritab*. Average particle size is 104 μm.
Solubility Soluble in chloroform, petroleum spirit, and hot propan-2-ol; practically insoluble in water.
Spectroscopy
IR spectra *see* Figure 1.

11 Stability and Storage Conditions

Hydrogenated vegetable oil type I is a stable material; typically it is assigned a 2-year shelf-life.

The bulk material should be stored in a well-closed container in a cool, dry place.

12 Incompatibilities

Incompatible with strong oxidizing agents.

13 Method of Manufacture

Hydrogenated vegetable oil type I is prepared from refined vegetable oils, which are hydrogenated using a catalyst.

14 Safety

Hydrogenated vegetable oil type I is used in food products and oral pharmaceutical formulations, and is generally regarded as a nontoxic and nonirritant excipient.

15 Handling Precautions

Observe normal precautions appropriate to the circumstances and quantity of material handled. Gloves, eye protection, and a dust mask are recommended when handling fine powder grades.

16 Regulatory Status

GRAS listed. Included in the FDA Inactive Ingredients Database (oral capsules and tablets; rectal and vaginal suppositories and topical preparations). Included in nonparenteral medicines licensed in the UK. Included in the Canadian Natural Health Products Ingredients Database.

17 Related Substances

Castor oil, hydrogenated; hydrogenated vegetable oil, type II; medium-chain triglycerides; suppository bases.

Hydrogenated vegetable oil, type II USP–NF

Comments Hydrogenated vegetable oil type II includes partially hydrogenated vegetable oils from different sources that have a wide range of applications. In general, type II materials have lower melting ranges and higher iodine values than type I materials. Many type II materials are prepared to meet specific customer requirements for use in cosmetics. Type II materials may also be used in the manufacture of suppositories. *See also* Section 9.

18 Comments

Products from different manufacturers may vary owing to differences in the source of the vegetable oil used for hydrogenation. Certain materials are made from mixed hydrogenated oils, e.g. hydrogenated soybean oil and hydrogenated castor oil (*Sterotex K*).

19 Specific References

1 Hölzer AW, Sjögren J. Evaluation of some lubricants by the comparison of friction coefficients and tablet properties. *Acta Pharm Suec* 1981; **18**: 139–148.
2 Staniforth JN. Use of hydrogenated vegetable oil as a tablet lubricant. *Drug Dev Ind Pharm* 1987; **13**(7): 1141–1158.
3 Lockwood PJ, *et al.* Influence of drug type and formulation variables on mechanisms of release from wax matrices. *Proc Int Symp Control Release Bioact Mater* 1987; **14**: 198–199.
4 Wang PY. Lipids as excipients in sustained release insulin implants. *Int J Pharm* 1989; **54**: 223–230.
5 Çiftçi K, *et al.* Formulation and *in vitro–in vivo* evaluation of sustained release lithium carbonate tablets. *Pharm Res* 1990; **7**: 359–363.
6 Watanbe Y, *et al.* Preparation and evaluation of enteric granules of aspirin prepared by acylglycerols. *Int J Pharm* 1990; **64**: 147–154.
7 Heng PW, *et al.* Investigation of agglomeration process with a hydrophobic binder in combination with sucrose stearate. *Eur J Pharma Sci* 2003; **19**: 381–393.
8 Dürr M, *et al.* [Dosing of liquids into liquid gelatin capsules at the production scale: development of compositions and procedures.] *Acta Pharm Technol* 1983; **29**(3): 245–251[in German].

20 General References

Banker GS *et al.*Tablet formulation and design. In: Lieberman HA, Lachman L, eds. *Pharmaceutical Dosage Forms: Tablets I*. New York: Marcel Dekker, 1989.
Bardon J, *et al.* [Temperature elevation undergone by mixtures of powders or granules during their transformation into tablets II: influence of nature and rate of lubricant.] *STP Pharma Sciences* 1985; **1**(6): 948–955[in French].
Miller TA, York P. Pharmaceutical tablet lubrication. *Int J Pharm* 1988; **41**: 1–19.
Staniforth JN, *et al.* Aspects of pharmaceutical tribology. *Drug Dev Ind Pharm* 1989; **15**: 2265–2294.

21 Author

RC Moreton.

22 Date of Revision

1 March 2012.

Vitamin E Polyethylene Glycol Succinate

1 Nonproprietary Names

USP–NF: Vitamin E Polyethylene Glycol Succinate

2 Synonyms

Speziol TPGS Pharma; tocofersolan; tocophersolan; tocopherol polyethylene glycol succinate; D-α-tocopheryl polyethylene glycol 1000 succinate; TPGS; vitamin E polyethylene glycol 1000 succinate; vitamin E TPGS; *VEGS*.

3 Chemical Name and CAS Registry Number

4-*O*-(2-Hydroxyethyl)-1-*O*-[2,5,7,8-tetramethyl-2-(4,8,12-trimethyltridecyl)-3,4-dihydrochromen-6-yl]butanedioate [9002-96-4] and [30999-06-5]

4 Empirical Formula and Molecular Weight

$C_{33}O_5H_{54}(CH_2CH_2O)_{20-22}$ ≈1 513

5 Structural Formula

n = 20-22

6 Functional Category

Antioxidant; emulsifying agent; nonionic surfactant; ointment base; solubilizing agent; suspending agent; tablet and capsule binder.

7 Applications in Pharmaceutical Formulation or Technology

Vitamin E polyethylene glycol succinate is an esterified vitamin E (tocopherol) derivative primarily used as a solubilizer or emulsifying agent because of its surfactant properties.[1] Structurally, it is amphipathic and hydrophilic, unlike the tocopherols, and therefore it is a water-soluble derivative that can be used in pharmaceutical formulations such as capsules,[2] tablets,[3] hot-melt extrusion,[4] microemulsions,[5] topical products,[6] and parenterals.[7] One of the most important applications is its use as a vehicle for lipid-based drug delivery formulations.

8 Description

Vitamin E polyethylene glycol succinate is a synthetic product. It is available as a white to light-brown, waxy solid and is practically tasteless. Chemically, it is a mixture composed principally of monoesterified polyethylene glycol 1000, the diesterified polyethylene glycol 1000, free polyethylene glycol 1000, and free tocopherol.[8]

9 Pharmacopeial Specifications

See Table I.

Table I: Pharmacopeial specifications for vitamin E polyethylene glycol succinate.

Test	USP35–NF30
Identification	+
Solubility in water	+
Acid value	+
Specific rotation	≥+24.0°
Assay (α-tocopherol)	≥25.0%

10 Typical Properties

Acid value ≤1.5[1]

Critical micelle concentration 0.02% by weight (37°C)[1]

HLB value ≈13.2[1]

Melting point 37–41°C[1]

Solubility Miscible in water in all parts.[1]

Specific gravity 1.06 (at 45°C)[1]

11 Stability and Storage Conditions

Vitamin E polyethylene glycol succinate is stable at ambient room temperature for up to 4 years. It reacts with alkalis and acids. Aqueous solutions of vitamin E polyethylene glycol succinate are stable over a pH range of 4.5–7.5 and can be further stabilized with propylene glycol.[1]

12 Incompatibilities

Vitamin E polyethylene glycol succinate is incompatible with strong acids and strong alkalis.

13 Method of Manufacture

Vitamin E polyethylene glycol succinate is prepared by esterification of the acid group of crystalline D-α-tocopheryl acid succinate by polyethylene glycol 1000.

14 Safety

Vitamin E polyethylene glycol succinate has been used at levels of 280 mg/capsule in the product *Agenerase* (amprenavir), which was dosed at 8 capsules (2240 mg vitamin E TPGS) per day.[2] An additional assessment of the safety of vitamin E polyethylene glycol succinate has been published, which includes a report showing no-observed-adverse-effect-level (NOAEL) in rats of 1000 mg/kg/day.[8]

15 Handling Precautions

Observe normal precautions appropriate to the circumstances and quantity of the material handled. Gloves and eye protection are recommended.

16 Regulatory Status

GRAS listed. Included in the FDA Inactive Ingredients Database (ophthalmic solution or drops; oral capsules, solution, tablet; topical solution or drops). Included in the Canadian Natural Health Products Ingredients Database.

17 Related Substances

Alpha tocopherol.

18 Comments

Vitamin E polyethylene glycol succinate has been characterized with respect to its mechanism of action and studied as a P-glycoprotein inhibitor.[9–12] It can also be used as a source of vitamin E.[1]

The PubChem Compound ID (CID) for vitamin E polyethylene glycol succinate is 71406.

19 Specific References

1 Wu H-WW, Hopkins WK. Characteristics of D-α-tocopheryl PEG 1000 succinate for applications as an absorption enhancer in drug delivery systems. *Pharm Technol* 1999; **23**: 52–68.
2 *Physician's Desk Reference*, 59th edn. Montvale, NJ: Medical Economics Company, 2005: 1396–1401.
3 Crowley MM, *et al.* Stability of polyethylene oxide in matrix tablets prepared by hot-melt extrusion. *Biomaterials* 2002; **23**(21): 4241–4248.
4 Repka MA, McGinity JW. Influence of vitamin E TPGS on the properties of hydrophilic films produced by hot-melt extrusion. *Int J Pharm* 2000; **202**(1–2): 63–70.
5 Suppasantorn P, *et al.* Microemulsions as topical delivery vehicles for the anti-melanoma prodrug, temozolomide hexyl ester (TMZA-HE). *J Pharm Pharmacol* 2007; **59**(6): 787–794.
6 Sheu MT, *et al.* Influence of micelle solubilization by tocopheryl polyethylene glycol succinate (TPGS) on solubility enhancement and percutaneous penetration of estradiol. *J Control Release* 2003; **88**(3): 355–368.
7 Constantinides PP, *et al.* Tocol emulsions for drug solubilization and perenteral delivery. *Adv Drug Deliv Res Rev* 2004; **56**: 1243–1255.
8 Opinion of the Scientific Panel on Food Additives, Flavourings, Processing Aids and Materials in Contact with Food on a request from the Commission related to D-alpha-tocopheryl polyethylene glycol 1000 succinate (TPGS) in use for food for particular nutritional purposes. *European Food Safety Authority (EFSA) Journal* 2007; **490**: 1–20.
9 Schulze JDR, *et al.* Impact of formulation excipients on human intestinal transit. *J Pharm Pharmacol* 2006; **38**(6): 821–825.
10 Collnot EM, *et al.* Mechanism of inhibition of P-glycoprotein mediated efflux by vitamin E TPGS: influence on ATPase activity and membrane fluidity. *Mol Pharm* 2007; **4**(3): 465–474.
11 Collnot E-M, *et al.* Influence of vitamin E TPGS poly(ethylene glycol) chain length on apical efflux transporters in Caco-2 cell monolayers. *J Control Release* 2006; **111**(1–2): 35–40.
12 Rege BD, *et al.* Effects of nonionic surfactants on membrane transporters in Caco-2 monolayers. *Eur J Pharm Sci* 2002; **16**(4–5): 237–246.

20 General References

Cognis. Product literature: *Speziol TPGS Pharma*, 2008.
Constantinides PP, *et al.* Advances in the use of tocols as drug delivery vehicles. *Pharm Res* 2006; **23**(2): 243–255.
Eastman Chemical Company. *Pharmaceutical Ingredients: Eastman Vitamin E TPGS NF. Applications and Properties* (Publication PCI-102B). Kingsport, Tennessee, 2005.
Eastman Chemical Company. *Eastman Vitamin E TPGS (D-Alpha-Tocopheryl Polyethylene Glycol-1000 Succinate): for nutritional supplements, food and beverage, personal care, and pharmaceutical applications* (publication V-4D). Kingsport, Tennessee, 2002.
Isochem. Material safety data sheet: *VEGS*, 2000.
Isochem. Product data sheet: *VEGS*, 2008.

21 Author

TC Dahl.

22 Date of Revision

1 March 2012.

Water

1 Nonproprietary Names

BP: Purified Water
JP: Purified Water
PhEur: Water, Purified
USP–NF: Purified Water
See also Sections 8 and 17.

2 Synonyms

Aqua; aqua purificata; hydrogen oxide.

3 Chemical Name and CAS Registry Number

Water [7732-18-5]

4 Empirical Formula and Molecular Weight

H_2O 18.02

5 Structural Formula

See Section 4.

6 Functional Category

Solvent.

7 Applications in Pharmaceutical Formulation or Technology

Water is widely used as a raw material, ingredient and solvent in the processing, formulation and manufacture of pharmaceutical products, active pharmaceutical ingredients (API) and intermediates, and analytical reagents. Specific grades of water are used for particular applications in concentrations up to 100%; *see* Table I.

Table I: Typical applications of specific grades of water.

Type	Use
Bacteriostatic water for injection	Diluent for ophthalmic and multiple-dose injections.
Potable water	Public supply suitable for drinking, the purity of which is unlikely to be suitable for use in the manufacture of pharmaceuticals.
Purified water	Vehicle and solvent for the manufacture of drug products and pharmaceutical preparations; not suitable for use in the manufacture of parenteral products.
Sterile water for inhalation	Diluent for inhalation therapy products.
Sterile water for injection	Diluent for injections.
Sterile water for irrigation	Diluent for internal irrigation therapy products.
Water for injections in bulk	Water for the bulk preparation of medicines for parenteral administration.

8 Description

The term 'water' is used to describe potable water that is freshly drawn direct from the public supply and is suitable for drinking. Water used in the pharmaceutical industry and related disciplines is classified as either drinking (potable) water, purified water, sterile purified water, water for injection (WFI), sterile water for injection, bacteriostatic water for injection, sterile water for irrigation, or sterile water for inhalation. Validation is required for all systems producing the water indicated, with the exception of potable water.

The chemical composition of potable water is variable, and the nature and concentrations of the impurities in it depend upon the source from which it is drawn. Water classified as potable water for applications such as some initial rinsing and API manufacturing operations, must meet the US Environmental Protection Agency's National Primary Drinking Water Regulations, or comparable regulations of the EU or Japan. For most pharmaceutical applications, potable water is purified by distillation, ion exchange treatment, reverse osmosis (RO), or some other suitable process to produce 'purified water'. For certain applications, water with pharmacopeial specifications differing from those of purified water should be used, e.g. WFI; *see* Sections 9 and 18.

Water is a clear, colorless, odorless, and tasteless liquid.

9 Pharmacopeial Specifications

See Table II. *See also* Section 17.

10 Typical Properties

Boiling point 100°C
Critical pressure 22.1 MPa (218.3 atm)
Critical temperature 374.2°C
Dielectric constant $D^{25} = 78.54$
Dipole moment
  1.76 in benzene at 25°C;
  1.86 in dioxane at 25°C.
Ionization constant 1.008×10^{-14} at 25°C.
Latent heat of fusion 6 kJ/mol (1.436 kcal/mol)
Latent heat of vaporization 40.7 kJ/mol (9.717 kcal/mol)
Melting point 0°C
Refractive index $n_D^{20} = 1.3330$
Solubility Miscible with most polar solvents.
Specific gravity 0.9971 at 25°C.
Specific heat (liquid) 4.184 J/g/°C (1.00 cal/g/°C) at 14°C.
Surface tension 71.97 mN/m (71.97 dynes/cm) at 25°C.
Vapor pressure 3.17 kPa (23.76 mmHg) at 25°C.
Viscosity (dynamic) 0.89 mPa s (0.89 cP) at 25°C.

11 Stability and Storage Conditions

Water is chemically stable in all physical states (ice, liquid, and vapor). Water leaving the pharmaceutical purification system and entering the storage tank must meet specific requirements. The goal when designing and operating the storage and distribution system is to keep the water from exceeding allowable limits during storage. In particular, the storage and distribution system must ensure that water is protected against ionic and organic contamination, which would lead to an increase in conductivity and total organic carbon, respectively. The system must also be protected against physical entry of foreign particles and microorganisms so that microbial growth is prevented or minimized. Water for specific purposes should be stored in appropriate containers; *see* Table III.

Table II: Pharmacopeial specifications of water for different pharmaceutical applications.

Test	Water JP XV	Purified water JP XV	Purified water in bulk PhEur 7.4	Purified water in containers PhEur 7.4	Purified water USP35–NF30	Water, highly purified PhEur 7.4	Sterile water for injection USP35–NF30	Bacteriostatic water for injection USP 35–NF30	Sterile water for inhalation USP35–NF30	Sterile water for irrigation USP35–NF30	Sterile purified water USP35–NF30	Water for injection[a] JP XV	Water for injection USP35–NF30	Water for injection (in bulk) PhEur 7.4	Sterile water for injection PhEur 7.4	Sterile purified water JP XV
Identification	–	–	–	–	–	–	–	–	–	–	–	–	–	–	–	–
Production	–	–	+	–	–	+	–	–	–	–	–	–	–	+	–	–
Characters	–	–	+	+	–	+	–	–	–	–	–	–	–	+	–	–
Appearance of solution	–	+	–	–	–	–	–	–	–	–	–	–	–	–	–	+
Odor and taste	–	+	–	–	–	–	–	–	–	–	–	–	–	–	–	+
pH	–	–	–	–	–	–	–	4.5–7.0	–	–	–	–	–	–	–	–
Acid or alkali	–	+	–	+	–	–	–	–	–	–	–	+	–	–	+	+
Cadmium	–	–	–	–	–	–	–	–	–	–	–	–	–	–	–	–
Chloride	–	+	–	+	–	–	+	–	–	–	–	–	+	–	+	+
Cyanide	–	–	–	–	–	–	–	–	–	–	–	–	–	–	–	–
Copper	–	–	–	–	–	–	–	–	–	–	–	–	–	–	–	–
Sulfate	–	+	–	+	–	–	+	+	–	–	–	+	–	–	+	+
Ammonium	≤0.05 mg/L	≤0.05 mg/L	–	≤0.2 ppm	–	–	+	–	–	–	–	+	–	–	≤0.2 ppm	≤0.05 mg/L
Iron	–	–	–	–	–	–	–	–	–	–	–	–	–	–	–	–
Calcium	–	–	–	+	–	–	+	+	–	–	–	–	–	–	+	–
Lead	–	–	–	–	–	–	–	–	–	–	–	–	–	–	–	–
Magnesium	–	–	–	+	–	–	–	–	–	–	–	–	–	–	+	–
Aluminum	–	–	≤10 ppb	–	–	≤10 ppb	–	–	–	–	–	–	–	≤10 ppb	≤10 ppb	–
Nitrate	–	–	≤0.2 ppm	–	–	≤0.2 ppm	–	–	–	–	–	–	–	≤0.2 ppm	≤0.2 ppm	+
Nitrogen from nitrate	–	+	–	–	–	–	–	–	–	–	–	+	–	–	–	+
Nitrogen from nitrite	–	+	–	–	–	–	–	–	–	–	–	+	–	–	–	+
Carbon dioxide	–	–	–	–	–	–	+	+	–	–	–	–	–	–	–	–
Heavy metals	–	+	≤0.1 ppm	–	–	–	–	–	–	–	–	+	–	–	–	+
Oxidizable substances	–	–	+	+	–	–	+	–	+	+	+	–	–	–	+	–
Potassium permanganate-reducing substances	–	+	–	–	–	–	–	–	–	–	–	+	–	–	–	+
Residue on evaporation	–	≤1.0 mg	–	≤0.001%	–	–	–	–	–	–	–	+	–	–	+	≤1.0 mg
Total organic carbon	–	–	+	–	+	≤0.5 mg/L	–	–	–	–	–	+[b]	+	≤0.5 mg/L	–	–
Total hardness	–	–	–	–	–	–	–	–	–	–	–	–	–	–	–	–
Conductivity	–	–	+	–	+	+	–	–				–	+	+	≤25 µS/cm for containers ≤10 mL, ≤5 µS/cm for containers ≥10 ml	–
Anionic surfactants	–	–	–	–	–	–	–	–	–	–	–	–	–	–	–	–
Antimicrobial agents	–	–	–	–	–	–	–	+	–	–	–	–	–	–	–	–
Sterility	–	–	–	–	–	–	+	+	+	+	+	+	–	–	+	+
Extractable volume	–	–	–	–	–	–	–	–	–	–	–	+	–	–	–	–
Particulate matter	–	–	–	–	–	–	+	+	–	–	–	–	–	–	+	–
Microbial contamination	–	–	–	$\leq 10^2$ cfu/mL	–	–	–	–	–	–	–	–	–	–	–	–
Bacterial endotoxins	–	–	≤0.25 IU/mL	–	–	≤0.25 IU/mL	≤0.25 EU/mL	<0.5 EU/mL	<0.5 EU/mL	≤0.25 EU/mL	–	≤0.25 EU/mL	≤0.25 EU/mL	≤0.25 IU/mL	<0.25 IU/mL	–

(a) For water for injection preserved in containers and sterilized, the JP XV provides separate tests for acid or alkali, chloride, ammonium, and residue on evaporation within the monograph.
(b) For water for injection prepared by reverse osmosis–ultrafiltration.

W

Table III: Storage requirements for different grades of water.

Type	Storage requirements[a]
Bacteriostatic water for injection	Preserve in single-dose and multiple-dose containers, preferably of Type I or Type II glass, not larger than 30 mL in size.
Potable water	Preserve in tightly sealed containers.
Purified water	Preserve in tightly sealed containers. If it is stored in bulk, the conditions of storage should be designed to limit the growth of microorganisms and avoid any other contamination.
Sterile water for inhalation	Preserve in single-dose containers, preferably of Type I or Type II glass.
Sterile water for injection	Preserve in single-dose containers, preferably of Type I or Type II glass, not more than 1000 mL in size.
Water for injection	Preserve in tightly sealed containers.
Water for injections in bulk	Collect and store in conditions designed to prevent growth of microorganisms and avoid any other contamination.

(a) To prevent evaporation and to maintain quality.

12 Incompatibilities

In pharmaceutical formulations, water can react with drugs and other excipients that are susceptible to hydrolysis (decomposition in the presence of water or moisture) at ambient and elevated temperatures.

Water can react violently with alkali metals and rapidly with alkaline metals and their oxides, such as calcium oxide and magnesium oxide. Water also reacts with anhydrous salts to form hydrates of various compositions, and with certain organic materials and calcium carbide.

13 Method of Manufacture

Unlike other excipients, water is not purchased from outside suppliers but is manufactured in-house by pharmaceutical companies. As naturally occurring water has a variety of contaminants, many treatment processes have been developed to remove these. A typical pharmaceutical water purification system contains several unit operations designed to remove various components. The selection of the most appropriate system and its overall design are crucial factors in ensuring that water of the correct quality is produced.[1,2]

To produce potable or drinking water, insoluble matter is first removed from a water supply by coagulation, settling (clarification), and filtering processes. Pathogenic microorganisms present are then destroyed by aeration, chlorination, or some other means. Water may also be rendered free of viable pathogenic microorganisms by active boiling for 15–20 minutes. Activated carbon filters are employed to remove chlorine and many dissolved organic materials found in water, although they may become a breeding ground for microorganisms. The palatability of the water is improved by aeration and charcoal filtration.

Purified water suitable for use in pharmaceutical formulations is usually prepared by purifying potable water by one of several processes, such as distillation, deionization, or RO.[1,3–8]

The quality attributes of WFI are stricter than those for purified water. Consequently, the preparation methods typically vary in the last stage to ensure good control of WFI quality. Methods for the production of WFI are the subject of current debate. The PhEur 7.4 indicates that only distillation would give assurance of consistent supply of the appropriate quality, but permits distillation, ion exchange, RO or any other suitable method that complies with regulations on water intended for human consumption laid down by the competent authority. The USP35–NF30 and the JP XV permit the use of RO in addition to distillation and ultrafiltration. In the past 10–15 years, RO has become the most common way to produce pharmaceutical purified water, either as a final treatment step or as a pretreatment step for the distillation stills.

Distillation Distillation is a process that involves the evaporation of water followed by the condensation of the resulting steam. While expensive, it allows removal of almost all organic and inorganic impurities and achieves very high quality water. It is also considered the safest method to avoid microbial and endotoxin contamination. To improve energy efficiency, distillation is usually conducted in multiple-effects stills designed to recover most of the energy spent on evaporating the water. A typical design consists of an evaporator, vapor separator, and compressor. The distilland (raw feed water) is heated in the evaporator to boiling and the vapor produced is separated from entrained distilland in the separator. The vapor then enters a compressor where the temperature of the vapors is raised to 107°C. Superheated vapors are then condensed on the outer surface of the tubes of the evaporator containing cool distilland circulating within.

Vapor compression stills of various sizes are commercially available and can be used to produce water of high purity when properly constructed. A high-quality distillate, such as WFI, can be obtained if the water is first deionized. The best stills are constructed from types 304 or 316 stainless steel and coated with pure tin, or are made from chemical-resistant glass.

Deionization An ionic exchange process is based on the ability of certain synthetic resins to selectively adsorb either cations or anions, and to release (exchange) other ions based on their relative activity. Cationic and anionic ion exchange resins are used to purify potable water by removing any dissolved ions. Dissolved gases are also removed, while chlorine, in the concentrations generally found in potable water, is destroyed by the resin itself. Some organics and colloidal particles are removed by adsorption and filtration. Resin beds may, however, foster microbial life and produce pyrogenic effluent unless adequate precautions are taken to prevent contamination. Another disadvantage is the type of chemicals required for resin regeneration. A continuous deionization system, which represents a combination of ion exchange and membrane separation technologies, uses an electrical current to continuously regenerate the ion exchange resin simultaneously with the water treatment process, eliminating the need to handle powerful chemicals. Ion exchange units are normally used today to treat raw feed water prior to distillation or RO processing.

Reverse osmosis Water is forced through a semipermeable membrane in the opposite direction to normal osmotic diffusion. Typically, membranes range between 1–10 Å and reject not only organic compounds, bacteria and viruses, but also 90–99% of all ions. It is common to use double-pass RO systems with two filtration stages connected in series. Such systems meet requirements for USP–NF purified water and WFI. However, EU regulations do not allow RO to be used as a final treatment step for the production of WFI.

Membrane filtration Membrane filters are surface-type filters, which stop particles larger than the pore size at the upstream surface of the polymeric membrane. Microfiltration uses membranes with pores in the 0.1–1.0 μm range, which can filter out particles of dust, activated carbon, ion exchange resin fines, and most microorganisms. Ultrafiltration uses membranes that reject not only solid particles but also dissolved matter with a high molecular weight. The 'molecular weight cut-off' point of such membranes varies in the range 10 000–100 000 Da, and bacteria, endotoxins, colloidal contaminants, and large organic molecules can be removed.

14 Safety

Water is the base for many biological life forms, and its safety in pharmaceutical formulations is unquestioned provided it meets standards of quality for potability[9] and microbial content; *see* Sections 9 and 18. Plain water is considered slightly more toxic upon injection into laboratory animals than physiological salt solutions such as normal saline or Ringer's solution.

Ingestion of excessive quantities of water can lead to water intoxication, with disturbances of the electrolyte balance.

WFI should be free from pyrogens.

LD_{50} (mouse, IP): 25 g/kg[10]

15 Handling Precautions

Observe normal precautions appropriate to the circumstances and quantity of material handled.

16 Regulatory Status

Included in nonparenteral and parenteral medicines licensed in the UK and US.

17 Related Substances

Bacteriostatic water for injection; carbon dioxide-free water; de-aerated water; hard water; soft water; sterile water for inhalation; sterile water for injection; sterile water for irrigation; water for injection (WFI).

Bacteriostatic water for injection

Comments The USP35–NF30 describes bacteriostatic water for injection as sterile water for injection that contains one or more suitable antimicrobial agents.

Carbon dioxide-free water

Comments Purified water that has been boiled vigorously for 5 minutes and allowed to cool while protecting it from absorption of atmospheric carbon dioxide.

De-aerated water

Comments Purified water that has been boiled vigorously for 5 minutes and cooled to reduce the air (oxygen) content.

Hard water

Comments Water containing the equivalent of not less than 120 mg/L and not more than 180 mg/L of calcium carbonate.

Soft water

Comments Water containing the equivalent of not more than 60 mg/L of calcium carbonate.

Sterile water for inhalation

Comments The USP35–NF30 describes sterile water for inhalation as WFI sterilized and suitably packaged. It contains no antimicrobial agents or other added substances, except where used in humidifiers or other similar devices, and where liable to contamination over a period of time.

Sterile water for injection

Comments The USP35–NF30 describes sterile water for injection as WFI sterilized and suitably packaged. It contains no antimicrobial agents or other substances.

Sterile water for injection is one of the materials that have been selected for harmonization by the Pharmacopeial Discussion Group. For further information see the General Information Chapter <1196> in the USP35–NF30, the General Chapter 5.8 in PhEur 7.4, along with the 'State of Work' document on the PhEur EDQM website, and also the General Information Chapter 8 in the JP XV.

Sterile water for irrigation

Comments The USP35–NF30 describes sterile water for irrigation as WFI sterilized and suitably packaged. It contains no antimicrobial agents or other substances.

Water for injection (WFI)

Comments The USP35–NF30 describes WFI as water purified by distillation or RO. It contains no added substances. The PhEur 7.4 title is 'water for injections' and comprises two parts: 'water for injections in bulk' and 'sterilized water for injection'. The PhEur 7.4 states that water for injections is produced by distillation.

18 Comments

In most pharmacopeias, the term 'water' now refers to purified or distilled water.

Without further purification, 'water' may be unsuitable for certain pharmaceutical applications; for example, the presence of calcium in water affects the viscosity and gel strength of algins and pectin dispersions, while the use of potable water affects the clarity and quality of cough mixtures, and the stability of antibiotic liquid preparations.

Water commonly contains salts of aluminum, calcium, iron, magnesium, potassium, sodium, and zinc. Toxic substances such as arsenic, barium, cadmium, chromium, cyanide, lead, mercury, and selenium may constitute a danger to health if present in excessive amounts. Ingestion of water containing high amounts of calcium and nitrate is also contraindicated. National standards generally specify the maximum limits for these inorganic substances in potable water. Limits have also been placed on microorganisms, detergents, phenolics, chlorinated phenolics, and other organic substances. The WHO[11] and national bodies have issued guidelines for water quality, although many countries have their own standards for water quality embodied in specific legislation.[12] *See* Table IV.

Control of microbiological contamination is critical for waters used in preparation of pharmaceuticals, as proliferation of microorganisms can potentially occur during all stages of manufacture, storage, or distribution. Suitable control is achieved by ensuring that the water system is well designed and well maintained. Purified water that is produced, stored, and circulated at ambient temperatures is susceptible to the establishment of biofilms; therefore, frequent monitoring, high usage, correct flow rate, and appropriate sanitization are all factors that require consideration to ensure that water is satisfactory.[13]

Monitoring of the whole system is essential in order to demonstrate that correct microbiological quality is achieved. For WFI, the recommended methodology is membrane filtration (0.45 μm) as a large sample size (100–300 mL) is required. For purified water, membrane filtration or plate count methods are typically used depending on the quality requirements of the system. It is important to set appropriate target, alert, and action limits to serve as an indication of action required to bring the quality of water back under control. It is recognized that limits are not intended as pass/fail criteria for water or product batches; however, an investigation regarding the implications should be conducted.[14]

Validation is conducted to provide a high level of assurance that the water production and distribution system will consistently produce water conforming to a defined quality specification. The validation process serves to qualify the design (DQ), installation (IQ), operation (OQ), and performance (PQ) of the system. The extent of monitoring data required should be defined, with consideration given to whether validation to FDA guidelines is required.[14] It is also important to have an ongoing control program with respect to maintenance, and periodic reviews of the performance of the water system.

The PubChem Compound ID (CID) for water is 962.

Table IV: Limits for inorganic substances in potable water (mg/L).

Contaminant	UK (mg/L)	WHO (mg/L)
Aluminum	0.2	0.2
Ammonium	0.5	–
Antimony	0.01	–
Arsenic	0.05	0.05
Barium	1.0	No limit
Beryllium	–	No limit
Boron	2.0	–
Cadmium	0.005	0.005
Calcium	250	–
Chloride	400	250
Chromium	0.05	0.05
Copper	3.0	1.0
Cyanide	0.05	0.1
Fluoride	1.5	1.5
Iron	0.2	0.3
Lead	0.05	0.05
Magnesium	50	–
Manganese	0.05	0.1
Mercury	0.001	0.001
Nickel	0.05	No limit
Nitrate (as N)	–	10
Nitrate (as NO_3)	50	–
Nitrite (as NO_2)	0.1	–
Phosphorus	2.2	–
Potassium	12	–
Selenium	0.01	0.01
Silver	0.01	No limit
Sodium	150	200
Sulfate	250	400
Zinc	5.0	5.0

19 Specific References

1 Thomas WH, Harvey H. Achieving purity in pharmaceutical water. *Manuf Chem Aerosol News* 1976; 47(10): 32, 36, 39, 40.
2 McWilliam AJ. High purity water distribution systems. *Pharm Eng* 1995; Sept/Oct: 54–71.
3 Honeyman T. Purified water for pharmaceuticals. *Manuf Chem* 1987; 58(3): 53, 54, 57, 59.
4 Cross J. Treating waters for the pharmaceutical industry. *Manuf Chem* 1988; 59(3): 34–35.
5 Cross J. Steam sterilisable ultrafiltration membranes. *Manuf Chem* 1989; 60(3): 25–27.
6 Horry JM, Cross JR. Purifying water for ophthalmic and injectable preparations. *Pharm J* 1989; **242**: 169–171.
7 Smith VC. Pure water. *Manuf Chem* 1990; **61**(3): 22–24.
8 Burrows WD, Nelson JH. IV fluidmakers: preparation of sterile water for injection in a field setting. *J Parenter Sci Technol* 1993; **47**(3): 124–129.
9 Walker A. Drinking water – doubts about quality. *Br Med J* 1992; **304**: 175–178.
10 Lewis RJ, ed. *Sax's Dangerous Properties of Industrial Materials*, 11th edn. New York: Wiley, 2004: 3692.
11 World Health Organization. *Guidelines for Drinking-water Quality*, vol. 1: *Recommendations*. Geneva: WHO, 1984.
12 Statutory Instrument 1147. The water supply (water quality) regulations 1989. London: HMSO, 1989. www.opsi.gov.uk/ (accessed 21 November 2011).
13 Riedewald F. Biofilms in pharmaceutical waters. *Pharm Eng* 1997; Nov/Dec: 8–18.
14 Food and Drug Administration. *Guide to Inspections of High Purity Water Systems*. Washington, DC: FDA, 1993. http://www.fda.gov/ICECI/Inspections/InspectionGuides/ucm074905.htm (accessed 21 November 2011).

20 General References

Collentro WV, ed. *Pharmaceutical Water: System Design, Operation and Validation*. Buffalo Grove, IL: Interpharm Press, 1999.

European Directorate for the Quality of Medicines and Healthcare (EDQM). European Pharmacopoeia – State Of Work Of International Harmonisation. *Pharmeuropa* 2011; 23(4): 714–715. www.edqm.eu/site/-614.html (accessed 21 November 2011).

Rössler R. Water and air, two important media in the manufacture of sterile pharmaceuticals, with regard to the GMP. *Drugs Made Ger* 1976; **19**: 130–136.

Santoro M, Maini C. Which water for pharmaceutical use? *Eur J Parenter Pharm Sci* 2003; 8: 15–20.

21 Authors

D Dubash, U Shah.

22 Date of Revision

1 March 2012.

Wax, Anionic Emulsifying

1 Nonproprietary Names

BP: Emulsifying Wax

PhEur: Cetostearyl Alcohol (Type A), Emulsifying

Cetostearyl Alcohol (Type B), Emulsifying

2 Synonyms

Collone HV; Crodex A; Cyclonette Wax; Lanette SX; Lanette W.

3 Chemical Name and CAS Registry Number

Anionic emulsifying wax [8014-38-8]

4 Empirical Formula and Molecular Weight

The PhEur 7.4 specifies that cetostearyl alcohol (type A), emulsifying contains a minimum of 80% cetostearyl alcohol and 7% sodium cetostearyl sulfate. Cetostearyl alcohol (type B), emulsifying contains a minimum of 80% cetostearyl alcohol and 7% sodium lauryl sulfate. A suitable buffer can be added to both.

The BP 2012 describes anionic emulsifying wax as containing cetostearyl alcohol, purified water, and either sodium lauryl sulfate or a sodium salt of a similar sulfated higher primary aliphatic alcohol. *See also* Section 18.

The BP 2012 specifies that the formula of anionic emulsifying wax is:

Cetostearyl alcohol 90 g
Sodium lauryl sulfate 10 g
Purified water 4 mL

5 Structural Formula

See Section 4.

6 Functional Categories

Emulsifying agent; solubilizing agent; stiffening agent.

7 Applications in Pharmaceutical Formulation or Technology

Anionic emulsifying wax is used in cosmetics and topical pharmaceutical formulations primarily as an emulsifying agent. The wax is added to fatty or paraffin bases to facilitate the production of oil-in-water emulsions that are nongreasy. In concentrations of about 2%, emulsions are pourable; stiffer emulsions, e.g. aqueous cream BP, may contain up to 10% of anionic emulsifying wax.

Creams should be adequately preserved and can usually be sterilized by autoclaving. A better-quality emulsion is produced by incorporating some alkali into the aqueous phase, although care should be taken not to use an excess.

Anionic emulsifying wax (3–30%) may also be mixed with soft and liquid paraffins to prepare anhydrous ointment bases such as emulsifying ointment BP. A preparation of 80% anionic emulsifying wax in white soft paraffin has been used as a soap substitute in the treatment of eczema.

In addition, anionic emulsifying wax (10%) has been added to theobroma oil (cocoa butter) to produce a suppository base with a melting point of 34°C.

8 Description

Anionic emulsifying wax occurs as an almost white or pale yellow colored, waxy solid or flakes, which when warmed become plastic before melting. It has a faint characteristic odor and a bland taste.

9 Pharmacopeial Specifications

See Table I.

Table I: Pharmacopeial specifications for anionic emulsifying wax.

Test	PhEur 7.4
Identification	+
Characters	+
Acid value	≤2.0
Iodine value	≤3.0
Saponification value	≤2.0
Water	≤3.0%

10 Typical Properties

Density 0.97 g/cm^3
Flash point >205°C
Melting range 49–54°C
Solubility Soluble in chloroform, ether, and, on warming, in fixed oils and mineral oil. The PhEur 7.4 specifies that cetostearyl alcohol, emulsifying (type A and type B) are soluble in hot water giving an opalescent solution, practically insoluble in cold water, and slightly soluble in ethanol (96%). The BP 2012 specifies that emulsifying wax is practically insoluble in water (forms an emulsion); partly soluble in ethanol (96%).
Spectroscopy
NIR spectra *see* Figure 1.

11 Stability and Storage Conditions

Solid anionic emulsifying wax is chemically stable and should be stored in a well-closed container in a cool, dry place.

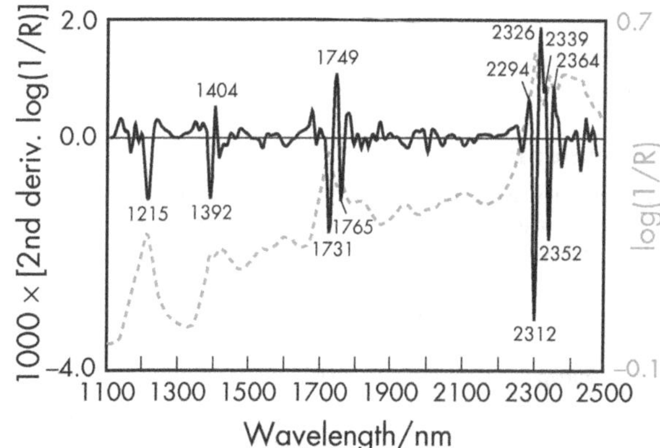

Figure 1: Near-infrared spectrum of anionic emulsifying wax measured by reflectance.

12 Incompatibilities

Incompatibilities of anionic emulsifying wax are essentially those of sodium alkyl sulfates and include cationic compounds (quaternary ammonium compounds, acriflavine, ephedrine hydrochloride, antihistamines, and other nitrogenous compounds), salts of polyvalent metals (aluminum, zinc, tin, and lead), and thioglycolates. Anionic emulsifying wax is compatible with most acids above pH 2.5. It is also compatible with alkalis and hard water.

Iron vessels should not be used when heating anionic emulsifying wax; stainless steel containers are satisfactory.

13 Method of Manufacture

Anionic emulsifying wax is prepared by melting cetostearyl alcohol and heating to about 95°C. Sodium lauryl sulfate, or some other suitable anionic surfactant, and purified water are then added. The mixture is heated to 115°C and, while this temperature is maintained, the mixture is stirred vigorously until any frothing ceases. The wax is then rapidly cooled.

14 Safety

Anionic emulsifying wax is used primarily in topical pharmaceutical formulations and is generally regarded as a nontoxic and nonirritant material. However, sodium lauryl sulfate, a constituent of anionic emulsifying wax, is known to be irritant to the skin at high concentrations; sodium cetyl sulfate is claimed to be less irritating.

Emulsifying ointment BP, which contains anionic emulsifying wax, has been found to have major sunscreen activity in clinically normal skin and should therefore not be used before phototherapy procedures.[1]

15 Handling Precautions

Observe normal precautions appropriate to the circumstances and quantity of material handled. Eye protection is recommended.

16 Regulatory Status

Included in the FDA Inactive Ingredients Database (rectal emulsions and aerosol foams; topical aerosols, emulsions, creams, lotions, and ointments). Included in nonparenteral medicines licensed in the UK. Included in the Canadian Natural Health Products Ingredients Database.

17 Related Substances

Cetostearyl alcohol; sodium lauryl sulfate; wax, nonionic emulsifying.

A number of emulsifying waxes are commercially available that contain different sodium alkyl sulfates and may not meet official compendial specifications. *See also* Section 18.

18 Comments

The nomenclature for emulsifying wax is confusing since there are three groups of emulsifying waxes, with different titles in Europe, the UK and US; *see* Table II.

Table II: Nomenclature for emulsifying wax.

	Europe	UK	US
Nonionic	—	Cetomacrogol emulsifying wax	Emulsifying wax
Anionic	Cetostearyl alcohol (type A), emulsifying Cetostearyl alcohol (type B), emulsifying	Emulsifying wax	—
Cationic	—	Cetrimide emulsifying wax	—

The waxes have similar physical properties but vary in the type of surfactant used, which, in turn, affects the range of compatibilities. Emulsifying wax BP and emulsifying wax USP–NF contain anionic and nonionic surfactants, respectively, and are therefore not interchangeable in formulations.

19 Specific References

1 Cox NH, Sharpe G. Emollients, salicylic acid, and ultraviolet erythema [letter]. *Lancet* 1990; **335**: 53–54.

20 General References

Eccleston GM. Properties of fatty alcohol mixed emulsifiers and emulsifying waxes. In: Florence AT, ed. *Materials Used in Pharmaceutical Formulation: Critical Reports on Applied Chemistry.* 6. Oxford: Blackwell Scientific, 1984; 124–156.

Eccleston GM. Functions of mixed emulsifiers and emulsifying waxes in dermatological lotions and creams. *Colloid Surface A* 1997; **123–124**: 169–182.

21 Author

G Moss.

22 Date of Revision

1 March 2012.

Wax, Carnauba

1 Nonproprietary Names

BP: Carnauba Wax

JP: Carnauba Wax

PhEur: Carnauba Wax

USP–NF: Carnauba Wax

2 Synonyms

Brazil wax; caranda wax; cera carnauba; E903.

3 Chemical Name and CAS Registry Number

Carnauba wax [8015-86-9]

4 Empirical Formula and Molecular Weight

Carnauba wax consists primarily of a complex mixture of esters of acids and hydroxy acids, mainly aliphatic esters, ω-hydroxy esters, *p*-methoxycinnamic aliphatic esters, and *p*-hydroxycinnamic aliphatic diesters composed of several chain lengths, in which C_{26} and C_{32} alcohols are the most prevalent.[1]

Also present are acids, oxypolyhydric alcohols, hydrocarbons, resinous matter, and water.

5 Structural Formula

See Section 4.

6 Functional Category

Coating agent.

7 Applications in Pharmaceutical Formulation or Technology

Carnauba wax is widely used in cosmetics, certain foods, and pharmaceutical formulations. Cosmetically, carnauba wax is commonly used in lip balms.[2]

Carnauba wax is the hardest and highest-melting of the waxes commonly used in pharmaceutical formulations and is used primarily as a 10% w/v aqueous emulsion to polish sugar-coated tablets. Aqueous emulsions may be prepared by mixing carnauba wax with an ethanolamine compound and oleic acid. The carnauba wax coating produces tablets of good luster without rubbing. Carnauba wax may also be used in powder form to polish sugar-coated tablets.

Carnauba wax (10–50% w/w) is also used alone or with other excipients such as hypromellose, hydroxypropyl cellulose, alginate/pectin-gelatin, polymethacrylates, and stearyl alcohol to produce sustained-release solid-dosage formulations.[3–10]

8 Description

Carnauba wax occurs as a light brown- to pale yellow-colored powder, flakes, or irregular lumps of a hard, brittle wax. It has a characteristic bland odor and practically no taste. It is free from rancidity. Various types and grades are available commercially.

9 Pharmacopeial Specifications

See Table I.

10 Typical Properties

Flash point 270–330°C

Refractive index $n_D^{90} = 1.450$

Table I: Pharmacopeial specifications for carnauba wax.

Test	JP XV	PhEur 7.4	USP35–NF30
Characters	+	+	—
Identification	—	+	—
Appearance of solution	—	+	—
Melting range	80–86°C	80–88°C	80–86°C
Acid value	≤10.0	2–7	2–7
Saponification value	78–95	78–95	78–95
Total ash	—	≤0.25%	≤0.25%
Heavy metals	—	—	≤20 ppm
Iodine value	5–14	—	—
Specific gravity	0.990–1.002	—	—

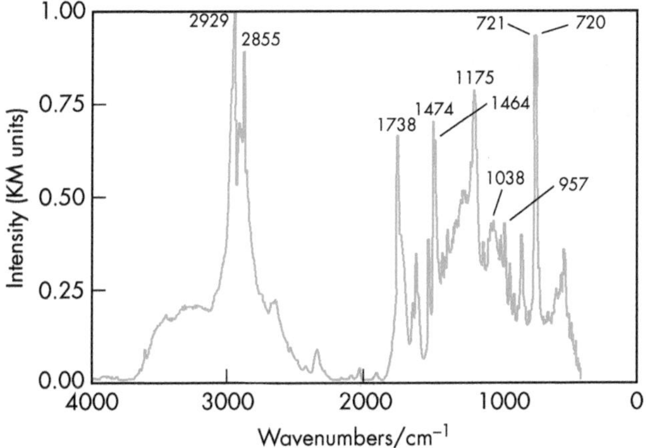

Figure 1: Infrared spectrum of carnauba wax measured by diffuse reflectance. Adapted with permission of Informa Healthcare.

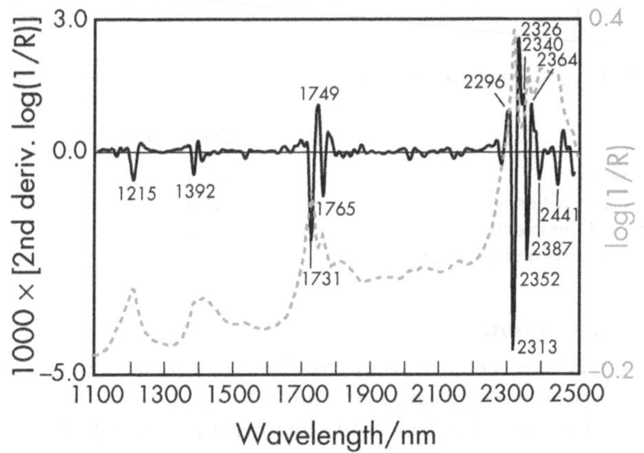

Figure 2: Near-infrared spectrum of carnauba wax measured by reflectance.

Solubility Freely soluble in warm benzene; soluble in warm chloroform and in warm toluene; slightly soluble in boiling ethanol (95%); practically insoluble in water.

Specific gravity 0.990–0.999 at 25°C

Spectroscopy

IR spectra *see* Figure 1.

NIR spectra *see* Figure 2.

Unsaponified matter 50–55%

11 Stability and Storage Conditions

Carnauba wax is stable and should be stored in a well-closed container, in a cool, dry place. Protect from light.

12 Incompatibilities

—

13 Method of Manufacture

Carnauba wax is obtained from the leaf buds and leaves of the Brazilian carnauba palm, *Copernicia cerifera*. The leaves are dried and shredded, and the wax is then removed by the addition of hot water.

14 Safety

Carnauba wax is widely used in oral pharmaceutical formulations, cosmetics, and certain food products. It is generally regarded as an essentially nontoxic and nonirritant material.[11–13] However, there have been reports of allergic contact dermatitis from carnauba wax in mascara.[14]

The WHO has established an acceptable daily intake of up to 7 mg/kg body-weight for carnauba wax.[15]

15 Handling Precautions

Observe normal precautions appropriate to the circumstances and quantity of material handled. When heated to decomposition it emits acrid smoke and irritating fumes.[16]

16 Regulatory Status

GRAS listed. Accepted for use as a food additive in Europe. Included in the FDA Inactive Ingredients Database (oral capsules and tablets). Included in nonparenteral medicines licensed in the UK. Included in the Canadian Natural Health Products Ingredients Database.

17 Related Substances

—

18 Comments

In cosmetics, carnauba wax is mainly used to increase the stiffness of formulations, e.g. lipsticks and mascaras. It has been experimentally investigated for use in producing microparticles in a novel hot air coating (HAC) process developed as an alternative to conventional spray-congealing techniques.[17] In addition, carnauba wax has been used to produce gel beads for intragastric floating drug delivery[18] and has been investigated for use in nanoparticulate sunscreen formulations.[19]

The EINECS number for carnauba wax is 232-399-4.

19 Specific References

1 Emås M, Nyqvist H. Methods of studying aging and stabilization of spray-congealed solid dispersions with carnauba wax. 1: Microcalorimetric investigation. *Int J Pharm* 2000; **197**: 117–127.

2 Marti-Mestres G, *et al.* Texture and sensory analysis in stick formulations. *STP Pharma Sci* 1999; **9**(4): 371–375.

3 Reza MS, *et al.* Comparative evaluation of plastic, hydrophobic and hydrophilic polymers as matrices for controlled-release drug delivery. *J Pharm Pharm Sci* 2003; **6**(2): 282–291.

4 Gioannola LI, *et al.* Carnauba wax microspheres loaded with valproic acid: preparation and evaluation of drug release. *Drug Dev Ind Pharm* 1995; **21**: 1563–1572.

5 Miyagawa Y, *et al.* Controlled-release of diclofenac sodium from wax matrix granule. *Int J Pharm* 1996; **138**(2): 215–224.

6 Aritomi H, *et al.* Development of sustained-release formulation of chlorpheniramine maleate using powder-coated microsponge prepared by dry impact blending method. *J Pharm Sci Tech Yakukzaigaku* 1996; **56**(1): 49–56.

7 Huang HP, *et al.* Mechanism of drug release from an acrylic polymer-wax matrix tablet. *J Pharm Sci* 1994; **83**(6): 795–797.

8 Joseph I, Venkataram S. Indomethacin sustained release from alginate-gelatin or pectin-gelatin coacervates. *Int J Pharm* 1995; **126**: 161–168.

9 Kumar K, *et al.* Sustained release tablet formulation of diethylcarbamazine citrate (Hetrazan). *Indian J Pharm* 1975; **37**: 57–59.
10 Dave SC, *et al.* Sustained release tablet formulation of diphenhydramine hydrochloride (Benadryl) - part II. *Indian J Pharm* 1974; **36**: 94–96.
11 Parent RA, *et al.* Subchronic feeding study of carnauba wax in beagle dogs. *Food Chem Toxicol* 1983; **21**(1): 85–87.
12 Parent RA, *et al.* Reproductive and subchronic feeding study of carnauba wax in rats. *Food Chem Toxicol* 1983; **21**(1): 89–93.
13 Rowland IR, *et al.* Short-term toxicity study of carnauba wax in rats. *Food Chem Toxicol* 1982; **20**(4): 467–471.
14 Chowdhury MM. Allergic contact dermatitis from prime yellow carnauba wax and coathylene in mascara. *Contact Dermatitis* 2002; **46**(6): 244.
15 FAO/WHO. Evaluation of certain food additives and naturally occurring toxicants. Thirty-ninth report of the joint FAO/WHO expert committee on food additives. *World Health Organ Tech Rep Ser* 1992; No. 828.
16 Lewis RJ, ed. *Sax's Dangerous Properties of Industrial Materials*, 11th edn. New York: Wiley, 2004.
17 Rodriguez L, *et al.* Hot air coating technique as a novel method to produce microparticles. *Drug Dev Ind Pharm* 2004; **30**(9): 913–923.
18 Sriamorsak P, *et al.* Wax-incorporated emulsion gel beads of calcium pectinate for intragastric floating drug delivery. *AAPS Pharm Sci Tech* 2008; **9**(2): 571–576.
19 Villalobos-Hernandez JR, Mueller-Goyman CC. *In vitro* erythemal UV-A protection factors of inorganic sunscreens distributed in aqueous media using carnauba wax-decyl oleate nanoparticles. *Eur J Pharm Biopharm* 2007; **65**(1): 122–125.

20 General References

—

21 Author

R Deanne.

22 Date of Revision

1 March 2012.

Wax, Cetyl Esters

1 Nonproprietary Names

USP–NF: Cetyl Esters Wax

2 Synonyms

Cera cetyla; *Crodamol SS*; *Cutina CP*; *Liponate SPS*; *Protachem MST*; *Ritaceti*; *Ritachol SS*; spermaceti wax replacement; *Starfol Wax CG*; *Synaceti 116*; synthetic spermaceti.

3 Chemical Name and CAS Registry Number

Cetyl esters wax [977067-67-6]

4 Empirical Formula and Molecular Weight

$C_nH_{2n}O_2$ ≈470–490 (where n = 26–38).

The USP35–NF30 describes cetyl esters wax as a mixture consisting primarily of esters of saturated fatty alcohols (C_{14}–C_{18}) and saturated fatty acids (C_{14}–C_{18}).

5 Structural Formula

See Section 4.

6 Functional Category

Emollient; stiffening agent.

7 Applications in Pharmaceutical Formulation or Technology

Cetyl esters wax is a stiffening agent and emollient used in creams and ointments as a replacement for naturally occurring spermaceti.

Cetyl esters wax is hydrophobic and has been proposed as a suitable component of an ophthalmic gelatin-based, controlled-release delivery matrix.[1]

The physical properties of cetyl esters wax vary greatly from manufacturer to manufacturer owing to differences between the mixtures of fatty acids and fatty alcohol esters that are used. Differences between products appear most obviously in the melting point, which can range from 43–47°C (USP35–NF30 range) to 51–55°C, depending on the mixture. Materials with a high melting point tend to contain predominantly cetyl and stearyl palmitates. *See* Table I.

Table I: Uses of cetyl esters wax.

Use	Concentration (%)
Cold cream	12.5
Rose water ointment	12.5
Spermaceti ointment	20.0
Topical creams and ointments	1–15

8 Description

Cetyl esters wax occurs as white to off-white, somewhat translucent flakes (typically in the range of 5 μm to several millimeters in the largest dimension), having a crystalline structure and a pearly luster when caked. It has a faint, aromatic odor and a bland, mild taste.

9 Pharmacopeial Specifications

See Table II.

Table II: Pharmacopeial specifications for cetyl esters wax.

Test	USP35–NF30
Melting range	43–47°C
Acid value	≤5
Iodine value	≤1
Saponification value	109–120
Paraffin and free acids	+

10 Typical Properties

Dielectric constant 6–18
Flash point >240°C
Peroxide value ≤0.5

Table III: Solubility of cetyl esters wax.

Solvent	Solubility at 20°C unless otherwise stated
Acetone	1 in 500
Chloroform	1 in 2.5
Dichloromethane	1 in 3
Ethanol	1 in 170
Ethanol (95%)	Practically insoluble 1 in 2.5 at 78°C
Ether	Soluble
Ethyl acetate	1 in 80
Fixed and volatile oils	Soluble
Hexane	1 in 8
Mineral oil	1 in 70
Water	Practically insoluble

Refractive index $n_D^{60} = 1.440$
Solubility High melting materials tend to be less soluble. *See* Table III.
Specific gravity 0.820–0.840 at 50°C
Viscosity (dynamic) 6.7–7.4 mPa s (6.7–7.4 cP) at 100°C

11 Stability and Storage Conditions

Store in a well-closed container in a cool, dry place. Avoid exposure to excessive heat (above 40°C).

12 Incompatibilities

Cetyl esters wax is incompatible with strong acids or bases.

13 Method of Manufacture

Cetyl esters wax is prepared by the direct esterification of the appropriate mixtures of fatty alcohols and fatty acids.

14 Safety

Cetyl esters wax is an innocuous material generally regarded as essentially nontoxic and nonirritant.

LD_{50} (rat, oral): >16 g/kg

15 Handling Precautions

Observe normal precautions appropriate to the circumstances and quantity of material handled.

16 Regulatory Status

Included in the FDA Inactive Ingredients Database (topical preparations). Included in nonparenteral medicines licensed in the UK. Included in the Canadian Natural Health Products Ingredients Database.

17 Related Substances

Spermaceti wax.

Spermaceti wax

CAS number [8002-23-1]
Appearance Spermaceti is a waxy substance obtained from the head of the sperm whale. It consists of a mixture of the cetyl esters of fatty acids (C_{12}–C_{18}) with cetyl laurate, cetyl myristate, cetyl palmitate, and cetyl stearate comprising at least 85% of the total esters. It occurs as white, translucent, slightly unctuous masses with a faint odor and mild, bland taste.
Iodine value 3.0–4.4
Melting point 44–52°C
Refractive index $n_D^{80} = 1.4330$
Saponification value 120–136
Solubility Soluble in chloroform, boiling ethanol (95%), ether, and fixed or volatile oils; practically insoluble in ethanol (95%) and water.
Specific gravity 0.938–0.944
Uses Spermaceti has been used in creams, ointments, and suppositories,[2] although it has largely been superseded in pharmaceutical and cosmetics formulation by the synthetic material, cetyl esters wax.
Comments The EINECS number for spermaceti wax is 232-302-5.

18 Comments

—

19 Specific References

1 Nadkarni SR, Yalkowsky SH. Controlled delivery of pilocarpine 1: *in vitro* characterization of Gelfoam matrices. *Pharm Res* 1993; **10**: 109–112.

2 Baichwal MR, Lohit TV. Medicament release from fatty suppository bases. *J Pharm Pharmacol* 1970; **22**: 427–432.

20 General References

Egan RR, Portwood O. Higher alcohols in skin lotions. *Cosmet Perfum* 1974; **89**(3): 39–42.

Holloway PJ. The chromatographic analysis of spermaceti. *J Pharm Pharmacol* 1968; **20**: 775–779.

Spencer GF, Kleiman R. Detection of spermaceti in a hand cream. *J Am Oil Chem Soc* 1978; **55**: 837–838.

21 Author

PJ Weller.

22 Date of Revision

1 March 2012.

Wax, Microcrystalline

1 Nonproprietary Names

USP–NF: Microcrystalline Wax

2 Synonyms

Amorphous wax; E907; petroleum ceresin; petroleum wax (microcrystalline).

3 Chemical Name and CAS Registry Number

Microcrystalline wax [63231-60-7]

4 Empirical Formula and Molecular Weight

Microcrystalline wax is composed of a mixture of straight-chain and randomly branched saturated alkanes obtained from petroleum. The carbon chain lengths range from C_{41} to C_{57}; cyclic hydrocarbons are also present.

5 Structural Formula

See Section 4.

6 Functional Category

Coating agent; modified-release agent; stiffening agent.

7 Applications in Pharmaceutical Formulation or Technology

Microcrystalline wax is used mainly as a stiffening agent in topical creams and ointments.

The wax is used to modify the crystal structure of other waxes (particularly paraffin wax) present in a mixture so that changes in crystal structure, usually exhibited over a period of time, do not occur. Microcrystalline wax also minimizes the sweating or bleeding of oils from blends of oils and waxes. Microcrystalline wax generally has a higher melting point than paraffin wax, and higher viscosity when molten, thereby increasing the consistency of creams and ointments when incorporated into such formulations.

Microcrystalline wax is also used in oral controlled-release matrix pellet formulations for various active compounds[1–3] and as a tablet- and capsule-coating agent. In controlled-release systems, microcrystalline wax coatings can also be used to affect the release of drug from ion-exchange resin beads.[4]

Microcrystalline wax is also used in confectionery, cosmetics, and food products.

8 Description

Microcrystalline wax occurs as odorless and tasteless waxy lumps or flakes containing small irregularly shaped crystals. It may vary in color from white to yellow, amber, brown, or black depending on the grade of material; pharmaceutical grades are usually white or yellow.

The USP35–NF30 describes microcrystalline wax as a mixture of straight-chain, branched-chain, and cyclic hydrocarbons, obtained by solvent fractionation of the still-bottom fraction of petroleum by suitable means of dewaxing or de-oiling.

9 Pharmacopeial Specifications

See Table I.

Table I: Pharmacopeial specifications for microcrystalline wax.

Test	USP35–NF30
Color	+
Melting range	54–102°C
Consistency	3–100
Acidity	+
Alkalinity	+
Residue on ignition	≤0.1%
Organic acids	+
Fixed oils, fats, and rosin	+

10 Typical Properties

Acid value 1.0

Density 0.928–0.941 g/cm^3

Freezing point 60.0–75.0°C

Refractive index n_D^{100} = 1.435–1.445

Saponification value 0.05–0.10

Solubility Soluble in benzene, chloroform, and ether; slightly soluble in ethanol; practically insoluble in water. When melted, microcrystalline wax is miscible with volatile oils and most warm fixed oils.

Spectroscopy

NIR spectra *see* Figure 1.

Viscosity (dynamic) 10.0–30.0 mPa s (10.0–30.0 cP) at 100°C.

11 Stability and Storage Conditions

Microcrystalline wax is stable in the presence of acids, alkalis, light, and air. The bulk material should be stored in a well-closed container in a cool, dry place.

12 Incompatibilities

—

13 Method of Manufacture

Microcrystalline wax is obtained by solvent fractionation of the still-bottom fraction of petroleum by suitable dewaxing or de-oiling.

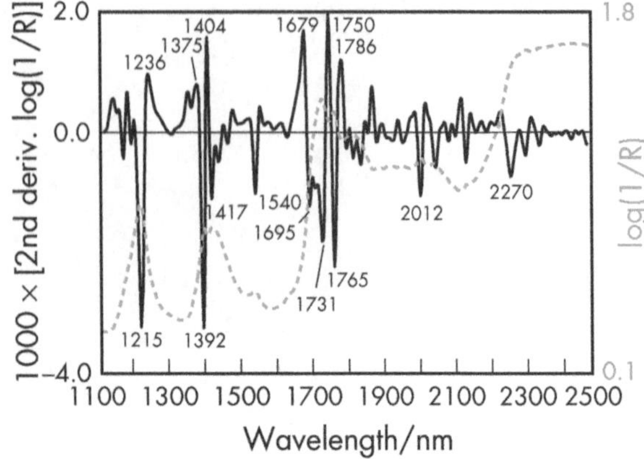

Figure 1: Near-infrared spectrum of microcrystalline wax measured by reflectance.

14 Safety

Microcrystalline wax is mainly used in topical pharmaceutical formulations but is also used in some oral products. It is generally regarded as a nontoxic and nonirritating material.

15 Handling Precautions

Observe normal precautions appropriate to the circumstances and quantity of material handled. Eye protection is recommended.

16 Regulatory Status

GRAS listed. Accepted for use as a food additive in Europe. Included in the FDA Inactive Ingredients Database (oral capsules; topical and vaginal preparations). Included in nonparenteral medicines licensed in the UK. Included in the Canadian Natural Health Products Ingredients Database.

17 Related Substances

Paraffin; wax, white; wax, yellow.

18 Comments

Rheological studies of a model ointment containing microcrystalline wax, white petroleum, and mineral oil showed that while the latter two substances control the rheology of the ointment, microcrystalline wax incorporates itself into the existing white petroleum structure and builds up the structure of the ointment.[5]

A specification for refined microcrystalline wax (petroleum wax) is contained in the *Food Chemicals Codex* (FCC).[6]

19 Specific References

1 De Brabander C, *et al.* Bioavailability of ibuprofen from matrix minitablets based on a mixture of starch and microcrystalline wax. *Int J Pharm* 2000; **208**: 81–86.

2 De Brabander C, *et al.* Matrix minitablets based on starch/microcrystalline wax mixtures. *Int J Pharm* 2000; **199**: 195–203.

3 Vergote GJ, *et al.* Oral controlled release matrix pellet formulation containing nanocrystalline ketoprofen. *Int J Pharm* 2001; **219**: 81–87.

4 Motycka S, Nairn J. Influence of wax coatings on release rate of anions from ion-exchange resin beads. *J Pharm Sci* 1978; **67**: 500–503.

5 Pena LE, *et al.* Structural rheology of a model ointment. *Pharm Res* 1994; **11**: 875–881.

6 *Food Chemicals Codex*, 7th edn. Bethesda, MD: United States Pharmacopeia, 2010: 776.

20 General References

Tennant DR. The usage, occurrences and dietary intakes of white mineral oils and waxes in Europe. *Food Chem Toxicol* 2004; **42**: 481–492.

Oauli-Bruns A, *et al.* Preparation of sustained-release matrix pellets by melt agglomeration in the fluidized bed: influence of formulation variables and modelling of agglomerate growth. *Eur J Pharm Biopharm* 2010; 74(3): 503–512.

21 Author

AH Kibbe.

22 Date of Revision

1 March 2012.

 Wax, Nonionic Emulsifying

1 Nonproprietary Names

BP: Cetomacrogol Emulsifying Wax
USP–NF: Emulsifying Wax

2 Synonyms

Collone NI; Crodex N; Emulgade 1000NI; Ester Wax NF; Lipowax P; Masurf Emulsifying Wax NF; Permulgin D; Polawax; Ritachol 2000; T-Wax.

3 Chemical Name and CAS Registry Number

See Section 4.

4 Empirical Formula and Molecular Weight

The USP35–NF30 designates nonionic emulsifying wax as emulsifying wax that is prepared from cetostearyl alcohol and contains a polyoxyethylene derivative of a fatty acid ester of sorbitan. However, the BP 2012 describes nonionic emulsifying wax as cetomacrogol emulsifying wax prepared from cetostearyl alcohol and macrogol cetostearyl ether (22). The UK and US materials are therefore constitutionally different. *See also* Section 18.

5 Structural Formula

See Section 4.

6 Functional Category

Emulsifying agent; solubilizing agent; stiffening agent.

7 Applications in Pharmaceutical Formulation or Technology

Nonionic emulsifying wax is used as an emulsifying agent in the production of oil-in-water emulsions that are unaffected by moderate concentrations of electrolytes and are stable over a wide pH range. The concentration of wax used alters the consistency of a product owing to its 'self-bodying action'; at concentrations up to about 5% a product is pourable.

Concentrations of about 15% of nonionic emulsifying wax are commonly used in creams, but concentrations as high as 25% may be employed, e.g. in chlorhexidine cream BP. Nonionic emulsifying wax is particularly recommended for use with salts of polyvalent metals and medicaments based on nitrogenous compounds. Creams are susceptible to microbial spoilage and should be adequately preserved.

Nonionic emulsifying wax is also used in nonaqueous ointment bases, such as cetomacrogol emulsifying ointment BP, and in barrier creams.

8 Description

Nonionic emulsifying wax occurs as a white or off-white waxy solid or flakes, which melt when heated to give a clear, almost colorless liquid. Nonionic emulsifying wax has a faint odor, which is characteristic of cetostearyl alcohol.

9 Pharmacopeial Specifications

See Table I.

Table I: Pharmacopeial specifications for nonionic emulsifying wax.

Test	BP 2012	USP35–NF30
Identification	+	—
Characters	+	—
Melting range	—	50–54°C
Solidifying point	45–53°C	—
pH (3% dispersion)	—	5.5–7.0
Alkalinity	+	—
Acid value	≤0.5	—
Hydroxyl value	175–192	178–192
Iodine value	—	≤3.5
Refractive index (at 60°C)	1.435–1.439	—
Saponification value	≤2.0	≤14
Sulfated ash	≤0.1%	—

10 Typical Properties

Density 0.94 g/cm^3

Flash point >150°C for *Masurf Emulsifying Wax NF*

Solubility The BP 2012 specifies that cetomagrocol emulsifying wax is practically insoluble in water, forming an emulsion, and is moderately soluble in ethanol (96%), and partly soluble in ether. The USP35–NF30 specifies that emulsifying wax is insoluble in water, soluble in alcohol and freely soluble in ether, chloroform, most hydrocarbon solvents, and aerosol propellants.

Spectroscopy

NIR spectra *see* Figure 1.

11 Stability and Storage Conditions

Nonionic emulsifying wax is stable and should be stored in a well-closed container in a cool, dry place.

12 Incompatibilities

Nonionic emulsifying wax is incompatible with tannin, phenol and phenolic materials, resorcinol, and benzocaine. It may reduce the antibacterial efficacy of quaternary ammonium compounds.

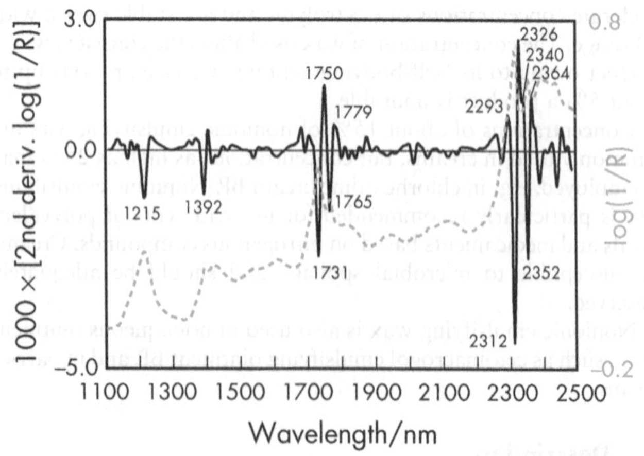

Figure 1: Near-infrared spectrum of nonionic emulsifying wax measured by reflectance.

13 Method of Manufacture

The BP 2012 specifies that cetomacrogol emulsifying wax (nonionic emulsifying wax) may be prepared by melting and mixing together 800 g of cetostearyl alcohol and 200 g of macrogol cetostearyl ether (22). The mixture is then stirred until cold.

The USP35–NF30 formula for nonionic emulsifying wax is a mixture of unstated proportions of cetostearyl alcohol and a polyoxyethylene derivative of a fatty acid ester of sorbitan.

14 Safety

Nonionic emulsifying wax is used in cosmetics and topical pharmaceutical formulations, and is generally regarded as a nontoxic and nonirritant material.

15 Handling Precautions

Observe normal precautions appropriate to the circumstances and quantity of material handled. Eye protection is recommended.

16 Regulatory Status

Included in the FDA Inactive Ingredients Database (rectal emulsions and aerosol foams; topical aerosols, emulsions, creams, lotions, and ointments). Included in nonparenteral medicines licensed in the UK. Included in the Canadian Natural Health Products Ingredients Database.

17 Related Substances

Cationic emulsifying wax; cetostearyl alcohol; polyoxyethylene alkyl ethers; wax, anionic emulsifying.

It should be noted that there are many similar nonionic emulsifying waxes composed of different nonionic surfactants and fatty alcohols.

Cationic emulsifying wax

Synonyms cetrimide emulsifying wax; *Crodex C.*

Method of manufacture Cetrimide emulsifying wax is prepared similarly to nonionic emulsifying wax and contains 90 g of cetostearyl alcohol and 10 g of cetrimide.

Comments Cationic emulsifying wax is claimed to be of particular value in cosmetic and pharmaceutical formulations when cationic characteristics are important. Thus it can be used in medicated creams, germicidal creams, ointments and lotions, hair conditioners, baby creams, and skin care products in which cationic compounds are included. Cationic emulsifying wax is compatible with cationic and nonionic materials, but is incompatible with anionic surfactants and drugs. Additional antimicrobial preservatives should be included in creams. Cetrimide may cause irritation to the eye; *see* Cetrimide.

18 Comments

The nomenclature for emulsifying wax is confusing since there are three groups of emulsifying waxes with different titles in Europe, the UK, and US; *see* Table II.

Table II: Nomenclature for emulsifying wax.

	Europe	UK	US
Nonionic	–	Cetomacrogol emulsifying wax	Emulsifying wax
Anionic	Cetostearyl alcohol (type A), emulsifying Cetostearyl alcohol (type B), emulsifying	Emulsifying wax	–
Cationic	–	Cetrimide emulsifying wax	–

The waxes have similar physical properties but vary in the type of surfactant used, which, in turn, affects the range of compatibilities. Emulsifying wax BP and emulsifying wax USP–NF contain anionic and nonionic surfactants, respectively, and are therefore not interchangeable in formulations.

19 Specific References

—

20 General References

Eccleston GM. Properties of fatty alcohol mixed emulsifiers and emulsifying waxes. In: Florence AT, ed. *Materials Used in Pharmaceutical Formulation: Critical Reports on Applied Chemistry*. 6. Oxford: Blackwell Scientific, 1984; 124–156.

Eccleston GM. Functions of mixed emulsifiers and emulsifying waxes in dermatological lotions and creams. *Colloids Surf A Physicochem Eng Asp* 1997; **123–124**: 169–182.

Hadgraft JW. The emulsifying properties of polyethyleneglycol ethers of cetostearyl alcohol. *J Pharm Pharmacol* 1954; **6**: 816–829.

Mason Chemical Company. Product literature: *Masurf Emulsifying Wax NF*, 2005.

21 Author

G Moss.

22 Date of Revision

1 March 2012.

Wax, White

1 Nonproprietary Names

BP: White Beeswax

JP: White Beeswax

PhEur: Beeswax, White

USP–NF: White Wax

2 Synonyms

Bleached wax; cera alba; E901.

3 Chemical Name and CAS Registry Number

White beeswax [8012-89-3]

4 Empirical Formula and Molecular Weight

White wax is the chemically bleached form of natural beeswax; *see* Section 13.

Beeswax consists of 70–75% of a mixture of various esters of straight-chain monohydric alcohols with even-numbered carbon chains from C_{24} to C_{36} esterified with straight-chain acids. These straight-chain acids also have even numbers of carbon atoms up to C_{36} together with some C_{18} hydroxy acids. The chief ester is myricyl palmitate. Also present are free acids (about 14%) and carbohydrates (about 12%) as well as approximately 1% free wax alcohols and stearic esters of fatty acids.

5 Structural Formula

See Section 4.

6 Functional Category

Coating agent; emulsion stabilizing agent; modified-release agent; stiffening agent.

7 Applications in Pharmaceutical Formulation or Technology

White wax is a chemically bleached form of yellow wax and is used in similar applications: for example, to increase the consistency of creams and ointments, and to stabilize water-in-oil emulsions. White wax is used to polish sugar-coated tablets and to adjust the melting point of suppositories.

White wax is also used as a film coating in sustained-release tablets.[1] White beeswax microspheres may be used in oral dosage forms to retard the absorption of an active ingredient from the stomach, allowing the majority of absorption to occur in the intestinal tract. Wax has been incorporated in dosage forms intended to float in the stomach fluids to prolong drug release, with mixed results.[2] Wax coatings can also be used to affect the release of drug from ion-exchange resin beads.[3–5]

See also Wax, Yellow.

8 Description

White wax consists of tasteless, white or slightly yellow-colored sheets or fine granules with some translucence. Its odor is similar to that of yellow wax but is less intense.

9 Pharmacopeial Specifications

See Table I.

Table I: Pharmacopeial specifications for white wax.

Test	JP XV	PhEur 7.4	USP35–NF30
Characters	+	+	—
Melting range	60–67°C	61–66°C	62–65°C
Relative density	—	≈0.960	—
Acid value	5–9 or 17–22	17–24	17–24
Ester value	—	70–80	72–79
Saponification value	80–100	87–104	—
Ceresin, paraffins, and certain other waxes	—	+	—
Purity	+	—	—
Glycerols and other polyols	—	+	—
Saponification cloud test	—	—	+
Fats or fatty acids, Japan wax, rosin, and soap	—	—	+

10 Typical Properties

Arsenic ≤3 ppm

Density 0.95–0.96 g/cm^3

Flash point 245–258°C

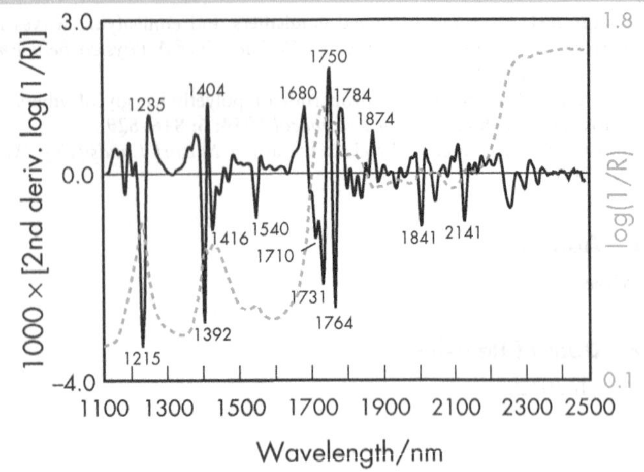

Figure 1: Near-infrared spectrum of white wax measured by reflectance.

Heavy metals ≤0.004%
Iodine number 8–11
Lead ≤10 ppm
Melting point 61–65°C
Peroxide value ≤8
Solubility Soluble in chloroform, ether, fixed oils, volatile oils, and warm carbon disulfide; sparingly soluble in ethanol (95%); practically insoluble in water.
Spectroscopy
NIR spectra *see* Figure 1.
Unsaponified matter 52–55%

11 Stability and Storage Conditions

When the wax is heated above 150°C, esterification occurs with a consequent lowering of acid value and elevation of melting point. White wax is stable when stored in a well-closed container, protected from light.

12 Incompatibilities

Incompatible with oxidizing agents.

13 Method of Manufacture

Yellow wax (beeswax) is obtained from the honeycomb of the bee (*Apis mellifera* Linné (Fam. Apidae)); *see* Wax, Yellow. Subsequent treatment with oxidizing agents bleaches the wax to yield white wax.

14 Safety

White wax is used in both topical and oral formulations, and is generally regarded as an essentially nontoxic and nonirritant material. However, although rare, hypersensitivity reactions to beeswax (attributed to contaminants in the wax) have been reported.[6,7]

15 Handling Precautions

Observe normal precautions appropriate to the circumstances and quantity of material handled.

16 Regulatory Status

GRAS listed. Accepted for use as a food additive in Europe. Included in the FDA Inactive Ingredients Database (oral capsules and tablets; rectal, topical, and vaginal preparations). Included in nonparenteral medicines licensed in the UK. Included in the Canadian Natural Health Products Ingredients Database.

17 Related Substances

Paraffin; wax, microcrystalline; wax, yellow.

18 Comments

A specification for white beeswax is contained in the *Food Chemicals Codex* (FCC).[8]

19 Specific References

1 Nughroho AK, Fudholi A. Comparison of mefenamic acid dissolution in sustained release tablets using hydroxypropyl methylcellulose and cera alba as film coating. *Indonesian J Pharm* 1999; **10**(2): 78–84.
2 Sriamornsak P, *et al.* Wax-incorporated emulsion gel beads of calcium pectinate for intragastric floating drug delivery. *AAPS PharmSciTech* 2008; 9(2): 571–576.
3 Giannola L, *et al.* White beeswax microspheres: a comparative *in vitro* evaluation of cumulative release of the anticancer agents fluorouracil and ftorafur. *Pharmazie* 1993; **48**: 123–126.
4 Giannola LI, *et al.* Preparation of white beeswax microspheres loaded with valproic acid and kinetic study of drug release. *Drug Dev Ind Pharm* 1995; **21**: 793–807.
5 Motycka S, Nairn J. Influence of wax coatings on release rate of anions from ion-exchange resin beads. *J Pharm Sci* 1978; **67**: 500–503.
6 Cronin E. Contact dermatitis from cosmetics. *J Soc Cosmet Chem* 1967; **18**: 681–691.
7 Rothenborg HW. Occupational dermatitis in beekeeper due to poplar resins in beeswax. *Arch Dermatol* 1967; **95**: 381–384.
8 *Food Chemicals Codex*, 7th edn. Bethesda, MD: United States Pharmacopeia, 2010: 83

20 General References

Puleo SL. Beeswax. *Cosmet Toilet* 1987; **102**(6): 57–58.
Tennant DR. The usage, occurrences and dietary intakes of white mineral oils and waxes in Europe. *Food Chem Toxicol* 2004; **42**: 481–492.

21 Author

AH Kibbe.

22 Date of Revision

1 March 2012.

W

Wax, Yellow

1 Nonproprietary Names

BP: Yellow Beeswax
JP: Yellow Beeswax
PhEur: Beeswax, Yellow
USP–NF: Yellow Wax

2 Synonyms

Apifil; cera flava; E901; refined wax.

3 Chemical Name and CAS Registry Number

Yellow beeswax [8012-89-3]

4 Empirical Formula and Molecular Weight

Yellow wax is naturally obtained beeswax; *see* Section 13.

Beeswax consists of 70–75% of a mixture of various esters of straight-chain monohydric alcohols with even-numbered carbon chains from C_{24} to C_{36} esterified with straight-chain acids. These straight-chain acids also have even numbers of carbon atoms up to C_{36} together with some C_{18} hydroxy acids. The chief ester is myricyl palmitate. Also present are free acids (about 14%) and carbohydrates (about 12%) as well as approximately 1% free wax alcohols and stearic esters of fatty acids.

5 Structural Formula

See Section 4.

6 Functional Category

Coating agent; emulsion stabilizing agent; modified-release agent; stiffening agent.

7 Applications in Pharmaceutical Formulation or Technology

Yellow wax is used in food, cosmetics, and confectionery products. Its main use is in topical pharmaceutical formulations, where it is used at a concentration of 5–20%, as a stiffening agent in ointments and creams. Yellow wax is also employed in emulsions because it enables water to be incorporated into water-in-oil emulsions.

In some oral formulations yellow wax is used as a polishing agent for sugar-coated tablets. It is also used in sustained-release formulations. Yellow wax coatings can be used to affect the release rate of drug from ion-exchange resin beads,(1) and yellow wax has also been used in multiparticulate controlled-release dosage forms of chlorphenamine maleate.(2)

Yellow wax forms a soap with borax.

8 Description

Yellow or light brown pieces or plates with a fine-grained matt, noncrystalline fracture and a faint characteristic odor. The wax becomes soft and pliable when warmed.

The PhEur 7.4 describes yellow wax as the wax obtained by melting the walls of the honeycomb made by the honeybee, *Apis mellifera*, with hot water and removing foreign matter.

9 Pharmacopeial Specifications

See Table I.

Table I: Pharmacopeial specifications for yellow wax.

Test	JP XV	PhEur 7.4	USP35–NF30
Characters	+	+	—
Melting range	60–67°C	61–66°C	62–65°C
Relative density	—	≈0.960	—
Acid value	5–9 or 17–22	17–22	17–24
Ester value	—	70–80	72–79
Saponification value	80–100	87–102	—
Ceresin, paraffins, and certain other waxes	—	+	—
Purity	+	—	—
Glycerol and other polyols (as glycerol)	—	≤0.5%	—
Saponification cloud test	—	—	+
Fats or fatty acids, Japan wax, rosin, and soap	—	—	+

10 Typical Properties

Acid value 20
Arsenic ≤3 ppm
Density 0.95–0.96 g/cm^3
Flash point 245–258°C
Heavy metals ≤0.004%
Iodine number 8–11
Lead ≤10 ppm
Melting point 61–65°C
Peroxide value ≤8
Solubility Soluble in chloroform, ether, fixed oils, volatile oils, and warm carbon disulfide; sparingly soluble in ethanol (95%); practically insoluble in water.
Spectroscopy
NIR spectra *see* Figure 1.
Unsaponified matter 52–55%
Viscosity (kinematic) 1470 mm^2/s (1470 cSt) at 99°C

11 Stability and Storage Conditions

When the wax is heated above 150°C esterification occurs with a consequent lowering of acid value and elevation of melting point. Yellow wax is stable when stored in a well-closed container, protected from light.

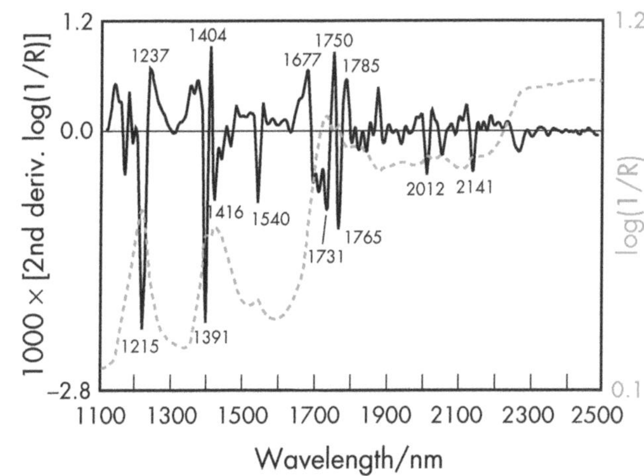

Figure 1: Near-infrared spectrum of yellow wax measured by reflectance.

W

12 Incompatibilities

Incompatible with oxidizing agents.

13 Method of Manufacture

Yellow wax is a natural secretion of bees (*Apis mellifera* Linné (Fam. Apidae)) and is obtained commercially from honeycombs. Honey is abstracted from combs either by draining or centrifugation, and water is added to the remaining wax to remove soluble impurities. Hot water is then added to form a floating melt, which is strained to remove foreign matter. The wax is then poured into flat dishes or molds to cool and harden.

14 Safety

Yellow wax is generally regarded as an essentially nontoxic and nonirritant material, and is used in both topical and oral formulations. However, hypersensitivity reactions attributed to contaminants in the wax, although rare, have been reported.[3,4]

15 Handling Precautions

Observe normal precautions appropriate to the circumstances and quantity of material handled.

16 Regulatory Status

GRAS listed. Accepted for use as a food additive in Europe. Included in the FDA Inactive Ingredients Database (oral capsules and tablets, and topical preparations). Included in nonparenteral medicines licensed in the UK. Included in the Canadian Natural Health Products Ingredients Database.

17 Related Substances

Paraffin; wax, microcrystalline; wax, white.

18 Comments

Studies have shown that yellow wax, when added to suppository formulations, increased the melting point of the preparation significantly and decreased the rate of release of the active substance.[5] *See also* White Wax.

A specification for yellow beeswax is contained in the *Food Chemicals Codex* (FCC).[6]

19 Specific References

1 Motycka S, Nairn J. Influence of wax coatings on release rate of anions from ion-exchange resin beads. *J Pharm Sci* 1978; **67**: 500–503.
2 Griffin EN, Niebergall PJ. Release kinetics of a controlled-release multiparticulate dosage form prepared using a hot-melt fluid bed coating method. *Pharm Dev Technol* 1999; **4**(1): 117–124.
3 Cronin E. Contact dermatitis from cosmetics. *J Soc Cosmet Chem* 1967; **18**: 681–691.
4 Rothenborg HW. Occupational dermatitis in beekeeper due to poplar resins in beeswax. *Arch Dermatol* 1967; **95**: 381–384.
5 Murrukmihadi M. Effect of cera flava on the release of sodium salicylate from suppository dosage form. *Indonesian J Pharm* 1999; **10**(3): 135–139.
6 *Food Chemicals Codex*, 7th edn. Bethesda, MD: United States Pharmacopeia, 2010: 84.

20 General References

Puleo SL. Beeswax. *Cosmet Toilet* 1987; **102**(6): 57–58.

21 Author

AH Kibbe.

22 Date of Revision

1 March 2012.

Xanthan Gum

1 Nonproprietary Names

BP: Xanthan Gum
PhEur: Xanthan Gum
USP–NF: Xanthan Gum

2 Synonyms

Corn sugar gum; E415; *Grindsted*; *Keldent*; *Keltrol*; polysaccharide B-1459; *Rhodicare S*; *Rhodigel*; *Rhodopol*; *Satiaxane U*; *Vanzan NF*; xanthani gummi; *Xantural*.

3 Chemical Name and CAS Registry Number

Xanthan gum [11138-66-2]

4 Empirical Formula and Molecular Weight

$(C_{35}H_{49}O_{29})_n$ approximately 1 000 000

The USP35–NF30 describes xanthan gum as a high molecular weight polysaccharide gum. It contains D-glucose and D-mannose as the dominant hexose units, along with D-glucuronic acid, and is prepared as the sodium, potassium, or calcium salt.

5 Structural Formula

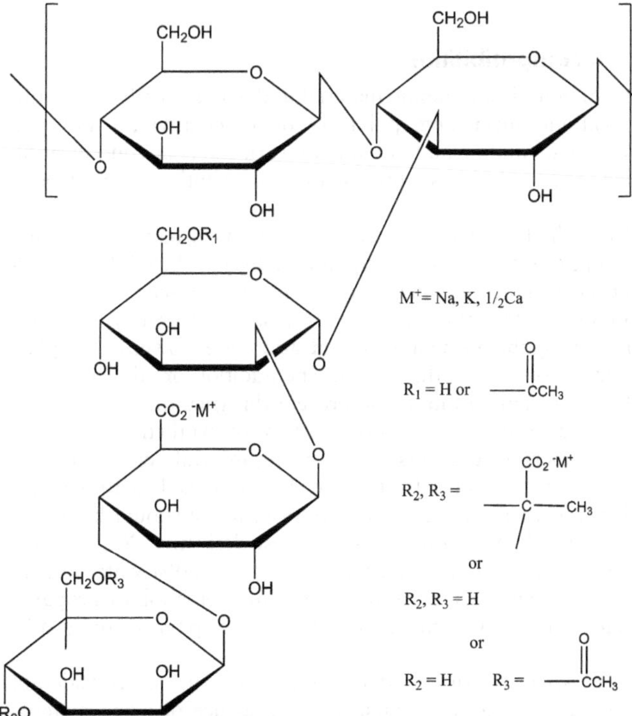

Each xanthan gum repeat unit contains five sugar residues: two glucose, two mannose, and one glucuronic acid. The polymer backbone consists of linear (1→4) linked β-D-glucose units, and is therefore identical in structure to cellulose. Trisaccharide side chains on alternating anhydroglucose units distinguish xanthan from cellulose. Each side chain comprises a glucuronic acid residue between two mannose units. Approximately 50% of terminal mannose residues are pyruvated and nonterminal residues usually carry an acetyl group at C-6. The glucuronic acid and pyruvic acid groups on the side chains give xanthan gum its anionic charge. The interaction of these anionic side chains with the polymer backbone and with each other determines the beneficial properties of xanthan gum solutions. The resulting stiff polymer chain may exist in solution as a single, double, or triple helix that interacts with other xanthan gum molecules to form complex, loosely bound networks.[1,2]

6 Functional Category

Emulsifying agent; gelling agent; modified-release agent; mucoadhesive; suspending agent; viscosity-increasing agent.

7 Applications in Pharmaceutical Formulation or Technology

Xanthan gum is widely used in oral and topical pharmaceutical formulations, cosmetics, and foods as a suspending agent, thickening agent and emulsifying agent.

Xanthan gum has been used as a suspending agent for conventional, dry and sustained-release suspensions.[3,4] Although primarily used as a suspending agent, xanthan gum has also been used to prepare sustained-release matrix tablets,[5–8] with chitosan,[9–11] guar gum,[6,12] galactomannan,[13] tragacanth,[14] locust bean gum,[15] and sodium alginate.[16]

A synergistic interaction occurs between xanthan gum and galactomannans such as guar gum, locust bean gum, and cassia gum, resulting in enhanced viscosity or gelation, which is valuable in applications where very high viscosity or gel formation is needed. Optimum synergistic effects are obtained with xanthan gum : guar gum ratios between 1 : 1 and 1 : 9. Similarly, mixtures of xanthan gum and magnesium aluminum silicate in ratios between 1 : 2 and 1 : 9 produce optimum synergistic rheological properties.[17]

Xanthan gum is also used as a hydrocolloid in the food industry, and in cosmetics it has been used as a thickening agent in shampoo.[18] Polyphosphate with xanthum gum in soft drinks is suggested to be effective at reducing erosion of enamel.[19,20]

See also Section 18.

8 Description

Xanthan gum occurs as a cream- or white-colored, odorless, free-flowing, fine powder.

9 Pharmacopeial Specifications

See Table I.

10 Typical Properties

Density 0.8 g/cm^3
Freezing point 0°C for a 1% w/v aqueous solution.
Heat of combustion 14.6 J/g (3.5 cal/g)
Melting point Chars at 270°C.
Particle size distribution Various grades with different particle sizes are available; *see* Table II.
Refractive index n_D^{20} = 1.333 (1% w/v aqueous solution).

Table I: Pharmacopeial specifications for xanthan gum.

Test	PhEur 7.4	USP35–NF30
Identification	+	+
Characters	+	–
pH	6.0–8.0	–
Viscosity	≥600 mPa s	≥600 mPa s
Propan-2-ol	≤750 ppm	≤0.075%
Other polysaccharides	+	–
Loss on drying	≤15.0%	≤15.0%
Total ash	6.5–16.0%	6.5–16.0%
Microbial contamination	+	+
Bacteria	$\leq 10^3$ cfu/g	–
Fungi	$\leq 10^2$ cfu/g	–
Pyruvic acid	≤1.5%	≤1.5%
Arsenic	–	≤3 μg/g
Lead	–	≤5 μg/g
Heavy metals	–	≤0.003%
Assay	–	91.0–108.0%

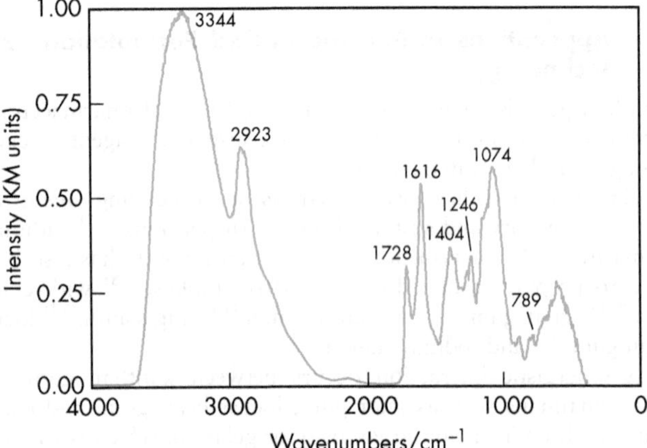

Figure 1: Infrared spectrum of xanthan gum measured by diffuse reflectance. Adapted with permission of Informa Healthcare.

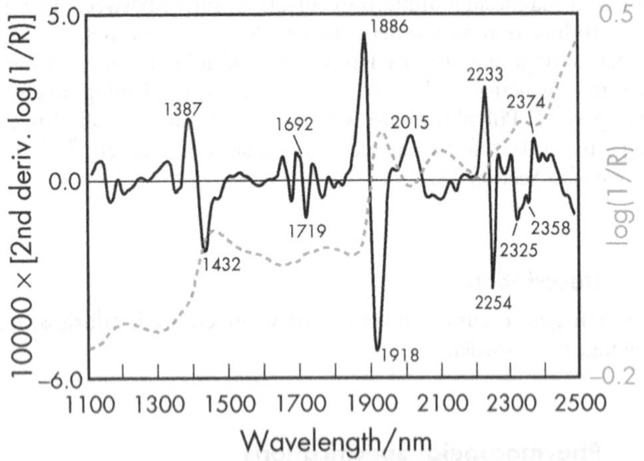

Figure 2: Near-infrared spectrum of xanthan gum measured by reflectance.

Solubility Soluble in water giving a highly viscous solution; practically insoluble in organic solvents.

Specific gravity 1.600 at 25°C

Spectroscopy

IR spectra *see* Figure 1.

NIR spectra *see* Figure 2.

Viscosity (dynamic) 1200–1600 mPa s (1200–1600 cP) for a 1% w/v aqueous solution at 25°C.

Table II: Particle size distribution of selected commercially available grades of xanthan gum.

Grade	Particle size (μm)
Vanzan NF	180
Vanzan NF-F	75
Vanzan NF-C	180
Xantural 180	180
Xantural 75	75
Xantural 11K	1100

11 Stability and Storage Conditions

Xanthan gum is a stable material. Aqueous solutions are stable over a wide pH range (pH 2–12), with slightly lower viscosity at extreme temperatures, although they demonstrate maximum stability at pH 4–10 and temperatures of 10–60°C. Xanthan gum solutions of less than 1% w/v concentration may be adversely affected by higher than ambient temperatures: for example, viscosity is reduced. Xanthan gum provides the same thickening, stabilizing, and suspending properties during long-term storage at elevated temperatures as it does at ambient conditions. In addition, it ensures excellent freeze–thaw stability. Solutions are also stable in the presence of enzymes, salts, acids, and bases.

Xanthan gum gels show pseudoplastic behavior, with the shear thinning being directly proportional to the shear rate. Viscosity returns to normal immediately on release of shear stress.

The bulk material should be stored in a well-closed container in a cool, dry place.

12 Incompatibilities

Xanthan gum is an anionic material and is not usually compatible with cationic surfactants, polymers, or preservatives, as precipitation occurs. Anionic and amphoteric surfactants at concentrations above 15% w/v cause precipitation of xanthan gum from a solution.

Under highly alkaline conditions, polyvalent metal ions such as calcium cause gelation or precipitation; this may be inhibited by the addition of a glucoheptonate sequestrant. The presence of low levels of borates (<300 ppm) can also cause gelation. This may be avoided by increasing the boron ion concentration or by lowering the pH of a formulation to less than pH 5. The addition of ethylene glycol, sorbitol, or mannitol may also prevent this gelation.

Xanthan gum is compatible with most synthetic and natural viscosity-increasing agents, many strong mineral acids, and up to 30% inorganic salts. If it is to be combined with cellulose derivatives, then xanthan gum free of cellulase should be used to prevent depolymerization of the cellulose derivative. Xanthan gum solutions are stable in the presence of up to 50% water-miscible organic solvents such as acetone, methanol, ethanol, or propan-2-ol. However, above this concentration precipitation or gelation occurs.

The viscosity of xanthan gum solutions is considerably increased, or gelation occurs, in the presence of some materials such as ceratonia, guar gum, and magnesium aluminum silicate.[17] This effect is most pronounced in deionized water and is reduced by the presence of salt. This interaction may be desirable in some instances and can be exploited to reduce the amount of xanthan gum used in a formulation; *see* Section 7.

Xanthan gum is incompatible with oxidizing agents such as persulfates, peroxides, and hypochlorites as they cause depolymerization, which is accelerated by heat and catalyzed by certain transition metals such as ferrous ions.

Xanthan gum interacts with some active ingredients such as amitriptyline, tamoxifen, and verapamil.[21]

13 Method of Manufacture

Xanthan gum is a high molecular weight exocellular polysaccharide derived from the bacterium *Xanthomonas campestris* using a natural, aerobic fermentation process. The process is conducted in a sterile environment where pH, oxygen content, and temperature are rigorously controlled. After fermentation is complete, the broth is sterilized and the gum is recovered by precipitation with isopropyl alcohol. It is then dried, milled, and packaged under aseptic conditions.

14 Safety

Xanthan gum is widely used in oral and topical pharmaceutical formulations, cosmetics, and food products, and is generally regarded as nontoxic and nonirritant at the levels employed as a pharmaceutical excipient.

Safety studies have indicated no adverse effects from ingestion of high doses of xanthan gum. The Joint FAO/WHO Expert Committee on Food Additives has established an acceptable daily intake (ADI) of'not specified' for xanthan gum.[22] An estimated acceptable daily intake for xanthan gum was previously set by the WHO at up to 10 mg/kg body-weight.[23]

LD_{50} (dog, oral): >20 g/kg[23]
LD_{50} (rat, oral): >45 g/kg
LD_{50} (mouse, oral): >1 g/kg[24]
LD_{50} (mouse, IP): >50 mg/kg[24]
LD_{50} (mouse, IV): 100–250 mg/kg

15 Handling Precautions

Observe normal precautions appropriate to the circumstances and quantity of material handled. Eye protection and gloves are recommended.

16 Regulatory Status

GRAS listed. Accepted for use as a food additive in Europe. Included in the FDA Inactive Ingredients Database (oral solutions, suspensions, and tablets; rectal and topical preparations). Included in nonparenteral medicines licensed in the UK. Included in the Canadian Natural Health Products Ingredients Database.

17 Related Substances

Ceratonia; guar gum.

18 Comments

Xanthan gum is available in several different grades that have varying particle sizes. Fine-mesh grades of xanthan gum are used in applications where high solubility is desirable since they dissolve rapidly in water. However, fine-mesh grades disperse more slowly than coarse grades and are best used dry blended with the other ingredients of a formulation. In general, it is preferable to dissolve xanthan gum in water first and then add the other ingredients of a formulation.

Xanthan gum in combination with pH-sensitive polymers[25] and with Konjac glucomannan[26,27] has been investigated for controlled colonic drug delivery. Xanthan gum with boswellia (3 : 1)[28] and guar gum (10 : 20)[29] have shown the best release profiles for the colon-specific compression coated systems of 5-fluorouracil for the treatment of colorectal cancer.

Xanthan gum has been studied with guar gum[30] and hypromellose[31–34] for the development of a floating drug delivery system. It has also been converted to sodium carboxymethyl xanthan gum and crosslinked with aluminum ions to prepare microparticles as a carrier for protein delivery.[35]

Xanthan gum increases gel strength and bioadhesion through interaction with mucin, with prolonged retention of the dosage form and increased drug bioavailability. A thermoreversible *in-situ* gel formulated in combination with mucoadhesive polymers such as xanthan gum has been studied for sustained ocular drug delivery.[36,37]

Xanthan gum has been found to increase the bioadhesive strength in vaginal formulations.[38,39] Xanthan gum alone or with carbopol 974P has been investigated as a mucoadhesive controlled-release excipient for buccal drug delivery.[40,41] Xanthan gum-based gels have been studied in the treatment of periodontal diseases.[42] Modified xanthan films, in combination with sodium alginate, have also been studied as a matrix system for transdermal delivery.[43,44]

Xanthan gum has been investigated as a gelling agent for topical formulations incorporating solid lipid nanoparticles of vitamin A[45] or microemulsions of ibuprofen.[46] A combined polymer system consisting of xanthan gum and carboxymethylcellulose with a polyvinyl pyrolidone-backboned polymer has also been investigated for relieving the symptoms of xerostomia.[47] Xanthan gum may also be used in spray-drying and freeze-drying processes for better results.[48,49]

Xanthan gum has been successfully used alone or in combination with agar for microbial culture media.[50]

Novel pH-sensitive hydrogel beads have been prepared using a copolymer of poly(acrylamide-g-xanthan) for targeting ketoprofen to the intestine.[51] These beads were able to retard drug release in the stomach, thus diminishing gastric side effects such as ulceration, hemorrhage and erosion of gastric mucosa.[51] Bioadhesive nasal inserts prepared from xanthan gum have a high potential as a new nasal dosage form for extended drug delivery.[52] Xanthan gum wafers have potential as drug delivery systems for suppurating wounds.[53,54]

The USP35–NF30 also includes a monograph for xanthan gum solution. A specification for xanthan gum is contained in the *Food Chemicals Codex* (FCC).[55]

The EINECS number for xanthan gum is 234-394-2. The PubChem Compound ID (CID) for xanthan gum is 7107.

19 Specific References

1 Jansson PE, *et al.* Structure of extracellular polysaccharide from *Xanthamonas campestris*. *Carbohydr Res* 1975; **45**: 275–282.
2 Melton LD, *et al.* Covalent structure of the polysaccharide from *Xanthamonas campestris*: evidence from partial hydrolysis studies. *Carbohydr Res* 1976; **46**: 245–257.
3 Junyaprasert VB, Manwiwattanakul G. Release profile comparison and stability of diltiazem-resin microcapsules in sustained release suspensions. *Int J Pharm* 2008; **352**(1–2): 81–91.
4 Gallardo LV, *et al.* Ondansetron: design and development of oral pharmaceutical suspensions. *Pharmazie* 2009; **64**(2): 90–93.
5 Mughal MA, *et al.* Guar gum, xanthan gum, and HPMC can define release mechanisms and sustain release of propranolol hydrochloride. *AAPS PharmSciTech* 2011; **12**(1): 77–87.
6 Hamza Yel-S, Aburahma MH. Design and *in vitro* evaluation of novel sustained-release double-layer tablets of lornoxicam: utility of cyclodextrin and xanthan gum combination. *AAPS PharmSciTech* 2009; **10**(4): 1357–1367.
7 Gohel MC, Bariya SH. Fabrication of triple-layer matrix tablets of venlafaxine hydrochloride using xanthan gum. *AAPS PharmSciTech* 2009; **10**(2): 624–630.
8 Sankalia JM, *et al.* Drug release and swelling kinetics of directly compressed glipizide sustained-release matrices: establishment of level A IVIVC. *J Control Release* 2008; **129**(1): 49–58.
9 Eftaiha AF, *et al.* Bioadhesive controlled metronidazole release matrix based on chitosan and xanthan gum. *Mar Drugs* 2010; **8**(5): 1716–1730.
10 Popa N, *et al.* Hydrogels based on chitosan-xanthan for controlled release of theophylline. *J Mater Sci Mater Med* 2010; **21**(4): 1241–1248.
11 Phaechamud T, Ritthidej GC. Formulation variables influencing drug release from layered matrix system comprising chitosan and xanthan gum. *AAPS PharmSciTech* 2008; **9**(3): 870–877.
12 Varshosaz J, *et al.* Use of natural gums and cellulose derivatives in production of sustained release metoprolol tablets. *Drug Deliv* 2006; **13**(2): 113–119.

13 Vendruscolo CW, *et al.* Xanthan and galactomannan (from *M. scabrella*) matrix tablets for oral controlled delivery of theophylline. *Int J Pharm* 2005; **296**(1–2): 1–11.
14 Rasul A, *et al.* Design, development and *in-vitro* evaluation of metoprolol tartrate tablets containing xanthan-tragacanth. *Acta Pol Pharm* 2010; **67**(5): 517–522.
15 Rajesh KS, *et al.* Effect of hydrophilic natural gums in formulation of oral-controlled release matrix tablets of propranolol hydrochloride. *Pak J Pharm Sci* 2009; **22**(2): 211–219.
16 Zeng WM. Oral controlled release formulation for highly water-soluble drugs: drug–sodium alginate–xanthan gum–zinc acetate matrix. *Drug Dev Ind Pharm* 2004; **30**(5): 491–495.
17 Kovacs P. Useful incompatibility of xanthan gum with galactomannans. *Food Technol* 1973; **27**(3): 26–30.
18 Howe AM, Flowers AE. Introduction to shampoo thickening. *Cosmet Toilet* 2000; **115**: 63–6668–69.
19 Hooper S, *et al.* A clinical study *in situ* to assess the effect of a food approved polymer on the erosion potential of drinks. *J Dent* 2007; **35**(6): 541–546.
20 Barbour ME, *et al.* An investigation of some food-approved polymers as agents to inhibit hydroxyapatite dissolution. *Eur J Oral Sci* 2005; **113**(6): 457–461.
21 Shanmugam S, *et al.* Natural polymers and their applications. *Natural Product Radiance* 2005; **4**(6): 478–481.
22 FAO/WHO. Safety Evaluation of certain food additives. Xanthan Gum (WHO Food Additives Series 21). http://www.inchem.org/documents/jecfa/jecmono/v21je13.htm (accessed 22 December 2011).
23 FAO/WHO. Evaluation of certain food additives and contaminants. Twenty-ninth report of the joint FAO/WHO expert committee on food additives. World Health Organ Tech Rep Ser 1986; No. 733.
24 Booth AN, *et al.* Physiologic effects of three microbial polysaccharides on rats. *Toxicol Appl Pharmacol* 1963; **5**: 478–484.
25 Asghar LF, *et al.* Colon specific delivery of indomethacin: effect of incorporating pH sensitive polymers in xanthan gum matrix bases. *AAPS PharmSciTech* 2009; **10**(2): 418–429.
26 Alvarez-Manceñido F, *et al.* Konjac glucomannan and konjac glucomannan/xanthan gum mixtures as excipients for controlled drug delivery systems. Diffusion of small drugs. *Int J Pharm* 2008; **349**(1–2): 11–18.
27 Alvarez-Manceñido F, *et al.* Konjac glucomannan/xanthan gum enzyme sensitive binary mixtures for colonic drug delivery. *Eur J Pharm Biopharm* 2008; **69**(2): 573–581.
28 Sinha VR, *et al.* Compression coated systems for colonic delivery of 5-fluorouracil. *J Pharm Pharmacol* 2007; **59**(3): 359–365.
29 Sinha VR, *et al.* Colonic drug delivery of 5-fluorouracil: an in vitro evaluation. *Int J Pharm* 2004; **269**(1): 101–108.
30 Patel VF, Patel NM. Statistical evaluation of influence of xanthan gum and guar gum blends on dipyridamole release from floating matrix tablets. *Drug Dev Ind Pharm* 2007; **33**(3): 327–334.
31 Jain S, *et al.* Development of a floating dosage form of ranitidine hydrochloride by statistical optimization technique. *J Young Pharm* 2010; 2(4): 342–349.
32 Bomma R, *et al.* Development and evaluation of gastroretentive norfloxacin floating tablets. *Acta Pharm* 2009; **59**(2): 211–221.
33 Jagdale SC, *et al.* Formulation and evaluation of gastroretentive drug delivery system of propranolol hydrochloride. *AAPS PharmSciTech* 2009; **10**(3): 1071–1079.
34 Patel A, *et al.* Development and *in vivo* floating behavior of verapamil HCl intragastric floating tablets. *AAPS PharmSciTech* 2009; **10**(1): 310–315.
35 Maiti S, *et al.* Controlled delivery of bovine serum albumin from carboxymethyl xanthan microparticles. *Pharm Dev Technol* 2009; **14**(2): 165–172.
36 Shastri DH, *et al.* Design and development of thermoreversible ophthalmic *in situ* hydrogel of moxifloxacin hydrochloride. *Curr Drug Deliv* 2010; 7(3): 238–243.
37 Shastri D, *et al.* Studies on *in situ* hydrogel: a smart way for safe and sustained ocular drug delivery. *J Young Pharm* 2010; **2**(2): 116–120.
38 Dobaria N, Mashru R. Design and *in vitro* evaluation of a novel bioadhesive vaginal drug delivery system for clindamycin phosphate. *Pharm Dev Technol* 2010; **15**(4): 405–414.
39 Ahmad FJ, *et al.* Development and *in vitro* evaluation of an acid buffering bioadhesive vaginal gel for mixed vaginal infections. *Acta Pharm* 2008; **58**(4): 407–419.
40 Sakeer K, *et al.* Use of xanthan and its binary blends with synthetic polymers to design controlled release formulations of buccoadhesive nystatin tablets. *Pharm Dev Technol* 2010; **15**(4): 360–368.
41 Singh S, *et al.* Preparation and evaluation of buccal bioadhesive tablets containing clotrimazole. *Curr Drug Deliv* 2008; **5**(2): 133–141.
42 Paolantonio M, *et al.* Clinical, microbiologic, and biochemical effects of subgingival administration of a xanthan-based chlorhexidine gel in the treatment of periodontitis: a randomized multicenter trial. *J Periodontol* 2009; **80**(9): 1479–1492.
43 Mundargi RC, *et al.* Evaluation and controlled release characteristics of modified xanthan films for transdermal delivery of atenolol. *Drug Dev Ind Pharm* 2007; **33**(1): 79–90.
44 Rajesh N, Siddaramaiah. Feasibility of xanthan gum-sodium alginate as a transdermal drug delivery system for domperidone. *J Mater Sci Mater Med* 2009; **20**(10): 2085–2089.
45 Pople PV, Singh KK. Development and evaluation of topical formulation containing solid lipid nanoparticles of vitamin A. *AAPS Pharm Sci Tech* 2006; 7(4): 91.
46 Chen H, *et al.* Microemulsion-based hydrogel formulation of ibuprofen for topical delivery. *Int J Pharm* 2006; **315**(1–2): 52–58.
47 Corcoran RA, *et al.* Evaluation of a combined polymer system for use in relieving the symptoms of xerostomia. *J Clin Dent* 2006; **17**(2): 34–38.
48 Patel N, *et al.* Spray-dried insulin particles retain biological activity in rapid *in-vitro* assay. *J Pharm Pharmacol* 2001; **53**(10): 1415–1418.
49 Corveleyn S, Remon JP. Stability of freeze-dried tablets at different relative humidities. *Drug Dev Ind Pharm* 1999; **25**(9): 1005–1013.
50 Babber SB, Jain R. Xanthan gum: an economical partial substitute for agar in microbial culture media. *Curr Microbiol* 2006; **52**(4): 287–292.
51 Kulkarni RV, Sa B. Enteric delivery of ketoprofen through functionally modified poly(acrylamide-grafted-xanthan)-based pH-sensitive hydrogel beads: preparation, in vitro and in vivo evaluation. *J Drug Target* 2008; **16**(2): 167–177.
52 Bertram U, Bodmeier R. In situ gelling, bioadhesive nasal inserts for extended drug delivery: in vitro characterization of a new nasal dosage form. *Eur J Pharm Sci* 2006; **27**(1): 62–71.
53 Matthews KH, *et al.* Gamma-irradiation of lyophilised wound healing wafers. *Int J Pharm* 2006; **313**(1–2): 78–86.
54 Matthews KH, *et al.* Lyophilised wafers as a drug delivery system for wound healing containing methylcellulose as a viscosity modifier. *Int J Pharm* 2005; **289**(1–2): 51–62.
55 *Food Chemicals Codex*, 7th edn. Bethesda, MD: United States Pharmacopeia, 2010.

20 General References

CP Kelco. Product literature: Xanthan Gum, 2008.
RT Vanderbilt Company Inc. Product literature: *Vanzan* xanthan gum. http://www.rtvanderbilt.com/vanzan.pdf (accessed 22 December 2011).
Sworn G. Xanthan Gum. In: Philips GO, Williams PA, eds. *Handbook of Hydrocolloids*. Cambridge, UK: Woodhead Publishing Ltd (also Boca Raton, FL: CRC Press LLC), 2000; 103-116.

21 Authors

HC Shah, KK Singh.

22 Date of Revision

1 March 2012.

X

Xylitol

1 Nonproprietary Names

BP: Xylitol
JP: Xylitol
PhEur: Xylitol
USP–NF: Xylitol

2 Synonyms

E967; *Klinit*; *meso*-xylitol; xilitol; *Xylifin*; *Xylisorb*; xylit; *Xylitab*; xylite; *Xylitolo*; xylitolum.

3 Chemical Name and CAS Registry Number

xylo-Pentane-1,2,3,4,5-pentol [87-99-0]

4 Empirical Formula and Molecular Weight

$C_5H_{12}O_5$ 152.15

5 Structural Formula

OH OH
HO OH
OH

6 Functional Category

Coating agent; emollient; flavor enhancer; humectant; sweetening agent; tablet and capsule diluent; taste-masking agent.

7 Applications in Pharmaceutical Formulation or Technology

Xylitol is used as a noncariogenic sweetening agent in a variety of pharmaceutical dosage forms, including tablets, syrups, and coatings. It is also widely used as an alternative to sucrose in foods and as a base for medicated confectionery.

Xylitol is finding increasing application in chewing gum,[1,2] mouthrinses,[3] and toothpastes[4] as an agent that decreases dental plaque and tooth decay (dental caries). Unlike sucrose, xylitol is not fermented into cariogenic acid end products[5] and it has been shown to reduce dental caries by inhibiting the growth of the cariogenic bacterium, *Streptococcus mutans*.[6,7] As xylitol has an equal sweetness intensity to sucrose, combined with a distinct cooling effect upon dissolution of the crystal, it is highly effective in enhancing the flavor of tablets and syrups and masking the unpleasant or bitter flavors associated with some pharmaceutical actives and excipients.

In topical cosmetic and toiletry applications, xylitol is used primarily for its humectant and emollient properties, although it has also been reported to enhance product stability through a combination of potentiation of preservatives and its own bacteriostatic and bactericidal properties.

Granulates of xylitol are used as diluents in tablet formulations, where they can provide chewable tablets with a desirable sweet taste and cooling sensation, without the 'chalky' texture experienced with some other tablet diluents. Xylitol solutions are employed in tablet-coating applications at concentrations in excess of 65% w/w. Xylitol coatings are stable and provide a sweet-tasting and durable hard coating.

In liquid preparations, xylitol is used as a sweetening agent and vehicle for sugar-free formulations. In syrups, it has a reduced tendency to 'cap-lock' by effectively preventing crystallization around the closures of bottles. Xylitol also has a lower water activity and a higher osmotic pressure than sucrose, therefore enhancing product stability and freshness. In addition, xylitol has also been demonstrated to exert certain specific bacteriostatic and bactericidal effects, particularly against common spoilage organisms.[8,9]

8 Description

Xylitol occurs as a white, granular solid comprising crystalline, equidimensional particles having a mean diameter of about 0.4–0.6 mm. It is odorless, with a sweet taste that imparts a cooling sensation. Xylitol is also commercially available in powdered form, and several granular, directly compressible forms.[10] *See also* Section 17.

9 Pharmacopeial Specifications

See Table I.

10 Typical Properties

Acidity/alkalinity pH = 5.0–7.0 (10% w/v aqueous solution).
Boiling point 215–217°C
Compressibility *see* Figure 1. Crystalline xylitol, under the same test conditions as illustrated in Figure 1, produces 12.5 mm tablets of 40 N hardness at 20 kN compression force.
Density (true) 1.52 g/cm^3
Density (bulk)
0.8–0.85 g/cm^3 for crystalline xylitol;
0.5–0.7 g/cm^3 for directly compressible granulated grades.
Flowability Flow characteristics vary depending upon the particle size of xylitol used. Fine-milled grades tend to be relatively poorly flowing, while granulated grades have good flow properties.

SEM 1: Excipient: xylitol (unsieved); magnification: 60×.

Table I: Pharmacopeial specifications for xylitol.

Test	JP XV	PhEur 7.4	USP35–NF30
Identification	+	+	+
Characters	+	+	—
Clarity and color of solution	+	+	—
Water	—	≤1.0%	≤0.5%
pH (50% w/w solution)	5.0–7.0	—	—
Melting point	93.0–95.0°C	92–96°C	—
Residue on ignition	≤0.1%	—	≤0.5%
Chloride	≤0.005%	—	—
Sulfate	≤0.006%	—	—
Nickel	+	≤1 ppm	—
Arsenic	≤1.3 ppm	—	—
Heavy metals	≤5 ppm	—	≤10 ppm
Reducing sugars (as dextrose)	+	≤0.2%	≤0.2%
Other polyols	—	—	≤2.0%
Related substances	—	≤2.0%	—
Lead	—	≤0.5 ppm	—
Bacterial endotoxins[(a)]	—	≤2.5 IU/g	—
Conductivity	—	≤20 μS cm^{-1}	—
Assay (anhydrous basis)	≥98.0%	98.0–102.0%	98.5–101.0%

(a) If intended for use in parenteral products.

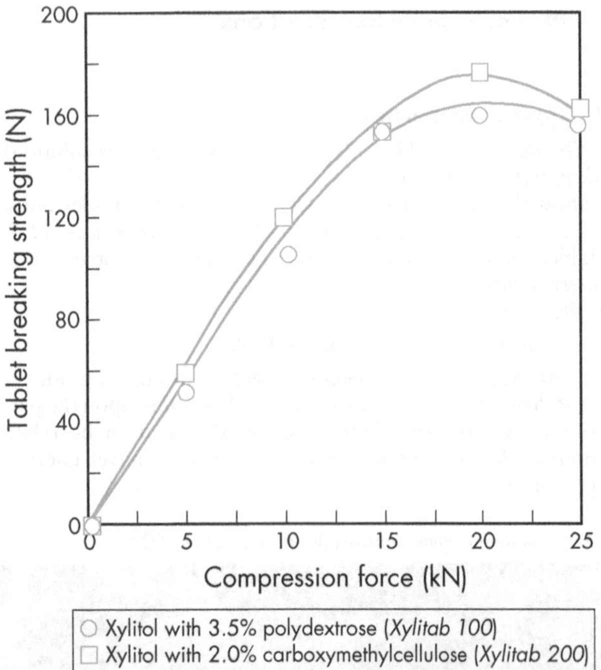

Figure 1: Compression characteristics of *Xylitab 100* and *Xylitab 200* (Danisco A/S).

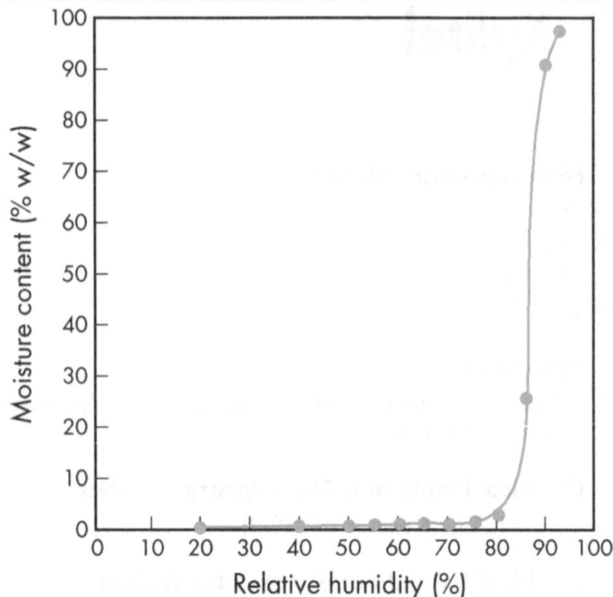

Figure 2: Moisture sorption isotherm of xylitol at 20°C.

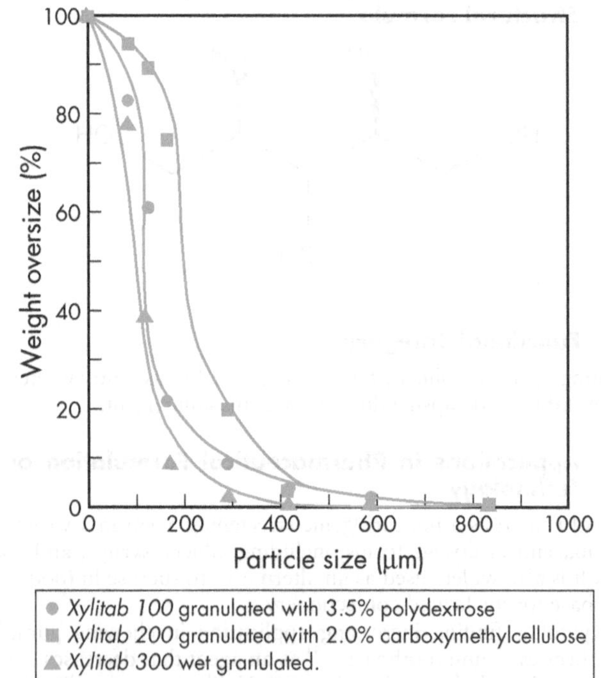

Figure 3: Particle size distribution of granulated xylitol (*Xylitab*, Danisco A/S).

Heat of solution −157.1 kJ/kg (−36.7 cal/g)

Melting point 92.0–96.0°C

Moisture content Xylitol is a moderately hygroscopic powder under normal conditions; *see also* Figure 2. At 20°C and 52% relative humidity, the equilibrium moisture content of xylitol is 0.1% w/w. After drying in a vacuum, over P_2O_5 at 80°C for 4 hours, xylitol loses less than 0.5% w/w water.

Osmolarity A 4.56% w/v aqueous solution is iso-osmotic with serum.

Particle size distribution The particle size distribution of xylitol depends upon the grade selected. Normal crystalline material typically has a mean particle size of 0.4–0.6 mm. Milled grades are commercially available that offer mean particle sizes as low as 50 μm. For particle size distributions of granulated xylitol, *see* Figure 3.

Solubility *see* Table II.

Specific rotation Not optically active.

Spectroscopy

NIR spectra *see* Figure 4.

Viscosity (dynamic) *see* Figure 5.

11 Stability and Storage Conditions

Xylitol is stable to heat but is marginally hygroscopic. Carameliza-tion can occur only if it is heated for several minutes near its boiling

Table II: Solubility of xylitol.

Solvent	Solubility at 20°C
Ethanol	1 in 80
Glycerin	Very slightly soluble
Methanol	1 in 16.7
Peanut oil	Very slightly soluble
Propan-2-ol	1 in 500
Propylene glycol	1 in 15
Pyridine	Soluble
Water	1 in 1.6

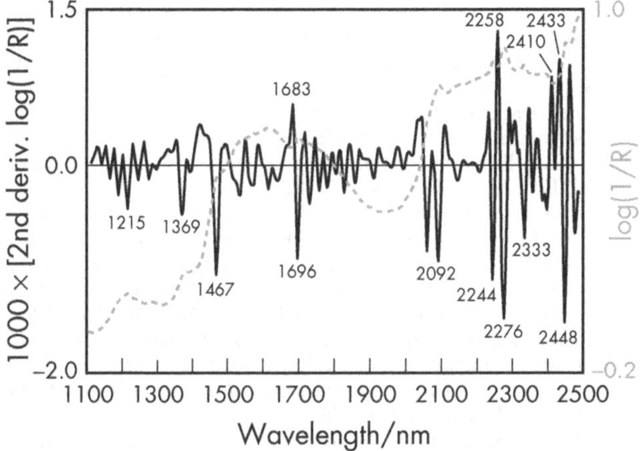

Figure 4: Near-infrared spectrum of xylitol measured by reflectance.

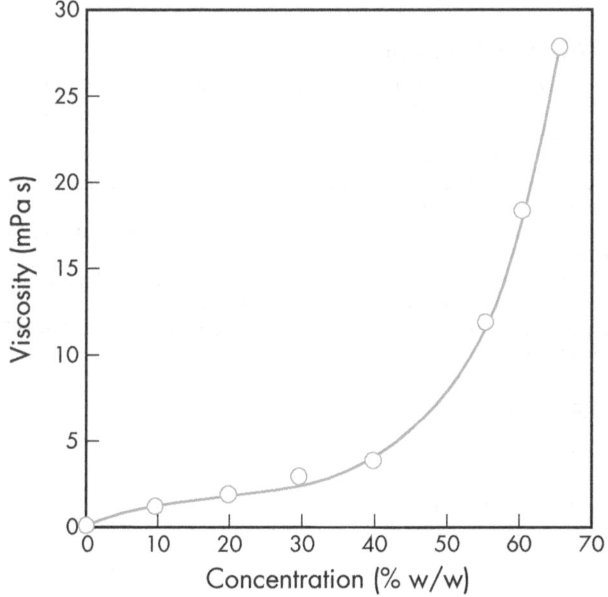

Figure 5: Viscosity of aqueous xylitol solutions at 20°C.

point. Crystalline material is stable for at least 3 years if stored at less than 65% relative humidity and 25°C. Milled and specialized granulated grades of xylitol have a tendency to cake and should therefore be used within 9 to 12 months. Aqueous xylitol solutions have been reported to be stable, even on prolonged heating and storage. Since xylitol is not utilized by most microorganisms, products made with xylitol are usually safe from fermentation and microbial spoilage.[8,9]

Xylitol should be stored in a well-closed container in a cool, dry place.

12 Incompatibilities

Xylitol is incompatible with oxidizing agents.

13 Method of Manufacture

Xylitol occurs naturally in many fruits and berries, although extraction from such sources is not considered to be commercially viable. Industrially, xylitol is most commonly derived from various types of hemicellulose obtained from such sources as wood, corn cobs, cane pulp, seed hulls, and shells. These materials typically contain 20–35% xylan, which is readily converted to xylose (wood sugar) by hydrolysis. This xylose is subsequently converted to xylitol via hydrogenation (reduction). Following the hydrogenation step, there are a number of separation and purification steps that ultimately yield high-purity xylitol crystals. The nature of this process, and the stringent purification procedures employed, result in a finished product with a very low impurity content. Potential impurities that may appear in small quantities are mannitol, sorbitol, galactitol, or arabitol.

Less commonly employed methods of xylitol manufacture include the conversion of glucose (dextrose) to xylose followed by hydrogenation to xylitol, and the microbiological conversion of xylose to xylitol.

14 Safety

Xylitol is used in oral pharmaceutical formulations, confectionery, and food products, and is generally regarded as an essentially nontoxic, nonallergenic, and nonirritant material.

Xylitol has an extremely low relative glycemic response and is metabolized independently of insulin. Following ingestion of xylitol, the blood glucose and serum insulin responses are significantly lower than following ingestion of glucose or sucrose. These factors make xylitol a suitable sweetener for use in diabetic or carbohydrate-controlled diets.[11]

Up to 100 g of xylitol in divided oral doses may be tolerated daily, although, as with other polyols, large doses may have a laxative effect. The laxative threshold depends on a number of factors, including individual sensitivity, mode of ingestion, daily diet, and previous adaptation to xylitol. Single doses of 20–30 g and daily doses of 0.5–1.0 g/kg body-weight are usually well tolerated by most individuals. Approximately 25–50% of the ingested xylitol is absorbed, with the remaining 50–75% passing to the lower gut, where it undergoes indirect metabolism via fermentative degradation by the intestinal flora.

An acceptable daily intake for xylitol of 'not specified' has been set by the WHO since the levels used in foods do not represent a hazard to health.[12]

LD_{50} (mouse, IP): 22.1 g/kg[13,14]
LD_{50} (mouse, IV): 12 g/kg
LD_{50} (mouse, oral): 12.5 g/kg
LD_{50} (rat, oral): 17.3 g/kg
LD_{50} (rat, IV): 10.8 g/kg
LD_{50} (rabbit, oral): 16.5 g/kg
LD_{50} (rabbit, IV): 4 g/kg

15 Handling Precautions

Observe normal precautions appropriate to the circumstances and quantity of material handled. Xylitol may cause transient gastrointestinal discomfort if ingested in large quantities; and may also be irritant to the eyes. Eye protection and gloves are recommended. Conventional dust-control practices should be employed. Xylitol is flammable, but does not ignite readily.

16 Regulatory Status

GRAS listed. Approved for use as a food additive in over 70 countries worldwide, including Europe, the USA and Japan.

Included in the FDA Inactive Ingredients Database (oral solution, chewing gum). Included in nonparenteral medicines licensed in the UK and USA. Included in the Canadian Natural Health Products Ingredients Database.

17 Related Substances

Various directly compressible forms of xylitol that contain other excipients are commercially available, e.g. *Xylitab 100*, which contains 3.5% polydextrose, and *Xylitab 200*, which contains 2.0% carboxymethylcellulose. A directly compressible form of pure xylitol called *Xylitab 300* is also available, which is produced via wet granulation.

Pyrogen-free grades of xylitol suitable for parenteral use are also commercially available.

18 Comments

The sweetening power of xylitol is approximately equal to that of sucrose, although it has been shown to be pH-, concentration-, and temperature-dependent; xylitol is approximately 2.5 times as sweet as mannitol.

Xylitol is highly chemically stable, meaning that it will not interact with pharmaceutical actives or excipients, and can be utilized over a wide pH range (pH 1–11).

Xylitol has a negative heat of solution that is far larger than that of other alternative sweetening agents; *see* Table III. Because of this, xylitol produces an intense cooling effect as the crystalline material dissolves. Xylitol's combination of sweetness and cooling can create product appeal while helping to mask the undesirable taste of many pharmaceutical actives or excipients.

Therapeutically, xylitol is used as an energy source for intravenous infusion therapy following trauma.[15]

A specification for xylitol is contained in the *Food Chemicals Codex* (FCC).[16]

The EINECS number for xylitol is 201-788-0. The PubChem Compound ID (CID) for xylitol is 6912.

Table III: Comparison of the heat of solution of selected sweetening agents.

Sweetening agent	Heat of solution (kJ/kg)
Lactitol (anhydrous)	−35.0
Maltitol	−69.2
Mannitol	−120.9
Sorbitol	−106.3
Sucrose	−23.0
Xylitol	−157.1

19 Specific References

1 Tanzer JM. Xylitol chewing gum and dental caries. *Int Dent J* 1995; 45(1): 65–76.

2 Soderling E, *et al.* Effects of xylitol, xylitol-sorbitol, and placebo chewing gums on the plaque of habitual xylitol consumers. *Eur J Oral Sci* 1997; **105**(2): 170–177.

3 Cobanera A, *et al.* Xylitol-sodium fluoride: effect on plaque. *J Dent Res* 1987; **66**: 814.

4 Sintes JL, *et al.* Enhanced anticaries efficacy of a 0.243% sodium fluoride/10% xylitol/silica dentifrice: 3-year clinical results. *Am J Dent* 1995; 8: 231–235.

5 Trahan L. Xylitol: a review of its action on mutans streptococci and dental plaque – its clinical significance. *Int Dent J* 1995; **45**(1): 77–92.

6 Hayes C. The effect of non-cariogenic sweeteners on the prevention of dental caries: a review of the evidence. *J Dent Educ* 2001; **65**(10): 1106–1109.

7 Makinen KK, *et al.* Properties of whole saliva and dental plaque in relation to 40-month consumption of chewing gums containing xylitol, sorbitol and sucrose. *Caries Res* 1996; **30**(3): 180–188.

8 Emodi A. Xylitol: its properties and food applications. *Food Technol* 1978; **Jan**: 28–32.

9 Makinen KK, Soderling E. Effect of xylitol on some food spoilage microorganisms. *J Food Sci* 1981; **46**(3): 950–951.

10 Garr JSM, Rubinstein MH. Direct compression characteristics of xylitol. *Int J Pharm* 1990; **64**: 223–226.

11 Natah SS, *et al.* Metabolic response to lactitol and xylitol in healthy men. *Am J Clin Nutr* 1997; **65**(4): 947–950.

12 FAO/WHO. Evaluation of certain food additives and contaminants. Twenty-seventh report of the joint FAO/WHO expert committee on food additives. *World Health Organ Tech Rep Ser* 1983; No. 696.

13 Sweet DV, ed. *Registry of Toxic Effects of Chemical Substances*. Cincinnati: US Department of Health, 1987: 5127–5128.

14 Lewis RJ, ed. *Sax's Dangerous Properties of Industrial Materials*, 11th edn. New York: Wiley, 2004: 3707.

15 Georgieff M, *et al.* Xylitol, an energy source for intravenous nutrition after trauma. *J Parenter Enteral Nutr* 1985; **9**: 199–209.

16 *Food Chemicals Codex*, 7th edn. Bethesda, MD: United States Pharmacopeia, 2010: 1093.

20 General References

Bond M, Dunning N. Xylitol. In: Mitchell H, ed. *Sweeteners and Sugar Alternatives in Food Technology*. Oxford: Blackwell Publishing, 2006: 295–324.

Counsell JN. *Xylitol*. London: Applied Science Publishers, 1978.

Danisco A/S. Product Information: *Xylitab*. http://www.danisco.com/ (accessed 15 June 2011).

O'Brien Nabors L, Gelardi RC, eds. *Alternative Sweeteners*, 2nd edn. New York: Marcel Dekker, 1991.

Thomas SE, *et al.* The use of xylitol as a carrier for liquid-filled hard-gelatin capsules. *Pharm Technol Int* 1991; 3(9): 36–40.

Ylikahri R. Metabolic and nutritional aspects of xylitol. *Adv Food Res* 1979; 25: 159–180.

21 Author

M Bond.

22 Date of Revision

1 March 2012.

Zein

1 Nonproprietary Names

USP–NF: Zein

2 Synonyms

—

3 Chemical Name and CAS Registry Number

Zein [9010-66-6]

4 Empirical Formula and Molecular Weight

Zein is a prolamin with a molecular weight of approximately 38 000.

5 Structural Formula

See Section 8.

6 Functional Category

Coating agent; modified-release agent; tablet and capsule binder.

7 Applications in Pharmaceutical Formulation or Technology

Zein is used as a tablet binder in wet-granulation processes or as a tablet-coating agent mainly as a replacement for shellac. It is used primarily as an enteric-coating agent or in extended-release oral tablet formulations.[1]

Zein is also used in food applications as a coating agent. *See* Table I.

Table I: Uses of zein.

Use	Concentration (%)
Tablet-coating agent	15
Tablet sealer	20
Wet granulation binder	30

8 Description

Zein is a prolamin obtained from corn (*Zea mays* Linné (Fam. Gramineae)). It occurs as a granular, straw- to pale yellow-colored amorphous powder or fine flakes and has a characteristic odor and bland taste.

Zein is a protein derivative that does not contain lysine or tryptophan. For the approximate amino acid content of zein, *see* Table II.

Table II: Approximate amino acid content of zein.

Alanine	8.3%	Leucine	19.3%
Arginine	1.8%	Methionine	2.0%
Asparagine	4.5%	Phenylalanine	6.8%
Cystine	0.8%	Proline	9.0%
Glutamic acid	1.5%	Serine	5.7%
Glutamine	21.4%	Threonine	2.7%
Glycine	0.7%	Tyrosine	5.1%
Histidine	1.1%	Valine	3.1%
Isoleucine	6.2%		

9 Pharmacopeial Specifications

See Table III.

Table III: Pharmacopeial specifications for zein.

Test	USP35–NF30
Identification	+
Microbial limits	≤1000 cfu/g
Loss on drying	≤8.0%
Residue on ignition	≤2.0%
Heavy metals	≤20 ppm
Protein content (dried basis)	81.9–100.0%

10 Typical Properties

Density 1.23 g/cm^3

Melting point When completely dry, zein may be heated to 200°C without visible signs of decomposition.

Particle size distribution 100% less than 840 μm in size.

Solubility Practically insoluble in acetone, ethanol, and water; soluble in aqueous alcohol solutions, aqueous acetone solutions (60–80% v/v), and glycols. Also soluble in aqueous alkaline solutions of pH 11.5 and above.

Spectroscopy

NIR spectra *see* Figure 1.

11 Stability and Storage Conditions

Zein should be stored in an airtight container, in a cool, dry place. It has not been reported to polymerize.[2,3]

12 Incompatibilities

Zein is incompatible with oxidizing agents.

13 Method of Manufacture

Zein is extracted from corn gluten meal with dilute propan-2-ol.

14 Safety

Zein is used in oral pharmaceutical formulations and food products, and is generally regarded as an essentially nontoxic and

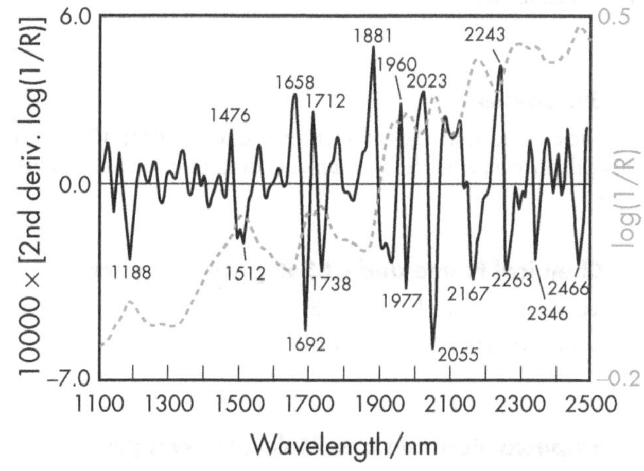

Figure 1: Near-infrared spectrum of zein measured by reflectance.

nonirritant material at the levels employed as an excipient. However, it may be harmful if ingested in large quantities. *See also* Section 18.

15 Handling Precautions

Observe normal precautions appropriate to the circumstances and quantity of material handled. Zein may be irritant to the eyes and may evolve toxic fumes on combustion. Eye-protection and gloves are recommended.

16 Regulatory Status

GRAS listed. Included in the FDA Inactive Ingredients Database (oral tablets). Included in nonparenteral medicines licensed in the UK. Included in the Canadian Natural Health Products Ingredients Database.

17 Related Substances

—

18 Comments

Zein may be safely consumed by persons sensitive to gluten.

A study has investigated the adjuvanticity and immunogenicity of zein microspheres being researched as drug and vaccine carriers.[4,5] Zein has been investigated in the preparation of microspheres for delivery of insulin and heparin, and in other delivery systems.[6–8]

A specification for zein is contained in the *Food Chemicals Codex* (FCC)[9] and the *Japanese Pharmaceutical Excipients* (JPE).[10]

The EINECS number for zein is 232-722-9.

19 Specific References

1 Katayama H, Kanke M. Drug release from directly compressed tablets containing zein. *Drug Dev Ind Pharm* 1992; **18**: 2173–2184.
2 Porter SC. Tablet coating. *Drug Cosmet Ind* 1996; **May**: 46–93.
3 Seitz JA *et al.* Tablet coating. In: Lachman L *et al.* eds. *The Theory and Practice of Industrial Pharmacy.* Philadelphia: Lea and Febiger, 1986: 346–373.
4 Hurtado-López P, Murdan S. An investigation into the adjuvanticity and immunogenicity of zein microspheres being researched as drug and vaccine carriers. *J Pharm Pharmacol* 2006; **58**: 769–774.
5 Hurtado-López P, Murdan S. Formulation and characterisation of zein microspheres as delivery vehicles. *J Drug Deliv Sci Tech* 2005; **15**(4): 267–272.
6 Wang HJ, *et al.* Heparin-loaded zein microsphere film and hemocompatibility. *J Control Release* 2005; **105**(1–2): 120–131.
7 Bernstein H *et al.* Protein microspheres and methods of using them, United States Patent 5,679,377; 1997.
8 Gao Z, *et al.* Study of a pingyangmycin delivery system: Zein/Zein-SAIB *in situ* gels. *Int J Pharm* 2007; **328**(1): 57–64.
9 *Food Chemicals Codex*, 7th edn. Bethesda, MD: United States Pharmacopeia, 2010: 1098.
10 Japan Pharmaceutical Excipients Council. *Japanese Pharmaceutical Excipients 2004*. Tokyo: Yakuji Nippo: 943-9844.

20 General References

Beck MI, *et al.* Physico-chemical characterization of zein as a film coating polymer: a direct comparison with ethyl cellulose. *Int J Pharm* 1996; **141**: 137–150.

21 Author

O AbuBaker.

22 Date of Revision

1 March 2012.

Zinc Acetate

1 Nonproprietary Names

BP: Zinc Acetate
PhEur: Zinc Acetate Dihydrate
USP–NF: Zinc Acetate

2 Synonyms

Acetic acid, zinc salt; dicarbomethoxy zinc; *Galzin*; *Wilzin*; zinc acetas dihydricus; zinc (II) acetate; zinc diacetate; zinc ethanoate.

3 Chemical Name and CAS Registry Number

Zinc acetate dihydrate [5970-45-6]
Zinc acetate anhydrous [557-34-6]

4 Empirical Formula and Molecular Weight

$C_4H_6O_4Zn{\cdot}2H_2O$ 219.50 (for dihydrate)
$C_4H_6O_4Zn$ 183.48 (for anhydrous)

5 Structural Formula

$H_3C-C(=O)O^-$ Zn^{2+} $^-O(O=)C-CH_3$

6 Functional Category

Emollient; emulsion stabilizing agent; gelling agent; opacifier.

7 Applications in Pharmaceutical Formulation or Technology

Zinc acetate has been used as an excipient in a variety of pharmaceutical formulations including topical gels, lotions and solutions, and subcutaneous injections. *See also* Section 18.

8 Description

Zinc acetate occurs as white crystalline, lustrous plates with a faint acetic odor and an astringent taste.

9 Pharmacopeial Specifications

See Table I. *See also* Section 18.

Table I: Pharmacopeial specifications for zinc acetate.

Test	PhEur 7.4	USP35–NF30
Identification	+	+
Characters	+	–
Appearance of solution	+	–
pH (5% w/v)	5.8–7.0	6.0–8.0
Reducing substances	+	–
Insoluble matter	–	+
Arsenic	≤2 ppm	≤3 ppm
Lead	≤10 ppm	≤0.002%
Chlorides	≤50 ppm	≤0.005%
Sulfates	≤100 ppm	≤0.010%
Aluminum	≤5 ppm	–
Cadmium	≤2 ppm	–
Copper	≤50 ppm	–
Iron	≤50 ppm	–
Alkalis and alkaline earths	–	≤0.2%
Assay	99.0–101.0%	98.0–102.0%

10 Typical Properties

Acidity/alkalinity pH = 6.0–8.0 (5% w/v aqueous solution of the dihydrate)

Boiling point Decomposes.

Melting point 237°C

Solubility For the dihydrate, *see* Table II.

Specific gravity 1.735

Table II: Solubility of zinc acetate dihydrate.

Solvent	Solubility at 20°C unless otherwise stated
Ethanol (95%)	1 in 30
	1 in 1 of boiling ethanol (95%)
Water	1 in 2.3
	1 in 1.6 at 100°C

11 Stability and Storage Conditions

Zinc acetate loses water of hydration above 100°C. Zinc acetate should be stored in an airtight container in a cool, dry place.

12 Incompatibilities

Zinc acetate is incompatible with oxidizing agents, zinc salts, alkalis and their carbonates, oxalates, phosphates, and sulfides.[1]

13 Method of Manufacture

Zinc acetate is synthesized by reacting zinc oxide with glacial acetic acid, with subsequent crystallization, separation by centrifugation, and drying and milling of the crystals. No organic solvents are used during the synthesis.

14 Safety

Zinc acetate is used in topical pharmaceutical formulations and subcutaneous injections, where it is generally regarded as relatively nontoxic and nonirritant when used as an excipient. However, zinc acetate is poisonous by intravenous and intraperitoneal routes; it is also moderately toxic following oral consumption.[1]

Zinc acetate:

LD_{50} (rat, oral): 2.510 g/kg[1]

LD_{50} (mouse, IP): 0.057 g/kg

Zinc acetate dihydrate:

LD_{50} (mouse, IP): 0.108 g/kg

LD_{50} (mouse, oral): 0.287 g/kg

LD_{50} (rat, IP): 0.162 g/kg

LD_{50} (rat, oral): 0.794 g/kg

15 Handling Precautions

Observe normal precautions appropriate to the circumstances and quantity of material handled. Eye protection and gloves are recommended. When heated to decomposition, zinc acetate emits toxic fumes of zinc oxide.

16 Regulatory Status

Included in the FDA Inactive Ingredients Database (SC injections; topical lotions and solutions). Included in medicines licensed in the UK.

17 Related Substances

—

18 Comments

Zinc acetate has been investigated for use in an oral controlled-release formulation for water-soluble drugs in combination with sodium alginate and xanthan gum.[2]

Therapeutically, zinc acetate is used as an astringent, and it has also been used in oral capsules for the treatment of Wilson's disease.[3] Zinc acetate has also demonstrated effectiveness as a spermicide in vaginal contraceptives.[4]

A specification for zinc acetate is included in the *Japanese Pharmaceutical Excipients* (JPE).[5]

The EINECS number for zinc acetate is 209-170-2.

19 Specific References

1 Lewis RJ, ed. *Sax's Dangerous Properties of Industrial Materials*, 11th edn. New York: Wiley, 2004; 3717–3718.

2 Zeng WM. Oral controlled-release formulation for highly water-soluble drugs: drug–sodium alginate–xanthan gum–zinc acetate matrix. *Drug Dev Ind Pharm* 2004; 30: 491–495.

3 Brewer GJ. Zinc acetate for the treatment of Wilson's disease. *Expert Opin Pharmacother* 2001; 2: 1473–1477.

4 Fahim MS, Wang M. Zinc acetate and lyophilized *Aloe barbadensis* as vaginal contraceptive. *Contraception* 1996; 53: 231–236.

5 Japan Pharmaceutical Excipients Council. *Japanese Pharmaceutical Excipients 2004*. Tokyo: Yakuji Nippo, 2004: 945–946.

20 General References

—

21 Author

BA Hanson.

22 Date of Revision

1 March 2012.

Zinc Oxide

1 Nonproprietary Names

BP: Zinc Oxide
JP: Zinc Oxide
PhEur: Zinc Oxide
USP–NF: Zinc Oxide

2 Synonyms

AZO; Chinese white; pigment white 4; zincite; zinci oxidum; zinc monoxide; zinc white.

3 Chemical Name and CAS Registry Number

Zinc oxide [1314-13-2]

4 Empirical Formula and Molecular Weight

ZnO 81.41

5 Structural Formula

See Section 4.

6 Functional Category

Antimicrobial preservative; coating agent; opacifier.

7 Applications in Pharmaceutical Formulation or Technology

Zinc oxide is widely used in creams, ointments, and lotions in pharmaceutical and cosmetic applications;[1,2] *see* Table I. It is most commonly used in baby products,[2,3] and in calamine lotion where it is blended with a small amount of iron oxide[4,5] and may be slightly pink depending on the amount of iron oxide present. *See also* Section 17.

Table I: Typical uses of zinc oxide.

Use	Concentration (%)
Baby cream	2–35[2]
Baby powder	10[3]
Cosmetic make-ups and creams	5–30[1]
Semisolid creams	5–13[4]

Zinc oxide completely absorbs UV light < 366 nm[6] and therefore is widely used as a sunscreen agent to block UVA and UVB.[7] Zinc oxide nanoparticles between 200 and 400 nm reflect and scatter light but 40–100 nm particles absorb and scatter UV and absorb visible wavelengths, making the screen transparent.[8]. It is used as an active 'physical' ingredient[9] since it physically blocks the UV from reaching the skin.

In addition, the ability of zinc oxide to scatter light makes it a useful opacifier for white creams. It can be used as a mild antibacterial or antifungicidal agent.[1,6] Zinc oxide is also used in small quantities in subcutaneous injections and suppositories.

8 Description

Zinc oxide occurs as a soft, white or yellowish-white odorless powder. It appears as loose aggregates consisting of amorphous submicron primary particles.

9 Pharmacopeial Specifications

See Table II.

Table II: Pharmacopeial specifications for zinc oxide.

Test	JP XV	PhEur 7.4	USP35–NF30
Identification	+	+	+
Characters	+	+	–
Alkalinity	+	+	+
Carbonate and color of solution	+	+	+
Arsenic	≤4 ppm	≤5 ppm	≤6 ppm
Cadmium	–	≤10 ppm	–
Lead	+	≤50 ppm	+
Loss on ignition	≤1.0%	≤1.0%	≤1.0%
Iron and other heavy metals	≤10 ppm	≤200 ppm	+
Sulfate	≤0.096%	–	–
Assay	≥99.0%	99.0–100.5%	99.0–100.5%

10 Typical Properties

Alkalinity/acidity pH 7.37 for pharmaceutical grades made by the 'French' process.
Density (bulk) 0.32–0.64 g/cm^3; *see also* Table III.

Table III: Physical properties of selected commercial grades of zinc oxide.

Grade	Specific surface area (m^2/g)	Particle size (μm)	Bulk density (g/cm^3)
AZO 66USP[11]	4.0–6.0	–	0.32–0.64
AZO 77USP[12]	6.0–9.0	–	0.32–0.64
Gold Seal[13]	3.5–6.7	–	–
USP-1[14]	9.0	0.12	0.48
USP-2[14]	3.2	0.33	0.64
USP-3[14]	5.0	0.21	0.56

Density (true) 5.67 g/cm^3[5]
Melting point > 1800°C[10]
Particle size distribution >99.9% passing through a #325 mesh; *see also* Table III.
Refractive index 2.015 for a 0.5 μm particle.[6]
Solubility Practically insoluble in water (0.42 mg/100 g water at 18°C);[6] soluble in dilute acids such as acetic acid, ammonia and alkali hydroxide solutions.[5]
Specific surface area 3.2–9.0 m^2/g depending on the grade;[11–14] *see* Table III.

11 Stability and Storage Conditions

Zinc oxide is stable and should be stored in a well-closed container.

12 Incompatibilities

Zinc oxide undergoes a violent reaction with magnesium or linseed oil.[10]

13 Method of Manufacture

Zinc oxide is manufactured by vaporizing high purity zinc metal at approx. 2000°C and mixing the vapor with air at controlled flow rates to produce zinc oxide. Pharmaceutical grades are made by this process, known as the 'French process'.[6] Varying the flow rates and time before cooling allows different crystal sizes and shapes to be formed.

14 Safety

Zinc oxide is generally considered a nontoxic material.[6] It is moderately toxic to humans by ingestion in its pure form.[10]

Zinc oxide is considered a nuisance powder[15] and can cause skin and eye irritation.[10] Inhalation of fresh fumes during zinc oxide manufacture may cause a temporary flu-like illness ('zinc chills')[6,10,16] but this does not apply to inhalation of zinc oxide powder.

The safety of zinc oxide nanoparticles is less well understood, particularly the possibility of dermal adsorption, and is an area of ongoing investigation.[9] There is some experimental evidence that zinc oxide may act as a mutagen or teratogen.[10]

LC_{50} (mouse, inhalation): 2.5 g/m^3[10]
LD_{50} (mouse, oral): 7.95 g/kg
LD_{50} (rat, IP): 0.24 g/kg
LD_{Lo}(human, oral): 0.5 g/kg
TC_{Lo} (human, inhalation): 0.6 g/m^3

15 Handling Precautions

Observe normal precautions appropriate to the circumstances and quantity of the material handled.

Pharmaceutical applications require standard inhalation protection appropriate to the level of potential exposure. In the US, the OSHA limit is 10 mg/m^3 for total dust, 5 mg/m^3 respirable fraction for zinc oxide.[10]

16 Regulatory Status

GRAS listed. Included in the FDA Inactive Ingredients Database (IM injections; SC injections and suspensions; suppositories). Approved by the FDA as a food additive and for externally applied drugs and cosmetics, including formulations applied to the eye area.[1] Included in the Canadian Natural Health Products Ingredients Database. Included in nonparenteral medicines licensed in the UK (creams, ointments, and suppositories).

17 Related Substances

Calamine.

Calamine

CAS number [8011-96-9]

Description Occurs as an odorless, preblended powder of white zinc oxide tinted with a small proportion of red iron oxide.

Pharmacopeial specifications see Table IV.

Solubility Soluble in dilute acetic acid or mineral acid, ammonia, ammonium carbonate, and fixed alkali hydroxide solution; insoluble in water.

Specific gravity 5.6–5.7

Storage Preserve in well-closed containers.

Comments Primarily used in traditional creams, ointment or lotions to treat a variety of skin conditions. Calamine is also available as a pre-blended powder,[4] and is used as a dusting powder.

18 Comments

Zinc oxide is approved by the FDA as a white pigment for external use in cosmetics, including the eye area. It is included in cosmetic face powders, rouges, eye make-up,[1] foundations, lipsticks, soaps and detergents, creams, deodorants, noncoloring shampoos and other products.[17]

Table IV: Pharmacopeial specifications for calamine.

Test	BP 2012[a]	USP35–NF30
Identification	+	+
Characters	+	–
Microbial limits	–	+
Loss on ignition	–	≤2.0%
Acid insoluble substances	–	≤2.0%
Alkaline substances	–	+
Arsenic	–	8 ppm
Calcium	+	+
Calcium or magnesium	–	+
Soluble barium salts	+	–
Lead	≤150 ppm	+
Chloride	+	–
Sulfate	+	–
Ethanol–soluble dyes	+	–
Matter insoluble in hydrochloric acid (dried residue)	≤10 mg	–
Water–soluble dyes	+	–
Residue on ignition	68.0–74.0%	–
Assay (ZnO)	–	98.0–100.5%

(a) Calamine BP contains basic zinc carbonate rather than zinc oxide.

Zinc oxide is also used as a food additive in dietary supplements and vitamins tablets,[6,18] and therapeutically as an antiseptic and mild astringent.[5]

A specification for zinc oxide is contained in the *Food Chemicals Codex* (FCC).[19]

The EINECS number for zinc oxide is 215-222-5. The PubChem Compound ID (CID) for zinc oxide is 24852301.

19 Specific References

1 Marmion DM, ed. Handbook of US Colorants: Foods, Drugs, Cosmetics and Medical Devices, 3rd edn. New York: Wiley & Sons; 1991.
2 Flick EW. *Cosmetic and Toiletry Formulations*, vol. 1, 2nd edn. New York: William Andrew Publishing/Noyes, 1989.
3 Niazi SK. Uncompressed solids formulations. In: *Handbook of Pharmaceutical Manufacturing Formulations*. London: Informa Healthcare, 2004.
4 Niazi SK. Part II: Formulations of semisolid drugs. In: *Handbook of Pharmaceutical Manufacturing Formulations*. London: Informa Healthcare, 2004.
5 O'Neil MJ. *The Merck Index – An Encyclopedia of Chemicals, Drugs, and Biologicals*, 14th edn. New Jersey: Merck Sharp & Dohme Corp, 2006: 2520.
6 Goodwin FE. Zinc compounds. In: *Kirk-Othmer Encyclopedia of Chemical Technology*, 5th edn. New York: John Wiley & Sons, 2004: 605–621.
7 Mitchnick MA, Fairhurst D, Pinnell SR. Microfine zinc oxide (Z-Cote) as a photostable UVA/UVB sunblock agent. *J Am Acad Dermatol* 1999; 40(1): 85–90.
8 Wolf R, *et al.* Sunscreens. *Clin Dermatol* 2001; **19**(4): 452–459.
9 Osmond MJ, McCall MJ. Zinc oxide nanoparticles in modern sunscreens: an analysis of potential exposure and hazard. *Nanotoxicology* 2010; 4(1): 15–41.
10 Lewis RJ, ed. *Sax's Dangerous Properties of Industrial Materials*, 11th edn. New York: Wiley, 2004: 3725.
11 US Zinc. Product Data Sheet: AZO 66USP, 2002.
12 US Zinc. Product Data Sheet: AZO 77USP, 2002.
13 IEQSA. Product Data Sheet: Gold Seal Zinc Oxide, September 2009.
14 Horsehead Corporation. Product Data Sheet: USP grade zinc oxides, June 2009.
15 Ash M, Ash I. *Handbook of Fillers, Extenders and Diluents*. Endicott NY: Synapse Information Resources Inc, 2007.
16 Mallinckrodt Inc. Material safety data sheet: Zinc oxide, 2009.
17 Cosmetic TaFA, Inc. International Cosmetic Ingredient Dictionary and Handbook, 11th edn. CFTA Inc, 2006.
18 Ash M, Ash I. *Handbook of Food Additives*, 3rd edn. Synapse Information Resources Inc, 2008: 497.

Z

19 *Food Chemicals Codex*, 7th edn (Suppl. 2). Bethesda, MD: United States Pharmacopeia, 2011: 1100.

20 General References

Morkoç H, Özgür Ü.;1; Zinc Oxide: Fundamentals, Materials and Device Technology. Weinheim: Wiley–VCH Verlag GmbH & KGaA, 2009.

21 Author

KP Hapgood.

22 Date of Revision

1 March 2012.

Zinc Stearate

1 Nonproprietary Names

BP: Zinc Stearate
PhEur: Zinc Stearate
USP–NF: Zinc Stearate

2 Synonyms

Cecavon; *Demarone*; dibasic zinc stearate; *HyQual*; *Kemilub*; *Metallac*; stearic acid zinc salt; *Synpro*; zinc distearate; zinc octadecanoate; zinc soap; zinci stearas.

3 Chemical Name and CAS Registry Number

Octadecanoic acid zinc salt [557-05-1]

4 Empirical Formula and Molecular Weight

$C_{36}H_{70}O_4Zn$ 632.33 (for pure material)

The USP35–NF30 describes zinc stearate as a compound of zinc with a mixture of solid organic acids obtained from fats, and consists chiefly of variable proportions of zinc stearate and zinc palmitate. It contains the equivalent of 12.5–14.0% of zinc oxide (ZnO).

The PhEur 7.4 states that zinc stearate $[(C_{17}H_{35}COO)_2Zn]$ may contain varying proportions of zinc palmitate $[(C_{15}H_{31}COO)_2Zn]$ and zinc oleate $[(C_{17}H_{33}COO)_2Zn]$. It contains not less than 10.0% and not more than 12.0% of zinc.

5 Structural Formula

See Section 4.

6 Functional Category

Tablet and capsule lubricant.

7 Applications in Pharmaceutical Formulation or Technology

Zinc stearate is primarily used in pharmaceutical formulations as a lubricant in tablet and capsule manufacture at concentrations up to 1.5% w/w. It has also been used as a thickening and opacifying agent in cosmetic and pharmaceutical creams, and as a dusting powder. *See* Table I.

Table I: Uses of zinc stearate.

Use	Concentration (%)
Tablet lubricant	0.5–1.5
Water-repellent ointments	2.5

8 Description

Zinc stearate occurs as a fine, white, bulky, hydrophobic powder, free from grittiness and with a faint characteristic odor.

9 Pharmacopeial Specifications

See Table II.

Table II: Pharmacopeial specifications for zinc stearate.

Test	PhEur 7.4	USP35-NF30
Identification	+	+
Characters	+	−
Acidity or alkalinity	+	−
Alkalis and alkaline earths	−	≤1.0%
Appearance of solution	+	−
Acid value of the fatty acids	195–210	−
Appearance of solution of fatty acids	+	−
Arsenic	−	≤1.5 ppm
Cadmium	≤5 ppm	−
Lead	≤25 ppm	≤10 ppm
Chlorides	≤250 ppm	−
Sulfates	≤0.6%	−
Assay (as Zn)	10.0–12.0%	−
Assay (as ZnO)	−	12.5–14.0%

SEM 1: Excipient: zinc stearate; magnification: 600×.

SEM 2: Excipient: zinc stearate; magnification: 2400×.

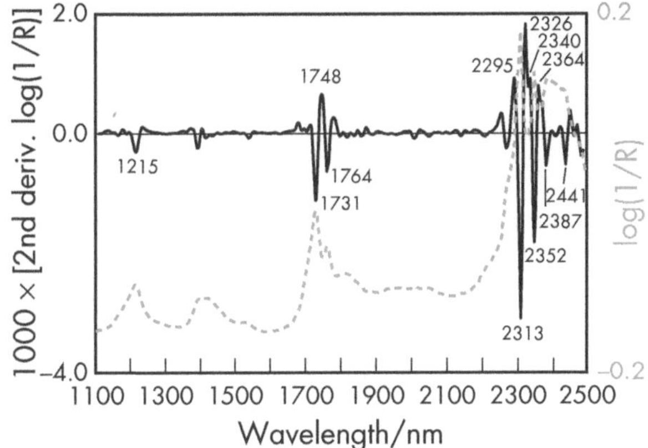

Figure 1: Near-infrared spectrum of zinc stearate measured by reflectance.

10 Typical Properties

Autoignition temperature 421°C
Density 1.09 g/cm^3
Flash point 277°C
Melting point 120–122°C; 130°C also reported.[1,2]
Particle size distribution 100% through a 44.5-μm sieve (#325 mesh).
Solubility Insoluble in ethanol (95%), ether, and water; practically insoluble in oxygenated solvents; soluble in acids, benzene, and other aromatic solvents.
Spectroscopy
NIR spectra *see* Figure 1.

11 Stability and Storage Conditions

Zinc stearate is stable and should be stored in a well-closed container in a cool, dry place.

12 Incompatibilities

Zinc stearate is decomposed by dilute acids. It is incompatible with strong oxidizing agents.

13 Method of Manufacture

An aqueous solution of zinc sulfate is added to sodium stearate solution to precipitate zinc stearate. The zinc stearate is then washed with water and dried. Zinc stearate may also be prepared from stearic acid and zinc chloride.

14 Safety

Zinc stearate is used in oral and topical pharmaceutical formulations, and is generally regarded as a nontoxic and nonirritant excipient. However, following inhalation, it has been associated with fatal pneumonitis, particularly in infants.[3] As a result, zinc stearate has now been removed from baby dusting powders.

LD_{50} (rat, IP): 0.25 g/kg

15 Handling Precautions

Observe normal precautions appropriate to the circumstances and quantity of material handled. Eye protection and gloves are recommended. Zinc stearate may be harmful on inhalation and should be used in a well-ventilated environment; a respirator is recommended. In the UK, the long-term (8-hour TWA) workplace exposure limit for zinc stearate is 10 mg/m^3 for total inhalable dust and 4 mg/m^3 for respirable dust. The short-term (15-minutes) workplace exposure limit for total inhalable dust is 20 mg/m^3.[4] In the US, the OSHA limit is 15 mg/m^3 for total dust, 5 mg/m^3 respirable fraction for zinc stearate.[5]

When heated to decomposition, zinc stearate emits acrid smoke and fumes of zinc oxide.

16 Regulatory Status

GRAS listed. Included in the FDA Inactive Ingredients Database (oral capsules and tablets). Included in nonparenteral medicines licensed in the UK. Included in the Canadian Natural Health Products Ingredients Database.

17 Related Substances

Calcium stearate; magnesium stearate; stearic acid.

18 Comments

Zinc stearate is neutral to moistened litmus paper.

The EINECS number for zinc stearate is 209-151-9. *See* Magnesium Stearate for further information and references.

19 Specific References

1 Sigma-Aldrich. Material safety data sheet: Zinc stearate, 2011.
2 ScienceLab.com, Inc. Material safety data sheet: Zinc stearate, 2008.
3 Ueda A. Experimental study on the pathological changes in lung tissue caused by zinc stearate dust. *Ind Health* 1984; **22**: 243–253.
4 Health and Safety Executive. *EH40/2005: Workplace Exposure Limits*. Sudbury: HSE Books, 2011. http://www.hse.gov.uk/pubns/priced/eh40.pdf (accessed 23 February 2012).
5 JT Baker Inc. Material safety data sheet: Zinc stearate, 2011. www.jtbaker.com/msds/englishhtml/z4275.htm (accessed 1 December 2011).

20 General References

Ferro Corp. Product information: *Synpro*. www.ferro.com (accessed 1 December 2011).

21 Author

LV Allen Jr.

22 Date of Revision

1 March 2012.

Appendix I: Suppliers Directory

EXCIPIENTS LIST

Acacia

UK
A and E Connock (Perfumery and Cosmetics) Ltd
AF Suter and Co Ltd
Blagden Specialty Chemicals Ltd
Colloides Naturels UK Ltd
Courtin & Warner Ltd
Lehvoss UK Ltd
Macron Chemicals UK
Thew, Arnott and Co Ltd

Other European
Alfa Aesar Johnson Matthey GmbH
Alland & Robert
Brenntag AG
Colloides Naturels International

USA
Advance Scientific & Chemical Inc
CarboMer Inc
Chart Corp Inc
Colloides Naturels Inc
Hawkins Inc
Hummel Croton Inc
KIC Chemicals Inc
Lipscomb Chemical Company
Macron Chemicals
Mitsubishi International Food Ingredients
Mutchler Inc
Penta Manufacturing Co
Premium Ingredients Ltd
Ruger Chemical Co Inc
Seidler Chemical Company Inc
Spectrum Chemicals & Laboratory Products Inc
TIC Gums
Voigt Global Distribution Inc

Acesulfame Potassium

UK
Sparkford Chemicals Ltd

Other European
Brenntag AG
Nutrinova Nutrition Specialities & Food Ingredients GmbH

USA
Advance Scientific & Chemical Inc
AerChem Inc
KIC Chemicals Inc
Nutrinova Inc
Premium Ingredients Ltd
Stadt Holdings Corporation

Others
Gangwal Chemicals Pvt Ltd
Kawarlal Excipients Pvt Ltd
Wintersun Chemical

Acetic Acid, Glacial

UK
Blagden Specialty Chemicals Ltd
BP plc
Eastman Company UK Ltd
Fisher Scientific UK Ltd
Macron Chemicals UK
Sigma-Aldrich Company Ltd
Tennants Distribution Ltd

Other European
Alfa Aesar Johnson Matthey GmbH
August Hedinger GmbH & Co KG
Brenntag GmbH
Impag GmbH
Panreac Quimica SAU

USA
Advance Scientific & Chemical Inc
Amresco Inc
Ashland Inc
Astro Chemicals Inc
BOC Sciences
BP Inc
Brenntag North America Inc
Brenntag Southwest
Celanese
Eastman Chemical Co
Fisher Scientific
Macron Chemicals
Penta Manufacturing Co
Ruger Chemical Co Inc
Spectrum Chemicals & Laboratory Products Inc
Thomas Scientific
Triple Crown America

Others
Boith Limited (China)
Priyanka Pharma/M K Ingredients & Specialities
SD Fine-Chem Ltd
Univar Canada Ltd
Wintersun Chemical

Acetone

UK
Leading Solvent Supplies Ltd
Macron Chemicals UK
Sigma-Aldrich Company Ltd

Other European
Alfa Aesar Johnson Matthey GmbH
August Hedinger GmbH & Co KG
Brenntag AG
Corcoran Chemicals Ltd
Dow Benelux NV
Panreac Quimica SAU

USA
Advance Scientific & Chemical Inc
AerChem Inc
Amresco Inc
Ashland Inc
Astro Chemicals Inc
Dow Chemical Co
Eastman Chemical Co
EMD Chemicals Inc
Fisher Scientific
Macron Chemicals
Penta Manufacturing Co
Ruger Chemical Co Inc
Seidler Chemical Company Inc
Spectrum Chemicals & Laboratory Products Inc
Thomas Scientific

Others
Boith Limited (China)
SD Fine-Chem Ltd
Univar Canada Ltd
Wintersun Chemical

Acetone Sodium Bisulfite

USA
Ruger Chemical Co Inc

Acetyltributyl Citrate

UK
Ubichem plc

Other European
Jungbunzlauer AG

USA
Jungbunzlauer Inc
Penta Manufacturing Co
Vertellus Specialties Inc

Acetyltriethyl Citrate

UK
Ubichem plc

Other European
Jungbunzlauer AG

USA
Jungbunzlauer Inc
Penta Manufacturing Co
Vertellus Specialties Inc

Adipic Acid

UK
Macron Chemicals UK

Other European
Alfa Aesar Johnson Matthey GmbH
Brenntag AG
Corcoran Chemicals Ltd
Panreac Quimica SAU

USA
Advance Scientific & Chemical Inc
DuPont
Macron Chemicals
Penta Manufacturing Co
Seidler Chemical Company Inc

Others
Boith Limited (China)
SD Fine-Chem Ltd
Wintersun Chemical

Agar

UK
AF Suter and Co Ltd
Mast Group Ltd
Sigma-Aldrich Company Ltd
Thew, Arnott and Co Ltd

Other European
Alfa Aesar Johnson Matthey GmbH
Panreac Quimica SAU

USA
Advance Scientific & Chemical Inc
Alfa Chem
Amresco Inc
Ashland Inc
CarboMer Inc

EMD Chemicals Inc
Penta Manufacturing Co
Premium Ingredients Ltd
Ruger Chemical Co Inc
Seidler Chemical Company Inc
Spectrum Chemicals & Laboratory Products Inc
TIC Gums
Voigt Global Distribution Inc

Others
Wintersun Chemical

Albumin

UK
AarhusKarlshamn UK Ltd
Bio Products Laboratory
Lehvoss UK Ltd

Other European
AarhusKarlshamn AB
Alfa Aesar Johnson Matthey GmbH

USA
AarhusKarlshamn USA Inc
Advance Scientific & Chemical Inc
AerChem Inc
Amresco Inc
CSL Behring
Seidler Chemical Company Inc
Voigt Global Distribution Inc

Alcohol

UK
Sigma-Aldrich Company Ltd
Tennants Distribution Ltd

Other European
Brenntag GmbH
Panreac Quimica SAU
Tate & Lyle

USA
Advance Scientific & Chemical Inc
Ashland Inc
Brenntag North America Inc
Brenntag Southwest
Dow Chemical Co
Grain Processing Corp
Penta Manufacturing Co
Seidler Chemical Company Inc

Alginic Acid

UK
Blagden Specialty Chemicals Ltd
Forum Products Ltd
IMCD UK Ltd
Rettenmaier UK Ltd

Other European
FMC Biopolymer
J Rettenmaier & Söhne GmbH and Co.KG

USA
Aceto Corp
Advance Scientific & Chemical Inc
Alfa Chem
American International Chemical Inc
CarboMer Inc
International Specialty Products Inc (ISP)
JRS Pharma LP
Mutchler Inc
Penta Manufacturing Co
Spectrum Chemicals & Laboratory Products Inc
Voigt Global Distribution Inc

Aliphatic Polyesters

UK
Purac Biochem (UK)

Other European
Evonik Industries AG

USA
Evonik Degussa Corp
Purac
SurModics Pharmaceuticals

Alitame

UK
Danisco Sweeteners Ltd

Almond Oil

UK
A and E Connock (Perfumery and Cosmetics) Ltd
AarhusKarlshamn UK Ltd
Alembic Products Ltd
Lehvoss UK Ltd
William Ransom & Son plc

USA
AarhusKarlshamn USA Inc
Advance Scientific & Chemical Inc
Arista Industries Inc
Charkit Chemical Corp
Chart Corp Inc
KIC Chemicals Inc
Penta Manufacturing Co
Pokonobe Industries Inc
Ruger Chemical Co Inc
Seidler Chemical Company Inc
Spectrum Chemicals & Laboratory Products Inc
Voigt Global Distribution Inc
Welch, Holme & Clark Co Inc

Alpha Tocopherol

UK
A and E Connock (Perfumery and Cosmetics) Ltd
Alembic Products Ltd
Cognis UK Ltd
Cornelius Group plc
Eastman Company UK Ltd
Ubichem plc

Other European
BASF Aktiengesellschaft
Cognis GmbH
Helm AG

USA
BASF Corp
Cognis Corp
Eastman Chemical Co
Spectrum Chemicals & Laboratory Products Inc
Takeda Pharmaceuticals North America Inc
Triple Crown America

Others
BASF Japan Ltd

Aluminum Hydroxide Adjuvant

UK
Reheis

Other European
Brenntag Nordic AS

USA
Advance Scientific & Chemical Inc
Alfa Chem
EM Sergeant Pulp & Chemical Co Inc
General Chemical Performance Products LLC
Penta Manufacturing Co
Reheis Inc
Ruger Chemical Co Inc
Seidler Chemical Company Inc

Others
Wintersun Chemical

Aluminum Monostearate

UK
Peter Greven UK

Other European
Magnesia GmbH

USA
Acme-Hardesty
AerChem Inc
Ashland Inc
Avantor Performance Materials
Ferro Corp
Penta Manufacturing Co
Ruger Chemical Co Inc
Spectrum Chemicals & Laboratory Products Inc

Others
Wintersun Chemical

Aluminum Oxide

UK
Fisher Scientific UK Ltd
Pumex (UK) Limited

Other European
Alfa Aesar Johnson Matthey GmbH
Brenntag GmbH
Evonik Industries AG

USA
EMD Chemicals Inc
Ruger Chemical Co Inc
Seidler Chemical Company Inc
SPI Pharma Inc

Others
Sumitomo Chemical
Wintersun Chemical

Aluminum Phosphate Adjuvant

UK
Reheis

Other European
Brenntag AG
Brenntag Nordic AS

USA
EM Sergeant Pulp & Chemical Co Inc
General Chemical Performance Products LLC
Reheis Inc

Ammonia Solution

UK
Macron Chemicals UK
Tennants Distribution Ltd
William Ransom & Son plc

Other European
Alfa Aesar Johnson Matthey GmbH
Brenntag GmbH
Panreac Quimica SAU

USA
Macron Chemicals
Ruger Chemical Co Inc
Seidler Chemical Company Inc
Spectrum Chemicals & Laboratory Products Inc
Triple Crown America

Others
SD Fine-Chem Ltd

Ammonium Alginate

USA
Penta Manufacturing Co

Ammonium Chloride

UK
Courtin & Warner Ltd
Peter Whiting (Chemicals) Ltd

Other European
Alfa Aesar Johnson Matthey GmbH
Merck KGaA
Panreac Quimica SAU

USA
AerChem Inc
Alfa Chem
BOC Sciences
Fisher Scientific
Penta Manufacturing Co
Ruger Chemical Co Inc
Seidler Chemical Company Inc

Others
SD Fine-Chem Ltd
Wintersun Chemical

Ascorbic Acid

UK
Fisher Scientific UK Ltd
Macron Chemicals UK
Peter Whiting (Chemicals) Ltd
Raught Ltd
Roche Products Ltd
Tennants Distribution Ltd

Other European
Alfa Aesar Johnson Matthey GmbH
BASF Aktiengesellschaft
Brenntag AG
Helm AG
Panreac Quimica SAU

USA
Aceto Corp
Advance Scientific & Chemical Inc
AerChem Inc
Alfa Chem
Allan Chemical Corp
American International Chemical Inc
Amresco Inc
Barrington Nutritionals
BASF Corp
Brenntag North America Inc
Brenntag Southwest
Charles Bowman & Co
Dastech International Inc
Fisher Scientific
George Uhe Co Inc
Glanbia Nutritionals Inc
Hawkins Inc
Helm New York Inc
Integra Chemical Co
KIC Chemicals Inc
Kraft Chemical Co
Macron Chemicals
Particle Dynamics Inc
Penta Manufacturing Co
Premium Ingredients Ltd
Ruger Chemical Co Inc
Seidler Chemical Company Inc
Spectrum Chemicals & Laboratory Products Inc
Takeda Pharmaceuticals North America Inc
Triple Crown America
Voigt Global Distribution Inc
Zhong Ya Chemical (USA) Ltd

Others
BASF Japan Ltd
Boith Limited (China)
CSPC Pharma
Univar Canada Ltd
Wintersun Chemical

Ascorbyl Palmitate

UK
A and E Connock (Perfumery and Cosmetics) Ltd
Roche Products Ltd

Other European
BASF Aktiengesellschaft

USA
Advance Scientific & Chemical Inc
American International Chemical Inc
BASF Corp
CarboMer Inc
George Uhe Co Inc
Hawkins Inc
Penta Manufacturing Co
RIA International
Ruger Chemical Co Inc
Spectrum Chemicals & Laboratory Products Inc
Voigt Global Distribution Inc

Others
BASF Japan Ltd
Gangwal Chemicals Pvt Ltd
Wintersun Chemical
Xinchem Co

Aspartame

UK
Blagden Specialty Chemicals Ltd
DSM UK Ltd
Tennants Distribution Ltd

Other European
Ajinomoto Sweeteners Europe SAS
Brenntag GmbH
Corcoran Chemicals Ltd
DSM Fine Chemicals
Helm AG

USA
Aceto Corp
Advance Scientific & Chemical Inc
AerChem Inc
Ashland Inc
BOC Sciences
Brenntag North America Inc
Brenntag Southwest
DSM Pharmaceuticals Inc
EM Sergeant Pulp & Chemical Co Inc
Hawkins Inc
Helm New York Inc
KIC Chemicals Inc
Merisant
NutraSweet Company, The
Penta Manufacturing Co
Premium Ingredients Ltd
Seidler Chemical Company Inc
Spectrum Chemicals & Laboratory Products Inc
Stadt Holdings Corporation
Triple Crown America
Voigt Global Distribution Inc
Zhong Ya Chemical (USA) Ltd

Others
Ajinomoto Co Inc
LS Raw Materials Ltd
Univar Canada Ltd
Wintersun Chemical

Attapulgite

USA
Advance Scientific & Chemical Inc
RIA International

Bentonite

UK
A and E Connock (Perfumery and Cosmetics) Ltd
Peter Whiting (Chemicals) Ltd
Tennants Distribution Ltd
Thew, Arnott and Co Ltd
Wilfrid Smith Ltd

Other European
Brenntag AG

USA
Advance Scientific & Chemical Inc
American Colloid Co
Brenntag Specialties Inc
Charles B Chrystal Co Inc
Farma International Inc
Kraft Chemical Co
Penta Manufacturing Co
Ruger Chemical Co Inc
Seidler Chemical Company Inc
Spectrum Chemicals & Laboratory Products Inc
Whittaker Clark, and Daniels Inc

Others
Signet Chemical Corporation Pvt Ltd

Benzalkonium Chloride

UK
Lonza Biologics Plc
Raught Ltd
Ubichem plc
White Sea and Baltic Company Ltd

Other European
Alfa Aesar Johnson Matthey GmbH
FeF Chemicals A/S
Lonza Ltd

USA
Advance Scientific & Chemical Inc
Brenntag North America Inc
Penta Manufacturing Co
RIA International
Ruger Chemical Co Inc
Sanofi-Aventis
Seidler Chemical Company Inc
Spectrum Chemicals & Laboratory Products Inc
Thomas Scientific
Triple Crown America

Others
Charles Tennant & Co (Canada) Ltd
Wintersun Chemical

Benzethonium Chloride

UK
Lonza Biologics Plc

Other European
Alfa Aesar Johnson Matthey GmbH
Lonza Ltd
Panreac Quimica SAU

USA
Advance Scientific & Chemical Inc
Penta Manufacturing Co
Ruger Chemical Co Inc
Spectrum Chemicals & Laboratory Products Inc

1

Benzoic Acid

UK
Clariant UK Ltd
Cornelius Group plc
Courtin & Warner Ltd
Dow Chemical Company (UK)
DSM UK Ltd
Fisher Scientific UK Ltd
Macron Chemicals UK
Sigma-Aldrich Company Ltd
Sparkford Chemicals Ltd
Tennants Distribution Ltd
Ubichem plc
White Sea and Baltic Company Ltd

Other European
Alfa Aesar Johnson Matthey GmbH
Brenntag GmbH
Clariant GmbH
DSM Fine Chemicals
Haltermann Products
Panreac Quimica SAU

USA
Aceto Corp
Advance Scientific & Chemical Inc
AerChem Inc
Alfa Chem
Amresco Inc
Ashland Inc
Brenntag North America Inc
Charkit Chemical Corp
Clariant Corp
Dastech International Inc
DSM Pharmaceuticals Inc
Fisher Scientific
KIC Chemicals Inc
Macron Chemicals
Parchem Trading Ltd
Penta Manufacturing Co
Premium Ingredients Ltd
RIA International
Ruger Chemical Co Inc
Seidler Chemical Company Inc
Spectrum Chemicals & Laboratory Products Inc
Triple Crown America
Voigt Global Distribution Inc
Zhong Ya Chemical (USA) Ltd

Others
Boith Limited (China)
Charles Tennant & Co (Canada) Ltd
LS Raw Materials Ltd
San Fu Chemical Company Ltd
Univar Canada Ltd
Wintersun Chemical

Benzyl Alcohol

UK
DSM UK Ltd
Fisher Scientific UK Ltd
Macron Chemicals UK
Symrise
Tennants Distribution Ltd
Ubichem plc

Other European
Alfa Aesar Johnson Matthey GmbH
DSM Fine Chemicals
Panreac Quimica SAU
Tessenderlo Chemie

USA
Advance Scientific & Chemical Inc
AerChem Inc
Alfa Chem
Astro Chemicals Inc
Brenntag North America Inc
Charkit Chemical Corp
DSM Pharmaceuticals Inc
Fisher Scientific
Hawkins Inc
Macron Chemicals
Penta Manufacturing Co
Premium Ingredients Ltd
Ruger Chemical Co Inc
Seidler Chemical Company Inc
Spectrum Chemicals & Laboratory Products Inc
Thomas Scientific
Voigt Global Distribution Inc

Others
Boith Limited (China)
LS Raw Materials Ltd
SD Fine-Chem Ltd
Univar Canada Ltd
Wintersun Chemical

Benzyl Benzoate

UK
Dow Chemical Company (UK)
Sigma-Aldrich Company Ltd
Symrise
White Sea and Baltic Company Ltd
William Ransom & Son plc

Other European
Alfa Aesar Johnson Matthey GmbH
Haltermann Products
Panreac Quimica SAU

USA
Advance Scientific & Chemical Inc
Alfa Chem
Penta Manufacturing Co
Premium Ingredients Ltd
Seidler Chemical Company Inc
Spectrum Chemicals & Laboratory Products Inc
Vertellus Specialties Inc
Voigt Global Distribution Inc

Others
Boith Limited (China)
LS Raw Materials Ltd

Boric Acid

UK
Courtin & Warner Ltd
Macron Chemicals UK
Peter Whiting (Chemicals) Ltd
Sigma-Aldrich Company Ltd

Other European
Brenntag AG
Panreac Quimica SAU

USA
Advance Scientific & Chemical Inc
AerChem Inc
Alfa Chem
Amresco Inc
Astro Chemicals Inc
BOC Sciences
Fisher Scientific
Integra Chemical Co
Macron Chemicals
Penta Manufacturing Co
Ruger Chemical Co Inc
Seidler Chemical Company Inc
Spectrum Chemicals & Laboratory Products Inc
Thomas Scientific

Others
Boith Limited (China)
LS Raw Materials Ltd
SD Fine-Chem Ltd

Bronopol

UK
IMCD UK Ltd
White Sea and Baltic Company Ltd

Other European
Alfa Aesar Johnson Matthey GmbH
BASF Aktiengesellschaft

USA
BASF Corp
Inolex Chemical Co
RIA International
Spectrum Chemicals & Laboratory Products Inc

Others
Cosmos Chemical Co Ltd
LS Raw Materials Ltd

Butylated Hydroxyanisole

UK
Eastman Company UK Ltd
IMCD UK Ltd
Sparkford Chemicals Ltd

Other European
Alfa Aesar Johnson Matthey GmbH
Brenntag GmbH

USA
Advance Scientific & Chemical Inc
Ashland Inc
Eastman Chemical Co
Kraft Chemical Co
Penta Manufacturing Co
Ruger Chemical Co Inc
Spectrum Chemicals & Laboratory Products Inc
Voigt Global Distribution Inc

Others
LS Raw Materials Ltd

Butylated Hydroxytoluene

UK
Eastman Company UK Ltd
IMCD UK Ltd
Raught Ltd
Sparkford Chemicals Ltd

Other European
August Hedinger GmbH & Co KG
Brenntag GmbH
Helm AG

USA
Advance Scientific & Chemical Inc
Ashland Inc
Brenntag North America Inc
Brenntag Southwest
Eastman Chemical Co
Kraft Chemical Co
Mutchler Inc
Penta Manufacturing Co
Ruger Chemical Co Inc
Spectrum Chemicals & Laboratory Products Inc
Voigt Global Distribution Inc

Others
LS Raw Materials Ltd
Wintersun Chemical

Butylene Glycol

Other European
Alfa Aesar Johnson Matthey GmbH

USA
Penta Manufacturing Co
Seidler Chemical Company Inc

Butylparaben

UK
Blagden Specialty Chemicals Ltd
Clariant UK Ltd
Macron Chemicals UK

Other European
Alfa Aesar Johnson Matthey GmbH
Clariant GmbH
Induchem AG

USA
Acme-Hardesty
Advance Scientific & Chemical Inc
Alfa Chem
Clariant Corp
Dastech International Inc
Hallstar Company, The
KIC Chemicals Inc
Kraft Chemical Co
Lipo Chemicals Inc
Macron Chemicals
Napp Technologies LLC
Penta Manufacturing Co
Protameen Chemicals Inc
Ruger Chemical Co Inc
Seidler Chemical Company Inc
Spectrum Chemicals & Laboratory Products Inc
Voigt Global Distribution Inc

Others
Charles Tennant & Co (Canada) Ltd
Wintersun Chemical

Butyl Stearate

UK
Cognis UK Ltd
Croda Europe Ltd
Macron Chemicals UK
Peter Whiting (Chemicals) Ltd

USA
Ashland Inc
Avatar Corp
Kraft Chemical Co
Macron Chemicals
Penta Manufacturing Co
Spectrum Chemicals & Laboratory Products Inc
Stepan Co
Universal Preserv-A-Chem Inc
Van Waters and Rogers Inc

Others
Oh Sung Chemical Ind Co Ltd

Calcium Acetate

UK
Macron Chemicals UK

Other European
Lehmann & Voss & Co

USA
Advance Scientific & Chemical Inc
AerChem Inc
Allan Chemical Corp
Charkit Chemical Corp
Macron Chemicals
Penta Manufacturing Co
Ruger Chemical Co Inc
Seidler Chemical Company Inc
Spectrum Chemicals & Laboratory Products Inc
Voigt Global Distribution Inc

Calcium Alginate

USA
Advance Scientific & Chemical Inc

Others
Kimica Corporation

Calcium Carbonate

UK
3M United Kingdom Plc
Allchem Performance
Blagden Specialty Chemicals Ltd
Fisher Scientific UK Ltd
Forum Products Ltd
Lehvoss UK Ltd
Macron Chemicals UK
Paroxite (London) Ltd
Peter Whiting (Chemicals) Ltd
Pumex (UK) Limited
Tennants Distribution Ltd
Thew, Arnott and Co Ltd

Other European
Alfa Aesar Johnson Matthey GmbH
Brenntag GmbH
Dr Paul Lohmann GmbH KG
Lehmann & Voss & Co
Magnesia GmbH
Schaefer Kalk GmbH & Co KG

USA
Aceto Corp
Advance Scientific & Chemical Inc
Asiamerica Ingredients Inc
Barrington Nutritionals
Brenntag North America Inc
Brenntag Southwest
Charles B Chrystal Co Inc
EM Sergeant Pulp & Chemical Co Inc
Fisher Scientific
Generichem Corp
Hawkins Inc
Hummel Croton Inc
Integra Chemical Co
Macron Chemicals
Mutchler Inc
Particle Dynamics Inc
Penta Manufacturing Co
Premium Ingredients Ltd
Ruger Chemical Co Inc
Seidler Chemical Company Inc
Spectrum Chemicals & Laboratory Products Inc
SPI Pharma Inc
Triple Crown America
Voigt Global Distribution Inc
Whittaker Clark, and Daniels Inc

Others
Boith Limited (China)
Signet Chemical Corporation Pvt Ltd
Wintersun Chemical

Calcium Chloride

UK
Peter Whiting (Chemicals) Ltd

Other European
Alfa Aesar Johnson Matthey GmbH
Brenntag AG
Lehmann & Voss & Co

USA
Advance Scientific & Chemical Inc
AerChem Inc
Alfa Chem
Astro Chemicals Inc
Penta Manufacturing Co
RIA International
Seidler Chemical Company Inc
Spectrum Chemicals & Laboratory Products Inc

Calcium Hydroxide

Other European
Alfa Aesar Johnson Matthey GmbH
Panreac Quimica SAU

USA
Advance Scientific & Chemical Inc
Fisher Scientific
Penta Manufacturing Co
Seidler Chemical Company Inc
Thomas Scientific

Others
Wintersun Chemical

Calcium Lactate

Other European
Brenntag NV
Lehmann & Voss & Co

USA
Advance Scientific & Chemical Inc
Alfa Chem
Kraft Chemical Co
Premium Ingredients Ltd
Purac
RIA International
Seidler Chemical Company Inc
Spectrum Chemicals & Laboratory Products Inc

Others
SD Fine-Chem Ltd
Wintersun Chemical

Calcium Phosphate, Dibasic Anhydrous

UK
Blagden Specialty Chemicals Ltd
Forum Products Ltd
Rettenmaier UK Ltd

Other European
Alfa Aesar Johnson Matthey GmbH
Brenntag GmbH
Chemische Fabrik Budenheim KG
IMCD Group BV
Lehmann & Voss & Co

USA
Advance Scientific & Chemical Inc
Alfa Chem
AnMar International
Asiamerica Ingredients Inc
Brenntag North America Inc
Budenheim USA Inc
Charkit Chemical Corp
Fuji Chemical Industries Health Science (USA) Inc
Innophos Inc
JRS Pharma LP
Mutchler Inc
Penta Manufacturing Co
Premium Ingredients Ltd
Ruger Chemical Co Inc
Seidler Chemical Company Inc
Spectrum Chemicals & Laboratory Products Inc
Triple Crown America
Zhong Ya Chemical (USA) Ltd

Others
Arihant Trading Co
Fuji Chemical Industry Co Ltd
Signet Chemical Corporation Pvt Ltd
Wintersun Chemical

Calcium Phosphate, Dibasic Dihydrate

UK
Blagden Specialty Chemicals Ltd
Fisher Scientific UK Ltd

1

Forum Products Ltd
Peter Whiting (Chemicals) Ltd
Rettenmaier UK Ltd

Other European
Alfa Aesar Johnson Matthey GmbH
Brenntag GmbH
Chemische Fabrik Budenheim KG
IMCD Group BV

USA
Aceto Corp
Advance Scientific & Chemical Inc
Asiamerica Ingredients Inc
Brenntag North America Inc
Budenheim USA Inc
Fisher Scientific
Innophos Inc
JRS Pharma LP
Mutchler Inc
Premium Ingredients Ltd
Ruger Chemical Co Inc
Spectrum Chemicals & Laboratory Products Inc
Triple Crown America

Others
Signet Chemical Corporation Pvt Ltd
Wintersun Chemical

Calcium Phosphate, Tribasic

UK
Fisher Scientific UK Ltd
Peter Whiting (Chemicals) Ltd

Other European
Alfa Aesar Johnson Matthey GmbH
Brenntag GmbH
Brenntag NV
Chemische Fabrik Budenheim KG
Panreac Quimica SAU

USA
Advance Scientific & Chemical Inc
American International Chemical Inc
Brenntag North America Inc
Budenheim USA Inc
CarboMer Inc
Fisher Scientific
Innophos Inc
Penta Manufacturing Co
Premium Ingredients Ltd
Ruger Chemical Co Inc
Seidler Chemical Company Inc
Spectrum Chemicals & Laboratory Products Inc
Triple Crown America
Voigt Global Distribution Inc
Zhong Ya Chemical (USA) Ltd

Others
Signet Chemical Corporation Pvt Ltd
Wintersun Chemical

Calcium Silicate

Other European
Alfa Aesar Johnson Matthey GmbH

USA
Advance Scientific & Chemical Inc
Celite Corporation
Mutchler Inc
Spectrum Chemicals & Laboratory Products Inc

Others
Wintersun Chemical

Calcium Stearate

UK
Allchem Performance
James M Brown Ltd
Lehvoss UK Ltd
Macron Chemicals UK
Peter Greven UK
Tennants Distribution Ltd

Other European
Alfa Aesar Johnson Matthey GmbH
CPI Chemicals
Dr Paul Lohmann GmbH KG
Lehmann & Voss & Co
Magnesia GmbH
Panreac Quimica SAU

USA
Aceto Corp
Acme-Hardesty
Advance Scientific & Chemical Inc
AerChem Inc
Allan Chemical Corp
American International Chemical Inc
Ashland Inc
Astro Chemicals Inc
Brenntag North America Inc
CarboMer Inc
Charkit Chemical Corp
Ferro Corp
Integra Chemical Co
KIC Chemicals Inc
Kraft Chemical Co
Macron Chemicals
Mutchler Inc
Parchem Trading Ltd
Penta Manufacturing Co
Premium Ingredients Ltd
Ruger Chemical Co Inc
Seidler Chemical Company Inc
Spectrum Chemicals & Laboratory Products Inc
Triple Crown America
Voigt Global Distribution Inc
Whittaker Clark, and Daniels Inc

Others
Charles Tennant & Co (Canada) Ltd
Signet Chemical Corporation Pvt Ltd
Wintersun Chemical

Calcium Sulfate

UK
Lehvoss UK Ltd
Peter Whiting (Chemicals) Ltd
Rettenmaier UK Ltd

Other European
Alfa Aesar Johnson Matthey GmbH
Dr Paul Lohmann GmbH KG
Merck KGaA

USA
Advance Scientific & Chemical Inc
Charles B Chrystal Co Inc
EM Sergeant Pulp & Chemical Co Inc
JRS Pharma LP
Particle Dynamics Inc
Penta Manufacturing Co
RIA International
Seidler Chemical Company Inc
Spectrum Chemicals & Laboratory Products Inc
Triple Crown America
Voigt Global Distribution Inc
Whittaker Clark, and Daniels Inc

Canola Oil

UK
AarhusKarlshamn UK Ltd
Adina Chemicals Ltd

Other European
AarhusKarlshamn AB

USA
AarhusKarlshamn USA Inc
Arista Industries Inc
Avatar Corp
Charkit Chemical Corp
KIC Chemicals Inc
Lipo Chemicals Inc
Penta Manufacturing Co
Pokonobe Industries Inc
Welch, Holme & Clark Co Inc

Carbomer

UK
A and E Connock (Perfumery and Cosmetics) Ltd
Evonik Goldschmidt UK Ltd

USA
Advance Scientific & Chemical Inc
Lubrizol Advanced Materials Inc.
Rita Corp
Ruger Chemical Co Inc
Seidler Chemical Company Inc
Spectrum Chemicals & Laboratory Products Inc
Voigt Global Distribution Inc

Others
Corel Pharma Chem
Kawarlal Excipients Pvt Ltd

Carbon Dioxide

UK
Air Liquide UK Ltd
Air Products (Gases) plc
BOC Gases

USA
Air Liquide America Corp
BOC Gases (USA)

Carboxymethylcellulose Calcium

UK
CP Kelco UK Ltd

USA
Advance Scientific & Chemical Inc
Ashland Inc
CP Kelco US Inc
Kraft Chemical Co

Carboxymethylcellulose Sodium

UK
CP Kelco UK Ltd
IMCD UK Ltd

Other European
Akzo Nobel Functional Chemicals bv
Alfa Aesar Johnson Matthey GmbH
Biogrund GmbH
Lehmann & Voss & Co
Panreac Quimica SAU

USA
American International Chemical Inc
Ashland Inc
Ashland Specialty Ingredients
Brenntag North America Inc
Brenntag Southwest
CarboMer Inc
CP Kelco US Inc
Dow Chemical Co
Kraft Chemical Co
Ruger Chemical Co Inc
Spectrum Chemicals & Laboratory Products Inc
Whittaker Clark, and Daniels Inc

Others
Arihant Trading Co

Ashland India Pvt Ltd
Kawarlal Excipients Pvt Ltd
Lucid Colloids Ltd
Nippon Paper Chemicals Co. Ltd
Signet Chemical Corporation Pvt Ltd

Carrageenan

UK
A and E Connock (Perfumery and Cosmetics) Ltd
AF Suter and Co Ltd
Blagden Specialty Chemicals Ltd
CP Kelco UK Ltd
Danisco UK Ltd
Lehvoss UK Ltd
Thew, Arnott and Co Ltd

Other European
Brenntag GmbH
Danisco A/S
FMC Biopolymer
IMCD Group BV

USA
Brenntag North America Inc
Brenntag Southwest
CarboMer Inc
CP Kelco US Inc
Danisco USA Inc
Seidler Chemical Company Inc
Spectrum Chemicals & Laboratory Products Inc
TIC Gums
Voigt Global Distribution Inc

Others
Priyanka Pharma/M K Ingredients & Specialities
Signet Chemical Corporation Pvt Ltd

Castor Oil

UK
A and E Connock (Perfumery and Cosmetics) Ltd
Adina Chemicals Ltd
Alembic Products Ltd
Blagden Specialty Chemicals Ltd
Croda Europe Ltd
Fisher Scientific UK Ltd
Kimpton Brothers Ltd
Lehvoss UK Ltd
Macron Chemicals UK
White Sea and Baltic Company Ltd
William Ransom & Son plc
WS Lloyd Ltd

Other European
Alfa Aesar Johnson Matthey GmbH
Brenntag GmbH
Panreac Quimica SAU

USA
Acme-Hardesty
Advance Scientific & Chemical Inc
Arista Industries Inc
Avatar Corp
Charkit Chemical Corp
Croda Inc
Fisher Scientific
KIC Chemicals Inc
Lipo Chemicals Inc
Lipscomb Chemical Company
Macron Chemicals
Mutchler Inc
Paddock Laboratories Inc
Penta Manufacturing Co
Pokonobe Industries Inc
Ruger Chemical Co Inc
Seidler Chemical Company Inc
Spectrum Chemicals & Laboratory Products Inc
Triple Crown America
Vertellus Specialties Inc
Voigt Global Distribution Inc
Welch, Holme & Clark Co Inc

Others
Univar Canada Ltd

Castor Oil, Hydrogenated

UK
Allchem Performance
Cognis UK Ltd
Cornelius Group plc
Croda Europe Ltd
Evonik Goldschmidt UK Ltd
Lehvoss UK Ltd
White Sea and Baltic Company Ltd

Other European
Arion & Delahaye
Cognis GmbH

USA
ABITEC Corp
Cognis Corp
Croda Inc
GR O'Shea Company
Mutchler Inc
Penta Manufacturing Co

Others
Signet Chemical Corporation Pvt Ltd
Wintersun Chemical

Cellaburate

UK
Eastman Company UK Ltd

Other European
Acros Organics BVBA

USA
Acros Organics
CarboMer Inc
Eastman Chemical Co

Cellulose, Microcrystalline

UK
Allchem Performance
Blagden Specialty Chemicals Ltd
Cornelius Group plc
DFE Pharma UK
Forum Products Ltd
IMCD UK Ltd
ISP Europe
Rettenmaier UK Ltd

Other European
Alfa Aesar Johnson Matthey GmbH
Biesterfeld Spezialchemie GmbH
DFE Pharma
FMC Biopolymer
IMCD Group BV
J Rettenmaier & Söhne GmbH and Co.KG
Lehmann & Voss & Co
NP Pharm
Pharmatrans Sanaq AG

USA
Alfa Chem
American International Chemical Inc
Ashland Inc
Asiamerica Ingredients Inc
Barrington Nutritionals
CarboMer Inc
Dastech International Inc
EM Sergeant Pulp & Chemical Co Inc
International Specialty Products Inc (ISP)
JRS Pharma LP
Mutchler Inc
Parchem
Spectrum Chemicals & Laboratory Products Inc
Voigt Global Distribution Inc

Others
Aastrid International
Arihant Trading Co
Asahi Kasei Corporation
Blanver
Charles Tennant & Co (Canada) Ltd
Glide Chem Pvt Ltd
Kawarlal Excipients Pvt Ltd
LS Raw Materials Ltd
NB Entrepreneurs
Signet Chemical Corporation Pvt Ltd
Wintersun Chemical

Cellulose, Microcrystalline and Carboxymethylcellulose Sodium

Other European
FMC Biopolymer
J Rettenmaier & Söhne GmbH and Co.KG

USA
FMC Biopolymer (USA)
JRS Pharma LP

Others
Signet Chemical Corporation Pvt Ltd

Cellulose, Powdered

UK
Allchem Performance

Other European
Blanver (Europe)
CFF GmbH and Co KG
J Rettenmaier & Söhne GmbH and Co.KG

USA
Blanver (USA)
EM Sergeant Pulp & Chemical Co Inc
International Fiber Corporation
Mutchler Inc
Triple Crown America
Voigt Global Distribution Inc

Others
Blanver Farmoquímica Ltda
NB Entrepreneurs
Nippon Paper Chemicals Co. Ltd

Cellulose, Silicified Microcrystalline

UK
Rettenmaier UK Ltd

Other European
J Rettenmaier & Söhne GmbH and Co.KG

USA
JRS Pharma LP
Mutchler Inc

Cellulose Acetate

UK
Eastman Company UK Ltd
IMCD UK Ltd

USA
CarboMer Inc
Eastman Chemical Co

Cellulose Acetate Phthalate

UK
Eastman Company UK Ltd
IMCD UK Ltd

1

Other European
FMC Biopolymer

USA
CarboMer Inc
Eastman Chemical Co

Ceratonia

USA
Ashland Inc

Ceresin

USA
Frank B Ross Co Inc
Koster Keunen LLC
Lambent Technologies
Mutchler Inc
Ruger Chemical Co Inc

Cetostearyl Alcohol

UK
A and E Connock (Perfumery and Cosmetics) Ltd
Allchem Performance
Cognis UK Ltd
Croda Europe Ltd
Efkay Chemicals Ltd
Evonik Goldschmidt UK Ltd
H Foster & Co (Stearines) Ltd
White Sea and Baltic Company Ltd

Other European
BASF Aktiengesellschaft
Berg + Schmidt (GmbH & Co.) KG
Cognis GmbH

USA
Acme-Hardesty
BASF Corp
Cognis Corp
Croda Inc
Evonik Degussa Corp
Penta Manufacturing Co
Rita Corp
Spectrum Chemicals & Laboratory Products Inc
Voigt Global Distribution Inc

Others
BASF Japan Ltd
LS Raw Materials Ltd
Signet Chemical Corporation Pvt Ltd

Cetrimide

UK
Peter Whiting (Chemicals) Ltd
Raught Ltd
White Sea and Baltic Company Ltd

Other European
FeF Chemicals A/S
Panreac Quimica SAU

USA
Aceto Corp
Alfa Chem
Triple Crown America

Others
LS Raw Materials Ltd

Cetyl Alcohol

UK
A and E Connock (Perfumery and Cosmetics) Ltd
AarhusKarlshamn UK Ltd
Adina Chemicals Ltd
Allchem Performance
Cognis UK Ltd
Croda Europe Ltd
Efkay Chemicals Ltd
Evonik Goldschmidt UK Ltd
Kimpton Brothers Ltd
Sasol UK Ltd
White Sea and Baltic Company Ltd

Other European
AarhusKarlshamn AB
Alfa Aesar Johnson Matthey GmbH
Berg + Schmidt (GmbH & Co.) KG
Brenntag GmbH
Chempri Oleochemicals (Europe)
Cognis GmbH
Corcoran Chemicals Ltd
Impag GmbH
Panreac Quimica SAU

USA
AarhusKarlshamn USA Inc
Acme-Hardesty
Advance Scientific & Chemical Inc
Brenntag North America Inc
CarboMer Inc
Cognis Corp
Croda Inc
Hawkins Inc
KIC Chemicals Inc
Kraft Chemical Co
Lipo Chemicals Inc
Mutchler Inc
P & G Chemicals
Penta Manufacturing Co
Protameen Chemicals Inc
Rita Corp
Ruger Chemical Co Inc
Sasol North America Inc
Seidler Chemical Company Inc
Spectrum Chemicals & Laboratory Products Inc

Others
Boith Limited (China)
LS Raw Materials Ltd
SD Fine-Chem Ltd
VVF Limited
Wintersun Chemical

Cetyl Palmitate

UK
Cornelius Group plc

Other European
Brenntag GmbH
Julius Hoesch GmbH & Co KG
SABO SPA
Union Derivan, SA (UNDESA)

USA
Ashland Inc
Croda Inc
Hallstar Company, The
Kraft Chemical Co
Penta Manufacturing Co
Ruger Chemical Co Inc
Spectrum Chemicals & Laboratory Products Inc
Stepan Co
Universal Preserv-A-Chem Inc

Cetylpyridinium Chloride

Other European
Alfa Aesar Johnson Matthey GmbH

USA
CarboMer Inc
Vertellus Specialties Inc

Chitosan

Other European
FMC Biopolymer
Kraeber GmbH & Co

USA
AerChem Inc
Alfa Chem
Fortitech Inc
Penta Manufacturing Co

Others
Boith Limited (China)
Wintersun Chemical

Chlorhexidine

USA
Alfa Chem
George Uhe Co Inc
Napp Technologies LLC

Others
LS Raw Materials Ltd
Wintersun Chemical

Chlorobutanol

UK
Blagden Specialty Chemicals Ltd
Courtin & Warner Ltd

USA
Allergan Inc
Carolina Biological Supply Co
Penta Manufacturing Co
Ruger Chemical Co Inc
Spectrum Chemicals & Laboratory Products Inc

Others
LS Raw Materials Ltd

Chlorocresol

USA
RIA International

Others
LS Raw Materials Ltd

Chlorofluorocarbons (CFC)

Other European
Honeywell Specialty Chemicals Seelze GmbH

Chloroxylenol

Other European
Alfa Aesar Johnson Matthey GmbH
Clariant GmbH

USA
Advance Scientific & Chemical Inc
Clariant Corp
Spectrum Chemicals & Laboratory Products Inc

Cholesterol

UK
A and E Connock (Perfumery and Cosmetics) Ltd
Macron Chemicals UK
Paroxite (London) Ltd
Ubichem plc

Other European
Alfa Aesar Johnson Matthey GmbH

USA
Aceto Corp
Advance Scientific & Chemical Inc
Alfa Chem
Amresco Inc

Avanti Polar Lipids Inc
BOC Sciences
Charles Bowman & Co
Penta Manufacturing Co
Rita Corp
Ruger Chemical Co Inc
Spectrum Chemicals & Laboratory Products Inc
Voigt Global Distribution Inc

Others
SD Fine-Chem Ltd

Citric Acid Monohydrate

UK
Blagden Specialty Chemicals Ltd
Cargill PLC
Courtin & Warner Ltd
Fisher Scientific UK Ltd
Macron Chemicals UK
Peter Whiting (Chemicals) Ltd
Raught Ltd
Roche Products Ltd
Sigma-Aldrich Company Ltd
Tate and Lyle plc
Tennants Distribution Ltd
Ubichem plc

Other European
Alfa Aesar Johnson Matthey GmbH
Arion & Delahaye
Brenntag GmbH
Cargill France SAS
Corcoran Chemicals Ltd
Dr Paul Lohmann GmbH KG
Jungbunzlauer AG
Panreac Quimica SAU

USA
Aceto Corp
Advance Scientific & Chemical Inc
Allan Chemical Corp
Amresco Inc
Ashland Inc
Avatar Corp
Brenntag North America Inc
Brenntag Southwest
Charkit Chemical Corp
Dastech International Inc
Fisher Scientific
George Uhe Co Inc
Hawkins Inc
Jungbunzlauer Inc
Kraft Chemical Co
Macron Chemicals
Mutchler Inc
Penta Manufacturing Co
RIA International
Spectrum Chemicals & Laboratory Products Inc
Triple Crown America
Voigt Global Distribution Inc

Others
Boith Limited (China)
LS Raw Materials Ltd
SD Fine-Chem Ltd
Univar Canada Ltd
Wintersun Chemical

Coconut Oil

UK
A and E Connock (Perfumery and Cosmetics) Ltd
Courtin & Warner Ltd
Sigma-Aldrich Company Ltd

USA
ABITEC Corp
Akzo Nobel Chemicals Inc
Arista Industries Inc
Avatar Corp
KIC Chemicals Inc
Kraft Chemical Co
Lambent Technologies
Pokonobe Industries Inc
Spectrum Chemicals & Laboratory Products Inc
Voigt Global Distribution Inc

Colloidal Silicon Dioxide

UK
Evonik Degussa Ltd
Grace Davison
Wacker Chemicals Ltd

Other European
Cabot GmbH
Evonik Industries AG
Wacker-Chemie GmbH

USA
Brenntag North America Inc
Cabot Corp
Evonik Degussa Corp
Ruger Chemical Co Inc
Spectrum Chemicals & Laboratory Products Inc
Wacker Chemical Corp

Colophony

USA
Penta Manufacturing Co
Spectrum Chemicals & Laboratory Products Inc

Others
Boith Limited (China)

Coloring Agents

UK
A and E Connock (Perfumery and Cosmetics) Ltd
Colorcon Ltd
DFE Pharma UK
Thew, Arnott and Co Ltd
White Sea and Baltic Company Ltd

Other European
DFE Pharma
IMCD Group BV

USA
Advance Scientific & Chemical Inc
AerChem Inc
Ashland Inc
Colorcon Inc
Sensient Technologies
Triple Crown America
Whittaker Clark, and Daniels Inc

Copovidone

UK
BASF Plc
ISP Europe

Other European
BASF Aktiengesellschaft
CPI Chemicals

USA
BASF Corp
International Specialty Products Inc (ISP)
Mutchler Inc

Others
BASF Japan Ltd
Signet Chemical Corporation Pvt Ltd

Corn Oil

UK
A and E Connock (Perfumery and Cosmetics) Ltd
AarhusKarlshamn UK Ltd
Alembic Products Ltd
Cargill PLC
Cognis UK Ltd

Other European
AarhusKarlshamn AB
Cargill France SAS
Cognis GmbH

USA
AarhusKarlshamn USA Inc
Arista Industries Inc
Avatar Corp
Cargill Inc
Charkit Chemical Corp
Cognis Corp
Grain Processing Corp
Integra Chemical Co
KIC Chemicals Inc
Penta Manufacturing Co
Pokonobe Industries Inc
Ruger Chemical Co Inc
Spectrum Chemicals & Laboratory Products Inc
Voigt Global Distribution Inc
Welch, Holme & Clark Co Inc

Corn Starch and Pregelatinized Starch

UK
Colorcon Ltd

USA
CarboMer Inc
Colorcon Inc

Corn Syrup Solids

UK
National Starch Personal Care

USA
Ashland Inc
Asiamerica Ingredients Inc
Cerestar USA Inc
Corn Products U.S.
Grain Processing Corp
National Starch Personal Care (USA)
Roquette America Inc

Cottonseed Oil

UK
A and E Connock (Perfumery and Cosmetics) Ltd
Blagden Specialty Chemicals Ltd
Fisher Scientific UK Ltd

USA
Advance Scientific & Chemical Inc
Arista Industries Inc
Avatar Corp
Charkit Chemical Corp
Fisher Scientific
Hawkins Inc
Parchem
Penta Manufacturing Co
Pokonobe Industries Inc
Ruger Chemical Co Inc
Spectrum Chemicals & Laboratory Products Inc
Voigt Global Distribution Inc
Welch, Holme & Clark Co Inc

1

Cresol

Other European
Alfa Aesar Johnson Matthey GmbH

USA
Advance Scientific & Chemical Inc
Amresco Inc
Penta Manufacturing Co
Seidler Chemical Company Inc
Spectrum Chemicals & Laboratory Products Inc

Croscarmellose Sodium

UK
Allchem Performance
Avebe UK Ltd
Blagden Specialty Chemicals Ltd
CP Kelco UK Ltd
DFE Pharma UK
IMCD UK Ltd

Other European
Avebe Group
DFE Pharma
FMC Biopolymer
IMCD Group BV
J Rettenmaier & Söhne GmbH and Co.KG
Lehmann & Voss & Co

USA
Advance Scientific & Chemical Inc
American International Chemical Inc
Avebe America Inc
CarboMer Inc
CP Kelco US Inc
EM Sergeant Pulp & Chemical Co Inc
Generichem Corp
Mutchler Inc
Parchem
RIA International
Spectrum Chemicals & Laboratory Products Inc
Voigt Global Distribution Inc

Others
Asahi Kasei Corporation
Blanver
Charles Tennant & Co (Canada) Ltd
Kawarlal Excipients Pvt Ltd
NB Entrepreneurs

Crospovidone

UK
BASF Plc
Blagden Specialty Chemicals Ltd
ISP Europe

Other European
August Hedinger GmbH & Co KG
BASF Aktiengesellschaft
IMCD Group BV
NP Pharm

USA
American International Chemical Inc
BASF Corp
Generichem Corp
International Specialty Products Inc (ISP)
Mutchler Inc

Others
BASF Japan Ltd
Glide Chem Pvt Ltd
Signet Chemical Corporation Pvt Ltd

Cyclodextrins

UK
Roquette (UK) Ltd
Wacker Chemicals Ltd

Other European
Alfa Aesar Johnson Matthey GmbH
Harke Pharma GmbH
Roquette Frères
Wacker-Chemie GmbH

USA
CarboMer Inc
CTD Inc
Mutchler Inc
Roquette America Inc
Spectrum Chemicals & Laboratory Products Inc
Wacker Chemical Corp

Others
Kawarlal Excipients Pvt Ltd

Cyclomethicone

UK
A and E Connock (Perfumery and Cosmetics) Ltd
Dow Corning (UK)

USA
Dow Corning

Decyl Oleate

UK
A and E Connock (Perfumery and Cosmetics) Ltd

Other European
Chempri Oleochemicals (Europe)
Mosselman

Others
Signet Chemical Corporation Pvt Ltd

Denatonium Benzoate

UK
A and E Connock (Perfumery and Cosmetics) Ltd

USA
Barrington Nutritionals
Burlington Bio-medical and Scientific Corp
Chart Corp Inc
RIA International
Spectrum Chemicals & Laboratory Products Inc

Dextran

UK
Sigma-Aldrich Company Ltd

USA
Advance Scientific & Chemical Inc
CarboMer Inc

Dextrates

UK
Rettenmaier UK Ltd

Other European
J Rettenmaier & Söhne GmbH and Co.KG

USA
Advance Scientific & Chemical Inc
JRS Pharma LP
Mutchler Inc
RIA International
Spectrum Chemicals & Laboratory Products Inc

Dextrin

UK
Avebe UK Ltd
Roquette (UK) Ltd

Other European
Alfa Aesar Johnson Matthey GmbH
Avebe Group
Roquette Frères

USA
Alfa Chem
Avebe America Inc
EM Sergeant Pulp & Chemical Co Inc
Hummel Croton Inc
Seidler Chemical Company Inc

Others
Signet Chemical Corporation Pvt Ltd

Dextrose

UK
Cargill PLC
Fisher Scientific UK Ltd
IMCD UK Ltd
Macron Chemicals UK
Pfanstiehl (Europe) Ltd
Raught Ltd
Roquette (UK) Ltd

Other European
Alfa Aesar Johnson Matthey GmbH
Biesterfeld Spezialchemie GmbH
Brenntag GmbH
Cargill France SAS
Corcoran Chemicals Ltd
Roquette Frères

USA
Advance Scientific & Chemical Inc
AerChem Inc
Ashland Inc
Brenntag North America Inc
Brenntag Southwest
CarboMer Inc
Cargill Inc
EM Sergeant Pulp & Chemical Co Inc
Ferro Corp
Fisher Scientific
Macron Chemicals
Mutchler Inc
Penta Manufacturing Co
Premium Ingredients Ltd
Roquette America Inc
Ruger Chemical Co Inc
Seidler Chemical Company Inc
Spectrum Chemicals & Laboratory Products Inc
Voigt Global Distribution Inc

Others
LS Raw Materials Ltd
SD Fine-Chem Ltd
Signet Chemical Corporation Pvt Ltd
Univar Canada Ltd

Dibutyl Phthalate

Other European
Panreac Quimica SAU

USA
Eastman Chemical Co
Penta Manufacturing Co
Seidler Chemical Company Inc

Others
Boith Limited (China)

Dibutyl Sebacate

UK
A and E Connock (Perfumery and Cosmetics) Ltd

USA
Acme-Hardesty

Mutchler Inc
Penta Manufacturing Co
Vertellus Specialties Inc

Others
Wintersun Chemical

Diethanolamine

UK
Sasol UK Ltd
Tennants Distribution Ltd
Ubichem plc

Other European
Alfa Aesar Johnson Matthey GmbH
Sasol Germany GmbH

USA
Advance Scientific & Chemical Inc
Amresco Inc
Brenntag North America Inc
Sasol North America Inc
Seidler Chemical Company Inc
Spectrum Chemicals & Laboratory Products Inc
Triple Crown America

Others
Wintersun Chemical

Diethylene Glycol Monoethyl Ether

Other European
Alfa Aesar Johnson Matthey GmbH
Gattefossé Group

USA
Alzo International Inc
Ashland Inc
Gattefossé Corp
HBCChem Inc
Penta Manufacturing Co
Spectrum Chemicals & Laboratory Products Inc

Others
Gattefossé Canada Inc.

Diethyl Phthalate

UK
BASF Plc
Eastman Company UK Ltd

Other European
BASF Aktiengesellschaft
Brenntag GmbH
Panreac Quimica SAU

USA
Advance Scientific & Chemical Inc
Allan Chemical Corp
BASF Corp
Brenntag North America Inc
Eastman Chemical Co
Penta Manufacturing Co
Seidler Chemical Company Inc
Spectrum Chemicals & Laboratory Products Inc

Others
BASF Japan Ltd

Diethyl Sebacate

USA
Penta Manufacturing Co

Difluoroethane (HFC)

Other European
DuPont de Nemours Int'l SA
Solvay Fluor GmbH

USA
Aeropres Corp

Dimethicone

UK
A and E Connock (Perfumery and Cosmetics) Ltd
Dow Corning (UK)
Evonik Goldschmidt UK Ltd
IMCD UK Ltd

Other European
Biesterfeld Spezialchemie GmbH
IMCD Group BV

USA
Chemtura Corporation
Dow Corning

Dimethyl Ether

UK
Air Liquide UK Ltd

Other European
DuPont de Nemours Int'l SA

USA
Aeropres Corp
Air Liquide America Corp

Dimethyl Phthalate

Other European
Alfa Aesar Johnson Matthey GmbH

USA
Eastman Chemical Co
Penta Manufacturing Co
Seidler Chemical Company Inc

Dimethyl Sulfoxide

Other European
Alfa Aesar Johnson Matthey GmbH
Panreac Quimica SAU

USA
Advance Scientific & Chemical Inc
Gaylord Chemical Company LLC
Penta Manufacturing Co
Seidler Chemical Company Inc
Spectrum Chemicals & Laboratory Products Inc

Others
Wintersun Chemical

Dimethylacetamide

UK
BASF Plc
Sigma-Aldrich Company Ltd

Other European
Alfa Aesar Johnson Matthey GmbH
BASF Aktiengesellschaft
DuPont de Nemours Int'l SA

USA
DuPont
Seidler Chemical Company Inc
Spectrum Chemicals & Laboratory Products Inc

Dipropylene Glycol

UK
Dow Chemical Company (UK)

USA
Dow Chemical Co
Penta Manufacturing Co
Seidler Chemical Company Inc

Disodium Edetate

UK
Macron Chemicals UK

Other European
August Hedinger GmbH & Co KG

USA
Macron Chemicals

Docusate Sodium

USA
Penta Manufacturing Co
Ruger Chemical Co Inc
Spectrum Chemicals & Laboratory Products Inc

Others
Signet Chemical Corporation Pvt Ltd

Edetic Acid

UK
Sigma-Aldrich Company Ltd

Other European
Akzo Nobel Functional Chemicals bv
Alfa Aesar Johnson Matthey GmbH

USA
Advance Scientific & Chemical Inc
AerChem Inc
Akzo Nobel Chemicals Inc
American International Chemical Inc
Brenntag North America Inc
Dow Chemical Co
Kraft Chemical Co
Penta Manufacturing Co
Seidler Chemical Company Inc
Spectrum Chemicals & Laboratory Products Inc
Thomas Scientific

Erythorbic Acid

Other European
Alfa Aesar Johnson Matthey GmbH
Brenntag AG

USA
Alfa Chem
Biddle Sawyer Corp
Brainerd Chemical Company Inc
KIC Chemicals Inc
Penta Manufacturing Co
Prinova
RIA International
Seidler Chemical Company Inc
Zhong Ya Chemical (USA) Ltd

Others
Wintersun Chemical

Erythritol

UK
Cargill PLC

Other European
Cargill France SAS

USA
Cargill Inc
EM Sergeant Pulp & Chemical Co Inc
Premium Ingredients Ltd
RIA International

Others
Cerestar Cargill Resources Maize Industry Co Ltd

Ethyl Acetate

UK
Eastman Company UK Ltd
Fisher Scientific UK Ltd
Leading Solvent Supplies Ltd
Tennants Distribution Ltd

Other European
Alfa Aesar Johnson Matthey GmbH
August Hedinger GmbH & Co KG
Corcoran Chemicals Ltd
Panreac Quimica SAU

USA
Advance Scientific & Chemical Inc
Astro Chemicals Inc
Dow Chemical Co
Eastman Chemical Co
Fisher Scientific
Penta Manufacturing Co
Ruger Chemical Co Inc
Seidler Chemical Company Inc
Spectrum Chemicals & Laboratory Products Inc
Thomas Scientific
Triple Crown America

Others
Aastrid International

Ethyl Lactate

UK
A and E Connock (Perfumery and Cosmetics) Ltd
White Sea and Baltic Company Ltd

Other European
Alfa Aesar Johnson Matthey GmbH

USA
Advance Scientific & Chemical Inc
Alzo International Inc
Penta Manufacturing Co

Others
SD Fine-Chem Ltd
Wintersun Chemical

Ethyl Maltol

USA
Advance Scientific & Chemical Inc
Chemtex USA Inc
Mitsubishi International Food Ingredients
Penta Manufacturing Co
Premium Ingredients Ltd

Others
Priyanka Pharma/M K Ingredients & Specialities

Ethyl Oleate

UK
A and E Connock (Perfumery and Cosmetics) Ltd
Croda Europe Ltd
Sigma-Aldrich Company Ltd

Other European
Alfa Aesar Johnson Matthey GmbH

USA
Croda Inc
Penta Manufacturing Co
Spectrum Chemicals & Laboratory Products Inc

Ethyl Vanillin

UK
Blagden Specialty Chemicals Ltd
Courtin & Warner Ltd
Rhodia Organic Fine Ltd

Other European
Alfa Aesar Johnson Matthey GmbH
Helm AG

USA
AerChem Inc
Ashland Inc
BOC Sciences
Brenntag North America Inc
Brenntag Southwest
Chart Corp Inc
Helm New York Inc
KIC Chemicals Inc
Penta Manufacturing Co
Premium Ingredients Ltd
Rhodia Pharma Solutions Inc
Spectrum Chemicals & Laboratory Products Inc
Voigt Global Distribution Inc

Others
Priyanka Pharma/M K Ingredients & Specialities
Wintersun Chemical

Ethylcellulose

UK
Hercules Ltd
IMCD UK Ltd

Other European
Alfa Aesar Johnson Matthey GmbH
FMC Biopolymer
Harke Pharma GmbH

USA
Alfa Chem
ASHA cellulose (Private Ltd)
Ashland Specialty Ingredients
Dow Chemical Co
Mutchler Inc
Spectrum Chemicals & Laboratory Products Inc

Others
Ashland India Pvt Ltd
Dow Chemical International Pvt Ltd
Glide Chem Pvt Ltd
Kawarlal Excipients Pvt Ltd
Signet Chemical Corporation Pvt Ltd
Univar Canada Ltd

Ethylene Glycol and Vinyl Alcohol Grafted Copolymer

UK
A and E Connock (Perfumery and Cosmetics) Ltd

Other European
BASF Aktiengesellschaft

Ethylene Glycol Stearates

USA
Acme-Hardesty
Seidler Chemical Company Inc

Ethylene Vinyl Acetate

UK
3M United Kingdom Plc

USA
3M Drug Delivery Systems

Ethylparaben

UK
Blagden Specialty Chemicals Ltd
Clariant UK Ltd
Evonik Degussa Ltd
Lanxess Ltd

Other European
Alfa Aesar Johnson Matthey GmbH
Clariant GmbH
Induchem AG

USA
Acme-Hardesty
Advance Scientific & Chemical Inc
Alfa Chem
BOC Sciences
Brenntag North America Inc
Clariant Corp
Dastech International Inc
Hallstar Company, The
KIC Chemicals Inc
Lanxess Corp
Lipo Chemicals Inc
Napp Technologies LLC
Protameen Chemicals Inc
Ruger Chemical Co Inc
Seidler Chemical Company Inc
Spectrum Chemicals & Laboratory Products Inc

Others
Charles Tennant & Co (Canada) Ltd
Univar Canada Ltd

Fructose

UK
Cargill PLC
Fisher Scientific UK Ltd
Forum Products Ltd
Pfanstiehl (Europe) Ltd
Raught Ltd

Other European
Alfa Aesar Johnson Matthey GmbH
Brenntag GmbH
Cargill France SAS
J Rettenmaier & Söhne GmbH and Co.KG
Lehmann & Voss & Co
Panreac Quimica SAU
SPI Pharma
Tate & Lyle

USA
Aceto Corp
Advance Scientific & Chemical Inc
Amresco Inc
Ashland Inc
Barrington Nutritionals
Brenntag North America Inc
CarboMer Inc
Cargill Inc
EM Sergeant Pulp & Chemical Co Inc
Evonik Degussa Corp
Ferro Corp
Fisher Scientific
JRS Pharma LP
Penta Manufacturing Co
Premium Ingredients Ltd
Spectrum Chemicals & Laboratory Products Inc
SPI Pharma Inc
Tate & Lyle (North America)
Thomas Scientific
Voigt Global Distribution Inc

Others
LS Raw Materials Ltd
SD Fine-Chem Ltd
SPI Pharma Inc India
Wintersun Chemical

Fructose and Pregelatinized Starch

USA
SPI Pharma Inc

Fumaric Acid

UK
DSM UK Ltd
Lonza Biologics Plc
Peter Whiting (Chemicals) Ltd
Sparkford Chemicals Ltd

Other European
Alfa Aesar Johnson Matthey GmbH
Brenntag GmbH
DSM Fine Chemicals
Lonza Ltd
Panreac Quimica SAU

USA
Aceto Corp
Advance Scientific & Chemical Inc
AerChem Inc
American International Chemical Inc
Ashland Inc
Astro Chemicals Inc
BOC Sciences
Brenntag North America Inc
Dastech International Inc
DSM Pharmaceuticals Inc
KIC Chemicals Inc
Penta Manufacturing Co
Premium Ingredients Ltd
Seidler Chemical Company Inc
Spectrum Chemicals & Laboratory Products Inc
Takeda Pharmaceuticals North America Inc
Tate & Lyle (North America)
Triple Crown America

Others
Aastrid International
BASF Japan Ltd
Univar Canada Ltd
Wintersun Chemical

Galactose

Other European
Danisco A/S
MeroPharm AG

USA
Advance Scientific & Chemical Inc
Aldrich
Amresco Inc
BioPure Healing Products LLC
CarboMer Inc
Danisco USA Inc
Ferro Corp
Macron Chemicals
Mutchler Inc
Penta Manufacturing Co
Ruger Chemical Co Inc
Seidler Chemical Company Inc
Spectrum Chemicals & Laboratory Products Inc

Others
Saccharides

Gelatin

UK
Croda Europe Ltd
Lehvoss UK Ltd
Macron Chemicals UK
PB Gelatins UK
Thew, Arnott and Co Ltd

Other European
Biogel AG
Ewald Gelatin GmbH
Gelita Europe
Harke Pharma GmbH
Italgelatine SpA
Juncá Gelatines S L SA
Lapi Gelatine SpA
Panreac Quimica SAU
PB Gelatins Belgium
Reinert Gruppe GmbH & Co KG
Rousselot SAS
Trobas Gelatine BV
Weishardt International, Europe

USA
Alfa Chem
Ashland Inc
Atlantic Gelatin/Kraft Foods Global Inc
Dastech International Inc
Eastman Gelatine Corp
Gelita USA
Gelnex
Glatech Productions LLC
Macron Chemicals
Nitta Gelatin NA Ltd
PB Leiner
Penta Manufacturing Co
Premium Ingredients Ltd
Rousselot Inc
Seidler Chemical Company Inc
Spectrum Chemicals & Laboratory Products Inc
Triple Crown America

Others
Gelco - Gelatinas de Colombia SA
Gelita Asia/Pacific/Africa
Gelita South America
Gelnex Indústria e Comércio Ltda
Geltech Co Ltd
Nippi Inc
Nitta Gelatin Inc
Nitta Gelatin India Ltd
PB Leiner Argentina SA
Progel Productora de Gelatina SA
Rousselot Argentina SA
Sammi Gelatin
Sterling Gelatin
The Rousselot (China) Gelatin Co Ltd
Weishardt International, Canada
Wintersun Chemical
Xiamen Xinwulong Gelatin Co Ltd

Glucose, Liquid

UK
Cargill PLC
Courtin & Warner Ltd
Raught Ltd
Roquette (UK) Ltd

Other European
Brenntag GmbH
Cargill Europe BVBA
Cargill France SAS
Corcoran Chemicals Ltd
Panreac Quimica SAU
Roquette Frères
Tate & Lyle

USA
Advance Scientific & Chemical Inc
Cargill Inc
Penta Manufacturing Co
Roquette America Inc
Thomas Scientific

Others
Saccharides

Glycerin

UK
Cognis UK Ltd
Courtin & Warner Ltd
Croda Europe Ltd
Fisher Scientific UK Ltd
H Foster & Co (Stearines) Ltd
Kimpton Brothers Ltd
Leading Solvent Supplies Ltd
Lonza Biologics Plc
Macron Chemicals UK
Peter Greven UK
Peter Whiting (Chemicals) Ltd
Raught Ltd
Sigma-Aldrich Company Ltd
Stan Chem International Ltd
Tennants Distribution Ltd
William Ransom & Son plc

Other European
August Hedinger GmbH & Co KG
Berg + Schmidt (GmbH & Co.) KG
Brenntag GmbH
Cognis GmbH
Corcoran Chemicals Ltd
Impag GmbH
Lonza Ltd
Panreac Quimica SAU

USA
Acme-Hardesty
Advance Scientific & Chemical Inc
AerChem Inc
Alfa Chem
American International Chemical Inc
Ashland Inc
Astro Chemicals Inc
Avatar Corp
Brenntag North America Inc
Brenntag Southwest
CarboMer Inc
Cognis Corp
Dow Chemical Co
Fisher Scientific
KIC Chemicals Inc
Kraft Chemical Co
Mitsubishi International Food Ingredients
Mutchler Inc
Paddock Laboratories Inc
Penta Manufacturing Co
Premium Ingredients Ltd
Protameen Chemicals Inc
Rita Corp
Ruger Chemical Co Inc
Seidler Chemical Company Inc
Spectrum Chemicals & Laboratory Products Inc
Thomas Scientific
Triple Crown America
Voigt Global Distribution Inc

Others
Charles Tennant & Co (Canada) Ltd
Gadot Petrochemical Industries Ltd
Kao Corporation
Kawarlal Excipients Pvt Ltd
Signet Chemical Corporation Pvt Ltd
Univar Canada Ltd
VVF Limited
Wintersun Chemical

Glyceryl Behenate

UK
A and E Connock (Perfumery and Cosmetics) Ltd
Alfa Chemicals Ltd/Gattefossé UK Ltd

Other European
Gattefossé SAS

USA
Gattefossé Corp
Lipscomb Chemical Company

Glyceryl Monooleate

UK
Alfa Chemicals Ltd/Gattefossé UK Ltd
Cognis UK Ltd
Croda Europe Ltd
Evonik Degussa Ltd
Evonik Goldschmidt UK Ltd
Peter Greven UK
Sasol UK Ltd

Other European
Cognis GmbH
Gattefossé SAS
Sasol Germany GmbH

USA
ABITEC Corp
Alfa Chem
Cognis Corp
Croda Inc
Gattefossé Corp
Hallstar Company, The
Lipscomb Chemical Company
Penta Manufacturing Co
Sasol North America Inc
Seidler Chemical Company Inc

Glyceryl Monostearate

UK
Alfa Chemicals Ltd
Alfa Chemicals Ltd/Gattefossé UK Ltd
Allchem Performance
Cognis UK Ltd
Croda Europe Ltd
Evonik Goldschmidt UK Ltd
H Foster & Co (Stearines) Ltd
IMCD UK Ltd
Lonza Biologics Plc
Peter Greven UK
Sasol UK Ltd

Other European
Cognis GmbH
Gattefossé SAS
Lonza Ltd
Sasol Germany GmbH

USA
ABITEC Corp
Allan Chemical Corp
Alzo International Inc
Brenntag North America Inc
Cognis Corp
Croda Inc
Gattefossé Corp
Hallstar Company, The
Kerry Bio-Science
Kraft Chemical Co
Lipo Chemicals Inc
Lonza Inc
Mutchler Inc
Penta Manufacturing Co
Protameen Chemicals Inc
Rita Corp
Sasol North America Inc
Seidler Chemical Company Inc

Others
Gattefossé Canada Inc.
Kao Corporation
LS Raw Materials Ltd
Univar Canada Ltd

Glyceryl Palmitostearate

UK
Alfa Chemicals Ltd/Gattefossé UK Ltd

Other European
Gattefossé SAS

USA
Gattefossé Corp
Lipscomb Chemical Company

Glycine

UK
A and E Connock (Perfumery and Cosmetics) Ltd
Macron Chemicals UK
Sigma-Aldrich Company Ltd

Other European
Alfa Aesar Johnson Matthey GmbH
Panreac Quimica SAU

USA
AerChem Inc
Alfa Chem
Amresco Inc
BOC Sciences
Budenheim USA Inc
Charkit Chemical Corp
Dastech International Inc
Fortitech Inc
Macron Chemicals
Mutchler Inc
Penta Manufacturing Co
Ruger Chemical Co Inc
Seidler Chemical Company Inc
Spectrum Chemicals & Laboratory Products Inc
Voigt Global Distribution Inc

Others
Boith Limited (China)

Guar Gum

UK
AF Suter and Co Ltd
Blagden Specialty Chemicals Ltd
Danisco UK Ltd
Stan Chem International Ltd
Thew, Arnott and Co Ltd

Other European
Brenntag GmbH
Corcoran Chemicals Ltd
Danisco A/S

USA
Advance Scientific & Chemical Inc
Ashland Inc
Ashland Specialty Ingredients
Barrington Nutritionals
Brenntag North America Inc
Brenntag Southwest
Chart Corp Inc
Danisco USA Inc
Penta Manufacturing Co
Premium Ingredients Ltd
Spectrum Chemicals & Laboratory Products Inc
TIC Gums
Triple Crown America
Voigt Global Distribution Inc

Others
Lucid Colloids Ltd
Priyanka Pharma/M K Ingredients & Specialities
Univar Canada Ltd

Hectorite

UK
Amcol Specialty Minerals
Fluorochem Ltd

Other European
ABCR GmbH & Co. KG

USA
BOC Sciences
International Laboratory Ltd
Spectrum Chemicals & Laboratory Products Inc

Others
Advanced Technology & Industrial Co Ltd

Heptafluoropropane (HFC)

Other European
DuPont de Nemours Int'l SA
Solvay Fluor GmbH

Hexetidine

Other European
Alfa Aesar Johnson Matthey GmbH

Hydrocarbons (HC)

UK
Air Products (Gases) plc

Other European
Chevron Lubricants

Hydrochloric Acid

UK
Macron Chemicals UK
Sigma-Aldrich Company Ltd
Tennants Distribution Ltd

Other European
Alfa Aesar Johnson Matthey GmbH
Brenntag GmbH
Panreac Quimica SAU

USA
Amresco Inc
Ashland Inc
Brenntag North America Inc
Brenntag Southwest
Fisher Scientific
Macron Chemicals
Seidler Chemical Company Inc
Spectrum Chemicals & Laboratory Products Inc
Triple Crown America

Others
Univar Canada Ltd
Wintersun Chemical

Hydrophobic Colloidal Silica

Other European
Evonik Industries AG
Wacker-Chemie GmbH

USA
Evonik Degussa Corp
Wacker Chemical Corp

Others
Wacker Chemicals (China) Co Ltd

Hydroxyethyl Cellulose

UK
A and E Connock (Perfumery and Cosmetics) Ltd
Hercules Ltd
IMCD UK Ltd
Lehvoss UK Ltd

Other European
SE Tylose GmbH & Co.KG

USA
Advance Scientific & Chemical Inc
Ashland Specialty Ingredients
Brenntag Southwest
CarboMer Inc
Dow Chemical Co
Spectrum Chemicals & Laboratory Products Inc
Voigt Global Distribution Inc

Others
Ashland India Pvt Ltd
Kawarlal Excipients Pvt Ltd

Hydroxyethylmethyl Cellulose

UK
Hercules Ltd

Other European
SE Tylose GmbH & Co.KG

USA
Ashland Specialty Ingredients
CarboMer Inc

Hydroxyethylpiperazine Ethane Sulfonic Acid

USA
Acros Organics
Fisher Scientific

Hydroxypropyl Betadex

Other European
Alfa Aesar Johnson Matthey GmbH
Wacker-Chemie GmbH

USA
Alfa Chem
Fitzgerald Industries International
Mutchler Inc
Roquette America Inc
Wacker Chemical Corp

Others
Gangwal Chemicals Pvt Ltd

Hydroxypropyl Cellulose

UK
Hercules Ltd
IMCD UK Ltd

Other European
Alfa Aesar Johnson Matthey GmbH
IMCD Group BV
Nippon Soda Co Ltd (Germany)

USA
Advance Scientific & Chemical Inc
Alfa Chem
Ashland Specialty Ingredients
CarboMer Inc
Mutchler Inc
Nippon Soda Co Ltd (USA)
Spectrum Chemicals & Laboratory Products Inc
Voigt Global Distribution Inc

Others
Arihant Trading Co
Ashland India Pvt Ltd
Kawarlal Excipients Pvt Ltd
Nippon Soda Co Ltd

Hydroxypropyl Cellulose, Low-substituted

UK
RW Unwin & Co Ltd
Shin-Etsu Chemical Co Ltd

Other European
Harke Pharma GmbH

USA
Biddle Sawyer Corp
Seppic Inc
Voigt Global Distribution Inc

Others
Arihant Trading Co
Signet Chemical Corporation Pvt Ltd

Hydroxypropyl Starch

UK
Cargill PLC

Hypromellose

UK
Colorcon Ltd
RW Unwin & Co Ltd
Shin-Etsu Chemical Co Ltd
Ubichem plc

Other European
Harke Pharma GmbH
IMCD Group BV
J Rettenmaier & Söhne GmbH and Co.KG
SE Tylose GmbH & Co.KG

USA
Advance Scientific & Chemical Inc
Ashland Inc
Biddle Sawyer Corp
CarboMer Inc
Colorcon Inc
Dow Chemical Co
Hawkins Inc
JRS Pharma LP
Mitsubishi International Food Ingredients
Sensient Technologies
Spectrum Chemicals & Laboratory Products Inc

Others
Arihant Trading Co
Glide Chem Pvt Ltd
Kawarlal Excipients Pvt Ltd
Signet Chemical Corporation Pvt Ltd
Univar Canada Ltd

Hypromellose Acetate Succinate

UK
RW Unwin & Co Ltd
Shin-Etsu Chemical Co Ltd

Other European
Harke Pharma GmbH

USA
CarboMer Inc

Others
Arihant Trading Co
Signet Chemical Corporation Pvt Ltd

Hypromellose Phthalate

UK
RW Unwin & Co Ltd
Shin-Etsu Chemical Co Ltd
Ubichem plc

Other European
Harke Pharma GmbH

USA
Biddle Sawyer Corp
CarboMer Inc
Mantrose-Haeuser Co Inc

Others
Arihant Trading Co
Kawarlal Excipients Pvt Ltd
Shasun
Signet Chemical Corporation Pvt Ltd

Imidurea

UK
ISP Europe

USA
Advance Scientific & Chemical Inc
International Specialty Products Inc (ISP)
Spectrum Chemicals & Laboratory Products Inc

Inulin

Other European
Alfa Aesar Johnson Matthey GmbH
BENEO-Orafti
Sensus

USA
Premium Ingredients Ltd
Sensus America Inc
TIC Gums

Iron Oxides

UK
Lanxess Ltd
PMC Chemicals Ltd

Other European
Alfa Aesar Johnson Matthey GmbH

USA
BOC Sciences
Lanxess Corp
Reade Advanced Materials Inc
Rockwood Pigments NA, Inc.

Others
LS Raw Materials Ltd
Priyanka Pharma/M K Ingredients & Specialities
Signet Chemical Corporation Pvt Ltd
Wintersun Chemical

Isomalt

Other European
Beneo-Palatinit GmbH
Cargill Europe BVBA

USA
American International Chemical Inc
Cargill Inc

Others
Cerestar Cargill Resources Maize Industry Co Ltd

Isopropyl Alcohol

UK
A and E Connock (Perfumery and Cosmetics) Ltd
IMCD UK Ltd
Macron Chemicals UK
Sasol UK Ltd
William Ransom & Son plc

Other European
Alfa Aesar Johnson Matthey GmbH
August Hedinger GmbH & Co KG
Corcoran Chemicals Ltd
Sasol Germany GmbH

USA
Advance Scientific & Chemical Inc
Amresco Inc
Astro Chemicals Inc
Brenntag North America Inc
Brenntag Southwest
Dow Chemical Co
Integra Chemical Co
Macron Chemicals
Paddock Laboratories Inc
Penta Manufacturing Co
Ruger Chemical Co Inc
Sasol North America Inc
Spectrum Chemicals & Laboratory Products Inc
Thomas Scientific
Triple Crown America

Others
SD Fine-Chem Ltd
Univar Canada Ltd
Wintersun Chemical

Isopropyl Myristate

UK
A and E Connock (Perfumery and Cosmetics) Ltd
Adina Chemicals Ltd
Cognis UK Ltd
Dow Chemical Company (UK)
Evonik Goldschmidt UK Ltd
Lehvoss UK Ltd
Peter Whiting (Chemicals) Ltd
White Sea and Baltic Company Ltd

Other European
Cognis GmbH
Haltermann Products

USA
Acme-Hardesty
Allan Chemical Corp
Alzo International Inc
Brenntag North America Inc
Brenntag Southwest
Cognis Corp
Croda Inc
Hallstar Company, The
Inolex Chemical Co
KIC Chemicals Inc
Kraft Chemical Co
Lipo Chemicals Inc
Parchem Trading Ltd
Penta Manufacturing Co
Rita Corp
Spectrum Chemicals & Laboratory Products Inc

Others
Charles Tennant & Co (Canada) Ltd
Croda Japan
Kao Corporation
LS Raw Materials Ltd
Signet Chemical Corporation Pvt Ltd

Isopropyl Palmitate

UK
A and E Connock (Perfumery and Cosmetics) Ltd
Adina Chemicals Ltd
Cognis UK Ltd
Dow Chemical Company (UK)
Evonik Goldschmidt UK Ltd
Lehvoss UK Ltd

Other European
Alfa Aesar Johnson Matthey GmbH
Cognis GmbH
Haltermann Products
Industrial Quimica Lasem, sa (IQL)

USA
Acme-Hardesty
Advance Scientific & Chemical Inc
Allan Chemical Corp
Alzo International Inc
Brenntag North America Inc
Cognis Corp
Inolex Chemical Co
KIC Chemicals Inc
Kraft Chemical Co
Lipo Chemicals Inc
Lubrizol Advanced Materials Inc.
Penta Manufacturing Co
Protameen Chemicals Inc
Rita Corp
Spectrum Chemicals & Laboratory Products Inc

Others
Charles Tennant & Co (Canada) Ltd
Choice Korea Co
Croda Japan
Kao Corporation
Pachem Distributions Inc

Kaolin

UK
Fisher Scientific UK Ltd
Lehvoss UK Ltd
Macron Chemicals UK
Peter Whiting (Chemicals) Ltd
Raught Ltd
Sigma-Aldrich Company Ltd
Tennants Distribution Ltd
Thew, Arnott and Co Ltd
William Ransom & Son plc

Other European
Alfa Aesar Johnson Matthey GmbH

USA
Advance Scientific & Chemical Inc
Charles B Chrystal Co Inc
Fisher Scientific
Macron Chemicals
Penta Manufacturing Co
Seidler Chemical Company Inc
Spectrum Chemicals & Laboratory Products Inc
Voigt Global Distribution Inc
Whittaker Clark, and Daniels Inc

Lactic Acid

UK
Fisher Scientific UK Ltd
Macron Chemicals UK
Peter Whiting (Chemicals) Ltd
Sigma-Aldrich Company Ltd
Tennants Distribution Ltd
White Sea and Baltic Company Ltd

Other European
Alfa Aesar Johnson Matthey GmbH
Arion & Delahaye
Brenntag GmbH
Corcoran Chemicals Ltd
Dr Paul Lohmann GmbH KG
Panreac Quimica SAU
Purac Biochem

USA
Advance Scientific & Chemical Inc
AerChem Inc
Amresco Inc
Astro Chemicals Inc
Brenntag North America Inc
Dastech International Inc
Fisher Scientific
Kraft Chemical Co
Macron Chemicals
Penta Manufacturing Co
Premium Ingredients Ltd
Purac
RIA International
Seidler Chemical Company Inc
Spectrum Chemicals & Laboratory Products Inc
Triple Crown America
Voigt Global Distribution Inc
Zhong Ya Chemical (USA) Ltd

Others
Boith Limited (China)
LS Raw Materials Ltd
Priyanka Pharma/M K Ingredients & Specialities
SD Fine-Chem Ltd
Univar Canada Ltd
Wintersun Chemical

Lactitol

UK
Danisco UK Ltd

Other European
Alfa Aesar Johnson Matthey GmbH
Danisco A/S

USA
Danisco USA Inc
Penta Manufacturing Co

Others
Saccharides

Lactose, Anhydrous

UK
DFE Pharma UK

Other European
Corcoran Chemicals Ltd
DFE Pharma
IMCD Group BV
Molkerei Meggle Wasserburg GmbH
Panreac Quimica SAU

USA
CarboMer Inc
Dastech International Inc
Hummel Croton Inc
Mitsubishi International Food Ingredients
Mutchler Inc
Parchem
Penta Manufacturing Co
RIA International
Seidler Chemical Company Inc
Sheffield Bio-Science
Spectrum Chemicals & Laboratory Products Inc
Thomas Scientific
Voigt Global Distribution Inc

Others
Charles Tennant & Co (Canada) Ltd
Glide Chem Pvt Ltd
Kawarlal Excipients Pvt Ltd
Signet Chemical Corporation Pvt Ltd

Lactose, Inhalation

UK
DFE Pharma UK

Other European
DFE Pharma
Molkerei Meggle Wasserburg GmbH

USA
Sheffield Bio-Science

Lactose, Monohydrate

UK
DFE Pharma UK
Forum Products Ltd
IMCD UK Ltd
Macron Chemicals UK

Other European
Alfa Aesar Johnson Matthey GmbH
Brenntag GmbH
DFE Pharma
IMCD Group BV
Molkerei Meggle Wasserburg GmbH

USA
Advance Scientific & Chemical Inc
Brenntag North America Inc
CarboMer Inc
EMD Chemicals Inc
Mutchler Inc
Penta Manufacturing Co
Sheffield Bio-Science
Spectrum Chemicals & Laboratory Products Inc
Voigt Global Distribution Inc

Others
Charles Tennant & Co (Canada) Ltd
LS Raw Materials Ltd

Lactose, Monohydrate and Corn Starch

Other European
Molkerei Meggle Wasserburg GmbH
Roquette Frères

Lactose, Monohydrate and Microcrystalline Cellulose

Other European
Molkerei Meggle Wasserburg GmbH

USA
Mutchler Inc

Lactose, Monohydrate and Povidone

UK
BASF Plc

Other European
BASF Aktiengesellschaft

USA
BASF Corp

Lactose, Monohydrate and Powdered Cellulose

Other European
Molkerei Meggle Wasserburg GmbH

USA
Mutchler Inc

Lactose, Spray-Dried

UK
DFE Pharma UK
Forum Products Ltd

Other European
DFE Pharma
IMCD Group BV
Molkerei Meggle Wasserburg GmbH

USA
Mutchler Inc
Sheffield Bio-Science
Spectrum Chemicals & Laboratory Products Inc

Others
Signet Chemical Corporation Pvt Ltd

Lanolin

UK
A and E Connock (Perfumery and Cosmetics) Ltd
Blagden Specialty Chemicals Ltd
Croda Europe Ltd
Fisher Scientific UK Ltd
Lehvoss UK Ltd
Macron Chemicals UK

Other European
Alfa Aesar Johnson Matthey GmbH
Corcoran Chemicals Ltd

USA
Advance Scientific & Chemical Inc
Brenntag North America Inc
Croda Inc
Fisher Scientific
Integra Chemical Co
Kraft Chemical Co
Paddock Laboratories Inc
Penta Manufacturing Co
Protameen Chemicals Inc
Rita Corp
Seidler Chemical Company Inc
Spectrum Chemicals & Laboratory Products Inc
Voigt Global Distribution Inc

Lanolin, Hydrous

UK
Adina Chemicals Ltd

USA
Lipo Chemicals Inc
Rita Corp
Spectrum Chemicals & Laboratory Products Inc

Lanolin Alcohols

UK
A and E Connock (Perfumery and Cosmetics) Ltd
Croda Europe Ltd
Lehvoss UK Ltd

USA
Charkit Chemical Corp
Croda Inc
Kraft Chemical Co
Rita Corp

Lauric Acid

UK
A and E Connock (Perfumery and Cosmetics) Ltd
Sigma-Aldrich Company Ltd
White Sea and Baltic Company Ltd

Other European
ABCR GmbH & Co. KG
Alfa Aesar Johnson Matthey GmbH
Corcoran Chemicals Ltd

USA
Acme-Hardesty
Avantor Performance Materials
BOC Sciences
Charkit Chemical Corp
KIC Chemicals Inc
Kraft Chemical Co
Penta Manufacturing Co
Pfaltz & Bauer Inc
Seidler Chemical Company Inc

Others
Advanced Technology & Industrial Co Ltd

Lecithin

UK
A and E Connock (Perfumery and Cosmetics) Ltd
AarhusKarlshamn UK Ltd
Alembic Products Ltd
Forum Products Ltd
Thew, Arnott and Co Ltd

Other European
AarhusKarlshamn AB
Alfa Aesar Johnson Matthey GmbH
Corcoran Chemicals Ltd
Imperial-Oel-Import
Lucas Meyer
Stern Wywiol Gruppe Holding GmbH & Co.KG

USA
AarhusKarlshamn USA Inc
Aceto Corp
Advance Scientific & Chemical Inc
Alfa Chem
American Lecithin Co
Ashland Inc
Avatar Corp
Brenntag North America Inc
CarboMer Inc
Fortitech Inc
KIC Chemicals Inc
Kraft Chemical Co
Lucas Meyer Inc
Parchem Trading Ltd
Penta Manufacturing Co
Premium Ingredients Ltd
Seidler Chemical Company Inc
Spectrum Chemicals & Laboratory Products Inc
Triple Crown America
Voigt Global Distribution Inc

Others
Wintersun Chemical

Leucine

UK
A and E Connock (Perfumery and Cosmetics) Ltd
Macron Chemicals UK
Sigma-Aldrich Company Ltd

Other European
Alfa Aesar Johnson Matthey GmbH
Corcoran Chemicals Ltd
Panreac Quimica SAU

USA
Advance Scientific & Chemical Inc
Alfa Chem
Amresco Inc
BOC Sciences
Fortitech Inc
Glanbia Nutritionals Inc
Lipscomb Chemical Company
Macron Chemicals
Penta Manufacturing Co
Premium Ingredients Ltd
Scandinavian Formulas Inc
Seidler Chemical Company Inc
Thomas Scientific

Others
LS Raw Materials Ltd
Wintersun Chemical

Linoleic Acid

Other European
Alfa Aesar Johnson Matthey GmbH

USA
Advance Scientific & Chemical Inc
AerChem Inc
Alfa Chem
Kraft Chemical Co
Loos & Dilworth Inc
Penta Manufacturing Co
Seidler Chemical Company Inc

Lysine Acetate

UK
Sigma-Aldrich Company Ltd

Lysine Hydrochloride

UK
Macron Chemicals UK

Other European
Panreac Quimica SAU

USA
Alfa Chem
American International Chemical Inc
Bachem Bioscience Inc
Barrington Chemical Corp
Functional Foods Corp

Helm New York Inc
Macron Chemicals
Penta Manufacturing Co
Ruger Chemical Co Inc
Spectrum Chemicals & Laboratory Products Inc
Triple Crown America
Voigt Global Distribution Inc

Others
Boith Limited (China)

Magnesium Aluminum Silicate

UK
Lehvoss UK Ltd

USA
American Colloid Co
Fuji Chemical Industries Health Science (USA) Inc
Kraft Chemical Co
Penta Manufacturing Co
RT Vanderbilt Company Inc
Spectrum Chemicals & Laboratory Products Inc
Whittaker Clark, and Daniels Inc

Others
Fuji Chemical Industry Co Ltd
Signet Chemical Corporation Pvt Ltd

Magnesium Carbonate

UK
Chance & Hunt
Courtin & Warner Ltd
Fisher Scientific UK Ltd
Intermag Co Ltd
Macron Chemicals UK
Paroxite (London) Ltd
Peter Whiting (Chemicals) Ltd
Tennants Distribution Ltd
William Ransom & Son plc

Other European
Alfa Aesar Johnson Matthey GmbH
Brenntag GmbH
Dr Paul Lohmann GmbH KG
Lehmann & Voss & Co
Magnesia GmbH

USA
Advance Scientific & Chemical Inc
AerChem Inc
Alfa Chem
Barrington Nutritionals
Brenntag North America Inc
Budenheim USA Inc
EM Sergeant Pulp & Chemical Co Inc
Fisher Scientific
Generichem Corp
Hummel Croton Inc
Kraft Chemical Co
Macron Chemicals
Particle Dynamics Inc
Penta Manufacturing Co
RIA International
Seidler Chemical Company Inc
Spectrum Chemicals & Laboratory Products Inc
Triple Crown America
Whittaker Clark, and Daniels Inc

Others
Signet Chemical Corporation Pvt Ltd
Univar Canada Ltd
Wintersun Chemical

Magnesium Oxide

UK
Blagden Specialty Chemicals Ltd
Fisher Scientific UK Ltd
Intermag Co Ltd
Lehvoss UK Ltd
Macron Chemicals UK
Tennants Distribution Ltd

Other European
Alfa Aesar Johnson Matthey GmbH
Brenntag GmbH
Dr Paul Lohmann GmbH KG
Magnesia GmbH
Panreac Quimica SAU

USA
Advance Scientific & Chemical Inc
AerChem Inc
Alfa Chem
Ashland Inc
Barrington Nutritionals
Brenntag North America Inc
Dastech International Inc
Fisher Scientific
Generichem Corp
Integra Chemical Co
Macron Chemicals
Particle Dynamics Inc
Penta Manufacturing Co
Premium Ingredients Ltd
RIA International
Ruger Chemical Co Inc
Seidler Chemical Company Inc
Spectrum Chemicals & Laboratory Products Inc
Whittaker Clark, and Daniels Inc

Others
LS Raw Materials Ltd
Signet Chemical Corporation Pvt Ltd
Univar Canada Ltd
Wintersun Chemical

Magnesium Silicate

Other European
Alfa Aesar Johnson Matthey GmbH

Others
Wintersun Chemical

Magnesium Stearate

UK
Allchem Performance
Blagden Specialty Chemicals Ltd
Fisher Scientific UK Ltd
Intermag Co Ltd
James M Brown Ltd
Lehvoss UK Ltd
Macron Chemicals UK
Peter Greven UK

Other European
Alfa Aesar Johnson Matthey GmbH
Biesterfeld Spezialchemie GmbH
CPI Chemicals
Dr Paul Lohmann GmbH KG
Lehmann & Voss & Co
Magnesia GmbH
Panreac Quimica SAU

USA
Aceto Corp
Acme-Hardesty
Advance Scientific & Chemical Inc
AerChem Inc
Alfa Chem
Allan Chemical Corp
Ashland Inc
Avatar Corp
Barrington Nutritionals
Brenntag North America Inc
CarboMer Inc
Dastech International Inc
EM Sergeant Pulp & Chemical Co Inc
Ferro Corp
Fisher Scientific
Generichem Corp
KIC Chemicals Inc
Kraft Chemical Co
Macron Chemicals
Mutchler Inc
Parchem Trading Ltd
Penta Manufacturing Co
Premium Ingredients Ltd
RIA International
Ruger Chemical Co Inc
Seidler Chemical Company Inc
Spectrum Chemicals & Laboratory Products Inc
Triple Crown America
Whittaker Clark, and Daniels Inc

Others
Charles Tennant & Co (Canada) Ltd
LS Raw Materials Ltd
Signet Chemical Corporation Pvt Ltd
Univar Canada Ltd
Wintersun Chemical

Magnesium Trisilicate

UK
Courtin & Warner Ltd
Intermag Co Ltd
William Ransom & Son plc

Other European
Alfa Aesar Johnson Matthey GmbH
Dr Paul Lohmann GmbH KG
Lehmann & Voss & Co
Magnesia GmbH

USA
Advance Scientific & Chemical Inc
Alfa Chem
American International Chemical Inc
Generichem Corp
Penta Manufacturing Co
RIA International
Ruger Chemical Co Inc
Spectrum Chemicals & Laboratory Products Inc

Others
LS Raw Materials Ltd

Maleic Acid

UK
Macron Chemicals UK
Sigma-Aldrich Company Ltd

Other European
Alfa Aesar Johnson Matthey GmbH
Corcoran Chemicals Ltd
Panreac Quimica SAU

USA
Alfa Chem
Amresco Inc
Hawkins Inc
Macron Chemicals
Premium Ingredients Ltd
Seidler Chemical Company Inc
Spectrum Chemicals & Laboratory Products Inc

Others
SD Fine-Chem Ltd
Wintersun Chemical
Xiamen Topusing Chemical Co Ltd

Malic Acid

UK
DSM UK Ltd
Lonza Biologics Plc

Peter Whiting (Chemicals) Ltd
Tennants Distribution Ltd
Ubichem plc

Other European
Alfa Aesar Johnson Matthey GmbH
Brenntag GmbH
Corcoran Chemicals Ltd
DSM Fine Chemicals
Lonza Ltd
Panreac Quimica SAU

USA
Advance Scientific & Chemical Inc
AerChem Inc
Alfa Chem
American International Chemical Inc
Amresco Inc
Ashland Inc
Astro Chemicals Inc
Brenntag North America Inc
DSM Pharmaceuticals Inc
Fortitech Inc
KIC Chemicals Inc
Kraft Chemical Co
Penta Manufacturing Co
Premium Ingredients Ltd
Seidler Chemical Company Inc
Spectrum Chemicals & Laboratory Products Inc
Triple Crown America
Voigt Global Distribution Inc
Zhong Ya Chemical (USA) Ltd

Others
Priyanka Pharma/M K Ingredients & Specialities
Univar Canada Ltd
Wintersun Chemical

Maltitol

UK
Cargill PLC
IMCD UK Ltd
Roquette (UK) Ltd

Other European
Alfa Aesar Johnson Matthey GmbH
Cargill France SAS
Corcoran Chemicals Ltd
Roquette Frères

USA
Advance Scientific & Chemical Inc
Ashland Inc
Cargill Inc
Premium Ingredients Ltd
Roquette America Inc
Spectrum Chemicals & Laboratory Products Inc

Others
Signet Chemical Corporation Pvt Ltd

Maltitol Solution

UK
Cargill PLC
Lonza Biologics Plc
Roquette (UK) Ltd

Other European
Cargill France SAS
Lonza Ltd
Roquette Frères

USA
Roquette America Inc

Maltodextrin

UK
Avebe UK Ltd
Cargill PLC
IMCD UK Ltd
Roquette (UK) Ltd

Other European
Avebe Group
Brenntag GmbH
Cargill Europe BVBA
Cargill France SAS
Corcoran Chemicals Ltd
Roquette Frères
Tate & Lyle

USA
Archer Daniels Midland Company (ADM)
Ashland Inc
Avebe America Inc
Brenntag North America Inc
CarboMer Inc
Cargill Inc
EM Sergeant Pulp & Chemical Co Inc
Generichem Corp
Grain Processing Corp
Premium Ingredients Ltd
Primera Foods
Roquette America Inc
Tate & Lyle (North America)
Voigt Global Distribution Inc

Others
Signet Chemical Corporation Pvt Ltd

Maltol

Other European
Alfa Aesar Johnson Matthey GmbH
Helm AG

USA
Ashland Inc
Helm New York Inc
Penta Manufacturing Co
Premium Ingredients Ltd

Others
Priyanka Pharma/M K Ingredients & Specialities

Maltose

UK
Cargill PLC
Forum Products Ltd
Pfanstiehl (Europe) Ltd

Other European
Alfa Aesar Johnson Matthey GmbH
Cargill France SAS
Lehmann & Voss & Co
SPI Pharma

USA
Cargill Inc
Ferro Corp
Penta Manufacturing Co
Seidler Chemical Company Inc
SPI Pharma Inc

Others
Hayashibara Co Ltd
SPI Pharma Inc India

Mannitol

UK
3M United Kingdom Plc
BASF Plc
Cargill PLC
Fisher Scientific UK Ltd
Forum Products Ltd
IMCD UK Ltd
Macron Chemicals UK
Pfanstiehl (Europe) Ltd
Roquette (UK) Ltd
Ubichem plc

Other European
Alfa Aesar Johnson Matthey GmbH
BASF Aktiengesellschaft
Brenntag GmbH
Cargill France SAS
Corcoran Chemicals Ltd
Helm AG
IMCD Group BV
Lehmann & Voss & Co
Merck KGaA
Panreac Quimica SAU
Roquette Frères
SPI Pharma
Ubichem Research Ltd

USA
3M Drug Delivery Systems
Aceto Corp
Advance Scientific & Chemical Inc
AerChem Inc
Alfa Chem
Amresco Inc
AnMar International
Asiamerica Ingredients Inc
BASF Corp
Brenntag North America Inc
Cargill Inc
Dastech International Inc
EM Sergeant Pulp & Chemical Co Inc
Ferro Corp
Fisher Scientific
George Uhe Co Inc
Macron Chemicals
Mutchler Inc
Penta Manufacturing Co
Premium Ingredients Ltd
RIA International
Roquette America Inc
Ruger Chemical Co Inc
Seidler Chemical Company Inc
Spectrum Chemicals & Laboratory Products Inc
SPI Pharma Inc
Thomas Scientific
Voigt Global Distribution Inc

Others
Arihant Trading Co
Kao Corporation
Kawarlal Excipients Pvt Ltd
LS Raw Materials Ltd
Saccharides
Signet Chemical Corporation Pvt Ltd
SPI Pharma Inc India
Univar Canada Ltd
Wintersun Chemical

Mannitol and Sorbitol

UK
Macron Chemicals UK
Sigma-Aldrich Company Ltd

Other European
SPI Pharma

USA
Aldrich
CarboMer Inc
Ferro Corp
Ferro Pfanstiehl Laboratories Inc
Macron Chemicals
Spectrum Chemicals & Laboratory Products Inc
SPI Pharma Inc

Others
Saccharides
SPI Pharma Inc India
Wintersun Chemical

Medium-chain Triglycerides

UK
Alfa Chemicals Ltd/Gattefossé UK Ltd
Blagden Specialty Chemicals Ltd
Cognis UK Ltd
Lonza Biologics Plc
Sasol UK Ltd
Thew, Arnott and Co Ltd

Other European
Berg + Schmidt (GmbH & Co.) KG
Cognis GmbH
Corcoran Chemicals Ltd
Gattefossé SAS
Lonza Ltd
Sasol Germany GmbH

USA
ABITEC Corp
Acme-Hardesty
AerChem Inc
Avatar Corp
Cognis Corp
Croda Inc
Gattefossé Corp
Lipscomb Chemical Company
Sasol North America Inc
Stepan Co

Others
Cognis Speciality Chemicals
Gangwal Chemicals Pvt Ltd
Kao Corporation
Signet Chemical Corporation Pvt Ltd

Meglumine

Other European
Alfa Aesar Johnson Matthey GmbH

USA
Spectrum Chemicals & Laboratory Products Inc

Menthol

UK
Courtin & Warner Ltd
Raught Ltd
Stan Chem International Ltd
Symrise
Thew, Arnott and Co Ltd

Other European
Alfa Aesar Johnson Matthey GmbH
Haarmann & Reimer GmbH
Helm AG
Panreac Quimica SAU

USA
Alfa Chem
CarboMer Inc
Charkit Chemical Corp
Chart Corp Inc
Helm New York Inc
KIC Chemicals Inc
Mutchler Inc
Penta Manufacturing Co
Premium Ingredients Ltd
RIA International
Seidler Chemical Company Inc
Spectrum Chemicals & Laboratory Products Inc
Voigt Global Distribution Inc

Others
LS Raw Materials Ltd
Wintersun Chemical

Methionine

UK
A and E Connock (Perfumery and Cosmetics) Ltd
Evonik Degussa Ltd
Macron Chemicals UK
Sigma-Aldrich Company Ltd

Other European
Alfa Aesar Johnson Matthey GmbH
Corcoran Chemicals Ltd
Panreac Quimica SAU

USA
Advance Scientific & Chemical Inc
AerChem Inc
Alfa Chem
Amresco Inc
BOC Sciences
Fortitech Inc
Macron Chemicals
Penta Manufacturing Co
Premium Ingredients Ltd
Seidler Chemical Company Inc
Spectrum Chemicals & Laboratory Products Inc
Thomas Scientific
Voigt Global Distribution Inc

Others
LS Raw Materials Ltd
Wintersun Chemical

Methylcellulose

UK
Colorcon Ltd
RW Unwin & Co Ltd
Shin-Etsu Chemical Co Ltd

Other European
Alfa Aesar Johnson Matthey GmbH
Harke Pharma GmbH
SE Tylose GmbH & Co.KG

USA
Advance Scientific & Chemical Inc
Alfa Chem
Ashland Specialty Ingredients
Biddle Sawyer Corp
Brenntag North America Inc
Colorcon Inc
Dow Chemical Co
Mitsubishi International Food Ingredients
Mutchler Inc
Parchem
RIA International
Spectrum Chemicals & Laboratory Products Inc

Others
Kawarlal Excipients Pvt Ltd
Signet Chemical Corporation Pvt Ltd

Methylparaben

UK
Blagden Specialty Chemicals Ltd
Clariant UK Ltd
Cornelius Group plc
Lanxess Ltd
White Sea and Baltic Company Ltd

Other European
Alfa Aesar Johnson Matthey GmbH
Clariant GmbH
Induchem AG

USA
Acme-Hardesty
Alfa Chem
Ashland Inc
Avatar Corp
Brenntag North America Inc
CarboMer Inc
Clariant Corp
Dastech International Inc
Hallstar Company, The
KIC Chemicals Inc
Kraft Chemical Co
Lanxess Corp
Lipo Chemicals Inc
Napp Technologies LLC
Penta Manufacturing Co
Premium Ingredients Ltd
Protameen Chemicals Inc
RIA International
Rita Corp
Ruger Chemical Co Inc
Seidler Chemical Company Inc
Spectrum Chemicals & Laboratory Products Inc
Thomas Scientific
Voigt Global Distribution Inc

Others
Charles Tennant & Co (Canada) Ltd
LS Raw Materials Ltd
San Fu Chemical Company Ltd
Univar Canada Ltd
Wintersun Chemical

Mineral Oil

UK
Fisher Scientific UK Ltd
Fuchs Lubricants (UK) plc
Macron Chemicals UK

Other European
Alfa Aesar Johnson Matthey GmbH
Brenntag GmbH
Chevron Lubricants
Corcoran Chemicals Ltd
Parafluid Mineraloelgesellschaft mbH
USOCO BV

USA
Advance Scientific & Chemical Inc
Astro Chemicals Inc
Avatar Corp
Brenntag North America Inc
Fisher Scientific
Macron Chemicals
Mutchler Inc
Paddock Laboratories Inc
Penta Manufacturing Co
Seidler Chemical Company Inc
Spectrum Chemicals & Laboratory Products Inc
Triple Crown America

Others
Univar Canada Ltd

Mineral Oil, Light

UK
Fisher Scientific UK Ltd
Fuchs Lubricants (UK) plc

Other European
Chevron Lubricants
Corcoran Chemicals Ltd
Parafluid Mineraloelgesellschaft mbH
USOCO BV

USA
Advance Scientific & Chemical Inc
Amresco Inc
Fisher Scientific
Mutchler Inc
Penta Manufacturing Co
Ruger Chemical Co Inc
Spectrum Chemicals & Laboratory Products Inc
Voigt Global Distribution Inc

Mineral Oil and Lanolin Alcohols

UK
Lehvoss UK Ltd

USA
Protameen Chemicals Inc
Rita Corp

Monoethanolamine

UK
Leading Solvent Supplies Ltd
Tennants Distribution Ltd

Other European
Alfa Aesar Johnson Matthey GmbH

USA
Advance Scientific & Chemical Inc
Brenntag North America Inc
Penta Manufacturing Co
Spectrum Chemicals & Laboratory Products Inc
Triple Crown America

Monosodium Glutamate

Other European
Alfa Aesar Johnson Matthey GmbH
Brenntag GmbH
Helm AG

USA
Advance Scientific & Chemical Inc
Ashland Inc
Brenntag North America Inc
Brenntag Southwest
Helm New York Inc
Penta Manufacturing Co
Premium Ingredients Ltd
Seidler Chemical Company Inc
Triple Crown America
Zhong Ya Chemical (USA) Ltd

Others
Wintersun Chemical

Monothioglycerol

USA
Penta Manufacturing Co

Myristic Acid

UK
A and E Connock (Perfumery and Cosmetics) Ltd
Brenntag (UK) Ltd
Sigma-Aldrich Company Ltd
White Sea and Baltic Company Ltd

Other European
Corcoran Chemicals Ltd

USA
Acme-Hardesty
Ashland Inc
Chemtura Corporation
Hallstar Company, The
KIC Chemicals Inc
Penta Manufacturing Co
Ruger Chemical Co Inc

Others
EPS Impex Co

Myristyl Alcohol

UK
Sasol UK Ltd
White Sea and Baltic Company Ltd

Other European
Alfa Aesar Johnson Matthey GmbH
Berg + Schmidt (GmbH & Co.) KG
Chempri Oleochemicals (Europe)
Sasol Germany GmbH

USA
Brown Chemical Co, Inc
Kraft Chemical Co
M Michel and Company Inc
P & G Chemicals
Parchem
Penta Manufacturing Co
Ruger Chemical Co Inc
Sasol North America Inc

Neohesperidin Dihydrochalcone

Other European
Exquim S.A.
Natura Internacional S.L.

USA
Penta Manufacturing Co

Neotame

USA
NutraSweet Company, The
Premium Ingredients Ltd

Nitrogen

UK
Air Liquide UK Ltd
Air Products (Gases) plc
BOC Gases

USA
Air Liquide America Corp
BOC Gases (USA)

Nitrous Oxide

UK
Air Liquide UK Ltd
BOC Gases

USA
Air Liquide America Corp
BOC Gases (USA)

Octyldodecanol

Other European
Cognis GmbH
Henkel AG & Co. KGaA

USA
Jarchem Industries Inc
Kraft Chemical Co

Others
Charles Tennant & Co (Canada) Ltd
Signet Chemical Corporation Pvt Ltd

Octyl Gallate

UK
White Sea and Baltic Company Ltd

Oleic Acid

UK
A and E Connock (Perfumery and Cosmetics) Ltd
Croda Europe Ltd
Fisher Scientific UK Ltd
H Foster & Co (Stearines) Ltd
Kimpton Brothers Ltd
Macron Chemicals UK
Peter Whiting (Chemicals) Ltd
Sigma-Aldrich Company Ltd
Tennants Distribution Ltd
White Sea and Baltic Company Ltd

Other European
Alfa Aesar Johnson Matthey GmbH
Panreac Quimica SAU

USA
Acme-Hardesty
Advance Scientific & Chemical Inc
AerChem Inc
Brenntag North America Inc
Brenntag Southwest
Croda Inc
Fisher Scientific
Kraft Chemical Co
Mutchler Inc
Penta Manufacturing Co
Ruger Chemical Co Inc
Seidler Chemical Company Inc
Spectrum Chemicals & Laboratory Products Inc
Triple Crown America
Voigt Global Distribution Inc
Welch, Holme & Clark Co Inc

Others
Boith Limited (China)
LS Raw Materials Ltd
Signet Chemical Corporation Pvt Ltd

Oleyl Alcohol

UK
A and E Connock (Perfumery and Cosmetics) Ltd
Croda Europe Ltd

Other European
Alfa Aesar Johnson Matthey GmbH
Chempri Oleochemicals (Europe)
Cognis GmbH

USA
Croda Inc
Penta Manufacturing Co

Others
Signet Chemical Corporation Pvt Ltd

Olive Oil

UK
A and E Connock (Perfumery and Cosmetics) Ltd
AarhusKarlshamn UK Ltd
Alembic Products Ltd
Lehvoss UK Ltd

Other European
AarhusKarlshamn AB

USA
AarhusKarlshamn USA Inc
Advance Scientific & Chemical Inc
Arista Industries Inc
Avatar Corp
Charkit Chemical Corp
Hawkins Inc
Mutchler Inc
Penta Manufacturing Co
Pokonobe Industries Inc
Ruger Chemical Co Inc
Spectrum Chemicals & Laboratory Products Inc
Triple Crown America
Voigt Global Distribution Inc
Welch, Holme & Clark Co Inc

Palmitic Acid

UK
A and E Connock (Perfumery and Cosmetics) Ltd
Sigma-Aldrich Company Ltd
White Sea and Baltic Company Ltd

Other European
Alfa Aesar Johnson Matthey GmbH
Cognis GmbH
Corcoran Chemicals Ltd

USA
Acme-Hardesty
Advance Scientific & Chemical Inc
Alzo International Inc
Ashland Inc
Chemtura Corporation
KIC Chemicals Inc
Penta Manufacturing Co
Ruger Chemical Co Inc

Others
Charles Tennant & Co (Canada) Ltd
Kao Corporation
NOF Corporation
Wintersun Chemical

Palm Kernel Oil

Other European
Corcoran Chemicals Ltd

USA
Pokonobe Industries Inc

Palm Oil

UK
AarhusKarlshamn UK Ltd

Other European
AarhusKarlshamn AB

USA
AarhusKarlshamn USA Inc
Arista Industries Inc
Penta Manufacturing Co
Seidler Chemical Company Inc

Paraffin

UK
AF Suter and Co Ltd
British Wax Refining Co Ltd
Cornelius Group plc
Fuchs Lubricants (UK) plc
Leading Solvent Supplies Ltd
Poth Hille & Co Ltd.
Thew, Arnott and Co Ltd
William Ransom & Son plc

Other European
Brenntag GmbH
Chevron Lubricants
Corcoran Chemicals Ltd
Panreac Quimica SAU
Parafluid Mineraloelgesellschaft mbH
USOCO BV

USA
Advance Scientific & Chemical Inc
Brenntag North America Inc
Brenntag Southwest
Koster Keunen LLC
Mutchler Inc
Penta Manufacturing Co
Seidler Chemical Company Inc
Spectrum Chemicals & Laboratory Products Inc
Strahl & Pitsch Inc
Voigt Global Distribution Inc

Others
Boith Limited (China)
LS Raw Materials Ltd
Sovereign Chemicals & Cosmetics

Peanut Oil

UK
A and E Connock (Perfumery and Cosmetics) Ltd
AarhusKarlshamn UK Ltd
Alembic Products Ltd
Allchem Performance

Other European
AarhusKarlshamn AB

USA
AarhusKarlshamn USA Inc
Advance Scientific & Chemical Inc
Arista Industries Inc
Charkit Chemical Corp
Croda Inc
Penta Manufacturing Co
Pokonobe Industries Inc
Ruger Chemical Co Inc
Spectrum Chemicals & Laboratory Products Inc
Voigt Global Distribution Inc
Welch, Holme & Clark Co Inc

Pectin

UK
Ingredients Consultancy Ltd, The
ISP Europe
Peter Whiting (Chemicals) Ltd

USA
Advance Scientific & Chemical Inc
CP Kelco US Inc
KIC Chemicals Inc
Parchem Trading Ltd
Penta Manufacturing Co
Ruger Chemical Co Inc
Seidler Chemical Company Inc
Spectrum Chemicals & Laboratory Products Inc
TIC Gums

Others
SD Fine-Chem Ltd

Pentetic Acid

UK
Dow Chemical Company (UK)

Other European
Alfa Aesar Johnson Matthey GmbH

USA
Dow Chemical Co

Petrolatum

UK
Fuchs Lubricants (UK) plc
Poth Hille & Co Ltd.

Other European
Parafluid Mineraloelgesellschaft mbH
Sasol Germany GmbH
USOCO BV

USA
Advance Scientific & Chemical Inc
Astro Chemicals Inc
Avatar Corp
Brenntag Southwest
Clarus Specialty Products
Integra Chemical Co
Mutchler Inc
Parchem Trading Ltd
Penta Manufacturing Co
Rita Corp
Ruger Chemical Co Inc
Seidler Chemical Company Inc
Spectrum Chemicals & Laboratory Products Inc
Voigt Global Distribution Inc

Petrolatum and Lanolin Alcohols

USA
Lubrizol Advanced Materials Inc.
Rita Corp

Phenol

UK
Chance & Hunt
Fisher Scientific UK Ltd
Macron Chemicals UK
Sigma-Aldrich Company Ltd
Tennants Distribution Ltd

Other European
Alfa Aesar Johnson Matthey GmbH
Brenntag GmbH
CHEMCO France

USA
Amresco Inc
Brenntag North America Inc
Dow Chemical Co
Fisher Scientific
Macron Chemicals
Penta Manufacturing Co
Seidler Chemical Company Inc
Spectrum Chemicals & Laboratory Products Inc
Voigt Global Distribution Inc

Others
Univar Canada Ltd

Phenoxyethanol

UK
Clariant UK Ltd
Dow Chemical Company (UK)
Lehvoss UK Ltd
Ubichem plc

Other European
Alfa Aesar Johnson Matthey GmbH
Clariant GmbH
Induchem AG

USA
Acme-Hardesty
Clariant Corp
Dow Chemical Co
Kraft Chemical Co
Penta Manufacturing Co
Spectrum Chemicals & Laboratory Products Inc

Phenylethyl Alcohol

USA
Alfa Chem
Penta Manufacturing Co
Spectrum Chemicals & Laboratory Products Inc

Phenylmercuric Acetate

Other European
Alfa Aesar Johnson Matthey GmbH
Panreac Quimica SAU

USA
Advance Scientific & Chemical Inc
George Uhe Co Inc
Penta Manufacturing Co
Ruger Chemical Co Inc
Spectrum Chemicals & Laboratory Products Inc

Phenylmercuric Borate

UK
Fluorochem Ltd

USA
Spectrum Chemicals & Laboratory Products Inc

Phenylmercuric Nitrate

Other European
Panreac Quimica SAU

USA
Advance Scientific & Chemical Inc
George Uhe Co Inc
Ruger Chemical Co Inc
Spectrum Chemicals & Laboratory Products Inc

Phospholipids

Other European
Lipoid GmbH

USA
Avanti Polar Lipids Inc

Others
Nippon Fine Chemical Co Ltd
NOF Corporation

Phosphoric Acid

UK
Macron Chemicals UK
Peter Whiting (Chemicals) Ltd

Other European
Brenntag GmbH

USA
Ashland Inc
Brenntag North America Inc
Brenntag Southwest
Macron Chemicals
Penta Manufacturing Co
Spectrum Chemicals & Laboratory Products Inc

Others
Boith Limited (China)
LS Raw Materials Ltd
SD Fine-Chem Ltd
Univar Canada Ltd

Polacrilin Potassium

UK
Rohm and Haas UK Ltd

USA
Rohm and Haas Co

Poloxamer

UK
BASF Plc

Other European
BASF Aktiengesellschaft

USA
Advance Scientific & Chemical Inc
BASF Corp
CarboMer Inc
Mutchler Inc
Spectrum Chemicals & Laboratory Products Inc
Voigt Global Distribution Inc

Others
BASF Japan Ltd
Signet Chemical Corporation Pvt Ltd

Polycarbophil

USA
Lubrizol Advanced Materials Inc.

Others
Arihant Trading Co
Lubrizol Advanced Materials, India Pvt Ltd

Polydextrose

UK
Danisco UK Ltd
Macron Chemicals UK

Other European
Danisco A/S
DuPont de Nemours Int'l SA

USA
Ashland Inc
Danisco USA Inc
Macron Chemicals
Tate & Lyle (North America)

Polyethylene Glycol

UK
Adina Chemicals Ltd
BASF Plc
Blagden Specialty Chemicals Ltd
Cornelius Group plc
Fisher Scientific UK Ltd
IMCD UK Ltd
Sasol UK Ltd
Tennants Distribution Ltd

Other European
Alfa Aesar Johnson Matthey GmbH
BASF Aktiengesellschaft
Brenntag GmbH
Corcoran Chemicals Ltd
Panreac Quimica SAU
Sasol Germany GmbH

USA
Advance Scientific & Chemical Inc
Alfa Chem
Ashland Inc
BASF Corp
Brenntag North America Inc
Dow Chemical Co
Fisher Scientific
Hawkins Inc
Lipo Chemicals Inc
Mutchler Inc
Paddock Laboratories Inc
Penta Manufacturing Co
Polysciences Inc
Protameen Chemicals Inc
Sasol North America Inc
Spectrum Chemicals & Laboratory Products Inc

Others
Aastrid International
BASF Japan Ltd
Dow Chemical International Pvt Ltd
Signet Chemical Corporation Pvt Ltd
Univar Canada Ltd
Wintersun Chemical

Polyethylene Oxide

UK
A and E Connock (Perfumery and Cosmetics) Ltd
Brenntag (UK) Ltd
Honeywill & Stein

Other European
Alfa Aesar Johnson Matthey GmbH
Degussa Hüls AG
Univar

USA
Ashland Inc
Avatar Corp
Aventis Behring LLC
BASF Corp
Dow Chemical Co
Integra Chemical Co
Mutchler Inc
Penta Manufacturing Co
Rhodia Inc
Spectrum Chemicals & Laboratory Products Inc

Polymethacrylates

UK
BASF Plc
Eastman Company UK Ltd
IMCD UK Ltd
Ubichem plc

Other European
BASF Aktiengesellschaft
Brenntag AG

USA
BASF Corp
CarboMer Inc
Eastman Chemical Co
Evonik Degussa Corp

Others
BASF Japan Ltd

Poly(methyl vinyl ether/maleic anhydride)

UK
Sigma-Aldrich Company Ltd

Other European
Matrix Marketing GmbH

USA
Fisher Scientific
International Specialty Products Inc (ISP)

Polyoxyethylene Alkyl Ethers

UK
Adina Chemicals Ltd
Cognis UK Ltd
Croda Europe Ltd
Lonza Biologics Plc
Sigma-Aldrich Company Ltd

Other European
Alfa Aesar Johnson Matthey GmbH
BASF Aktiengesellschaft
Brenntag NV
Cognis GmbH
Lonza Ltd

USA
Akzo Nobel Chemicals Inc
BASF Corp
Cognis Corp
Croda Inc
Global Seven Inc
Jeen International Corp
Lambent Technologies
Lipo Chemicals Inc
Protameen Chemicals Inc
Rita Corp

Others
BASF Japan Ltd
Kao Corporation
Nihon-Emulsion Co Ltd

Polyoxyethylene Castor Oil Derivatives

UK
Adina Chemicals Ltd
BASF Plc
Cognis UK Ltd
Lehvoss UK Ltd
White Sea and Baltic Company Ltd

Other European
BASF Aktiengesellschaft
Cognis GmbH

USA
ABITEC Corp
BASF Corp
Cognis Corp
Farma International Inc
Jeen International Corp
Lipo Chemicals Inc
Protameen Chemicals Inc

Others
BASF Japan Ltd
Nikko Chemicals Co Ltd

Polyoxyethylene Sorbitan Fatty Acid Esters

UK
Adina Chemicals Ltd
BASF Plc
Cognis UK Ltd
Croda Europe Ltd
Lonza Biologics Plc
Macron Chemicals UK

Other European
Alfa Aesar Johnson Matthey GmbH
BASF Aktiengesellschaft
Cognis GmbH
Lonza Ltd

USA
BASF Corp
Cognis Corp
Croda Inc
Hawkins Inc
Lipo Chemicals Inc
Protameen Chemicals Inc
Rita Corp

Others
Kao Corporation

Polyoxyethylene Stearates

UK
Adina Chemicals Ltd

USA
Alfa Chem
Lipo Chemicals Inc
Rita Corp

Polyoxyl 15 Hydroxystearate

UK
BASF Plc

USA
BASF Corp

Others
BASF Japan Ltd
Signet Chemical Corporation Pvt Ltd

Polyoxylglycerides

Other European
Gattefossé SAS

USA
Gattefossé Corp

Polyvinyl Acetate Dispersion

USA
CarboMer Inc

Others
Signet Chemical Corporation Pvt Ltd

Polyvinyl Acetate Phthalate

UK
Colorcon Ltd

USA
Colorcon Inc

Polyvinyl Alcohol

UK
Blagden Specialty Chemicals Ltd
IMCD UK Ltd
Nippon Gohsei (UK) Ltd

Other European
Alfa Aesar Johnson Matthey GmbH
DuPont de Nemours Int'l SA
Harke Pharma GmbH

USA
Alfa Chem
Astro Chemicals Inc
Celanese
DuPont
Mutchler Inc
Penta Manufacturing Co
Polysciences Inc
Ruger Chemical Co Inc
Seidler Chemical Company Inc
Spectrum Chemicals & Laboratory Products Inc

Others
Wintersun Chemical

Potassium Alginate

Other European
FMC Biopolymer

USA
International Specialty Products Inc (ISP)

Others
Kimica Corporation

Potassium Alum

Other European
Alfa Aesar Johnson Matthey GmbH

USA
Advance Scientific & Chemical Inc
Ruger Chemical Co Inc
Seidler Chemical Company Inc

Potassium Benzoate

UK
Dow Chemical Company (UK)
DSM UK Ltd
White Sea and Baltic Company Ltd

Other European
Alfa Aesar Johnson Matthey GmbH
Dr Paul Lohmann GmbH KG
DSM Fine Chemicals
Haltermann Products

USA
Advance Scientific & Chemical Inc
AerChem Inc
Ashland Inc
Brenntag North America Inc
Brenntag Southwest
DSM Pharmaceuticals Inc
Penta Manufacturing Co
Premium Ingredients Ltd
RIA International
Seidler Chemical Company Inc
Spectrum Chemicals & Laboratory Products Inc
Triple Crown America
Voigt Global Distribution Inc
Zhong Ya Chemical (USA) Ltd

Others
Univar Canada Ltd
Wintersun Chemical

Potassium Bicarbonate

UK
Peter Whiting (Chemicals) Ltd

Other European
Alfa Aesar Johnson Matthey GmbH
Corcoran Chemicals Ltd

USA
Advance Scientific & Chemical Inc
Amresco Inc
Fisher Scientific
Hummel Croton Inc
Mutchler Inc
Penta Manufacturing Co
Premium Ingredients Ltd
RIA International
Ruger Chemical Co Inc
Seidler Chemical Company Inc

Others
SD Fine-Chem Ltd
Wintersun Chemical

Potassium Chloride

UK
Fisher Scientific UK Ltd
ISP Europe
Macron Chemicals UK
Peter Whiting (Chemicals) Ltd
Sigma-Aldrich Company Ltd
Stan Chem International Ltd
Tennants Distribution Ltd

Other European
Alfa Aesar Johnson Matthey GmbH
Brenntag GmbH
Dr Paul Lohmann GmbH KG
Panreac Quimica SAU

USA
Advance Scientific & Chemical Inc
AerChem Inc
Alfa Chem
Amresco Inc
Brenntag North America Inc
Brenntag Southwest
Fisher Scientific
General Chemical Performance Products LLC
International Specialty Products Inc (ISP)
Macron Chemicals
Mutchler Inc
Penta Manufacturing Co
Reheis Inc
RIA International
Ruger Chemical Co Inc
Seidler Chemical Company Inc
Spectrum Chemicals & Laboratory Products Inc

Others
LS Raw Materials Ltd
Univar Canada Ltd
Wintersun Chemical

Potassium Citrate

UK
Courtin & Warner Ltd
Fisher Scientific UK Ltd
Peter Whiting (Chemicals) Ltd
Ubichem plc

Other European
Alfa Aesar Johnson Matthey GmbH
Brenntag GmbH
Dr Paul Lohmann GmbH KG
Lehmann & Voss & Co

USA
Advance Scientific & Chemical Inc
AerChem Inc
Alfa Chem
American International Chemical Inc
Ashland Inc
Brenntag North America Inc
Brenntag Southwest
Fisher Scientific
KIC Chemicals Inc
Kraft Chemical Co
Penta Manufacturing Co
RIA International
Seidler Chemical Company Inc
Spectrum Chemicals & Laboratory Products Inc
Tate & Lyle (North America)
Zhong Ya Chemical (USA) Ltd

Others
Boith Limited (China)
San Fu Chemical Company Ltd
Univar Canada Ltd
Wintersun Chemical

Potassium Hydroxide

UK
Fisher Scientific UK Ltd
Macron Chemicals UK
Peter Whiting (Chemicals) Ltd
Tennants Distribution Ltd
Ubichem plc

Other European
Alfa Aesar Johnson Matthey GmbH
Panreac Quimica SAU

USA
Advance Scientific & Chemical Inc
AerChem Inc
Amresco Inc
Brenntag North America Inc
Brenntag Southwest
Charkit Chemical Corp
Fisher Scientific
General Chemical Performance Products LLC
Macron Chemicals
Mutchler Inc
Penta Manufacturing Co
Ruger Chemical Co Inc
Seidler Chemical Company Inc
Spectrum Chemicals & Laboratory Products Inc
Voigt Global Distribution Inc

Others
Univar Canada Ltd
Wintersun Chemical

Potassium Metabisulfite

UK
Allchem Performance
Fisher Scientific UK Ltd
Ubichem plc

Other European
Alfa Aesar Johnson Matthey GmbH

USA
Advance Scientific & Chemical Inc
Brenntag North America Inc
Fisher Scientific
Penta Manufacturing Co
Premium Ingredients Ltd
Ruger Chemical Co Inc
Seidler Chemical Company Inc
Spectrum Chemicals & Laboratory Products Inc

Others
Univar Canada Ltd
Wintersun Chemical

Potassium Nitrate

USA
Seidler Chemical Company Inc

Others
Wintersun Chemical

Potassium Phosphate, Dibasic

Other European
Brenntag GmbH
FMC Biopolymer

USA
Alfa Chem
American International Chemical Inc
Amresco Inc
Brenntag Southeast
EMD Chemicals Inc
FMC Biopolymer (USA)
GFS Chemicals Inc
Kraft Chemical Co
Penta Manufacturing Co
Ruger Chemical Co Inc
Spectrum Chemicals & Laboratory Products Inc
Thomas Scientific
Universal Preserv-A-Chem Inc
Voigt Global Distribution Inc

Others
Wintersun Chemical

Potassium Sorbate

UK
Blagden Specialty Chemicals Ltd
Macron Chemicals UK
Peter Whiting (Chemicals) Ltd
Tennants Distribution Ltd
White Sea and Baltic Company Ltd

Other European
Alfa Aesar Johnson Matthey GmbH
Brenntag AG
Corcoran Chemicals Ltd
Helm AG
Panreac Quimica SAU

USA
Advance Scientific & Chemical Inc
AerChem Inc
Amresco Inc
Ashland Inc
Astro Chemicals Inc
Avatar Corp
Brenntag North America Inc
Brenntag Southwest
Charkit Chemical Corp
Dastech International Inc
Helm New York Inc
KIC Chemicals Inc
Macron Chemicals
Penta Manufacturing Co
Pfizer Inc
Premium Ingredients Ltd
Protameen Chemicals Inc
RIA International
Ruger Chemical Co Inc
Seidler Chemical Company Inc
Spectrum Chemicals & Laboratory Products Inc
Zhong Ya Chemical (USA) Ltd

Others
Boith Limited (China)
LS Raw Materials Ltd
Priyanka Pharma/M K Ingredients & Specialities
Univar Canada Ltd
Wintersun Chemical

Povidone

UK
BASF Plc
Blagden Specialty Chemicals Ltd
ISP Europe

Other European
August Hedinger GmbH & Co KG
BASF Aktiengesellschaft
IMCD Group BV
NP Pharm

USA
Advance Scientific & Chemical Inc
American International Chemical Inc
BASF Corp
EM Sergeant Pulp & Chemical Co Inc
Generichem Corp
International Specialty Products Inc (ISP)
Mutchler Inc
Napp Technologies LLC
Penta Manufacturing Co
Spectrum Chemicals & Laboratory Products Inc

Others
BASF Japan Ltd
Glide Chem Pvt Ltd
Kawarlal Excipients Pvt Ltd
Signet Chemical Corporation Pvt Ltd

Propionic Acid

UK
Tennants Distribution Ltd

Other European
Alfa Aesar Johnson Matthey GmbH

USA
Advance Scientific & Chemical Inc
Brenntag North America Inc
Brenntag Southwest
Dow Chemical Co
Penta Manufacturing Co
Seidler Chemical Company Inc
Spectrum Chemicals & Laboratory Products Inc

Others
Univar Canada Ltd

Propyl Gallate

UK
White Sea and Baltic Company Ltd

Other European
Alfa Aesar Johnson Matthey GmbH
Panreac Quimica SAU

USA
Aceto Corp
Advance Scientific & Chemical Inc
Alfa Chem
Penta Manufacturing Co
Premium Ingredients Ltd
Seidler Chemical Company Inc
Spectrum Chemicals & Laboratory Products Inc
Triple Crown America

Others
Wintersun Chemical

Propylene Carbonate

Other European
Alfa Aesar Johnson Matthey GmbH
Brenntag GmbH

USA
BOC Sciences
Brenntag North America Inc
Penta Manufacturing Co

Propylene Glycol

UK
Alfa Chemicals Ltd
BASF Plc
Eastman Company UK Ltd
Fisher Scientific UK Ltd
Macron Chemicals UK
Peter Whiting (Chemicals) Ltd
Sasol UK Ltd
Tennants Distribution Ltd

Other European
Alfa Aesar Johnson Matthey GmbH
August Hedinger GmbH & Co KG
BASF Aktiengesellschaft
Brenntag GmbH
Corcoran Chemicals Ltd
Gattefossé SAS
Lyondell Chemical Europe
Sasol Germany GmbH

USA
Acme-Hardesty
Advance Scientific & Chemical Inc
Amresco Inc
Ashland Inc
Avatar Corp
BASF Corp
Brenntag North America Inc
Brenntag Southwest
CarboMer Inc
Dow Chemical Co
Eastman Chemical Co
Fisher Scientific
Gattefossé Corp
KIC Chemicals Inc
Kraft Chemical Co
Lyondell Chemical Co
Macron Chemicals
Mitsubishi International Food Ingredients
Paddock Laboratories Inc
Penta Manufacturing Co
Premium Ingredients Ltd
Rita Corp
Ruger Chemical Co Inc
Sasol North America Inc
Seidler Chemical Company Inc
Spectrum Chemicals & Laboratory Products Inc
Voigt Global Distribution Inc

Others
Dow Chemical International Pvt Ltd
Gadot Petrochemical Industries Ltd
Univar Canada Ltd
Wintersun Chemical

Propylene Glycol Alginate

UK
A and E Connock (Perfumery and Cosmetics) Ltd

USA
CarboMer Inc
Integra Chemical Co
Penta Manufacturing Co
Spectrum Chemicals & Laboratory Products Inc

Others
Kimica Corporation

Propylene Glycol Dilaurate

UK
Connock

Propylene Glycol Monolaurate

UK
A and E Connock (Perfumery and Cosmetics) Ltd
Sasol UK Ltd

Other European
Gattefossé Group
Sasol Germany GmbH

USA
ABITEC Corp
Gattefossé Corp
Inolex Chemical Co
Lambent Technologies
Lipscomb Chemical Company
Lubrizol Advanced Materials Inc.
Sasol North America Inc
Stepan Co

Others
Nihon-Emulsion Co Ltd

Propylparaben

UK
Bayer plc
Blagden Specialty Chemicals Ltd
Clariant UK Ltd
Lanxess Ltd
White Sea and Baltic Company Ltd

Other European
Alfa Aesar Johnson Matthey GmbH
Clariant GmbH
Induchem AG

USA
Acme-Hardesty
Advance Scientific & Chemical Inc
Alfa Chem
Ashland Inc
Avatar Corp
Brenntag Southwest
Clariant Corp
Dastech International Inc
Hallstar Company, The
KIC Chemicals Inc
Kraft Chemical Co
Lanxess Corp
Napp Technologies LLC
Penta Manufacturing Co
Premium Ingredients Ltd
Protameen Chemicals Inc
RIA International
Rita Corp
Ruger Chemical Co Inc
Seidler Chemical Company Inc
Spectrum Chemicals & Laboratory Products Inc
Voigt Global Distribution Inc

Others
Charles Tennant & Co (Canada) Ltd
LS Raw Materials Ltd
San Fu Chemical Company Ltd
Univar Canada Ltd
Wintersun Chemical

Propylparaben Sodium

UK
Clariant UK Ltd

Other European
Clariant GmbH

USA
Clariant Corp

Pyroxylin

UK
Hercules Ltd

USA
CarboMer Inc
Hercules Inc
Penta Manufacturing Co
Ruger Chemical Co Inc
Spectrum Chemicals & Laboratory Products Inc
Van Waters and Rogers Inc
Voigt Global Distribution Inc

Pyrrolidone

UK
BASF Plc
ISP Europe

Other European
Alfa Aesar Johnson Matthey GmbH
BASF Aktiengesellschaft

USA
BASF Corp
EMD Chemicals Inc
International Specialty Products Inc (ISP)

Others
BASF Japan Ltd
Kawarlal Excipients Pvt Ltd
Signet Chemical Corporation Pvt Ltd

Raffinose

Other European
Alfa Aesar Johnson Matthey GmbH

USA
Advance Scientific & Chemical Inc

Saccharin

UK
Tennants Distribution Ltd

Other European
Alfa Aesar Johnson Matthey GmbH
Brenntag AG
Corcoran Chemicals Ltd
Helm AG
Hermes Sweeteners Ltd

USA
Aceto Corp
Advance Scientific & Chemical Inc
AerChem Inc
Ashland Inc
Brenntag North America Inc
Brenntag Southwest
Dastech International Inc
Helm New York Inc
Mutchler Inc
Penta Manufacturing Co
Pfaltz & Bauer Inc
PMC Specialities Group Inc
Seidler Chemical Company Inc
Spectrum Chemicals & Laboratory Products Inc
Triple Crown America
Voigt Global Distribution Inc

Others
LS Raw Materials Ltd
Univar Canada Ltd

Saccharin Sodium

UK
Blagden Specialty Chemicals Ltd
Fisher Scientific UK Ltd
Macron Chemicals UK

Other European
Alfa Aesar Johnson Matthey GmbH
Helm AG

USA
Advance Scientific & Chemical Inc
Alfa Chem
Fisher Scientific
George Uhe Co Inc
Helm New York Inc
Penta Manufacturing Co
Premium Ingredients Ltd
Seidler Chemical Company Inc
Spectrum Chemicals & Laboratory Products Inc
Voigt Global Distribution Inc

Safflower Oil

UK
A and E Connock (Perfumery and Cosmetics) Ltd

USA
Advance Scientific & Chemical Inc
Arista Industries Inc
Avatar Corp
KIC Chemicals Inc
Parchem
Penta Manufacturing Co
Pokonobe Industries Inc
Seidler Chemical Company Inc
Spectrum Chemicals & Laboratory Products Inc
Welch, Holme & Clark Co Inc

Saponite

USA
BOC Sciences
International Laboratory Ltd

Others
Advanced Technology & Industrial Co Ltd

Sesame Oil

UK
A and E Connock (Perfumery and Cosmetics) Ltd
AarhusKarlshamn UK Ltd
Adina Chemicals Ltd
Alembic Products Ltd
Croda Europe Ltd

Other European
AarhusKarlshamn AB

USA
AarhusKarlshamn USA Inc
AerChem Inc
Arista Industries Inc
Charkit Chemical Corp
Croda Inc
Hawkins Inc
KIC Chemicals Inc
Lipo Chemicals Inc
Penta Manufacturing Co
Pokonobe Industries Inc
Protameen Chemicals Inc
Spectrum Chemicals & Laboratory Products Inc
Voigt Global Distribution Inc
Welch, Holme & Clark Co Inc

Shellac

UK
AF Suter and Co Ltd
Colorcon Ltd
Kimpton Brothers Ltd
Mantrose (UK) Ltd
Paroxite (London) Ltd
Thew, Arnott and Co Ltd

Other European
Alland & Robert
Alltec Intertrade
Biogrund GmbH
Harke Pharma GmbH
Seppic SA
Stroever GmbH & Co. KG

USA
Colorcon Inc
Innovative Materials Technology (IMT)
Mantrose-Haeuser Co Inc
Seppic Inc
W Zinsser & Company Inc

Others
Arihant Trading Co
Excelacs Co. Ltd
Gifu Shellac Co Ltd

Simethicone

UK
A and E Connock (Perfumery and Cosmetics) Ltd
Dow Corning (UK)

USA
Advance Scientific & Chemical Inc
American International Chemical Inc
Dow Corning
Penta Manufacturing Co
Seidler Chemical Company Inc

Sodium Acetate

UK
Macron Chemicals UK
Sigma-Aldrich Company Ltd

Other European
Alfa Aesar Johnson Matthey GmbH
Brenntag GmbH
Panreac Quimica SAU

USA
Advance Scientific & Chemical Inc
Alfa Chem
Allan Chemical Corp
Amresco Inc
Astro Chemicals Inc
Macron Chemicals
Penta Manufacturing Co
Premium Ingredients Ltd
Seidler Chemical Company Inc
Spectrum Chemicals & Laboratory Products Inc
Zhong Ya Chemical (USA) Ltd

Others
Boith Limited (China)
SD Fine-Chem Ltd
Wintersun Chemical

Sodium Alginate

UK
Blagden Specialty Chemicals Ltd

Other European
FMC Biopolymer
NovaMatrix
Vion NV

USA
Advance Scientific & Chemical Inc
AerChem Inc
American International Chemical Inc
FMC Biopolymer (USA)
Spectrum Chemicals & Laboratory Products Inc
Voigt Global Distribution Inc

Others
Boith Limited (China)
Kimica Corporation

Sodium Ascorbate

UK
Peter Whiting (Chemicals) Ltd
Roche Products Ltd

Other European
Brenntag GmbH
Helm AG
Panreac Quimica SAU

USA
Advance Scientific & Chemical Inc
AerChem Inc
Brenntag North America Inc
Brenntag Southwest
Helm New York Inc
Integra Chemical Co
Penta Manufacturing Co
Premium Ingredients Ltd
RIA International
Seidler Chemical Company Inc
Spectrum Chemicals & Laboratory Products Inc
Takeda Pharmaceuticals North America Inc
Triple Crown America
Zhong Ya Chemical (USA) Ltd

Others
CSPC Pharma
LS Raw Materials Ltd
Wintersun Chemical

Sodium Benzoate

UK
Courtin & Warner Ltd
Dow Chemical Company (UK)
DSM UK Ltd
Fisher Scientific UK Ltd
Macron Chemicals UK
Peter Whiting (Chemicals) Ltd
Tennants Distribution Ltd
Ubichem plc
White Sea and Baltic Company Ltd

Other European
Alfa Aesar Johnson Matthey GmbH
Corcoran Chemicals Ltd
Dr Paul Lohmann GmbH KG
DSM Fine Chemicals
Haltermann Products
Panreac Quimica SAU

USA
Aceto Corp
Advance Scientific & Chemical Inc
AerChem Inc
Alfa Chem
Ashland Inc
Astro Chemicals Inc
Avatar Corp
Brenntag North America Inc
Brenntag Southwest
Dastech International Inc
DSM Pharmaceuticals Inc
Fisher Scientific
Hummel Croton Inc
KIC Chemicals Inc

Kraft Chemical Co
Macron Chemicals
Penta Manufacturing Co
Premium Ingredients Ltd
RIA International
Seidler Chemical Company Inc
Spectrum Chemicals & Laboratory Products Inc
Thomas Scientific
Triple Crown America
Zhong Ya Chemical (USA) Ltd

Others
Boith Limited (China)
LS Raw Materials Ltd
San Fu Chemical Company Ltd
Univar Canada Ltd
Wintersun Chemical

Sodium Bicarbonate

UK
Blagden Specialty Chemicals Ltd
Brunner Mond (UK) Ltd
Courtin & Warner Ltd
Fisher Scientific UK Ltd
Forum Products Ltd
Macron Chemicals UK
Peter Whiting (Chemicals) Ltd
Sigma-Aldrich Company Ltd
Tennants Distribution Ltd

Other European
Alfa Aesar Johnson Matthey GmbH
Brenntag GmbH
Corcoran Chemicals Ltd

USA
AerChem Inc
Alfa Chem
Astro Chemicals Inc
BOC Sciences
Brenntag North America Inc
Brenntag Southwest
Church and Dwight Co Inc
Fisher Scientific
Hummel Croton Inc
Integra Chemical Co
Macron Chemicals
Mutchler Inc
Penta Manufacturing Co
Premium Ingredients Ltd
RIA International
Seidler Chemical Company Inc
Spectrum Chemicals & Laboratory Products Inc
SPI Pharma Inc
Triple Crown America

Others
Univar Canada Ltd
Wintersun Chemical

Sodium Borate

UK
Borax Europe Ltd
Macron Chemicals UK
Sigma-Aldrich Company Ltd

Other European
Alfa Aesar Johnson Matthey GmbH
Brenntag AG

USA
Advance Scientific & Chemical Inc
EMD Chemicals Inc
Macron Chemicals
Mutchler Inc
Penta Manufacturing Co
Ruger Chemical Co Inc
Seidler Chemical Company Inc
Spectrum Chemicals & Laboratory Products Inc

Sodium Carbonate

UK
Sigma-Aldrich Company Ltd

Other European
Alfa Aesar Johnson Matthey GmbH
Panreac Quimica SAU

USA
Advance Scientific & Chemical Inc
Mutchler Inc
Penta Manufacturing Co
Premium Ingredients Ltd
Ruger Chemical Co Inc
Seidler Chemical Company Inc
Spectrum Chemicals & Laboratory Products Inc

Others
Boith Limited (China)
Wintersun Chemical

Sodium Chloride

UK
Courtin & Warner Ltd
Macron Chemicals UK
Sigma-Aldrich Company Ltd
Ubichem plc

Other European
Alfa Aesar Johnson Matthey GmbH
Brenntag AG
Panreac Quimica SAU

USA
Advance Scientific & Chemical Inc
Alfa Chem
Cargill Inc
Fisher Scientific
Hawkins Inc
Macron Chemicals
Penta Manufacturing Co
Seidler Chemical Company Inc
Spectrum Chemicals & Laboratory Products Inc
Triple Crown America

Others
SD Fine-Chem Ltd
Univar Canada Ltd
Wintersun Chemical

Sodium Citrate Dihydrate

UK
Cargill PLC
Courtin & Warner Ltd
Fisher Scientific UK Ltd
Macron Chemicals UK
Roche Products Ltd

Other European
Alfa Aesar Johnson Matthey GmbH
Cargill France SAS
Dr Paul Lohmann GmbH KG
Jungbunzlauer AG
Panreac Quimica SAU

USA
Advance Scientific & Chemical Inc
AerChem Inc
Cargill Inc
Fisher Scientific
KIC Chemicals Inc
Macron Chemicals
Penta Manufacturing Co
Spectrum Chemicals & Laboratory Products Inc
Tate & Lyle (North America)

Others
San Fu Chemical Company Ltd
Univar Canada Ltd

Sodium Cyclamate

UK
Blagden Specialty Chemicals Ltd

Other European
Alfa Aesar Johnson Matthey GmbH

USA
AerChem Inc
Premium Ingredients Ltd

Others
LS Raw Materials Ltd

Sodium Formaldehyde Sulfoxylate

Other European
Alfa Aesar Johnson Matthey GmbH
Panreac Quimica SAU

Others
Wintersun Chemical

Sodium Hyaluronate

Other European
Chemos GmbH
Contipro C a.s.
Kraeber GmbH & Co
Matrix Marketing GmbH
NovaMatrix

USA
RIA International

Others
Bio Chemicals Ltd
Kibun Food Chemifa Co Ltd
Signet Chemical Corporation Pvt Ltd

Sodium Hydroxide

UK
Fisher Scientific UK Ltd
Macron Chemicals UK
Sigma-Aldrich Company Ltd
Tennants Distribution Ltd
Ubichem plc

Other European
Alfa Aesar Johnson Matthey GmbH
Panreac Quimica SAU

USA
AerChem Inc
Brenntag North America Inc
Brenntag Southwest
Fisher Scientific
General Chemical Performance Products LLC
Hummel Croton Inc
Integra Chemical Co
Mutchler Inc
Penta Manufacturing Co
Seidler Chemical Company Inc
Spectrum Chemicals & Laboratory Products Inc
Triple Crown America

Others
LS Raw Materials Ltd
Univar Canada Ltd
Wintersun Chemical

Sodium Lactate

UK
White Sea and Baltic Company Ltd

Other European
Alfa Aesar Johnson Matthey GmbH

Dr Paul Lohmann GmbH KG
Panreac Quimica SAU
Purac Biochem

USA
Advance Scientific & Chemical Inc
Alfa Chem
American Ingredients Corp
Ashland Inc
EMD Chemicals Inc
Penta Manufacturing Co
Purac
Rita Corp
Ruger Chemical Co Inc
Seidler Chemical Company Inc
Spectrum Chemicals & Laboratory Products Inc
Zhong Ya Chemical (USA) Ltd

Others
Jiangxi Mosashino Bio-Chem Co Ltd
Signet Chemical Corporation Pvt Ltd
Wintersun Chemical

Sodium Lauryl Sulfate

UK
Allchem Performance
Cognis UK Ltd
Fisher Scientific UK Ltd
Sigma-Aldrich Company Ltd

Other European
Alfa Aesar Johnson Matthey GmbH
Cognis GmbH

USA
Acme-Hardesty
Advance Scientific & Chemical Inc
Akzo Nobel Chemicals Inc
Brenntag North America Inc
Brenntag Southwest
CarboMer Inc
Cognis Corp
Fisher Scientific
Kraft Chemical Co
Mutchler Inc
Penta Manufacturing Co
Seidler Chemical Company Inc
Spectrum Chemicals & Laboratory Products Inc

Others
Charles Tennant & Co (Canada) Ltd
Kawarlal Excipients Pvt Ltd
LS Raw Materials Ltd
SD Fine-Chem Ltd
Signet Chemical Corporation Pvt Ltd
Univar Canada Ltd
Wintersun Chemical

Sodium Metabisulfite

UK
Fisher Scientific UK Ltd
Peter Whiting (Chemicals) Ltd
Tennants Distribution Ltd
Ubichem plc

Other European
Alfa Aesar Johnson Matthey GmbH
Corcoran Chemicals Ltd

USA
Astro Chemicals Inc
Brenntag North America Inc
Brenntag Southwest
Fisher Scientific
Hawkins Inc
Penta Manufacturing Co
Seidler Chemical Company Inc
Spectrum Chemicals & Laboratory Products Inc
Triple Crown America

Others
LS Raw Materials Ltd
Wintersun Chemical

Sodium Phosphate, Dibasic

UK
Fisher Scientific UK Ltd
Macron Chemicals UK
Peter Whiting (Chemicals) Ltd
Tennants Distribution Ltd
Ubichem plc

Other European
Alfa Aesar Johnson Matthey GmbH
Brenntag GmbH

USA
Advance Scientific & Chemical Inc
AerChem Inc
Amresco Inc
Brenntag North America Inc
Brenntag Southwest
Budenheim USA Inc
Fisher Scientific
Mutchler Inc
Penta Manufacturing Co
Premium Ingredients Ltd
Spectrum Chemicals & Laboratory Products Inc
Triple Crown America
Zhong Ya Chemical (USA) Ltd

Others
Wintersun Chemical

Sodium Phosphate, Monobasic

UK
Fisher Scientific UK Ltd
Macron Chemicals UK
Peter Whiting (Chemicals) Ltd
Tennants Distribution Ltd
Ubichem plc

Other European
Brenntag GmbH
Thermphos International BV

USA
Advance Scientific & Chemical Inc
AerChem Inc
American International Chemical Inc
Brenntag North America Inc
Brenntag Southwest
Fisher Scientific
Mutchler Inc
Penta Manufacturing Co
Spectrum Chemicals & Laboratory Products Inc
Triple Crown America
Vertellus Specialties Inc
Zhong Ya Chemical (USA) Ltd

Sodium Propionate

UK
Ubichem plc

Other European
Alfa Aesar Johnson Matthey GmbH
Dr Paul Lohmann GmbH KG
Panreac Quimica SAU

USA
Advance Scientific & Chemical Inc
Brenntag North America Inc
Brenntag Southwest
Penta Manufacturing Co
Seidler Chemical Company Inc
Spectrum Chemicals & Laboratory Products Inc
Triple Crown America

Others
Wintersun Chemical

Sodium Starch Glycolate

UK
Allchem Performance
Avebe UK Ltd
Blagden Specialty Chemicals Ltd
DFE Pharma UK
Forum Products Ltd
Rettenmaier UK Ltd

Other European
Avebe Group
DFE Pharma
IMCD Group BV
J Rettenmaier & Söhne GmbH and Co.KG
Lehmann & Voss & Co
Pharmatrans Sanaq AG

USA
American International Chemical Inc
Avebe America Inc
Barrington Nutritionals
CarboMer Inc
Dastech International Inc
EM Sergeant Pulp & Chemical Co Inc
Generichem Corp
JRS Pharma LP
Mutchler Inc
Penta Manufacturing Co
RIA International
Spectrum Chemicals & Laboratory Products Inc

Others
Blanver
Charles Tennant & Co (Canada) Ltd
Kawarlal Excipients Pvt Ltd
NB Entrepreneurs
Signet Chemical Corporation Pvt Ltd

Sodium Stearate

UK
Peter Greven UK

Other European
Alfa Aesar Johnson Matthey GmbH

USA
Alfa Chem
Allan Chemical Corp
American International Chemical Inc
Ashland Inc
Avatar Corp
Kraft Chemical Co
Mutchler Inc
Parchem Trading Ltd
Penta Manufacturing Co
Ruger Chemical Co Inc
Spectrum Chemicals & Laboratory Products Inc
Universal Preserv-A-Chem Inc

Others
Boith Limited (China)
Viva Corporation
Wintersun Chemical

Sodium Stearyl Fumarate

UK
Allchem Performance
Blagden Specialty Chemicals Ltd
Forum Products Ltd
Rettenmaier UK Ltd

Other European
IMCD Group BV
J Rettenmaier & Söhne GmbH and Co.KG

USA
Aceto Corp
Advance Scientific & Chemical Inc
American International Chemical Inc
CarboMer Inc
JRS Pharma LP
Spectrum Chemicals & Laboratory Products Inc

Others
Arihant Trading Co
Signet Chemical Corporation Pvt Ltd

Sodium Sulfite

UK
BASF Plc
Macron Chemicals UK
Sigma-Aldrich Company Ltd

Other European
Alfa Aesar Johnson Matthey GmbH
Brenntag AG
Chemos GmbH
Panreac Quimica SAU

USA
Advance Scientific & Chemical Inc
Amresco Inc
Ashland Inc
Astro Chemicals Inc
Biddle Sawyer Corp
EMD Chemicals Inc
General Chemical Performance Products LLC
Macron Chemicals
Penta Manufacturing Co
Ruger Chemical Co Inc
Seidler Chemical Company Inc
Spectrum Chemicals & Laboratory Products Inc

Others
SD Fine-Chem Ltd
Wintersun Chemical

Sodium Thiosulfate

UK
Macron Chemicals UK
Sigma-Aldrich Company Ltd

Other European
Alfa Aesar Johnson Matthey GmbH
Evonik Industries AG

USA
Alfa Chem
Allan Chemical Corp
Charkit Chemical Corp
Fisher Scientific
Macron Chemicals
Penta Manufacturing Co
Ruger Chemical Co Inc
Seidler Chemical Company Inc
Spectrum Chemicals & Laboratory Products Inc
Thomas Scientific

Others
Boith Limited (China)
LS Raw Materials Ltd
Wintersun Chemical

Sorbic Acid

UK
Blagden Specialty Chemicals Ltd
Peter Whiting (Chemicals) Ltd
Tennants Distribution Ltd
White Sea and Baltic Company Ltd

Other European
Alfa Aesar Johnson Matthey GmbH
Brenntag GmbH
Panreac Quimica SAU

USA
Advance Scientific & Chemical Inc
AerChem Inc
Alfa Chem
Ashland Inc
Avatar Corp
BOC Sciences
Brenntag North America Inc
Brenntag Southwest
Dastech International Inc
KIC Chemicals Inc
Penta Manufacturing Co
Premium Ingredients Ltd
Protameen Chemicals Inc
Seidler Chemical Company Inc
Spectrum Chemicals & Laboratory Products Inc
Triple Crown America
Voigt Global Distribution Inc
Zhong Ya Chemical (USA) Ltd

Others
Boith Limited (China)
LS Raw Materials Ltd
Priyanka Pharma/M K Ingredients & Specialities
Univar Canada Ltd
Wintersun Chemical

Sorbitan Esters (Sorbitan Fatty Acid Esters)

UK
A and E Connock (Perfumery and Cosmetics) Ltd
Adina Chemicals Ltd
Cognis UK Ltd
Croda Europe Ltd
Esterchem Limited
Evonik Goldschmidt UK Ltd
Lonza Biologics Plc
White Sea and Baltic Company Ltd

Other European
Alfa Aesar Johnson Matthey GmbH
Cognis GmbH
Corcoran Chemicals Ltd
Lonza Ltd
Panreac Quimica SAU
Seppic SA

USA
Advance Scientific & Chemical Inc
Ashland Inc
Avatar Corp
Brenntag North America Inc
Brenntag Southwest
Cognis Corp
Croda Inc
Lipo Chemicals Inc
Penta Manufacturing Co
Protameen Chemicals Inc
Ruger Chemical Co Inc
Seidler Chemical Company Inc
Spectrum Chemicals & Laboratory Products Inc

Others
Univar Canada Ltd

Sorbitol

UK
Adina Chemicals Ltd
Cargill PLC
Colorcon Ltd
Forum Products Ltd
Lonza Biologics Plc
Pfanstiehl (Europe) Ltd
Roquette (UK) Ltd

Other European
Alfa Aesar Johnson Matthey GmbH
Biesterfeld Spezialchemie GmbH
Cargill France SAS
CPI Chemicals
IMCD Group BV
Lehmann & Voss & Co
Lonza Ltd
Roquette Frères
Tate & Lyle

USA
Advance Scientific & Chemical Inc
Alfa Chem
AnMar International
Ashland Inc
Asiamerica Ingredients Inc
Avatar Corp
Barrington Nutritionals
Brenntag North America Inc
Brenntag Southwest
CarboMer Inc
Cargill Inc
Corn Products U.S.
EM Sergeant Pulp & Chemical Co Inc
Ferro Corp
Kraft Chemical Co
Lipo Chemicals Inc
Mitsubishi International Food Ingredients
Mutchler Inc
Penta Manufacturing Co
Premium Ingredients Ltd
Roquette America Inc
Seidler Chemical Company Inc
Spectrum Chemicals & Laboratory Products Inc
SPI Pharma Inc
Thomas Scientific
Triple Crown America
Voigt Global Distribution Inc

Others
Signet Chemical Corporation Pvt Ltd
Univar Canada Ltd
Wintersun Chemical

Soybean Oil

UK
A and E Connock (Perfumery and Cosmetics) Ltd
AarhusKarlshamn UK Ltd
Croda Europe Ltd
Thew, Arnott and Co Ltd

Other European
AarhusKarlshamn AB
Corcoran Chemicals Ltd

USA
AarhusKarlshamn USA Inc
Advance Scientific & Chemical Inc
Arista Industries Inc
Avatar Corp
Charkit Chemical Corp
Croda Inc
KIC Chemicals Inc
Mutchler Inc
Parchem
Penta Manufacturing Co
Pokonobe Industries Inc
Seidler Chemical Company Inc
Spectrum Chemicals & Laboratory Products Inc
Welch, Holme & Clark Co Inc

Starch

UK
Avebe UK Ltd
Cargill PLC
Lehvoss UK Ltd

National Starch Personal Care
Roquette (UK) Ltd

Other European
Alfa Aesar Johnson Matthey GmbH
Avebe Group
Brenntag AG
Cargill France SAS
Corcoran Chemicals Ltd
IMCD Group BV
Roquette Frères
Tate & Lyle

USA
Advance Scientific & Chemical Inc
Alfa Chem
Ashland Inc
Avebe America Inc
Brenntag North America Inc
Brenntag Southwest
CarboMer Inc
Cargill Inc
EM Sergeant Pulp & Chemical Co Inc
Generichem Corp
Grain Processing Corp
Mutchler Inc
National Starch Personal Care (USA)
Penta Manufacturing Co
Roquette America Inc
Ruger Chemical Co Inc
Seidler Chemical Company Inc
Spectrum Chemicals & Laboratory Products Inc
Voigt Global Distribution Inc

Others
Colorcon Asia Pvt Ltd
NB Entrepreneurs
SD Fine-Chem Ltd
Wintersun Chemical

Starch, Modified

Others
Kawarlal Excipients Pvt Ltd

Starch, Pregelatinized

UK
Avebe UK Ltd
Cargill PLC
Colorcon Ltd
Lehvoss UK Ltd
Roquette (UK) Ltd

Other European
Avebe Group
Cargill Europe BVBA
Cargill France SAS
Roquette Frères

USA
Alfa Chem
Avebe America Inc
Cargill Inc
Colorcon Inc
Generichem Corp
Grain Processing Corp
Mutchler Inc
Roquette America Inc

Others
Asahi Kasei Corporation
Colorcon Asia Pvt Ltd
Kawarlal Excipients Pvt Ltd
Signet Chemical Corporation Pvt Ltd

Starch, Sterilizable Maize

UK
IMCD UK Ltd
Roquette (UK) Ltd

Other European
Roquette Frères
Tate & Lyle

USA
Roquette America Inc

Stearic Acid

UK
A and E Connock (Perfumery and Cosmetics) Ltd
Allchem Performance
Cognis UK Ltd
Croda Europe Ltd
H Foster & Co (Stearines) Ltd
James M Brown Ltd
Kimpton Brothers Ltd
Lehvoss UK Ltd
Macron Chemicals UK
Peter Whiting (Chemicals) Ltd
Poth Hille & Co Ltd.
Tennants Distribution Ltd
White Sea and Baltic Company Ltd

Other European
Alfa Aesar Johnson Matthey GmbH
Cognis GmbH
Corcoran Chemicals Ltd
Panreac Quimica SAU
Union Derivan, SA (UNDESA)

USA
Aceto Corp
Acme-Hardesty
Advance Scientific & Chemical Inc
Alzo International Inc
American International Chemical Inc
Ashland Inc
Astro Chemicals Inc
Brenntag North America Inc
Brenntag Southwest
Cognis Corp
EM Sergeant Pulp & Chemical Co Inc
Hummel Croton Inc
KIC Chemicals Inc
Kraft Chemical Co
Macron Chemicals
Mutchler Inc
Parchem Trading Ltd
Penta Manufacturing Co
Protameen Chemicals Inc
Rita Corp
Ruger Chemical Co Inc
Seidler Chemical Company Inc
Spectrum Chemicals & Laboratory Products Inc
Thomas Scientific
Triple Crown America

Others
Boith Limited (China)
Kawarlal Excipients Pvt Ltd
LS Raw Materials Ltd
Signet Chemical Corporation Pvt Ltd
Univar Canada Ltd
Wintersun Chemical

Stearyl Alcohol

UK
A and E Connock (Perfumery and Cosmetics) Ltd
AarhusKarlshamn UK Ltd
Adina Chemicals Ltd
Cognis UK Ltd
Croda Europe Ltd
Efkay Chemicals Ltd
Evonik Goldschmidt UK Ltd
Kimpton Brothers Ltd
Sasol UK Ltd
White Sea and Baltic Company Ltd

Other European
AarhusKarlshamn AB
Alfa Aesar Johnson Matthey GmbH
Berg + Schmidt (GmbH & Co.) KG
Chempri Oleochemicals (Europe)
Cognis GmbH
Corcoran Chemicals Ltd
Impag GmbH
Sasol Germany GmbH

USA
AarhusKarlshamn USA Inc
Acme-Hardesty
Brenntag North America Inc
Brenntag Southwest
Cognis Corp
Croda Inc
KIC Chemicals Inc
Kraft Chemical Co
Lipo Chemicals Inc
M Michel and Company Inc
P & G Chemicals
Penta Manufacturing Co
Protameen Chemicals Inc
Rita Corp
Sasol North America Inc
Seidler Chemical Company Inc
Spectrum Chemicals & Laboratory Products Inc

Others
Kao Corporation
VVF Limited

Sucralose

UK
Blagden Specialty Chemicals Ltd
Tate and Lyle plc

Other European
Corcoran Chemicals Ltd
Fusion Nutraceuticals Ltd

USA
BOC Sciences
McNeil Nutritionals LLC
Premium Ingredients Ltd
Spectrum Chemicals & Laboratory Products Inc

Others
Gangwal Chemicals Pvt Ltd
Kawarlal Excipients Pvt Ltd
Wintersun Chemical

Sucrose

UK
3M United Kingdom Plc
British Sugar Pharmaceutical Group
Fisher Scientific UK Ltd
IMCD UK Ltd
Macron Chemicals UK
Pfanstiehl (Europe) Ltd
Tate and Lyle plc

Other European
Alfa Aesar Johnson Matthey GmbH
Brenntag GmbH
IMCD Group BV

USA
3M Drug Delivery Systems
Advance Scientific & Chemical Inc
Brenntag North America Inc
Brenntag Southwest
Domino Foods, Inc.
EM Sergeant Pulp & Chemical Co Inc
Ferro Corp

1

Fisher Scientific
Macron Chemicals
Mutchler Inc
Penta Manufacturing Co
Ruger Chemical Co Inc
Seidler Chemical Company Inc
Spectrum Chemicals & Laboratory Products Inc
Tate & Lyle (North America)
Thomas Scientific
Voigt Global Distribution Inc

Others
Signet Chemical Corporation Pvt Ltd

Sucrose Octaacetate

USA
Advance Scientific & Chemical Inc
Penta Manufacturing Co

Sucrose Palmitate

USA
Mutchler Inc

Others
Mitsubishi-Kagaku Foods Corporation

Sucrose Stearate

UK
Croda Europe Ltd

Other European
Sisterna BV

USA
Croda Inc

Others
Croda Japan

Sugar, Compressible

UK
British Sugar Pharmaceutical Group
Forum Products Ltd
Wilfrid Smith Ltd

Other European
Tereos

USA
CarboMer Inc
Domino Foods, Inc.
EM Sergeant Pulp & Chemical Co Inc
Mutchler Inc

Others
Arihant Trading Co

Sugar, Confectioner's

USA
EM Sergeant Pulp & Chemical Co Inc
Hummel Croton Inc
Mutchler Inc

Sugar Spheres

UK
Forum Products Ltd
IMCD UK Ltd
Rettenmaier UK Ltd

Other European
J Rettenmaier & Söhne GmbH and Co.KG
NP Pharm

USA
JRS Pharma LP
Mutchler Inc

Others
Arihant Trading Co
Signet Chemical Corporation Pvt Ltd

Sulfobutylether β-Cyclodextrin

USA
CyDex Pharmaceuticals Inc

Sulfur Dioxide

Other European
Alfa Aesar Johnson Matthey GmbH
Brenntag GmbH

Sulfuric Acid

UK
Fisher Scientific UK Ltd
Macron Chemicals UK
Sigma-Aldrich Company Ltd
Tennants Distribution Ltd

Other European
Alfa Aesar Johnson Matthey GmbH
Brenntag GmbH
Panreac Quimica SAU

USA
Advance Scientific & Chemical Inc
Ashland Inc
Brenntag North America Inc
Brenntag Southwest
Fisher Scientific
Macron Chemicals
Seidler Chemical Company Inc
Spectrum Chemicals & Laboratory Products Inc
Thomas Scientific
Triple Crown America

Others
Univar Canada Ltd
Wintersun Chemical

Sunflower Oil

UK
A and E Connock (Perfumery and Cosmetics) Ltd
Thew, Arnott and Co Ltd

Other European
Corcoran Chemicals Ltd

USA
Arista Industries Inc
Avatar Corp
KIC Chemicals Inc
Mutchler Inc
Penta Manufacturing Co
Pokonobe Industries Inc
Seidler Chemical Company Inc

Suppository Bases, Hard Fat

UK
AarhusKarlshamn UK Ltd
Alfa Chemicals Ltd
Blagden Specialty Chemicals Ltd

Other European
AarhusKarlshamn AB
Gattefossé SAS

USA
AarhusKarlshamn USA Inc
Gattefossé Corp
Stepan Co
Voigt Global Distribution Inc

Tagatose

Other European
Alfa Aesar Johnson Matthey GmbH

USA
Spherix Incorporated

Others
Saccharides

Talc

UK
Baker Sillavan Barytes Ltd
Blagden Specialty Chemicals Ltd
Fisher Scientific UK Ltd
Lehvoss UK Ltd
Macron Chemicals UK
Pumex (UK) Limited
Tennants Distribution Ltd
Thew, Arnott and Co Ltd

Other European
August Hedinger GmbH & Co KG
Luzenac Europe
Panreac Quimica SAU

USA
Advance Scientific & Chemical Inc
Brenntag North America Inc
Charles B Chrystal Co Inc
Fisher Scientific
Luzenac America
Macron Chemicals
Mutchler Inc
Penta Manufacturing Co
Ruger Chemical Co Inc
Seidler Chemical Company Inc
Spectrum Chemicals & Laboratory Products Inc
Triple Crown America
Voigt Global Distribution Inc
Whittaker Clark, and Daniels Inc

Others
Arihant Trading Co
Signet Chemical Corporation Pvt Ltd
Univar Canada Ltd

Tartaric Acid

UK
Fisher Scientific UK Ltd
Macron Chemicals UK
Peter Whiting (Chemicals) Ltd
Sigma-Aldrich Company Ltd
Tennants Distribution Ltd
Ubichem plc
White Sea and Baltic Company Ltd

Other European
Alfa Aesar Johnson Matthey GmbH
Arion & Delahaye
Brenntag AG
Dr Paul Lohmann GmbH KG
Pahí SL
Panreac Quimica SAU

USA
Aceto Corp
Advance Scientific & Chemical Inc
AerChem Inc
American International Chemical Inc
Amresco Inc
Ashland Inc
Astro Chemicals Inc
BOC Sciences
Brenntag North America Inc
Brenntag Southwest
CarboMer Inc
Charkit Chemical Corp

Dastech International Inc
Fisher Scientific
George Uhe Co Inc
Macron Chemicals
Mutchler Inc
Penta Manufacturing Co
Premium Ingredients Ltd
Ruger Chemical Co Inc
Seidler Chemical Company Inc
Spectrum Chemicals & Laboratory Products Inc
Triple Crown America

Others
Arihant Trading Co
LS Raw Materials Ltd
Priyanka Pharma/M K Ingredients & Specialities
SD Fine-Chem Ltd
Univar Canada Ltd
Wintersun Chemical

Thaumatin

Other European
ABCR GmbH & Co. KG

USA
EM Sergeant Pulp & Chemical Co Inc
Penta Manufacturing Co
RFI Ingredients

Thimerosal

UK
Sigma-Aldrich Company Ltd
Ubichem plc

USA
Advance Scientific & Chemical Inc
AerChem Inc
Alfa Chem
George Uhe Co Inc
RIA International
Spectrum Chemicals & Laboratory Products Inc

Others
LS Raw Materials Ltd

Thymol

UK
Courtin & Warner Ltd
Sigma-Aldrich Company Ltd

Other European
Alfa Aesar Johnson Matthey GmbH

USA
Advance Scientific & Chemical Inc
EMD Chemicals Inc
Penta Manufacturing Co
Ruger Chemical Co Inc
Seidler Chemical Company Inc
Spectrum Chemicals & Laboratory Products Inc
Thomas Scientific

Others
Sarman Industries
Wintersun Chemical

Titanium Dioxide

UK
A and E Connock (Perfumery and Cosmetics) Ltd
BASF Plc
Cornelius Group plc
Kronos Ltd
Lehvoss UK Ltd
Peter Whiting (Chemicals) Ltd
Tennants Distribution Ltd
Thew, Arnott and Co Ltd
Tioxide Europe Ltd

Other European
Agrofert Holding a.s.
Alfa Aesar Johnson Matthey GmbH
BASF Aktiengesellschaft
Brenntag GmbH
DuPont de Nemours Int'l SA
Sachtleben Chemie GmbH

USA
Advance Scientific & Chemical Inc
AerChem Inc
Alfa Chem
American International Chemical Inc
Ashland Inc
BASF Corp
BOC Sciences
Brenntag North America Inc
Brenntag Southwest
Charles B Chrystal Co Inc
DuPont
Kraft Chemical Co
Lipscomb Chemical Company
Mutchler Inc
Penta Manufacturing Co
Premium Ingredients Ltd
Seidler Chemical Company Inc
Spectrum Chemicals & Laboratory Products Inc
Tioxide Americas Inc
Triple Crown America
Voigt Global Distribution Inc
Whittaker Clark, and Daniels Inc

Others
Arihant Trading Co
BASF Japan Ltd
Priyanka Pharma/M K Ingredients & Specialities
Signet Chemical Corporation Pvt Ltd
Univar Canada Ltd
Wintersun Chemical

Tragacanth

UK
AF Suter and Co Ltd
Blagden Specialty Chemicals Ltd
Fisher Scientific UK Ltd
Thew, Arnott and Co Ltd

Other European
Alfa Aesar Johnson Matthey GmbH

USA
Advance Scientific & Chemical Inc
Ashland Inc
Brenntag Southwest
CarboMer Inc
Chart Corp Inc
Fisher Scientific
Hummel Croton Inc
Lipscomb Chemical Company
Penta Manufacturing Co
Spectrum Chemicals & Laboratory Products Inc

Trehalose

UK
Cargill PLC
Hayashibara International

Other European
Alfa Aesar Johnson Matthey GmbH
Cargill France SAS

USA
AerChem Inc
Penta Manufacturing Co

Others
Asahi Kasei Corporation
Hayashibara Co Ltd

Triacetin

UK
Allchem Performance
Eastman Company UK Ltd
IMCD UK Ltd
Tennants Distribution Ltd
White Sea and Baltic Company Ltd

Other European
Alfa Aesar Johnson Matthey GmbH

USA
ABITEC Corp
Advance Scientific & Chemical Inc
Allan Chemical Corp
Eastman Chemical Co
Penta Manufacturing Co
Premium Ingredients Ltd
Seidler Chemical Company Inc
Spectrum Chemicals & Laboratory Products Inc

Others
Kawarlal Excipients Pvt Ltd
Priyanka Pharma/M K Ingredients & Specialities
Signet Chemical Corporation Pvt Ltd

Tributyl Citrate

UK
Ubichem plc

Other European
Jungbunzlauer AG

USA
Jungbunzlauer Inc
Penta Manufacturing Co
Vertellus Specialties Inc

Tricaprylin

UK
Sasol UK Ltd

Other European
Sasol Germany GmbH

USA
ABITEC Corp
Global Seven Inc
Sasol North America Inc

Others
Ecogreen Oleochemicals (S) Pte. Ltd.

Triethanolamine

UK
Fisher Scientific UK Ltd
Leading Solvent Supplies Ltd
Sigma-Aldrich Company Ltd
Tennants Distribution Ltd
Ubichem plc

Other European
Alfa Aesar Johnson Matthey GmbH
Panreac Quimica SAU

USA
Alfa Chem
Brenntag North America Inc
Fisher Scientific
Penta Manufacturing Co
Rita Corp
Seidler Chemical Company Inc
Spectrum Chemicals & Laboratory Products Inc
Triple Crown America

Others
Univar Canada Ltd

Triethyl Citrate

UK
Cognis UK Ltd
Ubichem plc

Other European
Alfa Aesar Johnson Matthey GmbH
Cognis GmbH
Jungbunzlauer AG

USA
Advance Scientific & Chemical Inc
Allan Chemical Corp
Cognis Corp
Jungbunzlauer Inc
Mutchler Inc
Penta Manufacturing Co
Spectrum Chemicals & Laboratory Products Inc
Vertellus Specialties Inc

Others
Arihant Trading Co

Tromethamine

Other European
Merck KGaA

USA
Aldrich
American International Chemical Inc
ANGUS Chemical Co
CarboMer Inc
EMD Chemicals Inc
GFS Chemicals Inc
Ruger Chemical Co Inc
Spectrum Chemicals & Laboratory Products Inc
Voigt Global Distribution Inc

Others
Xinchem Co

Vanillin

UK
Blagden Specialty Chemicals Ltd
Courtin & Warner Ltd
Fisher Scientific UK Ltd
Rhodia Organic Fine Ltd
Tennants Distribution Ltd
Ubichem plc

Other European
Alfa Aesar Johnson Matthey GmbH
Biesterfeld Spezialchemie GmbH
Helm AG
Panreac Quimica SAU

USA
AerChem Inc
Alfa Chem
Ashland Inc
BOC Sciences
Brenntag North America Inc
Brenntag Southwest
Charkit Chemical Corp
Chart Corp Inc
Fisher Scientific
Helm New York Inc
KIC Chemicals Inc
Mutchler Inc
Penta Manufacturing Co
Premium Ingredients Ltd
Rhodia Pharma Solutions Inc
Seidler Chemical Company Inc
Spectrum Chemicals & Laboratory Products Inc
Triple Crown America
Virginia Dare
Voigt Global Distribution Inc
Zhong Ya Chemical (USA) Ltd

Others
LS Raw Materials Ltd
Priyanka Pharma/M K Ingredients & Specialities
Wintersun Chemical

Vegetable Oil, Hydrogenated

UK
AarhusKarlshamn UK Ltd
Adina Chemicals Ltd
Forum Products Ltd
Rettenmaier UK Ltd

Other European
AarhusKarlshamn AB
Chevron Lubricants
J Rettenmaier & Söhne GmbH and Co.KG

USA
AarhusKarlshamn USA Inc
ABITEC Corp
Arista Industries Inc
JRS Pharma LP
Lipo Chemicals Inc
Mutchler Inc
Stepan Co

Vitamin E Polyethylene Glycol Succinate

UK
Cognis UK Ltd
Sigma-Aldrich Company Ltd

Other European
Cognis GmbH

USA
Cognis Corp

Others
HalloChem Pharma Co. Ltd
Signet Chemical Corporation Pvt Ltd

Water

UK
Fisher Scientific UK Ltd
Leading Solvent Supplies Ltd
Tennants Distribution Ltd

Other European
Alfa Aesar Johnson Matthey GmbH
August Hedinger GmbH & Co KG
Panreac Quimica SAU

USA
Advance Scientific & Chemical Inc
Fisher Scientific
Seidler Chemical Company Inc
Spectrum Chemicals & Laboratory Products Inc
Thomas Scientific

Wax, Anionic Emulsifying

UK
Adina Chemicals Ltd
British Wax Refining Co Ltd
Cognis UK Ltd
Croda Europe Ltd

Other European
Cognis GmbH

USA
Cognis Corp
Croda Inc
Lipo Chemicals Inc
Spectrum Chemicals & Laboratory Products Inc
Strahl & Pitsch Inc

Others
Sovereign Chemicals & Cosmetics

Wax, Carnauba

UK
AF Suter and Co Ltd
British Wax Refining Co Ltd
Cornelius Group plc
Kimpton Brothers Ltd
Lehvoss UK Ltd
Poth Hille & Co Ltd.
Tennants Distribution Ltd
Thew, Arnott and Co Ltd
Ubichem plc

Other European
Alfa Aesar Johnson Matthey GmbH

USA
Charkit Chemical Corp
Frank B Ross Co Inc
Koster Keunen LLC
Mutchler Inc
Penta Manufacturing Co
Ruger Chemical Co Inc
Strahl & Pitsch Inc
Whittaker Clark, and Daniels Inc

Others
Arihant Trading Co

Wax, Cetyl Esters

UK
A and E Connock (Perfumery and Cosmetics) Ltd
Cognis UK Ltd
Croda Europe Ltd

Other European
Cognis GmbH

USA
Cognis Corp
Croda Inc
Mutchler Inc
Rita Corp
Ruger Chemical Co Inc
Spectrum Chemicals & Laboratory Products Inc

Others
LS Raw Materials Ltd

Wax, Microcrystalline

UK
A and E Connock (Perfumery and Cosmetics) Ltd
AF Suter and Co Ltd
British Wax Refining Co Ltd
Cornelius Group plc
Kimpton Brothers Ltd
Paroxite (London) Ltd
Poth Hille & Co Ltd.
Thew, Arnott and Co Ltd

Other European
Chevron Lubricants
Parafluid Mineraloelgesellschaft mbH
USOCO BV

USA
Avatar Corp
Frank B Ross Co Inc
Koster Keunen LLC
Mutchler Inc
Strahl & Pitsch Inc
Voigt Global Distribution Inc
Whittaker Clark, and Daniels Inc

Others
Sovereign Chemicals & Cosmetics

Wax, Nonionic Emulsifying

UK
Adina Chemicals Ltd
Cognis UK Ltd
Croda Europe Ltd
Efkay Chemicals Ltd
Esterchem Limited
Lehvoss UK Ltd

Other European
Cognis GmbH

USA
Cognis Corp
Croda Inc
Koster Keunen LLC
Lipo Chemicals Inc
Mason Chemical Company
Rita Corp
Ruger Chemical Co Inc
Strahl & Pitsch Inc

Wax, White

UK
British Wax Refining Co Ltd
Cornelius Group plc
Fisher Scientific UK Ltd
Kimpton Brothers Ltd
Paroxite (London) Ltd
Poth Hille & Co Ltd.
Thew, Arnott and Co Ltd

Other European
Chevron Lubricants
USOCO BV

USA
Advance Scientific & Chemical Inc
Charkit Chemical Corp
Fisher Scientific
Koster Keunen LLC
Mutchler Inc
Penta Manufacturing Co
Rita Corp
Spectrum Chemicals & Laboratory Products Inc
Strahl & Pitsch Inc
Triple Crown America
Voigt Global Distribution Inc
Whittaker Clark, and Daniels Inc

Others
Charles Tennant & Co (Canada) Ltd
Sovereign Chemicals & Cosmetics

Wax, Yellow

UK
British Wax Refining Co Ltd
Cornelius Group plc
Fisher Scientific UK Ltd
Kimpton Brothers Ltd
Paroxite (London) Ltd
Poth Hille & Co Ltd.
Thew, Arnott and Co Ltd

Other European
USOCO BV

USA
Charkit Chemical Corp
Fisher Scientific
Koster Keunen LLC
Mutchler Inc
Penta Manufacturing Co
Rita Corp
Ruger Chemical Co Inc
Spectrum Chemicals & Laboratory Products Inc
Strahl & Pitsch Inc
Triple Crown America
Voigt Global Distribution Inc
Whittaker Clark, and Daniels Inc

Others
Charles Tennant & Co (Canada) Ltd
Sovereign Chemicals & Cosmetics

Xanthan Gum

UK
A and E Connock (Perfumery and Cosmetics) Ltd
AF Suter and Co Ltd
Blagden Specialty Chemicals Ltd
CP Kelco UK Ltd
Danisco UK Ltd
Thew, Arnott and Co Ltd

Other European
Biesterfeld Spezialchemie GmbH
Corcoran Chemicals Ltd
Danisco A/S
Jungbunzlauer AG
Lehmann & Voss & Co
Rhodia SA

USA
Advance Scientific & Chemical Inc
Ashland Inc
BOC Sciences
Brenntag North America Inc
Brenntag Southwest
CarboMer Inc
Cargill Inc
Chart Corp Inc
CP Kelco US Inc
Danisco USA Inc
Domino Foods, Inc.
Hawkins Inc
Jungbunzlauer Inc
KIC Chemicals Inc
Penta Manufacturing Co
Premium Ingredients Ltd
Rhodia Inc
Rhodia Pharma Solutions Inc
RT Vanderbilt Company Inc
Seidler Chemical Company Inc
Spectrum Chemicals & Laboratory Products Inc
TIC Gums
Voigt Global Distribution Inc
Zhong Ya Chemical (USA) Ltd

Others
Arihant Trading Co
Kawarlal Excipients Pvt Ltd
LS Raw Materials Ltd
Lucid Colloids Ltd
Priyanka Pharma/M K Ingredients & Specialities
Signet Chemical Corporation Pvt Ltd
Univar Canada Ltd
Wintersun Chemical

Xylitol

UK
Blagden Specialty Chemicals Ltd
Cargill PLC
Danisco UK Ltd
Forum Products Ltd
Roquette (UK) Ltd

Other European
Alfa Aesar Johnson Matthey GmbH
Arion & Delahaye
Cargill France SAS
Corcoran Chemicals Ltd
Danisco A/S
Helm AG
Roquette Frères

USA
Aceto Corp
Advance Scientific & Chemical Inc
AerChem Inc
Alfa Chem
American International Chemical Inc
Brenntag Southwest
Cargill Inc
Danisco USA Inc
EM Sergeant Pulp & Chemical Co Inc
George Uhe Co Inc
Hawkins Inc
Helm New York Inc
KIC Chemicals Inc
Penta Manufacturing Co
Premium Ingredients Ltd
RIA International
Roquette America Inc
Spectrum Chemicals & Laboratory Products Inc
Triple Crown America
Voigt Global Distribution Inc

Others
Signet Chemical Corporation Pvt Ltd
Wintersun Chemical

Zein

UK
Lehvoss UK Ltd
Ubichem plc

USA
Advance Scientific & Chemical Inc
Alfa Chem
Penta Manufacturing Co
Spectrum Chemicals & Laboratory Products Inc

Zinc Acetate

Other European
Alfa Aesar Johnson Matthey GmbH
Chemos GmbH
Honeywell Specialty Chemicals Seelze GmbH
Panreac Quimica SAU

USA
Advance Scientific & Chemical Inc
Amresco Inc
EMD Chemicals Inc
Penta Manufacturing Co
Ruger Chemical Co Inc
Seidler Chemical Company Inc
Thomas Scientific
Universal Preserv-A-Chem Inc
Vertellus Specialties Inc

Zinc Oxide

UK
Peter Whiting (Chemicals) Ltd

Other European
AAE Chemie NV
Alfa Aesar Johnson Matthey GmbH
Chemacon GmbH
Panreac Quimica SAU

USA
Aldrich
Alzo International Inc
American International Chemical Inc
Ashland Inc
Asiamerica Ingredients Inc
Barrington Nutritionals
BOC Sciences
EMD Chemicals Inc
Generichem Corp
Helm New York Inc
Horsehead Corporation

Kraft Chemical Co
Mutchler Inc
Particle Dynamics Inc
Penta Manufacturing Co
Premium Ingredients Ltd
Ruger Chemical Co Inc
Seidler Chemical Company Inc
Spectrum Chemicals & Laboratory Products Inc
Universal Preserv-A-Chem Inc
Voigt Global Distribution Inc

Others
Aastrid International
IEQSA
LS Raw Materials Ltd
Numinor
Rubamin Ltd
Wintersun Chemical

Zinc Stearate

UK
Allchem Performance
Fisher Scientific UK Ltd
James M Brown Ltd
Lehvoss UK Ltd
Macron Chemicals UK
Peter Greven UK
Sigma-Aldrich Company Ltd
Tennants Distribution Ltd

Other European
Alfa Aesar Johnson Matthey GmbH
Dr Paul Lohmann GmbH KG
Panreac Quimica SAU

USA
Aceto Corp
Acme-Hardesty
Advance Scientific & Chemical Inc
AerChem Inc
American International Chemical Inc
Astro Chemicals Inc
Brenntag North America Inc
CarboMer Inc
Fisher Scientific
George Uhe Co Inc
Hummel Croton Inc
KIC Chemicals Inc
Kraft Chemical Co
Macron Chemicals
Penta Manufacturing Co
RIA International
Seidler Chemical Company Inc
Spectrum Chemicals & Laboratory Products Inc
Triple Crown America
Voigt Global Distribution Inc
Whittaker Clark, and Daniels Inc

Others
Charles Tennant & Co (Canada) Ltd
Signet Chemical Corporation Pvt Ltd
Univar Canada Ltd
Wintersun Chemical

SUPPLIERS LIST: UK

3M United Kingdom Plc
3M Centre
Cain Road
Bracknell RG12 8HT
Tel: +44 (0)8705 360036
Web: www.3m.com
Trade names: CoTran.

A and E Connock (Perfumery and Cosmetics) Ltd
Alderholt Mill House
Fordingbridge
Hampshire SP6 1PU
Tel: +44 (0)142 565 3367
Fax: +44 (0)142 565 6041
E-mail: sales@connock.co.uk
Web: www.connock.co.uk

AarhusKarlshamn UK Ltd
King George Dock
Hull HU9 5PX
Tel: +44 1482 701271
Fax: +44 1482 709447
Web: www.aak.com
Trade names: Aextreff CT; Colzao CT; Cremao CS-34; Cremao CS-36; Hyfatol 16-95; Hyfatol 16-98; Shogun CT.

Aarhus United UK Ltd *see* AarhusKarlshamn UK Ltd

Acetex Chemicals Ltd *see* Celanese

Adina Chemicals Ltd
ACI House
8 Decimus Park
Kingstanding Way
Tunbridge Wells
Kent TN2 3GP
Tel: +44 (0)1892 517585
Fax: +44 (0)1892 517565
E-mail: sales@adina.co.uk
Web: www.cosmeticingredients.co.uk
Trade names: Lipocol C; Lipocol; Liponate IPP; Lipovol SES.

AF Suter and Co Ltd
Unit 1
Beckingham Business Park
Beckingham Road
Tolleshunt Major Essex
Tel: +44 (0)870 777 3952
Fax: +44 (0)870 777 3959
E-mail: afsuter@afsuter.com
Web: www.afsuter.com
Trade names: Swanlac.

Air Liquide UK Ltd
Cedar House
39 London Road
Reigate RH2 9QE
Tel: +44 (0)1675 462424
Fax: +44 (0)1675 467022
Web: www.airliquide.com

Air Products (Gases) plc
2 Millennium Gate
Westmere Drive
Crewe CW1 6AP
Tel: +44 (0)800 389 0202
Fax: +44 (0)1932 258502
Web: www.airproducts.co.uk

Air Products plc *see* Air Products (Gases) plc

Alembic Products Ltd
River Lane
Saltney
Chester CH4 8RQ
Tel: +44 (0)1244 680147
Fax: +44 (0)1244 680155
E-mail: sales@alembicproducts.co.uk
Web: www.alembicproducts.co.uk

Alfa Chemicals Ltd *see* Alfa Chemicals Ltd/Gattefossé UK

Alfa Chemicals Ltd/Gattefossé UK Ltd
Arc House
Terrace Road South
Binfield
Bracknell RG42 4PZ
Tel: +44 (0)1344 861800
Fax: +44 (0)1344 451400
E-mail: info@alfa-chemicals.co.uk
Web: www.alfa-chemicals.co.uk
Trade names: Labrafac Lipo; Precirol ATO 5.

Allchem Performance
22/23 Progress Business Centre
Whittle Parkway
Slough SL1 6DQ
Tel: +44 (0)1628 601 601
E-mail: info@acigroup.biz
Web: www.allchem.co.uk
Trade names: Vivapur; Vivasol; Vivastar P.

Alpha Therapeutic Europe Limited *see* Aarhus United UK Ltd

Amcol Specialty Minerals
Weaver Valley Road
Winsford
Cheshire CW7 3BU
Tel: +44 1606 868 200
Fax: +44 1606 868 268
Web: www.amcolspecialtyminerals.co.uk

Angus Chemie UK
Subsidiary of Dow Chemical Co.
Unit 7
Rotunda Business Centre
Thorncliffe Park Estate
Chapeltown
Sheffield S30 4PH
Tel: +44 114 257 1322
Fax: +44 114 257 1336

Avebe UK Ltd
Soff Lane
Butterswood
Goxhill
Barrow Upon Humber DN19 7NA
Tel: +44 (0)1469 532 222
Fax: +44 (0)1469 531 488
Web: www.avebe.com
Trade names: Paselli MD10 PH; Perfectamyl; Prejel; Primellose.

Baker *see* Mallinkrodt Baker UK

Baker Sillavan Barytes Ltd *see* Richard Baker Harrison
Trade names: Magsil Diamond.

BASF Plc
PO Box 4
Earl Road
Cheadle Hulme
Cheadle SK8 6QG
Tel: +44 (0)161 485 6222
Fax: +44 (0)161 486 0891
Web: www.basf.co.uk
Trade names: Cremophor A; Kollicoat MAE 100 P; Kollicoat MAE; Kollidon CL-M; Kollidon CL; Kollidon VA 64; Kollidon; Ludiflash; Ludipress LCE; Lutrol E; Luviskol VA; Pluriol E; Soluphor P; Solutol HS 15.

Bayer plc *see* Lanxess Ltd

Bio Products Laboratory
Dagger Lane
Elstree
Herts WD6 3BX
Web: www.bpl.co.uk
Trade names: Zenalb.

Blagden Specialty Chemicals Ltd
Osprey House
Black Eagle Square

Westerham
Kent TN16 1PA
Tel: +44 (0)1959 562000
Fax: +44 (0)1959 565111
E-mail: sales@blagden.co.uk
Web: www.blagden.co.uk
Trade names: MCC Sanaq.

BOC Gases
The Priestley Centre
10 Priestly Road
Surrey Research Park
Guildford GU2 7XY
Tel: +44 (0)800 111333
Fax: +44 (0)1483 505211
E-mail: customer.service@uk.gases.boc.com
Web: www.boc.com

Borax Europe Ltd
2 Eastbourne terrace
London W2 6LG
Tel: +44 (0)20 7781 1451
Fax: +44 (0)20 7781 1851

BP plc
International Headquaters
1 St James's Square
London SW1Y 4PD
Tel: +44 (0)20 7496 4000
Fax: +44 (0)20 7496 4630
Web: www.bp.com

Brenntag (UK) Ltd
Albion House, Rawdon Park
Green Lane
Yeadon
Leeds LS19 LXX
Tel: +44 (0)113 3879200
Fax: +44 (0)113 3879280
E-mail: enquiry@brenntag.co.uk
Web: www.brenntag.co.uk

British Sugar Pharmaceutical Group
British Sugar plc
Sugar Way
Peterborough
Cambridgeshire PE2 9AY
Tel: +44 (0) 1733 422555
E-mail: info@britishsugar.co.uk
Web: www.bspharma.co.uk

British Traders & Shippers Ltd *see* Nippon Gohsei (UK) Ltd

British Wax Refining Co Ltd
62 Holmethorpe Avenue
Holmethorpe Industrial Estate
Redhill
Surrey RH1 2NL
Tel: +44 (0)1737 761242
Fax: +44 (0)1737 761472
E-mail: wax@britishwax.com
Web: www.britishwax.com

Brunner Mond (UK) Ltd
Winnington Lane
Mond House
Northwich CW8 4DT
Tel: +44 (0)1606 724000
Fax: +44 (0)1606 781353
E-mail: sales@brunnermond.com
Web: www.brunnermond.com

Cargill PLC
Knowle Hill Park
Fairmile Lane
Cobham
Surrey KT11 2PD
Tel: +44 (0)1932 861000
Fax: +44 (0)1932 861200
Web: www.cargill.co.uk

Trade names: C*Dry MD; C*Pharm; C*PharmDex; C*PharmDry; C*PharmGel; C*PharmMaltidex; C*PharmMannidex; C*PharmSorbidex; C*PharmSweet; Satiaxane U; Treha; Zerose.

Chance & Hunt
Alexander House
Crown Gate
Runcorn WA7 2UP
Tel: +44 (0)1928 793000
Fax: +44 (0)1928 714351
Web: www.chance-hunt.com

Clariant UK Ltd
(Functional Chemicals Division)
Calverley Lane
Horsforth
Leeds LS18 4RP
Tel: +44 (0)113 258 4646
Fax: +44 (0)113 239 8473
Web: www.clariant.co.uk
Trade names: Nipabutyl; Nipacide PX; Nipagin A; Nipagin M; Nipanox BHA; Nipanox BHT; Nipantiox 1-F; Nipasol M Sodium; Nipasol M; Phenoxetol; Tylose CB.

Cognis UK Ltd
Charleston Road
Hythe
Hampshire SO45 3ZG
Tel: +44 (0)2380 894666
Fax: +44 (0)2380 243113
E-mail: anne.herbert@cognis.com
Web: www.uk.cognis.com
Trade names: Cegesoft; Cutina CP; Cutina GMS; Cutina HR; Emulgade 1000NI; Eumulgin; Hydagen CAT; Lanette O; Lanette SX; Lanette W; Monomuls 90-O18; Myritol; Novata; Speziol C16 Pharma; Speziol C16-18 Pharma; Speziol C18 Pharma; Speziol TPGS Pharma; Texapon K12P.

Colloides Naturels UK Ltd
The Triangle Business Centre
Exchange Square
Manchester M4 3TR
Tel: +44 (0)161 838 5744
Fax: +44 (0)161 838 5746
Web: www.cniworld.com

Colorcon Ltd
Flagship House
Victory Way
Crossways
Dartford
Kent DA2 6QD
Tel: +44 (0)1322 293000
Fax: +44 (0)1322 627200
Web: www.colorcon.com
Trade names: Methocel; Methocel; Opaglos R; Opaseal; Phthalavin; StarCap 1500; Starch 1500 G; Surelease; Sureteric.

Connock *see* A and E Connock (Perfumery and Cosmetics) Ltd

Cornelius Group plc
Cornelius House
Woodside
Dunmow Road
Bishop's Stortford
Hertfordshire CM23 5RG
Tel: +44 (0)1279 714 300
Fax: +44 (0)1279 714 320
E-mail: sales.dept@cornelius.co.uk
Web: www.cornelius.co.uk
Trade names: Pelemol CP; Tronox.

Courtin & Warner Ltd
19 Phoenix Place
Lewes
East Sussex BN7 1JX
Tel: +44 (0)1273 480611
Fax: +44 (0)1273 472249
E-mail: sales@courtinandwarner.com
Web: www.c-and-w.co.uk

Coventry Chemicals Ltd
Woodham's Road
Siskin Drive
Coventry CV3 4FX
Tel: +44 (0) 2476 639 739
Fax: +44 (0) 2476 639 717
E-mail: info@coventrychemicals.com
Web: www.coventrychemicals.com

CP Kelco UK Ltd
1 Cleeve Court
Cleeve Road
Leatherhead
Surrey KT22 7UD
Tel: +44 (0)1372 369 400
Fax: +44 (0)1372 369 401
Web: www.cpkelco.com
Trade names: Finnfix; Genu; Keltrol; Nymcel ZSB; Nymcel ZSX; Xantural.

Croda Europe Ltd
Cowick Hall
Snaith
Goole DN14 9AA
Tel: +44 (0)1405 860551
Fax: +44 (0)1405 861767
E-mail: healthcare-sales@croda-oleochemicals.com
Web: www.croda.com
Trade names: Brij; Byco; Cithrol; Crillet; Crodacol C70; Crodacol C90; Crodacol CS90; Crodacol S95; Crodamol EO; Crodamol IPIS; Crodamol SS; Crodex A; Crodex N; Croduret; Etocas; Novol; Polawax; Pricerine; Priolube 1408; Super Hartolan.

Danisco Sweeteners Ltd *see* Danisco UK Ltd

Danisco UK Ltd
Sweeteners Division
6 North Street
Beaminster DT8 3DZ
Tel: +44 (0)1308 862216
E-mail: info@danisco.com
Web: www.danisco.com
Trade names: Aclame; Grindsted; Grindsted; Litesse; Meyprodor; Xylitab.

Degussa Hüls Ltd *see* Evonik Degussa Ltd

DFE Pharma UK
PO Box 11
Teddington
Middlesex TW11 8YG
Tel: +44 (0)20 8943 5220
Fax: +44 (0)20 8943 5231
E-mail: nigel.roberts@dmv-fonterra-excipients.com
Trade names: Lactochem; Lactohale; Lactopress Anhydrous 250; Lactopress Anhydrous Crystals; Lactopress Anhydrous Fine Powder; Lactopress Anhydrous Microfine; Lactopress Anhydrous Powder; Lactopress Anhydrous; Lactopress Spray-Dried 250; Lactopress Spray-Dried 260; Lactopress Spray-Dried; Pharmacel; Pharmatose; Primellose; Primojel; Respitose; SuperTab 11SD; SuperTab 14SD; SuperTab 21AN; SuperTab 22AN; SuperTab 24AN; SuperTab 30GR.

DMV-Fonterra Excipients UK *see* DFE Pharma UK

Dow Chemical Company (UK)
Dow Chemical Company Ltd
Diamond House
Lotus Park
Kingsbury Crescent
Staines, Middlesex TW18 3AG
Tel: +44 (0)203 139 4000
Fax: +44 (0)203 139 4004
Web: www.dow.com
Trade names: Dowanol EPh; Polyox; Versenex.

Dow Corning (UK)
Center Northern Europe
Meriden Business Park
Copse Drive
Allesley
Coventry CV5 9RG
Tel: +44 (0)1676 528000
Fax: +44 (0)1676 528001
Web: www.dowcorning.com
Trade names: Dow Corning 245 Fluid; Dow Corning 246 Fluid; Dow Corning 345 Fluid; Dow Corning Q7-2243 LVA; Dow Corning Q7-2587; Dow Corning Q7-9120.

DSM UK Ltd
DSM House
Papermill Drive
Redditch B98 8QJ
Tel: +44 (0)1527 590552
Fax: +44 (0)1527 590555
Web: www.dsm.com

Eastman Company UK Ltd
European Technical Centre
Acornfield Road
Knowsley Industrial Park
Liverpool L33 7UF
Tel: +44 (0)151 547 2002
Fax: +44 (0)151 548 5100
Trade names: Eastacryl 30 D; Eastacryl; Tenox BHA; Tenox BHT; Tenox PG.

Edward Mendell *see* Rettenmaier UK Ltd

Efkay Chemicals Ltd
Allen House
The Maltings
Station Road
Sawbridgeworth CM21 9JX
Tel: +44 (0)1279 721 888
Fax: +44 (0)1279 722 261
E-mail: tricia@efkaychemicals.com

Esterchem Limited
Brooklands Way
Basford Lane Industrial Estate
Leekbrook
Nr. Leek
Staffordshire ST13 7QF
Tel: +44 (0)1538 383997
Fax: +44(0)1538 386855
E-mail: sales@esterchem.co.uk
Web: www.esterchem.co.uk
Trade names: Ester Wax NF.

Evonik Degussa Ltd
Tego House
Chippenham Drive
Kingston
Milton Keynes MK10 0AF
Tel: +44 (0)845 12895 77
Fax: +44 (0)845 12895 79
Web: www.evonik.com
Trade names: Aerosil; Tegin; Tegosept E.

Evonik Goldschmidt UK Ltd
Tego House
Chippenham Drive
Kingston
Milton Keynes MK10 OAF
Tel: +44 (0)845 1289577
Fax: +44 (0)845 1289579
Web: www.evonik.com
Trade names: ABIL; Tegin 4100; Tegin 515; Tegin M; Tegin; Tegin; Tego Carbomer; Tegosept E; Tegosoft DO; Tegosoft M; Tegosoft P.

Fisher Scientific UK Ltd
Part of Thermo Fisher Scientific
Bishop Meadow Road
Loughborough
Leicestershire LE11 5RG
Tel: +44 (0)1509 555500
E-mail: fsuk.sales@thermofisher.com
Web: www.fisher.co.uk

1

Fluorochem Ltd
Unit 14 Graphite Way
Rossington Park
Hadfield
Derbyshire SK13 1QH
Tel: +44 (0)1457 860 111
Fax: +44 (0)1457 892 799
E-mail: enquiries@fluorochem.co.uk
Web: www.fluorochem.net

Forum Biosciences Ltd *see* Forum Products Ltd

Forum Products Ltd
Betchworth House
57-65 Station Road
Redhill
Surrey RH1 1DL
Tel: +44 (0)1737 857700
E-mail: enquires@forumgroup.co.uk
Web: www.forum.co.uk
Trade names: Candex; Effer-Soda; Emcocel; Emcompress; Emdex; Explotab; Lubritab; Mannogem; ProSolv; Pruv; Sorbitab; Xylitab.

Foster & Co *see* H Foster & Co (Stearines) Ltd

Friesland Foods Domo UK Ltd
Riverside House
Brymau Three Estate
River Lane
Saltney
Chester CH4 8RQ
Tel: +44 (0)1244 680127
Fax: +44 (0)1244 671703
Web: www.domo.nl

Fuchs Lubricants (UK) plc
New Century Street
Hanley
Stoke-on-Trent
Staffordshire ST1 5HU
Tel: +44 (0)1782 203 700
Fax: +44 (0)1782 202072/3
E-mail: contact-uk@fuchs-oil.com
Web: www.fuchslubricants.com
Trade names: Silk; Sirius.

Global Ceramic Materials Ltd
Milton Works
Diamond Crescent
Milton
Stoke-on-Trent ST2 7PX
Tel: +44 (0) 1782 537297
E-mail: info@globalcm.co.uk
Web: www.globalcm.co.uk

Goldschmidt UK Ltd *see* Evonik Degussa UK Services Ltd
Trade names: DUB SIP.

Grace Davison
Oak Park Business Centre
Alington Road
Little Barford
St Neots
Cambs PE19 6WL
Tel: +44 (0)1480 324430
Web: www.grace.com

Haltermann Ltd *see* Dow Chemical Company (UK)

Hayashibara International
Thorncroft Manor
Thorncroft Drive
Leatherhead KT22 8JB
Tel: +44 (0)1372 700 756
E-mail: info@hayashibara-intl.com
Web: www.hayashibara-intl.com
Trade names: Treha.3M Drug Delivery Systems

Hercules Ltd *see* Ashland Aqualon Functional Ingredients
Trade names: Aqualon; Culminal MHEC; Klucel; Natrosol.

H Foster & Co (Stearines) Ltd
103 Kirkstall Road
Leeds
Yorkshire LS3 1JL
Tel: +44 (0)113 243 9016
Fax: +44 (0)113 242 2418
E-mail: sales@hfoster.co.uk
Web: www.hfoster.co.uk

Honeywill & Stein *see* IMCD UK Ltd

IMCD UK Ltd
Times House
Throwley Way
Sutton
Surrey SM1 4AF
Tel: +44 (0)208 770 7090
Fax: +44 (0)208 770 7295
Web: www.imcd.co.uk
Trade names: Ac-Di-Sol; Aquacoat cPD; Aquacoat ECD; Avicel PH; Blanose; Celphere; Klucel; Lycasin; Myvaplex 600P; Myvatex; Natrosol; Neosorb; Pearlitol; Protanal.

Ingredients Consultancy Ltd, The
PO Box 790
Upton upon Severn
Worcestershire WR8 0WF
Tel: +44 (0)1684 59 4949
Fax: +44 (0)1684 59 3692
E-mail: info@theingredients.co.uk
Web: www.theingredients.co.uk

Intermag Co Ltd
Bath Road
Felling Industrial Estate
Gateshead NE10 0LG
Tel: +44 (0)191 495 2220
Fax: +44 (0)191 438 4717
E-mail: sales@intermag.co.uk
Web: www.intermag.co.uk

ISP Europe
Waterfield
Tadworth KT20 5HQ
Tel: +44 17 37 37 7000
Web: www.ispcorp.com
Trade names: Celex; Germall 115; Pharmasolve; Plasdone S-630; Plasdone; Polyplasdone XL-10; Polyplasdone XL.

James M Brown Ltd
Napier Street
Fenton
Stoke-on-Trent ST4 4NX
Tel: +44 (0)1782 744171
Fax: +44 (0)1782 744473
E-mail: sales@jamesmbrown.co.uk
Web: www.jamesmbrown.co.uk

JT Baker UK *see* Mallinkrodt Baker UK

Karlshamns Ltd *see* AarhusKarlshamn UK Ltd
Trade names: Lipex 108; Lipex 204.

Kelco *see* CP Kelco UK Ltd

Kimpton Brothers Ltd
10–14 Hewett Street
London EC2A 3RL
Tel: +44 (0)20 7456 9999
Fax: +44 (0)20 7247 2784/7375 3584
E-mail: info@kimpton.co.uk
Web: www.kimpton.com

Kronos Ltd
Barons Court
Manchester Road
Wilmslow
Cheshire SK9 1BQ
Tel: +44 (0)1625 547200

Fax: +44 (0)1625 533123
E-mail: sales@kronosww.com
Web: www.kronostio2.com
Trade names: Kronos 1171.

Lanxess Ltd
Lichfield Road
Burton-Trent DE14 3WH
Tel: +44 (0)1283 714200
Fax: +44 (0)1283 714201
E-mail: pigment@lanxess.com
Web: www.lanxess.com
Trade names: Bayferrox 105M; Bayferrox 306; Bayferrox 920Z; Solbrol A; Solbrol M; Solbrol P.

Leading Solvent Supplies Ltd
Rudgate
Tockwith
York YO26 7QF
Tel: +44 (0)1423 358058
Fax: +44 (0)1423 358923
E-mail: sales@Leading-Solvent.co.uk
Web: www.Leading-Solvent.co.uk

Lehvoss UK Ltd
Paroxite Division
Office Unit 2
7 Dryden Court
Renfrew Road
Kennington
London SE11 4NH
Tel: +44 (0)20 7735 2425
Fax: +44 (0)20 7735 4408
E-mail: contact@lehvoss.co.uk
Web: www.lehvoss.co.uk
Trade names: EmCon CO; Fancol; Phenoxen; Pure-Dent B851; Pure-Dent; Spress B820.

Lloyd Ltd *see* WS Lloyd Ltd

Lonza Biologics Plc
228 Bath Road
Slough SL1 4DX
Tel: +44 (0)1753 777000
Fax: +44 (0)1753 777001
E-mail: contact.slough@lonza.com
Web: www.lonzagroup.com
Trade names: Ethosperse; Glycon G-100; Glycon; Hyamine 1622; Hyamine 3500.

Macron Chemicals UK
A division of Avantor *see* Avantor Performance Materials
Trade names: HyQual; Parlodion.

Mantrose (UK) Ltd
Unit 7B Northfield Farm
Great Shefford RG17 7BY
Tel: +44 (0)1488 648 988
Fax: +44 (0)1488 648 890
Web: www.mantrose.co.uk
Trade names: Crystalac; Mantrolac R-49; Mantrolac R-52.

Mast Group Ltd
Mast House
Derby Road
Bootle L20 1EA
Tel: +44 (0)151 9337277
Fax: +44 (0)151 9441332
Web: www.mastgrp.com

Messer UK Ltd *see* Air Liquide UK Ltd

National Starch & Chemical Ltd *see* National Starch Personal Care

National Starch Personal Care
(Division of Akzo Nobel)
Prestbury Court
Greencourts Business Park
333 Styal Road
Manchester M22 5LW
Tel: +44 (0)161 435 3200
Fax: +44 (0)161 435 3300
Web: www.personalcarepolymers.com
Trade names: Hylon; Purity 21; Uni-Pure.

Nipa Biocides
(Division of Clariant UK Ltd/Functional Chemicals Division)
Clariant UK Ltd

Nipa Laboratories Ltd *see* Clariant UK Ltd

Nippon Gohsei (UK) Ltd
Soarnol House
Kingston upon Hull HU12 8DS
Tel: +44 (0)1482 333320
Fax: +44 (0)1482 309332
E-mail: info@nippon-gohsei.com
Web: www.nippon-gohsei.com
Trade names: Gohsenol.

Noveon Inc *see* Lubrizol Advanced Materials Inc.

Nutrinova UK Ltd *see* Celanese

Paroxite (London) Ltd *see* Lehvoss UK

PB Gelatins UK
Building A6, Severn Road
Treforest
Industrial Estate
Pontypridd CF37 5SQ
Tel: +44 (0)1443 849300
Fax: +44 (0)1443 844209
E-mail: gelatin@tessenderlo.com
Web: www.tessenderlogroup.com
Trade names: Cryogel; Instagel; Solugel.

Penwest Ltd *see* Rettenmaier UK Ltd

Peter Greven UK
Albion Bridge Works
Vickers Street
Manchester M40 8EF
Tel: +44 (0)161 277 9877
Fax: +44 (0)161 203 4159
E-mail: sales@peter-greven.co.uk
Web: www.peter-greven.co.uk

Peter Whiting (Chemicals) Ltd
8 Barb Mews
Hammersmith
London W6 7PA
Tel: +44 (0)20 7605 7880
Fax: +44 (0)20 7603 3240
E-mail: sales@whiting-chemicals.co.uk
Web: www.whiting-chemicals.co.uk

Pfanstiehl (Europe) Ltd *see* Ferro Pfanstiehl Laboratories Inc

PMC Chemicals Ltd
12 Downham Chase
Timperley
Altrincham WA15 7TJ
Tel: +44 (0)161 904 0499
Fax: +44 (0)161 904 7080
E-mail: enquires@pmcchemicals.com
Web: www.pmcchemicals.com

Poth Hille & Co Ltd.
Unit 18 Easter Industrial Park
Ferry Lane South
Rainham RM13 9BP
Tel: +44 (0)1708 526 828
Fax: +44 (0)1708 525 695
Web: www.poth-hille.co.uk

Pumex (UK) Limited
Marsh Trees House
Marsh Parade
Newcastle-under-lyme ST1 1BT
Tel: +44 (0)1782 622 666
Fax: +44 (0)1782 622 655
E-mail: info@pumex.co.uk
Web: www.pumex.co.uk
Trade names: Magsil Osmanthus.

Purac Biochem (UK)
50–54 St Paul's Square
Birmingham B3 1QS
Tel: +44 (0)121 236 1828
Fax: +44 (0)121 236 1401
E-mail: puk@purac.com
Web: www.purac.com

Raught Ltd
117 The Drive
Ilford, Essex IG1 3JE
Tel: +44 (0)20 8554 9921
Fax: +44 (0)20 8554 8337
E-mail: technical@raught.co.uk
Web: www.raught.co.uk

Reheis *see* General Chemical LLC
Trade names: Rehydraphos.

Rettenmaier UK Ltd
Church House
48 Church Street
Reigate RH2 0SN
Tel: +44 (0)1737 222323
Fax: +44 (0)1737 222545
E-mail: techsales@jrspharma.co.uk
Web: www.jrspharma.com
Trade names: Compactrol; Emcocel; Emcompress Anhydrous; Emcompress; Emdex; Explotab; Lubritab; ProSolv; Pruv.

Rhodia Organic Fine Ltd
PO Box 46
St Andrews Road
Avonmouth
Bristol BS11 9YF
Tel: +44 (0)117 948 4242
Fax: +44 (0)117 948 4249
Trade names: Rhodiarome; Rhodopol; Rhovanil.

Richard Baker Harrison
253 Cranbrook Road
Ilford
Essex IG1 4TQ
Tel: +44 208 5540102
Fax: +44 208 5549282

Roche Products Ltd
6 Falcon Way
Shire Park
Hexagon Place
Welwyn Garden City AL7 1TW
Tel: +44 (0)170 736 6000
Fax: +44 (0)170 733 8297
Web: www.roche.com

Rohm and Haas UK Ltd
Rohm and Haas Europe Services ApS- UK branch
Heckmondwike Road
Dewsbury Moor
Dewsbury WF13 3NG
Tel: +44 (0)1924 403367
Fax: +44 (0)1824 405166
Web: www.rohmhaas.com/ionexchange
Trade names: Amberlite IRP-88.

Roquette (UK) Ltd
Sallow Road
Weldon Industrial Estate
Corby NN17 5JX
Tel: +44 (0)1536 273000
Fax: +44 (0)1536 263873
Web: www.roquette.com
Trade names: Flolys; Fluidamid R444P; Glucidex; Keoflo ADP; Kleptose; Lycadex PF; Lycasin 75/75; Lycasin 80/55; Lycasin 85/55; Lycasin HBC; Lycatab C; Lycatab DSH; Lycatab PGS; Neosorb; Pearlitol; Roclys; SweetPearl; Xylisorb.

RW Unwin & Co Ltd
Prospect Place
Welwyn AL6 9EW
Tel: +44 (0)1438 716441
Fax: +44 (0)1438 716067
E-mail: sales@rwunwin.co.uk
Web: www.rwunwin.co.uk
Trade names: Aqoat AS-HF/HG; Aqoat AS-LF/LG; Aqoat AS-MF/MG; Aqoat; HPMCP; L-HPC; Metolose; Metolose; Pharmacoat.

Sasol UK Ltd
No. 1 Hockley Court
2401 Stratford Road
Hockley Heath
Solihull B94 6NW
Tel: +44 (0)1564 783 060
Fax: +44 (0)1564 784 088
E-mail: keith.bernstone@uk.sasol.com
Web: www.sasol.com
Trade names: Imwitor 191; Imwitor 412; Imwitor 491; Imwitor 900K; Imwitor 948; Lipoxol; Miglyol 808; Miglyol 810; Miglyol 812; Nacol 14-95; Nacol 14-98; Nacol 16-95; Nacol 18-98; Nacol 18-98P; Witepsol.

Shin-Etsu Chemical Co Ltd *see* RW Unwin & Co Ltd
Trade names: Aqoat AS-HF/HG; Aqoat AS-LF/LG; Aqoat AS-MF/MG; Aqoat; HPMCP; L-HPC; Metolose; Metolose; Pharmacoat.

Sigma-Aldrich Company Ltd
Fancy Road
Poole BH12 4QH
Tel: +44 (0)1747 833000
Fax: +44 (0)1202 712239
E-mail: ukcustsv@europe.sial.com
Web: www.sigma-aldrich.com
Trade names: Brij; Thimerosal Sigmaultra; Trizma.

Sparkford Chemicals Ltd
58 The Avenue
Southampton SO17 2 1XS
Tel: +44 (0)23 8022 8747
E-mail: info@sparkford.co.uk
Web: www.sparkford.co.uk

Stan Chem International Ltd
4 Kings Road
Reading RG1 3AA
Tel: +44 (0)118 958 0247
Fax: +44 (0)118 958 9580
E-mail: info@stanchem.co.uk
Web: www.stanchem.co.uk

Symrise
Fieldhouse Lane
Marlow SL7 1TB
Tel: +44 (0)1628 646 017
Fax: +44 (0)1635 646 016
Web: www.symrise.com

Tate and Lyle plc
Head Office
Sugar Quay
Lower Thames Street
London EC3R 6DQ
Tel: +44 (0)20 7626 6525
Fax: +44 (0)20 7623 5213
Web: www.tate-lyle.co.uk
Trade names: Star-Dri.

Tennants Distribution Ltd
Hazelbottom Road
Cheetham
Manchester M8 0GR
Tel: +44 (0)161 2054454
Fax: +44 (0)161 2035985
E-mail: enquires@tennantsdistribution.com
Web: www.tennantsdistribution.com

Thew, Arnott and Co Ltd
270 London Road
Wallington, Surrey SM6 7DJ
Tel: +44 (0)20 8669 3131
Fax: +44 (0)20 8669 7747
E-mail: sales@thewarnott.co.uk
Web: www.thewarnott.co.uk

Tioxide Europe Ltd
(Huntsman Tioxide)
Tees Road
Hartlepool TS25 2DD
Tel: +44 (0)1642 546123
Fax: +44 (0)1642 546016
Web: www.huntsman.com
Trade names: Tioxide.

Ubichem plc
Mayflower Close
Chandlers Ford Industrial Estate
Eastleigh
Hampshire SO53 4AR
Tel: +44 (0)23 8026 3030
Fax: +44 (0)23 8026 3012
E-mail: info@ubichem.com
Web: www.ubichem.com

Uniqema *see* Croda Europe Ltd

Unwin *see* RW Unwin & Co Ltd

Wacker Chemicals Ltd
120 Bridge Road
Chertsey KT16 8LA
Tel: +44 (0)870 0480202
Fax: +44 (0)870 0480203
E-mail: info.uk@wacker.com
Web: www.wacker.com
Trade names: Cavamax W6 Pharma; Cavamax W7 Pharma; Cavamax W8 Pharma; HDK.

White Sea and Baltic Company Ltd
8 Kerry Hill
Horsforth
West Yorkshire LS18 4AY
Tel: +44 (0)113 259 0512
Fax: +44 (0)113 258 2393
E-mail: sales@whitesea.co.uk
Web: www.whitesea.co.uk

Whiting (Chemicals) Ltd *see* Peter Whiting (Chemicals) Ltd

Wilfrid Smith Ltd
Elm House
Medlicott Close
Oakley Hay
Corby NN18 9NF
Tel: +44 (0)1536 460020
Fax: +44 (0)1536 462400
E-mail: info@wilfred-smith.co.uk
Web: www.wilfrid-smith.co.uk

William Blythe Ltd
Church
Accrington
Lancashire BB5 4PD
Tel: +44 (0)1254 320000
Fax: +44 (0)1254 320001
Web: www.williamblythe.com

William Ransom & Son plc
Alexander House
40A Wilbury Way
Hitchin
Hertfordshire SG4 0AP
Tel: +44 (0)1462 437 615
Fax: +44 (0)1462 420 528
E-mail: info@williamransom.com
Web: www.williamransom.com

WS Lloyd Ltd
7 Redgrove House
Stonards Hill
Epping CM16 4QQ
Tel: +44 (0)1992 572670
Fax: +44 (0)1992 578074
E-mail: enquiries@wslloyd.com
Web: www.wslloyd.com

Xyrofin (UK) Ltd *see* Danisco Sweeteners Ltd

SUPPLIERS LIST: OTHER EUROPEAN

AAE Chemie NV
Duboisstraat 39
Antwerp 2060
Belgium
Tel: +32 3 568 1166
Fax: +32 3 568 0597
E-mail: info@aachemie.com
Web: www.aaechemie.com

AarhusKarlshamn AB
Skeppsgatan 19
Malmo SE 21119
Sweden
Tel: +46 40 627 83 00
Fax: +46 40 627 83 11
E-mail: info@aak.com
Web: www.aak.com
Trade names: Aextreff CT; Colzao CT; Cremao CS-34; Cremao CS-36; Hyfatol 16-95; Hyfatol 16-98; Shogun CT.

Aarhus Oliefabrik A/S *see* Aarhus United Denmark A/S

Aarhus United Denmark A/S *see* AarhusKarlshamn AB

ABCR GmbH & Co. KG
Im Schlehert 10
D-76187 Karlsruhe
Germany
Tel: +49 721 95061 0
Fax: +49 721 95061 80
E-mail: info@abcr.de
Web: www.abcr.de

Acetex Chimie SA *see* Celanese

Acros Organics BVBA
Janssen Pharmaceuticalaan 3a
2440 Geel
Belgium
Tel: +32 14 57 52 11
E-mail: info@acros.com
Web: www.acros.com

Agrofert Holding a.s.
Pyšelská 2327/2
149 00 Praha 4
Czech Republic
Tel: +420 272 192 111
E-mail: agrofert@agrofert.cz
Web: www.agrofert.cz
Trade names: Pretiox AV-01-FG.

Ajinomoto Sweeteners Europe SAS
Z.I.P des Huttes
Route de la Grande Hernesse
59820 Gravelines
France
Tel: +33 328 22 7400
Fax: +33 328 22 7500
Web: www.aji-aspartame.eu
Trade names: Pal Sweet Diet; Pal Sweet.

Akzo Nobel Functional Chemicals bv
Stationsstraat 77
3811 MH Amersfoort
PO Box 247
3800 AE Amersfoort
Netherlands
Tel: +31 33 467 6767
Fax: +31 33 467 6146
Web: www.akzonobel.com
Trade names: Akucell; Dissolvine.

Alfa Aesar Johnson Matthey GmbH
Postbox 11 07 65
76057 Karlsruhe
Germany
Tel: +49 721 84007 280
Fax: 49 721 84007 300

E-mail: EuroSales@alfa.com
Web: www.alfa-chemcat.com

Alland & Robert
9 rue de Saintonge
F-75003 Paris
France
Tel: +33 1 44 59 21 31
Fax: +33 1 42 72 54 38
E-mail: inforequest@allandetrobert.fr
Web: www.allandetrobert.fr

Alltec Intertrade
Friedensweg 16
Ganderkesee D-27777
Germany
Tel: +49 42 23925604
Fax: +49 42 23925605
E-mail: info@alltec-intertrade.de
Web: www.alltec-intertrade.de
Trade names: AT 10-1010.

Amylum Ibérica, SA *see* Tate & Lyle

Angus Chemie GmbH
Subsidiary of Dow Chemical Co.
Zeppelinstr. 30
49479 Ibbenbüren
Nordrhein-Westfalen
Germany
Tel: +49 5459 56 0
Fax: +49 5459 56 241
Web: www.dow.com/angus/index/htm

Arion & Delahaye
Kreglinger Europe NV
Grote Markt 7
B-2000 Antwerpen
Belgium
Tel: +32 (0)3 22 22 020
Fax: +32 (0)3 22 22 080
E-mail: info@kreglinger.com
Web: www.kreglinger-europe.com

August Hedinger GmbH & Co KG
Heiligenwiesen 26
D-70327 Stuttgart
Germany
Tel: +49 0711 402050
Fax: +49 0711 4020535
E-mail: info@hedinger.de
Web: www.hedinger.de

Avebe Group
PO Box 15
9640 AA Veendam
Netherlands
Tel: +31 598 66 91 11
Fax: +31 598 66 43 68
E-mail: info@avebe.com
Web: www.avebe.com
Trade names: Paselli MD10 PH; Perfectamyl; Prejel; Primellose.

Baerlocher GmbH
Freisinger Strasse 1
85716 Unterschleissheim
Germany
Tel: +49/89 14 37 30
Fax: +49/89 14 37 33 12
Web: www.baerlocher.com
Trade names: Ceasit PC.

BASF Aktiengesellschaft
Carl-Bosch-Strasse 38
D-67056 Ludwigshafen
Germany
Tel: +49 621 60 0
Fax: +49 621 60 42525
E-mail: global.info@basf.com
Web: www.basf.com
Trade names: Cremophor A; Kollicoat IR; Kollicoat MAE 100 P; Kollicoat MAE; Kollicoat SR 30 D; Kollidon CL-M; Kollidon CL; Kollidon VA 64; Kollidon; Ludiflash; Ludipress LCE; Lutrol E; Luviskol VA; Myacide; Palatinol A; Protectol; Soluphor P.

BENEO-Orafti
Aandorenstraat 1
B-3300 Tienen
Belgium
Tel: +32 (0)16 801 301
Fax: +32 (0)16 801 308
E-mail: info@BENEO-Orafti.com
Web: www.orafti.com
Trade names: Orafti.

Beneo-Palatinit GmbH
Gottlieb-Daimler Str 12
68165 Mannheim
Germany
Tel: +49 621 421 150
Fax: +49 621 421 160
E-mail: galenIQ@beneo-palatinit.com
Web: www.beneo-palatinit.com
Trade names: galenIQ; Palatinit.

BENEO-Remy NV
Remylaan 4
3018 Leuven-Wijgmaal
Belgium
Tel: +32 (0)16 24 85 11
Fax: +32 (0)16 44 01 44
E-mail: info@BENEO-Remy.com
Web: www.beneo-remy.com
Trade names: Remy.

Berg + Schmidt (GmbH & Co.) KG
An der Alster 81
Hamburg 20099
Germany
Tel: +49 (0)40 284 0390
Fax: +49 (0)40 284 03944
E-mail: info@berg-schmidt.de
Web: www.berg-schmidt.de
Trade names: Bergabest.

Biesterfeld Spezialchemie GmbH
Ferdinandstrasse 41
D-20095 Hamburg
Germany
Tel: +49 (0) 40 32 008 0
Fax: +49 (0) 40 32 008 443
E-mail: spezialchemie@biesterfeld.com
Web: www.biesterfeld-spezialchemie.com

Biogel AG
Haldenstr. 11
6006 Lucerne
Switzerland
Tel: +41 41 418 4050
Fax: +41 41 418 4049
Web: www.biogel.ch
Trade names: Vitagel.

Biogrund GmbH
Neukirchner Strasse 5
Hünstetten D-65510
Germany
Tel: +49 (0) 6126 952 63 0
Fax: +49 (0) 6126 952 63 33
E-mail: info@biogrund.com
Web: www.biogrund.com
Trade names: Walocel CRT.

Blanver (Europe)
Moia 1 - Tuset 3, 2nd Floor
08006 Barcelona
Spain
Tel: +34 93 241 3715
Fax: +34 93 414 7036
E-mail: pere@blanver.com

Web: www.blanver.com.br
Trade names: Microcel 3E-150.

Boehringer Ingelheim GmbH
Corporate Headquarters
Marketing & Sales Fine Chemicals
Binger Strasse 173
D-55216 Ingelheim
Germany
Tel: +49 6132 770
Fax: +49 6132 720
Web: www.boehringer-ingelheim.com/finechem

Brenntag AG *see* Brenntag GmbH

Brenntag GmbH
Stinnes-Platz 1
D-45472 Mülheim an der Ruhr
Germany
Tel: +49 208 7828 0
Web: www.brenntag-gmbh.de

Brenntag Nordic AS
Strandvejen 104A
2900 Hellerup
Denmark
Trade names: Alhydrogel.

Brenntag NV
Nijverheidslaan 38
B-8540 Deerlijk
Belgium
Tel: +32 56 77 69 44
E-mail: infor@brenntag.be
Web: www.brenntag.be
Trade names: Puracal; Tri-Cafos.

Cabot GmbH
Josef-Bautz-Strasse 15
D-63420 Hanau
Germany
Tel: +49 6181 505150
Fax: +49 6181 505201
Web: www.cabot-corp.com/cabosil
Trade names: Cab-O-Sil.

Cargill Europe BVBA
Bedrijvenlaan 9
2800 Mechelen
Belgium
Tel: +32 15 400 539
Fax: +32 15 400 554
Web: www.cargillexcipients.com
Trade names: C*Dry MD; C*PharmDry; C*PharmGel; C*PharmIsoMaltidex; C*PharmSweet; Cargill Dry; Satiaxane U.

Cargill France SAS
18/20 rue de Gaudines
78100 Saint-Germain-en-Laye
France
Tel: +33 (0)1 30 61 35 00
Web: www.cargill.fr
Trade names: C*Dry MD; C*Pharm; C*PharmDex; C*PharmDry; C*PharmGel; C*PharmMaltidex; C*PharmMannidex; C*PharmSorbidex; C*PharmSweet; Cargill Dry; Satiaxane U; Treha; Zerose.

CFF GmbH and Co KG
Arnstaedter Str.2
D-98708 Gehren
Germany
Tel: +49 (0) 36 78 38 82 0
Fax: +49 (0) 36 78 38 82 25 2
E-mail: customerservice@cff.de
Web: www.cff.de
Trade names: Sanacel Pharma; Sanacel Wheat; Sanacel.

Chemacon GmbH
Rossmarktstrasse 22
Speyer D-67346
Germany
Tel: +49 62 32 622 286
Fax: +49 62 32 622 287
E-mail: info@chemacon.de
Web: www.chemacon.de

CHEMCO France
Chemco France BP 4172
10 av Maurice Berteaux
F-78300 Poissy
France
Tel: +33 1 30 65 75 00
Fax: +33 1 30 65 74 94
E-mail: contact@chemco-france.com
Web: www.chemco-france.fr

Chemische Fabrik Budenheim KG
Rheinstrasse 27
D-55257 Budenheim
Germany
Tel: +49 6139 890
E-mail: info@budenheim.com
Web: www.budenheim.com
Trade names: Di-Cafos AN; Di-Cafos; Tri-Cafos.

Chemos GmbH
Werner-von-Siemensstr. 3
93128 Regenstauf
Germany
Tel: +49 9402 9336 0
Fax: +49 9402 9336 13
E-mail: sales@chemos-group.com
Web: www.chemos-group.com

Chempri Oleochemicals (Europe)
Ottergeerde 30
4941 VM Raamsdonksveer
Netherlands
Tel: +31 162 515550
Fax: +31 162 521594
E-mail: sales@chempri.com
Web: www.chempri.nl

Chevron Lubricants
Technologiepark Zwijnaarde 2
Gent B-9052
Belgium
Tel: +32 9 240 71 71
Fax: +32 9 240 71 95
E-mail: belux-lubeorders@chevron.com
Web: www.chevronlubricants.com

Chevron Texaco Global Lubricants Benelux *see* Chevron Lubricants

Chr. Hansen
Bøge Allé 10-12
Hørsholm DK-2970
Denmark
Tel: +45 45 74 74 74
Fax: +45 45 74 88 88
E-mail: dkinfo@chr-hansen.com
Web: www.chr-hansen.com
Trade names: Pharma-Carb.

Clariant GmbH
Industriepark Höchst
D-65926
Frankfurt/Main
Germany
Tel: +49 6930 518000
Fax: +49 6196 7578856
Web: www.clariant.de
Trade names: Nipabutyl; Nipacide PX; Nipagin A; Nipagin M; Nipasol M Sodium; Nipasol M; Phenoxetol; Tylose CB.

Cognis GmbH
Postfach 130164
D-40551 Düsseldorf
Germany
Tel: +49 211 7940 0
Fax: +49 211 798 2431
E-mail: info@cognis.de
Web: www.de.cognis.com

Trade names: Cegesoft; Cetiol V PH; Cutina CP; Cutina GMS; Cutina HR; Emulgade 1000NI; Eumulgin; Eutanol G PH; HD-Eutanol V PH; Hydagen CAT; Lanette 16; Lanette O; Lanette SX; Lanette W; Monomuls 90-O18; Myritol; Novata; Speziol C16 Pharma; Speziol C16-18 Pharma; Speziol C18 Pharma; Speziol TPGS Pharma; Texapon K12P.

Colloides Naturels International
129 Chemin de Croisset
PO Box 4151
F-76723 Rouen Cedex
France
Tel: +33 2 32 83 18 18
Fax: +33 2 32 83 19 19
E-mail: client-order@cniworld.com
Web: www.cniworld.com

Contipro C a.s.
Dolní Dobrouč 401
561 02 Dolní Dobrouč
Czech Republic
Tel: +420 465 520 035
Fax: +420 465 524 098
E-mail: sales@contipro.cz
Web: www.contipro.cz

Corcoran Chemicals Ltd
Kingbridge House
17-22 Parkgate Street
Dublin 8
Ireland
Tel: +353 1 63 30 400
Fax: +353 1 67 93 521
E-mail: info@corcoranchemicals.com
Web: www.corcoranchemicals.com

CPI Chemicals
Franz Josefs Kai 31/19
Vienna A-1010
Austria
Tel: +43 1 535 2612 0
Fax: +43 535 26 12 12
Web: www.cpichem.com

Danisco A/S
Headquarters
Langebrogade 1
PO Box 17
1001 Copenhagen K
Denmark
Tel: +45 3266 2000
Fax: +45 3266 2175
E-mail: info@danisco.com
Web: www.danisco.com
Trade names: Grindsted; Grindsted; Litesse; Meyprodor; Xylitab.

Degussa Hüls AG *see* Evonik Industries AG

DFE Pharma
Klever Strasse 187
P.O. Box 20 21 20
47574 Goch
Germany
Tel: +49 (0) 2823 9288 770
Fax: +49 (0) 2823 9288 7799
E-mail: pharma@dmv-fonterra-excipients.com
Web: www.dmv-fonterra-excipients.com
Trade names: Lactochem; Lactohale; Lactopress Anhydrous 250; Lactopress Anhydrous Crystals; Lactopress Anhydrous Fine Powder; Lactopress Anhydrous Microfine; Lactopress Anhydrous Powder; Lactopress Anhydrous; Lactopress Spray-Dried 250; Lactopress Spray-Dried 260; Lactopress Spray-Dried; Pharmacel; Pharmatose; Primellose; Primojel; Respitose; SuperTab 11SD; SuperTab 14SD; SuperTab 21AN; SuperTab 22AN; SuperTab 24AN; SuperTab 30GR.

DMV-Fonterra Excipients *see* DFE Pharma

Dow Benelux NV
Dow Belgium B.V.B.A
Prins Boudewijnlaan 41
2650 Edegem
Belgium
Tel: +32 (0)3 4502011
Fax: +32 (0) 3 4502913
Web: www.dow.com

Dr Paul Lohmann GmbH KG
Hauptstrasse 2
D-31860 Emmerthal
Germany
Tel: +49 5155 630
Fax: +49 5155 63134
E-mail: sales@lohmann-chemikalien.de
Web: www.lohmann4minerals.com

DSM Fine Chemicals
PO Box 43
NL-6130 AA Sittard
Netherlands
Tel: +31 46 477 3610
Fax: +31 46 4773605
E-mail: dfc.sales@dsm.com
Web: www.dsm.com

DSM Special Prods. BV
Po Box 601
6160 AP Geleen
Netherlands
Web: www.dsm.com

DuPont de Nemours Int'l SA
2, Chemin du Pavillion Box 50
CH-1218 Le Grand Saconnex
Geneva
Switzerland
Tel: + 41 22 717 5111
Fax: + 41 22 717 5109
Web: www.dupont.com
Trade names: Dymel 134a/P; Dymel 227 ea/P; Dymel A; Elvanol; TiPure.

Evonik Goldschmidt GmbH
Goldschmidtstrasse 100
45127 Essen
Germany
Tel: +49 201 173-01
Fax: +49 201 173-3000
Web: www.goldschmidt.com
Trade names: Tego Alkanol 16; Tegosoft PSE.

Evonik Industries AG
Rellinghauser Strasse 1-11
45128 Essen
Germany
Tel: +49 201 177 01
Fax: +49 201 17 3475
Web: www.evonik.com
Trade names: Aerosil; Aerosil R972; Eudragit; Tego Alkanol 16.

Evonik Röhm GmbH
Business Line Pharma Polymers
Kirschenallee
64293 Darmstadt
Germany
Tel: +49 61 51 18 4019
Fax: +49 61 51 18 3520
E-mail: eudragit.germany@evonik.com
Web: www.eudragit.com

Ewald Gelatin GmbH
Meddersheimer Straße 50
55566 Bad Sobernheim
Germany
Tel: +49 (0) 6751-86 0
Fax: +49 (0) 6751-86 49
Web: www.ewald-gelatine.de

Exquim S.A.
Edifici L'illa
Diagonal
549 5a Planta
Barcelona E-08029
Spain
Tel: 93 5044400

Fax: 93 5894502
E-mail: exquim@ferrergrupo.com
Trade names: Citrosa.

FeF Chemicals A/S
Københavnsvej 216
DK-4600 Køge
Denmark
Tel: +45 5667 1000
Fax: +45 5667 1001
E-mail: fefinfo@fefchemicals.com
Web: www.fefchemicals.com

FMC Biopolymer
Avenue Mounier 83
B-1200 Brussels
Belgium
Tel: +32 2 775 8311
Fax: +32 2 775 8300
Web: www.fmcbiopolymer.com
Trade names: Ac-Di-Sol; Aquacoat cPD; Aquacoat ECD; Avicel CL-611; Avicel PH; Avicel RC-501; Avicel RC-581; Avicel RC-591; Avicel RC/CL; Gelcarin; Kelcosol; Keltone; Marine Colloids; Protanal; Protanal; Protasan UP CL 113; SeaSpen PF; Viscarin.

FrieslandCampina Domo
Stationsplein 4
NL 3818 LE Amersfoort
PO Box 1551
NL3800 BN Amersfoort
Netherlands
Tel: +31 3 713 3333
Fax: +31 3 713 3334
E-mail: info.domo@frieslandcampina.com
Web: www.domo.nl

Friesland Foods Domo *see* FrieslandCampina Domo

Fusion Nutraceuticals Ltd
No 8 Block 1, Northwood Court
Northwood Business Campus
Santry
Dublin 9
Ireland
Tel: +353 1 842 8415
Fax: +353 1 842 8415
E-mail: info@fusionnutra.com
Web: www.fusionnutra.com
Trade names: SucraPlus.

Gattefossé Group
36, Chemin de GENAS
BP 603
69804 SAINT-PRIEST
France
Tel: +33 (0) 4 72 22 98 00
Fax: +33 (0) 4 78 90 45 67
Web: www.gattefosse.com
Trade names: Apifil; Lauroglycol 90.

Gattefossé SAS
36 Chemin de Genas
BP 603
F-69804 Saint Priest
France
Tel: +33 (0) 72 22 98 00
Fax: +33 (0) 78 90 45 67
E-mail: infopharma@gattefosse.com
Web: www.gattefosse.fr
Trade names: Compritol 888 ATO; Gelucire 44/14; Gelucire 50/13; Labrafil M1944CS; Labrafil M2125CS; Labrasol; Peceol; Precirol ATO 5; Transcutol HP; Transcutol P.

Gelita Europe
Uferstraße 7
Eberbach 69412
Germany
Tel: +49 (0) 62 71 / 01
Fax: +49 (0) 62 71 / 84 - 27 00

Gelnex Europe
Via dei Mille 3
IT-56029
Santa Croce sull'Arno (PI)
Italy
Tel: +39 (0) 571 387635
Fax: +39 (0) 571 386756
Web: www.gelnex.xom

Haarmann & Reimer GmbH *see* Symrise

Haltermann GmbH *see* Haltermann Products

Haltermann Products
(Division of Dow Chemical Co.)
Schlengendeich 17
D-21107 Hamburg
Germany
Tel: +49 40 333 180
Fax: +49 40 333 18 158
Web: www.dow.com/haltermann

Harke Pharma GmbH
Xantener Straße 1
45479 Mülheim an der Ruhr
Germany
Tel: +49 208 3069 2000
Fax: +49 208 3069 2300
E-mail: pharma@harke.com
Web: www.harke.com
Trade names: L-HPC.

Hedinger GmbH *see* August Hedinger GmbH & Co

Helm AG
Nordkanalstrasse 28
D-20097 Hamburg
Germany
Tel: +49 40 2375 0
Fax: +49 40 2375 1845
E-mail: info@helmag.com
Web: www.helmag.com

Henkel AG & Co. KGaA
Henkelstraße 67
Düsseldorf 40191
Germany
Tel: +49 211 797 0
Fax: +49 211 798 4008
Web: www.henkel.com
Trade names: Standamul G.

Hermes Sweeteners Ltd
Ankerstrasse 53
8026 Zurich
Switzerland
Tel: +41 (0)44 245 43 00
Fax: +41 (0)44 245 43 35
E-mail: info@hermesetas.com
Web: www.hermesetas.com
Trade names: Hermesetas.

Honeywell Specialty Chemicals Seelze GmbH
Wunstorfer Strasse 40
D-30926 Seelze
Germany
Tel: +49 (0) 5137 99 9 0
Fax: +49 (0) 5137 99 9 123
Web: www.honeywell.com
Trade names: Genetron.

IMCD Group BV
Wilhelminaplein 32
Rotterdam 3072DE
Netherlands
Tel: +31 10 290 8600
Fax: +31 10 290 8611
Web: www.imcdgroup.com

Impag GmbH
Impag Import GmbH

Fritz-Remy-Str.25
D-63071 Offenbach/Main
Germany
Tel: +49 69 8500080
Fax: +49 69 85000890
E-mail: impag@impag.de
Web: www.impag.de

Imperial-Oel-Import
Handelsgesellschaft mbH
Bergstrasse 11
Hamburg
D-20095
Germany
Tel: +49 40 3385 33 0
Fax: +49 40 3385 3385
E-mail: info@imperial-oel-import.de
Web: www.imperial-oel-import.de

Induchem AG
Industriestrasse 8a
CH-8604 Volketswil
Switzerland
Tel: +41 44 908 4333
Fax: +41 44 908 4330
E-mail: sales@induchem.com
Web: www.induchem.com
Trade names: Uniphen P-23; Uniphen P-23; Uniphen P-23; Uniphen P-23.

Industrial Quimica Lasem, sa (IQL)
Av. de la Industria, 7
Pol. Ind. Pla del Cami, s/n
08297 Castellgalí
Barcelona
Spain
Tel: +39 93 875 88 40
Fax: +39 93 875 88 41
E-mail: info.iql@lasem.com
Web: iql.lasem.com
Trade names: Waglinol 6016.

Interchim Austria GES.M.B.H
Brixentaler Strasse 67
A-6300 Wörgl
Austria
Tel: +43 5332 71947
Fax: +43 5332 75361
E-mail: office@interchim.at
Web: www.interchim.at

Italgelatine SpA
S.S. Alba-Bra
Nr. 201, 12060 Santa Vittoria d'Alba (Cuneo)
Italy
Tel: +39 0172-47 80 47
Fax: +39 0172-47 87 15
Web: www.italgelatine.com

J Rettenmaier & Söhne GmbH and Co.KG
Holzmühle 1
D-73494 Rosenberg
Germany
Tel: +49 7967 152 0
Fax: +49 7967 152 222
E-mail: info@jrs.de
Web: www.jrs.de
Trade names: Arbocel; Emcocel; Lubritab; ProSolv; Pruv; Vivapharm HPMC; Vivapur; Vivasol; Vivastar P.

Julius Hoesch GmbH & Co KG
Birkesdorfer Straße 5
Düren DE-52353
Germany
Tel: +49 (2421) / 807- 0
Fax: +49 (2421) / 807- 104
E-mail: info@julius-hoesch.de
Web: www.julius-hoesch.de
Trade names: Kessco CP.

Juncá Gelatines S L SA
Blanquers, 84-106
17820 Banyoles - Girona
Spain
Tel: +34 (0) 972-57 04 08
Fax: +34 (0) 972-57 33 54
Web: www.miqueljunca.com

Jungbunzlauer AG
St Alban-Vorstadt 90
CH-4002 Basel
Switzerland
Tel: +41 61 295 51 00
Fax: +41 61 295 51 08
Web: www.jungbunzlauer.com
Trade names: Citrofol AI; Citrofol AII; Citrofol BI.

Karlshamns AB *see* AarhusKarlshamn AB

Kraeber GmbH & Co
Pharmazeutische Rohstoffe
Waldhofstraße 14
Ellerbek 25474
Germany
Tel: +49 4101 3153 0
Fax: +49 4101 3153 90
E-mail: info@kraeber.de
Web: www.kraeber.de

Lapi Gelatine SpA
Via Lucchese 164
50053 Empoli-Firenze
Italy
Tel: +39 0571-84 03 1
Fax: +39 071-58 14 35
Web: www.lapigelatine.it

Lehmann & Voss & Co
Alsterufer 19
D-20354 Hamburg
Germany
Tel: +49 40 44197 0
Fax: +49 40 44197 219
E-mail: info@lehvoss.de
Web: www.lehvoss.de

Lipoid GmbH
Frigenstr. 4
Ludwigshafen D-67065
Germany
Tel: +49 621 53819 0
Fax: +49 621 553559
Web: www.lipoid.com
Trade names: Lipoid.

Lonza Ltd
Muenchensteinerstrasse 38
PO Box
CH-4002 Basel
Switzerland
Tel: +41 61 316 81 11
Fax: +41 61 316 91 11
E-mail: info@lonzagroup.com
Web: www.lonzagroup.com
Trade names: Ethosperse; Glycon G-100; Glycon; Hyamine 1622; Hyamine 3500; Lonzest GMS.

Lucas Meyer
Ausschläger Elbdeich 62
D-20539 Hamburg
Germany
Tel: +49 40 789 55 0
Fax: +49 40 789 83 29
E-mail: info@lucas-meyer.com

Luzenac Europe
(Division of Rio Tinto Minerals)
2 Place Edward Bouilleres
BP 33662
F-31036 Toulouse Cedex 1
France

Tel: +33 561 502 020
Fax: +33 561 400 623
Web: www.luzenac.com
Trade names: Luzenac Pharma.

Lyondell Chemical Europe
P.O Box 2416
3000 CK Rotterdam
Netherlands
Tel: +31 (10) 275 5500
Web: www.lyondell.com

Magnesia GmbH
PO Box 2168
D-21311 Lüneburg
Germany
Tel: +49 4131 8710 0
Fax: +49 4131 8710 55
E-mail: info@magnesia.de
Web: www.magnesia.de
Trade names: MagGran CC.

Matrix Marketing GmbH
Bahnweg Norg 35
CH-9475 Sevelen
Switzerland
Tel: +41 (0)81 740 5830
Fax: +41 (0)81 740 5831
E-mail: info@matrix-marketing.ch
Web: www.matrix-marketing.ch

Meggle Gmbh *see* Molkerei Meggle Wasserburg GmbH

Merck KGaA
Frankfurter Strasze 250
64293 Darmstadt
Germany
Tel: +49 (0) 6151 720
Fax: +49 (0) 6151 72 2000
E-mail: service@merck.de
Web: www.merck.de
Trade names: Emprove.

MeroPharm AG
Bahnhofstrasse 31
CH-8280 Kreuzlingen
Switzerland
Web: www.meropharm.com

Molkerei Meggle Wasserburg GmbH
Megglestr. 6–12
D-83512 Wasserburg
Germany
Tel: +49 80 71 73 476
Fax: +49 80 71 73 320
E-mail: service.pharma@meggle.de
Web: www.meggle-pharma.de
Trade names: CapsuLac; Cellactose 80; FlowLac 100; FlowLac 90; GranuLac; InhaLac; MicroceLac 100; PrismaLac; SacheLac; SorboLac; SpheroLac; StarLac; Tablettose.

Mosselman
Route de Wallonie, 4
B-7011 Ghlin
Belgium
Tel: +32 65 395 610
Fax: +32 65 395 612
E-mail: sales@mosselman.eu
Web: www.mosselman.be

Natura Internacional S.L.
Rio Guadalquivir 4
30130 Beniel
Spain
Tel: 34 96 6708283
Fax: 34 96 8975164
E-mail: naturainternacionalsalud@gmail.com
Web: www.naturainternacional.biz

Naturex SA
Site d'Agroparc - BP 1218
84911 AVIGNON cedex 9
France
Tel: 33 (0) 4 90 23 96 89
Fax: 33 (0) 4 90 23 73 40
E-mail: naturex@naturex.com
Web: www.naturex.com
Trade names: Talin.

Nippon Soda Co Ltd (Germany)
Nisso Chemical Europe GmbH
Berliner Allee 42
Dusseldorf D-40212
Germany
Tel: +49 211 1306686 0
E-mail: info@nisso-chem.de
Web: www.nisso-chem.de
Trade names: Nisso HPC.

NovaMatrix
FMC Biopolymer A.S.,d/b/a NovaMatrix
Industriveien 33
1337 Sandvika
Norway
Tel: +47 6781 5500
Fax: +47 6781 5510
E-mail: novamatrixinfo@fmc.com
Web: www.novamatrix.biz
Trade names: Pronova; Protasan.

Noviant *see* CP Kelco US Inc

NP Pharm
A subsidiary of Colorcon Inc
54 bis Impasse du Boeuf Couronné
F-78550 Bazainville
France
Tel: +33 134 877 897
Fax: +33 134 877 896
Web: www.colorcon.com
Trade names: Crospopharm; Ethispheres; Povipharm; Suglets.

Nutrinova Nutrition Specialities & Food Ingredients GmbH
(Division of Celanese)
Am Unisys-Park 1
65843 Sulzbach (Taunus)
Germany
Tel: +49 (0) 69 45009 0
Fax: +49 (0) 69 45009 50000
Web: www.celanese.com
Trade names: Sunett.

Pahí SL
64-66 Avenida Madrid
08028 Barcelona
Spain
Tel: +34 93 656 24 09; +34 93 656 23 51
Fax: +34 93 656 53 09
E-mail: pahi@tartaricchemicals.com
Web: www.tartaricchemicals.com

Palatinit Süßungsmittel GmbH *see* Beneo-Palatinit GmbH

Panreac Quimica SAU
C/ Garraf, 2,
Polígono Pla de la Bruguera
Barcelona
E-08211 Castellar del Vallès
Spain
Tel: +34-937-489 400
Fax: +34-937-489 401
E-mail: central@panreac.com
Web: www.panreac.es

Parafluid Mineraloelgesellschaft mbH
Export Department
PO Box 602060
Uberseering 9
D-22297 Hamburg
Germany

Tel: +49 406 3704 00
Fax: +49 406 3704 100
E-mail: export@parafluid.de
Web: www.parafluid.eu

PB Gelatins Belgium
(Division of Tessenderlo Chemie nv)
Marius Duchestraat 260
B-1800 Vilvoorde
Belgium
Tel: +32 2 255 62 21
Fax: +32 2 253 96 18
E-mail: gelatin@tessenderlo.com
Web: www.pbgelatins.com
Trade names: Cryogel; Instagel; Solugel.

Pharm Allergan GmbH
Pforzheimer Str.160
D-76275 Ettlingen
Germany
Tel: +49 (0)7243/501-0
Fax: +49 (0)7243/501-100
E-mail: receptionger-agn@allergan.com
Web: www.allergan.de
Trade names: Coliquifilm.

Pharmatrans Sanaq AG
Birsigstrasse 79
CH-4011 Basel
Switzerland
Tel: +41 61 225 9000
Fax: +41 61 225 9001
E-mail: info@pharmatrans-sanaq.com
Web: www.pharmatrans-sanaq.com
Trade names: Cellets; MCC Sanaq; SSG Sanaq.

Purac Biochem
PO Box 21
4200 AA Gorinchem
Netherlands
Tel: +31 183 695 695
Fax: +31 183 695 600
E-mail: pnl@purac.com
Web: www.purac.com

Reinert Gruppe GmbH & Co KG
Am Vogelgesang 3-5
50374 Erfstadt
Germany
Tel: +49 (0) 223-5/4 08 0
Fax: +49 (0) 223-5/4 08 77
Web: www.reinert-gruppe.de

Rhodia SA
Worldwide Headquarters
110, Esplanade Charles de Gaulle
La Défense 4
92931 Paris La Défense Cedex
France
Tel: +33 1 53 56 64 64
Fax: +33 1 53 56 64 00
Web: www.rhodia.com
Trade names: Rhodopol.

Roquette Frères
F-62080 Lestrem Cedex
France
Tel: +33 (0)3 21 63 36 00
Fax: +33 (0)3 21 63 38 50
Web: www.roquette.fr
Trade names: Flolys; Fluidamid R444P; Glucidex; Keoflo ADP; Kleptose; Lycadex PF; Lycasin 75/75; Lycasin 80/55; Lycasin 85/55; Lycasin HBC; Lycatab C; Lycatab DSH; Lycatab PGS; Neosorb; Pearlitol; Roclys; StarLac; SweetPearl; Xylisorb.

Rousselot SAS
10 Avenue de l'Arche
92419 Courbevoie Cedex
France
Tel: +33 (0) 1-46 67 87 20
Fax: +33 (0) 1-46 67 87 21

SABO SPA
Via Caravaggi
24040 Levate (BG)
Italy
Tel: 39 035 596 000
Fax: 39 035 594 400
Web: www.sabo.com
Trade names: Sabowax CP.

Sachtleben Chemie GmbH
Postfach 17 04 54
Duisburg
D-47184
Germany
Tel: +49 2066 22 0
Fax: +49 2066 22 2000
E-mail: info@sachtleben.de
Web: www.sachtleben.de
Trade names: Hombitan FF-Pharma.

Sasol Germany GmbH
Arthur-Imhausen-Str. 92
D-58453 Witten
Germany
Tel: +49 23 02 92 51 00
Fax: +49 23 02 92 55 00
Trade names: Imwitor 412; Imwitor 491; Imwitor 900K; Imwitor 948; Lipoxol; Merkur; Miglyol 808; Miglyol 810; Miglyol 812; Nacol 16-95; Witepsol.

Schaefer Kalk GmbH & Co KG
Louise Seher Strasse 6
D-65582 Diez
Germany
Tel: +49 6432 503 0
Fax: +49 6432 503 269
E-mail: info@schaeferkalk.de
Web: www.schaeferkalk.de

Sensus
Sensus Head Office
Borchwerf 3
4704 RG Roosendaal
Netherlands
Tel: +31 165 58 2500
E-mail: info@sensus.nl
Web: www.sensus.nl
Trade names: Frutafit.

Seppic SA
22 Terrasse Bellini
Paris La Défense
92806 Puteaux
France
Tel: +33 (0)1 42 91 40 00
Fax: +33 (0)1 42 91 41 41
Web: www.seppic.com

SE Tylose GmbH & Co.KG
(Division of Shin-Etsu)
Rheingaustr. 190-196
Wiesbaden 65203
Germany
Tel: +49 611 962 8571
Fax: +49 611 962 9267
Web: www.setylose.de
Trade names: Tylose H; Tylose MB; Tylose MH; Tylose MHB; Tylose.

Sisterna BV
Borchwerf 4M
4704RG Roosendaal
Netherlands
Tel: +31 (0) 165 524 730
Fax: +31 (0) 165 524 739
E-mail: info@sisterna.com
Web: www.sisterna.com
Trade names: Sisterna PS750-C.

SKW Biosystems *see* Sobel NV

Sobel NV *see* Vion NV

Solvay Fluor GmbH
Carl Ulrich Strasse 34
D-74206 AN Bad Wimpfen
Germany
Tel: +49 7063 510
Fax: +49 7063 512 55
Web: www.solvay-fluor.com

Solvay Fluor und Derivative *see* Solvay Fluor GmbH

SPI Pharma
Chemin du Vallan de Maire
13240 Septemes-Les Vallans
France
Tel: +33 4 91 96 36 000
Fax: +33 4 4212 4693
Web: www.spipharma.com
Trade names: Compressol S; Compressol SM; Compressol; Mannogem; Sorbitab SD 250; Sorbitol Special.

Stearinerie Dubois Fils
696, rue Yves Kermen
92658 Boulogne Billancourt
Cedex
France
Tel: +33 1 46 10 07 30
Fax: +33 1 49 10 99 48
Web: www.stearinerie-dubois.fr/pharmaciee.php
Trade names: DUB SE; DUB SE; Stelliesters SE 15S.

Stern Wywiol Gruppe Holding GmbH & Co.KG
An der Alster 81
D-20099 Hamburg
Germany
Tel: +49 (0)40284 0390
Fax: +49 (0)40284 03971
E-mail: info@stern-wywiol-gruppe.de
Web: www.stern-wywiol-gruppe.de

Stroever GmbH & Co. KG
Auf der Muggenburg 11
Bremen 28217
Germany
Tel: +49 (421) 386 13 0
Fax: +49 (421) 386 13 44
E-mail: info@stroever.de
Web: www.stroever.de
Trade names: SSB 55 Pharma; SSB 56 Pharma; SSB 57 Pharma.

Südzucker AG *see* Beneo-Palatinit GmbH

Suiker Unie
P.O. Box 100
4750 AC Oud Gastel
Noordzeedijk 113
Dinteloord 4671 TL
Netherlands
Tel: +31 165 525 252
Fax: +31 165 525 255
E-mail: suiker.info@suikerunie.com
Web: www.suikerunie.nl

Tate & Lyle
Parc Scientifique de la Haute Borne
2, Avenue de l'Horizon
Villeneuve d'Ascq
59650
France
Tel: +33 (0) 70 752 854
Fax: +33 (0) 76 707 517
Web: www.tateandlyle.com
Trade names: Fructamyl; Glucomalt; Glucosweet; Maldex G; Maldex; Maltosweet; Merigel; Meritena; Meritol; Mylose; Star-Dri.

Tereos
Tour Lilleurope
Parvis de Rotterdam
Lille F-59777
France
Tel: +33 (0) 3 28 38 79 30
Fax: + 33 (0) 3 28 38 79 31
E-mail: contact@tereos.com
Web: www.tereos.com
Trade names: Compressuc.

Tereos Syral SAS
A subsidiary of Tereos
Z. I. et Portuaire
B.P. 32
F–67 390 Marckolsheim
France
Tel: +33 (0) 3 88 58 60 60
Fax: +33 (0) 3 88 58 60 61
E-mail: sales-syral@tereos.com
Web: www.syral.com
Trade names: Glucodry; Maldex; Malta*Gran.

Tessenderlo Chemie
Troonstraat
Rue du Trône 130
BE-1050 Bruxelles
Belgium
Tel: +32 2 639 1811
Fax: +32 2 639 1999
E-mail: finechemicals@tessenderlo.com
Web: www.tessenderlo.com
Trade names: Cryogel; Instagel; Solugel.

Texas Global Products Benelux *see* Chevron Texaco Global Lubricants Benelux

The Orphan Pharmaceutical Company
70, avenue du Général de Gaulle
Paris La Défense 92058
France
Tel: +33 1 47 73 64 58
Fax: +33 1 49 00 18 00
E-mail: info@orphan-europe.com
Web: www.orphan-europe.com
Trade names: Wilzin.

Thermphos International BV
Netherlands
Tel: +44 (0)121 567 1006
Fax: +44 (0)121 567 1020
E-mail: Andrew.Mckeon@thermphos.com
Web: www.thermphos.com

Trobas Gelatine BV
PO Box 14
5100 AA Dongen
Netherlands
Tel: +31 (0) 162 314944
Fax: +31 (0) 162 317288
Web: www.trobas.nl

Ubichem Research Ltd
Illates ut 33
Budapest H-1097
Hungary
Tel: +36 1 347 5060
Fax: +36 1 347 5061
Web: www.ubichem.com

Union Derivan, SA (UNDESA)
Avda Generalitat 175-179
Barcelona
08840 Viladecans
Spain
Tel: +34 93 637 35 37
Fax: +34 93 659 19 02
E-mail: sales@undesa.com
Web: www.undesa.com
Trade names: Cristal G; Cristal S; Dervacid; Extra AS; Extra P; Extra S; Extra ST; Kemilub ES; Kemistab EC-F; Palmil C.

Univar
Rue de la petite Ile 4

Klein-Eiland 4
Anderlecht B-1070
Belgium
Tel: +32 02 525 05 11
Fax: +32 (0)2 520 17 51
E-mail: univar.benelux@univareurope.com
Web: www.univareurope.com

USOCO BV
Mandenmakerstraat 21
NL-2984 AS Ridderkerk
Netherlands
Tel: +31 0180 41 61 55
Fax: +31 0180 41 28 36
E-mail: info@usoco.nl
Web: www.usoco.nl

Vion NV
Netherlands
Web: www.vionfoodgroup.nl

Wacker-Chemie GmbH
Business Line Biotechnology
Johannes Hess Str. 24
D-84489 Burghausen
Germany
Tel: +49 8677 830
Fax: +49 8677 833 100
Web: www.wacker-biochem.com
Trade names: Cavamax W6 Pharma; Cavamax W7 Pharma; Cavamax W8 Pharma; Cavasol W7; HDK; HDK.

Weishardt International, Europe
BP 259
81305 Graulhet Cédex
France
Tel: +33 (0) 5-63 42 14 41
Fax: +33 (0) 5-63 42 35 18

SUPPLIERS LIST: USA

3M Drug Delivery Systems
3M Center
St Paul MN 55144-1000
Tel: +1 888 364 3577
Web: www.3m.com
Trade names: CoTran.

AarhusKarlshamn USA Inc
131 Marsh Street
Port Newark NJ 07114
Tel: +1 973 344 1300
Fax: +1 973 344 9049
Web: www.aak.com
Trade names: Aextreff CT; Colzao CT; Cremao CS-34; Cremao CS-36; Hyfatol 16-95; Hyfatol 16-98; Shogun CT.

Aarhus United USA Inc *see* AarhusKarlshamn USA Inc

Abbott Laboratories
U.S. Corporate Headquarters
100 Abbott Park Road
Abbott Park
Illinois 60064-3500
Tel: (847) 937-6100
Web: www.abbott.com
Trade names: Talatrol.

ABITEC Corp
501 West First Avenue
Columbus OH 43215
Tel: +1 614 429 6464
Fax: +1 614 299 8279
E-mail: sales@abiteccorp.com
Web: www.abiteccorp.com
Trade names: Acconon; Capmul GMO; Capmul GMS-50; Capmul PG-12; Captex 300; Captex 355; Captex 500; Captex 8000; Pureco 76; Sterotex HM; Sterotex.

Aceto Corp
One Hollow Lane
Lake Success NY 11042
Tel: +1 516 627 6000
E-mail: aceto@aceto.com
Web: www.aceto.com

Acme-Hardesty
450 Sentry Parkway
Suite 104
Blue Bell PA 19422
Tel: +1 215 591 3610
Fax: +1 215 591 3620
E-mail: sales@acme-hardesty.com
Web: www.acme-hardesty.com

Acros Organics
Part of Thermo Fisher Scientific
Tel: +1 800 227 6701
E-mail: chem.techinfo@thermofisher.com
Web: www.acros.com

Advance Scientific & Chemical Inc
2345 SW 34th Street
Fort Lauderdale FL 33312
Tel: +1 954 327 0900
Fax: +1 954 327 0903
E-mail: sales@advance-scientific.com
Web: www.advance-scientific.com

AerChem Inc
3935 W Roll Avenue
Bloomington IN 47403
Tel: +1 812 334 9996
Fax: +1 812 334 1960
E-mail: mmckean@aerchem.com
Web: www.aerchem.com

Aeropres Corp
Aeropres Headquarters
PO Box 78588
Shreveport, Louisiana
71137-8588
Tel: +1 318 221 6282
E-mail: jbowen@aeropres.com
Web: www.aeropres.com
Trade names: Aeropres 108; Aeropres 17; Aeropres 31.

AE Staley Mfg Co *see* Tate & Lyle

Air Liquide America Corp
2700 Post Oak Boulevard
Suite 1800
Houston TX 77056
Tel: +1 800 820 2522

Akzo Nobel Chemicals Inc
525 West Van Buren Street
Chicago IL 60607-3823
Tel: +1 312 544 7000
Fax: +1 312 544 7320
Web: www.akzonobel.com
Trade names: Brij; Dissolvine; Elfan 240.

Aldrich *see* Sigma-Aldrich Corp

Alfa Chem
2 Harbor Way
King's Point
NY 11024-2117
Tel: +1 516 504 0059
Fax: +1 516 504 0039
E-mail: alfachem@gmail.com
Web: www.alfachem1.com

Allan Chemical Corp
235 Margaret King Avenue
Ringwood NJ 07456
Tel: (973) 962 4014
Fax: (973) 962 6820
E-mail: info@allanchem.com
Web: www.allanchem.com

Allergan Inc
Corporate Headquarters
2525 Dupont Drive
Irvine CA 92612
Tel: +1 714 246 4500
Fax: +1 714 246 6987
Web: www.allergan.com
Trade names: Coliquifilm.

Alzo International Inc
650 Jernee Mill Road
Sayreville NJ 08872
Tel: +1 732 254 1901
Fax: +1 732 254 4423
E-mail: carolyn.zofchak@mail.alzointernational.com
Web: www.alzointernational.com
Trade names: Dermofat 4919; Dermol EL; Dermol IPM; Dermol IPP; Dermowax GMS; Wickenol 101; Wickenol 111; Wickenol 127; Wickenol 131.

American Colloid Co
2870 Forbs Avenue
Hoffman estates
IL 60192
Tel: +1 847 851 1700
Fax: +1 847 851 1799
Web: www.colloid.com
Trade names: Magnabrite; Polargel.

American Ingredients Corp
3947 Broadway
Kansas City MI 64111
Tel: +1 800 669 4092
E-mail: info@americaningredients.com
Web: www.patco-additives.com
Trade names: Patlac; Purasal.

American International Chemical Inc
135 Newbury Street
Framingham
Massachusetts 01701
Tel: +1 508 270 1800
Fax: +1 508 872 1566
E-mail: info@aicma.com
Web: www.aicma.com
Trade names: galenIQ; TriStar.

American Lecithin Co
115 Hurley Road
Unit 2B
Oxford CT 06478
Tel: +1 203 262 7100
Fax: +1 203 262 7101
Web: www.americanlecithin.com
Trade names: Phosal 53 MCT; Phospholipon 100 H.

Amresco Inc
6681 Cochran Road
Solon OH 44139
Tel: +1 440 349 1199
Fax: +1 440 349 3255
E-mail: info@amresco-inc.com
Web: www.amresco-inc.com

ANGUS Chemical Co
Subsidiary of Dow Chemical Co.
1500 East Lake Cook Road
Buffalo Grove
Illinois 60089
Tel: + 989 832 1560
Fax: + 989 832 1465
E-mail: dowcig@dow.com
Web: www.dow.com/angus/

AnMar International
540 Barnum Avenue
Bridgeport CT 06608
Tel: +1 203 336 8330
Fax: +1 203 336 5508
E-mail: Blanco@anmarint.com
Web: www.anmarinternational.com

Archer Daniels Midland Company (ADM)
Specialty Food Ingredients
4666 Faries Parkway
Decatur IL 62526
Tel: 800-637-5850
Fax: 217-451-4492
Web: www.adm.com
Trade names: Clintose.

Arista Industries Inc
557 Danbury Road
Wilton CT 06897
Tel: +1 800 255 6457
Fax: +1 203 761 4980
Web: www.aristaindustries.com

ASHA cellulose (Private Ltd)
Good Hope International Inc.
9807 Lackman Road
Lenexa KS 66219
Tel: +1 913 888 8088
Fax: +1 913 888 8075
E-mail: rshah@goodhopeinc.com
Web: www.ashacel.com
Trade names: Ashacel.

Ashland Inc
50 E. RiverCenter Blvd.
PO Box 391
Covington
KY 41012-0391
Tel: +1 859 815 3333
Web: www.ashland.com

Ashland Specialty Ingredients
(Division of Ashland)
Hercules Incorporated
1313 North Market Street
Wilmington
DE 19894-0001
Tel: +1 800 345 0447
Fax: +1 302 992 7507
Web: www.ashland.com
Trade names: Aqualon CMC; Aqualon; Benecel hypromellose; Benecel; Blanose; Culminal MC; Culminal MHEC; Galactosol; Klucel; Natrosol.

Asiamerica Ingredients Inc
245 Old Hook Road
Westwood
New Jersey 07675
Tel: +1 201 497 5993
Fax: +1 201 497 5994
E-mail: info@asiamericaingredients.com
Web: www.asiamericaingredients.com

Astro Chemicals Inc
126 Memorial drive
Springfield MA 01102
Tel: +1 413 781 7240
Fax: +1 413 781 7246
E-mail: sales@astrochemicals.com
Web: www.astrochemicals.com
Trade names: Drakeol; Hystrene; Industrene.

Atlantic Gelatin/Kraft Foods Global Inc
Hill Street
Woburn
Massachusetts 01801
Tel: +1 (781) 938-2200
Fax: +1 (781) 573-3876
E-mail: jmagnifico@kraft.com

Avanti Polar Lipids Inc
700 Industrial Park Drive
Alabaster
AL 35007-9105
Tel: +1 205 663 2494
Fax: +1 205 663 0756

E-mail: info@avantilipids.com
Web: www.avantilipids.com

Avantor Performance Materials
222 Red School Lane
Phillipsburg NJ 08865
Tel: 1-908-859-2151
Fax: 1-908-859-9318
E-mail: info@avantormaterials.com
Web: www.avantormaterials.com

Avatar Corp
500 Central Avenue
University Park IL 60484
Tel: +1 708 534 5511
Fax: +1 708 534 0123
E-mail: inquiries@avatarcorp.com
Web: www.avatarcorp.com
Trade names: Avatech; Citation; Pinnacle; ProKote 2855; ProKote LSC; Snow White.

Avebe America Inc
Princeton Corporate Center
4 Independence Way
Princeton NJ 08543-5307
Tel: +1 609 520 1400
Fax: +1 609 520 1473
Web: www.avebe.com
Trade names: Paselli MD10 PH; Perfectamyl; Prejel; Primellose.

Aventis Behring LLC *see* ZLB Behring

Bachem Bioscience Inc
3700 Horizon Dr.
King of Prussia
PA 19406
Tel: +1 610 239 0300
Fax: +1 610 239 0800
Web: bachem.com

Balchem Corp
52 Sunrise Park Road
New Hampton NY 10958
Tel: +1 845 326 5600
Fax: +1 845 326 5742
E-mail: nutrients@balchem.com
Web: www.balchem.com

Barrington Chemical Corp *see* Barrington Nutritionals Inc

Barrington Nutritionals
500 Mamaroneck Ave
Harrison NY 10528
Tel: +1 914 381 3500
Fax: +1 914 381 2232
E-mail: info@barringtonchem.com
Web: www.barringtonchem.com

BASF Corp
100 Campus Drive
Florham Park NJ 07932
Tel: +1 973 245 6000
Fax: +1 973 895 8002
Web: www.basf.com
Trade names: Cremophor A; Kollicoat MAE 100 P; Kollicoat MAE; Kollidon CL-M; Kollidon CL; Kollidon VA 64; Kollidon; Ludiflash; Ludipress LCE; Lutrol E; Luviskol VA; Myacide; Pluriol E; Soluphor P.

Bayer Corp *see* Lanxess Corp

BF Goodrich Speciality Chemicals *see* Noveon Inc

Biddle Sawyer Corp
21 Penn Plaza
360 West 31st Street
New York
NY 10001-2727
Tel: +1 212 736 1580
Fax: +1 212 239 1089
E-mail: BSC@biddlesawyer.com
Web: www.biddlesawyer.com
Trade names: L-HPC; Metolose; Metolose.

BioPure Healing Products LLC
P.O. Box 5023 Bellevue
WA 98009
Tel: 00 1 800 801 6187
E-mail: info@biopureus.com
Web: www.biopureus.com

Blanver (USA)
1515 South Federal Highway
Suite 105
Boca Raton FL 33432
Tel: +1 561 416 5513
Fax: +1 561 416 5663
E-mail: blanver@blanver.com
Web: www.blanver.com.br
Trade names: Microcel 3E-150.

BOC Gases (USA)
575 Mountain Avenue
Murray Hill
NJ 07974 2082
Tel: +1 908 464 8100
Fax: +1 410 749 4073
E-mail: USweb-inquiries@boc.com
Web: www.boc.com

BOC Sciences
45-16 Ramsey Road
Shirley NY 11967
Tel: +1 631 398 3562
Fax: +1 631 614 7828
E-mail: info@bocsci.com
Web: www.bocsci.com

Boehringer Ingelheim Chemicals Inc
2820 North Normandy Drive
Petersburg VA 23805
Tel: +1 804 50 48 600
Fax: +1 804 50 48 637
Web: www.boehringer-ingelheim.com/finechem

BP Inc
535 Madison Avenue
New York NY 10022
Tel: +1 212 421 5010
Web: www.bp.com

Brainerd Chemical Company Inc
1200 North Peoria
P.O Box 521150
Tulsa OK 74152-1150
Tel: +1 918 622 1214
Fax: +1 918 632 0851
E-mail: sales@brainerdchemical.com
Web: www.brainerdchemical.com

Brenntag Inc *see* Brenntag North America Inc

Brenntag North America Inc
North American Headquarters Office
5083 Pottsville Pike
Reading PA 19605
Tel: +1 610 926 6100
E-mail: brenntag@brenntag.com
Web: www.brenntagnorthamerica.com
Trade names: Sequestrene AA.

Brenntag Southeast
2000 East Pettigrew Street
Durham NC 27703
Tel: 00 1 800 849 700
Web: www.brenntagsoutheast.com

Brenntag Southwest *see* Brenntag Inc

Brenntag Specialties Inc
1000 Coolidge Street
South Plainfield NJ 07080
Tel: (908) 561 6100
Fax: (908) 757 3488
Web: www.brenntagspecialties.com
Trade names: Albagel.

Brown Chemical Co, Inc
302 West Oakland Ave
P.O. Box 440
Oakland
NJ 07436-0440
Tel: +1 201 337 0900
Fax: +1 201 337 9026
E-mail: sales@brownchem.com
Web: www.brownchem.com
Trade names: Lorol C14-95.

Budenheim USA Inc
2219 Westbrooke Drive
Columbus OH 43228
Tel: +1 614 345 2400
Web: www.budenheim.com

Burlington Bio-medical and Scientific Corp
71 Carolyn Boulevard
Farmingdale
NY 11735-1718
Tel: +1 631 694 4700
Fax: +1 631 694 9177
Trade names: Bitterguard.

Cabot Corp
2 Seaport Lane
Suite 1300
Boston MA 02210
Tel: +617 345 0100
Fax: +617 342 6103
Web: www.cabot-corp.com
Trade names: Cab-O-Sil.

CarboMer Inc
PO Box 261026-1026
San Diego
CA 92196-1026
Tel: +1 858 552 0992
Fax: +1 858 552 0999
E-mail: info@CarboMer.com
Web: www.carbomer.com

Cargill Corp *see* Cargill Inc
Trade names: C*Dry MD.

Cargill Inc
Cargill Office Center
PO Box 9300
Minneapolis
MN 55440-9300
Tel: +1 800 227 4455
Web: www.cargill.com
Trade names: C*Dry MD; C*PharmDry; C*PharmGel; Cargill Dry; Satiaxane U.

Carolina Biological Supply Co
2700 York Road
Burlington NC 27215
Tel: + 1 336 584 0381
Fax: + 1 800 222 7112
E-mail: carolina@carolina.com
Web: www.carolina.com

Celanese
Corporation headquaters
1601 West LBJ Freeway
Dallas
Texas 75234-4000
Tel: +1 972 443 4000
Web: www.celanese.com
Trade names: Celvol.

Celite Corporation
(Advanced Minerals Corporation)
130 Castilian Drive
Goleta CA 93117
Tel: +1 805 737 2460
Fax: +1 805 737 2466
E-mail: info@advancedminerals.com
Web: www.advancedminerals.com
Trade names: Micro-Cel.

Cerestar USA Inc *see* Cargill Corp

Charkit Chemical Corp
32 Haviland street
Norwalk
Connecticut 06854-4906
Tel: +1 203 299 3220
Fax: +1 203 299 1355
E-mail: sales@charkit.com
Web: www.charkit.com

Charles B Chrystal Co Inc
30 Vesey Street
New York NY 10007
Tel: +1 212 227 2151
Fax: +1 212 233 7916
E-mail: E-mail: info@cbchrystal.com
Web: www.cbchrystal.com
Trade names: Lion; Purtalc; Sim 90; Snow White.

Charles Bowman & Co
3328 John F. Donnelly Drive
Holland MI 49424
Tel: +1 616 786 4000
Fax: +1 616 786 2864
E-mail: cbc@charlesbowman.com
Web: www.charlesbowman.com

Chart Corp Inc
(Division of Naturex)
787 East 27th Street
Paterson NJ 07504
Tel: +1 201 345 5554
Fax: +1 201 345 2139

Chemtex USA Inc
150 River Road Unit G3B
Montville NJ 07045-9441
Tel: 011-973-335-2500
Fax: 011-973-335-2552
E-mail: contact@chemtexusa.com
Web: www.chemtexusa.com

Chemtura Corporation
Global Corporate Headquarters
199 Benson Road
Middlebury CT 06749
Tel: +1 203 573 2000
Web: www.chemtura.com
Trade names: Sentry.

Church and Dwight Co Inc
Performance Products Group
469 North Harrison Street
Princeton NJ 08543
Tel: +1 800 221 0453
Fax: +1 609 497 7176
E-mail: perfcustserv@churchdwight.com
Web: www.ahperformance.com

Clariant Corp
625 Catawba Avenue
Mount Holly (East)
NC 28120
Tel: +1 800 942 7239
Fax: +1 704 822 2204
E-mail: info@clariant.com
Web: www.ics.clariant.com
Trade names: Genapol; Hostacerin; Nipabutyl; Nipacide PX; Nipagin A; Nipagin M; Nipasol M Sodium; Nipasol M; Phenoxetol; Tylose CB.

Clarus Specialty Products
Tel: 1 866 400 0921
E-mail: info@clarussp.com
Web: www.clarussp.com

Cognis Corp
Care Chemicals
300 Brookside avenue

Ambler PA 19002
Tel: 800 531 0815
Fax: 215 628 1353
Web: www.na.cognis.com
Trade names: Cegesoft; Cutina CP; Cutina GMS; Cutina HR; Emulgade 1000NI; Eumulgin; Hydagen CAT; Lanette 16; Lanette O; Lanette SX; Lanette W; Monomuls 90-O18; Myritol; Novata; Speziol C16 Pharma; Speziol C16-18 Pharma; Speziol C18 Pharma; Speziol TPGS Pharma; Texapon K12P.

Colloides Naturels Inc
1140 US Highway 22 East
Center Point IV
Suite 102
Bridgewater NJ 08807
Tel: +1 908 707 9400
Fax: +1 908 707 9405

Colorcon Inc
415 Moyer Boulevard
West Point PA 19486
Tel: +1 215 699 7733
Fax: +1 215 661 2626
Web: www.colorcon.com
Trade names: Methocel; Methocel; Opaglos R; Opaseal; Phthalavin; StarCap 1500; Starch 1500 G; Surelease; Sureteric.

Corn Products U.S.
5 Westbrook Corporate Center
Westchester IL 60154
Tel: +1 708 551 2600
E-mail: info@cornproducts.com
Web: www.cornproducts.com
Trade names: Amidex; Globe Plus; Globe; Maltisweet 3145; Sorbogem.

CP Kelco US Inc
1000 Parkwood Circle
Suite 1000
Atlanta GA 30339
Tel: +1 678 247 7300
Fax: +1 678 247 2797
Web: www.cpkelco.com
Trade names: Finnfix; Genu; Genu; Keltrol; Nymcel ZSB; Nymcel ZSX; Xantural.

Croda Inc
300-A Columbus Circle
Edison NJ 08837
Tel: +1 732 417 0800
Fax: +1 732 417 0804
E-mail: marketing@crodausa.com
Web: www.crodausa.com
Trade names: Brij; Byco; Crillet; Crodacol C90; Crodacol CS90; Crodacol S95; Crodamol CP; Crodamol GTCC; Crodamol IPM; Crodamol IPP; Crodamol SS; Crodesta F; Crodex A; Crodex N; Croduret; Estol 3694; Etocas; Novol; Polawax; Priolube 1408; Super Hartolan.

CSL Behring
1020 First Avenue
PO Box 61501
King of Prussia
PA 19406 0901
Tel: +1 610 878 4000
Fax: +1 610 878 4009
Web: www.cslbehring.com

CTD Inc
27317 NW 78th Avenue
High Springs FL 32643
Tel: +1 386 454 0887
Fax: +1 386 454 8134
Web: www.cyclodex.com

Cultor Food Science *see* Danisco USA Inc

Cydex Inc *see* Cydex Pharmaceuticals Inc

CyDex Pharmaceuticals Inc
A Ligand Company
10513 W.84th Terrace
Lenexa KS 66214
Tel: +1 913 685 8850
Fax: +1 913 685 8856
E-mail: cdinfo@cydexpharma.com
Web: www.cydexpharma.com
Trade names: Captisol.

Danisco Cultor America Inc *see* Danisco USA Inc

Danisco USA Inc
Four New Century Parkway
New Century KS 66031
Tel: +1 913 764 8100
E-mail: usa.info@danisco.com
Web: www.danisco.com
Trade names: Grindsted; Grindsted; Litesse; MannoTab; Meyprodor.

Dastech International Inc
10 Cutter Mill Rd.
Great Neck
New York 11021
Tel: +1 516 466 7676
Fax: +1 516 466 7699
E-mail: info@dastech.com
Web: www.dastech.com

Degussa Hüls Corp *see* Evonik Degussa Corp

Domino Foods, Inc.
Domino Specialty Ingredients
One North Clematis Street
Suite 200
West Palm Beach FL 33401
Tel: +1 561 366 5150
Fax: +1 561 366 5158
Web: www.dominospecialtyingredients.com
Trade names: Di-Pac; Grindsted.

Dow Chemical Co
2030 Dow Center
Midland MI 48642
Tel: +1 989 636 1000
Fax: +1 989 636 3518
Web: www.dow.com
Trade names: Carbitol; Carbowax Sentry; Carbowax; Cellosize HEC; Dowanol EPh; Ethocel; Methocel; Methocel; Optim; Polyox; Versene Acid; Versenex; Walocel C.

Dow Corning
Corporate Center
PO Box 994
Midland MI 48686-0994
Tel: +1 989 496 4400
Fax: +1 989 496 6731
Web: www.dowcorning.com
Trade names: Dow Corning 245 Fluid; Dow Corning 246 Fluid; Dow Corning 345 Fluid; Dow Corning Q7-2243 LVA; Dow Corning Q7-2587; Dow Corning Q7-9120.

DSM Pharmaceuticals Inc
PO. Box 1887
Greenville
North Carolina
27835-1877
Tel: +1 (252) 758 34 36
Fax: +1 (252) 707 20 50
Web: www.dsm.com

DuPont
Packaging and Industrial Polymers
1007 Market Street
Wilmington DE 19898
Tel: +1 302 922 5225
Fax: +1 302 922 3495
Web: www.dupont.com
Trade names: Dymel 227 ea/P; Dymel A; Elvanol; TiPure.

DuPont (Packaging and Industrial Polymers) *see* DuPont

DURECT Corp
Corporate Headquarters
10260 Bubb Road

Cupertino
CA 95014-4166
Tel: 408 777 1417
Fax: 408 777 3577
Web: www.durect.com

Eastman Chemical Co
PO Box 431
Kingsport
TN 37662-5280
Tel: +1 423 229 2000
Fax: +1 423 229 1193
Web: www.eastman.com
Trade names: Eastacryl 30 D; Eastacryl; Eastman DBP; Eastman DEP; Pamolyn; Tenox BHA; Tenox BHT; Tenox PG.

Eastman Gelatine Corp
227 Washington Street
Peabidy
Massachusetts
01960-6998
Tel: +1 (978) 573-3700
Fax: +1 (978) 573-3876
Web: www.eastmangelatine.com

Edward Mendell Co *see* Rettenmaier UK Ltd

EMD Chemicals Inc
480 South Democrat Road
Gibbstown NJ 08027
Tel: +1 856 423 6300
Fax: +1 856 423 4389
E-mail: emdinfo@emdchemicals.com
Web: www.emdchemicals.com
Trade names: Sorbitol Instant.

EM Industries Inc *see* EMD Chemicals Inc

EM Sergeant Pulp & Chemical Co Inc
6 Chelsea Road
Clifton NJ 07012
Tel: +1 973 4729111
Fax: +1 973 472 5686
E-mail: info@sergeantchem.com
Web: www.sergeantchem.com

Evonik Degussa Corp
379 Interpace Parkway
Parsipanny NJ 07054
Tel: +1 973 541 8000
Fax: +1 973 541 8013
Web: www.evonik.com
Trade names: Aerosil ; Aerosil R972; Eudragit; Fructofin; Tego Alkanol 1618; Tego Alkanol 6855.

Farma International Inc
9501 Old South Dixie Highway
Miami FL 33156
Tel: +1 305 670 4416
Fax: +1 305 670 4417
E-mail: sales@farmainternational.com
Web: www.farmainternational.com
Trade names: Eumulgin; Veegum HS.

Ferro Corp
6060 Parkland Boulevard
Mayfield Heights OH 44124
Tel: +1 (216) 875-5600
Fax: +1 (216) 875-5627
E-mail: pfanstiehl-info@ferro.com
Web: www.ferro.com
Trade names: Synpro 90; Synpro; Synpro.

Ferro Pfanstiehl Laboratories Inc *see* Ferro Corp

Fisher Scientific
Part of Thermo Fisher Scientific
300 Industry Drive
Pittsburgh PA 15275
Tel: +1 800 766 7000
Fax: +1 800 926 1166
Web: www.fishersci.com

Fitzgerald Industries International
30 Sudbury Avenue
Suite 1A North
Acton MA 01720
Tel: +1 978 371 6446
Fax: +1 978 371 2266
E-mail: antibodies@fitzgerald-fii.com
Web: www.fitzgerald-fii.com

FMC Biopolymer (USA)
1735 Market Street
Philadelphia PA 19103
Tel: +1 800 526 3649
Fax: +1 215 299 6291
Web: www.fmcbiopolymer.com
Trade names: Avicel CL-611; Avicel RC-501; Avicel RC-581; Avicel RC-591; Avicel RC/CL; Kelcosol; Keltone; Profoam.

Foremost Farms USA
E10889 Penny Lane
PO Box 111
Baraboo WI 53913
Tel: +1 800 362 9196
Fax: +1 608 356 9005
E-mail: communications@foremostfarms.com
Web: www.foremostfarms.com

Fortitech Inc
Riverside Technology Park
2105 Technology Dr.
Schenectady
New York 12308
Tel: +1 518 372 5155
Fax: +1 518 372 5599
E-mail: info@fortitech.com
Web: www.fortitech.com

Frank B Ross Co Inc
970-H New Brunswick Avenue
Rahway NJ 07065
Tel: +1 732 669 0810
Fax: +1 732 669 0814
E-mail: techinfo@rosswaxes.com
Web: www.frankbross.com
Trade names: Ross Ceresine Wax.

Fuji Chemical Industries (USA) Inc *see* Fuji Chemical Industries Health Science (USA) Inc

Fuji Chemical Industries Health Science (USA) Inc
3 Terri Lane, Unit 12
Burlington NJ 08016
Tel: +1 609 386 3030
Fax: +1 609 386 3033
E-mail: contact@fujihealthscience.com
Web: www.fujihealthscience.com
Trade names: Fujicalin; Neusilin.

Functional Foods Corp
470 Rt. 9
Englishtown
New Jersey 07726
Tel: +1 732 972 2232
Fax: +1 732 536 9179
E-mail: yogi@functionalfoods.com
Web: www.functionalfoods.com

Gattefossé Corp
Main Office
Plaza I
115 West Century Road
Suit 340
Paramus NJ 07652
Tel: +1 201 265 4800
Fax: +1 201 265 4853
E-mail: info@gattefossecorp.com
Web: www.gattefossecorp.com
Trade names: Apifil; Compritol 888 ATO; Gelucire 44/14; Gelucire 50/13; Labrafac WL1349; Labrafil M1944CS; Labrafil M2125CS; Labrafil

M2130CS ; Labrasol; Lauroglycol 90; Peceol; Precirol ATO 5; Suppocire; Transcutol HP; Transcutol P.

Gaylord Chemical Company LLC
106 Galeria Blvd.
Slidell LA 70458
Tel: +1 800 426 6620
E-mail: info@gaylordchemical.com
Web: www.gaylordchemical.com
Trade names: Procipient.

Gelita USA
2445 Port Neal Industrial Rd
Sergeant Bluff IA 51054
Tel: +1 712 943 5516
Fax: +1 712 943 3372

Gelnex
1615 Northern Blvd
Suite 101
Manhasset NY 11030
Tel: +1 516 869 1623
Fax: +1 516 869 1057

General Chemical LLC *see* General Chemical Performance Products LLC

General Chemical Performance Products LLC
90 East Halsey Road
Parsippanny NJ 07054
Tel: +1 973 515 0900
Fax: +1 973 515 3232
E-mail: info@genchemcorp.com
Web: www.generalchemical.com
Trade names: Rehydragel; Rehydraphos.

Generichem Corp
5 Taft Road
PO Box 457
Totowa NJ 07511-0457
Tel: +1 973 256 9266
Fax: +1 973 256 0069
E-mail: info@generichem.com
Web: www.generichem.com
Trade names: Paselli MD10 PH; Prejel; Primellose; Primojel.

George Uhe Co Inc
219 River Drive
Garfield
New Jersey 07026
Tel: +1 800 850 4075
Fax: +1 201 843 7517
E-mail: global@uhe.com
Web: www.uhe.com

GFS Chemicals Inc
PO Box 245
Powell
OH 43065
Tel: +1 740 881 5501
Fax: +1 740 881 5989
E-mail: service@gfschemicals.com
Web: www.gfschemicals.com

Glanbia Nutritionals Inc
Regional Head Office
1603 Orrington
Suite 1000
Evanston IL 60201
Tel: +1 847 563 4100
Fax: +1 847 733 0216
E-mail: glanbianutritionalsna@glanbia.com
Web: www.glanbianutritionals.com

Glatech Productions LLC
325 Second Street
Lakewood NJ 08701
Tel: +1 (732) 364 8700
Fax: +1 (732) 370 0877
E-mail: info@koshergelatin.com
Web: www.koshergelatin.com
Trade names: Kolatin.

Global Seven Inc
6 Park Drive
Franklin NJ O7416
Tel: +1 (973) 209 7474
Fax: +1 (973) 209 6108
E-mail: sales@global-seven.com
Web: www.global-seven.com
Trade names: Hest TC; Hetoxol.

Grain Processing Corp
1600 Oregon Street
Muscatine IA 52761-1494
Tel: +1 563 264 4265
Fax: +1 563 264 4289
E-mail: sales@grainprocessing.com
Web: www.grainprocessing.com
Trade names: *Maltrin*; Maltrin QD; Maltrin; Pure-Dent; Spress B820.

GR O'Shea Company
650 East Devon Avenue
Suite 180
Itasca IL 60143-3142
Tel: +1 630 773 3223
Fax: +1 630 773 3553
E-mail: general@groshea.com
Web: www.groshea.com
Trade names: Castorwax MP 70; Castorwax MP 80; Castorwax.

Hallstar Company, The
120 South Riverside Plaza
Suite 1620
Chicago IL, 60606
Tel: +1 312 385 4494
E-mail: customerservice@hallstar.com
Web: www.hallstar.com
Trade names: CoSept B; CoSept E; CoSept M; CoSept P; Hallstar 653; HallStar CO-1695; HallStar GMO.

Hawkins Chemical Inc *see* Hawkins Inc

Hawkins Inc
Pharmaceutical Group
3000 East Hennepin Avenue
Minneapolis MN 55413
Tel: +1 612 617 6910
Fax: +1 612 617 0863
E-mail: pharmcs@hawkinsinc.com
Web: www.hawkinsinc.com

HBCChem Inc
2819 Whipple Road
Union City CA 94587
Tel: +1 510 219 6317
Fax: +1 650 486 1361
E-mail: sales@hbcchem-inc.com
Web: www.hbcchem-inc.com

Helm New York Inc
1110 Centennial Avenue
Piscataway NJ 08854-4169
Tel: +1 732 981 1160
Fax: +1 732 981 0965
E-mail: info@helmnewyork.com
Web: www.helmnewyork.com

Hercules Inc *see* Aqualon

Horsehead Corporation
4955 Steubenville Pike Suite 405
Pittsburgh PA 15205
Tel: 00 1 800 648 8897
E-mail: info@horseheadcorp.com
Web: www.horsehead.net

Hummel Croton Inc
10 Harmich Road
South Plainfield NJ 07080
Tel: +1 908 754 1800
Fax: +1 908 754 1815
E-mail: sales@hummelcroton.com
Web: www.hummelcroton.com

Huntsman Tioxide *see* Tioxide Americas Inc
Trade names: Surfonic.

Innophos Inc
259 Prospect Plains Rd
Cranbury NJ 08512
Tel: +1 609 495 2495
Web: www.innophos.com
Trade names: A-TAB; DI-TAB; TRI-TAB.

Innovative Materials Technology (IMT)
(A Division of Emerson Resources, Inc.)
600 Markley Street
Norristown PA 19401
Tel: +1 610 279 7450
Fax: +1 610 292 9722
E-mail: info@emersonresources.com
Web: www.emersonimt.com
Trade names: Marcoat 125.

Inolex Chemical Co
2101 South Swanson Street
Philadelphia PA 19148
Tel: +1 215 271 0800
Fax: +1 215 271 2621
E-mail: cheminfo@inolex.com
Web: www.inolex.com
Trade names: Lexol IPP-NF.

Integra Chemical Co
1216 6th Ave N
Kent
Washington 98032
Tel: +1 253 479 7000
Fax: +1 253 479 7079
E-mail: chemicals@integrachem.com
Web: www.integrachem.com

International Fiber Corporation
50 Bridge Street
North Tonawanda
NY 14120
Tel: +1 716 693 4040
Fax: +1 716 693 3528
E-mail: info@ifcfiber.com
Web: www.ifcfiber.com
Trade names: JustFiber; Solka-Floc.

International Laboratory Ltd
130 Produce Ave Ste F
South San Francisco
CA 94080
Tel: +1 650 278 9963
Fax: +1 650 745 2947
E-mail: admin@intlab.org
Web: www.intlab.org

International Specialty Products Inc (ISP)
1361 Alps Road
Wayne NJ 07470
Tel: +1 973 628 4000
Fax: +1 973 628 3311
E-mail: pharmaceuticalinfo@ispcorp.com
Web: www.ispcorp.com
Trade names: Celex; Ceraphyl 140; Gantrez; Germall 115; Pharmasolve; Plasdone S-630; Plasdone; Polyplasdone XL-10; Polyplasdone XL.

ISP *see* International Specialty Products

Jarchem Industries Inc
414 Wilson Avenue
Newark NJ 07105
Tel: +1 973 344 0600
Fax: +1 973 344 5743
E-mail: info@jarchem.com
Web: www.jarchem.com
Trade names: Jarcol 1-20.

Jeen International Corp
24 Madison Road
Fairfield NJ 07004
Tel: +1 800 771 5336
Fax: +1 973 439 1402
E-mail: info@jeen.com
Web: www.jeen.com
Trade names: Jeechem; Jeecol.

J Rettenmaier USA *see* JRS Pharma LP
Trade names: Vivapress Ca.

JRS Pharma LP
2981 Route 22, Suite 1
Patterson NY 12563-2359
Tel: +1 845 878 3414
Fax: +1 845 878 3484
E-mail: sales@jrspharma.com
Web: www.jrspharma.com
Trade names: Arbocel; Compactrol; Emcocel; Emcompress Anhydrous; Emcompress; Emdex; Explotab; Lubritab; ProSolv; Pruv; Vivapharm HPMC; Vivapress Ca; Vivapur MCG 591 PCG; Vivapur MCG 611 PCG; Vivapur; Vivasol; Vivastar P.

JT Baker Inc *see* Mallinkrodt Baker Inc

Jungbunzlauer Inc
7 Wells Avenue
Newton Centre
MA 02459
Tel: +1 617 969 0900
Fax: +1 617 964 2921
Web: www.jungbunzlauer.com
Trade names: Citrofol AI; Citrofol AII; Citrofol BI.

Kerry Bio-Science
158 State Highway
320 Norwich
NY 13815
Web: www.sheffield-products.com

KIC Chemicals Inc
87 South Ohioville Road
New Paltz NY 12561
Tel: +1 845 883 5306
Fax: +1 845 883 5326
E-mail: info@kicgroup.com
Web: www.kicgroup.com

Koster Keunen LLC
1021 Echo Lake Road
PO Box 69
Watertown
CT 06795-0069
Tel: +1 860 945 3333
Fax: +1 860 945 0330
E-mail: info@kosterkeunen.com
Web: www.kosterkeunen.com
Trade names: Koster Keunen Ceresine; Permulgin D.

Kraft Chemical Co
1975 N Hawthorne Avenue
Melrose Park IL 60160
Tel: +1 800 345 5200
Fax: +1 708 345 4005
E-mail: sales@kraftchemical.com
Web: www.kraftchemical.com

Lambent Technologies
3938 Porett Drive
Gurnee
Illinois 60031
Tel: +1 800 432 7187
Fax: +1 847 249 6792
E-mail: lambent@lambentcorp.com
Web: www.petroferm.com
Trade names: Cirashine CS; Lumulse.

Lanxess Corp
111, RIDC Park West Drive
Pittsburg PA 15275-1112
Tel: +1 800 526 9377
E-mail: info@LANXESS.com
Web: www.lanxess.com

1

Trade names: Bayferrox 105M; Bayferrox 306; Bayferrox 920Z; Solbrol A; Solbrol M; Solbrol P.

Lipo Chemicals Inc
207 19th Avenue
Paterson NJ 07504
Tel: +1 973 345 8600
Fax: +1 973 345 8365
E-mail: salesandmarketing@lipochemicals.com
Web: www.lipochemicals.com
Trade names: Lipo DGS; Lipocol C; Lipocol; Liponate IPP; Lipopeg 2-DEGS; Lipovol CAN; Lipovol CO; Lipovol SES; Uniphen P-23.

Lipscomb Chemical Company
4401 Atlantic Avenue, Suite 14
Long Beach, CA 90807
Tel: +1(562) 728 6321
Fax: +1(562) 728 9170
Web: www.lipscombchemical.com

Lonza Inc
90 Boroline Rd
Allendale NJ 07401
Tel: +1 201 316 9200
Fax: +1 201 785 9973
E-mail: contact.allendale@lonza.com
Web: www.lonza.com

Loos & Dilworth Inc
61 East Green Lane
Bristol PA 19007
Tel: +1 215 785 3591
Fax: +1 215 785 3597
E-mail: dtompkins@loosanddilworth.com
Web: www.loosanddilworth.com
Trade names: Pamolyn.

Lubrizol Advanced Materials Inc.
Headquaters (Noveon Inc)
9911 Brecksville Road
Cleveland OH 44141-3247
Tel: +1 216 447 5000
Fax: +1 216 447 5740
Web: www.pharma.lubrizol.com
Trade names: Carbopol; Noveon AA-1; Pemulen; Schercemol 318 Ester; Schercemol PGML; Schercemol; Vilvanolin CAB.

Lucas Meyer Inc
765 E Pythian Ave
Decatur IL 62526 2412
Tel: +1 217 8753660
Fax: +1 217 8775046
E-mail: lecithin@midwest.net

Luzenac America
(Division of Rio Tinto Minerals)
8051 E. Maplewood Ave
Building 4
Greenwood Village
CO 80111
Tel: +1 303 713 5000
Fax: +1 303 713 5769
Web: www.luzenac.com
Trade names: Imperial.

Lyondell Chemical Co
PO Box 3646
Houston TX 77253 3646
Tel: +1 713 652 7200
Web: www.lyondell.com

Macron Chemicals
A division of Avantor *see* Avantor Performance Materials
Trade names: Parlodion.

Mantrose Bradshaw Zinsser Group *see* Mantrose-Haeuser Co Inc

Mantrose-Haeuser Co Inc
1175 Post Road East
Westport CT 06880
Tel: +1 203 454 1800
Fax: +1 203 227 0558
E-mail: susan.coleman@mantrose.com
Web: www.mantrose.com
Trade names: Crystalac; Mantrocel HP-55; Mantrolac R-49; Mantrolac R-52.

Marroquin Organic International
303 Potrero St. Suite 18
Santa Cruz CA 95060
Tel: 831-423-3442
Fax: 831-423-3432
E-mail: info@marroquin-organics.com
Web: www.marroquin-organics.com
Trade names: Maisita; Stärkina.

Mason Chemical Company
721 West Algonquin Road
Arlington Heights IL 60005
Tel: +1 800 362 1855
Fax: +1 847 290 1625
E-mail: mason@maquat.com
Web: www.maquat.com
Trade names: Masurf Emulsifying Wax NF.

McNeil Nutritionals LLC
601 Office Center Drive
Fort Washington
PA 19034
Web: www.splenda.com
Trade names: Splenda.

Mendell *see* Penwest Pharmaceuticals Co

Merisant
33 North Dearborn street, Suite 200
Chicago IL 60602
Tel: +1 312 840 6000
Fax: +1 312 840 5400
Web: www.merisant.com
Trade names: Canderel; Equal.

Mitsubishi International Food Ingredients
Corporate, Sales & Marketing
5080 Tuttle Crossing Blvd
Suite 400
Dublin OH 43016
Tel: 800-628-3092
Fax: 614-798-8339
Web: www.mifiusa.com

M Michel and Company Inc
PO Box 788
Planetarium Station
New York NY 10024 0545
Tel: +1 212 344 3878
Fax: +1 212 344 3880
E-mail: corblok@aol.com
Web: www.mmichel.com
Trade names: Cachalot.

Morflex Inc *see* Vertellus Specialties Inc

Mutchler Inc
20 Elm Street
Harrington Park
NJ 07640
Tel: +1 201 768 1100
Fax: +1 201 768 9960
E-mail: info@mutchlerchem.com
Web: www.mutchlerchem.com

Nabi Biopharmaceuticals
Corporate Headquarters and Research and Development Facility
12270 Wilkins Avenue
Rockville MD 20852
Tel: +1 301 770 3099
Fax: +1 301 770 3097
Web: www.nabi.com

Napp Technologies LLC
401 Hackensack Ave
Hackensack NJ 0760
Tel: +1 201 843 4664

Fax: +1 201 843 4737
Web: www.napptech.com

National Starch & Chemical Co. *see* National Starch Personal Care (USA)

National Starch Personal Care (USA)
(Division of Akzo Nobel)
742 Grayson Street
Berkeley CA 94710 2677
Tel: +1 510 548 6722
Fax: +1 510 841 3150
Web: www.personalcarepolymers.com
Trade names: Hylon; Purity 21; Uni-Pure.

Naturex Inc
375 Huyler Street
South Hackensack
NJ 07606
Tel: +1 201 440 5000
Fax: +1 201 342 8000
E-mail: naturex.us@naturex.com
Web: www.naturex.com
Trade names: Talin.

Nipa Inc *see* Clariant Corp

Nippon Soda Co Ltd (USA)
Nisso America Inc
45 Broadway, Suite 2120
New York NY 10006
Tel: +1 212 490 0350
E-mail: info@nissoamerica.com
Web: www.nissoamerica.com
Trade names: Nisso HPC.

Nitta Gelatin NA Ltd
201 W, Passaic St.
Rochelle Park NJ 07662
Tel: +1-201-368-0071
Fax: +1-201-368-0282
Web: www.nitta-gelatin.com

NutraSweet Company, The
1762 Lovers Lane
PO Box 2387
Augusta GA 30903
Web: www.nutrasweet.com

Nutrinova Inc
(Division of Celanese)
1601 West LBJ Freeway
Dallas TX 75234-6034
Tel: +1 972 443 2055
Fax: +1 972 443 4950
E-mail: steven.kaufman@celanese.com
Trade names: Sunett.

O'Shea Company *see* GR O'Shea Company

P & G Chemicals
P & G Chemicals Americas
Sharon Woods Technical Centre
11530 Reed Hartman Highway
Cincinnati OH 45241
Tel: +1 800 477 8899
Fax: +1 513 626 3145
E-mail: chemicalsinfo.im@pg.com
Web: www.pgchemicals.com

Paddock Laboratories Inc
3940 Quebec Avenue North
Minneapolis MN 55427
Tel: +1 763 546 4676
Fax: +1 763 546 4842
E-mail: info@paddocklabs.com
Web: www.paddocklabs.com

Parchem
415 Huguenot street
New Rochelle NY 10801
Tel: +1 914 654 6800
E-mail: info@parchem.com
Web: www.parchem.com

Parchem Trading Ltd *see* Parchem

Particle Dynamics Inc
KV Pharmaceutical Co
2629 Hanley Road
St Louis MO 63144
Tel: +1 314 968 2376, +1 800 452 4682
Fax: +1 314 781 3354
E-mail: info@particledynamics.com
Web: www.particledynamics.com
Trade names: Calcipress; Descote; Destab; Destab; Destab; Destab.

PB Leiner
PO Box 645
Plainview NY 11803
Tel: +1 (563) 468 4567
Fax: +1 (563) 386 5755

Penta Manufacturing Co
50 Okner Parkway
Livingston NJ 07039
Tel: +1 973 740 2300
Fax: +1 973 740 1839
Web: www.pentamfg.com

Penwest Pharmaceuticals Co *see* JRS Pharma LP

Pfaltz & Bauer Inc
172 E. Aurora St
Waterbury CT 06708
Tel: +1 203 574 0075
Fax: +1 203 574 3181
E-mail: sales@pfaltzandbauer.com
Web: www.pfaltzandbauer.com
Trade names: Garantose.

Pfanstiehl Laboratories Inc *see* Ferro Pfanstiehl Laboratories Inc

Pfizer Inc
235 East 42nd Street
New York NY 10017
Tel: +1 212 573 2323
Fax: +1 212 573 7851
E-mail: info@pfizer.com
Web: www.pfizer.com

PMC Specialities Group Inc
501 Murray Road
Cincinnati OH 45217
Tel: +1 800 543 2466
Fax: +1 513 482 7373
Web: www.pmcsg.com
Trade names: Syncal.

Pokonobe Industries Inc
PO Box 1756
Santa Monica CA 90406
Tel: +1 310 392 1259
Fax: +1 310 392 3659
E-mail: info@pokonobe.com
Web: www.pokonobe.com

Polysciences Inc
400 Valley Road
Warrington PA 18976
Tel: +1 800 523 2575
Fax: +1 800 343 3291
E-mail: info@polysciences.com
Web: www.polysciences.com

Premium Ingredients Ltd *see* Prinova

Primera Foods
612 South 8th Street
P.O. Box 373
Cameron WI 54822
Tel: +1 715 458 4075
Web: www.primerafoods.com
Trade names: Malta*Gran; Rice*Trin; Tapi.

Prinova
285 East Fullerton Avenue
Carol Stream IL 60188
Tel: +1 630 868 0300
Fax: +1 630 868 0310
E-mail: info@prinovaUSA.com
Web: www.prinovagroup.com

Protameen Chemicals Inc
375 Minnisink Road
Totowa NJ 07511
Tel: +1 973 256 4374
Fax: +1 973 256 6764
E-mail: info@protameen.com
Web: www.protameen.com
Trade names: Procol; Protachem ; Protachem GMS-450; Protachem IPP; Protachem; Protalan anhydrous; Protalan M-16; Protalan M-26.

Purac
111 Barclay Boulevard
Lincolnshire Corporate Center
Lincolnshire IL 60069
Tel: +1 847 634 6330
Fax: +1 847 634 1992
E-mail: pam@purac.com
Web: www.purac.com
Trade names: Purac PF 90; Puracal.

Reade Advanced Materials Inc
Post Office Drawer 15039
850 Waterman Avenue
East Providence, Rhode Island
RI 02915-0039
Tel: +1 401 433 7000
Fax: + 401 433 7001
E-mail: sales@reade.com
Web: www.reade.com

Reheis Inc *see* General Chemical LLC

Rettenmaier *see* JRS Pharma LP

RFI Ingredients
300 Corporate Drive, Suite 14
Blauvelt NY 10913
Tel: +1 845 358 8600
Fax: +1 845 358 9003
E-mail: rfi@rfiingredients.com
Web: www.rfiingredients.com
Trade names: Talin.

Rhodia Inc *see* Rhodia Pharma Solutions Inc
Trade names: Rhodopol.

Rhodia Pharma Solutions Inc
259 Prospect Plains Road
CN 7500
Cranbury NJ 08512 7500
Tel: +1 609 860 3891
Fax: +1 609 860 1841
Web: www.rhodia.com
Trade names: Rhodiarome; Rhodopol; Rhovanil.

RIA International
11 Melanie Ln #17
East Hanover NJ 07936
Tel: +1 973 581 1282
Fax: +1 973 581 1283
E-mail: ria@riausa.com
Web: www.riausa.com

Rita Corp
PO Box 457
850 South Rt. 31
Crystal Lake
IL 60014-0457
Tel: +1 815 337 2500
Fax: +1 815 337 2522
E-mail: info@ritacorp.com
Web: www.ritacorp.com
Trade names: Acritamer; Forlan 500; Rita CA; RITA HA C-1-C; Rita IPM; Rita SA; Ritaceti; Ritachol 2000; Ritalac NAL; Ritawax; Ritoleth; Ritox; Tealan.

Rockwood Pigments NA, Inc.
7101 Muirkirk Road
Beltsville MD 20705-1333
Tel: +1 301 210 3400
Fax: +1 301 210 4967
E-mail: info.us@rpigments.com
Web: www.rockwoodpigments.com
Trade names: Ferroxide 212P; Ferroxide 226P; Ferroxide 505P; Ferroxide 510P; Ferroxide 78P; Ferroxide 88P; Sicovit B80; Sicovit B84; Sicovit B85; Sicovit R30; Sicovit Y10.

Rohm America Inc *see* Evonik Degussa Corp

Rohm and Haas Co
100 Independence Mall West
Philadelphia PA 19106
Tel: +1 215 592 3000
Fax: +1 219 592 3377
Web: www.rohmhaas.com
Trade names: Amberlite IRP-88.

Roquette America Inc
1417 Exchange Street
PO Box 6647
Keokuk IA 52632-6647
Tel: +1 319 524 5757
Fax: +1 319 526 2345
Web: www.roquette.com
Trade names: Flolys; Fluidamid R444P; Glucidex; Keoflo ADP; Kleptose HPB; Kleptose; Lycadex PF; Lycasin 75/75; Lycasin 80/55; Lycasin 85/55; Lycasin HBC; Lycatab C; Lycatab DSH; Lycatab PGS; Neosorb; Pearlitol; Roclys; SweetPearl; Xylisorb.

Rousselot Inc
1231 South Rochester Street
Mukwonago
Wisconsin 53149
Tel: +1 (262) 363-6051
Fax: +1 (262) 363-2789

RT Vanderbilt Company Inc
30 Winfield Street
Norwalk CT 06856-5150
Tel: +1 203 853 1400
Fax: +1 203 853 1452
Web: www.rtvanderbilt.com
Trade names: Vanzan NF; Veegum.

Ruger Chemical Co Inc
1515 West Blancke Street
Linden NJ 07036
Tel: +1 973 926 0331
Fax: +1 973 926 4921
Web: www.rugerchemical.com
Trade names: Patlac; Purasal.

Sanofi-Aventis
55 Corporate Drive
Bridgewater NJ 08807
Tel: +1 908 981 5000
Web: en.sanofi-aventis.com
Trade names: Zephiran.

Sasol North America Inc
900 Threadneedle, Suite 100
Houston TX 77079-2990
Tel: +1 281 588 3000
Fax: +1 281 588 3144
E-mail: info@us.sasol.com
Web: www.sasoltechdata.com
Trade names: Imwitor 191; Imwitor 412; Imwitor 491; Imwitor 900K; Imwitor 948; Lipoxol; Miglyol 808; Miglyol 810; Miglyol 812; Nacol 14-95; Nacol 14-98; Nacol 16-95; Nacol 18-98; Nacol 18-98P; Witepsol.

Scandinavian Formulas Inc
140 East Church St
Sellersville PA 18960

Tel: +1 215 453 2507
Fax: +1 215 257 9781
E-mail: info@scandinavianformulas.com
Web: www.scandinavianformulas.com

Seidler Chemical Company Inc
537 Raymond Blvd
Newark NJ 07105
Tel: +1 973 465 1122
Fax: +1 973 465 4469
E-mail: sales@sedielerchem.com
Web: www.seidlerchem.com

Seltzer Chemicals Inc *see* Glanbia Nutritionals Inc

Sensient Technologies
107 Wade Avenue
South Plainfield
NJ 07080 1311
Tel: +1 908 757 4500
Fax: +1 908 754 3222
E-mail: sensient-pharma@sensient-tech.com

Sensus America Inc
Princeton Coporate Plaza
1 Deer Park Drive, Suite J
Manmouth Junction NJ 08852
Tel: +1 646 452 6140
Fax: +1 646 452 6150
E-mail: Carol.Brown@Sensus.us
Web: www.sensus.us
Trade names: Frutafit.

Seppic Inc
(Subsidiary of Air Liquide Corp)
30 Two Bridges Road, Suite 210
Fairfield NJ 07004-1530
Tel: +1 973 882 5597
Fax: +1 973 882 5178
Web: www.seppic.com
Trade names: L-HPC; Sepifilm SN.

Sheffield Bio-Science
158 State Highway 320
Norwich NY 13815
Tel: +1 800 833 8308
Fax: +1 607 334 5022
Web: www.sheffield-products.com
Trade names: Anhydrous 120M; Anhydrous 60M; Anhydrous DT High Velocity; Anhydrous DT; Anhydrous Impalpable; Foremost Lactose 315; Foremost Lactose 316 Fast Flo; Monohydrate; Respitose.

Sheffield Pharma Ingredients *see* Sheffield Bio-Science

Spectrum Chemicals & Laboratory Products Inc
14422 South San Pedro Street
Gardena CA 90248 2027
Tel: +1 800 813 1514
Fax: +1 800 525 2299
E-mail: sales@spectrumchemical.com
Web: www.spectrumchemical.com

Spherix Incorporated
6430 Rockledge Drive #503
Bethesda MD 20817
Tel: +1 301 897 2540
Fax: +1 301 897 2567
E-mail: info@spherix.com
Web: www.spherix.com
Trade names: Naturlose.

SPI Pharma Inc
Rockwood Office Park
Suite 210
503 Carr Road
Wilmington DE 19809
Tel: +1 302 576 8600
Fax: +1 302 576 8567
Web: www.spipharma.com
Trade names: Advantose 100; Advantose FS 95; Advantose FS 95; Compressol S; Compressol SM; Effer-Soda; Lubripharm; Mannogem; Sorbitab SD 250; Sorbitab; Sorbitol Special.

Stadt Holdings Corporation
60 Flushing Avenue
Brooklyn NY 11205
Tel: +1 800 544 8610
Web: www.sweetone.com
Trade names: NatraTaste; Sweet One.

Staley Mfg Co *see* Tate & Lyle

Stepan Co
22 West Frontage Road
Northfield IL 60093
Tel: +1 847 446 7500
Fax: +1 847 501 2100
Web: www.stepan.com
Trade names: Neobee M5; Stepan 653; Wecobee.

Strahl & Pitsch Inc
230 Great East Neck Road
West Babylon NY 11704
Tel: +1 631 587 9000
Fax: +1 631 587 9120
E-mail: info@strahlpitsch.com
Web: www.spwax.com

SurModics Pharmaceuticals *see* Evonik Degussa Corp
Part of Evonik Degussa Corp
750 Lakeshore Parkway
Birmingham AL 35211
Tel: +1 205 917 2290
Fax: +1 205 917 2291
Web: www.surmodicsbiomaterials.com/

Takeda Pharmaceuticals North America Inc
One Takeda Parkway
Dearfield IL 60015
Tel: +1 224 554 6500
Web: www.tap.com

Tate & Lyle (North America)
Cereal Sweeteners
2200 E Eldorado Street
Decatur IL 62526
Tel: +1 800 526 5728
Fax: +1 217 421 3167
Web: www.tateandlyle.com
Trade names: Krystar; Maldex G; Maltosweet; STA-Lite; Star-Dri.

Thomas Scientific
PO Box 99
Swedesboro NJ 08085
Tel: +1 856 467 2000
Fax: +1 856 467 3087
E-mail: value@thomassci.com
Web: www.thomassci.com

Thornley Company
1 Innovation Way
Newark, Delaware
Tel: +1 302 224 8300
Fax: +1 302 224 8308
Web: www.thornleycompany.com

TIC Gums
10552 Philadelphia Rd
White Marsh MD 21162
Tel: +1 410 273 7300
Fax: +1 410 273 6469
E-mail: info@ticgums.com
Web: www.ticgums.com

Tioxide Americas Inc
(Huntsman Tioxide)
Huntsman Corporate Office
10003, Woodlock Forest Drive
The Woodlands TX 77380
Tel: +1 281 719 6000
Fax: +1 281 719 6054

Web: www.huntsman.com
Trade names: Tioxide.

Triple Crown America
13 North Seventh Street
Box 667
Perkasie PA 18944
Tel: +1 215 453 2500
Fax: +1 215 453 2508
E-mail: info@eurohealth.us
Web: www.triplecrownamerica.com

Universal Preserv-A-Chem Inc
33, Truman Drive South
Edison NJ 08817-2426
Tel: +1 732 777 7338
Fax: +1 732 777 7885

USG Corporation
550 West Adams Street
Chicago IL 60661-3676
Tel: + 1 312 436 4000
Web: www.usg.com
Web: www.gypsumsolutions.com
Trade names: Snow White; USG Terra Alba.

US Zinc
6020 Navigation Boulevard
Houston TX 77011-1132
Tel: 00 1 713 926 1705
Fax: 00 1 713 923 1783
Web: www.uszinc.com
Trade names: AZO.

Vanderbilt Company Inc *see* RT Vanderbilt Company Inc

Van Waters and Rogers Inc *see* Vopak USA Inc

Vertellus Specialties Inc
201 North Illinois Street
Suite 1800
Indianapolis IN 46204
Tel: +1 317 247 8141
Fax: +1 317 248 6472
Web: www.vertellus.com
Trade names: Castorwax; Citroflex 2; Citroflex 4; Citroflex A-2; Citroflex A-4; Crystal; Morflex DBS.

Virginia Dare
882 Third Avenue
Brooklyn NY 11232
Tel: +1 718 788 1776
Fax: +1 718 768 3978
E-mail: flavorinfo@virginiadare.com
Web: www.virginiadare.com

Voigt Global Distribution Inc
PO Box 1130
Lawrence KS 66044
Tel: +1 877 484 3552
Fax: +1 877 484 3554
E-mail: sales@VGDINC.com
Web: www.voigtglobal.com

Wacker Biochem Corp *see* Wacker Chemical Corp

Wacker Chemical Corp
1 Wacker Drive
Eddyville IA 52553
Tel: +1 515 969 4817
Fax: +1 515 969 4929
E-mail: info.usa@wacker.com
Web: www.wacker.com
Trade names: Cavamax W6 Pharma; Cavamax W7 Pharma; Cavamax W8 Pharma; Cavasol W7; HDK.

Welch, Holme & Clark Co Inc
7 Avenue L
Newark NJ 07105
Tel: +1 973 465 1200
Fax: +1 973 465 7332
Web: www.welch-holme-clark.com

Whittaker Clark, and Daniels Inc
1000 Coolidge St
S. Plainfield NJ 07080
Tel: +1 908 561 6100
Fax: +1 800 833 8139
E-mail: customerservice@wcdinc.com
Trade names: Albagel.

Witco Corp *see* Crompton Corp

W Zinsser & Company Inc
173 Belmont Drive
Somerset NJ 08875
Tel: +1 732 652 2000
Fax: +1 732 652 2491
Web: www.zinsser.com

Zhong Ya Chemical (USA) Ltd
140W Ethel Road, Unit V
Piscataway NJ 08854
Tel: +1 732 248 1008
Fax: +1 732 248 7676
E-mail: sales@zhongyachemical.com
Web: www.zhongyachemical.com

ZLB Behring *see* CSL Behring

SUPPLIERS LIST: OTHERS

Aastrid International
247-248 Udyog Bhavan
Sonawala Lane
Goregaon East
Mumbai 400 063
India
Tel: +91 22 26858570
Fax: +91 22 26859570
Web: www.aastrid.com

Advanced Technology & Industrial Co Ltd
Unit B, 1/F, Cheong Shing Building
17 Walnut Street
Tai Kok Tsui
Kln
Hong Kong
Tel: (852) 2390 2293
Fax: (852) 2394 5546
E-mail: cas@advtechind.com
Web: www.advtechind.com

Ajinomoto Co Inc
15-1, Kyobashi
1-chome, Chuo-ku
Tokyo 104-8315
Japan
Tel: +81 (3)5250 8111
E-mail: g-webmaster@ajinimoto.com
Web: www.ajinomoto.com
Trade names: Pal Sweet Diet; Pal Sweet.

Alembic Chemical Works Co Ltd
Prime Corporate Park, 2nd Floor, Behind ITC
Grand Maratha Sheraton
Sahar Road
Andheri East
Mumbai 400099
India
Tel: +91 22 30611698
Fax: +91 22 30611682
E-mail: info@alembic.co.in
Web: www.alembic-india.com

ANGUS Chemical (Singapore) Pte Ltd
438B, Alexandra Road
#08-12, Alexandra Technopark
119968 Singapore
Tel: +65 62751849
Fax: +65 63754005

Anzchem Pty Ltd
Mills Waterfront Estate

19/52 Holker Street
Silverwater NSW 2128
Australia
Tel: +61 2 9475 2200
Fax: +61 2 9475 2211
E-mail: info@anzchem.com.au
Web: www.anzchem.com.au
Trade names: L-HPC.

Arihant Trading Co
Head Office
119, Vasan Udyog Bhavan
1st Floor, Off Senapati Bapat Marg
Opp. Phoenix Mill, Lower Parel (West)
Mumbai 400013
India
Tel: +91 22 2495 4895
Fax: +91 22 2495 4894
E-mail: sales@arihantt.com
Web: www.arihantt.com
Trade names: Aqoat; Metolose; Pharmacoat.

Asahi Kasei Corporation
Jinbocho-Mitsui Building
1-105 Kanda Jinbocho
Chiyoda-ku
Tokyo 101-8101
Japan
Tel: +81 3 3296 3000
Fax: +81 3 3296 3161
Web: www.asahi-kasei.co.jp
Trade names: Celphere; Ceolus KG; Kiccolate.

Ashland India Pvt Ltd
Administrative Office
Office #601, 606, 607, 608
Plot No 17-18, Platinum Techno Park
Sector 30-A, Vashi
Navi Mumbai 400 705
India
Tel: +91 22 6148 9689
Fax: +91 22 6148 9616
E-mail: cjagtap@ashland.com
Web: www.ashland.com
Trade names: Aqualon; Blanose; Klucel; Natrosol.

Atul Ltd
310B, Veer Savarkar Marg (Cadell Road)
Adjacent to Prabhadevi Telephone Exchange
Opp. India United Mills (Dye Works)
Prabhadevi, Dadar (West)
Mumbai 400028
India
Tel: (91-22) 39876000
Fax: (91-22) 24376061 / 24386065
E-mail: mum@atul.co.in
Web: www.atul.co.in

BASF India Pvt Ltd
Corporate Office
1st Floor
Vibgyor Towers
Plot No. C-62
Bandra-Kurla Complex
Mumbai 400051
India
Tel: +91 22 6661 8000
Fax: +91 22 6758 2753
Web: www.india.basf.com

BASF Japan Ltd
Nanbu Bldg
3-3, Kioi-cho
102-8570 Chiyoda-ku
Tokyo
Japan
Tel: +81 33238 2500
Fax: +81 33238 2514
Trade names: Cremophor A; Kollicoat MAE; Kollidon CL-M; Kollidon CL; Kollidon VA 64; Kollidon; Lutrol E; Luviskol VA; Soluphor P.

Bio Chemicals Ltd
Biochem Division
Sanpeng Bridge
Baiguan
Shangyu 312351
China
Tel: +86 575 2210376
Fax: +86 575 2129555
Web: www.biochemicals.cn

Blanver *see* Blanver Farmoquímica Ltda
Trade names: Explosol; Microcel; Solutab; Tabulose.

Blanver Farmoquímica Ltda
Colorcon Do Brasil
Rua Ely, 76
06708-180 Cotia
Sao Paulo
Brazil
Tel: +55 11 4612 4262
Fax: +55 11 4612 3307
E-mail: blanver@blanver.com
Web: www.blanver.com.br
Trade names: Microcel 3E-150.

Boith Limited (China)
19/F, HuaMin Bldg.
No. 9 HuBin East Rd.
XiaMen
P.R.
China
Tel: +86 592 581 7016
Fax: +86 592 581 4960
E-mail: leeyuan@boith.com
Web: www.boith.com

Cerestar Cargill Resources Maize Industry Co Ltd
Jianguan Industry Park
Economy and Technology Development
138006 Songyuan Jilin Province
China
Tel: +86 (0)438 2181 101
Fax: +86 (0)438 2187 215
Web: www.cargillchina.com
Trade names: Zerose.

Charles Tennant & Co (Canada) Ltd
34 Clayson Road
Toronto ON M9M 2G8
Canada
Tel: +1 416 741 9264
Fax: +1 416 741 6642
Web: www.ctc.ca
Trade names: Jeecol ODD.

Chika Pvt Ltd
Industrial Assurance Building
5th Floor
Opposite Churchgate Station
Mumbai 400020
India
Tel: +91 22 22821382/22821386
Fax: +91 22 22830700/22882882
E-mail: chem@chika.co.in
Web: www.chika.co.in

Choice Korea Co
207 Shin Song Plaza 1423-2
Kwanyang-1 Dong
Donan-Ku
Anyang City
Kyunggi-do
South Korea
Tel: +82 314 240 212
Fax: +82 314 240 213
E-mail: choice4@kornet.net
Web: www.choicekorea.co.kr

Cognis Speciality Chemicals
B-103 Universal Business Park
Chanivali mam Road
off Saki Vihar Road
Andheri (East)
Mumbai 400072 Maharashtra
India
Tel: (22) 4033 7979/40026635
Web: www.cognis.com
Trade names: Chitopharm; Delios VF.

Colorcon Asia Pvt Ltd
Plot Nos. M14-M18
Verna Industrial Estate
Verna
Goa 403722
India
Tel: +91 832 288 3434
Fax: +91 832 288 3440
Web: www.colorcon.com
Trade names: Suglets.

Corel Pharma Chem
Corel House
Nr. Proton Business School
Sarkhej-Gandhinagar Highway
Gota, Ahmedabad - 382 481
Gujarat
India
Tel: + 91 79 27665556
Fax: + 91 - 79 27661754
E-mail: corel@corelpharmachem.com
Web: www.corelpharmachem.com
Trade names: Acrypol.

Cosmos Chemical Co Ltd
809-810 Longyin Plaza
217 North Zhongshan Road
Nanjing 210009
China
Tel: +86 25 3346885
Fax: +86 25 3346877
E-mail: cosmos@cosmoschem.com
Web: www.cosmoschem.com

Croda Chemicals (India) Pvt Ltd
Plot No 1/1 Part
TTC Industrial Area
Thane Belapur Road
Koparkhairne
Navi Mumbai 400710
India
Tel: +91 22 3094 8400/500/600
Fax: +91 22 2778 0007/8

Croda Japan
4-3 Hitotsubashi 2-chome
Chiyoda-ku
Tokyo
Japan
Tel: +81(0)3-3263-8270?
Fax: +81(0)3-3263-8277
E-mail: info-tec@croda.com
Web: www.croda.jp

Crystal Caschem India Ltd
Village: Boidra
Nr. Ashok Organic, Hansot Road
Ankleshwar 390 001
Gujarat
India
Tel: 0091 93281 70263
E-mail: info@crystalcaschem.com
Web: www.crystalcaschem.com
Trade names: Castorwax.

CSPC Pharma
276 Zhongshan West Road
Shijiazhuang
China
Tel: +86 311 7037015
Fax: +86 311 7039608
E-mail: zhangiv@mail.ecspc.com
Web: www.e-cspc.com

Dow Chemical International Pvt Ltd
Corporate Office
'G' Block, C/62, 2nd Floor
Vibgyor Towers
Bandra Kurla Complex
Bandar (E)
Mumbai 400051
India
Tel: +91 22 678 32 300
Fax: +91 22 265 24 217
E-mail: infoindia@dow.com
Web: www.dow.com
Trade names: Carbowax; Ethocel.

Ecogreen Oleochemicals (S) Pte. Ltd.
Headquaters
99 Bukit Timah Road
#03-01 Alfa Centre
229835 Singapore
Tel: +65 6337 7726
Fax: +65 6337 2110
E-mail: info@ecogreenoleo.com
Web: www.ecogreenoleo.com
Trade names: Rofetan GTC.

EPS Impex Co
PO Box 21904
Damai Plaza Luyang
88777 Kota Kinabalu
Malaysia
Tel: +60 88 316470
Fax: +60 88 316741
E-mail: patwary@streamyx.com
Web: www.epsimpex.com

Evonik Degussa India Pvt Ltd
12 A/A (13th Floor)
Bakhtawar
229 Nariman Point
Mumbai 400021
India
Tel: +91 22 67238866
Fax: +91 22 67238811
E-mail: gaurang.dave@evonik.com
Web: www.degussa-hpp.com

Excelacs Co. Ltd
29 Suanson 8
Ramkamhaeng 60 Rd
Huamark
Bangkapi
Bangkok 10240
Thailand
Tel: +66 (2) 3745 023
Fax: +66 (2) 3741 833
Web: www.shellacthailand.com
Trade names: Excelacs 3-Circles; Excelacs 3-Stars.

Fine Chemicals Corporation (Pty) Ltd
15 Hawkins Avenue
Epping Industria
Epping
Cape Town
South Africa
Tel: +27 21 531 6421
Web: www.fcc.co.za

Fuji Chemical Industry Co Ltd
55 Yokohoonji
Kamiichi-machi
Nakanikawa-gun
Toyama-Pref
Japan
Tel: +81 764 72 2323
Fax: +81 764 72 2330

Web: www.fujichemical.co.jp
Trade names: Fujicalin; Neusilin.

Gadot Petrochemical Industries Ltd
16 Habonim Street
P.O.B 8757
Netanya South Industrial Zone
42504 Israel
Israel
Tel: +972 9 892 9530
Fax: +972 9 865 3385
E-mail: gsales@gadot.com
Web: www.gadot.com

Gangwal Chemicals Pvt Ltd
306, Business Classic
Chincholi Bunder Road
Malad (West)
Mumbai 400064
India
Tel: +91 22 2888 9000
Fax: +91 22 883 5347
E-mail: sales@gangwalchem.com
Web: www.gangwalchem.com

Gattefossé Canada Inc.
170 Attwell Drive
Suite 580
Toronto ON M9W 5Z5
Canada
Tel: +1 416 243 5019
Fax: +1 416 243 8628
E-mail: service@gattefosse.ca
Web: www.gattefossecanada.ca
Trade names: Geleol.

Gattefosse India Pvt Ltd
Bombay College of Pharmacy
Ground Floor
C.S.T. Road, Kalina
Santacruz (East)
Mumbai 400098
India
Tel: +91 22 26656031 / 32 / 33
Web: www.gattefosse.com

Gelco - Gelatinas de Colombia SA
Carrera 42 No. 2B-100
Barranquilla
Colombia
Tel: +57 5 344 6444
Fax: +57 5 344 6261
Web: www.gelco-s-a.com

Gelita Asia/Pacific/Africa
Level 2, 2A Lord Street
Botany NSW 2019
Australia
Tel: +61 2 9578 7000
Fax: +61 2 9578 7050
Web: www.gelita.com

Gelita South America
Rua Phillip Leiner
200 - Bairro Rio Cotia
Cotia Sao Paulo
CEP 06714-285
Brazil
Tel: (011) 4612 8111
Fax: (011) 6845 2280

Gelnex Indústria e Comércio Ltda
Rod. SC 283 Km 25 Linho Rio Engano
CEP 897600-000 - Itá- SC
Caixa postal
400 - CEP 89700-000 Concórdia - SC
Brazil
Tel: (49) 3458 3500
Fax: (49) 3441 3501
Web: www.gelnex.com.br

Geltech Co Ltd
1520, Songjung-Dong
Kandsu-Gu
Busan
Korea (South)
Tel: +82 2-2217-4761
Fax: +82 2-2217-4762
Web: www.geltech.co.kr

Gifu Shellac Co Ltd
1-41 Higashiuzura
Gifu 500-8681
Japan
Tel: +81 58 272 0831
Fax: +81 58 272 0704
Trade names: Gifu Shellac GBN-PH; Gifu Shellac Pearl-811.

Glide Chem Pvt Ltd
Corporate Office
S-39 Rajouri Garden
New Delhi 110027
India
Tel: +91 11 414 4353 1/2
Fax: +91 11 2 511 1752/591 1962
E-mail: glidechem@gmail.com
Web: www.glidechem.com

HalloChem Pharma Co. Ltd
17F, Venus Science Incubate Center
No. 60 Xingguang Road
New North Zone
Chongqing 401121
China
Tel: +86 23 67030808
Fax: +86 23 67030809
E-mail: info@hallochem.com
Web: www.hallochem.com

Hayashibara Co Ltd
1-2-3 Shimoishii
Kitaku
Okayama 700-0907
Japan
Tel: +81 86 224 4311
Fax: +81 86 222 8942
Web: www.hayashibara.co.jp
Trade names: Maltose HH; Maltose PH; Treha.

Henley Chemicals
Canada

Highland International
India

Himedia
23, Vadhani Industrial Estate
LBS Marg
Mumbai 400086
India
Tel: +91 022 2500 1607 / 6116 9797
Fax: +91 022 2500 5764 / 2468 / 2286
E-mail: info@himedialabs.com

IEQSA
Av. Elmer Faucett No. 1920
Callao - Peru
PO BOX 4499 LIMA 100
Peru
Tel: (51) (1) 614 4300
Fax: (51) (1) 572 0118
E-mail: export@ieqsa.com.pe
Web: www.ieqsa.com.pe

Indchem International
3/20, 3rd Floor
Tardeo Air Conditioned Market 5
Tardeo
Mumbai 400034
India
Tel: +91 22 2351 0937/2351/0971
Fax: +91 22 2352 0939

JHD Lab
China
Tel: +86 20 89895988
Fax: +86 20 84343583
E-mail: jhdlab@jinhuada.com
Web: www.jhdlab.com
Trade names: Prodhygine.

Jiangxi Mosashino Bio-Chem Co Ltd
Xiaolan Industry Park of Nanchang
Jiangxi 330200
China
Tel: +86 791 576 1066
Fax: +86 791 576 1063
E-mail: admini@china-musashino.com
Web: www.china-musashino.com

Jiangxi Mosashino Co Ltd *see* Jiangxi Mosashino Bio-Chem Co Ltd

Kao Corporation
14-10, Nihonbashi Kayabacho 1-chome
Chuo-ku
Tokyo 103-8210
Japan
Web: chemical.kao.com
Trade names: Cetanol; Coconad; Emulgen; Exceparl IPM; Exceparl IPP; Lunac P-95; Rheodol MS-165V.

Kawarlal Excipients Pvt Ltd
27 Raghunayakula Street
Park Town
Chennai 600003
India
Tel: +91 044 2535 2767 / 5801 / 7534
Fax: +91 44 2534 0234
E-mail: info@kawarlal.com
Web: www.kawarlal.com
Trade names: Coatcel; Cyclocel; Deepcoat; Dynacel; Eratab; Fetocel; Kollicoat; Kollidon VA 64; Lubripharm; Ludipress LCE; Mannogem; NutraSweet; Primellose; Primojel; Rutocel A 55 RT; Rutocel; Speziol G; Speziol GTA; Speziol L2SM GF; Unisweet.

Kibun Food Chemifa Co Ltd
Sumitomo Irifune Building, Fifth Floor
2-1-1 Irifune, Chuo-ku
Tokyo 104-8553
Japan
Tel: +81 3 3206 0778
Fax: +81 3 3206 0788
Web: www.kibunfc.co.jp

Kimica Corporation
2-4-1 Yaesu
Chuo-ku
Tokyo 104-0028
Japan
Tel: 81 3 3548 1941
Fax: 81 3 3548 1942
E-mail: tokyo-office@kimica.jp
Web: www.kimica-alginate.com
Trade names: Kimiloid.

Lactose New Zealand
New Zealand

LS Raw Materials Ltd
Harav Kook
30/3 Petach Tikvah
49315Israel
Tel: 972 3 922 3966
Fax: 972 3 921 2647
E-mail: info@ls-rawmaterials.com
Web: www.ls-rawmaterials.com

Lubrizol Advanced Materials, India Pvt Ltd
5th Floor, Omega
Hiranandani Business Park
Powai
Mumbai-400076
India
Tel: +91 22 6602 7800 / 01
Fax: +91 22 6602 7888
Web: www.lubrizol.com

Lucid Colloids Ltd
401A Navbharat Estates
Zakaria Bunder Rd
Sweri West
Mumbai 400 015
India
Tel: +91 22 4158059
Fax: +91 22 4158074 / 75
E-mail: admin@lucidgroup.com
Web: www.lucidgroup.com

Matrix Laboratories Ltd
Plot No 564/A/22
Road No 92
Jubilee Hills
Hyderabad 500033
India
Tel: +91 40 30866666 / +91 40 23550543
Fax: +91 40 30866699
E-mail: matrix@matrixlabsindia.com
Web: www.matrixlabsindia.com

Mitsubishi-Kagaku Foods Corporation
2-11-1 Shiba-Koen
Minato-Ku
Tokyo 105-0011
Japan
Trade names: Ryoto; Surfhope SE Cosme; Surfhope SE Pharma; Surfhope SE.

NB Entrepreneurs
Uppalwadi
Near Industrial Estate
Kamptee Road
Nagpur 440 026
India
Tel: +91 (712) 2640062/63
Fax: +91 (712) 2641163
E-mail: nbent@dataone.in
Web: www.nbent.com
Trade names: Sancel-W.

Nihon-Emulsion Co Ltd
Japan
Tel: 81-3-3314-3211
Fax: 81-3-3312-7207
E-mail: trade@nihon-emulsion.co.jp
Web: www.nihon-emulsion.co.jp
Trade names: Emalex PG di-L; Emalex PGML; Emalex.

Nikko Chemicals Co Ltd
Nikko Chemicals Co Ltd
Chuo-ku
1-4-8, Nihonbashi-Bakurocho
Tokyo 103-0002
Japan
Tel: +81 3 3661 1677
Fax: +81 3 3664 8620
E-mail: info@nikkol.co.jp
Web: www.nikkol.co.jp
Trade names: DES-SP; Nikkol IPIS; Nikkol.

Nippi Inc
1-1-1 Senju Midori-cho
Adachi-ku
Tokyo 120-8601
Japan
Web: www.gelatin.nippi-inc.co.jp

Nippon Fine Chemical Co Ltd
Nippon Seika Bldg
2-49 Bingo-machi
Chuo-ku
Osaka 541-0051
Japan
Tel: +81 66231 4781
Fax: +81 66231 4787

E-mail: n-yoshida@nipponseika.com
Web: www.nipponseika.co.jp
Trade names: PhosphoLipid.

Nippon Paper Chemicals Co. Ltd
1-2-2, Hitotsubashi
Chiyoda-ku
Tokyo 100-0003
Japan
Tel: +81 3 5216 9111
Fax: +81 3 5216 8516
Web: www.npchem.co.jp
Trade names: KC Flock; Sunrose.

Nippon Soda Co Ltd
2-1 Otemachi 2-chome
Chiyoda-ku
Tokyo 100-8165
Japan
Tel: +81 3324 56159
E-mail: info@nippon-soda.co.jp
Web: www.nippon-soda.co.jp
Trade names: Nisso HPC.

Nitta Gelatin Inc
Tokyo Branch
08-12,2 Chome
Nihonbashi-Honchou
Cuou-Ku
Tokyo 103-0023
Japan
Tel: +81 3-6667-8252
Fax: +81 3-6667-8250
Web: www.nitta-gelatin.co.jp

Nitta Gelatin India Ltd
50/1002, SBT Avenue
Panampilly
Nagar
Cochin 682 036
India
Tel: +91 484-4099444 / 2317805
Fax: +91 484-2310568
Web: www.gelatin.in

NOF Corporation
Yebisu Garden Place Tower
20-3 Ebisu 4-chome
Shibuya-ku
Tokyo 150-6019
Japan
Tel: +81 3 5424 6771
Fax: +81 3 5424 6802
E-mail: ghonnls@nof.co.jp
Web: www.nof.co.jp
Trade names: Coatsome; NAA-160.

Numinor
PO Box 92, Maalot 24952
Israel
Tel: 972 4 9978 220
Fax: 972 4 9976 062
Web: www.numinor.com

Oh Sung Chemical Ind Co Ltd
645-10 Gojan-Dong
Namdong-gu
Incheon
Korea (South)
Tel: +82-32-547-3321
Fax: +82-32-547-3327
Web: www.oschem.co.kr

Pachem Distributions Inc
1800 Michelin Boulevard
Laval (Québec) H7L 4R3
Canada
Tel: +1 450 682 4044
Fax: +1 450 682 2044
E-mail: service@pachemdistribution.com
Web: www.pachemdistribution.com

PB Leiner Argentina SA
Parque Industrial Sauce Viejo
Ruta 11 - KM 455
CP 3017 - Sauce Viejo
Pcia. de Santa Fe
Argentina
Tel: +54 342 450 1101
Fax: +54 342 450 1108
Web: www.pbgelatins.com

Pfizer Ltd
Pfizer Centre
Patel Estate
S V Road
Jogeshwari (West)
Mumbai 400102
India
Tel: +91 22 6693 2000
Fax: +91 22 6693 2444
E-mail: contactus.india@pfizer.com
Web: www.pfizerindia.com

Priyanka Pharma/M K Ingredients & Specialities
No. 109-D
Mahendra Industrial Estate
1st Floor, Next to Vv. F. Ltd
Sion East
Mumbai 400022
India
Tel: +91 22 24091833

Progel Productora de Gelatina SA
Colombia 5026
Parque Industrial Juanchito
Apartado Aereo 1026
Manizales
Caldas
Colombia
Tel: +57 6 874 5599
Fax: +57 6 874 7630
Web: www.progel.com.co

Pro Lab Marketing Pvt Ltd
A-303
Ansal Chambers-1
3-Bhikaji Cama Place
New Delhi 110066
India
Tel: 011-6660 7725 / 6565 2166
Fax: 011-6660 7726
E-mail: info@prolabmarketing.com
Web: www.prolabmarketing.com

Raw Materials Ltd *see* LS Raw Materials Ltd

Rousselot Argentina SA
Av. Vergara, 2532
1688 Santos Tesei
Hurlingham
Buenos Aires
Argentina
Tel: +54 11 4489 8100
Fax: +54 11 4459 8101
Web: www.rousselot.com

Rubamin Ltd
Synergy House
Subhanpura
Vadodara 390 023
India
Tel: + 91 265 22 82 078
Fax: + 91 265 22 82 077
E-mail: info@rubamin.com
Web: www.rubamin.com

Saccharides
Division of Hofman International Inc
205, 259 Midpark Way SE

1

Calgary, Alberta T2X 1M2
Canada
Tel: +403 264 5150
Fax: +403 938 5150
E-mail: info@saccharides.net
Web: www.saccharides.net

Sammi Gelatin
222 Palgog Il-Dong
Ansan-City
Kyunggi-Do
Korea (South)
Tel: +82 31 437 0451
Fax: +82 31 437 0456
Web: www.sammi-gelatin.com

Samsung Fine Chemicals
Korea (South)
Web: www.sfc.samsung.co.kr
Trade names: Anycoat C; Mecellose.

San Fu Chemical Company Ltd
Member of the Fong Da group
Rm 1704, 17/F
Greenfield Tower, Concordia Plaza
1, Science Museum Road TST
Kowloon, Hong Kong
China
Tel: +1 852 2609 1138
Fax: +1 852 2609 0731
E-mail: info@fangda.com.hk
Web: www.fangda.com.hk

Sarman Industries
A-37 Gandhi Nagar
Moradabad 244001
India
Tel: +91 94 122 48444

SD Fine-Chem Ltd
315-317, T.V. Ind Estate
248 Worli Road
Mumbai 400030
India
Tel: +91 22 24959898 / 99
Fax: +91 22 24937232
E-mail: sales@sdfine.com

Shangyuchem *see* Bio Chemicals Ltd

Shasun
Shasun House
3 Doraiswamy Road
T-Nagar, Chennai-600017
Tamil Nadu
India
Tel: +91 44 24316700
Fax: +91 44 24348924
E-mail: shasun@shasun.com
Web: www.shasun.com

Shijiazhuang Pharmaceutical Group Co Ltd *see* CSPC Pharma

Sigma-Aldrich Corp
Bangalore
India
Tel: +91 80 6621 9400
Fax: +91 80 6621 9550
E-mail: india@sial.com
Web: www.sigmaaldrich.com

Signet Chemical Corporation Pvt Ltd
A-801
Crescenzo C/38-39
G-Block Bandra Kurla Complex
Mumbai 400 051
India
Tel: +91 22 61462725
Fax: +91 22 61462725
E-mail: sales@signetchem.com
Web: www.signetchem.com

Sovereign Chemicals & Cosmetics
4,Saibaba Market, Near Bank of Maharashtra
90 feet Road
Pantnagar
Ghatkopar (East)
Mumbai 400 075
India
Tel: 91 - 22- 64501221
Fax: 91 - 22 – 25157062
E-mail: response@sovereigngrp.net
Web: www.sovereigngrp.net
Web: www.petroleumjelly.co.in

SPI Pharma Inc India
21B Veerasandra Industrial Area
Hosur Road
Bangalore-560100
Karnataka
India
Tel: +91 8110 669 100
Fax: +91 8110 669 150
Trade names: Compressol S; Compressol SM; Compressol; Lubripharm; Mannogem; Sorbitab SD 250; Sorbitol Special.

Sterling Gelatin
Sandesara Estate
Padra Road
Atladra
Vadodara - 390012
Gujarat
India
Tel: +91 0265 2680730
Fax: +91 0265 2680257
Web: www.sterlinggelatin.com

Sumitomo Chemical
27-1 Shinkawa 2-chome
Chuo-ku
Tokyo 104-8260
Japan
Tel: + 81-3-5543-5102
Fax: + 81-3-5543-5901
Web: www.sumitomo-chem.co.jp

Tai'an Ruitai Cellulose Co Ltd
China
Tel: 86-538-3850703
Fax: 86-538-3852777
Web: www.ruitai.com
Trade names: Rutocel.

Teva Pharmaceutical Industries Ltd
Corporate Headquarters
5 Basel St
Petach
Tikva 49131
Israel
Tel: +972-3-9267267
Fax: +972-3-9234050
Web: www.tevapharm.com
Trade names: Galzin.

Thermo Fisher Scientific India Pvt Ltd
403-404, Delphi B Wing
Hiranandani Business Park
Powai
Mumbai 400 076
India
Tel: +91 22 6680 3000
Fax: +91 22 6680 3001/02

The Rousselot (China) Gelatin Co Ltd
25/A, 18 North CaoXi Road
Shanghai
China
Tel: 021-64272905
Fax: 021-64277336
Web: www.rousselotchina.com

Univar Canada Ltd
9800 Van Horne Way
Richmond BC V6X 1W5
Canada
Tel: +1 604 273 1441
Fax: +1 604 273 2046
Web: www.univarcanada.com

Univar USA *see* Univar Canada Ltd

Viva Corporation
India House No1
Kemps Corner
Bombay 400 026
India
Tel: +91 22 23001389
Fax: +91 22 23091237
E-mail: sales@sodiumstearate.com
Web: www.sodiumstearate.com

Voltas Ltd Chemical Division and Aqualon
Voltas House
'A' Block
Dr. Babasaheb Ambedkar Road
Chinchpokli
Mumbai 400033
India
Tel: +91 22 66656 666
Fax: +91 22 66656 311
Web: www.voltas.com

VVF Limited
Corporate Office
109, Sion (East)
Mumbai 400022
Maharashtra
India
Tel: +91 22 40282000
Fax: +91 22 24086698
E-mail: vvf.in.mumbai@vvfltd.com
Web: www.vvfltd.com
Trade names: Vegarol 1695; Vegarol 1698; Vegarol 1898.

Wacker Chemicals (China) Co Ltd
Room 1717, Western Tower
200 No.19, 4th Section Renmin South Road
Chengdu 200120
China
Tel: +86 28 8526-8320
Fax: +86 28 8526-8319
E-mail: info.china@wacker.com
Web: www.wacker.com

Weishardt International, Canada
895 Italia Street
Terrebonne
Quebec J6Y2C8
Canada
Tel: +1 450 621 2345
Fax: +1 450 621 3535

Weishardt International, Japan
Chalet Shibuya B-512
8-17 Sakuragaoka-cho
Shibuya-ku
Tokyo 150-0031
Japan
Tel: +81 (3) 3464-8042
Fax: +81 (3) 3464-8180
Web: www.weishardt.com

Wintersun Chemical
1150 S Mildred Ave
Ontario 91761
Canada
Tel: +1 909 930 1688
Fax: +1 909 947 1788
E-mail: sales@wintersunchem.com
Web: www.wintersunchem.com

Xiamen Topusing Chemical Co Ltd
7/H, Chang An Building
Lvling Road
Jiangtou
Xiamen 361009
China
Tel: +86 592 5538032 or +86 592 5538033
Fax: +86 592 5538092
E-mail: info@topusing.com
Web: www.topusing.com

Xiamen Xinwulong Gelatin Co Ltd
Unit E, F, 16/F
No. 2 East Hubin Road
361004 Xiamen
Fujian
China
Tel: +86-592-5505500
Fax: +86-592-5505505
Web: www.xwlgelatin.com

Xinchem Co
102/30, 3455 Chunshen Road
Shanghai 201100
China
Tel: +86 21 34123252
Fax: +86 21 54153973
E-mail: xinchem@xinchem.com
Web: www.xinchem.com

Yee Young Cerachem Ltd
Room 722
Hyundai Venture Ville
713 Suseo-Dong
Gangnam-Gu
Seoul
Korea (South)
Tel: +82 2 420 0331/0332/0433
Fax: +82 2 424 1877 or +82 2 6447 1877
E-mail: khchang@yeeyoung.co.kr
Web: www.yeeyoung.co.kr

Appendix II: List of Monographs by Functional Category

Appendix III: List of Related Substances

Acetic acid *see* Acetic Acid, Glacial
Activated attapulgite *see* Attapulgite
Aleuritic acid *see* Shellac
d-Alpha tocopherol *see* Alpha Tocopherol
d-Alpha tocopheryl acetate *see* Alpha Tocopherol
dl-Alpha tocopheryl acetate *see* Alpha Tocopherol
d-Alpha tocopheryl acid succinate *see* Alpha Tocopherol
dl-Alpha tocopheryl acid succinate *see* Alpha Tocopherol
Aluminum distearate *see* Aluminum Monostearate
Aluminum tristearate *see* Aluminum Monostearate
Anhydrous citric acid *see* Citric Acid Monohydrate
Anhydrous sodium citrate *see* Sodium Citrate Dihydrate
Anhydrous sodium propionate *see* Sodium Propionate
Aqueous shellac solution *see* Shellac
Artificial vinegar *see* Acetic Acid, Glacial
Aspartame acesulfame *see* Aspartame
Bacteriostatic water for injection *see* Water
Bentonite magma *see* Bentonite
Beta tocopherol *see* Alpha Tocopherol
Beta-carotene *see* Coloring Agents
n-Butyl lactate *see* Ethyl Lactate
Butylparaben sodium *see* Butylparaben
Calamine *see* Zinc Oxide
Calcium ascorbate *see* Sodium Ascorbate
Calcium cyclamate *see* Sodium Cyclamate
Calcium diorthosilicate *see* Calcium Silicate
Calcium polycarbophil *see* Polycarbophil
Calcium propionate *see* Sodium Propionate
Calcium sorbate *see* Sorbic Acid
Calcium sulfate hemihydrate *see* Calcium Sulfate
Calcium trisilicate *see* Calcium Silicate
Calcium trisodium pentetate *see* Pentetic Acid
Capric acid *see* Lauric Acid
Carbon dioxide-free water *see* Water
Cationic emulsifying wax *see* Wax, Nonionic Emulsifying
Ceratonia extract *see* Ceratonia
Cetylpyridinium bromide *see* Cetylpyridinium Chloride
Chlorhexidine acetate *see* Chlorhexidine
Chlorhexidine gluconate *see* Chlorhexidine
Chlorhexidine hydrochloride *see* Chlorhexidine
Chlorophenoxyethanol *see* Phenoxyethanol
Corn syrup solids *see* Maltodextrin
m-Cresol *see* Cresol
o-Cresol *see* Cresol
p-Cresol *see* Cresol
Crude olive-pomace oil *see* Olive Oil
Cyclamic acid *see* Sodium Cyclamate
De-aerated water *see* Water
Dehydrated alcohol *see* Alcohol
Delta tocopherol *see* Alpha Tocopherol
Denatured alcohol *see* Alcohol
Dextrose anhydrous *see* Dextrose
Diazolidinyl urea *see* Imidurea
Diethylene glycol monopalmitostearate *see* Ethylene Glycol Stearates
Dilute acetic acid *see* Acetic Acid, Glacial
Dilute alcohol *see* Alcohol
Dilute ammonia solution *see* Ammonia Solution
Dilute hydrochloric acid *see* Hydrochloric Acid
Dilute phosphoric acid *see* Phosphoric Acid
Dilute sulfuric acid *see* Sulfuric Acid
Dimethyl-β-cyclodextrin *see* Cyclodextrins
Dioctyl phthalate *see* Dibutyl Phthalate
Dipotassium edetate *see* Edetic Acid
Docusate calcium *see* Docusate Sodium
Docusate potassium *see* Docusate Sodium
Dodecyltrimethylammonium bromide *see* Cetrimide
Edetate calcium disodium *see* Edetic Acid
Eglumine *see* Meglumine
Ethyl gallate *see* Propyl Gallate
Ethyl linoleate *see* Linoleic Acid
Ethylene glycol monopalmitate *see* Ethylene Glycol Stearates
Ethylene glycol monostearate *see* Ethylene Glycol Stearates
Ethylparaben potassium *see* Ethylparaben
Ethylparaben sodium *see* Ethylparaben
Extra virgin olive oil *see* Olive Oil
Fine virgin olive oil *see* Olive Oil
Flexible collodion *see* Pyroxylin
Fuming sulfuric acid *see* Sulfuric Acid
Gamma tocopherol *see* Alpha Tocopherol
Glycerol ester of gum rosin *see* Colophony
Glyceryl triisooctanoate *see* Tricaprylin
Glycine hydrochloride *see* Glycine
Glycolic acid *see* Aliphatic Polyesters
Hard water *see* Water
Hesperidin *see* Neohesperidin Dihydrochalcone
Hexadecyltrimethylammonium bromide *see* Cetrimide
High-fructose syrup *see* Fructose
Hyaluronic acid *see* Sodium Hyaluronate
Hydrogenated lanolin *see* Lanolin
Hydrogenated rosin *see* Colophony
Hydrogenated vegetable oil, type II USP–NF *see* Vegetable Oil, Hydrogenated
2-Hydroxyethyl-β-cyclodextrin *see* Cyclodextrins
3-Hydroxypropyl-β-cyclodextrin *see* Hydroxypropyl Betadex
Indigo carmine *see* Coloring Agents
Invert sugar *see* Sucrose
Isodecyl oleate *see* Decyl Oleate
Isopropyl stearate *see* Isopropyl Isostearate
Isotrehalose *see* Trehalose
Laccaic acid B *see* Shellac
Lampante virgin olive oil *see* Olive Oil
Lanolin alcohols ointment *see* Petrolatum and Lanolin Alcohols
DL-Leucine *see* Leucine
Liquefied phenol *see* Phenol
Liquid fructose *see* Fructose
Lysine *see* Lysine Hydrochloride
Magnesium carbonate anhydrous *see* Magnesium Carbonate
Magnesium carbonate hydroxide *see* Magnesium Carbonate
Magnesium lauryl sulfate *see* Sodium Lauryl Sulfate
Magnesium metasilicate *see* Magnesium Silicate
Magnesium orthosilicate *see* Magnesium Silicate
Magnesium trisilicate anhydrous *see* Magnesium Trisilicate
D-Malic acid *see* Malic Acid
L-Malic acid *see* Malic Acid
d-Menthol *see* Menthol
l-Menthol *see* Menthol
D-Methionine *see* Methionine
DL-Methionine *see* Methionine
Methyl lactate *see* Ethyl Lactate
Methyl linoleate *see* Linoleic Acid
Methyl methacrylate *see* Polymethacrylates
Methyl oleate *see* Ethyl Oleate
Methylparaben potassium *see* Methylparaben
Methylparaben sodium *see* Methylparaben
N-Methylpyrrolidone *see* Pyrrolidone
Microcrystalline cellulose and carrageenan *see* Cellulose, Microcrystalline
Microcrystalline cellulose and dibasic calcium phosphate *see* Cellulose, Microcrystalline
Microcrystalline cellulose and guar gum *see* Cellulose, Microcrystalline
Microcrystalline cellulose and mannitol *see* Cellulose, Microcrystalline
Microcrystalline cellulose spheres *see* Sugar Spheres
Modified lanolin *see* Lanolin
Monobasic potassium phosphate *see* Potassium Phosphate, Dibasic
Monobasic potassium phosphate *see* Sodium Phosphate, Monobasic
Montmorillonite *see* Magnesium Aluminum Silicate
Neotrehalose *see* Trehalose

Normal magnesium carbonate *see* Magnesium Carbonate
NPTAB see Sugar Spheres
Octenyl succinic anhydride modified gum acacia *see* Acacia
Oleyl oleate *see* Oleyl Alcohol
Olive-pomace oil *see* Olive Oil
Palmitin *see* Palmitic Acid
Pentaerythritol ester of gum rosin *see* Colophony
Pentasodium pentetate *see* Pentetic Acid
Pharmaceutical glaze *see* Shellac
Phenoxypropanol *see* Phenoxyethanol
Polacrilin *see* Polacrilin Potassium
Poly(methyl methacrylate) *see* Polymethacrylates
Polymerized rosin *see* Colophony
Polyvinyl acetate *see* Polyvinyl Acetate Dispersion
Potassium bisulfite *see* Potassium Metabisulfite
Potassium myristate *see* Myristic Acid
Potassium propionate *see* Sodium Propionate
Powdered fructose *see* Fructose
Propan-1-ol *see* Isopropyl Alcohol
1,2-Propanediol, 1-laurate *see* Propylene Glycol Monolaurate
(*S*)-Propylene carbonate *see* Propylene Carbonate
Propylparaben potassium *see* Propylparaben
Purified bentonite *see* Bentonite
Purified stearic acid *see* Stearic Acid
Quaternium 18-hectorite *see* Hectorite
Rapeseed oil *see* Canola Oil
Recombinant albumin, human *see* Albumin
Refined almond oil *see* Almond Oil
Refined olive-pomace oil *see* Olive Oil
Saccharin ammonium *see* Saccharin
Saccharin calcium *see* Saccharin
Safflower glycerides *see* Safflower Oil
Self-emulsifying glyceryl monostearate *see* Glyceryl Monostearate
Sodium bisulfite *see* Sodium Metabisulfite
Sodium borate anhydrous *see* Sodium Borate
Sodium carbonate decahydrate *see* Sodium Carbonate
Sodium carbonate monohydrate *see* Sodium Carbonate
Sodium edetate *see* Edetic Acid
Sodium erythorbate *see* Erythorbic Acid
Sodium glycinate *see* Glycine
Sodium laurate *see* Lauric Acid
Sodium myristate *see* Myristic Acid
Sodium palmitate *see* Palmitic Acid
Sodium sorbate *see* Sorbic Acid
Sodium sulfite heptahydrate *see* Sodium Sulfite
Soft water *see* Water
Spermaceti wax *see* Wax, Cetyl Esters
Stearalkonium hectorite *see* Hectorite
Sterile water for inhalation *see* Water
Sterile water for injection *see* Water
Sterile water for irrigation *see* Water
Sucrose monolaurate *see* Sucrose Palmitate
Sugartab *see* Sugar, Compressible
Sunset yellow FCF *see* Coloring Agents
Synthetic paraffin *see* Paraffin
DL-Tagatose *see* Tagatose
L-Tagatose *see* Tagatose
Tartrazine *see* Coloring Agents
Theobroma oil *see* Suppository Bases, Hard Fat
Tocopherols excipient *see* Alpha Tocopherol
Tribasic sodium phosphate *see* Sodium Phosphate, Dibasic
Trimethyl-β-cyclodextrin *see* Cyclodextrins
Trimethyltetradecylammonium bromide *see* Cetrimide
Tripotassium phosphate *see* Potassium Phosphate, Dibasic
Trisodium edetate *see* Edetic Acid
Virgin olive oil *see* Olive Oil
Water for injection (WFI) *see* Water
White petrolatum *see* Petrolatum
Zinc formaldehyde sulfoxylate *see* Sodium Formaldehyde Sulfoxylate
Zinc propionate *see* Sodium Propionate
Zinc trisodium pentetate *see* Pentetic Acid

Appendix IV: List of Excipient 'E' Numbers

4

E Number	*Excipient*	
E100	Curcumin	206
E101	Riboflavin	206
E102	Tartrazine	206, 211
E104	Quinoline Yellow	206
E110	Sunset Yellow FCF	206, 210
E120	Carmine	206
E122	Carmoisine	206
E123	Amaranth	206
E124	Ponceau 4R	206
E127	Erythrosine	206
E129	Allura Red AC	206
E131	Patent Blue V	206
E132	Indigo Carmine	206, 209
E133	Brilliant Blue FCF	206
E140	Chlorophylls	206
E141	Copper Complexes of Chlorophylls and Chlorophyllins	206
E142	Green S	206
E150	Caramel	206
E151	Brilliant Black BN	206
E153	Vegetable Carbon	206
E160	Alpha-, Beta-, Gamma-Carotene	206
E160	Beta-apo-8′ Carotenal	206
E160	Capsanthin	206
E160	Capsorubin	206
E160	Ethyl Ester of Beta-apo-8′ Carotenoic Acid	206
E160	Lycopene	206
E160a	Beta-Carotene	209
E161	Canthaxanthin	206
E161	Lutein	206
E161	Xanthophylls	206
E162	Beetroot Red	206
E163	Anthocyanins	206
E163	Cyanidin	206
E163	Delphidin	206
E163	Malvidin	206
E163	Pelargonidin	206
E163	Peonidin	206
E163	Petunidin	206
E170	Calcium Carbonate	93, 206
E171	Titanium Dioxide	206, 851
E172	Iron Oxides	389
E172	Iron Oxides and Hydroxides	206
E173	Aluminum	206
E200	Sorbic Acid	769
E201	Sodium Sorbate	771
E202	Potassium Sorbate	659
E203	Calcium Sorbate	771
E210	Benzoic Acid	65
E211	Sodium Benzoate	718
E212	Potassium Benzoate	644
E214	Ethylparaben	306
E215	Ethylparaben Sodium	308
E216	Propylparaben	679
E217	Propylparaben Sodium	683
E218	Methylparaben	500
E219	Methylparaben Sodium	503
E220	Sulfur Dioxide	826
E221	Sodium Sulfite	765
E222	Sodium Bisulfite	748
E223	Sodium Metabisulfite	747
E224	Potassium Metabisulfite	653
E228	Potassium Bisulfite	654
E252	Potassium Nitrate	656
E260	Acetic Acid, Glacial	5
E262	Sodium Acetate	710
E263	Calcium Acetate	89

E Number	*Excipient*	
E270	Lactic Acid	406
E280	Propionic Acid	666
E281	Anhydrous Sodium Propionate	756
E281	Sodium Propionate	754
E282	Calcium Propionate	756
E283	Potassium Propionate	756
E284	Boric Acid	73
E285	Sodium Borate	725
E290	Carbon Dioxide	123
E296	Malic Acid	465
E297	Fumaric Acid	314
E300	Ascorbic Acid	47
E301	Sodium Ascorbate	715
E302	Calcium Ascorbate	717
E304	Ascorbyl Palmitate	50
E307	Alpha Tocopherol	34
E308	Gamma Tocopherol	36
E309	Delta Tocopherol	36
E310	Propyl Gallate	668
E311	Octyl Gallate	526
E312	Dodecyl Gallate	279
E315	Erythorbic Acid	284
E316	Sodium Erythorbate	285
E320	Butylated Hydroxyanisole	77
E321	Butylated Hydroxytoluene	80
E322	Lecithin	437
E325	Sodium Lactate	742
E327	Calcium Lactate	99
E330	Anhydrous Citric Acid	195
E330	Citric Acid Monohydrate	193
E331	Sodium Citrate Dihydrate	732
E332	Potassium Citrate	650
E334	Tartaric Acid	840
E338	Phosphoric Acid	569
E339	Sodium Phosphate, Dibasic	749
E339	Sodium Phosphate, Monobasic	752
E339	Tribasic Sodium Phosphate	751
E340	Monobasic Potassium Phosphate	754
E340	Potassium Phosphate, Dibasic	657
E341	Calcium Phosphate, Dibasic Anhydrous	101
E341	Calcium Phosphate, Dibasic Dihydrate	104
E341(iii)	Calcium Phosphate, Tribasic	107
E355	Adipic Acid	13
E385	Edetate Calcium Disodium	282
E400	Alginic Acid	23
E401	Sodium Alginate	712
E402	Potassium Alginate	641
E404	Ammonium Alginate	44
E404	Calcium Alginate	90
E405	Propylene Glycol Alginate	675
E406	Agar	15
E407	Carrageenan	130
E410	Ceratonia	157
E412	Guar Gum	341
E413	Tragacanth	854
E414	Acacia	1
E415	Xanthan Gum	897
E420	Sorbitol	776
E421	Mannitol	479
E422	Glycerin	324
E431	Polyoxyethylene Stearates	625
E432	Polyoxyethylene Sorbitan Fatty Acid Esters	620
E433	Polyoxyethylene Sorbitan Fatty Acid Esters	620
E434	Polyoxyethylene Sorbitan Fatty Acid Esters	620
E435	Polyoxyethylene Sorbitan Fatty Acid Esters	620
E436	Polyoxyethylene Sorbitan Fatty Acid Esters	620
E440	Pectin	543

E Number	Excipient	
E441	Gelatin	319
E452	Potassium Metaphosphate	655
E460	Cellulose, Microcrystalline	140
E460	Cellulose, Powdered	146
E461	Methylcellulose	496
E462	Ethylcellulose	297
E463	Hydroxypropyl Cellulose	363
E464	Hypromellose	373
E466	Carboxymethylcellulose Sodium	126
E471	Glyceryl Behenate	328
E-473	Sucrose Palmitate	813
E-473	Sucrose Stearate	815
E477	Propylene Glycol Dilaurate	676
E477	Propylene Glycol Monolaurate	678
E491	Sorbitan Esters (Sorbitan Fatty Acid Esters)	773
E492	Sorbitan Esters (Sorbitan Fatty Acid Esters)	773
E493	Sorbitan Esters (Sorbitan Fatty Acid Esters)	773
E494	Sorbitan Esters (Sorbitan Fatty Acid Esters)	773
E495	Sorbitan Esters (Sorbitan Fatty Acid Esters)	773
E500	Sodium Bicarbonate	721
E500i	Sodium Carbonate	726
E501	Potassium Bicarbonate	646
E504	Magnesium Carbonate	450
E504	Magnesium Carbonate Anhydrous	452
E504	Magnesium Carbonate Hydroxide	452
E504	Normal Magnesium Carbonate	453
E507	Hydrochloric Acid	351
E508	Potassium Chloride	647
E510	Ammonium Chloride	46
E513	Sulfuric Acid	828
E516	Calcium Sulfate	113
E516	Calcium Sulfate Hemihydrate	115
E524	Sodium Hydroxide	741
E525	Potassium Hydroxide	652
E526	Calcium Hydroxide	98
E530	Magnesium Oxide	454
E553a	Magnesium Silicate	456
E553a	Magnesium Trisilicate	462
E553b	Talc	837
E558	Bentonite	57

E Number	Excipient	
E559	Kaolin	403
E570	Stearic Acid	800
E621	Monosodium Glutamate	511
E640	Glycine	338
E642	Lysine Hydrochloride	445
E900	Dimethicone	262
E901	Wax, White	893
E901	Wax, Yellow	895
E903	Wax, Carnauba	886
E904	Shellac	704
E907	Wax, Microcrystalline	890
E913	Lanolin	429
E941	Nitrogen	521
E942	Nitrous Oxide	523
E943a	Hydrocarbons (HC)	349
E943b	Hydrocarbons (HC)	349
E944	Hydrocarbons (HC)	349
E950	Acesulfame Potassium	3
E951	Aspartame	52
E952	Calcium Cyclamate	736
E952	Cyclamic Acid	736
E952	Sodium Cyclamate	735
E953	Isomalt	391
E954	Saccharin	693, 695
E954	Saccharin Sodium	696, 698
E957	Thaumatin	844
E959	Neohesperidin Dihydrochalcone	518
E965	Maltitol	468
E965	Maltitol Solution	470
E966	Lactitol	408
E967	Xylitol	901
E968	Erythritol	285
E1200	Polydextrose	581
E1201	Povidone	661
E1202	Crospovidone	227
E1440	Hydroxypropyl Starch	371
E1505	Triethyl Citrate	867
E1518	Triacetin	859
E1520	Propylene Glycol	672

Appendix V: List of Excipient 'EINECS' Numbers

EINECS	*Excipient*	
200-018-0	Lactic Acid	407
200-061-5	Sorbitol	778
200-066-2	Ascorbic Acid	49
200-075-1	Dextrose	250
200-075-1	Glucose, Liquid	324
200-143-0	Bronopol	77
200-210-4	Thimerosal	847
200-238-7	Chlorhexidine	178
200-272-2	Glycine	339
200-289-5	Glycerin	327
200-302-4	Chlorhexidine Acetate	178
200-312-9	Palmitic Acid	536
200-313-4	Stearic Acid	802
200-317-6	Chlorobutanol	181
200-333-3	Fructose	312
200-334-9	Sucrose	810
200-338-0	Propylene Glycol	674
200-353-2	Cholesterol	193
200-412-2	D-Alpha Tocopherol	37
200-416-4	Galactose	318
200-431-6	Chlorocresol	184
200-432-1	DL-Methionine	495
200-449-4	Edetic Acid	283
200-456-2	Phenylethyl Alcohol	557
200-470-9	Linoleic Acid	442
200-522-0	Leucine	441
200-529-9	Edetate Calcium Disodium	283
200-532-5	Phenylmercuric Acetate	559
200-540-9	Calcium Acetate	90
200-559-2	Lactose, Anhydrous	412
200-559-2	Lactose, Monohydrate	420
200-562-9	L-Methionine	495
200-578-6	Alcohol	22
200-580-7	Acetic Acid, Glacial	7
200-618-2	Benzoic Acid	67
200-652-8	Pentetic Acid	546
200-661-7	Isopropyl Alcohol	396
200-662-2	Acetone	8
200-664-3	Dimethyl Sulfoxide	269
200-675-3	Sodium Citrate	734
200-711-8	Mannitol	482
200-716-5	Maltose	478
200-772-0	Sodium Lactate	743
201-064-4	Tromethamine	872
201-066-5	Acetyltriethyl Citrate	13
201-067-0	Acetyltributyl Citrate	11
201-069-1	Anhydrous Citric Acid	196
201-069-1	Citric Acid Monohydrate	196
201-070-7	Triethyl Citrate	868
201-071-2	Tributyl Citrate	862
201-176-3	Propionic Acid	667
201-321-0	Saccharin	695
201-550-6	Diethyl Phthalate	259
201-557-4	Dibutyl Phthalate	252
201-766-0	Tartaric Acid	841
201-772-3	Tagatose	837
201-788-0	Xylitol	904
201-793-8	Chloroxylenol	190
201-928-0	Erythorbic Acid	285
201-939-0	Menthol	493
201-944-8	Thymol	850
202-307-7	Propylparaben	682
202-318-7	Butylparaben	86
202-495-0	Monothioglycerol	514
202-598-0	Ethyl Lactate	291
202-601-5	Malic Acid	467
202-739-6	Trehalose	858
202-785-7	Methylparaben	503
202-859-9	Benzyl Alcohol	70
203-049-8	Triethanolamine	867
203-051-9	Triacetin	860
203-068-1	Phenylmercuric Borate	561
203-416-2	Dipropylene Glycol	274
203-529-7	Butylene Glycol	83
203-572-1	Propylene Carbonate	671
203-577-9	Cresol	223
203-599-9	Dipropylene Glycol	274
203-632-7	Phenol	553
203-672-5	Dibutyl Sebacate	254
203-742-5	Maleic Acid	465
203-743-0	Fumaric Acid	315
203-751-4	Isopropyl Myristate	400
203-764-5	Diethyl Sebacate	260
203-768-7	Sorbic Acid	771
203-821-4	Dipropylene Glycol	274
203-825-6	Squalane	784
203-868-0	Diethanolamine	256
203-889-5	Ethyl Oleate	294
203-919-7	Diethylene Glycol Monoethyl Ether	257
203-934-9	Isopropyl Stearate	398
203-993-0	Methyl Linoleate	442
204-000-3	Myristyl Alcohol	517
204-007-1	Oleic Acid	529
204-017-6	Stearyl Alcohol	804
204-065-8	Dimethyl Ether	265
204-214-7	Dioctyl Phthalate	252
204-271-8	Maltol	476
204-399-4	Ethylparaben	308
204-402-9	Benzyl Benzoate	72
204-464-7	Ethyl Vanillin	296
204-465-2	Vanillin	875
204-479-9	Benzethonium Chloride	64
204-498-2	Propyl Gallate	669
204-534-7	Triolein	870
204-589-7	Phenoxyethanol	555
204-593-9	Cetylpyridinium Chloride	171
204-648-7	Pyrrolidone	689
204-665-5	Butyl Stearate	87
204-673-3	Adipic Acid	15
204-696-9	Carbon Dioxide	124
204-772-1	Sucrose Octaacetate	812
204-823-8	Sodium Acetate	711
204-826-4	Dimethylacetamide	272
204-881-4	Butylated Hydroxytoluene	81
205-011-6	Dimethyl Phthalate	267
205-126-1	Sodium Ascorbate	717
205-290-4	Sodium Propionate	756
205-305-4	Ascorbyl Palmitate	52
205-316-4	Butyl Lactate	291
205-358-3	Disodium Edetate	276
205-391-3	Pentasodium Pentetate	546
205-483-3	Monoethanolamine	511
205-500-4	Ethyl Acetate	289
205-513-5	Hexetidine	348
205-538-1	Monosodium Glutamate	512
205-542-3	1,2-Propanediol, 1-laurate	679
205-571-1	Isopropyl Palmitate	401
205-582-1	Lauric Acid	435
205-597-3	Oleyl Alcohol	531
205-633-8	Sodium Bicarbonate	723
205-737-3	Erythritol	287
205-739-4	Sodium Formaldehyde Sulfoxylate	738
205-758-8	Trisodium Edetate	283
205-788-1	Sodium Lauryl Sulfate	746

EINECS	*Excipient*	
206-059-0	Potassium Bicarbonate	647
206-101-8	Aluminum Distearate	40
206-376-4	Capric Acid	435
206-483-6	D-Methionine	495
206-988-1	Sodium Palmitate	536
207-439-9	Calcium Carbonate	96
207-838-8	Sodium Carbonate	728
208-534-8	Sodium Benzoate	720
208-578-8	Aleuritic Acid	706
208-686-5	Tricaprylin	864
208-736-6	Cetyl Palmitate	169
208-761-2	Acetone Sodium Bisulfite	9
208-868-4	Ethyl Linoleate	442
208-875-2	Myristic Acid	515
208-915-9	Magnesium Carbonate	453
209-150-3	Magnesium Stearate	460
209-151-9	Zinc Stearate	911
209-170-2	Zinc Acetate	907
209-406-4	Docusate Sodium	279
209-481-3	Potassium Benzoate	645
209-566-5	Lactitol	409
209-567-0	Maltitol	469
211-082-4	Sodium Laurate	435
211-279-5	Aluminum Tristearate	40
211-519-9	Lysine Hydrochloride	446
212-406-7	Calcium Lactate	100
212-487-9	Sodium Myristate	515
212-490-5	Sodium Stearate	762
212-755-5	Potassium Citrate	651
212-828-1	*N*-Methylpyrrolidone	689
214-291-9	Cetrimide	165
214-620-6	Dodecyl Gallate	280
215-108-5	Bentonite	59
215-137-3	Calcium Hydroxide	99
215-168-2	Iron Oxides	390
215-171-9	Magnesium Oxide	455
215-181-3	Potassium Hydroxide	653
215-185-5	Sodium Hydroxide	742
215-222-5	Zinc Oxide	909
215-277-5	Iron Oxides	390
215-289-0	Saponite	702
215-478-8	Magnesium Aluminium Silicate	449
215-540-4	Sodium Borate Anhydrous	726
215-663-3	Sorbitan Esters (Sorbitan Fatty Acid Esters)	775
215-664-9	Sorbitan Esters (Sorbitan Fatty Acid Esters)	775
215-665-4	Sorbitan Esters (Sorbitan Fatty Acid Esters)	775
215-681-1	Magnesium Silicate	457
215-691-6	Aluminum Oxide	41
215-710-8	Calcium Silicate	110
215-798-8	Alpha Tocopherol	37
216-472-8	Calcium Stearate	112
217-895-0	Dipotassium Edetate	282
220-491-7	Sunset Yellow FCF	211
221-450-6	Magnesium Lauryl Sulfate	746
222-392-4	D-Mannose	485
222-981-6	Decyl Oleate	237
223-026-6	Chlorhexidine Hydrochloride	178
223-095-2	Denatonium Benzoate	238
223-781-1	Sodium Stearyl Fumarate	765
226-242-9	Octyldodecanol	526
227-407-8	Glycofurol	341
227-841-8	Glycine Hydrochloride	339
227-842-3	Sodium Glycinate	339
228-506-9	Meglumine	490
228-973-6	Sodium Erythorbate	285
230-325-5	Aluminum Monostearate	40
230-636-6	Beta-Carotene	209
230-907-9	Hydroxyethylpiperazine Ethane Sulfonic Acid	360
231-195-2	Sulfur Dioxide	827
231-211-8	Potassium Chloride	649
231-321-6	Calcium Sorbate	771
231-449-2	Monobasic Sodium Phosphate	754
231-493-2	Cyclodextrin	233
231-545-4	Colloidal Silicon Dioxide	201
231-595-7	Hydrochloric Acid	352
231-598-3	Sodium Chloride	731
231-633-2	Phosphoric Acid	570
231-635-3	Ammonia Solution	44
231-639-5	Sulfuric Acid	829
231-673-0	Sodium Metabisulfite	749
231-783-9	Nitrogen	522
231-818-8	Potassium Nitrate	657
231-819-3	Sodium Sorbate	771
231-821-4	Sodium Sulfite	766
231-837-1	Calcium Phosphate, Dibasic Anhydrous	103
231-837-1	Calcium Phosphate, Dibasic Dihydrate	106
231-837-1	Calcium Phosphate, Tribasic	108
231-867-5	Sodium Thiosulfate	768
231-900-3	Calcium Sulfate	115
231-913-4	Monobasic Potassium Phosphate	754
232-212-6	Potassium Metaphosphate	655
232-273-9	Sunflower Oil	831
232-276-5	Safflower Oil	700
232-280-7	Cottonseed Oil	221
232-281-2	Corn Oil	216
232-290-1	Ceresin	160
232-292-2	Castor Oil, Hydrogenated	138
232-293-8	Castor Oil	136
232-296-4	Peanut Oil	543
232-302-5	Spermaceti wax	889
232-307-2	Lecithin	439
232-313-5	Canola Oil	117
232-315-6	Paraffin	541
232-316-1	Palm Oil	538
232-348-6	Lanolin	431
232-360-1	Sorbitan Esters (Sorbitan Fatty Acid Esters)	775
232-373-2	Petrolatum	549
232-399-4	Wax, Carnauba	887
232-425-4	Palm Kernel Oil	537
232-430-1	Lanolin Alcohols	434
232-475-7	Colophony	204
232-479-9	Pentaerythritol Ester of Gum Rosin	203
232-482-5	Glycerol Ester of Gum Rosin	203
232-514-8	Glyceryl Palmitostearate	337
232-519-5	Acacia	2
232-524-2	Carrageenan	133
232-536-8	Guar Gum	343
232-541-5	Ceratonia	158
232-549-9	Shellac	707
232-553-0	Pectin	545
232-554-6	Gelatin	322
232-658-1	Agar	16
232-674-9	Cellulose, Powdered	148
232-674-9	Croscarmellose Sodium	226
232-675-4	Dextrin	247
232-678-0	Hyaluronic acid	740
232-678-0	Sodium Hyaluronate	740
232-679-6	Hydroxypropyl Starch	372
232-680-1	Alginic Acid	25
232-722-9	Zein	906
232-940-4	Maltodextrin	474
232-945-1	Pullulan	685
233-032-0	Nitrous Oxide	524
233-107-8	Calcium Diorthosilicate	110
233-139-2	Boric Acid	74
233-140-8	Calcium Chloride	97
233-141-3	Potassium Alum	644
233-466-0	DL-Alpha Tocopherol	37
234-394-2	Xanthan Gum	899
234-406-6	Quaternium 18-Hectorite	345
235-169-1	Calcium Trisodium Pentetate	546
235-186-4	Ammonium Chloride	47
235-192-7	Magnesium Carbonate Hydroxide	453
235-336-9	Calcium Trisilicate	110
235-340-0	Hectorite	345
236-550-5	Potassium Myristate	515
236-675-5	Titanium Dioxide	853
238-877-9	Talc	839
239-076-7	Magnesium Trisilicate	463
240-795-3	Potassium Metabisulfite	654

Index

Greek characters (α, β, γ etc.), numerical prefixes (50-, 1,2- etc.) and prefixes such as *para*, *ortho*, O-, N-, Z-, D-, L- etc. are excluded from alphabetization; page numbers in **bold** refer to monograph titles.

I